BERGEY'S MANUAL® OF
Systematic Bacteriology
Second Edition

Volume One
The *Archaea* and the Deeply Branching and Phototrophic *Bacteria*

Springer

New York
Berlin
Heidelberg
Barcelona
Hong Kong
London
Milan
Paris
Singapore
Tokyo

BERGEY'S MANUAL® OF
Systematic Bacteriology
Second Edition

Volume One
The *Archaea* and the Deeply Branching and Phototrophic *Bacteria*

David R. Boone
Richard W. Castenholz
EDITORS, VOLUME ONE

George M. Garrity
EDITOR-IN-CHIEF

EDITORIAL BOARD
James T. Staley, Chairman, **David R. Boone,** Vice Chairman,
Don J. Brenner, Richard W. Castenholz, George M. Garrity, Michael Goodfellow, Noel R. Krieg, Fred A. Rainey, Karl-Heinz Schleifer

WITH CONTRIBUTIONS FROM 105 COLLEAGUES

Springer

George M. Garrity
Department of Microbiology and Molecular Genetics
Bergey's Manual Trust
Michigan State University
East Lansing, MI 48824-1101
USA

Library of Congress Cataloging-in-Publication Data

Bergey's manual of systematic bacteriology / David R. Boone, Richard W. Castenholz,
editors, volume 1 ; George M. Garrity, editor-in-chief.—2nd ed.
   p. cm.
  Includes bibliographical references and index.
  Contents: v. 1. The archaea and the deeply branching and phototrophic bacteria.
  ISBN 0-387-98771-1 (alk. paper)
  1. Bacteria—Classification.   I. Title: Systematic bacteriology.   II. Boone, David R.
  III. Castenholz, Richard W.   IV. Garrity, George M.
  QR81.B46 2001
  579.3'01'2—dc21   2001020400

With 330 illustrations

The following proprietary names of products are used in this volume: Casamino™ acids; Vector NTI®;
XL10-Gold®; Gelrite®; Tryptone™; Phytagel™; bio-Trypcase™; Trypticase®; Oxoid® purified agar.

Printed on acid-free paper.

First edition published 1984–1989 by Bergey's Manual Trust and Williams & Wilkins, Baltimore.

Production coordinated by Impressions Book and Journal Services, Inc., and managed by Frederick
Bartlett, Theresa Kornak, and Catherine Lyons; manufacturing supervised by Jacqui Ashri.
Typeset by Impressions Book and Journal Services, Inc., Madison, WI.
Printed and bound by Maple-Vail Book Manufacturing Group, York, PA.
Printed in the United States of America.

9 8 7 6 5 4 3 2 1

ISBN 0-387-98771-1   SPIN 10711344

Springer-Verlag New York Berlin Heidelberg
A member of BertelsmannSpringer Science+Business Media GmbH

Preface to the Second Edition of *Bergey's Manual®* *of Systematic Bacteriology*

There is a long-standing tradition for the Editors of each successive edition of *Bergey's Manual* to open their respective volumes with the observation that the new edition is a departure from the earlier ones. We shall not waver from this tradition, as the very nature of our field compels us to make this pronouncement. Systematic bacteriology (or perhaps systematic procaryotic biology) is a dynamic field, driven by constant theoretical and methodological advances that will ultimately lead to a more perfect and useful classification scheme.

Since publication of the First Edition of the *Systematics Manual,* we have witnessed a major shift in how we view the relationships among *Bacteria* and *Archaea.* While the possibility of a universally applicable natural classification was evident as the First Edition was in preparation, it is only recently that the sequence databases became large enough, and the taxonomic coverage broad enough, to make such an arrangement feasible. We have relied heavily upon these data in organizing the contents of this edition of *Bergey's Manual of Systematic Bacteriology,* which will follow a phylogenetic framework based on analysis of the nucleotide sequence of the small ribosomal subunit RNA, rather than a phenotypic structure. This departs from the First Edition, as well as the Eighth and Ninth Editions of the *Determinative Manual.* While the rationale for presenting the content of this edition in such a manner should be evident to most readers, they should bear in mind that this edition, as have all preceding ones, represents a progress report rather than a final classification of procaryotes.

The Editors remind the readers that the *Systematics Manual* is a peer-reviewed collection of chapters, contributed by authors who were invited by the Trust to share their knowledge and expertise of specific taxa. Citation should refer to the author, the chapter title, and inclusive pages rather than to the Editors. The Trust is indebted to all of the contributors and reviewers, without whom this work would not be possible. The Editors are grateful for the time and effort that each expended on behalf of the entire scientific community. We also thank the authors for their good grace in accepting comments, criticisms, and editing of their manuscripts. We would also like to recognize the special efforts of Drs. Hans Trüper and Brian Tindall for their assistance on matters of nomenclature and etymology and Dr. Aharon Oren for his critical reading of large portions of the *Manual.*

We would like to express our thanks to the Department of Microbiology and Molecular Genetics at Michigan State University for housing our headquarters and editorial office and for providing a congenial and supportive environment for microbial systematics. We would also like to thank Connie Williams not only for her expert secretarial assistance, but also for unflagging dedication to the mission of Bergey's Manual Trust and Dr. Denise Searles for her editorial assistance and diligence in verifying countless pieces of critical information, along with Heather Everett, Alissa Wesche, and Mathew Winters for their assistance in fact-checking and compilation of the bibliography.

A project such as the *Systematics Manual* also requires the strong and continued support of a dedicated publisher, and we have been most fortunate in this regard. We would also like to express our gratitude to Springer-Verlag for supporting our efforts and for the development of the Bergey's Document Type Definition (DTD). We would especially like to thank our Executive Editor, Dr. Robert Badger, for his courage, patience, understanding, and support; Catherine Lyons for her expertise in designing and developing our DTD, and Terry Kornak and Fred Bartlett for their efforts during the pre-production and production phases. We would also like to acknowledge the support of ArborText, Inc., for providing us with state-of-the-art SGML development and editing tools at reduced cost. Lastly, I would like to express my personal thanks to my fellow trustees for providing me with the opportunity to participate in this effort, to Drs. David Boone and Richard Castenholz for their enormous efforts as volume editors and to my wife, Nancy, and daughter, Jane, for their patience, tolerance, and support.

Comments on this edition are welcomed and should be directed to Bergey's Manual Trust, Department of Microbiology and Molecular Genetics, Giltner Hall, Michigan State University, East Lansing, MI, USA 48824-1101. Email: garrity@pilot.msu.edu

George M. Garrity

Preface to the First Edition of *Bergey's Manual®* *of Systematic Bacteriology*

Many microbiologists advised the Trust that a new edition of the *Manual* was urgently needed. Of great concern to us was the steadily increasing time interval between editions; this interval reached a maximum of 17 years between the seventh and eighth editions. To be useful the *Manual* must reflect relatively recent information; a new edition is soon dated or obsolete in parts because of the nearly exponential rate at which new information accumulates. A new approach to publication was needed, and from this conviction came our plan to publish the *Manual* as a sequence of four subvolumes concerned with systematic bacteriology as it applies to taxonomy. The four subvolumes are divided roughly as follows: (a) the Gram-negatives of general, medical or industrial importance; (b) the Gram-positives other than actinomycetes; (c) the archaeobacteria, cyanobacteria and remaining Gram-negatives; and (d) the actinomycetes. The Trust believed that more attention and care could be given to preparation of the various descriptions within each subvolume, and also that each subvolume could be prepared, published, and revised as the area demanded, more rapidly than could be the case if the *Manual* were to remain as a single, comprehensive volume as in the past. Moreover, microbiologists would have the option of purchasing only that particular subvolume containing the organisms in which they were interested.

The Trust also believed that the scope of the *Manual* needed to be expanded to include more information of importance for systematic bacteriology and bring together information dealing with ecology, enrichment and isolation, descriptions of species and their determinative characters, maintenance and preservation, all focused on the illumination of bacterial taxonomy. To reflect this change in scope, the title of the *Manual* was changed and the primary publication becomes *Bergey's Manual of Systematic Bacteriology*. This contains not only determinative material such as diagnostic keys and tables useful for identification, but also all of the detailed descriptive information and taxonomic comments. Upon completion of each subvolume, the purely determinative information will be assembled for eventual incorporation into a much smaller publication which will continue the original name of the *Manual, Bergey's Manual of Determinative Bacteriology*, which will be a similar but improved version of the present *Shorter Bergey's Manual*. So, in the end there will be two publications, one systematic and one determinative in character.

An important task of the Trust was to decide which genera should be covered in the first and subsequent subvolumes. We were assisted in this decision by the recommendations of our Advisory Committees, composed of prominent taxonomic authorities to whom we are most grateful. Authors were chosen on the basis of constant surveillance of the literature of bacterial systematics and by recommendations from our Advisory Committees.

The activation of the 1976 Code had introduced some novel problems. We decided to include not only those genera that had been published in the Approved Lists of Bacterial Names in January 1980 or that had been subsequently validly published, but also certain genera whose names had no current standing in nomenclature. We also decided to include descriptions of certain organisms which had no formal taxonomic nomenclature, such as the endosymbionts of insects. Our goal was to omit no important group of cultivated bacteria and also to stimulate taxonomic research on "neglected" groups and on some groups of undoubted bacteria that have not yet been cultivated and subjected to conventional studies.

The invited authors were provided with instructions and exemplary chapters in June 1980 and, although the intended deadline for receipt of manuscripts was March 1981, all contributions were assembled in January 1982 for the final preparations. The *Manual* was forwarded to the publisher in June 1982.

Some readers will note the consistent use of the stem -var instead of -type in words such as biovar, serovar and pathovar. This is in keeping with the recommendations of the Bacteriological Code and was done against the wishes of some of the authors.

We have deleted much of the synonymy of scientific names which was contained in past editions. The adoption of the new starting date of January 1, 1980 and publication of the Approved Lists of Bacterial Names has made mention of past synonymy obsolete. We have included synonyms of a name only if they have been published since the new starting date, or if they were also on the Approved Lists and, in rare cases with certain pathogens, if the mention of an old name would help readers associate the organism with a clinical problem. If the reader is interested in tracing the history of a name we suggest he or she consult past editions of the *Manual* or the *Index Bergeyana* and its *Supplement*. In citations of names we have used the abbreviation AL to denote the inclusion of the name on the Approved Lists of Bacterial Names and VP to show the name has been validly published.

In the matter of citation of the *Manual* in the scientific literature we again stress the fact that the *Manual* is a collection of authored chapters and the citation should refer to the author, the chapter title and its inclusive pages, not the Editor.

To all contributors, the sincere thanks of the Trust is due; the Editor is especially grateful for the good grace with which the authors accepted comments, criticisms and editing of their manuscripts. It is only because of the voluntary and dedicated efforts of these authors that the *Manual* can continue to serve the science of bacteriology on an international basis.

A number of institutions and individuals deserve special acknowledgment from the Trust for their help in bringing about the publication of this volume. . .

Preface to the First Edition of *Bergey's Manual*®
of Determinative Bacteriology

The elaborate system of classification of the bacteria into families, tribes and genera by a Committee on Characterization and Classification of the Society of American Bacteriologists (1911, 1920) has made it very desirable to be able to place in the hands of students a more detailed key for the identification of species than any that is available at present. The valuable book on "Determinative Bacteriology" by Professor F. D. Chester, published in 1901, is now of very little assistance to the student, and all previous classifications are of still less value, especially as earlier systems of classification were based entirely on morphologic characters.

It is hoped that this manual will serve to stimulate efforts to perfect the classification of bacteria, especially by emphasizing the valuable features as well as the weaker points in the new system which the Committee of the Society of American Bacteriologists has promulgated. The Committee does not regard the classification of species offered here as in any sense final, but merely a progress report leading to more satisfactory classification in the future.

The Committee desires to express its appreciation and thanks to those members of the society who gave valuable aid in the compilation of material and the classification of certain species. . . .

The assistance of all bacteriologists is earnestly solicited in the correction of possible errors in the text; in the collection of descriptions of all bacteria that may have been omitted from the text; in supplying more detailed descriptions of such organisms as are described incompletely; and in furnishing complete descriptions of new organisms that may be discovered, or in directing the attention of the Committee to publications of such newly described bacteria.

David H. Bergey, *Chairman*
Francis C. Harrison
Robert S. Breed
Bernard W. Hammer
Frank M. Huntoon
Committee on Manual.
August, 1923.

Contents

DOMAIN *BACTERIA*

Contributors

Milton J. Allison
USDA, Agricultural Research Service—Midwest Area, National Animal Disease Center, Ames, IA 50010-0070, USA

Rudolf Amann
Nachwuchsgruppe Molekulare Ökologie, Max Planck Institute für Marine Mikrobiologie, Celsiusstrasse 1, D28359 Bremen, Germany

Chad C. Baker
Oregon Graduate Institute, P.O. Box 91000, Portland, OR 97291-1000, USA

John R. Battista
Department of Microbiology, Louisiana State University, Baton Rouge, LA 70803-0001, USA

Eberhard Bock
Inst. für Allgemeine Botanik und Botanischer Garten, Universität Hamburg, Ohnhorststrasse 18, Hamburg D-22609, Germany

David R. Boone
Department of Environmental Biology, Portland State University, Portland, OR 97207-0751, USA

Don J. Brenner
Meningitis & Special Pathogens Branch Laboratory Section, Centers for Disease Control & Prevention, Atlanta, GA 30333, USA

Frank Caccavo, Jr.
Department of Microbiology, University of New Hampshire, Rudman Hall/Spaulding, 46 College Road, Durham, NH 03824, USA

Richard W. Castenholz
Department of Biology, University of Oregon, Eugene, OR 97403-1210, USA

Song C. Chong
Department of Environmental Biology, Portland State University, Portland, OR 97207-0751, USA

Milton S. da Costa
Centro de Neurociências, Departamento de Zoologia, Universidade de Coimbra, Apartado 3126, P-3004-517, Coimbra, Portugal

Mary Ellen Davey
Microbiology Department, Dartmouth Medical School, Room 202, Vail Bldg., North College Street, Hanover, NH 03755, USA

Paul De Vos
Department of Biochemistry, Physiology & Microbiology (WE 10V), K.L. Ledeganckstraat 35, B-9000, Gent, Belgium

Wolfgang Eder
Lehrstuhl für Mikrobiologie, Universität Regensburg, Universitatsstrasse 31, Regensburg 93053, Germany

James G. Ferry
Department of Biochemistry & Molecular Biology, The Pennsylvania State University, University Park, PA 16802-4500, USA

Jean-Louis Garcia
Laboratoire de Microbiologie, Université de Provence, OR-STROM-ESIL-Case 925, 163, Avenue de Luminy, 13288 Marseille Cédex 9, France

George M. Garrity
Department of Microbiology and Molecular Genetics, Michigan State University, East Lansing, MI 48824-1101, USA

Jane Gibson
Section of Biochemistry, Molecular & Cell Biology, Division of Biological Sciences, Cornell University, Ithaca, NY 14853-0001, USA

Monique Gillis
Laboratorium voor Microbiologie en Microbiele Genetica (WE 10V), Rijksuniversiteit Gent, K.-L. Ledeganckstraat 35, B-9000 Gent, Belgium

Vladimir M. Gorlenko
Institute of Microbiology, Russian Academy of Sciences, Prospect 60-letiya, Octyabrya 7 k.2, Moscow 117811, Russia

William D. Grant
Microbiology & Immunology, Leicester University, University Road, Leicester LE1 9HN, England

Anthony C. Greene
School of Biomolecular and Biomedical Science, Academic 1 Building, Logan Campus, Griffith University, Meadowbrook, Queensland 4131, Australia

Doris Hafenbradl
10665 Sorrento Valley Road, San Diego, CA 92121, USA

E. Claude Hatchikian
IBSM-CNRS, Unité de Bioenergetique et Ingenierie des Proteines, 31, Chemin Joseph-Aiguier, 13402 Marseille Cedex 20, France

Michael Herdman
Physiologie Microbienne, Dept. B.M.G., CNRS-URA, 2172, Institut Pasteur, 28 Rue du Docteur Roux, F-75724 Paris Cedex 15, France

Lucien Hoffmann
Laboratoire d'Algologie, de Mycologie et de Systématique Expérimentale, Institut de Botanique, B. 22, Université de Liège, Sart Tilman, B-4000, Liège, Belgium

John G. Holt
Department of Microbiology and Molecular Genetics, Michigan State University, East Lansing, MI 48824-1101, USA

Gertrud Huber
Lehrstuhl für Mikrobiologie, Universität Regensburg, Universitätsstrasse 31, Regensburg, Germany

Harald Huber
Lehrstuhl für Mikrobiologie, Universität Regensburg, Universitätsstrasse 31, Regensburg, Germany

Robert Huber
Lehrstuhl für Mikrobiologie, Universität Regensburg, Universitätsstrasse 31, Regensburg, Germany

Yasuo Igarashi
Department of Biotechnology, University of Tokyo, 1-1-1 Yayoi, Bunkyo-ku, Tokyo 113-8657, Japan

Johannes F. Imhoff
Institut für Meereskunde, Universität Kiel, Abt. Marine Mikrobiologie, Düsternbrooker Weg 20, D-24105 Kiel, Germany

Masaharu Ishii
Department of Biotechnology, The University of Tokyo, 1-1-1 Yayoi, Bunkyo-ku, Tokyo 113-8657, Japan

Isabelle Iteman
Physiologie Microbienne, Dept. B.M.G., CNRS-URA 2172, Institut Pasteur, 28 Rue du Docteur Roux, F-75724 Paris Cedex 15, France

Takashi Itoh
Japan Collection of Microorganisms, The Institute of Physical and Chemical Research, Riken, Hirosawa, Wako-shi, Saitama 351-0198, Japan

Christian Jeanthon
Université de Bretagne Occidentale, Institut Universitaire Européen de la Mer, 29680 Plousané, France

D. Barrie Johnson
School of Biological Sciences, University of Wales, Bangor LL57 2UW, United Kingdom

Masahiro Kamekura
Noda Institute for Scientific Research, 399 Noda, Noda-shi, Chiba-ken 278, Japan

Toshiyuki Kawasumi
Department of Food and Nutrition, Faculty of Home Economics, Women's University, 2-8-1, Mejirodai, Bunkyo-ku, Tokyo, 112-8681, Japan

Olga I. Keppen
Department of Microbiology, Moscow State University, 119899 Moscow, Russia

Karel Kersters
Lab. voor Microbiologie, Rijksuniversiteit Gent, Vakgroep Biochemie, Fysiologie en Microbiologie, K.L. Ledeganckstraat 35, B-9000, Gent, Belgium

Jyoti Keswani
3157 Sylvan Circle, Morgantown, WV 26505, USA

Hans-Peter Klenk
VP Genomics, Epidauros Biotechnology Inc., Am Neuland 1, D-82347 Bernried, Germany

Tetsuo Kobayashi
Department of Applied Biological Sciences, Nagoya University, Lab. for Gene Regulation, School of Agricultural Sciences, Chikusa-ku, Nagoya-shi, Aichi 464-01, Japan

Yosuke Koga
Department of Chemistry, University of Occupational & Environmental Health, Fukuoka 807, Japan

Torsten Krafft
Am Grenzgraben 13, D-63067 Offenbach, Germany

Noel R. Krieg
Department of Biology, Virginia Polytechnic Institute & State University, Blacksburg, VA 24061-0406, USA

David P. Labeda
USDA, National Center for Agricultural Utilization Research, Microbial Properties Research, Peoria, IL 61604-3999, USA

Thomas A. Langworthy
Department of Microbiology, University of South Dakota School of Medicine, Vermillion, SD 57069-2390, USA

Stéphane L'Haridon
Université de Bretagne Occidentale, Institut Universitaire Européen de la Mer, 29680 Plousané, France

Wolfgang Ludwig
Lehrstuhl für Mikrobiologie, Technische Universität München, Am Hochanger 4, Freising, D-85350, Germany

Barbara J. MacGregor
Civil Engineering Department, Northwestern University, Evanston, IL 60208, USA

Joan M. Macy
Department of Microbiology, LaTrobe University, Bundoora Victoria 3083, Australia

Michael T. Madigan
Department of Microbiology, Southern Illinois University, Mail Stop 6508, Carbondale, IL 62901-4399, USA

Robert A. Mah
Division of Environmental Health Science, UCLA School of Public Health, Los Angeles, CA 90024-1772, USA

James S. Maki
Department of Biology, WEHR Life Science Building, Marquette University, Milwaukee, WI 53201-1881, USA

Terry J. McGenity
Department of Biological Sciences, University of Essex, Main Campus, Wivenhoe Park, Colchester, Essex CO4 3SQ, United Kingdom

Roy D. Meredith
Ringoes Wertsville R, Hopewell, NJ 08525, USA

Terry L. Miller
Wadsworth Centre for Lab. & Research, New York State Department of Health, Albany, NY 12201-0509, USA

Rafael Montalvo-Rodriguez
University of Nebraska, Lincoln, NB, USA

R.G.E. Murray
Department of Microbiology & Immunology, The University of Western Ontario, London, Ontario N6A 5C1, Canada

Takashi Nakase
Japan Collection of Microorganisms, The Institute of Physical and Chemical Research, Riken, Wako-shi, Saitama 351-0198, Japan

M. Fernanda Nobre
Departmento de Zoologia, Universidade de Coimbra, Apartado 3126, P-3000 Coimbra, Portugal

Norimichi Nomura
Laboratory of Marine Microbiology, Division of Applied Bioscience, Graduate School of Agriculture, Kyoto University, Kyoto 606-8502, Japan

Bernard Ollivier
Laboratoire de Microbiologie des Anaérobies, Université de Provence, CESB-ESIL ORSTOM, Case 921, 163 Avenue de Liminy, Marseille 13288 Cedex 9, France

Aharon Oren
Division of Microbial and Molecular Ecology, The Institute of Life Science, and the Moshe Shilo Minerva Center for Marine Biogeochemistry, The Hebrew University of Jerusalem, Givat Ram, Jerusalem 91904, Israel

Jörg Overmann
Institute für Chemie und Biologie des Meeres (ICBM), Universität Oldenburg, Carl-von-Ossietzky-Strasse 9-11, Postfach 25 03, D-26111 Oldenburg, Germany

Bharat K.C. Patel
School of Biomolecular & Biomedical Sciences, Faculty of Science & Technology, Griffith University, Nathan Campus, Brisbane, Queensland 4111, Australia

Girishchandra B. Patel
National Research Council of Canada, Institute for Biological Sciences, Ottawa, Ontario K1A OR6, Canada

Jerome J. Perry
3125 Eton Road, Raleigh, NC 27608-1113, USA

Norbert Pfennig
Primelweg 12, D-88662 Überlingen, Germany

Beverly K. Pierson
Department of Biological Sciences, University of Puget Sound, 1500 N. Warner, Jones Hall #007, Tacoma, WA 98416, USA

Fred A. Rainey
Department of Microbiology, Louisiana State University, Baton Rouge, LA 70803, USA

Anna-Louise Reysenbach
Department of Environmental Biology, Portland State University, Portland, OR 97207-0751, USA

Rosmarie Rippka
Physiologie Microbienne, Dept. B.M.G., CNRS-URA 2172, Institut Pasteur, 28 Rue du Docteur Roux, F-75724 Paris Cedex 15, France

James A. Romesser
Betz Dearborn, Inc. P.O. Box 4300, The Woodlands, TX 77380, USA

Yoshihiko Sako
Department of Applied Bioscience, Graduate School of Agriculture, Laboratory of Marine Microbiology, Kyoto University, Kyoto 606-8502, Japan

Priscilla C. Sanchez
Museum of Natural History, University of the Philippines, Los Baños, College, Laguna 4031, Philippines

Abigail A. Salyers
Department of Microbiology, University of Illinois-Urbana, Champaign, Urbana, IL 61801-3704, USA

Karl-Heinz Schleifer
Lehrstuhl für Mikrobiologie, Universität München, Am Hochanger 4, D-85350 Freising, Germany

Lindsay I. Sly
Centre for Bacterial Diversity and Identification, Department of Microbiology, University of Queensland, St. Lucia, Brisbane, Queensland 4072, Australia

Peter H.A. Sneath
Department of Microbiology & Immunology, School of Medicine, University of Leicester, P.O. Box 138, Leicester LE1 9HN, England

Kevin R. Sowers
Center for Marine Biotechnology, Maryland Biotechnology Institute, Baltimore, MD 21202, USA

Eva Spieck
Inst. für Allgemeine Botanik und Botanischer Garten, Universität Hamburg, Ohnhorststrasse 18, Hamburg D-22609, Germany

Stefan Spring
Deutsche Sammlung von Mikroorganismen und Zellkulturen, GmbH, Mascheroder Weg 1b, D-38124 Braunschweig, Germany

David A. Stahl
Department of Civil Engineering/Technology Institute, Northwestern University, Evanston, IL 60208-3109, USA

James T. Staley
Department of Microbiology, University of Washington, Seattle, WA 98195-0001, USA

Karl O. Stetter
Lehrstuhl für Mikrobiologie, Universität Regensburg, Universitätsstrasse 31, D-93053 Regensburg, Germany

Ken-ichiro Suzuki
Japan Collection of Microorganisms, The Institute of Physical and Chemical Research, Riken, Hirosawa, Wako-shi, Saitama 351-0198, Japan

Jean Swings
Laboratorium voor Microbiologie, Universiteit of Gent, Vakgroep WE 10V, Fysiologie en Microbiologie, K.L. Ledeganckstraat 35, B-9000, Gent, Belgium

Xinyu Tian
Institute of Microbiology, Academia Sinica, Beijing 100080, China

Brian J. Tindall
Deutsche Sammlung von Mikroorgenismen und Zellkulteren, GmbH, Mascheroder Weg 1b, D-38124 Braunschweig, Germany

Hans G. Trüper
Institute für Mikrobiologie und Biotechnologie, Rheinsche Friedrich-Wilhelms-Universität, Mechenheimer Allee 168, W-53115 Bonn, Germany

Peter Vandamme
Lab. voor Microbiologieen Microbiele Genetica, Universiteit of Gent, Faculteit Wetenschappen, K.L. Ledeganckstraat 35, B-9000 Gent, Belgium

Antonio Ventosa
Departamento de Microbiologia, Universidad de Sevilla y Parasitologia, Facultad de Farmacia, Apdo. 874, 41080 Sevilla, Spain

Russell H. Vreeland
Department of Biology, West Chester University, West Chester, PA 19383, USA

John B. Waterbury
Woods Hole Oceanographic Institute, Woods Hole, MA 02543, USA

William B. Whitman
Department of Microbiology, University of Georgia, Athens, GA 30602-2605, USA

Annick Wilmotte
Labo d'Algologique, Mycologie et Systematique Experimentale, Department de Botanique, B-22, Université de Liège, B-4000 Liège, Belgium

Yi Xu
Institute of Microbiology, Academia Sinica, Beijing 100080, China

Gerhard Zellner
Institute of Hydrology, GSF-National Research Center for Environment & Climate, Home address: Fuchsbergstrasse 7, D-85386 Eching, Germany

Tatjana N. Zhilina
Institute of Microbiology, Russian Academy of Sciences, Prospect 60-letja Oktyabrya 7a, Moscow 117312, Russia

Peijin Zhou
Institute of Microbiology, Academia Sinica, Beijing 100080, China

Wolfram Zillig
Max-Planck-Institut für Biochemie, Am Klopferspitz 18a, D-82152 Martinsried, Germany

Stephen H. Zinder
Department of Microbiology, Cornell University, Ithaca, NY 14853-0001, USA

The History of *Bergey's Manual*

R.G.E. Murray and John G. Holt

INTRODUCTION

Bergey's Manual of Determinative Bacteriology has been the major provider of an outline of bacterial systematics since it was initiated in 1923 and has provided a resource ever since to workers at the bench who need to identify bacterial isolates and recognize new species. It originated in the Society of American Bacteriologists (SAB) but it has since become a truly international enterprise directed by an independent Trust which was founded in 1936. It has gone through nine editions and has generated, as a more comprehensive resource, a unique compendium on bacterial systematics, *Bergey's Manual of Systematic Bacteriology* (Holt et al., 1984–1989), which now enters its second edition.

A number of dedicated bacteriologists (Table 1) have formed, guided the development of, and edited, each edition of *Bergey's Manual.* Many of these individuals have been well known for activity in their national societies and devotion to encouraging worldwide cooperation in bacteriology and particularly bacterial taxonomy. Some of them worked tirelessly on the international stage towards an effective consensus in taxonomy and common approaches to classification. This led to the formation in 1930 of an International Association of Microbiological Societies (IAMS) holding regular Congresses. The regulation of bacterial taxonomy became possible within IAMS through an International Committee on Systematic Bacteriology (ICSB), thus recognizing the need for international discussions of the problems involved in bacterial systematics. Eventually, the need for a Code of Nomenclature of Bacteria was recognized and was published in 1948 (Buchanan et al., 1948), and a Judicial Commission (JC) was formed by ICSB to adjudicate conflicts with the Rules. Despite these efforts, an enormous number of synonyms and illegitimate names had accumulated by the 1970s and were an evident and major problem for the Editor/Trustees of *Bergey's Manual* and for all bacteriologists (Buchanan et al., 1966; Gibbons et al., 1981). A mechanism for recognizing useful, and abandoning useless, names was accomplished by the ICSB and the JC largely due to the insistent arguments of V.B.D. Skerman. Lists were made based on the names included in the Eighth edition of *Bergey's Manual of Determinative Bacteriology* (Buchanan and Gibbons, 1974), because they had been selected by expert committees and individual author/experts, together with the recommendations of sub-committees of ICSB. The results were (1) the published Approved Lists of Bacterial Names (Skerman et al., 1980); (2) a new starting date for bacterial names of January 1, 1980 to replace those of May 1, 1753; (3) freeing of names not on the Approved Lists for use in the future; and (4) definition in the Bacteriological Code (1976 revision; Lapage et al., 1975)

of the valid and invalid publication of names. It is now evident that the care and thought of contributors to *Bergey's Manual* over the years played a major part in stimulating an orderly nomenclature for taxonomic purposes, in the development of a useful classification of bacteria often used as a basal reference, and in providing a continuing compendium of descriptions of known bacteria.

The *Manual* started as a somewhat idiosyncratic assembly of species and their descriptions following the interests and prejudices of the editor/authors of the early editions. Following the formation of the Bergey's Manual Trust in 1936 and the international discussions of the ICSB at Microbiological Congresses, the new editions became more and more the result of a consensus developed by advisory committees and specialist authors for each

TABLE 1. Members of the Board of Trustees

David H. Bergey	1923–1937
David R. Boone	1994–
Robert S. Breed	1923–1957 (Chairman 1937–1956)
Don J. Brenner	1979–
Marvin P. Bryant	1975–1986
R.E. Buchanan	1951–1973 (Chairman 1957–1973)
Richard W. Castenholz	1991–
Harold J. Conn	1948–1965
Samuel T. Cowan	late 1950s–1974
Geoffrey Edsall	late 1950s–1965
George M. Garrity	1997–
Norman E. Gibbons	1965–1976
Bernard W. Hammer	1923–1934
Francis C. Harrison	1923–1934
A. Parker Hitchens	1939–1950
John G. Holt	1973–
Frank M. Huntoon	1923–1934
Noel R. Krieg	1976–1991, 1996–
Stephen P. Lapage	1975–1978
Hans Lautrop	1974–1979
John Liston	1965–1976 (Chairman 1973–1976)
A.G. Lochhead	late 1950s–1960
James W. Moulder	1980–1989
E.G. D. Murray	1934–1964
R.G. E. Murray	1964–1990 (Chairman 1976–1990)
Charles F. Niven, Jr.	Late 1950s–1975
Norbert Pfennig	1978–1991
Arnold W. Ravin	1962–1980
Karl-Heinz Schleifer	1989–
Nathan R. Smith	1950–1964
Peter H.A. Sneath	1978–1994 (Chairman 1990–1994)
James T. Staley	1976–
Roger Y. Stanier	1965–1975
Joseph G. Tully	1991–1996
Jan Ursing	1991–1997
Stanley T. Williams	1989– (Chairman 1994–)

part or chapter of the volumes. This did not happen all at once; it developed out of practice and trials, and it is still developing as the basic sciences affecting taxonomy bring in new knowledge and new understanding of taxa and their relationships.

ANTECEDENTS OF *BERGEY'S MANUAL*

Classification of named species of bacteria did not arise quickly or easily (Buchanan, 1948). The Linnaean approach to naming life forms was adopted in the earliest of systems, such as Müller's use of *Vibrio* and *Monas* (Müller, 1773, 1786), for genera of what we would now consider bacteria. There were few observations, and there was insufficient discrimination in the characters available during most of the nineteenth century to allow any system, even the influential attempts by Ehrenberg (1838) and Cohn (1872, 1875), to provide more than a few names that still survive (e.g. *Spirillum*, *Spirochaeta*, and *Bacillus*). Most descriptions could rest only on shape, behavior, and habitat since microscopy was the major tool.

Müller's work was the beginning of the descriptive phase of bacteriology, which is still going on today because we now realize that the majority of bacteria in nature have not been grown or characterized. Early observations such as Müller's were made by cryptogamic botanists studying natural habitats, usually aquatic, and who usually gave Linnaean binomials to the objects they described microscopically. The mycologist H.F. Link (1809) described the first bacterium that we still recognize today, which he named *Polyangium vitellinum* and is now placed with the fruiting myxobacteria. Bizio (1823) attempted to explain the occurrence of red pigment formation on starchy foods such as polenta as the result of microbial growth and named the organism he found there *Serratia marcescens*, a name now associated with the prodigiosin-producing Gram-negative rod. Perhaps one of the most significant observers of infusoria in the early nineteenth century was C.G. Ehrenberg, who described many genera of algae and protozoa and, coincidentally, some bacteria (Ehrenberg, 1838). He named genera such as *Spirochaeta* and *Spirillum*, still recognized today, and *Bacterium*, which became a catch-all for rod-shaped cells, and was made *nomen rejiciendum* in 1947.

Logical classifications were attempted throughout the nineteenth century and that of Ferdinand Cohn (1872, 1875), with his attempts to classify the known bacteria, was most influential. In his 1872 paper Cohn recognized six genera of bacteria (*Micrococcus*, *Bacterium*, *Bacillus*, *Vibrio*, *Spirillum*, and *Spirochaeta*) and later (1875) expanded the classification to include the cyanobacteria while adding more bacterial genera (*Sarcina*, *Ascococcus*, *Leptothrix*, *Beggiatoa*, *Cladothrix*, *Crenothrix*, *Streptococcus* [not those recognized today], and *Streptothrix*). Buchanan (1925) suggested that Cohn's 1875 classification could be the starting date for bacterial nomenclature instead of Linnaeus' *Species Plantarum* of 1753 and discussed various ideas for the proper starting date for bacterial nomenclature, anticipating by a quarter of a century the actual change in starting date proposed in the revised Bacteriological Code (Lapage et al., 1975). The realization that cultivation was possible, and the development of pure culture techniques, extended enormously the capability to recognize and describe species by adding their growth characteristics and effects on growth media. The vague possibilities of pleomorphism gave way to a concept of fixity of species. All this was aided by the human preoccupation with health, the seriousness of infectious diseases, and the growing awareness of the association of particular kinds of bacteria with particular diseases. The result was a rapid increase in the number of taxonomic descriptions and the

recognition that similar but not identical species of bacteria were to be found both associated with higher life forms and more generally distributed in nature.

Between 1885 and 1910 there were repeated attempts at classification and arrangements based on perceived similarities, mostly morphological. There were genuine attempts to bring order out of chaos, and a preliminary publication often stimulated subsequent and repeated additions and revisions, but all these authors neglected the determinative requirements of bacteriology. Some notable examples were Zopf (1885), Flügge (1886), Schroeter (1886), and Trevisan (1887, 1889). Migula produced his first outline in 1890 and new versions in 1894, 1895, 1897, and 1900; others followed, notably Fischer (1895), and importantly, because of a degree of nomenclatural regularity, Lehmann and Neumann published their atlas in 1896. The latter was probably the most successful of the systems and was used in successive editions until 1930, especially in Europe. All these were important in their time. However, a major influence in the subsequent development of *Bergey's Manual* in the environment of the Society of American Bacteriologists (SAB) was the work of F.D. Chester, who produced reports in 1897 and 1898 of bacteria of interest in agriculture, to be followed in 1901 by his *Manual of Determinative Bacteriology*. Chester had recognized that the lack of an organized assembly of descriptions and a scheme of classification made the identification of isolates as known species and the recognition of new species an insurmountable task. Another classification provided by Orla-Jensen (1909, 1919) was influential because it represented an interpretation of "natural relationships", reflecting a more physiological approach to description based on his own studies of the lactic acid bacteria encountered in dairy bacteriology. He delimited genera and species on the basis of characteristics such as metabolic byproducts, fermentation of various sugars, and temperature ranges for growth, in addition to morphology. Most classifications to that time reflected the idiosyncrasies of the authors and their areas of experience. What was yet to come was the ordering of assemblies of all known bacteria, arranged with properties documented to facilitate determination and presenting continuing trials of hierarchical arrangements; it was in that format that *Bergey's Manual* started.

STEPS LEADING TO THE FIRST EDITION OF THE *MANUAL*

Bergey's Manual of Determinative Bacteriology arose from the interest and efforts of a group of colleagues in the Society of American Bacteriologists, who were fully aware of previous attempts to systematize the information available on bacterial species and who recognized that the determination of bacterial identity was difficult and required extensive experience. A committee was formed with C.-E.A. Winslow as chairman and J. Broadhurst, R.E. Buchanan, C. Krumweide Jr., L.A. Rogers, and G.H. Smith as members. Their discussions at the meetings of the SAB and their reports, which were published in the Journal of Bacteriology (Winslow et al., 1917, 1920), were signposts for future efforts in systematics. There were two "starters" for a *Manual*: R.E. Buchanan (Fig. 1a), a rising star in the bacteriological firmament, and President of the SAB in 1918, working at Iowa State College, and D.H. Bergey (Fig. 1b), a senior and respected bacteriologist and President of the SAB for 1915, working at the University of Pennsylvania.

Between 1916 and 1918 Buchanan wrote ten papers entitled "Studies on the nomenclature and classification of the bacteria" (Buchanan, 1916; 1917a, b, c; 1918a, b, c, d, e, f) which provided

FIGURE 1. *A*, Robert Earle Buchanan, 1883–1973; *B*, David Henricks Bergey, 1860–1937; *C*, Robert Stanley Breed, 1877–1956; *D*, Everitt G.D. Murray, 1890–1964. (Fig. 1C courtesy of American Society for Microbiology Archives Collection.)

substance for the Winslow Committee (Buchanan was a member), and was intended to be the basis of a systematic treatise. These papers were revolutionary, in the sense that they included all the bacteria (except the cyanobacteria) that were described at that time. Buchanan included, and named the higher groupings of, bacteria such as the actinomycetes, myxobacteria, phototrophs, and chemolithotrophs, along with the other bacteria included in the classifications of the day. This classification had a logical and aesthetic appeal that helped launch the systematic efforts that followed. No doubt Buchanan was driven by dissatisfaction with sloppy and confusing nomenclature as well as inadequate descriptions of "accepted" bacteria (indeed, much of his later work on *Bergey's Manual* and the *Index Bergeyana* reflected his preoccupation with names and illegitimacy, and had much to do with getting a bacteriological code of nomenclature started.) He must have known of Bergey's book and, perhaps because of increasing academic responsibilities, publication of his concepts in his *General Systematic Bacteriology* was delayed until 1925 (Buchanan, 1925). The book did not try to duplicate *Bergey's Manual*, but rather presented a history of bacterial classification and nomenclature, followed by a discussion of the history of all the bacterial genera and higher ranks, listed alphabetically.

R.E. Buchanan was the key player in the renewal of concern for a sensible (not necessarily "natural") classification of bacteria, with a well-regulated nomenclature, working continuously and firmly to those ends from 1916 to the end of his life. He was a man of his times developing his own priorities and prejudices, yet he recognized in the end that new science was needed for a significant phylogeny to develop. Furthermore, he was more influential in gaining support for the initiation and progress of the first few editions of *Bergey's Manual* under the slightly reluctant aegis of the SAB than is obvious in the *Manual's* pages and prefaces. He also played a dominant role in international efforts (representing the SAB) concerning the regulation and codification of classification and nomenclature. As a member of the "Winslow Committee" of the SAB directed to report on the classification of bacteria, he furnished much of the basis for discussion through his series of papers in the Journal of Bacteriology. He provided voluminous detailed suggestions for the revision of Dr. Winslow's drafts for their reports to the SAB (1917 and 1920). He was also in a powerful position to influence decisions, being elected President of SAB for 1918–1919 when critical discussions were taking place.

The Winslow Committee was engaged in protecting ("conserving") the generic names for well-established species by listing them as *genera conservanda*, together with type species for discussion at the 1918 SAB meeting. The intention was to provide a basis for recommendations for formal action at the next International Botanical Congress, since they were working under the general rules of the Botanical Code. They went further by classifying the genera within higher taxa and providing a key to assist recognition. They intended seeking formal approval of the whole report by the SAB. At this stage, R.S. Breed (Fig. 1c) wrote many letters of objection to having any society ratify the concepts involved in contriving a classification, because it would suggest that it is "official", and he attempted unsuccessfully to gain a postponement of the report's presentation. This polemical correspondence with Committee members, including Buchanan, ended in Breed's withdrawing his name from the report despite his evident interest in a workable classification and a more stable nomenclature. Winslow read the report to the SAB meeting on December 29, 1919. Although it emphasized that its listings were not to be considered as a standard or official classification, it did ask "that the names be accepted as definite and approved genera". The report was then published in the *Journal of Bacteriology*. The Committee was discharged and a new Committee on Taxonomy was appointed with R.E. Buchanan as Chairman. In 1920 Breed was added as a member of the new committee, with the responsibility of making the representations at the Botanical Congress because of his membership on the Botanical Code Revision Committee.

It was at this time and in this climate of opinion that Dr. David Bergey decided to put his own studies of bacteria together with the current views on their classification. To do this required more than one person and he assembled a like-minded group to form a Committee of the SAB for the production of a *Manual of Determinative Bacteriology* (F.C. Harrison, R.S. Breed, B.W. Hammer, and F.M. Huntoon). There is no direct evidence that Buchanan was ever asked to participate or, equally, that he raised any formal objections; it seems more likely that there could have been none of the formal encouragement to go ahead evident in 1921 and 1922 without his support. Indeed he seems to have thought it a good enterprise (Preface in his 1925 book). However, he did find it difficult to work with Breed (letter of January 8, 1951 to J.R. Porter) and in expressing this stated "I have ... always refused to become a member of the Editorial Board of the *Manual*". One wonders if his experiences with Breed between 1918 and 1951 ("Scarcely a month passes in which we do not have some disagreement ... but he has a good many excellent qualities") had kept him at arm's length but not out of touch with what was going on with the *Manual*.

The Winslow Committee had put before the SAB the possibility of a major compilation on bacterial systematics. No doubt Buchanan was in a position, as a Past President, to reinforce the value of that project in principle and David Bergey, likewise a Past President, must have been aware of all the discussions. At the time of the last report (Winslow et al., 1920) Bergey must have started on his book, because R.S. Breed reported to the 1922 SAB Council meeting that the work was approaching completion. A more formal proposal was made to the same Council meeting that Bergey's book be published under the aegis of the Society. The SAB agreed to this with the proviso that it go to a substantial publishing house and, following a discussion of the disposition of royalties, *Bergey's Manual of Determinative Bacteriology* was published in 1923 by the Williams & Wilkins Co., Baltimore (Bergey et al., 1923). It was a group effort from the start, with the authors listed as D.H. Bergey, F.C. Harrison, R.S. Breed, B.W. Hammer, and F.M. Huntoon, and there was an acknowledgment of the assistance of six other colleagues on special groups.

One can imagine that Buchanan was upset by this turn of events, for which the only evidence is his sending Bergey a long list of errors he found in the published book (personal communication). However, he was quite generous in his preface to his 1925 book, with his assessment of *Bergey's Manual* as a step towards reducing chaos and confusion in the classification, phylogeny, and naming of bacteria. He writes: "The most hopeful sign of importance in this respect probably has been the work of the committee on taxonomy of bacteria of the Society of American Bacteriologists under the chairmanship of Dr. Winslow and of the more recent work of a committee on classification of bacteria under the chairmanship of Dr. Bergey.... It is to be expected that, as a result of their work, eventually a practical system of nomenclature which will be satisfactory and applicable to all fields of bacteriology will be evolved" (Buchanan, 1925). Furthermore, he emphasized the differences between practical (medical) and academic attitudes towards individual species and the requirements of a classification. He was then, as later, concerned that bacterial nomenclature was not regulated by an appropriate Code. He writes: "It seems to be self-evident that until the bacteriologists can agree upon a code and follow it consistently, there is little hope or remedy for our present chaos". So it is not surprising that he contributed a section to the Fourth Edition (Bergey et al., 1934) discussing the International Botanical Code as a basis for a bacteriological code with modifications to make it more appropriate.

The committee that organized the First Edition stated that they did not regard their classification of species "as in any case final, but merely a progress report leading to more satisfactory classifications in the future". Clearly there was some feeling in the UK and Europe that this classification was an imposition on the part of the SAB*. As a counter, the Third Edition (Bergey et al., 1930) included a box opposite the title page which declares that it is *"Published at the direction of the Society"* which "disclaims any responsibility for the system of classification followed"; and states further that it "has not been formally approved by the Society and is in no sense official or standard" (italics are in the original). This shows that there had been, as indicated by the article by I.C. Hall in 1927 (Hall, 1927), some degree of contention among members of the SAB with the decisions of the Committee.

Hall's objections to the presentations of the Committee of the SAB on characterization and classification of bacterial types starts with the final report (Winslow et al., 1920) being "presented only to a small minority of the members of the Society who happened to return from lunch in time to attend a business session of the twenty-first annual meeting, which was held in Boston more than four months before the publication of the report". He regrets lack of opportunity for scientific consideration and "practically no discussion because only a few knew what was coming". He evidently objected to physiological criteria and believed that morphology should define genera, families, and orders; furthermore he disputed the validity of habitat and believed that serological characterization was futile. He was prepared to use cultural and physiological properties as criteria for species. He sought "unambiguous criteria". He quotes others who disagreed with the *Bergey's Manual* approach including W.W.C. Topley, who also expressed his distaste in his famous textbook ("Topley and Wilson") that was published in 1929.

Bergey's Manual was launched and successful enough for the publisher to encourage further editions with corrections and additions in 1925 and 1930, for which Bergey had the support of the same four co-authors. There were problems ahead. By 1930 Bergey was aging and becoming somewhat frail so that he was concerned about the *Manual*'s governance and future. He turned to Breed to an increasing degree for the overall editing and as a major contributor, but also to fight for financial support and for a degree of independence. The agreement co-signed by Bergey and Breed with the Society in 1922 had recommended that royalties " . . . be accumulated in a separate fund to be used to stimulate further work in this field" and Bergey himself felt that he had "donated" this fund to the Society for that purpose.

---

*As can be gathered from skeptical sentiments in the famous textbook by W.W.C Topley and G.S. Wilson, *Principles of Bacteriology and Immunity*, 1st ed. (1929), Edward Arnold Ltd., London, and continued in large part to the Fifth Edition (1964) but not thereafter.

THE STRUGGLE FOR FINANCIAL AND EDITORIAL INDEPENDENCE

Breed's correspondence after 1930 with the powerful Secretary-Treasurers of the SAB (J.M. Sherman 1923–1934; I.L. Baldwin 1935–1942) seeking funds to assist the business of producing new editions became increasingly sharp and argumentative because this assistance was almost uniformly refused. The royalties were small and the publisher did not pay any until the costs were covered; the result was that the Society felt they were exposed to risk with a property that they considered not likely to go on much longer. Sherman, in particular, strongly objected to Breed's rhetoric and proprietary attitude, yet he reluctantly agreed in 1933 to cede $900 (half the accumulated royalties) for Fourth Edition purposes. The Society felt that the funds were theirs (the contract was between the Society and Williams & Wilkins) and there might be others deserving of support from the fund. A request for funds by A.T. Henrici in 1935 brought the whole matter of ownership back into contention and into Baldwin's more diplomatic hands. At the same time Breed was asking for $1000 (essentially the remainder of royalties plus interest) and decisions had to be made during a flurry of correspondence with a repetitive *non placet obligato* from Sherman. There was also a *Bergey's Manual* Committee (Winslow, Buchanan and Breed) reporting to the Council in support of a mechanism for funding the *Manual*. In the end, and agreeably to all parties for different reasons, it was decided between Sherman and Baldwin that the SAB should cede the rights to the *Manual*, the royalties to come, and the accumulated fund to Dr. Bergey to do with as he would wish, and the Council agreed (December 28, 1935). In large part it was a gesture of respect for Dr. Bergey because both of them stated in letters that they did not expect the *Manual* to go through more editions, in which respect they were mistaken.

In preparation for the Fourth Edition, and recognizing that Bergey was not well and that Harrison, Hammer, and Huntoon would not stay for long, Breed added E.G.D. Murray (Fig. 1d) to his corps of editors/authors, so that with Harrison still enlisted there were two Canadian members. With the Fourth edition published in 1934, from late 1935 until early 1936 was a time of negotiation. It is clear that Bergey, Breed, and Murray wanted an independent entity, while Buchanan with his own ideas was presenting a plan to Baldwin involving sponsorship by the Society, and Breed was trying unsuccessfully to make peace with Buchanan. Bergey, for his part, was (January, 1936) consulting with the SAB and advisors in preparation for developing a deed of trust for the future development of the *Manual*, and asking that there be no further controversy. His feeling about the whole sad tale was voiced on January 29, 1936: "The arrangement I have made will be without hindrance from a group of persons who appear to have no kindly feeling toward advances in bacteriology in which they could not dictate every step". The *Bergey's Manual* Trust was indentured on January 2, 1936 in Philadelphia, Pennsylvania, and the Trustees were Bergey, Breed, and Murray. The only concession to the SAB, that continues to the ASM today, is that one of the Trustees is chosen as a representative who reports annually to the Society on the state of the Trust and its work.

Mr. R.S. Gill, the representative of the Williams & Wilkins Co., informed Breed in December, 1934 that copies of the Fourth Edition were exhausted and sought agreement for a new edition; Breed prevaricated because the situation was not yet clear. However, by 1937 he was seeking contributions from a number of colleagues for a future volume. Sadly, D.H. Bergey died on September 5, 1937 at age 77, but the trustees retained his name on the masthead of the Fifth Edition published in 1939. Breed was now Chairman of the Trust and remarked in a letter to E.B. Fred and I.L. Baldwin (January 26, 1938) that Dr. Bergey, who was so interested in seeing the *Manual* revised, would have liked "... to know how well his plans are developing and how ... interested specialists are cooperating with us in making this new edition much better than anything we have had before". So a new way of producing the *Manual* with many contributors was now in place for elaboration in future editions. The first printing of 2000 copies of the Fifth Edition (Bergey et al., 1939) was sold out before the end of the year and 1000 more copies were printed. It was obvious that the *Manual* was needed and served a useful purpose, vindicating the optimism Bergey and Breed had maintained in the face of opposition. Breed, Murray, and A.P. Hitchens (who was appointed to the Board of Trustees in 1939) had to organize a Sixth Edition, which needed to be completely revised and required much to be added. There were 1335 species descriptions in the Fifth Edition and the Sixth, when accomplished, would have 1630. They were faced not only with the need to make changes in the outline classification but also to make decisions about the inclusion or exclusion of large numbers of dubious and inadequately described bacteria. Furthermore, the exigencies of World War II took some of the trustees and many of their contributors out of contention for the duration. Nevertheless, the Sixth Edition was published in 1948 (Breed et al., 1948a) and acknowledged the assistance of 60 contributors. Some of the incompletely described species appeared in appendices following the listings in genera and the book included an index of sources and habitats as an attempt to be helpful. A novelty, and an approach not to be fully realized until 35 years later in the *Systematic Manual*, was a section on the *Myxobacterales* containing a preliminary discussion of the nomenclature and biological characteristics of members of that Order. For this, credit is given to J.M. Beebe, R.E. Buchanan, and R.Y. Stanier; it seems likely to those who knew all of them that this approach originated with Stanier. Additions to the Sixth Edition were sections on the classification of *Rickettsiales* prepared by I.A. Bengston and on the *Virales* or Filterable Viruses prepared by F.O. Holmes. The former was appropriate but the latter pleased very few, certainly preceded an adequate understanding that would have allowed for a rational classification, and never appeared again.

The original Board of Trustees went through changes due to death and the enlargement of the Board. H.J. Conn, a colleague of Breed's at Cornell, was added in 1948 to join Breed, Murray, and Hitchens. The next year A.P. Hitchens died and was replaced by N.R. Smith, an expert on *Bacillus* species. R.E. Buchanan was added as a member in 1951 and began to take an active role in the affairs of the Trust. In 1952 Breed expressed a desire to step down as Editor-in-Chief, he was 75, and the Board debated about his successor. Among those considered were E.G.D. Murray, who was about to retire from McGill University, L.S. McClung of Indiana University, and C.S. Pederson of Cornell, but no decision was made. In correspondence to Breed, Smith wrote that "No doubt, Dr. Buchanan would like to take over when you step aside . . . In fact one can read between the lines that 'no one besides Buchanan is capable of editing the *Manual* ". This change, however, did not come to pass as Breed stayed on until his death in 1956.

Breed pursued actively the production of a Seventh Edition in the 1950s with the active support of Murray and Smith (Breed et al., 1957). The task was no less formidable, and there were many new authorities mounting increasingly pointed discussions about shortcomings in bacterial taxonomy in the dinner sessions that Breed arranged at the annual SAB meetings. It was to be the last edition in which the bacteria are classified as Schizomycetes within a Division of the Plantae, the Protophyta, primordial plants. In fact, the Preface tells us, the opening statement describing the Schizomycetes as "typically unicellular plants", was hotly debated without attaining a change, yet there were some concessions to cytology in the rest of that description, particularly concerning nucleoids. Ten Orders were recognized, adding to the five in the Sixth Edition, and these now included *Mycoplasmatales* and considerable division of the Order *Eubacteriales*. The keys to the various taxa were improved for utility and, recognizing the many difficulties involved in determination, an inclusive key to the genera described in the book was devised by V.B.D. Skerman and appended. This key, which was referred to as a comprehensive key, was designed to lead the user by alternative routes to a diagnosis of a genus when a character might be variable. It proved to be extremely popular and useful with readers and was repeated as an updated version in the Eighth Edition. Overall, the substance of the Seventh Edition of the *Manual* was due to the efforts of 94 contributors from 14 different countries. The *Manual* was becoming an international effort; however, Breed complained that the slowness of communication between the USA and Europe hampered their efforts.

Breed did not see the fruits of his labors as Editor-in-Chief; he died February 10, 1956, with many of the contributions arranged and the form of the book decided, but leaving a serious problem of succession. The position of Chairman of the Board of Trustees and Editor-in-Chief was decided, appropriately, and given to R.E. Buchanan whose interest in bacterial nomenclature and taxonomy, with direct and indirect involvement in the *Manual*, dated back to its origins. There was the immediate problem of finishing the editorial work on the Seventh Edition after Breed's death. E.F. Lessel Jr. had been working as a graduate student with Breed in Geneva, NY on the *Manual*, but was called into military service before the job was finished, and was stationed at a camp in Texas. Upon taking over the Chairmanship, Buchanan contacted W. Stanhope Bayne-Jones, of the Army's Office of the Surgeon General and Lessel's superior, to ask that Lessel be assigned to work on the completion of the *Manual* while in the service. Bayne-Jones agreed and assigned Lessel to the Walter Reed Hospital in Washington, DC. Thus the last editorial polishing of the book could take place without undue delays. After his service commitments were fulfilled Lessel went to Iowa State and finished his Ph.D. under Buchanan's direction and acted on occasion as recording secretary for Trust meetings.

R.E. Buchanan for many years had held three important administrative posts at Iowa State (Bacteriology Department Head since 1912; Dean of Graduate College since 1919; and Director of the Agricultural Experiment Station since 1936), retiring from all three in 1948. After 1948 some of his energies went to compiling and annotating the text for the 1952 publication of the Bacteriological Code and starting the *International Bulletin of Bacteriological Nomenclature and Taxonomy*. The *International Bulletin* received its initial monetary start in 1950 with a $150 gift from the *Bergey's Manual* Trust, to which Murray objected, saying "the *Journal* would be ephemeral." Fortunately he was wrong because the *Bulletin* later changed its name to the *International Journal of Systematic Bacteriology* and is still being published by ICSB (IAMS) with about 1200 pages in the 1997 volume. When Buchanan became Editor-in-Chief of the *Manual*, he induced the Department of Bacteriology at Iowa State to provide him an office suite and the title of Research Professor, from which position he obtained grants from the National Library of Medicine to support the office. This support continued until his death in 1973 at the age of 89 years.

Buchanan's twenty-year involvement in the Trust was to see, near its end, the start of a new era, despite his many objections to change. The chief change to come arose from a growing lack of confidence in the sanctity of higher taxa, there being few and often no objective tests of correctness. In the production of the Seventh Edition, it was recognized that an expanding synonymy and the ever-growing list of species that were unrecognizable or inadequately described provided a burden that made for wasted space and unreasonably extensive appendices. The addition of Breed's collection of reprints to Buchanan's considerable collection formed an extensive taxonomic archive in the Trust headquarters. With this resource in mind the Trust decided that a separate publication was needed to assemble as complete a listing as possible of the names and references of all the taxa included in the *Manual*, as well as "species formerly found as appendices or indefinitely placed as *species incertae sedis*" that might or should have appeared in the *Manual*. These, together with an assessment of whether or not each name was validly published and legitimate, formed a monster book of nearly 1500 pages, published as *Index Bergeyana* (Buchanan et al., 1966). Each and every reference was checked for accuracy, for Buchanan rightly stated that there "was a lot of gossip about the description of each name." These labors were a personal interest of R.E. Buchanan, who directed several years of effort by J.G. Holt (then at Iowa State), E.F. Lessel Jr., and a number of graduate students and clerks in the undertaking. The lists served as a finder mechanism, an alphabetical listing of the names of the bacteria, and of special use as a reference after the new starting date for nomenclature, January 1, 1980, mandated by the revised Code (Lapage et al., 1975). Addenda were inevitable and more names were collected as a *Supplement to Index Bergeyana* published in 1981 under the direction of N.E. Gibbons, K.B. Pattee and J.G. Holt. These substantial reference works assisted the refining of the content of the Seventh and Eighth Editions of the *Manual* and allowed concentration on effectively described and legitimate taxa.

There were seemingly interminable discussions about what needed to be done for an effective new edition. This was particularly true in the period 1957–1964 after Breed's death, when the Trust membership changed and new ideas and new scientific approaches to taxonomy became available. In the late 1950s the Board of Trustees was enlarged with the addition of S.T. Cowan, C.F. Niven Jr., G. Edsall, and A.G. Lochhead (the record is unclear on the exact date of their appointment). The election of Cowan from the UK added a European member and continued the internationalization of the Board (Fig. 2). Each of these new members brought expertise in different areas of bacteriology and that policy of diversity of interest among members has continued to this day. Later, in 1962, Arnold Ravin, a bacterial geneticist, was added to replace the retiring Lochhead. Of primary concern in the late 1950s and early 1960s was the position of Editor-in-Chief and location of Trust headquarters. An arrangement with Iowa State University to have a candidate assume a professorship at the University and house the headquarters there was made. The position was offered to P.H.A. Sneath, who had gained renown with his invention of numerical taxonomy and production of a masterful monograph on the genus *Chromobacterium*. By 1963,

however, Sneath chose to stay in England and Buchanan stayed on as Editor-in-Chief. All other efforts to find a new editor failed until Buchanan's death in 1973. As he grew older, more difficult, and more autocratic, progress on a new edition slowed considerably. Even the replacement of E.G.D. Murray, Conn, Smith, and Edsall by R.G.E. Murray, J. Liston, R.Y. Stanier, and N.E. Gibbons did not change the speed of Board actions. It became a war of wills between Buchanan and the others on what was important and where progress could be made. Until decisions on the taxa to be included and their circumscriptions were made, there was slow progress in naming and putting to work the 20 or more advisory committees needed to direct the authors of the final texts on genera and species. One novel (to the Trust) approach for obtaining consensus on taxonomic matters was the organization of a conference of advisory committee members and trustees held in May, 1968 at Brook Lodge in Augusta, MI, under the auspices of the Upjohn Co. and chaired by R.G.E. Murray. Fifteen advisory committee members joined in discussions with the Trust to assess the status of current knowledge on the major groups of bacteria to be included in the Eighth Edition. Despite this helpful preliminary, it brought no agreement between Buchanan as Chairman, whose main focus was then on nomenclature, and the rest of the Trustees, whose interests mostly focused on biological, functional, and eco-physiological attributes. It was clear that many of the higher taxa rested on shaky ground and were hard to assess on strict taxonomic terms. Accordingly, there was a long argument over abandoning formal names above family level wherever possible, agreeing that a large number of genera were of uncertain affiliation or, at least, could only be related on the basis of some diagnostic characters, such as gliding motility, shape and Gram reaction, and methane production, all of which might or might not have phylogenetic significance. All former ideas about phylogeny and relationships were discarded. The Eighth Edition was planned as a book divided into "Parts", each with a vernacular descriptor. The Advisory Committee for each part (some needed more than one) was assigned a member of the Trust who was responsible for action and who, eventually, had to see that each genus had an assigned author (131 in the end) who was willing to write.

Molecular/genetic technology was well established by 1974 when the Eighth Edition was published, but was not yet widely applied to play a role in broad decisions in taxonomy. The procaryotic nature of bacteria and all cells related to them (i.e. including the Cyanobacteria) could be recognized and used to define the Kingdom Procaryotae. Monera was the old and partially applicable higher taxon but the description was not cytologically based. The molecular composition of DNA was useful for separating phenotypically similar but genetically distinct groups (e.g., *Micrococcus* and *Staphylococcus*) and many descriptions could include mol% G + C as a character. Genetic and subsequent biochemical-molecular data told us that species were only relatively "fixed" in their expressed characters. This concept needed to be addressed in the circumscriptions and aids to identification. Greater use was made of diagnostic tables and wherever possible there were indications regarding uncertainties and the percentages of positive or negative reactions for tests. The value of the Eighth Edition for identification purposes was increased by the emphasis of both the Trustees and the authors on refining descriptions (in terms as up-to-date as possible), tables, keys, and illustrations. As in previous editions, many old names of dubious or unrecognizable entities were discarded and synonymy was reduced to essentials; the old information and its location was not

FIGURE 2. Photograph of Trustees meeting at Iowa State University, Ames, November, 1960. *L. to R.*, G. Edsall, R.E. Buchanan, C.F. Niven, Jr., N.R. Smith, and E.G.D. Murray.

lost because it was available in the *Index Bergeyana* (Buchanan et al., 1966), or later in the *Supplement to Index Bergeyana* (Gibbons et al., 1981).

The Eighth Edition was a long time in gestation—17 years—but its success (40,000 copies over the next 10 years, and more than half outside of North America) was a testament to its necessity and utility. Most of the primary journals involved in publishing microbiological papers suggested or required the *Manual* as the nomenclatural resource for bacterial names, all this despite the treatment of some groups (e.g., the *Enterobacteriaceae*) not being universally accepted. But it was truly an international enterprise, with authors from 15 countries who could, at last, be named in literature citations as authors.

The editing of the Eighth Edition became a major operation requiring sharing of responsibilities and some redirection of effort. This was in part due to the age and increasing infirmities of R.E. Buchanan who had been both Chairman of the Trust and Editor, directing his efforts to nomenclature, synonymy, and etymology. It became evident that a Co-Editor was required and fortunately N.E. Gibbons, recently retired from the National Research Council of Canada, agreed to undertake the task. Shortly thereafter Gibbons became the *de facto* editor, due to Buchanan's illness and death in January, 1973, and did all the general technical editing from his home in Ottawa with help from his wife, Alice Gibbons (who handled the Index of Names), and a number of Trustees, especially S.T. Cowan. The book was published in 1974 (Buchanan and Gibbons, 1974).

With publication of the Eighth Edition the Board of Trustees went through another major change of membership, and over a period of two years Niven, Ravin, Liston, Gibbons, and Stanier left the Board. At the first meeting after Buchanan's death, held in October, 1973, J.G. Holt, who had served as Secretary to the Board from 1963–1966 and co-edited the *Index Bergeyana*, was elected member and Secretary. In 1974–1975, H. Lautrop, S. Lapage, and M. Bryant were added, and later in 1976 N.R. Krieg and J.T. Staley joined the Board. In 1975 Holt was appointed Editor-in-Chief. With the publication and healthy sales of the Eighth Edition and increasing international profile, it was de-

cided to meet at locations separate from the ASM venue and to meet every other year outside North America, and the 1975 meeting was held in Copenhagen, Denmark, at the Statens Seruminstitut. From then on a segment of each meeting was devoted to consultation with taxonomically inclined colleagues in that area.

The Trust had recognized, in the process of deciding the format of the book, that students and technologists were important users, with primary interests in identification and a lesser need for the extensive descriptions of individual species. An abridged edition of the Sixth Edition of the *Manual* had been produced (Breed et al., 1948b), but was only a modest success and not carried forward to the Seventh Edition. In 1974 the need seemed to be greater, so preparations were made to assemble an outline classification; the descriptions of genera, families and such higher taxa as were recognized; all the keys and tables for the identification of species; the glossary; all the illustrations; and two informative introductory chapters. It was recognized that there were both deletions and additions (new keys and synopses as well as new genera) to the material from the parent edition, so that at the most the abridged version would be considered an abstract of the work of the authors of the larger text. Therefore, citation could only be made to the complete Eighth Edition. It was published as *The Shorter Bergey's Manual of Determinative Bacteriology* in 1977 (Holt, 1977). It too was a great success, selling 20,000 copies over a span of 10 years. A few years later it was translated into Russian and sold throughout the USSR, with royalties accruing to the Trust.

The development of bacteriology, as we now appreciate, required the recognition and differentiation of the various groups of microbes as taxonomic entities. At the time that *Bergey's Manual* started, the nature of bacterial cells was not known. Bacteria were classified and named under the Botanical Code of Nomenclature as Schizomycetes and no one could then have substantiated present understanding that Cyanophyceae are really bacteria. The international discussions of bacterial classification were minimal and took place at Botanical Congresses, as befitted the view that the Schizomycetes and the Schizophyceae within the Phylum Schizophyta (later Protophyta) belonged in the plant kingdom.

This interpretation was maintained in *Bergey's Manuals* up to and including the Seventh Edition (1957); however, it was stated in an introductory chapter that E.G.D. Murray "... felt most strongly that the bacteria and related organisms are so different from plants and animals that they should be grouped in a kingdom equal in rank with these kingdoms". As expressed by Stanier and Van Niel (1962) in their seminal paper "The concept of a bacterium" it is "... intellectually distressing (for a biologist) to devote his life to the study of a group that cannot be readily and satisfactorily defined in biological terms ...". This marked the beginning of the useful and directive description of bacteria as cells of unique nature. With this approach it was clear that the cyanobacteria were included and there was, at last, a satisfactory unity. This was to be slowly elaborated in the next three decades by the recognition of phylogenetic information recorded in molecular sequences of highly conserved macromolecules, but in the meantime the Eighth Edition (1974) subscribed to the view based on cytological data that the bacteria (all the procaryotes) belong in a separate kingdom, the Procaryotae. This was not a surprising decision because two Trustees, Stanier and R.G.E. Murray, were then involved in the description of bacteria as cells with unique features.

INTERNATIONAL EFFORTS TO REGULATE TAXONOMY

The founders of *Bergey's Manual* were fully aware of the substratum of opinion, albeit not supported then by strong data, that the bacteria were a special form of life, requiring special methods and a different approach to classification, not necessarily the same as that required by the Botanical Code. In fact, between 1927 and 1930 there was a considerable international correspondence between bacteriologists interested in taxonomy in the varied fields of application in agriculture, medicine, soil science, etc, expressing their concerns. The correspondence also concerned what should be done about discussing bacteria at the forthcoming Botanical Congress to be held in Cambridge, England, in 1930, and about resolutions adopted by the Bacteriological Section of the Botanical Congress, of which J.M. Sherman had been Secretary, held in Ithaca, NY, in 1926. The resolutions were (1) exclusion of the requirement for a Latin diagnosis in bacteriological nomenclature; (2) greater emphasis on the "type concept"; (3) a special international and representative committee was needed to coordinate the special nomenclatural interests of bacteriologists; and (4) that a permanent International Commission on Bacteriological Nomenclature should be formed. Sherman, then Secretary-Treasurer of SAB, wrote to Prof. J. Briquet of the Permanent International Committee on Botanical Nomenclature pointing out that the past two Congresses had authorized a bacteriological committee on nomenclature, that it should be organized, and that the Bacteriological Section had prepared a distinguished list of nominations for membership. The list included three of the major contributors to discussions of systematics in the SAB (Buchanan, Breed, and Harrison) and two of them were intimately involved with *Bergey's Manual*.

A lively correspondence among the authorities resulted and much of it was stimulated by Breed writing to bacteriologists in Europe as well as America. He sums up an impression of the responses in a letter to the Secretary of the Botanical Congress, as follows: "... there is a general feeling that unless the Congress welcomes us into the ranks of botanists with the recognition of our peculiar and perplexing problems in the taxonomic field, we must organize an independent international group". At the same time he recognized the value of the work of Congresses in maintaining useful rules of nomenclature and reiterating the list of resolutions. The British correspondents were generally agreeable to bacteriological discussions but expressed sharp divisions as to associating or not with the botanists. Other players namely the newly formed International Society of Microbiology, and the Cambridge committee charged with organizing the bacteriological component of the 1930 Botanical Congress came on the scene in 1927. The former encouraged some thoughts of an independent base for microbiological congresses and taxonomy committees, while the latter questioned whether or not a Section of Bacteriology was desirable or even feasible, and asked H.R. Dean (Professor of Pathology at Cambridge University) to seek interest and act on it. Dean's correspondents in this matter were numerous and mostly British, but also included Breed, Buchanan, B. Issatchenko (USSR), and K.B. Lehmann (Germany) (letters regarding this information are now filed in The American Society for Microbiology Archives). The responses generally supported a Section at the Congress but the overall opinions on continuing association with the botanists varied from the enthusiastic (mostly general microbiologists) to outright contrary opinion (mostly medical bacteriologists). Paul Fildes wrote: "Personally I am of the opinion that bacteriology has nothing to gain by a close association with botany." And Sir John Ledingham, while agreeing with having general bacteriological discussions, thought in the future "If the botanists will not have us, maybe that is all to the good". J.W. McLeod wrote: "Frankly, I am not very enthusiastic about a Section of Bacteriology at an International Botanical Congress especially if we are going to have an International Association of Microbiology". Other views crept into letters such as one from F. Löhnis: "I know that there exists within ... (the SAB) ... a small but very active minority extremely eager to advance a scheme of classification and nomenclature that seems to me as to others quite contrary to international usage ... this minority has advanced its ideas in the U.S.A. and will probably try the same scheme at Cambridge in 1930 if there should be a separate Section of Bacteriology". Breed wrote Dean that there would be support in the SAB for a delegation and added a few remarks on differences with the botanists, including: "Our troubles, for example, do not concern type specimens kept in a herbarium. They are intimately concerned with the maintenance of type culture collections such as the English bacteriologists have been able to establish so splendidly at the Lister Institute". There were more meetings in 1929 of a subcommittee appointed to settle a program for the Bacteriology Section (Dean as Chairman, with Boycott, Topley, Ledingham, Paine, Thornton, Thaysen, and Murray) and charged to keep Briquet (Botanical Nomenclature Committee) informed of any discussion of bacteriological nomenclature that might take place

Attitudes to studying and naming bacteria were rather different in the UK and Europe in the 1920s than was evident in the USA and Canada. The influential members of the SAB involved in *Bergey's Manual* seemed to be able to muster support for their views and seek consensus even if there were rumblings of dissent (q.v. Hall, 1927). In Europe many, like Orla-Jensen, believed that individual bacteriologists of substance should prevail because they were the ones who knew their groups of bacteria and he objected to imposition from outside. Internationalism did not and does not come easily.

The International Society for Microbiology (ISM), formed during an international conference on rabies sponsored by the Institute Pasteur in April, 1927, elected Prof. J. Bordet as President and R. Kraus as Secretary-General. It was stated in the brochure that: "It will not only compose the Science of Bacteriology but all the sciences associated with Microbiology" and the concept was based on "the unanimous conviction that Science should unite Nations...". The idea that all Societies of Microbiology may join, and that National Committees may present individual microbiologists as members, was expressed. So, the concept of an international association was born in Europe without anyone from North America among the founding members from 14 countries. There was interest: Harrison wrote to Dean suggesting that contact should be established between the ISM and the Bacteriological Section meeting at the Botanical Congress. Ledingham wrote to Dean in June, 1928, to support a meeting of the Nomenclature Committee of the Pathological Society of Great Britain and Ireland with Breed and others who were visiting, "particularly with regard to joint action on this matter by the botanical bacteriologists and the new International Society for Microbiology. Possibly they might consent to turn the matter over entirely to the new International Society (if adequate guarantees given)". It is not clear what group meeting resulted although hints were made.

1930 was the year of change because the First International Congress for Microbiology was held in Paris and by a vote agreed to follow the rules of nomenclature accepted by the International Congresses of Botany and Zoology *"in so far as they may be applicable and appropriate"* (italics as given by Breed, 1943). This opened the doors for a dedicated committee which would be in action at the following Congress (1936, in London, England), and set in train the development of an International Committee for Systematic Bacteriology, the regulatory mechanisms that were to be so important to taxonomic decisions in years to come, and a bacteriological code of nomenclature. The Microbiology Congress and the Botanical Congress, prompted by its Bacteriology Section (and probably by a questionnaire circulated by Breed), both approved in plenary session that the starting date for bacteriological nomenclature should be May 1, 1753, the date of publication of *Species Plantarum* by Linnaeus.

No doubt, there was much going on behind the scenes and some degree of consensus about the ever contentious matters involved in bacterial taxonomy. However, it was clear that bacterial taxonomy would be a matter of international concern from then on.

THE ENLARGEMENT OF THE SCOPE OF THE *MANUAL*

In the period following the death of R.E. Buchanan, John Liston took over as Chairman until 1976 when he retired and was replaced by R.G.E. Murray. It was during this subsequent period, in the late 1970s, that plans were laid to expand the informational coverage of the *Manual*. What started as a discussion of a new edition of the determinative manual developed into a plan to include much more information on the systematics, biology, and cultivation of each genus covered. Hans Lautrop had analyzed the content of the Eighth Edition and suggested a format that would allow authors to expound on further descriptive information, isolation and maintenance, and taxonomic problems. Other planned departures from past editions included the profuse use of high quality illustrations and allowing publication of new names and combinations in the *Manual*. It was also decided to preface the book with essays on general aspects of bacterial systematics such as modern genetic techniques, culture collections, and nomenclature. This expanded coverage meant a large increase in the number of pages and it was decided to publish the book in four volumes, each containing a set of taxa divided

along somewhat practical lines. The final arrangement consisted of volumes covering the Gram-negatives of medical importance, the Gram-positives of medical importance, the other Gram-negatives (including the *Archaea* and, for the first time, the Cyanobacteria), and lastly, the Actinomycetes. This division allowed users to purchase separate volumes that suited their special professional requirements. This expansion demanded a more descriptive title and it was decided to call the book *Bergey's Manual of Systematic Bacteriology*. Production of each volume was set up on a cascading schedule with completion planned for the mid 1980s. Trust members were chosen to edit each sub-volume, with the final editing being done in the Ames office. Obviously, such an undertaking was an expensive endeavor, beyond royalty income, and extra funding was provided by a grant from the National Library of Medicine of the US National Institutes of Health for volumes 1 and 2, and an advance on royalties from the publisher. In the end the complete project cost around $400,000. Volume 1 was published in 1984 (Krieg and Holt, 1984), Volume 2 in 1986 (Sneath et al., 1986), and Volumes 3 and 4 in 1989 (Staley et al., 1989; Williams et al., 1989).

The book was a truly international project in which 290 scientists from 19 countries (and 6 continents) participated, and as much of a success as the Trust and its authors could have expected. Each of the volumes sold between 10 and 23 thousand copies in the 1984–1996 period and more than half of the sales were outside of the USA. The total royalties add up to in excess of $450,000, making the *Systematic Manual* both a scientific and business success. The challenge now is to find the finances, energies, and means to keep the *Manual* up to date, affordable and reasonably current.

One of the mandates of the Trust is to further bacterial taxonomy, and the modern Board of Trustees has taken other initiatives besides the publication of books to promote the field. There has been monetary support, however small, for worthwhile causes, such as the aforementioned gift to launch the *International Bulletin of Bacteriological Taxonomy and Nomenclature*. Also in 1980, the Trust contributed $3000 towards the publication of the Approved Lists of Bacterial Names (Skerman et al., 1980). Two ways have been found to honor people who have made important contributions to the field of bacterial systematics. In 1978 the Bergey Award was instituted as a joint effort by Williams & Wilkins and the Trust; the first award went to R.Y. Stanier and is an annual event. Table 2 lists the recipients of this award, which consists of $2,000 and expenses to allow travel to a meeting of the recipient's choice to receive the award. In the 1990s the Trust commissioned a medal, the Bergey Medal (Fig. 3), to be given to individuals who have made significant lifetime contributions to bacterial systematics and to recognize the service of Trustees (Table 3). In 1982, the Board of Trustees decided to stimulate the involvement of more people in the affairs of the Trust, beyond the legal limit of nine regular members set in the By-Laws. It instituted the appointment of *Bergey's Manual* Associates for five-year terms to contribute their scientific expertise to the needs of the *Manuals*, the Trust and its Editors (Table 4).

The *Systematic Manual* was produced during a time of significant advances in our understanding of relationships between bacterial taxa based on the comparison of molecular sequences in highly conserved nucleic acids and proteins. The work of Carl Woese and others dating from the 1970s began to provide solid, initially sparse but now burgeoning, information on the phylogenetic relationships of the bacteria and, indeed, all life forms. This new information had a potential impact on the organization

11

TABLE 2. Recipients of the Bergey Award

Roger Y. Stanier	1979
John L. Johnson	1980
Morrison Rogosa	1981
Otto Kandler	1982
Carl R. Woese	1983
W.E.C. Moore	1984
Jozef De Ley	1985
William H. Ewing	1986
Patrick A.D. Grimont	1987
Lawrence G. Wayne	1988
Hubert A. Lechevalier	1989
M. David Collins	1990
Erko Stackebrandt	1991
Wolfgang Ludwig	1992
Wesley E. Kloos	1993
Friedrich Widdel	1994
Michael Goodfellow	1995
Karel Kersters	1996
Rosmarie Rippka	1997
Barry Holmes	1998
David A. Stahl	1999
William B. Whitman	2000

TABLE 3. Recipients of the Bergey Medal

Eyvind A. Freundt	1994
R.G.E. Murray	1994
Riichi Sakazaki	1994
V.B.D. Skerman	1994
Dorothy Jones	1995
Norberto Palleroni	1995
Norbert Pfennig	1995
Thomas D. Brock	1996
Marvin P. Bryant	1996
John G. Holt	1996
Emilio Weiss	1996
Lillian H. Moore	1997
Ralph S. Wolfe	1997
George A. Zavarzin	1997
Kjell Bøvre	1998
Holger Jannasch	1998
Juluis P. Kreier	1998
Peter H.A. Sneath	1998
Wilhelm Frederiksen	1999
James W. Moulder	1999
Karl O. Stetter	1999
Hans G. Trüper	1999

FIGURE 3. Obverse view of the Bergey Medal, 3 in. diam., See Table 3 for a list of recipients.

TABLE 4. Past and Present Bergey's Manual Associates (1982–1999)

Martin Altwegg	Thomas McAdoo
Paul Baumann	W.E.C. Moore
David R. Boone	Aharon Oren
Richard W. Castenholz	Norberto J. Palleroni
Rita R. Colwell	Frederick A. Raincy
Gregory A. Dasch	Anna-Louise Reysenbach
Floyd E. Dewhirst	Morrison Rogosa
Paul De Vos	Abigail Salyers
Takayuki Ezaki	Juri Schindler
Monique Gillis	Karl-H. Schleifer
Michael Goodfellow	Haroun N. Shah
Peter Hirsch	Leslie I. Sly
Lillian Holdeman-Moore	Robert M. Smibert
Barry Holmes	Erko Stackebrandt
J. Michael Janda	Karl O. Stetter
Dorothy Jones	James M. Tiedje
Lev V. Kalakoutskii	Hans G. Trüper
Otto Kandler	Anne Vidaver
Karel Kersters	Lawrence G. Wayne
Helmut König	Robbin S. Weyant
Micah I. Krichevsky	Friedrich Widdel
L. David Kuykendall	Stanley T. Williams
David P. Labeda	George A. Zavarzin
Mary P. Lechevalier	

of the taxa in the *Manual*, however, the Trust and its advisors decided to continue to organize the book on phenotypic grounds. First, because the bench workers needing to identify isolates have to use these characters and, secondly, because the phylogenetic data were accumulating slowly during the early 1980s. The Trust decided to continue with a phenotypic arrangement and indicate, where appropriate and data were sufficient, the phylogenetic placement of the taxon being discussed. Finally, enough progress has been made in the last 20 years for this Second edition to be phylogenetically organized, although there are still gaps and uncertainties in our knowledge.

In the 1980s and early 1990s there was a large turnover in Board membership and leadership. New Board members included D.J. Brenner, J.W. Moulder, S.T. Williams, K.-H. Schleifer,

N. Pfennig, P.H.A. Sneath, R.W. Castenholz, J.G. Tully, and J. Ursing, some of whom have since retired (Table 1 and Fig. 4). In 1990 Board Chairman R.G.E. Murray retired after a long and fruitful tenure and was replaced by P.H.A. Sneath, who served until 1994 when S.T. Williams took over the helm. It should be explained that the Board of Trustees has a retirement age of 70 (members call it the "Buchanan Amendment"), which is no reflection on the quality of service of retired Board members. See Fig. 5 for the current membership of the Board of Trustees and Editors of sub-volumes of this Second Edition.

One important change in the Trust operations has been the establishment of a permanent headquarters. In the late 1980s the Trust decided to move from Iowa State University where it had resided since 1958, and set out to find a permanent home for the Editorial Office that was not tied to the tenure of the Editor. After an active search such a home was eventually found at Michigan State University which has a large, active Department

FIGURE 4. Trustees at their meeting in Stamford, England, September, 1985. *L. to R.*, D. Brenner, P. Sneath, N. Krieg, J. Holt, J. Moulder, N. Pfennig, J. Staley, S. Williams, M. Bryant, and R. Murray.

FIGURE 5. Current Trustees (with Emeritus Chairman P.H.A. Sneath) taken at Sun River, OR, August, 1997. *L. to R.*, J. Staley, S. Williams, G. Garrity, J. Holt, K. Schleifer, D. Brenner, N. Krieg, R. Castenholz, D. Boone, and P. Sneath.

of Microbiology and is the base for the NSF-funded Center for Microbial Ecology. In December, 1990, Holt and the Trust office and archives moved to East Lansing, Michigan. Holt subsequently retired as Editor-in-Chief in 1996 and a replacement was found who continued as a faculty member in the Department. The new Editor-in-Chief, George M. Garrity, assumed his duties in 1996.

All of these changes were accompanied by an increasingly active publishing program. After publication of the last two volumes of the *Systematic Manual* in 1989, plans were made to produce the Ninth Edition of the *Determinative Manual*. Based on a concept of N.R. Krieg, the format of the book was changed to a style between the Eighth Edition and the *Shorter Manual*; the species descriptions are summarized in extensive tables. It was published in 1994 (Holt et al., 1994) in softcover and contained the determinative information from the *Systematic* book plus descriptions of new genera and species named since publication of the larger book. This *Manual* is intended to be a prime resource for bench workers and all who are engaged in diagnostic bacteriology and the identification of isolates. The Trust published other books in the early 1990s, notably *Stedman's/Bergey's Bacteria Words* (Holt et al., 1992) (one of a series of wordbooks compiled for medical transcriptionist use), and provided the general editing of the Second Edition of the CDC manual on the *Identification of Unusual Pathogenic Gram-negative Aerobic and Facultatively Anaerobic Bacteria* (Weyant et al., 1996).

THE PUBLICATION PROCESS

It is no mean task to produce and get into print a taxonomic compendium; it is a major and complex project for authors, editors, and not least the publisher. The Williams & Wilkins Co. of Baltimore was the publisher of the *Manuals* from 1923 to 1998, and over those years there was an extraordinarily effective partnership between the Trust and the publisher which was mutually advantageous. The various editions of the *Determinative Manual* have been very successful in both the scientific and the commercial sense. The confidence of the publisher allowed them to provide financial support for the preparation of other ventures such as the *Systematic Manual*, which required some years of work and several editorial offices, adding to the up-front expenses. The great success of the published volumes vindicated and more than repaid the publisher's generous support of the enterprise. After major changes in the management of Williams & Wilkins and the merger of the company with another publisher, the Trust reexamined its publishing arrangements and entertained offers from other firms. In late 1998 a new publishing agreement was signed with Springer-Verlag of New York to publish this edition of the *Systematic Manual*, ushering in a new era of cooperation between the Trust, representing the microbiological community, and its publisher, who is committed to disseminating high-quality and useful books to that community.

Because of the number and complexity of the entries, the number of scientists involved in generating the text (or revising it, as is now more often the case), and the sheer number of indexable items, it has been obvious for years that some form of computer assistance would become essential. One of the long-term goals of the Trust and its publishers has been to produce an electronic version of the *Manual*. There were a number of objectives associated with this project. One was the obvious provision of a searchable CD-ROM version of the data contained in the *Manual*. The other, not so obvious, was the ability to streamline the process of updating new editions by supplying the phenotypic data of each taxon in a database that can be easily updated by authors and to which new information (which is accruing at an alarming rate) can be added. The Trust editorial office is now using the latest computer technology in producing this and subsequent versions of its manuals, utilizing the power of Standard Generalized Markup Language (SGML) to facilitate the storage, retrieval, typesetting, and presentation of the information in both print and electronic form. Planning for this new edition of the *Systematic Manual* has been underway for the past four years and two major problems have faced the Board and its Advisory Committees. One is the rapid rate of description of new taxa, many of which are not adequately differentiated by phenotypic characteristics. The other is the requirement that the book reflect the best of current science, including a phylogenetic classification based on semantides, particularly 16S rRNA. The phylogeny is incomplete but the gaps are being slowly filled. Problems occur when there is little correlation between the phylogenetic classification and the phenotypic groupings that prove essential to the initiation of identification. Therefore, broadly based and informational descriptions remain an essential feature of the *Manual* as well as a text that stimulates research.

We were most fortunate over the years to enjoy not only a cooperative and productive relationship with Williams & Wilkins, but also the friendly assistance of a series of liaison officers who have represented the Company and its interests and concerns. Among these most helpful people were Robert S. Gill, Dick Hoover, Sara Finnegan, and William Hensyl, whose abilities as facilitators and as interpreters of the disparate requirements of Trust and Publisher were essential. We look forward to our new relationship with Springer-Verlag which should be productive and benefit the entire microbiological community.

The concept of the *Bergey's Manuals*, i.e. encyclopedic taxonomic treatments of the procaryotic world that aid microbiologists at all levels and in all sub-disciplines, is alive and well. The vision of Bergey and Breed is being carried on by their successors and will continue well into the next millennium.

ACKNOWLEDGMENTS

A consideration of the history of the *Manuals* and publications of the Bergey's Manual Trust would be incomplete without an acknowledgment of the contributions of a large number of individuals. One such person that we wish to thank for assistance is The American Society of Microbiology Archivist, Jeff Karr, who sought and found correspondence and minutes that were of great use in preparing this manuscript. Many other people were often involved in complex operations going on in their place of work with no or limited formal recognition of their contribution, and frequently without recompense as in the case of wives of editors. Some individuals performed major tasks (e.g., Alice Gibbons, the whole index for the Eighth Edition). A long succession of helpers were involved over the 32 years the headquarters was at Iowa State, and their contributions were invaluable. Of special note was the long service to R.E. Buchanan of Elsa Zvirbulis, Mildred McConnell, and Vlasta Krakowska in Ames. J.G. Holt has been ably assisted by a series of excellent secretary/editorial assistants, especially Cynthia Pease in Ames, and Betty Caldwell and Constance Williams in East Lansing. Taxonomy and the production of useful compilations and classifications are "labors of love" involving both dedication and unremitting effort of those so inclined AND the people around them.

On Using the *Manual*

Noel R. Krieg and George M. Garrity

ARRANGEMENT OF THE MANUAL

One important goal of the *Manual* has always been to assist in the identification of procaryotes, but another goal, equally important, is to indicate the relatedness that exists among the various groups of procaryotes. This goal seemed elusive until the late 1950s and early 1960s, with the realization that the DNA of an organism makes it what it is. Initially, overall base compositions of DNAs (mol% G + C values) were used to compare procaryotic genomes, and organisms for which mol% G + C values differed markedly were obviously not of the same species. If, however, two organisms had the same mol% G + C value, they might or might not belong to the same species, and thus a much more precise method of comparison was needed. The development of DNA–DNA hybridization techniques fulfilled this need. The continua that often blurred the separation between groups defined by phenotypic characteristics did not usually occur with DNA–DNA hybridization. Organisms tended to be either closely related or not, because DNA–DNA duplex formation did not even occur if base pair mismatches exceeded 10–20%. Thus DNA–DNA hybridization solved many of the problems that had long plagued bacterial taxonomy at the species level of classification. It was almost useless, however, for estimating more distant relationships among procaryotes, i.e., at the generic level, family level, or above. An important development was the discovery by Doi and Igarashi (1965) and by Dubnau et al. (1965) that the ribosomal RNA (rRNA) cistrons in bacterial species were conserved (slower to change) to a greater extent than the bulk of the genome, probably because of their critical function for the life of a cell. This function would allow only slow changes in nucleotide sequence to occur over long periods of time, relative to other genes that were not so critical for the cell. This in turn led to the idea that rRNA–DNA hybridization might be useful for deducing the broader relationships that DNA–DNA hybridization could not reveal. For instance, in 1973 a monumental study by Palleroni et al. showed that the genus *Pseudomonas* consisted of five rRNA groups (tantamount to five different genera).

In the 1970s, the idea—based on cellular organization—that there were only two main groups of living organisms, the procaryotes and the eucaryotes, was challenged by Woese and Fox, who compared oligonucleotide catalogs of the 16S rRNA (and the analogous eucaryotic 18S rRNA) from a broad spectrum of living organisms. These comparisons indicated that there were two fundamentally different kinds of procaryotes: the *Archaea* (also called archaebacteria or archaeobacteria), and the *Bacteria* (also called eubacteria). Urcaryotes—i.e., that portion of eucaryotes exclusive of mitochondria or chloroplasts, these being

endosymbionts undoubtedly derived from procaryotes—differed from both the *Bacteria* and the *Archaea*. These findings led Woese, Kandler, and Wheelis (1990) to the view that the *Archaea*, the *Bacteria*, and the eucaryotes (now called the *Eucarya*) evolved by separate major evolutionary pathways from a common ancestral form, although just where the deepest branchings occur is still not clear.

Improvements in the methodologies of molecular biology have now made it possible to determine and compare sequences of the rDNA cistrons from a great number of procaryotes, and a comprehensive classification of procaryotes, based on relatedness deduced from 16S rDNA sequences, is underway. It is hoped that such a phylogenetic classification scheme will lead to more unifying concepts of bacterial taxa, to greater taxonomic stability and predictability, to the development of more reliable identification schemes, and to an understanding of how bacteria have evolved. However, sequencing of the complete genomes of a number of procaryotes and comparison of various genes among the organisms has led to some reservations about whether 16S rDNA sequence analyses are completely reliable for reconstructing evolutionary phylogenies. The study of genes other than rRNA genes has sometimes led to different phylogenetic arrangements. Some bacteria have been found to contain certain genes of the archaeal type, and some archaea have been reported to have certain genes of the bacterial type. These discrepancies might be due in part to a lateral transfer of genes by transformation, transduction, or conjugation from one present-day species to another, as distinguished from vertical transfer from ancestral forms. Thus the location of the deep evolutionary branchings deduced from 16S rRNA gene sequences may not be as firm as once thought. On the other hand, acquisition of an eclectic assortment of genes might have occurred in very primitive lifeforms that existed prior to the divergence of the three major evolutionary pathways. In any event, the present edition of the *Manual* provides the best available phylogenetic scheme based on 16S rRNA gene sequencing. Figure 1 shows the major groups of procaryotes and their relatedness to one another. The deeper branching points are not shown because of their uncertainty, and some crossover of branch points beneath the plane of projection is likely. Each branch is the equivalent of either a class or an order. The arrangement shown is reasonably firm and is unlikely to change very much as more information is gathered.

Within the *Archaea*, two phyla are recognized: "Crenarchaeota" and "Euryarchaeota". In the present classification, one class is accommodated in the "Crenarchaeota": "Thermoprotei", and seven in the "Euryarchaeota": "Methanobacteria", "Methanococci", "Halobacte-

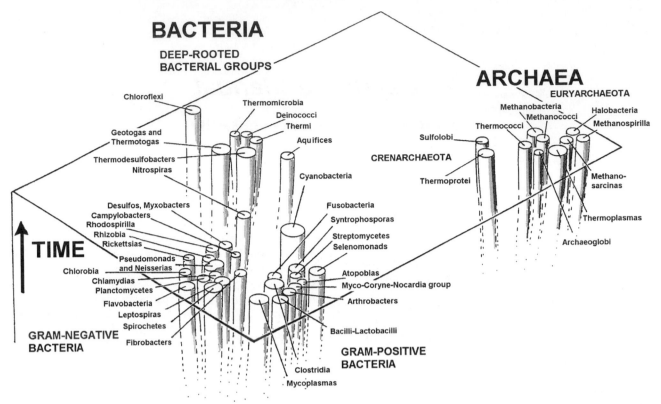

FIGURE 1. The major groups of procaryotes and their relatedness to one another. The relative size of the oval discs is an approximate indicator of the number of species in each group. The deeper origin of these groups, i.e., their evolution from more primitive forms, is still debatable and therefore is represented only by dashed lines. (Courtesy of Peter H.A. Sneath.)

ria", *"Thermoplasmata"*, *"Thermococci"*, *"Archaeoglobi"*, and *"Methanopyri"*.

The *Bacteria* have been grouped into 23 phyla, which are further subdivided into 28 classes. The deep-rooted *Bacteria* encompass nine phyla: the *"Aquificae"*, *"Thermotogae"*, *"Thermodesulfobacteria"*, the *"Deinococcus–Thermus"* phylum, the single-species phylum *"Chrysiogenetes"* (not shown), *"Chloroflexi"*, *"Thermomicrobia"*, *"Nitrospirae"*, and *"Deferribacteres"*.

Within the Gram-negatives, the *"Proteobacteria"* have been elevated to a phylum and subdivided into five classes corresponding to the *"Alphaproteobacteria"* (Rhodospirilla, Rhizobia, Rickettsias), the *"Betaproteobacteria"* (Neisserias), the *"Gammaproteobacteria"* (Pseudomonads), the *"Deltaproteobacteria"* (Desulfos and Myxos), and the *"Epsilonproteobacteria"* (Campylobacters); other Gram-negative phyla include the *"Planctomycetes"* (Planctomyces), the *"Chlamydiae"* (Chlamydias), the *"Chlorobi"* (Chlorobia), the *"Spirochaetes"* (Spirochetes and Leptospiras), the *"Fusobacteria"* (Fusiforms), the *"Verrucomicrobia"* (not shown), the *"Bacteroidetes"* (Bacteroides, Flavobacteria, Sphingobacteria), the *"Acidobacteria"* (not shown), *"Fibrobacteres"* (Fibrobacters), the *Cyanobacteria* (Cyanobacteria), and *"Dictyoglomi"* (not shown).

Traditionally, the Gram-positive bacteria have been separated on the basis of mol% G + C. The low G + C Gram-positives have been assigned to the phylum *"Firmicutes"* and include the Mycoplasmas, Clostridia, Bacilli–Lactobacilli, and Syntrophospora branches shown in Fig. 1. The high G + C Gram-positive bacteria have been assigned to the *"Actinobacteria"* and encompass the Arthrobacters, the Myco/Coryne/Nocardia group, the Atopobias, and the Streptomycetes. The *Cyanobacteria* represent the

last of the phyla depicted in the figure and consistently appear in close proximity to the Gram-positive bacteria, but represent a distinct phylogenetic lineage. These major groups are also shown in Fig. 2, in which boxes have been used to enclose the related groups. Only a few representative groups are shown in the figures because of space considerations.

THE PHYLA

The use of a phylogenetic schema for the organization and presentation of the contents of the *Manual* represents a departure from the first edition, in which genera were grouped together based on a few readily determined phenotypic criteria. All accepted genera have now been placed into a provisional taxonomic framework based upon the best available 16S rDNA sequence data (>1000 nts and <0.01% ambiguities). At the close of 1999, 16S rDNA sequences for approximately two-thirds of the validly published type strains were publicly available. In those instances where 16S rDNA data for the type species were not available, placement of a genus into the framework was based on either phenotypic characteristics or data derived from a closely related species. Some validly named genera are known to be either paraphyletic or polyphyletic. In such cases, allocation of the genus within the taxonomic framework was determined by the phylogenetic position of the type species. The order in which taxa are presented is based, in part, on the topology of the RDP tree (Release 6.01, November 1997) with some notable exceptions. This is discussed in more detail in the article entitled "Roadmap to the Manual".

It is generally agreed that procaryotes fall into two major lines

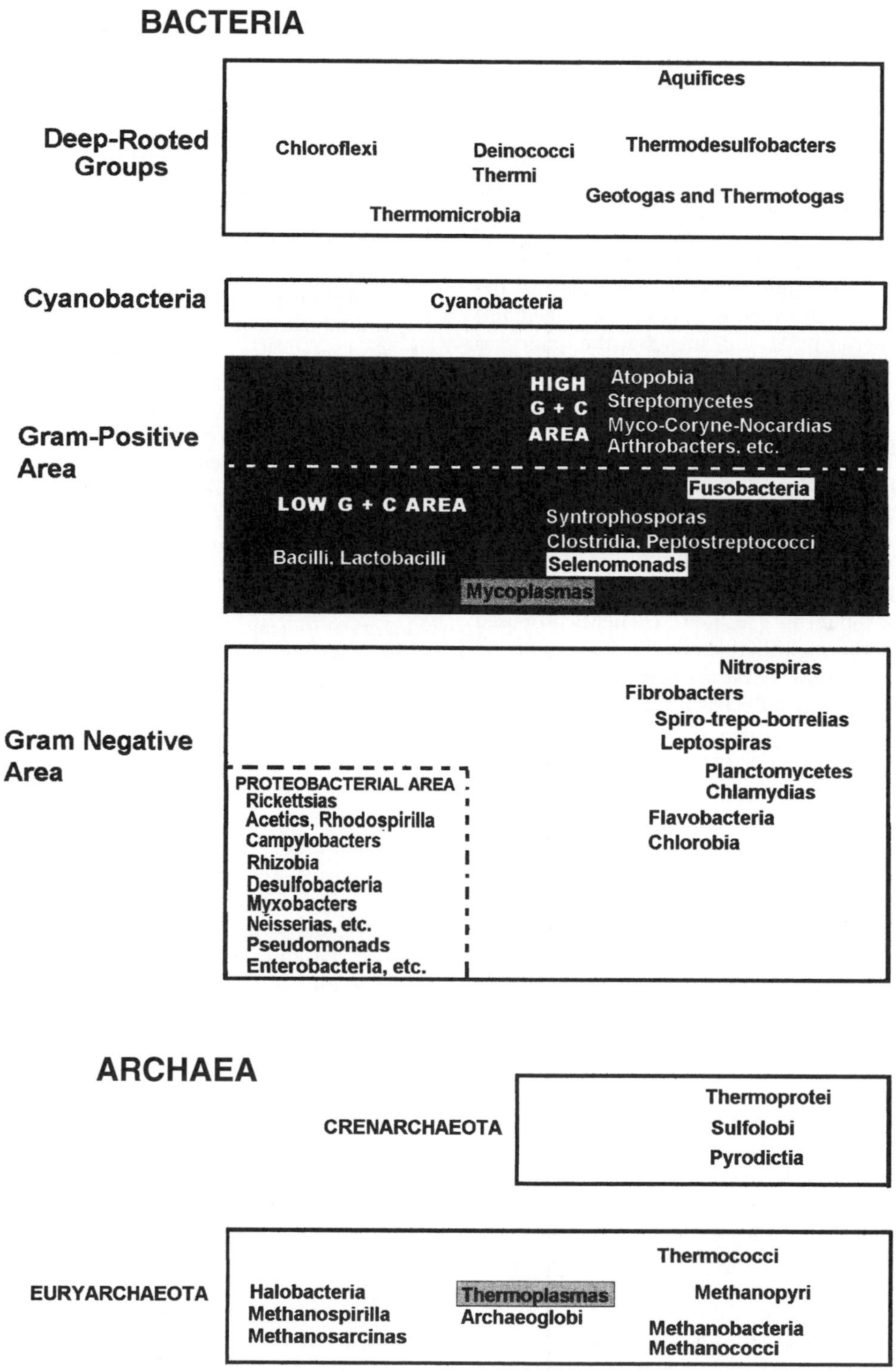

BACTERIA

Deep-Rooted Groups

Aquifices

Chloroflexi   Deinococci Thermi   Thermodesulfobacters

Thermomicrobia   Geotogas and Thermotogas

Cyanobacteria

Cyanobacteria

Gram-Positive Area

HIGH G + C AREA   Atopobia
Streptomycetes
Myco-Coryne-Nocardias
Arthrobacters, etc.

LOW G + C AREA   Fusobacteria

Syntrophosporas
Clostridia, Peptostreptococci

Bacilli, Lactobacilli   Selenomonads

Mycoplasmas

Gram Negative Area

Nitrospiras
Fibrobacters
Spiro-trepo-borrelias
Leptospiras
Planctomycetes
Chlamydias
Flavobacteria
Chlorobia

PROTEOBACTERIAL AREA
Rickettsias
Acetics, Rhodospirilla
Campylobacters
Rhizobia
Desulfobacteria
Myxobacters
Neisserias, etc.
Pseudomonads
Enterobacteria, etc.

ARCHAEA

CRENARCHAEOTA

Thermoprotei
Sulfolobi
Pyrodictia

EURYARCHAEOTA

Thermococci

Halobacteria   Thermoplasmas   Methanopyri
Methanospirilla   Archaeoglobi
Methanosarcinas   Methanobacteria
Methanococci

FIGURE 2. Simplified arrangement of the major groups of procaryotes. Blackened boxes in the *Bacteria* group indicate a Gram-positive staining reaction for members of *Bacteria*; shaded boxes indicate lack of any cell wall in members of *Archaea* and *Bacteria*.

of evolutionary descent: the **Archaea** and **Bacteria**. These will be dealt with as **domains**. The domains have been further subdivided into phyla that represent the major procaryotic lineages and will serve as the main organizational unit in this edition of the *Manual*. At present, the *Archaea* have been subdivided into two phyla and the *Bacteria* into 23 phyla. With the exception of the *Cyanobacteria* and the *"Actinobacteria"*, phyla are further subdivided into classes, orders, and families. In the case of the former, families are replaced by subsections, which may be further divided into subgroups. In addition, species generally do not appear as discrete entities. Rather, these are represented by strain designations. Some genera are also referred to as Form-genera. In the case of *"Actinobacteria"*, the taxonomic hierarchy is slightly modified and includes subclasses and suborders. Readers should note that names above the class level are not covered by the *Bacteriological Code* and should be regarded as informal or colloquial names. Furthermore, as additional sequence data become available, some phyla are likely to be combined while others may be split.

ARTICLES

Each article dealing with a bacterial genus is presented wherever possible in a definite sequence as follows:

a. *Name of the Genus.* Accepted names are in **boldface**, followed by the authority for the name, the year of the original description, and the page on which the taxon was named and described. The superscript AL indicates that the name was included on the Approved Lists of Bacterial Names, published in January 1980. The superscript VP indicates that the name, although not on the Approved Lists of Bacterial Names, was subsequently validly published in the *International Journal of Systematic Bacteriology*. Names given within quotation marks have no standing in nomenclature; as of the date of preparation of the *Manual* they had not been validly published in the *International Journal of Systematic Bacteriology*, although they had been "effectively published" elsewhere. Names followed by the term "gen. nov." are newly proposed but will not be validly published until they appear in the *International Journal of Systematic Bacteriology*; their proposal in the *Manual* constitutes only "effective publication", not valid publication.

b. *Name of Author(s).* The person or persons who prepared the Bergey article are indicated. The address of each author can be found in the list of Contributors at the beginning of the *Manual.*

c. *Synonyms.* In some instances a list of some synonyms used in the past for the same genus is given. Other synonyms can be found in the *Index Bergeyana* or the *Supplement to the Index Bergeyana.*

d. *Etymology of the Genus Name.* Etymologies are provided as in previous editions, and many (but undoubtedly not all) errors have been corrected. It is often difficult, however, to determine why a particular name was chosen, or the nuance intended, if the details were not provided in the original publication. Those authors who propose new names are urged to consult a Greek and Latin authority before publishing, in order to ensure grammatical correctness and also to ensure that the name means what it is intended to mean.

e. *Capsule Description.* This is a brief resume of the salient features of the genus. The most important characteristics are given in **boldface**. The name of the type species of the genus is also indicated.

f. *Further Descriptive Information.* This portion elaborates on the various features of the genus, particularly those features having significance for systematic bacteriology. The treatment serves to acquaint the reader with the overall biology of the organisms but is not meant to be a comprehensive review. The information is presented in a definite sequence, as follows:

  i. Colonial morphology and pigmentation
  ii. Growth conditions and nutrition
  iii. Physiology and metabolism
  iv. Genetics, plasmids, and bacteriophages
  v. Phylogenetic treatment
  vi. Antigenic structure

  vii. Pathogenicity
  viii. Ecology

g. *Enrichment and Isolation.* A few selected methods are presented, together with the pertinent media formulations.

h. *Maintenance Procedures.* Methods used for maintenance of stock cultures and preservation of strains are given.

i. *Procedures for Testing Special Characters.* This portion provides methodology for testing for unusual characteristics or performing tests of special importance.

j. *Differentiation of the Genus from Other Genera.* Those characteristics that are especially useful for distinguishing the genus from similar or related organisms are indicated here, usually in a tabular form.

k. *Taxonomic Comments.* This summarizes the available information related to taxonomic placement of the genus and indicates the justification for considering the genus a distinct taxon. Particular emphasis is given to the methods of molecular biology used to estimate the relatedness of the genus to other taxa, where such information is available. Taxonomic information regarding the arrangement and status of the various species within the genus follows. Where taxonomic controversy exists, the problems are delineated and the various alternative viewpoints are discussed.

l. *Further Reading.* A list of selected references, usually of a general nature, is given to enable the reader to gain access to additional sources of information about the genus.

m. *Differentiation of the Species of the Genus.* Those characteristics that are important for distinguishing the various species within the genus are presented, usually with reference to a table summarizing the information.

n. *List of the Species of the Genus.* The citation of each species is given, followed in some instances by a brief list of objective synonyms. The etymology of the specific epithet is indicated. Descriptive information for the species is usually presented in tabular form, but special information may be given in the text. Because of the emphasis on tabular data, the species descriptions are usually brief. The type strain of each species is indicated, together with the collection(s) in which it can be found. (Addresses of the various culture collections are given in the article entitled Culture Collections: An Essential Resource for Microbiology.) The 16S rRNA gene sequence used in phylogenetic analysis and for placement of the genus into the taxonomic framework is given, along with the GenBank accession number and RDP identifier for the aligned sequence. Additional comments may be provided to point the reader to other well-characterized strains of the species and any other known DNA sequences that may be relevant.

o. *Species Incertae Sedis.* The List of Species may be followed in some instances by a listing of additional species under the heading "Species Incertae Sedis". The taxonomic placement or status of such species is questionable and the reasons for the uncertainty are presented.

p. *Literature Cited.* All references given in the article are listed alphabetically at the end of the volume rather than at the end of each article.

TABLES

In each article dealing with a genus, there are generally three kinds of tables: (a) those that differentiate the genus from similar or related genera, (b) those that differentiate the species within the genus, and (c) those that provide additional information about the species (such information not being particularly useful for differentiation). Unless otherwise indicated, the meanings of symbols are as follows:

+ : 90% or more of the strains are positive
d: 11–89% of the strains are positive
− : 90% or more of the strains are negative
D: different reactions occur in different taxa (e.g., species of a genus or genera of a family)
v: strain instability (NOT equivalent to "d")

Exceptions to the use of these symbols, as well as the meaning of additional symbols, are clearly indicated in footnotes to the tables.

USE OF THE *MANUAL* FOR DETERMINATIVE PURPOSES

Each chapter has keys or tables for differentiation of the various taxa contained therein. Suggestions for identification may be found in the article on Polyphasic Taxonomy. For identification of species, it is important to read both the generic and species descriptions because characteristics listed in the generic descriptions are not usually repeated in the species descriptions.

The index is useful for locating the names of unfamiliar taxa and discovering what has been done with a particular taxon. Every bacterial name mentioned in the *Manual* is listed in the index. In addition, an up-to-date outline of the taxonomic framework, along with an alphabetized listing of genera, is provided in the Roadmap to the Manual. The table also provides the reader with an indication to which section a genus either was, or would have been, assigned in the first edition.

ERRORS, COMMENTS, SUGGESTIONS

As indicated in the Preface to the first edition of *Bergey's Manual of Determinative Bacteriology*, the assistance of all microbiologists in the correction of possible errors in the text is earnestly solicited. Comments on the presentation will also be welcomed, as well as suggestions for future editions. Correspondence should be addressed to:

Editorial Office,
Bergey's Manual Trust,
Michigan State University
East Lansing, MI 48824–1101
Telephone 517–432–2457;
fax 517–432–2458;
e-mail: garrity@msu.edu

Procaryotic Domains

Noel R. Krieg

Procaryotes can be described as follows:

Single cells or simple associations of similar cells (usually 0.2–10.0 μm in smallest dimension, although some are much larger) forming a group defined by cellular, not organismal, properties (i.e., by the structure and components of the cells of an organism rather than by the properties of the organism as a whole). The nucleoplasm (genophore) is, with a few exceptions, not separated from the cytoplasm by a unit-membrane system (nuclear membrane). Cell division is not accompanied by cyclical changes in the texture or staining properties of either nucleoplasm or cytoplasm; a microtubular (spindle) system is not formed. The plasma membrane (cytoplasmic membrane) is frequently complex in topology and forms vesicular, lamellar, or tubular intrusions into the cytoplasm; vacuoles and replicating cytoplasmic organelles independent of the plasma membrane system (chlorobium vesicles, gas vacuoles) are relatively rare and are enclosed by nonunit membranes. Respiratory and photosynthetic functions are associated with the plasma-membrane system in those members possessing these physiological attributes, although in the cyanobacteria there may be an independence of plasma and thylakoid membranes. Ribosomes of the 70S type (except for one group—the Archaea—with slightly higher S values) are dispersed in the cytoplasm; an endoplasmic reticulum with attached ribosomes is not present. The cytoplasm is immobile; cytoplasmic streaming, pseudopodial movement, endocytosis, and exocytosis are not observed. Nutrients are acquired in molecular form. Enclosure of the cell by a rigid wall is common but not universal. The cell may be nonmotile or may exhibit swimming motility (mediated by flagella of bacterial type) or gliding motility on surfaces.

In organismal terms, these ubiquitous inhabitants of moist environments are predominantly unicellular microorganisms, but filamentous, mycelial, or colonial forms also occur. Differentiation is limited in scope (holdfast structures, resting cells, and modifications in cell shape). Mechanisms of gene transfer and recombination occur, but these processes never involve gametogenesis and zygote formation.

Although procaryotic organisms can usually be readily differentiated from eucaryotic microorganisms, in some instances it may be difficult, especially with procaryotes that exhibit some attributes similar to those of microscopic eucaryotes. For instance, the hyphae formed by actinomycetes might be confused with the hyphae formed by molds; a fascicle of bacterial flagella could give the misleading impression of being a single eucaryotic flagellum; the ability of spirochetes to twist and contort their shape is suggestive of the flexibility exhibited by certain protozoa; some eucaryotic cells are as small as bacteria, and some bacteria are as large as eucaryotic cells (see footnote to Table 1). The most reliable approach is probably the demonstration of the absence of a nuclear membrane in procaryotes, but this involves electron microscopy of thin sections. Other procaryotic cell features range from those that are relatively easy to determine to the molecular characteristics that require sophisticated methods. Fluorescent- labeled gene probes that can easily distinguish between procaryotic and eucaryotic cells have been developed.

Some characteristics that may help to differentiate between procaryotes and eucaryotes are listed in Table 1.

ARCHAEA VS. BACTERIA

As shown in Figs. 1 and 2 in "On Using the Manual", the two fundamentally different groups (domains) that comprise the procaryotes are the *Bacteria* and the *Archaea*. Recent phylogenetic analyses of the *Bacteria*, *Archaea*, and *Eucarya* using conserved protein sequences have shown that the majority of trees support a closer relationship between the *Archaea* and *Eucarya* than between *Archaea* and *Bacteria*. Although the possibility of interdomain horizontal gene transfer complicates this picture, the apparent *Archaea–Eucarya* sisterhood raises interesting questions about the phylogenetic relationships between procaryotes and eucaryotes and the root of the universal tree of life. Table 2 provides some characteristics differentiating these two procaryotic groups. A general description of each group follows.

Bacteria For practical purposes the *Bacteria* may be divided into three phenotypic subgroups: (1) those that are Gram-negative and have a cell wall, (2) those that are Gram-positive and have a cell wall, and (3) those that lack any cell wall. (See chapters by Garrity and Holt and Ludwig and Klenk, this *Manual*, for a discussion of the phylogenetic relationships between Gram-positive and Gram-negative *Bacteria*.)

GRAM-NEGATIVE *BACTERIA* THAT HAVE A CELL WALL These have a Gram-negative type of cell-wall profile consisting of an outer membrane and an inner, relatively thin peptidoglycan layer (which contains muramic acid and is present in all but a few organisms that have lost this portion of wall; see footnote to Table 1) as well as a variable complement of other components outside or between these layers. They usually stain Gram-negative, although the presence of a thick exopolysaccharide layer around the outer membrane may result in a Gram-positive staining reaction, as seen in the cyst-like forms of some *Azospirillum* species. Cell shapes may be spheres, ovals, straight or curved rods, helices, or filaments; some of these forms may be sheathed or capsulated. Reproduction is by binary fission, but some groups show budding, and a rare group (Subsection II of the cyanobacteria) shows

TABLE 1. Some differential characteristics of procaryotes and eucaryotes[a]

Characteristic	Procaryotes	Eucaryotes
Cytological features		
Nucleoplasm (genophore, nucleoid) separated from the cytoplasm by a unit-membrane system (nuclear membrane)	−	+
Size of smallest dimension of cells (width or diameter):		
Usually 0.2–2.0 μm	+[b]	−
Usually 2.0 μm	−	+
Mitochondria present	−	Usually +
Chloroplasts present in phototrophs	−	+
Vacuoles, if present, enclosed by unit membranes	−	+
Gas vacuoles present[c]	D	−
Golgi apparatus present	−	D
Lysosomes present	−	D
Microtubular systems present	−[d]	D
Endoplasmic reticulum present	−	+
Ribosome location:		
Dispersed in the cytoplasm	+	−
Attached to endoplasmic reticulum	−	+
Cytoplasmic streaming, pseudopodial movement, endocytosis, and exocytosis	−	D
Cell division accompanied by cyclical changes in the texture or staining properties of either nucleoplasm or cytoplasm	−	+
Diameter of flagella, if present:		
0.01–0.02 μm	+	−
~0.2 μm	−	+
In cross-section, flagella have a characteristic "9 + 2" arrangement of microtubules	−	+
Endospores present[e]	D	−
Antibiotic susceptibility		
Susceptible to penicillin, streptomycin, or other antibiotics specific for procaryotes	D	−
Susceptible to cycloheximide or other antibiotics specific for eucaryotes	−	D
Features based on chemical analysis		
Poly-β-hydroxybutyrate present (as a storage compound in cytoplasmic inclusions)	D	−
Teichoic acids present (in cell walls)	D	−
Polyunsaturated fatty acids possibly present (in membranes)	Rare	Common
Branched-chain *iso*- or *anteiso*-fatty acids and cyclopropane fatty acids present (in membranes)	Common	Rare
Sterols present (in membranes)	−[f]	Common
Diaminopimelic acid present (in cell walls)	D[g]	−
Muramic acid present (in cell walls)	D[h]	−
Peptidoglycan (containing muramic acid) present in cell walls	D[h]	−
Nutrition		
Nutrients acquired by cells as soluble small molecules; to serve as sources of nutrients, particulate matter or large molecules must first be hydrolyzed to small molecules by enzymes external to the plasma membrane	+	D
Metabolic features		
Respiratory and photosynthetic functions and associated pigments and enzymes (e.g., chlorophylls, cytochromes), if present, are associated with the plasma membrane or invaginations thereof	+[i]	−
Chemolithotrophic type of metabolism occurs (inorganic compounds can be used as electron donors by organisms that derive energy from chemical compounds)	D	−
Ability to fix N_2	D	−
Ability to dissimilate NO_3^- to N_2O or N_2	D	−
Methanogenesis	D	−
Ability to carry out anoxygenic photosynthesis	D	−
Enzymic features		
Type of superoxide dismutase:		
Cu-Zn type	Rare	+
Mn and/or Fe type	+	−[j]
Reproductive features		
Cell division includes mitosis, and a microtubular (spindle) system is present	−	+
Meiosis occurs	−	D
Mechanisms of gene transfer and recombination, if they occur, involve gametogenesis and zygote formation	−	+
Molecular biological properties		
Number of chromosomes present per nucleoid	Usually 1	Usually 1
Chromosomes circular	+	−
Chromosomes linear	−[k]	+
Sedimentation constant of ribosomes:		
70S	+	−[l]
80S	−	+
Sedimentation constants of ribosomal RNA:		
16S, 23S, 5S	+	−
18S, 28S, 5.85S, 5S	−	+

(continued)

TABLE 1. (*continued*)

Characteristic	Procaryotes	Eucaryotes
First amino acid to initiate a polypeptide chain during protein synthesis:		
Methionine	D	+
N-Formylmethionine	D	−
Messenger-RNA binding site at AUCACCUCC at 3′ end of 16S or 18S ribosomal RNA	+	−

[a]Symbols: +, positive; −, negative; D, differs among organisms.

[b]A few bacteria (e.g., certain treponemes, mycoplasmas, *Haemobartonella*) may have a width as small as 0.1 µm; a few bacteria (e.g., *Achromatium, Macromonas*) may have a width greater than 10 µm. The largest known procaryote is a spherical, sulfur bacterium provisionally named *"Thiomargarita namibiensis"* and has a diameter of 100–750 µm. It is a member of the γ *Proteobacteria* and is related to the genus *Thioploca*. Its cytoplasm occurs as a narrow layer surounding a large, central, liquid vacuole that contains nitrate. The organism has not yet been isolated. Another large procaryote is *Epulopiscium fishelsoni*, a noncultured cigar-shaped bacterium that inhabits the intestinal tract of surgeonfish from the Red Sea; it can be larger than 80 × 600 µm. This organism is also viviparous, producing two live daughter cells within the mature cell.

[c]Gas vacuoles are not bounded by a unit membrane but by a protein. The vesicles composing the vacuoles can be caused to collapse by the sudden application of hydrostatic pressures, a feature essential to identify them. They can also be identified by electron microscopy.

[d]However, certain intracellular fibrils that may be microtubules have been reported in *Spiroplasma*, certain spirochetes, the cyanobacterium *Anabaena*, and in bacterial L forms.

[e]Bacterial endospores are usually resistant to a heat treatment of 80°C or more for 10 min, however, some types of endospores may be killed by this heat treatment and may require testing at lower temperatures.

[f]Except in membranes of most mycoplasmas.

[g]Present in virtually all Gram-negative bacteria and in many Gram-positive bacteria.

[h]Present in walled *Bacteria* except chlamydiae and planctomycetes; absent in *Archaea*.

[i]However, in cyanobacteria there may be an independence of cytoplasmic membrane and thylakoid membranes.

[j]Except in mitochondria, in which the Mn type occurs.

[k]A few bacteria such as some *Borrelia* species have linear chromosomes.

[l]Except in mitochondria and chloroplasts, which have 70S ribosomes.

multiple fission. Fruiting bodies and myxospores may be formed by the myxobacteria. Swimming motility, gliding motility, and nonmotility are commonly observed. Members may be phototrophic or nonphototrophic (both lithotrophic and heterotrophic) bacteria and include aerobic, anaerobic, facultatively anaerobic, and microaerophilic species; some members are obligate intracellular parasites.

GRAM-POSITIVE *BACTERIA* THAT HAVE A CELL WALL These *Bacteria* have a cell-wall profile of the Gram-positive type; there is no outer membrane and the peptidoglycan layer is relatively thick. Some members of the group have teichoic acids and/or neutral polysaccharides as components of the wall. A few members of the group have cell walls that contain mycolic acids. Reaction with Gram's stain is generally, but not always, positive; exceptions such as *Butyrivibrio*, which has an unusually thin wall and stains Gram-negative, may occur. Cells may be spheres, rods, or filaments; the rods and filaments may be nonbranching, but many show true branching. Cellular reproduction is generally by binary fission; some produce spores as resting forms (endospores or spores on hyphae). The members of this division include simple asporogenous and sporogenous bacteria, as well as the actinomycetes and their relatives. Gram-positive *Bacteria* are generally chemosynthetic heterotrophs and include aerobic, anaerobic, facultatively anaerobic, and microaerophilic species; some members are obligate intracellular parasites. Only one group, the heliobacteria, is photosynthetic and although these have a Gram-positive type of cell wall they nevertheless stain Gram-negative.

Table 3 provides some characteristics that help to differentiate the Gram-positive *Bacteria* from the Gram-negative *Bacteria*.

BACTERIA LACKING A CELL WALL These *Bacteria* are commonly called the mycoplasmas. They do not synthesize the precursors of peptidoglycan and are insensitive to β-lactam antibiotics or other antibiotics that inhibit cell wall synthesis. They are enclosed by a unit membrane, the plasma membrane. The cells are highly pleomorphic and range in size from large deformable vesicles to very small (0.2 µm), filterable elements. Filamentous forms with branching projections are common. Reproduction may be by budding, fragmentation, and/or binary fission. Some groups show a degree of regularity of form due to the placing of internal structures. Usually, they are nonmotile, but some species show a form of gliding motility. No resting forms are known. Cells stain Gram-negative. Most require complex media for growth (high-osmotic-pressure surroundings) and tend to penetrate the surface of solid media forming characteristic "fried egg" colonies. The organisms resemble the naked L-forms that can be generated from many species of bacteria (notably Gram-positive *Bacteria*) but differ in that the mycoplasmas are unable to revert and make cell walls. Most species are further distinguished by requiring both cholesterol and long-chain fatty acids for growth; unesterified cholesterol is a unique component of the membranes of both sterol-requiring and nonrequiring species if present in the medium. The mol% G + C content of rRNA is 43–48 (lower than the 50–54 mol% of walled Gram-negative and Gram-positive *Bacteria*); the mol% G + C content of the DNA is also relatively low, 23–46, and the genome size of the mycoplasmas at $0.5–1.0 \times 10^9$ Da is less than that of other procaryotes. The mycoplasmas may be saprophytic, parasitic, or pathogenic, and the pathogens cause diseases of animals, plants, and tissue cultures.

Archaea The *Archaea* are predominantly terrestrial and aquatic microbes, occurring in anaerobic, hypersaline, or hydrothermally and geothermally heated environments; some also occur as symbionts in animal digestive tracts. They consist of aerobes, anaerobes, and facultative anaerobes that grow chemolithoautotrophically, organotrophically, or facultatively organotrophically. *Archaea* may be mesophiles or thermophiles, with some species growing at temperatures up to 110°C.

The major groups of the *Archaea* include (a) the methanogenic *Archaea*, (b) the sulfate-reducing *Archaea*, (c) the extremely

TABLE 2. Some characteristics differentiating *Bacteria* from *Archaea*[a]

Characteristic	Bacteria	Archaea
General morphological and metabolic features		
Strict anaerobes that form methane as the predominant metabolic end product from H_2/CO_2, formate, acetate, methanol, methylamine or H_2/methanol. Cells exhibit a blue-green epifluorescence when excited at 420 nm	−	D
Strict anaerobes that form H_2S from sulfate by dissimilatory sulfate reduction. Extremely thermophilic (some grow up to 110°C). Exhibit blue-green epifluorescence when excited at 420 nm	−	D
Cells stain Gram negative or Gram positive and are aerobic or facultatively anaerobic chemoorganotrophs. Rods and regular to highly irregular cells occur. Cells require a high concentration of NaCl (1.5 M or above). Neutrophilic or alkaliphilic. Mesophilic or slightly thermophilic (up to 55°C). Some species contain the red-purple photoactive pigment bacteriorhodopsin and are able to use light for generating a proton motive force	−	D
Thermoacidophilic, aerobic, coccoid cells lacking a cell wall	−	D
Obligately thermophilic, aerobic, facultatively anaerobic, or strictly anaerobic Gram-negative rods, filaments, or cocci. Optimal growth temperature between 70°C and 105°C. Acidophiles and neutrophiles. Autotrophic or heterotrophic. Most species are sulfur metabolizers	−	D
Cell walls (if present)		
Contain muramic acid	+[b]	−
Antibiotic susceptibility		
Susceptible to β lactam antibiotics	D	−
Lipids		
Membrane phospholipids consist of:		
long chain alcohols (phytanols) that are ether linked to glycerol to form C_{20} diphytanyl glycerol diethers or C_{40} dibiphytanyl diglycerol tetraethers	−	+
Long chain aliphatic fatty acids that are ester linked to glycerol	+	−
Pathway used in formation of lipids:		
Mevalonate pathway	−	+
Malonate pathway	+	−
Molecular biological features		
Ribothymine is present in the "common arm" of the tRNAs	Usually +	−
Pseudouridine or 1-methylpseudouridine is present in the "common arm" of the tRNAs	−	+
First amino acid to initiate a polypeptide chain during protein synthesis:		
Methionine	−	+
N-Formylmethionine	+	−
Aminoacyl stem of the initiator tRNA terminates with the base pair "AU"	−	+
Protein synthesis by ribosomes inhibited by:		
Anisomycin	−	+
Kanamycin	+	−
Chloramphenicol	+	−
ADP-Ribosylation of the peptide elongation factor EF-2 is inhibited by diphtheria toxin	−	+
Elongation factor 2 (EF-2) contains the amino acid diphthamide	−	+
Some tRNA genes contain introns	−	+
DNA-dependent RNA polymerases are:		
Multicomponent enzymes	−	+
Inhibited by rifampicin and streptolydigin	+	−
Replicating DNA polymerases are inhibited by aphidicolin or butylphenyl-dGTP	+	−

[a] Symbols: +, positive; −, negative; D, differs among organisms.

[b] Except planctomycetes and chlamydiae, which have a protein wall.

halophilic *Archaea*, (d) the *Archaea* lacking cell walls, and (e) the extremely thermophilic S⁰-metabolizing *Archaea*.

A unique biochemical feature of all *Archaea* is the presence of glycerol isopranyl ether lipids. The lack of murein (peptidoglycan-containing muramic acid) in cell walls makes *Archaea* insensitive to β-lactam antibiotics. The "common arm" of the tRNAs contains pseudouridine or 1-methylpseudouridine instead of ribothymidine. The nucleotide sequences of 5S, 16S, and 23S rRNAs are very different from their counterparts in *Bacteria* and *Eucarya*.

Archaea share some molecular features with *Eucarya*: (a) the elongation factor 2 (EF-2) contains the amino acid diphthamide and is therefore ADP-ribosylable by diphtheria toxin, (b) amino acid sequences of the ribosomal "A" protein exhibit sequence homologies with the corresponding eucaryotic (L-7/L12) protein, (c) the methionyl initiator tRNA is not formylated, (d) some tRNA genes contain introns, (e) the aminoacyl stem of the initiator tRNA terminates with the base pair "AU," (f) the DNA-dependent RNA polymerases are multicomponent enzymes and are insensitive to the antibiotics rifampicin and streptolydigin, (g) like the α-DNA polymerases of eucaryotes, the replicating DNA polymerases of *Archaea* are not inhibited by aphidicolin or

butylphenyl-dGTP, and (h) protein synthesis is inhibited by anisomycin but not by chloramphenicol.

Autotrophic *Archaea* do not assimilate CO_2 via the Calvin cycle. In methanogens, autotrophic CO_2 is fixed via a pathway involving the unique coenzymes methanofuran, tetrahydromethanopterin, coenzyme F_{420}, HS-HTP, coenzyme M, HTP-SH, and coenzyme F_{430} whereas in *Acidianus* and *Thermoproteus*, autotrophic CO_2 is fixed via a reductive tricarboxylic acid pathway. Some methanogenic *Archaea* can fix N_2.

Gram stain results may be positive or negative within the same subgroup because of very different types of cell envelopes. Gram-positive-staining species possess pseudomurein, methanochondroitin, and heteropolysaccharide cell walls; Gram-negative-staining cells have (glyco-) protein surface layers. The cells may have a diversity of shapes, including spherical, spiral, plate or rod; unicellular and multicellular forms in filaments or aggregates also occur. The diameter of an individual cell may be 0.1–15 μm, and the length of the filaments can be up to 200 μm. Multiplication is by binary fission, budding, constriction, fragmentation, or unknown mechanisms. Colors of cell masses may be red, purple, pink, orange-brown, yellow, green, greenish black, gray, and white.

TABLE 3. Some characteristics differentiating Gram-positive bacteria having a cell wall from Gram-negative bacteria having a cell wall[a]

Characteristic	Gram-negative bacteria	Gram-positive bacteria
Cytological features		
An outer membrane is present (in the cell wall) in addition to the plasma (cytoplasmic) membrane	+	−
Acid-fast staining	−	D[b]
Endospores present	−[c]	D[d]
Filamentous growth with hyphae that show true branching	−	D
Locomotion		
Gliding motility occurs	D	−
Chemical features		
Percentage of the dry weight of the cell wall that is represented by lipid	Usually 11–22%	Usually 4%[e]
Teichoic or lipoteichoic acids present	−	D
Lipopolysaccharide (LPS)[f] occurs (in the outer membrane of the cell wall)	+	−
2-Keto-3-deoxyoctonate (KDO) present[g]	D	−
Percentage of the dry weight of the cell wall represented by peptidoglycan	Usually <10%	Usually 10%
Mycolic acids present	−	D[h]
Phosphatidylinositol mannosides present	−	D[i]
Phosphosphingolipids present	D[j]	−
Metabolic features		
Energy derived by the oxidation of inorganic iron, sulfur, or nitrogen compounds	−	D
Enzymic features		
Citrate synthases:		
Inhibited by reduced nicotinamide adenine dinucleotide (NADH)	Usually +[k]	Usually −
Molecular weight:		
~250,000	Usually +[l]	Usually −
~100,000	Usually −	Usually +
Succinate thiokinases, molecular weight of:		
70,000–75,000	Usually −	Usually +
140,000–150,000	Usually +	Usually −

[a] Symbols: +, positive; −, negative, D, differs among organisms.

[b] Acid-fast staining occurs in the genus *Mycobacterium*, and in some *Nocardia* species.

[c] An exception may be the genus *Coxiella*.

[d] Endospores occur in the genera *Bacillus, Clostridium, Desulfotomaculum, Sporosarcina, Thermoactinomyces, Sporomusa, Metabacterium,* and *Polyspora*.

[e] Except for *Mycobacterium, Corynebacterium, Nocardia,* and other genera whose walls contain mycolic acids.

[f] LPS consists of Lipid A (a β-linked D-glucosamine disaccharide to which phosphate residues are linked at positions 1 and 4 and fatty acids are linked to both the amino and hydroxyl groups of the glucosamines), a core polysaccharide (a short acidic heteropolysaccharide), and O antigens (side chains that are polysaccharide composed of repeating units).

[g] In many but not all Gram-negative bacteria, the core polysaccharide contains KDO which, if present, can serve as an indicator of the presence of LPS.

[h] Mycolic acids occur in *Corynebacterium, Nocardia, Mycobacterium, Bacterionema, Faenia,* and *Rhodococcus*.

[i] Present in certain actinomycete and coryneform bacteria.

[j] For example, in *Bacteroides*.

[k] Known exceptions include *Acetobacter, Thermus,* and cyanobacteria.

[l] One known exception is *Thermus*.

Classification of Procaryotic Organisms and the Concept of Bacterial Speciation

Don J. Brenner, James T. Staley and Noel R. Krieg

CLASSIFICATION NOMENCLATURE AND IDENTIFICATION

Taxonomy is the science of classification of organisms. Bacterial taxonomy consists of three separate, but interrelated areas: classification, nomenclature, and identification. Classification is the arrangement of organisms into groups (taxa) on the basis of similarities or relationships. Nomenclature is the assignment of names to the taxonomic groups according to international rules (*International Code of Nomenclature of Bacteria* [Sneath, 1992]). Identification is the practical use of a classification scheme to determine the identity of an isolate as a member of an established taxon or as a member of a previously unidentified species.

Some 4000 bacterial species thus far described (and the tens of thousands of postulated species that remain to be described) exhibit great diversity. In any endeavor aimed at an understanding of large numbers of entities it is practical, if not essential, to arrange, or classify, the objects into groups based upon their similarities. Thus classification has been used to organize the bewildering and seemingly chaotic array of individual bacteria into an orderly framework. Classification need not be scientific. Mandel said that "like cigars,... a good classification is one which satisfies" (Mandel, 1969). Cowan observed that classification is purpose oriented; thus, a successful classification is not necessarily a good one, and a good classification is not necessarily successful (Cowan, 1971, 1974).

Classification and adequate description of bacteria require knowledge of their morphologic, biochemical, physiological, and genetic characteristics. As a science, taxonomy is dynamic and subject to change on the basis of available data. New findings often necessitate changes in taxonomy, frequently resulting in changes in the existing classification, in nomenclature, in criteria for identification, and in the recognition of new species. The process of classification may be applied to existing, named taxa, or to newly described organisms. If the taxa have already been described, named, and classified, new characteristics may be added or existing characteristics may be reinterpreted to revise existing classification, update it, or formulate a new one. If the organism is new, i.e., cannot be identified as an existing taxon, it is named and described according to the rules of nomenclature and placed in an appropriate position in an existing classification, i.e., a new species in either an existing or a new genus.

Taxonomic ranks Several levels or ranks are used in bacterial classification. The highest rank is called a Domain. All procaryotic organisms (i.e., bacteria) are placed within two Domains, *Archaea* and *Bacteria*. Phylum, class, order, family, genus, species,

and subspecies are successively smaller, non-overlapping subsets of the Domain. The names of these subsets from class to subspecies are given formal recognition (have "standing in nomenclature"). An example is given in Table 1. At present, neither the kingdom nor division are used for *Bacteria*. In addition to these formal, hierarchical taxonomic categories, informal or vernacular groups that are defined by common descriptive names are often used; the names of such groups have no official standing in nomenclature. Examples of such groups are: the procaryotes, the spirochetes, dissimilatory sulfate- and sulfur-reducing bacteria, the methane-oxidizing bacteria, methanogens, etc.

SPECIES The basic and most important taxonomic group in bacterial systematics is the species. The concept of a bacterial species is less definitive than for higher organisms. This difference should not seem surprising, because bacteria, being procaryotic organisms, differ markedly from higher organisms. Sexuality, for example, is not used in bacterial species definitions because relatively few bacteria undergo conjugation. Likewise, morphologic features alone are usually of little classificatory significance because the relative morphologic simplicity of most procaryotic organisms does not provide much useful taxonomic information. Consequently, morphologic features are relegated to a less important role in bacterial taxonomy in comparison with the taxonomy of higher organisms.

The term "species" as applied to bacteria has been defined as a distinct group of strains that have certain distinguishing features and that generally bear a close resemblance to one another in the more essential features of organization. (A strain is made up of the descendants of a single isolation in pure culture, and usually is made up of a succession of cultures ultimately

TABLE 1. Taxonomic ranks

Formal rank	Example
Domain	*Bacteria*
Phylum	*Proteobacteria*
Class	*Alphaproteobacteria*
Order	*Legionellales*
Family	*Legionellaceae*
Genus	*Legionella*
Species	*Legionella pneumophila*
Subspecies	*Legionella pneumophila* subsp. *pneumophila*

derived from an initial single colony). Each species differs considerably and can be distinguished from all other species.

One strain of a species is designated as the type strain; this strain serves as the name-bearer strain of the species and is the permanent example of the species, i.e., the reference specimen for the name. (See the chapter on Nomenclature for more detailed information about nomenclatural types). The type strain has great importance for classification at the species level, because a species consists of the type strain and all other strains that are considered to be sufficiently similar to it as to warrant inclusion with it in the species. Any strain can be designated as the type strain, although, for new species, the first strain isolated is usually designated. The type strain need not be a typical strain.

The species definition given above is one that was loosely followed until the mid-1960s. Unfortunately, it is extremely subjective because one cannot accurately determine "a close resemblance", "essential features", or how many "distinguishing features" are sufficient to create a species. Species were often defined solely on the basis of relatively few phenotypic or morphologic characteristics, pathogenicity, and source of isolation. The choice of the characteristics used to define a species and the weight assigned to these characteristics frequently reflected the interests and prejudices of the investigators who described the species. These practices probably led Cowan to state that "taxonomy... is the most subjective branch of any biological science, and in many ways is more of an art than a science" (Cowan, 1965).

Edwards and Ewing (1962, 1986) were pioneers in establishing phenotypic principles for characterization, classification and identification of bacteria. They based classification and identification on the overall morphologic and biochemical pattern of a species, realizing that a single characteristic (e.g., pathogenicity, host range, or biochemical reaction) regardless of its importance was not a sufficient basis for speciation or identification. They employed a large number of biochemical tests, used a large and diverse strain sample, and expressed results as percentages. They also realized that atypical strains, when adequately studied, are often perfectly typical members of a given biogroup (biovar) within an existing species, or typical members of a new species.

Numerical taxonomic methods further improved the validity of phenotypic identification by further increasing the number of tests used, usually to 100–200, and by calculating coefficients of similarity between strains and species (Sneath and Sokal, 1973). Although there is no similarity value that defines a taxospecies (species determined by numerical taxonomy), 80% similarity is commonly seen among strains in a given taxospecies. Despite the additional tests and added sensitivity of numerical taxonomy, even a battery of 300 tests would assess only between 5–20% of the genetic potential of bacteria.

It has long been recognized that the most accurate basis for classification is phylogenetic. Kluyver and van Neil (1936) stated that "many systems of classification are almost entirely the outcome of purely practical considerations . . . (and) are often ultimately impractical . . ." They recognized that "taxonomic boundaries imposed by the intuition of investigators will always be somewhat arbitrary — especially at the ultimate systematic unit, the species. One must create as many species as there are organisms that differ in sufficiently fundamental characters" and they realized that "the only truly scientific foundation of classification is in appreciating the available facts from a phylogenetic view". The data necessary to develop a natural (phylogenetic)

species definition became available when DNA hybridization was utilized to determine relatedness among bacteria.

DNA hybridization is based upon the ability of native (double-stranded) DNA to reversibly dissociate or be denatured into its two complementary single strands. Dissociation is accomplished at high temperature. Denatured DNA will remain as single strands when it is quickly cooled to room temperature after denaturation. If it is then placed at a temperature between 25 and 30°C below its denaturation point, the complementary strains will reassociate to again form a double-stranded molecule that is extremely similar, if not identical, to native DNA (Marmur and Doty, 1961). Denatured DNA from a given bacterium can be incubated with denatured DNA (or RNA) from other bacteria and will form heteroduplexes with any complementary sequences present in the heterologous strand–DNA hybridization. This is the method used to determine DNA relatedness among bacteria.

Perfectly complementary sequences are not necessary for hybridization; the degree of complementary required for heteroduplex formation can be governed experimentally by changing the incubation temperature or the salt concentration. Increasing the incubation temperature and/or lowering the salt concentration in the incubation mixture increases the stringency of heteroduplex formation (fewer unpaired bases are tolerated), whereas decreasing the temperature and/or increasing the salt concentration decreases the stringency of heteroduplex formation. The percentage of unpaired bases within a heteroduplex is an indication of the degree of divergence present. One can approximate the amount of unpaired bases by comparing the thermal stability of the heteroduplex to the thermal stability of a homologous duplex. This is done by stepwise increases in temperature and measuring strand separation. The thermal stability is calculated as the temperature at which 50% of strand separation has occurred and is represented by the term "$T_{m(e)}$".

The ΔT_m values of heteroduplexes range from 0 (perfect pairing) to ~20°C, with each degree of instability indicative of approximately 1% divergence (unpaired bases). As DNA relatedness between two strains decreases, divergence usually increases.

A number of different DNA–DNA and DNA–RNA hybridization methods have been used to determine relatedness among bacteria (Johnson, 1985). Two of these, free solution reassociation with separation of single- and double-stranded DNA on hydroxyapatite (Brenner et al., 1982) and the S-1 endonuclease method (Crosa et al., 1973) are currently the most widely used for this purpose. These methods have been shown to be comparable (Grimont et al., 1980). An in-depth discussion of DNA hybridization methods has been presented by Grimont et al. (1980) and by Johnson (1985).

Experience with thousands of strains from several hundred well-established and new species led taxonomists to formulate a phylogenetic definition of a species (genomospecies) as "strains with approximately 70% or greater DNA–DNA relatedness and with 5°C or less ΔT_m. Both values must be considered" (Wayne et al., 1987). They further recommended that a genomospecies not be named if it cannot be differentiated from other genomospecies on the basis of some phenotypic property. DNA relatedness provides a single species definition that can be applied equally to all organisms and is not subject to phenotypic variation, mutations, or variations in metabolic or other plasmids. The major advantage of DNA relatedness is that it measures overall relatedness, and therefore the effects of atypical biochemical reactions, mutations, and plasmids are minimal since they affect only a very small percentage of the total DNA.

Once genomospecies have been established, it is simple to determine which variable biochemical reactions are species specific, and therefore to have an identification scheme that is compatible with the genetic concept of species. The technique is also extremely useful in determining the biochemical boundaries of a species, as exemplified for *Escherichia coli* in Table 2. The use of DNA relatedness and a variety of phenotypic characteristics in classifying bacteria has been called polyphasic taxonomy (Colwell, 1970), and seems to be the best approach to a valid description of species. DNA relatedness studies have now been carried out on more than 10,000 strains representing some 2000 species and hundreds of genera, with, to our knowledge, no instance where other data invalidated the genomospecies definition.

Stackebrandt and Goebel (1994) reviewed new species descriptions published in the *International Journal of Systematic Bacteriology*. In 1987, 60% of species descriptions included DNA relatedness studies, 10% were described on the basis of serologic tests, and 30% did not use these approaches. In 1993, 75% of species descriptions included DNA relatedness data, 8% used serology, and 3% used neither method. In the remaining 14%, 16S rRNA sequence analysis was the sole basis for speciation. As 16S rRNA sequence data have accumulated, the utility of this extremely powerful method for phylogenetic placement of bacteria has become evident (Woese, 1987; Ludwig et al., 1998b). The number of taxonomists using 16S rRNA sequencing is or soon will be greater than the number using DNA hybridization (Stackebrandt and Goebel, 1994), and many of them were creating species solely or largely on the basis of 16S rRNA sequence analysis. It soon became evident, however, that 16S rRNA sequence analysis was frequently not sensitive enough to differentiate between closely related species (Fox et al., 1992; Stackebrandt and Goebel, 1994). Stackebrandt and Goebel (1994) concluded that the genetic definition of 70% relatedness with 5%

or less divergence within related sequences continues to be the best means of creating species. They concluded that 16S rRNA sequence similarity of less than 97% between strains indicates that they represent different species, but at 97% or higher 16S rRNA sequence similarity, DNA relatedness must be used to determine whether strains belong to different species.

The validity and utility of the DNA relatedness based genetic definition of a species has been questioned (Maynard Smith, 1995; Vandamme et al., 1996a; Istock et al., 1996). These criticisms fall into several categories: (a) DNA relatedness (and any other current means of speciation) does not sufficiently sample bacterial diversity by employing large numbers of wild isolates from many different habitats; (b) it employs an arbitrary cutoff for a species whereas evolution is a continuum; (c) the DNA-relatedness based definition does not achieve standardization of species; (d) bacterial species are not real entities—named species are useful but not meaningful from an evolutionary standpoint; (e) DNA relatedness results are not comparable due to different methods; (f) DNA relatedness tests are too difficult and/or tedious to perform. In view of these perceived problems, it has been recommended that the best solution to the species problem in the absence of a "gold standard", which has not been provided by DNA relatedness, is a pragmatic polyphasic (consensus) taxonomy that integrates all available data.

Each of these criticisms has some merit; however each can be addressed, and none, in our opinion, represent fatal flaws nor significantly negate the usefulness of the DNA-relatedness based definition of a species. Large numbers of diverse strains (50–100) have been tested for DNA relatedness in a number of species including *E. coli*, *Legionella pneumophila*, *Enterobacter agglomerans*, *Klebsiella oxytoca*, *Yersinia enterocolitica*. In no case did the sample size or the diversity of sources and/or phenotypic characteristics change the results. For many other species only one or a few strains were tested—usually because that was the total number of strains available.

It is true that the 70% relatedness and 5% divergence values chosen to represent strains of a given species are arbitrary, and that there is a "gray area" around 70% for some species. Nonetheless, these values were chosen on the basis of results obtained from multiple strains, usually 10 or more, of some 600 species studied in a number of different reference laboratories. There are few, if any cases, in which the species defined in this manner have been shown to be incorrect.

The DNA relatedness approach has standardized the means of defining species by providing a single, universally applicable criterion. Since it has been successful, one must believe that it generates species that are compatible with the needs and beliefs of most bacteriologists. There are two areas in which genomospecies have actually or potentially caused problems. One of these is where two or more genomospecies cannot be separated phenotypically. In this case it has been recommended that these genomospecies not be formally named (Wayne et al., 1987). Alternatively, especially if a name already exists for one of the genomospecies, the others can be designated as subspecies. In this way there is no confusion at the species level and, one can, if one wishes, distinguish between the genomospecies using a genetic technique. The other "problem" is with nomenspecies that were split or lumped, usually on the basis of pathogenicity or phytopathogenic host range. These include species in the genera *Bordetella*, *Mycobacterium*, *Brucella*, *Shigella*, *Klebsiella*, *Neisseria*, *Yersinia*, *Vibrio*, *Clostridium*, and *Erwinia*. In some of these cases (*Klebsiella*, *Erwinia*) the classification has been changed and is now

TABLE 2. Classification of atypical strains that could be *E. coli*

Relatedness of biogroup to typical *E. coli*	Characteristic
80% or more	Urea positive and KCN positive
	Mannitol negative
	Inositol positive
	Adonitol positive
	H₂S positive or H₂S positive and yellow pigmented
	H₂S positive and citrate positive
	Citrate positive
	Phenylalanine deaminase positive
	Lysine and ornithine decarboxylase and arginine dihydrolase negative
	Indol negative
	Methyl red negative
	Methyl red negative and mannitol negative
	Urea positive and mannitol negative
	Anaerogenic, nonmotile, and lactose negative
60% or less	Yellow pigment, cellobiose positive, and KCN positive = *Escherichia hermannii*
	Urea positive, KCN positive, citrate positive, cellobiose positive = *Citrobacter amalonaticus*

accepted. In the others, changes have not yet been proposed or, as in the case of *Yersinia pestis* and *Yersinia pseudotuberculosis*, which are the same genomospecies, the change was rejected by the Judicial Commission because of possible danger to public health if there was confusion regarding *Y. pestis*, the plague bacillus.

If one agrees that a true species definition is not possible, the genomospecies definition is still useful in providing a single, universally applicable basis for designating species.

To criticize DNA relatedness because results obtained using different methods may not be totally comparable seems somewhat unjustified. When compared, the most frequently used methods have given similar results. Obviously, one should be careful in comparing data from various laboratories, especially when different methods are used. However, this is at least equally true for sequence data and phenotypic tests.

It is true that large amounts of DNA are required for the DNA relatedness protocols now used for taxonomic purposes, and that it is necessary to use radioactive isotopes. As for the difficulty involved and the limitations in strains that can be assayed (it is not uncommon to do 40–80 DNA relatedness comparisons daily), surely these are not credible reasons to stop using the method. Efforts can and should be made to automate the system, to miniaturize it, and to substitute nonradioactive compounds for the radioactive isotopes. With these improvements, the method will be available for use in virtually any laboratory. Even without them, one can argue that DNA hybridization is more affordable and practical than a consensus classification system in which several hundred tests must be done on each strain.

It is noteworthy that bacterial species can be compared to higher organisms on a molecular basis using mol% G + C range, DNA–DNA or DNA–rRNA relatedness, and similarity of 16S vs. 18S rDNA sequences (Staley, 1997, 1999). Thus, *E. coli* can be compared with its primate hosts based on the results of DNA–DNA hybridization. When this is done, it is apparent that the bacterial species is much broader than that of its hosts. For example, humans and our closest relative, the chimpanzee (*Pan troglodytes*), show 98.4% relatedness by this technique (Sibley and Ahlquist, 1987; Sibley et al., 1990). Indeed, even lemurs, which exhibit 78% DNA relatedness with humans, would be included in the same species as humans if the definition of a bacterial species was used. Furthermore, none of the primates would be considered to be threatened species using the bacterial definition. Likewise, the range of mol% G + C and the range of small subunit ribosomal RNA within *E. coli* strains shows a similar result, namely, that the bacterial species is much broader than that of animals (Staley, 1999).

One consequence of the broad bacterial species definition is that very few species have been described, fewer than 5000, compared with over a million animals. This has led some biologists to erroneously conclude that bacteria comprise only a minor part of the biological diversity on Earth (Mayr, 1998). In addition, with such a broad definition, not a single free-living bacterial species can be considered to be threatened with extinction (Staley, 1997). Therefore, biologists should realize, as mentioned earlier in this section, that the bacterial species is not at all equivalent to that of plants and animals.

In summary, the genetic definition of a species, if not perfect, appears to be both reliable and stable. DNA relatedness studies have already resolved many instances of confusion concerning which strains belong to a given species, as well as for resolving taxonomic problems at the species level. It has not been replaced as the current reference standard. It should remain the standard, at least until another approach has been compared to it and shown to be comparable or superior.

SUBSPECIES   A species may be divided into two or more subspecies based on consistent phenotypic variations or on genetically determined clusters of strains within the species. There is evidence that the subspecies concept is phylogenetically valid on the basis of frequency distribution of ΔT_m values. There are presently essentially no guidelines for the establishment of subspecies, which, although frequently useful, are usually designated at the pleasure of the investigator. Subspecies is the lowest taxonomic rank that is covered by the rules of nomenclature and has official standing in nomenclature.

INFRASUBSPECIFIC RANKS   Ranks below subspecies, such as biovars, serovars, phagovars, and pathovars, are often used to indicate groups of strains that can be distinguished by some special character, such as antigenic makeup, reactions to bacteriophage, etc. Such ranks have no official standing in nomenclature, but often have great practical usefulness. A list of some common infrasubspecific categories is given in Table 3.

GENUS   All species are assigned to a genus, which can be functionally defined as one or more species with the same general phenotypic characteristics, and which cluster together on the basis of 16S rRNA sequence. In this regard, bacteriologists conform to the binomial system of nomenclature of Linnaeus in which the organism is designated by its combined genus and species names. There is not, and perhaps never will be, a satisfactory definition of a genus, despite the fact that most new genera are designated substantially on the basis of 16S rRNA sequence analysis. In almost all cases, genera can be differentiated phenotypically, although a considerable degree of flexibility in genus descriptions is often needed. Considerable subjectivity continues to be involved in designating genera, and considerable reclassification, both lumping and splitting, is still occurring at the genus level. Indeed, what is perceived to be a single genus by one systematist may be perceived as multiple genera by another.

HIGHER TAXA   Classificatory relationships at the familial and higher levels are even less certain than those at the genus level, and descriptions of these taxa are usually much more general, if they exist at all. Families are composed of one or more genera that share phenotypic characteristics and that should be consistent from a phylogenetic standpoint (16S rRNA sequence clustering) as well as from a phenotypic basis.

MAJOR DEVELOPMENTS IN BACTERIAL CLASSIFICATION

A century elapsed between Antony van Leeuwenhoek's discovery of bacteria and Müller's initial acknowledgement of bacteria in

TABLE 3. Infrasubspecific designations

Preferred name	Synonym	Applied to strains having:
Biovar	Biotype	Special biochemical or physiologic properties
Serovar	Serotype	Distinctive antigenic properties
Pathovar	Pathotype	Pathogenic properties for certain hosts
Phagovar	Phage type	Ability to be lysed by certain bacteriophages
Morphovar	Morphotype	Special morphologic features

a classification scheme (Müller, 1786). Another century passed before techniques and procedures had advanced sufficiently to permit a fairly inclusive and meaningful classification of these organisms. For a comprehensive review of the early development of bacterial classification, readers should consult the introductory sections of the first, second, and third editions of *Bergey's Manual of Determinative Bacteriology*. A less detailed treatment of early classifications can be found in the sixth edition of the *Manual*, in which post-1923 developments were emphasized.

Two primary difficulties beset early bacterial classification systems. First, they relied heavily upon morphologic criteria. For example, cell shape was often considered to be an extremely important feature. Thus, the cocci were often classified together in one group (family or order). In contrast, contemporary schemes rely much more strongly on 16S rRNA sequence similarities and physiological characteristics. For example, the fermentative cocci are now separated from the photosynthetic cocci, which are separated from the methanogenic cocci, which are in turn separated from the nitrifying cocci, and so forth; with the 16S rRNA sequences of each group generally clustered together. Secondly, the pure culture technique which revolutionized bacteriology was not developed until the latter half of the 19th century. In addition to dispelling the concept of "polymorphism", this technical development of Robert Koch's laboratory had great impact on the development of modern procedures in bacterial systematics. Pure cultures are analogous to herbarium specimens in botany. However, pure cultures are much more useful because they can be (a) maintained in a viable state, (b) subcultured, (c) subjected indefinitely to experimental tests, and (d) shipped from one laboratory to another. A natural outgrowth of the pure culture technique was the establishment of type strains of species which are deposited in repositories referred to as "culture collections" (a more accurate term would be "strain collections"). These type strains can be obtained from culture collections and used as reference strains to duplicate and extend the observations of others, and for direct comparison with new isolates.

Before the development of computer-assisted numerical taxonomy and subsequent taxonomic methods based on molecular biology, the traditional method of classifying bacteria was to characterize them as thoroughly as possible and then to arrange them according to the intuitive judgment of the systematist. Although the subjective aspects of this method resulted in classifications that were often drastically revised by other systematists who were likely to make different intuitive judgments, many of the arrangements have survived to the present day, even under scrutiny by modern methods. One explanation for this is that the systematists usually knew their organisms thoroughly, and their intuitive judgments were based on a wealth of information. Their data, while not computer processed, were at least processed by an active mind to give fairly accurate impressions of the relationships existing between organisms. Moreover, some of the characteristics that were given great weight in classification were, in fact, highly correlated with many characteristics. This principle of correlation of characteristics appears to have started with Winslow and Winslow (1908), who noted that parasitic cocci tended to grow poorly on ordinary nutrient media, were strongly Gram-positive, and formed acid from sugars, in contrast to saprophytic cocci which grew abundantly on ordinary media, were generally only weakly Gram-positive and formed no acid. This division of the cocci studied by the Winslows (equivalent to the present genus *Micrococcus* (the saprophytes) and the genera *Staphylococcus* and *Streptococcus* (the parasites) has held up reasonably well even to the present day.

Other classifications have not been so fortunate. A classic example of one which has not is that of the genus "*Paracolobactrum*". This genus was proposed in 1944 and is described in the Seventh Edition of *Bergey's Manual* in 1957. It was created to contain certain lactose-negative members of the family *Enterobacteriaceae*. Because of the importance of a lactose-negative reaction in identification of enteric pathogens (i.e., *Salmonella* and *Shigella*), the reaction was mistakenly given great taxonomic weight in classification as well. However, for the organisms placed in "*Paracolobactrum*", the lactose reaction was not highly correlated with other characteristics. In fact, the organisms were merely lactose-negative variants of other lactose-positive species; for example "*Paracolobactrum coliform*" resembled *E. coli* in every way except in being lactose-negative. Absurd arrangements such as this eventually led to the development of more objective methods of classification, i.e., numerical taxonomy, in order to avoid giving great weight to any single characteristic.

Phylogenetic Classifications   We have already discussed the impact of DNA relatedness at the species level. Unfortunately, this method is of marginal value at the genus level and of no value above the genus level because the extent of divergence of total bacterial genomes is too great to allow accurate assessment of relatedness above the species level. At the genus level and above, phylogenetic classifications, especially as based on 16S rRNA sequence analysis, have revolutionized bacterial taxonomy (see Overview: A Phylogenetic Backbone and Taxonomic Framework for Procaryotic Systematics by Ludwig and Klenk).

Official Classifications   A significant number of bacteriologists have the impression that there is an "official classification" and that the classification presented in *Bergey's Manual* represents this "official classification". It is important to correct that misimpression. There is no "official classification" of bacteria. (This is in contrast to bacterial nomenclature, where each taxon has one [and usually only one] valid name, according to internationally agreed-upon rules, and judicial decisions are rendered in instances of controversy about the validity of a name.) The closest approximation to an "official classification" of bacteria would be one that is widely accepted by the community of microbiologists. A classification that is of little use to bacteriologists, regardless of how fine a scheme or who devised it, will soon be ignored or significantly modified. The editors of *Bergey's Manual* and the authors of each chapter make substantial efforts to provide a classification that is as accurate and up-to-date as possible, however it is not and cannot be "official".

It also seems worthwhile to emphasize something that has often been said before, viz. bacterial classifications are devised for microbiologists, not for the entities being classified. Bacteria show little interest in the matter of their classification. For the systematist, this is sometimes a very sobering thought!

ACKNOWLEDGMENTS

This chapter is dedicated to the memory of John L. Johnson, a consummate scientist, trusted colleague and friend, whose search for truth was uncompromising and unhindered by personal ego.

Identification of Procaryotes

Noel R. Krieg

THE NATURE OF IDENTIFICATION SCHEMES

Identification schemes are not classification schemes, although there may be a superficial similarity. An identification scheme for a group of organisms can be devised only **after** that group has first been classified (i.e., recognized as being different from other organisms). Identification of that group is based on one or more characteristics, or on a pattern of characteristics, which all the members of the group have and which other groups do not have.

The particular pattern of characteristics used for identifying a bacterial group should not be found in any other bacterial group. Following classification of the group, a relatively few characteristics which, taken together, are unique to that group are selected. The identifying characteristics may be phenotypic, such as cell shape and Gram reaction or the ability to ferment certain sugars, or they may be genotypic, such as a particular nucleotide sequence.

PURE CULTURES

Although it is possible to identify specific organisms, and even individual cells, in a mixed culture, pure cultures are usually used for identification. Moreover, in most laboratories identification is still being done mainly on the basis of the phenotypic characteristics of the culture, although it may be aided by commercial multitest identification systems, usually involving 96–well microtiter plates, that are capable of determining a variety of characteristics easily and quickly. Phenotypic identification systems work reasonably well with pure cultures, but if the culture is not pure the results will be a composite from all of the different organisms in the culture and thus can be very misleading.

In obtaining a pure culture, it is important to realize that the selection of a single colony from a plate does not necessarily assure purity. This is especially true if selective media are used; live but non-growing contaminants may often be present in or near a colony and can be subcultured along with the chosen organism. It is for this reason that non-selective media are preferred for final isolation, because they allow such contaminants to develop into visible colonies. Even with non-selective media, apparently well-isolated colonies should not be isolated too soon; some contaminants may be slow growing and may appear on the plate only after a longer incubation. Another difficulty occurs with bacteria that form extracellular slime or that grow as a network of chains or filaments; contaminants often become firmly embedded or entrapped in such matrixes and are difficult to remove. In the instance of cyanobacteria, contaminants frequently penetrate and live in the gelatinous sheaths that surround the cells, making pure cultures difficult to obtain.

In general, colonies from a pure culture that has been streaked on a solid medium are similar to one another, providing evidence of purity. Although this is generally true, there are exceptions, as in the case of S → R variation, capsular variants, pigmented or nonpigmented variants, etc., which may be selected by certain media, temperatures, or other growth conditions. Another criterion of purity is morphology: organisms from a pure culture generally exhibit a high degree of morphological similarity in stains or wet mounts. Again, there are exceptions, coccoid body formation, cyst formation, spore formation, pleomorphism, etc., depending on the age of the culture, the medium used, and other growth conditions. For example, examination of a broth culture of a marine spirillum after 2 or 3 days may lead one to believe the culture is highly contaminated with cocci, unless one is previously aware that following active growth such spirilla generally develop into thin-walled coccoid forms.

Universal Systems for Identifying a Pure Culture Although the goal of identification is merely to provide the name of an isolate, most identification systems depend on first determining a number of morphological, biochemical, cultural, antigenic, and other phenotypic characteristics of the isolate before the name can be assigned. An ideal universal system would be one that provides the name without having to determine these characteristics. In a sense, such a system would be a kind of "black box" into which the isolate, or an extract of it, is placed, to be followed some time later by a display of the name of the organism.

One system that has proven extremely useful is automated cellular fatty acid (CFA) analysis (Onderdonk and Sasser, 1995). The system depends on saponifying the fatty acids with sodium hydroxide, converting them to their volatile methyl esters, and then separating and quantifying each fatty acid by gas-liquid chromatography. A computer compares the resulting fatty acid profile with thousands of others in a huge database and calculates the best match or matches for the isolate. The computer can also indicate that an isolate does not closely match any other fatty acid profile, which can lead to discovery of new genera or species. The entire procedure is simple and takes about 2 h, and numerous specimens can be analyzed rapidly each day. One drawback is that the isolate must be cultured under highly standardized conditions of media and temperature in order to provide a valid basis of comparison with other fatty acid profiles. Another drawback is that the system may not be able to differentiate species that are very closely related by DNA–DNA hybridization, for example, *Escherichia coli* and *Shigella*. Still another drawback is that the system is extremely expensive to purchase or lease. Some commercial laboratories will perform the entire identification procedure on an isolate that is sent to them; this is helpful for one or a few isolates but becomes expensive if many isolates are to be identified.

A second universal system, and the one of choice at present, is one in which all or most of the nucleotide sequence of the 16s rRNA gene of an unknown isolate is determined. DNA is isolated from the strain and then universal primers are used to amplify the 16S rDNA by the polymerase chain reaction (PCR). The sequence of the PCR product is compared with other sequences stored in an enormous database. One such database is that used in the Ribosomal Database Project-II (RDP-II), which is a cooperative effort by scientists at Michigan State University and the Lawrence Berkeley National Laboratory. 16S sequencing is rapid enough to handle a large number of isolates in a short time; this service is provided by a number of institutions for medical and other types of isolates. Given a well-equipped sequencing lab, 16S sequences can be obtained and analyzed within 48 hours. The main drawback with sequence-based identification is that of the need for sophisticated equipment, which is present in relatively few microbiology labs. Another drawback is that although sequence-based identification is very effective for assigning an isolate to its most likely genus, it may not be able to identify an isolate to the species level if the sequences for two or more related species have greater than 97% similarity.

Traditional Identification Schemes for Identifying a Pure Culture **Phenotypic characteristics chosen for an identification scheme should be easily determinable by most microbiology laboratories.** Such characteristics should not be restricted to research laboratories or special facilities. Characteristics useful for identification are often not those that were involved in classification of the group. Classification might be based on a DNA–DNA hybridization study or on ribosomal RNA gene sequencing, whereas identification might be based on a few phenotypic characteristics that have been found to correlate well with the genetic information. Serological reactions, which generally have only limited value for classification, often have enormous value for identification. Slide agglutination tests, fluorescent antibody techniques, and other serological methods can be performed simply and rapidly and are usually highly specific; therefore, they offer a means for achieving quick, presumptive identification of bacteria. Their specificity is frequently not absolute, however, and confirmation of the identification by additional tests is usually required.

The goal of having easily determinable identifying characteristics may not always be possible, particularly with genera or species that are not susceptible to being identified by traditional phenotypic tests. For instance, the inability of *Campylobacter* species to use sugars makes phenotypic identification of species of this genus much more difficult than, say, the species of the family *Enterobacteriaceae*. In such instances one may need to resort to less common phenotypic characteristics such as the ability to grow at a specific temperature, antibiotic susceptibilities, and the ability to grow anaerobically with various electron acceptors such as trimethylamine oxide. There may even be a requirement for more sophisticated procedures, such as the use of cellular lipid patterns, DNA–DNA hybridization, or nucleic acid probes, in order to achieve an accurate identification. It may even be necessary to send the culture in question to a major reference facility that has the necessary equipment and technical expertise.

Identification of a strain should depend on a pattern of several characteristics, not merely one or a very few characteristics. If one feature is given great importance, it is possible that some strains may be mutants that do not exhibit that particular characteristic yet do have the other identifying features. For instance, hippurate hydrolysis was given great emphasis in differentiating *Campylobacter jejuni* from other *Campylobacter* species, but later it was discovered that hippurate-negative strains may occur. At first, these hippurate-negative strains were incorrectly thought to belong to a different species, *Campylobacter coli*, until DNA–DNA hybridization experiments showed that this was not correct.

Identification should rely on relatively few characteristics compared to classification schemes. Classification may involve hundreds of characteristics, as in a numerical taxonomy study but the prospect of inoculating hundreds of tubes of media in order to identify a strain is daunting. It may be possible, however, to use a large number of characteristics if they can be determined easily. To alleviate the need for inoculating large numbers of tubed media, a variety of convenient and rapid multitest systems have been devised and are commercially available for use in identifying particular groups of bacteria, particularly those of medical importance. A summary of some of these systems has been given by Smibert and Krieg (1993) and Miller and O'Hara (1995) but new systems are being developed continually. Each manufacturer provides charts, tables, coding systems, and characterization profiles for use with the particular multitest system being offered. It is important to realize that each system is for use in identifying only certain taxa and may not be applicable to other taxa. For instance, the commercial systems for identifying members of the family *Enterobacteriaceae* would give results that would be meaningless for identifying *Campylobacter* species.

Determination of the characteristics chosen for an identification scheme should be relatively inexpensive. Ordinary microbiology laboratories may not be able to afford expensive apparatus such as those required for cellular fatty acid profiles, 16S rRNA gene sequencing, or DNA probes. In regard to the latter, commercial kits for using DNA probes to identify particular taxa may be simple to use but may also be quite expensive. In general, such probes are best reserved for situations where it is essential to make a definitive identification because no other method will suffice.

The identification scheme should give results rapidly. This is especially true in clinical microbiology laboratories, where the treatment of a patient often depends on a rapid (but accurate) identification, and sometimes even a presumptive identification, of a pathogen. Serological methods have long been used for rapid detection of antigens associated with a particular species. For instance, a swab of the throat of a person with suspected case of streptococcal pharyngitis can be treated to extract the Lancefield Group A polysaccharide indicative of *Streptococcus pyogenes*. Anti-Group A antibodies can then be used in various ways, such as an ELISA test, to identify this antigen. Fluorescent antibodies can be used to obtain presumptive identification of individual cells in a mixture. For instance, cells of *Streptococcus pyogenes* can be seen in a swab from streptococcal pharyngitis by using fluorescent Lancefield Group A antiserum, and cells of *Vibrio cholerae* can be seen in diarrheic stools of cholera patients by using fluorescent O Group I antiserum. Antibodies are not always completely specific, however, and definitive identification usually requires isolation of the organism and determination of various identifying features.

Need for standardized test methods. One difficulty in devising identification schemes based on phenotypic characteristics is that the results of characterization tests may vary depending on the size of the inoculum, incubation temperature, length of the incubation period, composition of the medium, the surface-to-volume ratio of the medium, and the criteria used to define a "positive" or "negative" reaction. Therefore, the results of characterization tests obtained by one laboratory often do not match exactly those obtained by another laboratory, although the results within each laboratory may be quite consistent. The blind acceptance of an identification scheme without reference to the particular conditions employed by those who devised the scheme can lead to error (and, unfortunately, such conditions are not always specified). Ideally, it would be desirable to standardize the conditions used for testing various charactcristics, but this is easier said than done, especially on an international basis. The use of commercial multitest systems offers some hope of improving standardization among various laboratories because of the high degree of quality control exercised over the media and reagents, but no one system has yet been agreed on for universal use with any given taxon. **It is therefore advisable to always include strains whose identity has been firmly established** (type or reference strains, available from national culture collections) **for comparative purposes when making use of an identification scheme,** to make sure that the scheme is valid for the conditions employed in one's own laboratory.

Need for definitions of "positive" and "negative" reactions. Some tests may be found to be based on plasmid- or phage-mediated characteristics; such characteristics may be highly mutable and therefore unreliable for identification purposes. Even with immutable characteristics, certain tests are not well suited for use in identification schemes because they may not give highly reproducible results (e.g., the catalase test, oxidase test, Voges-Proskauer test, and gelatin liquefaction are notorious in this regard). Ideally, a test should give reproducible results that are clearly either positive or negative, without equivocal reactions. In fact, no such test exists. The Gram reaction of an organism may be "Gram variable," the presence of endospores in a strain that makes only a few spores may be very difficult to determine by staining or by heat-resistance tests, acid production from sugars may be difficult to distinguish from no acid production if only small amounts of acid are produced, and a weak growth response may not be clearly distinguishable from "no growth". A precise (although arbitrary) definition of what constitutes a "positive" and "negative" reaction is often important in order for a test to be useful for an identification scheme.

Sequence of tests used in identifying an isolate. In identifying an isolate, it is important to determine the **most general features first**. For instance, it would not be wise to begin by determining that melibiose is fermented, gelatin is liquefied, and that nitrate is reduced. Instead, it is better to begin with more general features such as the Gram staining reaction, morphology, and general type of metabolism. It is important to establish whether the new isolate is a chemolithotrophic autotroph, a photosynthetic organism, or a chemoheterotrophic organism. Living cells should be examined by phase-contrast microscopy and Gram-stained cells by light microscopy; other stains can be applied if this seems appropriate. If some outstanding morphological property, such as endospore production, sheaths, holdfasts, acid-fastness, cysts, stalks, fruiting bodies, budding division, or true branching, is obvious, then further efforts in identification can be confined to those groups having such a property. Whether or not the organisms are motile, and the type of motility (swimming, gliding) may be very helpful in restricting the range of possibilities. Gross

growth characteristics, such as pigmentation, mucoid colonies, swarming, or a minute size, may also provide valuable clues to identification. For example, a motile, Gram-negative rod that produces a water-soluble fluorescent pigment is likely to be a *Pseudomonas* species, whereas one that forms bioluminescent colonies is likely to belong to the family *Vibrionaceae*.

The **source** of the isolate can also help to narrow the field of possibilities. For example, a spirillum isolated from coastal sea water is likely to be an *Oceanospirillum* species, whereas Gram-positive cocci occurring in grape-like clusters and isolated from the human nasopharynx are likely to belong to the genus *Staphylococcus*.

The relationship of the isolate to oxygen (i.e., whether it is aerobic, anaerobic, facultatively anaerobic, or microaerophilic) is often of fundamental importance in identification. For example, a small microaerophilic vibrio isolated from a case of diarrhea is likely to be a *Campylobacter* species, whereas a Gram-negative anaerobic rod isolated from a wound infection may well be a member of the genera *Bacteroides, Prevotella, Porphrymomonas,* or *Fusobacterium*. Similarly, it is important to test the isolate for its ability to dissimilate glucose (or other simple sugars) to determine if the type of metabolism is oxidative or fermentative, or whether sugars are catabolized at all.

Above all, common sense should be used at each stage, as the possibilities are narrowed, in deciding what additional tests should be performed. There should be a reason for the selection of each test, in contrast to a "shotgun" type of approach where many tests are used but most provide little pertinent information for the particular isolate under investigation. As the category to which the isolate belongs becomes increasingly delineated, one should follow the specific tests indicated in the particular diagnostic tables or keys that apply to that category.

The following summary is taken from "The Mechanism of Identification" by S.T. Cowan and J. Liston in the eighth edition of the *Manual*, with some modifications:

1. Make sure that you have a pure culture.
2. Work from broad categories down to a smaller, specific category of organism.
3. Use all the information available to you in order to narrow the range of possibilities.
4. Apply common sense at each step.
5. Use the minimum number of tests to make the identification.
6. Compare your isolate to type or reference strains of the pertinent taxon to make sure the **identification** scheme being used is actually valid for the conditions in your particular laboratory.

If, as may well happen, you cannot identify your isolate from the information contained in the *Manual*, neither despair nor immediately assume that you have isolated a new genus or species; many of the problems of microbial classification are the result of people jumping to this conclusion prematurely. When you fail to identify your isolate, check (a) its **purity,** (b) that you have carried out the **appropriate tests,** (c) that your **methods are reliable,** and (d) that you have used correctly the various keys and tables of the *Manual*. It has been said that the most frequent cause of mistaken identity of bacteria is error in the determination of shape, Gram-staining reaction, and motility. In most cases, you should have little difficulty in placing your isolate into a genus; allocation to a species or subspecies may need the help of a specialized reference laboratory.

On the other hand, it is always possible that you have actually isolated a new genus or species. A comparison of the present edition of the *Manual* with the previous edition indicates that many new genera and species have been added. Undoubtedly, there exist in nature a great number of bacteria that have not yet been classified and therefore cannot yet be identified by existing schemes. However, before describing and naming a new taxon, one must be **very sure that it is really a new taxon** and not merely the result of an inadequate identification.

USE OF PROBES FOR FOR IDENTIFICATION OF A PARTICULAR SPECIES

DNA probes have made it possible to identify an isolate definitively without relying on phenotypic tests. A probe is a single-stranded DNA sequence that can be used to identify an organism by forming a "hybrid" with a unique complementary sequence on the DNA or rRNA of that organism. Using probes as a "shot-gun" approach to identification of an isolate, however, is costly and time-consuming. In general, probes are mainly used to verify the identification of an isolate after the microbiologist already has fairly good clues as to its identity.

Whether a probe consists of only a few nucleotides or many nucleotides, it must be specific for the particular species and must not bind to the DNA of other species. Also, the probe must have a label attached to it so that if it forms a hybrid duplex with a complementary sequence, that duplex can be readily detected.

Labeling can be accomplished by incorporating a radioactive isotope such as ^{32}P into the probe so that the hybrid duplex will be detectable by exposure to a photographic film. Because working with radioisotopes is dangerous and requires safe radioactive waste disposal, nonradioactive labeling of probes has become popular. One commonly used method is to chemically link digoxigenin to the probe. After the probe hybridizes to its target DNA an anti-digoxigenin antibody that has been chemically linked to alkaline phosphatase is used. After the antibody-enzyme conjugate binds to the digoxigenin on the probe, adamantyl-1,2-dioetane is applied as a substrate for the enzyme. The chemical reaction emits light which can be detected with photographic film.

Some of the more convenient procedures for identifying an isolate depend on the use of two probes for a particular organism, a detector (or reporter) probe and a capture probe, which bind to different regions of the same target DNA or RNA. (RNA is preferable because a bacterial cell has much more of it than DNA.) The detector probe has an antigenic group attached to it whereas the capture probe has a "tail" composed of a chain of similar nucleotides such as polyA or polyG. This tail allows the probe/target DNA hybrids to be removed by attachment to beads or plastic rods to which are bound the appropriate complementary chains of nucleotides (i.e., polyT or polyC). Detection of the removed hybrids is then done by means of an antibody/enzyme conjugate for the antigenic group on the detector probe, in which the enzyme attached to the antibody catalyzes a color-yielding reaction.

DNA probes can even be used to identify individual cells in mixed cultures under a microscope. A specific DNA probe is conjugated to a fluorescent dye (e.g., see DeLong et al., 1989; Amann et al., 1990; Angert et al., 1993) and applied to cells on a slide. If hybridization occurs between the probe DNA and the DNA or rRNA of an appropriate cell, the cell will become fluorescent when viewed under a fluorescence microscope. Methods have even been developed for rapid, nonradioactive, *in situ* hy-

bridization with bacteria in paraffin-embedded tissues (Barrett et al., 1998).

USE OF PROBES FOR IDENTIFICATION OF MULTIPLE SPECIES IN MIXED CULTURES

The methods of molecular biology have now made possible the definitive identification of many different organisms in a mixed culture, as in a sample of feces, soil, or water. The basis for this is the fact that approximately 70% of the 16S rRNA genes (i.e., 16S rDNA) of all procaryotes is highly conserved (identical in sequence) whereas other regions are unique to particular genera or species. This has made possible the construction of "universal primers" that can bind to any rDNA so that the various 16S rDNAs present in a mixed culture can be amplified by the polymerase chain reaction (PCR). The resulting PCR products are cloned and the unique rDNAs separated. These are sequenced and the corresponding organism is identified by comparing the sequence to a large database of 16S rDNA sequences (for examples, see Wise et al., 1997; Hugenholtz et al., 1998). The technique of denaturing gradient gel electrophoresis (DGGE) has been found useful for separating the PCR products derived from mixed cultures (e.g., see Teske et al., 1996; Fournier et al., 1998). When applied to mixed cultures from environmental sources such as soil and water, analysis of the 16S rDNA sequences has indicated that many of the sequences cannot be matched with those from any known organisms (i.e., are not identifiable as any cultured, described organism). The results indicate that even the present edition of the *Manual*, large as it is, probably describes less than 1% of existing procaryotic species.

IDENTIFICATION OF A PARTICULAR STRAIN OF A SPECIES

It is often necessary to identify one strain among the various strains of a species. One example is the need to identify a particular pathogenic strain so that the source of an outbreak of disease can be determined. For instance, one may wish to determine whether a strain of *Legionella pneumophila* isolated from an air conditioning system is the same strain as that isolated from a patient with Legionnaire's disease. As another example, in an ecological study one might be interested in learning whether a particular strain of *Bacillus sphaericus* that has been isolated from one soil sample is present in soil samples from other areas. The following are various methods for differentiating one bacterial strain from another. Some are traditional methods; others are DNA fingerprinting methods based on the techniques of molecular biology. DNA fingerprinting is the most specific way available to identify individual strains of a species.

Traditional methods

Antigenic Typing (Serotyping). Different strains of a species may have different antigens. The antigens present in a particular strain can be determined by the use of specific antisera. As examples, *Streptococcus pyogenes* is divided into >70 antigenic types based on M-proteins, *Streptococcus pneumoniae* is divided into >80 antigenic types based on capsular polysaccharides, and salmonellas are divided into >2000 serotypes based on O and H antigens.

Phage Typing. Strains of a bacterial species may be subject to attack and lysis by numerous bacteriophages. Some phages may attack a particular strain while others do not. The pattern of lysis by various bacteriophages constitutes the phage type of a strain. For example, *Salmonella typhi* can be divided into 33 phage types.

Antibiograms. Which of a large spectrum of antibiotics can inhibit growth of a strain and which cannot constitutes a specific identifying pattern.

DNA fingerprinting This method of identifying a bacterial strain can be done in various ways, as follows.

DNA fingerprinting using a probe and agarose gel electrophoresis. The DNA is treated with a restriction endonuclease to cleave it into many small pieces of differing molecular weight, which are then separated on an agarose gel according to their molecular weight. The gel is treated with an alkali to convert the double-stranded DNA fragments into single-stranded fragments. The pattern of DNA fragments on the gel is transferred to a nitrocellulose membrane and a labeled DNA probe is added. The probe binds only to DNA fragments containing a base sequence complementary to that of the probe. After removing any unbound probe, the location of the bound probe is determined by overlaying the membrane with photographic film, which will be exposed to either radiation or chemiluminescence.

Ribotyping. Ribotyping is a variation of the DNA fingerprinting method in which the DNA probe that is applied to the membrane is complementary to the gene for rRNA. A bacterial chromosome contains genes for three kinds of rRNA (23S, 16S, and 5S rRNA). These genes are transcribed from an *rrn* operon to yield a single large 30S precursor RNA molecule, which then undergoes a maturation process to yield the three different kinds of rRNA. Most operons in procaryotes occur only once on a chromosome but *rrn* operons occur more than once—from 2 to 14 per genome, depending on the species (Rainey et al., 1996). Ribotyping depends on the fact that the sequence of the DNA *between* the *rrn* operons varies from strain to strain in a bacterial species and, consequently, the sites for cleavage of this DNA by a restriction endonuclease will vary from one strain to another. If the DNA from each of two strains is treated with an appropriate endonuclease, the size of the resulting DNA fragments that contain an *rrn* operon will differ between the two strains and can be visualized by agarose gel electrophoresis. A universal DNA probe for 16S rDNA can be used to detect *only* the *rrn*-containing fragments on a membrane blot, and the pattern of these particular fragments will be unique for each strain.

DNA Fingerprinting Using Pulsed Field Gel Electrophoresis (PFGE). No probe is used in this method. When bacterial DNA is treated with a rare-cutting restriction endonuclease, many short fragments but only a few fragments of 1,000,000 bp or more are formed. Long DNA fragments cannot be separated on conventional agarose gel in the same way as short fragments but instead they "worm" their way through the matrix, as if they were going through a narrow, winding tube, and all migrate at a similar rate away from the cathode. Consequently, no banding pattern can be formed that could be used to characterize the bacterial strain. However, if the angle of the electric field suddenly changes, these DNAs must reorient their long axes along the new direction of the field before they can continue to migrate. The higher the molecular weight of the fragment, the longer the time it takes for this reorientation to occur. Thus the longer the fragment, the longer it takes to migrate through the gel. A PFGE apparatus causes a periodic switching of the angle of the electric field and thus allows the long fragments to become well separated

and form distinct bands. The bands can be visualized merely by soaking the gel in a solution of ethidium bromide, which binds to the fragments and fluoresces under ultraviolet light.

One problem with PFGE, however, is that the DNA must be treated *very gently* to avoid random mechanical breakage, because the only breakage must be that caused by the restriction endonuclease. Therefore, the intact bacterial cells are embedded in small blocks of low melting point agarose and lysed *in situ* before being treated with the restriction endonuclease. The blocks are then placed into a gel slab and subjected to pulsed-field gel electrophoresis at a low temperature for several hours.

Randomly amplified polymorphic DNA (RAPD) strain typing. RAPD strain identification is based on the PCR technique and the use of a single 10-base primer. Because the primer is short, there are usually many complementary sequences on the genomic DNA to which the primer will bind. DNA polymerase adds other bases to the primer, creating short pieces of double-stranded DNA. The PCR technique then creates millions of copies of these pieces. The various sizes of DNA pieces are then separated electrophoretically on an agarose gel and viewed by staining with ethidium bromide.

FURTHER READING

Amann, R.I., L. Krumholz and D.A. Stahl. 1990. Fluorescent-oligonucleotide probing of whole cells for determinative, phylogenetic, and environmental studies in microbiology. J. Bacteriol. *172*: 762–770.

Angert, E.R., K.D. Clements and N.R. Pace. 1993. The largest bacterium. Nature *362*: 239–241.

Barrett, D.M., D.O. Faigel, D.C. Metz, K. Montone and E.E. Furth. 1997. *In situ* hybridization for *Helicobacter pylori* in gastric mucosal biopsy specimens: quantitative evaluation of test performance in comparison with the CLO test and thiazine stain. J. Clin. Lab. Anal. *11*: 374–379.

Barrow, G.I. and R.K.A. Feltham (Editors). 1993. Cowan and Steel's Manual for the Identification of Medical Bacteria, 3rd Ed., Cambridge University Press, Cambridge.

Board, R.G., D. Jones and F.E. Skinner (Editors). 1992. Identification Methods in Applied and Environmental Microbiology, Blackwell Scientific Publications, Oxford.

DeLong, E.F., G.S. Wickham and N.R. Pace. 1989. Phylogenetic stains: ribosomal RNA-based probes for the identification of single cells. Science. *243*: 1360–1363.

Forbes, B.A., D.F. Sahn and A.S. Weissfeld. 1998. Bailey and Scott's Diagnostic Microbiology, 10th Ed., Mosby, St. Louis.

Fournier, D., R. Lemieux and D. Couillard. 1998. Genetic evidence for highly diversified bacterial populations in wastewater sludge during biological leaching of metals. Biotechnol. Lett. *20*: 27–31.

Hugenholtz, P., C. Pitulle, K.L. Hershberger and N.R. Pace. 1998. Novel division level bacterial diversity in a Yellowstone hot spring. J. Bacteriol. *180*: 366–376.

Logan, N.A. 1994. Bacterial systematics, Blackwell Scientific Publications, Oxford.

Miller, J.M. and C.M. O'Hara. 1995. Substrate utilization systems for the identification of bacteria and yeasts, *In* Murray, Baron, Pfaller, Tenover and Yolken (Editors), Manual of Clinical Microbiology, 6th Ed., American Society for Microbiology, Washington, D.C. pp. 103–109.

Murray, P.R., E.J. Baron, M.A. Pfallen, F.C. Tenover and R.H. Yolken (Editors). 1995. Manual of Clinical Microbiology, 6th Ed., American Society for Microbiology, Washington, D.C.

Onderdonk, A.B. and M. Sasser. 1995. Gas-liquid and high-performance chromatographic methods for the identification of micoorganisms, *In* Murray, Baron, Pfaller, Tenover and Yolken (Editors), Manual of Clinical Microbiology, 6th Ed., American Society for Microbiology, Washington, D.C. pp. 123–129.

Podzorski, R. and D.H. Persing. 1995. Molecular detection and identi-
fication of microorganisms, *In* Murray, Baron, Pfaller, Tenover and
Yolken (Editors), Manual of Clinical Microbiology, 6th Ed., American
Society for Microbiology, Washington, D.C. pp. 130–157.

Rainey, F.A., N.L. Ward-Rainey, P.H. Janssen, H. Hippe and E. Stackebrandt. 1996. *Clostridium paradoxum* DSM 7308^T contains multiple 16S
rRNA genes with heterogeneous intervening sequences. Microbiology
(Reading). *142*: 2087–2095.

Smibert, R.M. and N.R. Krieg. 1995. Phenotypic characterization, *In* Gerhardt, Murray, Wood and Krieg (Editors), Methods for General and
Molecular Bacteriology, American Society for Microbiology, Washington, D.C. pp. 607–654.

Teske, A., P. Sigalevich, Y. Cohen and G. Muyzer. 1996. Molecular identification of bacteria from a coculture by denaturing gradient gel
electrophoresis of 16S ribosomal DNA fragments as a tool for isolation
in pure cultures. Appl. Environ. Microbiol. *62*: 4210–4215.

Wise, M.G., TV. Matchers and LC. Shanties. 1997. Bacterial diversity of
a Carolina bay as determined by 16S rRNA gene analysis: confirmation
of novel taxa. Appl. Environ. Microbiol. *63*: 1505–1514.

Numerical Taxonomy

Peter H.A. Sneath

Numerical taxonomy (sometimes called **taxometrics**) developed in the late 1950s as part of multivariate analyses and in parallel with the development of computers. Its aim was to devise a consistent set of methods for classification of organisms. Much of the impetus in bacteriology came from the problem of handling the tables of data that result from examination of their physiological, biochemical, and other properties. Such tables of results are not readily analyzed by eye, in contrast to the elaborate morphological detail that is usually available from examination of higher plants and animals. There was thus a need for an objective method of taxonomic analyses, whose first aim was to sort individual strains of bacteria into homogeneous groups (conventionally species), and that would also assist in the arrangement of species into genera and higher groupings. Such numerical methods also promised to improve the exactitude in measuring taxonomic, phylogenetic, serological, and other forms of relationship, together with other benefits that can accrue from quantitation (such as improved methods for bacterial identification; see the discussion by Sneath of Numerical Identification in this *Manual*).

Numerical taxonomy has been broadly successful in most of these aims, particularly in defining homogeneous **clusters** of strains, and in integrating data of different kinds (morphological, physiological, antigenic). There are still problems in constructing satisfactory groups at high taxonomic levels, e.g., families and orders, although this may be due to inadequacies in the available data rather than any fundamental weakness in the numerical methods themselves.

The application of the concepts of numerical taxonomy was made possible only through the use of computers, because of the heavy load of routine calculations. However, the principles can easily be illustrated in hand-worked examples. In addition, two problems had to be solved: the first was to decide how to weight different variables or characters; the second was to analyze similarities so as to reveal the taxonomic structure of groups, species, or clusters. A full description of numerical taxonomic methods may be found in Sneath (1972) and Sneath and Sokal (1973). Briefer descriptions and illustrations in bacteriology are given by Skerman (1967), Lockhart and Liston (1970), Sneath (1978a), Priest and Austin (1993), and Logan (1994). A thorough review of applications to bacteria is that of Colwell (1973).

It is important to bear in mind certain definitions. Relationships between organisms can be of several kinds. Two broad classes are as follows.

Similarity on Observed Properties. Similarity, or resemblance, refers to the attributes that an organism possesses today, without reference to how those attributes arose. It is expressed as proportions of similarities and differences, for example, in existing attributes, and is called the **phenetic relationship**. This includes similarities both in phenotype (e.g., motility) and in genotype (e.g., DNA pairing).

Relationship by Ancestry, or Evolutionary Relationship. This refers to the **phylogeny** of organisms, and not necessarily to their present attributes. It is expressed as the time to a common ancestor, or the amount of change that has occurred in an evolutionary lineage. It is not expressed as a proportion of similar attributes, or as the amount of DNA pairing and the like, although evolutionary relationship may sometimes be deduced from phenetics on the assumption that evolution has indeed proceeded in some orderly and defined way. To give an analogy, individuals from different nations may occasionally look more similar than brothers or sisters of one family; their phenetic resemblance (in the properties observed) may be high though their evolutionary relationship is distant.

Numerical taxonomy is concerned primarily with phenetic relationships. It has in recent years been extended to phylogenetic work, by using rather different techniques; these seek to build upon the assumed regularities of evolution so as to give, from phenetic data, the most probable phylogenetic reconstructions. A review of the area is given by Sneath (1974).

The basic taxonomic category is the species. It is noted in the chapter on "Bacterial Nomenclature" that it is useful to distinguish a **taxospecies** (a cluster of strains of high mutual phenetic similarity) from a **genospecies** (a group of strains capable of gene exchange), and both of these from a **nomenspecies** (a group bearing a binomial name, whatever its status in other respects). Numerical taxonomy attempts to define taxospecies. Whether these are justified as genospecies or nomenspecies turns on other criteria. One may also distinguish a **genomospecies**, a group of strains that have high DNA–DNA relatedness. It should be emphasized that groups with high genomic similarity are not necessarily genospecies: genomic resemblance is included in phenetic resemblance; genospecies are defined by gene exchange.

Groups can be of two important types. In the first, the possession of certain invariant properties defines the group without permitting any exception. All triangles, for example, have three sides, not four. Such groupings are termed **monothetic**. Taxonomic groups, however, are not of this kind. Exceptions to the most invariant characters are always possible. Instead, taxa are **polythetic**, that is, they consist of assemblages whose members share a high proportion of common attributes, but not necessary any invariable set. Numerical taxonomy produces polythetic groups and thus permits the occasional exception on any character.

LOGICAL STEPS IN CLASSIFICATION

The steps in the process of classification are as follows:

1. Collection of data. The **bacterial strains** that are to be classified have to be chosen, and they must be examined for a number of relevant properties (**taxonomic characters**).
2. The data must be coded and scaled in an appropriate fashion.
3. The **similarity** or **resemblance** between the strains is calculated. This yields a table of similarities (**similarity matrix**) based on the chosen set of characters.
4. The similarities are analyzed for **taxonomic structure**, to yield the groups or clusters that are present, and the strains are arranged into **phenons** (phenetic groups), which are broadly equated with taxonomic groups (**taxa**).
5. The properties of the phenons can be tabulated for publication or further study, and the most appropriate characters (**diagnostic characters**) can be chosen on which to set up **identification systems** that will allow the best identification of additional strains.

It may be noted that those steps must be carried out in the above order. One cannot, for example, find diagnostic characters before finding the groups of which they are diagnostic. Furthermore, it is important to obtain complete data, determined under well-standardized conditions.

Data for numerical taxonomy The data needed for numerical taxonomy must be adequate in quantity and quality. It is a common experience that data from the literature are inadequate on both counts; most often it is necessary to examine bacterial strains afresh by an appropriate set of tests.

ORGANISMS   Most taxonomic work with bacteria consists of examining individual strains of bacteria. However, the entities that can be classified may be of various forms—strains, species, genera—for which no common term is available. These entities, *t* in number, are therefore called **operational taxonomic units** (**OTUs**). In most studies OTUs will be strains. A numerical taxonomic study, therefore, should contain a good selection of strains of the groups under study, together with type strains of the taxa and of related taxa. Where possible, recently isolated strains, and strains from different parts of the world, should be included.

CHARACTERS   A **character** is defined as any property that can vary between OTUs. The values it can assume are **character states**. Thus, "length of spore" is a character and "1.5 μm" is one of its states. It is obviously important to compare the same character in different organisms, and the recognition that characters are the same is called the **determination of homology**. This may sometimes pose problems, but in bacteriology these are seldom serious. A single character treated as independent of others is called a **unit character**. Sets of characters that are related in some way are called **character complexes**.

There are many kinds of characters that can be used in taxonomy. The descriptions in the *Manual* give many examples. For numerical taxonomy, the characters should cover a broad range of properties: morphological, physiological, biochemical. It should be noted that certain data are not characters in the above sense. Thus the degree of serological cross-reaction or the percent pairing of DNA are equivalent, not to character states, but to similarity measures.

NUMBER OF CHARACTERS   Although it is well to include a number of strains of each known species, numerical taxonomies are not greatly affected by having only a few strains of a species. This is not so, however, for characters. The similarity values should be thought of as estimates of values that would be obtained if one could include a very large number of phenotypic features. The accuracy of such estimates depends critically on having a reasonably large number of characters. The number, *n*, should be 50 or more. Several hundred are desirable, though the taxonomic gain falls off with very large numbers.

QUALITY OF DATA   The quality of the characters is also important. Microbiological data are prone to more experimental error than is commonly realized. The average difference in replicate tests on the same strain is commonly about 5%. Efforts should be made to keep this figure low, particularly by rigorous standardization of test methods. It is very difficult to obtain reasonably reproducible results with some tests, and they should be excluded from the analysis. As a check on the quality of the data, it is useful to reduplicate a few of the strains and carry them through as separate OTUs; the average test error is about half the percentage discrepancy in similarity of such replicates (e.g., 90% similarity implies about 5% experimental variation).

CODING OF THE RESULTS   The test reactions and character states now need coding for numerical analysis. There are several satisfactory ways of doing this, but for the present purposes of illustration only one common scheme will be described. This is the familiar process of coding the reactions or states into positive and negative form. The resulting table, therefore, contains entries + and − (or 1 and 0, which are more convenient for computation), for *t* OTUs scored for *n* characters. Naturally, there should be as few gaps as possible.

The question arises as to what weight should be given to each character relative to the rest. The usual practice in numerical taxonomy is to give each character equal weight. More specifically, it may be argued that unit characters should have unit weight, and if character complexes are broken into a number of unit characters (each carrying one unit of taxonomic information), it is logical to accord unit weight to each unit character. The difficulties of deciding what weight should be given *before* making a classification (and hence in a fashion that does not prejudge the taxonomy) are considerable. This philosophy derives from the opinions of the eighteenth-century botanist Adanson, and therefore numerical taxonomies are sometimes referred to as Adansonian.

Similarity The $n \times t$ table can then be analyzed to yield similarities between OTUs. The simplest way is to count, for any pair of OTUs, the number of characters in which they are identical (i.e., both are positive or both are negative). These **matches** can be expressed as a percentage or a proportion, symbolized as S_{SM} (for simple matching coefficient). This is the most common coefficient in bacteriology. Other coefficients are sometimes used because of particular advantages. Thus the Gower coefficient S_G accommodates both presence–absence characters and quantitative ones, the Jacquard coefficient S_J discounts matches between two negative results, and the Pattern coefficient S_P corrects for apparent differences that are caused solely by differences between strains in growth rate and hence metabolic vigor. These coefficients emphasize different aspects of the phenotype (as is quite legitimate in taxonomy) so one cannot regard one or another as necessarily the correct coefficient, but fortunately this makes little practical difference in most studies. Various special similarity coefficients can also be employed for electrophoretic and chemotaxonomic data.

The similarity values between all pairs of OTUs yields a checkerboard of entries, a square table of similarities known as a **similarity matrix** or **S matrix**. The entries are percentages, with 100% indicating identity and 0% indicating complete dissimilarity between OTUs. Such a table is symmetrical (the similarity of a to b is the same as that of b to a), so that usually only one half, the left lower triangle, is filled in.

These similarities can also be expressed in a complementary form, as dissimilarities. Dissimilarities can be treated as analogs of distances, when "taxonomic maps" of the OTUs are prepared, and it is a convenient property that the quantity $d = (1 - S_{SM})^{1/2}$ is equivalent geometrically to a distance between points representing the OTUs in a space of many dimensions (a **phenetic hyperspace**).

Taxonomic structure A table of similarities does not of itself make evident the **taxonomic structure** of the OTUs. The strains will be in an arbitrary order that will not reflect the species or other groups. These similarities therefore require further manipulation. It will be seen that a table of serological cross-reactions, if complete and expressed in quantitative terms, is analogous to a table of percentage similarities, and the same is true of a table of DNA pairing values. Such tables can be analyzed by the methods described below, though in serological and nucleic studies there are some particular difficulties on which further work is needed.

There are two main types of analyses to reveal the taxonomic structure: **cluster analysis** and **ordination**. The result of the former is a treelike diagram or **dendrogram** (more precisely a **phenogram**, because it expresses phenetic relationships), in which the tightest bunches of twigs represent clusters of very similar OTUs. The result of the latter is an **ordination diagram** or **taxonomic map**, in which closely similar OTUs are placed close together. The mathematical methods can be elaborate, so only a nontechnical account is given here.

In cluster analysis, the principle is to search the table of similarities for high values that indicate the most similar pairs of OTUs. These form the nuclei of the clusters and the computer searches for the next highest similarity values and adds the corresponding OTUs onto these cluster nuclei. Ultimately all OTUs fuse into one group, represented by the basal stem of the dendrogram. Lines drawn across the dendrogram at descending similarity levels define, in turn, phenons that correspond to a reasonable approximation to species, genera, etc. The most common cluster methods are the **unweighted pair group method with averages** (**UPGMA**) and **single linkage**.

In ordination, the similarities (or their mathematical equivalents) are analyzed so that the phenetic hyperspace is summarized in a space of only a few dimensions. In two dimensions this is a scattergram of the positions of OTUs from which one can recognize clusters by eye. Three-dimensional perspective drawings can also be made. The most common ordination methods are **principal components analysis** and **principal coordinates analysis**.

A number of other representations are also used. One example is a similarity matrix in which the OTUs have first been rearranged into the order given by a clustering method and then the cells of the matrix have been shaded, with the highest similarities shown in the darkest tone. In these "shaded S matrices", clusters are shown by dark triangles. Another representation is a table of the mean similarities between OTUs of the same cluster and of different clusters (**inter-** and **intragroup similarity table**);

if based on S_{SM} with UPGMA clustering, this table expresses the positions and radii of clusters (Sneath, 1979a) and consequently the distance between them and their probable overlap—properties of importance in numerical identification, as discussed later.

For general purposes, a dendrogram is the most useful representation, but the others can be very instructive, since each method emphasizes somewhat different aspects of the taxonomy.

The analysis for taxonomic structure should lead logically to the establishment or revision of taxonomic groups. We lack, at present, objective criteria for different taxonomic ranks, that is, one cannot automatically equate a phenon with a taxon. It is, however, commonly found that phenetic groups formed at about 80% *S* are equivalent to bacterial species. Similarly, we lack good tests for the statistical significance of clusters and for determining how much they overlap, though some progress is being made here (Sneath, 1977, 1979b). The fidelity with which the dendrogram summarizes the *S* matrix can be assessed by the **cophenetic correlation coefficient**, and similar statistics can be used to compare the **congruence** between two taxonomies if they are in quantitative form (e.g., phenetic and serological taxonomies). Good scientific judgment in the light of other knowledge is indispensable for interpreting the results of numerical taxonomy.

Descriptions of the groups can now be made by referring back to the original table of strain data. The better diagnostic characters can be chosen—those whose states are very constant within groups but vary between groups. It is better to give percentages or proportions than to use symbols such as +, (+), v, d, or − for varying percentages, because significant loss of statistical information can occur with these simplified schemes. It would, however, be superfluous to list percentages based on very few strains. As systematic bacteriology advances, it will be increasingly important to publish the actual data on individual strains or deposit it in archives; such data will show their full value when test methods become very highly standardized.

It is evident that numerical taxonomy and numerical identification place considerable demands on laboratory expertise. New test methods are continually being devised. New information is continually being accumulated. It is important that progress should be made toward agreed data bases (Krichevsky and Norton, 1974), as well as toward improvements in standardization of test methods in determinative bacteriology, if the full potential of numerical methods is to be achieved.

NUMERICAL IDENTIFICATION

The success of numerical taxonomy has in recent years led to the development of a new diagnostic method based upon it, called **numerical identification**. The rapidly growing field is well reviewed by Lapage et al. (1973), and Willcox et al. (1980). The essential principles can be illustrated geometrically (Sneath, 1978b) by considering the columns of percent positive test reactions in a new table, a table of *q* taxa for *m* diagnostic characters. If an object is scored for two variables, its position can be represented by a point on a scatter diagram. Use of three variables determines a position in a three-dimensional model. Objects that are very similar on the variables will be represented by clusters of points in the diagram or the model, and a circle or sphere can be drawn round each cluster so as to define its position and radius. The same principles can be extended to many variables or tests, which then represent a multidimensional space or "hyperspace". A column representing a species defines, in effect, a region in hyperspace, and it is useful to think of a species as being represented by a hypersphere in that space, whose position and radius are specified by the numerical values of these percentages. The tables form a reference library, or database, of properties of the taxa.

The operation of numerical identification is to compare an unknown strain with each column of the table in turn, and to calculate a distance (or its analog) to the center of each taxon hypersphere. If the unknown lies well within a hypersphere, this will identify it with that taxon. Further, such systems have important advantages over most other diagnostic systems. The numerical process allows a likelihood to be attached to an identification, so that one can know to some order of magnitude the certainty that the identity is correct. The results are not greatly affected by an occasional aberrant property of the unknown, or an occasional experimental mistake in performing the tests. Furthermore, the system is robust toward missing information, and quite good identifications can be obtained if only a moderate proportion of the tests have been performed.

Numerous applications of numerical identification are now being made. Most commercial testing kits or automatic instruments for microbial identification are based on these concepts, and they require the comparison of results on an unknown strain with a database using computer software or with printed material prepared by such means. Research sponsored by the Bergey's Manual® Trust (Feltham et al., 1984) shows that these concepts can be extended to a very wide range of genera.

FURTHER READING

Feltham, R.K.A., P.A. Wood and P.H.A. Sneath. 1984. A general-purpose system for characterizing medically important bacteria to genus level. J. Appl. Bacteriol. *57*: 279–290.

Lapage, S.P., S. Bascomb, W.R. Willcox and M.A. Curtis. 1973. Identification of bacteria by computer.I. General aspects and perspectives. J. Gen. Microbiol. *77*: 273–290.

Logan, N.A. 1994. Bacterial Systematics, Blackwell Scientific Publications, Oxford.

Sneath, P.H.A. 1978. Identification of microorganisms, *In* Norris and Richmond (Editors), Essays in Microbiology, John Wiley, Chichester. 10/1–10/32.

Willcox, W.R., S.P. Lapage and B. Holmes. 1980. A review of numerical methods in bacterial identification. Antonie van Leeuwenhoek J. Microbiol. Serol. *46*: 233–299.

Polyphasic Taxonomy

Monique Gillis, Peter Vandamme, Paul De Vos, Jean Swings, and Karel Kersters

INTRODUCTION

Bacterial taxonomy comprises the interrelated areas of classification, nomenclature, and identification and is supposed to reflect phylogeny and evolution. When looking back over the changes in bacterial systematics during the last 25 years, it is clear that the most spectacular changes occurred mainly in the areas of characterization and phylogeny. Characterization changed from simple procedures, in which a limited number of features of the bacterial cell (mainly morphological and physiological aspects) were studied, to a multidisciplinary approach using phenotypic, genotypic, and chemotaxonomic techniques. Determination of phylogenetic relationships (which is at this time essentially synonymous with 16S and/or 23S rRNA gene sequence similarities) became a routine procedure in bacterial taxonomy.

While the rules of bacterial nomenclature remain largely unchanged (Lapage et al., 1992; Stackebrandt and Goebel, 1994; Murray and Stackebrandt, 1995), the tools for identification diversified with the multidisciplinary approach to bacterial characterization. The names of bacteria, and certainly the number of named taxa, have also changed and/or increased drastically as a result of the application of this conceptual approach to bacterial taxonomy.

The term "polyphasic taxonomy" was introduced 30 years ago by Colwell (1970) to refer to a taxonomy that assembles and assimilates many levels of information, from molecular to ecological, and incorporates several distinct, and separable, portions of information extractable from a nonhomogeneous system to yield a multidimensional taxonomy. Nowadays, polyphasic taxonomy refers to a consensus type of taxonomy and aims to utilize all the available data in delineating consensus groups, decisive for the final conclusions.

The species is the basic unit of bacterial taxonomy, and the first recommendation for a polyphasic consensus delineation of a bacterial species is based on "the phylogenetic species definition" of Wayne et al. (1987). These authors defined a species as a group of strains, including the type strain, sharing at least 70% total genome DNA–DNA hybridization and less than 5°C ΔT_m.* Phenotypic features should agree with this genotypic definition and should override the "phylogenetic" concept of species only in a few exceptional cases. Total genome DNA–DNA hybridization values are the key parameter in this species delineation.

Considering the perception of a bacterial species, taxonomists

either sustain a coherent species definition without questioning if this corresponds with a biological reality, or they try to visualize bacterial species as condensed nodes in a cloudy and confluent taxonomic space. Genera and families represent mostly agglomerates of nodal species and internodal strains, and agglomerates of genera, respectively. Although in the present chapter most of the attention will be focused on the species level, the hierarchical structure of all current taxonomic classification requires us to consider higher taxa, such as genera and families, as well. Compared to the bacterial species, the higher taxa are much more difficult to delineate and phylogenetic divergence is not necessarily supported by phenotypic, chemotaxonomic, or polyphasic data. At present, no clear-cut genus definition is available and this has led to the creation of genera in which the genotypic and phenotypic divergence varies with the individual concepts of taxonomists. Therefore, delineation of genera by a consensus approach, including simple differential parameters and an accompanying polyphasic definition, is highly desirable if the present concept of bacterial classification is to be retained.

DIFFERENT TYPES OF INFORMATION USED IN POLYPHASIC TAXONOMY

In principle both genotypic and phenotypic information may be incorporated into polyphasic taxonomic studies. Sources of information and diverse techniques available to retrieve this information are represented schematically in Fig. 1. The ultimate characterization on the genomic level is the determination of the sequence and the organization of the total bacterial genome. As long as this cannot be performed routinely, the polyphasic approach is the most obvious strategy to collect a maximum amount of direct and indirect information about the total genome. It is not our intention to describe all the available techniques here. Our aim is to discuss the major categories of taxonomic techniques required to obtain a useful polyphasic characterization. Practical and theoretical aspects of the different techniques listed in Fig. 1 can be found in various papers (see Vandamme et al., 1996a) and handbooks. Of paramount importance is the level of taxonomic resolution of the different methods. Fig. 2 presents the discriminatory taxonomic power of the techniques summarized in Fig. 1. On the basis of this parameter, different categories of techniques can be distinguished: (i) those with a broad taxonomic resolution, of which the rRNA gene-based techniques are the best known for their impact on phylogenetic conclusions; (ii) those revealing differences on the species and/or genus level and (iii) various typing methods that are not necessarily relevant on the species level but can be used to screen for groups of similar strains. The various techniques differ

---

*T_m is the melting temperature of the hybrid as measured by stepwise denaturation; ΔT_m is the difference in T_m in degrees Celsius between the homologous and the heterologous hybrids formed under standard conditions.

GENOTYPIC INFORMATION

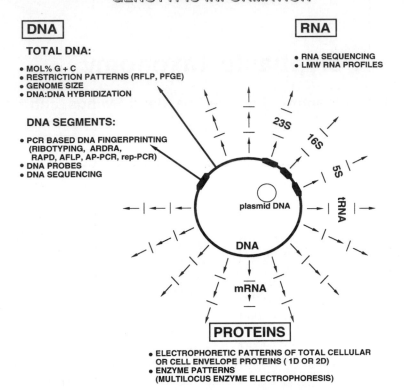

FIGURE 1. Schematic overview of various cellular components and techniques used in polyphasic bacterial taxonomy (adapted from Vandamme et al., 1996a). Abbreviations: AFLP, amplified fragment length polymorphism; AP-PCR, arbitrarily primed PCR; ARDRA, amplified rDNA restriction analysis; FAMEs, fatty acid methyl esters; LMW, low molecular weight; PFGE, pulsed-field gel electrophoresis; RAPD, randomly amplified polymorphic DNA; rep-PCR, repetitive element sequence-based PCR; RFLP, restriction fragment length polymorphism; 1D, 2D, one- and two-dimensional, respectively.

in the amount of effort required. Some have been automated, and some are relatively fast and cheap. It is obvious that fast, cost-effective, and preferentially automated methods with a fine taxonomic resolution (below the species level) are among those to be used for primary screening purposes, while total genome DNA–DNA hybridization, being a more laborious technique, can be restricted to a minimum number of strains representing groups defined using other appropriate methods. Techniques based on rRNA genes of representative strains are the most suitable to determine the phylogenetic position of bacterial groups.

In practice, it is nearly impossible to gather all the information that could possibly be used in a polyphasic study. The strategy in modern polyphasic taxonomy is to first estimate the different levels of taxonomic discrimination to be covered and then to choose the techniques accordingly. The total number of strains to be studied will also significantly affect the final choice.

The consensus polyphasic approach starts with making a choice of complementary techniques to be used simultaneously or stepwise in order to characterize and classify an individual strain or any group of strains. The goal is to evaluate all the results in relation to each other and to obtain a consensus view of the data with a minimum number of inconsistencies. Nomenclatural implications complete the evaluation, together with the search for adequate identification procedures. Each taxon should be described and, preferably, differentiated from related or similar taxa by its phenotypic, genotypic, and chemotaxonomic characteristics.

The minimal requirements for obtaining useful polyphasic data are: (i) a preliminary screening for groups of similar strains; (ii) determination of the phylogenetic placement of these groups; (iii) measurement of the relationships between the groups and their closest neighbors, and (iv) collection of various descriptive data, preferentially on different aspects of the cell.

POLYPHASIC STRATEGY

There is no universal strategy that can be employed in all polyphasic studies. The taxonomic levels to be covered vary with the

Technique

Restriction fragment length polymorphism (RFLP)
Low frequency restriction fragment analysis (PFGE)
Phage and bacteriocin typing
Serological (monoclonal, polyclonal) techniques
Ribotyping
DNA amplification (AFLP, AP-PCR, rep-PCR, RAPD, ...)
Zymograms (multilocus enzyme polymorphism)
Total cellular protein electrophoretic patterns
DNA–DNA hybridization
Mol% G+C
DNA amplification (ARDRA)
tDNA-PCR
Chemotaxonomic markers (polyamines, quinones, ...)
Cellular fatty acid fingerprinting (FAME)
Cell wall structure
Phenotype (classical, API, Biolog, ...)
rRNA sequencing
DNA probes
DNA sequencing

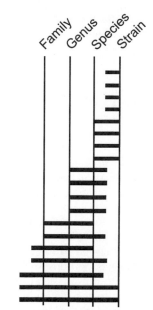

FIGURE 2. Taxonomic resolution of some of the currently used techniques in bacterial taxonomy (adapted from Vandamme et al., 1996a). Abbreviations: see legend of Figure 1.

objective of each study. The choice of the techniques to be used also depends on the number of strains to be studied. The more strains to be screened, the more one needs a fast and preferentially automated screening technique. On the other hand, the requirement for special analytical methods for a given technique is generally less important when only a few strains are under investigation. The taxonomic resolution of many techniques can differ depending on the bacterial group studied. The choice of methods can also be taxon-dependent when for any reason the preparation of particular cell constituents is very difficult or inefficient.

Development of a strategy for a polyphasic taxonomy can be illustrated by a theoretical example: suppose that one has to classify 50 bacterial isolates for which a minimal characterization (e.g., Gram reaction, origin, morphology, growth conditions) is available. According to the minimal requirements mentioned above, the 50 strains should initially be screened to identify groups of similar strains, preferably by at least two non-overlapping methods. The choice of techniques will certainly also be affected by the availability of the required instrumentation and the knowledge of each research group. For the delineation of groups, a thorough knowledge of the resolving power of each technique is necessary. An awareness of the limitations of the methods used to analyze and cluster the results is also essential. Armed with this knowledge, the main consensus groups obtained by the various techniques can then be determined. The second goal is to determine the phylogenetic position of the consensus groups by sequencing the 16S rRNA genes of representative strains.

Different theoretical possibilities for studying the relationships among the consensus groups and providing an emended or a new description of taxa exist. These possibilities are listed below:

Case 1 If the 16S rRNA gene sequence similarity between the representative strains under study and those found in GenBank for a particular genus exceeds 97%, it can be assumed that these strains are members of that genus. DNA–DNA hybrid-

izations can then be performed between several representative strains of each consensus group and all known species of that genus, to find out if the new consensus group belongs to one of the known species or constitutes a new species, as recommended by Stackebrandt and Goebel (1994). Members of a particular consensus group can be identified as a known species, the polyphasic consistency of the species can be verified, and, if needed, the description can be emended. When a new group is identified as belonging to a particular genus but not to one of its described species, the creation of a new species can be planned. Therefore a polyphasic description of the new taxon is required before a new species, with an appropriate name, can be proposed and described. Phenotypic characterization remains an indispensable part of the description allowing differentiation and description of the groups, but in the future genotypic and chemotaxonomic parameters should enhance the description of new taxa and assure the differentiation between the taxa.

Case 2 There are no clearcut recommendations for the delineation of bacterial genera or higher taxonomic levels. If the 16S rRNA gene sequence similarity between representative strains of a consensus group and those found in GenBank is less than 97%, it is often not straightforward to decide whether the particular group belongs to a new or existing genus. It is recommended to evaluate the stability of the phylogenetic position of the group in question and to compare its overall genotypic, chemotaxonomic, and phenotypic profile with that of its closest relatives. When both phenotypic and chemotaxonomic parameters support the phylogenetic group, the creation of a new genus can be considered.

Case 3 Strains not belonging to consensus groups must be further characterized to determine their exact taxonomic status.

POLYPHASIC TAXONOMY IN PRACTICE

Many examples of polyphasic taxonomic studies of diverse bacterial groups are available, and a general evaluation of the results shows that the various conclusions of these studies depend on the bacterial group(s) studied, on the techniques applied, and on the researcher involved (Vandamme et al., 1996a). In many cases the identification of consensus groups and the formulation of conclusions is not simple and may show various inconsistencies either with (i) earlier classification and nomenclature or (ii) within the new conclusions themselves. Some striking examples will be discussed briefly in Part A and the main problems will be summarized in Part B.

Part A

1. Polyphasic classification does not necessarily conform to special purpose classification practiced by specialists in the field, e.g., in the genera *Agrobacterium* (Kersters and De Ley, 1984) and *Xanthomonas* (Vauterin et al., 1995). In the former example the polyphasic groups did not at all correspond with the named species, of which the type species *A. tumefaciens* has a conserved status; this has led to the use of a "biovar" system to indicate the polyphasic groups. In *Xanthomonas* 14 new species corresponding to polyphasic groups have been created, partly replacing the former pathovar system.

2. Certain groups constitute very tight phylogenetic clusters that can be biochemically quite versatile e.g., *Bordetella* (Vandamme, 1998).

3. Occasionally members of very tight polyphasic groups are not classified accordingly e.g., *Escherichia* and *Shigella* species sharing more than 85% total DNA–DNA hybridization. For pragmatic reasons they remain classified in two separate genera (Brenner, 1984).

4. In contrast, biochemically restricted groups can be phylogenetically extremely heterogeneous. For example, *Campylobacter* (Vandamme et al., 1991) and *Capnocytophaga* (Vandamme et al., 1996b) originally included a large number of taxa characterized by a minimal set of common phenotypic features. The ability of various techniques to distinguish between members of these taxa at various taxonomic levels was reexamined leading to the development of molecular diagnostic tests.

5. For the lactic acid bacteria, traditionally applied phenotypic classification schemes do not correspond with the phylogenetic-based classification because of a large amount of sequence variation in the 16S rRNA genes (Vandamme et al., 1996a). The traditional phenotypic analysis remains important for identification purposes because the phylogenetic data have not yet been translated into new identification strategies.

6. In several bacterial lineages, such as *Comamonadaceae*, multiple subbranches (16S rRNA gene sequence similarities of 95–96%) have been identified, and some of them can be considered candidates for separate generic status (Willems et al, 1991a; Wen et al., 1999). However, as with species, it is important that genera exhibit some phenotypic coherence. This discrepancy has resulted either in the combination of multiple subbranches into a single genus or in the creation of separate genera for individual sub-branches.

7. In many examples genotypic groups could not be described phenotypically and therefore remain unnamed within species e.g., in *Comamonas terrigena* (Willems et al., 1991c) or within genera e.g., *Acinetobacter*, containing several unnamed genomic species (Bouvet and Jeanjean, 1989; Nishimura et al., 1987). Additional methods for unambiguous identification are required. It should be stressed that in such cases the requirement for phenotypic differentiation as formulated by

Wayne et al. (1987) makes it impossible to properly name genomic groups and can thus hinder the recognition of biological diversity.

8. In some cases it is very difficult to determine consensus groups because many strains occupy separate positions or the clusters are too narrow, due to the techniques used being too discriminating. Supplementary techniques, with slightly different levels of discriminating power, are recommended to improve the delineation of significant groups.

9. On occasion, a new genus has been created on the basis of significant phenotypic and physiological differences despite sharing more than 99% 16S rRNA gene similarity with an existing genus (Yurkov et al., 1997).

10. Molecular tools allow one to obtain various DNA sequences from diverse biotypes, providing an image of the total bacterial populations and consortia. Most uncultured or "unculturable" bacteria are only characterized by their 16S rRNA gene sequence, ITS, or 23S rRNA gene sequence, and many may represent new taxa that cannot yet be characterized polyphasically (Hugenholtz et al., 1998a). Identical sequences may originate from different cells of the same strain, from different strains of the same species, or from strains of closely related species. Therefore, it is not appropriate to propose to classify them like cultured organisms and to propose binomial species names. It is recommended to include such organisms temporarily in a new category, *Candidatus* as proposed by Murray and Schleifer (1994) and Murray and Stackebrandt (1995).

11. During the last years, many new species containing a single strain have been described. Likewise, there are numerous genera consisting of only one species. This does not correspond with an ideal definition nor with the reality of nature. However, if only a single strain representing a new taxon can be isolated, the creation of single strain species illustrates the breadth of bacterial diversity. Attempts to obtain additional isolates should be encouraged.

Part B   The major sources of conflict in practicing consensus taxonomy are: (i) characteristics expressing variability among organisms appear to be superior parameters for certain taxonomic ranks (DNA–DNA hybridization on the species level, rRNA gene sequencing on the genus and family level). However, the delineated groups cannot necessarily be revealed by the examination of phenotypic parameters. This has significant impact because of the need for a phenotypic description and differentiating phenotypic features; (ii) the lack of guidelines or minimal standards for description of a (polyphasic) genus.

1. **DNA–DNA hybridizations.** Classical hybridization techniques are laborious, require considerable amounts of DNA, and are consequently not suitable for large scale use. Moreover, several methods that do not necessarily give the same quantitative results are now in use, making quantitative comparisons difficult. New, more rapid, miniaturized, and standardized methods that will quickly delineate species (Adnan et al., 1993; Ezaki et al., 1989) are under development. The use of a single, rapid, and standardized method requiring small amounts of DNA is recommended. Regardless of the method used, it remains difficult to apply the 70% DNA–DNA hybridization rule for species definition because this rule was proposed mainly on the basis of differences among species in the family *Enterobacteriaceae*. These species are phenotypi-

cally well studied and exhibit a high degree of phenotypic heterogeneity that does not always correspond with genotypic heterogeneity. In many other bacterial families, members of a single species share DNA hybridization values of 40–100%, and evaluation of these lower values is often difficult.

2. **rRNA gene sequencing.** The comparison of 16S and 23S rRNA gene sequences is indispensable in polyphasic taxonomy and provides the phylogenetic framework for present day classification. The 16S rRNA gene sequence does not contain enough discriminating power to delineate species within certain groups, and additional DNA–DNA hybridizations are often required (Stackebrandt and Goebel, 1994). Moreover, all 16S rRNA gene sequences employed in classification should be used with care because high levels of sequence variation have been observed, even between strains of the same species. This variation is attributed to inter-operon differences, as well as to other differences (Clayton et al., 1995; Young and Haukka, 1996). For the reconstruction of phylogenetic trees, it is also important to include a wide range of related and unrelated reference organisms. Bootstrap analysis is highly recommended to determine the significance of the branching points.

3. **Phenotypic analysis.** According to Wayne et al. (1987), phenotypic data deserve special attention because of their impact on species delineation and because a description of new species requires a minimum number of phenotypic characteristics. Historically, phenotypic analysis was very important and many conventional tests have been used to describe and differentiate taxa. Taxonomic reports do not always provide much new phenotypic data, and data from older literature do not always reflect possible adaptations of strains or minor changes in test media and conditions. Nowadays few research groups perform extensive conventional phenotypic analysis because it is laborious and sometimes not reproducible. More and more commercialized, automated, and miniaturized methods, mostly conceived for particular bacterial groups are being used, resulting in the analysis of a restricted set of phenotypic properties and creating a dependence on the commercial dealer. A minimal phenotypic description may still be required in the long-term, but in the future genotypic and chemotaxonomic parameters should complete the description of new taxa and facilitate differentiation between the taxa. Phenotypic coherence at the species level does not usually represent a problem. However, at higher (genus) levels phenotypic coherence cannot always be found in the various phylogenetic groups, and clear differentiation of these groups is often doubtful or even impossible.

4. **Genus delineation.** Although there is a rather broad consensus among taxonomists that phylogenetic data are of superior value for the delineation of genera, the goal remains to define genera polyphasically, to describe them, and to differentiate them from their neighbors. However, there are no rules to delineate genera, except that it is generally accepted that genera should reflect phylogenetic relationships. The phylogenetic divergence within genera can differ with the bacterial groups under consideration, although most taxonomists do not accept very large (phylogenetically) heterogenous genera. The level of phylogenetic relatedness, as shown in an rRNA dendrogram, that corresponds to a given hierarchical line showing phenotypic coherence varies considerably (mostly between 4% and 10% 16S rRNA gene sequence difference). Phenotypic coherence does not always correspond to delineated phylogenetic groups and vice versa. Any new genus needs to be described. The goal of phenotypically differentiating genera from other closely related genera is regularly not fulfilled because the phenotypic data are often not complete or not comparable with results obtained from conventional tests described in the literature. We recommend, therefore, to also include chemotaxonomic and genomic data in order to improve the description and differentiation of taxa. Therefore, universal, comprehensive databases containing various kinds of molecular patterns, as well as phenotypic and chemotaxonomic data are required. To obtain reliable, reproducible, and exchangeable profiles which can be consulted, preferably on-line, the standardization of the experimental conditions and the use of tools to correct for inevitable small experimental aberrations becomes extremely important. Software programs for constructing and consulting such databases should be developed. For any new genus, a type species and an appropriate name must be proposed. Named taxa are necessary for the recognition of groups and for the practical use of bacterial classification.

We strongly recommend guidelines or minimal standards for delineating genera including (i) a phylogenetic parameter expressed as percentage 16S rRNA gene sequence similarity and a high bootstrap value for the relevant branching points; (ii) a polyphasic description including phenotypic, genomic, and chemotaxonomic data to provide a comprehensive description and allow differentiation.

CONCLUSION

The main conclusions concerning polyphasic taxonomy as it has been practiced widely during the last 20 years are as follows:

1. Replacing minimal numbers of characteristics by large numbers of features and characterizing different aspects of bacterial cells has resulted in more stable polyphasic classification systems.

2. Polyphasic species descriptions should (i) reflect phylogenetic relationships, (ii) be based on total genome DNA–DNA hybridization to determine the genomic relationships between representatives of groups and within these groups, and (iii) provide further descriptive genomic, phenotypic and chemotaxonomic information, as well as information on the infraspecific clonal structure as revealed by fine typing methodologies. In principle all methods studying a particular aspect of the cell can be useful as sources of information. In practice a choice of complementary methods has to be made to tackle any taxonomic problem.

3. The polyphasic bacterial species is more complex than the species defined by Wayne et al. (1987) because more aspects are considered. Polyphasic practice differs according to the groups studied, and the final impact of a particular category of characters may vary considerably. From polyphasic taxonomy studies, the bacterial species appears as an assemblage of isolates originating from a common ancestor population, in which the steady generation of genetic diversity has resulted in clones with different degrees of recombination. The polyphasic species is characterized by a degree of phenotypic consistency, by a significant degree of total genome DNA–DNA hybridization, and by over 97% 16S rRNA gene sequence similarity. Usually, a minimal phenotypic description

is required, supported by genomic and/or chemotaxonomic differentiating parameters.

4. Polyphasic taxonomy is purely empirical, follows no strict rules or guidelines, may integrate any type of information, and results in a consensus classification that reflects the phylogenetic relationships, a guarantee for its stability. The aim is to reflect as closely as possible the biological reality.

5. The usefulness of polyphasic characterization is to enable the selection of an appropriate technique to be used for a quick and accurate identification. For many bacteria encountered in routine diagnostic laboratories, monophasic, mostly phenotypic, identification will still be used, but other bacteria may require the utilization of more than one technique e.g., rRNA identification to determine the phylogenetic position and an appropriate fine technique (genomic, chemotaxonomic, and/or phenotypic tests) for the species level. Comparison with a standardized, accessible, universal database is a *conditio sine qua non*.

Perspectives

1. One of the most interesting perspectives in bacterial systematics is the technological progress to be expected in the near future and its enormous impact on polyphasic methodology. Data for large numbers of bacterial strains will be gathered even faster, but the challenge will be the processing of these enormous amounts of data into a classification system. Large sets of data can only be analyzed by computer- assisted techniques, and appropriate software programs are needed to agglomerate the most closely related strains and to represent the agglomerates. The application of fuzzy logic (Kosko, 1994), based on the idea that an isolate does not have to belong to a particular set of strains but can have a partial degree of membership in more than one set, may open new perspectives.

2. The accessibility of standardized genotypic, chemotaxonomic, and phenotypic features via universal databases is another goal, which can only be realized when complete standardization of all techniques is achieved. New methodologies that can fuse different databases are also required.

3. A further goal is the design of an accessible, cumulative, dynamic system allowing continuous recalculation of all existing information into "new synthetic taxonomies of the moment".

4. We will be dependent on these and other developments if we want to perform better in the discovery and description of bacterial biodiversity in nature. In such a system, non-clustered isolates and gene sequences of uncultured bacteria have their place and are available for comparison at the same level as recognized named taxa. In order to streamline labeling of taxa, the simplification of the actual nomenclatural practice might be considered.

5. Other macromolecules potentially useful for phylogenetic comparison e.g., β-subunit of ATPase and elongation factor Tu (Ludwig et al., 1993), chaperonin (Viale et al., 1994), various ribosomal proteins (Ochi, 1995), RNA polymerases (Zillig et al., 1989) and tRNAs (Höfle, 1990, 1991) should be further investigated to allow comparison of the results of phylogenetic analysis with the rRNA gene based dendrograms.

6. More whole-genome sequences and insights into genomic organization are available for a variety of bacterial organisms and will also become accessible to microbial taxonomists. It will be a formidable challenge in the next century to use this information to evaluate our present view on polyphasic classification.

7. Together with the comparison of sequences of particular genes or gene families, a better understanding of the evolution of bacterial genomes will become possible, shining a new light on the present (in)consistencies in bacterial systematics. If horizontal gene transfer is indeed not a marginal phenomenon but an important mechanism of procaryote evolution (Lake et al., 1999), complete genome sequencing may yield major revelations about the evolutionary tree of life.

Acknowledgments

We are indebted to the Fund for Scientific Research (Flanders, Belgium) for research and personnel grants (M.G., P.D.V., K.K. and J.S.) and for funding Postdoctoral Research Fellow (P.V.) and Research Director (P.D.V.) positions.

Overview: A Phylogenetic Backbone and Taxonomic Framework for Procaryotic Systematics

Wolfgang Ludwig and Hans-Peter Klenk

INTRODUCTION

Despite its relatively short history, microbial systematics has never been static but rather constantly subject to change. The evidence of this change is provided by many reclassifications in which bacterial taxa have been created, emended, or dissected, and organisms renamed or transferred. The development of a procaryotic systematics that reflects the natural relationships between microorganisms has always been a fundamental goal of taxonomists. However, the task of elucidating these relationships could not be addressed until the development of molecular methods (the analysis of macromolecules) that could be applied to bacterial identification and classification. Determination of genomic DNA G + C content, and chemotaxonomic methods such as analysis of cell wall and lipid composition, in many cases proved superior to classical methods based upon morphological and physiological traits. These tools provide information that can be used to differentiate taxa, but do not allow a comprehensive insight into the genetic and phylogenetic relationships of the organisms. DNA–DNA reassociation techniques provide data on genomic similarity and hence indirect phylogenetic information, but the resolution of this approach is limited to closely related strains. DNA–DNA hybridization is the method of choice for delimiting procaryotic species and estimating phylogeny at and below the species level. The current species concept is based on two organisms sharing a DNA–DNA hybridization value of greater than 70% (Wayne et al., 1987).

With improvement in molecular sequencing techniques, the idea of Zuckerkandl and Pauling (1965) to deduce the phylogenetic history of organisms by comparing the primary structures of macromolecules became applicable. The first molecules to be analyzed for this purpose were cytochromes and ferredoxins (Fitch and Marguliash, 1967). Subsequently, Carl Woese and co-workers demonstrated the usefulness of small subunit (SSU) rRNA as a universal phylogenetic marker (Fox et al., 1977). These studies suggested natural relationships between microorganisms on which a new procaryotic systematics could be based. The aims of this chapter are to provide a brief description of the methods used to reconstruct these phylogenetic relationships, to explore the phylogenetic relationships suggested by 16S rRNA and alternative molecular chronometers, and to present a justification for the use of the current 16S rRNA-based procaryotic systematics as a backbone for the structuring of the second edition of *Bergey's Manual of Systematic Bacteriology*.

RECONSTRUCTION AND INTERPRETATION OF PHYLOGENETIC TREES

Sequence alignment The critical initial step of sequence-based phylogenetic analyses is undoubtedly the alignment of primary structures. Alignment is necessary because only changes at positions with a common ancestry can be used to infer phylogenetic conclusions. These homologous positions have to be recognized and arranged in common columns to create an alignment, which then provides the basis for subsequent calculations and conclusions. Sequences such as SSU rRNA that contain a number of conserved sequence positions and stretches can be aligned using multiple sequence alignment software such as CLUSTAL W (Swofford et al., 1996). Furthermore, these conserved islands can be used a guide for arranging the intervening variable regions. The alignment of variable regions may remain difficult if deletions or insertions have occurred during the course of evolution. In addition, the homologous character of positions in variable regions is not necessarily indicated by sequence identity or similarity and hence can often not be reliably recognized. However, functional homology, if detectable or predictable, can be used to improve the alignment. In the case of rRNAs, functional pressure apparently dictates the evolutionary preservation of a common core of secondary or higher order structure which is manifested by the potential participation of 67% of the residues in helix formation by intramolecular base pairing. The majority of these structural elements are identical or similar with respect to their position within the molecule as well as number and position of paired bases, or internal and terminal loops. The primary structure sequence alignment can be evaluated and improved by checking for potential higher structure formation (Ludwig and Schleifer, 1994). Furthermore, the character of the base pairing, G–C versus non-G–C, Watson–Crick versus non-Watson–Crick, may be used to refine an alignment. The pairing is a byproduct of thermodynamic stability and consequently has an impact on function. Therefore, adjustments to the alignment appear rational from an evolutionary point of view. However, the recognition of homologous positions in regions which are highly variable with respect to primary as well as higher order structure may still be difficult or even impossible.

The principal problems of aligning rRNA sequences can be avoided by the routine user, if they take advantage of comprehensive databases of aligned sequences (including higher order structure information) that can be obtained from the Ribosomal Database Project (Maidak et al., 1999), the compilations of small

TABLE 1. Transformation of measured distances (lower triangle) into phylogenetic distances (upper triangle): applying the Jukes Cantor (Jukes and Cantor, 1969) transformation[a]

	Escherichia coli	Klebsiella pneumoniae	Proteus vulgaris	Pseudomonas aeruginosa	Bacillus subtilis	Thermus thermophilus	Geotoga subterranea
Escherichia coli		3.2	7	15.6	26	28.5	35.8
Klebsiella pneumoniae	3.1		7	15.1	25.8	28.2	36.4
Proteus vulgaris	6.7	6.7		17.6	26.6	29.9	37.8
Pseudomonas aeruginosa	14.1	13.7	15.7		23.5	29.2	34.3
Bacillus subtilis	22	21.8	22.4	20.2		27	30.4
Thermus thermophilus	23.7	23.5	24.7	24.2	22.6		32.4
Geotoga subterranea	28.5	28.8	29.7	27.6	25	26.3	

[a]The uncorrected distances were used for the reconstruction of the tree in Fig. 1. Given that the data are not ultrametric (see Swofford et al, 1996), they do not directly correlate with the branch lengths in the tree.

and large subunit rRNAs at the University of Antwerp (De Rijk et al., 1999; Van de Peer et al., 1999), or the ARB project as a guide to inserting new sequence data. The RDP offers alignment of submitted sequences as a service while the ARB program package contains tools for automated alignment, secondary structure check, and confidence test.*

Treeing methods The number and character of positional differences between aligned sequences are the basis for the inference of relationships. These primary data are then processed using treeing algorithms based on models of evolution. Usually, the phylogenetic analysis is refined by positional selection or weighting according to criteria such as variability or likelihood. The results of these analyses are usually visualized as additive trees. Terminal (the "organisms") and internal (the common "ancestors") nodes are connected by branches. The branching pattern indicates the path of evolution and the (additive) lengths of peripheral and internal branches connecting two terminal nodes indicate the phylogenetic distances between the respective organisms. There are two principal versions of presentation: radial trees or dendrograms (Fig. 1). The advantage of radial tree presentation is that phylogenetic relationships, especially of only moderately related groups, can usually be shown more clearly, and that all of the information is condensed into an area which can be inspected "at a glance". However, the number of terminal nodes (sequences, organisms, taxa) for which the relationships can be demonstrated is limited. This number is not limited in dendrograms.

A number of different treeing methods or algorithms based on sequence data have been developed. Most of them are based on models of evolution. These models describe assumed rules of the evolutionary process concerning parameters such as (overall) base frequencies or (number and weighting of) substitution types. A comprehensive review on methods for phylogenetic analyses, models of evolution, and the mathematical background is given by Swofford et al. (1996). The three most commonly used treeing methods, distance matrix, maximum parsimony, and maximum likelihood, operate by selecting trees which maximize the congruency of topology and branch lengths with the measured data under the criteria of a given model of evolution.

Distance treeing methods such as Neighbor Joining (Saitou and Nei, 1987) or the method of Fitch and Margoliash (Fitch and Margoliash, 1967) rely on matrices of distance values obtained by binary comparison of aligned sequences and calcula-

tion of the fraction of base differences. These treeing programs mostly perform modified cluster analyses by defining pairs and, subsequently, groups of sequences sharing the lowest distance values and connecting them into the framework of a growing tree. The tree topology is optimized by maximizing the congruence between the branch lengths in the tree and the corresponding inferred distances of the underlying matrix.

Before treeing, the measured differences are usually transformed into evolutionary distance values according to models of evolution. The underlying assumption is that the real number of evolutionary changes is underestimated by counting the detectable differences in present day sequences. For example, the Jukes Cantor transformation (Jukes and Cantor, 1969) accounts for this underestimation by superelevation of the measured distances (Table 1). Although the theoretical assumptions that provide the basis for transforming the measured distance values into phylogenetic distances are convincing with respect to overall branch lengths, there is a certain risk of misinterpretation or overestimation of local tree topologies. An intrinsic disadvantage of distance treeing methods is that only part of the phylogenetic information, the distances, is used, while the character of change is not taken into account. However, there are methods available to perform more sophisticated distance calculations than simply counting the differences (Felsenstein, 1982).

In contrast to distance methods, maximum parsimony-based treeing approaches use the original sequence data as input. According to maximum parsimony criteria, tree reconstruction and optimization is based on a model of evolution that assumes preservation to be more likely than change. Parsimony methods search for tree topologies that minimize the total tree length. That means the most parsimonious (Edgell et al., 1996) tree topology (topologies) require(s) the assumption of a minimum number of base changes to correlate the tree topology and the original sequence data. In principle, the problem of plesiomorphies (see below) can be handled more appropriately with parsimony than with distance methods, given that the most probable ancestor character state is estimated at any internal node of the tree. Long branch attraction is a disadvantage of the maximum parsimony approach. The parsimony approach does infer branching patterns but does not calculate branch lengths *per se*. To superimpose branch lengths on the most parsimonious tree topologies additional methods and criteria have to be applied. Both PAUP* and ARB parsimony tools are able to combine the

*Editorial Note: Software available from O. Strunk and W. Ludwig, Department of Microbiology, Technische Universität, München, Munich, Germany. ARB is a software environment for sequence data located at: www.mikro.biologie.tu-muenchen.de/pub/ARB.

*Editorial Note: Software available from David Swofford at the Laboratory of Molecular Systematics, National Museum of Natural History, Smithsonian Institution, Washington, D.C. Available from Sinauer Associates of Sunderland, MS at www.sinauer.com/formpurch.htm.

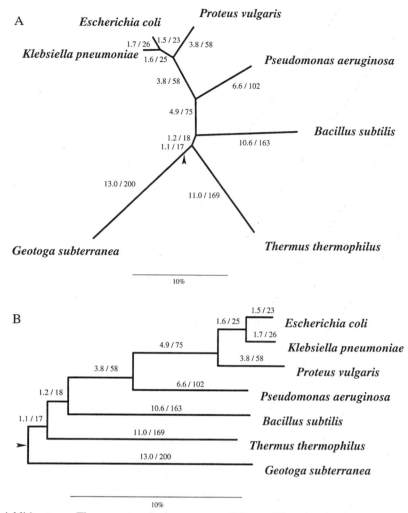

FIGURE 1. Additive trees. The same tree is shown as a radial tree (*A*) and a dendrogram (*B*). The tree was reconstructed by applying the neighbor joining method (Saitou and Nei, 1987) to a matrix of uncorrected binary 16S rRNA sequence differences for the organisms shown in the tree and a selection of archaeal sequences as outgroup references. *Arrowheads* indicate the branching of the archaeal reference sequences and the root of the trees. Bar = 10% sequence difference. The distance between two sequences (organisms) is the sum of all branch lengths directly connecting the respective terminal nodes or the sum of the corresponding horizontal branch lengths in the radial tree or the dendrogram, respectively. The numbers at the individual branches indicate overall percentage sequence divergence, followed by the number of different sequence positions (the length of the *E. coli* 16S rRNA sequence [1542 nucleotides] was used as reference in all calculations). Note: the tree topology was not evaluated by applying different methods and parameters.

reconstruction of topologies and the estimation of branch lengths.

The most sophisticated of the three independent phylogenetic treeing methods is maximum likelihood, where a tree topology is regarded as optimal if it reflects a path of evolution that, according to the criteria of given models of evolution, most likely resulted in the sequences of the contemporary organisms. The corresponding evolutionary models may include parameters such as transition/transversion ratio, positional variability, character state probability per position and many others. Given that the maximum likelihood approach utilizes more of the information content of the underlying sequences, it is considered to be superior to the other two treeing methods. An accompanying disadvantage is the need for expensive computing time and performance. Even if powerful computing facilities are accessible only a limited number of sequences can be handled within a

reasonable time. Rapid development in the field of computing hardware suggests that this powerful method may become applicable for larger data sets in the near future.

The use of filters   Most commonly used programs for phylogenetic treeing are capable of including filters or weighting masks that remove or weight down individual alignment columns while treeing, thus reducing the influence of highly variable positions. Conservation profiles can be calculated by simply determining the fraction of the most frequent character. More sophisticated approaches define positional variability, the rate of change, or the likelihood of a given character state, with respect to an underlying tree topology according to parsimony criteria, or by using a maximum likelihood approach. The choice of phylogenetic entities for which filters or masks should be generated depends on the group of organisms or the phylogenetic level

TABLE 2. Phylogenetic information content of procaryotic small subunit rRNA[a]

Intra-domain similarity	Bacteria &mt;67%				Archaea &mt;67%			
	Conserved		Variable		Conserved		Variable	
	Pos.	%	Pos.	%	Pos.	%	Pos.	%
Sequence conservation	568	36.8	974	63.2	571	37	971	63
Potential information (bits)			1948				1942	
Number of characters	1	2	3	4	1	2	3	4
Positional variability, %	36.8	23.2	13.5	26.5	37	28.3	15.2	19.5
Corrected information (bits)			1506				1385	

[a]The calculations were performed using the 16S rRNA sequence of *E. coli* (1542 nucleotides) as a reference. To avoid influences of sequencing, database, and alignment errors a 98% similarity criterion was applied to define "invariant" positions. Therefore, the term 'conserved' was used instead "invariant". Bits (of information) were calculated by multiplying the logarithm to the base two of the permissive character states (positional variability: different nucleotides per position) times the number of informative (variable) sites. Potential information was calculated as the maximum information content assuming positional variability of four. These values were corrected according measured positional variability.

(the corresponding area in a tree) of interest. Tools for the generation of profiles, masks, or filters are implemented in the ARB software package or available from other authors (Swofford et al., 1996; Maidak et al., 1999). The removal of positions also means loss of information; therefore it is recommended to perform treeing analyses of a given data set several times applying different filters. This helps to visualize the robustness or weakness of a specific tree topology and to estimate whether or not variable positions have had a substantial influence. Filters or masks should only be calculated using comprehensive data sets of full sequences; then these filters can also be applied to the analysis of partial sequences. The results of many years of tree reconstruction have shown that positions should only be removed up to 60% positional conservation, to avoid the loss of too much information. In most cases use of a 50% conservation filter is appropriate.

Confidence tests   Different treeing methods handle data according to particular assumptions and consequently may yield different results. The many inconsistencies of real sequence data also prevent easy and reliable phylogenetic inference; therefore the careful evaluation of tree topologies is to be recommended. Besides the application of filters and weighting masks and the use of different treeing approaches, resampling techniques can be used to evaluate the statistical significance of branching order. Bootstrapping or jackknifing (Swofford et al., 1996) are procedures that randomly sample or delete columns in sequence data (alignments) or distance values (distance matrix). Usually 100–1000 different artificial data sets are generated as inputs for treeing operations by these methods. For each data set the optimum tree topologies are inferred by the particular treeing method and, finally, a consensus tree topology is generated. In this consensus tree, bootstrap or jackknife values are assigned to the individual branches. These values indicate the number of treeing runs in which the subtree defined by the respective branch appeared as monophyletic with respect to all other groups. An example of a bootstrapped tree is shown in Fig. 2. Besides the bootstrap value, an area of low significance is indicated by circles centered on the individual (internal) nodes. These areas were estimated from the sampling values in relation to the corresponding (internal) branch lengths using the ARB software tools. No convincing significance can be expected if only a few residues provide information supporting the separation of branches or subtrees. Given that in most cases branch lengths indicate the degree of estimated sequence divergence, a subtree separated

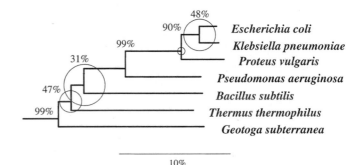

FIGURE 2. Confidence tests on tree topology. 1000 bootstrap operations were performed for evaluation of the tree in Figure 1B. The numbers at the furcations indicate the fraction of (1000 bootstrapped) trees which support the separation of the respective subtree (branches to the right of the particular furcation) from all other branches or groups in the tree. Circles indicating an area of "unsharpness" were calculated as a function of bootstrap values and branch lengths using ARB.

from the remainder of a phylogenetic tree by a short internal branch is highly unlikely to be assigned a high resampling value.

The resampling techniques can only be used to estimate the robustness of a tree reconstructed by applying a single treeing method and parameter set. Thus for reliable phylogenetic conclusions it is necessary to combine different treeing methods, as well as filters and weighting masks and resampling techniques. Even if appropriate software and powerful hardware are accessible, high quality tree evaluations may get rather expensive in working and computing time. An approach for estimating "upper bootstrap" limits without the need of expensive multiple treeings was developed and implemented in the ARB package. In the absence of resampling values, a critical "reading" of trees allows a rough estimate of the confidence of relative branching orders at a glance, assuming that a short branch length in most cases also indicates low significance of separation.

Why do trees differ?   Tree reconstruction can often be a frustrating experience, especially for researchers not familiar with the theoretical principles of phylogenetic treeing, when the application of different treeing methods or parameters to a single data set results in different tree topologies. This is not surprising since different treeing methods are based on different models of evolution, and therefore the data are processed in different ways. Consequently, a perfect match of tree topologies cannot necessarily be expected even if identical data sets are analyzed using identical parameters. None of the models reflect perfectly the reality of the evolutionary process. The assumption of independent evolution of different sequence positions, for example, does not hold true for the many functionally correlated residues such as base paired nucleotides in rRNAs. In addition, none of the treeing methods and software programs can really exhaustively test and optimize all possible tree topologies. For example, with only 20 sequences there would be 10^{20} possible tree topologies to be examined. Other factors, such as data selection (the organisms and sequence positions included in calculations), the order of data addition to the tree, and the presence of positions that have changed at a higher rate than the remainder of the data set, also influence tree topology. These instabilities do not usually concern the global tree topology but rather local branching patterns.

LIMITATIONS OF TREE RECONSTRUCTION

Information content of molecular chronometers The reconstruction of gene or organismal history, based upon the degree of divergence of present day sequences, relies on the number and character of detectable sequence changes that have accumulated during the course of evolution. Thus the maximum information content of molecules is defined by the number of characters (monomers), and the number of potential character states (different residues), per site. With real data, only a fraction of the sites are informative, as a reasonable degree of sequence conservation is needed to demonstrate the homologous character of molecules or genes and to recognize a phylogenetic marker as such. For example, there are 974 (63.2%) variable and hence informative positions in the 16S rRNA genes of members of the *Bacteria*, and 971 (63%) such positions in the *Archaea*. Given that the maximum information content per position is defined by the number of possible character states i.e., the four nucleotides (the potential fifth character state, deletion or insertion, is not considered here), there could be 1948 (*Bacteria*) or 1942 (*Archaea*) bits of information (logarithm to base 2 of the number of possible character states times the number of informative positions) in the SSU rRNA. However, due to functional constraints and evolutionary selective pressure, the number of allowed character states varies from position to position. As shown in Table 2, there are only 407 (26.4%; *Bacteria*) or 301 (19.5%; *Archaea*) positions in the investigated data set at which all four nucleotides are found, whereas only three different residues apparently are tolerated at 209 (13.6%; *Bacteria*) or 233 (15.2%; *Archaea*) positions, and only two character states are realized at 358 (23.2%; *Bacteria*) or 437 (28.3%; *Archaea*) positions. Thus the theoretical information content of 1984 (*Bacteria*) or 1938 (*Archaea*) bits in reality is reduced to 1506 (*Bacteria*) or 1385 (*Archaea*). The reduced information content draws attention to the need for careful sequence alignment and analysis.

The problem of plesiomorphy Any homologous residue in present day sequences can only report one evolutionary event. The higher the number of permitted characters at a particular position, the higher the probability that such an evolutionary event is directly detectable (by a difference). The majority of these events remain obscure since, especially at variable positions, identical residues are probably the result of multiple changes during the course of evolution, simulating an unchanged position (plesiomorphy). The effect of plesiomorphy on the topology of the resulting trees depends on the number of plesiomorphies supporting branch attraction and also on the treeing method used.

Such plesiomorphic sites may cause misleading branch attraction, as shown in Fig. 3, where a short stretch of aligned real 16S rRNA sequences is used to visualize branch attraction. Plesiomorphies may also be responsible for the observation that long "naked" branches represented by only one or a few highly similar sequences often "jump" in phylogenetic trees when the reference data set is changed or expanded. The positioning can usually be stabilized when further representatives of different phylogenetic levels of that branch become available. The rooting of trees may also be influenced by identities at plesiomorphic sites when single sequences are used as outgroup references. The influence of plesiomorphic positions can be reduced by using them at a lower weight for tree reconstruction, but is nevertheless still present.

Partial sequence data There are several convincing arguments for the use of only complete sequence data in the reconstruction of phylogenetic trees. These include the limited information content of the molecule, and the fact that different parts of the primary structure carry information for different phylogenetic levels (Ludwig et al., 1998b). Whenever partial sequences are added to a database of complete primary structures and phylogenetic treeing approaches are applied to the new data set, the new sequences may influence the overall tree topology. The inclusion of partial sequence data may impair phylogenetic trees or influence conclusions previously based on full data. Software which allows the addition of new data to a given data set, and placement of the new sequence according to optimality criteria in a validated tree without changing its topology, is now available. The ARB implementation of this software is capable of removing short partial sequences from a tree prior to the integration of a new highly similar but more complete sequence. Thus the more informative sequence is not "attracted" by a probably misplaced partial sequence. After finding the most similar sequences, the ARB tool compares the number of determined characters, removes the shorter version, and reinserts the data in the order of completeness. There are a number of recent publications presenting comprehensive trees based upon data sets which have been truncated to the regions comprised by included partial sequences. This procedure is not acceptable, given all the limitations of partial sequence data and of the methods of analysis.

Partial sequence data of appropriate regions of the gene may contain sufficient information for the identification of organisms. The determination and comparative analysis of partial sequences may be sufficient to reliably assign an organism to a phylogenetic group if the database contains sequences from closest relatives. A fraction of the 5′-terminal region of the SSU rRNA (*Escherichia coli* pos. 60–110) is one the most informative or discriminating regions for closely related organisms. Hence partial sequence data that include this region can be used to find the closest relative of an organism or to indicate a novel species. Short diagnostic regions (15–20 nucleotides) of partial sequences can also be used as targets for taxon-specific probes or PCR primers that are commonly used for the sensitive detection and identification of microorganisms (Schleifer et al., 1993; Amann et al., 1995; Ludwig et al., 1998a).

Bush-like trees The majority of names and definitions of major phylogenetic groups, such as the phylum *"Proteobacteria"* and the corresponding classes (*"Alphaproteobacteria"*, *"Betaproteobacteria"*, *"Gammaproteobacteria"*, *"Deltaproteobacteria"*, and *"Epsilonproteobacteria"*) originated in the early years of comparative rRNA sequence analysis. At that time phylogenetic clusters could easily be delimited, given that the trees contained many long "naked"

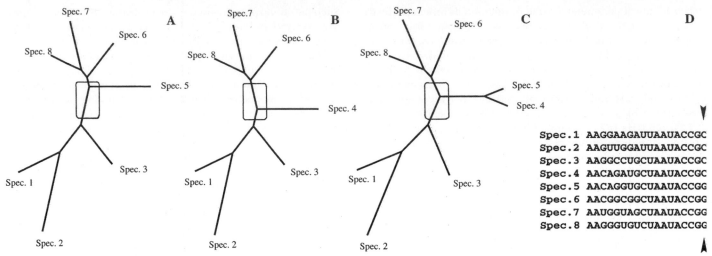

FIGURE 3. "Branch attraction". The trees show the effects of separate (*A*, *B*) or combined (*C*) inclusion of sequences Spec. 4 and Spec. 5 on the tree topology. The *rectangle* highlights the region of the tree where major changes can be seen. The trees were reconstructed using the neigh-bor joining method on the aligned 16S rRNA sequence fragments shown (*D*). The column of residues responsible for the attraction of Spec. 5 branch and the Spec. 6–8 subtree in *A* as well as the attraction of Spec. 4 branch and the Spec. 1–3 subtree in *B* is marked by arrows.

branches separating subtrees. This "phylogenetic clarity" was mainly an effect of the limited amount of available sequence data and has often been obscured by the rapid expansion in the number of sequence database entries. Most of the long "naked" branches have expanded and the tree-like topology changed to a bush-like topology. It is probably only a matter of time before the missing links will be found for the remaining "naked" branches such as the *Chlamydiales*, the *"Flexistipes"*, or branches assigned to cloned environmental sequences.

In bush-like areas of a tree, the probability that a given branch will exchange positions with a neighboring branch decreases with the distance between the two branches. This indicates that the relative order of closely neighboring branches cannot be reliably reconstructed, although their separation from more distantly located lineages remains robust. As a consequence, delimitation of taxa often cannot be based on individual local branching order. The use of criteria or additional data for the definition of taxonomic units remains the subjective decision of the taxonomist. In some cases, this leads to the definition of taxa that include paraphyletic groups.

PRESENTATION OF PHYLOGENETIC TREES

The main purpose of drawing trees is to visualize the phylogenetic relationships of the organisms or markers, and to allow the reader to recognize these relationships at a glance. It is often difficult to combine an easy-to-grasp presentation of phylogenetic relationships and associated information on the significance of branching patterns. There is no optimum solution to the problem of "correct" presentation of trees; however, ways of addressing this problem can be suggested.

One acceptable procedure would be to present all the trees (which may differ locally) obtained from the same data set by the application of different treeing methods and parameters. However, this may prove more confusing than helpful for readers not experienced in phylogenetic treeing. A more user-friendly solution is to present only one tree topology and to indicate the significance of the individual branches or nodes. However, showing multiple confidence values at individual nodes, or depicting

areas of confidence by shading or circles around the nodes, may make the tree unreadable, especially in areas of bush-like topology.

In many cases use of a consensus tree is advantageous. Some programs for consensus tree generation are able to present local topologies as multifurcations at which a relative branching order is not significantly supported by the results of tree evaluations. A fairly acceptable compromise is to use a consensus tree, and to visualize both a detailed branching pattern where stable topologies can be validated, and multifurcations that indicate inconsistencies or uncertainties. Such a multifurcation indicates missing information on that particular era of evolution rather than multiple events resulting in a high diversification within a narrow span of evolutionary time. This type of presentation is certainly more informative for the reader than a choice of various tree topologies, each showing low statistical significance (Ludwig et al., 1998b).

None of the modes of presentation described above can be applied to bush-like topologies and yield meaningful results. Although individual branches are likely to change their positions only locally within a large bush-like area, depending on the methods and parameters applied for treeing, there is no way to split up such an area by several multifurcations. No methods are currently available for the calculation of confidence values for the next, second, third and so on neighboring nodes, and highlighting areas of unsharpness makes the tree difficult to read. A legitimate solution is to base the calculations on the full data set but to hide some of the branches for presentation purposes, and show a tree topology containing a smaller number of significantly separated branches. Thus, while the tree would be based on all available information, only that part of the tree topology of interest for the particular phylogenetic problem is shown clearly laid out (Ludwig et al., 1998b).

16S RRNA: THE BENCHMARK MOLECULE FOR PROCARYOTE SYSTEMATICS

In principle, all the requirements of a phylogenetic marker molecule are fulfilled in SSU rRNAs to a greater extent than in almost all other described phylogenetic markers (Woese, 1987; Olsen and Woese, 1993; Olsen et al., 1994b; Ludwig and Schleifer, 1994; Ludwig et al., 1998b). Besides functional constancy, ubiquitous distribution, and large size (information content), genes coding for SSU rRNA exhibit both evolutionarily conserved regions and highly variable structural elements. The latter characteristic results from different functional selective pressures acting upon the independent structural elements. This varying degree of sequence conservation allows reconstruction of phylogenies for a broad range of relationships from the domain to the species level. A comprehensive SSU rRNA sequence data set (currently more than 16,000 entries) is available in public databases (Ludwig and Strunk, 1995*; Maidak et al., 1999; Van de Peer et al., 1999) in plain or processed (aligned) format, and is rapidly increasing in size. A significant fraction of validly described procaryotic species are represented by 16S rRNA sequences from type strains or closely related strains.

As with any new technique in the field of taxonomy, it took time to establish comparative sequencing of SSU rRNA (genes) as a powerful standard method for the identification of microorganisms and defining or restructuring procaryotic taxa according to their natural relationships. Rapid progress in sequencing and *in vitro* nucleic acid amplification technology led to the replacement of an expensive, sophisticated, and tedious methodology, available only to specialists, by rapid and easy-to-apply routine techniques. As a result, analysis of the genes coding for SSU rRNA is one of the most widely used classification techniques in procaryotic identification and systematics. It is widely accepted that SSU rRNA analysis should be integrated into a polyphasic approach for the new description of bacterial species or higher taxa.

SOME DRAWBACKS OF 16S rRNA GENE SEQUENCE ANALYSIS

Functional constraints Depending on functional importance, the individual structural elements of rRNAs cannot be freely changed. It is therefore assumed that sequence change in the rRNAs occurs in jumps rather than as a continuous process. The divergence of present day rRNA sequences may document the succession of common ancestors and their present day descendants, but a direct correlation to a time scale cannot be postulated.

Multiple genes It has been known since the early days of comparative rRNA sequence analysis that the genomes of microorganisms may contain multiple copies of some genes or operons. However, until recently it was commonly assumed that there are no remarkable differences between the rRNA gene sequences of a given organism. A significant degree of sequence divergence among multiple homologous genes within the same organism, such as has been found in *Clostridium paradoxum* (Rainey et al., 1996) and *Paenibacillus polymyxa* (Nübel et al., 1996), would call any sequence-based interorganism relationships into question. The underestimation of this problem may be attributed to the fact that such differences are not easy to recognize using sequencing techniques which depend on purified rRNA or amplified rDNA, and can be mistaken for artifacts. Only frame shifts resulting from inserted or deleted residues can be readily rec-

ognized. New techniques, such as denaturing gradient gel electrophoresis (DGGE) (Nübel et al., 1996), allow sequence variation in PCR-amplified rDNA fragments to be detected. The rapidly progressing genome sequencing projects have also provided detailed information on the topic of intraorganism rRNA heterogeneities. Different organisms vary with respect to the presence and degree of intercistron primary structure variation, and most differences concern variable positions and affect base-paired positions (Engel, 1999; Nübel et al., 1996). Although some projects to systematically investigate interoperon differences have been initiated, no comprehensive survey of the spectrum of microbial phyla has been performed. Current and future investigations will show whether regularities or hot spots for interoperon differences can be defined in general or in particular for certain phylogenetic groups. This knowledge can then be used to remove or weight such positions for phylogenetic reconstructions.

Interpretation of high 16S rRNA gene sequence similarity Organisms sharing identical SSU rRNA sequences may be more diverged at the whole genome level than others which contain rRNAs differing at a few variable positions. This has been shown by comparison of 16S rRNA sequence and genomic DNA–DNA hybridization data (Stackebrandt and Goebel, 1994). In the interpretation of phylogenetic trees, it is important to note that branching patterns at the periphery of the tree cannot reliably reflect phylogenetic reality. Given the low phylogenetic resolving power at these levels of close relatedness (above 97% similarity), it is highly recommended to support conclusions based on SSU rRNA sequence data analysis by genomic DNA reassociation studies (Stackebrandt and Goebel, 1994).

COMPARATIVE ANALYSES OF ALTERNATIVE PHYLOGENETIC MARKERS

Other genes have been investigated as potential alternative phylogenetic markers, to determine whether SSU rRNA-based phylogenetic conclusions can describe the relationships of the organisms, or merely reflect the evolutionary history of the respective genes. For sound testing of phylogenetic conclusions based on SSU rRNA data, the sequences used must originate from adequate phylogenetic markers. The principal requirements for such markers are ubiquitous distribution in the living world combined with functional constancy, sufficient information content, and a sequence database which represents diverse organisms, containing at least members of the major groups (phyla and lower taxa) as defined based upon SSU rRNA.

How many alternative phylogenetic markers are out there? Comparative analysis of the completed genome sequences suggests that there are only a limited number of genes that occur in all genomes and which also share sufficient sequence similarity to be recognized as ortho- or paralogous. Analysis of the first eight completely sequenced genomes (six *Bacteria*, one *Archaea*, and one yeast) showed that only 110 clusters of orthologous groups (COGs) were present in all genomes (Tatusov et al., 1997; Koonin et al., 1998; updated in www.ncbi.nlm.nih.gov/COG/) and only eight additional genes were ubiquitous in procaryotes. Another 126 COGs were found in the remaining five microbial genomes, excluding the mycoplasmas, which have a reduced genomic complement. The majority of the universally conserved COGs (65 out of 110) belong to the information storage and processing proteins, which appear to hold more promise for future phylogenetic analysis than the metabolic proteins. However, about half

---

Editorial Note: Software available from O. Strunk and W. Ludwig, Department of Microbiology, Technische Universität, München, Munich, Germany. ARB is a software environment for sequence data located at: www.mikro.biologie.tu-muenchen.de/pub/ARB.

of these information processing COGs contain ribosomal proteins, which are small and therefore not sufficiently informative for the inference of global phylogenies. This leaves us with about 40–100 genes that fulfill the basic requirements of useful phylogenetic markers.

It has been proposed that many genes involved in the processing of genetic information (components of the transcription and translation systems) exhibit concurrent evolution due to their housekeeping function (Olsen and Woese, 1997). It appears logical that these key systems would be optimized early and then conserved to confer maximum survival and evolutionary benefit on the organism.

Although the databases of alternative phylogenetic markers are small relative to that of the SSU rRNA, some of the other requirements for markers, including representation of phylogenetically diverse organisms, are met by, for example, LSU rRNA, elongation factor Tu/1α, the catalytic subunit of the proton translocating ATPase, recA, and the hsp 60 heat shock protein. For some other markers fulfillment of the ubiquity requirement can not be assessed because of the limited state of the sequence databases.

SOME DRAWBACKS OF ALTERNATIVE PHYLOGENETIC MARKERS

Lateral gene transfer and gene duplication Comparative analyses of the 18 published complete microbial genome sequences does not reveal a consensus picture of the root of the tree of life (Klenk et al., 1997b) or of the relative branching order of the early lineages within the domains. This contradicts the marked separation of the primary domains based on morphology, physiology, biochemical characteristics, and overall genome sequence data. A monophyletic origin of the domain *Archaea* has been put in question by some authors (Gupta, 1998), but genomic evidence for monophyly of this group has also been reported (Gaasterland and Ragan, 1999). This contradiction has led to the assumption that lateral gene transfer and/or gene duplications, often followed by the loss of one or more gene variants in different lineages, has occurred in some potential marker molecules, especially genes coding for proteins involved in central metabolism (Brown and Doolittle, 1997). Obviously, such genes or markers cannot be used for testing major phylogenies deduced from SSU rRNA data.

The usefulness of many proteins as potential phylogenetic markers is curtailed by the presence of duplicated genes in certain organisms. The degree of sequence divergence in these duplicated markers ranges from the interdomain level, as shown for the catalytic subunit of vacuolar and F_1F_0-ATPases of *Enterococcus hirae*, to the species level, exemplified by EF-Tu of *Streptomyces ramocissimus*. When conserved proteins are used as phylogenetic markers for inferring intradomain phylogenies, one has to take care that orthologous genes (common origin) rather than paralogous genes (descendants of duplications) are compared. The recognition of paralogous genes is a central problem in phylogenetic analyses, especially when only limited data sets are available as in the case of the catalytic subunit of the proton-translocating ATPase. Although the sequence similarities between bacterial F_1F_0 type, and archaeal and eucaryal vacuolar type, ATPase subunits are rather low (around 20%), it was initially assumed that the corresponding subunits (β and A or α and B) are homologous molecules (Iwabe et al., 1989; Ludwig et al., 1993). The presence of an F_1F_0 type ATPase β-subunit gene has been shown for all representatives of the domain *Bacteria* inves-

tigated thus far (Ludwig et al., 1993; Neumaier, 1996). However, the finding that *Thermus* and other members of the "*Deinococcus–Thermus*" phylum contain vacuolar type ATPases (Tsutsumi et al., 1991; Neumaier, 1996) threatened this ATPase-based phylogenetic picture. It was later found that genes for subunits of vacuolar type ATPases exist in many (but not all) bacterial species from different phyla in addition to the corresponding F_1F_0 type ATPase subunit genes (Kakinuma et al., 1991; Neumaier, 1996). It is commonly accepted that F_1F_0 type ATPase subunits α and β resulted from an early gene duplication and should be regarded as paralogous. The same is assumed for the vacuolar type ATPase subunits A and B. The findings described above suggest additional early gene duplications probably leading to the ancestors of F_1F_0 and vacuolar type ATPase (subunits). Whereas α and β, or A and B subunits, coexist in all cases investigated so far, this is not the case for the F_1F_0 and vacuolar type paralogs. The available data indicate that the former would have become the essential energy-gaining version in the bacterial domain, the latter in the archaeal and eucaryal domains. During the course of evolution, the other member of the duplicate pair apparently changed its function (Kakinuma et al., 1991) and may have lost its essential character. Therefore, the nonessential copy could have been lost by many (even closely related) organisms during the course of evolution. The "*Deinococcus–Thermus*" phylum, in which only vacuolar type ATPases have been found, might be an exception. Assuming an early diversification of the bacterial phyla, the functional diversification of the duplicated ATPases could have occurred during this era of evolution. The ancestor of the members of the "*Deinococcus–Thermus*" phylum may have lost the F_1F_0 version early in evolution. However, early lateral gene transfers as postulated by some authors (Hilario and Gogarten, 1993) cannot be excluded.

There are other examples of gene duplications and premature phylogenetic misinterpretations, as documented by the history of glyceraldehyde-3-phosphate dehydrogenase (GAPDH) based phylogenetic investigations (Martin and Cerff, 1986; Brinkmann et al., 1987; Martin et al., 1993; Henze et al., 1995). Besides these early gene duplications, there are also indications of more recent events, such as the EF-Tu of *Streptomyces*, *hsp60* of *Rhizobium*, or *recA* of *Myxococcus*. Paralogous genes occurring as a result of gene duplication or lateral gene transfer can only be recognized as such in organisms which have preserved more than one version of the (duplicated) gene. And even then it may remain difficult or even impossible to decide which genes can be regarded as orthologous. Obviously, only the orthologous gene, which represents the functionally essential compound, can be used for inferring or evaluating phylogenies. Thus, whenever new potential phylogenetic markers are investigated and major discrepancies with rRNA-based conclusions are found, a comprehensive data base should be established, accompanied by an extensive search for potential gene duplications.

Limited information content Based on currently available sequence data, the LSU rRNA is the only marker which carries more phylogenetic information than the small subunit rRNA. There are more than twice as many informative residues in the large subunit rRNA (Ludwig et al., 1998b). In the case of protein markers, the amino acid sequences are preferred over the coding gene sequences for phylogenetic analysis. Proteins provide the function, and consequently the amino acid sequences are the targets of evolutionary selective pressure. In contrast, the DNA sequence differences, especially at third base positions, are under pressure of the codon preferences of the particular organism. Most of the proteins recognized as useful phylogenetic markers comprise less informative primary structure sites than the rRNA markers. For example, EF-Tu/-1α and ATPase catalytic subunit protein primary structures contain 311 and 359 informative residues, respectively. This deficiency could be partly compensated for by the 20 possible character states (amino acids) per position. However, in real data the number of allowed character states—the positional variability—is reduced due to functional constraints. The current data sets (EF-Tu/-1α and ATPase catalytic subunit) do not contain positions at which more than 15 different amino acids occur, and the largest fractions of positions (18%–20%, 11%–12%, 9%–12%) are represented by positions with only 2, 3, or 4 different residues, respectively.

Conflicting tree topologies Identical tree topologies cannot be expected from phylogenetic analysis of different markers. Given the low phylogenetic information content of each of the markers, and the wide grid of resolution, it is unlikely that independently evolving markers have preserved information on the same eras of evolutionary time. In principle, one would expect that this missing phylogenetic information would yield reduced resolution but not change the topology of the tree. However, the latter is often the case, as shown in Fig. 4. A small stretch of aligned real 16S rRNA sequences was used to generate the tree in Fig. 4A. If it is assumed that this tree illustrates the phylogenetic truth and that the information for the common origin of Spec. 4 and Spec. 5 was lost during the course of evolution, one would expect a reduction in resolution. Removing the alignment column (marked by an arrowhead) responsible for this relationship should result in shortening or deleting of the common branch of Spec. 4 and Spec. 5, producing a multifurcation as shown in Fig. 4B. However, due to branch attraction by residues at other alignment positions the branches of Spec. 4 and Spec. 5 are separated as shown in Fig. 4C, misleadingly simulating a different history. Consequently, local differences of resolution and topology in trees derived from alternative phylogenetic markers do not necessarily indicate a different path of evolution.

ALTERNATIVE GENE TREES

Large subunit rRNA As alluded to above, the LSU rRNA may be the most informative alternative phylogenetic marker. The primary structure of this molecule is at least as conserved as that of the SSU rRNA, and it contains more and longer stretches of informative positions. The spectrum of the LSU rRNA database is superior to that of all other alternative (protein) markers. Given that both rRNAs are involved in the translation process, it can be assumed that a similar selective pressure has been exerted on both genes. Consequently, LSU rRNA should be more useful for supporting rather than evaluating SSU rRNA-based conclusions. The internal structure (branching orders of the major lineages) of the intradomain trees can also be evaluated, given the availability of representative data sets for both molecules. The overall topologies of trees based upon the sequences of small and large subunit rRNA genes are in good agreement (De Rijk et al., 1995; Ludwig et al., 1998b). A 23S rRNA-based bacterial phyla tree is shown in Fig. 5, the corresponding 16S rRNA-based tree in Fig. 6. Slight local differences between trees reconstructed from both genes with the same method and parameters have been documented (De Rijk et al., 1995; Ludwig et al., 1995). This finding does not really cast doubt on the SSU rRNA-based branching patterns but rather underlines the previously mentioned limitations of phylogenetic markers. The LSU rRNA might be the better phylogenetic marker, providing more information and greater resolution, but the major drawback of this molecule is the currently limited database. Unfortunately, this database has not grown as fast as that for the SSU rRNA.

Elongation factors The elongation factors are also intrinsic components of the translation process but are functionally different from the rRNAs. It is generally assumed that the different classes of elongation (and probably initiation) factors are paralogous molecules resulting from early gene duplications. At present, a reasonable data set is available for EF-Tu/1α. In general, EF-Tu/1α-based domain trees (Fig. 7) globally support rRNA-derived branching patterns (Ludwig et al., 1998b). However, some general problems of protein markers are also exhibited by EF-Tu/1α sequences. As with the rRNA markers, no significant relative branching order for the major intradomain lines of descent can be determined. No major contradictions, e.g., members of a given phylum defined by rRNA sequences clustering among representatives of another phylum, were seen between rRNA and EF-Tu/1α tree topologies. However, in detailed

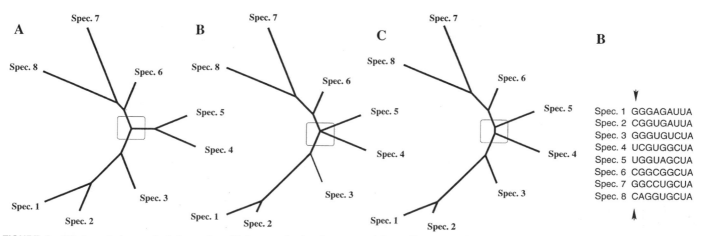

FIGURE 4. Missing phylogenetic information. If the tree in *A* reflects the true phylogeny, a tree topology showing a multifurcation for Spec. 4 and Spec. 5 as well as the other subtrees as shown in *B* would be correct if the phylogenetic information on the monophyletic origin of Spec. 4 and Spec. 5 was not preserved in present day sequences. This can be simulated by exclusion of the column marked by arrowheads in *D*. The loss of this information produces the misleading tree topology of *C* as a result of branch attraction.

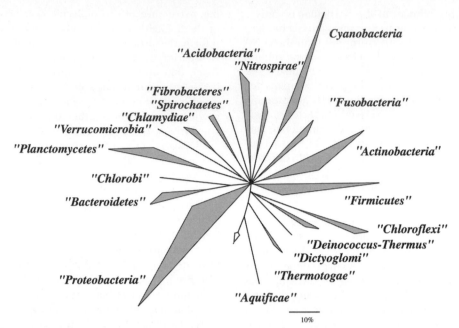

FIGURE 5. 23S rRNA based tree depicting the major bacterial phyla. The triangles indicate groups of related organisms, while the angle at the root of the group roughly indicates the number of sequences available and the edges represent the shortest and longest branch within the group. The tree was reconstructed, evaluated and optimized using the ARB parsimony tool. Only sequence positions sharing identical residues in at least 50% of all bacterial sequences were included in the calculations. All available almost complete homologous sequences from *Archaea* and *Eucarya* were used as outgroup references to root the tree (indicated by the *arrow*). Multifurcations indicate that a relative branching order could not be defined.

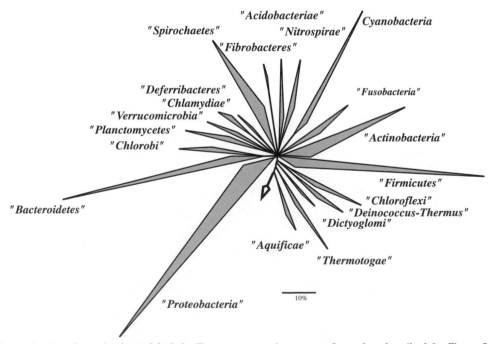

FIGURE 6. 16S rRNA based tree showing the major bacterial phyla. Tree reconstruction was performed as described for Figure 5. Tree layout of this and subsequent trees was according to the description for Figure 5.

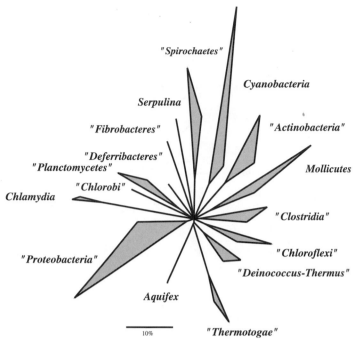

FIGURE 7. Elongation factor Tu based tree illustrating relationships among the major bacterial phyla. The tree was reconstructed from amino acid sequence data, and evaluated and optimized using the ARB parsimony tool. The tree is shown as unrooted, and only positions sharing identical residues in at least 30% of all sequences were included in the calculations.

trees local topological differences have been demonstrated (Ludwig et al., 1993). The reduced phylogenetic information content of EF-Tu (656 bits versus 1506 bits in the SSU rRNA; Ludwig et al., 1998b) may be responsible for the fact that the monophyletic status of some phyla such as the *"Proteobacteria"* is not supported by the protein-based trees. The separation of subgroups such as proteobacterial classes, however, is globally in agreement with the rRNA-based trees.

Interdomain sequence similarities for the rRNAs are 50% and higher, allowing the rooting and (at least to some extent) structuring of the lower branches for a given domain tree versus the other two. The interdomain protein similarities of the elongation factors are low (not more than 30%), making a reliable rooting or structuring of the bacterial tree difficult. The elongation factor database also contains examples of paralogy resulting from gene duplications or lateral gene transfer (Vijgenboom et al., 1994).

RNA polymerases The DNA-directed RNA polymerases (RNAPs) are essential components of the transcription process in all organisms, and the genes for the largest subunits (β and β′ in *Bacteria*; A′, A′′ and B in *"Crenarchaeota"*; B′ and B′′ in *"Euryarchaeota"*) are highly conserved and ubiquitous. The public databases contain RNAP sequences for about 40 species of *Bacteria* and 10 species of *Archaea*. The genes coding for RNAPs are located next to each other on the chromosomes of both *Bacteria* and *Archaea*, and contain 2300 (*Archaea*) to 2400 (*Bacteria*) amino acids that can be clearly aligned for phylogenetic purposes (Klenk et al., 1994). No paralogous genes are known for RNAPs. In general, for the *Bacteria* the intradomain topology of the trees derived from both RNAP large subunits supports the 16S rRNA-based tree in almost all details, with only one major discrepancy: the position of the root of the domain. Intensive rooting exper-

iments with a variety of archaeal and/or eucaryotic outgroups does not place the root of the *Bacteria* close to the extreme thermophiles (*Aquifex* or *Thermotoga* species) as in the rRNA tree, but next to *Mycoplasma* (Klenk et al., 1999). Since the placement of a root within a phylogenetic tree is not critical for most taxonomic purposes, it can be concluded that rRNAs and RNAPs in general support the same intradomain branching pattern for the *Bacteria*.

Proton translocating ATPase The catalytic subunit of proton-translocating ATPase is another example of a protein marker for which a reasonable data set is available, at least with respect to the spectrum of bacterial phyla (Ludwig et al., 1993, 1998b; Ludwig and Schleifer, 1994; Neumaier et al., 1996). This marker should be more appropriate than elongation factors or RNA polymerases for testing the validity of rRNA-based trees for organismal phylogeny, as the ATPase has nothing in common functionally with transcription or translation except its own synthesis.

In general, the F_1F_0 ATPase β-subunit data support the rRNA-based tree (Fig. 8), but the information content and resolving power is reduced. Again, local differences in branching patterns have been shown, and the monophyletic structure of some phyla, defined by rRNA analysis, is not supported (Ludwig et al., 1993).

A correct rooting of the ATPase β-subunit-based bacterial domain tree with the paralogous catalytic subunit of the vacuolar type ATPase (Hilario and Gogarten, 1998; Ludwig et al., 1998b) is not possible as the overall sequence similarities between the two paralogs are not higher than 23%.

There are not sufficient data available for the F_1F_0 ATPase α-subunit (most likely the paralogous pendant of the β-subunit) to allow effective comparison with the rRNA data. However, the currently available α-subunit data set does not indicate great dif-

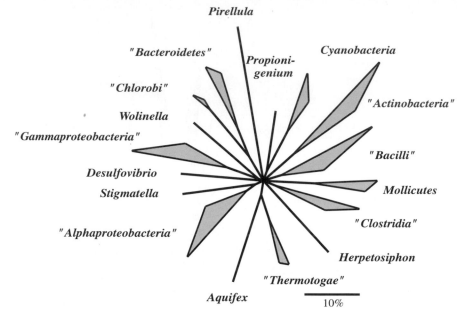

FIGURE 8. F_1F_0 ATPase β-subunit-based tree depicting the major bacterial phyla. The tree is shown as unrooted. Tree reconstruction was performed as described for Figure 7.

ferences in phylogenetic conclusions inferred from the two data sets. There are also insufficient data for the paralogous subunits A and B of the vacuolar type ATPase; however, a clear separation of the *Eucarya* from the *Bacteria* and *Archaea* is seen when the currently available data set is analyzed. The bacterial and archaeal lines appear intermixed at the lowest level of the corresponding subtree. At present, this intermixing cannot be proven or correctly interpreted (Neumaier, 1996). There is low significance for any branching pattern at this level of (potential) relatedness. Furthermore, only a few positions, which currently cannot be tested for plesiomorphy, are responsible for this intermixing. In addition, functional constancy can not be assumed for eucaryal and archaeal versus bacterial vacuolar type ATPases, and lateral gene transfer cannot be excluded (Hilario and Gogarten, 1993).

recA *protein* Most of the bacterial phyla are represented by one or a few sequences in the *recA* protein sequence data base (Wetmur et al., 1994; Eisen, 1995; Karlin et al., 1995). Comparative analysis of these data again supports the rRNA-based view of bacterial phylogeny. A homologous counterpart for the archaeal and eucaryal phyla has not yet been identified. A significant relative branching order of phyla cannot be defined. Although monophyly of the *"Proteobacteria"* or the Gram-positive bacteria with a low DNA G + C content is not observed, no major contradictions to the rRNA-based phylogeny have been reported. The higher phylogenetic groups (*"Proteobacteria"*, *Cyanobacteria*, *"Actinobacteria"*, *Chlamydiales*, *"Spirochaetes"*, *"Deinococcus–Thermus"*, *"Bacteroidetes"*, as well as *"Aquificae"*) are separated from each other as in the rRNA-derived phylogeny. However not surprisingly local differences in detailed branching patterns were found.

There is one major discrepancy: phylogenetic analysis of *Acidiphilium* using *recA* sequence data does not show it to cluster within the *"Alphaproteobacteria"* as is found with rRNA analyses. Two *recA* genes, which differ remarkably in sequence, have been found in *Myxococcus xanthus* and may indicate the occurrence of

gene duplications or lateral gene transfer. Therefore it is possible that such phenomena have occurred in the evolution of *recA* in *Acidiphilium*.

hsp60 heat shock proteins Sequences for hsp60 chaperonin have been determined for a wide spectrum of bacterial phyla (Viale et al., 1994; Gupta, 1996; 1998). A distant relationship has been postulated for hsp60, the eucaryotic TCP-1 complex, and the archaeal Tf-55 protein (Brown and Doolittle, 1997). However, given the low similarities, the homologous character of hsp60 and the TCP-1 complex or the Tf-55 protein cannot be demonstrated unambiguously.

Trees based upon the currently available hsp60 sequence data set support rRNA-based trees in that the different phyla are well separated from one another, and in cases where several sequences are available for a given phylum, subclusters resemble the rRNA-derived phylogeny. For example, the *"Gammaproteobacteria"* and *"Betaproteobacteria"* are more closely related to one another than to the *"Alphaproteobacteria"* sister group in both hsp60 and rRNA analyses. However, the use of hsp60 as a phylogenetic marker molecule is again complicated by the existence of duplicated genes as, for example, among *Rhizobium* species.

Other supporting and nonsupporting protein markers The hsp70 (70 kDa heat shock protein)-based tree globally supports rRNA-based clustering. The phyla appear to be separated and even the branching order of the classes of the *"Proteobacteria"* (*"Alphaproteobacteria"*, *"Gammaproteobacteria"*, *"Betaproteobacteria"*) is corroborated. The major concern associated with the hsp70-derived phylogeny is the intermixed rooting of bacterial and archaeal major lines of descent (Brown and Doolittle, 1997; Gupta, 1998). No significant branching order can be defined for the intermixed lines, and, as discussed above for the ATPase phylogeny, these findings may reflect missing resolution at the interdomain level.

At first glance, many other proteins (reviewed by Brown and Doolittle, 1997) seem to support the intradomain tree structures of rRNA-based phylogenies. However, meaningful comparative evaluation is difficult due to limitations in phylogenetic information content and/or databases that are insufficient in size and scope. Examples are provided by family B DNA polymerases which might represent useful markers for all three domains, aminoacyl-tRNA synthetases which differ in size and hence in potential information content, and ribosomal proteins which generally are short polypeptides and thus of very limited phylogenetic use (Brown and Doolittle, 1997). Among the enzymes involved in central metabolism, the usefulness of 3-phosphoglycerate kinase is also curtailed by a limited sequence database.

There are a number of potential protein markers for which deduced trees do not clearly support rRNA-based intradomain phylogenetic conclusions, including DNA gyrases and topoisomerases, some enzymes of the central metabolism, and of amino acid synthesis and degradation. However, as no comprehensive sequence databases are available, careful evaluation of the tree topologies is not possible.

RATIONALE FOR A 16S rRNA-DERIVED BACKBONE FOR *BERGEY'S MANUAL*

The introduction of comparative primary structure analysis of the SSU rRNA by Carl Woese and coworkers was undoubtedly a major milestone in the history of systematic biology. This approach opened the door to the elucidation of the evolutionary history of the procaryotes, and provided the first real opportunity to approach the ultimate goal in taxonomy i.e., systematics based upon the natural relationships between organisms. The rapid development of experimental procedures enabled the scientific community to characterize the majority of described species at the 16S rRNA level. During preparation of the new edition of *Bergey's Manual*, coordinated efforts to close the gaps and to investigate the missing species were initiated. There is a realistic prospect of completing the database with respect to all known validly described species in the near future.

Although the resolving power of the SSU rRNA approach has sometimes been overestimated, it has allowed a tremendous expansion in our knowledge of procaryotic relationships during recent years. This has been accompanied by the recognition of limitations in the existing procaryotic taxonomy, and efforts to redress these limitations. The taxonomic history of the pseudomonads is one impressive example of the "phylogenetic cleaning" of a genus that was phylogenetically heterogeneous in composition (Kersters et al., 1996).

It appears that the SSU rRNA is currently the most powerful phylogenetic marker, in terms of information content, depth of taxonomic resolution, and database size and scope. There is also good congruence between global tree topologies derived from different phylogenetic markers, indicating that SSU rRNA-based phylogenetic conclusions indeed reflect organismal evolution, at least at the global level. Local discrepancies in phylogenetic trees resulting from different information content, different rate or mode of change, or inadequate data analysis do not greatly compromise this general picture. The underlying cause of major tree discrepancies may in some cases be the analysis of paralogous genes, as indicated by multiple genes arising from duplication, loss, or lateral transfer of genes.

The logical consequence of these investigations and observations is to structure the present edition of *Bergey's Manual* according to our current (rRNA-based) concept of procaryotic phylogeny, using the global tree topology as a backbone, and to propose an emended framework of hierarchical taxa.

It should be considered that all phylogenetic conclusions and tree topologies presented here are models that represent the present, imperfect view of evolution. The information content of the SSU rRNA database is rather limited for representation of 3–4 billion years of evolution of cellular life. Furthermore, the methods of data analysis and the software and hardware for deciphering and visualizing this information are far from being optimal. For these reasons, the proposed backbone of the taxonomic scheme might be subject to change in the future. The introduction of new taxonomic tools and methods has always had a major impact on contemporaneous taxonomy. New sequence data and improved methods of data analysis may change our view of procaryotic phylogeny. Comparison of previous editions of *Bergey's Manual*, as well as updates of the Approved Lists of Bacterial Names (Skerman et al., 1980), indicates that the contemporary view of microbial taxonomy is determined mainly by the availability, applicability, and resolving power of the methods used to characterize organisms and elucidate their genetic and phylogenetic relationships.

THE SMALL SUBUNIT rRNA-BASED TREE

The global SSU rRNA-based intradomain phylogenetic relationships are discussed for the *Archaea* and *Bacteria* below. Given that the relationships of these organisms are described in detail in subsequent chapters, only higher phylogenetic levels are shown here. Reconstruction of general trees was performed using only sequences that were at least 90% complete (in relation to the *E. coli* 16S rRNA reference sequence). Lines of descent or phylogenetic groups containing a single or only a few sequences are (usually) not shown in these trees. Environmental sequences from organisms which have not yet been cultured were included in the calculations but are not depicted in the trees. The trees and discussions are based upon a comparative analysis of the current RDP (Maidak et al., 1999) and ARB trees. The RDP tree was reconstructed by applying a maximum likelihood method combined with resampling, whereas for the ARB tree a special maximum parsimony approach in combination with different optimization methods and upper bootstrap limit determination was used. The RDP tree contains the *Bacteria* and *Archaea*, while the ARB tree also includes the *Eucarya*. In both cases, the rooting and internal structuring of the domain trees was estimated using the full data set of the other domains. Although these trees were reconstructed using different methods, their global topologies are in good agreement.

A statistically significant relative branching order cannot be unambiguously determined for the majority of the phyla in the *Bacteria*, or for many of the intraphylum groups, as indicated by multifurcations within the trees. However, clustering tendencies are common to both trees. It should also be considered that most phyla were defined in the early days of comparative rRNA sequencing (Woese, 1987) when the data set was small and long "naked" branches facilitated clear-cut phylum delimitation. With the rapidly expanding database most of these "naked" branches expanded and in some cases it is no longer possible to demonstrate a monophyletic structure or to clearly delimit traditional phyla and other groups, as exemplified by the *"Proteobacteria"* and the low G + C Gram-positive bacteria (*"Bacilli"*, *"Clostridia"*, *Mollicutes*). The inter- and intra-genus relationships of each group are discussed in detail in subsequent chapters; described below is an overview of the phyla of the bacterial and archaeal domains and their major phylogenetic subclusters (Figs. 9, 10, 11, 12, 13, 14, 15, and 16).

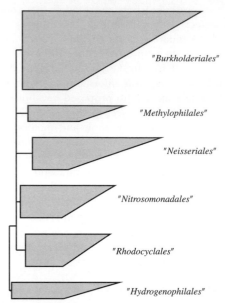

FIGURE 9. 16S rRNA-based tree showing the major phylogenetic groups of the *"Betaproteobacteria"*. Only groups represented by a reasonable number of almost complete sequences are shown. Tree topology is based on the ARB database of 16,000 sequences entries and was reconstructed, evaluated, and optimized using the ARB parsimony tool. A filter defining positions which share identical residues in at least 50% of all included sequences from *"Betaproteobacteria"* was used for reconstructing the tree. The topology was further evaluated by comparison with the current RDP tree, which was generated using a maximum likelihood approach in combination with resampling (Maidak et al., 1999). A relative branching order is shown if supported by both reference trees. Multifurcations indicate that a (statistically) significant relative branching order could not be determined or is not supported by both reference trees.

The Bacteria

THE *"Proteobacteria"* The traditional view of the *"Proteobacteria"* as a monophyletic phylum is not completely supported by careful analyses of the current 16S rRNA database. Although there is support for monophyly in the RDP tree, with the *"Deltaproteobacteria"* and *"Epsilonproteobacteria"* forming the deeper branches, a monophyletic structure that includes these two groups is not clearly supported by the ARB tree. Confidence analyses indicate that the significance of a relative branching order within the *"Proteobacteria"* is low in both trees. However, a closer relationship of the *"Gammaproteobacteria"* and *"Betaproteobacteria"*, as well as a common origin of these groups and the *"Alphaproteobacteria"*, is supported by the RDP as well as the ARB tree.

The *"Betaproteobacteria"* (Fig. 9) clearly represents a monophyletic group, comprising the described or proposed higher taxa *"Burkholderiales"*, *"Methylophilales"*, *"Nitrosomonadales"*, *"Neisseriales"*, and *"Rhodocyclales"*. A slightly deeper-branching group comprises the *"Hydrogenophilales"*.

The classical members of the *"Gammaproteobacteria"* (Fig. 10) represent a monophyletic group which includes the *"Betaproteobacteria"* as a major line of descent. In both reference trees the family *"Xanthomonadaceae"* appears to be the most likely sister group of the *"Betaproteobacteria"*. A common clustering of the families *Aeromonadaceae*, *"Alteromonadaceae"*, *Enterobacteriaceae*, *Pasteurellaceae*, and *Vibrionaceae* is supported in both trees. A relative branching order of this cluster and other major groups of the *"Gammaproteobacteria"* such as the families *Halomonadaceae*, *Legi-*

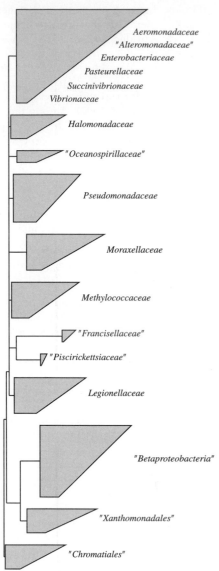

FIGURE 10. 16S rRNA-based tree depicting the major phylogenetic groups within the *"Gammaproteobacteria"*. Tree reconstruction and evaluation was performed as described for Figure 9 with the exception that a 50% filter calculated for the *"Gammaproteobacteria"* was used.

onellaceae, *Methylococcaceae*, *Moraxellaceae*, *"Oceanospirillaceae"*, *Pseudomonadaceae*, and the *"Francisellaceae"*-*"Piscirickettsiaceae"* group cannot be unambiguously determined. In both trees, these groups branch off higher than the *"Betaproteobacteria"*-*"Xanthomonadaceae"* branch, whereas the order *"Chromatiales"* forms a deeper branch. The phylogenetic position of the families *Moraxellaceae* and *Cardiobacteriaceae* relative to that of the *"Gammaproteobacteria"*-*"Xanthomonadaceae"* lineage depends on the treeing method used.

A closer relationship between the families *Rickettsiaceae* and *Ehrlichiaceae* within the *"Alphaproteobacteria"* (Fig. 11) can be seen in both reference trees. The results of tree evaluations indicate branching of this cluster followed by the families *"Sphingomonadaceae"* and the *"Rhodobacteraceae"*. The families *"Bradyrhizobiaceae"*, *Hyphomicrobiaceae*, *"Methylobacteriaceae"*, and *"Methylocystaceae"* represent another subcluster among the *"Alphaproteobacteria"*. A closer interrelated group is formed by the families *Bar-*

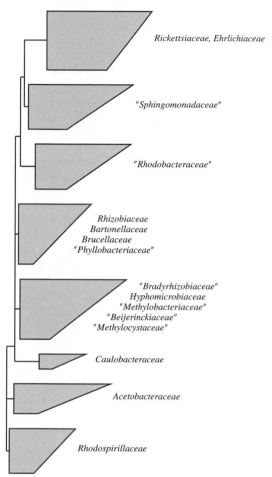

FIGURE 11. 16S rRNA-based tree showing the major phylogenetic groups within the *"Alphaproteobacteria"*. Tree reconstruction and evaluation was carried out as described for Figure 9 with the exception that a 50% filter calculated for the *"Alphaproteobacteria"* was used.

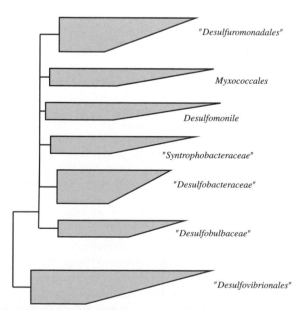

FIGURE 12. 16S rRNA-based tree illustrating the major phylogenetic groups within the *"Deltaproteobacteria"*. Tree reconstruction and evaluation was performed as described for Figure 9 with the exception that a 50% filter calculated for the *"Deltaproteobacteria"* was used.

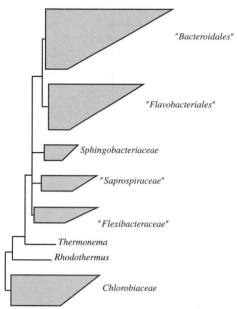

FIGURE 13. 16S rRNA-based tree depicting the major phylogenetic groups within the phylum *"Bacteroidetes"*. Tree reconstruction and evaluation was performed as described for Figure 9 with the exception that a 50% filter calculated for the *"Bacteroidetes"* phylum was used.

tonellaceae, Brucellaceae, Rhizobiaceae, and *"Phyllobacteriaceae"*. No reliable resolution of these major groups and the family *Caulobacteraceae* can be achieved, but it appears that a deeper branching of the families *Acetobacteraceae* and *Rhodospirillaceae* among the *"Alphaproteobacteria"* is indicated.

The order *"Desulfovibrionales"* currently represents the deepest branch of the *"Deltaproteobacteria"* (Fig. 12). Three other major subgroups comprise *Desulfomonile* and relatives, the *"Syntrophobacteraceae"*, as well as the *"Desulfobulbaceae"*. These subgroups are phylogenetically equivalent in depth to the lineages *"Desulfobacteraceae"*, *"Geobacteraceae"*, and *Myxococcales*.

The families *"Helicobacteraceae"* and *Campylobacteraceae* are the two major lines that form the *"Epsilonproteobacteria"*.

THE *"Spirochaetes"* The *"Spirochaetes"* phylum currently comprises three major subgroups: the sister groups of the families *Spirochaetaceae* and *"Serpulinaceae"*, as well as the deeper branching family *Leptospiraceae*.

"Deferribacteres" AND *"Acidobacteria"* PHYLA To date, the *"Deferribacteres"* phylum is represented by only two cultured species, while only three cultured species are found in the *"Acidobacteria"* phylum. However, a comprehensive data set of environmental sequences indicates a phylogenetic depth and diversity within the *"Acidobacteria"* comparable to that of the *"Proteobacteria"* (Ludwig et al., 1997).

THE *Cyanobacteria* The chloroplast organelles comprise a monophyletic subgroup within the *Cyanobacteria* phylum, which also contains a number of other major lines of descent. The current taxonomy of the cyanobacteria is far from being in accordance with the phylogenetic structure of the phylum.

"Verrucomicrobia", *"Chlamydiae"*, AND *"Planctomycetes"* The phylum *"Verrucomicrobia"* comprises a number of environmental sequences as well as a few cultured members of the genera *Verrucomicrobium* and *Prosthecobacter* (Hedlund et al., 1996). Both reference trees indicate a moderate degree of relationship be-

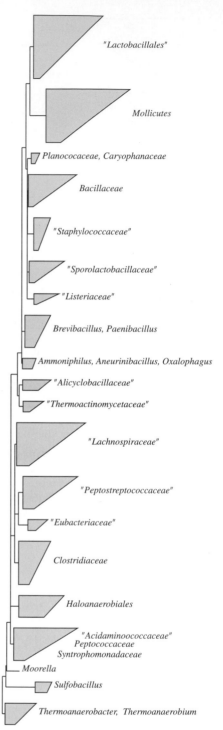

FIGURE 14. 16S rRNA-based tree showing the major phylogenetic groups of the *"Firmicutes"* (Gram-positive bacteria with a low DNA G + C content). Tree reconstruction and evaluation was carried out as described for Figure 9 with the exception that a 50% filter calculated for a core set of sequences (excluding the *Mycoplasmatales* and the deeper groups represented by *Moorella, Sulfobacillus, Thermoanaerobacter,* and *Thermoanaerobium*) was used.

tween the *"Verrucomicrobia"* and the *Chlamydiales* phylum. However, given the limited number of available sequences for the *"Verrucomicrobia"* and the long naked branch of the *Chlamydiales,* a sister group relationship between these two phyla should be regarded as tentative. A moderate relationship between these two

phyla and the *"Planctomycetes"* phylum is also indicated in both the ARB and RDP trees. However, the significance of this branching point is low, and their relationship may not be supported in the future by a growing database. The intraphylum structure of the *"Planctomycetes"* indicates two pairs of sister groups: *Pirellula/ Planctomyces* and *Isosphaera/ Gemmata.*

"Chlorobi" AND *"Bacteroidetes"* A monophyletic origin of the *"Chlorobi"* (containing the genera *Chlorobium, Pelodictyon, Prosthecochloris,* and some environmental sequences) and the *"Bacteroidetes"* (Gosink et al., 1998) phyla (Fig. 13) can be seen in both trees and is supported by alternative markers such as large subunit rRNA, and β-subunit of F_1F_0 ATPase. The thermophilic genera *Rhodothermus* and *Thermonema* represent the deepest branches of the phylum *"Bacteroidetes"*. A common root of the *"Bacteroidales"* and *"Flavobacteriales"* within the phylum is supported in both reference trees. This cluster seems to be phylogenetically equivalent to the other major groups i.e., the *Sphingobacteriaceae, "Saprospiraceae", "Flexibacteraceae", Flexithrix,* and *Hymenobacter.*

LOW G + C GRAM-POSITIVE BACTERIA Other than for the *"Proteobacteria"*, the most comprehensive 16S rRNA gene sequence database (with more than 1750 almost complete sequences) is available for the Gram-positive bacteria with a low DNA G + C content (*"Bacilli", "Clostridia", Mollicutes*). The common origin of the organisms classically assigned to this group is not significantly supported by all reference trees (see Fig. 14). The *Mollicutes,* comprising the families *Mycoplasmataceae, Acholeplasmataceae,* and their walled relatives, represent a monophyletic unit. The classical lactic acid bacteria are members of the families *"Aerococcaceae", "Carnobacteriaceae", "Enterococcaceae", Lactobacillaceae, "Leuconostocaceae",* and *Streptococcaceae,* and are unified in the order *"Lactobacillales"*. A clear resolution of the relationships between the families *Bacillaceae, Planococcaceae, "Staphylococcaceae", "Sporolactobacillaceae",* and *"Listeriaceae"* cannot be achieved. Two slightly deeper branching clusters comprise the genera groups of *Brevibacillus–Paenibacillus* and *Ammoniphilus–Aneurinibacillus–Oxalophagus.* The *"Alicyclobacillaceae"* and *Thermoactinomyces* groups represent a further deeper branch. Another major subbranch unifies the *"Eubacteriaceae", Clostridiaceae, "Lachnospiraceae",* and *"Peptostreptococcaceae"* . The *"Eubacteriaceae"* and *"Peptostreptococcaceae"* appear to be sister groups. The phylogenetic position of the order *Haloanaerobiales* is strongly influenced by the treeing method applied and should be regarded as tentative. The families *Haloanaerobiaceae* and *Halobacteroidaceae* constitute a well-defined phylogenetic unit in both reference trees. However, the assignment of this unit to the low G + C Gram-positive phylum is not clearly supported when different treeing methods are applied, suggesting that this group may represent its own phylum. A deeper rooting within the phylum is indicated for the *Peptococcaceae–Syntrophomonadaceae* cluster but the phylogenetic position of the genera *Moorella, Sulfobacillus, Thermoanaerobacter,* and *Thermoanaerobium* is uncertain. The latter two genera represent a phylogenetic unit, but this unit and each of the other genera probably represent additional phyla.

"Fusobacteria" PHYLUM The *"Fusobacteriaceae"* phylum so far comprises only three subclusters: *Fusobacterium, Propionigenium–Ilyobacter* and *Leptotrichia–Sebaldella.*

HIGH G + C GRAM POSITIVE BACTERIA (*"Actinobacteria"*) The phylum of the Gram-positive bacteria with a high G + C DNA content (the *"Actinobacteria"*) provides an example of a clearly defined and delimited major bacterial line of descent. As seen in Fig. 15, the families *Rubrobacteraceae* and *Coriobacteriaceae*

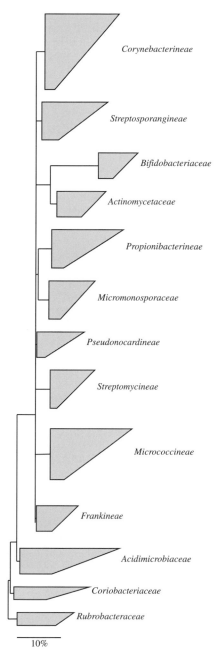

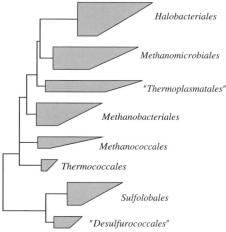

FIGURE 16. 16S rRNA-based tree showing the major phylogenetic groups within the *Archaea*. Tree reconstruction and evaluation was carried out as described for Figure 9 with the exception that tree optimization was performed independently for the *"Euryarchaeota"* and *"Crenarchaeota"* using a 50% filter in each case.

FIGURE 15. 16S rRNA-based tree depicting the major phylogenetic groups within the *"Actinobacteria"* (Gram-positive with a high DNA mol% G + C content). Tree reconstruction and evaluation was performed as described for Figure 9 with the exception that a 50% filter calculated for this phylum was used.

currently represent the deepest branches of the phylum, whereas the family *Acidimicrobiaceae* occupies an intermediate position between the former two and the remaining major subgroups of the phylum. There is some support for a common origin of the *Bifidobacteriaceae* and *Actinomycetaceae*, and for the clustering of the families *Propionibacteriaceae* and *Micromonosporaceae*. No significant or stable branching order for these and other subgroups such as *Corynebacteriaceae*, *Frankineae*, *Pseudonocardiaceae*, *Streptomycetaceae*, and *Streptosporangineae* could be achieved.

OTHER PHYLA   The *"Nitrospirae"* phylum contains a limited number of organisms, namely representatives of the genera *Ni-*

trospira, *Leptospirillum*, *Thermodesulfovibrio*, and *Magnetobacterium*. Similarly, only a limited number of organisms and environmental sequences represent the phylum of the green non-sulfur bacteria, which includes the families *"Chloroflexaceae"*, *"Herpetosiphonaceae"*, and *"Thermomicrobiaceae"*. Two major subgroups, the *Deinococcaceae* and the *"Thermaceae"*, have been identified within the *"Deinococcus–Thermus"* phylum. The orders *"Thermotogales"* and *"Aquificales"* constitute two of the deeper branching phyla within the bacterial domain.

The existence of additional phyla is suggested by the phylogenetic position of organisms such as *Dictyoglomus thermophilum* and *Desulfobacterium thermolithotrophum*, and of some environmental sequences. However, the phylum status of these lineages cannot be evaluated at this time, due to the paucity of available sequence data.

***The* Archaea**   Two major lines of descent (phyla) have been delineated within the *Archaea*: the *"Euryarchaeota"*, and the *"Crenarchaeota"*. Within the *"Euryarchaeota"*, the orders *Halobacteriales*, *Methanomicrobiales*, and *"Thermoplasmatales"* share a common root. A relationship between the first two orders is suggested in both reference trees (Fig. 16), and the order *Methanobacteriales* is indicated as the next deepest branch. A stable and significant tree topology resolving the relationship between these four orders and the orders *"Archaeoglobales"*, *Methanococcales*, *Thermococcales*, and *"Methanopyrales"* cannot be deduced from the current database.

The orders *Sulfolobales* and *"Desulfurococcales"* appear to be sister groups within the *"Crenarchaeota"*, while a monophyletic structure of the *Thermoproteales* is somewhat questionable. The genus *Thermofilum* tends to root outside the *Thermoproteales* group, however the significance of this branching is low and the database does not contain sufficient entries to allow careful evaluation of this outcome.

A third archaeal phylum, *"Korarchaeota"*, has been postulated on the basis of two partial environmental 16S rRNA sequences, but representatives of this lineage have not yet been isolated in pure culture. Consequently, the phylum status as well as phylogenetic position of the lineage can currently not be assessed.

Nucleic Acid Probes and Their Application in Environmental Microbiology

Rudolf Amann and Karl-Heinz Schleifer

I. INTRODUCTION

Microbiology has entered the molecular age. Almost every month another complete bacterial genome sequence is published (e.g., Fleischmann et al., 1995; Cole et al., 1998), and it is now routine to start the classification of a newly isolated microorganism with the determination and comparative analysis of at least one nucleic acid sequence. Clearly, the most commonly used molecule for this purpose is the ribonucleic acid of the small subunit of the ribosome, the 16S rRNA of *Bacteria* and *Archaea*. The high information content of nucleic acid sequences can, in principle, be accessed by two techniques. One is sequencing followed by comparative sequence analysis; the second is hybridization with nucleic acid probes.

As a first, rough definition, nucleic acid probes can be described as single-stranded pieces of nucleic acids that have the potential to bind specifically to their counterparts, complementary nucleic acid sequences. By this process, the so-called hybridization, probes facilitate the detection of their respective target molecules based on primary structure. In most cases, this is accomplished by conferring a detectable moiety, the label, to the target molecules (Fig. 1).

Nucleic acid hybridization techniques, such as DNA–DNA reassociation or DNA–rRNA hybridization, have been used by bacterial taxonomists at times when nucleic acid sequencing was still difficult and available only to specialists. Taxonomists adopted the methodology soon after the first hybridization experiments were performed in the early 1960s (Marmur and Lane, 1960; Hall and Spiegelman, 1961). *Bacteria* and *Archaea* generally lack the morphological diversity necessary for microscopic identification to be reliable. The traditional identification methods, based on the phenotypic characterization of pure culture isolates, were slow and often yielded unclear results because of the influ-

ence of exogenous and endogenous parameters on the expressed phenotype. Nucleic acid-based, genotypic methods promised a faster and more reliable identification based on stable genetic markers that were independent of the cultivation conditions. Furthermore, genotypic methods should allow bacterial taxonomists to transform an artificial classification system, suitable only for identification, into a more natural one reflecting the phylogeny of the bacteria. Indeed, studies using DNA–rRNA hybridization (Palleroni et al., 1973; De Ley and De Smedt, 1975) in the 1970s and 1980s yielded significant insights into the genotypic relationships of bacteria (Schleifer and Stackebrandt, 1983; De Ley, 1992). A DNA–DNA similarity of ≥70% continues to be used as an important determinant for placing bacterial strains into species. Meanwhile, nucleic acid probes have become a standard method of identifing fastidious, slow growing, or even hitherto uncultured bacteria (Amann et al., 1991).

The increasing availability of nucleic acid sequences in databases, the ease of synthesizing oligonucleotide probes, and numerous other methodological advances in molecular biology, such as the polymerase chain reaction (PCR; Saiki et al., 1988), have made nucleic acid probes a routinely used tool in microbiological laboratories. The use of probes is now so common that it has become impossible to review all applications. In this introductory chapter we will, therefore, focus on basic principles behind nucleic acid probing, describe the steps necessary for directed design and reliable application of nucleic acid probes, and discuss selected examples of the use of nucleic acid probes in environmental microbiology. Additional information may be obtained from two earlier reviews on the subject (Stahl and Amann, 1991; Schleifer et al., 1993) and a recently published book on "Molecular Approaches to Environmental Microbiology" (Pickup and Saunders, 1996). Examples for the application of

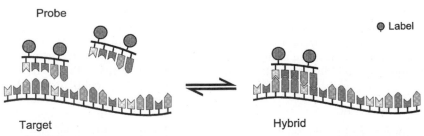

FIGURE 1. Specific hybridization of a nucleic acid probe to a target molecule.

nucleic acid probes in identification of bacteria are given by Amann et al. (1996a) and Schleifer et al. (1995).

II. BASIC PRINCIPLES OF NUCLEIC ACID PROBING

This part is intended to provide the groundwork for the rest of this chapter by explaining the principles of nucleic acid hybridization in a coherent way. It may also be used as a glossary for the specialized terminology.

In every cell there are two types of nucleic acids: deoxyribonucleic acid or DNA and ribonucleic acid or RNA. Whereas the former is the storage medium of genetic information, RNA molecules occur as ribosomal RNA (rRNA), messenger RNA (mRNA), or transfer RNA (tRNA) and are involved in the translation of genotypic information into the expressed characters. DNA usually forms a duplex of two antiparallel strands of polynucleotides that are fully complementary to each other. This means the base adenine (A) on one strand is opposing a thymine (T) on the other strand and cytosine (C) base pairs with guanine (G). The non-covalent bonding between the base pairs is mediated by hydrogen bonds which can be broken by physical or chemical means, resulting in the denaturation of the DNA molecule into single strands. The process is fully reversible and upon cooling or neutralization the DNA will reassociate to form the original duplex.

In RNA the base thymine is replaced by uracil (U). Internal base complementarities cause the single stranded RNA to fold into secondary structures that may, in addition to the canonical base pairs G–C and A–U, also be stabilized by non-canonical pairs such as G–U or G–A. It was first shown by Hall and Spiegelman (1961) that RNA may bind to or hybridize with denatured DNA, resulting in DNA–RNA duplex structures. The term hybridization is also used to describe the binding of a labeled single-stranded nucleic acid, the probe, to an unlabeled single-stranded nucleic acid, the target. When the degree of probe binding is plotted against temperature or the concentration of the naturing agent, the resulting profile is sigmoid (Fig. 2). Hybridization may also occur between two strands that are not fully complementary. In this case, canonical base pairs or matches stabilize a certain number of mismatches. The resulting hybrid will be less stable than a fully complementary hybrid and show a lower temperature of dissociation (T_d) of the probe as compared to the fully complementary target nucleic acid. The T_d is defined as that temperature at which 50% of the maximally bound probe has dissociated from the immobilized target. It is similar to the melting temperature, or T_m, of double stranded DNA. Parameters such as temperature and composition of the hybridization buffer have a strong influence on the kinetics and specificity of hybridization. The combined effect of these parameters is often referred to as the stringency of hybridization. As outlined in Fig. 2, determination of the optimal stringency of hybridization for a given probe is of utmost importance for the specificity of a hybridization assay and its capacity to discriminate between target and nontarget nucleic acids.

The optimum hybridization temperature is usually close to but below the T_d. If the stringency of hybridization is too low, the probe specificity may be compromised. If the stringency is too high, the sensitivity of the hybridization assay will be decreased.

III. DEVELOPMENT OF NUCLEIC ACID PROBES

The development of a new nucleic acid probe for the identification of a given strain, species, or a defined group of micro-

organisms can be accomplished by two different approaches. In addition to an approach that is based on comprehensive nucleic acid sequence collections, it is possible to generate nucleic acid probes for bacterial identification without prior knowledge of the target nucleic acid. In the following, the two approaches are discussed separately in the sections Empirical Probe Selection and Rational Probe Design.

A. Empirical probe selection   In a very simple hybridization format it is possible to use the total chromosomal DNA of a given strain after radioactive or nonradioactive labeling as a probe for the screening of total DNA. Such genomic DNA probes have been referred to as whole-cell probes (Stahl and Amann, 1991; Schleifer et al., 1993). Genomic DNA of the strain of interest is, e.g., labeled with photobiotin (Forster et al., 1985) and hybridized to DNA of reference strains that have been immobilized in microtiter plates (Ezaki et al., 1989). This assay is similar to, but much faster than, the traditional DNA–DNA hybridization, in which DNA must be released and purified from a large number of reference strains before it is subjected to pairwise hybridizations. Whole cell probes are, however, ill-defined and by their nature will always contain a fraction of sequences that are highly conserved, such as the rRNA genes. A considerable amount of nonspecific hybridization with less related species is therefore common (Grimont et al., 1985; Hyypiä et al., 1985) and often can not be prevented, even when highly stringent hybridization conditions are used (Ezaki et al., 1989).

Another straightforward possibility for using a natural nucleic acid fraction is the application of isolated RNA as a hybridization probe. This fraction consists mainly of rRNA, which when compared to the other RNAs (e.g., mRNA, tRNA) is more abundant and less rapidly degraded. It is important to clarify the essential difference between a chromosomal DNA–DNA hybridization and a DNA–(16S or 23S) rRNA hybridization. In the latter case, the probe is well known and characterized and therefore, *sensu stricto*, is not an example of empirical probe selection. With the saturation hybridization method (De Ley and De Smedt, 1975) the differences in the melting temperature of homologous and heterologous pairs of DNA and rRNA are determined as ΔT_m (e) values. These allow for the determination of intra- and intergeneric relationships of bacteria (De Ley, 1992). The limits of this widely accepted taxonomic method originate from the specific nature of the 16S and 23S rRNAs which are usually too conserved to allow for the differentiation of strains within one species or even closely related species. Furthermore, DNA–rRNA hybridization does not allow reconstruction of relationships above the level of families or orders (De Ley, 1992).

Random DNA fragments have the potential to be strain-specific. They can be selected from recombinant genomic DNA libraries by screening randomly chosen clones for specificity. Those clones that do not hybridize with closely related non-target strains are further evaluated. While this approach has been used frequently (for a review see e.g., Schleifer et al., 1993) it is rather time-consuming. Several strategies to enrich for DNA fragments with unique sequences have been developed. In the format of subtractive hybridization (Schmidhuber et al., 1988), DNA restriction fragments of the strain of interest are hybridized with biotinylated DNA fragments of a closely related strain. DNA that is similar in both strains is removed by binding to immobilized avidin, leaving behind a fraction that is enriched for DNA fragments unique to the strain of interest. Subtracter DNA may be obtained from one or several related organisms and is always

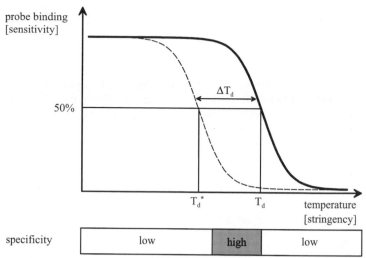

FIGURE 2. Theoretical dissociation profiles of a nucleic acid probe from a perfectly matched (*bold line*) and an imperfectly matched immobilized target nucleic acid (*broken line*). Probe binding (y-axis), which is directly proportional to the sensitivity, is shown over temperature (x-axis) which represents only one parameter that defines the hybridization stringency. Note that T_d, the temperature of dissociation of the perfect hybrid is higher than the T_d^* of the imperfect hybrid. It is the primary goal of probe design to maximize ΔT_d, the difference between the temperature of dissociation of a probe from target and non-target nucleic acid. The bar below the dissociation profiles indicates hybridization temperatures/stringencies with high and low specificity of discrimination of target and non-target nucleic acid.

used in excess. The limitations of this technique originate from its reliance on a single removal system which might not be sufficiently effective, and from problems with cloning the small quantities of DNA fragments resulting from this approach. Recently, an improved method that combines subtractive hybridization with PCR amplification and several removal systems and thereby largely overcomes these problems was described (Bjourson et al., 1992). A further simplification of the method was introduced by Wassill et al. (1998) and led to the successful identification of individual strains of lactococci.

After a random DNA probe has been selected for further use, the target sequence should be localized on either the chromosome or a plasmid, and in the latter case, the copy number of the plasmid should be established. Since plasmids are often mobile and might be lost, plasmid-targeted probes have been selected for identification of bacteria in only a few cases (e.g., Totten et al., 1983). Like probes targeted to species-specific repetitive sequences (e.g., Clark-Curtiss and Docherty, 1989), plasmid-targeted probes have the advantage of enhanced sensitivity, attributable to the increased number of target sites per cell. However, a clear disadvantage of randomly selected DNA fragments is that the biological function may remain unknown even after a sequence has been determined. Bacterial genome data clearly show that comparative sequence analysis still fails to assign defined functions to a substantial fraction of the sequences accumulated thus far (e.g., Fleischmann et al., 1995; Cole et al., 1998). Nevertheless, a complete sequence contains plenty of target sites for strain-specific probes. These probes are of importance in the monitoring of biotechnological strains such as starter cultures in food microbiology or production strains in amino acid fermentation.

B. Directed probe design The exponentially increasing number of nucleic acid sequences in various databases has prompted a move to directed probe design, which usually starts with the se-

lection of a target site. Good targets might, for example, be genes coding for well described virulence factors that allow for the differentiation of virulent and avirulent strains (Moseley et al., 1982). Other DNA probes have been targeted to genes specifying surface epitopes (Korolik et al., 1988) or antibiotic resistance (Groot Obbink et al., 1985). Because of rapid advances in molecular techniques, it is now easier and faster to screen for the presence of a gene of interest than to show its function. One has to realize, however, that detection of a gene by hybridization does not necessarily prove that it is present in a functional form, without deleterious mutations, or that it is expressed. The latter problem can be addressed by switching to mRNA as a target.

In the following, an example of a directed probe design that goes one step beyond the use of cloned DNA probes to target known molecules will be discussed in detail. The directed design of short oligonucleotide probes exploits defined signatures, e.g., single point mutations, that are initially detected in sequence databases by comparative analysis. This type of probe design is possible for any gene for which a reasonably large sequence database exists. By far the most commonly used target molecule, in this respect, is the 16S rRNA. The general steps in the design and optimization of an oligonucleotide probe are described in the following sections:

i. generation of a nucleic acid sequence database (see **Nucleic acid sequence databases**)

ii. design of probe (see **Computer-assisted probe design**)

iii. synthesis and labeling of the probe (see **Synthesis and labeling of oligonucleotide probes**)

iv. preparation of the target nucleic acid (see **Some common formats of oligonucleotide hybridization**)

v. optimization of the hybridization conditions (see **Optimizing the hybridization conditions**)

vi. evaluation of the probe specificity and sensitivity (see **Evaluation of probe specificity and sensitivity**)

Before starting the design of new probes, it is recommended that one checks whether suitable probes have already been developed and published. Rapid growth in the number of such probes precludes the provision of a compilation of available oligonucleotide probes, even if we restrict ourselves to rRNA-targeted probes as was done previously (Amann et al., 1995). Several probes databases are available, but are updated infrequently. Readers interested in rRNA-targeted probes might want to start with the Oligonucleotide Probe Database (OPD) which was accessible via the World Wide Web at the time of publication (http://www.cme.msu.edu/OPD; Alm et al., 1996). In addition to the probe sequences and references, this database also provides information on optimal hybridization conditions and T_d-values. Another special feature is the integration of OPD and the Ribosomal Database Project (RDP; Maidak et al., 1999), which allows the end user to reevaluate probe specificity against the constantly increasing rRNA sequence databases.

1. NUCLEIC ACID SEQUENCE DATABASES   Several large nucleic acid databases exist that are readily accessible via the Internet. For the design of 16S rRNA-targeted oligonucleotide probes specialized databases, offered by the Ribosomal Database Project (RDP) (Maidak et al., 1999) or the University of Antwerp (Van de Peer et al., 1999), are an ideal starting point. Both databases have collected more than 16,000 sequences of small subunit rRNA molecules. This includes the 16S rRNA of approximately 2750 of the validly described species of *Bacteria* and *Archaea* and numerous 18S rRNA sequences of *Eucarya*. It is likely that coverage of the 16S rRNA sequences of the cultured procaryotes will be almost complete in the near future. The value of these databases for the identification of microorganisms can not be overestimated.

For a scientist interested in the design of a new probe, the initial question is the availability of the target sequence. Is there a full or partial 16S rRNA sequence of the microbial strain of interest in the public databases? Have additional strains of the same and closely related species also been sequenced? If a complete detection system, consisting of multiple probes, is to be developed for a genus or an even wider taxonomic entity, how well do the available sequences cover this group? Are corresponding sequences for those organisms that must be discriminated against available? A critical examination of the database will frequently reveal a need to perform additional sequencing. Today, this is largely facilitated by direct sequencing of PCR products. Conserved primers for the 5′ and 3′ end of the 16S rRNA gene exist (Giovannoni, 1991), which enable amplification of

almost full length 16S rRNA genes from most, but not all (Marchesi et al., 1998), procaryotes. When starting from a pure culture, the resulting rDNA PCR product can be directly sequenced. Since a high-quality rRNA database is a prerequisite for reliable probe design, double-stranded sequencing should be performed on almost full length 16S rDNA sequences.*

Early in the design of a probe it is important to consider the intended application. If it is merely to screen isolates obtained from a specific set of samples, isolated on standardized media, the specificity requirements are more relaxed since one only needs to discriminate among those bacteria that are culturable under the selected conditions. However, if the probe is designed for *in situ* identification of a given microbial population in a complex environmental sample, it must be kept in mind that we have currently cultivated and described only a minority of the extant bacteria (Amann et al., 1995). It might, in such cases, be highly advisable to initially generate a 16S rRNA gene library of the community of interest to get at least a first impression of the natural diversity at the site of interest. Several environmental samples have already been investigated in this manner (e.g., Giovannoni et al., 1990, 1996; DeLong, 1992; Fuhrman et al., 1992; Barns et al., 1996; Snaidr et al., 1997; Hugenholtz et al., 1998b). The results not only indicate a huge microbial diversity but also add important 16S rRNA sequence information to the databases from organisms that have hitherto not been cultured. In the near future such "environmental" rRNA sequences (Barns et al., 1996) will outnumber sequences of well described, pure cultures and contribute significantly to our ability to perform directed probe design based on a reliable database.

2. COMPUTER-ASSISTED PROBE DESIGN   The principle behind directed design of oligonucleotide probes is the identification of sites at which all target sequences are identical and maximally different from all nontarget sequences. The process can be best described as a systematic search of a number of aligned sequences. In an alignment, sequences are arranged in such a way that homologous positions are written in columns. In a difference alignment, only those nucleotides that deviate from the uppermost sequence, which is usually the sequence of the target organism, are given as letters. A window of the width of the intended oligonucleotide probe is then shifted from left to right with the aim of identifying a region at which all nontarget sequences contain one or several mismatches (Fig. 3). If several such sites are identified, the number, quality, and location of mismatches provide the basis for a ranking of the potential target sites. The primary goal is to maximize the difference in the temperatures of dissociation of the probe from the target and the nontarget sequences (ΔT_d). In our example (Fig. 3) the third option would be the best. It contains not only more, but also stronger, mismatches. It has been shown by Ikuta et al. (1987) that for 19 base pair oligonucleotide–DNA duplexes containing different single mismatches, e.g., the destabilizing effect of A-A, T-T, C-A is more pronounced than that of the only slightly destabilizing G-T, G-A base pairs.

An important fine tuning of the ΔT_d may be achieved by shifting the probe target position in a way that the strong mismatches are located in the center. It has been shown previously that mismatches at the end of an oligonucleotide are generally less destabilizing than internal mismatches (Szostak et al., 1979). It is very difficult to differentiate a single terminal mismatch, whereas a strongly destabilizing A-A or T-U mismatch at position 11 of a 17mer results in a significant reduction of binding to

---

*The direct sequencing of a 16S rDNA PCR product of a pure culture assumes that it contains only one copy or, in the more frequent case of genomes with several rRNA operons (e.g., *E. coli* has seven, *B. subtilis* ten rRNA operons), identical copies of the gene coding for the 16S rRNA. This, however, is not always the case. Microheterogeneities between the different rRNA operons of bacteria exist (Nübel et al., 1996; Rainey et al., 1996). For the archaeon *Haloarcula marismortui*, which contains two rRNA operons, an exceptionally high sequence heterogeneity of 5% has been shown (Mylvaganam and Dennis, 1992). This clearly demonstrates that the "one organism-one rRNA sequence" hypothesis that applies for many organisms might be an oversimplification in some instances. This has various consequences, one being that a group of closely related sequences recovered from an environment may not represent a group of separate, phylogenetically highly related strains, but rather the sequence heterogeneity of the 16S rDNA contained within one strain. Such small sequence differences can be distinguished, e.g., by denaturing gradient gel electrophoresis (DGGE). Thus, DGGE of PCR-amplified 16S rRNA gene fragments from a single strain can result in multiple bands. Oligonucleotide probes targeted to the rRNA microheterogeneities allow one to analyze the expression of the different operons (Nübel et al., 1996).

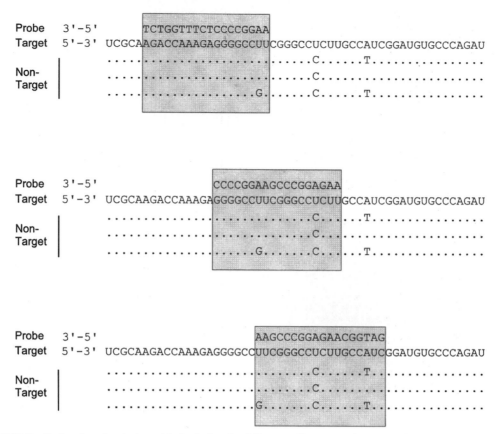

FIGURE 3. Rational probe design with the help of a difference alignment of target and non-target sequences. Note the differences in the number and the location of mismatches between the probe and the non-target sequences.

31% and 22%, respectively, of the binding to target rRNA without a mismatch (Manz et al., 1992). In the same study it was shown that the single mismatch discrimination could be significantly enhanced by competitor oligonucleotides. Competitors are unlabeled derivatives of the probe that are fully complementary to the known nontarget sequence. They are mixed with the labeled probe and efficiently prevent its hybridization to the nontarget sequence without significantly decreasing the homologous hybridization, thereby increasing probe specificity. The advantage of large databases is that such competitors can be designed in a directed way, preventing known potential unspecificities in advance. The variation in destabilizing effect of differently located mismatches during fluorescence *in situ* hybridization (FISH) with 16S rRNA-targeted oligonucleotide probes has also been evaluated (Neef et al., 1996).

Given the high number of 16S rRNA sequences now available, probe design must be performed with the aid of computer programs such as the PROBE_DESIGN tool of ARB* or the DESIGN_PROBE tool of the RDP (Maidak et al., 1999). Since rRNA sequences are patchworks of evolutionarily conserved regions, signature sites can be identified for any taxon between the level of the domains *Archaea, Bacteria, Eucarya*, and single species. In this respect, PROBE_DESIGN has the advantage that designation of target groups is done within a phylogenetic tree, assuring that

monophyletic assemblages are targeted. The relatively high degree of conservation of the rRNA molecules usually does not allow for the design of subspecies- and strain-specific rRNA-targeted probes. Modern probe design tools generate an ordered list of potential probe target sites that take into account the above mentioned key parameters, number, quality and location of mismatches. The ranking of target sites should be according to an estimated ΔT_d between the target and nontarget organisms. For group-specific probes, i.e., those targeted to families, orders or classes, it may be necessary to allow for incomplete coverage of the target group and few non-target hits.

It is in the very nature of the evolutionary process that mutations in a sequence site that is characteristic for a particular phylogenetic group may occur. A good signature site might be present in only 95% of all members of the group and, because of high microbial diversity, organisms might exist that are not members of the group but have the identical probe target site (Fig. 4A). A multiple probe approach was developed to address precisely such problems. By an intelligent combination of several probes, identifications can be made with a high degree of confidence. If possible, two or even three probes to separate signature sites of one target sequence are constructed (Amann, 1995a). If they bind to the same cells or colonies or to the same fragment or fraction of DNA, the possibility of false positives is virtually eliminated (Fig. 4B). Multiple probe systems can also be built from nested probe sets (Stahl, 1986) in which the first probe targets, e.g., a signature at the genus level, the second one at the species level, and the third is specific for selected strains

Editorial Note: Software available from O. Strunk and W. Ludwig: ARB: a software environment for sequence data. www.mikro.biologie.tu-muenchen.de (Department of Microbiology, Technische Universität, München, Munich, Germany.)

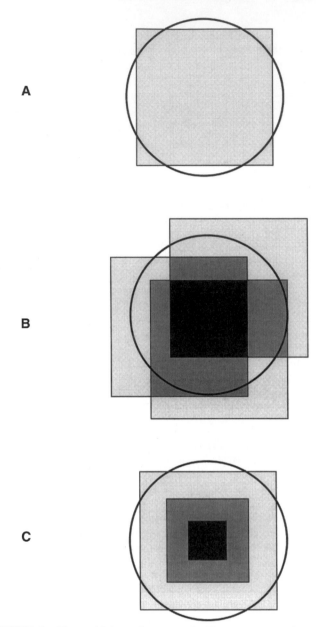

FIGURE 4. The multiple probe approach. The target group is represented by a circle. The specificities of the probes are described as squares. *A*, a probe detects most of the strains in the target group, but also some outside organisms. *B*, simultaneous application of three probes targeted to the same group. Note the different levels of gray indicating those subgroups that are detected by one, two, or three probes. *C*, nested set of probes focusing on increasingly narrow subgroups.

of this species (Fig 4C). If in agreement, the results of such a top- to-bottom approach (Amann et al., 1995) support each other, giving higher confidence to the final identification. For nested sets of probes used to quantify abundance of genera and species in mixed samples, it should also be possible to perform bookkeeping (Devereux et al., 1992) in the sense that the sum of all species-specific probes should be identical to the value obtained with the genus-specific probe. Here, the application of multiple probes allows for the identification of missing species within known genera. This directed screening for hitherto unknown bacteria that are genotypically different from, but related

to, known species, might be of considerable interest in biotechnology.

3. SYNTHESIS AND LABELING OF OLIGONUCLEOTIDE PROBES Based on the results of the probe design, oligonucleotides are chemically synthesized on a solid support. Today solid phase synthesis is a fast and reliable technique that is frequently performed by service units or private companies. The cost of oligonucleotide synthesis has become affordable, with a product of 0.2 μmol, sufficient for thousands of hybridization assays usually costing less than $50.

For standard assays the nucleic acid probe must be labeled prior to hybridization. There are two principal types of detection systems: (i) direct systems in which a label that can be directly detected is covalently attached to the probe and (ii) indirect systems in which the initial modification of the probe is detected via the secondary, non-covalent binding of a labeled reporter protein. This second step can, for example, be the specific detection of the vitamin biotin by labeled (strept)avidine (Langer et al., 1981) or an antigen-antibody reaction such as the detection of the hapten digoxigenin (Kessler, 1991). A rather complete compilation of labels and detection systems can be found elsewhere (Kessler, 1994).

The labeling of oligonucleotides during solid phase synthesis by the incorporation of modified nucleotides or the direct attachment of labels to the 5′ end using phosphoamidite chemistry is becoming more commonplace. Alternatively, labels or reporter molecules can be attached after the synthesis by enzymatic or chemical means. T4 polynucleotide kinase catalyzes, e.g., the ^{32}P- or ^{33}P-labeling of oligonucleotides at the 5′-end with γ-labeled nucleotide triphosphate (Maxam and Gilbert, 1980). Terminal transferase may be used to elongate the 3′ end with labeled nucleotide triphosphates (Ratcliff, 1981). Chemical labeling of oligonucleotides after synthesis can be accomplished via primary aliphatic amino groups incorporated into the oligonucleotides during synthesis. Detailed protocols are available for the attachment of activated fluorescent dye molecules such as fluorescein or rhodamine via 5′ aminolinkers (Amann, 1995b) or the covalent labeling of oligonucleotide probes with enzymes such as alkaline phosphatase (Jablonski et al., 1986) or horseradish peroxidase (Urdea et al., 1988).

The choice of detection system and the quality of labeling are of prime importance for the sensitivity of a hybridization assay. It is highly recommended to check any new batch of labeled oligonucleotides for expected length, homogeneity, and completeness of labeling by polyacrylamide gel electrophoresis (Sambrook et al., 1989). If, because of incomplete labeling and purification, a probe batch still contains an equimolar amount of unlabeled oligonucleotide of identical sequence, the resulting competition for target sites would reduce the hybridization signal by one half of the maximum. Another factor that strongly influences the sensitivity of a hybridization assay is, of course, the amount, purity, and accessibility of the target nucleic acid. Since the preparation of target nucleic acids is highly dependent on the hybridization format used, this aspect will be discussed subsequently (Section 4) in the context of some common formats of oligonucleotide hybridization.

4. OPTIMIZING THE HYBRIDIZATION CONDITIONS Unfavorable hybridization conditions may lead to the failure of even a well designed and highly purified probe. These conditions encompass the hybridization buffer, the temperature, and time of hybridization. The two main components of any hybridization

buffer are monovalent cations, added in the form of salts, and a buffer system. Hybridizations require a pH close to neutral. The monovalent cations are important for the speed of hybrid formation and the stability of the resulting duplexes. The time required for hybrid formation is also directly influenced by the complexity of the probe, of which probe length is a good indicator. The higher the probe complexity, the longer the time necessary for hybridization. As a rule of thumb, oligonucleotide hybridizations take 1 h, compared to 5 h or more for polynucleotide probes.

However, the most important characteristic that needs to be determined during the optimization of hybridization conditions is the melting point (T_m) of the probe-target hybrid, or in the case of an oligonucleotide probe, its temperature of dissociation (T_d). For long hybrids the melting point can be quite accurately estimated based on the mol% G + C content of the DNA (Stahl and Amann, 1991). There are also formulae for the estimation of T_d values, the simplest being that of Suggs et al. (1981), which applies for hybridization in 6X SSC (0.9 M sodium chloride; 0.09 M trisodium citrate; pH 7.0):

$$T_d = 4N_{G + C} + 2N_{A+T} \text{ (Suggs et al., 1981)}$$

Here $N_{G + C}$ and N_{A+T} are the numbers of G and C and of A and T which are assumed to add 4 and 2°C, respectively, to the thermal stability. Other more elaborate formulae, such as that of Lathe (1985), are available. They usually include the three parameters previously identified as key determinants of the T_d, the concentration of monovalent cations (M) in the hybridization buffer, and the length (n) and base composition (% G+C) of the oligonucleotide probe:

$$T_d = 81.5 + 16.6 \log M + 0.41 (\text{mol}\% \text{ G+C}) - 820/n \text{ (Lathe, 1985)}$$

All of these relationships have been empirically derived from experimental data. It must be noted that for oligonucleotides, the influence of the exact base sequence on the thermal stability is profound. The formulae should, therefore, only be used during probe design as an attempt to obtain a probe with a T_d within a certain range. As soon as the oligonucleotide has been synthesized, the T_d must be experimentally determined before the probe is used for the identification of microorganisms. For 16S rRNA-targeted oligonucleotide probes, several procedures have been described. In the original protocols, replicates of extracted, filter-immobilized total nucleic acid (Stahl et al., 1988; Raskin et al., 1994b) were hybridized at a relatively low temperature of 40°C with ^{32}P-labeled, rRNA-targeted oligonucleotide probes. The replicates were subsequently washed at successively higher temperatures (e.g., 40, 45, 50... 70°C) in 0.1% SDS–1X SSC for 30 min, and the amount of remaining radioactivity quantified. Alternatively, in the more economical elution technique, the same piece of membrane is transferred after hybridization through a series of washing steps at increasing temperature. Here, the amount of ^{32}P released in each washing step is quantified in scintillation vials and the total amount of released activity is plotted against temperature (Raskin et al., 1994b). In order to determine the optimum wash temperature, dissociation profiles for target and nontarget organisms need to be completed. Only then can conditions that fully discriminate nontarget nucleic acids and simultaneously yield good binding of the probe to the target nucleic acid be defined. Probes are specific only under certain conditions.

Changing the temperature of washing is, of course, not the only way to control the stringency of a hybridization assay. The temperature of hybridization is another obvious possibility, and, even at a constant temperature, the stringency of hybridization can be changed either by the addition of denaturing agents such as formamide or dimethylsulfoxide, or by varying the concentration of the duplex stabilizing monovalent cations. During hybridization of fixed whole microbial cells, high temperatures could have detrimental effects on the morphology. Therefore, formamide has been used to change the stringency of hybridization without altering the hybridization temperature of 46°C. This is done with the assumption that an addition of formamide of 1% is equivalent to an increase in hybridization temperature of 0.5°C (Wahl et al., 1987). The hybridization is followed by a slightly more stringent washing step at 48°C. In order to prevent production of excess amounts of potentially harmful waste, the stringency of the wash buffer is adjusted by lowering the concentration of monovalent cations rather than by the addition of formamide. This adjustment cannot be performed in the hybridization buffer since this would decrease the speed of hybridization. Based on the 0.5°C assumption of Wahl et al. (1987) and the salt term of the formula of Lathe (1985), Table 1 gives formamide concentrations and the salt concentrations that should yield comparable stringency.

By quantifying the fluorescence of the probe of interest after hybridization to selected target and nontarget cells, T_d values and optimum hybridization stringencies for whole cell hybridizations can be determined in a similar way as the optimum wash temperatures for immobilized extracted rRNA (Wagner et al., 1995; Neef et al., 1996). It has also been shown that the T_d values of a probe hybridized against extracted rRNA and against whole fixed cells are very similar (Amann et al., 1990b). One has to keep in mind, however, that the buffers used for T_d determinations are frequently very different. A T_d of 59°C in a buffer containing 2X SSC–0.1% SDS with a concentration of monovalent cations of roughly 390 mM is equivalent to a T_d in 1X SSC–0.1% SDS (195 mM) of 54°C. In the 900 mM NaCl buffer frequently used for *in situ* hybridization (Neef et al., 1996) the same T_d would be 65°C. If the temperature is kept at 46°C and the stringency is increased by adding formamide, half maximal binding would be at a concentration of 38% formamide (65 − 46 = 19°C equivalent to 38% formamide). Long term experience has shown that these correlations are quite robust. However, it should be stressed that it is best to determine T_d values in the

TABLE 1. Hybridization and washing buffers with corresponding stringencies for use in FISH

% Formamide [v/v] in a hybridization buffer containing 900 mM NaCl, 0.01% SDS, 10 mM Tris/HCl; pH 7.4	Concentration [mM] of monovalent cations in a wash buffer containing X mM NaCl, 0.01% SDS, 5 mM Na₂EDTA, 10 mM Tris/HCl; pH 7.4
0	900
5	636
10	450
15	318
20	225
25	159
30	112
35	80
40	56
45	40
50	28
55	20
60	14
65	10

actual format in which the probe is going to be used. If T_d values from separate determinations do not match, one should consider that the thermal stability of sequence identical duplexes increases from DNA–DNA to DNA–RNA and RNA–RNA (Saenger, 1984). As a rule, the T_d of a deoxyoligonucleotide hybridized against rRNA is about 2°C higher than against DNA.

Furthermore, even though each probe has a defined T_d, the optimal hybridization conditions are dependent on the hybridization format and the needs of the researcher. For slot blot hybridizations of total nucleic acids, the wash temperature frequently matches the T_d so as to reduce nonspecific binding. For FISH, in which discrete cells are stained, a weak nonspecific binding does not interfere, as long as it does not exceed the natural background fluorescence. In this format, which tends to be sensitivity-limited, the optimal stringency of hybridization is usually at the highest formamide concentration that still yields full fluorescence.

5. EVALUATION OF PROBE SPECIFICITY AND SENSITIVITY Specificity and sensitivity are key aspects of any identification method. In the process of generating a new probe, specificity has already been controlled on two levels: initially during probe design and subsequently in the optimization of hybridization conditions using selected target and nontarget reference strains. However, even when the hybridization conditions have been properly determined, it may still be too early to apply the newly designed probe for determinative purposes. Questions that should first be considered are whether all strains available for a given species are indeed detected by a species-specific probe. Since 16S rRNA sequences are not usually available for all strains of interest, the best approach is to check the newly designed probe against a panel of reference strains. Subsequently, one should consider which bacteria need to be discriminated and whether 16S rRNAs from those bacteria have been sequenced. If this is not the case, it must be demonstrated that these strains do not hybridize at the optimized hybridization conditions.

Finally, hybridization assays should always incorporate proper controls, including at the minimum a positive and a negative control to evaluate the specificity and sensitivity of hybridization.

IV. FORMATS OF OLIGONUCLEOTIDE HYBRIDIZATION

Numerous hybridization assays exist. A full coverage is beyond the scope of this article and readers interested in a more complete listing are referred to other recent reviews (e.g., Schleifer et al., 1993). We will restrict ourselves to hybridization with oligonucleotide probes and to a few commonly used assays that allow a rapid identification of microorganisms. For the scope of this chapter, it might be sufficient to discriminate, on one hand, standard from reverse formats and, on the other hand, assays that require extracted nucleic acids from those that detect target nucleic acids at their original location within microbial cells.

A. *Dot-blot/slot-blot and other membrane-based hybridization formats* These assays are all based on the immobilization of target nucleic acids that have been extracted from the samples of interest. Critical steps for these types of assays are the cell lysis, purification of the nucleic acids, denaturation, and immobilization of the target nucleic acid on nitrocellulose or nylon membranes.

1. QUANTITATIVE SLOT-BLOT HYBRIDIZATION Dot-blot and slot-blot refer to the technique of using a vacuum chamber with round (dot) or longitudinal (slot) holes for the defined application of target nucleic acid solutions to membranes (Kafatos et al., 1979). In contrast to simply spotting samples onto membranes, which is sufficient for qualitative screening of multiple organisms, blotting evenly immobilizes each target nucleic acid on the same, defined area, facilitating quantitation. One particular method, quantitative slot blot hybridization with rRNA-targeted oligonucleotide probes, was introduced to studies in microbial ecology by Stahl et al. (1988).

This assay was designed to be directly applicable to diverse environmental samples without the need to cultivate the populations of interest. The choice of rRNA as a target molecule allows the use of highly disruptive cell lysis methods, which would damage high molecular weight DNA. This is an important advantage, since little is known about the samples *a priori* and quantitation relies on the efficient and representative recovery of nucleic acids from physiologically diverse bacteria (e.g., thin-walled Gram-negative bacteria vs. thick-walled Gram-positive bacteria). DNA is much more sensitive to shearing than RNA, therefore many of the DNA-based methods that are dependent on the retrieval of relatively intact nucleic acids require the use of less harsh lysis protocols. Consequently, those methods might fail to recover certain groups of bacteria present in high abundance in the community under investigation. One such example was the complete absence of 16S rDNA clones of Gram-positive bacteria with a high DNA G + C content in a PCR based gene library obtained from municipal activated sludge known to contain significant numbers of bacteria belonging to this phylogenetic group (Snaidr et al., 1997). In this case the freeze-thaw lysis method applied might have been ineffective in releasing DNA from this important part of the natural microbial community.

For quantitative slot blot hybridizations, total nucleic acid is recovered from the sample of interest by mechanical disruption with zirconium beads. The lysis is performed at low pH in the presence of equilibrated phenol and sodium dodecylsulfate to minimize nucleic acid degradation. Subsequently, nucleic acids are further purified by sequential extraction with phenol/chloroform and chloroform followed by ethanol precipitation. After spectrophotometric quantitation, the RNA is denatured with 2% glutaraldehyde and applied to a nylon membrane using a slot blot device. Air drying and baking is used to further immobilize the nucleic acids. The membranes are prehybridized in a buffer containing Denhardt's solution (Denhardt, 1966) before a synthetic oligonucleotide probe (5′-end labeled with ^{32}P using polynucleotide kinase and [γ-^{32}P]ATP) is applied. Denhardt's solution saturates free nucleic acid binding sites on the membrane that would otherwise increase the background by nonspecifically binding the labeled probe. Membranes are usually hybridized in rotating cylinders to prevent drying during the 40°C incubation, which lasts for several hours. The subsequent 30 min washing step is then performed at, or close to, the T_d determined for each probe. The membranes are dried and the amount of radioactivity bound to each slot is quantified by phospor imaging or autoradiography combined with densitometry. Average signals obtained from triplicates of a particular sample (e.g., with a genus-specific probe) are normalized for differences in the total amount of immobilized rRNA by comparison to the average signal obtained from replicates of the very same sample with a universal probe that binds to the rRNA of all organisms. Several applications of this technique have been published (e.g., Stahl et al., 1988; Raskin et al., 1994a; b).

2. COLONY HYBRIDIZATION   In the special case of colony hybridization (Grunstein and Hogness, 1975), nucleic acids are released directly onto filters on which colonies were either directly grown or transferred by replica plating or filtration. This method was originally developed for the rapid screening of cloned DNA fragments to search for specific genes. Colony hybridization can also be used for the identification of culturable bacteria (e.g., screening of primary plate isolates obtained from environmental samples) (Sayler et al., 1985; Festl et al., 1986). It must, however, be considered that Gram-positive bacteria need considerably harsher lysis methods than Gram-negative bacteria. Gram-positive bacteria may be resistant to the frequently used alkaline lysis method and may therefore yield false negative results (Jain et al., 1988b). It has been shown that pretreatment of cells with 10% sodium dodecylsulfate improves the *in situ* lysis of a variety of Gram-positive bacteria (Betzl et al., 1990; Hertel et al., 1991). Under optimal conditions 1 in 10^6 colonies may be detected (Sayler et al., 1985). However, problems may arise from bacteria that show rapidly spreading growth, such as *Bacillus cereus* subsp. *mycoides*. In addition, only those bacteria that readily form colonies on the media employed can be identified. Media are always selective and allow the analysis of only a poorly defined subfraction of the microbial cells present in a given environment. These drawbacks are not so important if colony hybridization is used to follow the fate of defined, rapidly growing bacterial strains, so it can be used successfully for these applications. For the sake of brevity, we will provide only one example for several different areas: PCB-degrading bacteria were monitored in soil by detecting specific catabolic genes (Layton et al., 1994). Heavy metal resistant bacteria have been screened and enumerated (Barkay et al., 1985). More recently, colony hybridization was used to differentiate subspecies and biovars of *Lactococcus lactis* with a gene fragment from the histidine biosynthesis operon (Beimfohr et al., 1997). The survival of genetically modified microorganisms has been studied in aquatic environments (Amy and Hiatt, 1989) and in mammalian intestines (Brockmann et al., 1996), and colony hybridization has also been used to monitor the maintenance and transfer of genes (Jain et al., 1987b).

B. Reverse hybridization formats   In reverse hybridization formats the labeled target nucleic acid is analyzed using an array of immobilized probes. In contrast to the standard hybridization assays, multiple nucleic acid probes, rather than the target, are deposited or even synthesized on a support. Subsequently, the sample of interest, rather than the probe, is labeled and hybridized against the array. This approach was initially used by Saiki et al. (1989) for the genetic analysis of PCR-amplified DNA with immobilized sequence-specific oligonucleotide probes. Meanwhile, it has found numerous application for the identification of bacteria, e.g., *Listeria monocytogenes* (Bsat and Batt, 1993), clostridia (Galindo et al., 1993), and lactic acid bacteria (Ehrmann et al., 1994).

Reverse sample genome probing was introduced to environmental microbiology by Voordouw et al. (1991). This method follows the same concept as reverse hybridization, but uses immobilized genomic DNAs from reference strains to probe environmental DNA that is radioactively labeled by nick translation. Here, the quality of the identification obtained following the incubation of a labeled total chromosomal DNA probe is largely dependent on the number and types of bacterial standards spotted on the master filter, and is therefore restricted by our current ability to retrieve representative pure cultures. Furthermore,

since this assay represents essentially a massively parallel but classical DNA–DNA hybridization,the basic principle of this method must be considered. Because of potentially large differences in the mol% G + C content of DNA, which can range from 25–75%, a given incubation temperature might be optimal or relaxed for one type of DNA but highly stringent for another DNA. This directly influences the extent of binding and thereby the potential of this approach for accurate quantitation. The degree of DNA–DNA hybridization is high only between closely related species and quite low between less closely related species. As a consequence, the DNA of fairly closely related reference strains must be immobilized in order to detect even numerically abundant populations. So far the method has been restricted to the well characterized group of sulfate-reducing bacteria (e.g., screening of enrichments and isolates obtained from oil fields; Voordouw et al., 1992).

Supports for probe immobilization range from nylon membranes (e.g., Ehrmann et al., 1994) to microtiter plates (Galindo et al., 1993). Recently, Guschin and coworkers (1997) used oligonucleotide microchips as genosensors for determinative and environmental studies in microbiology. These microchips contain an array of deoxyribonucleotide oligomers that were immobilized after synthesis and purification within a polyacrylamide gel matrix bound to the surface of a glass slide. Oligonucleotide microchips were originally developed for rapid sequence analysis of genomic DNA by hybridization with oligonucleotides (Mirzabekov, 1994) and have proven to be suitable for analysis of mutations and gene polymorphisms (Yershov et al., 1996). Yet another fascinating possibility is the highly parallel synthesis of thousands of oligonucleotides. Here, photolithography is used to generate miniaturized arrays of densely packed oligonucleotides on a glass support (Fodor et al., 1991; Pease et al., 1994). These probe arrays, or DNA chips, are then used in hybridizations in which the analyzed nucleic acid is fluorescently labeled. Subsequently, the fluorescence arising from areas covered by the different oligonucleotides is quantified by laser scanning microscopy. Fluorescence signals from complementary probes were reported to be 5–35 times stronger than those arising from probes with one or two mismatches (Pease et al., 1994).

In the near future, it should be possible to immobilize thousands of species-specific probes or sets of nested probes tailored to the specific needs of microbiologists. Whereas light-generated DNA chips appear perfect for routine applications with a large commercial market (e.g., clinical microbiology), the postsynthesis loading of multiple oligonucleotide probes on suitable supports such as microchips or membranes could also be cost effective for more specialized applications. Along these lines, it is noteworthy that simultaneous transcriptional profiling on all open reading frames of the yeast *Saccharomyces cerevisiae* has been reported recently (Hauser et al., 1998).

C. In situ hybridization   *In situ* hybridization, defined in a strict sense, is a localization technique that identifies nucleic acids in cells that remain at the site where they live. In a somewhat wider definition, microbiologists are using the term to describe the detection of target nucleic acids within fixed whole cells, although early attempts were made to discriminate between true *in situ* and whole cell hybridization (Amann et al., 1990b). The *in situ* identification of fixed whole bacterial cells using fluorescently labeled, rRNA-targeted oligonucleotides originally described by DeLong and coworkers (1989) has, over the last decade, found numerous applications in microbiology (for review see Amann et al., 1995; Amann and Kühl, 1998). Ribosomal RNA is not the only target for *in situ* hybridization, but for obvious reasons the most common one. Its stability and high copy number makes rRNA a much easier target than, e.g., mRNA. This does not mean that *in situ* mRNA detection in single cells has not yet been achieved (e.g., Hahn et al., 1993; Hönerlage et al., 1995; Wagner et al., 1998), but that it has not yet been used for routine applications, as is the case for rRNA-targeted *in situ* hybridization probes.

The basic steps of fluorescence *in situ* hybridization are outlined in Fig. 5.

In principle, all the points that need to be considered for specific and sensitive detection of extracted target nucleic acids also apply to *in situ* hybridization. However, a couple of additional points are critical for *in situ* hybridization, especially for avoiding false negative results.

1. PERMEABILIZATION OF TARGET CELLS FOR NUCLEIC ACID PROBES   A prerequisite for successful *in situ* hybridization is that the probe molecules can get to the target molecules. For this, cell components such as the cell wall, membranes, and, if present, capsular material or other extracellular polymeric substances must be permeable for the probe molecules to enter. This is easier when smaller probes are used. Oligonucleotides are, in this regard, better than polynucleotides and small fluorescent labels with a molecular weight below 1 kDa are better than large enzyme labels such as horseradish peroxidase (Amann et al., 1992b; Schönhuber et al., 1997). Furthermore, since intact membranes are generally impermeable to standard oligonucleotides, a fixation step is required. Fixation is usually accomplished by treatment of the sample with crosslinking aldehyde solutions (paraformaldehyde, formalin) and/or denaturing alcohols (for detail see Amann, 1995b). This step also kills the cells. Even

though several fairly general fixation protocols have been described (Amann, 1995b; Amann et al., 1995), care should be taken to ensure that the procedure is optimized for the target cells, both so that their morphological integrity is not compromised and so that the cell walls do not become so strongly cross-linked that probe penetration is hindered. Thick-walled Gram-positive bacteria need different fixation protocols than Gram-negative bacteria (Roller et al., 1994; Erhart et al., 1997). Furthermore, diffusion of the probe requires a certain time, which is a function of the distance between the probe and the target. Therefore, larger aggregates need either to be dispersed, e.g., by sonication (Llobet-Brossa et al., 1998) or sectioned to preserve the natural organization (Ramsing et al., 1993; Schramm et al., 1996).

The impermeability of the thick peptidoglycan layer of many Gram-positive bacteria to horseradish peroxidase-labeled oligonucleotides has recently been exploited to estimate the state of the cell wall in individual bacteria using FISH (Bidnenko et al., 1998). The authors reasoned that the expression of intracellular, peptidoglycan-hydrolyzing enzymes, such as autolysins or phage-encoded lysins, should permeabilize the cell walls for probe entry and thereby make the cells detectable by this method. The concept worked for strains of *Lactococcus lactis* infected with the virulent bacteriophage bIL66. Whereas only few cells hybridized in an exponentially growing culture, after infection the frequency of hybridizing cells increased sharply to 90%. In contrast, FISH with peroxidase-labeled oligonucleotide probes cannot be used to estimate the state of the cell wall of the Gram-negative *E. coli*, which without further lysis is fully permeable for probes of that size.

2. *IN SITU* ACCESSIBILITY OF PROBE TARGET SITES   Ever since rRNA-targeted FISH was first performed it was obvious that some target sites yield stronger signals than others (Amann et al., 1995; Frischer et al., 1996). For denatured extracted nucleic acids, it is assumed that target molecules are completely single-stranded and that different target sites are equally accessible for different nucleic acid probes. This is not the case for *in situ* hybridization with rRNA-targeted oligonucleotides. Here, the target molecules, which are integral parts of the ribosome, remain in the cell. Consequently, both rRNA-protein and intramolecular rRNA–rRNA interactions may influence the accessibility of the target sites. It is therefore not surprising that 200 fluorescein-labeled oligonucleotides, targeting the 16S rRNA of *Escherichia coli* with a spacing of less than 10 nucleotides (Fuchs et al., 1998), showed large differences in their capacity to fluorescently stain the very same cells (Fig. 6).

A good choice of accessible target sites yielding bright fluorescent signals is of critical importance for the sensitivity of *in situ* identification. Since the higher-order structure of the rRNA molecules and the ribosome are quite conserved, the *in situ* accessibility map of the *E. coli* 16S rRNA should be helpful for the selection of target sites in other organisms. Nevertheless, variations in *in situ* accessibility between different species will exist (Fuchs et al., 1998). Therefore, in the event that a newly designed probe that works on extracted nucleic acid does not yield good signals *in situ*, it is recommended to use one of the well-established, strongly fluorescing, general probes (e.g., the EUB338; Amann et al., 1990a) to determine whether this problem is cell- or probe-related and could possibly be solved by switching to a different target site.

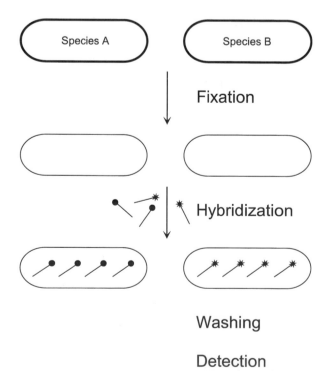

FIGURE 5. Principal steps of fluorescence *in situ* hybridization. The dots and asterisks indicate two different fluorescent dye molecules that are linked to two specific oligonucleotide probes.

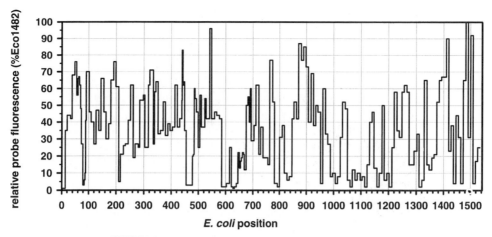

FIGURE 6. *In situ* accessibility of the 16S rRNA of *E. coli.*

3. Improving the sensitivity of *in situ* hybridization

When discussing the sensitivity of *in situ* hybridization, one has first to realize that even though an individual cell can be identified (Fig. 7), this cell first needs to be brought into the microscopic field of observation. In a marine sediment containing $>10^9$ cells/cm^3 and a very high fraction of autofluorescent particles, the detection limit might be no better than 0.1% or 10^6 cells/cm^3. However, given a relatively clean water sample and the right equipment, it should also be possible to detect <1 cell/cm^3.

Another frequently encountered problem is that bacterial cells from the environment have low signals after *in situ* hybridization with fluorescently monolabeled oligonucleotide probes (Amann et al., 1995). The fluorescence conferred by a rRNA-targeted probe will be sensitive to changes in the cellular rRNA content of the target cells. The linear relationship between the growth rate of *Salmonella typhimurium* (Schaechter et al., 1958) and cellular ribosome content is well known. This correlation also applies to other bacteria (Poulsen et al., 1993; Wallner et al., 1993) and might be the reason why small, starving cells with little to no growth are so difficult to detect by FISH with rRNA-targeted probes. On the other hand, if this correlation is really true for cells in the environment, then it should be possible to determine or, at least, to estimate *in situ* growth rates of individual cells based on quantitation of probe-conferred fluorescence. This has been attempted for sulfate-reducing bacteria in a biofilm using digital microscopy (Poulsen et al., 1993). However, there is a large difference between the highly controlled growth conditions in Schaechter's experiments (Schaechter et al., 1958) and those experienced by environmental bacteria which might have to cope with rapid changes in the physical and chemical environment. Since ribosome synthesis is energetically costly, ribosome degradation, as a rapid first response to the slowing of the growth rate, would be very wasteful. Indeed, during periods of starvation of up to several months, bacteria maintain cellular ribosome pools in excess of their current needs (Flärdh et al., 1992; Wagner et al., 1995). Consequently, in strongly fluctuating environments such as, e.g., sediments in the intertidal zone, the cellular ribosome content should not be used to estimate actual growth rates. Nevertheless, the FISH signal of a cell is ecologically meaningful since it reflects the potential of the cell to synthesize protein.

There have been several attempts to combine FISH with short term incubation of environmental samples with nutrients and/or antibiotics, with the aims of increasing the ribosome content of environmental cells or demonstrating viability. Oligotrophic biofilms in potable water were incubated for 8 h with a mixture of carbon sources and an antibiotic preventing cell division (Kalmbach et al., 1997) prior to FISH in a modification of the direct viable count technique (Kogure et al., 1979; 1984). The number of cells detectable by FISH increased from 50% to 80%, clearly demonstrating viability of the majority of the cells. In a similar approach, marine water samples were incubated for approximately one hour with chloramphenicol (Ouverney and Fuhrman, 1997). Again, an increase in detection yield from 75% to almost 100% was observed. It should be noted here that even though both studies described precautions taken to prevent changes in total cell number or microbial composition during the incubation of the samples, the treatments had effects, e.g., on the cellular ribosome content. These methods cannot therefore, in a strict sense, be regarded as *in situ* hybridizations, and it might, in any case, be helpful to also investigate parallel samples after direct fixation.

Recently, technical improvements that result in more sensitive FISH have been reported. These include the use of more sensitive fluorescent dye molecules such as CY3 (Glöckner et al., 1996), dual labeling of oligonucleotide probes (Wallner et al., 1993, Fuhrman and Ouverney, 1998), the application of tyramide signal amplification (Schönhuber et al., 1997), and detection of the probe-conferred fluorescence by highly sensitive cameras (e.g., Ramsing et al., 1996; Fuhrman and Ouverney, 1998). Still, a certain fraction of particles detected by binding of the DNA stain DAPI (Porter and Feig, 1980) and identified as cells by cell morphology cannot be detected by these improved methods. It has been suggested that some of these cells might represent "ghosts" lacking nucleoids (Zweifel and Hagström, 1995). Since many of these DAPI stained spots that remain undetected by FISH are at the limit of resolution of light microscopy, the possibility that they originate from large virus particles can also not be excluded. Furthermore, problems with probe penetration, and the possibility that even the most general 16S rRNA probe target sites contain mutations, should be considered.

Like any other method for the identification of microorganisms, FISH has specific limitations. In addition to those we have

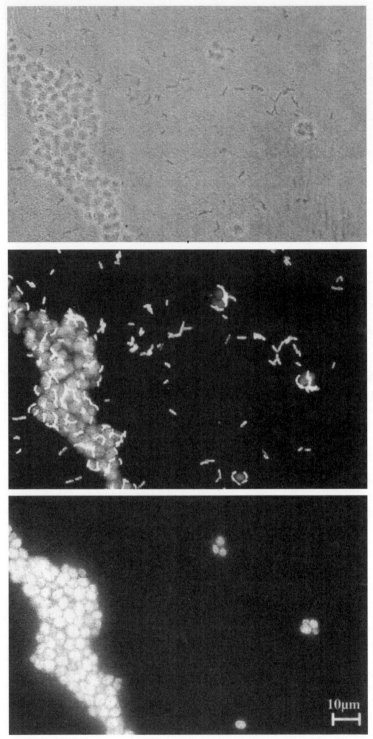

FIGURE 7. *In situ* identification of *E. coli* and the yeast *Saccharomyces cerevisiae* by *in situ* hybridization with two differently labeled rRNA-targeted oligonucleotide probes. Phase contrast and two epifluorescence images are shown for the same microscopic field.

already discussed, a further limitation is the lack of automation, as cells are frequently still counted manually. It should be stressed that FISH is the method of choice for determination of numbers of individual cells or localization of cells. However, for simple yes/no answers or rough estimates, this method may still be too complicated and time-consuming.

V. APPLICATIONS OF NUCLEIC ACID PROBING IN ENVIRONMENTAL MICROBIOLOGY

Nucleic acid probes have, in the last decade, revolutionized environmental microbiology. In each of the monthly issues of *Applied and Environmental Microbiology*, there are numerous examples in which hybridization or PCR assays are used to monitor defined strains or specific genes in different environments. This boom is due to the increasingly accepted view that traditional microbiological methods for analyzing environmental samples are less accurate and usually slower than molecular techniques. It also reflects the now-common idea that more than 100 years of scientific microbiology has, in terms of cultivation of representatives of extant microbial diversity, managed to describe only the tip of the iceberg. This does not imply that we should stop our efforts to isolate and study pure cultures. Most of what we know of microbiology and microorganisms is based on studies of pure cultures. Nevertheless, molecular techniques have become a faster and more accurate means to address questions concerning the exact size of a population or the composition of complex microbial communities. Before we review some examples, organized according to habitats, it should be noted that nucleic acid probes do not answer all our questions equally well, especially when the viability and physiology of certain bacteria is the focus of interest. In such instances, cultivation-based methods might still be the method of choice. For reasons of practicability, we will focus on applications of *in situ* hybridization in environmental microbiology. This, however, does not indicate that this method is superior to other assays. If one wants to analyze a relatively poorly studied habitat for microbial diversity or community composition, it is always wise to combine at least two different methods.

A. Soils and sediments Soils and sediments are among the most complex of all microbial habitats. The microbial diversity of soils has always been viewed as high, but the first attempts at quantifying this diversity were not reported until 1990. In their much-cited reassociation study of DNA isolated from a Norwegian forest soil, Torsvik et al. (1990a) suggest that the genetic diversity of DNA extracted from a bacterial cell fraction of a gram of soil corresponds to about 4000 completely different genomes of a size standard for soil bacteria. This was about 200 times the genetic diversity found in the strains isolated from the same soil sample (Torsvik et al., 1990b), which reflects the selectivity of cultivation based methods and the tendency of these methods to underestimate both the absolute number and the diversity of microorganisms already noticed before (Skinner et al., 1952; Sorheim et al., 1989). During the last years, molecular methods including the analysis of mol% G + C profiles (Holben and Harris, 1995; Nüsslein and Tiedje, 1998), the analysis of amplified rDNA by restriction analysis (ARDRA; e.g., Smith et al., 1997; Nüsslein and Tiedje, 1998), and denaturing or thermal gradient gel electrophoresis (Felske et al., 1997; Heuer et al., 1997) have increasingly been applied to soils.

With respect to nucleic acid probing of soils, colony hybridizations have often been performed (e.g., Sayler and Layton, 1990), whereas application of quantitative dot blot hybridization of rRNA–DNA is rather rare, suggesting difficulties with the extraction of good quality nucleic acids from soils (discussed, e.g., in Torsvik et al., 1990a; Holben and Harris, 1995). However, the few recent studies that have been reported (e.g., MacGregor et al., 1997; Rooney-Varga et al., 1997; Sahm et al., 1999) indicate that the method is applicable.

In situ monitoring of bacterial populations in soils and sediments by FISH has also long proven difficult (Hahn et al., 1992). Problems arise from autofluorescence of soil particles, irregular distribution of cells, and low detection yield. With the implementation of improved dyes (Zarda et al., 1997) and microscopic techniques such as confocal laser scanning microscopy (Assmus et al., 1995), the situation is much improved, leading to the application of FISH for studies of microbial community composition both in soil (Zarda et al., 1997; Ludwig et al., 1997; Chatzinotas et al., 1998) and in sediments (Llobet-Brossa et al., 1998; Rossello-Mora et al., 1999). As an example, Fig. 8 shows detection of a filament of the sulfate-reducing bacterium *Desulfonema* sp. in Wadden Sea sediments. Interestingly, quite unexpected groups of bacteria are found in high numbers in both habitats, e.g., the peptidoglycan-less planctomycetes (Zarda et al., 1997; Chatzinotas et al., 1998), as well as representatives of a thus far uncultured group of Gram-positive bacteria with a high DNA G + C content (Felske et al., 1997) and of the newly described phylum *Holophaga/Acidobacterium* (Ludwig et al., 1997). High abundance of members of the *Cytophaga/Flavobacterium* cluster has been reported for anoxic marine sediments (Llobet-Brossa et al., 1998; Rossello-Mora et al., 1999).

In the study of Chatzinotas et al. (1998), several molecular methods were compared for their potential to detect broad-scale differences in the microbial community composition of two pristine forest soils. This study highlights one of the many potential methodological pitfalls: FISH of dispersed soil slurries failed to detect Gram-positive bacteria with a high DNA G + C content, even though dot blot hybridization of extracted DNA indicated significant occurrence of members of this group. This was only in part a problem of cell permeability since care had been taken to ensure that at least the vegetative filaments of the actinomycetes were probe-permeable. Filaments could indeed be visualized in nondispersed soil samples. It appears that the filaments were destroyed by the methods used for soil dispersion, vortexing and sonication, which together with the physical effects of inorganic soil particles most likely resulted in the milling of actinomycete filaments. This again shows how important it is not to rely on a single technique for community composition analysis of the highly diverse soil microbiota.

B. In situ *hybridization of biofilms and aggregates* In many natural settings as well as in biotechnological wastewater treatment systems, immobilized communities of bacteria, the so-called biofilms, are the main mediators of biogeochemical reactions rather than free-living bacteria. Since, in addition to determination of community composition, *in situ* localization is of prime importance in the investigation of these systems, biofilms have been studied intensively with FISH. Thicker biofilms, in the mm to cm range, are known as microbial mats. Because of their size these can, unlike biofilms, also be studied with extraction-based molecular techniques, such as slot blot hybridization (e.g., Risatti et al., 1994) or DGGE analysis of amplified rDNA fragments (e.g., Ferris and Ward, 1997; Ferris et al., 1997).

The *in situ* visualization of defined bacterial biofilm populations was first achieved in an anaerobic fixed-bed reactor (Amann et al., 1992a). The initial colonization of a glass surface was monitored using FISH. Two morphologically distinct populations of Gram-negative sulfate reducing bacteria (a thick and a thin vibrio) could be assigned to 16S rDNA sequences related to *Desulfuromonas* and *Desulfovibrio* retrieved from the same reactor by oligonucleotide probing. One of the probes was later used to direct the enrichment and isolation of a sulfate-reducing strain representative for the *Desulfovibrio* sp. population (Kane et al., 1993).

Over the last few years, numerous FISH studies have been performed in systems such as activated sludge plants, trickling filters, or anaerobic sludge digesters (Harmsen et al., 1996a, b). Initial investigations of activated sludge targeted the higher level

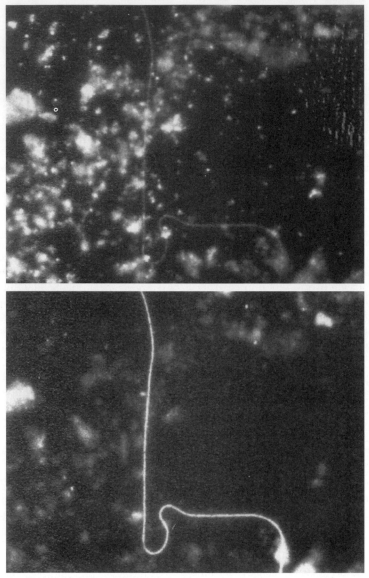

FIGURE 8. *In situ* visualization of *Desulfonema* sp. in Wadden Sea sediments. *Upper panel*, DAPI staining. *Lower panel*, FISH with *Desulfonema* probe.

bacterial taxa by applying, e.g., 16S or oligonucleotide probes for the α, β, and γ *Proteobacteria*, the *Cytophaga-Flavobacterium* cluster, or the *Actinobacteria* (Wagner et al., 1993; Manz et al., 1994; Kämpfer et al., 1996). Probes are now available for functionally important groups such as the ammonia- (Wagner et al., 1995, 1996; Mobarry et al., 1996) and nitrite-oxidizing bacteria (Wagner et al., 1996; Schramm et al., 1998), and for key genera and species in wastewater treatment such as *Acinetobacter* spp. (Wagner et al., 1994), *Zoogloea ramigera* (Rossello-Mora et al., 1995), and *Microthrix parvicella* (Erhart et al., 1997). The application of these probes has already resulted in some interesting findings. For instance, in contrast to textbook knowledge, *Acinetobacter* spp. seems to play no major role in biological phosphorus removal (Wagner et al., 1994). In the future, important processes such as floc formation and settling will be related to population sizes of defined bacteria. Based on previous indications that β-*Proteobacteria* are a major group in activated sludge, a high genetic diversity within this group was demonstrated by the simultaneous application of three oligonucleotide probes labeled with differ-

ent fluorochromes (Amann et al., 1996b; Snaidr et al., 1997). Here, the colocalization of two or even three probes within one fixed whole cell results in a better discrimination of closely related β-subclass *Proteobacteria*.

FISH can be combined with microsensor measurements to address both *in situ* structure and activity of biofilms. In the first example of such a study (Ramsing et al., 1993), a mature, thick trickling filter biofilm was first investigated with microsensors for oxygen, sulfide, and nitrate to quantify sulfate reduction before cryosectioning, and FISH was used for the *in situ* localization of SRB. Similarly, the structure/function correlation of nitrification has been analyzed both in a trickling filter treating the ammonia-rich effluent water of an eel farm (Schramm et al., 1996), and in a chemolithoautotrophic fluidized bed reactor (Schramm et al., 1998). Whereas the former contained, as expected, dense clusters of *Nitrosomonas* spp. (Fig. 9) and *Nitrobacter* spp., *Nitrosospira* spp. and *Nitrospira* spp. dominated the more oligotrophic fluidized bed. In another attempt to combine *in situ* activity measurement with *in situ* identification of individual cells, microau-

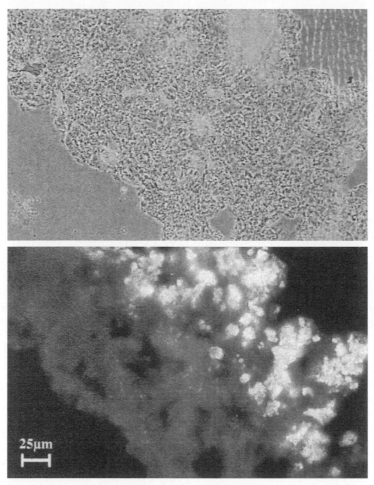

FIGURE 9. *In situ* visualization of clusters of *Nitrosomonas* sp. cells in a nitrifying biofilm. A phase contrast and an epifluorescence image are shown for the identical microscopic field.

toradiography has recently been used in conjunction with FISH (Nielsen et al., 1997). It should not be forgotten that the combination of traditional isolation and physiological characterization of pure cultures with FISH is a very powerful and straightforward way to correlate *in situ* distribution with particular metabolic properties. For example, it has been shown that *Paracoccus* is an important denitrifying genus in a methanol-fed sand filter (Neef et al., 1996). Traditional cultivation resulted in the isolation of numerous strains of *Paracoccus* that had the potential to denitrify using methanol as substrate. Genus-specific probes identified dense clusters of brightly stained paracocci accounting for about 3.5% of all cells in methanol-fed biofilms, whereas almost no cells were detected in a parallel filter that did not receive methanol and therefore showed no denitrification. Here, FISH allows for the assignment of functions studied in pure cultures to defined populations within a complex biofilm. A caveat of this approach may be the known metabolic plasticity of bacteria.

C. **In situ** *identification of planktonic bacteria in oligotrophic water samples* An early observation in the study of aquatic environments was that the free-living, planktonic bacteria were often more difficult to detect by FISH than those attached to the surfaces of the same water body (Manz et al., 1993). It was pointed out that these surfaces are enriched for nutrients so that immobilized bacteria are not as strongly nutrient-limited as planktonic bacteria. Indeed, the initial applications of FISH to bac-

terioplankton yielded good results only in highly eutrophic ponds (Hicks et al., 1992) or contaminated coastal water (Lee et al., 1993). The improvement of FISH detection yields by simultaneous application of multiple single labeled (Lee et al., 1993), brighter (Glöckner et al., 1996), or dual-labeled (Fuhrman and Ouverney, 1998) oligonucleotides in combination with image-intensifying CCD cameras (Ramsing et al., 1996; Fuhrman and Ouverney, 1998) indicates that the method remains sensitivity limited, and this is probably a function of small cell size and low cellular ribosome content. Some years ago, Edward DeLong, the pioneer of FISH, needed to apply quantitative dot blot hybridization to determine the abundance of archaeal rRNA in coastal Antarctic surface waters (DeLong et al., 1994). Assuming that rRNA abundance is a good indicator of biomass, it was suggested that as yet uncultured *Archaea* might constitute up to one-third of the total procaryotic biomass. FISH protocols have now been sufficiently improved to allow reliable *in situ* detection of greater than 50% of the bacteria in oligotrophic water samples (Glöckner et al., 1996; Fuhrman and Ouverney, 1998). The microbial community composition of the winter cover and pelagic zone of an Austrian high mountain lake has been described (Alfreider et al., 1996). A seasonal study of microbial community dynamics, that for the first time also included the monitoring of defined bacterial populations based on probes targeted to environmental 16S rDNA retrieved from the same lake, has also been reported (Pernthaler et al., 1998). The combined use of digital microscopy and FISH enabled the determination of biomasses and size distributions, an approach which has recently been extended to studies of morphological and compositional changes in a planktonic bacterial community in response to enhanced protozoan grazing (Pernthaler et al., 1997; Jürgens et al., 1999). While only the environmental applications of FISH have been discussed here, similar techniques may be applied to the analysis of waterborne pathogens such as *Legionella pneumophila* (Manz et al., 1995; Grimm et al., 1998) or bacterial endosymbionts (for a review see Amann et al., 1995).

D. Flow cytometry and fluorescence-activated cell sorting Flow cytometry is a technique for the rapid analysis and sorting of single cells. Up to 10^3 suspended cells per second can pass an observation point where the cells, aligned in a water jet like pearls on a string, interfere with one or several light sources, usually lasers. Several physical and chemical properties of each individual cell can be measured simultaneously on the basis of fluorescence emitted from specifically and stoichiometrically bound dyes, as well as from light scattering. Flow cytometry has a higher throughput than microscopic quantification of specifically stained microbial populations and can more readily be automated, which should facilitate more rapid and frequent monitoring of the composition of microbial communities. It has been demonstrated using pure cultures that FISH of fixed whole bacterial cells can be combined with flow cytometry (Amann et al., 1990a; Wallner et al., 1993). The approach can also be applied to environmental samples as has been shown with samples from a wastewater treatment plant (Wallner et al., 1995). Flow cytometric and microscopic counts were in general agreement, with some discrepancies found for those populations that occurred predominantly in flocs or chains.

Effective cell dispersion is a prerequisite for accurate counting and therefore flow cytometry is better suited for free-living cells than for immobilized microbial communities such as biofilms. Furthermore, application of flow cytometry in microbiology is frequently hindered by the small size and concomitant low scattering and fluorescence of microbial cells.

However, certain features of flow cytometry justify the effort required to change from microscopy to this approach for the analysis of bacteria. There is the above-mentioned high throughput and potential for automation, of which the study by Fuchs et al. (1998) on the quantitation of fluorescence conferred by 200 different 16S rRNA-targeted probes is a good example. An additional attractive feature is that many flow cytometers have a

sorting option. It was recently shown that bacteria could be sorted directly from environmental samples, without cultivation, based on differences in light scattering, DNA content, and affiliation to certain phylogenetic groups as revealed by FISH (Wallner et al., 1997). Microscopy of sorted cells showed that populations of originally low abundance could be strongly enriched (up to 1000-fold) by flow sorting (Snaidr et al., 1999). The ultimate purity of the sorted cells also depends on the sample analyzed and the original abundance, but in an optimal case can be close to 100% (Wallner et al., 1997). Gene fragments can subsequently be amplified from the sorted cells for further molecular analysis by PCR. In this way, the combination of flow sorting and FISH with probes targeted to 16S rRNA directly retrieved from the environment without cultivation allows selective access to the genetic information of microorganisms.

VI. OUTLOOK

The prospects for hybridization-based molecular methods in microbiology are bright. The future will likely bring not only even more sensitive methods and automation, but also the massively parallel application of user-friendly probe arrays. These will include nucleic acid probes for taxonomic identification at the species level, but may also allow strain level assignment. Probes for functional genes such as those coding for certain degradative pathways or virulence factors will also be available. It has been pointed out in this introductory chapter that probe design usually relies on knowledge of the target sequence. Therefore, in the near future we will see continued and most likely increased sequencing efforts both for specific genes, such as those coding for the 16S rRNA, and for full genomes. The only threat to the increased use of nucleic acid probes in the future might originate from the very same rapid development of nucleic acid sequencing technology. After all, probes are just tools for determining a short sequence that might in the future be determined as well or even faster by direct sequencing. We expect that in the near future the artificial boundaries between nucleic acid hybridization and sequencing will erode. Sequencing, rather than nucleic acid hybridization may become the standard approach to molecular identification, but this sequencing may be performed via hybridization to nucleic acid arrays.

ACKNOWLEDGMENTS

The original work of the authors has been supported by grants from the Deutsche Forschungsgemeinschaft, the Bundesministerium für Forschung und Technologie, the Fonds der Chemischen Industrie, the Max-Planck-Society, and the European Union. We would like to thank Bernhard Fuchs, Frank-Oliver Glöckner, and Wilhelm Schönhuber for artwork and critical reading of the manuscript.

Bacterial Nomenclature

Peter H.A. Sneath

SCOPE OF NOMENCLATURE

Nomenclature has been called the handmaid of taxonomy. The need for a stable set of names for living organisms, and rules to regulate them, has been recognized for over a century. The rules are embodied in international codes of nomenclature. There are separate codes for animals, noncultivated plants, cultivated plants, procaryotes, and viruses. But partly because the rules are framed in legalistic language (so as to avoid imprecision), they are often difficult to understand. Useful commentaries are found in Ainsworth and Sneath (1962), Cowan (1978), and Jeffrey (1977). There are proposals for a new universal code for living organisms (see the Proposed BioCode).

The nomenclature of the different kinds of living creatures falls into two parts: (a) informal or vernacular names, or very specialized and restricted names; and (b) scientific names of taxonomic groups (taxon, plural taxa).

Examples of the first are vernacular names from a disease, strain numbers, the symbols for antigenic variants, and the symbols for genetic variants. Thus one can have a vernacular name like the tubercle bacillus, a strain with the designation K12, a serological form with the antigenic formula Ia, and a genetic mutant requiring valine for growth labeled *val*. These names are usually not controlled by the codes of nomenclature, although the codes may recommend good practice for them.

Examples of scientific names are the names of species, genera, and higher ranks. Thus *Mycobacterium tuberculosis* is the scientific name of the tubercle bacillus, a species of bacterium.

These scientific names are regulated by the codes (with few exceptions) and have two things in common: (a) they are all Latinized in form so as to be easily recognized as scientific names, and (b) they possess definite positions in the taxonomic hierarchy. These names are international; thus microbiologists of all nations know what is meant by *Bacillus anthracis*, but few would know it under vernacular names like Milzbrandbacillus or Bactéridie de charbon.

The scientific names of procaryotes are regulated by the *International Code of Nomenclature of Bacteria*, which is also known as the *Revised Code* published in 1975 (Lapage et al., 1975). This edition authorized a new starting date for names of bacteria on January 1, 1980, and the starting document is the Approved Lists of Bacterial Names (Skerman et al., 1980), which contains all the scientific names of bacteria that retain their nomenclatural validity from the past. The operation of these Lists will be referred to later. The *Code* and the Lists are under the aegis of the International Committee on Systematic Bacteriology, which is a constituent part of the International Union of Microbiological Societies. The Committee is assisted by a number of Taxonomic Subcommittees on different groups of bacteria, and by the Judicial Commission, which considers amendments to the *Code* and any exceptions that may be needed to specific Rules. An updated edition of the *Revised Code* was published in 1992 (Lapage et al., 1992).

LATINIZATION

Since scientific names are in Latinized form, they obey the grammar of classic, medieval, or modern Latin (Neo-Latin). Fortunately, the necessary grammar is not very difficult, and the most common point to watch is that adjectives agree in gender with the substantives they qualify. Some examples are given later. The names of genera and species are normally printed in italics (or underlined in manuscripts to indicate italic font). For higher categories conventions vary: in Britain they are often in ordinary roman type, but in America they are usually in italics, which is preferable because this reminds the reader they are Latinized scientific names. Recent articles that deal with etymology and Latinization include that of MacAdoo (1993) and the accompanying article by Trüper on Etymology in Nomenclature of Procaryotes. The latter is particularly valuable because it clarifies the formation of names derived from names of persons.

TAXONOMIC HIERARCHY

The taxonomic hierarchy is a conventional arrangement. Each level above the basic level of species is increasingly inclusive. The names belong to successive **categories**, each of which possesses a position in the hierarchy called its **rank**. The lowest category ordinarily employed is that of species, though sometimes these are subdivided into subspecies. The main categories in decreasing rank, with their vernacular and Latin forms, and examples, are shown in Table 1.

Additional categories may sometimes be intercalated (e.g., subclass below class, and tribe below family). There is currently discussion on the best treatment for categories above kingdom; the BioCode (see later) uses the term, domain, above kingdom.

FORM OF NAMES

The form of Latinized names differs with the category. The species name consists of two parts. The first is the **genus name**. This is spelled with an initial capital letter, and is a Latinized substantive. The second is the **specific epithet**, and is spelled with a lower case initial letter. The epithet is a Latinized adjective in agreement with the gender of the genus name, or a Latin word in the genitive case, or occasionally a noun in apposition. Examples are given in the article by Trüper. Thus in *Mycobacterium*

TABLE 1. The ranking of taxonomic categories

Category	Example
Domain	*Bacteria*
Phylum in zoology or Division in botany and bacteriology	*Actinobacteria*
Class	*Actinobacteria*
Subclass	*Actinobacteridae*
Order	*Actinomycetales*
Suborder	*Actinomycineae*
Family	*Actinomycetaceae*
Genus	*Actinomyces*
Species	*Actinomyces bovis*

tuberculosis, the epithet tuberculosis means "of tubercle", so the species name means the mycobacterium of tuberculosis. The species name is called a **binominal name**, or **binomen**, because it has two parts. When subspecies names are used, a trinominal name results, with the addition of an extra **subspecific epithet**. An example is the subspecies of *Lactobacillus casei* that is called *Lactobacillus casei* subsp. *rhamnosus*. In this name, *casei* is the specific epithet and *rhamnosus* is the subspecific epithet. The existence of a subspecies such as *rhamnosus* implies the existence of another subspecies, in which the subspecific and specific epithets are identical, i.e., *Lactobacillus casei* subsp. *casei*.

One problem that frequently arises is the scientific status of a species. It may be difficult to know whether an entity differs from its neighbors in certain specified ways. A useful terminology was introduced by Ravin (1963). It may be believed, for example, that the entity can undergo genetic exchange with a nearby species, in which event they could be considered to belong to the same **genospecies**. It may be believed the entity is not phenotypically distinct from its neighbors, in which event they could be considered to belong to the same **taxospecies**. Yet, the conditions for genetic exchange may vary greatly with experimental conditions, and the criteria of distinctness may depend on what properties are considered, so that it may not be possible to make clear-cut decisions on these matters. Nevertheless, it may be convenient to give the entity a species name and to treat it in nomenclature as a separate species, a **nomenspecies**. It follows that all species in nomenclature should strictly be regarded as nomenspecies. They are, of course, usually also taxospecies.

Genus names, as mentioned above, are Latinized nouns, and so subgenus names (now rarely used) are conventionally written in parentheses after the genus name; e.g., *Bacillus* (*Aerobacillus*) indicates the subgenus *Aerobacillus* of the genus *Bacillus*. As in the case of subspecies, this implies the existence of a subgenus *Bacillus* (*Bacillus*).

Above the genus level most names are plural adjectives in the feminine gender, agreeing with the word *Procaryotae*, so that, for example, *Brucellaceae* means *Procaryotae Brucellaceae*.

PURPOSES OF THE CODES OF NOMENCLATURE

The codes have three main aims:

1. Names should be stable,
2. Names should be unambiguous,
3. Names should be necessary.

These three aims are sometimes contradictory, and the rules of nomenclature have to make provision for exceptions where they clash. The principles are implemented by three main devices: (a) priority of publication to assist stability, (b) establishment of nomenclatural types to ensure the names are not ambiguous, and (c) publication of descriptions to indicate that different names do refer to different entities. These are supported by subsidiary devices such as the Latinized forms of names, and the avoidance of synonyms for the same taxon (see Synonyms and Homonyms later in this section).

PRIORITY OF PUBLICATION

To achieve stability, the first name given to a taxon (provided the other rules are obeyed) is taken as the correct name. This is the **principle of priority**. But to be safeguarded in this way a name obviously has to be made known to the scientific community; one cannot use a name that has been kept secret. Therefore, names have to be published in the scientific literature, together with sufficient indication of what they refer to. This is called **valid publication**. If a name is merely published in the scientific literature, it is called **effective publication**; to be valid it also has to satisfy additional requirements, which are summarized later.

The earliest names that must be considered are those published after an official starting date. For many groups of organisms this is Linnaeus' *Species Plantarum* of 1753, but the difficulties of knowing to what the early descriptions refer, and of searching the voluminous and growing literature, have made the principle of priority increasingly hard to obey.

The *Code* of nomenclature for bacteria, therefore, established a new starting date of 1980, with a new starting document, the Approved Lists of Bacterial Names (Skerman et al., 1980). This list contains names of bacterial taxa that were recognizable and in current use. Names not on the lists lost standing in nomenclature on January 1, 1980, although there are provisions for reviving them if the taxa are subsequently rediscovered or need to be reestablished. To prevent the need to search the voluminous scientific literature, the new provisions for bacterial nomenclature require that for valid publication new names (including new names in patents) must be published in certain official publications. Alternatively, if the new names were effectively published in other scientific publications, they must be announced in the official publications to become validly published. Priority dates from the official publication concerned. At present the only official publication is the *International Journal of Systematic Bacteriology* (now the *International Journal of Systematic and Evolutionary Microbiology*).

NOMENCLATURAL TYPES

To make clear what names refer to, the taxa must be recognizable by other workers. In the past it was thought sufficient to publish a description of a taxon. This has been found over the years to be inadequate. Advances in techniques and in knowledge of the many undescribed species in nature have shown that old descriptions are usually insufficient. Therefore, an additional principle is employed, that of **nomenclatural types**. These are actual specimens (or names of subordinate taxa that ultimately relate to actual specimens). These type specimens are deposited in museums and other institutions. For procaryotes (like some other microorganisms that are classified according to their properties in artificial culture) instead of type specimens, **type strains** are employed. The type specimens or strains are intended to be typical specimens or strains that can be compared with other material when classification or identification is undertaken, hence the word "type". However, a moment's thought will show that if a type specimen has to be designated when a taxon is first described and named, this will be done at a time when little has yet been found out about the new group. Therefore, it is impossible to be sure that it is indeed a typical specimen. By the time a completely typical specimen can be chosen, the taxon may be so well known that a type specimen is unnecessary; no one would now bother to designate a type specimen of a bird so well known as the common house sparrow.

The word "type" thus does not mean it is typical, but simply that it is a **reference specimen for the name**. This use of the word "type" is a very understandable cause for confusion that may well repay attention by the taxonomists of the future. For this reason, the *Code* discourages the use of terms like serotype and recommends instead terms formed from -var, e.g., serovar.

In recent years other type concepts have been suggested. Numerical taxonomists have proposed the hypothetical median organism (Liston et al., 1963), or the centroid; these are mathematical abstractions, not actual organisms. The most typical strain in a collection is commonly taken to be the **centrotype** (Silvestri et al., 1962), which is broadly equivalent to the strain closest to the center (centroid) of a species cluster. Some workers have suggested that several type strains should be designated. Gordon (1967) refers to this as the "population concept". One strain, however, must be the official nomenclatural type in case the species must later be divided. Gibbons (1974b) proposed that the official type strain should be supplemented by reference strains that indicated the range of variation in the species, and that these strains could be termed the "type constellation". It may be noted that some of these concepts are intended to define not merely the center but, in some fashion, the limits of a species. Since these limits may well vary in different ways for different characters, or classes of characters, it will be appreciated that there may be difficulties in extending the type concept in this way. The centrotype, being a very typical strain, has often been chosen as the type strain, but otherwise these new ideas have not had much application to bacterial nomenclature.

Type strains are of the greatest importance for work on both classification and identification. These strains are preserved (by methods to minimize change to their properties) in culture collections from which they are available for study. They are obviously required for new classificatory work, so that the worker can determine if he has new species among his material. They are also needed in diagnostic microbiology, because one of the most important principles in attempting to identify a microorganism that presents difficulties is to compare it with authentic strains of known species. The drawback that the type strain may not be entirely typical is outweighed by the fact that the type strain is by definition authentic.

Not all microorganisms can be cultured, and for some the function of a type can be served by a preserved specimen, a photograph, or some other device. In such instances, these are the nomenclatural types, though it is commonly considered wise to replace them by type strains when this becomes possible. Molecular sequences are increasingly being used as important aspects of organisms, and sometimes they assume the functions of nomenclatural types, although they are not yet explicitly mentioned in the *Code*. Authors should, however, bear in mind the limitations of sequences for distinguishing very closely related organisms.

Sometimes types become lost, and new ones (**neotypes**) have to be set up to replace them; the procedure for this is described

in the *Code*. In the past it was necessary to define certain special classes of types, but most of these are now not needed.

Types of species and subspecies are type specimens or type strains. For categories above the species, the function of the type—to serve as a point of reference—is assumed by a *name*, e.g., that of a species or subspecies. The species or subspecies is tied to its type specimen or type strain.

Types of genera are **type species** (one of the included species) and types of higher names are usually **type genera** (one of the included genera). This principle applies up to and including the category, order. This can be illustrated by the types of an example of a taxonomic hierarchy shown in Table 2.

The type specimen or type strain must be considered a member of the species whatever other specimens or strains are excluded. Similarly, the **type species of a genus must be retained in the genus even if all other species are removed from it**. A type, therefore, is sometimes called a **nominifer** or **name bearer**; it is the reference point for the name in question.

DESCRIPTIONS

The publication of a name, with a designated type, does in a technical sense create a new taxon, insofar as it indicates that the author believes he has observations to support the recognition of a new taxonomic group. But this does not afford evidence that can be readily assessed from the bald facts of a name and designation of a type. From the earliest days of systematic biology, it was thought important to describe the new taxon for two reasons: (a) to show the evidence in support of a new taxon, and (b) to permit others to identify their own material with it—indeed this antedated the type concept (which was introduced later to resolve difficulties with descriptions alone).

It is, therefore, a requirement for valid publication that a description of a new taxon is needed. However, just how full the description should be, and what properties must be listed, is difficult to prescribe.

The codes of nomenclature recognize that the most important aspect of a description is to provide a list of properties that distinguish the new taxon from others that are very similar to it, and that consequently fulfill the two purposes of adducing evidence for a new group and allowing another worker to recognize it. Such a brief differential description is called a **diagnosis**, by analogy with the characteristics of diseases that are associated with the same word. Although it is difficult to legislate for adequate diagnoses, it is usually easy to provide an acceptable one; inability to do so is often because insufficient evidence has been obtained to support the establishment of the new taxon. It is generally unwise to propose a new taxon unless one can provide at least a few properties that distinguish it with good reliability from closely similar taxa.

The *Code* provides guidance on descriptions, in the form of recommendations. Failure to follow the recommendations does not of itself invalidate a name, though it may well lead later workers to dismiss the taxon as unrecognizable or trivial. The code for bacteria recommends that as soon as minimum stan-

TABLE 2. An example of taxonomic types

Category	Taxon	Type
Family	*Pseudomonadaceae*	*Pseudomonas*
Genus	*Pseudomonas*	*Pseudomonas aeruginosa*
Species	*Pseudomonas aeruginosa*	ATTC 10145

dards of description are prepared for various groups, workers should thereafter provide that minimum information; this is intended as a guide to good practice, and should do much to raise the quality of systematic bacteriology. For an example of minimum standards, see the report of the International Committee on Systematic Bacteriology Subcommittee on the Taxonomy of *Mollicutes* (1979).

CLASSIFICATION DETERMINES NOMENCLATURE

The student often asks how an organism can have two different names. The reason lies in the fact that a name implies acceptance of some taxonomy, and on occasion no taxonomy is generally agreed upon. Scientists are entitled to their own opinions on taxonomies; there are no rules to force the acceptance of a single classification.

Thus opinions may be divided on whether the bacterial genus *Pectobacterium* is sufficiently separate from the genus *Erwinia*. The soft-rot bacterium was originally called *Bacterium carotovorum* in the days when most bacteria were placed in a few large genera such as *Bacillus* and *Bacterium*. As it became clear that these unwieldy genera had to be divided into a number of smaller genera, which were more homogeneous and convenient, this bacterium was placed in the genus *Erwinia* (established for the bacterium of fireblight, *Erwinia amylovora*) as *Erwinia carotovora*. When further knowledge accumulated, it was considered by some workers that the soft-rot bacterium was sufficiently distinct to merit a new genus, *Pectobacterium*. The same organism, therefore, is also known as *Pectobacterium carotovorum*. Both names are correct in their respective positions. If one believes that two separate genera are justified, then the correct name for the soft-rot bacterium is *Pectobacterium carotovorum*. If one considers that *Pectobacterium* is not justified as a separate genus, the correct name is *Erwinia carotovora*.

Classification, therefore, determines nomenclature, not nomenclature classification. Although unprofitable or frivolous changes of name should be avoided, the freezing of classification in the form it had centuries ago is too high a price to pay for stability of names. Progress in classification must reflect progress in knowledge (e.g., no one now wants to classify all rod-shaped bacteria in *Bacillus*, as was popular a century ago). Changes in name must reflect progress in classification; some changes in name are thus inevitable.

CHANGES OF NAME

Most changes in name are due to moving species from one genus to another or dividing up older genera. Another cause, however, is the rejection of a commonly used name because it is incorrect under one or more of the Rules. A much-used name, for example, may not be the earliest, because the earliest name was published in some obscure journal and had been overlooked. Or there may already be another identical name for a different microorganism in the literature. Such problems are now rare because of the Approved Lists and the lists of new names in the *International Journal of Systematic Bacteriology* (see Proposal of New Names). Changes can be very inconvenient if a well-established name is found to be illegitimate (contrary to a Rule) because of a technicality. The codes of nomenclature therefore make provision to allow the organizations that are responsible for the codes to make exceptions if this seems necessary. A name thus retained by international agreement is called a **conserved name**, and when a name is conserved the type may be changed to a more suitable one.

When a species is moved from one genus into another, the specific epithet is retained (unless there is by chance an earlier name that forms the same combination, when some other epithet must be chosen), and this is done in the interests of stability. The new name is called a **new combination**. An example has been given above. When the original *Bacterium carotovorum* was moved to *Erwinia*, the species name became *Erwinia carotovora*. The gender of the species epithet becomes the same as that of the genus *Erwinia*, which is feminine, so the feminine ending, -*a*, is substituted for the neuter ending, -*um*.

NAMES SHOULD BE NECESSARY

The codes require that names should be necessary, i.e., there is only one correct name for a taxon in a given or implied taxonomy. This is sometimes expressed by the statement that an organism with a given position, rank, and circumscription can have only one correct name.

NAMES ARE LABELS, NOT DESCRIPTIONS

In the early days of biology, there was no regular system of names, and organisms were referred to by long Latin phrases that described them briefly, such as *Tulipa minor lutea italica folio latiore*, "the little yellow Italian tulip with broader leaves". The Swedish naturalist Linnaeus tried to reduce these to just two words for species, and in doing so he founded the present binominal system for species. This tulip might then become *Tulipa lutea*, just "the yellow tulip". Very soon it would be noted that a white variant sometimes occurred. Should it then still be named "the yellow tulip"? Why not change it to "the Italian tulip"? Then someone would find it in Greece and point out that the record from Italy was a mistake anyway. Twenty years later an orange or yellow form would be found in Italy after all. Soon the nomenclature would be confused again.

After a time it was realized that the original name had to be kept, even if it was not descriptive, just as a man keeps his name Fairchild Goldsmith as he grows older, and even if he becomes a farmer. The scientific names or organisms are today only labels, to provide a means of referring to taxa, just like personal names.

A change of name is therefore only rarely justified, even if it sometimes seems inappropriate. Provisions exist for replacement when the name causes great confusion.

CITATION OF NAMES

A scientific name is sometimes amplified by a citation, i.e., by adding after it the author who proposed it. Thus the bacterium that causes crown galls is *Agrobacterium tumefaciens* (Smith and Townsend) Conn. This indicates that the name refers to the organism first named by Smith and Townsend (as *Bacterium tumefaciens*, in fact, though this is not evident in the citation) and later moved to the genus *Agrobacterium* by Conn, who therefore created a new combination. Sometimes the citation is expanded to include the date (e.g., *Rhizobium*, Frank 1889), and more rarely to include also the publication, e.g., *Proteus morganii* Rauss 1936 *Journal of Pathology and Bacteriology* Vol. 42, p. 183.

It will be noted that citation is only necessary to provide a suitable reference to the literature or to distinguish between inadvertent duplication of names by different authors. A citation is not a means of giving credit to the author who described a taxon; the main functions of citation would be served by the bibliographic reference without mentioning the author's name. Citation of a name is to provide a **means of referring** to a name, just as a name is a means of referring to a taxon.

Synonyms and homonyms

A homonym is a name identical in spelling to another name but based on a different type, so they refer to different taxa under the same name. They are obviously a source of confusion, and the one that was published later is suppressed. The first published name is known as the **senior homonym**, and later published names are **junior homonyms**. Names of higher animals and plants that are the same as bacterial names are not treated as homonyms of names of bacteria, but to reduce confusion among microorganisms, bacterial names are suppressed if they are junior homonyms of names of fungi, algae, protozoa, or viruses.

A synonym is a name that refers to the same taxon under another scientific name. Synonyms thus come in pairs or even swarms. They are of two kinds:

1. **Objective synonyms** are names with the same nomenclatural type, so that there is no doubt that they refer to the same taxon. These are often called nomenclatural synonyms. An example is *Erwinia carotovora* and *Pectobacterium carotovorum*; they have the same type strain, American Type Culture Collection strain 15713.
2. **Subjective synonyms** are names that are believed to refer to the same taxon but that do not have the same type. They are matters of taxonomic opinion. Thus *Pseudomonas geniculata* is a subjective synonym of *Pseudomonas fluorescens* for a worker who believes that these taxa are sufficiently similar to be included in one species, *P. fluorescens*. They have different types, however (American Type Culture Collection strains #19374 and 13525, respectively), and another worker is entitled to treat them as separate species if he or she so wishes.

There are senior and junior synonyms, as for homonyms. The synonym that was first published is known as the **senior synonym**, and those published later are **junior synonyms**. Junior synonyms are normally suppressed.

Proposal of new names

The valid publication of a new taxon requires that it be named. The *Code* insists that authors should make up their minds about the new taxon; if they feel certain enough to propose a new taxon with a new name, then they should say they do so propose; if they are not sure enough to make a definite proposal, then the name of their taxon will not be afforded the protection of the *Code*. They cannot expect to suggest provisional names—or possible names, or names that one day might be justified—and then expect others to treat them as definite proposals at some unspecified future date. How can a reader possibly know when such vague conditions have been fulfilled?

If a taxon is too uncertain to receive a new name, it should remain with a vernacular designation (e.g., the marine form, group 12A). If it is already named, but its affinities are too uncertain to move it to another genus or family, it should be left where it is. There is one exception, and that is that a new species should be put into some genus even if it is not very certain which is the most appropriate, or if necessary a new genus should be created for it. Otherwise, it will not be validly published, it will be in limbo, and it will be generally overlooked, because no one else will know how to index it or how to seriously consider it. If it is misplaced, it can later be moved to a better genus. Names of procaryotic genera should not end in -myces, -phyces, -phyta, or -virus to avoid confusion with mycology, botany, or virology.

The formation of names is considered at length by Trüper in the accompanying section on Etymology in Nomenclature of Procaryotes. This gives advice on Latinization. He recommends that names should be short and easy to pronounce and should be formed from Latin or Greek roots where possible. He discusses the difficulties of forming names of taxa from the names of persons. Authors should refrain from naming taxa after themselves.

The basic needs for publication of a new taxon are four: (a) the publication should contain a new name in proper form that is not a homonym of an earlier name of bacteria, fungi, algae, protozoa, or viruses; (b) the taxon name should not be a synonym of an earlier taxon name; (c) a description or at least a diagnosis should be given; and (d) the type should be designated. A new species is indicated by adding the Latin abbreviation *sp. nov.*, a new genus by *gen. nov.*, and a new combination by *comb. nov.* The most troublesome part is the search of the literature to cover the first two points. This is now greatly simplified for bacteria, because the new starting date means that one need search only the Approved Lists of Bacterial Names and the issues of the *International Journal of Systematic Bacteriology* from January, 1980, onward for all validly published names that have to be considered. This task is made easier by the periodic cumulative updating of names in the *International Journal of Systematic Bacteriology* (e.g., Moore and Moore, 1989) and by the increasing availability of electronic online listings (e.g., Euzéby at Web site www-sv.cict.fr/bacterio/ and by the DSMZ, Braunsweig, Germany). However, the new name has to be published in that journal, with its description and designation of type, or, if published elsewhere, the name must be announced in that journal to render it validly published.

The proposed biocode

In recent years there has been growing awareness in botany and zoology of the problems for nomenclature from the huge numbers of new organisms that are being discovered. The different biological disciplines, therefore, have started the process of unifying the nomenclature of all living organisms, and a proposal for a universal BioCode is being actively pursued. A draft has been published (Greuter et al., 1998), which is now being studied by the organizations responsible for the codes for animals, plants, microorganisms, cultivated plants, and viruses. The aim of the BioCode is to introduce changes for names of taxa published at some date after January 1, 2000.

These proposals are at present only recommendations until the reforms are complete and widely accepted. The present codes of nomenclature will continue to operate in their own subject areas but will be revised to implement the provisions of the BioCode. The International Union of Microbiological Societies (which is the body ultimately responsible for the *Bacteriological Code*) is, in principle, in favor of this development, but the practical implementation will take some time. Nevertheless, it would be wise for microbiologists to take account of the main proposals.

Registration of new names for all organisms will be introduced by mechanisms similar to those in the *Bacteriological Code*. The main differences from that *Code* can be summarized as follows:

1. Phylum will replace division (the category below kingdom and above class).
2. Provision is made for numerous intercalations, with prefixes supra-, sub-, and infra-.
3. Nomenclature types will not be living specimens, although type strains in the form of viable but metabolically inactive organisms are acceptable.

4. Generic homonyms will be prohibited across all organisms. At present generic names of animals can be the same as those of plants (thus, *Pieris* is a genus of butterflies and a genus of ericaceous plants). Whether this is practicable remains to be seen. It will be easier to achieve when lists of genus names of plants and animals are more complete and are available in electronic form. The two serial publications, *Index Zoologicus* and *Index Nomina Genericorum Plantarum*, are widely available to check animal and plant genus names. The *Bacteriological Code* already prohibits homonyms among procaryotes, fungi, algae, protozoa, and viruses, as noted earlier.

5. There will be some complex rules on the use of synonyms extending above the genus to the rank of family. These are unfamiliar to bacteriologists, and it is not clear how readily they will be accepted.

6. There will be changes in the formal usage of certain terms. Thus, *effective publication* in bacteriology will become simply *publication* and *valid publication* will become *establishment by registration*. *Legitimate names* will become *acceptable names*. Synonyms will be *homotypic* and *heterotypic* instead of *objective* and *subjective*, respectively. *Priority* will become *precedence*, and senior and *junior* names will become *earlier* and *later* names.

7. Prohibition of genus names ending in -myces, -phyces, -phyta, and -virus has been mentioned earlier.

It is evident that revision of the *Bacteriological Code* will be required to achieve the aims of the BioCode, although it will often be possible to make exceptions for bacteriological work. It is to be hoped that such revision will ultimately lead to a version expressed in language familiar to bacteriologists and illustrated by examples from this discipline.

Etymology in Nomenclature of Procaryotes

Hans G. Trüper

I. INTRODUCTION

A. *Introductory remark* When I was invited to write this chapter I felt flattered. I have always been interested in names, in etymology and semantics. The invitation was probably due to more than 25 years of active membership in the International Committee for Systematic Bacteriology (ICSB) and in the Editorial Board of the *International Journal of Systematic Bacteriology*, and there especially my self-adopted task of watching the correctness of new Latin names by offering advice in etymology and questions of procaryote nomenclature. What I write hereafter is an outflow of the experiences I have gathered in these tasks including correspondence in etymological (often intertwined with nomenclatural) matters with hundreds of colleagues. Therefore, I shall try to write this chapter from the viewpoint of the microbiologist—as a user; for the user—rather than writing it *ex cathedra* as a classicist might want to do. Further, what I write here are my own opinions on these matters and they are not meant to offend anyone who has other or better insights.

B. *The Latin/Greek thesaurus of words and word elements* Scientific terminology, both in technical terminology and in nomenclature, has to fulfill requirements other than those of everyday language. These requirements have been excellently described by the late Fritz C. Werner (1972), a German zoologist.

The first requirement is that every term must unambiguously circumscribe a clearly conceivable idea and that every name stands for a special object or a special group of objects characterized by determined features.

The second requirement is that the total number of different words and word combinations must exceed the large number of discernible objects and abstract concepts, thereby ensuring that names are unambiguous. This is a real challenge as the number of objects, processes and concepts is continuously growing both in depth and breadth because of new scientific and social developments, and changes in nature due to human activities.

As more scientists from a wider range of nationalities participate in these developments, it is important that scientific terms and names fulfill a third requirement, namely universal comprehensibility.

These three requirements—unambiguousness, a large number of possible combinations, and universal usage—are met, to a high degree, by the fact that the terminology of natural sciences and medicine is largely based on the lexicon of classical Greek and Latin. The fact that these so called "dead" languages no longer undergo natural and living changes makes their word material a thesaurus that has been used and may be used further for contemporary needs. Consequently one has more or less

arbitrarily given these classical words and word elements certain new meanings. Using a living and constantly changing language in this way would promptly lead to problems and misunderstandings.

Firstly, the use of ancient word material allows the naming of the many new and—in their numbers—permanently increasing objects and concepts for which there are no respective words in contemporary spoken languages; even circumscriptions and combinations of words would hardly suffice. Latin and Greek offer a wealth of word elements and ways to form words that remain inexhausted thus far and are likely to serve our needs for a long time in the future, although scientists have not always been careful or reasonable in their "creations". By mixing Greek and Latin elements, by dropping syllables, repositioning letters, contracting words and creating arbitrary formations, the antique wealth of words has been changed, at times rather significantly. Furthermore, many other languages have contributed, and the names of scientists and other persons have been latinized.

What Werner (1972) did not emphasize was the fact that Latin remained the international language (*lingua franca*) of philosophy, religion, law, sciences, and politics throughout the European Middle Ages and the Renaissance and for philosophical and scientific publications up into the nineteenth century. Its usage, although limited to these circles, led to an enormous increase in vocabulary, usually adopted from other European or oriental languages (e.g., Arabic). It also needs to be mentioned here that Latin has remained the spoken language in the center of the Catholic Church, the Vatican, and is likely to be so into the future. This is particularly well documented by the fact that the *Libraria Editoria Vaticana* takes all efforts to integrate new Latin words coined for modern objects and concepts into the written and spoken Latin of the Vatican. The *Lexicon Recentis Latinitatis*, that appeared 1992 in Italian and 1998 in German, contains about 15,000 new Latin words, "from astronaut to zabaione", word combinations and circumscriptions of the fields of sciences, technology, religion, medicine, politics, sports, and even common idiomatic terms.

The thesaurus of words, enlarged this way, is thus no longer identical with that of either classical language but represents "something new" that has developed along historical lines and follows special contemporary laws of language.

All of the statements made by Werner (1972) apply to general scientific and medical terminology as well as to biological nomenclature. And they apply especially to the scientific nomenclature of procaryotes (eubacteria and archaebacteria) and viruses because these—in contrast to most animals, plants, and larger fungi—do not have popular or vernacular names in any living language because of their usual invisibility.

Nomenclature ("the system of names used in a branch of learning or activity") is an indispensable tool for correct information in our fast growing scientific world with its rapidly developing information networks. The binomial nomenclature used in biosystematics goes back to 1735 when the Swedish botanist Carolus Linnaeus (Karl von Linné, 1707–1778, ennobled 1757) published his famous "Systema Naturae" in Latin according to the scholarly habits of his times.

By introducing the species concept and the use of Latin and Greek for the names of living beings, Linnaeus laid down the principles of modern biological systematics as well as nomenclature. In our "age of informatics" one could certainly think of other ways to name the vast number of plants, fungi, animals, protists and procaryotes, perhaps by a number and/or letter code. For the human brain, however, names are still easier to memorize and work with as part of a system, as long as they are readable and pronounceable.

For the scientific names of procaryotes the *International Code of Nomenclature of Bacteria* (*ICNB, Bacteriological Code*), issue of 1992, is the compulsory compendium of governing Rules. It is the task of the accompanying chapter on nomenclature by P.H. A. Sneath to explain the *Bacteriological Code* (ICNB), whereas this chapter is intended to deal with etymology. Etymology means "origin and historical development of a word, as evidenced by study of its basic elements, earliest known use, and changes in form and meaning" or "the semantic derivation and evolution of a word". "Etymology" is derived from Greek etymon, "the truth" and thus aims at the true, the literal sense of a word.

Etymology is a necessary element in biological nomenclature as it explains the existing (i.e., so far given) names and helps to form new names. For the average microbiologist, "etymology" is that part of a species or genus description that stands first, describes the accentuation, origin and meaning of the name, contains a lot of strange abbreviations and is often considered as superfluous or nasty. I shall come to appropriate examples at the end of the chapter.

In 1993, the late professor of classical languages, Thomas Ozro MacAdoo of Blacksburg, VA, U.S.A., wrote a marvelous chapter on "Nomenclatural literacy" (MacAdoo, 1993) with the intention of helping bacteriologists form correct names. MacAdoo carefully described and examplified the five Latin declensions, the Greek alphabet and its Latin equivalents, the Greek declensions and their Latin equivalents, adjectives and participles, compounding in Latin and Greek, and the latinization of modern proper names. It cannot and will not be my task to equal this excellent and scholarly piece of work, as it contains an introduction to the two classical languages and requires a basic knowledge of, at least, Latin grammar. I highly recommend reading, or better studying, MacAdoo's paper. But I am afraid that I cannot agree with him on the way personal names should be latinized nowadays. (Additional literature recommended as etymological help for the formation of new bacterial names is marked by an asterisk in the further reading list.)

C. Pronunciation and accentuation   For many bacterial names the current common pronunciation differs from the pronunciation that is correct according to Latin rules (cf. common text books for Latin). It is unfortunately strongly influenced by the speaker's mother tongue, a clear indication that Latin is no longer the *lingua franca* of the scientific world. Whereas native speakers of languages that are written close to phonetics, such as Italian, Spanish, Portuguese, Dutch, or German, usually pronounce Latin close to its spelling, native speakers of French and especially of English (languages pronounced rather differently from their spelling) often pronounce Latin according to the pronunciation rules of their languages, i.e., further away from the written form. These differences in pronunciation are not generally that important as differences in spelling, because the name in question is often understood despite differences in pronunciation. Substantially helpful here, however, could be to pronounce at least the vowels as they are pronounced in Spanish and Italian, languages whose pronunciations stayed close to their Latin origin. International science will have to live with this problem until the day when all languages are written according to phonetic rules.

In many Central European high schools Latin pronunciation has gone back to the times of Caesar and Augustus when the Romans always pronounced the letter c as the sound k. As a consequence students pronounce, e.g., Caesar "Kaesar" (origin of the German word Kaiser which means emperor) or Cicero "Kikero". In bacteriology this leads to alternate pronunciations of *Acinetobacter, Acetobacter*, etc. (as akinetobakter, aketobakter, etc.) by some younger European microbiologists.

I consider it a pity that, for scientific terms used mainly in chemistry and physics, the writing of Greek k remained (keratin, kinetics) whereas in biological nomenclature it has usually, but not always, been latinized to c (*Triceratops, Acinetobacter*). Fortunately, classical Latin already introduced the Greek z for transliterated Greek words, and Medieval Latin introduced the letter j for the consonantic i. Meanwhile several names of bacteria starting with J have been proposed (e.g., *Janthinobacter* and the specific epithet *jejuni*). It makes sense to use the j in Latin names as the first letter of a word or word element when it is followed by a vowel.

One significant problem with pronunciation is that of some personal or geographical names used in generic names or specific epithets, e.g., the bacterial generic name *Buttiauxella*, named after the French microbiologist Buttiaux (pronounced: "buttio"). This generic name and specific epithets like *"bordeauxensis", "leicesterensis"*, or *"worcesterensis"* may be pronounced fully (as Latin would require) or pronounced as though they were spelled "buttioella", "bordoensis", "lesterensis", "woosterensis". I am afraid that we will have to leave the decision of pronunciation in such cases to the single scientist, as a rule for such "problems" seems rather difficult to conceive.

Frequently accentuation of Latin names appears to pose problems, especially when Greek word elements are involved. Here, the correct classical accentuation is often not used in bacterial names, e.g., the accepted accentuation of the name *Pseudomonas* is pseu-do-mo´-nas, whereas the classical Greeks would have accentuated the word pseu-do´-mo-nas. An almost universal guideline for accentuation of generic names is, that the syllable next to the last bears the accent. Although this holds for most specific epithets as well, we do tend to encounter other accentuations more often. The practical sense of natural scientists should prevail and the present common usage of accentuation in bacterial names should be the guideline.

II. FORMATION OF GENERIC NAMES AND SPECIFIC EPITHETS

Since Linnaeus, biological species bear binomial names, consisting of a *genus* (kind) and a *species* (appearance) name. The latter, if taken by itself, is called "specific epithet". A complete species name thus consists of the genus name and the specific epithet. In principle the language of biological nomenclatural names is Latin. In nomenclature, words of Greek origin as well as those of any other origin are handled as Latin, i.e., they have to be "latinized".

Only those bacterial names contained in the *Approved Lists of Names* (Skerman et al., 1980) and the *Validation Lists* that regularly appear in the *International Journal of Systematic Bacteriology* have standing in nomenclature. Regularly updated non-official lists of legitimate bacterial names (except for cyanobacteria described under the Botanical Code) are published by the German Culture Collection DSMZ, Braunschweig, Germany, twice a year. Dr.J.P. Euzéby, Toulouse, France, provides an even more detailed non-official list electronically on the Web site www- sv.cict.fr/bacterio/.

A. Compound names Compound names are formed by combining two or more words or word elements of Latin and/or Greek origin into one generic name or specific epithet. In most cases two word elements are used (e.g., *Thio/bacillus, thio/parus*), but up to four elements may be found (e.g., *Ecto/thio/rhodo/spira*).

In principle the formation of such combined or compound names is not at all difficult. There are four basic rules to be followed:

1. Except for the last word element, only the stems are to be used.
2. The connecting vowel is -o- when the preceding element is of Greek, it is -i- when the preceding element is of Latin origin.
3. A connecting vowel is dropped when the following element starts with a vowel.
4. Hyphens are not allowed.

In order to avoid later changes, these recommendations (cf. *Bacteriological Code*, Appendix 9 [Lapage et al., 1992]; Trüper, 1996) should be strictly followed, i.e., they should be considered as rules without exceptions.

The reader may protest here and mention, e.g., *Lactobacillus* as being against this ruling. *Lactibacillus* would indeed be the correct name, however, the name *Lactobacillus* is much older than the *Bacteriological Code* and has become a well established name. The ending *-phile* (or *-philic*) in English is often added to words of Latin origin connected by -o- (e.g., acidophile, francophile, anglophile, nucleophile, lactophile etc.). This is due to the meaning of -phile, "friendly to", which commands the dative case. In the most common Latin declension, the second, the dative is formed by adding an -o to the stem (acidophile, friendly to whom/what? friendly to acid). Therefore in bacteriology we have a number of older compound names of Latin origin with the connecting vowel -o-. By unknowingly taking over such originally dative-derived word elements ending on -o, names like *Lactobacillus* came into existence. Such cases prove that Appendix 9 of the *Bacteriological Code* (Lapage et al., 1992) does not have the power of a Rule yet. In the future new name formations of that kind should be avoided.

There are numerous mistakes with respect to compound names. Sometimes authors want to express that their new organism was isolated from a certain part of an animal's body, e.g., from the throat of a lion; throat is *pharynx* (Greek word stem: *pharyng-*), lion is *leo* (Latin word stem *leon-*). These stems may be correctly combined in two ways: "*pharyngoleonis*" or "*leonipharyngis*". Unfortunately the authors chose *leopharyngis*, which may

be corrected to the latter. This example demonstrates the different connecting vowels as well. Two more examples may emphasize the importance of word stems: so *Obesumbacterium* should be corrected to *Obesibacterium*, as the Latin stem of the first component is *obes-*, and the connecting vowel must be -i-. The generic name *Carbophilus* was formed the wrong way, because the stem of the first component is *carbon-*; the correct name would be *Carboniphilus*. For those scientists without training in Latin, a good Latin dictionary indicates the genitive of a noun thereby allowing them to identify the stem of a Latin noun. Typically, the genitive usually shows the stem (e.g., *carbo, carbonis*, the coal) well. MacAdoo (1993) gives a very useful overview on word stems and declensions for non-classicists. An excellent pocket book on word elements (stems) of Latin and Greek origin for usage in scientific terms and names was published by Werner (1972). However, it has only appeared in German to date. An English translation would be of great value for biologists world wide.

Other typical, yet well established misnomers whose connecting vowels were not dropped include *Acetoanaerobium, Cupriavidus, Haloanaerobacter, Haloanaerobium, Haloarcula, Pseudoalteromonas, Streptoalloteichus, Thermoactinomyces, Thermoanaerobacter, Thermoanaerobacterium*, not to speak of numerous equally malformed specific epithets.

B. Generic names The name of a genus (or subgenus) is a Latin noun (substantive) in the nominative case. If adjectives or participles are chosen to form generic names they have to be transformed into substantives (nouns) and handled as such.

Both Latin and Greek recognize three genders of nouns: masculine, feminine, and neuter. Adjectives associated with nouns follow these in gender. For the correct formation of specific epithets (as adjectives) it is therefore necessary to know the gender of the genus name or of its *last* component, respectively.

The more frequent last components in compound generic names of masculine gender are: *-arcus, -bacillus, -bacter, -coccus, -ferax, -fex, -ger, -globus, -myces, -oides, -philus, -planes, -sinus, -sipho, -vibrio*, and *-vorax*; of feminine gender: *-arcula, -bacca, -cystis, -ella, -ia, -illa, -ina, -musa, -monas, -opsis, -phaga, -pila, rhabdus* (*sic*), *-sarcina, -sphaera, -spira, -spina, -spora, -thrix* , and *-toga*; of neuter gender: *-bacterium, -bactrum, -baculum, -bium, - filamentum, -filum, -genium, -microbium, -nema, -plasma, -spirillum, -sporangium*, and *-tomaculum*.

C. Specific epithets As demanded by Rule 12c of the *Bacteriological Code*, the specific (or subspecific) epithet must be treated in one of the three following ways:

1. as an adjective that must agree in gender with the generic name.
2. as a substantive (noun) in apposition in the nominative case.
3. as a substantive (noun) in the genitive case.

Correct examples of these three ways are *Staphylococcus aureus* (adjective: "golden"), *Desulfovibrio gigas* (nominative noun: "the giant"), and *Escherichia coli* (genitive noun: "of the *colum*/colon"), respectively.

1. ADJECTIVES AND PARTICIPLES AS SPECIFIC EPITHETS Latin adjectives belong to the first, second, and third declension. Those of the first and second declension have different endings in the three genders, whereas in the third declension the situation is much more complicated, as there are adjectives that don't change with gender, others that do, and those that are identical in the masculine and feminine gender and different in the neu-

ter. Table 1 gives some representative examples. Note also that comparative adjectives are listed. I recommend always checking an adjective in the dictionary before using it in the formation of a name.

Participles are treated as if they were adjectives, i.e., they fall under Rule 12c, (2), of the *Bacteriological Code*. Infinitive (also named "present") participles in the singular do not change with gender. According to the four conjugations of Latin they end on *-ans* (e.g., *vorans* devouring, from *vorare* to devour), *-ens* (e.g., *delens* destroying, from *delere* to destroy, *deleo* I destroy), *-ens* (e.g., *legens* reading, from *legere* to read, *lego* I read), *-iens* (e.g., *capiens*, from *capere* to seize, *capio* I seize), *-iens* (e.g., *audiens*, from *audire* to listen, *audio* I listen). Note that the knowledge of the ending of the first person singular in the present is decisive!

Perfect participles change their endings with gender and are handled like adjectives of the first and second declension, e.g., *voratus, vorata, voratum* devoured, *deletus, deleta, deletum* destroyed, *lectus, lecta, lectum* (irregular) read, *captus, capta, captum* (irregular) seized, *auditus, audita, auditum*, listened/heard.

2. NOMINATIVE NOUNS IN APPOSITION AS SPECIFIC EPITHETS While the above mentioned first and third ways to form specific epithets are generally well understood and usually do not pose problems, the formation of epithets as substantives in apposition has obviously been misunderstood in several cases. So, for instance, when the name *Mycoplasma leocaptivus* was proposed for an isolate from a lion held in captivity, the authors, probably unintentionally, called their bacterium "the captive lion", whereas they wanted rather to explain the origin of their isolate "from a captive lion". Thus "*captivileonis*" would have been the correct epithet.

A nominative noun in apposition does not just mean that any nominative noun may be added to the generic name to automatically become its acceptable epithet. In grammar, apposition means "the placing of a word or expression beside another so that the second explains and has the same grammatical construction as the first"; i.e., the added nominative noun has an explanatory or specifying function for the generic name, like in general English usage "the Conqueror" has for "William" in "William, (called) the Conqueror". Thus *Desulfovibrio gigas* may be understood as *Desulfovibrio dictus gigas* and translated as "*Desul-*

fovibrio, called the giant", which, with reference to the unusual cell size of this species, makes sense.

Because all specific epithets ending with the Latin suffixes *-cola* (derived from *incola*, "the inhabitant, dweller") and *-cida* ("the killer") fulfill the above-mentioned requirement, they are to be considered correct.

Most legitimate specific epithets formed in bacteriology as nominative nouns in apposition so far have been mentioned and, where necessary, corrected recently (Trüper and de'Clari, 1997, 1998).

Although they are not explicitly ruled out by the *Bacteriological Code*, I have not yet encountered tautonyms, i.e., specific epithets identical with and repeating the genus name, in bacterial nomenclature (such as in zoology *Canis canis*, the dog). In order to avoid confusion, it would be wise to abstain from proposing such names.

3. GENITIVE NOUNS AS SPECIFIC EPITHETS   The formation of specific epithets as genitive nouns rarely poses problems, as the singular genitive of substantives (nouns) is usually given in the dictionaries.

If the plural genitive is preferred, as, e.g., in *Rhizobium leguminosarum* ("of legumes"), one has to find out the declension of the noun, as plural genitives are different in different declensions. This question will be addressed below.

D. *Formation of bacterial names from personal names*   Persons may be honored by using their name in forming a generic name or a specific epithet. This is an old custom in the whole area of biology. The *Bacteriological Code*, however, strongly recommends to refrain from naming genera (including subgenera) after persons quite unconnected with bacteriology or at least with natural science (Recommendation 10a) and in the case of specific epithets to ensure that, if taken from the name of a person, it recalls the name of one who discovered or described it, or was in some way connected with it (Recommendation 12c).

It is good style to ask the person to be honored by a scientific name for permission (as long as she/he is alive). Authors should refrain from naming bacteria after themselves or co-authors after each other in the same publication, as this is considered immodest by the majority of the scientific community.

The *Bacteriological Code* provides only two ways to form a ge-

TABLE 1. Examples of Latin adjectives

	Masculine	Feminine	Neuter	English translation
first and second declension:	bonus[a]	bona	bonum	good
	aureus[a]	aurea	aureum	golden
	miser	misera	miserum	wretched
	piger	pigra	pigrum	fat, lazy
	ruber	rubra	rubrum	red
	pulcher	pulchra	pulchrum	beautiful
third declension:	puter	putris	putre	rotten
	celer	celeris	celere	rapid
	facilis[a]	facilis	facile	easy
	facilior	facilior	facilius	easier
	maior	maior	maius	more
	minor	minor	minus	less
	simplex	simplex	simplex	simple
	egens	egens	egens	needy

[a]Most common types.

neric name from a personal name, either directly or as a diminutive: Both are always in feminine gender.

Appendix 9 of the ICBN recommends how such names should be formed. Appendix 9 has, however, not the power of the Rules.

The application of the classical Roman rules for name-giving, as was done by MacAdoo (1993), does not make sense as modern names worldwide follow different and various rules and regulations. A differentiation in *prenomina, nomina,* and *cognomina* is therefore no longer applicable and should not be used as a basis for latinization of names nowadays. Principally, modern family names are either *nomina* or *cognomina* in the classical sense. Continuing latinization of names as practiced in ancient Rome would have the advantage that the practice would not change over time. Rather, it would remain fixed. Therefore MacAdoo (1993) would have preferred to establish a uniform rule for latinization of names. But attention must be paid to the fact that since classical times throughout the Middle Ages up into the nineteenth century, (usually learned) people of others than the Roman nation have latinized their names, and thus several varieties of latinization have developed and must be considered as historically evolved. Thus, if such names are not incorrect, they cannot be denied or refused under the *Bacteriological Code* (Appendix 9). I have therefore tried to give the recommended rulings of Appendix 9 (adopted as editorial policy by the *Bergey's Manual* Trust) a simpler and clearer wording and have given examples according to those latinizations that have historically precedence (Trüper, 1996). The results were revised and are compiled in Table 2.

Some personal names in Europe were already latinized before 1800 and kept since then. If they end on *-us,* replace the ending by *-a* or *-ella* (diminutive) respectively (e.g., the name Bucerius would result in "*Buceria*" or "*Buceriella*"). Beware, however, of Lithuanian names like Didlaukus, Zeikus etc.! These are not latinized but genuine forms and would receive the ending *-ia* according to Table 2.

No more than one person can be honored in a given generic name or epithet. In the case of the Brazilian microbiologist Henrique da Rocha Lima, the generic name *Rochalimaea* was formed by dropping the particle *da* and combining his two family names. Combinations of the names of two or more persons cannot be constructed under this aspect. Here the only possibility would be the provision of the *Bacteriological Code* for forming "arbitrary names". These are treated below.

If an organism is named after a person, the name cannot be shortened, e.g., "*Wigglesia*" after Wigglesworth, "*Stackia*" after Stackebrandt or "*Goodfellia*" after Goodfellow etc., but must fully appear. Certainly titles (*Sir, Lord, Duke, Baron, Graf, Conte,* etc.) and particles (*de, da, af, van, von,* etc.) indicating nobility or local origin of the family should not be included in bacterial names, although they may belong to the name according to the laws of the respective country.

Rarely, generic names or specific epithets have been formed from forenames (first names, given names, Christian names), i.e., not from the family name, so the genus *Erwinia* was named after the American microbiologist Erwin F. Smith. The first name Elizabeth appears in *Bartonella* (formerly *Rochalimaea*) *elizabethae.* One could imagine that, in avoiding the usually long Thai family names first names should be chosen in respective cases. Also unusually long double (hyphenated) names like the (hypothetical) Basingstoke-Thistlethwaite or Saporoshnikov-Shindlefrink hopefully do not occur so often among microbiologists as to be honored by a bacterial name (hyphens are not allowed, anyhow!).

One could think of a simplified standard procedure to ease formation of generic names from personal names:

1. All names ending on consonants or *-a* receive the ending *-ia,* all others the ending *-a.*
2. Diminutive formation: All names ending on consonants receive the ending *-ella,* all names ending on vowels receive the ending *-nella.*

This simplified scheme should perhaps be recommended by the *Bacteriological Code* as an optional alternative to Appendix 9. Such a ruling should, however, not be introduced with retroactive power as Principle 1 of the *Bacteriological Code* aims at constancy of names.

TABLE 2. Ways to form generic names from personal names (names in quotation marks are hypothetical)

Personal name ending on	Add ending	Person	Example (direct formation)	Diminuitive ending	Example (diminutive formation)
-a	-ea	da Rocha Lima	Rochalimaea	drop a, add -ella	"Rochalimella"
	-a	Benecke	Beneckea	-lla	"Beneckella"
-e	-ia	Burke	Burkeia	-lla	"Burkella"
-i	-a	Nevski	Nevskia	-ella	"Nevskiella"
	-a	Beggiato	Beggiatoa	-nella	"Beggiatonella"
-o	-nia	Cato	"Catonia"	-nella	Catonella
-u	-ia	Manescu	"Manescuia"	-ella	"Manescuella"
-y	-a	Deley	Deleya	-ella	"Deleyella"
	-a	Buchner	Buchnera	-ella	"Buchnerella"
-er	-ia	Lister	Listeria	-iella	"Listeriella"
any consonant	-ia	Cabot	"Cabotia"	-(i)ella	"Cabot(i)ella"
		Wang	"Wangia"	-(i)ella	"Wang(i)ella"
		Salmon	"Salmonia"	-ella	Salmonella
		Escherich	Escherichia	-(i)ella	"Escherich(i)ella"
		Zeikus[a]	"Zeikusia"	-(i)ella	"Zeikus(i)ella"

[a]This name of Lithuanian origin is not a genuine latinized name. If it were so, the genus names "Zeikia" or "Zeik(i)ella" might have been possible.

To form specific epithets from personal names there are, in principle, two possibilities: the adjective form and the genitive noun form. The adjective form has no means of recognizing the sex of the honored person, which, in principle is not necessary for nomenclatural purposes. The personal names receive appropriate endings according to the gender of the generic name as indicated in Table 3. Thus an adjective epithet is formed that has the meaning of "pertaining/belonging to . . . (the person)".

When the genitive of a latinized personal name is formed for a specific epithet, the sex of the person to be honored may be taken into consideration as indicated in Table 4.

On the basis of classical, medieval, and modern usage any of the forms of latinization listed in Table 4 may be chosen. As evident from Table 4 the formation of specific epithets from personal names as genitive nouns poses certain problems only with names ending on -a and -o.

Classical Roman names of male persons like Agrippa, Caligula, Caracalla, Galba, Seneca, etc. (predominantly *cognomina*) were used in the first declension like the masculine nouns *poeta* (the poet), *nauta* (the sailor), or *agricola* (the land dweller, farmer), regardless of the fact that most of the nouns in this declension are of feminine gender. If bacteria would have been named after these gentlemen, their specific epithets were *agrippae*, *caligulae*, *caracallae*, *galbae*, and *senecae*, respectively. I think that Volta, Migula, and Komagata are dignified successors in this row.

If authors consider it necessary to indicate the sex of the person to be honored, there are several choices, in the following exemplified by the Japanese name Nakamura:

1. Mr. Nakamura is latinized to Nakamuraus, resulting in a specific epithet "*nakamurai*".
2. Mr. Nakamura is latinized to Nakamuraeus (like Linnaeus or my ancestors Nissaeus and Molinaeus), resulting in a specific epithet "*nakamuraei*".
3. Respectively, Ms. Nakamura may be latinized to Nakamuraea resulting in a specific epithet "*nakamuraeae*".
4. Mr. Nakamura is latinized to Nakamuraius, as in MacAdoo's opinion it should be normative (MacAdoo, 1993), resulting an a specific epithet "*nakamuraii*".
5. Respectively, Ms. Nakamura is latinized to Nakamuraia, resulting in a specific epithet "*nakamuraiae*".

By now the reader will understand that possibilities 2–5, although permissible or even recommended by MacAdoo (1993), look and sound rather awkward and are likely to produce numerous misspellings. Therefore I strongly suggest to use the classical version and version 1 only.

Roman names ending on -*o* usually followed the third declension, i.e., the genitive is formed by adding the ending -*nis*, which also reveals that such words have stems ending on n, e.g., Nero/Neronis, Cicero/Ciceronis, or the noun *leo/leonis* (the lion). Medieval Latin followed this custom. So, for the medieval German emperors named Otto the genitive Ottonis was used in writing, which was all in Latin at that time. Therefore it makes

TABLE 3. Formation of specific epithets from personal names in the adjectival form (examples given are hypothetical)

Ending of name	Example: family name	Add the endings for gender		
		masculine	feminine	neuter
consonant	Grant	-ianus	-iana	-ianum
-a	Kondratieva	-nus	-na	-num
-e	Lee	-anus	-ana	-anum
-i	Bianchi	-anus	-ana	-anum
-o	Guerrero	-anus	-ana	-anum
-u	Manescu	-anus	-ana	-anum
-y	Bergey	-anus	-ana	-anum

TABLE 4. Formation of specific epithets from personal names as genitive nouns (hypothetical epithets in quotation marks)

Ending of name	Add for female	Example (female person)	Add for male	Example (male person)
-a	-e (first declension)	Catarina, "catarinae"	-e (classic)	Komagata, komagatae Volta, voltae
			-i	Thomalla, "thomallai"
	-ea	Julia, "juliaeae"	-ei	Poralla, "porallaei"
	-iae	Mateka, "matekaiae"	-ii	Ventosa, "ventosaii"
-e	-ae	Hesse, "hesseae"	-i	Stille, "stillei"
-i	-ae	Kinski, "kinskiae"	-i	Suzuki, "suzukii"
-o	-niae	Cleo, "cleoniae"	-nii	Guerrero, "guerreronii"
			-nis	Otto, "ottonis"
-u	-iae	Feresu, "feresuiae"	-ii	Manescu, "manescuii"
-y	-ae	Macy, "macyae"	-i	Deley, deleyi
-er	-ae	Miller, "millerae"	-i	Stutzer, stutzeri Stanier, stanieri
any other letter	-iae	Gordon, "gordoniae"	-ii	Pfennig, pfennigii Zeikus, "zeikusii"

sense to treat Spanish, Italian, Portuguese, Japanese, Chinese, Ukrainian, Indonesian, as well as all other names that end on -o the same way.

Several European names are derived from classical Greek and end on -as, such as Thomas, Andreas, Aeneas, Cosmas, etc.. In their genitive form, they receive the ending -ae: Thomae, Andreae, Aeneae, Cosmae, etc. Although one could argue for a Latinization to Thomasius, Andreasius, etc., to form the specific epithets *thomasii*, *andreasii*, etc., I would recommend the use of the classical ending -ae.

E. Formation of bacterial names from geographical names Authors often consider it necessary to indicate the geographical origin, provenance, or occurrence of their isolates in the respective specific epithets.

Such epithets are simply constructed by adding the ending -ensis (masculine or feminine gender) or -ense (neuter gender) to the geographical name in agreement with the latter's gender. If the name of the locality ends on -a or -e or -en these letters are dropped before adding -ensis/-ense (e.g., *jenensis* from Jena, *hallensis* from Halle, *bremensis* from Bremen). Sometimes authors make the mistake of adding iensis/-iense. This is only correct if the locality's name ends on -ia (e.g., California leads to *californiensis*). The advice given above guarantees that such mistakes will not happen.

Specific local landscape names such as tundra, taiga, puszta, prairie, jungle (from Sanskrit *jangala*), steppe and savanna may be dealt with in the same way (*tundrensis, taigensis, pusztensis, prairiensis, jangalensis, steppensis* and *savannensis*, respectively).

Epithets on the basis of geographical names may not be formed as substantives in the genitive case, as if they were derived from personal names (e.g., the city of Austin, Texas, cannot lead to "*austinii*" but must lead to "*austinensis*").

Quite a number of localities in the Old World (Europe, Asia, Africa) have classical Greek, Latin, and medieval Latin names and adjectives derived from these: *europaeus, aegyptius, africanus, asiaticus, ibericus, italicus, romanus* (Rome), *germanicus, britannicus, hibernicus* (Ireland), *indicus* (India), *arabicus* (Arabia), *gallicus* (France), *polonicus, hungaricus, graecus* (Greece), *hellenicus* (Hellas, classical Greece), *hispanicus* (Spain), *rhenanus* (Rhineland), *frisius* (Friesland), *saxonicus* (Saxony), *bavaricus* (Bavaria), *bretonicus* (Brittany), *balticus* (Baltic Sea), *mediterraneus* (Mediterranean Sea), etc.

Since the discovery of the other parts of the world by European sailors and travelers, European geographers have continued to give Latin names to "new" continents and countries, so adjectives like *americanus, cubanus, mexicanus*, etc. were introduced. Wherever older adjectives exist they may be used as specific epithets to indicate geographical origins.

European and Mediterranean cities and places of classical times may have had very different names than those in current useage: e.g., *Lucentum* (Alicante, Spain), *Argentoratum* (Strasbourg, France), *Lutetium* (Paris, France), *Traiectum* (Utrecht, Netherlands), *Ratisbona* (Regensburg, Germany), *Eboracum* (York, U.K.), *Londinium* (London, U.K.), *Hafnia* (Copenhagen, Denmark). Microbiologists are free to demonstrate their knowledge of these ancient names but may use epithets derived from the present names of such places, e.g., *alicantensis, strasburgensis, parisensis, utrechtensis, yorkensis, regensburgensis* (MacAdoo, 1993).

Many localities (mostly lakes, rivers, seas, valleys, islands, capes, rocks or mountains, but also some towns or cities) have names that consist of two words, usually an adjective and a substantive (noun), e.g., Deep Lake, Black Sea, Dead Sea, Red River, Rio Grande, Rio Tinto, Long Island, Blue Mountain, Baton Rouge etc., or of two substantives, e.g., Death Valley, Lake Windermere, Loch Ness, Martha's Vineyard, Ayers Rock, Woods Hole, Cape Cod etc. Although such epithets would be correct in the sense of the *Bacteriological Code*, formation of specific epithets from such localities' names may pose a problem, because the use of the adjectival suffix -ensis, -ense may lead to rather strange looking or awkward constructions, such as "deeplakensis" or "bluemountainense". If the name of a locality lends itself to translation into Latin, specific epithets may alternatively well be formed as genitive substantives by forming the genitives of the two components and concatenating them without hyphenation, e.g., like the existing ones *lacusprofundi* (of Deep Lake), *marisnigri* (of the Black Sea), *marismortui* (of the Dead Sea), or (of two nouns) *vallismortis* (of Death Valley). Note that in Latin the basic noun comes first, the determining word (adjective or noun) second. If possible one should avoid the inclusion of articles such as the, el, il, le, la, de, den, het, der, die, das, or their plurals los, les, las, ils, gli, le, de, die, etc. as they are used for locations in several languages, e.g., La Jolla, La Paz, El Ferrol, El Alamein, Le Havre, The Netherlands, Die Schweiz, Den Haag, Los Angeles, etc. Articles would unnecessarily elongate names without adding information.

F. Formation of names for bacteria living in association or symbiosis with other biota An enormous reservoir of bacteria for future research is the microflora that is more or less tightly associated with other biota. I predict that at least two million new species (Trüper, 1992) will be described for the gut flora of various animal species.

Also the plant microfloras have so far been mainly investigated with respect to nitrogen fixation and diseases of economically important plants. To date, little has been done to investigate the phytopathogens that attack economically unimportant plants or weeds.

It is to be expected that microbiologists working in these fields will want to give new isolates names that relate to their hosts or associates.i.e., Latin nomenclatural names of animals, fungi, plants, and protists have been, and to a much larger extent, will be used.

This area of bacterial name-giving is unfortunately full of traps. Clearly, naming a bacterium after a host animal bearing a tautonym (such as *Picus picus*, the woodpecker) is easier than having to choose between generic name and a different specific epithet of the host. It is therefore important to know what these mean and how they were formed (adjective, substantive in genitive, etc.), in order to avoid nasty, ridiculous, or embarrassing mistakes.

The following example may demonstrate this situation: Certainly a bacterium isolated from the common house fly *Musca domestica* should not receive the epithet *domesticus, -a, -um* ("pertaining to the house"); its epithet should rather be *muscae* (of the fly) or *muscicola* (dwelling in/on the fly) the latter being a nominative noun in apposition.

The *domestica* associated with *Musca* is an adjective. If we theoretically consider it an independent noun meaning "the one pertaining to the house" one could, of course, form the genitive from it and thus produce a bacterial epithet *domesticae*. In this example, however, that would not make much sense as too many things "pertain to a house". But formally it would not violate the Rules of the *Bacteriological Code*.

The easiest way of forming such specific epithets is the use of the genitive case of the generic name of the eucaryote in question, e.g., *suis, equi, bovis, muscae, muris, aquilae, falconis, gypis, elephantis* (of the pig, horse, cow, fly, mouse, eagle, falcon, vulture, elephant), or: *fagi, quercus* (fourth declension genitive, spoken with long u), *castaneae, aesculi, rosae, liliae* (of the beech, the oak, chestnut, horse chestnut, rose, lily).

Alternatively the genitive of the plural is recommended, especially if several species of the eucaryotic genus house the bacterial species in question. The formation of the plural genitive needs the knowledge of the stem and declension of the word. The following examples may be of some principal assistance:

1. First declension: *-arum* (*muscarum*, of flies; *rosarum*, of roses)
2. Second declension: *-orum* (*equorum*, of horses; *pinorum*, of pines)
3. Third declension: *-um* (*leonum*, of lions; *canum*, of dogs)
4. Fourth declension: *-um* (*quercuum*, of oaks)
5. Fifth declension: *-rum* (*scabierum*, of different forms of scabies, a skin disease)

Be aware of irregular forms such as *bos* (the cow), genitive: *bovis*, plural genitive *boum*! Use dictionaries and look up the declension in MacAdoo (1993)!

G. Names taken from languages other than Latin or Greek

Besides names of persons or localities, many words from languages other than Latin or Greek have been used in bacterial names and certainly will be in the future. Here a few examples may suffice to demonstrate the width and variety of such cases:

During late medieval and renaissance times alchemy became rather fashionable among European scientists and many Arabic words entered into the terminology that would eventually be used in chemistry. One of these, which is often used in bacterial names, is "alkali" (Arabic *al-qaliy*, the ashes of saltwort) from which the element kalium (K, English: potassium) received its name. As the *-i* at the end of the word belongs to the stem it is wrong to speak and write of al<u>ca</u>lophilic instead of al<u>ka</u>liphilic microbes. Latinized names of bacteria containing this stem should therefore be corrected to, e.g., *Alkaligenes, alkaliphilus*, etc., and new ones should be formed correctly!

A rather common mistake occurs with the English suffix -philic (e.g., hydrophilic—friendly to water, water-loving). This is clearly an English transformation of the Latin *-philus, -a, -um* (originating from Greek *philos*, friendly). All names formed thus far ending on *-philicus, -a, -um* are wrong and should, in my opinion, be changed to *-philus, -a, -um* as soon as possible. Here, however, Rule 57a (accordance with the rules of Latin) would have to be weighed against Rule 61 (retaining the original spelling) of the *Bacteriological Code*.

National foods or fermentation products often do not have equivalent Latin names and if typical microorganisms found in them or causing their fermentations are described, they have been (and may be) named after them, e.g., sake, tofu, miso, yogurt, kvas, kefir, pombe, pulque, aiva, etc. However, these names cannot be used unaltered as specific epithets in the form of nominative substantives in apposition (Trüper and de'Clari, 1997). They must be properly Latinized. The best way to do so is to form a neuter substantive from them by adding *-um* (e.g., *sakeum, tofuum, kefirum, pombeum*, etc.) and use the genitive of that (ending: *-i*) in the specific epithet (e.g., *sakei, tofui, kefiri, pombei*, etc.)

Another point worth mentioning is the "unnecessary" usage of words from languages other than Greek or Latin. For instance, the formation of the epithet *simbae* from the East African Swahili word *simba*, lion, for a *Mycoplasma* species was not necessary because in this genus the corresponding Latin epithet *leonis* (of the lion) had not been used before.

H. Formation of bacterial names from names of elements and compounds used in chemistry and pharmacy

The almost unlimited biochemical capacities of bacteria is another rather inexhaustible source for new names. Many generic names, as well as specific epithets, have been formed from names of chemical elements, compounds and even pharmaceutical and chemical products or their registered or unregistered trade names.

The late Robert E. Buchanan (1960, reprinted 1994) listed numerous examples of such generic names and specific epithets. Based on the classical Latin/Greek thesaurus and enriched by numerous Arabic words, the pharmaceutical sciences have, since the Middle Ages, developed a Neo-Latin terminology for chemicals of all categories.

The vast majority of names of chemicals are latinized as neuter nouns of the second declension with nominatives ending *-um*, genitives in *-i*. The following groups belong in this category:

1. Most of the chemical elements with the exception of carbon (L. *carbo, carbonis*), phosphorus (L. *phosphorus, phosphori*), and sulfur (L. *sulfur, sulfuris*) have the ending *-(i)um*; nitrogen may also be called *azotum* besides *nitrogenium*, calcium may also be called *calx* (genitive: *calcis*).
2. Chemical and biochemical compounds ending on *-ide* (anions), *-in, -ane, -ene, -one, -ol* (only non-alcoholic compounds), *-ose* (sugars), *-an* (polysaccharides), *-ase* (enzymes) (*-um* is added, or the *-e* at the end is replaced by *-um*, respectively).
3. Acids are named by *acidum* (L. neuter noun, acid), followed by a descriptive neuter adjective, e.g., sulfurous acid *acidum sulfurosum*, sulfuric acid *acidum sulfuricum*, acetic acid *acidum aceticum*.

The second largest category of chemicals are treated as neuter nouns of the third declension: these end on *-ol* (the alcohols), *-al* (aldehydes), *-er* (ethers, esters), and *-yl* (organic radicals); latinization does not change their names at the end, whereas the genitive is formed by adding *-is*.

Anions ending in *-ite* and *-ate* are treated as masculine nouns of the third declension. The English ending *-ite* is latinized to *-is*, with the genitive *-itis*, e.g., nitrite becomes *nitris, nitritis*. The English ending *-ate* is latinized to *-as*, with the genitive *-atis*, e.g., nitrate becomes *nitras, nitratis*.

Only few chemicals have names that are latinized in the first declension as feminine nouns, ending on *-a* with a genitive on *-ae*. Besides chemicals that always had names ending on *-a* (like urea), these are drugs found in classical and medieval Latin, such as gentian (*gentiana*) and camphor (*camphora*), further modern drugs, whose Latin names were formed by adding *-a*, like the French ergot becoming *ergota* in Latin.

The most important group of this category are alkaloids and other organic bases, such as nucleic acid bases and amino acids with English names on *-ine*. In Neo-Latin this ending is *-ina*, with the genitive *-inae*, e.g., *betaina, -ae; atropina, -ae; adenina, -ae; alanina, -ae*; etc.

For their use in bacterial generic names and specific epithets word stems and genitives of latinized chemical names are the basis. In principle they are then treated like any other word elements.

I. Arbitrary names Either genus names or specific epithets "may be taken from any source and may even be composed in an arbitrary manner" (*Bacteriological Code*, Rule 10a and Rule 12c). They must, however, be treated as Latin. These "rubber" paragraphs open up a box of unlimited possibilities for people whose Latin is exhausted. But in view of the million names that will have to be formed in the future they are a simple necessity, whether Latin purists like them or not.

Examples for arbitrary generic names are *Cedecea*, *Afipia*, and in the near future "*Vipia*" and "*Desemzia*", that were derived from the abbreviations CDC (Center for Disease Control), AFIP, VPI (Virginia Polytechnical Institute), and DSMZ (Deutsche Sammlung von Mikroorganismen und Zellkulturen), respectively. Examples for arbitrary specific epithets are, e.g., (*Salmonella*) *etousae*, derived from the abbreviation ETOUSA (European Theater of Operations of the U.S. Army), and (*Bacteroides*) *thetaiotaomicron*, formed from the three Greek letter names *theta*, *iota*, and *omicron*.

More recently, the new genus *Simkania* was described. The name is a latinized contraction of the first and the family name of the microbiologist Simona Kahane. Certainly an arbitrary name, short, elegant and easy to pronounce, points to future possibilities of bacterial name-giving. Authors should aim at such easily spelled and pronounced short names, when they take advantage of arbitrary name-giving.

III. SOME CASE HISTORIES OF MALFORMED NAMES

From the viewpoint of classical Latin many of the existing bacterial names are, plainly said, lousy in their grammar and etymology. However, under the Rules of the *Bacteriological Code* they are acceptable. A few case histories of wrong bacterial names are worth mentioning in a chapter on etymology because of their scurrility.

Acetobacter xylinus: This specific epithet goes back to Brown 1886, who described a *Bacterium xylinum*. Several subsequent changes of the genus (Trevisan 1889, *Bacillus xylinus*; Ludwig 1898, *Acetobacterium xylinum*; Pribram 1933, *Ulvina xylina*) prove by the change in gender that the epithet is an adjective. Because before 1951 (*Bacteriological Code*, Opinion 3), the gender of names ending in -*bacter* was not fixed as masculine, *Acetobacter xylinum* (Holland 1920 and Bergey et al. 1925) (all names and dates before 1950 cited were taken from *Index Bergeyana*, Buchanan et al., 1966) was not wrong either. As a consequence of Opinion 3 the species should be named *Acetobacter xylinus*. The *Approved Lists* of names (Skerman et al., 1980), however, listed the organism as *Acetobacter aceti* subspecies *xylinum*! Yamada (1983) revived the species status and correctly called it *A. xylinus*. The compiler of Validation List 14 (*International Journal of Systematic Bacteriology*, 1984) incorrectly put a *sic* after *xylinus* and changed it to the neuter form *xylinum*! (The Latin expression *sic* is used to point out a mistake or other peculiarity.) Unexpectedly the previous authors obeyed this falsifying change and even tried to give the neuter epithet justification by explaining it as a nominative noun in apposition (*xylum*, M.L. neut.n. cotton). "*Acetobacter*, called the cotton" makes little sense and certainly does not meet the requirements of a nominative noun in apposition (cf. Trüper and de'Clari, 1997), Finally, Euzéby (1997) corrected the name to *A. xylinus*.

Methanobrevibacter arboriphilus: In 1975 the new species *Methanobacterium arbophilicum* was described. The organism was isolated from rotting trees and the authors wanted to express "friendly to trees" by the epithet. In Latin, tree is *arbor*, genitive *arboris*, ie., the stem is clearly *arbor*-, not *arbo*-. The second error

was that the English ending -*philic* was latinized to -*philicum* instead of correctly to -*philum*. Although this was first pointed out to the authors in 1976, they did not correct the epithet themselves. Then, in a review paper, Balch et al. (1979) rearranged the methanogenic procaryotes and transferred the species to the genus *Methanobrevibacter* as *M. arboriphilus* (the correct form of the epithet). It was again the compiler of the Validation List No. 6 (*International Journal of Systematic Bacteriology*, 1981), who created a new wrong form of the epithet, *arboriphilicus*! Although immediately informed of his error, he did not correct it. And so this wrong epithet still occurred in *Bergey's Manual of Systematic Bacteriology*, Vol. 3 (1989). To my knowledge it has not been corrected!*

Some time ago an author wanted to create the specific epithet "*nakupumuans*" and explained this word as derived from the Maori word *nakupumua*, breaking protein down to fragments. Becoming informed that there was neither need to use another language than Latin, nor any specific connection between the Maori and protein degradation the author decided to call the isolate *proteoclasticum*. Accepting such name formations in procaryote nomenclature would mean giving up Latin as the basic language of biological nomenclature. As long as names can be formed from the Latin/Greek thesaurus at our hands, names from other languages should be avoided.

In another instance, an author wanted to propose a specific epithet in honor of a colleague and formed an epithet ending in -*icus*. As this is not within the Rules, I advised him to choose either an epithet ending on -*ii* (genitive noun) or on -*ianus* (adjective). His answer was that he did not like the former and felt that the latter sounded like an insult to the colleague to be honored!

Another colleague correctly formed the generic name *Acidianus* (accentuation: a.cid.ia'nus) from the Latin neuter noun *acidum*, acid and the Latin masculine noun *Ianus*, the Roman god with the two faces, by which he wanted to indicate the ability of the organism to both oxidize and reduce elemental sulfur. With this spelling the epithet promptly became mispronounced (a.ci.di.a'nus) suggesting a different meaning and causing suggestive jokes. Here the use of the consonantic i, (i.e., j) would have sufficed to suppress the misinterpretation: *Acidijanus* would be the choice.

These examples also show that nobody is free from making mistakes. During my work in this field I have made several, and sometimes even given wrong advice, quite to my embarrassment afterwards.

IV. PRACTICAL ETYMOLOGY IN DESCRIPTIONS OF GENERA AND SPECIES

As mentioned before, for the average microbiologist "etymology" is a kind of nasty linguistic exercise necessary for the description of a new genus or species. In reality he/she has to "create" a new name; the organism has been isolated and determined by the author, not "created"! The better and more modest wording would be, to "propose" a new name.

On the basis of six examples of such "etymologies" I shall try to explain how these are composed.

1. *Escherichia coli*: Esch.er.i'chi.a (better: E.sche.ri'chi.a) M.L. fem.n. *Escherichia*, named after Theodor Escherich, who iso-

---

Editorial Note: As of January 2000, this name still appears on the Approved List. No action to correct the name has been taken.

lated the type species of the genus. co'li Gr.n. *colon* large intestine, colon; M.L. gen.n. *coli* of the colon.

2. *Rhodospirillum rubrum*: Rho.do.spi.ril'lum Gr.n. *rhodon*, the rose; M.L. dim neut.n. *Spirillum*, a bacterial genus; M.L. neut n. *Rhodospirillum*, a red *Spirillum*. (Etymology of the latter: Gr. n. *spira*, spiral, M.L. dim. neut n. *Spirillum*, a small spiral.) rub'rum.L. neut. adj. *rubrum*, red.

3. *Azotobacter paspali*: A.zo.to.bac'ter French n. *azote*, nitrogen; M.L. masc.N. *bacter*, the equivalent of Gr. neut.n. *bactrum*, a rod or staff.M.L. masc n. *Azotobacter*, nitrogen rod. pas.pal'i (better: pas.pa'li). M.L. gen n. *paspali*, named for *Paspalum*, generic name of a grass.

4. *Pseudomonas fluorescens*: Pseu.do.mo'nas (seldom: Pseu.do'mo.nas). Gr. adj. *pseudos*, false; Gr.n. *monas*, a unit; M.L. fem.n. *Pseudomonas*, false monad. flu.o.res'cens.M.L. v. *fluorescere* (*fluoresco*), fluoresce; M.L. part adj. *fluorescens*, fluorescing.

5. *Desulfovibrio gigas*: De.sul.fo.vi'brio (or: De.sul.fo.vib'rio). L. pref. *de*, from; L.n. *sulfur*, sulfur; L.v. *vibrare*, vibrate; M.L. masc.n. *Vibrio*, that which vibrates, a bacterial generic name; M.L. masc.n. *Desulfovibrio*, a vibrio that reduces sulfur compounds. (Note: If we were meticulous, the name should either be "*Desulfativibrio*" referring to sulfate, or "*Desulfurivibrio*" referring to sulfur. As *Desulfo-* may cover both, in this case it is certainly the best name for the genus!) gi'gas.L. nom.n. *gigas*, the giant.

6. *Thermoanaerobium aotearoense*: Ther.mo.an.ae.ro'bi.um. Gr. adj. *thermos*, hot; Gr. pref. *an-*, without; Gr.n. *aer*, air; Gr.n. *bios*, life; M.L. neut.n. *Thermoanaerobium*, life in heat without air. a.o.te.a.ro.en'se. Maori n. *Aotearoa*, New Zealand; L. neut. suffix *-ense*, indicating provenance; M.L. neut. adj. *aotearoense*, from or pertaining to Aotearoa (New Zealand).

From these examples several regularities can be deduced:

1. After the name or epithet the "etymology" starts with an indication of accentuation. The word is broken into a row of syllables interrupted by periods. The accent-bearing syllable is indicated by an accent sign behind it (note: never before it!) instead of a period. The classical Latin language did not develop explicit rules about breaking up words into syllables; the Romans broke written words the way they were spoken, and logically split compound words between compounds. As the rules for breaking words into syllables are different for different modern languages, in my opinion, one should continue to follow the Roman custom rather than the rules for any modern language.

2. The accentuation is followed by the etymology proper of the name. The abbreviations commonly in use indicate the language of origin (Gr. classical Greek, L. classical Latin, M.L. modern Latin), the type of word or word element (adj. adjective, n. noun/substantive, v. verb, part. adj. participle used as adjective, dim. diminutive, pref. prefix, suff. suffix), the case (gen. genitive, nom. nominative, the latter being seldom indicated) and the gender of nouns or adjectives (fem. feminine, masc. masculine, neut. neuter).

3. The word elements are explained in the sequence they occur in the name. Then, like a summary, the language, gender, and the word type of the complete name or epithet is given, followed by the Latin name and its translation.

The abbreviation M.L. is very often misunderstood as medieval Latin. I personally would therefore prefer a ruling that M.L. would really mean medieval Latin and that modern Latin, better Neo-Latin, would be abbreviated N.L.

V. RECOMMENDATIONS (FROM THE VIEWPOINT OF LANGUAGE) FOR FUTURE EMENDATIONS OF THE *BACTERIOLOGICAL CODE*

We should not aim for pure classical Latin in biological nomenclature but rather develop the current Latin/Greek thesaurus further by following the Rules of the ICNB or the respective codes of nomenclature applicable to other fields of biology. This is in reality what has happened since Linnaeus' time. In my opinion the ICNB has excellent provisions to do so. This is already documented by the low number of Opinions that had to be issued by the Judicial Commission of the ICSB during the last ten years.

For several years the development of a uniform code of nomenclature for all biological taxa has been underway, enlisting the participation of well-known taxonomists from bacteriology, botany, mycology, phycology, protozoology, virology, and zoology. This effort has received the support of the International Unions of Biological and Microbiological societies, IUBS and IUMS (Hawksworth and NcNeill, 1998). These activities reflect the general scientific need to assess the total extent of biodiversity on Earth, in order to facilitate conservation and, perhaps, prevent further extinction of the biota. For this purpose a unified system of biological names has been considered indispensable. Drafts of the future universal "BioCode" have been published, the latest (fourth) draft by Greuter et al. (1998). As soon as the BioCode is accepted by the taxonomic committees of the different biological disciplines involved, the *Bacteriological Code* will have to be revised to conform with any new recommendation. Changes in etymological rulings should be expected. Unfortunately the recommendations for latinization (Articles 37–39) are not yet formalized, therefore comments and recommendations cannot be offered at this time.

Besides the cases mentioned in the text above, where certain changes or simplifications have been recommended, there are a few other points where, in my opinion, the Rules need further development with respect to etymology:

1. Stronger emphasis should be put on short and easily pronounceable names.

2. Words from languages other than Latin or Greek should be banned as long as an equivalent exists in Greek or Latin or can be constructed by combining word elements from these two languages, and as far as they are not derived from names of geographical localities or local foods or drinks (e.g., sake, kefir, kvas, pombe, tofu, miso, yogurt, etc.), for which no Latin/Greek names exist.

3. Formation of bacterial names on the basis of latinized names of chemical compounds should be regulated under the Code. Here the recommendations of Buchanan (1994), as explained above, should be the basis.

4. The principal ban on ordinal numbers (adjectives) for the formation of bacterial names (ICNB, Rule 52, -2-) only makes sense for those numbers above ten because of their length. Therefore, this part of Rule 52 should be abandoned.

5. In the transliteration of the Greek letter k to the Latin letter c the k sound is lost when the vowels e, i, or y follow. Instead the c is pronounced as a sharp s as in English. Therefore, to preserve the k sound before e, i, and y, the letter K should be kept even in the Latin transliteration (example: *Akinetobacter* as in kinetics instead of *Acinetobacter*).

6. Authors should refrain from naming bacteria after themselves or coauthors after each other in the same publication, as this is considered immodest by the majority of the scientific community.

7. Generic names and specific epithets formed from personal names can only contain the name of one person, not a combination or contraction of the names of two or more persons.

8. In the future, bacteriologists (including those that work on archaebacteria and cyanobacteria) should avoid names that end on -myces or -phyces in order to avoid confusion with mycology and phycology, i.e., with eucaryote nomenclature. Articles 25–28 of the future BioCode (Greuter et al., 1998) will forbid procaryote names ending in -myces, -phyta, -phyces, etc. or in -virus.

9. In the etymology given with the description of a taxon, there should be an indication whether a Latin name is from classical Latin ("L.") or Greek ("G."), from a medieval Latin ("M.L.") source or formed as Neo-Latin ("N.L."). This will save time for those who want to look up such names and words in dictionaries, and it will end ambiguous interpretation of M.L. as either "modern" Latin or medieval Latin. Already Buchanan (1960, reprinted 1994) prefered "Neo-Latin" over "modern" Latin.

ACKNOWLEDGMENTS

I wish to thank Eckhard Bast (Bonn), Jean P. Euzéby (Toulouse), Lanfranco de'Clari (Lugano), Roy Moore (Ulster), and Bernhard Schink (Konstanz), for their extremely helpful correspondence and discussions on etymology of bacterial names, and the Fonds der Chemischen Industrie for financial support.

FURTHER READING

*Bailly, A. (Editor).1950. Dictionnaire Grec-Francais, Hachette, Paris.

Balch, W.E., G.E. Fox, L.J. Magrum, C.R. Woese and R.S. Wolfe. 1979. Methanogens: reevaluation of a unique biological group. Microbiol. Rev. *43*: 260–296.

*Brown, W.B. 1956. Composition of Scientific Words - A Manual of Methods and a Lexicon of Materials for the Practice of Logotechnics, Brown, pp. 882

*Buchanan, R.E. 1994. Chemical terminology and microbiological nomenclature. Int. J. Syst. Bacteriol. *44*: 588–590.

Buchanan, R.E., J.G. Holt and E.F. Lessel (Editors). 1966. Index Bergeyana, The Williams & Wilkins Co., Baltimore.

*Calonghi-Badellino, G. (Editor).1966. Dizionario della lingua latina, Rosenberg and Sallier, Torino.

*Diefenback, L. 1857. Glossarium Latino-Germanicum Mediae et Infirmae Aetatis, Frankfurt.

*Egger, C. (Editor).1992. Lexicon Recentis Latinitatis (Italian/Latin), Libraria Editoria Vaticana, Rome.

Euzéby, J.P. 1997. Revised nomenclature of specific or subspecific epithets that do not agree in gender with generic names that end in -bacter. Int. J. Syst. Bacteriol. *47*: 585.

*Farr, E.R., J.A. Leussink and F.A. Stafleu. 1979. Index Nominum Genericorum (Plantarum), Sheltema and Holkema, Utrecht.

*Feihl, S., C. Grau, H. Offen and A. Panella (Editors). 1998. Neues Latein Lexikon, translated from Italian, Libraria Editoria Vaticana, Bonn.

Greuter, W., D.L. Hawksworth, J. McNeill, M.A. Mayo, A. Minelli, P.H.A. Sneath, B.J. Tindall, P. Trehane and P. Tubbs. 1998. Draft Biocode (1997): the prospective international rules for the scientific naming of organisms. Taxon. *47*: 129–150.

*Habel, E. and F. Gröbel. 1989. Mittellateinisches Glossar, Schöningh Verlag, Paderborn.

Hawksworth, D.L. and J. McNeill. 1998. The International Committee on Bionomenclature (ICB), the draft BioCode (1997), and the IUBS resolution on bionomenclature. Taxon. *47*: 123–136.

*Lapage, S.P., P.H.A. Sneath, E.F. Lessel, V.B.D. Skerman, H.P.R. Seeliger and W.A. Clark (Editors). 1992. International Code of Nomenclature of Bacteria (1990) Revision. Bacteriological Code, American Society for Microbiology, Washington, D.C.

*Lewis, C.T. and C. Short (Editors). 1907. A New Latin Dictionary, American Book Company, New York.

*Liddell, H.G., R. Scott, H.S. Jones and R. McKenzie (Editors). 1968. A Greek- English Lexicon, Oxford University Press, Oxford.

*MacAdoo, T.O. 1993. Nomenclatural literacy, In Goodfellow, M. and A.G. O'Donnell (Editors), Handbook of New Bacterial Systematics, Academic Press Ltd., London. pp. 339–360.

*Noel, F. (Editor).1833. Dictionarium Latino-Gallicum, Le Normant, Paris.

*Simpson, D.P. (Editor).1959. Cassell's New Latin Dictionary, Funk and Wagnalls, New York.

Skerman, V.B.D., V. McGowan and P.H.A. Sneath. 1980. Approved lists of bacterial names. Int. J. Syst. Bacteriol. *30*: 225–420.

*Stearn, W.T. (Editor).1983. Botanical Latin, David and Charles, Devon.

Trüper, H.G. 1992. Prokaryotes: an overview with respect to biodiversity and environmental importance. Biodivers. Conserv. *1*: 227–236.

*Trüper, H.G. 1996. Help! Latin! How to avoid the most common mistakes while giving Latin names to newly discovered prokaryotes. Microbiologia *12*: 473–475.

Trüper, H.G. and L. de'Clari. 1997. Taxonomic note: Necessary correction of specific epithets formed as substantives (nouns) "in apposition". Int. J. Syst. Bacteriol. *47*: 908–909.

Trüper, H.G. and L. de'Clari. 1998. Taxonomic note: erratum and correction of further specific epithets formed as substantives (nouns) "in apposition". Int. J. Syst. Bacteriol. *48*: 615.

*Werner, F.C. (Editor).1972. Wortelemente Lateinisch-Griechischer Fachausdrücke in den Biologischen Wissenschaften, 3rd ed., Suhrkamp Taschenbuch Verlag, Berlin.

*Woodhouse, S.C. (Editor).1979. English-Greek Dictionary: A Vocabulary of the Attic Language, Routledge and Kegan Paul, London.

Yamada, Y. 1983. *Acetobacter xylinus* sp. nov., nom. rev., for the cellulose-forming and cellulose-less acetate-oxidising acetic acid bacteria with the Q-10 system. J. Gen. Appl. Microbiol. *29*: 417–420.

*Yancey, P.H. 1945. Origin from mythology of biological names and terms. Bios. *16*: 7–19.

Microbial Ecology—New Directions, New Importance

Stephen H. Zinder and Abigail A. Salyers

INTRODUCTION: MICROBIAL ECOLOGY—THE CORE THAT LINKS ALL BRANCHES OF MICROBIOLOGY

Microbial ecology is the study of microorganisms in their natural habitats. In these habitats, they are rarely in pure culture and are usually interacting with other microorganisms, are sometimes interacting with host organisms, and are always interacting with their physicochemical environment. These conditions are usually very different from those used to grow microorganisms in pure culture in the laboratory. Since *Bergey's Manual* is a compendium of properties of pure cultures of procaryotes, it might appear that a discussion of microbial ecology is inappropriate. However, ecological studies have a profound effect on our understanding of pure cultures, and this impact will become more important in the twenty-first century. This chapter will not give a comprehensive overview of microbial ecology, but will, instead, discuss the relevance of microbial ecology to pure culture studies and vice versa.

Microbial ecology has a long history that reaches back to Antony van Leeuwenhoek's microscopic observations of microbial populations in various habitats including rainwater, dental plaque, and feces. Until the late nineteenth century, when the techniques developed by Louis Pasteur and Robert Koch allowed new approaches to be taken, the microscope was essentially the only tool available to study microorganisms, and only natural populations of microorganisms could be studied. Many of Pasteur's early studies on fermentations, spontaneous generation, and the distribution of microorganisms in air had an ecophysiological bent, describing phenomena such as the effect of oxygen on species composition and metabolism. Pasteur eventually used the techniques and concepts he developed in these studies to investigate pathogenesis.

Pure culture microbiology began in the late nineteenth century with the development of isolation techniques, particularly the use of semisolid agar media by Robert Koch. Koch's postulates demanded isolation as an essential step in proving microbial causation of a disease. Koch and his followers, the "microbe hunters", took center stage in microbiology in the first half of the twentieth-century. They isolated and characterized nearly all of the important pathogens, leading to the almost complete elimination of infectious diseases from the Western world through better sanitation and use of vaccines and antibiotics, an achievement that is certainly one of the great triumphs of twentieth-century science.

In the latter half of the twentieth century, molecular biologists took center stage in microbiology, working mainly with *Escherichia coli*. They defined genes and operons, mapped their positions on the chromosome, and studied the regulation of their expression. These studies culminated in the recent determination of the entire genome sequence of *E. coli* K12, as well as those for over 25 other microorganisms at the time this volume went to press, with many more microbial genome sequences in the offing. The molecular characterization of *E. coli* and other bacteria is also a major landmark of twentieth-century science.

Thus, the twentieth century could be considered the "Age of the Pure Culture". Working in a reductionist style with pure cultures was extremely successful and therefore very seductive. Many microbiologists came to believe that only work with pure cultures or with macromolecules could be good science, and forgot the communities from which their microbe had been taken.

During the twentieth century, most microbial ecologists worked mainly at agricultural and technical schools. Their work was often directed towards applied areas such as soil microbiology related to agriculture or the environment. The dearth of researchers, the prevalence of applied rather than fundamental research, relatively low levels of funding, and, as will be described presently, formidable technical difficulties in studying microbial ecology, all contributed to this field lagging behind pure culture microbiology.

Starting in the 1980s, a series of unpleasant surprises on the disease front brought clinical microbiologists face to face with the fact that microbial ecology was indeed central to their interests. The emergence of new diseases such as Lyme disease, AIDS, and ehrlichosis, and the reemergence of old diseases such as cholera in South America and tuberculosis in the United States, demonstrated that changing human practices (human ecology) could create new windows of opportunity for microbes and that understanding the way that they moved into new niches (microbial ecology) was critical for controlling further spread of diseases. Large outbreaks of salmonellosis and *E. coli* O157:H7 raised anew questions concerning the factors that control colonization of animals by these pathogens, and whether the normal microbiota of animals and plants could be manipulated to decrease colonization opportunities. Increased concern about antibiotic resistance, the study of which had long been dominated by molecular biologists, led to the realization that all of the practical questions about how to control the spread of resistance were centered instead on the ecology of resistance—how genes were spreading in various microbial communities. In general, the realization that it was better to prevent disease than intervene after the disease had established a foothold gave new impetus to understanding where disease-causing organisms are normally found, how they fit into their normal ecological niches, and how they adapt to new niches that in some cases were quite different from their usual ones.

Another humbling finding was that, despite decades of intensive research on *E. coli* K12, a function could not be ascribed to 38% of the genes in its genome (Blattner et al., 1997). The unknown genes probably encode functions that help *E. coli* to live in habitats as diverse as the human intestinal tract and freshwater creeks and to make a living under conditions more demanding than those experienced growing in Luria broth or even in a chemostat. Moreover, evidence has been obtained that a considerable fraction (at least 18%) of the genes in the *E. coli* K12 genome were transferred from other organisms (Lawrence and Ochman, 1998). Finally, several *E. coli* strains have genomes several hundred million bases larger than that of strain K12 (Bergthorsson and Ochman, 1995), indicating that we have much to learn about this species which we thought we knew so well. Thus, functional genomics also leads us into microbial ecology.

Meanwhile, environmental microbiologists, who had identified themselves all along as microbial ecologists, began to turn away from characterizing steps in pathways using biochemical analysis of pure cultures, and returned to asking questions about how such pathways operated in nature. At one time such questions would have seemed futile because of the complexity of microbial communities and the suspicion that there remained many uncultivated microbes. Microbial ecologists were also beginning to look at familiar environments in new ways. Bacteriophages began to be recognized as important predators of bacteria in some settings. Horizontal gene transfer assumed new prominence as ecologists began to realize that bacteria in a complex community could interact sexually as well as metabolically, and that gene transfer can even occur between eubacteria and archaea (Doolittle and Logsdon, 1998). The discovery of syntrophic interactions, in which two microbes work together to carry out a reaction that is thermodynamically impossible for one organism (Schink, 1997), opened up a new dimension in metabolic interactions. Indeed, the paper originally describing the resolution of *"Methanobacillus omelianskii"* into two syntrophic organisms (Bryant et al., 1967) was considered to be one of the 100 most important in twentieth-century microbiology (Joklik, 1999). Report after report appeared of microbes that could carry out reactions previously thought to be improbable, if not impossible: anaerobic breakdown of aromatic (Evans et al., 1991) and aliphatic (Aeckersberg et al., 1998) hydrocarbons, "fermentation" of inorganic sulfur compounds (Bak and Pfennig, 1987), utilization of chlorinated organics as respiratory electron acceptors (Mohn and Tiedje, 1992), and methanogenesis in aerobic methane-oxidizing bacteria, using enzymes from the methanogenesis pathway of archaea in the reverse direction (Chistoserdova et al., 1998).

Microbiologists from many different areas have begun to rediscover microbial communities and to recall that the conditions under which microbes normally live are very different from those used to grow them in the laboratory. The pendulum began to swing back to a position where pure culture studies were declared by some to be unscientific and inappropriate, and community analysis became the imperative (Caldwell, 1994). While this represents an extreme position that few would advocate, microbial communities have been neglected too long, and the time is ripe for their study. Moreover, as new technologies have been introduced, another need has become evident—the need for more sophisticated models and theories about community structure

and interactions among members of the community. Just as the availability of genome sequences has challenged scientists working on individual microbes to find creative new ways to use this information, the availability of molecular tools for analyzing microbial communities calls for conceptual advances that will make maximal use of new technologies.

CLASSICAL MICROBIAL ECOLOGY

Microbial ecology, as practiced through most of the twentieth century, employed nonmolecular biological tools to study natural microbial populations. These consist mainly of activity measurements, biomass measurements, microscopy, and cultivation techniques (Atlas and Bartha, 1993). When applied to procaryotes, all of these techniques suffer from limitations that are mainly due to the small size of these organisms and the complexity of their environments. For example, a single 1-mm crumb of soil contains microhabitats that are aerobic, anaerobic, wet, dry, organic-rich, organic-poor, acidic, and basic. Thus methodological problems arising from the microenvironment are particularly formidable.

As an example of this, consider some commonly used ways of measuring various activities in microbial populations. A compound, sometimes isotopically labeled, is added to the environment. If the compound is at its natural concentration, a chemical transformation can be measured to estimate the rate of that process. Alternatively, the metabolic potential for that process can be determined if a higher concentration is used. A problem with these methods when applied to microbial populations is that they are essentially bulk measurements. For example, one can measure the rate of sulfate reduction in a sample, but several different populations of sulfate reducers may be contributing to that rate. Kinetic measurements and analysis or inhibition studies may provide more fine structure information about the process, and clever use of microelectrodes (Fossing et al., 1995) or microautoradiography (Krumholz et al., 1997) can give spatial information on an appropriate scale for that process. Still, more often than not, the information we obtain from these studies is of low resolution. Moreover, natural microbial populations are often perturbed during sampling, by such processes as mixing or simply placing them in a vial, so that delicate spatial relationships are destroyed.

Microbial biomass can be estimated by a variety of bulk techniques in which some cell constituent such as organic matter, protein, or chlorophyll is extracted and quantified from microbial populations. Measurement of amount of ATP and other nucleotides can give an estimation of the active biomass (Karl, 1980). More specific methods based on quantifying lipids, including those considered "signatures" of various microbial groups such as archaea and eubacterial methylotrophs, have been developed (Hedrick et al., 1991), but their suitability for application to complex natural microbial habitats is uncertain.

Microscopy remains an extremely important technique in microbial ecology, especially since it brings the researcher down to the scale of the microenvironment. Particularly useful are fluorescent microscopical methods, such as staining with nucleic acid specific stains such as acridine orange or DAPI (Amann et al., 1995). A problem with microscopic observation of procaryotic cells is that they are, with certain exceptions, morphologically nondescript to our eyes, so that we cannot simply identify them by looking at them the way we can plants and animals. Even using electron microscopy, many procaryotes are indistinguishable. In some cases, fluorescent antibodies have been useful in identifying microbes *in situ*, but the antibody specificity must be carefully assessed to correlate serotype with taxonomic group (Macario et al., 1991b).

The culture of microorganisms is a cornerstone of microbial ecology. Enrichment culture techniques were developed by Beijerinck and Winogradsky at the beginning of the twentieth century and used by them and their followers to cultivate a variety of metabolically diverse organisms from natural habitats. In a manner similar to application of Koch's postulates, organisms that carried out in pure culture processes detected in natural habitats such as nitrogen fixation, pesticide degradation, or pyrite oxidation were isolated.

Despite success in applying cultural techniques to the study of microbial ecosystems, it has long been known that the number of organisms obtained from most natural habitats using cultural techniques is usually one to several orders of magnitude lower than that seen under the microscope, a phenomenon termed the "great plate count anomaly" (Staley and Konopka, 1985). The potential causes of this anomaly will be discussed below, but it should be mentioned here that it was not clear at the time whether the relatively low viable counts were mainly a matter of poor recovery of known organisms, or whether there were entire microbial groups which were not being cultured.

Two sets of classical microbial ecological studies in the twentieth century are particularly notable: those of Robert Hungate on gastrointestinal habitats and those of Thomas Brock on hot springs. Hungate studied the microbiota of the termite gut and the animal rumen from the 1940s until the 1980s (Hungate, 1979). He developed a novel set of anaerobic culture techniques for growing fastidious anaerobes, and enunciated the concept that growth media must simulate the microbial habitat to promote growth of the organisms; for example, by adding sterilized rumen fluid as a nutrient supplement and assuring that the medium, as closely as possible, matched the physicochemical characteristics of the rumen. These studies led to the isolation and characterization of a large variety of previously unknown fastidious anaerobes, most of which had complex nutritional requirements, indicating nutritional interdependence of rumen populations. Once isolated, organisms in the rumen fluid, such as cellulose degraders were enumerated, to determine whether their numbers were sufficiently high to account for a significant fraction of the activity measured directly in the rumen. Hungate and his disciples made the rumen an example of how a microbial habitat can be studied quantitatively. Indeed, in some studies of rumen and colon populations, nearly half of the directly counted organisms were cultured.

Brock studied microbial populations in hot springs, mainly in Yellowstone National Park, in the 1960s to 1970s (Brock, 1995). In the initial isolation studies, he also applied principles of habitat simulation by using a medium containing relatively low concentrations of organic nutrients, a mineral composition similar to that of the hot spring, and an incubation temperature of 70–75°C. Whereas many of his predecessors used rich media, incubated at temperatures near 60°C, and invariably obtained *Bacillus stearothermophilus* and its relatives, Brock and colleagues obtained *Thermus aquaticus*, which grew at temperatures up to 78°C, and produced a thermostable DNA polymerase that made possible automation of the polymerase chain reaction. *T. aquaticus* was the beginning of a flood of thermophiles isolated by Brock and later by Karl Stetter (Stetter, 1995) and others. In other studies, Brock and colleagues applied several physiological and microscopy techniques to study microbial populations in the hot springs. These included studies in which the effects of tempera-

ture on processes such as CO_2 fixation were examined in natural populations, and studies in which microautoradiography was employed to examine growth and metabolism of individual cells. Growth of organisms in boiling water was demonstrated by examining colonization of microscope cover slips *in situ*, a technique used by Arthur Henrici in the 1930s to culture (but not isolate) organisms such as *Caulobacter* from aquatic habitats. Cover slips which were repeatedly treated with germicidal ultraviolet radiation had considerably fewer organisms present, demonstrating that the organisms were mainly growing on the slides rather than passively attaching to the cover slips. Electron micrographs of these populations showed unusual ultrastructures that today are easily recognized as archaeal. These boiling water organisms eluded culture until it was realized that most were anaerobes or microaerophiles (Stetter, 1995).

THE WOESEAN REVOLUTION AND ITS IMPACT ON MICROBIAL ECOLOGY

As is amply described elsewhere in this volume, Carl Woese's studies on molecular phylogeny had a revolutionary effect on microbial systematics. In terms of understanding microbial diversity and evolution, Woese's phylogeny exposed the fallacy of assuming that the number of named species is any indication of diversity, a notion put forth by Mayr (1998); for example, that there are millions of animal species and only thousands of bacterial ones. This fallacy should have already been evident considering that if macrobiologists created species by the same criteria of genetic relatedness as microbiologists (two members of the same species have at least 70% DNA–DNA hybridization and 5°C difference in melting temperature of heteroduplexes, equivalent to 4–5% sequence divergence [Stackebrandt and Goebel, 1994]), humans and chimpanzees (~1.6% sequence divergence) would be members of the same species (Sibley et al., 1990; Staley, 1997, 1999). Early microbiologists were well aware of the metabolic diversity of the microbial world, but this sense of microbial diversity had been lost during the era when molecular biology first came to dominate microbiology. Woese's phylogeny made the extent of diversity in the microbial world more apparent by displaying genetic diversity in a way that had high visual impact and showed clearly how much diversity exists in the microbial world.

About ten years after the Woesean revolution in microbial systematics began, it started to have an equally profound effect on microbial ecology. The development, mainly by Norman Pace and his colleagues (Pace, 1997; Hugenholtz, et al., 1998a), of techniques to retrieve rRNA gene sequences from nature has enabled researchers to identify organisms in natural habitats without the need for culturing them.

Many of the features of 16S rRNA that make it a good taxonomic tool, especially its universal distribution and the fact that it contains regions with various degrees of sequence conservation, also make it a powerful tool for ecological analysis. In one of the most straightforward methods using 16S rRNA (Fig. 1), the DNA is extracted from a mixed microbial population, and primers directed at universally conserved regions of the 16S rRNA gene (rDNA) are used to amplify these genes using the polymerase chain reaction (PCR). The resulting population of rDNAs are then cloned and sequenced (Fig. 1). The different 16S rDNA clones can be analyzed phylogenetically by comparison to the databases of known 16S rRNA genes. Thus, a semiquantitative census or community analysis of the organisms present in a habitat can be obtained without culturing them. It should

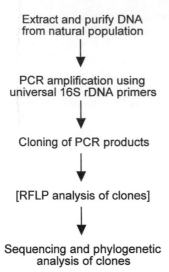

Extract and purify DNA
from natural population

↓

PCR amplification using
universal 16S rDNA primers

↓

Cloning of PCR products

↓

[RFLP analysis of clones]

↓

Sequencing and phylogenetic
analysis of clones

FIGURE 1. Community analysis of 16S rRNA from a natural microbial community. Restriction fragment length polymorphism (RFLP) analysis of clones can be performed to identify potentially identical clones and minimize the amount of sequencing done. However, there can be subtle sequence differences between clones with identical RFLP patterns.

be mentioned that there are biases at each step of these procedures, so that care must be taken in applying and interpreting the results of such analyses (Wilson, 1997; von Wintzingerode et al., 1997; Polz and Cavanaugh, 1998; Suzuki et al., 1998). For example, any organism from which DNA is not extracted by the procedure will not be included in the census.

The nearly universal conclusion obtained from applying this technique and its many variants to microbial habitats is that the diversity of uncultured organisms far exceeds, both in number and in kind, the diversity of those cultured. The studies by Pace and colleagues of a single hot spring, Obsidian Pool, in Yellowstone National Park (Fig. 2) are illustrative. One study (Barns et al., 1996) focusing on *Archaea* demonstrated several new branches of the *Crenarchaeota*, a group that on the basis of the small number of representatives cultured, was not considered to be very diverse. Moreover, two archaeal sequences did not cluster in either the *Crenarchaeota* or the *Euryarchaeota*, and were considered a new archaeal group tentatively named the *Korarchaeota*. Other studies have demonstrated that *Crenarchaeota*, all presently cultured members of which are thermophilic, can be found in moderate temperature habitats such as soils, the surface of a sponge (Preston et al., 1996), and even in Antarctic waters (DeLong et al., 1994). A study of the *Bacteria* in Obsidian Pool sediments (Hugenholtz et al., 1998b) revealed several novel phylum-level branches (called division-level branches in the original publication). This and other studies, including those using culturing approaches, greatly increased the diversity of the eubacteria from 11 divisions in 1987 (Woese, 1987) to over 30 (Fig. 3). Finally, PCR amplification of rDNA is being used to re-examine certain chronic diseases which may be caused by an uncultured microorganism, as was the case for ulcer causation by *Helicobacter pylori* (Relman, 1999).

Soil seems to harbor a particularly great diversity of organisms. In one study (Borneman and Triplett, 1997), samples were taken from two Amazonian soils, 16S rDNA clone libraries were generated, and 50 clones from each soil were sequenced. No 2 sequences of the resulting 100 were identical with each other, nor

were there any exact matches between the soil clones and sequences of cultured organisms in the database (Table 1). While certain biases in the PCR procedure may have overemphasized diversity (Polz and Cavanaugh, 1998; Suzuki et al., 1998), it is still the equivalent of pulling out 100 jelly beans from a bag, and finding each to be a novel color and a different color from all of the others. Moreover, many of the classical cultural studies of soil have led to the impression that Gram-positive bacteria and *Proteobacteria* are the dominant procaryote groups. However, molecular studies have shown that less well characterized groups, such as the phyla *Verrucomicrobia* and *Acidobacteria*, are apparently of equal quantitative importance.

Molecular ecological studies using 16S rDNA have led to the recognition of many novel phylum-level branches of the procaryotic phylogenetic tree from which only a few or even no organisms have been cultured (Fig. 4). On the basis of number and diversity of sequences, some of these phylum-level branches have phylogenetic depth comparable to that of some of the better characterized phyla such as the *Proteobacteria*, yet our knowledge of these organisms at best is scant. From the ecological perspective, such studies are useful but represent only a promising beginning. The real challenge is to determine how these microbes interact with their environment and each other.

A suite of other molecular techniques has been developed to further characterize microbial populations in natural habitats. Because 16S rRNA has regions that change at different rates, one can design oligonucleotide probes and PCR primers of various specificities such that some are species specific, whereas others cover broader phylogenetic groupings such as a genus, the *Proteobacteria* (a phylum), or all *Bacteria* (a domain). One can use these probes to measure the amount of rRNA or rDNA from various microbial groups in a natural habitat, usually by filter hybridization. Whereas PCR amplification studies are semiquantitative at best, quantitative information can be derived from hybridization reactions. For example, in one study of an anaerobic bioreactor (Raskin et al., 1994b) it was demonstrated that the sum of the hybridizations to probes specific for various methanogenic groups was roughly equal to that of a probe for all *Archaea*, indicating that all significant archaeal groups had been accounted for.

One can also obtain an index of various phylogenetic groups in a population using techniques such as denaturing gradient gel electrophoresis (DGGE) and thermal gradient gel electrophoresis techniques which separate 16S rDNA PCR products on the basis of their mol% G + C (Muyzer and Smalla, 1998). Another indexing method is terminal restriction length polymorphism (T-RFLP) analysis in which one of the PCR primers is end-labeled with a fluorochrome and the resulting PCR products are subjected to restriction enzyme digestion and electrophoretic product analysis (Liu et al., 1997). These and similar methods give a characteristic pattern of either bands or peaks for a given microbial population, and this pattern can then be compared with patterns of other populations. The effects of a change in environmental conditions, such as a temperature shift, on microbial populations can also be examined. The mobility of the bands and peaks can be compared with those of known standards. They can be identified directly by sequencing, so that specific populations can be studied using these techniques. However, since different sequences can exhibit similar migration and behavior using any of these methods, identification is not always conclusive.

In the future, as DNA sequencing technology continues to

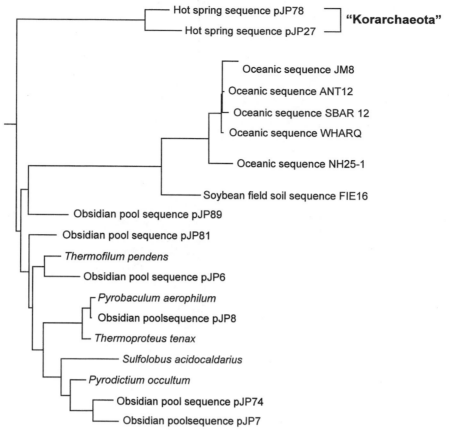

FIGURE 2. Phylogenetic tree showing crenarchaeal sequences including several derived from natural habitats. Also included are the two sequences considered to be members of the new archaeal subdomain, the *"Korarchaeota"*.

improve, it may become feasible to sequence the "genome" of an entire microbial community (Rondon et al, 1999). Information from genome sequencing would suggest hypotheses about the activities of different members of the community, hypotheses which could then be tested biochemically. It is important to note that this exciting possibility requires not just rapid and cheap DNA sequencing technology, but an increased knowledge of bacterial physiology and gene sequences obtained from the study of pure cultures.

An interesting development in modern microbial ecology has been the return to favor of an old ally, the microscope. Microbiologists should not have abandoned their microscopes in the first place, but many of them did. It has taken some fancy new technology to awaken microbiology to the tremendous amount of information they can obtain from microscopic examination of an environmental sample. Fluorescent *in situ* hybridization (FISH), described below, may have stimulated the move back to microscopy, but scientists are rediscovering that a lot can be learned from microscopic examination without resorting to molecular stains. Some bacteria such as cyanobacteria have very distinctive morphologies and naturally fluoresce red when illuminated with light of the appropriate wavelength. The blue-green fluorescence of factor F_{420} is characteristic for many methanogens. A novel use of molecular technology is to insert a gene for green fluorescent protein in an appropriate place in an organism's genome and then use fluorescence microscopy to follow the fate of introduced labeled cells in a habitat. Confocal microscopy provides a three dimensional view of a community. Finally, types of motility and potentially interesting associations can be identified. For example, observation of a cyanobacterium gliding along surrounded by a layer of motile bacteria that contain sulfur granules suggests possible interactions that might be missed by taking a less dynamic view of the population.

FISH is a particularly powerful molecular technique that allows visual identification of phylogenetic groups in natural microbial populations. In its common usage in microbial ecology, a fluorescently labeled oligonucleotide is added to a sample of permeabilized cells, so that it can hybridize to rRNA in the cells and make them fluorescent. Those cells are then viewed using a fluorescence microscope. The specificity of the probe can be adjusted as described above to include all organisms (universal probes), all organisms in a particular phylum, or a single species. Variations of this technique use multiple probes, each with a different fluorescent label (Amann et al., 1995), so that different populations can be visualized in a single sample. FISH is a good way to determine how well a 16S rDNA community analysis reflects the actual composition of the microbial community.

Figure 5 demonstrates the use of FISH to visualize the iron and manganese oxidizing filamentous procaryote *Leptothrix discophora* in a natural aquatic sample. Another study demonstrated that the dominant components of a microbial population in an aerobic sewage digestor were members of the β-*Proteobacteria*, whereas cultural studies indicated that members of the α-*Proteobacteria* were more abundant (Amann et al., 1995). This technique has its own biases and artifacts. It depends upon reliable permeabilization of the target cell populations and on the sample

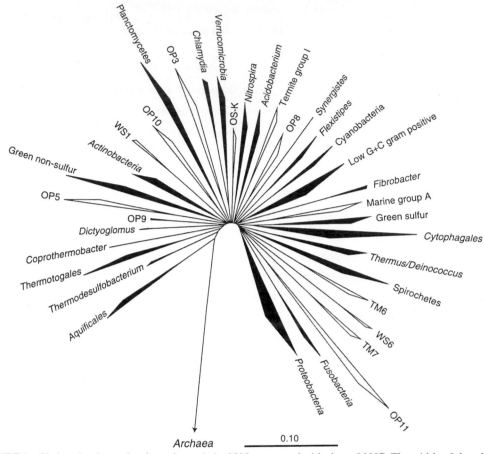

FIGURE 3. Phylum level tree for the eubacteria in 1999 compared with that of 1987. The width of the phylum lines is proportional to the phylogenetic depth within them, and unfilled lines represent uncultured phyla. From Hugenholtz et al. (1998a).

TABLE 1. Phylogenetic assignments of clones derived from PCR amplification of SSU rDNA from two soil samples taken from the Amazon river region[a]

Taxon (Phylum)	Mature forest	Pasture
Crenarchaeota	1	1
Chloroplasts	0	1
Verrucomicrobia	7	6
Planctomycetes	1	0
Low mol% G + C Gram-positive (*Firmicutes*)	11	8
Bacillus	2	8
Clostridium	9	0
High mol% G + C Gram-positive (*Actinobacteria*)	0	3
Cytophaga-Flavobacterium-Bacteroides	3	4
Acidobacteria	9	10
Proteobacteria	6	10
Alpha subdivision	2	2
Beta subdivision	1	0
Delta subdivision	1	4
Epsilon subdivision	2	4
Unclassified	12	7

[a]Although the primers used were universal, no eucaryotes were detected. Data from Borneman and Triplett (1997).

having a low background fluorescence. Moreover, the target cells must contain at least 10,000 ribosomes, a condition that may not be met by slow growing populations often encountered in natural

habitats. Finally, hybridization conditions must be optimized to give the desired specificity.

A part of the molecular revolution that remains relatively underdeveloped is the use of molecules other than nucleic acids to characterize microbial communities. For example, clinical microbiologists have long used antibodies to identify pathogens. This approach might be useful for detecting enzyme expression. If the enzyme is encoded by a regulated gene, its expression indicates that the microbes are experiencing a certain set of conditions.

Some microbial ecologists are beginning to exploit another type of technology for monitoring gene expression: tagging a gene with a reporter group such as the gene encoding green fluorescent protein, and monitoring expression of the gene by measuring fluorescence emitted by the microbe in its natural environment. A drawback to this approach is that the organism whose gene expression is to be monitored must be genetically manipulable in order for the reporter gene to be introduced into the bacterial genome. Moreover, addition of the reporter gene to the organism's natural complement of genes may reduce its fitness in the environment. Nonetheless, the reporter gene approach looks very promising for the future. There are also natural "reporter" traits. For example if an organism growing in pure culture changes shape during starvation, its shape in an environmental sample can give some indication of the degree of starvation it is experiencing. As previously mentioned, the

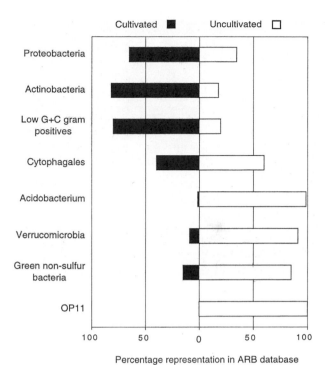

Cultivated ■   Uncultivated □

Percentage representation in ARB database

FIGURE 4. Relative representation of selected bacterial phyla of 16S rDNA sequences from cultivated and uncultivated organisms. From Hugenholtz et al. (1998a).

strength of fluorescence from *in situ* hybridization probes may reflect the number of ribosomes and thus the nutritional state of the microbe (Amann et al., 1994). Methods for direct PCR amplification of genes from bacteria that have been fixed on a slide and permeabilized may soon be sufficiently sensitive to allow single copy genes to be detected, providing important clues about the microbe's metabolic potential. For example, if a bacterium tested positive with primers designed to amplify ribulose bis-phosphate carboxylase, the microbe probably is capable of autotrophic growth, fixing carbon dioxide by the Calvin cycle. As more and more genome sequences appear, it should become easier to identify highly conserved regions of genes associated with a particular microbial activity and to design primers to amplify these genes on slides. By incorporating fluorescent nucleotides in the reaction mixture, the organism containing the gene of interest becomes visible under a fluorescence microscope.

THE CHALLENGE OF UNCULTURED SPECIES

It is clear that we have cultured only a very small proportion of the procaryotic species in nature. These studies have left us with thousands of what Howard Gest (1999) has termed "virtual bacteria", sequences without a corresponding isolated organism. The increase in number of these sequences shows no signs of abating. Clearly, we would know much more from isolates of these organisms than from their 16S rDNA sequences alone. It has been suggested by some that many of these "uncultured" organisms are "unculturable". There are reports that in natural habitats some microorganisms enter a state, often upon starvation, called "viable, but non-culturable". That is, they cannot be grown using standard culture media but are still viable in their environment or, in the case of pathogens, are still able to cause disease (Colwell, 1993). This concept is controversial (Bogosian et al., 1998),

but that controversy is not relevant to this discussion. Even assuming that the concept is completely valid, there is still no reason to equate uncultured with unculturable. We believe that a free-living uncultured organism is culturable until proven otherwise. Our concern with this semantic point is that equating the terms uncultured and unculturable legitimizes not attempting to culture organisms represented by novel rDNA sequences, since if they are nonculturable anyway, why bother trying?

The idea of unculturability is bolstered by the previously mentioned "great plate count anomaly", since direct microscopic counts are so often much higher than viable counts. There are several other reasonable explanations for this phenomenon, however. No single growth medium, even if well designed, can necessarily culture from a sample all of the organisms present, that could include aerobes, anaerobes, heterotrophs, chemolithotrophs, and phototrophs. Moreover, microorganisms in natural habitats are often associated with particles, microcolonies, or biofilms, which must be dispersed to provide accurate counts.

In all too many studies, inappropriate growth media have been used to culture organisms from natural habitats. In the most common examples, overly rich media, often designed for culturing microorganisms from the human body, are used in attempts to culture organisms from soil and water. In addition, most microbiologists are accustomed to growing pure cultures on high concentrations of soluble substrates in batch culture to obtain high organism densities. In many natural habitats, soluble organic compounds are found at vanishingly small steady-state concentrations, and exposure to high nutrient concentrations can be toxic to these starved organisms. Better results can be obtained using appropriately habitat-simulating media. For example, one study (Button et al., 1993) showed that viable counts representing up to 60% of direct microscopic counts could be obtained from seawater if a growth medium consisting of sterilized seawater without organic amendment was used for a most-probable-number enumeration technique in which a fluorescence-activated cell sorter with a detection limit of 10^4 cells/ml was used for detection of tubes positive for growth. Addition of as little as 5 mg/l Casamino acids led to complete inhibition of growth of the natural population. Eventually a hydrocarbon oxidizing organism, less susceptible to this inhibition once in culture, was isolated from these and similar samples (Dyksterhouse et al., 1995; Button et al., 1998). Organic contaminants in agar, especially in "bacteriological grade agars" which are brown from contaminants, are inhibitory to some organisms. Better results can be obtained with more purified agar preparations, other gelling agents such as bacterial polysaccharides, or silica gel, which was used by Winogradski to isolate nitrifiers.

Many organisms require organic nutrients in their growth media, and supplying and identifying those nutrients can pose a challenge. Typical nutrients added to media include vitamins, amino acids, and nucleic acid precursors, or complex nutrient sources such as yeast extract (which, incidentally, lacks vitamin B_{12}). Examples of unconventional nutrients required by microorganisms include polyamines such as putrescine (Cote and Gherna, 1994), the peptidoglycan precursor diaminopimelic acid (Cote and Gherna, 1994), cresol (Stupperich and Eisinger, 1989), and the methanogenic cofactors coenzyme M and coenzyme B (Kuhner et al., 1991). Some organisms require lipids or lipid precursors such as mevalonic acid, an isoprenoid precursor. Certain rumen anaerobes require branched-chain volatile fatty acids as precursors of branched-chain amino acids, but corresponding amino acids sometimes do not support growth because

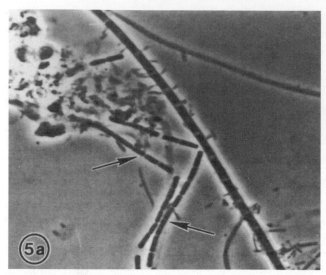
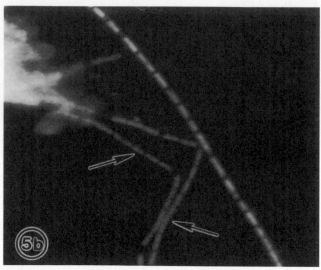

FIGURE 5. Phase contrast (left) and fluorescence *in situ* hybridization (FISH) (right) micrographs of natural populations from an iron/manganese layer in swamp water. The probe used was specific for *Leptothrix discophora*, a sheathed rod with a morphology similar to the cells indicated with arrows. Note that a much thicker filamentous rod is also stained while many other organisms present are not. When the hybridization was done at higher temperature (greater stringency), the thicker rods were not stained. From Ghiorse et al. (1996).

they are not transported into the cell. Often, good sources of unknown nutrients are extracts or supernatants of natural populations or mixed laboratory cultures containing that organism. For example, rumen fluid, a supernatant of rumen contents, is used to cultivate rumen bacteria, and an extract of a mixed culture was used to isolate an organism which utilizes chlorinated ethenes (Maymó-Gatell et al., 1997). These extracts and supernatants help simulate the habitat from which the organism was derived, although the steady state concentrations of essential nutrients in those preparations may not support significant growth in batch culture. Care must also be taken to simulate the physicochemical environment of an organism (Breznak and Costilow, 1994), which includes factors such as pH, oxidation-reduction potential, and concentrations of various solutes. Finally, more sophisticated factors, such as cytokines (Mukamolova et al., 1998), may be needed for the culture of a given microorganism.

Enrichment culture is a powerful tool for isolation of novel metabolic types of organisms, especially ones that carry out a given process rapidly. However, enrichment may not always yield the most numerous organism carrying out that process. As an example, a thermophilic cyanobacterium named *Synechococcus lividus* was routinely enriched from undiluted samples of hot spring photosynthetic mats in Yellowstone National Park. *S. lividus* was morphologically identical to the cyanobacteria seen in microscopic examinations of hot spring mat material. However, molecular community analysis showed that several other cyanobacteria with 16S rRNA sequences highly divergent from that of *S. lividus* were present in the mat material (Ward et al, 1998). Moreover, according to hybridization and DGGE analyses (Ferris et al., 1996), *S. lividus* was only a minor component of the microbial community. Dilutions of Octopus Spring mat material, containing ~10^{10} cyanobacterial cells per ml, into liquid growth medium showed that *S. lividus* dominated in low dilutions, whereas in dilutions which received 25–1000 cells, much slower growing strains predominated. Two strains were isolated from these high dilutions: one of these had a 16S rDNA sequence which matched one of the major sequences from the mat, while the other had a sequence not found in the mat material, an instance of culture

techniques finding an organism not found using the molecular approach. Thus, one conclusion from these studies was that *S. lividus* was a "weed", able to grow rapidly in low dilutions in enrichments, and thereby overtake the slower growing strains which predominated in the mat material. Simply diluting the material prior to enrichment yielded some more ecologically important strains.

Molecular techniques can provide information that aids in the culturing of target organisms. One can sometimes infer the physiology of the target organism from that of its relatives. For example, early in Brock's studies of Yellowstone National Park hot springs, large pink tufts of a filamentous bacterium were noted in the outflow of Octopus Spring, but the organism eluded culture. Subsequent molecular ecological studies (Reysenbach et al., 1994), including FISH analysis, showed that the 16S rDNA sequence of the dominant organism fell within the *Aquifex–Hydrogenobacter* group, members of which can grow microaerophilically on hydrogen. This information was used to isolate *Thermocrinis ruber* (Huber et al., 1998c). One can also use molecular techniques to determine optimal culture conditions for growth of an organism. Do its numbers increase under aerobic or anaerobic conditions, with one nutrient present or with another? FISH can allow visualization of an organism, for example, one member of the *Korarchaeota* was shown to be a long crooked rod (Burggraf et al., 1997a). In another case, FISH analysis showed that a large coccoid organism in an anaerobic enrichment culture was associated with a particular archaeal rDNA sequence. Although the organism was not numerically dominant, laser tweezers were used to move single cells of this organism into sterile microcapillary tubes from which they were then transferred to culture medium and isolated (Huber et al., 1995a).

It should be mentioned that knowing the 16S rDNA sequence of a microorganism may not always allow one to predict whether it possesses a particular phenotype. Thus, the fact that an uncultivated microbe is found to be closely related to a cultivated photosynthetic organism at the rDNA sequence level does not mean that the uncultivated microbe is necessarily photosynthetic. Yet, as demonstrated in the case of *Thermocrinis*, sequence information about genetic relatedness helps to narrow the possibilities and to suggest cultivation strategies that might not otherwise have been tried.

We conclude this section with a plea for greater future efforts to culture uncultured organisms. Whereas much intellectual energy has been expended on developing and optimizing molecular techniques to study microbial populations, considerably less work has gone into culturing members of those populations. Novel approaches and techniques need to be developed. Ideally, the molecular revolution will stimulate new efforts to cultivate the currently uncultivated microorganisms in two ways. First, as already mentioned, a knowledge of what types of microorganisms might be present may suggest the type of media that would be most successful. Second, once scientists know that a particular organism is present in the site, they are less likely to give up after a few halfhearted attempts at cultivation. If a significant fraction of the procaryotic species detected using molecular techniques can be cultured, the next edition of *Bergey's Manual* promises to be greatly expanded and much more indicative of the true diversity of microorganisms in nature.

RELEVANCE OF PURE CULTURE STUDIES TO MICROBIAL ECOLOGY

There are those who have claimed that the study of pure cultures of microorganisms is irrelevant to natural habitats in which microorganisms are interacting with each other and experiencing conditions quite different from those encountered in laboratory media. Indeed, studying a microorganism in pure culture is akin to studying animal behavior at the zoo, and one must be careful about extrapolating results from pure cultures to natural settings. However, to completely reject pure culture results as irrelevant to nature is throwing out the baby with the bathwater. For example, *E. coli* geneticists are baffled by the *lac* operon, especially in view of the fact that lactose is mostly absorbed before it reaches the locations where *E. coli* tends to flourish. If they were more aware of the intestinal milieu, however, they would realize that the ability of *E. coli* to cleave β-galactoside linkages and numerous other sugar linkages would be of great value to a bacterium that is grazing on the mucin layer covering the mucosa of the small and large intestine. The important point, whether the above is the case or not, is that a thorough knowledge of an organism's physiology and the characteristics of its normal location can put a scientist at an enormous advantage in interpreting the results of conventional genetic analysis. Even more important, a microbial ecologist with an eye firmly fixed on the environment understands that the genetic analysis is not an end in itself, but is merely a hypothesis-generating step that generates the hypothesis and the reagents to test this hypothesis in a real environmental sample.

We will provide a single example of the usefulness of pure culture results in the study of natural populations. In the 1970s, it was found that the marine luminescent organism *Vibrio fischeri* was bioluminescent only in late stages of culture (for a recent perspective, see Hastings and Greenberg, 1999). It was eventually determined that the concentration in those cultures of a soluble

acylhomoserine lactone (AHSL) autoinducer built up to a critical level, thereby turning on expression of the luminescence genes. Because of this autoinduction phenomenon, the bacteria were luminescent in the light organ of a marine organism, where they were present in high density, but not when in seawater. Eventually, these AHSLs were found in a variety of other *Proteobacteria*, including pathogens, usually regulating some function which is optimal at high organism densities, such as polymer hydrolysis or conjugation. This form of microbial cell-to-cell communication was termed quorum sensing (Fuqua et al., 1994). A similar phenomenon is mediated by small peptides in Gram- positive bacteria and cyanobacteria. It has been demonstrated that *Pseudomonas aeruginosa* production of an AHSL is essential for proper formation of biofilms (Davies et al., 1998), one of the quintessential forms of microbial communities in natural habitats. Moreover, there are now signs of AHSL-mediated communication among different species (Fuqua and Greenberg, 1998) and evidence of their presence in natural aquatic microbial biofilms attached to submerged rocks (McLean et al., 1997). Microbial cell-to-cell communication will be a fertile and fascinating area of study in microbial ecology, and our present level of understanding of this phenomenon is the result of pure culture studies.

Thus, pure culture work can suggest hypotheses, and in some cases give answers, that provide insights into how a natural community operates. In the future, molecules identified in pure culture studies may prove to be valuable indicators of microbial activities in a particular setting. Furthermore, pure culture studies can lead to the development of more sophisticated hypotheses about metabolic interactions between microorganisms in a natural environment.

FINALE

The prospects for microbial ecology in the twenty-first century are promising indeed. The application of molecular techniques in conjunction with more classical ones will provide a wealth of information about the natural habits of the most fundamental and arguably the most dominant form of life on Earth (Gould, 1996; Whitman et al, 1998). Most importantly, future editions of *Bergey's Manual* are guaranteed to have a much more complete description of the diversity of procaryotes. We conclude with a quote from the closing chapter of naturalist E.O. Wilson's autobiography (Wilson, 1994) in which he describes what he would do if he were beginning his scientific career again.

"If I could do it all over again, and relive my vision in the twenty-first century, I would be a microbial ecologist. Ten billion bacteria live in a gram of ordinary soil, a mere pinch held between thumb and forefinger. They represent thousands of species, almost none of which are known to science. Into that world I would go with the aid of modern microscopy and molecular analysis. I would cut my way through clonal forests sprawled across grains of sand, travel in an imagined submarine through drops of water proportionately the size of lakes, and track predators and prey in order to discover new life ways and alien food webs."

Culture Collections: An Essential Resource for Microbiology

David P. Labeda

Collections of microbial strains have existed since bacteriologists were first able to isolate and cultivate pure cultures of microorganisms, and have always been an important aspect of microbiology, whether used as a source of strains for teaching purposes or as an archive of reference material for research, taxonomic, or patent purposes. Although the field of microbiology is assuming an increasingly molecular emphasis, the need for culture collections has not diminished. Culture collections still provide a significant degree of continuity with the past through the preservation and distribution of microbial strains described or cited in publications, and collections often maintain novel microorganisms awaiting future exploitation by biotechnology.

HISTORICAL PERSPECTIVE

The history of microbial culture collections has been reviewed in detail by numerous authors over the years (Porter, 1976; Malik and Claus, 1987), but there is consensus that the first culture collection established specifically to preserve and distribute strains to other researchers was that of Professor František Král in Prague during the 1880s (Martinec et al., 1966). Upon the death of Professor Král in 1911, this collection was acquired by Professor Ernst Pribham, who transferred it to the University of Vienna and issued several catalogs listing the holdings of the collection. Part of this collection was brought to Loyola University in Chicago by Pribham in the 1930s. Many of these strains were subsequently transferred to the American Type Culture Collection upon Pribham's death, but others remained in the collection at Loyola University (Porter, 1976). The Vienna portion of the Pribham Collection was apparently lost during World War II (Martinec et al., 1966). The next oldest culture collection, the Centraalbureau voor Schimmelcultures (CBS), was founded in 1906 and is still active in Baarn and Delft, The Netherlands (Malik and Claus, 1987). It has served to rescue the collections of many European laboratories. The large number of filamentous fungi collected by Charles Thom at USDA in Washington, DC, in the early part of this century formed the nucleus of the collections of these microorganisms at both the American Type Culture Collection and the Agricultural Research Service (NRRL) Culture Collection in Peoria, Illinois, when the strains were transferred in the 1940s (Kurtzman, 1986). Both of these collections have also served as repositories for many other orphaned culture collections from individuals and institutions, such as the N.R. Smith *Bacillus* Collection or the U.S. Army Quartermaster Collection.

ROLE OF CULTURE COLLECTIONS

The major culture collections throughout the world, some of which are listed in Table 1, have as their primary commission the preservation and distribution of germplasm that has been demonstrated to have significance to the microbiology community. The importance of a particular strain may be as a reference for medical or taxonomic research, as an assay organism for testing or screening, or as an essential component of a patent application for a product or process in which it is involved. Alternatively, the strain may be placed in a collection with reference to the publication in which it was cited as part of the investigation. This latter form deposition is essential on account of the inherent transience of researchers and their research programs, making it possible for later investigators to repeat or advance published research that would be impossible in the absence of the strains involved. As mentioned above, the many national reference and service collections have succeeded in preserving, for later generations of microbiologists, many of the private and specialized collections of microorganisms that may represent an entire career of one microbiologist. In other cases, however, the acquisition of a collection of strains may well result from a change in the direction of the research program in a scientist's laboratory.

Active culture collections represent centers of expertise in the methods of preservation of microbial germplasm and collection management practices, by virtue of their day-to-day activities in these areas. As such, they are an invaluable resource for training others in these important activities.

The major culture collections of the world also serve as centers for excellence in research in systematics and taxonomy. In large part the identification and characterization of strains is an integral function of collections, and the availability of a large collection of strains is essential for this type of research. Culture collections that have contract identification services are also continually searching for faster and more reliable methods to characterize unknown strains for their clients. In many cases, the strains maintained in any collection will directly reflect the taxonomic interests of the curators, in terms of the depth and breadth of particular taxonomic groups.

ROLE IN PRESERVATION OF MICROBIAL BIODIVERSITY

The Convention on Biodiversity, also known as the Rio Treaty (Convention on Biological Diversity Secretariat, 1992), has resulted from the recent global emphasis on conservation of biodiversity and, although dealing more specifically with higher or-

TABLE 1. Some of the world's major bacterial culture collections

Collection	Address
ATCC	American Type Culture Collection 10801 University Boulevard, Manassas, VA 20110–2209 USA Telephone: 703–365–2700; Fax: 703–365–2701 Web site: www.atcc.org/
BCCM®/LMG	Belgian Coordinated Collections of Microorganisms Laboratorium voor Microbiologie, Universiteit Gent (RUG) K.L. Ledeganckstraat 35, B-9000 Gent BELGIUM Telephone: 32-9-264 51 08 ; Fax: 32-9-264 53 46 E-mail: bccm.lmg@rug.ac.be Web site: www.belspo.be/bccm/
DSMZ	Deutsche Sammlung von Mikroorganismen und Zellkulturen GmbH (German Collection of Microorganisms and Cell Cultures) Mascheroder Weg 1b, D-38124 Braunschweig GERMANY Telephone: 49-531–2616 Ext. 0; Fax: 49 531–2616 Ext. 418 E-mail: help@dsmz.de Web site: www.dsmz.de
IFO	Institute for Fermentation, Osaka 17–85 Juso-Honmachi 2-chome, Yodogawa-ku, Osaka, 532 JAPAN Telephone: 81–6-300–6555; Fax: 81–6-300–6814 Web site: wwwsoc.nacsis.ac.jp./ifo/index.html
NRRL	Agricultural Research Service Culture Collection National Center for Agricultural Utilization Research 1815 North University Street, Peoria, IL 61604 USA Telephone: 309–681–6560; Fax: 309–681–6672 E-mail: nrrl@mail.ncaur.usda.gov Web site: nrrl.ncaur.usda.gov
JCM	Japan Collection of Microorganisms RIKEN Hirosawa, Wako-shi, Saitama, 351–01 JAPAN Telephone: 81–48–462–1111; Fax: 81–48–462–4617 Web site: www.jcm.riken.go.jp
NCIMB	National Collections of Industrial and Marine Bacteria, Ltd. 23 St. Machar Drive, Aberdeen, AB24 3RY Scotland, UNITED KINGDOM Telephone: 44–0-1224 273332; Fax: 44–0-1224 487658 E-mail: ncimb@abdn.ac.uk Web site: www.ncimb.co.uk
NCTC	National Collection of Type Cultures PHLS Central public Health Laboratory 61 Colindale Avenue, London, NW9 5HT UNITED KINGDOM Telephone: 44–181–2004400; Fax: 44–181–2007874 Web site: www.ukncc.co.uk

ganisms such as plants and animals, suggests *in situ* conservation of genetic resources through the establishment of protected habitats. The very nature of microorganisms makes this concept somewhat untenable, and thus culture collections should play a major role in the cataloging and *ex situ* preservation of microbial germplasm. Moreover, although the Rio Treaty encouraged the establishment of means of conserving genetic resources in the country of origin, this may not be economically or technologically feasible because of the costs and training involved in the *de novo* establishment of a culture collection (Kirsop, 1996). The established national culture collections are staffed with experienced personnel well versed in the preservation of microorganisms. The relative shortage worldwide of trained microbial taxonomists magnifies this problem, and since the large established collections are centers of excellence in systematics, this is additionally supportive of their potential role in conservation of microbial biodiversity.

TYPE STRAINS

A mission critical function of the culture collections is the preservation and distribution of type strains as a primary reference for taxonomic research. The importance of type strains to microbial systematics has been reiterated in virtually every edition of *Bergey's Manual.*

Type strains represent the primary reference for taxonomic characterization, whether it is identification of unknown strains, re-characterization of known taxa, or description of new taxa. The advent of molecular phylogenetic characterization and analysis based on sequence determination of the 16S ribosomal RNA gene, or other conserved genes, does not diminish the importance of culture collections; for after all, type strains represent a significant part of the "foliage" on the procaryotic tree of life. Sequence databases have largely been constructed using the type strains of microorganisms held in the international reference and service culture collections. Phylogenetic trees from 16S rRNA gene sequences serve as an indication of the evolutionary relationship among strains, but may underestimate the actual differences between strains. Type strains are thus still necessary for evaluation of subtle phenotypic differences between strains and are essential if other gene sequences are to be determined.

The deposition of type strains of new taxa in one or more of the internationally recognized permanent culture collections, in conjunction with description and valid publication, was a recommendation under Rule 30 of the *Bacteriological Code* through the 1992 Revision (Sneath, 1992). The International Committee on Systematic Bacteriology, upon the recommendation of the Judicial Commission, emended Rule 30 of the code to change this recommendation to an absolute requirement (International Committee of Systematic Bacteriology, 1997). Under this revised rule, a taxon cannot be considered validly published and hence a valid name unless it has been deposited and is available for distribution from a recognized culture collection. Moreover, descriptions of new taxa are not accepted for publication in the *International Journal of Systematic and Evolutionary Microbiology*, the official organ of the International Committee on Systematic Bacteriology, unless type strains have been deposited. Thus, it is the role and responsibility of the permanent culture collections to preserve and distribute to the scientific community type material for all of the validly published taxa. The skill in strain preservation in the major culture collections is such that, barring a major disaster, it is unlikely that type strains held there will be lost, and the frequent distribution and replication of type material among permanent collections is another form of protection against such a loss. Should the type strain of a taxon be lost for any reason, however, the procedure for defining a neotype strain is outlined in the *Bacteriological Code*, and a culture of this strain must be deposited in one or several of the permanent collections.

NETWORKING AND DATABASES

With the advent of the Internet, networking of culture collection information has become commonplace. The first compilation of information regarding the culture collections of the world, based on an international survey, was provided in the first edition of the *World Directory of Culture Collections* (Martin and Skerman, 1972). This was replaced ten years later by the updated second edition (McGowan and Skerman, 1982). A computerized database of this information was maintained as the *World Data Centre for Culture Collections of Microorganisms* (WDCM) at the University of Queensland until 1986, when WDCM was moved to RIKEN, Japan, the site of the Japan Collection for Microorganisms. The WDCM was subsequently moved again in 1996 to the National Institute of Genetics in Japan. Currently 498 culture collections are registered in the WDCM database. The culture collection and strain information compiled and held at the WDCM have been available to microbiologists throughout the world via the World Wide Web since 1994, with approximately 30,000 average accesses per month (H. Sugawara, personal communication). There also has been an explosion in the number of on-line cat-alogs for culture collections now available on the Internet (Canhos et al., 1996). The WDCM website (wdcm.nig.ac.jp) provides a useful starting point on the Internet to begin a search for this on-line culture collection information. The collection database at WDCM is useful for deciphering the siglas (e.g., ATCC, DSMZ, JCM, NRRL, etc.) used by collections and identifying the location of the collection and contact information. The STRAINS database allows searching for taxonomic names and provides an indication of which culture collections throughout the world have a strain or strains available for a particular species.

The global interest in the study of microbial diversity and biotechnological utilization of microorganisms has greatly accelerated the placement and interrelating of the collection data with databases related to genomics and physiological properties of microorganisms. The efforts toward the total integration of microbial data using the Internet has been well reviewed by Canhos et al. (1996), Larsen et al. (1997), and Sugawara et al. (1996), and so will not be discussed here.

Intellectual Property of Procaryotes

Roy D. Meredith

Procaryotes and their macromolecular components are protectable as intellectual property, which is a composite legal field of mostly federal laws on patents, trademarks, and copyrights. Patents cover scientific inventions evidencing practical application, and provide exclusive rights for a limited period. Trademarks, as well as tradenames and trade dress, are labels designed to identify to the public particular goods or services, and function to preserve the reputation of a business and to prevent confusing similarity. Copyrights protect original works fixed in any tangible medium of expression, and may be applicable to nucleotide or amino acid sequences. All three of these kinds of intellectual property possess the common characteristic of enabling the owner to obtain an injunction against unlicensed use, and to seek monetary damages. Except where specifically noted, the present essay covers only federal laws of the United States.

PATENTS

An invention is patentable if it is new, useful, and not an obvious variation of what is known. What is held to be new under the law is roughly any invention without its anticipation existing in the public domain, i.e., there is no closely similar invention by another, whether published or publicly known. An invention must be useful to be patented, and this requirement of utility includes some practical application with at least some initial evidence that the invention will work as stipulated, e.g., a DNA sequence capable of expressing a structural protein of medical value with an experiment showing such expression in one host cell. Applications for perpetual motion machines are deemed incredible and lack such utility. Other features of the requirement of utility relate to statutory subject matter, and prevent patenting of mathematical equations, methods of doing business, evolutionary trees, and the like. Finally, a patentable discovery must not be an obvious variant of what is known, the standard of obviousness being defined with reference to a person of ordinary skill in the art. What is an obvious discovery and therefore not patentable under the law is similar in scope to a balanced expert's view of what is obvious in his or her field of expertise.

The patent law on procaryotes in the United States has undergone rapid development ever since a well publicized decision of the U.S. Supreme Court in 1980. This decision was partly responsible for substantial increases in investment and business development in the commercial application of recombinant DNA methods. In *Diamond* v. *Chakrabarty*, a patent claim to a microorganism *per se* was held to be patentable as appropriate statutory subject matter (No. 79–1464, 1980). The U.S. Patent and Trademark Office (USPTO) had rejected a patent claim to a recombinant *Pseudomonas* species capable of metabolizing camphor and octane, two components of oil. The practical application of the microorganism for the clearance of oil spills was not an issue, but the USPTO held the patent claim to be inappropriate statutory subject matter. On appeal after intermediate appellate review, the U.S. Supreme Court held that the patent law permits patenting of "anything under the sun that is made by man." The decision was split 5–4, a hint of potential weakness as binding precedent for future legal decisions of the U.S. Supreme Court. However, no challenges have since been made to overturn or substantially modify *Diamond* v. *Chakrabarty*.

Patenting of the macromolecular components of a procaryotic cell, or of any other cell, largely follow classic guidelines and case law of chemical entities. Since *Diamond* v. *Chakrabarty*, the patent practitioner has available an increasing body of case law relating to polynucleotides or proteins of defined amino acid sequence, vectors, plasmids, and so on. This development is largely consistent with older case law on defined synthetic molecules of an organic or inorganic nature. Patent practitioners can now provide advice of a more certain nature to inventors, providing more opportunities for business development.

Patent law in the United States and Europe differs in various respects. First, a patent claim filed in the United States on a living organism *per se* cannot be properly rejected on moral or ethical grounds, unless the subject matter is repugnant, e.g., claims to a virulent strain of *Yersinia pestis* intended for biological warfare. In contrast, the laws in Europe may prevent patenting of recombinantly altered living organisms wholly confined to the laboratory (e.g., transgenic mice having exclusive uses related to research and drug development). At the time of this writing, the European Union has not resolved the issues, so such ethical concerns may continue as long standing impediments to obtaining coverage in Europe for patent claims to certain kinds of organisms *per se*. The European Patent Convention now permits patent claims to microorganisms alone.

A second important difference is the effect of an inventor's own publication. In the United States, there is a one-year grace period for filing a patent application after the publication date of the invention in a scientific paper, or abstract. In contrast, Europe, Japan, and many other countries have the rule of absolute novelty, which requires the filing of a patent application before publishing. For valuable inventions, an inventor is well advised to follow the absolute novelty rule to obtain non-U.S. patent protection.

A third important difference relates to priority of invention. In the United States, priority depends on the first to invent, not the first to file. In Europe, Japan, Canada, Australia, and many other countries, priority depends on the first to file, prompting a rush to the patent office. The consequence in practice allows an inventor filing in the United States to prove a date of invention earlier than the filing date, in a patent interference proceeding. Notwithstanding these differences, most patent applications having substantial commercial value should be filed diligently after the invention is actually reduced to practice, so that the inventor is both the first to invent and the first to file (Meredith, 1997).

Enforcement of a patent arises after allowance and issue by the USPTO, when the invention is patented. The patent owner possesses the legal right to prevent any third party from making, using, selling, or offering for sale his or her patented invention in the United States. Alternatively, the patent owner may license the invention on terms and conditions acceptable to the patent owner and his or her licensee(s). To enforce the patent, the patent owner files an action of patent infringement in federal court, which has substantially uniform procedure throughout the United States. Such litigation is often expensive, time-consuming, and not prone to settlement. However, when there is valid patent infringement, the owner may be adequately compensated for lost profits.

The ethics of patenting living things primarily relates to several concerns. First, patenting living things may be inherently unethical because it involves ownership of a living thing. This view ignores long standing law allowing ownership of living things, even bailment law on cattle from medieval English common law. Second, patenting living things may encourage inappropriate commercial exploitation. This second view has merit to the extent that it is morally wrong to commercially exploit a particular technology. For example, there is a consensus that it is unethical to promote the commercial exploitation of methods for altering the human germ line by in vitro recombinant DNA. Perhaps it is worth pointing out that the patentee does not seek to patent a living thing with the intention to inflict cruel and unusual punishment. Patents also have limited terms, now 20 years after filing, in most countries including the United States. Third, patenting living things discourages research by suppressing publication, and therefore it may be unethical. This third view correctly points out that delay of scientific publications sometimes occurs before the filing of a patent application, but proper planning typically avoids this situation. These ethical concerns are not likely to be settled in the near future.

TRADEMARKS AND RELATED CONCEPTS

Many commercial products and kits for laboratory use are trademarked, and are well known to any laboratory scientist. Examples include the SORVALL® centrifuge, Vector NTI™ molecular biology software, or Epicurian Coli® XL10-Gold™ cells for high transformation frequency. Trademarks distinguish particular goods or services from others in the marketplace. They function to preserve business reputation by preventing copying of the trademark in the sale of similar goods or services.

Trademarking usually occurs years after filing of a patent application, e.g., just before commercial launch. In the drug industry, it is common practice to trademark the same drug with several different trademarks in the course of commercial development.

A trademark is a word, symbol, or device used to identify the goods of a manufacturer or distributor and to distinguish them from the goods or services of others. The mark must not be confusingly similar to those already registered or in use. Tradenames and trade dress are related concepts, and are also important intellectual property rights. A tradename is the name of a corporation, business or other organization, e.g., Genome-SystemsInc™. Trade dress involves the total image of a product and may include features such as size, shape, color, graphics, and so on.

To obtain enforceable rights, the corporate sponsor is well advised to undertake a trademark search, and seek professional advice from legal counsel. It is not uncommon to wait too long, with the result that a mark, name, or trade dress already in use by internal marketing personnel must be substituted with a new one. If advertising investment is made to establish an exclusive association of a unique mark with a particular product, the long term value of the mark will be substantially greater.

Enforcement against unlicensed use is carried out in federal court, in much the same way as patent enforcement. The owner may obtain an injunction and monetary damages. The laws of trademarks, tradenames, and trade dress are a mosaic of federal and state law, as well as numerous additional iterations in foreign countries.

COPYRIGHTS

There is some scholarly authority that DNA sequences, and by implication amino acid and RNA sequences, are copyrightable (Kayton, 1982). Others disagree (Cooper, 1997). Copyrighted subject matter gives the author or owner exclusive rights to reproduce the copyrighted work or its derivative. Courts have recently extended copyright protection to computer software. By analogy, it seems reasonable to extend copyright law to natural or synthetic sequences having biological significance, particularly synthetic sequences created by the experimental scientist.

Copyrightable subject matter is defined as "original works of authorship fixed in any tangible medium of expression" (17 U.S C Section 102(a)). Original includes any new collection and assembling of preexisting materials, e.g., nucleotides or amino acids to be polymerized into a sequence. The work must be fixed in a tangible medium. Court cases find that floppy disks, CD-ROM, and RAM are examples of a tangible medium, suggesting that a DNA strand or protein would be similarly held to be a tangible medium. An idea, e.g., a mathematical formula, is not copyrightable, just like patent law.

One advantage of copyrighting polynucleotide or amino acid sequences is the ease of registration. Mere registration in the U.S. Copyright Office may be sufficient to confer an enforceable right. There is no lengthy examination of the copyright, although registration may be refused. By contrast, patent claims are not enforceable until after examination, allowance, and issue by the USPTO. Examination, allowance, and issue of a patent application often takes more than several years.

CONCLUSION

Procaryotes and their macromolecular components are protectable as intellectual property. An explosion of commercial interest in this scientific area over the past two decades has created a body of case law allowing legal counsel to give better defined advice to prospective inventors, universities, and corporate sponsors. However, many issues in the intellectual property of cells remain to be clarified by the courts, legislative bodies, and regulatory agencies.

This sketch outlines the current state of the intellectual property of procaryotes in 1998. It will be appreciated that, as a general matter, the law is an ever evolving collection of rules that are sometimes subject to substantial modification, even reversal. In this respect, the law differs fundamentally from scientific advances, which are timeless and always true. The microbiologist seeking to bring an invention to the marketplace is well advised to seek legal counsel early in the process of business and commercial development, particularly for patent protection.

ACKNOWLEDGMENTS

The Board would like to acknowledge the wise counsel and friendship of Rick Meredith who passed away in 1998.

FURTHER READING

No. 79–1464. Diamond v. Chakrabarty, 447 U.S. 303, 65 L.Ed.2d 144 (Supreme Court).

Cooper, I.P. 1997. Biotechnology and the Law, Clark Boardman Callaghan Co., New York.

Goldbach, K., H. Vogelsang-Wenke and F.-J. Zimmer. 1997. Protection of Biotechnological Matter under European and German Law: A Handbook for Applicants, VCH-Law Books, Wenheim.

Kayton, I. 1982. Copyright in living genetically engineered works. Geo. Wash.L. Rev. *50*: 191–218.

Meredith, R. 1997. Winning the Race To Invent. Nat. Biotechnol. *15*: 283–284.

Meredith, R. 1995. Good news for inventors: A lawyer's view. Soc. Ind. Microbiol. *45*: 73–74.

Miller, A.R. 1993. Copyright protection for computer programs, databases, and computer-generated works: Is anything new since CONTU? Harv.L. Rev. *106*: 978–1073.

The Road Map to the *Manual*

George M. Garrity and John G. Holt

INTRODUCTION

The Second Edition of *Bergey's Manual of Systematic Bacteriology* represents a major departure from the First Edition, as well as from the Eighth and Ninth Editions of the *Manual of Determinative Bacteriology*, in that the organization of content follows a phylogenetic framework, based on analysis of the nucleotide sequence of the ribosomal small subunit RNA, rather than a phenotypic structure. The Eighth and Ninth Editions of the *Determinative Manual* and the First Edition of the *Systematic Manual* were organized in a non-hierarchical scheme because information about higher taxa was insufficient for construction of a formal hierarchical classification, such as those used in all the previous editions. Instead, the genera were organized into phenotypic groupings, e.g., the Gram-positive cocci, and these groupings were called Parts, Sections, or Groups. By the Ninth Edition of the *Determinative Manual*, which was based on the information included in the *Systematic Manual*, there were 35 separate phenotypic groupings.

As early as the 1970s, molecular comparisons suggested a natural phylogenetic classification eventually would be possible. Some sections of both manuals were organized into cogent phylogenetic groups, but the editors were not able to place the bulk of genera into a scientifically sound hierarchy based on phylogenetic relationships. By the end of the 1990s the number of 16S rDNA sequences became large enough, and the taxonomic coverage sufficiently broad, to justify the organization of this edition of the *Systematic Manual* along phylogenetic lines. The rationale for presenting the content of this edition, using the continually evolving natural classification scheme based on 16S rDNA sequence data, should be clear to most readers.

It was predicted as early as the 1960s that a phylogenetic arrangement of taxa might not be fully compatible with the known phenotypes of the subject taxa (Sokal and Sneath, 1963). While there are a number of clear-cut examples of coincidence between phenotype and phylogeny, (e.g., the spirochetes, green-sulfur bacteria [*Chloroflexi*], and the actinomycetes), the picture is much less clear for the many Gram-negative and low G + C Gram-positive genera, which represent a significant proportion of the known, cultivable taxa. Many seemingly contradictory groupings exist, and since there has been relatively little effort to systematically correlate phenotype with phylogenetic grouping we are left with a dilemma in attempting to present what many bacteriologists require: an organizational scheme that also contains determinative information other than the 16S rDNA sequence. This situation is expected to change dramatically, as additional full genome sequences become available. However, it is unclear as to whether those data will clarify or obscure the picture.

We have also learned, at least at first pass, that an incomplete phylogenetic scheme presents some unique challenges when used as the basis for organizing printed material, especially in the case of the first volume of a multivolume work. The phylogenetic framework presents fewer mnemonic devices than the earlier scheme and no unique order of appearance. In fact, in many regions within the large scale phylogenetic trees, the precise location of a given species, genus or even higher taxon may be uncertain (see Ludwig and Klenk in this volume* for a detailed discussion of this issue). Consequently, readers might have difficulty in intuitively determining the precise location of a given taxon, especially when searching for less familiar ones. We believe this will be a transient problem for most readers and, with some guidance, they should be able to rapidly assimilate the "new" taxonomy presented in this edition.

To expedite this process, the Editorial Board decided that this edition required a "road map" to help the reader find his or her way through the *Manual*. This task fell to the past and present Editors-in-Chief, as they were principally responsible for creating and maintaining the taxonomic hierarchy as well as the overall organization of the content of this edition. A reader may need to know where a certain genus fits into the overall classification and where it will be found in the book; he or she may need to know the identity of an isolate; or he or she may need to know the phenotypic characteristics of the closest relatives of an isolate that has been identified by molecular means. Thus, it has also been our responsibility to act as a guide for the readership, to ensure that their needs would be met.

Each volume of this edition will have an updated "road map" chapter (at the current rate of description of new taxa, the number of new genera between this volume and the fifth and last could number well over 250). Also, there will continue to be refinements in the phylogenetic classification and, we hope, the discovery of new phenotypic features to further describe the taxa.

THE TAXONOMIC HIERARCHY

The outline classification presented here is a work in progress. It was started in the early 1990s in the Trust editorial office. The principal objective was to devise a classification that reflected the phylogeny of procaryotes based upon 16S rDNA sequence analysis and to place all validly named taxa into the classification at a single point, based on the sequence data derived from the type

---

*Overview: A Phylogenetic Backbone and Taxonomic Framework for Procaryotic Systematics.

strain, type species, or type genus. We acknowledge that some workers may raise objections to such an approach, as there are a number of existing genera (e.g., *Clostridium*) that are paraphyletic, with species appearing in multiple locations within phylogenetic trees. Such instances indicate the need for taxonomic revision, as the species appearing in clades apart from the type strain are clearly misclassified. Authors of individual treatments have been requested to provide readers with detailed discussions of the relevant taxonomic and phylogenetic issues, and resolutions or proposals as to how such matters are best resolved where they arise.

Initially, the RDP (Ribosomal Database Project) tree was used to guide the placement of genera within the outline. However, 16S rDNA sequences for the majority of type strains were not available until recently. As a result, provisional placement, based on phenotypic similarity to sequenced strains, was the only option. As new sequence data became available for existing species, placement in the outline was changed accordingly. Also, as new taxa were described they were added to the classification. During this time P.H.A. Sneath and R.G.E. Murray were charged with the task of devising a hierarchical classification based upon the phylogenetic trees. All these efforts were combined into the outline presented in this edition.

In October, 1997, in collaboration with the Center for Microbial Ecology at Michigan State University, the Trust hosted a two-day meeting to discuss progress on updating the RDP tree and to compare the evolving classification with the ARB tree maintained by Ludwig and Strunk at the Technical University of Munich. A panel of 16 internationally recognized experts in procaryotic phylogeny and taxonomy was assembled to discuss known problems within these two large-scale phylogenetic trees and the ramifications of those problems on the further development of a natural classification for *Bacteria* and *Archaea*. Placement of taxa within the two phylogenetic trees was thoroughly reviewed and areas of uncertainty and discordance were highlighted. In addition, a number of other technical issues, having direct or indirect bearing on the development of a workable taxonomy, were raised. These included a lack of control over the quality and authenticity of some sequences, the actual identity of the organisms from which the sequences were obtained, a lack of published documentation on the calculations and algorithms used in the construction of the RDP tree, and the impact of sequence alignment methods on the resulting phylogenies. This effort lead to a significant, albeit slow, improvement in the number and quality of sequences in the trees. More recently, it has also led to experimentation in alternative methods of visualizing very large sets of sequence data.

There were, of course, nomenclatural problems that needed to be addressed. First of all, the Bacteriological Code (Lapage et al., 1992) does not cover taxa above the rank of Class, so we have had to follow other Codes for naming these higher ranks. Secondly, the trees are of little help in determining the limits of ranks above Order or Class. Thirdly, there is no recognition of rank above Kingdom, yet most phylogeneticists state that the living world is contained in three groups in a rank above Kingdom (variously called Domain or Empire), namely the *Archaea*, *Bacteria*, and *Eucarya*. After considerable deliberation, the Trust has concluded that the rank of Domain should be incorporated into the hierarchy. Furthermore, the rank of Kingdom would not be used to avoid possible conflicts with other codes. Within the classification, the *Archaea* and *Bacteria* are divided into Phyla. The phyla are, in turn, successively divided into classes, orders (except

for the *Cyanobacteria* which use the rank of subdivision), families, and genera. In the *Actinobacteria*, subclasses and suborders are also recognized. At the close of 1999, 4314 validly named procaryotic species had appeared either in the Approved List of Bacterial Names (Skerman et al., 1980) or in Validation Lists 1–71,* In addition, 910 synonymies had been recorded as a result of taxonomic emendments. These are summarized in Table 1.

Adoption of a hierarchical classification presents several other issues. By definition, each species must be a member of successively higher ranks (six of which will be recognized for the majority of taxa in this edition of the *Systematics Manual*). Yet, there is considerable reluctance among many workers to place new species and genera into higher taxa, especially at the intermediate levels (family, order, and class). In compiling the outline we have had to deal with situations where new species were variously assigned to a class or domain without being ascribed membership in any of the intervening taxa. This may be attributed to a lack of clear rules for delineating higher taxa. It may also reflect the inherent limitations of the 16S rRNA gene for defining higher taxonomic structure, especially when contemporary phylogenetic techniques, that rely solely on tree graphs, are used to analyze small and inherently biased data sets. We have also observed a general lack of consistency in defining the boundaries of genera based on 16S rDNA sequence analysis. This is particularly problematic in "bushy" areas of the phylogenetic trees where uncertainty of branching order is high and clear demarcation of taxonomic groups is impossible in the absence of other supporting data.

In dealing with such problems, we have "filled- in" the missing taxa so that the hierarchy is complete. Names of higher taxa are based largely on priority, except in instances where such a strategy might lead to unnecessary confusion (e.g., *Helicobacteraceae* rather than *Wolinellaceae*). Each of these higher taxa has also been scrutinized for phylogenetic coherence so as to avoid paraphyletic or polyphyletic taxa, wherever possible. However, since 16S rDNA sequences are not yet available for all validly named species, some such instances are likely to occur. Considerable effort has been spent in confirming the placement of genera within higher taxa. We also acknowledge that some existing taxa are "problematic" and contain misidentified species. There are also areas within the phylogenetic trees that are ambiguous. While corrections have been made to taxa appearing in Volume 1 of this edition, further corrections will have to await publication of subsequent volumes, when authors address these issues in detail.

Despite these limitations, use of the well established phylogeny based on the 16S rRNA gene provides a marked improvement over the earlier artificial classifications. The technique (16S rDNA sequencing) is universal in applicability and will soon provide a single type of data that will be available for all validly named species. Given the rapid advancements in sequencing technology, we expect that other gene sequences (e.g., 23S rRNA gene) will follow in the near future and help in placement of "problem taxa". Therefore, readers must recognize that the current classification is fluid and as each new "road map" is published there

---

*Lists 1–71 were published in the International Journal of Systematic Bacteriology *27* (1977) 306; *29* (1979) 79, 436; *30* (1980) 601, 676; *31* (1981) 215, 382; *32* (1982) 266, 384; *33* (1983) 438, 672, 896; *34* (1984) 91, 270, 355, 503; *35* (1985) 223, 375, 535; *36* (1986) 354, 489, 573; *37* (1987) 179; *38* (1988) 136, 220, 328, 449; *39* (1989) 93, 205, 371, 495; *40* (1990) 105, 212, 320, 470; *41* (1991) 178, 331, 456, 580; *42* (1992) 191, 327, 511, 656; *43* (1993) 188, 398, 624, 864; *44* (1994) 182, 370, 595, 852; *45* (1995) 197, 418, 619, 879; *46* (1996) 362, 625, 836, 1189; *47* (1997) 242, 601, 915, 1274; *48* (1998) 327, 627, 631, 1083; *49* (1999) 1, 341, 935, 1325.

TABLE 1. Summary of taxonomic scheme employed in the Second Edition of *Bergey's Manual of Systematic Bacteriology*

Taxonomic rank[a]	Total	Archaea	Bacteria
Domains	2	1	1
Phyla	25	2	23
Classes	40	8	32
Subclasses	5	0	5
Order/subsection	89	12	77
Suborders	14	0	14
Families	203	21	182
Genera	941	69	871
Species	5224	217	5007

[a]At the close of 1999 there were 4314 validly named species and 849 validly named genera of *Bacteria* and *Archaea*. The increased numbers that will appear in the second edition of *Bergey's Manual of Systematic Bacteriology* are attributed to inclusion of *Cyanobacteria*, which are covered by the Botanical Code, and synonymies (objective and subjective).

will be changes in the placement of some taxa. If it is true that we have described only about 10% of the extant procaryotes, then it is highly likely that this current classification will expand and change. We are in a period of rapid isolation and description of new procaryotic taxa which rivals the expansion of the field in the late 1800s. We hope that this current "natural history" approach will provide the basis for a more meaningful and predictive classification of procaryotes in the future.

MAPPING THE TAXONOMIC SPACE

One of the greatest difficulties we experienced while constructing and updating the outline was to easily visualize the higher taxonomic structure of the procaryotes based on 16S rDNA sequence analysis. While it was clear that such structure should exist, based on similarity in the topology of many regions of the ARB and RDP trees (e.g., the separation of *Archaea* and *Bacteria* and the consistent presence of deeply branching taxa in the *Bacteria*), the validity of many of the intermediate taxa appearing in the outline was less obvious. Summary trees, drawn in various ways (Barnes et al., 1996; Hugenholtz et al., 1998a; Ludwig and Klenk), suggested the presence of 25–40 major lineages within the procaryotes. However, such trees yield relatively little information about either the number of member taxa or their relatedness. On the other hand, while these relationships can be examined in larger trees, such trees obscure the spatial relationships among the taxa, especially when the groups of interest may be separated by 10s or even 100s of pages. To that end, we sought alternative methods of exploring the sequence data for evidence of taxonomic structure and to independently confirm placement of genera within the taxonomic outline.

Following the 1997 meeting on phylogenetic trees, P.H.A. Sneath used principal coordinate analysis to prepare a two-dimensional projection of the major procaryotic groups. The analysis was based upon branch lengths (evolutionary distances) of type strains appearing in Version 6.01 of the RDP tree (See Fig. 1 in Krieg and Garrity's "On Using the *Manual*"). His analysis supported the clear separation of the procaryotic domains. It also supported separation of the deeply branching *Bacteria* and oxygenic phototrophic *Bacteria* from the Gram-negative and Gram-positive *Bacteria*. There was, however, proportionately less separation among the many phyla of the Gram-positive *Bacteria* than was observed among the Gram-negative *Bacteria*. Although it is unclear how one can use such a plot to assert evolutionary relationships, it seems quite reasonable to infer that points that reside close together in these plots are more likely to have a recent common ancestry than those that do not plot closely to-

gether, assuming that the plot of those two principal coordinates accounts for a significant portion of the variance within the data. To show the relationship between the planar projection of evolutionary data and phylogenetic trees, Sneath drew imaginary branches below the plane, along a third axis, time.

Sneath's analysis suggested that ordination techniques might provide a useful alternative to phylogenetic trees, especially for uncovering higher order taxonomic structure within very large sets of sequences (>1000). To that end, a series of experiments were conducted by Garrity and Lillburn using principal component analysis as a means of exploring aligned 16S rDNA sequence data. While their approach differed from Sneath's in several ways, the results showed remarkable consistency.

Garrity and Lilburn's strategy was to calculate evolutionary distances for aligned, full-length 16S rDNA sequences (a subset of sequences contained in Version 7.1 of the RDP) to 184 reference sequences representing type strains on which families were based in the *Bergey's* outline. The rationale for this approach was to use these reference sequences as benchmarks, much like those employed in the production and validation of topographical maps. The sequences ($n = 4502$, length > 1350 nts, ambiguities ≥ 3%) were masked to exclude positions that were not conserved and matrices of evolutionary distances were calculated in PAUP (Version 4.02b, Swofford, 1999; Sinauer Associates, Massachusetts) using maximum likelihood methods, as that approach was found to yield the best separation of taxa without significantly distorting either close or distant relationships. The parameters used in the maximum likelihood estimates were comparable to the F84 model used in fastDNAml or DNAml (Tateno et al., 1994b; Kishino and Hasegawa, 1989). The resulting matrix of vectors was then subjected to a principal component analysis (PCA) in S-Plus 2000 (Mathsoft, Seattle).

Like principal coordinate analysis, PCA provides a means of visualizing high dimension data in a lower dimensional space by finding the uncorrelated single linear combinations of the original variables that explain most of the underlying variability within the data (Mardia et al., 1979; Venables and Ripley, 1994). PCA has been widely used in taxonomic and ecological studies in the past (Sneath and Sokal, 1973; Dunn and Everitt, 1982), it is a well understood method, and is suited to answering questions about higher-order structure within data and uncovering outliers (e.g., misclassified taxa). PCA also offers several advantages. Unlike many "treeing" algorithms used in phylogenetic analysis, PCA is a computationally efficient method, allowing the rapid analysis of data sets with thousands of taxa. The reliability of PCA scatter plots (created by plotting scores for each principal component)

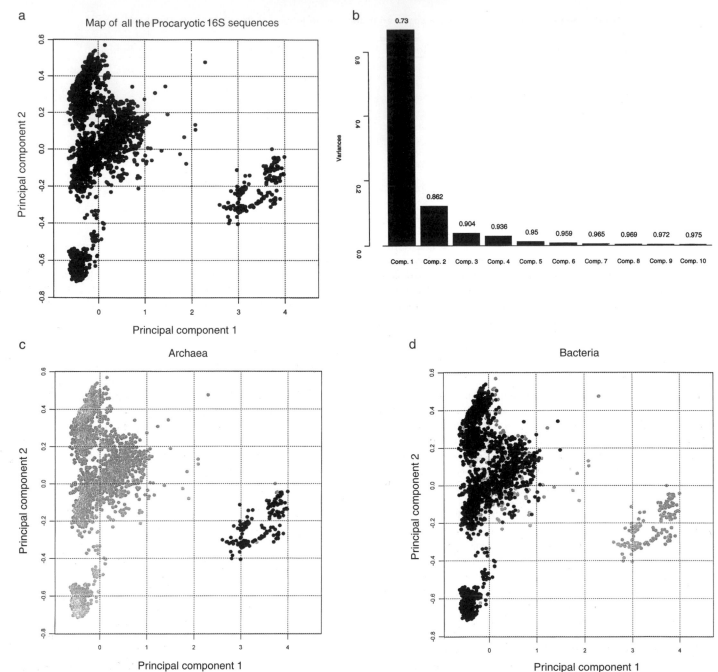

FIGURE 1. *a*, Map of the procaryotes based on a principal component analysis of evolutionary distances derived from full-length 16S rDNA sequences in the RDP (Release 7.01). Evolutionary distances for aligned, full-length 16S rDNA sequences (*n* = 4502, length >1350 nts, ambiguities ≤3%) were estimated, using maximum likelihood methods, to 184 reference strains on which families were based in the Bergey's outline. The resulting matrix was then subjected to a principal component analysis and the principal component scores plotted for principal component 1 vs. principal component 2. *b*, Screeplot of principal component analysis data reveals that the first two principal components account for 86.2% of the total variance within the evolutionary data matrix used to compute the principal components. *c*, Location of type strains of validly named species of the *Archaea* within the map of the procaryotes. The selected strains are highlighted in black, whereas non-type and bacterial strains are shaded. This scheme of presentation will be used throughout to highlight different groups of interest. *d*, Location of the *Bacteria* within the map of the procaryotes.

can be easily tested by estimating the cumulative residual variance explained by the principal components; ranked in descending order. If the first two or three account for >85% of the total variance (Mardia et al., 1979; Venables and Ripley, 1994), the plot is generally considered a good depiction of the relationships among the taxa. Furthermore, if the underlying principal component scores are available for further analysis, one can readily create plots identifying the location of subsets of the original taxa projected back into the original coordinate system. This allows one to work in a fixed space of constant dimension and orientation, overcoming one of the more common problems of PCA: recomputation of principal components for subsets, leading to different views of the data. Despite the obvious utility of PCA and other ordination techniques, it does not appear that these methods have been applied previously to the exploration of evolutionary data.

Proof of principle A PCA plot of all of the sequence data is presented in Fig. 1a. A screeplot (Mardia et al., 1979; Venebles and Ripley, 1994) of the cumulative variance revealed that the first component accounted for >73.0% of the total variance, with second and third components accounting for 13.2% and 4.2%, respectively (Fig. 1b). These results were quite surprising as they indicate that the dimensionality of the evolutionary data, derived from the 16S rDNA sequences, could be significantly reduced with very little loss of information. This finding was confirmed in a series of experiments using successively smaller numbers of randomly selected reference sequences ("benchmarks") in the data vectors. Essentially no impact on the overall topology in PCA plots was observed, until the number of reference points (benchmarks species) dropped below 10. Subsequently, the underlying data were modified, adding the names of the higher taxa to which each species (sequence) belonged and the phenotypic group in which it best fit. These names and categories were then used to create different views, in which subsets could be visualized against the background of all taxa and compared to known placement within the RDP and ARB trees. This tool was used to validate the taxonomic outline and highlight misidentified species (sequences). The locations of the domains *Archaea* and *Bacteria* are shown in Fig. 1c and 1d. Points that correspond to sequences from type strains appear in black. Those which correspond to nomenspecies or environspecies are shaded.

PHENOTYPIC GROUPS WITHIN THE PROCARYOTES

In the Ninth Edition of the *Determinative Manual*, approximately 590 genera were subdivided into 35 major phenotypic groups. These phenotypic groups were based upon readily recognizable characters that could be used for the presumptive identification of species that are routinely encountered in a wide variety of ecological niches. As stated earlier, the objective of the Ninth Edition of the *Manual of Determinative Bacteriology* was utilitarian, and readers were advised that no attempt at creating a natural classification had been made.

Since publication of the *Determinative Manual* in 1994, use of molecular methods of identification has become increasingly common, often to the exclusion of traditional methods of phenotypic profiling. Molecular probes, based largely on conserved regions of the genome (principally 16S and 23S rRNA), have come into routine use and provide a universally applicable technique to aid in the detection and identification of bacteria in virtually any sample, without the need to cultivate. However, despite these advances, there remains a need for incorporation of both phenotypic and genotypic data in formal descriptions and in identification schemes for procaryotes. This is discussed in considerable detail by Gillis et al. in their introductory essay, "Polyphasic Taxonomy," in this volume. Phenotypic information can also play a role in the separation of closely related taxa, especially in ambiguous regions of phylogenetic trees where 16S rDNA sequence data proves inadequate for resolution of such taxa (Ludwig and Klenk). A prime example is provided by the "*Acidaminococcaceae*", a well defined group of cocci possessing a Gram-negative cell wall that is found within the low G + C Gram-positive phylum *Firmicutes*.

To that end, we have undertaken the task of mapping each of the genera into broad phenotypic groups based on those that

were used in the last edition of the *Determinative Manual*. We recognize that some of the original categories have proven to be unnecessarily broad and now include Gram-negative and Gram-positive staining phyla or *Archaea* and *Bacteria*. To address these shortcomings, we have subdivided those groups along phylogenetic lines. We have also added several categories to accommodate newly recognized phenotypes. Assignment of a phenotype to each genus was based upon information provided either in this volume or from the original published descriptions. In instances where phyla will appear in subsequent volumes, the phenotypic categories should be viewed as "working descriptions" of higher taxa. Fig. 2 summarizes the relationships between phenotypic groupings and the 25 phyla defined above. Table 2 provides the reader with an alphabetical listing of the genera that will be included in this edition of the *Systematics Manual*, the phylum and class to which each genus has been (or likely will be) assigned, the phenotypic group, and the volume in which a detailed treatment will appear.

THE PROCARYOTIC PHYLA

As indicated in Table 1, the 2 procaryotic domains have been subdivided into 25 phyla, 2 of which occur within the *Archaea*. The remaining 23 phyla are ascribed to the *Bacteria*. The following are brief, working descriptions that are intended to provide readers with some indication of the relationship between the taxonomic outline and phylogenies proposed by the RDP and ARB trees, along with known problems. It also provides the readers with an indication of where each taxonomic group will appear in this or subsequent volumes of the *Systematics Manual*. Although the contents will be presented in a phylogenetic context, some practical considerations were necessary in the final layout of the individual volumes. Twelve phyla will be presented in this volume and 10 in Volume 5, while Volumes 2, 3, and 4 will each cover a single phylum (the *Proteobacteria*, the *Firmicutes*, and the *Actinobacteria*, respectively). Furthermore, Volume 1 will deviate slightly from the phylogenetic model as the phototrophic species will be presented together, as a phenotypically coherent group. Phototrophic species within the *Proteobacteria* and *Firmicutes* will also be presented (and updated) in Volumes 2 and 3, respectively, in their proper phylogenetic context.

In addition, readers are cautioned that the numbering scheme used in the outline is, to some extent, arbitrary, especially at the lower levels. As the branching order of species within genera is oftentimes ambiguous and the data set is known to be incomplete, the use of phylogenetic trees as a guide to the appearance of taxa in the *Systematics Manual* proved to be untenable. Thus, we have had to adopt a more workable and all-inclusive strategy. The type genus will always appear first within a family, and all other genera within each family (if more than one genus is included) will appear in alphabetical order. Readers are reminded that the numbering of subordinate taxa will be subject to change as new taxa are described and existing taxa split apart. Likewise, the fact that the *Archaea* and deeply branching *Bacteria* are presented first is based largely on the earlier versions of the RDP tree. There will be some deviation from the RDP ordering of taxa in subsequent volumes for a variety of practical reasons.

Phylum A1 **Crenarchaeota** *phy. nov.* Garrity and Holt. *Cren.ar.ch.ae.o'ta* M.L. fem. pl. n. *Crenarchaeota* from the Kingdom *Crenarchaeota* (Woese, Kandler and Wheelis 1990, 4579).*

Editorial Note: names of phyla are given the feminine gender and plural number, in conformance with Rule 7 of the *International Code of Nomenclature of Bacteria* (1990 revision).

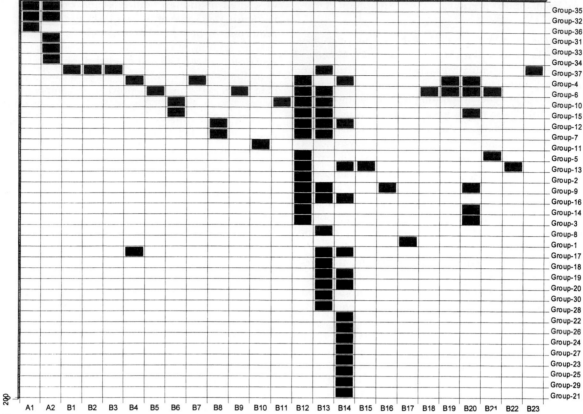

FIGURE 2. Occurrence of major phenotypic groups within the 25 procaryotic phyla. The group number refers to the phenotypic group used in the Ninth Edition of *Bergey's Manual of Determinative Bacteriology* (Holt et al., 1994). Phylum A1-*Crenarchaeota*, A2-*Euryarchaeota*, B1-*Aquificae*, B2-*Thermotogae*, B3-*Thermodesulfobacteria*, B4-*"Deinococcus-Thermus"*, B5-*Chrysiogenetes*, B6-*Chloroflexi*, B7-*Thermomicrobia*, B8-*Nitrospirae*, B9-*Deferribacteres*, B10-*Cyanobacteria*, B11-*Chlorobi*, B12-*Proteobacteria*, B13-*Firmicutes*, B14-*Actinobacteria*, B15-*Planctomycetes*, B16-*Chlamydiae*, B17-*Spirochaetes*, B18-*Fibrobacteres*, B19-*Acidobacteria*, B20-*Bacteroidetes*, B21-*Fusobacteria*, B22-*Verrucomicrobia*, B23-*Dictyoglomi*. Phenotypic Group 1-Spirochetes, Group 2-Aerobic/microaerophilic, motile, helical/vibrioid Gram-negative *Bacteria*, Group 3-Nonmotile or rarely motile, curved Gram-negative *Bacteria*, Group 4-Gram-negative aerobic/microaerophilic rods and cocci, Group 5-Facultatively anaerobic Gram-negative rods, Group 6-Anaerobic, straight, curved, and helical Gram-negative rods, Group 7-Dissimilatory sulfate- or sulfite-reducing *Bacteria*, Group 8-Anaerobic Gram-negative cocci, Group 9-Symbiotic and parasitic *Bacteria* of vertebrate and invertebrate species, Group 10-Anoxygenic phototrophic *Bacteria*, Group 11-Oxygenic photo-trophic *Bacteria*, Group 12-Aerobic chemolithotropic *Bacteria* and associated genera, Group 13-Budding and/or appendaged *Bacteria*, Group 14-Sheathed *Bacteria*, Group 15-Nonphotosynthetic, nonfruiting, gliding *Bacteria*, Group 16-Fruiting gliding *Bacteria*: The myxobacteria, Group 17-Gram-positive cocci, Group 18-Endospore-forming Gram-positive rods and cocci, Group 19-Regular, nonsporulating, Gram-positive rods, Group 20-Irregular, nonsporulating, Gram-positive rods, Group 21-Mycobacteria, Group 22-Nocardioform actinomycetes, Group 23-Actinomycetes with multilocular sporangia, Group 24-Actinoplanetes, Group 25-*Streptomyces* and related genera, Group 26-Maduromycetes, Group 27-*Thermomonospora* and related genera, Group 28-*Thermoactinomyces*, Group 29-Other actinomycete genera, Group 30-Mycoplasmas, Group 31-The methanogens, Group 32-Archaeal sulfate reducers, Group 33-Extremely halophilic *Archaea*, Group 34-*Archaea* lacking a cell wall, Group 35-Extremely thermophilic and hyper thermophilic S⁰-metabolizers, Group 36-Hyperthermophilic non-S⁰ metabolizing *Archaea*, Group 37-Thermophilic and hyperthermophilic *Bacteria*.

The phylum consists of a single class, the *Thermoprotei*, which is well supported by 16S rDNA sequence data. It is subdivided into three orders: the *Thermoproteales*, the *Desulfurococcales*, and the *Sulfolobales*. At present, Ludwig and Klenk indicate good support for the *Sulfolobales* and *Desulfurococcales*. In PCA plots (Fig. 3), member species for which high quality sequences were available map to a position intermediate between the major subgroups within the *Euryarchaeota*.

Morphologically diverse, including rods, cocci, filamentous forms, and disk-shaped cells. Stain Gram-negative. Motility observed in some genera. Obligately thermophilic, with growth occurring at temperatures ranging from 70 to 113°C. Acidophilic. Aerobic, facultatively anaerobic, or strictly anaerobic chemolithoautotrophs or chemoheterotrophs. Most metabolize S⁰. Chemoheterotrophs may grow by sulfur respiration. RNA polymerase of the BAC type. In the *Determinative Manual*, the known crenarchaeotes were assigned to Group 36.

Phylum A2 **Euryarchaeota** *phy. nov.* Garrity and Holt.
Eur.y.arch.ae.o'ta M.L. fem. pl. n. *Euryarchaeota* from the Kingdom *Euryarchaeota* (Woese, Kandler and Wheelis 1990, 4579).

The phylum currently consists of seven classes: the *Methanobacteria*, the *Methanococci*, the *Halobacteria*, the *Thermoplasmata*, the *Thermococci*, the *Archaeoglobi*, and the *Methanopyri*. With the sole exception of the *Methanococci*, which is subdivided into three orders, each class contains a single order. In phylogenetic analyses, Ludwig and Klenk indicate that the *Methanobacteria*, *Methanomicrobiales*, *Halobacteria*, and *Thermoplasmata* share a common root. The relationships among the remaining classes are more ambiguous. In PCA plots, the *Euryarchaeota* split into four sub-

Crenarchaeota

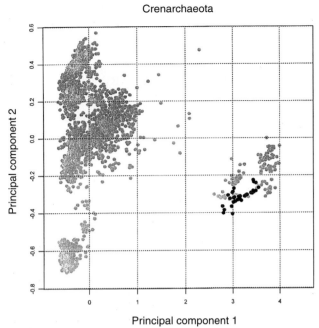

Euryarchaeota

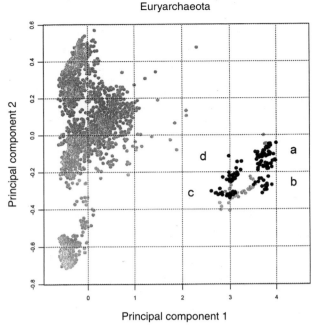

FIGURE 3. Location of Phylum *Crenarchaeota* within the map of the procaryotes.

FIGURE 4. Location of Phylum *Euryarchaeota* within the map of the procaryotes. Region a contains the *Halobacteriales*, *Thermoplasmatales*, and *Methanomicrobiales*. Region b contains the *Methanosarcinales*. Region c contains the *Thermococci*, *Archaeoglobi*, *Methanopyri*, and *Methanothermaceae*. Region d contains the *Methanobacteriaceae* and some members of the *Methanococcaceae*, which appears to be polyphyletic.

clusters, separated by the *Crenarchaeota*. The *Halobacteria* and *Thermoplasmata* form distinct clusters that are found in region *a*, as indicated in Fig. 4. The *Thermococci*, *Archaeoglobi*, and *Methanopyri* also form coherent clusters, located in region *c*. As currently defined in the outline, the *Methanobacteria* and *Methanococci* may be problematic. While the *Methanobacteriaceae* form a discrete and coherent group in region *d*, *Methanothermus* is located in region *c*. The *Methanococci* are split into four discrete clusters, with the *Methanococcus* species falling into two tight clusters in regions c and d; the *Methanomicrobiales* clustering with the *Halobacteria* in region *a*, and the *Methanosarcinales* forming a discrete group within region *b*, at the distal end of the *Crenarchaeota* cluster.

The *Euryarchaeota* are morphologically diverse and occur as rods, cocci, irregular cocci, lancet-shaped, spiral-shaped, disk shaped, triangular, or square cells. Cells stain Gram-positive or Gram-negative based on the presence or absence of pseudomurein in cell walls. In some classes, cell walls consist entirely of protein or may be completely absent (*Thermoplasmata*). Five major physiological groups have been described previously.

The methanogenic Archaea Strict anaerobes that are able to form methane as the principal metabolic end product. H_2/CO_2, formate, acetate, methanol, methylamines, or H_2/methanol can serve as substrates. S^0 may be reduced to H_2S without the gain of energy. Blue-green epifluorescence when excited at 420 nm. Cells possess coenzyme M, factor 420, factor 430, and methanopterin. RNA polymerases are of the AB′B″ type. Genera exhibiting this phenotype are found in the *Methanobacteria*, the *Methanococci*, and the *Methanopyri*. (BMDB9 Group 31).

The extremely halophilic Archaea Cells stain Gram-negative or Gram-positive, aerobic or facultatively anaerobic chemoorganotrophs. Rods and regular to highly irregular cells occur. Cells require a high concentration of NaCl (1.5 M or above). Neutrophilic or alkaliphilic. Mesophilic or slightly thermophilic (up to 55°C). RNA polymerase is of the AB′B″C type. Some species

contain bacteriorhodopsin and use light for ATP synthesis. All known genera with this phenotype are members of a single class, the *Halobacteriales*. (BMDB9 Group 33).

Archaea lacking a cell wall Thermoacidophilic, aerobic, coccoid cells lacking a cell envelope. Cytoplasmic membrane contains a mannose-rich glycoprotein and a lipoglycan. RNA polymerase is of the BAC type. Only a single genus within the *Thermoplasmata* is currently known to exhibit this phenotype. (BMDB9 Group 34).

Sulfate reducing Archaea Cells are strict anaerobes that are able to form H_2S from sulfate by dissimilatory sulfate reduction. Traces of methane are also formed. Extremely thermophilic (growth up to 92°C). Exhibit blue-green fluorescence at 420 nm by fluorescence microscopy. Cells possess factor 420 and methanopterin, but no coenzyme M and no factor 430. RNA polymerase is of the (A+C)B′B″ type. Within the *Euryarchaeota*, this phenotype is restricted to the *Archaeoglobi*.

Extremely thermophilic S^0 metabolizers Obligately thermophilic, strictly anaerobic Gram-negative irregular-shaped cocci. Optimal growth temperature between 67 and 103°C. Heterotrophic growth. S^0 required for growth of some species. Within the *Euryarchaeota*, this phenotype is unique to the *Thermococci*. (BMDB9 Group 35).

Phylum B1 Aquificae The phylum *Aquificae* Reysenbach 2001, 359 consists of a single class and order. In phylogenetic analysis of 16S and 23S rDNA sequence data, the *Aquificae* are generally considered to be one of the deepest and earliest branching groups within the *Bacteria*. However, phylogenetic analyses of other protein and gene sequences show the placement with respect to other *Bacteria* to be variable (Huber and Stetter, Genus

Aquifex). One genus, *Desulfurobacterium*, is provisionally placed within the phylum and may represent another, currently undefined phylum (Ludwig and Klenk; Reysenbach, Phylum *Aquificae*). In PCA plots (Fig. 5), the *Aquificae* are represented by a single sequence from the type strain, *Aquifex pyrophilus*. Readily identifiable, high quality sequences were not publicly available for members of the other genera at the end of 1999. All members of the *Aquificae* are Gram-negative nonsporulating rods or filaments. Thermophilic, with optimum growth in the range of between 65 and 85°C. Chemolithoautotrophic or chemoorganotrophic growth. Many species grow anaerobically and are capable of nitrate reduction. However, both microaerophilic and aerobic species also occur. (BMDB9 Group 36*)

Phylum B2 Thermotogae The phylum *Thermotogae* Reysenbach 2001, 369 consists of a single class and order. In phylogenetic analysis, this phylum consistently branches deeply, along with the *Aquificae*. In PCA plots (Fig. 6), the type strains for member species fall in the same general vicinity as the *Aquificae* and are widely spaced, suggesting a high level of variability. All *Thermotogae* are Gram-negative, nonsporulating, rod-shaped bacteria that possess a characteristic sheath-like outer layer or "toga". *meso*-Diaminopimelic acid not present in peptidoglycan. Strictly anaerobic heterotrophs, utilizing a broad range of organic compounds for growth. Thiosulfate and/or S⁰ reduced. Growth inhibited by H₂. (BMDB9 Group 36).

Phylum B3 Thermodesulfobacteria *phy. nov.* Garrity and Holt. Ther' mo.de' sul.fo.bac' ter.ia M.L. fem. pl. n. *Thermodesulfobacteriales* type order of the phylum; dropping the ending to denote a phylum; M.L. fem. pl. n. *Thermodesulfobacteria* the phylum of *Thermodesulfobacteriales*.

The phylum *Thermodesulfobacteria* is currently represented by a single species which branches deeply in the ARB tree. High quality sequence data were not available from the RDP at the time of publication; therefore, the location of this phylum in PCA plots is unkown. Gram-negative, rod-shaped cells possessing an outer membrane layer which forms protrusions. Thermophilic, strictly anaerobic, chemoheterotrophs exhibiting a dissimilatory sulfate-reducing metabolism. (BMDB9 Group 36).

Phylum B4 "Deinococcus–Thermus" The *"Deinococcus–Thermus"* phylum represents a deep branching line of descent that is defined largely on the basis of 16S rDNA signature nucleotides (Battista and Rainey, Family *Deinococcaceae*, and da Costa, Nobre and Rainey, Genus *Thermus*). The phylum is subdivided into two orders, each of which contain a single family. In PCA plots (Fig. 7), member species group together into a coherent, elongated cluster, with *Deinococcus* and *Meiothermus* species overlapping. Two major phenotypes, which are consistent with the phylogenetic branching, are observed.

Deinococcales Gram-positive, spherical to rod shaped cells with a complex, multi-layered cell wall. Mesophilic to thermophilic. Nonmotile. Chemoorganotrophic with a respiratory metabolism. Resistant to ionizing radiation. (BMDB9 Group 17).

Thermales Gram-negative straight rods or filaments. Nonmotile, nonsporulating. Thermophilic. Aerobic heterotrophs or chemoheterotrophs with a strictly respiratory metabolism. Some species capable of utilizing nitrate or nitrite as a terminal electron acceptor. (BMDB9 Group 4).

Phylum B5 Chrysiogenetes *phy. nov.* Garrity and Holt. Chry.si.o' ge.ne.tes M.L. fem. pl. n. *Chrysiogenales* type order of the phylum; dropping the ending to denote a phylum; M.L. fem. pl. n. *Chrysiogenetes* the phylum of *Chrysiogenales*.

The phylum *Chrysiogenetes* is currently represented by a single species, which was reportedly distinct from members of other phyla. Exact placement is currently uncertain, but a distant relationship to *Deferribacteres* is likely. In PCA plots (Fig. 8), *Chry-*

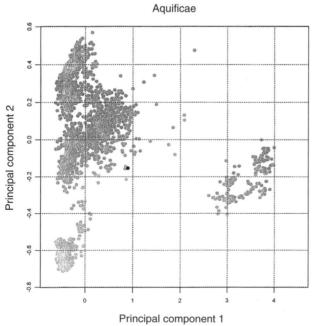

FIGURE 5. Location of Phylum *Aquificae* within the map of the procaryotes.

Editorial Note: Neither *Aquifex* nor any other genus within the phylum appears in the ninth edition of *Bergey's Manual of Determinative Bacteriology*.

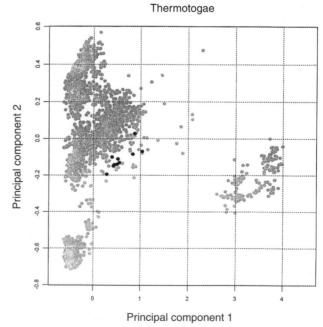

FIGURE 6. Location of Phylum *Thermotogae* within the map of the procaryotes.

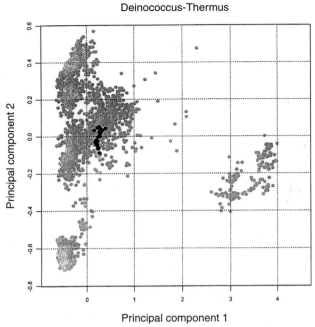

FIGURE 7. Location of Phylum *"Deinococcus-Thermus"* within the map of the procaryotes.

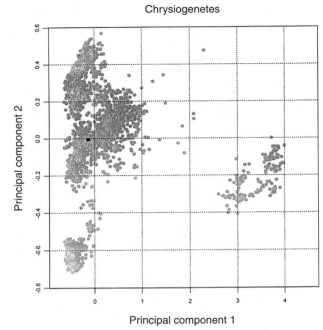

FIGURE 8. Location of Phylum *Chrysiogenetes* within the map of the procaryotes.

siogenes is located in a densely populated region, in close proximity to *Geovibrio* as well as the *"Clostridia"*, *"Bacilli"*, and the *"Deltaproteobacteria"*. Gram-negative, motile, curved, rod-shaped cells. Mesophilic, exhibiting anaerobic respiration in which arsenate serves as the electron acceptor. (BMDB9 Group 6).

Phylum B6 **Chloroflexi** *phy. nov.* Garrity and Holt.
Chlo.ro.flex′ i M.L. masc. n. *Chloroflexus* genus of the phylum; dropping the ending to denote a phylum; M.L. fem. pl. n. *Chloroflexi* phylum of *Chloroflexus*.

The phylum *Chloroflexi* is a deep branching lineage of *Bacteria*. The single class within *Chloroflexi* subdivides into two orders: the *"Chloroflexales"* and the *"Herpetosiphonales"*. In phylogenetic trees, the *Chloroflexi* tend to group with the *Thermomicrobia*; however, PCA plots (Fig. 9) show a clear separation, in keeping with the marked differences in phenotype. Gram-negative, filamentous *Bacteria* exhibiting gliding motility. Peptidoglycan contains L-ornithine as the diamino acid. Lipopolysaccharide-containing outer membrane not present.

"Chloroflexales" Contain bacteriochlorophyll and carotenoids. Obligate or facultative anoxygenic phototrophs. Photosynthetic pigments borne in chlorosomes. All genera are facultatively aerobic and preferentially utilize organic substrates in both phototrophic and chemotrophic metabolism. (BMDB9 Group 10).

"Herpetosiphonales" Lack bacteriochlorophylls, but carotenoids present. Aerobic, chemoheterotrophic metabolism. (BMDB9 Group 15).

Phylum B7 **Thermomicrobia** *phy. nov.* Garrity and Holt.
Ther.mo.mi.cro′ bi.a M.L. fem. pl. n. *Thermomicrobiales* type order of the phylum; dropping the ending to denote a phylum; M.L. fem. pl. n. *Thermomicrobia* the phylum of *Thermomicrobiales*.

The phylum *Thermomicrobia* consists of a single known representative that branches deeply in the RDP and ARB trees and is distantly related to the *Chloroflexi*. Although Ludwig and Klenk

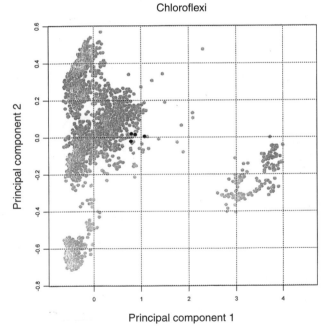

FIGURE 9. Location of Phylum *Chloroflexi* within the map of the procaryotes.

argue that the *Thermomicrobia* are monophyletic with the *Chloroflexi*, we have provisionally placed them into a separate phylum. Relatively little work has been done with this organism since the First Edition of the *Systematics Manual* was published, and no new relatives have been reported. Gram-negative, short, irregularly shaped nonmotile rods. In PCA plots (Fig. 10), the type strain maps to a sparsely populated region in which other deeply branching taxa and environspecies are found. Nonsporulating. No diamino acid present in peptidoglycan in significant amount.

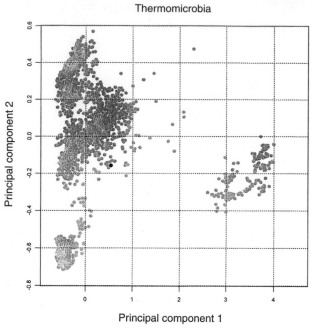

FIGURE 10. Location of Phylum *Thermomicrobia* within the map of the procaryotes.

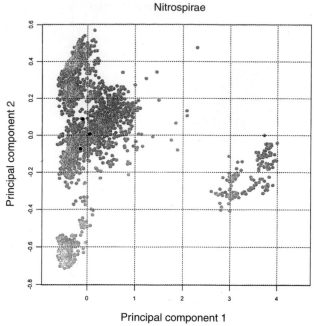

FIGURE 11. Location of Phylum *Nitrospirae* within the map of the procaryotes.

Hyperthermophilic, optimum growth temperature 70–75°C. Obligately aerobic and chemoorganotrophic. (BMDB9 Group 4a)

Phylum B8 **Nitrospirae** *phy. nov.* Garrity and Holt.
Ni.tro.spi' rae M.L. fem. n. *Nitrospira* genus of the phylum; dropping the ending to denote a phylum; M.L. fem. pl. n. *Nitrospirae* the phylum of *Nitrospira.*

The phylum *Nitrospirae* is based mainly on phylogenetic grounds. At present, it consists of a single class, order and family of *Bacteria* and environtaxa that branch deeply in the ARB and RDP trees; member taxa consistently group together. In PCA plots (Fig. 11) the *Nitrospirae* cluster in a densely populated region in close proximity to the *"Clostridia"*, *"Bacilli"*, and the *"Deltaproteobacteria"*. Gram-negative, curved, vibrioid or spiral-shaped cells. Metabolically diverse, most genera are aerobic chemolithotrophs including nitrifiers, dissimilatory sulfate reducers, and magnetotactic forms. One genus (*Thermodesulfovibrio*) is thermophilic, and obligately acidophilic and anaerobic. (BMDB9 Groups 12 and 7).

Phylum B9 **Deferribacteres** *phy. nov.* Garrity and Holt.
De.fer.ri.bac' te.res M.L. fem. pl. n. *Deferribacterales* type order of the phylum; dropping the ending to denote a phylum; M.L. fem. pl. n. *Deferribacteres* the phylum of *Deferribacterales.*

The phylum *Deferribacteres* is a distinct lineage within the *Bacteria* based on phylogenetic analysis of 16S rDNA sequences. At present the members of this phylum are organized into a single class, order, and family. The relationships within the phylum may, however, be more distant and warrant further subdivision in the future. In PCA plots (Fig. 12), the member species are widely separated along the Y axis and are in close proximity to the *"Deltaproteobacteria"* and *Nitrospirae*. While such separations are consistent for members of a phylum, it is quite probable that the individual strains belong to different classes and/or orders. Further subdivision will await the inclusion of additional sequence data into the models. Chemoorganotrophic heterotrophs that respire anaerobically with terminal electron acceptors including

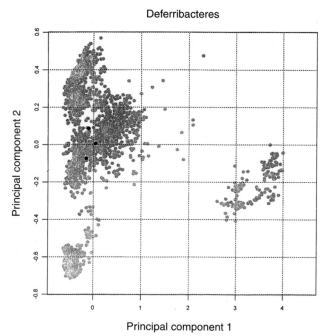

FIGURE 12. Location of Phylum *Deferribacteres* within the map of the procaryotes.

Fe(II), Mn(IV), S[0], Co(III), and nitrate. Placement of one genus, *Synergistes*, is provisional. (BMDB9 Group 6).

Phylum B10 **Cyanobacteria** Although the *Cyanobacteria* represent a major lineage within the *Bacteria* and include the chloroplasts as a distinct and highly diverse subgroup, the nomenclature of taxa ascribed to this phylum is governed by the *Botanical Code of Nomenclature* rather than the *Bacteriological Code.* The phylum consistently branches close to the low G + C Gram-positive *Bacteria* (the *Firmicutes*). In PCA plots (Fig. 13a), the

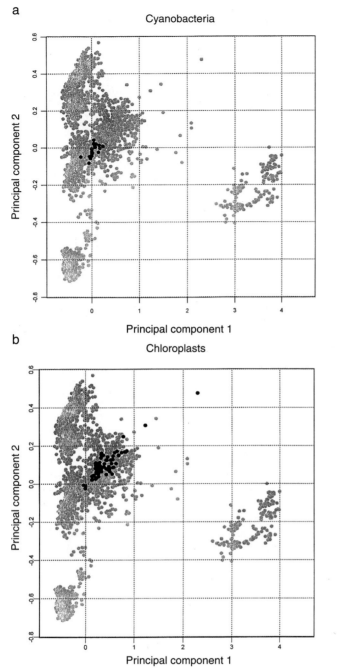

FIGURE 13. *a,* Location of Phylum *Cyanobacteria* within the map of the procaryotes. *b,* Location of a selected set of chloropasts in reference to procaryotic 16S rDNA sequences.

Cyanobacteria are located in a region densely populated by the *Mollicutes.* Chloroplasts (Fig. 13b) also map to the same general location, but span a much broader area and include some of the most distant outliers, presumably because of unique stem and loop structures (B. Maidak, personal communication). At present, the taxonomy of this group is in a state of flux. In a separate chapter in this volume, Wilmotte and Herdman (see "Phylogenetic Relationships Amongst the Cyanobacteria Based on 16S rRNA Sequences" in this volume) discuss the manifold discrepancies between the current 16S rDNA-based phylogeny and the distinctive phenotypes of these bacteria. Gram-negative unicellular, colonial, or filamentous oxygenic and photosynthetic *Bac-*

teria exhibiting complex morphology and life cycles. The principal characters that define all members of this phylum are the presence of two photosystems (PSII and PSI) and the use of H_2O as the photoreductant in photosynthesis. Although facultative photoheterotrophy or chemoheterotrophy may occur in some species or strains, all known members are capable of photoautotrophy (using CO_2 as the primary source of cell carbon). Lipopolysaccharide outer membrane present. Thick peptidoglycan layer (2–200 nm). Contain chlorophyll *a.* Phycobiliproteins (allophycocyanin, phycocyanin, and sometimes phycoerythrin) may or may not be present. The *Cyanobacteria* are subdivided into five subsections that are equivalent to orders. The current classification scheme is structured according to phenotypic characteristics rather than the 16S rDNA phylogeny as such a scheme is not yet possible. (BMDB9 Group 11).

SUBSECTION I   Unicellular or nonfilamentous aggregates of cells held together by outer walls or a gel-like matrix (colonies). Reproduction by binary fission in one, two, or three planes, symmetric or asymmetric; or by budding.

SUBSECTION II   Unicellular or nonfilamentous aggregates of cells held together by outer walls or a gel-like matrix (colonies). Reproduction by internal multiple fissions with production of daughter cells smaller than the parent; or by multiple plus binary fission.

SUBSECTION III   Filamentous; trichome of cells, branched or unbranched, uniseriate or multiseriate. Reproduction by binary fission in one plane only, giving rise to uniseriate unbranched trichomes, although false branching may occur. Trichomes composed of cells, which do not differentiate into heterocysts or akinetes.

SUBSECTION IV   Filamentous; trichome of cells, branched or unbranched, uniseriate or multiseriate. Reproduction by binary fission in one plane only, giving rise to uniseriate unbranched trichomes, although false branching may occur. One or more cells of trichomes differentiate into heterocysts when the concentration of ammonium or nitrate in the medium is low; some also produce akinetes.

SUBSECTION V   Filamentous; trichome of cells, branched or unbranched, uniseriate or multiseriate. Reproduction by binary fission periodically or commonly in more than one plane, giving rise to multiseriate trichomes or trichomes with various types of side branches or both.

Phylum B11   Chlorobi *phy. nov.* Garrity and Holt.
Chlo.ro′ bi M.L. neut. n. *Chlorobium* genus of the phylum; dropping the ending to denote a phylum; M.L. fem. pl. n. *Chlorobi* phylum of *Chlorobium.*

The *Chlorobi* share a common root with *Bacteroidetes* in both the RDP and ARB trees. In PCA plots (Fig. 14), the *Chlorobi* are found in closest proximity to the "*Flavobacteria*", a class within the *Bacteroidetes.* At present, the phylum contains a single class, order, and family. Gram-negative, spherical, ovoid, straight, or curved rod-shaped *Bacteria.* Strictly anaerobic, obligately phototrophic. Cells grow preferentially by photoassimilation of simple organic compounds. Some species may utilize sulfide or thiosulfate as an electron donor for CO_2 accumulation. Sulfur globules accumulate on the outside of the cells when grown in the presence of sulfide and light, and sulfur is rarely oxidized further to sulfate. Most genera are able to grow as chemoheterotrophs under microaerobic or aerobic conditions. Ammonia and di-

Chlorobi

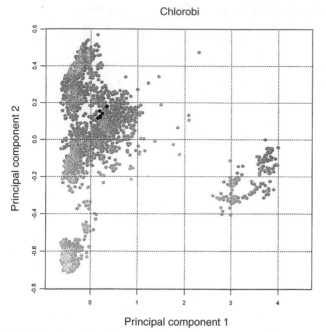

FIGURE 14. Location of Phylum *Chlorobi* within the map of the procaryotes.

nitrogen used as the nitrogen source. Most genera require one or more growth factors; the most common are biotin, thiamine, niacin, and *p*-aminobenzoic acid.

Phylum B12 **Proteobacteria** *phy. nov.* Garrity and Holt.
Pro.te.o.bac.te' ri.a M.L. fem. pl. n. *Proteobacteria* class *Proteobacteria* elevated to phylum.

Encompassing over 1300 validly named species, most of which are cultivable, and assigned to 384 genera, the *Proteobacteria* currently represent the largest, phylogenetically coherent group within the *Bacteria*. The group was originally proposed as the Class *Proteobacteria* by Stackebrandt et al. (1988) and contained four informally named subclasses; we have elevated the class to the rank of phylum. At the time Stackebrandt et al. proposed the class *Proteobacteria*, it was generally agreed that the group had undergone a rapid evolution and generated numerous branches in which physiologically and morphologically diverse forms grouped together. Despite this incongruity, the *Proteobacteria* were generally thought to be monophyletic. More recently, the consensus has changed somewhat, largely as a result of additional sequences having been added to the phylogenetic models. At present, five major lines of descent are recognized within the phylum. However, it now appears that the *Proteobacteria* may not be monophyletic (Ludwig and Klenk) and the separation between the *"Betaproteobacteria"* and *"Gammaproteobacteria"* is much less clear, with the former appearing as a subgroup of the latter. The *"Alphaproteobacteria"*, the *"Deltaproteobacteria"*, and the *"Epsilonproteobacteria"* remain distinct lineages. In PCA plots (Fig. 15a–f), the overlap among the *"Betaproteobacteria"* and *"Gammaproteobacteria"* is also apparent, as is the separation of the *"Alphaproteobacteria"*, the *"Deltaproteobacteria"*, and the *"Epsilonproteobacteria"*.

The *"Alphaproteobacteria"* Within our classification, the *"Alphaproteobacteria"* have been further subdivided into six orders: *Rhodospirillales, Rickettsiales, "Rhodobacterales", "Sphingomonadales",*

Caulobacterales, and *"Rhizobiales"*. This differs slightly from the ARB tree, which shows a separation of the families *Rhodospirillaceae* and *Acetobacteraceae* into separate clades; we have treated these as families of the same order. Within the *Rickettsiales*, sequences from species of two genera, *Wolbachia* (a) and *Polynucleobacter* (b), appear as outliers, separated from the other members of the order, which cluster tightly. These taxa map into the region occupied by the *"Betaproteobacteria"* and *"Gammaproteobacteria"* and might be misplaced or misidentified strains. A similar pattern is observed with one species of *Azotobacter* (c) within the *Rhodospirillales*. Otherwise, the orders form discrete clusters that together form an essentially complete coverage of a region of the map that overlaps only slightly with the other four classes of *Proteobacteria*.

The *"Betaproteobacteria"* The *"Betaproteobacteria"* have also been subdivided into six orders: *"Burkholderiales", "Hydrogenophilales", "Methylophilales", "Neisseriales", "Nitrosomonadales",* and *"Rhodocyclales"*. Within the *"Burkholderiales"*, one species of *Duganella* (a) appears to be misidentified or misplaced. The *"Hydrogenophilales"* also appear to be somewhat problematic, as all of the member species appear as outliers in the PCA plots. Ludwig and Klenk indicate that this group branches more deeply within the ARB tree.

The *"Gammaproteobacteria"* The *"Gammaproteobacteria"* are subdivided into 13 orders within our classification: *"Chromatiales", "Xanthomonadales", "Cardiobacteriales", "Thiotrichales", "Legionellales", "Methylococcales", "Oceanospirillales", Pseudomonadales, "Alteromonadales", "Vibrionales", "Aeromonadales", "Enterobacteriales",* and *"Pasteurellales"*. While this is generally in good agreement with the ARB and RDP trees, Ludwig and Klenk note a common clustering of the last five orders and show a deeper division between the *Pasteurellaceae* and *Moraxellaceae*, equivalent to separate orders. In PCA plots, the *Pasteurellaceae* and *Moraxellaceae* appear in adjacent regions of a densely populated area, proximal to the *"Alteromonadales", "Vibrionales", "Aeromonadales", Enterobacteriales,* and *"Pasteurellales"*. These authors also note that the relative branching order within the *"Gammaproteobacteria"* cannot be unambiguously determined. We believe that this is consistent with our observation of nearly confluent coverage of taxonomic space. Within the *"Gammaproteobacteria"*, the orders appear relatively coherent in our classification. Some outliers exist, notably *Thiomicrospira* (a) which appears in the region of the *"Epsilonproteobacteria"*, and *Nitrosococcus* and *Halorhodospira* species (b) which are somewhat removed from the main cluster. Some species assigned to the genus *Pseudomonas* also appear to be misplaced. The order *"Cardiobacteriales"* may also be problematic as it appears to be polyphyletic.

The *"Deltaproteobacteria"* The taxonomy of the *"Deltaproteobacteria"* is currently in a state of flux and some rearrangement within this class prior to publication of the second volume of this edition is likely. At present, the class is subdivided into seven orders: *"Desulfurellales", "Desulfovibrionales", "Desulfobacterales", "Desulfuromonadales", "Syntrophobacterales", "Bdellovibrionales",* and *Myxococcales*. The *"Desulfobacterales"* may be paraphyletic, with the *"Desulfoarculaceae"* (Family III) appearing on a separate branch in the ARB tree. This family contains the genus *Nitrospina*, which appears in different locations in the ARB and RDP trees. In PCA plots, the *"Desulfobacterales"* appear as a discontinuous group. One additional order, the *"Bdellovibrionales"*, may be misplaced. At present, this group (which is represented by a single sequence

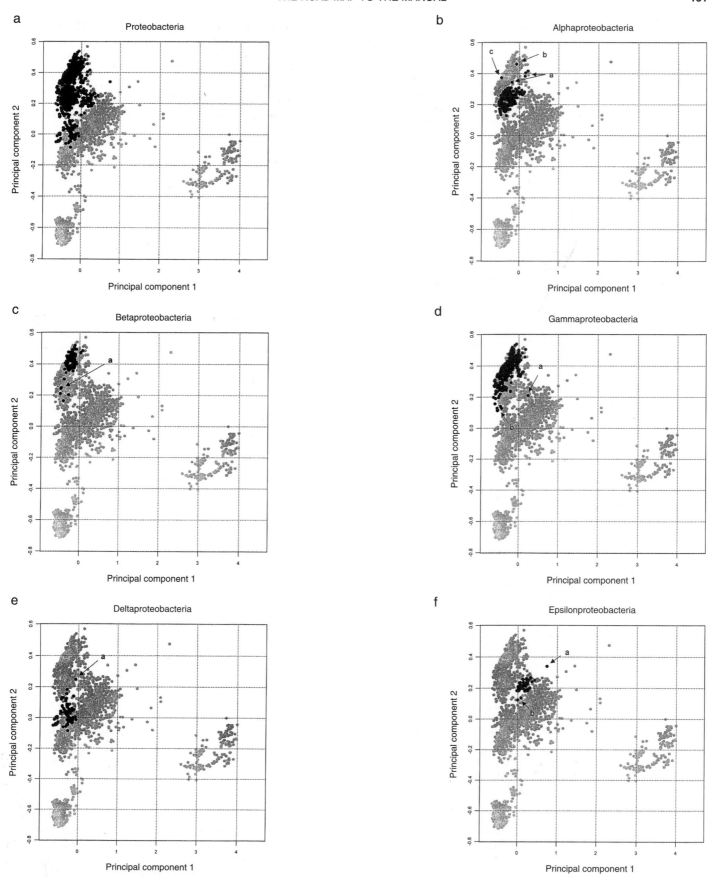

FIGURE 15. *a*, Location of Phylum *Proteobacteria* within the map of the procaryotes. *b*, Class *"Alphaproteobacteria"*—Outliers are identified as members of the genera *Wolbachia* (a), *Polynucleobacteria* (b), and *Azotobacter* (c). *c*, Class *"Betaproteobacteria"*—Two of the outliers (a) are identified as members of the genus *Duganella*. The remaining outliers are members of the *Hydrogenophilales*. *d*, Class *"Gammaproteobacteria"*—Outliers are identified as members of the genera *Thiomicrospira* (a), *Nitrococcus* and *Halorhodospira* (b). *e*, Class *"Deltaproteobacteria"*—Outlier identified as a member of the genus *Bdellovibrio* (a). *f*, Class *"Epsilonproteobacteria"*—Outliers identified as members of the genera *Helicobacteria* (a) and *Wolinella* (b).

appears within the region occupied by the *"Epsilonproteobacteria"* (a).

The "Epsilonproteobacteria" The *"Epsilonproteobacteria"* represent a more recently recognized line of descent within the *Proteobacteria* and encompass two families within a single order. The group is well supported in phylogenetic trees and appears as a well separated and tightly clustered group in PCA plots. Three outliers appear in the plot. One (a) is a potentially misidentified species ascribed to the genus *Helicobacter*, while the other two represent strains of *Wolinella* (b).

***Phenotypic characteristics of the* Proteobacteria**   As noted above, the *Proteobacteria* are phenotypically diverse and the five classes are heterogeneous in this regard. In the Ninth Edition of the *Determinative Manual*, the *Proteobacteria* were distributed among 12 of the phenotypic groups, and only 3 of those groups are unique to any one of the current classes.

GRAM-NEGATIVE AEROBIC/MICROAEROPHILIC RODS AND COCCI Predominantly chemoorganotrophic heterotrophs, but some may grow autotrophically by using H_2 as an electron donor. Capable of growth under an air atmosphere and have a strictly respiratory type of metabolism with O_2 as the terminal electron acceptor. Some genera are also capable of anaerobic respiration using alternate terminal electron acceptors. Occur in soil, fresh water, or marine environments, within plant roots, or in the reproductive organs, intestinal tract, and oral cavity of humans and animals. Some are pathogenic for animals or humans. Two major subtypes occur within this group. The first includes aerobic rods and cocci that can grow in the presence of air (21% oxygen). Some genera are microaerophilic under nitrogen-fixing conditions but grow as aerobes when supplied with a source of fixed nitrogen. The second subtype consists of microaerophilic straight rods that cannot grow in the presence of air (21% oxygen). Oxygen-dependent growth exhibited at reduced O_2 concentrations and in the absence of alternate electron acceptors. Some genera can respire anaerobically using nitrate, fumarate, or other electron acceptors. Common among the *"Alphaproteobacteria"*, *"Betaproteobacteria"*, *"Gammaproteobacteria"*, and *"Deltaproteobacteria"*. (BMDB9 Group 4).

ANAEROBIC STRAIGHT, CURVED, AND HELICAL GRAM-NEGATIVE RODS   Chemoorganotrophic heterotrophs. Obtain energy by anaerobic respiration or by fermentation. Do not include genera that respire anaerobically using sulfate, other oxidized sulfur compounds, or S^0 as electron acceptors. This phenotype is quite common in the *"Betaproteobacteria"*, the *"Gammaproteobacteria"*, and the *"Deltaproteobacteria"*. (BMDB9 Group 6).

ANOXYGENIC PHOTOTROPHIC BACTERIA   Bacteria that contain bacteriochlorophyll and can use light as an energy source. When growing under illumination the organisms are anaerobic and do not evolve O_2 during photosynthesis. Some are also capable of growing in the dark by respiring with oxygen. Do not contain phycobiliproteins. Three subtypes occur within the *Proteobacteria*. The first is characterized by cells able to grow with sulfide and sulfur as the sole electron donor for photosynthetic CO_2 assimilation. Grow well under photoautotrophic conditions. Sulfur globules accumulate inside the cells when grown in the presence of sulfide and light and may be further oxidized to sulfate. Contain bacteriochlorophyll *a* or *b* and carotenoids. May require vitamin B_{12} for growth. This subtype is restricted to the *"Gammaproteobacteria"*. (BMDB9 Group 10).

The second subtype is characterized by cells that grow preferentially by photoassimilation of simple organic compounds. Some species may utilize sulfide or thiosulfate as an electron donor for CO_2 accumulation. Sulfur globules form only on the outside of the cells when grown in the presence of sulfide and light and sulfur is rarely oxidized further to sulfate. Most genera are able to grow as chemoheterotrophs under microaerobic or aerobic conditions. Ammonia and dinitrogen used as the nitrogen source. Most genera require one or more growth factors; the most common are biotin, thiamine, niacin, and *p*-aminobenzoic acid. This subtype occurs in the *"Alphaproteobacteria"* and *"Betaproteobacteria"*.

The third subtype is defined by cells that grow chemoheterotrophically under aerobic conditions. No growth occurs under anaerobic conditions in the light. Metabolism is predominantly respiratory. Cells contain bacteriochlorophyll *a* and carotenoids. This subtype is restricted to some species within the *"Alphaproteobacteria"*.

NON-PHOTOSYNTHETIC, NON-FRUITING GLIDING BACTERIA Non-phototrophic rods or filaments that lack flagella but which can glide across solid surfaces. Simple life cycles. Sheaths may occur in some genera. Myxospores produced by some genera. Four subtypes are observed. Subtype one consists of single-celled, rod-shaped, gliding bacteria that are often pleomorphic. Chemoorganotrophic, aerobic, and facultatively or obligately anaerobic. Restricted to some members of the *"Gammaproteobacteria"*. Subtype two consists of cells that occur as flattened, filamentous, gliding bacteria. Chemoorganotrophic, aerobic, saccharolytic. Found in oral cavity of warm-blooded vertebrates. Restricted to the *"Betaproteobacteria"*. Subtype three consists of sulfur oxidizing, gliding bacteria. Cell size varies widely. Most are filamentous. Sulfur deposited internally in the presence of H_2S and often from thiosulfate. Chemoorganotrophic and chemolithotrophic nutrition known; possibly mixotrophic nutrition for some genera. Respiratory metabolism. This subtype is restricted to the *"Gammaproteobacteria"*. The fourth subtype consists of other gliding forms that do not fall into the three subtypes described above. Such forms occur in small number within the *"Betaproteobacteria"* and *"Gammaproteobacteria"*. (BMDB9 Group 15).

AEROBIC CHEMOLITHOTROPHIC BACTERIA AND ASSOCIATED GENERA   Nonphototrophic. Three subtypes found within the *Proteobacteria*: nitrifiers, which utilize reduced inorganic nitrogen compounds (ammonia and nitrite) as energy sources for growth; sulfur oxidizers, which either oxidize or utilize reduced inorganic sulfur compounds as the sole source of energy; and obligate hydrogen oxidizers, which utilize hydrogen gas as the energy source for growth rather than organic sources of carbon. Nitrifiers occur in the *"Betaproteobacteria"*, the *"Gammaproteobacteria"*, and the *"Epsilonproteobacteria"*. Sulfate oxidizers are found solely within the *"Betaproteobacteria"* and obligate hydrogen oxidizers occur in all but the *"Epsilonproteobacteria"*. (BMDB9 Group 12).

FACULTATIVELY ANAEROBIC GRAM-NEGATIVE RODS   Chemoorganotrophic heterotrophs, but some may grow autotrophically by using H_2 as an electron donor. Capable of growing under an air atmosphere by a respiratory type of metabolism; also capable of growing anaerobically by fermentation. Occur either free-living or in association with animal, human, or plant hosts. Some are pathogenic. Initially, this phenotype was restricted to three families generically defined as enteric bacteria. However, the phenotype occurs more broadly and is found in all but the *"Epsilon-proteobacteria"*. (BMDB9 Group 5).

BUDDING AND/OR APPENDAGED NONPHOTOTROPHIC BACTERIA This phenotype is further subdivided into two subtypes, based on the presence of absence of prosthecae and the mode of cell division. Both forms occur within the *"Alphaproteobacteria"*, while only nonprosthecate forms occur within the *"Gammaproteobacteria"*. (BMDB9 Group 13).

AEROBIC/MICROAEROPHILIC, MOTILE, HELICAL/VIBRIOID GRAM-NEGATIVE BACTERIA Vibrioid or helical cells. Motility by polar flagella exhibiting a characteristic corkscrew-like motion when swimming. Aerobic or microaerophilic. Respiratory type of metabolism with O_2 as the terminal electron acceptor. Anaerobic respiration may occur with fumarate or nitrate as alternate electron acceptor. Most genera are chemoorganotrophic but some may grow autotrophically with H_2 as the electron donor. Occur in soil, fresh water, and marine environments, in association with plant roots, or in the reproductive organs, intestinal tract, and oral cavity of humans. Some species are predatory on other microorganisms. This phenotype occurs in all classes of the *Proteobacteria*. (BMDB9 Group 2).

SYMBIOTIC AND PARASITIC BACTERIA OF VERTEBRATE AND INVERTEBRATE SPECIES Obligate intracellular parasites of eucaryotic hosts (vertebrates or arthropods). May be rod-shaped, coccoid, or pleomorphic. Multiply by binary fission. Cell walls contain muramic acid. Some are pathogenic to humans or other vertebrate and invertebrate hosts. This phenotype is found among members of the *"Alphaproteobacteria"* and the *"Gammaproteobacteria"*. (BMDB9 Group 9).

FRUITING GLIDING BACTERIA Chemoorganotrophic, strictly aerobic bacteria that lack flagella but can glide across solid surfaces. Under conditions of nutrient deprivation, cells aggregate to form fruiting bodies, composed of modified slime and cells, which are often brightly colored and of macroscopic dimensions. Fruiting bodies vary in complexity, from simple mounds to complex structures consisting of sporangia of characteristic shape and dimensions which may be sessile or borne singly or in groups on simple or branched stalks. Within the fruiting bodies, myxospores or microcysts are produced. A narrowly defined phenotype that occurs within a single order of the *"Deltaproteobacteria"*, the *Myxococcales*. (BMDB9 Group 16).

SHEATHED BACTERIA Nonphototrophic. Aerobic. Do not exhibit gliding motility. Characterized by a filamentous pattern of growth in which the cells are enclosed with a transparent sheath. Sheaths may appear yellow to dark brown, owing to the deposition of iron and manganese oxides. Single cells may be motile by polar or subpolar flagella or they may be nonmotile. An uncommon phenotype that is found in two genera of *"Betaproteobacteria"*. (BMDB9 Group 14).

NONMOTILE OR RARELY MOTILE, CURVED GRAM-NEGATIVE BACTERIA Chemoorganotrophic heterotrophs. Saprophytic. Four morphotypes were originally defined, one of which is observed in a single genus, *Ancylobacter*, within the *"Alphaproteobacteria"*. Curved or C-shaped cells that may form rings by overlapping cell ends. Coils and helices may occur. Gas vacuoles may occur. Aerobic. (BMDB9 Group 3).

Phylum B13 **Firmicutes** *phy. nov.* Gibbons and Murray 1978,[AL] emend. Garrity and Holt.

Fir.mi.cu' tes M.L. fem. pl. n. *Firmicutes* named for the division *Firmicutes*.

Originally described by Gibbons and Murray (1978) as the division *Firmicutes* and encompassing all of the Gram-positive *Bacteria*; we propose conservation of the name for the phylum containing the Gram-positive *Bacteria* with a low DNA mol% G + C content. At the end of 1999, the phylum included 154 genera which have been subdivided into three classes: the *"Clostridia"*, the *Mollicutes*, and the *"Bacilli"*. Ludwig and Klenk note that while the *Mollicutes* appear to be a monophyletic group in reference trees, a common origin of the classical "low G + C Gram positives" is not significantly supported by either the ARB or RDP trees. Within PCA plots (Fig. 16a–d), we find that the *Firmicutes* occupy a large area between the *Proteobacteria* and the *Actinobacteria* along the second principal component, and between the *Proteobacteria* and *Archaea* along the first principal component. The *"Clostridia"* and *"Bacilli"* fall into a densely populated region and overlap significantly. There is also a slight overlap with the *"Deltaproteobacteria"* in the large scale "maps" but the groups can be clearly separated by recalculation of principal components for the subset of sequences limited to the *Proteobacteria* and *Firmicutes*. The *Mollicutes* fall into a separate region of the plots, overlapping slightly with the *Cyanobacteria*. The phylum is phenotypically diverse and has been the subject of a number of recent rearrangements. While the current taxonomic scheme appears to be relatively stable, it is expected that a number of rearrangements and refinements will occur prior to publication of the third volume of this series, which will deal with the *Firmicutes* exclusively. Examination of PCA plots at the order, family, and genus level reveals that most taxa form coherent clusters. There are, however, a number of instances in which outliers occur at the genus and family level. This is typically indicative of misplaced or misnamed taxa or sequences.

THE *"CLOSTRIDIA"* The class *"Clostridia"* is subdivided into three orders: the *Clostridiales*, the *"Thermoanaerobacteriales"*, and the *Haloanaerobiales*. The *Clostridiales* are coherent in PCA plots, with only a few outliers. One of the outliers is a *Clostridium* species that is most likely misidentified (a) and appears adjacent to a *Eubacterium* species that is also potentially classified in error (b). Two other outliers were clearly in error and are left uncorrected in these figures for demonstration purposes. Both were "identified" as *Eubacterium* species (d), on the basis of sequence annotations, but clearly mapped into the region occupied by deeply branching *Actinobacteria*. The annotations were, however, out of date and did not reflect reassignment to other genera. The *"Thermoanaerobacteriales"* appears to be polyphyletic. The type genus stands alone. *Coprothermobacter* (c) plots into a sparsely populated region occupied by deeply branching phyla and the remaining genera fall into a distinctly different region. The *Haloanaerobiales* is also problematic in that members of the genus *Acetohalobium* map into three distinct places within the PCA plots. We suspect that the genus is either malformed or contains unidentified strains or sequences.

THE *MOLLICUTES* The *Mollicutes* represent a well formed class that is subdivided into four orders, in which the current taxonomy and phylogeny are largely in good agreement. The *Mollicutes* form a broad and essentially coherent cluster with only three outliers. Two of the outliers are species of *Haemobartonella* (a) and are particularly noteworthy as they fall into a sparsely populated region of the map which is occupied by a few sequences derived from environ taxa and atypical chloroplasts. *Hae-*

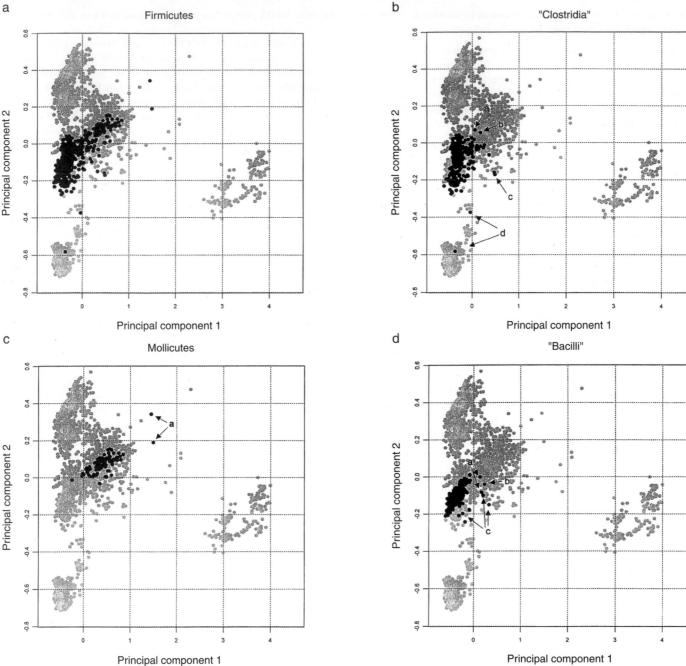

FIGURE 16. *a,* Location of Phylum *Firmicutes* within the map of the pro-caryotes. *b,* Class *"Clostridia"*. Outliers identified as members of the genera *Clostridium* (a), *Eubacterium* (b, d), and *Coprothermobacter* (c). The outliers at position (d) provide an example of two reclassified species for which the annotation had not been updated in the sequence data bases. The upper strain is *Atopobium fossor* (*Eubacterium fossor*) and the lower strain is *Eggerthella lenta* (*Eubacterium lentum*). These were retained in the plots for demonstration purposes, as both species have been transferred to the *Actinobacteria* as a result of the reclassification. *c,* Class *Mollicutes.* Outliers are members of the genus *Haemobartonella* (a). *d,* Class *"Bacilli"*. Outliers are members of the genera *Lactobacillus* (a) and *Oenococcus* (b). Those at positions (c) are members of the *"Alicyclobacillaceae"*.

mobartonella species are obligately parasitic bacteria and the un-usual location of these species in the PCA plots may be indicative of a highly unusual primary and/or secondary 16S rRNA struc-ture resulting from reductive evolution. Several other parasitic species also fall along the periphery of the major clusters and behave like deeply branching taxa.

THE *"BACILLI"* The *"Bacilli"* also form a coherent class which is currently subdivided into two orders: *Bacillales* and *"Lactoba-cillales"*. Ludwig and Klenk indicate that the five families of lactic

acid bacteria that form the *"Lactobacillales"* constitute a phylo-genetically coherent group, whereas four of the nine families in the *Bacillales* branch more deeply than the remaining five. In PCA plots we find nine outliers, representing potentially mis-classified taxa. Two are currently identified as members of the genus *Lactobacillus* (a), one belongs to the genus *Oenococcus* (b). The remaining six are all assigned to the *"Alicyclobacillaceae"*. The two *Alicyclobacillus* species map within the region occupied by *"Clostridia"*, as do two of the four *Sulfobacillus* species. The remaining two *Sulfobacillus* sequences map into a sparsely populated region occupied by deeply branching taxa and environtaxa.

***Phenotypic characteristics of the* Firmicutes**   The *Firmicutes* are phenotypically diverse and the three classes within the phylum can be ascribed to 13 phenotypic groups. As compared to the *Proteobacteria*, the level of phenotypic uniqueness is significantly higher, with 8 of the 14 phenotypes occurring in only one class within the phylum. There is, however, phenotypic overlap with other phyla, including the *Proteobacteria*.

THERMOPHILIC AND HYPERTHERMOPHILIC BACTERIA   Growth occurring at temperatures over 65°C. This characteristic is limited to two genera within the "*Clostridia*": *Coprothermobacter* and *Carboxydothermus*. (BMDB9 Group 36).

ANAEROBIC STRAIGHT, CURVED, AND HELICAL GRAM-NEGATIVE RODS   Chemoorganotrophic heterotrophs. Obtain energy by anaerobic respiration or by fermentation. Does not include genera that respire anaerobically with sulfate, other oxidized sulfur compounds, or S^0 as electron acceptors. Phenotype occurs widely within all three orders of "*Clostridia*" as well as in a single genus within "*Bacilli*": *Acetoanaerobium*. (BMDB9 Group 6).

ANOXYGENIC PHOTOTROPHIC BACTERIA   Cells grow by photoassimilation of simple organic substrates. Strictly anaerobic and photoheterotrophic. Reduced sulfur compounds are not utilized. Ammonia and dinitrogen are used as nitrogen sources. Internal membrane systems and chlorosomes are absent. Cells contain bacteriochlorophyll *g* and carotenoids. Cell walls lack lipopolysaccharide. Phenotype limited to a single family within the "*Clostridia*", the "*Heliobacteriaceae*". This family will be covered in volume one as well as volume three. (BMDB9 Group 10).

NONPHOTOSYNTHETIC, NONFRUITING, GLIDING BACTERIA   Nonphototrophic rods or filaments that lack flagella but which can glide across solid surfaces. Simple life cycles. Occurs within two genera within the "*Bacilli*": *Filifactor* and *Agitococcus*. (BMDB9 Group 15).

AEROBIC, NONPHOTOTROPHIC, CHEMOLITHOTROPHIC BACTERIA   Reduced inorganic sulfur compounds can be oxidized. Occurs in a single genus within the "*Bacilli*": *Sulfobacillus*. (BMDB9 Group 12).

DISSIMILATORY SULFATE- OR SULFITE-REDUCING BACTERIA   Chemoorganotrophic heterotrophs. Respire anaerobically with either sulfate and other oxidized sulfur compounds, or with S^0, as terminal electron acceptors. Occurs in a single genus within the "*Clostridia*": *Desulfotomaculum*. (BMDM9 Group 7).

SYMBIOTIC AND PARASITIC BACTERIA OF VERTEBRATE AND INVERTEBRATE SPECIES   Obligate intracellular parasites of eucaryotic hosts (vertebrates or arthropods). May be rod-shaped, coccoid, or pleomorphic. Many species are pathogenic. Phenotype occurs within the *Mollicutes* and "*Bacilli*". (BMDB9 Group 9).

ANAEROBIC GRAM-NEGATIVE COCCI   Chemoorganotrophic heterotrophs having a strictly fermentative type of metabolism. This phenotype is restricted to some members of the family "*Acidaminococcaceae*" within the "*Clostridia*". (BMDB9 Group 8).

GRAM-POSITIVE COCCI   Chemoorganotrophic, mesophilic, nonsporeforming cocci that stain Gram-positive. Three physiological subtypes: Type one consists of aerobic cocci that occur in pairs, clusters, or tetrads. Catalase positive. Cytochromes present. Teichoic acids not present in cell walls. Acid production from

carbohydrates may often be negative or weak. Type two consists of facultatively anaerobic or microaerophilic cocci that occur in pairs, chains, clusters, or tetrads. The presence of catalase, cytochromes, and cell wall teichoic acids varies. Cytochromes may or may not be present. Type three consists of strictly anaerobic cocci that occur in pairs, chains, tetrads, or cuboidal packets. Cytochromes are absent in those genera that have been tested. The catalase reaction is usually negative, although in some instances there is a weak or pseudocatalase reaction. Commonly occurring in members of the "*Clostridia*" (types 2 and 3) and "*Bacilli*" (types 1, 2, and 3). (BMDB9 Group 17).

ENDOSPORE-FORMING GRAM-POSITIVE RODS AND COCCI   Bacteria that produce heat-resistant endospores. Mostly motile rods or filaments, however, one genus (*Sporosarcina*) contains motile cocci (in tetrads or cuboidal packets). Most stain Gram positive, at least in young cultures; however, one genus stains Gram negative (*Sporohalobacter*). Strict aerobes, facultative anaerobes or microaerophiles, or strict anaerobes. (BMDM9 Group 18).

REGULAR, NONSPORULATING GRAM-POSITIVE RODS   Rod-shaped cells (coccoid to elongated rods or filaments), Gram-positive, nonsporulating, generally nonpigmented, mesophilic, chemoorganotrophic, and grow only in complex media. Some are pathogens of animals. Three major physiological subtypes are recognized. Type one consists of fermentative, saccharolytic microaerophiles that do not possess heme-containing catalase, cytochromes or menaquinones, and which utilize oxygen only via flavin-containing oxidases and peroxidases. Type two consists of aerobes or facultative anaerobes that possess cofactors and enzymes for respiration. These organisms are also able to ferment sugars, mainly to lactic acid, under oxygen-limited or anaerobic conditions. Type three consists of strict aerobes which neither utilize glucose as a carbon or energy source nor ferment sugars to organic acids. Variations of this phenotype occur in all three classes of the *Firmicutes*. (BMDM9 Group 19).

IRREGULAR, NONSPORULATING GRAM POSITIVE RODS   The majority are irregular rods which stain Gram positive or Gram negative to Gram variable. Grow in the presence of air and do not produce endospores. Oxygen requirements range from strictly aerobic, facultatively anaerobic to microaerophilic, or strictly anaerobic; may be pathogenic to animals or plants. A common phenotype occurring in all three classes of the *Firmicutes*. (BMDB9 Group 20).

MYCOPLASMAS   Pleomorphic cells devoid of cell walls. Growth on agar exhibits characteristic "fried egg" appearance. Some require sterols for growth. May show gliding motility. Facultatively anaerobic to obligately anaerobic. Have low mol% G + C of DNA of 23–46. This phenotype is restricted to members of the class *Mollicutes*. (BMDB9 Group 30).

THERMOACTINOMYCES   Gram-positive *Bacteria*. Stable filaments produce aerial growth. Endospores produced in both aerial and vegetative filaments. Thermophilic. Cell walls contain meso-DAP. Phenotype restricted to a single genus, *Thermoactinomyces*, within the "*Bacilli*". (BMDB9 Group 28).

***Phylum B14*   Actinobacteria** *phy. nov.* Garrity and Holt.
Ac.ti.no.bac.te' ri.a M.L. fem. pl. n. *Actinobacteria* class of the phylum; M.L. fem. pl. n. the phylum of *Actinobacteria*.

We propose elevation of the class *Actinobacteria* (Stackebrandt et al., 1997) to the rank of phylum, recognizing that the phylogenetic depth represented in this lineage is equivalent to that of existing phyla and that the group shows clear separation from the *Firmicutes*. Within the phylum, we recognize a single class, *Actinobacteria*, and preserve the complete hierarchical structure of Stackebrandt et al (1997), including the five subclasses and ten suborders.

In their analysis of the reference trees, Ludwig and Klenk indicate that the phylum is clearly defined and delimited, with the *Rubrobacterales* and *Coriobacteriales* representing the deepest lineages, and the *Acidimicrobiales* occupying a position of intermediate depth. This is consistent with the current classification of *Actinobacteria*. No mention was made of *Sphaerobacterales*, which is included in the current classification. The sequence contained within the RDP is known to be problematic and the sole member of the class is misplaced in the RDP tree. These authors also indicate that neither a significant nor stable branching order could be established for the families within the order *Actinomycetales* using 16S rDNA, 23S rDNA, or the β subunit of F_1F_0 ATP-ase.

Within the PCA plots (Fig. 17) we find that the *Actinobacteria* map to a location completely removed from the *Firmicutes*, further confirming the likelihood that this group represents a separate line of evolutionary descent. Consistent with the reference trees, we find that the two *Rubrobacterales* sequences (a) map into the region of *Clostridia*, while the *Coriobacteriales* (b) and *Acidimicrobiales* (c) map into the sparsely populated region between the *Firmicutes* and the main cluster of *Actinobacteria*. This is the same region in which two misidentified *Eubacterium* species were located; these were reassigned to the genera *Atopobium* (Kageyama et al., 1999) and *Eggerthella* (Wade et al., 1999) (members of the *Coriobacteriales*). We also find that the *Bifidobacteriales* (e) are a separate group lying adjacent to, but removed from, the major lineages within the *Actinomycetales*, consistent with the published phylogenetic model of Stackebrandt et al. This differs slightly from the Ludwig and Klenk subtree, in which the *Bifi-*

dobacteriales could not be resolved. In addition, sequences from three *Actinomyces* species (e) fall outside of the main cluster in which the remainder of the genus maps.

The major cluster of *Actinobacteria* contains over 500 data points, representing type strains for more than 85% of the validly named genera. We find the compactness of the cluster to be remarkable, and to suggest a possible explanation as to why it may be impossible to determine stable or significant branching order within the *Actinomycetales*. It is quite likely that the level of sequence variability, using the current alignment, is simply too low to yield a degree of separation comparable to that found for other phyla.

While the 16S rDNA sequence diversity might appear somewhat lower than that found with some other phyla, the *Actinobacteria* have long been recognized for a very high level of morphological, physiological, and genomic diversity. During the past 30 years, considerable effort has been spent in developing a polyphasic approach to the classification and identification of the *Actinobacteria*, and most of the characteristics (especially the molecular and chemotaxonomic) correspond with the current phylogenetic classification. The level of congruence with morphology and conventional biochemical approaches is lower. Despite this potential shortcoming, and the need for specialized microscopy techniques, morphological characteristics are still of value, especially in the preliminary classification and identification of many genera of arthrospore-forming actinobacteria.

UNICELLULAR *ACTINOBACTERIA*

Gram-negative aerobic rods and cocci Chemoorganotrophic heterotrophs, capable of growth under an air atmosphere (21% oxygen) and have a strictly respiratory type of metabolism with O_2 as the terminal electron acceptor. May grow microaerophilically under nitrogen-fixing conditions but grow as aerobes when supplied with a source of fixed nitrogen. Phenotype occurs within a restricted number of genera belonging to the *Acidothermaceae* and the *Genera Incertae Sedis*. (BMDB9 Group 4).

Aerobic sulfur oxidizers Nonphototrophic. Reduced inorganic sulfur compounds can be oxidized, and used as a sole source of energy. Phenotype currently occurs in a single genus, *Acidimicrobium*. (BMDB9 Group 12b).

Budding and/or appendaged bacteria Nonphototrophic. Restricted to a single genus, *Blastococcus*, within the suborder *Frankineae*. (BMDB9 Group 13).

Gram-positive cocci Chemoorganotrophic, mesophilic, nonsporeforming cocci that stain Gram positive. Strictly anaerobic. Occur in pairs, chains, tetrads, or cuboidal packets. Cytochromes are absent in those genera that have been tested. The catalase reaction is usually negative, although in some instances there is a weak or pseudocatalase reaction. Phenotype occurs in a number of genera within the *Micrococcineae*, *Propionibacterineae*, and the "*Kineosporiaceae*". (BMDB9 Group 17).

Regular, nonsporulating Gram-positive rods Rod-shaped cells (coccoid to elongated rods or filaments), generally nonpigmented, mesophilic, chemoorganotrophic. Grow only in complex media. Aerobic or facultatively anaerobic. Possess cofactors and enzymes for respiration. (BMDB9 Group 19).

Irregular, nonsporulating Gram-positive rods The majority are irregular rods which stain Gram positive, grow in the presence of air, and do not produce endospores. Some may exhibit club-shaped forms, branched filamentous elements, mixtures of rods

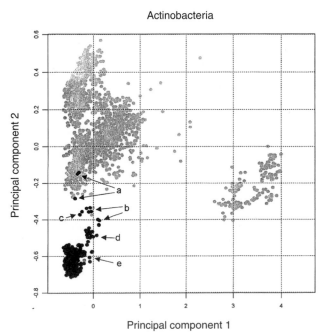

FIGURE 17. Location of Phylum *Actinobacteria* within the map of the procaryotes. Outliers are members of the orders *Rubrobacterales* (a), *Coriobacteriales* (b), *Acidimicrobiales* (c), and *Bifidobacteriales* (d). Those at position (e) are atypical members of the genus *Actinomyces*.

or filamentous and coccoid forms, and some may have a rod-coccus cycle. May stain Gram negative to Gram variable. Oxygen requirements include strictly aerobic, facultatively anaerobic to microaerophilic, or strictly anaerobic. Some are pathogens of animals or plants. (BMDB9 Group 20).

Mycobacteria Aerobic, nonmotile, nonsporing, slow-growing, rod-shaped *Bacteria* that are characteristically acid-fast. Gram stain reaction is weak. Branched filaments are formed occasionally. No aerial mycelium is formed. Only one genus was ascribed to this group: *Mycobacterium*. Some species are pathogenic to animals and humans. (BMDB9 Group 21).

THE SPOROACTINOMYCETES

Nocardioform actinomycetes Gram-positive *Bacteria*. Morphologically diverse. Cells of some genera occur as straight to curved rods. Others produce a mycelium with rudimentary to extensive branching that fragments into bacillary or coccoid elements. Some genera may produce an aerial mycelium and chains of arthrospores. Genera are distinguished mainly on the basis of chemotaxonomic markers. Aerobic or facultatively anaerobic. Chemoorganotrophic. Most are free-living saprophytes. Some genera are pathogenic to animals and humans. Four subtypes are recognized. Type one consists of *Corynebacterineae*, exclusive of *Mycobacterium*. Type two consists of *Pseudonocardineae*. Type three consists of *Nocardioidaceae*. Type four consists of *Promicromonosporaceae*. (BMDB9 Group 22).

Actinomycetes with multilocular sporangia Gram-positive *Bacteria*. Filaments divide by formation of longitudinal and transverse septa to produce a thallus composed of coccoid elements. Spores may be motile or nonmotile. Restricted to the *Frankineae* and *Dermatophilaceae*. (BMDB9 Group 23).

Actinoplanetes Gram-positive *Bacteria*. Chemoorganotrophic. Stable filaments formed with little or no aerial growth. Some genera produce motile spores which are borne in sporangia. Others may produce nonmotile spores either singly or in short chains. Cell walls contain meso-DAP and glycine. Arabinose and xylose present in whole cell hydrolysates. This phenotype is limited to members of the *Micromonosporaceae* other than the type genus. (BMDB9 Group 24).

Streptomyces and related genera Gram-positive *Bacteria*. Most are strict aerobes. Chemoheterotrophic. Filamentous growth, often accompanied by extensive aerial growth with long chains of spores. Cell walls contain L-DAP and glycine. Although the most commonly encountered phenotype because of the shear abundance of *Streptomyces* species in culture and in nature, it is limited to a few genera. (BMDM9 Group 25).

Maduromycetes Gram-positive *Bacteria*. Chemoheterotrophic. Aerobic. Stable filaments formed with varying amounts of aerial growth. Arthrospores borne on aerial mycelia in chains of varying length or in sporangia. Cell walls contain meso-DAP and cell hydrolysates contain madurose. Two subtypes recognized. Type one contains actinomycetes that form a branched, nonfragmenting, vegetative mycelium that bears an aerial mycelium that can differentiate into two or more arthrospores or sporangia bearing one to many spores. Spores may be motile or nonmotile. Type two consists of actinomycetes bearing extensively branched, nonfragmenting hyphae that form a dense vegetative mycelium. Aerial mycelium may be moderate or absent. Chains of arthrospores of varying length arise from the aerial mycelium and may

be straight, flexuous, or form irregular spirals. (BMDM9 Group 26).

Thermomonospora and related genera Gram-positive *Bacteria*. Produce stable filaments with varying amounts of aerial growth. Spores borne singly, in chains or in sporangium-like structures. Cell walls contain meso-DAP. (BMDM9 Group 26).

Other sporoactinomycete genera Gram-positive *Bacteria*. Chemoorganotrophic. Exhibit filamentous growth, produce long chains of arthrospores on the aerial mycelium and do not produce mycolic acids. (BMDM9 Group 29).

Phylum B15 **Planctomycetes** *phy. nov.* Garrity and Holt. Planc.to.my.ce′tes M.L. fem. pl. n. *Planctomycetales* type order of the phylum, ending to denote a phylum; M.L. fem. pl. n. *Planctomycetes* phylum of *Planctomycetales*.

The *Planctomycetes* phylum branches deeply within the bacterial radiation in the ARB and RDP trees and has consistently shown a distant relationship to the *Chlamydiae*. While the precise location within either the ARB or RDP trees remains uncertain, Ludwig and Klenk note that the phylum consistently subdivides into two sister groups, one consisting of *Planctomyces* and *Pirellula*, and the second containing of *Gemmata* and *Isosphaera*. Both groups are currently ascribed to the family *Planctomycetaceae* within the order *Planctomycetales* (Schlesner and Stackebrandt, 1986). Within PCA plots (Fig. 18), *Planctomyces* and *Pirellula* cluster together closely and both split into two subgroups that each overlap. The *Gemmata* (a) species map very close to *Planctomyces/Pirellula* whereas *Isosphaera* (b) maps to a sparsely populated region some distance from the other genera. The *Planctomycetes* are Gram-negative *Bacteria* that reproduce by budding. Cells are spherical to ovoid or bulbiform. Cells may produce one or more multifibrillar appendages that may terminate in holdfasts. Cell envelope lacks peptidoglycan. Some members of the *Planctomy-*

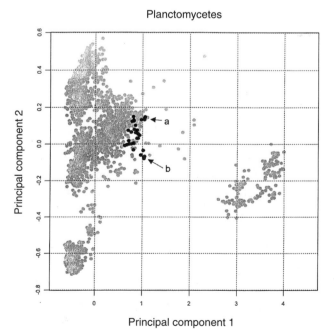

FIGURE 18. Location of Phylum *Planctomycetes* within the map of the procaryotes. Outliers are members of the genera *Isosphaera* (a) and *Gemmata* (b).

cetes exhibit a membrane enclosed nucleoid (*Gemmata* and *Pirellula*). (BMDB9 Group 13).

Phylum B16 Chlamydiae *phy. nov.* Garrity and Holt.
Chla.my′ di.ae M.L. fem. pl. n. *Chlamydiales* type order of the phylum, dropping the ending to denote a phylum; M.L. fem. pl. n. *Chlamydiae* phylum of *Chlamydiales.*

The phylum *Chlamydiae* is defined on the basis of 16S rDNA sequence data as a separate evolutionary lineage within the *Bacteria.* Ludwig and Klenk regard the *Chlamydiae* as a sister group of the *Verrucomicrobia,* but caution that the relationship is tentative. More recently, Everett et al. (1999) proposed an emendation of the *Chlamydiales,* subdividing the order into four families: *Chlamydiaceae, Parachlamydiaceae, Simkaniaceae,* and *Waddliaceae.* We have incorporated this scheme into the current version of the outline and have added the class "*Chlamydiae*" and the phylum *Chlamydiae* to complete the hierarchy. In PCA plots (Fig. 19), the *Chlamydiaceae, Parachlamydiaceae,* and *Waddliaceae* form a tightly clustered group that is clearly separated from the *Planctomycetes* and *Verrucomicrobia,* with which the *Chlamydiae* have often been grouped. The *Simkaniaceae* are clearly separate from the remaining members of the phylum, an observation that is consistent with the published trees (Everett et al., 1999), suggesting a possible misplacement. All members of the phylum are nonmotile, obligately parasitic, coccoid bacteria that multiply within membrane-bound vacuoles in the cytoplasm of cells of mammalian and avian origin. Gram-negative or Gram-variable (*Parachlamydia*). Multiplication occurs by means of a complex life cycle. Pathogenic. Cell walls do not contain muramic acid or only a trace. (BMDB9 Group 9).

Phylum B17 Spirochaetes *phy. nov.* Garrity and Holt.
Spi.ro.chae′ tes M.L. fem. pl. n. *Spirochaetales* type order of the phylum, ending to denote a phylum; M.L. fem. pl. n. *Spriochaetes* phylum of *Spirochaetales.*

The phylum *Spirochaetes* represents a separate line of evolutionary descent in the bacterial domain. Three subgroups are

clearly evident within the reference trees: *Spirochaetaceae,* "*Serpulinaceae*", and *Leptospiraceae,* the last branching more deeply. A similar pattern is observed in PCA plots (Fig. 20) in which the leptospiras (a) form a discrete and well-separated cluster. The *Spirochaetes* remain remarkably uniform in morphology. All validly named species are Gram-negative, helically shaped, highly flexible cells. Motility by periplasmic flagella. Chemoorganotrophic. Anaerobic, microaerophilic, facultatively anaerobic, or aerobic. Free-living or associated with host animals (arthropods, mollusks, and mammals, including humans). Some species are pathogenic. (BMDB9 Group 1).

Phylum B18 Fibrobacteres *phy. nov.* Garrity and Holt.
Fi.bro.bac′ te.res M.L. masc. n. *Fibrobacter* genus of the phylum, ending to denote a phylum; M.L. fem. pl. n. *Fibrobacteres* phylum of *Fibrobacter.*

The phylum *Fibrobacteres* is another lineage currently represented by a single genus, *Fibrobacter,* and was first proposed as such by Montgomery et al. (1988). The *Fibrobacteres* tend to branch in the general region of the *Spirochaetes* and *Chlamydiae* in the ARB and RDP trees. In PCA plots (Fig. 21), we find them mapping to a similar location. Gram-negative, obligately anaerobic, nonsporulating, rod-shaped or pleiomorphic *Bacteria.* Capnophilic. Chemoorganotrophic heterotrophs. Associated with the digestive tracts of various herbivorous mammals. (BMDM9 Group 6).

Phylum B19 Acidobacteria *phy. nov.* Garrity and Holt.
A.ci.do.bac.te′ ri.a M.L. neut. n. *Acidobacterium* genus of the phylum, dropping the ending to denote a phylum; M.L. fem. pl. n. *Acidobacteria* phylum of *Acidobacterium.*

The phylum *Acidobacteria* represents a lineage for which relatively few isolates have been cultivated to date, however, Ludwig and Klenk note that sequences of environspecies related to the *Acidobacteria* are abundant. At present, the phylum includes three genera; quality sequences were available in the RDP for two of

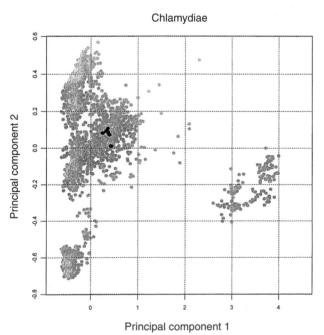

FIGURE 19. Location of Phylum *Chlamydiae* within the map of the procaryotes. Members of the family *Simkaniaceae* lie outside the main cluster.

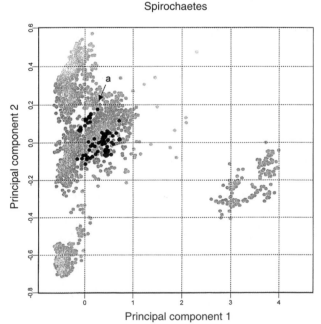

FIGURE 20. Location of Phylum *Spirochaetes* within the map of the procaryotes. The upper cluster at position (a) consists of the family *Leptospiraceae.*

Fibrobacteres

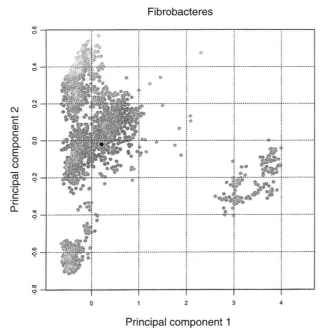

FIGURE 21. Location of Phylum *Fibrobacteres* within the map of the procaryotes.

Acidobacteria

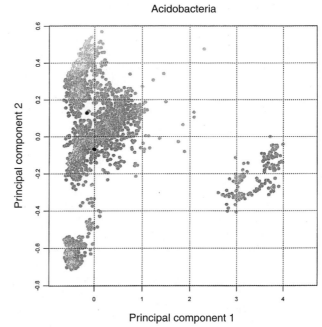

FIGURE 22. Location of Phylum *Acidobacteria* within the map of the procaryotes. The type genus of the phylum is the lower point. The upper point is the type strain of *Holophaga*.

these. PCA plots (Fig. 22) reveal that *Acidobacterium capsulatum* maps to a region in close proximity to *Sporohalobacter*, within the *Halobacteroidaceae* (*Firmicutes*). *Geothrix*, the other genus for which a 16S rDNA sequence was available, mapped to a position removed from *Acidobacterium*, in a sparsely populated region between the *"Deltaproteobacteria"* and the *"Gammaproteobacteria"*. Although the members of the phylum are currently ascribed to a single class, order, and family, it is possible that further subdivision is warranted once additional data becomes available for inclusion into the models. Two phenotypic groups are currently recognized in genera assigned to this phylum. Type one consists of Gram-negative, aerobic, nonsporeforming, rod-shaped *Bacteria*. Chemoorganotrophic heterotrophs. Mesophilic. Acid tolerant. The second group consists of Gram-negative, anaerobic, rod-shaped *Bacteria* which obtain energy by anaerobic respiration or by fermentation. Chemoorganotrophic heterotrophs. (BMDB9 Groups 4 and 6).

Phylum B20 **Bacteroidetes** *phy. nov.* Garrity and Holt.
Bac.te.roi.de' tes M.L. fem. pl. n. *Bacteroidaceae* family of the phylum, ending to denote a phylum; M.L. fem. pl. n. *Bacteroidetes* phylum of *Bacteroidaceae*.

As discussed above, the *Bacteroidetes* share a common root with the *Chlorobi* in the ARB and RDP trees. We have opted to treat these as separate phyla at present, as the branching occurs at a depth that is equivalent to that observed for several other phyla. Within the *Bacteroidetes* there are three distinct lineages that have been accorded the rank of class: the *"Bacteroidetes"*, the *"Flavobacteria"*, and the *"Sphingobacteria"*. In PCA plots (Fig. 23), the class *"Bacteroidetes"* forms a relatively coherent group with only three outliers. One is a species of *Bacteroides* (a) and is likely misidentified, as it maps into the region of the *Clostridiales*. A second is a species of *Porphyromonas* (b), which maps into the sparsely populated region where deeply branching taxa are generally located. The *"Flavobacteria"* map to the same general vicinity, with two outliers, which are misidentified strains of *Flavobac-*

Bacteroidetes

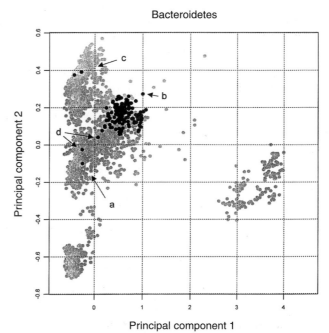

FIGURE 23. Location of Phylum *Bacteroidetes* within the map of the procaryotes. Outliers are identified as members of the genera *Bacteroides* (a), *Porphyromonas* (b), *Flavobacterium* and *Flectobacillus* (c), and *Rhodothermus* (d).

terium (c), appearing in the region of the *"Betaproteobacteria"* and *"Gammaproteobacteria"*. A second cluster of outliers is also observed and consists of all of the species of *Blattabacterium*. *Blattabacterium* species are insect symbionts and we note that symbionts and parasites often map some distance away from phylogenetically related, nonsymbiotic or nonparasitic species in PCA plots. The *"Sphingobacteria"* appear to be more heterogeneous and cluster more loosely. Three outliers also occur within the class. One is likely to be a misidentified species of *Flectobacillus* (c) as it maps into the region of the *"Betaproteobacteria"* and *"Gammaproteobacteria"*. The other two outliers are species of *Rhodothermus* (d), which Ludwig and Klenk indicate is one of the two deepest branching genera within the phylum.

The *Bacteroidetes* are phenotypically diverse, and overlap significantly with members of other phyla. Member species can be ascribed to the following broad categories.

GRAM-NEGATIVE AEROBIC/MICROAEROPHILIC RODS Predominantly chemoorganotrophic heterotrophs, but some may grow autotrophically by using H_2 as an electron donor. Capable of growth under an air atmosphere and have a strictly respiratory type of metabolism with O_2 as the terminal electron acceptor. Some genera are also capable of anaerobic respiration using alternate terminal electron acceptors. Occur in soil, fresh water, or marine environments, within plant roots, or in the reproductive organs, intestinal tract, and oral cavity of humans and animals. Some are pathogenic for animals or humans. Two subtypes observed. Subtype one consists of aerobic rods that can grow in the presence of air (21% oxygen). Some genera are microaerophilic under nitrogen-fixing conditions but grow as aerobes when supplied with a source of fixed nitrogen. Subtype two consists of microaerophilic straight rods that cannot grow in the presence of air (21% oxygen). Oxygen-dependent growth exhibited at reduced O_2 concentrations and in the absence of alternate electron acceptors. Some genera can respire anaerobically using nitrate, fumarate, or other electron acceptors. Occurs predominantly in the *"Flavobacteria"* and *"Sphingobacteria"*. (BMDB9 Group 4).

ANAEROBIC GRAM-NEGATIVE RODS Chemoorganotrophic heterotrophs. Obtain energy by anaerobic respiration or by fermentation. Do not include genera that respire anaerobically with sulfate, other oxidized sulfur compounds, or S^0 as electron acceptors. The phenotype occurs mainly within the *Bacteroidetes*. One genus of *"Flavobacteria"*, *Psychromonas*, also exhibits this phenotype. (BMDM9 Group 6).

NONPHOTOSYNTHETIC, NONFRUITING, GLIDING *BACTERIA* Gram-negative, single-celled, rod-shaped gliding *Bacteria*. Often pleomorphic. Lack flagella, but motile by gliding across solid surfaces or nonmotile. Chemoorganotrophic, aerobic, and facultatively or obligately anaerobic. Nonphototrophic. Simple life cycles. Found mainly within species ascribed to the *"Sphingobacteria"*, however also occurs with some species of *Bacteroides* and *"Flavobacteria"*. (BMDM9 Group 15).

SYMBIOTIC *BACTERIA* OF INVERTEBRATE SPECIES Obligate intracellular parasites of eucaryotic hosts (vertebrates or arthropods). Gram-negative, rod-shaped. Restricted to a single genus of *Bacteroidetes*, *Blattabacterium*. (BMDB9 Group 9).

SHEATHED *BACTERIA* Nonphototrophic. Aerobic. Do not exhibit gliding motility. Characterized by growing as filaments, the cells of which are enclosed with a transparent sheath. Sheaths may appear yellow to dark brown, owing to the deposition of iron and manganese oxides. Single cells may be motile by polar or subpolar flagella or they may be nonmotile. Phenotype restricted to a small number of genera within the *"Sphingobacteria"*. (BMDB9 Group 14).

NONMOTILE OR RARELY MOTILE, CURVED, GRAM-NEGATIVE *BACTERIA* Chemoorganotrophic heterotrophs. Saprophytic. Morphologically distinctive. Curved or C-shaped cells that may form rings by overlapping cell ends. Coils and helices may occur. Gas vacuoles do not occur. Aerobic, aerotolerant, or anaerobic. Occur in soil, fresh water, or marine environments. (BMDB9 Group 3).

***Phylum B21* Fusobacteria** *phy. nov.* Garrity and Holt.
Fu.so.bac.te' ri.a M.L. neut. n. *Fusobacterium* genus of the phylum, dropping the ending to denote a phylum; M.L. fem. pl. n. *Fusobacteria* phylum of *Fusobacterium*.

The *Fusobacteria* represent a line of descent within the ARB and RDP trees that, according to Ludwig and Klenk, contains three subclusters. Our PCA analysis (Fig. 24) shows the group to be quite heterogeneous, with a main cluster located in a region adjacent to the *Leptospiraceae*, but the remaining genera widely dispersed across the region occupied by the *Firmicutes* and *Bacteroidetes*. It is unclear at this time whether this is attributable to the inclusion of sequences derived from misidentified or misplaced species in our model. At present, we recognize a single class, order, and family. Phenotypically, *Fusobacteria* are homogeneous and are characterized as anaerobic, Gram-negative rods with a chemoorganotrophic heterotrophic metabolism. (BMDB9 Group 6).

***Phylum B22* Verrucomicrobia** *phy. nov.* Hedlund, Gosink and Staley 1997, 35, emend. Garrity and Holt
Ver.ru.co.mi.cro' bi.a M.L. fem. pl. n. *Verrucomicrobiales* type order of the phylum, dropping ending to denote a phylum; M.L. fem. pl. n. *Verrucomicrobia* phylum of *Verrucomicrobiales*.

The phylum *Verrucomicrobia* was recently proposed by Hedlund et al. (1997) as a new division within the bacterial domain. We will retain this taxon, however, it will be referred to as a phylum rather than a division. The *Verrucomicrobia* represent another distinct lineage within the phylogenetic reference trees and contains a number of environ species as well as a small number of cultured species assigned to two genera: *Verrucomicrobium* and *Prosthecobacter*, which we ascribe to the family *Verrucomicrobiaceae*. In most

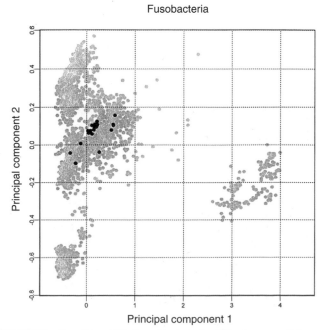

FIGURE 24. Location of Phylum *Fusobacteria* within the map of the procaryotes.

Verrucomicrobia

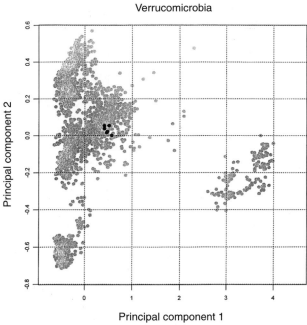

FIGURE 25. Location of Phylum *Verrucomicrobia* within the map of the procaryotes.

Dictyoglomi

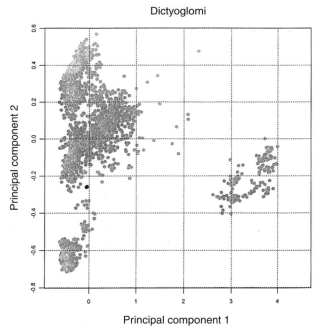

FIGURE 26. Location of Phylum *Dictyoglomi* within the map of the procaryotes.

instances, *Verrucomicrobia* have shown a moderate degree of relationship to the *Planctomycetes* and *Chlamydiae*; however, significance of the common branching is generally low and the relationships among these three phyla are likely to change as additional species are characterized. Within PCA plots (Fig. 25), the *Verrucomicrobia* map to a region adjacent to the boundaries of the *Planctomycetes*. Within the same region, we also find a single representative of the *Chlamydiae*, *Simkania negevensis*, which is an outlier of that phylum.

Phenotypically, members of the *Verrucomicrobia* are Gram-negative *Bacteria* that possess peptidoglycan containing diaminopimelic acid. Some species are capable of producing prosthecae and fimbriae. Aerobic or facultatively aerobic. Chemoheterotrophic metabolism. Mesophilic. Multiply by binary fission or asymmetrically by budding. Buds may be produced at the tip of a prostheca or on the cell surface. (BMDB9 Group 13).

Phylum B23 **Dictyoglomi** *phy. nov.* Garrity and Holt.
Dic.ty.o.glo′ mi L. n. *Dictyoglumus* genus of the phylum, dropping the ending to denote a phylum; M.L. fem. pl. n. *Dictyoglomi* phylum of *Dictyoglomus.*

The phylum *Dictyoglomi* is currently represented by a single species. In the ARB and RDP reference trees, *Dictyoglomi* behaves as a deeply branching group and is found in the vicinity of the *Thermomicrobia* and *"Deinococcus–Thermus"* phylum. In PCA

plots (Fig. 26), *Dictyoglomi* maps to the sparsely populated region beneath the *Firmicutes*. Within this region we also find members of the *"Acidaminococcaceae"*, with *Succiniclasticum* appearing as the nearest neighbor. Gram-negative, rod-shaped, extremely thermophilic *Bacteria*. Nonsporulating. Obligately anaerobic, chemoorganoheterotrophic with a fermentative metabolism. Single cells may aggregate into spherical, membrane bound structures of up to several hundred cells.

ACKNOWLEDGMENTS

We extend our thanks to Dr. Timothy Lilburn of the Ribosomal Database Project, and Peter Sneath of the University of Leichester, who provided invaluable assistance in building and testing evolutionary models and statistical concepts that were vital in compilation of this chapter. We also would like to acknowledge Drs. Wolfgang Ludwig and Karl-Heinz Schleifer of the Technical University of Munich for their numerous and helpful discussions on resolving discrepancies between the Bergey's outline and the phylogenetic reference trees. Special thanks go to Dr. Denise Searles of Bergey's Manual Trust for her assistance in building and maintaining the *Bergey's* outline and researching the history of each type strain, and to Matthew Winters for his assistance in testing and refining our SGML implementation of the outline. We would also like to extend our thanks to Drs. Fred Rainey, Joseph Tully, Floyd Dewhirst, Philip Hugenholtz, Thomas Schmidt, Norman Pace, Bonnie Maidak, David Boone, Erko Stackebrandt, Ross Overbeek, Niels Larsen, Oliver Strunk, and Gary Olsen for their helpful discussions and constructive comments.

TABLE 2. Phenotypic grouping of the procaryotic phyla

Genus	Phylum	Class	Volume	BMDB9
Abiotrophia	Firmicutes	"Bacilli"	3	Group-17
Acetitomaculum	Firmicutes	"Clostridia"	3	Group-20
Acetivibrio	Firmicutes	"Clostridia"	3	Group-6
Acetoanaerobium	Firmicutes	"Bacilli"	3	Group-6
Acetobacter	Proteobacteria	"Alphaproteobacteria"	2	Group-4a
Acetobacterium	Firmicutes	"Clostridia"	3	Group-20
Acetofilamentum	Bacteroidetes	"Bacteroidetes"	5	Group-6
Acetogenium	Firmicutes	"Clostridia"	3	Group-6
Acetohalobium	Firmicutes	"Clostridia"	3	Group-6
Acetomicrobium	Bacteroidetes	"Bacteroidetes"	5	Group-6
Acetonema	Firmicutes	"Clostridia"	3	Group-6
Acetothermus	Bacteroidetes	"Bacteroidetes"	5	Group-6
Acholeplasma	Firmicutes	Mollicutes	3	Group-30
Achromatium	Proteobacteria	"Gammaproteobacteria"	2	Group-15.3
Achromobacter	Proteobacteria	"Betaproteobacteria"	2	Group-4a
Acidaminobacter	Firmicutes	"Clostridia"	3	Group-6
Acidaminococcus	Firmicutes	"Clostridia"	3	Group-8
Acidianus	Crenarchaeota	Thermoprotei	1	Group-35.1
Acidimicrobium	Actinobacteria	Actinobacteria	4	Group-12.2
Acidiphilium	Proteobacteria	"Alphaproteobacteria"	2	Group-4a
Acidobacterium	Acidobacteria	"Acidobacteria"	5	Group-4a
Acidocella	Proteobacteria	"Alphaproteobacteria"	2	Group-4a
Acidomonas	Proteobacteria	"Alphaproteobacteria"	2	Group-4a
Acidothermus	Actinobacteria	Actinobacteria	4	Group-4a
Acidovorax	Proteobacteria	"Betaproteobacteria"	2	Group-4a
Acinetobacter	Proteobacteria	"Gammaproteobacteria"	2	Group-4a
Actinobacillus	Proteobacteria	"Gammaproteobacteria"	2	Group-5.3
Actinobaculum	Actinobacteria	Actinobacteria	4	Group-20
Actinobispora	Actinobacteria	Actinobacteria	4	Group-22.2
Actinocorallia	Actinobacteria	Actinobacteria	4	Group-27
Actinokineospora	Actinobacteria	Actinobacteria	4	Group-22.2
Actinomadura	Actinobacteria	Actinobacteria	4	Group-26.2
Actinomyces	Actinobacteria	Actinobacteria	4	Group-20
Actinoplanes	Actinobacteria	Actinobacteria	4	Group-24
Actinopolyspora	Actinobacteria	Actinobacteria	4	Group-22.2
Actinosynnema	Actinobacteria	Actinobacteria	4	Group-27
Aegyptianella	Proteobacteria	"Alphaproteobacteria"	2	Group-9
Aerococcus	Firmicutes	"Bacilli"	3	Group-17
Aeromicrobium	Actinobacteria	Actinobacteria	4	Group-20
Aeromonas	Proteobacteria	"Gammaproteobacteria"	2	Group-5.2
Aeropyrum	Crenarchaeota	Thermoprotei	1	Group-35.1
Afipia	Proteobacteria	"Alphaproteobacteria"	2	Group-4a
Agitococcus	Firmicutes	"Bacilli"	3	Group-15.5
Agrobacterium	Proteobacteria	"Alphaproteobacteria"	2	Group-4a
Agrococcus	Actinobacteria	Actinobacteria	4	Group-17
Agromonas	Proteobacteria	"Alphaproteobacteria"	2	Group-4a
Agromyces	Actinobacteria	Actinobacteria	4	Group-20
Ahrensia	Proteobacteria	"Alphaproteobacteria"	2	Group-4a
Alcaligenes	Proteobacteria	"Betaproteobacteria"	2	Group-4a
Alcanivorax	Proteobacteria	"Gammaproteobacteria"	2	Group-4a
Alicyclobacillus	Firmicutes	"Bacilli"	3	Group-18
Allochromatium	Proteobacteria	"Gammaproteobacteria"	2	Group-10.1
Alloiococcus	Firmicutes	"Bacilli"	3	Group-17
Allomonas	Proteobacteria	"Gammaproteobacteria"	2	Group-5
Allorhizobium	Proteobacteria	"Alphaproteobacteria"	2	Group-4a
Alterococcus	Proteobacteria	"Gammaproteobacteria"	2	Group-5.1
Alteromonas	Proteobacteria	"Gammaproteobacteria"	2	Group-4a
Alysiella	Proteobacteria	"Betaproteobacteria"	2	Group-15.2
Amaricoccus	Proteobacteria	"Alphaproteobacteria"	2	Group-4a
Aminobacter	Proteobacteria	"Alphaproteobacteria"	2	Group-4a
Aminobacterium	Firmicutes	"Clostridia"	3	Group-6
Aminomonas	Firmicutes	"Clostridia"	3	Group-6
Ammonifex	Firmicutes	"Clostridia"	3	Group-6
Ammoniphilus	Firmicutes	"Bacilli"	3	Group-18
Amoebobacter	Proteobacteria	"Gammaproteobacteria"	2	Group-10.1
Amphibacillus	Firmicutes	"Bacilli"	3	Group-18
Amycolatopsis	Actinobacteria	Actinobacteria	4	Group-22.2
Anabaena	Cyanobacteria	"Cyanobacteria"	1	Group-11.4a
Anabaenopsis	Cyanobacteria	"Cyanobacteria"	1	Group-11
Anaeroarcus	Firmicutes	"Clostridia"	4	Group-6

(continued)

TABLE 2. *(continued)*

Genus	Phylum	Class	Volume	BMDB9
Anaerobacter	*Firmicutes*	*"Clostridia"*	3	Group-18
Anaerobaculum	*Firmicutes*	*"Clostridia"*	3	Group-6
Anaerobiospirillum	*Proteobacteria*	*"Gammaproteobacteria"*	2	Group-6
Anaerobranca	*Firmicutes*	*"Clostridia"*	3	Group-20
Anaerofilum	*Firmicutes*	*"Clostridia"*	3	Group-19
Anaeromusa	*Firmicutes*	*"Clostridia"*	3	Group-6
Anaeroplasma	*Firmicutes*	*Mollicutes*	3	Group-30
Anaerorhabdus	*Bacteroidetes*	*"Bacteroidetes"*	5	Group-6
Anaerosinus	*Firmicutes*	*"Clostridia"*	3	Group-6
Anaerovibrio	*Firmicutes*	*"Clostridia"*	3	Group-6
Anaplasma	*Proteobacteria*	*"Alphaproteobacteria"*	2	Group-9
Ancalochloris	*Chlorobi*	*"Chlorobia"*	1	Group-10.5
Ancalomicrobium	*Proteobacteria*	*"Alphaproteobacteria"*	2	Group-13.1
Ancylobacter	*Proteobacteria*	*"Alphaproteobacteria"*	2	Group-3
Aneurinibacillus	*Firmicutes*	*"Bacilli"*	3	Group-18
Angiococcus	*Proteobacteria*	*"Deltaproteobacteria"*	2	Group-16
Angulomicrobium	*Proteobacteria*	*"Alphaproteobacteria"*	2	Group-13.3
Antarctobacter	*Proteobacteria*	*"Alphaproteobacteria"*	2	Group-13.3
Aphanizomenon	*Cyanobacteria*	*"Cyanobacteria"*	1	Group-11.4a
Aquabacter	*Proteobacteria*	*"Alphaproteobacteria"*	2	Group-4a
Aquabacterium	*Proteobacteria*	*"Betaproteobacteria"*	2	Group-4a
Aquamicrobium	*Proteobacteria*	*"Alphaproteobacteria"*	2	Group-4a
Aquaspirillum	*Proteobacteria*	*"Betaproteobacteria"*	2	Group-2
Aquifex	*Aquificae*	*Aquificae*	1	Group-36
Arcanobacterium	*Actinobacteria*	*Actinobacteria*	4	Group-20
Archaeoglobus	*Euryarchaeota*	*Archaeoglobi*	1	Group-32
Archangium	*Proteobacteria*	*"Deltaproteobacteria"*	2	Group-16
Arcobacter	*Proteobacteria*	*"Epsilonproteobacteria"*	2	Group-2
Arhodomonas	*Proteobacteria*	*"Gammaproteobacteria"*	2	Group-4a
Arsenophonus	*Proteobacteria*	*"Gammaproteobacteria"*	2	Group-5.1
Arthrobacter	*Actinobacteria*	*Actinobacteria*	4	Group-20
Arthrospira	*Cyanobacteria*	*"Cyanobacteria"*	1	Group-11.3
Asteroleplasma	*Firmicutes*	*Mollicutes*	3	Group-30
Asticcacaulis	*Proteobacteria*	*"Alphaproteobacteria"*	2	Group-13.1
Atopobium	*Actinobacteria*	*Actinobacteria*	4	Group-19
Aureobacterium	*Actinobacteria*	*Actinobacteria*	4	Group-20
Azoarcus	*Proteobacteria*	*"Betaproteobacteria"*	2	Group-4a
Azomonas	*Proteobacteria*	*"Gammaproteobacteria"*	2	Group-4a
Azorhizobium	*Proteobacteria*	*"Alphaproteobacteria"*	2	Group-4a
Azospirillum	*Proteobacteria*	*"Alphaproteobacteria"*	2	Group-2
Azotobacter	*Proteobacteria*	*"Gammaproteobacteria"*	2	Group-4a
Bacillus	*Firmicutes*	*"Bacilli"*	3	Group-18
Bacteroides	*Bacteroidetes*	*"Bacteroidetes"*	5	Group-4b
Balneatrix	*Proteobacteria*	*"Gammaproteobacteria"*	2	Group-2
Bartonella	*Proteobacteria*	*"Alphaproteobacteria"*	2	Group-4a
Bdellovibrio	*Proteobacteria*	*"Deltaproteobacteria"*	2	Group-2
Beggiatoa	*Proteobacteria*	*"Gammaproteobacteria"*	2	Group-15.3
Beijerinckia	*Proteobacteria*	*"Alphaproteobacteria"*	2	Group-4a
Bergeyella	*Bacteroidetes*	*"Flavobacteria"*	5	Group-4a
Beutenbergia	*Actinobacteria*	*Actinobacteria*	4	Group-17
Bifidobacterium	*Actinobacteria*	*Actinobacteria*	4	Group-20
Bilophila	*Proteobacteria*	*"Deltaproteobacteria"*	2	Group-6
Blastobacter	*Proteobacteria*	*"Alphaproteobacteria"*	2	Group-13.3
Blastochloris	*Proteobacteria*	*"Alphaproteobacteria"*	2	Group-10.3
Blastococcus	*Actinobacteria*	*Actinobacteria*	4	Group-13.3
Blastomonas	*Proteobacteria*	*"Alphaproteobacteria"*	2	Group-13.3
Blattabacterium	*Bacteroidetes*	*"Flavobacteria"*	5	Group 9
Bogoriella	*Actinobacteria*	*Actinobacteria*	4	Group-20
Bordetella	*Proteobacteria*	*"Betaproteobacteria"*	2	Group-4a
Borrelia	*Spirochaetes*	*"Spirochaetes"*	5	Group-1
Borzia	*Cyanobacteria*	*"Cyanobacteria"*	1	Group-11.3
Bosea	*Proteobacteria*	*"Alphaproteobacteria"*	2	Group-4a
Brachybacterium	*Actinobacteria*	*Actinobacteria*	4	Group-20
Brachymonas	*Proteobacteria*	*"Betaproteobacteria"*	2	Group-4a
Brachyspira	*Spirochaetes*	*"Spirochaetes"*	5	Group-1
Bradyrhizobium	*Proteobacteria*	*"Alphaproteobacteria"*	2	Group-4a
Brenneria	*Proteobacteria*	*"Gammaproteobacteria"*	2	Group-5.1
Brevibacillus	*Firmicutes*	*"Bacilli"*	3	Group-18
Brevibacterium	*Actinobacteria*	*Actinobacteria*	4	Group-20
Brevinema	*Spirochaetes*	*"Spirochaetes"*	5	Group-1

(continued)

TABLE 2. *(continued)*

Genus	Phylum	Class	Volume	BMDB9
Brevundimonas	*Proteobacteria*	"*Alphaproteobacteria*"	2	Group-4a
Brochothrix	*Firmicutes*	"*Bacilli*"	3	Group-19
Brucella	*Proteobacteria*	"*Alphaproteobacteria*"	2	Group-4a
Buchnera	*Proteobacteria*	"*Gammaproteobacteria*"	2	Group-4a
Budvicia	*Proteobacteria*	"*Gammaproteobacteria*"	2	Group-5.1
Burkholderia	*Proteobacteria*	"*Betaproteobacteria*"	2	Group-4a
Buttiauxella	*Proteobacteria*	"*Gammaproteobacteria*"	2	Group-5.1
Butyrivibrio	*Firmicutes*	"*Clostridia*"	3	Group-6
Caedibacter	*Proteobacteria*	"*Alphaproteobacteria*"	2	Group 9
Calderobacterium	*Aquificae*	*Aquificae*	1	Group-36
Caldicellulosiruptor	*Firmicutes*	"*Clostridia*"	3	Group-6
Caldivirga	*Crenarchaeota*	*Thermoprotei*	1	Group-35.2
Caloramator	*Firmicutes*	"*Clostridia*"	3	Group-18
Calothrix	*Cyanobacteria*	"*Cyanobacteria*"	1	Group-11.4c
Calymmatobacterium	*Proteobacteria*	"*Gammaproteobacteria*"	2	Group-5.4
Campylobacter	*Proteobacteria*	"*Epsilonproteobacteria*"	2	Group-2
Capnocytophaga	*Bacteroidetes*	"*Flavobacteria*"	5	Group-15.1
Carbophilus	*Proteobacteria*	"*Alphaproteobacteria*"	2	Group-4a
Carboxydothermus	*Firmicutes*	"*Clostridia*"	3	Group-36
Cardiobacterium	*Proteobacteria*	"*Gammaproteobacteria*"	2	Group-5.4
Carnimonas	*Proteobacteria*	"*Gammaproteobacteria*"	2	Group-4a
Carnobacterium	*Firmicutes*	"*Bacilli*"	3	Group-19
Caryophanon	*Firmicutes*	"*Bacilli*"	3	Group-19
Catellatospora	*Actinobacteria*	*Actinobacteria*	4	Group-24
Catenococcus	*Proteobacteria*	"*Betaproteobacteria*"	2	Group-4
Catenuloplanes	*Actinobacteria*	*Actinobacteria*	4	Group-24
Catonella	*Firmicutes*	"*Clostridia*"	3	Group-6
Caulobacter	*Proteobacteria*	"*Alphaproteobacteria*"	2	Group-13.1
Cedecea	*Proteobacteria*	"*Gammaproteobacteria*"	2	Group-5.1
Cellulomonas	*Actinobacteria*	*Actinobacteria*	4	Group-20
Cellulophaga	*Bacteroidetes*	"*Flavobacteria*"	5	Group-15.1
Cellvibrio	*Proteobacteria*	"*Gammaproteobacteria*"	2	Group-2
Centipeda	*Firmicutes*	"*Clostridia*"	3	Group-6
Cetobacterium	*Fusobacteria*	"*Fusobacteria*"	5	Group-6
Chamaesiphon	*Cyanobacteria*	"*Cyanobacteria*"	1	Group-11.1
Chelatobacter	*Proteobacteria*	"*Alphaproteobacteria*"	2	Group-4a
Chelatococcus	*Proteobacteria*	"*Alphaproteobacteria*"	2	Group-4a
Chitinophaga	*Bacteroidetes*	"*Sphingobacteria*"	5	Group-15.1
Chlamydia	*Chlamydiae*	"*Chlamydiae*"	5	Group-9.2
Chlamydophila	*Chlamydiae*	"*Chlamydiae*"	5	Group-9.2
Chlorobium	*Chlorobi*	"*Chlorobia*"	1	Group-10.5
Chloroflexus	*Chloroflexi*	"*Chloroflexi*"	1	Group-10.6
Chlorogloeopsis	*Cyanobacteria*	"*Cyanobacteria*"	1	Group-11.5
Chloroherpeton	*Chlorobi*	"*Chlorobia*"	1	Group-10.5
Chloronema	*Chloroflexi*	"*Chloroflexi*"	1	Group-10.6
Chondromyces	*Proteobacteria*	"*Deltaproteobacteria*"	2	Group-16
Chromatium	*Proteobacteria*	"*Gammaproteobacteria*"	2	Group-10.1
Chromobacterium	*Proteobacteria*	"*Betaproteobacteria*"	2	Group-5.4
Chromohalobacter	*Proteobacteria*	"*Gammaproteobacteria*"	2	Group-4a
Chroococcidiopsis	*Cyanobacteria*	"*Cyanobacteria*"	1	Group-11.2
Chroococcus	*Cyanobacteria*	"*Cyanobacteria*"	1	Group-11.1
Chryseobacterium	*Bacteroidetes*	"*Flavobacteria*"	5	Group-4a
Chryseomonas	*Proteobacteria*	"*Gammaproteobacteria*"	2	Group-4a
Chrysiogenes	*Chrysiogenetes*	*Chrysiogenetes*	1	Group-6
Citrobacter	*Proteobacteria*	"*Gammaproteobacteria*"	2	Group-5.1
Clavibacter	*Actinobacteria*	*Actinobacteria*	4	Group-20
Clevelandina	*Spirochaetes*	"*Spirochaetes*"	5	Group-1
Clostridium	*Firmicutes*	"*Clostridia*"	3	Group-18
Collinsella	*Actinobacteria*	*Actinobacteria*	4	Group-20
Colwellia	*Proteobacteria*	"*Gammaproteobacteria*"	2	Group-5.2
Comamonas	*Proteobacteria*	"*Betaproteobacteria*"	2	Group-4a
Coprococcus	*Firmicutes*	"*Clostridia*"	3	Group-16
Coprothermobacter	*Firmicutes*	"*Clostridia*"	3	Group-36
Coriobacterium	*Actinobacteria*	*Actinobacteria*	4	Group-20
Corynebacterium	*Actinobacteria*	*Actinobacteria*	4	Group-20
Couchioplanes	*Actinobacteria*	*Actinobacteria*	4	Group-24
Cowdria	*Proteobacteria*	"*Alphaproteobacteria*"	2	Group-9.1
Coxiella	*Proteobacteria*	"*Gammaproteobacteria*"	2	Group-9.1
Craurococcus	*Proteobacteria*	"*Alphaproteobacteria*"	2	Group-4a
Crenothrix	*Bacteroidetes*	"*Sphingobacteria*"	5	Group-14

(continued)

TABLE 2. *(continued)*

Genus	Phylum	Class	Volume	BMDB9
Crinalium	*Cyanobacteria*	"*Cyanobacteria*"	1	Group-11.3
Cristispira	*Spirochaetes*	"*Spirochaetes*"	5	Group-1
Cryobacterium	*Actinobacteria*	*Actinobacteria*	4	Group-20
Cryptobacterium	*Actinobacteria*	*Actinobacteria*	4	Group-19
Cryptosporangium	*Actinobacteria*	*Actinobacteria*	4	Group-23
Cupriavidus	*Proteobacteria*	"*Betaproteobacteria*"	2	Group-4a
Curtobacterium	*Actinobacteria*	*Actinobacteria*	4	Group-20
Cyanobacterium	*Cyanobacteria*	"*Cyanobacteria*"	1	Group-11
Cyanobium	*Cyanobacteria*	"*Cyanobacteria*"	1	Group-11
Cyanocystis	*Cyanobacteria*	"*Cyanobacteria*"	1	Group-11
Cyanospira	*Cyanobacteria*	"*Cyanobacteria*"	1	Group-11
Cyanothece	*Cyanobacteria*	"*Cyanobacteria*"	1	Group-11.1
Cyclobacterium	*Bacteroidetes*	"*Sphingobacteria*"	5	Group-3
Cycloclasticus	*Proteobacteria*	"*Gammaproteobacteria*"	2	Group-4a
Cylindrospermopsis	*Cyanobacteria*	"*Cyanobacteria*"	1	Group-11
Cylindrospermum	*Cyanobacteria*	"*Cyanobacteria*"	1	Group-11.4a
Cystobacter	*Proteobacteria*	"*Deltaproteobacteria*"	2	Group-16
Cytophaga	*Bacteroidetes*	"*Sphingobacteria*"	5	Group-15.1
Dactylococcopsis	*Cyanobacteria*	"*Cyanobacteria*"	1	Group-11.1
Dactylosporangium	*Actinobacteria*	*Actinobacteria*	4	Group-24
Deferribacter	*Deferribacteres*	*Deferribacteres*	1	Group-6
Defluvibacter	*Proteobacteria*	"*Alphaproteobacteria*"	2	Group-4a
Dehalobacter	*Firmicutes*	"*Clostridia*"	3	Group-6
Deinococcus	"*Deinococcus–Thermus*"	*Deinococci*	1	Group-17
Deleya	*Proteobacteria*	"*Gammaproteobacteria*"	2	Group-4a
Delftia	*Proteobacteria*	"*Betaproteobacteria*"	2	Group-4a
Demetria	*Actinobacteria*	*Actinobacteria*	4	Group-23
Dendrosporobacter	*Firmicutes*	"*Clostridia*"	3	Group-18
Dermabacter	*Actinobacteria*	*Actinobacteria*	4	Group-20
Dermacoccus	*Actinobacteria*	*Actinobacteria*	4	Group-17
Dermatophilus	*Actinobacteria*	*Actinobacteria*	4	Group-23
Dermocarpella	*Cyanobacteria*	"*Cyanobacteria*"	1	Group-11.2
Derxia	*Proteobacteria*	"*Alphaproteobacteria*"	2	Group-4a
Desemzia	*Firmicutes*	"*Bacilli*"	3	Group-19
Desulfacinum	*Proteobacteria*	"*Deltaproteobacteria*"	2	Group-7.3
Desulfitobacterium	*Firmicutes*	"*Clostridia*"	3	Group-19
Desulfoarculus	*Proteobacteria*	"*Deltaproteobacteria*"	2	Group-7.3
Desulfobacca	*Proteobacteria*	"*Deltaproteobacteria*"	2	Group-7.3
Desulfobacter	*Proteobacteria*	"*Deltaproteobacteria*"	2	Group-7.3
Desulfobacterium	*Proteobacteria*	"*Deltaproteobacteria*"	2	Group-7.3
Desulfobacula	*Proteobacteria*	"*Deltaproteobacteria*"	2	Group-7.3
Desulfobotulus	*Proteobacteria*	"*Deltaproteobacteria*"	2	Group-7.3
Desulfobulbus	*Proteobacteria*	"*Deltaproteobacteria*"	2	Group-7.2
Desulfocapsa	*Proteobacteria*	"*Deltaproteobacteria*"	2	Group-7.2
Desulfocella	*Proteobacteria*	"*Deltaproteobacteria*"	2	Group-7.3
Desulfococcus	*Proteobacteria*	"*Deltaproteobacteria*"	2	Group-7.3
Desulfofaba	*Proteobacteria*	"*Deltaproteobacteria*"	2	Group-7.2
Desulfofrigus	*Proteobacteria*	"*Deltaproteobacteria*"	2	Group-7.3
Desulfofustis	*Proteobacteria*	"*Deltaproteobacteria*"	2	Group-7.2
Desulfohalobium	*Proteobacteria*	"*Deltaproteobacteria*"	2	Group-7.3
Desulfomicrobium	*Proteobacteria*	"*Deltaproteobacteria*"	2	Group-7.2
Desulfomonas	*Proteobacteria*	"*Deltaproteobacteria*"	2	Group-7.2
Desulfomonile	*Proteobacteria*	"*Deltaproteobacteria*"	2	Group-7.3
Desulfonatronovibrio	*Proteobacteria*	"*Deltaproteobacteria*"	2	Group-7.3
Desulfonema	*Proteobacteria*	"*Deltaproteobacteria*"	2	Group-7.3
Desulfonispora	*Firmicutes*	"*Clostridia*"	3	Group-18
Desulforhabdus	*Proteobacteria*	"*Deltaproteobacteria*"	2	Group-7.3
Desulforhopalus	*Proteobacteria*	"*Deltaproteobacteria*"	2	Group-7.2
Desulfosarcina	*Proteobacteria*	"*Deltaproteobacteria*"	2	Group-7.3
Desulfospira	*Proteobacteria*	"*Deltaproteobacteria*"	2	Group-7.3
Desulfosporosinus	*Firmicutes*	"*Clostridia*"	3	Group-18
Desulfotalea	*Proteobacteria*	"*Deltaproteobacteria*"	2	Group-7.2
Desulfotomaculum	*Firmicutes*	"*Clostridia*"	3	Group-7.1
Desulfovibrio	*Proteobacteria*	"*Deltaproteobacteria*"	2	Group-7.2
Desulfurella	*Proteobacteria*	"*Deltaproteobacteria*"	2	Group-7.4
Desulfurobacterium	*Aquificae*	*Aquificae*	1	Group-36
Desulfurococcus	*Crenarchaeota*	*Thermoprotei*	1	Group-35.3
Desulfuromonas	*Proteobacteria*	"*Deltaproteobacteria*"	2	Group-7.4
Desulfuromusa	*Proteobacteria*	"*Deltaproteobacteria*"	2	Group-7.3
Dethiosulfovibrio	*Firmicutes*	"*Clostridia*"	3	Group-6

(continued)

TABLE 2. *(continued)*

Genus	Phylum	Class	Volume	BMDB9
Devosia	*Proteobacteria*	"Alphaproteobacteria"	2	Group-4a
Dialister	*Firmicutes*	"Clostridia"	3	Group-6
Dichelobacter	*Proteobacteria*	"Gammaproteobacteria"	2	Group-6
Dichotomicrobium	*Proteobacteria*	"Alphaproteobacteria"	2	Group-13.1
Dictyoglomus	*Dictyoglomi*	"Dictyoglomi"	5	Group-36
Dietzia	*Actinobacteria*	Actinobacteria	4	Group-22.1
Diplocalyx	*Spirochaetes*	"Spirochaetes"	5	Group-1
Dolosicoccus	*Firmicutes*	"Bacilli"	3	Group-17
Dolosigranulum	*Firmicutes*	"Bacilli"	3	Group-19
Duganella	*Proteobacteria*	"Betaproteobacteria"	2	Group-4a
Ectothiorhodospira	*Proteobacteria*	"Gammaproteobacteria"	2	Group-10.2
Edwardsiella	*Proteobacteria*	"Gammaproteobacteria"	2	Group-5.1
Eggerthella	*Actinobacteria*	Actinobacteria	4	Group-19
Ehrlichia	*Proteobacteria*	"Alphaproteobacteria"	2	Group-9.1
Eikenella	*Proteobacteria*	"Betaproteobacteria"	2	Group-5.4
Empedobacter	*Bacteroidetes*	"Flavobacteria"	5	Group-4a
Enhydrobacter	*Proteobacteria*	"Gammaproteobacteria"	2	Group-5.2
Ensifer	*Proteobacteria*	"Alphaproteobacteria"	2	Group-4a
Enterobacter	*Proteobacteria*	"Gammaproteobacteria"	2	Group-5.1
Enterococcus	*Firmicutes*	"Bacilli"	3	Group-17
Entomoplasma	*Firmicutes*	Mollicutes	3	Group-30
Eperythrozoon	*Firmicutes*	Mollicutes	3	Group-9.1
Eremococcus	*Firmicutes*	"Bacilli"	3	Group-17
Erwinia	*Proteobacteria*	"Gammaproteobacteria"	2	Group-5.1
Erysipelothrix	*Firmicutes*	Mollicutes	3	Group-19
Erythrobacter	*Proteobacteria*	"Alphaproteobacteria"	2	Group-4a
Erythromicrobium	*Proteobacteria*	"Alphaproteobacteria"	2	Group-10.7
Erythromonas	*Proteobacteria*	"Alphaproteobacteria"	2	Group-10.7
Escherichia	*Proteobacteria*	"Gammaproteobacteria"	2	Group-5.1
Eubacterium	*Firmicutes*	"Clostridia"	3	Group-20
Ewingella	*Proteobacteria*	"Gammaproteobacteria"	2	Group-5.1
Excellospora	*Actinobacteria*	Actinobacteria	4	Group-26
Exiguobacterium	*Firmicutes*	"Bacilli"	3	Group-20
Facklamia	*Firmicutes*	"Bacilli"	3	Group-17
Falcivibrio	*Actinobacteria*	Actinobacteria	4	Group-20
Ferrimonas	*Proteobacteria*	"Gammaproteobacteria"	2	Group-5.2
Ferroglobus	*Euryarchaeota*	Archaeoglobi	1	Group-32
Fervidobacterium	*Thermotogae*	Thermotogae	1	Group-36
Fibrobacter	*Fibrobacteres*	"Fibrobacteres"	5	Group-6
Filibacter	*Firmicutes*	"Bacilli"	3	Group-15
Filifactor	*Firmicutes*	"Clostridia"	3	Group-18
Filomicrobium	*Proteobacteria*	"Alphaproteobacteria"	2	Group-13.1
Fischerella	*Cyanobacteria*	"Cyanobacteria"	1	Group-11.5
Flammeovirga	*Bacteroidetes*	"Sphingobacteria"	5	Group-15.1
Flavimonas	*Proteobacteria*	"Gammaproteobacteria"	2	Group-4a
Flavobacterium	*Bacteroidetes*	"Flavobacteria"	5	Group-4a
Flectobacillus	*Bacteroidetes*	"Sphingobacteria"	5	Group-3
Flexibacter	*Bacteroidetes*	"Sphingobacteria"	5	Group-15.1
Flexistipes	*Deferribacteres*	Deferribacteres	1	Group-6
Flexithrix	*Bacteroidetes*	"Sphingobacteria"	5	Group-15.1
Formivibrio	*Proteobacteria*	"Betaproteobacteria"	5	Group-6
Francisella	*Proteobacteria*	"Gammaproteobacteria"	2	Group-4a
Frankia	*Actinobacteria*	Actinobacteria	4	Group-23
Frateuria	*Proteobacteria*	"Gammaproteobacteria"	2	Group-4a
Friedmanniella	*Actinobacteria*	Actinobacteria	4	Group-22.3
Frigoribacterium	*Actinobacteria*	Actinobacteria	4	Group-20
Fundibacter	*Proteobacteria*	"Gammaproteobacteria"	2	Group-2
Fusibacter	*Firmicutes*	"Clostridia"	3	Group-19
Fusobacterium	*Fusobacteria*	"Fusobacteria"	5	Group-6
Gallionella	*Proteobacteria*	"Betaproteobacteria"	2	Group-12.2
Gardnerella	*Actinobacteria*	Actinobacteria	4	Group-20
Geitleria	*Cyanobacteria*	"Cyanobacteria"	1	Group-11.5
Geitlerinema	*Cyanobacteria*	"Cyanobacteria"	1	Group-11.3
Gelidibacter	*Bacteroidetes*	"Flavobacteria"	5	Group-4a
Gemella	*Firmicutes*	"Bacilli"	3	Group-17
Gemmata	*Planctomycetes*	"Planctomycetacia"	5	Group-13.2
Gemmiger	*Proteobacteria*	"Alphaproteobacteria"	2	Group-13.3
Gemmobacter	*Proteobacteria*	"Alphaproteobacteria"	2	Group-4a
Geobacter	*Proteobacteria*	"Deltaproteobacteria"	2	Group-6
Geodermatophilus	*Actinobacteria*	Actinobacteria	4	Group-23

(continued)

TABLE 2. *(continued)*

Genus	Phylum	Class	Volume	BMDB9
Geothrix	*Acidobacteria*	"*Acidobacteria*"	5	Group-6
Geotoga	*Thermotogae*	*Thermotogae*	1	Group-36
Geovibrio	*Deferribacteres*	*Deferribacteres*	1	Group-6
Glaciecola	*Proteobacteria*	"*Gammaproteobacteria*"	2	Group-5.2
Globicatella	*Firmicutes*	"*Bacilli*"	3	Group-19
Gloeobacter	*Cyanobacteria*	"*Cyanobacteria*"	1	Group-11.1
Gloeocapsa	*Cyanobacteria*	"*Cyanobacteria*"	1	Group-11.1
Gloeothece	*Cyanobacteria*	"*Cyanobacteria*"	1	Group-11.1
Gluconacetobacter	*Proteobacteria*	"*Alphaproteobacteria*"	2	Group-4a
Gluconobacter	*Proteobacteria*	"*Alphaproteobacteria*"	2	Group-4a
Glycomyces	*Actinobacteria*	*Actinobacteria*	4	Group-29
Gordonia	*Actinobacteria*	*Actinobacteria*	4	Group-22.1
Gracilibacillus	*Firmicutes*	"*Bacilli*"	3	Group-18
Haemobartonella	*Firmicutes*	*Mollicutes*	3	Group-9.1
Haemophilus	*Proteobacteria*	"*Gammaproteobacteria*"	2	Group-5.3
Hafnia	*Proteobacteria*	"*Gammaproteobacteria*"	2	Group-5.1
Haliscomenobacter	*Bacteroidetes*	"*Sphingobacteria*"	5	Group-14
Hallella	*Bacteroidetes*	"*Bacteroidetes*"	5	Group-6
Haloanaerobacter	*Firmicutes*	"*Clostridia*"	3	Group-6
Haloanaerobium	*Firmicutes*	"*Clostridia*"	3	Group-6
Haloarcula	*Euryarchaeota*	*Halobacteria*	1	Group-33
Halobacillus	*Firmicutes*	"*Bacilli*"	3	Group-18
Halobacterium	*Euryarchaeota*	*Halobacteria*	1	Group-33
Halobacteroides	*Firmicutes*	"*Clostridia*"	3	Group-6
Halobaculum	*Euryarchaeota*	*Halobacteria*	1	Group-34
Halocella	*Firmicutes*	"*Clostridia*"	3	Group-6
Halochromatium	*Proteobacteria*	"*Gammaproteobacteria*"	2	Group-10.1
Halococcus	*Euryarchaeota*	*Halobacteria*	1	Group-33
Haloferax	*Euryarchaeota*	*Halobacteria*	1	Group-33
Halogeometricum	*Euryarchaeota*	*Halobacteria*	1	Group-33
Halomonas	*Proteobacteria*	"*Gammaproteobacteria*"	2	Group-4a
Halorhabdus	*Euryarchaeota*	*Halobacteria*	1	Group-33
Halorhodospira	*Proteobacteria*	"*Gammaproteobacteria*"	2	Group-10.3
Halorubrum	*Euryarchaeota*	*Halobacteria*	1	Group-33
Haloterrigena	*Euryarchaeota*	*Halobacteria*	1	Group-33
Halothermothrix	*Firmicutes*	"*Clostridia*"	3	Group-6
Helcococcus	*Firmicutes*	"*Clostridia*"	3	Group-17
Helicobacter	*Proteobacteria*	"*Epsilonproteobacteria*"	2	Group-2
Heliobacillus	*Firmicutes*	"*Clostridia*"	3	Group-10.4
Heliobacterium	*Firmicutes*	"*Clostridia*"	3	Group-10.4
Heliophilum	*Firmicutes*	"*Clostridia*"	3	Group-10.4
Heliothrix	*Chloroflexi*	"*Chloroflexi*"	1	Group-10.6
Herbaspirillum	*Proteobacteria*	"*Betaproteobacteria*"	2	Group-2
Herbidospora	*Actinobacteria*	*Actinobacteria*	4	Group-26
Herpetosiphon	*Chloroflexi*	"*Chloroflexi*"	1	Group-15.5
Hippea	*Proteobacteria*	"*Deltaproteobacteria*"	2	Group-7.4
Hirschia	*Proteobacteria*	"*Alphaproteobacteria*"	2	Group-13.1
Holdemania	*Firmicutes*	*Mollicutes*	3	Group-20
Hollandina	*Spirochaetes*	"*Spirochaetes*"	5	Group-1
Holophaga	*Acidobacteria*	"*Acidobacteria*"	5	Group-6
Holospora	*Proteobacteria*	"*Alphaproteobacteria*"	2	Group-9
Hydrogenobacter	*Aquificae*	*Aquificae*	1	Group-36
Hydrogenophaga	*Proteobacteria*	"*Betaproteobacteria*"	2	Group-4a
Hydrogenophilus	*Proteobacteria*	"*Betaproteobacteria*"	2	Group-12.1b
Hydrogenovibrio	*Proteobacteria*	"*Gammaproteobacteria*"	2	Group-4a
Hymenobacter	*Bacteroidetes*	"*Sphingobacteria*"	5	Group-4a
Hyperthermus	*Crenarchaeota*	*Thermoprotei*	1	Group-35.3
Hyphomicrobium	*Proteobacteria*	"*Alphaproteobacteria*"	2	Group-13.1
Hyphomonas	*Proteobacteria*	"*Alphaproteobacteria*"	2	Group-13.1
Ideonella	*Proteobacteria*	"*Betaproteobacteria*"	2	Group-4a
Ignavigranum	*Firmicutes*	"*Bacilli*"	3	Group-17.2
Igneococcus	*Crenarchaeota*	*Thermoprotei*	1	Group-35.1
Ilyobacter	*Fusobacteria*	"*Fusobacteria*"	5	Group-6
Intrasporangium	*Actinobacteria*	*Actinobacteria*	4	Group-25
Iodobacter	*Proteobacteria*	"*Betaproteobacteria*"	2	Group-5.4
Isochromatium	*Proteobacteria*	"*Gammaproteobacteria*"	2	Group-10.1
Isosphaera	*Planctomycetes*	"*Planctomycetacia*"	5	Group-13.2
Iyengariella	*Cyanobacteria*	"*Cyanobacteria*"	1	Group-11
Janibacter	*Actinobacteria*	*Actinobacteria*	4	Group-19
Janthinobacterium	*Proteobacteria*	"*Betaproteobacteria*"	2	Group-4a

(continued)

TABLE 2. *(continued)*

Genus	Phylum	Class	Volume	BMDB9
Johnsonella	*Firmicutes*	"*Clostridia*"	3	Group-6
Jonesia	*Actinobacteria*	*Actinobacteria*	4	Group-20
Kibdelosporangium	*Actinobacteria*	*Actinobacteria*	4	Group-22.2
Kineococcus	*Actinobacteria*	*Actinobacteria*	4	Group-17
Kineosporia	*Actinobacteria*	*Actinobacteria*	4	Group-25
Kingella	*Proteobacteria*	"*Betaproteobacteria*"	2	Group-4a
Kitasatospora	*Actinobacteria*	*Actinobacteria*	4	Group-25
Klebsiella	*Proteobacteria*	"*Gammaproteobacteria*"	2	Group-5.1
Kluyvera	*Proteobacteria*	"*Gammaproteobacteria*"	2	Group-5.1
Kocuria	*Actinobacteria*	*Actinobacteria*	4	Group-17
Kribbella	*Actinobacteria*	*Actinobacteria*	4	Group-25
Kurthia	*Firmicutes*	"*Bacilli*"	3	Group-19
Kutzneria	*Actinobacteria*	*Actinobacteria*	4	Group-29
Kytococcus	*Actinobacteria*	*Actinobacteria*	4	Group-17
Labrys	*Proteobacteria*	"*Alphaproteobacteria*"	2	Group-13.1
Lachnospira	*Firmicutes*	"*Clostridia*"	3	Group-6
Lactobacillus	*Firmicutes*	"*Bacilli*"	3	Group-19
Lactococcus	*Firmicutes*	"*Bacilli*"	3	Group-17
Lactosphaera	*Firmicutes*	"*Bacilli*"	3	Group-17
Lamprobacter	*Proteobacteria*	"*Gammaproteobacteria*"	2	Group-10.1
Lamprocystis	*Proteobacteria*	"*Gammaproteobacteria*"	2	Group-10.1
Lampropedia	*Proteobacteria*	"*Gammaproteobacteria*"	2	Group-4a
Lautropia	*Proteobacteria*	"*Betaproteobacteria*"	2	Group-4a
Lawsonia	*Proteobacteria*	"*Deltaproteobacteria*"	2	Group-4a
Leclercia	*Proteobacteria*	"*Gammaproteobacteria*"	2	Group-5.1
Legionella	*Proteobacteria*	"*Gammaproteobacteria*"	2	Group-4a
Leifsonia	*Actinobacteria*	*Actinobacteria*	4	Group-20
Leminorella	*Proteobacteria*	"*Gammaproteobacteria*"	2	Group-5.1
Lentzea	*Actinobacteria*	*Actinobacteria*	4	Group-27
Leptolyngbya	*Cyanobacteria*	"*Cyanobacteria*"	1	Group-11.3
Leptonema	*Spirochaetes*	"*Spirochaetes*"	5	Group-1
Leptospira	*Spirochaetes*	"*Spirochaetes*"	5	Group-1
Leptospirillum	*Nitrospirae*	"*Nitrospira*"	1	Group-12.2
Leptothrix	*Proteobacteria*	"*Betaproteobacteria*"	2	Group-14
Leptotrichia	*Fusobacteria*	"*Fusobacteria*"	5	Group-6
Leucobacter	*Actinobacteria*	*Actinobacteria*	4	Group-19
Leuconostoc	*Firmicutes*	"*Bacilli*"	3	Group-17
Leucothrix	*Proteobacteria*	"*Gammaproteobacteria*"	2	Group-15.5
Lewinella	*Bacteroidetes*	"*Sphingobacteria*"	5	Group-14
Limnothrix	*Cyanobacteria*	"*Cyanobacteria*"	1	Group-11.3
Listeria	*Firmicutes*	"*Bacilli*"	3	Group-19
Listonella	*Proteobacteria*	"*Gammaproteobacteria*"	2	Group-5.2
Lonepinella	*Proteobacteria*	"*Gammaproteobacteria*"	2	Group-5.3
Luteococcus	*Actinobacteria*	*Actinobacteria*	4	Group-17
Lyngbya	*Cyanobacteria*	"*Cyanobacteria*"	1	Group-11.3
Lysobacter	*Proteobacteria*	"*Gammaproteobacteria*"	2	Group-15.1
Lyticum	*Proteobacteria*	"*Alphaproteobacteria*"	2	Group-9
Macrococcus	*Firmicutes*	"*Bacilli*"	3	Group-17
Macromonas	*Proteobacteria*	"*Gammaproteobacteria*"	2	Group-12.1a
Magnetobacterium	*Nitrospirae*	"*Nitrospira*"	1	Group-12.5
Magnetospirillum	*Proteobacteria*	"*Alphaproteobacteria*"	2	Group-2
Malonomonas	*Proteobacteria*	"*Deltaproteobacteria*"	5	Group-6
Mannheimia	*Proteobacteria*	"*Gammaproteobacteria*"	2	Group-5.3
Maricaulis	*Proteobacteria*	"*Alphaproteobacteria*"	2	Group-13.1
Marichromatium	*Proteobacteria*	"*Gammaproteobacteria*"	2	Group-10.1
Marinilabilia	*Bacteroidetes*	"*Bacteroidetes*"	5	Group-15.1
Marinobacter	*Proteobacteria*	"*Gammaproteobacteria*"	2	Group-4a
Marinobacterium	*Proteobacteria*	"*Gammaproteobacteria*"	2	Group-4a
Marinococcus	*Firmicutes*	"*Bacilli*"	3	Group-17
Marinomonas	*Proteobacteria*	"*Gammaproteobacteria*"	2	Group-2
Marinospirillum	*Proteobacteria*	"*Gammaproteobacteria*"	2	Group-2
Megamonas	*Bacteroidetes*	"*Bacteroidetes*"	5	Group-6
Megasphaera	*Firmicutes*	"*Clostridia*"	3	Group-8
Meiothermus	"*Deinococcus–Thermus*"	*Deinococci*	1	Group-4a
Melissococcus	*Firmicutes*	"*Bacilli*"	3	Group-17
Melittangium	*Proteobacteria*	"*Deltaproteobacteria*"	2	Group-16
Meniscus	*Bacteroidetes*	"*Sphingobacteria*"	5	Group-3
Mesophilobacter	*Proteobacteria*	"*Gammaproteobacteria*"	2	Group-4a
Mesoplasma	*Firmicutes*	*Mollicutes*	3	Group-30
Mesorhizobium	*Proteobacteria*	"*Alphaproteobacteria*"	2	Group-4a

(continued)

TABLE 2. *(continued)*

Genus	Phylum	Class	Volume	BMDB9
Metallosphaera	*Crenarchaeota*	*Thermoprotei*	1	Group-35.1
Methanobacterium	*Euryarchaeota*	*Methanobacteria*	1	Group-31.1
Methanobrevibacter	*Euryarchaeota*	*Methanobacteria*	1	Group-31.1
Methanocalculus	*Euryarchaeota*	*Methanococci*	1	Group-31
Methanocaldococcus	*Euryarchaeota*	*Methanococci*	1	Group-31
Methanococcoides	*Euryarchaeota*	*Methanococci*	1	Group-31.3
Methanococcus	*Euryarchaeota*	*Methanococci*	1	Group-31.2
Methanocorpusculum	*Euryarchaeota*	*Methanococci*	1	Group-31.2
Methanoculleus	*Euryarchaeota*	*Methanococci*	1	Group-31.2
Methanofollis	*Euryarchaeota*	*Methanococci*	1	Group-31.2
Methanogenium	*Euryarchaeota*	*Methanococci*	1	Group-31.2
Methanohalobium	*Euryarchaeota*	*Methanococci*	1	Group-31.3
Methanohalophilus	*Euryarchaeota*	*Methanococci*	1	Group-31.3
Methanolacinia	*Euryarchaeota*	*Methanococci*	1	Group-31.2
Methanolobus	*Euryarchaeota*	*Methanococci*	1	Group-31.3
Methanomicrobium	*Euryarchaeota*	*Methanococci*	1	Group-31.2
Methanoplanus	*Euryarchaeota*	*Methanococci*	1	Group-31.2
Methanopyrus	*Euryarchaeota*	*Methanopyri*	1	Group-31
Methanosaeta	*Euryarchaeota*	*Methanococci*	1	Group-31
Methanosalsum	*Euryarchaeota*	*Methanococci*	1	Group-31
Methanosarcina	*Euryarchaeota*	*Methanococci*	1	Group-31.3
Methanosphaera	*Euryarchaeota*	*Methanobacteria*	1	Group-31.1
Methanospirillum	*Euryarchaeota*	*Methanococci*	1	Group-31.2
Methanothermobacter	*Euryarchaeota*	*Methanobacteria*	1	Group-31
Methanothermococcus	*Euryarchaeota*	*Methanococci*	1	Group-31
Methanothermus	*Euryarchaeota*	*Methanobacteria*	1	Group-31.1
Methanotorris	*Euryarchaeota*	*Methanococci*	1	Group-31
Methylobacillus	*Proteobacteria*	"*Betaproteobacteria*"	2	Group-4a
Methylobacter	*Proteobacteria*	"*Gammaproteobacteria*"	2	Group-4a
Methylobacterium	*Proteobacteria*	"*Alphaproteobacteria*"	2	Group-4a
Methylocaldum	*Proteobacteria*	"*Gammaproteobacteria*"	2	Group-4a
Methylococcus	*Proteobacteria*	"*Gammaproteobacteria*"	2	Group-4a
Methylocystis	*Proteobacteria*	"*Alphaproteobacteria*"	2	Group-4a
Methylomicrobium	*Proteobacteria*	"*Gammaproteobacteria*"	2	Group-4a
Methylomonas	*Proteobacteria*	"*Gammaproteobacteria*"	2	Group-4a
Methylophaga	*Proteobacteria*	"*Gammaproteobacteria*"	2	Group-4a
Methylophilus	*Proteobacteria*	"*Betaproteobacteria*"	2	Group-4a
Methylopila	*Proteobacteria*	"*Alphaproteobacteria*"	2	Group-4a
Methylorhabdus	*Proteobacteria*	"*Alphaproteobacteria*"	2	Group-4a
Methylosinus	*Proteobacteria*	"*Alphaproteobacteria*"	2	Group-4a
Methylosphaera	*Proteobacteria*	"*Gammaproteobacteria*"	2	Group-4a
Methylovorus	*Proteobacteria*	"*Betaproteobacteria*"	2	Group-4a
Micavibrio	*Proteobacteria*	"*Deltaproteobacteria*"	2	Group-2
Microbacterium	*Actinobacteria*	*Actinobacteria*	4	Group-20
Microbispora	*Actinobacteria*	*Actinobacteria*	4	Group-26.1
Microbulbifer	*Proteobacteria*	"*Gammaproteobacteria*"	2	Group-4a
Micrococcus	*Actinobacteria*	*Actinobacteria*	4	Group-17
Microcoleus	*Cyanobacteria*	"*Cyanobacteria*"	1	Group-11.3
Microcystis	*Cyanobacteria*	"*Cyanobacteria*"	1	Group-11.1
Microlunatus	*Actinobacteria*	*Actinobacteria*	4	Group-17
Micromonospora	*Actinobacteria*	*Actinobacteria*	4	Group-24
Micropruina	*Actinobacteria*	*Actinobacteria*	4	Group 17.1
Microscilla	*Bacteroidetes*	"*Sphingobacteria*"	5	Group-15.1
Microsphaera	*Actinobacteria*	*Actinobacteria*	4	Group-17
Microtetraspora	*Actinobacteria*	*Actinobacteria*	4	Group-26.1
Microvirgula	*Proteobacteria*	"*Betaproteobacteria*"	2	Group-4a
Mitsuokella	*Firmicutes*	"*Clostridia*"	3	Group-6
Mobiluncus	*Actinobacteria*	*Actinobacteria*	4	Group-20
Modestobacter	*Actinobacteria*	*Actinobacteria*	4	Group-20
Moellerella	*Proteobacteria*	"*Gammaproteobacteria*"	2	Group-5.1
Moorella	*Firmicutes*	"*Clostridia*"	3	Group-18
Moraxella	*Proteobacteria*	"*Gammaproteobacteria*"	2	Group-4a
Morganella	*Proteobacteria*	"*Gammaproteobacteria*"	2	Group-5.1
Moritella	*Proteobacteria*	"*Gammaproteobacteria*"	2	Group-4a
Morococcus	*Proteobacteria*	"*Gammaproteobacteria*"	2	Group-4a
Mycobacterium	*Actinobacteria*	*Actinobacteria*	4	Group-21
Mycoplana	*Proteobacteria*	"*Alphaproteobacteria*"	2	Group-4a
Mycoplasma	*Firmicutes*	*Mollicutes*	3	Group-30
Myroides	*Bacteroidetes*	"*Flavobacteria*"	5	Group-4a
Myxococcus	*Proteobacteria*	"*Deltaproteobacteria*"	2	Group-16

(continued)

TABLE 2. *(continued)*

Genus	Phylum	Class	Volume	BMDB9
Myxosarcina	*Cyanobacteria*	*"Cyanobacteria"*	1	Group-11.2
Nannocystis	*Proteobacteria*	*"Deltaproteobacteria"*	2	Group-16
Natrialba	*Euryarchaeota*	*Halobacteria*	1	Group-33
Natrinema	*Euryarchaeota*	*Halobacteria*	1	Group-33
Natroniella	*Firmicutes*	*"Clostridia"*	3	Group-6
Natronincola	*Firmicutes*	*"Clostridia"*	3	Group-18
Natronobacterium	*Euryarchaeota*	*Halobacteria*	1	Group-33
Natronococcus	*Euryarchaeota*	*Halobacteria*	1	Group-33
Natronomonas	*Euryarchaeota*	*Halobacteria*	1	Group-33
Natronorubrum	*Euryarchaeota*	*Halobacteria*	1	Group-33
Neisseria	*Proteobacteria*	*"Betaproteobacteria"*	2	Group-4a
Neorickettsia	*Proteobacteria*	*"Alphaproteobacteria"*	2	Group-9
Neptunomonas	*Proteobacteria*	*"Gammaproteobacteria"*	2	Group-4a
Nesterenkonia	*Actinobacteria*	*Actinobacteria*	4	Group-17
Nevskia	*Proteobacteria*	*"Gammaproteobacteria"*	2	Group-13.3
Nitrobacter	*Proteobacteria*	*"Alphaproteobacteria"*	2	Group-12.3a
Nitrococcus	*Proteobacteria*	*"Gammaproteobacteria"*	2	Group-12.3a
Nitrosococcus	*Proteobacteria*	*"Gammaproteobacteria"*	2	Group-12.3b
Nitrosomonas	*Proteobacteria*	*"Betaproteobacteria"*	2	Group-12.3b
Nitrosospira	*Proteobacteria*	*"Betaproteobacteria"*	2	Group-12.3b
Nitrospina	*Proteobacteria*	*"Deltaproteobacteria"*	2	Group-12.3a
Nitrospira	*Nitrospirae*	*"Nitrospira"*	1	Group-12.3a
Nocardia	*Actinobacteria*	*Actinobacteria*	4	Group-22.1
Nocardioides	*Actinobacteria*	*Actinobacteria*	4	Group-22.3
Nocardiopsis	*Actinobacteria*	*Actinobacteria*	4	Group-27
Nodularia	*Cyanobacteria*	*"Cyanobacteria"*	1	Group-11.4a
Nonomuria	*Actinobacteria*	*Actinobacteria*	4	Group-26
Nostoc	*Cyanobacteria*	*"Cyanobacteria"*	1	Group-11.4a
Nostochopsis	*Cyanobacteria*	*"Cyanobacteria"*	1	Group-11.3
Obesumbacterium	*Proteobacteria*	*"Gammaproteobacteria"*	2	Group-5.1
Oceanospirillum	*Proteobacteria*	*"Gammaproteobacteria"*	2	Group-2
Ochrobactrum	*Proteobacteria*	*"Alphaproteobacteria"*	2	Group-4a
Octadecabacter	*Proteobacteria*	*"Alphaproteobacteria"*	2	Group-4a
Oenococcus	*Firmicutes*	*"Bacilli"*	3	Group-17
Oerskovia	*Actinobacteria*	*Actinobacteria*	4	Group-22.4
Oligella	*Proteobacteria*	*"Gammaproteobacteria"*	2	Group-4a
Oligotropha	*Proteobacteria*	*"Alphaproteobacteria"*	2	Group-4a
Orenia	*Firmicutes*	*"Clostridia"*	3	Group-6
Orientia	*Proteobacteria*	*"Alphaproteobacteria"*	2	Group-9
Ornithinicoccus	*Actinobacteria*	*Actinobacteria*	4	Group-20
Ornithobacterium	*Bacteroidetes*	*"Flavobacteria"*	5	Group-4a
Oscillatoria	*Cyanobacteria*	*"Cyanobacteria"*	1	Group-11.3
Oscillochloris	*Chloroflexi*	*"Chloroflexi"*	1	Group-10.6
Oscillospira	*Firmicutes*	*"Bacilli"*	3	Group-18
Oxalobacter	*Proteobacteria*	*"Betaproteobacteria"*	2	Group-6
Oxalophagus	*Firmicutes*	*"Bacilli"*	3	Group-18
Oxobacter	*Firmicutes*	*"Clostridia"*	3	Group-18
Paenibacillus	*Firmicutes*	*"Bacilli"*	3	Group-18
Pantoea	*Proteobacteria*	*"Gammaproteobacteria"*	2	Group-5.1
Parachlamydia	*Chlamydiae*	*"Chlamydiae"*	5	Group-9
Paracoccus	*Proteobacteria*	*"Alphaproteobacteria"*	2	Group-4a
Paracraurococcus	*Proteobacteria*	*"Alphaproteobacteria"*	2	Group-4a
Pasteurella	*Proteobacteria*	*"Gammaproteobacteria"*	2	Group-5.3
Pasteuria	*Firmicutes*	*"Bacilli"*	3	Group-9
Pectinatus	*Firmicutes*	*"Clostridia"*	3	Group-6
Pectobacterium	*Proteobacteria*	*"Gammaproteobacteria"*	2	Group-5a
Pediococcus	*Firmicutes*	*"Bacilli"*	3	Group-17
Pedobacter	*Bacteroidetes*	*"Sphingobacteria"*	5	Group-4a
Pedomicrobium	*Proteobacteria*	*"Alphaproteobacteria"*	2	Group-13.1
Pelczaria	*Actinobacteria*	*Actinobacteria*	4	Group-4a
Pelistega	*Proteobacteria*	*"Betaproteobacteria"*	2	Group-4a
Pelobacter	*Proteobacteria*	*"Deltaproteobacteria"*	2	Group-6
Pelodictyon	*Chlorobi*	*"Chlorobia"*	1	Group-10.5
Peptococcus	*Firmicutes*	*"Clostridia"*	3	Group-17
Peptostreptococcus	*Firmicutes*	*"Clostridia"*	3	Group-17
Persicobacter	*Bacteroidetes*	*"Sphingobacteria"*	5	Group-15.1
Petrotoga	*Thermotogae*	*Thermotogae*	1	Group-36
Pfennigia	*Proteobacteria*	*"Gammaproteobacteria"*	2	Group-10.1
Phaeospirillum	*Proteobacteria*	*"Alphaproteobacteria"*	2	Group-10.3
Phascolarctobacterium	*Firmicutes*	*"Clostridia"*	3	Group-6

(continued)

TABLE 2. *(continued)*

Genus	Phylum	Class	Volume	BMDB9
Phenylobacterium	*Proteobacteria*	"*Alphaproteobacteria*"	2	Group-4a
Phocoenobacter	*Proteobacteria*	"*Gammaproteobacteria*"	2	Group-5.3
Photobacterium	*Proteobacteria*	"*Gammaproteobacteria*"	2	Group-5.2
Photorhabdus	*Proteobacteria*	"*Gammaproteobacteria*"	2	Group-5.1
Phyllobacterium	*Proteobacteria*	"*Alphaproteobacteria*"	2	Group-4a
Picrophilus	*Euryarchaeota*	*Thermoplasmata*	1	Group-35.3
Pilimelia	*Actinobacteria*	*Actinobacteria*	4	Group-24
Pillotina	*Spirochaetes*	"*Spirochaetes*"	5	Group-1
Pirellula	*Planctomycetes*	"*Planctomycetacia*"	5	Group-13.2
Piscirickettsia	*Proteobacteria*	"*Gammaproteobacteria*"	2	Group-9
Planctomyces	*Planctomycetes*	"*Planctomycetacia*"	5	Group-13.2
Planktothrix	*Cyanobacteria*	"*Cyanobacteria*"	1	Group-11.3
Planobispora	*Actinobacteria*	*Actinobacteria*	4	Group-26.1
Planococcus	*Firmicutes*	"*Bacilli*"	3	Group-17
Planomonospora	*Actinobacteria*	*Actinobacteria*	4	Group-26.1
Planopolyspora	*Actinobacteria*	*Actinobacteria*	4	Group-26
Planotetraspora	*Actinobacteria*	*Actinobacteria*	4	Group-26.1
Plesiomonas	*Proteobacteria*	"*Gammaproteobacteria*"	2	Group-5.2
Pleurocapsa	*Cyanobacteria*	"*Cyanobacteria*"	1	Group-11.2
Polaribacter	*Bacteroidetes*	"*Flavobacteria*"	5	Group-4a
Polaromonas	*Proteobacteria*	"*Betaproteobacteria*"	2	Group-4a
Polyangium	*Proteobacteria*	"*Deltaproteobacteria*"	2	Group-16
Polynucleobacter	*Proteobacteria*	"*Alphaproteobacteria*"	2	Group-9
Porphyrobacter	*Proteobacteria*	"*Alphaproteobacteria*"	2	Group-4a
Porphyromonas	*Bacteroidetes*	"*Bacteroidetes*"	5	Group-6
Pragia	*Proteobacteria*	"*Gammaproteobacteria*"	2	Group-5.1
Prauserella	*Actinobacteria*	*Actinobacteria*	4	Group-22.2
Prevotella	*Bacteroidetes*	"*Bacteroidetes*"	5	Group-6
Prochlorococcus	*Cyanobacteria*	"*Cyanobacteria*"	1	Group-11.3
Prochloron	*Cyanobacteria*	"*Cyanobacteria*"	1	Group-11.6
Prochlorothrix	*Cyanobacteria*	"*Cyanobacteria*"	1	Group-11.6
Prolinoborus	*Proteobacteria*	"*Betaproteobacteria*"	2	Group-4a
Promicromonospora	*Actinobacteria*	*Actinobacteria*	4	Group-22.4
Propionibacter	*Proteobacteria*	"*Betaproteobacteria*"	2	Group-6
Propionibacterium	*Actinobacteria*	*Actinobacteria*	4	Group-20
Propioniferax	*Actinobacteria*	*Actinobacteria*	4	Group-20
Propionigenium	*Fusobacteria*	"*Fusobacteria*"	5	Group-6
Propionispira	*Firmicutes*	"*Clostridia*"	3	Group-6
Propionivibrio	*Proteobacteria*	"*Betaproteobacteria*"	5	Group-6
Prosthecobacter	*Verrucomicrobia*	*Verrucomicrobiae*	5	Group-13.1
Prosthecochloris	*Chlorobi*	"*Chlorobia*"	1	Group-10.5
Prosthecomicrobium	*Proteobacteria*	"*Alphaproteobacteria*"	2	Group-13.1
Proteus	*Proteobacteria*	"*Gammaproteobacteria*"	2	Group-5.1
Protomonas	*Proteobacteria*	"*Alphaproteobacteria*"	2	Group-4a
Providencia	*Proteobacteria*	"*Gammaproteobacteria*"	2	Group-5.1
Pseudaminobacter	*Proteobacteria*	"*Alphaproteobacteria*"	2	Group-4a
Pseudoalteromonas	*Proteobacteria*	"*Gammaproteobacteria*"	2	Group-4a
Pseudobutyrivibrio	*Firmicutes*	"*Clostridia*"	3	Group-6
Pseudocaedibacter	*Proteobacteria*	"*Alphaproteobacteria*"	2	Group-9
Pseudomonas	*Proteobacteria*	"*Gammaproteobacteria*"	2	Group-4a
Pseudonocardia	*Actinobacteria*	*Actinobacteria*	4	Group-22.2
Pseudoramibacter	*Firmicutes*	"*Clostridia*"	3	Group-19
Pseudoxanthomonas	*Proteobacteria*	"*Gammaproteobacteria*"	2	Group-4a
Pseudanabaena	*Cyanobacteria*	"*Cyanobacteria*"	1	Group-11.6
Psychrobacter	*Proteobacteria*	"*Gammaproteobacteria*"	2	Group-4a
Psychroflexus	*Bacteroidetes*	"*Flavobacteria*"	5	Group-15.1
Psychromonas	*Bacteroidetes*	"*Flavobacteria*"	5	Group-6
Psychroserpens	*Bacteroidetes*	"*Flavobacteria*"	5	Group-3
Pyrobaculum	*Crenarchaeota*	*Thermoprotei*	1	Group-35.2
Pyrococcus	*Euryarchaeota*	*Thermococci*	1	Group-35.3
Pyrodictium	*Crenarchaeota*	*Thermoprotei*	1	Group-35.3
Pyrolobus	*Crenarchaeota*	*Thermoprotei*	1	Group-35.3
Quinella	*Firmicutes*	"*Clostridia*"	3	Group-8
Rahnella	*Proteobacteria*	"*Gammaproteobacteria*"	2	Group-5.1
Ralstonia	*Proteobacteria*	"*Betaproteobacteria*"	2	Group-4a
Rarobacter	*Actinobacteria*	*Actinobacteria*	4	Group-20
Rathayibacter	*Actinobacteria*	*Actinobacteria*	4	Group-20
Renibacterium	*Actinobacteria*	*Actinobacteria*	4	Group-19
Rhabdochromatium	*Proteobacteria*	"*Gammaproteobacteria*"	2	Group-10.1
Rhizobacter	*Proteobacteria*	"*Gammaproteobacteria*"	2	Group-4a

(continued)

TABLE 2. *(continued)*

Genus	Phylum	Class	Volume	BMDB9
Rhizobium	*Proteobacteria*	"*Alphaproteobacteria*"	2	Group-4a
Rhizomonas	*Proteobacteria*	"*Alphaproteobacteria*"	2	Group-4a
Rhodanobacter	*Proteobacteria*	"*Gammaproteobacteria*"	2	Group-4a
Rhodobacter	*Proteobacteria*	"*Alphaproteobacteria*"	2	Group-10.3
Rhodobium	*Proteobacteria*	"*Alphaproteobacteria*"	2	Group-10.3
Rhodocista	*Proteobacteria*	"*Alphaproteobacteria*"	2	Group-10.3
Rhodococcus	*Actinobacteria*	*Actinobacteria*	4	Group-22.1
Rhodocyclus	*Proteobacteria*	"*Betaproteobacteria*"	2	Group-10.3
Rhodoferax	*Proteobacteria*	"*Betaproteobacteria*"	2	Group-10.3
Rhodomicrobium	*Proteobacteria*	"*Alphaproteobacteria*"	2	Group-10.3
Rhodopila	*Proteobacteria*	"*Alphaproteobacteria*"	2	Group-10.3
Rhodoplanes	*Proteobacteria*	"*Alphaproteobacteria*"	2	Group-10.3
Rhodopseudomonas	*Proteobacteria*	"*Alphaproteobacteria*"	2	Group-10.3
Rhodospira	*Proteobacteria*	"*Alphaproteobacteria*"	2	Group-10.3
Rhodospirillum	*Proteobacteria*	"*Alphaproteobacteria*"	2	Group-10.3
Rhodothalassium	*Proteobacteria*	"*Alphaproteobacteria*"	2	Group-10.3
Rhodothermus	*Bacteroidetes*	"*Sphingobacteria*"	5	Group-4a
Rhodovibrio	*Proteobacteria*	"*Alphaproteobacteria*"	2	Group-10.3
Rhodovulum	*Proteobacteria*	"*Alphaproteobacteria*"	2	Group-10.3
Rickettsia	*Proteobacteria*	"*Alphaproteobacteria*"	2	Group-9
Rickettsiella	*Proteobacteria*	"*Gammaproteobacteria*"	2	Group-9
Riemerella	*Bacteroidetes*	"*Flavobacteria*"	5	Group-4a
Rikenella	*Bacteroidetes*	"*Bacteroidetes*"	5	Group-6
Rivularia	*Cyanobacteria*	"*Cyanobacteria*"	1	Group-11
Roseateles	*Proteobacteria*	"*Betaproteobacteria*"	2	Group-10.3
Roseburia	*Firmicutes*	"*Clostridia*"	3	Group-6
Roseivivax	*Proteobacteria*	"*Alphaproteobacteria*"	2	Group-4a
Roseobacter	*Proteobacteria*	"*Alphaproteobacteria*"	2	Group-4a
Roseococcus	*Proteobacteria*	"*Alphaproteobacteria*"	2	Group-4a
Roseomonas	*Proteobacteria*	"*Alphaproteobacteria*"	2	Group-4a
Roseospira	*Proteobacteria*	"*Alphaproteobacteria*"	2	Group-10.3
Roseovarius	*Proteobacteria*	"*Alphaproteobacteria*"	2	Group-13.3
Rothia	*Actinobacteria*	*Actinobacteria*	4	Group-20
Rubrimonas	*Proteobacteria*	"*Alphaproteobacteria*"	2	Group-4a
Rubrivivax	*Proteobacteria*	"*Betaproteobacteria*"	2	Group-10.3
Rubrobacter	*Actinobacteria*	*Actinobacteria*	4	Group-20
Ruegeria	*Proteobacteria*	"*Alphaproteobacteria*"	2	Group-4a
Rugamonas	*Proteobacteria*	"*Gammaproteobacteria*"	2	Group-4a
Ruminobacter	*Proteobacteria*	"*Gammaproteobacteria*"	2	Group-6
Ruminococcus	*Firmicutes*	"*Clostridia*"	3	Group-17
Runella	*Bacteroidetes*	"*Sphingobacteria*"	5	Group-3
Saccharobacter	*Proteobacteria*	"*Gammaproteobacteria*"	2	Group-5.1
Saccharococcus	*Firmicutes*	"*Bacilli*"	3	Group-17
Saccharomonospora	*Actinobacteria*	*Actinobacteria*	4	Group-22.2
Saccharopolyspora	*Actinobacteria*	*Actinobacteria*	4	Group-22.2
Saccharothrix	*Actinobacteria*	*Actinobacteria*	4	Group-29
Sagittula	*Proteobacteria*	"*Alphaproteobacteria*"	2	Group-4a
Salibacillus	*Firmicutes*	"*Bacilli*"	3	Group-18
Salinicoccus	*Firmicutes*	"*Bacilli*"	3	Group-17
Salinivibrio	*Proteobacteria*	"*Gammaproteobacteria*"	2	Group-5.2
Salmonella	*Proteobacteria*	"*Gammaproteobacteria*"	2	Group-5.1
Sandaracinobacter	*Proteobacteria*	"*Alphaproteobacteria*"	2	Group-10.7
Sanguibacter	*Actinobacteria*	*Actinobacteria*	4	Group-20
Saprospira	*Bacteroidetes*	"*Sphingobacteria*"	5	Group-15.5
Sarcina	*Firmicutes*	"*Clostridia*"	3	Group-17
Schwartzia	*Firmicutes*	"*Clostridia*"	3	Group-6
Scytonema	*Cyanobacteria*	"*Cyanobacteria*"	1	Group-11.4b
Sebaldella	*Fusobacteria*	"*Fusobacteria*"	5	Group-6
Selenomonas	*Firmicutes*	"*Clostridia*"	3	Group-6
Seliberia	*Proteobacteria*	"*Alphaproteobacteria*"	2	Group-13.3
Serpens	*Proteobacteria*	"*Gammaproteobacteria*"	2	Group-4a
Serpulina	*Spirochaetes*	"*Spirochaetes*"	5	Group-1
Serratia	*Proteobacteria*	"*Gammaproteobacteria*"	2	Group-5.1
Shewanella	*Proteobacteria*	"*Gammaproteobacteria*"	2	Group-5.2
Shigella	*Proteobacteria*	"*Gammaproteobacteria*"	2	Group-5.1
Simkania	*Chlamydiae*	"*Chlamydiae*"	5	Group-9
Simonsiella	*Proteobacteria*	"*Betaproteobacteria*"	2	Group-15.2
Sinorhizobium	*Proteobacteria*	"*Alphaproteobacteria*"	2	Group-4a
Skermanella	*Proteobacteria*	"*Alphaproteobacteria*"	2	Group-10.3
Skermania	*Actinobacteria*	*Actinobacteria*	4	Group-22.2

(continued)

TABLE 2. *(continued)*

Genus	Phylum	Class	Volume	BMDB9
Slackia	*Actinobacteria*	*Actinobacteria*	4	Group-20
Smithella	*Proteobacteria*	*"Deltaproteobacteria"*	2	Group-6
Sodalis	*Proteobacteria*	*"Gammaproteobacteria"*	2	Group-4a
Sphaerobacter	*Actinobacteria*	*Actinobacteria*	4	Group-20
Sphaerotilus	*Proteobacteria*	*"Betaproteobacteria"*	2	Group-14
Sphingobacterium	*Bacteroidetes*	*"Sphingobacteria"*	5	Group-4a
Sphingomonas	*Proteobacteria*	*"Alphaproteobacteria"*	2	Group-4a
Spirilliplanes	*Actinobacteria*	*Actinobacteria*	4	Group-24
Spirillospora	*Actinobacteria*	*Actinobacteria*	4	Group-26.1
Spirillum	*Proteobacteria*	*"Betaproteobacteria"*	2	Group-2
Spirochaeta	*Spirochaetes*	*"Spirochaetes"*	5	Group-1
Spiroplasma	*Firmicutes*	*Mollicutes*	3	Group-30
Spirosoma	*Bacteroidetes*	*"Sphingobacteria"*	5	Group-3
Spirulina	*Cyanobacteria*	*"Cyanobacteria"*	1	Group-11.3
Sporichthya	*Actinobacteria*	*Actinobacteria*	4	Group-25
Sporobacter	*Firmicutes*	*"Clostridia"*	3	Group-18
Sporobacterium	*Firmicutes*	*"Clostridia"*	3	Group-18
Sporocytophaga	*Bacteroidetes*	*"Sphingobacteria"*	5	Group-15.1
Sporohalobacter	*Firmicutes*	*"Clostridia"*	3	Group-18
Sporolactobacillus	*Firmicutes*	*"Bacilli"*	3	Group-18
Sporomusa	*Firmicutes*	*"Clostridia"*	3	Group-6
Sporosarcina	*Firmicutes*	*"Bacilli"*	3	Group-18
Sporotomaculum	*Firmicutes*	*"Clostridia"*	3	Group-18
Staleya	*Proteobacteria*	*"Alphaproteobacteria"*	2	Group-4a
Stanieria	*Cyanobacteria*	*"Cyanobacteria"*	1	Group-11.3
Staphylococcus	*Firmicutes*	*"Bacilli"*	3	Group-17
Staphylothermus	*Crenarchaeota*	*Thermoprotei*	1	Group-35.3
Stappia	*Proteobacteria*	*"Alphaproteobacteria"*	2	Group-4a
Starria	*Cyanobacteria*	*"Cyanobacteria"*	1	Group-11.3
Stella	*Proteobacteria*	*"Alphaproteobacteria"*	2	Group-13.1
Stenotrophomonas	*Proteobacteria*	*"Gammaproteobacteria"*	2	Group-4a
Stetteria	*Crenarchaeota*	*Thermoprotei*	1	Group-32
Stigmatella	*Proteobacteria*	*"Deltaproteobacteria"*	2	Group-16
Stigonema	*Cyanobacteria*	*"Cyanobacteria"*	1	Group-11.5
Stomatococcus	*Actinobacteria*	*Actinobacteria*	4	Group-17
Streptoalloteichus	*Actinobacteria*	*Actinobacteria*	4	Group-27
Streptobacillus	*Fusobacteria*	*"Fusobacteria"*	5	Group-5.4
Streptococcus	*Firmicutes*	*"Bacilli"*	3	Group-17
Streptomyces	*Actinobacteria*	*Actinobacteria*	4	Group-25
Streptosporangium	*Actinobacteria*	*Actinobacteria*	4	Group-26.1
Streptoverticillium	*Actinobacteria*	*Actinobacteria*	4	Group-25
Stygiolobus	*Crenarchaeota*	*Thermoprotei*	1	Group-35.1
Succiniclasticum	*Firmicutes*	*"Clostridia"*	3	Group-6
Succinimonas	*Proteobacteria*	*"Gammaproteobacteria"*	5	Group-6
Succinispira	*Firmicutes*	*"Clostridia"*	3	Group-6
Succinivibrio	*Proteobacteria*	*"Gammaproteobacteria"*	2	Group-6
Sulfitobacter	*Proteobacteria*	*"Alphaproteobacteria"*	2	Group-4a
Sulfobacillus	*Firmicutes*	*"Bacilli"*	3	Group-12.2
Sulfolobus	*Crenarchaeota*	*Thermoprotei*	1	Group-35.1
Sulfophobococcus	*Crenarchaeota*	*Thermoprotei*	1	Group-36
Sulfurisphaera	*Crenarchaeota*	*Thermoprotei*	1	Group-35.1
Sulfurococcus	*Crenarchaeota*	*Thermoprotei*	1	Group-35.1
Sulfurospirillum	*Proteobacteria*	*"Epsilonproteobacteria"*	2	Group-2
Sutterella	*Proteobacteria*	*"Betaproteobacteria"*	2	Group-6
Suttonella	*Proteobacteria*	*"Gammaproteobacteria"*	2	Group-4a
Symbiotes	*Proteobacteria*	*"Alphaproteobacteria"*	2	Group-9
Symploca	*Cyanobacteria*	*"Cyanobacteria"*	1	Group-11.3
Synechococcus	*Cyanobacteria*	*"Cyanobacteria"*	1	Group-11.1
Synechocystis	*Cyanobacteria*	*"Cyanobacteria"*	1	Group-11.1
Synergistes	*Deferribacteres*	*Deferribacteres*	1	Group-6
Syntrophobacter	*Proteobacteria*	*"Deltaproteobacteria"*	2	Group-6
Syntrophobotulus	*Firmicutes*	*"Clostridia"*	3	Group-6
Syntrophococcus	*Firmicutes*	*"Bacilli"*	3	Group-8
Syntrophomonas	*Firmicutes*	*"Clostridia"*	3	Group-6
Syntrophospora	*Firmicutes*	*"Clostridia"*	3	Group-18
Syntrophus	*Proteobacteria*	*"Deltaproteobacteria"*	2	Group-6
Tatumella	*Proteobacteria*	*"Gammaproteobacteria"*	2	Group-5.1
Taylorella	*Proteobacteria*	*"Betaproteobacteria"*	2	Group-4b
Tectibacter	*Proteobacteria*	*"Alphaproteobacteria"*	2	Group-9
Telluria	*Proteobacteria*	*"Betaproteobacteria"*	2	Group-4a

(continued)

TABLE 2. *(continued)*

Genus	Phylum	Class	Volume	BMDB9
Terrabacter	*Actinobacteria*	*Actinobacteria*	4	Group-20
Terracoccus	*Actinobacteria*	*Actinobacteria*	4	Group-17
Tessaracoccus	*Actinobacteria*	*Actinobacteria*	4	Group-17.2
Tetragenococcus	*Firmicutes*	"*Bacilli*"	3	Group-17
Thauera	*Proteobacteria*	"*Betaproteobacteria*"	2	Group-4a
Thermaerobacter	*Firmicutes*	"*Clostridia*"	3	Group-36/37
Thermoactinomyces	*Firmicutes*	"*Bacilli*"	3	Group-28
Thermoanaerobacter	*Firmicutes*	"*Clostridia*"	3	Group-20
Thermoanaerobacterium	*Firmicutes*	"*Clostridia*"	3	Group-6
Thermoanaerobium	*Firmicutes*	"*Clostridia*"	3	Group-20
Thermobacillus	*Firmicutes*	"*Bacilli*"	3	Group-4a
Thermobifida	*Actinobacteria*	*Actinobacteria*	4	Group-27
Thermobispora	*Actinobacteria*	*Actinobacteria*	4	Group-26.1
Thermobrachium	*Firmicutes*	"*Clostridia*"	3	Group-19
Thermochromatium	*Proteobacteria*	"*Gammaproteobacteria*"	2	Group-10.1
Thermocladium	*Crenarchaeota*	*Thermoprotei*	1	Group-35.2
Thermococcus	*Euryarchaeota*	*Thermococci*	1	Group-35.3
Thermocrinis	*Aquificae*	*Aquificae*	1	Group-36
Thermocrispum	*Actinobacteria*	*Actinobacteria*	4	Group-22.2
Thermodesulfobacterium	*Thermodesulfobacteria*	*Thermodesulfobacteria*	1	Group-36
Thermodesulforhabdus	*Proteobacteria*	"*Deltaproteobacteria*"	2	Group-7.3
Thermodesulfovibrio	*Nitrospirae*	"*Nitrospira*"	1	Group-7.2
Thermodiscus	*Crenarchaeota*	*Thermoprotei*	1	Group-35.3
Thermofilum	*Crenarchaeota*	*Thermoprotei*	1	Group-35.2
Thermohydrogenium	*Firmicutes*	"*Clostridia*"	3	Group-20
Thermoleophilum	*Proteobacteria*	"*Gammaproteobacteria*"	2	Group-4a
Thermomicrobium	*Thermomicrobia*	*Thermomicrobia*	1	Group-4a
Thermomonospora	*Actinobacteria*	*Actinobacteria*	4	Group-27
Thermonema	*Bacteroidetes*	"*Sphingobacteria*"	5	Group-15.1
Thermoplasma	*Euryarchaeota*	*Thermoplasmata*	1	Group-34
Thermoproteus	*Crenarchaeota*	*Thermoprotei*	1	Group-35.2
Thermosipho	*Thermotogae*	*Thermotogae*	1	Group-36
Thermosphaera	*Crenarchaeota*	*Thermoprotei*	1	Group-35.3
Thermosyntropha	*Firmicutes*	"*Clostridia*"	3	Group-6
Thermoterrabacterium	*Firmicutes*	"*Clostridia*"	3	Group-19
Thermothrix	*Proteobacteria*	"*Betaproteobacteria*"	2	Group-12.1b
Thermotoga	*Thermotogae*	*Thermotogae*	1	Group-36
Thermus	"*Deinococcus–Thermus*"	*Deinococci*	1	Group-4a
Thiobacillus	*Proteobacteria*	"*Betaproteobacteria*"	2	Group-12.1b
Thiobacterium	*Proteobacteria*	"*Gammaproteobacteria*"	2	Group-12.1a
Thiocapsa	*Proteobacteria*	"*Gammaproteobacteria*"	2	Group-10.1
Thiococcus	*Proteobacteria*	"*Gammaproteobacteria*"	2	Group-10.1
Thiocystis	*Proteobacteria*	"*Gammaproteobacteria*"	2	Group-10.1
Thiodictyon	*Proteobacteria*	"*Gammaproteobacteria*"	2	Group-10.1
Thiohalocapsa	*Proteobacteria*	"*Gammaproteobacteria*"	2	Group-10.1
Thiolamprovum	*Proteobacteria*	"*Gammaproteobacteria*"	2	Group-10.1
Thiomargarita	*Proteobacteria*	"*Gammaproteobacteria*"	2	Group-12.1a
Thiomicrospira	*Proteobacteria*	"*Gammaproteobacteria*"	2	Group-12.1b
Thiomonas	*Proteobacteria*	"*Betaproteobacteria*"	2	Group-12.1b
Thiopedia	*Proteobacteria*	"*Gammaproteobacteria*"	2	Group-10.1
Thioploca	*Proteobacteria*	"*Gammaproteobacteria*"	2	Group-15.3
Thiorhodococcus	*Proteobacteria*	"*Gammaproteobacteria*"	2	Group-10.1
Thiorhodospira	*Proteobacteria*	"*Gammaproteobacteria*"	2	Group-12.3a
Thiorhodovibrio	*Proteobacteria*	"*Gammaproteobacteria*"	2	Group-10.1
Thiospira	*Proteobacteria*	"*Gammaproteobacteria*"	2	Group-12.1a
Thiospirillum	*Proteobacteria*	"*Gammaproteobacteria*"	2	Group-10.1
Thiothrix	*Proteobacteria*	"*Gammaproteobacteria*"	2	Group-15.3
Thiovulum	*Proteobacteria*	"*Epsilonproteobacteria*"	2	Group-12.1a
Tindallia	*Firmicutes*	"*Clostridia*"	3	Group-19.1
Tissierella	*Firmicutes*	"*Clostridia*"	3	Group-6
Tolumonas	*Proteobacteria*	"*Gammaproteobacteria*"	2	Group-5.1
Tolypothrix	*Cyanobacteria*	"*Cyanobacteria*"	1	Group-11.3
Toxothrix	*Bacteroidetes*	"*Sphingobacteria*"	5	Group-15.5
Trabulsiella	*Proteobacteria*	"*Gammaproteobacteria*"	2	Group-5.1
Treponema	*Spirochaetes*	"*Spirochaetes*"	5	Group-1
Trichococcus	*Firmicutes*	"*Bacilli*"	3	Group-17
Trichodesmium	*Cyanobacteria*	"*Cyanobacteria*"	1	Group-11.3
Tsukamurella	*Actinobacteria*	*Actinobacteria*	4	Group-22.1
Turicella	*Actinobacteria*	*Actinobacteria*	4	Group-20
Tychonema	*Cyanobacteria*	"*Cyanobacteria*"	1	Group-11.3

(continued)

TABLE 2. *(continued)*

Genus	Phylum	Class	Volume	BMDB9
Ureaplasma	*Firmicutes*	*Mollicutes*	3	Group-30
Vagococcus	*Firmicutes*	"*Bacilli*"	3	Group-17
Vampirovibrio	*Proteobacteria*	"*Deltaproteobacteria*"	2	Group-2
Variovorax	*Proteobacteria*	"*Betaproteobacteria*"	2	Group-4a
Veillonella	*Firmicutes*	"*Clostridia*"	3	Group-8
Verrucomicrobium	*Verrucomicrobia*	*Verrucomicrobiae*	5	Group-13.1
Verrucosispora	*Actinobacteria*	*Actinobacteria*	4	Group-24
Vibrio	*Proteobacteria*	"*Gammaproteobacteria*"	2	Group-5.2
Virgibacillus	*Firmicutes*	"*Bacilli*"	3	Group-18
Vitreoscilla	*Proteobacteria*	"*Betaproteobacteria*"	2	Group-15.5
Vogesella	*Proteobacteria*	"*Betaproteobacteria*"	2	Group-4a
Waddlia	*Chlamydiae*	"*Chlamydiae*"	5	Group-9
Weeksella	*Bacteroidetes*	"*Flavobacteria*"	5	Group-4a
Weissella	*Firmicutes*	"*Bacilli*"	3	Group-17
Wigglesworthia	*Proteobacteria*	"*Gammaproteobacteria*"	2	Group-5.1
Williamsia	*Actinobacteria*	*Actinobacteria*	4	Group-20
Wolbachia	*Proteobacteria*	"*Alphaproteobacteria*"	2	Group-9
Wolinella	*Proteobacteria*	"*Epsilonproteobacteria*"	2	Group-2
Xanthobacter	*Proteobacteria*	"*Alphaproteobacteria*"	2	Group-4a
Xanthomonas	*Proteobacteria*	"*Gammaproteobacteria*"	2	Group-4a
Xenococcus	*Cyanobacteria*	"*Cyanobacteria*"	1	Group-11.2
Xenorhabdus	*Proteobacteria*	"*Gammaproteobacteria*"	2	Group-5.1
Xylella	*Proteobacteria*	"*Gammaproteobacteria*"	2	Group-4a
Xylophilus	*Proteobacteria*	"*Gammaproteobacteria*"	2	Group-4a
Yersinia	*Proteobacteria*	"*Gammaproteobacteria*"	2	Group-5.1
Yokenella	*Proteobacteria*	"*Gammaproteobacteria*"	2	Group-5.1
Zavarzinia	*Proteobacteria*	"*Alphaproteobacteria*"	2	Group-4a
Zoogloea	*Proteobacteria*	"*Betaproteobacteria*"	2	Group-4a
Zymobacter	*Proteobacteria*	"*Gammaproteobacteria*"	2	Group-5.1
Zymomonas	*Proteobacteria*	"*Alphaproteobacteria*"	2	Group-5.4
Zymophilus	*Firmicutes*	"*Clostridia*"	3	Group-6

Taxonomic Outline of the *Archaea* and *Bacteria*

Readers are advised that the taxonomic scheme presented here is a work-in-progress and is based on data available in March 2000. Some rearrangement and emendment is expected to occur as new data become available and subsequent volumes go to press.

Domain *Archaea*
  Phylum AI. *Crenarchaeota* phy. nov.
    Class I. *Thermoprotei* class. nov.
      Order I. *Thermoproteales*[VP (T)]
        Family I. *Thermoproteaceae*[VP]
          Genus I. *Thermoproteus*[VP (T)]
          Genus II. *Caldivirga*[VP]
          Genus III. *Pyrobaculum*[VP]
          Genus IV. *Thermocladium*[VP]
        Family II. *Thermofilaceae*[VP]
          Genus I. *Thermofilum*[VP (T)]
      Order II. *Desulfurococcales* ord. nov.
        Family I. *Desulfurococcaceae*[VP]
          Genus I. *Desulfurococcus*[VP (T)]
          Genus II. *Aeropyrum*[VP]
          Genus III. *Ignicoccus*[VP]
          Genus IV. *Staphylothermus*[VP]
          Genus V. *Stetteria*[VP]
          Genus VI. *Sulfophobococcus*[VP]
          Genus VII. *Thermodiscus* gen. nov.
          Genus VIII. *Thermosphaera*[VP]
        Family II. *Pyrodictiaceae*[VP]
          Genus I. *Pyrodictium*[VP (T)]
          Genus II. *Hyperthermus*[VP]

          Genus III. *Pyrolobus*[VP]
      Order III. *Sulfolobales*[VP]
        Family I. *Sulfolobaceae*[VP]
          Genus I. *Sulfolobus*[AL (T)]
          Genus II. *Acidianus*[VP]
          Genus III. *Metallosphaera*[VP]
          Genus IV. *Stygiolobus*[VP]
          Genus V. *Sulfurisphaera*[VP]
          Genus VI. *Sulfurococcus*[VP]
  Phylum AII. *Euryarchaeota* phy. nov.
    Class I. *Methanobacteria* class. nov.
      Order I. *Methanobacteriales*[VP (T)]
        Family I. *Methanobacteriaceae*[AL]
          Genus I. *Methanobacterium*[AL (T)]
          Genus II. *Methanobrevibacter*[VP]
          Genus III. *Methanosphaera*[VP]
          Genus IV. *Methanothermobacter*[VP]
        Family II. *Methanothermaceae*[VP]
          Genus I. *Methanothermus*[VP (T)]
    Class II. *Methanococci* class. nov.
      Order I. *Methanococcales*[VP (T)]
        Family I. *Methanococcaceae*[VP]
          Genus I. *Methanococcus*[AL (T)]
          Genus II. *Methanothermococcus* gen. nov.

Family II.   *Methanocaldococcaceae* fam. nov.
     Genus I.   *Methanocaldococcus* gen. nov. [T]
     Genus II.   *Methanotorris* gen. nov.
Order II.   *Methanomicrobiales* [VP]
   Family I.   *Methanomicrobiaceae* [VP]
     Genus I.   *Methanomicrobium* [VP (T)]
     Genus II.   *Methanoculleus* [VP]
     Genus III.   *Methanofollis* [VP]
     Genus IV.   *Methanogenium* [VP]
     Genus V.   *Methanolacinia* [VP]
     Genus VI.   *Methanoplanus* [VP]
   Family II.   *Methanocorpusculaceae* [VP]
     Genus I.   *Methanocorpusculum* [VP (T)]
   Family III.   *Methanospirillaceae* fam. nov.
     Genus I.   *Methanospirillum* [AL (T)]
   Genera incertae sedis
     Genus I.   *Methanocalculus* [VP]
Order III.   *Methanosarcinales* ord. nov.
   Family I.   *Methanosarcinaceae* [VP]
     Genus I.   *Methanosarcina* [AL (T)]
     Genus II.   *Methanococcoides* [VP]
     Genus III.   *Methanohalobium* [VP]
     Genus IV.   *Methanohalophilus* [VP]
     Genus V.   *Methanolobus* [VP]
     Genus VI.   *Methanosalsum* gen. nov.
   Family II.   *Methanosaetaceae* fam. nov.
     Genus I.   *Methanosaeta* [VP (T)]
Class III.   *Halobacteria* class. nov.
   Order I.   *Halobacteriales* [VP (T)]
   Family I.   *Halobacteriaceae* [AL]
     Genus I.   *Halobacterium* [AL (T)]
     Genus II.   *Haloarcula* [VP]
     Genus III.   *Halobaculum* [VP]
     Genus IV.   *Halococcus* [AL]
     Genus V.   *Haloferax* [VP]
     Genus VI.   *Halogeometricum* [VP]
     Genus VII.   *Halorhabdus* [VP]
     Genus VIII.   *Halorubrum* [VP]
     Genus IX.   *Haloterrigena* [VP]
     Genus X.   *Natrialba* [VP]
     Genus XI.   *Natrinema* [VP]
     Genus XII.   *Natronobacterium* [VP]
     Genus XIII.   *Natronococcus* [VP]
     Genus XIV.   *Natronomonas* [VP]
     Genus XV.   *Natronorubrum* [VP]
Class IV.   *Thermoplasmata* class. nov.
   Order I.   *Thermoplasmatales* ord. nov. [T]
   Family I.   *Thermoplasmataceae* fam. nov.
     Genus I.   *Thermoplasma* [AL (T)]
   Family II.   *Picrophilaceae* [VP]
     Genus I.   *Picrophilus* [VP (T)]
Class V.   *Thermococci* class. nov.
   Order I.   *Thermococcales* [VP (T)]
   Family I.   *Thermococcaceae* [VP]
     Genus I.   *Thermococcus* [VP (T)]
     Genus II.   *Pyrococcus* [VP]
Class VI.   *Archaeoglobi* class. nov.
   Order I.   *Archaeoglobales* ord. nov. [T]
   Family I.   *Archaeoglobaceae* fam. nov.
     Genus I.   *Archaeoglobus* [VP (T)]
     Genus II.   *Ferroglobus* [VP]
Class VII.   *Methanopyri* class. nov.

Order I.   *Methanopyrales* ord. nov. [T]
   Family I.   *Methanopyraceae* fam. nov.
     Genus I.   *Methanopyrus* [VP (T)]
Domain *Bacteria*
   Phylum BI.   *Aquificae* phy. nov.
   Class I.   *Aquificae* class. nov.
     Order I.   *Aquificales* ord. nov. [T]
     Family I.   *Aquificaceae* fam. nov.
       Genus I.   *Aquifex* [VP (T)]
       Genus II.   *Calderobacterium* [VP]
       Genus III.   *Hydrogenobacter* [VP]
       Genus IV.   *Thermocrinis* [VP]
     Genera incertae sedis
       Genus I.   *Desulfurobacterium* [VP]
   Phylum BII.   *Thermotogae* phy. nov.
   Class I.   *Thermotogae* class. nov.
     Order I.   *Thermotogales* ord. nov. [T]
     Family I.   *Thermotogaceae* fam. nov.
       Genus I.   *Thermotoga* [VP (T)]
       Genus II.   *Fervidobacterium* [VP]
       Genus III.   *Geotoga* [VP]
       Genus IV.   *Petrotoga* [VP]
       Genus V.   *Thermosipho* [VP]
   Phylum BIII.   *Thermodesulfobacteria* phy. nov.
   Class I.   *Thermodesulfobacteria* class. nov.
     Order I.   *Thermodesulfobacteriales* ord. nov. [T]
     Family I.   *Thermodesulfobacteriaceae* fam. nov.
       Genus I.   *Thermodesulfobacterium* [VP (T)]
   Phylum BIV.   "Deinococcus-Thermus"
   Class I.   *Deinococci* class. nov.
     Order I.   *Deinococcales* [VP]
     Family I.   *Deinococcaceae* [VP]
       Genus I.   *Deinococcus* [VP (T)]
     Order II.   *Thermales* ord. nov.
     Family I.   *Thermaceae* fam. nov.
       Genus I.   *Thermus* [AL (T)]
       Genus II.   *Meiothermus* [VP]
   Phylum BV.   *Chrysiogenetes* phy. nov.
   Class I.   *Chrysiogenetes* class. nov.
     Order I.   *Chrysiogenales* ord. nov. [T]
     Family I.   *Chrysiogenaceae* fam. nov.
       Genus I.   *Chrysiogenes* [VP (T)]
   Phylum BVI.   *Chloroflexi* phy. nov.
   Class I.   "Chloroflexi"
     Order I.   "Chloroflexales"
     Family I.   "Chloroflexaceae"
       Genus I.   *Chloroflexus* [AL]
       Genus II.   *Chloronema* [AL]
       Genus III.   *Heliothrix* [VP]
       Genus IV.   *Oscillochloris* [VP]
     Order II.   "Herpetosiphonales"
     Family I.   "Herpetosiphonaceae"
       Genus I.   *Herpetosiphon* [AL]
   Phylum BVII.   *Thermomicrobia* phy. nov.
   Class I.   *Thermomicrobia* class. nov.
     Order I.   *Thermomicrobiales* ord. nov. [T]
     Family I.   *Thermomicrobiaceae* fam. nov.
       Genus I.   *Thermomicrobium* [AL (T)]
   Phylum BVIII.   *Nitrospirae* phy. nov.
   Class I.   "Nitrospira"
     Order I.   "Nitrospirales"
     Family I.   "Nitrospiraceae"

Genus I.   *Nitrospira*[VP]
Genus II.   *Leptospirillum*[VP]
Genus III.   *Magnetobacterium*[VP]
Genus IV.   *Thermodesulfovibrio*[VP]
Phylum BIX.   *Deferribacteres* phy. nov.
  Class I.   *Deferribacteres* class. nov.
    Order I.   *Deferribacterales* ord. nov.[(T)]
      Family I.   *Deferribacteraceae* fam. nov.
      Genus I.   *Deferribacter*[VP (T)]
      Genus II.   "*Flexistipes*"
      Genus III.   *Geovibrio*[VP]
    Genera incertae sedis
      Genus I.   *Synergistes*[VP]
Phylum BX.   *Cyanobacteria*
  Class I.   "*Cyanobacteria*"
    Subsection I.
      Form Genus I.   *Chamaesiphon*
      Form Genus II.   *Chroococcus*
      Form Genus III.   *Cyanobacterium*
      Form Genus IV.   *Cyanobium*
      Form genus V.   *Cyanothece*
      Form Genus VI.   *Dactylococcopsis*
      Form Genus VII.   *Gloeobacter*
      Form Genus VIII. *Gloeocapsa*
      Form Genus IX.   *Gloeothece*
      Form Genus X.   *Microcystis*
      Form Genus XI.   *Prochlorococcus*
      Form Genus XII.   *Prochloron*
      Form genus XIII. *Synechococcus*
      Form genus XIV.   *Synechocystis*
    Subsection II.
      Subgroup I.
        Form Genus I.   *Cyanocystis*
        Form Genus II.   *Dermocarpella*
        Form Genus III.   *Stanieria*
        Form Genus IV.   *Xenococcus*
      Subgroup II.
        Form Genus I.   *Chroococcidiopsis*
        Form Genus II.   *Myxosarcina*
        Form genus III.   *Pleurocapsa*
    Subsection III.
      Form Genus I.   *Arthrospira*
      Form Genus II.   *Borzia*
      Form Genus III.   *Crinalium*
      Form genus IV.   *Geitlerinema*
      Form genus V.   *Leptolyngbya*
      Form Genus VI.   *Limnothrix*
      Form Genus VII.   *Lyngbya*
      Form genus VIII.   *Microcoleus*
      Form Genus IX.   *Oscillatoria*
      Form genus X.   *Planktothrix*
      Form Genus XI.   *Prochlorothrix*
      Form Genus XII.   *Pseudanabaena*
      Form Genus XIII.   *Spirulina*
      Form Genus XIV.   *Starria*
      Form genus XV.   *Symploca*
      Form Genus XVI.   *Trichodesmium*
      Form Genus XVII.   *Tychonema*
    Subsection IV.
      Subgroup I.
        Form Genus I.   *Anabaena*
        Form Genus II.   *Anabaenopsis*

Form Genus III.   *Aphanizomenon*
Form Genus IV.   *Cyanospira*
Form Genus V.   *Cylindrospermopsis*
Form Genus VI.   *Cylindrospermum*
Form Genus VII.   *Nodularia*
Form Genus VIII.   *Nostoc*
Form Genus IX.   *Scytonema*
  Subgroup II.
    Form Genus I.   *Calothrix*
    Form Genus II.   *Rivularia*
    Form Genus III.   *Tolypothrix*
Subsection V.
  Family I.
    Form Genus I.   *Chlorogloeopsis*
    Form Genus II.   *Fischerella*
    Form Genus III.   *Geitleria*
    Form Genus IV.   *Iyengariella*
    Form Genus V.   *Nostochopsis*
    Form Genus VI.   *Stigonema*
Phylum BXI.   *Chlorobi* phy. nov.
  Class I.   "*Chlorobia*"
    Order I.   *Chlorobiales*[AL]
      Family I.   *Chlorobiaceae*[AL]
      Genus I.   *Chlorobium*[AL]
      Genus II.   *Ancalochloris*[AL]
      Genus III.   *Chloroherpeton*[VP]
      Genus IV.   *Pelodictyon*[AL]
      Genus V.   *Prosthecochloris*[AL]
Phylum BXII.   *Proteobacteria* phy. nov.
  Class I.   "*Alphaproteobacteria*"
    Order I.   *Rhodospirillales*[AL]
      Family I.   *Rhodospirillaceae*[AL]
      Genus I.   *Rhodospirillum*[AL]
      Genus II.   *Azospirillum*[AL]
      Genus III.   *Magnetospirillum*[VP]
      Genus IV.   *Phaeospirillum*[VP]
      Genus V.   *Rhodocista*[VP]
      Genus VI.   *Rhodospira*[VP]
      Genus VII.   *Rhodothalassium*[VP]
      Genus VIII.   *Rhodovibrio*[VP]
      Genus IX.   *Roseospira*[VP]
      Genus X.   *Skermanella*[VP]
      Family II.   *Acetobacteraceae*[VP]
      Genus I.   *Acetobacter*[AL]
      Genus II.   *Acidiphilium*[VP]
      Genus III.   *Acidocella*[VP]
      Genus IV.   *Acidomonas*[VP]
      Genus V.   *Craurococcus*[VP]
      Genus VI.   *Gluconacetobacter*[VP]
      Genus VII.   *Gluconobacter*[AL]
      Genus VIII.   *Paracraurococcus*[VP]
      Genus IX.   *Rhodopila*[VP]
      Genus X.   *Roseococcus*[VP]
      Genus XI.   *Stella*[VP]
      Genus XII.   *Zavarzinia*[VP]
    Order II.   *Rickettsiales*[AL]
      Family I.   *Rickettsiaceae*[AL]
      Genus I.   *Rickettsia*[AL]
      Genus II.   *Orientia*[VP]
      Genus III.   *Wolbachia*[AL]
      Family II.   *Ehrlichiaceae*[AL]
      Genus I.   *Ehrlichia*[AL]

Genus II.   *Aegyptianella*[AL]
Genus III.   *Anaplasma*[AL]
Genus IV.   *Cowdria*[AL]
Genus V.   *Neorickettsia*[AL]
Family III.   "Holosporaceae"
  Genus I.   *Holospora*[VP]
  Genus II.   *Caedibacter*[VP]
  Genus III.   *Lyticum*[VP]
  Genus IV.   *Polynucleobacter*[VP]
  Genus V.   *Pseudocaedibacter*[VP]
  Genus VI.   *Symbiotes*[AL]
  Genus VII.   *Tectibacter*[VP]
Order III.   "Rhodobacterales"
  Family I.   "Rhodobacteraceae"
    Genus I.   *Rhodobacter*[VP]
    Genus II.   *Ahrensia*[VP]
    Genus III.   *Amaricoccus*[VP]
    Genus IV.   *Antarctobacter*[VP]
    Genus V.   *Gemmobacter*[VP]
    Genus VI.   *Hirschia*[VP]
    Genus VII.   *Hyphomonas*[VP]
    Genus VIII.   *Maricaulis*[VP]
    Genus IX.   *Octadecabacter*[VP]
    Genus X.   *Paracoccus*[AL]
    Genus XI.   *Rhodovulum*[VP]
    Genus XII.   *Roseivivax*[VP]
    Genus XIII.   *Roseobacter*[VP]
    Genus XIV.   *Roseovarius*[VP]
    Genus XV.   *Rubrimonas*[VP]
    Genus XVI.   *Ruegeria*[VP]
    Genus XVII.   *Sagittula*[VP]
    Genus XVIII.   *Staleya*[VP]
    Genus XIX.   *Stappia*[VP]
    Genus XX.   *Sulfitobacter*[VP]
Order IV.   "Sphingomonadales"
  Family I.   "Sphingomonadaceae"
    Genus I.   *Sphingomonas*[VP]
    Genus II.   *Blastomonas*[VP]
    Genus III.   *Erythrobacter*[VP]
    Genus IV.   *Erythromicrobium*[VP]
    Genus V.   *Erythromonas*[VP]
    Genus VI.   *Porphyrobacter*[VP]
    Genus VII.   *Rhizomonas*[VP]
    Genus VIII.   *Sandaracinobacter*[VP]
    Genus IX.   *Zymomonas*[AL]
Order V.   *Caulobacterales*[AL]
  Family I.   *Caulobacteraceae*[AL]
    Genus I.   *Caulobacter*[AL]
    Genus II.   *Asticcacaulis*[AL]
    Genus III.   *Brevundimonas*[VP]
    Genus IV.   *Phenylobacterium*[VP]
Order VI.   "Rhizobiales"
  Family I.   *Rhizobiaceae*[AL]
    Genus I.   *Rhizobium*[AL]
    Genus II.   *Agrobacterium*[AL]
    Genus III.   *Allorhizobium*[VP]
    Genus IV.   *Carbophilus*[VP]
    Genus V.   *Chelatobacter*[VP]
    Genus VI.   *Ensifer*[VP]
    Genus VII.   *Sinorhizobium*[VP]
  Family II.   *Bartonellaceae*[AL]
    Genus I.   *Bartonella*[AL]

Family III.   *Brucellaceae*[AL]
  Genus I.   *Brucella*[AL]
  Genus II.   *Mycoplana*[AL]
  Genus III.   *Ochrobactrum*[VP]
Family IV.   "Phyllobacteriaceae"
  Genus I.   *Phyllobacterium*[VP]
  Genus II.   *Aminobacter*[VP]
  Genus III.   *Aquamicrobium*[VP]
  Genus IV.   *Defluvibacter*[VP]
  Genus V.   *Mesorhizobium*[VP]
  Genus VI.   *Pseudaminobacter*[VP]
Family V.   "Methylocystaceae"
  Genus I.   *Methylocystis*[VP]
  Genus II.   *Methylopila*[VP]
  Genus III.   *Methylosinus*[VP]
Family VI.   "Beijerinckiaceae"
  Genus I.   *Beijerinckia*[AL]
  Genus II.   *Chelatococcus*[VP]
  Genus III.   *Derxia*[AL]
Family VII.   "Bradyrhizobiaceae"
  Genus I.   *Bradyrhizobium*[VP]
  Genus II.   *Afipia*[VP]
  Genus III.   *Agromonas*[VP]
  Genus IV.   *Blastobacter*[AL]
  Genus V.   *Bosea*[VP]
  Genus VI.   *Nitrobacter*[AL]
  Genus VII.   *Oligotropha*[VP]
  Genus VIII.   *Rhodopseudomonas*[AL]
Family VIII.   *Hyphomicrobiaceae*[AL]
  Genus I.   *Hyphomicrobium*[AL]
  Genus II.   *Ancalomicrobium*[AL]
  Genus III.   *Ancylobacter*[VP]
  Genus IV.   *Angulomicrobium*[VP]
  Genus V.   *Aquabacter*[VP]
  Genus VI.   *Azorhizobium*[VP]
  Genus VII.   *Blastochloris*[VP]
  Genus VIII.   *Devosia*[VP]
  Genus IX.   *Dichotomicrobium*[VP]
  Genus X.   *Filomicrobium*[VP]
  Genus XI.   *Gemmiger*[AL]
  Genus XII.   *Labrys*[VP]
  Genus XIII.   *Methylorhabdus*[VP]
  Genus XIV.   *Pedomicrobium*[AL]
  Genus XV.   *Prosthecomicrobium*[AL]
  Genus XVI.   *Rhodomicrobium*[AL]
  Genus XVII.   *Rhodoplanes*[VP]
  Genus XVIII.   *Seliberia*[AL]
  Genus XIX.   *Xanthobacter*[AL]
Family IX.   "Methylobacteriaceae"
  Genus I.   *Methylobacterium*[AL]
  Genus II.   *Protomonas*[VP]
  Genus III.   *Roseomonas*[VP]
Family X.   "Rhodobiaceae"
  Genus I.   *Rhodobium*[VP]
Class II.   "Betaproteobacteria"
  Order I.   "Burkholderiales"
    Family I.   "Burkholderiaceae"
      Genus I.   *Burkholderia*[VP]
      Genus II.   *Cupriavidus*[VP]
      Genus III.   *Lautropia*[VP]
      Genus IV.   *Thermothrix*[VP]
    Family II.   "Ralstoniaceae"

Genus I.   *Ralstonia*[VP]
Family III.   "Oxalobacteraceae"
Genus I.   *Oxalobacter*[VP]
Genus II.   *Duganella*[VP]
Genus III.   *Herbaspirillum*[VP]
Genus IV.   *Janthinobacterium*[AL]
Genus V.   *Telluria*[VP]
Family IV.   *Alcaligenaceae*[VP]
Genus I.   *Alcaligenes*[AL]
Genus II.   *Achromobacter*[VP]
Genus III.   *Bordetella*[AL]
Genus IV.   *Pelistega*[VP]
Genus V.   *Sutterella*[VP]
Genus VI.   *Taylorella*[VP]
Family V.   *Comamonadaceae*[VP]
Genus I.   *Comamonas*[VP]
Genus II.   *Acidovorax*[VP]
Genus III.   *Aquabacterium*[VP]
Genus IV.   *Brachymonas*[VP]
Genus V.   *Delftia*[VP]
Genus VI.   *Hydrogenophaga*[VP]
Genus VII.   *Ideonella*[VP]
Genus VIII.   *Leptothrix*[AL]
Genus IX.   *Polaromonas*[VP]
Genus X.   *Rhodoferax*[VP]
Genus XI.   *Roseateles*[VP]
Genus XII.   *Rubrivivax*[VP]
Genus XIII.   *Sphaerotilus*[AL]
Genus XIV.   *Thiomonas*[VP]
Genus XV.   *Variovorax*[VP]
Order II.   "Hydrogenophilales"
Family I.   "Hydrogenophilaceae"
Genus I.   *Hydrogenophilus*[VP]
Genus II.   *Thiobacillus*[AL]
Order III.   "Methylophilales"
Family I.   "Methylophilaceae"
Genus I.   *Methylophilus*[VP]
Genus II.   *Methylobacillus*[AL]
Genus III.   *Methylovorus*[VP]
Order IV.   "Neisseriales"
Family I.   *Neisseriaceae*[AL]
Genus I.   *Neisseria*[AL]
Genus II.   *Alysiella*[AL]
Genus III.   *Aquaspirillum*[AL]
Genus IV.   *Catenococcus*[VP]
Genus V.   *Chromobacterium*[AL]
Genus VI.   *Eikenella*[AL]
Genus VII.   *Formivibrio*[VP]
Genus VIII.   *Iodobacter*[VP]
Genus IX.   *Kingella*[AL]
Genus X.   *Microvirgula*[VP]
Genus XI.   *Prolinoborus*[VP]
Genus XII.   *Simonsiella*[AL]
Genus XIII.   *Vitreoscilla*[AL]
Genus XIV.   *Vogesella*[VP]
Order V.   "Nitrosomonadales"
Family I.   "Nitrosomonadaceae"
Genus I.   *Nitrosomonas*[AL]
Genus II.   *Nitrosospira*[AL]
Family II.   *Spirillaceae*[AL]
Genus I.   *Spirillum*[AL]
Family III.   *Gallionellaceae*[AL]

Genus I.   *Gallionella*[AL]
Order VI.   "Rhodocyclales"
Family I.   "Rhodocyclaceae"
Genus I.   *Rhodocyclus*[AL]
Genus II.   *Azoarcus*[VP]
Genus III.   *Propionibacter*[VP]
Genus IV.   *Propionivibrio*[VP]
Genus V.   *Thauera*[VP]
Genus VI.   *Zoogloea*[AL]
Class III.   "Gammaproteobacteria"
Order I.   "Chromatiales"
Family I.   *Chromatiaceae*[AL]
Genus I.   *Chromatium*[AL]
Genus II.   *Allochromatium*[VP]
Genus III.   *Amoebobacter*[AL]
Genus IV.   *Halochromatium*[VP]
Genus V.   *Isochromatium*[VP]
Genus VI.   *Lamprobacter*[VP]
Genus VII.   *Lamprocystis*[AL]
Genus VIII.   *Marichromatium*[VP]
Genus IX.   *Nitrosococcus*[AL]
Genus X.   *Pfennigia*[VP]
Genus XI.   *Rhabdochromatium*[VP]
Genus XII.   *Thermochromatium*[VP]
Genus XIII.   *Thiocapsa*[AL]
Genus XIV.   *Thiococcus*[VP]
Genus XV.   *Thiocystis*[AL]
Genus XVI.   *Thiodictyon*[AL]
Genus XVII.   *Thiohalocapsa*[VP]
Genus XVIII.   *Thiolamprovum*[VP]
Genus XIX.   *Thiopedia*[AL]
Genus XX.   *Thiorhodococcus*[VP]
Genus XXI.   *Thiorhodovibrio*[VP]
Genus XXII.   *Thiospirillum*[AL]
Family II.   *Ectothiorhodospiraceae*[VP]
Genus I.   *Ectothiorhodospira*[AL]
Genus II.   *Arhodomonas*[VP]
Genus III.   *Halorhodospira*[VP]
Genus IV.   *Nitrococcus*[AL]
Genus V.   *Thiorhodospira*[VP]
Order II.   "Xanthomonadales"
Family I.   "Xanthomonadaceae"
Genus I.   *Xanthomonas*[AL]
Genus II.   *Frateuria*[VP]
Genus III.   *Lysobacter*[AL]
Genus IV.   *Nevskia*[AL]
Genus V.   *Pseudoxanthomonas*[VP]
Genus VI.   *Rhodanobacter*[VP]
Genus VII.   *Stenotrophomonas*[VP]
Genus VIII.   *Xylella*[VP]
Order III.   "Cardiobacteriales"
Family I.   *Cardiobacteriaceae*[VP]
Genus I.   *Cardiobacterium*[AL]
Genus II.   *Dichelobacter*[VP]
Genus III.   *Suttonella*[VP]
Order IV.   "Thiotrichales"
Family I.   "Thiotrichaceae"
Genus I.   *Thiothrix*[AL]
Genus II.   *Achromatium*[AL]
Genus III.   *Beggiatoa*[AL]
Genus IV.   *Leucothrix*[AL]
Genus V.   *Macromonas*[AL]

Genus VI. *Thiobacterium*[VP]
Genus VII. *Thiomargarita*[VP]
Genus VIII. *Thioploca*[AL]
Genus IX. *Thiospira*[AL]
Family II. "Piscirickettsiaceae"
Genus I. *Piscirickettsia*[VP]
Genus II. *Cycloclasticus*[VP]
Genus III. *Hydrogenovibrio*[VP]
Genus IV. *Methylophaga*[VP]
Genus V. *Thiomicrospira*[AL]
Family III. "Francisellaceae"
Genus I. *Francisella*[AL]
Order V. "Legionellales"
Family I. *Legionellaceae*[AL]
Genus I. *Legionella*[AL]
Family II. "Coxiellaceae"
Genus I. *Coxiella*[AL]
Genus II. *Rickettsiella*[AL]
Order VI. "Methylococcales"
Family I. *Methylococcaceae*[VP]
Genus I. *Methylococcus*[AL]
Genus II. *Methylobacter*[VP]
Genus III. *Methylocaldum*[VP]
Genus IV. *Methylomicrobium*[VP]
Genus V. *Methylomonas*[VP]
Genus VI. *Methylosphaera*[VP]
Order VII. "Oceanospirillales"
Family I. "Oceanospirillaceae"
Genus I. *Oceanospirillum*[AL]
Genus II. *Balneatrix*[VP]
Genus III. *Fundibacter*[VP]
Genus IV. *Marinomonas*[VP]
Genus V. *Marinospirillum*[VP]
Genus VI. *Neptunomonas*[VP]
Family II. *Halomonadaceae*[VP]
Genus I. *Halomonas*[VP]
Genus II. *Alcanivorax*[VP]
Genus III. *Carnimonas*[VP]
Genus IV. *Chromohalobacter*[VP]
Genus V. *Deleya*[VP]
Genus VI. *Zymobacter*[VP]
Order VIII. *Pseudomonadales*[AL]
Family I. *Pseudomonadaceae*[AL]
Genus I. *Pseudomonas*[AL]
Genus II. *Azomonas*[AL]
Genus III. *Azotobacter*[AL]
Genus IV. *Cellvibrio*[VP]
Genus V. *Chryseomonas*[VP]
Genus VI. *Flavimonas*[VP]
Genus VII. *Lampropedia*[AL]
Genus VIII. *Mesophilobacter*[VP]
Genus IX. *Morococcus*[VP]
Genus X. *Oligella*[VP]
Genus XI. *Rhizobacter*[VP]
Genus XII. *Rugamonas*[VP]
Genus XIII. *Serpens*[AL]
Genus XIV. *Thermoleophilum*[VP]
Genus XV. *Xylophilus*[VP]
Family II. *Moraxellaceae*[VP]
Genus I. *Moraxella*[AL]
Genus II. *Acinetobacter*[AL]
Genus III. *Psychrobacter*[VP]

Order IX. "Alteromonadales"
Family I. "Alteromonadaceae"
Genus I. *Alteromonas*[AL]
Genus II. *Colwellia*[VP]
Genus III. *Ferrimonas*[VP]
Genus IV. *Glaciecola*[VP]
Genus V. *Marinobacter*[VP]
Genus VI. *Marinobacterium*[VP]
Genus VII. *Microbulbifer*[VP]
Genus VIII. *Moritella*[VP]
Genus IX. *Pseudoalteromonas*[VP]
Genus X. *Psychromonas*[VP]
Genus XI. *Shewanella*[VP]
Order X. "Vibrionales"
Family I. *Vibrionaceae*[AL]
Genus I. *Vibrio*[AL]
Genus II. *Allomonas*[VP]
Genus III. *Enhydrobacter*[VP]
Genus IV. *Listonella*[VP]
Genus V. *Photobacterium*[AL]
Genus VI. *Salinivibrio*[VP]
Order XI. "Aeromonadales"
Family I. *Aeromonadaceae*[VP]
Genus I. *Aeromonas*[AL]
Genus II. *Tolumonas*[VP]
Family II. *Succinivibrionaceae*[VP]
Genus I. *Succinivibrio*[AL]
Genus II. *Anaerobiospirillum*[AL]
Genus III. *Ruminobacter*[VP]
Genus IV. *Succinimonas*[VP]
Order XII. "Enterobacteriales"
Family I. *Enterobacteriaceae*[AL]
Genus I. *Escherichia*[AL]
Genus II. *Alterococcus*[VP]
Genus III. *Arsenophonus*[VP]
Genus IV. *Brenneria*[VP]
Genus V. *Buchnera*[VP]
Genus VI. *Budvicia*[VP]
Genus VII. *Buttiauxella*[VP]
Genus VIII. *Calymmatobacterium*[VP]
Genus IX. *Cedecea*[VP]
Genus X. *Citrobacter*[AL]
Genus XI. *Edwardsiella*[AL]
Genus XII. *Enterobacter*[AL]
Genus XIII. *Erwinia*[AL]
Genus XIV. *Ewingella*[VP]
Genus XV. *Hafnia*[AL]
Genus XVI. *Klebsiella*[AL]
Genus XVII. *Kluyvera*[VP]
Genus XVIII. *Leclercia*[VP]
Genus XIX. *Leminorella*[VP]
Genus XX. *Moellerella*[VP]
Genus XXI. *Morganella*[AL]
Genus XXII. *Obesumbacterium*[AL]
Genus XXIII. *Pantoea*[VP]
Genus XXIV. *Pectobacterium*[AL]
Genus XXV. *Photorhabdus*[VP]
Genus XXVI. *Plesiomonas*[AL]
Genus XXVII. *Pragia*[VP]
Genus XXVIII. *Proteus*[AL]
Genus XXIX. *Providencia*[AL]
Genus XXX. *Rahnella*[VP]

Genus XXXI.   *Saccharobacter*[VP]
Genus XXXII.   *Salmonella*[AL]
Genus XXXIII.   *Serratia*[AL]
Genus XXXIV.   *Shigella*[AL]
Genus XXXV.   *Sodalis*[VP]
Genus XXXVI.   *Tatumella*[VP]
Genus XXXVII.   *Trabulsiella*[VP]
Genus XXXVIII.   *Wigglesworthia*[VP]
Genus XXXIX.   *Xenorhabdus*[AL]
Genus XL.   *Yersinia*[AL]
Genus XLI.   *Yokenella*[VP]
Order XIII.   "Pasteurellales"
Family I.   *Pasteurellaceae*[VP]
Genus I.   *Pasteurella*[AL]
Genus II.   *Actinobacillus*[AL]
Genus III.   *Haemophilus*[AL]
Genus IV.   *Lonepinella*[VP]
Genus V.   *Mannheimia*[VP]
Genus VI.   *Phocoenobacter*[VP]
Class IV.   "Deltaproteobacteria"
Order I.   "Desulfurellales"
Family I.   "Desulfurellaceae"
Genus I.   *Desulfurella*[VP]
Genus II.   *Hippea*[VP]
Order II.   "Desulfovibrionales"
Family I.   "Desulfovibrionaceae"
Genus I.   *Desulfovibrio*[AL]
Genus II.   *Bilophila*[VP]
Genus III.   *Lawsonia*[VP]
Family II.   "Desulfomicrobiaceae"
Genus I.   *Desulfomicrobium*[VP]
Family III.   "Desulfohalobiaceae"
Genus I.   *Desulfohalobium*[VP]
Genus II.   *Desulfomonas*[AL]
Genus III.   *Desulfonatronovibrio*[VP]
Order III.   "Desulfobacterales"
Family I.   "Desulfobacteraceae"
Genus I.   *Desulfobacter*[VP]
Genus II.   *Desulfobacterium*[VP]
Genus III.   "Desulfobacula"
Genus IV.   "Desulfobotulus"
Genus V.   *Desulfocella*[VP]
Genus VI.   *Desulfococcus*[VP]
Genus VII.   *Desulfofaba*[VP]
Genus VIII.   *Desulfofrigus*[VP]
Genus IX.   *Desulfonema*[VP]
Genus X.   *Desulfosarcina*[VP]
Genus XI.   *Desulfospira*[VP]
Genus XII.   *Desulfotalea*[VP]
Family II.   "Desulfobulbaceae"
Genus I.   *Desulfobulbus*[VP]
Genus II.   *Desulfocapsa*[VP]
Genus III.   *Desulfofustis*[VP]
Genus IV.   *Desulforhopalus*[VP]
Family III.   "Desulfoarculaceae"
Genus I.   "Desulfoarculus"
Genus II.   *Nitrospina*[AL]
Genus III.   *Desulfobacca*[VP]
Genus IV.   *Desulfomonile*[VP]
Order IV.   "Desulfuromonadales"
Family I.   "Desulfuromonadaceae"
Genus I.   *Desulfuromonas*[AL]

Genus II.   *Desulfuromusa*[VP]
Family II.   "Geobacteraceae"
Genus I.   *Geobacter*[VP]
Family III.   "Pelobacteraceae"
Genus I.   *Pelobacter*[VP]
Genus II.   *Malonomonas*[VP]
Order V.   "Syntrophobacterales"
Family I.   "Syntrophobacteraceae"
Genus I.   *Syntrophobacter*[VP]
Genus II.   *Desulfacinum*[VP]
Genus III.   *Desulforhabdus*[VP]
Genus IV.   *Thermodesulforhabdus*[VP]
Family II.   "Syntrophaceae"
Genus I.   *Syntrophus*[VP]
Genus II.   *Smithella*[VP]
Order VI.   "Bdellovibrionales"
Family I.   "Bdellovibrionaceae"
Genus I.   *Bdellovibrio*[AL]
Genus II.   *Micavibrio*[VP]
Genus III.   *Vampirovibrio*[VP]
Order VII.   *Myxococcales*[AL]
Family I.   *Myxococcaceae*[AL]
Genus I.   *Myxococcus*[AL]
Genus II.   *Angiococcus*[VP]
Family II.   *Archangiaceae*[AL]
Genus I.   *Archangium*[AL]
Family III.   *Cystobacteraceae*[AL]
Genus I.   *Cystobacter*[AL]
Genus II.   *Melittangium*[AL]
Genus III.   *Stigmatella*[AL]
Family IV.   *Polyangiaceae*[AL]
Genus I.   *Polyangium*[AL]
Genus II.   *Chondromyces*[AL]
Genus III.   *Nannocystis*[VP]
Class V.   "Epsilonproteobacteria"
Order I.   "Campylobacterales"
Family I.   *Campylobacteraceae*[VP]
Genus I.   *Campylobacter*[AL]
Genus II.   *Arcobacter*[VP]
Genus III.   *Sulfurospirillum*[VP]
Genus IV.   *Thiovulum*[AL]
Family II.   "Helicobacteraceae"
Genus I.   *Helicobacter*[VP]
Genus II.   *Wolinella*[VP]
Phylum BXIII.   *Firmicutes* phy. nov.
Class I.   "Clostridia"
Order I.   *Clostridiales*[AL]
Family I.   *Clostridiaceae*[AL]
Genus I.   *Clostridium*[AL]
Genus II.   *Acetivibrio*[VP]
Genus III.   *Acidaminobacter*[VP]
Genus IV.   *Anaerobacter*[VP]
Genus V.   *Caloramator*[VP]
Genus VI.   *Natronincola*[VP]
Genus VII.   *Oxobacter*[VP]
Genus VIII.   *Sarcina*[AL]
Genus IX.   *Sporobacter*[VP]
Genus X.   *Thermobrachium*[VP]
Genus XI.   *Tindallia*[VP]
Family II.   "Lachnospiraceae"
Genus I.   *Lachnospira*[AL]
Genus II.   *Acetitomaculum*[VP]

Genus III.   *Anaerofilum*[VP]
Genus IV.   *Butyrivibrio*[AL]
Genus V.   *Catonella*[VP]
Genus VI.   *Coprococcus*[AL]
Genus VII.   *Johnsonella*[VP]
Genus VIII.   *Pseudobutyrivibrio*[VP]
Genus IX.   *Roseburia*[VP]
Genus X.   *Ruminococcus*[AL]
Genus XI.   *Sporobacterium*[VP]
Family III.   "Peptostreptococcaceae"
Genus I.   *Peptostreptococcus*[AL]
Genus II.   *Filifactor*[VP]
Genus III.   *Fusibacter*[VP]
Genus IV.   *Helcococcus*[VP]
Genus V.   *Tissierella*[VP]
Family IV.   "Eubacteriaceae"
Genus I.   *Eubacterium*[AL]
Genus II.   *Acetobacterium*[AL]
Genus III.   *Pseudoramibacter*[VP]
Family V.   *Peptococcaceae*[AL]
Genus I.   *Peptococcus*[AL]
Genus II.   *Anaeroarcus*[VP]
Genus III.   *Anaerosinus*[VP]
Genus IV.   *Anaerovibrio*[AL]
Genus V.   *Carboxydothermus*[VP]
Genus VI.   *Centipeda*[VP]
Genus VII.   *Dehalobacter*[VP]
Genus VIII.   *Dendrosporobacter*[VP]
Genus IX.   *Desulfitobacterium*[VP]
Genus X.   *Desulfonispora*[VP]
Genus XI.   *Desulfosporosinus*[VP]
Genus XII.   *Desulfotomaculum*[AL]
Genus XIII.   *Mitsuokella*[VP]
Genus XIV.   *Propionispira*[VP]
Genus XV.   *Succinispira*[VP]
Genus XVI.   *Syntrophobotulus*[VP]
Genus XVII.   *Thermoterrabacterium*[VP]
Family VI.   "Heliobacteriaceae"
Genus I.   *Heliobacterium*[VP]
Genus II.   *Heliobacillus*[VP]
Genus III.   *Heliophilum*[VP]
Family VII.   "Acidaminococcaceae"
Genus I.   *Acidaminococcus*[AL]
Genus II.   *Acetonema*[VP]
Genus III.   *Anaeromusa*[VP]
Genus IV.   *Dialister*[VP]
Genus V.   *Megasphaera*[AL]
Genus VI.   *Pectinatus*[AL]
Genus VII.   *Phascolarctobacterium*[VP]
Genus VIII.   *Quinella*[VP]
Genus IX.   *Schwartzia*[VP]
Genus X.   *Selenomonas*[AL]
Genus XI.   *Sporomusa*[VP]
Genus XII.   *Succiniclasticum*[VP]
Genus XIII.   *Veillonella*[AL]
Genus XIV.   *Zymophilus*[VP]
Family VIII.   *Syntrophomonadaceae*[VP]
Genus I.   *Syntrophomonas*[VP]
Genus II.   *Acetogenium*[VP]
Genus III.   *Aminobacterium*[VP]
Genus IV.   *Aminomonas*[VP]
Genus V.   *Anaerobaculum*[VP]

Genus VI.   *Anaerobranca*[VP]
Genus VII.   *Caldicellulosiruptor*[VP]
Genus VIII.   *Dethiosulfovibrio*[VP]
Genus IX.   *Syntrophospora*[VP]
Genus X.   *Thermaerobacter*[VP]
Genus XI.   *Thermanaerovibrio*[VP]
Genus XII.   *Thermohydrogenium*[VP]
Genus XIII.   *Thermosyntropha*[VP]
Order II.   "Thermoanaerobacteriales"
Family I.   "Thermoanaerobacteriaceae"
Genus I.   *Thermoanaerobacterium*[VP]
Genus II.   *Ammonifex*[VP]
Genus III.   *Coprothermobacter*[VP]
Genus IV.   *Moorella*[VP]
Genus V.   *Sporotomaculum*[VP]
Genus VI.   *Thermoanaerobacter*[VP]
Genus VII.   *Thermoanaerobium*[VP]
Order III.   *Haloanaerobiales*[VP]
Family I.   *Haloanaerobiaceae*[VP]
Genus I.   *Haloanaerobium*[VP]
Genus II.   *Halocella*[VP]
Genus III.   *Halothermothrix*[VP]
Genus IV.   *Natroniella*[VP]
Family II.   *Halobacteroidaceae*[VP]
Genus I.   *Halobacteroides*[VP]
Genus II.   *Acetohalobium*[VP]
Genus III.   *Haloanaerobacter*[VP]
Genus IV.   *Orenia*[VP]
Genus V.   *Sporohalobacter*[VP]
Class II.   *Mollicutes*[AL]
Order I.   *Mycoplasmatales*[AL]
Family I.   *Mycoplasmataceae*[AL]
Genus I.   *Mycoplasma*[AL]
Genus II.   *Eperythrozoon*[AL]
Genus III.   *Haemobartonella*[AL]
Genus IV.   *Ureaplasma*[AL]
Order II.   *Entomoplasmatales*[VP]
Family I.   *Entomoplasmataceae*[VP]
Genus I.   *Entomoplasma*[VP]
Genus II.   *Mesoplasma*[VP]
Family II.   *Spiroplasmataceae*[VP]
Genus I.   *Spiroplasma*[AL]
Order III.   *Acholeplasmatales*[VP]
Family I.   *Acholeplasmataceae*[AL]
Genus I.   *Acholeplasma*[AL]
Order IV.   *Anaeroplasmatales*[VP]
Family I.   *Anaeroplasmataceae*[VP]
Genus I.   *Anaeroplasma*[AL]
Genus II.   *Asteroleplasma*[VP]
Genera incertae sedis
Family I.   "Erysipelotrichaceae"
Genus I.   *Erysipelothrix*[AL] *
Genus II.   *Holdemania*[VP] *
Class III.   "Bacilli"
Order I.   *Bacillales*[AL]
Family I.   *Bacillaceae*[AL]
Genus I.   *Bacillus*[AL]
Genus II.   *Amphibacillus*[VP]
Genus III.   *Exiguobacterium*[VP]
Genus IV.   *Gracilibacillus*[VP]
Genus V.   *Halobacillus*[VP]
Genus VI.   *Saccharococcus*[VP]

Genus VII.   *Salibacillus*[VP]
Genus VIII.   *Virgibacillus*[VP]
Family II.   *Planococcaceae*[AL]
Genus I.   *Planococcus*[AL]
Genus II.   *Filibacter*[VP]
Genus III.   *Kurthia*[AL]
Genus IV.   *Sporosarcina*[AL]
Family III.   *Caryophanaceae*[AL]
Genus I.   *Caryophanon*[AL]
Family IV.   "Listeriaceae"
Genus I.   *Listeria*[AL]
Genus II.   *Brochothrix*[AL]
Family V.   "Staphylococcaceae"
Genus I.   *Staphylococcus*[AL]
Genus II.   *Gemella*[AL]
Genus III.   *Macrococcus*[VP]
Genus IV.   *Salinicoccus*[VP]
Family VI.   "Sporolactobacillaceae"
Genus I.   *Sporolactobacillus*[AL]
Genus II.   *Marinococcus*[VP]
Family VII.   "Paenibacillaceae"
Genus I.   *Paenibacillus*[VP]
Genus II.   *Ammoniphilus*[VP]
Genus III.   *Aneurinibacillus*[VP]
Genus IV.   *Brevibacillus*[VP]
Genus V.   *Oxalophagus*[VP]
Genus VI.   *Thermobacillus*[VP]
Family VIII.   "Alicyclobacillaceae"
Genus I.   *Alicyclobacillus*[VP]
Genus II.   *Pasteuria*[AL]
Genus III.   *Sulfobacillus*[VP]
Family IX.   "Thermoactinomycetaceae"
Genus I.   *Thermoactinomyces*[AL]
Order II.   "Lactobacillales"
Family I.   *Lactobacillaceae*[AL]
Genus I.   *Lactobacillus*[AL]
Genus II.   *Pediococcus*[AL]
Family II.   "Aerococcaceae"
Genus I.   *Aerococcus*[AL]
Genus II.   *Abiotrophia*[VP]
Genus III.   *Dolosicoccus*[VP]
Genus IV.   *Eremococcus*[VP]
Genus V.   *Facklamia*[VP]
Genus VI.   *Globicatella*[VP]
Genus VII.   *Ignavigranum*[VP]
Family III.   "Carnobacteriaceae"
Genus I.   *Carnobacterium*[VP]
Genus II.   *Agitococcus*[VP]
Genus III.   *Alloiococcus*[VP]
Genus IV.   *Desemzia*[VP]
Genus V.   *Dolosigranulum*[VP]
Genus VI.   *Lactosphaera*[VP]
Genus VII.   *Trichococcus*[VP]
Family IV.   "Enterococcaceae"
Genus I.   *Enterococcus*[VP]
Genus II.   *Melissococcus*[VP]
Genus III.   *Tetragenococcus*[VP]
Genus IV.   *Vagococcus*[VP]
Family V.   "Leuconostocaceae"
Genus I.   *Leuconostoc*[AL]
Genus II.   *Oenococcus*[VP]
Genus III.   *Weissella*[VP]

Family VI.   *Streptococcaceae*[AL]
Genus I.   *Streptococcus*[AL]
Genus II.   *Lactococcus*[VP]
.Genera incertae sedis
Genus I.   *Acetoanaerobium*[VP]
Genus II.   *Oscillospira*[AL]
Genus III.   *Syntrophococcus*[VP]
Phylum BXIV.   *Actinobacteria* phy. nov.
Class I.   *Actinobacteria*[VP]
Subclass I.   *Acidimicrobidae*[VP]
Order I.   *Acidimicrobiales*[VP]
Suborder I.   "Acidimicrobineae"
Family I.   *Acidimicrobiaceae*[VP]
Genus I.   *Acidimicrobium*[VP]
Subclass II.   *Rubrobacteridae*[VP]
Order I.   *Rubrobacterales*[VP]
Suborder I.   "Rubrobacterineae"
Family I.   *Rubrobacteraceae*[VP]
Genus I.   *Rubrobacter*[VP]
Subclass III.   *Coriobacteridae*[VP]
Order I.   *Coriobacteriales*[VP]
Suborder I.   "Coriobacterineae"
Family I.   *Coriobacteriaceae*[VP]
Genus I.   *Coriobacterium*[VP]
Genus II.   *Atopobium*[VP]
Genus III.   *Collinsella*[VP]
Genus IV.   *Cryptobacterium*[VP]
Genus V.   *Eggerthella*[VP]
Genus VI.   *Slackia*[VP]
Subclass IV.   *Sphaerobacteridae*[VP]
Order I.   *Sphaerobacterales*[VP]
Suborder I.   "Sphaerobacterineae"
Family I.   *Sphaerobacteraceae*[VP]
Genus I.   *Sphaerobacter*[VP]
Subclass V.   *Actinobacteridae*[VP]
Order I.   *Actinomycetales*[AL]
Suborder I.   *Actinomycineae*[VP]
Family I.   *Actinomycetaceae*[AL]
Genus I.   *Actinomyces*[AL]
Genus II.   *Actinobaculum*[VP]
Genus III.   *Arcanobacterium*[VP]
Genus IV.   *Mobiluncus*[VP]
Suborder VI.   *Micrococcineae*[VP]
Family I.   *Micrococcaceae*[AL]
Genus I.   *Micrococcus*[AL]
Genus II.   *Arthrobacter*[AL]
Genus III.   *Bogoriella*[VP]
Genus IV.   *Demetria*[VP]
Genus V.   *Kocuria*[VP]
Genus VI.   *Leucobacter*[VP]
Genus VII.   *Nesterenkonia*[VP]
Genus VIII.   *Renibacterium*[VP]
Genus IX.   *Rothia*[AL]
Genus X.   *Stomatococcus*[VP]
Genus XI.   *Terracoccus*[VP]
Family II.   *Brevibacteriaceae*[AL]
Genus I.   *Brevibacterium*[AL]
Family III.   *Cellulomonadaceae*[VP]
Genus I.   *Cellulomonas*[AL]
Genus II.   *Oerskovia*[AL]
Genus III.   *Rarobacter*[VP]
Family IV.   *Dermabacteraceae*[VP]

Genus I.   *Dermabacter*[VP]
Genus II.   *Brachybacterium*[VP]
Family V.   *Dermatophilaceae*[AL]
    Genus I.   *Dermatophilus*[AL]
    Genus II.   *Dermacoccus*[VP]
    Genus III.   *Kytococcus*[VP]
Family VI.   *Intrasporangiaceae*[VP]
    Genus I.   *Intrasporangium*[AL]
    Genus II.   *Janibacter*[VP]
    Genus III.   *Ornithinicoccus*[VP]
    Genus IV.   *Sanguibacter*[VP]
    Genus V.   *Terrabacter*[VP]
Family VII.   *Jonesiaceae*[VP]
    Genus I.   *Jonesia*[VP]
Family VIII.   *Microbacteriaceae*[VP]
    Genus I.   *Microbacterium*[AL]
    Genus II.   *Agrococcus*[VP]
    Genus III.   *Agromyces*[AL]
    Genus IV.   *Aureobacterium*[VP]
    Genus V.   *Clavibacter*[VP]
    Genus VI.   *Cryobacterium*[VP]
    Genus VII.   *Curtobacterium*[AL]
    Genus VIII.   *Frigoribacterium*[VP]
    Genus IX.   *Leifsonia*[VP]
    Genus X.   *Rathayibacter*[VP]
Family IX.   "Beutenbergiaceae"
    Genus I.   *Beutenbergia*[VP]
Family X.   *Promicromonosporaceae*[VP]
    Genus I.   *Promicromonospora*[AL]
Suborder VII.   *Corynebacterineae*[VP]
Family I.   *Corynebacteriaceae*[AL]
    Genus I.   *Corynebacterium*[AL]
Family II.   *Dietziaceae*[VP]
    Genus I.   *Dietzia*[VP]
Family III.   *Gordoniaceae*[VP]
    Genus I.   *Gordonia*[VP]
    Genus II.   *Skermania*[VP]
Family IV.   *Mycobacteriaceae*[AL]
    Genus I.   *Mycobacterium*[AL]
Family V.   *Nocardiaceae*[AL]
    Genus I.   *Nocardia*[AL]
    Genus II.   *Rhodococcus*[AL]
Family VI.   *Tsukamurellaceae*[VP]
    Genus I.   *Tsukamurella*[VP]
Family VII.   "Williamsiaceae"
    Genus I.   *Williamsia*[VP]
Suborder VIII.   *Micromonosporineae*[VP]
Family I.   *Micromonosporaceae*[AL]
    Genus I.   *Micromonospora*[AL]
    Genus II.   *Actinoplanes*[AL]
    Genus III.   *Catellatospora*[VP]
    Genus IV.   *Catenuloplanes*[VP]
    Genus V.   *Couchioplanes*[VP]
    Genus VI.   *Dactylosporangium*[AL]
    Genus VII.   *Pilimelia*[AL]
    Genus VIII.   *Spirilliplanes*[VP]
    Genus IX.   *Verrucosispora*[VP]
Suborder IX.   *Propionibacterineae*[VP]
Family I.   *Propionibacteriaceae*[AL]
    Genus I.   *Propionibacterium*[AL]
    Genus II.   *Luteococcus*[VP]
    Genus III.   *Microlunatus*[VP]

    Genus IV.   *Propioniferax*[VP]
    Genus V.   *Tessaracoccus*[VP]
Family II.   *Nocardioidaceae*[VP]
    Genus I.   *Nocardioides*[AL]
    Genus II.   *Aeromicrobium*[VP]
    Genus III.   *Friedmanniella*[VP]
    Genus IV.   *Kribbella*[VP]
    Genus V.   *Micropruina*[VP]
Suborder X.   *Pseudonocardineae*[VP]
Family I.   *Pseudonocardiaceae*[VP]
    Genus I.   *Pseudonocardia*[AL]
    Genus II.   *Actinopolyspora*[AL]
    Genus III.   *Amycolatopsis*[VP]
    Genus IV.   *Kibdelosporangium*[VP]
    Genus V.   *Kutzneria*[VP]
    Genus VI.   *Prauserella*[VP]
    Genus VII.   *Saccharomonospora*[AL]
    Genus VIII.   *Saccharopolyspora*[AL]
    Genus IX.   *Streptoalloteichus*[VP]
    Genus X.   *Thermobispora*[VP]
    Genus XI.   *Thermocrispum*[VP]
Family II.   *Actinosynnemataceae*[VP]
    Genus I.   *Actinosynnema*[AL]
    Genus II.   *Actinokineospora*[VP]
    Genus III.   *Lentzea*[VP]
    Genus IV.   *Saccharothrix*[VP]
Suborder XI.   *Streptomycineae*[VP]
Family I.   *Streptomycetaceae*[AL]
    Genus I.   *Streptomyces*[AL]
    Genus II.   *Kitasatospora*[VP]
    Genus III.   *Streptoverticillium*[AL]
Suborder XII.   *Streptosporangineae*[VP]
Family I.   *Streptosporangiaceae*[VP]
    Genus I.   *Streptosporangium*[AL]
    Genus II.   *Herbidospora*[VP]
    Genus III.   *Microbispora*[AL]
    Genus IV.   *Microtetraspora*[AL]
    Genus V.   *Nonomuraea*[VP]
    Genus VI.   *Planobispora*[AL]
    Genus VII.   *Planomonospora*[AL]
    Genus VIII.   *Planopolyspora*[VP]
    Genus IX.   *Planotetraspora*[VP]
Family II.   *Nocardiopsaceae*[VP]
    Genus I.   *Nocardiopsis*[AL]
    Genus II.   *Thermobifida*[VP]
Family III.   *Thermomonosporaceae*[VP]
    Genus I.   *Thermomonospora*[AL]
    Genus II.   *Actinomadura*[AL]
    Genus III.   *Spirillospora*[AL]
Suborder XIII.   *Frankineae*[VP]
Family I.   *Frankiaceae*[AL]
    Genus I.   *Frankia*[AL]
Family II.   *Geodermatophilaceae*[VP]
    Genus I.   *Geodermatophilus*[AL]
    Genus II.   *Blastococcus*[AL]
    Genus III.   *Modestobacter*[VP]
Family III.   *Microsphaeraceae*[VP]
    Genus I.   *Microsphaera*[VP]
Family IV.   *Sporichthyaceae*[VP]
    Genus I.   *Sporichthya*[AL]
Family V.   *Acidothermaceae*[VP]
    Genus I.   *Acidothermus*[VP]

Family VI.   "Kineosporiaceae"
   Genus I.   *Kineosporia*[AL]
   Genus II.   *Cryptosporangium*[VP]
   Genus III.   *Kineococcus*[VP]
Suborder XIV.   *Glycomycineae*[VP]
   Family I.   *Glycomycetaceae*[VP]
      Genus I.   *Glycomyces*[VP]
Order II.   *Bifidobacteriales*[VP]
   Family I.   *Bifidobacteriaceae*[VP]
      Genus I.   *Bifidobacterium*[AL]
      Genus II.   *Falcivibrio*[VP]
      Genus III.   *Gardnerella*[VP]
   Genera Incertae sedis
      Genus I.   *Actinobispora*[VP]
      Genus II.   *Actinocorallia*[VP]
      Genus III.   *Excellospora*[AL]
      Genus IV.   *Pelczaria*[VP]
      Genus V.   *Turicella*[VP]
Phylum BXV.   *Planctomycetes* phy. nov.
   Class I.   "Planctomycetacia"
      Order I.   *Planctomycetales*[VP]
         Family I.   *Planctomycetaceae*[VP]
            Genus I.   *Planctomyces*[AL]
            Genus II.   *Gemmata*[VP]
            Genus III.   *Isosphaera*[VP]
            Genus IV.   *Pirellula*[VP]
Phylum BXVI.   *Chlamydiae* phy. nov.
   Class I.   "Chlamydiae"
      Order I.   *Chlamydiales*[AL]
         Family I.   *Chlamydiaceae*[AL]
            Genus I.   *Chlamydia*[AL]
            Genus II.   *Chlamydophila*[VP]
         Family II.   *Parachlamydiaceae*[VP]
            Genus I.   *Parachlamydia*[VP]
         Family III.   *Simkaniaceae*[VP]
            Genus I.   *Simkania*[VP]
         Family IV.   *Waddliaceae*[VP]
            Genus I.   *Waddlia*[VP]
Phylum BXVII.   *Spirochaetes* phy. nov.
   Class I.   "Spirochaetes"
      Order I.   *Spirochaetales*[AL]
         Family I.   *Spirochaetaceae*[AL]
            Genus I.   *Spirochaeta*[AL]
            Genus II.   *Borrelia*[AL]
            Genus III.   *Brevinema*[VP]
            Genus IV.   *Clevelandina*[VP]
            Genus V.   *Cristispira*[AL]
            Genus VI.   *Diplocalyx*[VP]
            Genus VII.   *Hollandina*[VP]
            Genus VIII.   *Pillotina*[VP]
            Genus IX.   *Treponema*[AL]
         Family II.   "Serpulinaceae"
            Genus I.   *Brachyspira*[VP]
            Genus II.   *Serpulina*[VP]
         Family III.   *Leptospiraceae*[AL]
            Genus I.   *Leptonema*[VP]
            Genus II.   *Leptospira*[AL]
Phylum BXVIII.   *Fibrobacteres* phy. nov.
   Class I.   "Fibrobacteres"
      Order I.   "Fibrobacterales"
         Family I.   "Fibrobacteraceae"
            Genus I.   *Fibrobacter*[VP]

Phylum BXIX.   *Acidobacteria* phy. nov.
   Class I.   "Acidobacteria"
      Order I.   "Acidobacteriales"
         Family I.   "Acidobacteriaceae"
            Genus I.   *Acidobacterium*[VP]
            Genus II.   *Geothrix*[VP]
            Genus III.   *Holophaga*[VP]
Phylum BXX.   *Bacteroidetes* phy. nov.
   Class I.   "Bacteroides"
      Order I.   "Bacteroidales"
         Family I.   *Bacteroidaceae*[AL]
            Genus I.   *Bacteroides*[AL]
            Genus II.   *Acetofilamentum*[VP]
            Genus III.   *Acetomicrobium*[VP]
            Genus IV.   *Acetothermus*[VP]
            Genus V.   *Anaerorhabdus*[VP]
            Genus VI.   *Megamonas*[VP]
         Family II.   "Rikenellaceae"
            Genus I.   *Rikenella*[VP]
            Genus II.   *Marinilabilia*[VP]
         Family III.   "Porphyromonadaceae"
            Genus I.   *Porphyromonas*[VP]
         Family IV.   "Prevotellaceae"
            Genus I.   *Prevotella*[VP]
   Class II.   "Flavobacteria"
      Order I.   "Flavobacteriales"
         Family I.   *Flavobacteriaceae*[VP]
            Genus I.   *Flavobacterium*[AL]
            Genus II.   *Bergeyella*[VP]
            Genus III.   *Capnocytophaga*[VP]
            Genus IV.   *Cellulophaga*[VP]
            Genus V.   *Chryseobacterium*[VP]
            Genus VI.   *Coenonia*[VP]
            Genus VII.   *Empedobacter*[VP]
            Genus VIII.   *Gelidibacter*[VP]
            Genus IX.   *Ornithobacterium*[VP]
            Genus X.   *Polaribacter*[VP]
            Genus XI.   *Psychroflexus*[VP]
            Genus XII.   *Psychroserpens*[VP]
            Genus XIII.   *Riemerella*[VP]
            Genus XIV.   *Weeksella*[VP]
         Family II.   "Myroidaceae"
            Genus I.   *Myroides*[VP]
         Family III.   "Blattabacteriaceae"
            Genus I.   *Blattabacterium*[AL]
   Class III.   "Sphingobacteria"
      Order I.   "Sphingobacteriales"
         Family I.   *Sphingobacteriaceae*[VP]
            Genus I.   *Sphingobacterium*[VP]
            Genus II.   *Pedobacter*[VP]
         Family II.   "Saprospiraceae"
            Genus I.   *Saprospira*[AL]
            Genus II.   *Haliscomenobacter*[AL]
            Genus III.   *Lewinella*[VP]
         Family III.   "Flexibacteraceae"
            Genus I.   *Flexibacter*[AL]
            Genus II.   *Cyclobacterium*[VP]
            Genus III.   *Cytophaga*[AL]
            Genus IV.   *Flectobacillus*[AL]
            Genus V.   *Hymenobacter*[VP]
            Genus VI.   *Meniscus*[AL]
            Genus VII.   *Microscilla*[AL]

Genus VIII. *Runella*[AL]
Genus IX. *Spirosoma*[AL]
Genus X. *Sporocytophaga*[AL]
Family IV. "Flammeovirgaceae"
    Genus I. *Flammeovirga*[VP]
    Genus II. *Flexithrix*[AL]
    Genus III. *Persicobacter*[VP]
    Genus IV. *Thermonema*[VP]
Family V. *Crenotrichaceae*[AL]
    Genus I. *Crenothrix*[AL]
    Genus II. *Chitinophaga*[VP]
    Genus III. *Rhodothermus*[VP]
    Genus IV. *Toxothrix*[AL]
Phylum BXXI. *Fusobacteria* phy. nov.
Class I. "Fusobacteria"
Order I. "Fusobacteriales"
Family I. "Fusobacteriaceae"
    Genus I. *Fusobacterium*[AL]

Genus II. *Ilyobacter*[VP]
Genus III. *Leptotrichia*[AL]
Genus IV. *Propionigenium*[VP]
Genus V. *Sebaldella*[VP]
Genus VI. *Streptobacillus*[AL]
Family II. Incertae sedis
    Genus I. *Cetobacterium*[VP]
Phylum BXXII. *Verrucomicrobia* phy. nov.
Class I. *Verrucomicrobiae*[VP]
Order I. *Verrucomicrobiales*[VP]
Family I. *Verrucomicrobiaceae*[VP]
    Genus I. *Verrucomicrobium*[VP]
    Genus II. *Prosthecobacter*[VP]
Phylum BXXIII. *Dictyoglomi* phy. nov.
Class I. "Dictyoglomi"
Order I. "Dictyoglomales"
Family I. "Dictyoglomaceae"
    Genus I. *Dictyoglomus*[VP]

DOMAIN *ARCHAEA*

Phylum AI. Crenarchaeota *phy. nov.*

GEORGE M. GARRITY AND JOHN G. HOLT

Cren.arch.ae.o' ta. M.L. fem. pl. n. *Crenarchaeota* from the Kingdom *Crenarchaeota* (Woese, Kandler and Wheelis 1990, 4579).

The phylum consists of a single class, the *Thermoprotei*, which is well supported by 16S rDNA sequence data. It is subdivided into three orders: the *Thermoproteales*, the *Desulfurococcales*, and the *Sulfolobales*. Morphologically diverse, including rods, cocci, filamentous forms, and disk-shaped cells. Stain Gram-negative. Motility observed in some genera. Obligately thermophilic, with growth occurring at temperatures ranging from 70 to 113°C.

Acidophilic. Aerobic, facultatively anaerobic, or strictly anaerobic-chemolithoautotrophs or chemoheterotrophs. Most metabolize S^0. Chemoheterotrophs, may grow by sulfur respiration. RNA polymerase of the BAC type.

Type order: **Thermoproteales** Zillig and Stetter 1982, 267, emend. Burggraf, Huber and Stetter 1997b, 659 (Effective publication: Zillig and Stetter *in* Zillig, Stetter, Schäfer, Janekovic, Wunderl, Holz and Palm 1981, 224).

Class I. **Thermoprotei** *class. nov.*

ANNA-LOUISE REYSENBACH

Ther.mo.pro' te.i. M.L. fem. pl. n. *Thermoproteales* type order of the class; dropping the ending to denote a class; M.L. fem. pl. n. *Thermoprotei* the class of *Thermoproteales.*

Rod, disc, or spherical cells of varied size and shape; rods are from ~0.1 to 0.5 × 1 to ~100 μm long. Coccoid forms are from ~0.5 to 5 μm in diameter and may be pleomorphic. Gram-negative and thermophilic. Grows best from 65 to >100°C. Obligate anaerobes, facultative aerobes, or obligate aerobes. Chemolithoautotrophs, facultative chemolithoautotrophs, or organotrophs. Sulfur reduction or oxidation, or sulfur respiration of various organic substrates. Occur worldwide in acidic thermal springs and mudpots, and in submarine thermal environments.

Type order: **Thermoproteales** Zillig and Stetter 1982, 267, emend. Burggraf, Huber and Stetter 1997b, 659 (Effective publication: Zillig and Stetter *in* Zillig, Stetter, Schäfer, Janekovic, Wunderl, Holz and Palm 1981, 224).

Key to orders of the class Thermoprotei

1. Rods and filaments, varying in diameter from 0.15 to 0.6 μm and up to 100 μm long. Often swollen spherical protrusions at ends of cells. Best growth at 75–100°C. Moderately acidophilic. Optimum pH 4.5–7.

   Order I. *Thermoproteales*, p. 170

2. Irregular disc-shaped or coccoid cells ranging from 0.3 to 2.5 μm in diameter. Hyperthermophilic, best growth above 85°C. Moderately acidophilic. Optimum pH 4.5–7.

   Order II. *Desulfurococcales*, p. 179

3. Irregular coccoid cells, vary from 0.8 to 2.0 μm in diameter. Obligately or facultatively chemolithoautotrophic. Aerobic, or facultatively or obligately anaerobic. Grows best at 65–85°C. Acidophilic. Optimum pH 2–4.5

   Order III. *Sulfolobales*, p. 198

Order I. **Thermoproteales** Zillig and Stetter 1982, 267,[VP] emend. Burggraf, Huber and Stetter 1997b, 659 (Effective publication: Zillig and Stetter *in* Zillig, Stetter, Schäfer, Janekovic, Wunderl, Holz and Palm 1981, 224)

HARALD HUBER AND KARL O. STETTER

Ther.mo.pro.te.a' les. M.L. masc. n. *Thermoproteus* type genus of the order; *-ales* ending to denote an order; M.L. fem. pl. n. *Thermoproteales* the order of *Thermoproteus*.

Cells are rods about 0.1–0.5 μm in diameter and 1 to almost 100 μm in length. Septa have not been encountered. Terminal spheres ("golf clubs") observed under normal (aerobic) phase contrast microscopy (Zillig et al., 1981). "Golf clubs" cannot be seen under anaerobic conditions at growth temperatures (Horn et al., 1999). **Gram-negative. Anaerobic to facultatively anaerobic. Hyperthermophilic** (optimal growth temperature 75–100°C). **Grow either chemolithoautotrophically** by gaining energy from the reaction $H_2 + S^0 \rightarrow H_2S$ using CO_2 as sole carbon source **or by sulfur respiration of various organic substrates** yielding CO_2 and H_2S. Some genera are able to gain energy by respiration using O_2, nitrate, or nitrite as electron acceptors. Cell envelope S-layer composed of protein or glycoprotein subunits in dense packing hexagonal arrangement, devoid of muramic acid. Lipids contain glycerol ethers of polyisoprenoid C_{40} and, in lesser amounts, C_{20} alcohols. Transcription is resistant to rifampicin and streptolydigin. RNA polymerases exhibit a complex BAC-component pattern.

16S rRNA sequence analysis (Burggraf et al., 1997b) revealed that the *Thermoproteales*, along with the *Sulfolobales* and the *Desulfurococcales*, comprise the three orders within the crenarchaeal branch of the *Archaea*.

Widely distributed in solfataric hot springs and submarine hydrothermal systems.

The order comprises two validly described families, the *Thermoproteaceae* and the *Thermofilaceae*.

The mol% G + C of the DNA is: 46–57.

Type genus: **Thermoproteus** Zillig and Stetter 1982, 267 (Effective publication: Zillig and Stetter *in* Zillig, Stetter, Schäfer, Janekovic, Wunderl, Holz and Palm 1981, 225).

Key to the families of the order Thermoproteales

1. Rigid rods 0.4–0.5 μm in diameter and 1.5–20 μm in length. H_2/S^0 autotrophy.
   Family I. *Thermoproteaceae.*
2. Thin rods 0.15–0.35 μm in diameter and 1–100 μm in length. No H_2/S^0 respiration.
   Family II. *Thermofilaceae*, p. 178

Family I. **Thermoproteaceae** Zillig and Stetter 1982, 267,[VP] emend. Burggraf, Huber and Stetter 1997b, 659 (Effective publication: Zillig and Stetter *in* Zillig, Stetter, Schäfer, Janekovic, Wunderl, Holz and Palm 1981, 224)

HARALD HUBER AND KARL O. STETTER

Ther.mo.pro.te.a' ce.ae. M.L. masc. n. *Thermoproteus* type genus of the family; *-aceae* ending to denote a family; M.L. fem. pl. n. *Thermoproteaceae* the family of *Thermoproteus*.

Rigid rods, 0.4–0.5 × 1.5–20 μm. Energy produced chemolithoautotrophically by reduction of elemental sulfur with H_2 forming H_2S and using CO_2 as sole carbon source. **Organotrophic growth by sulfur respiration utilizing various organic substrates or respiration with O_2 or nitrate as electron acceptors.** See also description of the order for features.

The family harbors four genera: *Thermoproteus, Caldivirga, Pyrobaculum,* and *Thermocladium.*

Type genus: **Thermoproteus** Zillig and Stetter 1982, 267 (Effective publication: Zillig and Stetter *in* Zillig, Stetter, Schäfer, Janekovic, Wunderl, Holz and Palm 1981, 225).

Key to the genera of the family Thermoproteaceae

1. Obligately anaerobic; sulfur respiration with complex organic substrates. Temperature optimum 85–90°C; temperature maximum: 97°C.
   Genus I. *Thermoproteus*, p. 171
2. Grows anaerobically or microaerobically. Optimum growth at 85°C and pH 3.7–4.2.
   Genus II. *Caldivirga*, p. 173
3. Anaerobic or facultatively anaerobic; growth on organic substrates by sulfur, nitrate, nitrite, or oxygen respiration. Temperature optimum 100°C; temperature maximum 104°C.
   Genus III. *Pyrobaculum*, p. 174
4. Anaerobic or microaerobic; growth on organic substrates by fermentation. Sulfur or thiosulfate required. Temperature optimum around 75°C.
   Genus IV. *Thermocladium*, p. 177

Genus I. **Thermoproteus** *Zillig and Setter 1982, 267^VP (Effective publication: Zillig and Stetter in Zillig, Stetter, Schäfer, Janekovic, Wunderl, Holz and Palm 1981, 225)*

WOLFRAM ZILLIG AND ANNA-LOUISE REYSENBACH

Ther.mo.pro' te.us. Gr. fem. n. *therme* heat; Gr. masc. n. *proteus* a mythical figure able to assume different forms; M.L. masc. n. *Thermoproteus* the genus of thermophilic bacteria of various forms.

Rigid rods of ~0.4 μm in diameter and from <1 to ~100 μm in length. No cell septa. Otherwise, the description is the same as that for the family.

Recovered from acidic hot springs and water holes in Iceland, Italy, North America, New Zealand, the Azores, and Indonesia at pH values 1.7–6.5 and temperatures up to 100°C.

Type species: **Thermoproteus tenax** Zillig and Stetter 1982, 267 (Effective publication: Zillig and Stetter *in* Zillig, Stetter, Shäfer, Janekovic, Wunderl, Holz and Palm 1981, 225).

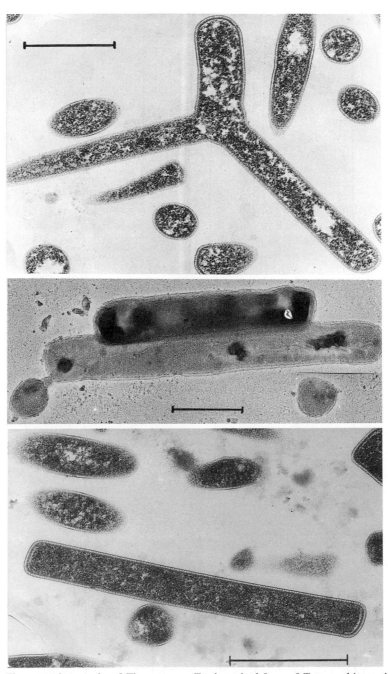

FIGURE A1.1. Electron micrographs of *Thermoproteus*. *Top*, branched form of *T. tenax*, thin section. *Middle*, golf club and released spheroid of *Thermoproteus* species isolate H3, platinum-shadowed. *Bottom*, normal rod of *T. tenax*, thin section. Bars = 1 μm. (Micrographs courtesy of D. Janekovic.)

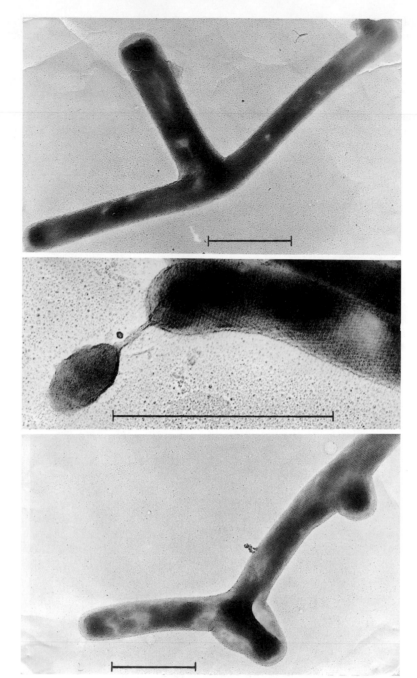

FIGURE A1.2. *Top*, branched form of *T. tenax*. *Middle*, golf club of *T. tenax*. *Bottom*, budding form of *T. tenax*. All platinum-shadowed. Bars = 1 um. (Micrographs courtesy of D. Janekovic.)

Key to the species of the genus *Thermoproteus*

1. Grows best at pH 5. Host to four rod-shaped DS–DNA bacteriophage.

    T. tenax.

2. Grows best at pH 6.5. Resistant to *T. tenax* bacteriophage.

    T. neutrophilus.

3. Grows best at pH 5.6.

    "*T. uzoniensis*".

List of species of the genus Thermoproteus

1. **Thermoproteus tenax** Zillig and Stetter 1982, 267[VP] (Effective publication: Zillig and Stetter *in* Zillig, Stetter, Shäfer, Jane-kovic, Wunderl, Holz and Palm 1981, 225).

    te'nax. L. masc. adj. *tenax* tenacious, resistant.

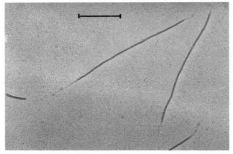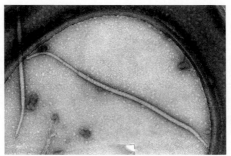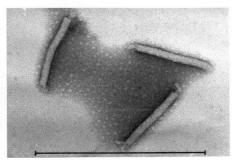

FIGURE A1.3. Electron micrographs of viruses of *T. tenax*. *Left* to *right*, TTV3, TTV2 and TTV1. Bar on the *left* = 1 μm. Bar on the *right* = 1 μm for the middle and right sections of the figure. Negatively stained. (Micrographs courtesy of D. Janekovic.)

Cell shapes typical of genus (Figs. A1.1 and A1.2). Nonmotile. Pili attached laterally and/or terminally. Grows chemolithoautotrophically with H_2 and S^0 as energy sources and CO_2 as carbon source, or by sulfur respiration of different compounds serving as carbon sources, including glucose, starch, glycogen, fumarate, and amino acids, but not lactate, acetate, and pyruvate. Malate replaces sulfur as electron acceptor. Growth on organic substrates stimulated by yeast extract, at concentrations of 0.2–0.5 g/l. Forms CO_2. Fermentation pathways are unknown.

Grows best at about pH 5 and 90°C but grows even up to 96°C. The DNA-dependent RNA polymerase is stable at 95°C.

Strain Kra 1 is from Krafla, Iceland, and is host to four different rod-shaped viruses (Fig. A1.3) of a previously unknown type containing double-stranded DNA (Janekovic et al., 1983; Zillig et al., 1986a).

The mol% G + C of the DNA is: 55.5.

Type strain: Kra 1, ATCC 35583, DSMZ 2076.

2. **Thermoproteus neutrophilus** Zillig 1989b, 496^VP (Effective publication: Zillig 1989a, 2241).

neu.tro.phi' lus. L. masc. adj. *neuter* neutral; Gr. adj. *philos* loving, preferring; M.L. masc. adj. *neutrophilus* preferring neutral pH.

Facultative chemolithotroph, which can utilize acetate instead of CO_2 as carbon source, gaining energy by the formation of H_2S from H_2 and S^0 (Schäfer et al., 1986). Grows optimally at pH 6.5 but also at a pH as high as pH 7.0. In other respects resembles *T. tenax* but not as a host to its viruses.

The mol% G + C of the DNA is: not available.

Type strain: DSMZ 2338.

GenBank accession number (16S rRNA): AB009618.

3. **"Thermoproteus uzoniensis"** Bonch-Osmolovskaya, Miroshnichenko, Kostrikina, Chernych and Zavarzin 1990, 559.

u.zo.ni' en.sis. M.L. masc. adj. *uzoniensis* inhabiting hot springs and soil of the Uzon caldera.

The nonmotile cells are straight or slightly curved rods 0.3–0.4 × 1–20 μm, occasionally branching or with spherical bulbs. The cell envelope consists of hexagonal subunits, covered with a layer of less well-defined structure. The isolate grows anaerobically from 74 to 102°C (optimum temperature: 90°C) and at a pH of 4.6–6.8 (optimum pH 5.6). Cells utilize peptides, and can simultaneously reduce S^0. Fermentation products include acetate, isobutyrate, and isovalerate.

The mol% G + C of the DNA is: 56.5 (T_m).

Deposited strain: Z-605, DSMZ 5263.

Genus II. **Caldivirga** *Itoh, Suzuki, Sanchez and Nakase 1999, 1162^VP*

TAKASHI ITOH, KEN-ICHIRO SUZUKI, PRISCILLA C. SANCHEZ AND TAKASHI NAKASE

Cal.di.vir' ga. L. adj. *caldus* hot; L. n. *virga* rod; M.L. fem. n. *Caldivirga* a hot rod.

Straight or slightly curved rods, occasionally bent or branching, either singly or extensively. Most cells are 0.4–0.7 × 3–20 μm. Often, **spherical bodies protrude at the end of the cells, or laterally.** Cells multiply by branching or budding, and divide by constriction. Pili attach to cells terminally or laterally. Nonmotile. **Grows anaerobically or microaerobically. Optimum growth at high temperature (85°C) and weakly acidic conditions (pH 3.7–4.2).** Resistant to chloramphenicol, kanamycin, oleandomycin, and streptomycin. Sensitive to erythromycin, novobiocin, and rifampicin. Possesses at least five cyclized glycerol-bisphytanyl-glycerol tetraethers. The 16S rDNA is typical of a chrenarchaeote in the signature sequence analysis. Phylogenetically represents an independent lineage in the family *Thermoproteaceae*. Isolated from acidic hot springs. Monospecific at present.

The mol% G + C of the DNA is: 43.

Type species: **Caldivirga maquilingensis** Itoh, Suzuki, Sanchez and Nakase 1999, 1162.

ENRICHMENT AND ISOLATION PROCEDURES

Caldivirga can be enriched using the medium for *Sulfolobus*, supplemented with yeast extract (0.2 g/l), resazurin (1 mg/l), the vitamin mixture of Balch et al. (1979) (10 ml/l), and sulfur powder (5.0 g/l). The pH should be adjusted to 4.0–5.0. The medium is reduced with $Na_2S \cdot 9H_2O$. Growth is observed after one week at 85°C under a N_2 atmosphere. Strains are then isolated by serial dilutions as described by Itoh et al. (1998).

DIFFERENTIATION OF THE GENUS *CALDIVIRGA* FROM OTHER GENERA

Caldivirga is assigned to the family *Thermoproteaceae* on the basis of 16S rRNA sequence analysis, cell morphology, and the presence of cyclic glycerol diphytanyl tetraether lipids. *Caldivirga maquilingensis* can be distinguished from other species in this family by a lower mol% G + C content (43.0%). It grows at more acidic pH than *Thermoproteus* or *Pyrobaculum* species and at a higher temperature than *Thermocladium modestius*.

List of species of the genus Caldivirga

1. **Caldivirga maquilingensis** Itoh, Suzuki, Sanchez and Nakase 1999, 1162.[VP]

ma.qui.lin.gen' sis. L. adj. suff. *-ensis* belonging to; M.L. adj. pertaining to Mount Maquiling, a volcano in the Phillippines.

Description is the same as the genus.

Anaerobic or microaerobic. Heterotrophic. Growth is stimulated significantly by archaeal cell extracts or a vitamin mixture. Strain IC-167[T] grows optimally at 85°C and pH 3.7–4.2, with a doubling time of 8 h. Growth occurs from 60 to 92°C, and between pH 2.3 and 5.9 when buffered with 10 mM trisodium citrate. Alternatively, grows up to pH 6.4 by using 10 mM MES as buffer agent. Grows at low salinity (0–0.75% NaCl). Utilizes glycogen, gelatin, beef extract, peptone, tryptone, and yeast extract as carbon sources. Requires sulfur, thiosulfate, or sulfate as possible electron acceptors. Produces sulfide during growth. The 16S rDNA may contain introns. The type strain was isolated from a hot spring at Mt. Maquiling, Laguna, the Philippines.

The mol% G + C of the DNA is: 43.

Type strain: IC-167, ANMR 0178, JCM 10307, MCC-UPLB 1200.

GenBank accession number (16S rRNA): AB013926.

Genus III. **Pyrobaculum** *Stetter, Huber and Kristjansson 1988, 221,[VP] emend. Huber, Völkl and Stetter in Völkl, Huber, Drobner, Rachel, Burggraf, Trincone and Stetter 1993, 2924 (Effective publication: Stetter, Huber and Kristjansson in Huber, Kristjansson and Stetter 1987b, 100)*

ROBERT HUBER AND KARL O. STETTER

Py.ro.ba' cu.lum. Gr. neut. n. *pyr* fire; L. neut. n. *baculum* stick; M.L. neut. n. *Pyrobaculum* the fire stick.

Rod-shaped, almost rectangular cells. During growth, the rods form aggregates arranged in a V-shape with various angles. X- and raft-shaped aggregates occur. **Rods form terminal spheres**. Cells usually 0.5–0.6 × 1.5–8 µm. Gram negative. **No murein present**. No spores formed. **No cell septa**. Cells surrounded by **a single or double S-layer** of protein subunits, **arranged on a p6 lattice**. **Isoprenyl ether lipids present** in the membrane. Motile due to flagellation. Facultatively aerobic or strictly anaerobic. **Hyperthermophilic**, optimum growth temperature: 100°C; maximum: 104°C; minimum: 74°C. Elongation factor G is ADP-ribosylated. Facultative or obligate heterotrophs.

Based on 16S rDNA analysis, *Pyrobaculum* represents a genus within the order *Thermoproteales*.

The mol% G + C of the DNA is: 45–52.

Type species: **Pyrobaculum islandicum** Stetter, Huber and Kristjansson 1988, 221 (Effective publication: Stetter, Huber and Kristjansson *in* Huber, Kristjansson and Stetter 1987b, 100).

FURTHER DESCRIPTIVE INFORMATION

Within the 16S rDNA phylogenetic tree, the genus *Pyrobaculum* belongs to the *Crenarchaeota* within the *Archaea* domain (Huber et al., 1987b; Woese et al., 1990; Huber and Stetter, 1992b; Burggraf et al., 1993; Völkl et al., 1993). *Pyrobaculum* belongs to the *Thermoproteales* with closest relationship to members of the genus *Thermoproteus* (Zillig et al., 1981; Burggraf et al., 1997b).

Cells of *Pyrobaculum* are cylinder-shaped, rigid rods with nearly rectangular ends (Figs. A1.4 and A1.5). At the end of the exponential growth phase, about 1–10% of the rods form terminal spherical bodies, similar to the "golf clubs" of *Thermoproteus* (Zillig et al., 1981). In the stationary phase, the terminal spheres of *Pyrobaculum aerophilum* tend to enlarge and several rods convert completely into spheres during further incubation. At high nitrate concentrations or pH values ≥8.0, cultures of *P. aerophilum* contain about 50% coccoid cells. *Pyrobaculum* species show motility, which can be strongly enhanced by heating a microscopic slide to about 90°C.

The cell envelope of *Pyrobaculum organotrophum* is composed of two distinct hexagonally arrayed crystalline protein layers. Between the two layers, fibrils appear to be sandwiched. The outer layer has a p6 symmetry and a lattice spacing of 20.6 nm (Phipps et al., 1991b). In contrast, the cell wall of *Pyrobaculum islandicum* and *P. aerophilum* consists of one S-layer with a p6 symmetry and is connected via a spacer to the cytoplasmic membrane (Phipps et al., 1990; Völkl et al., 1993). An additional surface coat of fibrillar material covers the S-layer (Phipps et al., 1990; Rieger et al., 1997). Thin sections of *P. aerophilum* revealed a novel type of cell-to-cell connection in the form of thin tubes approximately 15–20 nm in diameter (Rieger et al., 1997). The complex lipids of the *Pyrobaculum* species consist mainly of phosphoglycolipids type I and type II. Aminophospholipids are absent (Trincone et al., 1992; Völkl et al., 1993). The core lipids are mainly composed of acyclic and cyclic glycerol diphytanyl glycerol tetraethers with

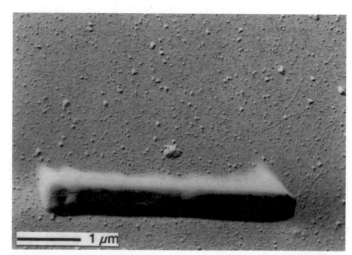

FIGURE A1.4. Platinum-shadowed cell showing flagella of *Pyrobaculum aerophilum.*

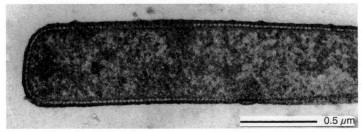

FIGURE A1.5. Ultrathin-section of *Pyrobaculum aerophilum.*

one to four pentacyclic rings. In addition, traces of glycerol di-phytanyl glycerol diethers are present (Trincone et al., 1992; Völkl et al., 1993). In *P. islandicum* and *P. organotrophum,* fully saturated menaquinones have been identified (Tindall, 1989; Tindall et al., 1991b).

P. organotrophum and *P. islandicum* are strictly anaerobic. Both species can grow heterotrophically, using sulfur or sulfur compounds as electron acceptor, forming H_2S and CO_2. The complete oxidation of organic compounds to CO_2 proceeds via the citric acid cycle (Selig and Schönheit, 1994). In contrast to the strictly heterotrophic, *P. organotrophum, P. islandicum* can grow chemolithoautotrophically with H_2, CO_2, and sulfur. *P. aerophilum* is facultatively anaerobic, growing by aerobic respiration or dissimilatory nitrate reduction, using organic or inorganic compounds as substrates (Table A1.1). During autotrophic growth under aerobic conditions, hydrogen or thiosulfate serves as electron donor (Table A1.1). The optimal oxygen concentrations for aerobic growth during autotrophic and organotrophic nutrition are around 0.6% and 1% O_2; the upper limit is between 3 and 5% O_2. Under anaerobic conditions, *P. aerophilum* grows by nitrate respiration, forming nitrite and traces of NO. Hydrogen or thiosulfate serves as electron donor during autotrophic growth with nitrate as electron acceptor (Table A1.1). Furthermore, *P. aerophilum* grows in the presence of organic substrates by nitrite reduction, forming N_2 and traces of N_2O and NO as final products. N_2O is not used as an alternative electron acceptor. Growth of *P. aerophilum* is inhibited on plates or in liquid medium, when S^0 is added (Table A1.1). For growth of the *Pyrobaculum* species under anaerobic conditions, the anaerobic technique according to Balch and Wolfe (1976) is used. The physiological growth conditions of the different *Pyrobaculum* species are listed in Table A1.2. For detailed growth conditions see also Huber and Stetter (1992b). High cell densities of *P. aerophilum* are obtained when the organism is cultivated in cellulose capillary tubes (Rieger et al., 1997).

A fosmid-based genomic map was constructed from the genome of *P. aerophilum.* So far, 95% of the genome sequence has been determined and 474 putative genes have been identified (Fitz-Gibbon et al., 1997). The sequencing of the entire 2.22 Mb genome of *P. aerophilum* by three groups under the leadership of the California Institute of Technology is in progress and the data should be available in 2000.

Within the single 23S rRNA-encoding gene of *P. organotrophum,* two introns occur. The RNA products circularize after excision from the 23S rRNA and are stable in the cell (Dalgaard and Garret, 1992). Furthermore, a 713-bp intron identified in the 16S rRNA gene of *P. aerophilum* also circularizes upon excision. The intron contains an open reading frame, whose protein translation shows no certain homology with any known protein sequence (Burggraf et al., 1993).

TABLE A1.1. Substrate utilization of *Pyrobaculum aerophilum*[a]

Substrate (0.05%, final concentration)	Electron acceptor		
	O_2	KNO_3	KNO_2
Yeast extract	+ + +	+ + +	+ + +
Meat extract	+ + +	+ + +	+ +
Tryptone	+ + +	+ + +	+ +
Peptone	+ + +	+ + +	+ +
Gelatin	+	+ +	nd
Casein	−	+	nd
Casamino Acids	+	+ +	nd
Starch	−	−	nd
Sucrose	−	−	−
Glucose	−	−	−
Lactose	−	−	−
Ribose	−	−	−
Pyruvate	−	−	−
Propionate	+	+	+
Acetate	+	+	+
Sodium thiosulfate	+ +	+	nd
Sulfite	−	−	nd
Sulfur	−	−	−
Sulfur–yeast extract	−	−	−
Hydrogen	+	+	nd
None	−	−	−

[a]nd, not determined; + + +, 10^8 or more cells/ml; + +, 5×10^7–9×10^7 cells/ ml; +, 1×10^7–4×10^7 cells/ml; −, no growth.

TABLE A1.2. Diagnostic and descriptive features of the *Pyrobaculum* species[a]

Characteristic	1. *Pyrobaculum islandicum*	2. *Pyrobaculum aerophilum*	3. *Pyrobaculum organotrophum*
pH range	5.0–7.0	5.8–9.0	5.0–7.0
pH optimum	6	7	6
Temperature range, °C	74–102	75–104	78–102
Temperature optimum, °C	100	100	102
NaCl range, %	<0.9	0–3.6	<0.6
NaCl optimum, %	nd	1.5	nd
Optimal doubling time, min	150	180	690

[a]nd, not determined.

Cells of *P. islandicum* and *P. organotrophum* are resistant to streptomycin, phosphomycin, chloramphenicol, penicillin G, and vancomycin (100 μg/ml). Growth of *P. islandicum* and *P. organotrophum* is reversibly inhibited in the presence of rifampicin (100 μg/ml).

So far, members of *Pyrobaculum* have been obtained mainly from continental solfataras with low salinity. *P. organotrophum* was originally isolated from a strongly gassed spring hole at Grandalur (Hveragerthi, Iceland), while *P. islandicum* was obtained from an outflow of superheated water of an overpressure valve at the Krafla geothermal power plant (Iceland). Further members of *Pyrobaculum* were isolated from Pisciarelli Solfatara (Italy) and from the Ribeira Quente solfatara field (Azores). As an exception, the marine *P. aerophilum* was isolated from a hot, strongly gassed marine water hole at the tidal zone of Maronti Beach in Italy, Naples.

ENRICHMENT AND ISOLATION PROCEDURES

Due to their extremely high growth temperatures, the three *Pyrobaculum* species can be selectively enriched, using an incubation temperature of 100°C. For the enrichment and isolation of the *Pyrobaculum* species, the anaerobic technique of Balch and Wolfe (1976) is suitable. For the enrichment of relatives of *P. organotrophum* and *P. islandicum*, a low salt medium with pH 6, containing sulfur, yeast extract, and peptone, and a gas phase of 300 kPa H_2/CO_2 (80:20; v/v) is recommended under conditions described in Table A1.2 (Huber et al., 1987b; Huber and Stetter, 1992b). A still more selective enrichment of *P. islandicum* can be achieved when the organic compounds are omitted from the medium. Isolates of *P. islandicum* can be obtained by plating under anaerobic conditions, using a stainless-steel anaerobic jar (Balch et al., 1979) and plates, solidified by 20% starch, containing thiosulfate instead of sulfur. Tiny colonies will be formed after anaerobic incubation for about 7 days at 85°C. Colonies of *P. islandicum* are greenish black. For the selective enrichment of *P. aerophilum*, half strength seawater medium, pH 7, supplemented with nitrate and a gas phase of 300 kPa H_2/CO_2 (80:20; v/v) can be used under conditions described in Table A1.2 (Völkl et al., 1993). Isolates are obtained, using plates solidified by 0.6% Gelrite and a stainless-steel anaerobic jar for incubation at 85°C (Balch et al., 1979). After about 2 weeks, round grayish-yellow colonies will be visible. High plating efficiencies approaching 100% can be obtained on Gelrite plates at 92°C under aerobic

conditions and an incubation time of only 4 days (Völkl et al., 1993). *Pyrobaculum* species can be also isolated by the "selected cell cultivation" technique, using a strongly focused infrared laser beam as "optical tweezers" (Huber et al., 1995a; Beck and Huber, 1997).

MAINTENANCE PROCEDURES

Pyrobaculum species can be stored for at least 3 months at 4°C under anaerobic conditions. Storage in liquid nitrogen at −130°C in the presence of 5% DMSO is possible for long-term preservation. No loss of cell viability of *Pyrobaculum* was observed after storage over a period of more than 4 years.

DIFFERENTIATION OF THE GENUS *PYROBACULUM* FROM OTHER GENERA

Pyrobaculum can be phylogenetically differentiated from other genera within the *Thermoproteales*, by comparison of the 16S rDNA sequences. *Pyrobaculum* can be distinguished from *Thermoproteus* by its inability to form filaments, its growth at 100°C, negligible DNA similarity, and a 10% lower mol% G + C content in DNA (*P. islandicum* and *P. organotrophum*). *P. aerophilum* can be further differentiated from *Thermoproteus* (and from *P. islandicum* and *P. organotrophum*) in its physiological and metabolic properties.

ACKNOWLEDGMENTS

We would like to thank R. Rachel for electron microscopy.

FURTHER READING

Huber, R. and K.O. Stetter. 1992. The order *Thermoproteales. In* Balows, Trüper, Dworkin, Harder and Schleifer (Editors), The Prokaryotes. A Handbook of Bacteria: Ecophysiology, Isolation, Identification, Applications, 2nd ed., Vol. I, Springer-Verlag, New York, pp. 677–683.
Stetter, K.O. 2000. Volcanoes, hydrothermal venting, and the origin of life. *In* Marit and Ernst (Editors), Volcanoes and the Environment, Cambridge University Press, Cambridge, in press.

DIFFERENTIATION OF THE SPECIES OF THE GENUS *PYROBACULUM*

For general characteristics of the species see generic description. Other characteristics of the *Pyrobaculum* species are listed in Tables A1.1 and A1.2.

List of species of the genus Pyrobaculum

1. **Pyrobaculum islandicum** Stetter, Huber and Kristjansson 1988, 221[VP] (Effective publication: Stetter, Huber and Kristjansson *in* Huber, Kristjansson and Stetter 1987b, 100).
   is.lan' di.cum. M.L. neut. adj. *islandicum* of Iceland, describing the place of its first isolation.

   Facultative chemolithoautotroph. Chemolithoautotrophic growth on H_2, CO_2, and sulfur, forming H_2S. S^0 strictly required as electron acceptor for autotrophic growth. Heterotrophic growth on yeast extract, peptone, meat extract, and cell homogenates of *Bacteria* (e.g., *Thermotoga maritima*) or *Archaea* (e.g., *Staphylothermus marinus*) in the presence of sulfur, sulfite, thiosulfate, L(−)-cystine, or oxidized glutathione. DL-Lanthionine, fumarate, tetrathionate, dimethyl sulfone, and sulfate are not used as electron acceptors. No growth on galactose, glucose, maltose, starch, glycogen, formamide, methanol, ethanol, Casamino acids, formate, malate, propionate, L(+)-lactate, or acetate.

   Highly thermostable glutamate dehydrogenase present, which requires NAD^+ as a coenzyme for L-glutamate deamination (Kujo and Ohshima, 1998). A dissimilatory siro-

heme-sulfite-reductase-type protein has been identified, which has an $\alpha_2\beta_2$ structure and contains high-spin siroheme, nonheme iron and acid-labile sulfide (Molitor et al., 1998).

   Contains two new tRNA nucleosides, 3-hydroxy-*N*-[[9-β-D-ribofuranosyl-9*H*-purin-6-yl amino]carbony]-norvaline and 3-hydroxy-*N*-[[9-β-D-ribofuranosyl-9*H*-2-methyl-thiopurin-6-yl)amino]-carbonyl]-norvaline (Reddy et al., 1992).

   65% DNA similarity to *P. organotrophum*.
   The mol% G + C of the DNA is: 45 (T_m) and 47 (HPLC).
   Type strain: GEO3, DSMZ 4184.
   GenBank accession number (16S rRNA): L07511.

2. **Pyrobaculum aerophilum** Völkl, Huber and Stetter 1996, 836[VP] (Effective publication: Völkl, Huber and Stetter *in* Völkl, Huber, Drobner, Rachel, Burggraf, Trincone and Stetter 1993, 2924).
   a.e.ro' phi.lum. Gr. n. *aero* air; Gr. masc. adj. *philos* loving; M.L. neut. adj. *aerophilum* air-loving, because of its ability to use oxygen for growth in contrast to all other described species of the genus *Pyrobaculum*.

Nutrition is described in the generic description and Table A1.1.

Anaerobically grown cell masses exhibit a deep green color; aerobically grown cell masses show a brownish yellow color and exhibit a catalase-positive reaction.

A subtilisin-type protease ("aerolysin") with activity between 80 and 130°C was identified (Völkl et al., 1994). Cytochrome oxidases with novel prenylated hemes as cofactors are present (Lübben and Morand, 1994).

The mol% G + C of the DNA is: 52 (HPLC, T_m).

Type strain: IM2, DSMZ 7523.

GenBank accession number (16S rRNA): L07510.

3. **Pyrobaculum organotrophum** Stetter, Huber and Kristjansson 1988, 221[VP] (Effective publication: Stetter, Huber and Kristjansson *in* Huber, Kristjansson and Stetter 1987b, 101.)

or.ga.no.tro'phum. Gr. neut. n. *organon* tool (pertaining to organic chemical compounds in M.L.); Gr. masc. n. *trophos* one who feeds; M.L. neut. adj. *organotrophum* feeding on organic material.

Strictly heterotrophic. Growth on peptone, yeast extract, meat extract, and cell homogenates of *Bacteria* (e.g., *Thermotoga maritima*) or *Archaea* (e.g., *Staphylothermus marinus*) in the presence of sulfur, L(−)-cystine or oxidized glutathione. DL-Lanthionine, tetrathionate, thiosulfate, sulfite, dimethyl sulfone, fumarate, or sulfate is not used as electron acceptors. No growth on glycogen, starch, maltose, galactose, glucose, Casamino acids, formamide, methanol, ethanol, formate, malate, propionate, L(+)-lactate, or acetate.

65% DNA similarity to *P. islandicum*.

The mol% G + C of the DNA is: 46 (T_m) and 48 (HPLC).

Type strain: H10, DSMZ 4185.

Genus IV. **Thermocladium** *Itoh, Suzuki and Nakase 1998, 886*[VP]

ANNA-LOUISE REYSENBACH

Ther.mo.cla' di.um. Gr. adj. *thermos* hot; Gr. dim. n. *cladion* small branch; M.L. neut. n. *Thermocladium* a hot twig, indicating branching cells in a hot environment.

Cells are **straight or slightly curved rods about 0.5 × 5–20 μm**. Spherical swollen protrusions (1–3 μm diameter) may exist at the ends of cells. **Nonmotile**. Multiply by **branching and budding** and can divide by constriction. **Gram-negative. Anaerobic or microaerobic heterotrophic thermophile** growing between **60 and 80°C and pH 3.0 and 5.9**. Best growth occurs at 75°C at pH 4.2. Possesses at least five tetraether core lipids with cyclopentane rings in the isopranyl chains. Sensitive to rifampicin but not to chloramphenicol, kanamycin, or oleandomycin.

The mol% G + C of the DNA is: 52.

Type species: **Thermocladium modestius** Itoh, Suzuki and Nakase 1998, 886.

FURTHER DESCRIPTIVE INFORMATION

The cells of *Thermocladium* resemble those of *Thermoproteus* in shape and size. The branching and "golf club" appearance produced by the spherical bodies at the ends of the cells are consistent with other members of the *Thermoproteales*. The cells are very variable in length and frequently can reach lengths of 50 μm. In early stationary growth phase, up to 20% of the cells may be branched. The cell membranes contain no fatty acids, but large amounts of at least five different types of tetraether lipids and small amounts of diether lipids.

Thermocladium is an obligate heterotroph that grows best on yeast extract and gelatin. The isolate grows moderately on glycogen, starch, beef extract, Casamino acids, malt extract, peptone, and tryptone. Sulfur, thiosulfate, and L-cystine are good electron acceptors, and little growth is observed with sulfate, nitrate, or FeCl$_2$. A vitamin solution or archaeal extract enhances growth. The organism can grow at reduced oxygen concentration.

Strains of the organism were isolated from thermal springs, solfataric soils, and muds from nine different solfataric areas in Japan, including Noji-onsen, Fukushima; Kawayu-onsen, Hokkaido; and Tamagawa-onsen, Akita.

ENRICHMENT AND ISOLATION PROCEDURES

Thermocladium can be isolated using the medium for *Sulfolobus*, supplemented with 0.02% yeast extract and 0.5% sulfur and at pH 5. An H$_2$/CO$_2$ (4:1 v/v) gas headspace is used and the medium is reduced with Na$_2$S·9H$_2$0. Growth is observed after a week at 70°C and the enrichments are often mixed rods and cocci. Initial purifications of the strains were achieved by serial dilutions.

DIFFERENTIATION OF THE GENUS *THERMOCLADIUM* FROM OTHER GENERA

Based on DNA–rRNA cross-hybridization, *Thermocladium* is phylogenetically distinct from its relative *Thermoproteus tenax*. In addition, phylogenetic analysis using 16S rRNA sequences groups *Thermocladium* with *Thermoproteus* and *Pyrobaculum*, but as a distinct lineage. The organism branches more frequently and it has a lower optimal temperature for growth (75°C) than the other members of the *Thermoproteaceae*.

TAXONOMIC COMMENTS

Although *Thermocladium* has many features in common with the members of the *Thermoproteaceae*, such as branching and cell size, its phylogenetic position places it as a distinct lineage, much like the *Thermofilaceae*. As more isolates are obtained it is possible *Thermocladium* may be reclassified into a family of its own, although DNA–DNA hybridization data do not support such a distinction.

DIFFERENTIATION OF THE SPECIES OF THE GENUS *THERMOCLADIUM*

Only one species is described, the type species *Thermocladium modestius*.

1. **Thermocladium modestius** Itoh, Suzuki and Nakase 1998, 886.[VP]

mo.des' ti.us. L. comp. adj. *modestior* (neut. *modestius*) referring to relatively modest temperature growth range.

See generic description of features.

Grows optimally at 75°C, pH 4.2, and under H_2/CO_2 gas phase. Utilizes wide range of carbon sources, but grows best with 0.1–0.2% yeast extract. Sulfur, thiosulfate, L-cystine, or sulfate can be used as electron acceptor and sulfide is produced during growth. Growth stimulated by CO_2 in the gas phase and by archaeal cell extract or a vitamin mixture.

The mol% G + C of the DNA is: 52 (HPLC).

Type strain: IC-125, JCM 10088.

GenBank accession number (16S rRNA): AB005296.

Additional Remarks: Two additional strains are JCM 10089 and 10090.

Family II. **Thermofilaceae** Burggraf, Huber and Stetter 1997b, 659[VP]

HARALD HUBER AND KARL O. STETTER

Ther.mo.fi.la' ce.ae. M.L. neut. n. *Thermofilum* type genus of the family; *-aceae* ending to denote a family; M.L. fem. pl. n. *Thermofilaceae* the *Thermofilum* family.

Thin rods, 0.15–0.35 μm in diameter and 1 to >100 μm in length. Obligately anaerobic. Gram negative. **Temperature optimum 85–90°C. Sulfur respiration with complex organic substrates.**

See also description of the order for features.

Type genus: **Thermofilum** Zillig and Gierl 1983, 673 (Effective publication: Zillig and Gierl *in* Zillig, Gierl, Schreiber, Wunderl, Janekovic, Stetter and Klenk 1983a, 86).

Genus I. **Thermofilum** *Zillig and Gierl 1983, 673[VP] (Effective publication: Zillig and Gierl in Zillig, Gierl, Schreiber, Wunderl, Janekovic, Stetter and Klenk 1983a, 86)*

WOLFRAM ZILLIG AND ANNA-LOUISE REYSENBACH

Ther.mo.fi' lum. Gr. fem. n. *therme* heat; L. neut. n. *filum* thread, filament; M.L. neut. n. *Thermofilum* filament existing in a hot environment.

Cells 0.15–0.35 μm in diameter and 1 to >100 μm in length with terminal pili. Only **rarely branched or with sharp bends**; often with terminal spherical protrusions on both ends; sometimes with swollen sections; ghosts become spiral shaped. Gram negative.

Anaerobic. Thermoacidophilic. Utilizes peptides by sulfur respiration.

Lives in solfataric hot springs at pH 2.8–6.7 and temperatures up to 100°C. Optimum pH for growth is ~5. Grows best at 85–90°C.

Type species: **Thermofilum pendens** Zillig and Gierl 1983, 673 (Effective publication: Zillig and Gierl *in* Zillig, Gierl, Schreiber, Wunderl, Janekovic, Stetter and Klenk 1983a, 86).

1. **Thermofilum pendens** Zillig and Gierl 1983, 673[VP] (Effective publication: Zillig and Gierl *in* Zillig, Gierl, Schreiber, Wunderl, Janekovic, Stetter and Klenk 1983a, 86).

pen' dens. L. neut. part. adj. *pendens* depending; growth depends on a factor from *Thermoproteus* species.

Description is the same as for the genus, except that it requires a polar lipid component of *Thermoproteus tenax* for growth. See Fig. A1.6.

The mol% G + C of the DNA is: 57.6.

Type strain: Hrk 5, ATCC 35544, DSMZ 2475.

GenBank accession number (16S rRNA): X14835.

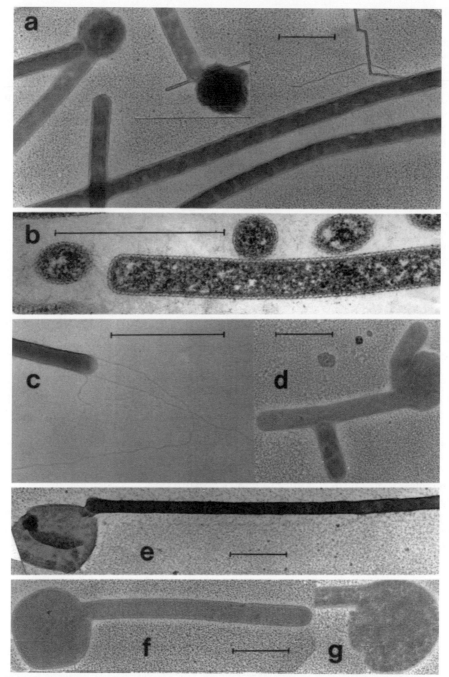

FIGURE A1.6. Electron micrographs of *Thermofilum pendens*. *a*, normal rods and "golf club" forms, rotary-shadowed with platinum. *b*, normal rods, longitudinal sections, and cross-sections, doubly contrasted with lead citrate and uranyl acetate. *c*, end of rod with pili. *d*, branching or budding. *e*, golf club form. *f* and *g*, spiralized forms. *c-g*, rotary shadowed with platinum. Bars = 1 μm. (Micrographs courtesy of D. Janekovic.)

Order II. **Desulfurococcales** *ord. nov.*

HARALD HUBER AND KARL O. STETTER

De.sul.fu.ro.coc.ca' les. M.L. masc. n. *Desulfurococcus* type genus of the order; *-ales* ending to denote an order; M.L. fem. pl. n. *Desulfurococcales* the order of *Desulfurococcus.*

Cells are regular to irregular coccoid to disc-shaped with 0.2–5 μm diameters. Septa have not been encountered. **Gram-negative.** Anaerobic, facultatively anaerobic or aerobic. **Hyperthermophilic** (optimal growth temperature from 85 to 106°C). **Growth either** chemolithoautotrophically by H_2 oxidation with S^0, nitrate, nitrite, or thiosulfate, and CO_2 as sole carbon source, **or organotrophically by sulfur respiration of various organic substrates or fermentation**. Cell envelope composed of an S-layer of protein or glycoprotein subunits; devoid of muramic acid. The major core lipids are glycerol dialkylglycerol tetraethers and 2,3-di-*O*-phytanyl-*sn*-glycerol. Transcription is resistant to rifampicin and streptolydigin. The DNA-dependent RNA polymerases are of the BAC type.

16S rRNA sequence analysis (Burggraf et al., 1997b) revealed that the *Desulfurococcales* are one of three orders within the *Crenarchaeota* (Fig. A1.7).

Widely distributed mainly in submarine hydrothermal systems but also in solfataric hot springs. The order comprises two validly published families, the *Desulfurococcaceae* and the *Pyrodictiaceae*.
The mol% G + C of the DNA is: 35–67.

Type genus: **Desulfurococcus** Zillig and Stetter 1983, 438 (Effective publication: Zillig and Stetter *in* Zillig, Stetter, Prangishvili, Schäfer, Wunderl, Janekovic, Holz and Palm 1982, 315.)

Key to the families of the order Desulfurococcales

1. Coccoid to disc-shaped cells; optimal growth temperature between 85 and 95°C, temperature maximum up to 102°C.
   Family I. *Desulfurococcaceae*, p. 180
2. Coccoid to disc-shaped cells, sometimes within a network of hollow tubules (cannulae); optimal growth temperature between 95 and 106°C, temperature maximum 108–113°C.
   Family II. *Pyrodictiaceae*, p. 191

Family I. **Desulfurococcaceae** Zillig and Stetter 1983, 438,[VP] emend. Burggraf, Huber and Stetter 1997b, 659 (Effective publication: Zillig and Stetter *in* Zillig, Stetter, Prangishvili, Schäfer, Wunderl, Janekovic, Holz and Palm 1982, 315)

HARALD HUBER AND KARL O. STETTER

De.sul.fu.ro.coc.ca' ce.ae. M.L. masc. n. *Desulfurococcus* type genus of the family; *-aceae* ending to denote a family; M.L. fem. pl. n. *Desulfurococcaceae* the family of *Desulfurococcus*.

Coccoid to disc-shaped cells. Hyperthermophilic, optimal growth temperature between 85 and 95°C, maximal growth temperature up to 102°C. Anaerobic or aerobic (only one genus). **Chemolithoautotrophic growth by sulfur reduction to H_2S using CO_2 as sole carbon source. Heterotrophic growth by sulfur respiration of various organic substrates or by fermentation**. See also description of the order for features.

So far, the family harbors six validly published genera: *Desul-furococcus, Staphylothermus, Aeropyrum, Thermosphaera, Stetteria*, and *Sulfophobococcus*. Two further genera also belong to this family, *Thermodiscus* and *Ignicoccus* (Stetter, 1986a; Burggraf et al., 1997b).

Type genus: **Desulfurococcus** Zillig and Stetter 1983, 438 (Effective publication: Zillig and Stetter *in* Zillig, Stetter, Prangishvili, Schäfer, Wunderl, Janekovic, Holz and Palm 1982, 315).

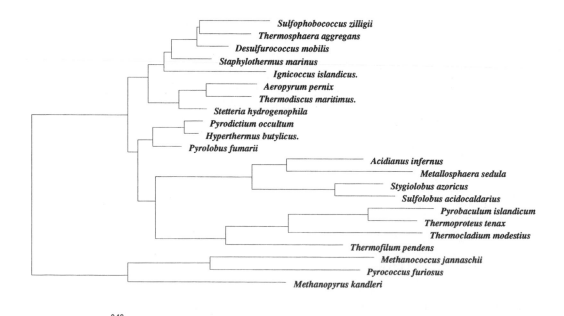

FIGURE A1.7. Phylogenetic tree of the type species of all crenarchaeotal genera. The tree was calculated by the maximum parsimony method. Bar = 10 substitutions per hundred nucleotides.

Key to the genera of the family Desulfurococcaceae

1. Cells are regular cocci. Obligately anaerobic; sulfur respiration with complex organic substrates, fermentation of peptides. Temperature optimum 85–90°C; temperature maximum 97°C. The mol% G + C of genomic DNA is 42–51. Thrives in continental solfataric areas.

   Genus I. *Desulfurococcus*, p. 181

2. Cells are irregular cocci. Obligately aerobic; respiration of complex organic substrates with oxygen. Temperature optimum 90–95°C; temperature maximum 100°C. The mol% G + C of genomic DNA is 67. Lives in marine habitats.

   Genus II. *Aeropyrum*, p. 183

3. Cells are regular cocci. Anaerobic; grow only chemolithotrophically by S^0/H_2 autotrophy. Temperature optimum 90°C; temperature maximum 100°C. The mol% G + C of genomic DNA is from 41 to 45. Marine.

   Genus III. *Ignicoccus*, p. 184

4. Cells are cocci in aggregates. Obligately anaerobic; organotrophic growth by fermentation of peptides. Temperature optimum 92°C; temperature maximum 98°C. The mol% G + C of genomic DNA is 35. Lives in marine environments.

   Genus IV. *Staphylothermus*, p. 186

5. Cells are irregular to disc-shaped cocci. Obligately anaerobic; sulfur respiration with complex organic substrates, H_2 obligately required. Temperature optimum 95°C; temperature maximum 102°C. The mol% G + C of genomic DNA is 65. Marine.

   Genus V. *Stetteria*, p. 187

6. Cells are regular cocci occurring singly or in aggregates. Anaerobic; growth on organic substrates by fermentation. Inhibition by elemental sulfur. Temperature optimum 85°C; temperature maximum 95°C. The mol% G + C of genomic DNA is around 55. Terrestrial.

   Genus VI. *Sulfophobococcus*, p. 188

7. Cells are disc-shaped. Obligately anaerobic; sulfur respiration with complex organic substrates. Temperature optimum 88°C; temperature maximum 98°C. The mol% G + C of genomic DNA is 49. Marine.

   Genus VII. *Thermodiscus*, p. 189

8. Cells are cocci, occurring singly, in pairs, short chains, or grapelike aggregates. Anaerobic; growth on organic substrates by fermentation. Inhibition by elemental sulfur or molecular hydrogen. Temperature optimum 85°C; temperature maximum 90°C. The mol% G + C of genomic DNA is 46. Lives in solfataric hot springs.

   Genus VIII. *Thermosphaera*, p. 190

Genus I. **Desulfurococcus** *Zillig and Stetter 1983, 438^VP (Effective publication: Zillig and Stetter* in *Zillig, Stetter, Prangishvili, Schäfer, Wunderl, Janekovic, Holz and Palm 1982, 316)*

WOLFRAM ZILLIG

De.sul.fu.ro.coc' cus. L. pref. *de* from; L. n. *sulfur* sulfur; Gr. n. *koccos* berry; M.L. masc. n. *Desulfurococcus* sulfur-reducing coccus.

Cells cocci. Stains Gram-negative. See Fig. A1.8 for appearance of cells.

Anaerobic. Utilizes protein, peptides, or carbohydrates facultatively by sulfur respiration or fermentation.

Cell envelope flexible, composed of subunits. Lipids, RNA polymerase, and resistance to antibiotics are the same as those described for the order.

Type species: **Desulfurococcus mucosus** Zillig and Stetter 1983, 438 (Effective publication: Zillig and Stetter *in* Zillig, Stetter, Prangishvili, Schäfer, Wunderl, Janekovic, Holz and Palm 1982, 316).

List of species of the genus Desulfurococcus

1. **Desulfurococcus mucosus** Zillig and Stetter 1983, 438^VP (Effective publication: Zillig and Stetter *in* Zillig, Stetter, Prangishvili, Schäfer, Wunderl, Janekovic, Holz and Palm 1982, 316).

   mu.co' sus. L. masc. adj. *mucosus* slimy.

   The description is the same as that for the genus with the addition that it forms a strongly smelling unknown product and produces a slimy polymer attached to its surface. Nonmotile. Optimum pH for growth: 6. Optimum temperature for growth: 85°C. Sugars are not used.

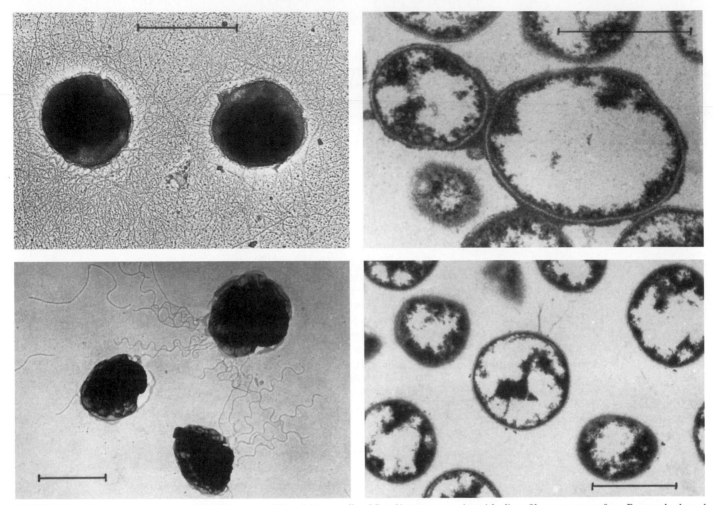

FIGURE A1.8. Electron micrographs of *Desulfurococcus*. *Upper left*, two cells of *Desulfurococcus* species with slime filaments on surface. Rotary-shadowed with platinum. *Upper right*, thin section of *D. mucosus* with solid slime layer attached to envelope. *Lower left*, three cells of *D. mobilis*, with one showing monopolar polytrichous flagellation. Rotary-shadowed with platinum. *Lower right*, thin section of *D. mobilis* showing apparatus to which flagella are attached within the cell. Bars = 1 μm. (Micrographs courtesy of D. Janekovic.)

Lives in solfataric hot springs at pH 2.2–6.5 and up to 97°C.

The mol% G + C of the DNA is: 51.3.

Type strain: 07, ATCC 35584, DSMZ 2162.

2. **Desulfurococcus mobilis** Zillig and Stetter 1983, 438^{VP} (Effective publication: Zillig and Stetter *in* Zillig, Stetter, Prangishvili, Schäfer, Wunderl, Janekovic, Holz and Palm 1982, 316).

mo'bi.lis. L. masc. adj. *mobilis* motile.

Description is identical with *D. mucosus*, except that cells are motile and do not produce slime. In addition, the 23S rRNA gene contains an intron (Larsen et al., 1986). Sugars are not used.

The mol% G + C of the DNA is: 50.

Type strain: Hvv3, ATCC 35582, DSMZ 2161.

3. **"Desulfurococcus amylolyticus"** Bonch-Osmolovskaya, Slesarev, Miroshnichenko, Svetlichnaya and Alekseev 1988, 100.

a.my.lo.ly'ti.cus. L. masc. adj. *amylolyticus* starch degrading.

Irregular cocci, 0.7–1.5 μm in diameter with noncellular appendages ~1 μm long. Hyperthermophile, grows best at 90–92°C. Obligately anaerobic heterotroph. Utilizes peptides, amino acids, starch, glycogen, and pectin. Growth stimulated by S^0. Produces sulfide. Isolated from Kamchatka and Kunashir Island, Russia.

The mol% G + C of the DNA is: 41.2 (HPLC and T_m).

Deposited strain: Z-533, DSMZ 3822.

4. **"Desulfurococcus saccharovorans"** Stetter 1986a, 59.

Description is identical with *D. mucosus* except that sugars are utilized (e.g., 0.5% glucose in the presence of 0.02% yeast extract).

Genus II. **Aeropyrum** Sako, Nomura, Uchida, Ishida, Morii, Koga, Hoaki and Maruyama 1996, 1075[VP]

YOSHIHIKO SAKO AND NORIMICHI NOMURA

Ae.ro.py'rum. Gr. n. *aer* air; Gr. neut. n. *pyr* fire; M.L. masc. n. *Aeropyrum* air fire, referring to the hyperthermophilic respirative character of the organism.

Cells are **coccoid**, and highly irregular in shape. Cells are usually about 0.8–1.0 μm in diameter. Cells are surrounded by a cell envelope (S-layer-like structure) about 25 nm wide covering the cytoplasmic membrane (Fig. A1.9). Gram negative. **Highly motile**. Cells are frequently surrounded by pililike appendages. **Strictly aerobic**, with molecular oxygen used as the terminal electron acceptor. **Heterotrophic hyperthermoneutrophiles**. Growth occurs between 70 and 100°C (optimum, 90–95°C), at pH 5–9 (optimum, pH 7.0), and at 1.8–7.0% salinity (optimum, 3.5% salinity). Various proteinaceous complex compounds serve as substrates during aerobic respiration. On the basis of 16S rRNA analysis, the genus *Aeropyrum* is closely related to the members of the families *Pyrodictiaceae* and *Desulfurococcaceae*.

The mol% G + C of the DNA is: 56.

Type species: **Aeropyrum pernix** Sako, Nomura, Uchida, Ishida, Morii, Koga, Hoaki and Maruyama 1996, 1076.

FURTHER DESCRIPTIVE INFORMATION

Cells are resistant to ampicillin, vancomycin, and cycloserine and sensitive to chloramphenicol. Only di-sesterterpanyl ($C_{25}C_{25}$) glycerol ether lipids are present in the cell membrane. No fatty acid is detected. The 23S rRNA of *Aeropyrum pernix* shares a secondary structural feature with those of the members in the *Pyrodictiaceae* and *Desulfurococcaceae* that might be a defining signature. The helix at positions 1445–1465 (*E. coli* numbering) in the 23S rRNA secondary structure characterized by a G bulge six nucleotides 5′ from the capping loop and an AG bulge six nucleotides 3′ from the capping loop. This motif was not found elsewhere in the *Archaea* domain. The complete genome sequence has been determined (Kawarabayashi et al., 1999). *A. pernix* possesses two family B DNA polymerases (Cann et al., 1999).

ENRICHMENT AND ISOLATION PROCEDURES

Enrichment of *Aeropyrum pernix* can be achieved by using "JX" medium* (Sako et al., 1996). After 2 days of incubation at 90°C, coccoid organisms become visible. After the cell density of an enrichment culture reaches approximately 10^7 cells/ml, pure cultures can be obtained by a dilution-to-extinction technique in "JXT" medium, prepared by adding 0.1% (w/v) $Na_2S_2O_3 \cdot 5H_2O$ to JX medium.

MAINTENANCE PROCEDURES

Cultures of *Aeropyrum pernix* are routinely grown for 3 days at 90°C and then transferred into fresh JXT medium. Stock cultures can be stored at −80°C for at least 6 months, without transfer.

DIFFERENTIATION OF THE GENUS *AEROPYRUM* FROM OTHER GENERA

On the basis of its hyperthermophily and strict aerobic mode of life, *Aeropyrum* is distinct from other archaea. Several members within the *Sulfolobales* and *Thermoplasmatales* are able to grow aerobically, depending on the culture conditions. *Sulfolobus* spp. are irregular coccoid shape, and capable of growing under aerobic conditions, either heterotrophically, or chemolithoautotrophically by the oxidation of S^0 (Segerer and Stetter, 1989a). Among them, *Sulfolobus solfataricus* DSMZ 1616[T] grows optimally at 87°C (Zillig et al., 1980a). *Acidianus* and *Desulfurolobus* spp. are

---

*"JX" medium consists of (per liter of artificial seawater or filtrated natural seawater): yeast extract, 1g; and peptone, 1g; pH 7.0–7.2 (adjusted with HCl at room temperature).

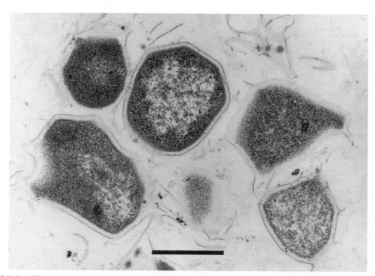

FIGURE A1.9. Electron micrograph of an ultrathin section of *A. pernix* strain K1. Bar = 0.5 μm.

facultatively aerobic S^0 oxidizers (Zillig et al., 1986b; Segerer and Stetter, 1989b). *Metallosphaera sedula* is a strict aerobe, either by heterotrophy or chemolithoautotrophy (Huber et al., 1989a). *Thermoplasma* spp. are facultative aerobes (Langworthy and Smith, 1989). However, all of them have distinctly lower mol% G+C contents, which range from 31 to 46, and grow optimally below pH 4.0, whereas *Aeropyrum pernix* has a mol% G + C content of 56 and grows within a pH range from 5.0 to 9.0. Furthermore, this genus is clearly differentiated from any known member of the *Archaea* by the idiosyncratic lipid composition and the low level of similarity of the 16S rRNA sequence.

ACKNOWLEDGMENTS

We wish to thank Tadaaki Yoshida, Kureha Chemical Industry Co. Ltd., Japan, for his help in ultrathin sectioning.

FURTHER READING

Sako, Y., N. Nomura, A. Uchida, Y. Ishida, H. Morii, Y. Koga, T. Hoaki and T. Maruyama. 1996. *Aeropyrum pernix* gen. nov., sp. nov., a novel aerobic hyperthermophilic archaeon growing at temperatures up to 100°C. Int. J. Syst. Bacteriol. *46*: 1070–1077.

List of species of the genus Aeropyrum

1. **Aeropyrum pernix** Sako, Nomura, Uchida, Ishida, Morii, Koga, Hoaki and Maruyama 1996, 1076.[VP]

   per′ nix. L. masc. adj. *pernix* nimble, indicating high motility in microscopic inspection.

   Cells are irregular, coccoid. Vigorous motility is evident by light microscopy either at room temperature or at 90°C. Packed cells are brown. Growth occurs between 70 and 100°C (optimum, 90–95°C), at pH 5–9 (optimum, pH 7.0), and at 1.8–7.0% salinity (optimum, 3.5% salinity). Cells lyse by low osmotic shock (below 1.5% salinity). Batch cultures are grown in cotton-plugged 2-liter Erlenmeyer flasks containing 500 ml of JXT medium and incubated at 90°C in an air bath rotary shaker with vigorous shaking (180 rpm). Thiosulfate stimulates growth. Oxygen serves as a possible electron acceptor. During growth, H_2S is not produced either in the presence or absence of thiosulfate. By 16S rRNA sequence comparisons, the evolutionary distances (estimated changes per 100 nucleotides) to *Pyrodictium occultum* and *Desulfurococcus mobilis* are 7.9 and 11.0, respectively. The single copy of the rRNA operon (16S-23S) of *Aeropyrum pernix* strain K1 contains three introns (Nomura et al., 1998). The core lipids consist solely of di-sesterterpanyl ($C_{25}C_{25}$) glycerol ether lipids. The polar lipids consist of five components, but only two are predominant: phosphoglycolipid (archaetidyl [glucosyl] inositol; 91%) containing inositol and glucose, and phospholipid (archaetidyl inositol; 9%) (Morii et al., 1999).

   The mol% G + C of the DNA is: 56 (genome sequencing).
   Type strain: K1, JCM 9820.
   GenBank accession number (16S rRNA): AB008745, D83259.

Genus III. **Ignicoccus** Huber, Burggraf and Stetter 2000, 2098[VP]

HARALD HUBER AND KARL O. STETTER

I′g.ni.coc′ cus. L. masc. n. *ignis* fire; Gr. masc. n. *koccos* berry; M.L. masc. n. *Ignicoccus* the fireball.

Irregular cocci, about 1–3 µm in diameter; monopolar polytrichous flagella. Gram negative. **Occurring singly and in pairs**. Cell envelope consists of cytoplasmic membrane, a periplasm (20–350 nm wide), and a sheath resembling an outer membrane. Cells contain phytanyl di- and tetraether lipids. No rigid sacculus. Murein, pseudomurein, or S-layer absent. **Strictly anaerobic. Chemolithoautotrophic growth in the presence of H_2 and CO_2 with sulfur as electron acceptor**. Sulfate, sulfite, thiosulfate, tetrathionate, nitrate, and oxygen are not used as electron acceptors. **H_2S formed during growth. Growth between 70 and 98°C**, pH optimum ~6, NaCl optimum ~2%. Ampicillin, rifampicin, and vancomycin resistant. Growth stimulated by addition of meat extract or tryptone (0.1% w/v).

Based on 16S rDNA sequence comparison, member of the family *Desulfurococcaceae*, within the phylum *Crenarchaeota* (Burggraf et al., 1997b).

The mol% G + C of the DNA is: 41–45.
Type species: **Ignicoccus islandicus** Huber, Burggraf and Stetter 2000, 2099.

FURTHER DESCRIPTIVE INFORMATION

Cells are irregular in shape. The cytoplasm is densely packed and surrounded by a membrane (~6–8 nm wide). The periplasm has a variable width, ranging from 20 to 350 nm. It contains numerous round or elongated vesicles, 50–60 nm in diameter and up to 300 nm in length, each surrounded by a membrane. In sections, the outermost part of the cell envelope occasionally has a weak double-layer appearance. This "sheath" is frequently found to be fractured into two leaflets in freeze–etch/freeze–fracture experiments and, therefore, most likely represents a (lipid) membrane. A two-dimensional crystalline arrangement, as observed in S-layer sheets of most *Archaea* and many *Bacteria*, is not observed. The core lipids of isolate Kol8 are acyclic 2,3-di-O-phytanyl-*sn*-glycerol and glyceroldialkyl glycerol tetraether in a relative ratio of about 1:1.

The organisms are obligate anaerobes, growing chemolithoautotrophically by sulfur reduction using molecular hydrogen as an electron donor. In the presence of H_2 and S^0 growth is stimulated by the addition of meat extract, tryptone (each 0.1%), and glucose (0.05%). No growth is observed on organic substrates such as meat extract, yeast extract, peptone, Casamino acids, gelatin, starch, formate, acetate, and glucose, when cultures are pressurized with hydrogen-free gas (N_2/CO_2; 80:20; 200 kPa). Sulfur cannot be replaced by oxygen, thiosulfate, tetrathionate, sulfite, sulfate, or nitrate as electron acceptor. Cells lyse at NaCl concentrations in the medium above 6.0%.

Ignicoccus was isolated from hot sediments at the Kolbeinsey Ridge north of Iceland and from active black smoker walls from the East Pacific.

ENRICHMENT AND ISOLATION PROCEDURES

Ignicoccus can be enriched anaerobically in ½ SME medium (Stetter et al., 1983; Pley et al., 1991) at pH 5.5 using elemental sulfur and a H_2/CO_2 atmosphere (80:20; 300 kPa) at 90°C. After 2–3 days irregular cocci appear and high amounts of H_2S can be detected.

Ignicoccus can be isolated by serial dilution or by using optical tweezers (Huber et al., 1995a).

MAINTENANCE PROCEDURES

For long-term preservation (at least 3 years), cell cultures containing 5% DMSO should be stored over liquid nitrogen ($-140°C$) or freeze dried.

DIFFERENTIATION OF THE GENUS *IGNICOCCUS* FROM OTHER GENERA

Ignicoccus belongs to the archaeal family of the *Desulfurococcaceae* due to the coccoid cell shape, the lack of a cell sacculus, the negative Gram reaction, the optimal growth temperature of around 90°C, and its phylogenetic position based on analysis of 16S rDNA sequences. They can be differentiated from the other members of the *Desulfurococcaceae* by their unique cell wall architecture (Baumeister and Lembcke, 1992) and the extremely limited spectrum of utilized electron donors and acceptors: molecular hydrogen is the only electron donor, and no acceptors other than S^0 can be used. Therefore, the organisms are obligate hydrogen/sulfur autotrophs, producing H_2S. Some organic components stimulate growth by decreasing the shortest doubling times and yielding higher final cell densities. However, they cannot be used as sole energy sources. In contrast, all other anaerobic members of the *Desulfurococcaceae* known so far grow either by fermentation of sugars or complex organic compounds (e.g., *Staphylothermus*, *Sulfophobococcus*, and *Thermosphaera*) or by sulfur respiration of organic compounds, yielding H_2S in addition to organic acids or alcohols (e.g. *Desulfurococcus*, *Thermodiscus*, or *Stetteria*) (Huber and Stetter, in press*). This physiological separateness is confirmed by 16S rDNA sequence analyses, where the new isolates exhibit phylogenetic distances of at least 6% to any other representative of the *Desulfurococcaceae*.

TAXONOMIC COMMENTS

On the basis of 16S rDNA sequence analysis, the *Ignicoccus* species represent a deep branch within the *Desulfurococcaceae*. They exhibit phylogenetic distances of between 6% and 10% to any of the other members of this family (Huber et al., in press).

FURTHER READING

Burggraf, S., H. Huber and K.O. Stetter. 1997. Reclassification of the crenarchaeal orders and families in accordance with 16S rRNA sequence data. Int. J. Syst. Bacteriol. *47*: 657–660.

Huber, H., S. Burggraf, T. Mayer, I. Wyschkony, M. Biebl, R. Rachel and K.O. Stetter. 2000. *Ignicoccus* gen. nov., a novel genus of hyperthermophilic, chemolithoautotrophic *Archaea*, represented by two new species *Ignicoccus islandicus* sp. nov. and *Ignicoccus pacificus* sp. nov. Int. J. Syst. Evol. Microbiol. *50*: (in press).

DIFFERENTIATION OF THE SPECIES OF THE GENUS *IGNICOCCUS*

Two species of the genus *Ignicoccus* exist: *I. islandicus*, and *I. pacificus*.

Key to the species of the genus *Ignicoccus*

1. Growth temperature from 70°C and 98°C; at NaCl concentrations from 0.3% to 5.5%. The mol% G + C content of the DNA is 41.

   I. islandicus.

2. Growth temperature between 75°C to 98°C; at NaCl concentrations from 1.0% to 5.0%. The mol% G + C content of the DNA is 45.

   I. pacificus.

List of species of the genus Ignicoccus

1. **Ignicoccus islandicus** Huber, Burggraf and Stetter 2000, 2099.[VP]

   is.la' n.di.cus. M.L. masc. adj. *islandicus* Icelandic, describing the location of its first isolation.

   Slightly irregular cocci, about 1.2–3 μm in diameter; monopolar polytrichous flagella. Gram negative. Occurring singly and in pairs. No rigid sacculus; murein or pseudomurein and S-layer absent. Growth from 70 to 98°C, pH 3.8–6.5, and 0.3–5.5% NaCl; optima: 90°C, pH 5.8, 2% NaCl. Strictly anaerobic. Chemolithoautotrophic; growth in the presence of H_2 and CO_2 with sulfur as an electron acceptor. No chemoorganotrophic growth on meat extract, yeast extract, peptone, Casamino acids, gelatin, starch, formate, acetate, and glucose. Sulfate, sulfite, thiosulfate, nitrate, and oxygen are not used as electron acceptors. H_2S (up to 25 μmol/ml) is formed during growth. Growth stimulated by addition of 0.1% (w/v) meat extract, tryptone, or glucose (0.05%, w/v). No significant DNA–DNA similarity to *Ignicoccus pacificus*. Isolated from the Kolbeinsey Ridge, north of Iceland.

   See also generic description for features.

   The mol% G + C of the DNA is: 41 (T_m and direct analysis).

   Type strain: Kol8, DSMZ 13165.

   GenBank accession number (16S rRNA): X99562.

2. **Ignicoccus pacificus** Huber and Stetter 2000, 2099.[VP]

   pa.ci' fi.cus. M.L. masc. adj. *pacificus* from the Pacific, describing the site of its first isolation.

   Slightly irregular cocci, ~1–2 μm in diameter; monopolar polytrichous flagella. Gram negative. Occurring singly and in pairs. No rigid sacculus; murein or pseudomurein and S-layer absent. Growth from 75 to 98°C, pH 4.5–7.0, and 1.0–5.0% NaCl; optima: 90°C, pH 6, 2% NaCl. Strictly anaerobic. Chemolithoautotrophic growth in the presence of H_2 and CO_2 with sulfur as an electron acceptor. No chemoorganotrophic growth on meat extract, yeast extract, peptone, Casamino acids, gelatin, starch, formate, acetate, and glucose.

Editorial Note: Encyclopedia of Life Sciences will be published in 2001. The chapter on *Crenarchaeota* may be viewed at www.els.net.

Sulfate, sulfite, thiosulfate, tetrathionate, and nitrate are not used as electron acceptors. H₂S formed during growth. Ampicillin, rifampicin, and vancomycin resistant. No significant DNA–DNA similarity to *Ignicoccus islandicus*. Isolated from Black Smoker samples at 9°N and 104°W in the Pacific.

See also generic description for features.
The mol% G + C of the DNA is: 45 (T_m).
Type strain: LPC33, DSMZ 13166.
GenBank accession number (16S rRNA): AJ271794.

Genus IV. **Staphylothermus** Stetter and Fiala 1986, 573[VP] (Effective publication: Stetter and Fiala in Fiala, Stetter, Jannasch, Langworthy and Madon 1986, 112)

KARL O. STETTER

Sta.phy.lo.ther'mus. Gr. fem. n. *staphyle* bunch of grapes; Gr. fem. n. *therme* heat; M.L. masc. n. *Staphylothermus* grape (-forming) thermophile.

Cells slightly irregular cocci occurring singly, in pairs, as short chains, and as aggregates of up to 100 cells. Width about 0.5–15 μm, depending on the culture conditions (Fig. A1.10). **Strictly anaerobic. Heterotrophic growth in the presence of sulfur on peptone, tryptone, yeast extract, meat extract, and extracts of bacteria or archaea. H₂S, CO₂, acetate, and isovalerate are formed as metabolic products. Hyperthermophilic.** Optimum nutrition (0.1% yeast extract [w/v] + 0.5% peptone [w/v]), optimum temperature is 92°C and maximum, 98°C. In minimal medium (0.02% yeast extract [w/v] + 0.004% *Methanosarcina* extract [v/v]), optimum temperature is 85°C and maximum, 92°C. Resistant to vancomycin, kanamycin, streptomycin, and chloramphenicol. Gram negative. Cell envelope strongly stained by ruthenium red. Elongation factor 2 is ADP-ribosylated by diphtheria toxin. Phytanyl glycerol diether, dibiphytanyl diglycerol tetraether and a yet unknown ether lipid component are present.

The mol% G + C of the DNA is: 35.

Type species: **Staphylothermus marinus** Stetter and Fiala 1986, 573 (Effective publication: Stetter and Fiala in Fiala, Stetter, Jannasch, Langworthy and Madon 1986, 112.)

FURTHER DESCRIPTIVE INFORMATION

At low substrate concentrations (0.005–0.02% yeast extract [w/v]), large aggregates are mainly formed; at high substrate concentrations (0.1% yeast extract [w/v] + 0.5% peptone [w/v]), single cells, pairs and, very rarely, aggregates of up to 5 cells can be seen. Cells are 0.5–1 μm in diameter. At high yeast extract concentrations (0.2%), giant cells with diameters of up to 15 μm appear. Each contains one to a few highly contrasted dark granules (seen by phase-contrast microscopy) which most likely consist of glycogen, due to their brownish staining reaction with the Lugol reagent (J-KJ solution). The S-layer is formed by a poorly ordered meshwork of branched, filiform morphological subunits resembling dandelion seed heads ("tetrabrachion") (Peters et al., 1995).

In the presence of oxygen, the titer of viable cells is reduced by 50% within 7 min at 85°C; at 4°C the half-life of survival was about 20 h, thus indicating that the organism is extremely sensitive to oxygen during growth.

S. marinus was isolated from anaerobic samples taken from

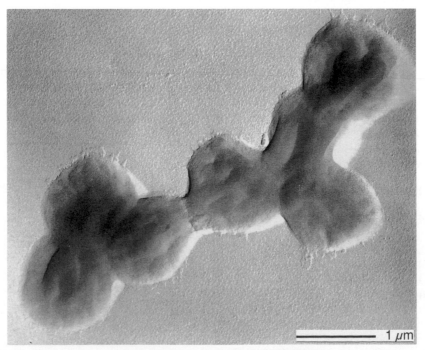

FIGURE A1.10. Electron micrograph of an aggregate of cells of *Staphylothermus marinus*. Platinum shadowed.

the beach of Vulcano, Italy, and from an abyssal "black smoker" vent at the East Pacific Rise.

ENRICHMENT AND ISOLATION PROCEDURES

S. marinus can be enriched anaerobically at 85°C in SME* medium (Stetter et al., 1983) supplemented with yeast extract and peptone. After about one week, coccoid cells in aggregates often appear. Purification by repeated serial dilutions in fresh medium or from single cells selected under the laser microscope ("optical tweezers") (Huber et al., 1995a).

MAINTENANCE PROCEDURES

Cultures of *S. marinus* are grown for 3 days at 85°C. Storage is possible anaerobically at 4°C, conditions at which cells remain viable for at least 2 years. For long-term storage, in the presence of 5% DMSO, 20 μl portions of cultures are kept frozen in capillaries over liquid nitrogen at −140°C.

DIFFERENTIATION OF THE GENUS *STAPHYLOTHERMUS* FROM OTHER GENERA

The genus *Staphylothermus* differs from all other genera of the *Desulfurococcaceae* (Burggraf et al., 1997b) by its 16S rRNA sequence and its very low mol% G + C content of 35. From the other genera adapted to marine environments (*Aeropyrum, Stetteria, Thermodiscus, Ignicoccus*) and from the terrestrial genus *Desulfurococcus*, *Staphylothermus* is differentiated by its growth in ag-

gregates. From the terrestrial *Sulfophobococcus* and *Thermosphaera*, in addition to its 16S rRNA sequence, *Staphylothermus* is differentiated by its sulfur requirement.

TAXONOMIC COMMENTS

By 16S rRNA total sequencing, *S. marinus* is a member of the *Desulfurococcaceae* and represents a deeply branching lineage that presumably is not rapidly evolving based on the short branch length (Kjems et al., 1992; Burggraf et al., 1997b).

FURTHER READING

Burggraf, S., H. Huber and K.O. Stetter. 1997. Reclassification of the crenarchaeal orders and families in accordance with 16S rRNA sequence data. Int. J. Syst. Bacteriol. *47*: 657–660.

Fiala, G., K.O. Stetter, H.W. Jannasch, T.A. Langworthy and J. Madon. 1986. *Staphylothermus marinus*, sp. nov., represents a novel genus of extremely thermophilic submarine heterotrophic archaebacteria growing up to 98°C. Syst. Appl. Microbiol. *8*: 106–113.

Kjems, J., N. Larsen, J.Z. Dalgaard, R.A. Garrett and K.O. Stetter. 1992. Phylogenetic relationships amongst the hyperthermophilic archaea determined from partial 23S rRNA gene sequences. Syst. Appl. Microbiol. *15*: 203–208.

Peters, J., M. Nitsch, B. Kühlmorgen, R. Golbik, A. Lupas, J. Kellermann, H. Engelhardt, J.P. Pfander, S. Müller, K. Goldie, A. Engel, K.O. Stetter and W. Baumeister. 1995. Tetrabrachion: a filamentous archaebacterial surface protein assembly of unusual structure end extreme stability. J. Mol. Biol. *245*: 385–401.

DIFFERENTIATION OF THE SPECIES OF THE GENUS *STAPHYLOTHERMUS*

Only one species, *S. marinus*, is currently known.

List of species of the genus Staphylothermus

1. **Staphylothermus marinus** Stetter and Fiala 1986, 573[VP] (Effective publication: Stetter and Fiala *in* Fiala, Stetter, Jannasch, Langworthy and Madon 1986, 112.)
ma.ri' nus. L. masc. adj. *marinus* of the sea.

See the generic description for the features.
The mol% G + C of the DNA is: 35 (T_m).
Type strain: F1, ATCC 49053, DSMZ 3639.
GenBank accession number (16S rRNA): X99560.

Genus V. **Stetteria** Jochimsen, Peinemann-Simon, Völker, Stüben, Botz, Stoffers, Dando and Thomm 1998, 328[VP] (Effective publication: Jochimsen, Peinemann-Simon, Völker, Stüben, Botz, Stoffers, Dando and Thomm 1997, 72)

ANNA-LOUISE REYSENBACH

Stet.te' ria. M.L. fem. *stetteria* of Stetter, named after Karl Otto Stetter.

Gram-negative irregular and disc-shaped cocci about 0.5–1.5 μm in diameter. Motile with a single flagellum. S-layer with protein subunits arranged in hexagonal array. Heterotrophic growth by oxidizing complex organics (yeast extract, peptone, tryptone, Casamino acids, and cell-free extracts of *Thermococcus celer* or *Pyrococcus furiosus*) in the presence of H_2 and S^0 or thiosulfate. H_2S is produced during growth with S^0. CO_2 stimulates growth. Growth between 68 and 102°C and at pH 1.0–5.8, best at 95°C, pH 6. The optimal NaCl concentration for growth is 2–3.5% (range 0.5–6%). Isolated from shallow marine submarine hydrothermal springs at Paleohori Bay in Milos, Greece.
The mol% G + C of the DNA is: 65.

Type species: **Stetteria hydrogenophila** Jochimsen, Peinemann-Simon, Völker, Stüben, Botz, Stoffers, Dando and Thomm 1998, 328 (Effective publication: Jochimsen, Peinemann-Simon, Völker, Stüben, Botz, Stoffers, Dando and Thomm 1997, 72).

FURTHER DESCRIPTIVE INFORMATION

Cells of *Stetteria* form cytoplasmic protrusions during stationary growth phase, when grown on low sulfur (0.5%), or if sulfur is replaced by thiosulfate. Growth of the isolate is obligately dependent upon S^0 and hydrogen. Produces H_2S. The isolate grows mixotrophically on peptide substrates in the presence of H_2. Acetate and ethanol are among the products of metabolism when grown in SME* medium. In the absence of CO_2 in the medium, the isolate forms CO_2 during growth.

Stetteria has only been isolated from the geothermally active area at Paleohori Bay at Milos along the South Aegean volcanic arc.

---

*"SME" medium consists of (per liter): NaCl, 13.85 g; MgSO₄·2H₂O, 3.5 g; MgCl₂·6H₂O, 2.75 g; CaCl₂·2H₂0, 1.0 g; KCl, 0.325 g; NaBr, 0.05g; H₃BO₃, 0.01 5 g; SrCl₂·6H₂0, 0.0075 g; KH₂PO₄, 0.5 g; KI, 0.025 mg; NiNH₄SO₄, 0.002 g; sulfur, 25 g; resazurine, 0.001 g; and trace minerals (Balch et al., 1979), 10 ml.

ENRICHMENT AND ISOLATION PROCEDURES

Stetteria can be enriched from geothermally heated sediments in the brine seep area of Paleohori Bay at a depth of 10 m. Half-strength SME-medium (Stetter et al., 1983) was supplemented with 2% sulfur and the salt concentration adjusted to the concentration measured for the upper layer of the brine seep. The enrichment was incubated at 95°C for 2 weeks. Pure cultures were isolated by serial dilution.

DIFFERENTIATION OF THE GENUS *STETTERIA* FROM OTHER GENERA

Stetteria can be distinguished from other members of the *Desulfurococcaceae* primarily based on its 16S rRNA sequence, DNA–DNA hybridization, and its obligate requirement for hydrogen for growth. The isolate differs from its closest relative, *Staphylothermus*, which has a mol% G + C content of 35, whereas *Stetteria* has a mol% G + C content of 65. Additionally, unlike *Stetteria*,

Staphylothermus forms cell aggregates of up to 100 cells and it does not require hydrogen for growth. *Stetteria* has a similar mol% G + C to that of *Pyrodictium*; however, they differ in physiology and morphology. *Pyrodictium* can grow in the absence of hydrogen, and it is able to grow on carbohydrates, acetate, or formate. *Pyrodictium* cells form characteristic ultrathin fibers that are absent in *Stetteria*. The genera *Hyperthermus*, *Thermodiscus*, and *Desulfurococcus* differ from *Stetteria* in their ability to grow in the absence of hydrogen, and in their mol% G + C content.

TAXONOMIC COMMENTS

The DNA–DNA hybridization and 16S rRNA phylogeny clearly place this isolate within the *Desulfurococcaceae* as a new genus.

FURTHER READING

Stetter, K.O., H. König and E. Stackebrandt. 1983. *Pyrodictium* gen. nov., a new genus of submarine disc-shaped sulphur reducing archaebacteria growing optimally at 105°C. Syst. Appl. Microbiol. *4*: 535–551.

List of species of the genus Stetteria

1. **Stetteria hydrogenophila** Jochimsen, Peinemann-Simon, Völker, Stüben, Botz, Stoffers, Dando and Thomm 1998, 328[VP] (Effective publication: Jochimsen, Peinemann-Simon, Völker, Stüben, Botz, Stoffers, Dando and Thomm 1997, 72). *hy.dro.ge.no.phi' la.* M.L. neut. n. *hydrogenium* hydrogen; Gr. adj.

philos loving; M.L. fem. adj. *hydrogenophila* like hydrogen since growth depends upon on hydrogen.

The description is the same as that of the genus.
The mol% G + C of the DNA is: 65 (HPLC).
Type strain: 4ABC, DSMZ 11227.
GenBank accession number (16S rRNA): Y07784.

Genus VI. **Sulfophobococcus** Hensel, Matussek, Michalke, Tacke, Tindall, Kohlhoff, Siebers and Dielenschneider 1997b, 915[VP] (Effective publication: Hensel, Matussek, Michalke, Tacke, Tindall, Kohlhoff, Siebers and Dielenschneider 1997a, 109)

ANNA-LOUISE REYSENBACH

Sul.fo.pho.bo.coc' cus. L. neut. n. *sulfur* sulfur; Gr. v. *phobein* to fear, to avoid; Gr. masc. n. *koccos* grain; *Sulfophobococcus* the sulfur fearing coccus.

Cells are **regular or slightly irregular cocci ~3–5 μm in diameter** with a **tuft of filaments approximately 40 μm in length**. Occur **singly or as aggregates**. No murein present in the cell wall. **Strict heterotrophic anaerobe** that grows between **70 and 95°C at pH 6.5–8.5**. Growth is inhibited by sulfur. Membrane lipids are only dibiphytanylglycerol tetraethers.

The mol% G + C of the DNA is: 54–56.

Type species: **Sulfophobococcus zilligii** Hensel, Matussek, Michalke, Tacke, Tindall, Kohlhoff, Siebers and Dielenschneider 1997b, 915 (Effective publication: Hensel, Matussek, Michalke, Tacke, Tindall, Kohlhoff, Siebers and Dielenschneider 1997a, 109).

FURTHER DESCRIPTIVE INFORMATION

The cells often occur singly or as aggregates. The role of the tufts of filaments remains unanswered. These filaments may play a role in motility or adhesion to particles.

Under heterotrophic conditions the cells are completely inhibited by S^0. Sulfite, thiosulfate, diothionate and L(-)-cysteine do not affect growth rates. Sulfate partially supports growth. The organism can grow on yeast extract, peptone, Casamino acids, glucose, ethanol, acetate, xylan, and cellulose. At the optimal growth temperature (87°C) and pH (pH 7.5), the doubling time is 4.5 h.

The isolate was obtained from hot alkaline springs in Iceland.

Antibodies raised to whole cells of the isolate were used to determine the distribution of *Sulfophobococcus* in 52 different hot springs in Iceland. Maximal cell titers were obtained in springs of pH 8.0–9.5 and at temperatures of 85–90°C.

ENRICHMENT AND ISOLATION PROCEDURES

Sulfophobococcus can be isolated in a mineral medium supplemented with 0.1% yeast extract. Isolates are obtained by growth of successive serial dilutions of the enrichment in the same medium.

DIFFERENTIATION OF THE GENUS *SULFOPHOBOCOCCUS* FROM OTHER GENERA

Based on 16S rRNA phylogeny, *Sulfophobococcus* groups with *Pyrodictium* and *Desulfurococcus* (differs by 7.0 and 4.5%, respectively). However, the isolate differs from the described *Desulfurococcus* species in that *Sulfophobococcus* only uses yeast extract as an organic substrate and cannot use tryptone or casein. Additionally, growth of all *Desulfurococcus* isolates is stimulated by S^0, whereas *Sulfophobococcus* growth is inhibited by sulfur. The desulfurococci are inhibited by rifampicin, whereas *Sulfophobococcus* is not sensitive to this antibiotic. *Sulfophobococcus* has a higher pH optimum for growth (pH 7.5) than *Desulfurococcus* (pH 6.0). Furthermore, chemotaxonomic features distinguish this organism from other *Crenarchaeota*. The isolate does not produce any respiratory lipoquinones (present in genera of *Thermoproteus* and *Pyrobaculum*, and members of the *Sulfolobales*). *Sulfophobococcus* contains predominantly tetraether lipids, with no evidence of pentacyclic rings associated with the tetraethers.

TAXONOMIC COMMENTS

The 16S rRNA phylogeny groups *Sulfophobococcus* with *Desulfurococcus*. A 95.5% similarity in sequence suggests that the isolate should be incorporated as a member of the *Desulfurococcales*. However, the distinctive physiological and chemotaxonomic differences of these two genera will require a reassessment of the description of the *Desulfurococcales*. Furthermore, the *Sulfophobococcus* 16S rRNA contains a number of deviations from the known crenarchaeotal sequences at positions where all *Crenarchaeota* have conserved bases. Examples are at positions 45/470, 1017/1028, and 1052/1088.

DIFFERENTIATION OF THE SPECIES OF THE GENUS *SULFOPHOBOCOCCUS*

Only one species is described; the type species *Sulfophobococcus zilligii*.

List of species of the genus Sulfophobococcus

1. **Sulfophobococcus zilligii** Hensel, Matussek, Michalke, Tacke, Tindall, Kohlhoff, Siebers and Dielenschneider 1997b, 915^VP (Effective publication: Hensel, Matussek, Michalke, Tacke, Tindall, Kohlhoff, Siebers and Dielenschneider 1997a, 109). *zil.lig′ i.i.* M.L. gen. n. *zilligii* of Zillig, in honor of Wolfram Zillig.

    See generic description of features.

    Regular to slightly irregular cocci (3–5 μm in diameter) with tufts of filaments. Forms aggregates and adsorbs to amorphous particles in the medium. Optimal growth at 85°C and pH 7. Strictly anaerobic fermentative growth on yeast extract. No growth on starch, peptone, Casamino acids, glucose, ethanol, acetate, xylan, or cellulose. Growth inhibited by S^0.

    The mol% G + C of the DNA is: 54.7 ± 0.8 (T_m).

    Type strain: K1, DSMZ 11193, JCM 10309.

    GenBank accession number (16S rRNA): X98064.

Genus VII. Thermodiscus gen. nov.

KARL O. STETTER

Ther.mo.dis′ cus. Gr. fem. n. *therme* heat; L. masc. n. *discus* disc; M.L. masc. n. *Thermodiscus* the hot disc.

Cells are irregular dish- to disc-shaped cocci, varying in diameter from about 0.3 to 3 μm (Fig. A1.11). The discs show a **thickness of only about 0.1–0.2 μm**. Cells occur singly. Gram negative. No cannulae formed. Pili-like structures about 0.01 μm in diameter and up to 15 μm long present, sometimes connecting two individual cells. Flagella and motility absent. Cell envelope composed of protein subunits in hexagonal array, about 33 nm in diameter. Isoprenoid ether-linked lipids are present (Langworthy, personal communication). **Strictly anaerobic. Optimum growth temperature: 90°C**; maximum: 98°C; minimum: 75°C.

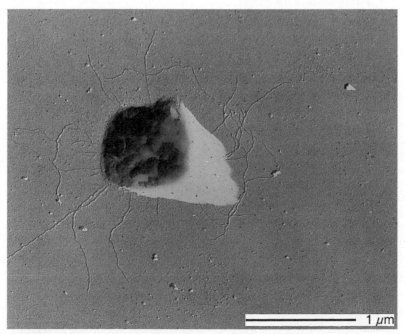

FIGURE A1.11. Electron micrograph of a *Thermodiscus* cell. Platinum shadowed.

Growth occurs at pH 5–7; optimum pH ~5.5. Optimum NaCl concentration: ~2%. **Heterotrophic. H₂S may be formed by S⁰ reduction.**

> *The mol% G + C of the DNA is:* 49.
>
> *Type species:* **Thermodiscus maritimus** sp. nov.

FURTHER DESCRIPTIVE INFORMATION

Growth by sulfur respiration in artificial seawater (Stetter et al., 1983) in the presence of S^0 and 0.02% yeast extract. H_2 may stimulate growth (in some isolates). Sometimes growth by fermentation of yeast extract, in the presence and absence of S^0 (Stetter, 1986a). The cell envelope contains large amounts of a periodate-Schiff-positive protein with an apparent molecular weight of about 84,000 kDa. Crude extracts carry a protein which is ADP-ribosylated by diphtheria toxin (Stetter, 1986a).

ENRICHMENT AND ISOLATION PROCEDURES

Thermodiscus can be enriched by using SME medium* supplemented with 0.02% yeast extract and S^0 (Stetter et al., 1983) under strict anaerobic conditions. Incubation at pH 5.5 at 90°C under shaking (gas phase: H_2/CO_2; 80%:20%; 3 bar). *Thermodiscus* can be isolated by repeated serial dilution. Alternatively, pure cultures can be obtained from single cells using optical tweezers (Huber et al., 1995a).

MAINTENANCE PROCEDURES

Cultures of *Thermodiscus* can be stored anaerobically at 4°C for at least one year. For long-term storage, cells are suspended in 5% DMSO, 20 µl portions are kept frozen at −140°C in capillaries over liquid nitrogen.

DIFFERENTIATION OF THE GENUS *THERMODISCUS* FROM OTHER GENERA

Thermodiscus belongs to the *Desulfurococcaceae* based on its 16S rRNA sequence and metabolic properties. The genus *Thermodiscus* differs from all other genera of the *Desulfurococcaceae* by its 16S rRNA sequence (Burggraf et al., 1997b) and its disk-shaped cell morphology (Stetter, 1986a).

TAXONOMIC COMMENTS

By 16S rRNA sequence comparisons, *Thermodiscus* is among the *Desulfurococcaceae* and is specifically related to *Aeropyrum pernix* (Burggraf et al., 1997b).

FURTHER READING

Burggraf, S., H. Huber and K.O. Stetter. 1997. Reclassification of the crenarchaeal orders and families in accordance with 16S rRNA sequence data. Int. J. Syst. Bacteriol. *47:* 657–660.

Stetter, K.O. 1986. Diversity of extremely thermophilic archaebacteria. *In* Brock (Editor), Thermophiles: General, Molecular and Applied Microbiology, John Wiley & Sons, New York, pp. 39–74.

DIFFERENTIATION OF THE SPECIES OF THE GENUS *THERMODISCUS*

Only one species, *Thermodiscus maritimus*, is currently known.

List of species of the genus Thermodiscus

1. **Thermodiscus maritimus** sp. nov.

   ma.ri′ ti.mus. L. masc. adj. *maritimus* belonging to the sea; describing its habitat.

   See the generic description for the features.

   Isolated from a hot marine sediment at the beach of Vulcano Island, Italy.

   The mol% G + C of the DNA is: 49 (T_m and direct analysis).

   Deposited strain: S2.

   GenBank accession number (16S rRNA): X99554.

Genus VIII. **Thermosphaera** *Huber, Dyba, Huber, Burggraf and Rachel 1998b, 36*[VP]

ROBERT HUBER AND KARL O. STETTER

Ther.mo.sphae′ ra. Gr. fem. n. *therme* heat; L. fem. n. *sphaera* sphere; L. fem. n. *Thermosphaera* the hot sphere.

Coccoid cells with a diameter of 0.2–0.8 µm. They occur singly, in pairs, in short chains and in **grapelike cell aggregates** up to 10 µm in diameter with several 100 individuals (Fig. A1.12). Cell envelope composed of a cytoplasmic membrane covered by an amorphous layer. **No surface layer protein detectable. Acyclic and cyclic glycerol diphytanyl tetraethers** in the membrane. No spores formed. Flagella present. **Obligate anaerobe. Hyperthermophilic**, optimum growth temperature: 85°C; maximum: 90°C; minimum: 67°C. Optimal pH: 6.5–7.2. **Heterotrophic growth** on yeast extract, peptone, gelatin, amino acids, heat-treated xylan, amino acids, and glucose. **H₂, CO₂, isovalerate, and acetate formed as fermentation products**. Molecular hydrogen or **sulfur inhibits growth**.

By 16S rDNA analysis, *Thermosphaera* represents a new genus within the *Desulfurococcaceae*.

The mol% G + C of the DNA is: 46.

Type species: **Thermosphaera aggregans** Huber, Dyba, Huber, Burggraf and Rachel 1998b, 36.

FURTHER DESCRIPTIVE INFORMATION

By 16S rDNA sequence comparisons, *Thermosphaera aggregans* was identified as a member of the *Crenarchaeota* within the *Archaea* domain (Woese et al., 1990). Within this phylogenetic branch, *T. aggregans* belongs to the *Desulfurococcaceae* (Huber et al., 1995a; Burggraf et al., 1997b) with closest relationships to the genera *Desulfurococcus* (Zillig et al., 1982), *Staphylothermus* (Fiala et al., 1986) and *Sulfophobococcus* (Hensel et al., 1997a).

The cell aggregates of *T. aggregans* are very rigid and cannot be disintegrated into single cells by enzyme treatment (cellulase, proteinase K, trypsin), supersonic treatment, or mechanical stirring. The aggregates exhibit a weak bluish-green fluorescence at 420 nm, fading rapidly under UV radiation.

---

*"SME" medium consists of (per liter): NaCl, 13.85 g; MgSO₄·2H₂O, 3.5 g; MgCl₂·6H₂O, 2.75 g; CaCl₂·2H₂O, 1.0 g; KCl, 0.325 g; NaBr, 0.05g; H₃BO₃, 0.01 5 g; SrCl₂·6H₂O, 0.0075 g; KH₂PO₄, 0.5 g; KI, 0.025 mg; NiNH₄SO₄, 0.002 g; sulfur, 25 g; resazurine, 0.001 g; and trace minerals (Balch et al., 1979), 10 ml.

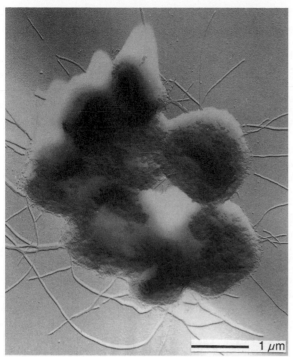

FIGURE A1.12. Platinum-shadowed cell aggregate of *Thermosphaera aggregans.*

T. aggregans grows optimally under anaerobic conditions at 85°C, pH 6.5, and in the absence of exogenous sodium chloride. The optimal doubling time at 85°C is 110 min. The apparent activation energy for growth is about 149 kJ^{-1}. No growth on meat extract, bovine heart infusion, peptone, amylose, glycogen, cellulose, cellobiose, maltose, raffinose, pyruvate, and acetate.

T. aggregans was isolated from Obsidian Pool, a terrestrial hot spring in Yellowstone National Park, WY, USA.

ENRICHMENT AND ISOLATION PROCEDURES

T. aggregans was originally enriched and obtained in pure culture by a newly developed procedure, which allowed the isolation of a 16S rDNA sequence-predicted, hyperthermophilic archaeum from a natural environment for the first time. This procedure is a combination of *in situ* 16S rDNA sequence analysis, specific cell hybridization within enrichment cultures, and "selected cell cultivation" by the use of a laser microscope ("optical tweezers"; Barns et al., 1994; Huber et al., 1995a; Beck and Huber, 1997).

MAINTENANCE PROCEDURES

T. aggregans can be stored in liquid nitrogen at −140°C in the presence of 5% DMSO.

DIFFERENTIATION OF THE GENUS *THERMOSPHAERA* FROM OTHER GENERA

Based on 16S rDNA sequence data, *T. aggregans* can be distinguished from the genera *Staphylothermus*, *Desulfurococcus*, and *Sulfophobococcus*. *T. aggregans* can be further distinguished from *Sulfophobococcus* on the basis of different conserved bases in the 16S rDNA sequence. *T. aggregans* differs from *Desulfurococcus* and *Staphylothermus* by the lack of significant DNA similarity, the presence of cyclic tetraether lipids in its membrane and the absence of a regular cell surface lattice.

FURTHER READING

Huber, R., S. Burggraf, T. Mayer, S.M. Barns, P. Rossnagel and K.O. Stetter. 1995. Isolation of a hyperthermophilic archaeum predicted by *in situ* RNA analysis. Nature (Lond.) *376*: 57–58.

Stetter, K.O. 2000. Volcanoes, hydrothermal venting, and the origin of life. *In* Marit and Ernst (Editors), Volcanoes and the Environment, Cambridge University Press, Cambridge, in press.

List of species of the genus Thermosphaera

1. **Thermosphaera aggregans** Huber, Dyba, Huber, Burggraf and Rachel 1998b, 36.[VP]
   ag'gre.gans. L. v. *aggregare* referring to the ability of the cells to form grapelike aggregates.

   Description is the same as for the genus.
   The mol% G + C of the DNA is: 46 (T_m).
   Type strain: M11TL, DSMZ 11486.
   GenBank accession number (16S rRNA): X99556.

Family II. **Pyrodictiaceae** Burggraf, Huber and Stetter 1997b, 659[VP]

HARALD HUBER AND KARL O. STETTER

Pyr.o.dic' ti.a.ce.ae. M.L. neut. n. *Pyrodictium* type genus of the family; -*aceae* ending to denote a family; M.L. fem. pl. n. *Pyrodictiaceae* the *Pyrodictium* family.

Coccoid to disc-shaped cells, *Pyrodictium* **species form a network of cannulae. Hyperthermophilic,** maximal growth temperature between 108 and 113°C. **Grows either chemolithoautotrophically by gaining energy from the reduction of S⁰ or thiosulfate to H₂S** using CO_2 as sole carbon source **or by fermentation.** Some genera gain energy by respiration using O_2 or nitrate as electron acceptors. Three genera are described: *Pyrodictium*, *Hyperthermus* and *Pyrolobus*.

Type genus: **Pyrodictium** Stetter, König and Stackebrandt 1984, 270, emend. Pley and Stetter *in* Pley, Schipka, Gambacorta, Jannasch, Fricke, Rachel and Stetter 1991, 251 (Effective publication: Stetter, König and Stackebrandt 1983, 549).

Key to the genera of the family Pyrodictiaceae

1. Cells are discs within a network of cannulae. Obligately anaerobic; H₂/S⁰ autotrophy and sulfur respiration with complex organic substrates. Temperature optimum: 105°C; temperature maximum: 110°C.

   Genus I. *Pyrodictium.*

2. Cells are irregular cocci. Obligately anaerobic; growth by fermentation of complex organic material. Temperature optimum between 95 and 106°C; temperature maximum 108°C.

Genus II. *Hyperthermus*, p. 195

3. Cells are irregular cocci. Facultatively anaerobic; obligately chemolithoautotrophic; growth by H_2 oxidation with S^0, thiosulfate or (low concentrations of) oxygen. Temperature optimum 106°C; temperature maximum 113°C.

Genus III. *Pyrolobus*, p. 196

Genus I. **Pyrodictium** *Stetter, König and Stackebrandt 1984, 270,[VP] emend. Pley and Stetter in Pley, Schipka, Gambacorta, Jannasch, Fricke, Rachel and Stetter 1991, 251 (Effective publication: Stetter, König and Stackebrandt 1983, 549)*

KARL O. STETTER

Pyr.o.dic′ti.um. Gr. neut. n. *pyr* fire; Gr. neut. n. *diktyon* network; M.L. neut. n. *Pyrodictium* fire-loving network.

Cells disk- to dish-shaped, highly variable in diameter, ranging from 0.3 to 2.5 µm, frequently with ultraflat areas, about 0.1–0.2 µm thick (Fig. A1.13). **Produce tubule-shaped structures ("cannulae")**, 0.025 µm thick, **which form networks connecting the cells** (Fig. A1.14). Strictly and facultatively **chemolithotrophic**. H_2S is formed from molecular sulfur and hydrogen. No mobility observed. Cell envelope composed of protein subunits (Fig. A1.15) in hexagonal array. Isopranoid di- and tetraether lipids present (Langworthy and Pond, 1986a). Elongation factor 2 is ADP-ribosylated by diphtheria toxin (F. Klink, personal communication). **Strictly anaerobic. Optimum temperature: 97–105°C; maximum: 110°C**; minimum: 80–82°C. Growth occurs at pH 5–7; optimum pH: ~5.5. Optimum NaCl concentration: ~1.5%. Vigorous shaking prevents cell growth.

The mol% G + C of the DNA is: 60–62.

Type species: **Pyrodictium occultum** Stetter, König and Stackebrandt 1984, 270, emend. Pley and Stetter *in* Pley, Schipka, Gambacorta, Jannasch, Fricke, Rachel and Stetter 1991, 251 (Effective publication: Stetter, König and Stackebrandt 1983, 549).

FURTHER DESCRIPTIVE INFORMATION

Cells are fragile and irregular in shape. They occur singly. Cell division by binary fission (Horn et al., 1999). The S-layer consists of subunits 21 nm in diameter. Cells can be disintegrated by sodium dodecyl sulfate (SDS, 1% w/v). The cell envelope of *Pyrodictium* contains a major protein that stains positive with the periodate-Schiff reagent and therefore is most likely a glycoprotein. The networks composed of cannulae and cells can be seen in the light microscope with high-intensity dark field illumination and in the electron microscope (Stetter, 1982a). The cannulae often form plectenchyme-like bundles. Cannulae are hollow cylinders with a diameter of about 25 nm, composed of glycoprotein subunits in helical array (Rieger et al., 1995). The thickness of the wall is 2–3 nm. A triple-layered unit membrane was not found. The cannulae are up to 150 µm in length (Horn et al., 1999). They cannot be disintegrated by SDS. Granules of sulfur are seen frequently, sticking to the surface of the cannulae. Precipitations of zinc and sulfur (Stetter et al., 1983), most likely zinc sulfide, can be seen on the cannulae, the same as on the cell envelope.

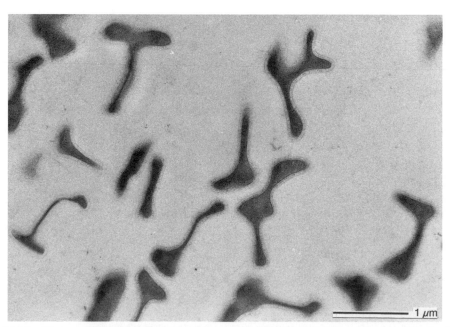

FIGURE A1.13. Electron micrograph of an ultrathin section of *Pyrodictium* cells, cultivated inside the lumen of cellulose capillary tubes (prepared by high-pressure freezing and freeze-substitution).

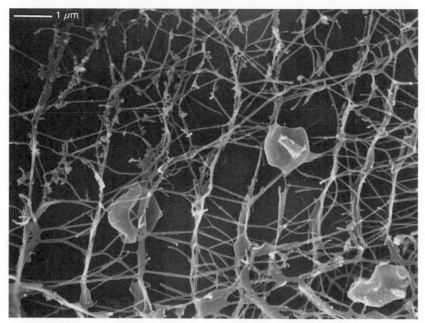

FIGURE A1.14.   Scanning electron micrograph of the network of *Pyrodictium* cells with their extracellular cannulae.

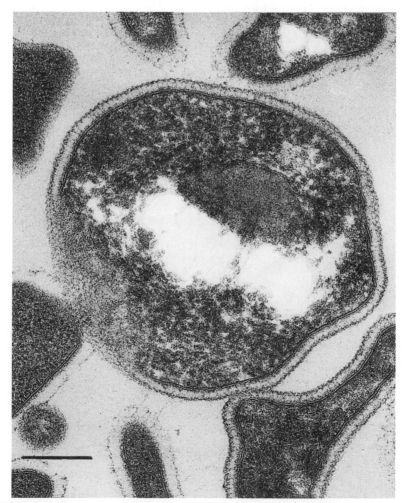

FIGURE A1.15.   Electron micrograph of a thin section of *P. occultum* strain Pl-19. Bar = 0.2 μm.

In addition to the cannulae, flagella-like appendages up to 5 μm long and 10 nm in diameter, protruding from the disks, can be seen (König et al., 1988; Rieger et al., 1995). Fine structures are well preserved when grown in cellulose capillary tubes (Rieger et al., 1997).

Pyrodictium possesses a heat-inducible molecular chaperone named thermosome (Phipps et al., 1993). During growth at 108°C it amounts to about 80% of the soluble protein in *Pyrodictium* (Phipps et al., 1991a). *Pyrodictium abyssi* contains a heat stable protein-serine/threonine phosphatase which is active between 40–110°C (Mai et al., 1998). An extremely thermostable membrane-bound sulfur-reducing enzyme complex has been characterized from *Pyrodictium abyssi* (Dirmeier et al., 1998).

Cell growth is extremely sensitive to oxygen. At 4°C, tolerance to oxygen is variable, depending on unknown factors. Growth of chemolithoautotrophic strains is slightly stimulated by yeast extract (0.02%, w/v) and citric acid (0.001%, w/v). In the fermentor, pyrite is formed during growth (Stetter et al., 1983). Significant growth of facultatively heterotrophic strains occurs on yeast extract, meat extract, peptone, casein, and glycogen as single energy source.

Pyrodictium was isolated from anaerobic sediment and rock samples taken from the hot seafloor and from the beach at Vulcano Island, Italy, and from abyssal black smoker vents at the Mid Atlantic Ridge and East Pacific Rise.

ENRICHMENT AND ISOLATION PROCEDURES

Pyrodictium can be enriched in SME medium* (Stetter et al., 1983; Pley et al., 1991) supplemented with sulfur and pressurized with H_2/CO_2 (80%:20%; 300 kPa) and incubated anaerobically at 100°C without shaking. After 1–3 days, a fluffy layer, strongly reminiscent of mold growth, appears above the sulfur. This layer sticks only very loosely to the sulfur and can be removed easily by gentle shaking. The enrichment cultures can be purified by thrice-repeated serial dilution in SME medium (Stetter et al., 1983) after vigorous shaking of the enrichment culture. No colonies are formed on medium solidified by agar, polysilicate, or starch, because of wetting of the surface during incubation. Pure cultures can be obtained from single cells isolated using optical tweezers (Huber et al., 1995a). Packed cells, originally grown in liquid medium, are white to gray. For growth in fermentors, the steel parts have to be protected, e.g., with enamel. Alternatively, titanium fermentors may be used (Stetter, unpublished).

MAINTENANCE PROCEDURES

Stock cultures of *Pyrodictium* can be stored anaerobically at 4°C for at least 2 years without transfer. Long-term storage is possible by suspending cells in 5% DMSO and aliquoting in 20-μl portions into capillaries at −140°C over liquid nitrogen. Cultures are transferred routinely into SME medium and incubated at 105°C.

DIFFERENTIATION OF THE GENUS *PYRODICTIUM* FROM OTHER GENERA

The genus *Pyrodictium* belongs to the *Pyrodictiaceae* within the *Archaea*, where it is unique because of its morphology, cannulae matrix, and mol% G + C content.

TAXONOMIC COMMENTS

The phylogenetic relationship of *P. occultum* has been determined by sequence analysis of the 16S and 23S rRNAs (Burggraf et al., 1997b). *Pyrodictium* groups with *Pyrolobus* and *Hyperthermus*, forming the shortest branch within the hyperthermophilic *Crenarchaeota* (Fig. A1.7).

FURTHER READING

Pley, U., J. Schipka, A. Gambacorta, H.W. Jannasch, H. Fricke, R. Rachel and K.O. Stetter. 1991. *Pyrodictium abyssi*, sp. nov. represents a novel heterotrophic marine archaeal hyperthermophile growing at 110°C. Syst. Appl. Microbiol. *14*: 245–253.

Stetter, K.O. 1982. Ultrathin mycelia-forming organisms from submarine volcanic areas having an optimum growth temperature of 105°C. Nature (Lond.) *300*: 258–260.

Stetter, K.O., H. König and E. Stackebrandt. 1983. *Pyrodictium* gen. nov., a new genus of submarine disc-shaped sulphur reducing archaebacteria growing optimally at 105°C. Syst. Appl. Microbiol. *4*: 535–551.

DIFFERENTIATION OF THE SPECIES OF THE GENUS *PYRODICTIUM*

Three species, *P. occultum*, *P. brockii*, and *P. abyssi* are currently described.

Key to the species of the genus *Pyrodictium*

1. Growth strictly H_2-dependent. Optimal growth temperature 105°C. The cell envelope contains a major periodate-Schiff-positive protein with a molecular weight of 172,000 kDa.

      P. occultum.

2. Growth heterotrophic. H_2 stimulates growth. Optimal growth temperature 97°C. The cell envelope contains a glycoprotein with a molecular weight of 126,000 kDa.

      P. abyssi.

3. Growth strictly H_2-dependent. Optimal growth temperature 105°C. The cell envelope contains a dominant periodate-Schiff-negative (or slightly positive) protein with a molecular weight of 150,000 kDa.

      P. brockii.

List of species of the genus Pyrodictium

1. **Pyrodictium occultum** Stetter, König and Stackebrandt 1984, 270,[VP] emend. Pley and Stetter *in* Pley, Schipka, Gambacorta, Jannasch, Fricke, Rachel and Stetter 1991, 251 (Effective publication: Stetter, König and Stackebrandt 1983, 549).

oc.cul′tum. L. neut. adj. *occultum* hidden, indicating the invisibility of the network in the phase-contrast microscope.

Cells polymorphous disks and dishes, about 0.3–2.5 μm in diameter and usually 0.2 μm thick. Networks of cannulae

*"SME" medium consists of (per liter): NaCl, 13.85 g; $MgSO_4·2H_2O$, 3.5 g; $MgCl_2·6H_2O$, 2.75 g; $CaCl_2·2H_2O$, 1.0 g; KCl, 0.325 g; NaBr, 0.05g; H_3BO_3, 0.01 5 g; $SrCl_2·6H_2O$, 0.0075 g; KH_2PO_4, 0.5 g; KI, 0.025 mg; $NiNH_4SO_4$, 0.002 g; sulfur, 25 g; resazurine, 0.001 g; and trace minerals (Balch et al., 1979), 10 ml.

formed. Growth by hydrogen–sulfur autotrophy or, in the presence of 0.02% yeast extract, on H_2/CO_2 and $S_2O_3^{2-}$. Optimal growth temperature around 105°C. Growth between 0.2 and 12% NaCl and pH 4.5 and 7.2. The cell envelope contains a major periodate-Schiff-positive protein with a molecular weight of 172,000 kDa.

The mol% G + C of the DNA is: about 62 (T_m and direct analysis).

Type strain: PL-19, DSMZ 2709.

GenBank accession number (16S rRNA): M21087.

2. **Pyrodictium abyssi** Pley and Stetter 1991, 580^{VP} (Effective publication: Pley and Stetter *in* Pley, Schipka, Gambacorta, Jannasch, Fricke, Rachel and Stetter 1991, 252).

ab.y'ss.i. L. gen. n. *abyssi* immense depth, living at great depth within the oceans.

Cells are polymorphous disk-shaped, mainly about 1–2 μm in diameter with flat protrusions. Growth in liquid cultures in colonies ("tiny white balls"), about 1 mm in diameter, consisting of cannulae (about 0.025 μm in width) and cells. Grayish flakes formed in the late growth phase. Cell envelope composed of regularly arrayed subunits, about 0.015 μm in diameter. The cell envelope contains a glycoprotein with a molecular weight of 126,000 kDa. Growth from 80 to 110°C (optimum: 97° C), 0.7 to 4.2% NaCl (optimum: ~2% NaCl), and pH 4.7 to 7.1 (optimum: pH 5.5). Strictly heterotrophic. Proteins, carbohydrates, cell extracts, acetate, and formate fermented. H_2 stimulates growth. Fermentation products (on yeast + meat extract + H_2) are isovalerate, isobutyrate, butanol, and CO_2. H_2S formed from S^0 or $S_2O_3^{2-}$ in the pres-

ence of H_2. Core lipids consist of glycerol-dibisphytanyl-glyceroltetraether and $C_{20}C_{20}$-diether. Isolated from the wall of an abyssal active "black smoker" off Guaymas, Mexico and from a submarine hot vent at the Kolbeinsey ridge north of Iceland.

The mol% G + C of the DNA is: about 60 (T_m and direct analysis).

Type strain: AV2, DSMZ 6158.

GenBank accession number (16S rRNA): X99559.

3. **Pyrodictium brockii** Stetter, König and Stackebrandt 1984, 270,^{VP} emend. Pley and Stetter *in* Pley, Schipka, Gambacorta, Jannasch, Fricke, Rachel and Stetter 1991, 252 (Effective publication: Stetter, König and Stackebrandt 1983, 549).

brock'i.i. M.L. gen. n. *brockii* of Brock; named for T. D. Brock for his pioneering work on the extreme thermophiles.

Cells polymorphous disks and dishes, about 0.3–2.5 μm in diameter and usually 0.2 μm thick. Networks of cannulae are formed. Growth by hydrogen-sulfur-autotrophy or, in the presence of 0.02% yeast extract, on H_2/CO_2 and SO_3^{2-}. No growth with $S_2O_3^{2-}$ as electron acceptor. In the presence of S^0 and H_2/CO_2 the growth yield is significantly stimulated by 0.2% yeast extract. Optimal growth temperature around 105°C. Growth at 0.2–12% NaCl and pH 4.5–7.2. The cell envelope contains a periodate-Schiff-negative or slightly positive protein, with a molecular weight of about 150,000 kDa as the dominant component.

The mol% G + C of the DNA is: about 62 (T_m and direct analysis).

Type strain: S1, DSMZ 2708.

Genus II. **Hyperthermus** *Zillig, Holz and Wunderl 1991, 170*^{VP}

ANNA-LOUISE REYSENBACH AND WOLFRAM ZILLIG

Hy.per.ther' mus. Gr. prep. *hyper* above; Gr. fem. n. *therme* heat; M.L. masc. n. *Hyperthermus* an organism existing in a very hot environment.

Cells are **irregular cocci** about 1.5 μm in diameter. Reproduce by constriction. Gram negative. Envelope is a highly regular hexagonal S-layer. Projections that resemble pili are present. **Obligately anaerobic heterotrophic hyperthermophile growing between 72 and 108°C.** Best growth occurs between 95 and 106°C at pH 7.0 and in the presence of 1.7% NaCl. Utilizes a mixture of peptides as carbon and energy sources.

The mol% G + C of the DNA is: 55.6.

Type species: **Hyperthermus butylicus** Zillig, Holz and Wunderl 1991, 170.

FURTHER DESCRIPTIVE INFORMATION

The cells of *Hyperthermus* resemble those of *Sulfolobus* in shape, but are slightly larger (1.5 μm). Clumping of the cells occurs if the cultures are not agitated. Many pili-like projections are present on the cell surface. The S-layer has a distinct hexagonal pattern. At high temperatures the cells often contain vacuoles, sometimes immediately below the S-layer. The boundary between the vacuoles and the cytoplasm is sharp, but is not a typical bilayered membrane. The predominant lipids are tetraethers. The RNA polymerase of *Hyperthermus* resembles that of other crenarchaeotes.

Hyperthermus is an obligate heterotroph that grows by fermentation of peptide mixtures including tryptone, Trypticase,

peptone, and gelatin. Growth on tryptone is stimulated in the presence of S^0 and H_2 with the resultant formation of H_2S. Fermentation products from growth on tryptone are CO_2, *n*-butanol, acetic acid, propionic acid, phenylacetic acid, and hydroxyphenylacetic acid and in low amounts propylbenzene, acetophenone, and hydroxyacetophenone.

The organism was isolated from a sandy, hydrothermally heated and gassed seafloor with temperatures up to 114°C in 10 m depth, off the coast of São Miguel Island in the Azores.

ENRICHMENT AND ISOLATION PROCEDURES

Hyperthermus has been enriched using the medium employed to isolate *Pyrodictium* (Stetter et al., 1983), with the addition of trace minerals (Balch and Wolfe, 1976) and 2.5 mg KCl, 2 mg $NiSO_4·6H_2O$, and 0.5g NH_4Cl. Tryptone (6 g/l) was added as the carbon source. The medium was reduced with H_2S and the headspace was 800 kPa CO_2, with or without 200 kPa of H_2. The organism can be isolated using K9A40 gellan gum (Merck). Eight grams of gum are dissolved in a liter of boiling water. The gum is poured into glass Petri dishes after 1 g $CaSO_4$ is added. After the medium is solidified, the gum is soaked with a saturated solution of S^0 in 1 M $(NH_4)_2S$. Colloidal sulfur is precipitated with 1M H_2SO_4, and after washing with water, the plates are equilibrated overnight with the culture medium. The gels are dried at 37°C and stored in anaerobic incubation vessels in the presence of H_2S. The enrichments are plated and incubated at 99°C for 40–60 h at which time single, small amber colonies become visible.

DIFFERENTIATION OF THE GENUS *HYPERTHERMUS* FROM OTHER GENERA

Based on DNA–rRNA cross-hybridizations *Hyperthermus* is phylogenetically distinct from its relatives in the *Sulfolobales* and from *Thermococcales* and *Thermoproteales*. Additional phylogenetic analysis using 16S rRNA sequences has placed *Hyperthermus* close to *Pyrodictium* into the *Pyrodictiaceae* (Burggraf et al., 1997b) as also indicated by the nature of the lipids (see species description). The distinct hexagonal S-layer of *Hyperthermus* differs significantly from *Desulfurococcus* sp. (tetragonal symmetry) and from *Pyrodictium* (different lattice constant and structure). Additionally, the mol% G + C of *Hyperthermus* is 4% lower than that of *Pyrodictium*. The RNA polymerase of *Hyperthermus* resembles that of the *Thermoproteales*, *Sulfolobales*, *Thermococcales*, and *Thermoplasmatales* in containing an unsplit B component.

DIFFERENTIATION OF THE SPECIES OF THE GENUS *HYPERTHERMUS*

Only one species is described.

List of species of the genus Hyperthermus

1. **Hyperthermus butylicus** Zillig, Holz and Wunderl 1991, 170.[VP]

   bu.ty'li.cus. M.L. masc. adj. *butylicus* butylic, referring to the production of butanol.

   See generic description of features.

   Grows by fermentation of proteolysis products. The products are CO_2, *n*-butanol, acetic acid, propionic acid, and phenylacetic acid. Growth stimulated by S^0 and H_2. Optimal growth with 0.6% tryptone, 1.7% NaCl, pH 7, and between 95 and 107°C. No growth below 75°C and very slow growth occurs at 108°C. The membrane lipids are mainly bisisopranyl tetraether lipds containing two or more cyclopentane rings in the isopranyl chains.

   The mol% G + C of the DNA is: 55.6 (HPLC).
   Type strain: DSMZ 5456.
   GenBank accession number (16S rRNA): X99553.

Genus III. **Pyrolobus** *Blöchl, Rachel, Burggraf, Hafenbradl, Jannasch and Stetter 1999, 1325[VP] (Effective publication: Blöchl, Rachel, Burggraf, Hafenbradl, Jannasch and Stetter 1997, 20)*

KARL O. STETTER

Pyr.o.lo'bus. Gr. neut. n. *pyr* fire; Gr. masc. n. *lobos* lobe; M.L. masc. n. *Pyrolobus* the "fire lobe" (the hyperthermophilic lobe).

Cells regularly to irregularly lobed cocci, about 0.7–2.5 μm in diameter (Fig. A1.16). **Gram-negative**. No spores formed. **Immotile. S-layer of the cell envelope** consists of protein subunits arranged in a p4 lattice. **Membrane contains glycerol-dialkyl-glycerol-tetraether. Growth between 90 and 113°C** (optimum 106°C), at pH 4.0–6.5 (optimum pH 5.5) and at 1–4% NaCl (optimum 1.7%). Exponential cultures survive a one-hour autoclaving at 121°C. Facultatively aerobic. Obligately chemolithoautotrophic. Growth by anaerobic and microaerophilic H_2 oxidation, with NO_3^-, $S_2O_3^{2-}$, and O_2 as electron acceptors. NO_3^- is reduced to NH_4^+. Growth inhibition by S^0.

By 16S rRNA sequence analysis, specifically related to the *Pyrodictiaceae*.

The mol% G + C of the DNA is: 53.

Type species: **Pyrolobus fumarii** Blöchl, Rachel, Burggraf, Hafenbradl, Jannasch and Stetter 1999, 1325 (Effective publication: Blöchl, Rachel, Burggraf, Hafenbradl, Jannasch and Stetter 1997, 20).

FURTHER DESCRIPTIVE INFORMATION

Cells occurring singly and in short chains. White colonies, about 1 mm in diameter, formed on plates solidified by Gelrite. Center-to-center spacing of neighboring S-layer protein complexes of 18.5 nm. Nonhydrolyzed lipids contain a main spot on TLC staining blue (instead of violet) by anisaldehyde. Most extreme hyperthermophile known so far. No growth at 85°C or below. Pressure of 25,000 kPa tolerated. No stimulation of growth by organic compounds. Growth inhibition by acetate, pyruvate, glucose, and starch. Membrane-associated hydrogenase with an optimal reaction temperature of 119°C present in cells grown on H_2/NO_3^-. Crude extracts of cells contain a heat-shock protein that serologically cross-reacts strongly with the thermosome of *Pyrodictium occultum*.

No significant DNA–DNA similarity to *Pyrodictium*.

Pyrolobus was isolated from abyssal submarine hydrothermal systems like the walls of active black smokers at the "TAG" site, Mid Atlantic Ridge, at the East Pacific Rise and at the Guaymas vents.

ENRICHMENT AND ISOLATION PROCEDURES

Pyrolobus can be enriched anaerobically at 110°C in half strength SME* medium (½ SME) in the presence of $NaNO_3$ and with H_2/CO_2 as the gas phase (Blöchl et al., 1997). *Pyrolobus* can be isolated by plating on ½ SME solidified by Gelrite and incubated anaerobically at 102°C. White colonies formed after 7 days. In addition, pure cultures of *Pyrolobus* can be obtained from single cells isolated with optical tweezers (Huber et al., 1995a).

---

*"SME" medium consists of (per liter): NaCl, 13.85 g; $MgSO_4 \cdot 2H_2O$, 3.5 g; $MgCl_2 \cdot 6H_2O$, 2.75 g; $CaCl_2 \cdot 2H_2O$, 1.0 g; KCl, 0.325 g; NaBr, 0.05g; H_3BO_3, 0.01 5 g; $SrCl_2 \cdot 6H_2O$, 0.0075 g; KH_2PO_4, 0.5 g; KI, 0.025 mg; $NiNH_4SO_4$, 0.002 g; sulfur, 25 g; resazurine, 0.001 g; and trace minerals (Balch et al., 1979), 10 ml.

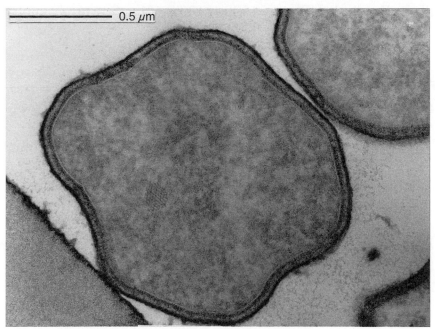

FIGURE A1.16. Electron micrograph of an ultrathin section of *Pyrolobus fumarii*. Contrasted with lead citrate and uranyl acetate.

MAINTENANCE PROCEDURES

Pyrolobus can be stored at room temperature for more than six months. For long-term preservation, cells can be suspended in medium containing 5% DMSO, dispensed into capillaries in 20-l portions at $-140°C$ over liquid nitrogen.

DIFFERENTIATION OF THE GENUS *PYROLOBUS* FROM OTHER GENERA

By 16S rRNA sequence data, *Pyrolobus* is specifically related to *Pyrodictium*. *Pyrolobus* is differentiated from *Pyrodictium* by insignificant DNA similarity, lower mol% G + C content (9% lower than *Pyrodictium*), the lack of a cannulae matrix, inability to grow by S^0 respiration, and energy conservation by nitrate and oxygen respiration.

TAXONOMIC COMMENTS

By 16S rRNA sequencing, *Pyrolobus* and *Pyrodictium* belong to the *Pyrodictiaceae* lineage, which is one of the shortest and deepest branches close to the root of the *Crenarchaeota* within the *Archaea*.

FURTHER READING

Blöchl, E., R. Rachel, S. Burggraf, D. Hafenbradl, H.W. Jannasch and K.O. Stetter. 1997. *Pyrolobus fumarii*, gen. and sp. nov., represents a novel group of archaea, extending the upper temperature limit for life to 113°C. Extremophiles *1*: 14–21.

Burggraf, S., H. Huber and K.O. Stetter. 1997. Reclassification of the crenarchaeal orders and families in accordance with 16S rRNA sequence data. Int. J. Syst. Bacteriol. *47*: 657–660.

Huber, R., S. Burggraf, T. Mayer, S.M. Barns, P. Rossnagel and K.O. Stetter. 1995. Isolation of a hyperthermophilic archaeum predicted by *in situ* RNA analysis. Nature (Lond.) *376*: 57–58.

DIFFERENTIATION OF THE SPECIES OF THE GENUS *PYROLOBUS*

Only one species of *Pyrolobus* is presently described.

List of species of the genus Pyrolobus

1. **Pyrolobus fumarii** Blöchl, Rachel, Burggraf, Hafenbradl, Jannasch and Stetter 1999, 1325[VP] (Effective publication: Blöchl, Rachel, Burggraf, Hafenbradl, Jannasch and Stetter 1997, 20).
   fum.a′ri.i. L. gen. n. *fumarii* of the chimney, referring to its black smoker habitat.

Description is the same as for the genus.
The mol% G + C of the DNA is: 53 (T_m and direct analysis).
Type strain: 1A, DSMZ 11204.
GenBank accession number (16S rRNA): X99555.

Order III. **Sulfolobales** Stetter 1989d, 496[VP] (Effective publication: Stetter 1989c, 2250)

HARALD HUBER AND KARL O. STETTER

Sul.fo.lo.ba' les. M.L. masc. n. *Sulfolobus* type genus of the order; *-ales* ending to denote an order; M.L. fem. pl. n. *Sulfolobales* the order of the *Sulfolobus.*

Cells coccoid, irregularly lobed. Extreme thermoacidophiles: optimum pH for growth around 2, temperature optima between 60 and 90°C. Aerobic, facultatively anaerobic, or obligately anaerobic. **Facultatively or obligately chemolithoautotrophic S⁰ metabolizers**.

Type genus: **Sulfolobus** Brock, Brock, Belly and Weiss 1972, 66.

Family I. **Sulfolobaceae** Stetter 1989d, 496[VP] (Effective publication: Stetter 1989c, 2250)

HARALD HUBER AND KARL O. STETTER

Sul.fo.lo.ba' ce.ae. M.L. masc. n. *Sulfolobus* type genus of the family; *-aceae* ending to denote a family; M.L. fem. pl. n. *Sulfolobaceae* the *Sulfolobus* family.

Description of the family is the same as for the order.

Type genus: **Sulfolobus** Brock, Brock, Belly and Weiss 1972, 66.

Key to the genera of the family Sulfolobaceae

1. Strictly aerobic growth. Cells can oxidize H_2S to S^0; S^0 or tetrathionate to sulfuric acid. Some strains grow on sulfidic ores, producing soluble metal sulfates or oxidizing ferrous iron. Most strains are facultative heterotrophs. The mol% G + C of the DNA is between 34 and 42.

    Genus I. *Sulfolobus*, p. 198

2. Cells are able to grow under aerobic conditions by oxidation of H_2S or S^0 producing H_2SO_4. Furthermore, H_2 or ferrous iron can be oxidized. Anaerobically, cells reduce S^0 to H_2S with H_2 as electron donor (H_2/S^0 autotrophy). Some strains are facultative heterotrophs. The mol% G + C of the DNA is around 31.

    Genus II. *Acidianus*, p. 202

3. Cells are obligate aerobes and oxidize H_2S, S^0, and tetrathionate to sulfuric acid, and H_2 to H_2O. Soluble metal sulfates are produced by oxidation of sulfidic ores. All strains are facultative heterotrophs. Organotrophic growth on complex organic substrates. The mol% G + C of the DNA is around 45.

    Genus III. *Metallosphaera*, p. 204

4. Cells are strict anaerobes and reduce S^0 to H_2S with H_2 as electron donor (H_2/S^0 autotrophy). Obligate chemolithoautotrophs. The mol% G + C of the DNA is around 38.

    Genus IV. *Stygiolobus*, p. 207

5. Cells are facultative anaerobes. Aerobic growth on complex organic substrates (no data on sulfur oxidation available). Anaerobic growth with S^0 and H_2 (production of H_2S). The mol% G + C of the DNA is 33.

    Genus V. *Sulfurisphaera*, p. 208

6. Cells are obligate aerobes and oxidize S^0 and sulfidic ores to sulfuric acid. Soluble metal sulfates are produced by oxidation of sulfidic ores. All strains are facultative heterotrophs, organotrophic growth on complex organic substrates and sugars. The mol% G + C of the DNA is around 45.

    Genus VI. *Sulfurococcus*, p. 209

Genus I. **Sulfolobus** Brock, Brock, Belly and Weiss 1972, 66[AL]

HARALD HUBER AND KARL O. STETTER

Sul.fo.lo' bus. L. neut. n. *sulfur* sulfur; Gr. masc. n. *lobos* lobe; M.L. masc. n. *Sulfolobus* lobed sulfur-oxidizing organism.

Cells coccoid, highly irregular with a diameter from about 0.7 to 2 μm, **usually occurring singly**. Gram negative. Immotile or motile by possession of one or more flagella. **Obligate aerobe. Facultatively chemolithoautotrophic. Lithotrophic growth via oxidation of sulfidic ores, sulfide, S⁰, or tetrathionate** resulting in the production of sulfuric acid. No growth by anaerobic sulfur reduction. Aerobic hydrogen oxidation by some species. **Organotrophic growth on complex organic material** (e.g., beef extract, peptone, or yeast extract), sugars, or amino acids. **Optimal growth temperatures between 65 and 85°C. Growth between pH 1 and 5.5** (Brock et al., 1972; Brock, 1978). Based on 16S rRNA sequence data, the *Sulfolobus* species are distributed over several branches within the *Sulfolobales* (*Sulfolobaceae*), indicating that they do not represent one cluster (genus) within this order (family) (Fuchs et al., 1996a).

The mol% G + C of the DNA is: 34–42.

Type species: **Sulfolobus acidocaldarius** Brock, Brock, Belly and Weiss 1972, 66.

FURTHER DESCRIPTIVE INFORMATION

Cells are highly irregular in shape, often lobed, but occasionally spherical. No septum formation visible. Cell envelope composed of glycoprotein subunits with p3 symmetry. Cell membrane consists of isopranyl ether lipids, 95% tetraethers with cyclopentane rings (glycerol-dialkylglycerol tetraethers, GDGT and glycerol-dialkyl-nonitol tetraethers, GDNT) and 5% diethers. In addition, unique quinones are found like caldariellaquinone and sulfolobusquinone. Flagella, pilus-like, and pseudopodium-like structures are described.

Several species and strains of the genus *Sulfolobus* are mixotrophic or obligately heterotrophic. Sulfur can be replaced by Fe^{2+} ions or by molecular hydrogen as an electron donor. In the absence of oxygen, sulfur oxidation occurs by using Fe^{3+} ions or MoO_4^{2-} ions as electron acceptors. In contrast to *Acidianus*, *Sulfurisphaera*, or *Stygiolobus*, *Sulfolobus* does not grow anaerobically by S^0 reduction.

DNA-dependent RNA polymerase shows the "BAC" type with incomplete immunochemical cross-reaction between the species of the genus and with enzymes of *Metallosphaera* and *Acidianus*. ADP-ribosylation of an elongation factor II-like protein by diphtheria toxin. No significant DNA–DNA relatedness to representatives of the genera *Metallosphaera*, *Acidianus*, or *Stygiolobus*.

Sulfolobus was isolated from acidic continental solfatara fields including Yellowstone National Park (Wyoming, USA), New Mexico, Solfatara Crater and Pisciarelli Solfatara, Naples, Italy, Dominica, El Salvador, New Zealand, Iceland, Japan, the Azores, and Sumatra. It has also been isolated from hot deposits from heaps at the open cast mining area near Ronneburg, Germany, which indicates a worldwide distribution. On the surface of boiling mudholes, oily glimmering films contain large amounts of *Sulfolobus*-like cells (up to 10^8/ml).

Several viruses have been isolated from different strains, like the lemon-shaped SSV1 (representing the *Fuselloviridae*) from *Sulfolobus shibatae* (Grogan et al., 1990). It can be induced by UV light and contains double-stranded DNA of 15 kb length. Further viruses (SIRV1 and SIRV2, belonging to the *Rudiviridae*) have been found in "*S. islandicus*" (not yet described) which is closely related to *S. solfataricus* and *S. shibatae* (Zillig et al., 1998).

ENRICHMENT AND ISOLATION PROCEDURES

Sulfolobus can be enriched in a modified Allen medium (Allen, 1959; Brock et al., 1972) using S^0, sulfidic ores, complex organic substrates, or molecular hydrogen as energy sources. Incubation should be carried out under aerobic conditions at pH 2 at temperatures around 75°C. *Sulfolobus* can be isolated by plating under microaerobic conditions on Gelrite, starch or polysilicate plates or by using optical tweezers (Huber et al., 1995a).

MAINTENANCE PROCEDURES

Sulfolobus cultures can be stored at 4°C or at room temperature for a few weeks. It is recommended that the culture media be neutralized before storage by the additon of $CaCO_3$. For longer periods (up to at least 5 years) cultures can be stored over liquid nitrogen (-140°C) in 5% DMSO.

DIFFERENTIATION OF THE GENUS *SULFOLOBUS* FROM OTHER GENERA

Sulfolobus is assigned to the order *Sulfolobales* on the basis of the cell morphology, an optimal growth temperature between 70 and 80°C and a pH optimum around 2. In contrast to *Acidianus*, *Sulfurisphaera*, and *Stygiolobus*, growth does not occur under anaerobic conditions. The mol% G + C content of 35–38 separates *Sulfolobus* from *Metallosphaera* and *Sulfurococcus*. *Thermoplasma*, another group of thermoacidophilic *Archaea*, differs by the lack of a protein cell envelope, a mol% G + C content of 46, and its inability to grow by sulfur oxidation.

TAXONOMIC COMMENTS

By 16S rRNA sequencing, the species of the genus *Sulfolobus* are spread over several branches within the *Sulfolobales*, indicating that phylogenetically they do not represent a genus (Fig. A1.17) (Fuchs et al., 1996a). A close relationship was determined only between *Sulfolobus solfataricus* and *Sulfolobus shibatae* (0.4% phylogenetic distance), but these organisms show no significant DNA–DNA similarity. The closest relative of *Sulfolobus acidocaldarius* is *Stygiolobus azoricus*, an obligate anaerobe. *Sulfolobus hakonensis* is related to both *Metallosphaera* species, although the mol% G + C content of its DNA is significantly lower than that of *Metallosphaera*. *Sulfolobus metallicus* represents a separate lineage within the *Sulfolobales* with at least 9% phylogenetic distance to all other *Sulfolobales* species.

FURTHER READING

Brock, T.D. 1978. Thermophilic Microorganisms and Life at High Temperatures, Springer-Verlag, Heidelberg.

Segerer, A. and K.O. Stetter. 1992. The order *Sulfolobales*. *In* Balows, Trüper, Dworkin, Harder and Schleifer (Editors), The Prokaryotes: A handbook of Bacteria: Ecophysiology, Isolation, Identification, Applications, 2nd Ed., Springer-Verlag, New York, pp 684–701.

Fuchs, T., H. Huber, S. Burggraf and K.O. Stetter. 1996. 16S rDNA-based phylogeny of the archaeal order *Sulfolobales* and reclassification of *Desulfurolobus ambivalens* as *Acidianus ambivalens* comb. nov. Syst. Appl. Microbiol. *19*: 56–60.

DIFFERENTIATION OF THE SPECIES OF THE GENUS *SULFOLOBUS*

Six species of the genus *Sulfolobus* are described: *S. acidocaldarius*, *S. hakonensis*, *S. metallicus*, *S. shibatae*, *S. solfataricus*, and *S. yangmingensis*. Two further species are so far not validly published: "*S. islandicus*" (Zillig et al., 1994) and "*S. thuringiensis*" (Fuchs et al., 1996a).

Key to the species of the genus *Sulfolobus*

1. Optimum temperature for growth: 70–75°C, maximum 85°C. Growth on glucose, sucrose, mannose, tryptophan, and glutamic acid. No growth on galactose and lactose. DNA-dependent RNA polymerase stable up to 75°C *in vitro*.

S. acidocaldarius.

2. Optimum temperature for growth 70°C, maximum 80°C. Facultative chemolithotrophic. Growth on S⁰, ferrous sulfide, tetrathionate, and yeast extract, weak growth on maltose, glutamic acid, or tryptophan.

S. hakonensis.

3. Optimum temperature for growth: 65°C, maximum 75°C. Obligate chemolithoautotrophic. Growth on sulfidic ores or S⁰. No growth by the "Knallgas"-reaction.

S. metallicus.

4. Optimum temperature for growth: 81°C. Growth on galactose, lactose, glucose, sucrose. Cannot readily be distinguished from *S. solfataricus* by physiological or chemical methods.

S. shibatae.

5. Optimum temperature for growth: 85°C. Growth on galactose, lactose, glucose, sucrose, and glutamic acid. DNA-dependent RNA polymerase stable up to 85°C *in vitro.*

S. solfataricus.

6. Optimum temperature for growth: 80°C; maximum 95°C. Facultative chemolithoautotrophic. Utilization of sugars and amino acids as sole carbon sources.

S. yangmingensis.

List of species of the genus Sulfolobus

1. **Sulfolobus acidocaldarius** Brock, Brock, Belly and Weiss 1972, 66.[AL]

a.ci.do.cal.dar′i.us. M.L. neut. n. *acidum* acid; L. masc. adj. *caldarius* pertaining to warm or hot; M.L. masc. adj. *acidocaldarius* organism living in acid-hot environments.

Cell diameter 0.8–1.0 μm (Fig. A1.18). Growth temperature: 55–85°C; optimum: 70–75°C. pH for growth: 1–6; optimum pH: 2–3. Cells lyse at pH >7.5. Aerobic. Growth on complex organic substrates like yeast extract, tryptone, and Casamino acids and on sugars like glucose, ribose, sucrose, and xylose. No growth on lactose and mannose (Huber, 1987; Takayanagi et al., 1996). Poor growth by oxidation of molecular hydrogen ("Knallgas"-reaction) (Huber et al., 1992a). DNA-dependent RNA polymerase consists of 10 subunits of different molecular weights. The molecular weights of the subunits are slightly different from the corresponding ones of *Sulfolobus solfataricus* (Zillig et al., 1980a). The enzyme is activated by Mg²⁺ ions (optimum: 20–50 mM Mg²⁺ at 30 mM NH₄⁺) and is stable up to 75°C *in vitro.* About 60% DNA–DNA similarity to "*S. thuringiensis*", no significant DNA–DNA similarities to other representatives of this genus. The type strain has lost its ability to oxidize S⁰.

See also generic description for features.

The mol% G + C of the DNA is: ~37 (direct analysis, T_m).

Type strain: 98-3, ATCC 33909, DSMZ 639.

GenBank accession number (16S rRNA): D14053, D14876, U32320, and Z21972.

2. **Sulfolobus hakonensis** Takayanagi, Kawasaki, Sugimori, Yamada, Sugai, Ito, Yamasato and Shioda 1996, 381.[VP]

ha.ko.nen′ sis. M.L. masc. adj. *hakonensis* pertaining to Hakone

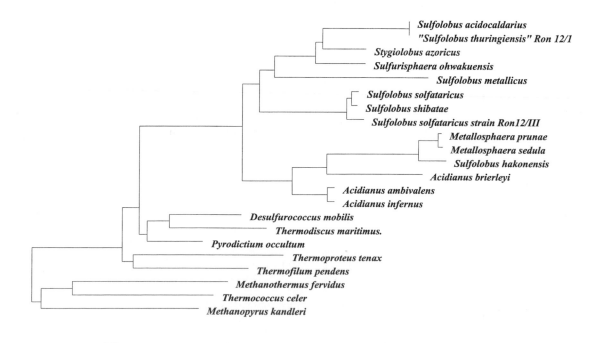

FIGURE A1.17. Phylogenetic tree of the members of the *Sulfolobales.* The tree was calculated by the maximum parsimony method. Bar = 10 substitutions per hundred nucleotides.

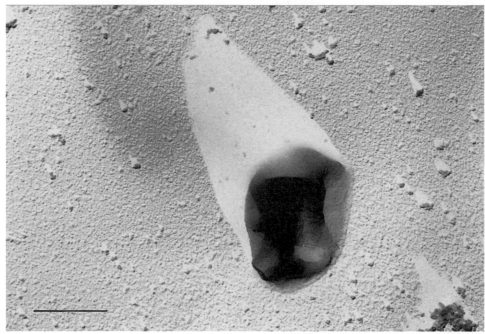

FIGURE A1.18. Electron micrograph of a cell of *Sulfolobus acidocaldarius*. Platinum-shadowed. Bar = 0.5 μm.

National Park, Japan, the location of the hot spring from which the organism was isolated.

Lobed cocci, 0.9–1.1 μm in diameter. Gram negative. Cell wall consists of regularly arranged structures. Growth temperature: 50–80°C, optimum: 70°C; pH 1.0–4.0, optimum: 3.0. Aerobic and facultatively chemolithoautotrophic. Lithotrophic growth occurs on S[0], ferrous sulfide, sodium tetrathionate, and hydrogen sulfide. Weak growth in the presence of 1.0% maltose, glutamic acid, or tryptophan as a sole carbon and energy source. The main cellular lipids are calditoglycero-caldiarchaeol (CGTE). The levels of similarity between the 16S rRNA of *S. hakonensis* and the 16S rRNAs of *S. acidocaldarius* DSMZ 639, *S. solfataricus* IFO 15331, and *S. shibatae* DSMZ 5389 are 89.8, 89.3, and 89.4%, respectively.

See also generic description for features.

The mol% G + C of the DNA is: 38.4 (Bd).

Type strain: HO1-1, ATCC 51241, DSMZ 7519, IAM 14250, JCM 8857.

GenBank accession number (16S rRNA): D14052.

3. **Sulfolobus metallicus** Huber and Stetter 1992a, 191[VP] (Effective publication: Huber and Stetter 1991, 377).

me.tal′ li.cus. L. masc. n. *metallicus* the miner.

Irregular cocci, ~1.5 μm in diameter. Growth between 50 and 75°C, at pH 1.0–4.5 and at 0–3% NaCl. Optimum at 65°C. Obligate chemolithoautotrophic growth on sulfidic ores like pyrite, sphalerite, and chalcopyrite and on S[0] with formation of sulfuric acid. Cell envelope consists of an S-layer of glycoproteins. Isopranyl ether lipids and caldariellaquinone present. An elongation factor II-like protein is sensitive to diphtheria toxin. No DNA–DNA similarity to other strains of the genus. "BAC"-type DNA-dependent RNA polymerase with molecular masses of the polypeptides of 130, 103, 45, 34.5, 33.5, 27, 20.5, 14.5, 13, and 11 kDa (determined by SDS polyacrylamide gel electrophoresis). Antibodies against the native RNA polymerase of strain Kra23 show incomplete cross

reaction with RNA polymerases of the type strains of *Acidianus infernus*, *Metallosphaera sedula*, and *Sulfolobus acidocaldarius*. By 16S rRNA sequencing, at least 9% phylogenetic distance to all other members of the *Sulfolobales*. Isolated from continental solfataric fields in Iceland.

See also generic description for features.

The mol% G + C of the DNA is: 38 (direct analysis, T_m).

Type strain: Kra 23, DSMZ 6482.

GenBank accession number (16S rRNA): D85519, X90479.

4. **Sulfolobus shibatae** Grogan, Palm and Zillig 1991, 457[VP] (Effective publication: Grogan, Palm and Zillig 1990, 598).

shi.ba′ tae. M.L. gen. n. *shibatae* of the family name of Masaru Shibata who enabled us to obtain the samples from which the organism was isolated.

Cells coccoid, 0.7–1.5 μm in diameter, but may have several flattened-to-deeply indented faces; occur singly. Colonies are pale tan, translucent, smooth, and convex with entire margins. Aerobic. Weakly motile, depending on culture conditions. Growth inhibited by phosphate concentrations above ~10 mM. Optimum growth temperature 81°C, maximum 86°C. D-Arabinose, D-galactose, D-glucose, D-mannose, lactose, lactulose, maltose, sucrose, raffinose, starch, and tryptone utilized. Facultatively chemolithoautotrophic growth by sulfur oxidation. DNA-dependent RNA polymerase component pattern similar to that of *S. solfataricus* and different from that of *S. acidocaldarius*. According to DNA–DNA cross-hybridization phylogenetically close to *S. solfataricus*. Isolated from acidic geothermal spring.

See also generic description for features.

The mol% G + C of the DNA is: 35 (direct analysis, T_m).

Type strain: B12, DSMZ 5389.

GenBank accession number (16S rRNA): D26490, X90478.

5. **Sulfolobus solfataricus** Zillig, Stetter, Wunderl, Schulz, Priess and Scholz 1980b, 676[VP] (Effective publication: Zillig, Stetter, Wunderl, Schulz, Priess and Scholz 1980a, 268).

sol.fa.ta' ri.cus. M.L. masc. adj. *solfataricus* living in solfatara habitats.

Cells are 0.8–2.0 μm in diameter. Temperature for growth: 50–87°C; optimum: 85°C. pH for growth: 2–5.5; optimum pH: ~4.5. Cells lyse at pH >7.5. Organotrophic growth on complex substrates like yeast extract, tryptone, and Casamino acids and on sugars like galactose, but also on glucose, ribose, sucrose, lactose, mannose, and xylose (Huber, 1987; Takayanagi et al., 1996). Poor growth by oxidation of molecular hydrogen ("Knallgas"-reaction) (Huber et al., 1992a). DNA-dependent RNA polymerase consists of 11 subunits of different molecular weights. The molecular weights of the subunits are slightly different from the corresponding ones of *S. acidocaldarius* (Zillig et al., 1980a). The enzyme is activated by Mg^{2+} ions (optimum: 3 mM Mg^{2+} at 30 mM NH_4^+) and is stable up to 85°C *in vitro*. Low DNA similarity to *S. shibatae* (around 20–25%). Type strain has lost its ability to oxidize S^0.

See also generic description for features.

The mol% G + C of the DNA is: ~35 (direct analysis, T_m).

Type strain: ATCC 35091, DSMZ 1616.
GenBank accession number (16S rRNA): D26490, X90478.

6. **Sulfolobus yangmingensis** Jan, Wu, Chaw, Tsai and Tsen 1999, 1815.[VP]

yang.ming.en' sis. M.L. adj. *yangmingensis* pertaining to the Yang-Ming National Park, Taiwan, Republic of China, from where the organism was isolated.

Lobed, spherical cells, 1.1 ± 0.2 μm in diameter. Thermoacidophilic. Optimum pH 4.0 (range: 2.0–6.0). Optimum temperature 80°C (range: 65–95°C). Growth on amino acids (except cysteine), D-arabinose, D-xylose, D-fructose, D-galactose, D-glucose, D-mannose, L-rhamnose, lactose, maltose, sucrose, raffinose, sorbitol, cellobiose, and trehalose. Weak growth on D-ribose. The major cellular lipids are calditoglycerocaldarchaeol and caldarchaeol.

Isolated from an acidic, muddy hot spring.
The mol% G + C of the DNA is: 42.
Type strain: YM1.
GenBank accession number (16S rRNA): AB010957.

Genus II. **Acidianus** *Segerer, Neuner, Kristjansson and Stetter 1986, 561*[VP]

HARALD HUBER AND KARL O. STETTER

A.cid.i.a' nus. L. masc. adj. *acidus* acid; L. masc. n. *Ianus* a mythical Roman figure with two faces looking in opposite directions; M.L. masc. n. *Acidianus* acidic bifaced (bacterium).

Cells coccoid, highly irregular, about 0.5–2 μm in diameter, **occurring singly almost exclusively.** Gram negative. Neither motility nor flagella was detected. Cell envelope composed of protein subunits with p3 symmetry. Isopranyl ether lipids, calditol, caldariellaquinone and sulfolobusquinone present. Elongation factor EF-G is ADP-ribosylated by diphtheria toxin. **Facultatively anaerobic. Lithotrophic growth aerobically via S^0 oxidation** to sulfuric acid or by H_2 oxidation ("Knallgas" reaction). **Under anaerobic conditions reduction of S^0 with H_2 to H_2S. Thermoacidophilic, growth above 45°C at pH 1–6.** Optimal NaCl concentration for growth in culture media around 0.2%, maximum around 4%.

Based on 16S rRNA sequence data, *Acidianus* clusters with the genus *Metallosphaera* (Fuchs et al., 1996a).

The mol% G + C of the DNA is: 31.

Type species: **Acidianus infernus** Segerer, Neuner, Kristjansson and Stetter 1986, 562.

FURTHER DESCRIPTIVE INFORMATION

Cells are highly irregular in shape: lobed, disk- to dish-shaped or appearing as polyhedrons; occasionally spherical (Figs. A1.19 and A1.20). No septa formation visible. Cell envelope consists of protein subunits about 25 nm in width with p3 symmetry.

The organisms are autotrophic and mixotrophic. Some strains are able to grow heterotrophically on complex organic substrates (e.g., yeast extract, peptone) in the presence of oxygen. No growth on sugars or amino acids. Optimal O_2 concentration for hydrogen oxidation around 4%. Anaerobic growth via S^0 oxidation takes place in the presence of molybdate. No growth at very low redox potential (around -300 mV) in the absence of either sulfur or hydrogen or both. Anaerobically grown cells are significantly more resistant to storage and disintegration procedures than are aerobically grown cells.

The DNA-dependent RNA polymerase exhibits incomplete immunochemical cross-reaction with antibodies against the enzyme of *Sulfolobus acidocaldarius*. No significant DNA–DNA similarity to representatives of the genera *Sulfolobus*, *Metallosphaera*, and *Stygiolobus*. The genome sizes of the different species range from 1.83 to 1.88 Mb.

Acidianus was isolated from aerobic and anaerobic samples taken from acidic solfatara springs and mudholes at Solfatara Crater, Naples, Italy, Yellowstone National Park, Wyoming, USA and from solfatara fields in Iceland, The Azores and Java (Indonesia). It also occurs in geothermally heated acidic marine environments at the beach of Vulcano Island, Italy.

ENRICHMENT AND ISOLATION PROCEDURES

Acidianus can be enriched anaerobically in a modified Allen medium (Allen, 1959; Brock et al., 1972) at pH 2 using S^0 and an H_2/CO_2 (80%:20%) atmosphere (300 kPa) at the appropriate temperature. After 2–3 days mature cultures are transferred into aerobic medium and further incubated.

Acidianus can be isolated by plating under aerobic or anaerobic conditions on plates with 0.5–1% colloidal sulfur containing 0.8% Gelrite or 10% starch or by using optical tweezers (Huber et al., 1995a).

MAINTENANCE PROCEDURES

Since anaerobically grown cells are more resistant to storage procedures than are aerobically grown cells, stock cultures of *Acidianus* can be routinely stored anaerobically at 4°C for a few months after raising the pH of the medium to 5.5 with sterile $CaCO_3$. For longer periods (at least 2 years) cultures should be stored over liquid nitrogen (-140°C) or freeze dried.

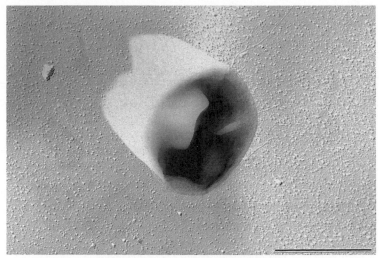

FIGURE A1.19. Electron micrograph of a cell of *Acidianus infernus.* Platinum-shadowed. Bar = 1 μm

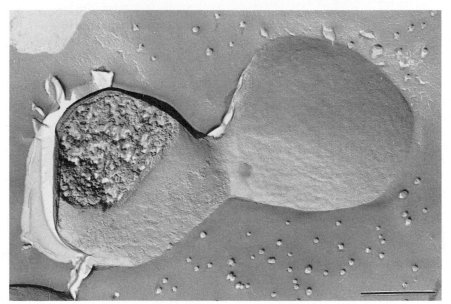

FIGURE A1.20. Electron micrograph of freeze-etched dividing cell of *Acidianus infernus.* Bar = 0.5 μm.

DIFFERENTIATION OF THE GENUS *ACIDIANUS* FROM OTHER GENERA

Acidianus belongs to the archaeal order of the *Sulfolobales* based on its morphology, the thermoacidophilic mode of life, the low mol% G + C content of its genomic DNA, the mode of aerobic energy conservation by S^0 oxidation producing sulfuric acid, and the presence of caldariellaquinone and calditol. In contrast to the representatives of the genera *Sulfolobus* and *Metallosphaera*, growth occurs also under anaerobic conditions via reduction of S^0 with molecular hydrogen producing H_2S. Aerobic growth by some species on sulfidic ores results in metal-mobilization. The mol% G + C content of 31 separates *Acidianus* from all other genera of the *Sulfolobales*.

TAXONOMIC COMMENTS

By 16S rRNA sequencing, the closest relatives of the *Acidianus* species are the representatives of *Metallosphaera*. The 16S rRNA sequences of *A. infernus* and *A. ambivalens* (formally *Desulfurolobus ambivalens*) exhibit a phylogenetic distance of only 0.7% (Fuchs

et al., 1996a). In agreement with this result a similarity of about 50% was determined in DNA–DNA hybridization experiments (Huber et al., 1987a). Therefore, *Desulfurolobus ambivalens* was reclassified as *Acidianus ambivalens* (Fuchs et al., 1996a).

FURTHER READING

Fuchs, T., H. Huber, S. Burggraf and K.O. Stetter. 1996. 16S rDNA-based phylogeny of the archaeal order *Sulfolobales* and reclassification of *Desulfurolobus ambivalens* as *Acidianus ambivalens* comb. nov. Syst. Appl. Microbiol. *19*: 56–60.

Segerer, A., A. Neuner, J.K. Kristjansson and K.O. Stetter. 1986. *Acidianus infernus* gen. nov., sp. nov., and *Acidianus brierleyi* comb. nov.: facultatively aerobic, extremely acidophilic thermophilic sulfur-metabolizing archaebacteria. Int. J. Syst. Bacteriol. *36*: 559–564.

Segerer, A., K.O. Stetter and F. Klink. 1985. Two contrary modes of chemolithotrophy in the same archaebacterium. Nature (Lond.) *313*: 787–789.

Zillig, W., S. Yeats, I. Holz, A. Böck, M. Rettenberger, F. Gropp and G. Simon. 1986. *Desulfurolobus ambivalens*, gen. nov., sp. nov., an autotrophic archaebacterium facultatively oxidizing or reducing sulfur. Syst. Appl. Microbiol. *8*: 197–203.

DIFFERENTIATION OF THE SPECIES OF THE GENUS *ACIDIANUS*

Three species of the genus *Acidianus* are described: *A. infernus,*
A. brierleyi, and *A. ambivalens.* The latter was previously positioned
in the genus *Desulfurolobus* (which no longer exists).

Key to the species of the genus *Acidianus*

1. Optimum temperature for growth 85–90°C; maximum 96°C. Aerobic growth strictly lithotrophic
   by sulfur oxidation.

                                     A. infernus.

2. Optimum temperature for growth 80°C; maximum 87°C. Aerobic growth strictly lithoautotrophic
   by sulfur oxidation.

                                    A. ambivalens.

3. Optimum temperature for growth 70°C; maximum 75°C. Aerobic growth lithotrophically by oxidation of S^0 or organotrophically by oxidation of complex organic substrates (e.g., yeast extract).

                                     A. brierleyi.

List of species of the genus Acidianus

1. **Acidianus infernus** Segerer, Neuner, Kristjansson and Stetter
   1986, 562.[VP]

   in.fer′ nus. L. masc. adj. *infernus* emerged from Hades, referring to the locus typicus at the Solfatara Crater, where Dante
   placed the gate to hell in his *Divina Commedia.*

   Growth between 65 and 96°C, optimum around 90°C; pH
   for growth 1–5.5; optimum around 2. Optimal doubling time
   around 2.5 hours. Obligately chemolithoautotrophic. Aerobic hydrogen oxidation, optimal O_2 concentration around
   4% (v/v). No organotrophic growth on yeast extract, peptone, tryptone, Casamino acids, and beef extract. The cell
   width depends on the culture conditions, with aerobically
   grown cells being about 1.5 times larger than anaerobically
   grown ones of the same growth phase. Cells lyse at pH values
   >8.5. Genome size 1.829 Mb. About 50% DNA similarity to
   A. ambivalens.

   See also generic description for features.

   The mol% G + C of the DNA is: ~31 (T_m and direct
   analysis).

   Type strain: So4a, DSMZ 3191.

   GenBank accession number (16S rRNA): D38773, D85505,
   and X89852.

2. **Acidianus ambivalens** (Zillig and Böck *in* Zillig, Yeats, Holz,
   Böck, Rettenberger, Gropp, and Simon 1986b) Fuchs, Huber,
   Burggraf and Stetter 1996b, 836[VP] (Effective publication:
   Fuchs, Huber, Burggraf and Stetter 1996a, 59) (*Desulfurolobus
   ambivalens* Zillig and Böck *in* Zillig, Yeats, Holz, Böck, Ret-
   tenberger, Gropp and Simon 1986b, 202).

   am.bi.va′ lens. L. masc. adj. *ambo* both; L. part. *valens* being
   able; M.L. part. adj. *ambivalens* ambivalent.

   Growth up to 87°C, optimum around 80°C; pH for growth
   1–3.5.; optimum around 2.5. Optimal doubling time around

4 h. Obligate chemolithoautotrophic. Genome size 1.855 Mb.
About 50% DNA similarity to *A. infernus.* Cells contain 3
plasmids. The major plasmid, 7.7 kb in size, amplified upon
reductive growth with the formation of virus-like particles
resembling the *Sulfolobus* virus-like particles SSV1.

See also generic description for features.

The mol% G + C of the DNA is: ~32.7 (T_m and direct
analysis).

Type strain: Lei1, DSMZ 3772.

GenBank accession number (16S rRNA): D38774, D85506,
and X90484.

3. **Acidianus brierleyi** (Zillig, Stetter, Wunderl, Schulz, Priess
   and Scholz 1980a) Segerer, Neuner, Kristjansson and Stetter
   1986, 562[VP] (*Sulfolobus brierleyi* Zillig, Stetter, Wunderl, Schulz,
   Priess and Scholz 1980a, 268).

   brier′ ley.i. M.L. gen. n. *brierleyi* of Brierley, named for J. A.
   Brierley who isolated the organism.

   Temperature for growth between 45 and 75°C, optimum
   around 70°C; pH for growth 1–6; optimum 1.5–2. Facultatively chemolithoautotrophic. Aerobic hydrogen oxidation.
   Organotrophic growth on yeast extract, peptone, tryptone,
   Casamino acids, and beef extract. Cells lyse at pH values >8.5.
   DNA-dependent RNA polymerase requires Mg^{2+} ions, consists of nine subunits of different molecular weight, and is
   resistant to rifampicin and streptolydigin. Genome size 1.88
   Mb. No significant DNA similarity to *A. ambivalens* and *A.
   infernus.*

   See also generic description for features.

   The mol% G + C of the DNA is: ~31 (T_m and direct
   analysis).

   Type strain: DSMZ 1651.

   GenBank accession number (16S rRNA): X90477.

Genus III. **Metallosphaera** *Huber, Spinnler, Gambacorta and Stetter 1989b, 496[VP] (Effective
publication: Huber, Spinnler, Gambacorta and Stetter 1989a, 45)*

HARALD HUBER AND KARL O. STETTER

Me.tal.lo.sphae′ ra. L. neut. n. *metallum* ore; L. fem. n. *sphaera* sphere; M.L. fem. n. *Metallosphaera* the
metal-mobilizing sphere.

Cells regular to irregular cocci with a diameter of 0.6–1.5 μm
(Fig. A1.21). Gram negative. Motile by possession of one or more

flagella. **Obligate aerobic. Facultatively chemolithoautotrophic.**
Lithotrophic growth on sulfidic ores like pyrite, sphalerite, and

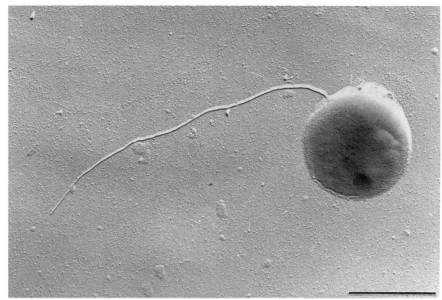

FIGURE A1.21.   Electron micrograph of a cell of *Metallosphaera sedula* with a flagellum. Platinum-shadowed. Bar = 1 μm.

chalcopyrite, and **on S⁰** resulting in the production of sulfuric acid. Aerobic hydrogen oxidation. **Organotrophic growth on beef extract, peptone, and yeast extract**. No growth by anaerobic sulfur reduction. **Optimal growth temperature around 75°C between pH 1 and 4.5**. Based on 16S rRNA sequence data, the *Metallosphaera* species are members of the family *Sulfolobaceae* and form a cluster with the representatives of the genus *Acidianus* (Fuchs et al., 1996a).

The mol% G + C of the DNA is: 45.

Type species: **Metallosphaera sedula** Huber, Spinnler, Gambacorta and Stetter 1989b, 496 (Effective publication: Huber, Spinnler, Gambacorta and Stetter 1989a, 46).

FURTHER DESCRIPTIVE INFORMATION

Cells are regular or irregular in shape, sometimes lobed. No septum formation visible. Cell envelope composed of protein subunits with p3 symmetry (Fig. A1.22). Cell membrane consists of isopranyl ether lipids. Flagella exhibit diameters between 10–13 nm. RNA polymerase shows the "BAC" type with incomplete immunochemical cross-reaction with antibodies against the enzyme of *Sulfolobus acidocaldarius*. ADP-ribosylation of an elongation factor II-like protein by diphtheria toxin. Genome sizes from 1.879 to 1.89 Mb. No significant DNA–DNA hybridization to representatives of the genera *Sulfolobus, Acidianus,* and *Stygiolobus*. *Metallosphaera* was isolated from acidic solfatara springs and mudholes at Pisciarelli Solfatara, Naples, Italy, and from hot deposits from heaps at the open cast mining area near Ronneburg, Germany.

ENRICHMENT AND ISOLATION PROCEDURES

Metallosphaera can be enriched in a modified Allen medium (Allen, 1959; Brock et al., 1972) using S⁰, sulfidic ores (highly selective), complex organic substrates, or molecular hydrogen as energy sources. Incubation should be carried out under aerobic conditions at pH 2 and a temperature around 75°C. *Metallosphaera* can be isolated by plating on media containing 0.8% Gel-

rite, or polysilicate or by using optical tweezers (Huber et al., 1995a).

MAINTENANCE PROCEDURES

Metallosphaera can be stored at room temperature or at 4°C for a few weeks. For longer periods (up to at least 2 years) cultures should be stored over liquid nitrogen (−140°C). It is not necessary to neutralize the culture media before storage.

DIFFERENTIATION OF THE GENUS *METALLOSPHAERA* FROM OTHER GENERA

Metallosphaera belongs to the *Sulfolobales* by its morphology, the pH optimum around 2, and an optimal growth temperature around 75°C. In contrast to the representatives of the genera *Acidianus* and *Stygiolobus*, growth does not occur under anaerobic conditions. The mol% G + C content of 45 separates *Metallosphaera* from the genus *Sulfolobus*. Efficient metal-mobilization from sulfidic ores.

TAXONOMIC COMMENTS

By 16S rRNA sequencing, the *Metallosphaera* species are related to the representatives of the genus *Acidianus* (around 9.5% phylogenetic distance). The two species of *Metallosphaera* are closely related (only 0.1% phylogenetic distance) but show no significant DNA–DNA similarity.

FURTHER READING

Fuchs, T., H. Huber, S. Burggraf and K.O. Stetter. 1996. 16S rDNA-based phylogeny of the archaeal order *Sulfolobales* and reclassification of *Desulfurolobus ambivalens* as *Acidianus ambivalens* comb. nov. Syst. Appl. Microbiol. *19*: 56–60.

Fuchs, T., H. Huber, T. Teiner, S. Burggraf and K.O. Stetter. 1995. *Metallosphaera prunae*, sp. nov., a novel metal-mobilizing, thermoacidophilic archaeum, isolated from a uranium mine in Germany. Syst. Appl. Microbiol. *18*: 560–566.

Huber, G., C. Spinnler, A. Gambacorta and K.O. Stetter. 1989. *Metallosphaera sedula* gen. and sp. nov. represents a new genus of aerobic, metal-mobilizing, thermoacidophilic archaebacteria. Syst. Appl. Microbiol. *12*: 38–47.

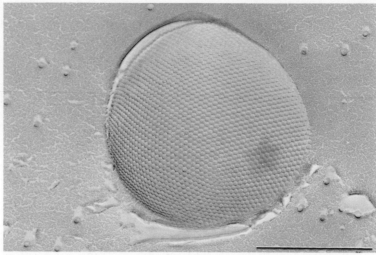

FIGURE A1.22. Electron micrograph of a freeze-etched cell of *Metallosphaera sedula* with surface layer. Bar = 0.5 μm.

DIFFERENTIATION OF THE SPECIES OF THE GENUS *METALLOSPHAERA*

Only two species of the genus *Metallosphaera* are described: *M. sedula* and *M. prunae*.

Key to the species of the genus *Metallosphaera*

1. Optimal O_2 concentration for growth on molecular hydrogen 0.5%. No more than one flagellum per cell. No fibrillar surface coat.
   M. sedula.
2. Optimal O_2 concentration for growth on molecular hydrogen 8–12%. Up to six flagella per cell. Possession of a fibrillar surface coat most likely consisting of carbohydrates.
   M. prunae.

List of species of the genus Metallosphaera

1. **Metallosphaera sedula** Huber, Spinnler, Gambacorta and Stetter 1989b, 496[VP] (Effective publication: Huber, Spinnler, Gambacorta and Stetter 1989a, 46).

   se' du.la. L. fem. adj. *sedula* busy, describing the efficient metal mobilization.

   Growth between 50 and 80°C, optimum around 75°C and pH 1.0–4.5. Optimal doubling time around 5 hours. Aerobic; facultatively chemolithoautotrophic growth on sulfidic ores and S[0]. Aerobic hydrogen oxidation, optimal O_2 concentration around 0.5% (v/v). No sulfur reduction in the presence of molecular hydrogen. Organotrophic growth on yeast extract, peptone, tryptone, Casamino acids, and beef extract. Cell envelope composed of protein subunits with p3 symmetry. RNA polymerase exhibits the "BAC" type and is composed of subunits with molecular weights of 130, 104, 42, 33.5, 32, 24.5, 15.8, 14.5, 13, 11.8, and 11.3 kDa. Genome size 1.89 Mb.

   See also generic description for features.

   The mol% G + C of the DNA is: ~45 (T_m and direct analysis).

   Type strain: TH2, DSMZ 5348.
   GenBank accession number (16S rRNA): X90481.

2. **Metallosphaera prunae** Fuchs, Huber, Teiner, Burggraf and Stetter 1996c, 625[VP] (Effective publication: Fuchs, Huber, Teiner, Burggraf and Stetter 1995, 565).

   pru' nae. L. fem. gen. n. *pruna* of the glowing coal.

   Growth between 55 and 80°C, optimum around 75°C and pH 1.0–4.5. Optimal doubling time 3.5 h. Aerobic, facultatively chemolithoautotrophic growth on sulfidic ores and S[0]. Aerobic hydrogen oxidation, optimal O_2 concentration around 12% (v/v). Organotrophic growth on yeast extract, peptone, and beef extract. No sulfur reduction anaerobically by molecular hydrogen. Cell envelope composed of protein subunits with p3 symmetry and, depending on the growth conditions, of a fibrillar surface coat. Genome size 1.879 Mb.

   See also generic description for features.

   The mol% G + C of the DNA is: ~46 (T_m and direct analysis).

   Type strain: Ron 12/II, DSMZ 10039.
   GenBank accession number (16S rRNA): X90482.

Genus IV. **Stygiolobus** Segerer, Trincone, Gahrtz and Stetter 1991, 497[VP]

HARALD HUBER AND KARL O. STETTER

Sty.gi.o.lo'bus. L. masc. n. *stygius* from hell; Gr. masc. n. *lobos* lobe; M.L. masc. n. *Stygiolobus* lobed organism from Hades (its habitat was the gate to hell in Dante's *Divina Commedia*).

Cells coccoid, highly irregular with a diameter from about 0.5 to 2 μm. Gram negative. Motile by possession of flagella. **Obligately anaerobic, chemolithoautotrophic. Growth by reduction of S⁰ with molecular hydrogen**, production of H₂S. **Optimal growth temperature around 80°C between pH 1 and 5.5.** Based on 16S rRNA sequence data, *Stygiolobus* clusters with *Sulfolobus acidocaldarius* (Fuchs et al., 1996a).

The mol% G + C of the DNA is: 38.

Type species: **Stygiolobus azoricus** Segerer, Trincone, Gahrtz and Stetter 1991, 497.

FURTHER DESCRIPTIVE INFORMATION

Cells are highly irregular in shape, either strongly lobed or with sharp edges and bends (Fig. A1.23). No septum formation visible. Cell envelope composed of protein subunits with p3 symmetry. Neither muramic acid nor *meso*-diaminopimelic acid present. Cell membrane contains glycerol-dibisphytanyl-nonitol tetraether lipids and sulfolobusquinone, but no caldariellaquinone. Incomplete cross-reaction of protein fractions with polyclonal antibodies against the RNA polymerase of *Sulfolobus acidocaldarius* and the native (NiFe) hydrogen uptake hydrogenase from *Acidianus brierleyi*. Complete immunochemical cross-reactions with polyclonal antibodies against histone-like proteins HSNP C′ and DBNP B from *Sulfolobus acidocaldarius*. ADP-ribosylation of the elongation factor EF-G sensitive to diphtheria toxin. Genome size only about 1.55 Mb. No significant DNA–DNA hybridization to representatives of the genera *Sulfolobus*, *Acidianus*, and *Metallosphaera*. *Stygiolobus* was isolated from hot acidic solfataric springs, mud and soil near Furnas and Ribeira Grande on São Miguel Island, Azores.

ENRICHMENT AND ISOLATION PROCEDURES

Stygiolobus can be enriched in a modified Allen medium (Allen, 1959; Brock et al., 1972) under strict anaerobic conditions (Balch et al., 1979) using S⁰ as electron acceptor and an atmosphere consisting of H₂/CO₂ (e.g., 80:20, v/v; 300 kPa). Incubation should be carried out at pH 2 and a temperature around 80°C. However, a parallel enrichment of *Acidianus infernus* or *Acidianus ambivalens* can not be excluded. *Stygiolobus* can be isolated by repeated end point dilution.

MAINTENANCE PROCEDURES

Stygiolobus can be stored at 4°C in anaerobic media for at least one year after the pH is increased to pH 5.5 by adding sterile CaCO₃. For longer periods cultures should be stored over liquid nitrogen (−140°C).

DIFFERENTIATION OF THE GENUS *STYGIOLOBUS* FROM OTHER GENERA

The genus *Stygiolobus* belongs to the *Sulfolobales* based on its morphology, the pH optimum of ~2 and an optimal growth temperature around 80°C. In contrast to the representatives of the other genera of the *Sulfolobales*, it grows obligately anaerobic. Sulfur cannot be oxidized to sulfuric acid but is reduced with molecular hydrogen to H₂S (S⁰/H₂ lithotrophy). The mol% G + C content of 38 separates *Stygiolobus* from the genus *Acidianus*. No significant DNA–DNA similarity is observed to the other thermoacidophilic genera.

TAXONOMIC COMMENTS

By 16S rRNA sequencing, the closest relative of *Stygiolobus* is *Sulfolobus acidocaldarius* with a phylogenetic distance of 5%. However, due to the great distances of the representatives of the genus *Sulfolobus* to each other (see also Genus *Sulfolobus*), no closer relationship is obvious to the other *Sulfolobus* species (Fuchs et al., 1996a).

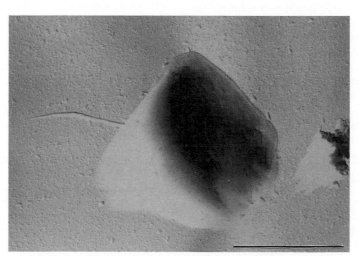

FIGURE A1.23. Electron micrograph of a cell of *Stygiolobus azoricus* with a single flagellum. Platinum-shadowed. Bar = 1 μm.

Further Reading

Fuchs, T., H. Huber, S. Burggraf and K.O. Stetter. 1996. 16S rDNA-based phylogeny of the archaeal order *Sulfolobales* and reclassification of *Desulfurolobus ambivalens* as *Acidianus ambivalens* comb. nov. Syst. Appl. Microbiol. *19*: 56–60.

Segerer, A.H., A. Trincone, M. Gahrtz and K.O. Stetter. 1991. *Stygiolobus azoricus* gen. nov., sp. nov., represents a novel genus of anaerobic, extremely thermoacidophilic archaebacteria of the order *Sulfolobales*. Int. J. Syst. Bacteriol. *41*: 495–501.

DIFFERENTIATION OF THE SPECIES OF THE GENUS *STYGIOLOBUS*

Only one species, *Stygiolobus azoricus*, is currently known.

List of species of the genus Stygiolobus

1. **Stygiolobus azoricus** Segerer, Trincone, Gahrtz and Stetter 1991, 497.[VP]

   a.zo'ri.cus. M.L. masc. adj. *azoricus* from the Azores, referring to the place of isolation.

   Exponentially growing cells are about 0.5–1.8 μm wide and occur singly almost exclusively. At elevated pH of ≥4, aggregates of 2–50 cells occur. Cells are frequently surrounded by pilus- or fimbria-like appendages. Growth between 57 and 89°C (optimum ~80°C) and between pH 1.0 and 5.5 (optimum pH 2.5–3). Optimal doubling time around 4.5 h. Obligately chemolithoautotrophic growth by means of S^0/H_2 lithoautotrophy. Low amounts of yeast extract (0.005–0.02%, w/v) stimulate growth. Genome size 1.543 Mb.

   See also generic description for features.

   The mol% G + C of the DNA is: ~38 (T_m).

   Type strain: FC6, DSMZ 6296.

   GenBank accession number (16S rRNA): X90480.

Genus V. **Sulfurisphaera** Kurosawa, Itoh, Iwai, Sugai, Uda, Kimura, Horiuchi and Itoh 1998, 455[VP]

ANNA-LOUISE REYSENBACH

Sul.fu.ri.sphae'ra. L. neut. n. *sulfur* sulfur; L. fem. n. *sphaera* sphere; M.L. fem. n. *Sulfurisphaera* sulfur-metabolizing spherical cells from sulfataric fields.

Cells are **irregular cocci, 1.2–1.5 μm in diameter. Gram-negative. Facultatively anaerobic heterotroph. Envelope** surrounding the cell membrane. Cells contain **caldarchaeol and calditoglycerocaldarchaeol** as the core lipids. No pili or flagella present. Growth occurs between **63 and 92°C and pH 1.0 and 5.0.** The organisms occur in acidic solfataras.

The mol% G + C of the DNA is: 32.9.

Type species: **Sulfurisphaera ohwakuensis** Kurosawa, Itoh, Iwai, Sugai, Uda, Kimura, Horiuchi and Itoh 1998, 455.

FURTHER DESCRIPTIVE INFORMATION

Transmission electron microscopy revealed a distinctive 24 nm thick envelope surrounding the cell membrane. This structure is similar to that reported for *Acidianus infernus*, *Stygiolobus azoricus*, *Metallosphaera sedula*, and *Sulfolobus acidocaldarius*.

The isolates can be grown both in liquid and solid media, however growth on 0.6% gellan gum plates is much slower. The organism grows optimally at 84°C and pH 2.0 with a doubling time of 5.9 h. *Sulfurisphaera* is able grow on proteinaceous, complex substrates such as yeast extract or tryptone, however, growth does not occur on simple sugars or amino acids. Poor growth does occur under anaerobic conditions by oxidation or reduction of S^0.

Sulfurisphaera grows in hot acidic (about 70–80°C, pH 3) springs in Ohwaku Valley, Hakone, Japan.

ENRICHMENT AND ISOLATION PROCEDURES

Sulfurisphaera can be enriched using a medium containing $(NH_4)_2SO_4$, KH_2PO_4, $MgSO_4$, $CaCl_2$ and yeast extract, pH 3.0 and incubation at 80°C for up to 3 days. Isolated colonies can be obtained by inoculating a liquid culture onto gellan gum plates and incubating the plates for about 8 days.

MAINTENANCE PROCEDURES

Aerobic cultures can be maintained at −80°C in 20% glycerol in basal medium for at least 10 months.

DIFFERENTIATION OF THE GENUS *SULFURISPHAERA* FROM OTHER GENERA

DNA–DNA hybridization and 16S rRNA analysis places *Sulfurisphaera* as a distinct genus from its physiological relatives, the facultative anaerobic acidophiles like *Acidianus*.

TAXONOMIC COMMENTS

Based on 16S rRNA sequence similarity, *Sulfurisphaera* is 87–90% similar to members of the genus *Acidianus*, and about 94% to its closest phylogenetic relatives *Sulfolobus acidocaldarius* and *Stygiolobus azoricus*. However, the mol% G + C of *Sulfurisphaera* most closely resembles other facultative anaerobic isolates of the *Sulfolobales*, namely *Acidianus* spp.

DIFFERENTIATION OF THE SPECIES OF THE GENUS *SULFURISPHAERA*

Only one species has been described; *Sulfurisphaera ohwakuensis*.

List of species of the genus Sulfurisphaera

1. **Sulfurisphaera ohwakuensis** Kurosawa, Itoh, Iwai, Sugai, Uda, Kimura, Horiuchi and Itoh 1998, 455.[VP]

   oh.wa.ku.en'sis. M.L. fem. adj. *ohwakuensis* from Ohwaku Valley, referring to the place of isolation.

   Irregular cocci are 1.2–1.5 μm in diameter. Cell envelope about 24 nm wide covers the cell membrane. Facultative anaerobe. Organotrophic growth on yeast extract or tryptone.

Cells grow at 63–92°C; optimum temperature, 84°C. The pH range for growth is 1.0–5.0; optimum pH ~2.0. Isolate was obtained from muddy water from an acidic hot spring in Ohwaku Valley, Hakone, Japan.

The mol% G + C of the DNA is: 32.9 (HPLC).
Type strain: TA-1, IFO 15161.
GenBank accession number (16S rRNA): D85507.

Genus VI. **Sulfurococcus** *Golovacheva, Val'ekho-Roman and Troitski 1995, 880[VP] (Effective publication: Golovacheva, Val'ekho-Roman and Troitski 1987, 91)*

ANNA-LOUISE REYSENBACH

Sul.fu.ro.coc' cus. L. neut. n. *sulfur* sulfur; Gr. n. *koccos* berry; M.L. masc. n. *Sulfurococcus* sulfur-oxidizing coccus.

Cells coccoid, occurring singly or in pairs, 0.8–2.5 μm in diameter. Reproduce by binary fission and budding. Gram-negative. Cell wall lacks murein. S-layer with protein subunits arranged in hexagonal array. **Pili and capsule present. Aerobic. Facultative lithotroph.** Lithotrophy via oxidation of sulfur to sulfuric acid. Heterotrophic growth by oxidizing complex organics (e.g., yeast extract, peptone, sugars, amino acids). **Thermoacidophilic, growing between 40 and 85°C and at pH 1.0 and 5.8.**

The mol% G + C of the DNA is: 43–46.

Type species: **Sulfurococcus mirabilis** Golovacheva, Val'ekho-Roman and Troitski 1995, 880 (Effective publication: Golovacheva, Val'ekho-Roman and Troitski 1987, 91).

FURTHER DESCRIPTIVE INFORMATION

Regardless of growth conditions the cells are nonmotile cocci, varying in size between 0.8 μm and 1.2 μm. Occasional large (3 μm) cells may be present and the cells may acquire an angular shape after three days of incubation. The cell wall consists of a 7.5–8 nm thick S-layer. The six proteins are arranged in a hexagonal structure. When grown on a medium with S^0, *S. yellowstonensis* cells are surrounded by a 0.2–0.4 μm capsule of uncharacterized composition. Pili (about 10 nm) occur in growing cultures. Complex ester bonds are absent in the cytoplasmic membrane. Large electron dense polyphosphate granules and electron transparent granules of unknown function are present in the cytoplasm of *S. yellowstonensis* cells.

The distinguishing characteristic of *Sulfurococcus* is its ability to reproduce by binary fission and budding. The mode of budding is reminiscent of that found in *Thermoproteus, Thermofilum, Pyrodictium* and *Thermodiscus*. One to five buds develop on the cell surface or on hyphae-like appendages that are about 10 nm in diameter and can be 3–4 μm in length. The cells that arise are small (~0.3 μm in diameter). A similar mode of reproduction has not been reported for the close relative, *Sulfolobus*.

S. mirabilis forms punctate, spherical, and transparent colonies when grown on agar media.

Sulfurococcus is a strict aerobe, capable of oxidizing S^0 to sulfuric acid under autotrophic and mixotrophic conditions. In addition, *S. yellowstonensis* is able to oxidize Fe^{2+} and sulfide minerals such as pyrite, chalcopyrite, and sphalerite. *Sulfurococcus* grows well on many different organic substrates at concentrations of 0.1–0.2%. Preferred substrates include fructose, sucrose, glucose, peptone, yeast extract, casein hydrolyzate, and phenylalanine.

The mol% G + C of the DNA of the two species of *Sulfurococcus* is distinct from *Sulfolobus* (mol% G + C is 36–38) and closer to *Metallosphaera sedula* (mol% G + C is 43.7–46.7). The DNA–DNA similarity (1–7%) confirms the closer relationship of *Sulfurococcus yellowstonensis* with *M. sedula* strain TH-2. The 5S rRNA sequence comparisons place the two species of *Sulfurococcus* within the same clade of the *Sulfolobales* lineage. The 5S rRNA sequences of *Sulfurococcus yellowstonensis* and *Sulfurococcus mirabilis* are 95.4% similar, and the sequence similarity between *Sulfurococcus* and *Sulfolobus solfataricus* is 75–79%.

Sulfurococcus has been isolated from acidic continental geothermal springs in Yellowstone National Park, USA, and in the crater of the Uzon volcano in Kamchatka, Russia.

Glucose uptake and respiration in *S. mirabilis* shared the temperatures for optimal and maximal growth; however, uptake and respiration continue to occur until 30°C (15°C lower than the minimum temperature for growth). Temperatures lower than 30°C appear to destabilize the intracellular pH, and can cause cell death (Golavacheva et al., 1991).

ENRICHMENT AND ISOLATION PROCEDURES

Sulfurococcus can be isolated in a modified AB medium of Allen and Brock (Allen, 1959; Brock et al., 1972) containing 0.01–0.02% yeast extract and under either heterotrophic or autotrophic conditions. Enrichment cultures are agitated in the presence of air and incubated between 60 and 80°C. Pure cultures can be obtained on agar plates with modified AB medium.

DIFFERENTIATION OF THE GENUS *SULFUROCOCCUS* FROM OTHER GENERA

Sulfurococcus differs from the related genus *Sulfolobus* by its higher mol% G + C content, 5S rRNA nucleotide sequence, and its ability to reproduce by budding.

TAXONOMIC COMMENTS

The 5S rRNA sequence comparisons places *Sulfurococcus* within the *Sulfolobales*. From both 5S rRNA data and DNA–DNA hybridization results, *Sulfurococcus* is a distinct genus from *Sulfolobus*. However, the physiological characteristics of the two genera are very similar, and confirmation of separate genera using 16S rRNA sequences may be required. Additionally, the primary characteristic used to distinguish *Sulfurococcus* from other members of the *Sulfolobales* was its ability to reproduce by budding. Although there is no report of budding in *Sulfolobus*, it is possible that this aspect of the morphology of this genus has been overlooked, and that budding is not restricted to *Sulfurococcus*. The two species of *Sulfurococcus* are more clearly defined, based on their ability to oxidize sulfide minerals and Fe^{2+}, different 5S rRNA sequences, low DNA–DNA homologies, and different optimum temperatures for growth.

FURTHER READING

Golovacheva, R.S., K.M. Val'ekho-Roman and A.V. Troitskii. 1987. *Sulfurococcus mirabilis* gen. nov., sp. nov., a new thermophilic archaebacterium with the ability to oxidze sulfur. Mikrobiologiya *56*: 100–107.

Golovacheva, R.S., N.V. Zhilina, L.O. Severina and A.G. Dorofeev. 1991. Effect of external conditions on the behavior of *Sulfurococcus* B. Mikrobiologiya *60*: 628–636.

Karavaiko, G.I., O.V. Golyshina, A.V. Troitskii, K.M. Val'ekho-Roman, R.S. Golovacheva and T.A. Pivovarova. 1994. *Sulfurococcus yellowstonii* sp. nov., a new species of iron- and sulfur-oxidizing thermoacidophilic archaebacteria. Microbiology *63*: 379–387.

Severina, L.O., N.V. Zhilina and V.K. Plakunov. 1991. Some characteristics of glucose transport system in extreme thermoacidophilic archaebacteria belonging to the genus *Sulfurococcus*. Microbiology *60*: 285–289.

DIFFERENTIATION OF THE SPECIES OF THE GENUS *SULFUROCOCCUS*

Only two species are described; *Sulfurococcus mirabilis* and *Sulfurococcus yellowstonensis*. These have been distinguished on the basis of their 5S rRNA structure, DNA–DNA hybridization results, and ability to oxidize Fe^{2+}.

List of species of the genus Sulfurococcus

1. **Sulfurococcus mirabilis** Golovacheva, Val'ekho-Roman and Troitski 1995, 880[VP] (Effective publication: Golovacheva, Val'ekho-Roman and Troitski 1987, 91).

    mi.ra.bi' lis. L. masc. adj. *mirabilis* wonderful.

    Cells occur singly or in pairs, 0.8–1.2 μm, some cells up to 2.5 μm. Oxidizes S^0 to sulfuric acid. Temperature for growth: 50–86°C; optimum 70–75°C. pH for growth: 1–5.8; optimum: pH 2.0–2.6, cells tolerate pH up to 6.2 and lyse at pH 7.0. 5S rRNA differs from *Sulfolobus solfataricus* at 23 nucleotide positions.

    See also generic description of features.

    The mol% G + C of the DNA is: 43–46 (Bd, HPLC).

    Type strain: INMI AT-59.

2. **Sulfurococcus yellowstonensis** Karavaiko, Golyshina, Troitski, Val'ekho-Roman, Golovacheva and Pivovarova 1995, 880[VP] (Effective publication: Karavaiko, Golyshina, Troitski, Val'ekho-Roman, Golovacheva and Pivovarova 1994, 386).

    yel.low.sto.nen' sis. M.L. masc. gen. n. *yellowstonensis* of Yellowstone, named after Yellowstone National Park, where organism was isolated.

    Spherical cells 0.8–1 μm, occur singly, in pairs, chains of 3–5 cells or grapelike microcolonies. Lithotrophic growth on Fe^{2+}, S^0, sulfide minerals. Temperature for growth: 40–80°C; optimum: 60°C. pH for growth: 1.0–5.5; optimum: 2–2.6. 12% DNA–DNA hybridization with *S. mirabilis*. 5S rRNA differs from *S. mirabilis* in 7 nucleotide positions.

    See also generic description of features.

    The mol% G + C of the DNA is: 44.6 (HPLC).

    Type strain: Str6kar.

Phylum AII. Euryarchaeota *phy. nov.*

GEORGE M. GARRITY AND JOHN G. HOLT

Eur.y.arch.ae.o' ta. M.L. fem. pl. n. *Euryarchaeota* from the Kingdom *Euryarchaeota* (Woese, Kandler and Wheelis 1990, 4579).

The phylum currently consists of seven classes: the *Methanobacteria*, the *Methanococci*, the *Halobacteria*, the *Thermoplasmata*, the *Thermococci*, the *Archaeoglobi*, and the *Methanopyri*. With the sole exception of the *Methanococci*, which is subdivided into three orders, each class contains a single order. The *Euryarchaeota* are morphologically diverse and occur as rods, cocci, irregular cocci, lancet-shaped, spiral-shaped, disk-shaped, triangular, or square cells. Cells stain Gram-positive or Gram-negative based on the presence or absence of pseudomurein in cell walls. In some classes, cell walls consist entirely of protein or may be completely absent (*Thermoplasmata*). Five major physiological groups have been described previously: the methanogenic *Archaea*, the extremely halophilic *Archaea*, *Archaea* lacking a cell wall, sulfate reducing *Archaea*, and the extremely thermophilic S^0 metabolizers.

Type order: **Methanobacteriales** Balch and Wolfe 1981, 216 (Effective publication Balch and Wolfe *in* Balch, Fox, Magrum, Woese and Wolfe 1979, 268).

Taxonomy of Methanogenic *Archaea*

WILLIAM B. WHITMAN, DAVID R. BOONE, YOSUKE KOGA AND JYOTI KESWANI

The taxonomy of methanogenic *Archaea* presented in the following chapters is an attempt to form taxa of consistent phylogenetic depth with taxa of equal ranks used in the taxonomy of other procaryotes (Boone et al., 1993b). Consistency within the taxonomy between major procaryotic groups is necessary to allow investigators from many different backgrounds to infer the relative closeness of relationships based upon their experiences with other procaryotic groups. Historically, the taxonomy of methanogens has not reflected the extent of their phylogenetic diversity. This situation developed in part because the last major revision of the taxonomy of methanogens created taxa whose phylogenetic depth was much greater than common usage in other groups of procaryotes (Balch et al., 1979). While this strategy may have been justifiable considering the small number of strains known at that time and the general absence of quantitative phylogenetic data for other procaryotes, in the last two decades a large number of new taxa of methanogens have been described and a wealth of phylogenetic data has been obtained for the procaryotes. Because many of the new taxa of methanogens were created following conventions commonplace in other procaryotic groups, the resulting taxonomy was no longer uniform.

For instance, many genera of methanogens were much more diverse than most bacterial genera. Within the *Methanococcales*, only one genus was recognized for hyperthermophilic, extremely thermophilic, and mesophilic species (Balch et al., 1979). Although the *Methanococcales* have few phenotypic characteristics to distinguish among species, lumping these species into a single genus masks a tremendous phylogenetic diversity that is readily apparent by molecular and biochemical analyses. On the basis of sequence comparisons of 16S rRNA as well as a variety of proteins (Keswani et al., 1996, and unpublished data), the hyperthermophilic and mesophilic methanococci are about as related as *Escherichia coli* and *Haemophilus influenzae*, which are classified in separate orders. In contrast, in the *Methanomicrobiales*, where phenotypic differences among species were more readily recognizable, the usage of genera has been more typical. The phylogenetic diversity, on the basis of 16S rRNA sequence similarity, was comparable to that found in other procaryotic genera.

In the absence of clear quantitative rules to determine taxonomic rank from phylogenetic or genetic relatedness, consistency can only be approximate. In the following chapters, the following guidelines have been adopted. For species-level distinctions, the recommendations of the International Committee on Systematic Bacteriology are followed (Wayne et al., 1987). Thus, organisms are distinguished as separate species if their DNA reassociation is less than 70%, the change in the melting temperature of their hybrid DNA is greater than 5°C, and substantial phenotypic differences exist. When 16S rRNA sequence data were available, a sequence similarity of <98% was considered evidence for a separate species (Devereux et al., 1990). However, sequence similarities of >98% were not considered sufficient evidence that two organisms belonged to the same species (Stackebrandt and Goebel, 1994). For genera, interspecies DNA reassociation values of less than 20–30% were considered evidence for separate genera (Johnson, 1984). When 16S rRNA sequence data were available, a similarity of less than about 93–95% was considered evidence for separate genera (Devereux et al., 1990). The usage of the rank of family has varied greatly in methanogen taxonomy (Balch et al., 1979; Zellner et al., 1989c). In general, this situation reflects the ambiguous usage in other procaryotic groups. For the taxonomy proposed here, the rank of family was used to indicate closely related genera with low DNA reassociation values and 16S rRNA sequence similarities greater than 88–93%. In some cases, families within the same order also showed differences in the components of their cellular lipids (Koga et al., 1998). This usage was chosen to avoid underrepresentation of the phylogenetic diversity even though it leads to recognition of some families for which only a single species has been described. As emphasized by others, phylogenetic family definitions must also be consistent with phenotypic differences and hence must be fairly flexible (Fox and Stackebrandt, 1987). The rank of order was used to recognize deep phylogenetic differences between families that were also indicative of well-characterized phenotypic differences including cellular lipid and wall components and substrates for methanogenesis. Although the usefulness of higher taxa is less well supported, the rank of class is proposed to distinguish major groups of methanogens. Hopefully, this working hypothesis will be tested in future studies that elucidate the deep phylogenetic relationships among the euryarchaeotes.

In the chapters that follow, five orders of methanogens are recognized: *Methanobacteriales*, *Methanococcales*, *Methanomicrobiales*, *Methanosarcinales*, and *Methanopyrales*. Organisms from distinct orders have 16S rRNA sequence similarities of no more than 82%.

In addition, the orders are readily distinguished on the basis of cell envelope structure, lipid composition, and substrate utilization.

The order *Methanobacteriales* contains species that possess pseudomurein as a major component of the cellular envelope and lipids composed of caldarchaeol and *myo*-inositol. These methanogens reduce CO_2 (with one exception) as the major substrate for methanogenesis. The order is composed of two families, *Methanobacteriaceae* and *Methanothermaceae*. The families are distinguished by 16S rRNA sequence similarities below 90% and differences in cell wall structure. In addition to pseudomurein, the *Methanothermaceae* possess a protein surface layer. The families are further distinguished by the presence of *N*-acetylglucosamine in the lipids of the *Methanothermaceae* and serine in the lipids of the *Methanobacteriaceae*.

Members of the order *Methanopyrales* also possess pseudomurein and reduce CO_2 as a substrate for methanogenesis, but their lipids contain galactose and mannose and lack caldarchaeol. Currently, this order only contains one hyperthermophilic species.

The order *Methanococcales* lacks pseudomurein, and the cell envelope is composed of a protein surface layer. All species form methane by CO_2 reduction. It contains two families, the *Methanococcaceae* and *Methanocaldococcaceae*, that are distinguished by 16S rRNA sequence similarities of less than 93%. The *Methanocaldococcaceae* are also entirely hyperthermophilic, while the *Methanococcaceae* are extremely thermophilic and mesophilic. The two families are further distinguished by the presence of cyclic archaeol in the *Methanocaldococcaceae*. In contrast to the other families proposed, 16S rRNA phylogenetic trees of the *Methano-*

Taxonomy of methanogens

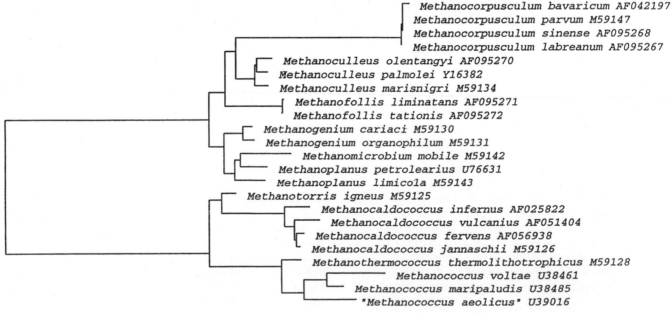

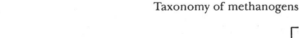
10% difference

FIGURE A2.1. Phylogenetic tree of the orders *Methanococcales* and *Methanomicrobiales*, prepared using the weighted least squares method implemented in PAUP*, with inverse-squared weighting. Distances were calculated using 1182 positions and corrected according to the HKY85 model. (Courtesy of the Ribosomal Database Project.)

caldococcaceae suggest that this taxon is not monophyletic, and *Methanotorris (Methanococcus) igneus* appears to form a lineage separate from the other species (see Fig. A2.1). This distinction is also supported by the presence of hydroxyarchaeol and the absence of caldarchaeol in this group. While the classification of *Methanotorris* within a new family is difficult to justify without additional strains, phenotypic differences and phylogenetic distance suggest that the creation of a new family for members of this genus may be necessary in the future.

The order *Methanomicrobiales* is composed of three families, *Methanomicrobiaceae*, *Methanocorpusculaceae*, and *Methanospirillaceae*. Members of the order all form methane by CO_2 reduction and possess cell envelopes containing a glycoprotein surface layer. The lipids all contain galactose, aminopentanetetrols (which are unique to this taxon), and glycerol. The families are distinguished by 16S rRNA sequence similarities of less than 89%. The *Methanospirillaceae* are further distinguished by their spirillum morphology and complex cellular envelope. The *Methanocorpusculaceae* are further distinguished by their very small size and the presence of diaminopropane as the major polyamine in the species tested.

Lastly, the *Methanosarcinales* are composed of two families, *Methanosarcinaceae* and *Methanosaetaceae*. Members of this order contain lipids composed of hydroxyarchaeol (with one exception), *myo*-inositol, and ethanolamine. The families are distinguished by 16S rRNA sequence similarities below 88%. The family *Methanosarcinaceae* produces methane from methyl group containing compounds, although many species can also reduce CO_2 to methane and split acetate to methane and CO_2 in the aceticlastic reaction. The cell envelope is formed of a surface layer protein, although complex heteropolysaccharides may also be present. Lipids also contain glycerol. The *Methanosaetaceae* utilize only acetate as a substrate for methanogenesis and possess a sheath. Galactose is a component of their lipids.

Key to the Methanogenic *Archaea*

1. Rod-shaped, lancet-shaped, or coccoid methanogens that reduce CO_2 or sometimes methyl compounds with H_2, formate, or secondary alcohols as the electron donors; cell walls contain pseudomurein; lipids contain caldarchaeol and *myo*-inositol in the absence of glycerol as a polar group of the lipid.

   Order *Methanobacteriales*, p. 214

2. Coccoid methanogens from marine environments that grow by catabolizing H_2/CO_2 or formate; cell walls are protein S- layers; lipids contain *N*-acetylglucosamine and hydroxyarchaeol, caldarchaeol, or cyclic archaeol in the absence of *myo*- inositol.

   Order *Methanococcales*, p. 236

3. Coccoid, rod-, or spiral-shaped methanogens that catabolize H_2/CO_2, formate, and sometimes secondary alcohols; cell walls are glycoprotein S-layers, sometimes with an exterior sheath; lipids contain aminopentanetetrols, galactose, and glycerol as polar groups.

   Methanomicrobiales, p. 246

4. Hyperthermophilic rod-shaped methanogens; cell walls contain pseudomurein; caldarchaeol, hydroxyarchaeol, and cyclic archaeol lipids are absent.

   Methanopyrales, p. 353

5. Pseudosarcinal, coccoid, or sheathed rod-shaped methanogens which catabolize methyl groups or acetate; sometimes H_2/CO_2 is catabolized, but never formate; cell walls contain an S-layer and sometimes other material; lipids usually contain hydroxyarchaeol and contain glycerol or galactose as polar groups in the absence of aminopentanetetrols.

   Methanosarcinales, p. 268

Class I. **Methanobacteria** *class. nov.*

DAVID R. BOONE

Me.tha.no.bac.te' ri.a. M.L. fem. pl. n. *Methanobacteriales* type order of the class; dropping the ending to denote a class; M.L. fem. pl. n. *Methanobacteria* the class of *Methanobacteriales*.

Description is the same as that of *Methanobacteriales*, the sole order of the class.

The mol% G + C of the DNA is: 23–61.

Type order: **Methanobacteriales** Balch and Wolfe 1981, 216 (Effective publication Balch and Wolfe *in* Balch, Fox, Magrum, Woese and Wolfe 1979, 268).

Order I. **Methanobacteriales** Balch and Wolfe 1981, 216[VP] (Effective publication Balch and Wolfe *in* Balch, Fox, Magrum, Woese and Wolfe 1979, 268)

DAVID R. BOONE, WILLIAM B. WHITMAN AND YOSUKE KOGA

Me.tha.no.bac.te.ri.a' les. M.L. neut. n. *Methanobacterium* type genus of the order; *-ales* ending to denote an order; M.L. fem. pl. n. *Methanobacteriales* the order of *Methanobacterium.*

Cells are short, lancet-shaped cocci to long, filamentous **rods**. **Nonmotile**. Cells typically stain as Gram-positive, although some strains are reported to be Gram-variable. Cell wall structure may appear to be typically Gram-positive when viewed by electron microscopy of thin sections, but the wall does not contain muramic acid. **Pseudomurein is the predominant peptidoglycan polymer. Cell membranes are composed of lipids containing caldarchaeol and *myo*-inositol.** The *Methanobacteriales* are very **strict anaerobes**, and they all **grow by oxidizing H$_2$**. Formate, secondary alcohols, or CO may also be oxidized. CO$_2$ is the normal electron acceptor, but some cells reduce methanol instead; sulfur may also be reduced to sulfide, but sulfide production does not lead

to growth. They form a highly specialized physiological group, which **does not catabolize carbohydrates, proteinaceous material, or organic compounds other than methanol, some secondary alcohols, formate, or CO**. They are widely distributed in nature, being found in anaerobic habitats such as aquatic sediments, soil, anaerobic sewage digestors, and the gastrointestinal tracts of animals, and in ecosystems where geothermally produced H$_2$ accumulates.

The mol% G + C of the DNA is: 23–61.

Type genus: **Methanobacterium** Kluyver and van Niel 1936, 399, emend Balch, Fox, Magrum, Woese and Wolfe 1979, 284.

Key to the families of the order Methanobacteriales

I. Little or no growth at or above 70°C.
  Family I. *Methanobacteriaceae.*
II. No growth below 60°C; fastest growth above 70°C.
  Family II. *Methanothermaceae,* p. 233

Family I. **Methanobacteriaceae** Barker 1956, 15[AL], emend. Balch and Wolfe *in* Balch, Fox, Magrum, Woese and Wolfe 1979, 267

DAVID R. BOONE, WILLIAM B. WHITMAN AND YOSUKE KOGA

Me.tha.no.bac.te.ri.a' ce.ae. M.L. neut. n. *Methanobacterium* type genus of the family; *-aceae* ending to denote a family; M.L. fem. pl. n. *Methanobacteriaceae* the family of *Methanobacterium.*

Cells are short, lancet-shaped cocci to long, filamentous **rods**. **Nonmotile**. Cells typically stain as Gram-positive, although some strains are reported to be Gram-variable. Cell wall structure may appear to be typically Gram-positive when viewed by electron microscopy of thin sections, but the wall does not contain muramic acid. **Pseudomurein is the predominant peptidoglycan polymer. Cell membranes are composed of archaeol, caldarchaeol, and (in one genus) hydroxyarchaeol. Phospholipid head groups include glucose, *myo*-inositol, serine, and (in some genera) ethanolamine.** The *Methanobacteriaceae* are very **strict anaerobes**, and all **grow by oxidizing H$_2$**. Formate or CO may also be oxidized. CO$_2$ is the normal electron acceptor, but some cells reduce methanol instead; sulfur may also be reduced to sulfide, but sulfide

production does not lead to growth. Cells contain coenzyme M, factor F$_{420}$, factor F$_{430}$, and methanopterin. They form a highly specialized physiological group that **does not catabolize carbohydrates, proteinaceous material, or organic compounds other than methanol, some secondary alcohols, formate, or CO. Little or no growth occurs at temperatures above 70°C**. Widely distributed in nature, being found in anaerobic habitats such as aquatic sediments, soil, anaerobic sewage digestors, and the gastrointestinal tracts of animals, and in ecosystems where geothermally produced H$_2$ accumulates.

The mol% G + C of the DNA is: 23–61.

Type genus: **Methanobacterium** Kluyver and van Niel 1936, 399, emend Balch, Fox, Magrum, Woese and Wolfe 1979, 284.

Key to the genera of the family Methanobacteriaceae

I. Cells are short to long rods or filaments; no growth above 60°C.
  Genus I. *Methanobacterium,* p. 215
II. Cells are short rods and lancet-shaped; grow by reducing CO$_2$.
  Genus II. *Methanobrevibacter,* p. 218
III. Cells are cocci, growing by reducing methanol with H$_2$.
  Genus III. *Methanosphaera,* p. 226
IV. Cells are rods or filaments growing above 60°C.
  Genus IV. *Methanothermobacter,* p. 230

Genus I. **Methanobacterium** *Kluyver and van Niel 1936, 399,[AL] emend. Balch and Wolfe* in *Balch, Fox, Magrum, Woese and Wolfe 1979, 284*

DAVID R. BOONE

Me.tha.no.bac.te'ri.um. M.L. masc. n. *methanum* methane; Gr. masc. n. *bakterion* a small rod; M.L. neut. n. *Methanobacterium* methane (-producing) rod.

Curved, crooked, or straight rods, long to filamentous, about 0.5–1.0 µm in width. Endospores not formed. Usually stain Gram-positive, and cell wall structure appears as Gram-positive in electron micrographs, but cell wall differs chemically from peptidoglycan. Nonmotile. Fimbriae may be present. Strictly anaerobic. Optimum growth temperatures 37–45°C. Energy metabolism by reduction of CO_2 to CH_4; H_2 is electron donor for this reduction, and some strains can also use formate, secondary alcohols, and CO. Ammonia, or for some strains dinitrogen, can serve as sole nitrogen source, and sulfide can serve as sulfur source.

The mol% G + C of the DNA is: 32–61.

Type species: **Methanobacterium formicicum** Schnellen 1947, 85.

FURTHER DESCRIPTIVE INFORMATION

Rods are long and form chains or often filaments. The cell wall is a single, thick layer external to the cytoplasmic membrane and composed of pseudomurein (Kandler and König, 1978; König et al., 1982). The members of this genus generally stain Gram-positive, although results from *Methanobacterium formicicum* and *Methanobacterium bryantii* are variable. Intracytoplasmic membranous bodies in thin-section electron micrographs (Zeikus and Wolfe, 1972; Zeikus and Bowen, 1975a) may be artifacts of the fixation process (Aldrich et al., 1987). Cells possess fimbriae (Doddema et al., 1979).

Growth occurs by CO_2 reduction coupled to H_2 oxidation. Some strains can use formate (Benstead et al., 1991), secondary alcohols, or CO (Kluyver and Schnellen, 1947). Formate and H_2 can be interconverted by formate-using strains (Wu et al., 1993). Elemental sulfur may be reduced to sulfide, but this reaction does not support cell growth (Stetter and Gaag, 1983). Several species are autotrophic, and the mechanism of CO_2 fixation is probably reduction of two CO_2 molecules to acetate by a modified Ljungdahl-Wood pathway (Simpson and Whitman, 1993). Neither the reductive pentose cycle, the serine pathway, nor the hexose-phosphate pathway can be demonstrated (Ferry et al., 1976; Taylor et al., 1976; Daniels and Zeikus, 1978; Stupperich and Fuchs, 1981). An incomplete reductive tricarboxylic acid cycle may be present (Sprott et al., 1993), and gluconeogenesis from acetate occurs via pyruvate.

Agar colonies are tannish white or grayish white, round, diffuse, and somewhat filamentous.

Members of the genus *Methanobacterium* may be found in methanogenic environments such as anaerobic digestors, freshwater sediments, marshy soils, and the rumen of cattle or sheep. H_2/CO_2 or formate is used as a methanogenic substrate, even in environments in which the concentrations of H_2 and formate are very low. Nevertheless, H_2 and formate may turn over very rapidly and thus be quantitatively important substrates.

ENRICHMENT AND ISOLATION PROCEDURES

Members of the genus *Methanobacterium* may be isolated directly from their methanogenic environments by inoculation of roll tube media. Enrichment techniques may increase the numbers

of H_2-using methanogens such as *Methanobacterium* species relative to organisms unable to grow on H_2. However, some nonmethanogens such as acetogenic bacteria can grow under anoxic conditions with H_2 as a substrate, and these would also be enriched. When enrichment cultures are transferred several times, the high H_2 concentrations in such cultures may allow *Methanobacterium* strains to be outcompeted by these faster-growing nonmethanogens. The incorporation of antibiotics such as penicillin or cycloserine may be necessary for isolation of methanogens from environments or enrichment cultures that contain large numbers of nonmethanogenic H_2-using bacteria. Formate may be used to enrich or isolate formate-using strains of *Methanobacterium* such as *M. formicicum*. Some strains can also grow on secondary alcohols (Widdel, 1986; Zellner and Winter, 1987b; Bleicher et al., 1989).

Because species of *Methanobacterium* are autotrophic or have limited growth factor requirements, media for isolation or enrichment need have little or no organic additions. Extracts from the natural habitat or rumen fluid additions to mineral media may meet all the nutritional requirements. Unusual metal ions such as nickel, cobalt, tungsten, or molybdenum are required. Ammonia should be included as a nitrogen source and sulfide as a sulfur source. The most successful culture techniques involve the use of roll tubes, although plating methods may be used (Edwards and McBride, 1975; Balch and Wolfe, 1976). Agar plates are prepared and inoculated within an anaerobic chamber and then sealed in anaerobic jars for incubation.

MAINTENANCE PROCEDURES

MS medium (Boone et al., 1989) with H_2 added as a catabolic substrate is suitable for maintenance of cultures of *Methanobacterium*. This may be preferable to MS mineral medium because the presence of yeast extract and peptones allows turbid growth of nonmethanogenic contaminants, should they occur, making their presence more readily discernible. *Methanobacterium* may be stored after lyophilization (Hippe, 1984), as aqueous suspensions at liquid nitrogen temperatures (Hippe, 1984; Boone, 1995; Tumbula et al., 1995), or cultures can be maintained by regular subculturing (Boone, 1995).

DIFFERENTIATION OF THE GENUS *METHANOBACTERIUM* FROM OTHER GENERA

Members of the genus *Methanobacterium* can be distinguished from other genera of H_2-utilizing methanogens by their morphology, their lack of motility, and their inability to grow at high temperatures. The other rod-shaped, H_2-utilizing methanogens are in the genera *Methanothermobacter, Methanothermus, Methanopyrus, Methanobrevibacter, Methanospirillum, Methanomicrobium,* and *Methanolacinia. Methanothermobacter, Methanothermus,* and *Methanopyrus* are thermophiles or hyperthermophiles and can be distinguished from *Methanobacterium* by their higher growth temperatures. *Methanobrevibacter, Methanomicrobium,* and *Methanolacinia* rods are generally much shorter than those of *Methanobacterium. Methanospirillum* grows in long chains enclosed within a sheath, but the regularity of the gentle spiral curvature is easy to distinguish from the irregularly crooked *Methanobacterium* morphology. Also, *Methanospirillum*, *Methanomicrobium*, and *Methanolacinia* are motile, whereas *Methanobacterium* is not.

Fimbriae occur and have been found only in this genus and in *Methanothermobacter* (Doddema et al., 1979).

TAXONOMIC COMMENTS

The taxonomy set forth in this *Manual* proposes the transfer of all the thermophilic species formerly in *Methanobacterium* into

the new genus *Methanothermobacter*. This proposal was prompted by analysis of 16S rDNA sequences that indicate a deep phylogenetic separation of the former mesophilic and thermophilic species of *Methanobacterium* (Boone et al., 1993b; Wasserfallen et al., 2000).

FURTHER READING

Balch, W.E., G.E. Fox, L.J. Magrum, C.R. Woese and R.S. Wolfe. 1979. Methanogens: reevaluation of a unique biological group. Microbiol. Rev. *43*: 260–296.

DIFFERENTIATION OF THE SPECIES OF THE GENUS *METHANOBACTERIUM*

Strains of *Methanobacterium* cannot be accurately assigned to species based on phenotype. The species descriptions below are based on a limited number of strains, often a single strain. Thus, the constancy of the phenotypes is not known, and the phenotypic differences among the species indicated in Table A2.1 may not be dependable for assignment of new strains to species.

List of species of the genus Methanobacterium

1. **Methanobacterium formicicum** Schnellen 1947, 85.[AL]
   for.mi'ci.cum. M.L. neut. n. *acidum formicum* M.L. neut. adj. *formicicum* pertaining to formic acid.

   Original description supplemented by material from Mylroie and Hungate (1954), Smith (1966), Langenberg et al. (1968), Zeikus and Bowen (1975a), and Bryant and Boone (1987b).

   Cells are long, slender, crooked rods with blunt, rounded ends, often forming chains or filaments. Cells are 0.4–0.8 × 2–15 μm. Each strain grows with a relatively constant diameter but with a length that may vary. Fimbriae present (Doddema et al., 1979).

   Surface colonies are white to gray, flat, and filamentous. Deep colonies are profusely filamented spheroids. In roll tubes with H_2/CO_2, colonies appear after 3–5 d, attaining a diameter up to 5 mm within two weeks. Appearance of liquid cultures depends on the strain: cultures may be uniformly turbid or growth may occur as highly granular clumps which do not break up even with vigorous agitation (Langenberg et al., 1968).

   Some strains may be autotrophic; acetate and cysteine may be stimulatory. The maximum specific growth rate is 0.053/h, and maximum H_2 uptake rate is 0.13 $mol \cdot h^{-1} \cdot g^{-1}$ (Lundbäck et al., 1990).

   Habitat: may be numerous in anaerobic digestors or anaerobic freshwater sediments; may be present in low numbers in the rumen of cattle or as endosymbionts in anaerobic protozoa (van Bruggen et al., 1984; Magingo and Stumm, 1991).

   Isolated from a sewage sludge digestor.

   The mol% G + C of the DNA is: 41–42 (Bd).

   Type strain: MF, ATCC 33274, DSMZ 1535, OCM 55.

   Additional Remarks: Described strains appearing to belong in this species include JF-1 (DSMZ 2639, OCM 105) (Schauer and Ferry, 1980) and DSMZ 3636, which was not named in the original publication (van Bruggen et al., 1984) but was later referred to as strain MS-1.

2. **Methanobacterium alcaliphilum** Worakit, Boone, Mah, Abdel-Samie and El-Halwagi 1986, 381.[VP]

   al.ca.li'phil.um. M.L. n. *alcali* (from Arabic *al* end; *qaliy* soda ash); Gr. adj. *philos* loving; M.L. neut. adj. *alcaliphilum* liking alkaline environments.

   Original material supplemented by Boone et al. (1986).

   Cells are long rods, 0.5–0.6 × 2–25 μm, occurring individually or in pairs, more rarely in short chains or filaments. Cells stain Gram-negative. Nonmotile. Colonies grow to 1.0–1.5 mm in diameter.

   Deep colonies in roll tubes are cream colored, small (diameter, 0.2 mm), irregular spheroids. Surface colonies are yellowish to cream colored, smooth, opaque, raised, convex, and circular with entire margins.

   H_2/CO_2 is the sole substrate for growth and methanogenesis. Growth factors present in peptone or yeast extract are required for growth. Acetate, when present, is assimilated.

   Habitat: alkaline lake sediments.

   Optimal growth occurs between pH 8.1 and 9.1 and near 37°C.

   Isolated from an alkaline lake of the Wadi el Natrun in Egypt.

   The mol% G + C of the DNA is: 57 (Bd).

   Type strain: WeN4, DSMZ 3387, OCM 11.

   Additional Remarks: Described strains classified in this species include WeN1 (ATCC 43376, DSMZ 3457, OCM 8), WeN2 (ATCC 43377, DSMZ 3458, OCM 9), and WeN3 (ATCC 43378, DSMZ 3459, OCM 10) (Boone, et al., 1986).

3. **Methanobacterium bryantii** Boone 1987, 173[VP] (ex Balch and Wolfe *in* Balch, Fox, Magrum, Woese and Wolfe 1979, 284). *bry.an'ti.i.* M.L. gen. n. *bryantii* of Bryant; named for M.P. Bryant for his pioneering work on methanogens and for the separation and characterization of this organism from the "*Methanobacillus omelianskii*" syntrophic coculture.

   Original description of the type culture by Bryant et al. (1967); description supplemented with material of Bryant et al. (1971) and Langenberg et al. (1968).

   Cells are slender rods with blunt, rounded ends. They often form chains and filaments which are irregularly crooked. Cells are 0.5–1.0 μm wide, and chains and filaments are 10–15 μm long. Gram stain results are variable. Nonmotile. Fimbriae are present.

TABLE A2.1. Morphological and physiological characteristics of the species of the genus *Methanobacterium*[a,b]

Characteristic	1. *M. formicicum*	2. *M. alcaliphilum*	3. *M. bryantii*	4. *M. espanolense*	5. *M. ivanovii*	6. *M. palustre*	7. *M. subterraneum*	8. *M. uliginosum*
Morphology:								
Long rods	+	+	+	+	+	+	−	+
Filaments	+	+	+	−	−	+	−	−
Ultrastructure appears as Gram-positive	+	+	+	+	+	+	+	+
Gram stain results	v	−	v	+	+	+	+	+
Motility	−	−	−	−	−	−	nd	−
Fimbriae	+	nd	+	nd	nd	nd	nd	nd
Catabolic substrates:								
H_2/CO_2	+	+	+	+	+	+	+	+
Formate	+	−	−	−	−	+	+	−
Secondary alcohols	−	nd	NG[c]	NG[c]	−	+	−	−
Methanol or methyl amines	−	−	−	−	−	−	−	−
Acetate	−	−	−	−	−	−	−	−
Growth requirements:								
Chemoautotrophic	+	−	+	NT[d]	+	+	+	+
Acetate stimulates growth	+	−	+	nd	+	nd	−	+
Compounds stimulating growth	−	YE, P	Vit	Vit	−	nd	−	YE, P
Conditions supporting most rapid growth:								
Growth temperature 30–45°C	+	+	+	+	+	+	+	+
Growth pH 5–6.5	−	−	−	+	−	−	−	+
Growth pH 6.5–7.0	−	−	+	−	−	+	−	+
Growth pH 7.0–7.5	+	−	+	−	+	+	+	+
Growth pH 8–9	−	+	−	−	−	−	+	−
Mol% G + C	38–42	57	33–38	34	37	34	54.5	30–34

[a]For symbols see standard definitions; YE, yeast extract; P, peptones; Vit, vitamins.

[b]nd, not determined.

[c]NG, alcohols are oxidized, but do not result in growth.

[d]NT, not tested; cells grew in vitamin-free medium that contained acetate.

Surface colonies, which can reach 1–5 mm in diameter, are flat with diffuse to filamentous edges and have a characteristic gray to light gray-green appearance. Deep colonies are rounded and filamentous. Cells tend to clump in liquid culture.

Ammonia is used as a nitrogen source. Acetate, cysteine, and B vitamins are highly stimulatory for growth.

Habitat: anaerobic digestors and sediments.

Isolated from the coculture "*Methanobacillus omelianskii*", which was originally isolated from an anaerobic digestor.

The mol% G + C of the DNA is: 33–38 (Bd).

Type strain: M.o.H., ATCC 33274, DSMZ 863, OCM 110.

GenBank accession number (16S rRNA): M59124.

Additional Remarks: The type strain is an established neotype (Boone, 1987).

4. **Methanobacterium espanolense** corrig. Patel, Sprott and Fein 1990, 17[VP] (*Methanobacterium espanolae* (sic) Patel, Sprott and Fein 1990, 17).

es.pa.no.len' se. M.L. neut. adj. *espanolense* from Espanola, Ontario, Canada.

Rods, 0.8 × 3–9 μm, often in chains and with some filaments up to 22 μm long. Stain Gram-positive. Nonmotile. Flagella absent.

Surface colonies are circular, slightly raised, creamy to light yellow, with entire margins and 0.5–1.0 mm in diameter after 7–14 d of incubation.

Originally enriched and isolated from primary sludge of a bleach-craft mill.

The mol% G + C of the DNA is: 34 (T_m).

Type strain: GP9, OCM 178.

5. **Methanobacterium ivanovii** Jain, Thompson, Conway de Macario and Zeikus 1988a, 136[VP] (Effective publication: Jain, Thompson, Conway de Macario and Zeikus 1987a, 81).

i.va.no' vi.i. M.L. gen. n. *ivanovii* of Ivanov, after the Russian scientist M.V. Ivanov for his pioneering interest in environmental biogeochemistry and microbial methanogenesis in deep subsurface environments.

Original material supplemented by material from Belyaev et al. (1987, 1983) and Bhatnagar et al. (1984, 1986).

Cells are slightly curved rods with rounded ends, 0.5–0.8 × 1–15 μm, occurring singly or in short chains. Stain Gram-positive. Nonmotile.

Surface colonies yellow-greenish, smooth, and round, 3–6 mm in diameter after 14 d of incubation. Grows with ammonia or dinitrogen (Magot et al., 1986; Souillard and Sibold, 1986) as a nitrogen source. Sulfide, cysteine, or methionine serves as a sulfur source.

The type strain was originally designated strain 31 (Belyaev et al., 1987) and strain Ivanov (Belyaev et al., 1987; Jain et al., 1987a) and was isolated from an oil-bearing sandstone rock core obtained from 1650 m deep in the Bondyuzshkoe oil field on the Kama River of the Tartar Republic of the former Soviet Union.

The mol% G + C of the DNA is: 37 (T_m).

Type strain: 31, DSMZ 2611, OCM 140.

6. **Methanobacterium palustre** Zellner, Bleicher, Braun, Kneifel, Tindall, Conway de Macario and Winter 1990a, 470[VP] (Effective publication: Zellner, Bleicher, Braun, Kneifel, Tindall, Conway de Macario and Winter 1989a, 8).

pa.lus' tre. M.L. neut. adj. *palustre* muddy, indicating its occurrence in a muddy environment.

Cells are rods 0.5 × 2.5–5 µm; occasionally form filaments up to 65 µm long. Cells stain Gram-positive. Nonmotile.

H_2/CO_2, formate, 2-propanol/CO_2, and 2-butanol/CO_2 are catabolic substrates. No organic growth factors are required. Ammonia serves as sole nitrogen source; sulfide serves as sole sulfur source.

Isolated from marshy soil.

The mol% G + C of the DNA is: 34 (T_m).

Type strain: F, DSMZ 3108, OCM 238.

7. **Methanobacterium subterraneum** Kotelnikova, Macario and Pedersen 1998, 365.[VP]

sub.ter.ra' ne.um. L. adj. neut. *subterraneum* underground, below the earth/soil surface.

Small, thin rods, 0.1–0.15 × 0.6–1.2 µm, often in aggregates but not in chains. Nonmotile. Substrates used for growth and methane production include H_2/CO_2 and formate, but not methylamines, acetate, pyruvate, dimethyl sulfide, methanol or other alcohols plus CO_2. Grows autotrophically in mineral medium without any organic additions. Growth inhibited by yeast extract (2 g/l), Casamino acids (1 g/l), isobutyric acid (5 mg/l), n-butyric acid (5 mg/l), Na_2SeO_3 (2 mg/l), $ZnCl_2$ (2 mg/l), $CoCl_2$ (2 mg/l), $NiCl_2$ (20 mg/l), and $MnCl_2$ (20 mg/l). Vitamins are not essential for growth. Optimum growth temperature 20–40°C (range: 3.6–45°C). Optimum pH 7.8–8.8 (range: 6.5–9.2). Optimum NaCl concentration 0.2 M (range: 0.2–1.4M). Isolated from

granitic rock groundwater from the Äspö hard rock laboratory tunnel, south-eastern Sweden. Strain A8p[T] was isolated from granitic groundwater at a depth of 68 m. Reference strains 3067 and C2BIS (DSM 11075) were isolated from granitic groundwater at depths of 409 and 420 m, respectively.

The mol% G + C of the DNA is: 54.5 ± 0.5 (T_m)

Type strain: A8p, DSM 11074.

GenBank accession number (16S rRNA): X99044.

8. **Methanobacterium uliginosum** König 1985, 375[VP] (Effective publication: König 1984, 1480).

u.li.gi.no' sum. M.L. neut. adj. *uliginosum* occurring wet, since it occurs in marshy soil.

Cells are rods 0.2–0.6 × 1.9–3.8 µm; some spherical cells may be produced at the ends of the rods, and they may remain attached or be released. Cells stain Gram- positive. Nonmotile.

H_2/CO_2 is the sole catabolic substrate. Ammonia serves as sole nitrogen source; sulfide probably serves as sole sulfur source, although this was not tested in the absence of L-cysteine.

Isolated by inoculation of enrichment cultures with marshy soil followed by treatment with antibiotics and purification by dilution in liquid medium.

The mol% G + C of the DNA is: 29.4 (T_m) or 33.8 (determined by direct nucleotide analysis).

Type strain: P2St, ATCC 35997, DSMZ 2956, OCM 176.

Genus II. **Methanobrevibacter** *Balch and Wolfe 1981, 216[VP] (Effective publication: Balch and Wolfe in Balch, Fox, Magrum, Woese and Wolfe 1979, 284)*

TERRY L. MILLER

Me.tha.no.bre.vi.bac' ter. M.L. neut. n. *methanum* methane; L. masc. adj. *brevis* short; M.L. masc. n. *bacter* equivalent of Gr. neut. n. *bakterion* rod, staff; M.L. masc. n. *Methanobrevibacter* short methane (-producing) rod.

Oval rods or cocci to short rods, usually occurring in pairs or chains; about 0.5–0.7 µm in width and 0.8–1.4 µm in length. Rarely, filaments are formed. Nonsporing, Gram-positive. **Cell walls are composed of pseudomurein.** Nonmotile. Strict anaerobes. Optimum temperature, 37–40°C; maximum, ~45°C; minimum, ~30°C.

Energy for growth is obtained by reduction of CO_2 to CH_4 by using H_2 and sometimes formate as the electron donor. Acetate, methanol, methylamines, or other organic compounds are not used as electron donors for CH_4 formation. NH_4^+ is a major source of cell nitrogen. One or more B-complex vitamins are required for growth. Acetate may be a major source of cell carbon.

The mol% G + C of the DNA is: 27.5–31.6.

Type species: **Methanobrevibacter ruminantium** (Smith and Hungate 1958) Balch and Wolfe 1981, 216 (Effective publication: Balch and Wolfe *in* Balch, Fox, Magrum, Woese and Wolfe, 1979, 284) (*Methanobacterium ruminantium* Smith and Hungate 1958, 717).

FURTHER DESCRIPTIVE INFORMATION

M. ruminantium, Methanobrevibacter smithii, and *Methanobrevibacter arboriphilus* were established on the basis of differences in 16S rRNA oligonucleotide catalog values (Balch et al., 1979). The 16S rRNA gene sequences of six of the seven species of the genus

have been determined. A comparison of the sequence similarities of six species is shown in Table A2.2.

Cells are coccobacillary with tapered ends to short rods with rounded ends (Fig. A2.2). They occur singly but more often in pairs or short chains and may appear in long chains. Cell walls are composed of pseudomurein (König et al., 1982). Pseudomurein is composed of N-acetyl amino sugars, L-amino acids, and neutral sugars. The glycan moieties of *M. ruminantium* and *M. smithii* contain D-glucosamine, D-galactosamine, and L-talosaminuronic acid. *M. arboriphilus* (strains DH1 and AZ) contains only D-galactosamine and D-talosaminuronic acid. The peptide moiety

TABLE A2.2. 16S rRNA gene sequence similarity values for *Methanobrevibacter* species[a]

| Species | % Sequence Similarity | | | | |
	DH1	RFM–2	RFM–1	RFM–3	PS
M. ruminantium M1	94.5	93.9	93.9	94.4	94.5
M. arboriphilus DH1		95.8	97	95.7	95
M. curvatus RFM–2			95.9	95.7	94.1
M. cuticularis RFM–1				95.2	94.2
M. filiformis RFM–3					93.6
M. smithii PS					

[a]Values are based on the percent differences among 1164 unambiguously aligned nucleotides (Leadbetter and Breznak,1996; J.R. Leadbetter and J.A. Breznak, personal communication).

of all species contains L-alanine, L-glutamate, and L-lysine; however, L-threonine can partially or completely replace L-alanine in the wall of *M. ruminantium*. *M. smithii* (strain PS) cell wall contains ornithine as an additional component of the peptide moiety. *M. ruminantium* and *M. arboriphilus* (DH1) have high phosphate levels in their cell walls. The lipid composition of strains of *M. ruminantium*, *M. smithii*, and *M. arboriphilus* have been examined (Tornabene and Langworthy, 1979; Tornabene et al., 1979; Morii et al., 1988). The lipids of the species differ primarily in the composition of the isoprenoid hydrocarbon neutral lipid.

Morphological and ultrastructural features of the type strains of six species are shown in Figs. A2.2, A2.3, and A2.4.

M. ruminantium and *M. smithii* require acetate as a major source of cell carbon. Acetate is a precursor of 60% of the cell carbon of *M. ruminantium* (Bryant et al., 1971). CO_2 can serve as the sole carbon source of *M. arboriphilus*, but one or more B-complex vitamins are required for growth. *M. ruminantium* requires 2-methylbutyrate, 2-mercaptoethanesulfonic acid (coenzyme M), and a mixture of amino acids (Bryant et al., 1971; Taylor et al., 1974). Trace metal requirements have not been determined, although *M. smithii* was shown to require nickel (Diekert et al., 1981). All of the species grow well with H_2/CO_2 as energy sources. *M. ruminantium* and *M. smithii* can use formate as an energy source, but growth is usually slow and cultures do not grow to the extent observed with H_2 and CO_2. *M. arboriphilus* does not usually grow with formate, although a sewage sludge isolate and a rice paddy isolate that are morphologically and immunologically similar to *M. arboriphilus* grow with formate as the sole energy source (Morii et al., 1983; Asakawa et al., 1993). *Methanobrevibacter curvatus*, *Methanobrevibacter cuticularis*, and *Methanobrevibacter filiformis* do not use formate as an energy source (Leadbetter and Breznak, 1996; Leadbetter et al., 1998a).

Some features of the biosynthetic capabilities of *Methanobrevibacter* have been studied by assessing the incorporation of ^{13}C-labeled precursors into cell macromolecules. *M. smithii* incorporates acetate and CO_2 into cell carbon by reductive carboxylation of acetate to form pyruvate (Choquet et al., 1994b). *M. smithii* and *M. arboriphilus* use the reductive TCA cycle for synthesis of the glutamate family of amino acids (Sprott et al., 1993). *M. smithii* synthesizes ribose via the oxidative branch of the pentose phosphate pathway (Choquet et al., 1994b). *M. smithii* synthesizes hexose by the reverse of the Embden–Meyerhof–Parnas pathway (Chouquet et al., 1994b). *M. smithii* incorporates either CO_2 or formate into the C-2 and C-8 of purines (Choquet et al., 1994a). The ^{13}C-labeling studies indicate that *Methanobrevibacter* synthesizes amino acids by conventional pathways.

M. smithii and *M. arboriphilus* are resistant to many antibiotics that inhibit eubacterial membrane function or cell wall, RNA, or protein synthesis (Hilpert et al., 1981; Pecher and Böck, 1981).

Methanobrevibacter species occur in ruminant, human, and other animal gastrointestinal tracts, termite hindgut, human oral cavity, municipal sewage sludges, decaying woody tissues, and rice paddy soil.

ENRICHMENT AND ISOLATION PROCEDURES

All enrichments and isolations must be carried out under strictly anaerobic conditions. Isolation and cultivation procedures are based on the techniques developed by Hungate (1969). Serum bottle modifications of the Hungate technique (Miller and Wolin, 1974; Balch and Wolfe, 1976) allow the use of syringes for additions and incubation under elevated gas pressures to increase the availability of the energy sources, H_2/CO_2. Liquid cultures

with H_2 and CO_2 are incubated with rotation or shaking. Anaerobic glove boxes based on the design of Aranki and Freter (1972) facilitate the isolation and handling of pure cultures.

M. smithii can be enumerated and isolated semiselectively from human fecal samples by plating on medium 1 of Balch et al. (1979) supplemented with 0.1% additional NH_4Cl, 10% rumen fluid, 2% agar (modified medium 1), and with clindamycin and cephalothin (Miller and Wolin, 1982). Many eubacteria are inhibited by these antibiotics. Methanogens are insensitive, owing to their unique macromolecular properties. The roll tubes are incubated statically at 37°C under 80% H_2 and 20% CO_2 gas phase. The presence of methanogens is confirmed by gas chromatographic analysis for CH_4 in the headspaces of the roll tubes. A single methanogenic colony can produce detectable CH_4. Colonies are picked from the roll tubes having the most dilute inocula and detectable methane and are subcultured in liquid medium. The purity of cultures is established by noting that all cells show F_{420} fluorescence when viewed with epifluorescence microscopy and show lack of growth in complex media with energy sources other than H_2 and CO_2 or formate. The above-mentioned antibiotic medium may also be used to enrich or enumerate *Methanobrevibacter* species in rumen contents and animal feces (Miller et al., 1986a, b). The antibiotic medium is not specific for *Methanobrevibacter* and may be useful for enriching and/or enumerating other methanogens in other ecosystems.

MAINTENANCE PROCEDURES

M. ruminantium, *M. smithii*, and *M. arboriphilus* are maintained for short periods (weeks) on agar slants of modified medium 1 without antibiotics (see above). A broth culture (24–48 h, 0.1–0.3 ml) is inoculated into a tube containing 10 ml of reduced agar medium and 1 ml of reduced liquid medium. Inoculated tubes are regassed and pressurized to 2 atm with 80% H_2 and 20% CO_2 and incubated statically and horizontally at 37°C. After growth, the head space is regassed with H_2/CO_2, and the culture is stored at 4°C. Cultures are directly transferred to fresh slant tubes every two weeks.

M. ruminantium, *M. smithii*, and *M. arboriphilus* are preserved for longer periods (months) by preparing agar–liquid cultures on the basis of biphasic culture techniques (Krieg and Gerhardt, 1981; Miller and Wolin, 1985a; Miller et al., 1986b). Double-strength modified medium 1 without antibiotics and containing 3% Difco agar is prepared and dispensed into serum bottles under an atmosphere of 80% N_2 and 20% CO_2. After autoclaving, double-strength reducing agent is added, the headspace of the bottle is replaced with 80% H_2 and 20% CO_2, and the bottle is laid on its side. When the agar solidifies, an amount of reduced single-strength broth medium is added in the ratio of 1 volume of liquid medium to 3 volumes of solid medium. An inoculum equivalent to 10% of the liquid volume is added, and the bottles are pressurized to 2 atm with 80% H_2 and 20% CO_2. The cultures are incubated at 37°C with gentle rocking and regassed and pressurized 1–2 times daily until an OD (optical density) of >2.0 (1-cm cuvette) is obtained. After outgrowth, biphasic culture bottles are regassed and repressurized with 80% H_2 and 20% CO_2, precooled for 1 h at 4°C, and stored at −76°C.

Cultures are removed every 6–12 months, rapidly thawed under warm running water, and transferred by using a 10% inoculum into reduced single-strength broth medium. Cells of *M. smithii* remain viable after 1–2 years of storage at −76°C. Addition of sterile glycerol (20%, v/v) to biphasic cultures of *Methanobrevibacter* species did not enhance viability and resulted in growth

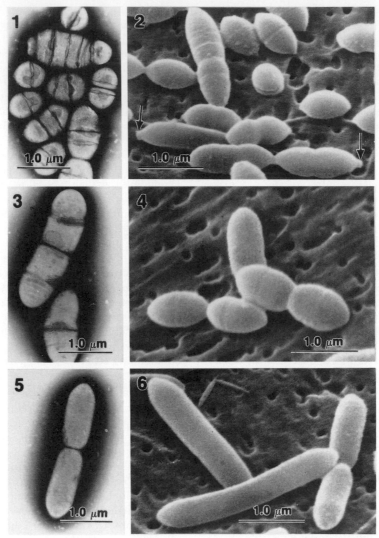

FIGURE A2.2. Morphology of *Methanobrevibacter ruminantium*, *M. smithii*, and *M. arboriphilus*. The three species were grown in liquid modified medium I (see text) with 2 atm H_2/CO_2 (80%:20%) at 37°C. Cultures were regassed and repressurized 1–2 times daily for 2–4 days. Final OD_{660} (d = 1 cm) was *M. ruminantium*, 1.3 (72 h); *M. smithii*, 3.6 (96 h); *M. arboriphilus*, 1.2 (48 h). *Part 1.* Negative stain preparation, 2% sodium phosphotungstate, pH 7 (2% NAPTA) of *M. ruminantium* M1. The multiple septa of the cells are penetrated by the stain. *Part 2.* Scanning electron micrograph (SEM) of *M. ruminantium* M1. Cells from culture fluid were collected on a filter with mild vacuum and fixed *in situ* with 2% glutaraldehyde in 0.09 N sodium cacodylate, pH 7.2, for 30 min followed by rinsing with the same buffer and fixing in 1% osmium in veronal acetate buffer (Kellenberger et al., 1958) for 1 h. After rinsing in double distilled water and dehydration in a graded ethanol series, the material was critical-point-dried from liquid CO_2 and sputter-coated with gold. The division septa are apparent; cell ends appear more tapered than those seen by negative staining (part 1), and remnants of wall material from a recent cell division are observed (*arrows*). *Part 3.* Negative stain (2% NAPTA) of *M. smithii*, strain PS. Multiple septa are penetrated by the stain. *Part 4.* SEM of *M. smithii*, strain PS (prepared as for Part 2). Some cell septa are visible, and the cell ends are rounder than *M. ruminantium* (part 2). *Part 5.* Negative stain (2% NAPTA) of *M. arboriphilus* DH1. The cell surface is smooth. *Part 6.* SEM of *M. arboriphilus*, strain DH1. Septa are not present. The ends of the cells are slightly truncated.

lags of 1–2 d when transfers were made into a liquid medium. This method of culture storage is routinely used by the author to stock a wide variety of other genera and species of methanogens, including thermophilic species. Biphasic cultures have also been used to obtain sufficient cells for cell wall and DNA analyses (Miller and Wolin, 1985a; Miller et al., 1986b; König, 1986).

DIFFERENTIATION OF THE GENUS *METHANOBREVIBACTER* FROM OTHER GENERA

In 1979, the genus *Methanobrevibacter* was phylogenetically differentiated from other genera by comparison of oligonucleotide catalog values of 16S rRNA (Balch et al., 1979). The genus is currently phylogenetically differentiated from other genera on the basis of differences in 16S rRNA gene sequence. *Methano-*

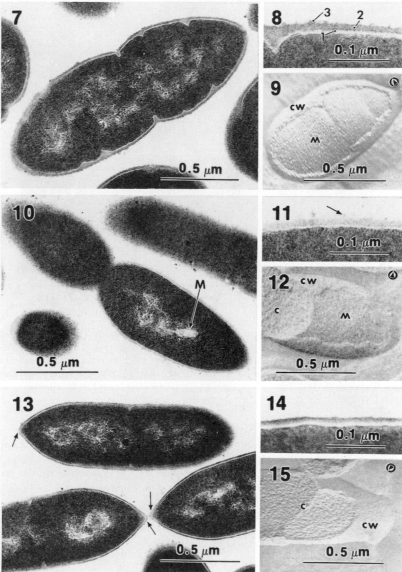

FIGURE A2.3. Ultrastructure features of *Methanobrevibacter ruminantium*, *M. smithii*, and *M. arboriphilus*. Cultures were grown as described in the legend to Fig. A2.2. *Part 7*. Thin section of *M. ruminantium* M1, prepared as described by Samsonoff et al. (1970). The thick cell wall is invaginated at multiple septum sites. *Part 8*. Thin section through the cell wall of *M. ruminantium* M1. The cell wall is composed of three layers as previously reported by Zeikus and Bowen (1975a): a thin electron dense inner layer (*1*), a thicker less electron dense middle layer (*2*), and a rough irregular outer layer (*3*). *Part 9*. Freeze-etched preparation of *M. ruminantium* M1. Cells were harvested by centrifugation, frozen in freon-22, stored in liquid N_2, and fractured in a Balzers 360M freeze-fracture device. After 1 min of etching at 110°C, the samples were shadowed with carbon platinum. No organized structural patterns can be seen in the cell wall (*cw*). Fractures through the cytoplasmic membrane (*M*) revealing the protoplasmic face (Branton et al., 1975) are frequent. The encircled arrow indicates the direction of the shadow. *Part 10*. Thin section of *M. smithii* PS (prepared as for part 7). Membranous structures (*M*) are frequently seen near the nucleoid and in some instances extend to the cytoplasmic membrane (not shown). *Part 11*. Thin section through the cell wall of *M. smithii* PS. The wall appears as a single thick electron-dense layer with a rough irregular outer surface (*arrow*). *Part 12*. Freeze-etched preparation of *M. smithii* PS (prepared as for part 9). No organized structural pattern is seen in the cell wall (*cw*). Most fractures are through the cytoplasm (*c*) and only occasionally occur through the cytoplasmic membrane (*M*), revealing the protoplasmic face. The encircled arrow indicates the direction of the shadow. *Part 13*. Thin section of *M. arboriphilus* DH1 (prepared as for part 7). The cell ends are slightly truncated (*arrows*). *Part 14*. Thin section through the cell wall of *M. arboriphilus* DH1. The cell wall appears as a single layer which is more electron dense toward the outer surface. *Part 15*. Freeze-etched preparation of *M. arboriphilus* DH1 (prepared as for part 9). The cell wall (*cw*) has no apparent organized structural pattern. Fractures occur frequently through the cytoplasm (*c*) but not through the cytoplasmic membrane. The encircled arrow indicates the direction of the shadow.

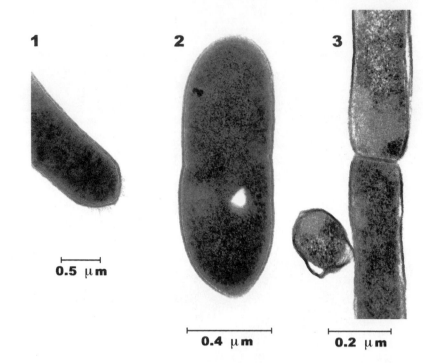

FIGURE A2.4. Morphology of *Methanobrevibacter curvatus (Part 1)*, *M. cuticularis (Part 2)*, and *M. filiformis (Part 3)*. The three species were grown in liquid JF1 medium with 0.1% yeast extract with H_2/CO_2(80%:20%) until late exponential phase (J. Breznak, personal communication). (Parts 1 and 2 are reprinted with permission from J.R. Leadbetter and J.A. Breznak, Applied and Environmental Microbiology *62*: 3620–3631, (1996), ©American Society for Microbiology, Washington, D.C. Part 3 is reprinted with permission from J.R. Leadbetter et al., Archives of Microbiology *169*: 287–292, 1998, ©Springer-Verlag, Berlin.)

brevibacter gives a positive Gram reaction, and all other genera, except *Methanosarcina*, *Methanobacterium*, and *Methanosphaera*, are Gram-negative. *Methanobrevibacter* is distinguished from *Methanosarcina* on the basis of morphology, energy sources for growth, and the presence of pseudomurein in *Methanobrevibacter* cell walls. *Methanobrevibacter* is differentiated from *Methanobacterium* on the basis of morphology. *Methanobrevibacter* is differentiated from *Methanosphaera* on the basis of morphology and energy sources for growth.

The differentiation of species of *Methanobrevibacter* on the basis of phenotypic differences is unsatisfactory because of the lack of distinguishing morphological, biochemical, and physiological characteristics. There are phenotypic differences among strains, for example, bile sensitivity, formate utilization, and requirements for acetate, coenzyme M, 2-methylbutyrate, and probably amino acids. However, the limited number of markers and the lack of information about their distribution among strains of *Methanobrevibacter* species require the use of more powerful molecular tools for establishing phylogenetic relationships. Sequence analysis of the 16S rRNA gene may reveal phylogenetic relationships of isolates to existing species. However, *Methanobrevibacter* isolates from different animal intestinal ecosystems share >97% 16S rRNA gene sequence similarity but little genomic DNA similarity (Lin and Miller, 1998). In such cases, genomic DNA reassociation studies are essential for differentiating new organisms at the species level (Lin and Miller, 1998).

Acknowledgments

I thank W.A. Samsonoff for electron microscopic analyses of *M. ruminantium*, *M. smithii*, and *M. arboriphilus* and for the electron micrographs presented in this description.

Further Reading

Balch, W.E., G.E. Fox, L.J. Magrum, C.R. Woese and R.S. Wolfe. 1979. Methanogens: reevaluation of a unique biological group. Microbiol. Rev. *43*: 260–296.

Ferrari, A., T. Brusa, A. Rutili, E. Canzi and B. Biavati. 1994. Isolation and characterization of *Methanobrevibacter oralis* sp. nov. Curr. Microbiol. *29*: 7–12.

Leadbetter, J.R. and J.A. Breznak. 1996. Physiological ecology of *Methanobrevibacter cuticularis* sp. nov. and *Methanobrevibacter curvatus* sp. nov., isolated from the hindgut of the termite *Reticulitermes flavipes*. Appl. Environ. Microbiol. *62*: 3620–3631.

Zeikus, J.G. and D.L. Henning. 1975. *Methanobacterium arbophilum* sp. nov. an obligate anaerobe isolated from wetwood of living trees. Antonie Leeuwenhoek *41*: 543–552.

List of species of the genus Methanobrevibacter

1. **Methanobrevibacter ruminantium** (Smith and Hungate 1958) Balch and Wolfe 1981, 216[VP] (Effective publication: Balch and Wolfe *in* Balch, Fox, Magrum, Woese and Wolfe 1979, 284) (*Methanobacterium ruminantium* Smith and Hungate 1958, 717[AL]).

ru.mi.nan′ti.um. L. part. adj. *ruminans, ruminantis* ruminating; M.L. neut. pl. n. *ruminantia* ruminants; M.L. pl. gen. *ruminantium* of ruminants.

Only one strain of the species (M1 or DSMZ 1093), isolated from bovine rumen contents by Bryant (1965), has been phylogenetically characterized. The following features are based on studies of this strain.

Short oval rod or coccobacillus with tapered ends 0.7 μm in width and 0.8–1.7 μm in length. Cells occur predominantly in pairs in young cultures and in chains in older cultures. Strong Gram-positive reaction, even in relatively old cultures. Nonmotile. Flagella appear to be absent by negative staining or freeze fracture procedures (W.A. Samsonoff, personal communication). Langenberg et al. (1968) first described the coccoid appearance and the presence of large numbers of cross- walls, presumably because the cells are constantly dividing (Fig. A2.2, parts 1 and 2). The cell wall is composed of three layers (Fig. A2.3, part 8).

Surface colonies on organically complex agar medium in roll tube cultures with H_2/CO_2 gas phase are translucent, convex, and circular with entire margins and are frequently light yellow. They may be visible after ~3 d of incubation at 37°C and can reach a diameter of 3–4 mm, depending on the number of colonies and the availability of the energy source. Colonies in deep agar are lenticular.

M. ruminantium requires acetate as a major source (60%) of cell carbon (Bryant et al., 1971). In addition to one or more B vitamins, it requires 2-mercaptoethanesulfonic acid (coenzyme M), 2-methylbutyric acid, and a mixture of amino acids (Bryant, 1965; Bryant et al., 1971; Taylor et al., 1974). Coenzyme M transport is energy dependent and is inhibited by 2-bromoethanesulfonic acid (Balch and Wolfe, 1979b). Radioisotopic studies indicate that 2-methylbutyric acid is a precursor of isoleucine via a reductive carboxylation pathway (Robinson and Allison, 1969). The amino acids cannot replace NH_4^+ as the major source of cell nitrogen. Growth is inhibited in modified medium 1 containing 2% oxgall and 0.1% sodium deoxycholate (T.L. Miller, unpublished data). H_2 and CO_2 serve as energy sources. Cells contain methanofuran and tetrahydromethanopterin (Jones et al., 1985). Growth with formate as an energy source is slow, and cultures do not grow to the same optical density as with H_2 and CO_2. Formate is probably first converted to H_2 and CO_2 via a formate dehydrogenase coupled to a hydrogenase. A magnesium- and ATP-dependent methyl-coenzyme M methylreductase system has been demonstrated in cell-free extracts (Gunsalus and Wolfe, 1978; Romesser and Wolfe, 1982).

Rabbit antisera and the corresponding antigen preparations of strain M1 do not cross-react with antigens and antisera, respectively, of other species of the genus or other members of the family (Conway de Macario et al., 1982c).

There is limited information on the antibiotic sensitivity of the strain. Clindamycin (2 μg/ml) and cephalothin (8 μg/ml) are not lethal in liquid or solid media. Bacitracin (100 μg/ml) inhibits growth in liquid medium (T.L. Miller, unpublished data). Growth and CH_4 production are inhibited by 2-bromoethanesulfonate unless its molar concentration is exceeded by that of coenzyme M (Balch and Wolfe, 1979a).

The strain isolated by Smith and Hungate (1958) was present in rumen contents in concentrations of 10^6–10^8 per ml of rumen contents in grass- and/or alfalfa-fed steers. It is no longer in extant culture. Both coenzyme M-requiring and coenzyme M-nonrequiring *Methanobrevibacter* strains are present in high concentrations in bovine rumen contents (Lovley et al., 1984 and Miller et al., 1986a). The taxonomic relationship of the *Methanobrevibacter* species isolated in these

studies to the *M. ruminantium* type strain or to each other has not yet been clearly established (Miller et al., 1986a).

The mol% G + C of the DNA is: 30.6 (Bd).

Type strain: M1, DSMZ 1093.

Additional Remarks: The 16S rRNA sequence has not been deposited with GenBank. The Ribosomal Database Project designation for the 16S rRNA gene sequence is Mbb.rumina.

2. **Methanobrevibacter arboriphilus** corrig. (Zeikus and Henning 1975) Balch and Wolfe 1981, 216[VP] (Effective publication: Balch and Wolfe *in* Balch, Fox, Magrum, Woese and Wolfe 1979, 284) (*Methanobrevibacter arboriphilicus* [sic] Balch and Wolfe 1981, 216) (*Methanobacterium arbophilicum* Zeikus and Henning 1975, 550[AL]).

ar.bo.ri.phi' lus. L. gen. n. *arbor* tree; Gr. adj. *philos* loving; M.L. masc. adj. *arboriphilus* tree-loving.

Five strains of *M. arboriphilus* are presently in pure culture: strains DH1 (DSMZ 1125), DC (DSMZ 1536), AZ (DSMZ 744), A2 (DSMZ 2462), and SA (DSMZ 7056). Strains DH1 and DC have identical 16S rRNA oligonucleotide catalog similarity values (Balch et al., 1979). The 16S rRNA oligonucleotide catalog of strain AZ has a similarity index value of 0.84 with strains DH1 and DC. The 16S rRNA gene sequence of strain DH1 has been determined. Strain AZ has <70% genomic DNA reassociation values with the other strains of the species and may represent a different species.

Some physiological features of the strains are summarized in Table A2.4.

Cells of DH1 grown in liquid culture are short rods with rounded ends, 0.5 μm in width and 1.2–1.4 μm in length. Some cells may have a slightly truncated end (Fig. A2.3, part 13). They occur singly or in pairs. In agar medium, cells are elongated, often as much as 12 times the length of cells grown in liquid medium (Zeikus and Henning, 1975). Cells in liquid culture tend to clump together and are not easily dispersed by vigorous shaking or vortexing. Gram positive. Nonmotile. Cells of strain AZ were reported to have a single polar flagellum (Doddema et al., 1979). Flagella appear to be absent from cells of strain DH1 by negative stain or freeze fracture procedures (W.A. Samsonoff, personal communication). Multiple division septa are not usually present (Fig. A2.2, parts 5 and 6). The cell wall appears as a single electron-dense layer (Fig. A2.3, part 14).

Surface colonies in organically complex agar medium in roll tubes with H_2 and CO_2 gas phase are roughly round, diffuse, or filamentous, and creamy white to yellow or dark brown. Mature colonies do not exceed 5 mm in diameter.

Carbon dioxide is the major and possibly the sole source of cell carbon; however, one or more B vitamins are required for good growth. Growth is stimulated by Trypticase pep-

TABLE A2.3. Genomic DNA reassociation of *Methanobrevibacter smithii* strains[a]

Strain	% Reassociation with labeled DNA from		
	PS	ALI	B181
PS	100	112	110
ALI	88	100	113
B181	78	91	100

[a]Data are from Miller and Wolin (1986). DNA reassociation was measured using the direct-binding nitrocellulose membrane method (Johnson, 1984).

tones, yeast extract, and rumen fluid. Growth of strains DH1, DC, and AZ is inhibited in modified medium 1 containing 2% oxgall and 0.1% sodium deoxycholate (T.L. Miller, unpublished data). Strain SA grows in media containing 0 to 0.1 M NaCl. H_2 and CO_2 may be the sole or preferred energy sources. Methane formation from H_2 and CO_2 by cell suspensions of strain AZ is dependent on sodium ions (Perski et al., 1982). Growth with formate as a sole energy source has been reported for strains A2 and SA (Morii et al., 1983; Asakawa et al., 1993). The other strains do not use formate as an energy source. Extracts of DH1 oxidize CO with reduction of benzyl viologen, but CO cannot substitute for H_2 as the electron donor for CO_2 reduction to CH_4 (Daniels et al., 1977). Strains DH1 and AZ synthesize coenzyme M (Balch and Wolfe, 1979a). Cell extracts of strain AZ have formyl-methanofuran dehydrogenase, methylene-tetrahydromethanopterin (H4MPT) dehydrogenase, methylene-H4MPT reductase, and heterodisulfide reductase activities (Schwörer and Thauer, 1991). Strain DH1 has a reductive tricarboxylic acid pathway (Sprott et al., 1993). Strain DH1 contains corrinoids (Krzycki and Zeikus, 1980). Strain AZ has a low polyamine content (Scherer and Kneifel, 1983). Strains A2 and SA have the tetraether type of lipids (Asakawa et al., 1993).

The lack of detectable immunologic cross-reactivity between reciprocal rabbit antisera and antigens between strains DH1, DC, and AZ and the existence of distinct immunovars indicate strain differences (Table A2.4; Conway de Macario et al., 1982a). Strain A2 weakly cross-reacts with strain DC antiserum but not with strain AZ or DH1 antisera (Morii et al., 1983). Cells of strain SA cross-react with strain A2 antiserum (Asakawa et al., 1993).

Strain DH1 is inhibited (2 μg/ml) by chloramphenicol and ansiomycin (Pecher and Böck, 1981). The following antibiotics produce zones of inhibition (13–23 mm) with strain AZ: bacitracin, gardimycin, enduracidin, chloramphenicol, gentamicin, and lasalocid (Hilpert et al., 1981).

The genomic DNA reassociation values of strains DH1, DC, AZ, A2, and SA are shown in Table A2.5. The low degree of genomic DNA reassociation of strain AZ with other strains of the species suggests that it is a member of a different species (Asakawa et al., 1993).

The type strain was isolated from enrichments of decaying cottonwood tissue (Zeikus and Henning, 1975). Strains AZ

and A2 were isolated from enrichments of anaerobic sewage sludge (Zehnder and Wuhrmann, 1977; Morii et al., 1983). Strain DC was isolated in the laboratory of R.S. Wolfe from an anaerobic sewage sludge enrichment provided by D. Castignetti (D. Castignetti, personal communication). Strain SA was isolated from rice paddy soil (Asakawa et al., 1993).

The mol% G + C of the DNA is: 25.5–31.6 (Bd or T_m).

Type strain: DH1, DSMZ 1125.

Additional Remarks: Reference strains include DC (DSMZ 1536), AZ (DSMZ 744), A2 (DSMZ 2462), and SA (DSMZ 7056). The 16S rRNA sequence has not been deposited with GenBank. The Ribosomal Database Project designation for the 16S rRNA gene sequence is Mbb.arbori.

3. **Methanobrevibacter curvatus** Leadbetter and Breznak 1997, 601[VP] (Effective publication: Leadbetter and Breznak 1996, 3629).

cur.va' tus. L. particip. *curvatus* bent, referring to the curved shape of the cell.

Curved rods with slightly tapered ends, 0.34 by 1.6 μm in size, occurring singly or in pairs. Nonmotile. Gram-positive-like by staining and cell wall ultrastructure. Cells have polar fibers of 3 by 300 nm. No endospores formed.

Strict anaerobe. Catalase positive, oxidase negative. Metabolizes H_2 and CO_2, yielding CH_4 as the sole product. Methanol, methanol plus H_2, CO, formate, acetate, ethanol, isopropanol, trimethylamine, dimethylamine, theobromine, theophylline, trimethoxybenzoate, lactate, pyruvate, and glucose are not metabolized.

Optimum temperature is 30°C (range 10–30°C). Optimum pH is 7.1–7.2 (range 6.5–8.5). Complex nutritional supplements, e.g., 40% (v/v) clarified rumen fluid and nutrient broth (Difco) are required for growth.

Strain RFM-2 was isolated from hindgut contents of the termite *Reticulitermes flavipes* (Kollar) (Rhinotermitidae).

The mol% G + C of the DNA is: not known.

Type strain: RFM-2, DSMZ 11111.

GenBank accession number (16S rRNA): U62533.

4. **Methanobrevibacter cuticularis** Leadbetter and Breznak 1997, 601[VP] (Effective publication: Leadbetter and Breznak 1996, 3629).

cu.ti.cu.la' ris. L. fem. n. *cuticula* dim. skin; L. fem. adj. *cuticularis* referring to the cuticular surface of the termite hindgut epithelium, which is colonized by this organism.

Straight short rods with slightly tapered ends, 0.4–1.2 μm in size, occurring singly, in pairs, or in short chains. Non-

TABLE A2.4. Physiological features of *Methanobrevibacter arboriphilus*[a]

Characteristic	Strain				
	DH1	DC	AZ	A2	SA
Mol% G + C (Bd)	27.5	27.7	31.6	29.6	nd
Mol% G + C (T_m)	25.8	26.2	25.5	25.9	26.4
Serology:[b]					
DH1 antiserum	4+	−	−	−	nd
DC antiserum	−	4+	−	1+	nd
AZ antiserum	−	−	4+	−	nd
A2 antiserum	nd	nd	nd	4+	3+
16S rRNA Sab[c]	1	1	0.84	nd	nd
Energy source:					
H_2/CO_2	Yes	Yes	Yes	Yes	Yes
Formate	No	No	No	Yes	Yes

[a] Symbols: −, no reaction; nd, not determined.

[b] Data from Conway de Macario et al. (1982b) and Asakawa et al. (1993).

[c] Relative to DH1 (See Balch et al., 1979).

TABLE A2.5. Genomic DNA reassociation of *Methanobrevibacter arboriphilus* strains[a]

Strain	% Reassociation with labeled DNA from				
	DH1	DC	AZ	A2	SA
DH1	100	66 (105)	39 (72)	76	76
DC	74 (60)	100	42 (70)	86	92
AZ	31 (38)	34 (54)	100	40	35
A2	60	80	35	100	88
SA	73	79	37	108	100

[a] Data are from Asakawa et al. (1993) and (in parentheses) from Miller and Wolin (1986). DNA reassociation was measured using the direct-binding nitrocellulose membrane method (Johnson, 1984).

motile. Cells stain Gram-positive. TEM of thin sections shows that the cell wall lacks an outer membrane and resembles that of Gram-positive members of the *Bacteria*. No endospores are formed.

Strict anaerobe. Oxygen tolerant. Catalase positive, oxidase negative. H_2 and CO_2 are the preferred energy sources. Formate is a poor substrate for methanogenesis. Methanol, methanol plus H_2, CO, acetate, ethanol, isopropanol, trimethylamine, dimethylamine, theobromine, theophylline, trimethoxybenzoate, lactate, pyruvate, and glucose are not metabolized.

Optimum temperature is 37°C (range 10–37°C). Optimum pH is 7.7 (range 6.5–8.5). Yeast extract, a source of amino acids, and ~2.0% clarified rumen fluid are markedly stimulatory to growth.

Strain RFM-1 is resistant to rifamycin SV and cephalothin (10 µg/ml each).

Strain RFM-1 was isolated from hindgut contents of the termite *Reticulitermes flavipes* (Kollar) (Rhinotermitdae).

The mol% G + C of the DNA is: not known.

Type strain: RFM-1, DSMZ 11139.

GenBank accession number (16S rRNA): U41095.

5. **Methanobrevibacter filiformis** Leadbetter, Crosby and Breznak 1998b, 1083[VP] (Effective publication: Leadbetter, Crosby and Breznak 1998a, 291).

fi.li.for' mis. L. neut. n. *filum* a thread; L. fem. n. *forma* shape; M.L. masc. adj. *filiformis* thread shaped.

Filament-forming rods with slightly tapered ends, 0.23–0.28 µm in width by up to several hundred µm in length. Septation within filaments typically occurs at ~4-µm intervals. Rarely occurs as single 4-µm-long cells. Nonmotile. Gram-positive-like by staining and by cell wall ultrastructure. No endospores are formed.

Strict anaerobe. Catalase positive. Metabolizes H_2 and CO_2 to CH_4. Methanol, methanol/H_2, formate, CO, acetate, ethanol, isopropanol, trimethylamine, dimethylamine, theobromine, theophylline, trimethoxybenzoate, lactate, pyruvate, and glucose are not metabolized.

Optimum temperature is 30°C (range 10–33.5°C). Optimum pH is 7.0–7.2 (range 6.0–7.5). Yeast extract (>0.01%) is required for growth. Growth is inhibited in media with 1 mM cysteine or sulfide as a reducing agent, but not by 1 mM dithiothreitol.

The type strain was isolated from hindgut contents of the termite *Reticulitermes flavipes* (Kollar) (Rhinotermitdae) collected in Woods Hole, Massachusetts, USA. It was not part of the hindgut flora of *R. flavipes* collected in Dansville, Michigan.

The mol% G + C of the DNA is: not known.

Type strain: RFM-3, DSMZ 11501.

GenBank accession number (16S rRNA): U82322.

6. **Methanobrevibacter oralis** Ferrari, Brusa, Rutili, Canzi and Biavati 1995, 880[VP] (Effective publication: Ferrari, Brusa, Rutili, Canzi and Biavati 1994, 11).

o.ra' lis. M.L. masc. adj. *oralis* of the mouth.

The phylogenetic relationship of this species to the other species of the genus is not known.

Cells are short, oval rods with tapered ends, 0.4–0.5 µm in width and 0.7–1.2 µm in length, occurring most frequently in pairs or short chains. Cells give a Gram-positive reaction

when less than 4 d old. Ultrathin sections show a tristratified cell wall that is highly invaginated. Nonmotile.

Surface colonies are 0.5–1.0 mm in diameter, have entire margins, and are creamy to light yellow in color.

Strict anaerobe. Growth occurs with H_2 and CO_2. Formate, methanol, and acetate are not used as substrates for methanogenesis. The optimum sodium chloride concentration is between 0.01 and 0.1 M. There is no growth above 0.2 M. Fecal extract is required for growth, and a branched chain volatile fatty acid mixture is highly stimulatory.

Optimum growth temperature is 35–38°C (range 25–39°C). Optimum pH is 6.9–7.4 (range 6.2–8.0).

The type strain was isolated from human subgingival plaque. Belay et al. (1988a) reported the isolation of methanogens from dental plaque that were antigenically similar to *M. smithii*. The description of the cell wall of strain ZR (Ferrari et al., 1994) is similar to that of *M. ruminantium* (Fig. A2.3, part 7 and part 8). Strain ZR has not been examined for its antigenic reactivity with antisera against other members of the genus. Genomic DNA of type strain ZR was reported not to hybridize with genomic DNA from *M. ruminantium* strain M1, *M. smithii* strain PS, or *M. arboriphilus* strain DH1 in a dot blot assay with nonradioactive genomic DNA (Ferrari et al., 1994). Quantitative genomic DNA reassociation studies of strain ZR with species of the genus has not been examined.

The mol% G + C of the DNA is: 28 (T_m).

Type strain: ZR, DSMZ 7256.

7. **Methanobrevibacter smithii** Balch and Wolfe 1981, 216[VP] (Effective publication: Balch and Wolfe *in* Balch, Fox, Magrum, Woese and Wolfe 1979, 284).

smith' i.i. M.L. gen. n. *smithii* of Smith; named after P.H. Smith, who isolated the type strain.

The species description is based on characteristics of the type strain PS and two strains from human feces. The two strains, B181 and ALI, share a single nucleotide base difference in the 16S rRNA gene sequence with strain PS and are identical to each other (Lin and Miller, 1998). All three strains share greater than 90% genomic DNA similarity as measured by DNA reassociation studies (Miller and Wolin, 1986; Lin and Miller, 1998).

Cells are short oval rods or coccobacilli with tapered ends, 0.6–0.7 µm in width and ~1.0 µm in length. Cells occur most frequently in pairs or in chains of 4–6 cells. Gram positive. Nonmotile. A strain cited as "PS1", but confirmed to have been strain PS, was reported to have a single polar flagellum (Doddema et al., 1979; H.D. Doddema and G.D. Vogels, personal communication). Other investigators have found that flagella are absent from cells of strain PS or human fecal strains that are morphologically, physiologically, and immunologically indistinguishable from strain PS (Miller et al., 1982; M. Edwards and W.A. Samsonoff, personal communication). Multiple septa are frequently observed on the cell surface (Fig. A2.2, parts 3 and 4), but septum formation is not as extensive as that observed with *M. ruminantium* cells. The cell wall appears as a single electron-dense thick layer (Fig. A2.4, part 11).

Surface colonies in roll tube cultures on complex rumen fluid-containing medium are translucent, effuse to low convex, usually circular or elliptical with entire margins, and light to dark tan, often with a tiny brown center. They can reach a diameter of 2–3 mm in roll tubes with few colonies and excess energy source.

One or more B vitamins are required for stimulatory to growth, and acetate is required as a major source of cell carbon (Bryant et al., 1971). Nickel is required for growth (Diekert et al., 1981). Other trace metal requirements have not been investigated. Growth of the type strain or human fecal strains is not inhibited in modified medium 1 containing 2% oxgall and 0.1% sodium deoxycholate, with H_2 and CO_2 as energy sources (Miller et al., 1982). NH_4^+ is the sole source of cell nitrogen, and H_2S may serve as the sole source of cell sulfur (Bryant et al., 1971). H_2 and CO_2 are the preferred energy sources. Growth on formate is poor. Cell extracts do not have CO dehydrogenase activity (Bott et al., 1985). A F_{420}- dependent formate dehydrogenase oxidizes formate to CO and reduced F_{420} (Tzeng et al., 1975a). A hydrogenase is F_{420}- linked, and biosynthetic reducing power may be generated via F_{420}/NADPH oxidoreductase (Tzeng et al., 1975b). These enzymatic reactions were the first demonstration of the function of F_{420} in electron transfer reactions in methanogens. *M. smithii* also contains the cofactors 2-mercaptoethanesulfonic acid (Balch and Wolfe, 1979a) and factor 430 (Diekert et al., 1981). Cells have corrinoids (Krzycki and Zeikus, 1980). Strain PS also has methanofuran and tetrahydromethanopterin (Jones et al., 1985). The polyamine content is low (Scherer and Kneifel, 1983). Some features of the metabolic pathways of strain PS were determined by analysis of the incorporation of ^{13}C-precursors into cellular components (Sprott et al., 1993; Choquet et al., 1994a, b). Strain PS synthesizes α-ketoglutarate from oxalacetic acid via succinate by the reducing reactions of the incomplete tricarboxylic acid pathway. *M. smithii* incorporates acetate and CO_2 into cell carbon by reductive carboxylation of acetate to form pyruvate. It synthesizes ribose via the oxidative branch of the pentose phosphate pathway. Hexose is formed by the reverse of the Embden–Meyerhof– Parnas pathway.

The pseudomurein cell wall of strain PS contains ornithine. However, the cell wall of strain ALI lacks ornithine, indicating that ornithine is not a reliable marker of the species (König, 1986). Antisera and the corresponding antigen preparations of strain PS do not cross-react with antigens and antisera, respectively, of other species in the genus (Conway de Macario et al., 1982c). PS antisera strongly cross-react with human fecal strains (Conway de Macario et al., 1982b; Miller and Wolin, 1982; Miller et al., 1982).

The following antibiotics produce zones of inhibition (20– 40 mm): bacitracin, gardimycin, enduracidin, chloramphenicol, and lasalocid (Hilpert et al., 1981). In rumen fluid medium, monensin causes a delayed growth response (Chen and Wolin, 1979). Bacitracin (10 µg/ml) completely inhibits growth in liquid modified medium I (T.L. Miller, unpublished data).

The type strain was isolated from an anaerobic sewage sludge enrichment with formate as the exogenously added energy source (Smith, 1961). *M. smithii* is the dominant methanogen in feces of humans who harbor methanogens in their large bowels (Nottingham and Hungate, 1968; Miller and Wolin, 1982). Concentrations range from extremely low numbers (a few cells per gram of dry feces) to as high as 10^{10} per gram of dry fecal matter and in some individuals can be equal to 10% of the total concentration of viable anaerobic bacteria (Weaver et al., 1986). *Methanobrevibacter* species have been isolated from feces of several different animals, but to date, *M. smithii* appears to be unique to the human large bowel ecosystem (Miller and Wolin, 1986; Miller et al., 1986b; Weaver et al., 1986; Lin and Miller, 1998).

The range of mol% G + C of three human fecal isolates is 28.8–29.5 (T_m) (Miller et al., 1986b; T.L. Miller, unpublished data). DNA reassociation studies show a high level of similarity (>94%) between the two fecal strains and the type strain (Table A2.3; Lin and Miller, 1998).

The mol% G + C of the DNA is: 30.0–31.0 (T_m, Bd)

Type strain: PS, DSMZ 861.

GenBank accession number (16S rRNA): U55233.

Additional Remarks: Reference strains include ALI (DSMZ 2375) and B181 (DSMZ 11975).

Genus III. **Methanosphaera** *Miller and Wolin 1985b, 535[VP] (Effective publication: Miller and Wolin 1985a, 121)*

TERRY L. MILLER

Me.tha.no.sphae'ra. M.L. neut. n. *methanum* methane; L. fem. n. *sphaera* a sphere; M.L. fem. n. *Methanosphaera* methane-producing sphere.

Round cells, usually occurring in pairs, tetrads, and clusters, about 1.0 µm in diameter. Resting cells, such as spores, are not known. Gram positive. Nonmotile. Very strict anaerobe. **Cell walls are composed of pseudomurein.** Optimum temperature: near 37°C. Optimum pH: 6.5–6.9. **Chemoorganotrophic.**

Energy for growth is obtained by using 1 mol of H_2 to reduce 1 mol of methanol to 1 mol of CH_4. Methane is not produced from methanol in the absence of H_2. Carbon dioxide, carbon monoxide, sulfate, fumarate, choline, and nitrate do not substitute for methanol. Methane is not produced from acetate, methylamines, or formate, with or without H_2. No growth or methane is obtained with ethanol and H_2. **Easily visible pigments are not produced, and cytochromes are absent.** Corrinoids are present.

Carbon dioxide and acetate are required for growth. NH_4^+ and one or more amino acids may be major sources of cell nitrogen. One or more B vitamins may be required for, or stimulatory to, growth.

The mol% G + C of the DNA is: 23–26.

Type species: **Methanosphaera stadtmanae** Miller and Wolin 1985b, 535 (Effective publication: Miller and Wolin 1985a, 121.)

FURTHER DESCRIPTIVE INFORMATION

The genus is currently represented by two species, *Methanosphaera stadtmanae* and *Methanosphaera cuniculi*. They are distinguished from each other based on a lack of genomic DNA reassociation.

ENRICHMENT AND ISOLATION PROCEDURES

All enrichments and isolations require stringent anaerobic conditions (Miller and Wolin, 1974). Selective enrichment procedures have not yet been developed. The type species was present in enrichments of human fecal material with methanol as the exogenously added methanogenic substrate and an initial gas phase of 80% N_2 and 20% CO_2 and was subsequently isolated from subcultures of the enrichment with methanol and 80% H_2 and 20% CO_2 (Miller and Wolin, 1983). *M. stadtmanae* was isolated from an individual who harbored *Methanobrevibacter smithii* as the numerically dominant methanogen morphotype (Miller and Wolin, 1983, 1985a). *Methanosphaera* probably uses methanol produced by other organisms that degrade pectin in the intestinal habitat.

MAINTENANCE PROCEDURES

M. stadtmanae is maintained in anaerobic agar/liquid biphasic culture in the phosphate-buffered, complex rumen fluid medium described by Miller and Wolin (1985a) with a ratio of 1 volume of single-strength liquid medium to 4 volumes of double-strength agar medium. The methanol concentration is 1.2% in the agar phase and 0.6% in the liquid phase with a gas phase of 203 kPa H_2/CO_2 (80:20). An inoculum (OD $\geq$0.7, diameter = 1.8 cm) equivalent to 10% of the liquid phase is added, and the bottle is incubated at 37°C with gentle shaking. After outgrowth, the biphasic cultures are regassed and pressurized to 203 kPa with H_2/CO_2, prechilled for 1 h at 4°C, and frozen at −76°C. Biphasic cultures preserved by this method remain viable for up to one year of storage in the frozen state.

DIFFERENTIATION OF THE GENUS *METHANOSPHAERA* FROM OTHER GENERA

Methanosphaera is currently the only described methanogen that is restricted to methanol and H_2 as its sole source of methanogenic substrates (Miller and Wolin, 1983, 1985a). *Methanosarcina barkeri* (strain Fusaro, DSMZ 804) grows and produces methane by direct reduction of 1 mol of methanol with 1 mol of H_2 (Müller et al., 1986). However, *Methanosarcina* is not restricted to this mechanism of methanogenesis and can use a variety of substrates for methanogenesis and growth, including methanol without H_2, methylamines, acetate, and H_2 and CO_2.

In 1985, *Methanosphaera* was phylogenetically differentiated from *Methanosarcina* and other genera of methanogens by the oligonucleotide catalog sequence of its 16S rRNA (Miller and Wolin, 1985a; C. Woese, personal communication). Several characteristics indicate that *M. stadtmanae* is closely related to the family *Methanobacteriaceae*. The 16S rRNA oligonucleotide catalog relationships showed a familial relationship to the *Methanobacteriaceae*, although the S_{AB} index (0.45) indicated that it did not belong to either of the recognized genera of the family, *Methanobacterium* or *Methanobrevibacter*. *M. stadtmanae* shares 91% 16S rRNA gene sequence similarity with *Methanobacterium formicicum* and *Methanobrevibacter ruminantium* (based on data in the Ribosomal Database Project; Maidak et al., 1997). The latter two organisms are the type species of the genera of the *Methanobacteriaceae*. The cell envelope of *Methanosphaera* contains pseudomurein, the major polymer of the cell envelopes of the family (König et al., 1982; König, 1986). Cell envelopes of other families of methanogens do not contain pseudomurein. The amino acid composition of the pseudomurein of *Methanosphaera* is distinguished from that of other members of the family by the presence of serine (König, 1986). The subunit pattern of the methyl reductase of *M. stadtmanae* is similar to that of *Methanothermobacter thermautotrophicus* (*Methanobacterium thermoautotrophicum*), *Methanobacterium formicicum*, and *Methanobrevibacter smithii* (Rouvière and Wolfe, 1987). Immunological fingerprinting showed a relationship between *M. stadtmanae* and *Methanothermobacter thermautotrophicus* but no relationship to any other member of the *Methanobacteriaceae* or any member of the other families of methanogens (Conway de Macario and Macario, 1986).

ACKNOWLEDGMENTS

I thank Dr. W.A. Samsonoff of the Wadsworth Center for providing the electron microscopic analysis.

FURTHER READING

Biavati, B., M. Vasta and J.G. Ferry. 1988. Isolation and characterization of *Methanosphaera cuniculi* sp. nov. Appl. Environ. Microbiol. *54*: 768–771.

Miller, T.L. and M.J. Wolin. 1983. Oxidation of hydrogen and reduction of methanol to methane is the sole energy source for a methanogen isolated from human feces. J. Bacteriol. *153*: 1051–1055.

Miller, T.L. and M.J. Wolin. 1985. *Methanosphaera stadtmaniae*, gen. nov., sp. nov.: a species that forms methane by reducing methanol with hydrogen. Arch. Microbiol. *141*: 116–122.

List of species of the genus Methanosphaera

1. **Methanosphaera stadtmanae** Miller and Wolin 1985b, 535[VP] (Effective publication: Miller and Wolin 1985a, 121) (*Methanosphaera stadtmaniae* [sic] Miller and Wolin 1985a, 121). stadt.man′ae. M.L. gen. n. *stadtmanae* of Stadtman; named in honor of T.C. Stadtman for her important contributions to the microbiology and biochemistry of methanogenesis.

   Morphological characteristics are shown in Fig. A2.5. A distinctive cleavage furrow is observed in dividing cells. Numerous electron-dense inclusions that are stable in the electron beam are usually, but not always, located near the cell wall (Fig. A2.5, parts c–e). The cell envelope consists of a single (18–20-nm-thick) electron-dense layer (Fig. A2.5, parts c and e); its surface is smooth and lacks any organized structure. The peptide moiety of the pseudomurein contains L-glutamate, L-lysine, and L-serine. Flagella are absent from

cells treated with negative stains or by freeze fracture procedures (W.A. Samsonoff, personal communication). The lipid composition was investigated by Jones and Holzer (1991). The major component of the neutral lipids is a tail-to-tail linked C_{20} isoprenoid, 2,6,11,15-tetramethylhexadecane, the first finding of a tail-to-tail-linked C_{20} isoprenoid hydrocarbon in any biological system. Smaller amounts of squalene, dihydrosqualene, and tricyclic terpenes are also found. The polar lipids consist primarily of diphytanyl diglycerol tetraether and traces of dibiphytanyl diglycerol tetraether. *M. stadtmanae* contains large quantities and a broad distribution of C_{14} to C_{30} fatty acids, including an unusual component, 1-hydroxy-nonadecanoic acid.

Surface colonies in complex rumen fluid medium are opaque to translucent, effuse, circular, or elliptical with entire margins and are light tan. Colonies in uncrowded roll tubes

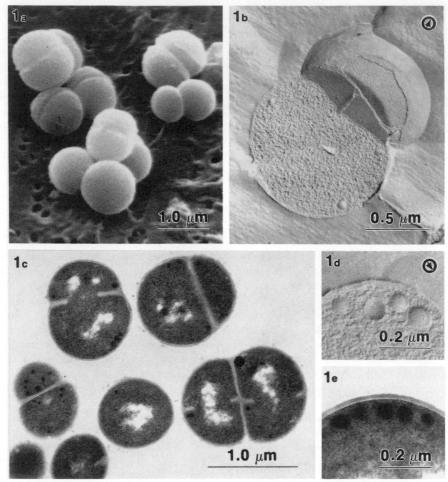

FIGURE A2.5. Morphology and ultrastructure of *Methanosphaera stadtmanae*. The organism was grown in a liquid semi-defined medium with 1.0% methanol and 202 kPa H_2/CO_2 (80%:20%) and cells were prepared for electron microscopy as described by Miller and Wolin, (1985a). (*a*) Scanning electron micrograph illustrating the spherical cellular morphology and the cleavage furrow seen in rapidly dividing cells. (*b*) Freeze–fracture through dividing cells; the outer surface of the wall appears smooth; the encircled arrow indicates the direction of the shadow. (*c*) Thin section through dividing cells reveals multiple septa and numerous electron-dense bodies. (*d*) High-magnification freeze fracture through inclusion bodies illustrating their presence in unfixed material. (*e*) High-magnification thin section through the electron-dense inclusion bodies near the cell wall, which consists of a single electron dense layer. (Reprinted with permission from T.L. Miller and M.J. Wolin, Archives of Microbiology *141:* 116–122, 1985, ©Springer-Verlag, Berlin).

with excess methanol and hydrogen may reach a diameter of 2 mm.

Methanol and hydrogen are stoichiometrically converted to methane (Miller and Wolin, 1983). Growth in a complex medium, with H_2 in excess, increases as the methanol concentration is increased, up to ~0.45 M (1.4%). Higher concentrations of methanol are inhibitory. The growth yield is 4 g (dry weight) of cells/mol of methane. No growth or methane production occurs with CO alone or with H_2. CO inhibits growth and methane production from methanol and H_2.

The optimum pH is 6.5–6.9. The temperature range is 30–40°C. No growth occurs at 25°C or at 45°C. Thiamine is required for growth, and biotin is stimulatory. Coenzyme M, vitamin B_{12}, or other B vitamins are not required for growth.

Acetate and CO_2 are required for growth. Radioisotopic incorporation studies indicate ~50% of the cell carbon is derived from CO_2 and about 50% is derived from acetate (Miller and Wolin, 1983; Miller et al., 1995). Methanol and formate are not significant sources of cell carbon, although small amounts can be incorporated into cell carbon (Miller et al., 1995). Growth is not inhibited in a complex rumen fluid medium containing 2% oxgall and 0.1% sodium deoxycholate. NH_4^+ and isoleucine are essential for growth, and leucine is stimulatory to growth. Leucine is transported into the cell by a Na^+-dependent active transport system (Sparling et al., 1993a). Isoleucine competitively inhibited [^{14}C]leucine uptake, suggesting that isoleucine is transported into the cell by the same mechanism. 2-Methylbutyric and isovaleric acids do not substitute for isoleucine and leucine, respectively. Sulfur and trace metal requirements have not been determined. The organism can be grown in a chemically defined medium containing acetate, CO_2, thiamine, leucine, isoleucine, NH_4^+, minerals, and trace metals (Miller et al., 1995).

The biochemistry of methanogenesis from methanol has been examined in *M. stadtmanae*. The inability of this organism to reduce CO_2 to methane is explained by the lack of methanofuran, the C1-carrier cofactor involved in the first two reductive steps in the pathway of CO_2 methanogenesis (van de Wijngaard et al., 1991). In addition, the enzymes responsible for reduction of CO_2 (formylmethanofuran dehydrogenase, methylenetetrahydromethanopterin dehydrogenase, and methylenetetrahydromethanopterin reductase) have very low specific activities in *M. stadtmanae* (Schwörer and Thauer, 1991; van de Wijngaard et al., 1991). The specific activity of heterodisulfide reductase is sufficient to account for the rates of *in vivo* methane formation from methanol (Schwörer and Thauer, 1991). *M. stadtmanae* contains high concentrations of methyl-coenzyme M reductase (Rouvière and Wolfe, 1987). Van de Wijngaard et al. (1991) provided evidence that a methylated enzyme-bound corrinoid was identical to the Co-methyl derivative of 5-hydroxybenzimidazolylcobamide. The results suggest that the intermediates in methanol reduction are an enzyme-bound methylated corrinoid and methyl-coenzyme M. Washed cell suspensions of *M. stadtmanae* coupled the reduction of methanol to ATP synthesis by a proton motive force (Sparling et al., 1993b). The mechanism by which H_2 reduces the bound methyl group to methane is not completely understood. Membrane-bound F_{420}-dependent hydrogenase activity was absent, but methyl viologen- dependent hydrogenase activity was present in the membranes (Deppenmeier et al., 1989). *M. stadtmanae* also has corrinoid-containing membrane proteins (Dangel et al., 1987).

The pathways of biosynthesis of cellular monomers were established by using ^{13}C-enriched precursors of cellular carbon (Choquet et al., 1994a, b; Miller et al., 1995). Acetate and CO_2 are assimilated into cellular carbon by reductive carboxylation of acetate to form pyruvate. The ^{13}C-labeling studies provide evidence that the organism synthesizes hexose by the reverse reactions of the Embden–Meyerhof–Parnas pathway. Pentoses are synthesized by the oxidative branch of the pentose-phosphate pathway. The carbon skeletons of alanine, serine, and cysteine arise from pyruvate. α-Ketoglutarate is formed from succinate via the symmetrical intermediate succinate which is formed by the reducing reactions of the incomplete tricarboxylic acid pathway. Histidine is formed from 5-phosphoryl-1-pyrophosphate, ATP, and glutamine. All other amino acids are formed from pyruvate and CO_2 by conventional biosynthetic pathways. The methyl group of acetate is the precursor of the hydroxymethyl of serine, the methyl of methionine, C7 of histidine, and C2 and C8 of the purine ring. Serine is the direct donor of C1 units for biosynthetic reactions in *M. stadtmanae*. The labeling of the carbon atoms of the purine ring and pyridines are consistent with the pathways found in eubacteria. Solid NMR spectroscopy of ^{13}C-labeled whole cells of *M. stadtmanae* provided indirect evidence that the methyl of thymidine is derived from the methyl of acetate via the hydroxymethyl group of serine (Bank et al., 1996). The chemical nature of the C-1-carrier cofactor in *M. stadtmanae* is not known. Typical derivatives of tetrahydromethanopterin are not found in detectable levels in extracts (Keltjens and Vogels, 1988; van de Wijngaard et al., 1991).

The following antibiotics do not inhibit growth in liquid culture medium (μg/ml): trimethoprim (10), methotrexate (10), sulfanilamide (500), cephalothin (1.7), and clindamycin (6.7). Metronidazole (1 μg/ml) and bacitracin (10 μg/ml) completely inhibit growth. Monensin (10 μg/ml) and bromoethanesulfonate (1 mM) cause growth lags of ~5 d, after which cultures grow to turbidities similar to those obtained in controls without added inhibitors (Miller and Wolin, 1985a).

The type strain was isolated from enrichments of human feces (Miller and Wolin, 1983, 1985a). Similar morphotypes have been observed in human and some animal feces but are usually far outnumbered by the numerically dominant H_2/CO_2-using *Methanobrevibacter* morphotype (Miller and Wolin, 1986; Miller et al., 1986b; T.L. Miller, unpublished data). In the intestinal habitat, the organism probably uses methanol produced by other intestinal organisms that degrade pectin.

The mol% G + C of the DNA is: 25.8 (T_m).

Type strain: MCB3, DSMZ 3091.

GenBank accession number (16S rRNA): M59139.

2. **Methanosphaera cuniculi** Biavati, Vasta and Ferry 1990, 470[VP] (Effective publication: Biavati, Vasta and Ferry 1988, 768).

cu.ni' cu.li. L. masc. n. *cuniculus* rabbit; L. gen. n. *cuniculi* of the rabbit.

The species is currently represented by one isolate, strain 1R7. The following description is based on the study by Biavati et al. (1988).

Cells are round, 0.6 to 1.2 μm in diameter, Gram-positive, nonmotile, and usually occur in pairs. Cell walls are composed of pseudomurein and contain serine. The optimum temperature is 35–40°C, and the optimum pH is 6.8. No growth occurs at 25 or 45°C. The organism is anaerobic. Surface colonies are circular or elliptical, yellowish, and have entire edges. The diameters of colonies are ~1 mm after 10 d of incubation.

Both hydrogen and methanol are required for growth. No growth occurs with H_2/CO_2 alone or under a N_2/CO_2 atmosphere in media containing methanol or acetate. No growth occurs under a H_2/CO_2 or N_2/CO_2 atmosphere in media containing ethanol, isopropanol, methylamines, or formate. Acetate is required for growth. Vitamin and other nutritional requirements are not known.

Strain 1R7 was isolated from enrichments containing vancomycin (100 μg/ml) that were sequentially subcultured to media containing clindamycin (6.7 μg/ml) and cephalothin (1.7 μg/ml), and then penicillin (40 μg/ml). Cells of this strain react strongly with polyvalent rabbit antisera raised against *M. stadtmanae* MCB3 and show no reaction with anti-*Methanothermobacter thermautotrophicus* LH and GC1 polyvalent antisera at the S probe concentration. Using the S1 procedure described by Johnson (1985), the average genomic DNA similarity between strain 1R7 and *M. stadtmanae* MCB3 was 29% (Biavati et al., 1988).

The type strain was isolated from enrichments of the contents of rabbit rectum.

The mol% G + C of the DNA is: 23 (T_m).

Type strain: 1R7, DSMZ 4103.

Genus IV. **Methanothermobacter** *Wasserfallen, Nölling, Pfister, Reeve and Conway de Macario 2000, 51*[VP]

DAVID R. BOONE

Me.tha.no.ther.mo.bac' ter. M.L. neut. n. *methanum* methane; Gr. adj. *thermos* hot; M.L. masc. n. *bacter* equivalent of Gr. neut. n. *bakterion* rod, staff; M.L. masc. n. *Methanothermobacter* thermophilic methane rod.

Curved or crooked slender rods, moderately long to filamentous, 0.3–0.5 μm wide. Endospores not formed. **Cells stain Gram-positive**, and ultrastructure appears typically Gram-positive, but cell walls are composed of pseudomurein. **Nonmotile. Cells produce fimbriae. Strictly anaerobic. Fastest growth between 55°C and 65°C. Energy metabolism by reduction of CO_2 to CH_4, with H_2 as electron donor**; some cells can also use formate as an electron donor. Sulfur is reduced to sulfide, but this reaction does not yield energy for growth. Ammonia is the sole nitrogen source, and sulfide may serve as sulfur source.

The mol% G + C of the DNA is: 32–61.

Type species: **Methanothermobacter thermautotrophicus** (Zeikus and Wolfe 1972) Wasserfallen, Nölling, Pfister, Reeve and Conway de Macario 2000, 51 (*Methanobacterium thermoautotrophicus* Zeikus and Wolfe 1972, 713).

FURTHER DESCRIPTIVE INFORMATION

Methanothermobacter is a phylogenetically coherent genus of methanogens formerly classified in the genus *Methanobacterium* (Fig. A2.6). The phylogenetic separation of *Methanothermobacter* from *Methanobacterium* is mirrored by the difference in optimum temperature (see Taxonomic Comments).

Rods are usually long, often forming filaments and sometimes chains. However, *Methanothermobacter wolfeii* cultures may occur as coccoid cells. The cell wall is a single thick layer external to the cytoplasmic membrane, morphologically similar to that of Gram-positive bacteria but composed of pseudomurein rather than peptidoglycan (Kandler and König, 1985). Gram-stained preparations appear to be Gram-positive, although results for some strains are variable. Cells are not motile and do not possess flagella, but they possess fimbriae (Doddema et al., 1979) that are 4–5 nm in diameter and resistant to proteases, lipase, and nucleases. Intracytoplasmic membranous bodies have been observed in electron micrographs (Zeikus and Bowen, 1975a), but these may be artifacts of sample preparation (Aldrich et al., 1987).

Lipids are largely diether lipids characteristic of *Archaea*, but membranes contain some fatty acids, mainly palmitic and stearic but with smaller amounts of myristic and 18:1 acids (Pugh and Kates, 1994). *Methanothermobacter thermautotrophicus* membrane lipids contain gentiobiosylarchaeol (a diether glycolipid), and serine-, inositol-, and ethanolamine-containing diether and tetraether phospholipids and phosphoglycolipids (archaetidyl-L- serine, caldarchaetidyl-L-serine, gentiobiosylcaldarchaetidyl-L-serine, 1D-1-(archaetidyl)-*myo*-inositol, 1D-1(caldarchaetidyl)-*myo*-inositol, 1D-1-(gentiobiosylcaldarchaetidyl)-*myo*-inositol, and archaetidylethanolamine) (Nishihara et al., 1987, 1989).

Cells obtain energy for growth by the reduction of CO_2 to CH_4; H_2 or sometimes formate serves as an electron donor. The F_{420}-reducing hydrogenase is membrane bound (Braks et al., 1994). *Methanothermobacter marburgensis* produces a flavoprotein during growth with iron limitation (Wasserfallen et al., 1995). Two distinct coenzyme M reductases are present (Rospert et al., 1990; Brenner et al., 1992). One of the isozymes (*mcr*II) is produced only early during batch growth (Pihl et al., 1994) or in continuous cultures with high concentrations of H_2, and the other (*mcr*I) is produced during growth with low H_2 concentration (Morgan et al., 1997). Growth yields (Y_{CH4}) were higher during growth with low H_2 concentration (*mcr*I) (Morgan et al., 1997). The methyl reductase of *Methanothermobacter thermautotrophicus* catalyzes reductive dechlorination of 1,2-dichloroethane (Holliger et al., 1992a).

Elemental sulfur may also be reduced to sulfide, but this reaction does not appear to support growth (Stetter and Gaag, 1983). Most strains are autotrophic, but *M. thermophilus* requires coenzyme M for growth (Laurinavichus et al., 1987). Acetate is synthesized by the Ljungdahl-Wood pathway first described for homacetigenic clostridia (Simpson and Whitman, 1993). Tryptophan synthesis is regulated at the transcriptional level (Gast et al., 1994). Ammonia serves as a sole nitrogen source, but ammonia can be inhibitory at concentrations above 3–4 g/l (Hen-

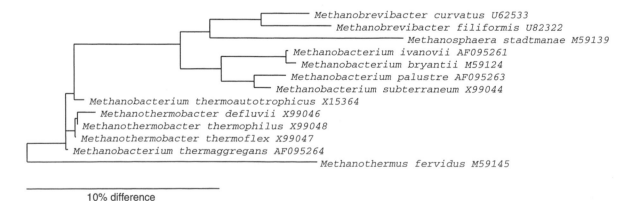

10% difference

FIGURE A2.6. Phylogenetic tree of *Methanobacterium* and *Methanothermobacter* based on analysis of their 16S rRNA gene sequences. (Courtesy of the Ribosomal Database Project.)

driksen and Ahring, 1991). Sulfide can serve as a sulfur source. Cells are strictly anaerobic, although *M. marburgensis* contains superoxide dismutase (Meile et al., 1995). *M. thermautotrophicus* and *M. marburgensis* each can adapt to salinities up to 0.5 M with little effect on growth rate; they synthesize organic solutes, including cyclic-2,3-diphosphoglycerate and glutamate (Ciulla et al., 1994b). *M. marburgensis* also forms 1,3,4,6-tetracarboxyhexane (Gorkovenko et al., 1994) and 1-α-glutamate at low salinities and forms a novel solute at salinities greater than 0.5 M (Ciulla et al., 1994b). Halotolerance of *M. marburgensis*, but not *M. thermautotrophicus*, is stimulated by yeast extract (Ciulla et al., 1994b).

Molybdate is not required by *M. wolfeii* and *Methanothermobacter marburgensis*, which have a formylmethanofuran dehydrogenase isozyme that can use tungstate rather than molybdate (Schmitz et al., 1992a, b; Bertram et al., 1994). *Methanosarcina barkeri* and other methanogens that do not possess this isozyme require molybdate (Schmitz et al., 1992b, 1994).

The physical maps of *M. wolfeii* and *M. marburgensis* are very similar, although the restriction patterns showed major discrepancies (Stettler and Leisinger, 1992). Mobile genetic elements are carried as plasmids or integrated into the chromosome (Nölling et al., 1993a). The *M. marburgensis* strains Marburg and ZH3 contain closely related plasmids (pME2001 and pME2200, respectively) (Stettler et al., 1994). Phage ψM1 transduces *M. marburgensis* at a rate of about 10^{-5} (Meile et al., 1990), and a related prophage is present in *M. wolfeii* (Stettler et al., 1995). Virulent phages φF1 and φF3 infect some *M. thermautotrophicus* strains, but not strain THF. These do not infect *M. marburgensis* or *M. wolfeii* (strains CB12, SF-4, and HN4) (Nölling et al., 1993a). Phage ψM1 is virulent for *M. marburgensis* but not for other species of *Methanothermobacter* (Meile et al., 1989).

ENRICHMENT AND ISOLATION PROCEDURES

Methanothermobacter is easily enriched or isolated in mineral media with H_2/CO_2 as a catabolic substrate. MS mineral medium with H_2 added is suitable for enrichment of *Methanothermobacter*, though growth is faster with 3 mM acetate added as an anabolic substrate. MS mineral medium is MS medium (Boone et al., 1989) with peptones, yeast extract, and coenzyme M omitted and with the concentration of $Na_2S \cdot 9H_2O$ increased to 0.5 g/l. Few other organisms can grow in such media at 60°C, so enrichment steps may not be necessary.

MAINTENANCE PROCEDURES

Methanothermobacter grows well in MS medium (Boone et al., 1989) with H_2 added as an electron donor for catabolic CO_2 reduction or in MS mineral medium with H_2 added. *Methanothermobacter* cultures can be maintained by biweekly transfer to fresh medium, by lyophilization, or by freezing the cultures in liquid nitrogen (Boone, 1995).

DIFFERENTIATION OF THE GENUS *METHANOTHERMOBACTER* FROM OTHER GENERA

Methanothermobacter is the only genus of rod-shaped methanogens that grows on H_2/CO_2 optimally at 60–65°C. Other thermophilic rod-shaped methanogens are *Methanosaeta thermophila* (which grows on acetate but not on H_2/CO_2) and *Methanothermus* (which grows optimally at temperatures above 80°C).

TAXONOMIC COMMENTS

The thermophilic species that until recently belonged to the genus *Methanobacterium* are, as a group, phylogenetically coherent

and distinct from the other members of the genus and should be considered as members of a separate genus (Boone et al., 1993b; Wasserfallen et al., 2000). The first edition of *Bergey's Manual of Systematic Bacteriology* described four thermophilic species of *Methanobacterium* (Boone and Mah, 1989). Two of these, *Methanobacterium thermoautotrophicum* and *Methanobacterium wolfei*, were transferred to *Methanothermobacter* as *Methanothermobacter thermautotrophicus* and *Methanothermobacter wolfeii*, respectively (Wasserfallen et al., 2000). The two other thermophilic *Methanobacterium* species described in the previous edition (*Methanobacterium thermoformicicum* and *Methanobacterium thermalcaliphilum*) are subjective synonyms of *Methanothermobacter thermautotrophicus* (*Methanobacterium thermoautotrophicum*) (Touzel et al., 1992; Kotelnikova et al., 1993a; Wasserfallen et al., 2000).

Since the publication of the previous edition, four new species of thermophilic *Methanobacterium* have been described. Three of these (*Methanobacterium defluvii*, *Methanobacterium thermophilum*, and *Methanobacterium thermoflexum*) were not transferred to *Methanothermobacter* because Wasserfallen et al. (2000) considered them to be junior subjective synonyms of *M. thermautotrophicus*. The basis for this synonymy was similarity of phenotypes and of 16S rDNA sequences. However, DNA–DNA hybridization data of Kotelnikova et al. (1993b) suggest that these species are phylogenetically distinct (Table A2.6) and should be considered as separate species, and the 16S rRNA data of Wasserfallen et al. (2000) indicate that they are sufficiently similar to be placed in the same genus, if not species. Therefore, it is proposed that these species be transferred into the genus *Methanothermobacter*. DNA–DNA hybridization values between *Methanobacterium thermoflexum* or *Methanobacterium defluvii* and other strains of thermophilic *Methanobacterium* and *Methanothermobacter* (including the type strain of *Methanothermobacter thermautotrophicus*) are low (22–47%), supporting the status of these two species as separate from *M. thermautotrophicus*. Although this DNA–DNA hybridization study (Kotelnikova et al., 1993b) included no direct comparison between *Methanobacterium thermophilum* and *M. thermautotrophicus*, DNA from *M. thermautotrophicus* gave significantly different results than the equivalent experiments with DNA from *Methanobacterium thermophilum* (Kotelnikova et al., 1993b), indirectly suggesting significant differences between the DNA sequences of *M. thermautotrophicus* and those of *Methanobacterium thermophilum*.

A fourth new thermophilic species, *Methanobacterium thermaggregans* (Blotevogel and Fischer, 1985), has a type strain obtained from cattle manure by dilution. The highest dilution of manure into culture medium containing penicillin G (50 mg/l) was repeatedly grown and transferred into the same medium until the culture appeared to be axenic. This culture is not derived from a single cell or colony-forming unit, so the data from the reported

TABLE A2.6. DNA hybridization of *Methanothermobacter* strains[a]

Strain	DNA sequence similarity (%)	
	M. defluvii	*M. thermoflexus*
M. defluvii ADZ	100	46
M. thermoflexus IDZ	46	100
M. thermautotrophicus H	39	43
M. thermautotrophicus Z–245	47	44
M. thermautotrophicus DSMZ 3266	42	33
M. thermautotrophicus AC60	29	35
M. wolfeii DSMZ 2790	26	22
M. thermophilum M	23	10

[a]Data from Kotelnikova et al., 1993a.

tests are difficult to interpret. Therefore, *M. thermaggregans* was not transferred to the genus *Methanothermobacter* and is not listed as a species in this *Manual.*

Phylogenetic and phenotypic analysis (Wasserfallen et al., 2000) indicates that thermophilic strains formerly classified in *Methanobacterium* fall into three monophyletic groups represented by *M. thermautotrophicus*, *M. wolfeii*, and *M. marburgensis*. The *M. thermautotrophicus* group also includes *Methanobacterium thermophilum*, *Methanobacterium defluvii*, and *Methanobacterium thermoflexum*.

DIFFERENTIATION OF THE SPECIES OF THE GENUS *METHANOTHERMOBACTER*

The differential characteristics of the species of the genus *Methanothermobacter* are indicated in Table A2.7.

List of species of the genus Methanothermobacter

1. **Methanothermobacter thermautotrophicus** (Zeikus and Wolfe 1972) Wasserfallen, Nölling, Pfister, Reeve and Conway de Macario 2000, 51[VP] (*Methanobacterium thermoautotrophicus* Zeikus and Wolfe 1972, 713).

ther.mau.to.tro'phi.cus. Gr. adj. *thermos* hot; Gr. pref. *auto* self; Gr. adj. *trophikos* one who feeds; M.L. masc. adj. *thermautotrophicus* thermophilic and autotrophic.

Cells are slender, cylindrical, irregularly crooked rods, 0.35–0.6 × 3–7 μm, with frequent filaments 10–120 μm in length. Deep colonies in roll tubes are tannish white, roughly spherical, diffuse, and filamentous.

At least two strains (THF and Z-245) have plasmids with a significant amount of similarity (Nölling et al., 1992), including a methyltransferase gene (Nölling and de Vos, 1992).

Habitat: thermophilic, anaerobic, sewage sludge digestors.

Isolated from an anaerobic sewage sludge digestor.

Described strains classified within this species: strain YT (ATCC 29183, DSMZ 1850); strain YTB (ATCC 35610) (Zeikus et al., 1980); strain Z-245 (DSMZ 3720) (Zhilina and Ilarionov, 1984); strain THF (Lee and Zinder, 1988); strain FTF (DSMZ 3012); strain FF1 (Nölling et al., 1991); strain FF3 (Nölling et al., 1991); and strain JW510 (DSMZ 1910). Strain Marburg (DSMZ 2133, OCM 82) has often been classified as *M. thermautotrophicus* (*Methanobacterium thermoautotrophicum*) in spite of the fact that phylogenetic analysis demonstrates that strain Marburg is not closely related to the type strain (strain ΔH) (Brandis et al., 1981; Wasserfallen et al.,

2000). Strain Marburg is now the type strain of *M. marburgensis*.

The mol% G + C of the DNA is: 50 or 52 (Bd) or 49 (T_m).

Type strain: ΔH, ATCC 29096, DSMZ 1053.

GenBank accession number (16S rRNA): X68720.

2. **Methanothermobacter defluvii** comb. nov. (*Methanobacterium defluvii* Kotelnikova, Obraztsova, Gongadze and Laurinavichus 1993b, 434).

de.flu'vi.i. L. neut. gen. n. *defluvii* of sewage.

Cells are curved or crooked rods 0.4 × 3–6 μm.

Habitat: anaerobic sewage sludge digestors.

The mol% G + C of the DNA is: 62.2 (T_m).

Type strain: ADZ, DSMZ 7466, OCM 570, VKM B- 1962.

GenBank accession number (16S rRNA): X99046.

3. **Methanothermobacter marburgensis** Wasserfallen, Nölling, Pfister, Reeve and Conway de Macario 2000, 52.[VP]

mar.bur.gen'sis. M.L. masc. adj. *marburgensis* from Marburg.

Cells are slender, cylindrical, straight or curved rods, 0.4–0.6 × 3–6 μm, and frequently occur in pairs or chains up to 30 μm long. Colonies on agar plates are 1–4 mm in diameter, white or slightly yellowish, and convex. One plasmid is present, and some strains are infected by lytic phages.

Habitat: mesophilic, sewage sludge digestors.

Isolated from a mesophilic anaerobic sewage sludge digestor.

TABLE A2.7. Diagnostic and descriptive features of *Methanothermobacter* species[a]

Characteristic	1. *M. thermophilus*	2. *M. defluvii*	3. *M. marburgensis*	4. *M. thermautotrophicus*	5. *M. thermoflexus*	6. *M. wolfeii*
Morphology:						
Rods	+	+	+	+	+	+
Filaments	+	−	+	+	+	+
Gram stain results	−	−	+	+	+	+
Catabolic substrates:						
H$_2$/CO$_2$	+	+	+	+	+	+
Formate	−	+	−	d	+	−
Growth requirements:						
Autotrophic	+	−	+	+	+	+
Growth factors:						
Acetate		−	−	−	−	−
Yeast extract		−	−	−	−	
Coenzyme M	+	+	−	−	+	
Peptones		−	−	−	−	
pH range 7–8	+	+	+	+	+	+

[a]For symbols see standard definitions.

Described strains classified in this species include Hveragerdi (DSMZ 3590) (Roth et al., 1986) and strain ZH3 (Stettler et al., 1994).

The mol% G + C of the DNA is: 47.6 (T_m).

Type strain: DSMZ 2133, OCM 82.

GenBank accession number (16S rRNA): X15364, X05482.

4. **Methanothermobacter thermoflexus** (Kotelnikova, Obraztsova, Gongadze and Laurinavichus 1993b) comb nov. (*Methanobacterium thermoflexum* Kotelnikova, Obraztova, Gongadze and Laurinavichus 1993b, 434).

ther.mo.fle' xus. Gr. adj. *thermos* hot; L. masc. pred. adj. *flexus* bent or crooked; M.L. masc. adj. *thermoflexus* heat-(loving) and bent or crooked.

Cells are crooked rods, 0.4 × 7–20 μm, forming chains. Colonies in roll tubes are yellowish, 0.5–1.0 mm in diameter, with entire margins.

Habitat: anaerobic, sewage sludge digestors.

The mol% G + C of the DNA is: 55 (T_m).

Type strain: IDZ, DSMZ 7268, OCM 571, VKM B- 1963.

GenBank accession number (16S rRNA): X99047.

5. **Methanothermobacter thermophilus** comb. nov. (*Methanobacterium thermophilum* Laurinavichus, Kotelnikova and Obraztsova 1987, 320).

ther.mo' phi.lus. Gr. adj. *thermos* hot; Gr. adj. *philos* loving; M.L. masc. adj. *thermophilus* thermophilic.

Cells are slender, cylindrical, irregularly crooked rods, 0.36 × 1.4–6.5 μm, with filaments up to 30 μm in length.

(F_{420}) fluorescence is absent or weak.

Habitat: thermophilic, anaerobic sewage sludge digestors.

The mol% G + C of the DNA is: 44.7 (T_m).

Type strain: M, DSMZ 6529, VKM B-1786, OCM 231.

GenBank accession number (16S rRNA): X99048.

6. **Methanothermobacter wolfeii** (Winter and Lerp *in* Winter, Lerp, Zabel, Wildenauer, König and Schindler 1984) Wasserfallen, Nölling, Pfister, Reeve and Conway de Macario 2000, 51[VP] (*Methanobacterium wolfei* Winter and Lerp *in* Winter, Lerp, Zabel, Wildenauer, König and Schindler 1984, 465).

wol' fei.i. M.L. gen. n. *wolfeii* of Wolfe; named for R.S. Wolfe for his pioneering research on the biochemistry of methanogenesis.

Cells are rods, 0.4–0.6 × 2.5–6 μm, with rounded ends; some filaments occur. Cells occur singly, in pairs, or in chains. Surface colonies are 0.5–2 mm in diameter after 5–6 d, convex or raised with entire margins, smooth, moist, translucent, and white to yellowish gray.

Isolated from a mixture of sewage sludge and river sediment.

Described strains classified in this species: strain CB12 (ATCC 43574, DSMZ 3664, OCM 36) (Zhao et al., 1986), strain SF-4, and strain HN4 (Nölling et al., 1993b).

The mol% G + C of the DNA is: is 61 (T_m).

Type strain: ATCC 43096, DSMZ 2970, OCM 154.

GenBank accession number (16S rRNA): X89406.

Family II. **Methanothermaceae** Stetter 1982b, 266[VP] (Effective publication: Stetter *in* Stetter, Thomm, Winter, Wildgruber, Huber, Zillig, Janekovic, König, Palm and Wunderl 1981, 176)

KARL O. STETTER

Me.tha.no.ther.ma' ce.ae. M.L. masc. n. *Methanothermus* type genus of the family; *-aceae* ending to denote a family; M.L. fem. pl. n. *Methanothermaceae* the *Methanothermus* family.

Rod-shaped cells, 0.3–0.4 × 1–3 μm, occurring singly and in short chains. Gram positive. The cell envelope contains pseudomurein and an outer S-layer. Two tufts of polar flagella. Strictly anaerobic. Chemolithotrophic. Utilize H_2 and CO_2 to form methane. S^0 is reduced to H_2S (Stetter and Gaag, 1983). No growth below 60°C.

Free-living in hot anaerobic environments within solfatara fields.

Comparison with members of the *Methanobacteriaceae* by means

of 16S rRNA sequencing revealed that they are phylogenetically distant (Burggraf et al., 1991; Burggraf et al., 1997b).

So far, one genus with two species is described (Stetter et al., 1981; Lauerer et al., 1986).

The mol% G + C of the DNA is: 33.

Type genus: **Methanothermus** Stetter 1982b, 267 (Effective publication: Stetter *in* Stetter, Thomm, Winter, Wildgruber, Huber, Zillig, Janekovic, König, Palm and Wunderl 1981, 177).

Genus I. **Methanothermus** Stetter 1982b, 267[VP] (Effective publication: Stetter *in* Stetter, Thomm, Winter, Wildgruber, Huber, Zillig, Janekovic, König, Palm and Wunderl 1981, 177)

KARL O. STETTER

Me.tha.no.ther' mus. M.L. neut. n. *methanum* methane; Gr. fem. n. *therme* heat; M.L. masc. n. *Methanothermus* methane- (producing) thermophile.

Straight to slightly curved rods, usually 0.3–0.4 × 1–3 μm, **occurring singly and in short chains**. Motile by bipolar polytrichous flagellation. The **cell envelope consists of a double layer, an inner layer of pseudomurein and an outer S-layer** (Sleytr and Messner,

1983) **of protein** subunits (Fig. A2.7). At the poles, flagellum-containing channels are visible, leading radially through the pseudomurein. Gram positive. Strictly anaerobic. **Optimum temperature: 80–88°C**; maximum, ~97°C; minimum, 55–60°C.

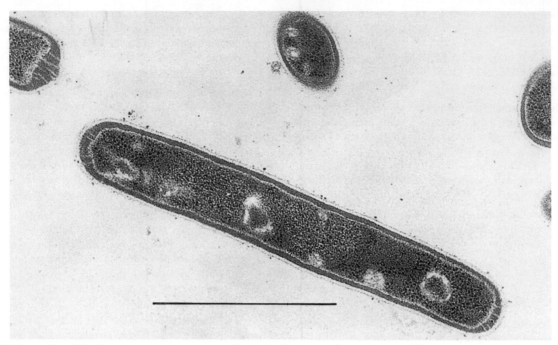

FIGURE A2.7. Thin section of *Methanothermus fervidus* contrasted with lead citrate and uranyl acetate. Electron micrograph. Bar = 1 μm. (Reprinted with permission from K.O. Stetter, 1984. Microbial growth on C₁ compounds. *In* Proceedings of the Fourth International Symposium, ©American Society for Microbiology, Washington, D.C., p. 178.)

Chemolithotrophic, growing on hydrogen and CO_2. The end product is methane. Formate, acetate, methanol, and methylamines cannot serve as substrates (Stetter et al., 1981). In the presence of sulfur, hydrogen, and CO_2, large amounts of H_2S are formed, and sulfur granules attached to the cells are visible (Stetter and Gaag, 1983).

The mol% G + C of the DNA is: 33.

Type species: **Methanothermus fervidus** Stetter 1982b, 267 (Effective publication: Stetter *in* Stetter, Thomm, Winter, Wildgruber, Huber, Zillig, Janekovic, König, Palm and Wunderl 1981, 177).

FURTHER DESCRIPTIVE INFORMATION

Cells are rigid. They occur singly, in pairs, and in short chains. At 97°C, irregular giant cells with large diameters (up to ~2 μm) appear. At 65°C and below, the rods become curly and occur in clumps. Cells divide by binary fission. The S-layer of the envelope can be removed by pronase or sodium dodecyl sulfate. The pseudomurein sacculus consists of *N*-acetylglucosamine, *N*-acetylgalactosamine, *N*-acetyltalosaminuronic acid, glutamic acid, alanine, and lysine (Stetter et al., 1981). Cells of *Methanobacterium formicicum* show a slight serological cross-reaction with *Methanothermus fervidus* cells (Macario and Conway de Macario, 1983). The RNA polymerase shows a subunit pattern with a spacing similar to that of *Methanothermobacter thermautotrophicus* (strains ΔH and W; Stetter et al., 1980). However, no serological cross-reaction could be observed in the Ouchterlony immunodiffusion test (Stetter et al., 1981).

Methanothermus species could be isolated only from samples of Icelandic solfatara fields at Kerlingarfjoll and Hveragerthi with a pH of 6.5 and a temperature of 85°C.

ENRICHMENT AND ISOLATION PROCEDURES

M. fervidus can be enriched anaerobically in medium 1 (Balch et al., 1979) at pH 6.5 in stoppered serum bottles with a gas phase of H_2/CO_2 (80%:20%) at ~70°C. No growth occurs above pH 7. *M. fervidus* can be isolated by plating on medium 1 solidified by polysilicate and by incubation with H_2/CO_2 in a pressure cylinder (Balch et al., 1979) at 85°C. After 3 d, round, smooth, opaque, slightly grayish colonies 1–3 mm in diameter appeared. No growth on agar (Stetter et al., 1981).

MAINTENANCE PROCEDURES

Stock cultures of *Methanothermus*, stable for at least 4 months without transfer, were obtained by 6 h of growth (5% inoculation) followed by renewing the gas phase of 200 kPa of H_2/CO_2 (80%:20%) and storage at 4°C. For longer periods cultures should be stored over liquid nitrogen (−140°C) in the presence of 5% dimethylsulfoxide.

DIFFERENTIATION OF THE GENUS *METHANOTHERMUS* FROM OTHER GENERA

Methanothermus belongs to the order *Methanobacteriales* due to its positive Gram reaction, the presence of a pseudomurein cell wall, and its rod shape. It differs from the genera *Methanobacterium* and *Methanobrevibacter* by the lack of serological cross-reaction of its RNA polymerases, the high temperature optimum, and the existence of an S-layer (Sleytr and Messner, 1983) outside of the cell wall. It differs from the genus *Methanopyrus* by its lower growth temperature and the lack of 2,3-di-*O*-geranylgeranyl-*sn*-glycerol (Hafenbradl et al., 1993).

TAXONOMIC COMMENTS

By 16S rRNA sequencing, the *Methanothermus* species are related to members of the genus *Methanobacterium* within the order *Methanobacteriales* (Burggraf et al., 1991).

FURTHER READING

Burggraf, S., K.O. Stetter, P. Rouvière and C.R. Woese. 1991. *Methanopyrus kandleri*: an archaeal methanogen unrelated to all other known methanogens. Syst. Appl. Microbiol. *14*: 346–351.

Lauerer, G., J.K. Kristjansson, T.A. Langworthy, H. König and K.O. Stetter. 1986. *Methanothermus sociabilis*, sp. nov., a second species within the *Methanothermaceae* growing at 97°C. Syst. Appl. Microbiol. *8*: 100–105.

Stetter, K.O., M. Thomm, J. Winter, G. Wildgruber, H. Huber, W. Zillig, D. Janecovic, H. König, P. Palm and S. Wunderl. 1981. *Methanothermus fervidus*, sp. nov., a novel extremely thermophilic methanogen isolated from an icelandic hot spring. Zentbl. Bakteriol. Mikrobiol. Hyg. *C2*: 166–178.

DIFFERENTIATION OF THE SPECIES OF THE GENUS *METHANOTHERMUS*

Only two species of *Methanothermus* are presently known, *M. fervidus* and *Methanothermus sociabilis*.

Key to the species of the genus *Methanothermus*

1. Cells occur singly or in short chains. Temperature optimum 80–85°C.
   M. fervidus.
2. Cells grow in large clusters. Optimum growth at 88°C.
   M. sociabilis.

List of species of the genus Methanothermus

1. **Methanothermus fervidus** Stetter 1982b, 267[VP] (Effective publication: Stetter *in* Stetter, Thomm, Winter, Wildgruber, Huber, Zillig, Janekovic, König, Palm and Wunderl 1981, 177).

fer' vi.dus. L. masc. adj. *fervidus* fervent, because of its growth in almost-boiling water.

See the generic description for features.
The mol% G + C of the DNA is: 33 (T_m and direct analysis).
Type strain: V24S, DSMZ 2088, JCM 10308.
GenBank accession number (16S rRNA): M59145.

2. **Methanothermus sociabilis** Stetter 1986b, 573[VP] (Effective publication: Stetter *in* Lauerer, Kristjansson, Langworthy, König and Stetter 1986, 104).

so.ci.a' bi.lis. L. masc. adj. *sociabilis* social, because of its growth in large aggregates.

Cells grow in large clusters. Optimum growth: ~88°C. The DNA–DNA similarity with *M. fervidus* is ~63%. The lipid pattern is different from that of *M. fervidus*.
The mol% G + C of the DNA is: 33 (T_m and direct analysis).
Type strain: Kf1-F1, DSMZ 3496.
GenBank accession number (16S rRNA): AF095273.

Class II. **Methanococci** *class. nov.*

DAVID R. BOONE

Me.tha.no.coc' ci. M.L. fem. pl. n. *Methanococcales* type order of the class; dropping the ending to denote a class; M.L. fem. pl. n. *Methanococci* the class of *Methanococcales.*

Cell shapes include cocci and many varieties of **coccoid shapes, rods, and sheathed rods** including spirals. Most cells have a protein cell wall, some walls contain an acidic heteropolysaccharide, and some cells are surrounded by a sheath. Peptidoglycan and pseudomurein are absent. Lipids are largely isoprenoid hydrocarbons ether-linked to glycerol. Cells are **strictly anaerobic**. They obtain energy by formation of methane. Many form methane by CO_2 reduction, with H_2, formate, or alcohols as electron donors. Some form methane by reduction or dismutation of methyl compounds including methyl amines, methyl sulfides, or methanol, and some split acetate to CH_4 and CO_2. Some strains reduce S^0, but this reaction does not appear to support cell growth. Cells occupy a wide range of anoxic habitats including aquatic sediments, anaerobic sewage digestors, and the gastrointestinal tracts of animals.

Type order: **Methanococcales** Balch and Wolfe 1981, 216 (Effective publication: Balch and Wolfe *in* Balch, Fox, Magrum, Woese and Wolfe 1979, 285).

Key to the orders of the class Methanococci

I. Coccoid organisms from marine environments; protein cell walls; Lipids contain N-acetylglucosamine and hydroxyarchaeol or cyclic archaeol. Cells grow autotrophically by using H_2 or formate to reduce CO_2.

II. Coccoids, rods, or sheathed rods that use H_2, formate, or alcohols to reduce CO_2. Lipids contain aminopetanetetrols, galactose, and glycerol.

III. Coccoids, pseudosarcinae, or sheathed rods that catabolize methyl groups (methanol, methyl amines, or methyl sulfides); some cells can grow on acetate or on H_2/CO_2. Lipids contain hydroxyarchaeol or cyclic archael and glycerol or galactose in the absence of aminopentanetetrols.
Order III. *Methanosarcinales*, p. 268

Order I. **Methanococcales** Balch and Wolfe 1981, 216[VP] (Effective publication: Balch and Wolfe *in* Balch, Fox, Magrum, Woese and Wolfe 1979, 285)

WILLIAM B. WHITMAN, DAVID R. BOONE AND YOSUKE KOGA

Me.tha.no.coc.ca' les. M.L. masc. n. *Methanococcus* type genus of the order; *-ales* ending to denote an order; M.L. fem. pl. n. *Methanococcales* the *Methanococcus* order.

Strictly anaerobic methane-producing archaeon. Hydrogen or formate may be electron donors. The cell wall is composed of protein. The order includes two families, *Methanococcaceae* Balch and Wolfe 1981, and *Methanocaldococcaceae*.

Type genus: **Methanococcus** Kluyver and van Niel 1936, 400, emend. Barker 1936, 430, Mah and Kuhn 1984b, 264 (Nom. Cons., Opinion 62 of the Jud. Comm. 1986a, 491) emend.

Key to families of the order Methanococcales

I. Temperature optimum for growth 70°C and below.
Family I. *Methanococcaceae.*
II. Temperature optimum for growth above 70°C.
Family II. *Methanocaldococcaceae*, p. 242

Family I. **Methanococcaceae** Balch and Wolfe 1981, 216[VP] (Effective publication: Balch and Wolfe *in* Balch, Fox, Magrum, Woese and Wolfe 1979, 285) emend.

WILLIAM B. WHITMAN, DAVID R. BOONE AND YOSUKE KOGA

Me.tha.no.coc.ca' ce.ae. M.L. masc. n. *Methanococcus* type genus of the family; *-aceae* ending to denote a family; M.L. fem. pl. n. *Methanococcaceae* the *Methanococcus* family.

Irregular cocci with a protein cell wall. Motile by means of flagella. Obligately anaerobic methane-producing archaeon which uses carbon dioxide as an electron acceptor. Hydrogen and formate are electron donors. Temperature range varies from mesophilic to thermophilic. Selenium is stimulatory to growth. The genera are *Methanococcus* (Kluyver and van Niel 1936, emend. Barker 1936) Mah and Kuhn 1984b and *Methanothermococcus*.

Type genus: **Methanococcus** Kluyver and van Niel 1936, 400, emend. Barker 1936, 430, Mah and Kuhn 1984b, 264 (Nom. Cons., Opinion 62 of the Jud. Comm. 1986a, 491) emend.

Key to the genera of the family Methanococcaceae

I. Temperature optimum for growth below 50°C.
Genus I. *Methanococcus.*
II. Temperature optimum for growth 50°C and above.
Genus II. *Methanothermococcus*, p. 241

Genus I. **Methanococcus** Kluyver and van Niel 1936, 400, emend. Barker 1936, 430,[AL] Mah and Kuhn 1984b, 264 (Nom. Cons., Opinion 62 of the Jud. Comm. 1986a, 491) emend.

WILLIAM B. WHITMAN

Me.tha.no.coc' cus. M.L. neut. n. *methanum* methane; Gr. n. *kokkos* a spherical cell; M.L. masc. n. *Methanococcus* methane coccus.

Irregular cocci, 1.0–2.0 µm in diameter during balanced growth. Cells from older cultures or colonies are extremely irregular. Cells lose integrity during Gram staining and lyse completely within 10 s when suspended in either distilled water or 0.01% sodium dodecylsulfate. **Motile, by means of polar tufts of flagella. Obligately anaerobic.** Mesophilic (temperature optima: 35–40°C). pH optimum between 6 and 8. NaCl required for growth, optimal concentrations 0.5–4% (w/v). **Obligately methanogenic;**

H₂ and formate serve as electron donors. Acetate, methanol, and methylamines are not substrates for methanogenesis. Alcohols, including isopropanol, are not substrates for methanogenesis. With one exception, all species grow autotrophically in mineral medium. Amino acids and acetate are stimulatory for some species. Nitrogen sources include ammonium, N_2 gas and alanine. Storage materials include glycogen. Organisms are found in **salt-marsh, marine, and estuarine sediments**.

The mol% G + C of the DNA is: 29–34.

Type species: **Methanococcus vannielii** Stadtman and Barker 1951b, 269 (Nom. Cons., Opinion 62 of the Jud. Comm. 1986a, 491).

FURTHER DESCRIPTIVE INFORMATION

Cell morphology In growing cultures, cells are slightly irregular and uniform in size, between 1 and 2 μm in diameter. Pairs of cells are common. In stationary cultures, colonies or enrichment cultures, cell shape is very irregular, and large cells up to 10 μm in diameter are observed (Jones et al., 1977). In wet mounts, a few cells in a preparation may slowly swell and burst (Ward, 1970). Cells found on the edge of a slide, where drying may occur, are much larger, more regular, and more transparent than cells found in the center of the slide. Cells from older cultures are mechanically fragile and rupture during vigorous stirring or upon harvesting by some continuous centrifugation devices. Cells are also osmotically fragile and lyse rapidly in distilled water. Cell integrity is maintained in 2% NaCl (w/v). Cells lyse rapidly in 0.01% sodium dodecylsulfate and contain a protein cell wall or S-layer. In *M. vannielii* and *M. voltae*, the outer cell surface is composed of hexagonally ordered structures (Jarrell and Koval, 1989).

The pattern of flagellation has been described. *M. vannielii* contains two tufts or bundles of flagella at the same pole.

Colony morphology Colonies are round, entire, and slightly convex. The surface is smooth or shiny. Color varies from pale green to yellow, depending on the species and the growth conditions. Colony size is 1–2 mm after 5 d. With transmitted light, colonies are translucent with either irregular or circular dark centers. However, because the colony morphology depends greatly on the culture conditions, it is not a reliable characteristic for classification.

Growth conditions All species grow rapidly on H_2/CO_2 in pressurized culture tubes (Balch and Wolfe, 1976). Under optimum conditions, generation times are <3 h. All species grow with formate. Acetate, methanol, and methylamines are not substrates for methanogenesis. All strains that have been tested are unable to utilize alcohols such as ethanol, isopropanol, isobu-

tanol, and cyclohexanol as electron donors for CO_2 reduction (Zellner and Winter, 1987b). An unnamed isolate of *Methanococcus* has also been reported to utilize methylfurfural compounds as a substrate for methanogenesis (Boopathy, 1996). However, Boopathy (1996) found that the tested *Methanococcus maripaludis* strain did not utilize methylfurfural. Other growth conditions are described in Table A2.8.

Mineral requirements In addition to NaCl, high concentrations of magnesium salts are stimulatory or required by the methanococci (Whitman et al., 1982, 1986; Corder et al., 1983; Jones et al., 1983b). Calcium is required by *M. voltae* (Whitman et al., 1982). Selenium is stimulatory to all species which have been tested (Jones and Stadtman, 1977; Whitman et al., 1982; Jones et al., 1983b). Iron, nickel, and cobalt are required or stimulatory for *M. voltae* (Whitman et al., 1982), and tungsten and nickel are required or stimulatory for *M. vannielii* (Jones and Stadtman, 1977; Diekert et al., 1981).

Carbon sources Except for *M. voltae*, the methanococci will grow in mineral medium with sulfide as the sole reducing agent and carbon dioxide as the sole carbon source (Whitman et al., 1986). Under these conditions, acetate is stimulatory to *M. maripaludis* but not to *M. vannielii* and "*M. aeolicus.*" *M. voltae* requires acetate, isoleucine, and leucine for growth (Whitman et al., 1982). Isovalerate and 2-methylbutyrate can substitute for leucine and isoleucine, respectively. Pantoyllactone and pantoic acid, which are formed from pantothenate during autoclaving, are also stimulatory to growth. Glycogen has been identified as a storage product in most methanococci (König et al., 1985; Yu et al., 1994).

Sulfur and nitrogen sources Sulfide is sufficient as a sulfur source for all methanococci. Elemental sulfur is also reduced to sulfide (Stetter and Gaag, 1983; Whitman, unpublished data). Cysteine, dithiothreitol, and sulfate do not substitute for sulfide (Whitman et al., 1982, 1987). Some strains of *M. maripaludis* utilize thiosulfate as a sulfur source (Rajagopal and Daniels, 1986).

Ammonium is sufficient as a nitrogen source for all methanococci and is required by *M. voltae* even during growth with amino acids (Whitman et al., 1982). N_2 gas and alanine are additional nitrogen sources for *M. maripaludis*. *M. vannielii* cannot utilize N_2 gas or amino acids as nitrogen sources, but it will utilize purines (DeMoll and Tsai, 1986; Whitman, 1989).

Lipid, polyamine, and compatible solute content A number of compounds are useful chemotaxonomic markers. The core lipid of the methanococci is composed of archaeol and hydroxyarchaeol (Koga et al., 1998). The polar head groups are composed of glucose, N-acetylglucosamine, serine, and ethanolamine

TABLE A2.8. Diagnostic characteristics of the species of the genus *Methanococcus*[a]

Characteristic	1. *M. vannielii*	2. *M. maripaludis*	3. *M. voltae*	4. "*M. aeolicus*"
Cell diameter:				
≤1.4 μm	+	+	−	−
>1.4 μm	−	−	+	+
Autotrophic growth	−	−	+	−
Growth with N₂ as sole N source	−	+	−	+
Mol% G + C	32.5	33–35	29–32	32

[a]Symbols: +, property of the species; −, not a property of the species; nd, not determined.

(in some species). The most abundant polyamine is spermidine (Kneifel et al., 1986). The compatible solute common in the methanothermococci, β-glutamate, has not been detected (Robertson et al., 1990b).

Immunological data Rabbit antisera to formalin-treated whole cells of methanococci cross-react weakly with other members of the genus but not other methane-producing bacteria (Conway de Macario et al., 1981; Jones et al., 1983b; Bryniok and Trosch, 1989). In a similar procedure, goat antisera to the type strain seldom cross-reacted to members of other species. However, it also did not cross-react to every member of the species (Keswani et al., 1996), and care must be taken during interpretation of immunological data.

Antibiotics and other inhibitors Like other methanogens, methanococci are generally resistant to low concentrations of many common antibiotics (Jones et al., 1977). Some antibiotics that are inhibitory at low concentrations are: adriamycin, chloramphenicol, efrapeptin, leucinostatin, metonidazole, monensin, pleuromutilin, pyrollnitrin, and virginiamycin (Elhardt and Böck, 1982; Böck and Kandler, 1985). Methanococci are also sensitive to low concentrations of organic tin-containing compounds such as: phenyltin, tripropyltin, and triethyltin (Boopathy and Daniels, 1991).

Pathogenicity The methanococci have not been reported to be associated with disease.

Habitats To date, methanococci have only been isolated from marine environments. *M. vannielii* was isolated from the shore of the San Francisco Bay (Stadtman and Barker, 1951b). The type strain of *M. voltae* was isolated from sediments from the mouth of the Waccasassa estuary in Florida (Ward, 1970). The type strain of *M. maripaludis* was isolated from salt-marsh sediments near Pawley's Island, South Carolina (Jones et al., 1983b). Additional strains of both *M. voltae* and *M. maripaludis* have been isolated from salt-marsh sediments in Georgia and Florida (Whitman et al., 1986; Keswani et al., 1996). An unnamed methanococcal isolate has also been obtained from the biofilm of a ship hull (Boopathy and Daniels, 1992).

ENRICHMENT AND ISOLATION PROCEDURES

Methanococci may be easily isolated after enrichment under H_2/CO_2 (80:20) in pressurized tubes or bottles (Miller and Wolin, 1974; Balch and Wolfe, 1976). Because of their rapid growth, methanococci frequently outgrow other H_2-utilizing methanogens found in marine sediments. Therefore, this enrichment is somewhat specific, and enrichment cultures that take longer than 5 d to develop seldom contain methanococci. The enrichments are transferred to medium containing antibiotics (0.2 mg/ml penicillin G, erythromycin, and streptomycin sulfate) before plating on agar plates or roll tubes (Jones et al., 1983b, c). In some cases, it is necessary to include antibiotics in the solid medium to prevent growth of spreading bacteria over colonies of methanococci. Isolated colonies are picked with a syringe needle and transferred to liquid medium. Purity may be demonstrated by microscopic examination, by restreaking on agar medium and by the absence of growth in mineral medium supplemented with 1% yeast extract under N_2/CO_2 (80:20).

MAINTENANCE PROCEDURES

A useful medium for rapid growth of methanococci consists of (in g/l): KCl, 0.34; NaCl, 22; $MgSO_4 \cdot 7H_2O$, 3.5; $MgCl_2 \cdot 6H_2O$,

2.8; NH_4Cl, 0.5; $CaCl_2 \cdot 2H_2O$, 0.14; K_2HPO_4, 0.14; $NaHCO_3$, 5; $Fe(NH_4)_2(SO_4)_2 \cdot 6H_2O$, 0.01; and 10 ml/l of trace metal solution (in g/l): nitrilotriacetic acid, neutralized with KOH, 1.5; $MnSO_4 \cdot 2H_2O$, 0.1; $Fe(NH_4)_2(SO_4)_2 \cdot 6H_2O$, 0.2; $CoCl_2 \cdot 6H_2O$, 0.1; $ZnSO_4 \cdot 7H_2O$, 0.1; $CuSO_4 \cdot 5H_2O$, 0.01; $NiCl_2 \cdot 6H_2O$, 0.025; Na_2SeO_3, 0.2; $Na_2MoO_4 \cdot 2H_2O$, 0.1; and $Na_2WO_4 \cdot 2H_2O$, 0.1. An oxygen indicator, resazurin (1 mg/l), may also be added prior to boiling the medium under N_2/CO_2 (80:20). After boiling, 2-mercaptoethanesulfonate, 0.5 g/l, is added to reduce the medium. After dispensing the medium anaerobically and sterilizing, the medium may be stored for several months in an anaerobic chamber (Coy Laboratories, Ann Arbor, Michigan). Just prior to inoculation, one part of sterile 2.5% $Na_2S \cdot 9H_2O$ (w/v) is added to 50 parts of medium. For *M. voltae*, the mineral medium must be supplemented with either yeast extract, 2 g/l, or $NaCH_3COO \cdot 3H_2O$, 1.4 g/l, L-isoleucine, 0.5 g/l, L-leucine, 0.5 g/l, and pantoyllactone, 1.3 mg/l (Whitman et al., 1986).

Because the methanococci lyse shortly after the cessation of growth in liquid media, stock cultures are grown below the temperature optimum (30°C) and stored at room temperature for up to 3 weeks. Strains of all mesophilic methanococci have been stored with little loss in viability for up to 3 years in 25% glycerol at $-70°C$ (Whitman et al., 1986; Tumbula et al., 1995). Cultures of 10 years or older have been routinely revived. Tube cultures are first concentrated by centrifugation and resuspended in a one-fifth volume of medium containing yeast extract and 25% glycerol (v/v). Portions of the cell suspension are transferred to sterile 1-ml screw-top glass vials in the anaerobic chamber. The vials are then stored at $-70°C$ without anaerobic precautions. To revive the cultures, 0.2 ml of the cell suspension are allowed to thaw in the anaerobic chamber and transferred to fresh medium. Methanococci have also been stored by freeze-drying (Hippe, 1984) and freezing in glycerol (Winter, 1983).

DIFFERENTIATION OF THE GENUS *METHANOCOCCUS* FROM OTHER GENERA

Several morphologically and nutritionally similar species of *Methanogenium* and *Methanomicrobium* have also been obtained from marine environments (Romesser et al., 1979; Rivard and Smith, 1982; Ferguson and Mah, 1983; Rivard et al., 1983). Methanococci may be distinguished from these other species by a faster growth rate, a requirement for higher concentrations of NaCl for optimal growth, the lack of organic growth requirements (except *M. voltae*) and lower mol% G + C of the DNA. However, identification of isolates based solely on morphology and growth characteristics is equivocal, and the use of salt or mineral requirements has been particularly deceptive (see below). Thus, antigenic cross-reactivity (Conway de Macario et al., 1981) and 16S rRNA sequencing (Keswani et al., 1996) are helpful for final identification of new isolates. Species of methanococci may also be distinguished by one-dimensional sodium dodecyl sulfate-polyacrylamide gel electrophoresis (SDS-PAGE) of cellular proteins (Whitman, 1989; Keswani et al., 1996). Cultures are grown to an absorbance of 1.0 cm^{-1} at 600 nm, and 5-ml cultures are harvested by centrifugation. The cells are resuspended in 0.1 ml of mineral medium or a salt solution prepared without reducing agents. This cell suspension may be stored at $-20°C$ prior to electrophoresis. After thawing, the suspension is vortexed until evenly dispersed, and 15 μl are added to 60 μl of sample buffer containing sodium dodecyl sulfate and 2-mercaptoethanol. A 35-μl portion of this mixture is electrophoresed on a 12% polyacrylamide gel. The protein profile on SDS–PAGE is sufficiently distinctive to distinguish species of mesophilic methanococci from each other or from other methanogenic bacteria like *Methanogenium* species (data not shown).

TAXONOMIC COMMENTS

In the 8th Edition of *Bergey's Manual* (Bryant, 1974), the genus *Methanococcus* included methane-producing cocci which did not form regular packets (i.e., *Methanosarcina*) or chains (i.e., some species of *Methanobacterium*, now *Methanobrevibacter*). At that time, the type species, *Methanococcus mazei*, was not available in pure culture, and only one other species, *M. vannielii*, was known. Upon analysis of the partial sequence of the 16S rRNAs of methanogens (Balch et al., 1979) and isolation of an archaeon with the phenotype of *M. mazei* (Mah, 1980), it became apparent that *M. mazei* was related to the *Methanosarcinaceae* and that *M. vannielii* was related to a new species of *Methanococcus*, *M. voltae*. Thus, it was proposed that *M. vannielii* become the new type species for the genus and *M. mazei* be reclassified as *Methanosarcina mazeii* (Balch et al., 1979; Mah and Kuhn, 1984b; Judicial Commission, 1986a). This proposal was adopted in the last edition of *Bergey's Manual of Systematic Bacteriology*, and the genus *Methanococcus* included five phenotypically and phylogenetically related species (Whitman, 1989).

Likewise, species that more closely resemble *M. mazei* were placed in the *Methanosarcinaceae*. Thus, *Methanococcus halophilus* Zhilina 1984, which utilizes methylamines for methane synthesis, was not classified with the *Methanococcaceae* (Zhilina, 1983). Similarly, *Methanococcus frisius* Blotevogel, Fischer and Lüpkes 1986, resembles *Methanosarcina mazeii* by nutritional and morphological criteria (Blotevogel et al., 1986) and was also classified with the *Methanosarcinaceae*.

Although the remaining methanococcal species were more closely related to each other than other methane-producing bacteria, they were not a closely-knit group (Whitman, 1989). In particular, the 16S rRNA of *M. jannaschii* was different enough from the other methanococci to justify creation of a new genus (Jones et al., 1983a). Because the description of *M. jannaschii* was based on a single isolate and its phenotype was similar to other methanococci, the creation of a new genus was not undertaken at that time. Subsequently, the isolation of additional hyperthermophilic methanococci and the nearly complete sequencing of the 16S rRNAs of many methanococci has led to the recommendation that the *Methanococcus* genus be further subdivided into four genera, thus reducing the genetic diversity within the group (Boone et al., 1993b). This informal proposal would leave the mesophilic species in the genus *Methanococcus* and place the thermophiles and hyperthermophiles in novel genera. The 16S rRNA sequence similarities (>91%) and DNA hybridizations (>3%) among the remaining, mesophilic species were still somewhat lower than found in many other bacterial and archaeal genera, suggesting that organisms in this group were still rather diverse (Keswani et al., 1996). However, in the absence of additional species, little benefit in subdividing this genus exists at this time.

The present descriptions of some methanococci are based on single isolates. Therefore, the described phenotypic characteristics may not be truly representative of the species. This difficulty is illustrated by a description of methanococci isolated from sediments of salt marshes (Whitman et al., 1986). It was not possible to identify natural isolates by their sodium or magnesium requirements, which were somewhat variable. Although strains related to *M. voltae* were recognized by their requirement for isoleucine, leucine, and acetate, the larger number of autotrophic isolates could not be identified by their nutritional properties or their mol% G + C. Subsequent comparison of these isolates by SDS–PAGE proved very useful (Keswani et al., 1996). Moreover, these results indicate that *Methanococcus maripaludis* and *Methanococcus deltae*, which was largely described on the basis of differences in salt requirements with *M. maripaludis* (Corder et al., 1983), are more similar to each other than are other species in this genus and probably are subjective synonyms.

By 16S rRNA sequencing and other criteria, the methanococci can be separated into four species, only three of which have been validly published (Keswani et al., 1996). The deepest branching taxon is "*Methanococcus aeolicus*", an isolate which has been described parenthetically in the literature and appears to warrant species status (Schmid et al., 1984; Whitman et al., 1986; Whitman, 1989). "*Methanococcus aeolicus*" is a mesophilic, coccal-shaped methane-producing archaeon which readily lyses in 0.01% sodium dodecyl sulfate or distilled water. Cells are 1.5–2.0 μm in diameter. Growth is autotrophic with hydrogen or formate as the electron donors. Ammonium and N$_2$ gas can serve as nitrogen sources. Sodium and magnesium salts are required for growth. "*M. aeolicus*" may be readily distinguished from other mesophilic methanococci by SDS–PAGE.

The next deepest branch is *M. voltae*. Described strains include a cluster closely related to the type strain and one additional genospecies related to strain A3 (Keswani et al., 1996). Although the DNA hybridization between the type strain PS and strain A3 was 35–38%, the phenotypic characteristics were insufficient to distinguish them from each other. Similarly, the described strains of *M. maripaludis* include three genospecies represented by strains C5, C6, and C7. These genospecies possess 61–69% DNA hybridization to strain JJ, the type strain, and 64–81% DNA hybridization to each other. In part because of its high DNA hybridization (83%) with strain JJ, *M. deltae* ΔRC is considered a subjective synonym of *M. maripaludis* and is named *M. maripaludis* ΔRC.

FURTHER READING

Boone, D.R., W.B. Whitman and P. Rouvière. 1993. Diversity and taxonomy of methanogens. *In* Ferry (Editor), Methanogenesis: Ecology, Physiology, Biochemistry, and Genetics, Chapman & Hall, New York, pp. 35–80.

Jarrell, K.F. and S.F. Koval. 1989. Ultrastructure and biochemistry of *Methanococcus voltae*. Crit. Rev. Microbiol. *17*: 53–87.

Whitman, W.B 1985. Methanogenic bacteria. *In* Woese and Wolfe (Editors), The Bacteria, Vol. VIII: Archaebacteria, Academic Press, Orlando, pp. 3–84.

List of species of the genus Methanococcus

1. **Methanococcus vannielii** Stadtman and Barker 1951b, 269.[AL]
   van.nie' li.i. M.L. gen. n. *vannielii* of van Niel; named for C.B. van Niel, the bacteriologist who developed the carbon dioxide reduction theory of methane formation.

Cell morphology, nutrition, and physiology are described in Tables A2.8 and A2.9 and in the generic description.

The mol% G + C of the DNA is: 31–33 (Bd and liquid chromatography).

TABLE A2.9. Descriptive characteristics of the species of the genus *Methanococcus*[a]

Characteristic	1. *M. vannielii*	2. *M. maripaludis*	3. *M. voltae*	4. "*M. aeolicus*"
Irregular coccus	+	+	+	+
Cell diameter, μm	1.3	0.9–1.3	1.3–1.7	1.7
Substrates for methane synthesis:				
H$_2$/CO$_2$ (80:20)	+	+	+	+
Formate	+	+	+	+
Acetate	−	−	−	−
Methylamines	−	−	−	−
Growth requirement:				
Acetate	−	−	+	−
Isoleucine	−	−	+	−
Leucine	−	−	+	−
Ca^{2+} (1 mM)	−	−	+	−
Growth stimulatory:				
Selenium	+	+	+	+
Pantoyllactone	−	−	+	−
Acetate	−	+	na	−
Amino acids	−	+	na	−
Sulfur sources:				
Sulfide	+	+	+	+
Elemental sulfur	+	+	+	+
Thiosulfate	−	d	−	−
Sulfite	−	−	−	−
Sulfate	−	−	−	−
Nitrogen sources:				
NH$_3$	+	+	+	+
N$_2$	−	+	−	+
Alanine	−	+	−	−
Temperature range, °C	≤20–45	≤20–45	≤20–45	≤20–45
pH range	6.5–8	6.5–8	6.5–8	6.5–8
NaCl optimum, % (w/v)	0.6–2	0.6–2	1–2	1–2
NaCl range, % (w/v)	0.3–5	0.3–5	0.3–5	1 to >5
Mol% G + C	32.5	33–35	29–32	32

[a]Symbols: +, property of the species; −, not a property of the species; d, property of some strains of the species; na, not applicable.

Type strain: SB, ATCC 35089, DSMZ 1224.

GenBank accession number (16S rRNA): M36507.

Additional Remarks: The type strain was not named in the publication in which it was originally described. Subsequently, it has become known as strain SB.

2. **Methanococcus maripaludis** Jones, Paynter and Gupta 1984, 270[VP] (Effective publication: Jones, Paynter and Gupta 1983b, 91).

ma.ri.pa.lu' dis. L. n. *mare* sea; L. n. *palus* marsh; M.L. pl. gen. n. *maripaludis* of the sea marsh.

Cell morphology, nutrition, and physiology are described in Tables A2.8 and A2.9 and in the generic description.

The mol% G + C of the DNA is: 33 (Bd and liquid chromatography).

Type strain: JJ, DSMZ 2067.

GenBank accession number (16S rRNA): U38484.

3. **Methanococcus voltae** Balch and Wolfe 1981, 216[VP] (Effective publication: Balch and Wolfe *in* Balch, Fox, Magrum, Woese and Wolfe 1979, 285).

vol' tae. M.L. gen. n. *voltae* of Volta; named for the Italian physicist A. Volta for discovery of the combustible nature of gas from anaerobic sediments.

Cell morphology, nutrition, and physiology are described in Tables A2.8 and A2.9 and in the generic description.

The mol% G + C of the DNA is: 30–31 (Bd and liquid chromatography).

Type strain: PS, ATCC 33273, DSMZ 1537.

GenBank accession number (16S rRNA): M59290.

Other Organisms

1. *"Methanococcus aeolicus"*

Mentioned parenthetically in Schmid et al. (1984), this bacterium has not been fully described. However, DNA hybridization data and 16S rRNA sequence analysis suggest that it represents a new species of *Methanococcus* (Keswani et al., 1996). The SDS–PAGE pattern is distinctive. Nutritional properties closely resemble other autotrophic methanococci (Whitman et al., 1986). It is capable of autotrophic growth and is not stimulated by acetate or amino acids. Hydrogen gas and formate are substrates for methanogenesis; acetate, methylamines, and methanol are not. Nitrogen sources include ammonium and N$_2$ gas. The type strain has not been designated.

2. *Methanococcus deltae* Corder, Hook, Larkin and Frea 1988, 221[VP] (Effective publication: Corder, Hook, Larkin and Frea 1983, 32).

del' tae. M.L. gen. n. *deltae* from the Greek *delta*.

The cell morphological, nutritional, and other properties closely resemble *M. maripaludis* JJ. Considered as a subjective synonym of *M. maripaludis* (Keswani et al., 1996).

Genus II. **Methanothermococcus** gen. nov.

WILLIAM B. WHITMAN

Me.tha.no.ther.mo.coc'cus. M.L. neut. n. *methanum* methane; Gr. adj. *thermos* hot; Gr. n. *kokkos* a spherical cell; M.L. masc. n. *Methanothermococcus* a coccus producing methane at thermophilic growth temperatures.

Regular to irregular coccus, 1.5 μm in diameter. Cells stain Gram-negative and lyse rapidly in dilute solutions of sodium dodecylsulfate. **Motile, by means of a polar tuft of flagella. Obligately anaerobic. Thermophilic** (temperature optimum: 60–65°C). pH optimum between 5.1 and 7.5. NaCl required for growth, optimal concentration 2–4% (w/v). **Obligately methanogenic; H_2 and formate serve as electron donors. Acetate, methanol, and methylamines are not substrates for methanogenesis.** Alcohols, including isopropanol, are not substrates for methanogenesis. Grows autotrophically in mineral medium. Organic carbon sources are not stimulatory. Nitrogen sources include ammonium, nitrate and N_2 gas. Sulfur sources include sulfide, elemental sulfur, and other oxidized sulfur-containing compounds such as thiosulfate, sulfite, and sulfate. Isolated from shallow, sandy, geothermally heated marine sediments and marine oil-reservoir water.

The mol% G + C of the DNA is: 31–34.

Type species: **Methanothermococcus thermolithotrophicus** comb. nov. (*Methanococcus thermolithotrophicus* Huber, Thomm, König, Thies and Stetter 1982, 50).

FURTHER DESCRIPTIVE INFORMATION

Cell morphology In growing cultures, cells are slightly irregular and uniform in size, between 1 and 2 μm in diameter. Pairs of cells are common. Cells on the edge of a slide, where drying may occur, are much larger, more regular, and more transparent than cells from the center of the slide. Cells contain a protein cell wall or S-layer and a single tuft of about 20 flagella.

Colonial morphology Colonies are circular, smooth, and yellowish in color. Colony size is 1 mm.

Growth conditions Grows rapidly on H_2/CO_2 (80:20) in pressurized culture tubes or on formate. For the two strains described, the growth properties varied somewhat (Huber et al., 1982; Nilsen and Torsvik, 1996). The temperature range for growth is 17–70°C, and the temperature optimum is 60–65°C. An increase in the hydrostatic pressure increases the growth rate slightly, but does not increase the temperature range (Bernhardt et al., 1988). The pH range for growth is 4.9–9.8. The range of NaCl concentrations for growth is 0.6–9.4%, and the optimum concentration is 2–4%. Under optimal conditions, generation times are <1 h. Other growth conditions are described in Table A2.10.

Sulfur and nitrogen sources Sulfide is sufficient as a sulfur source for methanothermococci. Elemental sulfur is reduced to sulfide with inhibition of methanogenesis (Stetter and Gaag, 1983). Oxidized sulfur compounds including thiosulfate, sulfite, and sulfate are also sulfur sources (Daniels et al., 1986). In the presence of very low concentrations of sulfide, organic sulfur sources such as methanethiol, ethanethiol, propanethiol, butanethiol, methylsulfide, ethylsulfide, dimethylsulfide, and carbon disulfide serve as supplemental sulfur sources (Rajagopal and Daniels, 1986).

Ammonium is sufficient as a nitrogen source. N_2 gas and nitrate are also assimilated as nitrogen sources (Belay et al., 1984, 1990).

Lipid, polyamine, and compatible solute content Several compounds are useful chemotaxonomic markers. The core lipid of the methanothermococci is composed of archaeol, caldarchaeol, and hydroxyarchaeol (Koga et al., 1998). The polar head groups are composed of glucose, *N*-acetylglucosamine, and serine. The most abundant polyamine is spermidine, and small amounts of putrescine and spermine are also present (Kneifel et al., 1986; Hamana et al., 1998). Although the abundance varies with salt concentration in the growth medium, methanothermococci contain glutamate, β-glutamate, and *N*-acetyl-β-lysine as compatible solutes (Robertson et al., 1990b, 1992b).

Sensitivity to antibiotics and other inhibitors Like other methanogens, methanothermococci are generally resistant to low concentrations of many common antibiotics. Organotin compounds including phenyltin, tripropyltin, and triethyltin are inhibitory at low concentrations (IC$_{50}$ = 5 μM; Boopathy and Daniels, 1991).

Pathogenicity The methanothermococci have not been reported to be associated with disease.

Habitats Methanothermococci have been isolated from shallow, geothermally heated marine sediments and oil-field reservoir water (Nilsen and Torsvik, 1996).

ENRICHMENT AND ISOLATION PROCEDURES

Methanothermococci may be isolated after enrichment under H_2/CO_2 (80:20) in pressurized tubes or bottles at 50–60°C (Huber et al., 1982; Nilsen and Torsvik, 1996). The antibiotics

TABLE A2.10. Descriptive characteristics of the species of the genus *Methanothermococcus*[a]

Characteristic	*M. thermolithotrophus*
Irregular coccus	+
Cell diameter, μm	1.5
Substrates for methane synthesis:	
H_2/CO_2 (80:20)	+
Formate	+
Acetate	−
Methylamines	−
Autotrophic growth	+
Sulfur sources:	
Sulfide	+
Elemental sulfur	+
Thiosulfate	+
Sulfite	+
Sulfate	+
Nitrogen sources:	
NO$_3^-$	+
NH$_3$	+
N$_2$	+
Temperature optimum, °C	60–65
Temperature range, °C	17–70
pH optimum	5.1–7.5
pH range	4.9–9.8
NaCl optimum, % (w/v)	2–4
NaCl range, % (w/v)	0.6–9.4
Mol% G + C	32.5

[a]Symbols: +, property of the species; −, not a property of the species.

(30 µg/ml penicillin G and 50 µg/ml vancomycin) may be included to facilitate isolation. Plating can be performed at elevated temperature on 2% agar or Gelrite gellan gum.

MAINTENANCE PROCEDURES

A useful medium for rapid growth of methanothermococci consists of (in g/l): KCl, 0.34; NaCl, 22; MgSO$_4$·7H$_2$O, 3.5; MgCl$_2$·6H$_2$O, 2.8; NH$_4$Cl, 0.5; CaCl$_2$·2H$_2$O, 0.14; K$_2$HPO$_4$, 0.14; NaHCO$_3$, 5; Fe(NH$_4$)$_2$(SO$_4$)$_2$·6H$_2$O, 0.01; and 10 ml/l of trace metal solution (in g/l): nitrilotriacetic acid, neutralized with KOH, 1.5; MnSO$_4$·2H$_2$O, 0.1; Fe(NH$_4$)$_2$(SO$_4$)$_2$·6H$_2$0, 0.2; CoCl$_2$·6H$_2$O, 0.1; ZnSO$_4$·7H$_2$O, 0.1; CuSO$_4$·5H$_2$O, 0.01; NiCl$_2$·6H$_2$O, 0.025; Na$_2$SeO$_3$, 0.2; Na$_2$MoO$_4$·2H$_2$O, 0.1; and Na$_2$WO$_4$·2H$_2$O, 0.1. An oxygen indicator, resazurin (1 mg/l), may also be added prior to boiling the medium under N$_2$/CO$_2$ (80:20). After boiling, 2-mercaptoethanesulfonate, 0.5 g/l, is added to reduce the medium. After dispensing the medium anaerobically and sterilizing, the medium may be stored for several months in an anaerobic chamber (Coy Laboratories, Ann Arbor, Michigan). Just prior to inoculation, one part of sterile 2.5% Na$_2$S·9H$_2$O (w/v) is added to 50 parts of medium. The pH of the medium prior to growth is 6.8.

Because the methanothermococci lyse rapidly shortly after the cessation of growth in liquid media, stock cultures are grown below the optimum temperature. Cultures can be stored at 4°C for six months.

DIFFERENTIATION OF THE GENUS *METHANOTHERMOCOCCUS* FROM OTHER GENERA

Methanothermococcus is distinguished from the mesophiles *Methanococcus* and the hyperthermophiles *Methanocaldococcus* and *Methanotorris* by its moderate thermophily. It differs from most other moderately thermophilic methanogens (such as *Methanothermobacter*) by its coccoid morphology. So far, *Methanoculleus thermophilicus* is the only other moderately thermophilic coccus

described (Rivard and Smith, 1982; Ferguson and Mah, 1983). *M. thermophilicus* requires acetate for growth, grows in low concentrations of NaCl, has a mol% G + C of 56–60, and is easily distinguishable by these properties. Antigenic cross-reactivity (Conway de Macario et al., 1981; Bryniok and Trosch, 1989; Nilsen and Torsvik, 1996), 16S rRNA sequencing (Keswani et al., 1996), and sodium dodecyl sulfate-polyacrylamide gel electrophoresis (SDS–PAGE) of cellular proteins (Nilsen and Torsvik, 1996) are also helpful for final identification of new isolates.

TAXONOMIC COMMENTS

In the first edition of *Bergey's Manual of Systematic Bacteriology*, the type species of *Methanothermococcus*, *M. thermolithotrophicus*, was included in the genus *Methanococcus* on the basis of its morphology, 16S rRNA sequence, simple wall structure, and simple nutritional properties (Whitman, 1989). This classification, which was based in part on rRNA catalogs, did not take into account the quantitative differences in 16S rRNA sequences between mesophilic, thermophilic, and hyperthermophilic species. With the availability of nearly complete 16S rRNA sequences, it was recommended that the genus *Methanococcus* be further subdivided into four genera to reduce the genetic diversity within the group (Boone et al., 1993b). The genus *Methanococcus* was restricted to mesophiles. The moderate thermophiles with 16S rRNA sequence similarities of <91% to the type strain of *Methanococcus* were placed in new genera. The moderate thermophilic genus, *Methanothermococcus*, was further distinguished from the hyperthermophilic *Methanocaldococcus* and *Methanotorris* genera based on its low 16S rRNA sequence similarity (<91%) and difference in growth temperature.

FURTHER READING

Boone, D.R., W.B. Whitman and P. Rouvière. 1993. Diversity and taxonomy of methanogens. *In* Ferry (Editor), Methanogenesis: Ecology, Physiology, Biochemistry, and Genetics, Chapman & Hall, New York, pp. 35–80.

List of species of the genus Methanothermococcus

1. **Methanothermococcus thermolithotrophicus** comb. nov. (*Methanococcus thermolithotrophicus* Huber, Thomm, Kōnig, Thies and Stetter 1982, 50).
   ther.mo.li.tho.tro'phi.cus. Gr. fem. n. *therme* heat; Gr. masc. n. *lithos* stone; Gr. masc. n. *trophos* one who feeds; M.L. masc. adj. *thermolithotrophicus* grows lithotrophically at elevated temperatures.

   Cell morphology, nutrition, and physiology are described in Table A2.10 and in the generic description.
   The mol% G + C of the DNA is: 31–34 (T_m and liquid chromatography).
   Type strain: SN-1, DSMZ 2095.
   GenBank accession number (16S rRNA): M59128.

Family II. **Methanocaldococcaceae** *fam. nov.*

WILLIAM B. WHITMAN, DAVID R. BOONE AND YOSUKE KOGA

Me.tha.no.cal.do.coc.ca'ce.ae. M.L. masc. n. *Methanocaldococcus* type genus of the family; -aceae ending to denote family; M.L. fem. pl. n. *Methanocaldococcaceae* the *Methanocaldococcus* family.

Irregular coccus with a protein cell wall. Obligately anaerobic methane-producing archaeon which uses carbon dioxide as an electron acceptor. Hydrogen is the electron donor. Hyperther-

mophilic. Selenium may be stimulatory to growth. The genera are *Methanocaldococcus* and *Methanotorris*.
Type genus: **Methanocaldococcus** gen. nov.

Key to the genera of the family Methanocaldococcaceae

I. Motile and selenium required or stimulatory for growth.
Genus I. *Methanocaldococcus.*
II. Nonmotile and selenium not required for growth.
Genus II. *Methanotorris*, p. 245

Genus I. **Methanocaldococcus** gen. nov.

WILLIAM B. WHITMAN

Me.tha.no.cal.do.coc' cus. M.L. neut. n. *methanum* methane; L. adj. *caldus* hot; Gr. n. *kokkos* a spherical cell; M.L. masc. n. *Methanocaldococcus* a coccus producing methane at hyperthermophilic growth temperatures.

Regular to irregular coccus, 1–3 μm in diameter. Cells stain Gram-negative and lyse rapidly in dilute solutions of sodium dodecyl-sulfate. **Motile, by means of polar tufts of flagella. Obligately anaerobic. Hyperthermophilic** (temperature optima: 80–85°C). pH optimum between 5.2 and 7.6. NaCl required for growth, optimal concentration 2–4% (w/v). **Obligately methanogenic; H_2 serves as the electron donor. Formate, acetate, methanol, and methylamines are not substrates for methanogenesis.** Grows autotrophically in mineral medium; complex carbon sources are sometimes stimulatory. Nitrogen sources include ammonium, nitrate, and N_2 gas. Sulfur sources include sulfide and S^0. Isolated from deep-sea hydrothermal vents and surrounding sediments.

The mol% G + C of the DNA is: 31–33.

Type species: **Methanocaldococcus jannaschii** comb. nov. (*Methanococcus jannaschii* Jones, Leigh, Mayer, Woese and Wolfe 1983a, 260).

FURTHER DESCRIPTIVE INFORMATION

Cell morphology Irregular cocci, 1–3 μm in diameter. Pairs of cells are common. Cells contain a protein cell wall or S-layer with hexagonal spacing and two or three flagellar tufts.

Growth conditions Grows rapidly on H_2/CO_2 (80:20) in pressurized culture tubes (Jones et al., 1983a; Zhao et al., 1988; Jeanthon et al., 1998, 1999b). The temperature range for growth is 48–91°C, and the temperature optimum is 80–85°C. The pH range for growth is 5.2–7.6. The optimal pH for growth is 6.0–6.5. The range of NaCl concentrations for growth is 0.5–5.6%, and the optimum concentration is near 2–3%. Under optimal conditions, generation times are <1 h. Other growth conditions are described in Table A2.11.

Sulfur and nitrogen sources Sulfide is sufficient as a sulfur source for methanocaldococci. Elemental sulfur is reduced to sulfide with inhibition of methanogenesis (Jones et al., 1983a; Zhao et al., 1988; Jeanthon et al., 1998, 1999b). Thiosulfate and sulfate are not utilized as sulfur sources. Ammonium is sufficient as a nitrogen source, and nitrates are also assimilated in the species tested.

Lipid, polyamine, and compatible solute contents So far, only *M. jannaschii* has been examined for these compounds. Its core lipid is composed of archaeol, caldarchaeol, and cyclic archaeol (Koga et al., 1998). The polar head groups are composed of glucose, *N*-acetylglucosamine, ethanolamine, and serine. The composition of the lipid also varies with the growth temperature; caldarchaeol, cyclic archaeol, and phospholipids become more abundant at higher temperatures (Sprott et al., 1991). The most abundant polyamine is spermine, and small amounts of sper-midine and traces of putrescine are present (Hamana et al., 1998). The most abundant compatible solute is β-glutamate (Robertson et al., 1990b).

Sensitivity to antibiotics and other inhibitors Like other methanogens, methanocaldococci are generally resistant to low concentrations of many common antibiotics. In particular, resistance to ampicillin and kanamycin have been noted. Sensitivity to chloramphenicol (75 μg/ml) and rifampicin (50 μg/ml) has also been observed for all species except *M. fervens* (Jeanthon et al., 1999b).

Pathogenicity The methanocaldococci have not been reported to be associated with disease.

Habitats Methanocaldococci have been isolated from chimney material and sediments from deep-sea hydrothermal vents in both the Pacific and Atlantic Oceans.

ENRICHMENT AND ISOLATION PROCEDURES

Methanocaldococci may be isolated after enrichment under H_2/CO_2 (80:20) in pressurized tubes or bottles at 80°C (Jeanthon et al., 1998). Isolated colonies are obtained by plating at elevated temperature on 0.8% Phytagel (Sigma Chemical, Co., St. Louis, MO).

MAINTENANCE PROCEDURES

Cultures can be stored at 4°C in culture medium or at −80°C in medium plus 20% glycerol (Jeanthon et al., 1998). A typical medium consists of (per liter of distilled water): Sea salts, 30 g; NH_4Cl, 1 g; KH_2PO_4, 0.35 g; $NaHCO_3$, 1 g; PIPES, 3.46 g; yeast extract, 2 g; cysteine-HCl, 0.5 g; trace element solution (Widdel and Bak, 1992), 1 ml; tungstate, 30 mg; selenate, 0.5 mg; vitamin solution (Widdel and Bak, 1992), 1 ml; thiamine solution (Widdel and Bak, 1992), 1 ml; vitamin B_{12}, 0.05 mg; growth-stimulating factors (Pfennig et al., 1981), 1 ml; and resazurin, 1 mg. The pH is adjusted to 6.5 before autoclaving, and the medium is reduced by the addition of $Na_2S\cdot9H_2O$ to a final concentration of 0.05% (w/v). The medium is then pressurized to 200 kPa with H_2/CO_2 (80:20 v/v). Strictly anaerobic procedures are used throughout the preparation of the medium and cultivation.

DIFFERENTIATION OF THE GENUS *METHANOCALDOCOCCUS* FROM OTHER GENERA

Methanocaldococcus is distinguished from the mesophile *Methanococcus* and the thermophile *Methanothermococcus* by its hyperthermophily. It is distinguished from the other hyperthermophilic coccus, *Methanotorris*, by its growth requirement for selenium, the presence of flagellar tufts, and its 16S rRNA sequence.

TABLE A2.11. Descriptive characteristics of the species of the genus *Methanocaldococcus*[a]

Characteristic	1. *M. jannaschii*	2. *M. fervens*	3. *M. infernus*	4. *M. vulcanius*
Irregular coccus	+	+	+	+
Cell diameter, μm	1.5	1–2	1–3	1–3
Number of flagellar tufts	2	nd	3	3
Substrates for methane synthesis:				
H_2/CO_2	+	+	+	+
Formate	−	−	−	−
Acetate	−	−	−	−
Methylamine	−	−	−	−
Autotrophic growth	+	+	+	+
Growth requirement:				
Selenium	+	+	+	+
Tungsten	nd	+	+	+
Growth stimulatory:				
Yeast extract	−	+	+	+
Sulfur sources:				
Sulfide	+	+	+	+
Elemental sulfur	+	+	+	+
Thiosulfate	nd	nd	−	−
Sulfate	−	nd	−	−
Nitrogen sources:				
NH_3	+	+	+	+
NO_3^-	nd	nd	+	+
Temperature optimum, °C	85	85	85	80
Temperature range, °C	50–86	48–92	49–89	55–91
pH optimum	6	6.5	6.5	6.5
pH range	5.2–7	5.5–7.6	5.2–7	5.2–7
NaCl optimum, %	3	3	2.5	2.5
NaCl range, %	1.0–5.0	0.5–5.0	1.2–5.0	0.6–5.6
Mol% G + C	31	33	33	31

[a]Symbols: +, property of the species; −, not a property of the species; nd, not determined.

Restriction fragment length polymorphism of the gene encoding the 16S rRNA has proven useful for distinguishing species of *Methanocaldococcus* from each other as well as from strains of *Methanotorris* (Jeanthon et al., 1999a). It is difficult to distinguish species of *Methanocaldococcus* on phenotypic properties alone. Of the four species currently described, only *M. fervens* is resistant to rifampicin, and only *M. jannaschii* is not stimulated by yeast extract.

TAXONOMIC COMMENTS

In the first edition of *Bergey's Manual of Systematic Bacteriology* the type species of *Methanocaldococcus*, *M. jannaschii*, was included in the genus *Methanococcus* on the basis of its morphology, 16S rRNA sequence, simple wall structure, and simple nutritional properties (Whitman, 1989). This classification, which was based in part on rRNA catalogs, did not take into account the quantitative differences in 16S rRNA sequences between mesophilic, ther-

mophilic, and hyperthermophilic species. With the availability of nearly complete 16S rRNA sequences, Boone et al. (1993b) recommended that the genus *Methanococcus* be further subdivided into four genera to reduce the genetic diversity within the group. The genus *Methanococcus* was thus restricted to mesophiles. The thermophile *M. thermolithotrophicus* was placed in a new genus *Methanothermococcus*. The hyperthermophiles were placed in two new genera, *Methanocaldococcus* and "*Methanoignis*" (called *Methanotorris* here, see below), on the basis of their low 16S rRNA sequence similarity (<93%) to each other and their differences in nutritional and other phenotypic properties.

FURTHER READING

Boone, D.R., W.B. Whitman and P. Rouvière. 1993. Diversity and taxonomy of methanogens. *In* Ferry (Editor), Methanogenesis: Ecology, Physiology, Biochemistry, and Genetics, Chapman & Hall, New York, pp. 35–80.

List of species of the genus Methanocaldococcus

1. **Methanocaldococcus jannaschii** comb. nov. (*Methanococcus jannaschii* Jones, Leigh, Mayer, Woese and Wolfe 1983a, 260). *jan.na' schi.i.* M.L. gen. n. *jannaschii* of Jannasch; named for the marine microbiologist H.W. Jannasch.

   Cell morphology, nutrition, and physiology are described in Table A2.11 and in the generic description.
   The mol% G + C of the DNA is: 31(Bd).
   Type strain: JAL-1, DSMZ 2661.
   GenBank accession number (16S rRNA): M59126.

2. **Methanocaldococcus fervens** comb. nov. (*Methanococcus fervens* Jeanthon, L'Haridon, Reysenbach, Corre, Vernet, Messner, Sleytr and Prieur 1999b, 588.)

fer' vens. L. part. adj. *fervens* boiling hot, referring to its high growth temperature.

   Cell morphology, nutrition, and physiology are described in Table A2.11 and in the generic description.
   The mol% G + C of the DNA is: 33 (T_m).
   Type strain: AG86, DSMZ 4213.
   GenBank accession number (16S rRNA): AF056938.

3. **Methanocaldococcus infernus** comb. nov. (*Methanococcus infernus* Jeanthon, L'Haridon, Reysenbach, Vernet, Messner, Sleytr and Prieur 1998, 917).
   in.fer' nus. L. masc. adj. *infernus* referring to the place of isolation, deep-sea hydrothermal vents.

Cell morphology, nutrition, and physiology are described in Table A2.11 and in the generic description.

The mol% G + C of the DNA is: 33 (T_m).

Type strain: ME, DSMZ 11812.

GenBank accession number (16S rRNA): AF025822.

4. **Methanocaldococcus vulcanius** comb. nov. (*Methanococcus vulcanius* Jeanthon, L'Haridon, Reysenbach, Corre, Vernet, Messner, Sleytr and Prieur 1999b, 588).

vul.ca' ni.us. L. masc. adj. *vulcanius* referring to Vulcanus, the Roman fire god, and to the place of isolation, the deep-sea hydrothermal vents.

Cell morphology, nutrition, and physiology are described in Table A2.11 and in the generic description.

The mol% G + C of the DNA is: 31(T_m).

Type strain: M7, DSMZ 12094.

GenBank accession number (16S rRNA): AF051404.

Genus II. **Methanotorris** gen. nov.

WILLIAM B. WHITMAN

Me.tha.no.tor' ris. M.L. neut. n. *methanum* methane; L. masc. n. *torris* fire; M.L. masc. n. *Methanotorris* the methane-producer at fiery temperatures.

Regular to irregular coccus, 1.3–1.8 μm in diameter. **Nonmotile. Obligate anaerobic. Hyperthermophilic** (temperature optimum: 88°C). pH optimum 5.7. NaCl required for growth, optimal concentration 1.8% (w/v). **Obligate methanogen; H_2 serves as the electron donor. Formate, acetate, methanol, and methylamines are not substrates for methanogenesis.** Grows autotrophically in mineral medium. Complex carbon sources are not stimulatory. Nitrogen sources include ammonium. Sulfur sources include sulfide and elemental sulfur. Isolated from deep-sea and shallow marine hydrothermal systems.

The mol% G + C of the DNA is: 31.

Type species: **Methanotorris igneus** comb. nov. (*Methanococcus igneus* Burggraf, Fricke, Neuner, Kristjansson, Rouvière, Mandelco, Woese and Stetter 1990a, 268).

FURTHER DESCRIPTIVE INFORMATION

Cell morphology   Irregular cocci, 1.3–1.8 μm in diameter. Pairs of cells are common. Cells contain a protein cell wall or S-layer. Although nonmotile, a few flagellar-like structures are observed by electron microscopy.

Growth conditions   Grows rapidly on H_2/CO_2 (80:20) in pressurized culture tubes (Burggraf et al., 1990a). The temperature range for growth is 45–91°C, and the temperature optimum is 88°C. The pH range for growth is 5.0–7.5. The optimal pH for growth is 5.7. The range of NaCl concentrations for growth is 0.9–5.4%, and the concentration optimum is near 1.8%. Under optimal conditions, generation times are about 0.5 h. Other growth conditions are described in Table A2.12.

Sulfur and nitrogen sources   Sulfide is sufficient as a sulfur source for methanothermococci. Elemental sulfur is reduced to sulfide with inhibition of methanogenesis (Burggraf et al., 1990a). Ammonium is sufficient as a nitrogen source. Nitrate and other potential nitrogen sources have not been tested.

Lipid and compatible solute content   The core lipid is composed of archaeol, caldarchaeol, and cyclic archaeol (Koga et al., 1998). The polar head groups are composed of glucose, N-acetylglucosamine, ethanolamine, and serine. The compatible solutes include glutamate, β-glutamate, and di-*myo*-inositol-1,1'-phosphate (Robertson et al., 1990b; Ciulla et al., 1994a).

Sensitivity to antibiotics and other inhibitors   Like other methanogens, *Methanotorris* is generally resistant to low concentrations of many common antibiotics. In particular, resistance to ampicillin, kanamycin, and rifampicin has been noted. Sensitivity to chloramphenicol (75 μg/ml) has also been observed (Jeanthon et al., 1999b).

Pathogenicity   The genus *Methanotorris* has not been reported to be associated with disease.

Habitats   *Methanotorris* was originally isolated from a shallow marine hydrothermal system near Iceland (Burggraf et al., 1990a). Additional strains have been isolated from chimney material and sediments from deep-sea hydrothermal vents in the Guaymas Basin and the Mid-Atlantic Ridge (Jeanthon et al., 1999a).

ENRICHMENT AND ISOLATION PROCEDURES

Methanotorris may be isolated after enrichment under H_2/CO_2 (80:20) in pressurized tubes or bottles at 80°C (Jeanthon et al., 1998). Isolated colonies are obtained by plating at elevated temperature on 0.8% Phytagel (Sigma Chemical, Co., St. Louis, MO).

TABLE A2.12.   Descriptive characteristics of the species of the genus *Methanotorris*[a]

Characteristic	M. igneus
Irregular coccus	+
Cell diameter, 1–2 μm	+
Substrates for methane synthesis:	
H_2/CO_2 (80:20)	+
Formate	−
Acetate	−
Methylamines	−
Autotrophic growth	+
Growth requirement:	
Selenium	−
Tungsten	−
Sulfur sources:	
Sulfide	+
Elemental sulfur	+
Thiosulfate	−
Sulfate	−
Nitrogen sources:	
NH_3	+
Temperature optimum, °C	88
Temperature range, °C	45–91
pH optimum	5.7
pH range	5.0–7.5
NaCl optimum, % (w/v)	1.8
NaCl range, % (w/v)	0.9–5.4
Mol% G + C	31

[a]Symbols: +, property of the species; −, not a property of the species.

MAINTENANCE PROCEDURES

Cultures can be stored at 4°C in culture medium or at -80°C in medium plus 20% glycerol (Jeanthon et al., 1998). Typically, a medium consists of: Sea salts, 30 g; NH$_4$Cl, 1 g; KH$_2$PO$_4$, 0.35 g; NaHCO$_3$, 1 g; PIPES, 3.46 g; yeast extract, 2 g; cysteine-HCl, 0.5 g; trace element solution (Widdel and Bak, 1992), 1 ml; tungstate, 30 mg; selenate, 0.5 mg; vitamin solution (Widdel and Bak, 1992), 1 ml; thiamine solution (Widdel and Bak, 1992), 1 ml; vitamin B$_{12}$, 0.05 mg; growth-stimulating factors (Pfennig et al., 1981), 1 ml; resazurin, 1 mg; and distilled water to yield one liter final volume. The pH is adjusted to 6.5 before autoclaving, and the medium is reduced by the addition of Na$_2$S·9H$_2$O to a final concentration of 0.05% (w/v). The medium is then pressurized to 200 kPa with H$_2$/CO$_2$ (80:20 v/v). Strict anaerobic procedures are used throughout the preparation of the medium and cultivation.

DIFFERENTIATION OF THE GENUS *METHANOTORRIS* FROM OTHER GENERA

Methanotorris is distinguished from the mesophile *Methanococcus* and the thermophile *Methanothermococcus* by its hyperthermophily. It is distinguished from the other hyperthermophilic coccus, *Methanocaldococcus*, by the absence of a growth requirement for selenium, the absence of motility and flagellar tufts, and its 16S rRNA sequence. Restriction fragment length polymorphism

of the gene encoding the 16S rRNA has proven useful for distinguishing *Methanotorris* from species of *Methanocaldococcus* (Jeanthon et al., 1999a).

TAXONOMIC COMMENTS

The genus *Methanotorris* was created to recognize the genetic diversity within the hyperthermophilic *Methanocaldococcaceae*. Although the 16S rRNA sequence similarity to species of *Methanocaldococcus* is in the range of 92–93%, most phylogenetic treeing algorithms suggest that *Methanotorris* represents an independent clade. Within the genus *Methanocaldococcus*, the 16S rRNA sequence similarity is greater than 95%. Therefore, placement of *M. igneus* in a separate genus distinguishes it from these closely related species. This treatment follows the recommendation of Boone et al. (1993b), who subdivided the genus *Methanococcus* into four genera. However, the proposed genus name "*Methanoignis*" was incorrectly formed, the correct latinization being "*Methanignis*". To preserve the prefix "*methano-*", the genus name *Methanotorris* is proposed here.

FURTHER READING

Boone, D.R., W.B. Whitman and P. Rouvière. 1993. Diversity and taxonomy of methanogens. *In* Ferry (Editor), Methanogenesis: Ecology, Physiology, Biochemistry, and Genetics, Chapman & Hall, New York, pp. 35–80.

List of species of the genus Methanotorris

1. **Methanotorris igneus** comb. nov. (*Methanococcus igneus* Burggraf, Fricke, Neuner, Kristjansson, Rouvière, Mandelco, Woese and Stetter 1990a, 268).
   ig' ne.us. L. masc. adj. *igneus* belonging to the fire, denoting its high growth temperature.

Cell morphology, nutrition, and physiology are described in Table A2.12 and in the generic description.
The mol% G + C of the DNA is: 31 (T_m).
Type strain: Kol 5, DSMZ 5666.
GenBank accession number (16S rRNA): M59125.

Order II. **Methanomicrobiales** Balch and Wolfe 1981, 216[VP] (Effective publication: Balch and Wolfe *in* Balch, Fox, Magrum, Woese and Wolfe 1979, 286).

DAVID R. BOONE, WILLIAM B. WHITMAN AND YOSUKE KOGA

Me.tha.no.mic.ro.bi.a' les. M.L. neut. n. *Methanomicrobium* type genus of the order; *-ales* the ending to denote an order; M.L. fem. pl. n. *Methanomicrobiales* the order of *Methanomicrobium*.

Cell shapes include cocci and many varieties of **coccoid shapes, rods, and sheathed rods** including spirals. Most cells have a protein cell wall, and some cells are surrounded by a sheath. Peptidoglycan and pseudomurein are absent. Lipids contain galactose, aminopentanetetrols, and glycerol as polar head groups. Cells are **strictly anaerobic**. They obtain energy by formation of methane. Many form methane by CO$_2$ reduction, with H$_2$, formate, or alcohols as electron donors. Some strains reduce S^0, but

this reaction does not appear to support cell growth. Cells occupy a wide range of anoxic habitats including aquatic sediments, anaerobic sewage digestors, and the gastrointestinal tracts of animals.

Type genus: **Methanomicrobium** Kluyver and van Niel 1936, 399, emend. Balch and Wolfe *in* Balch, Fox, Magrum, Woese and Wolfe 1979, 284.

Key to the families of the order Methanomicrobiales

1. Coccoid organisms that grow by using H$_2$, formate, or alcohols to reduce CO$_2$; difficult to distinguish from *Methanocorpusculaceae* (below) except by phylogenetic analysis.
   Family I. *Methanomicrobiaceae*, p. 247
2. Coccoid organisms that grow by using H$_2$, formate, or sometimes secondary alcohols to reduce CO$_2$; difficult to distinguish from *Methanomicrobiaceae* (above) except by phylogenetic analysis.
   Family II. *Methanocorpusculaceae*, p. 262

3. Sheathed rods forming a gentle helix; use H$_2$, formate, or alcohols to reduce CO$_2$.
Family III. *Methanospirillaceae*, p. 264

Family I. **Methanomicrobiaceae** Barker 1956, 15,[AL] emend. Balch and Wolfe *in* Balch, Balch and Wolfe *in* Balch, Fox, Magrum, Woese and Wolfe 1979, 268.

DAVID R. BOONE, WILLIAM B. WHITMAN AND YOSUKE KOGA

Me.tha.no.mic.ro.bi.a' ce.ae. M.L. neut. n. *Methanomicrobium* type genus of the family; *-aceae* the ending to denote a family; M.L. fem. pl. n. *Methanomicrobiaceae* the family of *Methanomicrobium.*

Cells are **coccoids, short rods, or plate-shaped**. Cell walls are protein. Peptidoglycan and pseudomurein are absent. **Strictly anaerobic. Energy by CO$_2$ reduction to form methane, with H$_2$, formate, or sometimes secondary alcohols** as electron donor. Some strains reduce S^0, but this reaction does not appear to support cell growth. Acetate is required; some species have additional growth factors. Many species are marine, and most grow fastest with NaCl at 100–500 mM. Cells occupy a wide range of anoxic habitats including marine sediments and anaerobic sewage digestors.

The mol% G + C of the DNA is: 38–61.

Type genus: **Methanomicrobium** Kluyver and van Niel 1936, 399, emend. Balch and Wolfe *in* Balch, Fox, Magrum, Woese and Wolfe 1979, 284.

TAXONOMIC COMMENTS

The genera of *Methanomicrobiaceae* share many phenotypic features, and it is impossible to provide a key of common phenotypic features that clearly divides the genera. In many cases assignments of organisms to species, and sometimes even assignment to genera, cannot be done with confidence except with phylogenetic analysis.

Key to the genera of the family Methanomicrobiaceae

I. Rod-shaped cells require acetate and 7-mercaptoheptanoylthreonine phosphate for growth.
Genus I. *Methanomicrobium.*
II. Coccoid cells 0.5–2 µm in diameter, mesophilic or thermophilic.
Genus II. *Methanoculleus*, p. 251
III. Coccoid cells, 1.5–3 µm in diameter, mesophilic.
Genus III. *Methanofollis*, p. 253
IV. Coccoid cells, 0.5–2.5 µm in diameter, marine.
Genus IV. *Methanogenium*, p. 256
V. Rods, 0.6 × 1.5–2.5 µm, marine.
Genus V. *Methanolacinia*, p. 258
VI. Disc-shaped cells, 0.08–0.3 µm thick × 1–3.5 µm wide and long.
Genus VI. *Methanoplanus*, p. 259

Genus I. **Methanomicrobium** Balch and Wolfe 1981, 216[VP] (Effective publication: Balch and Wolfe in Balch, Fox, Magrum, Woese and Wolfe 1979, 286)

GERHARD ZELLNER

Me.tha.no.mi.cro' bi.um. M.L. neut. n. *methanum* methane; Gr. adj. *micros* small; Gr. n. *bios* life; M.L. neut. n. *Methanomicrobium* methane (-producing) small life (-form).

Short, straight to slightly curved, irregular rods, 0.6–0.7 µm wide and 1.5–2.5 µm long, with rounded ends. Morphology may be influenced by substrate availability, becoming coccoid (0.5–1.0 µm in diameter) under starvation conditions. **Cells occur singly and in pairs but not in chains**. Capsules are not formed. Endospores are not produced. Gram negative. Species may be motile, with monotrichous polar flagellation, or nonmotile. Strictly anaerobic. Growth between 25 and 45°C with an optimum at 40°C. Growth and methanogenesis occur from pH 5.9 to 7.7 with an optimum between 6.1 and 7.0. **H$_2$/CO$_2$ and formate serve as substrates for methanogenesis and growth**. Acetate is a required nutrient. Species may have no other requirements for organic nutrients or may have several. A heat-stable cofactor present in

clarified rumen fluid and in cell extracts of *Methanothermobacter thermautotrophicus* is required for growth. This requirement could be satisfied by addition of 7-mercaptoheptanoylthreonine phosphate.

The mol% G + C of the DNA is: 48.8.

Type species: **Methanomicrobium mobile** (Paynter and Hungate 1968) Balch and Wolfe 1981, 216 (Effective publication: Balch and Wolfe *in* Balch, Fox, Magrum, Woese and Wolfe 1979, 286) (*Methanobacterium mobilis* Paynter and Hungate 1968, 1943).

FURTHER DESCRIPTIVE INFORMATION

The genus currently is represented by one species, *Methanomicrobium mobile*. Specific information regarding the species is given in the species description. Another strain previously described as *Methanomicrobium paynteri* (Rivard et al., 1983) was reclassified as member of a new genus, *Methanolacinia paynteri* (Zellner et al., 1989b).

Cells of *M. mobile* are rod-shaped and motile by means of a single polar flagellum (Fig. A2.8A), but cells may become pleo-

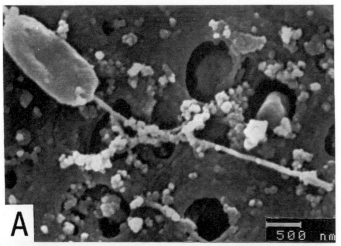

FIGURE A2.8. *Methanomicrobium mobile* BP. Scanning electron micrographs of (*A*) rod-shaped cell of *M. mobile* BP with single polar flagellum and (*B* and *C*) variety of pleomorphic, irregularly coccoid cells of *M. mobile* BP. (Photo courtesy of H. Diekmann).

morphic and irregularly coccoid and lose their flagella (Fig. A2.8B) depending upon nutritional status. The pleomorphic cells appear in media containing suboptimal concentrations of required nutrients or when starvation occurs, and they have a tendency to lyse (Tanner, 1982; Rivard et al., 1983). A rigid cell wall is lacking for *M. mobile*. Muramic acid is absent, and murein and pseudomurein are not present (Kandler and König, 1978). The cell envelope is sensitive to sodium dodecyl sulfate (SDS) and trypsin (Kandler and König, 1978). The cell envelope of *M. mobile* consists of a cytoplasmic membrane and an S-layer of hexagonally arranged glycoprotein subunits (Fig. A2.9a and b). The center- to-center spacing of the morphological units of the S-layer lattice of *M. mobile* is ~14.5 nm. The apparent M_r of the glycoprotein subunits, as indicated by a positive PAS-staining reaction, is 160,000 (Fig. A2.9C).

Lipophilic core components of polar lipids extracted from *M. mobile* cells include diphytanylglycerol diether, bisdiphytanyldiglycerol tetraether, and several unidentified novel core lipids (Grant et al., 1985; Zellner et al., 1989b; Koga et al., 1993). *M. mobile* has a complex pattern of core ether lipids with $C_{20}C_{20}$, $C_{40}C_{40}$, and unidentified ether core lipids (Grant et al., 1985).

Motility of *M. mobile* may be difficult to observe. Brief exposure to air, or growth under less than optimum nutritional conditions (M.J.B. Paynter, unpublished), results in very weak tumbling motility which is difficult to distinguish from Brownian motion.

A heat stable factor is required for growth and methanogenesis. It is present in bovine rumen fluid, mixed rumen bacteria, and *Methanothermobacter thermautotrophicus*, from which it has been partially purified (Tanner, 1982). The mobile factor was isolated and chemically characterized (Tanner and Wolfe, 1988), and it was demonstrated that 7-mercaptoheptanoylthreonine phosphate could satisfy the growth requirement of *M. mobile* for heat-stable factor (mobile factor) (Kuhner et al., 1991).

Colonies of *M. mobile* grown on H_2/CO_2 in a medium containing clarified rumen fluid are barely visible after 4 d of incubation at 39°C. Growth for 15 d results in surface colonies that are 0.7–1.0 mm in diameter, with entire edges, translucent, colorless to pale yellow, smooth, and convex. Deep colonies are lenticular.

Rabbit antiserum prepared against freeze-dried cells of *M. mobile* does not react in the indirect immunofluorescence assay with methanogens of other families or with *Methanospirillum hungateii*. The anti-*M. mobile* antibody probe shows a weak positive reaction with *Methanogenium cariaci* (Macario and Conway de Macario, 1985). Cells of *Methanolacinia paynteri* also have a weak positive reaction with the anti-*M. mobile* antibody probe but not with antisera to methanogens belonging to other families (Rivard et al., 1983).

Methanomicrobium has been isolated from bovine rumen (Paynter and Hungate, 1968). Methanogens antigenically related to *M. mobile* strain BP were also detected with anti-*M. mobile* antibody probe in anaerobic methane-producing wastewater bioreactors (Macario et al., 1991a; Zellner et al., 1996, 1997). In a wood-fermenting anaerobic bioreactor, 89–97% of all immunologically detectable methanogens were weakly related (antigenically) to *M. mobile* BP (Macario et al., 1991a).

M. mobile was found to be light sensitive (Olson et al., 1991). *M. mobile* reached a final optical density OD_{660} of 0.21 in the absence of light, while no growth was detected in the presence of light (General Electric blue fluorescent, 8 W/m²). The inhibition was attributed to a wavelength of light between 370 and 430 nm.

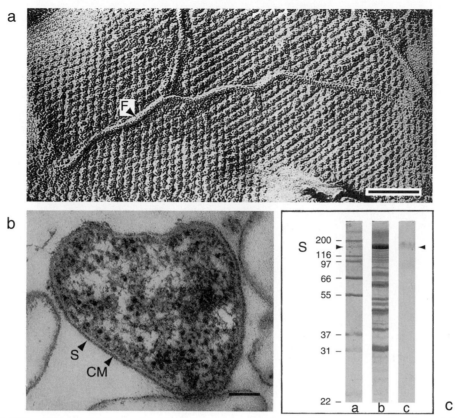

FIGURE A2.9. *Methanomicrobium mobile* BP. Electron micrographs of (*a*) freeze-fractured and metal-shadowed intact cells of *M. mobile* showing the hexagonal S-layer and flagella and (*b*) an ultrathin section of a cell. S, S-layer; CM, cytoplasmic membrane; F, flagella. Bar = 100 nm. SDS-PAGE (*c*) of SDS-solubilized whole-cell extracts. *Lane a*, molecular mass standard; *lane b*, Coomassie blue staining; *lane c*, periodic acid-Schiff staining. (Photo courtesy of P. Messner).

ENRICHMENT AND ISOLATION PROCEDURES

Because of the strict anaerobic nature of this organism, all manipulations must be performed using anaerobic techniques. Cultivation requires that media be both anoxic and at a negative redox potential. This is achieved for tube and flask cultures by using reducing agents and the anaerobic procedures of Hungate (1969) with modifications (Balch et al., 1979). Anaerobiosis can also be achieved by using an anaerobic glove box. This allows cultivation on reduced agar media contained in Petri plates.

M. mobile is isolated by serial dilution of bovine rumen contents on a clarified rumen fluid medium* with H_2/CO_2 (80:20) as the gas phase (Paynter and Hungate, 1968). Growth in closed vessels is improved if the atmosphere is pressurized to 300 kPa or if the gas phase is periodically renewed.

Purification is accomplished by repeated serial dilution on the enrichment or isolation medium.

MAINTENANCE PROCEDURES

M. mobile requires transfer every 2–3 weeks when grown in H_2/CO_2 rumen fluid broth. Colonies in rumen fluid agar roll tubes survive for several months under dry ice storage (M.J.B. Paynter, unpublished data). Cultures are preserved in liquid nitrogen.

DIFFERENTIATION OF THE GENUS *METHANOMICROBIUM* FROM OTHER GENERA

Distinctive features are presented in Table A2.13.

The mol% G + C for the DNA of *Methanomicrobium* is 48.8 (Bd) and is similar to that of irregular cocci of *Methanogenium* (*Methanogenium organophilum*, *Methanogenium frittonii*), *Methanoplanus* (*Methanoplanus limicola*), and *Methanolacinia* (*Methanolacinia paynteri*), which need to be distinguished because of their similar morphologies of irregular cocci.

Complete 16S rRNA sequencing and phylogenetic analysis revealed that *M. mobile* forms a phylogenetic cluster within the *Methanogenium* group comprising *Methanogenium cariaci*, *Methanogenium organophilum*, *Methanoplanus limicola* and *Methanomicrobium mobile* (Rouvière et al., 1992). The complete 16S rRNA sequence of *M. mobile* was deposited in the NCBI database under accession number M59142. In addition, partial sequences of the 16S rRNA genes of unpublished *Methanomicrobium* sp. strains BRM9 and BRM16 were deposited in the NCBI database under the accession

---

*Clarified rumen fluid contains the following (final percentage composition [w/v]): rumen contents centrifuged at 25,000 × g for 15 min, 30 (v/v); KH_2PO_4, 0.05; K_2HPO_4, 0.05; NaCl, 0.1; $(NH_4)_2SO_4$, 0.05; $MgSO_4$, 0.01; $CaCl_2$, 0.01; resazurin, 0.0001; $NaHCO_3$, 0.5; cysteine·HCl, 0.03; Na_2S, 0.03; 80% H_2 and 20% CO_2 gas phase.

TABLE A2.13. Differential characteristics of genera of the family *Methanomicrobiaceae*[a]

Characteristic	*Methanomicrobium*	*Methanoculleus*	*Methanofollis*	*Methanogenium*	*Methanolacinia*	*Methanoplanus*
Cell morphology:						
Short, straight rods	+	−	−	−	+	−
Irregular coccus	+	+	+	+	+	+
Ring-shaped cells	−	−	+	−	−	−
Disk-shaped cells	−	−	−	−	−	+
Cell diameter, μm:						
1.0		+	−	+	+	−
1.5	+	+	+	+	+	+
2.0	+	+	+	+	+	+
3.0	−	−	+	+	−	+
M_r of S-layer glycoprotein subunit	160,000	101,000–138,000	118,000–120,000	106,000–117,000	155,000 (135,000)	110,000–143,000
Substrate:						
H_2/CO_2	+	+	+	+	+	+
Formate	+	+	+	D	+	+
2–Propanol/CO_2	−	D	D	D	D	−
2–Butanol/CO_2	nd	D	D	D	D	−
Cyclopentanol/CO_2	nd	D	D	D	D	−
Growth requirements:						
Yeast extract	+	D	D	D	+	+
Peptone	+	D	D	−	D	D
Clarified rumen fluid	+	−	−	−	−	−
Acetate	+	+	+	+	+	+
0.1 M NaCl	−	D	D	D	−	+
Growth promotion by tungstate	−	+	+	+	−	D
Optimal growth temperature, °C, (min/max)	25/45	10/<70	20/45	0/62	20/45	16/41
Mol% G + C	49[b]	49–62	54–60	47–52	38/45	39–48

[a]For symbols, see standard definitions; nd, not determined.
[b]bd.

numbers X99138 and X99139, respectively. Both strains were isolated from rumen of Southern Hemisphere cattle (G.N. Jaris, E.R.B. Moore and K.N. Joblin, unpublished).

FURTHER READING

Conway de Macario, E., M.J. Wolin and A.J.L. Macario. 1981. Immunology of archaebacteria that produce methane gas. Science *214*: 74–75.

Paynter, M.J. and R.E. Hungate. 1968. Characterization of *Methanobac-* *terium mobilis*, sp. nov., isolated from the bovine rumen. J. Bacteriol. *95*: 1943–1951.

Rouvière, P.F., L. Mandelco, S. Winker and C.R. Woese. 1992. A detailed phylogeny for the *Methanomicrobiales*. Syst. Appl. Microbiol. *15*: 363–371.

Zellner, G., P. Messner, H. Kneifel, B.J. Tindall, J. Winter and E. Stacke-brandt. 1989. *Methanolacinia* gen. nov., incorporating *Methanomicrobium paynteri* as *Methanolacinia paynteri* comb. nov. J. Gen. Appl. Microbiol. *35*: 185–202.

List of species of the genus Methanomicrobium

1. **Methanomicrobium mobile** (Paynter and Hungate 1968) Balch and Wolfe 1981, 216[VP] (Effective publication: Balch and Wolfe *in* Balch, Fox, Magrum, Woese and Wolfe 1979, 286) (*Methanobacterium mobilis* Paynter and Hungate 1968, 1943).

mo′bi.le. L. neut. adj. *mobile* motile, movable.

Characteristics are as described for the genus. Cells are 0.7 × 1.5–2.0 μm. The polar flagellum is 81 nm wide and up to 12.5 μm long. Lipophilic core components of polar lipids extracted from *M. mobile* cells include diphytanylglycerol diether, bisdiphytanyldiglycerol tetraether, and several unidentified novel core lipids. *M. mobile* has a complex pattern of core ether lipids with $C_{20}C_{20}$, $C_{40}C_{40}$, and unidentified ether core lipids.

Substrate for growth and methanogenesis is H_2/CO_2 or formate. H_2 or formate can serve as an electron donor for reduction of CO_2 to CH_4. Acetate, propionate, butyrate, isobutyrate, valerate, isovalerate, caproate, succinate, glucose, pyruvate, methanol, ethanol, propanol, 2-propanol, and butanol are not substrates.

Acetate, isobutyrate, 2-methylbutyrate, isovalerate, tryptophan, pyridoxine, thiamine, and biotin are required for growth; *p*-aminobenzoate is not required but is stimulatory.

A heat-stable factor is required for growth and methanogenesis, which is present in bovine rumen fluid, extract of mixed rumen bacteria, and *Methanothermobacter thermautotrophicus*. 7-Mercaptoheptanoylthreonine phosphate could fulfill the growth requirement of *M. mobile* for heat-stable factor (mobile factor).

Growth occurs from 30 to 45°C with an optimum at 40°C; no growth at 28 or 50°C. The pH range is 5.9–7.7 with an optimum at 6.1–6.9 in bicarbonate or acetate buffers. Trishydrochloride (0.1 M) buffer is completely inhibitory, and phosphate (0.1 M) reduces methanogenesis by 50%. The organism was isolated from bovine rumen.

The mol% G + C of the DNA is: 48.8 (Bd).

Type strain: BP, ATCC 35094, DSMZ 1539.

GenBank accession number (16S rRNA): M59142.

Genus II. **Methanoculleus** *Maestrojuán, Boone, Xun, Mah and Zhang 1990, 121*[VP]

SONG C. CHONG AND DAVID R. BOONE

Me.tha.no.cul′le.us. M.L. neut. n. *methanum* methane; L. n. *culleus* bag; M.L. masc. n. *Methanoculleus* methane producing.

Cells are irregular coccoids, 0.5–2.0 μm in diameter, occurring singly or pairs. Stain Gram-negative. **Does not form spores.** Cell wall is protein (S-layer of glycoprotein subunits), sensitive to lysis by detergents. **Thermophilic or mesophilic.** Rapid growth at NaCl concentration of 0.1–0.2 M. **Chemolithotrophic, strictly anaerobic, utilizing H₂/CO₂.** All species but *Methanoculleus olentangyi* grow on formate, and some grow on secondary alcohols/CO₂, with **methane as a catabolic product.** Nonmotile, but flagella and fimbriae may occur. Most rapid growth between 20 and 45°C and between pH 6.2 and 8.0. Habitat is anaerobic digestors, anoxic lake sediments, or sea sediments.

The mol% G + C of the DNA is: 49–61.

Type species: **Methanoculleus bourgensis** corrig. (Ollivier, Mah, Garcia and Boone 1986) Maestrojuán, Boone, Xun, Mah and Zhang 1990, 121[VP] (*Methanogenium bourgense* Ollivier, Mah, Garcia and Boone 1986, 300).

FURTHER DESCRIPTIVE INFORMATION

Cells are irregular coccoids occurring singly or in pairs. Although some species (*Methanoculleus marisnigri* and *Methanoculleus palmolei*) have flagella, motility is undetectable. Colonies are white, yellow, sometimes with a shiny surface, circular, and convex. The cell wall is a glycosylated S-layer with hexagonal symmetry, and cells are very sensitive to lysis by physical and chemical treatment, such as 0.02% sodium dodecyl sulfate. The S-layer of protein subunits (138,000 Da [Zabel et al., 1984]) contains sugar moieties of only 0.8% galactose, 0.3% glucose, and 0.1% mannose on a dry weight basis (Romesser et al., 1979).

All species can grow on H₂/CO₂ or formate as a catabolic substrate, and only *M. palmolei* can use secondary alcohols (2-propanol, 2-butanol, 2-pentanol, and cyclopentanol). Methylotrophic substrates are not used. Several species (*M. olentangyi*, *M. marisnigri*, and *M. oldenburgensis*) can grow in mineral medium but are stimulated by yeast extract, peptones, or vitamins. The other species of the genus require one or more of these organic additions.

ENRICHMENT AND ISOLATION PROCEDURES

Methanoculleus can be isolated from anaerobic digestors and fresh or marine anoxic sediments. Enrichment cultures of *Methanoculleus* are obtained by the serum tube modifications (Sowers and Noll, 1995) of the anaerobic techniques of Hungate (1969). Small amounts of salt in the medium (2.5 g of NaCl per liter or higher concentrations for enrichment from marine samples) often improve the growth rate. MS medium, a bicarbonate-buffered medium, has been modified for good growth of these species by adding 2.5 g/l NaCl and 5 mmol/l sodium acetate (MG medium) (Maestrojuán et al., 1990). Cultures are isolated by streaking on anaerobic agar plates (Sowers and Noll, 1995) or more commonly by inoculation of roll tube media.

MAINTENANCE PROCEDURES

Cultures can be maintained by periodic subculturing or storage as liquid suspensions in liquid nitrogen (Hippe, 1984; Boone, 1995).

DIFFERENTIATION OF THE GENUS *METHANOCULLEUS* FROM OTHER GENERA

Coccoid methanogens that are mesophilic or moderately thermophilic and grow on H₂/CO₂ but not trimethylamine may be found in the following genera: *Methanococcus*, *Methanocorpusculum*, *Methanoculleus*, *Methanogenium*, and *Methanoplanus*. These genera are morphologically very similar, although members of the genus *Methanoplanus* are described as plate shaped. However, this distinction is very difficult to discern. Likewise, there are differences in cell diameter among these genera, but this feature also is a difficult one to use for taxonomic separation. Cells of *Methanoculleus* (0.5–2.0 μm in diameter) are somewhat larger than those of *Methanocorpusculum* (less than 1 μm in diameter), but the ranges of size overlap. However, DNA hybridization studies indicate sufficiently deep divisions between these two groups to justify placement as separate genera (Maestrojuán et al., 1990). Species of *Methanoculleus* also have higher mol% G + C of their DNA (49–61) than *Methanococcus* (31–41) and *Methanogenium* (47–52).

List of species of the genus Methanoculleus

1. **Methanoculleus bourgensis** corrig. (Ollivier, Mah, Garcia and Boone 1986) Maestrojuán, Boone, Xun, Mah and Zhang 1990, 121[VP] (*Methanogenium bourgense* Ollivier, Mah, Garcia and Boone 1986, 300).
   bour.gen′sis. M.L. masc. adj. *bourgensis* from Bourg-en-Bresse, France, name of a locality in France.

   See Table A2.14 and the generic description for many characteristics.

   Cells are irregular coccoids, nonmotile, 1–2 μm in diameter. Colonies are white or yellow, circular and convex. Methane produced from H₂/CO₂ and formate. Yeast extract and Trypticase peptones (BBL) stimulate growth. Fastest growth at mesophilic temperatures, at neutral pH, and in media with 0.17 M NaCl. The most rapid growth at 37°C. pH range for growth is 5.5–8.0 (fastest growth at pH 6.7).

   Isolated from a digestor fermenting tannery by-products, which was originally inoculated with digested sewage sludge from Bourg-en-Bresse, France. Strain LX1 (OCM 24) is a reference strain.

   The mol% G + C of the DNA is: 59.

   Type strain: MS2, DSMZ 3045, OCM 15.

   GenBank accession number (16S rRNA): AF095269.

2. **Methanoculleus marisnigri** (Romesser, Wolfe, Mayer, Spiess and Walther-Mauruschat 1979) Maestrojuán, Boone, Xun, Mah and Zhang 1990, 121[VP] (*Methanogenium marisnigri* Romesser, Wolfe, Mayer, Spiess and Walther-Mauruschat 1979, 152).
   ma.ris.ni′gri. L. gen. n. *maris* of the sea; L. masc. adj. *niger* black; M.L. neut. n. *marisnigri* of the Black Sea.

TABLE A2.14. Morphological and physiological characteristics of the species of the genus *Methanoculleus*[a]

Characteristic	1. *M. bourgensis*	2. *M. marisnigri*	3. *M. oldenburgensis*	4. *M. olentangyi*	5. *M. palmolei*	6. *M. thermophilicus*
Fimbriae	−	−	+	−	−	−
Flagella	−	+	−	−	+	D
Pili	−	−	−	−	−	D
Motility	−	−	−	−	−	D
Requires acetate	+	−	+	+	+	−
Growth on formate	+	+	+	−	+	+
Growth at 60°C	−	−	−	−	−	+
Growth at 15°C	−	+	−	−	−	−
Mol% G + C	59	61	49	54	59–60	59

[a]For symbols see standard definitions.

See Table A2.14 and the generic description for many characteristics.

Cells up to 1.3 µm in diameter. Colonies are circular, convex, and yellow with entire edges and shiny surfaces. Trypticase peptones (BBL) required for growth. Fastest growth at 20–25°C, at pH values of 6.2–6.6, and in media with 0.1 M NaCl.

Isolated from sediments of the Black Sea. Strain AN8 (DSMZ 4552, OCM 51) is a reference strain.

The mol% G + C of the DNA is: 61 (Bd).

Type strain: JR1, DSMZ 1498, OCM 56.

GenBank accession number (16S rRNA): M59134.

3. **Methanoculleus oldenburgensis** Blotevogel, Gahl-Janssen, Fischer, Pilz, Auling, Macario and Tindall 1998, 327[VP] (Effective publication: Blotevogel, Gahl-Janssen, Fischer, Pilz, Auling, Macario and Tindall 1991, 58).

ol.den.bur.gen' sis. M.L. masc. adj. *oldenburgensis* from Oldenburg (Germany).

See Table A2.14 and the generic description for many characteristics.

Irregular cocci 0.5–1.5 µm in diameter, occurring singly, seldom in pairs. Nonmotile. Possess fimbriae.

H_2/CO_2 and formate serve as energy substrates, but alcohols do not. Acetate is required, and other organic additions such as yeast extract, Casamino acids, and peptones do not stimulate growth.

Growth occurred at temperatures of 25–50°C, with most rapid growth at 45°C. The pH range for growth is 6.5–8.5, with fastest growth at pH 7.5–8.0. Fastest growth in media with 0.04 to 0.17 M NaCl; growth was inhibited with 0.26 M NaCl or more.

Isolated from the Hunte River sediment near Oldenburg, Germany.

The mol% G + C of the DNA is: 48.6 (T_m).

Type strain: CB-1, DSMZ 6216.

4. **Methanoculleus olentangyi** (Corder, Hook, Larkin and Frea 1983) Maestrojuán, Boone, Xun, Mah and Zhang 1990, 121[VP] (*Methanogenium olentangyi* Corder, Hook, Larkin and Frea 1983, 32).

o.len.tan' gy.i. M.L. gen. n. *olentangyi* of the Olentangy River.

See Table A2.14 and the generic description for many characteristics.

Growth occurs at 30–45°C, with most rapid growth at 37°C. Growth is fastest in media with 0.17 M NaCl.

Isolated from the Olentangy River, Ohio.

The mol% G + C of the DNA is: 54 (Bd).

Type strain: RC/ER, OCM 52.

GenBank accession number (16S rRNA): AF095270.

5. **Methanoculleus palmolei** Zellner, Messner, Winter and Stackebrandt 1998, 1115.[VP]

pal.mo.le' i. L. fem. n. *palma* palm; L. masc. n. *oleum* oil; M.L. gen. n. *palmolei* of palm oil.

See Table A2.14 and the generic description for many characteristics.

Irregular cocci, 1.25–2.0 µm in diameter. Stains Gram-negative. Flagella present, but motility was not observed. Growth by CO_2 reduction with H_2, formate, 2-propanol, 2-butanol, and cyclopentanol as electron donors. Ethanol, 1-propanol, cyclohexanol, acetoin, and 2,3-butanediol not used. Acetate is required, and growth is stimulated by potassium and tungstate. Growth occurs at 22–50°C, with most rapid growth at 40°C. Growth at pH values between 6.5 and 8.0, with most rapid growth at pH 6.9–7.5.

Habitat: anaerobic digestors at mesophilic temperatures.

Isolated from an anaerobic digestor treating wastewater of a palm oil mill in North Sumatra, Indonesia.

The mol% G + C of the DNA is: 59 (T_m) or 59.5 (HPLC).

Type strain: INSLUZ, DSMZ 4273.

GenBank accession number (16S rRNA): Y16382.

6. **Methanoculleus thermophilicus** (Rivard and Smith 1982) Maestrojuán, Boone, Xun, Mah and Zhang 1990, 121[VP] (*Methanogenium thermophilicum* Rivard and Smith 1982, 436).

ther.mo.phi' li.cus. Gr. n. *therme* heat; Gr. adj. *philos* loving; M.L. adj. *thermophilicus* heat loving.

See Table A2.14 and the generic description for many characteristics.

Cells are irregular cocci, 1.0–1.3 µm in diameter. Colonies are beige and circular with entire edges.

Trypticase peptones (BBL) and vitamins required. Fastest growth at 55°C (growth range 37–65°C), at pH 7, and in medium with 0.20 M NaCl.

The type strain was isolated from sediment underlying a high temperature effluent channel from a nuclear power plant. Reference strains are strain Los Angeles (Ferguson and Mah, 1983) (DSMZ 2624, OCM 6), Ratisbona (Zabel et al., 1985) (DSMZ 2640, OCM 71), and TCI (Widdel et al., 1988) (DSMZ 391).

The mol% G + C of the DNA is: 59 (Bd) and 57–60 (T_m) for other strains of the species.

Type strain: CR-1, DSMZ 2373, OCM 174.

Genus III. **Methanofollis** Zellner, Boone, Keswani, Whitman, Woese, Hagelstein, Tindall and Stackebrandt 1999, 253[VP]

GERHARD ZELLNER AND DAVID R. BOONE

Me.tha.no.fol' lis. M.L. masc. n. *methanum* methane; L. masc. n. *follis* bag; M.L. masc. n. *Methanofollis* methane-producing bag.

Irregular cocci, occurring singly, 1.5–3.0 µm in diameter. Cells lysed by the addition of 1% SDS (w/v) because of its proteinaceous S-layer composed of hexagonally arranged glycoprotein subunits with an apparent molecular mass of 118–120 kDa. Gramnegative. **Some strains are motile.** Capsules are not formed. Endospores are not produced. Obligately anaerobic, no microaerophilic or aerobic growth. Mesophilic (range 20–44°C, optimal temperature 37–40°C). Optimum pH for growth is about 7.0 (pH range 6.3–8.8). Chemolithotrophic; substrates for growth and methane production are H_2/CO_2 or formate; some species may use 2-propanol/CO_2, 2-butanol/CO_2, and cyclopentanol/CO_2. Sulfate is not reduced. Acetate is a required organic carbon source. Species may have no other complex nutrient requirements or may have several (usually fulfilled by yeast extract). NaCl may be required by strains or may be stimulatory for growth. Growth of cells is usually promoted by supplementation of medium with 1 to 2 µM tungstate.

The habitats are anaerobic high-rate wastewater bioreactors or solfataric fields.

The mol% G + C of the DNA is: 54–60 (T_m).

Type species: **Methanofollis tationis** (Zabel, König and Winter 1984) Zellner, Boone, Keswani, Whitman, Woese, Hagelstein, Tindall and Stackebrandt 1999, 253 (*Methanogenium tatii* Zabel, König and Winter 1984, 313).

FURTHER DESCRIPTIVE INFORMATION

The original description was supplemented with material on the utilization of secondary and cyclic alcohols from Zellner and Winter (1987b) and Bleicher et al. (1989). *Methanofollis liminatans* contains a secondary alcohol dehydrogenase (sADH) activity which is devoid of zinc and dependent on factor F_{420} (Bleicher and Winter, 1991), while on the other hand, the sADH of *Methanobacterium palustre* is dependent on $NADP^+$ (Bleicher and Winter, 1991). The pterin of *Methanofollis tationis* has a structural modification not found in other methanogens and is called tatiopterin. Tatiopterin is lacking a 7-methyl substituent but has both a glutamyl and an aspartyl residue conjugated to the phosphoglutaryl moiety (Raemakers-Franken et al., 1989) and can be distinguished from methanopterin and sarcinapterin. Methanopterin [7-methylpterin (2-amino-4-hydroxy-7-methylpteridine)] is detected in *Methanothermobacter thermautotrophicus* and *Methanolobus tindarius* (Keltjens et al., 1983, DiMarco et al., 1990), while sarcinapterin, a methanopterin derivative containing an additional glutamyl moiety conjugated to the α-carboxylate of the α-hydroxyglutarate, is detected in *Methanosarcina* spp. and *Methanococcus voltae.*

The ether lipids comprise diphytanyl diether and dibiphytanyl tetraethers. Other derivatives including the pentacyclic rings are not detected. The major polar lipids of *M. tationis* and *M. liminatans* are phospholipids, glycolipids, and phosphoglycolipids. Among these the predominant compounds are the ether derivatives of phosphatidyl glycerol, diglycosyl glycolipid, phosphopentanetetrol amine, phosphopentanetetrol trimethylamine, and a phosphoglycolipid. Minor components include a glycolipid, a phospholipid, an aminophospholipid, a phosphoglyco-

lipid, and an aminophosphoglycolipid. The major glycolipid present is a diglycosyl ether lipid; the nature of the sugars and their mode of linkage has not been determined. A single major phosphoglycolipid is detected. The presence of the amino-pentanetetrol containing phospholipids is a characteristic feature also found for other members of the family *Methanomicrobiaceae.*

Respiratory lipoquinones were not detected in cell extracts of *Methanofollis* strains (Zellner et al., 1999).

Methanofollis liminatans is antigenically unrelated to other methanogenic reference strains, including *Methanogenium cariaci* JR1, *Methanoculleus marisnigri* JR1, *Methanoplanus limicola* M3, and *Methanomicrobium mobile* BP (Zellner et al., 1990c).

Analyses of 16S rRNA and 16S rDNA sequences demonstrated the phylogenetic relationship of *Methanofollis tationis* strain DSMZ 2702, *Methanofollis liminatans* strains GKZPZ and BM1 to other members of the family *Methanomicrobiaceae.* The three species form a phylogenetically coherent genus almost equidistant from *Methanoculleus* and *Methanogenium* (Fig. A2.10)

ENRICHMENT AND ISOLATION PROCEDURES

Methanofollis has been isolated from solfataric fields and anaerobic, methane-producing wastewater bioreactors. Because of the strictly anaerobic nature of the organisms, all manipulations must be performed with strictly anaerobic techniques as described by Balch et al. (1979). Cultivation requires that media be both anoxic and at a negative redox potential. This is achieved for tube and serum bottle cultures by using reducing agents and the anaerobic procedures of Hungate (1969) and by the use of an anaerobic glove box, stainless steel anaerobic jars for incubation of plates with reduced agar media under 300 kPa H_2/CO_2 (4:1, v/v), and pressurized serum bottles with 300 kPa H_2/CO_2 or N_2/CO_2 (4:1, v/v) as described previously (Balch et al., 1979).

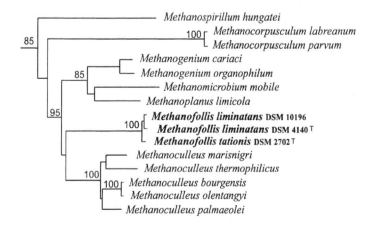

10%

FIGURE A2.10. Dendrogram showing the phylogenetic position of *M. tationis* DSMZ 2702, *M. liminatans* GKZPZ[T] (DSMZ 4140), and strain BM1 (DSMZ 10196) within the family *Methanomicrobiaceae.* Bar represents 10% sequence divergence. (Courtesy of E. Stackebrandt).

M. tationis was isolated by agar plating and picking a well separated colony. Growth in closed vessels is improved if the atmosphere is pressurized to 300 kPa or if the gas phase is periodically flushed with the appropriate gas mixture.

Purification of *M. liminatans* GKZPZ was accomplished by repeated serial dilutions of a sample from the Biohochreaktor of the wastewater treatment plant of Hoechst company in Kelsterbach (Frankfurt, Germany) on enrichment/isolation medium (Zellner et al., 1990c). The reference strain BM1 was isolated from a sample of a butyrate-degrading fluidized-bed reactor (Zellner et al., 1991) by picking a well-separated colony on an agar plate (tungsten-supplemented, acetate-containing defined medium with 2.5% [w/v] agar, grown in a stainless steel anaerobic jar under a gas atmosphere of 300 kPa H_2/CO_2 [4:1, v/v] at 37°C) (Zellner et al., 1999).

Maintenance Procedures

M. tationis requires transfers every four weeks when grown on H_2/CO_2 in complex medium. Cultures can be stored at 4°C for ~4–6 weeks when the gas phase is replaced by H_2/CO_2. *M. liminatans* strains require regular transfers every two weeks when grown on H_2/CO_2 in synthetic medium, and they lyse quickly when substrate is consumed. Exponentially grown cultures can be stored at 4°C for approximately four weeks if the gas phase is replaced by H_2/CO_2. However, cultures stored in this manner will lyse gradually.

Colonies on agar plates or agar roll tubes survive for several months under H_2/CO_2 at 4°C. Cultures may be preserved in lyophilized form in skim milk or in liquid nitrogen for archival purposes.

Differentiation of the genus *Methanofollis* from other genera

Distinctive features are presented in Table A2.13.

With the exception of *M. liminatans*, which may appear as ring-shaped cells by phase-contrast microscopy (Zellner et al., 1990c), cells of *Methanofollis* are irregular cocci and morphologically indistinguishable from those of *Methanoculleus* and *Methanogenium* (Boone et al., 1993b). However, some differences in cell envelope composition have been observed (Table A2.13). All irregularly coccoid methanogens within the *Methanomicrobiales* possess hexagonal S-layer lattices consisting of glycoprotein subunits with relative molecular masses (M_r) ranging from 90,000

to 155,000. However, only methanogens belonging to the family *Methanocorpusculaceae* have an S-layer glycoprotein with M_r ranging between 90,000 and 94,000. The M_r of S-layer glycoprotein subunits of genera of the family *Methanomicrobiaceae* ranged from 101,000 to 155,000, but for *Methanofollis* species M_r of 118,000 (*M. liminatans*) and 120,000 (*M. tationis*) were found.

The cells could not be distinguished from those species of *Methanogenium, Methanoculleus, Methanoplanus, Methanolacinia,* and *Methanocorpusculum* on the basis of substrate spectrum (Table A2.13).

The mol% G + C of *Methanofollis tationis* is similar to the mol% G + C of *Methanoculleus olentangyi* and *Methanoculleus thermophilicus* TCI; the mol% G + C of *Methanofollis liminatans* is similar to that of *Methanoculleus marisnigri, Methanoculleus thermophilicus* strains, and *Methanoculleus palmolei* (Zellner et al., 1998). The mol% G + C of both *Methanofollis* species is clearly distinct from that of *Methanogenium, Methanoplanus, Methanolacinia,* and *Methanomicrobium* species (Table A2.13).

Antigenic fingerprinting of *Methanofollis liminatans* GKZPZ demonstrated that it is antigenically unrelated to other methanogenic reference strains, including *Methanogenium cariaci* JR1, *Methanoculleus marisnigri* JR1, *Methanoplanus limicola* M3, and *Methanomicrobium mobile* BP of the family *Methanomicrobiaceae* (Zellner et al., 1990c).

Further Reading

Boone, D.R., W.B. Whitman and P. Rouvière. 1993. Diversity and taxonomy of methanogens. *In* Ferry (Editor), Methanogenesis: Ecology, Physiology, Biochemistry, and Genetics, Chapman & Hall, New York, pp. 35–80.

Zabel, H.P., H. König and J. Winter. 1984. Isolation and characterization of a new coccoid methanogen, *Methanogenium tatii*, new species from a solfataric field on Mount Tatio (Chile). Arch. Microbiol. *137*: 308–315.

Zellner, G., D.R. Boone, J. Keswani, W.B. Whitman, C.R. Woese, A. Hagelstein, B.J. Tindall and E. Stackebrandt. 1999. Reclassification of *Methanogenium tationis* and *Methanogenium liminatans* as *Methanofollis tationis* gen. nov., comb. nov. and *Methanofollis liminatans* comb. nov. and description of a new strain of *Methanofollis liminatans*. Int. J. Syst. Bacteriol. *49*: 247–255.

Zellner, G., U.B. Sleytr, P. Messner, H. Kneifel and J. Winter. 1990. *Methanogenium liminatans*, spec. nov., a new coccoid, mesophilic methanogen able to oxidize secondary alcohols. Arch. Microbiol. *153*: 287–293.

Differentiation of the species of the genus *Methanofollis*

Table A2.15 provides information to distinguish among species of *Methanofollis*.

List of species of the genus Methanofollis

1. **Methanofollis tationis** (Zabel, König and Winter 1984) Zellner, Boone, Keswani, Whitman, Woese, Hagelstein, Tindall and Stackebrandt 1999, 253[VP] (*Methanogenium tatii* Zabel, König and Winter 1984, 313).

   ta.ti.o' nis. M.L. gen. n. *tationis* of Tatio, to indicate the source of the sample used for isolation (Mount Tatio, Chile).

   See Tables A2.13 and A2.15 and the generic description for many characteristics.

   Irregular cocci, 1.5–3 μm in diameter, with a proteinaceous, SDS-sensitive S-layer. Stain Gram-negative. Cells are motile or nonmotile, and peritrichously flagellated. Flagella have a length of 4–10 μm and a diameter of 8.3 ± 0.5 nm.

   Obligately anaerobic, no microaerophilic or aerobic growth. Strains are mesophilic (range 25–44°C, optimum 40–44°C). Optimal pH for growth is about 7.0 (range 6.3–8.8). Substrates for growth and methane production are H_2/CO_2, and formate. Methanol, ethanol, 1-propanol, 1-butanol, 2-propanol, 2-butanol, 2-pentanol, cyclopentanol, cyclohexanol, methylamines, and acetate are not utilized as catabolic substrates. No chemolithoautotrophic growth. Yeast extract, peptone, acetate, and 0.1 M NaCl required for optimal growth. Lipid composition same as that described for the genus.

   Colonies are yellow.

TABLE A2.15. Differential characteristics of the species of the genus *Methanofollis*[a]

Characteristic	1. *M. tationis*[b]	2. *M. liminatans*
Cell morphology:		
Irregular coccus	+	+
Ring-shaped cells	−	+
Cell diameter, μm:		
1.5	−	+
2.0	+	+
3.0	+	−
M_r of S-layer glycoprotein subunit	120,000	118,000
Substrates:		
H_2/CO_2	+	+
Formate	+	+
2-Propanol/CO_2	−	+
2-Butanol/CO_2	−	+
Cyclopentanol/CO_2	−	+
Growth requirements:		
Yeast extract	+	−
Peptone	+	−
Acetate	+	+
0.1 M NaCl	+	−
Optimal growth temperature, °C	40–44	37–40
Mol% G + C	54	59.3–60

[a]For symbols see standard definitions.

[b]The properties of *M. tationis* are based upon the features of a single strain; the properties of *M. liminatans* are based upon the features of two strains.

Habitat: solfataric fields.

Type strain isolated from a mud sample in 0.5 m depth of a small, moderately thermophilic solfataric pool (50°C, pH 6.1) close to oasis San Pedro di Atacama on Mount Tatio, Atacama desert, Northern Chile.

The mol% G + C of the DNA is: 54 (T_m).

Type strain: Chile 9, DSMZ 2702, OCM 43.

GenBank accession number (16S rRNA): AF095272.

2. **Methanofollis liminatans** (Zellner, Sleytr, Messner, Kneifel and Winter 1990c) Zellner, Boone, Keswani, Whitman, Woese, Hagelstein, Tindall and Stackebrandt 1999, 253[VP]

(*Methanogenium liminatans* Zellner, Sleytr, Messner, Kneifel and Winter 1990c, 292).

li.mi.na' tans. L. masc. n. *limus* sludge; L. parti. *natans* swimming; M.L. masc. adj. *liminatans* a methane- producing bag swimming in sludge.

See Tables A2.13 and A2.15 and the generic description for many characteristics.

Irregularly coccoid or ring-shaped cells, 1.25–2.0 μm in diameter with a proteinaceous, SDS-sensitive S-layer of hexagonally arranged glycoprotein subunits (Fig. A2.11). Stain Gram-negative. Some strains may be motile. Endospores not formed.

Obligately anaerobic, no microaerophilic or aerobic growth.

Cells are mesophilic (range ≥15–44°C) with an optimum at 40°C (strain GKZPZ) or 37–40°C (strain BM1). The optimal pH is about 7. Acetate is required for growth.

Substrates for growth and methane production are H_2/CO_2, formate, 2-propanol/CO_2, 2-butanol/CO_2, and cyclopentanol/CO_2. Secondary and cyclic alcohols are oxidized to the respective ketones. Methanol, ethanol, 1-propanol, 1-butanol, 2-pentanol, methylamines, and acetate are not utilized as catabolic substrates.

Lipid composition same as that described for the genus.

Habitat: anaerobic industrial wastewater bioreactors at moderate temperatures.

Type strain isolated from Biohochreaktor, Hoechst company, Kelsterbach, Frankfurt, Germany. Reference strain (strain BM1) isolated from butyrate-degrading fluidized bed reactor inoculated with granular sludge of upflow anaerobic sludge blanket (UASB) reactor of wastewater treatment plant of sugar refinery at Brühl, Germany.

The mol% G + C of the DNA is: 60 (T_m) (Zellner et al., 1990c) and that of the reference strain BM1 is 60 (HPLC) (Zellner et al., 1999).

Type strain: GKZPZ, DSMZ 4140.

GenBank accession number (16S rRNA): AF095271.

FIGURE A2.11. Electron micrograph of freeze-etched preparation of *Methanofollis liminatans* GKZPZ (DSMZ 4140) showing the S-layer architecture of hexagonally arranged glycoprotein subunits and flagella. Bar = 0.1 μm. (Reprinted with permission from G. Zellner et al., Archives of Microbiology *153*: 287–293, 1990, ©Springer-Verlag, Berlin.)

Genus IV. **Methanogenium** Romesser, Wolfe, Mayer, Spiess and Walther-Mauruschat 1981, 216ᵛᴾ (Effective publication: Romesser, Wolfe, Mayer, Spiess and Walther-Mauruschat 1979, 152)

JAMES A. ROMESSER

Me.tha.no.ge'ni.um. M.L. neut. n. *methanum* methane; Gr. v. suff. *genes* producing; M.L. neut. n. *Methanogenium* methane producing.

Irregular cocci occurring singly or in pairs, usually 0.5–2.6 µm in diameter. **Gram-negative.** Does not form spores. No motility observed, although some strains flagellated. Strictly anaerobic. **Chemolithotrophic, utilizing hydrogen or sometimes formate or primary or secondary alcohols as the electron donor, reducing carbon dioxide to methane. Growth factors required. NaCl generally required or stimulatory for growth.** Optimal temperature range 15–57°C; optimal pH 6.4–7.9.

The mol% G + C of the DNA is: 47–52.

Type species: **Methanogenium cariaci** Romesser, Wolfe, Mayer, Spiess and Walther-Mauruschat 1981, 216 (Effective publication: Romesser, Wolfe, Mayer, Spiess and Walther-Mauruschat 1979, 152).

FURTHER DESCRIPTIVE INFORMATION

Cells are often highly irregular (almost raisin-shaped) cocci. The cells can be made to round up into more regular cocci by lowering the salt concentration, but a reduction in cell growth rate is generally observed. Cells lyse when suspended in distilled water or when 0.01% sodium dodecyl sulfate is present. Although the cells stain Gram-negative, micrographs of ultrathin sections of *M. cariaci* do not show an outer membrane (Romesser et al., 1979). The cell walls appear to be constructed of protein subunits and lack muramic acid or amino sugars (Kandler and König, 1978). Freeze-etch analysis has shown the cell walls to have a periodic surface pattern consisting of structural units about 14 nm in diameter (Romesser et al., 1979). *M. cariaci* has peritrichously arranged flagella but has not been observed to be motile. *M. cariaci* and *Methanogenium frittonii* bear long, thin pili or fimbriae.

Colonies are shiny, yellow, and circular and have entire edges. Colonies of *Methanogenium frittonii* are convex; those of *M. cariaci* are umbonate. *M. frigidum* has not been successfully cultured on agar.

All strains require or are stimulated by organic growth factors, and most either require or are stimulated by NaCl. All strains utilize H_2/CO_2 or formate for growth. Psychrophilic, mesophilic, and thermophilic strains have been described.

The cells occur in anaerobic marine or freshwater sediments, where they presumably function as hydrogen scavengers.

ENRICHMENT AND ISOLATION PROCEDURES

The organisms may be cultivated in 125-ml serum vials in sterile growth medium prepared under a strictly anaerobic growth atmosphere (H_2/CO_2, 80:20 v/v) by a modification of the Hungate technique (1950), as described by Bryant and Robinson (1961)

and revised by Balch and Wolfe (1976). Enrichment medium contains the following constituents in a mixture of 30% distilled water and 70% seawater at the indicated final concentration (%, w/v): ammonium acetate, 0.05; sodium formate, 0.1; Trypticase (BBL), 0.2; yeast extract, 0.1; Na_2CO_3, 0.1; $NaHCO_3$, 0.2; cysteine hydrochloride, 0.05; $Na_2S \cdot 9H_2O$, 0.05; resazurin, 0.0001. The pH of the medium after equilibration with a H_2/CO_2 (80%:20%) gas atmosphere at 137 kPa is 7.4. Enrichments for *M. cariaci* are incubated at 20°C, those for *M. organophilum* at 37°C, those for *M. frittonii* at 57°C, and those for *M. frigidum* at 15°C. Methane formation is measured periodically by gas chromatography.

Isolation of these methanogens (except *M. frigidum*) is achieved by streaking enrichment cultures on agar plates in a Freter type anaerobic hood (Aranki and Freter, 1972) as described by Romesser et al. (1979). Alternatively, agar roll tubes may be used. Colonies are yellow and exhibit a dull greenish-blue fluorescence when illuminated with long wave UV light. Wet mounts of cells exhibit osmotic fragility. Isolation of *M. frigidum* is achieved by dilution to extinction (Franzmann et al., 1997).

MAINTENANCE PROCEDURES

Cultures are maintained by serial transfer in broth culture or on agar slants. Incubation at suboptimal growth temperatures is recommended so that transfers need not be made as frequently. Cells may be susceptible to freezing and thawing lysis, and storage procedures requiring freeze-thaw are not recommended.

DIFFERENTIATION OF THE GENUS *METHANOGENIUM* FROM OTHER GENERA

Since the publication of the first edition of *Bergey's Manual of Systematic Bacteriology*, all species originally described in the genus *Methanogenium*, except the type strain, have been reclassified as members of other genera of *Methanomicrobiales*. Analysis of rRNA and rDNA sequences has resulted in reclassification of *Methanogenium aggregans* to the genus *Methanocorpusculum* (Xun et al., 1989), *Methanogenium bourgense*, *Methanogenium marisnigri*, *Methanogenium olentangyi*, and *Methanogenium thermophilicum* to the genus *Methanoculleus* (Maestrojuán et al., 1990), and *Methanogenium tationis* to the genus *Methanofollis* (Zellner et al., 1999). Precise determination of phylogenetic relationships of the genus *Methanogenium* to other genera of methanogens will continue to depend heavily on studies of relatedness of nucleic acids, especially 16S rDNA.

FURTHER READING

Rouvière, P., L. Mandelco, S. Winker and C.R. Woese. 1992. A detailed phylogeny for the *Methanomicrobiales*. Syst. Appl. Microbiol. *15*: 363–371.

DIFFERENTIATION OF THE SPECIES OF THE GENUS *METHANOGENIUM*

The differential characteristics of the species of *Methanogenium* are indicated in Table A2.16.

List of species of the genus Methanogenium

1. **Methanogenium cariaci** Romesser, Wolfe, Mayer, Spiess and Walther-Mauruschat 1981, 216[VP] (Effective publication: Romesser, Wolfe, Mayer, Spiess and Walther-Mauruschat 1979, 152).

car.i.a' ci. M.L. gen. n. *cariaci* of Cariaco.

See Table A2.16, Fig. A2.12, and the generic description for many characteristics.

Cells up to 2.6 μm in diameter. Yeast extract required for growth. Optimal growth in 0.46 M NaCl. Colonies are circular, umbonate, and greenish-yellow with entire edges and a shiny surface.

Type strain isolated from sediment from the Cariaco Trench.

The mol% G + C of the DNA is: 52 (Bd).

Type strain: JR1, ATCC 35093, DSMZ 1497, OCM 49.

GenBank accession number (16S rRNA): M59130.

2. **Methanogenium frigidum** Franzmann, Liu, Balkwill, Aldrich, Conway de Macario and Boone 1997, 1071.[VP]

fri' gi.dum. L. neut. adj. *frigidum* cold, referring to the low growth temperatures of the species.

TABLE A2.16. Characteristics differentiating the species of the genus *Methanogenium*[a]

Characteristic	1. *M. cariaci*	2. *M. frigidum*	3. *M. frittonii*	4. *M. organophilum*
Requires acetate	+	+	−	+
Pili/fimbriae	+	−	+	−
Temperature optimum, °C	20–25	15	57	30–35
Flagella	+	−	−	−
Ethanol as hydrogen donor	−	+	−	−

[a]For symbols see standard definitions.

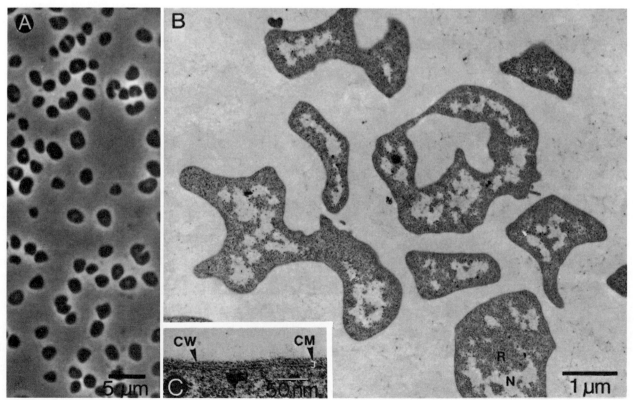

FIGURE A2.12. *M. cariaci. A*, phase-contrast photomicrograph of living cells. *B*, ultrathin section. *C*, section showing cell wall (*CW*) and cytoplasmic membrane (*CM*). (Reprinted with permission from J.A. Romesser et al., Archives of Microbiology *121*: 147–153, 1979, ©Springer-Verlag, Berlin).

See Table A2.16 and the generic description for many characteristics.

Cells from 1.2 to 2.5 μm in diameter. Acetate required as growth factor. Peptones and yeast extract stimulatory. Optimal growth requires 350–600 mM Na$^+$.

A partial 16S rDNA sequence has been determined. The sequence was most similar to that of *M. organophilum* (level of similarity, 98.6%) and *M. cariaci* (level of similarity, 98.2%). The similarities to sequences of members of the order *Methanomicrobiales* other than *Methanogenium* strains were between 89.5 and 94.0%, and the similarities to other members of the *Archaea* that do not belong to the order *Methanomicrobiales* were between 75.9 and 80.6%.

Type strain isolated from anoxic waters of Ace Lake in Antarctica.

The mol% G + C of the DNA is: not determined.

Type strain: Ace-2, OCM 469.

GenBank accession number (16S rRNA): AF009219.

3. **Methanogenium frittonii** Harris, Pinn and Davis 1996, 625^{VP} (Effective publication: Harris, Pinn and Davis 1984, 1127).

frit.ton' i.i. M.L. gen. n. *frittonii* of the Fritton.

See Table A2.16 and the generic description for many characteristics.

Cells from 1.0 to 2.5 μm in diameter. Yeast extract, tryptose, and Casamino acids are stimulatory. NaCl is not required for growth and is inhibitory above 2% (w/v). Colonies are circular, convex, and dark yellow with entire edges and shiny surfaces.

Type strain isolated from sediment of Fritton Lake, England.

The mol% G + C of the DNA is: 49.2 (Bd).

Type strain: FR-4, DSMZ 2832, OCM 200.

4. **Methanogenium organophilum** Widdel, Rouvière and Wolfe 1989, 93^{VP} (Effective publication: Widdel, Rouvière and Wolfe 1988, 480).

or.ga.no' phi.lum. M.L. pref. *organo* pertaining to organic substance; Gr. n. *organon* organ; Gr. adj. *philos* loving; M.L. neut. adj. *organophilum* that likes organic compounds.

See Table A2.16 and the generic description for many characteristics.

Cells from 0.5 to 1.5 μm in diameter. Biotin, 4-aminobenzoate, and vitamin B_{12} required as growth factors. Tungstate required as a trace mineral. Acetate required as a carbon source for growth on substrates other than ethanol. Optimal growth requires the addition of 2.0% NaCl and 0.3% $MgCl_2 \cdot 6H_2O$ (w/v) to growth media.

Colonies are smooth and yellowish.

H_2, formate, 2-propanol, 2-butanol, ethanol, and 1-propanol serve as hydrogen donors for CO_2 reduction to CH_4. Secondary alcohols are oxidized to ketones, primary alcohols to monocarboxylic acids.

The complete 16S rDNA sequence has been determined and is most closely related to that determined for *M. cariaci*.

Type strain isolated from marine mud.

The mol% G + C of the DNA is: 46.7 (T_m).

Type strain: CV, DSMZ 3596, OCM 72.

GenBank accession number (16S rRNA): M59131.

Genus V. **Methanolacinia** *Zellner, Messner, Kneifel, Tindall, Winter and Stackebrandt 1990b, 470^{VP} (Effective publication: Zellner, Messner, Kneifel, Tindall, Winter and Stackebrandt 1989b, 199)*

Song C. Chong and David R. Boone

Me.tha.no.la.ci' ni.a. M.L. neut. n. *Methanum* methane; L. fem. n. *lacinia* a lobe, to indicate the irregular lobed shape of the organism.

Small, **irregular rods**, 0.6 × 1.5–2.5 μm, occurring singly. Colonies are translucent and off-white. **Endospores not formed**. Stain Gram-negative. **Susceptible to lysis by detergent**. Nonmotile, although flagella are present. **Strictly anaerobic**. The cells grow fastest at a sodium ion concentration of 0.15 M (growth range 0–0.8 M). Growth is fastest at pH 7.0 (range 6.6–7.3) and mesophilic temperature (40°C). **Energy metabolism is by reduction of CO_2 to CH_4 with H_2 (or secondary alcohols) as electron donor**. Acetate, formate, methylamines, methanol, and methanol/H_2 are not catabolized. Ammonia can serve as the sole nitrogen source, and sulfide can serve as a sulfur source; acetate is required as a carbon source.

The mol% G + C of the DNA is: 38–44.9.

Type species: **Methanolacinia paynteri** (Rivard, Henson, Thomas and Smith 1983) Zellner, Messner, Kneifel, Tindall, Winter and Stackebrandt 1990b, 470 (Effective publication: Zellner, Messner, Kneifel, Tindall, Winter and Stackebrandt 1989b, 199) (*Methanomicrobium paynteri* Rivard, Henson, Thomas and Smith 1983, 486).

FURTHER DESCRIPTIVE INFORMATION

Cells undergo a morphological change during substrate deprivation, becoming very irregularly coccoid. Cells have an S-layer of hexagonal glycoprotein subunits that have a lattice constant of 15.3 nm. Cells are easily lysed by treatment with 0.1% sodium dodecyl sulfate, sodium desoxycholate, or Triton X-100.

Cells catabolize 2-propanol and 2-butanol but not ethanol, methanol, methylamine, dimethylamine, trimethylamine, formate, acetate, pyruvate, propionate, glutamate, or glucose. Yeast extract and peptones stimulate growth but are not required. The temperature range for growth is 25–42°C, with no growth at 20°C or 45°C.

The species *Methanomicrobium paynteri* was transferred by Zellner et al. (1989b) into this genus based on its phenotypic and phylogenetic differences from *Methanomicrobium*.

ENRICHMENT AND ISOLATION PROCEDURES

Cells are strictly anaerobic, and their cultivation requires the use of Hungate techniques (Hungate, 1969) or an equivalent for preparation of media and manipulation of cultures. Enrichment cultures from suitable sediments are made in media with H_2/CO_2 as an energy source. After incubation at 37°C on a shaker until growth is complete (approximately 1 week), cultures may be isolated from the enrichment culture by serial dilution and inoculation into roll tube media.

MAINTENANCE PROCEDURES

Anaerobic cultures may be stored at room temperature for as long as 6 months (Rivard et al., 1983), or cultures can be preserved in liquid nitrogen (Hippe, 1984; Boone, 1995).

List of species of the genus Methanolacinia

1. **Methanolacinia paynteri** (Rivard, Henson, Thomas and Smith 1983) Zellner, Messner, Kneifel, Tindall, Winter and Stackebrandt 1990b, 470[VP] (Effective publication: Zellner, Messner, Kneifel, Tindall, Winter and Stackebrandt 1989b, 199) (*Methanomicrobium paynteri* Rivard, Henson, Thomas and Smith 1983, 486).

payn' te.ri. M.L. gen. n. *paynteri* of Paynter; named after M.J.B. Paynter, who first isolated a species of the genus.

Small, irregular rods, 0.6 × 1.5–2.5 µm, occurring singly. Colonies are translucent and off-white. Endospores not formed. Stain Gram-negative. Susceptible to lysis by detergent. Nonmotile, although flagella are present. Strictly anaerobic. The cells grow fastest at a sodium ion concentration

of 0.15 M (growth range 0–0.8 M). Growth is fastest at pH 7.0 (range 6.6–7.3) and mesophilic temperature (40°C). Energy metabolism is by reduction of CO_2 to CH_4 with H_2 (or secondary alcohols) as an electron donor. Acetate, formate, methylamines, methanol, and methanol/H_2 are not catabolized. Ammonia can serve as the sole nitrogen source and sulfide can serve as sulfur source; acetate is required as a carbon source.

Isolated from marine sediment collected from a mangrove swamp, Grand Cayman, British West Indies.

The mol% G + C of the DNA is: 44.9 (Bd) or 38 (T_m).

Type strain: G2000, ATCC 33997, DSMZ 2545, OCM 124.

Genus VI. **Methanoplanus** Wildgruber, Thomm and Stetter 1984, 270[VP] (Effective publication: Wildgruber, Thomm and Stetter in Wildgruber, Thomm, König, Ober, Ricchiuto and Stetter 1982, 36)

HARALD HUBER, GERTRUD HUBER AND KARL O. STETTER

Me.tha.no.pla' nus. M.L. neut. n. *methanum* methane; L. masc. adj. *planus* flat; M.L. masc. n. *Methanoplanus* the methane (-producing) plate.

Cells are angular, crystal-like plates or disc shaped. They are 1–3.5 µm long, 1–2 µm wide, and only 0.07–0.3 µm thick (Figs. A2.13 and A2.14). **They occur singly or in pairs and are some-** times branched, without septa. Stain Gram-negative. Flagella or pilus-like structures are found. **Strictly anaerobic. Chemolithotrophic growth on H_2 and CO_2 or formate.** One species can use

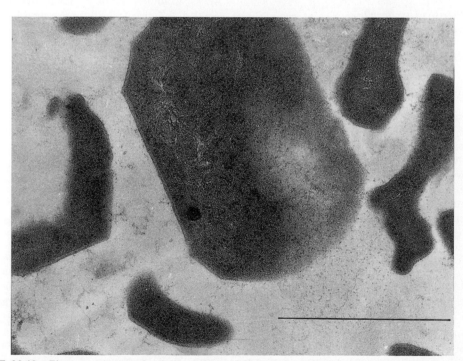

FIGURE A2.13. Electron micrograph of thin section of *Methanoplanus limicola.* Contrast with lead citrate and uranyl acetate. Bar = 1 µm.

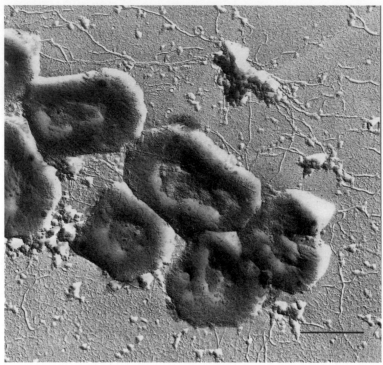

FIGURE A2.14. Electron micrograph of cells of *Methanoplanus limicola*. Platinum shadowed. Bar = 1.0 µm.

CO_2 and 2-propanol as an energy source for growth and **methane formation**. No growth on methanol or methylamines. On molecular hydrogen in the presence of S^0, H_2S is formed in addition to methane. The organisms are **mesophilic** with optimum growth temperatures between 32 and 40°C at neutral pH. Based on 16S rRNA sequence data, *Methanoplanus* belongs to the family *Methanomicrobiaceae*.

The mol% G + C of the DNA is: 39–50.

Type species: **Methanoplanus limicola** Wildgruber, Thomm and Stetter 1984, 270 (Effective publication: Wildgruber, Thomm and Stetter *in* Wildgruber, Thomm, König, Ober, Ricciuto and Stetter 1982, 36).

FURTHER DESCRIPTIVE INFORMATION

Cells are osmotically fragile. They can be easily lysed by detergents, e.g., SDS (0.001%) and Triton (0.01%). The cells are sometimes branched; no septum formation visible. Cytoplasmic inclusions, e.g., grana of polyphosphate, can be seen within the cells. The cell envelope shows a hexagonal surface layer consisting of proteins or glycoproteins. Acetate or tungsten is required for growth. Growth is stimulated by organic material, e.g., yeast extract. Cells are resistant to vancomycin, penicillin, kanamycin, and tetracycline. *Methanoplanus* has been isolated from an anaerobic environment within a swamp, from an oil-producing well, and from a marine sapropelic ciliate.

ENRICHMENT AND ISOLATION PROCEDURES

Methanoplanus limicola can be enriched anaerobically in medium 3 (Balch et al., 1979) in pressurized (200 kPa H_2/CO_2 [80:20]) serum bottles at 30°C in the presence of vancomycin, penicillin, tetracycline, and kanamycin (each 150 µg/ml). It can be isolated by streaking onto polysilicate plates containing medium 3 (Balch et al., 1979). Round, smooth, bright ocher-colored colonies about

2 mm in diameter are visible after 3 months at 30°C. No growth on agar (Wildgruber et al., 1982).

Methanoplanus endosymbiosus can be isolated from homogenized cells of *Metopus contortus* by plating the homogenate on solid media (van Bruggen et al., 1986a) containing penicillin (10^3 IU/ml) or lysozyme (1 mg/ml). Incubate at 30°C for 3 weeks. Colonies, about 2 mm in diameter, are whitish yellow, convex, and circular with entire margins.

Methanoplanus petrolearius can be enriched anaerobically in a mineral medium supplemented with yeast extract (0.1%) and bio-Trypticase (Ollivier et al., 1997) after one week of incubation at 37°C. It was isolated in pure culture after 1 month of incubation at 37°C in agar roll tubes (Ollivier et al., 1997). Colonies were circular and about 2 mm in diameter.

MAINTENANCE PROCEDURES

Stock cultures of *M. limicola* can be stored anaerobically at −20°C for several months after the gas phase is renewed. Cultures can also be stored over liquid nitrogen (−140°C) after the addition of 5% DMSO.

DIFFERENTIATION OF THE GENUS *METHANOPLANUS* FROM OTHER GENERA

Methanoplanus belongs to the family *Methanomicrobiaceae* on the basis of the utilization of H_2/CO_2 or formate as an energy source, the relatively high G + C content of the DNA, and the lack of a rigid cell wall. The requirement for acetate in combination with the inability to use acetate, methanol, and methylamines as energy sources is a further characteristic of this group. *Methanoplanus* differs from the other genera of the *Methanomicrobiaceae* by the flat cell shape and the existence of flagella. 16S rRNA data demonstrate that members of the genera *Methanogenium* and

Methanomicrobium are the closest relatives (Ollivier et al., 1997; Huber and Stetter, in press.*).

TAXONOMIC COMMENTS

Initial 16S rRNA–DNA hybridization experiments indicated that *Methanoplanus* should be placed as a separate family, the *Methanoplanaceae*, within the order *Methanomicrobiales* (Tu et al., 1982). However, later 16S rRNA sequence comparisons indicated that *Methanoplanus* species are closely related to members of the genera *Methanomicrobium*, *Methanogenium*, and *Methanoculleus*, which all belong to the family *Methanomicrobiaceae* (Boone et al., 1993b; Ollivier et al., 1997; Huber and Stetter, in press*). Therefore, the genus *Methanoplanus* should be placed within this family and the family *Methanoplanaceae* should no longer be recognized.

FURTHER READING

Ollivier, B.M., J.L. Cayol, B.K.C. Patel, M. Magot, M.L. Fardeau and J.L. Garcia. 1997. *Methanoplanus petrolearius* sp. nov., a novel methanogenic bacterium from an oil-producing well. FEMS Microbiol. Lett. *147*: 51–56.

van Bruggen, J.J.A., K.B. Zwart, J.G.F. Hermans, E.M. van Hove, C.K. Stumm and G.D. Vogels. 1986. Isolation and characterization of *Methanoplanus endosymbiosus*, sp. nov., an endosymbiont of the marine sapropelic ciliate *Metopus contortus* Quennerstedt. Arch. Microbiol. *144*: 367–374.

Wildgruber, G., M. Thomm, H. König, K. Ober, T. Ricchiuto and K.O. Stetter. 1982. *Methanoplanus limicola*, a plate-shaped methanogen representing a novel family, the *Methanoplanaceae*. Arch. Microbiol. *132*: 31–36.

DIFFERENTIATION OF THE SPECIES OF THE GENUS *METHANOPLANUS*

Differential characteristics of the speicies of the genus *Methanoplanus* are indicated in Table A2.17.

List of species of the genus Methanoplanus

1. **Methanoplanus limicola** Wildgruber, Thomm and Stetter 1984, 270[VP] (Effective publication: Wildgruber, Thomm and Stetter *in* Wildgruber, Thomm, König, Ober, Ricciuto and Stetter 1982, 36).

li.mi' co.la. L. masc. n. *limicola* inhabitant of a swamp.

Chemolithotrophic, growing on H_2/CO_2 or formate. Growth from 17 to 41°C, optimum at 40°C with a doubling time of 7 h. Growth at NaCl concentration from 0.4 to 5.4%, optimum around 1%. Acetate required. Cell envelope most likely consists of a glycoprotein with an M_r of 143,000. Tuft of flagella. Occurring in swamps.

The mol% G + C of the DNA is: 47.5 (T_m).

Type strain: M3, DSMZ 2279.

GenBank accession number (16S rRNA): M59143.

2. **Methanoplanus endosymbiosus** van Bruggen, Zwart, Herman, van Howe, Stumm and Vogels 1986b, 573[VP] (Effective publication: van Bruggen, Zwart, Herman, van Howe, Stumm and Vogels 1986a, 373).

en.do.sym.bi.o' sus. Gr. adj. *endo* inside; M.L. masc. adj. *symbiosus* living together; M.L. masc. adj. *endosymbiosus* living symbiotically inside of another organism.

Endosymbiont of the sapropelic marine ciliate *Metopus contortus*. Chemolithotrophic, growing on H_2 and CO_2 or formate. Growth from 16 to 36°C, optimum at 32°C with a doubling time of 7 h. Growth at NaCl concentration from 0 to 0.75 M, optimum around 0.25 M. Acetate not required. Tungsten (0.1 μM) required. Cell envelope most likely consists of a glycoprotein with an M_r of 110,000. Peritrichous flagellation.

The mol% G + C of the DNA is: 38.7 (T_m).

Type strain: MC1, DSMZ 3599.

GenBank accession number (16S rRNA): Z29435.

3. **Methanoplanus petrolearius** Ollivier, Cayol, Patel, Magot, Fardeau and Garcia 1998a, 1083[VP] (Effective publication: Ollivier, Cayol, Patel, Magot, Fardeau and Garcia 1997, 55).

pe.tro.le.a' ri.us. L. fem. n. *petra* rock; L. masc. adj. *olearius* related to vegetal oil; M.L. masc. adj. *petrolearius* related to mineral oil.

Chemolithotrophic, growing on H_2/CO_2, formate, and 2-propanol/CO_2; acetate required. No growth at 25–45°C, optimum between 35 and 40°C with a doubling time of 10 h. Growth at NaCl concentrations between 0 and 5%, optimum between 1 and 3%. Isolated from an offshore oil field.

The mol% G + C of the DNA is: 50 (HPLC).

Type strain: 4847, CM 486, DSMZ 11571.

GenBank accession number (16S rRNA): U76631.

Editorial Note: Encyclopedia of Life Sciences will be published in 2001. The chapter on *Euryarchaeota* may be viewed at www.els.net.

TABLE A2.17. Key to the species of the genus *Methanoplanus*[a]

Characteristic	1. *M. limicola*	2. *M. endosymbiosus*	3. *M. petrolearius*
Acetate required	+	−	+
Tungsten (0.1 μM) required	−	+	−
Growth on CO_2/2-propanol	−	−	+
Optimum growth temperature, °C	40	32	35–40
Mol% G + C	47.5	38.7	50

[a]For symbols see standard definitions.

Family II. **Methanocorpusculaceae** Zellner, Stackebrandt, Messner, Tindall, Conway de Macario, Kneifel, Sleyter and Winter 1989d, 371[VP] (Effective publication: Zellner, Stackebrandt, Messner, Tindall, Conway de Macario, Kneifel, Sleyter and Winter 1989c, 388)

DAVID R. BOONE, WILLIAM B. WHITMAN AND YOSUKE KOGA

Me.tha.no.cor.pus.cu.la' ce.ae. M.L. neut. n. *Methanocorpusculum* type genus of the family; *-aceae* ending to denote a family; M.L. fem. pl. n. *Methanocorpusculaceae* the family of *Methanocorpusculum*.

Description is the same as for the sole genus of the family, *Methanocorpusculum*.

Type genus: **Methanocorpusculum** Zellner, Alten, Stacke-brandt, Conway de Macario and Winter 1988, 136 (Effective publication: Zellner, Alten, Stackebrandt, Conway de Macario and Winter 1987, 18) emend. Xun, Boone and Mah 1989, 110.

Genus I. **Methanocorpusculum** Zellner, Alten, Stackebrandt, Conway de Macario and Winter 1988, 136[VP] (Effective publication: Zellner, Alten, Stackebrandt, Conway de Macario and Winter 1987, 18) emend. Xun, Boone and Mah 1989, 110

SONG C. CHONG AND DAVID R. BOONE

Me.tha.no.cor.pus' cu.lum. M.L. neut. n. *methanum* relating to methane; L. neut. n. *corpusculum* a particle; M.L. neut. n. *Methanocorpusculum* a methane-producing particle.

Small, irregular cocci, <2 μm in diameter. Endospores not formed. Stain Gram-negative. Lysed by detergent or hypotonic shock. **Nonmotile or very weakly motile. Strictly anaerobic. Fastest growth at 30–40°C, neutral pH, and NaCl concentration of 0.1–0.25 M.** Energy production by reduction of CO_2 to CH_4, with H_2, formate, and sometimes secondary alcohols as electron donors; acetate or methylamines are not catabolized. **Acetate and either yeast extract, peptones, or rumen fluid are required** as carbon and nitrogen sources. Sulfide serves as a sulfur source. May be found in anaerobic digestors or anoxic lake sediments.

Type species: **Methanocorpusculum parvum** Zellner, Alten, Stackebrandt, Conway de Macario and Winter 1988, 136 (Effective publication: Zellner, Alten, Stackebrandt, Conway de Macario and Winter 1987, 18).

FURTHER DESCRIPTIVE INFORMATION

Cells are small and coccoid, occurring singly, in pairs, or in aggregates. Motility is weak or absent, although most strains (other than *Methanocorpusculum labreanum*) possess flagella. The cell wall is an S-layer of hexagonally arranged units with lattice constants of 14.3 nm for *Methanocorpusculum parvum*, 15.8 nm for *Methanocorpusculum sinense*, and 16.0 nm for *Methanocorpusculum bavaricum* (Zellner et al., 1989c). The S-layer plays an important role in the lobed cell morphology of these *Archaea* (Pum et al., 1991).

ENRICHMENT AND ISOLATION PROCEDURES

Methanocorpusculum strains are enriched from anaerobic digestors or sediments by inoculating dilutions into minimal medium containing mineral salts (including 1 mM sodium tungstate), peptones, and rumen fluid, and with H_2 and CO_2 as the catabolic substrate (Winter, 1983; Zhao et al., 1989). Acetogenic bacteria grow in this medium more rapidly than do methanogens. However, if fewer acetogens than methanogens are present in inocula, acetogens may be avoided in enrichments by acquiring enrichments from high dilutions of the inocula. Otherwise, antibiotics such as D-cycloserine and penicillin can aid in the enrichment

of methanogens. Cells are strictly anaerobic, so anaerobic techniques such as those of Hungate (1969) or adaptations of those techniques (Balch et al., 1979; Sowers and Noll, 1995) must be used. Roll tube methods may be used for isolation.

MAINTENANCE PROCEDURES

Culture can be maintained by periodic subculturing, or they can be stored as cell suspensions at liquid nitrogen temperatures (Hippe, 1984; Boone, 1995).

DIFFERENTIATION OF THE GENUS *METHANOCORPUSCULUM* FROM OTHER GENERA

Coccoid methanogens able to use H_2 or formate as as electron donor for CO_2 reduction may be found in several genera other than *Methanocorpusculum*, including *Methanococcus*, *Methanoculleus*, *Methanofollis*, and *Methanogenium*. These genera are phenotypically similar, and certain differentiation may require phylogenetic tests. Many of these coccoid methanogens are found in marine or estuarine environments, so the source of an isolate may be a clue to its classification. *Methanocorpusculum* has not been found in marine or estuarine environments, the habitat of *Methanococcus* and many *Methanogenium* strains. The most evident feature of many *Methanocorpusculum* species is their small size (<1 μm), whereas *Methanoculleus*, *Methanofollis*, and *Methanogenium* are generally larger (1–3 μm in diameter). Further, *Methanogenium* and *Methanoculleus* are nonmotile.

TAXONOMIC COMMENTS

When *Methanogenium aggregans* Ollivier, Mah, Garcia and Robinson 1985 was transferred into the genus *Methanocorpusculum* (as *Methanocorpusculum aggregans*) it was considered as a separate species closely related to *M. parvum* (Xun et al., 1989). These two species have 71% sequence similarity of their genomic DNA (Xun et al., 1989), indicating a phylogenetic distance near the recommended minimum limit that should separate species of procaryotes. Later, noting that DNA reassociation studies which use S1 nuclease tend to give lower values than membrane-filter methods, Boone et al. (1993b) suggested that *M. aggregans* should be considered a junior subjective synonym of *M. parvum*. We have retained that view here, although future studies may find sufficient phenotypic and phylogenetic distinction to consider these as separate species.

M. labreanum is phylogenetically distinct from *M. parvum*, having only 31% sequence similarity to the type strain (strain XII)

and 48% similarity to *M. parvum* strain MSt (the type strain of *M. aggregans*), as determined by the S1 nuclease procedure (Xun et al., 1989). DNA sequence similarity between these strains and the two newer species of the genus (*M. bavaricum* and *M. sinense*) has not been determined, but whole-cell protein electrophoresis and other characteristics (Zellner et al., 1989c) suggest that these two species are distinct from each other and from *M. parvum*.

DIFFERENTIATION OF THE SPECIES OF THE GENUS *METHANOCORPUSCULUM*

Differential characteristics of the species of the genus *Methanocorpusculum* are indicated in Table A2.18.

List of species of the genus Methanocorpusculum

1. **Methanocorpusculum parvum** Zellner, Alten, Stackebrandt, Conway de Macario and Winter 1988, 136[VP] (Effective publication: Zellner, Alten, Stackebrandt, Conway de Macario and Winter 1987, 18).

par'vum. L. neut. adj. *parvum* small.

Small, irregular coccoid, occurring singly or in pairs, up to 1 μm in diameter. Weakly motile, with a single flagellum. Stains Gram-negative. Chemoorganotrophic and strictly anaerobic. Optimum growth temperature 37°C (range 20–45°C) and pH 6.8–7.5. Good growth with NaCl at 0–47 g/l. Energy metabolism by reduction of CO_2 to CH_4 with the following as electron donors: H_2, formate, 2-propanol, or 2-butanol. No growth with acetate, methylamines, or methanol. Acetate, yeast extract, and tungsten (1 mmol/l) are required for best growth, and rumen fluid is stimulatory.

Habitat: sewage sludge. The type strain (XII) was isolated from an anaerobic sour whey digestor inoculated with sewage sludge.

The mol% G + C of the DNA is: 48.5 (T_m).

Type strain: XII, DSMZ 3823, OCM 63.

GenBank accession number (16S rRNA): M59147.

Additional Remarks: The reference strain is MSt (DSMZ 3027, OCM 21).

2. **Methanocorpusculum bavaricum** Zellner, Stackebrandt, Messner, Tindall, Conway de Macario, Kneifel, Sleyter and Winter 1989d, 371[VP] (Effective publication: Zellner, Stackebrandt, Messner, Tindall, Conway de Macario, Kneifel, Sleyter and Winter 1989c, 388).

ba.va'ri.cum. M.L. neut. gen. n. *bavaricum* of Bavaria, to indicate the original habitat of the type strain of the species which was isolated from anaerobic sediment of a wastewater treatment pond at a sugar factory in Regensburg, Germany.

Small, irregular cocci, occurring singly. Less than 1 μm in diameter. Weakly motile. Stains Gram-negative. Strictly anaerobic, chemoorganotrophic. S-layer cell wall. Rapid growth at 37°C (range 15–45°C) and pH 7.0. Growth by CO_2 reduction with H_2, formate, 2-propanol, and 2-butanol as electron donors. Rumen fluid required for growth.

The mol% G + C of the DNA is: 47.7 (T_m) or 51 (HPLC).

Type strain: SZSXXZ, DSMZ 4179, OCM 127.

GenBank accession number (16S rRNA): AF042197, AF095266.

3. **Methanocorpusculum labreanum** Zhao, Boone, Mah, Boone and Xun 1989, 12.[VP]

lab.re.a'num. M.L. neut. adj. *labreanum* isolated from surface sediments of the La Brea Tar Pits, Los Angeles, California.

Irregular cocci, 0.4 to 2.0 μm in diameter. Stains Gram-negative. Nonmotile. Surface colonies are tan, circular, clear, and convex with entire edges. Growth is most rapid at pH 7.0 (range 6.5–7.5), 37°C, and in the presence of 0 to 15 g/l of NaCl. Catabolic substrate is H_2/CO_2 or formate. Protein cell wall, sensitive to lysis by detergent. Trypticase peptones or yeast extract is required, and acetate may be stimulatory.

The mol% G + C of the DNA is: 50 (Bd).

Type strain: Z, DSMZ 4855, OCM 1.

GenBank accession number (16S rRNA): AF095267.

4. **Methanocorpusculum sinense** Zellner, Stackebrandt, Messner, Tindall, Conway de Macario, Kneifel, Sleyter and Winter 1989d, 371[VP] (Effective publication: Zellner, Stackebrandt, Messner, Tindall, Conway de Macario, Kneifel, Sleyter and Winter 1989c, 388).

si.nen'se. M.L. neut. adj. *sinense* of China, indicating the

TABLE A2.18. Characteristics differentiating the species of the genus *Methanocorpusculum*[a]

Characteristic	1. *M. parvum*	2. *M. bavaricum*	3. *M. labreanum*	4. *M. sinense*
Flagella	+	+	−	+
Motility	+	+	−	+
Secondary alcohols/CO_2	+	+		−
Optimum temperature range, °C:				
30–35	−	−	−	+
35–40	−	+	+	−
Mol% G + C	48.5[b]	48[b], 51[c]	50[d]	52[b], 50[c]

[a]For symbols see standard definitions.
[b]T_m.
[c]HPLC.
[d]Bd.

source of the type strain which was isolated from a pilot plant for treatment of distillery wastewater in Chengdu, China.

Small, irregular coccoids, occurring singly, less than 1 μm in diameter. Weakly motile. Stain Gram-negative. Cell wall is an S-layer protein. Obligately anaerobic and chemoorganotrophic. The most rapid growth occurs at 30°C (range 15–45°C) and pH 7.0. Energy metabolism by CO_2 reduction to methane with H_2 or formate as an electron donor. Rumen fluid is required for growth under laboratory conditions.

The mol% G + C of the DNA is: 52.0 (T_m) or 50.0 (HPLC).
Type strain: China Z, DSMZ 4274, OCM 128.
GenBank accession number (16S rRNA): AF095268.

Family III. **Methanospirillaceae** *fam. nov.*

DAVID R. BOONE, WILLIAM B. WHITMAN AND YOSUKE KOGA

Me.tha.no.spi.ril.la' ce.ae. M.L. neut. n. *Methanospirillum* type genus of the family; *-aceae* the ending to denote a family; M.L. fem. pl. n. *Methanospirillaceae* the family of *Methanospirillum*.

Description is the same as for the sole genus of the family, *Methanospirillum*.

Type genus: **Methanospirillum** Ferry, Smith and Wolfe 1974, 469.

Genus I. **Methanospirillum** *Ferry, Smith and Wolfe 1974, 469*[AL]

JAMES G. FERRY AND DAVID R. BOONE

Me.tha.no.spi.ril' lum. M.L. neut. n. *methanum* methane; Gr. n. *speira* a spiral; M.L. neut. n. *Methanospirillum* methane (-producing) spiral.

Symmetrically curved rods 0.4–0.5 × 7.4–10 μm that often form wavy filaments from 15 to several hundred μm in length. Stain Gram-negative. Progressively **motile** by means of polar, tufted flagella. **Strictly anaerobic.** Fastest growth at 30–37°C. Fixes N_2. Energy metabolism by reduction of CO_2 to CH_4 with formate or H_2 as electron donors. Capable of autotrophic growth, but exogenous organic compounds may be used when present.

The mol% G + C of the DNA is: 45–49.5.

Type species: **Methanospirillum hungateii** Ferry, Smith and Wolfe 1974, 469 (*Methanospirillum hungatii* [sic] Ferry, Smith and Wolfe 1974, 469).

FURTHER DESCRIPTIVE INFORMATION

The gentle α-helical curvature of *Methanospirillum hungateii* is readily observed by phase-contrast microscopy (Fig. A2.15). The cell envelope consists of a rigid paracrystalline outer sheath encasing the walls and membranes of one or more cells (Ferry et al., 1974) (Fig. A2.16).

The flexible cell wall sacculus is an S-layer composed of non-glycosylated polypeptides of 110 and 114 kDa (Firtel et al., 1993). Funnel-shaped protein channels 4.0– and 2.0–nm wide may allow passage of a small (<10 kDa) sheath precursor (Firtel et al., 1993). This protein layer is involved in septum formation (Sprott et al., 1979; Sprott and McKellar, 1980).

The sheath, which can be purified after lysis of the cells, contains a mix of polypeptides ranging from 10 to 40 kDa that have been analyzed immunochemically (Southam and Beveridge, 1991). Hydrolysis of the sheath yields a wide variety of amino acids, no amino sugars, and only a trace of neutral sugars. The asymmetry between the two faces of the sheath is correlated to distinct asymmetry in the distribution of exposed polypeptides. Novel phenol-soluble polypeptides are sandwiched between polypeptides that can be solubilized by SDS and β-mercaptoethanol, forming a trilaminar structure (Southam and Beveridge, 1992a). The detergent-soluble polypeptides are rich in cysteine, but most of the organosulfur groups are not exposed, suggesting these

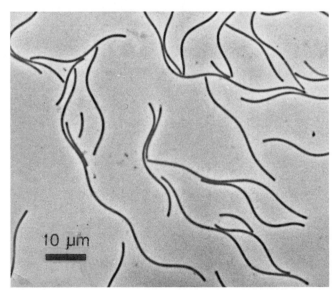

FIGURE A2.15. Phase-contrast micrograph of *Methanospirillum hungateii* showing single cells and filaments. Wet mounts of cultures were prepared on glass slides that had been coated with 2% washed Noble agar (Difco).

polypeptides confer some rigidity to the sheath. However, the phenol-soluble polypeptides confer most of the rigidity (Southam and Beveridge, 1992b). Scanning tunneling microscopy reveals that the sheath has a paracrystalline structure with small pores impervious to solutes with a hydrated radius of greater than 0.3 nm (Beveridge et al., 1990; Blackford et al., 1994). Thus, Gram stain procedures stain the termini as Gram-positive, whereas the regions between the termini appear Gram-negative (Beveridge et al., 1991). The sheath expands with pressure and may act as a pressure regulator by allowing methane to escape only after a certain pressure is attained (Xu et al., 1996). The sheath has a distinct fibrillar surface pattern (Kandler and König, 1978; Sprott and McKellar, 1980; Southam and Beveridge, 1991), and negatively stained cells show surface striations (Fig. A2.17) which result from the presence of circumferential rings likened to the hoops of a barrel (Sprott et al., 1986; Beveridge et al., 1990).

A disc-shaped partition (sometimes called a cell spacer or plug) occurs between the ends of adjacent cells (Zeikus and Bowen, 1975b; Southam et al., 1993) (Fig. A2.16). It consists of four paracrystalline layers of two types (Southam et al., 1993; Firtel et al., 1994), plus amorphous coatings (Firtel et al., 1994). The pores in the layers are 6-fold larger than the sheath pores, permitting the diffusion of solutes into the cell (Beveridge et al., 1991).

Cells typically grow as single cells or short chains within their sheath (10–30 μm long); they are motile by means of two polar tufts of flagella that allow smooth swimming for several minutes in the same direction at 3–10 μm per second. The flagella transect the multilayered terminal plug (Southam et al., 1990). The fla-

gellar wavelength is 2 μm (± 0.2 μm) with an amplitude of 0.34 μm and a width of 9.45 nm (thin section) to 9.9 nm (negative stain). The flagella are stable up to 80°C and at pH 4–10 (Faguy et al., 1994). They are composed of two major flagellins of 24 and 26 kDa (Cruden et al., 1989) that are independent gene products (Faguy et al., 1994). The basal structure has been reported to have two rings with a morphology similar to those of Gram-positive bacteria (Cruden et al., 1989) or alternatively as a simple knob without apparent ring or hook structures (Faguy et al., 1994). During growth at suboptimum temperatures (Faguy et al., 1993) or in medium high in phosphate (Southam et al., 1990) or low in Ca^{2+} (Faguy et al., 1993), the flagella are lost. Also, in low-calcium media cells may grow as long cells or filaments (up to 900 μm) with little or no curvature (Faguy et al., 1993).

Surface colonies of the type strain of *M. hungateii* (1–3 mm in diameter) appear yellow, circular, and convex with lobate margins (Ferry et al., 1974). Microscopic examination of agar surface

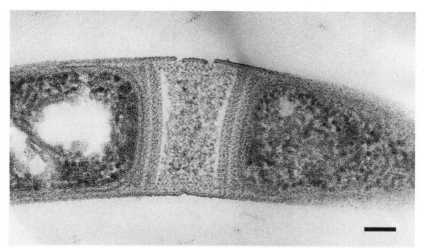

FIGURE A2.16. Thin section of *M. hungateii* JF1 showing the lamellae of the end plates of two adjacent cells of a filament, with an amorphous matrix (spacer) separating the end plates. Bar = 100 μm. (Reprinted with permission from Beveridge et al., Canadian Journal of Microbiology *33*:725–732, 1987, ©National Research Council.)

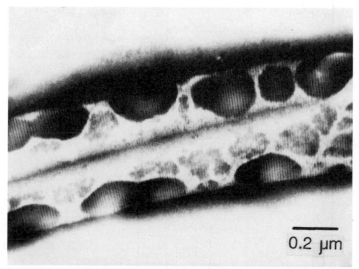

0.2 μm

FIGURE A2.17. Negatively stained whole cells of *Methanospirillum hungateii* strain JF-1. Cells were stained with an aqueous solution of 0.2 g of phosphotungstic acid per liter adjusted to pH 7.0 with KOH.

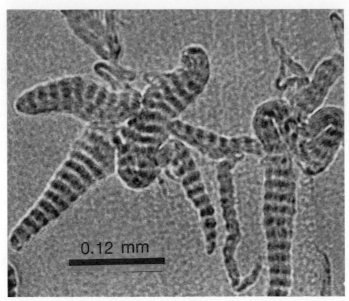

FIGURE A2.18. Colony of *Methanospirillum hungateii* developing on the surface of an agar plate.

colonies shows a unique optical pattern of regular light and dark striations approximately two cell lengths (16 μm) apart (Fig. A2.18). Surface colonies of strain GP1 are light blue with serrated edges (Patel et al., 1976).

The energy-yielding metabolism of *M. hungateii* is the reduction of CO_2 to methane, with H_2 or formate as an electron donor (Ferry and Wolfe, 1977; Sprott et al., 1983). Some strains can also use 2-propanol as an electron donor (Widdel et al., 1988).

Nitrogen is fixed (Belay et al., 1988b). *M. hungateii* JF-1 grows well in mineral salts medium with an H_2/CO_2 (4:1) atmosphere (Ferry and Wolfe, 1977), but strain GP1 requires acetate as a carbon source (Ekiel et al., 1983). Strain JF-1 shows positive chemotaxis to acetate (Migas et al., 1989). The addition of B vitamins or yeast extract and peptones stimulates the growth of strain JF-1, but yeast extract and peptones inhibit the growth of strain GP1 (Breuil and Patel, 1980). Other differences between strains JF-1 and GP1 include the amino acid composition of the outer sheath and the DNA base composition (Patel et al., 1976).

Among 28 tested antibiotics, only enduracidin, chloramphenicol, gramicidin S, and monensin inhibit growth (Hilpert et al., 1981). A decline in the growth rate and intracellular ATP concentration occurs when cultures are exposed to dicyclohexylcarbodiimide (Sprott and Jarrell, 1982).

Strains of *M. hungateii* have been isolated from sewage sludge and pear waste digestors. A salt-tolerant strain has been isolated from marine sediments by K.R. Sowers. *M. hungateii* is the dominant H_2-utilizing organism associated with benzoate- and fatty-acid-oxidizing organisms in enrichment cultures from nonmarine habitats (Ferry et al., 1976; Boone and Bryant, 1980; McInerney et al., 1981a; Mountfort and Bryant, 1982). Formate is metabolized at low concentrations, which implies that this substrate may be directly utilized in some habitats (Schauer et al., 1982).

ENRICHMENT AND ISOLATION PROCEDURES

All procedures for the preparation of culture media and transfer of cultures are done under an O_2-free atmosphere of H_2/CO_2

(Balch and Wolfe, 1976; Balch et al., 1979). Media for enrichment of *M. hungateii* contain minerals, vitamins, sodium formate, sodium acetate, and a cysteine-sulfide reducing agent. A CO_2-bicarbonate buffer system is used to maintain the pH between 6.8 and 7.2 (Ferry et al., 1974; Patel et al., 1976). Additions of acetate and effluent from the habitat of the inoculum are recommended for enrichment and isolation of new strains. Isolation media may include additions of yeast extract and peptones. Colonies are most easily identified in deep agar roll tubes as diffuse light-yellow colonies that develop in correlation with the formation of methane.

MAINTENANCE PROCEDURES

Methanospirillum strains are maintained by biweekly transfer of liquid cultures or agar slants. Cells grow well in MS medium (Boone et al., 1989) with H_2 or formate added as a catabolic substrate. Cultures can be frozen and stored in liquid nitrogen (Boone, 1995).

DIFFERENTIATION OF THE GENUS *METHANOSPIRILLUM* FROM OTHER GENERA

The genus *Methanospirillum* is distinguished from other genera of methanogenic bacteria primarily by its spiral shape and double-layered cell envelope. Immunological fingerprinting (Conway de Macario et al., 1982a, c) and 16S rRNA sequences show the genus *Methanospirillum* to be most closely related to the family *Methanomicrobiaceae*. The protein sheath of *M. hungateii* and the protein cell walls of other genera contain similar distributions of amino acids; however, unlike the protein cell walls, the sheath is resistant to 2% SDS or 4 M urea at alkaline pH (Kandler, 1982b).

The membrane lipids of *M. hungateii* contain dibiphytanyl diglycerol tetraethers common to other genera of methanogenic bacteria, but the fraction is unique in that one of the free hydroxyl groups is esterified with glycerophosphoric acid and the other is glycosidically linked to a disaccharide (Kushwaha et al., 1981).

List of species of the genus Methanospirillum

1. **Methanospirillum hungateii** Ferry, Smith and Wolfe 1974, 469^AL (*Methanospirillum hungatii* [sic] Ferry, Smith and Wolfe 1974, 469).

hun.gat' ei.i. M.L. gen. n. *hungateii* of Hungate; named for R. E. Hungate who has made many contributions to the ecological study of methanogenic bacteria.

The description of the species is the same as that given for the genus.

The mol% G + C of the DNA is: 45 (Bd).

Type strain: JF-1, DSMZ 864, OCM 16.

Additional Remarks: Other strains classified in this species are GP1 (DSMZ 1101, NRC Canada 2239) (Patel et al., 1976) and SK (DSMZ 35950 (Widdel et al., 1988).

Genus incertae sedis I. **Methanocalculus** *Ollivier, Fardeau, Cayol, Magot, Patel, Prensier and Garcia 1998b, 826^VP*

BERNARD OLLIVIER

Me.tha.no.cal' cu.lus. M.L. neut. n. *methanum* methane; M.L. masc. n. *calculus* pebble, gravel; M.L. masc. n. *Methanocalculus* methane (-producing) pebble-shaped organism.

Irregular cocci with peritrichous flagella. Endospores not formed. Strictly anaerobic. **Mesophilic. Growth at pH values of 7.0–8.4 and salinities up to about 125 g/l. Energy metabolism by reduction of CO_2 to CH_4; H_2 and formate are electron donors for this reduction.**

The mol% G + C of the DNA is: 55.

Type species: **Methanocalculus halotolerans** Ollivier, Fardeau, Cayol, Magot, Patel, Prensier and Garcia 1998b, 826.

ENRICHMENT AND ISOLATION PROCEDURES

Members of the genus *Methanocalculus* may be isolated directly from their methanogenic environments, either directly by inoculation of roll tube medium or after enrichment. Culture medium should be reduced and completely free of oxygen. It should contain CO_2 as an electron acceptor and either H_2 or formate as an electron donor; yeast extract and peptones should be included in the medium.

MAINTENANCE PROCEDURES

MSH medium (Boone et al., 1989) with H_2 added as a catabolic substrate is suitable for maintenance of cultures of *Methanocalculus*. Cultures may be stored after lyophilization (Hippe, 1984) as aqueous suspensions at liquid nitrogen temperatures (Hippe,

1984; Boone, 1995; Tumbula et al., 1995), or cultures can be maintained by regular subculturing (Boone, 1995).

DIFFERENTIATION OF THE GENUS *METHANOCALCULUS* FROM OTHER GENERA

Members of the genus *Methanocalculus* can be distinguished from other genera of H_2-utilizing methanogens by their tolerance of NaCl concentrations up to 125 g/l. Other coccoid H_2-using methanogens that are less halotolerant can be found in the order *Methanococcales* and in the genera *Methanogenium, Methanocorpusculum,* and *Methanoculleus.*

TAXONOMIC COMMENTS

Phylogenetic analysis indicates that *Methanocalculus* belongs to the order *Methanomicrobiales,* but the assignment of *Methanocalculus* to a family is not so clear. Phylogenetic analysis of the 16S rRNA sequence of the sole strain of *Methanocalculus* indicates its nearest neighbors are the families *Methanocorpusculaceae* and *Methanospirillaceae.* Although *Methanocalculus* may be slightly closer to the former family, it is not clear whether this genus should be placed in one of these or into a new family. Thus, *Methanocalculus* is listed here as a *genus incertae sedis* within the order *Methanomicrobiales.*

List of species of the genus Methanocalculus

1. **Methanocalculus halotolerans** Ollivier, Fardeau, Cayol, Magot, Patel, Prensier and Garcia 1998b, 826.^VP

ha.lo.to' le.rans. Gr. n. *hals, halos* salt; L. pres. part. *tolerans* tolerating; M.L. part. *halotolerans* salt tolerating.

Irregular cocci 0.8–1.0 μm in diameter, occurring singly or in pairs. Peritrichous flagella. Endospores not formed. Strictly anaerobic. Mesophilic, with fastest growth at 38°C. Growth at pH values of 7.0–8.4 and salinities up to about 125 g/l. Doubling time is 12 h under optimal conditions. Energy metabolism by reduction of CO_2 to CH_4; H_2 and formate are

electron donors for this reduction. Secondary alcohols are not catabolized. Acetate is required, and yeast extract stimulates growth.

Colonies in roll tube media appear after 10 weeks at 37°C and are round and 1 mm in diameter.

The only known strain was isolated from an oil-producing well.

The mol% G + C of the DNA is: 55 (HPLC).

Type strain: SEBR 4845, OCM 470.

GenBank accession number (16S rRNA): AF033672.

Order III. **Methanosarcinales** *ord. nov.*

DAVID R. BOONE, WILLIAM B. WHITMAN AND YOSUKE KOGA

Me.tha.no.sar.ci.na' les. M.L. fem. n. *Methanosarcina* type genus of the order; *-ales* the ending to denote an order; M.L. fem. pl. n. *Methanosarcinales* the order of *Methanosarcina*.

Cells are **coccoids, pseudosarcinae, or sheathed rods**. Most cells have a protein cell wall, and some are surrounded by a sheath or acidic heteropolysaccharide. Peptidoglycan and pseudomurein are absent. Lipids usually contain hydroxyarchaeol and contain glycerol or galactose as polar head groups in the absence of aminopentanetetrols. Cells are **strictly anaerobic**. They obtain energy by formation of methane. All **can grow by dismutating methyl compounds (methanol, methyl amines, or methyl sulfides) or by splitting acetate**. Some can form methane by reduction of CO$_2$ or methyl compounds, by using H$_2$, but never formate, as electron donor. Some strains reduce S^0, but this reaction does not appear to support cell growth. Cells occupy a wide range of anoxic habitats including aquatic sediments, anaerobic sewage digestors, and the gastrointestinal tracts of animals.

Type genus: **Methanosarcina** Kluyver and van Niel 1936, 400, emend. Mah and Kuhn 1984a, 266 (Nom. Cons., Opinion 63 of the Jud. Comm. 1986b, 492).

Key to the families of the order Methanosarcinales

I. Coccoids, pseudosarcinae, or "cysts" capable of growth by dismutating methyl compounds; growth may also occur by reduction of methyl compounds with H$_2$ or by splitting of acetate.
Family I. *Methanosarcinaceae*.

II. Rod-shaped cells contained in a sheath, with sole source of energy by splitting of acetate.
Family II. *Methanosaetaceae*, p. 289

Family I. **Methanosarcinaceae** Balch and Wolfe 1981, 216[VP], emend. Sowers, Johnson and Ferry 1984b, 448 (Effective publication: Balch and Wolfe *in* Balch, Fox, Magrum, Woese and Wolfe 1979, 288)

DAVID R. BOONE, WILLIAM B. WHITMAN AND YOSUKE KOGA

Me.tha.no.sar.ci.na' ce.ae. M.L. fem. n. *Methanosarcina* type genus of the family; *-aceae* the ending to denote a family; M.L. fem. pl. n. *Methanosarcinaceae* the family of *Methanosarcina*.

Cell morphology is **coccoidal or pseudosarcinal**. Most cells have a protein cell wall, and some cells are surrounded by a sheath or acidic heteropolysaccharide. Peptidoglycan and pseudomurein are absent. Lipids contain *myo*-inositol, ethanolamine, and glycerol as polar head groups. Cells are **strictly anaerobic**. They obtain energy by formation of methane. All **can grow by dismutating methyl compounds (methanol, methyl amines, or methyl sulfides)**. Some can use acetate, and some can form methane by CO$_2$ reduction, with H$_2$, but never formate, as electron donor. Some strains reduce S^0, but this reaction does not appear to support cell growth. Cells occupy a wide range of anoxic habitats including aquatic sediments, anaerobic sewage digestors, and the gastrointestinal tracts of animals.

Type genus: **Methanosarcina** Kluyver and van Niel 1936, 400, emend. Mah and Kuhn 1984a, 266 (Nom. Cons., Opinion 63 of the Jud. Comm. 1986b, 492).

Key to the genera of the family Methanosarcinaceae

I. Cells are coccoid bodies, pseudosarcinae, or "cysts", growing by dismutating methyl compounds; some may reduce CO$_2$ with H$_2$ or grow by splitting acetate; fastest growth usually occurs at salinities below 0.5 M NaCl.
Genus I. *Methanosarcina*, p. 269

II. Coccoid cells growing only by dismutating methyl compounds; fastest growth in media with about 0.5 M NaCl, and little or no growth with 1.5 M NaCl; may require phylogenetic analysis to distinguish from the genus *Methanolobus* (below).
Genus II. *Methanococcoides*, p. 276

III. Coccoid cells growing only by dismutating methyl compounds; fastest growth in media with at least 1.5 M NaCl.
Genus III. *Methanohalobium*, p. 279

IV. Coccoid cells growing only by dismutating methyl compounds; fastest growth in media with 0.5–1.5 M NaCl.
Genus IV. *Methanohalophilus*, p. 281

V. Coccoid cells growing only by dismutating methyl compounds; fastest growth in media with about 0.5 M NaCl, and little or no growth with 1.5 M NaCl; may require phylogenetic analysis to distinguish from the genus *Methanococcoides* (above).

Genus V. *Methanolobus*, p. 283

VI. Coccoid cells growing only by dismutating methyl compounds; fastest growth in media with 0.5–1.5 M NaCl and pH 8–10.

Genus VI. *Methanosalsum*, p. 287

Genus I. **Methanosarcina** *Kluyver and van Niel 1936, 400,[AL] emend. Mah and Kuhn 1984a, 266 (Nom. Cons., Opinion 63 of the Jud. Comm. 1986b, 492)*

DAVID R. BOONE AND ROBERT A. MAH

Me.tha.no.sar' ci.na. M.L. neut. n. *methanum* methane; L. fem. n. *sarcina* a package, bundle; M.L. fem. n. *Methanosarcina* methane sarcina.

Irregular **spheroid bodies** (1–3 μm in diameter), occurring alone or typically in **aggregates of cells** (aggregates up to 1000 μm in diameter). Sometimes occur as **large cysts** with a common outer wall surrounding individual coccoid cells. Refer to Figs. A2.19, A2.20, and A2.21 for typical morphologies. Endospores not formed. Gram stain results are variable. **Nonmotile.** May contain gas vesicles (Fig. A2.22). **Strictly anaerobic.** Optimum growth temperatures are 30–40°C for mesophilic species and 50–55°C for thermophiles. Energy metabolism via formation of **methane from acetate, methanol, monomethylamine, dimethylamine, trimethylamine, H_2/CO_2, and CO.** Some strains do not use H_2/CO_2 as the sole energy substrate. May grow very slowly on pyruvate. N_2 may be fixed.

The mol% G + C of the DNA is: 36–43.

Type species: **Methanosarcina barkeri** Schnellen 1947, 73 (Nom. Cons., Opinion 63 of the Jud. Comm. 1986b, 492).

FURTHER DESCRIPTIVE INFORMATION

Aggregates are small to large spheroid bodies comprising many irregular subunits (Zhilina, 1971, 1976; Zeikus and Bowen, 1975a). The multilocular nature is visible only by transmission electron microscopy of thin sections, although phase-contrast microscopy reveals some surface indentations (Figs. A2.19 and A2.20). The spheroid bodies may exist as (a) small coccoid

shape(s) 1–3 μm in diameter, with a tendency to irregularity. Surface indentations are not visible in these bodies, but the bodies may be subdivided by irregular cross walls, which are apparent only when thin sections are examined by transmission electron microscopy; (b) larger coccoid bodies 5–10 μm in diameter, occurring in clusters of 5–10 or more. Surface indentations are visible; and (c) large spheroid bodies 20–100 μm or more in diameter. Surface indentations are visible, giving an appearance similar to mulberries (plant genus *Morus*). The larger clusters are always subdivided by irregular cross walls, and surface indentations are always visible under phase-contrast microscopy. Occasionally, aggregates may form large rafts 1000 μm or more across. A cyst with a common outer layer may enclose myriad smaller irregular coccoid elements. Such cysts may be ruptured by applying external pressure or may disaggregate enzymatically (Liu et al., 1985; Xun et al., 1990) to release the coccoid elements. The disaggregatase is localized in the cell walls or membranes of strains of *Methanosarcina mazeii* (Conway de Macario et al., 1993). In natural environments, the outer wall may serve as a point of attachment for other microbes, whose catabolism may provide acetate to the *Methanosarcina* cells (Tatton et al., 1989). Some species undergo a life cycle involving several of these forms, or a single species may exist as only one of the forms. The outer layer is often lost when cultures are grown in saline medium,

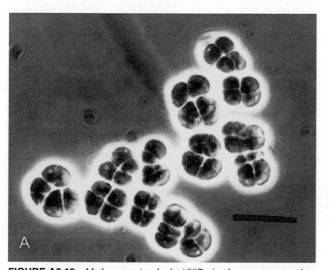
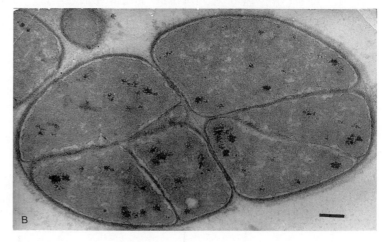

FIGURE A2.19. *Methanosarcina barkeri* 227. *A*, phase-contrast micrograph of typical pseudosarcinae. Bar = 10 μm. *B*, thin-section electron micrograph. Bar = 200 nm. (Courtesy of J. Pangborn, Facility for Advanced Instrumentation, University of California, Davis.)

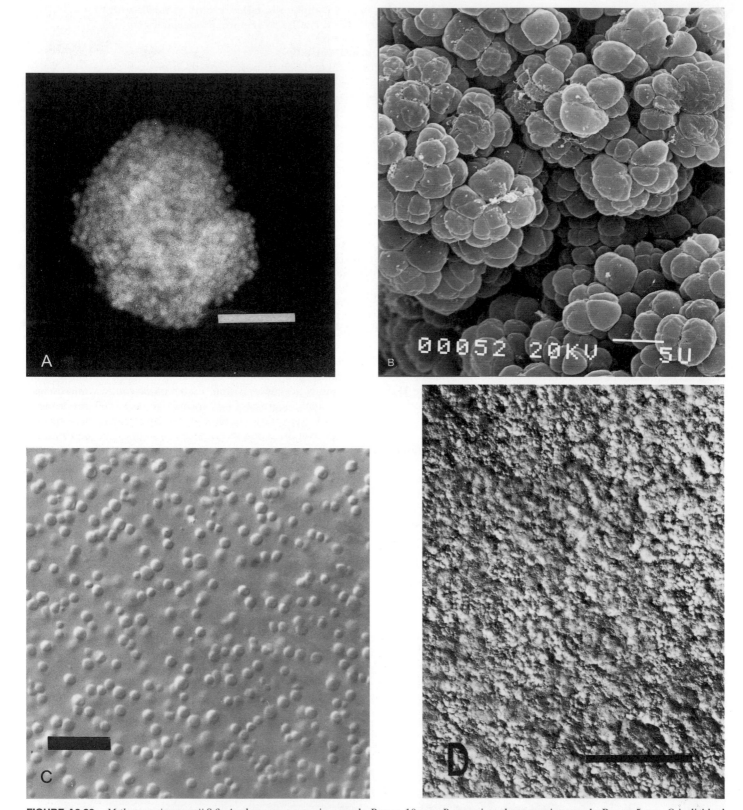

FIGURE A2.20. *Methanosarcina mazeii* S-6. *A*, phase-contrast micrograph. Bar = 10 μm. *B*, scanning electron micrograph. Bar = 5 μm. *C*, individual coccoid units viewed by Nomarski optics. Bar = 5 μm. *D*, lamina viewed by differential interference contrast. Bar = 1.2 μm. A, B, and C, courtesy of Ralph Robinson, Department of Microbiology, University of California, Los Angeles; D, courtesy of Drs. Everly Conway de Macario and Alberto J.L. Macario.

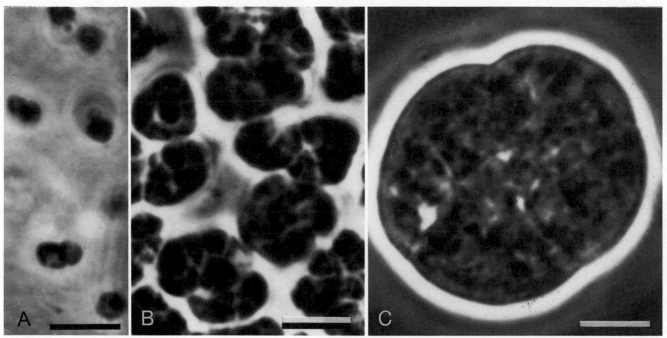

FIGURE A2.21. Phase-contrast photomicrographs of *Methanosarcina acetivorans* C2A. Bars = 5 μm. *A*, individual coccoid units. *B*, small aggregates of cells. *C*, a mature cyst. (Courtesy of K. Sowers, Department of Anaerobic Bacteriology, Virginia Polytechnic Institute and State University, Blackburg.) (Reprinted with permission from K.R. Sowers et al., Applied and Environmental Microbiology, *47*: 971–978, 1984, ©American Society for Microbiology, Washington, D.C.)

resulting in a coccoid morphology (Sowers et al., 1993). However, elevated concentrations of divalent cations rather than sodium ions trigger the loss of the outer wall (Boone and Mah, 1987; Ahring et al., 1991a).

The cell wall of *Methanosarcina* contains a protein layer adjacent to the cell membrane. External to this layer, there may often be an outer layer composed of heteropolysaccharide. This outer layer is absent in *Methanosarcina acetivorans*, which is sensitive to lysis by sodium dodecyl sulfate (SDS). The dissolution of the heteropolysaccharide outer layer of other *Methanosarcina* strains, which may occur during growth, gives rise to individual SDS-sensitive coccoid units. The heteropolysaccharide outer layer is composed mainly of galactosamine, glucose, mannose, and glucuronic or galacturonic acid. Sulfate is not a major component of this layer, but a small amount (5%) of protein is present. SDS-extracted cell wall material of *M. mazeii* contains five amino acids (molar amino acid concentrations relative to glycine are given in parentheses): serine (1.8), glycine (1.0), lysine (0.9), ornithine (0.7), and alanine (0.2) (Robinson et al., 1985). These amino acids are apparently not contaminants from the inner, protein cell wall, since the concentrations of other detectable amino acids are less than 1/10 the molar amount of glycine.

Antigenic analysis with polyclonal antibody probes and whole bacterial cells as antigens indicates a high degree of antigenic relatedness among the species of *Methanosarcina* and no cross-reactivity with methanogens outside the family (Conway de Macario et al., 1982b; Macario and Conway de Macario, 1982). These findings augment the morphological and physiological factors that distinguish the species of *Methanosarcina* from other genera of methanogens. Gas vesicles may occur. *M. mazeii* accumulates polyphosphates up to 14% of cell dry weight (Rudnick et al., 1990).

All species of *Methanosarcina* use methanol and methylamines as catabolic substrates. Most species can also use H_2/CO_2 and acetate. Stationary-phase, H_2/CO_2-grown cells may require a period of adaptation for growth on acetate (Boone et al., 1987). However, if cells are grown with both H_2/CO_2 and acetate present, H_2/CO_2 is used first, then acetate without a long lag (Mah et al., 1981). Acetate is degraded by the aceticlastic reaction, with the methyl group reduced to CH_4 and the carboxyl group oxidized to CO_2 (Mah et al., 1978). Growth on CO is accomplished by oxidation to CO_2 and H_2. Carbon monoxide dehydrogenase is a cytosolic enzyme (Gokhale et al., 1993), but CO oxidation is coupled to the proton translocation (Bott and Thauer, 1989). Cells grow very slowly on pyruvate (Böck et al., 1994; Böck and Schönheit, 1995). In acetate-grown cells, electrons from oxidation of the carboxyl group apparently are passed down an electron transport chain including cytochrome *b* and then to the heterodisulfide via a soluble reductase (Peer et al., 1994). *Methanosarcina thermophila* contains a ferredoxin (Clements and Ferry, 1992; Clements et al., 1994). Some catabolic enzymes are regulated: F_{420}-dependent N^5N^{10}-methyleneH$_4$MPT dehydrogenase is stimulated during growth on H_2 or methanol but not acetate (when it would not be needed) (Mukhopadhyay et al., 1993). *Methanosarcina barkeri* produces carbonic anhydrase when grown on acetate but not when grown on methanol or H_2 (Karrasch et al., 1989).

M. thermophila TM-1 does not produce nor consume CO during growth on acetate, but *M. thermophila* CALS-1 does (Zinder and Anguish, 1992). *Methanosarcina* strains produce a small amount (40 Pa) of H_2 during growth on acetate and consume H_2 very slowly when its partial pressure is higher than this (Ahring et al., 1991b). H_2 at high partial pressures inhibits acetate degradation (Ahring et al., 1991b).

Nitrogen fixation by *M. barkeri* requires molybdenum, although vanadium can substitute (Scherer, 1989). The *nifD* gene

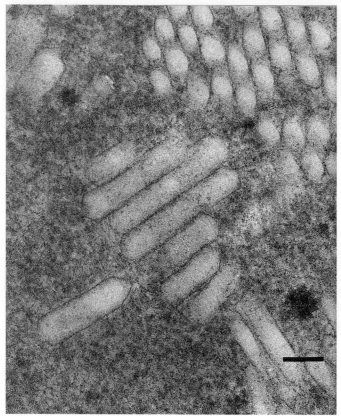

FIGURE A2.22. Thin-section electron micrograph of gas vesicles of *Methanosarcina vacuolata* W. Bar = 100 nm. (Courtesy of J. Pangborn.) (Reprinted with permission from R.A. Mah and M.R. Smith, *In* Starr, Stolp, Trüper, Balows, and Schlegel (editors), The Prokaryotes. A Handbook on Habitats, Isolation, and Identification of Bacteria. ©Springer-Verlag, Berlin, 1981.)

is evolutionarily related to that of *Clostridium pasteurianum* (Chien and Zinder, 1994).

Methanosarcina sp. strain DCM and *M. mazeii* S-6 dechlorinate chloroform, but they form $^{14}CO_2$ rather than $^{14}CH_4$, suggesting that the mechanism is not reductive (Mikesell and Boyd, 1990). However, *Methanosarcina* has also been shown to dechlorinate reductively (Krone et al., 1989; Jablonski and Ferry, 1992; Krone and Thauer, 1992). Reduced corrinoids or F_{430} alone, from *M. barkeri* Jülich, can reductively dechlorinate 1,2-dichloroethane, whether F_{430} is reduced enzymatically or abiotically with Ti(III) citrate (Holliger et al., 1992b).

Acetate catabolism by *M. barkeri* 227 shows isotopic discrimination of $-21.3‰$ for the [2-^{13}C]acetate and a similar value for [1-^{13}C]acetate (Gelwicks et al., 1994).

M. barkeri and *Methanosarcina vacuolata* are slightly acidophilic, or at least acid tolerant (Maestrojuán and Boone, 1991), growing well at pH values as low as 5. Cells growing on acetate are more sensitive to low pH (Maestrojuán and Boone, 1991), possibly because of acetate toxicity at low environmental pH. Most species are slightly halotolerant, but *Methanosarcina siciliae* is halophilic. In *M. thermophila*, increased environmental salinities up to about 0.4 M NaCl are balanced by cytosolic accumulation of glutamate, but at salinities above 0.4 M $N^{ε}$-acetyl-β-lysine accumulates (Sowers et al., 1990). Other solutes that accumulate include glycine betaine and K^+ (Sowers and Gunsalus, 1995).

Members of the genus *Methanosarcina* may be isolated from freshwater and marine environments where acetate or methylamines are degraded. These include anaerobic digestors, freshwater and marine sediments, and the rumen of cattle or sheep. They may also be found in the digestive tracts of animals fed diets with fishmeal that contains trimethylamine oxide (which in anoxic environments is reduced to trimethylamine). *Methanosarcina* species in pure culture also grow by using energy substrates other than acetate or methylamines, but these may be the major natural substrates. Existing *Methanosarcina* strains have a higher K_m for H_2 than other H_2-oxidizing methanogens and may not be competitive for H_2 in mixed-culture environments (Boone et al., 1987).

In environments where organic matter is completely mineralized to CH_4 and CO_2, acetate is the major source of methane. Acetate is degraded by the aceticlastic reaction, in which the methyl group is cleaved with its hydrogens intact. Members of two genera of methanogens, *Methanosarcina* and *Methanosaeta*, are capable of carrying out this reaction. *Methanosarcina* strains have higher maximum growth rates and may be more tolerant of low pH than *Methanosaeta* (Maestrojuán and Boone, 1991), but *Methanosaeta* has a lower K_m for acetate and can consume acetate at a lower concentration than can *Methanosarcina* strains (Min and Zinder, 1989; Westermann et al., 1989b; Fukuzaki et al., 1990). When acetate concentrations are very low, *Methanosarcina* strains cannot consume acetate but rather produce small amounts of it (Westermann et al., 1989a). Thus, *Methanosarcina* may be expected to dominate in digestors and other environments with pH values below 7 or with rapid dilution rates and higher acetate concentrations. *Methanosaeta* may dominate in sediments and digestors with higher pH values or those operated with longer retention times of solids.

ENRICHMENT AND ISOLATION PROCEDURES

Enrichment cultures of *Methanosarcina* may be initiated in media containing acetate, methanol, or methylamines as a substrate. NH_3 may serve as a nitrogen source, and sulfide may serve as a sulfur source. Some strains require organic growth factors, and some may require unusual metal ions, such as nickel or cobalt. Trimethylamine is the substrate of choice for rapid enrichment of *Methanosarcina*. Although many halophilic methanogens, including *Methanolobus tindarius* (König and Stetter, 1982), *Methanococcoides methylutens* (Sowers and Ferry, 1983), and others (Zhilina, 1983, 1986; Mathrani and Boone, 1985; Paterek and Smith, 1985; Zhilina and Zavarzin, 1987a, 1990) may also use trimethylamine (or methanol), these strains do not grow in freshwater environments. Methanol may also be used as an enrichment substrate, but nonmethanogens, such as *Eubacterium limosum*, may predominate (Sharak Genthner et al., 1981). Growth of nonmethanogens may be inhibited by antibiotics, such as penicillin, cycloserine, or vancomycin. Acetate is perhaps the most selective substrate for *Methanosarcina*, but growth is slower than with trimethylamine and not all *Methanosarcina* species can use acetate. Also, acetate enrichment cultures may yield *Methanosaeta* instead of *Methanosarcina*, especially at pH values of 7.0 or above. Lowering the pH below 7.0 usually leads to the predominance of *Methanosarcina* in enrichment cultures when these organisms are present in the inoculum.

M. thermophila normally occurs only in large aggregates; in some media with high salt content, it may form individual coccoid units (Sowers and Gunsalus, 1988a). Because of its existence in large aggregates, the numbers of colony-forming units may be small compared with its metabolic importance in mixed-culture systems. Colonies of *M. thermophila* may occur in roll tubes inoculated with 100 μl or more of sample compared with an equivalent mass-density of *M. mazeii*, whose individual coccoid cells may form colonies when as little as 0.1 μl is inoculated. To isolate *M. thermophila* from roll tubes at low dilutions, it may be necessary to incorporate antibiotics into the medium to inhibit nonmethanogenic species of bacteria (Zinder and Mah, 1979).

MAINTENANCE PROCEDURES

Species of *Methanosarcina* may be stored after lyophilization or as liquid suspensions at liquid nitrogen temperatures (Boone, 1995), or cultures can be maintained by regular subculturing (Hippe, 1984).

DIFFERENTIATION OF THE GENUS *METHANOSARCINA* FROM OTHER GENERA

Members of the genus *Methanosarcina* can be distinguished from most genera by morphology and by their ability to produce CH_4 as a metabolic product from most of the following substrates: H_2/CO_2, acetate, monomethylamine, dimethylamine, trimethylamine, or methanol. Only four other genera of methanogens (*Methanolobus*, *Methanococcoides*, *Methanohalobium*, and *Methanohalophilus*) use methylamines or methanol in the absence of added H_2; these organisms are unable to use H_2/CO_2 or acetate as a sole catabolic substrate. *Methanosphaera stadtmanae* (Miller and Wolin, 1983, 1985a) uses methanol only by reducing it with H_2 to methane. The only other methanogenic genus that can catabolize acetate, *Methanosaeta*, cannot use trimethylamine. *Methanosaeta* can be distinguished easily by its square-ended rods occurring in long chains, in contrast to *Methanosarcina*, which typically grows in large, irregular, coccoid aggregates. The genus *Methanosarcina* may often be differentiated from other species of methanogens by its morphology alone. However, the life cycles of *M. mazeii* and *M. acetivorans* include individual coccoid units that are difficult to distinguish from other coccus-shaped or coccoid methanogenic bacteria. By using epifluorescence microscopy, it is usually possible to identify the typical *Methanosarcina* morphology (large, irregular, coccoid, fluorescent aggregates) by microscopic examination alone.

TAXONOMIC COMMENTS

The species *Methanococcus frisius* Blotevogel, Fischer and Lüpkes 1986[VP] (Blotevogel et al., 1986) is antigenically related to *Methanosarcina* species (Blotevogel and Macario, 1989) and has a 16S rRNA sequence similar to that of *M. mazeii* Göl (Eggen et al., 1992). Therefore, it was transferred to the genus *Methanosarcina* as *Methanosarcina frisia*[VP] Blotevogel and Fischer 1989, 91 (Blotevogel and Fischer, 1989). *M. frisia* is considered here to be a junior subjective synonym of *M. mazeii* because the DNA sequences of the type strains of these species are 77% similar. Also, these two species have similar physiologies (Maestrojuán et al., 1992). However, some may consider *M. frisia* to be a separate species. An analysis of partial sequences of the *mcrA* gene shows differences between *M. mazeii* and *M. frisia* (Springer et al., 1995) that are consistent with the latter view.

Further Comments The unusual morphologies exhibited by *Methanosarcina* species may be helpful in the recognition and classification of the species, but they have led to some confusion of terms. The term "sarcina" was used to describe methanosarcinae because of the superficial resemblance of these methan-

ogens to true sarcinae. Methanosarcinae, however, exhibit non-perpendicular division planes, and numbers of cells or bodies within an aggregate are not usually a power of two. Therefore, the term "pseudosarcina" ("pseudo-sarcine") was suggested by Mazé (1903) to describe the unordered appearance. Several other genera of microorganisms also exhibit this unusual pseudosarcina morphology, viz., *Geodermatophilus* (Ishiguro and Wolfe, 1970), *Cyanobacteria* (Stanier et al., 1971), and a sulfate reducer (Widdel, 1980).

The term "cyst" describes another unusual morphology exhibited by *M. mazeii* and *M. acetivorans*. The methanosarcina cyst is a life cycle stage in which a common wall surrounds a tightly packed mass of loose coccoid cells (Fig. A2.20). This form has been called "lamina" (Mayerhofer et al., 1992). This type of cyst differs from others because it is neither acellular, thick walled, resistant to desiccation, nor connected to a fruiting structure. The cyst of *M. mazeii* is a large body 100–1000 μm or more in diameter containing tens of thousands of coccoid units. In *M. acetivorans* the cysts may be smaller. Cysts of *Methanosarcina* have a surface layer that appears to be composed of moribund cells surrounding a myriad of (or many, in the case of *M. acetivorans*) coccoid elements. The cysts may be physically disrupted, releasing viable coccoid cells, or rupture of the cysts may occur spontaneously by the action of a disaggregating enzyme (Liu et al., 1985).

The separation of *M. vacuolata* from other species of *Methanosarcina* is based on the characteristics of a single isolate (Zhilina and Zavarzin, 1979a). Although several other vacuolated strains of *Methanosarcina* have been described (Mah et al., 1977; Archer and King, 1983), it has not been demonstrated that these strains share the characteristics of the type strain of *M. vacuolata* on which the species description was based, and these other strains may be *M. barkeri* rather than *M. vacuolata*. Further, vacuolated *Methanosarcina* strains may lose their ability to form vacuoles for many months and regain them for no apparent reason (R.A. Mah, unpublished data). Thus, some strains currently classified as *M. barkeri* may in fact be phylogenetically closer to *M. vacuolata*, and molecular data may be necessary to distinguish these species. DNA–DNA hybridization studies (Zhilina and Zavarzin, 1979a), membrane lipid analysis (Osipov et al., 1984), 16S rRNA cataloging (Balch et al., 1979) or sequencing, ribosomal protein analysis (Douglas et al., 1980) and immunological analysis (Zhilina and Zavarzin, 1987b) indicate significant phylogenetic differences between *M. vacuolata* and *M. barkeri*. Other gas-vacuolated strains may appear to be *M. vacuolata*, based on their physiologies and morphologies, but may be unrelated phylogenetically. For instance, it is not clear that strain W (Mah et al., 1977) and strain FR-1 (Archer and King, 1983), two other gas-vacuolated strains, are *M. vacuolata*. Strain W forms white colonies (unpublished data), whereas those of *M. vacuolata* are yellow. The mol% G + C of the DNA of strain W is 40.5 (Balch et al., 1979), and that of strain FR-1 is 40.7 (Archer and King, 1983), but that of the type strain of *M. vacuolata*, strain Z-761, is 36. Distinction between *M. vacuolata* and *M. barkeri* by phenotypic traits, if possible, will require the characterization and comparison of additional strains.

FURTHER READING

Balch, W.E., G.E. Fox, L.J. Magrum, C.R. Woese and R.S. Wolfe. 1979. Methanogens: reevaluation of a unique biological group. Microbiol. Rev. *43*: 260–296.

Ferry, J.G. 1993. Methanogenesis: Ecology, Physiology, Biochemistry, and Genetics, Chapman & Hall, New York.

Mah, R.A. 1980. Isolation and characterization of *Methanococcus mazei*. Curr. Microbiol. *3*: 321–326.

Robinson, R.W. 1986. Life cycles in the methanogenic archaebacterium *Methanosarcina mazei*. Appl. Environ. Microbiol. *52*: 17–27.

Sowers, K.R., S.F. Baron and J.G. Ferry. 1984. *Methanosarcina acetivorans*, sp. nov., an acetotrophic methane-producing bacterium isolated from marine sediments. Appl. Environ. Microbiol. *47*: 971–978.

Zinder, S.H. and R.A. Mah. 1979. Isolation and characterization of a thermophilic strain of *Methanosarcina* unable to use H$_2$–CO$_2$ for methanogenesis. Appl. Environ. Microbiol. *38*: 996–1008.

DIFFERENTIATION OF THE SPECIES OF THE GENUS *METHANOSARCINA*

Some differential characteristics of the species of the genus *Methanosarcina* are shown in Table A2.19.

List of species of the genus Methanosarcina

1. **Methanosarcina barkeri** Schnellen 1947, 73,[AL] emend. Bryant and Boone 1987a, 169.

bar′ ke.ri. M.L. gen. masc. n. *barkeri* of Barker; named for H.A. Barker, who made many definitive studies of this and other methanogenic bacteria.

Original material supplemented with material from Kluyver and Schnellen (1947), Stadtman and Barker (1951a), Mah et al. (1978), Weimer and Zeikus (1978), Hippe et al. (1979), Murray and Zinder (1984), Lobo and Zinder (1988, 1990), Maestrojuán and Boone (1991), and Maestrojuán et al. (1992).

Coccoid bodies, 1.5–2.0 μm in diameter, occurring mostly in irregular aggregates ranging from several to several hundred micrometers in size. Membranes contain C$_{25}$ isoprenoids as the major neutral lipid but no C$_{30}$ isoprenoids (Langworthy et al., 1982). Nonmotile. Stain Gram-positive. Not lysed by SDS.

Deep colonies in methanol agar with inorganic salts are whitish to light yellow and 0.5–1.0 mm in diameter. In liquid medium, growth may occur as sediment with active gas formation.

Energy-yielding metabolism involves methane production. H$_2$/CO$_2$; methanol; monomethylamine, dimethylamine, and trimethylamine; acetate; or CO may be used as substrates. The methyl group of methanol or acetate is reduced to CH$_4$ without intermediate oxidation to CO$_2$. Cells contain similar amounts of F$_{420}$ regardless of whether they are grown on methanol or H$_2$/CO$_2$ (Peck, 1989). Carbohydrates, amino acids, formate, ethanol, propionate, and butyrate are not fermented.

Ammonia or N$_2$ serves as a nitrogen source, and sulfide serves as a sulfur source. Growth and CH$_4$ formation are more rapid in medium with H$_2$/CO$_2$ or methanol than with acetate. Cells contain cytochromes *b* and *c* (Kamlage and Blaut, 1992) and cytochrome *bc* (Kumazawa et al., 1994). Optimum growth is obtained at pH 7.0 and at 30–40°C. Strictly anaerobic.

Habitat: freshwater and marine mud, rumens of ungu-

TABLE A2.19. Diagnostic and descriptive features of *Methanosarcina* species[a]

Characteristic	1. *M. barkeri*	2. *M. acetivorans*	3. *M. mazeii*	4. *M. siciliae*	5. *M. thermophila*	6. *M. vacuolata*
Morphology:						
Coccoid cells or small aggregates	+	+	+	+	−	+
Large aggregates	−	+	+	+	+	−
Cysts	−	+	+	−	−	−
Life cycle	−	+	+	−	−	−
Gram stain	+	−	+	−	+	+
Gas vacuoles	−	−	−	−	−	+
Substrates:						
H$_2$/CO$_2$	+	+	d	−	−	+
Acetate	+	+	d	−	+	+
Methylamines	+	+	+	+	+	+
Methanol	+	+	+	+	+	+
Methyl sulfides				+		
Growth requirements:						
Chemoautotrophic	+	+	+	+	−	+
Compounds stimulating growth	Vitamins	AA, YE	P, YE	YE	PABA	P, YE
Conditions supporting most rapid growth:						
Temperature 30–40°C	+	+	+	+	−	+
Temperature 55–60°C	−	−	−	−	+	−
pH 5–6[b]	+	−	−	−	−	+
pH 6–7	+	+	+	+	+	+
pH 7.5	+	+	+	−	+	−
0.0 M NaCl	+	+	+	−	+	+
0.3 M NaCl	+	+	+	+	+	+
0.5 M NaCl	+	+	+	+	+	−
Mol% G + C	39–44	41	42	41–43	42	36

[a]For symbols see standard definitions; AA, amino acids, YE, yeast extrct, P, peptones, PABA, *p*-aminobenzoate.

[b]When grown on methyl substrates.

lates, animal waste lagoons, and sludge from anaerobic sewage sludge digestors.

Described strains classified in this species: 227 (ATCC 43567, DSMZ 1538, OCM 35) (Mah et al., 1978), S-3 (OCM 37) (Smith, 1966), Fusaro (ATCC 29787, DSMZ 804, OCM 83) (Kandler and Hippe, 1977), 3 (ATCC 29786, DSMZ 805, OCM 84) (Kandler and Hippe, 1977), Jülich (DSMZ 2948, OCM 86) (Scherer and Bochem, 1983), Weismoor (OCM 89) (Scherer and Bochem, 1983), FR-1 (DSMZ 2256, OCM 93) (Archer and King, 1983; Maestrojuán et al., 1992), and NIH- 1 (purified from strain NIH) (OCM 266). Electrophoretic analysis of whole-cell proteins (Maestrojuán et al., 1992) suggests that strain UBS (DSMZ 1311, OCM 27) should be classified as *M. mazeii*.

The mol% G + C of the DNA is: is 39–44 (Bd).

Type strain: MS, DSMZ 800, OCM 38.

GenBank accession number (16S rRNA): AJ012094.

Additional Remarks: Strain MS is the established neotype strain; isolated from a butyrate enrichment derived from an anaerobic sewage sludge digestor (Bryant and Boone, 1987a).

2. **Methanosarcina acetivorans** Sowers, Baron and Ferry 1986, 355[VP] (Effective publication: Sowers, Baron and Ferry 1984a, 977).

a.ce.ti′vo.rans. L. neut. n. *acetum* vinegar; L. part. adj. *vorans* consuming; M.L. fem. part. adj. *acetivorans* consuming acetic acid.

Individual, irregular coccoid cells 1.5–2.5 µm in diameter. During growth on acetate, septate cell aggregates form. These small aggregates develop into cysts that contain individual coccoid elements within a common wall. Nonmotile. Stain Gram-negative. Cell walls are thin (10 nm), osmotically fragile, and composed of protein. Cells lysed by SDS.

Colonies in roll tubes are pale yellow and are 0.5 mm in diameter after 14 d. Surface colonies are smooth, circular, and convex with entire edges.

Acetate, methanol, monomethylamine, dimethylamine, and trimethylamine are converted to methane. H_2/CO_2 and formate are not used. No organic growth factors are required. NaCl (optimum: 0.2 M) and Mg^{2+} (optimum: 50–100 mM) are required for growth.

Optimum growth is obtained at 35–40°C and pH 6.5–7.0. Strictly anaerobic.

Minimum doubling time occurs with methanol as a substrate (5.2 h). Monomethylamine, dimethylamine, and trimethylamine give doubling times of 6.7, 7.8, and 7.3 h, respectively. Doubling time on acetate is 24.1 h.

Habitat: littoral marine sediments.

The mol% G + C of the DNA is: 41 (T_m).

Type strain: C2A, ATCC 35395, DSMZ Z2834, OCM 95.

GenBank accession number (16S rRNA): M59137.

Additional Remarks: The type strain was isolated from marine sediment (Sowers et al., 1984a).

3. **Methanosarcina mazeii** (Barker 1936) Mah and Kuhn 1984b, 263[VP] (*Methanosarcina mazei* (sic) Mah and Kuhn 1984b, 263) (*Methanococcus mazei* Barker 1936, 433).

ma′zei.i. M.L. gen. n. *mazeii* of Mazé; named for P. Mazé, the French bacteriologist who first studied the organism.

Original description is supplemented with material from Mah (1980), Maestrojuán and Boone (1991), and Maestrojuán et al. (1992).

Individual, irregular coccoid cells 1.0–3.0 µm in diameter form irregular clumps 20–100 µm or more in diameter. These irregular clumps exhibit surface indentations and may cluster together in large rafts 1000 µm or more in diameter. The clumps later may become cysts, which can give rise to individual coccoid cells. A life cycle involving these morphological stages may occur (Zhilina and Zavarzin, 1979b; Liu et al., 1985; Robinson et al., 1985; Robinson, 1986). The organism is nonmotile. Stain Gram-negative. Isolated coccoid units, but not aggregates or cysts, are sensitive to lysis by SDS.

Colonies in roll tubes are buff white to tannish yellow, with a grainy appearance when young (<7 d). Older surface colonies are smooth, circular, transparent, glistening, mucoid, and pulvinate with entire margins. Appearance of colonies in roll tubes may depend on whether cells are aggregated and on growth conditions, which may cause strains to grow in a disaggregated state; colonies of disaggregated cells may appear transparent. Occasionally, gas is trapped in the mucoid colony surface, fulminating into clusters of bubbles.

Methanol, methylamine, dimethylamine, and trimethylamine are converted to methane. Either acetate or H_2/CO_2 may be used. H_2/CO_2 may be used concurrently with trimethylamine or methanol. Butyrate, ethanol, butanol, and acetone are not methanogenic substrates.

Methanogenic substrates are utilized as a sole energy source in the presence of yeast extract and Trypticase Peptone® (BBL Microbiology Systems). The organism is stimulated by sludge supernatant fluid. Ammonia may serve as a nitrogen source.

Grows well at 30–40°C and pH 5.5–8.0; optimum growth is obtained at pH 6.8–7.2. Strictly anaerobic.

During exponential methane formation, the generation times on various substrates are as follows: acetate, 17 h; methanol, methylamine, or trimethylamine, 7–15 h; and H_2/CO_2, 9 h. Not all strains are capable of rapid growth on H_2/CO_2 or acetate.

Habitat: decaying leaves, garden soil, sewage sludge digestors, black mud, and feces of herbivorous animals; also isolated from urban solid waste and various sewage and animal waste digestors and lagoons.

Described strains classified in this species: Z-558 (biotype 3, DSMZ 2244, OCM 92) (Zhilina and Zavarzin, 1979b), CW3A (OCM 20) (McInerney et al., 1981b), UBS (DSMZ 1311, OCM 27) (Zeikus and Winfrey, 1976), LYC (ATCC 43573, DSMZ 4556, OCM 34) (Liu et al., 1985), Göl (DSMZ 3647, OCM 88) (Jussofie et al., 1986), G1 (DSMZ 3338, OCM 87) (Hippe et al., 1979), MC3 (DSMZ 2907) (Touzel and Albagnac, 1983), MC3 (DSMZ 2907, OCM 97) (Touzel and Albagnac, 1983), C 16 (ATCC 43340, DSMZ 3318, OCM 98), and nine strains from anaerobic fixed-film reactors (Equi [OCM 61], GA4 [OCM 76], GA5 [OCM 77], GA6 [OCM 78], GA8 [OCM 79], GA12 [OCM 80], and GA15 [OCM 81]) (Maestrojuán et al., 1992).

The mol% G + C of the DNA is: 42 (Bd).

Type strain: S-6, DSMZ 2053, OCM 26.

GenBank accession number (16S rRNA): AJ012095, U20151.

Additional Remarks: The type strains was isolated from an anaerobic sewage sludge digestor.

4. **Methanosarcina thermophila** Zinder, Sowers and Ferry 1985, 523.[VP]

ther.mo'phi.la. Gr. adj. *thermos* hot; Gr. adj. *philos* loving; M.L. fem. adj. *thermophila* heat loving.

Irregular aggregates, 100 μm or more across, are comprised of coccoid bodies. These bodies appear to be individual cells; many nonperpendicular division planes are evident. Tetrads do not occur, and individual cell bodies are rare. Nonmotile. Stain Gram-positive. Not lysed by SDS.

Surface colonies in agar roll tubes are yellow to brown and have a rough, granular appearance. Growth occurs on acetate, methanol, monomethylamine, dimethylamine, and trimethylamine. H_2/CO_2 and formate are not used, although H_2 may be used during growth on methanol. NH_3 serves as a nitrogen source; N_2 is not fixed. Growth occurs in defined medium with p-aminobenzoate and methanol and acetate as catabolic substrates and sole carbon and energy source (Murray and Zinder, 1985).

Optimum growth temperature, 50–55°C; range, <35–55°C. Good growth occurs over a broad range of pH values (5.5–8.0), with an optimum at 6–7. Maximum specific growth rate on acetate is about 0.05/h (Clarens and Moletta, 1990). Strictly anaerobic.

Habitat: thermophilic anaerobic digestors.

Described strains classified in this species: CHTI-55 (ATCC 43170, DSMZ 2906, OCM 90) (Touzel et al., 1985), MP (DSMZ 2980, OCM 91) (Ollivier et al., 1984), and MST-A1 (MST; Thomas et al., 1986), DSMZ 2905, OCM 96).

The mol% G + C of the DNA is: 42 (Bd).

Type strain: TM-1, DSMZ 1825, OCM 12.

GenBank accession number (16S rRNA): M59140.

Additional Remarks: The type strain was isolated from a 55°C anaerobic digestor (Zinder and Mah, 1979.)

5. **Methanosarcina vacuolata** Zhilina and Zavarzin 1987b, 283.[VP]

va.cu.o.la'ta. L. adj. *vacuus* empty; M.L. fem. part. adj. *vacuolata* equipped with gas vacuoles.

Original material (Zhilina and Zavarzin, 1979a) supplemented with material from Zhilina (1971, 1976, 1978), Archer and King (1983, 1984), and Lysenko and Zhilina (1985).

Coccoid bodies, 1–2 μm in diameter, occurring sometimes as individual cells but mostly in irregular, rounded aggregates ranging from several to several hundred micrometers in size; individual cells within aggregates may not be distinguished by light microscopy. May contain gas vesicles. Membranes contain C_{30} isoprenoids (Osipov et al., 1984). Nonmotile. Stain Gram-positive. Not lysed by SDS.

Deep colonies in agar medium are light yellow, angular, granular, and 0.5–1.0 mm in diameter. In liquid medium, growth may occur as a light yellow, easily dispersed sediment with clear supernatant medium.

Energy-yielding metabolism involves methane production. H_2/CO_2, methanol, methylamines, and acetate may be used as substrates. Formate is not used. Growth on methanol faster than on acetate. Carbohydrates, amino acids, formate, ethanol, propionate, and butyrate are not fermented.

Ammonia serves as a nitrogen source, and sulfide serves as a sulfur source. Growth of some strains may be stimulated by addition of organic compounds, but all can grow autotrophically. Optimum growth is obtained at pH 7.5 and 37–40°C. Strictly anaerobic.

Habitat: soil, freshwater mud, and sludge from anaerobic sewage sludge digestors.

The mol% G + C of the DNA is: is 36.3 (T_m) (Lysenko and Zhilina, 1985).

Type strain: Z-761, biotype-2, strain Z, ATCC 35090, DSMZ 1232, OCM 85.

GenBank accession number (16S rRNA): U20150.

Additional Remarks: The type strain was isolated from a methanogenic digestor (Zhilina and Zavarzin, 1979b).

Genus II. **Methanococcoides** *Sowers and Ferry 1985b, 223*[VP] *(Effective publication: Sowers and Ferry 1983, 688)*

KEVIN R. SOWERS

Me.tha.no.coc.co'i.des. Gr. adj. suff. *-oides* similar to; M.L. neut. n. *Methanococcoides* organism similar to *Methanococcus.*

Irregular cocci, 0.8–1.8 μm in diameter; occurring singly or in pairs. If motility is observed, it is by a monotrichous flagellum. **Strictly anaerobic.** Temperature range: 1.7–35°C; optimum: 23–35°C. **Slightly halophilic,** with optimal growth near 0.2 M NaCl. Mg^{2+} is required for growth, as $MgSO_4$ or $MgCl_2$ at concentrations greater than 0.01 M. **Trimethylamine, dimethylamine, methylamine, and methanol are substrates for growth** and methanogenesis; **acetate, dimethylsulfide, formate, and H_2/CO_2 are not.**

The mol% G + C of the DNA is: 40–42.

Type species: **Methanococcoides methylutens** Sowers and Ferry 1985b, 223 (Effective publication: Sowers and Ferry 1983, 688).

FURTHER DESCRIPTIVE INFORMATION

The irregularly shaped cells (Fig. A2.23 and A2.24) become spherical as the NaCl or Mg^{2+} concentration is decreased and lyse when either is eliminated (Sowers and Ferry, 1983; Franz-

mann et al., 1992). Whole cells are immediately lysed by 0.01% sodium dodecyl sulfate or 0.001% Triton X-100. Electron microscopy of thin sections shows a monolayered cell wall ~10 nm thick (Fig. A2.25). Acid hydrolysates of isolated cell walls yield a variety of amino acids, which indicates that the walls are protein. Aspartic and glutamic acids are predominant. Amino sugars are not detected. The membrane polar lipid fraction consists of 2,3-diphytanyl glycerol ethers; dibiphytanyl diglycerol tetraethers are not detected. Electron-dense inclusion bodies occur occasionally in cytoplasm.

Surface colonies (0.5–1.5 mm) are yellow, circular, and convex with entire edges. Colonies fluoresce blue-green under UV light.

Medium that contains seawater, mineral salts, and growth substrate prepared anaerobically under a N_2/CO_2 (4:1) atmosphere is required for growth. Organic growth factors may be required (Sowers and Ferry, 1985a). A mixture of NaCl, $MgSO_4$, KCl, and

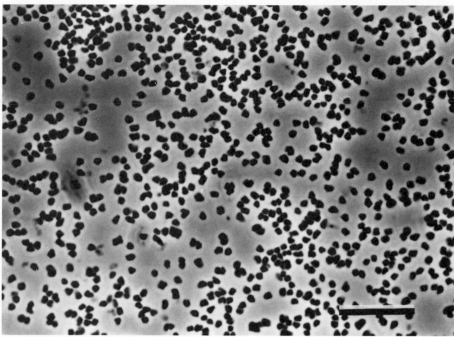

FIGURE A2.23. Phase-contrast photomicrograph of *Methanococcoides methylutens*. Bar = 10 μm.

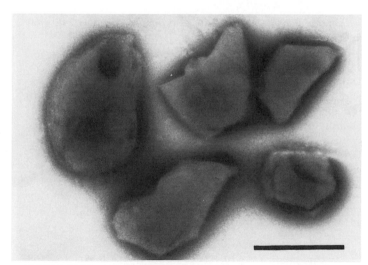

FIGURE A2.24. Transmission electron photomicrograph of *Methanococcoides methylutens* negatively stained with ammonium molybdate. Bar = 1 μm.

$CaCl_2$ may be substituted for seawater. Yeast extract, Trypticase peptones, or rumen fluid stimulate growth. Essential trace metals include nickel, iron, and cobalt (Sowers and Ferry, 1985a). *Methanococcoides* will grow in the presence of vancomycin (100 mg/ 1).

ENRICHMENT AND ISOLATION PROCEDURES

Culture media are prepared anaerobically under a N_2/CO_2 (4:1) atmosphere (Sowers and Noll, 1995). Enrichment medium contains a solution of 80% artificial seawater diluted with demineralized water plus trace metals, vitamins, cysteine sulfide reducing agent, and trimethylamine-HCl (Sowers and Ferry, 1983). Resazurin is added as an E_h indicator. The pH is maintained at 7.2 by a CO_2-bicarbonate buffer system. Alternatively, MGM me-

dium (Franzmann et al., 1992) or disaggregating medium (Sowers et al., 1993) may be used with the addition of trimethylamine-HCl (20 mM) and vancomycin (100 mg/ml). Enrichment cultures are incubated at 20–30°C (depending on the ambient temperature of the inoculum source) and assayed for turbidity and methane production. Isolation medium is the same as enrichment medium, with the addition of 2% purified agar (BBL Microbiology Systems) or Noble agar (Difco). Inoculum is streaked onto agar-solidified medium, or serial dilutions are added to molten agar and poured into Petri plates (Apolinario and Sowers, 1996) or roll tubes (Hungate, 1969). Yellow colonies are observed within 5 d and are differentiated from nonmethanogens by their blue-green fluorescence in long-wave UV light (Mink and Dugan, 1977).

MAINTENANCE PROCEDURES

Strains are maintained by transfer every 3 months on agar slants or in liquid medium stored at room temperature in the dark. For long-term storage, cultures are maintained by freezing in liquid growth medium and glycerol (3:1) as described by Tumbula et al. (1995). Cultures stored in glycerol at −70°C have remained viable for over 10 years.

DIFFERENTIATION OF THE GENUS *METHANOCOCCOIDES* FROM OTHER GENERA

The genus *Methanococcoides* is distinguished from *Methanosarcina* based on its inability to grow on acetate or H_2/CO_2 and a 16S rRNA sequence similarity of 92% (Franzmann et al., 1992). In addition, the thick (≤400–nm) methanochondroitin outer-wall layer that is synthesized by several *Methanosarcina* species (Kreisl and Kandler, 1986) is not formed by either described species of *Methanococcoides*. Among the obligately methylotrophic genera, *Methanococcoides* and *Methanolobus* are slightly halophilic, growing optimally in 0.2 M NaCl, which distinguishes them from moderately halophilic *Methanohalophilus* and extremely halophilic *Methanohalobium*, which grow optimally in 0.5–2.5 M and >2 M

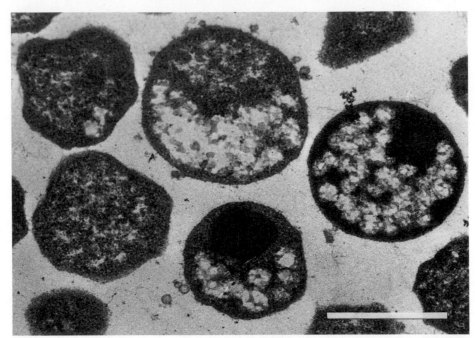

FIGURE A2.25. Thin-section transmission electron photomicrograph of *Methanococcoides methylutens* stained with uranyl acetate. Bar = 1 μm.

NaCl, respectively (Zhilina and Zavarzin, 1987a; Paterek and Smith, 1988). *Methanococcoides* is distinguished from *Methanolobus* by its lower temperature range for growth (König and Stetter, 1982), differences in polar lipid patterns based on thin-layer-chromatographic mobilities (Franzmann et al., 1992), and a 16S rRNA sequence similarity of ≤94%. The genera show no similarity by DNA–DNA reassociation. Immunological fingerprinting by indirect immunofluorescence (S probe) shows no cross-reactivity between *Methanococcoides burtonii* and *Methanolobus tindarius* Tindari 3 and only slight cross-reactivity between *Methanococcoides*

methylutens TMA-10 and *Methanolobus tindarius* Tindari 3 (Sowers and Ferry, 1983; Franzmann et al., 1992).

Taxonomic Comments

Differentiation of the slightly halophilic genera *Methanococcoides* and *Methanolobus* by phenotypic traits is currently based on only two species of *Methanococcoides* and five species of *Methanolobus*. Definitive phylogeny of new strains in either genus should be confirmed by comparative sequence analysis of 16S rRNA and DNA reassociation until a greater number of representative species are described.

List of species of the genus Methanococcoides

1. **Methanococcoides methylutens** Sowers and Ferry 1985b, 223[VP] (Effective publication: Sowers and Ferry 1983, 688). *me.thyl.u′ tens.* M.L. n. *methylum* methyl; L. part. adj. *utens* using; *methylutens* using methyl.

    See Table A2.20 and generic description for characteristics.

TABLE A2.20. Characteristics differentiating species of *Methanococcoides*

Characteristic	1. *M. methylutens*	2. *M. burtonii*
Growth temperature, °C:		
Optimum	35	23.4
Maximum	35	29.5
Minimum	15	1.7[a]
Optimum pH	7.0–7.5	7.7
NaCl range	0.1–1.0 M	0.1–0.5 M
Growth factors	Biotin	None
Mol% G + C (T_m)	42	39.6

[a]Lowest demonstrated growth; predicted T_{min} is −2.5°C. For growth at temperatures of <5.6°C in freshly inoculated medium, cultures must be preincubated at a higher temperature (e.g., 20°C).

The mol% G + C of the DNA is: 42 (T_m).
Type strain: TMA-10, ATCC 33938, DSMZ 2657, OCM 158.
GenBank accession number (16S rRNA): M59127.
Additional Remarks: The type strain was isolated from a submarine canyon that contained large deposits of organic material.

2. **Methanococcoides burtonii** Franzmann, Springer, Ludwig, Conway de Macario and Rhode 1993, 398[VP] (Effective publication: Franzmann, Springer, Ludwig, Conway de Macario and Rhode 1992, 579). *bur.ton′ i.i.* M.L. gen. n. *burtonii* of Burton; named after Harry R. Burton, a limnologist who discovered methane in Ace Lake.

    See Table 2.20 and generic description for characteristics.
    The mol% G + C of the DNA is: 39.6 (T_m).
    Type strain: DSMZ 6242, OCM 468.
    GenBank accession number (16S rRNA): X65537.
    Additional Remarks: The type strain was isolated from the anoxic hypolimnion of Ace Lake, Antarctica.

Genus III. **Methanohalobium** Zhilina and Zavarzin 1988, 136[VP] (Effective publication: Zhilina and Zavarzin 1987a, 467)

TATJANA N. ZHILINA

Me.tha.no.ha.lo'bi.um. M.L. neut. n. *methanum* methane; Gr. n. *hals, halos* salt; Gr. n. *bios* life; M.L. neut. n. *Methanohalobium* methane-producing organism living in salt.

Cells flat, polygonal, or irregular spheroid bodies in small aggregates (5–10 μm) or single (0.2–2 μm; average, 1 μm). Gram stain variable. **Nonmotile. Strictly anaerobic. Extremely halophilic**: optimum NaCl concentration, 4.3 M, with a range of 2.6–5.1 M. **Moderately thermophilic**. Optimum temperature, 40–55°C; maximum, ~60°C; minimum, 35–37°C. Grows between pH 6.0 and 8.3; optimum pH, 7.0–7.5. **Strictly methylotrophic. Trimethylamine, dimethylamine, and methylamine are substrates** for growth and methanogenesis; **acetate, formate, and H$_2$/CO$_2$ are not utilized**; methanol is utilized at very low concentrations (<5 mM).

The mol% G + C of the DNA is: 37.

Type species: **Methanohalobium evestigatum** Zhilina and Zavarzin 1988, 136 (Effective publication: Zhilina and Zavarzin 1987a, 467) (*Methanohalobium evestigatus* (sic) Zhilina and Zavarzin 1987a, 467).

FURTHER DESCRIPTIVE INFORMATION

Phylogenetic treatment *Methanohalobium* belongs to the cluster of methylotrophic halophilic methanogenic euryarchaeota of the order *Methanosarcinales* and family *Methanosarcinaceae* (Boone et al., 1993b). The genus includes a single species, *M. evestigatum*. Specific information regarding the species is given in the species description below.

ENRICHMENT AND ISOLATION PROCEDURES

Methanohalobium can be enriched from highly saline environments (e.g., salterns) anaerobically in mineral medium with 4.3 M NaCl (Zhilina and Zavarzin, 1987a), trimethylamine (0.3–0.5%), and ampicillin and penicillin (100 mg/l). Cultures are incubated at 50–55°C and isolated on the same medium by serial dilution in liquid medium.

MAINTENANCE PROCEDURES

M. evestigatum is maintained by regular subculturing in liquid medium and stored anaerobically at 4°C; no lysis occurs for up to a year. Cultures may be stored in liquid nitrogen (H. Hippe, personal communication).

DIFFERENTIATION OF THE GENUS METHANOHALOBIUM FROM OTHER GENERA

The genus *Methanohalobium* is differentiated from other genera of the *Methanosarcinaceae* as an extremely halophilic methanogen.

It differs from *Methanosarcina*, to which it is somewhat similar in morphology of cells and cell aggregates, by obligate and extreme halophily, and an inability to utilize either H$_2$/CO$_2$ or acetate. It differs from the obligately halophilic methylotrophic methanogens *Methanohalophilus*, *Methanolobus*, and *Methanococcoides* by the higher range and optimal salinity, and temperature optima; morphology, cell wall structure, and lipid profile; resistance to sodium dodecyl sulfate (SDS) lytic action; and immunology. It differs from alkaliphilic *Methanosalsum* by morphology, the higher range of salinity, inability to grow at high pH, and inability to utilize dimethylsulfide as an energy source.

TAXONOMIC COMMENTS

The phylogenetic relationship of *Methanohalobium* to the *Methanosarcinaceae* is indicated by sequence analysis of 16S rRNA (Boone et al., 1993b) and partial sequences of the methyl-coenzyme M reductase (*mcrI*) gene, which is a phylogenetic tool for this family (Springer et. al., 1995), and by the shared ability to utilize methylamines. Its status as a separate genus is based on 5S rRNA (Chumakov et al., 1987) and 16S rRNA phylogenies (Springer et al., 1995), DNA–DNA similarity (Wilharm et al., 1991), and immunological differences (Bezrukova et al., 1989).

ACKNOWLEDGMENTS

I thank Prof.G. A. Zavarzin for his contribution.

FURTHER READING

Chumakov, K.M., T.N. Zhilina, I.S. Zvyagintseva, A.L. Tarasov and G.A. Zavarzin. 1987. 5S rRNA in archaebacteria. Zh. Obshch. Biol. *48*: 167–181.

Wilharm, T., T.N. Zhilina and P. Hummel. 1991. DNA–DNA hybridization of methylotrophic halophilic methanogenic bacteria and transfer of *Methanococcus halophilus*[VP] to the genus *Methanohalophilus* as *Methanohalophilus halophilus*, comb. nov. Int. J. Syst. Bacteriol. *41*: 558– 562.

Zhilina, T.N. and T.P. Svetlichnaya. 1989. The ultrafine structure of *Methanohalobium evestigatus*, an extremely halophilic methanogenic bacterium. Mikrobiologiya *58*: 312–318.

Zhilina, T.N. and G.A. Zavarzin. 1987. *Methanohalobium evestigatus*, gen. nov. sp. nov., the extremely halophilic methanogenic archaebacterium. Dokl. Akad. Nauk. SSSR *293*: 464–468.

Zhilina, T.N. and G.A. Zavarzin. 1990. Extremely halophilic, methylotrophic, anaerobic bacteria. FEMS Microbiol. Rev. *87*: 315–321.

List of species of the genus Methanohalobium

1. **Methanohalobium evestigatum** Zhilina and Zavarzin 1988, 136[VP] (Effective publication: Zhilina and Zavarzin 1987a, 467) (*Methanohalobium evestigatus* (sic) Zhilina and Zavarzin 1987a, 467).

e.ves.ti.ga' tum. L. neut. adj. *evestigatum* found.

Cell morphology is strongly influenced by the concentration of divalent cations. In media containing 1.5 mM Ca^{2+},

18 mM Mg^{2+}, and 4.3 M NaCl, the cells are irregular spheroid bodies united in aggregates by the common capsulae and resemble *Methanosarcina* (Fig. A2.26a). In older cultures or in the presence of 4.3 M sea salt with 4.4 mM Ca^{2+} and 46 mM Mg^{2+}, the cells are flattened and irregularly shaped and have a "broken-glass" appearance (Fig. A2.26b). Oxidized cells have strong autofluorescence, as do many other meth-

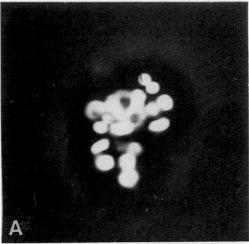

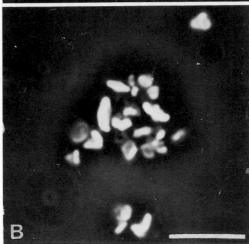

FIGURE A2.26. *Methanohalobium evestigatum* appearance in the light microscope (Anoptral contrast). Bar = 10 μm. (*A*) Aggregates of irregular spheroid cells in 25% (w/v) NaCl medium. (*B*) Aggregates of polygonal flat cells in 25% (w/v) seasalt medium.

anogens. *M. evestigatum* is thermotolerant and growth was observed after 8 min at 100°C.

Fine structure differs for the irregular spheroid and flat cells. Rounded cells have strong cell walls 30–50 nm thick with distinct microcapsulae. Multiplication occurs by multiple fission as for methanosarcinae, causing variability in cell size. The cell wall inserts as a diaphragm (Fig. A2.27a). Flat cells are polygonal, with sharp edges; layers of cell walls produce sheets with regular arrangements of definite subunits of 12.5-nm dimensions assembled in rows at 25-nm distance. Grana of polyphosphate and glycogen are present (Fig. A2.27b) (Zhilina and Svetlichnaya, 1989). The overall morphology of the cells and aggregates resembles that of *Methanosarcina mazeii*.

The cell wall is SDS resistant. The lipid profile of *M. eves-*

tigatum has only isoprenes $C_{25:5}$, $C_{25:3}$, and $C_{25:4}$, thus differing from *Methanohalophilus*, which contains squalenes $C_{30:5}$ and $C_{30:6}$ but no isoprenes (Osipov et al., 1988).

Energy-yielding metabolism involves methane and ammonia production from methylamines; trimethylamine is the preferred substrate. Hydrogen is formed during growth, with the final concentration up to 200 ppm with trimethylamine as a substrate and 1000 ppm with methylamine. There is no growth on methanol at 123 mM (Zhilina and Zavarzin, 1987a) or at 20 mM (Boone et al., 1993a), but growth at 5 mM was reported (Boone et al., 1993a). In spite of the absence of growth and methanogenesis at high concentrations of methanol, it could be the source for production of up to 400 ppm of H_2 (Bodnar et al., 1987).

Medium that contains 2.6–5.1 M NaCl, mineral salts, vitamins, and substrate with an N_2/CO_2 (9:1) atmosphere is sufficient for growth of *M. evestigatum* (Zhilina and Zavarzin, 1987a). No organic compounds are needed for growth; yeast extract at 500 mg/l is inhibitory, but at 50–100 mg/l, it replaces vitamins. Trimethylamine may serve as a sole carbon and energy source. Ammonia serves as a nitrogen source, and sulfide serves as a sulfur source. Mg^{2+} in the range of 0.31–200 mM does not affect growth.

By DNA–DNA hybridization, *Methanohalobium* demonstrated no relatedness with the species of genus *Methanohalophilus* (*M. mahii*, *M. halophilus*, and *M. portucalensis*) or with *Methanosalsum zhilinae*, including two other extremely halophilic strains, Z-7403 and Z-7408 (Wilharm et al., 1991; Boone et al., 1993a). Immunologically slight resemblance was found between *Methanohalobium* and *Methanosarcina* species, but not with other halophilic methanogens, such as *Methanolobus tindarius*, *Methanococcoides methylutens*, and *Methanohalophilus halophilus* (Bezrukova et al., 1989).

The organism grows in the presence of ampicillin and penicillin (100 mg/l).

M. evestigatum is a causative agent in a "noncompetitive pathway" of methanogenesis in an anaerobic microbial community at a high salinity range. In coculture with the haloanaerobe *Acetohalobium arabaticum*, it produces methane from betaine, a common osmoregulatory compound of halophilic proteobacteria (Zhilina and Zavarzin, 1990). It inhabits lagoons of high salinity heated by the sun, which might be the reason for its broad temperature optima and thermotolerance.

The type strain of *M. evestigatum* was isolated from the sediments of saline lagoons in Sivash, covered by a dense cyanobacterial mat of *Microcoleus chthonoplastes*.

The mol% G + C of the DNA is: 37 (T_m) (Wilharm et al., 1991).

Type strain: Z-7303, DSMZ 3721, OCM 161.

GenBank accession number (16S rRNA): U20149.

Additional Remarks: GenBank accession number for 5S rRNA sequence: X62861; methyl-coenzyme M reductase (*mcrI*) gene sequence: U22236.

Other Organisms

1. *"Methanohalobium strain SD-1"*

Irregular cocci 1 μm in diameter, single, in pairs or small regular tetragonal clumps. Gram negative. Cells fluoresce in epifluorescence microscopy, but the fluorescence fades after

about 30 s. Optimum salinity, 1–3.5 M NaCl (Boone et al., 1993a). Cells are sensitive to lysis by osmotic shock or in the presence of SDS (50 mg/l). Strain SD-1 utilizes mono-, di-, and trimethylamines and methanol as catabolic substrates.

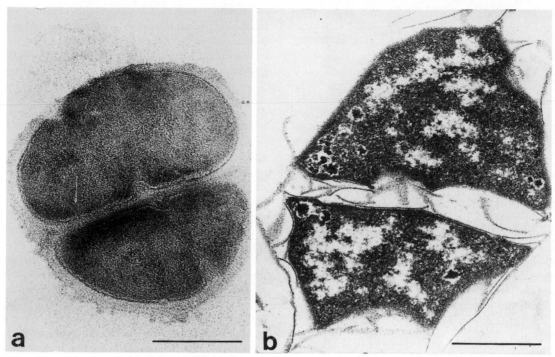

FIGURE A2.27. Ultrathin sections of *Methanohalobium evestigatum*. (*a*) Small aggregate of irregular spheroid cells in NaCl medium. Thick cell walls and the type of fission are seen. Bar = 0.5 μm. (*b*) Small aggregate of the flat polygonal cells in seasalt medium. The sheets of cell envelope surrounding cells are obviously seen. Bar = 1 μm.

With trimethylamine as a substrate, the cells grew fastest at 42°C and pH 7.8, with 0.9–3.5 M Na$^+$. Cells grew rapidly (specific growth rate was 0.015 h^{-1}) in mineral medium with no organic compound other than trimethylamine added; vitamins did not stimulate growth. Chloramphenicol inhibited the growth of cells but did not kill them. Tetracycline, ampicillin, carbenicillin, cycloserine, erythromycin, and penicillin had no effect at 100 mg/l. Strain SD-1 was isolated from a saltern in San Diego, California. (Boone et al., 1993a.)

DNA hybridization and whole-cell protein analysis show that strain SD-1 is phylogenetically distant from the type species of *Methanohalophilus* and phenotypically different from the type strain *Methanohalobium* (Boone et al., 1993a). However, in the *mcr* sequence tree it is phylogenetically related to the genus *Methanohalobium* (Springer et al., 1995).

The mol% G + C of the DNA is: 41.

Deposited strain: OCM 134

Additional Remarks: GenBank accession number for methyl-coenzyme M reductase (*mcrI*) gene sequence: U22256.

2. Other reported *Methanohalobium*-like strains

During the evaporation of lagoons, a series of moderately and extremely halophilic methylotrophic methanogens, together with *Methanohalobium*, were isolated from the brines (Zhilina, 1986). Among moderately halophilic strains Z-7301 (DSMZ 5700), Z-7302 (DSMZ 5701), Z-7401 (DSMZ 5702), Z-7404 (DSMZ 5699), and Z-7405 (DSMZ 5703), some were identified by DNA–DNA hybridization as representatives of the new species within the genus *Methanohalophilus* (Wilharm et al., 1991), and strains Z-7401, Z-7302, Z-7405, and Z-7404 were ascribed to *Methanohalophilus portucalensis* (Boone et al., 1993a). The extremely halophilic strains Z-7403 and Z-7408 (DSMZ 5814) are phenotypically similar to *Methanohalobium evestigatum*; however, they differ significantly in DNA–DNA similarity (Wilharm et al., 1991).

Genus IV. **Methanohalophilus** *Paterek and Smith 1988, 122*[VP]

DAVID R. BOONE

Me.tha.no.ha.lo'phi.lus. M.L. neut. n. *methanum* methane; Gr. masc. n. *hals, halos* salt; Gr. adj. *philos* loving; M.L. masc. n. *Methanohalophilus* salt-loving methanogen.

Irregular cocci, ~1 μm in diameter, occurring singly or in small clumps. Nonmotile. Does not form endospores. Stain Gram-negative. **Lysed by detergents or hypotonic shock.** Surface colonies are cream to pale yellow, circular, with entire edges.

Cells grow on methyl substrates, including methylamines and methanol, producing methane and carbon dioxide (and ammonia, in the case of methylamines). Methanol is toxic at high

(40 mM) concentrations. **H$_2$, formate, secondary alcohols, and acetate are not catabolized. Strictly anaerobic. Moderately halophilic** (fastest growth with 1.0–2.5 M NaCl). Fastest growth at pH values near neutral and temperatures 35–40°C. Some strains have no requirement for organic growth factors, but others require vitamins.

Habitat: anoxic sediments of saline lakes and evaporation ponds.

Type species: **Methanohalophilus mahii** Paterek and Smith 1988, 122.

FURTHER DESCRIPTIVE INFORMATION

Methanohalophilus is an obligate halophile that balances the osmolarity of its cytosol with that of the environment by accumulating or synthesizing organic solutes, such as glycine betaine, glutamate (Robertson et al., 1990a, b), N^ε-acetyl-β-lysine (Robertson et al., 1990a, b), and β-glutamine (Lai et al., 1991) but not β-glutamate (Robertson et al., 1990b). Glycine betaine and N^ε-acetyl-β-lysine may be the dominant osmoregulatory compounds (Robertson et al., 1992a).

ENRICHMENT AND ISOLATION PROCEDURES

Enrichment and isolation of *Methanohalophilus* may require only a habitat-simulating medium with trimethylamine added as a catabolic substrate. Good growth occurs in MH medium with 20 mM trimethylamine. MH medium is the same as MS medium (Boone et al., 1989) but with the NaCl concentration increased to 1.5 M, with $MgCl_2 \cdot 6H_2O$ increased to 6 g/l, and with 1.5 g of KCl added per liter.

MAINTENANCE PROCEDURES

Cultures grow well in MH medium with 20 mM trimethylamine. Cultures can survive weeks or even months after growth at room temperature, as long as they remain anoxic. They can be maintained by monthly transfer into fresh medium. They can also be stored as frozen stocks by adding 5% anoxic glycerol as a cryoprotectant before anoxically distributing them to vials and freezing them at 1°C/min to −40°C (Boone, 1995).

DIFFERENTIATION OF THE GENUS *METHANOHALOPHILUS* FROM OTHER GENERA

Methanohalophilus is the only recognized genus of moderately halophilic methanogens. All methylotrophic halophiles are classified within the family *Methanosarcinaceae*. Genera in this family other than *Methanohalophilus* are either nonhalophilic or slightly halophilic (*Methanosarcina*, *Methanolobus*, *Methanococcoides*, and *Methanosalsum* grow fastest at salinities less than 0.5 M NaCl) or extremely halophilic (*Methanohalobium* grows fastest at salinities greater than 2.5 M NaCl).

TAXONOMIC COMMENTS

Methanohalophilus oregonensis, previously classified in this genus, is phylogenetically distinct from the genus and should be transferred into the genus *Methanolobus* as *Methanolobus oregonensis* (Boone et al., 1993b). *Methanohalophilus zhilinae*, also previously in this genus, is likewise phylogenetically distinct and should be moved to a new genus (Boone et al., 1993b). In accordance with these recommendations, we have classified *M. oregonensis* as *Methanolobus oregonensis* (comb. nov.) and *M. zhilinae* in the new genus *Methanosalsum* (gen. nov.) as *Methanosalsum zhilinae* (comb. nov.).

With the removal of these two species, the three species remaining in the genus *Methanohalophilus* are very similar morphologically and physiologically. In fact, they cannot be differentiated by their phenotypic characteristics. Although there are considerable phylogenetic distances among these species (Boone et al., 1993b), this is not reflected in dependably distinct phenotypes. Nucleic acid analysis or one-dimensional denaturing gel analysis of whole-cell proteins can distinguish the species (Boone et al., 1993a).

Another genus of moderately halophilic, methylotrophic methanogens has been proposed (*Halomethanococcus* Yu and Kawamura 1988, 328[VP]), but this genus may be considered as a junior subjective synonym of *Methanohalophilus*. Based on its published phenotypic description, *Halomethanococcus* cannot be distinguished from *Methanohalophilus*, and *Halomethanococcus doii*, the only species in that genus, is not distinguishable from any of the three species of *Methanohalophilus*. The type strain of *H. doii* has been lost, so it will never be possible to identify phenotypic traits to differentiate this species, nor will it be possible to demonstrate that the species has significant phylogenetic differences from species of *Methanohalophilus*.

DIFFERENTIATION OF THE SPECIES OF THE GENUS *METHANOHALOPHILUS*

Differential characteristics of the species of the genus *Methanohalophilus* are indicated in Table A2.21.

List of species of the genus Methanohalophilus

1. **Methanohalophilus mahii** Paterek and Smith 1988, 122.[VP]
   ma' hi.i. M.L. masc. gen. n. *mahii* of Mah, in honor of Professor Robert A. Mah for his noteworthy research in anaerobic microbiology and on methanogenic bacteria.

   Irregular cocci about 1 μm in diameter. Surface colonies are cream to pale yellow and circular with entire margins.
   The mol% G + C of the DNA is: 48.5 (Bd).
   Type strain: SLP, ATCC 35705, DSMZ 5219, OCM 68.
   GenBank accession number (16S rRNA): M59133.
   Additional Remarks: The type strain was isolated from sediments of the Great Salt Lake.

2. **Methanohalophilus halophilus** (Zhilina 1983) Wilharm, Zhilina and Hummel 1991, 561[VP] (*Methanococcus halophilus* Zhilina 1983, 290).

   ha.lo' phi.lus. Gr. masc. n. *hals, halos* salt; Gr. adj. *philos* loving; M.L. masc. adj. *halophilus* salt loving.

   Irregular cocci, 0.5–2.0 μm in diameter, occurring singly, in pairs, and as irregular aggregates. Cell wall is an S-layer, surrounded by a slime layer.

   Growth in liquid medium is dispersed. No growth factors are required.

   Contains no plasmids (Sowers and Gunsalus, 1988b).
   The mol% G + C of the DNA is: 41–44 (T_m).
   Type strain: Z-7982, DSMZ 3094, OCM 160.
   Additional Remarks: The type strain was isolated from a cyanobacterial mat and bottom deposits at Hamelin Pool, Shark Bay, northwestern Australia.

3. **Methanohalophilus portucalensis** Boone, Mathrani, Liu, Menaia, Mah and Boone 1993a, 436.[VP]
   por.tu.ca.len' sis. M.L. masc. adj. *portucalensis* from Portugal.

TABLE A2.21. Diagnostic and descriptive features of *Methanohalophilus* species[a]

Characteristic	1. *M. halophilus*	2. *M. mahii*	3. *M. portucalensis*
Morphology:			
Coccoid cells	+	+	+
Motile	–	–	–
Substrates:			
H_2/CO_2	–	–	–
Acetate	–	–	–
Methylamines	+	+	+
Methanol	+	+	+
Methyl sulfides			
Growth requirements:			
Growth in mineral medium plus catabolic substrate	+	–	–
Stimulated by other organic components in medium	+	Biotin, thiamine	Biotin
Environmental conditions supporting rapid growth:			
Temp. 30–40°C	+	+	+
Temp. 55–60°C	–	–	–
pH 5	–	–	–
pH 6	–	–	–
pH 7	+	–	+
pH 8	+	+	+
pH 9	–	+	–
0 M NaCl	–	–	–
0.3–1.0 M NaCl	+	+	+
Mol% G + C	39	41	41

[a]For symbols see standard definitions.

Irregular cocci, 0.6–2 µm in diameter, occurring singly, in pairs, and in irregular clumps. Surface colonies are 0.5 mm in diameter within 7 d of incubation; they are circular, tannish yellow, convex, shiny, and opaque or grainy with entire margins.

The mol% G + C of the DNA is: 43–44.
Type strain: FDF-1, OCM 59.
Additional Remarks: The type strain was isolated from sediments of a solar saltern in Figiera da Foz, Portugal.

Genus V. **Methanolobus** *König and Stetter 1983, 439ᵛᴾ (Effective publication: König and Stetter 1982, 488)*

David R. Boone

Me.tha.no.lo' bus. M.L. neut. n. *methanum* methane; Gr. masc. n. *lobos* lobe; M.L. masc. n. *Methanolobus* methane (-producing) lobe.

Irregular cocci, 0.8–1.25 µm in diameter, sometimes forming loose aggregates (Fig. A2.28). Cells are surrounded by a unit membrane and a protein S-layer (Fig. A2.29). **Flagella, when present, are monotrichous** (Fig. A2.30). Stain Gram-negative. **Strictly anaerobic.** Optimum temperature, 37°C; maximum, 40–45°C; minimum, 10–15°C. Fastest growth occurs at marine salinity. **Catabolic substrates are methanol, methylamines, and sometimes methyl sulfides.** No growth on H_2/CO_2, formate, acetate, or alcohols other than methanol (König and Stetter, 1982; Zellner and Winter, 1987a). In the presence of molecular sulfur, H_2S, in addition to methane and CO_2, is formed from methanol (Stetter and Gaag, 1983). Tungstate is required (Zellner and Winter, 1987b).

The mol% G + C of the DNA is: 39–41.

Type species: **Methanolobus tindarius** König and Stetter 1983, 439 (Effective publication: König and Stetter 1982, 488).

Further descriptive information

Since the previous edition of *Bergey's Manual of Systematic Bacteriology*, one species was transferred out of *Methanolobus* and an-

other species was transferred into the genus (see Taxonomic Comments), leaving the remaining species as a coherent group based on phylogenetic analysis of 16S rRNA sequences. The RNA polymerase of *M. vulcani* shows serological cross-reaction with that of *Methanosarcina barkeri* (Thomm and Stetter, unpublished observation). Antibodies prepared against whole cells of *Methanosarcina* species show a slight cross-reaction with *M. tindarius*. On the other hand, there is no serological cross- reaction of antibodies prepared against *M. tindarius* with members of any other genera of methanogens except *Methanococcoides* (Macario and Conway de Macario, 1983; König and Stetter, unpublished observation).

Cells stain Gram-negative, but there is no outer membrane. The cell wall morphology is an S-layer (König and Stetter, 1986), the major protein of which, in the case of *M. tindarius*, is a glycoprotein with apparent molecular mass of 156 kDa (Sleytr and Messner, 1983). Cells are lysed by 1% SDS. Grana of polyphosphate and glycogen granules may be present (König et al., 1985).

Growth occurs throughout the pH range of 6.8–9, and cells

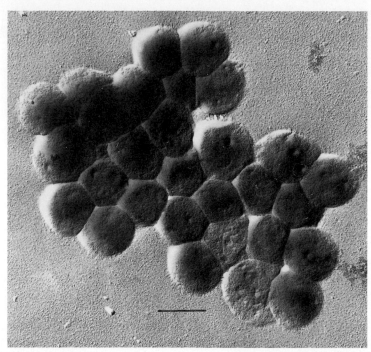

FIGURE A2.28. Platinum-shadowed electron micrograph of a *Methanolobus vulcani* aggregate. Bar = 1 μm.

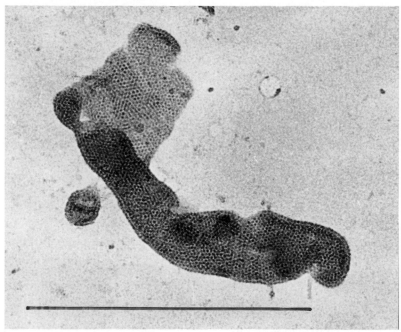

FIGURE A2.29. Cell envelope of *Methanolobus tindarius*. Negative staining with uranyl formate. Bar = 1 μm. (Reprinted with permission from O. Kandler and H. König, 1985. The Bacteria: A Treatise on Structure and Function. ©Academic Press, New York.)

maintain a reversed pH gradient, with their cytosol more acidic than the surrounding environment (Ni et al., 1994a; Kadam and Boone, 1996). *M. tindarius* fixes dinitrogen (König et al., 1985) and contains *nif* genes (Magot et al., 1986).

Methanolobus is commonly found in anaerobic sea sediments with *in situ* temperatures between 10 and 45°C and pH values between 5 and 8. *M. oregonensis* has been found in alkaline aquifers.

ENRICHMENT AND ISOLATION PROCEDURES

Methanolobus may be enriched in habitat-simulating mineral medium to which trimethylamine has been added as a catabolic substrate. Methanol can be used in place of trimethylamine. Antibiotics, such as ampicillin, penicillin, kanamycin, and tetracycline, may be employed to inhibit the growth of nonmethanogenic bacteria. Solidified agar in roll tubes or Petri dishes can be used for isolation under anaerobic conditions. After 5–10 d,

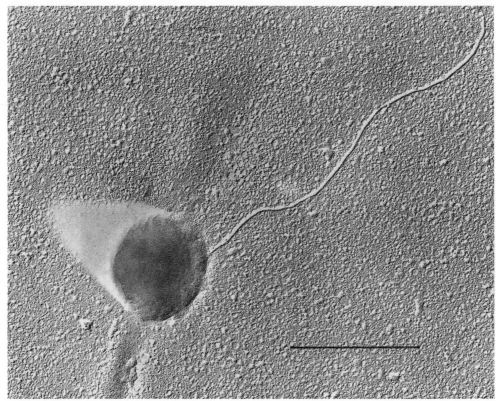

FIGURE A2.30. A platinum-shadowed electron micrograph of a single cell of *Methanolobus tindarius* showing a flagellum. Bar = 1 μm.

visible colonies 1–3 mm in diameter form; they are circular, transparent to translucent, greenish white, ocher, or colorless with entire margins (König and Stetter, 1982; Kadam et al., 1994).

MAINTENANCE PROCEDURES

Methanolobus strains may be grown in medium of marine salinity, such as MSH medium (Ni, 1991), with trimethylamine as a catabolic substrate. For storage, cultures may be frozen by adding cryoprotectant (5% anoxic glycerol), dispensing them to anoxic vials, and cooling them at 1°C/min to −40°C; then they are stored at a lower temperature. Cultures may also be maintained by monthly transfer to fresh medium.

DIFFERENTIATION OF THE GENUS *METHANOLOBUS* FROM OTHER GENERA

The genus *Methanolobus* may be differentiated from all genera outside the family *Methanosarcinaceae* by the ability to form methane from methanol and methyl amines in the absence of added H_2. Within the family *Methanosarcinaceae*, *Methanolobus* can be distinguished from most other genera by being slightly halophilic. The genera *Methanohalophilus*, *Methanohalobium*, and *Methanosalsum* are moderately or extremely halophilic. The genus *Methanosarcina* contains mainly halotolerant (but not halophilic) species, although *Methanosarcina acetivorans* (Sowers et al., 1984a) is a slight halophile. However, *M. acetivorans* differs from *Methanolobus* by using acetate and CO_2/H_2 as an energy source.

The only other slightly halophilic genus in *Methanosarcinaceae* is *Methanococcoides* (Sowers and Ferry, 1983), which shares many features with *Methanolobus*. It is difficult to distinguish these two genera except by phylogenetic means.

TAXONOMIC COMMENTS

Methanolobus siciliae is phylogenetically more closely related to the genus *Methanosarcina* than it is to the genus *Methanolobus*; therefore, it was transferred by Ni et al. (1994b) as *Methanosarcina siciliae*.

Phylogenetic analysis of the type strain of *Methanohalophilus oregonensis* showed that it is more closely related to *Methanolobus* than to *Methanohalophilus* (Boone et al., 1993b), so it is herein transferred to *Methanolobus* as *Methanolobus oregonensis* comb. nov.

FURTHER READING

König, H. and K.O. Stetter. 1982. Isolation and characterization of *Methanolobus tindarius*, sp. nov., a coccoid methanogen growing only on methanol and methylamines. Zentbl. Bakteriol. Mikrobiol. Hyg. 1 Abt Orig. C. *3*: 478–490.

Sowers, K.R. and J.G. Ferry. 1983. Isolation and characterization of a methylotrophic marine methanogen, *Methanococcoides methylutens*, gen. nov., sp. nov. Appl. Environ. Microbiol. *45*: 684–690.

Sowers, K.R., J.L. Johnson and J.G. Ferry. 1984. Phylogenetic relationships among the methylotrophic methane-producing bacteria and emendation of the family *Methanosarcinaceae*. Int. J. Syst. Bacteriol. *34*: 444–450.

DIFFERENTIATION OF THE SPECIES OF THE GENUS *METHANOLOBUS*

Differential characteristics of the species of the genus *Methanolobus* are indicated in Table A2.22.

List of species of the genus Methanolobus

1. **Methanolobus tindarius** König and Stetter 1983, 439[VP] (Effective publication: König and Stetter 1982, 488).

   tin.da'ri.us. M.L. masc. adj. *tindarius* from Tindari, the place of isolation in Sicily, Italy.

   Lacks plasmids (Sowers and Gunsalus, 1988b).

   The mol% G + C of the DNA is: 40 (T_m).

   Type strain: Tindari3, strain T3, ATCC 35996, DSMZ 2278, OCM 150.

   GenBank accession number (16S rRNA): M59135.

2. **Methanolobus bombayensis** Kadam, Ranade, Mandelco and Boone 1994, 606.[VP]

   bom.ba.yen'sis. M.L. masc. adj. *bombayensis* from Bombay, named for its source in the sediments of the Bay of Bombay.

   Cells are irregular, nonmotile coccoids (diameter, 1–1.5 µm) occurring singly. Stain Gram-negative. Lysed by 0.1 g/l SDS. Surface colonies are circular, convex, translucent, and colorless with entire margins and have diameters of 2–3 mm after 7–10 d of incubation. The habitat is marine sediments.

   The mol% G + C of the DNA is: 39.2 (HPLC).

   Type strain: B-1, OCM 438.

   GenBank accession number (16S rRNA): U20148.

   Additional Remarks: Type strain was isolated from Arabian Sea sediments obtained near Bombay, India.

3. **Methanolobus oregonensis** comb. nov. (*Methanohalophilus oregonensis* Liu, Boone and Choy 1990, 115).

   o.re.go.nen'sis. M.L. masc. adj. *oregonensis* from Oregon, named for the state in which it was isolated.

   Irregular cocci, 1.0–1.5 µm in diameter, occurring indi-

vidually or in aggregates. Stain Gram-negative. Nonmotile. Sensitive to lysis by detergents. Cells grow with catabolic substrate plus either vitamins or peptones as sole organic additions to the medium, but yeast extract stimulates growth.

Surface colonies are ~1 mm in diameter after 10 d. They are circular, entire, smooth, and tan.

The mol% G + C of the DNA is: 40.9 (determined chromatographically).

Type strain: WAL-1, DSMZ 5435, OCM 99.

GenBank accession number (16S rRNA): U20152.

Additional Remarks: Enriched and isolated from a slurry of subsurface solids from a saline, alkaline aquifer.

4. **Methanolobus taylorii** Oremland and Boone 1994, 574.[VP]

   tay.lo'ri.i. M.L. gen. n. *taylorii* of Taylor, named in honor of Barrie F. Taylor for his contributions to our understanding of the marine organosulfur cycle and his mentorship of many students and postdoctoral associates in the field of marine microbiology.

   Original material supplemented with material from Oremland et al. (1989) and Kiene et al. (1986).

   Cells are coccoid bodies 0.5–1.0 µm in diameter, sometimes growing in large clumps. Lysed by detergents. Stain Gram-negative. Nonmotile. Biotin and catabolic substrate are the only organic requirements for growth.

   Cells grow and form methane from methylamines, methylsulfides, and methanol. Cells form methane but do not grow on dimethyl selenide and methylmercury; methane is formed from diethyl sulfide.

   Habitat: estuarine sediments.

TABLE A2.22. Diagnostic and descriptive features of *Methanolobus* species[a]

Characteristic	1. *M. tindarius*	2. *M. bombayensis*	3. *M. oregonensis*	4. *M. taylorii*	5. *M. vulcani*
Morphology:					
Coccoid cells	+	+	+	+	+
Loose aggregates	+	−	+	+	+
Motile (monotrichous)	+	−	−	−	−
Substrate:					
Methyl sulfides	−	+	+	+	−
Growth requirements:					
Growth in mineral medium plus catabolic substrate	+	+	−	−	−
Stimulated by other organic components in medium		+	Biotin, thiamine	Biotin	Biotin
Environmental conditions supporting rapid growth:					
Temperature 30–40°C	+	+	+	+	+
pH 5	−	−	−	−	−
pH 6	+	−	−	−	+
pH 7	+	+	−	+	+
pH 8	+	+	+	+	+
pH 9	−	−	+	−	−
0 M NaCl	−	−	+	−	−
0.3–1.0 M NaCl	+	+	+	+	+
Plasmids	−				mMP1
Mol% G + C	46	39	41	41	39

[a]For symbols see standard definitions.

The mol% G + C of the DNA is: 40.8 (determined chromatographically).

Type strain: GS-16, DSMZ 9005, OCM 58.

GenBank accession number (16S rRNA): U20154.

5. **Methanolobus vulcani** Stetter, König and Thomm 1989, 496,[VP]emend. Kadam and Boone 1995, 401 (Effective publication: Stetter 1989a, 2207).

vul.ca' ni. L. gen. n. *vulcani* from Vulcan, the god after whom Vulcano Island was named (Insula Vulcani).

Irregular cocci. Stain Gram-negative. Nonmotile. Lysed by detergents. Biotin and catabolic substrate are only organic requirements for growth, but yeast extract and peptones stimulate growth.

A plasmid (mMP1) with a molecular weight of 4.6×10^6 is present in *M. vulcani* (Thomm et al., 1983).

The mol% G + C of the DNA is: 39 (Bd).

Type strain: PL-12/M, DSMZ 3029, OCM 157.

GenBank accession number (16S rRNA): U20155.

Additional Remarks: Type strain was isolated from sea sediments near Vulcano Island, Italy.

Genus VI. Methanosalsum gen. nov.

DAVID R. BOONE AND CHAD C. BAKER

Me.tha.no.sal' sum. M.L. neut. n. *methanum* methane; L. neut. part. adj. *salsum* salted, salty; M.L. neut. n. *Methanosalsum* the salty methane (bacterium).

The generic description is based on that of the type species, *Methanosalsum zhilinae* (*Methanohalophilus zhilinae* Mathrani, Boone, Mah, Fox and Lau, 1988, 141). Additional material supplied by Kevbrin et al. (1997).

Irregular, angular cocci occurring individually or occasionally in **clumps and tetrads**. Stain Gram-negative. Cell wall is an S-layer. Slowly **motile** by one or two flagella (Fig. A2.31A). Strictly anaerobic. Grows most rapidly at 35–45°C (growth range, 20–50°C), 0.4–0.7 M Na$^+$ (growth range, 0.2–2.1 M Na$^+$), and pH 8.7–9.5 (**growth range, 8–10**). Bicarbonate is required. Energy metabolism is by the formation of **methane from methylamines, methanol, or dimethylsulfide**; acetate, formate, H$_2$/CO$_2$, tetramethylammonium, pyruvate, and alcohols other than methanol are not used.

The mol% G + C of the DNA is: 38–39.5.

Type species: **Methanosalsum zhilinae** comb. nov. (*Methanohalophilus zhilinae* Mathrani, Boone, Mah, Fox and Lau 1988, 141).

FURTHER DESCRIPTIVE INFORMATION

Cells are generally coccoid but very irregular and angular (Fig. A2.31B). They appear to divide by pinching off daughter cells (Fig. A2.31C). The cell wall is an S-layer (Fig. A2.31D) that is easily disrupted by ionic detergents, causing cell lysis. Cells grow in mineral medium, but growth is stimulated by medium enriched with Trypticase Peptone (BBL), yeast extract, or rumen fluid. Stimulation by yeast extract is in part due to the glycine betaine it contains, which is used by *M. zhilinae* as an osmoregulatory compound (Robertson et al., 1990a). However, growth is stimulated by yeast extract even in the presence of glycine betaine. Trimethylamine may serve as a sole carbon and energy source. Dimethylsulfide (5 mM) supports growth and methanogenesis as a sole catabolic substrate; however, higher concentrations (20 mM) are inhibitory.

M. zhilinae may be very resistant to ammonia toxicity, even at high pH (Kadam and Boone, 1996): at pH 9.5, the growth rate of the type strain is only 50% inhibited by 45 mM un-ionized ammonia. Strain Z-7936 is inhibited (methanogenic rate decreased by <50%) by ammonia or methylamine concentrations of 20 mM.

Growth is inhibited by antibiotics that affect ribosomes (chloramphenicol and tetracycline) but not by cell wall active agents, such as penicillin, carbenicillin, and cycloserine.

The sources of the type strain are alkaline and saline sediments from a lake of the Wadi el Natrun in Egypt, and a second strain (Z-7936) was isolated from Lake Magadi, another alkaline soda lake.

ENRICHMENT AND ISOLATION PROCEDURES

Methanosalsum may be enriched from suitable habitats by using culture medium (pH 8–9.5) containing methylamines as a catabolic substrate. NH$_3$ may serve as a nitrogen source, and sulfide may serve as a sulfur source. Isolation can be accomplished by using the same medium in roll tubes with agar added.

MAINTENANCE PROCEDURES

Species of *Methanosalsum* may be stored for several weeks as liquid suspensions, or suspensions can be frozen in the presence of cryoprotectant (5% glycerol) by cooling at 1°C/min and stored at liquid nitrogen temperatures (Boone, 1995); cultures can also be maintained by regular subculturing (Hippe, 1984).

DIFFERENTIATION OF THE GENUS METHANOSALSUM FROM OTHER GENERA

The genus *Methanosalsum* can be separated from genera outside the family *Methanosarcinaceae* by the ability to grow by catabolism of methylamines with the formation of methane. The only methylotrophic methanogen outside *Methanosarcinaceae* is *Methanosphaera*, which differs from *Methanosalsum* by growing only by catabolism of methanol plus H$_2$.

Methanosalsum can be distinguished from most other species of *Methanosarcinaceae* by its environmental requirements of salinity and pH. *Methanosalsum* is moderately halophilic, whereas *Methanohalobium* is extremely halophilic and *Methanosarcina* is halotolerant (Table A2.23). The salinity requirement of *Methanosalsum* is similar to those of *Methanohalophilus, Methanococcoides,* and *Methanolobus*, but these genera generally grow at much lower pH values than does *Methanosalsum*. *Methanosalsum* is most similar in its environmental requirements of pH and salinity to *Methanolobus oregonensis*. *M. oregonensis* has a lower temperature optimum (35–37°C) and a lower maximum salinity (1.7 M Na$^+$) than *Methanosalsum* (45°C and 2 M Na$^+$). Further, *M. oregonensis* (but not *Methanosalsum*) requires thiamin as a growth substrate.

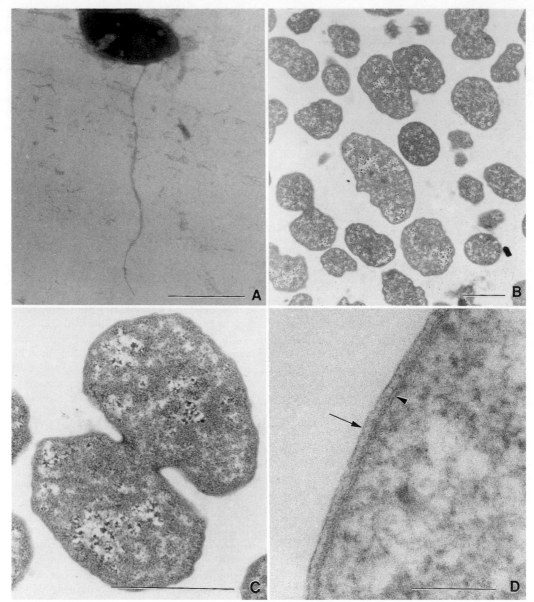

FIGURE A2.31. Transmission electron micrographs of *Methanosalsum zhilinae*. *A*, negatively stained cell showing a single flagellum. *B*, thin section at low magnification showing variation in cell shapes. *C*, thin section of a single cell nearing completion of division. *D*, thin section at high magnification showing plasma membrane (*arrowhead*) and external S-layer (*arrow, left*). In frames *A*, *B*, and *C* the scale markers represent 1 µm; in frame *D* the marker represents 0.1 µm. (Courtesy of Henry C. Aldrich, University of Florida.)

List of species of the genus Methanosalsum

1. **Methanosalsum zhilinae** comb. nov. (*Methanohalophilus zhilinae* Mathrani, Boone, Mah, Fox, and Lau 1988, 141).
   zhi′ li.nae. M.L. gen.n. *zhilinae* of Zhilina; named for Tatjana Zhilina, who made many definitive studies of halophilic methanogens.

   Original material supplemented with material from Boone et al. (1986), Mathrani et al. (1988), Kadam and Boone (1996), and Kevbrin et al. (1997).

   Coccoid cells, 0.8–1.5 µm in diameter, occurring singly, in pairs, and in tetrads. Colonies in roll tubes after 7 d are 0.2–0.4 mm in diameter, yellowish tan, smooth, convex,

opaque, and circular with entire margins. The type strain has the highest resistance to un-ionized ammonia of any known organism (Kadam and Boone, 1996).

   The mol% G + C of the DNA is: 38 (Bd) or 39.5 (T_m); that of strain Z-7938 is 40.1 (T_m).

   Type strain: WeN5, DSMZ 4017, OCM 62.

   Additional Remarks: Type strain was isolated from sediments of a lake in the Wadi el Natrun in Egypt. Strain Z-7936 is a reference strain isolated from sediments of Lake Magadi.

TABLE A2.23. Diagnostic and descriptive features of the genera of *Methanosarcinales*[a]

Characteristic	Methanococcoides	Methanohalobium	Methanohalophilus	Methanolobus	Methanosaeta	Methanosalsum	Methanosarcina
Morphology:							
Coccoid cells	+	+	+	+	–	+	+
Aggregates	+	+	+	+	–	+	+
Cysts	+	–	–	+	–	–	+
Sheathed rods	–	–	–	–	+	–	–
Motile	–	–	–	–	–	–	–
Substrates:							
H_2/CO_2	–	–	–	–	–	–	D
Acetate	–	–	–	–	+	–	+
Methylamines	+	+	+	+	–	+	+
Methanol	+	+	+	+	–	+	+
Methyl sulfides				D		+	
Organic growth factors:							
None	–	D	D	D	D	+	D
Biotin	+	D	D	D	D	–	D
Thiamine	–	D	D	D	–	–	–
Conditions supporting rapid growth:							
10–20°C	D	–	–	–	–	–	–
30–40°C	D	+	+	+	D	+	D
55–60°C	–	–	–	–	D	–	D
pH 5–6	–	–	–	–	–	–	+
pH 6–7	–	–	–	D	D	–	+
pH 7–8	+	+	+	D	D	–	–
pH 8–9	–	–	–	D	D	+	–
pH 9–10	–	–	–	D	–	+	–
0 M NaCl	–	–	–	–	+	–	D
0.3–1.0 M NaCl	+	–	–	+	–	+	D
1–2 M NaCl	–	–	+	–	–	+	–
2–4 M NaCl	–	+	–	–	–	–	–
Mol% G + C	42	37	39–41	39–46	48–54	38	36–43

[a]For symbols see standard definitions.

Family II. **Methanosaetaceae** *fam. nov.*

DAVID R. BOONE, WILLIAM B. WHITMAN AND YOSUKE KOGA

Me.tha.no.sae.ta' ce.ae. M.L. fem. n. *Methanosaeta* type genus of the family; *-aceae* the ending to denote a family; M.L. fem. pl. n. *Methanosaetaceae* the family of *Methanosaeta*.

Sheathed rods. Nonmotile. Stain Gram-negative. **Sole energy source is acetate**, which is split into methane and CO_2. **Strict anaerobes.** Lipids contain *myo*-inositol, ethanolamine, and galactose as polar head groups.

Habitat: Anaerobic sediments and anaerobic sewage sludge digestors.

Type genus: **Methanosaeta** Patel and Sprott 1990, 80.

Genus I. **Methanosaeta** *Patel and Sprott 1990, 80*[VP]

GIRISHCHANDRA B. PATEL

Me.tha.no.sae' ta. M.L. neut. n. *methanum* methane; Gr. n. *saeta* bristle; M.L. fem. n. *Methanosaeta* methane (-producing) bristle.

Straight rods with flat ends; single cells are usually 0.8–1.3 μm wide by 2.0–7.0 μm long and are enclosed within a tubular sheath structure. Forms short (~5–25 μm) to long (average of up to 150 μm and longer) flexible chains of cells within the sheath that may resemble filaments (Figs. A2.32 and A2.33). Gas vacuoles are generally observed in thermophilic strains but have so far not been observed in mesophilic strains. **Obligate anaerobe.** Nonmotile; stain Gram-negative. Optimum temperature for mesophilic strains, 35–40°C with a range of 10–45°C; for thermophilic strains, 55–60°C with a range of 30–70°C. Optimum pH: 6.5–7.5 with a range of 5.5–8.4.

Acetate is the only substrate used as an energy source and is almost stoichiometrically converted to methane and carbon dioxide.

Acetate and CO_2 serve as carbon sources.

The mol% G + C of the DNA is: 48–54.

Type species: **Methanosaeta concilii** (Patel 1984) Patel and Sprott 1990, 80 (*Methanothrix concilii* Patel 1984, 1394).

FURTHER DESCRIPTIVE INFORMATION

The structural architecture of *Methanosaeta* is unique. The tubular sheath enclosing the cells has regular striations (Patel et al., 1986; Zinder et al., 1987; Kamagata et al., 1992) and is made of annular hoops stacked together (Beveridge et al., 1986b; Patel et al., 1986), somewhat similar to the sheath of *Methanospirillum hungateii* (Zeikus and Bowen, 1975b). The individual cells within the continuous sheath have a thin granular "wall" layer (Beveridge et al., 1986b; Zinder et al., 1987) and are separated from

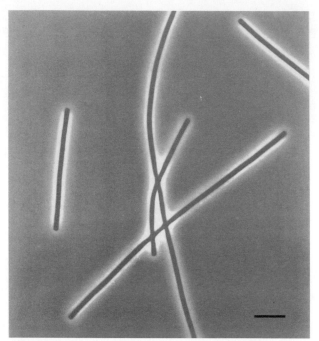

FIGURE A2.32. *Methanosaeta concilii* culture observed under phase contrast microscopy. Bar = 5 μm.

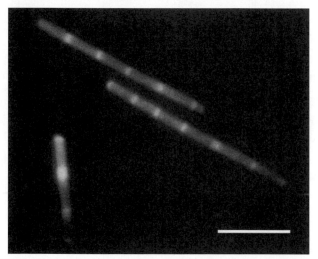

FIGURE A2.33. Visualization of spacer plugs in *Methanosaeta concilii* under UV fluorescence after staining with the fluorescent dye Calcofluor White M2R. The individual cells in the filament, as judged from the distance between the spacer plugs, are about 2.2 μm long. Bar = 5 μm. (Reprinted with permission from Patel, Canadian Journal of Microbiology *30:* 1383–1396, 1984, ©National Research Council.)

one another by structures referred to as spacer plugs or cross walls (Patel, 1984; Beveridge et al., 1986b; Zinder et al., 1987) (Fig. A2.34A and B). Each end of the sheath, and hence the cell filament, is also blocked by a spacer plug (Beveridge et al., 1986b). The spacer plugs of *Methanosaeta* consist of a series of concentric rings (Fig. A2.35A), and they are not as complex as those of *Methanospirillum* (Fig. A2.35B), whose spacer plugs are multilayered and possess a hexagonal symmetry (Beveridge et al., 1986b; Patel et al., 1986). The sheaths of both *Methanosaeta* and *Methanospirillum* are predominantly proteinaceous and con-

tain carbohydrates, but both sheath types have distinct chemical compositions (Kandler and König, 1978; Patel et al., 1986). Although the sheaths and cell spacers of *Methanosaeta* and *Methanospirillum* are quite resilient to physical and chemical stressors, those of *Methanosaeta* were the most resilient, requiring harsher alkali treatment for lysis and sheath isolation (Patel et al., 1986).

Replication in mesophilic species of *Methanosaeta* requires both cell division and breakage of the sheath, is unlike that described for any eubacterium, and represents a new form of procaryotic division (Beveridge et al., 1986a). Although the mechanism of replication in thermophilic species of *Methanosaeta* has not been studied, based on the similarities in the gross ultrastructures of mesophilic and thermophilic species (Patel, 1984; Beveridge et al., 1986a, b; Zinder et al., 1987), it would be reasonable to suggest a similar mechanism.

The polar membrane lipids of mesophilic strain GP6 of *Methanosaeta* have been studied extensively and have been shown to contain core lipid structures based on $C_{20}C_{20}$-diether (archaeol; 2,3-di-O-phytanyl-sn-glycerol) and its 3'-hydroxy derivative (Ferrante et al., 1988a, b, 1989).

Methanosaeta can be cultivated in a defined mineral salts medium containing vitamins and acetate as the sole organic compounds (Patel, 1984; Patel et al., 1988; Zinder et al., 1987). Energy for growth is derived from the aceticlastic reaction (Patel, 1984; Zinder et al., 1987; Kamagata et al., 1992), which results in the almost stoichiometric formation of methane and carbon dioxide from acetate. Methane is almost exclusively derived from the methyl carbon and carbon dioxide from the carboxyl carbon of acetate (Patel, 1984; Ekiel et al., 1985; Zinder et al., 1987). Both carbons of acetate, as well as carbon from the bicarbonate in the medium, are incorporated into cell carbon (Patel, 1984; Ekiel et al., 1985).

Acetate is the only substrate that supports growth and methane production. Formate, H_2/CO_2, methanol, and methylamines do not serve as substrates for growth or methane production (Patel, 1984; Zinder et al., 1987; Kamagata and Mikami, 1991). In the presence of acetate, formate may be split into H_2 and CO_2 by some mesophilic strains, but this is not converted to methane (Patel, 1984). Formate is not split by thermophilic strains isolated to date (Zinder et al., 1987; Kamagata and Mikami, 1991). H_2 and CO_2 do not inhibit growth or methanogenesis from acetate (Patel, 1984; Zinder et al., 1987; Kamagata and Mikami, 1991). *Methanosaeta*, similar to *Methanosarcina*, uses an incomplete tricarboxylic acid pathway, operating in the oxidative direction, for the synthesis of glutamate and amino acids derived from glutamate (Ekiel et al., 1985). This is contrary to the incomplete TCA pathway operating in the reductive direction in most other methanogens (Ekiel et al., 1985; Sprott et al., 1993). The F_{420} contents in *Methanosaeta* species vary, and they are much lower than those reported in other methanogen species (Kamagata and Mikami, 1991), which may help explain the weak to no observed autofluorescence when cells are excited with UV light at 350-nm wavelength (Patel, 1984; Zinder et al., 1987).

Nutritional requirements and factors affecting methanogenesis have been most extensively studied for the mesophilic strain GP6 (Patel, 1984; Patel et al., 1988, 1991b; Patel and Sprott, 1991; Sprott and Patel, 1986; Baudet et al., 1988) and less so for the thermophilic strains P_T and CALS-1 (Zinder et al., 1987; Kamagata and Mikami, 1991). Vitamins are required for growth and/or are stimulatory for some strains. Sludge fluid may be stimulatory, and yeast extract may be inhibitory, for some strains.

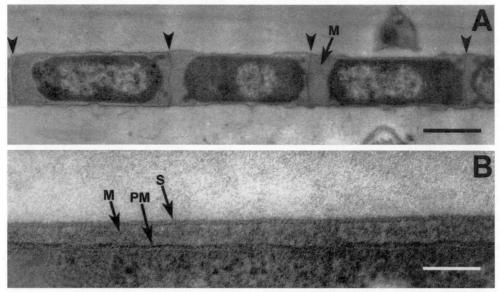

FIGURE A2.34. Ultrastructure of *Methanosaeta concilii*. *A*, thin section of three cells within a filament. The entire filament is bounded by a tubular sheath and each cell is separated from the next by a spacer plug (*arrowheads*). Surrounding each cell is an amorphous granular matrix (*M*). Bar = 1 µm. *B*, higher magnification showing the enveloping layers in thin section. The plasma membrane (*PM*) has an unusually thick and electron-dense surface leaflet. *M*, matrix; *S*, sheath. Bar = 100 nm. (Reprinted with permission from Beveridge et al., Canadian Journal of Microbiology *32*: 703–710, 1986, ©National Research Council.)

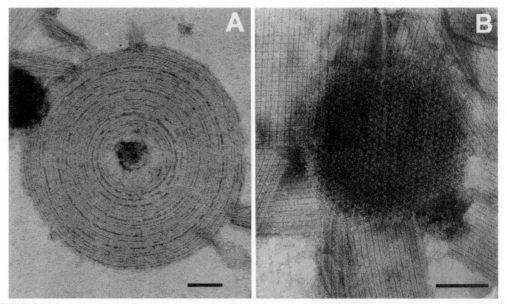

FIGURE A2.35. *A*, negatively stained spacer plug from *Methanosaeta concilii* showing the concentric rings which help make up its fabric. Bar = 100 nm. *B*, negatively stained spacer plug from *Methanospirillum hungateii* GP1 showing the hexagonal arrangement of subunits and the moiré effect of the multiple layers. Fragments of sheath surround the spacer plug. Bar = 100 nm. (Reprinted with permission from Patel et al., Canadian Journal of Microbiology *32*: 623–631, 1986, ©National Research Council.)

Sodium sulfide has been shown to serve as the sulfur source for growth. Comparison of the kinetics of acetate utilization in thermophilic strains of *Methanosarcina* and *Methanosaeta* supports the model in which higher acetate concentrations in the growth environment would favor growth of the former and lower concentrations (below 1–3 mM) would favor the growth of the *Methanosaeta* (Min and Zinder, 1989). Comparison of kinetics of acetate utilization in mesophilic strains of *Methanosarcina* (Smith and Mah, 1978) and *Methanosaeta* (Patel, 1984) also suggest that the former would be favored at higher acetate concentrations.

The protein patterns of whole cells of the thermophilic strains P_T and CALS-1 of *Methanosaeta* were almost identical and were distinct from that of the mesophilic strain GP6 (Kamagata et al., 1992). The alignment of the partial 16S rRNA sequences obtained from the mesophilic strain GP6 and the thermophilic strains P$_T$ and CALS-1 indicated 99% similarity among the thermophilic strains and 94–95% similarity between the two thermophilic strains and the mesophilic strain (Kamagata et al., 1992). Immunological analyses of the mesophilic strain GP6 and the thermophilic strain CALS-1 of *Methanosaeta*, using poly- and monoclonal antibody probes for 29 reference methanogens, indicated that although the two strains were sufficiently interrelated to form an immunologically cohesive group, they were antigenically distinct from one another and also from the rest of the methanogens (Macario and Conway de Macario, 1987). *Methanosaeta* strains exhibited a weak immunologic reaction to *Methanosarcina* alone, but were nonreactive with other methanogens, including the sheathed methanogen *Methanospirillum hungateii* (Patel, 1984; Macario and Conway de Macario, 1987). Comparison of small-subunit rRNA sequences of 20 methanogens in the order *Methanomicrobiales* showed that strain CALS-1 was in the "methanosarcina" cluster of the phylogenetic tree (Rouvière et al., 1992). However, the genera *Methanosaeta* and *Methanosarcina* are morphologically distinct and, unlike some *Methanosarcina* species which can grow and produce methane using the aceticlastic reaction in addition to growth on other methanogenic substrates, *Methanosaeta* strains can only grow by the aceticlastic reaction. These data support the placement of the mesophilic and thermophilic strains of aceticlastic rod-shaped methanogens in one distinct genus, *Methanosaeta*, but as two separate species.

Methanosaeta can be isolated from mesophilic and thermophilic anaerobic digestors treating domestic wastes; from granular sludge, anaerobic fixed-bed reactors, and other types of anaerobic systems; and from acetate enrichments inoculated with cultures from anaerobic waste treatment digestors. *Methanosaeta*-like methanogens have been observed in, or enriched from, anoxic mud from a river (Westermann et al., 1989b) and from thermophilic environments, such as Icelandic hot springs (Sonne-Hansen and Ahring, 1997) and thermal lakes (Nozhevnikova and Chudina, 1985).

ENRICHMENT AND ISOLATION PROCEDURES

Methanosaeta can be enriched, using appropriate inoculum from a natural habitat, in anaerobic (reduced with cysteine-HCl/Na$_2$S and under a N$_2$/CO$_2$ headspace atmosphere) defined mineral salts media (pH, 7.0 ± 0.5) containing acetate as the main organic source of carbon and energy for growth (Patel, 1984). Vitamins should be included, since some strains have shown a requirement and others are stimulated. Yeast extract should be avoided, since it may be inhibitory. Axenic culture is usually successful when highly enriched cultures (from batch or continuous culture) are serially diluted in the presence of antibiotics, such as vancomycin, which are effective in eliminating nonmethanogenic contaminants but do not inhibit the methanogens (Patel, 1984; Zinder et al., 1987; Kamagata and Mikami, 1991). Purity of culture maintained in antibiotic-free medium for several transfers should be checked in anaerobic, complex growth media containing substrates (in addition to acetate) that would be expected to support the growth of contaminating bacteria. Under phase-contrast microscopic observation of such test cultures, it is easy to distinguish any contaminants from the distinct morphology of *Methanosaeta* cells. All axenic cultures available to date, from recognized, open culture collections, have been obtained by the serial dilution technique. Attempts to isolate *Methanosaeta* from colony growth on solid media, as well as to obtain viable

colonies from the currently available axenic cultures, have so far been unsuccessful.

MAINTENANCE PROCEDURES

Working cultures of *Methanosaeta* can be maintained by transfer into appropriate broth media at regular intervals of 1–4 weeks, depending on the strain. Often large inocula (10–50%) are required. Longer-term preservation of stock cultures can be accomplished by using traditional lyophilization techniques, but under strictly anaerobic conditions.

TAXONOMIC COMMENTS

In the first edition of *Bergey's Manual of Systematic Bacteriology*, aceticlastic, rod-shaped methanogens were described under the genus *Methanothrix*, with *Methanothrix soehngenii* as the type species for the genus (Zehnder, 1989). The original description of *Methanothrix* and its type species, *Methanothrix soehngenii*, was based on strain Opfikon (Zehnder et al., 1980; Huser et al., 1982). The strain was probably impure, making both the genus and species illegitimate (Patel and Sprott, 1990; Boone, 1991; Patel, 1992). Also, DSMZ 2139 (generally assumed to represent strain Opfikon) and DSMZ 3013 (referred to as strain FE) cultures of *Methanothrix soehngenii* quoted in the last edition of *Bergey's Manual of Systematic Bacteriology* are not pure cultures, and strain VNBF was never deposited in a culture collection (Touzel et al., 1988; Boone, 1991; Patel, 1992). An axenic culture of mesophilic, aceticlastic, rod-shaped methanogens was described by Patel, (1984, 1985) as *Methanothrix concilii* based on the description of strain GP6 as the type strain for the species, and this culture was deposited in several open, permanently established culture collections (ATCC 35969, DSMZ 3671, OCM 69, NRC 2989), where its axenic nature has been verified (Patel, 1992). Later, Patel and Sprott (1990) assigned strain GP6 to a new genus, *Methanosaeta*, with *Methanosaeta concilii* as the type species and strain GP6 as the type strain for the species. However, based on comparative studies of strain GP6 (DSMZ 3671) and DSMZ 2139 culture (strain Opfikon), it was suggested that these were subjective synonyms (see discussion in Boone, 1991; Patel, 1992). In response to a Request for Opinion (Boone, 1991), the Judicial Commission of the International Committee on Systematic Bacteriology denied the request to assign strain GP6 as a neotype for *Methanothrix soehngenii* (Wayne, 1994). This action reaffirmed strain GP6 as the type strain of *Methanosaeta concilii* and reaffirmed *M. concilii* as the type species of the genus *Methanosaeta*.

Thermophilic strains of aceticlastic rod-shaped methanogens were described as *Methanothrix thermophila*, with strain P$_T$ as the type strain for the species (DSMZ 6194) and strain CALS-1 (DSMZ 3870), isolated by Zinder et al. (1987), as another strain of *Methanothrix thermophila* (Kamagata et al., 1992). Kamagata et al. (1992) also suggested that the name "*Methanothrix thermoacetophila*" ("*Methanosaeta thermoacetophila*") be rejected because its description (Nozhevnikova and Chudina, 1985) was based on the impure culture strain Z-517. In their paper, Nozhevnikova and Chudina (1985) stated that a pure culture was not available. Although an axenic culture of Z-517 was later deposited with the DSMZ as DSMZ 4774 (see personal communication in Patel and Sprott, 1990), the species description is not valid because it was based on an impure culture referred to as strain Z-517 (Kamagata et al., 1992; Boone and Kamagata, 1998). It has recently been proposed that *M. thermophila* be transferred to the genus *Methanosaeta* as *Methanosaeta thermophila* comb. nov. (Boone and Kamagata, 1998). Boone and Kamagata (1998) have also submitted a Request for Opinion to reject the species *Methanothrix soehngenii* and the genus *Methanothrix* as *nomina confusa* and to place these two names on the list of rejected names (*nomen rejiciendum*). On the basis of the proposal of Boone and Kamagata (1998), thermophilic strains originally validly described as *Methanothrix thermophila* (Kamagata et al., 1992) are included in the current chapter as *Methanosaeta thermophila*. The Judicial Commission's decision on the Request for Opinion regarding rejection of the names *Methanothrix soehngenii* and *Methanothrix* should not have

a bearing on the proposal to transfer *Methanothrix thermophila* to the genus *Methanosaeta* as *Methanosaeta thermophila* comb. nov. Descriptions of aceticlastic, rod-shaped, sheathed methanogens that were based on impure cultures and/or on cultures that were never deposited in an open, permanently established culture collection (irrespective of what they had been called in the past) have been omitted in the current chapter. Hence, to date, *Methanosaeta* includes two species, *M. concilii* (the type species) and *M. thermophila*.

DIFFERENTIATION OF THE SPECIES OF THE GENUS *METHANOSAETA*

Some differential characteristics of the species of *Methanosaeta* are shown in Table A2.24.

List of species of the genus Methanosaeta

1. **Methanosaeta concilii** (Patel 1984) Patel and Sprott 1990, 80[VP] (*Methanothrix concilii* Patel 1984, 1394).
   con.ci' li.i. L. gen. n. *concilii* of a council, named after the National Research Council of Canada, in whose laboratories the type strain was isolated.

   Cells are 0.8 μm wide by 2–6 μm long, usually growing as chains in long filaments up to 150 μm long, especially in stationary cultures in which pH is not controlled. Much shorter filaments (5–30 μm) are seen in fermenter cultures in which optimum pH and acetate levels are maintained to prevent growth limitation (personal observations). Growth in stationary broth cultures occurs as loose or fluffy sediment or mat forms, both of which can be easily dispersed by gentle shaking. Acetate is the only substrate that supports growth and methanogenesis; formate, acetate, and CO_2 can serve as carbon sources for growth. In cultures started from small inocula (<10% [v/v]), CO_2 or sodium carbonate or bicarbonate is required for initiation of growth. Ammonium serves as the sole nitrogen source; organic sources, such as cysteine, alanine, phenylalanine, glutamate, aspartate, or methylamine, do not. Sodium sulfide, cysteine-HCl, or glutathione can serve as the sole sulfur source. An unspecified vitamin(s) is required for growth. Growth is stimulated by biotin, thiamine-HCl, *p*-aminobenzoic acid, and sludge fluid. Calcium and magnesium are essential; cobalt, nickel, zinc, iron, and

manganese stimulate growth. There seems to be a specific system for nickel uptake; labeled nickel is found to be associated with two soluble proteins having carbon monoxide dehydrogenase and methyl-coenzyme M reductase activities. Methanogenesis from acetate, in thoroughly washed cell suspensions, requires the presence of both sodium chloride and cobaltous chloride (lithium chloride and methylcobalamin can be substituted for NaCl and $CoCl_2$, respectively). In short-term incubation studies (up to 80 min), severe inhibition of methanogenesis by >30 mM ammonium (NH_4Cl) is significantly reversed by $CaCl_2$ (20–40 mM) or $MgCl_2$ (5 mM). Longer-term growth studies (21 d) indicate that 120 mM NH_4Cl causes ~80% inhibition of growth and methanogenesis. As little as 0.05% (w/v) yeast extract causes severe inhibition of growth and methane production. Methyl or benzyl viologen (5 μM), chloroform (50 μM), potassium cyanide (100 μM), 2-bromoethanesulfonic acid (10 μM), D-cycloserine (0.1 mg/ml), and 1.0 mg of vancomycin or kanamycin/ml causes severe or complete inhibition of growth. Penicillin G, bacitracin (1 mg/ml), and ampicillin (0.09 mg/ml) cause minimal or no inhibition. Toxicity studies with eight benzene ring compounds indicate that relatively low concentrations of abietic acid and pentachlorophenol cause severe inhibition of growth and methane production.

The metabolism of 1 mol of acetate results in the production of ~1 mol each of methane and CO_2 and 1.13–1.16 g (dry weight) of cells. K_S for acetate is 1.2 mM. The 2-mercaptoethanesulfonic acid (HSCoM) content in GP6 cells is ~0.45 nmol/mg (dry weight) of cells.

The mol% G + C of the DNA is: 49.0 ± 1.25% (T_m).

Type strain: GP6, ATCC 35969, DSMZ 3671, NRC 2989, OCM 69.

2. **Methanosaeta thermophila** (Kamagata, Kawasaki, Oyaizu, Nakamura, Mikami, Endo, Koga and Yamasato 1992) Boone and Kamagata 1998, 1080[VP] (*Methanothrix thermophila* Kamagata, Kawasaki, Oyaizu, Nakamura, Mikami, Endo, Koga and Yamasato 1992, 465).
   ther.mo.phi' la. Gr. adj. *thermos* hot; Gr. adj. *philos* loving; M.L. fem. adj. *thermophila* heat loving

   Single cells are 0.8–1.3 μm wide and 2–6 μm long. Sheathed chains of cells are usually 10–100 μm long and sometimes more than 100 μm long. Strain CALS-1 cells are usually 2–5 cell units long; longer filaments are rare. Filament length may be related to growth conditions, as seen for *M. concilii*. Broth cultures grow as loose sediment and exhibit

TABLE A2.24. Differential characteristics of the species of the genus *Methanosaeta*[a, b]

Characteristic	1. *M. concilii*	2. *M. thermophila*
Growth temperature, °C:		
Optimal	35–40	55–60
Range	>10 to ≤45	>30 to ≤70
Growth pH:		
Optimal	7.1– 7.5	6.5–6.7
Range	≤6.6 to >7.8	>5.5 to ≤8.4
Mass doubling time (h)	65–70[c]	24–36
Gas vesicles	NR	+
Formate split into H_2/CO_2	+	−
Inhibition by 0.1% yeast extract (w/v)	+	±
Requires vitamin(s)	+	±
Mol% G + C	49.0 ± 1.25[c] (T_m)	52.7–54.2 (HPLC)

[a]Symbols: +, positive; −, negative; ±, some strains reported to possess the trait; NR, not reported to date.

[b]Data for *M. concilii* are based on strain GP6 and are from Patel (1984), except as indicated otherwise; data for *M. thermophila* are based on strains P_T (Kamagata and Mikami, 1991; Kamagata et al., 1992) and CALS-1 (Zinder et al., 1987).

[c]Data from Patel and Sprott (1990).

opalescent turbidity upon gentle shaking. Strain P_T shows blue-green autofluorescence under epifluorescence microscopy; autofluorescence is not seen with strain CALS-1 at 420 or 350 nm. Gas vesicles have been observed in both strains. Acetate is the only substrate used for growth and methane production; H_2/CO_2, formate, methanol, methylamines, and methanol do not support growth or methane production. Growth rate is enhanced by HSCoM, but in the presence of sodium sulfide and cysteine, HSCoM and cysteine appeared to serve primarily as reducing agents rather than as sulfur sources for growth of strain CALS- 1. Sodium sulfide serves as the sulfur source for strain CALS- 1; >1 mM sulfide is inhibitory. Biotin is required for growth of strain CALS-1. Growth of strain CALS-1 does not occur in the presence of 1 mM of 2-mercaptoethanol. Yeast extract (0.1%) is inhibitory to strain P_T. Chloramphenicol (5 μg/ml), bacitracin, neomycin, cycloserine (10 μg/ml), tetracycline, and kanamycin (100 μg/ml) severely inhibit growth of strain P_T; streptomycin, vancomycin (100 μg/ml), and penicillin G (500 μg/ml) cause little or no inhibition. Strain CALS-1 does not grow below pH 7 when bicarbonate-CO_2 buffer is replaced with 20 mM phosphate buffer, and this was presumed to be due to phosphate toxicity at lower pH values.

Utilization of acetate by strain CALS-1 was concentration independent down to 0.1–0.2 mM acetate, and the threshold values were as low as 12–21 μM. The growth yield of strain CALS-1 is 1.1 ± 0.1 g (dry weight) per mole of methane produced from acetate.

Comparison of the N-terminal amino acid sequences of the two major subunits (67 and 52 kDa) of membrane ATPase of strain P_T, as well as other assays, indicated similarities to the ATPase of *Methanosarcina barkeri*, as well as to other archaeal ATPases which belong to the V-type ATPase family (Inatomi et al., 1993).

The mol% G + C of the DNA is: 52–54 (HPLC).

Type strain: PT, DSMZ 6194, OCM 780.

Additional Remarks: Additional strains include CALS-1, DSMZ 3870.

Class III. Halobacteria class. nov.

WILLIAM D. GRANT, MASAHIRO KAMEKURA, TERRY J. McGENITY AND ANTONIO VENTOSA

Ha.lo.bac.te' ri.a. M.L. fem. pl. n. *Halobacteriales* type order of the class; dropping the ending to denote a class; M.L. fem. pl. n. *Halobacteria* the class of *Halobacteriales*.

Description is the same as that of *Halobacteriales*, the sole order of the class.

The mol% G + C of the DNA is: 59–71 (major component) and 51–59 (minor component).

Type order: **Halobacteriales** Grant and Larsen 1989b, 495[VP] (Effective publication: Grant and Larsen 1989a, 2216).

Order I. Halobacteriales Grant and Larsen 1989b, 495[VP] (Effective publication: Grant and Larsen 1989a, 2216)

WILLIAM D. GRANT, MASAHIRO KAMEKURA, TERRY J. McGENITY AND ANTONIO VENTOSA

Ha.lo.bac.te.ri.a' les. M.L. neut. n. *Halobacterium* type genus of the order; *-ales* ending to denote an order; M.L. fem. pl. n. *Halobacteriales* the *Halobacterium* order.

Rods, cocci, a multitude of pleomorphic forms including flat disks, triangles, and squares. Resting stages are not known, although there are reports of structures referred to as halocysts in some strains. **Non-motile or motile by tufts of flagella.** Stain Gram-negative or Gram-positive (after fixation in 2% (w/v) acetic acid). **Aerobic or facultatively anaerobic** with or without nitrate. Some strains have a fermentative mode of growth on arginine.

Require at least 1.5 M NaCl for growth. Most strains grow best at 3.5–4.5 M NaCl. **Colonies of most strains are various shades of red due to the presence of C_{50} carotenoids** (bacterioruberins) that impart red or pink coloration to mass developments in the natural environment. Retinal-based pigments capable of producing light-dependent movements of ions across the cell membrane or functioning as photosensors are present in many strains. One of these pigments, bacteriorhodopsin, acts as a proton pump driven by light energy. Optimum temperature for growth: 35–50°C. Chemoorganotrophic, using amino acids or carbohydrates as carbon source. Have an RNA polymerase of the ABB''C type.

Occur ubiquitously in nature where the salt concentration is high, i.e., in salt lakes, soda lakes, and salterns. Are common in crude solar salts and proteinaceous products (fish and hides) heavily salted with solar salt. Also present in subterranean salt deposits.

The cell's DNA commonly comprises a major and a minor component, with the minor component making up to 10–30% of the total DNA. Many strains harbor large plasmids (>100 kb).

The mol% G + C of the DNA is: 59–71 (major component) and 51–59 (minor component).

Type genus: **Halobacterium** Elazari-Volcani 1957, 207, emend. Larsen and Grant 1989, 2222.

FURTHER DESCRIPTIVE INFORMATION

The organisms are known by the trivial name "halobacteria". Also sometimes known as "haloarchaea", although it should be remembered that there are also moderately halophilic methanogenic archaea, and one genus (*Methanohalobium*) is extremely halophilic, growing most rapidly from 3 M NaCl to saturation (Zhilina and Zavarzin, 1990).

The halobacteria can be unequivocally distinguished from other extremely halophilic procaryotes by their archaeal characteristics, particularly the possession of ether-linked phosphoglycerides readily detectable by the procedures of Ross et al. (1985), Torreblanca et al. (1986a), and Kamekura and Dyall-Smith (1995). The lipids of all halobacteria examined to date contain phytanyl ether analogs of phosphatidyl glycerol and phosphatidyl glycerol phosphate methyl ester. Many strains also contain phosphatidyl glycerol sulfate. One or more glycolipids and sulfated glycolipids are also present in most strains including a sulfated tetraglycosyl diether, triglycosyl diethers, and diglycosyl diethers. Structures are detailed in Kates (1993) and typical thin layer chromatographic properties are shown in Kamekura and Dyall-Smith (1995) and Holt et al. (1994). All halobacteria have diphytanyl ($C_{20}C_{20}$) glycerol ether core lipids, but some strains have additional phytanyl-sesterterpanyl ($C_{20}C_{25}$) glycerol ether core lipids (De Rosa et al., 1982; Kamekura and Dyall-Smith, 1995), and certain strains of haloalkaliphiles have di-sesterterpanyl ($C_{25}C_{25}$) glycerol ether lipids (De Rosa et al., 1983). Bidiphytanyl tetraethers ($C_{40}C_{40}$) have not been detected in any halobacteria to date. Isoprenoid quinones are of the menaquinone type (MK-8 and MK-8(H_2); Collins et al., 1981), not the ubiquinone type.

In subsequent chapters describing the different genera of halobacteria, the following polar lipid abbreviations have been used: PG, phosphatidyl glycerol; PGP-Me, phosphatidyl glycerol phosphate methyl ester; PGS, phosphatidyl glycerol sulphate; DGD, diglycosyl diether; TGD, trigylcosyl diether; TGD-1, galactosyl mannosyl glucosyl diether; TGD-2, glucosyl mannosyl glucosyl diether; S-DGD, sulfated diglycosyl diether; S-TGD, sulfated triglycosyl diether; S- TeGD, sulfated tetraglycosyl diether; S2-DGD, disulfated diglycosyl diether.

The most striking feature of the halobacteria is their absolute requirement for high concentrations of NaCl. Although some strains may grow at salt concentrations as low as 1.5 M, most of the strains grow best at concentrations of 3.5–4.5 M and grow well in saturated NaCl (5.2 M). To compensate for the high salt concentrations in the environment, the organisms accumulate mainly KCl, up to 5 M (Matheson et al., 1976) and may be growth limited by the amount of KCl in media. NMR studies are conflicting over whether most of the KCl is in the free state (Shporer and Civan, 1977), or bound within the cells (Ginzburg, 1978). The functional and structural units of the cells are adapted to these high salt concentrations. Most intracellular enzymes have a requirement for high levels of KCl (Kushner, 1985) although normally NaCl will substitute. However, NaCl is required for the stabilization of the cell walls of noncoccoid types and KCl will not substitute. The molecular basis of the salt requirement for protein stability and function is not fully understood, although halobacterial proteins are considerably more acidic overall than non-halophilic counterparts. Detailed structural investigations of a few highly purified halobacterial proteins (e.g., malate dehydrogenase, dihydrofolate dehydrogenase, dihydrolipoamide dehydrogenase, and ferredoxin) indicate that the high concentrations of negative charges in such proteins is associated with separate and distinct domains within these proteins. Denatured halobacterial proteins lose the ability to retain large amounts of water and salt (Werber et al., 1986), and undergo a remarkable

loss of secondary and tertiary structure. The main stabilization mechanism consists of the formation of co-operative hydrate bonds between the network of these surface acidic groups and hydrated salt ions maintaining these stable and soluble in conditions where non-halophilic proteins would precipitate. Halobacterial proteins interact much more strongly with water than their non-halophilic counterparts (Dym et al., 1995; Madern et al., 1995; Pieper et al., 1998).

Noncoccoid strains assume a variety of forms from rods to flat irregular discs (see Grant and Larsen, 1989a). Indeed, some species of *Haloarcula* have flat triangular cells (Takashina et al., 1990) and sheets of flat square microorganisms (presumably halobacteria), first observed in hypersaline brines by Walsby (1980), have since been found in a variety of saturated brines (Oren, 1994). The lack of turgor pressure within halobacterial cells presumably enables the cells to tolerate the formation of corners. The unifying feature of these halobacteria is their flatness, the function of which remains unknown, though it is conceivable that enhanced surface area facilitates oxygen transfer in oxygen-depleted brines.

Whereas the cell envelopes of the coccoid halobacteria (genera *Halococcus* and *Natronococcus*) are stable in the absence of salts, those of noncoccoid isolates maintain their integrity only in the presence of high concentrations of NaCl or KCl. The surface of the cell envelope of noncoccoid species has a hexagonal pattern due to the regular packing of glycoprotein subunits which are held together only in the presence of salt. Because of the proteinaceous nature of the cell surface, noncoccoid halobacterial cells are susceptible to the attack by proteolytic enzymes, thus leading to the lysis of the cells (Kamekura and Seno, 1989). Upon gradual dilution of the growth medium with water, the cells change their shape, through irregular forms to spheres which undergo lysis. The composition of the growth medium, particularly Mg^{2+}, affects the point at which lysis occurs. There are thus reports of noncoccoid isolates being relatively stable in sea water (Torreblanca et al., 1986a) which has a relatively high Mg^{2+} concentration, whereas these isolates lyse in NaCl solutions at sea-water concentration. Some species of halobacteria are reported as forming halocysts and thallus-like structures within a common capsule that may be stable to hyposaline conditions (Kostrikina et al., 1991; Cline and Doolittle, 1992). There have also been isolations of noncoccoid types from relatively low salt environments (McGenity et al., 1998) and recovery of halobacterial 16S rRNA gene clones from apparently low salt environments (Jurgens et al., 1997; Munson et al., 1997).

Coccoid halobacteria have very different cell wall structures from those of the other halobacteria. Cells of *Halococcus morrhuae* are surrounded by a thick (50 nm) cell wall. This structure is unique, consisting of a complex highly sulfated heteropolysaccharide arranged in three domains (Schleifer et al., 1982). *H. morrhuae* remains intact at very low salt concentrations due to the rigidity of its cell wall. Novel cell wall structures have been found in the alkaliphilic coccus *Natronococcus occultus*, and the neutrophilic coccus originally named *Halococcus turkmenicus*, since renamed *Haloterrigena turkmenica* (see Genus VIII) (Kandler, 1994). The cell walls of both species contain glutamic acid, acetate groups, and glucosamine; *Natronococcus occultus* additionally contains uronic acid. The structural arrangement of this unusual glucosaminoglycan remains to be elucidated. *Natronococcus* spp. also remain intact at low salt concentrations although some leakage of the cell contents occurs. Many halobacteria are motile, and some flat cells seem to flex, possibly due to the presence of actin or tubulin-like components (Grant and Larsen, 1989a; Nishiyama et al., 1992). Halobacteria have right-handed helical flagella arranged in bundles of several filaments which slip against each other when changing rotational directions (Alam and Oesterhelt, 1984). The constituting flagellins are glycoproteins carrying the same sulfated oligosaccharide moieties as the cell surface glycoprotein (Wieland et al., 1985).

The majority of halobacteria examined to date have retinal-based pigments capable of the light-mediated translocation of ions across the cell membrane. The best known of these is bacteriorhodopsin, an outwardly-directed proton pump involved in energy conservation, the only non-chlorophyll-mediated light energy transducing system. Other retinal-based pigments include halorhodopsin, an inwardly directed chloride pump that is involved in osmotic homeostasis, and two other rhodopsins involved in the phototactic response of cells (Mukohata, 1994; Spudich et al., 1995). It is probable that all halobacteria possess halorhodopsin and the photoreceptor pigments, but not all possess bacteriorhodopsin. Because halobacteria grow perfectly well in the dark, the systems are not essential. There is no doubt that the possession of bacteriorhodopsin confers survival value on illuminated cells otherwise starved of energy (Brock and Petersen, 1976; Hartmann et al., 1980; Rodriguez-Valera et al., 1983b). Reviews by Lanyi should be consulted for details of the mechanisms by which ions move across membranes (Lanyi, 1991, 1993, 1995). There is controversy over whether halobacteria are capable of photoautotrophic growth. Oren (1983) speculated that halobacteria might contribute significantly to light-dependent CO_2 fixation in the Dead Sea. CO_2 fixation by halobacteria has been demonstrated on a number of occasions, but is believed to be due to several different anaplerotic mechanisms (Danon and Caplan, 1977; Javor, 1988; Rajagopalan and Altekar, 1991). Ribulose bisphosphate carboxylase/oxygenase has, however, been detected in several heterotrophically-grown *Haloferax* and *Haloarcula* strains (Altekar and Rajagopalan, 1990) but the levels of ribulose bisphosphate carboxylase/oxygenase are at best only a few percent of those typically found in autotrophic bacteria. The enzyme has now been purified and characterized (Rajagopalan and Altekar, 1994). To date, there is still no convincing evidence for photoautotrophic growth.

Some strains contain gas vacuoles, which make colonies appear pink or even white on agar media. Upon cultivation in the laboratory, strains have a tendency to lose the ability to produce vacuoles at high frequency, and sectoring may occur in gas-vacuolate colonies where reversion to non-gas vacuole forms has taken place. How gas vesicles form and the control of expression has been the subject of a considerable body of work notably by Pfeifer, DasSarma and co-workers (DasSarma and Arora, 1997; Pfeifer et al., 1997).

Most strains grow well in nutrient-rich complex media with yeast extract, peptone, or Casamino acids as carbon and energy sources. Difco's Bacto-Peptone, however, has been known to cause lysis of noncoccoid halobacteria due to the presence of high concentration of taurine conjugate of cholic acid (Kamekura et al., 1988). Some strains grow in defined media (Rodriguez-Valera et al., 1980) with glucose or sucrose as carbon sources and ammonia or glutamate as nitrogen sources. Anaerobic growth occurs either by an uncharacterized fermentative mode, linked to arginine utilization (Hartmann et al., 1980), or by using alternative electron acceptors such as nitrate (Tomlinson et al., 1986), dimethyl sulfoxide or trimethylamine N-oxide (Oren and Trüper, 1990), or fumarate (Oren, 1991).

Species are mostly insensitive to (eu)bacterial inhibitors such as penicillin, chloramphenicol, streptomycin, erythromycin, and tetracycline, but usually sensitive to anisomycin, aphidicolin, bacitracin, novobiocin, and vibriostat reagent (0/129).

Physical and genetic maps of *Haloferax* and *Halobacterium* spp., are available (Charlebois, 1995a, b). There has also been a considerable body of work on plasmids (DasSarma, 1995) and halophages (Nuttall and Dyall-Smith, 1993b, 1995; DasSarma and Stolt, 1995). Functional transformation/mating systems have been demonstrated for halobacteria (Pfeifer et al., 1994; Tchelet and Mevarech, 1994) and shuttle vectors are available for halobacteria and *E. coli* (Holmes et al., 1991, 1994; Cline and Doolittle, 1992) plus composite transposons (Dyall-Smith and Doolittle, 1994) with selectable marker genes for the rapid isolation and analysis of halobacterial genes. *The Halophile Laboratory Manual* (DasSarma and Fleischmann, 1995) should be consulted for details of plasmids and plasmid vectors. Halocins (halobacterial bacteriocins) are also produced by many halobacteria (Torreblanca et al., 1994).

Halobacteria are the most halophilic organisms known, and form the dominant microbial population when hypersaline waters approach saturation (Rodriguez-Valera et al., 1981), frequently imparting a red or pink coloration to the brines. Indeed, very old Chinese literature noted the red coloration of saturated salterns, presumably a consequence of halobacterial blooms (see Baas-Becking, 1931). The importance of this reddening in promoting rapid precipitation of sea salt has also been documented. It is now known that the carotenoid pigments of halobacteria trap solar radiation, increasing the ambient temperature and evaporation rates in salterns. Neutral salt lakes and solar salterns may contain 10^7–10^8 cells/ml as judged by microscopic examination (Post, 1977; Oren, 1993, 1994), although the recovery of colony-forming units is usually two or more orders of magnitude lower. Saline soda lakes support comparable blooms of alkaliphilic types (Jones et al., 1998), and brines within ancient evaporite deposits also support populations of halobacteria whose origins are unclear (Grant et al., 1998a). The addition of archaea-specific inhibitors to Dead Sea brines virtually abolished the incorporation of amino acids, whereas the addition of (eu)bacterial inhibitors had little effect, giving support to the view that halobacteria make up the overwhelmingly dominant population in environments with salt concentrations exceeding 250 g/l (Oren, 1994). Dead Sea brines stored for more than 50 years still contained a diverse population of halobacteria (Arahal et al., 1996). The characteristic red bacterioruberins possessed by most naturally occurring isolates seem to play a protective role against the strong sunlight where these organisms are found. These carotenoid pigments have been shown to protect the cells against photooxidative damage (Wu et al., 1983). Colorless strains are rarely reported, but have been isolated from some sites such as beach sands (Onishi et al., 1985).

Saturated brines, crude solar salt, and salted products are potential sources of halobacteria. The material can be spread directly on the surface of agar media containing penicillin to prevent the growth of halotolerant and halophilic (eu)bacteria. Wais (1988) has noted that the plating efficiency of halobacteria is markedly improved by the incorporation of halobacterial extracts into the agar media. The plates are wrapped in plastic bags to prevent drying and are incubated at 37–40°C. Red colonies appear after 4–21 d. Enrichments may be obtained in liquid media vigorously shaken to produce good aeration. Selective procedures that are specific for most of the individual genera or

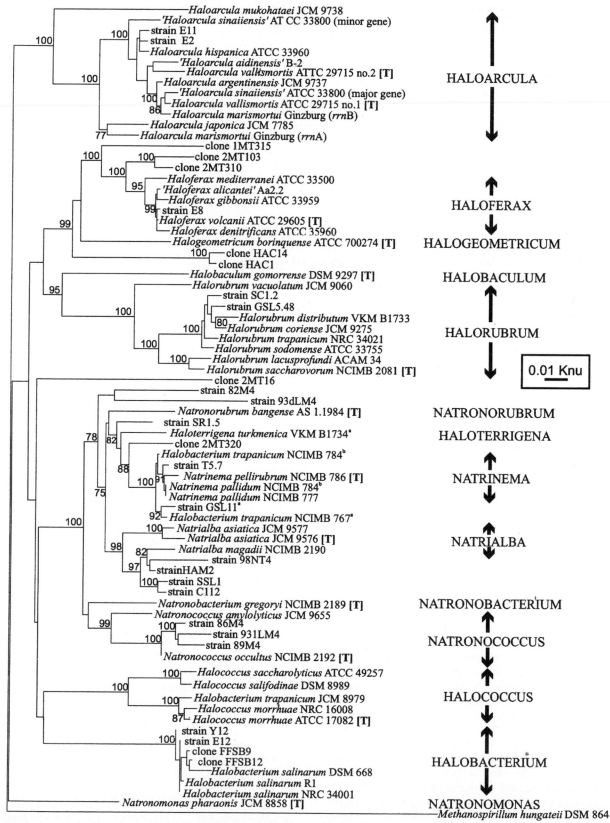

FIGURE A2.36. Phylogeny of the order *Halobacteriales*. The least-squares algorithm of Fitch and Margoliash (1967) was used to construct the tree from evolutionary distances (Jukes and Cantor, 1969) obtained from an alignment of 16S rRNA sequences (nucleotides 18–1534, *E. coli* numbering). Bootstrap percentages greater than 75% are indicated at nodes. The extent of each genus is illustrated by arrows to the right of the tree. [T] indicates the type species. (a) indicates three strains which were designated by Ventosa et al. (1999b) as one species, *Haloterrigena*

turkmenica, but which require further consideration in light of a preceding publication (McGenity et al., 1998). (b) indicates two nonidentical sequences from strain NCIMB 784. Bar represents 0.01 Knu (1 nucleotide substitution per 100 bases). For sequence accession numbers, see the species descriptions, McGenity et al. (1998), Montalvo-Rodríguez et al. (1998), Ventosa et al. (1999b), and Xu et al. (1999). Aligned sequences are available electronically from the authors.

TABLE A2.25. Signature 16S rDNA sequences of the genera of the order *Halobacteriales*

Genus	Signature sequence A[a]	Signature sequence B	Signature sequence C
Halobacterium	[137]CTGTGGACGGGAATACTC[154]	[468]GTGGTGCACGAATAAGGAC[503]	[1436]GCCGGCATGCGCTGGT[1465]
Haloarcula	[168]GCCAATAGCGGATA[181]	[717]CACCAATGGGGAAACCAC[734]	
Haloferax	[168]GCTAATAGTTCATACGGGAGTC[183 + b]	[593]GAAGGTTCATCGGGAAATCCGCCA[616]	[1246]AGACAATGGGTTGCTATCTCGAAAG[1270]
Halococcus	[183 +]GCTTTCATGTTGGAA[190]	[635]CGGCGGAAACCAGTCGG[650]	[1306]ATTGCGGGTTGAAACCCACCCGC[1328]
Halorubrum	[199]CTCCGGCGCCGAAGGAT[229]	[656]GACCGGAAGGCGCGACG[672]	[1242]GTCGAGACAAAGGGTTCC[1259c]
Halobaculum	[181]ACTGTCTCCTACGCTTGAACACC[193]	[656]GACAGGGAGACTCGAAGG[673]	[1273]GACGGTAATCTCAGAAACCTGG[1294]
Halogeometricum	[137]CTGCAGACACGGACAAC[153]	[635]CGGCGGAAACTTCGTGG[650]	[1283]TCCTAAATCCGGTC[1296]
Natrialba			[1460]CCGGGTCGAATCTGG[1474]
Natrinema[d]	[190]GTGGCACGAACGCGAAACGTTACG[211]		[1438]CGACGCAACGTCGGT[1465]
Natronomonas	[40]CGGTGTCCGATTTAGCCATGCGA[60]	[730]AGCACCTTGGAAGGACGG[747]	[1271]GTGAAGCTAAATCTCCTAAACTTGG[1294]
Natronobacterium	[138]TACAGAGCGGGATAAC[153]	[609]ATCTCCTCGCTCAACGAGGAG[628]	[1307]TTGCAGGCTGAAACTCGCCTGC[1328]
Natronococcus	[127]GCCAAACTACCCTCTGGAG[144]		[1435]GCGCTTCCGGCGGT[1465]

[a]A, signature sequences at the 5′ end of the 16S rRNA gene; B, signature sequences in the middle of the 16S rRNA gene; C, signature sequences at the 3′ end of the 16S rRNA gene. Numbers above the sequence represent their position based on the sequence of *E. coli*.

[b]183 + indicates a region in which there are gaps in the sequence of *E. coli* when aligned with halobacteria.

[c]This signature has a single base mismatch with *Halorubrum vacuolatum*.

[d]Includes *Halobacterium trapanicum* NCIMB 767 and strain GSL11.

species have not been devised. However, a specific enrichment for the genus *Halobacterium* has now been devised, based on arginine fermentation (Oren and Litchfield, 1999). Generation times are commonly 3–6 h in complex media, depending on the isolate, whereas generation times of natural populations measured indirectly in brines *in situ* are 54–120 h (Oren, 1994).

Growth on agar slants in screw-capped tubes can be kept in the refrigerator for 4–5 months. Lyophilization does not give reliable results for some strains. L-drying, drying from liquid using a pre-dried, sterile skim milk sponge, has been used successfully at IFO (Sakane et al., 1992). Preservation in liquid nitrogen has been used for many years at DSMZ (Hippe, 1991; Tindall, 1991). Preparations of cells suspended in growth medium supplemented with 15% (w/v) glycerol may be stored at −80°C using the bead procedure of Feltham et al. (1978).

TAXONOMIC COMMENTS

In the first edition of *Bergey's Manual of Systematic Bacteriology*, Grant and Larsen (1989a) created the order *Halobacteriales* to accommodate the halobacteria, and proposed a new taxonomic structure largely based on chemotaxonomy, notably the possession or absence of particular glycolipids and phospholipids. At that time, relatively few 16S rRNA gene sequences were available, but in general, the phylogenetic placements were consistent with the chemotaxonomic groupings. A total of six genera were described, with, in addition, a considerable number of *species incertae sedis*. Many halobacteria that do not precisely match existing taxa have been described since, which, together with reassessment of existing taxa using phylogenetic methods, has led to the taxonomy of the group being in a state of flux. For example, Kamekura et al. (1997), on the basis of 16S rRNA gene sequence comparisons, showed that the haloalkaliphiles are much more diverse than had previously been imagined. In all, there are presently fourteen validly described genera of halobacteria (Kamekura et al., 1997; Oren et al., 1997b; McGenity et al., 1998; Montalvo-Rodríguez et al., 1998; Ventosa et al., 1999b; Xu et al., 1999).

In general, halobacterial taxonomy, originally largely based on polar lipid profiles, has continued to be remarkably consistent with the phylogenetic data derived from 16S rRNA gene sequence comparisons as sequences become available for type spe-

cies. In addition, *species incertae sedis* have subsequently been assigned to phylogenetically coherent groups (McGenity and Grant, 1995; McGenity et al., 1998). However, the alkaliphilic halobacteria, which are phenotypically and chemotaxonomically

TABLE A2.26. Signature 16S rDNA bases specific for *Halobacteriales*

Genus	Signature bases[a]
Halobacterium	103C(T), 110A(Y), 131G(A):231C(T), 139G(M):224C(K), 153T(C), 435A(G), 1154T(R), 1426T(C)
Haloarcula[b]	253T(T): 273A(G), 721A(G), 726G(C): 731C(G), 934+1C(T),[c] 988T(C):1217A(G), 1044+29C(T), 1044+33T(C), 1219C(T), 1246T(R), 1515A(G)
Haloferax	110A(Y), 1153A(G)
Halococcus	30T(C):553A(G), 50C(T), 116C(A), 218C(G), 229C(T), 233A(G), 557A(G), 1219C(T), 1289C(A), 1314T(C)
Halorubrum[d]	139T(M):224A(K), 192A(S), 294C(K), 295A(G), 667G(Y): 739C(R), 847G(M), 1149G(Y), 1145G(T), 1165G(Y):1171C(R), 1252A(T), 1317C(A)
Halobaculum	131T(A), 218A(S), 231A(Y), 758C(G), 762A(G), 833C(T)
Halogeometricum	645T(R), 825A(G), 875T(C), 1119A(Y), 1247A(G), 1290T(C)
Natrialba	560G(H)
Natrinema	174G(H), 193+2A(B), 209A(C)
Natronomonas	43T(R), 59G(T), 111T(A), 134G(A), 294C(K), 295A(K)
Natronobacterium and *Natronococcus*	1355C(R), 1356G(Y):1366C(G)

[a]The numbering of signature nucleotides is based on an alignment with *E. coli* 16S rRNA sequence. Nucleotides in parentheses are those of other genera. Occasionally, two genera share the same signature base at a particular position, e.g., *Halobacterium/Haloferax* 110A(Y). A colon separates complementary nucleotides.

[b]Bases possessed by all species except the outlier *Haloarcula mukohataei* are underlined.

[c]934+1C etc., indicates 1 base after *E. coli* base 934. This occurs when there are gaps in the sequence of *E. coli* when aligned with halobacteria.

[d]Bases possessed by all species except the outlier *Halorubrum vacuolatum* are underlined.

extremely similar, originally classified in the genera *Natronobacterium* and *Natronococcus*, have turned out to be more phylogenetically diverse than one would have predicted (Kamekura et al., 1997). The review by Kamekura (1998a) should be consulted for details of how the systematics of the group has evolved since 1980. Halobacteria constitute a monophyletic group with the most distantly related species having 83.2% 16S rRNA gene sequence similarity. The methanogens are their closest relatives (Olsen et al., 1994b), but still have less than 80% 16S rRNA gene sequence similarity to halobacteria. Two halobacterial genomes are currently in the process of being sequenced (see genus I *Halobacterium*).

It is clear that halobacterial diversity extends beyond the fourteen formally described genera, when uncultivated halobacteria are considered. For example, phylogenetically distinct 16S rRNA clones have been obtained from both neutral (Benlloch et al., 1995) and alkaline crystallizer ponds (Grant et al., 1999) and Munson et al. (1997) obtained several distinct halobacterial clones from salt marshes. Phylogenetic tree construction (Fig. A2.36) from all the available sequences at the time of writing (including uncultivated clones) indicates that the generic groupings are consistent, albeit with a number of monospecific genera (*Halobaculum*, *Natronomonas*, *Natronobacterium*). It should be noted that all members of the *Haloarcula* taxon examined to date possess at least two heterogeneous copies of the 16S rRNA gene, differing in sequence similarity by up to 5% (Dennis et al., 1998; Gemmell et al., 1998). Nevertheless, *Haloarcula* spp. still comprise a monophyletic group although *Haloarcula mukohataei* is an outlier. Details of the organisms/clones/sequences in Fig. A2.36 are given in McGenity et al. (1998), Montalvo-Rodríguez et al. (1998), Ventosa et al. (1999b), and Xu et al., 1999. Aligned sequences are available on request. The major change from the treatment in the previous *Manual* is the definition of the genus *Halorubrum* from a number of *species incertae sedis*, the addition of *Halobaculum* and a reworking of existing alkaliphilic halobacteria, together with *species incertae sedis* and a number of new isolates, to comprise the new genera *Natrinema*, *Natrialba*, *Natronomonas*, and *Haloterrigena*. It is probable that the genus *Natrinema* should be extended to include the representatives indicated by the dashed line in Fig. A2.36. There is evidence that unnamed soda lake isolates 82 M4 and 93d LM4 may constitute a new taxon. Signature sequences and signature bases for the genera are listed in Tables A2.25 and A2.26. Signature sequences have not yet been tested as identification probes.

There have been problems over the years regarding strain identities which have led to considerable confusion. The limited value of much of the earlier phenotypic data has resulted in many of the deposited strains retaining their classification in the oldest genus *Halobacterium*. Tindall (1992) has addressed some of these problems. Particular problems include the loss of the type species of *Halobacterium*, the type strain of *Haloarcula marismortui* and differences in the type strains of *Halorubrum trapanicum* held by different culture collections (McGenity and Grant, 1995). Attempts have been made to redress the situation by deposition of neotypes (Grant et al., 1998b; Oren et al., 1990) and recent comparative studies (Ventosa et al., 1999b). In addition, the specific epithets *Halobacterium halobium*, *Halobacterium cutirubrum*, and *Halobacterium salinarium* have, on occasion, been given to isolates with nearly 100% DNA/DNA relatedness (Grant and Larsen, 1989a; Tindall, 1992). These almost identical strains have now been subsumed into the more correctly named species *Halobacterium salinarum* (Ventosa and Oren, 1996). On the contrary, some isolates with the above epithets, e.g., *Halobacterium halobium* NCIMB 777 and *Halobacterium cutirubrum* NCIMB 763, should not be assigned to the genus *Halobacterium*, and 16S rRNA gene sequence data are beginning to clarify their taxonomic position (e.g., McGenity et al., 1998).

Minimal standards for description of new taxa of the order *Halobacteriales* are now available (Oren et al., 1997b).

Family I. **Halobacteriaceae** Gibbons 1974a, 269[AL]

WILLIAM D. GRANT, MASAHIRO KAMEKURA, TERRY J. MCGENITY AND ANTONIO VENTOSA

Ha.lo.bac.ter.i.a' ce.ae. M.L. neut. n. *Halobacterium* type genus of the family; *-aceae* ending to denote a family; M.L. pl. fem. n. *Halobacteriaceae* the *Halobacterium* family.

Description of the family is the same as for the order.

The reader should remember that some members of the halophilic (eu)bacteria have generic names with the prefix Halo-, such as *Haloanaerobium*, *Halobacteroides*, *Halocella*, *Halochromatium*, *Haloincola*, *Halomonas*, *Halothermothrix*, and *Halovibrio*.

At present the following fourteen genera have been validated in the family *Halobacteriaceae*: *Halobacterium*, *Haloarcula*, *Haloferax*, *Halococcus*, *Halorubrum*, *Natrialba*, *Halobaculum*, *Halogeometricum*, *Haloterrigena*, *Natrinema*, *Natronomonas*, *Natronobacterium*, *Natronococcus*, and *Natronorubrum*.

Type genus: **Halobacterium** Elazari-Volcani 1957, 207, emend. Larsen and Grant 1989, 2222.

Key to the genera of the Halobacteriaceae

1. Cells are rods of varying length under optimal conditions in young liquid culture. Pleomorphic and coccoid forms may be present in old liquid culture and in agar-grown cultures. Cells lyse in distilled water. Cells are motile and uniformly stain Gram-negative. Mg^{2+} requirement moderate (5–50 mM). Amino acids are required for growth. Optimum salt concentration for growth: 3.5–4.5 M NaCl. The pH range for growth is 5–8. Characteristic sulfated triglycosyl and tetraglycosyl diethers present.

Genus I. *Halobacterium*, p. 301

2. Cells are extremely pleomorphic. Under optimal conditions in liquid media, flat triangles, rectangles, and irregular disks are commonly observed. Cells lyse in distilled water. Motile or nonmotile and uniformly stain Gram-negative. Mg^{2+} requirement moderate (5–50 mM).

Amino acids not required for growth. Optimum salt concentration for growth: 2–3 M NaCl. The pH range for growth is 5–8. Characteristic triglycosyl diether lipid present.

Genus II. *Haloarcula*, p. 305

3. Cells are rods of varying length under optimal conditions in young liquid cultures. Pleomorphic and coccoid forms may be present in old liquid cultures and in agar-grown cultures. Cells lyse in distilled water. Cells are motile and uniformly stain Gram-negative. Mg^{2+} requirement moderate (5–50 mM). Amino acids are required for growth. Optimum salt concentration for growth: 3.5–4.5 M NaCl. The pH range for growth is 5–8. Sulfated diglycosyl diether present. Phosphatidyl glycerol sulfate absent.

Genus III. *Halobaculum*, p. 309

4. Cells are coccoid under all conditions of growth, occurring in pairs, tetrads, sarcinae, or irregular clusters. Cells are nonmotile and do not lyse in distilled water, and some cells at least stain Gram-positive. Optimum salt concentration for growth: 3.5–4.5 M NaCl. The pH range for growth is 5–8. Possess both $C_{20}C_{20}$ and $C_{20}C_{25}$ diether core lipids. A sulfated diglycosyl diether is present.

Genus IV. *Halococcus*, p. 311

5. Cells are extremely pleomorphic under optimal conditions in liquid media, pleomorphic rods and flat disks most commonly present. Cells lyse in distilled water. Cells are motile or non-motile and uniformly stain Gram-negative. Mg^{2+} requirement high (20–50 mM). Amino acids are not required for growth. Optimum salt concentration for growth: 2–3 M NaCl. The pH range for growth is 5–8. Characteristic sulfated diglycosyl diether lipid present. Phosphatidyl glycerol sulfate absent.

Genus V. *Haloferax*, p. 315

6. Cells are extremely pleomorphic, including short and long rods, squares, triangles, and ovals. Cells lyse in distilled water. Cells are motile and uniformly stain Gram- negative. Mg^{2+} requirement high (40–80 mM). Optimum salt concentration for growth: 3.5–4 M NaCl. The pH range for growth is 6–8. Contains a yet unidentified non-sulfate-containing glycolipid. Phosphatidyl glycerol sulfate absent.

Genus VI. *Halogeometricum*, p. 318

7. Cells are rods of varying length under optimal conditions in young liquid cultures. Pleomorphic and coccoid forms may be present in old liquid cultures and in agar-grown cultures. Cells lyse in distilled water. Some strains are motile. Cells stain Gram-negative. Mg^{2+} requirement moderate (5–50 mM) except for one species with low requirement (<1 mM). Optimum salt concentration for growth: 2.5–4.5 M NaCl. The pH range for growth is usually 5–8, except for one alkaliphilic species with pH range 8.5–11.0. Sulfated diglycosyl diethers present, except in the alkaliphilic species.

Genus VII. *Halorubrum*, p. 320

8. Cells are rods or coccoid, nonmotile. Cells are uniformly Gram-negative. Amino acids are required for growth. Mg^{2+} requirement moderate (5–50 mM). Optimum salt concentration for growth: 3–4 M NaCl. The pH range for growth is 6–8.5. Possess both $C_{20}C_{20}$ and $C_{20}C_{25}$ diether core lipids. Characteristic disulfated diglycosyl diether lipid present. Phosphatidyl glycerol sulfate absent.

Genus VIII. *Haloterrigena*, p. 324

9. Cells are rods of varying length under optimal conditions in young liquid cultures. Pleomorphic and coccoid forms may be present in old liquid cultures and in agar-grown cultures. Cells lyse in distilled water. Cells are motile and uniformly stain Gram-negative. Amino acids are required for growth. Mg^{2+} requirement moderate (5–50 mM) except for alkaliphilic species (<1 mM). Optimum salt concentration for growth: 3.5–4.5 M NaCl. The pH range for growth is 5–10 or 6–8 depending on species. Possess both $C_{20}C_{20}$ and $C_{20}C_{25}$ diether core lipids. In one species, disulfated diglycosyl diethers present. Unidentified phospholipids are present in alkaliphilic species, glycolipids are absent in alkaliphilic species.

Genus IX. *Natrialba*, p. 325

10. Cells are rods of varying lengths under optimal conditions. Pleomorphic and coccoid forms may be present in old liquid cultures and in agar-grown cultures. Cells lyse in distilled water. Cells are uniformly Gram-negative. Mg^{2+} requirement moderate (5–50 mM). Amino acids are required for growth. Optimum salt concentration for growth: 3.4–4.3 M NaCl. The pH range for growth is 5–8.5. Possess both $C_{20}C_{20}$ and $C_{20}C_{25}$ diether core lipids. Possess several unidentified glycolipids.

Genus X. *Natrinema*, p. 327

11. Cells are rods of varying lengths in young liquid cultures but become coccoid as they age. Cells from agar cultures are coccoid. Cells lyse in distilled water. Motile or nonmotile and uniformly stain Gram-negative. Optimum salt concentration for growth: 3.5–4.5 M NaCl. The pH range for growth is pH 8.5–11.0, with a very low Mg^{2+} requirement (<1 mM). Possess both $C_{20}C_{20}$ and

$C_{20}C_{25}$ diether core lipids. Unidentified phospholipids present, glycolipids absent. Phosphatidyl glycerol sulfate absent.

Genus XI. *Natronobacterium*, p. 329

12. Cells are coccoid under all conditions of growth, occurring singly or in pairs, tetrads, and irregular clusters. Cells suspended in distilled water show some leakage of contents as judged by an increase in viscosity of the suspension, but they appear intact when examined microscopically. At least some cells stain Gram-positive. Nonmotile. Optimum salt concentration for growth: 3.0–4.0 M NaCl. The pH range for growth is pH 8.5–11.0, with a very low Mg^{2+} requirement (<1 mM). Possess both $C_{20}C_{20}$ and $C_{20}C_{25}$ diether core lipids. Unidentified phospholipids present, glycolipids absent. Phosphatidyl glycerol sulfate absent.

Genus XII. *Natronococcus*, p. 330

13. Cells are rods of varying length under optimal conditions in young liquid cultures. Pleomorphic and coccoid forms may be present in old liquid cultures and in agar-grown cultures. Cells lyse in distilled water. Cells are motile and uniformly stain Gram-negative. Amino acids are required for growth. Optimum salt concentration for growth: 3.5–4.5 M NaCl. The pH range for growth is pH 8.5–11.0, with a very low Mg^{2+} requirement (<1 mM). Possess both $C_{20}C_{20}$ and $C_{20}C_{25}$ diether core lipids. Unidentified phospholipids present, glycolipids absent. Phosphatidyl glycerol sulfate absent.

Genus XIII. *Natronomonas*, p. 332

14. Cells are pleomorphic rods. Cells lyse in distilled water. Cells are nonmotile and stain Gram-negative. Amino acids are required for growth. Optimum salt concentration for growth: 3.4–3.8 M NaCl. The pH range for growth is 8.0–10.0, with a very low Mg^{2+} requirement. Possesses both $C_{20}C_{20}$ and $C_{20}C_{25}$ diether lipids. Unidentified phospholipids present, glycolipids absent. Phosphatidyl glycerol sulfate absent.

Genus XIV. *Natronorubrum*, p. 333

ACKNOWLEDGMENTS

We are grateful to Aharon Oren for many comments and suggestions during the preparation of this overview.

FURTHER READING

DasSarma, S. and E.M. Fleischmann (Editors). 1995. Archaea: A Laboratory Manual, Cold Spring Harbor Laboratory Press, Plainview, NY.

Grant, W.D. and H. Larsen. 1989. Extremely halophilic archaeobacteria. *In* Staley, Bryant, Pfennig and Holt (Editors), Bergey's Manual of Systematic Bacteriology, 1st ed., Vol. 3, The Williams & Wilkins Co., Baltimore, pp. 2216–2219.

Kamekura, M. 1998. Diversity of extremely halophilic bacteria. Extremophiles 2: 289–296.

Kates, M. 1993. Membrane lipids of extreme halophiles: biosynthesis, function and evolutionary significance. Experientia (Basel) 49: 1027–1036.

Kushner, D.J. 1985. The *Halobacteriaceae. In* Woese and Wolfe (Editors), The Bacteria: A Treatise on Structure and Function, Vol. VIII. The Archaebacteria, Academic Press, New York, pp. 171–214.

Olsen, G.J., C.R. Woese and R. Overbeek. 1994. The winds of (evolutionary) change: breathing new life into microbiology. J. Bacteriol. 176: 1–6.

Oren, A. 1994. The ecology of the extremely halophilic archaea. FEMS Microbiol. Rev. 13: 415–440.

Oren, A., A. Ventosa and W.D. Grant. 1997. Proposed minimal standards for description of new taxa in the order *Halobacteriales*. Int. J. Syst. Bacteriol. 47: 233–238.

Tindall, B.J. 1992. The family *Halobacteriaceae. In* Balows, Trüper, Dworkin, Harder and Schleifer (Editors), The Prokaryotes. A Handbook of Bacteria: Ecophysiology, Isolation, Identification, Applications, 2nd ed., Vol. 1, Springer-Verlag, New York, pp. 768–808.

Genus I. **Halobacterium** *Elazari-Volcani 1957, 207,*[AL] *emend. Larsen and Grant 1989, 2222*

WILLIAM D. GRANT

Ha.lo.bac.te'ri.um. Gr. n. *hals, halos* the sea, salt; Gr. n. *bakterion* a small rod; M.L. neut. n. *Halobacterium* salt bacterium.

Under optimum growth conditions cells are rod shaped (0.5–1.2 × 1.0–6.0 μm). Pleomorphic forms common (bent and swollen rods, flat ribbons, clubs, spheres). Cells divide by constriction. No known resting stages. Stain Gram-negative but no outer membrane is present. Motile by tufts of polar flagella. Some strains have gas vacuoles. Most strains are strict aerobes, but some exhibit facultatively anaerobic growth. Oxidase and catalase positive. Extremely halophilic with growth occurring in media containing 3.0–5.2 M NaCl. Most strains grow best at 3.5–4.5 M NaCl. Optimum temperature: 35–50°C; maximum, 55°C; minimum, 15–20°C. pH range, 5.5–8.5. Chemoorganotrophic. Amino acids required for growth. Most strains are proteolytic.

The DNA is usually composed of a major component and a minor component. The latter makes up 10–30% of the total DNA (Bd). Strains having a minor component usually harbor a large plasmid (144 kb). 16S rDNA signature sequence(s) are listed in Table A2.25 in Order *Halobacteriales*.

The mol% G + C of the DNA is: 67.1–71.2 (major component) and 57–60 (minor component).

Type species: **Halobacterium salinarum** corrig. (Harrison and Kennedy 1922) Elazari-Volcani 1957, 208; emend. Larsen 1984, 262; emend. Larsen and Grant 1989, 2222; emend. Ventosa and Oren 1996, 347 (*Pseudomonas salinaria* Harrison and Kennedy 1922, 120; *Halobacterium halobium* (Petter 1931) Elazari-Volcani 1957, 210; *Halobacterium cutirubrum* (Lochhead 1934) Elazari-Volcani 1957, 209.)

FURTHER DESCRIPTIVE INFORMATION

The Genus *Halobacterium* is phylogenetically well defined, with representatives having 16S rRNA genes with >98.5% similarity. *Halobacterium* strains have 16S rRNA genes with less than 88.2% similarity to those of other genera of halobacteria.

Pleomorphism is a distinctive feature of the halobacteria, including members of the genus *Halobacterium*. In some strains the cells are regular slender rods when grown at moderate temperature under optimal conditions, but they display a multitude of involution forms when in the stationary phase of growth or when grown at inappropriate temperatures or NaCl concentrations. Upon gradual dilution of the medium with water, the cells change shape from rods through irregular transition forms to spheres, and the spheres undergo lysis. The composition of the growth medium may affect the point at which lysis occurs, since in media whose ion composition (with the exception of NaCl) is based on seawater (Torreblanca et al., 1986a), many strains are reported to be stable in 2% (w/v) total salts, whereas in other media lysis is widely reported to occur between 5–10% (w/v) NaCl (Kushner, 1985). The composition of the medium may also affect the minimum Mg^{2+} requirement for growth, originally considered to be 0.1–0.5 M in complex media (Larsen, 1984) but reported as much lower in seawater-based medium (Torreblanca et al., 1986a).

The lysis phenomenon in hypotonic solutions is not primarily due to an osmotic effect but rather to the need for high concentrations of salt (NaCl) to maintain the cell envelope. The cell envelope of members of the genus *Halobacterium* have an outer layer (Steensland and Larsen, 1969) which is largely composed of a characteristic eucaryotic-like sulfated glycoprotein of high molecular weight (Wieland et al., 1982) responsible for maintaining the structural integrity of the cell wall. The surface of the cell envelope has a hexagonal pattern due to the regular packing of glycoprotein subunits (Fig. A2.37) (Larsen and Grant, 1989). The proteinaceous subunits of the cell envelope are held together only in the presence of NaCl; KCl will not substitute.

Some members of the genus *Halobacterium* (and other halobacteria), when grown under low oxygen tension, form patches of a special pigmented protein, bacteriorhodopsin in the cell membrane. The reaction center of the chromophore contains retinal; the membrane patches are referred to as the purple membrane. Bacteriorhodopsin acts as a proton pump driven by light energy absorbed by the pigment. Important energy-requiring phenomena in the cell utilize the energy made available by the proton gradient thus produced. A number of other retinal-based pigments are also present; they include halorhodopsin, an inwardly directed Cl^- pump similar in structure to bacteriorhodopsin, and sensory retinal chromophores that appear to function as signal transducers that mediate the phototactic responses of cells (see *Halobacteriales*, Further Descriptive Information).

Most strains grow well in complex media with yeast extract or Casamino acids as carbon and energy sources. Many variations of complex media have been used over the years. Tindall (1992), Kushner (1993), and DasSarma et al. (1995) detail a number of these. The most commonly used complex medium is based on the formulation of Gibbons and his collaborators (Sehgal and Gibbons, 1960).* Balanced mixtures of sea salts supplemented with NaCl are sometimes used for the salt components (Rodriguez-Valera et al., 1980). Defined media have also been developed for certain isolates (Rodriguez-Valera et al., 1980; Larsen, 1981). However, not all isolates grow in these media. Most strains have complex nutrition requirements and preferentially utilize amino acids as a source of carbon and energy. Many strains are proteolytic. Ducharme et al. (1972) found that arginine was rapidly removed from the medium by *Halobacterium cutirubrum (salinarum)*. Arginine is degraded via the arginine deaminase pathway, and the resulting citrulline is converted to ornithine with the production of carbamoyl phosphate in a reaction catalyzed by ornithine carbamoyl transferase (Larsen, 1967; Hartmann et al., 1980). The utilization of aromatic amino acids by *Halobacterium salinarum* has been studied by Lobyreva et al. (1987).

NMR experiments with *H. salinarum* indicate the presence of a glycolytic pathway in addition to a modified Entner-Doudoroff pathway (Sonawat et al., 1990). *H. salinarum* possesses a complete oxidative citric acid cycle to generate energy and key intermediates for biosynthesis and all the enzymes required for gluconeogenesis have been detected in *H. salinarum* (Danson, 1988; Danson and Hough, 1992; Ghosh and Sonawat, 1998).

Several strains of *H. salinarum* can use dimethylsulfoxide as an electron acceptor, reducing it to dimethylsulfide and also reducing trimethylamine *N*-oxide to trimethylamine and fumarate to succinate (Oren and Trüper, 1990; Oren, 1991).

Even under aerobic conditions, growth is relatively slow; generation times of 3 h are the best that have been recorded, but 6–7 h is more common.

Some strains contain gas vacuoles, which make colonies appear pink or even white on agar media. Upon cultivation in the laboratory, the strains have a tendency to lose the ability to produce vacuoles at high frequency, and sectoring may occur in gas vacuolate colonies where reversion to non-gas vacuole forms has taken place (DasSarma, 1993). Observations such as this have led to the realization that certain strains of *H. salinarum* have considerable genetic instability due to the presence of multiple

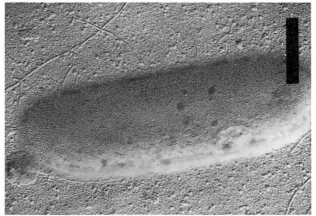

FIGURE A2.37. *Halobacterium salinarum* 1 (ATCC 19700, DSMZ 668), grown at 30°C in a medium of tap water containing the following (g/l): crude solar salt, 250.0; MgSO₄·7H₂O, 20.0; KCl, 5.0; CaCl₂·6H₂O, 0.2; yeast extract (Difco), 5.0; and tryptone (Difco), 5.0; pH 6.8. Replica of the surface of a cell from a culture in late exponential phase. Bar = 0.5 µm. (Courtesy of V. Mohr.)

*The medium contains (g/l): NaCl, 200.0; MgSO₄·7H₂O, 20.0; KCl, 2.0; trisodium citrate, 3.0; yeast extract, 10.0; and Casamino acids, 7.5. Fe²⁺ and Mn²⁺ are added as trace elements at 10 and 0.1 ppm, respectively. The NaCl is sterilized separately from the rest of the components. The final pH should be between 7.2–7.6. Addition of starch to the growth media at 1% (w/v) often favorably affects colony recovery (Oren, 1990). Addition of brine from the site and addition of cell-free extracts of halobacteria are also recorded as markedly improving recovery (Wais, 1988; Oren, 1994).

copies of different insertion elements which inactivate genes where they integrate into DNA. The genome can be divided into two fractions on the basis of mol% G + C content (Ebert and Goebel, 1985; Pfeifer, 1995). The FI fraction contains ~68% G + C and is believed to be the chromosome. The FII fraction contains ~58% G + C and corresponds to one or more large plasmids and AT-rich islands of the chromosome. The insertion sequences are mainly clustered in the lower percentage of G + C component of the DNA, which also contains the hypermutable gas vacuole operon (Pfeifer et al., 1989; Charlebois and Doolittle, 1989). In a model for the genome structure, the plasmids represent a hypermutable domain where new functions are created by transpositional events and advantageous mutations are rendered stable by transfer to the chromosome (Hackett et al., 1994).

A number of phages for *H. salinarum* strains have now been characterized (halophages). Most of these resemble bacteriophage λ in morphology but require high salt concentrations to maintain their infectivity. In some cases, lysogeny is established. The best studied is φH, which has a 59 kb double-stranded genome that has been partially sequenced (Stolt and Zillig, 1994). A variety of natural plasmids have also been characterized from *H. salinarum* strains, and shuttle vectors based on them that replicate in *Escherichia coli* have been constructed (DasSarma, 1995).

The polar lipids of *Halobacterium* species are based on 2,3-di-*O*-phytanyl-*sn*-glycerol ($C_{20}C_{20}$) ethers. All strains have $C_{20}C_{20}$ forms of PG, PGP-Me, PGS, TGD-1, S-TGD-1, and S-TeGD.* S-TGD-1 and S-TeGD are characteristic of the genus (Ross and Grant, 1985; Larsen and Grant, 1989; Kates, 1993).

Isolates are usually insensitive to penicillin, chloramphenicol, streptomycin, erythromycin, and tetracycline but sensitive to anisomycin, aphidicolin, bacitracin, novobiocin, and vibriostat reagent (0/129) (Hilpert et al., 1981; Bonelo et al., 1984; Ross and Grant, 1985).

Members of the genus *Halobacterium* make up a variable fraction of the microbial population of neutral salt lakes and salterns, particularly thallasohaline examples. They are also present in subterranean salt deposits. Members of this genus are responsible for the spoilage of proteinaceous products that have been heavily salted with solar salt, such as fish and hides, reflecting their proteolytic capabilities. The organisms have also been isolated from salted fermented products such as fish sauce (Thongthai et al., 1992).

ENRICHMENT AND ISOLATION PROCEDURES

Concentrated neutral brines, crude solar salt, and proteinaceous salted products are potential sources of members of the genus *Halobacterium*. The material can be spread directly on the surface of agar media. The plates are wrapped in plastic bags to prevent drying and are incubated at 37–40°C. Red colonies appear after 4–21 d. Enrichments may be obtained in liquid media vigorously shaken to produce good aeration. Members of the genus *Halobacterium* and other halobacteria are obtained by these procedures. A selective procedure for the enrichment and isolation of *Halobacterium* has now been described, based on arginine screening (Oren and Litchfield, 1999). Proteinaceous salted products might be expected to have very high numbers of members of the genus *Halobacterium* rather than other halobacteria.

---

*See *Halobacteriales* for definitions of lipid abbreviations.

MAINTENANCE PROCEDURES

See under Order *Halobacteriales*, Further descriptive information.

DIFFERENTIATION OF THE GENUS *HALOBACTERIUM* FROM OTHER GENERA

The key to the family *Halobacteriaceae* indicates characteristics useful in distinguishing *Halobacterium* species from other genera. In particular, the polar lipid composition of members of the genus *Halobacterium* is characterized by the presence of S-TGD-1 and S-TeGD.

TAXONOMIC COMMENTS

Halobacteria (usually members of the genus *Halobacterium*) are commonly associated with the spoilage of salted fish, meats, and hides. The first recorded isolations of the organisms are from these sources, notably salted fish, and Larsen (1984) was of the opinion that the first reliable description of rod-shaped forms from these sources was given by Klebahn (1919), who referred to them as *"Bacillus halobius ruber"*. Later work by Harrison and Kennedy (1922) proposed the name *"Pseudomonas salinaria"*. Petter (1931), again working on salted fish, isolated a number of strains described as *"Bacterium halobium"*, some of which still survive. Schoop (1935b) was probably the first to propose the use of the generic designation *Halobacterium*, although the credit for the designation is given to Elazari-Volcani (1957), who first recognized the unusual and unifying properties of the isolates. Sources of other isolates included salted hides (Lochhead, 1934) and the Dead Sea (Elazari-Volcani, 1940). It is not possible to be certain how diverse the collection of these organisms was, although it is likely that those from salted proteinaceous products were *Halobacterium sensu strictu*. Many of the isolates were then brought together by the National Research Council (NRC) Laboratories in Ottawa (see Colwell et al., 1979). Unfortunately, the NRC no longer maintains a culture collection and many of the isolates have now been lost. Tindall (1992) has exhaustively reviewed the history of the halobacteria, pointing out the inadequacies of much of the taxonomic work since those early days. The accounts of Larsen and Grant (1989) and Kamekura (1998a) should be consulted together with that of Tindall (1992), for detailed information about the uncertain state of halobacterial systematics, particularly relating to the genus *Halobacterium*, that has persisted until fairly recently.

Many of the halobacteria are biochemically unreactive, and thus limited phenotypic data, in the absence of much chemosystematic and phylogenetic information, has resulted in many of the noncoccoid halobacteria being assigned in the past to the genus *Halobacterium*. Larsen and Grant (1989) attempted to clarify the situation, proposing on the basis of chemotaxonomy, notably polar lipid signatures, and limited 16S rRNA gene sequence information that there should be only one species in the genus *Halobacterium*, *H. salinarium*, pointing out that there were a considerable number of *species incertae sedis* and other organisms that awaited taxonomic clarification. That view has turned out to be largely correct as further sequence data have become available, and the isolates of uncertain taxonomic standing, with the exception of *H. cutirubrum* NCIMB 763, have now been assigned to other genera or have had new genera created to accommodate them. There were also nomenclatural problems, particularly relating to the use of the different specific epithets *halobium*, *salinarium*, and *cutirubrum*, for a number of strains that appeared to be largely identical. Larsen and Grant (1989) considered that *H.*

cutirubrum along with *Halobacterium halobium* should be subsumed into *H. salinarium*, the last specific epithet having historical precedence. This has now been accepted, having been validly published by Ventosa and Oren (1996), who further changed the epithet *salinarium* to the more correct *salinarum*.

The nomenclatural and taxonomic fates of the *species incertae sedis* and other misclassified *Halobacterium* species referred to in the first edition of *Bergey's Manual of Systematic Bacteriology* (Larsen and Grant, 1989) are as follows:

H. saccharovorum ATCC 29252 → *Halorubrum saccharovorum*
H. trapanicum NRC 34021 (neotype NCIMB 13488) → *Halorubrum trapanicum*
H. sodomense ATCC 33755 → *Halorubrum sodomense*
H. denitrificans ATCC 35960 → *Haloferax denitrificans*
H. trapanicum NCIMB 784 → *Natrinema pallidum*
H. salinarium (cutirubrum) NCIMB 786 → *Natrinema pellirubrum*
H. halobium NCIMB 777 → *Natrinema pellirubrum*

The following species described after 1989 and listed in *Bergey's Manual of Determinative Bacteriology* (9th ed.) (Holt et al., 1994) have also been reclassified:

H. lacusprofundi ACAM 34 → *Halorubrum lacusprofundi*
H. distributum VKMB 1733 → *Halorubrum distributum*

H. cutirubrum NCIMB 763 still awaits further investigation but probably constitutes a distinct taxon.

Genus V, *Haloferax*, Genus VII, *Halorubrum*, and Genus X, *Natrinema*, should be consulted for further details.

At present, following from the reworking of the taxonomy described by Larsen and Grant (1989), *H. salinarum* is still the only species considered to belong to the genus *Halobacterium*. Problems associated with lack of availability of type strains of the organisms variously called *H. cutirubrum* and *H. halobium*, now subsumed into *H. salinarum*, have now largely receded, since, quoting from Tindall (1992), "strains of *H. salinarium (salinarum)* have been deposited in a variety of culture collections and there seems to be no problem with this organism". Unfortunately, although there are a number of 16S rRNA gene sequences of strains of *H. salinarum* available, largely due to previous confusions over *halobium/salinarium/cutirubrum* identities, the type strain has never been subjected to phylogenetic analysis. However, various comparative studies have provided strong indirect evidence that the organisms that have been sequenced (*H. halobium* DSMZ 671 strain R1, *H. cutirubrum* NRC 34001, and *H. salinarum* DSMZ 668) are likely to be closely related to the type strain as judged by DNA–DNA hybridization (Ross and Grant, 1985; Larsen and Grant, 1989). The cluster of sequences shown in Fig. A2.36 (including unnamed strains Yl2 and G12 plus cloned sequences) indicates a very high sequence similarity with no obvious relationship with any other group of halobacteria. Accordingly, there is no reason not to believe that the organisms, including the type strain, constitute a well-defined monospecific genus.

In view of the large numbers of largely uncharacterized halobacteria in culture collections with specific epithets of *halobium*, *salinarium*, or *cutirubrum*, it is unwise to assume that an organism so named is *H. salinarum sensu stricto*. Only the organisms that have been characterized by chemotaxonomic procedures, or phylogenetic, or DNA–DNA hybridization analysis can definitely be ascribed to this species. These include *H. cutirubrum* CCM 2088, *H. cutirubrum* NRC 34001 (DSMZ 669), *H. halobium* CCM 2090, *H. halobium* NCIMB 736 (NRC 34007), *H. halobium* DSMZ 670

(NRC 34020), *H. halobium* NCIMB 2080 (DSMZ 671), *H. salinarium* CCM 2148, *H. salinarium* CCM 2084, and *H. salinarium* ATCC 19700 (DSMZ 668) (Ross and Grant, 1985).

The R1 strain (DSMZ 671) extensively used in genetic and biochemical studies is a spontaneous gas vacuole-minus mutant of a wild-type strain that is held as *H. halobium* DSMZ 670 (see below). The isolate referred to as *H. halobium* NRC 817 has also been extensively used in genetic studies (Pfeifer, 1988; Krüger and Pfeifer, 1996). The designation 817 is a consequence of a misreading of the DSMZ catalog, and the strain is probably also *H. halobium* DSMZ 670.

There are two genome sequencing projects currently under way involving *H. salinarum* strains. One of these is at The Max-Planck Institute for Biochemistry at Martinsried, Germany, where the strain R1 is being sequenced. The other is a joint venture between the Universities of Massachusetts and Washington in the United States. In this case the strain is NRC-1, whose origin is less clear. NRC unfortunately no longer maintains a culture collection, and the histories of some of their strains that have yielded valuable biochemical and genetic information are difficult to establish. The paper by Stoeckenius and Kunau (1968) indicates that *H. halobium* R1 was isolated as a spontaneous mutant from a plate inoculated with *H. halobium* NRL, obtained from the NRC, Ottawa, Canada. In the paper of Sapienza and Doolittle (1982) it is said that "*H. halobium* strain R1 was isolated as a spontaneously occurring, gas vacuole-deficient variant of the 'wild-type' strain NRC-1 in 1969 (W. Stoeckenius, personal communication)". The DSMZ catalog refers to *H. halobium* DSMZ 670 as the wild-type parent strain of strain R1 received from Stoeckenius, with the NRC as the origin. Accordingly, in all probability, NRL and NRC-1 are one and the same and held as *H. halobium* DSMZ 670. Where and how the strain was isolated is not clear. NRC-1 has a 2000-kb chromosome and an additional 600-kb plasmid fraction. With the exception of a domain of ~210 kb subject to transpositional rearrangements, the two strains have extremely similar chromosome maps (Hackett et al., 1994), supporting the view that they are closely related. Details of these sequencing projects can be accessed at the TIGR Microbial Database site.*

An ordered cosmid library covering 99% of the genome of what is referred to as *Halobacterium* strain GRB has also been constructed (St. Jean et al., 1994; Charlebois, 1995a). The organism was originally isolated from Guerrero Negro in Baja California (Ebert et al., 1984) and has been considered as a suitable candidate for genetic analysis, since it lacks insertion elements (Soppa and Oesterhelt, 1989). Whether this strain is a *Halobacterium* sp. *sensu stricto* has not been established, although the similarity of the restriction map to those of the R1 and NRC1 strains suggest that it is likely to be so (Hackett et al., 1994).

FURTHER READING

Grant, W.D. and H. Larsen. 1989. Extremely halophilic archaeobacteria. *In* Staley, Bryant, Pfennig and Holt (Editors), Bergey's Manual of Systematic Bacteriology, 1st ed., Vol. 3, The Williams & Wilkins Co., Baltimore, pp. 2216–2219.

Hackett, N.R., Y. Bobovnikova and N. Heyrovska. 1994. Conservation of chromosomal arrangement among three strains of the genetically unstable archaeon *Halobacterium salinarium*. J. Bacteriol. *176*: 7711–7718.

Editorial Note: At the time of publication, the TIGR Microbial Database could be accessed at the following website: http://www.tigr.org/tdb/mdb/mdb.html. The genome sequence of NRC-1 is now complete (Ng et al., 2000).

Kamekura, M. 1998. Diversity of extremely halophilic bacteria. Extremophiles 2: 289–296.

Kates, M. 1993. Biology of halophilic bacteria, Part II: Membrane lipids of extreme halophiles: Biosynthesis, function and evolutionary significance. Experientia (Basel) 49: 1027–1036.

Ross, H.N.M. and W.D. Grant. 1985. Nucleic acid studies on halophilic archaebacteria. J. Gen. Microbiol. 131: 165–174.

Tindall, B.J. 1992. The family Halobacteriaceae. In Balows, Trüper, Dworkin, Harder and Schleifer (Editors), The Prokaryotes. A Handbook of Bacteria: Ecophysiology, Isolation, Identification, Applications, 2nd ed., Vol. 1, Springer-Verlag, New York, pp. 768–808.

List of species of the genus Halobacterium

1. **Halobacterium salinarum** corrig. (Harrison and Kennedy 1922) Elazari-Volcani 1957, 208;[AL] emend. Larsen 1984, 262; emend. Larsen and Grant 1989, 2222; emend. Ventosa and Oren 1996, 347 (*Pseudomonas salinaria* Harrison and Kennedy 1922, 120; *Halobacterium halobium* (Petter 1931) Elazari-Volcani 1957, 210; *Halobacterium cutirubrum* (Lochhead 1934) Elazari-Volcani 1957, 209.)

sa.li.na'rum. L. neut. adj. *salinarum* belonging or pertaining to salt works.

Rod-shaped ($0.5–1.0 \times 1.0–6.0$ μm or more in length) but displaying a multitude of pleomorphic forms, especially in deficient media or at elevated temperatures. Some strains contain gas vacuoles. Motile by tufts of polar flagella. Generally aerobic, but may grow anaerobically in the presence of dimethylsulfoxide, trimethylamine *N*-oxide, or fumarate, or fermentatively in the presence of arginine.

Fastest growth occurs at 3.5–4.5 M NaCl; good growth occurs in saturated NaCl (~5.2 M); no growth occurs below 3 M NaCl. Temperature range for growth: 20–55°C; optimum temperature: 50°C. Mg^{2+} requirement: 0.05–0.1 M. pH range: 5.5–8.0.

Chemoorganotrophic. Amino acids are required for growth. Carbohydrates are not utilized, but stimulation of growth is observed in the presence of glycerol. Starch is not hydrolyzed. Indole positive. Gelatin is hydrolyzed. Media become alkaline as a result of the deamination or decarboxylation of amino acids. Urease negative. Arginine dehydrolase present.

Commonly found in neutral salt lakes, marine salterns, and proteinaceous products heavily salted with crude solar salt. Also found in ancient evaporite deposits.

Other characteristics are the same as those described for the genus.

The DNA is composed of a major component and a minor component. No 16S rDNA sequence is available for the type strain.

Other sequences: DSMZ 668 (X92978), DSMZ 669 (K02971), and DSMZ 671 (M38280).

The mol% G + C of the DNA is: 66–70.9 and 57–60 (Bd) (minor component).

Type strain: ATCC 33171, NCMB 764, NRC 34002.

Genus II. **Haloarcula** Torreblanca, Rodriguez-Valera, Juez, Ventosa, Kamekura and Kates 1986b, 573[VP] (Effective publication: Torreblanca, Rodriguez-Valera, Juez, Ventosa, Kamekura and Kates 1986a, 98.)

ANTONIO VENTOSA

Ha.lo.ar'cu.la. Gr. n. *hals, halos* salt; L. fem. n. *arcula* small box; M.L. fem. n. *Haloarcula* salt (-requiring) small box.

Cells may be rods in liquid culture but are extremely pleomorphic, usually flat, 1–3 μm across with a range of shapes from triangles and squares to irregular disks. Resting stages are not known. Gram negative. Motile or nonmotile. **Aerobic or facultatively anaerobic**. Oxidase and catalase positive. **Extremely halophilic, with growth occurring in media containing 2.0–5.2 M NaCl**; optimum concentration, 2.5–3 M NaCl. Temperature range for growth, 30–55°C; optimum temperature, 40–45°C. pH range for growth: 5–8; optimum pH, 7–7.5. Minimal Mg^{2+} requirement for growth, 0.005–0.05 M. Chemoorganotrophic, utilizing many substrates as sources of carbon and energy. **Acids are produced from sugars. Amino acids are not required. The polar lipids are characterized by the presence of $C_{20}C_{20}$ derivatives of TGD-2**.

The mol% G + C of the DNA is: 60.1–65. A minor DNA component may be present.

Type species: **Haloarcula vallismortis** (Gonzalez, Gutiérrez and Ramirez 1978) Torreblanca, Rodriguez-Valera, Juez, Ventosa, Kamekura and Kates 1986b, 573 (Effective publication: Torreblanca, Rodriguez-Valera, Juez, Ventosa, Kamekura and Kates 1986a, 98) (*Halobacterium vallismortis* Gonzalez, Gutiérrez and Ramirez 1978, 710.)

FURTHER DESCRIPTIVE INFORMATION

Cells contain bacterioruberins and retinal pigments. The main polar lipids are $C_{20}C_{20}$ derivatives of PG, PGP-Me, PGS, and TGD-2 (Evans et al., 1980; Ross and Grant, 1985; Torreblanca et al., 1986a). Torreblanca et al. (1986a) indicate that small amounts of DGD-2 are also present. It should be noted that Ross and Grant (1985) and Ross et al. (1985) misnamed TGD-1 as TGD-2 and TGD-2 as TGD-1, respectively.*

Insensitive to penicillin, chloramphenicol, streptomycin, and tetracycline and sensitive to anisomycin, bacitracin, novobiocin, and vibriostat (0/129). Some species are very susceptible to heavy metals (Nieto et al., 1987).

Organisms like these have been found in different thalassohaline and athalassohaline environments, including the Dead Sea, salt pools in Death Valley, and water and soil of salterns. Members of the genus *Haloarcula*, closely related to *H. hispanica*, have been isolated from 57-year-old enrichments prepared from Dead Sea water samples (Arahal et al., 1996).

---

*See *Halobacteriales* for definitions of lipid abbreviations.

ENRICHMENT AND ISOLATION PROCEDURES

These organisms grow well on many of the standard media used for culturing halobacteria. There are no known ways of specifically enriching for organisms of this group.

DIFFERENTIATION OF THE GENUS HALOARCULA FROM OTHER GENERA

The key to the family *Halobacteriaceae* indicates those characteristics useful for distinguishing *Haloarcula* from the other genera of the family. The pleomorphic nature of the cells, the use of a wide range of carbon sources and the relatively low Mg^{2+} requirement for growth are not in themselves sufficient to distinguish these organisms from other pleomorphic types, and they can be reliably separated from other halobacteria only by chemotaxonomic or molecular procedures. The polar lipids of these organisms, with the exception of *Haloarcula mukohataei*, characteristically contain the triglycosyl glycolipid TGD-2, and this is the simplest way of distinguishing the genus *Haloarcula*. They can also be differentiated by comparative analysis of their 16S rRNA sequences.

TAXONOMIC COMMENTS

The definition of the genus *Haloarcula* was based on the numerical taxonomic study of Torreblanca et al. (1986a), based on phenotypic features, and on the polar lipid composition of the strains, adopting the generic name first suggested by Javor et al. (1982) for a variety of pleomorphic isolates. Later, DNA–DNA hybridization experiments of representative strains of the *Haloarcula* phenons (Gutiérrez et al., 1990) supported this study. The type species of the genus, *Haloarcula vallismortis*, was originally classified in the genus *Halobacterium* (Gonzalez et al., 1978).

A common feature of the representative strains of the genus *Haloarcula* is the presence of at least two heterogeneous 16S rRNA genes. This property has been reported in the species *Haloarcula vallismortis*, *H. marismortui*, *H. quadrata*, and *"H. sinaiiensis"* (Myl-

vaganam and Dennis, 1992; Kamekura, 1998b; Oren et al., 1999). The 16S rRNA sequence comparison shows that the species of the genus *Haloarcula*, with the exception of *H. mukohataei*, constitute a consistent monophyletic group (Fig. A2.36 of *Halobacteriales* overview). Two 16S rRNA signature sequences, as well as several 16S rRNA signature bases specific for *Haloarcula* have been proposed (Tables A2.25 and A2.26 of *Halobacteriales* overview).

The recently described species *H. mukohataei* was placed in the genus *Haloarcula* on the basis of the similarity of its 16S rRNA gene sequence to the 16S rRNA sequences of the other species of this genus. However, its different polar lipid composition as well as important differences in the 16S rRNA signature bases (Table A2.26 of *Halobacteriales* overview) could justify the transfer of this species to a new genus.

ACKNOWLEDGMENTS

I am grateful to W.D. Grant, M. Kamekura, and A. Oren for their comments and suggestions during the preparation of this article.

FURTHER READING

Kamekura, M. 1999. Diversity of members of the family *Halobacteriaceae*. *In* Oren (Editor), Microbiology and Biogeochemistry of Hypersaline Environments, CRC Press, Boca Raton, pp. 13–25.

Oren, A. 1999. The enigma of square and triangular halophilic bacteria. *In* Seckbach (Editor), Enigmatic Microorganisms and Life in Extreme Environments, Kluwer Academic Publishers, Dordrecht, pp. 339–355.

Tindall, B.J. 1992. The family *Halobacteriaceae*. *In* Balows, Trüper, Dworkin, Harder and Schleifer (Editors), The Prokaryotes. A Handbook of Bacteria: Ecophysiology, Isolation, Identification, Applications, 2nd ed., Vol. 1., Springer-Verlag, New York, pp. 768–808.

Torreblanca, M., F. Rodriguez-Valera, G. Juez, A. Ventosa, M. Kamekura and M. Kates. 1986. Classification of non-alkaliphilic halobacteria based on numerical taxonomy and polar lipid composition, and description of *Haloarcula* gen. nov. and *Haloferax* gen. nov. Syst. Appl. Microbiol. 8: 89–99.

DIFFERENTIATION OF THE SPECIES OF THE GENUS HALOARCULA FROM OTHER GENERA

Some differential features of the species of the genus *Haloarcula* are given in Table A2.27.

TABLE A2.27. Differential characteristics of the species of the genus *Haloarcula*[a]

Characteristic	1. *H. vallismortis*	2. *H. argentinensis*	3. *H. hispanica*	4. *H. japonica*	5. *H. marismortui*	6. *H. mukohataei*	7. *H. quadrata*
Acid from:							
Fructose	+	+	+	+	+	−	−
Galactose	−	+	+	+	nd	+	+
Mannose	−	+	+	−	−	+	−
Starch hydrolysis	+	nd	+	−	−	nd	+
Gelatin hydrolysis	−	nd	+	−	−	nd	−
Indole production	+	nd	d	+	−	nd	−
TGD-2 as glycolipid	+	+	+	+	+	−	+
Mol% G + C	64.7	62	62.7	63.3	62	65	60.1

[a]For symbols see standard definitions; nd, not determined.

List of species of the genus Haloarcula

1. **Haloarcula vallismortis** (Gonzalez, Gutiérrez and Ramirez 1978) Torreblanca, Rodriguez-Valera, Juez, Ventosa, Kamekura and Kates 1986b, 573[VP] (Effective publication: Torreblanca, Rodriguez-Valera, Juez, Ventosa, Kamekura and Kates 1986a, 98) (*Halobacterium vallismortis* Gonzalez, Gutiérrez and Ramirez 1978, 710.)

val.lis.mor' tis. L. gen. n. *vallis* of the valley; L. gen. n. *mortis* of death; M.L. fem. n. *vallismortis* of the valley of death; named after Death Valley, California.

See Table A2.27 and the generic description for many features.

Cells are pleomorphic rods, 0.6–1.0 × 3–5 μm. Motile.

Optimal growth in media containing 4.3 M NaCl; no growth below 2.5 M. Minimal Mg^{2+} concentration, 0.005 M (Torreblanca et al., 1986a). Temperature range for growth: 20–45°C; optimal growth at 40°C. The pH range for growth is 5.5–8.5. Optimum pH, 7.4–7.5.

Possibly facultatively anaerobic in the presence of nitrate. Able to use a variety of compounds as sole carbon and energy sources (glucose, fructose, galactose, sucrose, maltose, trehalose, glycerol, and gluconate). Acid is produced from sugars. Indole is produced. Nitrate is reduced, producing gas. Starch is hydrolyzed. H$_2$S is produced from cysteine.

Isolated from salt pools in Death Valley.

A minor DNA component has not been detected, although large plasmids are present (Gutiérrez et al., 1986).

The mol% G + C of the DNA is: 64.7 (Bd).

Type strain: J.F. 54, ATCC 29715, DSMZ 3756.

GenBank accession number (16S rRNA): U17593.

2. **Haloarcula argentinensis** Ihara, Watanabe and Tamura 1997, 76.[VP]

ar.gen.ti.nen' sis. M.L. fem. adj. *argentinensis* Argentinian.

See Table A2.27 and the generic description for many features.

Cells are triangular and flat, 0.3 × 1.0 μm. Motile by polar flagella. Colonies are orange-red and mucoid on old plates.

Chemoorganotrophic. Strictly aerobic. Acid is produced from glucose and other sugars (sucrose, maltose, galactose, mannose, ribose, glycerol, and fructose). The fructose-bisphosphate aldolase is class I (EDTA insensitive).

Growth occurs in media containing 2.0–4.5 M NaCl. Optimum growth at 2.5–3.0 M NaCl at 40°C. Minimal Mg^{2+} concentration, 0.03 M. Optimum Mg^{2+} concentration, 0.1 M. Optimum growth temperature: 40°C.

Contains a retinal protein, designated cruxrhodopsin-1 (Tateno et al., 1994a).

Isolated from soil from salterns in Argentina.

The mol% G + C of the DNA is: 62 (T$_m$).

Type strain: arg-1, DSMZ 12282, JCM 9737.

GenBank accession number (16S rRNA): D50849.

3. **Haloarcula hispanica** Juez, Rodriguez-Valera, Ventosa and Kushner 1986b, 573[VP] (Effective publication: Juez, Rodriguez-Valera, Ventosa and Kushner 1986a, 78.)

his.pa' ni.ca. L. fem. adj. *hispanica* Spanish.

See Table A2.27 and generic description for many features.

Cells are small pleomorphic rods, 0.5–1.0 × 0.3 μm. Motile by polar flagella.

Growth occurs in media containing 2.5–5.2 M NaCl. Optimum NaCl concentration: 3.5–4.2 M. Minimal Mg^{2+} concentration: 0.005 M. Temperature range for growth: 25–50°C. Optimal growth at 35–40°C. The pH range for growth is 6–8. Optimum pH, 7.0.

Possibly facultatively anaerobic in the presence of nitrate. Utilizes a variety of compounds as sole carbon and energy sources (glucose, lactose, sucrose, glycerol, mannitol, sorbitol, acetate, citrate, lactate, malate, pyruvate, succinate, arginine, glutamine, and lysine). Acid is produced from sugars. Starch is hydrolyzed. Some strains hydrolyze casein. Indole is variable. Nitrate is reduced with gas production. Susceptible to only high concentration of bacitracin (40 μg/ml).

Isolated from marine salterns in Spain.

The mol% G + C of the DNA is: 62.7 (T$_m$).

Type strain: Y27, ATCC 33960, DSMZ 4426.

GenBank accession number (16S rRNA): U68541.

4. **Haloarcula japonica** Takashina, Hamamoto, Otozai, Grant and Horikoshi 1991, 178[VP] (Effective publication: Takashina, Hamamoto, Otozai, Grant and Horikoshi 1990, 180.)

ja.po' ni.ca. M.L. fem. adj. *japonica* Japanese.

See Table A2.27 and generic description for many features.

Cells are pleomorphic, triangular, and flat, 0.2–0.5 × 2.0–5.0 μm. Gas vacuoles not produced. Motile.

Growth occurs in media containing 2.5–5 M NaCl. Optimum NaCl concentration, 3.5 M. Minimal Mg^{2+} concentration, 0.04 M. Optimal Mg^{2+} concentration, 0.08 M. Temperature range for growth, 24–45°C. Optimal growth at 42–45°C. The pH range for growth is 6–8. Optimum pH, 7–7.5.

Chemoorganotrophic; able to use a wide variety of compounds as sole carbon and energy sources (glucose, fructose, galactose, arabinose, xylose, rhamnose, sucrose, glycerol, mannitol, and sorbitol). Utilization of sugars produces acidification of the medium. Nitrate and nitrite are reduced with gas formation. Casein, gelatin, and starch are not hydrolyzed. Indole positive. H$_2$S is produced from thiosulfate.

A minor DNA component or plasmid of ~145 kb has been detected.

Isolated from saltern soil in Japan.

The mol% G + C of the DNA is: 63.3 (Bd).

Type strain: TR-1, DSMZ 6131, JCM 7785.

GenBank accession number (16S rRNA): D28872.

5. **Haloarcula marismortui** (Elazari-Volcani 1957) Oren, Ginzburg, Ginzburg, Hochstein and Volcani 1990, 210[VP] (*Halobacterium marismortui* Elazari-Volcani 1957, 210.)

ma.ris.mor' tu.i. L. gen. n. *maris* of the sea; L. masc. adj. *mortuus* dead; M.L. gen. n. *marismortui* of the Dead Sea.

See Table A2.27 and the generic description for many features.

Cells are pleomorphic, flat, and disk shaped (1.0–2.0 × 2.0–3.0 μm). Motile; motility often difficult to observe.

Growth occurs in media containing 1.7–5.1 M NaCl. Optimum NaCl concentration, 3.4–3.9 M. Optimum growth temperature, 40–50°C.

Chemoorganotrophic; able to use a wide variety of compounds as sole carbon and energy sources (glucose, fructose, sucrose, glycerol, acetate, succinate, and malate). Acid is produced from glucose, fructose, ribose, xylose, maltose, sucrose, mannitol, sorbitol, and glycerol. Utilization of nitrate occurs under anaerobic conditions. Starch is very slowly hydrolyzed.

Isolated from the Dead Sea.

The mol% G + C of the DNA is: 55 (Bd).

Type strain: ATCC 43049, DSMZ 3752.

GenBank accession number (16S rRNA): X61688 (*rrn*A gene), X61689 (*rrn*B gene).

6. **Haloarcula mukohataei** Ihara, Watanabe and Tamura 1997, 76.[VP]

mu.ko.ha' tae.i. M.L. adj. *mukohataei* of Mukohata; named after Japanese biochemist and biophysicist Yasuo Mukohata.

See Table A2.27 and the generic description for many features.

Cells are short rods (0.5 × 2.0 μm) in actively growing cultures and pleomorphic in stationary cultures.

Growth occurs in media containing 2.5–4.5 M NaCl. Optimum growth at 3.0–3.5 M NaCl at 40°C. Minimal Mg^{2+} concentration: 0.003 M. Tolerant to 0.3 M Mg^{2+}. Optimum growth temperature, 40°C.

Chemoorganotrophic. Strictly aerobic. Acid is produced from glucose and other sugars (sucrose, maltose, galactose, mannose, ribose, and glycerol). The fructose-bisphosphate aldolase is class I (EDTA insensitive).

Contains a retinal protein, designated cruxrhodopsin-2 (Sugiyama et al., 1994).

Isolated from soil of salterns in Argentina.

The mol% G + C of the DNA is: 65 (T_m).

Type strain: arg-2, DSMZ 12286, JCM 9738.

GenBank accession number (16S rRNA): D50850.

7. **Haloarcula quadrata** Oren, Ventosa, Gutiérrez and Kamekura 1999, 1154.[VP]

qua.dra' ta. L. fem. adj. *quadrata* square.

See Table A2.27 and generic description for many features.

Cells are pleomorphic, predominantly flat, and square (2–3 μm) (Fig. A2.38). Motile by one or more flagella.

Growth occurs in media containing 2.7–4.3 M NaCl. Optimum NaCl concentration, 3.4–4.3 M. Minimal Mg^{2+} concentration, 0.05 M. Optimum Mg^{2+} concentration, 0.1–0.5 M. Optimum growth temperature, 53°C. The pH range for growth is 5.9–8.0. Optimum pH, 6.5–7.0.

Chemoorganotrophic; able to use glucose, sucrose, and maltose as single carbon and energy sources. Acid is pro-

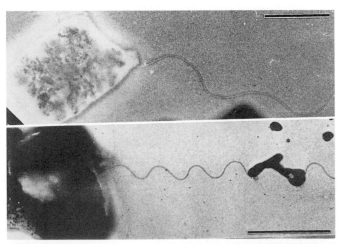

FIGURE A2.38. Electron micrographs of monotrichously flagellated cells of *Haloarcula quadrata*. Bar = 1 μm. (Reprinted with permission from Alam et al., EMBO Journal *3:* 2899–2903, 1984, ©Oxford University Press.)

duced from glucose, galactose, sucrose, xylose, and ribose. Anaerobic growth in the presence of nitrate with formation of nitrite and gas. Starch is hydrolyzed. Gelatin, casein, Tween-20, and Tween-80 are not hydrolyzed. Indole is not produced.

Isolated from a brine pool in the Sinai Peninsula (Egypt).

The mol% G + C of the DNA is: 60.1 (T_m).

Type strain: 801030/1, DSMZ 11927.

GenBank accession number (16S rRNA): AB010964 (*rrn*A gene), AB010965 (*rrn*B gene).

8. **"Haloarcula californiae"** Javor, Requadt and Stoeckenius 1982, 1532.

The taxonomic status of this organism is uncertain because of a lack of detailed descriptive information.

Originally described largely on the basis of its unusual angular pleomorphic flat cells, with rectangular and square forms being common. Torreblanca et al. (1986a) described the organism as being nonproteolytic but lipolytic, producing H_2S from thiosulfate, reducing nitrate, and being amylase negative. Amino acids are not required for growth, and acid is produced from sugars. The polar lipid composition is similar to that reported for the genus *Haloarcula*. DNA–DNA hybridization studies show the closest relationship to *H. marismortui* (Gutiérrez et al., 1990) with low similarity values with other species of *Haloarcula*.

Isolated from a Californian saltern.

The mol% G + C of the DNA is: 59.8.

Deposited strain: BJGN-2, ATCC 33799.

9. **"Haloarcula sinaiiensis"** Javor, Requadt and Stoeckenius 1982, 1532.

The taxonomic status of this organism is uncertain because of a lack of detailed descriptive information. Stated as being different from *"H. californiae"* on the basis of restriction endonuclease digest patterns. Does not group with any other in the numerical taxonomic analysis of Torreblanca et al. (1986a). The polar lipid pattern is similar to that of the species of the genus *Haloarcula*. Two heterologous 16S rRNA genes have been described. The 16S rRNA sequence comparison reveals a closer relationship with *H. vallismortis* than with other species of *Haloarcula*.

Isolated from a Red Sea sabkha.

The mol% G + C of the DNA is: 59.7.

Deposited strain: BJSG-2, ATCC 33800.

GenBank accession number (16S rRNA): D14129 (major gene) and D14130 (minor gene).

10. **"Haloarcula aidinensis"** Zhou, Xu, Xiao, Ma and Liu 1994, 89.

The taxonomic status of this organism is uncertain because of a lack of detailed descriptive information. The description of this species is based on the phenotypic features and polar lipid content of three strains (A5, B2, and B-B2) isolated from salt lakes of Xinjiang, China.

Deposited strain: A5, AS 1.2042.

GenBank accession number (16S rRNA): AB000563 (strain B2).

Species Incertae Sedis

1. Walsby's square bacterium Walsby 1980, 69.

Square bacteria were originally seen in brine from a sabkha in the southern Sinai (Walsby, 1980), and since then they have been seen in a variety of other saline habitats in Israel, California, Spain, and Senegal (Oren, 1999). The bacteria are characterized as flat, perfectly square boxes (Fig. A2.39) with sides 2–5 μm long and an overall thickness of 0.1–0.2 μm. The best analogy for the shape and form of these organisms is that of postage stamps, and sheets containing up to 64 have been described (Kessel and Cohen, 1982). The organisms are nonmotile and gas vacuolate. To date, these gas vacuole-containing square bacteria have not been grown in culture, although other halobacteria can be readily isolated from such habitats. It is possible that the shape is determined by environmental factors. There is no doubt that these organisms are halobacteria, since they contain bacterioruberins and retinal pigments (Stoeckenius, 1981) and their polar lipid pattern is similar to those of the halobacteria (Fredrickson et al., 1989; Oren et al., 1996). They also have a hexagonal pattern of subunits on the cell surface similar to those found in other halobacteria (Kessel et al., 1985).

These organisms are difficult to see with the light microscope but are probably common in saline environments throughout the world. In the absence of any convincing laboratory cultures, it is not possible to tell whether more than one type exists. There is no *a priori* reason that these should be associated with the genus *Haloarcula* other than the observation that the *Haloarcula* species do occasionally produce squares in culture, although the sheets of cells characteristic of Walsby's organisms have not been seen in any of the laboratory cultures of pleomorphic organisms. Polar lipid analysis of a brine sample from a saltern in Eilat, Israel, in which the majority of the cells showed a square to trapezoidal flat shape with gas vacuoles, showed a glycolipid pattern similar to that of *Haloferax*, while PGS was present (Oren et al., 1996), and it can be assumed that square bacteria are not necessarily related to the genus *Haloarcula*.

FIGURE A2.39. Phase-contrast micrographs of square bacteria after concentration of brine. The gas vacuoles were collapsed by the centrifugation and are not visible. (Reproduced with permission from W. Stoeckenius, *Journal of Bacteriology 148:* 352–360, 1981, ©American Society for Microbiology.)

Genus III. **Halobaculum** Oren, Gurevich, Gemmell and Teske 1995, 752[VP]

AHARON OREN

Ha.lo.ba' cu.lum. Gr. n. *hals, halos* sea, salt; L. neut. n. *baculum* stick; M.L. neut. n. *Halobaculum* salt stick.

Gram-negative rods. Extremely halophilic, with optimum growth occurring in media containing 1–2.5 M NaCl and 0.6–1 M MgCl₂. Optimum temperature 40°C. Colonies small, round, convex, entire, translucent orange-red. Oxidase and catalase positive. Chemoorganotrophic; **aerobic**. Nitrate reduced to nitrite. No growth occurs anaerobically with nitrate or with arginine. Yeast extract and Casamino acids in low concentrations are good sources of organic nutrients. Growth on single carbon sources not observed. Certain carbohydrates stimulate growth with acid production. Polar lipids are characterized by the presence of C₂₀C₂₀ derivatives of sulfated diglycosyl glycolipid (S-DGD-1) and the absence of phosphatidyl glycerol sulfate (PGS).

The mol% G + C of the DNA is: 70.

Type species: **Halobaculum gomorrense** Oren, Gurevich, Gemmell and Teske 1995, 753.

FURTHER DESCRIPTIVE INFORMATION

Cells of *Halobaculum gomorrense*, the only *Halobaculum* species described or isolated thus far, are rod shaped, measuring 5–10 × 0.5–1 μm, and nonmotile. The strain lacks the typical pleomorphic flattened shape of *Haloferax* species. However, when first isolated, the strain was pleomorphic, and it acquired its rod shape upon subculturing. Gas vesicles have not been detected. Cells stain Gram-negative.

High salt concentrations required for structural integrity. Both in media high in magnesium and in suspensions in NaCl, at least 15% salt is needed to maintain the native rod shape. At lower concentrations cells become spherical, and below 5% salt lysis occurs.

Culture requires at least 1 M NaCl for growth (in the presence of 0.8 M MgCl₂). Optimum NaCl concentration range: 1.5–2.5 M at 35–40°C; optimum MgCl₂ concentration, 0.6–1 M (in the presence of 2.1 M NaCl). Optimum temperature, 40°C (in medium containing 2.1 M NaCl and 0.8 M MgCl₂). Optimum pH, 6–7.

Growth on single carbon sources was never observed. Glucose, maltose, sucrose, galactose, xylose, trehalose, starch, and glycerol stimulate growth with acid production. Growth is stimulated by lactate. Yeast extract and Casamino acids at a concentration of 0.5% are inhibitory. Including starch in the medium proved highly stimulatory; starch was hydrolyzed. Gelatin is not hydrolyzed. Indole is not produced.

Cells contain bacterioruberins but may not contain bacteriorhodopsin (Oren et al., 1995). The main polar lipids are PG, PGP-Me, and the sulfated diglycosyl glycolipid (S-DGD-1).*

Sensitive to novobiocin, bacitracin, anisomycin, vibriostat (0/ 129), taurocholate, and deoxycholate and insensitive to penicillin, ampicillin, kanamycin, chloramphenicol, streptomycin, and neomycin.

ENRICHMENT AND ISOLATION PROCEDURES

The only isolate of the only species described thus far was obtained from a sample collected in July, 1992, from a depth of 4 m at the deepest part of the Dead Sea. It developed in the highest positive tube of a dilution series in a medium rich in magnesium, resembling that designed for *Halorubrum sodomense* (Oren, 1983), containing (in g/l): NaCl, 125; $MgCl_2 \cdot 6H_2O$, 160; K_2SO_4, 5.0; $CaCl_2 \cdot 2H_2O$, 0.1; yeast extract, 1.0; Casamino acids, 1.0; and starch, 2.0; pH 7.

In many properties, *Halobaculum gomorrense* resembles *Halorubrum sodomense*. Thus, very high magnesium concentrations are required for optimal growth, making them both well adapted to the extremely high magnesium concentrations found in the Dead Sea. The isolation of *Halobaculum* species has never been reported from other environments.

The polar lipid composition of the biomass collected from the Dead Sea during an archaeal bloom in 1992 showed the presence of PG, PGP-Me, and S-DGD-1, and a lack of PGS. This was first considered as evidence that *Haloferax*-like organisms dominated the archaeal community (Oren and Gurevich, 1993). However, the discovery of *H. gomorrense*, with its much higher magnesium tolerance than most *Haloferax* species, made it feasible that *Halobaculum*-type organisms were in fact dominant. As long as no selective growth and enrichment procedures for *Halo-*

baculum exist, and as long as the main method to differentiate it from other halophilic archaea is based on comparison of 16S rRNA sequences, there is no simple way to assess the importance of *Halobaculum* in the Dead Sea and in other environments inhabited by halophilic archaea.

MAINTENANCE PROCEDURES

Growth on agar slants in screw-cap tubes can be kept in the refrigerator for 4–5 months. Lyophilization gives satisfactory results. No experience has been obtained with frozen storage.

DIFFERENTIATION OF THE GENUS *HALOBACULUM* FROM OTHER GENERA

The genus *Halobaculum* can be differentiated from most other archaea by the presence of S-DGD-1 and the lack of PGS. Differentiation between *Halobaculum* and *Haloferax*, which possess identical polar lipid compositions (Tindall, 1992), can only be based on comparison of 16S rRNA nucleotide sequences (Oren et al., 1995). *H. gomorrense* shares with *Halorubrum sodomense* a relatively low requirement for sodium ions and a very high requirement for magnesium for optimal growth (Oren, 1983). The two can be differentiated by the type of glycolipid present and the presence or absence of PGS.

TAXONOMIC COMMENTS

The definition of the genus *Halobaculum* is based on the characterization of polar lipids (enabling differentiation from most other genera of the *Halobacteriaceae*, with the exception of *Haloferax*) and 16S rRNA analysis, showing only remote phylogenetic relationship with the other recognized genera within the *Halobacteriaceae*. Sequence similarities of 89.0–89.2% and 88.8–89.4% were found with representatives of the genera *Haloferax* and *Halorubrum*, respectively. The intermediate position of *Halobaculum*, with respect to *Halorubrum* and *Haloferax*, is suggested by the 16S rRNA sequence data (Fig. A2.36), is thus confirmed by its physiological characteristics (Table A2.28).

---

*See *Halobacteriales* for definitions of lipid abbreviations.

TABLE A2.28. Characteristics differentiating *Halobaculum*, *Haloferax*, and *Halorubrum* species[a]

Characteristic	*Halobaculum gomorrense*	*Haloferax denitrificans*	*Haloferax gibbonsii*	*Haloferax mediterranei*	*Haloferax volcanii*	*Halorubrum sodomense*
Cell morphology	Rod	Pleomorphic	Pleomorphic	Pleomorphic	Pleomorphic	Rod
Motility	−	−	+	d	d	+
Gas vesicles	−	−	−	+	−	−
Bacteriorhodopsin	−	−	−	−	−	+
Lipids:						
S-DGD-1	+	+	+	+	+	−
PGS	−	−	−	−	−	+
Optimum Mg^{2+} conc.	>0.5 M	≤0.5 M	≤0.5 M	≤0.5 M	≤0.5 M	>0.5 M
Indole from tryptophan	−	−	+	+	+	−
Nitrite from nitrate	+	+	d	+	+	+
Gas from nitrate	−	+	−	+	−	−
Acid from:						
Mannose	−	−	+	nd	nd	−
Galactose	+	+	+	nd	+	−
Fructose	−	+	+	+	+	+
Lactose	−	−	nd	+	+	−
Arabinose	−	nd	+	+	nd	−
Starch hydrolyzed	+	−	−	+	−	+
Mol% G + C	70	64	62–64	59–62	63–66	68

[a]For symbols see standard definitions; nd, not determined.

List of species of the genus Halobaculum

1. **Halobaculum gomorrense** Oren, Gurevich, Gemmell and Teske 1995, 747.[VP]

go.mor.ren' se. M.L. neut. adj. *gomorrense* pertaining to Gomorra, a biblical city near the Dead Sea.

Cells are rod shaped (5–10 × 0.5–1 μm). Nonmotile. Gas vesicles lacking.

Requires at least 1 M NaCl (in the presence of 0.8 M MgCl$_2$). Optimal growth in 1.5–2.5 M NaCl and 0.6–1 M MgCl$_2$; optimum temperature 40°C.

Aerobic; no growth occurs anaerobically with nitrate or with arginine; nitrate reduced to nitrite.

No indole produced from tryptophan.

Susceptible to novobiocin, bacitracin, anisomycin, vibriostatic agent 0/129, taurocholate, and deoxycholate; insensitive to penicillin, ampicillin, kanamycin, chloramphenicol, streptomycin, and neomycin.

Glucose, maltose, sucrose, galactose, xylose, trehalose, starch, and glycerol stimulate growth with acid production. Starch hydrolyzed. No acid produced from mannose, fructose, ribose, lactose, arabinose, mannitol, or sorbitol. Growth stimulated by lactate. No stimulation observed by acetate, citrate, propionate, succinate, glycine, alanine, or glutamate.

Found in the Dead Sea.

Other characteristics are the same as those described for the genus.

The mol% G + C of the DNA is: 70 (HPLC).

Type strain: Strain DS2807, DSMZ 9297.

GenBank accession number (16S rRNA): L37444.

Genus IV. **Halococcus** Schoop 1935a, 817[AL]

WILLIAM D. GRANT

Ha.lo.coc' cus. Gr. n. *hals, halos* the sea, salt; Gr. n. *kokkos* a berry; M.L. masc. n. *Halococcus* salt-coccus.

Cocci, 0.8–1.5 μm in diameter, occurring in pairs, tetrads, sarcina packets, or irregular clusters. Divide by septation. Does not form spores. **Stain mainly Gram-negative with at least some cells Gram-positive. No lysis occurs in hypotonic solutions or in the presence of** *N*-**laurylsarcosine. Nonmotile. Strictly aerobic.** Oxidase and catalase positive. **Extremely halophilic, requiring at least 2.5 M NaCl for growth and 3.5–4.5 M NaCl for best growth.** Optimum temperature: 30–40°C; maximum 50°C; minimum 28°C. pH range for growth 6–9.5. Chemoorganotrophic.

16S rRNA signature sequences are listed in Table A2.25 in Order *Halobacteriales.*

The mol% G + C of the DNA is: 59.5–67. Some strains contain satellite DNA (~30% of total DNA) of lower mol% G + C compared to the major DNA component.

TABLE A2.29. Characteristics differentiating *Halococcus morrhuae, Halococcus saccharolyticus,* and *Halococcus salifodinae*[a]

Characteristic	1. *H. morrhuae*	2. *H. saccharolyticus*	3. *H. salifodinae*
Acid production from glucose	−	+	+
Tween-80 hydrolysis	+	−	nd
Utilization of compounds as carbon and energy sources:			
Amygdalin	−	+	nd
Lactose	−	+	nd
D-salicin	−	+	nd
D-xylose	d	−	+
Erythritol	−	+	−
L-arabinose	−	+	+
L-rhamnose	−	−	+
Adonitol	−	+	nd
Ethanol	−	+	nd
Erythritol	−	+	−
Glycerol	d	+	+
cis-aconitate	−	+	nd
D-gluconate	−	+	nd
D-glucuronate	−	+	nd
Hippurate	−	+	nd
DL-lactate	−	+	nd
Propionate	−	+	nd
Quinate	−	+	nd
D-saccharate	−	+	nd
L-methionine	−	+	nd
Sensitivity to antibiotics:			
Chloramphenicol	−	−	+
Tetracycline	−	−	+
Novobiocin	+	−	+
Growth at pH 9.5	−	−	+
Mg^{2+} requirement	5 mM	40 mM	≤ 0.1 mM

[a]For symbols see standard definitions; nd, information not available.

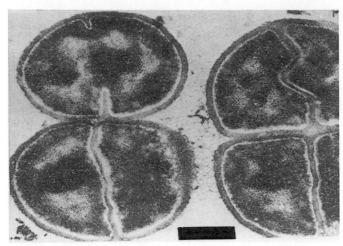

FIGURE A2.40. *Halococcus morrhuae* 24 (DSMZ 1309, CCM 2226). Thin section of cells at the end of the exponential growth phase. Bar = 1 µm. (Reprinted with permission from H. Steensland and H. Larsen, Kgl. Norske Vidensk. Selsk. Skrifter No. 8, Universitetsforlaget, pp. 1–5, 1971,©The Norwegian Research Council for Science and the Humanities, Oslo.)

Type species: **Halococcus morrhuae** (Klebahn 1919) Kocur and Hodgkiss 1973, 154 (Not *Sarcina litoralis* Poulsen 1879, 254; not *Sarcina morrhuae* Farlow 1880, 974) (*Sarcina morrhuae* Klebahn 1919, 38; *Micrococcus (Diplococcus) morrhuae* Klebahn 1919, 42; *Halococcus litoralis* Schoop 1935a, 817.)

FURTHER DESCRIPTIVE INFORMATION

Red, extremely halophilic cocci have a thick (30–60 nm) rigid cell wall and form septa, not constrictions, during division (Fig. A2.40). The wall material contains, in addition to simple sugars, amino sugars, uronic acids, and glycine, forming a complex heteroglycan which is highly sulfated, in some cases sulfonated, and apparently responsible for the rigid structure of the wall and the resistance to lysis in hypotonic solutions and lysis by *N*-laurylsarcosine (Reistad, 1975; Schleifer et al., 1982; Kamekura, 1998a).

The main lipids are both $C_{20}C_{20}$ and $C_{20}C_{25}$ derivatives of PG, PGP-Me, and the sulfated diglycosyl glycolipid S-DGD-1.* The presence of another glycolipid in *H. morrhuae*, provisionally identified as TGD-2, was originally reported (Ross and Grant, 1985), but the precise identity of this glycolipid has not subsequently been confirmed. Unidentified glycolipids are also present in the other species (Montero et al., 1989; Denner et al., 1994). Phosphatidyl glycerol sulfate is absent (Ross and Grant, 1985; Kamekura, 1998b; Kates, 1993). The cells contain carotenoids which are of the C_{50} bacterioruberin type, menaquinones (MK-8 and MK-8(H_2)) but not ubiquinones, and retinal, as reported for other halobacteria (Kushwaha et al., 1974; Collins et al., 1981).

As in the case of other halobacteria, the proteins are acidic (Reistad, 1970; Denner et al., 1994), and enzymes are activated by salt (Larsen, 1967). Although the cell wall is resistant to lysis in hypotonic solutions, the cell membrane disintegrates (Reistad, 1970).

Some *H. morrhuae* strains have been reported to contain satellite DNA (~30% of total DNA), with the mol% G + C of the major DNA component being 67% and that of the minor component being 59% (Moore and McCarthy, 1969). Ross and Grant (1985) also reported that the DNA of the type strain of *H. morrhuae* contained a major and a minor component. Other species have not been investigated for the presence of a minor DNA component.

A yeast extract-salt medium similar to that described for other halobacteria (see footnote in Genus I. *Halobacterium*) is a suitable growth medium. *H. saccharolyticus* requires a higher concentration of Mg^{2+} than other halococci (Montero et al., 1989). Good aeration must be provided. Growth is slow even under optimum conditions, and a generation time of ~14 h is the best that has been reported. A number of amino acids, purines, and pyrimidines are required for growth (Onishi et al., 1965). *H. morrhuae* strains have generally been regarded as being unreactive in typical substrate utilization tests (Kocur and Hodgkiss, 1973). However, in the course of a numerical taxonomy study of around 100 strains of halococci isolated from marine salterns, Montero et al. (1988) showed that many of the 80 strains in the phenon that contained the reference strains of *H. morrhuae* were nutritionally quite versatile, being able to use a variety of compounds as sole carbon, nitrogen, and energy sources. *Halococcus saccharolyticus*, defined originally by the numerical taxonomy study of Montero et al. (1988), is much more metabolically versatile, growing on a wide range of compounds as sole carbon and energy sources (Montero et al., 1989). *Halococcus salifodinae* grows in a minimal medium containing 0.1% yeast extract supplemented with various sugars (Denner et al., 1994). The slow growth on laboratory media may be the reason why these halobacteria seem to be less frequently encountered in natural samples. They are, however, found in the same places as other halobacteria, in strongly saline lakes and marine salterns, and have been shown to cause spoilage of salted fish and hides. They have also been isolated from seawater (Rodriguez-Valera et al., 1979), hypersaline soils (Quesada et al., 1982), salterns (Montero et al., 1988), and salt mines (Norton et al., 1993; Denner et al., 1994).

Like other halobacteria, they are largely insensitive to antibiotics such as penicillin, chloramphenicol, streptomycin, and tetracycline, but they are usually sensitive to bacitracin, aphidicolin, anisomycin, and novobiocin (Hilpert et al., 1981; Bonelo et al., 1984).

ENRICHMENT AND ISOLATION PROCEDURES

The methods are the same as those given for *Halobacterium*. Addition of sugars to media containing >4% (w/v) $MgCl_2$ might be expected to select for *H. saccharolyticus*. *H. salifodinae* grows in a minimal medium (0.1% (w/v) yeast extract) supplemented with glucose as a carbon source.

MAINTENANCE PROCEDURES

See under Order *Halobacteriales* — Further descriptive information.

DIFFERENTIATION OF THE GENUS *HALOCOCCUS* FROM OTHER GENERA

The key to the family *Halobacteriaceae* indicates characteristics useful in distinguishing *Halococcus* spp. from members of other

---

*See *Halobacteriales* for definitions of lipid abbreviations.

genera. In particular, the invariant coccoid morphology and the stability of the cells on suspension in distilled water are useful diagnostic features. The only other coccoid halobacteria (*Natronococcus* spp.) are obligately alkaliphilic.

Differential features of the three currently recognized species *H. morrhuae*, *H. saccharolyticus*, and *H. salifodinae* are given in Table A2.29. For an exhaustive listing of phenotypic differences between numerical taxonomy phenons containing *H. morrhuae* and *H. saccharolyticus* strains see Montero et al. (1988).

TAXONOMIC COMMENTS

The history of the genus *Halococcus* includes a considerable amount of controversy. The genus *Halococcus*, as created by Schoop (1935a), comprises the red, extremely halophilic cocci. Prior to 1935, such organisms were commonly classified as either *Micrococcus litoralis* (Poulsen 1879) Kellerman 1915 (*Sarcina litoralis*) or as *Micrococcus morrhuae* (Farlow 1880) Klebahn 1919 (*Sarcina morrhuae*). Unfortunately, the types of these species were descriptions rather than isolated cultures, and neither of the descriptions includes the red, extremely halophilic cocci currently known as *Halococcus*. Farlow's (1880) description of *Sarcina morrhuae* was of material taken from salt fish appearing microscopically as tetrad-forming cocci. However, Farlow explicitly described the coccoid material as "colorless." Moreover, the cocci measured 5–8 μm in diameter, and the tetrads were "always surrounded by a thin hyaline envelope." Farlow (1886) later used *Sarcina litoralis* Poulsen 1879 as a senior subjective synonym, but Poulsen's description also was not that of red, extremely halophilic cocci. The first to describe the red, extremely halophilic cocci in a satisfactory manner was Klebahn (1919), who isolated and studied several strains from salted fish. Perhaps because of ambiguities in the published descriptions, Klebahn (1919) recognized both *Micrococcus litoralis* and *Sarcina morrhuae* as species of red, extremely halophilic cocci, transferring the latter to *Micrococcus* as *Micrococcus morrhuae*. Petter (1932) later argued that the red, extremely halophilic cocci were somewhat variable in morphology in that they may form irregular packets or regular arrangements and vary in size depending on the growth conditions, and she proposed that only one species should be retained, *Sarcina morrhuae* Klebahn, giving recognition to Klebahn as the discoverer of these organisms.

Schoop (1935a) recognized that the red, extremely halophilic cocci should be classified in a separate genus, and transferred *Sarcina litoralis* to the new genus *Halococcus* as *H. litoralis*, the type and sole species of the genus. However, Schoop's opinion that the original description of *Sarcina litoralis* Poulsen included red, extremely halophilic cocci was in error (Kocur and Hodgkiss, 1973). Although red, extremely halophilic cocci were commonly classified as *Sarcina litoralis* prior to that time, the original description that constitutes the type of this species does not include such organisms. Therefore, Kocur and Hodgkiss (1973) transferred *Micrococcus morrhuae* (Farlow 1880) Klebahn 1919 to the genus *Halococcus* as *H. morrhuae*. Kocur and Hodgkiss (1973) emended the genus to exclude *H. litoralis* (Poulsen) Schoop 1935a, making *H. morrhuae* (Farlow 1880) Kocur and Hodgkiss 1973 the sole species and the new type species of the genus. Unfortunately, like *H. litoralis*, the original circumscription of *H. morrhuae* does not include the red, extremely halophilic cocci generally considered to belong to this genus. Regardless, Kocur and Hodgkiss (1973) designated a red, extremely halophilic coccus as the neotype strain (ATCC 17082). This taxonomy was adopted in the 8th Edition of the *Manual* (Gibbons, 1974a). Although this neotype strain is not consistent with the original description of the species, it is included in the Approved Lists 1980 as the type strain of *H. morrhuae* (Farlow 1880) Kocur and Hodgkiss 1973.

The first edition of *Bergey's Manual of Systematic Bacteriology* continued to list this as the sole species of *Halococcus* (Larsen, 1989). However, the species description was attributed to Klebahn as *H. morrhuae* Klebahn 1919. Although Klebahn (1919) was probably the first to adequately describe this group of organisms, he did not propose a new species.

Since the publication of the first edition, two new species of *Halococcus* have been described. Numerical taxonomy of halococci isolated from salterns (Montero et al., 1988) revealed a new species of nutritionally versatile strains, *H. saccharolyticus* (Montero et al., 1989). The phylogenetic distinction of these strains from *H. morrhuae* was demonstrated by DNA–rRNA hybridization (Montero et al., 1993). A coccoid isolate was then obtained by Denner et al. (1994) from an Austrian salt mine, which, on the basis of whole cell protein patterns, partial 16S rDNA sequence, lipid composition, and pH range for growth, was proposed as a new species, *H. salifodinae*. 16S rRNA gene sequences of strains of the type species, *H. morrhuae*, have been deposited, together with sequences for the type strains of *H. salifodinae* and *H. saccharolyticus* (Ventosa et al., 1999b). Analysis of these sequences clearly shows that the genus *Halococcus* represents a distinct group within the halobacteria, with the different species having >94% 16S rDNA sequence similarity and the nearest halobacterial neighbor sequences <89.5% similar (Fig. A2.36).

It should be noted that a number of different culture collection strains masquerading under the name *Halobacterium trapanicum* (NCIMB 767, JCM 897, and ATCC 43102) were suspected to be halococci in view of their morphologies (Tindall, 1992). 16S rRNA gene sequence determinations have now confirmed that view as correct (Ventosa et al., 1999b) (see Genus *Halorubrum* below). The organism named *Halococcus turkmenicus* (Zvyagintseva and Tarasov, 1987), which was noted in the 9th Edition of *Bergey's Manual of Determinative Bacteriology* (Holt et al., 1994), is not a halococcus and has been renamed *Haloterrigena turkmenica* (Ventosa et al., 1999b) (see Genus *Natrinema*). In particular, this organism has S2-DGD rather than S-DGD-1 (Kamekura, 1998b; Ventosa et al., 1999b) and is lysed with *N*-laurylsarcosine, whereas halococci *sensu strictu* are resistant (Kamekura, 1998a).

FURTHER READING

Denner, E.B.M., T.J. McGenity, H.J. Busse, W.D. Grant, G. Wanner and H. Stan-Lotter. 1994. *Halococcus salifodinae* sp. nov., an archaeal isolate from an Austrian salt mine. Int. J. Syst. Bacteriol. *44*: 774–780.

Larsen, H. 1989. Genus VI. *Halococcus*. *In* Staley, Bryant, Pfennig and Holt (Editors), Bergey's Manual of Systematic Bacteriology, 1st ed., Vol. 3, The Williams & Wilkins Co, Baltimore, pp. 2228–2230.

Montero, C.G., A. Ventosa, F. Rodriguez-Valera, M. Kates, N. Moldoveanu and F. Ruiz-Berraquero. 1989. *Halococcus saccharolyticus* sp. nov., a new species of extremely halophilic non-alkaliphilic cocci. Syst. Appl. Microbiol. *12*: 167–171.

Montero, C.G., A. Ventosa, F. Rodriguez-Valera and F. Ruiz-Berraquero. 1988. Taxonomic study of non-alkaliphilic halococci. J. Gen. Microbiol. *134*: 725–732.

Tindall, B.J. 1992. The family *Halobacteriaceae*. *In* Balows, Trüper, Dworkin, Harder and Schleifer (Editors), The Prokaryotes. A Handbook of Bacteria: Ecophysiology, Isolation, Identification, Applications, 2nd ed., Vol. 1, Springer-Verlag, New York, pp. 768–808.

Ventosa, A., M.C. Gutiérrez, M. Kamekura and M.L. Dyall-Smith. 1999. Proposal for the transfer of *Halococcus turkmenicus*, *Halobacterium tra-* *panicum* JCM 9743 and strain GSL 11 to *Haloterrigena turkmenica* gen. nov., comb. nov. Int. J. Syst. Bacteriol. *49*: 131–136.

List of species of the genus Halococcus

1. **Halococcus morrhuae** (Klebahn 1919) Kocur and Hodgkiss 1973, 154[AL] (Not *Sarcina litoralis* Poulsen 1879, 254; not *Sarcina morrhuae* Farlow 1880, 974) (*Sarcina morrhuae* Klebahn 1919, 38; *Micrococcus (Diplococcus) morrhuae* Klebahn 1919, 42; (*Halococcus litoralis* Schoop 1935a, 817.)
   morr.hu' ae. M.L. n. *morrhuae* from the specific epithet of the codfish, *Gadus morhua* L. (often misspelled *morrhua*); M.L. gen. n. *morrhuae* of the codfish.

   The morphology is the same as that described for the genus.

   Strictly aerobic. Amino acids are needed for growth and are also used as a source of energy. Glucose is not used as a main source of carbon for growth, and acid and gas are not produced from glucose when tested for by the usual methods. Nitrate is reduced to nitrite without production of gas. Catalase and oxidase positive. H_2S is usually produced from thiosulfate and frequently from cysteine. Urease negative. Some strains produce indole and hydrolyze gelatin and starch.

   The following compounds are used as sole carbon and energy sources: trehalose, *N*-acetyl glucosamine, and butyrate. The following compounds are used as sole carbon, nitrogen, and energy sources: L-glutamine, L-histidine, L-proline, and L- valine. The following compounds are not utilized as sole carbon, nitrogen, and energy sources: allantoin, L-aspartic acid, creatinine, ethionine, glycine, putrescine, and sarcosine.

   Optimum pH for growth 7.2. Some growth occurs at pH 5.5; no growth occurs at pH 8.

   Isolated from seawater, saline lakes, salterns, and salted products.

   The mol% G + C of the DNA is: 61–66 (T_m). Appreciable amounts of satellite DNA may be present (mol% G + C 57–59 [Bd]).

   Type strain: L.D.3.1, ATCC 17082, CCM 537, DSMZ 1307, NCIMB 787, NRC 16008.

   GenBank accession number (16S rRNA): D11106.

2. **Halococcus saccharolyticus** Montero, Ventosa, Rodriguez-Valera, Kates, Moldoveanu and Ruiz-Berraquero 1990, 105[VP] (Effective publication: Montero, Ventosa, Rodriguez-Valera, Kates, Moldoveanu and Ruiz-Berraquero 1989, 169.)
   sac.cha.ro.ly' ti.cus. Gr. n. *sacchar* sugar; Gr. adj. *lytikos* dissolving; M.L. masc. adj. *saccharolyticus* producing lysis of sugar.

   The morphology is the same as that described for the genus.

   Strictly aerobic. Catalase and oxidase positive. Acid is produced from D-glucose. Acid is not produced from lactose, D-mannitol, or sucrose. Starch, Tween-80, casein, esculin, and tyrosine are not hydrolyzed. Gelatin hydrolysis is variable. Selenite is reduced. Urease and phenylalanine deaminase are not produced.

   Indole positive. H_2S is produced from cysteine but not from thiosulfate. Nitrate is reduced to nitrite; reduction of nitrite variable.

   Insensitive to novobiocin.

   The following compounds are utilized as sole carbon and energy sources: *N*-acetylglucosamine, amygdalin, D-fructose, D-galactose, D-glucose, lactose, maltose, D-mannose, D-salicin, trehalose, adonitol, ethanol, erythritol, glycerol, *myo*-inositol, D-mannitol, propanol, D-sorbitol, acetate, *cis*-aconitate, δ-aminovalerate, butyrate, fumarate, D-gluconate, D-glucuronate, hippurate, DL-lactate, DL-malate, pyruvate, propionate, quinate, D-saccharate, and succinate. The following compounds are not utilized as sole carbon and energy sources: esculin, L-arabinose, D-galactosamine, D-glucosamine, inulin, D-melibiose, D-raffinose, L-rhamnose, D-ribose, sucrose, D-xylose, dulcitol, aminobutyrate, benzoate, caprylate, *p*-hydroxybenzoate, malonate, oxalate, salicylate, suberate, or D-tartrate. The following amino acids are utilized as sole carbon, nitrogen, and energy sources: L-alanine, L-arginine, L-asparagine, L-glutamic acid, L-glutamine, L-histidine, L-isoleucine, L-leucine, L-lysine, L-methionine, L-ornithine, L-proline, DL- phenylalanine, L-serine, L-threonine, L-tryptophan, and L-valine. The following compounds are not utilized as sole carbon, nitrogen, and energy sources: allantoin, L-aspartic acid, creatinine, ethionine, glycine, putrescine, or sarcosine.

   Unidentified phospholipids and glycolipids are present. pH range for growth 6–8; optimum 7.2. Temperature range for growth 28–42°C; optimum 37°C. Requires at least 4% (w/v) $MgCl_2$ for growth.

   Isolated from marine salterns.

   The mol% G + C of the DNA is: 59.5 (T_m).

   Type strain: ATCC 49257, CCM 4147, DSMZ 5350.

   GenBank accession number (16S rRNA): AB004876.

3. **Halococcus salifodinae** Denner, McGenity, Busse, Grant, Wanner and Stan-Lotter 1994, 779.[VP]
   sa.li.fo' di.nae. L. gen. n. *salifodinae* of a salt mine, referring to the source of the isolate.

   The morphology is the same as that described for the genus.

   Strictly aerobic. Oxidase and catalase positive. Nitrate is reduced to nitrite. Gelatin is liquefied.

   Slightly sensitive to chloramphenicol and tetracycline.

   Grows in minimal medium containing 0.1% (w/v) yeast extract supplemented with D-galactose, L-rhamnose, D-xylose, D-arabinose, D-glucose, D-fructose, D-trehalose, D-raffinose, or glycerol as a carbon source. Broad pH range for optimum growth at 6.8–9.5, with no growth at pH 10.5 or pH 6. Colonies on complex medium at neutral pH are pink, whereas colonies are nonpigmented at pH 9.5. Temperature range for growth 28–50°C, optimum 40°C. Unidentified glycolipids are present.

   Isolated from an Austrian salt mine.

   The mol% G + C of the DNA is: 62 ± 1 (HPLC).

   Type strain: ATCC 51437, DSMZ 8989.

   GenBank accession number (16S rRNA): AB004877 and Z28387.

Genus V. **Haloferax** Torreblanca, Rodriguez-Valera, Juez, Ventosa, Kamekura and Kates 1986b, 573ᵛᴾ (Effective publication: Torreblanca, Rodriguez-Valera, Juez, Ventosa, Kamekura and Kates 1986a, 98)

ANTONIO VENTOSA

Ha.lo.fe'rax. Gr. n. *hals, halos* salt; L. neut. adj. *ferax* fertile; M.L. neut. n. *Haloferax* salt (requiring) and fertile.

Cells extremely pleomorphic, most commonly flattened disks or cups (1–3 × 2–3 μm). Resting stages are not known. Gram negative. **Nonmotile or motile; motility often difficult to observe.** One species is gas vacuolate. Colonies have a mucoid appearance. Strictly aerobic. Nitrate can be reduced under anaerobic conditions. Oxidase and catalase positive. **Extremely halophilic, with growth occurring in media containing 1.5–4.5 M NaCl**; optimum concentration: 2.5 M NaCl. Temperature range for growth: 30–55°C; optimum temperature: ~35°C. pH range for growth: 5–8; optimum pH: 7. Minimal Mg^{2+} requirement for growth: 0.01–0.04 M. Chemoorganotrophic, utilizing many substrates as sources of carbon and energy. **Acid is produced from sugars. Amino acids are not required. Polyhydroxyalkanoates are accumulated under certain conditions. The polar lipids are characterized by the presence of $C_{20}C_{20}$ derivatives of S-DGD-1 and the absence of PGS.**

The mol% G + C of the DNA is: 59.1–65.5. A minor DNA component may be present.

Type species: **Haloferax volcanii** (Mullakhanbhai and Larsen 1975) Torreblanca, Rodriguez-Valera, Juez, Ventosa, Kamekura and Kates 1986b, 573 (Effective publication: Torreblanca, Rodriguez-Valera, Juez, Ventosa, Kamekura and Kates 1986a, 98) (*Halobacterium volcanii* Mullakhanbhai and Larsen 1975, 213.)

FURTHER DESCRIPTIVE INFORMATION

Rod-shaped cells are rare even under optimal conditions. In common with other halobacteria, a large glycoprotein can be extracted from cell envelopes (Rodriguez-Valera et al., 1983a; Juez et al., 1986a). Cells lyse in hypotonic solutions but are more stable at low salt than other halobacterial cells, remaining intact in 0.5 M NaCl.

Cells contain bacterioruberins and retinal pigments but may not contain bacteriorhodopsin (Kamekura et al., 1998). The main polar lipids are $C_{20}C_{20}$ derivatives of PG, PGP-Me, DGD-1, and S-DGD-1 (Ross and Grant, 1985; Torreblanca et al., 1986a).

All examples are insensitive to penicillin, chloramphenicol, streptomycin, and tetracycline and sensitive to anisomycin, bacitracin, novobiocin, and vibriostatic reagent (0/129). Species of *Haloferax* are more metal tolerant than other halobacteria, especially to arsenate and, in the case of *Haloferax mediterranei*, to cadmium (Nieto et al., 1987).

Many genetic and molecular biology studies on halobacteria have been based on *Haloferax volcanii* and *H. mediterranei* (Schalkwyk, 1993; DasSarma and Fleischmann, 1995; López-García et al., 1995; Mojica et al., 1995).

Organisms like these constitute a fraction of the microbial population of the Dead Sea and are presumably to be found in other high-Mg^{2+} environments. However, they are also found in thallasohaline salterns. A recent study of 22 halobacteria isolated from 57-year-old enrichments prepared from Dead Sea water samples showed that some isolates were phenotypically and phylogenetically related to *H. volcanii* (with a 16S rRNA similarity of 99.7%) (Arahal et al., 1996; Ventosa et al., 1999a).

ENRICHMENT AND ISOLATION PROCEDURES

Organisms of this group have been isolated from the Dead Sea and from salterns at the Spanish Mediterranean coast and in San Francisco, California. *H. volcanii* was originally enriched in a complex medium containing 0.2 M Mg^{2+} and 4 M NaCl (Mullakhanbhai and Larsen, 1975), but many other halobacteria will grow in this medium. No systematic attempts have been made to devise a specific enrichment medium, but in view of the high Mg^{2+} tolerance (>1 M), the relatively low NaCl requirement (<1 M), and utilization of carbohydrates and other compounds as carbon sources by these microorganisms, such an enrichment medium should not be difficult to design. Rodriguez-Valera et al. (1980) and Torreblanca et al. (1986a) used media with simple carbon sources whose composition was based on a mixture of salts (at 25% or 15%) corresponding in proportions to those found in seawater. Samples may be directly spread on agar media, with incubated plates in sealed plastic bags to prevent desiccation. Colonies appear after 3–14 d of incubation at 37°C.

DIFFERENTIATION OF THE GENUS *HALOFERAX* FROM OTHER GENERA

The key to the family *Halobacteriaceae* indicates those characteristics useful for distinguishing *Haloferax* from the other genera of the family. Although *Haloferax* species have a relatively high Mg^{2+} requirement, they can be reliably separated from other halobacteria by comparative analysis of the 16S rRNA sequences as well as by chemotaxonomic procedures. The polar lipids of these microorganisms characteristically contain S-DGD-1 and its unsulfated derivative and lack PGS; this, together with the observation of extremely pleomorphic cells, is the simplest way of distinguishing the genus.

TAXONOMIC COMMENTS

The genus *Haloferax* was proposed by Torreblanca et al. (1986a) on the basis of a numerical taxonomic study of a large number of phenotypic features and the polar lipid composition. The name (*ferax*, fertile) suggests that species of this genus have very high growth rates (doubling times of about 2–3 h) in comparison with other halobacteria and are physiologically very versatile, using a high number of compounds as sources of carbon and energy. Further DNA–DNA hybridization studies supported the placement of these halobacteria in different species (Ross and Grant, 1985; Gutiérrez et al., 1989). Three of the four species in the genus, *H. volcanii, H. mediterranei,* and *H. denitrificans,* were previously classified in the genus *Halobacterium. H. volcanii* and *H. mediterranei* were placed in the genus *Haloferax* by Torreblanca et al. (1986a), and subsequently, Tindall et al. (1989) transferred *Halobacterium denitrificans* to the genus *Haloferax.*

The 16S rRNA sequence comparison studies showed that the species of *Haloferax* constitute a phylogenetically coherent genus (Fig. A2.36), and signature 16S rRNA sequences specific for this genus have been reported (Table A2.25 of *Halobacteriales* overview).

ACKNOWLEDGMENTS

I am grateful to W.D. Grant, M. Kamekura, and A. Oren for their comments and suggestions during the preparation of this article.

FURTHER READING

DasSarma, S. and E.M. Fleischmann. 1995. Archaea: A Laboratory Manual, Cold Spring Harbor Laboratory Press, Plainview, NY.
Kamekura, M. 1999. Diversity of members of the family *Halobacteriaceae*. *In* Oren (Editor), Microbiology and Biogeochemistry of Hypersaline Environments, CRC Press, Boca Raton, pp. 13–25.

Tindall, B.J. 1992. The family *Halobacteriaceae*. *In* Balows, Trüper, Dworkin, Harder and Schleifer (Editors), The Prokaryotes. A Handbook of Bacteria: Ecophysiology, Isolation, Identification, Applications, 2nd ed., Vol. 1, Springer-Verlag, New York, pp. 768–808.
Torreblanca, M., F. Rodriguez-Valera, G. Juez, A. Ventosa, M. Kamekura and M. Kates. 1986. Classification of non-alkaliphilic halobacteria based on numerical taxonomy and polar lipid composition, and description of *Haloarcula* gen. nov. and *Haloferax* gen. nov. Syst. Appl. Microbiol. *8*: 89–99.

DIFFERENTIATION OF THE SPECIES OF THE GENUS *HALOFERAX*

Some differential features of the species of the genus *Haloferax* are given in Table A2.30.

List of species of the genus Haloferax

1. **Haloferax volcanii** (Mullakhanbhai and Larsen 1975) Torreblanca, Rodriguez-Valera, Juez, Ventosa, Kamekura and Kates 1986b, 573[VP] (Effective publication: Torreblanca, Rodriguez-Valera, Juez, Ventosa, Kamekura and Kates 1986a, 98) (*Halobacterium volcanii* Mullakhanbhai and Larsen 1975, 213.) *vol.ca′ ni.i.* M.L. gen. n. *volcanii* of Volcani; named after Israeli microbiologist B.E. Volcani, discoverer of life in the Dead Sea.

   See Table A2.30 and the generic description for many features.

   Cells are extremely pleomorphic, frequently flat, disk shaped, or cup shaped even under optimal conditions (1–2 × 2–3 µm) (Fig. A2.41). In young cultures, many elongated disks occur, some of which may rotate around their long axis. Flagella have not been demonstrated, and the movement is different from that of the polarly flagellated strains of halobacteria.

   Best growth occurs in media containing 1.5–2.5 M NaCl; 5 M NaCl is inhibitory. Minimal Mg^{2+} concentration, 0.02 M (Torreblanca et al., 1986a); tolerant to 1.5 M Mg^{2+}. Optimum growth temperature, 45°C.

   Chemoorganotrophic. Growth occurs on sugars as the sole source of carbon; acid is produced from glucose and other sugars. Indole is produced. H_2S is produced from thiosulfate. Nitrate is reduced to nitrite.

   Found in the Dead Sea.

   The DNA is composed of a major component and a minor component. Plasmids are present (Pfeifer et al., 1981; Charlebois et al., 1991; Charlebois, 1995b).

   The mol% G + C of the DNA is: 63.4 (T_m) and 66.5 (Bd), and 55.3 (Bd) (minor component).

   Type strain: DS 2, ATCC 29605, CCM 2852, DSMZ 3757, NCIMB 2012.

   GenBank accession number (16S rRNA): K00421.

2. **Haloferax denitrificans** (Tomlinson, Jahnke and Hochstein 1986) Tindall, Tomlinson and Hochstein 1989, 360[VP] (*Halobacterium denitrificans* Tomlinson, Jahnke and Hochstein 1986, 69.) *de.ni.tri′ fi.cans.* M.L. part. adj. *denitrificans* denitrifying.

   See Table A2.30 and the generic description for many features.

   Colonies are orange-red. Cells occur singly and exhibit a range of morphological forms, including disk-shaped cells (0.8–1.0 × 2.0–3.0 µm). Gas vacuoles not formed. Nonmotile.

   Growth occurs in media containing 1.5–4.5 M NaCl. Temperature range for growth, 30–55°C. The pH range for growth is 6–8, optimum pH, 6.7 (at 37°C).

   Chemoorganotrophic; able to use a wide variety of compounds as carbon and energy sources (glucose, galactose, fructose, maltose, sucrose, acetate, citrate, fumarate, glycerol, lactate, α-ketoglutarate, malate, succinate, and pyruvate). The utilization of sugars produces acidification of the medium. Utilization of nitrate or nitrite occurs under anaerobic conditions. Tween-40 and gelatin are hydrolyzed, but urea, starch, and Tween-80 are not hydrolyzed. Sulfide is produced from thiosulfate. Indole negative.

TABLE A2.30. Differential characteristics of the species of the genus *Haloferax*[a]

Characteristic	1. *H. volcanii*	2. *H. denitrificans*	3. *H. gibbonsii*	4. *H. mediterranei*
Presence of gas vacuoles	−	−	−	+
Gelatin hydrolysis	−	+	−	+
Starch hydrolysis	−	−	−	+
Nitrate reduction	+	+	−	+
Tween-80 hydrolysis	−	−	+	+
H_2S from thiosulfate	+	+	+	−
Indole production	+	−	+	+
Mol% G + C	63.4	64.2	61.8	60

[a]For symbols see standard definitions.

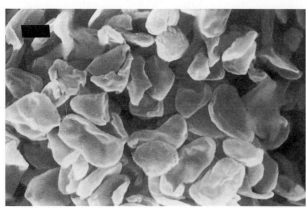

FIGURE A2.41. *Haloferax volcanii* DS2 (NCIMB 2012), grown at 30°C in a medium consisting of (g/l of distilled water): NaCl, 125.0; $MgCl_2·6H_2O$, 50.0; K_2SO_4, 5.0; $CaCl_2·6H_2O$, 0.2; yeast extract (Difco), 5.0; and tryptone (Difco), 5.0; pH 6.8. The culture was photographed at the end of the exponential growth phase. Bar = 1 μm. (Courtesy of G. Bentzen; photographed by the Laboratory of Clinical Electron Microscopy, University of Bergen.)

The major menaquinones present are MK8 and MK8(VIII-H_2).

Isolated from salterns in San Francisco, California.

The mol% G + C of the DNA is: 64.2 (T_m).

Type strain: S1, ATCC 35960, DSMZ 4425.

GenBank accession number (16S rRNA): D14128.

3. **Haloferax gibbonsii** Juez, Rodriguez-Valera, Ventosa and Kushner 1986b, 573^VP (Effective publication: Juez, Rodriguez-Valera, Ventosa and Kushner 1986a, 78.)

gib.bon' si.i. M.L. gen. n. *gibbonsii* of Gibbons; named for the Canadian microbiologist N.E. Gibbons, one of the pioneers in the study of halobacteria.

See Table A2.30 and generic description for many features.

Cells are short pleomorphic rods (0.4 × 0.5–2.5 μm). Motile by tufts of polar flagella.

Growth occurs in media containing 1.5–5.2 M NaCl; optimum concentration, 3–4 M NaCl at 40°C and 2–3 M NaCl at 30°C. Temperature range for growth, 25–55°C; optimum temperature, 35–40°C. Mg^{2+} requirement for growth, 0.2 M. pH range for growth, 5–8; optimum pH, 6.5–7.0.

Chemoorganotrophic; acid is produced from a wide range of sugars (arabinose, fructose, galactose, glucose, maltose, mannose, sucrose, and xylose). Indole positive. H_2S is produced from thiosulfate. Starch is not hydrolyzed. Tween-20, -40, -60 and -80 are hydrolyzed. A few strains hydrolyze gelatin and reduce nitrate.

Produces halocins (Meseguer et al., 1986; Torreblanca et al., 1994), of which halocin H6 has been characterized (Torreblanca et al., 1989; Meseguer et al., 1995).

Isolated from marine salterns in Spain.

The mol% G + C of the DNA is: 61.8 (T_m).

Type strain: Strain Ma 2.38, ATCC 33959, DSMZ 4427, NCIMB 2188.

GenBank accession number (16S rRNA): D13378.

4. **Haloferax mediterranei** (Rodriguez-Valera, Juez and Kushner 1983a) Torreblanca, Rodriguez-Valera, Juez, Ventosa, Kamekura and Kates 1987, 179^VP (Effective publication: Torreblanca, Rodriguez-Valera, Juez, Ventosa, Kamekura and Kates 1986a, 97) (*Halobacterium mediterranei* Rodriguez-Valera, Juez and Kushner 1983a, 379.)

me.di.ter.ra' ne.i. L. gen. n. *mediterranei* of the Mediterranean Sea.

See Table A2.30 and the generic description for many features.

Colonies are pink and highly mucoid after 5–7 d at 37°C. Cells are pleomorphic rods (0.5 × 2 μm). During the late exponential and stationary phases, gas vacuoles are produced. Motile by a tuft of polar flagella.

Growth occurs in media containing 1–5.2 M NaCl. Minimal Mg^{2+} concentration, 0.02 M (Torreblanca et al., 1986a). Temperature range for growth, 20–55°C, optimum temperature, 47–54°C (complex medium), 43–51°C (defined medium) (Shand and Perez, 1999). Optimum pH 6.5.

Chemoorganotrophic; able to use a wide variety of compounds as carbon and energy sources (glucose, sucrose, fructose, lactose, glycerol, maltose, mannitol, sorbitol, xylose, arabinose, succinate, malate, lactate, pyruvate, citrate, acetate, glutamate, lysine, and arginine). Utilization of sugars (glucose, fructose, maltose, sucrose) causes acidification of the medium. Nitrate and nitrite are reduced. Indole positive. H_2S is not produced from thiosulfate. Starch, gelatin, and Tween-80 are hydrolyzed.

Produces bacteriocins (halocins) active against a range of other halobacteria (Rodriguez-Valera et al., 1982; Meseguer et al., 1986; Torreblanca et al., 1994; Shand et al., 1999). Produces halocin H4, the first halocin to be characterized (Meseguer and Rodriguez-Valera, 1985, 1986; Cheung et al., 1997). Resistant to relatively high level of bacitracin (40 mg/ml).

A minor DNA component has not been detected, but large plasmids are present (Gutiérrez et al., 1986; López-García et al., 1992; Anton et al., 1995).

Isolated from marine salterns in Spain.

The mol% G + C of the DNA is: 60.0 (T_m) and 62.2 (Bd).

Type strain: R-4, ATCC 33500, CCM 3361, DSMZ 1411, NCIMB 2177.

GenBank accession number (16S rRNA): D11107.

5. **"Haloferax alicantei"** Holmes, Scopes, Moritz, Simpson, Englert, Pfeifer and Dyall-Smith 1997.

This organism was originally isolated from Spanish salterns and phenotypically characterized by Torreblanca et al. (1986a), showing a polar lipid profile typical of the genus *Haloferax*. The 16S rRNA sequence comparison studies support its placement within this genus (Fig. A2.36 of *Halobacteriales* overview). DNA–DNA hybridization studies confirm its placement as a different species of this genus (A. Ventosa et al., unpublished).

This strain was largely used for genetic studies (Holmes and Dyall-Smith, 1990, 1991, 1999; Holmes et al., 1991, 1997). Bacteriophage HF1 infecting this and other halobacteria has been reported (Nuttall and Dyall-Smith, 1993b).

Deposited strain: Aa 2.2.

GenBank accession number (16S rRNA): M33803, M33804, M33805.

6. *Haloferax* sp. strain D1227 Emerson, Chauhan, Oriel and Breznak 1994, 445.

Isolated from soil contaminated with highly saline oil brine near Grand Rapids, Michigan; can utilize a variety of aromatic compounds (benzoic acid, cinnamic acid, 3-phenylpropionic acid, and 3-hydroxybenzoic acid) as sole carbon and energy sources. The degradation of 3-phenylpropionic acid in this strain has been studied (Fu and Oriel, 1998, 1999). Requires 2 M NaCl for optimal growth. The sequence comparison of its 16S rDNA (GenBank accession number AF069950) confirms its placement in the genus *Haloferax*, phylogenetically related to *H. volcanii* and *H. mediterranei* (Fu and Oriel, 1999).

Genus VI. **Halogeometricum** Montalvo-Rodríguez, Vreeland, Oren, Kessel, Betancourt and López-Garriga 1998, 1310[VP]

AHARON OREN, RAFAEL MONTALVO-RODRÍGUEZ AND RUSSELL H. VREELAND

Ha.lo.ge.o.me'tri.cum. Gr. n. *hals, halos* the sea, salt; L. adj. *geometricum* geometrical; M.L. neut. n. *Halogeometricum* salty geometrical shape.

Gram-negative pleomorphic cells, including short and long rods, squares, triangles, and ovals. **Extremely halophilic**, with optimal growth occurring in media containing 3.5–4 M NaCl and 0.04–0.08 M $MgCl_2$. Optimum temperature 40°C. Colonies small, mucoid, pink. Oxidase and catalase positive. Chemoorganotrophic; **aerobic**. Grows anaerobically with nitrate with formation of nitrite and gaseous products. No growth occurs anaerobically with arginine. Casamino acids and yeast extract are good sources of organic nutrients. Several sugars can be used as carbon sources in the presence of low concentrations of yeast extract. Polar lipids are characterized by the presence of a yet-to-be characterized nonsulfated diether glycolipid and the absence of PGS.

The mol% G + C of the DNA is: 59–60.5.

Type species: **Halogeometricum borinquense** Montalvo-Rodríguez, Vreeland, Oren, Kessel, Betancourt and López-Garriga 1998, 1310.

FURTHER DESCRIPTIVE INFORMATION

Cells of *Halogeometricum borinquense*, the only *Halogeometricum* species described thus far, are extremely pleomorphic, showing a variety of short and longer rods, squares, triangles, and oval cells, measuring 1–3 × 1–2 µm. Cells are motile by peritrichous flagella. Gas vesicles are produced, and in nonshaken liquid cultures cells float to the air-medium interface. Cells stain Gram-negative.

High salt concentrations required for structural integrity. Requires at least 1.4 M NaCl for growth. Optimal NaCl concentration range: 3.5–4 M at 40°C; optimal $MgCl_2$ concentration: 0.04–0.08 M. At NaCl concentrations below 1 M cells become spherical, and cells lyse in the absence of salt. Grows between 22 and 50°C (optimum temperature: 40°C) and in the pH range 6–8.

A variety of sugars and related compounds (glucose, maltose, fructose, xylose, trehalose, cellobiose, raffinose, and glycerol) stimulate growth, in many cases with acid production. Starch is not hydrolyzed. Gelatin is strongly hydrolyzed. Indole is produced.

Cells are colored pink due to the presence of bacterioruberins. The main polar lipids are PG, PGP-Me, and a yet unidentified, nonsulfate-containing glycolipid. PGS is absent.

Sensitive to novobiocin, bacitracin, and sulfamethazone plus trimethoprim and insensitive to penicillin, ampicillin, kanamycin, chloramphenicol, vancomycin, tetracycline, and erythromycin.

ENRICHMENT AND ISOLATION PROCEDURES

Three isolates of the only species yet described were obtained from samples collected in May and August 1994 from the solar salterns in Cabo Rojo, Puerto Rico. These developed as colonies on agar plates with Sehgal-Gibbons medium (Sehgal and Gibbons, 1960) containing (in g/l): NaCl, 250; $MgSO_4 \cdot 7H_2O$, 20; KCl, 2; trisodium-citrate, 3; yeast extract, 10; and Casamino acids, 7.5 (pH 7.5–7.8) or with glycerol solar salt medium (in g/l): solar salt, 250; glycerol, 10; and Casamino acids, 1.

No selective growth and enrichment procedure for *Halogeometricum* exists.

TABLE A2.31. Characteristics differentiating *Halogeometricum borinquense* from other pleomorphic types of halophilic *Archaea*[a]

Characteristic	Halogeometricum borinquense	Haloferax volcanii	Haloferax mediterranei	Haloferax denitrificans	Haloferax gibbonsii	Haloarcula species	Haloarcula mukohataei
Motility	+	−	v	−	+	D	+
Gas vesicles	−	−	+	−	−	−	−
Lipids:							
Sulfated glycolipid	+	−	−	−	−	−	−
TGD-2	−	−	−	−	−	+	−
S-DGD-1	−	+	+	+	+	−	+
PGS	−	−	−	−	−	+	+
Indole from tryptophan	+	+	+	−	+	D	nd
Nitrite from nitrate	+	+	+	+	d	+	nd
Gas from nitrate	+	−	+	+	nd	+	nd
Starch hydrolyzed	−	−	+	−	−	+	nd
Mol% G + C	59.1	63.4	59.1–62.2	64.2	61.8	60.1–64.7	65

[a]For symbols see standard definitions; nd, not tested.

MAINTENANCE PROCEDURES

Growth on agar slants in screw-cap tubes can be kept in the refrigerator for 4–5 months. Lyophilization gives satisfactory results. Cells may also be stored on ceramic beads at −80°C or under liquid nitrogen.

DIFFERENTIATION OF THE GENUS *HALOGEOMETRICUM* FROM OTHER GENERA

The genus *Halogeometricum* can be differentiated from the other halophilic archaeal genera on the basis of the diverse cell morphology, polar lipid composition, and comparison of 16S rRNA sequences (Table A2.31). The pleomorphic cell shape, content of gas vesicles, and motility are highly characteristic. The absence of PGS and the presence of a single glycolipid chromatographically similar, but not identical, to TGD-2, the glycolipid of the genus *Haloarcula* (a genus that does contain PGS), can also be used as a diagnostic trait. Comparison of 16S rRNA nucleotide sequences (Fig A2.42) also differentiates the genus *Halogeometri-cum* from the other recognized genera within the family *Halobacteriaceae*.

TAXONOMIC COMMENTS

The definition of the genus *Halogeometricum* is based on cell morphology, characterization of polar lipids, and 16S rRNA analysis, showing only remote phylogenetic relationship to the other recognized genera within the *Halobacteriaceae*. Sequence similarities of 93.7% and 86.2–87.4% were found with *Haloferax volcanii* and *Haloarcula* species, respectively.

FURTHER READING

Montalvo-Rodríguez, R., R.H. Vreeland, A. Oren, M. Kessel, C. Betancourt and J. López-Garriga. 1998. *Halogeometricum borinquense* gen. nov., sp. nov., a novel halophilic archaeon from Puerto Rico. Int. J. Syst. Bacteriol. *48*: 1305–1312.

Sehgal, S.N. and N.E. Gibbons. 1960. Effect of some metal ions on the growth of *Halobacterium cutirubrum*. Can. J. Microbiol. *6*: 165–169.

List of species of the genus Halogeometricum

1. **Halogeometricum borinquense** Montalvo-Rodríguez, Vreeland, Oren, Kessel, Betancourt and López-Garriga 1998, 1310.[VP]

bo.rin.quen' se. M.L. neut. adj. *borinquense* of Borinquen, the native Indian name for Puerto Rico.

Cells are extremely pleomorphic, 1–3 × 1–2 μm. Motile by peritrichous flagella. Gas vesicles present. Requires at least 1.4 M NaCl. Optimal growth in 3–3.5 M NaCl and 0.04–0.08 M MgCl₂; optimum temperature 40°C.

Aerobic; grows anaerobically with nitrate; does not grow anaerobically with arginine. Indole is produced from tryptophan. Glucose, maltose, fructose, xylose, trehalose, cellobiose, raffinose, and glycerol stimulate growth. Starch not hydrolyzed. Acid produced from fructose, ribose, lactose, arabinose, xylose, and sucrose.

Susceptible to novobiocin, bacitracin, and sulfamethazone plus trimethoprim; insensitive to penicillin, ampicillin, kanamycin, chloramphenicol, vancomycin, tetracycline, and erythromycin.

Glucose, maltose, fructose, xylose, trehalose, cellobiose, raffinose, and glycerol stimulate growth. Starch not hydrolyzed. Acid produced from fructose, ribose, lactose, arabinose, xylose, and sucrose.

Found in solar salterns in Puerto Rico.

Other characteristics are the same as those described for the genus.

The mol% G + C of the DNA is: 59.1 (HPLC).

Type strain: PR3, ATCC 700274.

GenBank accession number (16S rRNA): AF002984.

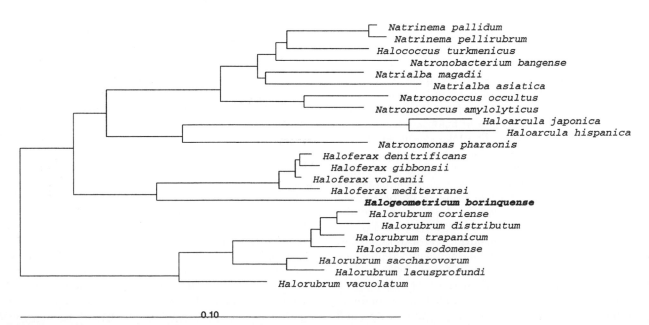

FIGURE A2.42. Phylogenetic tree showing the position of *Halogeometricum borinquense*. Bar = 0.10 fixed mutation per nucleotide position.

Genus VII. **Halorubrum** McGenity and Grant 1996, 362[VP] (Effective publication: McGenity and Grant 1995, 241)

TERRY J. MCGENITY AND WILLIAM D. GRANT

Ha.lo.ru'brum. Gr. n. *hals, halos* salt; L. neut. adj. *rubrum* red; M.L. neut. n. *Halorubrum* salt (requiring) and red.

Cells rod shaped 1–12 μm × 0.5–1.2 μm or pleomorphic. Does not form spores. Colonies orange-red, smooth, circular, convex, or bright pink in gas-vacuolate species. Cells stain Gram-negative or variable. Some strains motile. **Strict aerobes. Chemoorganotrophic, using many substrates, including sugars, as sources of carbon and energy.** Oxidase and catalase positive. **Produces nitrite from nitrate. Growth in medium between 1.5 and 5.2 M NaCl.** Temperature range for growth 4–56°C. A minor DNA component may be present. Forms a phylogenetic cluster with more than 92.8% 16S rRNA sequence similarity. Has three signature sequences (Table A2.25).

The mol% G + C of the DNA is: 62.7–71.2.

Type species: **Halorubrum saccharovorum** (Tomlinson and Hochstein 1976) McGenity and Grant 1996, 362 (Effective publication: McGenity and Grant 1995, 241) (*Halobacterium saccharovorum* Tomlinson and Hochstein 1976, 587; *Halorubrobacterium saccharovorum* (Tomlinson and Hochstein 1976) Kamekura and Dyall-Smith 1995, 345.)

FURTHER DESCRIPTIVE INFORMATION

Halorubrum species commonly appear as flattened rods, but when grown under suboptimal conditions, they may exhibit a variety of forms. For example, *Halorubrum vacuolatum* is spherical on solid media and in the stationary phase in liquid media (Mwatha and Grant, 1993). A detailed study of *Halorubrum distributum* by Kostrikina et al. (1991) demonstrated four morphotypes dominated by rods. *Halorubrum coriense* is pleomorphic even when actively growing (Nuttall and Dyall-Smith, 1993a).

The genus *Halorubrum* is phylogenetically quite diverse, with three main clusters supported by 100% bootstrap analysis (Fig. A2.36) containing 1) *Halorubrum saccharovorum* and *Halorubrum lacusprofundi* (98.3% 16S rRNA sequence similarity); 2) *Halorubrum sodomense, Halorubrum trapanicum, H. coriense,* and *H. distributum* (>97.7% 16S rRNA sequence similarity); and 3) *H. vacuolatum,* the most deeply-branching species.

All species except *H. vacuolatum* are neutrophilic, and *H. vacuolatum* is the only representative of the genus with gas vacuoles.

Spectra typical of C_{50} carotenoids have been observed in four species of the genus and can be assumed to be present in the others, given the similar pigmentation of the cells and the widespread occurrence of these pigments in halobacteria.

The polar lipids of the six neutrophilic species are $C_{20}C_{20}$ derivatives of phosphatidyl glycerol, phosphatidyl glycerol phosphate methyl ester, phosphatidyl glycerol sulfate, and sulfated mannosyl-glucosyl-glycerol diether (Ross et al., 1985; Kamekura and Dyall-Smith, 1995). The detailed structure of this sulfated glycolipid is discussed under the section "Differentiation of the Species of *Halorubrum*". *H. vacuolatum,* however, has both $C_{20}C_{20}$ and $C_{20}C_{25}$ derivatives of phosphatidyl glycerol, phosphatidyl glycerol phosphate methyl ester, and two unidentified phospholipids but has no glycolipid (Mwatha and Grant, 1993).

The DNA is *dam* methylated in all species analyzed (Lodwick et al., 1986; Nuttall and Dyall-Smith, 1993a), a feature not found in any of the other hybridization similarity groups of Ross and Grant (1985).

Halorubrum species have been isolated from hypersaline environments varying widely in chemical and physical properties, e.g., salterns, salt mines, Antarctic salt lakes, the Dead Sea, Lake Magadi, and solonchak soil (Tomlinson and Hochstein, 1976; Oren, 1983; Rodriguez-Valera et al., 1985; Zvyagintseva and Tarasov, 1987; Franzmann et al., 1988; Mwatha and Grant, 1993; Norton et al., 1993; Nuttall and Dyall-Smith, 1993a).

ENRICHMENT AND ISOLATION PROCEDURES

The methods are the same as those given for the genus *Halobacterium* for neutrophiles and *Natronobacterium* for the alkaliphilic species.

MAINTENANCE PROCEDURES

See under Order *Halobacteriales.*

DIFFERENTIATION OF THE GENUS *HALORUBRUM* FROM OTHER GENERA

Several characteristics given in the description of the genus differentiate *Halorubrum* species from other genera, in particular phylogenetic analysis of 16S rRNA sequences together with signature sequence analysis (Table A2.25). Their utilization of sugars differentiates them from some halobacterial genera, e.g., *Halobacterium.* Polar lipids of neutrophilic strains serve to distinguish them from other neutrophilic genera.

TAXONOMIC COMMENTS

The genus *Halorubrum* was formally proposed in 1995 to accommodate *Halorubrum saccharovorum, Halorubrum sodomense, Halorubrum trapanicum* NRC 34021, and *Halorubrum lacusprofundi,* which had been classified previously in the genus *Halobacterium* (McGenity and Grant, 1995). The generic name *Halorubrum* appeared in Validation List number 56 (McGenity and Grant, 1996). Almost at the same time Kamekura and Dyall-Smith (1995) proposed the genus *Halorubrobacterium* to accommodate *Halorubrobacterium saccharovorum, Halorubrobacterium sodomense, Halorubrobacterium lacusprofundi, Halorubrobacterium distributum* (previously *Halobacterium distributum*), and *Halorubrobacterium coriense* sp. nov. The generic name *Halorubrobacterium* appeared in Validation List number 57 (Kamekura and Dyall-Smith, 1996) and was treated as a later synonym of *Halorubrum* (Oren and Ventosa, 1996).

Kamekura et al. (1997) transferred *Natronobacterium vacuolatum* into the genus *Halorubrum* based largely on phylogenetic analysis but recognized that *H. vacuolatum* is relatively deep branching (Fig. A2.36). The combination of the factors, alkaliphily, presence of gas vacuoles, and different polar lipids, indicates that *H. vacuolatum* is an outlier within the genus. However, the assignment of this species into *Halorubrum* by Kamekura et al. (1997) is supported by the isolation of two gas-vacuolate strains (R.T. Gemmell, T.J. McGenity and W.D. Grant, unpublished) which branch more deeply than *H. vacuolatum,* but which are neutrophilic and possess polar lipids typical of the genus *Halorubrum.* This indicates that polar lipids, which are normally good chemotaxonomic markers in the *Halobacteriaceae*, may not be as useful in alkaliphilic species.

It had been recognized for a decade before the formal proposal of the genus *Halorubrum* that several species within the genus *Halobacterium* were distinct from *Halobacterium salinarum*. Grant and Ross (1986) suggested the genus name *Halorubrum* for *H. saccharovorum*, *H. sodomense*, and *H. trapanicum* NRC 34021 largely on the basis of 16S rRNA–DNA hybridization and polar lipid analysis (Ross and Grant, 1985) and *dam* methylation of DNA (Lodwick et al., 1986). Other early studies, mostly of *H. saccharovorum*, confirmed that this species did not belong to the genus *Halobacterium*. For example, *H. saccharovorum* is immunologically distinct from other halobacteria (Conway de Macario et al., 1986) and, unlike many other halobacteria, has class I fructose bisphosphate aldolases (Dhar and Altekar, 1986). Torreblanca et al. (1986a) proposed that *H. saccharovorum* and *H. sodomense* should remain in the genus *Halobacterium*, based on phenotypic and chemotaxonomic criteria. However, an anomalous strain of *H. saccharovorum* (strain ATCC 29252) was used in the study of Torreblanca et al. (1986a) as determined by lipid analyses (Tindall, 1990). In addition, *H. sodomense* was not shown to be closely related to other halobacteria, but this was explained by its slow growth rate affecting interpretation of the data (Torreblanca et al., 1986a). Gutiérrez et al. (1989) also used strain ATCC 29252 to determine the mol% G + C of 64.3 (T_m), which differs considerably from the value for strain NCIMB 2081 of 71.2% (Bd) (Ross and Grant, 1985).

There have been several minor changes to some specific epithets in the genus *Halorubrum* for gender agreements, as noted in the species descriptions.

A complication has arisen because *Halobacterium trapanicum* NCIMB 767, *Halobacterium trapanicum* JCM 8979, and *Halorubrum trapanicum* NRC 34021 are supposedly derived from the strain described as "the orange rod of the 'Trapani' salt" isolated by Petter (1931). These three strains fall within three different phylogenetic lineages (Fig. A2.36), which indicates problems of transfer between culture collections in the past. The situation is currently being clarified by the deposition of *Halorubrum trapanicum* NCIMB 13488 as a proposed neotype, a strain derived from the original NRC 34021 strain held in the laboratory of W.D. Grant (Grant et al., 1998b).

FURTHER READING

Kamekura, M. 1999. Diversity of members of the family *Halobacteriaceae*. *In* Oren (Editor), Microbiology and Biogeochemistry of Hypersaline Environments, CRC Press, Boca Raton, pp. 13–25.

Kamekura, M. and M.L. Dyall-Smith. 1995. Taxonomy of the family *Halobacteriaceae* and the description of two new genera *Halorubrobacterium* and *Natrialba*. J. Gen. Appl. Microbiol. *41*: 333–350.

McGenity, T.J. and W.D. Grant. 1995. Transfer of *Halobacterium saccharovorum*, *Halobacterium sodomense*, *Halobacterium trapanicum* NRC 34021, and *Halobacterium lacusprofundi* to the genus *Halorubrum* gen. nov., as *Halorubrum saccharovorum* comb. nov., *Halorubrum sodomense* comb. nov., *Halorubrum trapanicum* comb. nov., and *Halorubrum lacusprofundi* comb. nov. Syst. Appl. Microbiol. *18*: 237–243.

DIFFERENTIATION OF THE SPECIES OF THE GENUS *HALORUBRUM*

A genus-wide matrix of DNA–DNA hybridizations has not been determined, although a few comparisons have been made: *H. saccharovorum* has DNA–DNA similarities of 45% and 52% to *H. sodomense* and *H. trapanicum* NRC 34021, respectively, which supports separation at species level (Ross and Grant, 1985). No pair of species within *Halorubrum* has more than 98.8% 16S rRNA sequence similarity, which by itself is not strong evidence for delineation of species. However, Norton et al. (1993) showed that strain 54R had 48% DNA similarity to *H. saccharovorum* yet had 99.7% 16S rRNA sequence similarity. Therefore, it is probable that the DNA relatedness of each species pair would be below the 60–70% level which delineates species.

Physiological characteristics which distinguish the seven species often reflect their habitats. For example, *H. sodomense*, isolated from the Dead Sea, in which the Mg^{2+} concentration is often >1 M, grows optimally in media with 0.6–1.2 M Mg^{2+} (Oren, 1983). *H. lacusprofundi*, isolated from an Antarctic saline lake, is the only halobacterium able to grow, albeit slowly, at 4°C (Franzmann et al., 1988). *H. vacuolatum*, from Lake Magadi, an alkaline saline lake, is the only member of the genus that can grow at pH 10.5 (Mwatha and Grant, 1993).

H. vacuolatum has cylinder-shaped gas vacuoles throughout its growth cycle in media containing 2.5–4.2 M NaCl but lacks gas vacuoles when grown in media containing 2.2 M NaCl (Mayr and Pfeifer, 1997).

H. vacuolatum has both $C_{20}C_{20}$ and $C_{20}C_{25}$ diether core lipids differentiating it from the other six neutrophilic species, which have only $C_{20}C_{20}$ diether core lipids. Furthermore, *H. vacuolatum* has different polar lipids, lacking phosphatidyl glycerol sulfate and a characteristic sulfated glycolipid found in the other species (Mwatha and Grant, 1993).

The characteristic glycolipid of the six neutrophilic species is a sulfated mannosyl-glucosyl-glycerol diether (S-DGD, sulfated diglycosyl diether). There has been considerable confusion over the nomenclature and structure of this glycolipid. Kamekura (1998b) observed that the mobility by thin-layer chromatography of S-DGD is the same in all species of *Halorubrum* and is different from that of the similar glycolipids in other genera. In that respect, it serves as a useful marker for the genus. However, differences have been found in the detailed structure, which has been determined from *H. saccharovorum* (Lanzotti et al., 1988), *H. sodomense* (Trincone et al., 1990), and *H. trapanicum* NRC 34021 (Trincone et al., 1993). In *H. sodomense*, the linkage between the sugars of S-DGD named S-DGD-3 by Kamekura (1998b) is 1–4 and sulfate is on C-2 of the mannose residue (Trincone et al., 1990). In *H. trapanicum* NRC 34021, the linkage is 1–2, and sulfate is on C-2 of the mannose residue (Trincone et al., 1993), and the S-DGD was named S-DGD-5 by Kamekura (1998b). In *H. saccharovorum*, the linkage was reported as 1–2, with sulfate on C-6 of the mannose residue (Lanzotti et al., 1988) and the S-DGD was named S-DGD-1 by Kamekura (1998b). Kamekura (1998b) named the glycolipid in other neutrophilic *Halorubrum* spp. S-DGD-3 on the basis of comparative mobility. The structure of this glycolipid in *Halorubrum* spp. needs to be re-examined.

A selection of characteristics useful in distinguishing representatives of the genus *Halorubrum* is given in Table A2.32. There are some conflicting reports on carbon-source requirements and production of acid from sugars, largely due to tests having been carried out in different laboratories using different minimal media, etc.

List of species of the genus Halorubrum

1. **Halorubrum saccharovorum** (Tomlinson and Hochstein 1976) McGenity and Grant 1996, 362[VP] (Effective publication: McGenity and Grant 1995, 241) (*Halobacterium saccharovorum* Tomlinson and Hochstein 1976, 587; *Halorubrobacterium saccharovorum* (Tomlinson and Hochstein 1976) Kamekura and Dyall-Smith 1995, 345.)

sac.cha.ro.vo′rum. L. n. *saccharum* sugar; L. v. *voro* to devour; M.L. neut. adj. *saccharovorum* sugar-devouring.

Cells are rod shaped, 0.6–1.2 μm × 2.5 μm, but become pleomorphic under unfavorable conditions. Gram negative. Gas vacuoles are absent. Motile. Colonies orange-red. Strictly aerobic. NaCl concentration for growth: 1.5–5.2 M; optimum, 3.5–4.5 M. Mg^{2+} requirement, 0.005 M. Temperature range for growth, 30–56°C, optimum 50°C. Oxidase and catalase positive. Gelatinase, caseinase, urease, and amylase negative. Phosphatase positive. Indole not produced. Xanthine and Tween not degraded. Hydrogen sulfide produced in small quantities from sodium thiosulfate but not from cysteine. Amino acids required for growth. Chemoorganotrophic. Glucose is metabolized by a modified Entner-Doudoroff pathway. Many sugars or sugar derivatives utilized with production of acid: arabinose, ribose, xylose, rhamnose, fructose, galactose, glucose, mannose, lactose, maltose, sucrose, and glycerol. Production of bacteriorhodopsin not detected. The polar lipids are $C_{20}C_{20}$ derivatives of phosphatidyl glycerol, phosphatidyl glycerol phosphate methyl ester, phosphatidyl glycerol sulfate, and sulfated mannosyl-glucosyl-glycerol diether. Susceptible to infection by halophage HF2. Isolated from marine salterns. A minor DNA component has not been detected, though a plasmid has been observed.

Two papers were published in 1995 proposing the transfer of *Halobacterium saccharovorum*, *Halobacterium lacusprofundi*, and *Halobacterium sodomense* to a new genus. McGenity and Grant (1995) proposed the genus *Halorubrum*, which appeared in Validation List number 56 (McGenity and Grant, 1996), while Kamekura and Dyall-Smith (1995) proposed the genus *Halorubrobacterium*, which appeared in Validation List number 57 (Kamekura and Dyall-Smith, 1996). Therefore, *Halorubrobacterium* was treated as a later synonym of *Halorubrum* (Oren and Ventosa, 1996). *H. saccharovorum* was considered *species incertae sedis* in the last edition of the *Manual* (Larsen and Grant, 1989).

The mol% G + C of the DNA is: 71.2 (Bd).

Type strain: M6, ATCC 29252, DSMZ 1137, NCIMB 2081.

GenBank accession number (16S rRNA): U17364, X82167.

2. **Halorubrum coriense** (Kamekura and Dyall-Smith 1995) Oren and Ventosa 1996, 1180[VP] (*Halorubrobacterium coriensis* Kamekura and Dyall-Smith 1995, 347.)

co.ri.en′se. M.L. neut. adj. *coriense* pertaining to Corio, a saltern along Corio Bay, Geelong, Australia, where the organism was originally isolated.

Cells are pleomorphic. Actively growing cultures consist of short rods and cup-shaped cells, while stationary-phase cultures contain predominantly long rods (0.5–5 μm). Stain Gram-negative. Gas vacuoles are absent. Motile. Colonies orange-red. Strictly aerobic. NaCl concentration for growth, 2–5.2 M; optimum, 2.2–2.7 M. Mg^{2+} requirement, 0.005 M. Temperature range for growth: 30–56°C; optimum 50°C. Oxidase and catalase positive. Indole not produced. Amino acids not required for growth. Chemoorganotrophic. Four compounds tested all showed enhanced growth: galactose, glucose, lactose, and glycerol. The polar lipids are $C_{20}C_{20}$ derivatives of phosphatidyl glycerol, phosphatidyl glycerol phos-

TABLE A2.32. Characteristics differentiating representatives of the genus *Halorubrum*[a]

Characteristic	1. H. saccharovorum	2. H. coriense	3. H. distributum	4. H. lacusprofundi	5. H. sodomense	6. H. trapanicum	7. H. vacuolatum
$C_{20}C_{20}$ and $C_{20}C_{25}$ core lipids	−	−	−	−	−	−	+
Presence of PGS	+	+	+	+	+	+	−
Presence of S-DGD[b]	+	+	+	+	+	+	−
Motile	+	+	+	d	+	−	−
Gas vacuoles	−	−	−	−	−	−	+
Growth at 2 M NaCl	+	−	+	+	+	−	−
Growth at 10°C	−	−	−	+	−	−	−
Growth at pH 10.5	−	−	−	−	−	−	+
Starch hydrolyzed	−	nd	−	−	+	−	nd
Xanthine degraded	−	nd	nd	+	nd	nd	nd
Amino acids required	+	−	nd	+	+	+	nd
Glucose used	+	+	−	+	+	+	+
Sucrose used	+	+	−	nd	+	+	+
Lactose used	+	+	nd	+	+	−	nd
Galactose used	+	nd	−	+	−	+	+
Acid from glucose	+	nd	+	−	+	+	nd

[a]For symbols see standard definitions; nd, not determined.

[b]See text for details of the glycolipid S-DGD.

phate methyl ester, phosphatidyl glycerol sulfate, and sulfated mannosyl-glucosyl-glycerol diether. Susceptible to infection by halophage HF2. Isolated from marine salterns.

Nuttall and Dyall-Smith (1993a) gave a detailed description of strain Ch2, but did not propose a name because the taxonomy of related species had not been clarified.

The mol% G + C of the DNA is: not known.

Type strain: Ch2, ACM 3911, DSMZ 10284, JCM 9275.

GenBank accession number (16S rRNA): L00922.

3. **Halorubrum distributum** (Zvyagintseva and Tarasov 1987) Oren and Ventosa 1996, 1180[VP] (*Halobacterium distributum* Zvyagintseva and Tarasov 1987, 667; *Halorubrobacterium distributum* (Zvyagintseva and Tarasov 1987) Kamekura and Dyall-Smith 1995, 346.)

dis.tri.bu' tum. L. neut. adj. *distributum* distributed (widely).

Cells are pleomorphic, but predominantly rod shaped 0.8– 1 μm × 2.5–7 μm. Gas vacuoles are absent. Motile. Colonies orange-red. Strictly aerobic. NaCl concentration for growth; 1.7–5.2 M; optimum: 2.5–4.3 M. Temperature range for growth, 26–50°C; optimum 37–45°C. Oxidase and catalase positive. Gelatinase and amylase negative. Indole not produced. Hydrogen sulfide produced. Sugars (galactose, glucose, and sucrose) did not stimulate growth but were metabolized with the formation of acid. Acidification of the medium was observed in the presence of glycerol. The polar lipids are $C_{20}C_{20}$ derivatives of phosphatidyl glycerol, phosphatidyl glycerol phosphate methyl ester, phosphatidyl glycerol sulfate, and sulfated mannosyl-glucosyl-glycerol diether. Isolated from marine salterns. A minor DNA component has been detected according to Tindall (1992).

Zvyagintseva and Tarasov (1987) originally used the specific epithet "*distributus*", which was subsequently corrected. Also Zvyagintseva et al. (1996) proposed that strain 1M (VKM B-1733) should be deposed as the type strain and moved to a different species. Oren et al. (1997a) stated that this was contrary to the rules of the International Code of Nomenclature of Bacteria and that strain 1M should remain as the type strain. Several other strains, including strain 4p (VKM B- 1739), have been referred to as *Halorubrum (Halobacterium) distributum*, but they should be considered as a new species in the genus *Halorubrum*.

The mol% G + C of the DNA is: 63.6 (major component) and 54.6 (minor component) (Tarasov et al., 1987).

Type strain: 1M, ATTC 51197, JCM 9100, NCIMB 13203, VKM B-1733.

GenBank accession number (16S rRNA): D63572.

4. **Halorubrum lacusprofundi** (Franzmann, Stackebrandt, Sanderson, Volkman, Cameron, Stevenson, McMeekin and Burton 1988) McGenity and Grant 1996, 362[VP] (Effective publication: McGenity and Grant 1995, 242) (*Halobacterium lacusprofundi* Franzmann, Stackebrandt, Sanderson, Volkman, Cameron, Stevenson, McMeekin and Burton 1988, 26; *Halorubrobacterium lacusprofundi* (Franzmann, Stackebrandt, Sanderson, Volkman, Cameron, Stevenson, McMeekin and Burton 1988) Kamekura and Dyall-Smith 1995, 345.)

la.cus.pro.fun' di. L. masc. n. *lacus* lake; L. masc. adj. *profundus* deep; M.L. gen. n. *lacusprofundi* of the deep lake, referring to Deep Lake, Antarctica, where the organism was originally isolated.

Cells are thin rods up to 12 μm long but become pleomorphic under unfavorable conditions. Gas vacuoles are absent. The type strain is motile, but a second strain is not. Colonies orange-red. Strictly aerobic. NaCl concentration for growth, 1.5–5.2 M; optimum, 2.5–3.5 M. Mg^{2+} requirement, 0.005 M. Temperature range for growth, 4–40°C, optimum: 31–36°C. Oxidase and catalase positive. Gelatinase, caseinase, urease, and amylase negative. Phosphatase positive. Indole not produced. Xanthine degraded. Tween not hydrolyzed. Hydrogen sulfide not produced from cysteine. Amino acids required for growth. Chemoorganotrophic. Many sugars utilized without acid production: ribose, galactose, glucose, mannose, and lactose. Other carbon sources include: formate, acetate, propionate, and ethanol. Glycerol and lactate are not used. The polar lipids are $C_{20}C_{20}$ derivatives of phosphatidyl glycerol, phosphatidyl glycerol phosphate methyl ester, phosphatidyl glycerol sulfate, and sulfated mannosyl-glucosyl-glycerol diether. Isolated from Deep Lake, Antarctica.

For further taxonomic comments, see last paragraph in *H. saccharovorum* description.

The mol% G + C of the DNA is: 65.3–65.8 (Bd) (major component) and 54.8–56.5 (Bd) (minor component).

Type strain: ACAM 34, ATCC 49239, DSMZ 5036, JCM 8891, NCIMB 12997, UQM 3107.

GenBank accession number (16S rRNA): X82170.

5. **Halorubrum sodomense** (Oren 1983) McGenity and Grant 1996, 362[VP] (Effective publication: McGenity and Grant 1995, 242) (*Halobacterium sodomense* Oren 1983, 385; *Halorubrobacterium sodomense* (Oren 1983) Kamekura and Dyall- Smith 1995, 345.)

so.do.men' se. M.L. neut. adj. *sodomense* pertaining to Sodom, near the Dead Sea, where the organism was originally isolated.

Cells are rod shaped 0.5 μm × 2.5–5 μm, but become pleomorphic under unfavorable conditions. Stain Gram-negative. Gas vacuoles are absent. Motile by a tuft of polar flagella. Colonies orange-red. Strictly aerobic. Growth occurs in media containing between 0.5 (in the presence of 1.5–2 M Mg^{2+}) and 4.3 M NaCl; optimum, 1.7–2.5 M (in the presence of 0.6– 1.2 M Mg^{2+}). Mg^{2+} can be partially replaced by Ca^{2+}. Mg^{2+} requirement, 0.005 M; optimum, 0.6–1.2 M. Temperature range for growth, 20–50°C; optimum 40°C. Oxidase and catalase positive. Gelatinase negative. Amylase positive. Indole not produced. Amino acids required for growth. Chemoorganotrophic. Certain sugars (xylose, fructose, glucose, maltose, sucrose) and glycerol utilized with production of acid. Little or no acid produced from galactose, mannose, lactose, arabinose, rhamnose, ribose, or mannitol. Bacteriorhodopsin is produced in the light under reduced oxygen tensions. The polar lipids are $C_{20}C_{20}$ derivatives of phosphatidyl glycerol, phosphatidyl glycerol phosphate methyl ester, phosphatidyl glycerol sulfate, and sulfated mannosyl-glucosyl-glycerol diether. Isolated from the Dead Sea. A minor DNA component has not been detected, though a plasmid has been observed.

H. sodomense was considered *species incertae sedis* in the first edition of the *Manual* (Larsen and Grant, 1989). For further taxonomic comments, see last paragraph in *H. saccharovorum* description.

The mol% G + C of the DNA is: 67.4 (Bd), 68 (Bd), 66 (T_m).

Type strain: RD 26, ATCC 33755, DSMZ 3755, NCIMB 2197.

GenBank accession number (16S rRNA): D13379, X82169.

6. **Halorubrum trapanicum** (Petter 1931) McGenity and Grant 1996, 362[VP] (Effective publication: McGenity and Grant 1995, 242) (*Bacterium trapanicum* Petter 1931, 1419; *Halobacterium trapanicum* (Petter 1931) Elazari-Volcani 1957, 211.)

tra.pa' ni.cum. M.L. neut. adj. *trapanicum* belonging to Trapani, Sicily, the site of the salt from which the organism was originally isolated.

Cells generally rod shaped 0.7–1 μm × 1.5–3 μm but often pleomorphic. Stain Gram-negative. Gas vacuoles are absent. Nonmotile. Cells contain low levels of carotenoids and so colonies are often pale orange, almost colorless. Strictly aerobic. NaCl concentration for growth: 2.5–5.2 M. Temperature optimum 37°C. Oxidase and catalase positive. Gelatinase and amylase negative. Indole not produced. Gas produced from nitrate. Amino acids required for growth. Chemoorganotrophic. Many sugars or sugar derivatives utilized with production of acid: fructose, galactose, glucose, mannose, maltose, sucrose, trehalose, and glycerol. No growth stimulation or acid production was observed on rhamnose, arabinose, sorbose, melezitose, ribose, xylose, lactose, raffinose, dextrin, inulin, starch, mannitol, or salicin. The polar lipids are $C_{20}C_{20}$ derivatives of phosphatidyl glycerol, phosphatidyl glycerol phosphate methyl ester, phosphatidyl glycerol sulfate, and sulfated mannosyl-glucosyl-glycerol diether. Isolated from Trapani salt from a cannery in Bergen (Norway). A minor DNA component has not been detected.

The mol% G + C of the DNA is: 64.3 (Bd).

Type strain: DSMZ 5647.

GenBank accession number (16S rRNA): X82168.

Additional Remarks: Strains from four culture collections, supposedly derived from the original strain isolated by Petter (1931), are all different (Tindall, 1992). Strain JCM 8979 is phylogenetically and morphologically like *Halococcus* sp. and does not match the original description. Strain ATCC 43102 is also a coccus (Grant et al., 1998b). Strain NCIMB 767 is phylogenetically related to the genus *Natrinema* (McGenity et al., 1998). The strain deposited by N.E. Gibbons as NRC 34021 survives only in the laboratory of W.D. Grant. This strain was used in the study in which *Halobacterium trapanicum* was transferred to the genus *Halorubrum*. The properties of strain NRC 34021 generally agree with the original description, and so it has been deposited in the National Collection of Industrial and Marine Bacteria, Aberdeen, United Kingdom, as strain NCIMB 13488 and proposed as the neotype (Grant et al., 1998b). *Halorubrum trapanicum* was considered *species incertae sedis* in the first edition of the *Manual* (Larsen and Grant, 1989).

7. **Halorubrum vacuolatum** (Mwatha and Grant 1993) Kamekura, Dyall-Smith, Upsani, Ventosa and Kates 1997, 857[VP] (*Natronobacterium vacuolatum* Mwatha and Grant 1993, 403.)

va.cu.o.la' tum. L. neut. adj. *vacuolatum* vacuolate (containing gas vacuoles).

Cells are short rods 0.5–0.7 μm × 1.5–3.0 μm in the exponential phase of growth in liquid medium and spherical in the stationary phase and on solid medium (1–1.5 μm diameter). Variable Gram stain. Cells produce large gas vacuoles in the stationary phase and on solid medium. Nonmotile. Colonies bright pink. Strictly aerobic. NaCl concentration for growth, 2.5–5.2 M; optimum, 3.5 M. Obligately alkaliphilic, requiring sodium carbonate, with a pH range for growth of 8.5–10.5; optimum pH 9.5. Temperature range for growth: 20–50°C, optimum 35–40°C. Oxidase and catalase positive. Gelatinase and amylase negative. Many sugars, sugar derivatives, and amino acids utilized: galactose, glucose, sucrose, fumarate, succinate, citrate, acetate, proline, and lysine. Fructose, mannitol, and pyruvate are not used. Possesses menaquinone MK-8. The polar lipids are both $C_{20}C_{20}$ and $C_{20}C_{25}$ derivatives of phosphatidyl glycerol and phosphatidyl glycerol phosphate methyl ester with no phosphatidyl glycerol sulfate or glycolipids. Isolated from Lake Magadi, Kenya.

Mwatha and Grant (1993) originally used the specific epithet "*vacuolata*" which was subsequently corrected.

The mol% G + C of the DNA is: 62.7 (T_m).

Type strain: M 24, ATTC 51376, DSMZ 8800, JCM 9060.

GenBank accession number (16S rRNA): D87972.

Genus VIII. **Haloterrigena** *Ventosa, Gutiérrez, Kamekura and Dyall-Smith 1999b, 135[VP]*

ANTONIO VENTOSA AND MASAHIRO KAMEKURA

Ha.lo.ter.ri.ge' na. Gr. n. *hals, halos* the sea, salt; L. fem. adj. *terrigena* born from the earth; M.L. fem. adj. *Haloterrigena* the salt born from the earth.

Gram-negative rods or oval cells. Nonmotile. Oxidase and catalase positive. **Extremely halophilic**, requiring at least 2 M NaCl. Optimum temperature 45°C. Colonies are pigmented red or light pink due to the presence of C_{50}-carotenoids. Chemoorganotrophic; **aerobic**. Polar lipids are glycerol-diether analogs of PG, PGP-Me, and some glycolipids (S2-DGD)*.

The mol% G + C of the DNA is: 59.2–60.2.

Type species: **Haloterrigena turkmenica** (Zvyagintseva and Tarasov 1987) Ventosa, Gutiérrez, Kamekura and Dyall-Smith 1999b, 135 (*Halococcus turkmenicus* Zvyagintseva and Tarasov 1987, 843.)

FURTHER DESCRIPTIVE INFORMATION

The genus is currently represented by a single species, *H. turkmenica*, which was isolated from a saline soil (Zvyagintseva and Tarasov, 1987). The genus was established to remove *Halococcus turkmenicus* from the other species of the genus *Halococcus*, with which it is phylogenetically unrelated (Ventosa et al., 1999b). The taxonomic description is not very complete and few studies on this organism have been reported (Tarasov et al., 1987, 1989; Kandler, 1994).

Haloterrigena turkmenica contains the bis-sulfated glycolipid S2-DGD (mannose-2,6 disulfate 1 → 2 glucose-glycerol diether) as its sole glycolipid; this glycolipid was found before only in

---

*See *Halobacteriales* for definitions of lipid abbreviations.

Natrialba asiatica. Besides, both diphytanyl moieties ($C_{20}C_{20}$) and phytanyl-sesterterpanyl moieties ($C_{20}C_{25}$) of PG and PGP-Me are present (Ventosa et al., 1999b). The cell wall polymer of *H. turkmenica* resembles the L-glutaminyl glycan of *Natronococcus occultus* (Niemetz et al., 1997). The hydrolyzate of the cell wall contains L-glutamic acid, acetic acid, and D-glucosamine (Kandler, 1994).

ENRICHMENT AND ISOLATION PROCEDURES

See *Halobacteriales* and the genus *Halobacterium.*

MAINTENANCE PROCEDURES

See *Halobacteriales*—Further Descriptive Information.

DIFFERENTIATION OF THE GENUS *HALOTERRIGENA* FROM OTHER GENERA

The key to the family *Halobacteriaceae* indicates those characteristics useful for distinguishing *Haloterrigena* from the other genera of the family. This genus can be differentiated from other related genera by the presence of S2-DGD, as well as by comparative analysis of the 16S rRNA sequences.

TAXONOMIC COMMENTS

The genus *Haloterrigena* was created to accommodate the single species *Halococcus turkmenicus* and some other strains (*Halobacterium trapanicum* JCM 9743 and strain GSL-11, a Great Salt Lake isolate [Post and Al-Harjan, 1988]) (Ventosa et al., 1999b). McGenity et al. (1998) proposed the placement of some misclassified halobacterial strains in the genus *Natrinema*: *Halobacterium salinarum* NCIMB 786 as *Natrinema pellirubrum*, and *Halo-*

bacterium trapanicum NCIMB 784 and *Halobacterium halobium* NCIMB 777 as *Natrinema pallidum.* These halobacteria are phylogenetically closely related to *Haloterrigena* and it is suggested that *Halobacterium trapanicum* JCM 9743 (originally derived from NCIMB 767) and strain GSL-11 could be included in the genus *Natrinema* instead of *Haloterrigena.* Further studies on these strains would help to clarify their correct taxonomic placement.

An isolate, strain PR5, obtained from a saltern in Puerto Rico, with a similar polar lipid pattern to *H. turkmenica* and able to grow optimally at 50°C (Montalvo-Rodríguez, 1996) could represent a second species of the genus *Haloterrigena, H. thermotolerans* (Montalvo-Rodríguez et al., in press).

ACKNOWLEDGMENTS

We are grateful to M.C. Gutiérrez for the critical reading of this article.

FURTHER READING

Kamekura, M. 1999. Diversity of members of the family *Halobacteriaceae.* In Oren (Editor), Microbiology and Biogeochemistry of Hypersaline Environments, CRC Press, Boca Raton, pp. 13–25.

McGenity, T.J., R.T. Gemmell and W.D. Grant. 1998. Proposal of a new halobacterial genus *Natrinema* gen. nov., with two species *Natrinema pellirubrum* nom. nov. and *Natrinema pallidum* nom. nov. Int. J. Syst. Bacteriol. 48: 1187–1196.

Ventosa, A., M.C. Gutiérrez, M. Kamekura and M.L. Dyall-Smith. 1999. Proposal for the transfer of *Halococcus turkmenicus, Halobacterium trapanicum* JCM 9743 and strain GSL 11 to *Haloterrigena turkmenica* gen. nov., comb. nov. Int. J. Syst. Bacteriol. 49: 131–136.

Zvyagintseva, I.S. and A.L. Tarasov. 1987. Extreme halophilic bacteria from saline soils. Mikrobiologiya 56: 839–844.

List of species of the genus Haloterrigena

1. **Haloterrigena turkmenica** (Zvyagintseva and Tarasov 1987) Ventosa, Gutiérrez, Kamekura and Dyall-Smith 1999b, 135[VP] (*Halococcus turkmenicus* Zvyagintseva and Tarasov 1987, 843.) turk.me.ni.ca. M.L. fem. adj. *turkmenica* of Turkmen (Turkmenistan), from where the bacterium was originally isolated.

Cells are oval or coccoid, 1.5–2.0 μm in diameter. Nonmotile.

Optimal growth in media containing 15–25% NaCl. Optimal growth temperature 45°C.

Acid is produced from glucose, fructose, mannose, ribose, and sucrose. Indole is not produced. Nitrate is reduced without production of gas. Starch or gelatin is not hydrolyzed. H_2S is not produced.

Isolated from sulfate saline soil in Turkmen (Turkmenistan).

The mol% G + C of the DNA is: 59.8 (T_m).

Type strain: 4K, ATCC 51198, DSMZ 5511, NCIMB 13204, VKM B-1734.

GenBank accession number (16S rRNA): AB004978.

Genus IX. **Natrialba** *Kamekura and Dyall-Smith 1996, 625[VP] (Effective publication: Kamekura and Dyall-Smith 1995, 347)*

MASAHIRO KAMEKURA

Na.tri.al' ba. M.L. n. *natrium* sodium; M.L. fem. adj. *alba* white; M.L. fem. n. *Natrialba* sodium white.

Cells are rods when grown under optimum conditions, 0.5–1.0 × 1.0–5.0 μm. Cells stain Gram-negative. Motile.

Strictly aerobic. Oxidase and catalase positive. **Growth occurs in media containing between 2.0 M and 5.3 M (saturated) NaCl.** Isolated from hypersaline habitats (salt ponds, beach sands, saline soda lakes, etc.). **All strains of this genus possess $C_{20}C_{25}$ core diether lipids,** as easily detected by the double spots of PG (phosphatidyl glycerol) and PGP-Me (phosphatidyl glycerol phosphate methyl ester) on thin-layer chromatograms or by mass spectrometric analyses of the isolated polar lipid components (Matsu-

bara et al., 1994). There is one signature base (Table A2.26 in overview) (see Kamekura, 1998b).

The mol% G + C of the DNA is: 60.3–63.1.

Type species: **Natrialba asiatica** Kamekura and Dyall-Smith 1996, 625 (Effective publication: Kamekura and Dyall-Smith 1995, 348.)

FURTHER DESCRIPTIVE INFORMATION

Some strains are neutrophilic, but some are alkaliphilic. Cells of some strains are not pigmented, while others contain red-pink pigments.

ENRICHMENT AND ISOLATION PROCEDURES

See under *Halobacteriales* and the genus *Halobacterium*.

TAXONOMIC COMMENTS

The genus *Natrialba* was created to accommodate two neutrophilic strains, 172P1 and B1T, isolated in Japan and Taiwan, respectively, as *Natrialba asiatica* (Kamekura and Dyall-Smith, 1995). These strains are not red pigmented, unlike other members of the existing genera, and were shown to contain a glycolipid, S2-DGD-1 (disulfated diglycosyl diether), not detected in members of other genera.

The second species of the genus, *N. magadii*, was transferred from the genus *Natronobacterium* because its 16S rDNA sequence is closer to that of *Natrialba asiatica* (93–94%) than to that of the type species of *Natronobacterium* (*Natronobacterium gregoryi*) (Kamekura et al., 1997). This transfer is supported by maximum likelihood analysis (Fig. A2.36 of the overview). On the other hand, the lipid analyses do not support the close 16S rRNA sequence similarities. No glycolipid has been detected in *N. magadii*, while *N. asiatica* has a glycolipid, S2-DGD-1 (mannose-2,6 disulfate 1→2 glucose-glycerol diether) (Matsubara et al., 1994). The structures of these lipids suggest that the enzymes involved in the biosynthesis are different in the two species. However, one of the two strains of *N. asiatica*, strain B1T, is able to grow at alkaline pH (pH 5–10) (Kamekura and Dyall-Smith, 1995). Furthermore, *N. magadii* shares the signature base of the genus *Natrialba*, 560G (see Table A2.26 of the overview of *Halobacteriales*). Moreover, 626A is confined to *N. asiatica* strain B1T, *N. magadii*, and five other strains which belong to this genus, mentioned in

"Other Organisms". Altogether, these data (high similarities in 16S rRNA genes, signature bases, and growth at alkaline pH) seem to support the view that *N. magadii* is a member of the genus *Natrialba*.

The genus *Natrialba* is the first genus of the family *Halobacteriaceae* which comprises both neutrophilic and alkaliphilic strains. There remains a possibility, however, that a more divergent survey of alkaliphilic, as well as neutrophilic, extreme halophiles may split the genus in the future.

Recently a variety of extreme halophiles were isolated from hypersaline ponds in Northern China by a group led by P. Zhou in Beijing. Some neutrophilic strains, OF8 and QX1, have been suggested to belong to the genus *Natrialba* based on the presence of have the glycolipid S2-DGD-1 (M. Kamekura, T. Ito and P. Zhou, unpublished data). The haloalkaliphilic strains "*Natronobacterium wudunaoensis*" Y21, "*Natronobacterium innermongoliae*" HAM-2, and "*Natronobacterium chahannaoensis*" C112 have also been isolated (Y. Xu, P. Zhou and D. Wang, personal communication). These strains might belong to the genus *Natrialba* as judged from the 16S rDNA phylogenetic tree reconstruction (Fig. A2.36) and the fact that they share the signature base specific for this genus (Table A2.26 of the overview). The GenBank/EMBL/DDBJ accession numbers of 16S rRNA gene sequences of the last three strains are: AJ001376, AF009601, and AJ004806, respectively. Strain 98NT4 isolated from Lake Natron, which straddles the Kenyan-Tanzanian border (Duckworth et al., 1996), is also a member of the genus *Natrialba*. (The GenBank/EMBL/DDBJ accession number of the 16S rRNA gene sequence is X92174.)

DIFFERENTIATION OF THE SPECIES OF THE GENUS *NATRIALBA*

Differential characteristics of the species of the genus *Natrialba* are indicated in Table A2.33.

List of species of the genus Natrialba

1. **Natrialba asiatica** Kamekura and Dyall-Smith 1996, 625[VP] (Effective publication: Kamekura and Dyall-Smith 1995, 348.) *a.si.a' ti.ca*. L. fem. adj. *asiatica* from Asia referring to the geographical region from which these organisms were isolated.

Gram-negative rods, 0.5–1.0 × 1.0–5.0 μm. Motile. Gas vacuoles not present. Colonies are round and smooth. Does not form visible amounts of bacterioruberins, the red-pink pigments, in the whole range of NaCl concentrations, temperature, and illumination in various growth media. Content of bacterioruberins was less than 0.1% of that of *Halobacterium salinarum* NRC 34001.

Growth occurs in media containing 2.0–5.3 M (saturated) NaCl. Isolated from hypersaline habitats, salt ponds, beach sands.

Structure of the glycolipid (S2-DGD-1) is 2,3-di-*O*-phytanyl- or phytanyl-sesterterpanyl-[2,6-(HSO$_3$)2-α-mannopyranosyl- (1→2)-α-glucopyranosyl]-*sn*-glycerol (Matsubara et al., 1994).

The strain 172P1 does not require vitamins for growth at 37°C. The requirement for NaCl is partly replaceable by KCl; growth is obtained in a medium containing 6% NaCl and 12% KCl, but no growth occurs in 20% KCl.

The strain 172P1, isolated in Japan from beach sands with granular salts attached, produces an extracellular halophilic protease (Kamekura et al., 1992), and strain B1T was isolated from solar salts produced in Taiwan. Neither strain produces bacteriorhodopsin-like proteins in the membranes (Kamekura et al., 1998).

DNA–DNA hybridization experiments showed that strains 172P1 and B1T have more than 80% similarity, indicating they constitute a single species.

The mol% G + C of the DNA is: 60.3–63.1 (T$_m$).
Type strain: 172P1, DSMZ 12278, JCM 9576.
GenBank accession number (16S rRNA): D14123.

2. **Natrialba magadii** (Tindall, Ross and Grant 1984a) Kamekura and Dyall-Smith; Upasani, Ventosa and Kates 1997, 857[VP] (*Natronobacterium magadii* Tindall, Ross and Grant 1984a, 41.) *ma.ga' di.i*. M.L. gen. n. *magadii* of Magadi; named for Lake Magadi, a saline soda lake in Kenya.

Colonies are 2–3 mm in diameter after 5–7 d at 37°C, wrinkled, friable, and orange-red. In liquid culture, cells are

TABLE A2.33. Differential characteristics of the species of the genus *Natrialba*

Characteristic	1. *N. asiatica*	2. *N. magadii*
Membrane glycolipid	S2-DGD-1	not detected
Growth at pH 7	+	−

short rods 0.7–0.9 × 2–4 μm, spherical on solid medium (1–1.5 μm in diameter). Motile by a tuft of polar flagella.

Growth occurs in media containing 2–5.2 M NaCl with an optimum at 3.5 M. Growth occurs at 20–50°C; optimum temperature, 37–40°C. Alkaliphilic with obligate requirement for sodium carbonate. pH range for growth, 8.5–11.0; optimum 9.5.

Sugars are not utilized, although acetate stimulates growth. Casamino acids are used as a nitrogen source. Gelatin is liquefied; sulfide is produced from thiosulfate; does not hydrolyze starch.

The DNA is composed of a major component and a minor component. A large plasmid (~144 kb) is present.

The mol% G + C of the DNA is: 63.0 (major component) and 49.5 (Bd) (minor component).

Type strain: MS3, ATCC 43099, CCM 3739, DSMZ 3394, NCIMB 2190.

GenBank accession number (16S rRNA): X72495.

Other Organisms

1. "*Natronobacterium chahanensis*" Wang and Tang 1989, 72.

The taxonomic status of this organism is uncertain because of a lack of detailed descriptive information. This organism was originally isolated from Chahannao soda lake in the Inner Mongolia Autonomous Region. The cells are rod shaped, 0.8–1.2 × 1.5–3.0 μm. Gram negative. Grows well in 12–30% NaCl, with optimum growth at 20% and 37°C. Spherical cells are present in lower concentrations of NaCl. Alkaliphilic, with optimum pH 8.7 and no growth at pH below 8.5. Motile with flagella. Orange pigments are produced under illumination.

$C_{20}C_{20}$ and $C_{20}C_{25}$ ether lipids are present. As with all *Archaea*, muramic acid and diaminopimelic acid are absent. Sensitive to erythromycin, and rifampicin, but insensitive to penicillin, streptomycin, and neomycin. Utilizes fructose, glucose, maltose, and glycerol. Produces extracellular amylase with optimal activity at 16% NaCl and pH 8.0.

The strain C212 has a 16S rRNA gene sequence very similar to that of *Natrialba magadii* (P. Zhou, T. Ito and M. Kamekura, unpublished data), suggesting that it may belong to that species.

The mol% G + C of the DNA is: 59.5 (T_m).

2. *Natrialba* sp. strain SSL1 (Kamekura, Dyall-Smith, Upasani, Ventosa and Kates 1997.)

The taxonomic status of this organism is uncertain because of a lack of detailed descriptive information. This strain was originally isolated from an alkaline saline brine from Sambhar Salt Lake, India (Upasani and Desai, 1990). In liquid culture, cells are motile, long rods 0.5–0.8 × 4–10 μm. On solid media cells are spherical, 1–1.5 μm in diameter. Obligate alkaliphilic halophile with optimum growth at pH 9.5 and 17.5% NaCl (w/v). On agar plates colonies are pale pink and opaque due to the presence of gas vacuoles.

Cell membranes contain a minor amount of glycolipid DGD-4 (glucose 1→6 glucose-glycerol diether) (Upasani et al., 1994). No bacteriorhodopsin produced (Kamekura et al., 1998). The 16S rRNA gene sequence strongly suggests that this isolate is a member of the genus *Natrialba*; the similarity between SSL1 and *N. asiatica* 172P1, and that between SSL1 and *N. asiatica* B1T are 95.9% and 96.6%, respectively (Kamekura et al., 1997).

The mol% G + C of the DNA is: 60.2 ± 0.4 (T_m).

Deposited strain: SSL1, ATCC 43988.

GenBank accession number (16S rRNA): D88256.

Genus X. **Natrinema** *McGenity, Gemmell and Grant 1998, 1194*[VP]

TERRY J. MCGENITY, WILLIAM D. GRANT AND MASAHIRO KAMEKURA

Na.tri.ne'ma. M.L. n. *natrium* sodium; Gr. n. *nema* a thread; M.L. neut. n. *Natrinema* sodium (-requiring) thread.

Cells rod shaped, 1–5 × 0.6–1.0 μm, but pleomorphic under unfavorable conditions. Colonies light orange-red or pale orange, 1–2 mm in diameter, smooth, circular, convex. Cells stain Gram-negative. Strict aerobes. Chemoorganotrophs. Nitrogen sources: Casamino acids. Carbon sources, Casamino acids and certain sugars. Optimum growth concentration 3.4–4.3 M NaCl. Optimum pH 7.0–7.6. Gelatin liquefied, starch not hydrolyzed, sulfide and indole not produced. Possess menaquinones MK-8 and MK-8(H_2) and both $C_{20}C_{20}$ and $C_{20}C_{25}$ diether core lipids. Possess several unidentified but characteristic glycolipids. Phylogenetically tightly clustered genus (99.1% 16S rRNA sequence similarity between the two species), with two signature sequences (Table A2.25).

The mol% G + C of the DNA is: 69.9 (major component) and 60.0 (minor component) but has been identified only for the type species.

Type species: **Natrinema pellirubrum** McGenity, Gemmell and Grant 1998, 1194.

FURTHER DESCRIPTIVE INFORMATION

Both species of *Natrinema* possess the phospholipids phosphatidyl glycerol and phosphatidyl glycerol phosphate methyl ester, but *Natrinema pellirubrum* has a very small amount of phosphatidyl glycerol sulfate compared with *Natrinema pallidum*. Several glycolipids are characteristic of the genus but remain unidentified (Ross et al., 1985).

A large plasmid of ~144 kb has been detected in *N. pellirubrum*, but has not been tested for in *N. pallidum*, contrary to a previous report (McGenity et al., 1998).

Some strains grow slowly at pH 8.6. Both species are sensitive to anisomycin, bacitracin, novobiocin, vibriostatic agent (0/129), and sulfa-methoxazole/trimethoprim. Sanz et al. (1993) revealed that *N. pallidum* NCIMB 777, which has three rRNA operons (Sanz et al., 1988), was distinct from other halobacteria in terms of the sensitivity of its protein synthesis systems to antibiotics.

Members of the genus *Natrinema* were isolated from spoiled, salted fish and hides (Formisano, 1962), while similar strains have been isolated from salt lakes and salterns (Post and Al-Harjan, 1988; McGenity et al., 1998). Cells lyse in less than 1.5 M NaCl (McGenity et al., 1998), but Formisano (1962) noted that *N. pallidum* (formerly *Halobacterium trapanicum*) NCIMB 784 not only survived but grew moderately at 0.7 M NaCl, which is at odds with the requirement of 1.8 M NaCl for growth reported by McGenity et al. (1998). The ability to withstand low salt concentrations may be lost upon repeated subculture of laboratory strains at high salt concentrations. It should be noted that strains related to *Natrinema* spp. have been isolated from former salterns in which the salinity is relatively low. For example, strain T5.7 (Fig. A2.36) was isolated from a saltern containing 0.7 M NaCl (McGenity et al., 1998).

ENRICHMENT AND ISOLATION PROCEDURES

The methods are the same as those given for the genus *Halobacterium*.

MAINTENANCE PROCEDURES

See under *Halobacteriales*.

DIFFERENTIATION OF THE GENUS *NATRINEMA* FROM OTHER GENERA

Several characteristics given in the description of the genus differentiate *Natrinema* from other genera, in particular, phylogenetic analysis of 16S rRNA sequences together with signature sequences and bases (Tables A2.25 and A2.26). All representatives of the genus *Natrinema* are neutrophilic, which differentiates them from phylogenetically related genera. Polar lipid profiles may be used to distinguish *Natrinema* spp. from phylogenetically related, neutrophilic, rod-shaped halobacteria (e.g., *Natrialba asiatica*).

TAXONOMIC COMMENTS

The genus *Natrinema* was created to accommodate three well-characterized but misclassified strains (McGenity et al., 1998). In the last edition of the *Manual*, *Halobacterium salinarium* NCIMB 786 (now the type species, *Natrinema pellirubrum*), together with *Halobacterium trapanicum* NCIMB 784 and *Halobacterium halobium* NCIMB 777 (both now *N. pallidum*), were listed as "other organisms" within the genus *Halobacterium* (Larsen and Grant, 1989). Larsen and Grant (1989) stated that "these three organisms together constitute a taxon distinct from other halobacteria" on the basis of 16S rRNA-DNA hybridization (Ross and Grant, 1985) and polar lipid analysis (Ross et al., 1985). This has been supported by 16S rRNA sequence analysis (Fig. A2.36) and other

taxonomic data (McGenity et al., 1998). Grant and Ross (1986) originally suggested the genus name *"Halonema"* for these three species, but this was never validly published.

There has been some confusion over the naming of strain NCIMB 786—although it has been referred to widely as *Halobacterium salinarium* (or *Halobacterium salinarium* subsp. *proteolyticum*), it was originally described as *Halobacterium cutirubrum* var. *proteolyticum* (Formisano, 1962).

It has recently become clear that several other misnamed halobacteria may be related to *Natrinema* species (Kamekura and Dyall-Smith, 1995; McGenity et al., 1998; Ventosa et al., 1999b). Pending the valid publication of *Natrinema* gen. nov. (McGenity et al., 1998), Ventosa et al. (1999b) independently proposed the creation of a new genus, *Haloterrigena*, with a single species, *Haloterrigena turkmenica*, to accommodate the organisms *Halococcus turkmenicus* VKM B1734, *Halobacterium trapanicum* NCIMB 767 (redeposited by M. Kamekura as JCM 9743), and strain GSL11 on the basis of high (>70%) DNA–DNA reassociation. Strain VKM B1734, however, shares only 95.6% 16S rRNA sequence similarity with strain GSL11. Such high DNA–DNA reassociation between strains NCIMB 767, GSL11, and VKM B1734 may have been caused in part by lateral transfer of large extrachromosomal elements — a hypothesis that warrants further investigation.

Polar lipid analysis, normally a good indicator of relatedness for neutrophilic halobacteria, suggests that strain VKM B1734 might be considered distinct from strains GSL11 and NCIMB 767. Strain VKM B1734 has $C_{20}C_{20}$ and $C_{20}C_{25}$ diether core lipids and the phospholipids phosphatidyl glycerol and phosphatidyl glycerol phosphate methyl ester, but it also has the glycolipid S2-DGD (mannose-2,6 disulfate 1→2 glucose-glycerol diether), which is present in *Natrialba asiatica* but absent from *Natrinema* spp. and strains GSL11 and NCIMB 767 (Ventosa et al., 1999b). From phylogenetic analysis (Fig. A2.36) and some phenotypic studies, it appears that strains GSL11 and NCIMB 767 probably belong to the genus *Natrinema* (McGenity et al., 1998; Ventosa et al., 1999b), but more detailed characterization is required to clarify the taxonomic position of the organism *Haloterrigena turkmenica*, formerly referred to as *Halococcus turkmenicus*.

FURTHER READING

McGenity, T.J., R.T. Gemmell and W.D. Grant. 1998. Proposal of a new halobacterial genus *Natrinema* gen. nov., with two species *Natrinema pellirubrum* nom. nov. and *Natrinema pallidum* nom. nov. Int. J. Syst. Bacteriol. *48*: 1187–1196.
Ventosa, A., M.C. Gutiérrez, M. Kamekura and M.L. Dyall-Smith. 1999. Proposal for the transfer of *Halococcus turkmenicus*, *Halobacterium trapanicum* JCM 9743 and strain GSL 11 to *Haloterrigena turkmenica* gen. nov., comb. nov. Int. J. Syst. Bacteriol. *49*: 131–136.

DIFFERENTIATION OF THE SPECIES OF THE GENUS *NATRINEMA*

N. pallidum NCIMB 784 and NCIMB 777 have 52% and 50% DNA similarity, respectively, to *N. pellirubrum* NCIMB 786 (Ross and Grant, 1985). Other features distinguishing the two species, in addition to those listed in Table A2.34 include 1) 99.0–99.1% 16S rRNA sequence similarity; 2) different ribotype patterns when digested with *Sal*I, *Kpn*I, and *Eco*RI; 3) different whole-cell protein profiles (75% similarity measured by the Pearson product-moment coefficient); and 4) different polar lipid composition — *N. pellirubrum* has an additional glycolipid and only amounts of phosphatidyl glycerol sulfate and lacks one of three glycolipids which are present in *N. pallidum* and other closely

related strains (Ross et al., 1985; Kamekura and Dyall-Smith, 1995).

TABLE A2.34. Characteristics differentiating *Natrinema* species

Characteristic	1. *N. pellirubrum*	2. *N. pallidum*
Presence of phosphatidyl glycerol sulfate	Trace	+
Ability to use ribose	+	−
Sensitivity to rifampicin (50 mg/disc)	−	Slight

List of species of the genus Natrinema

1. **Natrinema pellirubrum** McGenity, Gemmell and Grant 1998, 1194.[VP]

pel.li.rub′rum. L. fem. n. *pellis* skin or hide; L. neut. adj. *rubrum* red; M.L. neut. n. *pellirubrum* red-hide.

Cells are rod shaped (1–4 μm × 0.6–1.0 μm) but are pleomorphic under unfavorable conditions. Cells contain carotenoids, making colonies light red or orange. Cells lyse in less than 1.5 M NaCl. Growth in 2 M NaCl; optimum 3.4–4.3 M. Temperature range for growth 20–45°C. pH range for growth 6.0–8.6; optimum 7.2–7.8. Chemoorganotrophic. Strictly aerobic. Nitrogen sources, Casamino acids. Carbon sources, Casamino acids and the sugars fructose, glucose, lactose, and ribose. Gelatinase positive; nitrate reduced to nitrite, nitrite reduced but no gas produced, end product unknown. Does not produce indole, hydrolyze starch, or produce sulfide from cysteine. Possesses $C_{20}C_{20}$ and $C_{20}C_{25}$ diether core lipids and several unidentified glycolipids. Possesses phospholipids: phosphatidyl glycerol, phosphatidyl glycerol phosphate methyl ester, but only trace amounts of phosphatidyl glycerol sulfate. Sensitive to anisomycin, bacitracin, novobiocin, vibriostatic agent (0/129), and sulfamethoxazole/trimethoprim. Insensitive to ampicillin, chloramphenicol, ketoconzaole, flavomycin, rifampicin, streptomycin, and tetracycline. Isolated from salted hides. Possesses a large plasmid of ∼144 kb.

The mol% G + C of the DNA is: 69.9 (major component) and 60.0 (minor component) (Bd).

Type strain: NCIMB 786.

GenBank accession number (16S rRNA): AJ002947.

Additional Remarks: NCIMB 786 was originally described (but not formally proposed) as *Halobacterium cutirubrum* var. *proteolyticum* by Formisano (1962), but the deposited strain has been referred to throughout the literature as *Halobacterium salinarium* or *Halobacterium salinarium* subsp. *proteolyticum.*

2. **Natrinema pallidum** McGenity, Gemmell and Grant 1998, 1194.[VP]

pal′li.dum. L. neut. adj. *pallidum* pale.

Cells are rod shaped (1.5–6.0 μm × 0.7–1.0 μm) but become pleomorphic under unfavorable conditions. Requires at least 1.7 M NaCl for growth, and cells lyse in less than 1.5 M NaCl. Optimum NaCl concentration for growth 3.4–4.3 M. Cells contain only low levels of carotenoids and hence colonies are pale orange, beige, or almost colorless. Colonies are circular, entire, convex, translucent, 1.0–2.0 mm in diameter. Isolated from salted fish and hides. Chemoorganotrophic; strict aerobe. Nitrogen sources, Casamino acids. Carbon sources: Casamino acids, glucose, fructose, lactose, but not ribose. Reduces nitrate to nitrite and nitrite to an unknown end product. Liquifies gelatin; does not produce sulfide from cysteine. Indole and amylase negative. Possesses $C_{20}C_{20}$ and $C_{20}C_{25}$ diether core lipids and menaquinones MK-8 and MK-8(H$_2$). Possesses phospholipids: phosphatidyl glycerol, phosphatidyl glycerol phosphate methyl ester, significant amounts of phosphatidyl glycerol sulfate, and several unidentified glycolipids. Grows in pH range 6.0–8.4; optimum 7.2–7.6. Temperature optimum 37–40°C. Sensitive to anisomycin, bacitracin, novobiocin, vibriostatic agent (0/129), and sulfa-methoxazole/trimethoprim. Slightly sensitive to rifampicin. Insensitive to amphotericin B, ampicillin, chloramphenicol, erythromycin, flavomycin, streptomycin, and tetracycline.

NCIMB 777 was deposited as *Halobacterium halobium* by D. J. Kushner. Strain NCIMB 784, deposited as *Halobacterium trapanicum* by Formisano (1962), is regarded as synonymous with NCIMB 777 based on identical nutritional requirements and antibiotic sensitivity, very similar whole-cell protein profiles (85% similarity, calculated using the Pearson product-moment coefficient), and identical ribotype patterns when digested with *Sal*I, *Taq*I, *Eco*RI, and *Xho*I (McGenity et al., 1998); and this is supported by their identical 16S rRNA sequences and polar lipids (Ross et al., 1985).

The mol% G + C of the DNA is: not known.

Type strain: NCIMB 777.

GenBank accession number (16S rRNA): AJ002949.

Genus XI. **Natronobacterium** *Tindall, Ross and Grant 1984b, 355*[VP] *(Effective publication: Tindall, Ross and Grant 1984a, 41) emend.*

BRIAN J. TINDALL

Na.tro.no.bac.te′ri.um. arbitrarily derived from the Arabic n. *natrun* soda, salt; Gr. n. *bakterion* a small rod; M.L. neut. n. *Natronobacterium* soda rod.

Rods in liquid culture in the exponential phase of growth, usually **0.5–1.0 × 2–15 μm. Cells become shorter** in older liquid cultures, while cells are **short rods** to **coccobacilli** or **coccoid** on agar plates (Fig. A2.43). Resting stages are not known. **Gram-negative. Cells lyse rapidly in distilled water. Nonmotile. Aerobic. Growth stimulated by carbohydrates.** Nitrate and nitrite are not reduced. Gelatin hydrolyzed. Starch is not hydrolyzed. Sulfide produced from thiosulfate. **Alkaliphilic,** requiring a pH of at least 8.5 for growth, with an **optimum at pH 9.5–10.0.** Grows at 2.0–5.2 M NaCl, with an optimum at 3 M. **Requires only low concentrations of magnesium** (<10 mM), elevated concentrations becoming toxic. Temperature range for growth 25–40°C, with an optimum at 37°C.

$C_{20}C_{20}$ and $C_{20}C_{25}$ diethers are present, the relative composition of which has been shown to vary under different growth conditions and during the growth phase (Tindall et al., 1984a, 1991a; Tindall, 1985; Morth and Tindall, 1985b). The predominant polar lipids present are the diether derivatives of phosphatidyl glycerol and methyl-phosphatidyl glycerophosphate, together with small amounts of unidentified phospholipids; glycolipids are not present in significant quantities (Tindall et al., 1984a; Morth and Tindall, 1985a). Menaquinones are the only respiratory lipoquinones present. Members of the genus *Natronobacterium* are unique in that they produce MK-8, MK-8(VIII-H$_2$), MMK-8, MMK-8(VIII-H$_2$), DMMK-8, and DMMK-8(VIII-H$_2$) (Collins and Tindall, 1987).

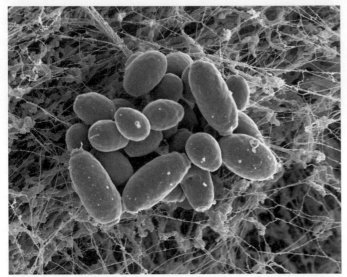

FIGURE A2.43. Scanning electron micrograph of liquid culture of *Natronobacterium gregoryi* (DSMZ 3393) in the late exponential/stationary phase. Note the presence of short rods and coccobacilli. (Courtesy of Dr. M. Rohde, GBF, Braunschweig.)

Type species: **Natronobacterium gregoryi** Tindall, Ross and Grant 1984b, 355 (Effective publication: Tindall, Ross and Grant 1984a, 41.)

FURTHER DESCRIPTIVE INFORMATION

DNA–RNA hybridization data indicate a specific relationship between *Natronobacterium gregoryi*, *Natronomonas pharaonis*, and *Natrialba magadii* (Ross and Grant, 1985), whereas 16S rDNA se-

quence analysis indicates a specific relationship to members of the genus *Natronococcus* (see Kamekura et al., 1997).

Members of the genus *Natronobacterium* are currently represented by a single species, *N. gregoryi*, which was isolated from the salterns at Lake Magadi, Kenya. No other isolates of this genus are known.

ENRICHMENT AND ISOLATION PROCEDURES

Members of the genus *Natronobacterium* have been isolated only from the solar salt pans at Lake Magadi (Tindall, 1980; Tindall et al., 1984a). There are no reports that members of this genus occur in other alkaline, highly saline environments. The method of isolation employed a medium containing high levels of NaCl, and low levels of magnesium and calcium, together with sufficient Na_2CO_3 to maintain the pH between 9.0 and 10.0 (Tindall et al., 1984a). The method of isolation does not appear to be specific for members of this genus, since members of the species *Natronomonas pharaonis* (basonym: *Halobacterium pharaonis*; *Natronobacterium pharaonis*) and *Natrialba magadii* (basonym: *Natronobacterium magadii*), together with a large number of uncharacterized alkaliphilic halobacteria were isolated concurrently with *Natronobacterium* (Tindall, 1980; Tindall et al., 1984a).

MAINTENANCE PROCEDURES

Members of the genus *Natronobacterium* may be maintained on the media of Tindall et al. (1984a) and Tindall (1985). Long-term storage of members of this genus is best achieved by drying or by storing in liquid nitrogen as outlined by Tindall (1991). A critical aspect in the drying of members of this genus is avoiding the prior freezing of the cells. Methods described as freeze drying result in rapid loss of viability, whereas methods employing some form of liquid drying result in good recovery.

List of species of the genus Natronobacterium

1. **Natronobacterium gregoryi** Tindall, Ross and Grant 1984b, 355[VP] (Effective publication: Tindall, Ross and Grant 1984a, 41.)

   gre.go′ry.i. M.L. gen. n. *gregoryi* of Gregory; named for J.W. Gregory, Scottish geologist who first described the geology of the rift valley.

   The species description is identical with that of the genus with the following additions. Growth stimulated by ribose, fructose, glucose, mannose, and sucrose. Resistant to ampi-

   cillin (25 μg/disk), carbenicillin (100 μg/disk), and tetracycline (50 μg/disk) but sensitive to novobiocin (10 μg/disk), bacitracin (5 U/disk), anisomycin (5 μg/disk), ciprofloxacin (MIC, 15 μg/ml;, IC50 5 μg/ml), perfloxacin (IC50, 60 μg/ml), norfloxacin (IC50, 25 μg/ml), and coumermycin (IC50, 0.2 μg/ml).

   The mol% G + C of the DNA is: 65.0 (T_m); 63.2–65.0 (Bd).

   Type strain: SP2, ATCC 43098, CCM 3738, DSMZ 3393, JCM 8860, NCIMB 2189, VKM B-1750.

   GenBank accession number (16S rRNA): D87970.

Genus XII. **Natronococcus** *Tindall, Ross and Grant 1984b, 355[VP] (Effective publication: Tindall, Ross and Grant, 1984a, 41) emend.*

BRIAN J. TINDALL

Na.tro.no.coc′cus. arbitrarily derived from the Arabic n. *natrun* soda, salt; Gr. n. *kokkos* a berry, small rod; M.L. masc. n. *Natronococcus* soda berry.

Cocci in liquid culture and on plates usually 1.0 × 2.0 μm. **Cells occur in irregular clusters, although they may also occur in pairs or singly.** Resting stages are not known. **Gram variable. Cells do not lyse rapidly in distilled water.** Nonmotile. Aerobic. Colonies on plates smooth, entire, and **pale brown to red-orange. Growth may be stimulated by carbohydrates**. Nitrate may be reduced,

although there is no evidence that this is coupled to growth. Gelatin and starch may be hydrolyzed. Sulfide may be produced from thiosulfate. **Alkaliphilic**, requiring a pH of at least 8.5 for growth, with an **optimum at pH 9.5–10.0**. Grows at 1.4–5.2 M NaCl, with the optimum in the range 2.5–3 M. **Requires only low concentrations of magnesium** (<10 mM), elevated concentrations becoming toxic. Temperature range for growth 22–50°C, with an optimum in the range 35–45°C.

$C_{20}C_{20}$ and $C_{20}C_{25}$ diethers are present. The predominant polar lipids present are the diether derivatives of phosphatidyl glycerol and methyl-phosphatidyl glycerophosphate, together with small amounts of unidentified phospholipids; significant quantities of glycolipids are not present (Tindall et al., 1984a; Morth and Tindall, 1985a). A phosphatidylglycero-(cyclo)-phosphate may also be present (Lanzotti et al., 1989). Menaquinones are the only respiratory lipoquinones present, comprising MK-8 and MK-8(VIII-H$_2$) (Tindall et al., 1984a).

Type species: **Natronococcus occultus** Tindall, Ross and Grant 1984b, 355 (Effective publication: Tindall, Ross and Grant 1884a, 41.)

FURTHER DESCRIPTIVE INFORMATION

DNA–RNA hybridization data indicate that the genus *Natronococcus* is a distinct lineage within the family *Halobacteriaceae* (Ross and Grant, 1985), whereas 16S rDNA sequence analysis indicates a specific relationship to members of the genus *Natronobacterium* (see Kamekura et al., 1997). It should be noted that the position of *Natronococcus occultus* in the DNA–RNA dendrogram appears to correspond to the position of *Natronomonas pharaonis* in the 16S rDNA sequence based dendrogram, and the position of *Natronomonas pharaonis* in the DNA–RNA dendrogram appears to correspond to the position of *N. occultus* in the 16S rDNA sequence based dendrogram.

Members of the genus *Natronococcus* are currently represented by two species, *N. occultus* and *Natronococcus amylolyticus*, both of which were isolated from Lake Magadi, Kenya. No other species in this genus are known.

ENRICHMENT AND ISOLATION PROCEDURES

Members of the genus *Natronococcus* have been isolated from Lake Magadi (Tindall et al., 1984a; Kanai et al., 1995). There are no reports that members of this genus occur in other alkaline, highly saline lakes. The methods of isolation employed media containing high levels of NaCl and low levels of magnesium and calcium, together with sufficient Na$_2$CO$_3$ to maintain the pH between 9.0 and 10.0 (Tindall et al., 1984a; Kanai et al., 1995). Although the addition of nitrate resulted in the enrichment of *Natronococcus occultus* and the addition of starch to the medium resulted in the isolation of *Natronococcus amylolyticus* there are numerous reports of studies using similar alkaline, highly saline media that have resulted in the isolation of alkaliphilic halobacteria not belonging to the genus *Natronococcus* (Soliman and Trüper, 1982; Tindall, 1986, 1988; Xu et al., 1999).

MAINTENANCE PROCEDURES

Members of the genus *Natronococcus* may be maintained on the media of Tindall et al. (1984a), Kanai et al. (1995), and Tindall (1985). Long-term storage of members of this genus is best achieved by drying or by storing in liquid nitrogen as outlined by Tindall (1991). A critical aspect in the drying of members of this genus is avoiding the prior freezing of the cells. Methods described as freeze drying result in rapid loss of viability, whereas methods employing some form of liquid drying result in good recovery.

List of species of the genus Natronococcus

1. **Natronococcus occultus** Tindall, Ross and Grant 1984b, 355[VP] (Effective publication: Tindall, Ross and Grant 1884a, 41.) *oc.cul′ tus.* L. masc. adj. *occultus* hidden, the hidden *Natronococcus.*

    Cocci in liquid culture and on plates usually 1.0 × 2.0 μm. Cells occur in irregular clusters, but may also occur in pairs or singly (Fig. A2.44). Resting stages are not known. Gram variable. Nonmotile. Aerobic. Colonies on plates smooth, entire, and pale brown. Growth stimulated by ribose, glucose, and sucrose. Nitrate is reduced. Gelatin hydrolyzed. Starch is not hydrolyzed. Sulfide produced from thiosulfate. Grows at 1.4–5.2 M NaCl, with an optimum in the range 3.5–3.6 M. Requires only low concentrations of magnesium (<10 mM), elevated concentrations becoming toxic. Temperature range for growth 25–50°C, with optimum in the range 35–40°C. Resistant to ampicillin (25 μg/disk), carbenicillin (100 μg/disk), chloramphenicol (400 μg/disk), polymixin B (300 IU/disk), streptomycin (25 μg/disk), and tetracycline (50 μg/disk) but sensitive to novobiocin (10 μg/ disk), bacitracin (5 U/disk), and anisomycin (5 μg/disk). *N. occultus* has been shown to be relatively resistant to potassium tellurite (MIC 10 mM) and possesses inducible tellurite reductase activity (Pearion and Jablonski, 1999). The cell wall polymer of *N. occultus* has been shown to consist of L-glutamate, N-acetyl-D-glucosamine, N-acetyl-D-galactosamine, D-galacturonic acid, D-glucuronic acid, and D-glucose in a ratio of 5:7:1:8:0.5:0.3 (Niemetz et al., 1997).

    The polar lipids are based on $C_{20}C_{20}$ and $C_{20}C_{25}$ diether lipids (Tindall et al., 1984a; Tindall, 1985), the relative percentages of which may change with growth conditions (Morth and Tindall, 1985b). A phosphatidylglycero-(cyclo)-phosphate is present (Morth and Tindall, 1985a; Lanzotti et al., 1989).

    The mol% G + C of the DNA is: 61.2 (major component) (Bd) and 51.9 (minor component) (Bd).

    Type strain: SP4, ATCC 43101, DSMZ 3396, JCM 8859, NCIMB 2192, VKM B-1752.

    GenBank accession number (16S rRNA): Z28378.

2. **Natronococcus amylolyticus** Kanai, Kobayashi, Aono and Kudo 1995, 765.[VP] *a.my.lo.ly′ ti.cus.* L. neut. n. *amylum* starch; Gr. adj. *lytikos* dissolving; M.L. masc. adj. *amylolyticus* starch dissolving.

    Cells are coccoid, 1–2 μm in diameter, usually occurring in irregular clusters. Colonies are circular, entire, and orange-red. The optimum NaCl concentration for growth is 15–20% NaCl, but growth occurs over the range 8–30% NaCl. Temperature range for growth is 22–55°C, with an optimum between 40 and 45°C. Growth occurs over the pH range from 8.0 to 10.0, with an optimum at about 9.0. Oxidase and catalase positive. Obligately aerobic. Chemoorganotrophic. Nitrate and nitrite are reduced. Starch is hydrolyzed, but gelatin is not liquefied. No inhibition of growth in the presence of ampicillin, chloramphenicol, polymixin B, or streptomycin. Sensitive to anisomycin, bacitracin, erythromycin, novobiocin, and tetracycline.

    The major polar lipids present are diether derivatives of phosphatidyl glycerol and methyl-phosphatidyl glycerol phosphate. The phosphatidylglycero-(cyclo)-phosphate detected

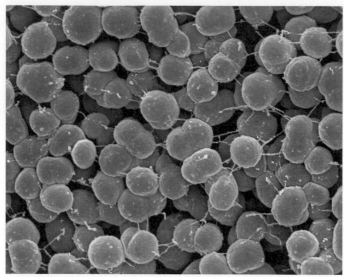

FIGURE A2.44. Scanning electron micrograph of liquid cultures of *Natronococcus occultus* (DSMZ 3396) in the late exponential/stationary phase. Note the presence of diplococci and cocci. (Courtesy of Dr. M. Rohde, GBF, Braunschweig.)

in *Natronococcus occultus* is not present, serving to distinguish the two species. Both $C_{20}C_{20}$ and $C_{20}C_{25}$ diether lipids are present.

Studies of the α-amylase from strain Ah-36 have shown that it produces a maltotriose from starch, amylose, and amylopectin, with slower hydrolysis of glycogen (Kobayashi et al., 1992). The gene coding for this enzyme can be cloned and expressed in *Haloferax volcanii* (strain WFD11, DSMZ 5716) (Kobayashi et al., 1994a).

The mol% G + C of the DNA is: 63.5 (method of determination not specified).

Type strain: Ah-36, ATCC 51971, DSMZ 10542, JCM 9655.
GenBank accession number (16S rRNA): D43628.

Genus XIII. **Natronomonas** *Kamekura, Dyall-Smith, Upasani, Ventosa and Kates 1997, 856[VP]*

MASAHIRO KAMEKURA

Na.tro.no.mo' nas. Arbitrarily derived from the Arabic n. *natrun* soda (sodium carbonate); Gr. n. *monas* unit; M.L. fem. n. *Natronomonas* the soda unit.

Cells are short rods $0.8 \times 1-3$ μm in liquid culture. Cells stain Gram-negative. Motile by a tuft of polar flagella. Chemoorganotrophic and **strictly aerobic**. Halophilic; **growth occurs in media containing 2–5.2 M NaCl**, with optimum concentration of 3.5 M. **Alkaliphilic, growing between pH 7 and 10**, with optimum pH of 8.5–9.5. Growth occurs at 25–50°C. There are six signature bases (Table A2.25).

The mol% G + C of the DNA is: 61.2–64.3.

Type species: **Natronomonas pharaonis** (Soliman and Trüper 1982) Kamekura, Dyall-Smith, Upasani, Ventosa and Kates 1997, 856 [VP] (*Halobacterium pharaonis* Soliman and Trüper 1982, 327; *Natronobacterium pharaonis* Tindall, Ross and Grant 1984a, 55.)

FURTHER DESCRIPTIVE INFORMATION

Sugars are not utilized. Casamino acids and glutamate are used as nitrogen sources. Gelatin is liquefied by most strains. Sulfide is formed from thiosulfate. Indole is formed from tryptophan. Starch and casein are not hydrolyzed. Nitrate is reduced to nitrite. Colonies are pigmented red because of the presence of C_{50} carotenoids.

Polar lipids are glycerol-diether analogs of phosphatidyl glycerol (PG), methyl ester of phosphatidyl glycerol phosphate (PGP-Me), phosphatidic acid (PA), and an unknown phospholipid PL1. Both diphytanyl moieties ($C_{20}C_{20}$) and phytanyl-sesterterpanyl moieties ($C_{20}C_{25}$) detected.

ENRICHMENT AND ISOLATION PROCEDURES

See under *Halobacteriales* and the genus *Halobacterium*.

TAXONOMIC COMMENTS

Phylogenetically, *Natronomonas pharaonis* is distantly related to the other taxa in the family *Halobacteriaceae*, including other alkaliphilic members. Similarities of the 16S rRNA gene sequence with those of other genera are less than 90% (Kamekura et al., 1997).

List of species of the genus Natronomonas

1. **Natronomonas pharaonis** (Soliman and Trüper 1982) Kamekura, Dyall-Smith, Upasani, Ventosa and Kates 1997, 856[VP] (*Halobacterium pharaonis* Soliman and Trüper 1982, 327; *Natronobacterium pharaonis* Tindall, Ross and Grant 1984a, 55.) *pha.ra.o' nis.* M.L. gen. n. *pharaonis* of Pharaoh, title of the kings of ancient Egypt.

Colonies are 1–2 mm in diameter after 5–7 d at 37°C, translucent, and red. The first alkaliphilic extreme halophile described. Pleomorphic motile rods, originally isolated from the alkaline brines of eutrophic desert lakes of Wadi Natrun, Egypt. Optimum temperature for growth 45°C. See generic description for many features.

Strain SP1, which showed 96% similarity to DSMZ 2160[T] in DNA–DNA hybridization, was isolated from a Kenyan soda lake, Lake Magadi, and deposited as NCIMB 2191 (Tindall et al., 1984a). NCIMB 2191 contains a small amount of disesterterpanyl moiety ($C_{25}C_{25}$) lipids, whereas DSMZ 2160[T] does not (Tindall, 1985).

Cell membranes of *Natronomonas pharaonis* contain two retinal-containing proteins, pharaonis halorhodopsin and pharaonis phoborhodopsin, which have chloride-pumping activity and negative phototactic activity, respectively (Tomioka and Sasabe, 1995). No bacteriorhodopsin has been detected (Kamekura et al., 1998).

The DNA is composed of a major component and a minor component. A large plasmid (~144 kb) is present.

The mol% G + C of the DNA is: 64.3 (T_m) and 61.2 (Bd) (major component) and 51.9 (Bd) (minor component).

Type strain: Gabara, ATCC 35678, DSMZ 2160, NCIMB 2260.

GenBank accession number (16S rRNA): D87971.

Genus XIV. Natronorubrum *Xu, Zhou and Tian 1999, 261[VP]*

YI XU, PEIJIN ZHOU, XINYU TIAN AND AHARON OREN

Na.tro.no.ru' brum. Arabic n. *natron* soda; L. neut. adj. *rubrum* red; M.L. neut. n. *Natronorubrum* the red of soda.

Gram-negative pleomorphic flat-shaped cells. Extremely halophilic and alkaliphilic, requiring at least 2 M NaCl and pH values between 8.0 and 11.0. Optimum pH 9.0–9.5. Optimum temperature 45°C. Colonies red. Oxidase and catalase positive. Chemoorganotrophic; **aerobic.** Yeast extract and Casamino acids are good sources of organic nutrients. Certain carbohydrates stimulate growth, sometimes with acid production. Polar lipids are characterized by the presence of $C_{20}C_{20}$ and $C_{20}C_{25}$ derivatives of phosphatidyl glycerol, phosphatidyl glycerophosphate methyl ester, and unidentified minor phospholipids.

The mol% G + C of the DNA is: 59.9–60.1.

Type species: **Natronorubrum bangense** Xu, Zhou and Tian 1999, 261.

FURTHER DESCRIPTIVE INFORMATION

Cells of *Natronorubrum bangense* and *Natronorubrum tibetense*, the only *Natronorubrum* species described thus far, are pleomorphic, flat, triangular, square, disk, or polygonal shaped, and 1–3 μm in diameter. Cells are nonmotile. Cells stain Gram-negative.

Natronorubrum species are haloalkaliphiles that require both high salt concentrations and high pH values for growth. Cells lyse in distilled water. Optimal NaCl concentration range: 3.4–3.8 M. Optimum temperature: 45°C. Optimum pH: 9.0–9.5.

Growth on single carbon sources has not been tested. Several sugars (glucose, maltose, fructose, sucrose, and lactose) stimulate growth, in some cases with the production of acids. Growth is also stimulated by acetate. Starch is not hydrolyzed. Indole is produced.

The main polar lipids are the $C_{20}C_{20}$ and $C_{20}C_{25}$ derivatives of PG and PGP-Me. Several minor, yet to be identified phospholipids are also present. As is the case for most other alkaliphilic halophilic *Archaea*, glycolipids are absent.

ENRICHMENT AND ISOLATION PROCEDURES

The two species of the genus *Natronorubrum* were isolated from sediments of the Bange hypersaline alkaline lake in Tibet (pH 10), using a high-salt, high pH medium as described by Tindall et al. (1980). No specific enrichment or isolation procedure favoring development of *Natronorubrum* species over that of other haloalkaliphilic *Archaea* has yet been described, and no information is available on the distribution of the two *Natronorubrum* species beyond the soda lake in Tibet from which the first strains were isolated.

MAINTENANCE PROCEDURES

No specific information on suitable procedures for the maintenance of *Natronorubrum* species is available. It may be assumed that procedures satisfactory for other haloalkaliphiles (e.g., *Natronobacterium gregoryi*, *Natronomonas pharaonis*, *Halorubrum vacuolatum*, and *Natrialba magadii*) will also be suitable for the maintenance of *Natronorubrum* isolates.

DIFFERENTIATION OF THE GENUS *NATRONORUBRUM* FROM OTHER GENERA

Differentiation of the genus *Natronorubrum* from other haloalkaliphilic noncoccoid *Archaea* is based on cell morphology and on 16S rRNA gene sequence analysis. The two *Natronorubrum* species are highly pleomorphic, while other haloalkaliphiles are rod-shaped (*Natronobacterium gregoryi*, *Natronomonas pharaonis*, *Halorubrum vacuolatum*, *Natrialba magadii*) or coccoid (*Natronococcus*). Other properties useful in the differentiation of *Natronorubrum* from other haloalkaliphiles are the formation of sulfide from thiosulfate, the reduction of nitrate, and the relatively low mol% G + C of the DNA. Table A2.35 presents the phenotypic

TABLE A2.35. Characteristics differentiating *Natronorubrum bangense* and *Natronorubrum tibetense* from each other and from other noncoccoid haloalkaliphililc members of the *Halobacteriaceae*[a]

Characteristic	*Natronorubrum bangense*	*Natronorubrum tibetense*	*Natronobacterium gregoryi*	*Natronomonas pharaonis*	*Natrialba magadii*	*Halorubrum vacuolatum*
Cell morphology	Pleomorphic	Pleomorphic	Long rod	Rod	Rod	Rod-spherical
Motility	−	−	−	+	+	−
Gas vesicles	−	−	−	−	−	+
Sulfide from thiosulfate	−	−	+	+	+	nd
Nitrite from nitrate	−	−	−	−	−	+
Casein hydrolysis	−	−	−	−	+	nd
Gelatin liquefaction	−	+	+	+	+	−
Tween hydrolysis:						
Tween 40	−	+	nd	−	nd	nd
Tween 60	−	+	nd	−	nd	nd
Tween 80	−	−	nd	−	nd	nd
Growth stimulated by:						
Acetate	+	+	+	−	+	+
Glucose	+	+	+	−	−	+
Fructose	+	+	+	−	−	−
Sucrose	+	+	+	−	−	+
Mannitol	−	−	+	−	−	−
Sensitivity to erythromycin	+	+	+	−	+	−
Mol% G + C	59.9 (T_m)	60.1 (T_m)	65.0 (Bd)	64.3 (T_m); 61.2 (major) and 51.9 (minor) (Bd)	63.0 (major) and 49.5 (minor) (Bd)	62.7 (T_m)

[a]For symbols see standard definitions; nd, not determined. Data were derived from Xu et al. (1999) and from the original species descriptions.

features allowing differentiation of *Natronorubrum* species from other validly described noncoccoid haloalkaliphiles.

TAXONOMIC COMMENTS

Based on 16S rRNA gene sequence data, the closest relatives of *Natronorubrum* are the genera *Natrialba*, *Haloterrigena*, and *Natrinema*.

FURTHER READING

Tindall, B.J., A.A. Mills and W.D. Grant. 1980. An alkalophilic red halophilic bacterium with a low magnesium requirement from a Kenyan soda lake. J. Gen. Microbiol. *116*: 257–260.

Xu, Y., P. Zhou and X. Tian. 1999. Characterization of two novel haloalkaliphilic archaea *Natronorubrum bangense* gen. nov., sp. nov. and *Natronorubrum tibetense* gen. nov., sp. nov. Int. J. Syst. Bacteriol. *49*: 261–266.

List of species of the genus Natronorubrum

1. **Natronorubrum bangense** Xu, Zhou and Tian 1999, 261.[VP]
   ban.gen' se. M.L. neut. adj. *bangense* from Bange, China.

   Cells are pleomorphic, flat-shaped, nonmotile, 1–3 μm in diameter.

   Requires 2.1–4.3 M NaCl and pH 8.0–11.0. Optimal growth in 3.8 M NaCl and pH 9.5.

   Aerobic. Indole produced from tryptophan. Glucose, maltose, fructose, sucrose, lactose, and acetate stimulate growth. No stimulation by mannitol observed. Starch, casein, and Tween 40, 60, and 80 are not hydrolyzed. Gelatin is not liquefied. Nitrate is not reduced.

   Found in the Bange hypersaline alkaline lake in Tibet.

   Other characteristics are the same as those described for the genus.

   The mol% G + C of the DNA is: 59.9 (T_m).

   Type strain: AS 1.1984 (strain A33).

   GenBank accession number (16S rRNA): Y14028.

2. **Natronorubrum tibetense** Xu, Zhou and Tian 1999, 261.[VP]
   ti.bet.en' se. M.L. neut. adj. *tibetense* from Tibet.

   Cells are pleomorphic, flat-shaped, nonmotile, 1–3 μm in diamter.

   Requires 2.1–5.1 M NaCl and pH 8.5–11.0. Optimal growth in 3.4 M NaCl and pH 9.0.

   Aerobic. Indole produced from tryptophan. Glucose, maltose, fructose, sucrose, lactose, and acetate stimulate growth. No stimulation by mannitol observed. Tween 40 and 60 are hydrolyzed. Starch, casein, and Tween 80 are not hydrolyzed. Gelatin is liquefied. Nitrate is not reduced.

   Found in the Bange hypersaline alkaline lake in Tibet.

   Other characteristics are the same as those described for the genus.

   The mol% G + C of the DNA is: 60.1 (T_m).

   Type strain: AS 1.2123 (strain GA33).

   GenBank accession number (16S rRNA): AB005656.

Class IV. **Thermoplasmata** *class. nov.*

ANNA-LOUISE REYSENBACH

Ther.mo.plas.ma' ta. M.L. fem. pl. n. *Thermoplasmatales* the type order of the class; dropping the ending to denote a class; M.L. fem. pl. n. *Thermoplasma* the class of *Thermoplasmatales.*

The description for this class is the same as that for the order *Thermoplasmatales.*

Type order: **Thermoplasmatales** ord. nov.

Order I. **Thermoplasmatales** *ord. nov.*

ANNA-LOUISE REYSENBACH

Ther.mo.plas.ma.ta' les. M.L. neut. n. *Thermoplasma* the type genus of the order; *-ales* the ending to denote an order; M.L. fem. pl. n. *Thermoplasmatales* the order of *Thermoplasma.*

This order includes the only wall-less representative in the *Archaea, Thermoplasma,* and the extremely acidophilic thermophile *Picrophilus.* All members are thermophilic acidophiles, growing best around 60°C and at pH ≤2. Recently, nonthermophilic members of this order have been described (Edwards et al., 2000).

Type genus: **Thermoplasma** Darland, Brock, Samsonoff and Conti 1970, 1418.

Key to the families of Thermoplasmatales

1. Irregular spherical cells varying in shape from 0.5 to 5.0 μm in diameter. Lacks true cell wall. Facultative aerobe and thermoacidophilic heterotroph. Grows best at pH 2.0. Motile.
   Family I. *Thermoplasmataceae.*
2. Irregular cocci about 1 μm in diameter. Heterotrophic extremely thermoacidophilic aerobe. Grows best at pH 0.7. Nonmotile. Cell wall present.
   Family II. *Picrophilaceae,* p. 339

Family I. **Thermoplasmataceae** *fam. nov.*

ANNA-LOUISE REYSENBACH

Ther.mo.plas.ma.ta' ce.ae. M.L. neut. n. *Thermoplasma* the type genus of the family; *-aceae* the ending to denote a family; M.L. fem. pl. n. *Thermoplasmataceae* the family of *Thermoplasma.*

The *Thermoplasmataceae* can be distinguished from *Picrophilaceae* based on 16S rRNA sequence comparisons and the absence of a true cell wall in member taxa. This family is represented by a single genus, *Thermoplasma.*

Type genus: **Thermoplasma** Darland, Brock, Samsonoff and Conti 1970, 1418.

Key to the genera of Thermoplasmataceae

1. Wall-less pleomorphic spherical cells, 0.1–0.5 μm in diameter. Facultative aerobe and obligate heterotroph. Growth from 33 to 67°C (optimum ~60°C) and pH 1.0–4.0 (optimum pH 2.0). Motile. The mol% G + C is between 38 and 46. Found in self-heating coal refuse piles and terrestrial solfataras.
   Genus I. *Thermoplasma.*

Genus I. **Thermoplasma** *Darland, Brock, Samsonoff and Conti 1970, 1418[AL]*

THOMAS A. LANGWORTHY

Ther.mo.plas' ma. Gr. n. *therme* heat; Gr. fem. n. *plasma* something formed or molded, a form; M.L. neut. n. *Thermoplasma* heat (-loving) mycoplasma.

Pleomorphic, varying in shape from spherical (0.1–5.0 μm) to filamentous structures. **Cells lack a true cell wall** and are surrounded by a single triple-layer membrane ~5–10 nm thick. Membrane contains ether lipids based on 40-carbon, isopranoid-branched **diglycerol tetraethers**. Resting stages not known. Stain Gram-negative. Cells may be motile via monopolar monotrichous flagellation. **Facultatively aerobic**. Anaerobic growth is enhanced by elemental sulfur, which is reduced to H_2S via respiration. **Obligate thermoacidophile**. Optimum growth occurs at 55–59°C and pH 1–2. Cells undergo lysis near neutrality. On agar at pH 2, colonies attain a diameter of about 0.3 mm and are dark brown in color, flat and coarsely granular, and some exhibit a typical "fried egg" appearance with a translucent peripheral zone. Biochemical and nutritional characteristics relatively poorly defined. Do not require cholesterol. **Obligately heterotrophic** with growth on extracts of yeast, meat, and *Bacteria* or *Archaea*. **Occur free-living in self-heating coal refuse piles and acidic solfatara fields**.

The mol% G + C of the DNA is: 46.

Type species: **Thermoplasma acidophilum** Darland, Brock, Samsonoff and Conti 1970, 1418 (*Thermoplasma acidophila* [sic] Darland, Brock, Samsonoff and Conti 1970, 1418.)

FURTHER DESCRIPTIVE INFORMATION

When grown either aerobically or anaerobically in liquid medium* consisting of Allen's basal salts solution (Allen, 1959) adjusted to pH 2 and supplemented with 0.1% yeast extract and 1.0% glucose, *Thermoplasma* exhibits a typical mycoplasmal morphology by light and phase microscopy. Cells appear as pleomorphic spheres varying in size from 0.3 to 2.0 μm and occasionally, large cells, up to 5 μm in diameter. Filamentous structures exhibiting budding characteristics are also common, particularly in young cultures.

A cell wall is absent, evidenced by electron micrographs of thin-sectioned cells (Darland et al., 1970; Belly et al., 1973; Langworthy, 1979; Segerer et al., 1988a) (Fig. A2.45). The surrounding cytoplasmic membrane averages about 7 nm in thickness. Membrane lipids lack fatty acid ester residues. The lipids are ether-linked C_{40} biphytanyl diglycerol tetraethers along with small amounts of C_{20} phytanyl glycerol diethers characteristic of the *Archaea* (Langworthy, 1977, 1979, 1985; Langworthy et al., 1982; Langworthy and Pond, 1986a, b; Kates, 1978; Tornabene and Langworthy, 1979). No internal membranes or organelles are present.

Cells are generally motile, typically with a singular polar flagellum (Black et al., 1979; Segerer et al., 1988a; Faguy et al., 1996).

Growth on an agar surface is unreliable and difficult to achieve due to drying out at high temperature and hydrolysis of the agar in the presence of acid. These difficulties can sometimes be overcome by combining double-strength liquid medium and agar after cooling to 45–50°C followed by incubation in a humidified atmosphere (see *Thermoplasma* medium*). Alternatively 10% starch may be employed in place of agar. When growth can be initiated, colonies are small (about 0.3 mm diameter), and some show a fried egg appearance (Belly et al., 1973). Colonies are typically flat, coarsely granular, and dark brown. Scanning electron microscopy shows individual cells to have an imbricate surface texture characteristic of cells which lack a cell wall (Mayberry-Carson et al., 1974). Optimum temperature for growth: 59°C. Growth between 33 and 67°C. Growth is slow to slight at the extremes. Optimum pH: 2.0. Growth occurs between the pH limits of 0.5–4, but growth is very slow at the extremes.

Hydrogen ions are specifically required for maintenance of cellular stability. Cells undergo lysis at neutral pH (Belly and Brock, 1972; Smith et al., 1973). Other monovalent cations, divalent cations, or osmotic stabilizers do not substitute for the required hydrogen ion. The phenomenon is analogous to the sodium ion requirement of certain *Halobacteriaceae* for the maintenance of cellular integrity. The intracellular hydrogen ion concentration of *Thermoplasma*, however, is not in equilibrium with the external environment, but the internal pH is near neutrality (Hsung and Haug, 1975; Searcy, 1976).

Thermoplasma is a facultative aerobe. It possesses cytochromes and menaquinone-7, suggesting the presence of a complete respiratory chain (Belly et al., 1973; Holländer, 1978). Aerobic growth is stimulated by slight aeration, but excessive aeration inhibits growth (Smith et al., 1973). Because the amount of dissolved oxygen at 59°C is low, *Thermoplasma* might be considered microaerophilic. All isolates are able to grow well under strictly anaerobic conditions by sulfur respiration (Segerer et al., 1988a).

Nutritionally, *Thermoplasma* is an obligate heterotroph. It requires yeast extract, but some isolates can display some growth on meat extract and extracts from eubacterial or archaebacterial cells (Belly et al., 1973; Smith et al., 1975; Segerer et al., 1988a). At yeast extract concentrations below 0.025%, there is little or no growth. At concentrations higher than 0.25%, growth is inhibited. Between the limiting concentrations, growth is proportional to the concentration of yeast extract used. Growth rates and yields can vary, depending upon the manufacturer and the lot of yeast extract employed. The component or components supplied by yeast extract for growth appear to be basic oligopeptide(s) (Smith et al., 1975). No growth occurs aerobically on elemental sulfur or ferrous iron.

Cell yields are influenced by the inoculum size. Total cell yields decrease with inoculum sizes of <5% (v/v). Under optimal conditions, *Thermoplasma* has a generation time of about 5 h (Belly et al., 1973; Smith et al., 1973; Segerer et al., 1988a). Cell numbers increase with optical density (540 nm) to the stationary phase, reaching about 1×10^9 cells/ml, at which point there is a drastic loss in viability although there is no great reduction in optical density.

Sucrose, glucose, galactose, mannose, and fructose, when added in 0.1% concentration to the basal medium containing the growth-limiting concentration of yeast extract (0.025%), appear to stimulate growth of *Thermoplasma* (Belly et al., 1973).

No lysis occurs when cells are suspended in distilled water, heated to 100°C for 30 min, or treated with EDTA, primary alcohols, digitonin, lysozyme, trypsin, or Pronase. Cells are rapidly

---

*Thermoplasma basal medium contains the following ingredients (in g/l of distilled water): KH_2PO_4, 3.0; $MgSO_4$, 0.5; $CaCl_2·2H_2O$, 0.25; $(NH_4)_2SO_4$, 0.2; yeast extract (Difco), 1.0; glucose 10.0; and trace element solution, 10 ml (see below). Adjust to pH 2 with 10 N H_2SO_4. After the medium is autoclaved, add separately sterilized stock solutions of 10 g of glucose (25 ml of a 40% glucose solution) to give a final concentration of 1.0% and a 10% w/v yeast extract to give a final concentration of 0.1%. For agar medium, mix equal volumes of separately sterilized double-strength liquid medium and 5.6% Ionagar no. 2 (Consolidated Laboratories, Inc.), after cooling to 45°C, to give a final agar concentration of 2.8%. Incubate in a sealed and humidified atmosphere. Alternatively 10% starch may be used in place of agar. For element solution, add the following (per liter of distilled water): $FeCl_3·6H_2O$, 1.930 g; $MnCl_2·4H_2O$, 0.180 g; $Na_2B_4O_7·10H_2O$, 0.450 g; $ZnSO_4·7H_2O$, 22.0 mg; $CuCl_2·2H_2O$, 5.0 mg; $Na_2MoO_4·2H_2O$, 3.0 mg; $VOSO_4·5H_2O$, 3.8 mg; and $CoSO_4·7H_2O$, 2.0 mg. For anaerobic growth; the acidified basal medium is supplemented with sulfur at the rate of 4.0 g/l. The medium is poured into 100-ml serum bottles (20 ml/bottle) under a gas mixture of N_2/CO_2 (80:20) at 200 kPa and then sterilized by Tyndallization. Yeast extract and glucose are added separately. For strict anaerobiosis, the bottles are further supplemented with a separately sterilized solution of $Na_2S·9H_2O$ and resazurine (to a final concentration of 0.05% and 0.0001%, respectively) and the final pH is adjusted to 2.0.

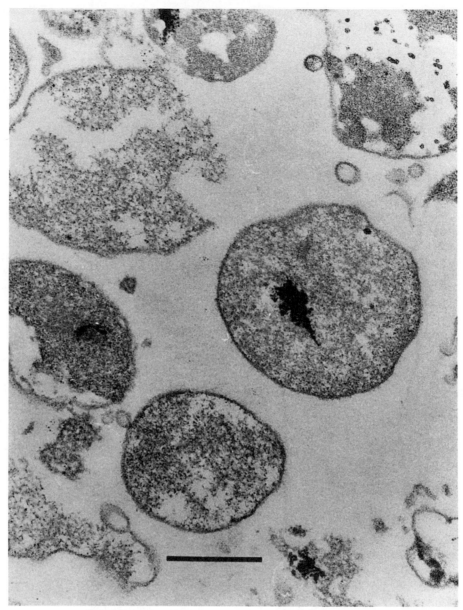

FIGURE A2.45. Thin section of *Thermoplasma*. Bar = 0.5 μm.

lysed by sodium dodecyl sulfate and more slowly by cetyl tri-methylammonium bromide (Belly and Brock, 1972; Smith et al., 1973).

Thermoplasma is resistant to the cell wall inhibitors vancomycin, ampicillin, and ristocetin and to novobiocin, rifampicin, strep-tomycin, and chloramphenicol in concentrations of 150 μg/ml (Darland et al., 1970; Belly et al., 1973; Brock, 1978; Segerer et al., 1988a).

Molecular characteristics which further distinguish *Thermoplasma* include the following: a membrane-associated linear li-poglycan containing 24 mannose residues and 1 glucose residue (Smith, 1980); a mannosyl membrane glycoprotein (Yang and Haug, 1979); a small genome size of about 8.4×10^8 to 1×10^9 Da (Christiansen et al., 1975; Searcy and Doyle, 1975); histone-like proteins associated with the DNA (DeLang et al., 1981; Segerer et al., 1988a); a seven-subunit DNA-dependent RNA poly-merase that is resistant to rifampicin, streptolydigin, and α-aman-

itine (Sturm et al., 1980); an unusual modification pattern in tRNAs (Gupta and Woese, 1980; Kuchino et al., 1982); an unusual nucleotide sequence in the 16S rRNA (Woese et al., 1980); a 5S rRNA secondary structure which does not conform to the usual models employed for either procaryotic or eucaryotic 5S rRNAs (Luehrsen et al., 1981); occurrence of diphthamide (Kessel and Klink, 1980); an unusual superoxide dismutase (Searcy and Searcy, 1981); coenzyme F_{420} (Lin and White, 1986); possession of proteosomes (Nitsch et al., 1997); and a cytoplasmic mem-brane which exists as a lipid monolayer rather than a lipid bilayer and which accounts for the characteristic cross-fracture rather than tangential fracture of freeze-etched cells (Langworthy, 1979, 1985; Langworthy et al., 1982; Langworthy and Pond, 1986a, b).

Thermoplasma occurs in self-heating coal refuse piles in south-ern Indiana and western Pennsylvania, which was originally its only known habitat (Belly et al., 1973). It has now been found to be relatively widely distributed as a common inhabitant of moderately hot acidic solfatara fields and geothermal areas found throughout the world (Segerer et al., 1988a).

ENRICHMENT AND ISOLATION PROCEDURES

Thermoplasma was originally isolated aerobically from a coal refuse pile at the Friar Tuck mine in southwestern Indiana by incubating 20 ml of *Thermoplasma* medium with 1.0 g of coal refuse (Darland et al., 1970). The isolation medium has been modified to include the acid-stable antibiotic vancomycin at a concentration of 1225 µg/ml to inhibit the growth of rod-shaped bacteria such as *Bacillus acidocaldarius* (Belly et al., 1973; Segerer et al., 1988a). *Thermoplasma* has also been isolated from water samples by filtration through a membrane filter (0.45-µm pore size) followed by passage of the filtrate through a second filter (0.22-µm pore size), with subsequent incubation in culture medium (Belly et al., 1973). Isolation samples are incubated at 55°C for 4–6 weeks or until the development of visible turbidity. The presence of *Thermoplasma* is confirmed by microscopic examination. Cultures are purified by dilution in liquid medium, since reproducible growth of colonies on agar is generally difficult to obtain. The above-mentioned procedures are selective for the isolation of *Thermoplasma*, with the possible exception of *Sulfolobus*, *Acidianus*, and *Picrophilus*, which can be distinguished by morphology and physiological characteristics. Anaerobic isolation employing anaerobic media can result in the specific selection of *Thermoplasma* from its habitat, with the possible exception of *Acidianus*, which can easily be distinguished morphologically and physiologically.

MAINTENANCE PROCEDURES

The most reliable procedure to maintain aerobically grown cultures is by continuous passage after 2–3 d of incubation. A 10–20% inoculum should be used, and multiple culture tubes should be incubated, since growth is sometimes spurious. Glassware should be free of any trace of detergent or soap residue, which will kill the cells. Aerobically grown cells remain viable at room temperature for 10–15 d but die upon refrigeration. In contrast, anaerobically grown cultures are much more stable and can be stored for at least one year without loss of viability (Segerer et al., 1988a). Sometimes cells can be recovered from the frozen state, but sufficient time (1–2 weeks) is required for development of visible turbidity. Cells are killed by lyophilization, and neutralization of cultures prior to preservation is precluded by cell lysis.

DIFFERENTIATION OF THE GENUS *THERMOPLASMA* FROM OTHER GENERA

Thermoplasma is distinguished from other genera of mycoplasmas by its stability, by its requirement for hot acid, and by its molecular features. Table A2.36 provides the primary characteristics of *Thermoplasma* that distinguish it from other genera of morphologically or physiologically similar thermoacidophilic bacteria.

TAXONOMIC COMMENTS

Thermoplasma is, by definition, a mycoplasma by virtue of being a free-living organism which morphologically lacks a cell wall. Biochemical characteristics, however, also indicate that *Thermoplasma* is a thermoacidophilic member of the *Archaea*. Although lacking a cell wall, *Thermoplasma* shares the basic common archaeal features of isopranoid (phytanyl-based) ether lipids, unusual but similar RNA polymerase structure and certain nucleotide sequences in the 16S rRNA, 5S rRNA, and tRNA (Fox et al., 1980; Kandler, 1982a; Kandler and Zillig, 1986; Stetter and Zillig, 1985). The lack of DNA–DNA sequence similarity (<0.25%) indicates the absence of a close genetic relationship between *Thermoplasma* and *Sulfolobus* and *Acidianus*; i.e., *Thermoplasma* is not merely a stable L-form derived from *Sulfolobus* (Christiansen et al., 1981). Thus, on biochemical grounds, *Thermoplasma* is an archaeon. The most closely related relative appears to be *Picrophilus*, which is found in the same habitat but can be distinguished by its strictly aerobic growth and presence of a filigreed surface S-layer external to the cellular membrane (Schleper et al., 1995).

FURTHER READING

Balows, A., H.G. Trüper, M. Dworkin, W. Harder and K.H. Schleifer (Editors). 1992. The Prokaryotes. A Handbook of Bacteria: Ecophysiology, Isolation, Identification, Applications, Springer-Verlag, New York. 4126 pp.

Belly, R.T., B.B. Bohlool and T.D. Brock. 1973. The genus *Thermoplasma*. Ann. N. Y. Acad. Sci. *225*: 94–107.

Brock, T.D. 1978. Thermophilic Microorganisms and Life at High Temperatures, Springer–Verlag, Heidelberg.

Brock, T.D. (Editor). 1986. Thermophiles: General, Molecular, and Applied Microbiology, John Wiley & Sons, New York.

Kandler, O. and W. Zillig (Editors). 1986. Archaebacteria '85, Gustav Fischer Verlag, Stuttgart.

Schleper, C., G. Pühler, I. Holz, A. Gambacorta, D. Janekovic, U. Santarius, H.P. Klenk and W. Zillig. 1995. *Picrophilus* gen. nov., fam. nov.: a novel aerobic, heterotrophic, thermoacidophilic genus and family comprising archaea capable of growth around pH 0. J. Bacteriol. *177*: 7050–7059.

TABLE A2.36. Differential characteristics of the genus *Thermoplasma* and other genera of thermoacidophilic bacteria[a]

Characteristic	*Thermoplasma*	*Picrophilus*	*Sulfolobus*	*Bacillus acidocaldarius*
Shape:				
Pleomorphic spheres	+	+	−	−
Lobed spheres	−	−	+	−
Rods, filaments		−	−	+
S-layer present	−	+	+	+
Ether lipids present	+	+	+	−
Endospores formed	−	−	−	+
Aerobic growth	+	+	+	+
Anaerobic growth	+	−	−	−
Nutrition:				
Requires yeast extract	+	+	−	−
Facultative autotroph	−	−	+	−
Lysis at neutral pH	+	+	−	−
Mol% G + C	46	36	40	60–64

[a]Symbols: +, positive for all strains; −, negative for all strains.

Segerer, A., T.A. Langworthy and K.O. Stetter. 1988. *Thermoplasma acidophilum* and *Thermoplasma volcanium*, new species from solfatara fields. Syst. Appl. Microbiol. *10*: 161–171.

Woese and Wolfe (Editors). 1985. The Bacteria, Academic Press, New York.

Woese, C.R. 1987. Bacterial evolution. Microbiol. Rev. *51*: 221–271.

DIFFERENTIATION OF THE SPECIES OF THE GENUS *THERMOPLASMA*

Two distinct species of *Thermoplasma* have been established, *Thermoplasma acidophilum* and *Thermoplasma volcanium*. *T. acidophilum* isolates demonstrate serological diversity and can be differentiated into five antigenic groups by immunofluorescence and immunodiffusion analysis (Belly et al., 1973; Bohlool and Brock, 1974). *T. volcanium* isolates display diversity based upon DNA–DNA similarities which can be divided into three DNA similarity groups (Segerer et al., 1988a).

List of species of the genus Thermoplasma

1. **Thermoplasma acidophilum** Darland, Brock, Samsonoff and Conti 1970, 1418[AL] (*Thermoplasma acidophila* [sic] Darland, Brock, Samsonoff and Conti 1970, 1418.)

   a.ci.do'phil.um. L. neut. adj. *acidum* acid; Gr. adj. *philos* loving; M.L. neut. adj. *acidophilum* acid-loving.

   Morphology and nutritional requirements are as described for the genus. Growth between 45 and 63°C (optimum 59°C) and pH values of 0.5–4 (optimum pH 2.0). No significant DNA–DNA similarity to *T. volcanium*.

   Occur free-living in warm acidic solfatara fields and self-heating coal refuse piles.

   Reference strains: ATCC 27657 (isolate 3–24), ATCC 27658 (isolate 122–1B3), and ATCC 27656 (isolate 124–1).

   The mol% G + C of the DNA is: 46 (T_m, Bd, and direct analysis).

   Type strain: 122–1B2, AMRC C-165, ATCC 25905, DSMZ 1728.

   GenBank accession number (16S rRNA): M38637.

2. **Thermoplasma volcanium** Segerer, Langworthy and Stetter 1988b, 379[VP] (Effective publication: Segerer, Langworthy and Stetter 1988a, 169.)

   vol.ca'ni.um. M.L. neut. adj. *volcanium* belonging to Volcanus, the Roman god of fire, who lived in a volcano, the origin of isolation of the type strain.

   Morphology and nutritional requirements are as described for the genus. Growth between 33 and 67°C (optimum 60°C) and pH values of 1.0–4.0 (optimum pH 2.0). Three different DNA–DNA similarity groups not hybridizing closely with each other or with *T. acidophilum*. The three groups are represented by isolates GSS1 (group 1), KD3 (group 2), and KO2 (group 3).

   Occur free-living in continental and submarine solfataras in Italy (group 1), continental solfataras and a tropical swamp in Java (group 2), and continental solfataras in Iceland and the United States (group 3).

   Reference strains: DSMZ 4300 (isolate KD3, DNA–DNA group 2), DSMZ 4301(isolate KO2, DNA–DNA group 3).

   The mol% G + C of the DNA is: 38 (T_m and direct analysis).

   Type strain: GSS1 (DNA–DNA group 1), ATCC 51530, DSMZ 4299.

Family II. **Picrophilaceae** Schleper, Zillig and Pülher *in* Schleper, Pühler, Klenk and Zillig 1996, 814[VP]

ANNA-LOUISE REYSENBACH AND WOLFRAM ZILLIG

Pi.cro.phi.la'ce.ae. M.L. masc. n. *Picrophilus* the type genus of the family; *-aceae* ending to denote a family; M.L. fem. pl. n. *Picrophilaceae* the family of *Picrophilus*.

The *Picrophilaceae* is another family in the order *Thermoplasmatales*, the acidophilic thermophiles. The taxa are separated based on their differences in 16S rRNA sequence (>9.3%), differences in DNA-dependent RNA polymerase, and the presence of an S-layer in taxa of the *Picrophilaceae*. Only one genus exists.

Type genus: **Picrophilus** Schleper, Zillig and Pülher *in* Schleper, Pühler, Klenk and Zillig 1996, 814.

Genus I. **Picrophilus** Schleper, Zillig and Pülher *in* Schleper, Pühler, Klenk and Zillig 1996, 814[VP]

ANNA-LOUISE REYSENBACH AND WOLFRAM ZILLIG

Pi.cro.phi'lus. Gr. adj. *pikros* acidic; Gr. adj. *philos* loving; M.L. masc. n. *picrophilus* acid-loving organism.

Thermophilic, hyperacidophilic, obligately aerobic irregular cocci about 1 μm in diameter. **Nonmotile.** Divides by constriction. Filigreed tetragonal S-layer present. **Grows heterotrophically** in the presence of 0.1–0.5% yeast extract. No growth occurs on organic substrates in the absence of yeast extract. **Growth occurs between 47 and 60°C and between pH 0 and 3.5.**

The mol% G + C of the DNA is: 36.

Type species: **Picrophilus oshimae** Schleper, Zillig and Pülher *in* Schleper, Pühler, Klenk and Zillig 1996, 815.

FURTHER DESCRIPTIVE INFORMATION

The cells are irregular cocci, 1–1.5 μm in diameter. During exponential growth the major fraction of cells is present in incompletely divided division forms of two or three nascent cells. Transmission electron microscopy revealed areas of low electron density, resembling vacuoles, but no membranes were present to separate these areas from the rest of the cytoplasm. A distinctive 40 nm-thick S-layer is present, with a dense outer stratum and an inner stratum consisting of widely spaced pillars that anchor the surface layer in the membrane. The S-layer has a very regular tetragonal symmetry. No pili or flagella have been noted. The lipids of *Thermoplasma* and *Picrophilus* are similar; however, unlike *Thermoplasma*, *Picrophilus* has a β-glycosyl residue in the major lipid component.

The isolates can be grown both in liquid and solid media. When grown on 12.5% starch agar plates at pH 1, the convex shiny colonies are 2–5 mm in diameter and appear whitish-yellow. The best growth occurs at pH 0.7 and 60°C, although *Picrophilus* can grow between 45 and 65°C and between pH 0 and 3.5. The organism grows heterotrophically on 0.1–0.5% yeast extract and cannot grow by fermentation or chemolithoautotrophically by formation of H_2S from sulfur and hydrogen or by sulfur respiration. Higher cell densities, but slower growth, are achieved when sugars (1% glucose, sucrose, or lactose) are added to the medium in the presence of 0.2% yeast extract. No growth occurs if sugars, starch, or Casamino acids are added in the absence of yeast extract. Very poor growth occurs on tryptone. Growth is inhibited by 0.2 M sodium chloride.

The organism has a DNA-dependent RNA polymerase that has a unique subunit pattern and cross-reacts with the antibody against the RNA polymerase of *T. acidophilum*. Two different plasmids, one of 8.3 kb and one of 8.8 kb, have been isolated from strains of *Picrophilus oshimae*, but not from *Picrophilus torridus*.

Picrophilus grows in hot (about 55°C) geothermal solfatara soils (pH <0.5) and springs (pH 2.2) in Hokkaido, Japan. As the organism is inhibited by 0.2 M NaCl, it is likely restricted to terrestrial geothermal environments.

ENRICHMENT AND ISOLATION PROCEDURES

Picrophilus can be enriched in the medium described by Smith et al. (1975) (according to E.A. Freundt) and amended with 0.2%

yeast extract as the carbon source. The pH is adjusted with sulfuric acid, and the cultures are incubated at 60°C with moderate shaking. Colonies can be obtained using 12.5% starch plates containing the same medium at pH 1.

MAINTENANCE PROCEDURES

Cultures can be maintained at −70°C in 20% glycerol in a basal salt medium of pH 4.5. Cells lyse above pH 5.

DIFFERENTIATION OF THE GENUS *PICROPHILUS* FROM OTHER GENERA

Picrophilus differs from its close relative *Thermoplasma* in a number of ways: *Picrophilus* has an optimum pH for growth of 0.7, whereas *Thermoplasma* grows optimally at pH 2. *Picrophilus* has a lower mol% G + C (36) than *Thermoplasma* (46 for *T. acidophilum* and 38–40 for *T. volcanium*). *Thermoplasma* can grow anaerobically via sulfur respiration, whereas *Picrophilus* is a strict aerobe. In contrast to the archaeal mycoplasmas *Thermoplasma* spp., both *Picrophilus* species have a surface layer. Antibodies to the RNA polymerases of the two genera do not cross-react. The 16S rRNA sequences of the two genera are sufficiently dissimilar (>9.3%) to support their placement in separate genera.

TAXONOMIC COMMENTS

Picrophilus is a close relative of *Thermoplasma acidophilum*, although comparison of the 16S rRNA sequences indicated they are 9.3–11.5% different; sufficient to be considered different genera. Additionally, antisera directed against the DNA-dependent RNA polymerase from *Sulfolobus acidocaldarius* and *Thermoplasma acidophilum* did not react with the RNA polymerase isolated from *P. oshimae*. Unlike *Thermoplasma*, *Picrophilus* cannot grow anaerobically with or without sulfur. Additionally, *Picrophilus* has a surface cell layer and a distinctive S-layer.

DIFFERENTIATION OF THE SPECIES OF THE GENUS *PICROPHILUS*

Only two species have been described, *P. oshimae* and *P. torridus*. The species were distinguished from one another based on their 16S rRNA sequences (3% difference of 250 nucleotides compared). Additionally, *P. torridus* has a higher growth rate than *P. oshimae*.

List of species of the genus Picrophilus

1. **Picrophilus oshimae** Schleper, Zillig and Pülher in Schleper, Pühler, Klenk and Zillig 1996, 815.[VP]

   o.shi' mae. M.L. gen. n. *oshimae* of Oshima, referring to the Japanese biochemist Tairo Oshima.

   Cell morphology is as described for the genus. The bisphytanyl tetraethers are dominated by a single phosphoglycolipid component. Type *b* cytochromes present.

   P. oshimae strains contain plasmids either 8.3 or 8.8 kb long.

   The mol% G + C of the DNA is: 36 (HPLC).

   Type strain: KAW 2/2, DSMZ 9789.

   GenBank accession number (16S rRNA): X84901.

2. **Picrophilus torridus** Zillig, Schleper and Pülher *in* Schleper, Pühler, Klenk and Zillig 1996, 816.[VP]

   tor' ri.dus. L. part. *torridus* dried, burned, referring to dry hot soil where the strain was isolated.

   DNA restriction digests and 16S rRNA phylogeny distinguish *P. torridus* from *P. oshimae. P. torridus* grows significantly faster than *P. oshimae*. No plasmids are present. Comparison of 250 nucleotides of the 16S rRNA of *P. torridus* with *P. oshimae* revealed a 3% difference.

   The mol% G + C of the DNA is: not available.

   Type strain: KAW 2/3, DSMZ 9790.

Class IV. **Thermococci** *class. nov.*

WOLFRAM ZILLIG AND ANNA-LOUISE REYSENBACH

Ther.mo.coc' ci. M.L. fem. pl. n. *Thermococcales* the type order of the class; dropping the ending to denote a class; M.L. fem. pl. n. *Thermococci* the class of *Thermococcales.*

Irregular coccoid cells, vary in size from ~1–4 μm in diameter. Hyperthermophilic heterotrophs, most members require sulfur for growth. Optimal growth from 75 to 100°C. Occur widely at deep-sea and shallow marine hydrothermal vents and have been isolated from terrestrial thermal springs in New Zealand. Additional characteristics are described below.

Type order: **Thermococcales** Zillig 1988, 136 (Effective publication: Zillig *in* Zillig, Holz, Klenk, Trent, Wunderl, Janekovic, Imsel and Haas 1987, 69.)

Key to Orders of the class Thermococci

I. Irregular coccoid cells varying in size from 1 to 4 μm in diameter. Hyperthermophilic heterotrophs, capable of growing on a wide range of peptides and carbohydrates. Sulfur is usually required for growth. Best growth between 75 and 100°C.

Order I. *Thermococcales.*

Order I. **Thermococcales** Zillig 1988, 136 [VP](Effective publication: Zillig *in* Zillig, Holz, Klenk, Trent, Wunderl, Janekovic, Imsel and Haas 1987, 69)

WOLFRAM ZILLIG AND ANNA-LOUISE REYSENBACH

Ther.mo.coc.ca' les. M.L. masc. n. *Thermococcus* type genus of the order; *-ales* ending to denote an order; M.L. fem. pl. n. *Thermococcales* the *Thermococcus* order.

Cells spherical and sometimes pleomorphic, about 1 μm in diameter. Often occur as diplococci (cell division by constriction) or as clusters of up to 30 cells. Cytoplasm is electron-dense in contrast to *Desulfurococcus*. Envelope S-layer is composed of subunits.

Strictly anaerobic hyperthermophilic heterotrophs. Utilize a variety of carbon substrates such as peptides and/or carbohydrates, generally by sulfur respiration (forming H_2S from S^0). Other electron acceptor may be used. Members produce strong smelling sulfur-based products such as mercaptans. Lipids are dominated by a bisphytanyl phospholipid component. Members produce glycogen.

Occur in shallow marine and deep-sea hydrothermal environments at neutral pH and at temperatures about 80–103°C. Recently new species have been described from terrestrial thermal springs in New Zealand.

Type genus: **Thermococcus** Zillig 1983, 673 (Effective publication: Zillig *in* Zillig, Holz, Janekovic, Schäfer and Reiter 1983b, 93.)

Family I. **Thermococcaceae** Zillig 1988, 136 [VP](Effective publication: Zillig *in* Zillig, Holz, Klenk, Trent, Wunderl, Janekovic, Imsel and Haas 1987, 69)

WOLFRAM ZILLIG AND ANNA-LOUISE REYSENBACH

Ther.mo.coc.ca' ce.ae. M.L. masc. n. *Thermococcus* type genus of the family; *-aceae* ending to denote family; M.L. fem. pl. n. *Thermococcaceae* the *Thermococcus* family.

Cell appearance, metabolism, and general characteristics are the same as those described for the order.

Type genus: **Thermococcus** Zillig 1983, 673 (Effective publication: Zillig *in* Zillig, Holz, Janekovic, Schäfer and Reiter 1983b, 93.)

Key to the genera of the family Thermococcaceae

I. Extremely thermophilic (optimum growth 75–88°C), obligate and facultative sulfur respirers, isolated from marine and terrestrial thermal environments. The mol% G + C of the DNA ranges from 38 to 60.

Genus I. *Thermococcus*, p. 342

II. Hyperthermophilic chemoorganotrophs, facultatively sulfidogenic, that are distinguished from *Thermococcus* primarily due to their higher temperature optimum for growth (about 100°C) and the clustering of their 16S rRNA sequences as a separate clade within the *Thermococcales*. The mol% G + C of the DNA ranges from 38 to 44.

Genus II. *Pyrococcus*, p. 346

Genus I. **Thermococcus** Zillig 1983, 673[VP] (Effective publication: Zillig in Zillig, Holz, Janekovic, Schäfer and Reiter 1983b, 93)

TETSUO KOBAYASHI

Ther.mo.coc′cus. Gr. fem. n. *therme* heat; Gr. masc. n. *kokkos* berry; M.L. masc. n. *Thermococcus* berry existing in hot environment.

Regular to irregular cocci occurring singly and in pairs, 0.7–2.5 μm in diameter (Fig. A2.46). **Cell division by constriction**. Nonmotile or **motile with polar tufts of flagella. Strictly anaerobic. Extremely thermophilic with optimal temperatures for growth between 75 and 88°C**; maximum 103°C; minimum 50°C. Optimum pH: 6–8; maximum 9; minimum 3. **NaCl is required for growth, with optimal concentrations between 2 and 4%**; maximum 8; minimum 0.8. **Obligately heterotrophic growth on peptides**. Some species also utilize carbohydrates. **Elemental sulfur significantly stimulates growth rates with production of H_2S**. Some species require elemental sulfur for growth.

The mol% G + C of the DNA is 46.5–58 (Bd, T_m, and HPLC).

Type species: **Thermococcus celer** Zillig 1983, 673 (Effective publication: Zillig *in* Zillig, Holz, Janekovic, Schäfer and Reiter 1983b, 93.)

FURTHER DESCRIPTIVE INFORMATION

Cells are surrounded by a single- or double-layered envelope based on thin sections. All species utilize peptides as energy substrates, most likely by fermentation. Other energy substrates utilized depend on strains and include amino acids, casein, pyruvate, maltose, starch, pectin, and chitin. No growth on glucose. Produce a strong-smelling product. In the absence of elemental sulfur, H_2 is produced, and usually both growth rate and cell yield decrease dramatically. Polar lipids are mainly glycerol diethers and/or tetraethers.

Resistant to antibiotics such as ampicillin, chloramphenicol, streptomycin, and vancomycin. However, most species are sensitive to rifampicin at 100 μg/ml, probably because of interference with the integrity of the cytoplasmic membrane but not because of inhibition of RNA synthesis (Huber et al., 1987b). Inhabit coastal and submarine solfataras, including deep-sea hy-

drothermal vents. Have been isolated from New Zealand terrestrial thermal springs.

ENRICHMENT AND ISOLATION PROCEDURES

Samples have been taken from shallow or submarine solfataras. Research submersibles have been used to obtain deep-sea samples. Media used for the enrichment and isolation have been basically artificial seawater supplemented with peptides such as peptone or tryptone, yeast extract, and elemental sulfur. Na_2S (0.5 g/l) is added to provide anoxic conditions, and resazurine (1 mg/l) is often added as a redox indicator. *Thermococcus* strains can be purified by conventional serial dilution methods or by colonization on agar or Gelrite plates supplemented with polysulfide or colloidal sulfur instead of elemental sulfur.

MAINTENANCE PROCEDURES

Thermococcus species can be stored at −80°C or in liquid nitrogen for years.

DIFFERENTIATION OF THE GENUS *THERMOCOCCUS* FROM OTHER GENERA

Thermococcus species are mainly differentiated from members in the orders *Thermoproteales* and *Sulfolobales* by 16S rRNA sequence comparison.

The genus *Pyrococcus* in the order *Thermococcales* has higher optimum temperatures for growth (around 100°C) and lower mol% G + C (38 for the type species, *Pyrococcus furiosus*) than those of the genus *Thermococcus*.

FURTHER READING

Zillig, W. 1991. The Order *Thermococcales*. *In* Balows, Trüper, Dworkin, Harder and Schleifer (Editors), The Prokaryotes, Springer-Verlag, New York, pp. 702–706.

DIFFERENTIATION OF THE SPECIES OF THE GENUS *THERMOCOCCUS*

Some differentiating features of the species of the genus *Thermococcus* are listed in Table A2.37. Since they are morphologically and physiologically so similar to each other, DNA–DNA hybridization analysis may be required for differentiation.

List of species of the genus Thermococcus

1. **Thermococcus celer** Zillig 1983, 673[VP] (Effective publication: Zillig *in* Zillig, Holz, Janekovic, Schäfer and Reiter 1983b, 93.) *ce′ler*. L. masc. adj. *celer* fast, due to high growth rate.

See Table A2.37, Fig. A2.46, and the generic description. Contains diether lipids with a trace amount of tetraether lipids in the membrane. The size of the chromosome was

estimated to be 1890 ± 27 kb (Noll, 1989). Isolated from solfataric marine water holes of Vulcano, Italy.

The mol% G + C of the DNA is: 56.6 (T_m).
Type strain: Vu13, ATCC 35543, DSMZ 2476.
GenBank accession number (16S rRNA): M21529.

2. **Thermococcus aggregans** Canganella, Jones, Gambacorta and Antranikian 1998, 1183.[VP]

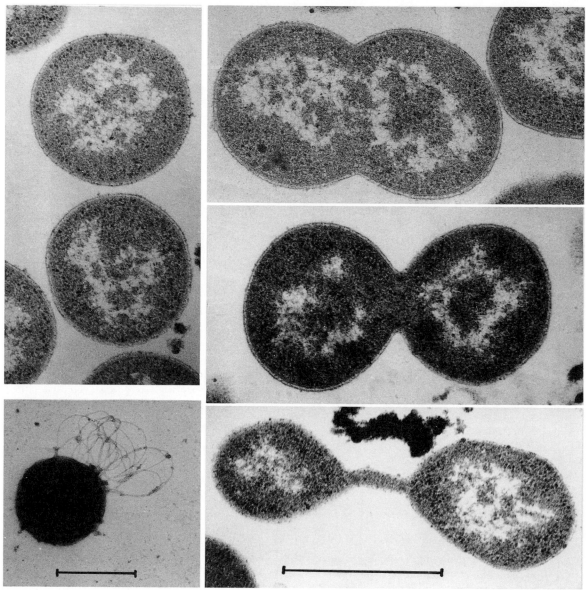

FIGURE A2.46. Electron micrographs of *T. celer*. *Upper left,* thin section of two "normal" cells of *T. celer*. *Right, top to bottom,* thin sections of diploid forms of *T. celer* showing increasing constriction. *Lower left,* cell of *T. celer* showing monopolar polytrichous flagellation. Negatively stained. Bars = 1 μm. (Micrographs courtesy of D. Janekovic.)

ag′gre.gans. L. v. *aggregare* to stick together, to aggregate; L. pres. part. *aggregans* assembling, aggregating.

See Table 2.37 and the generic description. Nonmotile. Occur as single cells, diploid forms, or (particularly when cultivated on yeast extract/tryptone plus sulfur) aggregates of many cells (up to 50). Colonies are circular, convex, and pale gray, ranging from 0.5 to 1.0 mm after 48–72 h at 70°C under strictly anoxic conditions. Grows on proteinaceous substrates such as yeast extract or Trypticase, casein, maltose, and starch. Glucose is also utilized. When grown on yeast extract plus Trypticase and sulfur, produces acetate, propionate, isobutyrate, isovalerate, CO_2, and H_2S; when grown on glucose or maltose plus sulfur, produces only acetate, CO_2, and H_2S. Lipids are mainly represented by one phospholipid and some minor components; core lipids are characteristically archaeol and caldarchaeol compounds in equal amounts. Isolated from marine sediments at the Guaymas Basin hydrothermal vent site (2000 m depth) in the Gulf of California.

The mol% G + C of the DNA is: 42.0 (HPLC).
Type strain: TY, DMS 10597, JCM 10137.
GenBank accession number (16S rRNA): Y08384.

3. **Thermococcus alcaliphilus** Keller, Braun, Dirmeier, Hafenbradl, Burggraf, Rachel and Stetter 1997, 601[VP] (Effective publication: Keller, Braun, Dirmeier, Hafenbradl, Burggraf, Rachel and Stetter 1995, 394.)

al.ca.li′phi.lus. M.L. neut. n. *alcalum* alkali from the Arabic *al* of the end; *quali* soda ash; Gr. adj. *philos* loving; M.L. masc. adj. *alcaliphilus* liking alkaline media.

See Table A2.37 and the generic description. Cells are surrounded by a regular surface layer protein with p6 symmetry based on freeze etching. Possesses a single flagellum.

TABLE A2.37. Diagnostic and descriptive characteristics of *Thermococcus* species[a]

Characteristic	1. *T. celer*	2. *T. aggregans*	3. *T. alcaliphilus*	4. *T. barophilus*	5. *T. chitonophagus*	6. *T. fumicolans*	7. *T. gorgonarius*	8. *T. guaymasensis*	9. *T. hydrothermalis*	10. *T. pacificus*	11. *T. peptonophilus*	12. *T. profundus*	13. *T. stetteri*	14. *T. zilligii*	15. *"T. litoralis"*
Cell diameter, μm	1	1.0–1.5	0.8–1.3	0.8–2.0	1.2–2.5	0.8–2.0	0.3–1.2	1.0–3.0	0.8–2	0.7–1.0	0.7–2	1–2	1–2	0.5–2.0	0.5–3.0
Flagella	+	Non-motile	+ Single	+	+	+	+	Non-motile	+	+	+	+	Non-motile	+	–
Temperature, °C															
Maximum	93	94	90	995	93	103	95	90	100	95	100	90	85	85	95
Optimum	88	88	85	85	85	85	80–88	88	80–90	80–88	85	80	75	75–80	88
Minimum	NR[b]	60	56	48	60	73	68	56	55	70	60	50	60	55	65
pH:															
Maximum	NR	7.9	10.5	9.5	9	9.5	8.5	8.1	9.5	8	8	8.5	7.2	9.2	8.5
Optimum	5.8	7	9	7	6.7	8	6.5–7.2	7.2	5.5–6.5	6.5	6	7.5	6.5	7.4	7.2
Minimum	NR	5.6	6.5	4.5	3.5	4.5	5.8	5.6	3.5	6	4	4.5	5.7	5.4	6.2
NaCl, g/l:															
Maximum	NR	30	60	40	80	40	50	50	52	60	50	60	40	11.7	65
Optimum	40	25	20–30	20–30	20	13–26	20–35	30	20–26	20–35	30	20	25	2.9	25
Minimum	lysis below 35	10	10	10	8	6	10	10	13	10	10	10	10	0.6	18
Sulfur effect[c]	S	S	S	S	S	S	R	S	S	R	S	R	R	R	S
Energy substrate:															
Peptides	+	+	+	+	+	+	+	+	+	+	+	+	+	+	+
Casein	+	+	+	+	NR	+	NR	+	+	NR	+	+	NR	+	+
Amino acids	NR	–	–	+	NR	+	–	–	+	–	–	NR	–	–	–
Starch	NR	+	–	–	NR	–	–	+	–	+	NR	+	+	–	–
Pectin	NR	NR	NR	NR	–	NR	NR	NR	NR	NR	NR	NR	+	NR	NR
Chitin	NR	–	NR	NR	+	–	NR	–	NR	NR	NR	NR	NR	NR	NR
Maltose	NR	+	–	–	–	+	–	+	+	–	–	+	–	–	–
Pyruvate	NR	NR	NR	weak	NR	+	weak	NR	+	–	–	+	NR	NR	+
Mol% G + C	56.6	42	42	37.1	46.5	54–55	50.6	46	58	53.3	52	52.5	50.2	46.2	38

[a]Characteristics of the type strains.

[b]NR; not reported.

[c]S = stimulatory, R = required.

Grows at the highest pH range among *Thermococcus* species (optimum around pH 9.0). Fermentation products (on yeast extract with sulfur) are isovalerate, isobutyrate, propionate, acetate, CO_2, NH_3, and H_2S. Isolated from a shallow marine hydrothermal system at Vulcano Island, Italy.

The mol% G + C of the DNA is: 42 (T_m, HPLC).

Type strain: AEDII12, DSMZ 10322.

4. **Thermococcus barophilus** Marteinsson, Birrien, Reysenbach, Vernet, Marie, Gambacorta, Messner, Sleytr and Prieur 1999, 357.[VP]

bar.o' phi.lus. Gr. neut. n. *baros* weight; Gr. adj. *philos* loving; M.L. adj. *barophilus* weight lover, referring to the weight of the water column.

See Table A2.37 and the generic description. Cell envelope consists of a hexagonal S-layer lattice (spacing 28.4 nm). At atmospheric pressure, growth occurs at 48–95°C; optimum growth at 85°C under 0.3 MPa hydrostatic pressure. High hydrostatic pressure (40 MPa) stimulates growth at 75, 80, 90, and 95°C, with an optimum growth temperature of 85°C. Obligately barophilic between 95 and 100°C and requires 15.0– 17.5 MPa for growth. Sulfur is not necessary for growth, but greatly enhances growth and can be replaced by cystine. Membrane lipids are composed of one major phospholipid and an archaeol component. Isolated from a wall of a thermal vent, under 40 MPa hydrostatic pressure at 95°C, from the Mid-Atlantic Ridge, Snakepit at depth of 3550 m.

The mol% G + C of the DNA is: 37.1 (T_m).

Type strain: MP, CNCM I-1946.

GenBank accession number (16S rRNA): U82237.

5. **Thermococcus chitonophagus** Huber and Stetter 1996, 836[VP] (Effective publication: Huber and Stetter *in* Huber, Stöhr, Hohenhaus, Rachel, Burggraf, Jannasch and Stetter 1995b, 262.)

chi.to.no.pha' gus. Gr. masc. n. *chiton* sheath; Gr. masc. n. *phagos* degrader; M.L. masc. n. *chitonophagus* the sheath degrader.

See Table A2.37 and the generic description. Cell envelope composed of periplasmic space and a protein surface layer exhibiting p6 symmetry, based on ultrathin section of a freeze- substituted cell.

Isolated using chitin as a sole carbon and energy source. Cells stick to chitin particles during exponential growth. Chitin is degraded to H_2, CO_2, NH_3, acetate, and formate. Chitonoclastic enzyme system is cell associated, oxygen stable, and inducible. Growth occurs on chitin, yeast extract, and meat extract. In the presence of elemental sulfur, casein peptone and meat peptone are also utilized. Core lipids consist of glycerol diphytanyl diethers and acyclic and cyclic glycerol diphytanyl tetraethers with one pentacyclic ring.

Isolated from a deep-sea hydrothermal vent environment (depth, 2600 m) at Guaymas hot vent area.

The mol% G + C of the DNA is: 46.5 (T_m, HPLC).

Type strain: GC74, DSMZ 10152.

GenBank accession number (16S rRNA): X99570.

6. **Thermococcus fumicolans** Godfroy and Meunier *in* Godfroy, Meunier, Guezennec, Lesongeur, Raguénès, Rimbault and Barbier 1996, 1118.[VP]

fu.mi.co' lans. L. masc. n. *fumus* smoke, referring to the smokers or chimneys of deep-sea vents; L. part. pres. *colans* living on; M.L. adj. *fumicolans* living on smoke.

See Table A2.37 and the generic description. Cells are surrounded by two-layered cell envelope. Grows on proteolysis products and a mixture of amino acids and produces organic acids such as acetate, phenylacetate, lactate, and 2-methylpropionate as the major fermentation products. Isolated from an active chimney wall fragment obtained at a hydrothermal site in the North Fiji Basin in the southwest Pacific Ocean.

The mol% G + C of the DNA is: 54–55 (T_m, HPLC).

Type strain: ST557, CIP 104690.

GenBank accession number (16S rRNA): Z70250.

7. **Thermococcus gorgonarius** Miroshnichenko, Gongadze, Rainey, Kostyukova, Lysenko, Chernyh and Bonch-Osmolovskaya 1998, 28.[VP]

gor.go.na' ri.us. M.L. masc. adj. *gorgonarius* pertaining to Gorgon, a Greek mythological creature with snakes for hair.

See Table A2.37 and the generic description. Cell envelope consists of one layer of subunits. The major cell wall protein has an Mr of 75,000. Protrusions of different kinds: prostheca- like, chains of bubbles, fimbriae network. Grows on peptides and weakly on pyruvate. In closed vessels growth is dependent on the presence of S^0 which is reduced to H_2S. Produces growth products on peptone are CO_2, H_2S, acetate, propionate, isobutyrate, and isovalerate. Isolated from sand from the shore of Whale Island, New Zealand.

The mol% G + C of the DNA is: 50.6 (T_m).

Type strain: W-12, DSM 10395.

GenBank accession number (16S rRNA): Y16226.

8. **Thermococcus guaymasensis** Canganella, Jones, Gambacorta and Antranikian 1998, 1184.[VP]

gua.i.mas' ensis. L. gen. n. *guaymasensis* named after the Guaymas Basin, the name of the deep-sea site where the species was isolated.

See Table A2.37 and the generic description. Nonmotile. Colonies are circular, convex, and pale gray, ranging from 0.5 to 1.0 mm after 48–72 h at 70°C under strictly anoxic conditions. Grows on proteinaceous substrates such as yeast extract or Trypticase, casein, maltose, and starch. Glucose is also utilized. When grown on yeast extract plus Trypticase and sulfur, produces acetate, propionate, isobutyrate, isovalerate, CO_2, H_2S; when grown on glucose or maltose plus sulfur, produces only acetate, CO_2, and H_2S. Lipids are mainly represented by one phospholipid and some minor components; core lipids are characteristically archaeol and caldarchaeol compounds in equal amounts with traces of tricyclized caldarchaeol. Isolated from marine sediments at the Guaymas Basin hydrothermal vent site (2000 m depth) in the Gulf of California.

The mol% G + C of the DNA is: 46.0 (HPLC).

Type strain: TYS, DMS 11113, JCM 10136.

GenBank accession number (16S rRNA): Y08385.

9. **Thermococcus hydrothermalis** Godfroy, Lesongeur, Ragué nès, Quérellou, Antoine, Meunier, Guezennec and Barbier 1997, 626.[VP]

hy.dro.ther.ma' lis. M.L. masc. adj. *hydrothermalis* pertaining to a hydrothermal vent.

See Table A2.37 and the generic description. Cell envelope consists of two layers. Grows on peptides and a mixture of amino acids and produces acetate, phenylacetate, lactate, and 2-methylpropionate as the major fermentation products. Isolated from an active chimney wall fragment recovered from a hydrothermal site on the East Pacific Rise at latitude 21°N.

The mol% G + C of the DNA is: 58 (T_m).

Type strain: AL662, CNCMI 1319.

GenBank accession number (16S rRNA): Z70244.

10. **Thermococcus pacificus** Miroshnichenko, Gongadze, Rainey, Kostyukova, Lysenko, Chernyh and Bonch-Osmolovskaya 1998, 28.[VP]

pa.ci' fi.cus. L. masc. adj. for Pacific Ocean, from the southwest part of which the new organism was isolated.

See Table A2.37 and the generic description. Cell envelope consists of one layer of subunits. The major cell wall protein has an Mr or 56,000. Grows on peptide and starch. In closed vessels, growth is dependent on the presence of S^0 which is reduced to H_2S. Growth products on peptone are CO_2, H_2S, acetate, propionate, isobutyrate, and isovalerate. Isolated from bottom deposits of the Bay of Plenty, New Zealand.

The mol% G + C of the DNA is: 53.3 (T_m).

Type strain: P-4, DMS 10394.

GenBank accession number (16S rRNA): Y16227.

11. **Thermococcus peptonophilus** González, Kato and Horikoshi 1996, 625[VP] (Effective publication: González, Kato and Horikoshi 1995, 164.)

pep.to.no' phi.lus. Gr. adj. *peptos* cooked; Gr. adj. *philos* loving; M.L. adj. *peptonophilus* peptone-loving, indicating that it grows only on peptides as a carbon source.

See Table A2.37 and the generic description. The cell envelope of the type strain, OG-1, consists of two layers, the external one being electron dense, while strain SM-2 has a 10–15-nm single layer of low electron density. Isolated from deep-sea hydrothermal vents at Izu-Bonin (OG-1) and South Mariana Trough (SM-2) areas in the western Pacific Ocean.

The mol% G + C of the DNA is: 52 (HPLC).

Type strain: OG-1, JCM 9653.

GenBank accession number (16S rRNA): D37982.

12. **Thermococcus profundus** Kobayashi and Horikoshi 1995, 418[VP] (Effective publication: Kobayashi and Horikoshi *in* Kobayashi, Kwak, Akiba, Kudo and Horikoshi 1994b, 236.)

pro.fun' dus. L. masc. adj. *profundus* deep, living within the depths of the oceans.

See Table A2.37 and the generic description. The cell envelope consists of a 40–45-nm-thick layer of low electron density. Shows very good growth on peptides, pyruvate, maltose, and starch. Possesses an extremely large amount of glutamate dehydrogenase and excretes L-alanine in the culture broth (Kobayashi et al., 1995), as observed in *Pyrococcus fu-*

riosus (Consalvi et al., 1991; Kengen and Stams, 1994). These organisms probably share similar nitrogen metabolic pathways. Isolated from a deep-sea hydrothermal vent on Iheya Small Ridge at the Mid-Okinawa Trough.

The mol% G + C of the DNA is: 52.5 (HPLC).
Type strain: DT5432, JCM 9378.
GenBank accession number (16S rRNA): Z75233.

13. **Thermococcus stetteri** Miroshnichenko 1990, 321[VP] (Effective publication: Miroshnichenko *in* Miroshnichenko, Bonch-Osmolovskaya, Neuner, Kostrikina, Chernych and Alekseev 1989, 261.)

stet' te.ri. M.L. gen. n. *stetteri* of Stetter; named after K.O. Stetter because of his important contribution to the knowledge of extreme thermophiles.

See Table A2.37 and the generic description. The cell envelope consists of two layers of subunits (20 nm thick [each]) with a thin electron-dense layer between them (10 nm thick). Produces H_2S, CO_2, acetate, isobutyrate, and isovalerate. Strains K-15 and K-17, which show high DNA–DNA hybridization to K-3 (the type strain), are also members of *Thermococcus stetteri*. The type strain, K-3, is nonmotile, while K-15 and K-17 are motile with a polar tuft of flagella. Temperature range for growth of K-15 and K-17 is much higher than that of K-3. They grow between 75 and 98°C, with optima at 88°C. K-15 and K-17 do not utilize starch and pectin. Isolated from marine hydrothermal vents in Kraternaya cove (Northern Kurils).

The mol% G + C of the DNA is: 50.2 (T_m).
Type strain: K-3, DSMZ 5262.
GenBank accession number (16S rRNA): Z75240.

14. **Thermococcus zilligii** Ronimus, Reysenbach, Musgrave and Morgan 1999, 935[VP] (Effective publication: Ronimus, Reysenbach, Musgrave and Morgan 1997, 247).

zil.lig' i.i. M.L. gen. n. *zilligii* named after Wolfram Zillig for his contribution to the understanding of transcription in the *Archaea* and the characterization of hyperthermophilic *Archaea* and of their viruses.

See Table A2.37 and the generic description. Optimum NaCl concentration is 50 mM (0.29%), with no growth above 200 mM. Absolutely requires either Na^+ or Li^+ for growth. Sulfur (either as S^0, cystine, or oxidized glutathione, with the sulfur being reduced to H_2S) is also a requirement. Grows on peptone, casein, or yeast extract. Inclusion bodies often formed when grown in the presence of S^0. When grown on Trypticase peptone, produces acetate, propionate, isobutyrate, mercapto-based compounds, isovalerate/2-methylbutyrate, H_2, H_2S, CO_2, and NH_4^+. Isolated from a terrestrial, fresh-water geothermal pool (Rt1) in Kuirau Park, Rotorua, New Zealand.

The mol% G + C of the DNA is: 46.2 (T_m).
Type strain: AN1, DSM 2770.
GenBank accession number (16S rRNA): U76534.

15. **"Thermococcus litoralis"** Neuner, Jannasch, Belkin and Stetter 1990, 207.

li.to.ra' lis. L. masc. adj. *litoralis* pertaining to the shore.

Cells are not flagellated. Strain A3 (DSMZ 5474) shows a growth temperature range and optimum slightly different from those of the proposed type species. It grows between 50 and 96°C, with an optimum at 85°C. Isolated from shallow submarine solfataras at Lucrino/Naples and Porto di Levente/Vulcano. The mol% G + C is identical to that of *Pyrococcus furiosus*.

The mol% G + C of the DNA is: 38 (T_m, HPLC).
Deposited strain: NS-C, DSMZ 5473.
GenBank accession number (16S rRNA): Z70252.

Genus II. **Pyrococcus** *Fiala and Stetter 1986b, 573*[VP] *(Effective publication: Fiala and Stetter 1986a, 60)*

KARL O. STETTER AND HARALD HUBER

Pyr.o.coc' cus. Gr. neut. n. *pyr* fire; Gr. masc. n. *kokkos* berry; M.L. masc. n. *Pyrococcus* fireball.

Cells slightly irregular cocci, 0.8–2.5 μm in width, **occurring singly or in pairs**. Monopolar polytrichous flagellated. Strictly anaerobic. **Heterotrophic** growth on peptone, tryptone, yeast extract, meat extract, extracts of *Bacteria* and *Archaea*, casein, starch, maltose, and Casamino acids. **Temperature for growth, 67–103°C; optimum, around 100°C.** pH for growth, 5–9; optimum, ~7. Optimum NaCl concentration: 2–3%; maximum, 5%; minimum, 0.5%. Shortest doubling time, about 35 min under optimum conditions. Gram negative. Isopranyl ether lipids present. ADP ribosylation of elongation factor 2 by diphtheria toxin.

The mol% G + C of the DNA is: 38–47.

Type species: **Pyrococcus furiosus** Fiala and Stetter 1986b, 573 (Effective publication: Fiala and Stetter 1986a, 60.)

FURTHER DESCRIPTIVE INFORMATION

Cells are surrounded by an envelope about 50 nm thick (Fig. A2.47) which can be differentiated into an inner part of high contrast, measuring about 16 nm thick, and a weakly contrasted outer region. Cells are stained by the periodate-Schiff reagent

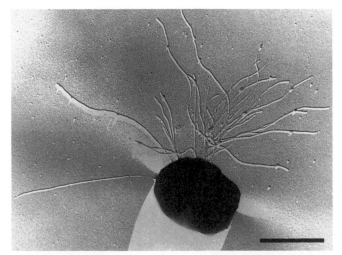

FIGURE A2.47. Electron micrograph of a *Pyrococcus furiosus* cell. Platinum shadowed. Bar = 1 μm.

and are therefore most likely glycoproteins. Maltose is fermented *Pyrococcus furiosus* to acetate, alanine, H_2, and CO_2 (Kengen and Stams, 1994; Schönheit and Schäfer, 1995). *Pyrococcus* contains many heat stable enzymes (Leuschner and Antranikian, 1995), including ADP-dependent kinases in glucose fermentation in *Pyrococcus furiosus* (Kengen et al., 1996) and tungsten-containing aldehyde ferredoxin oxidoreductases in *P. furiosus* (Adams and Kletzin, 1996).

ENRICHMENT AND ISOLATION PROCEDURES

Enrichment of *P. furiosus* can be achieved by using "SME" medium* (Stetter et al., 1983) supplemented with yeast extract and peptone (Fiala and Stetter, 1986a). After 1 d incubation at 100°C, coccoid organisms become visible and can be purified by repeated serial dilutions (due to a very short doubling time, even the 10^{-8} dilutions usually grow overnight). Purification can be achieved alternatively by streaking onto agar-solidified medium. Round, smooth, white colonies about 0.5 mm in diameter appear after 1 week of anaerobic incubation at 70°C.

MAINTENANCE PROCEDURES

Cultures of *Pyrococcus furiosus* are routinely grown for 10–15 h at 100°C and then transferred into fresh medium. Stock cultures can be stored anaerobically at 4°C for at least 1 year without transfer.

DIFFERENTIATION OF THE GENUS *PYROCOCCUS* FROM OTHER GENERA

The genus *Pyrococcus* belongs to the *Thermococcales* by its 16S rRNA sequence (Huber and Stetter, in press[†]). It can be distinguished from *Thermococcus* (Zillig et al., 1983b) by its lower mol% G + C content (10% lower than *Thermococcus*) and its higher optimum and maximum growth temperatures.

TAXONOMIC COMMENTS

By 16S rRNA sequencing, *Pyrococcus* is closely related to *Thermococcus* (Huber and Stetter, in press[†]). The total genome of *Pyrococcus horikoshii* was published recently (Kawarabayasi et al., 1998). Furthermore the total genome of *"Pyrococcus abyssi"* and parts of the genome sequence of *Pyrococcus furiosus* are available from the websites of Genoscope and the University of Maryland Biotechnology Institute Center of Marine Biotechnology.[‡]

FURTHER READING

Erauso, G., A.L. Reysenbach, A. Godfroy, J.R. Meunier, B. Crump, F. Partensky, J.A. Baross, V. Marteinsson, G. Barbier, N.R. Pace and D. Prieur. 1993. *Pyrococcus abyssi* sp. nov., a new hyperthermophilic archaeon isolated from a deep-sea hydrothermal vent. Arch. Microbiol. *160*: 338–349.

Fiala, G. and K.O. Stetter. 1986. *Pyrococcus furiosus*, sp. nov. represents a novel genus of marine heterotrophic archaebacteria growing optimally at 100°C. Arch. Microbiol. *145*: 56–61.

Zillig, W., I. Holz, H.-P. Klenk, J. Trent, S. Wunderl, D. Janekovic, E. Imsel and B. Haas. 1987. *Pyrococcus woesei*, sp. nov., an ultra-thermophilic marine archaebacterium, representing a novel order, *Thermococcales*. Syst. Appl. Microbiol. *9*: 62–70.

DIFFERENTIATION OF THE SPECIES OF THE GENUS *PYROCOCCUS*

Five species have been described and are mainly distinguished on the basis of their anaerobic utilization of organic carbon sources. However, only four species (*P. furiosus*, *P. glycovorans*, *P. horikoshii*, and *P. woesei*) have been validly published. Furthermore, several *Pyrococcus* species appear in the literature without being described in detail, e.g., *"P. kodadaraensis"*, *"P. endeavori"*, or *"P. hellenicus"*.

Key to the species of the genus *Pyrococcus*

1. Growth by fermentation of peptones, yeast extract, and polysaccharides. Salt optimum 2%. Periplasmic space in cell envelope. The mol% G + C of the DNA is 38.

   P. furiosus.

2. Growth by fermentation of proteinaceous substrates and carbohydrates. Glucose used as carbon source. The mol% G + C of the DNA is 47.

   P. glycovorans.

3. No growth by fermentation of sugars and polysaccharides. Requires peptides for growth in addition to tryptophan. Periplasmic space in cell envelope. The mol% G + C of the DNA is 42–44.

   P. horikoshii.

4. Does not ferment peptones, yeast extract, or polysaccharides but uses elemental sulfur as an electron acceptor for these substrates, forming hydrogen sulfide as an end product. The mol% G + C of the DNA is 37.5.

   P. woesei.

5. Obligate organotrophic growth by fermentation of peptones, yeast extract and polysaccharides. Salt optimum 3%. Periplasmic space absent; cell envelopes consist of a double S-layer. The mol% G + C of the DNA is 45.

   "P. abyssi".

---

*"SME" medium consists of (per liter): NaCl, 13.85 g; $MgSO_4$·$7H_2O$, 3.5 g; $MgCl_2$·$6H_2O$, 2.75 g; $CaCl_2$·$2H_2O$, 1.0 g; KCl, 0.325 g; NaBr, 0.05 g; H_3BO_3, 0.015 g; $SrCl_2$·$6H_2O$, 0.0075 g; KH_2PO_4, 0.5 g; KI, 0.025 mg; $NiNH_4SO_4$, 0.002 g; sulfur, 25 g; resazurine, 0.001 g; and trace minerals (Balch et al., 1979), 10 ml.

[†]*Editorial Note: Encyclopedia of Life Sciences* will be published in 2001. The chapter on *Euryarchaeota* may be viewed at www.els.net.

[‡]At the time of publication, Genoscope and the Center of Marine Biotechnology could be accessed at the following websites: www.genoscope.cns.fr./Pab/ and combdna.umbi.umd.edu/bags.html.

List of species of the genus Pyrococcus

1. **Pyrococcus furiosus** Fiala and Stetter 1986b, 573[VP] (Effective publication: Fiala and Stetter 1986a, 60.)

fu.ri.o' sus. L. masc. adj. *furiosus* raging.

Cells are highly motile due to a bundle of about 50 flagella, each about 7 µm long and 7 nm thick. Close to the origin of the flagella an electron-dense granum-like body of up to 1 µm in width can often be seen. H_2 and CO_2 are formed as metabolic products from yeast extract and peptone. In the presence of sulfur, the H_2 concentration is diminished, H_2S is formed, and final cell concentrations are about five times higher than without S^0, indicating a detoxification of H_2 by H_2S formation.

The mol% G + C of the DNA is: 38 (T_m).

Type strain: Vc1, DSMZ 3638.

GenBank accession number (16S rRNA): U20163.

2. **Pyrococcus glycovorans** Barbier, Godfroy, Meunier, Quérellou, Canbon, Lesongeur, Grimont and Raguénès 1999, 1835.[VP]

gly.co.vo' rans. Gr. adj. *glykos* sweet, referring to glucose; L. part. pres. *vorans* eating, devouring; M.L. part. adj. *glycovorans* eating glucose.

Cocci with a mean diameter of 1 µm (range 0.5–1.5 µm). Motile with polar flagella. Cell division occurs by constriction. Obligately anaerobic. Optimum growth at 30 g/l sea salt, pH 7.5. At atmospheric pressure, growth occurs between 75 and 104°C. Obligately chemoorganotrophic. Ferments protein, proteolytic products, and carbohydrates, especially glucose. Sulfur not required for growth but significantly enhances final cell concentrations when grown in closed culture vessels. S^0 reduced to H_2S. At low cell concentration, under optimal conditions and at atmospheric pressure, doubling time is 0.5 h and maximal concentrations at the beginning of the stationary phase can reach 5×10^8 cells/ml. The type strain was isolated from pieces of a deep-sea smoker and the surrounding alvinellid tubes collected at 13°N at a depth of 2650 m on the East Pacific Rise.

The mol% G + C of the DNA is: 47.

Type strain: AL585, CNCM I-2120.

GenBank accession number (16S rRNA): Z70247.

3. **Pyrococcus horikoshii** González, Robb and Kato 1999, 1325[VP] (Effective publication: González, Robb and Kato *in* González, Masuchi, Robb, Ammermann, Maeder, Yanagibayashi, Tamaoka and Kato 1998, 129.)

ho.ri.ko' shi.i. M.L. gen. n. *horikoshii* named to honor Dr.K. Horikoshi.

Cells are slightly irregular cocci, 0.8–2.0 µm in diameter, showing a polar tuft of flagella. The cell envelope consists of a complete S-layer enclosing a periplasmic space around the cytoplasmic membrane. Obligate anaerobe. Growth occurs at temperatures between 80 and 102°C, with an optimum at 98°C. Optimum pH for growth is 7.0, and growth occurs from pH 5–8, with no growth observed at pH 8.5. NaCl concentration allowing growth ranges from 1 to 5%; optimum 2.4%. The shortest doubling time is 32 min. Elemental sulfur is not essential for growth, but it greatly stimulates growth and results in higher growth yield. Heterotrophic growth occurs on complex proteinaceous substrates (peptone, yeast extract, beef extract, tryptone, casein); on Casamino acids, supplemented with tryptophan in the presence of vitamins; and on a mixture of 21 amino acids when vitamins are present. Tryptophan is required for growth. Inhabits marine hydrothermal vents in the western Pacific Ocean. Isolated from the Okinawa Trough.

The mol% G + C of the DNA is: 44 (T_m) or 42 (direct analysis).

Type strain: OT3, JCM 9974.

GenBank accession number (16S rRNA): D45214.

Additional Remarks: The GenBank accession number for 16S rRNA of strain JA-1 (JCM 9975) is D87344.

4. **Pyrococcus woesei** Zillig 1988, 136[VP] (Effective publication: Zillig *in* Zillig, Holz, Klenk, Trent, Wunderl, Janekovic, Imsel and Haas 1987, 69.)

woe' se.i. M.L. gen. n. *woesei* of Woese; named for C.R. Woese, who recognized *Archaea* and their testimony for phylogeny.

Roundly spherical to elongated, often constricted cells of 0.5–2 µm in diameter, frequently linked into doublets by short, thin threads. Organism stains Gram-negative with large bundles of smoothly bent filaments (flagella?) attached to one pole when cells grow on solid supports.

Hyperthermophilic anaerobes. Exist in marine solfataras at ~100–103°C in the presence of 3% NaCl. Upper temperature limit is 104.8°C.

Fermentation not observed. Yeast extract or peptones (e.g., Bacto tryptone) or polysaccharides are used by sulfur respiration. Polysaccharides are used well only in the presence of H_2. Yeast extract may also be used in the absence of elemental sulfur, apparently using an endogenous electron acceptor. Isolated from 102 to 103°C sediments in vents of marine solfataras of Vulcano Island, Italy.

The mol% G + C of the DNA is: DNA is 37.5.

Type strain: DSMZ 3773.

5. **"Pyrococcus abyssi"** Erauso, Reysenbach, Pace and Prieur *in* Erauso, Reysenbach, Godfroy, Meunier, Crump, Partensky, Baross, Marteinsson, Barbier, Pace and Prieur 1993, 347.

a.bys' si. L. gen. n. *abyssi* bottomless, immense depth, living within the ocean depths.

Cells are Gram-negative, slightly irregular cocci, variable in size (usually 0.8–2 µm), and shape depends upon growth conditions. Type strain (GE5) is highly motile by means of a polar tuft of flagella. Cell envelope consists of a double S-layer. Cell division occurs by constriction. Obligate anaerobe. Grows optimally at 2–3% NaCl and pH around 7.0. At atmospheric pressure, growth occurs between 67 and 102°C (optimum 96°C corresponding to 33-min doubling time). Facultatively barophilic: elevated hydrostatic pressures significantly extend upper growth temperature and stimulate growth. Barotolerant to facultatively barophilic.

Obligate chemoorganotroph, fermenting proteolysis products (e.g., tryptones and gelatin) and mixtures of amino acids and pyruvate in presence of vitamins in H_2, CO_2, and volatile fatty acids. H_2 produced inhibits growth, but this can be alleviated by addition of S^0, cystine, or polysulfides. In this case, H_2S and mercaptans are produced.

Membrane lipids are composed mainly of tetraether lipids with 0–4 cyclopentane rings in the isoprenoid chains.

Isolated from hot fluid from an active chimney in the North Fiji Basin (Southwest Pacific).

The mol% G + C of the DNA is: 45 $(T_m$ and direct analysis).

Deposited strain: GE5, CNCM I-1302.

GenBank accession number (16S rRNA): L19921.

Class VI. **Archaeoglobi** *class. nov.*

GEORGE M. GARRITY AND JOHN G. HOLT

Arch.ae.o.glo'bi. M.L. fem. pl. n. *Archaeoglobales* type order of the class; dropping the ending to denote a class; M.L. fem. pl. n. *Archaeoglobi* the class of *Archaeoglobales.*

Description of the class is the same as that for the order *Archaeoglobales.*

Type order: **Archaeoglobales** ord. nov. Stetter 1989, 2216.

Order I. **Archaeoglobales** *ord. nov.* Stetter 1989, 2216

HARALD HUBER AND KARL O. STETTER

Ar.chae.o.glo.ba'les. M.L. masc. n. *Archaeoglobus* type genus of the order; *-ales* ending to denote an order; M.L. fem. pl. n. *Archaeoglobales* the order of *Archaeoglobus.*

Regular to highly irregular cocci occurring singly or in pairs. **Strictly anaerobic. Hyperthermophilic,** temperature optimum around 80°C. **Neutrophilic.** Cell envelope consisting of protein subunits. **Blue-green fluorescence at 420 nm. Sulfate, sulfite, thiosulfate and/or nitrate serve as electron acceptors,** but S^0 inhibits growth. Only one family is presently known.

The mol% G + C of the DNA is: 40–46.

Type genus: **Archaeoglobus** Stetter 1988b, 328, emend. Huber, Jannasch, Rachel, Fuchs and Stetter 1997, 379 (Effective publication: Stetter 1988a, 172.)

Family I. **Archaeoglobaceae** *fam. nov.* Stetter 1989, 2216

HARALD HUBER AND KARL O. STETTER

Ar.chae.o.glo.ba'ce.ae. M.L. masc. n. *Archaeoglobus* type genus of the family; *-aceae* ending to denote a family; M.L. fem. pl. n. *Archaeoglobaceae* the family of *Archaeoglobus.*

Description of the family is the same as that for the order.

Type genus: **Archaeoglobus** Stetter 1988b, 328, emend. Huber, Jannasch, Rachel, Fuchs and Stetter 1997, 379 (Effective publication: Stetter 1988a, 172).

Key to the genera of the family Archaeoglobaceae

1. Sulfate or sulfite and thiosulfate used as electron acceptors. H$_2$S formed as end product. Facultative chemolithoautotrophic or mixotrophic.
   Genus I. *Archaeoglobus.*
2. Fe (II), sulfide or molecular hydrogen used as electron donor and nitrate as electron acceptor. Formation of nitrite and nitrous gases. Reduction of thiosulfate in the presence of H$_2$ leads to H$_2$S.
   Genus II. *Ferroglobus,* p. 352

Genus I. **Archaeoglobus** *Stetter 1988b, 328,*[VP] *emend. Huber, Jannasch, Rachel, Fuchs and Stetter 1997, 379 (Effective publication: Stetter 1988a, 172)*

HARALD HUBER AND KARL O. STETTER

Arch.ae.o.glo'bus. Gr. masc. adj. *archaios* ancient; L. masc. n. *globus* sphere; M.L. masc. n. *Archaeoglobus* the ancient sphere.

Cells coccoid, regular to highly irregular, 0.4–1.2 μm in width, occurring singly and in pairs. Gram negative. Monopolar polytrichous flagella. **Blue-green fluorescence at 420 nm. Growth between 60 and 95°C and pH 5.5–7.5. Strictly anaerobic. Sulfate or sulfite and thiosulfate used as electron acceptors. Chemolithoautotrophic growth in the presence of H$_2$/CO$_2$. Chemoorganotrophic growth on formate, formamide, D(−)- and L(+)-lac**tate, glucose, starch, Casamino acids, peptone, gelatin, casein, meat extract, yeast extract, and extracts of bacterial and archaeal cells. S^0 not used as electron acceptor; inhibits growth. Cell envelope consists of an S-layer composed of subunits in hexagonal array containing a periodate-Schiff-positive polypeptide. Cells contain phytanyl ether lipids. Rifampicin and streptolydigin resistant.

By 16S rRNA total sequence comparisons, *Archaeoglobus*, together with *Ferroglobus*, represents a separate lineage within the euryarchaeotal branch of the *Archaea*.

The mol% G + C of the DNA is: 41–46.

Type species: **Archaeoglobus fulgidus** Stetter 1988b, 328 (Effective publication: Stetter 1988a, 172.)

FURTHER DESCRIPTIVE INFORMATION

Cells are often highly irregular cocci to triangular plate-shaped lobes (Fig. A2.48). No septum formation visible. Greenish-black smooth colonies, 1–2 mm in diameter. Nitrate and nitrite not used as electron acceptors. The organisms are autotrophic and/or organotrophic sulfate/sulfite respirer. H_2S (up to 6 μmol/ml) and methane (up to 0.1 μmol/ml) formed during growth. Acetate, pyruvate, fumarate, isopropanol, ethanol, and extracts of bacterial and archaeal cells may be used as substrates. No rigid sacculus. Murein and pseudomurein absent. ADP ribosylable protein present in crude extracts. F_{420}- and methanopterin-like compounds present, but no coenzyme M (CoM) and no factor 430. Methyl-CoM reductase genes absent. Adenosine triphosphate sulfurylase, adenylylsulfate reductase, and bisulfite reductase activities present in cell extract. Enzymes of the reductive carbon monoxide dehydrogenase pathway for autotrophic CO_2 fixation present (in *"Archaeoglobus lithotrophicus"* and, with the exception of CO dehydrogenase, also in *Archaeoglobus profundus*) (Vorholt et al., 1995). DNA polymerase sensitive to aphidicolin.

Archaeoglobus has been isolated from anaerobic samples of geothermally heated (shallow) marine sediments at Vulcano Island and Naples, Italy, at Ribeira Quente, Azores, and at the Kolbeinsey Ridge north of Iceland. Also isolated from deep-sea hydrothermal systems off Guaymas, Mexico, black smokers at the Mid-Atlantic Ridge, the crater of an active Polynesian submarine volcano, and production fluids of oil reservoirs at the East Shetland Basin (North Sea) and below the permafrost surface of the North Slope of Alaska (Stetter, 1992; Stetter et al., 1993; Huber et al., 1997).

ENRICHMENT AND ISOLATION PROCEDURES

Archaeoglobus can be obtained anaerobically in sulfate, sulfite, or thiosulfate-containing media (MGG medium, [Balch et al., 1979; Huber et al., 1982]) at temperatures around 85°C. In order to enrich *Archaeoglobus fulgidus*, lactate has been used as sole carbon source (Stetter et al., 1987). *A. profundus* can be enriched on H_2/CO_2 in the presence of acetate (Burggraf et al., 1990b); for *Archaeoglobus veneficus*, sulfite or thiosulfate is necessary as an electron acceptor, in combination with H_2/CO_2 as an electron donor (Huber et at., 1997); and *"A. lithotrophicus"* was enriched with sulfate in combination with H_2/CO_2 (Stetter et al., 1993).

Archaeoglobus can be isolated by serial dilutions, by plating on media containing 1.5% agar (only *A. fulgidus*) or 0.6% Gelrite, or by using optical tweezers (Huber et al., 1995a).

MAINTENANCE PROCEDURES

Cultures of *Archaeoglobus* can be stored anaerobically at 4°C for a few weeks. For longer periods (at least up to five years) cultures should be stored over liquid nitrogen ($-140°C$) after the addition of 5% DMSO.

DIFFERENTIATION OF THE GENUS *ARCHAEOGLOBUS* FROM OTHER GENERA

Archaeoglobus belongs to the archaeal order *Archaeoglobales* due to its morphology, the anaerobic mode of life, the G + C content of its genomic DNA, the blue-green fluorescence at 420 nm, the possession of phytanyl ether lipids in the membrane, the S-layer envelope, and its 16S rRNA sequence. In contrast to *Ferroglobus*, growth occurs via reduction of sulfate or sulfite and thiosulfate by production of H_2S. Nitrate cannot be used as an electron acceptor.

TAXONOMIC COMMENTS

By 16S rRNA sequencing *Archaeoglobus* branches near the base of the euryarchaeotal side of the archaeal tree. However, this

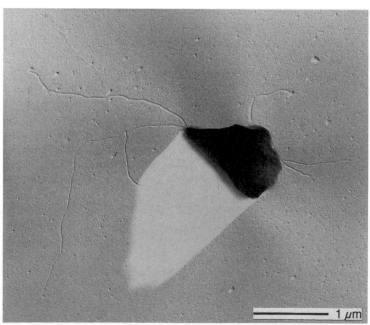

FIGURE A2.48. Electron micrograph of a cell of *Archaeoglobus fulgidus*. Platinum-shadowed.

originally reported deep branching turned out to be incorrect. Reexamination of the phylogenetic position using both 16S and 23S rRNA sequences (Woese et al., 1991) demonstrated that the correct branch off can be achieved only by using transversion analysis. The investigation revealed that *Archaeoglobus* branches between the lineages of the *Methanobacteriales* and the *Methanomicrobiales*/extreme halophiles. Furthermore, 16S rRNA sequencing demonstrated that the closest relative of *Archaeoglobus* is the genus *Ferroglobus*, represented so far by only one species, *Ferroglobus placidus* (Hafenbradl et al., 1996).

The complete genome of *Archaeoglobus fulgidus* has been sequenced (Klenk et al., 1997a).

FURTHER READING

Burggraf, S., H.W. Jannasch, B. Nicolaus and K.O. Stetter. 1990. *Archaeoglobus profundus*, sp. nov., represents a new species within the sulfate-reducing archaebacteria. Syst. Appl. Microbiol. *13*: 24–28.

Huber, H., H. Jannasch, R. Rachel, T. Fuchs and K.O. Stetter. 1997. *Archaeoglobus veneficus* sp. nov., a novel facultative chemolithoautotrophic hyperthermophilic sulfite reducer, isolated from abyssal black smokers. Syst. Appl. Microbiol. *20*: 374–380.

Stetter, K.O., G. Lauerer, M. Thomm and A. Neuner. 1987. Isolation of extremely thermophilic sulfate reducers: evidence for a novel branch of archaebacteria. Science (Wash., D.C.) *236*: 822–824.

Woese, C.R., L. Achenbach, P. Rouvière and L. Mandelco. 1991. Archaeal phylogeny: reexamination of the phylogenetic position of *Archaeoglobus fulgidus* in light of certain composition-induced artifacts. Syst. Appl. Microbiol. *14*: 364–371.

DIFFERENTIATION OF THE SPECIES OF THE GENUS *ARCHAEOGLOBUS*

Three species of the genus *Archaeoglobus* are described: *A. fulgidus*, *A. profundus*, and *A. veneficus*. "*A. lithotrophicus*", a strain isolated from oil reservoirs below the bed of the North Sea, is so far not validly published (Stetter et al., 1993).

Key to the species of the genus *Archaeoglobus*

1. Dissimilatory sulfate (sulfite) and thiosulfate reduction; facultatively chemolithoautotrophic; the mol% G + C of the DNA is 46.

   A. fulgidus.

2. Dissimilatory sulfate and thiosulfate reduction; obligately chemolithoheterotrophic, requiring H_2 and an organic carbon source; the mol% G + C of the DNA is 41.

   A. profundus.

3. Dissimilatory sulfite and thiosulfate reduction; no growth with sulfate as electron acceptor; facultatively chemolithoautotrophic; the mol% G + C of the DNA is 45.

   A. veneficus.

4. Dissimilatory sulfate (sulfite) and thiosulfate reduction; obligately chemolithotrophic, addition of complex organic material results in slow growth and low final cell densities; the mol% G + C of the DNA is 40.

   "*A. lithotrophicus*".

List of species of the genus Archaeoglobus

1. **Archaeoglobus fulgidus** Stetter 1988b, 328[VP] (Effective publication: Stetter 1988a, 172.)

   ful′gi.dus. L. masc. adj. *fulgidus* shining, on account of its fluorescence under the UV microscope.

   Description same as for the genus. Optimum growth at 83°C. Isolated from the geothermally heated sea floor at Vulcano, Italy.

   The mol% G + C of the DNA is: 46 (T_m).
   Type strain: VC-16, DSMZ 4304.
   GenBank accession number (16S rRNA): X05567, Y00275.

2. **Archaeoglobus profundus** Burggraf, Jannasch, Nicolaus and Stetter 1990c, 320[VP] (Effective publication: Burggraf, Jannasch, Nicolaus and Stetter 1990b, 27.)

   pro.fun′dus. L. masc. adj. *profundus* deep, living within the depths of the oceans.

   Highly irregular cocci, sometimes triangular. Occur singly and in pairs. Gram negative. Motility and flagella not observed. Blue-green fluorescence at 420 nm. Temperature for growth: 65–90°C (optimum ~82°C); pH 4.5–7.5; 0.9–3.6% NaCl. Strictly anaerobic. Mixotrophic. H_2 obligately required. Acetate, lactate, pyruvate, yeast extract, meat extract, peptone, and (acetate-containing) crude oil may serve as organic carbon sources. Sulfate, thiosulfate, and sulfite serve as electron acceptors for growth. S^0 is inhibitory to growth. Cell envelope consists of an S-layer. Resistant to rifampicin and streptolydigin. No significant DNA–DNA hybridization with *Archaeoglobus fulgidus*. Isolated from the deep sea hydrothermal system off Guaymas, Mexico.

   The mol% G + C of the DNA is: ~41 (T_m).
   Type strain: AV 18, DSMZ 5631.

3. **Archaeoglobus veneficus** Huber, Jannasch, Rachel, Fuchs and Stetter 1998a, 327[VP] (Effective publication Huber, Jannasch, Rachel, Fuchs and Stetter 1997, 379.)

   ve.ne′fi.cus. L. masc. adj. *veneficus* the poison mixer, to give credit to its ability to convert one toxic compound (sulfite) into another (sulfide).

   Cells are highly irregular cocci to triangular plate-shaped lobes, 0.5–1.2 μm in diameter, occurring singly or in pairs. Gram negative. Motile by polar flagella. Blue-green fluorescence at 420 nm. Growth temperature, 65–85°C (optimum, 80°C); pH 6.5–8.0 (optimum, pH 7); 0.5–4% NaCl (optimum, 2% NaCl). Shortest doubling time, 1 h. Obligately anaerobic. Chemolithoautotrophic growth with molecular hydrogen using sulfite or thiosulfate as electron acceptors. Organotrophic sulfite respiration with formate, acetate, pyruvate, isopropanol, ethanol, fumarate, and glucose. H_2S (up to 6 μmol/ml) and methane (up to 2 nmol/ml) are formed during growth. Sulfate, nitrate, or nitrite cannot serve as electron acceptors.

Elemental sulfur inhibits growth. Resistant to ampicillin, vancomycin, rifampicin, streptolydigin. Cell envelope consists of a surface layer composed of hexagonally arranged complexes (center-to-center distance, 19 nm). Insignificant DNA–DNA hybridization to *A. fulgidus* and *A. profundus*. Based on 16S rRNA sequence comparison 1.9% phylogenetic distance to *A. fulgidus*. Isolated from a black smoker wall, "Snake Pit" site, Mid-Atlantic Ridge.

The mol% G + C of the DNA is: 45 (T_m and direct analysis).

Type strain: SNP6, DSMZ 11195.

GenBank accession number (16S rRNA): Y10011.

4. **"Archaeoglobus lithotrophicus"** Stetter, Huber, Blöchl, Kurr, Eden, Fiedler, Cash and Vance 1993.

li.tho.tro' phi.cus. Gr. masc. n. *lithos* stone; Gr. masc. n. *trophos* one who feeds; M.L. masc. adj. *lithotrophicus* grows lithotrophically.

Regular to highly irregular cocci, cell diameter 0.4–1.2 µm. Occur singly and in pairs. Gram negative. Peritrichous flagellated. Blue-green fluorescence at 420 nm. Temperature for growth, 55–87°C (optimum, ~80°C); pH 6.0–8.5; 0.6–4.8% NaCl (optimum, 1.2% NaCl). Strictly anaerobic. Obligately chemolithoautotrophic. Sulfate, thiosulfate, and, for a few strains, sulfite serve as electron acceptors. Growth significantly decreased by the addition of yeast extract, meat extract, peptone, or Casamino acids. S^0 is inhibitory to growth. Cell envelope consists of an S-layer in hexagonal array. No significant DNA–DNA hybridization with *A. fulgidus* and *A. profundus*. Isolated from production fluids of oil reservoirs at the East Shetland Basin (North Sea).

The mol% G + C of the DNA is: ~40 (T_m).

Deposited strain: TF2.

Genus II. **Ferroglobus** *Hafenbradl, Keller, Dirmeier, Rachel, Rossnagel, Burggraf, Huber and Stetter 1997, 601^{VP} (Effective publication: Hafenbradl, Keller, Dirmeier, Rachel, Rossnagel, Burggraf, Huber and Stetter 1996, 312)*

DORIS HAFENBRADL AND KARL O. STETTER

Fer.ro.glo' bus. L. masc. n. *ferrum* iron; L. masc. n. *globus* ball; M.L. masc. n. *Ferroglobus* the iron ball.

Cells irregular coccoid, 0.7–1.3 µm in diameter. Motile by monopolar flagella. Weak **blue-green fluorescence at 420 nm. Optimal growth at ~85°C, pH 7.0**, and 2% NaCl. **Strictly anaerobic. Facultatively chemolithoautotrophic.** Growth by **oxidation of Fe(II)**, S^{2-}, and H_2; **nitrate used as electron acceptor**, and, in the presence of H_2, also with $S_2O_3^{2-}$ as electron acceptor. S-layer envelope.

The mol% G + C of the DNA is: ~43.

Type species: **Ferroglobus placidus** Hafenbradl, Keller, Dirmeier, Rachel, Rossnagel, Burggraf, Huber and Stetter 1997, 601 (Effective publication: Hafenbradl, Keller, Dirmeier, Rachel, Rossnagel, Burggraf, Huber and Stetter 1996, 313.)

FURTHER DESCRIPTIVE INFORMATION

Sulfate cannot be used as an electron acceptor. Inhibition of growth by elemental sulfur. Cell membrane contains phytanyl ether lipids. CO_2 fixation during autotrophic growth occurs via the carbon monoxide dehydrogenase pathway (Vorholt et al., 1997). *Ferroglobus* has a denitrifying capacity. By phylogenetic analysis of the 16S rRNA sequence, related to the genus *Archaeoglobus* and a member of the *Archaeoglobaceae* (phylogenetic distance to *Archaeoglobus fulgidus*, 4.2%).

Ferroglobus has so far been isolated only from a shallow marine hydrothermal system at Vulcano Island, Italy.

ENRICHMENT AND ISOLATION PROCEDURES

Selective enrichment of *Ferroglobus* can be achieved by using anaerobic "FM" medium (Hafenbradl et al., 1996) with nitrate as an electron acceptor and Fe(II) as an energy source at high temperatures (85°C). After about 4 d growth of irregular cocci can be observed. The organisms can be purified by repeated serial dilutions or alternatively by using the "optical tweezer" technique (Huber et al., 1995a).

MAINTENANCE PROCEDURES

Cultures of *Ferroglobus* are routinely grown at 85°C and then transferred into fresh medium. They can be stored at room temperature but have to be transferred every 2 d. Stock cultures containing 5% dimethylsulfoxide can be stored at −140°C over liquid nitrogen for longer periods.

DIFFERENTIATION OF THE GENUS *FERROGLOBUS* FROM OTHER GENERA

Ferroglobus belongs to the archaeal order of the *Archaeoglobales* due to its morphology, the anaerobic mode of life, the G + C content of its genomic DNA, the blue-green fluorescence at 420 nm, the possession of phytanyl ether lipids in the membrane, the S-layer envelope, and its 16S rRNA sequence. In contrast to the representatives of the genus *Archaeoglobus*, growth occurs via oxidation of ferrous iron, molecular hydrogen, and sulfide with nitrate as an electron acceptor. With thiosulfate and molecular hydrogen, H_2S is produced. No growth occurs with sulfate or sulfite as an electron acceptor.

TAXONOMIC COMMENTS

By 16S rRNA sequencing, the closest relatives of *Ferroglobus* are the representatives of *Archaeoglobus*. For the correct position in the 16S rRNA-based phylogenetic tree the sequence data have to be calculated by transversion analysis, which is also true for members of the genus *Archaeoglobus* (Woese et al., 1991). These calculations reveal that *Ferroglobus* branches between the *Methanobacteriales* and the *Methanomicrobiales* and extreme halophiles (Hafenbradl et al., 1996).

FURTHER READING

Hafenbradl, D., M. Keller, R. Dirmeier, R. Rachel, P. Rossnagel, S. Burggraf, H. Huber and K.O. Stetter. 1996. *Ferroglobus placidus* gen. nov., sp. nov., a novel hyperthermophilic archaeum that oxidized Fe^{2+} at neutral pH under anoxic conditions. Arch. Microbiol. *166*: 308–314.

DIFFERENTIATION OF THE SPECIES OF THE GENUS *FERROGLOBUS*

Only one species exists so far: *Ferroglobus placidus.*

List of species of the genus Ferroglobus

1. **Ferroglobus placidus** Hafenbradl, Keller, Dirmeier, Rachel, Rossnagel, Burggraf, Huber and Stetter 1997, 601[VP] (Effective publication: Hafenbradl, Keller, Dirmeier, Rachel, Rossnagel, Burggraf, Huber and Stetter 1996, 313.)

pla'ci.dus. L. masc. adj. *placidus* peace loving (because of reduction of nitrate, a component of gun powder).

Cells are irregular cocci, about 0.7–1.3 µm in diameter, occurring singly and in pairs, sometimes in aggregates. Growth between 65°C and 95°C (optimum, 85°C), 0.5–4.5% NaCl (optimum, 1.8–2.0% NaCl), and pH 6.0–8.5 (optimum, pH 7.0). Strictly lithotrophic, strictly anaerobic. Fe(II), H_2, and sulfide serve as electron donors. Reduction of nitrate leads to formation of nitrite and nitrous gases. Reduction of thiosulfate in the presence of H_2 by formation of H_2S. Sulfur is inhibitory to growth. Pyruvate, acetate, and yeast extact do not serve as electron donors for nitrate reduction but stimulate growth. Core lipids consist of 2,3-di-*O*-phytanyl-*sn*-glycerol and glycerol-dialkyl-glycerol.

See the generic description for features.

The mol% G + C of the DNA is: 43 (direct analysis).

Type strain: AEDII12DO, DSMZ 10642.

GenBank accession number (16S rRNA): X99565.

Class VII. **Methanopyri** *class. nov.*

GEORGE M. GARRITY AND JOHN G. HOLT

Me.tha.no.py'ri. M.L. fem. pl. n. Methanopyrales type order of the class; dropping the ending to denote a class; M.L. fem. pl. n. Methanopyri the class of Methanopyrales.

Description of the class is the same as that for the order *Methanopyrales.*

Type order: **Methanopyrales** ord. nov.

Order I. **Methanopyrales** *ord. nov.*

ROBERT HUBER AND KARL O. STETTER

Me.tha.no.py.ra'les. M.L. masc. n. Methanopyrus type genus of the order; -ales the ending to denote an order; M.L. fem. pl. n. Methanopyrales the order of Methanopyrus.

Description is the same as that of *Methanopyraceae.*

Type genus: **Methanopyrus** Kurr, Huber, König, Jannasch, Fricke, Trincone, Kristjansson and Stetter 1992, 327 (Effective publication Kurr, Huber, König, Jannasch, Fricke, Trincone, Kristjansson and Stetter 1991, 245.)

Family I. **Methanopyraceae** *fam. nov.*

ROBERT HUBER AND KARL O. STETTER

Me.tha.no.py.ra'ce.ae. M.L. masc. n. Methanopyrus type genus of the family; -aceae the ending to denote a family; M.L. fem. pl. n. Methanopyraceae the family of Methanopyrus.

Rod-shaped cells with a **Gram-positive cell wall structure**. Strictly anaerobic. **Cell walls composed of pseudomurein. Growth is chemoautotrophic with H_2/CO_2 converted to methane as energy forming reaction. Growth at 100°C and above, but no growth below 80°C.** Dominant core lipid is 2,3-di-*O*-geranylgeranyl-*sn*-glycerol (Hafenbradl et al., 1993).

Type genus: **Methanopyrus** Kurr, Huber, König, Jannasch, Fricke, Trincone, Kristjansson and Stetter 1992, 327 (Effective publication Kurr, Huber, König, Jannasch, Fricke, Trincone, Kristjansson and Stetter 1991, 245.)

Genus I. **Methanopyrus** *Kurr, Huber, König, Jannasch, Fricke, Trincone, Kristjansson and Stetter 1992, 327^VP (Effective publication: Kurr, Huber, König, Jannasch, Fricke, Trincone, Kristjansson and Stetter 1991, 245)*

ROBERT HUBER AND KARL O. STETTER

Me.tha.no.py' rus. M.L. neut. n. *methanum* methane; Gr. neut. n. *pyr* fire; M.L. masc. n. *Methanopyrus* the "methane fire" (the hyperthermophilic methanogen).

Rod-shaped cells, usually occurring singly and in chains, about 2–14 µm long and 0.5 µm in diameter (Figs. A2.49, A2.50). Nonsporulating. **Cell division by septum formation**. Double-layered surface coat. Cell wall composed of pseudomurein. **AB′B″C subunit pattern of the DNA-dependent RNA polymerase. 2,3-di-*O*-phytanyl-*sn*-glycerol and 2,3-di-*O*-geranylgeranyl-*sn*-glycerol present in the lipids.** Gram positive. Motile. **Obligately anaerobic. Hyperthermophilic**, optimum growth temperature, 98°C; maximum, 110°C; minimum, 84°C. No growth at 80°C or below. Growth between 0.2 and 4% NaCl; optimum, 2% NaCl. Growth between pH 5.5 and 7; optimum, pH 6.5. Chemolithoautotrophic. **Energy for growth is obtained by reduction of CO₂ to CH₄ by using H₂ as the electron donor. Formate, acetate, methanol, methylamine, propanol, L(+)-lactate, and glycerol are not used as electron donors for CH₄ formation.** Epifluorescence at 420 nm. Cyclic 2,3-diphosphoglycerate present in high concentra-

tions. Sulfur reduced to H₂S, causing cell lysis. Based on 16S rDNA analysis, a separate and very deeply branching lineage within the euryarchaeal side of the archaeal tree.

The mol% G + C of the DNA is: 59–60.

Type species: **Methanopyrus kandleri** Kurr, Huber, König, Jannasch, Fricke, Trincone, Kristjansson and Stetter 1992, 327 (Effective publication: Kurr, Huber, König, Jannasch, Fricke, Trincone, Kristjansson and Stetter 1991, 245.)

FURTHER DESCRIPTIVE INFORMATION

Phylogenetic studies by 16S rDNA sequence comparisons showed that *Methanopyrus kandleri* represents a very deep branch-off within the *Euryarchaeota* and is unrelated to any other methanogen known (Huber et al., 1989c; Burggraf et al., 1991). On the basis of the phylogenetic divergence, the new order *Methanopyrales* was proposed (Kurr et al., 1991). The uniqueness of *M.*

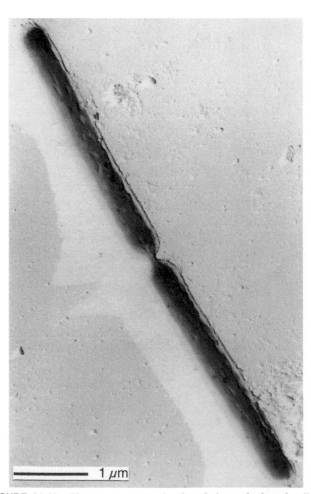

FIGURE A2.49. Electron micrograph of a platinum-shadowed cell of *Methanopyrus kandleri.*

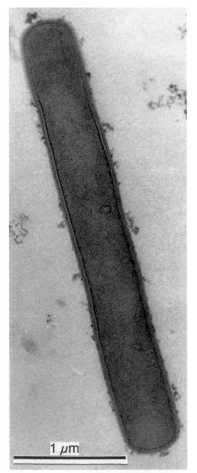

FIGURE A2.50. Ultrathin section of a single cell of *Methanopyrus kandleri.*

kandleri is further supported by (1) a phylogenetic analysis of the methyl coenzyme M reductase operons (Nölling et al., 1996); (2) a different posttranscriptional modification pattern of the tRNAs (Palmer et al., 1992); (3) the presence of a terpenoid lipid, which is considered to be a rather primitive feature from the viewpoint of evolution (Hafenbradl et al., 1993); and (4) the novel DNA topoisomerase V, a relative of the eucaryotic topoisomerase I (Slesarev et al., 1993 1994). Common to all methanogens is the pathway of CO_2 reduction to CH_4, which in *M. kandleri* has been shown to be identical to that used in other methanogens (Klein et al., 1993; Thauer et al., 1993).

Cells of *M. kandleri* are rod shaped, growing singly or in long chains with up to 70 individuals. Under shaking, flocs and string-like aggregates are formed, consisting of millions of cells in raft-like array. Cells are very rigid and do not lyse in the presence of Brij 58, sodium desoxycholate, or proteinase K.

The pseudomurein of *M. kandleri* is composed of L-glutamic acid, L-alanine, L-lysine, L-ornithine, *N*-acetylgalactosamine, and talosaminuronic acid. *N*-acetylglucosamine is not present. The core lipids consist of 2,3-di-*O*-phytanyl-*sn*-glycerol and of a novel unsaturated glycerol ether lipid. This lipid was identified as 2,3-di-*O*-geranylgeranyl-*sn*-glycerol and is the first unsaturated ether lipid found in *Archaea* (Hafenbradl et al., 1993). The surface coat is composed of two layers (total width, ~10–15 nm). On the surfaces of many cells, a "fuzzy" coat with an unknown composition can be seen. There is no evidence that the protein on the cell surface is regularly arrayed on a two-dimensional lattice and, therefore, could *bona fide* be called a "S-layer" (Rachel, 1999). Molecular weights of RNA polymerase subunits are: A, 107,000 Da; B′, 76,000 Da; B′′, 62,500 Da; and C, 59,000 Da.

Greenish, irregular colonies with a diameter of about 1 mm are formed after 7 d of anaerobic incubation at 98°C.

The pathway of CO_2 reduction to CH_4 is identical to that used in other methanogens (Klein et al., 1993; Thauer et al., 1993). Most of the corresponding enzymes are dependent on the presence of high concentrations of lyotrophic salts, e.g., $(NH_4)_2SO_4$, for activity and/or thermostability (Klein et al., 1993; Thauer et al., 1993; Ermler et al., 1997; Shima et al., 1998b). *M. kandleri* contains high intracellular concentrations of cyclic 2,3-diphospho-glycerate (1.1 M cDPG) (Huber at al., 1989c). Recently, it was shown that potassium cDPG has a function in the activation and thermostabilization of methanogenic enzymes (Shima et al., 1998a).

A thermosome was isolated from *M. kandleri* which shows high amino acid sequence similarity with the members of the group II chaperonins. It has barrel-shaped symmetry, and the complex appears to be a homo-oligomer composed of two rings with 8-fold symmetry (Andrä et al., 1996; Minuth et al., 1999).

M. kandleri was originally isolated from samples taken at the Guaymas hot vents at a depth of 2000 m. Further *Methanopyrus* isolates were obtained from the Kolbeinsey Ridge north of Iceland (depth; 106 m) and recently from hydrothermally heated black smoker walls and sediments from the TAG site and Snake Pite site at the Mid-Atlantic Ridge (depth; 3650 m) (R. Huber and K.O. Stetter, unpublished).

ENRICHMENT AND ISOLATION PROCEDURES

M. kandleri can be selectively enriched at an incubation temperature of 100°C in anaerobic seawater medium (Kurr et al., 1991) at pH 6.5 with a gas phase consisting of 300 kPa H_2/CO_2 (80:20 [v/v]). The isolates are obtained by plating under anaerobic conditions, using a stainless steel anaerobic jar (Balch et al., 1979) and plates solidified with Gelrite. The incubation temperature is 98°C, and colonies appear within 7 d. *Methanopyrus* species can be also isolated by a newly developed plating-independent isolation method (i.e., "selected cell cultivation" technique), using a strongly focused infrared laser beam as "optical tweezers" (Huber et al., 1995a; Beck and Huber, 1997).

MAINTENANCE PROCEDURES

Long term storage of *M. kandleri* in liquid nitrogen at −140°C in the presence of 5% dimethylsulfoxide is possible.

DIFFERENTIATION OF THE GENUS *METHANOPYRUS* FROM OTHER GENERA

M. kandleri can be differentiated from most members of the rod-shaped *Methanobacteriales* by its growth only at temperatures above 84°C (Balch et al., 1979). *M. kandleri* can be distinguished from the hyperthermophilic family *Methanothermaceae* within the *Methanobacteriales* by a much higher G + C content (33 mol% versus 60 mol%), insignificant DNA similarity, the lack of cyclic dibisphytanyl tetraethers, the lack of *N*-acetylglucosamine in the pseudomurein, and its high growth temperature (up to 110°C). For further differentiation, see the introductory chapter, Overview: A Phylogenetic Backbone and Taxonomic Framework for Procaryotic Systematics by Wolfgang Ludwig and Hans-Peter Klenk.

FURTHER READING

Thauer, R.K. 1998. Biochemistry of methanogenesis: a tribute to Marjory Stephenson. Microbiology *144*: 2377–2406.

Stetter, K.O 2000. Volcanoes, hydrothermal venting, and the origin of life. *In* Marit and Ernst (Editors), Volcanoes and the Environment, Cambridge University Press, Cambridge, in press.

List of species of the genus Methanopyrus

1. **Methanopyrus kandleri** Kurr, Huber, König, Jannasch, Fricke, Trincone, Kristjansson and Stetter 1992, 327[VP] (Effective publication: Kurr, Huber, König, Jannasch, Fricke, Trincone, Kristjansson and Stetter 1991, 245.)

   kand′le.ri. M.L. gen. n. *kandleri* honoring the microbiologist and botanist Otto Kandler.

   Description is the same as for the genus.
   The mol% G + C of the DNA is: 59 (T_m) and 60 (HPLC).
   Type strain: AV19, DSMZ 6324.
   GenBank accession number (16S rRNA): M59932.

DOMAIN *BACTERIA*

Phylum BI. Aquificae *phy. nov.*

ANNA-LOUISE REYSENBACH

A.qui.fi' cae. M.L. fem. pl. n. *Aquificales* type order of the phylum; dropping the ending to denote a phylum; M.L. fem. pl. n. *Aquificae* the phylum of *Aquificales.*

Rod-shaped, moderately thermophilic to hyperthermophilic. Isolated from marine and terrestrial hydrothermal springs. Includes microaerophilic chemolithotrophic hydrogen oxidizers, and chemoorganotrophs. Most members have been reported to be motile. All members, with the exception of *Hydrogenobacter acidophilum*, grow best between pH 6.0 and 8.0.

Additional characterization of the novel lineage represented by the genus *Desulfurobacterium* may result in the placement of this genus as a representative of a new phylum. This latter genus forms a distinct 16S rRNA lineage between the *Thermotogales* and *Aquificales*, and it is the only example of a strict chemolithoautotrophic, anaerobic, thermophilic sulfur reducing member of the domain *Bacteria.*

Type order: **Aquificales** ord. nov.

Class I. **Aquificae** *class. nov.*

ANNA-LOUISE REYSENBACH

A.qui.fi' cae. M.L. fem. pl. n. *Aquificales* type order of the class; dropping the ending to denote a class; M.L. fem. pl. n. *Aquificae* the class of *Aquificales.*

The description is the same as that of the phylum *Aquificae.*

Type order: **Aquificales** ord. nov.

Order I. **Aquificales** *ord. nov.*

ANNA-LOUISE REYSENBACH

A.qui.fi.ca' les. M.L. masc. n. *Aquifex* type genus of the order; *-ales* ending to denote an order; M.L. fem. pl. n. *Aquificales* the order of *Aquifex.*

Thermophilic motile and nonmotile rods that vary from 0.2 to 6 µm in length. Gram negative. Spores not formed. Long filamentous forms may develop under some growth conditions. All members are capable of chemolithotrophic microaerophilic growth using H_2, O_2 and CO_2. All isolates grow best at 70°C or above and are found in terrestrial and shallow marine thermal springs.

Type genus: **Aquifex** Huber and Stetter 1992e, 656 (Effective publication: Huber and Stetter *in* Huber, Wilharm, Huber, Trincone, Burggraf, König, Rachel, Rockinger, Fricke and Stetter 1992b, 349).

Key to the families of the Order Aquificales

1. Rods or filaments ~0.5 µm in diameter and generally up to 3.0 µm in length. Chemolithotrophic and microaerophilic growth best at 70°C or above, on CO_2, H_2, and O_2. Mol% G + C is 35–48.
Family I. *Aquificaceae.*

Family I. **Aquificaceae** *fam. nov.*

ANNA-LOUISE REYSENBACH

A.qui.fi.ca' ce.ae. M.L. masc. n. *Aquifex* type genus of the family; *-aceae* ending to denote a family; M.L. fem. pl. n. *Aquificaceae* the family of *Aquifex.*

Description is the same as for the order.

Type genus: **Aquifex** Huber and Stetter 1992e, 656 (Effective publication: Huber and Stetter *in* Huber, Wilharm, Huber, Trin- cone, Burggraf, König, Rachel, Rockinger, Fricke and Stetter 1992b, 349).

Key to the genera of the Family Aquificaceae

1. Motile rods up to 6 μm in length. Marine hyperthermophile. Grows best at 85°C. Mol% G + C is ~47.

    Genus I. *Aquifex.*

2. Nonmotile or motile rods up to 3.0 μm long. Terrestrial thermophile grows best at 74–76°C. Mol% G + C is ~40.

    Genus II. *Calderobacterium*, p. 362

3. Motile or nonmotile rod. Thermophile grows best at 65–75°C. Mol% G + C is ~35–46.

    Genus III. *Hydrogenobacter*, p. 363

4. Motile, nonmarine rod up to 3.0 μm in length. Can form filaments. Grows best at 80°C. Mol% G + C is ~47–48. Cannot grow anaerobically by nitrate reduction. Can grow chemoorgano- trophically on formate and formamide.

    Genus IV. *Thermocrinis*, p. 364

Genus I. **Aquifex** *Huber and Stetter 1992e, 656^VP (Effective publication: Huber and Stetter* in *Huber, Wilharm, Huber, Trincone, Burggraf, König, Rachel, Rockinger, Fricke and Stetter 1992b, 349)*

ROBERT HUBER AND KARL O. STETTER

A' qui.fex. L. fem. n. *aqua* water; L. v. *facere* to make; M.L. masc. n. *Aquifex* the water maker.

Rod-shaped cells with rounded ends, about 2–6 μm long and 0.4–0.5 μm in diameter. Cells occur singly, in pairs, and in ag- gregates of up to about 100 individuals (Figs. B1.1 and B1.2). Wedge-shaped refractile areas in the cells are formed during growth. Nonsporulating. Complex cell envelope, consisting of a **peptidoglycan layer (murein type A1γ)**, an outer membrane, and a surface layer protein (hexagonal lattice, center-to-center spac- ing 18 nm). **Diaminopimelic acid present** in the murein. The main components of the complex lipids are an aminophospho- lipid, a glycolipid, and a phospholipid. The **core lipids consist mainly of alkyl ethers of glycerol**. Traces of fatty acids present,

phytanyl-ether lipids absent. **Gram-negative**. Motile by polytri- chous flagellation. **Facultatively aerobic. Microaerophilic. Hy- perthermophilic**, optimum growth temperature: 85°C; maxi- mum: 95°C; minimum: 67°C. Growth at 1–5% NaCl; optimum: 3%. Growth at pH 5.4–7.5; optimum: pH 6.8. **Strictly chemolith- oautotrophic. CO_2 is fixed via the reductive citric acid cycle** (Beh et al., 1993; Deckert et al., 1998). **Molecular hydrogen, thiosul- fate, and S^0 serve as electron donor and oxygen and nitrate as electron acceptors**. Sulfuric acid is formed from sulfur and thio- sulfate. In the presence of sulfur and hydrogen, H_2S is also pro- duced. From nitrate, nitrite and dinitrogen are formed.

Based on 16S rDNA sequence analysis, *Aquifex* represents a deep phylogenetic branch within the bacterial domain.

The mol% G + C of the DNA is: 47.3–46.9.

Type species: **Aquifex pyrophilus** Huber and Stetter 1992e, 656 (Effective publication: Huber and Stetter *in* Huber, Wilharm, Huber, Trincone, Burggraf, König, Rachel, Rockinger, Fricke and Stetter 1992b, 349.)

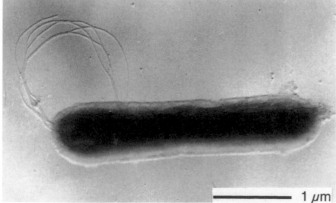

FIGURE B1.1.   Electron micrograph of a platinum-shadowed cell of *Aqui- fex pyrophilus.*

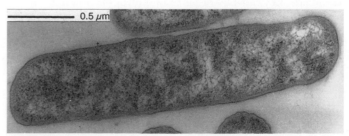

FIGURE B1.2.   Ultrathin section of *Aquifex pyrophilus.*

FURTHER DESCRIPTIVE INFORMATION

By 16S rDNA sequence comparisons, it was shown that *Aquifex pyrophilus* represents a very deep phylogenetic branch within the bacterial domain (Burggraf et al., 1992; Huber et al., 1992b). However, the analysis of individual protein sequences shows that the placement of *A. pyrophilus* is variable relative to other bacterial groups in the 16S rRNA phylogenetic tree (Acca et al., 1994; Klenk et al., 1994; Wetmur et al., 1994; Bocchetta et al., 1995; Brown and Dolittle, 1995; Baldauf et al., 1996; Pennisi, 1998). So far, the analysis of different genes from the completely sequenced genome gives no constant picture of the phylogeny of *Aquifex* (Deckert et al., 1998; Pennisi, 1998). Comparison of 16S rDNA sequences showed that the genera *Calderobacterium* and *Hydrogenobacter* (Kryukov et al., 1983; Kawasumi et al., 1984; Pitulle et al., 1994) and *Thermocrinis* (Huber et al., 1998c) also belong to the family of the *Aquificaceae* within the *Aquificales* order (Fig. B1.3) (Huber et al., 1992b). Together with representatives of the order *Thermotogales* (Huber and Stetter, 1992c), members of the *Aquificales* represent the bacteria with the highest growth temperatures known so far.

Under the UV microscope at 412 nm, *A. pyrophilus* exhibits a weak bluish green fluorescence, fading rapidly under UV radiation. The cells have polytrichous flagella and are highly motile, even after storage for two weeks at 4°C. The 19-nm-diameter flagellar filaments show prominent helical arrays of subunits and are composed of a 54 kDa flagellin monomer, which is highly thermostable (Behammer et al., 1995).

Colonies are round and brownish yellow; the colony size is about 1 mm in diameter.

From genomic DNA of *A. pyrophilus*, a physical map was established, and a random sequence analysis was performed (Shao et al., 1994; Choi et al., 1997). From *"Aquifex aeolicus"*, a second species within the genus *Aquifex*, the complete genome sequence was recently determined (Deckert et al., 1998). The genome contains 1,551,335 bp (G + C content: 43.4%) and is densely packed: most genes are apparently expressed in polycistronic operons. In addition, many convergently transcribed genes overlap slightly. Many genes of *"A. aeolicus"* are found dispersed throughout the genome or appear in novel operons. This is in contrast to other organisms, in which many genes are functionally grouped in operons. Genes coding for introns or inteins were not detected in the genome. A single, extrachromosomal element with a length of 39,456 bp (G + C content: 36.4%) was identified, present at roughly twice the copy number of the chromosome (Deckert et al., 1998).

So far, members of the genus *Aquifex* have been exclusively isolated from marine hydrothermal systems. From a marine hydrothermal vent system at the Kolbeinsey Ridge north of Iceland (depth: 106 m), *A. pyrophilus* was isolated (Huber et al., 1992b). *"A. aeolicus"* was obtained from a shallow marine hydrothermal system at Vulcano, Italy, sharing its biotope with *A. pyrophilus*. Recently, novel *Aquifex* relatives were isolated from anhydrite smokers with temperatures of up to 250°C, detected at a depth of 400 m near Grimsey Island, north of Iceland.

ENRICHMENT AND ISOLATION PROCEDURES

Members of the genus *Aquifex* can be selectively enriched at an incubation temperature of 90°C in seawater medium under microaerophilic culture conditions (Huber et al., 1992b). A pH ~6.5 and a gas phase consisting of 300 kPa $H_2/CO_2/O_2$ (79.75:19.75:0.5; by vol) is recommended. Addition of 0.05% sulfur or 0.2% sodium thiosulfate to the medium may enhance growth of *Aquifex* relatives. The enrichment bottles are incubated and are examined periodically over a period of 2 weeks by phase microscopy. When rod-shaped cells become visible, the enrichments are serially diluted. The isolates are obtained by plating under microaerophilic conditions, using a stainless steel anaerobic jar (Balch et al., 1979) and plates, solidified by 1.5% Gelrite. The incubation temperature is 85°C, and colonies with a diameter of 1–2 mm will appear within 5–10 days. *Aquifex* species grow well on Gelrite-plates and plating efficiencies of up to 100% were obtained with *"A. aeolicus"*. From liquid cultures, *Aquifex* species can be also isolated by a plating-independent, newly developed isolation method ("selected cell cultivation" technique), using a strongly focused infrared laser beam as "optical tweezers" (Huber et al., 1995a; Beck and Huber, 1997).

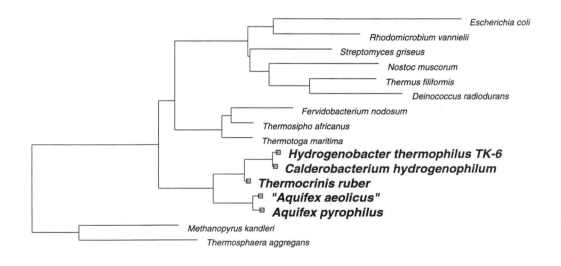

FIGURE B1.3. Phylogenetic position of members of the genus *Aquifex* to related bacteria. The tree was constructed by neighbor-joining analysis of 16S rDNA sequences. Bar = 0.10 fixed mutation per nucleotide position.

MAINTENANCE PROCEDURES

Storage of *A. pyrophilus* for several months at 4°C is possible. Storage in liquid nitrogen at −140°C in the presence of 5% DMSO is recommended for long-term preservation. No loss of cell viability of *A. pyrophilus* was observed after a storage over a period of 6 years.

DIFFERENTIATION OF THE GENUS *AQUIFEX* FROM OTHER GENERA

A. pyrophilus can be phylogenetically differentiated from *Calderobacterium*, *Hydrogenobacter*, and *Thermocrinis* by 16S rDNA sequence comparison. *A. pyrophilus* may also be differentiated from these genera by a higher maximal growth temperature and by the ability to grow by nitrate reduction under anaerobic conditions. *A. pyrophilus* can be further distinguished from *Hydroge-* *nobacter* and *Calderobacterium* by (a) its motility and presence of flagella, (b) the formation of H_2S from sulfur, (c) the presence of diaminopimelic acid in the murein, and (d) the composition of the lipids. *A. pyrophilus* can be further differentiated from *Thermocrinis* by its growth at higher salinity and by the presence of a regular surface lattice.

FURTHER READING

Deckert, G., P.V. Warren, T. Gaasterland, W.G. Young, A.L. Lenox, D.E. Graham, R. Overbeek, M.A. Snead, M. Keller, M. Aujay, R. Huber, R.A. Feldman, J.M. Short, G.J. Olsen and R.V. Swanson. 1998. The complete genome of the hyperthermophilic bacterium *Aquifex aeolicus*. Nature (Lond.) *392*: 353–358.

Stetter, K.O. 1998. Volcanoes, hydrothermal venting, and the origin of life. *In* Marit and Ernst (Editors), Volcanoes and the Environment, Cambridge University Press, Cambridge, in press.

List of species of the genus Aquifex

1. **Aquifex pyrophilus** Huber and Stetter 1992e, 656[VP] (Effective publication: Huber and Stetter *in* Huber, Wilharm, Huber, Trincone, Burggraf, König, Rachel, Rockinger, Fricke and Stetter 1992b, 349.)

    pyr.o'phi.lus. Gr. neut. n. *pyr* fire; Gr. adj. *philos* loving; M.L. adj. *pyrophilus* fire loving.

    Description is the same as for the species.

    The mol% G + C of the DNA is: 47.3 (HPLC) and 46.9 (T_m).

    Type strain: Kol5a, DSMZ 6858.

    GenBank accession number (16S rRNA): M83548.

2. **"Aquifex aeolicus"**

    "Aquifex aeolicus" can be distinguished from *Aquifex pyrophilus* by: (a) its obligate hydrogen requirement, (b) its inability to grow anaerobically by nitrate reduction, (c) insignificant DNA–DNA hybridization, and (d) about 4% lower G + C content of the DNA.

    The mol% G + C of the DNA is: 43.4 by total genome sequencing (Deckert et al., 1998).

    Deposited strain: VF5.

Genus II. **Calderobacterium** *Kryukov, Savel'eva and Pusheva 1984, 270*[VP] *(Effective publication: Kryukov, Savel'eva and Pusheva 1983, 787)*

ANNA-LOUISE REYSENBACH

Cal.de.ro.bac.te' ri.um. M.L. fem. n. *caldera,* from the Portuguese n. *caldera* cauldron; Gr. neut. n. *bakterion* a small rod; *Calderobacterium* a rod living in a cauldron.

Cells are **long rods about 0.35–0.5 × 2–8 μm. Nonmotile**, stain **Gram-negative. Cell division occurs by septation. Obligate chemolithotrophic thermophile** growing between **50 and 82°C.** Best growth occurs between **74 and 76°C at pH 6.0–7.0.** Cell walls contain peptidoglycan.

The mol% G + C of the DNA is: 39–41.

Type species: **Calderobacterium hydrogenophilum** Kryukov, Savel'eva and Pusheva 1984, 270 (Effective publication: Kryukov, Savel'eva and Pusheva 1983, 787.)

FURTHER DESCRIPTIVE INFORMATION

Liquid cultures appear grayish-white to opalescent, and darken to brown when old. Better growth is achieved by agitation. On solid agar media, the colonies are small, convex, and light yellow. In liquid medium, with H_2, O_2, and CO_2, the generation time is 2 h.

A constitutive, soluble, membrane-bound, NAD-specific, and thermostable hydrogenase was isolated from *Calderobacterium hydrogenophilum*. Maximum specific activity coincides with the early stationary growth phase.

The organism was isolated from bacterial mats in the thermal springs in Geyser valley in and caldera of the Uzon volcano, both in Kamchatka, Russia.

ENRICHMENT AND ISOLATION PROCEDURES

Calderobacterium can be isolated from terrestrial thermal springs using Schelgel mineral medium, under an atmosphere of H_2/ O_2/CO_2 (7:2:1) and incubated at 64–74°C. Colonies can be obtained on 3% agar plates.

DIFFERENTIATION OF THE GENUS *CALDEROBACTERIUM* FROM OTHER GENERA

Phylogenetic analysis using 16S rRNA sequences places *Calderobacterium* apart from *Aquifex*, but very close to *Hydrogenobacter thermophilus* strain TK-6 (98% similar). In addition, *Calderobacterium* has a slightly lower mol% G + C than *Hydrogenobacter thermophilus*, although other *Hydrogenobacter* spp. such as *H. acidophilus* and the proposed *"H. halophilus"* have mol% G + C values of 35 and 46, respectively.

TAXONOMIC COMMENTS

There is some discrepancy from the initial isolation report (Savel'eva et al., 1982), where the isolate is reported to also grow organotrophically on some fatty acids and amino acids.

FURTHER READING

Savel'eva, N.D., V.R. Kryukov and M.A. Pusheva. 1982. An obligate thermophilic hydrogen bacterium. Mikrobiologiya *51*: 765–769.

DIFFERENTIATION OF THE SPECIES OF THE GENUS *CALDEROBACTERIUM*

Only one species is described, the type species *Calderobacterium hydrogenophilum*.

List of species of the genus Calderobacterium

1. **Calderobacterium hydrogenophilum** Kryukov, Savel'eva and Pusheva 1984, 270[VP] (Effective publication: Kryukov, Savel'eva and Pusheva 1983, 787.)

hy.dro.ge.no.phi' lum. M.L. neut. n. *hydrogenium* hydrogen; Gr. adj. *philos* loving; M.L. neut. adj. *hydrogenophilum* hydrogen loving.

LSee generic description of features.

Nonspore-forming long rod. Neither carotenoid pigments nor poly-β-hydroxybutyric acid are produced. Does not re-quire vitamins for growth. Obligate chemolithotroph using H_2 as electron donor and CO_2 as carbon source. Ammonium salts and nitrates are used as nitrogen sources. Grows in bacterial mats in thermal springs in Kamchatka at temperatures ranging from 54 to 75°C.

The mol% G + C of the DNA is: 41 (HPLC).

Type strain: DSMZ 2913, INMI Z-829.

GenBank accession number (16S rRNA): Z30242.

Genus III. **Hydrogenobacter** *Kawasumi, Igarashi, Kodama and Minoda 1984, 9[VP]*

MASAHARU ISHII, TOSHIYUKI KAWASUMI, YASUO IGARASHI AND ANNA-LOUISE REYSENBACH

Hy.dro.ge.no.bac' ter. M.L. neut. n. *hydrogenium* hydrogen, that which produces water; M.L. n. *bacter* masc. form of Gr. neut. n. *bakterion* a rod; M.L. masc. n. *hydrogenobacter* hydrogen rod.

Straight rods, 0.3–0.5 × 2.0–3.0 μm, occurring singly or in pairs. **Gram-negative. Nonmotile**; cells do not possess flagella. Endospores absent. **Aerobic. Optimum temperature: 70–75°C**; maximum temperature: ~80°C. The optimum pH for growth is around neutrality. **Chemolithotrophic, using H_2 and thiosulfate as electron donor and CO_2 as carbon source.** Chemoorganotrophic growth has not been observed. Growth factors are not required.

A straight-chain saturated $C_{18:0}$ acid and a straight-chain unsaturated $C_{20:1}$ acid are the major components of the cellular fatty acids. 2-Methylthio-3-VI,VII-tetrahydromultiprenyl[7]-1,4-naphthoquinone (methionaquinone) is the major component of the quinone system.

The mol% G + C of the DNA is: 43.5–43.9.

Type species: **Hydrogenobacter thermophilus** Kawasumi, Igarashi, Kodama and Minoda 1984, 9.

FURTHER DESCRIPTIVE INFORMATION

No filamentous bacterial form is observed regardless of the growth temperature or phase.

All members of the genus can be easily cultivated on basal inorganic medium in an atmosphere consisting of $H_2/O_2/CO_2$ (75:15:10). Liquid cultivation is routinely used. The specific growth rate under autotrophic conditions at the optimum temperature is 0.33–0.42 (h[−1]). Ammonium and nitrate ions are used as nitrogen source, but urea and gaseous nitrogen are not. Nitrite inhibits growth. Assimilatory nitrate reduction and peroxidase are positive. Urease is negative.

None of the following organic compounds (0.1%) or media is used as sole sources of energy and carbon: glucose, fructose, galactose, maltose, sucrose, xylose, raffinose, L-rhamnose, D-mannose, D-trehalose, mannitol, starch, formate, acetate, propionate, pyruvate, succinate, malate, citrate, fumarate, maleate, glycolate, gluconate, DL-lactate, α-ketoglutarate, *p*-hydroxybenzoate, DL-β-hydroxybutyrate, betaine, methanol, ethanol, methylamine, dimethylamine, trimethylamine, glycine, L-glutamate, L-aspartate, L-serine, L-leucine, L-valine, L-tryptophan, L-histidine, L-alanine, L-lysine, L-proline, L-arginine, nutrient broth, yeast extract-malt extract medium, and brain-heart infusion. Under an atmosphere containing $CO/CO_2/O_2$ (90:5:5) no growth in the strains has been observed.

Strain TK-6 can grow anaerobically when H_2 and nitrate are used as an electron donor and an electron acceptor, respectively.

No oxygen uptake by intact cells of strain TK-6 has been detected with use of the following organic compounds (1 mM): glucose, fructose, acetate, pyruvate, citrate, α-ketoglutarate, succinate, fumarate, malate, oxalacetate, and glutamate.

Strain TK-6 fixes CO_2 by using a reductive tricarboxylic acid cycle (Fig. B1.4). The following enzymes and proteins involved in the cycle have been purified and characterized; ATP:citrate lyase (Ishii et al., 1989), ferredoxin (Ishii et al., 1996), 2-oxoglutarate:ferredoxin oxidoreductase (Yoon et al., 1996a), and pyruvate:ferredoxin oxidoreductase (Yoon et al., 1997). NADH:ferredoxin oxidoreductase, whose activity was detected in strain TK-6 (Yoon et al., 1996b), is believed to reduce ferredoxin *in vivo*.

Membrane-bound hydrogenase activity and NAD-dependent soluble hydrogenase activity have been observed in strain TK-6.

Type *b* and *c* cytochromes have been found in cell-free extracts of all the strains of the genus. In strain TK-6, both types of cytochrome and methionaquinone are reduced by membrane-bound hydrogenase reaction (Ishii et al., 1991).

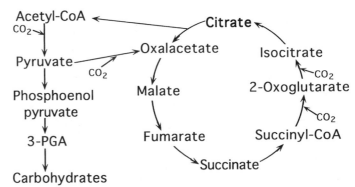

FIGURE B1.4. Reductive tricarboxylic acid cycle in *Hydrogenobacter thermophilus*.

The electrophoretic patterns of the total soluble proteins are essentially the same among all strains of the genus, although there are differences in the relative mobilities of several proteins.

All of the *Hydrogenobacter* strains described so far were isolated from hot (>45°C) water-containing soils from various hot springs in Japan.

ENRICHMENT AND ISOLATION PROCEDURES

H. thermophilus can be enriched using inorganic agar medium in an atmosphere containing $H_2/O_2/CO_2$ (85:5:10) at 65°C. Colonies that develop on plates are inoculated into the same liquid medium and gas mixture. After repeated liquid cultivation, *H. thermophilus* can be isolated in pure culture by the dilution method (Kawasumi et al., 1980). Once purified, none of the strains could grow as colonies on agar medium. This can be explained by the fact that the organism is sensitive to agar. However, strain TK-6 grows as a colony on Gelrite medium (Ishii et al., 1987).

MAINTENANCE PROCEDURES

Hydrogenobacter strains can be preserved for about 1 year by lyophilization using 0.1 M phosphate buffer (pH 7.0) containing 3% sodium glutamate as the suspending medium.

TAXONOMIC COMMENTS

Studies of the 16S rDNA sequences showed that the *Aquifex–Hydrogenobacter* complex belongs to a very deep branching order, the *Aquificales* (Pitulle et al., 1994).

List of species of the genus Hydrogenobacter

1. **Hydrogenobacter thermophilus** Kawasumi, Igarashi, Kodama and Minoda 1984, 9.[VP]

    ther.mo' phi.lus. Gr. n. *therme* heat; Gr. adj. *philos* loving; M.L. masc. adj. *thermophilus* heat-loving.

    The description of this species is the same as that of the genus.

    The mol% G + C of the DNA is: 43.5–43.9 (T_m).
    Type strain: TK-6, DSMZ 6534, IAM 12695.
    GenBank accession number (16S rRNA): Z30214.

2. **Hydrogenobacter acidophilus** Shima and Suzuki 1993, 707.[VP]

    a.ci.do' phi.lus. L. neut. adj. *acidum* acid; Gr. adj. *philos* loving; M.L. masc. adj. *acidophilus* acid loving.

    Short (1.3–1.8 μm long), motile, Gram-negative, non-sporulating rods. Grows best at 65°C and pH 3–4. Chemolithotrophic, using hydrogen and reduced sulfur compounds as electron donors and O_2 as the electron acceptor. Requires S^0 or thiosulfate for growth. Both soluble and membrane-bound hydrogenases present. CO_2 is fixed by the reductive TCA cycle. Methionaquinone is the major component of the quinone system. Straight chain saturated $C_{18:0}$ acid and straight chain unsaturated $C_{18:1}$ and $C_{20:1}$ acids are the major fatty acids. 16S rRNA phylogenetic analysis places the new species as a deep branch within the *Hydrogenobacter* lineage.

    The mol% G + C of the DNA is: 35 (HPLC).
    Type strain: 3H-1, JCM8795.
    GenBank accession number (16S rRNA): D16296.

3. **"Hydrogenobacter halophilus"** Nishihara, Igarashi and Kodama 1990, 297.

    hy.lo' phil.lus. Gr. n. *hals, halos* salt; Gr. adj. *philos* loving; M.L. masc. adj. *halophilus* salt-loving.

    This species grows optimally at around 0.3–0.5 M NaCl, while *H. thermophilus* cannot grow at such NaCl concentration levels. Furthermore, *"H. halophilus"* differs from *H. thermophilus* in the following characteristics: cell dimension: 0.3–0.5 × 1.0–2.5 μm; major components of the cellular fatty acids: $C_{18:0}$, $C_{18:1}$, and $C_{20:1}$; alternative energy sources: thiosulfate and S^0; distribution of hydrogenase: membrane-bound hydrogenase only; nitrogen source: ammonium ions only.

    The mol% G + C of the DNA is: 46.0 (HPLC).
    Deposited strain: TH-112.

Genus IV. **Thermocrinis** *Huber, Eder, Heldwein, Wanner, Huber, Rachel and Stetter 1999, 341*[VP] *(Effective publication: Huber, Eder, Heldwein, Wanner, Huber, Rachel and Stetter 1998c, 3581)*

ROBERT HUBER, WOLFGANG EDER AND KARL O. STETTER

Ther.mo.cri' nis. Gr. fem. n. *therme* heat; L. masc. n. *crinis* hair; M.L. masc. n. *Thermocrinis* hot hair.

Rod-shaped cells with rounded ends, about 1–3 × 0.4–0.5 μm. The rods occur singly, in pairs, in aggregates or as **long filaments**. Double-layered cell envelope. **Diaminopimelic acid present**. Non-sporulating. **Gram negative**. Motile by monopolar polytrichous flagellation. **Microaerophilic**. **Hyperthermophilic**, optimum growth temperature, 80°C; maximum 89°C; minimum, 44°C. Growth at 0–0.4% NaCl and at pH 7 and 8.5. Growth occurs on solid surfaces. **Chemolithoautotrophic or chemoorganoheterotrophic metabolism**.

Based on 16S rDNA analysis, *Thermocrinis* represents a deep phylogenetic branch within the *Bacteria*.

The mol% G + C of the DNA is: 47.2–47.8.

Type species: **Thermocrinis ruber** Huber, Eder, Heldwein, Wanner, Huber, Rachel and Stetter 1999, 341 (Effective publication: Huber, Eder, Heldwein, Wanner, Huber, Rachel and Stetter 1998c, 3582.)

FURTHER DESCRIPTIVE INFORMATION

Phylogenetic studies by 16S rDNA sequence comparisons showed that the genus *Thermocrinis* (Huber et al., 1998c) belongs, together with the genera *Aquifex* (Huber et al., 1992b), *Hydrogenobacter*, and *Calderobacterium* (Kryukov et al., 1983; Kawasumi et al., 1984; Pitulle et al., 1994), to the family *Aquificaceae* within the order *Aquificales* (Huber et al., 1992b). For further taxonomic comments see also Genus *Aquifex* in this *Manual*).

The morphology of *Thermocrinis ruber* depends strongly on the culture conditions. When grown in closed serum tubes, cells of *T. ruber* are rod-shaped, growing singly or in pairs during the logarithmic growth phase. In addition, cell aggregates composed of four to several hundred individuals can be observed in the late-logarithmic growth phase. The cells are able to attach on solid surfaces such as silicon-coated cover glasses or muscovite glimmer. After growth, the surfaces of these materials are tightly covered with rods, which occur singly or in pairs (Fig. B1.5). However, in a permanent flow of medium under exposure to air, *T. ruber* grows predominantly as long filaments, forming a pink network of cells (Fig. B1.6).

T. ruber cells have a double-layered cell envelope covering the membrane. The cell wall consists of a periplasmic space (most likely containing peptidoglycan) and an outer membrane. There is no evidence of large, well-ordered crystalline arrays indicative of S-layer proteins.

Brownish red, round colonies with a diameter of about 1 mm are formed after 7 days of microaerophilic incubation at 80°C on 1.5% Gelrite plates. The gas phase consists of N_2/O_2 (97:3; v/v), and thiosulfate serves as the electron donor.

T. ruber is able to grow chemolithoautotrophically or chemoorganoheterotrophically under microaerophilic culture conditions (maximum oxygen concentration: 6%, v/v). Molecular hydrogen, thiosulfate, S^0, formate, and formamide are used as electron donors, and oxygen serves as the electron acceptor. CO_2, formate, and formamide are used as single carbon sources. Ammonia is formed from formamide. Formaldehyde, methanol, L(+)-lactate, citrate, fumarate, acetate, propionate, pyruvate, succinate, glycolate, DL-malate, peptone, yeast extract, meat extract, and brain heart infusion are not used as growth substrates with an N_2/O_2 gas phase (97:3, v/v). At oxygen tensions below 3% (v/v) S^0 is reduced to H_2S in the presence of H_2. Cell masses cultivated in the presence of 3% oxygen (v/v) in the dark or under exposure to light exhibit a comparable pink color. Catalase activity is present.

T. ruber was isolated from biomass consisting of pink filaments from the upper outflow channel near the source pool of Octopus Spring (formerly Pool A) (Reysenbach et al., 1994; Brock, 1995), a hot spring of the thermal features of Yellowstone National Park, Wyoming. The upper end of the filament growth was about 1.5 m from the actual lip of the pool. Further *Thermocrinis* isolates, most likely representing new species within this genus, were obtained from terrestrial hot springs in Hveragerthi (Iceland) and from the Uzon Valley, Kamchatka (Russia).

ENRICHMENT AND ISOLATION PROCEDURES

T. ruber can be enriched at an incubation temperature of 80–85°C in microaerophilic low salt media (Huber et al., 1998c) at pH 7 with a gas phase consisting of 300 kPa $N_2/H_2/O_2$ (96:3:1; v/v) or 300 kPa N_2/O_2 (99.95:0.05; v/v) in the presence of formate or formamide. Enrichment cultures of relatives of *T. ruber* can be obtained from filamentous streamers thriving in hot springs and from hot ponds not containing visible streamers. *Thermocrinis* species can be isolated by a plating-independent, newly developed isolation method ("selected cell cultivation" technique), using a strongly focussed infrared laser beam as "optical tweezers" (Huber et al., 1995a; Beck and Huber, 1997).

MAINTENANCE PROCEDURES

For short-term storage, *T. ruber* can be maintained at room temperature or at 4°C for about 2 months. Long-term storage of *T. ruber* in liquid nitrogen at −140°C in the presence of 5% DMSO is possible.

DIFFERENTIATION OF THE GENUS *THERMOCRINIS* FROM OTHER GENERA

T. ruber can be phylogenetically differentiated from the other genera within the *Aquificales* by comparison of the 16S rDNA sequences. *T. ruber* differs from *Hydrogenobacter/Calderobacterium* by (a) the formation of H_2S from S^0, (b) a positive catalase reaction, and (c) the presence of diaminopimelic acid in the murein. In contrast to members of *Aquifex*, *T. ruber* is not able to grow at high salinity levels or anaerobically by nitrate reduction. *T. ruber* differs from *Aquifex*, *Hydrogenobacter*, and *Calderobacterium*, based on its ability to form streamers, by its lack of a regular surface lattice and by its chemoorganotrophic growth on formate and formamide.

FURTHER READING

Brock, T.D. 1967. Life at high temperatures. Science *158*: 1012–1019.

Brock, T.D. 1978. Thermophilic Microorganisms and Life at High Temperatures, Springer–Verlag, Heidelberg.

Stetter, K.O. 2000. Volcanoes, hydrothermal venting, and the origin of life. *In* Marit and Ernst (Editors), Volcanoes and the Environment, Cambridge University Press, in press.

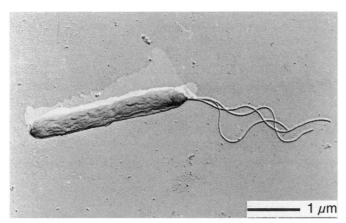

FIGURE B1.5. Platinum-shadowed, single, flagellated cell of *Thermocrinis ruber*.

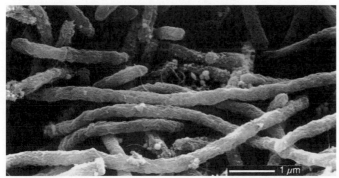

FIGURE B1.6. Scanning electron micrograph of *Thermocrinis ruber*.

1. **Thermocrinis ruber** Huber, Eder, Heldwein, Wanner, Huber, Rachel and Stetter 1999, 341^{VP} (Effective publication: Huber, Eder, Heldwein, Wanner, Huber, Rachel and Stetter 1998c, 3582.)

ru′ber. L. masc.adj. *ruber* referring to the cell color.

Description is the same as for the genus.

The mol% G + C of the DNA is: 47.2 (HPLC) and 47.8 (T_m).

Type strain: OC 1/4, DSMZ 12173.

GenBank accession number (16S rRNA): AJ005640.

Genus *Incertae Sedis* I. **Desulfurobacterium** *L'Haridon, Cilia, Messner, Raguénès, Gambacorta, Sleytr, Prieur and Jeanthon 1998, 709^{VP}*

STÉPHANE L'HARIDON AND CHRISTIAN JEANTHON

De.sul.fu.ro.bac.te′ri.um. L. pref. *de* from; L. n. *sulfur* sulfur; Gr. neut. dim. n. *bakterion* a small rod; M.L. neut. n. *Desulfurobacterium* sulfur-reducing rod-shaped bacterium.

Rod-shaped cells, about 0.4–0.5 × 1–2 μm. The rods occur singly or in pairs during the logarithmic growth phase. Double-layered cell envelope. Spores are not produced. **Gram-negative.** Highly motile with monopolar flagellation.

Strictly anaerobic. Thermophilic; temperature range for growth, 40–75°C; optimum growth at 70°C. pH range is 4.4–8.0; optimum pH 6.0–6.5. **NaCl dependent**; growth occurs between 1.5 and 7% sea salts. **Strict chemolithoautotrophic metabolism** by reduction of thiosulfate, S^0, or sulfite.

Based on 16S rDNA analysis, *Desulfurobacterium* represents a deep phylogenetic branch within the *Bacteria* and a line of descent that is equidistant from the previously described deeply branching thermophilic members of the orders *Thermotogales* and *Aquificales.*

Habitat: walls of a deep-sea hydrothermal vent chimney at 23°N (3500 m depth, mid-Atlantic ridge).

The mol% G + C of the DNA is: 35.0.

Type species: **Desulfurobacterium thermolithotrophum** L'Haridon, Cilia, Messner, Raguénès, Gambacorta, Sleytr, Prieur and Jeanthon 1998, 709.

FURTHER DESCRIPTIVE INFORMATION

Whitish, round colonies with a diameter of about 1 mm are formed after 3 days incubation at 65°C on 0.7% Phytagel plates (a gellan gum from Sigma). The gas phase consists of H$_2$/CO$_2$ (80:20, v/v; 200 kPa) and thiosulfate and polysulfides serve as the electron acceptors. Packed cell masses exhibit a pink color.

Under the microscope, cells of *Desulfurobacterium thermolithotrophum* appear to be highly motile, and up to three flagella can be observed by negative staining (Fig. B1.7). In the stationary growth phase, some rods become spherical. Freeze fracturing of intact cells show that the organisms are completely covered with an oblique S-layer lattice with center-to-center spacings of approximately 11.3 and 6.3 nm and an angle γ between the lattice vectors of approximately 80°. Ultrathin sections clearly demonstrate that *D. thermolithotrophum* cells possess the typical cell envelope profile of Gram-negative bacteria with a cytoplasmic membrane and an outer membrane. The S-layer on top of the outer membrane very frequently peels off, forming large loops.

D. thermolithotrophum grows anaerobically and chemolithoautotrophically. Thiosulfate, S^0, sulfite, and polysulfides are used as electron acceptors in the presence of H$_2$. Exponential H$_2$S production parallels growth. Sulfate, cystine, nitrate, and nitrite are not used. CO$_2$ is used as the sole carbon source. Acetate, formate,

methanol, monomethylamine, and yeast extract are not used as growth substrates in the presence or absence of sulfur with a N$_2$/CO$_2$ (80:20, v/v) or H$_2$ (100%) gas phase. The organism is unable to grow in the presence of oxygen even at low concentrations. Ammonium, nitrate, tryptone, and yeast extract are used as nitrogen sources. Catalase is not produced.

D. thermolithotrophum was isolated from a deep-sea hydrothermal vent chimney located at mid-Atlantic ridge (23°N). Additional *Desulfurobacterium* isolates, most likely representing new species within this genus, were obtained from deep-sea hydrothermal vent samples collected at 13°N (East Pacific Rise), 14°N (mid- Atlantic ridge), and Guaymas Basin (Gulf of California). *In situ* hybridization experiments demonstrated that this species and other phylogenetically closely related organisms represented up to 40% of the bacterial population inhabiting hydrothermal vent chimneys, suggesting that they may play a significant role within marine hydrothermal environments (Harmsen et al., 1997).

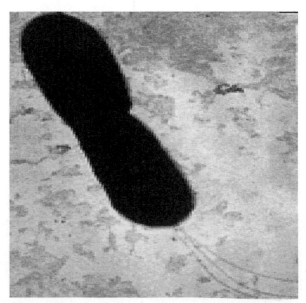

FIGURE B1.7. Negatively stained dividing cell of *Desulfurobacterium thermolithotrophum* showing polar flagella. (Reprinted with permission from L'Haridon et al., International Journal of Systematic Bacteriology *48:* 701–711, 1998, ©International Union of Microbiological Societies.)

ENRICHMENT AND ISOLATION PROCEDURES

D. thermolithotrophum can be enriched at 65–70°C in anaerobic marine medium (L'Haridon et al., 1998) at pH 6.5 with a gas phase consisting of 200 kPa H_2/CO_2 (80:20, v/v) in the presence of sulfur compounds.

MAINTENANCE PROCEDURES

For short-term storage, *D. thermolithotrophum* can be maintained at 4°C for about 2 months. Long-term storage of *D. thermolithotrophum* in liquid nitrogen is possible in the presence of 10% DMSO.

DIFFERENTIATION OF THE GENUS *DESULFUROBACTERIUM* FROM OTHER GENERA

D. thermolithotrophum can be phylogenetically differentiated from genera within the orders *Thermotogales* and *Aquificales* by comparison of the 16S rDNA sequences. Phenotypically, *D. thermo-*

lithotrophum differs from members of the order *Thermotogales*, which always produce a "sheath" and are strict heterotrophs (Huber and Stetter, 1992c). Although *D. thermolithotrophum* shares a diagnostic 16S rRNA secondary structural feature with members of the order *Aquificales* (Burggraf et al., 1992), it differs from the *Aquificales* in its inability to grow in microaerophilic conditions. Based on the 16S rRNA sequences, this novel lineage might be considered as a novel family.

FURTHER READING

Huber, R., T. Wilharm, D. Huber, A. Trincone, S. Burggraf, H. König, R. Rachel, I. Rockinger, H. Fricke and K.O. Stetter. 1992. *Aquifex pyrophilus* gen. nov. sp. nov., represents a novel group of marine hyperthermophilic hydrogen-oxidizing bacteria. Syst. Appl. Microbiol. *15*: 340–351.

Pitulle, C., Y. Yang, M. Marchiani, E.R.B. Moore, J.L. Siefert, M. Aragno, P. Jurtshuk, Jr. and G.E. Fox. 1994. Phylogenetic position of the genus *Hydrogenobacter*. Int. J. Syst. Bacteriol. *44*: 620–626.

List of species of the genus Desulfurobacterium

1. **Desulfurobacterium thermolithotrophum** L'Haridon, Cilia, Messner, Raguénès, Gambacorta, Sleytr, Prieur and Jeanthon 1998, 709.[VP]

   ther.mo.li.tho.tro'phum. Gr. adj. *thermos* hot; Gr. masc. n. *lithos* stone; Gr. masc. n. *trophos* one who feeds; M.L. neut. adj. *thermolithotrophum* referring to its thermophilic way of life and lithotrophic metabolism.

   Description is the same as for the genus.
   The mol% G + C of the DNA is: 35 (T_m).
   Type strain: BSA, DSMZ 11699.
   GenBank accession number (16S rRNA): AJ001049.

Phylum BII. Thermotogae *phy. nov.*

ANNA-LOUISE REYSENBACH

Ther.mo.to' gae. M.L. fem. pl. n. *Thermotogales* type order of the phylum; dropping the ending to denote a phylum; M.L. fem. pl. n. *Thermotogae* the phylum of *Thermotogales.*

Extremely thermophilic rod-shaped bacteria, non-sporeforming; Gram negative with an outer sheath-like envelope of "toga". The members are all anaerobic and fermentative. The phylum is represented by a single class and a single order.

Type order: **Thermotogales** ord. nov.

Class I. **Thermotogae** *class. nov.*

ANNA-LOUISE REYSENBACH

Ther.mo.to' gae. M.L. fem. pl. n. *Thermotogales* type order of the class; dropping the ending to denote a class; M.L. fem. pl. n. *Thermotogae* the class of *Thermotogales.*

The description is the same as that of the phylum *Thermotogae.*

Type order: **Thermotogales** ord. nov.

Order I. **Thermotogales** *ord. nov.* Huber and Stetter 1992c, 3809

ANNA-LOUISE REYSENBACH

Ther.mo.to.ga' les. M.L. fem. n. *Thermotoga* type genus of the order; *-ales* ending to denote an order; M.L. fem. pl. n. *Thermotogales* the order of *Thermotoga.*

Thermophilic Gram-negative rods, with sheath-like structure that balloons at the ends. Does not form spores. Meso-diaminopimelic acid not present in the peptidoglycan. Anaerobic heterotrophs that ferment a range of substrates including complex organics such as yeast extract, Trypticase, and sugars such as glucose and xylose. Hydrogen inhibits growth. Thiosulfate is reduced. Sensitive to lysozyme. Unusual long chain dicarboxylic fatty acids present in lipids.

Type genus: **Thermotoga** Stetter and Huber 1986, 575 (Effective publication: Stetter and Huber *in* Huber, Langworthy, König, Thomm, Woese, Sleytr and Stetter 1986, 332.)

Key to the families of the order "Thermotogales"

1. Gram-negative, rod-shaped organisms surrounded by a characteristic sheath-like outer structure ("toga") with terminal protuberances at the cell ends. Heterotrophic growth. Acetate, CO_2, and H_2 are products of glucose fermentation. Best growth at pH 6.5–7.5. Moderately thermophilic to hyperthermophilic.

Family I. *Thermotogaceae.*

369

Family I. **Thermotogaceae** *fam. nov.*

ANNA-LOUISE REYSENBACH

Ther.mo.to.ga' ce.ae. M.L. fem. n. *Thermotoga* type genus of the family; *-aceae* ending to denote a family; M.L. fem. pl. n. *Thermotogaceae* the family of *Thermotoga.*

Description is the same as for the order *Thermotogales.*

Type genus: **Thermotoga** Stetter and Huber 1986, 575 (Effective publication: Stetter and Huber *in* Huber, Langworthy, König, Thomm, Woese, Sleytr and Stetter 1986, 332.)

Key to the genera of the family Thermotogaceae

1. Rod-shaped cells, 1.5–11 μm in length. Best growth at 66–80°C. Found in marine, terrestrial hydrothermal systems and in oil reservoirs. 15,16-dimethyl-30-glyceroloxytriacontanoic acid is present. Mol% G + C is ranges from 39.6 to 50.
   Genus I. *Thermotoga.*
2. Rod-shaped cells, 1.0–40 μm in length. Very large (1–4 μm) terminal "bleb" or spheroid at on end of the cell. Best growth at 65–70°C. Maximum salt tolerance varies (0.55–6.0%). RNA polymerase is unique. Mol% G + C is ranges from 34 to 41.
   Genus II. *Fervidobacterium*, p. 375
3. Rods 3–20 μm in length. Best growth around 50°C. Grows in a broad salt range from 0.5 to 10.0% NaCl (optimum 3.0–4.0% NaCl). Together with *Petrotoga*, forms a fast evolving 16S rRNA lineage. Mol% G + C is 30.
   Genus III. *Geotoga*, p. 377
4. Rods vary in length from 2 to 7.5 μm. Best growth around 55°C. High salt tolerance (optimum 2.0% NaCl). Distinguished from *Geotoga* by 16S rRNA sequence. Mol% G + C is 40.
   Genus IV. *Petrotoga*, p. 382
5. Rods 1–4 μm, can form chains of up to 12 cells surrounded by a sheath. Grows best at 70–75°C. Maximum salt tolerance, 3% NaCl. RNA polymerase does not cross-react with that of *Thermotoga* and *Fervidobacterium*. Mol% G + C is 30.
   Genus V. *Thermosipho*, p. 385

Genus I. **Thermotoga** *Stetter and Huber 1986, 575VP (Effective publication: Stetter and Huber in Huber, Langworthy, König, Thomm, Woese, Sleytr and Stetter 1986, 332)*

ROBERT HUBER AND KARL O. STETTER

Ther.mo.to' ga. Gr. fem. n. *therme* heat; L. fem. n. *toga* Roman outer garment; M.L. fem. n. *Thermotoga* the hot outer garment.

Rod-shaped cells, occurring singly and in pairs, about 1.5–11 × 0.5–1 μm. The rods are surrounded by a typical, **sheath-like outer structure** ("toga"), ballooning over the ends (0.6 × 3.5– 14 μm). Resting stages are not known. **Gram negative.** Motile by flagellation. **Obligately anaerobic. Thermophilic or hyperthermophilic**, optimum growth temperature: 66–80°C. Optimal pH: 6.5–7.5. **Heterotrophic growth. Acetate, CO$_2$, and H$_2$ are metabolites from glucose fermentation. Thiosulfate is reduced to H$_2$S.** Molecular hydrogen inhibits growth. Sensitive to inhibitors of cell-wall, protein, and nucleic acid biosynthesis.

Based on 16S rDNA analysis, deep phylogenetic branch within the bacterial domain.

The mol% G + C of the DNA is: 39.6–50.

Type species: **Thermotoga maritima** Stetter and Huber 1986, 575 (Effective publication: Stetter and Huber *in* Huber, Langworthy, König, Thomm, Woese, Sleytr and Stetter 1986, 332.)

FURTHER DESCRIPTIVE INFORMATION

By 16S rDNA sequence comparisons, it was shown that *Thermotoga maritima* represents a slowly evolving lineage and a deep phylogenetic branch within the domain *Bacteria* (Fig. B2.1) (Huber et al., 1986; Huber and Stetter, 1992c; Woese, 1987). Based on these results, a thermophilic origin of all bacteria was suggested (Achenbach-Richter et al., 1987; Huber and Stetter, 1992d). By comparative analysis of such unrelated macromolecules as 23S rRNA, elongation factor Tu and G, β-subunit of the ATPase, *fus* gene, and ferredoxins, the phylogenetic placement was confirmed (Bachleitner et al., 1989; Schleifer and Ludwig, 1989; Tiboni et al., 1991; Ludwig et al., 1993; Blamey at al., 1994; Darimont and Sterner, 1994). One report, derived from sequence comparisons of bacterial RNA-polymerase large subunits, placed *T. maritima* next to the chloroplasts (Palm et al., 1993). Based on 16S rDNA sequences, the genera *Thermosipho* (Huber et al., 1989d), *Fervidobacterium* (Patel et al., 1985a; Huber et al., 1990), *Geotoga*, and *Petrotoga* (Davey et al., 1993a) also belong to the *Thermotoga* branch, representing the new order *Thermotogales* with the family of the *Thermotogaceae* (Fig. B2.1; Huber and Stetter, 1992c,d). Members of this order represent, together with members of the order *Aquificales*, the bacteria with the highest growth temperatures known so far (Huber and Stetter, 1992d; Huber et al., 1992b).

Cells of *Thermotoga* are Gram-negative and rod-shaped; they form an outer sheath, ballooning over the ends ("toga"), visible in all phases of growth (Figs. B2.2 and B2.3). In the stationary phase, the rods become spheres. After lysozyme, penicillin G, and proteinase treatment, cells of *T. maritima* round up, while cell lysis occurs in the presence of SDS. Most *Thermotoga* species possess flagella and are motile (Table B2.1). Motility can be

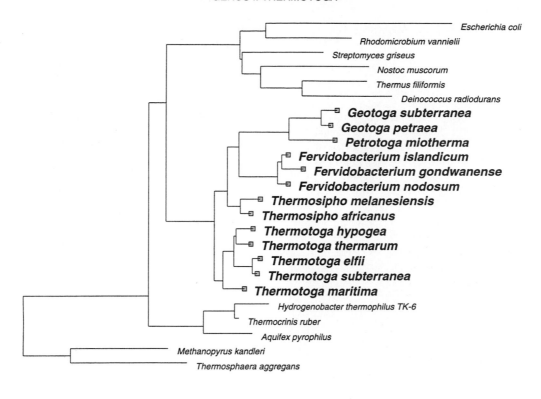

Escherichia coli
Rhodomicrobium vannielii
Streptomyces griseus
Nostoc muscorum
Thermus filiformis
Deinococcus radiodurans
Geotoga subterranea
Geotoga petraea
Petrotoga miotherma
Fervidobacterium islandicum
Fervidobacterium gondwanense
Fervidobacterium nodosum
Thermosipho melanesiensis
Thermosipho africanus
Thermotoga hypogea
Thermotoga thermarum
Thermotoga elfii
Thermotoga subterranea
Thermotoga maritima
Hydrogenobacter thermophilus TK-6
Thermocrinis ruber
Aquifex pyrophilus
Methanopyrus kandleri
Thermosphaera aggregans

0.10

FIGURE B2.1. Phylogenetic position of members of the genus *Thermotoga* to related bacteria. The tree was constructed by neighbor-joining analysis of 16S rRNA sequences. Bar = 0.10 fixed mutation per nucleotide position.

strongly enhanced by heating the microscopic slide to 80°C. For *T. maritima*, a thermotactic response to temporal temperature changes was reported, and thermostable chemotaxis proteins have been identified (Gluch et al., 1995; Swanson et al., 1996).

The outer membrane of *T. maritima* is composed of two major proteins. Ompα appears as a rod-shaped spacer that spans the periplasm, connecting the outer membrane to the inner body (Engel et al., 1992; Lupas et al., 1995). Ompβ, identified as a true porin, is the regularly arrayed outer membrane protein (Engel et al., 1993). The subunits are hexagonally arrayed, having a center-to-center spacing of approximately 12.4 nm. The murein of *T. maritima* is composed of muramic acid, *N*-acetylglucosamine, glutamic acid, alanine, and lysine (molar ratio = 0.41:0.69:1.00:1.43:0.89). D- and L-lysine are present, not found so far in Gram-negative bacteria. About 50% of the total polar lipids are two amphipathic monopolar glycolipids with a very rare α-(1–4)-diglucosyl structure (Manca et al., 1992). The core lipids from different *Thermotoga* species are composed of fatty acid methyl esters, unusual long-chain dicarboxylic fatty acids, and a new ether core lipid, identified as 15,16-dimethyl-30-glyceroloxytriacontanoic acid (Huber et al., 1986; De Rosa et al., 1988). So far, a total of 37 different fatty acids have been identified in *T. maritima*, including the novel 13,14-dimethyloctacosanedioic acid (Carballeira et al., 1997).

Colonies are round and white. The colony size is about 1 mm in diameter after 2–14 days incubation.

For the cultivation of the strictly anaerobic *Thermotoga* species, the anaerobic technique according to Balch and Wolfe (1976) is used. Physiological growth conditions for the different *Thermotoga* species are listed in Table B2.1. *T. maritima* and *Thermotoga*

neapolitana grow at a maximum temperature of 90°C and at an optimum around 80°C (Table B2.1) and exhibit the highest growth temperatures within this genus (Huber et al., 1986; Huber and Stetter, 1992d). Under optimal conditions, the generation time range from 0.75 to 4.75 h. *Thermotoga* relatives are obligate heterotrophs, preferentially fermenting carbohydrates, or complex organic matter (for details,: see species description). For detailed culture conditions see also Huber and Stetter (1992c).

Glucose is mainly catabolized by *T. maritima* via the Embden-Meyerhoff glycolytic pathway and, to a lesser extent, via the Entner-Doudoroff pathway (Selig et al., 1997). Final products are L(+)-lactate, acetate, ethanol, L-alanine, CO_2, and H_2. L-Alanine production from sugar fermentation was proposed to be a remnant of an ancestral metabolism (Ravot et al., 1996a). H_2 is a potent inhibitor of growth. Depending on the species, this inhibition can be overcome by the addition of sulfur or inorganic sulfur-containing compounds. Under these culture conditions, H_2S is formed as final product. However, growth of some species is even inhibited in the presence of sulfur. When grown in the presence of thiosulfate, all *Thermotoga* species produce H_2S (Ravot et al., 1995b). The activity of various glycolytic enzymes vary over a wide range and have high heat stability (Jaenicke et al., 1996; Selig et al., 1997). Several metabolic enzymes have already been expressed in the active form in a mesophilic host, *Escherichia coli* (Jaenicke et al., 1996; Bibel et al., 1998). D-glucose is taken up via an active transport system, energized by an ion gradient. The gradient is generated by ATP, derived from substrate-level phosphorylation (Galperin et al., 1996). The arginine biosynthesis pathway proceeds via *N*-acetylated intermediates in an eight-step pathway (van de Casteele et al., 1990). Compatible

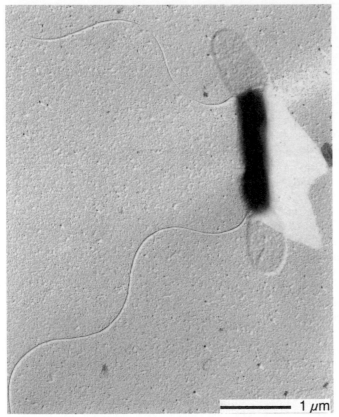

FIGURE B2.2. Platinum-shadowed cell of *Thermotoga maritima* with a sheath-like outer membrane ("toga"), overballooning at both ends.

solutes including the newly described compounds 2-*O*-β-mannosyl-di-*myo*-inositol-1,3′-phosphate and di-*myo*-inositol-1,3′-phosphate, were identified in *T. maritima* and *T. neapolitana* (Martins et al., 1996).

The genome of *T. maritima* has been studied by random sequencing of cDNA and genomic libraries. To date, 175 unique clones were analyzed. At least 52 clones contained significant amino acid sequence similarity to known proteins in other organisms (Kim et al., 1993). Sequencing of the *T. maritima* genome (1.8 Mb) was done at the Institute for Genomic Research (TIGR; Rockville, MD, USA). Comparison of the genome sequences of *Thermotoga maritima*, *Aquifex pyrophilus*, and *Archaeoglobus fulgidus* suggests that lateral gene transfer occurred between hyperthermophilic members of the *Bacteria* and *Archaea* domains (Nelson et al., 1999).

Auxotrophic and antimetabolite-resistant mutants of *T. neapolitana* have been isolated by the use of mutagenic agents (Vargas and Noll, 1994). So far, the only plasmid discovered was found in *Thermotoga* sp. RQ7, an isolate, obtained from the hot seafloor of Ribeira Quente, the Azores (Huber et al., 1986). The plasmid is the smallest natural replicon so far described, with only 846 bp, replicating by the rolling-circle mechanism (Harriott et al., 1994; Yu and Noll, 1997).

Growth of *T. maritima*, *T. neapolitana*, and *"Thermotoga subterranea"* is inhibited by 100 µg/ml of streptomycin and 100 µg/ml chloramphenicol. *T. maritima* is also sensitive to penicillin G, ampicillin, vancomycin, cycloserine, and phosphomycin (100 µg/ml), while penicillin G and ampicillin (100 µg/ml) do not inhibit growth of *"T. subterranea"*. While *Thermotoga thermarum* is sensitive

to low concentrations of rifampicin (1 µg/ml), *T. maritima*, *T. neapolitana*, and *"T. subterranea"* are resistant to 100 µg rifampicin/ml. The purified RNA polymerase of *T. maritima* is resistant to 1 µg rifampicin/ml, and only about 80% of its activity is inhibited by 200 µg/ml (Huber et al., 1986). Furthermore, growth of *T. maritima* is not inhibited by 10 µg/ml of aminoglycoside antibiotics and the purified ribosomes are resistant to these antibiotics as well (Londei et al., 1988).

Members of *Thermotoga* are cosmopolitan. They thrive within shallow and deep-sea marine hydrothermal systems, continental solfatara springs of low salinity, and marine or continental oil reservoirs (Table B2.1) (Huber and Stetter, 1992d; Stetter et al., 1993; Grassia et al., 1996).

ENRICHMENT AND ISOLATION PROCEDURES

Depending on the salt requirement of the *Thermotoga* species, either media with low ionic strength or half-strength seawater media are used (Table B2.1) (Huber and Stetter, 1992c). A pH of ~7 and a gas phase consisting of 300 kPa N_2 or 300 kPa N_2/CO_2 (80:20; v/v) is recommended. The media, supplemented with complex organic material (e.g., yeast extract and peptone) and a defined carbon source (e.g., starch, maltose), are inoculated with 0.5–1 ml of original sampling material. The enrichment bottles are incubated at the appropriate temperatures (Table B2.1) and examined periodically over a period of two weeks by phase microscopy. *T. maritima* or *T. neapolitana* can be selectively enriched with starch as the only carbon and energy source and an incubation temperature of 85°C. When rod-shaped cells with an outer sheath-like structure become visible, the enrichments are serially diluted. The isolates are obtained by plating under anaerobic conditions, using either a stainless steel anaerobic jar (Balch et al., 1979) and plates, solidified with 0.8–3% agar or 0.7–1.0% Gelrite, agar shake tubes (2%), or phytagel roll tubes (4%). The incubation temperature is between 60 and 75°C, and colonies appear within 2–14 days. For *T. neapolitana*, mean plating efficiencies of up to 84% on 0.7% Gelrite were reported (Childers et al., 1992). *Thermotoga* species can be also isolated by a plating-independent, newly developed isolation method ("selected cell cultivation" technique), using a strongly focused infrared laser beam as "optical tweezers" (Huber et al., 1995a ; Beck and Huber, 1997).

MAINTENANCE PROCEDURES

Thermotoga can be stored for several months at 4°C under anaerobic conditions. Storage in liquid nitrogen at −140°C in the presence of 5% DMSO is recommended for long-term preservation. No loss of cell viability of *T. maritima* was observed after storage over a period of ten years.

DIFFERENTIATION OF THE GENUS *THERMOTOGA* FROM OTHER GENERA

Thermotoga can be phylogenetically differentiated from the other genera within the *Thermotogales* by comparison of the 16S rDNA sequences. Furthermore, *Thermotoga* differs from all other genera by the presence of the new ether core lipid 15,16-dimethyl-30-glyceroloxytriacontanoic acid. *Fervidobacterium* can be distinguished morphologically from *Thermotoga* by the formation of spheroids, occurring terminally. *Fervidobacterium* and *Thermosipho* differ from *Thermotoga* by the presence of RNA polymerases, which exhibit no serological cross-reaction with the same enzyme from *T. maritima*. *Thermosipho* grows in chains (up to 12 rods) surrounded by a sheath and is therefore morphologically differ-

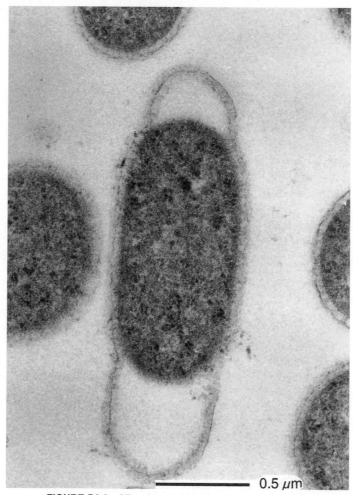

FIGURE B2.3. Ultrathin section of *Thermotoga maritima*.

TABLE B2.1. Diagnostic and descriptive features of the species of *Thermotoga*[a]

Characteristic	1. *Thermotoga maritima*	2. *Thermotoga elfii*	3. *Thermotoga hypogea*	4. *Thermotoga neapolitana*	5. *Thermotoga thermarum*	6. "*Thermotoga subterranea*"
Habitat	Geothermal heated seafloor, Vulcano, Italy	African oil well	African oil well	Shallow submarine hot spring, Naples, Italy	Hot solfataric spring, Lac Abbé, Djibouti, Africa	Paris oil well
pH range	5.5–9.0	5.5–8.7	6.1–9.1	5.5–9.0	6.0–9.0	6.0–8.5
pH optimum	6.5	7.5	7.3–7.4	7	7	7
Temperature range, °C	55–90	50–72	56–90	55–90	55–84	50–75
Temperature optimum, °C	80	66	70	80	70	70
NaCl range, %	0.25–6.0	0–2.4	0–1.5	0.25–6.0	0.2–0.55	0–2.4
NaCl optimum,%	2.7	1	0–0.2	2	0.35	1.2
Flagella	single subpolar flagellum	peritrichous flagella	laterally	–	laterally	nd

[a]For symbols see standard definitions; nd, not determined.

ent from *Thermotoga*. Furthermore, *Thermosipho* can be distinguished from *Thermotoga* by its ability to grow at much lower temperatures (35–77°C) and by a mol% G + C content of only 30. *Thermotoga* is clearly different from *Petrotoga* and *Geotoga* on the basis of its sodium chloride requirement and its temperature range for growth.

ACKNOWLEDGMENTS

We wish to thank R. Rachel for electron microscopy.

FURTHER READING

Adams, M.W.W. and R.M. Kelly. 1995. Enzymes from microorganisms in extreme environments. Chem. Eng. News *18*: 32–42.

Huber, R. and K.O. Stetter. 1992. The order *Thermotogales. In* Balows, Trüper, Dworkin, Harder and Schleifer (Editors), The Prokaryotes: A Handbook of Bacteria: Ecophysiology, Isolation, Identification, Applications, 2nd ed., Springer-Verlag, New York, pp. 3809–3815.

Jannasch, H.W. 1997. Small is powerful: Recollections of a microbiologist and oceanographer. Annu. Rev. Microbiol. *52*: 1–45.

Leuschner, C. and G. Antranikian. 1995. Heat-stable enzymes from extremely thermophilic and hyperthermophilic microorganisms. World J. Microbiol. Biotechnol. *11*: 95–114.

Stetter, K.O. 2000. Volcanoes, hydrothermal venting, and the origin of

life. *In* Marit and Ernst (Editors), Volcanoes and the Environment, Cambridge University Press, in press.

Woese, C.R. 1987. Bacterial evolution. Microbiol. Rev. *51*: 221–271.

DIFFERENTIATION OF THE SPECIES OF THE GENUS *THERMOTOGA*

For general characteristics of the species see generic description. Other characteristics of the species of *Thermotoga* are listed in Table B2.1.

List of species of the genus Thermotoga

1. **Thermotoga maritima** Stetter and Huber 1986, 575[VP] (Effective publication: Stetter and Huber *in* Huber, Langworthy, König, Thomm, Woese, Sleytr and Stetter 1986, 332.)

   ma.ri′ti.ma. L. fem. adj. *maritima* belonging to the sea, describing its biotope.

   Growth on glucose, ribose, xylose, xylan, galactose, sucrose, raffinose, lactose, salicin, maltose, cellulose, mannan, starch, glycogen, yeast extract, whole cell extracts of bacteria (e.g., *Lactobacillus bavaricus*), and archaea (e.g., *Pyrodictium occultum*). No growth on acetate, lactate, formate, pyruvate, propionate, ethanol, methanol, glycerol, glutamate, glycine, and Casamino acids. Glucose is fermented to L(+)-lactate, acetate, CO_2, H_2, and two minor unidentified compounds. Hydrogen inhibition is reversible in the presence of sulfur. *T. maritima* is able to fix molecular nitrogen and to use it as a nitrogen source (Huber and Stetter, 1992c). Elongation factor 2 is resistant to ADP ribosylation by diphtheria toxin. Contains two new tRNA nucleosides, 3-hydroxy-*N*-[[9-β-D-ribofuranosyl-9*H*-purin-6-yl)amino]carbonyl] norvaline, and 3-hydroxy-*N*-[[9-β-D-ribofuranosyl-9*H*-2-methyl-thiopurin-6-yl)amino]carbonyl]norvaline (Reddy et al., 1992).

   The DNA-dependent RNA polymerase exhibits a bacterial subunit pattern. The core enzyme consists of subunits with molecular weights of 184,000, 141,000, and 45,000.

   The mol% G + C of the DNA is: 46 (HPLC, T_m).

   Type strain: MSB8, DSMZ 3109.

   GenBank accession number (16S rRNA): M21774.

2. **Thermotoga elfii** Ravot, Magot, Fardeau, Patel, Prensier, Egan, Garcia and Ollivier 1995a, 312.[VP]

   el.fi′i. L. gen. n. *elfii* named after ELF-Aquitaine.

   Requires yeast extract and bio-Trypcase (tryptic digested casein-peptone; bioMérieux, Graponne, France) for growth. In the presence of thiosulfate, oxidation of D-glucose, D-arabinose, D-fructose, lactose, maltose, D-mannose, D-ribose, sucrose, and D-xylose, but not L-arabitol, D-mannitol, L-rhamnose, L-sorbose, L-xylose, acetate, butyrate, lactate, or propionate, occurs. Yeast extract cannot be replaced by a vitamin solution, Casamino acids or a mixture of both. Bio-Trypcase is fermented in the presence of yeast extract. Glucose (in the presence of yeast extract/bio-Trypcase) is fermented to acetate, CO_2 and H_2. Addition of thiosulfate enhances glucose utilization. Sulfur prevents growth.

   The mol% G + C of the DNA is: 39.6 (HPLC).

   Type strain: SEBR 6459, DSMZ 9442.

   GenBank accession number (16S rRNA): X80790.

3. **Thermotoga hypogea** Fardeau, Ollivier, Patel, Magot, Thomas, Rimbault, Rocchiccioli and Garcia 1997, 1018.[VP]

   hy.po.ge′ a. Gr. pron. *hypo* under; Gr. n. *ge* earth; M.L. fem. adj. *hypogea* under the earth, referring to the site of isolation.

   Ferments yeast extract and bio-Trypcase. Yeast extract or

bio-Trypcase required for growth on carbohydrates. Yeast extract cannot be replaced by a vitamin solution, Casamino acids or a mixture of Casamino acids and vitamins. Uses D-glucose, DL-fructose, D-galactose, DL-lactose, DL-maltose, D-mannose, D-sucrose, D-xylose, and xylan, but not D-arabinose, D-ribose, L-sorbose, L-xylose, acetate, butyrate, lactate, or propionate. Glucose and xylose are fermented to acetate, CO_2, H_2, L-alanine, and traces of ethanol in the absence or presence of thiosulfate. Addition of thiosulfate enhances glucose utilization and final cell densities. S^0 does not inhibit growth and is not used as an electron acceptor.

The mol% G + C of the DNA is: 50 (HPLC).

Type strain: SEBR 7054, DSMZ 11164.

GenBank accession number (16S rRNA): U89768.

4. **Thermotoga neapolitana** Jannasch, Huber, Belkin and Stetter 1989, 93[VP] (Effective publication: Jannasch, Huber, Belkin and Stetter 1988, 104.)

   ne.a.po.li.ta′ na. M.L. fem. adj. *neapolitana* having been isolated near Naples.

   Growth on ribose, glucose, sucrose, maltose, lactose, galactose, starch, and glycogen. Growth on yeast extract and weak growth on peptone or tryptone (Belkin et al., 1986). Addition of sulfur to yeast extract-glucose medium enhances final cell yields up to fourfold (Belkin et al., 1986). No growth on acetate, lactate, formate, pyruvate, propionate, ethanol, methanol, glycerol, glutamate, glycine, and Casamino acids. In the presence of sulfur or cystine, hydrogen inhibition can be overcome.

   No significant DNA–DNA similarity to *T. maritima*.

   The mol% G + C of the DNA is: 41 (T_m).

   Type strain: NS-E, ATCC 49049, DSMZ 4359.

5. **Thermotoga thermarum** Windberger, Huber and Stetter 1992, 327[VP] (Effective publication: Windberger, Huber and Stetter *in* Windberger, Huber, Trincone, Fricke and Stetter 1989, 511.)

   ther.ma′ rum. L. gen. n. *thermarum* living in hot continental springs with low ionic strength.

   Yeast extract required for growth. Addition of glucose, maltose, and starch stimulate growth. Sulfur prevents growth.

   The molecular weights of the RNA polymerase core enzyme subunits are 170,000, 135,000, and 41,000. Incomplete serological cross-reaction with *T. maritima* RNA polymerase antiserum. No significant DNA–DNA similarity to *T. maritima* and *T. neapolitana*.

   The mol% G + C of the DNA is: 40 (HPLC, T_m).

   Type strain: LA3, DSMZ 5069.

   GenBank accession number (16S rRNA): M21774.

6. **"Thermotoga subterranea"** Jeanthon, Reysenbach, L'Haridon, Gambacorta, Pace, Glénat and Prieur 1995, 96.

sub.terr.a.ne.a. L. fem. adj. *subterranea* under the earth, describing its site of isolation.

Growth on peptone, tryptone, casein, and yeast extract. Addition of glucose or maltose to yeast extract containing medium enhances final cell yields. No growth on glucose, maltose, lactose, ribose, raffinose, lactate, and acetate. Growth inhibition in the presence of sulfur. Hydrogen inhibition reversible in the presence of cystine or thiosulfate.

The mol% G + C of the DNA is: 40 (T_m).

Deposited strain: SL1, DSMZ 9912.

GenBank accession number (16S rRNA): U22664.

Genus II. **Fervidobacterium** *Patel, Morgan and Daniel 1985b, 535*[VP] *(Effective publication: Patel, Morgan and Daniel 1985a, 68)*

ROBERT HUBER AND KARL O. STETTER

Fer.vi.do.bac.te′ri.um. L. adj. *fervidus* hot; Gr. n. *bakterion* a small rod; M.L. neut. n. *Fervidobacterium* rods which grow at higher temperatures.

Rod-shaped cells, generally occurring singly or in pairs, about 0.5–0.6 × 1–40 µm. **Gram-negative**. The majority of the rods produce terminal **protuberances on one end of the cell** ("spheroids"). No resting stages known. Obligately anaerobic. Extremely thermophilic, optimum growth temperature: 65–70°C. Optimal pH: 6.5–7.2. **Heterotrophic growth. Products from glucose fermentation are acetate, CO₂, and H₂.** Sulfur is reduced to H₂S. Molecular hydrogen inhibits growth. Sensitive to inhibitors of cell-wall, protein, and nucleic acid biosynthesis.

Deep phylogenetic branch within the bacteria by 16S rDNA analysis.

The mol% G + C of the DNA is: 33.7–41.

Type species: **Fervidobacterium nodosum** Patel, Morgan and Daniel 1985b, 535 (Effective publication: Patel, Morgan and Daniel 1985a, 68.)

FURTHER DESCRIPTIVE INFORMATION

Based on 16S rDNA sequence comparison, it was shown that the genus *Fervidobacterium* (Patel et al., 1985a; Huber et al., 1990) belongs, together with the genera *Thermotoga* (Huber et al., 1986), *Petrotoga, Geotoga* (Davey et al., 1993a), and *Thermosipho* (Huber at al., 1989d; Ravot et al., 1996b), to the *Thermotogales* order (Huber and Stetter, 1992c,d). For further taxonomic comments see also Genus *Thermotoga* in this *Manual.*

Cells of *Fervidobacterium* are Gram-negative rods, occurring predominantly singly and in pairs (Fig. B2.4). The majority of the rods form a characteristic, terminal bleb ("spheroid") on one end of the cells, which is visible during all stages of growth (diameter: 1–4 µm; Fig. B2.4). Furthermore, growth in short chains was observed and *Fervidobacterium islandicum* frequently forms aggregates of up to about 50 cells. After lysozyme treatment of *F. islandicum* (1 mg/ml), the rods turn into spheres (3–5 min), while the outer sheath with the terminal bleb retains its shape. Besides spheroids, *Fervidobacterium* species form giant spheres ("rotund bodies"; diameter: 5–8 µm), membrane-bound structures containing one to seven individual cells. Motility was reported for *F. islandicum* and *F. nodosum.*

Electron microscopic studies showed that the cell envelope of *Fervidobacterium* is composed of two layers. The outer layer protrudes to form the "spheroids" (Patel et al., 1985a). In *F. nodosum*, the two layers have an irregular convoluted structure, are connected with regular junctions and the outer layer is susceptible to lysis by SDS. The cells are able to grow and multiply within the "spheroid envelope", thus forming "rotund bodies" (Patel et al., 1985a). In *F. islandicum*, diaminopimelic acid, typical for Gram-negative bacteria, is absent. The lipids of *Fervidobacterium* are composed of fatty acid methyl esters (Huber et al.,

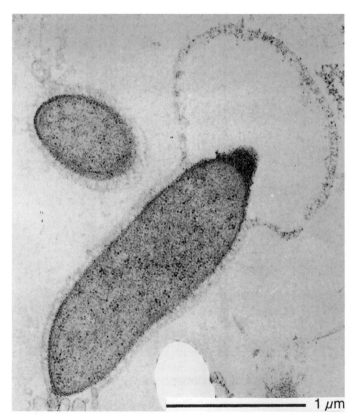

FIGURE B2.4. Ultrathin section of *Fervidobacterium islandicum.*

1990; Patel et al., 1991). In addition, unusual C_{28} to C_{34}-dicarboxylic fatty acids are present (Huber et al., 1990).

Fervidobacterium species form milky or creamy white, round colonies, which become visible after two to five days of anaerobic incubation.

Fervidobacterium species must be cultivated under anaerobic conditions. For the preparation of the cultivation media, different anaerobic techniques can be used (e.g., Balch and Wolfe, 1976; Patel et al., 1985a). Table B2.2 summarizes the physiological growth conditions. Under optimal conditions, the doubling time ranges from 79 to 150 min. *Fervidobacterium* relatives are obligate heterotrophs, preferentially using carbohydrates for growth (for details see species description). The presence of complex organic matter for growth (e.g., yeast extract) is necessary. For detailed culture conditions see also Huber and Stetter (1992c).

TABLE B2.2. Diagnostic and descriptive features of *Fervidobacterium* species[a]

Characteristic	1. *Fervidobacterium nodosum*	2. *Fervidobacterium gondwanense*	3. *Fervidobacterium islandicum*	4. *Fervidobacterium pennivorans*
Habitat	hot spring, New Zealand	runoff channel of the Great Artesian Basin, Australia	continental solfatara field, Iceland	hot spring, Azores
pH range	6.0–8.0	6.0–8.0	6.0–8.0	5.5–8.0
pH optimum	7	7	7	6.5
Temperature range, °C	41–79	45 to <80	50–80	50–80
Temperature optimum, °C	70	65–68	65	70
NaCl range, %	nd	nd	nd	0–4.0
NaCl optimum, %	<1.0	<0.2	<0.7	0.4

[a]nd, not determined.

During growth on glucose, L(+)-lactate, acetate, ethanol, butyric acid, valeric acid, L-alanine, CO_2, and H_2 were identified as final products. H_2 is a potent inhibitor of growth for different *Fervidobacterium* species. Hydrogen inhibition of *F. pennivorans* may be prevented by the addition of thiosulfate or S^0. *F. nodosum* and *F. islandicum* can reduce sulfur, forming H_2S. However, in contrast to *F. pennivorans*, neither of these species is able to grow in the presence of hydrogen, when sulfur is added to the medium.

Members of the genus *Fervidobacterium* seem to be restricted to habitats of low salinity. While most members of *Fervidobacterium* were isolated from volcanic hot springs, *F. gondwanense* was obtained from a nonvolcanic, geothermally heated water source (Table B2.2).

ENRICHMENT AND ISOLATION PROCEDURES

Enrichments should be performed under anaerobic conditions (e.g., Balch and Wolfe, 1976) at neutral pH in a low salt medium (Table B2.2), supplemented with yeast extract, Trypticase peptone, and glucose. Keratinophilic (feather-degrading) *Fervidobacterium* species can be enriched in yeast extract/tryptone medium with native feathers from chickens, ducks, or geese. The media, inoculated with 0.1–1 ml of the original samples, are incubated at 70°C and are examined for growth with a phase-contrast microscope over a period of two weeks. When rod-shaped cells with a single terminal swelling become visible, the positive enrichments should be subcultured at least three times. From the enrichments, pure cultures of *F. islandicum* and *F. pennivorans* are obtained by plating under anaerobic conditions, using agar plates (1.5–2% agar) and a stainless steel anaerobic jar (Balch et al., 1979). In the case of *F. pennivorans*, the plates are supplemented with 0.05% skim milk instead of feathers. Enrichment cultures of *F. nodosum* and *F. gondwanense* can be purified by using the end-point dilution technique and anaerobic medium, supplemented with 2% agar. Single colonies, which serve as an inoculum for liquid cultures, will be visible within 2–

6 days of incubation. *Fervidobacterium* species can be also isolated by the "selected cell cultivation" technique, using a strongly focused infrared laser beam as "optical tweezers" (Huber et al., 1995a; Beck and Huber, 1997).

MAINTENANCE PROCEDURES

Stock cultures of *F. nodosum* remain viable for more than two years at room temperature, and stock cultures of *F. islandicum* and *F. pennivorans* for one year at 4°C. For long-term preservation, *F. islandicum* and *F. nodosum* can be stored in liquid nitrogen at −140°C in the presence of 5% DMSO. No loss of cell viability was observed after storage over a period of six years.

DIFFERENTIATION OF THE GENUS *FERVIDOBACTERIUM* FROM OTHER GENERA

Fervidobacterium can be distinguished from the other genera within the *Thermotogales* by 16S rDNA sequence comparison. Within this order, *Fervidobacterium* is morphologically unique by the formation of characteristic blebs, formed terminally on one end of the rod-shaped cells. Furthermore, *Thermotoga* and *Thermosipho* can be differentiated from *Fervidobacterium* by the presence of RNA polymerases that do not cross-react serologically with the same enzyme from *Fervidobacterium*. From *Thermotoga*, *Fervidobacterium* is further differentiated by the chemical nature of the lipids. The inability of *Fervidobacterium* to grow at NaCl concentrations above 1% also serves to distinguish it from *Thermosipho*, *Geotoga*, and *Petrotoga*.

FURTHER READING

Huber, R. and K.O. Stetter. 1992. The order *Thermotogales*. *In* Balows, Trüper, Dworkin, Harder and Schleifer (Editors), The Prokaryotes: A Handbook of Bacteria: Ecophysiology, Isolation, Identification, Applications, 2nd ed., Springer-Verlag, New York, pp. 3809–3815.

Stetter, K.O. 2000. Volcanoes, hydrothermal venting, and the origin of life. *In* Marit and Ernst (Editors), Volcanoes and the Environment, University Press, Cambridge, in press.

List of species of the genus Fervidobacterium

1. **Fervidobacterium nodosum** Patel, Morgan and Daniel 1985b, 535[VP] (Effective publication: Patel, Morgan and Daniel 1985a, 68.)

no.do'sum. L. neut. adj. *nodosum* knotty or swollen.

In the presence of yeast extract, glucose, raffinose, galactose, mannose, fructose, arabinose, sorbitol, maltose, sucrose, pectin, glycerol, and to a minor extent, rhamnose and pyruvate, are used. Cellulose is not degraded, and no growth is observed on acetate, lactate, dextrin, xanthan, citrate, ser-

ine, glycine, glutamine, arginine, succinate, formate, fumarate, xylitol, carragenan, inulin, karaya gum, or oxalate. Main products of glucose fermentation are acetate, ethanol, lactate, CO_2 and H_2 and traces of *n*-butyric and *n*-valeric acid.

F. nodosum is sensitive to penicillin, chloramphenicol, and tetracycline (10 µg/ml), but resistant to D-cycloserine (100 µg/ml).

No significant DNA–DNA hybridization with *F. islandicum*. The mol% G + C of the DNA is: 33.7 (T_m).

Type strain: Rt17-B, ATCC 35602, DSMZ 5306.
GenBank accession number (16S rRNA): M59177.

2. **Fervidobacterium gondwanense** Andrews and Patel 1996, 268.[VP]

gond.wa.nen' se. M.L. neut. adj. *gondwanense* pertaining to the large land mass known as Gondwana, which included Australia, Africa, India, and South America before they separated.

In the presence of Trypticase and yeast extract, glucose, maltose, mannose, starch, amylopectin, cellobiose, carboxymethyl cellulose, fructose, xylose, galactose, lactose, dextrin, and pyruvate are used, but not Casamino acids, sorbose, ribose, raffinose, arabinose, dextran, gelatin, cellulose, xylan, or chitin. During fermentation of glucose, lactate, ethanol, acetate, H_2, and CO_2 are produced. Sulfite, sulfate, and thiosulfate are not reduced to sulfide.

Sensitive to penicillin, streptomycin, novobiocin, phosphomycin, tetracycline, and vancomycin (10 μg/ml). Growth is also inhibited by chloramphenicol, polymyxin B, and rifampicin (100 μg/ml). Resistant to D-cycloserine (100 μg/ml).

The mol% G + C of the DNA is: 35 (T_m).
Type strain: AB39 (ACM 5017).
GenBank accession number (16S rRNA): Z49117.

3. **Fervidobacterium islandicum** Huber, Woese, Langworthy, Kristjansson and Stetter 1991, 178[VP] (Effective publication: Huber, Woese, Langworthy, Kristjansson and Stetter 1990, 109.)

is.lan' di.cum. M.L. neut. adj. *islandicum* from Island, describing its place of isolation.

Yeast extract required for growth. In the presence of yeast extract, pyruvate, ribose, glucose, maltose, raffinose, starch, and cellulose stimulate growth. In the presence of glucose,

L(+)-lactate, acetate, ethanol, H_2, CO_2, and traces of isobutyric and isovaleric acid are formed.

An enriched RNA polymerase fraction does not cross-react with antibodies against the purified RNA polymerase of *Thermotoga maritima.*

Growth is inhibited in the presence of vancomycin, streptomycin, chloramphenicol, ampicillin, and rifampicin (10 μg/ml).

No significant DNA–DNA hybridization with *F. nodosum.*
The mol% G + C of the DNA is: 41 (T_m) and 40 (HPLC).
Type strain: H 21, DSMZ 5733.
GenBank accession number (16S rRNA): M59176.

4. **Fervidobacterium pennivorans** Friedrich and Antranikian 1999, 1[VP] (Effective publication: *Fervidobacterium pennavorans* (sic) Friedrich and Antranikian 1996, 2879.)

pen' ni.vo' rans. L. n. *penna* feather, wing; L. v. *voro* to eat, devour; L. part. adj. *vorans* devouring; M.L. part. adj. *pennivorans* feather-devouring (referring to the ability of the organism to degrade keratin, a protein found in feathers).

In the presence of yeast extract, starch, glycogen, pullulan, glucose, fructose, maltose, and xylose are used as growth substrates. Lactose, amylose, arabinose, casein, collagen, succinate, or Tween 80 were not used. As fermentation products, ethanol and acetate were identified with glucose or starch as growth substrates. Degradation of unprocessed feathers at high temperature. Cell-bound keratinase, classified as a serine protease. The purified enzyme, with a molecular mass of 130 kDa and an isoelectric point of 3.8, converts native feather meal to peptides. The keratinase is SDS-resistant and active between pH 6.0 and 10.5 (optimum pH: 10.0) and 50–100°C (optimum: 80°C).

The mol% G + C of the DNA is: 40 (HPLC).
Type strain: Ven 5, DSMZ 9078.

Genus III. **Geotoga** Davey, Wood, Key, Nakamura and Stahl 1993b, 864[VP] (Effective Publication: Davey, Wood, Key, Nakamura and Stahl 1993a, 198)

MARY ELLEN DAVEY, BARBARA J. MACGREGOR AND DAVID A. STAHL

Ge.o.to' ga. Gr. fem. n. *ge* earth; L. fem. n. *toga* Roman outer garment; M.L. fem. n. *Geotoga* the earth outer garment.

Rods 3–20 × 0.5–0.7 μm surrounded by an outer sheath-like structure, which is clearly visible under phase microscopy. The cells usually appear singly or in pairs within the sheath, however they may have as many as five cells per sheath. Gram-negative, **fermentative** bacteria. They are **strictly anaerobic**. Form enlarged spherical bodies in stationary phase. Cultures remain viable for up to six months when stored at 4°C, even after the medium becomes oxidized. Colonies are circular, convex, with entire margins, and whitish in color. Cells are **motile**. Sensitive to rifampicin. Growth occurs at pH 5.5–9.0 and at **0.5–10.0% NaCl**, with an optimum around 3.5%. No growth occurs at 0.05% and 12.5% NaCl. The isolates do not form endospores. They are **moderately thermophilic** with growth occurring at 30–60°C and an optimum

of around 50°C. Yeast extract is required for growth. Hydrogen inhibits growth of some strains. S^0 reduced to H_2S.

Sequences of 16S rRNA indicate that this genus represents a distinct branch within the order *Thermotogales* in the domain *Bacteria.*

The mol% G + C of the DNA is: 30.

Type species: **Geotoga petraea** Davey, Wood, Key, Nakamura and Stahl 1993b, 864 (Effective publication: Davey, Wood, Key, Nakamura and Stahl 1993a, 198.)

FURTHER DESCRIPTIVE INFORMATION

Geotoga is closely related to the genus *Petrotoga* (84% 16S rRNA similarity). Together, these genera constitute a distinct lineage within the *Thermotogales*. This lineage appears to be faster evolving, as indicated by the longer branch length relative to other lineages within the order. This rate difference likely contributes to different topologies inferred using distance and maximum likelihood methods of tree construction. Using distance methods, the *Geotoga/Petrotoga* lineage defines the point of earliest divergence within the *Thermotogales*. By contrast, maximum likelihood shows divergence within the order. Since distance methods tend to place fast-evolving sequences inappropriately deep, we present only the maximum likelihood topology (Fig. B2.5). They share phenotypic characteristics with many members of the *Thermotogales*. Cells are rod shaped with a characteristic outer sheath-like structure, or "toga" (Fig. B2.6). The sheath is readily visible under phase microscopy, and one or two cells per sheath are routinely observed, but occasionally one species (*Geotoga petraea*) has up to five cells in one sheath (Fig. B2.7). In stationary growth phase, they form enlarged spherical bodies surrounded by the sheath.

Geotoga are moderate thermophiles. *Geotoga subterranea* grows at 30–60°C with an optimum of around 45°C, and *Geotoga petraea* grows at 30–55°C, with an optimum of around 50°C. They are able to tolerate a wide range of salt concentrations. Growth was observed at 0.5–10.0% (w/v NaCl) with an optimum of around

3.0% for *Geotoga petraea*, and 4.0% for *Geotoga subterranea*. No growth occurred at 0.05% and 12.5% NaCl. They contain C_{12} to C_{18} fatty acids which are typical for microorganisms in the bacterial domain. The presence of long-chain fatty acids (C_{22} and greater), which are characteristic of thermophilic *Bacteria*, has not yet been determined.

On agar medium, colonies are circular, convex, with entire margins, and whitish in color. They require yeast extract (0.01%) for growth and this requirement cannot be replaced with a vitamin mixture or Casamino acids. Growth of *Geotoga subterranea* is enhanced by the addition of tryptone (0.05%). Stock cultures stored at 4°C will remain viable for up to six months even after the medium has become oxidized as indicated by the redox indicator (resazurin) in the medium. This characteristic not only indicates a tolerance to oxygen exposure, but also an ability to persist in culture.

Geotoga are able to use mannose, starch, maltodextrins, glucose, lactose, sucrose, galactose, and maltose as sole carbon and energy sources. In contrast to *Petrotoga miotherma*, *Geotoga* spp. do not use xylose. They are not able to use raffinose, ethanol, lactate, Casamino acids, acetate, or formate. In addition, amylase activity was found to be extracellular (unpublished data, Davey, M.E., 1992), in contrast to the amylases of *Thermotoga maritima* which

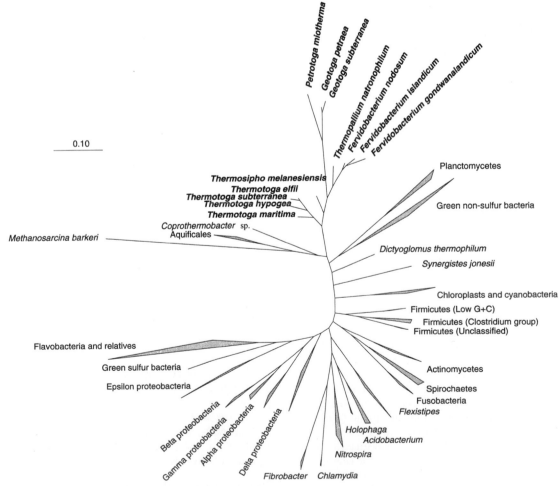

FIGURE B2.5. This maximum likelihood tree was calculated using FastDNAml algorithm (Felsenstein, 1992) as implemented in the ARB sequence analysis program (Software available from O. Strunk and W. Ludwig: ARB: a software environment for sequence data. www.mikro.biologie.tu-muenchende).

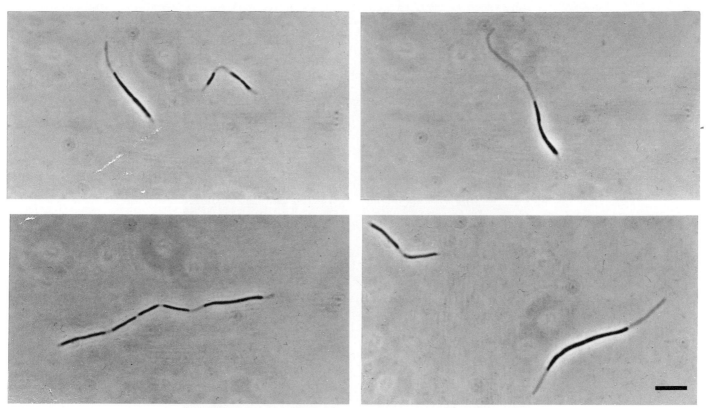

FIGURE B2.6. Scanning electron micrographs of *Geotoga subterranea*. Bar = 10 μm. (Reprinted with permission from M. Davey et al., Systematic and Applied Microbiology *16*: 191–200, 1993, ©Urban and Fischer Verlag.)

are exclusively associated with the "toga" (Schumann et al., 1991). The fermentation products after incubation at the optimal temperature for one week in the presence of glucose are H_2, CO_2, acetate, and ethanol.

Hydrogen inhibits growth on glucose of *Geotoga petraea*. No growth occurs in cultures with a headspace of H_2/CO_2 (80:20); however, this effect can be overcome by gassing the headspace with N_2. This inhibition by hydrogen can also be eliminated by the addition of S^0 (0.1%) which results in the production of H_2S; however, the addition of S^0 does not increase the growth yield. *Geotoga subterranea* is not inhibited by this concentration of hydrogen and also has the ability to reduce S^0 to H_2S. Members of the *Thermotogales* have also been reported to form H_2S with no concomitant increase in growth (Huber et al., 1986). Huber and colleagues. proposed that it is a means of relieving the inhibitory effect of hydrogen. Therefore, this ability may provide these bacteria with a mechanism to grow in environments with elevated levels of hydrogen which they would not otherwise be able to tolerate.

Growth is inhibited by the addition of vancomycin, chloramphenicol, and rifampicin; but not by streptomycin, when added to the culture prior to incubation at optimal temperatures (100 mg/ml; final concentration).

Geotoga were originally isolated from brines obtained from petroleum reservoirs located in Oklahoma and Texas. Other strains of "sheathed" bacteria have been isolated from a variety of petroleum reservoir sites in the same region, as well as from marine sediments (Davey, unpublished); however, these isolates have not been characterized.

Geotoga spp. have the potential to persist in a variety of environments. This ability is indicated by their tolerance to a broad range of temperatures and salt concentrations, as well as exposure to oxygen. Their apparent persistence in petroleum reservoirs also suggests a tolerance to a number of other parameters, i.e., high heavy metal concentrations, pressure, and petroleum. In addition, the ability to reduce S^0 in order to mitigate the inhibitory affect of hydrogen indicates another means of tolerance.

There are no known functions ascribed to the "toga". Investigations into the structure of the "toga" of *Thermotoga maritima* (Rachel et al., 1988, 1990) have determined that the main protein constituent is related to the outer membrane proteins of other bacteria. Since all these bacteria were isolated from unusual environments, it is conceivable that this structure could play a role in their survival.

ENRICHMENT AND ISOLATION PROCEDURES

Geotoga can be isolated from petroleum field brines, which are collected anaerobically at production and injection sites in pre-reduced, sterile bottles. The brine samples should be maintained at an ambient temperature and placed in an anaerobic chamber within 24 h of sampling. These samples are used both as a source of inoculum and as a base for the culture medium. Enrichments should be designed to mimic the *in situ* conditions (temperature, pH, E_h, and salinity) of the petroleum reservoir from which the brine samples are collected. This can be accomplished by inoculating 100-ml serum bottles (pre-reduced in a Coy anaerobic chamber with an atmosphere of $N_2/H_2/CO_2$; 85:10:5) containing

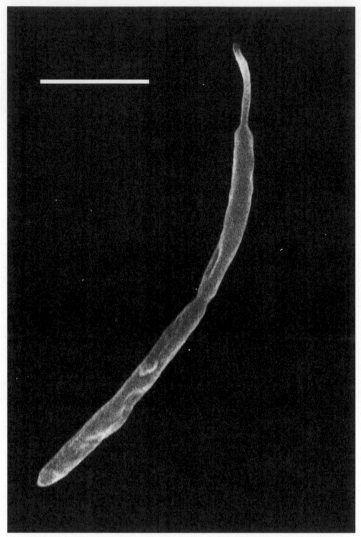

FIGURE B2.7. Phase contrast micrographs of cells of *Geotoga petraea*. Bar = 2 μm. (Reprinted with permission from M. Davey et al., *Systematic and Applied Microbiology 16:* 191–200, 1993, ©Urban and Fischer Verlag.)

0.1 g starch and 0.04 g α-glycerophosphate with 100 ml of field brine. This procedure results in minimal alteration of the concentration of salts and the pH of the sample. Starch or maltodextrins are provided as the preferred carbon and energy source for *Geotoga*. The α-glycerophosphate is added as a source of phosphate instead of inorganic phosphate as a result of precipitation of calcium-phosphate in many brines. If salt concentrations are low, inorganic phosphates can be used. Nitrogen is not generally a limiting nutrient in petroleum brines; however, the samples should be tested. This can be done by preparing enrichments without nitrogen and determining if growth is limited. The enrichments are then incubated anaerobically (initial E_h approximately -200 mV) at the reservoir temperature. The enrichments will grow to a concentration of 10^8 cells/ml within 10 days, at which time the sheathed organisms are readily visible by phase microscopy or with an epifluorescent microscope when stained with acridine orange (Hobbie et al., 1977).

The cultures can be monitored by phase microscopy for the appearance of "sheathed" bacteria and then subcultured for iso-

lation. The strains can be isolated on agar plates of Mel's Anaerobic Medium (MAM)*. They should be incubated in an anaerobic chamber with an atmosphere of $N_2/H_2/CO_2$ (85:10:5). The concentrations of inorganic compounds should be varied to match the particular brine being used in the enrichment, and the temperature for incubation should be the same as the reservoir from which the brines were collected. Circular, convex, white colonies will appear on the plates after one week of incubation and should be predominantly "sheathed" bacteria. Subsequently, four additional single colony transfers should be done to ensure purity.

---

*Mel's Anaerobic Medium (MAM): starch, 1.0 g/l; yeast extract, 0.5 g/l; NaCl, 33.3 g/l; CaCl$_2$·2H$_2$O, 7.6 g/l; MgCl$_2$·6H$_2$O, 2.5 g/l; PIPES, 6.8 g/l; NH$_4$Cl, 0.5 g/l; trace minerals (Widdel and Pfennig, 1984), 1 ml/l; resazurin, 1 mg/l; α-glycerophosphate, 0.4 g/l; ascorbic acid, 0.02%; cysteine-HCl, 0.01%; agar (Difco), 15.0 g/l; pH 6.8.

MAINTENANCE PROCEDURES

Lyophilization in skim milk (20%) is recommended for long term storage. Frozen stocks in skim milk (10%) and DMSO (5%) are best for stocks that are used routinely to start a new culture (Gherna, 1994). Working stock cultures can best be maintained at 4°C for one month in MAM with reduced salt concentrations (NaCl, 16.6 g; CaCl$_2$·2H$_2$O, 3.8 g). Even though the stocks do remain viable for several months longer, significant changes in morphology, such as enlarged, rounded, and bifurcated cells, are often observed in these older stock cultures, indicative of possible mutations; therefore, the use of older stocks is not recommended. MAM with reduced salt concentration can also be used to reconstitute lyophilized cultures and as a solid medium for plating. Colonies form in one week when plates are incubated at 45°C in an anaerobic chamber with an atmosphere of N$_2$/H$_2$/ CO$_2$ (85:10:5).

DIFFERENTIATION OF THE GENUS *GEOTOGA* FROM OTHER GENERA

Geotoga is morphologically and physiologically similar to other members of the order *Thermotogales*. Differentiation from *Petrotoga* is based primarily on 16S rRNA sequence similarity. Other distinguishing characteristics are shown in Table B2.3. Three characteristics (salt tolerance, temperature range for growth, and 16S rRNA sequence) can be used to differentiate *Geotoga* from the other members of the *Thermotogales*. *Geotoga*, like *Petrotoga*, tolerate a broad range of salt concentrations, which is not characteristic of the other genera. It has been determined that *Petrotoga miotherma* has the ability to make a variety of compatible solutes, such as trehalose, glycine betaine, and β-mannosylglycerate (Martins et al., 1996). This characteristic has not been examined in *Geotoga*. The *Thermotogales* are characteristically thermophilic. Two genera within the order *Thermotogales*, *Thermosipho* (Huber et al., 1989d) and *Fervidobacterium* (Huber et al., 1990), include strains whose temperature range for growth is more moderate and similar to that of *Geotoga*; however, all their temperature optima (*Fervidobacterium islandicum*, 65°C; and *Thermosipho africanus*, 75°C) are higher than that of the *Geotoga* isolates (*G. subterranea*, 45°C; *G. petraea* 50°C). Furthermore, *Geotoga* 16S ribosomal sequences are only distantly related to sequences from other characterized members of this order.

TAXONOMIC COMMENTS

Geotoga is a newly defined genus consisting of two species. Phenotypic and genotypic characteristics of the two isolates clearly place this genus within the *Thermotogales*; however, further analysis of these strains, as well as the isolation and characterization of more species that are representative of this genus, is required to better resolve phenotypic and genotypic diversity within the genus.

ACKNOWLEDGMENTS

This research was supported by Phillips Petroleum Co., Bartlesville, Oklahoma; as well as research agreement N00014–91-J4083 from the office of Naval Research; and research agreement CR815285 from the Environmental Protection Agency to D.A.S.) (Davey et al., 1993a).

DIFFERENTIATION OF THE SPECIES OF THE GENUS *GEOTOGA*

Table B2.3 presents the main differential characteristics of the two species.

List of species of the genus Geotoga

1. **Geotoga petraea** Davey, Wood, Key, Nakamura and Stahl 1993b, 864[VP] (Effective publication: Davey, Wood, Key, Nakamura and Stahl 1993a, 198.)

   pe.trae' a. L. fem. adj. *petraea* rocky, stony, dwelling on/in rocks, describing its isolation from a rock formation.

   The characteristics are as described for the genus and as indicated in Table B2.3.
   The mol% G + C of the DNA is: 30.
   Type strain: T5, ATCC 51226.
   GenBank accession number (16S rRNA): L10658.

2. **Geotoga subterranea** Davey, Wood, Key, Nakamura and Stahl 1993b, 864[VP] (Effective publication: Davey, Wood, Key, Nakamura and Stahl 1993a, 198.)

   sub.ter.ra' ne.a. L. fem. adj. *subterranea* under the earth, describing its site of isolation.

   The characteristics are as described for the genus and as indicated in Table B2.3.
   The mol% G + C of the DNA is: 30.
   Type strain: CC-1, ATCC 51225.
   GenBank accession number (16S rRNA): L10659.

TABLE B2.3. Characteristics differentiating *Petrotoga miotherma*, *Geotoga subterranea*, and *G. petraea*[a]

Characteristic	P. miotherma	P. mobilis	G. petraea	G. subterranea
Mol% G + C of DNA (HPLC)	34	34	30	30
Motility	−	+	+	+
Xylose used	+	+	−	−
Salt range (optimum)	0.25–10% (2%)	0.5–9% (3–4%)	0.25–10% (3%)	0.25–10% (4%)
Temperature range, °C (optimum)	35–65 (55)	40–65 (58–60)	30–55 (50)	30–60 (45)
Growth inhibited by hydrogen	+	+	+	−
Growth enhanced with tryptone (0.01%)	−	nd	+	−

[a]For symbols see standard definitions; nd, not determined.

Genus IV. **Petrotoga** Davey, Wood, Key, Nakamura and Stahl 1993b, 864[VP] (Effective publication: Davey, Wood, Key, Nakamura and Stahl 1993a, 199)

MARY ELLEN DAVEY, BARBARA J. MACGREGOR AND DAVID A. STAHL

Pe.tro.to'ga. Gr. and L. fem. n. *petra* rock, stone; L. fem. n. *toga* Roman outer garment; M.L. fem. n. *Petrotoga* the stone outer garment.

Rods 1–50 × 0.5–1.5 μm surrounded by an outer sheath-like structure which is clearly visible under phase microscopy. The cells frequently appear singly, in pairs, or in chains within the sheath. They are Gram-negative **fermentative** bacteria and **strictly anaerobic**. The cultures remain viable for up to six months when stored at 4°C, even after the medium becomes oxidized. Colonies are circular, convex, with entire margins and whitish in color. Cells are sensitive to rifampicin. Growth occurs at pH 5.5–9.0 with an optimum of 6.5–7.0 and at **0.5–10.0% NaCl** with an optimum of around 2.0–4.0%. No growth occurs at 0.05% and 12.5% NaCl. They are **moderately thermophilic,** with growth occurring at 35–65°C (optimum around 55–60°C). Endospores are not formed. Yeast extract is required for growth in culture. The cells are able to **reduce S[0] to hydrogen sulfide**.

The 16S rRNA sequence indicates that this genus represents a distinct branch within the order *Thermotogales* in the bacterial lineage.

The mol% G + C of the DNA is: 32.0–34.2.

Type species: **Petrotoga miotherma** Davey, Wood, Key, Nakamura and Stahl 1993b, 864 (Effective publication: Davey, Wood, Key, Nakamura and Stahl 1993a, 199.)

FURTHER DESCRIPTIVE INFORMATION

Petrotoga is closely related to the genus *Geotoga* (84% 16S rRNA similarity). Together, these genera constitute a distinct lineage within the *Thermotogales*. This lineage appears to be faster evolving, as indicated by the longer branch length relative to other lineages within the order. This rate difference likely contributes to different topologies inferred using distance and maximum likelihood methods of tree construction. Using distance methods the *Geotoga/Petrotoga* lineage defines the point of earliest divergence within the *Thermotogales*. By contrast, maximum likelihood shows divergence within the order. Since distance methods tend to place fast-evolving sequences inappropriately deep, we

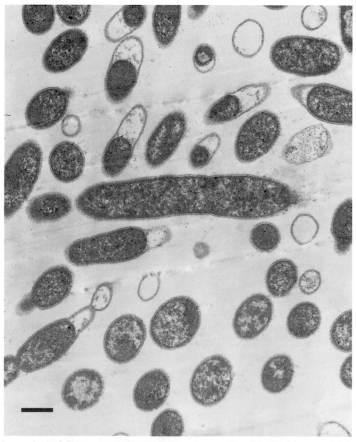

FIGURE B2.8. Thin section of *Petrotoga miotherma*. Bar = 1 μm. (Reprinted with permission from M. Davey et al., Systematic and Applied Microbiology *16*: 191–200, 1993, ©Urban and Fischer Verlag.)

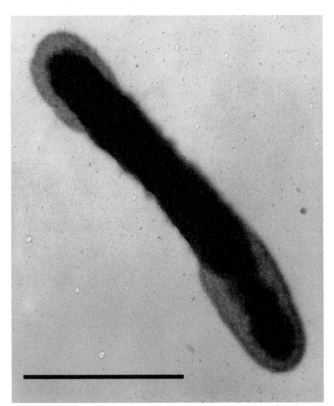

FIGURE B2.9. Negatively stained cell of *Petrotoga miotherma.* EM-micrograph. Bar = 1 μm. (Reprinted with permission from M. Davey et al., *Systematic and Applied Microbiology* *16:* 191–200, 1993, ©Urban and Fischer Verlag.)

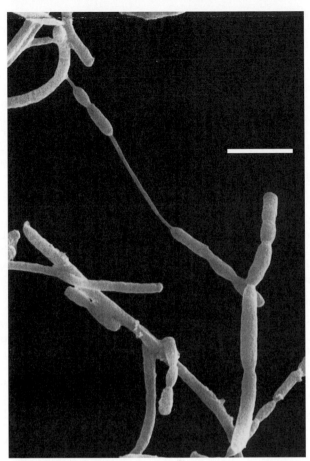

FIGURE B2.10. Scanning electron micrographs of *Petrotoga miotherma.* Bar = 5 μm. (Reprinted with permission from M. Davey et al., *Systematic and Applied Microbiology* *16:* 191–200, 1993, ©Urban and Fischer Verlag.)

present only the maximum likelihood topology (Fig. B2.5). They share phenotypic characteristics with many members of the *Thermotogales*; however, their 16S ribosomal RNA indicates that this genus is not closely related to other characterized members of this order. The distinct microscopic appearance of *Petrotoga* is highly similar to the morphology of other members of the *Thermotogales*. Cells are rod shaped with a characteristic outer sheath-like structure or "toga", as illustrated in Figs. B2.8, B2.9, and B2.10. The sheath is readily visible under phase microscopy, and one or two cells per sheath are routinely observed. In stationary growth phase, the cells form enlarged round spheres surrounded by the sheath.

Petrotoga are moderately thermophilic, growing at 35–65°C, and are able to tolerate a broad range of salt concentrations (0.5–10% NaCl). They contain C_{12} to C_{18} fatty acids which are typical for microorganisms in the bacterial domain. The presence of long-chain fatty acids (C_{22} and greater) which are characteristic of thermophiles has not yet been determined.

On agar medium, colonies are circular, convex, with entire margins, and whitish in color. Yeast extract (0.01%) is required for growth and cannot be replaced with a vitamin mixture or Casamino acids. Stock cultures stored at 4°C will remain viable for up to six months even after the medium becomes oxidized, as indicated by the redox indicator (resazurin). This characteristic indicates not only their ability to persist in culture, but also their ability to tolerate exposure to oxygen. It is also noteworthy that *P. miotherma* produce pink colonies and a pink pellet in the presence of resazurin, indicating an affinity for resazurin by the sheath.

Petrotoga are able to use xylose, mannose, starch, maltodextrins, glucose, lactose, sucrose, galactose, maltose, cellobiose, and ribose as sole energy and carbon sources. *P. mobilis* is also able to ferment xylan. They are not able to use raffinose, ethanol, lactate, Casamino acids, acetate, or formate. In addition, amylase activity was found to be exclusively extracellular in *P. miotherma* (Davey, unpublished data), in contrast to the amylases of *Thermotoga maritima*, which are exclusively associated with the "toga" (Schumann et al., 1991). The fermentation products after incubation at the optimal temperature for one week in the presence of glucose are H_2, CO_2, acetate, and ethanol.

Hydrogen inhibits growth on glucose. No growth occurs in cultures with a headspace of H_2/CO_2 (80:20); however, this effect can be overcome by gassing the headspace with N_2. This inhibition by hydrogen can also be eliminated by the addition of S^0 (0.1%), which results in the production of H_2S; however, the addition of S^0 does not increase the growth yield or change the stoichiometry of fermentation end products. Members of the *Thermotogales* have also been reported to form H_2S in the presence of hydrogen and S^0, which does not accompany an increase in growth (Huber et al., 1986). Huber and colleagues proposed that this is a means of relieving the inhibitory effect of hydrogen. Therefore, this ability may provide these bacteria with a mechanism to grow in environments with elevated levels of hydrogen that they would not otherwise be able to tolerate.

Growth is inhibited by the addition of vancomycin, chloramphenicol, and rifampicin, and by streptomycin when added to the culture incubation at their optimal temperatures (100 µg/ml; final concentration).

Petrotoga have been isolated from North Sea oil wells and petroleum reservoirs located in Shidler, Oklahoma.

Petrotoga can potentially inhabit a number of distinct environments. This is indicated by their ability to tolerate a broad range of temperatures and salt concentrations. Their apparent persistence in petroleum reservoirs also suggests a tolerance to a number of other parameters, i.e., high heavy metal concentrations, pressure, and petroleum.

Investigations into the structure of the "toga" of *Thermotoga maritima* (Rachel et al., 1988, 1990) have determined that the main protein constituent is related to the outer membrane proteins of other bacteria. There are no known functions ascribed to the outer sheath-like structure; however, since all these bacteria were isolated from "atypical" environments, it is conceivable that this structure could play a role in their survival.

ENRICHMENT AND ISOLATION PROCEDURES

Petrotoga can be isolated from petroleum field brines, which are collected anaerobically at production and injection sites in prereduced, sterile bottles. The brine samples should be maintained at ambient temperature and placed in an anaerobic chamber within 24 h of sampling. These samples are used both as a source of inoculum and as a base for the culture medium. Enrichments are designed to mimic the *in situ* conditions (temperature, pH, E_h, and salinity) of the petroleum reservoir from which the brine samples are collected. This can be accomplished by inoculating a 100 ml serum bottle (pre-reduced in a Coy anaerobic chamber with an atmosphere of $N_2/H_2/CO_2$; 85:10:5) containing 0.1 g starch and 0.04 g α-glycerophosphate with 100 ml of field brine. This procedure results in minimal alteration of the concentration of salts and the pH of the sample. Starch or maltodextrins are provided as the preferred carbon and energy source for *Petrotoga*. The α-glycerophosphate is added as a source of phosphate instead of inorganic phosphate precipitation of calcium-phosphate in many brines. If salt concentrations are low, inorganic phosphates can be used. Nitrogen is often not a limiting nutrient in petroleum brines; however, the samples should be tested. This can be done by preparing enrichments without nitrogen and determining if growth is limited. The enrichments are then incubated anaerobically (initial E_h approximately -200 mV) at the reservoir temperature. The enrichments will grow to a concentration of 10^8 cells/ml within 10 days, at which time the sheathed organisms are readily visible with an epifluorescent microscope when stained with acridine orange or by phase microscopy (Hobbie et al., 1977).

The cultures can be monitored by phase microscopy for the appearance of "sheathed" bacteria and then subcultured for isolation. The strains can be isolated on agar plates of Mel's Anaerobic Medium (MAM)*. They should be incubated in an anaerobic chamber with an atmosphere of $N_2/H_2/CO_2$ (85:10:5). The concentrations of inorganic compounds should be varied

to match the particular brine being used in the enrichment and the temperature for incubation should be the temperature of the reservoir from which the brines were collected. Circular, convex, white colonies will appear on the plates after one week of incubation and they should be predominantly "sheathed" bacteria. Subsequently, four additional single colony transfers should be done to ensure purity.

MAINTENANCE PROCEDURES

Lyophilization in skim milk (20%) is recommended for long-term storage. Frozen stocks in a solution of milk (10%) and DMSO (5%) are best for stocks that are used routinely to start a new culture (Gherna, 1994). Stock cultures can best be maintained at 4°C for one month in MAM with reduced salt concentrations (NaCl, 16.6 g; $CaCl_2 \cdot 2H_2O$, 3.8 g). Even though the stocks do remain viable for several months longer, significant changes in morphology, such as enlarged, rounded, and bifurcated cells, are often observed in these older stock cultures, possibly indicative of mutations; therefore, the use of older stocks is not recommended. MAM with reduced salt concentration can also be used to reconstitute lyophilized cultures and as a solid medium for plating. Colonies form in one week when plates are incubated at 45°C in an anaerobic chamber with an atmosphere of $N_2/H_2/CO_2$ (85:10:5).

DIFFERENTIATION OF THE GENUS *PETROTOGA* FROM OTHER GENERA

Petrotoga are morphologically and physiologically similar to other members of the order *Thermotogales*. *Petrotoga* is represented by two species, *P. miotherma* and *P. mobilis*. Differentiation from *Geotoga* is based primarily on 16S rDNA sequence similarity. Other distinguishing characteristics are shown in Table B2.3. Three characteristics (salt tolerance, temperature range for growth, and 16S rRNA sequence) can be used to differentiate *Petrotoga* from the *Thermotogales*. They differ from the other *Thermotogales* by their tolerance to higher salt concentrations; attributed, in part, to their ability to make a variety of compatible solutes, such as trehalose, glycine betaine, and β-mannosylglycerate (Martins et al., 1996). The *Thermotogales* are characteristically thermophilic. Two genera within the order of *Thermotogales*, *Thermosipho* (Huber et al., 1989d), and *Fervidobacterium* (Huber et al., 1990), include strains whose temperature range for growth is slightly more moderate, like *Geotoga*. However, their temperature optimum for growth is higher than that for *Petrotoga* (55–60°C), i.e., *Thermosipho africanus*, 75°C, and *Fervidobacterium islandicum*, 65°C. Furthermore, *Petrotoga* 16S rDNA sequences are only distantly related to sequences from other characterized members of this order.

TAXONOMIC COMMENTS

Petrotoga is a newly defined genus consisting of two species (*P. miotherma* and *P. mobilis*). Phenotypic and genotypic characteristics clearly place it within the *Thermotogales*; however, further analysis of these strains, as well as the isolation and characterization of more species, which are representative of this genus, is required to confirm the classification of this genus.

ACKNOWLEDGMENTS

This research was supported by Phillips Petroleum Co., Bartlesville, Oklahoma; as well as research agreement N00014–91-J4083 from the Office of Naval Research; and research agreement CR815285 from the Environmental Protection Agency to D.A.S.

---

*Mel's Anaerobic Medium (MAM) medium: starch, 1.0 g/l; yeast extract, 0.5 g/l; NaCl, 33.3 g/l; $CaCl_2 \cdot 2H_2O$, 7.6 g/l; $MgCl_2 \cdot 6H_2O$, 2.5 g/l; PIPES, 6.8 g/l; NH_4Cl, 0.5 g/l; trace minerals (Widdel and Pfennig, 1984), 1 ml/l; resazurin, 1 mg/l; α-glycerophosphate, 0.4 g/l; ascorbic acid, 0.02%; cysteine-HCl, 0.01%; agar (Difco), 15.0 g/l; pH 6.8.

List of species of the genus Petrotoga

1. **Petrotoga miotherma** Davey, Wood, Key, Nakamura and Stahl 1993b, 864[VP] (Effective publication: Davey, Wood, Key, Nakamura and Stahl 1993a, 199.)

mi.o.ther′ma. M.L. fem. adj. *miotherma* from Gr. adj. *meiothermos* Gr. fem. adj. less hot, referring to the optimum temperature for growth.

The characteristics are as described for the genus and as indicated in Table B2.3.

The mol% G + C of the DNA is: 32.0–34.2.

Type strain: 42–6, ATCC 51224.

GenBank accession number (16S rRNA): L10657.

2. **Petrotoga mobilis** Lien, Madsen, Rainey and Birkeland 1998, 1012.[VP]

mo′bi.lis. L. fem. adj. *mobilis* motile.

Rod-shaped cells, 0.5–1.0 × 1.0–5.0 µm, with a sheath-like outer structure. Occurs as single cells, in pairs, or in chains, sometimes with more than 20 cells within the sheath. Gram-negative. Obligately anaerobic. Heterotrophic, able to ferment a broad spectrum of carbohydrates including xylan. Frequently motile with subpolar flagellation. Neither endospores nor enlarged spherical bodies in stationary phase are formed. Growth occurs at pH 5.5–8.5, 40–65°C, and at NaCl concentrations of 0.5–9.0% NaCl (optimum 3–4%). Several vitamins are required and yeast (extract) stimulates growth. Isolated from hot oilfield water from an oil reservoir in the North Sea.

The mol% G + C of the DNA is: 31 and 34 (T_m and HPLC, respectively).

Type strain: SJ95, DMS 10674.

GenBank accession number (16S rRNA): Y15479.

Genus V. **Thermosipho** *Huber, Woese, Langworthy, Fricke and Stetter 1989e, 496[VP]*
(Effective publication: Huber, Woese, Langworthy, Fricke and Stetter 1989d, 36), emend. Ravot, Ollivier, Patel, Magot and Garcia 1996b, 322

ROBERT HUBER AND KARL O. STETTER

Ther.mo.si′pho. Gr. fem. n. *therme* heat; L. masc. n. *sipho* tube; M.L. masc. n. *Thermosipho* the hot tube, due to the sheath surrounding the bacteria.

Rod-shaped cells, 1–4 × 0.4–0.6 µm, occurring singly, in pairs and in short chains. The rods form an **outer sheath**, visible in all phases of growth, ballooning over the ends. No resting stages known. **Gram-negative. Obligately anaerobic. Hyperthermophilic**, optimum growth temperature: 70–75°C. Optimal pH: 6.5–7.5. **Heterotrophic growth. Products from glucose fermentation are acetate, CO$_2$, and H$_2$**. Sulfur is reduced to H$_2$S. Inhibition of growth by molecular hydrogen. Sensitive to inhibitors of cell-wall, protein, and nucleic acid biosynthesis.

Deep phylogenetic branch within the bacteria, based on 16S rDNA sequence analysis.

The mol% G + C of the DNA is: 30–30.5.

Type species: **Thermosipho africanus** Huber, Woese, Langworthy, Fricke and Stetter 1989e, 496 (Effective publication: Huber, Woese, Langworthy, Fricke and Stetter 1989d, 36), emend. Ravot, Ollivier, Patel, Magot and Garcia 1996b, 322.

FURTHER DESCRIPTIVE INFORMATION

16S rDNA sequence comparison showed that the genus *Thermosipho* (Huber et al., 1989d; Ravot et al., 1996b), together with the genera *Thermotoga* (Huber et al., 1986), *Petrotoga*, *Geotoga* (Davey et al., 1993a), and *Fervidobacterium* (Patel et al., 1985a), belongs to the order *Thermotogales* (Huber and Stetter, 1992c,d). For further taxonomic comments see also Genus *Thermotoga* in this *Manual*.

Thermosipho cells occur singly, in pairs, and in short chains up to 12 cells in length, depending on the growth phase. The cells are surrounded by a sheath-like structure, ballooning over the ends (Fig. B2.11). In the late-stationary growth phase, *T. africanus* cells tend to become large spheres.

Cells of *T. africanus* are sensitive to lysozyme, indicating the presence of a cell wall. They show a Gram-negative staining reaction, but meso-diaminopimelic acid is absent. The lipids are composed of fatty acid methyl esters (Huber et al., 1989d; Antoine et al., 1997). In addition, unusual C$_{28}$- to C$_{34}$- dicarboxylic fatty acids were identified in *T. africanus* (Huber et al., 1989d).

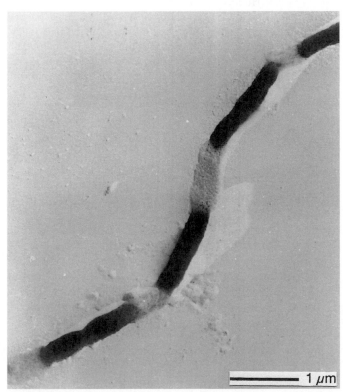

FIGURE B2.11. Four rod-shaped cells of *Thermosipho africanus* in a tube-like sheath; platinum-shadowed.

After five days of anaerobic incubation at 65°C, *T. africanus* forms round and colorless colonies, about 1 mm in diameter.

For the growth of *Thermosipho*, the anaerobic technique of Balch and Wolfe (1976) is recommended. Table B2.4 summarizes the physiological conditions for growth. *Thermosipho* species are obligate heterotrophs, preferentially using carbohydrates for growth (for details see species description). For heterotrophic growth, complex organic material such as yeast extract is necessary. Detailed growth conditions were described recently by Huber and Stetter (1992c).

During glucose fermentation, lactate, acetate, ethanol, butyric acid, valeric acid, L-alanine, CO_2, and H_2 are formed as final products. In the presence of thiosulfate, growth of *T. africanus* is improved and acetate production from sugars is increased (Ravot et al., 1995b). H_2 inhibits growth of *Thermosipho*. This hydrogen inhibition can be eliminated by the addition of S^0, which is reduced to H_2S.

Thermosipho species are sensitive to different antibiotics, which typically inhibit bacterial growth. Rifampicin is inhibitory only at high concentration ≥100 μg/ml.

The type species *T. africanus* was originally isolated from marine hydrothermal springs and sandy sediments of the Gulf of Tadjoura southwest of Obock (Djibouti, Africa) (Huber et al., 1989d). *T. melanesiensis*, a second species within this genus, was isolated recently from the gills of the deep-sea hydrothermal vent mussel *Bathymodiolus brevior*. The invertebrate was collected at the bottom of a black smoker in the Lau Basin (depth: 1832 m) of the southwest Pacific Ocean (Antoine et al., 1997). A new, thus far undescribed *Thermosipho* species was obtained from a deep oil reservoir, located in the East Shetland Basin of the North Sea (Huber and Stetter, unpublished).

ENRICHMENT AND ISOLATION PROCEDURES

A selective enrichment procedure for *Thermosipho* is not known. In general, enrichments of *Thermosipho* relatives must be performed under anaerobic culture conditions (Balch and Wolfe, 1976). Half-strength seawater media at neutral pH (Table B2.4), supplemented with yeast extract and an additional carbon source like starch or glucose is suitable. The media are inoculated with the original sampling material and are incubated at 70°C with a headspace of N_2/CO_2 (80:20; v/v). The enrichment bottles are examined over a period of two weeks for growth by the use of a phase-contrast microscope. Cultures predominated by chains of rods covered with an outer, sheath-like structure are chosen for further work. The enrichments are serially diluted three times in sequence. Afterward, pure cultures of *Thermosipho* can be obtained under anaerobic conditions by plating, using 1.5% agar plates or Gelrite plates and a stainless steel anaerobic jar (Balch

TABLE B2.4. Diagnostic and descriptive features of the *Thermosipho* species[a]

Characteristic	1. *Thermosipho africanus*	2. *Thermosipho melanesiensis*
pH range	6.0–8.0	4.5–8.5
pH optimum	7.2	6.5–7.5
Temperature range, °C	35–77	50–75
Temperature optimum, °C	75	70
NaCl range, %	0.11–3.6	1.0–6.0
NaCl optimum, %	nd	3
Optimal doubling time, min	35	100

[a]nd, not determined.

et al., 1979). Single colonies will become visible within 3–5 days of incubation at 60–65°C and can serve as an inoculum for liquid media. *Thermosipho* species can be also isolated by the new "selected cell cultivation" technique, using a strongly focused infrared laser beam as "optical tweezers" (Huber et al., 1995a; Beck and Huber, 1997).

MAINTENANCE PROCEDURES

T. africanus can be stored at −140°C in liquid nitrogen in the presence of 5% DMSO. No loss of cell viability was observed after storage over a period of six years.

DIFFERENTIATION OF THE GENUS *THERMOSIPHO* FROM OTHER GENERA

Phylogenetically, *Thermosipho* can be differentiated from the other *Thermotogales* genera by comparison of the 16S rDNA sequences. *Thermosipho* may also be differentiated from these genera by cellular morphology: growth in chains of up to 12 rods, surrounded by a sheath. In addition, *Thermotoga* and *Fervidobacterium* can be differentiated from *Thermosipho* by the presence of RNA polymerases, which show no serological cross reaction with the same enzyme of *Thermosipho*. *Thermosipho* can be further distinguished from *Thermotoga* by a different lipid composition and a mol% G + C content of only 30. *Thermosipho* may be differentiated from *Geotoga* and *Petrotoga* by differences in its sodium chloride requirement and in its higher temperature optimum and temperature maximum for growth.

FURTHER READING

Huber, R. and K.O. Stetter. 1992. The order *Thermotogales*. *In* Balows, Trüper, Dworkin, Harder and Schleifer (Editors), The Prokaryotes: A Handbook of Bacteria: Ecophysiology, Isolation, Identification, Applications, 2nd Ed., Springer-Verlag, New York, pp. 3809–3815.

Stetter, K.O. 2000. Volcanoes, hydrothermal venting, and the origin of life. *In* Marit and Ernst (Editors), Volcanoes and the Environment, Cambridge University Press, in press.

List of species of the genus Thermosipho

1. **Thermosipho africanus** Huber, Woese, Langworthy, Fricke and Stetter 1989e, 496[VP] (Effective publication: Huber, Woese, Langworthy, Fricke and Stetter 1989d, 36), emend. Ravot, Ollivier, Patel, Magot and Garcia 1996b, 322.
   af.ri.ca' nus. L. masc. adj. *africanus* belonging to Africa, describing its place of isolation.

   For heterotrophic growth, complex organic material such as yeast extract, peptone or tryptone is necessary. D-Glucose, starch, D-ribose, and maltose used. D-Galactose, fructose, and sucrose poorly used. Enhancement of growth on all these

sugars in the presence of thiosulfate. No growth observed on cellulose, dulcitol, L-arabinose, lactose, L-xylose, D-xylose, L-sorbose, L-rhamnose, and D-mannitol. Main products during glucose fermentation are acetate, CO_2, H_2, and small amounts of ethanol and lactate. Sulfur and thiosulfate reduced to H_2S. No growth inhibition in the presence of 10 μg/ml rifampicin. Growth is inhibited in the presence of streptomycin, vancomycin, chloramphenicol, and phosphomycin (10 μg/ml) and rifampicin (100 μg/ml). Sensitive to lysozyme. No serological cross-reaction of an enriched RNA polymerase fraction with antibodies against the purified RNA polymerase of *Thermotoga maritima*.

No significant DNA–DNA hybridization with *T. melanesiensis*.

The mol% G + C of the DNA is: 29 (T_m) and 30 (HPLC).
Type strain: Ob7, DSMZ 5309.
GenBank accession number (16S rRNA): M24022.

2. **Thermosipho melanesiensis** Antoine, Cilia, Meunier, Guezennec, Lesongeur and Barbier 1997, 1122.[VP]

me.la.ne.si.en′ sis. L. masc. adj. *melanesiensis* originating from Melanesia, describing its site of isolation.

For heterotrophic growth, complex organic material such as yeast extract or brain heart infusion is necessary. Growth is supported by malt extract, tryptone, and carbohydrates supplemented by yeast extract. In the presence of yeast extract, decreasing growth yields are obtained with sucrose and starch, glucose, maltose, lactose, cellobiose, and galactose. In combination with yeast extract, meat extract, xylose, fructose, ethanol, mannitol, sorbitol, glycine, glycogen, formate, acetate, and Casamino acids do not support growth. Sulfur is used as electron acceptor, but neither thiosulfate nor sulfate is used. The presence of sulfur increases the growth yield. Growth inhibited in the presence of rifampicin, chloramphenicol, and streptomycin (100 µg/ml).

No significant DNA–DNA hybridization with *T. africanus*.
The mol% G + C of the DNA is: 30.5 (T_m).
Type strain: BI429, CIP 104789, DSMZ 12029.
GenBank accession number (16S rRNA): Z70248.

Phylum BIII. Thermodesulfobacteria *phy. nov.*

George M. Garrity and John G. Holt

Ther.mo.de.sul.fo.bac.te' ri.a. M.L. fem. pl. n. *Thermodesulfobacteriales* type order of the phylum; dropping the ending to denote a phylum; M.L. fem. pl. n. *Thermodesulfobacteria* the phylum of *Thermodesulfobacteriales.*

The phylum *Thermodesulfobacteria* is currently represented by a single genus which branches deeply in the major reference trees. Gram-negative, rod-shaped cells possessing an outer membrane layer which forms protrusions. Thermophilic, strictly anaerobic, chemoheterotrophs exhibiting a dissimilatory sulfate-reducing metabolism.

Type order: **Thermodesulfobacteriales** ord. nov.

Class I. **Thermodesulfobacteria** *class. nov.*

E. Claude Hatchikian , Bernard Ollivier and Jean-Louis Garcia

Ther.mo.de.sul.fo.bac.te' ri.a. M.L. fem. pl. n. *Thermodesulfobacteriales* type order of the class; dropping the ending to denote an class; M.L. fem. pl. n. *Thermodesulfobacteria* the class of *Thermodesulfobacteriales.*

Rod-shaped cells, 0.3–0.6 × 0.9–2.0 µm; occur singly, in pairs, or in chains in young cultures; sometimes pleomorphic in old cultures. Possess an outer wall membrane layer. **Gram negative**. Does not form spores. **Usually nonmotile**, but motility might be observed in some species. **Thermophilic**. Neutrophilic. Chemoorganotrophic, **strict anaerobe**, ferments pyruvate. The principal fermentation end products are acetate, CO_2, and hydrogen. Sulfate and thiosulfate are used as electron acceptor for growth.

Lactate and pyruvate used as electron donors for growth. In the presence of sulfate, lactate and pyruvate are incompletely oxidized to acetate. *Thermodesulfobacterium* forms a distinct lineage within the domain *Bacteria* (Fig. B3.1). Occur in thermal environments, including thermophilic digestors, hot springs, and hot oil reservoirs.

The mol% G + C of the DNA is: 34–38.

Type order: **Thermodesulfobacteriales** ord. nov.

Order I. **Thermodesulfobacteriales** *ord. nov.*

E. Claude Hatchikian , Bernard Ollivier and Jean-Louis Garcia

Ther.mo.de.sul.fo.bac.te.ri.a' les. M.L. neut. n. *Thermodesulfobacterium* type genus of the order; *-ales* ending to denote an order; M.L. fem. pl. n. *Thermodesulfobacteriales* the order of *Thermodesulfobacterium.*

Only one order, *Thermodesulfobacteriales*, is accepted in the class *Thermodesulfobacteria*; the description of the order is therefore the same as for the class.

Type genus: **Thermodesulfobacterium** Zeikus, Dawson,

Thompson, Ingvorsen and Hatchikian 1995, 197[VP] (Effective publication: Zeikus, Dawson, Thompson, Ingvorsen and Hatchikian 1983, 1167.)

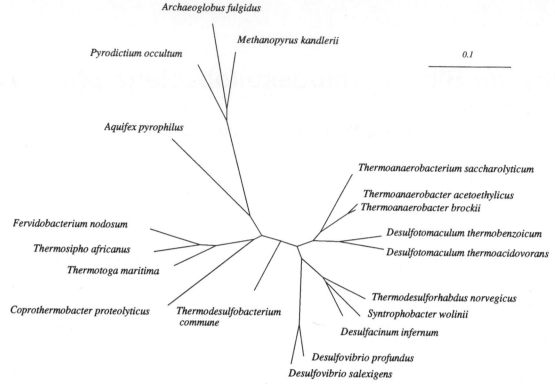

FIGURE B3.1. Unrooted phylogenetic tree based on 16S rRNA gene sequence comparison, and obtained by a neighbor-joining algorithm (Phylip package), indicating the position of *Thermodesulfobacteria* class among members of low branches of the tree of life. Bar = 10 nucleotide changes per 100 nucleotides.

Family I. **Thermodesulfobacteriaceae** *fam. nov.*

E. CLAUDE HATCHIKIAN , BERNARD OLLIVIER AND JEAN-LOUIS GARCIA

Ther.mo.de.sul.fo.bac.te.ri.a' ce.ae. M.L. neut. n. *Thermodesulfobacterium* type genus of the family; *-aceae* ending to denote a family; M.L. fem. pl. n. *Thermodesulfobacteriaceae* the family of *Thermodesulfobacterium.*

Only one family, *Thermodesulfobacteriaceae*, is accepted in the order *Thermodesulfobacteriales*; the description of the family is therefore the same as for the class. Only one genus is accepted in the family *Thermodesulfobacteriaceae*.

Type genus: **Thermodesulfobacterium** Zeikus, Dawson, Thompson, Ingvorsen and Hatchikian 1995, 197[VP] (Effective publication: Zeikus, Dawson, Thompson, Ingvorsen and Hatchikian 1983, 1167.)

Genus I. **Thermodesulfobacterium** *Zeikus, Dawson, Thompson, Ingvorsen and Hatchikian 1995, 197[VP] (Effective publication: Zeikus, Dawson, Thompson, Ingvorsen and Hatchikian 1983, 1167)*

E. CLAUDE HATCHIKIAN AND BERNARD OLLIVIER

Ther.mo.de.sul.fo.bac.te' ri.um. Gr. masc. n. *thermos* heat; L. pref. *de* from; L. neut. n. *sulfur* sulfur; Gr. n. *bakterion* a small rod; M.L. neut. n. *Thermodesulfobacterium* a thermophilic rod reducing sulfate.

Straight rod-shaped cells, 0.3–0.6 × 0.9–2.0 µm; occur singly, in pairs, or in chains. Possess an outer wall membrane layer. Gram-negative. Cellular extrusions or blebs form next to the outer membranous layer. Contain non-phytanyl ether-linked lipids, cytochrome c_3, desulfofuscidin, but no desulfoviridin. **Thermophilic**. Chemoorganotrophic, strict anaerobe, and dissimilatory sulfate-reducing metabolism. Lactate and pyruvate are used as electron donors, and sulfate or thiosulfate are used as electron acceptors for growth. In the presence of sulfate, lactate and pyruvate are incompletely oxidized to acetate. Occur in thermal environments including hot springs and hot oil reservoirs.

The mol% G + C of the DNA is: 31–38.

Type species: **Thermodesulfobacterium commune** Zeikus, Dawson, Thompson, Ingvorsen and Hatchikian 1995, 197 (Effective publication: Zeikus, Dawson, Thompson, Ingvorsen and Hatchikian 1983, 1168.)

FURTHER DESCRIPTIVE INFORMATION

In contrast to *Thermodesulfobacterium commune*, cells of *Thermodesulfobacterium mobile* (formerly *Desulfovibrio thermophilus*) are motile by a single polar flagellum and may occur in long chains (Rozanova and Khudyakova, 1974). Cells of the latter tend to be the largest in diameter and in length. Both species possess a typical Gram-negative cell wall architecture as observed under thin-section electron microscopy. Cell material resembling the "blebs" results from extrusions near the outer membranous layer and cell division in *T. commune*. Large quantities of polyribosomes are found in the cytoplasm of *T. mobile*. They all contain cytochrome c_3-type (Hatchikian et al., 1984; Fauque et al., 1990) and desulfofuscidin as specific dissimilatory sulfite reductase (Hatchikian and Zeikus, 1983; Fauque et al., 1990; Hatchikian, 1994). The two strains possess MK-7 as a major menaquinone (Collins and Widdel, 1986), and *T. commune* was reported to contain a two-[4Fe-4S] cluster ferredoxin involved in the phosphoroclastic reaction (Guigliarelli et al., 1985; Papavassiliou and Hatchikian, 1985). Immunocytochemical localization of desulfofuscidin and APS reductase in *T. mobile* have shown that the former enzyme was located in the cytoplasm, whereas the latter was associated with the cytoplasmic membrane, in contrast to *Desulfovibrio* species (Kremer et al., 1988). The main polyamine synthesized by *T. commune* is N^4-*bis*(aminopropyl)spermidine (Hamana et al., 1996).

The two *Thermodesulfobacterium* species oxidize a limited range of substrates, including hydrogen, lactate, and pyruvate in the presence of sulfate as electron acceptor; lactate and pyruvate are incompletely oxidized to acetate and CO_2, and sulfate is reduced into sulfide. A similar bacterium, that used only H_2 as energy source, was isolated from hot oil field water (Cord-Ruwisch et al., 1986). Pyruvate is fermented to acetate, CO_2, and H_2 by the two species, whereas lactate is only slightly fermented by *T. mobile* with production of small amounts of H_2. However, H_2 was also reported as a trace gas formed during lactate oxidation by *T. commune* grown in the presence of sulfate. In contrast to *T. mobile*, *T. commune* is not known to use sulfite as electron acceptor (Zeikus et al., 1983), but in the case of *T. commune*, the sulfite concentration tested (25 mM) might have been highly toxic for growth, since this electron acceptor is usually used at concentrations of around 2 mM. The lipids of *T. commune* comprise unique non-isoprenoid branched glycerol diethers and monoethers (Langworthy et al., 1983). The lipid fraction also contains some fatty acids and hydrocarbons (Langworthy et al., 1983).

Bacteria similar to *T. commune* are found in thermal environments such as Yellowstone National Park Springs, where organic matter is actively decomposed (i.e., the Octopus Spring algal-bacterial mat) or in waters and sediments that contain geothermal H_2 (i.e., the Ink Pot Hot Spring environment). This suggests that in these environments, *T. commune* grows preferentially on H_2 resulting from fermentation processes or geothermal reactions (Widdel, 1992) and that acetate is available as a carbon source for sustainable growth on H_2. In addition, *T. commune* was also isolated from a hot oil reservoir and cultivated on the oil-water phase, suggesting its involvement in the oxidation of aliphatic and aromatic hydrocarbons (L'Haridon et al., 1995). Interestingly, *T. mobile* is also found in oil field ecosystems. The type

strain 7 (VKM B-1128) was isolated from stratal water of the oil deposit of the Apsheron Peninsula on the Caspian Sea (Rozanova and Khudyakova, 1974), whereas *T. mobile* strain GFA1 was immunomagnetically captured from a North Sea oil field water (Christensen et al., 1992). Based on these findings, Christensen et al. (1992) concluded that *T. mobile* might be indigenous to several oil-bearing strata throughout the world. These results finally suggest that *Thermodesulfobacterium* species have not only terrestrial, but also subterrestrial, thermal origins. Nevertheless, all known *Thermodesulfobacterium* strains so far do not display a pronounced salt requirement or a significant salt tolerance (Widdel, 1992; L'Haridon et al., 1995).

ENRICHMENT AND ISOLATION PROCEDURES

Enrichment of *Thermodesulfobacterium* species can be obtained on different media, including medium 77 of Postgate (1969), and classical media for sulfate-reducing bacteria (Widdel, 1980; Widdel and Pfennig, 1984). A NaCl content lower than 2% is preferable. Enrichments can be performed in the presence of lactate as electron donor and sulfate as electron acceptor. When H_2 is used as the energy source, acetate (around 2 mM) is required as a carbon source. Incubations ranging from 60 to 70°C are recommended. Positive growth is indicated by the production of sulfide and acetate when lactate is used as a primary source of energy. At least three subcultures in the same growth conditions are necessary before isolation. If classical enrichment procedures with consecutive transfers to selective media are unsuccessful, immunomagnetic enrichment using specific antisera as capture agents can be used (for more details, see Christensen et al., 1992).

After several transfers, the enrichment cultures are serially diluted by using the method of Hungate (1969) with roll tubes containing the basal medium and purified agar at 2–4%. At least two colonies are picked up, and the process of serial dilution in roll tubes is repeated to purify the cultures. Purity can be checked by microscopic examination after growth on a complex rich medium containing sugars to allow the emergence of classical heterotrophs. Isolation is also possible via serial dilutions in liquid medium (Widdel, 1992).

MAINTENANCE PROCEDURES

Stock cultures can be maintained on Medium 77, as described by Postgate (1969), and transferred monthly. Liquid cultures retained viability after several months at room temperature. Cultures can be also refrigerated (Widdel, 1992). Because of easy death/lysis of cells when cultures are incubated longer at high temperature, the preferred method of preparing stock cultures is to harvest cells at the end of the exponential growth phase.

For long-term preservation, 5–10% (v/v) DMSO is added to cultures kept in liquid nitrogen (Widdel, 1992).

DIFFERENTIATION OF THE GENUS *THERMODESULFOBACTERIUM* FROM OTHER GENERA

Differentiation of the genus from other taxa with which it might be confused is indicated in Table B3.1.

TAXONOMIC COMMENTS

The genus *Thermodesulfobacterium* was first proposed by Zeikus et al. (1983) for a nonsporulating thermophilic sulfate-reducing rod-shaped bacterium, *Thermodesulfobacterium commune*. Thereafter, based on phenotypic and genotypic characteristics, another thermophilic bacterium, previously named *Desulfovibrio thermo-*

TABLE B3.1. Differentiation of the genus *Thermodesulfobacterium* from other nonspore-forming Gram-negative thermophilic sulfate-reducing genera[a]

Characteristic	*Thermodesulfobacterium*	*Thermodesulforhabdus*	*Desulfacinum*	*Thermodesulfovibrio*
Straight rods	+	+	−	−
Motility	D	+	−	+
Growth at temperatures 80°C	+	−	−	−
Autotrophic growth	−	−	+	−
Sulfite reductase: desulfofuscidin	+	−	nd	+
Fermentation of pyruvate	+	−	nd	+
Electron acceptors:				
Sulfate	+	+	+	+
Thiosulfate	+	−	+	+
Electron donors:				
Acetate	−	+	+	−
Malate	−	+	+	−
Ethanol	−	+	+	−
Fatty acids C_4–C_{10}, C_{13}–C_{18}	−	+	+	−
Oxidation of organic substrates	incomplete	complete	complete	incomplete

[a]For symbols see standard defintions; nd, not determined.

philus (Rozanova and Khudyakova, 1974; Widdel and Pfennig, 1984), was tentatively renamed *Thermodesulfobacterium mobile* (Rozanova and Pivovarova, 1988, 1991). The placement of this species in *Thermodesulfobacterium* was confirmed by analysis of 16S rRNA sequences (Widdel, 1992). The genus *Thermodesulfobacterium* was described long ago but was only recently validated together with the species *T. commune* (Zeikus et al., 1995). As recommended by Henry et al. (1994a) and Tao et al. (1996), the renaming of *Desulfovibrio thermophilus* as *Thermodesulfobacterium mobile* was not in accordance with the international rules of nomenclature. The correct reassignment should be *Thermodesulfobacterium thermophilus*, in accordance with Rule 41a of the Bacteriological Code.

In a recent study, a new thermophilic sulfate-reducing bacterium named *Thermodesulfovibrio yellowstonii* was described. Like *Thermodesulfobacterium commune*, it was also isolated from a thermal pool at Yellowstone National Park. Both microorganisms represent a lineage that branches deeply within the *Bacteria* domain. This lineage is clearly distinct from the previously defined phylogenetic lines of sulfate-reducing bacteria represented by the Gram-negative mesophilic species of the proteobacteria, the Gram-positive species, and the hyperthermophilic archaeon, *Archaeoglobus* sp. (Henry et al., 1994a).

ACKNOWLEDGMENTS

The authors wish to thank J.L. Cayol for drawing the phylogenetic tree. We also thank M.L. Fardeau and P.A. Roger for their helpful discussions.

FURTHER READING

Henry, E.A., R. Devereux, J.S. Maki, C.C. Gilmour, C.R. Woese, L. Mandelco, R. Schauder, C.C. Remsen and R. Mitchell. 1994a. Characterization of a new thermophilic sulfate-reducing bacterium *Thermodesulfovibrio yellowstonii*, gen. nov. and sp. nov.: its phylogenetic relationship to *Thermodesulfobacterium commune* and their origins deep within the bacterial domain. Arch. Microbiol. *161*: 62–69.

Rozanova, E.P. and A.I. Khudyakova. 1974. A new nonspore-forming thermophilic sulfate-reducing organism, *Desulfovibrio thermophilus* nov. sp. Mikrobiologiya *43*: 1069–1075.

Rozanova, E.P. and T.A. Pivovarova. 1988. Reclassification of *Desulfovibrio thermophilus* (Rozanova, Khudyakova, 1974). Mikrobiologiya *57*: 102–106.

Zeikus, J.G., M.A. Dawson, T.E. Thompson, K. Ingvorsen and E.C. Hatchikian. 1983. Microbial ecology of volcanic sulphidogenesis: isolation and characterization of *Thermodesulfobacterium commune* gen. nov. and sp. nov. J. Gen. Microbiol. *129*: 1159–1169.

DIFFERENTIATION OF THE SPECIES OF THE GENUS *THERMODESULFOBACTERIUM*

Differential characteristics of the species of *Thermodesulfobacterium* are indicated in Table B3.2.

List of species of the genus Thermodesulfobacterium

1. **Thermodesulfobacterium commune** Zeikus, Dawson, Thompson, Ingvorsen and Hatchikian 1995, 197[VP] (Effective publication: Zeikus, Dawson, Thompson, Ingvorsen and Hatchikian 1983, 1168.)

com.mu' ne. L. neut. adj. *commune* widespread.

-DSmall rods, 0.3 × 0.9 µm, occurring singly or in pairs. Possess a Gram-negative cell wall (Fig. B3.2). Cells often form blebs at their ends. Motility not observed. Lactate, pyruvate, and hydrogen are used as electron donors, and sulfate or thiosulfate are used as electron acceptors for growth. In the presence of sulfate, lactate is incompletely oxidized to acetate, CO_2, and sulfate is reduced into H_2S. Pyruvate is fermented to H_2, CO_2, and acetate as end products. Growth with hydrogen as sole electron donor for energy metabolism, acetate being required as carbon source. Fumarate and nitrate are not reduced. Does not use glucose, malate, and ethanol.

Temperature range: 45–82°C; optimum temperature for growth: 70°C. pH range: 6.0–8.0; optimum pH for growth: 7.0. Desulfofuscidin is present, but not desulfoviridin or desulforubidin. Does not require NaCl. Growth inhibited by 2% NaCl, 250 µg/ml sodium azide, and 100 µg/ml penicillin, cycloserine, chloramphenicol, and tetracycline.

Habitat: anaerobic niches associated with volcanic thermal environments. Isolated from Ink Pot Spring, Yellowstone National Park, USA.

TABLE B3.2. Characteristics differentiating *Thermodesulfobacterium commune* and *Thermodesulfobacterium mobile*[a]

Characteristic	1. *Thermodesulfobacterium commune*	2. *Thermodesulfobacterium mobile*
Dimensions, μm	0.3 × 0.9	0.6 × 2.0
Presence of cells in chains	−	+
Motility	−	+
Optimum temperature, °C	70	65
Mol% G + C	34	38–31[b]
Lactate fermentation	−	(+)
Electron acceptors:		
Sulfate	+	+
Thiosulfate	+	+
Sulfite	(−)[c]	+

[a]For symbols see standard definitions; (+), slight growth.

[b]Data from Henry et al. (1994a).

[c]Tested at 25 mM.

The mol% G + C of the DNA is: 34.

Type strain: YSRA-1, ATCC 33708, DSMZ 2178.

GenBank accession number (16S rRNA): L10662.

2. **Thermodesulfobacterium mobile** (Rozanova and Khudyakova 1974) Rozanova and Pivovarova 1991, 179[VP] (Effective publication: Rozanova and Pivovarova 1988, 88) (*Desulfovibrio thermophilus* Rozanova and Khudyakova 1974, 912.)

mo.bi.le. L. neut. adj. *mobile* motile.

Rods with rounded ends, average size 2.0 × 0.5 μm with a cell wall structure typical of Gram-negative bacteria. Cells occur in pairs or in chains. Small vesicular evaginates of the cell membrane are sometimes located between cells or around them. Cells are motile by a single polar flagellum. Lactate and pyruvate promote growth in the presence of sulfate as electron acceptor and are incompletely oxidized to acetate. Thiosulfate or sulfite can replace sulfate. Ferments pyruvate and lactate (slightly), with the evolution of small amounts of H_2. Hydrogen and formate are used as energy sources in the presence of sulfate as electron acceptor, but only when the medium is supplemented with acetate or yeast extract. Does not use methanol, ethanol, butanol, isobutanol, acetate, propionate, butyrate, oxalate, glucose, lactose, and choline.

Temperature range: 45–85°C; optimum temperature: 65°C. Does not require NaCl; grows rapidly in media with a NaCl content of 0.05–1% and more slowly with 2% NaCl. Desulfofuscidin is present, but not desulfoviridin.

Habitat: stratal water of the oil deposit on the Apsheron Peninsula on the Caspian Sea with a temperature of 84°C.

The mol% G + C of the DNA is: 38.

Type strain: VKMV-1128, DSMZ 1276.

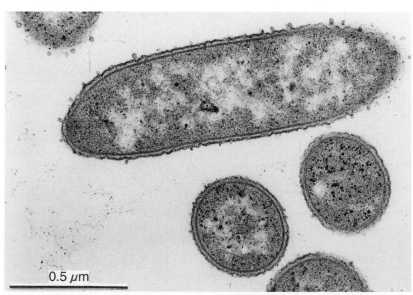

FIGURE B3.2. Ultrathin section of *Thermodesulfobacterium commune* showing a cell wall structure typical of Gram-negative bacteria.

Phylum BIV. "Deinococcus–Thermus"

Class I. **Deinococci** *class. nov.*

GEORGE M. GARRITY AND JOHN G. HOLT

Dei.no.coc' ci. M.L. fem. pl. n. *Deinococcales* type order of the class; dropping the ending to denote a class; M.L. fem. pl. n. *Deinococci* the class of *Deinococcales*.

The description of the class is the same as for the order *Deino-coccales*.

Type order: **Deinococcales** Rainey, Nobre, Schumann, Stacke-brandt and da Costa 1997, 513.

Order I. **Deinococcales** Rainey, Nobre, Schumann, Stackebrandt and da Costa 1997, 513[VP]

JOHN R. BATTISTA AND FRED A. RAINEY

Dei.no.coc.ca' les. M.L. masc. n. *Deinococcus* type genus of the order; *-ales* ending to denote an order; M.L. fem. pl. n. *Deinococcales* the *Deinococcus* order.

The order *Deinococcales* was described with the family *Deinococ-caceae* as the type family (Rainey et al., 1997), and the descriptive details are given under *Deinococcus*. The order is described as a **phylogenetic lineage** that contains all species of the genus *Dei-*

nococcus as a coherent group (Fig. B4.1). A set of 16S RNA–rDNA **signature nucleotides** defines the order (Table B4.1).

Type genus: **Deinococcus** Brooks and Murray 1981, 354, emend. Rainey, Nobre, Schumann, Stackebrandt and da Costa 1997, 513.

Family I. **Deinococcaceae** Brooks and Murray 1981, 356,[VP] emend. Rainey, Nobre, Schumann, Stackebrandt and da Costa 1997, 513

JOHN R. BATTISTA AND FRED A. RAINEY

Dei.no.coc.ca' ce.ae. M.L. masc. n. *Deinococcus* type genus of the family; *-aceae* ending to denote family; M.L. fem. pl. n. *Deinococcaceae* the *Deinococcus* family.

The family *Deinococcaceae* was described with the genus *Deinococcus* as the type genus (Brooks and Murray, 1981), and the descriptive details are given under *Deinococcus*. Strains are **nonmotile**, without differentiated resting stages. All strains examined are **chemoor-ganotrophic**, **aerobic**, and **catalase positive**. Metabolism is res-piratory. Strains may be **mesophilic** or **thermophilic** with opti-mum growth temperatures of 25–35°C and 45–50°C, respectively. The phospholipids do not include phosphatidylglycerol, diphos-phatidylglycerol, or their derivatives. Fatty acids are saturated or monounsaturated. Unbranched 15, 16, and 17 carbon acids pre-dominate in mesophilic species, branched 16 and 17 carbon acids in thermophilic species.

Type genus: **Deinococcus** Brooks and Murray 1981, 354, emend. Rainey, Nobre, Schumann, Stackebrandt and da Costa 1997, 513.

FURTHER DESCRIPTIVE INFORMATION

The genus *Deinococcus* was described for *"Micrococcus radiodurans"* and other radiation resistant cocci by Brooks and Murray (1981) following the study of 55 red-pigmented micrococci. Five species were placed in the new genus: *Deinococcus radiodurans, Deinococcus*

radiophilus, Deinococcus proteolyticus, Deinococcus radiopugnans, and *Deinococcus erythromyxa.* Subsequent chemotaxonomic analysis in-dicated that *D. erythromyxa* was not a *Deinococcus* (Murray and Brooks, 1986), and recent 16S rRNA gene sequence based phy-logenetic analyses has established that *D. erythromyxa* is an acti-nomycete of the genus *Kocuria* (Rainey et al., 1997). Despite a morphological similarity, these deinococci were differentiated from the genus *Micrococcus* and other genera of Gram-positive bacteria by a layered cell wall, the presence of L-ornithine in the peptidoglycan, a lack of teichoic acid, and extreme resistance to UV and ionizing radiation.

Oyaizu et al. (1987) described the genus *Deinobacter* and added it to the family *Deinococcaceae.* The genus *Deinobacter* was created based on the morphology of the isolate and *Deinobacter grandis* designated as the type species. Subsequent 16S rRNA gene se-quence comparisons demonstrated that *D. grandis* and all validly described *Deinococcus* species formed a distinct phylogenetic line-age (Rainey et al., 1997). On the basis of the close relationship between *D. grandis* and other members of the genus *Deinococcus,* *Deinobacter grandis* was transferred to the genus *Deinococcus* as *Deinococcus grandis* comb. nov. (Rainey et al., 1997). Two recent

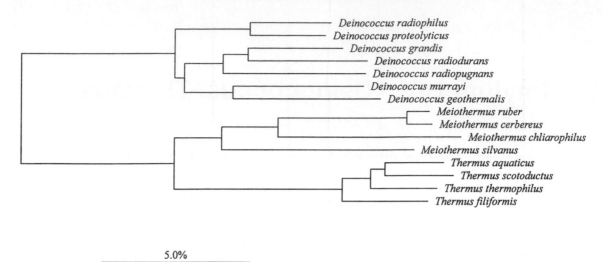

5.0%

FIGURE B4.1. Phylogenetic dendrogram based on 16S rDNA–rRNA sequence comparisons. The dendrogram was constructed by using the neighbor-joining method (Saitou and Nei, 1987) from evolutionary distances. Bar = 5 inferred nucleotide substitutions per 100 nucleotides.

TABLE B4.1. 16S rRNA gene sequence signature nucleotide positions that distinguish members of the genera *Deinococcus* from members of other phylogenetic lineages within the domain *Bacteria*

Positions	Nucleotides in:	
	Deinococcus lineage	Other bacteria
584–757	C–U	G–U
657–749	C–G	U/G–A/C/U
1050–1208	G–U	G–C
1421–1479	A–C/U	U/G/C–C/G/A
1429–1471	G–C	A/C–G/U

additions to the genus *Deinococcus*, *Deinococcus geothermalis* and *Deinococcus murrayi*, are thermophiles; 16S rRNA gene sequence comparisons, chemotaxonomic and morphological characteristics, and gamma radiation resistance showed that these species are part of the *Deinococcaceae* (Ferreira et al., 1997).

16S rRNA catalogs revealed the deinococci as a distinct phylogenetic lineage in the domain *Bacteria* (Woese et al., 1985b;

Woese, 1987), and established that the genera *Deinococcus* and *Thermus* were related (Hensel et al., 1986). Analyses of a nearly complete 16S rRNA sequence of *D. radiodurans* strain (UWO 298) with other complete 16S rRNA sequences established the *Deinococcus–Thermus* phylum (Weisburg et al., 1989). Recent 16S rRNA gene sequence based phylogenetic analyses of all species of the genus *Deinococcus* have demonstrated that all species fall within a distinct phylogenetic lineage (Ferreira et al., 1997; Rainey et al., 1997), and set of signature 16S rRNA gene sequence nucleotides (Table B4.1) was defined (Rainey et al., 1997) for the deinococci. All veritable species cluster together, the levels of 16S rRNA gene sequence similarity being between 88.1 and 93.7%. The levels of 16S rRNA gene sequence similarity between the *Deinococcus* group and the *Thermus-Meiothermus* group ranged from 77.5 to 81%, confirming the phylogenetic relatedness of these genera. A phylogenetic dendrogram, constructed using the neighbor joining method (Saitou and Nei, 1987), is shown in Fig. B4.1.

Genus I. **Deinococcus** *Brooks and Murray 1981, 354,[VP] emend. Rainey, Nobre, Schumann, Stackebrandt and da Costa 1997, 513*

JOHN R. BATTISTA AND FRED A. RAINEY

Dei.no.coc′ cus. Gr. adj. *deinos* strange or unusual; Gr. n. *kokkos* a grain or berry; M.L. masc. n. *Deinococcus* unusual coccus.

Deinococcus strains form either **spherical cells** 0.5–3.5 μm in diameter, or **rods** 0.6–1.2 μm wide × 1.5–4.0 μm long. Spherical cells form pairs or tetrads and appear larger than other cocci. Species are **mesophilic** or **thermophilic**, with optimum growth temperatures of 25–35°C or 45–50°C, respectively. With the exception of *D. grandis*, all species stain Gram-positive. **Nonmotile**, without a differentiated resting stage. **Aerobic, catalase is produced.** Several distinct cell wall layers are visible by electron microscopy of sections and an outer membrane is included. The peptidoglycan is type A3β and contains **L-ornithine. Menaquinone 8** (MK-8) is the predominant respiratory quinone. The phospholipids do not include phosphatidylglycerol, diphosphatidylglycerol, or derivatives. Fatty acids are saturated or monounsaturated. Unbranched 15, 16, and 17 carbon acids predominate in mesophilic species, branched 16 and 17 carbon acids in thermophilic species. **Chemoorganotrophic**; metabolism is respiratory. May be proteolytic. All natural isolates are **ionizing-radiation resistant.** The habitat of the mesophilic species is not defined, thermophilic species are found in hot springs.

The mol% G + C of the DNA is: 62–70.

Type species: **Deinococcus radiodurans** Brooks and Murray 1981, 354.

FURTHER DESCRIPTIVE INFORMATION

Cells divide alternately in two planes, exhibiting pairs and tetrads when grown in rich liquid media. In exponential phase, approximately 95% of *D. radiodurans* R1 cells exist as diplococci, the remainder being tetracocci (Hansen, 1978). In defined media* cells occur singly and as pairs in a ratio of about 1:2 (Park and Battista, unpublished observations). The fine structure of cell division is unusual (Thornley et al., 1965; Murray et al., 1983). A new cell wall septum appears to bud from opposite sides of the cell, establishing a plane of division (Fig. B4.2A). As the septum forms, each side grows toward the other, like curtains being closed. Near the center of the cell, the halves of the new cell wall turn 90° in opposite directions (Fig. B4.2B), and grow to meet a second septum forming perpendicular to the first septum (Fig. B4.2C). Eventually, all septa join, resulting in four individual cells. With the exception of rounded inclusion bodies (Fig. B4.3), the cytoplasm is devoid of distinctive structures. The inclusions are reported to be composed of polysaccharide (Murray and Brooks, 1986). The genome of *D. radiodurans* R1 is 3.3 Mbp (Hansen, 1978; Grimsley et al., 1991) and has been sequenced in its entirety (White, et al., 1999). The genome is segmented, consisting of a 2.64-Mb chromosome, a 0.41-Mb chromosome (designated chromosome I and II, respectively), a 0.18-Mb megaplasmid, and a 0.045-Mbp plasmid. Cells are multi-genomic (Hansen, 1978; Harsojo et al., 1981). There is a minimum of four identical copies of the chromosome per stationary phase cell (Hansen, 1978) when grown in rich media. In exponentially growing cells the number of chromosome copies increases to as many as 10 per cell (Harsojo et al., 1981). Naturally occurring plasmids have been found in the type strain of *D. radiophilus* and strains R1 and SARK of *D. radiodurans* (MacKay et al., 1985).

The cell envelope of *D. radiodurans* is a complex multilayered structure (Fig. B4.4), 50 to 60 nm thick and reminiscent of the cell walls of Gram-negative organisms (Thornley et al., 1965; Work and Griffiths, 1968; Brooks et al., 1980; Embley et al., 1987). The plasma and outer membranes are separated by a 14–20 nm peptidoglycan layer and an uncharacterized "compartmentalized layer". Many, but not all, strains of *D. radiodurans* have a hexagonally arrayed layer of proteins (HPI layer) associated with the surface of the outer membrane (Kubler and Baumeister, 1978; Thompson and Murray, 1982a). A few strains exhibit a dense carbohydrate coat (Baumeister et al., 1981).

The plasma and outer membranes appear to have the same lipid composition (Thompson and Murray, 1981). Conventional phospholipids (phosphatidyl-glycerol, di-phosphatidyl-glycerol, phosphatidyl-ethanolamine, phosphatidyl-serine, phosphatidyl-choline, phosphatidyl-inositol) are absent, being replaced by an unusual mixture of glycolipids and phosphoglycolipids (Thompson et al., 1980; Thompson and Murray, 1981; Counsell and Murray, 1986; Melin et al., 1986). Twelve polar lipids have been identified by thin layer chromatography, but most remain uncharacterized (Counsell and Murray, 1986). Forty-three percent of the total lipid content of *D. radiodurans* R1 is composed of two related phosphoglycolipids: 2′-*O*-(1,2-diacyl-*sn*-glycerol-3-phospho)-3′-*O*-(α-galactosyl)-*N*-D-glyceroyl-alkylamine, and 2′-*O*-(1,2-diacyl-*sn*-glycerol-3-phospho)-3′-*O*-(α-*N*-acetylglucosaminyl)-*N*-D-glyceroyl-alkylamine (Anderson and Hansen, 1985; Huang and Anderson, 1989, 1991). These lipids appear to be derived from the same precursor, a novel phosphatidylglycerolalkylamine that,

when glycosylated with galactose or glucosamine, forms these lipids. Although glucosamine-containing lipids have been found in other species, notably members of *Thermus* (Pask-Hughes and Shaw, 1982), these phosphoglycolipids are, at present, considered unique to *D. radiodurans*. There is no evidence of lipopolysaccharide. Attempts to find hydroxy fatty acids, lipid A, and heptoses have been unsuccessful (Murray and Brooks, 1986; Murray, 1992).

The fatty acid composition of *D. radiodurans* is distinctive (Ferreira et al., 1997). A mixture of 15, 16, 17, and 18 carbon saturated and monounsaturated acids are present. Polyunsaturated, cyclopropyl, and branched chain fatty acids are not present at detectable levels. *cis*-7-Hexadecenoic acid ($16:1_{\omega 7c}$) predominates in *D. radiodurans*, comprising approximately 40% of all fatty acids, followed by hexadecanoic acid (16:0) and *cis*-8-heptadecenoic acid ($17:1_{\omega 8c}$) at 14% and 11%, respectively. Significant levels of positional isomers ($16:1_{\omega 9c}$, $17:1_{\omega 6c}$) of the major monounsaturated fatty acids are also present.

The reddish pigmentation of *D. radiodurans* is due to carotenoids present in the membranes of this species (Brooks et al., 1980). Six colored components, designated X_1–X_6, have been identified in chloroform/methanol extracts of *D. radiodurans* R1, and components X_4 and X_5 were characterized as xanthophylls (Carbonneau et al., 1989). Detailed structural characterization of these compounds has not been reported.

The MK-8 menaquinone system of *D. radiodurans* R1 is unusual (Yamada et al., 1977). These quinones consist of eight isoprene units and are fully unsaturated. The isolated menaquinones show characteristic absorption peaks at 247, 270, and 332 nm in ethanol.

In electron micrographs, the peptidoglycan appears as a thick dense layer (Fig. B4.4). Depending on the technique used to stain thin sections, the peptidoglycan may appear "holely" or fenestrated (Thornley et al., 1965; Work and Griffiths, 1968; Thompson and Murray, 1982b). This phenomenon is related to the differential reactivity of peptidoglycan toward lead (Murray, 1992), and has no known physiological significance. There is no evidence that the peptidoglycan is associated with any other polymers. Teichoic acids have not been detected in *D. radiodurans*.

The chemical structure of the peptidoglycan of *D. radiodurans* SARK has been elucidated using plasma desorption time-of-flight mass spectrometry (Quintela et al., 1999). The monomeric subunit is *N*-acetylglucosamine-*N*-acetylmuramic acid-L-Ala-L-Glu-(δ)-L-Orn-[(γ)Gly-Gly]-D-Ala-D-Ala, consistent with the A3β classification given to *D. radiodurans* (Scheilfer and Kandler, 1972; Brooks et al., 1980; Sleytr and Glauert, 1982). Glycan strands terminate in (1→6) anhydromuramic acid. $(Gly)_2$ bridges form the cross-links. Cross-linkage was measured at 47.4% with 8 mol% trimers.

The peptidoglycan fine structure of *D. radiodurans* strain SARK reflects the phylogenetic proximity of *Deinococcus* and *Thermus*. *Thermus thermophilus* HB8 (Quintela et al., 1995) has an A3β murein chemotype, and its peptidoglycan is built from the same monomeric subunit [(γ)Gly-Gly-D-Ala-D-Ala] as *D. radiodurans* strain SARK. The composition of the terminal dipeptides differs between the genera. *T. thermophilus* HB8 is enriched in D-Ala-D-Ala and D-Ala-Gly, whereas *D. radiodurans* strain SARK is abundant in D-Ala-D-Glu. Also, *D. radiodurans* strain SARK does not have the phenyl acetate containing muropeptides characteristic of *T. thermophilus* HB8.

Strains of *D. radiodurans* are highly resistant to ultraviolet (UV) light and ionizing radiation. Exponential phase cultures will sur-

---

*Defined medium for *D. radiodurans* consists of (per liter 10 mM phosphate buffer, pH 7.2): glucose, 5 g; niacin, 0.5 mg; biotin, 0.5 mg; L-glutamic acid, 100 mg; L-methionine, 100 mg; $(NH_4)_2SO_4$, 0.33 mg; $CaCl_2$, 10 mg; $FeSO_4 \cdot 7H_2O$, 2.5 mg; $MgCl_2 \cdot 6H_2O$, 100 mg; $CuSO_4 \cdot 5H_2O$, 0.5 μg; $MnCl_2 \cdot 4H_2O$, 10 μg; $ZnSO_4 \cdot 7H_2O$, 200 μg; $CoCl_2$, 20 μg.

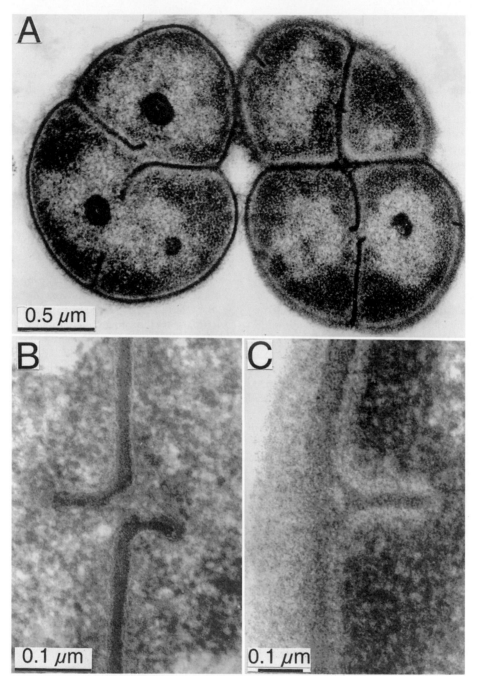

FIGURE B4.2. *A*, electron micrograph of *D. radiodurans* SARK showing a dividing cell. *B*, electron micrograph of *D. radiodurans* R1 illustrating how the advancing septa avert just before the advancing edges meet. *C*, electron micrograph of *D. radiodurans* R1 illustrating an early septum forming in the second division plane. (*A* reprinted with permission from R.G. Murray et al., Canadian Journal of Microbiology *29*: 1412–1423, 1983, ©National Research Council of Canada. *B* and *C*, courtesy of P.A. O'Cain, M.C. Henk, and J.R. Battista.)

vive 500 J/m² UV and 5000 Gray (Gy) γ radiation without loss of viability (Moseley and Mattingly, 1971; Mattimore et al., 1995). This property is often used to facilitate their isolation from natural microflora.

The strains of *Deinococcus* species are typically grown on a rich medium composed of 5 g tryptone, 1 g glucose, and 3 g yeast extract (TGY), and 1.5% agar if solid media is required. Glucose is not necessary, but facilitates growth. There are reports that the addition of either methionine or glutamate (1 g/l) to TGY also enhances growth. *D. radiodurans, D. radiophilus, D. proteolyticus, D.*

radiopugnans, and *D. grandis* are mesophiles, growing optimally at 30°C. Growth for these species does not proceed at temperatures above 45°C (Bridges et al., 1969a, b). Even though the generation time of *D. radiodurans* strain R1 under optimal conditions is about 1 h, it requires 36–48 h for an inoculum of 10³ CFU to reach stationary phase when grown in 10 ml of well-aerated TGY broth at 30°C (Battista, unpublished observations). The lag phase for this strain appears to be very long, especially if the inoculum is obtained from a frozen cell stock.

On TGY agar, colonies are smooth, convex, and pink to red

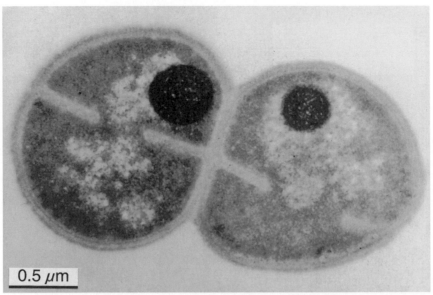

FIGURE B4.3. Electron micrograph of *D. radiodurans* R1 in exponential phase growth showing the characteristic inclusion bodies. (Courtesy of P.A. O'Cain, M.C. Henk, and J.R. Battista.)

in color. Color varies depending on the composition of the medium, colonies being redder when a greater concentration of yeast extract is added to the media (Murray and Brooks, 1986).

The defined medium will permit growth of *D. radiodurans*, *D. radiophilus*, *D. proteolyticus*, and *D. radiopugnans* (Manuel and Battista, unpublished observations). The vitamins niacin and biotin must be present. The addition of methionine and glutamate enhance the growth rate, but are not absolute requirements. Growth on this defined medium is slow, generation times being approximately 18 h at 30°C. This medium has been used successfully to assist in the isolation and characterization of auxotrophic strains of *D. radiodurans* strain R1 (Curnow et al., 1998), and should assist in defining additional biochemical characteristics that can be used to distinguish strains of the mesophilic species of *Deinococcus*. Other more complex defined media have also been described (Raj et al., 1960; Shapiro et al., 1977).

D. geothermalis and *D. murrayi* are thermophiles and grow optimally between 45 and 50°C on *Thermus* or Degryse medium 162 (Ferreira et al., 1997). In Degryse media, *D. geothermalis* shows a pH optimum of 6.8, whereas *D. murrayi* grows best at pH 8.0.

ENRICHMENT AND ISOLATION PROCEDURES

The natural habitat of the mesophilic *Deinococcus* species has not been defined. Only two ecological studies of *Deinococcus* species have been reported (Krabbenhoft et al., 1965; Masters et al., 1991), and even though these studies were limited to relatively small geographical areas, each concluded that the deinococci were rare but widely distributed soil organisms. Isolates have most frequently been obtained by selecting for ionizing radiation resistance. All described species survive 5000 Gy γ radiation without loss of viability, and some survivors can be found at doses as high as 15,000 Gy. Doses of ionizing radiation this high eradicate all but a few bacterial species, making isolation of *Deinococcus* strains from mixed bacterial flora relatively straightforward. *Deinococcus* strains have been isolated following γ-irradiation of soil (Krabbenhoft et al., 1965; Christensen and Kristensen, 1981; Kristensen and Christensen, 1981), animal feces (Oyaizu et al., 1987), sewage (Ito et al., 1983), sawdust (Ito, 1977), processed meats (Davis

et al., 1963; Grant and Patterson, 1989), dried foods (Lewis, 1971), room dust (Christensen and Kristensen, 1981), medical instruments (Christensen and Kristensen, 1981), and textiles (Kristensen and Christensen, 1981).

Mesophilic *Deinococcus* strains have also been isolated without the use of γ-irradiation as an air contaminant (Murray and Brooks, 1986) and from a stream (Krabbenhoft et al., 1965) near the site where *D. radiodurans* strain R1 was first isolated (Anderson et al., 1956). In each case, samples obtained by filtration were plated, and plates screened for pink colonies that were subsequently identified as a *Deinococcus* species.

It is possible to identify ionizing radiation-resistant bacteria from natural microflora by selecting for desiccation resistance (Sanders and Maxcy, 1979). *D. radiodurans* is exceptionally resistant to desiccation. Cultures dried onto a glass slide and held for 2 years in a desiccator at 5% humidity exhibit approximately 80% viability (Mattimore and Battista, unpublished observations). The desiccation resistance of *D. radiodurans* and other *Deinococcus* species may provide some insight into the ecology of this genus. Perhaps the dry but viable organism becomes airborne, and is distributed randomly within an environment.

The thermophilic species, *D. geothermalis* and *D. murrayi*, were isolated without irradiation from hot springs in Italy and Portugal (Ferreira et al., 1997), and it is assumed that this is their natural habitat. *D. geothermalis* has also been isolated as an accidental contaminant of printing paper machines running at 45–50°C (Vaisanen et al., 1998). The source of this contamination was not determined.

MAINTENANCE PROCEDURES

Colonies remain viable on sealed TGY agar plates at room temperature for at least 2 weeks. Storage at −80°C in 15% glycerol or 15% DMSO is perhaps the best general method for long-term preservation. Lyophilization is also very effective, but should be used with caution when preserving strains that are ionizing-radiation sensitive, because of the link made between the sensitivity of *D. radiodurans* to ionizing radiation and its sensitivity to desiccation (Mattimore and Battista, 1996). Agar slant cultures re-

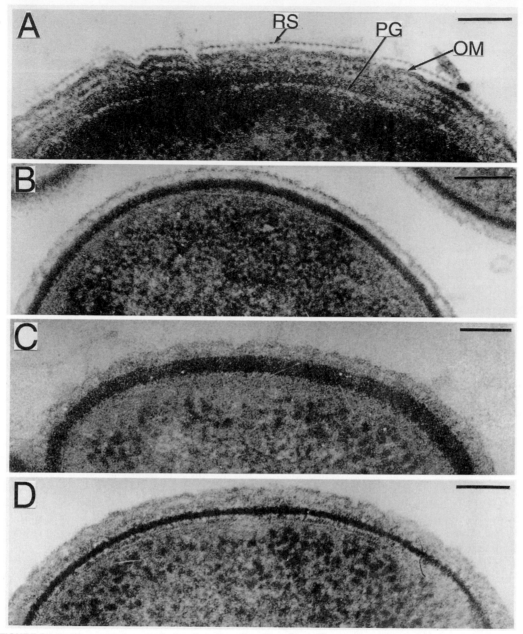

FIGURE B4.4. Electron micrograph showing the cell wall profiles of four species of *Deinococcus*: A, *D. radiodurans*; B, *D. radiophilus*; C, *D. proteolyticus*; D, *D. radiopugnans*. The sections are stained with uranyl acetate and lead citrate. *RS*, regular surface protein array; *PG*, peptidoglycan layer; *OM*, outer membrane. Bar = 100 nm.

main viable for long periods at room temperature, but reports of failures suggest that frequent subculturing is prudent (Murray and Brooks, 1986).

TAXONOMIC COMMENTS

All species of the genus *Deinococcus* cluster together within a single lineage based on 16S rRNA gene sequence-based phylogenetic analyses. It was the distinctness of this lineage that was used to justify the description of the order *Deinococcales* (Rainey et al., 1997; Fig. B4.1). Three clusters exist within the *Deinococcus* lineage, one containing the thermophiles *D. geothermalis* and *D. murrayi* (Ferreira et al., 1997), a lineage consisting of *D. proteolyticus* and *D. radiophilus* (Rainey et al., 1997), and a lineage composed of *D. radiodurans*, *D. radiopugnans*, and *D. grandis* (Rainey et al., 1997). The levels of 16S rRNA gene sequence similarity within the *Deinococcus* lineage range from 88.3 to 95.5%. There are no simple diagnostic tests that reliably differentiate the mesophilic *Deinococcus* species from one another (Table B4.2). Obviously, optimal growth temperature and habitat distinguish the thermophilic deinococci from the mesophilic species, and the morphology and Gram-stain reaction of *D. grandis* set it apart from the other described members of the genus, but these characteristics have little phylogenetic significance and cannot be considered definitive methods of identification and differentiation. All species of the genus *Deinococcus* can be easily identified and differentiated by comparative 16S rRNA gene sequence analyses

TABLE B4.2. Characteristics of the type strains of the species of the genus *Deinococcus*[a]

Characteristic	1. *D. radiodurans*[b]	2. *D. geothermalis*[c]	3. *D. grandis*[d]	4. *D. murrayi*[c]	5. *D. proteolyticus*[b]	6. *D. radiophilus*[b]	7. *D. radiopugnans*[b]
Morphology	Spherical	Spherical	Rod	Spherical	Spherical	Spherical	Spherical
Cell size, μm	1.5–3.0	1.2–2	0.6–1.2 × 1.5–4	1.2–2	1.0–2.0	1.0–2.0	1.0–2.0
Catalase	+	+	+	+	+	+	+
Esculin hydrolysis	−	nd	+	nd	+	−	−
β-Galactosidase	−	+	nd	+	−	−	−
Growth at 37°C	+	+	+	+	+	+	+
Resistance to 10 kGy gamma radiation	+	+	+	+	+	+	+
Colony color	Red	Orange	Pink/Red	Orange	Pink/Red	Pink/Red	Orange/red
Peptidoglycan type	A3β (L-Orn-Gly$_{2-3}$)	A3β (L-Orn-Gly$_2$)	A3β (L-Orn-Gly$_2$)	A3β (L-Orn-Gly$_2$)	A3β (L-Orn-Gly$_2$)	A3β (L-Orn-Gly$_2$)	A3β (L-Orn-Gly$_2$)
Gram reaction	Positive	Positive	Negative	Positive	Positive	Positive	Positive
Predominant fatty acid	$C_{16:1\omega7c}$	$C_{17:0\ iso}$	$C_{15:1\omega6c}$	$C_{17:1\omega9c\ iso}$	$C_{16:1\omega7c}$ $C_{17:1\omega8c}$	$C_{16:0}$	$C_{17:1\omega9c\ iso}$
Branched chain fatty acid	nd	$C_{13:0\ iso}$, $C_{14:0\ iso}$, $C_{15:0\ iso}$, $C_{16:0\ iso}$, $C_{17:0\ iso}$	$C_{15:0\ iso}$, $C_{16:0\ iso}$, $C_{17:0\ iso}$	$C_{15:0\ iso}$, $C_{17:0\ iso}$, $C_{15:1\ iso\ F}$, $C_{17:1\omega9c\ iso}$	$C_{17:0\ iso}$, $C_{17:1\omega9c\ iso}$	nd	$C_{13:0\ iso}$, $C_{15:0\ iso}$, $C_{15:1\ iso\ F}$, $C_{17:1\omega9c\ iso}$, $C_{17:0\ iso}$
Nitrate reduction	18/24	−	+	−	−	−	−
Mol% G + C	67	65.9	69	69.9	65	62	70
Menaquinone system	MK-8	MK-8	MK-8	MK-8	MK-8	MK-8	MK-8
Colonial morphology	Smooth and convex with regular edge	Smooth and convex with regular edge	Smooth and convex with regular edge	Smooth and convex with regular edge	Smooth and convex with regular edge	Smooth and convex with regular edge	Smooth and convex with regular edge

[a]For symbols see standard definitions; nd, not determined.

[b]Data taken from Brooks and Murray (1981).

[c]Data taken from Ferreira et al. (1997).

[d]Data taken from Oyaizu et al. (1987).

(Ferriera et al., 1997; Rainey et al., 1997). This method is the only feasible way to identify novel species from environmental isolates. Mean fatty acid composition may also be useful. Ferriera et al. (1997) showed that the type strain of each species displays a different fatty acids profile (Table B4.2).

Several biochemical parameters (Table B4.3) vary between the thermophilic strains of *Deinococcus* (Ferreira et al., 1997). These differences are detailed in the species description.

ACKNOWLEDGMENTS

The support of our research by the National Science Foundation is gratefully acknowledged.

FURTHER READING

Battista, J.R. 1997. Against all odds: The survival strategies of *Deinococcus radiodurans*. Annu. Rev. Microbiol. *51*: 203–224.

Ferreira, A.C., M.F. Nobre, F.A. Rainey, M.T. Silva, R. Wait, J. Burghardt, A.P. Chung and M.S. Da Costa. 1997. *Deinococcus geothermalis* sp. nov. and *Deinococcus murrayi* sp. nov., two extremely radiation-resistant and slightly thermophilic species from hot springs. Int. J. Syst. Bacteriol. *47*: 939–947.

Moseley, B.E. 1983. Photobiology and radiobiology of *Micrococcus* (*Deinococcus*) *radiodurans*. Photochem. Photobiol. Rev. *7*: 223–274.

Murray, R.G.E. 1992. The family *Deinococcaceae*. *In* Ballows, Trüper, Dworkin, Harder and Schleifer (Editors), The Prokaryotes: A Handbook of Bacteria: Ecophysiology, Isolation, Identification, Applications, Vol. 4, Springer-Verlag, New York, pp. 3732–3744.

TABLE B4.3. Biochemical characteristics the distinguish *D. geothermalis* and *D. murrayi*[a]

Characteristic	*D. geothermalis*[b]	*D. murrayi*[b]
Reduction of tellurite (0.5%)	+	−
Hydrolysis of arbutin	+	−
Presence of:		
α-Galactosidase	−	+
β-Galactosidase	w	+
Utilization of:		
D-Cellobiose	+	−
Lactose	+	−
D-Galactose	+	−
L-Rhamnose	+	−
D-Xylose	+	−
D-Mannitol	+	−
L-Sorbitol	+	−
L-Arginine	−	+
Malate	+	−
Succinate	+	−

[a]For symbols see standard definitions.

[b]Data taken from Ferreira et al., 1997.

Rainey, F.A., M.F. Nobre, P. Schumann, E. Stackebrandt and M.S. da Costa. 1997. Phylogenetic diversity of the deinococci as determined by 16S ribosomal DNA sequence comparison. Int. J. Syst. Bacteriol. *47*: 510–514.

List of species of the genus Deinococcus

1. **Deinococcus radiodurans** Brooks and Murray 1981, 354.[VP]
ra.di.o.du' rans. M.L. prefix *radio* radiation; L. part. adj. *durans* enduring; M.L. part. adj. *radiodurans* resisting radiation.

The strains of this species form spherical, nonmotile cells, 0.5–3.5 μm in diameter, that occur singly and in pairs, dividing in two planes. Cells do not have differentiated resting stage. Colonies are usually pink to brick red after 48 h growth on TGY medium at 30°C. Cells are Gram-positive, but have several distinct cell wall layers including an outer membrane. Teichoic acids are absent. Some strains have a distinctive hexagonally arrayed S layer. Cells are aerobic, and catalase is produced. All natural isolates are resistant to 1.5 kGy gamma radiation and to exposure to 1500 J/m² of ultraviolet radiation. Palmitoleate ($16:1_{\omega7c}$) is the predominant fatty acid, accounting for approximately 40% of the total fatty acid. Branched chain and cyclopropyl fatty acids are not detected. *D. radiodurans* strains have been isolated from ground pork and beef.

The mol% G + C of the DNA is: 67 (T_m).
Type strain: ATCC 13939, DSMZ 20539, UWO 288.
GenBank accession number (16S rRNA): Y11332.

2. **Deinococcus geothermalis** Ferreira, Nobre, Rainey, Silva, Wait, Burghardt, Chung and da Costa 1997, 945.[VP]
ge.o.ther.ma' lis. Gr. n. *ge* earth; Gr. adj. *thermos* hot; M.L. adj. *geothermalis* of hot springs.

Deinococcus geothermalis strains form spherical nonmotile cells 1.2–2.0 μm in diameter. Cells do not have a differentiated resting stage. Cells stain Gram-positive. Colonies are orange-pigmented after 72 h on Degryse medium 162. Growth occurs between 30°C and 55°C in Degryse medium 162; the optimum growth temperature for the type strain is about 47°C. Optimum pH is 6.5; growth does not occur at pH 4.0 or pH 9.0. Yeast extract is not required for growth. The major fatty acids are $C_{17:0\,iso}$ and $C_{15:0\,iso}$; unsaturated fatty acids are present in low relative proportions. Oxidase and catalase positive. Nitrate is not reduced to nitrite, α-galactosidase is negative, β-galactosidase is positive. The type strain degrades casein, gelatin, hide powder azure, hippurate, arbutin, and starch. Some strains do not degrade starch. The type strain utilizes D-cellobiose, D-trehalose, lactose, maltose, D-fructose, D-galactose, D-glucose, D-mannose, L-rhamnose, sucrose, D-xylose, D-mannitol, D-sorbitol, glycerol, L-asparagine, L-glutamate, L-glutamine, L-proline, L-serine, malate, pyruvate, and succinate. Strains of *D. geothermalis* do not utilize D-raffinose, D-melibiose, L-arabinose, *myo*-inositol, ribitol, L-arginine, and citrate. The type strain is resistant to 1.5 Mrad γ radiation. The three described isolates of this species are from hot springs; the type strain AG-3a and strain AG-5a are from Agnano, Naples, Italy, whereas strain RSPS-2a is from São Pedro do Sul, Portugal (Ferreira et al., 1997). Additional strains have been isolated from paper machines (Vaisanen et al., 1998) and from the soil near man-made steam vents (Rainey et al., unpublished data).

The mol% G + C of the DNA is: 65.9 (T_m).
Type strain: AG-3a, DSMZ 11300.
GenBank accession number (16S rRNA): Y13038.

3. **Deinococcus grandis** (Oyaizu, Stackebrandt, Schleifer, Ludwig, Pohla, Ito, Hirata, Oyaizu, Komagata 1987) Rainey, Nobre, Schumann, Stackebrandt and da Costa 1997, 513[VP] (*Deinobacter grandis* Oyaizu, Stackebrandt, Schleifer, Ludwig, Pohla, Ito, Hirata, Oyaizu, Komagata 1987, 66.)
gran' dis. L. masc. adj. *grandis* large.

The strains of this species form rod-shaped, nonmotile cells, ranging from 0.6 to 1.2 μm × 1.5 to 4.0 μm. Cells do not have a differentiated resting stage. Colonies are usually pink to red after 48 h growth on TGY medium at 30°C. Cells stain Gram-negative, but have the distinct cell wall layers characteristic of the deinococci. The peptidoglycan layer is approximately 10 nm, thinner than other *Deinococcus* species. Teichoic acids are absent. Cells are aerobic, and catalase is produced. Most strains hydrolyze gelatin, starch, and esculin. Tween 80 is not hydrolyzed. Nitrate is reduced to nitrite. Isolates are resistant to 1.5 kGy γ radiation. The 15 carbon fatty acids ($C_{15:0}$, and $C_{15:1\,\omega6c}$) predominate, accounting for about 50% of the mean fatty acid composition. Branched chain, saturated fatty acids of $C_{15:0\,iso}$, $C_{16:0\,iso}$, and $C_{17:0\,iso}$ are found, composing approximately 10% of the total fatty acid. Strain KS 0485 was isolated from irradiated *Elephas maximas* feces.

The mol% G + C of the DNA is: 69 (T_m).
Type strain: KS 0485, DSMZ 3963, IAM 13005.
GenBank accession number (16S rRNA): Y11329.

4. **Deinococcus murrayi** Ferreira, Nobre, Rainey, Silva, Wait, Burghardt, Chung and da Costa 1997, 945.[VP]
mur' ray.i. L. gen. n. *murrayi* of Murray, named after the Canadian microbiologist R.G.E. Murray in recognition of his research.

Morphological, biochemical, and chemotaxonomic characteristics of *D. murrayi* are similar to those described for *D. geothermalis*. Growth occurs between 30°C and 52.5°C in Degryse medium 162, the optimum growth temperature for the type strain is about 47°C. The optimum pH is about 8.0; growth does not occur at pH 5.0 or pH 10.5. The major fatty acids are $C_{17:1\,\omega9c\,iso}$ and $C_{17:0\,iso}$. Oxidase and catalase are positive. Nitrate is not reduced to nitrite; α-galactosidase and β-galactosidase are positive in the type strain, but these activities were absent in some reference strains. Casein, gelatin, hide powder azure, hippurate, and starch are degraded. The type strain utilizes D-trehalose, maltose, D-fructose, D-glucose, D-mannose, sucrose, glycerol, L-arginine, L-asparagine, L-glutamate, L-glutamine, L-proline, L-serine, and pyruvate. Strains may not utilize D-mannose and L-glutamine. None of the strains utilize D-raffinose, D-cellobiose, D-melibiose, lactose, D- galactose, L-rhamnose, D-xylose, L-arabinose, D-mannitol, D- sorbitol, *myo*-inositol, ribitol, citrate, malate, and succinate. The type strain is resistant to 1.5 Mrad γ radiation. The three described isolates of this species are from hot springs; the type strain ALT-1b is from Alcafache, Portugal, whereas strains RSPS-7a and RSG-1.2 are from São Pedro do Sul and São Gemil, Portugal, respectively (Ferreira et al., 1997).

The mol% G + C of the DNA is: 69.9 (T_m).
Type strain: ALT-1b, DSMZ 11303.
GenBank accession number (16S rRNA): Y13041.

5. **Deinococcus proteolyticus** Brooks and Murray 1981, 357.[VP]
pro.te.o.ly' ti.cus. M.L. n. adj. *proteolyticus* proteolytic.

Morphological, biochemical, and chemotaxonomic characteristics of *D. proteolyticus* are almost identical to those described for *D. radiodurans*. The strains of this species are typically 1.0–2.0 μm in diameter. *cis*-Hexadecenoic ($C_{16:1\omega7c}$) and *cis*-heptadecenoic acids ($C_{17:1\omega8c}$) are the predominant fatty acids, each accounting for approximately 27% of the total fatty acid composition. The branched chain fatty acids ($C_{17:0\ iso}$, $C_{17:1\omega9c\ iso}$) are found, differentiating *D. proteolyticus* from *D. radiodurans* and *D. radiophilus*. The specific epithet implies that this species peptonizes proteins (milk, soy, and gelatin) as was observed by the original isolators (Kobatake et al., 1973). The other mesophilic *Deinococcus* species also show some activity but this species is the most active. Strain CCM 2703 was isolated from irradiated *Llama glama* feces.

The mol% G + C of the DNA is: 65 (T_m).

Type strain: ATCC 35074, CCM 2703, DSMZ 20540, UWO 1056.

GenBank accession number (16S rRNA): Y11331.

6. **Deinococcus radiophilus** Brooks and Murray 1981, 357.[VP]
ra.di.o'phil.us. M.L. prefix *radio* radiation; Gr. adj. *philos* loving; M.L. adj. *radiophilus* radiation loving.

Morphological, biochemical, and chemotaxonomic characteristics of *D. radiophilus* are almost identical to those described for *D. radiodurans*. The strains of this species are typically 1.0–2.0 μm in diameter. Hexadecanoic (16:0) and *cis*-hexadecenoic acids ($16:1\omega7c$) are the predominant fatty acids. Branched chain and cyclopropyl fatty acids are not detected.

D. radiophilus is distinguished from *D. radiodurans* by presence of significant levels of 12, 13, and 14 carbon saturated fatty acids. Strain UWO 1055 was isolated from irradiated Bombay duck (*Harpodon nehereus*).

The mol% G + C of the DNA is: 62 (T_m).

Type strain: ATCC 27603, DSMZ 20551, UWO 1055.

GenBank accession number (16S rRNA): Y11333.

7. **Deinococcus radiopugnans** Brooks and Murray 1981, 358.[VP]
ra.di.o.pug'nans. M.L. prefix *radio* radiation; L. part. adj. *pugnans* fighting or resisting; M.L. adj. *radiopugnans* radiation resisting.

Morphological, biochemical, and chemotaxonomic characteristics of *D. radiopugnans* are almost identical to those described for *D. radiodurans*. The strains of this species are typically 1.0–2.0 μm in diameter. The branched chain fatty acid ($C_{17:1\omega9c\ iso}$) predominates. $C_{13:0\ iso c}$, $C_{15:1\ iso F}$, $C_{15:0 iso}$, and $C_{17:0\ iso}$ are also present, differentiating *D. radiopugnans* from other mesophilic deinococci. Smooth and rough variants as well as variants with less pigment may occur. The 15-carbon, saturated, branched chain fatty acid component may be absent. Nitrate is reduced to nitrite. Strain UWO 293 was isolated from irradiated haddock.

The mol% G + C of the DNA is: 70 (T_m).

Type strain: ATCC 19172, UWO 293.

GenBank accession number (16S rRNA): Y11334.

Order II. **Thermales** *ord. nov.*

FRED A. RAINEY AND MILTON S. DA COSTA

Ther.ma'les. M.L. masc. n. *Thermus* type genus of the order, *-ales* ending to denote an order; M.L. fem. pl. n. *Thermales* the order of *Thermus*.

Cells are straight rods of variable length, filaments are also present. **Nonmotile**; flagella are not present. Gram negative. Endospores are not observed. Most strains form **yellow- or red-pigmented colonies**; some strains are nonpigmented. Aerobic with a strictly respiratory type of metabolism, but some strains grow anaerobically with nitrate and nitrite as terminal electron acceptors. Oxidase positive; most strains are catalase positive. **Thermophilic**, with an optimum growth temperature range of ~50–75°C. **Menaquinone 8** is the predominant respiratory quinone; **ornithine** is the principal diamino acid of the peptidoglycan. One major phospholipid is present in all strains, one or two major glycolipids are also present. Fatty acids are predominantly *iso*- and *anteiso*-branched; branched chain 2-hydroxy and/or 3-hydroxy fatty acids are present in many strains. **Heterotrophic**, some strains may be chemolithoheterotrophic oxidizing sulfur compounds. Isolated from and detected in **hydrothermal areas** with neutral to alkaline pH, also commonly isolated from man-made thermal environments.

The mol% G + C of the DNA is: 57–70.

Type genus: **Thermus** Brock and Freeze 1969, 295, emend. Nobre, Trüper and da Costa 1996b, 605.

Family I. **Thermaceae** *fam. nov.*

MILTON S. DA COSTA AND FRED A. RAINEY

Ther.ma'ce.ae. M.L. masc. n. *Thermus* type genus of the family; *-aceae* ending to denote a family; M.L. fem. pl. n. *Thermaceae* the *Thermus* family.

Cells are straight rods of variable length, filaments are also present. **Nonmotile**; flagella are not present. Gram negative. Endospores are not observed. Most strains form **yellow or red-pigmented colonies**, some strains are nonpigmented. **Aerobic** with a strictly respiratory type of metabolism, but some strains grow anaerobically with nitrate and nitrite as terminal electron acceptors. Oxidase positive; most strains are catalase positive. **Ther**mophilic, with an optimum growth temperature range of ~50–75°C. **Menaquinone 8** is the predominant respiratory quinone; **ornithine** is the principal diamino acid of the peptidoglycan. One major phospholipid is present in all strains, one or two major glycolipids are also present. Fatty acids are predominantly *iso*- and *anteiso*-branched; some strains of the genus *Thermus* possess 3-hydroxy fatty acids; all strains of the genus *Meiothermus* possess

2-hydroxy fatty acids, some strains also possess 3-hydroxy fatty acids. **Heterotrophic**, some strains may be chemolithoheterotrophic oxidizing sulfur compounds. Isolated from and detected in **hydrothermal areas** with neutral to alkaline pH, also commonly isolated from man-made thermal environments.

The mol% G + C of the DNA is: 57–70.

Type genus: **Thermus** Brock and Freeze 1969, 295, emend. Nobre, Trüper and da Costa 1996b, 605.

Genus I. **Thermus** *Brock and Freeze 1969, 295,*[AL] *emend. Nobre, Trüper and da Costa 1996b, 605*

MILTON S. DA COSTA, M. FERNANDA NOBRE AND FRED A. RAINEY

Ther' mus. Gr. adj. *thermos* hot; M.L. masc. n. *Thermus* to indicate an organism living in hot places.

Straight rods, 0.5–0.8 μm in diameter; the cell length is variable. Short filaments are also formed under some culture conditions. Some strains have a stable filamentous morphology. **Nonmotile**; do not possess flagella. Endospores are not observed. Stain Gram-negative. Most strains form **yellow-pigmented colonies**, some strains are nonpigmented. **Aerobic** with a strictly respiratory type of metabolism, but some strains grow anaerobically with nitrate and nitrite as terminal electron acceptors. Oxidase positive and catalase positive. **Thermophilic**, with an optimum growth temperature of about 70–75°C; most strains have a maximum growth temperature below 80°C, but some strains grow at higher temperatures. The optimum pH is about 7.8. **Menaquinone 8** is the predominant respiratory quinone; **ornithine** is the principal diamino acid of the peptidoglycan. One major phospholipid and one major glycolipid dominate the polar lipid pattern on thin-layer chromatography. Additional phospholipids and glycolipids are minor components. Fatty acids are predominantly **iso- and anteiso-branched**; branched chain 3-hydroxy fatty acids are present in some strains. Proteins and peptides are hydrolyzed by all strains. Starch is hydrolyzed by some strains. Monosaccharides, disaccharides, amino acids, and organic acids are used as sole carbon and energy sources. The utilization of pentoses and polyols is very rare. Most strains require yeast extract or cofactors for growth. Found in **hydrothermal areas** with neutral to alkaline pH, also commonly isolated from man-made thermal environments.

The mol% G + C of the DNA is: 57–65.

Type species: **Thermus aquaticus** Brock and Freeze 1969, 295.

FURTHER DESCRIPTIVE INFORMATION

A number of early 5S rRNA studies of *T. aquaticus* and *T. thermophilus* indicated that these species had no clear phylogenetic relationship to any of the known bacterial groups for which 5S rRNA sequences were available for comparison (Erdmann et al., 1984; Pace et al., 1985; Vandenberghe et al., 1985). The later studies of Bakeeva et al. (1986) and Chumakov (1987) demonstrated a relationship between the genera *Thermus* and *Deinococcus* by 5S rRNA sequence analyses.

The first 16S rRNA studies of members of the genus *Thermus* were in the form of incomplete oligonucleotide catalogs (Hensel et al., 1986) but demonstrated a clear phylogenetic relationship between the genera *Thermus* and *Deinococcus* and concluded that *Thermus* was a member of the *Deinococcus* division as defined by Stackebrandt and Woese (1981). The degree of sequence similarity between full 16S rRNA sequences of *T. aquaticus* and *D. radiodurans* was first reported by Woese (1987) to be 81%, and it was suggested that this value was high enough to place these taxa in the same phylum, but in different subdivisions of the phylum *"Deinococcus–Thermus"*. The same author also indicated

that the *Deinococcus–Thermus* lineage was the third deepest eubacterial branch at that time. The deep branching phylogenetic position of the genus *Thermus* was further investigated by Hartmann et al. (1989), who determined the full 16S rRNA sequence of *T. thermophilus* HB8. As new deep branching procaryotic lineages have been discovered, both as organisms isolated in pure culture or as environtaxa detected by molecular approaches in the environment, the order of branching within the procaryotic phylogenetic tree has changed, but the *Deinococcus–Thermus* lineage still represents a deep branching phylum (Rainey et al., 1997). The relationship of the genus *Thermus* and the genus *Deinococcus* was also demonstrated by Embley et al. (1993), and the study provided additional complete 16S rDNA sequences for *Thermus* strains VI-7a (erroneously designated Vi17) (EMBL Accession No. Z15061), YS38 (EMBL Accession No. Z15062), SPS-14 (EMBL Accession No. Z15060), and *Meiothermus (Thermus) ruber* ATCC 35948 (EMBL Accession No. Z15059).

The fine structure of the strains of the genus *Thermus* shows that the cells have an envelope consisting of a cytoplasmic membrane with a simple outline, a cell wall with an inner, electron-dense thin layer presumably representing the peptidoglycan connected by irregularly spaced invaginations to an outer corrugated "cobble-stone" layer (Brock and Edwards, 1970; Pask-Hughes and Williams, 1975; Kristjánsson et al., 1994) (Fig. B4.5). Unusual morphological structures, commonly called "rotund bodies", are occasionally seen in many strains by phase-contrast and transmission electron microscopy. These structures consist of several cells bound longitudinally by a common external layer of the cell envelope enclosing a large space between the cells (Brock and Edwards, 1970; Kraepelin and Gravenstein, 1980; Becker and Starzyk, 1984). The type strain of *Thermus filiformis* and one unclassified strain isolated from New Zealand (D. Cowan, personal communication) have, in contrast to other strains, a stable filamentous morphology and do not form rod-shaped cells on solid or in liquid media. *T. filiformis* forms septate cells within a continuous outer sheath, although published electron micrographs do not show if the cells separate completely or not. There also appears to be an additional layer surrounding the envelope of the type strain of *T. filiformis* (Hudson et al., 1987b). Several other strains form long filamentous cells resembling those of the type strain of *T. filiformis* when specific D-amino acids are added to the culture medium (Janssen et al., 1991). Most of these strains originated from New Zealand and some belong to *T. filiformis*; however, *T. scotoductus* strain NH also formed filamentous cultures, but *T. aquaticus* YT-1 and *T. thermophilus* HB8 did not.

A crystalline surface protein layer (S-layer) has been identified in *T. thermophilus* strains HB8 and HB-27. The S-layer of strain HB8 is composed of a major protein with a molecular weight of 100 kDa, designated P100 (Berenguer et al., 1988; Castón et al.,

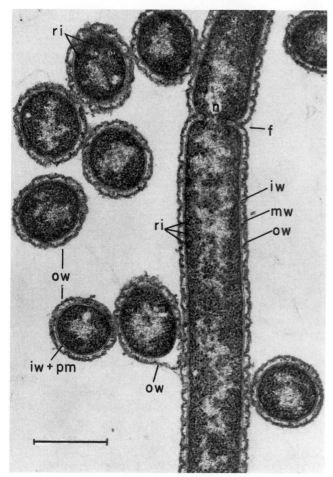

FIGURE B4.5. Transmission electron micrograph of *Thermus aquaticus* YT-1 showing the nucleoplasm (*n*) surrounded by numerous ribosomes (*ri*). The cell envelope comprises the plasma membrane (*pm*) and wall exhibiting an outer dense layer (*ow*), a middle light zone (*mw*), and a thin inner dense layer (*iw*). Note cell division furrow (*f*). Bar = 0.5 μm. (Reprinted with permission from T.D. Brock and M. Edwards, Journal of Bacteriology *104:* 509–517, 1970, ©American Society of Microbiology.)

1988; Faraldo et al., 1988). This protein forms oligomeric complexes stabilized by Ca^{2+} that interact with the peptidoglycan. The gene that codes for this protein has been cloned and sequenced, and mutants unable to produce P100 have been produced (Faraldo et al., 1992; Olabarría et al., 1996b). Strain HB-27 has a similar S-layer, but the major protein has a molecular weight of 95 kDa (Fernández-Herrero et al., 1995). Mutants defective for the production of P100 and P95 grow slower than the wildtype strains and produce cells with altered morphology. For example, these mutants frequently form structures composed of groups of cells bound by a common envelope reminiscent of the "rotund bodies" seen in many wildtype strains of the genus, but no conclusions are yet possible on the relationship between both structures. Other outer envelope proteins also form an S-layer-like array in strain HB8 and may be minor S-layer components, but little is yet known about the function of these proteins (Olabarría et al., 1996a; Fernández-Herrero et al., 1997).

The peptidoglycan of the strains of the genus *Thermus* contain L-ornithine as the diamino acid and glycylglycine as the interpeptide bridge (Merkel et al., 1978a; Pask-Hughes and Williams, 1978); this peptidoglycan composition is consistent with the A3β

murein type of Schleifer and Kandler (1972) that is also found in the genera *Meiothermus* and *Deinococcus* (Hensel et al., 1986; Embley et al., 1987; Sharp and Williams, 1988). The structure of the peptidoglycan of *T. thermophilus* HB8 was recently examined by Quintela et al. (1995), who found that the N-terminal glycine was substituted by phenylacetic acid in a significant proportion of the interpeptide bridges.

The major respiratory quinone of strains of the genus *Thermus* is menaquinone 8 (MK-8) (Collins and Jones, 1981; Hensel et al., 1986; Williams, 1989). The presence of ornithine and MK-8 corroborate the phylogenetic interpretation that the genera *Thermus*, *Meiothermus*, and *Deinococcus* are related to each other, although the species of the *Thermus/Meiothermus* line of descent have few other characteristics in common with the species of the genus *Deinococcus*.

The polar lipid composition of the species of *Thermus* consists of one major phospholipid, designated phospholipid 2 (PL-2), and one major glycolipid, designated glycolipid 1 (GL-1), which comprise between 80% and 95% of the total polar lipid phosphorus and carbohydrate. Other minor polar lipids, namely phospholipid 1 (PL-1) and glycolipid 2 (GL-2), are also detected by thin-layer chromatography in most strains of the genus *Thermus* (Pask-Hughes and Shaw, 1982; Prado et al., 1988; Donato et al., 1990). The major glycolipid of several strains has been identified as a diglycosyl-(*N*-acyl)glycosaminyl-glucosyldiacylglycerol, which contains three hexose residues and one *N*-acylated hexosamine, giving a hexose/hexosamine/glycerol ratio of approximately 3:1:1. Depending on the strain, the polar head group of GL-1 contains *N*-acylglucosamine or *N*-acylgalactosamine, three glucose residues, two glucose residues plus one galactose, or one glucose plus two galactose residues (Oshima and Yamakawa, 1974; Pask-Hughes and Shaw, 1982; Prado et al., 1988; Wait et al., 1997). The innermost hexose bound to glycerol always appears to be glucose. The terminal galactose, present in GL-1 of several strains, such as *T. thermophilus* strain HB8, is in the rare furanose configuration instead of the more common pyranose configuration (Oshima and Yamakawa, 1974; Wait et al., 1997). In contrast to the usual polar head group, GL-1 from *Thermus aquaticus* strain YS 004 was recently shown to have *N*-acetylgalactosamine in place of the subterminal hexose residue resulting in a hexose/hexosamine/glycerol ratio of 2:2:1 (Carreto et al., 1996).

One exception to the canonical polar lipid pattern of the strains of the genus *Thermus* is found in one colony variant of *Thermus scotoductus* strain X-1 (ATCC 27978). This strain produces two colony types, designated t1 and t2 (Tenreiro et al., 1995b). The polar lipid composition of colony type t2 is typical of most *Thermus* strains, consisting of the major phospholipid (PL-2), the major glycolipid (GL-1), and traces of a minor glycolipid (GL-2), whereas in colony type t1 GL-2 is the major glycolipid and only trace amounts of GL-1 are detected (Tenreiro et al., 1995b). Glycolipid 2 (GL-2) of this strain, and the same minor glycolipid of *T. oshimai* strain SPS-11, was identified as a truncated version of GL-1 lacking the terminal hexose (Wait et al., 1997). The structure of the major phospholipid (PL-2) of the genus *Thermus* has never been identified, but recently it was shown to be identical to the major phospholipid found in *Deinococcus radiodurans*, reinforcing the phylogenetic interpretation that the *Thermus/Meiothermus* line of descent is related to the species of the genus *Deinococcus* (Hensel et al., 1986; Huang and Anderson, 1989; Rainey et al., 1997; Wait et al., unpublished results).

Unexpectedly, terminally and subterminally branched long chain diols, identified as 16-methylheptadecane-1,2-diol and 15-methylheptadecane-1,2-diol, were detected as major components of GL-1 and GL-2 of *T. scotoductus* X-1 and *T. filiformis* Tok A4 (Wait et al., 1997). Long chain diols had only been detected in *Thermomicrobium roseum* where they appeared to be the exclusive backbone structure of the polar lipids, which apparently lack the normal glycerol-based lipids (Pond et al., 1986; Pond and Langworthy, 1987). In the species of *Thermus*, long chain diol-based lipids never completely replace the normal glycerolipids, and the polar head group of both structural types of glycolipids is identical. The levels of diols vary with the strain, and some strains have only vestigial amounts. Long chain diols have also been identified in the glycolipids of one of the four species of the genus *Meiothermus* (Ferreira et al., 1999). The presence of long chain diols in *Thermomicrobium roseum*, *Thermus* and *Meiothermus* species, in conjunction with 16S rDNA phylogenies, leads to the hypothesis that these lipid backbones reflect a distant, but definite relationship between the *Deinococcus–Thermus* phylum and the Green Non-sulfur Bacteria (Weisburg et al., 1989; van de Peer et al., 1994).

Iso- and anteiso-branched $C_{15:0}$ and $C_{17:0}$ fatty acids are the predominant acyl chains of the strains of the genus *Thermus*. Straight chain saturated fatty acids and unsaturated branched chain fatty acids are minor components at the optimum growth temperature in the vast majority of the strains (Ray et al., 1971; Donato et al., 1990; Nobre et al., 1996a); nevertheless, straight chain $C_{16:0}$ reaches levels of about 20% of the total fatty acids in *T. thermophilus* AT-62 (Nobre et al., 1996a). *Iso*-branched fatty acids predominate over anteiso-branched fatty acids in the vast majority of the strains at the optimum growth temperature (Pask-Hughes and Shaw, 1982; Prado et al., 1988; Nobre et al., 1996a), although the type strain of *T. filiformis* possesses between 60 and 70% anteiso-fatty acids at all growth temperatures examined (Donato et al., 1990). Some strains of the genus *Thermus*, namely most strains of *T. aquaticus* and the type strain of *T. filiformis*, also contain moderate levels of branched chain 3-hydroxy fatty acids. 3-Hydroxy fatty acids are exclusively amide-linked to the galactosamine present in the glycolipids, but are never present in the strains where glucosamine replaces galactosamine (Carreto et al., 1996; Ferreira et al., unpublished results).

The majority of the isolates of the species of the genus *Thermus* form yellow-pigmented colonies, although the color varies considerably from deep yellow to very pale yellow. Many strains, isolated primarily from man-made environments that are maintained dark, are nonpigmented, although yellow-pigmented strains can also be isolated from these environments. Several nonpigmented *Thermus* strains have been isolated from abyssal hot springs, but even here, some isolates are yellow-pigmented (Marteinsson et al., 1995). Several novel carotenoids, designated thermozeaxanthins and thermobiszeaxanthins, were identified in *Thermus* strain HB-27 by Yokoyama et al. (1995, 1996), who also established the biosynthetic pathway of these compounds. In some strains, pigmentation appears to be an unstable characteristic because spontaneous nonpigmented mutants are frequently produced that never revert to yellow pigmentation. Moreover, the consistent isolation of nonpigmented strains from dark environments leads to the hypothesis that the yellow pigmentation of *Thermus* is favored in natural thermal areas exposed to sunlight where carotenoids would protect the cells from sunlight, whereas nonpigmented strains would have a selective advantage in non-illuminated environments because the production of ca-rotenoids would be energetically expensive. This hypothesis has its roots in the observation that the nonpigmented strain X-1 had a higher growth rate than *T. aquaticus* YT-1 (Ramaley and Hixson, 1970). Recently, a gene cluster involved in the synthesis of carotenoids in strain HB-27 was located on a large plasmid, designated pTT27. The same authors reported that carotenoid over-producing mutants were more resistant to UV-irradiation than either the wild-type strain or the carotenoid underproducing mutants (Hoshino et al., 1994; Tabata et al., 1994). Nevertheless, the carotenoid overproducing mutants grew slower at supraoptimal temperatures than the wild-type strain. These results could explain why nonpigmented strains predominate in dark environments over pigmented strains, whereas pigmented strains constitute the predominant isolates in hot springs exposed to sunlight. The presence of carotenogenic genes on plasmids could also explain the high frequency of pigmentation loss in some strains due to curing of the plasmids under laboratory conditions.

The cardinal growth temperatures of the species of the genus *Thermus* range between about 45°C and 82–83°C. However, only a few strains, all closely related to *T. thermophilus* HB8, are capable of growth at 80°C or above (Manaia et al., 1994). The majority of the strains of the genus *Thermus* have, in fact, a maximum growth temperature slightly below 80°C (Brock and Freeze, 1969; Hudson et al., 1989; Santos et al., 1989; Manaia and da Costa, 1991). The optimum growth temperature of all strains is in the vicinity of 70°C, but in some strains the optimum growth temperature could be as high as 75°C. The type strain of *T. scotoductus*, however, was reported to have an optimum growth temperature of 65–70°C, and a maximum growth temperature of 73°C (Kristjánsson et al., 1994).

The strains of the species of the genus *Thermus* have a respiratory metabolism and are aerobic. Many strains are capable of growth under anaerobic conditions using nitrate as the electron acceptor; some strains also reduce nitrite (Munster et al., 1986; Hudson et al., 1989; Santos et al., 1989). None of the strains, however, appears to be capable of carrying out fermentation. The strains of the genus *Thermus* possess phosphofructokinase (Yoshida et al., 1971; Yoshida, 1972), fructose-1,6-bisphosphate aldolase (Freeze and Brock, 1970), glyceraldehyde-3-phosphate dehydrogenase (Fujita et al., 1976; Harris et al., 1980; Tanner et al., 1996), phosphoenolpyruvate carboxylase (Sundaram and Bridger, 1979), enolase (Barnes and Stellwagen, 1973), and lactate dehydrogenase (Machida et al., 1985), and it is presumed that these species have a complete Embden-Meyerhoff pathway for the initial catabolism of hexoses. Several enzymes of the tricarboxylic acid cycle (Nishiyama et al., 1986; Miyazaki, 1996), the glyoxylate bypass (Degryse and Glansdorff, 1976), and several components of the respiratory chain, such as menaquinone 8, NADH-quinone oxidoreductase (Yano et al., 1997), cytochromes ba_3 (Zimmermann et al., 1988; Oertling et al., 1994; Keightley et al., 1995), and cytochrome c_{552} (Soulimane et al., 1997), have been identified and characterized in these organisms. A V-type ATPase, found in *Archaea*, eucaryotic endomembrane systems, and some bacteria, was initially detected in *T. thermophilus* HB8 by Yokoyama et al. (1990), but later Radax et al. (1998) found that some strains, namely the type strains of *T. filiformis* and *T. scotoductus*, possessed an F-type ATPase found in other bacteria, mitochondria, and chloroplasts. Conversely, *T. aquaticus* TY-1, *T. thermophilus* HB-27, and *Meiothermus chliarophilus* possess V-type ATPase. The presence of two different types of ATPases in species of the same genus is not clearly understood, although the authors speculate that horizontal gene transfer from other organisms may be responsible for the presence of the V-type ATPase in some strains. Alternatively, a common ancestor may have had both enzymes and only one or the other was retained during evolution.

All strains examined are catalase positive, in contrast to the strains of several species of the genus *Meiothermus* that do not have this enzyme under the conditions examined (Tenreiro et al., 1995b; Chung et al., 1997). A manganese superoxide dismutase has also been identified in *T. thermophilus* (Stallings et al., 1985; Lah et al., 1995). All strains are chemoorganotrophic and are capable of growth on amino acids, peptides and proteins, organic acids, and simple and complex carbohydrates (Table B4.4). With the exception of some strains of *T. thermophilus*, most other strains are not able to assimilate pentoses (Williams and da Costa, 1992). Recent results show that many of the strains appear to be able to hydrolyze Tween 20 to Tween 80, as well (Chung et al., unpublished results).

Thermus medium and Degryse medium 162 are the most frequently used media for growth of the strains of this genus (Degryse et al., 1978; Brock, 1984; Williams and da Costa, 1992). Both media contain yeast extract, and most strains require some cofactors for growth provided by this supplement, although yeast extract can be replaced by a complex mixture of amino acids, vitamins, and nucleotides (Sharp and Williams, 1988). Some strains, namely strain HB8, have been grown in a minimal salts medium containing a carbon source, biotin, and thiamine (Tanaka et al., 1981); strain YT-1 was grown in a minimal medium containing a carbon source, biotin, thiamine, and nicotinic acid (Yeh and Trela, 1976), whereas strain ZO5 has been grown in an inorganic medium without co-factors and containing pyruvate as the carbon source (van de Casteele et al., 1997). Some caution is necessary when attempting to examine the growth of *Thermus* strains in these media, because it can be very difficult to reproduce.

To date, the genomic structure of the strains of the genus *Thermus* has not been extensively studied. A physical map of the chromosome of strain HB8 has been constructed using pulsed-field gel electrophoresis (PFGE) of macrorestriction fragments, and several genes have been located on the chromosome of this strain that is estimated to be about 1.74 Mb (Borges and Bergquist, 1993); the physical map of strain HB-27 was also constructed using several restriction endonucleases and was estimated at 1.82 Mb (Tabata et al., 1993). The chromosome size of several other strains of the genus *Thermus* was later estimated to range from about 1.8 to 2.5 Mb using PFGE (Rodrigo et al., 1994; Moreira et al., 1997).

The first plasmids from strains of the genus *Thermus* were isolated by Hishinuma et al. (1978), but plasmids have now been detected from the majority of the strains examined. For example, Munster et al. (1985) found plasmids in about 60% of the isolates from Yellowstone National Park, and Moreira et al. (1995) detected plasmids in about 80% of the strains examined. Plasmid pTT1 (plasmid pTT8 of Hishinuma et al., 1978) was the first to be characterized (Eberhard et al., 1981). Two other plasmids, one from *T. thermophilus* AT-62 and the other from *T. thermophilus* HB8, respectively designated pTF62 and pVV8, were also characterized by restriction endonucleases (Vasquez et al., 1981, 1983). Plasmid pVV8 is associated with cell aggregation of strain HB8 in rich medium (Mather and Fee, 1990), whereas, as stated above, plasmid pTT27 encodes carotenogenic genes in strain HB-

27 (Hoshino et al., 1994; Tabata et al., 1994). All other plasmids appear to be cryptic at this time (Raven, 1995).

The discovery that *Thermus* strains are naturally transformable, coupled to the ease of growth of these organisms in solid and liquid media, renders them excellent candidates for the development of thermophilic host–vectors systems (Koyama et al., 1986). In particular, strain HB-27, devoid of the Taq1 restriction system found in many strains, has been used extensively as host for genetic manipulation (Koyama et al., 1986; Koyama and Furukawa 1990; Lasa et al., 1992b; Mather and Fee, 1992; Fernández-Herrero et al., 1995). Other strains, such as SPS-7 and SPS-10 are also naturally transformable (Peist, Marugg and da Costa, unpublished results). The plasmid vector pYK105 described by Koyama et al. (1989) constituted the first of a series of *Thermus*-specific and *Thermus*–*E. coli* shuttle vectors that have been constructed by several groups using *trpB* (Koyama and Furukawa, 1990), β-galactosidase (Koyama et al., 1990), or kanamycin resistance as selection markers (Lasa et al., 1992a; Mather and Fee, 1992; Wayne and Xu, 1997). For their replication in thermophiles, these vectors rely on cryptic *Thermus* vectors or on entire cryptic plasmids. Integration vectors have also been developed that have proved useful for stable expression and gene analysis (Lasa et al., 1992a; Fernández-Herrero et al., 1995; Weber et al., 1995; Tamakoshi et al., 1997).

Bacteriophage have also been found in strains of the genus *Thermus*, but have not been investigated in detail (Raven, 1995). The first *Thermus* phages were isolated by Sakaki and Oshima (1975) from *T. thermophilus* HB8 and other Japanese strains. Phage φYS40 grew within the temperature range of *T. thermophilus*, where it formed clear plaques indicative of complete lysis of the culture. This phage has an icosahedral head and a tail terminated by a plate and tail fibers. The genetic material is double-stranded DNA of about 175 kb, and the mol% G + C of the DNA was estimated at 35%, unlike the mol% G + C value of the *T. thermophilus* DNA, which is about 64%. Other bacteriophage from strains of the genus *Thermus*, namely φYB10, also have low G + C content, and some type of unknown thermostabilization mechanism must be necessary to prevent strand separation of the DNA of these viruses at high temperatures (Raven, 1995).

The exploitation of thermostable enzymes is a major goal of biotechnology; however, the main thrust is directed to the utilization of enzymes from organisms that grow at or near the boiling point of water, leaving out organisms in the growth range of the strains of the genus *Thermus*. Despite all the interest in the biotechnological exploitation of hyperthermophilic organisms, *T. aquaticus* YT-1 produces one of the most valuable enzymes in scientific and economic terms. This enzyme is, of course, Taq polymerase. This enzyme made the polymerase chain reaction (PCR) possible, and even other polymerases from hyperthermophiles, namely from *Thermococcus litoralis*, have not replaced the Taq polymerase.

Other useful enzymes have been isolated from *Thermus* strains, and some have been cloned in *E. coli*, but these have never really found a niche in the enzyme market because other, more stable enzymes are becoming available. Several proteases, such as Aqualysin I, Caldolysin, Caldolase, and PreTaq, have been purified and characterized, but there is very little biotechnological interest in them (Cowan and Daniel, 1982a, b; Matsuzawa et al., 1988; Saravani et al., 1989; Peek et al., 1992). An amylase has also been characterized from *T. filiformis* strain Ork A2 (Egas et al., 1998), but other more thermostable amylases from *Archaea* are known

TABLE B4.4. Characteristics of the type strains of the species of the genus *Thermus*[a]

Characteristic	1. *T. aquaticus* YT–1[b]	2. *T. brockianus* YS38[c]	3. *T. filiformis* Wai33 A1[b]	4. *T. oshimai* SPS–17[b]	5. *T. scotoductus* SE–1[d]	6. *T. thermophilus* HB8[b]
Pigmentation	Deep yellow	Light yellow	Deep yellow	Light yellow	Colorless	Yellow
Colonies	Compact	Spreading	Compact	Compact	Compact	Compact
Optimum temperature, °C	70	70	70	70	65–70	70
Growth at 80°C	–	–	–	–	–	+
Growth in 1% NaCl	+	nd	–	–	nd	+
Growth in 3% NaCl	–	–	–	–	–	+
Anaerobic growth with NO_3^-	–	+	–	+	+	–
Presence of:						
Oxidase	+	+	+	+	+	+
Catalase	+	+	+	+	+	+
DNase	+	nd	–	–	nd	+
α-Galactosidase	+	nd	+	+	nd	+
β-Galactosidase	–	nd	+	+	nd	+
Hydrolysis of:						
Elastin	+	nd	–	+	nd	–
Fibrin	+	nd	–	+	nd	–
Casein	+	–	–	+	nd	+
Gelatin	+	–	+	+	–	+
Starch	+	–	–	–	–	–
Arbutin	–	nd	+	+	nd	+
Esculin	–	nd	+	+	nd	–
Utilization of:						
D-Glucose	+	+	+	+	+	+
D-Fructose	+	+	+	+	+	+
L-Rhamnose	–	nd	–	–	nd	–
L-Arabinose	–	nd	–	–	nd	–
D-Xylose	–	–	–	–	–	+
D-Galactose	+	+	+	+	nd	–
D-Mannose	+	nd	–	+	nd	+
D-Cellobiose	+	nd	–	–	nd	+
Lactose	+	+	+	+	nd	–
Sucrose	+	nd	+	+	nd	+
D-Trehalose	+	nd	+	+	nd	+
D-Melibiose	+	+	+	+	nd	+
D-Raffinose	+	nd	+	+	nd	–
Dextrin	–	nd	–	–	nd	–
Salicin	–	nd	+	+	nd	–
D-Mannitol	–	nd	–	–	nd	–
D-Sorbitol	–	nd	–	–	nd	–
myo-Inositol	–	nd	–	–	nd	–
Glycerol	–	nd	–	–	+	–
Pyruvate	+	+	+	+	+	+
Acetate	+	nd	+	+	–	+
Citrate	–	nd	–	–	nd	–
Lactate	–	nd	–	–	nd	–
Malate	+	nd	+	+	nd	–
Succinate	+	nd	+	+	–	–
Arginine	+	nd	+	+	nd	–
Proline	+	+	+	+	+	+
Ornithine	+	nd	+	+	nd	–
Serine	+	nd	–	–	nd	+
Acetamide	–	nd	–	–	nd	–
Mol% G + C	64	63	65	63	65	65

[a]Symbols: +, positive result; –, negative result or no growth; nd, not determined.
[b]Results from Manaia and da Costa (1991).
[c]Results from Manaia et al. (1994).
[d]Results from Kristjánsson et al. (1994).

and are being examined for the conversion of starch to high glucose syrups.

Strains of the genus *Thermus* are ubiquitous in natural hydrothermal areas with neutral to alkaline pH; the first isolates of the genus *Thermus* were obtained from hydrothermal areas in Yellowstone National Park and Pacheteau's Calistoga in California by Brock and Freeze (1969). Isolates were later recovered from many inland hydrothermal areas in Japan (Yoshida and Oshima, 1971; Saiki et al., 1972; Taguchi et al., 1982), Iceland (Pask-Hughes and Williams, 1977; Kristjánsson and Alfredsson 1983; Alfredsson et al., 1985; Hudson et al., 1987a), New Zealand (Hudson et al., 1986), New Mexico (Hudson et al., 1989), the island of São Miguel in the Azores, continental Portugal (Santos et al., 1989; Manaia and da Costa, 1991), the Australian Artesian Basin

(Denman et al., 1991), and the Kamchakta Peninsula (R. Sharp, personal communication). Furthermore, Yellowstone National Park continues to be the source of many more *Thermus* isolates (Munster et al., 1986; Hudson et al., 1989). In addition to continental hydrothermal areas where the concentration of sodium chloride is generally very low, strains of the genus *Thermus* have also been isolated from shallow marine hot springs off the coast of Iceland (Kristjánsson et al., 1986), on beaches on the island. of Fiji (Hudson et al., 1989), and the island of São Miguel in the Azores (Manaia and da Costa, 1991), along the coast north of Naples, Italy (da Costa, unpublished results), and the island of Monserrat in the Caribbean (N. Raven, personal communication). Recently, isolates of the genus *Thermus* were also obtained from the most unique type of thermal environment— abyssal geothermal areas in the Mid-Atlantic Ridge and in the Guaymas Basin, Gulf of California, at depths of 3500 and 2000 m, respectively (Marteinsson et al., 1995), some of which belong to *T. thermophilus* (Marteinsson et al., 1999).

The isolation of strains of this genus is not restricted to natural hydrothermal areas. In fact, the isolation of strains from manmade environments was simultaneous with the description of *Thermus aquaticus* by Brock and Freeze (1969), who included yellow-pigmented strains isolated from cold and hot water taps at Indiana University, where no hot springs exist. Later, nonpigmented *Thermus* strains were isolated from hot water taps (Pask-Hughes and Williams, 1975; Stramer and Starzyk, 1981), domestic and industrial hot water systems (Brock and Boylen, 1973), and thermally polluted streams (Ramaley and Hixson, 1970; Brock and Yoder, 1971; Ramaley and Bitzinger, 1975; Degryse et al., 1978). In addition to these environments, and perhaps unexpectedly, strains of the genus *Thermus* have also been isolated from self-heating (thermogenic) compost piles in Switzerland and Germany (Beffa et al., 1996; G. Antranikian, personal communication).

In natural environments, these organisms are generally isolated from hydrothermal areas where the water temperature ranges from 55 to 70°C and the pH ranges from 5.1 to 10.5 (Kristjánsson and Alfredsson, 1983; Munster et al., 1986; Hudson et al., 1989; Santos et al., 1989). However, isolates have occasionally been recovered from water with temperatures as high as 95°C, and pH values as low as 3.9 (Kristjánsson and Alfredsson, 1983; Hudson et al., 1986; Munster et al., 1986). The recovery of strains of the genus *Thermus* from geothermal sites with low pH and very high temperatures, as well as from cold water sources, is believed to be due to dispersal from environments that these strains actually colonize (Williams and da Costa, 1992; Alfredsson and Kristjánsson, 1995).

Most of the isolates of the genus *Thermus* originate from terrestrial hot springs venting fresh water because these are more common than marine geothermal areas and because they are sampled more frequently. Some shallow marine hot springs contain less total salts than the surrounding sea (Kristjánsson et al., 1986), but other marine hot springs contain concentrations of salts similar to those of sea water, and all isolates from marine hot springs are, to our knowledge, halotolerant (Manaia and da Costa, 1991; Tenreiro et al., 1997). These organisms have higher growth rates in *Thermus* medium without added NaCl, but grow in this medium containing 3–4% NaCl, whereas the vast majority of the strains of *Thermus* isolated from inland hydrothermal areas will not grow at salinities above 1% NaCl (Kristjánsson et al., 1986; Hudson et al., 1989; Santos et al., 1989; Manaia and da Costa, 1991; Manaia et al., 1994). The halotolerant strains from

inland hot springs, namely strains HB8, AT-62, GK-24 from Japan, strain B from Iceland, and strain RQ-1 from the island of São Miguel in the Azores, are closely related to one another and to the isolates from marine hot springs (Manaia et al., 1994). However, the inland sites in Japan from which halotolerant strains were isolated are reported to be saline, and inland saline hot springs are also found in Iceland (Waring, 1965; Alfredsson and Kristjánsson, 1995). Nevertheless, the site on the island of São Miguel from which halotolerant strains RQ-1 and RQ-3 were isolated has a low salinity (Santos et al., 1989; Veríssimo et al., 1991), but is only about 5 km from shallow marine hot springs that have yielded similar halotolerant *Thermus* strains and *Rhodothermus marinus* (Manaia and da Costa, 1991; Nunes et al., 1992). On the other hand, none of the strains isolated from the Furnas area, about 1 km further inland, was halotolerant. These observations lead to the view that halotolerant isolates of *Thermus* are, as expected, primarily marine organisms that may occasionally colonize inland hot springs of low salinity. Moreover, all the halotolerant strains studied belong to the species *T. thermophilus*, or to other, closely related species that remain unclassified (Manaia and da Costa, 1991; Chung et al., unpublished results).

Studies based on whole DNA–DNA hybridization values and 16S rDNA sequence similarities of a large number of isolates show that some of the species appear to have a wide distribution in geothermal and man-made environments, whereas others have a restricted distribution. For example, *T. brockianus* strains have been isolated from Yellowstone National Park, and Iceland (Hudson et al., 1987a; Saul et al., 1993; Williams et al., 1995; Chung et al., unpublished results); strains belonging to *T. thermophilus*, or to species closely related to *T. thermophilus*, have been isolated from marine and inland hot springs in Japan, Iceland, the island of São Miguel, the island of Fiji, Naples, abyssal black smokers, and thermogenic composts; and strains of *T. scotoductus* have been isolated from Iceland, the Azores, continental Portugal, a thermally polluted stream in the United States, hot springs in New Mexico, and hot tap water in London. However, strains of *T. aquaticus* have only been isolated from Yellowstone National Park, and *T. filiformis* has been isolated only from New Zealand. Strains of *T. brockianus* and *T. aquaticus* are easily isolated from the same springs in Yellowstone National Park, and the absence of isolates of the latter species from other hydrothermal areas cannot be due to the difficulty in isolating strains of *T. aquaticus*. Moreover, many strains of *T. filiformis*, based on high DNA–DNA hybridization values and 16S rDNA sequence similarity, are known, but all originate from New Zealand (Georganta et al., 1993; Saul et al., 1993; Chung et al., unpublished results). In fact, with the exception of strains closely related to *T. thermophilus* from the South Island of New Zealand, strains belonging to other species have not been isolated from New Zealand. These results indicate that some species of the genus *Thermus* appear to be restrcted to a limited number of sites or geothermal areas, whereas others are frequently isolated from widely dispersed hydrothermal areas. The lack of extensive sampling and characterization of isolates, and the lack of culture-independent phylogenetic studies of samples, is the most likely explanation for the inability to recover strains of some species from widely separated geothermal areas, but it is equally possible that physical, chemical, and biological parameters of the hydrothermal areas restrict the distribution of the strains of some species. It is interesting to note that only strains closely related to *T. brockianus* and *T. aquaticus* continue to be found in Yellowstone National Park. For example, Nold and Ward (1995) performed 16S rDNA sequence analysis on a number of isolates from Octopus Spring in Yellowstone National Park; one isolate belonged to *T. brockianus*, whereas the other strain was closely related to *T. aquaticus* YT-1 and to strain YSPID isolated several years earlier from the same area by Hudson et al. (1989). More intriguing are the recent results of Hugenholtz et al. (1998b), who used culture-independent phylogenetic methods to analyze the microbial community of Obsidian Pool in Yellowstone National Park. The only *Thermus* rRNA gene types found were all closely related to strain YSPID. Sequences closely related to *T. aquaticus* YT-1 or *T. brockianus* YS38 were not found. These results indicate that, at least in some hot springs, only strains of one species can be detected, which somehow may be related to unknown factors governing colonization of the geothermal sites.

ENRICHMENT AND ISOLATION PROCEDURES

Brock and Freeze (1969) isolated the original strains of *Thermus aquaticus* on Castenholz D basal salts medium (Castenholz, 1969) supplemented with yeast extract (1.0 g/l) and tryptone (1.0 g/l). This medium has been used in most studies to isolate and to grow strains of the genus *Thermus*, and has been called Castenholz medium D for *Thermus* (Hudson et al., 1986) or simply *Thermus* medium (Munster et al., 1986; Williams and da Costa, 1992). Medium 162 of Degryse et al. (1978) is also commonly used to isolate and grow these organisms. The basal salts medium 162 is slightly different from Castenholz D medium, and generally 2.5 g/l of yeast extract and 2.5 g/l of tryptone are used. Both media are, nevertheless, adequate for the growth of all known strains of this genus.

The growth of the majority of the strains of *Thermus* is inhibited by levels of organic nutrients higher than about 1.0%. Hexoses are particularly inhibitory, apparently because of acidification of the medium. Some strains, particularly those closely related to *T. thermophilus* HB8, are more resistant to organic nutrients of the culture medium and are frequently grown in a medium containing (per liter water) Trypticase or polypeptone, 8.0 g; yeast extract, 4.0 g; and NaCl, 2.0 g (Oshima and Imahori, 1974). The review by Sharp et al. (1995) gives an extensive list of media, and their composition, used to grow the strains of this genus for several purposes, and should be consulted.

Most isolates of the genus *Thermus* have been obtained by enrichment in *Thermus* medium or in medium 162. Water or biofilm samples are inoculated into liquid medium and incubated at 70–75°C for 2–3 days. Turbid cultures are spread on the same medium solidified with agar (2–3%) and incubated at the same temperature until yellow or nonpigmented colonies appear and can be isolated. Alternatively, samples are directly spread on solid media (Hudson et al., 1986, 1989; Munster et al., 1986; Santos et al., 1989). Membrane filtration methods have also been used extensively for the isolation of strains of the genus *Thermus* and offer the advantage of recovering a larger number of different colonial types and minor populations, as compared to liquid enrichments that tend to select clones that grow better in the media used or that constitute the major populations of the samples. The membrane filtration method can be used with water or biofilms that have been macerated and shaken vigorously with small amounts of water from the same hot spring or phosphate-buffered saline (Kristjánsson and Alfredsson, 1983; Kristjánsson et al., 1986; Manaia and da Costa, 1991). Adequate volumes of the samples, or dilutions, are filtered through cellulose nitrate or acetate membrane filters (47 mm dia., 0.22–0.45 μm pore

size). The filters are placed onto the surface of plates of *Thermus* medium, or a similar, low nutrient medium solidified with 2–3% agar. The plates are then inverted, wrapped in plastic film, and incubated for 2–7 days at temperatures ranging from 70 to 75°C. Yellow or colorless colonies can easily be observed and picked for further purification. Other organisms may also be isolated under these conditions, namely aerobic spore-formers, although few of these are isolated at temperatures above 70°C. Identification of isolates of the genus *Thermus* can be easily accomplished by assessing the presence of cytochrome oxidase and catalase, cell morphology, fatty acid composition, and the characteristic polar lipid composition (Prado et al., 1988; Donato et al., 1990; Nobre et al., 1996a).

MAINTENANCE PROCEDURES

All strains grow well on *Thermus* medium or Degryse medium 162 solidified with 2% or 3% agar. Other solidifying agents such as Gelrite are not necessary at temperatures of 70–75°C. Nevertheless, glass plates may be useful at temperatures above 70°C because some plastic Petri plates can become deformed at high temperatures. Sometimes colonies of thermophilic spore-formers derived from the agar appear during cultivation of organisms at 70°C. Correct autoclaving procedures must be followed to insure that the medium is thoroughly brought to autoclaving temperature for the specified time. In our experience, it is really not necessary to sterilize the media at temperatures higher than 121°C. Sometimes it may be convenient to incubate uninoculated plates at high temperatures to insure that spore-forming strains did not survive sterilization. During incubation, the Petri plates should be wrapped in plastic film to prevent evaporation. Cultures on solid medium can be maintained for a few weeks in the dark at room temperature. Cultures can be stored for longer times by freezing at −70°C in cryotubes containing liquid media, mentioned above, supplemented with glycerol to yield a final concentration of 15% (v/v). Cultures have been maintained for several years without loss of viability by freeze-drying or by storage in liquid nitrogen.

TAXONOMIC COMMENTS

At this time, the genus *Thermus* consists of the species *T. aquaticus* (Brock and Freeze, 1969), *T. thermophilus* (Oshima and Imahori, 1974; Manaia et al., 1994), *T. filiformis* (Hudson et al., 1987b), *T. scotoductus* (Kristjánsson et al., 1994), *T. brockianus* (Williams et al., 1995), and *T. oshimai* (Williams et al., 1996). The study of Saul et al. (1993) provided the first insight into the intrageneric phylogenetic relationships of the genus *Thermus*. The 20 strains for which the 16S rDNA sequence was determined in that study included the three species that were validly described at that time, namely *T. aquaticus* YT-1 (L09663), *T. filiformis* Wai33 A1 (L09667), and *T. ruber* (L09672) (later transferred to the genus *Meiothermus* [Nobre et al., 1996b]), the then-invalid species *T. thermophilus* HB8 (L09659) and "*T. flavus*" AT-62 (L09660), plus 14 strains from Iceland, New Zealand, and United States. This study demonstrated that the species of the genus *Thermus* could be differentiated using 16S rDNA sequence comparisons and provided some information on the biogeographical distribution of the *Thermus* species.

After the study of Saul et al. (1993), three new species of the genus *Thermus* were described, namely *T. brockianus* (Williams et al., 1995), *T. oshimai* (Williams et al., 1996), and *T. scotoductus* (Kristjánsson et al., 1994). The descriptions of these species did not include phylogenetic analyses of 16S rDNA sequence data nor did the authors demonstrate the phylogenetic position of these species within the radiation of the genus *Thermus*. Until recently (Chung et al., unpublished results), 16S rDNA sequences were not available for the type strains of *T. oshimai* and *T. scotoductus*.

Comparison of the 16S rDNA sequences of the type strains of each of the six validly described species of the genus *Thermus* shows the 16S rDNA sequence similarities within the genus *Thermus* to be in the range 91.2–96.4%. *Thermus oshimai* is the most distantly related of the species of the genus *Thermus* and this is reflected in the 16S rDNA sequence similarity values. The sequence of the type strain of *T. oshimai* has similarity values to *T. aquaticus* of 92.2%, to *T. brockianus* of 91.7%, to *T. filiformis* of 91.2%, to *T. scotoductus* of 91.9% and to *T. thermophilus* of 93.0%. The latter five species of the genus *Thermus* have 16S rDNA sequence similarities in the range 94.1–96.4%.

Within each species the 16S rDNA similarity values are in the range 98.9–99.7% for *T. aquaticus*, 99.9–100% for *T. brockianus*, 99.2–99.9% for *T. filiformis*, 99.8–100% for *T. oshimai*, 98.7–99.9% for *T. scotoductus*, and 99.4–100% for *T. thermophilus*. These values are based on the comparison of all published and a large number of unpublished full 16S rDNA sequences for strains of species of the genus *Thermus* (Fig. B4.6). These values clearly demonstrate the usefulness of 16S rDNA sequence data in the identification of new *Thermus* isolates or in determining their novelty and relatedness to the validly described *Thermus* species. Such data, however, are of little use in the differentiation of strains within a *Thermus* species.

The species *Meiothermus ruber* (Loginova et al., 1984), *M. silvanus*, and *M. chliarophilus* (Tenreiro et al., 1995a) were initially included in the genus *Thermus* despite their lower cardinal growth temperatures, because these species had many characteristics in common with the high-temperature strains and were phylogenetically closely related to them (Brock, 1984; Hensel et al., 1986; Sharp and Williams, 1988; Williams and da Costa, 1992; Embley et al., 1993). The description of *T. silvanus* and *T. chliarophilus* with higher 16S rDNA sequence similarity to *T. ruber* than to the high-temperature species of *Thermus*, coupled with the lower optimum growth temperature range and the presence of moderate levels of 2-OH fatty acids, led to the proposal of the genus *Meiothermus* and the emendation of the genus *Thermus*, which retained only the species with optimal growth temperatures around 70°C (Nobre et al., 1996b). 16S rDNA sequence similarity values of 84.9–86.7% are found between the species of the genera *Thermus* and *Meiothermus*, demonstrating the clear distinction between these genera. The large differences between

the 16S rDNA sequences of members of the genera *Thermus* and *Meiothermus* clearly allow differentiation of species of these two genera.

Two yellow-pigmented strains from Japan, designated *"Thermus flavus"* for strain AT-62 (Saiki et al., 1972) and *"Thermus caldophilus"* for strain GK-24 (Taguchi et al., 1982), were never validly described, and, in fact, have been shown to belong to the species *T. thermophilus* (Manaia et al., 1994; Williams et al., 1995). One strain, named *"Thermus lacteus"*, is a patent strain of unknown affinity deposited with the American Type Culture Collection (ATCC 31557).

Due to extremely variable biochemical and physiological characteristics and fatty acid composition, it is very difficult to define most of the species of the genus *Thermus* (Tables B4.4 and B4.5). The variability of biochemical and physiological parameters was noticed in early studies involving numerical taxonomy and has constituted one of the great hurdles for an adequate classification of most of the isolates of this genus (Cometta et al., 1982; Alfredsson et al., 1985; Hudson et al., 1986, 1987a, 1989; Munster et al., 1986; Santos et al., 1989). The phenotypic variability may be due to natural diversity within each species, but there is the possibility that it could also be the result of technical difficulties in assessing phenotypic characteristics. For example, it has been reported that glucose is not assimilated by many strains, but lowering the growth temperature to about 65°C results in assimilation of this carbon source by the majority of the strains (Manaia and da Costa, 1991). In an initial numerical classification of Portuguese and Azorean strains, the species later named *T. oshimai* formed two distinct clusters, yet all strains had high DNA–DNA hybridization values (Santos et al., 1989; Williams et al., 1995). In another study based primarily on strains isolated from Yellowstone National Park, one phenetic cluster contained most of the strains of *T. aquaticus*, whereas the other major cluster contained most of the strains of *T. brockianus*. However, the latter cluster also contained, for example, strains of *T. thermophilus* (Munster et al., 1986).

The same type of variation was found in the fatty acid composition of a large number of strains belonging to all of the species of the genus; the species *T. thermophilus*, for example, had extremely variable fatty acid compositions, even though many strains share very high DNA–DNA hybridization values (Manaia et al., 1994; Nobre et al., 1996a).

To further complicate matters, most species have been described on the basis of a small number of isolates that have not been extensively characterized, so that interspecific diversity of phenotypic characteristics has not been assessed. For example, the species *T. filiformis* was described on the basis of one strain from New Zealand with a stable filamentous morphology (Hudson et al., 1987b). Other strains from New Zealand belong to this species on the basis of DNA–DNA hybridization values, but are not filamentous (Georganta et al., 1993). Moreover, the type strain of *T. filiformis* possesses very high levels of anteiso-fatty acids as well as 3-OH fatty acids, whereas the other strains have high levels of iso-fatty acids and lack 3-OH fatty acids (Ferraz et al., 1994; Nobre et al., 1996a). The biochemical and physiological diversity of *T. filiformis* strains that share high DNA–DNA hybridization values makes it difficult, if not impossible, to define a distinct phenotype for this species (Hudson et al., 1987b, 1989; Georganta et al., 1993).

The description of the species *T. scotoductus* was based on strains isolated from hydrothermally fed hot water taps in Iceland, as well as strain X-1 isolated from a thermally polluted

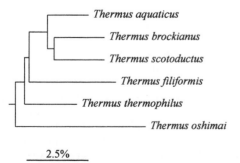

FIGURE B4.6. Phylogenetic dendrogram based on 16S rDNA sequence comparison of the type strains of the genus *Thermus*. Bar = 2.5 inferred nucleotide substitutions per 100 nucleotides.

TABLE B4.5. Fatty acid composition of the type strains of the genus *Thermus* grown at 70°C

| | % of total in:[a] | | | | | |
Fatty acid	1. *T. aquaticus* YT–1	2. *T. brockianus* YS38	3. *T. filiformis* Wai33 A1	4. *T. oshimai* SPS–17	5. *T. scotoductus* SE–1	6. *T. thermophilus* HB8
$C_{14:0 \text{ iso}}$	0.9	1.6	0.9	0.5	—	0.8
$C_{15:0 \text{ iso}}$	17.6	33.5	4.1	37.7	15.8	32.4
$C_{15:0 \text{ anteiso}}$	1.9	3.1	18.9	3.8	16.1	4.7
$C_{15:0}$	—	0.6	—	3.7	1.1	—
$C_{16:0 \text{ iso}}$	13.0	12.1	8.4	3.1	3.8	5.8
$C_{16:0}$	16.3	9.2	3.5	7.2	8.0	8.0
UN[b]	—	—	2.9	—	2.5	—
$C_{15:0 \text{ iso 3-OH}}$	3.2	—	0.6	—	—	—
$C_{15:0 \text{ anteiso 3-OH}}$	—	—	1.1	—	—	—
$C_{17:0 \text{ iso}}$	27.0	34.1	6.6	36.1	25.3	41.9
$C_{17:0 \text{ anteiso}}$	2.7	3.3	36.7	3.7	25.8	5.7
$C_{17:0}$	—	—	—	2.1	1.2	—
$C_{16:0 \text{ iso 3-OH}}$	2.4	—	0.9	—	—	—
$C_{16:0 \text{ 3-OH}}$	2.5	—	—	—	—	—
$C_{18:0 \text{ iso}}$	0.6	0.5	0.9	—	—	—
$C_{18:0}$	0.9	—	—	—	—	—
$C_{17:0 \text{ iso 3-OH}}$	7.6	—	2.4	—	—	—
$C_{17:0 \text{ anteiso 3-OH}}$	0.7	—	8.8	—	—	—

[a]Values for fatty acids present at levels of less than 0.5% in all strains are not shown. Under the conditions used to determine the fatty acid composition, diols were detected in low levels and were not included in the table.

[b]Unknown fatty acid or alcohol with an equivalent chain length of 16.090.

stream in the United States (Ramaley and Hixson, 1970; Kristjánsson et al., 1994). The lack of pigmentation was one of the characteristics used to discriminate this species from other species of the genus, although all these strains came from nonilluminated or artificial environments, where nonpigmented strains would be expected to constitute the dominant populations. Two other nonpigmented strains, designated NH and DI, isolated from hot tap water in London, were later found to be practically identical to the type strain from Iceland (Pask-Hughes and Williams, 1975; Tenreiro et al., 1995b). Other strains from the hot spring at the end of a nonilluminated tunnel at Vizela in northern Portugal were found to be closely related to the type strain of *Thermus scotoductus* (Tenreiro et al., 1995b), and were, not surprisingly, colorless. However, yellow-pigmented strains from hot springs exposed to sunlight in the Azores were also found to be very closely related to the Vizela strains and to *T. scotoductus* (Santos et al., 1989; Tenreiro et al., 1995b; Williams et al., 1996). Moreover, only the strains from Iceland and from London have identical fatty acid compositions (Nobre et al., 1996a). These results also appear to leave the species *T. scotoductus* without distinct phenotypic characteristics.

The species *T. aquaticus* can be easily distinguished from *T. brockianus*, but all of the strains extensively characterized originate from Yellowstone National Park and variability, therefore is limited (Munster et al., 1986; Williams et al., 1995). However, when biochemical and physiological characteristics of these organisms are compared to those of other species, the distinctiveness of *T. brockianus* and *T. aquaticus* breaks down, even though there is no doubt that these two species constitute distinct genomic species. It is interesting to note that all of the *T. brockianus* strains isolated from different hot springs at Yellowstone National Park have an identical fatty acid composition (Nobre et al., 1996a), and an identical genomic structure based on PFGE of large DNA fragments, leading us to believe that they constitute one clone (Moreira et al., 1997), whereas the *T. aquaticus* isolates have more variable fatty acid compositions and PFGE profiles. It is still possible to distinguish the strains of *T. aquaticus* from all other strains of the genus *Thermus* due to the presence of high

levels of iso-fatty acids coupled with the presence of 3-OH fatty acids (Nobre et al., 1996a). However, one strain from Yellowstone National Park that belongs to the species *T. aquaticus* by DNA–DNA hybridization values (Williams et al., 1996) does not possess 3-OH fatty acids or galactosamine in GL-1 (Ferreira et al., unpublished results) and may prove, after further examination, to constitute a nasty little exception to the classification and identification of *T. aquaticus* by fatty acid analysis. The description of the species *T. oshimai* was based on strains isolated from continental Portugal, the Azores, and Iceland, and appears to have a fairly homogeneous phenotype despite the formation of two phenotypic clusters in the original characterization of these strains (Santos et al., 1989; Nobre et al., 1996a; Williams et al., 1996).

Two characteristics, growth of the strains at 80°C or higher and halotolerance, appear to distinguish *T. thermophilus* from other species of this genus (Manaia et al., 1994). Several strains are very closely related to strain HB8 on the basis of DNA–DNA hybridization results and phylogenetic analysis, and can clearly be assigned to *T. thermophilus*. These are strains AT-62, HB-27, B, GK-24, RQ-1, and RQ-3. Other strains isolated from shallow marine hot springs are also closely related to strain HB8, sharing about 60% DNA–DNA hybridization values. Several strains recovered from shallow marine hot springs were also assigned to *T. thermophilus* because they could not be distinguished from this species (Manaia et al., 1994), but we are now of the opinion that the taxonomic status of strains such as Fiji A3 should be reassessed (Manaia et al., 1994; da Costa, unpublished results).

The species *T. thermophilus* was validly described by Oshima and Imahori (1974) based on strain HB8, previously named *"Flavobacterium thermophilum"* (Yoshida and Oshima, 1971). However, this species was not included in the Approved Lists of Bacterial Names (Skerman et al., 1980) because it could not be distinguished from *T. aquaticus*. Williams (1989) showed that strain HB8 had high DNA–DNA hybridization values with strains AT-62, GK-24, and B, but did not attempt to revive the name *T. thermophilus* due to the lack of phenotypic characteristics that could distinguish this species from the other species of the genus. The isolation of halotolerant strains from marine hot springs on the island of São Miguel in the Azores led to a detailed phenotypic characterization, and DNA–DNA hybridization study of strains belonging to this species (Manaia and da Costa, 1991; Manaia et al., 1994). The results showed that several marine and inland strains shared the ability to grow at temperatures above 80°C and were halotolerant. These characteristics, coupled with high DNA–DNA hybridization values, led to the revival of the species *T. thermophilus*. The revival of the name *T. thermophilus* by Manaia et al. (1994) appeared in the Validation List several pages after the publication of Williams et al. (1995) where the revival of the name *T. thermophilus* was also proposed. However, Williams et al. (1995) proposed the name *T. thermophilus* as valid only in the abstract of the publication without a formal proposal in the text, and did not present phenotypic results showing that the species could be distinguished from other species of the genus

Thermus. For these reasons the publication by Manaia et al. (1994) is considered the effective publication for the revival of the species *T. thermophilus* (H. Trüper, B. Tindall, N. Weiss, personal communication).

Acknowledgments

Until Brock and Freeze described *Thermus aquaticus* in 1969, the interest in thermophilic bacteria was negligible. Other thermophiles were known at that time, but the lack of an ecological perspective diverted our minds from the possibility that many organisms could inhabit natural environments with extremely high temperatures. It was the insight of Thomas Brock on life at high temperatures that led to the intense research now conducted on thermophilic bacteria, and it was *Thermus aquaticus* that led the way. For this reason, and many others, we wish to dedicate this chapter to Thomas D. Brock.

List of species of the genus Thermus

1. **Thermus aquaticus** Brock and Freeze 1969, 295.[AL]
a.qua′ti.cus. L.masc. adj. *aquaticus* living in water.

The strains of this species form rod-shaped cells 0.5–0.8 μm in diameter with variable length. Short filaments are present. Colonies are about 1 mm in diameter after 48 h growth on *Thermus* medium at 70°C and are bright yellow. The strains of this species generally hydrolyze casein, gelatin, and starch; the strains do not utilize lactose or melibiose, and do not reduce nitrate or nitrite. With one possible exception, all other strains examined have iso 3-OH fatty acids. Strains YS 004, YS 013, YS 025, YS 031, and YS 041 of Munster et al. (1986) can be assigned to this species (Williams, 1989; Williams et al., 1995; Nobre et al., 1996a). These strains have mol% G + C of the DNA between about 60% and 64%. The strains of this species have only been isolated from Yellowstone National Park.

The mol% G + C of the DNA is: 64 (T_m) (type strain).
Type strain: YT-1, ATCC 25104, DSMZ 625.
GenBank accession number (16S rRNA): L09663.

2. **Thermus brockianus** Williams, Smith, Welch, Micallef and Sharp 1995, 498.[VP]
brock′i.a.nus. M.L. masc. adj. *brockianus* named after Thomas D. Brock.

Strains of this species form rod-shaped cells and short filaments. The colonies of the strains from Yellowstone National Park are pale yellow, and spread on *Thermus* agar. These strains do not hydrolyze casein, and only a few hydrolyze gelatin or starch; nitrate and nitrite are reduced; lactose, trehalose and melibiose are utilized as single carbon sources. The strains do not possess 3-hydroxy fatty acids. All isolates from Yellowstone National Park are very closely related, and appear to constitute one clone. Several strains of phenetic group 2 of Munster et al. (1986), namely strains YS 07, YS 11, YS 19, YS 30, YS 40, and YS 44, belong to this species (Williams et al., 1995; Nobre et al., 1996a). Strain ZHG1 A1 from Iceland (Hudson et al., 1986) also belongs to this species (Williams et al., 1995).

The mol% G + C of the DNA is: 63 (T_m).
Type strain: YS38, NCIB 12676.
GenBank accession number (16S rRNA): Z15062.

3. **Thermus filiformis** Hudson, Morgan and Daniel 1987b, 435.[VP]

fi.li.for′ mis. L. neut. n. *filum* thread; L. fem. n. *forma* shape; L. masc. adj. *filiformis* thread shaped.

The strains of this species form bright yellow colonies. The type strain of this species has a stable filamentous morphology, but other strains assigned to this species produce rod-shaped cells. The type strain possesses very high levels of anteiso-fatty acids and anteiso 3-hydroxy fatty acids. Anteiso fatty acids are present in low levels in other strains and 3-hydroxy fatty acids are absent. It appears that this species has no phenotypic characteristics that distinguishes it from other species of the genus *Thermus*. Several isolates of Hudson et al. (1986), namely strains T351, Rt358, Tok22, and Rt6 A1 belong to this species due to high DNA–DNA hybridization values (Georganta et al., 1993). Strains of this species have been isolated only from hot springs in New Zealand.

The mol% G + C of the DNA is: 65 (T_m).
Type strain: Wai33 A1, ATCC 43280, DSMZ 4687.
GenBank accession number (16S rRNA): L09667, X58345.

4. **Thermus oshimai** Williams, Smith, Welch and Micallef 1996, 406.[VP]
o.shi′ma.i. M.L.gen. n. *oshimai* named after Tairo Oshima.

Strains of this species form rod-shaped cells and short filaments. The colonies of most strains are pale yellow; some strains are not pigmented. The strains of this species hydrolyze casein and fibrin. Most strains reduce nitrate and nitrite. The strains of this species utilize sucrose, maltose, trehalose, lactose, and melibiose, and possess α- and β-galactosidase. This species includes strains of phenetic clusters E and F of Santos et al. (1989) isolated from the hot spring at São Pedro do Sul, continental Portugal, and the island of São Miguel, the Azores, and strains JK-66, JK-90, and JK-91 from Iceland (Williams et al., 1996).

The mol% G + C of the DNA is: 63 (T_m).
Type strain: SPS-17, ATCC 700435, NCIB 13400.

5. **Thermus scotoductus** Kristjánsson, Hjörleifsdóttir, Marteinsson and Alfredsson 1995, 418[VP] (Effective publication: Kristjánsson, Hjörleifsdóttir, Marteinsson and Alfredsson 1994, 49.)
sco.to.duc′ tus. Gr. n. *scotos* darkness; L. masc. n. *ductus* Roman water duct; M.L. masc. n. *scotoductus* living in pipes and producing a dark pigment.

Cells are about 0.5 μm × 1.5 μm; filaments are present.

Old cultures of some strains produce a dark water-soluble pigment. The type strain and strains X-1 (ATCC 27978), NH (NCIB 11245) and DI (NCIB 11246), are nonpigmented. The optimum growth temperature of the type strain is between 65 and 70°C, but strain X-1 has a higher optimum growth temperature. Strain X-1 is composed of two colony types; in colony type t1 glycolipid 2 (GL-2) is the major glycolipid, and glycolipid 1 (GL-1) is not detected or is present in vestigial concentrations. Strains of this species have been isolated from hot water systems in Iceland and London, and a thermally polluted stream in the United States. Nonpigmented strains from Vizela, continental Portugal, yellow-pigmented strains of phenetic groups A and B from the island of São Miguel in the Azores (Santos et al., 1989), and strain NMX2 A1 from New Mexico (Hudson et al., 1989) probably belong to this species.

The mol% G + C of the DNA is: 65 (T_m).

Type strain: strain SE-1, ATCC 51532, DSMZ 8553.

GenBank accession number (16S rRNA): AF032127.

6. **Thermus thermophilus** (ex Oshima and Imahori 1974) Manaia, Hoste, Gutierrez, Gillis, Ventosa, Kersters and da Costa 1995, 619[VP] (Effective publication: Manaia, Hoste, Gutierrez, Gillis, Ventosa, Kersters and da Costa 1994, 530.)

ther.mo'phi.lus. Gr. n. *therme* heat; Gr. adj. *philos* loving; M.L. masc. adj. *thermophilus* heat-loving.

The cells are rod shaped, and short filaments are also formed. The strains of this species form light yellow nonspreading colonies on *Thermus* medium. All the strains can grow at temperatures as high as 80–82°C and in *Thermus* medium containing 3.0% NaCl. Strains B (NCIB 11247), AT-62 (ATCC 33923), RQ-1 (DSMZ 9247), GK-24 (Taguchi et al., 1982), IB-21, and HB-27 can be assigned to this species (Manaia et al., 1994; Williams et al., 1995). In the absence of other distinguishing characteristics, strains Fiji3, A1 (Hudson et al., 1989), and several halotolerant strains from marine hot springs on the island of São Miguel also should be included in this species (Manaia et al., 1994). This species has been isolated from inland and marine hot springs Japan, Iceland, and the island of São Miguel, Azores.

The mol% G + C of the DNA is: 65 (T_m).

Type strain: HB8, ATCC 27634, DSMZ 579, NCIB 11244.

GenBank accession number (16S rRNA): M26923, X07998, X58342.

Genus II. **Meiothermus** Nobre, Trüper and da Costa 1996b, 605[VP]

M. Fernanda Nobre and Milton S. da Costa

Mei.o.ther'mus. Gr. prefix *meio-* less; Gr. adj. *thermos* hot; M.L. masc. n. *Meiothermus* to indicate an organism in a less hot place.

Straight rods, 0.5–0.8 μm in diameter; the cell length is variable. Long filaments are also formed under some culture conditions. **Nonmotile**; do not possess flagella. Endospores are not observed. Stain Gram-negative. Most strains form **red- or orange-pigmented colonies**, some strains are bright **yellow**. Aerobic with a strictly respiratory type of metabolism, but some strains use nitrate as terminal electron acceptor. Oxidase positive; the strains of one species are catalase positive, whereas the strains of the other species are catalase negative. **Slightly thermophilic**, with optimum growth temperatures of 50–65°C; strains do not grow at 70°C. The optimum pH ranges from 7.5 to 8.0. **Menaquinone 8** is the predominant respiratory quinone; **ornithine** is present in the peptidoglycan. One major phospholipid and two prominent glycolipids migrating close to each other dominate the polar lipid pattern. Additional phospholipids and glycolipids are minor components. Fatty acids are predominantly **iso- and anteiso-branched**. **Branched-chain 2-hydroxy fatty acids** are present in all strains. Proteins and peptides are hydrolyzed by all strains. Starch is hydrolyzed by some strains. Hexoses, a few pentoses and a few polyols, disaccharides, amino acids, and organic acids are used as sole carbon and energy sources. Most strains require yeast extract or cofactors for growth. Found in **hydrothermal areas** with neutral to alkaline pH, also isolated from fermentors.

The mol% G + C of the DNA is: 59–70.

Type species: **Meiothermus ruber** (Loginova, Egorova, Golovacheva and Seregina 1984) Nobre, Trüper and da Costa 1996b, 605 (*Thermus ruber* Loginova, Egorova, Golovacheva, Seregina 1984, 498.)

Further descriptive information

Phylogenetic analysis based on 16S rDNA sequence analysis shows that the species of the genus *Meiothermus* form a sister line of descent with the species of the genus *Thermus* with which they share only about 86% sequence similarity (Nobre et al., 1996b). These two closely related genera constitute the order *Thermales*, which, along with the distantly related species of the order *Deinococcales*, constitute the *Deinococcus–Thermus* phylum within the domain *Bacteria* (Weisburg et al., 1989; Embley et al., 1993; Nobre et al., 1996b; Rainey et al., 1997). It had been noted for some time that the red-pigmented "low-temperature" species designated *Thermus ruber* formed a separate line of descent from the "high-temperature" species of the genus *Thermus* (Weisburg et al., 1989; Bateson et al., 1990; Embley et al., 1993), but the species was nevertheless maintained in this genus. The description of two new, slightly thermophilic species that, based on 16S rRNA sequence analysis and chemotaxonomic parameters, were more closely related to *Thermus ruber* than to the other species of the genus *Thermus* led to the proposal of the genus *Meiothermus* for the species with low growth temperatures (Tenreiro et al., 1995a; Nobre et al., 1996b).

Transmission electron microscopy shows that the cells of the genus *Meiothermus* have an envelope consisting of a cytoplasmic membrane with a simple outline, a cell wall with an inner, electron-dense thin layer, presumably representing the peptidoglycan connected to an outer corrugated "cobble-stone" layer by irregularly spaced invaginations. The species of the genus *Meiothermus* are morphologically indistinguishable from the species of *Ther-*

mus, except that "rotund bodies" frequently seen in the species of the latter genus have not, to our knowledge, been observed in the species of *Meiothermus* (Brock and Edwards, 1970; Hensel et al., 1986; Tenreiro et al., 1995a).

The peptidoglycan of several strains of *M. ruber* examined contains L-ornithine as the diamino acid and glycylglycine as the interpeptide bridge, and is consistent with the A3β murein type of Schleifer and Kandler (1972). This type of peptidoglycan is also found in the genera *Thermus* and *Deinococcus* (Schleifer and Kandler 1972; Hensel et al., 1986; Embley et al., 1987; Sharp and Williams, 1988). Likewise, the major respiratory quinone of the strains of the genus *Meiothermus* is also menaquinone 8 (MK-8) (Hensel et al., 1986; Nobre et al., 1996b). These two chemical parameters corroborate the phylogenetic interpretation that the three genera are related to each other, although the species of the *Thermus/Meiothermus* line of descent have few other characteristics in common with the species of the genus *Deinococcus*. The polar lipid composition of the species of *Meiothermus* consists of one major phospholipid, designated phospholipid 1 (PL-1), and two prominent glycolipids, designated glycolipid 1a (GL-1a) and glycolipid 1b (GL-1b), migrating very close to each other and to the origin (Donato et al., 1991; Tenreiro et al., 1995a; Chung et al., 1997) (Fig. B4.7). Iso- and anteiso-branched C_{15} and C_{17} fatty acids are the predominant acyl chains of the strains of the genus *Meiothermus*. Straight-chain saturated fatty acids and unsaturated branched-chain fatty acids are found in minor concentrations. On the other hand, branched-chain 2-hydroxy fatty acids, at concentrations that vary between 7 and 13% of the total fatty acids, are present in all strains (Tenreiro et al., 1995a; Nobre et al., 1996a; Chung et al., 1997). The strains of *M. ruber* and *M.*

cerbereus also contain appreciable amounts of branched-chain 3-hydroxy fatty acids, but these are practically absent in *M. chliarophilus* and *M. silvanus* (Nobre et al., 1996a; Chung et al., 1997). The formation of two separate glycolipids, by thin layer chromatography, in the strains of the genus *Meiothermus*, is due to the differential binding of fatty acids to the hexosamine of the polar head group; 2-hydroxy fatty acids are exclusively amide-linked to the hexosamine of GL-1a, whereas 3-hydroxy and/or saturated branched-chain fatty acids are amide-linked to the hexosamine of GL-1b. The type strain of *M. silvanus*, like some strains of *Thermus*, has a mixture of glycolipids formed by long chain 1,2-diols and glycerolipids (Wait et al., 1997; Ferreira et al., 1999).

The strains of three of the four species of the genus *Meiothermus* form reddish-pigmented colonies, although the color varies considerably from orange-red to deep red. The differences in pigmentation appear unrelated to the site of isolation nor to the taxonomic status of these strains, because isolates recovered from the same geothermal area and belonging to the same species can have different pigmentation (Hensel et al., 1986; Sharp and Williams, 1988; Chung et al., 1997). The strains of *M. chliarophilus* form bright yellow-pigmented colonies (Tenreiro et al., 1995a), but this species has been recovered only from one hot spring in central Portugal and the isolates may belong to one clone. It is possible, therefore, that strains of *M. chliarophilus* originating from other geothermal areas could also produce red-pigmented cells. The red or yellow pigmentation of the strains of *Meiothermus* is due to a complex mixture of carotenoids that have not been characterized (Hensel et al., 1986; Sharp and Williams, 1988).

The cardinal growth temperatures of the species of the genus *Meiothermus* are lower than those of the species of the genus *Thermus*. *M. ruber* has an optimum growth temperature of about 60–65°C, a maximum growth temperature just below 70°C, and a minimum growth temperature of about 35–40°C. The species *M. silvanus* and *M. cerbereus* have an optimum growth temperature around 55°C, a maximum growth temperature of 60–65°C, and a minimum growth temperature of about 35°C, whereas *M. chliarophilus* is the least thermophilic of the known species of this genus with an optimum growth temperature of about 50°C and a maximum growth temperature of about 60°C (Loginova et al., 1984; Tenreiro et al., 1995a; Chung et al., 1997).

The strains of the species of the genus *Meiothermus* are aerobic, but some strains are capable of growth using nitrate as electron acceptor; none of the strains appears to be capable of carrying out fermentation. On the other hand, all strains of the species of *Meiothermus*, with the exception of the strains of *M. ruber*, are catalase negative. This observation came quite as a surprise, but no explanation for the absence of catalase has been given (Tenreiro et al., 1995a; Chung et al., 1997). All strains are chemoorganotrophs capable of growth on amino acids, peptides and proteins, and simple and complex carbohydrates (Table B4.6). However, none of the strains appears to be able to hydrolyze lipids. *Thermus* medium and Degryse medium 162 are the most commonly used media for growth of the strains of this genus (Degryse et al., 1978; Williams and da Costa, 1992). Both media contain yeast extract, although this supplement can be replaced by a complex mixture of amino acids, vitamins, and nucleotides (Sharp and Williams, 1988).

The strains of *M. cerbereus* and some strains of *M. silvanus* were found to require reduced sulfur compounds, such as thiosulfate, cysteine, or thioglycolate, for growth in *Thermus* liquid medium, but not in the corresponding medium solidified with agar. Under these conditions sulfate is not produced during growth on thiosulfate. The inability to detect thiosulfate utilization by these

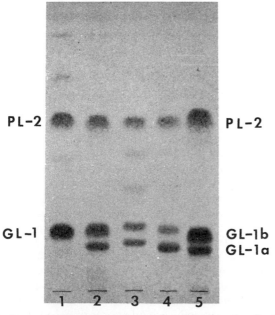

FIGURE B4.7. Single-dimension thin-layer chromatography of polar lipids of *Meiothermus* and *Thermus* strains. Lanes: 1, *T. oshimai*ᵀ; 2, *M. ruber*ᵀ; 3, *M. chliarophilus*ᵀ; 4, *M. silvanus*ᵀ; 5, *M. cerbereus*ᵀ. PL-2, phospholipid 2; GL-1, glycolipid 1; GL-1a, glycolipid 1a; GL-1b, glycolipid 1b.

TABLE B4.6. Characteristics of the type strains of the species of the genus *Meiothermus*[a,b]

Characteristic	1. *M. ruber*	2. *M. cerbereus*	3. *M. chliarophilus*	4. *M. silvanus*
Pigmentation	Red	Red	Yellow	Red
Optimum growth, °C	60	55	50	55
Presence of:				
Oxidase	+	+	+	+
Catalase	+	−	−	−
α-Galactosidase	+	+	+	−
DNase	+	+	+	+
β-Galactosidase	+	+	+	+
Hydrolysis of:				
Elastin	+	+	+	−
Fibrin	+	+	+	+
Starch	−	−	+	−
Casein	+	+	+	+
Gelatin	+	+	+	+
Arbutin	+	+	+	+
Esculin	+	+	+	−
Xylan	−	−	−	−
Reduction of nitrate	−	−	+	+
Utilization of:				
D-Glucose	+	+	+	+
D-Fructose	+	+	+	+
D-Mannose	+	+	+	+
D-Galactose	+	+	+	+
D-Melibiose	+	+	+	+
Maltose	+	+	+	+
Lactose	+	+	+	+
D-Trehalose	+	+	+	−
Sucrose	+	+	+	+
D-Cellobiose	+	+	+	−
D-Raffinose	+	−	+	−
D-Arabinose	−	−	−	−
L-Rhamnose	−	−	−	−
D-Xylose	+	−	+	+
Ribitol	−	−	−	−
D-Mannitol	+	−	+	−
D-Sorbitol	+	−	+	−
Glycerol	+	−	+	−
Citrate	−	−	−	−
Pyruvate	−	+	+	+
Succinate	+	−	−	−
Malate	+	−	−	−
myo-Inositol	+	−	+	−
L-Asparagine	+	−	+	+
L-Glutamate	+	+	+	+
L-Glutamine	+	−	+	+
L-Serine	+	−	+	+
L-Proline	+	+	+	−
L-Arginine	+	−	+	+
Menaquinone (MK)	MK-8	MK-8	MK-8	MK-8
Mol% G + C of DNA	66.0	60.9	69.9	63.6

[a]Symbols: +, positive result or growth; −, negative result or no growth.

[b]*M. chliarophilus* was grown at 50°C; other strains were grown at 55°C.

strains may be due to the small amounts used for growth, leading to the hypothesis that these isolates are unable to reduce sulfate by the assimilatory sulfate reduction pathway. On the other hand, the type strain of *M. ruber* oxidizes thiosulfate to sulfate, although this strain shows no requirement for reduced sulfur compounds, and growth is not improved by their addition to the growth medium. The oxidation of thiosulfate to sulfate by the type strain of *M. ruber* may be analogous to the so-called gratuitous oxidation of sulfur compounds found in other heterotrophic bacteria (Schook and Berk, 1978; Das et al., 1996). In *M. ruber*, however, only about half of the thiosulfate utilized was accounted for in the form of sulfate, and sulfur crystals were not observed by phase contrast microscopy (Chung et al., 1997).

The species *M. ruber*, *M. silvanus*, *M. chliarophilus*, and *M.*

cerbereus have been validly described (Loginova et al., 1984; Tenreiro et al., 1995a; Nobre et al., 1996b; Chung et al., 1997). The phylogenetic analysis of the species of the genus *Meiothermus* shows that *M. ruber* and *M. cerbereus* are closely related to each other, sharing about 98% 16S rDNA sequence similarity and a whole DNA–DNA hybridization value of about 50% (Chung et al., 1997). On the other hand, *M. silvanus* and *M. chliarophilus* are more distantly related to the *M. ruber*/ *M. cerbereus* clade. The phylogenetic analysis also shows that *M. silvanus* and *M. chliarophilus* are quite unrelated, sharing no more than about 90% sequence similarity with each other or with the other two species of *Meiothermus* (Tenreiro et al., 1995a; Chung et al., 1997). On this basis, it could be argued that *M. chliarophilus* and *M. silvanus* should be placed in two separate genera. However, it is prudent to maintain these species in the genus *Meiothermus* because of the lack of physiological and chemical characteristics that justify their transfer to a new genus (Fig. B4.8).

The strains of the genus *Meiothermus* have been isolated from a large number of geothermal areas extending from Yellowstone National Park, the Island of S. Miguel in the Azores, several hot springs in continental Portugal, the Hveragerthi and Geysir areas of Iceland, the Kamchatka Peninsula, to the North Island of New Zealand (Loginova et al., 1984; Sharp and Williams, 1988; Donato et al., 1991; Ruffett et al., 1992; Tenreiro et al., 1995a; Chung et al., 1997). Isolates resembling strains of *Meiothermus* were also recovered from soil in the Togo Republic, but so little information was provided that it is impossible to know the origin of the soil or to discuss the characteristics of the strains (Loginova et al., 1978). The species *M. ruber* has also been isolated from fermentors fed with yeast factory wastewater and maintained at 60°C (Hensel et al., 1986). In natural environments, these organisms are generally isolated from areas where the water temperature ranges from 40 to 70°C and the pH is neutral to alkaline. Unlike strains belonging to *Thermus thermophilus*, or other closely related halotolerant species of the genus *Thermus*, strains of *Meiothermus* have not been isolated from marine hot springs, and no halotolerant or slightly halophilic strains are known.

The most commonly isolated strains of this genus, and perhaps the most widespread strains, resemble the type strain of *M. ruber*, having been isolated from all of the geothermal areas mentioned above. The strains of *M. silvanus* have been isolated from two hot springs in northern Portugal and from the Geysir area of Iceland. Strains of *M. cerbereus* have been isolated from the Geysir area of Iceland, and closely related strains that have not been extensively characterized have also been isolated from Yellowstone National Park (Nold and Ward, 1995). Strains of *M. chliarophilus* have been isolated only from the hot spring at Alcafache in central Portugal with a vent temperature of 50.5°C and a pH of 8.1. On the other hand, strains of the genus *Meiothermus* could not be isolated from the hot spring at S. Gemil only a few kilometers away with a vent temperature of 48°C and a pH of 7.9. Furthermore, only strains closely related to *M. ruber* have been consistently isolated from the hot spring and hot spring runoff at S. Pedro do Sul located about 20 km from Alcafache (Donato et al., 1991; Nobre et al., 1996b). One factor leading to the frequent isolation of *M. ruber*-like strains from several geothermal areas around the world may be the high temperature used for enrichment of water and sediment samples for

these organisms. Enrichments performed in liquid media at 60°C or higher invariably lead to the isolation of strains with optimum growth temperatures of about 60–65°C that appear to be closely related to *M. ruber*. The other species with optimum growth temperatures of about 50–55°C and maximum growth temperatures of about 60°C will not be isolated under these enrichment conditions. There could also be physicochemical or biological characteristics of the hot springs that could limit the distribution of some of the species, although these are not known. The isolation of *M. ruber* from fermentors shows that these organisms can also colonize artificial thermal environments (Hensel et al., 1986). This is not surprising since strains of the genus *Thermus*, with higher growth temperature ranges, have also been isolated from man-made thermal habitats (Brock and Boylen, 1973; Degryse et al., 1978).

ENRICHMENT AND ISOLATION PROCEDURES

Loginova et al. (1984) isolated the original strains of *Meiothermus ruber* on a potato–yeast extract–peptone medium, and Hensel et al. (1986) used wastewater from a yeast factory diluted 10-fold and supplemented with KCl to enrich strains that grew in a fermentor containing the original concentrated wastewater. However, all other isolates of the genus *Meiothermus* have been obtained with *Thermus* medium (Ramaley and Hixson, 1970; Brock, 1978; Williams and da Costa, 1992), although other low nutrient media such as medium 162 (Degryse et al., 1978) should also yield adequate recovery of these organisms. Many of the strains of *M. ruber* have been obtained after enrichment of water or biofilm samples in liquid *Thermus* medium at 60°C for 2–3 days. After the enrichment in liquid medium, small amounts of the cultures are spread on the same medium solidified with agar and incubated at 60°C until red-pigmented colonies appear and can be isolated (Sharp and Williams, 1988). Instead of enrichments in liquid medium it is, perhaps, preferable to use a membrane filtration method. This method can be used with water or biofilms that have been macerated and shaken vigorously with small amounts of water from the same hot spring, or phosphate buffered-saline. Adequate volumes of the samples, or dilutions, are filtered through sterile 47-mm-diameter cellulose nitrate or acetate membrane filters with pore sizes of 0.22 or 0.45 μm. The filters are placed on the surface of *Thermus* agar plates, or a similar low nutrient medium, the plates inverted, and incubated for 2–7 days at temperatures ranging from 50 to 65°C. Plates are wrapped in plastic film to prevent evaporation. Red, yellow, or

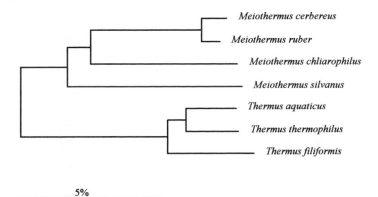

5%

FIGURE B4.8. 16S rDNA-based phylogenetic dendrogram of the type strains of the species of the genus *Meiothermus* (Chung et al., 1997). Bar = 5 inferred nucleotide substitution per 100 nucleotides.

colorless colonies can easily be observed and picked for further purification. Since many of the species of the genus *Meiothermus* have lower growth temperatures than *M. ruber*, it is preferable to incubate the plates or liquid enrichment tubes simultaneously at several temperatures. It is also important to isolate colonies other than red-pigmented varieties, since the known strains of *M. chliarophilus* are yellow and some spontaneous nonpigmented variants of red-pigmented strains have been observed (unpublished results). Other organisms unrelated to the genus *Meiothermus* will also grow under these isolation conditions, namely spore-forming Gram-positive rods. At this time there is no reliable selective medium for *Meiothermus* strains. Identification of isolates of the genus *Meiothermus* can be easily assessed by the characteristic polar lipid composition, or by the presence of branched-chain C_{15} and C_{17} fatty acids together with branched-chain 2-hydroxy fatty acids (Nobre et al., 1996a).

MAINTENANCE PROCEDURES

Cultures of *Meiothermus* are very easy to keep. All strains grow well on *Thermus* medium (Williams and da Costa, 1992) or Degryse medium 162 (Degryse et al., 1978) solidified with 2% or 3% agar, though it may be preferable to add 0.1 g/l of thiosulfate or cysteine to the medium to insure growth of all strains. During incubation, the Petri plates should be wrapped in plastic film to prevent evaporation. Long-term cultures can be maintained frozen at −70°C in plastic tubes containing the liquid media, men-tioned above, supplemented with 15% (v/v) glycerol. Cultures have been maintained for several years without loss of viability by lyophilization or frozen in liquid nitrogen storage.

TAXONOMIC COMMENTS

The first strains of the genus *Meiothermus* were isolated by Loginova and Egorova (1975) from the Kamchatka Peninsula and were designated *Thermus ruber*, but the species was not included in the Approved Lists of Bacterial Names (Skerman et al., 1980). The name was later revived by Loginova et al. (1984) on the basis of the strain described in the original publication. The species was retained in the genus *Thermus*, because it had biochemical and physiological characteristics similar to the high temperature strains of the genus *Thermus* in spite of the lower cardinal growth temperatures. In the 1st Edition of *Bergey's Manual of Systematic Bacteriology* (1984), Thomas Brock suggested that in the absence of further distinguishing characteristics, the red-pigmented strains belonged to *Thermus aquaticus*. It was, however, subsequently shown that *Thermus ruber* was distinct from *Thermus aquaticus* on the basis of DNA–DNA hybridization studies, biochemical characteristics, and 16S rDNA sequence analysis (Hensel et al., 1986; Sharp and Williams, 1988). The description of *T. silvanus* and *T. chliarophilus* with higher 16S rDNA sequence

TABLE B4.7. Differential characteristics of the species of the genus *Meiothermus*[a,b,c]

Characteristic	1. *M. ruber*[T]	2. *M. cerbereus*[T]	3. *M. chliarophilus*[T]	4. *M. silvanus*[T]
Pigmentation	Red	Red	Yellow	Red
Optimum growth, °C	60	55	50	55
Presence of:				
Catalase	+	−	−	−
α-Galactosidase	+	+	+	−
Hydrolysis of:				
Elastin	+	+	+	−
Starch	−	−	+	−
Esculin	+	+	+	−
Reduction of nitrate	−	−	+	+
Utilization of:				
D-Cellobiose	+	+	+	−
D-Raffinose	+	−	+	−
D-Mannitol	+	−	+	−
D-Sorbitol	+	−	+	−
D-Xylose	+	−	+	+
D-Trehalose	+	+	+	−
Glycerol	+	−	+	−
Pyruvate	−	+	+	+
Succinate	+	−	−	−
Malate	+	−	−	−
myo-Inositol	+	−	+	−
L-Asparagine	+	−	+	+
L-Glutamine	+	−	+	+
L-Serine	+	−	+	+
L-Proline	+	+	+	−
L-Arginine	+	−	+	+

[a]Symbols: T, Type strain; +, positive result or growth; −, negative result or no growth.

[b]*M. chliarophilus* was grown at 50°C; other strains were grown at 55°C.

[c]All species tested positive for oxidase, DNase, β-galactosidase, hydrolysis of fibrin, casein, gelatin, and arbutin; utilization of D-glucose, D-fructose, D-mannose, D-galactose, D-melibiose, maltose, lactose, sucrose, and L-glutamate; menaquinone, MK-8. All species tested negative for utilization of L-rhamnose, citrate, D-arabinose, ribitol, and xylan.

TABLE B4.8. Mean fatty acid composition of the strains examined after growth at 50°C

Fatty acid	1. *M. ruber* (13)[a]	2. *M. cerbereus* (6)	3. *M. chliarophilus* (3)	4. *M. silvanus* (5)
	\% of total in:			
$C_{13:0\ iso}$	0.6	1.4	1.5	0.8
$C_{14:0\ iso}$	0.7	2.7	1.7	0.7
$C_{14:0}$	0.6	−	0.7	0.3
$C_{13:0\ iso\ 3OH}$	0.4	−	−	0.8
$C_{15:1\ iso\ F}$[b]	2.7	3.8	−	−
$C_{15:0\ iso}$	33.0	34.6	42.1	25.6
$C_{15:0\ anteiso}$	5.5	11.1	8.1	26.3
$C_{15:0}$	1.8	1.6	2.1	0.4
$C_{16:1\ \omega7t\ alcohol}$	0.8	1.9	−	−
$C_{16:0\ iso}$	2.9	4.0	2.5	1.5
$C_{16:1\ \omega9c}$	0.4	−	−	−
$C_{16:1\ \omega7c}$	0.7	−	0.5	−
$C_{15:0\ iso\ 2OH}$	−	−	−	0.9
$C_{16:0}$	7.6	4.5	9.1	6.4
UN 1[c]	1.1	−	0.7	2.5
$C_{15:0\ 3OH}$	−	−	1.0	−
$C_{15:0\ 2OH}$	0.4	−	0.3	−
$C_{iso\ 17:1\ \omega9c}$	6.5	4.8	−	−
$C_{anteiso\ 17:1\ \omega9c}$	1.0	−	−	−
$C_{17:0\ iso}$	13.3	5.8	16.4	10.0
$C_{17:0\ anteiso}$	3.7	2.5	2.7	6.4
$C_{17:1\ \omega8c}$	0.8	−	−	−
$C_{17:1\ \omega6c}$	0.8	−	0.7	1.3
$C_{17:0}$	0.8	−	1.2	0.3
$C_{16:0\ 2OH}$	0.6	−	0.4	0.5
$C_{17:0\ iso\ 2OH}$	7.8	3.3	7.3	9.6
$C_{17:0\ anteiso\ 2OH}$	0.4	−	0.6	3.0
$C_{17:0\ iso\ 3OH}$	1.1	4.7	−	−
$C_{17:0\ 2OH}$	0.5	−	0.2	−
$C_{17:0\ anteiso\ 3OH}$	0.6	1.4	−	−
$C_{1,2-diol\ 18:0\ iso}$	−	−	−	1.6

[a]Number of strains examined.

[b]The double bond position of this fatty acid is unknown.

[c]Unknown fatty acid and/or alcohol with an equivalent chain length of 16.090.

similarity to *T. ruber* than to the high-temperature species of *Thermus*, coupled with the lower optimum growth temperature range and the presence of moderate levels of 2-hydroxy fatty acids, led to the proposal of the genus *Meiothermus* (Nobre et al., 1996b).

The species *M. ruber*, *M. silvanus*, *M. chliarophilus*, and *M. cerbereus* can be identified on the basis of biochemical and physiological characteristics (Table B4.7), but it is perhaps most convenient to use the fatty acid composition to identify and distinguish the species of *Meiothermus* from one another (Table B4.8).

The type and reference strains of *M. cerbereus* isolated from Iceland require reduced sulfur compounds for growth in liquid medium. Moreover, several isolates that are very closely related to *M. silvanus* and come from the Geysir area of Iceland and from the hot spring at Chaves in Northern Portugal also have this requirement. Another isolate, also very closely related to *M. silvanus*, requires ascorbate for growth in liquid medium. Therefore, the requirement for reduced sulfur compounds by *M. cerbereus* does not appear to be restricted to this species and cannot be considered a valid distinguishing characteristic of *M. cerbereus*. Based on 16S rDNA sequence similarity, two strains isolated from Yellowstone National Park (ac-2 and ac-17) appear to be very closely related to the strains of *M. cerbereus* (Nold and Ward,

1995). The lack of extensive characterization, namely of the requirement for reduced sulfur compounds as well as DNA–DNA hybridization results, makes it difficult to place these strains in the species *M. cerbereus* and to assess the distribution of the reduced sulfur requirement.

Whole DNA–DNA hybridization studies, coupled to 16S rDNA sequencing, show that several red-pigmented isolates from Iceland, Portugal, the Azores, Yellowstone National Park, and New Zealand are very closely related to the type strain of *M. ruber* and can be assigned to this species (Sharp and Williams, 1988; Chung et al., unpublished results). The lack of molecular genetic studies on many other isolates that resemble *M. ruber*, and the large diversity of physiological and biochemical parameters of these strains (Ruffett et al., 1992), makes it difficult to assign any of these strains conclusively to this species. However, many strains tentatively assigned to *M. ruber* have a fatty acid composition that distinguishes these strains from those of other three species of this genus (Tenreiro et al., 1995a; Nobre et al., 1996a; Chung et al., 1997). One strain, designated *"T. rubens"*, has been deposited in the American Type Culture Collection (ATCC 31556) as a patent strain, but was never validly described and is believed to belong to *M. ruber* (Ado et al., 1982).

List of species of the genus Meiothermus

1. **Meiothermus ruber** (Loginova, Egorova, Golovacheva and Seregina 1984) Nobre, Trüper and da Costa 1996b, 605[VP] (*Thermus ruber* Loginova, Egorova, Golovacheva and Seregina 1984, 498.)

   ru′ber. M.L. masc. adj. *ruber* red.

   The cells of the strains of this species, as in all other species of this genus, are 0.5–0.8 μm in diameter and variable in cell length. The colonies are 1.0–2.0 mm in diameter and the color varies from deep-red to pink. This species, unlike the other species of this genus, is catalase positive. The strains of this species can easily be distinguished from the other species of this genus by the high optimum growth temperature and the fatty acid composition. The type strain of this species, unlike the type strains of the other species of the genus *Meiothermus*, utilizes succinate, malate, and *myo*-inositol. Strains of this species have been isolated from hot springs in the Kamchatka Peninsula, and strains similar to the type strain have been isolated from hot springs and runoffs in other geographical areas as well as from fermentors. The characteristics of this species are as described in Tables B4.6, B4.7, and B4.8.

   The mol% G + C of the DNA is: 66.0 (T_m).

   Type strain: Loginova 21, ATCC 35948, DSMZ 1279, VKMB 1258.

   GenBank accession number (16S rRNA): Z15059.

2. **Meiothermus cerbereus** Chung, Rainey, Nobre, Burghardt and da Costa 1997, 1229.[VP]

   cer.be.re′ us. L. masc. adj. *cerbereus* named for Cerberus, the monster guarding the mythological Greek Hades.

   The colony dimensions are similar to those of the other species of the genus, and the colonies are red-orange pigmented. Catalase negative. The type strain and the other isolates examined, all of which originate from Geysir, Iceland, require reduced sulfur compounds, such as cysteine, thiosulfate, or thioglycolate, for growth in liquid *Thermus* medium

or liquid medium 162, but this requirement may not be a characteristic of this species. The levels of 3-hydroxy fatty acids are higher than the levels of 2-hydroxy fatty acids. The isolates of the species assimilate few single carbon sources; sugars appear to be the main carbon sources for growth, although pyruvate, glutamate, and proline are also utilized. Isolated from the Geysir geothermal area, Iceland. Closely related strains isolated from Yellowstone National Park, United States. The characteristics of this species are as described in Tables B4.6, B4.7, and B4.8.

   The mol% G + C of the DNA is: 60.9 (T_m).

   Type strain: GY-1, DSMZ 11376.

   GenBank accession number (16S rRNA): Y13594.

   Additional Remarks: GenBank accession number (16S rRNA) for reference strain GY-5 (DSMZ 11377) is Y13595.

3. **Meiothermus chliarophilus** (Tenreiro, Nobre and da Costa 1995a) Nobre, Trüper and da Costa 1996b, 605[VP] (*Thermus chliarophilus* Tenreiro, Nobre and da Costa 1995a, 638.)

   chli.a.ro′ phi.lus. Gr. adj. *chliaros* warm; Gr. adj. *philos* loving; M.L. masc. adj. *chliarophilus* warmth loving.

   The cell and colony dimensions are as described for the other species of this genus. All known isolates form bright yellow colonies. Catalase negative. This species has the lowest growth temperature range of the species of this genus. Has been isolated only from the hot spring at Alcafache in Central Portugal. The characteristics of this species are as described in Tables B4.6, B4.7, and B4.8.

   The mol% G + C of the DNA is: 69.9 (T_m).

   Type strain: ALT-8, DSMZ 9957.

   GenBank accession number (16S rRNA): X84212.

4. **Meiothermus silvanus** (Tenreiro, Nobre and da Costa 1995a) Nobre, Trüper and da Costa 1996b, 605[VP] (*Thermus silvanus* Tenreiro, Nobre and da Costa 1995a, 637.)

   sil.va′ nus. L. adj. *silvanus* named after Manuel T. Silva, Portuguese microbiologist and immunologist.

The colony dimensions are similar to those of the other species of the genus, and the colonies are red pigmented. The optimum growth temperature is about 55°C. The strains of this species are catalase negative, α-galactosidase negative, and do not assimilate proline. The strains of this species have high relative proportions of both iso- and anteiso-C_{15} fatty acids, the latter of which predominates. This species possesses higher levels of 2-hydroxy anteiso-fatty acids than the other species of the genus *Meiothermus*. The glycolipids of this species contain a mixture of long chain 1,2-diols and glycerolipids. Strains of this species have been isolated from the hot spring and runoffs at Vizela in northern Portugal, and the Geysir geothermal area in Iceland. The characteristics of this species are as described in Tables B4.6, B4.7, and B4.8.

The mol% G + C of the DNA is: 63.6 (T_m).

Type strain: VI-R2, DSMZ 9946.

GenBank accession number (16S rRNA): X84211.

Phylum BV. Chrysiogenetes *phy. nov.*

GEORGE M. GARRITY AND JOHN G. HOLT

Chry.si.o.ge.ne' tes. M.L. fem. pl. n. *Chrysiogenales* type order of the phylum; dropping ending to denote a phylum; M.L. fem. pl. n. *Chrysiogenetes* the phylum of *Chrysiogenales.*

The phylum *Chrysiogenetes* is currently represented by a single species, which was reportedly distinct from members of other phyla. Exact placement is currently uncertain, but a distant relationship to *Deferribacteres* is likely. Gram-negative, motile, curved, rod-shaped cells. Mesophilic, exhibiting anaerobic respiration in which arsenate serves as the electron acceptor.

Type order: **Chrysiogenales** ord. nov.

Class I. **Chrysiogenetes** *class. nov.*

GEORGE M. GARRITY AND JOHN G. HOLT

Chry.si.o.ge.ne' tes. M.L. fem. pl. n. *Chrysiogenales* type order of the class; dropping ending to denote a class; M.L. fem. pl. n. *Chrysiogenetes* the class of *Chrysiogenales.*

Description is the same as for the genus *Chrysiogenes.*

Type order: **Chrysiogenales** ord. nov.

Order I. **Chrysiogenales** *ord. nov.*

GEORGE M. GARRITY AND JOHN G. HOLT

Chry.si.o.gen' ales. M.L. masc. n. *Chrysiogenes* type genus of the order; *-ales* ending to denote an order; M.L. fem. pl. n. *Chrysiogenales* the order of *Chrysiogenes.*

Description is the same as for the genus *Chrysiogenes.*

Type genus: **Chrysiogenes** Macy, Nunan, Hagen, Dixon, Harbour, Cahill and Sly 1996, 1156.

Family I. **Chrysiogenaceae** *fam. nov.*

GEORGE M. GARRITY AND JOHN G. HOLT

Chry.si.o.ge.na' ce.ae. M.L. masc. n. *Chrysiogenes* type genus of the family; *-aceae* ending to denote a family; M.L. fem. pl. n. *Chrysiogenaceae* the family of *Chrysiogenes.*

Description is the same as for the genus *Chrysiogenes.*

Type genus: **Chrysiogenes** Macy, Nunan, Hagen, Dixon, Harbour, Cahill and Sly 1996, 1156.

Genus I. **Chrysiogenes** *Macy, Nunan, Hagen, Dixon, Harbour, Cahill and Sly 1996, 1156*[VP]

JOAN M. MACY , TORSTEN KRAFFT AND LINDSAY I. SLY

Chry.si.o'ge.nes. Gr. n. *chryseion* gold mine; Gr. patronymic suff. *-genes* sprung from, born from; M.L. masc. n. *Chrysiogenes* sprung from a gold mine.

Curved, rod-shaped cells with rounded ends, 1.0–2.0 μm in length and 0.50–0.75 μm in diameter. **Gram-negative**. Each cell is motile by a **single polar flagellum. Strictly anaerobic**. In agar, colonies are small, round-lenticular with entire edges, convex and white. In liquid medium, cells form swarms of motile bacteria. The optimum temperature for growth is between 25 and 30°C. Cells grow in defined minimal medium by **anaerobic respiration** with **acetate as the electron donor and carbon source** and **arsenate as the electron acceptor**. Acetate is oxidized to CO_2. Arsenate is reduced to arsenite. Phylogenetically, the genus is the first representative of a new deeply branching lineage of the *Bacteria*.

The mol% G + C of the DNA is: 49.

Type species: **Chrysiogenes arsenatis** Macy, Nunan, Hagen, Dixon, Harbour, Cahill and Sly 1996, 1156.

FURTHER DESCRIPTIVE INFORMATION

At present, the genus contains only a single species, *C. arsenatis*. The bacteria of this species appear as single, curved, rod-shaped cells which occur as either S-shaped pairs (Fig. B5.1) or W-shaped pairs prior to cell separation. The organism is extremely motile via a single polar flagellum (Fig. B5.1).

In liquid medium, most of the cells form swarms of bacteria in which they are very close together, without making physical contact. The sizes and shapes of the swarms vary, with some appearing as circles, teardrops, bananas, cigars, or triangles with rounded ends. Single swarms can move as a whole, and upon coming into contact with another swarm usually form a unified larger swarm.

Small, white-tan, convex lenticular-shaped colonies are formed in the agar of Hungate roll tubes. Growth on agar plates has not yet been established.

The bacteria are usually grown anaerobically at 25–30°C in a minimal medium containing: NaCl, 20 mM; KCl, 4 mM; NH_4Cl, 2.8 mM; KH_2PO_4, 1.5 mM; Na_2SO_4, 0.2 mM; $MgCl_2 \cdot 6H_2O$, 2 mM; $CaCl_2 \cdot 2H_2O$, 1 mM; $NaHCO_3$, 0.05% (w/v); 1 ml/l trace element solution ($Na_2EDTA \cdot 2H_2O$, 5.2 g/l; $FeCl_2 \cdot 4H_2O$, 1.5 g/l; $ZnCl_2$, 70 mg/l; $MnCl_2 \cdot 2H_2O$, 100 mg/l; H_3BO_3, 62 mg/l; $CoCl_2 \cdot 6H_2O$, 190 mg/l; $CuCl_2 \cdot 2H_2O$, 17 mg/l; $NiCl_2 \cdot 6H_2O$, 24 mg/l; and $Na_2MoO_4 \cdot 2H_2O$, 36 mg/l) and 5 ml/l vitamin solution (in g/l: folic acid, 0.01; riboflavin, 0.04; thiamine, 0.02; pantothenic acid, 0.12; niacinamide, 0.2; vitamin B_{12}, 0.04; biotin, 0.2; pyridoxine, 0.02; and *p*-aminobenzoic acid, 0.02). Sodium acetate (5 mM) is used as the electron donor/carbon source and sodium arsenate

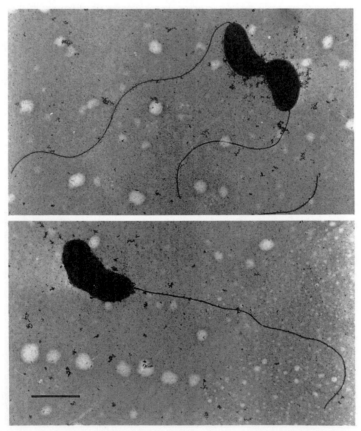

FIGURE B5.1. Electron micrograph of *Chrysiogenes arsenatis*. Bar = 1 μm. (Reprinted with permission from J.M. Macy et al., International Journal of Systematic Bacteriology *46:* 1153–1157, 1996, ©International Union of Microbiological Societies.)

(5 mM) as the terminal electron acceptor. The initial pH of the medium is 7.0. Growth occurs according to the following equation: $CH_3COO^- + 2HAsO_4^{2-} + 2H_2AsO_4^- + 5H^+ \rightarrow 2HCO_3^- + 4H_3AsO_3$. During growth, while acetate is being oxidized to CO_2 and arsenate is being reduced to arsenite, the pH rises to approximately 9.4. The doubling time is 4 h. No growth occurs when either acetate or arsenate is absent from the medium. In addition to using the soluble form of arsenate the bacterium can also grow using arsenate bound to iron oxide. Since *C. arsenatis* is able to grow in a minimal medium with the respiratory substrate acetate as the electron donor and arsenate as the electron acceptor, the organism must conserve energy using arsenate respiration.

With acetate (5 mM) as the electron donor and carbon source, arsenate can be replaced as the terminal electron acceptor by nitrate (5 mM) or nitrite (5 mM). Nitrate is reduced via nitrite to ammonia. The final cell density with nitrite is not as high as with arsenate or nitrate as electron acceptors. Sulfate, thiosulfate, selenate, Fe(III), and oxygen cannot serve as terminal electron acceptors.

In the presence of either arsenate (5 mM) or nitrate (5 mM) as terminal electron acceptors, acetate can be replaced as the electron donor and carbon source by the following compounds: pyruvate, L- and D-lactate (growth with D-lactate is better than with L-lactate), fumarate, succinate, and malate. Growth does not occur in the absence of arsenate or nitrate with any of these substrates. Hydrogen, formate, citrate, glutamate, threonine, aspartate, serine, phenylalanine, arginine, valine, histidine, methanol, glucose, fructose, sucrose, ribose, xylose, and benzoate do not support growth when arsenate or nitrate is present as the electron acceptor.

The presence of yeast extract (0.1%, w/v) stimulates the rate and extent of growth on acetate plus arsenate or nitrate. Molecular hydrogen stimulates growth on acetate only when arsenate, not nitrate, is the terminal electron acceptor.

The respiratory arsenate reductase, which catalyzes the reduction of arsenate to arsenite during growth, has been purified and characterized (Krafft and Macy, 1998). It is located in the periplasm of the cells and the synthesis of the protein is induced by the presence of arsenate. The enzyme activity can be measured with reduced benzyl viologen as the artificial electron donor. Cytochrome c_{551} is also present in the periplasm. The existence of a cytoplasmic malate dehydrogenase supports the conclusion that *C. arsenatis* uses the tricarboxylic acid cycle to oxidize electron donors such as acetate, pyruvate, and succinate.

ENRICHMENT AND ISOLATION PROCEDURES

C. arsenatis was originally isolated from arsenic-contaminated mud of a reed bed which is part of a waste water purification scheme associated with Ballarat Goldfields in Ballarat, Australia. Organisms have also been found in similar environments in Bendigo, Australia.

The organisms are enriched by inoculating arsenic-contaminated mud (approximately 0.1 g), into 10 ml of the anoxic minimal medium described above, containing 10 mM sodium arsenate and 10 mM sodium acetate. Enrichments are incubated at 28°C for up to 6 days. After a few days, as arsenate is reduced and acetate is oxidized, the pH of the cultures should begin to increase. When the pH has increased to above 8.0, the medium will become slightly turbid; the turbidity is due primarily to the presence of a precipitate. If the pH is maintained at less than 7.8, a precipitate does not form, and the optical density (600

TABLE B5.1. Comparison of *Chrysiogenes arsenatis* with other arsenate-respirers[a]

Characteristic	Chrysiogenes arsenatis	MIT-13	SES-3	Desulfotomaculum auripigmentum
Morphology	curved rod-shaped	vibrio-shaped	vibrio	slightly curved rod
Length, µm	1.0–2.0	1.0	1.5	2.5
Diameter, µm	0.5–0.75	0.1	0.3	0.4
Motility	+	+	+	−
Gram reaction	negative	negative	negative	positive
Temperature optimum, °C	25–30	nd	nd	25–30
Relationship to oxygen	strictly anaerobic	strictly anaerobic	microaerophilic	strictly anaerobic
Doubling time on arsenate, h	4	~14	5	~20
Electron donor:				
Hydrogen	−	−	−	+
Formate	−	nd	+[b]	−
Acetate	+	−	−	−
Pyruvate	+	+	nd	+
Lactate	+	+	+	+
Fumarate	+	+	nd	−
Succinate	+	nd	+	−
Malate	+	nd	nd	+
Citrate	−	nd	+	−
Benzoate	−	−	nd	−
Methanol	−	nd	nd	−
Glucose	−	nd	nd	−
Electron acceptor:				
Arsenate	+	+	+	+
Nitrate	+	+	+	+
Sulfate	−	−	−	+
Thiosulfate	−	nd	+	+
Fe(III)	−	nd	+	−
Selenate	−	−	+	−

[a]For symbols see standard definitions; nd, not determined.

[b]Growth occurs only in the presence of acetate.

nm) of the culture reaches approximately 0.14. When viewed microscopically, usually only a single bacterial type (described above), will be present — *C. arsenatis*.

The organisms are isolated by serial dilution into agar roll tubes (Hungate, 1969) containing the minimal medium described above, 1.5% Oxoid purified agar, 5 or 10 mM sodium acetate, and sodium arsenate. Colonies that develop in the agar (lenticular) should be removed under a stream of nitrogen gas (Hungate, 1969) and subcultured into agar medium at least twice. Organisms can normally tolerate up to at least 30 mM arsenate, although not all of this arsenate is reduced to arsenite, as the organism is usually inhibited when the concentration of arsenite reaches 10 mM.

MAINTENANCE PROCEDURES

C. arsenatis is normally cultured in anaerobic Balch tubes (Bellco) with the minimal medium described above. After inoculation the cells are incubated at 25–30°C for approximately 16 h to allow growth. The grown cultures can be maintained at 4°C for at least one week.

C. arsenatis can be stored at −70°C, as a 20× concentrated suspension of a grown culture in anoxic minimal salts medium containing 25% (v/v) glycerol.

DIFFERENTIATION OF THE GENUS *CHRYSIOGENES* FROM OTHER GENERA

The nearest relative of *C. arsenatis* is *Flexistipes sinusarabici*, which is a strictly anaerobic, Gram-negative, flexible, rod-shaped, thermophilic bacterium that was isolated from brine water samples of the Red Sea at a depth of 2000 m (Fiala et al., 1990). Except for the fact that both of these organisms are Gram-negative and strictly anaerobic, *C. arsenatis* shares no other characteristics with the thermophilic *F. sinusarabici*. Both organisms are physiologically different.

Besides *C. arsenatis*, three other bacterial species are known which have the ability to reduce arsenate to arsenite during growth. These include *Sulfurospirillum arsenophilum* MIT-13 (Ahmann et al., 1994), *Sulfurospirillum barnesii* SES-3 (Oremland et al., 1994; Laverman et al., 1995), and *Desulfotomaculum auripigmentum* (Newman et al., 1997). These organisms differ from *C. arsenatis* in that they are not able to use the respiratory substrate acetate as the electron donor during anaerobic growth with arsenate as the terminal electron acceptor. Instead they use the non-respiratory substrate lactate as the electron donor. Comparisons of these organisms to *C. arsenatis* are presented in Table B5.1. Phylogenetic analysis has placed *Desulfotomaculum auripig-*

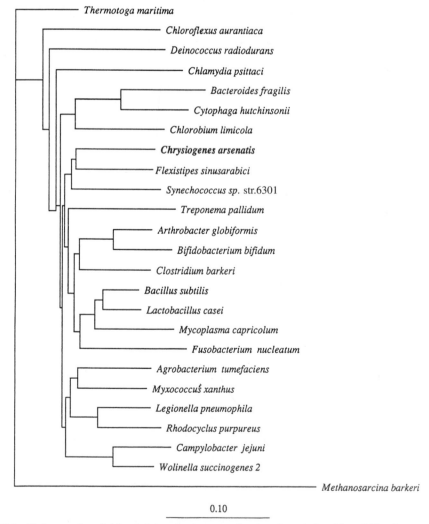

FIGURE B5.2. Phylogenetic neighbor-joining dendrogram showing the relationships of *Chrysiogenes arsenatis* with representatives of the phyla of *Bacteria*. Bar = 0.1% sequence difference.

mentum in the low G + C Gram-positive phylum (*Firmicutes*), and *Sulfurospirillum arsenophilum* and *Sulfurospirillum barnesii* within the "*Epsilonproteobacteria*" (Newman et al., 1997, Stolz et al., 1999). The organisms are, therefore, phylogenetically unrelated to *C. arsenatis*.

Escherichia coli and other *Enterobacteriaceae*, *Staphylococcus aureus*, and *Staphylococcus xylosus* are also known to reduce arsenate to arsenite (Silver, 1996). In contrast to *C. arsenatis* and the other arsenate-respiring bacteria mentioned above, energy is not conserved during this reduction. Instead arsenate reduction is only part of an arsenic resistance mechanism which is conferred by similar proteins encoded by *ars*-operons (Silver, 1996).

TAXONOMIC COMMENTS

Circumscription of *C. arsenatis* is based on the characteristics of a single isolate. Characterization of more isolates will be required before the phenotypic diversity of the genus and species can be evaluated.

The species is the only known member of a deep phylogenetic lineage (Fig. B5.2). In an analysis of 16S rRNA sequences (Macy et al., 1996) *C. arsenatis* branches deeply with *Flexistipes sinusarabici* and *Synechococcus* sp. strain 6301. The analysis of binary sequence similarity showed that *C. arsenatis* has only 74.8–81.8% similarity with the sequences of representatives of all other major phylogenetic groups within the *Bacteria*. The highest similarities are with *Lactobacillus casei* (81.8%), *Flexistipes sinusarabici* (81.7%), *Bacillus subtilis* (81.7%), and *Synechococcus* (81%). Bootstrap analysis showed that some deep branching points were statistically unstable. When the neighbor-joining method was used, the branch including *C. arsenatis*, *F. sinusarabici*, and *Synechococcus* sp., and that joining *C. arsenatis* and *F. sinusarabici* occurred in only 36% and 50% of trees, respectively. Comparison of the 16S rRNA sequence of *C. arsenatis* with the eubacterial consensus sequence (conserved in 95% or more of eubacterial 16S rRNAs) (Lane, 1991) showed that the sequence of *C. arsenatis* differed from the consensus sequence at four positions (the base in the eubacterial sequence is given first): position 607 (A versus U); position 678

TABLE B5.2. Characteristics of the species of the genus *Chrysiogenes*[a]

Characteristic	C. arsenatis
Morphology	curved, rod-shaped
Length, µm	1.0–2.0
Diameter, µm	0.5–0.75
Gram reaction	negative
Motility	+
Relationship to oxygen	strictly anaerobic
Colonies:	
Shape	lenticular
Color	white-tan
Swarm forming ability	+
Temperature optimum, °C	25–30
pH range	7.0–9.4
Doubling time on arsenate, h	4
Electron donor:	
Acetate, pyruvate, lactate, fumarate, succinate, malate	+
Hydrogen, formate, citrate, glutamate, threonine, aspartate, serine, phenylalanine, arginine, valine, histidine, methanol, glucose, fructose, sucrose, ribose, xylose, benzoate	–
Electron acceptor:	
Arsenate, nitrate, nitrite	+
Sulfate, thiosulfate, Fe(III), selenate, oxygen	–

[a]For symbols see standard definitions.

(U versus G); position 712 (A versus U); and position 1159 (U versus A). Positions 678 and 712 are complementary, so that a U–A pair is replaced by G–U.

ACKNOWLEDGMENTS

The authors would like to thank Dr. Aidan C. Parte, Managing Editor of the International Journal of Systematic Bacteriology, for the permission to use the electron micrograph of *C. arsenatis* (Fig. B5.1).

This chapter is dedicated to Professor Joan M. Macy, a microbial physiologist of outstanding ability and one whose true worth has yet to be appreciated.

List of species of the genus Chrysiogenes

1. **Chrysiogenes arsenatis** Macy, Nunan, Hagen, Dixon, Harbour, Cahill and Sly 1996, 1156.[VP]

   ar.se.na' tis. M.L. gen. n. *arsenatis* of arsenate, referring to the ability of the organism to reduce arsenate to arsenite.

   The characteristics of the species are listed in Table B5.2, depicted in Fig. B5.1, and given in the description for the genus.

   Isolated from mud which was obtained from a reed bed near the Ballarat Goldfields in Ballarat, Australia.

   The mol% G + C of the DNA is: 49 (T_m).

   Type strain: DSMZ 11915, ATCC 700172.

   GenBank accession number (16S rRNA): X81319.

Phylum BVI. Chloroflexi *phy. nov.*

GEORGE M. GARRITY AND JOHN G. HOLT

Chlo.ro.flex' i. M.L. masc. n. *Chloroflexus* genus of the phylum; dropping ending to denote a phylum; M.L. fem. pl. n. *Chloroflexi* phylum of the genus *Chloroflexus.*

The phylum *Chloroflexi* is a deep branching lineage of *Bacteria.* The single class within *Chloroflexi* subdivides into two orders: the *"Chloroflexales"* and the *"Herpetosiphonales".* Gram-negative, filamentous *Bacteria* exhibiting gliding motility. Peptidoglycan contains L-ornithine as the diamino acid. Lipopolysaccharide-containing outer membrane not present.

Class I. "Chloroflexi"

RICHARD W. CASTENHOLZ

The phylum/class *"Chloroflexi"* is represented by a group of filamentous *Bacteria* phylogenetically related (by 16S rDNA sequence analyses) and contains at present two orders (*"Chloroflexales"* and *"Herpetosiphonales"*). The *"Chloroflexales"* are obligate or facultative phototrophs or at least filamentous *Bacteria* that contain some form of bacteriochlorophyll. The *"Herpetosiphonales"* are "flexibacteria" that lack bacteriochlorophylls and thus lack phototrophic capability. Phenotypic characters that should be considered a consistent feature of this phylum are as follows: the lack of the LPS outer membrane, and the presence of a polysaccharide, peptidoglycan complex in which meso-diaminopimelic acid is replaced by L-ornithine. Gliding motility is also a consistent character, but it is well known that this ability of gliding bacteria can commonly be lost in culture.

Order I. "Chloroflexales"

Family I. "Chloroflexaceae"

Filamentous Anoxygenic Phototrophic Bacteria

BEVERLY K. PIERSON

Four genera of filamentous gliding bacteria capable of anoxygenic photosynthesis with bacteriochlorophylls in the light are grouped together as "Filamentous Anoxygenic Phototrophic Bacteria" or the "Photosynthetic Flexibacteria". This grouping is a natural one based on morphological and phylogenetic data. The cells of all genera are uniseriately arranged in multicellular filaments that are capable of gliding motility. Three of the genera, *Chloroflexus, Chloronema,* and *Oscillochloris* contain chlorosomes, structural elements that are attached to cellular membranes and contain the light-harvesting bacteriochlorophylls *c* and *d.* One genus, *Heliothrix,* lacks chlorosomes and bacteriochlorophylls *c* or *d.* All four genera contain bacteriochlorophyll *a.*

In the previous edition of this manual and in many other publications, this group of organisms has been defined as the filamentous green bacteria or the green non-sulfur bacteria implying some level of association with the green sulfur bacteria. The latter group is a distinctive and phylogenetically cohesive family (*Chlorobiaceae*) of anoxygenic strictly anaerobic photoautotrophs that also contain chlorosomes housing the light-harvesting bacteriochlorophylls *c, d,* or *e.* Unlike the green sulfur bacteria, however, most (but not all) of the strains of filamentous phototrophic bacteria described in this section are facultatively aerobic and preferentially utilize organic substances in their phototrophic or chemotrophic metabolism. For this reason they have been referred to as the "green non-sulfur bacteria" by analogy to the purple sulfur vs. non-sulfur bacteria. At first this analogy seemed appropriate. As more strains of these recently described filamentous organisms have been studied *in situ* and in culture, however, more occurrences of autotrophic and sulfide-dependent metabolism with the absence of aerobic metabolism have been detected. Those species that have been the most intensely studied (*Chloroflexus aurantiacus* and *C. aggregans*) are now known to be phylogenetically closely related to each other (based on 16S rRNA) but are phylogenetically distant from the green sulfur bacteria. Thus the two groups of "green" bacteria do not have the close phylogenetic relationship that exists among the purple sulfur and non-sulfur bacteria. Several important photosynthetic characteristics of the cells of the filamentous phototrophs such as the membrane-bound light-harvesting bacteriochlorophyll *a* complexes and reaction centers are clearly more closely related to those of the purple bacteria than to those of the green bacteria, further eroding the usefulness of the analogy. Finally, the genus *Heliothrix*, shown to be phylogenetically in the same phylum as *Chloroflexus*, lacks chlorosomes altogether and is not green at all. Consequently this group of organisms has been renamed to reflect those features that its members have in common — anoxygenic photosynthesis and filamentous gliding morphology.

Based on 16S rRNA sequences, the genus *Chloroflexus* and its associated phototrophs are not closely related to the green sulfur bacteria or to any other photosynthetic bacteria. They belong in one of the deepest phyla (called the green non-sulfur bacteria phylum) of the *Bacteria* forming the only phototroph-containing branch occurring prior to the nearly simultaneous radiation of all the other bacterial phyla (Woese, 1987; Stackebrandt et al., 1996). Trees based on 16S rRNA within this phylum show that the nearest genetic relatives to *Chloroflexus* are *Oscillochloris* and

Heliothrix as well as the chemoorganotrophic flexibacterium *Herpetosiphon* and other chemotrophic bacteria (Fig. B6.1) (see also Weller et al., 1992; Hanada et al., 1995a; Sly et al., 1998; Ward et al., 1998). This supports the view of Castenholz (1973) who considers the genera *Chloroflexus* and *Heliothrix* as phototrophic flexibacteria. The sequence of 5S rRNA was the first evidence to support the relationship of *Heliothrix* to *Chloroflexus* and *Herpetosiphon* (Pierson et al., 1985), and was the first evidence to show that *Oscillochloris* was also more closely related to this group of bacteria than to any other phototrophs (Keppen et al., 1994). The 16S rRNA (Fig. B6.1) confirmed these relationships. So far we have no phylogenetic data to support the placement of the genus *Chloronema* in this group. Although the filamentous phototrophs cluster together as the deepest division of photosynthetic bacteria, their phylogenetic relationships to each other need further clarification and it may be premature to build a higher order taxonomic system around them.*

The filamentous anoxygenic phototrophs belong to a phylogenetically cohesive group that also contains non-photosyn-

---

Editorial Note: As this volume was in press, Keppen et al. proposed the family *Oscillochlorodaceae* based upon differences in 16S rDNA sequences between the genera *Oscillochloris* and *Chloroflexus* (Keppen, O.I., T.P. Tourova, B.B. Kuznetsov, R.N. Ivanovsky and V.M. Gorlenko. 2000. Proposal of *Oscillochlorodaceae* fam. nov. on the basis of a phylogenetic analysis of the filamentous anoxygenic phototrophic bacteria, and emended description of *Oscillochloris* and *Oscillochloris trichoides* in comparison with further new isolates. Int. J. Syst. Evol. Microbiol. *50*: 1529–1537.

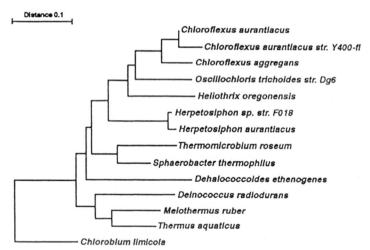

FIGURE B6.1. Phylogenetic tree including the filamentous phototrophs *Chloroflexus aurantiacus*, *Chloroflexus aggregans*, and *Oscillochloris trichoides*. *Heliothrix oregonensis* was not included in this tree because of incomplete sequence data. Other trees have been published that include *Heliothrix* (Weller et al., 1992; Ward et al., 1998).

TABLE B6.1. Differential characteristics of the genera of gliding filamentous anoxygenic phototrophs[a]

Genus	*Chloroflexus*	*Heliothrix*	*Oscillochloris*	*Chloronema*
Bacteriochlorophyll	*a,c*	*a*	*a,c*	*a,d*
Carotenoid	β- + γ-carotene and OH-γ-carotene glucosides	Oxo-γ-carotene and Glucosides	β- + γ-carotene	nd
Color	Orange-green	Orange	Green	Green
Cell diameter, μm	0.5–1.5	1.5	0.8–1.4; 4.5–5.5[b]	2.0–2.5
Sheath	±	−	±	+
Gas vesicles	−	−	±	+
Chemoheterotrophy	±	nd	±	nd

[a]Symbols: see standard definitions; nd, not determined; ±, some strains positive under some conditions.

[b]There are two species in this genus with very different cell diameter ranges.

thetic members, some of which are filamentous (*Herpetosiphon*) and some of which are not (Fig. B6.1). Recent analyses have revealed novel 16S rRNA gene sequences that cluster within this group from diverse environments including activated sludge (Bradford et al., 1996), open ocean (Giovannoni et al., 1996), and a high temperature thermal spring (Hugenholtz et al., 1998b). Clearly we have much to learn about the phenotypic range and phylogenetic diversity of the filamentous phototrophs and their non-photosynthetic relatives.

The major differentiating characteristics of the four genera of the filamentous anoxygenic phototrophic bacteria are presented in Table B6.1.

Genus I. **Chloroflexus** *Pierson and Castenholz 1974a, 7[AL]*

BEVERLY K. PIERSON AND RICHARD W. CASTENHOLZ

Chlo.ro.flex' us. Gr. adj. *chloros* green; L. masc. n. *flexus* a bending; M.L. masc. n. *Chloroflexus* green bending.

Filaments of indefinite length; cells 0.5–1.5 × 2–6 μm; none differentiated (Fig. B6.2). Cell division by fission, no branching. No internal proliferations of cell membrane, except mesosomes; **chlorosomes present when anaerobically grown. Motile by gliding**. Thin sheath may be present. Stain Gram-negative, but cell wall not typical of Gram-negative bacteria. No flagella.

Anaerobic and facultatively aerobic (some). **Primarily photoheterotrophic**, secondarily photoautotrophic (probably not all strains) and chemoheterotrophic (not all strains). Several carbon sources utilized: e.g., acetate, glycerol, glucose, pyruvate, and glutamate. **Bacteriochlorophylls *a* and *c* present under anaerobic conditions; major carotenoids include β- and γ-carotene and hydroxy-γ-carotene glucoside ester**.

The mol% G + C of the DNA is: 53.1–54.9 or 56.7–57.1.

Type species: **Chloroflexus aurantiacus** Pierson and Castenholz 1974a, 7.

FURTHER DESCRIPTIVE INFORMATION

Phylogenetic Treatment The 16S rRNA sequences of the species of *Chloroflexus* set this genus apart from all other phototrophs except *Oscillochloris* and *Heliothrix*. *Chloroflexus* is only distantly related to the groups of non-filamentous phototrophs based not only on the 16S rRNA sequences but also on sequences of specific photosynthetic genes such as those for the reaction center and light-harvesting polypeptides. Among all phototrophs the genus *Chloroflexus* and its filamentous relatives form the deepest phylum on the Bacterial branch of the tree of life. On the basis of 5S and 16S rRNA sequences, the filamentous phototroph, *Heliothrix oregonensis*, is related to *Chloroflexus* (Pierson et al., 1985; Weller et al., 1992), as is the genus *Oscillochloris* (Keppen et al., 1994).

Chloroflexus is by far the best studied of all four genera of anoxygenic filamentous phototrophs with two species and several strains available in pure culture. Except for one published paper on a mesophilic variety of *Chloroflexus* (Pivovarova and Gorlenko, 1977), all work has been done on thermophilic strains of the two species, *C. aurantiacus* and *C. aggregans*.

Cell Morphology, Fine Structure, and Cell Wall Composition In thermophilic *C. aurantiacus* (Pierson and Castenholz, 1974a), *C. aurantiacus* var. *mesophilus* (Pivovarova and Gorlenko, 1977), and *C. aggregans* (Hanada et al., 1995a), transmission electron microscopy indicates a relatively simple cell interior with no invaginations of the cell membrane except for mesosomes which were more common and conspicuous in the mesophilic variety. The DNA area is centrally located; elongated chlorosomes are closely appressed to the cell membrane (Pierson and Castenholz, 1974a; Hanada et al., 1995a) (Fig. B6.3). Extensive work on the

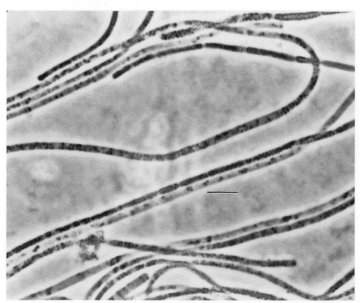

FIGURE B6.2. *C. aurantiacus* strain OK-70-fl. The transverse septa are not apparent. Phase-contrast photomicrograph. Bar = 6 μm.

composition and development of these pigment-bearing sacs has been done. See Blankenship et al. (1995) and Olson (1998) for recent reviews of chlorosome structure and function.

Poly-β-hydroxybutyrate granules, polyphosphate granules and, possibly, glycogen bodies may be seen by use of transmission electron microscopy (Pierson and Castenholz, 1974a) (Fig. B6.3). Polyglucose was present in cells under certain conditions (Sirevåg and Castenholz, 1979).

Motility is by gliding. A wide range of gliding rates is found within the genus. *C. aurantiacus* was reported to glide at 0.01– 0.04 μm/s (Pierson and Castenholz, 1974a) while *C. aggregans* was much faster (1–3 μm/s) (Hanada et al., 1995a).

Although *Chloroflexus* stains Gram-negative, analysis of the wall of *C. aurantiacus* revealed characteristics that were Gram-positive in nature. The peptidoglycan contains ornithine and is possibly phosphodiester-linked to polysaccharide *via* the muramic acid (Jürgens et al., 1987). Furthermore, *Chloroflexus* lacks peptido-glycan-linked lipoprotein. Meissner et al. (1988a) found no evidence for a lipopolysaccharide layer or any of its constituents in either of two strains of *C. aurantiacus*.

Pigments and Lipids   The bulk bacteriochlorophyll *a* has *in vivo* absorbance maxima near 802 and 865 nm, unlike the single maximum at about 810 nm in the chlorosome-containing green sulfur bacteria (Pierson and Castenholz, 1974a). The *in vivo* absorbance maximum for bacteriochlorophyll *c* in *Chloroflexus* is near 740 nm. Gloe and Risch (1978) found that the bacterio-

chlorophyll *c* from *Chloroflexus aurantiacus* (four thermophilic strains) was esterified to stearol. Takaichi et al. (1995) more recently reported that bacteriochlorophyll *c* in two strains of *C. aurantiacus* (J-10-fl and OK-70-fl) existed as mixtures of stearyl, phytyl, geranylgeranyl, and cetyl/deyl esters with phytyl and stearyl esters dominating under low light growth conditions and phytyl and geranylgeranyl esters dominating under higher light growth conditions. A "baseplate" bacteriochlorophyll a_{792} is located in the chlorosome adjacent to the point of attachment to the cell membrane (Schmidt et al., 1980; Betti et al., 1982).

As in the case of other anoxygenic photosynthetic bacteria, the synthesis of bacteriochlorophyll *a* is controlled by both light intensity and oxygen. Bacteriochlorophyll *c* synthesis is also controlled by these factors (Lehmann et al., 1994; Hanada et al., 1995a; Ma et al., 1996). Bacteriochlorophyll synthesis decreased in the presence of high light or oxygen but the responses were differential, suggesting independent regulation of the two pigments by light and oxygen (Pierson and Castenholz, 1974b; Oelze, 1992).

Carotenoids of phototrophic cells consisted largely of monocyclic γ-carotene, OH-γ-carotene, and OH-γ-carotene glucoside fatty acid ester (Halfen et al., 1972; Schmidt et al., 1980; Hanada et al., 1995a; Takaichi et al., 1995). Although the bicyclic β-carotene is a major constituent of *C. aurantiacus* (Halfen et al., 1972; Schmidt et al., 1980; Takaichi et al., 1995), it is present in only small amounts in *C. aggregans* (Hanada et al., 1995a; Takaichi et al., 1995). Aerobic, heterotrophic cells of *C. aurantiacus* con-

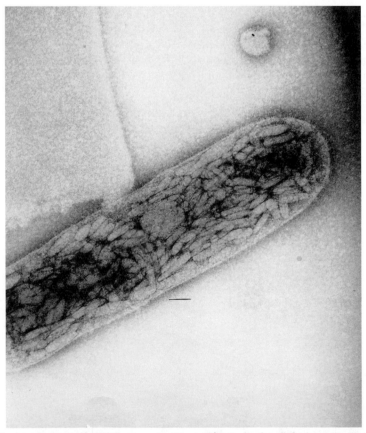

FIGURE B6.3.   *C. aurantiacus* strain J-10-fl grown at about 45°C under low light intensity (incandescent, ~320 lux). Samples were negatively stained with 1% phosphotungstic acid on a carbon-stabilized Formvar grid and viewed with a Siemens Elmiskop operating at 80 kV. Numerous chlorosomes may be seen. The large granules are probably poly-β-hydroxybutyrate. Transmission electron micrograph. Bar = 0.1 μm. (Courtesy of M. Broch-Due.)

tained largely 4-oxo-β-carotene (echinenone) and 4-oxo-γ-carotene glucoside (myxobactone) (K. Schmidt, personal communication). Unlike the weak response or lack of carotenoid synthesis under aerobic conditions in the purple bacteria, *Chloroflexus* shows enhanced synthesis of some carotenoids (Pierson and Castenholz, 1974b; Hanada et al., 1995a).

The reaction center of *Chloroflexus* was initially characterized independently by two research groups (Pierson and Thornber, 1983; Pierson et al., 1983; Bruce et al., 1982). It is similar to the reaction center of the purple bacteria but differs in having three bacteriochlorophyll *a* and three bacteriopheophytin *a* molecules as opposed to the four and two found in purple bacteria. The reaction center of *Chloroflexus* contains only two protein subunits (L and M) lacking the H subunit. For a recent review of the *Chloroflexus* reaction center see Feick et al. (1995). The *C. aurantiacus* reaction center has recently been crystallized (Feick et al., 1996). The pigment protein interactions differ from those of the purple bacteria (Ivancich et al., 1996).

Three strains of *C. aurantiacus* possessed both monogalactosyl- and digalactosyl-diglyceride lipids (Kenyon and Gray, 1974). Phosphatidylethanolamine and cardiolipin were absent (Knudsen et al., 1982). Phosphatidylglycerol and phosphatidylinositol were also present in substantial amounts, as were single wax esters (unusual in bacteria) (Knudsen et al., 1982). Waxes ranged from C_{28} to C_{38}, with fully saturated C_{36} being the major species. Longer chain fatty acids (C_{17} and C_{16-20}) predominated with a high degree of saturation, although monoenoic (16:1, 17:1, 18:1, 19:1, 20:1) acids occurred (Kenyon and Gray, 1974; Knudsen et al., 1982). *C. aurantiacus* also contains four unsaturated fatty alcohols (C_{16-19}) (Knudsen et al., 1982). The aminoglycosphingolipid found in green sulfur bacteria was absent in *Chloroflexus* (Jensen et al., 1991). The absence of hopanoid triterpenes and steroids from *Chloroflexus* and the presence of a unique verrucosan-2β-ol and verrucosane point to a unique pathway of lipid biosynthesis (Hefter et al., 1993).

Nutrition and Growth Conditions *Chloroflexus aurantiacus* can use various carbon sources but growth in complex medium is superior in terms of both rate and yield. Yeast extract, sometimes with added Casamino acids, produces the most rapid growth rates and greatest yields under photoheterotrophic or aerobic chemoheterotrophic conditions. Pierson and Castenholz (1974a) found that in strain J-10-fl, 80–90% of the expected yeast extract yield could be achieved with glycerol and 60–70% could be achieved with acetate if either was supplemented with "growth factor" concentrations of yeast extract. Lower yields occurred with glucose, pyruvate, glutamate, or lactate with the same yeast extract supplement. Ethanol, succinate, malate, and butyrate were not adequate substrates. Madigan et al. (1974) had both similar and different results. Without supplemental yeast extract but with a vitamin mixture, Madigan et al. (1974) found consistently good growth with glucose in all strains and good growth with glycerol, pyruvate, glutamate, or aspartate in a few strains only—all under anaerobic conditions in the light. Only moderate to low yields were achieved under these conditions with acetate, lactate, succinate, malate, butyrate, citrate, ribose, galactose, ethanol, mannitol, or glycylglycine. Aerobic dark growth occurred with most of the substrates already mentioned, but most strains would not grow on ethanol and citrate. No strains grew on glycylglycine. Inorganic carbon was present under all conditions described. Løken and Sirevåg (1982) found that acetate supported substantial growth of strain OK-70-fl in the absence of CO_2.

Both strains of *C. aggregans* also grew best photoheterotrophically on yeast extract or Casamino acids (Hanada et al., 1995a). Both strains grew, although much less vigorously on aspartate, glutamate, lactate, succinate, malate, and glucose. Growth was not obtained with glycylglycine, acetate, or glycerol, although faint growth was observed for strain MD-66 on acetate and glycerol.

Because of the variations in preferences for carbon sources observed among the various strains of both species of *Chloroflexus*, and because of the substantially improved growth obtained with the complex organic substrates (yeast extract and Casamino acids), carbon sources are not useful distinguishing characteristics for this genus or its species.

Some strains of *C. aurantiacus* lack the ability to grow aerobically in darkness (Giovannoni et al., 1987a). Although O_2 is tolerated (and even consumed), growth does not resume until light is applied (if bacteriochlorophyll is still present). If pigment is low, semianaerobic conditions are necessary for synthesis to occur before growth resumes.

CO_2 may serve as sole carbon source in at least one strain of *C. aurantiacus* (OK-70-fl) (Madigan et al., 1974; Madigan and Brock, 1975, 1977). Sulfide-dependent photoautotrophic growth is slow in the laboratory (Castenholz, 1973; Madigan and Brock, 1977). It is apparent, however, that this mode of growth operates under some field conditions to the probable exclusion of photoheterotrophy (Castenholz, 1973; Giovannoni et al., 1987a). Photoautotrophic growth with sulfide or thiosulfate was not observed in *C. aggregans* (Hanada et al., 1995a). *C. aurantiacus* strain OK-70-fl grows autotrophically on hydrogen using a pathway of CO_2 fixation unique to *Chloroflexus* (see section on metabolic pathways).

Growth factor requirements have not been thoroughly investigated for all strains. Madigan (1976) reported that folic acid and thiamine were required by the two strains of *C. aurantiacus* tested. These vitamins were also required by both strains of *C. aggregans* (Hanada et al., 1995a).

Ammonium and some amino acids will serve as nitrogen sources; nitrate will not (Madigan, 1976; Brock, 1978). No evidence for nitrogen fixation has been found in *Chloroflexus*, and the genes for nitrogenase appear to be absent (Madigan, 1995).

The optimum growth temperature (52–60°C) is known for strains of *C. aurantiacus* and *C. aggregans*. The strains originally isolated and those used in the original genus description were all thermophiles from hot springs. The upper limit was 65–70°C in the laboratory (Pierson and Casteanholz, 1974a). The lower limit was 30–35°C. At temperatures below 40°C, the filaments formed aggregates and often adhered to culture vessel walls (Pierson and Castenholz, 1974a). In strain J-10-fl, these observations were more systematically confirmed by Oelze and Fuller (1983). The optimum pH of the holotype (J-10-fl) was 7.6–8.4, and growth was severely limited below pH 7.0 (Pierson and Castenholz, 1974a). *C. aggregans* grew over a pH range of 7.0–9.0 (Hanada et al., 1995a).

The mesophilic variety *C. aurantiacus* var. *mesophilus* (Pivovarova and Gorlenko, 1977) had a temperature range for growth of 10–40°C, with an optimum at about 20–25°C (Gorlenko, 1975). The optimum pH (7.0–7.2) was also lower for the mesophilic strains than for the thermophilic strains in culture. It appears that the mesophilic strains are no longer available, at least as axenic cultures (V. M. Gorlenko, personal communication).

Metabolism and Metabolic Pathways A recent review (Sirevåg, 1995) of carbon metabolism in *Chloroflexus* evaluated two proposed CO_2 fixation pathways. In strain OK-70-fl, evidence has been presented for the unique 3-hydroxypropionate cycle (Holo, 1989; Strauss et al., 1992; Eisenreich et al., 1993; Strauss and Fuchs, 1993). The 3-hydroxypropionate pathway carboxylates acetyl-CoA and forms 3-hydroxypropionate as an important intermediate. A second carboxylation of propionyl-CoA results eventually in the production of glyoxylate. An alternative autotrophic pathway for *C. aurantiacus* strain B3 was proposed by Ivanovsky et al. (1993). In this pathway, acetyl-CoA is carboxylated to pyruvate. A second carboxylation of phosphoenolpyruvate to oxaloacetate leads finally to glyoxylate.

Evidence for operation of a TCA cycle under both aerobic and anaerobic conditions was found (Sirevåg and Castenholz, 1979). Cells of *Chloroflexus* grown photoheterotrophically on acetate as sole organic carbon source have enhanced activities of key enzymes of the glyoxylate cycle: isocitrate lyase and malate synthase (Løken and Sirevåg, 1982).

The electron transport pathways for photosynthesis and respiration in *Chloroflexus* were recently reviewed (Zannoni, 1995). Several features distinguish the pathways in this genus. Both anaerobically and aerobically grown *Chloroflexus* lack the soluble cytochrome c_2 that characteristically feeds electrons to the reaction center or terminal oxidase in most purple bacteria. Photosynthetic cells have a tetraheme Cyt*c*-554 and a novel peripheral copper protein, auracyanin, (Trost et al., 1988) not present in aerobically grown cells. A Cyt*bc*$_1$ complex appears to function in both photosynthesis and respiration. Although aerobically grown cells lack the tetraheme Cyt*c*-554 and auracyanin, they have multiple membrane-bound *c*-type cytochromes.

The terminal oxidase for aerobic respiration appears to vary (Zannoni, 1995). Cyt*aa*$_3$ is present in cells grown at less than saturated oxygen levels and a Cyt*a*, is present in cells grown at higher oxygen levels. *Chloroflexus* has menaquinone rather than ubiquinone in its electron transfer pathways. Analysis of quinones in two strains of *C. aurantiacus* and *C. aggregans* (Hanada et al., 1995b) showed MK-10 was the dominant quinone (96% of the total quinone in *C. aurantiacus* and 91% in *C. aggregans*). Trace amounts of MK-8, -9, and -11 were found in all strains. The two strains of *C. aggregans* also contained small but significant amounts (2.5–4.5 mol%) of MK-4.

Several enzymes of specialized interest have been purified from *Chloroflexus*. A maltotetraose- and maltotriose-producing amylase was purified and characterized from *C. aurantiacus* J-10-fl (Ratanakhanokchai et al., 1992). ATPase (Yanyushin, 1988) and hydrogenase (Serebryakova et al., 1989) have been isolated. Relatively few metabolic pathways have been fully characterized, however. Sirevåg (1995) recently reviewed what is known of carbon metabolism in *Chloroflexus*. *Chloroflexus* operates both TCA and glyoxylate cycles when grown under anaerobic conditions in the light. Glyoxylate is also the product of the CO_2-fixation pathway in *Chloroflexus*. Synstad et al. (1996) studied and sequenced the malate dehydrogenase in *C. aurantiacus*. They found that the enzyme was thermostable and had sequence similarity to other bacterial malate dehydrogenases and the typical MDH sequences at the substrate binding site. The sequence of amino acids in the coenzyme binding domain was more similar to lactate dehydrogenase. Synthesis of Bchl *c* is via the c-5 glutamate pathway (Swanson and Smith, 1990).

Genetics and Molecular Data   There is relatively little known of the genetics of *Chloroflexus*. So far no plasmids have been reported and no phages (specific or nonspecific) are known. A number of mutants have been obtained using chemical and UV

mutagenesis. These are strains lacking colored carotenoid pigments or deficient in bacteriochlorophyll *c* as a result of mutagenesis (Pierson et al., 1984b). All strains showed frequent reversion of the latter mutation. *C. aurantiacus* is quite resistant to UV radiation, growing photoheterotrophically with continuous exposure to levels of UV-C that were immediately lethal to *E. coli* (Pierson et al., 1993).

Chloroflexus aurantiacus strain J-10-fl is sensitive to penicillin and ampicillin but was resistant to carbenicillin up to 100 μg/ml (Pierson et al., 1984b).

The presence of endogenous restriction endonucleases and the nature of DNA restriction in *C. aurantiacus* were recently reviewed by Shiozawa (1995).

Hanada et al. (1995b) used a dot blot hybridization assay with labeled DNA from *C. aurantiacus* J-10-fl. Strain OK-70-fl had 98% hybridization. Three of their new isolates of *C. aurantiacus* had hybridization values of 76–87% with J-10-fl. The two isolates assigned to the new species, *C. aggregans*, had 9 and 10% DNA hybridization with *C. aurantiacus* (J-10-fl). Hanada et al. (1995a) resequenced the 16S rRNA gene from *C. aurantiacus* J-10-fl and compared it with *C. aggregans* MD-66. Comparing sequences corresponding to positions 28 to 1491 in the *E. coli* sequence representing 95% of the entire 16S rRNA gene, they found 92.8% similarity. *C. aurantiacus* was the closest relative to *C. aggregans* in a tree constructed on the basis of distance matrix data (Hanada et al., 1995a). *C. aggregans* could have been designated a separate genus from *Chloroflexus* on the basis of this low 16S rRNA sequence identity. However, since very few phenotypic characters could be used for generic separation, it was assigned as a new species instead (Hanada et al., 1995a).

Considerable work has been done in sequencing photosynthesis genes and studying the organization and expression of the *Chloroflexus* genome (see Shiozawa, 1995). Most significantly the photosynthesis genes show substantial differences from the corresponding genes in other phototrophs. *Chloroflexus* appears to lack an H subunit for its reaction center, and the L and M subunit genes appear to be homologous to those in the purple bacteria (Shiozawa et al., 1987, 1989). Unlike the purple bacterial photosynthetic genes, however, the *Chloroflexus* reaction center genes are organized in a separate operon from the genes for the light-harvesting polypeptides. The genes for the B806-866 light-harvesting complex and the tetraheme Cyt*c*-554 in *C. aurantiacus* form an operon that lies at least 3 kb from the reaction center operon (Watanabe et al., 1995).

The sequence identity is low between the light harvesting bacteriochlorophyll *a* polypeptides of *C. aurantiacus* and those of the purple bacteria (Wechsler et al., 1985, 1987). There is even less similarity between the chlorosome polypeptides of *C. aurantiacus* and those of the green bacteria (Shiozawa, 1995). Watanabe et al. (1995) suggest that the low sequence identities and differences in chromosomal organization of the photosynthetic genes in *Chloroflexus* support the phylogenetic distance from other phototrophs predicted by the 16S rRNA sequences.

Ecology   Populations of thermophilic *Chloroflexus* in their native hot spring habitats have been studied extensively (reviewed by Brock, 1978; Castenholz, 1984a, b; Pierson and Castenholz, 1992; Castenholz and Pierson, 1995). Gorlenko (1975) suggests that mesophilic *Chloroflexus* populations are common in the anaerobic portions of freshwater lakes. Likewise, mesophilic *Chloroflexus*-like organisms from shallow, marine microbial mats have been reported, but the necessary culture isolations from both freshwater and marine mesophilic environments have not been made or maintained (see Castenholz, 1984b; Mack and Pierson, 1988; Pierson et al., 1994).

Confirmed and massive *Chloroflexus* populations occur worldwide in neutral to alkaline hot springs forming gel-like, orange-pink-reddish mats generally under a thin (1 mm or less) surface layer of cyanobacteria. *Chloroflexus* trichomes may be seen microscopically mixed with the cyanobacteria up to about 70–72°C in North America. The filaments form conspicuous undermats at temperatures less than 68–69°C. Such mats may be found at temperatures as low as 30–40°C in hot spring effluents. Hanada et al. (1995b) cultured *Chloroflexus* from about 80% of the hot springs they sampled with water between pH 6.4–8.2 over a range of 50–70°C. They detected no *Chloroflexus* in samples from extremely saline or high iron springs in this range.

The masses of filaments forming the conspicuous "*Chloroflexus*" undermat in hot springs may not be identical to the cultivated *Chloroflexus* strains. In fact, due to culturing bias, the cultivated strains may represent only a small fraction of the total bacterial population within the mat. Using a *Chloroflexus*-specific antiserum prepared against a purified strain isolated from Octopus Spring mat, Tayne et al. (1987) showed that 29–59% of the widest filaments in a mat sample bound to the antiserum and hence were likely to be *Chloroflexus*. Abundant narrower filaments within the mat (16–36% of the wide filament population) did not react with the antiserum at all. Since the antiserum reacted positively with 14 isolated strains of *Chloroflexus* from a world-wide geographic distribution, it seems likely that the other nonbinding filaments in the mat were not *Chloroflexus* (Tayne et al., 1987).

Evidence for the presence of specific and novel sequences of 16S rRNA in microbial mats has been used to study the distribution of particular species in the mats and to determine the relative abundance and diversity of known and novel species in mats. Weller et al. (1992) retrieved two *Chloroflexus*-like sequences from Octopus Spring mat in Yellowstone. Neither was identical with *C. aurantiacus* or *H. oregonensis*. In fact, no sequences identical to *C. aurantiacus* (J-10-fl) or another *C. aurantiacus* strain cultivated from the same mat were obtained (Ward et al., 1990; Weller et al., 1992). These studies show that extrapolation from selectively cultured strains to the "mass" population in the native hot spring mat must be done cautiously. Species diversity within the mat is much greater than previously thought (Ward et al., 1990).

Different *Chloroflexus* strains or species may be adapted to specific niches within the hot spring mats. This would explain why *C. aurantiacus* Y-400-fl rRNA was detected only in mats sampled at higher temperatures (61–70°C) or when samples from lower temperature sites were preincubated at higher temperatures in hot spring mats (Ruff-Roberts et al., 1994). *Chloroflexus*-like 16S rRNAs were found in highest abundance in the top mm of the mat. None was detected below 3.5 mm (Ruff-Roberts et al., 1994).

The natural hot spring populations of *Chloroflexus* appear to be primarily (a) heterotrophic or (b) photoheterotrophic if the usual anaerobic conditions of undermat have allowed bacteriochlorophyll synthesis. The organic carbon is ultimately derived from the cyanobacteria (Ward et al., 1984). The organic substrates for heterotrophic growth of *Chloroflexus* in hot springs are not totally known, but fermentation products in the undermat (e.g., acetate, lactate, propionate, and butyrate) from decomposing cyanobacteria may be used (Ward et al., 1984; Anderson et al., 1987; Bateson and Ward, 1988). Being dependent on cy-

anobacteria restricts major development of heterotrophic populations to the temperature range of the cyanobacteria. In western North America, cyanobacteria commonly can be found in temperatures up to 73–74°C. In most of the rest of the world, upper limits for cyanobacteria are lower (about 63–64°C), thus limiting the distribution of *Chloroflexus* to this lower temperature (see Pierson and Castenholz, 1992, and Castenholz and Pierson, 1995, for more extensive discussion of the temperature effects on the distribution of *Chloroflexus* in hot springs).

Whenever sufficient sulfide (H_2S, HS^-, S^{2-}) is present in source waters, however, the dependence of *Chloroflexus* on cyanobacteria is relieved, and the appearance of sulfide-dependent photoautotrophy occurs. In this case, *Chloroflexus* can commonly be found in temperatures up to about 66°C, forming essentially pure mats devoid of cyanobacteria (see Castenholz, 1973; Pierson and Castenholz, 1992; Castenholz and Pierson, 1995). This is seen even in North America, since sulfide inhibits and excludes the cyanobacterial species otherwise found above 57°C in nonsulfide waters (Castenholz, 1973; Revsbech and Ward, 1984). The capacity for sulfide-dependent photoautotrophy in natural populations of *Chloroflexus* has been demonstrated in such restricted habitats in hot springs in Yellowstone National Park (Giovannoni et al., 1987a; Ward et al., 1989). In the total absence of oxygen, these mats comprised primarily of *Chloroflexus* developed over a temperature range of 50–66°C in the presence of sulfide (30–1000 µM). This *Chloroflexus* strain (GCF) might represent a new species since it could not be grown in the presence of oxygen in the laboratory (Giovannoni et al., 1987a). In the Bol'shaya River hot springs near Lake Baikal, however, cyanobacteria and *Chloroflexus* were both absent at temperatures above 54°C (Yurkov et al., 1991). The authors attributed this unusual absence of *Chloroflexus* to high pH (above 9.0) and the simultaneous occurrence of high sulfide (around 10 mg/l) and oxygen (0.8–2.8 mg/l). At lower temperatures (45–54°C) *Chloroflexus* was abundant (Yurkov et al., 1991).

In an Icelandic high sulfide thermal environment *Chloroflexus* was the dominant phototroph above 60°C in the presence of 60 µM H_2S (Jørgensen and Nelson, 1988). At lower temperatures cyanobacteria dominated. In the high temperature *Chloroflexus* mat, a thin layer of the oxygen-producing cyanobacterium *Mastigocladus* was sandwiched within the *Chloroflexus* mat. Jørgensen and Nelson (1988) suggested that cyanobacterial oxygenic photosynthesis could occur within the *Chloroflexus* mat only when *Chloroflexus* had consumed sufficient sulfide by photosynthesis.

Pentecost (1995) studied the ecology of photosynthetic bacteria including *Chloroflexus* in travertine-depositing springs. *Chloroflexus* was the dominant photosynthetic bacterium in these springs over a temperature range of 40–60°C and sulfide concentration of 0–100 µM. *Chloroflexus* was isolated from 39.5 to 63.4°C (pH: 6.32–7.01; E_h: −300 to +80 mV). The *Chloroflexus* isolations were most abundant from 55 to 60°C. The *Chloroflexus* was found in hotter and more sulfide rich waters than cyanobacteria. The *Chloroflexus* filaments formed extensive frondose growths at the travertine surface and upon encrustation with aragonite produced a delicate bush-like microfabric readily identifiable in older deposits (Pentecost, 1995).

ENRICHMENT AND ISOLATION PROCEDURES

Gorlenko (1975) reported a successful enrichment of a freshwater, mesophilic variety of *C. aurantiacus* in lake-bottom water and sediment columns to which mineral salts, trace elements, Na_2SO_4 (0.6 g/l), $Na_2S\cdot9H_2O$ (0.1 g/l), yeast extract (0.025 g/

l), and casein hydrolysate (0.025 g/l) had been added. He used a low light intensity (<3000 lux) and a temperature of 25–30°C (Pierson and Castenholz, 1992).

Early attempts at the enrichment of thermophilic varieties of *C. aurantiacus* did not meet with great success due to the large number of contaminating chemoheterotrophs, although enrichment could be obtained (see Pierson and Castenholz, 1992). However, co-enrichment with cyanobacteria proved to be a simple and usually successful method of obtaining the common physiological types of *Chloroflexus*, i.e., those that will grow under aerobic or semiaerobic as well as anaerobic conditions. Cyanobacterial samples from hot spring mats (pH 5–11) will almost always contain *Chloroflexus*, even if it is not obvious. The cyanobacteria (one or more strains) may be enriched for simply by inoculating a small sample into a common cyanobacterial, liquid medium and incubating this at an appropriate temperature (e.g., 45–55°C) under the low intensity light (1000–3000 lux) provided by fluorescent or incandescent bulbs (Pierson and Castenholz, 1992). A mixed culture or a monoculture of cyanobacteria will develop, and trichomes of *Chloroflexus* and other heterotrophs will be carried along. It is essentially impossible to rid this enrichment of *Chloroflexus*.

Thermophilic *Chloroflexus* may be isolated from a cyanobacterial enrichment or directly from hot spring mats by placing small (4–16 mm²) pieces of cyanobacteria or sample on agar-solidified (1.5% w/v) cyanobacterial medium and incubating for one to a few days. Commonly, some of the cyanobacteria will spread extensively by gliding motility. *Chloroflexus*, a much slower glider, may be carried along to some extent but will form its own wisps of gliding trichomes which will commonly spread away from central or radiating masses of less motile or nonmotile cyanobacteria (Fig. B6.4). These wisps of *Chloroflexus* may then be picked up by cutting out the minute piece of agar on which they rest, inverting it on a new plate which contains some yeast extract (e.g., 0.02–0.2 g/l), and streaking, smearing, or dragging the trichomes over the sterile agar. By repeatedly picking trichomes from the tips of spreading wisps, an axenic culture can be established. These cultures can be grown in liquid or agar-solidified modified DG medium*. More details of this procedure are given by Pierson and Castenholz (1974a) and Pierson and Castenholz (1992).

Clones of the 1.0-µm-diameter trichomes may be established by the manual picking of individual trichomes spread on an agar surface. Presumptive clones of these and the narrower diameter trichomes (0.5–0.7 µm) may be established by streaking. Units <20 µm long can be prepared by short term (10–20s) ultrasonic disruption, and then plated in a thin overlay of 0.7% (w/v) agar medium (Pierson et al., 1984b).

For thermophilic varieties of *Chloroflexus* that do not grow in the continued presence of oxygen, such as the GCF strains, anaerobic techniques need to be applied (see Giovannoni et al., 1987a and Pierson and Castenholz, 1992 for more details). However, the other aspects of the procedure are similar. *Chlo-*

roflexus samples directly from hot spring mats are placed or streaked on agar plates which have been stored previously under anoxic conditions. For samples that contain cyanobacteria, approximately 10^{-5} M 3-(3,4,-dichlorophenyl)-1,1-dimethylurea (DCMU) should be used in the medium to prevent oxygenic photosynthesis. Incubation and growth in gas packs containing H_2/CO_2 have resulted in strains that are obligately phototrophic but that can tolerate O_2 without lethal effects for at least several hours (Giovannoni et al., 1987a).

Hanada et al. (1995b) reported a new medium and methods for enrichment and isolation of *Chloroflexus* strains that gave improved efficiency and reproducibility of recovery. The medium for isolation and enrichment was designated PE medium**. Samples of bacterial mats from hot springs are suspended and streaked on plates of PE medium (1.5% agar). The plates are incubated at 55°C under incandescent light (2000 lux). Colonies (green or orange) appear within a few days. As described above, cells are picked from gliding advancing edges of the colonies and restreaked. Plates are incubated under alternating aerobic and anaerobic conditions to eliminate fermentative and obligately aerobic bacteria. By comparing the yield of colonies from identical inocula on DG and PE isolation plates Hanada et al., (1995b) determined that colonies appeared three to four days faster on PE plates incubated aerobically at 55°C and that ten times as many colonies appeared as on DG plates.

Malik (1996a) reported improved growth of several *Chloroflexus* strains including *C. aurantiacus* (J-10-fl) and *C. aggregans* (MD-66) in a slightly modified version of the DG medium used for *Chloroflexus* enrichment. By máintaining strict anaerobic conditions with added sulfide and gassing with nitrogen, it was claimed that photoautotrophic growth was obtained. These claims have not been reconciled with earlier reports of the absence of sulfide-dependent photoautotrophy in these specific strains. Since the modified medium contained yeast extract (Malik 1996a), it is possible that cultures were growing photoheterotrophially though perhaps benefiting from the lowered redox potential produced by the added sulfide.

MAINTENANCE PROCEDURES

Cultures may be maintained under anaerobic or semianaerobic conditions in liquid or agar-solidified medium. Presumably, a variety of media would suffice, but the authors and several others have used modified versions of medium DG with the chelator nitrilotriacetic acid (0.1 g/l) and the buffer glycylglycine (0.8 g/l) (e.g., Pierson and Castenholz, 1992). Yeast extract (2–3 g/l) or 1 g yeast extract plus Casamino acids (2 g/l) are used as principal substrates. Other labs have successfully altered various components of the medium to meet experimental needs. Specifically glycylglycine buffer can be replaced with 10 mM Tris hydrochloride (pH 8.0) (Rolstad et al., 1988) or with 0.1% w/w potassium-phosphate (pH 8.0) (Ivanovsky et al., 1993). For

---

*Modified DG medium contains (per liter distilled H₂O): nitrilotriacetic acid, 0.1 g; micronutrient solution, 0.5 ml; FeCl₃ solution (0.29 g/l), 1.0 ml; CaSO₄·2H₂O, 0.06 g; MgSO₄·7H₂O, 0.10 g; NaCl, 0.008 g; Na₂HPO₄, 0.07 g; KH₂PO₄, 0.036 g; NH₄Cl, 0.2 g; glycylglycine, 0.8 g; yeast extract (Difco), 2.0 g or yeast extract (Difco), 1.0 g, and vitamin-free Casamino acids (Difco), 2.0 g. Micronutrient solution contains (per liter distilled H₂O): H₂SO₄, 0.5 ml; MnSO₄·H₂O, 2.28 g; ZnSO₄·7H₂O, 0.5 g; H₃BO₃, 0.5 g; CuSO₄·5H₂O, 0.025 g; Na₂MoO₄·2H₂O, 0.025 g; CoCl₂·6H₂O, 0.045 g. After all additions and before autoclaving, pH is adjusted to 8.2. Agar (15 g/l) is added after pH adjustment.

**PE medium contains (per liter distilled H₂O): sodium glutamate, 0.5 g; sodium succinate, 0.5 g; sodium acetate, 0.5 g; yeast extract (Difco), 0.5 g; Casamino acids (Difco), 0.5 g; Na₂S₂O₃, 0.5 g; KH₂PO₄, 0.38 g; K₂HPO₄, 0.39 g; (NH₄)₂SO₄, 0.5 g; vitamin solution, 1.0 ml; basal salt solution, 5.0 ml. Vitamin solution contains (per 100 ml distilled H₂O): nicotinic acid, 100 mg; thiamine HCl, 100 mg; biotin, 5 mg; *p*-aminobenzoic acid, 50 mg; vitamin B₁₂, 1 mg; calcium pantothenate, 50 mg; pyridoxine HCl, 50 mg; folic acid, 50 mg. Basal salt solution contains (per liter distilled H₂O): trisodium EDTA, 4.53 g; FeSO₄·7H₂O, 1.11 g; MgSO₄·7H₂O, 24.65 g; CaCl₂·2H₂O, 2.94 g; NaCl, 23.4 g; MnSO₄·4H₂O, 111 mg; ZnSO₄·7H₂O, 28.8 mg; Co(NO₃)₂·6H₂O, 29.2 mg; CuSO₄·5H₂O, 25.2 mg; Na₂MoO₄·2H₂O, 24.2 mg; H₃BO₃, 31.0 mg. The pH is adjusted to 7.5.

FIGURE B6.4. *C. aurantiacus* strain J-10-fl grown on agar plate in co-culture with the unicellular cyanobacterium *Synechococcus* species seen as dark streaks (*white arrow*). The growth temperature was 45°C. Bar = ~50 μm.

photoautotrophic growth of Strain B-3, the latter authors used an atmosphere of H_2 and added $NaHCO_3$ (0.2%) as sole carbon source. For photoheterotrophic growth, screw-capped tubes or flasks nearly filled with sterile complex organic medium provide sufficiently anaerobic conditions for massive pigment synthesis as well as luxuriant and rapid growth. Incandescent lamps provide good growth at low intensities (1000–3000 lux). Cultures may be maintained easily at 45–60°C, however, cultures must be transferred every 7–14 days (Pierson and Castenholz, 1992).

Pigment-containing trichomes of the common forms of *Chloroflexus* will develop on agar plates in incubators (45–55°C) with or without an overlay procedure. Plate cultures richer in bacteriochlorophylls require incubation in anaerobic conditions. These cultures must be transferred every 7–14 days. Agar shake or stab cultures may also be used.

Another maintenance system, which requires less frequent transferring, is the use of a coculture with a known pure strain of cyanobacterium (in this case a thermophilic unicell, *Synechococcus lividus*, strain OH-53-s). In such cultures, no yeast extract or Casamino acids are used, and organic carbon for *Chloroflexus* is provided by the cyanobacterium. Although this process is essentially aerobic, under low light intensity, bacteriochlorophyll will be present in the coculture. When pure *Chloroflexus* is required, separate wisps (Fig. B6.4) may be manually removed and transferred to organic medium. In the coculture presently used, the cyanobacterium has a lower temperature tolerance than does the *Chloroflexus* and can be eliminated by raising the temperature in liquid cultures to 60–62°C for 4–5 days.

Hanada et al. (1995b) reported successful maintenance of *Chloroflexus* cultures in PE medium. Viability in liquid anaerobic culture under low light (less than 1000 lux) at 55°C was a few months for most strains.

Lowering both the light intensity and the temperature contributes to increasing the survival time of cultures. Malik (1996b) developed conditions for long-term culture maintenance in which viability was retained by cultures of *C. aurantiacus* and *C. aggregans* for 12 months or more. The same modified medium (Malik, 1996a) was used for long-term maintenance with added sulfide and anaerobic conditions. Cultures were maintained un-

der "slow growing" conditions at low temperature (37–40°C) in dim light (100–200 lux) and were fed yeast extract monthly. The addition of activated charcoal after an initial growth period apparently helped to maintain viability by adsorbing waste products (Malik, 1996b).

Chloroflexus may be kept in cryo-ampules at −196°C (liquid N_2) for a few years, but recovery results have been variable. Dense, but not aged, cultures should be used, suspended in their own medium. Recovery from freeze-drying storage has been variable (Malik, 1996b).

More details on culture maintenance are given by Pierson and Castenholz (1992), Hanada et al. (1995b), and Malik (1996b).

DIFFERENTIATION OF THE GENUS *CHLOROFLEXUS* FROM OTHER GENERA

Chloroflexus may not be easily distinguished by morphological criteria from the other three genera considered to be related (Table B6.1). Ultrastructural and biochemical characteristics are required. All are filamentous. The larger trichome diameter and the abundance of chlorosomes on complete and incomplete septa seem sufficient to separate *Oscillochloris chrysea* (Gorlenko and Pivovarova, 1977) from *Chloroflexus*, although this species of *Oscillochloris* has not been isolated in axenic culture. The narrower *Oscillochloris trichoides* would be difficult to distinguish from *Chloroflexus* without the use of electron microscopy and biochemical analysis. The possession of gas vesicles, a characteristic so erratically spread through the *Bacteria* and *Archaea*, should probably not be used as a character to distinguish genera. Characteristics of the genus *Chloronema* (Dubinina and Gorlenko, 1975), i.e., gas vesicles, trichome diameters of 2.0–2.5 μm (*Chloronema giganteum*) or 1.5–2.0 μm (*Chloronema spiroideum*), and the presence of a conspicuous sheath, are not necessarily adequate for erecting a genus separate from *Chloroflexus*. Again, neither species of *Chloronema* has been grown in culture, other than as enrichments. The probable presence of bacteriochlorophyll *d* rather than of bacteriochlorophyll *c* in *Chloronema* is also not a good distinguishing characteristic, since bacteriochlorophylls *c* and *d* occur in different species of *Chlorobium* and even in a single strain

(Broch-Due and Ormerod, 1978) and are insufficient for a generic distinction. Recent detailed phylogenetic studies of the green sulfur bacteria confirm that gas vesicles and pigmentation are not useful characters for making generic distinctions (Overmann and Tuschak, 1997).

Heliothrix is a thermophilic, filamentous bacterium that falls within the range of cell diameters for species of *Chloroflexus* (1.5 µm), but *Heliothrix* possesses neither chlorosomes nor the accessory bacteriochlorophyll *c* or *d* that would be associated with these structures (Pierson et al., 1984a, 1985) (see under *Heliothrix* in this section).

Herpetosiphon and other aerobic filamentous bacteria are of dimensions and color similar to orange *Chloroflexus* but lack chlorophylls of any type (Soriano, 1973). Some carotenoids (e.g., γ-carotene, and carotenoid glucosides) are similar to those of *Chloroflexus* (Kleinig and Reichenbach, 1977).

TAXONOMIC COMMENTS

The species *Chloroflexus aggregans* has been designated with an invalid generic name, *"Chlorocrinus aggregans"*, in at least one publication (Ward et al., 1998). As noted here, *Chloroflexus aggregans* has been described as a valid species of the genus *Chloroflexus* (Hanada et al., 1995a).

Chloroflexus was originally placed in a family (*"Chloroflexaceae"*) defined by the possession of chlorosomes, accessory bacteriochlorophylls *c* or *d*, filamentous morphology, and gliding motility (Trüper, 1976). Pfennig (1989) grouped the apparently related filamentous phototrophic bacteria into a group designated the

"Multicellular filamentous green bacteria" in the previous edition of this *Manual*. Woese (1987) affirmed the phylogenetic group containing *Chloroflexus* in the domain *Bacteria* as the "green non-sulfur" bacteria while cautioning that there was little phenotypic cohesion to the group. All of these terms have been widely used. Since sulfide-dependent autotrophy occurs in this group, Trüper (1987) objected to the "green non-sulfur bacteria" designation. It is misleading. Trüper (1987) further pointed out that if *Heliothrix oregonensis* indeed was phylogenetically related to *Chloroflexus*, then the family definition would have to be emended to remove the requirement for chlorosomes and accessory bacteriochlorophylls. Pierson and Castenholz (1992 and 1995) have discussed the taxonomic problems of the filamentous phototrophs thoroughly and recommended designating them as a group called the "filamentous anoxygenic phototrophs" or the "photosynthetic flexibacteria" as originally proposed by Castenholz (1973). These terms would accurately describe all the phototrophs currently found in the same phylogenetic group as several non-phototrophs. Since both 5S (Pierson et al., 1985) and 16S rRNA (Weller et al., 1992) sequences show that *Heliothrix oregonensis* is phylogenetically associated with *Chloroflexus*, such recognition of the inappropriateness of the "green" designation is needed.

ACKNOWLEDGMENTS

R. W. Castenholz wrote the section on *Chloroflexus* in the previous edition of this *Manual* (1989). Much has been retained from that version including the figures.

List of species of the genus Chloroflexus

1. **Chloroflexus aurantiacus** Pierson and Castenholz 1974a, 7.[AL]
au.ran.ti'a.cus. M.L. neut. n. *aurantium* specific name of the orange; M.L. masc. adj. *aurantiacus* orange-colored.

The characteristics are the same as those described for the genus. Filaments of most strains are less than 1.5 µm in diameter. Filament diameter of the type strain is less than 1.0 µm. The type species is thermophilic and grows in the temperature range of 40–66°C with an optimum at about 52–60°C. Optimum pH: 7.6–8.4.

The mesophilic variety of *C. aurantiacus* (var. *"mesophilus"*; Gorlenko, 1975) grows in the temperature range of 10–40°C with an optimum at about 20–25°C. Optimum pH: 7.0–7.2.

The mol% G + C of the DNA is: 54.9 (Bd), 56.9 (HPLC).
Type strain: J-10-fl, ATCC 29366, DSMZ 635.
GenBank accession number (16S rRNA): D38365.

2. **Chloroflexus aggregans** Hanada, Hiraishi, Shimada and Matsuura 1995a, 680.[VP]
ag'gre.gans. L. v. *aggregare* to flock or band together; L. pres. part. *aggregans* assembling, aggregating.

The characteristics are the same as those described for the genus. Cell diameter is 1.0–1.5 µm and filaments are clearly septate. Cultures grow by forming mat-like aggregates in liquid medium that have the appearance of green balls. Cell aggregates reform rapidly whenever a growing culture is dispersed to form a uniform suspension. Optimal growth temperature is 55°C.

The mol% G + C of the DNA is: 56.7–57.0 (HPLC).
Type strain: MD-66, DSMZ 9485.
GenBank accession number (16S rRNA): D32255.

Other Organisms

Mesophilic marine and hypersaline strains of organisms identical in structure and apparent pigmentation to *Chloroflexus* have been observed by several investigators. They tend to be abundant in marine intertidal or solar saltern microbial mats where they are found often in association with cyanobacteria and purple sulfur bacteria in the presence of sulfide. They are so abundant that they are easily detected by routine pigment analysis or microscopic examination of mat samples. *Chloroflexus*-like filaments have been observed in several different locations in hypersaline mats in Baja, California, Mexico (Stolz, 1983, 1984, 1990; Palmisano et al., 1988; D'Amelio et al., 1989; Lopez-Cortez, 1990; Pierson, 1994; Pierson et al., 1994). They have been observed in

mats from Solar Lake, Sinai, Egypt (Cohen, 1984; D'Amelio et al., 1989), in the Ebre Delta, Spain (Villanueva et al., 1994), in mats from the lagoons of Arabat near the Sea of Azov (Venetskaya and Gerasimenko, 1988), and in the intertidal mats of Great Sippewissett Salt Marsh, MA, USA (Mack and Pierson, 1988; Pierson et al., 1994). Mesophilic *Chloroflexus*-like filaments have been identified by transmission electron microscopy (i.e., chlorosome-containing filaments). Migrating mats of masses of *Chloroflexus*-like organisms have been observed in marine intertidal locations (Mack and Pierson, 1988; Pierson et al., 1994). The ultrastructure (Pierson et al., 1994) and pigmentation and physiology (Mack and Pierson, 1988) were similar to *C. aurantiacus*. The freshwater mesophilic strains (Gorlenko, 1975) and marine mesophilic strains have been unexpectedly difficult to culture. The mesophilic strains isolated by Gorlenko are no longer available in pure culture (Gorlenko, personal communication). Mack succeeded in maintaining a marine strain from Mellum Island (Pierson et al., 1994) in mixed culture for several years.

Genus II. **Chloronema** Dubinina and Gorlenko 1975, 515[AL]

VLADIMIR M. GORLENKO AND BEVERLY K. PIERSON

Chlo.ro.ne'ma. Gr. adj. *chloros* green; Gr. n. *nema* thread; M.L. neut. n. *Chloronema* green filament.

Cells cylindrical, combined in a trichome surrounded by a sheath. Trichomes **straight or spiral**, appear **yellow-green** (Fig. B6.5). Cells multiply by separation of sections of trichomes (hormogonia) of varying length. **Gliding motility**. The major pigment is **bacteriochlorophyll** *d*; also contain bacteriochlorophyll *c*. The primary pigment-bearing structures are **chlorosomes**. Cells are capable of **anoxygenic photosynthesis**.

The mol% G + C of the DNA has not been studied.

Type species: **Chloronema giganteum** Dubinina and Gorlenko 1975, 515.

FURTHER DESCRIPTIVE INFORMATION

Descriptive information is based on the study of *Chloronema* in natural water bodies (Dubinina and Gorlenko, 1975; Dubinina and Kuznetsov, 1976; Gorlenko and Lokk, 1979). Cells of *C. giganteum*, 2.0–2.5 × 3.5–4.5 μm, form straight or spiral trichomes 75–250 μm long (Fig. B6.5). Trichomes are surrounded by a slimy sheath of varying thickness. In lakes containing iron salts, sheaths of *Chloronema* contain iron oxides, occasionally in considerable amounts. In this case, sheaths appear brown. Bacteria multiply by separation of a short end section of trichomes or of an individual cell, which separate from the sheath by gliding. The rate of gliding is about 10 μm/s. *Chloronema* organisms rapidly attach themselves to solid surfaces, e.g., glass slides, and move by gliding.

The sheath is filamentous. The trichome and the sheath are separated by a vacant space. The cell wall appears in electron micrographs of thin sections to be of the Gram-negative type. Bacteria stain Gram-negative. The cytoplasmic membrane forms multiple invaginations of the mesosome type. Chlorosomes are located predominantly in the periphery of cells but also occur along the cell septa and are associated with the membrane invaginations. During division of cells, the cytoplasmic membrane extends to the inner portion of the cell, and then a cell septum is formed which is similar to the cell wall. The central part of each cell is filled with elongated gas vesicles (Fig. B6.6).

Analysis of natural samples from zones of mass development of *C. giganteum* revealed the presence of bacteriochlorophyll *d* (*in vivo* absorption maximum: 720 nm). *Chloronema* occurs in dimictic freshwater lakes of the mesohumus type, with high ferrous iron content and only traces of H_2S. *C. giganteum* occurs predominantly below the chemocline under anoxic conditions.

Observations at natural habitats indicate that *Chloronema* is capable of anoxygenic photosynthesis. *C. giganteum* exhibits positive aerotaxis in the dark, possibly indicating a capacity for aerobic chemotrophic metabolism.

ENRICHMENT AND ISOLATION PROCEDURES

Chloronema requires low light intensity and low temperature (4–15°C). Trichomes may accumulate on glass slides immersed in water samples rich in *Chloronema* filaments. Growth has not been obtained in liquid or on solid media.

DIFFERENTIATION OF THE GENUS *CHLORONEMA* FROM OTHER GENERA

The major differentiating characteristics of *Chloronema* in comparison with the other genera of multicellular filamentous green bacteria are given in Table B6.1.

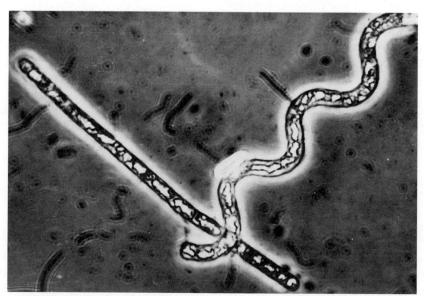

FIGURE B6.5. *C. giganteum* from a water sample of dimictic Lake Lesnaya Lamba. Phase-contrast light micrograph (×2000).

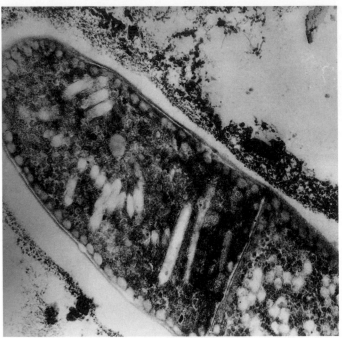

FIGURE B6.6. *C. giganteum* from a water sample of dimictic Lake Lesnaya Lamba. Electron micrograph of ultrathin section fixed with OsO$_4$ and glutaraldehyde ($\times$18,000).

TAXONOMIC COMMENTS

The lack of cultures with appropriate physiological data and the lack of any molecular phylogenetic data compromise our understanding of the validity of the genus and species designations of these organisms. The characters that distinguish them from the other chlorosome-containing genera (*Chloroflexus* and *Oscillochloris*) are few and variable enough to warrant caution in the taxonomy of this group. Its large size, presence of gas vacuoles, bacteriochlorophyll *d*, and planktonic distribution lead to its recognition *in situ*. As in the case of the other prominent uncultured filamentous phototrophs, the presence of *Chloronema* is conspicuous in the environments where it is found, and numerous investigators have seen it and studied it *in situ* (see Steenbergen and Korthals, 1982; Steenbergen et al., 1989; Eichler and Pfennig, 1990; Abella and Garcia-Gil, 1992 for a few examples). Abella and Garcia-Gil (1992) studied the distribution of the straight and spiral forms in several lakes documenting the importance of gas vacuoles and morphology to the migration of *Chloronema*. It is a major constituent of the plankton in lakes and hence may have a significant although as yet undetermined environmental impact. Because of its widespread occurrence and apparent importance, the genus should probably stand as is until culture or molecular data allow clarification of its taxonomy and phylogeny. In its present form, the description allows the recognition of the organisms *in situ* which in turn assures their continued study.

List of species of the genus Chloronema

1. **Chloronema giganteum** Dubinina and Gorlenko 1975, 515.[AL]
gi.gan.te′ um. M.L. neut. adj. *giganteum* gigantic.

Cells are cylindrical, 2–2.5 $\times$ 3.5–4.5 μm, and contain large centrally located gas vacuoles. Cells are combined in trichomes surrounded by a sheath (Fig. B6.6). Iron oxides may be accumulated in the sheath, sometimes in considerable amounts. Trichomes, together with the sheath, are 3–4 μm wide and up to 250 μm long. Waves in spiral trichomes are 15–20 μm long.

Motile by gliding. Multiplication of trichomes by separation of a part of the trichome (hormogonium) from the mother filament. Hormogonia actively separate themselves from the sheath. Bacteria are capable of phototaxis and chemotaxis.

Filaments appear yellowish green. Contain bacteriochlorophyll *d*. The photosynthetic structures are chlorosomes which are underlying and attached to the cytoplasmic membrane (Fig. B6.6).

Habitat: the metalimnion and upper hypolimnion of stratified fresh-water lakes with high ferrous iron content and no or low concentrations of H$_2$S. This species occurs together with many other species of the purple and green sulfur bacteria.

The mol% G + C of the DNA is: unknown.
Type strain: none.

Other Organisms

1. *Chloronema spiroideum* Dubinina and Gorlenko 1975, 516[AL].
Placement with the green filamentous anoxygenic bacteria is questionable as the presence or absence of chlorosomes and bacteriochlorophylls was not determined.

Genus III. **Heliothrix** *Pierson, Giovannoni, Stahl and Castenholz 1986, 354*[VP] *(Effective publication: Pierson, Giovannoni, Stahl and Castenholz 1985, 164).*

BEVERLY K. PIERSON AND RICHARD W. CASTENHOLZ

He′ li.o.thrix. Gr. masc. n. *helios* sun; Gr. fem. n. *thrix* hair; M.L. fem. n. *Heliothrix* sun hair.

Clearly septate filaments of indefinite length, unbranched and undifferentiated. Cells about 1.5 μm in diameter, much longer than broad (Fig. B6.7). Thin sheath present or not. Cell division by fission, in one plane only. **Gliding motility, rapid aggregation of suspended filaments into tight clumps, no flagella.** Cells granular, poly-β-hydroxybutyrate inclusions present; gas vacuoles unknown. Cells stain Gram-negative. **Bacteriochlorophyll *a* present, but no accessory bacteriochlorophylls or chlorosomes. Filaments orange; carotenoids abundant.** Internal membranous structures unknown.

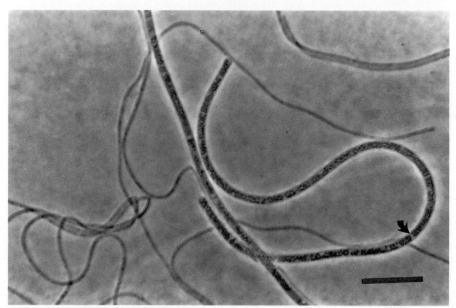

FIGURE B6.7. *H. oregonensis* from a population collected at Kah-nee-ta Hot Springs, Warm Springs, Oregon (U.S.A.). A cross-wall is indicated by an arrow. The narrower filaments are *C. aurantiacus* (J-10-fl) added to the collection for comparison. Phase-contrast micrograph. Bar = 10 µm. (Reprinted with permission from B.K. Pierson et al., Archives of Microbiology *142:* 164–167, 1984, ©Springer-Verlag.)

Aerotolerant; anaerobic growth uncertain; probably mainly aerobic **photoheterotrophic metabolism**, light-dependent uptake of acetate. Temperature range: 35–56°C; optimum temperature: 40–55°C (for acetate uptake).

Type species: **Heliothrix oregonensis** Pierson, Giovannoni, Stahl and Castenholz 1986, 354 (Effective publication: Pierson, Giovannoni, Stahl and Castenholz 1985, 164).

FURTHER DESCRIPTIVE INFORMATION

Only the type species has been described. The gliding rate is 0.1–0.4 µm/s (Pierson and Castenholz, 1971). Although this rate is 10 times faster than that of *C. aurantiacus*, it is still about 10 times slower than that of *C. aggregans*. Electron microscopy revealed no conspicuous or identifiable internal inclusions except granules of poly-β-hydroxybutyrate (Pierson et al., 1984a). En masse, *H. oregonensis* is bright orange. Bacteriochlorophyll *a* was identified by characteristic absorption spectra of cavitated cells and of methanolic extracts (Pierson and Castenholz, 1971; Pierson et al., 1985). Oxygenated derivatives of γ-carotene and glycosides of γ-carotene are the predominant carotenoids (K. Schmidt, personal communication; Pierson et al., 1984a).

H. oregonensis grows in a relatively few hot springs in Oregon where it forms a conspicuous orange mat, a few millimeters in vertical thickness, which generally overlies a greenish layer composed of several cyanobacteria (Castenholz, 1984a, b; Pierson et al., 1984a). The pH is about 8.5. In these springs, the population constitutes a spectacular enrichment, of which *H. oregonensis* may comprise >95%. The masses of filaments are commonly found as a "puffy" upper layer. An apparently identical organism is also common in several alkaline hot springs of Yellowstone National Park.

ENRICHMENT AND ISOLATION PROCEDURES

Filaments from the puffy upper layer can easily be collected by a syringe with a 17- or 18-gauge needle. Since the filaments also aggregate rapidly, the collected material can be washed and al-

lowed to aggregate repeatedly, thus further purifying the material by washing out unicellular cyanobacteria and other bacteria (Pierson et al., 1984a). The filaments are large enough to be seen as individuals at low power with a compound microscope or at high power with a dissecting microscope thus permitting "picking" individual filaments.

H. oregonensis has been brought into coculture with *Isosphaera pallida*, an aerobic, chemoheterotrophic bacterium (Giovannoni et al., 1987b). A single, axenic filament of *H. oregonensis* is manually removed from an agar plate, on which filaments have been gliding, and transferred to the edge of a colony of axenic *I. pallida* growing in the light at 45°C on IMC medium (Pierson et al., 1984a; Giovannoni et al., 1987b). Numerous attempts to grow *H. oregonensis* axenically have failed.

MAINTENANCE PROCEDURES

Cocultures of *H. oregonensis* and *I. pallida*, as well as axenic cultures of *I. pallida*, have been maintained by monthly transfers on slants or plates of agar-solidified IMC medium (pH 8.1). They are incubated at 45°C and under a moderate light intensity (3000 lux, cool white fluorescent lamps) in an atmosphere of 95% air and 5% CO_2. It is not known whether the coculture survives lyophilization or storage at the temperature of liquid nitrogen.

DIFFERENTIATION OF THE GENUS *HELIOTHRIX* FROM OTHER GENERA

The genus *Heliothrix* may be distinguished from the other filamentous phototrophic genera since *Heliothrix* possesses bacteriochlorophyll *a* only and no other accessory light-harvesting bacteriochlorophyll or chlorosomes. In the one known species of *Heliothrix*, the filament diameter is similar to the largest *Chloroflexus* filaments (1.5 µm) and the septa (cross-walls) may be seen easily in *Heliothrix* with phase-contrast microscopy (Fig. B6.7). The morphology at the level of light microscopy, coupled with its rapid gliding and clumping behavior make it superficially similar to *C. aggregans*. *Herpetosiphon* may resemble *Heliothrix* morphologically (Reichenbach and Golecki, 1975), but lacks bacteriochlorophyll.

TAXONOMIC COMMENTS

The complete nucleotide sequence of the 5S rRNA of *H. oregonensis* indicated a similarity of 0.767 compared with that of *Chloroflexus aurantiacus* (see Pierson et al., 1985). Although these two organisms are not very closely related, they are more closely related to each other than either is to any other phototrophs (see Pierson et al., 1985). The partial sequence of the 16S rRNA of *Heliothrix oregonensis* showed that it was a different genus from *Chloroflexus*, and was more closely related than *Herpetosiphon aurantiacus* to *Chloroflexus* (Weller et al., 1992; Ward et al., 1998). *Chloroflexus*-specific antibodies that reacted substantially with 13 different strains of *Chloroflexus* isolated from all over the world showed no cross-reactivity with *H. oregonensis* (Tayne et al., 1987).

List of species of the genus Heliothrix

1. **Heliothrix oregonensis** Pierson, Giovannoni, Stahl and Castenholz 1986, 354[VP] (Effective publication: Pierson, Giovannoni, Stahl and Castenholz 1985, 164).

    o.re.gon.en' sis. M.L. fem. adj. *oregonensis* pertaining to Oregon, a state in the U.S.A.

    The characteristics are the same as those described for the genus.

    Habitat: orange top layer over cyanobacterial mats in relatively few hot springs in Oregon (U.S.A.).

    The mol% G + C of the DNA is: unknown.

    Type strain: IS/F-1.

    GenBank accession number (16S rRNA): L04675.

    Additional Remarks: Readers are cautioned that the 16S rRNA sequence noted above is less than 1000 nucleotides in length.

Genus IV. **Oscillochloris** Gorlenko and Pivovarova 1989, 496[VP] (Effective publication: Gorlenko and Pivovarova 1977, 406)

OLGA I. KEPPEN, VLADIMIR M. GORLENKO AND BEVERLY K. PIERSON

Os.cil' lo.chlo' ris. L. n. *oscillans* oscillating; Gr. adj. *chloros* green; M.L. fem. n. *Oscillochloris* oscillating green (bacterium).

Cells arranged in uniseriately multicellular flexible filaments with gliding motility. Trichomes uniformly wide throughout their length, length variable, **nonbranching**, with or without sheath. Trichomes multiply by separation of short segments or single cells from the parent trichome. Cells usually contain gas vacuoles. Cells stain Gram-positive or Gram-negative. **Trichomes appear green or yellow-green. Contain bacteriochlorophylls *c* and *a* as well as carotenoids.** The light-harvesting structures of the photosynthetic apparatus are **chlorosomes. Photolithoheterotrophic or photolithoautotrophic under anaerobic conditions**; capable of photosynthesis in the presence of hydrogen sulfide during which they deposit elemental sulfur globules outside the cells. Some strains are capable of growth under microaerobic conditions.

Type species: **Oscillochloris chrysea** (Gicklhorn 1921) Gorlenko and Pivovarova 1989, 496 (Effective publication: Gorlenko and Pivovarova 1977, 406) (*Oscillatoria coerulescens* Gicklhorn 1921).

FURTHER DESCRIPTIVE INFORMATION

Morphology and Fine Structure   Width of trichomes varies from 0.8 to 1.4 μm in *O. trichoides* (Fig. B6.8). Individual cell lengths are 2.0–5.0 μm (Fig. B6.9). *O. chrysea* cells are wider (4.5–5.5 μm) (Fig. B6.10). Cell walls are formed by diaphragm-like narrowing and closing of cell septa and subsequent formation of the other components of the cell wall. Multiple incomplete septa are typical in *O. chrysea* (Fig. B6.11). Formation of porous thick cell septa results in the breakage of the trichome and its multiplication. Short sections of trichomes may separate at the end of a filament, or the entire filament may break up into segments resembling hormogonia. Long trichomes as well as "hormogonia" possess gliding motility at rates of 0.1–0.2 μm/s in *O. trichoides* and up to 7 μm/s in *O. chrysea*. The cell wall consists of multiple layers and appears similar to the Gram-negative type in *O. trichoides*, although analysis is not yet complete. Cells of *O.*

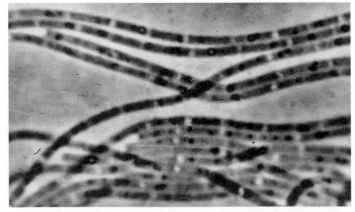

FIGURE B6.8. *O. trichoides* SR-1. Phase-contrast light micrograph (×2500). (Reprinted with permission from V.M. Gorlenko and S.A. Kozotkov, Izvestiya Akademii Nauk S.S.S.R. Seriya Biologicheskaya (Moskva) *6:* 848–857, 1979.)

chrysea stain Gram-positive. Cells usually contain gas vesicles localized along cell septa. Cells also contain poly-β-hydroxybutyrate granules. Globules and crystals of sulfur are deposited outside the cells in the immediate proximity of the trichomes.

Pigments Cells contain bacteriochlorophyll *c* as the predominant pigment and small amounts of bacteriochlorophyll *a*. The major carotenoids in *O. trichoides* are β- and γ-carotene. Colonies of the original isolate of *O. trichoides* (monoculture SR-1) grown at low oxygen concentrations either in dark or in the light appeared orange or pink suggesting that oxygen suppressed the synthesis of bacteriochlorophyll *c*, while the synthesis of carotenoids was unaffected. *O. trichoides* strain DG-6, however, is an obligate anaerobe. Under anaerobic phototrophic conditions, cell suspensions of all strains appear green, yellow-green, or olivegreen.

Metabolism, Nutrition, and Growth Conditions Cells of *O. trichoides* are capable of photolithoheterotrophic (preferred) and photolithoautotrophic growth under anaerobic conditions. Photoautotrophic growth utilizing H_2S or H_2 as electron donor is poor. The optimal concentration of sulfide is 500–700 mg/l (as $Na_2S \cdot 9H_2O$) but cells tolerate sulfide concentrations as high as 1.5 g/l. Sulfide is oxidized to S^0 only and deposited externally. Growth in the dark under aerobic (microaerobic) conditions was possible for monoculture SR-1 but not for strain DG-6. CO_2 is fixed by the Calvin Cycle in strain DG-6 (ribulosebisphosphate carboxylase and phosphoribulokinase activities) (Ivanovsky et al., 1999).

Cells of strain DG-6 grow best on media containing sulfide and bicarbonate as well as acetate, pyruvate, casein hydrolysate, or yeast extract. Formate, propionate, butyrate, malate, succinate, fumarate, citrate, and benzoate do not support growth. Tricarboxylic acid cycle and glyoxylate pathway do not function in *O. trichoides* DG-6. Nitrogen sources are ammonium salts, urea, and amino acids. N_2 did not support good growth but acetylene reduction assays with *O. trichoides* DG-6 were positive for N_2 fixation. All strains are mesophiles. The optimum temperature for growth of *O. trichoides* DG-6 is 28–30°C. Growth occurs from pH 6.8 to 9.0 but is optimal from 7.5 to 8.5.

Genetics and Molecular Data Keppen et al. (1994) determined that *O. trichoides* strain DG-6 had only 5% DNA–DNA hybridization with *C. aurantiacus* strain B-3. 5S rRNA sequence data were used to determine phylogenetic relationships with other bacteria and showed that *O. trichoides* was more closely related to *C. aurantiacus* than to any other phototrophs or other groups within the *Bacteria*. The 16S rRNA analysis showed an 82.5–86.5% sequence similarity with *Chloroflexus* (see Fig. B6.1).

Ecology Members of the genus *Oscillochloris* are typically benthic freshwater forms, occurring in algobacterial mats formed on the surface of sulfide-containing anaerobic mud sediments. *O. trichoides* may occur in cold and warm springs with temperatures up to 45°C (Lenkoran [Azerbaijan], Dagestan [Caucasus], Bailak region [Siberia], Kamchatka), and in shallow freshwater pools and brackish soda lakes in areas with high H_2S content or in microzones below purple sulfur bacteria. *Oscillochloris* occurs together with other anoxygenic phototrophs and filamentous cyanobacteria. *O. chrysea* develops in water bodies with high content of organic matter (e.g., domestic effluents).

ENRICHMENT AND ISOLATION PROCEDURES

The original monoculture of *O. trichoides* (SR-1) was isolated from algobacterial mats by serial dilutions in test tubes with a modified agar medium used for green sulfur bacteria (Pfennig, 1989) which contained the following: additional anhydrous Na_2SO_4, 330 mg/l; Difco yeast extract, 25 mg/l; tryptone, 25 mg/l; vitamin B_{12}, 20 µg/l; trace elements; agar, 0.2–0.4%; and $Na_2S \cdot 9H_2O$, 300 mg/l; at pH 7.5. Inocula were incubated at a light intensity of 2000 lux and at a temperature of 20–35°C. Colonies (green or yellow-green, of irregular fuzzy shape) were isolated from the highest dilutions and purified by further dilution series. Thus, associated phototrophic bacteria (mainly purple non-sulfur bacteria) were removed after 3 to 4 dilution series. Monocultures still contained (a) facultative anaerobic fermenting bacteria producing H_2 from lactate and (b) sulfur- and sulfate-reducing bacteria. Good growth of cultures free of other phototrophic bacteria was obtained in the medium described above.

The monoculture SR-1 of *O. trichoides* was subsequently lost and another strain (DG-6) was obtained in pure culture and studied by Keppen et al. (1993, 1994). Isolation and culture methods for *O. trichoides* strain DG-6 were reported by Keppen et al. (1993). *O. trichoides* DG-6 was isolated from a cyanobacterial mat at the outlet of the Talgy spring (Dagestan). The spring water salinity was 2 g/l and H_2S concentration was 5 mg/l. The pH was 8.0, Eh was +70 mV, and temperature was 25°C. The pure cultures were isolated by serial dilutions in tubes of DGN medium (Pierson and Castenholz, 1992) containing 0.8% agar. Nitrates were omitted from the DGN medium, and it was supplemented with 1 g/l KH_2PO_4 and the vitamin mix used by Madigan et al.

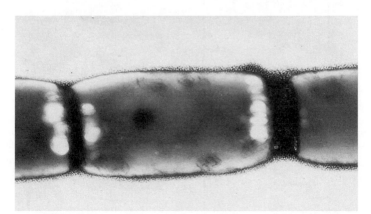

FIGURE B6.9. *O. trichoides* SR-1. Electron micrograph of section stained with phosphotungstic acid (×25,000). (Reprinted with permission from V.M. Gorlenko and S.A. Kozotkov, Izvestiya Akademii Nauk S.S.S.R. Seriya Biologicheskaya (Moskva) *6:* 848–857, 1979.)

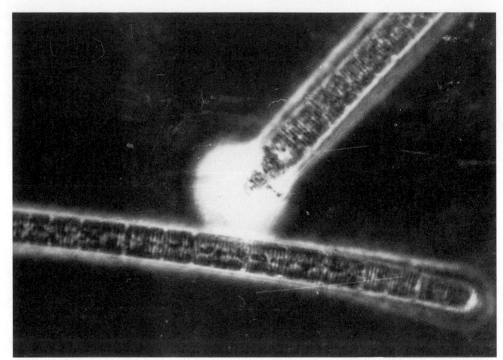

FIGURE B6.10. *O. chrysea* from a mud-water sample from a creek contaminated with domestic effluents. Phase-contrast light micrograph of sample with India ink (×2700). (Reprinted with permission from V.M. Gorlenko and N.N. Pivovarova, Izvestiya Akademii Nauk S.S.S.R. Seriya Biologicheskaya (Moskva) *23:* 396–409, 1977.)

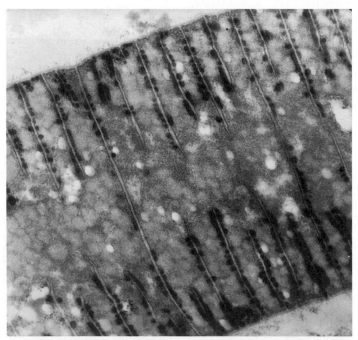

FIGURE B6.11. *O. chrysea* from a mud-water sample from a creek contaminated with domestic effluents. Electron micrograph of ultrathin section fixed with $KMnO_4$ (×18,000). (Reprinted with permission from V.M. Gorlenko and N.N. Pivovarova, Izvestiya Akademii Nauk S.S.S.R. Seriya Biologicheskaya (Moskva) *23:* 396–409, 1977.)

(1974). The medium also contained 0.2% bicarbonate, 0.02–0.1% sodium sulfide, and was adjusted to pH 8.2–8.4. Cultures were grown under anaerobic conditions at 28°C with illumination of 1000 lux. Larger yields were obtained by supplementing the medium with 0.2% acetate. *O. chrysea* did not grow in semisolid agar media. Therefore, it was not possible to obtain pure cultures.

MAINTENANCE PROCEDURES

Cultures for experimental studies and maintenance are grown in a medium modified from DGN and described by Keppen et al. (1994).* Conditions for anaerobic growth were obtained in completely filled vessels ranging from 50 to 1000 ml capacity or in 100-ml bottles containing 50 ml of medium and atmospheres of pure Ar, H_2, or N_2 (Keppen et al., 1994).

Cultures of *O. trichoides* may be stored at 4°C in liquid medium for 1–2 months at low light intensity.

DIFFERENTIATION OF THE GENUS *OSCILLOCHLORIS* FROM OTHER GENERA

The major differentiating characteristics of *Oscillochloris* in comparison with the other genera of multicellular filamentous bacteria are given in Table B6.1.

TAXONOMIC COMMENTS

Mesophilic representatives of the multicellular filamentous bacteria were first studied in monoculture, mixed culture, or in samples from the mass development in nature (Lauterborn, 1915; Gicklhorn, 1921; Gorlenko and Pivovarova, 1977; Gorlenko and Korotkov, 1979). The original description of the genus *Oscillochloris* was based on field observations of *O. chrysea*, which was never in culture. Culture SR-1, upon which the original description of *O. trichoides* was based, was not an axenic culture and was subsequently lost. *O. trichoides* strain DG-6 is the first strain to be successfully grown as an axenic culture and the first to be subjected to molecular analysis. The genus description provided here includes additional information gained from the study of the pure culture of *O. trichoides*, strain DG-6 (Keppen et al., 1993, 1994; Ivanovsky et al., 1999).

The availability of sequence data (5S and 16S rRNA) links *O. trichoides* DG-6 to the *Chloroflexus* and *Heliothrix* group of organisms (Fig. B6.1) (Keppen et al., 1994). The low DNA hybridization results reported by Keppen et al. (1994) with *Chloroflexus*, the higher mol% G + C content in *Oscillochloris*, and the low (82.5–86.5%) sequence similarity in 16S rRNA between *Chloroflexus* and *Oscillochloris* (see Fig. B6.1) indicate that different families are probably warranted for these two genera.

Although the original reference material for *O. chrysea* and *O. trichoides* culture SR-1 are no longer available, *O. trichoides* strain DG-6 is available in pure culture. *O. chrysea* has been validated as the type species for *Oscillochloris*. Culture SR-1 was not validated as the type strain for *O. trichoides*, and since it has been lost, the new strain DG-6 is now the type strain for this species (Rule 18f).

List of species of the genus Oscillochloris

1. **Oscillochloris chrysea** (Gicklhorn 1921) Gorlenko and Pivovarova 1989, 496[VP] (Effective publication: Gorlenko and Pivovarova 1977, 406) (*Oscillatoria coerulescens* Gicklhorn 1921).

   chry.se' a. L. fem. adj. *chrysea* gold-shining.

   The description of *O. chrysea* is based on natural collections. Since it did not grow on semisolid agar media, it was not possible to obtain pure cultures. Trichomes yellow-green, multicellular, flexible, 4.5–5.5 μm wide and up to 2.5 mm long. Individual cells vary in length from 3.5 to 7.0 μm but may be shorter (Figs. B6.10 and B6.11). Trichomes carry multiple transverse striae because of complete and incomplete cell septa (Fig. B6.11). The terminal cells of trichomes are smaller in diameter and possess a thicker cell envelope. Occasionally, the ends of trichomes carry slimy dome-shaped structures used for the attachment to solid particles (Fig. B6.10). Division and formation of septa occur by centripetal infolding of the cytoplasmic membrane and inner peptidoglycan layer of the cell wall. The cell wall has no pronounced outer membrane and contains a thicker nonuniform peptidoglycan layer (up to 25.2 nm). Trichomes have no sheath but are sometimes surrounded by a microcapsule 25–30 nm thick. Stain Gram-positive. Motile by gliding at a rate of about 7 μm/s. Multiplication of trichomes by hormogonia. More frequently, the end section of trichomes 15 μm or more long is separated, or trichomes break up in the middle. Less frequently, trichomes break up into short sections of 15–30 μm. Some cells contain a few gas vacuoles localized along the cell septa.

   Individual trichomes appear yellow-green in the presence or absence of oxygen. Cell suspension appears dark green. Trichomes appear blue if observed under the microscope in reflected light.

   Chlorosomes, the photosynthetic structures, are localized on the cytoplasmic membrane along complete and incomplete septa (fig. B6.11). Chlorosomes are absent from the cytoplasmic membrane parallel to the long axis of the filaments.

   The major photosynthetic pigment is bacteriochlorophyll *c*; bacteriochlorophyll *a* is present in smaller amounts. Carotenoid pigments are present.

   Physiology: phototrophic, tolerant of hydrogen sulfide (500 mg/l $Na_2S \cdot 9H_2O$), probably capable of aerobic dark metabolism. Storage materials: poly-β-hydroxybutyrate, glycogen, polymetaphosphates. Habitat: surface of hydrogen sulfide-containing mud of freshwater bodies. Development at pH 7.5–8.5 and at 10–20°C. The following other bacteria were observed in the same habitat: *Oscillatoria* sp., flexibacteria, *Beggiatoa alba* and, occasionally, *Amoebobacter roseus*, *Thiodictyon bacillosus*, and *Chlorobium limicola*.

   The mol% G + C of the DNA is: unknown.

   Type strain: no pure culture.

2. **Oscillochloris trichoides** (Lauterborn 1915) Gorlenko and Korotkov 1989, 496[VP] (Effective publication: Gorlenko and Korotkov 1979, 855) (*Oscillatoria trichoides* (Szafer) Lauterborn 1915, 436).

   tri.cho' i.des. Gr. n. *thrix* hair; Gr. suffix adj. *-oides*; M.L. fem. adj. *trichoides* hairlike (bacteria).

   The characteristics are the same as described for the genus with some variations between monoculture SR-1 and strain DG-6. Trichomes straight or wavy, yellow-green, and 0.8–1.0

---

*Medium for growth of *Oscillochloris trichoides* contains (per liter distilled H_2O): EDTA, 0.02 g; $CaSO_4 \cdot 2H_2O$, 0.06 g; $MgSO_4 \cdot 7H_2O$, 0.10 g; NaCl, 0.008 g; Na_2HPO_4, 0.11 g; NH_4Cl, 0.2 g; $FeCl_3$ solution (100 mg/l), 3.0 ml; micronutrient solution, 0.5 ml; glycylglycine, 0.8 g; $NaHCO_3$, 1.0 g; acetate, 1.0 g; $Na_2S \cdot 9H_2O$, 0.5 g; vitamin solution, 1.0 ml. Micronutrient solution contains (per liter distilled H_2O): H_2SO_4, 0.5 ml; $MnSO_4 \cdot H_2O$, 2.28 g; $ZnSO_4 \cdot 7H_2O$, 0.5 g; H_3BO_3, 0.5 g; $CuSO_4 \cdot 5H_2O$, 0.025 g; $Na_2MoO_4 \cdot 2H_2O$, 0.025 g; $CoCl_2 \cdot 6H_2O$, 0.045 g. Vitamin solution contains (per 100 ml distilled H_2O): thiamine HCl, 10.0 mg; biotin, 0.5 mg; riboflavin, 3.0 mg; folic acid, 5.0 mg.

μm wide (DG-6) or 1.2–1.4 μm wide (SR-1) (Figs. B6.8 and B6.9). Individual cells in a filament vary in length from 2.0 to 5.0 μm. There may or may not be a thin sheath, or trichomes are surrounded by a thin slime layer (microcapsule). Cells are motile by slow gliding at a rate of 0.2 μm/s (SR-1) and multiply by separation of a short section of the filament (hormogonium) or of an individual cell from the mother trichome. In older cultures, gas vacuoles are formed that are localized along the cell septa (Fig. B6.9). The cell wall appears to be of the Gram-negative type. The primary pigment-bearing structures are chlorosomes. Trichomes deposit poly-β-hydroxybutyrate inside the cells. Sulfur is not accumulated inside the cells. Globules and crystals of sulfur are present outside the cells in the immediate proximity of the trichomes. Cell suspensions appear dark green when grown under anaerobic conditions in the light and appear orange if grown under microaerobic conditions (SR-1 only). In stab cultures grown in agar medium in the light, dark green colonies are formed; colonies in the upper part of the tube may appear orange (SR-1). Strain DG-6 does not produce colonies in the upper part of the tube. The *in vivo* absorption spectrum of cell suspensions (strain DG-6) has maxima at 333, 436, 748, and 852 nm. The major photosynthetic pigment is bacteriochlorophyll *c*; a small amount of bacteriochlorophyll *a* is also present. The major carotenoids are β- and γ-carotene.

Cultures of strain DG-6 grow only in the light under anaerobic conditions. Photoheterotrophic, but capable of slow photoautotrophic development with hydrogen or sulfide (optimum initial concentration of sulfide of 500–700 mg $Na_2S \cdot 9H_2O/l$). Sulfide oxidized to sulfur and deposited externally. Grow best on media containing bicarbonate and sulfide and supplemented with acetate, pyruvate, casein hydrolysate, or yeast extract. Sources of sulfur include sulfide, cysteine, cystine, but not sulfate. Nitrogen sources include ammonium salts, urea, glycine, glutamate, glutamine, asparagine, and caseine hydrolysate. Slow growth with N_2. Optimum development at pH 7.5–8.5. Mesophilic. Growth factor requirements are satisfied by yeast extract or a mixture of vitamins (thiamine, biotin, riboflavin, and folic acid).

Habitat: mats on the surface of mud containing hydrogen sulfide in freshwater pools or springs. These bacteria occur together with the following other phototrophic organisms: *Chlorobium* sp., purple sulfur bacteria species, cyanobacteria (*Oscillatoria* sp. and *Pseudanabaena* sp.), and with colorless sulfur bacteria.

The mol% G + C of the DNA is: 57.3–59.2 (melting point).

Type strain: DG-6.

GenBank accession number (16S rRNA): AF093427.

Additional Remarks: Monoculture SR-1 was not in pure culture and was subsequently lost after publication of original description. Strain DG-6 is in pure culture in Moscow State University Culture Collection (KM MGU 327) and VKM B-2210 in Russian Collection of Microorganisms, Pushchino, Moscow Region.

Other Organisms

Mesophilic organisms bearing strong resemblance to the genus *Oscillochloris* and in particular the species *O. chrysea* have been observed repeatedly in hypersaline microbial mats containing sulfide. Stolz (1990) published electron micrographs of sections of hypersaline microbial mats from Laguna Figueroa, Baja California N., Mexico, showing *Oscillochloris*-like cells with abundant internal membranes lined with chlorosomes. The filaments were observed in mats with other phototrophs such as cyanobacteria, purple sulfur bacteria, and *Chloroflexus*-like organisms. Organisms nearly identical to the ones described by Stolz were reported by Pierson (1994) and Pierson et al. (1994) in electron micrographs of enrichment cultures from hypersaline mats from Guerrero Negro, Baja California, Mexico. The mats from which the enrichments came contained other phototrophs such as cyanobacteria and purple sulfur bacteria. Pierson et al. (1994) studied the pigmentation and physiology as well as the ultrastructure of these microbes and found them very tolerant of high sulfide and capable of photoautotrophy. Pierson maintained two of these organisms in mixed cultures for several years and currently has one viable mixed culture. It remains puzzling why these mesophilic filamentous phototrophs are so difficult to grow in pure culture.

Order II. "Herpetosiphonales"

RICHARD W. CASTENHOLZ

The order "*Herpetosiphonales*" and family "*Herpetosiphonaceae*", under the phylum *Chloroflexi*, are represented by one genus (*Herpetosiphon*). Other genera that may eventually be added to this order are "flexibacteria" that also lack bacteriochlorophylls, are aerobic chemoheterotrophs, and are organisms that show a close phylogenetic relationship to members of the phototrophic *Chloroflexus* group of the order "*Chloroflexales*"/family "*Chloroflexaceae*" based on 16S rDNA sequence comparisons and by the use of sequence analyses of other molecules. Other principal characters that have been a consistent feature of this order/family and the "*Chloroflexales*"/ "*Chloroflexaceae*" are as follows: the lack of the LPS outer membrane, the presence of a polysaccharide, peptidoglycan complex in which *meso*-diaminopimelic acid is replaced by L-ornithine, and usually by gliding motility. It is likely that these aerobic heterotrophic organisms have evolved, by loss of a photosystem from the "*Chloroflexaceae*", which also show in some strains the potential for facultative aerobic chemoheterotrophic growth.

Family I. "Herpetosiphonaceae"

Genus I. Herpetosiphon *Holt and Lewin 1968, 2408*[AL]

JOHN G. HOLT AND RICHARD W. CASTENHOLZ

Her.pet.o'si'phon. Gr. n. *herpeton* gliding animal, reptile; Gr. masc. n. *siphon* tube or cylinder; M.L. masc. n. *Herpetosiphon* gliding cylinder or tube.

Herpetosiphon is represented currently by two species that are **unbranched, flexible filaments or rods**, 0.5–1.5 × 5–150 µm or more (to several mm), consisting of individual cells 2–3 µm long. Appearance of short, transparent "sheaths" extending from ends of some filaments are termed sleeves by Reichenbach (1992) and are characteristic of *Herpetosiphon*. They may not be true sheaths, but rather the "ghosts" of necridial (dead) cells of the filament (Reichenbach, 1992). Since gliding motility of whole filaments occurs, it is unlikely that a true sheath could exist at that point. Resting stages are unknown. Gram-negative. Cells contain **peptidoglycan in which *meso*-diaminopimelic acid is replaced by L-ornithine. In addition, the outer membrane is lacking (together with its lipopolysaccharide component). Motile by gliding.**

The mol% G + C of the DNA is: 48–50.

Type species: **Herpetosiphon aurantiacus** Holt and Lewin 1968, 2408.

FURTHER DESCRIPTIVE INFORMATION

The two species are aerobic chemoheterotrophs with O_2 as terminal electron acceptor in respiration. Colonies or cell masses are usually yellow to orange in color. The predominant pigment is a 4-oxo-α-carotene with a hydroxyl at C-1' to which a disaccharide (mainly glucose) is attached via a glycosidic bond (Reichenbach, 1992). Phylogenetic analyses are required for identification. Genus *Herpetosiphon* shows relationship to *Chloroflexus* and relatives, and future inclusions into this genus or to future genera in this order/family should agree. The two species represented here are both of freshwater origins. Three additional species included in the 1989 edition are marine, but have been removed to the genus *Lewinella* (see below).

ENRICHMENT AND ISOLATION PROCEDURES

Herpetosiphon may be isolated from freshwater and soil habitats, and presumably from decaying organic matter (e.g., rotting wood, compost, dung), and from activated sludge of sewage plants, and from lake surfaces with proximal soil or sewage input (see Reichenbach, 1992). The species transferred to *Lewinella* are marine and will not be discussed here.

Isolation is primarily through self-isolation of gliding filaments from a central inoculum or repeated isolation of "swarms" of filaments on agar-solidified media (see Reichenbach, 1992). Plain water agar (1.5%) may be used with the addition of 0.1% $CaCl_2·2H_2O$; pH adjusted to 7.2. Filter-sterilized cycloheximide (25 µm/ml in medium) is added after autoclaving. Since media

with low organic content are preferred, BG-11 medium may be used. Another commonly used medium is VY/2 (Baker's yeast [fresh weight of yeast cake], 0.5%; $CaCl_2·2H_2O$, 0.1%; vitamin B_{12}, 0.5 µg/ml; agar, 1.5%; pH adjusted to 7.2 [see Reichenbach, 1992]).

TAXONOMIC COMMENTS

The genus *Herpetosiphon* in the 1st edition of *Bergey's Manual of Systematic Bacteriology* included five species, all of which had been validated. The original description and naming by J. Holt and R.A. Lewin in 1968 was for *H. aurantiacus* strain ATCC 23779. Subsequently Lewin (1970) published a description of the four other species of this genus, all of which had phenotypic similarities to *H. aurantiacus*. However, 16S rDNA sequence analyses (>1417 nucleotides) by Sly et al. (1998) have shown that only *H. geysericola* Lewin (1970) shows a close relationship to the type species, *H. aurantiacus* (96.4% similarity), indicative of separate species status within this genus. The other three species included in the 1st edition of the *Manual* are closely related to each other but not to the *H. aurantiacus/H. geysericola* clade. These three have been proposed as species of a new genus, *Lewinella* gen. nov., that is allied with the *Flexibacter/Bacteroides/Cytophaga* (FBC) phylum (Sly et al., 1998). These former *Herpetosiphon* species [i.e., *H. cohaerens* Lewin (1970), *H. persicus* Lewin (1970), and *H. nigricans* Lewin (1970)] have now been renamed as species (comb. nov.) of the newly described genus *Lewinella* Sly, Taghavit and Fegan 1998. They will not be included in the following discussion of *Herpetosiphon*. Sequence analyses of 16S rRNA has consistently shown that the type species (*Herpetosiphon aurantiacus* strain ATCC 23779) is closely related to *Chloroflexus aurantiacus* (see phylogenetic tree under "*Chloroflexales*"). This relationship is supported by similar biochemical and structural properties, such as the lack of an LPD outer membrane, and the replacement of *meso*-diaminopimelic acid (*meso*-DAP) by ornithine in both groups. This *Chloroflexus/Herpetosiphon* line is one of the first evolutionary branches of the *Bacteria*, and consistently shows itself to be the earliest branch with photosynthetic members. Other photosynthetic genera of this clade have been described earlier. The first described species of *Chloroflexus* (*C. aurantiacus* Pierson and Castenholz 1974a) is a true thermophile, but several other mesophilic, phototrophic types have now been discovered, so that confusion between *Herpetosiphon* and *Chloroflexus* is very likely if only partial phenotypic characteristics are used for identification.

List of species of the genus Herpetosiphon

1. **Herpetosiphon aurantiacus** Holt and Lewin 1968, 2408.[AL]
   au.ran.ti'a.cus. M.L. neut. n. *aurantium* specific name of the orange; M.L. masc. adj. *aurantiacus* orange-colored.

   Cells, 1.0–1.5 × 5–10 µm, in unbranched, flexible filaments which may exceed 500 µm in length and usually have

sheath-like "sleeves" (see Reichenbach, 1992). Gliding motility occurs.

Starch is hydrolyzed, cellulose is not. A crystalline suspension of tyrosine is degraded, with formation of a reddish-brown pigment. Catalase positive. Growth has apparently not been accomplished in defined medium (Reichenbach, 1992).

The definition was based on three isolates of E.E. Jeffers obtained from a slime coat of *Chara* sp. (Phylum: Charophyta) from Birch Lake, Minnesota.

The mol% G + C of the DNA is: 48.1 (Bd).

Type strain: ATCC 23779, DSMZ 785.

GenBank accession number (16S rRNA): M34117.

2. **Herpetosiphon geysericola** Lewin 1970, 517.[AL]

gey.ser.i' co.la. Icelandic n. *geysir* geyser; L. n. *cola* dweller; M.L. n. *geysericola* hot spring dweller.

Flexible filaments or rods 0.5 × 10–150 µm (or more) with sheaths or "sleeves". Starch is hydrolyzed, cellulose is digested. Carboxymethyl cellulose is depolymerized. Gliding motility occurs.

Isolated in the immediate vicinity of a hot spring in Baja California.

The mol% G + C of the DNA is: 48.5 (Bd).

Type strain: ATCC 23076, DSMZ 7119.

GenBank accession number (16S rRNA): AF039293.

Additional Remarks: The original description of this species assumed synonymy with *Phormidium geysericola* Copeland 1936, a presumed cyanobacterium. The reason for assuming *H. geysericola* to be the organism that Copeland described was the morphological similarities with field material alone. The only culture strain of *H. geysericola* was not from a hot spring at 60–80°C, but from the surroundings of a hot spring. There is no evidence that it is thermotolerant or thermophilic. The ATCC lists a growth temperature of 30°C, which is essentially impossible for any bacterium capable of growth at 60°C or higher. The upper limit for *H. geysericola* is 38–40°C (see Reichenbach, 1992). Therefore, the synonym with *Phormidium geysericola* Copeland should be dropped, since the true identity of Copeland's organism is not known. A good possibility, however, is that it was a *Chloroflexus* sp. that has very similar morphological properties. *Chloroflexus* is well known from hot springs in the range of 60 to ~70°C (see above).

Phylum BVII. Thermomicrobia *phy. nov.*

GEORGE M. GARRITY AND JOHN G. HOLT

Ther.mo.mi.cro'bi.a. M.L. fem. pl. n. *Thermomicrobiales* type order of the phylum; dropping ending to denote a phylum; M.L. fem. pl. n. *Thermomicrobia* the phylum of *Thermomicrobiales.*

The phylum *Thermomicrobia* consists of a single known representative that branches deeply in the major reference trees and is distantly related to the *Chloroflexi.* Gram negative, short, irregularly shaped nonmotile rods. Nonsporulating. No diamino acid present in peptidoglycan in significant amount. Hyperthermophilic, optimum growth temperature 70–75°C. Obligately aerobic and chemoorganotrophic.

Type order: **Thermomicrobiales** ord. nov.

Class I. **Thermomicrobia** *class nov.*

GEORGE M. GARRITY AND JOHN G. HOLT

Ther.mo.mi.cro'bi.a. M.L. fem. pl. n. *Thermomicrobiales* type order of the class; dropping ending to denote a class; M.L. fem. pl. n. *Thermomicrobia* the class of *Thermomicrobiales.*

Description is the same as for the genus *Thermomicrobium.*

Type order: **Thermomicrobiales** ord. nov.

Order I. **Thermomicrobiales** *ord. nov.*

GEORGE M. GARRITY AND JOHN G. HOLT

Ther.mo.mi.cro.bia'les. M.L. neut. n. *Thermomicrobium* type genus of the order; *-ales* ending to denote an order; M.L. fem. pl. n. *Thermomicrobiales* the order of *Thermomicrobium.*

Description is the same as for the genus *Thermomicrobium.*

Type genus: **Thermomicrobium** Jackson, Ramaley and Meinschein 1973, 34.

Family I. **Thermomicrobiaceae** *fam. nov.*

GEORGE M. GARRITY AND JOHN G. HOLT

Ther.mo.mi.cro.bi.a'ce.ae. M.L. neut. n. *Thermomicrobium* type genus of the family; *-aceae* ending to denote a family; M.L. fem. pl. n. *Thermomicrobiaceae* the family of *Thermomicrobium.*

Description is the same as for the genus *Thermomicrobium.*

Type genus: **Thermomicrobium** Jackson, Ramaley and Meinschein 1973, 34.

Genus I. **Thermomicrobium** *Jackson, Ramaley and Meinschein 1973, 34*[AL]

JEROME J. PERRY

Ther.mo.mi.cro' bi.um. Gr. n. *therme* heat; Gr. adj. *micros* small; Gr. n. *bios* life; M.L. neut. n. *Thermomicrobium* indicates a small organism living in hot environments.

Short, irregularly shaped rods, 1.3–1.8 µm in diameter and 3.0–6.0 µm in length. The pleomorphic forms are dumbbell shaped or appear irregular in diameter and occur singly or in pairs. **Neither resting stages nor endospores are formed.** Gram negative. **No peptidoglycan diamino acid occurs in the cell walls in significant amounts.** Nonmotile. Obligately **aerobic. Optimum temperature for growth, 70–75°C**; maximum, 80°C, minimum, 45°C. Optimal pH, 8.2–8.5, but good growth occurs between 7.5 and 8.7. Chemoorganotrophic, having a strictly respiratory type of metabolism with oxygen as the terminal electron acceptor. Catalase positive. Colonies have a rose-pink color. Maximum growth occurs on a medium consisting of yeast extract and peptone (0.5% each). **Growth does not occur on glucose.** *n*-Alkanes **are not utilized. Generation time, 5.5 h.** Isolated from a hot spring in Yellowstone National Park, U.S.A.

The mol% G + C of the DNA is: 64.

Type species: **Thermomicrobium roseum** Jackson, Ramaley and Meinschein 1973, 34.

FURTHER DESCRIPTIVE INFORMATION

Although the original description of the species noted the appearance of possibly motile cells, there is no evidence of flagella in electron micrographs and motility is not presently included in the description.

Electron micrographs of longitudinal sections of the organism indicate that it has a layered cell wall similar to that of Gram-negative bacteria. The outermost layer appears to be a repeating structure covering the cell surface in a regular mosaic pattern (Ramaley et al., 1978). Attempts to isolate a peptidoglycan from

T. roseum by various techniques (Merkel et al., 1980) revealed that the cell wall material obtained was unlike that from other Gram-negative thermophilic bacteria (Merkel et al., 1978a). The purified wall fraction from *T. roseum* is composed mainly of a protein with a monomeric molecular weight of 75,000. The amino acid composition of the cell wall fraction is shown in Table B7.1, column A, and more closely resembles the amino acid composition of subunit cell wall polymers (Thornley, 1975; Mescher and Strominger, 1976) than the typical cell wall of Gram-negative thermophilic organisms (Merkel et al., 1978a). The major cell wall protein has been purified electrophoretically and the amino acid composition (Table B7.1, column B) indicates high concentrations of proline, glutamic acid, glycine, and alanine. There are some minor differences in composition and in the molar ratios of the amino acids when compared to the purified cell wall. The obvious similarities do indicate that the monomeric protein occurs as a major component of the outer envelope. The role of this protein in the stability of the organism is not known at the present time. The atypical nature of the cell wall of *T. roseum* is a distinct feature and probably of major taxonomic significance. Other Gram-negative, obligately thermophilic bacteria studied appear to have a more uniform peptidoglycan composition, in which pentaamines and hexamine are present (Hamana et al., 1990).

Nutritional studies have been accomplished with Allen's salts medium* (Allen, 1959) and in a salts medium devised by Castenholz (1969) with nitrate as a source of inorganic nitrogen.† *T. roseum* grew better with increasing concentrations of yeast extract from 0.1 to 0.7% with an equal amount of added tryptone. Total growth decreased at concentrations above 0.7%. In media with defined carbon sources, e.g., glycerol or succinate, the organism did not grow unless glutamate was present, and then solely in Castenholz's medium. When complex media were employed, better growth was attained with Allen's salts. A possible vitamin requirement has not been completely assessed.

The pink pigment of *T. roseum* is a carotenoid with absorbance maxima similar to torulene and 3,4-dehydrolycopene (Jackson et al., 1973).

ENRICHMENT AND ISOLATION PROCEDURES

Only one strain of *T. roseum* has been isolated. Isolation was accomplished from Toadstool Spring in Yellowstone National Park; a sample of mat and water taken near the source of the

TABLE B7.1. Amino acid analysis of the cell wall and cell wall protein of *Thermomicrobium roseum*[a]

Amino Acid	Cell Wall (A)		Cell Wall Protein (B)	
Threonine	8.1[b]	0.5[c]	3.6[b]	0.3[c]
Serine	6.2	0.4	4.0	0.3
Proline	8.1	0.5	14.0	1.2
Muramic acid	3.1	0.2	0	0
Glucosamine	0	0	0	0
Glutamic acid	12.8	0.8	11.9	1.0
Glycine	19.3	1.2	33.8	2.9
Alanine	16.8	1.0	11.5	1.0
Valine	4.0	0.2	1.8	0.2
Diaminopimelic acid	1.2	0.1	0	0
Leucine	6.2	0.4	3.2	0.3
Isoleucine	0	0	1.8	0.2
Tyrosine	4.7	0.3	2.2	0.2
Galactosamine	3.1	0.2	0	0
Histidine	1.2	0.1	3.2	0.3
Lysine	0	0	Tr	Tr
Arginine	3.7	0.2	Tr	Tr
Phenylalanine	0	0	1.8	0.2
Ornithine	0	0	7.2	0.6
Tryptophan	nd	nd	nd	nd

[a]Symbols: Tr, trace; nd, not determined. (Reproduced with permission from G.J. Merkel, D.R. Durham and J.J. Perry, Canadian Journal of Microbiology 26: 556–559, 1980, © National Research Council of Canada.)

[b]Percentage of total micromoles of amino acid.

[c]Molar ratio with alanine equal to 1.

*Allen's basal salts (mg/l deionized distilled water): $(NH_4)_2SO_4$, 1300; KH_2PO_4, 280; $MgSO_4 \cdot 7H_2O$, 247; $CaCl_2 \cdot 2H_2O$, 74; $FeCl_3 \cdot 6H_2O$, 19; $MnCl_2 \cdot 4H_2O$, 1.8; $Na_2B_4O_7 \cdot 10H_2O$, 4.4; $ZnSO_4 \cdot 7H_2O$, 0.22; $CuCl_2 \cdot H_2O$, 0.05; $Na_2MoO_4 \cdot 2H_2O$, 0.03; and VCl_2, 0.03; pH adjusted to 2.0 with H_2SO_4 during storage to prevent precipitation.

†Castenholz's basal salts, × 10 stock solution (per liter of distilled water): nitrilotriacetic acid, 1.0 g; $CaSO_4 \cdot 2H_2O$, 0.06 g; $MgSO_4 \cdot 7H_2O$, 1.0 g; NaCl, 0.08 g; KNO_3, 1.03 g; $NaNO_3$, 6.89 g; Na_2HPO_4, 1.11 g; $FeCl_3$ solution (0.28 g/l of distilled water), 10.0 ml; and micronutrient solution, 10.0 ml. The micronutrient solution contains (per liter of distilled water): H_2SO_4, 0.05 ml; $MnSO_4 \cdot H_2O$, 2.2 g; $ZnSO_4 \cdot 7H_2O$, 0.5 g; H_3BO_3, 0.5 g; $CuSO_4$, 0.016 g; $Na_2MoO_4 \cdot 2H_2O$, 0.025 g; and $CoCl_2 \cdot 6H_2O$, 0.046 g. The Castenholz × 10 basal salts stock is adjusted to pH 8.2 with 1 N NaOH and autoclaved.

TABLE B7.2. Differential characteristics of *Thermus* sp., *Thermomicrobium roseum*, and unclassified, Gram-negative, nonsporulating thermophilic rods[a]

Characteristic	*Thermus* Strains	*Thermomicrobium roseum*	Unclassified Strains[b]	
			Group A	Group B
Peptidoglycan diamino acid present in significant amounts	Ornithine	None	Diaminopimelic acid (DAP)	DAP plus lysine or ornithine
n-Alkane utilization	−	−	+	+
Growth on glucose	+	−	+	−
Growth on complex media	+	+	+	−
Generation time	20–60 min	5.5 h	1–2 h[c]	5–6 h
Optimum growth temperature, °C	60–70	70–75	55–60[c]	60
Mol% G + C of DNA	61–71	64	52–58	68–72

[a]For symbols see standard definitions.

[b]Merkel et al., 1978a. Also see Other Organisms at end of article.

[c]Glucose as substrate.

spring (pH 8.9) at 74°C yielded compact pink colonies after 1 week incubation on plates at 70°C, using a medium consisting of 0.1% yeast extract, 0.1% tryptone, and the mineral salts mixture described by Allen (1959). The other colonies that grew on primary isolation were of the genus *Thermus*. A pure culture was obtained by continued restreaking and incubation for 5 days.

DIFFERENTIATION OF THE GENUS *THERMOMICROBIUM* FROM OTHER GENERA

Under phase-contrast microscopy, *Thermomicrobium* cells are quite small and pleomorphic, whereas *Thermus* cells appear as long, thin, regular rods. *Thermomicrobium* can also be distinguished from the genus *Thermus* on the basis of generation time (5–6 h for *Thermomicrobium*, 1 h for *Thermus*), nutrition and cell wall composition. *Thermus* strains have few growth factor requirements and grow on a wide array of sugars and organic acids as carbon sources. The genus *Thermomicrobium* requires glutamate (and possibly other factors) and is more limited in substrate range. For details, see Tables B7.2 and B7.3.

TAXONOMIC COMMENTS

The only species currently included in this genus is *T. roseum*, and this species is represented by only a single strain. Another species, *T. fosteri* represented by ATCC strain 29033, was initially placed in the genus (Phillips and Perry, 1976), but subsequent study of *T. roseum* (Merkel et al., 1980) has indicated an atypical cell wall composition which *T. fosteri* does not have. *T. fosteri* is more closely related to other hydrocarbon-utilizing thermophiles (Merkel et al., 1978a) than it is to *T. roseum*.

TABLE B7.3. Physiological characteristics of *Thermomicrobium roseum*[a]

Characteristic	Reaction or Result
Pink pigmentation	+
Growth on *n*-heptadecane	−
Substrate utilization[b]:	
In Castenholz's salts medium	
D-Fructose, D-glucose, glycerol, sodium succinate, mannitol, sucrose, sodium acetate, sodium citrate, peptone, brain heart infusion, Trypticase soy broth, or tryptone	−
In Castenholz's salts with 0.2% glutamate	
Glycerol, sucrose, nutrient broth, or yeast extract	+
Sodium glutamate, or casein hydrolysate	Weak
In Allen's salts	
Peptone, casein hydrolysate, brain heart infusion, nutrient broth, Trypticase soy broth, tryptone, yeast extract	+
Susceptible to the following antibiotics (amt/disk):	
Chloramphenicol, 30 µg; erythromycin, 15 µg; kanamycin, 5 µg; neomycin, 5 µg; novobiocin, 5 µg; penicillin 2 U; streptomycin, 2 µg; tetracycline, 5 µg	+

[a]For symbols see standard definitions.

[b]Substrate added at 0.2%.

A number of phylogenetic studies confirm the placement of *Thermomicrobium* as a deeply branched, green, non- sulfur bacterium (Van den Eynde et al., 1990; Gupta et al., 1997; Haas and Brown, 1998).

List of species of the genus Thermomicrobium

1. **Thermomicrobium roseum** Jackson, Ramaley and Meinschein 1973, 34.[AL]

    ro' se.um M.L. neut. adj *roseum* rose colored.

    The description is as given for the genus. The cell wall composition is given in Table B7.1.

Isolated from a hot spring in Yellowstone National Park.

The mol% G + C of the DNA is: 64 (T_m).

Type strain: ATCC 27502.

GenBank accession number (16S rRNA): M34115.

Other Organisms

Several hydrocarbon-utilizing, obligately thermophilic, Gram-negative nonsporulating rods have been isolated, including ATCC 29033 which was initially named *Thermomicrobium fosteri*. All of these organisms possess a typical peptidoglycan and therefore differ from *Thermomicrobium roseum*. They are presently assigned to the genus *Thermoleophilum* (Zarilla and Perry, 1984).

All of the strains are capable of growth on normal alkanes with a carbon chain length from C_{13} to C_{20}. One group of strains contains diaminopimelic acid (DAP) as the major diamino acid in the peptidoglycan, has a mol% G + C from 52 to 58, forms nonpigmented colonies, has a generation time of 1.8–3.7 h on n-heptadecane or 0.7–1.8 h on glucose, grows on complex media, and has an optimum growth temperature of 55–65°C. A second group, including the strain previously named *T. fosteri*, contains DAP plus lysine or ornithine as a peptidoglycan constituent, has a generation time of 4–6 h on n-heptadecane and an optimum temperature for growth at 60°C. Members of this group cannot utilize sugars, fatty acids, or any of the carbon sources tested except the C_{13} to C_{20} normal alkanes. Complex media also fail to support growth. The mol% G + C ranges from 68 to 72 and only one strain (*T. fosteri*) forms pigmented colonies (light pink).

For further descriptive information, see the articles by Merkel et al., 1978a, b.

Phylum BVIII. Nitrospirae *phy. nov.*

GEORGE M. GARRITY AND JOHN G. HOLT

Ni.tro.spi' rae. M.L. fem. n. *Nitrospira* genus of the phylum; ending to denote a phylum; M.L. fem. pl. n. *Nitrospirae* the phylum of *Nitrospira.*

The phylum *Nitrospirae* is based mainly on phylogenetic grounds. At present, it consists of a single class, order, and family of *Bacteria* and environtaxa that branch deeply in the major reference trees; member taxa consistently group together. Gram-negative, curved, vibrioid or spiral-shaped cells. Metabolically diverse, most genera are aerobic chemolithotrophs including nitrifiers, dissimilatory sulfate reducers, and magnetotactic forms. One genus (*Thermodesulfovibrio*) is thermophilic, and obligately acidophilic and anaerobic.

Class I. "Nitrospira"

Order I. "Nitrospirales"

Family I. "Nitrospiraceae"

Genus I. **Nitrospira** Watson, Bock, Valois, Waterbury and Schlosser 1986b, 489[VP] (Effective publication: Watson, Bock, Valois, Waterbury and Schlosser 1986a, 6)

EVA SPIECK AND EBERHARD BOCK

Ni.tro.spi' ra. L. n. *nitrum* nitrate; Gr. n. *spira* a coil, spiral; M.L. fem. n. *Nitrospira* nitrate spiral.

Vibrio-like to spiral-shaped rods, 0.2–0.4 × 0.9–2.2 μm. Cells reproduce by binary fission. **Intracytoplasmic membranes are lacking.** Gram negative. Usually **nonmotile. Aerobic.** The major source of energy and reducing power is from the **oxidation of nitrite to nitrate. Lithoautotrophs,** but cells can also grow mixotrophically. Cells were isolated from ocean environments and heating systems. Also occurs in soil samples, freshwater, and activated sludge.

The mol% G + C of the DNA is: 50.0–56.9.

Type species: **Nitrospira marina** Watson, Bock, Valois, Waterbury and Schlosser 1986b, 489 (Effective publication: Watson, Bock, Valois, Waterbury and Schlosser 1986a, 6).

FURTHER DESCRIPTIVE INFORMATION

Nitrospira was shown recently to branch at a deep level from the *Proteobacteria* (Ehrich et al., 1995). The species *Nitrospira marina* and *"Nitrospira moscoviensis"* share 16S rRNA primary structure similarity of 88.9%. Comparative analysis revealed a moderate phylogenetic relationship to *Leptospirillum ferrooxidans, Thermodesulfovibrio yellowstonii* and *"Candidatus* Magnetobacterium bavaricum".* These organisms represent a new phylum in the domain *Bacteria. Nitrospira* is so far the only nitrifier which does not belong to the *Proteobacteria.* This genus is phylogenetically distinct from the genera *Nitrobacter, Nitrococcus,* and *Nitrospina,* which are affiliated with different subclasses of the *Proteobacteria.* A more generalized description of the nitrite-oxidizing bacteria, including additional figures of *Nitrospira,* as well as detailed information about isolation, physiology, and ecology, will be published in Volume 2 of the *Manual.*

Cells of *Nitrospira* are curved rods which occur as tightly to loosely wound helices with 1–12 turns (Fig. B8.1). One important feature is the enlarged, electron-dense periplasmic space up to 30–40 nm wide (Watson et al., 1986a). The cytoplasmic membrane is asymmetric with an electron dense layer on the periplasmic side. The membrane-associated nitrite-oxidizing system (NOS) of *"Nitrospira moscoviensis"* was isolated from heat-treated membranes. The four major proteins of the enzyme fraction had apparent molecular masses of 130, 62, 46, and 29 kDa. The 130 kDa protein is proposed to be the α-subunit, whereas the 46 kDa protein was identified as the β-NOS by the use of monoclonal antibodies (Spieck et al., 1998). By immuno-electron microscopy this protein was shown to be located in the periplasmic space (Fig. B8.1), where a periodic arrangement of membrane-associated particles of 13–15 nm was found in the form of a hexagonal pattern (Fig. B8.2). They were less electron dense in the middle and composed of several smaller particles. It is supposed that these particles represent the NOS in *Nitrospira*. Electron microscopy of the isolated enzyme revealed uniform particles with a size of 7 × 9 nm. Cytochromes of the *b*- and *c*-type but not of

the *a*-type were detected. No carboxysomes were found in *Nitrospira*. Cells contain glycogen-like deposits. Optimum growth occurs in mineral medium with 2–3 mM nitrite. Optimum pH range: 7.6–8.0. High concentrations of nitrite are toxic. No heterotrophic growth has been observed. The minimum generation times for lithoautotrophic growth are 12 to 90 h. Species were isolated from marine environments and heating systems. This genus may be the most important nitrite oxidizer in marine sediments and marine waters rich in organic matter (Watson et al., 1989). *Nitrospira*-like bacteria were also shown to be the dominant nitrite oxidizers in activated sludge (Juretschko et al., 1998), freshwater aquaria (Hovanec et al., 1998), and microbial aggregates in a fluidized bed reactor (Schramm et al. 1998). Using the rRNA approach, several specific oligonucleotide probes were developed that indicated the existence of uncultured species. Isolation was hindered by EPS (extracellular polymeric substances). In addition, genus typical microcolonies were observed in enrichment cultures of soil from Elbmarsh and permafrost of Siberia (Hartwig, personal communication). Here, cells could be detected immunologically with monoclonal antibodies recognizing the β-NOS (Bartosch et al., 1999).

DIFFERENTIATION OF THE GENUS *NITROSPIRA* FROM OTHER GENERA

Table B8.1 lists characteristic features of *Nitrospira* which can be used to differentiate cells from those of the other genera of nitrite oxidizers. Besides morphology, *Nitrospira* can be separated from *Nitrobacter* by the substrate range as well as biochemical and immunological investigations (Table B8.2). Analysis of the fatty acids of the genera *Nitrobacter*, *Nitrococcus*, *Nitrospina*, and *Nitrospira* showed genus-specific profiles suitable for chemotaxonomic classification. *"Nitrospira moscoviensis"* was characterized by the new fatty acid 11-methyl branched palmitate (Lipski, personal communication).

TAXONOMIC COMMENTS

Classification of *Nitrospira* as a distinct genus of nitrite oxidizers is based primarily on its unique morphology and ultrastructure. The separation from other genera was confirmed by phylogenetic investigations. Both species differ in growth characteristics and mol% G + C content of the DNA. Cluster analysis of the fatty acid profiles gave first evidence for an undescribed species of *Nitrospira* to be present in two cultures growing optimal at temperatures of 42 or 47°C, respectively (Lipski, personal communication).

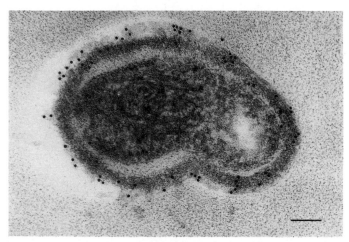

FIGURE B8.1. Ultrathin section of *"Nitrospira moscoviensis"*. The location of the β-NOS in the periplasmic space was shown by immunogold labeling. (Reproduced with permission from E. Spieck, S. Ehrich, J. Aamand, and E. Bock, Archives of Microbiology *169*: 225–230, 1998, ©Springer-Verlag.) Bar = 100 nm.

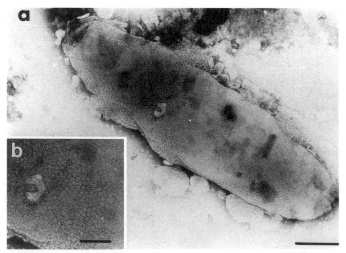

FIGURE B8.2. Hexagonal pattern of particles on the periplasmic side of the cytoplasmic membrane of *"Nitrospira moscoviensis"*. *a*, Electron micrograph of a partly destroyed cell in negative contrast. Bar = 0.3 μm. *b*, The regular arrangement of particles is shown in detail. Bar = 0.1 μm. (Reproduced with permission from E. Spieck, S. Ehrich, J. Aamand, and E. Bock, Archives of Microbiology *169*: 225–230, 1998, ©Springer-Verlag.)

FURTHER READING

Bartosch, S., I. Wolgast, E. Spieck and E. Bock. 1999. Identification of nitrite-oxidizing bacteria with monoclonal antibodies recognizing the nitrite oxidoreductase. Appl. Environ. Microbiol. *65*: 4126–4133.

Ehrich, S., D. Behrens, E. Lebedeva, W. Ludwig and E. Bock. 1995. A new obligately chemolithoautotrophic, nitrite-oxidizing bacterium, *Nitrospira moscoviensis* sp. nov. and its phylogenetic relationship. Arch. Microbiol. *164*: 16–23.

Juretschko, S., G. Timmermann, M. Schmid, K.H. Schleifer, A. Pommerening-Röser, H.P. Koops and M. Wagner. 1998. Combined molecular and conventional analyses of nitrifying bacterium diversity in activated sludge: *Nitrosococcus mobilis* and *Nitrospira*-like bacteria as dominant populations. Appl. Environ. Microbiol. *64*: 3042–3051.

Spieck, E., S. Ehrich, J. Aamand and E. Bock. 1998. Isolation and immunocytochemical location of the nitrite-oxidizing system in *Nitrospira moscoviensis*. Arch. Microbiol. *169*: 225–230.

Watson, S.W., E. Bock, H. Harms, H.P. Koops and A.B. Hooper. 1989.

TABLE B8.1. Differentiation of the four genera of nitrite-oxidizing bacteria

Characteristic	*Nitrobacter*	*Nitrococcus*	*Nitrospina*	*Nitrospira*
Cell shape	pleomorphic rod	coccus	straight rod	vibrio to helical-shaped cells
Intracytoplasmic membranes	polar stacks of lamellar membranes	randomly arranged tubes	none[b]	none
Main cytochrome types[a]	a,c	a,c	c	b,c
Mol% G + C	59–62	61	58	50–57

[a]Lithoautotrophic growth.

[b]Occasional invaginations of the plasma membrane.

TABLE B8.2. Specific reactions of three monoclonal antibodies (mABs) with the different genera in correlation with their phylogeny (Bartosch et al., 1999) and type of nitrite-oxidizing system (NOS)

Characteristic	*Nitrobacter*			*Nitrococcus*		*Nitrospina*	*Nitrospira*
Phylogenetic position	*Alphaproteobacteria* (Phylum *Proteobacteria*)			*Gammaproteobacteria* (Phylum *Proteobacteria*)		*Deltaproteobacteria* (Phylum *Proteobacteria*)	Phylum *Nitrospirae*
Location of the NOS	inner side of the cytoplasmic membrane			inner side of the cytoplasmic membrane		periplasmic side of the cytoplasmic membrane	periplasmic side of the cytoplasmic membrane
two-dimensional structure	rows of particle dimers			particles in rows		hexagonal pattern of particles	hexagonal pattern of particles
Molecular masses of the NOS, KDa	65	130	65	65	65	48	46
mABs	153–1	153–2	153–3	153–1	153–3	153–3	153–3

Nitrifying bacteria. *In* Staley, Bryant, Pfennig and Holt (Editors), Bergey's Manual of Systematic Bacteriology, 1st ed., Vol. 3, The Williams & Wilkins Co., Baltimore, pp. 1808–1834.

Watson, S.W., E. Bock, F.W. Valois, J.B. Waterbury and U. Schlosser. 1986. *Nitrospira marina*, gen. nov. sp. nov.: a chemolithotrophic nitrite-oxidizing bacterium. Arch. Microbiol. *144*: 1–7.

List of species of the genus Nitrospira

1. **Nitrospira marina** Watson, Bock, Valois, Waterbury and Schlosser 1986b, 489[VP] (Effective publication: Watson, Bock, Valois, Waterbury and Schlosser 1986a, 6).

   ma.ri′na. L. fem. adj. *marina* of the sea, marine.

   Helical to vibrio-shaped cells with a width of 0.3–0.4 μm and a spiral amplitude of 0.8–1.0 μm. When grown lithotrophically, 1–12 turns were observed. Under mixotrophic conditions, most cells have only one turn. Growth occurs in medium containing 70–100% seawater enriched with nitrite and other inorganic salts. The organisms grow better mixotrophically than lithoautotrophically with minimum generation times of 23 h and 90 h, respectively. Cells can use pyruvate or glycerol as carbon source and yeast extract or peptone as nitrogen source. High concentrations of organics or nitrite are inhibitory. Optimum temperature range for growth is 20–30°C. Membranes possess six major proteins with molecular masses of 130, 85, 75, 62, 55, and 46 kDa (Ehrich et al., 1995). Cell suspensions and cell-free extracts show characteristic (dithionite-reduced oxidized minus) absorption peaks at 416, 550 (shoulder), and 558 nm. The strain was isolated from Gulf of Maine water sample.

   The mol% G + C of the DNA is: 50.0 (T_m).

   Type strain: ATCC 43039.

2. **"Nitrospira moscoviensis"** Ehrich, Behrens, Lebedeva, Ludwig and Bock 1995, 22.

   mos.co.vien′sis. L. adj. of Moscow, named after the place where this organism was first isolated.

   Mesophilic, curved rods with 1–3 turns. Logarithmically growing cells are irregularly shaped. The doubling time was 12 h in a mineral medium with nitrite. Growth was not supported by organic matter. Optimum growth occurs at 39°C with a range of 33–40°C. After consumption of nitrite, flocs of 1–2 mm in diameter appeared due to the production of extracellular polymeric substances. In nitrite-oxidizing membranes eight major proteins were present with molecular masses of 130, 85, 66, 62, 58, 46, 34, and 29 kDa. Cell-free extracts show absorption peaks characteristic for *b*- and *c*-type cytochromes (maxima at 418, 524, 559 nm, shoulders at 531, 550 nm). Cytochromes of the *a*-type were not detectable. Isolated from an iron pipe of municipal heating systems.

   The mol% G + C of the DNA is: 56.9 (T_m).

   Deposited strain: M-1, DSMZ 10035.

Genus II. **Leptospirillum** Hippe 2000, 503[VP] (ex Markosyan 1972, 26)

D. BARRIE JOHNSON

Lep.to.spi.ril′lum. Gr. adj. *lepto* thin, narrow, fine; Gr. n. *spira* a spiral; M.L. neut. n. *Leptospirillum* a thin spiral.

Small, **Gram-negative, vibrioid or spiral-shaped** cells (0.9–2.0 × 0.2–0.5 μm). **Motile** by means of a **single** (in most cases) **polar** flagellum. No known resting stages. Obligately chemolithotrophic, using the **oxidation of ferrous iron** (or ferrous iron-containing sulfide minerals, such as pyrite) as sole energy source, and **fixing carbon dioxide** by the Benson-Calvin cycle. **Obligately aerobic and acidophilic**; pH optima for growth is generally 1.3–2.0. Mesophilic isolates and a single moderately thermophilic isolate have been described. *Leptospirillum* forms a distinct lineage within the proposed *Nitrospira* phylum (Ehrich et al., 1995). Widely distributed in natural and industrial environments where the accelerated oxidation of sulfide ores creates acidic, metal-rich ecosystems.

The mol% G + C of the DNA is: 51–56.

Type species: **Leptospirillum ferrooxidans** Hippe 2000, 503 (ex Markosyan 1972, 26).

FURTHER DESCRIPTIVE INFORMATION

Two species of *Leptospirillum* have been described; the type species *L. ferrooxidans* and *L. thermoferrooxidans*. However, only the type species has been studied in detail. When the 16S rRNA sequences of *L. ferrooxidans* isolates were determined by Lane et al. (1992), they were noted to be distinct (<80% sequence identity) from those of all other bacteria (some 350 available at the time) and could not be affiliated with any of the known higher level taxa. This included the *"Spirillaceae"* family, within which the genus had originally been included. More recent analysis indicates that *L. ferrooxidans* forms a distinct lineage within the proposed *Nitrospira* phylum. Several nonacidophilic bacteria, including *"Candidatus* Magnetobacterium bavaricum", *Thermodesulfovibrio yellowstonii, Nitrospira marina,* and *"Nitrospira moscoviensis"* also fall within this phylum (Ehrich et al., 1995).

There is mounting evidence that mesophilic iron-oxidizing bacteria identified as *L. ferrooxidans* comprise more than a single species. Data from analysis of DNA base composition grouped isolates with relatively low (~51%) and relatively high (55–56%) mol% G + C contents, while DNA–DNA hybridization studies of the same six *L. ferrooxidans*-like isolates identified at least two hybridization groups (Harrison and Norris, 1985). Similar results were reported by Hallmann et al. (1992) using different strains of *L. ferrooxidans*. The pioneering work of Lane et al. (1992) indicated that the original *L. ferrooxidans* isolate (variously designated as Z-2, Z-1, or L15) shared only 94% 16S rRNA sequence similarity to two other strains (BU-1 and LfLa); Goebel and Stackebrandt (1995) later found that three clones of *L. ferrooxidans* isolated from a mineral leachate sample were more closely related to strains BU-1 and LfLa than to strain L15, and concluded that the original *L. ferrooxidans* isolate was possibly an atypical strain (particularly in the context of mineral leaching) and that the mesophilic *Leptospirillum* cluster comprises at least two species, or even genera. Chemotaxonomic and physiological variations between *L. ferrooxidans* strain Z-2 and other isolates have also been noted, including differences in their respiratory lipoquinones (Goebel and Stackebrandt, 1995) and a requirement for zinc by the original isolate (Norris, 1989). In this discussion, most references are made to data obtained from *L. ferrooxidans* strain Z-2, and therefore may not necessarily hold for other *Leptospirillum*-like strains.

Goebel and Stackebrandt (1995) have successfully screened PCR-generated 16S rDNA clone libraries using a *Leptospirillum*-specific primer designated "Lf 176" and other strain-dependent primers designated LfI 460, LfII 459, and LfIII 473. It was noted that the original isolate (strain Z-2) of *L. ferrooxidans* did not, however, give a positive reaction with the Lf 176 primer in PCR analysis. De Wulf-Durand and co-workers (1997) have subsequently described a PCR primer, designated LEPTO679R, that matches all *L. ferrooxidans* strains for which the 16S rRNA gene has been sequenced. Edwards et al. (1999) used fluorescent *in situ* hybridization using two gene probes (designated LC206 and LF581) to examine the distribution of *L. ferrooxidans* in acidic drainage waters at a mine site (Iron Mountain) in California. Their results indicated that *L. ferrooxidans* was the dominant iron-oxidizing bacterium present in extremely low pH (~0.5) and higher temperature (~40°C) sites within the mine.

Bacteria of the genus *Leptospirillum* occur as curved rods (vibrios), 0.2–0.5 × 0.9–2.0 µm (Figs. B8.3 and B8.4). Helical forms (generally comprising 2 to 5 turns, though helixes of over 20 turns have been observed) are common under some growth conditions. In other circumstances (e.g., when grown at extremely low pH), coccoid and pseudococcoid (tightly coiled vibrios) forms have been observed. Cells are highly motile by means of a single polar flagellum, 18–22 nm in diameter in mesophilic

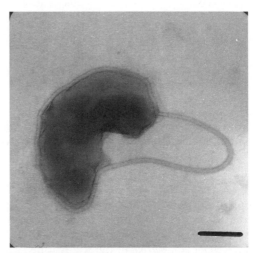

FIGURE B8.3. Transmission electron micrograph of phosphotungstate-stained *Leptospirillum ferrooxidans* (strain Z-2). Bar = 200 nm.

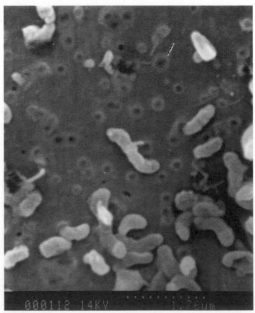

FIGURE B8.4. Scanning electron micrograph of *Leptospirillum ferrooxidans* (strain Z-2).

strains but somewhat larger (~25 nm) in the single moderately thermophilic isolate that has been described. In some isolates, the flagellar insertion may be sub-polar rather than polar and may be sheathed. Occasionally cells may have two flagella (Goebel, 1997). Cell division precedes with septum formation and is generally uniform; occasionally, nonuniform cell division may occur, producing a small (500–800 nm diameter) daughter cell and a larger, curved, mother cell (Pivovarova et al., 1981). The cell wall structure is similar to that of other Gram-negative bacteria. The cell membrane comprises two electron-dense layers (0.6–1.0 and 0.35–0.6 nm thick) with an electron-translucent layer (0.7–0.8 μm) between. The periplasmic space between the cell wall and cell membrane spans ca. 0.8–1.5 nm. Nuclear structures are visible in the center of cells, but intracellular membrane structures are less obvious (Pivovarova et al., 1981). Large numbers of polyribosomes may occur in the cytoplasm, but β-hydroxybutyrate reserves have not been observed.

L. ferrooxidans-like isolates often produce copious amounts of extracellular slime, particularly when grown at relatively low (<25°C) temperatures. When grown in ferrous sulfate liquid medium, this may result in the formation of bacterial aggregates, or flocs, ~1–2 mm in diameter. Alternatively, when grown in a "mineral" medium containing fine-grain pyrite, bacteria/mineral aggregates of ~1–3 mm may be observed. When grown in mixed cultures with heterotrophic *Acidiphilium* spp., flocs (intergrowths of autotrophic and heterotrophic bacteria) tend to be significantly larger. The exocellular slime material is rich in polysaccharide, and apparently differs in composition between ferrous sulfate- and pyrite-grown cells (Gehrke et al., 1995). The exopolymer has a significant ferric iron content.

Growth of *Leptospirillum* spp. occurs only with ferrous iron as the electron donor. This may be provided as ferrous sulfate or (in the case of *L. ferrooxidans*) in mineral-form, such as pyrite (FeS$_2$). Oxidation of pyrite proceeds via an "indirect" mechanism, in which ferric iron acts as a chemical oxidant of the mineral, thereby being reduced to ferrous iron which is, in turn, reoxidized by *L. ferrooxidans*. In this scenario, intimate contact between the bacterium and mineral is not required. Alternatively,

cells that are attached to sulfides may accelerate mineral oxidation by a direct mechanism. This mechanism is less well understood, although recent evidence indicates that the iron associated with the exopolymeric materials plays a central role in acting as an electron shuttle between the bacteria and the mineral (Sand et al., 1995). The affinity of *L. ferrooxidans* for ferrous iron (K_m ~0.25 mM) is significantly greater than that of the better-known iron-oxidizing acidophile, *Thiobacillus ferrooxidans**(K_m ~1.35 mM) (Norris et al., 1988). Minimum culture doubling times of *L. ferrooxidans* in ferrous iron medium is strain dependent, varying from 10 to over 20 hours.

L. ferrooxidans grows very poorly or not at all on most solid media (such as agar-gelled liquid media). This appears to be related to the sensitivity of the autotroph to organic materials, such as those produced by acid hydrolysis of the gelling agent during plate incubation. However, use of an "overlayer" approach, in which acidophilic heterotrophic bacteria are incorporated into the lower zone of a bilayered medium, circumvents this problem, allowing the plating efficiency of *L. ferrooxidans* to exceed 90% for the strains tested (Johnson, 1995). Colonies are first visible after 7–14 days incubation at 30°C, and develop into small (1–2 mm), round, entire, orange to light brown colored, iron-encrusted forms. Some variation in colony size and morphology occurs with different strains of *L. ferrooxidans*.

Leptospirillum spp. are obligate aerobes and acidophiles, growing optimally at pH 1.3–2.0, with a lower pH limit of 1.1 (Battaglia et al., 1994). Acidity results from the hydrolysis of ferric iron and, when grown in mixed culture with thionic bacteria, from the production of sulfuric acid. Of the two species described, *L. ferrooxidans* is a mesophilic chemolithotroph, but is notably more tolerant of higher temperatures (35°C) and less tolerant of lower (<25°C) temperatures than *T. ferrooxidans*; *L. thermoferrooxidans* is a moderate thermophile (Table B8.3). Leptospirilla are highly sensitive to many organic materials (such as organic acids), and to many inorganic anions, with the notable exception of sulfate.

*Editorial note: Thiobacillus ferrooxidans has been renamed Acidithiobacillus ferrooxidans (Kelly and Wood, 2000).

TABLE B8.3. Phenotypic characteristics of acidophilic iron-oxidizing bacteria[a]

Characterisitic	*Leptospirillum ferrooxidans*	*Leptospirillum thermoferrooxidans*	*Thiobacillus ferrooxidans*	"*Thiobacillus ferrooxidans*" m-1	*Thiobacillus prosperus*	"*Ferromicrobium acidophilus*"	*Sulfobacillus thermosulfidooxidans*	*Sulfobacillus acidophilus*	*Acidimicrobium ferrooxidans*
Cell morphology	vibrios 1 μm, spirilla	vibrios 1.5–2 μm, spirilla	rods 1–2 μm	rods 1–1.5 μm	rods 3–4 μm	rods 1–3 μm	rods 2–6 μm	rods 3–5 μm	rods 1–1.5 μm, filaments
Endospores	−	−	−	−	−	−	+	+	−
Motility	+ +	+ +	(+)	+	+	+	(+)	(+)	+
S^0 oxidation	−	−	+	−	+	−	+	+	−
Utilization of yeast extract	−	−	−	−	−	+	+	+	+
Growth at 50°C	−	+	−	−	−	−	+	+	+
Mol% G + C	51–56	56	58–59	65	63–64	52–55	48–50	55–57	67–68.5

[a]Symbols: +, positive; −, negative; + +, strongly motile; (+), limited motility.

As a general rule, *L. ferrooxidans* strains are more tolerant to ferric iron than is *T. ferrooxidans*, though less so to some other metal cations, such as copper.

The electron transfer chain involved in the transfer of electrons from ferrous iron to molecular oxygen in *Leptospirillum* spp. has not been fully resolved, though thermodynamic constraints dictate that it is necessarily short. Iron-grown cells of *L. ferrooxidans* produce large amounts of a novel red-colored, acid-stable, and acid-soluble cytochrome with a distinctive (reduced minus oxidized) peak at 579 nm, that is rapidly oxidized by ferrous iron in cell-free extracts, implying that this protein is a key component in the respiratory chain (Blake and Shute, 1997). The E^o of the ferrous/ferric couple ($+770$ mV) dictates that the reduction of NAD(P) by *Leptospirillum* spp. requires the consumption of ATP. *Leptospirillum* spp. are obligately autotrophic. CO_2 is assimilated via the Benson-Calvin cycle; active ribulose bisphosphate carboxylase has been found in cell-free extracts of *L. ferrooxidans*. It has also been reported that *L. ferrooxidans* is capable of fixing dinitrogen (Norris et al., 1995).

Leptospirillum spp. typically inhabit highly acidic, metal-rich environments associated with the oxidation of sulfide minerals. These include metal and coal mines, mine spoils, and the streams ("acid mine drainage") that drain from them. *L. ferrooxidans* is also a major component of the microbial microflora of ore leach liquors in "biomining" operations, including both tank and heap leaching operations. Isolates have also been obtained from acidic geothermal environments. Early attempts at estimating indigenous acidophilic microflora have tended to discriminate against *L. ferrooxidans* in favor of the more easily isolated and well-studied iron-oxidizing chemolithotroph *T. ferrooxidans*. More recent data indicates that *L. ferrooxidans* is, quite frequently, the dominant iron-oxidizing bacterium in extremely acidic environments, particularly where the pH is <2, and the ferrous iron/ferric iron ratio is low (Rawlings et al., 1999).

ENRICHMENT AND ISOLATION PROCEDURES

In the environment, and in industrial (biomining) operations, *L. ferrooxidans* generally exists in association with a variety of other chemolithotrophic, iron- and/or sulfur-oxidizing bacteria, as well as with acidophilic heterotrophs and some eucaryotic life forms. Ferrous iron overlay solid medium (Johnson, 1995) is suitable for the simultaneous isolation of *L. ferrooxidans* and *T. ferrooxidans*-like bacteria, using a spread plate technique. After 15–25 days incubation at 30°C colonies of *L. ferrooxidans* are generally small (1–2 mm) round, entire, and orange to light brown in color, and contrast with those of *T. ferrooxidans*, which tend to show considerable (strain dependent) variation in size, color, and morphology. When crushed, and viewed under a phase contrast microscope, the cellular characteristics of *L. ferrooxidans* (highly motile vibrios and spirilli) can aid further differentiation from *T. ferrooxidans* (variably-motile straight rods). Confirmation of *L. ferrooxidans* may be obtained by screening for physiological traits (Table B8.3) or by the application of molecular approaches, e.g., 16S rDNA sequence analysis. Pure cultures of *L. ferrooxidans* may also be obtained by enrichment and serial dilution. Ferrous sulfate liquid medium is inappropriate for this purpose, as the faster growth rate of *T. ferrooxidans* invariably selects for the latter iron-oxidizing acidophile. A liquid medium containing pyrite (1–2%, w/v) and basal salts, adjusted initially to ~pH 1.8 will, with prolonged incubation, select for *L. ferrooxidans* rather than *T. ferrooxidans*. Cultures, incubated at 30–40°C, should be examined regularly until highly motile, vibrioid cells are observed as the dominant microorganism present (this may take 2–3 weeks) and

serial dilutions inoculated into ferrous sulfate (or pyrite) liquid media, or else spread onto ferrous iron-containing overlay medium (as above). Given the propensity of *L. ferrooxidans* to grow in coculture with heterotrophic and sulfur-oxidizing acidophiles, the purity of liquid cultures of *L. ferrooxidans* should be monitored regularly using solid (or liquid) media containing a suitable organic substrate (e.g., yeast extract) or S^0.

MAINTENANCE PROCEDURES

The mortality rates of *L. ferrooxidans* in oxidized ferrous sulfate medium tend to be considerably greater than those of *T. ferrooxidans*, and difficulty may be experienced in recovering viable cells less than 20 days after depletion of ferrous iron. Liquid media containing pyrite (1–2%) are more suitable for longer-term maintenance of active cultures of *L. ferrooxidans*. Viable cultures may be stored for at least 6–12 months at <10°C by addition of a sterile suspension of pyrite to stationary-phase ferrous iron-grown cultures. Readers are cautioned that some strains have been reported to lose viability more rapidly when stored at 4°C than at 17°C (Hallmann et al., 1992). *L. ferrooxidans* has also been reported not to survive lyophilization, though cultures may be preserved in liquid nitrogen.

The long-term viability of pure cultures of *L. ferrooxidans* in both oxidized ferrous sulfate and pyrite media is often considerably less than corresponding mixed cultures containing acidophilic heterotrophic bacteria, such as *Acidiphilium* and *"Ferromicrobium"* spp. This is presumably related to the sensitivity of *L. ferrooxidans* to organic materials, which accumulate during culture growth (via cell leakage and lysis), and which are metabolized by heterotrophic "contaminant" organisms. Short-term or medium-term storage of defined mixed cultures may therefore be more successful and convenient from which pure cultures of *L. ferrooxidans* may be obtained by streaking onto overlaid solid medium and single colony isolation, as noted above.

DIFFERENTIATION OF THE GENUS *LEPTOSPIRILLUM* FROM OTHER GENERA

The key phenotypic characteristics that may be used to differentiate mesophilic and moderately thermophilic *Leptospirillum* spp. from other iron-oxidizing acidophiles are listed in Table B8.3. Particular note should be made of the iron-oxidizing acidophile strain m-1, which has several key traits in common with *L. ferrooxidans* rather than *T. ferrooxidans* (with which it has been included); these include the inability to oxidize sulfur and a relatively high K_i (inhibition constant) for ferric iron.

TAXONOMIC COMMENTS

The genus *Leptospirillum* was first proposed over 25 years ago but has been validated only recently (Hippe, 2000). Though there is now a considerable data bank of physiological and (phylo)genetic information relating to the type species, *L. ferrooxidans*. In contrast, the phylogenetic position of the moderate thermophile *L. thermoferrooxidans* is unknown; this acidophile has been isolated only once, and has since been lost prior to being deposited in any culture collection. *L. ferrooxidans*-like isolates that have been sequenced (16S rRNA or rDNA) form a distinct cluster, as noted above, and it appears that these bacteria may represent more than a single species and perhaps multiple genera.

ACKNOWLEDGMENTS

The author is grateful for the suggestions and advice from Brett Goebel (Australian Magnesium Co., Queensland, Australia).

FURTHER READING

De Wulf-Durand, P., L.J. Bryant and L.I. Sly. 1997. PCR-mediated detection of acidophilic, bioleaching-associated bacteria. Appl. Environ. Microbiol. 63: 2944–2948.

Edwards, K.J., T.M. Gihring and J.F. Banfield. 1999. Seasonal variations in microbial populations and environmental conditions in an extreme acid mine drainage environment. Appl. Environ. Microbiol. 65: 3627–3632.

Goebel, B.M., and E. Stackebrandt. 1995. Molecular analysis of the microbial biodiversity in a natural acidic environment. In Jerez, Vargas, Toledo and Weirtz (Editors), Biohydrometallurgical Processing II, University of Chile, Santiago, pp. 43–52.

Golovacheva, R.S., O.V. Golyshina, G.I. Karavaiko, A.G. Dorofeev, T.A. Pivovarova and N.A. Chernykh. 1992. A new iron-oxidizing bacterium, Leptospirillum thermoferrooxidans sp. nov. Mikrobiologiya 61: 1056–1065.

Hippe, H. 2000. Leptospirillum gen. nov. (ex Markosyan 1972), nom. rev., including Leptospirillum ferrooxidans sp. nov. (ex Markosyan 1972), nom. rev. and Leptospirillum thermoferrooxidans sp. nov. (Golovacheva et al. 1992). Int. J. Syst. Evol. Microbiol. 50: 501–503.

Johnson, D.B. 1995. Selective solid media for isolating and enumerating acidophilic bacteria. J. Microbiol. Methods 23: 205–218.

Lane, D.J., A.P. Harrison, Jr., D. Stahl, B. Pace, S.J. Giovannoni, G.J. Olsen and N.R. Pace. 1992. Evolutionary relationships among sulfur- and iron-oxidizing eubacteria. J. Bacteriol. 174: 269–278.

Markosyan, G.E. 1972. A new iron-oxidizing bacterium, Leptospirillum ferrooxidans gen. et sp. nov.. Biol. Zh. Arm. 25: 26.

Pivovarova, T.A., G.E. Markosyan and G.I. Karavaiko. 1981. Morphogenesis and fine structure of Leptospirillum ferrooxidans. Mikrobiologiya 50: 482–486.

Rawlings, D.E., H. Tributsch and G.S. Hansford. 1999. Reasons why "Leptospirillum"-like species rather than Thiobacillus ferrooxidans are the dominant iron-oxidizing bacteria in many commercial processes for the biooxidation of pyrite and related ores. Microbiology 145: 5–13.

List of species of the genus Leptospirillum

1. **Leptospirillum ferrooxidans** Hippe 2000, 503^{VP} (ex Markosyan 1972, 26).

fer.ro.ox′i.dans. L. n. ferrum iron; M.L. v. oxido oxidize; M.L. part. adj. iron-oxidizing.

Vibrios (0.9–1.1 μm) and spirilli (of variable length) are most common, though occasional cocci and pseudococci occur. Motile by means of a single (in most cases) polar flagellum. Obligately chemolithotrophic and autotrophic. Strictly aerobic. Utilizes ferrous iron or iron-containing sulfide minerals as sole energy source. Optimum temperature 30–37°C; growth range <10–45°C. Optimum pH 1.3–2.0; minimum pH: 1.1. Widely distributed in extremely acidic (pH <3) metal- rich environments and in industrial mineral leaching operations.

The mol% G + C of the DNA is: 51–56 (T_m).

Type strain: L15, ATCC 29047, DSMZ 2705.

GenBank accession number (16S rRNA): X86776.

2. **Leptospirillum thermoferrooxidans** Hippe 2000, 503^{VP} (Effective publication: Golovacheva, Golyshina, Karavaiko, Dorofeev, Pivovarova and Chernykh 1992, 749).

ther′mo.fer.ro.ox′i.dans. Gr. adj. thermos hot; L. n. ferrum iron; M.L. v. oxido oxidize; M.L. part. adj. thermoferrooxidans hot iron-oxidizing.

Vibrios (1.5–2.0 × 0.2–0.5 μm) and spiral (2.0–3.0 × 0.2–0.5 μm) forms. Motile by means of a single polar flagellum. Obligately chemolithotrophic and autotrophic. Strictly aerobic. Utilizes ferrous iron as sole energy source, but not iron-containing sulfide minerals. Optimum temperature 45–50°C; growth range 30–55°C. Optimum pH 1.65–1.90; minimum pH: 1.3. DNA–DNA relatedness to the type strain of L. ferrooxidans is 26.7%. A single strain, isolated from acidic, hydrothermal springs on Kunashir (one of the Kuril Islands, located north of Japan), has since been lost.

The mol% G + C of the DNA is: 56 (T_m).

Type strain: L-88.

---

*Genus III. "**Candidatus** Magnetobacterium" Spring, Amann, Ludwig, Schleifer, van Gemerden and Petersen 1993, 2398*

STEFAN SPRING AND KARL-HEINZ SCHLEIFER

Mag.ne′to.bac.te′ri.um. Gr. n. magnes magnet, comb. form magneto-; Gr. n. bakterion a small rod; M.L. neut. n. Magnetobacterium a magnetic rod.

Rod-shaped, large, magnetic cells occurring in freshwater sediments. The original description is based on bacteria enriched from the littoral sediment of a freshwater lake in Southern Germany (Chiemsee); similar types of bacteria were also observed in sediments of other freshwater lakes and ponds in Southern Germany and Brazil. Counts of active bacteria in different vertical layers of Chiemsee sediment indicated that "Candidatus M. bavaricum" is a typical gradient organism, particularly adapted to zones characterized by low levels of oxygen, where it reaches the highest abundance (Spring et al., 1993). A minor fraction of these bacteria was also found in the anoxic zone, whereas high concentrations of oxygen can apparently not be tolerated by this bacterium over longer periods of time.

Type species: "**Candidatus Magnetobacterium bavaricum**" Spring, Amann, Ludwig, Schleifer, van Gemerden and Petersen 1993, 2398.

FURTHER DESCRIPTIVE INFORMATION

Phylogenetic analyses of the 16S rRNA gene of "Candidatus M. bavaricum" revealed only a distant relationship to all hitherto identified magnetic bacteria and most other known representatives of the domain Bacteria. The closest related cultivated species is currently Thermodesulfovibrio yellowstonii, a thermophilic sulfate-reducing bacterium, sharing a sequence similarity of 86.8% with the 16S rRNA gene of "Candidatus M. bavaricum". The nearest neighbors in reconstructed phylogenetic trees besides T. yellowstonii are members of the genera Nitrospira and Leptospirillum. These bacteria together with "Candidatus M. bavaricum" form a coherent phylogenetic group that is not related to other major lineages of the domain Bacteria and was therefore tentatively named Nitrospira-group after the first validly described species within this lineage (Ehrich et al., 1995).

Cells of *"Candidatus* M. bavaricum" can be phenotypically distinguished from other magnetic bacteria by their unique morphology. The rod-shaped cells are Gram-negative having dimensions of 1–1.5 × 6–9 μm (determined by phase-contrast microscopy) and are motile by a polar tuft of flagella. They contain up to a thousand magnetosomes arranged in three to five rope-shaped bundles of chains parallel to the long axis of the cell and large sulfur inclusions (Fig. B8.5). The magnetosomes have a characteristic hook-shaped morphology, consist of magnetite (Fe_3O_4), and have a length of 110–150 nm (Fig. B8.6).

The phylogenetic identity of the morphovar *"Candidatus* M. bavaricum" could be shown by *in situ* hybridization with a specific oligonucleotide probe (Fig. B8.7). The oligonucleotide is directed against a 16S rRNA gene retrieved from large, rod-shaped magnetic bacteria, which were obtained from a mixture of magnetically separated microorganism by flow cytometry. The sequence of the oligonucleotide probe is 5′-GCCATCCCCTCGCTTACT-3′ and the almost complete sequence of the 16S rRNA gene is deposited under the accession number X71838.

Under the influence of a magnetic field, cells of *"Candidatus*

M. bavaricum" show a characteristic motility caused by the intracellular magnetosomes, which confer a stable magnetic moment to these bacteria. The total magnetic moment per cell was determined as 13–64 × 10^{-15} A/m^2 using the rotating field method (Steinberger et al., 1994). Cells are oriented along the geomagnetic field lines and swim unidirectionally forward (toward the North pole) with an average speed of 40 μm/s. In contrast to the random walk motility of a chemotactic response, forward swimming is not disrupted by tumbling.

Cells respond, however, to chemical gradients by reversing their swimming direction for short periods of time. When a cell is swimming forward, the flagella are wound around the rotating cell; when a cell is swimming in the opposite direction, the flagellar tuft is reversed and propels the cell.

The morphological characteristics and preferred microhabitat at the oxic-anoxic transition zone allow some speculation about the potential metabolism of *"Candidatus* M. bavaricum". It seems probable that this bacterium oxidizes reduced sulfur compounds, thereby depositing S^0 intracellularly as intermediate product like many other chemolithotrophic sulfur-oxidizers (e.g.,

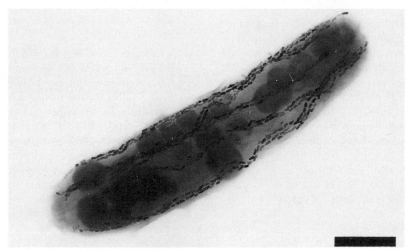

FIGURE B8.5. Electron micrograph of a whole cell of *"Candidatus* M. bavaricum" showing magnetosome bundles and large sulfur inclusions. Bar = 1 μm. (Reprinted with permission from M. Hanzlik.)

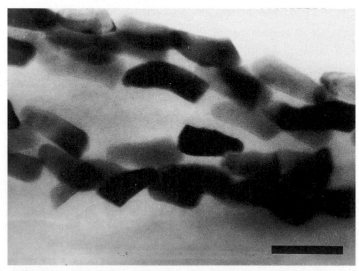

FIGURE B8.6. Hook-shaped magnetosomes typical for *"Candidatus* M. bavaricum". Bar = 100 nm. (Reprinted with permission from M. Hanzlik.)

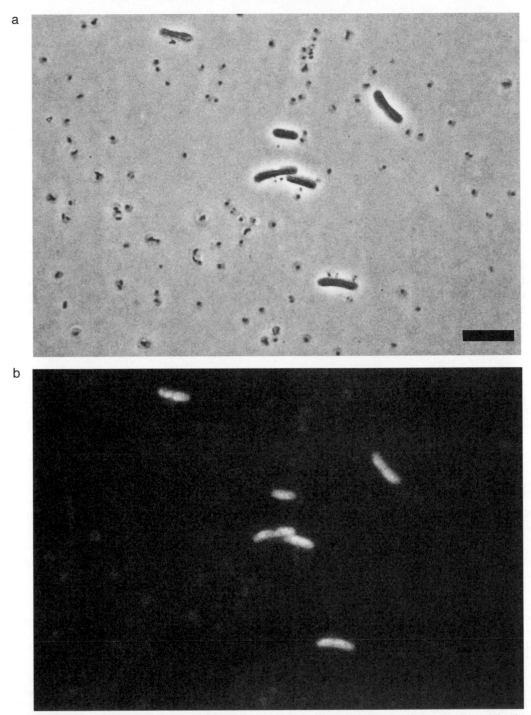

FIGURE B8.7. *In situ* hybridization of *"Candidatus* M. bavaricum" with a specific oligonucleotide probe. Bar = 10 μm. Mixed sample of magnetically separated bacteria viewed by phase-contrast (*a*) and epifluorescence microscopy (*b*). Only cells of the morphovar *"Candidatus* M. bavaricum" bound the fluorescently labeled oligonucleotide probe.

Beggiatoa spp.). Possible electron acceptors could be oxygen, nitrate, sulfate, or iron oxides. If the latter were used, the accumulation of large quantities of magnetite could be explained by the redox-cycling of iron.

Although the natural habitat of this bacterium is at a constant low temperature of 4–10°C, this organism does not seem to be psychrophilic, growing well at an ambient temperature of around 25°C.

ENRICHMENT AND ISOLATION PROCEDURES

To date, all attempts to take *"Candidatus* M. bavaricum" in axenic culture have failed. An enrichment of this magnetic bacterium, however, is often observed after storing freshwater sediment over longer periods of time in an aquarium at room temperature in dim light. Due to "magnetotactic" behavior, this microorganism can be easily separated from sediment along with other magnetic bacteria by imposing an artificial magnetic field.

List of species of the genus

1. **"*Candidatus* Magnetobacterium bavaricum"** Spring, Amann,
   Ludwig, Schleifer, van Gemerden and Petersen 1993, 2398.
   ba.var' i.cum. M.L. neut. adj. *bavaricum* bavarian.

See genus description.
Type strain: none available.
GenBank accession number (16S rRNA): X71838.

Genus IV. **Thermodesulfovibrio** Henry, Devereux, Maki, Gilmour, Woese, Mandelco, Schauder, Remsen and Mitchell 1994b, 595^{VP} (Effective publication: Henry, Devereux, Maki, Gilmour, Woese, Mandelco, Schauder, Remsen and Mitchell 1994a, 68)

JAMES S. MAKI

Ther.mo.de.sul.fo.vib.ri.o. Gr. adj. *thermos* warm, hot; L. pref. *de* from; L. n. *sulfur* sulfur; L. v. *vibrare* to vibrate; M.L. masc. n. *Vibrio* that which vibrates; M.L. masc. n. *Thermodesulfovibrio* a thermophilic curved bacterium that reduces sulfur.

Curved or vibrioid rod-shaped single cells or in chains, 0.3 × 1.7 µm, spore formation not observed. Gram negative. **Motile by means of single polar flagellum. Strict anaerobe**. Growth between 40 and 70°C, with **optimal growth at 65°C**. pH optimum 6.8–7.0. Chemoorganotroph, fermentative and **dissimilatory sulfate-reducing metabolism**. Contains desulfofuscidin, does not contain desulfoviridin. Sulfate and thiosulfate used as electron acceptors for growth. Sulfite and nitrate may also be electron acceptors. Organic substrates oxidized to acetate. The genus is among the deeply brancing *Bacteria* based upon 16S rRNA sequence analysis.

The mol% G + C of the DNA is: 29.5–38.

Type species: **Thermodesulfovibrio yellowstonii** Henry, Devereux, Maki, Gilmour, Woese, Mandelco, Schauder, Remsen and Mitchell 1994b, 595 (Effective publication: Henry, Devereux, Maki, Gilmour, Woese, Mandelco, Schauder, Remsen and Mitchell 1994a, 68).

FURTHER DESCRIPTIVE INFORMATION

Based on 16S rDNA sequence analysis, the genus *Thermodesulfovibrio* has been shown to be a member of the group which branches at a deep level within the *Bacteria* (Ehrich et al., 1995), forming the phylum *Nitrospirae* (Fig. B8.8). Other members of this group include *Nitrospira marina* (Watson et al., 1986), "*Nitrospira moscoviensis*" (Ehrich et al., 1995), *Leptospirillum ferrooxidans* (Balashova et al., 1974; Pivovarova et al., 1981), and "*Candidatus* Magnetobacterium bavaricum" (Spring et al., 1993). The latter has not been isolated in pure culture. Morphologically the cells of *Thermodesulfovibrio* are curved rods (Fig. B8.9), a feature that is shared by other members of the group, with the exception of "*Candidatus* M. bavaricum". However, species of *Nitrospira* and *Leptospirillum* form spiral or helical cells that have not been observed in *Thermodesulfovibrio*, although an observation indicates that a recently isolated species of the latter may form chains

(Sonne-Hansen and Ahring, 1999). Electron microscopy shows the cells to have a Gram-negative cell wall with a plasma membrane and an outer cell wall separated by a periplasmic space (Fig. B8.9). No intracytoplasmic membranes are visible. Cells are motile and have a single polar flagellum (Fig. B8.10).

Small colonies (1 mm in diameter) may take 4–5 weeks to develop on lactate/thioglycollate agar medium (Table B8.4; Henry et al., 1994a) when incubated at 60°C under N_2 gas in an anaerobe jar. Colonies may develop more readily using an agar dilution series with a bicarbonate-buffered and sulfide-reduced medium (Table B8.4). In the latter case, lens-shaped colonies are golden brown. In liquid culture, the cells of *T. yellowstonii* may attach to iron sulfide particles or may become loosely packed in elongated aggregates. Both adhesion and aggregation appear to be mediated by the polar flagellum. The cells of "*T. islandicus*" during growth may form chains of two to three cells that may increase in length as the cultures get older.

Optimal growth in various media (Table B8.4 and Sonne-Hansen and Ahring, 1999) occurs at 65°C, although growth can occur between 40 and 70°C. The pH of the medium should be between 6.8 and 7.0 for optimal growth. Growth may occur between a pH of 6.5 and 7.7. NaCl concentrations below 5 g/l result in favorable growth. Vitamins are not required. *Thermodesulfovibrio* can use sulfate, thiosulfate, and/or sulfite and nitrate as electron acceptors (Table B8.5). Electron donors include hydrogen plus acetate, lactate, and pyruvate. Growth with the addition of lactate and sulfate results in lactate oxidized stoichiometrically to acetate rather than CO_2. This is coupled with sulfate reduction to sulfide.

The type strain for this genus, *T. yellowstonii*, was isolated from hydrothermal vent water in Yellowstone Lake, Wyoming, while "*T. islandicus*" was isolated from a microbial mat sample taken from an alkaline hot spring in Iceland. This indicates that the

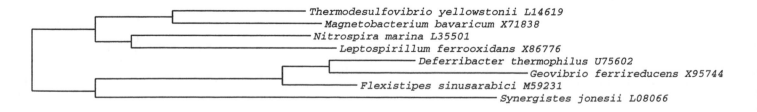

10% difference

FIGURE B8.8. Phylogenetic tree of the Nitrospiras and Deferribacters.

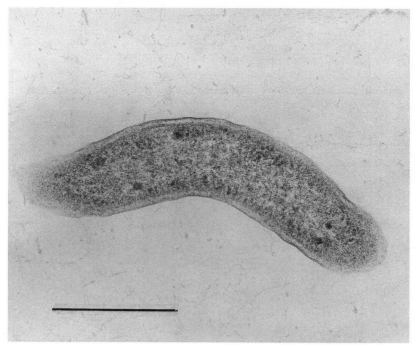

FIGURE B8.9. Morphology of *Thermodesulfovibrio yellowstonii* showing curved rod-shaped cells. Fine structure of cell shows Gram-negative cell wall. Bar = 0.5 μm. (Courtesy of J. S. Maki.)

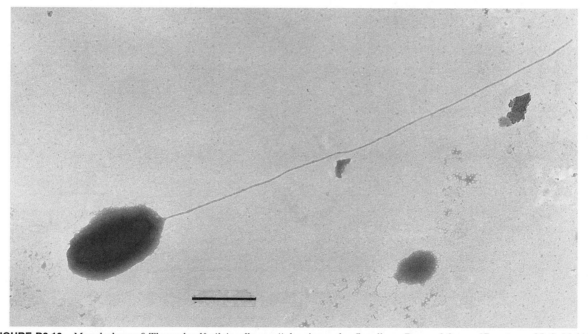

FIGURE B8.10. Morphology of *Thermodesulfovibrio yellowstonii* showing polar flagellum. Bar = 1.0 μm. (Courtesy of J. S. Maki.)

habitat of this genus is a freshwater anaerobic one associated with hydrothermal vents or other thermal features.

ENRICHMENT AND ISOLATION PROCEDURES

Enrichment of *Thermodesulfovibrio* was accomplished by inoculating either lactate/thioglycollate medium (Table B8.4) or Postgate medium 3 (Postgate, 1963), supplemented with 0.5 mg/l resazurin and 1 ml/l trace element solution (Angelidaki et al., 1990), with a subsample of thermal water or microbial mat sample and incubating at 55–60°C until the liquid became turbid and the precipitate turned black, indicating production of hydrogen sulfide. Subsequent isolation by colony formation was accomplished by using the same medium with 3% agar as a hardening agent and incubating plates under N_2 gas in an anaerobe jar at 60°C (Henry et al., 1994a) or in anaerobic roll-tubes containing 1.1% Phytagel as the solidifying agent (Sonne-Hansen and Ahring, 1999).

MAINTENANCE PROCEDURES

Cultures can be maintained on either solid or liquid media (Table B8.4). Cultures are maintained either on plates under N_2 gas in anaerobic jars or in agar dilution series as described above. Cultures grow in liquid media at 60°C without shaking. Cultures were observed to survive for as long as a year in liquid culture at room temperature, although growth was not observed (Henry et al., 1994a). Placement of these cultures into a 60°C incubator without additional electron donors or acceptors allowed growth to recommence. The bacterium can also be freeze-dried and is supplied in this manner by the American Type Culture Collection.

Characteristics for *Thermodesulfovibrio* are listed in Table B8.5. All methodology for analyses of these characteristics are presented in Henry et al. (1994a) and/or Sonne-Hansen and Ahring (1999).

DIFFERENTIATION OF THE GENUS *THERMODESULFOVIBRIO* FROM OTHER GENERA

In the *"Nitrospira"* group (Fig. B8.8), *Thermodesulfovibrio* is the only thermophilic sulfate reducing genus. It does not form the spiral or helical cells of other members of this group (*Nitrospira*

and *Leptospirillum*) nor does it contain magnetosomes as does *"Candidatus* Magnetobacterium bavaricum". Comparison with other Gram-negative thermophilic sulfate reducers (Table B8.5) show it to have a unique morphology as a curved rod (vibrio) and have a lower mol% G + C than the other species. In addition, its phylogeny based on 16S rRNA sequence places it in a lineage distinct from previously defined phylogenetic lines of other mesophilic sulfate-reducing bacteria, hyperthermophilic *Archaea* (Henry et al., 1994a), and other thermophilic sulfate-reducing bacteria (Table B8.5).

TAXONOMIC COMMENTS

Classification of *Thermodesulfovibrio* as a distinct genus of thermophilic sulfate-reducing bacteria is based on its unique morphological and other phenotypic and genotypic characteristics, as well as its 16S rRNA sequence.

FURTHER READING

Henry, E.A., R. Devereux, J.S. Maki, C.C. Gilmour, C.R. Woese, L. Mandelco, R. Schauder, C.C. Remsen and R. Mitchell. 1994. Characterization of a new thermophilic sulfate-reducing bacterium *Thermodesulfovibrio yellowstonii*, gen. nov. and sp. nov.: its phylogenetic relation-

TABLE B8.4. Composition of media for enrichment and cultivation of *Thermodesulfovibrio*

Component	Enrichment[a]	Cultivation[b]	Cultivation[c]
NaCl	1	1	
KCl			0.5
Na_2SO_4		3	4
$MgSO_4 \cdot 7H_2O$	2		
$MgCl_2 \cdot 6H_2O$		0.4	0.4
NH_4Cl	1	0.3	0.25
$CaCl_2 \cdot 2H_2O$	0.1	0.15	0.15
KH_2PO_4	0.015	0.2	
Na_2HPO_4			0.2
MOPS	2.1		
$FeSO_4 \cdot 7H_2O$	0.05		
Yeast extract	0.05		0.5
Sodium lactate	3 ml of 60% solution	20 mM	3 ml of 60% solution
Trace element solution	1 ml (SL7)[b]	1 ml (SL10)[e]	10 ml[c]
Vitamin solution	1 ml[f]	1 ml[g]	10 ml[c]
Resazurin			1.0 mg
Sodium thioglycollate[d]	0.1		0.2
Sodium ascorbate[d]	0.1		0.2
Bicarbonate		30 ml of solution[h]	1.3
Sulfide solution[i]		3 ml	
Selenite-tungstate solution[j]		1 ml	

[a]From Henry et al. (1994a). In g/l unless otherwise indicated. The medium should be adjusted to pH 7.0 with 1 M HCl, boiled under N_2, dispensed into serum vials, sealed with black butyl rubber stoppers, gassed with nitrogen, and autoclaved.

[b]From Widdel and Pfennig (1981). In g/l unless otherwise indicated. Modified by changing the trace element solution, the vitamin solution, and the selenite-tungstate solution.

[c]Medium 1895 from American Type Culture Collection (1996). In g/l unless otherwise indicated. Combine the first 11 components and adjust pH to 7.5. Heat to boiling and cool under 100% N_2. Add remaining components. Seal tube anaerobically under same gas phase and autoclave.

[d]Sodium thioglycollate and sodium ascorbate used as reducing agents should be aseptically filtered (0.2 μm pore size), at a final concentration of 0.1 g/l of medium and added after the medium has been autoclaved and cooled.

[e]From Widdel et al. (1983).

[f]Vitamin solution of Widdel and Pfennig (1981). In final concentration: biotin (10 ng/ml), calcium pantothenate (10 ng/ml), thiamine (20 ng/ml), p-aminobenzoic acid (50 ng/ml), nicotinic acid (100 ng/ml), and pyridoxamine (250 ng/ml) supplemented with cyanocobalamin, 10 mg/l of vitamin solution. The vitamin solution is added aseptically filtered (0.2 μm pore size) after the medium has been autoclaved and cooled.

[g]Vitamin solution of Pfennig (1978). In final concentration: biotin (10 ng/ml), calcium pantothenate (25 ng/ml), thiamine (50 ng/ml), p-aminobenzoic acid (50 ng/ml), nicotinic acid (100 ng/ml), and pyridoxamine (250 ng/ml). In addition, 1 ml of cyanocobalamin, 5 mg/100 ml, was added to the medium.

[h]$NaHCO_3$, 84 g/l, CO_2 saturated and autoclaved under CO_2 atmosphere.

[i]$Na_2S \cdot 9H_2O$, 120 g/l, autoclaved under N_2 atmosphere.

[j]$Na_2SeO_3 \cdot 5H_2O$, 6 mg/l; $Na_2WO_4 \cdot 2H_2O$, 8 mg/l; NaOH, 0.5 g/l.

TABLE B8.5. Comparison of characteristics of the Gram-negative thermophilic sulfate reducing bacteria[a]

Characteristic	Thermodesulfovibrio yellowstonii and "T. islandicus"[b]	Thermodesulfobacterium commune[c]	Thermodesulfobacterium mobile[d]	Desulfacinum infernum[e]	Thermodesulforhabdus norvegicus[f]
Morphology	vibrio	rod	rod	oval	rod
Dimensions (μm)	0.3 × 1.5 [0.4 × 1.7]	0.3 × 0.9	0.6 × 2.0	1.5 × 2.5–3.0	1 × 2.5
Motility/Flagellation	+/single polar	−	+/single polar	−	+/single polar
Mol% G + C	29.5 [38]	34	38(31)[g]	64	51
16s rRNA Phylogeny (Phylum)	"Nitrospiraceae" (Nitrospirae)	Thermodesulfo-bacteriaceae (Thermodesulfobacteria)	Unknown	"Desulfobulbaceae" (Proteobacteria)	"Desulfobulbaceae" (Proteobacteria)
Sulfite reductases:					
Desulfoviridin	−	−	−	−	−
Desulfofuscidin	+ [unknown]	+	+		−
Cytochromes	c [unknown]	c_3	c		c
Temperature optimum, °C	65	70	65	60	60
pH optimum	6.8–7.0	near 7.0		7.1–7.5	6.9
Fermentation of:					
Lactate	−	−	(+)		
Pyruvate	+	+	+		
Electron acceptors:					
Sulfate	+	+	+	+	+
Thiosulfate	+	+	+	+	−
Sulfite	+ [−]	−	+	+	+
Sulfur	−		−	−	−
Nitrate	− [+]	−		−	−
Fumarate	−	−			
Incomplete oxidation	+	+	+		
Electron donors:					
Hydrogen + acetate	+ [H₂ alone, +]	+	+		−
Formate	(+) [+]		+	+	−
Lactate	+	+	+	+	+
Pyruvate	+	+	+	+	+
Acetate	−	−	−	+	+
Propionate	−			+	−
Butyrate	−			+	+
Ethanol	−	−	−	+	+
Malate	−	−	−	+	+

[a]Only one strain has been described for each species. For symbols see standard definitions; (+), slight growth.

[b]Data from Henry et al. (1994a) for *T. yellowstonii* and, when *"T. islandicus"* is different, data from Sonne-Hansen and Ahring (1999) appears in brackets.

[c]Data from Zeikus et al. (1983).

[d]Data from Rozanova and Pivovarova (1988).

[e]Data from Rees et al. (1995).

[f]Data from Beeder et al. (1995).

[g]Mol% G+C of 31 determined by Henry et al. (1994a).

ship to *Thermodesulfobacterium commune* and their origins deep within the bacterial domain. Arch. Microbiol. *161*: 62–69.

Sonne-Hansen, J. and B.K. Ahring. 1999. *Thermodesulfobacterium hverager-dense* sp. nov., and *Thermodesulfovibrio islandicus* sp. nov., two thermophilic sulfate reducing bacteria isolated from a Icelandic hot spring. Syst. Appl. Microbiol. *22*: 559–564.

List of species of the genus Thermodesulfovibrio

1. **Thermodesulfovibrio yellowstonii** Henry, Devereux, Maki, Gilmour, Woese, Mandelco, Schauder, Remsen and Mitchell 1994b, 595[VP] (Effective publication: Henry, Devereux, Maki, Gilmour, Woese, Mandelco, Schauder, Remsen and Mitchell 1994a, 68).

   yel.low.sto.ni.i. M.L. gen. n. *yellowstonii* of Yellowstone.

   Morphological, cultural, and biochemical characteristics are the same as those described for the genus. Figs. B8.9 and B8.10 illustrate morphological features. Other characteristics listed in Table B8.5.

   Originally isolated from hydrothermal vent water in Yellowstone Lake, Yellowstone National Park, Wyoming. Found in anaerobic niches associated with freshwater hydrothermal vent water.

   The mol% G + C of the DNA is: 29.5 ± 1.0 (T_m).
   Type strain: YP87, ATCC 51303, DSMZ 11347.
   GenBank accession number (16S rRNA): L14619.

2. **"Thermodesulfovibrio islandicus"** Sonne-Hansen and Ahring 1999, 563.

   is.lan.di.cus. M.L. masc. adj. *islandicus* Icelandic, pertaining to Iceland.

   Morphological, cultural, and biochemical characteristics are the same as those described for the genus and found in Table B8.5. Differences between *"T. islandicus"* and *T. yellowstonii* include the mol% G + C DNA composition (38% and 29.5%, respectively), the ability to use nitrate and not sulfite as an electron acceptor, and growth of cells in chains.

Furthermore, DNA–DNA hybridization shows a 69% relatedness between the two species.

Originally isolated from a microbial mat in a slightly alkaline thermal spring in Iceland. Found in anaerobic niches associated with thermal springs.

The mol% G + C of the DNA is: 38 (HPLC).
Deposited strain: DSMZ 12570.
GenBank accession number (16S rRNA): X96726.

Phylum BIX. Deferribacteres *phy. nov.*

GEORGE M. GARRITY AND JOHN M. HOLT

De.fer.ri.bac' te.res. M.L. fem. pl. n. *Deferribacterales* type order of the phylum; dropping ending to denote a phylum; M.L. fem. pl. n. *Deferribacteres* the phylum of *Deferribacterales.*

The phylum *Deferribacteres* is a distinct lineage within the *Bacteria* based on phylogenetic analysis of 16S rDNA sequences. At present the members of this phylum are organized into a single class, order, and family. The relationships within the phylum may, however, be more distant and warrant further subdivision in the future. Chemoorganotrophic heterotrophs that respire anaerobically with terminal electron acceptors including Fe(II), Mn(IV), S^0, Co(III), and nitrate. Placement of one genus, *Synergistes*, is provisional.

Type order: **Deferribacterales** ord. nov.

Class I. **Deferribacteres** *class. nov.*

HARALD HUBER AND KARL O. STETTER

De.fer.ri.bac' te.res. M.L. fem. pl. n. *Deferribacterales* type order of the class; dropping ending to denote a class; M.L. fem. pl. n. *Deferribacteres* the class of *Deferribacterales.*

The description is the same as for the family *Deferribacteraceae*. Only one order known.

Type order: **Deferribacterales** ord. nov.

Order I. **Deferribacterales** *ord. nov.*

HARALD HUBER AND KARL O. STETTER

De.fer.ri.bac.ter.a' les. M.L. fem. pl. n. *Deferribacter* type genus of order; *-ales* ending to denote an order; M.L. fem. pl. n. *Deferribacterales* the order of *Deferribacter.*

The description is the same as for the family *Deferribacteraceae*. Only one family known.

Type genus: **Deferribacter** Greene, Patel and Sheehy 1997, 508

Family I. **Deferribacteraceae** *fam. nov.*

HARALD HUBER AND KARL O. STETTER

De.fer.ri.bac.ter.a' ce.ae. M.L. masc. n. *Deferribacter* type genus of the family; *-aceae* ending to denote a family; M.L. fem. pl. n. *Deferribacteraceae* the family of *Deferribacter.*

Straight to bent and flexible rods or vibrio-shaped cells. Gram-negative. No spores formed. Mesophilic to thermophilic, optimal growth temperature between 35 and 60°C. **Anaerobic respiration of different organic substrates by using Fe (III), Mn (IV), or nitrate as electron acceptors or heterotrophic growth by fermen-**tation. Although the common physiological features are limited, the three genera *Deferribacter*, *Flexistipes*, and *Geovibrio* form one separate lineage on the 16S rRNA based phylogenetic tree with similarities around 89%. Each genus is so far represented by only one species, therefore their positioning in one family may change, after further representatives of this group have been isolated.

So far, the family harbors three genera: *Deferribacter, Flexistipes,* and *Geovibrio.* The genus *Synergistes* (Allison et al., 1992) does not belong to this group, it exhibits sequence differences of about 19% to the other three representatives of the family.

The mol% G + C of the DNA is: 34–43.

Type genus: **Deferribacter** Greene, Patel and Sheehy 1997, 508.

Key to the genera of the family Deferribacteraceae

I. Cells are straight to bent rods. Anaerobic growth with Fe (III), Mn (IV), or nitrate as electron acceptor. Complex organic substrates and organic acids serve as electron donors. No fermentation. Temperature optimum 60°C; salinity range 0–6% NaCl. The mol% G + C of genomic DNA is 34. Thrives in petroleum reservoirs.

Genus I. *Deferribacter.*

II. Cells are flexible rods. Anaerobic growth by fermentation of complex organic compounds. Temperature optimum 45–50°C; salinity range 3–18% NaCl. The mol% G + C of genomic DNA is 39. Lives in brine waters of the red sea.

Genus II. *Flexistipes,* p. 468

III. Cells are vibroid. Anaerobic respiration of complex organic substrates, organic acids or proline with Fe (III). Mesophilic. Salinity range 0–2 NaCl%. The mol% G + C of genomic DNA is 43. Lives in surface sediments.

Genus III. *Geovibrio,* p. 468

Genus I. **Deferribacter** *Greene, Patel and Sheehy 1997, 508*[VP]

ANTHONY C. GREENE AND BHARAT K. C. PATEL

De.fer.ri.bac′ter. L. pref. *de* from; L. n. *ferrum* iron; Gr. hyp. masc. n. *bacter* rod; M.L. masc. n. *Deferribacter* rod that reduces iron.

Gram-negative rods. Nonsporulating. Nonmotile. Anaerobic. Fe(III), Mn(IV), and nitrate utilized as electron acceptors. Complex organic extracts and organic acids used as electron donors. No fermentation occurs.

Type species: **Deferribacter thermophilus** Greene, Patel and Sheehy 1997, 509.

ENRICHMENT AND ISOLATION PROCEDURES

An enrichment culture of *Deferribacter* can be obtained after 3–5 d incubation in MR medium* at 60°C. Pure cultures can be isolated using the agar shake dilution technique. The process involves serially diluting enrichment cultures in MR medium amended with 20 mM NaNO₃ in place of MnO₂ and fortified with purified agar at a concentration of 2%. Several colonies are selected and tested for the ability to reduce Mn(IV) and Fe(III). The purity of the cultures are checked microscopically. Liquid cultures retain viability for several months at room temperature or when lyophilized.

DIFFERENTIATION OF THE GENUS *DEFERRIBACTER* FROM OTHER GENERA

Phylogenetic analysis places *Deferribacter* in a phylum with *Flexistipes* and *Geovibrio* (Fig. B9.1). *D. thermophilus* grows at a higher temperature and has a lower mol% G + C DNA content than either *Geovibrio ferrireducens* or *Flexistipes sinusarabici.* Furthermore, it differs in cell morphology from *G. ferrireducens* and grows at lower salinity than *F. sinusarabici.* Unlike *D. thermophilus, G. ferrireducens* and *F. sinusarabici* are unable to reduce Mn(IV) and *F. sinusarabici* does not reduce Fe(III). *D. thermophilus* and *G. ferrireducens* respire anaerobically, while *F. sinusarabici* has a fermentative metabolism. *D. thermophilus* is thermophilic, while *F. sinusarabici* is only moderately thermophilic and *G. ferrireducens* is mesophilic.

FURTHER READING

Greene, A.C., B.K.C. Patel and A.J. Sheehy. 1997. *Deferribacter thermophilus* gen. nov., sp. nov., a novel thermophilic manganese- and iron-reducing baterium isolated from a petroleum reservoir. Int. J. Syst. Bacteriol. *47:* 505–509.

List of species of the genus Deferribacter

1. **Deferribacter thermophilus** Greene, Patel and Sheehy 1997, 509.[VP]

   ther.mo′phil.us. Gr. adj. *thermus* warm, hot; Gr. adj. *philos* loving; M.L. masc. adj. *thermophilus* heat loving.

Gram-negative rods, 0.3–0.5 × 1–5 μm. Nonsporulating. Motility is not evident. Anaerobic. Thermophilic. Optimum temperature ~60°C (range 50–65°C); optimum NaCl 20g/l (range 0–50 g/l); pH 6.5 (range pH 5–8). Fe(III), Mn(IV), and nitrate used as electron acceptors. Yeast extract, peptone, Casamino acids, tryptone, hydrogen, acetate, malate, citrate, pyruvate, lactate, succinate, and valerate used as electron donors. Sensitive to penicillin, vancomycin, streptomycin, and cycloserine. Resistant to tetracycline.

Strain BMA was isolated from the produced formation

---

*MR medium contains (per liter distilled H₂O): NH₄Cl, 1 g; K₂HPO₄ ·3H₂O, 0.08 g; MgCl₂ ·6H₂O, 4.5 g; CaCl₂ ·2H₂O, 0.375 g; NaCl, 32 g; NaHCO₃, 3.6 g; yeast extract, 2 g; MnO₂, 1.3 g. After pH is adjusted to 7.1, medium is boiled, cooled under a stream of N₂ and dispensed into serum bottles under N₂/CO₂ (80:20). The bottles are capped with butyl rubber septa and autoclaved.

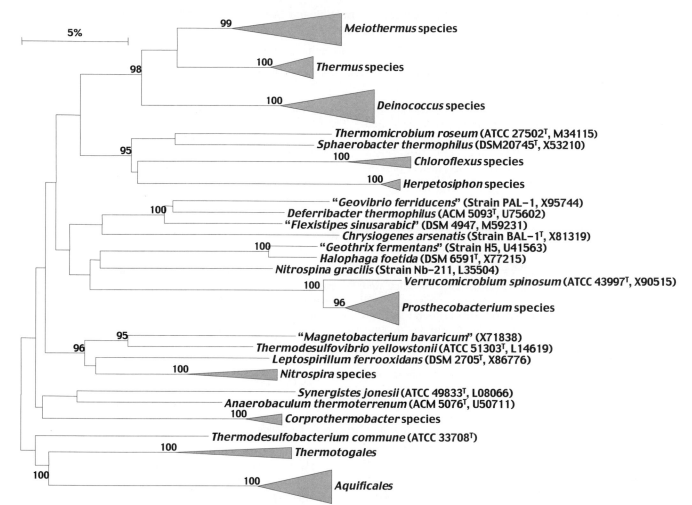

FIGURE B9.1. The position of *Deferribacter thermophilus* (ACM 5093[T]) in relation to its phylogenetic relatives is shown. The strain numbers from culture collections and their corresponding 16S rRNA gene sequence accession numbers extracted from GenBank/EMBL are shown in brackets. Various members of the same order/genera are shown as clusters and are represented as filled triangles. The members for the *Aquificales* cluster include *Aquifex pyrophilus* (DSM 6858[T], M83548), *Calderobacterium hydrogenophilum* (INMI Z-829[T], Z30242), *Hydrogenobacter thermophilus* (JCM 8795[T], Z30189), and *Thermocrinis ruber* (DSM 12173[T], AJ005640). The *Thermotogales* cluster includes *Fervidobacterium nodosum* (ATCC 35602[T], M59177), *Thermotoga elfii* (DSM 9442[T], X80790), and *Thermosipho africanus* (DSM 5309[T], M83140). The *Coprothermobacter* cluster includes *Coprothermobacter platensis* (DSM 11748[T], Y08935) and *Coprothermobacter proteolyticus* (ATCC 35245[T], X69335). The *Chloroflexus* cluster includes *Chloroflexus aurantiacus* (ATCC 29366[T],D38365) and *Chloroflexus aggregans* (DSM 9485[T], D32255). The *Herpetosiphon* cluster includes *Herpetosiphon aurantiacus* (ATCC 23779[T], M34117) and *Herpetosiphon geysericola* (ATCC 23076[T], AF039293). The *Thermus* cluster includes *Thermus thermophilus* (ATCC 27634[T], X07998), *Thermus aquaticus* (ATCC 25104[T], L09663), and *Thermus scotoductus* (DSM

8553[T], AF032127). The *Meiothermus* cluster includes *Meiothermus cerbereus* (DSM 11376[T], Y13594), *Meiothermus ruber* (ATCC 35948[T], Z15059), *Meiothermus chliarophilus* (DSM 9957[T], X84212), and *Meiothermus silvanus* (DSM 9946[T], X84211). The *Deinococcus* cluster includes *Deinococcus proteolyticus* (DSM 20540[T], Y11331), *Deinococcus radiodurans* (DSM 20539[T], Y11332), *Deinococcus radiopugnans* (ATCC 19172[T], Y11334), and *Deinococcus grandis* (DSM 3963[T], Y11329). The *Nitrospira* cluster includes *Nitrospira marina* (ATCC 43039[T], L35501) and *"Nitrospira moscoviensis"* (DSM 10035, X82558). The *Prosthecobacter* cluster includes *Prosthecobacter dejongeii* (ATCC 27091[T], U60012), *Prosthecobacter debontii* (ATCC 70020[T], U60014), *Prosthecobacter fusiformis* (DSM 8960[T],U60015), and *Prosthecobacter vanneervenii* (ATCC 70019[T], U60013). Phylogenetic analysis was performed on 933 unambiguous nucleotides using dnadist and neighbor-joining programs which form part of the PHYLIP suit of software. Bar 5 nucleotide changes per 100 nucleotides. The following abbreviations have been used: T, Type Culture; ATCC, American Type Culture Collection; DSM, Deutsche Sammlung von Mikroorganismen und Zellkulturen GmbH; ACM, Australian Collection of Microorganisms; INMI, Culture Collection of the Institute of Microbiology; JCM, Japan Collection of Microorganisms.

water of Beatrice oil field, a high-temperature, seawater-flooded petroleum reservoir located in the British sector of the North Sea.

The mol% G + C of the DNA is: 34 (T_m).

Type strain: BMA, ACM 5093.

GenBank accession number (16S rRNA): U75602.

Genus II. **Flexistipes** Fiala, Woese, Langworthy and Stetter 2000, 1415[VP] (Effective publication: Fiala, Woese, Langworthy and Stetter 1990, 125)

KARL O. STETTER AND HARALD HUBER

Fle.xi.sti'pes. M.L. masc. n. *flexus* winding; M.L. masc. n. *stipes* stick; M.L. masc. n. *Flexistipes* the flexible stick.

Flexible rods, nonmotile. No spores formed. Gram negative. **Heterotrophic growth under anaerobic conditions. Halophilic, neutrophilic, and slightly thermophilic.** Muramic acid and mesodiaminopimelic acid present. Ester lipids in the cytoplasmic membrane. Based on 16S rRNA analyses, *Flexistipes* **clusters with the genera** *Deferribacter* **and** *Geovibrio*.

The mol% G + C of the DNA is: ~39

Type species: **Flexistipes sinusarabici** Fiala, Woese, Langworthy and Stetter 2000, 1415 (Effective publication: Fiala, Woese, Langworthy and Stetter 1990, 125).

ENRICHMENT AND ISOLATION PROCEDURES

Flexistipes can be enriched using the procedure of Balch and Wolfe (1976) for cultivation of strict anaerobes. Yeast extract (1 g/l) and sulfur (25 g/l) were added to filter-sterilized water and enrichment cultures were incubated at 46°C without shaking. Isolates were obtained by serial dilution. For defined growth experiments, a synthetic medium (3 × MB) was prepared based upon ZoBells marine broth (ZoBell, 1941), but containing three times the amount of inorganic compounds and supplemented with 1 g/l yeast extract.

MAINTENANCE PROCEDURES

Cultures remain viable at 4°C for at least one month. Long term storage at −140°C over liquid nitrogen is possible after suspending cells in 5% DMSO.

DIFFERENTIATION OF THE GENUS *FLEXISTIPES* FROM OTHER GENERA

Phylogenetic analysis places *Flexistipes* in a family with *Deferribacter* and *Geovibrio*. *F. sinusarabici* can be distinguished from *D. thermophilus* and *G. ferrireducens* by growth at higher salinity and by having fermentative metabolism rather than respiratory metabolism.

FURTHER READING

Fiala, G., C.R. Woese, T.A. Langworthy and K.O. Stetter. 1990. *Flexistipes sinusarabici,* a novel genus and species of eubacteria occurring in the Atlantis II Deep brines of the Red Sea. Arch. Microbiol. *154*: 120–126.

List of species of the genus Flexistipes

1. **Flexistipes sinusarabici** Fiala, Woese, Langworthy and Stetter 2000, 1415[VP] (Effective publication: Fiala, Woese, Langworthy and Stetter 1990, 125).

   sinus.a.ra'bi.ci. L. masc. g. *sinus* gulf; L. masc. adj. *arabicus* arabic; L. masc. g. *sinusarabici* of the Red Sea; describing the place of isolation.

   Flexible rods, 0.3 × 5–50 μm. Nonmotile. Optimum temperature 45–50°C (range 30–53°C); pH 6–8; 3–10% NaCl. Doubling time of 8.5 h under optimal conditions. Hetero- trophic growth on yeast extract, meat extract, peptone, and tryptone. Weak growth on acetate and Casamino acids. Strictly anaerobic. Lipids contain C_{16}–C_{18} fatty acids as principal components. Sensitive to penicillin, ampicillin, vancomycin, and streptomycin. Resistant to tetracycline and rifampicin. Isolated from Atlantic II deep brines of the Red Sea (depth 2000 m).

   The mol% G + C of the DNA is: 39 (T_m).

   Type strain: MAS 10, DSMZ 4947.

   GenBank accession number (16S rRNA): M59231.

Genus III. **Geovibrio** Caccavo, Coates, Rossello-Mora, Ludwig, Schleifer, Lovley and McInerney 2000, 1415[VP] (Effective publication: Caccavo, Coates, Rossello-Mora, Ludwig, Schleifer, Lovley and McInerney 1996, 375)

FRANK CACCAVO, JR.

Ge.o.vi'bri.o. Gr. n. *ge* the earth; L. hyp. masc. v. *vibrare* to vibrate; M.L. masc. n. *Geovibrio* vibrating from the earth.

Vibrioid Gram-negative cells 0.5 × 2–3 μm and motile by a single polar flagellum. Spore formation is not observed. **Cells occur singly or in chains.** Strictly anaerobic, obligately respiratory chemoorganotroph that oxidizes acetate with Fe(III), S⁰, or Co(III) as an electron acceptor. Hydrogen, **proline**, lactate, propionate, **Casamino acids**, succinate, fumarate, pyruvate, and yeast extract are also used as electron donors for Fe(III) reduction, while other carboxylic acids, sugars, alcohols, amino acids, and aromatic hydrocarbons are not. Optimum temperature, 35°C. Cells contain *c*-type cytochromes. Grows best in freshwater medium, but will grow in up to 2% NaCl. 16S rRNA sequence analysis shows that *Geovibrio* currently **forms a separate line of descent within the Bacteria.**

The mol% G + C of the DNA is: 42.84 ± 1.51.

Type species: **Geovibrio ferrireducens** Caccavo, Coates, Rossello-Mora, Ludwig, Schleifer, Lovley and McInerney 2000, 1415 (Effective publication: Caccavo, Coates, Rossello-Mora, Ludwig, Schleifer, Lovley and McInerney 1996, 375.)

FURTHER DESCRIPTIVE INFORMATION

Comparative 16S rRNA sequence analysis of the genus *Geovibrio* shows that it is phylogenetically distinct, with only a moderate relationship to the genus *Flexistipes* (Fig. B9.2). These two genera form a separate phylum among the known major bacterial lines of descent.

Geovibrio are obligately anaerobic respiratory chemoorgano-

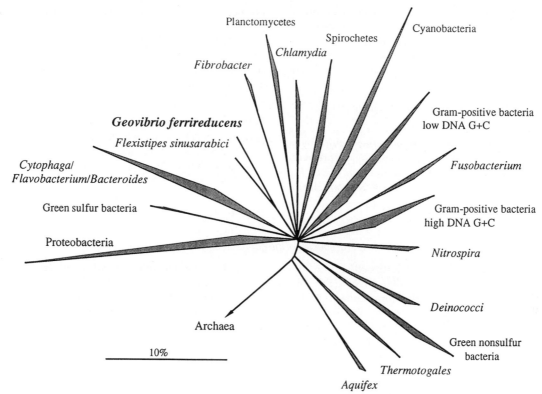

FIGURE B9.2. Phylogenetic tree showing the relationship between *Geovibrio ferrireducens* and representatives of the major bacterial lines of descent. The tree is based on distance matrix and parsimony analyses of about 2800 16S rRNA sequences from bacteria. Homologous archaeal sequences were included to root the tree. The triangles indicate major bacterial phylogenetic groups. Multifurcations indicate that a relative branching order could not be unambiguously determined. Bar = 10% estimated sequence divergence. (Reprinted with permission from Caccavo, Jr. et al., *Archives of Microbiology 165:*370–376, 1996, ©Springer-Verlag.)

trophs that use acetate as an electron donor for the dissimilatory reduction of ferric pyrophosphate, ferric oxyhydroxide, ferric citrate, Co(III)-EDTA, or S⁰. The oxidation of proline, hydrogen, lactate, propionate, succinate, fumarate, pyruvate, Casamino acids, or yeast extract can also be coupled to Fe(III) reduction. *Geovibrio* is the first organism known to couple the oxidation of an amino acid to Fe(III) reduction. Other carboxylic acids, sugars, alcohols, amino acids, and aromatic hydrocarbons are not used as electron donors. *Geovibrio* does not use oxygen, Mn(IV), U(VI), Cr(VI), nitrate, sulfate, sulfite, or thiosulfate as an electron acceptor.

Geovibrio can be cultured in a defined anaerobic basal medium* with various electron donors and acceptors. The optimum temperature for growth is 35°C. Optimal growth occurs in freshwater medium, although the genus will grow in up to 2% NaCl.

Difference spectra of whole cells show the presence of *c*-type cytochromes. The reduced cytochromes have absorbance peaks at 553 nm, 523 nm, and 422 nm.

The type strain of *Geovibrio* was isolated from a hydrocarbon-contaminated ditch.

ENRICHMENT AND ISOLATION PROCEDURES

Geovibrio is selectively enriched using the defined anaerobic basal medium with proline as an electron donor and amorphous ferric oxyhydroxide as a sole electron acceptor. Bottles or tubes (100-, 50- or 20-ml capacity) can serve as culture vessels. The medium is inoculated with anaerobic sediment; 0.5–1% of the total culture volume should be added. Enrichments are incubated in the dark at 25–35°C. After the ferric oxyhydroxide has been reduced to a dark, partially magnetic mineral (12–21 days), transfers to fresh medium are made (10% inoculum). The enrichments should be mixed well before subculturing. A series of six transfers should be made. The sixth transfer is streaked for isolation on anoxic ferric pyrophosphate agar slants, which are made by adding a final concentration of 1.5% (w/v) purified agar to the anaerobic basal medium with proline as the electron donor and soluble ferric pyrophosphate as the electron acceptor. The pinpoint colonies and the surrounding medium turn white as the Fe(III) in the medium is reduced. Colonies are restreaked onto agar slants until a pure culture of vibrioid cells is obtained.

MAINTENANCE PROCEDURES

Stock cultures may be maintained in liquid anaerobic basal medium or on anaerobic agar slants using acetate as an electron

---

*The defined anaerobic basal medium has the following composition (in g/l of distilled water): NaHCO₃, 2.5; NH₄, Cl 1.5; KH₂PO₄, 0.6; KCl, 0.1; vitamins, 10 ml; and trace minerals, 10 ml. The medium is boiled and cooled under a constant stream of N₂/CO₂ (80:20, v/v), dispensed into aluminum-sealed culture tubes under the same gas phase, capped with butyl rubber stoppers, and sterilized by autoclaving (121°C, 20 min.). The pH of the medium is 7.0. The vitamin mix contains (in mg/l of distilled water): biotin, 2; folic acid, 2; pyridoxine hydrochloride, 10; riboflavin, 5; thiamine, 5; nicotinic acid, 5; pantothenic acid, 5; vitamin B₁₂, 0.1; *p*-aminobenzoic acid, 5; and thiotic acid, 5. The mineral mix contains (in g/l of distilled water): N(CH₂COOH)₃, 1.5; MgSO₄·6H₂O, 3.0; MnSO₄·H₂O, 0.5; NaCl, 1.0; FeSO₄·7H₂O, 0.1; CaCl₂·2H₂O, 0.1; ZnCl, 0.13; CuSO₄, 0.01; AlK(SO₄)·12H₂O, 0.01; H₃BO₂, 0.01; Na₂MnoO₄·2H₂O, 0.025; NiCl₂·6H₂O, 0.024; Na₂WO₄, 0.025.

donor and soluble ferric pyrophosphate as an electron acceptor. Culture bottles or tubes are sealed anaerobically and stored in the dark at 30°C. Stock cultures are transferred monthly.

Geovibrio may be preserved indefinitely be suspending cells in anaerobic basal medium containing 5% DMSO and storing in liquid nitrogen.

DIFFERENTIATION OF THE GENUS *GEOVIBRIO* FROM OTHER GENERA

Geovibrio differs from other anaerobic iron-reducing bacteria by its vibrioid cell morphology. A characteristic physiological property of *Geovibrio* is its growth with proline as an electron donor for Fe(III) reduction. The 16S rRNA sequence of *Geovibrio* distin-

guishes it phylogenetically from all known iron-reducing bacteria.

TAXONOMIC COMMENTS

The 16S rRNA sequence (1545 nts) of the type species of *Geovibrio* showed 83.1% identical residues with that of *Flexistipes sinusarabici*. These two genera form a separate line of descent within the *Bacteria*.

FURTHER READING

Caccavo, F., Jr., J.D. Coates, R.A. Rossello-Mora, W. Ludwig, K.H. Schleifer, D.R. Lovley and M.J. McInerney. 1996. *Geovibrio ferrireducens*, a phylogenetically distinct dissimilatory Fe(III)-reducing bacterium. Arch. Microbiol. *165*: 370–376.

List of species of the genus Geovibrio

1. **Geovibrio ferrireducens** Caccavo, Coates, Rossello-Mora, Ludwig, Schleifer, Lovley and McInerney 2000, 1415[VP] (Effective publication: Caccavo, Coates, Rossello-Mora, Ludwig, Schleifer, Lovley and McInerney 1996, 375.)

   fer.ri.re.du' cens. L. n. *ferrum* iron; L. part. adj. *reducens* converting to a different state; M.L. adj. *ferrireducens* reducing iron.

The characteristics are as described for the genus.

Occur in surface sediments of a hydrocarbon-contaminated ditch in Norman, Oklahoma, U.S.A.

The mol% G + C of the DNA is: 42.84 ± 1.51 (HPLC).

Type strain: PAL-1, ATCC 51996.

GenBank accession number (16S rRNA): X95744.

Genus Incertae Sedis I. **Synergistes** *Allison, Mayberry, McSweeney and Stahl 1993, 398*[VP]
(Effective publication: Allison, Mayberry, McSweeney and Stahl 1992, 528)

MILTON J. ALLISON, BARBARA J. MACGREGOR AND DAVID A. STAHL

Sy.ner.gis' tes. Arbitrarily derived from English n. *synergist*; M.L. masc. n. *synergistes* a co-worker.

Gram-negative, nonmotile rods. Nonsporing, **strictly anaerobic** chemoorganotroph. Growth occurs in media that contain yeast extract plus any of a variety of peptones. **Certain amino acids and/ or pyridinediols are fermented** and products formed from peptones include formate, acetate, propionate, H₂, and ammonia. **Carbohydrates are not fermented**. The 16S rRNA is affiliated with the bacterial domain. No sequence similarities greater than ∼84% were identified among near-full length 16S rRNA sequences in the 7.1 release of the RDP (Maidak et al., 1999) or GenBank (as of March 2000). The 16S rDNA is most closely related (80–83% sequence similarity) to two recently described amino acid fermenters, *Aminomonas paucivorans* and *Thermoanaerovibrio acidaminovorans*, and also demonstrates a distant relationship to *Dethiosulfovibrio peptidovorans* by both maximum likelihood (Fig. B9.3) and neighbor-joining analysis. The only known habitat is the rumen but other gastrointestinal habitats are likely.

The mol% G + C of the DNA is: 57–59.

Type species: **Synergistes jonesii** Allison, Mayberry, McSweeney and Stahl 1993, 398 (Effective publication: Allison, Mayberry, McSweeney and Stahl 1992, 528).

FURTHER DESCRIPTIVE INFORMATION

All strains are obligate anaerobes that use arginine, histidine, or 2,3-DHP or 3,4-DHP as growth substrates. Morphologic or physiologic differences between the strains were not detected but differences between strains were observed by comparisons of cel-

lular proteins (PAGE) or profiles of cellular fatty acids (Allison et al., 1992).

ENRICHMENT AND ISOLATION PROCEDURES

Isolates able to degrade 3-hydroxy-4(1H)-pyridone (3,4-DHP) and its isomer 2,3-DHP were obtained from enrichment cultures of mixed rumen populations from a goat in Hawaii. Enrichments were in media containing 3,4-DHP or 2,3-DHP. Isolation success using anaerobic roll tube methods was enhanced by selecting black colonies that formed in roll tube cultures in media containing 3,4-DHP and FeCl₃.

MAINTENANCE PROCEDURES

Cultures may be recovered from freeze-dried preparations stored at 5°C, or cultures plus glycerol (10%) stored at −70°C.

DIFFERENTIATION OF THE GENUS *SYNERGISTES* FROM OTHER GENERA

The most convincing evidence for uniqueness of the genus is based on an analysis of the 16S rDNA sequence. No sequence similarities greater than ∼84% were identified among near-full length 16S rDNA sequences in the 7.1 release of the RDP or GenBank (as of March 2000). The 16S rDNA is most closely related to two recently described amino acid fermenters, *Aminomonas paucivorans* and *Thermoanaerovibrio acidaminovorans*. No other anaerobic, Gram-negative bacteria able to degrade the 2,3- or 3,4-pyridinediols are known.

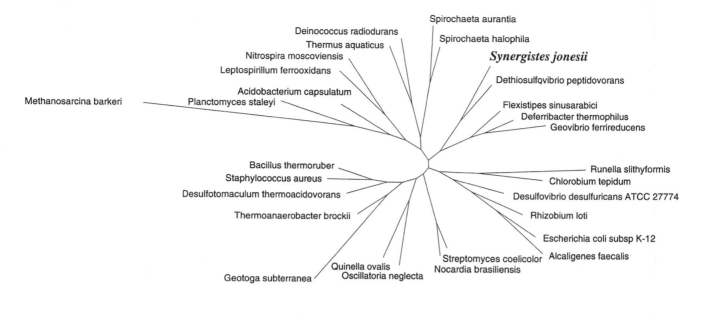

Spirochaeta aurantia
Spirochaeta halophila
Synergistes jonesii
Dethiosulfovibrio peptidovorans
Flexistipes sinusarabici
Deferribacter thermophilus
Geovibrio ferrireducens
Deinococcus radiodurans
Thermus aquaticus
Nitrospira moscoviensis
Leptospirillum ferrooxidans
Acidobacterium capsulatum
Planctomyces staleyi
Methanosarcina barkeri
Runella slithyformis
Chlorobium tepidum
Desulfovibrio desulfuricans ATCC 27774
Rhizobium loti
Escherichia coli subsp K-12
Alcaligenes faecalis
Bacillus thermoruber
Staphylococcus aureus
Desulfotomaculum thermoacidovorans
Thermoanaerobacter brockii
Streptomyces coelicolor
Nocardia brasiliensis
Quinella ovalis
Oscillatoria neglecta
Geotoga subterranea

0.10

FIGURE B9.3. 16S rRNA phylogeny of *Synergistes jonesii*. The maximum likelihood phylogenetic tree was constructed using fast DNAml (Olsen et al., 1994a), as implemented in the ARB sequence analysis package. Bar = 0.1 base changes per position.

List of species of the genus Synergistes

1. **Synergistes jonesii** Allison, Mayberry, McSweeney and Stahl 1993, 398[VP] (Effective publication: Allison, Mayberry, McSweeney and Stahl 1992, 528).

jonesii. M.L. gen. n. *jonesii* in honor of Raymond J. Jones, the Australian scientist who identified the activity of this bacterium in detoxification of 3,4-DHP and inoculated the rumens of cattle with this organism to solve animal intoxication problems when *Leucaena leucocephala* is grazed.

Cells are oval rods (0.6–0.8 μm × 1.2–1.8 μm), Gram-negative, and nonmotile. Spores are not formed.

Strictly anaerobic chemoorganotroph. Growth occurs in media that contain yeast extract and any of a variety of peptones. Histidine, arginine, glycine, and smaller amounts of tyrosine and phenylalanine are utilized during growth in complex media containing peptones. Arginine is metabolized by the arginine deaminase pathway and products include ornithine, citrulline, CO_2, acetate, butyrate, and ammonia. Citrulline is not produced but formate and propionate are additional products when histidine is metabolized (McSweeney et al., 1993a). Capacity to utilize pyridinediols (both 2,3- and 3,4-DHP) is regulated and is repressed by arginine and induced by the specific pyridinediols. Hydrogen is produced during growth, but indole or hydrogen sulfide are not produced and nitrate is not reduced. Urease, catalase, starch hydrolysis, and gelatin liquefaction were not detected. Carbohydrates are not fermented. Cellular fatty acid profiles are dominated by straight and branched-chain, saturated and unsaturated odd-carbon fatty acids, including a relatively high abundance of

$C_{17:1}$ and $C_{19:1}$ fatty acids, C_{20} cyclopropane acid and $C_{15\ 3-OH}$. Sensitive to tetracycline, penicillin, clindamycin, cephalothin, ampicillin, and deoxycycline, but resistant to a large number of other antibiotics (Allison et al., 1992).

The impetus for isolation of these bacteria arose from the observation that ruminants in Hawaii grazed *Leucaena leucocephala* (leucaena) without becoming poisoned by 3,4- DHP, which is a toxic compound that is produced in the rumen from mimosine (β-[N-(3-hydroxy-4- pyridone)]-α-aminopropionic acid). The latter is a toxic amino acid that is found in leucaena. Bacteria with DHP-degrading capabilities were not detected in ruminal samples from animals in parts of the world where animals are vulnerable to intoxication when they are fed leucaena. Ruminal inoculation with *S. jonesii* provides protection for cattle from leucaena toxicity (Jones and Megarrity, 1986; Hammond et al., 1989). Detection and quantitation of these bacteria in mixed rumen populations using oligonucleotide probes targeting specific rRNA sequences was demonstrated (McSweeney et al., 1993b). This is the first known example of geographic limits to the distribution of important functional ruminal bacteria.

Anaerobic bacterium that degrades toxic pyridinediols.

Found in the rumen of animals in certain, but not all, parts of the world.

The mol% G + C of the DNA is: 57–59 (T_m).

Type strain: 78–1, ATCC 49833.

GenBank accession number (16S rRNA): L08066.

Phylum BX. Cyanobacteria
Oxygenic Photosynthetic Bacteria

Richard W. Castenholz

The oxygenic photosynthetic procaryotes comprise a single taxonomic and phylogenetic group (see master phylogenetic tree of the *Bacteria*). In the last edition of the *Manual*, two separate groups were described, but it is now apparent that members of the *Prochlorales* simply represent different, unrelated genera which fall into the main cluster of the *Cyanobacteria* (see Oxygenic Photosynthetic Bacteria, below). The principal character that defines all of these oxygenic photosynthetic procaryotes is the presence of two photosystems (PSII and PSI) and the use of H_2O as the photoreductant in photosynthesis. Although facultative photo- or chemo-heterotrophy may occur in some species or strains, all known members are capable of photoautotrophy (using CO_2 as the primary source of cell carbon).

A special glossary for these organisms has been included.

GLOSSARY TO THE OXYGENIC PHOTOSYNTHETIC BACTERIA

Akinete a thick-walled "resting" cell or "spore" that differentiates from a vegetative cell.

Baeocyte a small cell formed internally by multiple fissions in parent cells of Subsection II.

Binary fission the division of one cell into two, followed by the enlargement of at least one, and usually both, daughter cells.

Budding the asymmetric binary fission of a parental cell, in which one daughter cell is smaller than the other.

False branch a branch formed by a slipping to one side of a section or loop of trichome through the sheath; a branch not formed by lateral division of a cell (see Figs. B10.1 and B10.45).

Filament a chain of cells in which cell separation does not include an intercalating sheath, although a linear sheath may be present.

Gas vacuole an area in cells that refracts light, during observation under light microscopy, because of its gaseous nature, often appearing red; composed of numerous gas vesicles.

Gas vesicle a subunit of a gas vacuole; each subunit is a cylindrical structure with conical ends (as seen with transmission electron microscopy), composed of a single protein envelope ($\sim$2 nm thick), impervious to water and containing gas.

Geminate to become doubled or paired; arranged in pairs.

Heterocyst a specialized cell that functions in N_2-fixation; differentiates from vegetative cell.

Hormocyst a hormogoniumlike short chain of cells enveloped or encapsulated by a thick outer envelope or sheath.

Hormogonium a short segment of trichome that is motile and apparently functions as a disseminule.

Lectotype an element subsequently designated or selected from among syntypes to serve as the definitive type.

Meristem a localized region of cell division.

Multiple fission the internal fission of a parental cell into several smaller cells which do not enlarge to parental size until after release.

Multiserate arranged with more than one row of cells, as a result of cells in trichome dividing in more than one plane.

Necridial cell a localized dead or dead-appearing cell of a trichome; the possible result of regulated differentiation and lysis; the cell at which trichome fragmentation occurs.

Pseudofilament a chain of cells in which all cells in the chain are separated from each other by sheath material or a fibrous layer.

Syntype any of two or more elements used as types by the author of a name, whether or not it is designated as such by the author.

Thallus a plantlike body composed of many cells or filaments; the thallus may have a definite shape but lacks internal differentiation.

Thylakoid the intracellular membrane system that includes the reaction center, chlorophyll *a* (and *b* in some), and some carotenoid pigments.

Trichome a chain of cells without an investing sheath (in cyanobacteria).

True branch a branch formed by lateral or oblique division of a cell in a trichome.

Tychoplankton floating or free-living organisms in shallow water of a lake, intermingled with attached vegetation and periphyton, usually near shore.

Uniserate arranged in a single row or series of cells, such as in a branched or unbranched trichome.

Vegetative cells the normal cells that grow and divide by binary or multiple fission.

Oxygenic Photosynthetic Bacteria

RICHARD W. CASTENHOLZ

Introduction The oxygenic photosynthetic procaryotes comprise a single phylogenetic branch within the domain *Bacteria* (Eubacteria). On the basis of the most recent 16S rDNA sequence analyses, this branch or cluster is most closely related to the Gram-positive bacteria (see master phylogenetic tree), particularly the low G + C group which includes *Clostridium innocuum* and *Bacillus subtilis*.

At the time of the 1989 edition, it was thought that the order *Prochlorales* included in the phylum *"Prochlorophyta"* (the oxygenic photosynthetic bacteria that lack phycobilisomes and possess chl *b* in addition to chl *a*) constituted an evolutionary branch distinct from the cyanobacteria. This hypothesis has now been shown to be incorrect and has been rejected (Palenik and Swift, 1996). The three described genera of the chl *a/b*-containing oxygenic procaryotes (*Prochloron*, *Prochlorothrix*, and *Prochlorococcus*) fall within three distinct polyphyletic branches of the cyanobacterial tree according to 16S rDNA sequence analyses, and are, therefore, more closely related to a number of typical cyanobacteria than to each other (Wilmotte, 1994; Turner, 1997). This conclu-sion is corroborated by the identification of functional genes for the *a* and *b* subunits of a type of phycoerythrin in one strain of *Prochlorococcus* and the occurrence of a small amount of a PE-like pigment in some deep water populations of this genus (Hess et al., 1996). Furthermore, in a traditional cyanobacterial strain (*Synechococcus* PCC 7942) a chl *a/b* binding protein, similar to that in chloroplasts of eucaryotes, has been found (Dolganov et al., 1995). These discoveries give further support to the theory that cyanobacteria and "prochlorophytes" evolved from ancient oxygenic ancestors that possessed chl *a* and chl *b* together with phycobiliproteins. A unicellular cyanobacterium that synthesizes chl *d* as a major pigment in addition to small amounts of chl *a* and phycobilins, a combination similar to the chloroplasts of some species of eucaryotic red algae, has been recently discovered (Miyashita et al., 1997). Thus, the synthesis of a second type of light-harvesting chlorophyll, or the lack of phycobiliproteins or phycobilisomes within the cyanobacteria, should not be considered as a major phylogenetic divergence.

General Characteristics of the Cyanobacteria

RICHARD W. CASTENHOLZ

Besides sharing the basic cellular features of other *Bacteria*, the cyanobacteria possess unique and diagnostic characteristics that will be briefly described here, but less extensively than in the 1989 edition of *Bergey's Manual of Systematic Bacteriology* (Staley et al., 1989). Guidelines for characterization of cyanobacteria are given in Table B10.1. Several other reviews (e.g., Stanier and Cohen-Bazire, 1977; Stewart, 1980; Fay, 1983) are available as well as the books, *Biology of Cyanobacteria* (Carr and Whitton, 1982), *Photosynthetic Picoplankton* (Platt and Li, 1986), *The Cyanobacteria* (Fay and Van Baalen, 1987), *Cyanobacteria* (Packer and Glazer, 1988), *Molecular Biology of Cyanobacteria* (Bryant, 1994), *Bergey's Manual of Determinative Bacteriology* (Holt et al., 1994), and *Ecology of Cyanobacteria: Their Diversity in Time and Space* (Whitton and Potts, 2000).

Cell Envelope The cell wall in cyanobacteria is of a Gram-negative type, but the structural peptidoglycan layer is often considerably thicker than in the Gram-negative *Proteobacteria*. This layer is usually 1–10 nm thick, but reaches 200 nm in the case of *Oscillatoria princeps*. Thicker walls, as found in this species, are often partially perforated by numerous incomplete pore pits extending from the cell interior and carrying the cytoplasmic membrane outward to contact the lipopolysaccharide (LPS) outer membrane. Small-diameter pores (5–13 nm) are present in regular or scattered order in the walls of all cyanobacteria, but the arrangement varies greatly. For example, in filamentous forms the peripheral (longitudinal) wall may be perforated by pores covering most of the surface or alternatively by rings of pores subtending the cross-walls (junctional pores). The cross-wall (transverse septum) may be perforated by a single minute central pore or by numerous pores (microplasmodesmata).

Numerous unicellular, colonial, and filamentous cyanobacteria possess an "envelope" outside of the outer membrane. This is variously called the sheath, glycocalyx, or capsule or, depending on the consistency, may be referred to as gel, mucilage, or slime. Many sheaths show a microfibrillar substructure. The sheaths of cyanobacteria are predominantly composed of polysaccharide, but in some strains >20% of the weight may consist of polypeptides. In the firm sheaths of many colonial and filamentous cyanobacteria, yellow, red, or blue pigments may accumulate and mask the color of the cells. The typical yellow-brown pigment has now been characterized as a UV-absorbing, protective pigment (scytonemin) (see Garcia-Pichel and Castenholz, 1991; Castenholz and Garcia-Pichel, 2000). In *Nostoc commune*, and potentially in other cyanobacteria, UV-A/B-absorbing mycosporine-like amino acids also occur in the sheath (Ehling-Schulz et al., 1997), rather than in the cytoplasm as found in other cyanobacteria investigated (Garcia-Pichel and Castenholz, 1993).

Cell Division Most unicellular and colonial cyanobacteria and some filamentous forms undergo binary fission by a constrictive type of division in which all envelope layers (often including sheath) grow inward until cell separation is complete or nearly complete. In others, particularly the "oscillatorian" types, which lack severe constrictions at the cross-walls, the outer membrane (and sheath if present) is continuous and not involved in division. Instead, the cytoplasmic membrane invaginates and a peptidoglycan layer is synthesized between the two membranes. In *Chamaesiphon*, however, reproduction occurs by a process similar to budding. In other cases, what appears to be budding is merely an asymmetric constrictive division of a polarly differentiated unicell or "exospore". In a number of unicellular and pseudofilamentous forms, reproduction occurs by internal multiple fissions, often in addition to binary fission in some vegetative cells. The minute cells resulting from these multiple divisions (baeocytes or microcytes; formerly called "endospores") may in

TABLE B10.1. Guidelines for characterization of the cyanobacteria

CELL MORPHOLOGY
Cell shape, polarity
Cell dimensions (diameter, length)
Number and regularity of planes of fission
Color of cells and cell suspension
If present:
Sheath or glycocalyx description
Gas vacuoles with location
Motility (speed, rotation, smoothness)
Baeocyte ("endospore") description
Heterocyst and akinete description
Photomicrographs: large forms: bright field and Nomarski; small
    forms: phase contrast
ULTRASTRUCTURE
Thylakoid arrangement
Cell wall structural appearance
Internal inclusions, storage granules
Sheath structure
COLONY/TRICHOME MORPHOLOGY
Colony or thallus shape, symmetry, etc.
Trichome type: tapered, straight, helical, septal constriction, shape of
    terminal cell, terminal hair, type of hormogonia, type of false or
    "true" branching
Location and pattern of heterocysts and akinetes
Morphological responses to nutrient deficiencies
GENETIC CHARACTERS
DNA base composition (mol% G + C)
Complete sequence analysis of 16S rDNA and of other components,
    including various gene products
DNA–DNA hybridization data (with other spp.)
Lipid profile
PHYSIOLOGY/BIOCHEMISTRY
Absorption spectrum (*in vivo*)
Types of phycobilin pigments present
Capacity for chromatic adaptation
Temperature optimum and upper limit
Capacity for dark chemoheterotrophy (aerobic vs. anaerobic)
Photoheterotrophy (with DCMU)
Growth and/or acetylene reduction, aerobic and anaerobic, free of
    combined N
Salinity tolerance
pH range and tolerance
Vitamin requirements
Sulfide sensitivity; sulfide-dependent anoxygenic photosynthetic
    capacity
$CaCO_3$-deposition; boring ability in $CaCO_3$
CULTURE CONDITIONS
Specifics of isolation, growth medium, light intensity; conservation of
    cultures including type or reference strain; ability to grow on humid
    or dry agar
HABITAT/ECOLOGY
General and specific description of habitat: marine, freshwater,
    brackish, terrestrial; trophic level; flowing or static water,
    temperature, depth, light intensity regime; associated organisms of
    community; microenvironment, substrate, etc.

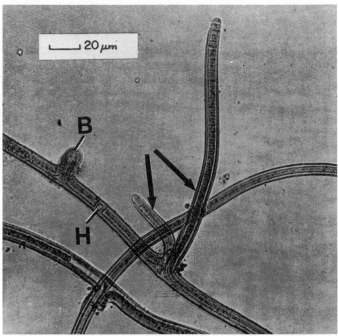

FIGURE B10.1. *Scytonema* species showing geminate (*double*) false branching. An initial bulge (*B*) breaks, with the two ends then growing out separately (*arrows*). A heterocyst (*H*) is indicated. (Reproduced with permission from G.E. Fogg, W.D.P. Stewart, P. Fay and A.E. Walsby. 1973. The Blue-Green Algae. Academic Press, London, p. 18.)

some cases show slow gliding motility after release from the mother cell.

Binary fission in cyanobacteria results in unicellular populations when constrictive separation is complete and when no sheath holds daughter cells together. When fission regularly alternates the plane of division in two or three planes, and when a sheath or gel retains cells together, orderly or disorderly colonies of many cells result. When fission occurs in only one plane and separation is incomplete, short or long chains of cells (trichomes) occur. In sheathed trichomes (trichome + sheath = filament), local weaknesses in the sheath may allow the bulging out of a trichome as a loop which eventually breaks (geminate or double false branching, Fig. B10.1) or the breaking out of a broken single trichome from a sheath (single false branching) (Fig. B10.2). In some filamentous genera, true branches of sheathed trichomes occur when the plane of division shifts about 90° in some cells of the trichome (Fig. B10.3). Other variations of branching may occur (see Figs. B10.85, B10.86, and B10.87 in Subsection V).

Cell Exterior and Motility   Fimbriae (or pili) occur abundantly with diverse patterns in many cyanobacteria. Although procaryotic flagella have never been demonstrated in cyanobacteria, swimming motility in small unicellular types has been described (Waterbury et al., 1985). No organelles of propulsion have been identified. A gliding (sliding) type of motility is known in a large variety of cyanobacteria but mainly in filamentous forms. In this form of propulsion, contact with a solid or semisolid material is required. The mechanism is not yet clear, but, in the case of oscillatorian cyanobacteria, continuous, proteinaceous, microfibrillar bands associated with the wall or periplasmic space may be responsible (Adams et al., 1999), although propulsion by mucilage extrusion through the junctional pores has also been claimed as a mechanism (Hoiczyk and Baumeister, 1998). Gliding motility in other non-photosynthetic groups of procaryotes may be a result of different mechanisms (see Castenholz, 1982).

Cell Interior   Although not all thylakoids in cyanobacteria appear to be invaginations of the cytoplasmic membrane, there are orderly "attachment points" or "thylakoid centers" associated with the periphery of the cytoplasm or the cytoplasmic membrane. The thylakoids of cyanobacteria, whether arranged concentrically or radially, or seemingly interspersed in another manner, do not show the accordion-like pleating or the obvious, frequent cytoplasmic membrane connections so characteristic of many phototrophic members of the *Proteobacteria* (i.e., photo-

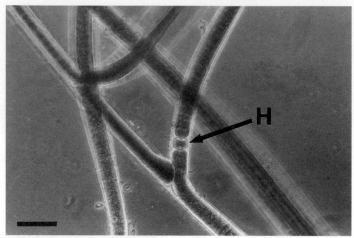

FIGURE B10.2. *Scytonema* cf. *polycystum* from tropical marine mangrove "lake". *H*, heterocyst above single false branch. Bar = 40 µm.

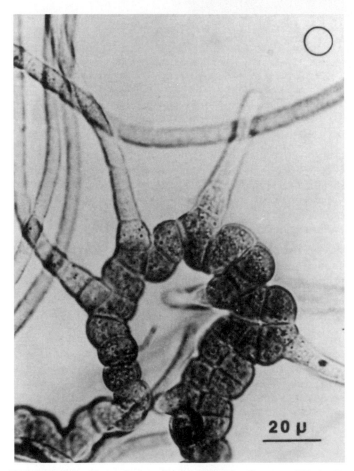

FIGURE B10.3. Photomicrograph of multiseriate filament of *Fischerella* with true branches. (Reproduced with permission from E.L. Thurston and L.O. Ingram, ©Journal of Phycology 7: 203–210, 1971.)

trophic purple bacteria). Fingerlike tubular thylakoids and open vesicular types are absent in cyanobacteria, at least in healthy cells. Most cyanobacteria possess upright hemidiscoidal or hemispherical phycobilisomes, which are complex protein-pigment aggregates arranged in orderly rows on both surfaces of the thylakoids (Fig. B10.4). The size of these structures (20–70 nm in diameter) is greater than the thickness of the double thylakoid membrane. The phycobilisome comprises the major light-harvesting complex of most cyanobacteria. The chl *a/b*-containing oxygenic cyanobacteria do not possess these structures, and they are also absent in all anoxygenic bacteria. *Gloeobacter*, a unicellular cyanobacterium, lacks thylakoids, and the cytoplasmic membrane serves as the chlorophyll-bearing vehicle. Instead of discrete individual phycobilisomes, *Gloeobacter* possesses rows of phycobilisome-type rods standing perpendicular to the cytoplasmic membrane. The pigments contained in the typical phycobilisome are allophycocyanin (APC) in the core, which abuts on the surface of the thylakoid membrane, usually directly in contact with the light-harvesting chlorophyll *a* of PSII, and phycocyanin (PC) in the rods. Phycoerythrocyanin (PEC) or phycoerythrin (PE), if present, occur as outer disks of the rods (Fig. B10.5). For further information on phycobilisomes and thylakoids in cyanobacteria see various chapters in Bryant, 1994, and Grossman et al., 1993.

In the cytoplasm of cyanobacteria there are many other components and "inclusions", most of which can be visualized readily using various preparative techniques for transmission electron microscopy. They include:

1. Glycogen (polyglucose) granules which are either ovoid or elongate and rod-shaped, and usually located between the thylakoids. Polyhydroxybutyrate granules are present in a few cyanobacteria.
2. Cyanophycin granules, which are polymers of arginine and aspartic acid. These "structured granules" are recognizable by a radiating substructure pattern and are often large enough to be detectable by light microscopy (500 nm diameter). They are apparently unique to cyanobacteria, although some species lack them. Functionally, they serve as reserves of nitrogen.
3. Carboxysomes (polyhedral bodies) which are large angular structures composed largely of ribulose bisphosphate carboxylase/oxygenase (RUBISCO), apparently serving as reserves of this carboxylating enzyme.
4. Polyphosphate (volutin) granules reaching 100–300 nm in diameter. With transmission electron microscopy they appear as spherical, electron-dense or porous structures, depending on fixation and staining methods.
5. Gas vacuoles which are composed of many elongate cylindrical gas vesicles bearing a pointed cap at each end (Fig.

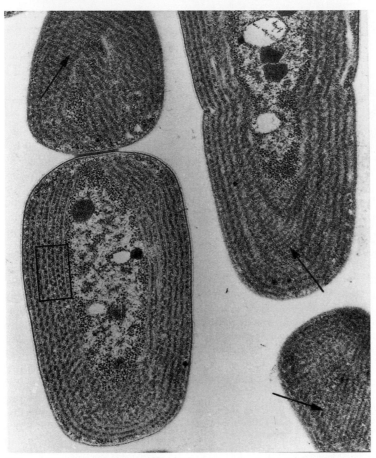

FIGURE B10.4. Thin section of *Pseudanabaena* PCC 7408. The rows of phycobilisomes are seen in cross-section, longitudinal section, and tangential section. In tangential sections, the rows of closely packed phycobilisomes appear as electron-dense cords running at an angle to the long axis of the cells (*arrows*). When the rows of phycobilisomes are cut in cross-section, they are seen in face view (e.g., in *box*). They alternate with one another on the apposing stromal surfaces of thylakoids. (Reproduced with permission from G. Cohen-Bazire and D.A. Bryant. 1982. Phycobilisomes: composition and structure. *in* Carr and Whitton [eds.], The Biology of Cyanobacteria. Blackwell, Oxford, p. 147.)

FIGURE B10.5. Model of hemidiscoidal phycobilisome of *Mastigocladus laminosus* (*Fischerella*). Two disks of phycoerythrocyanin (*gray*) lie at the distal end of each of the six rods. Each rod also contains five disks of phycocyanin (*dark*) which lie proximal to the allophycocyanin core (*light*). Each disk symbolizes a trimeric biliprotein aggregate including the uncolored proteins. (Reproduced with permission from M. Nies and W. Wehrmeyer, Archives of Microbiology *129*: 374–379, 1981, ©Springer-Verlag.)

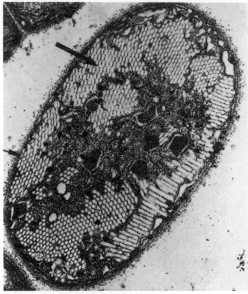

FIGURE B10.6. Electron micrograph of a section of a vegetative cell of *Anabaena circinalis*, a planktonic form, with most of the cell volume occupied by gas vesicles (*arrow*). (Reproduced with permission from M.J. Daft and W.D.P. Stewart, New Phytologist *72*: 799–808, 1973, ©Blackwell Science.)

B10.6). Gas vesicles are constructed of a 2 nm thick protein coat, contain gases, and are completely impervious to water. They occur mainly in planktonic cyanobacteria (see Walsby, 1994).

6. The "nucleoid region" of cyanobacterial cells, often called the "centroplasm", appears as a lighter pigmented area using light microscopy, or as a structure of lesser density when viewed with transmission electron microscopy. The DNA fibrils of cyanobacteria occur in a complex, folded arrangement, but each is circular when unfolded. The total molecular weight of cyanobacterial genomes ranges between 1.6 and 8.6×10^9 daltons (Herdman et al., 1979a). Multiple or "excess" DNA occurs in the more complex, filamentous types. Unicellular forms, however, usually have genome sizes below 3.6×10^9 daltons, similar to those known for procaryotes other than cyanobacteria. For example, *Escherichia coli* has a genome size of 2.5×10^9 daltons (Carbon et al., 1979) while the genome of *Bacillus subtilis* is 2.3×10^9 daltons in size (Ash et al., 1991). The complete genome of *Synechocystis* PCC 6803 has been sequenced (Kaneko et al., 1996); its genome size is 2.1×10^9 daltons. The genome size of two types of "cyanelles" (cyanobacterial derivatives serving as "chloroplasts" in unicellular eucaryotic hosts) is very small ($<1 \times 10^9$ daltons), much smaller than that of any free-living cyanobacterium.

Specialized Cells and Differentiation

Heterocysts and N₂-fixation   The cyanobacteria of Subsections IV and V produce heterocysts in trichomes at intervals or at their termini through the differentiation of vegetative cells, usually only after the concentration of combined inorganic nitrogen (especially ammonia) in the surrounding medium has decreased considerably. Heterocysts are unique to cyanobacteria. The structure, development, and function of heterocysts have been studied extensively (see Wolk et al., 1994). In brief, vegetative cells differentiate into less granular cells with additional wall layers and

modified thylakoids that lack functional photosystem II. As an anoxygenic PS I-functioning cell that develops an anoxic interior, nitrogenase is synthesized, and N₂-fixation proceeds, especially in the light with ample ATP provided by the cyclic electron transport of PS I. The regulation of heterocyst spacing and numbers of heterocysts relative to oxygenic photosynthetic cells is mediated through the intercellular transport of a 17 amino acid peptide, recently discovered in *Anabaena* PCC 7120 (Yoon and Golden, 1998).

N₂-fixation also occurs in several cyanobacteria lacking heterocysts (Bergman et al., 1997). Some of these species fix N₂ primarily during the dark period when anoxic conditions return, within the confines of tight associations such as microbial mats (Stal, 1995). In some planktonic forms (e.g., *Trichodesmium*) the cellular or environmental conditions allowing N₂-fixation are still uncertain, but a circadian rhythm of nitrogenase activity has been demonstrated (Chen et al., 1996, 1998).

Akinetes   Akinetes ("resting spores") are produced by many cyanobacteria in Subsection IV. Many (but not all) of the heterocystous cyanobacteria produce akinetes, particularly under conditions of nutrient deficiency and/or light limitation. Akinetes differentiate from vegetative cells, grow, and acquire a thick wall surrounding the old wall. Akinetes accumulate cyanophycin, glycogen, lipids, and carotenoid pigments, but polyphosphates disappear. The pattern of akinete formation is generally related to the location of heterocysts, with akinetes found either adjacent to or always distant from these structures. These cells tolerate drying, freezing, and long-term storage in anoxic sediments, and apparently do not require a resting period before germination (see Herdman, 1988; van Dok and Hart, 1996, 1997).

Hormogonia   Although a hormogonium is simply a short chain of cells, the cells usually differ somewhat from the vegetative cells of the trichome. Hormogonia are often defined as chains of 5–15 cells with cell diameters less than those of the vegetative trichomes. Hormogonia are generally the migrating (dispersal) phase, in which gliding motility or flotation of gas vacuolate forms into the plankton occurs. Cell division does not occur until the hormogonial stage ends its dispersal period. Hormogonia may also form in specific regions of the parent organism. The formation and liberation of hormogonia appears to be a timed process associated with environmental conditions (e.g., phosphorus repletion) or with particular stages of a morphogenetic cycle (see Herdman and Rippka, 1988; Tandeau de Marsac, 1994).

Termini   In some filamentous cyanobacteria the terminal cell may be highly differentiated into various tapered shapes, some conical and others long, tapered, and hooked. The entire end of the trichome, including several subterminal cells, may be tapered. Some filamentous types, such as *Calothrix* (Subsection IV), develop an extended terminal taper of cells (multicellular hairs) in which most of the thylakoids and pigments have been lost.

Physiology and Biochemistry   The chief physiological/biochemical characteristic of cyanobacteria, distinguishing them from all other procaryotes, is the dual photosystem that allows the use of H₂O as photoreductant with the consequent liberation of O₂. Chlorophyll *a* serves as the reaction center pigment and is also involved in light harvesting for all cyanobacteria, as in the chloroplasts of eucaryotic algae and plants. Chlorophyll *a* and *b* (or vinyl chl *a/b*) occur in the cyanobacteria termed "prochlorophytes" and chlorophyll *d* has been reported in one strain or species of this group (see Introduction).

The phycobiliproteins making up the phycobilisomes constitute the major light-harvesting pigments in almost all cyanobacteria, except under some conditions of stress, such as exposure to high solar irradiance. CO_2 reduction mainly utilizes the reductive pentose phosphate cycle.

The oxygenic photosynthesis of some cyanobacteria can be altered in an adaptive response to the presence of free sulfide. PS II is partly or wholly inhibited and electrons derived from sulfide enter the photosynthetic electron transport system closer to PSI and result in CO_2 reduction (anoxygenic photosynthesis) (Cohen et al., 1986). Not all cyanobacteria are capable of tolerating or adapting to sulfide in this way. Some combine oxygenic and anoxygenic photosynthesis or simply protect PS II from inhibition by sulfide (Castenholz and Utkilen, 1984).

Most cyanobacteria are obligate photoautotrophs, since dark catabolic rates are often not increased by external substrates, either because of negligible uptake rates or because of constitutive rates of NADP reduction or respiration that are not enhanced by external substrates. A relatively small number of cyanobacteria, however, can grow as aerobic dark heterotrophs at much slower rates than as photoautotrophs, mainly by the utilization of glucose, sucrose, or fructose. Photoheterotrophy may be artificially forced in some cyanobacteria, with DCMU used to block electron transport from PSII. Respiration is normally greatly reduced under photosynthetic light conditions, since a part of the photosynthetic electron chain of the thylakoids is also used in respiration. Anaerobic metabolism in the dark is limited to fermentation, and serves mainly for maintenance (see Stal, 1995).

Cyanobacterial doubling times during exponential growth are often > 12 h and usually 24 h or much longer, even in culture with unlimited nutrients and a saturating intensity of continuous light. There are at least a few unicellular and oscillatorian strains in culture that have doubling times of < 6 h.

Although growth and reproduction in cyanobacteria occurs principally by cell division (binary fission), some species undergo a complex morphogenetic life cycle which may involve nongrowth hormogonial dispersal phases, an aseriate stage, and the differential growth of certain portions of filaments (e.g., Whitton, 1987). Phases of some cycles are controlled by light through one or more photoreversible phytochrome-like pigments (Yeh et al., 1997; Pepper, 1998). The photobiology of cyanobacteria, in addition to photosynthesis and photomorphogenesis, includes the control of phycobilin synthesis (chromatic adaptation), again via a reversible photoreceptor pigment. For example, many cyanobacteria cease synthesis of phycoerythrin when light is enhanced in the red wavelengths rather than in the green portion of the spectrum (Tandeau de Marsac, 1977).

Phototaxis (positive and negative) and photophobic reactions (step-up and step-down) are also well documented in the cyanobacteria. Many cyanobacteria move towards a light source (positive phototaxis) because of a cessation of reversals when trichomes become oriented parallel to the light gradient (see Castenholz, 1982; Häder, 1987), while others are quite capable of steering or turning movements when light is unidirectional (Nultsch and Wenderoth, 1983; Mrozek, 1990). Photophobic reactions are known in some of the same species that exhibit phototaxis. Step-down responses have been documented more commonly than step-up responses; these result in the retention of filamentous cyanobacteria in illuminated niches. Recently, however, step-up responses have also been documented; these probably allow avoidance of high inhibitory intensities of solar radiation that occur as a steep gradient in some communities such as intertidal mud flats or microbial mats (Kruschel and Castenholz, 1998; Castenholz and Garcia-Pichel, 2000).

Endogenous rhythms, previously unknown in procaryotes, have been demonstrated in cyanobacteria for nitrogen fixation (Grobbelaar et al., 1986; Chen et al., 1998) and for general rhythmicity (Kondo and Ishiura, 1994; Kondo et al., 1997).

Ecology The occurrence and predominance of cyanobacteria in a vast array of habitats is a result of several general characteristics and of some features characterizing certain cyanobacterial species clusters. Many species are generalists and will tolerate a great range of environmental conditions, including extremes that usually or often exclude eucaryotic algae. It is presumed that cyanobacteria per se evolved in the Precambrian well before the Paleozoic boundary, and this is borne out by the existence of microfossils of the middle and late Proterozoic that are nearly identical morphologically to some living cyanobacteria (see Knoll, 1985; Schopf, 1996).

Planktonic dominance by several filamentous forms is probably strongly related to the possession of gas vesicles and the ability to regulate buoyancy or simply to be buoyant (see Walsby, 1994). In eutrophic lakes, N_2-fixation capability following depletion of combined nitrogen by blooms of other algae and cyanobacteria and/or the ability to efficiently utilize a very low photon flux density is thought to enhance the success of some cyanobacteria (Gibson and Smith, 1982; Van Liere and Walsby, 1982; Mur, 1983). The very efficient use of CO_2 at low concentrations or the ability to use HCO_3^- at high pH have also been proposed as factors which promote cyanobacterial dominance under some conditions (Shapiro, 1997). Other hypotheses have been suggested for specific situations (see Vincent, 1987). In oligotrophic marine or fresh waters, the N_2-fixing ability may be of primary importance (e.g., for *Trichodesmium* species). The minute picoplanktonic unicells of *Synechococcus* and *Prochlorococcus* may thrive in deep-mixing oligotrophic waters because of their efficient absorption of photons under a low flux density (Wood, 1985; Glover, 1986). It should also be noted that some phycoerythrin-containing cyanobacteria achieve large population densities in the metalimnia or upper hypolimnia of lakes where dim green light is the only photosynthetically active radiation available.

It has been suggested that siderochrome (trihydroxymate) excretion by planktonic cyanobacteria in iron-poor waters may aid in sequestering iron. Extracellular excretion of these substances (e.g., hydroxymate) may also inhibit the growth of potential competitors, and has been suggested as contributing to cyanobacterial success (see Gibson and Smith, 1982). In addition, several planktonic cyanobacteria produce potent toxins of two types: alkaloid neurotoxins and peptide hepatotoxins (Skulberg et al., 1984; Namikoshi and Rinehart, 1996; Willén and Mattsson, 1997; Chorus and Bartram, 1999). Cyanobacterin, an effective inhibitor of other cyanobacteria and some eucaryotic algae, is a diaryl-substituted lactone with a chlorine substitution on one of the aromatic rings. It is produced by the nonplanktonic cyanobacterium *Scytonema hofmanni* (Gleason and Baxa, 1986). The size of cyanobacterial populations is sometimes controlled by attacks of specific cyanophages, lytic myxobacters, or aquatic fungi (Garza and Suttle, 1998).

The temperature optimum for growth of many cyanobacteria is higher by at least several degrees than for most eucaryotic algae. This characteristic may play an important role in the noted summertime dominance of cyanobacteria in temperate latitudes. Some species tolerate even higher temperatures, inhabiting hot springs (up to 74°C for at least one species; Castenholz, 1984a), terrestrial rock surfaces, and hot desert soils (Garcia-Pichel and Belnap, 1996), habitats where eucaryotic phototrophs may be inhibited or excluded. Cyanobacteria also predominate at low and freezing temperatures in freshwater ponds of polar regions, probably due to the exclusion of most other phototrophs, but in these cases the actual temperature optimum for most of these organisms may be considerably higher (see Vincent, 1988; Tang et al., 1997). Extensive freshwater and terrestrial microbial mats of the Antarctic are composed mainly of cyanobacteria. The dominance of cyanobacteria may be achieved through the ability to tolerate alternating freezing and thawing or slow freeze-drying.

The ability to withstand high salinity (Borowitzka, 1986) allows cyanobacteria to dominate many hypersaline marine lagoons and inland saline lakes (Bauld, 1981; Javor and Castenholz, 1981; Stal, 1995). Again, many of these organisms grow better at salinities lower than that found in their most common habitats (Garcia-Pichel et al., 1998). Cyanobacteria may also be especially tolerant of specific substances at higher concentrations. Free sulfide is tolerated by many cyanobacteria at levels much higher than those which eucaryotic algae can withstand (Castenholz, 1976; Cohen et al., 1986).

Some cyanobacteria are especially desiccation-resistant, hence the prevalence of extensive cyanobacterial mats in ephermal hypersaline bodies of water, desert and tropical soils, as epilithic crusts, as cryptoendoliths in cold and hot desert sandstones, and in terrestrial or subaerial habitats in numerous tropical or subtropical regions (Potts, 1994). Cyanolichen associations are very common in many climatic regions. Nitrogen-fixing cyanobacteria may be the sole photosynthetic partner with the fungus or constitute the nitrogen-fixing partner in a tripartite association where a green alga provides the photosynthate (see Nash, 1996). The specificity of the cyanobiont varies (Paulsrud et al., 1998).

Other symbioses include cyanobacteria that serve as functional "chloroplasts" in a variety of eucaryotic "hosts" or simply as nitrogen-fixing "factories" in several unrelated types of green plants (Bergman et al., 1992). Most of these are intercellular and capable of independent growth outside of the host.

"Cyanelles" are functional intracellular chloroplasts found in a few restricted unicellular host eucaryotes (Bhattacharya and Schmidt, 1997). They are thought to be derived from unicellular cyanobacteria that, through a long-term symbiosis in the host cell, have lost the potential for autonomy (Trench, 1982). The loss of the cyanobacterial wall is nearly complete, but remnants of peptidoglycan are retained (Scott et al., 1984). However, current phylogenetic analyses indicate that cyanelles investigated so far are more closely related to extant chloroplasts than to any free-living cyanobacteria (Bhattacharya and Schmidt, 1997).

An extensive review of cyanobacterial ecology is available: *Ecology of Cyanobacteria: Their Diversity in Time and Space* (Whitton and Potts, 2000).

Taxa of the Cyanobacteria Two courses for the delimitation of genera have been followed by cyanobacterial specialists: (a) to retain "small" genera (Anagnostidis and Komárek, 1985) or (b) to gather many species into fewer genera (Bourrelly, 1985). It appears that the first course may be the more practical of the two and that small clusters of species within more unambiguously defined genera will eventually be most acceptable. However, we believe that there is still insufficient information, based on cultural characteristics and genetic information, to pursue this course. Therefore, the "larger" genera used here are, in most cases, intended to serve as temporary vehicles or form genera

for various groups of species, ecotypes, or strains. It is now obvious that many generic names used in the past (e.g., *Synechococcus*, *Oscillatoria*, *Leptolyngbya*, *Pseudanabaena*) will eventually be split into several new genera. This process has already begun, and even in the nonexhaustive treatment provided in this volume, the reader will find several new genera not appearing in the 1989 and 1994 *Bergey's Manuals*. Many of the "botanically classified" cyanobacteria (e.g., Geitler, 1932; Desikachary, 1959) have been recently revised, based mainly on morphological criteria and usually without the use of cultures, although morphological analyses are quite detailed (e.g., Komárek and Anagnostidis, 1986, 1989, 1998; Anagnostidis and Komárek, 1988, 1990). New arrangements and new generic names have been proposed and used.*

The wisdom of not attempting a major revision of the cyanobacterial genera in the present volume is borne out by everchanging phylogenetic interpretations based not only on sequences of 16S rDNA or rRNA, but also using other genes (see Turner, 1997; Wilmotte and Herdman, this volume).

At this time, a modest revision of the temporary scheme published by Rippka et al. (1979) and in *Bergey's Manual of Systematic Bacteriology* (Vol. 3, 1989) will be used. Except in reference to older literature or to publications of other authors, species epithets will not be used. Instead, generic names alone with the addition when needed of culture identification numbers in place of the species epithets will be used. It is much too early to characterize and catalog species of cyanobacteria, even though this has been done by authors who continue to use mainly phenotypic characteristics (e.g., Komárek and Anagnostidis, 1998). All of these classifications will have to be revised almost completely when genetic comparisons become available. The subsections (I–V) used here are equivalent to the orders established earlier in the botanical literature.

The key to the subsections of cyanobacteria which follows is simple and is merely a modification of that used in the first edition of the *Manual*. A phylogenetic grouping of genera is not used; such an attempt would be premature (see Wilmotte and Herdman, this volume). As before, the traditional "subsections", based mainly on morphology, are used. In some cases these appear to agree with the clusters found in phylogenetic trees. When this is the case, such congruence will be indicated. The generic descriptions included in each subsection are based largely on cultured material. It should be understood that only a small number of genera of cyanobacteria are represented in axenic and/or clonal culture and many of these have not been studied extensively enough to be used for characterization. Nevertheless, the number of cyanobacteria in culture is considerably greater than at the time of the first edition 10 years ago. Many of the cultured strains have not been maintained under conditions which elicit the various morphological features generally ascribed to them from observation of environmentaly samples. Consequently, the degree of detail used in generic descriptions will vary considerably, depending on the thoroughness of recent studies. The classification system and genera included here will continue to undergo a long process of revision and expansion.

Editorial Note: In this edition of *Bergey's Manual of Systematic Bacteriology*, the authors will use only a few of these names proposed during the past 10–11 years. For the most part the new generic names added since the 1989 edition are those that have been characterized and therefore delimited to some extent in established and easily accessed culture collections, particularly the PCC (for a list of culture collections containing cyanobacteria, see Table B10.2). A few well characterized genera, not in culture but morphologically complex (e.g., with distinctive branch patterns), are also included in this edition.

TABLE B10.2. Partial listing of culture collections[a]

Canada
Univ.Toronto Culture Collection of Algae and Cyanobacteria (UTCC)
Department of Botany, Univ. Toronto, 25 Willcocks St., Toronto,
Ontario
Canada M5S 3B2
Phone: +416–978–3641
Fax: +416–978–5878
E-mail: jacreman@botany.utoronto.ca
Internet: www.botany.utoronto.ca/utcc/

Europe
Culture Collection of Algae
Correspondent: Georg Gartner
Univ. Innsbruck, Inst. Botanik, Sternwartestr 15,A-6020 Innsbruck
Austria
Phone: +512–5075939
Fax: +512–293439

Cyanobacterial Culture Collection
Correspondent: George Schmetterer
Univ. Vienna
Austria
E-mail: rbo10@dmswwula.bitnet

Collection of Phytoplankton, Benthos and Epilithon
Correspondent: Christine Cocquyt
Lab Plantkunde, K.L. Ledeganckstraat 35, 9000 Gent
Belgium
Phone: +32–9-2645063
Fax: +32–9-2645334

Cyanobacteria Culture Collection
Correspondent: Lucien Hoffman
Univ. Liége, 4000 Liége
Belgium
E-mail: u213101@bliulg11.bitnet

Cyanobacterial Culture Collection
Correspondent: Bart Nelissen
Univ. Instelling, Antwerpen
Belgium
E-mail: rrna@ccv.uia.ac.be

Microalgal Culture Collection
Correspondent: René Matagne
Univ. Liége, Genetics of Microorganisms, Department of Botany, 4000
Liége
Belgium
Phone: +3241–663820/27
Fax: +3241–663840
E-mail: chlabel@vm1.ulg.ac.be

Univ. Gent Herbarium
Dept. Morphology, Systematics and Ecology, K.L. Ledeganckstraat 151,
9000 Gent
Belgium
Phone: +32–9-2645064
Fax: +32–9-2645334
E-mail: koen.sabbe@rug.ac.be
Notes: North Sea, Mediterranean, Indian Ocean species

CCALA–Culture Collection of Autotrophic Organisms
Curator: Jaromir Lukavsky
Fax: +42–333–2391
Czech Acad Sciences, Inst. Botany, Dukelska 145, Cs-379 82 Trebon
Czech Republic
Phone: +42–333–721156
Fax: +42–333–721136
E-mail: hauser@omega.jh.jcu.cz

USB–Microalgal Culture Collection
Correspondent: David Kaftan
Univ. South Bohemia, Fac. Biological Sciences, Department of
Ecological and Systematical Biology
Czech Republic
Phone: +42–38–817629
Fax: +42–38–45985
E-mail: kaftan@bio.bf.jcu.cz

(continued)

TABLE B10.2. *(continued)*

Botanical Museum Copenhagen
Gothersgade 120, 1123 Copenhagen
Denmark
Phone: +35–8-35–322185
Fax: +35–8-35–322185
E-mail: ruth@bot.ku.dk
Notes: Species over the world

SCCAP–Scandinavian Culture Center for Algae and Protozoa
Correspondent: Niels Larsen
Univ. Copenhagen, Botanical Inst., Oster Farimagsgade 2d, 1353
Copenhagen
Denmark
Fax: +45–35–322321
E-mail: sccap@bot.ku.dk

Universitetsparken
Inst. Biology, Bygning 137, 8000 Aarhus
Denmark
Phone: +45–89–423188
Fax: +45–86–139326

Microalgal Culture Collection
Correspondent: Juhani Lehtonen
Univ. Turku, Department of Biology, 21500 Turku
Finland
Phone: +358–21–3335562
Fax: +358–21–3335549
E-mail: juanle@sara.cc.utu.fi

CEA–Cadarche
Correspondent: C. Gudin
St. Paul les Durance, 13108 Cadarche
France
Phone: +33–42254384
Fax: +33–42–253020

CN–Centre de Nantes
Correspondent: P. Larssus
Lab Phycotoxines et Nuisances, bp1105, 44311 Nantes Cedex 03
France
Phone: +33–40–374130
Fax: +33–40–374073
E-mail: plassus@ifremer.fr

COM–Centre d'Oceanologie Maritime
Correspondent: B. Berland
CNRS, URA 41, Statione Marine d'Endoumé, Rue de la Batterié des
Lions, 13007 Marseilles
France
Phone: +33- 91–041638
Fax: +33–91–041635
E-mail: berland@com.univ-mars.fr

Herbarium Plouzane
BP 70, 29280 Plouzane
France
Phone: +33–98–224319
Fax: +33–98–224548
E-mail: belsher@ifremer.fr
Notes: Mediterranean, English Channel species

INRA
Correspondent: J.-F. Humbert
BP 511, 74203 Thonon les Bains Cedex
France
Phone: +33–50–714955
Fax: +33–50–260760

INSERM
Correspondent: D. Pesando
Station Marine, 06230 Villefranche-Sur-Mer
France
Phone: +33–93–763870
Fax: +33–93–763–877

(continued)

Microalgal Culture Collection
Correspondent: Dominique Grizeau
Lab Biotechnology Marine Cnam, Digue de Collignon, 50110
  Tourlaville
France
Phone: +33–33–203765
Fax: +33–33–205608

Microalgal Culture Collection
Correspondent: Daniel Vaulot
Station Biologique, BP74, 29682 Roscoff
France
Phone: +33–98–292334
Fax: +33–98–292324
E-mail: vaulot@sb-roscoff.fr

Microalgal Culture Collection–Univ. Caen
Correspondent: Benoit Veron
Univ. Caen, Biologie et Biotechnologie Marines, 14032 Caen
France
Phone: +33–31–455837
Fax: +33–31–455346
E-mail: billard@criuc.unicaen.fr

Museum National d'Histoire Naturelle–Cryptogamie
Correspondent: Alain Couté
12 Rue Buffon, 75005 Paris
France
Phone: +33–1–40793196
Fax: +33–1–40793594

PCC–Pasteur Culture Collection Cyanobacterial Strains
Correspondent: R. Rippka, M. Herdman
Unité de Physiologie Microbienne, Inst. Pasteur, 28 Rue du Docteur
  Roux, 75724 Paris Cedex 15
France
Phone: +33–1–45688416
Fax: +33–1–40613042
E-mail: rrippka@pasteur.fr; mherdman@pasteur.fr
Internet: www.pasteur.fr/recherche/banques/PCC/

Akung
Biologische Station Hiddensee, Schwedenhagen 6, 18656 Kloster-
  Hiddensee
Germany
Phone: +49–38300–6440
Fax: +49–38300–64444

Alfred Wegener Inst. for Polar and Marine Research
Box 120161, 27515 Bremerhaven
Germany
Phone: +49–471–4831530
Fax: +49–471–4831425
E-mail: rcrawford@awi-bremerhaven.de
Notes: over the world

Biologische Anstalt Helgoland
Correspondent: Elbraechter Malte
Wattenmeerstation Sylt, Hafenstr 43, 25992 List
Germany
Phone: +49–4651–956135
Fax: +49–4651–956200

Heinrich Heine Univ.
Correspondent: Klaus Kowallik
Universitätstr 1, 40225 Düsseldorf
Germany
Phone: +49–211–8112455
Fax: +49–211–8112455
E-mail: kowallik@uni-duesseldorf.de

Igv. Inst. für Getreideverarbeitung
Correspondent: Otto Pulz
GmbH, Arthur-Scheunert Allee 40–41, 14558 Bergholz-Rehbrucke
Germany
Phone: +49–33–20089151
Fax: +49–33–20089220

(*continued*)

Inst Freshwater and Fish Ecology
Correspondent: Lothar Krienitz
16775 Neuglobsow
Germany
Phone: +49–33081/238
Fax: +49–33081/238

Inst für Pflanzenphysiologie
Correspondent: Wolfgang Gross
Univ. Berlin, Königin-Luise Str 12–16a, 14195 Berlin
Germany
Phone: +49–30–8383111
Fax: +49–30–8384313

Liste der Sammlung von Conjugaten-Kulturen (SVCK)
Correspondent: M. Engels
E-mail: engels@botanik.uni-hamburg.de
am Inst. für Allgemeine Botanik der Univ. Hamburg, Ohnhorststr. 18
  D-22609 Hamburg, Germany
Internet: www.itz,uni-hamburg.de

Max Planck Inst. für Limnologie
Correspondent: Rainer Suetfeld
Box 165, 24302 Plön
Germany
Phone: +49–4522–763287
Fax: +49–4522–763310
E-mail: suetfeld@mpil-ploen.mpg.d400.de

Sammlung von Algenkulturen (SAG)
Leitung: Prof. Dr. Thomas Friedl Mitarbeiter
Fax: 49–551–397871
Inst Pflanzenphysiologie, Univ. Göttingen, 18 Nikolausbergerweg, 340
  Göttingen
Germany
Phone: +49–351–26160
Fax: +49–531–261648
Internet: www.gwdg.de/~botanik/phykologia/epsag.html

Culture Collection–Univ. Thessaloniki
Correspondent: Eleni Tryfon
Aristotle Univ. Thessaloniki, Inst. Botany, School Biology
Greece
Phone: +31–31–998266
Fax: +31–31–998389
E-mail: tryfon@olymp.ccf.auth.gr

CNR Culture Collection
Correspondent: Mario Tredici
CNR, Centro di Studio dei Microorganismi Autotrofi, P.LE delle
  Cascine 27, 50144 Firenze
Italy
Phone: +39–55–328806
Fax: +39–55–330431
E-mail: tredici@csma.fi.cnr.it

Culture Collection Naples
Correspondent: Raffaella Casotti
Stazionne Zoologica a. Dohrn, Naples
Italy
Phone: +39–81–5833211
Fax: +39–81–7641355
E-mail: raffa@alpha.szn.it

Univ. Roma Tor Vergata Culture Collection
Correspondent: Patricia Albertano
Dept. Biologia, Via della Ricerca Scientifica, 00133 Roma
Italy
Phone: +39–6–72594345
Fax: +39–6–2023500
E-mail: albertano@towx1.ccd.utorvrm.it

Culture Collection Vilnius
Correspondent: Jurate Kasperoviciene
Inst. Botany, Zaliuju Ezeru 47, 2021 Vilnius
Lithuania
Phone: +370–2-736251
Fax: +370–2-729950

(*continued*)

TABLE B10.2. (*continued*)

NICMM Culture Collection
Correspondent: Louis Peperzak
Box 8039, 4330 EA Middleburg
Netherlands
Phone: +31–118–672306
Fax: +31–118–616500
E-mail: peperzak@kikz.rws.minvenw.nl

Rijksherbarium
Research Inst., Rijksherbarium/Hortus Botanicus, Box 9514, 2300
  Leiden
Netherlands
Phone: +31–71–5274729
Fax: +31–71–5273511
E-mail: prudhomme@rulrhb.leidenuniv.nl
Notes: species over the world

NIVA–Culture Collection of Algae
Correspondent: O.M. Skulberg
Norwegian Inst. for Water Research, Box 173 Kjelsas, 0411 Oslo
Norway
Phone: +47–22–185100
Fax: +47–22–185200

ACOI Univ. Coimbra
Correspondent: M.F. Santos
Univ. Coimbra, Dept. Botanica, Calçada Maritim de Freitas, 3049
  Coimbra
Portugal
Phone: +351–39–22897
Fax: +351–39–20780
E-mail: mfsantos@ci.uc.pt
Internet: www.uc.pt/botanica/ACOI.htm

Univ. Babes–Bolyai Culture Collection
Correspondent: Leontin Stefan Peterfi
Dept. Plant Biology, Republicii St 42, 3400 Cluj-Napoca
Romania
Phone: +40–64–192152
Fax: +40–64–161238

Bispsu–Biological Inst. St. Petersburg State Univ.
Correspondent: S. Karpov
Univ. St. Petersburg, Oranienbaumskoye 2, Stary Peterhof, 198904 St.
  Petersburg
Russia
Phone: +7–812–4279669
Fax: +7–812–4286649

DMMSU–Culture Collection of Microbiology Department
Moscow State Univ., Mgu Vorobiovy Gory, 119899 Moscow
Russia
Phone: +7–95–9393033
Fax: +7–95–9390126

IPPAS–Collection of Microalgae
Correspondent: Elena.S. Kuptsova
Russian Acad. Sciences, Inst. Plant Physiology, 35, Botanischeskaya St.,
  127276 Moscow
Russia
Phone: +7–95–4824491; 7–95–9039380; 7–95–9039346
Fax: +7–95–4821685
Internet: panizzi.shef.ac.uk/msdn/ippas.html

Slovak Acad. Sciences
Correspondent: Frantisek Hindak
Inst. Botany, Ubravska Cesta 14, 84223 Bratislava
Slovakia
Phone: +42–7–3782505
Fax: +42–7–371948
E-mail: botuhin@savba.savba.sk

TABLE B10.2. (*continued*)

BCA–Banco Canario de Alagas
Correspondent: Antera Martel
Inst. Applied Algology, Muelle de Taliarte S/N, 35214 Telde, las
  Palmas de Gran Canaria
Spain
Phone: +34–28–133290
Fax: +34–28–132950
E-mail: martel@amq1.ext.ulpgc.es

ICMA–Marine Microalgae Culture Collection
Correspondent: Luis M. Lubian Chaichio
Inst. Ciencias Marinas de Andalucia, Poligono Rio San Pedro S/N,
  11501, Puerto Real, Cadiz
Spain
Phone: +34–56–832612
Fax: +34–56–834701

UAM–Culture Collection
Correspondent: Antonio Quesada
Dept. Biologia, Univ. Autonoma de Madrid, 28049 Madrid
Spain
Phone: +34–1–3978181
Fax: +34–1–3978344
E-mail: quesada@vm1.sdi.uam.es

UCM–Culture Collection
Correspondent: Eduardo Costas
Univ. Complutense, Fac. Veterinaria, 28040 Madrid
Spain
Phone: +34–1–3943769
Fax: +34–1–3943778
E-mail: vlrodas@eucmax.sim.ucm.es

Avd for Marine Ekology
Correspondent: L. Gisselson
Dept. Marine Ecology, Lund Univ., Lund
Sweden
Phone: +46–46–08366
Fax: +46–46–104003

IBASU–A Collection of Algal Strains
Correspondent: V.P. Yunger
Division of Spore Plantae, Inst. Botany, Terenschevskoskaya 2, 252601
  Kiev
Ukraine
Phone: +380–224–5157
Fax: +380–224–1064

CCAP–Culture Collection of Algae and Protozoa (freshwater)
The Librarian: I. McCulloch
Freshwater Biological Association, The Ferry House, Ambleside,
  Cumbria LA22 0LP
United Kingdom
Phone: +44–1394–42468
Fax: +44–1394–46914
E-mail: ccap@ife.ac.uk

CCAP–Culture Collection of Algae and Protozoa (marine)
Correspondent: Michael Turner
Dunstaffnage Marine Lab, Box 3, Oban, Argyl
United Kingdom
Phone: +44–1631–562244
Fax: +44–1631–565518
E-mail: ccapn@dml.ac.uk

Culture Collection Cardiff
Correspondent: Ingrid Jüttner
Univ. Wales, School of Pure and Applied Biology, Box 915, Cardiff
United Kingdom
Phone: +44–1222–874000, EXT. 29
Fax: +44–1222–874305
E-mail: juttner@cardiff.ac.uk

(*continued*)   (*continued*)

TABLE B10.2. (*continued*)

ECCO–European Culture Collection's Organization
Correspondent: A. Doyle
ECACC/CAMR Poton Dwon, Salsbury
United Kingdom
Phone: +44–1980–612512
Fax: +44–1980–611315
E-mail: email.ecacc@ecacc.demon.co.uk
Notes: Africa species

PLY–Plymouth Marine Lab Algal Culture Collection
Correspondent: J. Green
Marine Biological Association, The Citadel, Plymouth
United Kingdom
Phone: +44–1752–222772
Fax: +44–1752–226865

SARU–Swansea Algal Research Unit
Correspondent: M.J. Merrett
Univ. Wales, School Biological Sciences, Swansea
United Kingdom
Phone: +44–1792–295362
Fax: +44–1792–295447
E-mail: m.j.merrett@swansea.ac.uk

Univ. Dundee Culture Collection
Correspondent: G.A. Codd
Univ. Dundee, Dept. Biological Sciences, Dundee
United Kingdom
Phone: +44–1382–344272
Fax: +44–1382–322318
E-mail: g.a. codd@dundee.ac.uk

Univ. Wales Culture Collection
Correspondent: D.H. Jewson
Freshwater Lab, Univ. Ulster, Traad Point, Ballyronan, Magherafelt, C.
  Derry
United Kingdom
Phone: +44–1648–417264
Fax: +44–1648–417777
E-mail: d.jewson@ulst.ac.uk

UW–Univ. Westminster Algal Collection
Correspondent: Jane Lewis
Univ. Westminster, 115 New Cavendish St, London WIM 8JS
United Kingdom
Phone: +44–171–9115000, EXT. 581
Fax: +44–171–9115087
E-mail: lewisjm@westminster.ac.uk

Israel
NCIM National Collection Industrial Microorganisms
Correspondent: A. Ben-Amotz
Iolr Israel Oceanographic Limnological Research Ltd., POB 8030,
  31080 Haifa
Israel
Phone: +972–4-515202
Fax: +972–4-511911

Japan
IAM Culture Collection
Head: Akira Yokota
Center for Cellular and Molecular Research
Institute of Molecular and Cellular Biosciences
Univ. Tokyo
Yayoi 1–1-1, Bunkyo-ku, Tokyo 113–0032
Japan
Phone: +81–35841–7828 (general, fungi and yeasts)
Phone: +81–35841–7867 (bacteria and microalgae)
Fax: +81–3-5841–8490
E-mail: iamcc@iam.u-tokyo.ac.jp
Internet: imcbns.iam.u-tokyo.ac.jp/misyst/ColleBOX/
  IAMcollection.html

(continued)

TABLE B10.2. (*continued*)

National Institute for Environmental Study (NIES)
16–2, Onogawa, Tsukuba, Ibaraki 305
Japan
Phone: +81–298–50–2471
Fax: +81–298–50–2427

United States
American Type Culture Collection (ATCC)
10801 University Blvd.Manassas, Virginia 20110–2209
USA
Phone: +703–365–2700
E-mail: news@atcc.org
Internet: www.atcc.org/

Culture Collection of Microorganisms from Extreme Environments
  (CCMEE)
Correspondent: Richard W. Castenholz
Department of Biology, Univ. Oregon, Eugene, Oregon 97403
USA
Phone: +541–346–4530
Fax: +541–346–2364
E-mail: rcasten@darkwing.uoregon.edu

Provasoli-Guillard National Center for Cultures of Marine
  Phytoplankton (CCMP)
Bigelow Laboratory, McKown Point, West Boothbay Harbor, Maine
  04575
USA
Phone: +207–633–9630
Fax: +207–633–9641
Internet:ccmp@bigelow.org

Univ. and Jepson Herbaria
1001 Valley Life Sciences Bldg., # 2465
Univ. of California, Berkeley CA 94720–2465
Phone: UC: +510–642–2465
Phone: JEPS: +510–643–5390
E-mail: herbaria@ucjeps.herb.berkeley.edu

UTEX (Univ. Texas Culture Collection of Algae)
The UTEX Director: Jerry Brand
MCDB, School of Biological Sciences, Univ. Texas at Austin, Austin,
  Texas 78712
USA
Phone: +512–471–4019
Fax: +512–471–0354
E-mail: UTAlgae@uts.cc.utexas.edu
Internet: www.bio.utexas.edu/research/utex/

[a]These and others are listed in Cyanosite: (bilbo.bio.purdue.edu/www-cyanosite/
collec.html). Most collections have or will have Web sites in the near future.
Besides the above, other local culture collections may be found. The principal
collection of cyanobacteria is that of the Institut Pasteur (PCC) (Unité de Physiologie Microbienne, Dept. Biochimie et Génétique Moléculaire, 75724 Paris
Cedex 15, France). Many of the strains were included in the taxonomic analysis of Rippka et al. (1979). A more recent listing is in Rippka and Herdman
(1992).

TABLE B10.3. Composition of freshwater media for *Cyanobacteria*[a]

Ingredient	Chu no. 10 (modified)	Gerloff et al.	BG-11[b]	D medium[c]	Allen and Arnon	Kratz and Myers
Disodium EDTA	–	–	1[e]	–	4[j]	–
Nitrilotriacetic acid (NTA)	–	–	–	100	–	–
Citric acid	3	3	6	–	–	165[k]
$NaNO_3$	–	41	1500[f]	700	–	–
KNO_3	–	–	–	100	2020	1000
$Ca(NO_3)_2 \cdot 4H2O$	40–60	–	–	–	–	25
$K_2HPO_4 \cdot 3H_2O$	13	–	40	–	456	1000
Na_2HPO_4	–	8	–	110	–	–
$MgSO_4 \cdot 7H_2O$	25	15	75	100	246	250
$CaSO \cdot 2H_2O$	–	–	–	60	–	–
$MgCl_2 \cdot 6H_2O$	–	21	–	–	–	–
$CaCl_2 \cdot 2H_2O$	–	36	36	–	74	–
KCl	–	9	–	–	–	–
NaCl	–	–	–	8	232	–
$Na_2CO_3(H_2O)$	20	20	20	–	–	20 (optional)
Ferric ammonium citrate	–	–	6	–	–	–
Ferric citrate	3	3	–	–	–	–
$FeCl_3$	3	–	–	0.3[h]	–	–
$Fe_2(SO_4)_3 \cdot 6H_2O$	–	–	–	–	–	4
Micronutrients	–[d]	–[d]	1 ml[g]	0.5 ml[i]	–[d]	1 ml[g]
Vitamin mix	–[d]	–[d]	–[d]	–[d]	–[d]	–[d]

[a]Unless indicated, concentrations are in mg/l of double-distilled or deionized water; see Castenholz, 1988a, for details.

[b]pH 7.4 after cooling.

[c]Prepared as a 20-fold concentrated stock, stored at 4°C. Micronutrients and $FeCl_3$ are included in the stock. pH is adjusted to 8.2 with NaOH before autoclaving. After cooling and clearing, pH is about 7.5. Several variations of this medium are described in Castenholz, 1988a.

[d]Micronutrients and vitamins optional. If used, 0.5–1.0 ml of any mixture in Table B10.5.

[e]Disodium-magnesium EDTA is generally used.

[f]The nitrate concentration is often lowered.

[g]The medium generally uses A_5 + Co (Table B10.5).

[h]Sometimes, 2–4 times this amount is used. A stock solution of 0.29 g/l is kept at 4°C.

[i]D micro (Table B10.5).

[j]13% ferric-sodium EDTA.

[k]Trisodium citrate dihydrate.

Diagnostic guide to the Subsections ("Orders") of Cyanobacteria

A. 1. Unicellular or nonfilamentous aggregates of cells held together by outer walls or a gel-like matrix (colonies). B

 2. Filamentous; trichome of cells branched or unbranched, uniseriate or multiseriate. C

B. 1. Reproduction by binary fission in one, two, or three planes, symmetric or asymmetric; or by budding.

Subsection I, p. 493

 2. Reproduction by internal multiple fissions with production of daughter cells smaller than the parent; or by multiple plus binary fission.

Subsection II, p. 514

C. 1. Reproduction by binary fission in one plane only giving rise to uniseriate unbranched trichomes, although false branching may occur (see "General Characteristics of the Cyanobacteria"). D

 2. Reproduction by binary fission periodically or commonly in more than one plane, giving rise to multiseriate trichomes or trichomes with various types of "true" side branches or both (see "General Characteristics of the Cyanobacteria").

Subsection V, p. 589

D.   1. Trichomes composed of cells which do not differentiate into heterocysts or akinetes.
            Subsection III, p. 539
      2. One or more cells of trichomes differentiate into heterocysts, at least when the concentration
        of ammonium or nitrate in the medium is low; some also produce akinetes.
            Subsection IV, p. 562

Methods of Isolating and Culturing Cyanobacteria   The various methods for the isolation and cultivation of cyanobacteria will not be described here. These may be found in the 1989 edition and in the 1994 *Manual*, but also in more detail in Castenholz (1988a, b) and in Rippka (1988a). However, recipes for culture medium are provided here in Tables B10.3, B10.4, and B10.5.

TABLE B10.4. Composition of Marine and Hypersaline Media for Cyanobacteria[a]

Ingredient	Grund[b]	F/2[b]	MN[c]	Ong et al.[d]	ASP-M[e]	Aquil[f]	Erdschreiber's[b]	Yopp et al.[g]
Disodium EDTA	2	10[j]	0.5	5	0.8[j]	–	10[j]	5[j]
Citric acid	–	–	3	–	–	–	–	–
NaNO$_3$	40	90	750	750	40–70	8.5	150	–
Ca(NO$_3$)$_2$·4H$_2$O	–	–	–	–	–	–	–	1000
K$_2$HPO$_4$·3H$_2$O	–	–	–	–	–	–	–	–
Na$_2$HPO$_4$	4	–	–	–	–	–	40	–
NaH$_2$PO$_4$·H$_2$O	–	5–20	20	15	7–14	0.5	–	65
MgSO$_4$·7H$_2$O	–	–	38	–	4920	–	–	10,000
MgCl$_2$·6H$_2$O	–	–	–	–	4040	11030	–	10,680
CaCl$_2$·2H$_2$O	10	–	18	–	1270	1000–1350	–	–
Na$_2$CO$_3$·H$_2$O	–	–	20	–	–	–	–	–
NaHCO$_3$	–	–	–	–	168	200	–	–
Na$_2$SO$_4$	–	–	–	–	–	4090	–	–
NaCl	–	–	–	–	23,200	24,360	–	117,000
KCl	–	–	–	–	740	695	–	2000
KBr	–	–	–	–	–	10	–	–
NaF	–	–	–	–	–	3	–	–
Fe$_2$(SO$_4$)$_3$·6H$_2$O	0.2	–	–	–	–	–	–	–
Na$_2$SeO$_4$ (0.01 mM stock)	1 ml	–	–	–	–	–	–	–
NiSO$_4$(NH$_4$)$_2$SO$_4$·6H$_2$O (0.1 mM stock)	1 ml	–	–	–	–	–	–	–
Micronutrients	0.2 ml[h]	1 ml[k]	1 ml[l]	1 ml	1 ml[m]	0.5 ml[o]	–	1 ml[r]
Ferric ammonium citrate	–	–	3	–	–	–	–	–
Vitamin mix	0.5 ml[i]	1 ml[i]	–	1–2 ml	1 ml[n]	0.5–1.0 ml[p]	0.5 ml[p]	–
Natural or artificial seawater	1000 ml	1000 ml	750 ml	877 ml	–	–	950 ml	–
Distilled/deionized water	–	–	250 ml	120 ml	1000 ml	1000 ml	–	1000 ml
Soil extract	–	–	–	–	–	–	50 ml[q]	–

[a]Unless indicated, concentrations are in mg/l of natural or artificial seawater or distilled water; see Castenholz, 1988a, for details.

[b]McLachlan, 1973.

[c]Rippka et al., 1979.

[d]Ong and Glazer, 1984.

[e]See McLachlan, 1973; in addition to ingredients listed, 660–1320 mg/l glycylglycine may be added as buffer for axenic cultures.

[f]See Castenholz, 1988a, for preparation details; 30 mg/l H$_3$BO$_3$ and 17 mg/l SrCl$_2$·6H$_2$O is also added.

[g]See Waterbury and Stanier, 1981; 500 mg/l glycylglycine is used as buffer.

[h]A$_5$ + Co or D Micro generally used (Table B10.5).

[i]Optional: mix of Ong and Glazer, 1984, would generally be adequate (Table B10.5).

[j]13% ferric-sodium EDTA.

[k]Optional, but if used eliminate ferric EDTA of original formula.

[l]A$_5$ + Co (Table B10.5).

[m]See McLachlan, 1973, for suggested micronutrient addition; however, F/2 should be adequate (Table B10.5).

[n]S-3 mix (Table B10.5).

[o]Use PIV for F/2 (Table B10.5).

[p]Use Ong and Glazer, 1984 (Table B10.5).

[q]See Castenholz, 1988a, for preparation.

[r]Sheridan and Castenholz solution (Table B10.5).

TABLE B10.5. Composition of micronutrient solutions and vitamin mixes

Ingredient	Micronutrients (g/l)					Ingredient	Vitamins (mg/ml)[f]		
	A5 + Co[a]	D Micro[b]	Sheridan and Castenholz[c]	PIV[d]	F/2[e]		DN[g]	S-3[e]	Ong et al.[h]
H_2SO_4 (conc.)	–	0.5 ml	–	–	–	Nicotinic acid	0.100	0.1	–
HCl (conc.)	–	–	3 ml	–	–	PABA	0.010	0.10	–
H_3BO_3	2.86	0.5	0.5	–	–	Biotin	0.001	0.001	0.001
$MnSO_4 \cdot H_2O$	–	2.28	–	–	–	Thiamin	0.200	0.5	2.0
$MnCl_2 \cdot 4H_2O$	1.81	–	2.0	0.041	0.177	Cyanocobalamin	0.001	0.001	0.001
$ZnNO_3 \cdot 6H_2O$	–	–	0.5	–	–	Folic Acid	0.001	0.002	–
$ZnSO_4 \cdot 7H_2O$	0.22	0.5	–	–	0.018	myo-Inositol	0.001	5.0	–
$ZnCl_2$	–	–	–	0.005	–	Thymine	–	3.0	–
$CuCl_2 \cdot 2H_2O$	–	–	0.025	–	–	Calcium pantothenate	0.100	0.10	–
		–							
$CuSO_4 \cdot 5H_2O$	0.08	0.025	–	–	0.010				
$NaMoO_4 \cdot 2H_2O$	0.39	0.025	0.025	0.004	0.007				
$Co(NO_3)_2 \cdot 6H_2O$	0.049	–	0.025	–	–				
$CoCl_2 \cdot 6H_2O$	–	0.045	–	0.002	0.011				
$VOSO_4 \cdot 6H_2O$	–	–	0.025	–	–				
$FeCl_3 \cdot 6H_2O$	–	–	–	0.097	1.90				
$NiSO_4(NH_4)_2SO_4 \cdot 6H_2O$	–	0.019	–	–	–				
Na_2SeO_4	–	0.004	–	–	–				
Disodium EDTA	–	–	–	0.75 (add first)	4.35				

[a]Rippka et al., 1979.

[b]Castenholz, 1981.

[c]Waterbury and Stanier, 1981. Ni and Se have recently been added by Castenholz.

[d]Starr, 1978.

[e]McLachlan, 1973.

[f]Concentrations are usually designed for additions of 1 ml/l. Vitamin mixes are generally filter sterilized and added after the medium is autoclaved. No cyanobacteria have been shown to have a complex vitamin requirement; most have none at all.

[g]Nelson et al., 1982.

[h]Ong and Glazer, 1984.

Phylogenetic Relationships Among the Cyanobacteria Based on 16S rRNA Sequences

ANNICK WILMOTTE AND MICHAEL HERDMAN

The purpose of this chapter is to identify the major cyanobacterial lineages by phylogenetic analysis, and to show why it is premature, for this edition of the *Manual*, to treat the taxonomy of cyanobacteria on a phylogenetic basis. Using the Maximum Likelihood method (fastDNAml, Olsen et al., 1994a) and the Neighbor-joining method with several corrections (Jukes and Cantor, Tajima and Nei, Kimura, Transversion analysis, Galtier and Gouy, and the substitution rate calibration method) as well as a bootstrap analysis involving 500 resamplings, as implemented in the software TREECON (Van de Peer and de Wachter, 1994; www.uia.ac.be/u/yvdp/), we have compared the different trees obtained. Two examples of tree topologies, constructed using the same 123 cyanobacterial strains, are given in Figs. B10.7 and B10.8. In Fig. B10.7, the Neighbor-joining method was applied to a distance matrix calculated using the substitution rate calibration where all positions are weighted according to their rate of variability (Van de Peer et al., 1993), whereas Fig. B10.8 shows the tree obtained using the fastDNAml software. We observe that the branching orders are quite variable, especially at the base of the tree. This makes a hierarchical arrangement of taxa, based on these trees, quite impossible. However, we also observe, at the tip of the branchings, groups of sequences which are related and consistently cluster in the same lineage. These groupings have a true phylogenetic significance and will be discussed below in more detail.

Preliminary warnings Scientists interested in taxonomic or other studies should not assume that the strains they use have been correctly identified, and many culture collections contain misidentified strains (Komárek, 1994). Therefore, they should either use well characterized strains or make the effort themselves to carry out a polyphasic taxonomic study including a morphological description. In this chapter, the interpretation of phylogenetic trees based on the 16S rRNA sequences is complicated by the presence of strains of unknown morphology, because no descriptions were given. Where the latter appear closely related phylogenetically to another organism, it is impossible to know whether they share additional similarities and whether taxonomic inferences can be made about this grouping. In the case of "*Microcoleus* 10 mfx" (*Geitlerinema* PCC 9452) which appeared very closely related to *Oscillatoria* PCC 7105 (Wilmotte, 1994), it took several years before a cultivated strain was available to the present authors. We concluded that the morphology of the two strains was extremely similar and that both could better be assigned to *Geitlerinema*. Another example of confusion involves two nonidentical 16S rRNA sequences that were submitted for each of the "*Microcystis*" strains NIES 42 and NIES 43. These strains occupy different positions in the trees, depending on the source of the sequence. This situation may be explained either by a high degree of sequencing errors in one set of sequences, mislabeling of strains in one of the laboratories, or the presence of two different genotypes in each culture. Moreover, both strains are incorrectly identified and do not belong to the genus *Microcystis* as defined in this *Manual*.

The trees based on cyanobacterial 16S rRNA sequences One striking feature of the 16S rRNA trees constructed with Maximum Likelihood or distance methods by us and others (e.g., Turner, 1997) is that all branches at the base diverge in a short interval of evolutionary distance, in a "fan-like" manner as observed originally by Giovannoni et al. (1988). As hypothesized by the latter authors, this topology may reflect the revolutionary invention of oxygenic photosynthesis which allowed an explosive radiation of the cyanobacteria in a short time span.

The trees (Figs. B10.7 and B10.8) also show that a number of sequences have no close relatives, and could be called "loner" sequences. Except for their isolated position, no taxonomic inference can be made for these strains, and sequences of related strains are needed. Unfortunately, these loner sequences tend to be unstable, grouping with others in a rather erratic manner, depending on the positions, strains, and tree building method used. This instability is reflected in the low bootstrap values for these groupings. The origin of these artefactual attractions is probably a combination of different factors and is linked to the fact that false identities might arise by chance between groups without real phylogenetic affinities; the absence of organisms at intermediate levels of relatedness does not allow us to infer which identities are real and which are due to chance (Ludwig et al., 1998b). We also deplore the presence in the present dataset of many partial sequences, which further complicates the phylogenetic analyses by restricting the number of positions present in all sequences which can be used for the calculations. Our observations show that the addition of new complete sequences tends to reduce the number of artifactual groupings and we have therefore used unpublished sequences in the trees of Figs. B10.7 and B10.8. As the number of published cyanobacterial sequences has increased greatly during the past year, we expect that the phylogenetic trees will become much more useful for taxonomic inferences. However, we are more skeptical about the potential for distinguishing a well structured hierarchy of different taxonomic levels, due to the explosive radiation of equivalent lineages in a short domain of evolutionary distance.

A number of clusters can be identified in the trees of Figs. B10.7 and B10.8, and are discussed below:

I. **The heterocystous cluster** – As observed by Giovannoni et al. (1988), this cluster corresponds to a homogenous genotypic lineage well supported by the bootstrap analysis. However, within this cluster, strains presently assigned to different genera, such as *Nostoc* PCC 7120 and *Cylindrospermum* PCC 7417 or *Nodularia* PCC 73104 and *Anabaena cylindrica* PCC 7122, are situated in the same lineage. This highlights the need to study the genotypic relationships of these strains in more detail. The three strains which divide in more than one plane, *Fischerella* PCC 7414, *Chlorogloeopsis fritschii* PCC 6718, and *Chlorogloeopsis* HTF ("*Mastigocladus* HTF") PCC 7518, are equally distant from each other in 16S rRNA similarity and represent three different branches, probably related at the genus level. The latter strain no longer produces heterocysts, probably as the re-

sult of a mutation (Wilmotte et al., 1993). It should be noted that although strains assigned to these genera are treated in this *Manual* as a distinct Subsection (Subsection V), their separation from the other heterocystous cyanobacteria of Subsection IV is not justified on phylogenetic grounds. *Scytonema* PCC 7110 has no close relatives in the phylogenetic trees. With the recent addition of new sequences, however, some interesting relationships are emerging. Three typical *Calothrix* strains (PCC 6303, PCC 7102, PCC 7709) cluster together and may represent a single species, since the latter two strains showed about 70% relative binding in DNA–DNA hybridization studies (Lachance, 1981). These strains, isolated from terrestrial or freshwater habitats, are related to the marine *Rivularia* PCC 7116 (though with bootstrap support of less than 50% in Fig. B10.7). However, a fourth *Calothrix* strain (PCC 7507) is not related to this cluster and clearly represents a different genus; this strain produces akinetes (Rippka et al., 1979) and is phylogenetically related to another akinete-former, *Cylindrospermum* PCC 7417. A cluster of closely related *Tolypothrix* strains is well separated from *Calothrix*, in accordance with their treatment in Subsection IV of this *Manual*. Five *Nostoc* strains, including the type strain *N. punctiforme* PCC 73102, can be considered to be members of a single species, whereas *Nostoc* PCC 7120 is only distantly related to this group. Finally, the planktonic gas-vacuolate organisms *Nodularia*, *Anabaenopsis*, *Cyanospira*, *Aphanizomenon*, and *Anabaena flos-aquae* are closely related, but this group also includes *Anabaena cylindrica* PCC 7122; although the latter was isolated from an aquatic environment, it does not produce gas vesicles.

II. **The lineage containing *Prochlorococcus marinus*, *Synechococcus*, *Cyanobium*, and sequences from the Sargasso Sea** – Four 16S rRNA sequences directly retrieved from water samples of the Sargasso Sea, and for which the phenotype is unknown, appear closely related to *P. marinus* and marine picoplanktonic *Synechococcus*. A more complete phylogenetic tree of these organisms is given by Urbach et al. (1998), see also the description of organisms of Subsection I in this *Manual*. It is surprising that two *Cyanobium* strains (*C. gracile* PCC 6307 and *C. marinum* PCC 7001) that are of freshwater and marine origins, respectively, and identified as *Synechococcus* in Rippka et al., 1979, cluster with the picoplanktonic organisms, regardless of major differences in mol% G + C (32% for *P. marinus* PCC 9511 [Herdman and Rippka, unpublished] and 70% for the *Cyanobium* strains). Included in this lineage is a pair of sequences of "*Microcystis holsatica*" and "*M. elabens*" discussed earlier. Sequences carrying the same strain designations but obtained in a different laboratory fall outside this cluster. As described in the chapter on Subsection I of this *Manual*, the Geitlerian identification of such organisms as *Microcystis* requires revision.

III. **The *Prochlorothrix hollandica* strains** – This chlorophyll *a*/*b* containing organism was first isolated and described in detail by Burger-Wiersma et al. (1989). The axenic type strain is PCC 9006. Although the morphological similarity of the second strain (Zwart et al., 1998) to PCC 9006 is not documented, both strains are phylogenetically closely related. In the previous edition of the *Manual*, *Prochlorothrix* was placed in an order separate from the cyanobacteria. In this edition it is treated as a member of Subsection III,

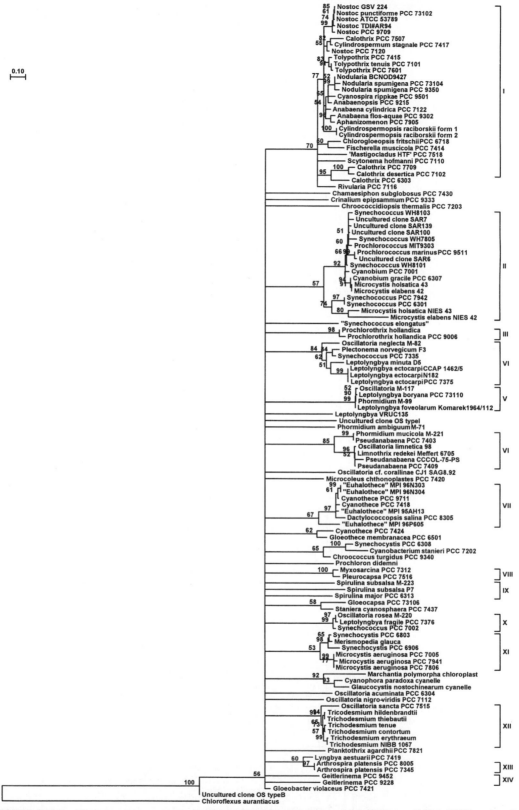

FIGURE B10.7. Distance tree constructed with the Neighbor-joining method (Saitou and Nei, 1987), using the rate substitution calibration (Van de Peer et al., 1993) to calculate the evolutionary distances between pairs of sequences, as implemented in the software package TREECON for Windows (Van de Peer and de Wachter, 1994). The parameter *p* was 0.34. A bootstrap analysis involving 500 resamplings was performed, and bootstrap values higher than 50% are given in front of the concerned node. The nodes not supported by at least 50% of bootstrap replicates are drawn as unresolved. 123 cyanobacterial sequences of cultivated and uncultivated organisms were included, and *Chloroflexus aurantiacus* was used as outgroup. 820 positions, common to all the sequences, were used and indels were not taken into account. The scale given on top corresponds to an evolutionary distance of 0.1.

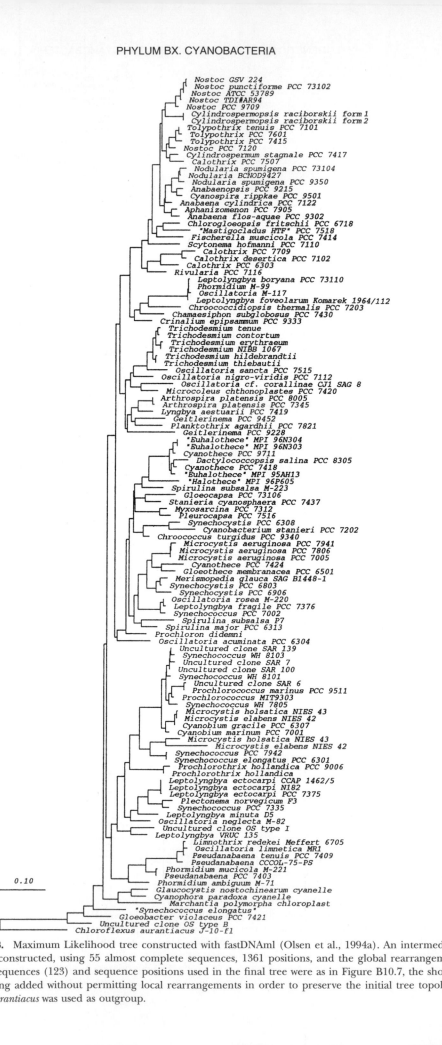

FIGURE B10.8. Maximum Likelihood tree constructed with fastDNAml (Olsen et al., 1994a). An intermediate tree was first constructed, using 55 almost complete sequences, 1361 positions, and the global rearrangement option. The sequences (123) and sequence positions used in the final tree were as in Figure B10.7, the shorter sequences being added without permitting local rearrangements in order to preserve the initial tree topology. *Chloroflexus aurantiacus* was used as outgroup.

since its exclusion from the cyanobacteria is not justified phylogenetically.

IV. **The marine *Leptolyngbya* lineage** – This lineage contains only marine phycoerythrin-containing strains with narrow filaments, corresponding to the genus *Leptolyngbya*. Three strains identified as *Leptolyngbya ectocarpi* and having almost identical 16S rRNA sequences were isolated from different parts of the world: Australia, East Coast of the USA, and Corsica (France, Europe). This shows a wide geographical distribution of strains with a similar genotype. A sequence divergence of 6% separates the cluster of these three strains from strain D5, assigned to *Leptolyngbya minuta* and having deeper constrictions at the cross-walls (Wilmotte et al., 1992, 1997). The marine strain *Plectonema norvegicum* F3 is distinguished from the *Leptolyngbya* strains by more flexuous trichomes, more rounded cells which cannot become much longer than wide, a more irregular sheath and the presence of false branching (Wilmotte, 1991). It is still questionable whether the genus *Plectonema* should be retained or merged with *Leptolyngbya*. The morphology of *Oscillatoria neglecta* M-82 corresponds to a *Leptolyngbya*-type (R. Rippka, personal communication). The marine unicellular *Synechococcus* PCC 7335 is also, unexpectedly, a member of this lineage. It is one of the two cases (the other being *Synechococcus* PCC 7002) where filamentous and unicellular cyanobacteria are very closely related.

V. **A freshwater "*Leptolyngbya*" lineage** – A lineage containing strains which can be assigned to the genus *Leptolyngbya* emerges. *L. foveolarum* Komárek 1964/112 and *L. boryana* PCC 73110 have almost identical 16S rRNA sequences and share a common morphology, though the latter strain is distinguished by the presence of sacrificial cells or necridia (Albertano and Kovacik, 1994; Nelissen et al., 1996). The morphology of strains *Phormidium* M-99 and *Oscillatoria* M-117 (Ishida et al., 1997) is not documented, although their sequences are virtually indistinguishable from those of the *Leptolyngbya* strains.

VI. **The *Pseudanabaena* cluster** – The strains *Pseudanabaena* PCC 7403, PCC 7409, *Pseudanabaena galeata* CCCOL-75-PS, and *Limnothrix redekei* Meffert 6705 ("*Oscillatoria redekei*"), as well as *Oscillatoria limnetica* MR1 (Zwart et al., 1998) and *Phormidium mucicola* M-221, belong to the same lineage. *Pseudanabaena* PCC 6903 also belongs to this lineage (not shown). The morphology of *O. limnetica* MR1 is similar to *L. redekei* without gas vesicles but there is no description of strain *P. mucicola* M-221 (Ishida et al., 1997). If this strain corresponds to the original description of the species, all of the seven strains share a number of morphological similarities: cell diameter between 1 and 3 μm and absence of sheath. Gas vesicles are present at the cross-walls, except in PCC 7409 and MR1, which have lost them, and *Phormidium mucicola*, where they were not explicitly reported in the original description. The constrictions at the cross-walls are consistently quite deep, except for in *Limnothrix redekei* Meffert 6705. Meffert (1987) showed that in this strain the constrictions varied from inconspicuous, when gas vesicles were abundant, to quite deep when the volume occupied by the gas vesicles was reduced under particular culture conditions. The lineage containing these seven strains could correspond to the definition of the genus *Pseudanabaena* if a certain variation in the degree of constrictions at the cross-walls as a function of the environmental conditions is admitted.

VII. **The halotolerant unicellular strains** – The morphology and physiology of these marine, euryhaline, and moderately thermophilic "Euhalothece" and "Halothece" strains were studied by Garcia-Pichel et al. (1998), and the phylogenetic position determined by analysis of the 16S rRNA sequences. They observed that the morphology was very diverse (cell widths from 2.8 to 10.3 μm, single-celled to colonial habit, round to needle-like cells, baeocyte formation or binary division) and that the morphological variability was conspicuous, depending on culture conditions. In contrast, the physiological characters were quite similar and the sequences grouped into one phylogenetic lineage. This study is a nice example of a polyphasic approach to the taxonomy of cyanobacteria.

VIII. **The lineage containing the baeocyte-forming strains** – There is a cluster grouping the marine strains *Pleurocapsa* PCC 7516 and *Myxosarcina* PCC 7312, but another freshwater tropical baeocyte forming strain, *Stanieria* PCC 7437, and the terrestrial *Chroococcidiopsis thermalis* PCC 7203 are "loner" strains according to our criteria. Sequences of more baeocyte-formers are therefore needed to understand the phylogenetic relationships of strains assigned to Subsection II.

IX. **The *Spirulina* strains** – In the distance tree of Fig. B10.7, the three *Spirulina major* and *S. subsalsa* strains are grouped together but with a bootstrap value lower than 50%. In the Maximum Likelihood tree of Fig. B10.8, *Spirulina subsalsa* M-223 (Ishida et al., 1997) does not cluster with the others. In the case of strains P7 and PCC 6313, the trichome diameters are similar, though there is a distinct space between the spirals of PCC 6313 whereas the spirals of P7 are contiguous (Rippka et al., 1979; Wilmotte, 1991). Like *S. subsalsa* P7, strain M-223 exhibits the morphology typical of this species (R. Rippka, personal communication). This points to the existence of morphological and genotypic divergence within the genus *Spirulina*, which should be studied in more detail.

X. ***Synechococcus* PCC 7002 and *Leptolyngbya fragile* PCC 7376** – As in the marine *Leptolyngbya* lineage (above), a unicellular and a filamentous strain of similar marine origin and cell diameter are clustered in the same lineage, but with a 16S rRNA sequence similarity (97.4%) which is a bit too low for conspecificity. *Oscillatoria rosea* M-220, of unknown origin and morphology (Ishida et al., 1997), is also a member of this lineage. If the presence of both morphotypes in the same phylogenetic lineage is hypothesized to result from mutations, the impact of such phenotypic plasticity on the taxonomic system of the cyanobacteria cannot be determined because the extent to which the underlying mutation occurs is not known. It would probably not require many genetic changes to allow a filamentous form to break up into single cells or for single cells to evolve into filaments by remaining attached after cell division.

XI. **The *Synechocystis* PCC 6906–*Microcystis* lineage** – There is a moderately well supported grouping of three unicellular strains assigned to each of the *Synechocystis* and *Microcystis* genera. *Microcystis* and the high GC cluster of *Synechocystis* are very similar morphologically and in mol% G + C (Waterbury and Rippka, 1989). *Microcystis* is distinguished by the presence of gas vesicles and often forms colonies in nature. However, these two characters can be lost in culture, which would lead to an identification as *Synechocystis*. Indeed, the non gas-vacuolate *Microcystis* PCC 7005 was originally classified as *Synechocystis* (Rippka et al., 1979) but its correct identity was revealed by DNA–DNA hybridization studies: PCC 7005 shows 75% relative binding to PCC 7806 and PCC 7941, but no similarity to *Synechocystis* PCC 6803 (Rippka and Herdman, unpublished). Noteworthy is that the other *Synechocystis* strain present in the trees (PCC 6906) does not seem closely related to PCC 6803, despite their similar morphology and mol% G + C content (Rippka et al., 1979). The inclusion within the *Synechocystis* group of a strain identified as *Merismopedia glauca* is not surprising, as discussed in the description of strains of Subsection I in this *Manual.* As described below, a further *Synechocystis* isolate, PCC 6308, is unrelated to the above isolates; this strain differs from the others by about 12 mol% G + C (Rippka et al., 1979) and clearly deserves a separate generic assignment.

XII. **The *Trichodesmium–Oscillatoria* PCC 7515 lineage** – It is remarkable that all the *Trichodesmium* sequences, of which most come from natural populations, cluster tightly together. This is one case where the morphological and ecological similarities are truly consistent with the 16S rRNA sequence analysis. The 16S rRNA sequence of the cultivated strain *Trichodesmium* NIBB1067 shares 94.9% sequence similarity with freshwater *Oscillatoria sancta* PCC 7515. Both strains are situated in the same lineage in the trees of Figs. B10.7 and B10.8. In view of these data, the conspicuous difference of 9% in the G + C content of the two strains (Herdman et al., 1979b; Zehr et al., 1991), and the differences in morphology and ecology, we support the retention of a separate genus *Trichodesmium*.

XIII. **The *Arthrospira* lineage** – The two *Arthrospira* strains, of different geographical origins, have almost identical 16S rRNA sequences and show no close genotypic relationship to the *Spirulina* strains. The ITS sequences of 37 *Arthrospira* strains, including PCC 7345 and PCC 8005, from four continents fall into only two separate clusters differing by 44 out of 478 nucleotide positions. This indicates a remarkable degree of similarity for geographically distant strains (Nelissen et al., 1994; Baurain and Wilmotte, unpublished). In Fig. B10.9, the two *Arthrospira* strains are also closely related to strain NIVA-CYA 136/2, of Kenyan origin. Only 479 nucleotides were determined for this strain (Rudi et al, 1997). It is noteworthy that the *Arthrospira* strains always group with *Lyngbya aestuarii* PCC 7419, which has a quite distinct and different morphology, but this might be due to the absence of any close relative to this large-celled (15 μm wide) marine *Lyngbya* in the phylogenetic data set.

XIV. **The *Geitlerinema* lineage** – The grouping of the two *Geitlerinema* strains is not well supported in the trees of Figs. B10.7 and B10.8, but stabilizes when more *Planktothrix* strains are added prior to tree constructions (Fig. B10.9). This is an example of the difficulties experienced in interpreting phylogenetic trees with the data set currently available.

Other groups   With the exception of *Planktothrix agardhii* PCC 7821 (ex *Oscillatoria agardhii*), the available sequences for members of *Planktothrix* and *Tychonema* genera are short (less than 500 nucleotides) and our phylogenetic interpretation (Fig. B10.9) should thus be treated as preliminary. Nevertheless, it appears that all of the *Planktothrix* isolates cluster so closely that their division into four species is not justified. Similarly, it appears that the two species of *Tychonema* should be combined. The taxonomy of these genera is explained in more detail in the description of Subsection III.

Three other groupings are consistently present in our trees, but with quite low bootstrap support in the distance tree (Fig. B10.7) and long branches in the Maximum Likelihood tree of Fig. B10.8. They are the groups of *Cyanothece* PCC 7424 and *Gloeothece membranacea* PCC 6501, of *Gloeocapsa* PCC 73106 and *Stanieria cyanosphaera* PCC 7437, and of *Synechocystis* PCC 6308, *Cyanobacterium stanieri* PCC 7202, and *Chroococcus turgidus* PCC 9340. We cannot find significant similarities in other phenotypic or genotypic characters to support these groupings, and the availability of additional sequences of related strains might help determine whether the groupings represent artifacts or true relationships. However, the separation of *Cyanothece* PCC 7424 from the other strains of this genus (PCC 7418 and PCC 9711) is consistent with the fact that PCC 7424 is a freshwater isolate with low salt tolerance, whereas the others are moderate halophiles, and indicates that this genus should be divided into two. The same is true for the genus *Synechocystis*, since the low GC member (PCC 6308) is clearly separated from the high GC strains (PCC 6803 and PCC 6906).

A filamentous strain isolated from Roman frescoes, *Leptolyngbya* VRUC 135, has some degree of genotypic affinity to the 16S rRNA sequence OS-VI-L16 retrieved from a hot spring in Yellowstone and for which the corresponding morphology is currently unknown (Weller et al., 1992).

Conclusions and recommendations   Since publication of the last edition of this *Manual,* it has become increasingly clear that

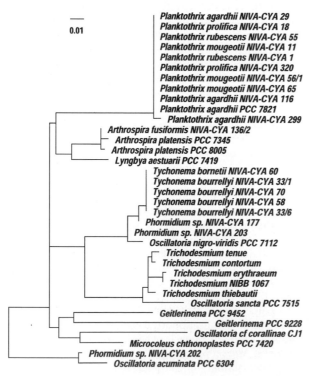

FIGURE B10.9. A clade extracted from the Maximum Likelihood tree of Figure B10.8, to which incomplete (less than 500 nucleotides) sequences of strains of *Planktothrix*, *Tychonema*, *Arthrospira*, and *Phormidium*, all from the NIVA culture collection, were added. Bar = 0.01 fixed mutation per nucleotide position.

members of the orders *Chroococcales*, *Pleurocapsales*, and *Oscilla-toriales* do not form coherent phylogenetic lineages but instead are dispersed throughout the tree. In addition, members of the *Prochlorales* which, with the exception of the more recently discovered *Prochlorococcus*, were treated as a separate order in the last edition, are not phylogenetically distinct from other cyanobacteria. For practical purposes, pending further studies, we recommend that the organisms be arranged in five major groups ("Sections"*, Rippka et al., 1979) as before, but that two of these (Subsections I and III) accommodate unicellular and filamentous "prochloralian" members, respectively.

In spite of the limited number of sequences available, useful information on the genotypic relationships of cyanobacteria has been gained through 16S rRNA sequence analysis. For a meaningful taxonomic interpretation of these results, it is necessary to link the molecular and phenotypic data using a polyphasic approach.

Acknowledgments   Permission to use unpublished sequences was kindly given by Thérèse Coursin, Isabelle Iteman, Wilhelm Schönhuber from the Institut Pasteur (Paris, France), and Filip Haes and Bart Nelissen (University of Antwerp, Belgium).

Subsection I. (Formerly **Chroococcales** Wettstein 1924, emend. Rippka, Deruelles, Waterbury, Herdman and Stanier 1979)

MICHAEL HERDMAN, RICHARD W. CASTENHOLZ, ISABELLE ITEMAN, JOHN B. WATERBURY AND ROSMARIE RIPPKA

Analysis of 16S rRNA gene sequence data (see Turner, 1997; Wilmotte and Herdman, this volume) shows that the unicellular cyanobacteria included in this Subsection do not form a coherent phylogenetic clade but are widely dispersed among strains of other Subsections. Furthermore, with the exception of *Cyanobium* and *Microcystis*, all of the genera for which multiple sequence data are available are polyphyletic. Further examples of polyphyly may arise when more sequences are available for genera that are currently represented by only a single sequence. It is presently impossible to classify the "chroococcalean" cyanobacteria on a phylogenetic basis. However, the data emerging from phylogenetic studies support many taxonomic proposals and will help in the subdivision of some of the form-genera discussed here.

Members of Subsection I are **unicellular cyanobacteria that reproduce by equal binary fission or by budding. Cells are spherical, ellipsoidal, or rod-shaped** and vary in size from **0.5–30 μm** in width or diameter. **Fission occurs in one, two, or three successive planes** at right angles to one another or in irregular planes, resulting in **cells occurring singly or in aggregates of varying size.** Colonial forms are held together by mucilage or multilaminated sheath layers, and cell aggregate form depends on the number of planes and regularity of division. **Planktonic forms generally contain gas vesicles.** With one exception, *Gloeobacter*, all members contain thylakoids. **Most synthesize phycobiliproteins**, light-harvesting proteins organized in the phycobilisome; **two genera** (*Prochloron* and *Prochlorococcus*) **lack phycobilisomes** and synthesize **chlorophyll *b* in addition to chlorophyll *a***, or derivatives of one or both of these photosynthetic pigments. Members of this Subsection range in mean DNA base composition from 31 to 71 mol% G + C and in genomic complexity from 1.3 to 5.2 Gdal.

Unless otherwise stated, phenotypic and genotypic data used here are taken from: Stanier et al., 1971; Kenyon, 1972; Waterbury and Stanier, 1977; Herdman et al., 1979a, b; Rippka et al., 1979; Rippka and Cohen-Bazire, 1983; Bryant, 1982; Waterbury

et al., 1986; Waterbury and Rippka, 1989; Komárek et al., 1999; Rippka et al., 2000; and our unpublished results.

TAXONOMIC COMMENTS

Subsection I here largely corresponds to section I of Rippka et al., 1979 and to the order *Chroococcales* described in the last edition (Waterbury and Rippka, 1989). However, it differs by inclusion of the genera *Chroococcus* (previously a member of the *Gloeocapsa* group), *Cyanobacterium* and *Cyanobium* (both ex *Synechococcus* group), *Dactylococcopsis*, *Microcystis* (previously a member of the *Synechocystis* group), and two "prochlorophytes" (*Prochloron* and *Prochlorococcus*) that were previously treated in a separate order, *Prochlorales*. Subsection I differs from traditional botanical definitions of the order *Chroococcales* by including these "prochlorophytes" and unicellular cyanobacteria that divide by budding (*Chamaesiphon*). The budding cyanobacteria have traditionally been included in the order *Dermocarpales* (synonym, *Chamaesiphonales*) together with the unicellular cyanobacteria that reproduce by multiple fission.

In field-based botanical taxonomic treatments of the unicellular cyanobacteria, genera have been defined almost exclusively by structural characteristics such as cell size and shape, planes of division, and the presence or absence of a well-defined sheath. Stanier et al. (1971) recognized that use of such features often resulted in heterogeneous assemblages of strains that could not be considered true genera because they exhibited large variations in DNA base composition. Use of additional phenotypic characteristics, such as ultrastructural properties and physiological and chemical characteristics, has in many cases provided sufficient additional information to enable genera to be more clearly defined (Rippka et al., 1979).

Two major structural and developmental properties of unicellular cyanobacteria merit particular comment: the presence or absence of cell aggregates and the number and regularity of the planes of cell division. These are major characteristics used to delineate genera of unicellular cyanobacteria in culture and in the traditional botanical literature. Cell aggregates in members of Subsection I are held together either by multilaminated sheath material or by amorphous slime or capsular material. The possession of extracellular sheath layers has proven to be a stable

---

Editorial Note: Sections referred to by Rippka et al. (1979) will be referred to as Subsections in the *Manual.*

feature of many cyanobacterial groups in culture, and it is a primary character used here in the description of several unicellular genera (*Chroococcus, Gloeobacter, Gloeocapsa,* and *Gloeothece*). Other unicellular cyanobacteria, primarily some members of the form-genus *Synechocystis,* occur in cell aggregates that are held together by amorphous slime or capsular material. Slime production has proven to be an unreliable taxonomic characteristic because its production in culture is affected by the growth phase of the cyanobacteria and the conditions under which they are grown. Binary fission occurs in one, two, or three successive planes at right angles to one another or in irregular planes. The number and regularity of the successive planes of division are stable features of individual cyanobacteria that should, in principle, be readily determinable from cultured material. In practice it is often difficult to determine the number of planes of division, the distinction between division in two or three planes being especially problematical. Examination of batch cultures growing either in liquid or on solid media is usually insufficient to unequivocally provide this information. Ideally, it should be possible to grow individual strains in slide culture and to document the planes of division by time lapse photography and light microscopy. This has not been done for many of the strains currently in pure culture.

Most traditional botanical taxonomic treatments of the *Chroococcales* (Fritsch, 1945; Desikachary, 1959; Bourrelly, 1970, 1985) follow the system of Geitler (1925, 1942, 1960) where the order contains two families. The first, *Chroococcaceae,* conforms to Subsection I as defined here, except that Geitler did not include *Chamaesiphon* (that reproduces by bud formation) and the "prochlorophytes" that had not been discovered at that time. The second family, the *Entophysalidaceae,* is a poorly defined group whose members undergo binary fission in irregular planes to form vegetative cell aggregates that often resemble those produced by members of the *Pleurocapsales* (Subsection II in this volume). The *Entophysalidaceae* are distinguished from members of Subsection II by the absence of baeocyte formation. Future studies will probably show that many entophysalidacean forms are actually members of the *Pleurocapsales*.

Drouet and Daily (1956), in their revision of the coccoid *Myxophyceae* (unicellular cyanobacteria), drastically reduced the number of chroococcalean genera and species. Their system, based primarily on the examination of many preserved specimens, has not proven successful for the identification of unicellular cyanobacteria in the field or in pure culture (Stanier et al., 1971).

Komárek and Anagnostidis (1986, 1998) made a significant departure from the Geitlerian system and placed all the chroococcalean and pleurocapsalean cyanobacteria in one order (the *Chroococcales*) containing seven families, the unicellular forms (here Subsection I) being assigned to three families. However, phylogenetic analyses of 16S rRNA gene sequences support neither the unification of these organisms into a single order, nor this proposed family concept (see Wilmotte and Herdman, this volume).

Key to the form-genera of Subsection I

I. Phycobilisomes absent, phycobiliproteins absent or present only in trace amounts.
    A. Free-living, marine, cell diameter 0.5–0.8 µm.
                Form-genus *Prochlorococcus,* p. 506
    B. Symbiotic on didemnids, marine, cell diameter 8–14 µm.
                Form-genus *Prochloron,* p. 507
II. Phycobiliproteins and phycobilisomes present on the plasma membrane, thylakoids absent.
                Form-genus *Gloeobacter,* p. 502
III. Phycobiliproteins and phycobilisomes present on the thylakoids.
    A. Reproduction exclusively by asymmetrical binary fission (budding) at the apical pole of ovoid, club- shaped, or pyriform cells.
                Form-genus *Chamaesiphon,* p. 495
    B. Reproduction by symmetrical transverse binary fission in a single plane, structured sheath absent.
       1. Cells spherical to rod-shaped, 1.7–2.3 µm in diameter; thylakoids dispersed throughout the cell; mean DNA base composition 39–41 mol% G + C.
                Form-genus *Cyanobacterium,* p. 497
       2. Cells spherical to rod-shaped, 0.8–1.4 µm in diameter; contain peripheral thylakoids; mean DNA base composition 66–71 mol% G + C.
                Form-genus *Cyanobium,* p. 498
       3. Cells spherical to rod-shaped, 3 µm or greater in width; mean DNA base composition 41–49 mol% G + C.
                Form-genus *Cyanothece,* p. 499
       4. Cells spherical to rod-shaped, 0.6–2.1 µm in diameter; contain peripheral thylakoids; may be united into colonial aggregates by mucilage formation; mean DNA base composition 47–66 mol% G + C.
                Form-genus *Synechococcus,* p. 508
       5. Cells fusiform in shape, 4–8 µm in width, 35–80 µm in length (dividing or paired cells may be up to 135 µm long); gas vesicles present along cell margins; mean DNA base composition (determined for only 1 strain) 43 mol% G + C.
                Form-genus *Dactylococcopsis,* p. 501
    C. Reproduction by symmetrical transverse binary fission in a single plane, structured sheath present.

1. Cells spherical to rod-shaped, 4–6 μm in width, united into aggregates by well defined, often laminated, sheath layers; thylakoids distributed throughout the cell; capable of N_2 fixation under aerobic conditions; mean DNA base composition 40–43 mol% G + C.

Form-genus *Gloeothece*, p. 504

D. Reproduction by transverse binary fission in two or three planes, structured sheath absent.

  1. Cells spherical, 3–4 μm in diameter, gas vesicles normally present; mean DNA base composition around 42 mol% G + C.

Form-genus *Microcystis*, p. 505

  2. Cells spherical, 2–6 μm in diameter that typically occur singly or in pairs in culture; in nature (but rarely in culture) some may occur in square or cubical aggregates held together by amorphous slime material; gas vesicles absent; mean DNA base composition 35–48 mol% G + C.

Form-genus *Synechocystis*, p. 512

E. Reproduction by transverse binary fission in two planes, structured sheath present.

  1. Cells 22–32 μm in diameter, occurring as aggregates of generally 4 cells; reproducing by binary fission in two successive planes at right angles; cells are initially spherical, but become hemispherical and quadrantal after the first and second cell divisions due to the constraints imposed by their tightly appressed sheath layers; mean DNA base composition (determined for only 1 strain) 38 mol% G + C.

Form-genus *Chroococcus*, p. 496

  2. Cells spherical, 3–10 μm in diameter occurring in aggregates; mean DNA base composition 40–46 mol% G + C.

Form-genus *Gloeocapsa*, p. 503

Form-genus I. **Chamaesiphon** Braun and Grunow 1865, emend. Geitler 1925

MICHAEL HERDMAN, RICHARD W. CASTENHOLZ, JOHN B. WATERBURY AND ROSMARIE RIPPKA

Cha.mae.si' phon. Gr. adv. *chamai* dwarf; Gr. masc. n. *siphon* tube; M.L. masc. n. *Chamaesiphon* microbial tube.

Unicellular cyanobacteria that reproduce exclusively by successive unequal binary fission (budding) at the apical pole of ovoid, club-shaped, or pyriform cells. The small spherical buds are produced in succession at one pole of the mother cell. This mode of reproduction confers an intrinsic heteropolarity on the cell.

The mol% G + C of the DNA is: 47.

FURTHER DESCRIPTIVE INFORMATION

In the developmental cycle shown schematically in Fig. B10.10, the bud enlarges and elongates during *period A*. During *period B*, unequal binary fission (budding) produces an apical bud (traditionally termed exospore) and a larger basal cell. The second generation bud forms from the reproductive pole (*r*) of the primary basal cell during *period C*. The future reproductive pole (*r′*) of the bud has been demonstrated for only one strain (PCC 6605). In many natural samples, cells are partly surrounded by a structured sheath ("pseudovagina" in the botanical literature) which is discontinuous at the apical (reproductive) pole. This feature is lacking in the strains currently in the PCC.

In nature, members of the genus *Chamaesiphon* typically develop as epiphytes and epiliths in freshwater streams and lakes. The cells are attached to the substrate by the nonreproductive (basal) pole and are often partly enclosed by a sheath layer. Newly formed buds may be liberated shortly after the completion of division or may remain attached to one another, forming a short chain of spherical cells that adhere to the reproductive (apical) pole of the mother cell.

The three strains (PCC 6605, PCC 7430, and PCC 8308) carried as *Chamaesiphon* in the PCC differ with respect to size and cell shape: two (PCC 7430 and PCC 8308) are slightly larger (diameter of buds about 3–4 μm; mother cells about 5–7 μm) and clearly pyriform prior to bud formation (Fig. B10.11), whereas strain PCC 6605 exhibits smaller cells (diameter of buds

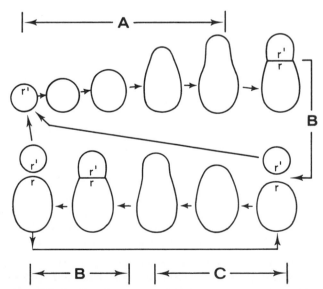

FIGURE B10.10. Schematic representation of the development cycle of *Chamaesiphon* strains PC 6605 and PCC 7430. See text for explanation and description. (Reprinted with permission from J.B. Waterbury and R.Y. Stanier, Archives of Microbiology *115*: 249–257, 1977, ©Springer-Verlag.)

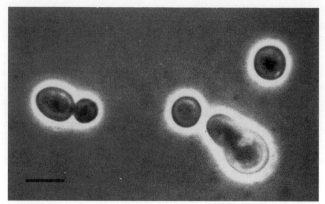

FIGURE B10.11. Light micrograph of *Chamaesiphon* strain PCC 7430, grown in liquid medium BG-11. Phase contrast. Bar = 5.0 μm. (Reproduced with permission from R. Rippka et al., Journal of General Microbiology *111*: 1–61, 1979, ©Society for General Microbiology.)

about 1.5–2 μm; mother cells about 3 μm) that remain spherical to ovoid even after the onset of asymmetric binary fission. They are all nonmotile and do not produce gas vesicles. None produce a structured sheath. They all synthesize C-phycoerythrin and undergo complementary chromatic adaptation (of types III and I for PCC 6605 and PCC 7430, respectively). They have a low salt tolerance, reflecting their origin from freshwater habitats. They differ in nutritional properties, strain PCC 7430 being a facultative photoheterotroph. Strains PCC 6605 and PCC 7430 have a similar mean DNA base composition (46.9 and 46.7 mol% G + C, respectively) and genome complexity (3.8 and 3.6 Gdal, respectively), and neither are able to synthesize nitrogenase (even at low oxygen concentrations).

Reference strains

PCC 6605 (co-identity ATCC 27169) isolated from stream, Berkeley, California, 1966 by M.M. Allen; mol% G + C is 46.9; genome size is 3.8 Gdal. Corresponds most closely to the botanical description of the species *Chamaesiphon minutus* sensu Kann 1972 (Waterbury and Rippka, 1989).

PCC 7430 (co-identity ATCC 29397; SAG 5.82) isolated from a freshwater stream, Sarka Valley near Prague, Czechoslovakia, 1963 by J. Komárek and F. Hindák; mol% G + C is 46.7; genome size is 3.6 Gdal. Corresponds most closely to botanical description of *Chamaesiphon subglobosus* sensu Kann 1972 (Waterbury and Rippka, 1989).

TAXONOMIC COMMENTS

Chamaesiphon as defined here conforms to the botanical description of the genus, except for differences in terminology. In the botanical literature the smaller daughter cell (bud) resulting from unequal binary fission is termed an "exospore", and the basal cell the "sporangium" (Geitler, 1932). The one *Chamaesiphon* strain for which the 16S rDNA sequence has been determined (PCC 7430) clusters phylogenetically most closely with filamentous genera, such as *Leptolyngbya* strains (see Turner, 1997, Honda et al., 1999, Wilmotte and Herdman, this volume) and is sufficiently distinct from other genera of Subsection I to warrant independent generic status.

FURTHER READING

Kann, E. 1972. Zur Systematik und Ökologie der Gattung *Chamaesiphon* (Cyanophyceae). I. Systematik. Arch. Hydrobiol. Suppl. 41, Algological Studies 7: 117–171.
Waterbury, J.B. and R.Y. Stanier. 1977. Two unicellular cyanobacteria which reproduce by budding. Arch. Microbiol *115*: 249–257.

Form-genus II. **Chroococcus** *Nägeli 1849*

ROSMARIE RIPPKA, RICHARD W. CASTENHOLZ, ISABELLE ITEMAN AND MICHAEL HERDMAN

Chro.o.coc'cus. Gr. fem. n. *chroa* skin; Gr. n. *kokkos* grain or kernel; M.L. masc. n. *Chroococcus* kernel in skin.

Large cells (22–32 μm in diameter) that reproduce by binary fission in two successive planes at right angles; the cells occur in aggregates, generally of four cells, held together by multilaminated sheath material. Due to the constraints imposed by the tightly appressed sheath layers, a single, spherical, predivisional cell gives rise to hemispherical and quadrantal cells after the first and second cell divisions, respectively.

FURTHER DESCRIPTIVE INFORMATION

RFLP analysis of PCR amplicons of the 16S rRNA–23S rRNA ITS region of five strains in the PCC show that the two freshwater isolates form a cluster that is distinct from three marine isolates (K. Romari, I. Iteman and R. Rippka, unpublished data).

Cluster 1 This contains two strains (PCC 7946 and PCC 9442), isolated from a peat bog in Switzerland and a flooded limestone quarry (Bloomington, Indiana), respectively. Both have a **low salt tolerance**, do not contain gas vesicles, and **synthesize C-phycoerythrin**; strain PCC 7946 is known to undergo complementary chromatic adaptation of type I. They differ in cell diameter (25–30 μm and up to 50 μm, respectively).

Reference strain PCC 7946; isolated from a peat bog near Kastanienbaum, Switzerland by J.B. Waterbury; is the strain most similar to the "botanical" *C. turgidus* above.

Cluster 2 This cluster contains three strains (PCC 8987, PCC 9106, and PCC 9340) that differ from members of cluster 1 in being true marine isolates, having an **obligate requirement for elevated concentrations of Na$^+$, Mg^{2+}, and Ca^{2+}**. The first two strains were isolated from a salt marsh on the Mediterranean coast and the intertidal region, Gulf of Elat, respectively; PCC 9106 is strain S24 of Potts et al. (1983). PCC 9340 (Fig. B10.12) originated as a cryptoendolith in a desert rock (strain N41, Potts et al., 1983). The three strains are similar in size (25–30 μm diameter) and exhibit gliding motility. They do not synthesize gas vesicles, and **none produce C-phycoerythrin**. Strain PCC 9106 is a facultative photoheterotroph using only sucrose (among 10 organic compounds tested) as organic carbon source (Potts et al., 1983) and like PCC 9340 is capable of nitrogen fixation under anoxic conditions (Potts et al., 1983). These two strains were reported to have mean mol% G + C DNA base compositions of 47.1 and 48.9, respectively. However, reexamination of strain PCC

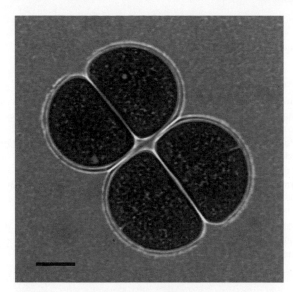

FIGURE B10.12. Light micrograph of *Chroococcus* strain PCC 9340, grown in liquid medium ASN-III. Bright field. Bar = 10.0 μm.

9340 (N41) gave a value of 38.1 mol% G + C (R. Rippka and M. Herdman, unpublished data).

Reference strain   PCC 9340 (co-identity CCMEE N-41); isolated as a cryptoendolith from a sandstone rock, Negev Desert, Israel, by R. Ocampo-Friedmann; mol% G + C is 38.1, genome size is 4.1 Gdal. Despite its origin, this is a marine strain, having an

obligate requirement for elevated concentrations of Na^+, Mg^{2+}, and Ca^{2+}.

TAXONOMIC COMMENTS

Traditionally, *Chroococcus* and *Gloeocapsa* were separated on the basis of sheath consistency, being tight in the former and loose in the latter genus. This was not accepted by Rippka et al. (1979). Waterbury and Rippka (1989) used the term "*Gloeocapsa* group" to accommodate both the genus *Gloeocapsa* (sensu Rippka et al., 1979) and the large-celled *Chroococcus* isolates. None of the *Gloeocapsa* species described in Geitler (1932) have cell diameters over 15 μm and the majority are much smaller, whereas many of the *Chroococcus* species are larger than this, particularly *Chroococcus turgidus*, the type species (Geitler, 1942). Further information is given under "Taxonomic Comments" for the genus *Gloeocapsa*. Two strains, *Chroococcus* PCC 9106 and PCC 9340 (*Chroococcus* S24 and N41; Potts et al., 1983), assigned to the "*Gloeocapsa* group" in the last edition, were reassigned to the genus *Chroococcus* (Rippka and Herdman, 1992) in agreement with their original designation. The present separation of these genera, based on size alone, is confirmed by phylogenetic analyses of 16S rRNA gene sequences: *Chroococcus* PCC 9340 (N-41) is not closely related to *Gloeocapsa* PCC 73106 (see Wilmotte and Herdman, this volume).

FURTHER READING

Potts, M., R. Ocampo-Friedmann, M.A. Bowman and B. Tözün. 1983. *Chroococcus* S24 and *Chroococcus* N41 (cyanobacteria): morphological, biochemical and genetic characterization and effects of water stress on ultrastructure. Arch. Microbiol *135:* 81–90.

Form-genus III. **Cyanobacterium** *Rippka and Cohen-Bazire 1983*

ROSMARIE RIPPKA, RICHARD W. CASTENHOLZ AND MICHAEL HERDMAN

Cy.ano'bact.er.ium. Gr. adj. *kuanos* blue (color); Gr. n. *bakterion* (diminutive of *baktron*) small rod; M.L. neut. n. *Cyanobacterium* small blue-(green) rod.

Unicellular cyanobacteria that reproduce by transverse binary fission in a single plane. Cells are spherical to rod-shaped, 1.7–2.3 μm in diameter. Thylakoids are parallel and located throughout the cell.

The mol% G + C of the DNA is: 39–41.

Type species: (Botanical Code): **Cyanobacterium stanieri** Rippka and Cohen-Bazire 1983.

FURTHER DESCRIPTIVE INFORMATION

This genus is currently represented by two strains in the PCC, PCC 7202^T (Fig. B10.13) and PCC 7502. They were isolated from an alkaline pond in Chad and from a sphagnum bog, Switzerland, respectively. Both are nonmotile, do not synthesize C-phycoerythrin, and are obligate photoautotrophs. They have similar mean DNA base compositions (39 and 40.5 mol% G + C, respectively). They differ from each other with respect to their salt tolerance (PCC 7202^T being euryhaline), reflecting differences in original habitats. They probably represent different species of the same genus.

Holotype   PCC 7202^T (co-identity ATCC 29140), proposed as type strain of the type species, *Cyanobacterium stanieri* (Rippka and Cohen-Bazire, 1983); isolated from an alkaline pond, Chad, 1963 by M. Lefévre; mol% G + C is 39.0; genome size is 2.7 Gdal.

TAXONOMIC COMMENTS

The two strains assigned to this genus were, like the members of the genus *Cyanobium* described below, included in the *Syne-*

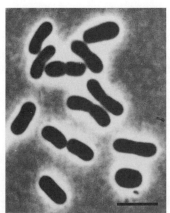

FIGURE B10.13. Light micrograph of *Cyanobacterium stanieri* strain PCC 7202^T, grown in liquid medium BG-11. Phase contrast. Bar = 5.0 μm. (Reproduced with permission from R. Rippka and G. Cohen-Bazire, Annales de Microbiologie (Institut Pasteur) *134B:* 21–36, 1983, ©Masson Editeur.)

chococcus "group" in the last edition (see also Rippka et al., 1979). Rippka and Cohen-Bazire (1983) proposed the creation of the genus *Cyanobacterium* to accommodate unicellular cyanobacteria that reproduce by binary fission in a single plane, and can be distinguished from other members of the form-genus *Synechococcus* by their low mean DNA base composition. Although based only on strain PCC 7202[T], the separation of *Cyanobacterium* from both *Cyanobium* and the form-genus *Synechococcus* is justified by phylogenetic evidence (see Wilmotte and Herdman, this volume) and by the different arrangement of the thylakoids, which seem to be entirely peripheral and parallel to the cell wall in members of the latter two genera (Komárek et al., 1999). Further support for the separation of *Cyanobacterium* from *Synechococcus* is provided by the low relative binding (13%) observed in DNA–DNA hybridization of *Synechococcus* PCC 6301 with *Cyanobacterium* PCC 7502 (Wilmotte and Stam, 1984). Rippka and Cohen-Bazire (1983) proposed strain PCC 7202[T] as the holotype of *Cyanobac-* *terium stanieri*, the type species of this genus. This proposal has been validated under the Botanical Code of Nomenclature (Komárek et al., 1999).

FURTHER READING

Komárek, J., J. Kopecký and V. Cepák. 1999. Generic characters of the simplest cyanoprokaryotes *Cyanobium*, *Cyanobacterium* and *Synechococcus*. Cryptogam. Algol. *20*: 209–222.

Rippka, R. and G. Cohen-Bazire. 1983. The *Cyanobacteriales*: a legitimate order based on the type strain *Cyanobacterium stanieri*? Ann. Microbiol. (Institut Pasteur) *134B*: 21–36.

Rippka, R., J. Deruelles, J.B. Waterbury, M. Herdman and R.Y. Stanier. 1979. Generic assignments, strain histories and properties of pure cultures of cyanobacteria. J. Gen. Microbiol. *111*: 1–61.

Wilmotte, A.M.R. and W.T. Stam. 1984. Genetic relationships among cyanobacterial strains originally designated as '*Anacystis nidulans*' and some other *Synechococcus* strains. J. Gen. Microbiol. *130*: 2737–2740.

Form-genus IV. **Cyanobium** *Rippka and Cohen-Bazire 1983*

ROSMARIE RIPPKA, RICHARD W. CASTENHOLZ AND MICHAEL HERDMAN

Cy.ano' bi.um. Gr. adj. *kuanos* blue color; Gr. masc. n. *bios* life; M.L. neut. n. *Cyanobium* blue-(green) life-(form).

Unicellular cyanobacteria that reproduce by transverse binary fission in a single plane. Cells are spherical to rod-shaped, 0.8–1.4 μm in diameter and contain peripheral thylakoids.

The mol% G + C of the DNA is: 66–71.

Type species: (Botanical Code) **Cyanobium gracile** Rippka and Cohen-Bazire 1983.

FURTHER DESCRIPTIVE INFORMATION

Seven of the 12 strains assigned to this genus in the PCC were previously included in the *Synechococcus* "group" as the *Cyanobium* cluster, and one strain (PCC 7001) was included in the *Synechococcus* "group" marine-cluster B. Although variable with respect to size (range 0.8–1.4 μm in width), the 12 strains are mostly slightly smaller than members of the form-genus *Synechococcus*, (see Fig. B10.14 for strain PCC 6307[T]) from which they are distinguished by their high mean DNA base composition. All strains are nonmotile; they do not synthesize C-phycoerythrin. Where tested, they do not synthesize nitrogenase in anaerobiosis (PCC 8905, PCC 8939, PCC 8966, and PCC 8975 have not been examined). All strains except PCC 8975 are obligate photoautotrophs. On the basis of their salt tolerance, members of this genus can be assigned to two clusters.

Cluster 1 Strains of this cluster (PCC 6307[T], PCC 6710, PCC 6713, PCC 6904, PCC 6907, PCC 6911, PCC 7009, PCC 8905, PCC 8966, and PCC 8975) were isolated either from freshwater habitats or from brackish water, and **none are capable of sustained growth in a marine medium**. Strains PCC 6307[T], PCC 6713, PCC 6904, PCC 6907, and PCC 6911 are hosts to the cyanophage SM-1 (Safferman et al., 1969; R. Rippka, unpublished data). Given the apparent host specificity of this phage (R. Rippka, unpublished data), these five strains most likely represent independent isolates of the same species. Strains PCC 6710 and PCC 7009 are not hosts to cyanophage SM-1, and thus may represent members of a different species. The remaining strains have not yet been tested for phage sensitivity.

Holotype PCC 6307[T] (co-identity ATCC 27147), proposed as type strain of the type species, *Cyanobium gracile* (Rippka and Cohen-Bazire, 1983); isolated from lake water, Wisconsin, USA, 1949 by G.C. Gerloff (Gerloff et al., 1950); mol% G + C is 69.7; genome size is 2.6 Gdal.

Cluster 2 The two strains (PCC 7001 and PCC 8939) representing this cluster **differ from those included in cluster 1 by their higher salt tolerance**; they are euryhaline, growing equally well in freshwater and marine medium. They are similar in mean DNA base composition (69.5 and 69.9 mol% G + C, respectively) and in genome size (2.0 Gdal).

Reference strain PCC 7001 (co-identity ATCC 27194); isolated from intertidal mud, City Island, New York, USA, 1961 by C. van Baalen (van Baalen, 1962); mol% G + C is 69.5; genome size is 2.0 Gdal.

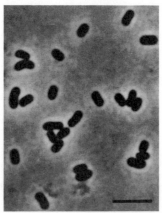

FIGURE B10.14. Light micrograph of *Cyanobium gracile* strain PCC 6307[T], grown in liquid medium BG-11. Phase contrast. Bar = 5.0 μm. (Reproduced with permission from R. Rippka and G. Cohen-Bazire, Annales de Microbiologie (Institut Pasteur) *134B*: 21–36, 1983, ©Masson Editeur.)

TAXONOMIC COMMENTS

As evidenced by their high mean DNA base composition (mol% G + C 66–71), strains now assigned to *Cyanobium* are genetically distinct from members of the form-genus *Synechococcus*. For this reason, Rippka and Cohen-Bazire (1983) proposed the creation of a new genus and designated strain PCC 6307^T as the neotype strain of the new combination, *Cyanobium gracile*. This genus and type species have been validated under the Botanical Code of Nomenclature by Komárek et al. (1999). Separation of *Cyanobium* from *Synechococcus* is supported by the low relative binding (13%) observed in DNA–DNA hybridization of *Synechococcus* PCC 6301 with *Cyanobium* PCC 6307^T (Wilmotte and Stam, 1984). Strains PCC 6307^T and PCC 7001 cluster relatively closely in phylogenetic trees inferred from 16S rRNA gene sequences (Wilmotte and Herdman, this volume), suggesting that the two clusters of this genus represent different species.

Strain PCC 6307^T (cluster 1) was identified prior to entering the Collection as *Coccochloris peniocystis*. Recognition of these epithets, as an alternative to creation of the new combination *Cyanobium gracile*, was rejected by Rippka and Cohen-Bazire (1983) on the grounds that they were based on the disputed revisions of Drouet and Daily (1956). Strain PCC 6907 (UTEX 563) was originally identified as *Synechococcus elongatus*, but should, on the basis of its dimensions, be assigned to *Synechococcus leopoliensis* (Stam and Holleman, 1979). Thus, although the specific name gracile was proposed for the neotype of *Cyanobium* (Rippka and Cohen-Bazire, 1983), the epithet leopoliensis would have been equally valid. Strain PCC 6911 was identified prior to entering the PCC as *Microcystis aeruginosa* NRC-1. However, this was clearly a misidentification, or the result of mislabeling of cultures.

Strain PCC 7001 (cluster 2) was previously described as *Anacystis marina* (van Baalen, 1962). Since this identification is also based on Drouet and Daily (1956), it seems wise to reject these epithets. In contrast, since phylogenetic evidence (see Wilmotte and Herdman, this volume) shows that PCC 7001 merits separation at the specific level from other members of the genus, the combination *Cyanobium oceanicum* (Komárek et al., 1999) would seem appropriate, since Komárek (1976) suggested that PCC 7001 corresponds in morphology and ecological habitat to *Synechococcus oceanicus*.

Strain PCC 6603, included in the *Cyanobium* cluster in the last edition, was previously reported to have a mean DNA base composition of 63 and 65.7 mol% G + C by Edelman et al. (1967) and Stanier et al. (1971), respectively. Since this is not in agreement with later results (48.5 mol% G + C, M. Herdman, unpublished data), it is possible that strain PCC 6603 no longer corresponds to the organism previously described under this number. It is therefore not included in this genus in the PCC.

FURTHER READING

Komárek, J., J. Kopecký and V. Cepák. 1999. Generic characters of the simplest cyanoprokaryotes *Cyanobium*, *Cyanobacterium* and *Synechococcus*. Cryptogam. Algol. *20*: 209–222.

Rippka, R. and G. Cohen-Bazire. 1983. The *Cyanobacteriales*: a legitimate order based on the type strain *Cyanobacterium stanieri*? Ann. Microbiol. (Institut Pasteur) *134B*: 21–36.

Wilmotte, A.M.R. and W.T. Stam. 1984. Genetic relationships among cyanobacterial strains originally designated as '*Anacystis nidulans*' and some other *Synechococcus* strains. J. Gen. Microbiol. *130*: 2737–2740.

Form-genus V. Cyanothece

ROSMARIE RIPPKA, RICHARD W. CASTENHOLZ, JOHN B. WATERBURY AND MICHAEL HERDMAN

Cy.an.o.the' ce. Gr. adj. *kuanos* blue color; Gr. n. *theke* case, envelope; M.L. fem. n. *Cyanothece* blue (green) sheath.

The form-genus *Cyanothece* is a provisional assemblage of strains loosely defined as unicellular coccoid to rod-shaped cyanobacteria that divide by transverse binary fission in a single plane. The cells are ~3 µm or greater in width and lack structured (e.g., laminated) sheaths.

The mol% G + C of the DNA is: 41–49.

FURTHER DESCRIPTIVE INFORMATION

The form-genus *Cyanothece* is superficially distinguished from the form-genus *Synechococcus* by cell size, and from *Gloeothece* and *Gloeobacter* by the absence of well-defined sheath layers.

In addition to morphological similarities, the members of this form-genus that have been examined are either capable of dinitrogen fixation at atmospheric oxygen concentrations, or capable of synthesizing the enzyme nitrogenase under anaerobic conditions. By contrast, only one strain (PCC 7335) in the large form-genus *Synechococcus* is capable of nitrogenase synthesis under anaerobic conditions.

In other respects the strains included in the form-genus *Cyanothece* are a diverse assemblage. Members of clusters 1 and 2 were isolated from rice paddies and from various other freshwater sources. Strains assigned to cluster 3 were isolated from marine coastal algal mats, or from coastal solar evaporation ponds. In all cases, their major ionic requirements for growth reflect their origins.

On the basis of salt tolerance, cell size (Figs. B10.15 and B10.16), and, where studied, their mean DNA base composition and phylogenetic affinities, 22 strains carried in the PCC, ATCC, and MPI collections and here assigned to the form-genus *Cyanothece* may be divided into three clusters, each probably representing a separate genus.

Cluster 1 This cluster comprises three strains (PCC 7424, PCC 7822, and PCC 9224) of similar size (**4–6 µm** in width), all of which were isolated from **freshwater habitats** and have a **low salt tolerance**. They were originally characterized by abundant mucilage production (particularly in the nonaxenic state), but this property is no longer expressed by strains PCC 7424 (Fig. B10.15) and PCC 7822. Two of the strains (PCC 7424 and PCC 7822) produce C-phycoerythrin irrespective of the spectral quality of light provided for growth, and are capable of aerobic nitrogen fixation (Singh, 1973; Kallas et al., 1985). The latter property was lost by strain PCC 7424, but a revertant has been isolated (R. Rippka, unpublished data). Strain PCC 7424, the only isolate of this cluster examined in more detail, has a cell diameter of 5–6 µm, is an obligate photoautotroph, and has a mean DNA base composition of 41.2 mol% G + C and a relatively large genome complexity (4.2 Gdal). Strain PCC 9224 differs from

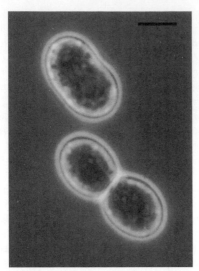

FIGURE B10.15. Light micrograph of strain PCC 7424, reference strain for cluster 1 of the form-genus *Cyanothece*, grown in liquid medium BG-11. Phase contrast. Bar = 5.0 μm. (Reproduced with permission from R. Rippka et al., Journal of General Microbiology *111:* 1–61, 1979, ©Society for General Microbology.)

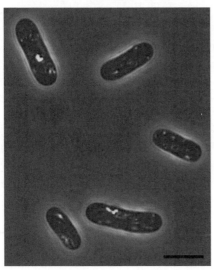

FIGURE B10.16. Light micrograph of strain PCC 7418, reference strain for cluster 3 of the form-genus *Cyanothece*, grown in liquid artificial seawater medium (ASN-III in 2× Turks salts). Phase contrast. Bar = 5.0 μm.

strains PCC 7424 and PCC 7822 by its bright green color, and most likely does not produce a phycoerythrinoid pigment. Furthermore, it does not grow at the expense of atmospheric nitrogen in the presence of air.

Reference strain   PCC 7424 (co-identity ATCC 29155); isolated from a rice field, Senegal, 1972 by R. Roger; mol% G + C is 41.2; genome size is 4.2 Gdal.

Cluster 2   This cluster is comprised of eight strains isolated from **freshwater habitats** (PCC 6910, PCC 7425, PCC 8303, PCC 8801, PCC 8802, PCC 8955, PCC 9308, and PCC 9318) that are characterized by a **smaller cell size (3–4 μm in width) than the members of cluster 1**. The two strains examined in more detail,

PCC 6910 and 7425, are obligate photoautotrophs, have a mean DNA base composition of 47.5 and 48.6 mol% G + C and small genomes of 2.5 and 3.1 Gdal, respectively. Their mean DNA base composition is significantly higher than that of the two other clusters of the form-genus *Cyanothece*. Two strains (PCC 8801 and PCC 8802), isolated from rice fields in Taiwan, were first described as *Synechococcus* RF-1 and RF-2 and are capable of nitrogen fixation in the presence of atmospheric concentrations of oxygen (Huang and Chow, 1988). Strain PCC 8801 is the only member of this cluster that is able to synthesize C-phycoerythrin. Strain PCC 7425, isolated from a rice field in Senegal, fixes nitrogen under anoxic conditions (Rippka and Waterbury, 1977). Five strains (PCC 7425, PCC 8303, PCC 8955, PCC 9308, and PCC 9318), isolated from diverse geographical regions, contain prominent light refractile inclusions that, on the basis of three strains examined, are most likely composed of polyalkanoates (Porta et al., 2000). Strain PCC 6910 was identified prior to entering the PCC as *Gloeocapsa alpicola* and was included in the last edition as a member of the *Synechococcus* "group".

Reference strain   PCC 7425 (co-identity ATCC 29141); isolated from a rice field, Senegal, 1972, by R. Roger; mol% G + C is 48.6; genome size is 3.1 Gdal.

Cluster 3   This cluster contains two strains, PCC 7418 (Fig. B10.16) and PCC 9711 (MPI 95AH10), which differ from the members of clusters 1 and 2 in growing best at **salt concentrations greater than that of seawater (3% NaCl)** and tolerating salinities **as high as 20% NaCl**, reflecting their original habitat, solar evaporation ponds (Garlick et al., 1977; Garcia-Pichel et al., 1998). A further seven strains (MPI 95AH11, MPI 95AH13, MPI 96AL06, MPI 96P402, MPI 96P408, MPI 96AL03, and Syn C1 P22) of the culture collection of the Max-Planck Institut (Bremen), and strain *Aphanothece halophytica* ATCC 43922, can be assigned to this form-genus, since they form a relatively tight cluster (showing around 95% 16S rRNA gene sequence identity, see Garcia-Pichel et al., 1998) with strains PCC 7418 and MPI 95AH10. These strains were all assigned to the "Euhalothece subcluster" by Garcia-Pichel et al. (1998). They were all isolated from solar evaporation ponds, except strain Syn C1 P22 (isolated from a benthic microbial mat). They vary in cell width from **3–10 μm**, have **a minimum requirement of 1.5–6% NaCl and a maximum tolerance of 16–25%**; the two strains examined (PCC 7418 and ATCC 43922), like *Dactylococcopsis* PCC 8305, produce quaternary ammonium compounds as osmolytes (see Garcia-Pichel et al., 1998). None of the strains of cluster 3 synthesize C-phycoerythrin, and only PCC 7418 produces gas vesicles, distributed throughout the cell (Garlick et al., 1977). This strain is a facultative anoxyphototroph, using Na_2S or H_2 as the electron donor (Belkin and Padan, 1978), has a mean DNA base composition of 42.4 mol% G + C, a genome complexity of 3.1 Gdal, and a cell diameter of 4–5 μm. One additional strain, MPI 96P605, also isolated from a solar evaporation pond, is distantly related to the other strains of cluster 3 (Garcia-Pichel et al., 1998; Wilmotte and Herdman, this volume), and would appear to represent a different genus. Garcia-Pichel et al. (1998) described this entire group of halophilic cyanobacteria as the "Halothece cluster". Interestingly, *Dactylococcopsis* PCC 8305, also halophilic and isolated from the same habitat, groups within the "Halothece cluster" (Garcia-Pichel et al., 1998). The freshwater isolate PCC 7424 is clearly separated from the halophiles on the basis of 16S rRNA gene sequence analyses (Garcia-Pichel et al., 1998; Wilmotte and Herdman, this volume), justifying a generic separation of the respective clusters.

Reference strain PCC 7418 (co-identity ATCC 29534); isolated from Solar Lake, Sinai, 1972 by Y. Cohen as *Aphanothece halophytica*; mol% G + C is 42.4; genome size is 3.1 Gdal.

Strains unclassified at present Three strains (WH 8501, WH 8502, and WH 8503) were isolated from the tropical Atlantic Ocean, two (WH 8501 and WH 8502) having been included as the "marine cluster" in the *Synechocystis* "group" in the last edition, since they were thought to divide in two planes. The cells are spherical, **2.5–4 μm in diameter**, occur singly or in pairs, divide in one plane, and are nonmotile. Neither cell aggregates nor amorphous capsular material are produced. The strains are obligate photoautotrophs incapable of using organic compounds as sole sources of cell carbon. They are capable of fixing N_2 under atmospheric concentrations of oxygen. N_2 fixation occurs in the dark at the expense of glycogen reserves accumulated in the light. The photosynthetic thylakoids are dispersed throughout the cytoplasm. The cells contain phycoerythrin high in phycourobilin content as their primary light-harvesting pigment and are incapable of chromatic adaptation. The temperarature range is restricted, with growth occurring between 26°C and 32°C, which restricts their distribution to the equatorial oceans. They are **obligately marine with elevated requirements for Na^+, Cl^-, Mg^{2+}, and Ca^{2+}** for growth. The small cell size and **low mean DNA base composition (30–32 mol% G + C)** preclude inclusion of these strains as a cluster of this form-genus. The three strains are very similar and will be assigned to a new genus and species (J.B. Waterbury, unpublished data). The reference strain for this group is WH 8501, isolated from the tropical Atlantic Ocean (28°S, 43°W), 1984 by S.W. Watson and F.W. Valois; mol% G + C is 30.5.

TAXONOMIC COMMENTS

This genus was created by Komárek (1976) and is based on the botanical type species *Cyanothece aeruginosa* (synonym *Synechococcus aeruginosa* Nägeli 1849). Although none of the strains here assigned to *Cyanothece* conform perfectly to the "botanical" description of the type species, which is very wide celled (7–16 μm) (see Geitler, 1932, 1942) the genus nomen is used here as a form-genus to temporarily accommodate rod-shaped unicellular cyanobacteria that reproduce by equal binary fission in a single plane, and that are significantly larger (>3 μm in width) than their morphological counterparts assigned to the form-genus *Synechococcus* (0.5–2 μm in width). The strains assigned to clusters 1 and 2 were identified as *Aphanothece* prior to their entry into the PCC. In botanical taxonomic treatises (Geitler, 1932, 1942; Bourrelly, 1970; Komárek and Anagnostidis, 1986) this genus is reserved for cyanobacteria possessing the morphological and division properties of *Synechococcus* or *Cyanothece*, but that can be distinguished by the production of very mucilaginous colonies. However, this property is influenced by environmental conditions and may not be expressed (or is lost) in culture. For this reason, and in agreement with Rippka et al. (1979) and Waterbury and Rippka (1989), the genus *Aphanothece* is not accepted here. Based on the size range of its cells (3–4 μm), Komárek (1976) concluded that strain PCC 6910 of cluster 2 corresponds to *Cyanothece cedrorum*.

FURTHER READING

Huang, T.C. and T.J. Chow. 1988. New type of N_2-fixing cyanobacterium (blue-green algae). FEMS Microbiol. Lett. *36*: 109–110.

Komárek, J. 1976. Taxonomic reviews of the genera *Synechocystis* Sauv. 1892; *Synechococcus* Näg, 1849, and *Cyanothece* gen. nov. (Cyanophyceae). Arch. Protistenkd. *118*: 119–179.

Komárek, J. and K. Anagnostidis. 1998. Cyanoprokaryota 1. Teil Chroococcales, G. Fischer, Jena.

Mitsui, A., S. Kumazawa, A. Takahashi, H. Ikemoto, S. Cao and T Arai. 1986. Strategy by which nitrogen-fixing unicellular cyanobacteria grow photoautotrophically. Nature *323*: 730–732.

Yopp, J.H., D.R. Tindall, D.M. Miller and W.E. Schmid. 1978. Isolation, purification and evidence of the obligate halophilic nature of the blue-green alga *Aphanothece halophytica* Frenig (*Chroococcales*). Phycologia *17*: 172–177.

Form-genus VI. **Dactylococcopsis** Hansgirg 1888 (sensu Komárek 1969).

ROSMARIE RIPPKA, RICHARD W. CASTENHOLZ AND MICHAEL HERDMAN

Dac.ty.lo.coc.cop'sis. Gr. n. *dactylos* finger; Gr. n. *kokkos* grain or kernel; Gr. fem. n. *opsis* appearance; M.L. fem. n. *Dactylococcopsis* appearing as finger-shaped grain.

Unicellular cyanobacteria that are fusiform in shape. The cells are 4–8 × 35–80 μm. Dividing or paired cells may be up to 135 μm long. A well-defined sheath is absent; gas vacuoles present along cell margins, but are most abundant at the mature (pointed) ends of the cell. The only axenic isolate (PCC 8305) is halophilic and halotolerant.

FURTHER DESCRIPTIVE INFORMATION

The single axenic isolate, strain PCC 8305 (Fig. B10.17), undergoes transverse binary fission; the newly-created extremities of the cells are hemispherical rather than pointed. Depending on culture conditions (e.g., salinity) cells may take more or less elongate shapes, and asymmetric cell division may occur (see Garcia-Pichel et al., 1998). This strain will not grow at salinities below that of seawater (see Walsby et al., 1983). Optimal growth was obtained at total salinities between 5% and 20%, but growth was possible up to about 25% (Garcia-Pichel et al., 1998). A greater salinity tolerance was shown at 38°C than at 25°C. Phylogenetically strain PCC 8305 clusters very closely with the halophilic strains of *Cyanothece*/Euhalothece (Garcia-Pichel et al., 1998; see Wilmotte and Herdman, this volume).

Reference strain PCC 8305; isolated from hypersaline Solar Lake, Sinai, 1979 by A.E. Walsby; mean DNA base composition is 43.4 mol% G + C; genome size is 2.4 Gdal. *Dactylococcopsis salina* proposed as type species (Walsby et al., 1983).

TAXONOMIC COMMENTS

The generic nomen *Dactylococcopsis* has a complex history, since it may have been used originally for a green alga (*Ankistrodesmus*) (Komárek, 1969; 1983). Two alternative botanical genera for cyanobacteria exhibiting the morphological properties of strain PCC 8305 have been described: *Myxobactron* Schmidle 1904 and *Rhabdogloea* Schröder 1917 (see Komárek and Anagnostidis, 1986). However, the type material of *Myxobactron* was reported to contain only diatoms (see Komárek, 1983; Komárek and An-

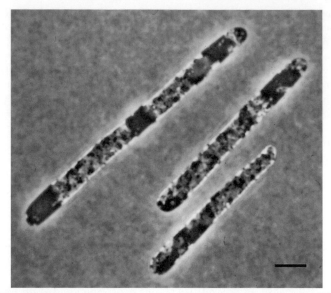

agnostidis, 1986) and is thus of doubtful validity. Komárek (1983) proposed to transfer several species (*Dactylococcopsis smithii* R. et F. Chaudat 1925, *Dactylococcopsis linearis* Geitler 1935, and *Dactylococcopsis planctonica* Teiling 1942) to the genus *Rhabdogloea*, even though the authenticity of the type species (*Rhabdogloea ellipsoidea* Schröder 1917) had not been reconfirmed. Disregarding this proposal, Komárek and Anagnostidis (1986) subsequently concluded that the isolate of Walsby et al. (1983) should be assigned to *Myxobactron salinum*. We here retain the original epithets of Walsby et al., *Dactylococcopsis salina*, since the name *Dactylococcopsis* antedates that of *Myxobactron* and *Rhabdogloea* and thus has priority, and the three genera were considered to be synonymous (Geitler, 1942).

FURTHER READING

Walsby, A.E., J. van Rijn and Y. Cohen. 1983. The biology of a new gas vacuolate cyanobacterium *Dactylococcopsis salina*, a new species in Solar Lake, Sinae, Egypt. Proc.R. Soc. Lond. Ser.B. Biol. Sci. *217*: 417–448.

FIGURE B10.17. Light micrograph of *Dactylococcopsis salina* strain PCC 8305, grown in liquid artificial sea-water medium (ASN-III in 2× Turks salts). Bar = 5.0 μm.

Form-genus VII. **Gloeobacter** *Rippka, Waterbury and Cohen-Bazire 1974*

MICHAEL HERDMAN, RICHARD W. CASTENHOLZ AND ROSMARIE RIPPKA

Gloe.o.bac′ ter. Gr. adj. *gloios* sticky; Gr. n. *bakterion* (dim. of *baktron*) small rod; M.L. n. *Gloeobacter* sticky bacterium.

Unicellular cyanobacteria that possess oval- to rod-shaped cells that divide by transverse binary fission in a single plane. Cells occur in irregular aggregates held together by sheath material. Intracellular photosynthetic thylakoids are lacking; the light-harvesting and photosynthetic apparatus is associated with the cytoplasmic membrane.

The mol% G + C of the DNA is: 64.

Type species: **Gloeobacter violaceus** Rippka, Waterbury and Cohen-Bazire 1974.

FURTHER DESCRIPTIVE INFORMATION

The three strains currently in culture (PCC 7421[T], PCC 8105, and PCC 9601) were isolated from rocks in mountainous regions of Switzerland. The holotype strain PCC 7421[T] (Fig. B10.18) is characterized by relatively small cells (1–1.5 μm wide) that contain unusually large polyphosphate granules, typically located near each cell pole (Rippka et al., 1974; Waterbury and Rippka, 1989). The phycobiliproteins are present in a cortical layer 50–70 nm thick, in the form of cylindrical elements on the inner surface of the cytoplasmic membrane (Cohen-Bazire and Bryant, 1982). Healthy colonies or cell aggregates are generally violet in color due to a low chlorophyll *a* content, the presence of phycoerythrin (which is synthesized irrespective of the spectral quality of light provided for growth) and a deep red carotenoid glycoside (Rippka et al., 1974). Cultures may turn bluish gray under suboptimal conditions (senescent cultures or those grown under high light intensity). The phycoerythrin of PCC 7421[T] is unique for non-marine cyanobacteria in having two types of chromophores, phycoerythrobilin (PEB) and phycourobilin (PUB) (Bryant et al., 1981) that absorb at 564 and 498 nm, respectively, whereas the more widely occurring C-phycoerythrin, analyzed for a large number of strains (Bryant, 1982), contains only PEB.

Strain PCC 7421[T] is an obligate photoautotroph, does not synthesize nitrogenase (even under anaerobic conditions), and has a high mean DNA base composition (64 mol% G + C). Growth in liquid cultures results in a mixture of single cells and aggregates of widely varying size, resulting in a clumpy and heterogeneous appearance. Growth rates are slow, a mean doubling time of 73 h (strain PCC 7421[T]) being achieved in continuous light (~500 lux) at 25°C. Growth does not occur at 37°C.

Holotype *Gloeobacter violaceus* Rippka et al. 1974 strain PCC 7421[T]; isolated from a calcareous rock, near Kastanienbaum, Vierwaldstättersee, Switzerland, 1972 by R. Rippka; mol% G + C is 64.0; genome size is 2.7 Gdal.

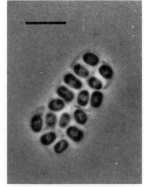

FIGURE B10.18. Light micrograph of *Gloeobacter violaceus* strain PCC 7421[T], grown in liquid medium BG-11. Phase contrast. Bar = 5.0 μm. (Reproduced with permission from R. Rippka et al., Journal of General Microbiology *111*: 1–61, 1979, ©Society for General Microbiology.)

Taxonomic Comments

This genus is currently represented by a single species, *Gloeobacter violaceus*. The genus *Gloeobacter* can be differentiated from all other genera of unicellular cyanobacteria by its lack of photosynthetic thylakoids. This is a characteristic not discernible in the field but readily recognizable in thin sections of fixed material examined by transmission electron microscopy. Before it was known that these cyanobacteria lack thylakoids, they were placed in the genus *Gloeothece*. *Gloeobacter violaceus* conforms to the botanical species *Gloeothece coerulea* Geitler 1927, but with minor discrepancies in color. However, the current PCC strains of *Gloeothece* possess thylakoids and have a lower mol% G + C content of 40–44, in contrast to the 64 mol% of *Gloeobacter violaceus* PCC 7421[T]. In addition, 16S rRNA gene sequence analysis of PCC 7421[T] shows that it falls at the base of the cyanobacterial tree and is sufficiently distinct (see Turner, 1977; Wilmotte and Herdman, this volume) from all other chroococcalean cyanobacteria (including *Gloeothece* PCC 6501) to deserve independent generic status.

Further Reading

Bryant, D.A., G. Cohen-Bazire and A.N. Glazer. 1981. Characterization of the biliproteins of *Gloeobacter violaceus*: chromophore content of a cyanobacterial phycoerythrin carrying phycourobilin chromophore. Arch. Microbiol. *129*: 190–198.

Rippka, R., J.B. Waterbury and G. Cohen-Bazire. 1974. A cyanobacterium which lacks thylakoids. Arch. Microbiol. *100*: 419–436.

Form-genus VIII. Gloeocapsa *Kützing 1843*

MICHAEL HERDMAN, RICHARD W. CASTENHOLZ, ISABELLE ITEMAN AND ROSMARIE RIPPKA

Gloe.o.cap' sa. Gr. adj. *gloios* sticky; L. n. *capsa* box; M.L. fem. n. *Gloeocapsa* sticky box.

Unicellular cyanobacteria with spherical cells 3–10 μm in diameter that divide by binary fission in two or three successive planes at right angles to one another and occur in aggregates that are held together by a multilaminated sheath.

The mol% G + C of the DNA is: 40–46.

Further descriptive information

On the basis of cell size and thermotolerance, the nine members assigned to the genus *Gloeocapsa* in the PCC can be divided into three clusters. The separation of clusters 2 and 3 is further justified by RFLP analysis of the 16S rRNA–23S rRNA ITS region (K. Romari, I. Iteman and R. Rippka, unpublished data).

Cluster 1 This cluster accommodates three strains (PCC 7007, PCC 73106, and PCC 7501) that have **cells 6–10 μm in diameter**. They all have low salt tolerance and synthesize C-PE constitutively. Strains PCC 73106 (Fig. B10.19A) and PCC 7501 exhibit gliding motility on solid substrates when exposed to unidirectional light. These two strains have very similar DNA base compositions (41.7 and 40.9 mol% G + C, respectively) and genome size (3.0 and 3.4 Gdal, respectively); neither is capable of N₂ fixation even under anoxic conditions. However, they vary with respect to a few properties: PCC 73106 has a firmer sheath consistency than PCC 7501, and is an obligate photoautotroph, in contrast to PCC 7501 which is a facultative photo- and chemoheterotroph using glucose as a substrate. Slide culture studies with strain PCC 73106 have shown that binary fission occurs in three planes at right angles to each other, but due to compression by the sheath material after several rounds of division, reproduction may appear to occur at more irregular angles (R. Rippka, unpublished data).

The 16S rRNA gene sequence analysis of strain PCC 73106 shows that this cyanobacterium forms a moderately tight cluster with *Stanieria* PCC 7437 and several other members of Subsection II (see "Phylogenetic relationships amongst the *Cyanobacteria*," p. 487) and is genetically distinct from *Chroococcus* PCC 9340.

Reference strain PCC 73106 (co-identity ATCC 27928), isolated from an acidic bog, Switzerland, 1971 by R. Rippka; mol% G + C is 41.7; genome size is 3.0 Gdal.

Cluster 2 This contains strains PCC 7512 and PCC 7701, characterized by **significantly smaller cell dimensions of the postdivisional cells (3–3.5 μm in diameter)**. They both have a low salt tolerance. PCC 7512 is an obligate photoautotroph, synthesizes C-phycoerythrin constitutively, has a mean DNA base composition of 39.8 mol% G + C, and a genome size of 2.9 Gdal. PCC

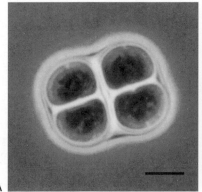
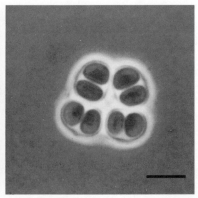

FIGURE B10.19. Light micrograph of *Gloeocapsa* strains PCC 73106 (*A*) and PCC 7428 (*B*), grown in liquid medium BG-11. Phase contrast. Bar = 5.0 μm. (Reproduced with permission from R. Rippka et al., Journal of General Microbiology *111*: 1–61, 1979, ©Society for General Microbiology.)

7701 is a facultative photoheterotroph and does not synthesize C-phycoerythrin. Neither are capable of N₂-fixation under oxic or anoxic conditions.

Reference strain   PCC 7512 (co-identity ATCC 29115); isolated from a rock scraping, Pont Neuf, Paris, 1975 by J.B. Waterbury; mol% G + C is 39.8; genome size is 2.9 Gdal.

Cluster 3   This cluster accommodates four strains (PCC 7428, PCC 8702, PCC 9817, and PCC 9820) that, like the strains of cluster 2, possess small post-divisional cells (3–3.5 μm in diameter). However, they are **thermotolerant, capable of growth at temperatures up to 50°C** (K. Romari and R. Rippka, unpublished data), and were **isolated from moderate hot springs**. None exhibit gliding motility. Strain PCC 7428 (Fig. B10.19B), the only isolate characterized in detail, has a high salt tolerance, is a facultative photo- and chemoheterotroph (using glucose, fructose, ribose, or sucrose), produces no phycoerythrinoid pigment and is incapable of N₂-fixation under oxic or anoxic conditions. Strain PCC 7428 has a mol% G + C content of 46.0 and a genome size of 3.5 Gdal. DNA–DNA hybridization studies with all strains except PCC 8702 have shown that they are members of the same species (K. Romari, R. Rippka and M. Herdman, unpublished data).

Reference strain   PCC 7428 (co-identity ATCC 29159); isolated from a moderate temperature hot spring, Maha Oya, Sri Lanka, 1973 by A. Neilson; mol% G + C is 46; genome size is 3.5 Gdal.

TAXONOMIC COMMENTS

This genus corresponds in part to the "*Gloeocapsa* group" of the last edition. However, two strains, *Chroococcus* PCC 9106 and PCC 9340 (*Chroococcus* S24 and N41; Potts et al., 1983), were assigned to this "group" in the last edition, as explained in the Taxonomic Comments for *Chroococcus*. These strains were reassigned to the genus *Chroococcus* (Rippka and Herdman, 1992) in agreement with their original designation, and are included in *Chroococcus* here. This separation is confirmed by phylogenetic analyses of the 16S rRNA gene sequences (see Wilmotte and Herdman, this volume).

The strains currently assigned to *Gloeocapsa* in the PCC remain a somewhat heterogeneous assemblage that in the traditional botanical literature would be assigned to either *Gloeocapsa* or *Chroococcus*. Cell shape and the characteristics of the extracellular sheath layers are the primary characteristics used in the botanical literature to distinguish between the two genera. Immediately following division, *Chroococcus* possesses hemispherical cells that are surrounded by tightly appressed sheath layers. *Gloeocapsa* has spherical cells that are held loosely together by their surrounding sheath layers. However, distinction between the two genera in field material and in laboratory cultures with reference to the two primary structural characteristics can be problematic due to transitional stages. Sheath color, used as an additional taxonomic trait (Geitler, 1932), is not a valid characteristic, since the presence or intensity of yellow-brown color (scytonemin) or red to violet color (gloeocapsin) is likely to be determined in both cases by the intensity or dosage of UV radiation (see Garcia-Pichel and Castenholz, 1991). In the present treatment, cell size is used as the major characteristic for distinguishing the two genera, members of *Chroococcus* exhibiting cell diameters of more than 20 μm (compare Fig. B10.19B of *Gloeocapsa* PCC 7428 with Fig. B10.12 of *Chroococcus* PCC 9340).

Form-genus IX. **Gloeothece** *Nägeli 1849 (sensu Rippka, Deruelles, Waterbury, Herdman and Stanier 1979)*

ROSMARIE RIPPKA, RICHARD W. CASTENHOLZ, JOHN B. WATERBURY AND MICHAEL HERDMAN

Gloe.o.the' ce. Gr. adj. *gloios* sticky; Gr. fem. n. *theke* case, envelope; M.L. fem. n. *Gloeothece* gelatinous sheath.

Unicellular rod-shaped cyanobacteria that divide by transverse binary fission in a single plane, the cells being united into aggregates by well defined, often laminated, sheath layers.

The mol% G + C of the DNA is: 40–43.

FURTHER DESCRIPTIVE INFORMATION

Three characteristics are used here to distinguish members of the genus *Gloeothece* from other unicellular cyanobacteria. Possession of well-defined sheath layers separates *Gloeothece* from members of both form-genera *Synechococcus* and *Cyanothece*. Division by binary fission in a single plane distinguishes *Gloeothece* from members of *Gloeocapsa*, and possession of photosynthetic thylakoids dispersed throughout the cytoplasm of the cell distinguishes *Gloeothece* from *Gloeobacter*. However, a misidentification of what we are here calling *Gloeothece* can easily be made with microscopic observation alone.

Cell orientation within the aggregates of *Gloeothece* can be misleading. The juxtaposition of the cells gives the impression that division has occurred in three planes at right angles to one another, an interpretation that led to the early identification of pure cultures of this cyanobacterium as members of the genus *Gloeocapsa* (Stanier et al., 1971). In reality, the cell axes, not the planes of division, of the daughter cells shift following cell division (Fig. B10.20). This shift is the result of cell movement caused by tension exerted on the cells by their surrounding sheath layers (R. Rippka and J. Deruelles, unpublished data). Another problem is that during exponential growth the cells may be almost spherical.

Gloeothece is represented by eight strains in the PCC (PCC 6501, PCC 6909, PCC 7109, PCC 73107, PCC 73108, PCC 8302, PCC 8803, and PCC 8804). This genus has a wide geographic distribution and has been isolated from freshwater, acid bogs, and alkaline caves. All strains are of similar cell size (4–6 μm wide), and are capable of fixing N₂ under aerobic conditions. Their mean DNA base composition is 40.4–42.7 mol% G + C and their genome size is 5.0–5.2 Gdal. All strains examined are obligate photoautotrophs and have a low salt tolerance. All but one strain (PCC 73107) produce C-phycoerythrin, but complementary chromatic adaptation does not occur. Phycourobilin does not occur as a chromophore, as incorrectly reported by Waterbury and Rippka (1989). Individual cells, colonies on solid media, and liquid cultures appear dark olive-green as a result of the presence of both C-phycocyanin and C-phycoerythrin.

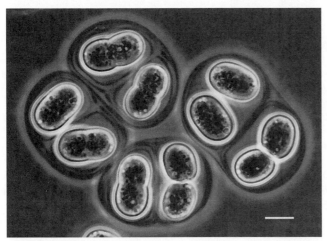

FIGURE B10.20. Light micrograph of *Gloeothece membranacea* strain PCC 6501, grown in liquid medium BG-11. Phase contrast. Bar = 5.0 μm.

Reference strain PCC 6501 (co-identity ATCC 27151); isolated from freshwater, California, U.S.A., 1965 by M.M. Allen; proposed as type strain of the neotype species, *Gloeothece membranacea* (see Waterbury and Rippka, 1989); mol% G + C is 41.7; genome size is 2.31 Gdal.

TAXONOMIC COMMENTS

Gloeothece as defined here conforms to the traditional botanical descriptions of the genus (Geitler, 1932; Desikachary, 1959; Bourrelly, 1970; 1985; Komárek and Anagnostidis, 1986, 1998). The treatise of Geitler (1932) contains 16 species in the genus, including *G. coerulea* which is now placed in the genus *Gloeobacter*. Although from diverse habitats, the eight strains presently in the PCC as *Gloeothece* are phenotypically similar and could be assigned to a single botanical species, *Gloeothece membranacea* Bornet 1892. Reference strain PCC 6501 is suggested as the neotype strain for this botanical species. The botanical type species, *Gloeothece linearis* Nägeli 1849, has much smaller cells (1.5–2.5 μm wide), has not been isolated, and is of doubtful validity. The reference strain PCC 6501 clusters relatively closely with *Cyanothece* PCC 7424 in phylogenetic trees based on 16S rRNA gene sequences (Turner et al., 1999; Wilmotte and Herdman, this volume).

Form-genus X. **Microcystis** *Kützing ex Lemmermann 1907 nom. cons.*

MICHAEL HERDMAN, RICHARD W. CASTENHOLZ, ISABELLE ITEMAN AND ROSMARIE RIPPKA

Mi.cro.cys' tis. Gr. adj. *mikros* small; Gr. n. *cystis* bladder, bag; M.L. fem. n. *Microcystis* small bag.

Planktonic unicellular cyanobacteria that reproduce by binary fission in two (or possibly three?) planes at right angles to one another. Cells are spherical, 3–6 μm in diameter. May be united into colonial aggregates by mucilage, but are never enclosed by multilaminated sheath material. Gas vesicles are normally present.

The mol% G + C of the DNA is: 42.

FURTHER DESCRIPTIVE INFORMATION

Five strains (PCC 7005, PCC 7806, PCC 7813, PCC 7820, and PCC 7941) assigned to the genus *Microcystis* by Rippka and Herdman (1992) were included in the *Synechocystis* "group" in the last edition (see also Rippka, 1988b). They resemble members of cluster 1 of the latter form-genus in cell size, and members of cluster 2 in mean DNA base composition. The majority are easily **distinguished from members of the form-genus *Synechocystis* by the production of gas vesicles** (Fig. B10.21). They all possess spherical cells 3–4 μm in diameter, not surrounded by a mucilaginous sheath, and show low salt tolerance. They synthesize neither C-phycoerythrin nor phycoerythrocyanin. They are similar in mean DNA base composition (41.6–42.5 mol% G + C) and genome size (3.1–3.5 Gdal). They were all isolated from freshwater lakes in either Canada, The Netherlands, Scotland, or the United States.

Three of the five strains (PCC 7806, PCC 7820, and PCC 7941) are easily distinguished from *Synechocystis* by their ability to produce gas vesicles, which give the cells a typical light refractile appearance when viewed by phase contrast microscopy. They produce the hepatotoxin microcystin (Codd and Carmichael, 1982; Birk et al., 1989), a toxin associated with many waterblooms (Carmichael, 1988). Although strain PCC 7813 does not produce gas vesicles, it does contain the genes encoding the gas vesicle

structural proteins (Damerval et al., 1989) and produces microcystin (G.A. Codd, personal communication). PCC 7813 was isolated from the same primary sample as PCC 7820, to which it shows 100% similarity by DNA–DNA hybridization (R. Rippka and M. Herdman, unpublished data); PCC 7813 and PCC 7820 are also unique among the five strains in being facultative photo- and chemoheterotrophs, utilizing glucose. PCC 7813 is thus most likely a spontaneous mutant of PCC 7820, incapable of producing gas vesicles. Such spontaneous mutation is frequently observed in gas-vacuolated cyanobacteria (R. Rippka, unpublished data), and would appear to have also occurred in strain PCC 7005,

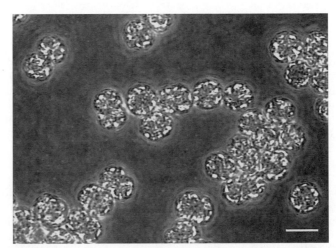

FIGURE B10.21. Light micrograph of *Microcystis aeruginosa* strain PCC 7941, grown in liquid medium BG-11 containing NaNO₃ (2 mM), NaHCO₃ (10 mM). Phase contrast. Bar = 5.0 μm.

originally identified as a member of the genus *Synechocystis* (Rippka et al., 1979). This strain, unable to produce a hepatotoxin (Eloff, 1981), has been shown by DNA–DNA hybridization studies (R. Rippka and M. Herdman, unpublished data) and by RFLP analysis of the 16S rRNA–23S rRNA ITS region (I. Iteman and J. Filée, unpublished data) to be closely related to the other strains here identified as *Microcystis*. In contrast, it shows no significant genetic relatedness to *Synechocystis* PCC 6803, with which it shares a close morphological resemblance.

Strains PCC 7005 and PCC 7806 release large amounts of β-cyclocitral, a feature that distinguishes them from all other cyanobacteria examined (Jüttner, 1984).

A further 19 strains of *Microcystis* have recently entered the PCC. They are currently under study and are not listed individually here. They were isolated from freshwater in Africa, Australia, Europe, and North America. Their inclusion does not significantly change the descriptive information above, which is based on only five strains: six possess slightly larger cells (up to 6 μm diameter); with one exception they all produce gas vesicles; seven possess mucilage; they all show low salt tolerance; 10 are capable of facultative photo- and chemoheterotrophic growth at the expense of glucose; none synthesize C-phycoerythrin.

Reference strain Although strain PCC 7820 was proposed in the last edition as the reference strain for the "*Microcystis* cluster", Rippka and Herdman (1992) designated *Microcystis* PCC 7941, isolated and elegantly described by Gorham and collaborators (Hughes et al., 1958) more than 20 years earlier than PCC 7820, as the type strain of the type species, *Microcystis aeruginosa*.

PCC 7941 (co-identity CCAP 1450/4, SAG 14.85, UTEX 2385), proposed as type strain of the type species, *Microcystis aeruginosa*, is most likely *Microcystis aeruginosa* NRC-1 (SS-17) (P.R. Gorham, personal communication); isolated by P.R. Gorham and W.W. Carmichael, Little Rideau Lake, Ontario, Canada, 1954 (Hughes et al., 1958; Zehnder and Gorham, 1960); mol% G + C is 42.5; genome size is 3.1 Gdal.

TAXONOMIC COMMENTS

Five strains available from the PCC, previously assigned to the genus *Synechocystis* (Rippka, 1988b) or to the "*Microcystis* cluster" of the *Synechocystis* "group" in the last edition, were recognized by Rippka and Herdman (1992) as members of the genus *Microcystis*. This separation is justified genetically by DNA–DNA hybridization studies (R. Rippka and M. Herdman, unpublished data) and by phylogenetic analysis (see Wilmotte and Herdman, this volume).

DNA–DNA hybridization studies also demonstrated that the five strains available from the PCC are probably members of the same nomenspecies, since they show a minimum of 70% relative binding (R. Rippka and M. Herdman, unpublished data). Furthermore, they form a close group in phylogenetic trees inferred from 16S rRNA gene sequences (see Wilmotte and Herdman, this volume). RFLP analysis of the 16S rRNA–23S rRNA ITS region of all 24 PCC strains has shown that they form a tight cluster, divisible into six subclusters whose significance requires further study (I. Iteman, J. Filée and R. Rippka, unpublished data). From their habitat and cellular morphology, these strains all conform to *Microcystis aeruginosa* Kützing ex Lemmermann 1907. Under the Botanical Code of Nomenclature the genus *Microcystis*, as defined by Geitler (1932) and Desikachary (1959), harbors both spherical and rod-shaped unicellular cyanobacteria, many of which are planktonic and contain gas vesicles. Species are primarily defined by their type of "colony" formation ("clathrate", "lobed", or "lenticular") as observed in field samples. However, strains identified as *M. aeruginosa*, *M. viridis*, and *M. wesenbergii* were not separable by phylogenetic analysis of the 16S rRNA gene (Neilan et al., 1997a) or by RFLP analysis of the 16S rRNA–23S rRNA ITS region (Neilan et al., 1997b). Further isolates assigned to these three species, together with strains named as *M. ichthyoblabe* and *M. novacekii*, formed a tight cluster in phylogenetic trees based on the sequence of the ITS region (Otsuka et al., 1999). The separation of these strains into five species is therefore not supported by genetic evidence.

Form-genus XI. **Prochlorococcus** *Chisholm, Frankel, Goericke, Olson, Palenik, Waterbury, West-Johnsrud and Zettler 1992, 299*

ROSMARIE RIPPKA, RICHARD W. CASTENHOLZ AND MICHAEL HERDMAN

Pro.chlor′o.coc′cus. Gr. prefix *pro* primitive; Gr. adj. *chloros* green; Gr. n. *kokkos* grain or kernel; M.L. masc. n. *Prochlorococcus* primitive green kernel (cell).

Marine planktonic cyanobacteria that possess spherical to slightly rod-shaped cells (0.5–0.8 × 0.7–1.6 μm) and divide by binary fission in a single plane. They lack phycobilisomes and contain divinyl-chlorophyll *a* and *b* (*a₂* and *b₂*).

Type species: **Prochlorococcus marinus** Chisholm, Frankel, Goericke, Olson, Palenik, Waterbury, West-Johnsrud and Zettler 1992, 299.

FURTHER DESCRIPTIVE INFORMATION

The members of this genus (Fig. B10.22) range in size from 0.5 to 0.8 μm in width and 1.0–1.6 μm in length (predivisional cells). Although they show a superficial resemblance to various slightly larger strains of picoplanktonic marine *Synechococcus* (see Fig. B10.24F), all isolates of *Prochlorococcus* are distinguished by their lack of phycobilisomes and phycobiliproteins (except small quantities of PE III in the case of strain CCMP 1375^T; see Hess et al.,

1999, and possibly other strains). As a light-harvesting-antenna, members of the genus *Prochlorococcus* synthesize divinyl-chlorophyll *a* and *b* (*a₂* and *b₂*) (Goericke and Repeta, 1993). Thus, the principal characteristic distinguishing this genus is the replacement of phycobilisomes by a chlorophyll *a/b* complex analogous to that of green algal and green plant chloroplasts. This pigment composition confers a yellow-green color to the cultures.

Some strains of *Prochlorococcus* that synthesize divinyl chlorophyll *a* and *b* may contain monovinyl-chlorophyll *b* (*b₁*), but always lack monovinyl-chlorophyll *a*. Minor quantities of a chlorophyll *c*-like pigment (Mg, 3–8 divinyl phaeoporphyrin *a₅*) and α-carotene, rather than β-carotene, are additional characteristic features. Axenic strain PCC 9511 uses ammonia or urea as a nitrogen source; nitrate is not utilized (Rippka et al., 2000). Several organic phosphorus compounds efficiently replace phosphate, indicating ecto-phosphohydrolase activity.

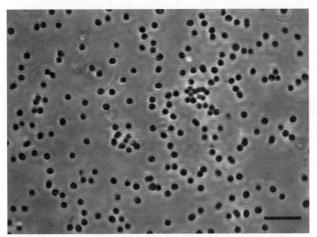

FIGURE B10.22. Light micrograph of *Prochlorococcus* strain PCC 9511, grown in liquid artificial sea-water medium. Phase contrast. Bar = 5.0 μm.

The holotype, strain CCMP 1375[T], originated from a depth of 120 m near the bottom of the euphotic zone in the Sargasso Sea, whereas strain CCMP 1378 was isolated from surface water (5 m) of the northwestern Mediterranean Sea (Chisholm et al., 1992; Partensky et al., 1993). Representatives of this group of oxygenic phototrophs are apparently ubiquitous in the water column (0–200 m) of most tropical and temperate oceans. They generally dominate the lower portion of the euphotic zone (100–200 m) in clear oceanic waters and coexist in various proportions with phycoerythrin-rich unicellular small cyanobacteria assigned to the genus *Synechococcus* (see Partensky et al., 1999).

Nonaxenic *Prochlorococcus* was initially cultured in medium K/10-Cu (Chisholm et al., 1992), but showed poor growth in this medium. A medium termed PCR-S11 was used successfully for the culturing of axenic and nonaxenic strains (see Rippka et al.,

2000). The axenic strain (PCC 9511, Fig. B10.22) was isolated fortuitously by serial dilution of the primary culture.

Holotype *Prochlorococcus marinus* Chisholm et al. 1992 (strain CCMP 1375[T]), isolated from 120 m depth in Sargasso Sea.

Reference strains

CCMP 1375[T] type strain; isolated from a depth of 120 m, Sargasso Sea (Chisholm et al. 1992).

PCC 9511 axenic strain isolated from primary culture (SARG) containing CCMP 1375[T], also Sargasso Sea, 1995 by R. Rippka; mol% G + C is 32; genome size is 1.3 Gdal (2 Mbp). The genome size is the smallest known for any oxygenic procaryotic phototroph.

TAXONOMIC COMMENTS

The genus *Prochlorococcus* was created for two nonaxenic isolates, both of which were assigned to the single species *P. marinus* Chisholm et al. 1992. The formal description was based on the nonaxenic nomenclatural type, strain CCMP 1375[T]. However, strain PCC 9511 was derived and purified to axenic status from the same primary culture (SARG) from which the holotype was obtained (Rippka et al., 2000). *Prochlorococcus* PCC 9511 differs from CCMP 1375[T] in possessing horseshoe-shaped thylakoids, exhibiting a low chlorophyll b_2 content, and lacking phycoerythrin (as in strain CCMP 1378). Analyses of 16S rRNA gene sequences (Urbach et al., 1998; see Wilmotte and Herdman, this volume) confirm that axenic strain PCC 9511 is not co-identical with the nomenclatural type, but is located in a clade that includes all representatives of *Prochlorococcus*, such as strain MIT 9303 (a distant relative), together with marine strains of *Synechococcus* such as WH 7805. Rippka et al. (2000) have proposed that strain PCC 9511 be considered a subspecies of *P. marinus* (subsp. *pastoris*).

Form-genus XII. **Prochloron** *Lewin 1977, 216*[VP]

MICHAEL HERDMAN, RICHARD W. CASTENHOLZ AND ROSMARIE RIPPKA

Pro.chlo′ ron. Gr. prefix *pro* primitive; Gr. adj. *chloros* green; M.L. neut. n. *Prochloron* primitive green (cell).

Unicellular, spherical, without evident mucilaginous sheath. Reproduction by binary fission. Almost exclusively extracellular symbionts of colonial ascidians (chiefly didemnids) on subtropical marine shores. Form chlorophylls *a* and *b* and lack phycobiliproteins.

FURTHER DESCRIPTIVE INFORMATION

The type species (Fig. B10.23) was described on the basis of cells with diameters of 8–14 μm, growing as facultative symbionts on colonies of *Didemnum*. Cells up to 30 μm in diameter are also found as obligate symbionts inside colonies of other genera and species of didemnids. A tabulation of cell sizes in various host

associations has been published by Lewin et al. (1984). Whether these are conspecific has not been established, since they have not been grown in laboratory culture. A thorough description of this genus was given in the last edition (Griffiths and Thinh, 1989; Lewin, 1989).

The thylakoids tend to be paired or stacked; phycobilisomes (and phycobiliproteins) are absent. The predominant green pigments are chlorophylls *a* and *b* associated with β-carotene, and a variety of xanthophylls including echinenone, β-cryptoxanthin, mutachrome, and isocryptoxanthin. The cell walls are composed of peptidoglycan and muramic acid. The lipids comprise glycerides with low proportions of polyunsaturated fatty acids, phosphatidylglycerol (but no phosphatidylcholine or phosphatidyl-

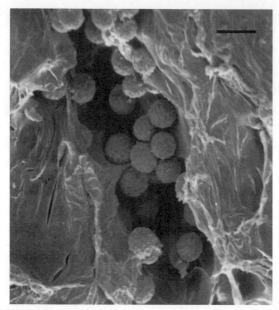

FIGURE B10.23. Scanning electron micrograph of *Prochloron* cells, 8 μm in diameter, in the perioral groove on the upper surface of a colony of *Didemnum candidum*. Bar = 10.0 μm. (Reproduced with permission from L. Cheng.)

ethanolamine), and small amounts of steroids. Carbohydrates, including branched and unbranched 1,4-polyglucose, have been demonstrated, but no cyanophycin [poly (*N*-arginyl-aspartic acid)] or poly-β-hydroxybutyric acid. Gas vesicles have not been observed. Based on one collected sample, the mol% G + C of the DNA is 41, the genome size is 3.6 Gdal. 16S rRNA gene sequence data provide firm evidence for the inclusion of this "prochlorophyte" within the cyanobacteria (see Wilmotte and Herdman, this volume).

Reference strain None (this organism has not yet been cultured).

FURTHER READING

Lewin, R.A. 1975. A marine *Synechocystis* (Cyanophyta, *Chroococcales*) epizoic on ascidians. Phycologia *14*: 153–160.

Lewin, R.A. 1977. *Prochloron*, a type genus of the Prochlorophyta. Phycologia *16*: 216.

Lewin, R.A., L. Cheng and R.S. Alberte. 1984. *Prochloron*-ascidian symbioses: photosynthetic potential and productivity. Micronesia *19*: 165–170.

Form-genus XIII. Synechococcus

MICHAEL HERDMAN, RICHARD W. CASTENHOLZ, JOHN B. WATERBURY AND ROSMARIE RIPPKA

Syn.e.cho.coc′ cus. Gr. n. *synechia* continuity; Gr. n. *kokkos* grain or kernel; M.L. masc. n. *Synechococcus* contiguous cells.

This form-genus serves as a provisional repository for unicellular cyanobacteria that share the morphological properties of members assignable to the botanical genus *Synechococcus* Nägeli 1849. Cells are spherical to rod shaped, 0.6–2.1 μm in diameter and contain photosynthetic thylakoids located peripherally. Reproduction occurs by transverse binary fission in a single plane. The cells may be united into colonial aggregates by mucilage formation, but never produce well defined sheath layers.

The mol% G + C of the DNA is: 47–66.

FURTHER DESCRIPTIVE INFORMATION

The representatives in culture (Fig. B10.24) were isolated from a wide range of habitats (freshwater, hot springs, brackish water, or the marine environment) and exhibit a variable tolerance to, or requirement for, salts. Photosynthetic pigment content, ability to grow photoautotrophically, and motility are equally variable. They may be divided into five clusters by their morphological, physiological, and genetic properties. These clusters are well supported by phylogenetic studies and are equivalent to genera. However, these have not been formalized because they are based on an incomplete data set, with only a relatively limited number of strains and phenotypic properties available. Two clusters are further divided into subclusters, which may be related at the specific level.

Cluster 1 Strains assigned to this cluster are 1–1.5 μm in diameter, nonmotile, obligate photoautotrophs and do not synthesize C-phycoerythrin. They exhibit a **low salt tolerance**, being incapable of sustained growth in marine media. Their mean DNA base composition ranges from 50 to 56 mol% G + C and their genome size from 1.6 to 2.1 Gdal. They are divided into two subclusters on the basis of cell size and mean DNA base composition.

Cluster 1.1 The five strains of this cluster, PCC 6301 (Fig. B10.24A), PCC 6311, PCC 6908, PCC 7942, and PCC 7943, share similar cell dimensions (**1–1.2 μm in width**) and are **hosts to the cyanophage AS-1M** (Sherman and Connelly, 1976; Rippka and Cohen-Bazire, 1983). They have a mean DNA base composition of **55–56 mol% G + C** (not determined for PCC 7943). The high degree of DNA similarity (94–97% relative binding) between strains PCC 6301, PCC 7942, and PCC 7943 (Wilmotte and Stam, 1984), and the close phylogenetic relationship of PCC 6301 and PCC 7942 (see "Phylogenetic relationships amongst the *Cyanobacteria*," p. 487), show that these strains are assignable to the same nomenspecies.

Reference strain PCC 6301 (co-identity ATCC 27144, for other identities see Rippka and Herdman, 1992); isolated from freshwater (Waller Creek), Texas, United States, 1952 by W.A. Kratz (Kratz and Myers, 1955; Rippka et al., 1979); mol% G + C is 55.1; genome size is 2.1 Gdal. Proposed as type strain of the species *Synechococcus elongatus* (Rippka and Cohen-Bazire, 1983).

Cluster 1.2 This "cluster" contains only PCC 6312, which differs from members of cluster 1.1 by a larger cell diameter (**1.3–1.5 μm**), a lower mean DNA base composition (**50.2 mol% G + C**) and in being **insensitive to cyanophage AS1-M**. The genome size is 2.0 Gdal. Based on the DNA–DNA hybridization results of Wilmotte and Stam (1984), this strain is moderately related (31% relative binding, $\Delta T_{m(e)}$ 5.6°C) to *Synechococcus elongatus* PCC 6301 (cluster 1.1). It may thus be assignable to *Synechococcus*, but clearly represents a different species.

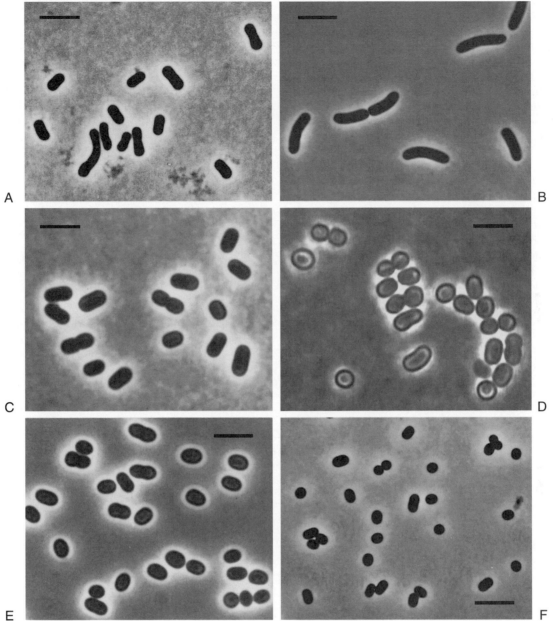

FIGURE B10.24. *A*, Light micrograph of strain PCC 6301, reference strain for cluster 1.1 of the form-genus *Synechococcus*, grown in liquid medium BG-11. Phase contrast. (Reproduced with permission from R. Rippka and G. Cohen-Bazire, Annales de Microbiologie [Institut Pasteur] *134B*: 21–36, 1983, ©Masson Editeur.) *B*, Light micrograph of strain PCC 6716, a thermophilic strain of cluster 2 of the form-genus *Synechococcus*, grown in liquid medium BG-11. Phase contrast. *C*, Light micrograph of strain PCC 7002, a euryhaline member of cluster 3 of the form-genus *Synechococcus*, grown in liquid medium ASN-III. Phase contrast. *D*, Light micrograph of strain PCC 7003, a marine member of cluster 3 of the form-genus *Synechococcus*, grown in liquid medium ASN-III. Phase contrast. *E*, Light micrograph of strain PCC 7335, reference strain of cluster 4 of the form-genus *Synechococcus*, grown in liquid medium ASN-III. Phase contrast. (Reproduced with permission from R. Rippka et al., Journal of General Microbiology *111*: 1–61, 1979, ©Society for General Microbiology.) *F*, Light micrograph of strain WH 5701, reference strain of cluster 5.2 of the form-genus *Synechococcus*, grown in liquid medium ASN-III. Phase contrast. Bars = 5.0 µm.

Reference strain PCC 6312 (co-identity ATCC 27167); isolated from freshwater, California, United States, 1963 by M.M. Allen (Rippka et al., 1979); mol% G + C is 50.2; genome size is 2.0 Gdal.

Cluster 2 The three strains included in this cluster, PCC 6715, PCC 6716 (Fig. B10.24B), and PCC 6717, are of similar size (**1.2–1.4 µm** in width) and have a mean DNA base composition of **52–53.6 mol% G + C**. They are nonmotile, exhibit a **low salt tolerance**, are obligate photoautotrophs and do not synthesize C-phycoerythrin. They were **isolated from hot springs and are thermophilic**, growth occurring at 53°C or higher (Stanier et al., 1971).

Reference strain PCC 6715; isolated from a hot spring, Yellowstone National Park, United States, 1961 by D.L. Dyer (Dyer and Gafford, 1961; Rippka et al., 1979); mol% G + C is 53.6.

Cluster 3 This cluster corresponds to marine cluster C in the last edition, with PCC 7335 removed. The four strains in-

are obligate photoautotrophs, incapable of using organic compounds as sole sources of cell carbon, and none are capable of synthesizing nitrogenase in anaerobiosis. Members of the two subclusters differ primarily in their ability to synthesize C-phycoerythrin and in distinct, though overlapping, ranges of mean DNA base composition.

Cluster 5.1 This cluster is equivalent to marine cluster A in the last edition and is represented by 15 strains isolated from both coastal waters and the open ocean: WH 6501, WH 7802, WH 7803, WH 7805, WH 8003, WH 8005, WH 8010, WH 8011, WH 8018, WH 8102, WH 8103, WH 8105, WH 8110, WH 8112, and WH 8113. Cells range in diameter from 0.6 to 1.7 µm and are either nonmotile or capable of swimming (WH 8011, WH 8103, WH 8112, and WH 8113). This novel form of motility has not been reported in any other group of cyanobacteria but is common in the open ocean isolates (Waterbury et al., 1986). All the strains of this cluster **contain C-phycoerythrin** as their major light- harvesting pigment but are incapable of complementary chromatic adaptation. There is a considerable spectral diversity of phycoerythrins, resulting primarily from the presence or absence of phycourobilin and the ratio of phycoerythrobilin to phycourobilin chromophores. All the strains have elevated growth requirements for Na^+, Mg^{2+}, and Ca^{2+} that reflect the chemistry of seawater. The mean DNA base composition ranges from **55– 62 mol% G + C**.

Reference strain WH8103 (co-identity ATCC 53061); isolated from surface waters, Sargasso Sea (28°N, 67°W), 1981 by J.B. Waterbury (Waterbury et al., 1986); mol% G + C is 59.

Cluster 5.2 This cluster contains three strains, WH 5701 (Fig. B10.24F), WH 8007, and WH 8101, isolated from coastal marine samples. It is equivalent to marine cluster B of the last edition, except that strain PCC 7001 has been removed. Cells range in diameter from 0.9 to 1.4 µm and are nonmotile. **None contain C- phycoerythrin.** One strain (WH 8007) has elevated salt requirements for growth, whereas the others are merely halotolerant. The mean DNA base composition ranges from **63–66 mol% G + C**.

Reference strain WH 5701; isolated from Long Island Sound, United States, 1957 by R.R.L. Guillard (Waterbury et al., 1986); mol% G + C is 66.

TAXONOMIC COMMENTS

Previous studies on axenic cultures revealed that, in spite of their morphological simplicity and similarity, strains traditionally assignable to *Synechococcus* differed greatly, both in size and in physiological and biochemical properties (Rippka et al., 1979). Furthermore, they were characterized by a wide span of mean DNA base composition (39–71 mol% G + C; Herdman et al., 1979b). It was thus clear in the last edition that the *Synechococcus* "group" required subdivision into several genera and species. Rippka and Cohen-Bazire (1983) accepted the genus *Cyanothece* Komárek 1976 for the larger members (more than 3 µm in width). In addition, they proposed two new genera, *Cyanobacterium* and *Cyanobium*, the descriptions of which were based on the morphological and genetic properties of two strains, PCC 7202^T and PCC 6307^T, for which they suggested the specific epithets *Cyanobacterium stanieri* and *Cyanobium gracile*, respectively. The existence of the three genera *Cyanothece*, *Cyanobacterium*, and *Cyanobium* is well supported by phylogenetic studies (see Wilmotte and Herdman, this volume). The new genera proposed by Rippka and

Cohen-Bazire (1983) have now been validated under the Botanical Code of Nomenclature (Komárek et al., 1999), and are accepted here.

As a result of the above changes, the form-genus *Synechococcus* is now more restricted and differs substantially from its description in the last edition. However, further taxonomic revision of this form-genus must be anticipated, as is evident from the dispersal of its members throughout the phylogenetic tree (see Wilmotte and Herdman, this volume, and Fig. B10.25). Members of the five clusters fall into distinct clades and clearly represent different genera.

Cluster 1 is of major importance for the taxonomy of the form-genus *Synechococcus*, since it harbors a number of strains, in particular PCC 6301, which have been recognized by botanical authorities (Padmaja and Desikachary, 1968; Komárek, 1976) as representative of *Synechococcus elongatus* Nägeli 1849, the botanical type species of *Synechococcus* (even though some inconsistencies remain: compare Komárek, 1970; Komárek, 1976). For this reason, and in agreement with an earlier proposal by Stanier et al. (1971), Rippka and Cohen-Bazire (1983) suggested that strain PCC 6301 be designated the living neotype of *Synechococcus elongatus*. The majority of the strains included in cluster 1 were previously identified as *Anacystis nidulans*, based on the taxonomic revisions of Drouet and Daily (1956). However, these epithets should be avoided, since these revisions are highly disputed by the botanical community (Anagnostidis and Komárek, 1985). Furthermore, the epithet *Anacystis* has a long and controversial history and its rejection has been proposed (Padmaja and Desikachary, 1968). Strain PCC 6908, although a typical member of *Synechococcus elongatus*, was identified prior to entering the PCC as *Synechococcus cedrorum*. This was clearly a misidentification, since members of the latter species are characterized by much larger cell dimensions (3–5 µm in width) and are here included in the genus *Cyanothece* Komárek 1976.

The thermophilic strains (PCC 6715, PCC 6716, and PCC 6717) currently placed in cluster 2 of the form-genus *Synechococcus* are phylogenetically distant from other members of this group and clearly represent a distinct genus. They are all 100% identical to two other thermophilic isolates, *Synechococcus lividus* Y-7c-s and *Synechococcus* isolate C1 (ATCC 700243), in the hypervariable V9 region of the 16S rRNA (Ferris et al., 1996: PCC 6716 and PCC 6717 were listed as ATCC 27179 and ATCC 27180, respectively), and all five strains clearly represent independent isolates of the same species. Phylogenetic analysis (Fig. B10.25) reveals an additional member of this species, "*S. elongatus*" Toray strain), which is closely related to isolate C1, the only other strain for which an almost complete sequence is available. However, these isolates are only moderately related (Fig. B10.25, see also Turner et al., 1999) to other thermophilic strains assigned to *Synechococcus* (isolates C9 = ATCC 700244, P1 = ATCC 700245, and P2 = ATCC 700246), and the latter strains may deserve separate generic status. Recent 16S rDNA sequence analyses of isolates from an Oregon hot spring have shown that several strains collected at temperatures of about 45°C and that grow up to only about 55–57°C cluster together. However, they are only distantly related to three groups of strains, originating from higher temperatures, that are able to grow, in the case of the highest temperature isolates, at up to about 70°C (S.R. Miller and R.W. Castenholz, unpublished data). It appears that the strains of the lower temperature group will fall into cluster 2 of this volume. Further analyses are required to properly place the members of the higher temperature group.

According to Komárek (1976), strains of cluster 3 are most

likely representatives of *Synechococcus bacillaris*. However, the phylogenetic data suggest that members of this cluster require assignment to a separate genus, since all four strains are closely related to each other and distinct from all other members of the form-genus *Synechococcus* (Fig. B10.25). This cluster contains two strains (PCC 7002 and PCC 73109) that were described as *Agmenellum quadruplicatum* (van Baalen, 1962). However, as pointed out by Komárek and Anagnostidis (1986), these epithets are incorrect: *Agmenellum* is synonymous with *Merismopedia*, both genera having been described more or less simultaneously in 1839 for unicellular cyanobacteria that divide in two successive planes at right angles to one another and thus form colonies composed of rectangular tablets of cells. According to the definition of the form-genus *Synechococcus*, the strains included in this cluster divide in a single plane, precluding their assignment to *A. quadruplicatum*. Strain PCC 7003 was originally identified by F. Drouet (van Baalen, 1962) as a species of *Coccochloris*, an older synonym of the genus *Aphanothece* Nägeli 1849 (Desikachary, 1959). The latter genus was defined by the fundamental properties (morphology and division patterns) characteristic of the form-genus

Synechococcus, but was restricted to mucilage-producing members (Geitler, 1932; Desikachary, 1959). However, it included both large "species" that now have been transferred into the form-genus *Cyanothece*, and small-celled representatives of different ecological origin. Thus, the epithet *Coccochloris* should be avoided.

Members of clusters 5.1 and 5.2 (marine clusters A and B in the last edition) differ primarily only in their phycobiliprotein content. Phylogenetic analysis (see "Phylogenetic relationships amongst the *Cyanobacteria*," p. 487, and Fig. B10.25) places three members of cluster 5.1 (WH 7803, WH 7805, and WH 8103) into the same clade as WH 8101 (cluster 5.2). This clade of marine cyanobacteria also contains organisms assigned to the genera *Cyanobium* and *Prochlorococcus*. Komárek et al. (1999) suggested that strains of cluster 5 be assigned to *Cyanobium*. However, despite the extensive range of properties exhibited by members of this clade, the phylogenetic distances separating them are remarkably short, and further studies are required in order to establish whether specific or generic separation is required for clusters 5.1 and 5.2.

Form-genus XIV. Synechocystis

MICHAEL HERDMAN, RICHARD W. CASTENHOLZ, ISABELLE ITEMAN AND ROSMARIE RIPPKA

Syn.e.cho.cys' tis. Gr. n. *synechia* continuity; Gr. n. *kystis* bladder or bag; M.L. fem. n. *Synechocystis* contiguous bladder-like cells.

Cyanobacteria of this provisional form-genus are unicellular strains that possess spherical cells 2–6 μm in diameter and reproduce by binary fission in two or three successive planes at right angles. Cells typically occur singly or in pairs in culture; but in nature (and rarely in culture) may occur in aggregates held together by amorphous slime material; well-defined sheath layers are never produced. Gas vacuoles are absent.

The mol% G + C of the DNA is: 35–48.

FURTHER DESCRIPTIVE INFORMATION

Strains assigned to the form-genus *Synechocystis* have a wide range of mean DNA base composition (**35–48 mol% G + C**) and are certainly representative of several genera and species. Phylogenetic analysis (see "Phylogenetic relationships amongst the *Cyanobacteria*," p. 487), although restricted to three strains (PCC 6308, PCC 6803, and PCC 6906), and a more extensive analysis of the intergenic transcribed spacer region (ITS) of the rRNA operon (see descriptions of clusters 1, 2, and 3), give firm support for further generic separation. However, these genera have not yet been formalized because characterization was performed on a limited number of strains, and only a few phenotypic properties were examined.

On the basis of their size, physiological properties, and mean DNA base composition, the strains assigned to the form-genus *Synechocystis* (Figs. B10.26 and B10.27) are subdivided into three clusters. None of the PCC strains fix N₂, even under anoxic conditions. It should be noted that determination of the size range for spherical cells is a matter of interpretation, and estimates may vary extensively depending on whether only postdivisional cells, or both post and predivisional cells, are measured. Our own measurements are made only on postdivisional cells, and thus some of the values quoted may not be in entire agreement with previous publications.

Cluster 1 This cluster, equivalent to the "low GC cluster" in the last edition, comprises five strains (PCC 6308, PCC 6701, PCC 6711, PCC 6804, and PCC 6808) of cell diameter **4–6 μm**, that have a narrow range of mean DNA base composition (**34.7-37.0 mol% G + C**) and are similar in genome size (2.1–2.5 Gdal). All strains divide in two successive planes at right angles and, under certain conditions, produce rectangular colonies that traditionally define the genus *Merismopedia* (Stanier et al., 1971; R. Rippka, unpublished data). They were all isolated from freshwater, show **low salt tolerance**, are **obligate photoautotrophs**, and have a **low content of polyunsaturated fatty acids**. They are all mesophiles, their temperature maxima for growth being in the range 37–39°C. However, the five strains differ in size, strains

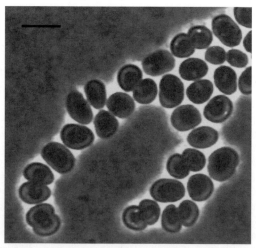

FIGURE B10.26. Light micrograph of strain PCC 6308, reference strain for cluster 1 of the form-genus *Synechocystis*, grown in liquid medium BG-11. Phase contrast. Bar = 5.0 μm.

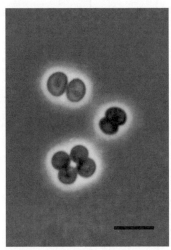

FIGURE B10.27. Light micrograph of strain PCC 6803, reference strain for cluster 2 of the form-genus *Synechocystis*, grown in liquid medium BG-11. Phase contrast. Bar = 5.0 μm.

PCC 6804 and PCC 6808 exhibiting an average cell diameter of 6 μm, compared to 4 μm for strains PCC 6308 (Fig. B10.26), PCC 6701, and PCC 6711. This cluster may therefore contain two species. Both large and small members are heterogeneous with respect to three properties: strains PCC 6711 and PCC 6804 produce amorphous capsular material; strains PCC 6701, PCC 6711, and PCC 6808 synthesize C-phycoerythrin and are capable of complementary chromatic adaptation of type II (Bryant, 1982), whereas PCC 6308 and PCC 6804 lack the latter pigment; PCC 6711, PCC 6804, and PCC 6808 exhibit gliding motility, the remaining strains being nonmotile. Three additional strains (PCC 8010, PCC 8981, and PCC 9027), of undetermined mean DNA base composition, are assigned to cluster 1 on the basis of their close relationship to the other members, revealed by RFLP analysis of the ITS region (I. Iteman and J. Filée, unpublished data); the latter two strains synthesize C-phycoerythrin and exhibit gliding motility.

Reference strain   PCC 6308 (co-identity ATCC 27150, for other identities see Rippka and Herdman, 1992); isolated from lake water, Wisconsin, United States, 1949 by G.C. Gerloff (Gerloff et al., 1950); mol% G + C is 34.7. This strain was identified prior to entering the PCC as *Gloeocapsa alpicola*. Since it does not produce well-defined sheath layers, this was clearly a misidentification.

Cluster 2   This cluster was part of the "high GC cluster" in the last edition. It comprises 15 strains with a mean DNA base composition in the range of **46–48 mol% G + C**, and probably represents several different species. Although precise measurements are lacking for some of these strains, comparative microscopy established that all possess relatively small cells (**2–4 μm** in diameter). They all divide in at least two planes, but we have no clear evidence of division in three planes. None synthesize phycoerythrin. On the basis of physiological properties they can be assigned to two subclusters.

Cluster 2.1   The 13 strains of known mean DNA base composition of this subcluster (PCC 6702, PCC 6714, PCC 6803, PCC 6805, PCC 6806, PCC 6905, PCC 6906, PCC 7201, PCC 8906, PCC 8912, PCC 8915, PCC 8962, and PCC 8990) were isolated from diverse habitats (terrestrial, freshwater, brackish water, or from a saline environment), geographically widely separated. In

spite of this, they may all be representatives of a single species. They are similar in size (**2.3–2.6 μm** in diameter, Fig. B10.27), they have a very narrow range of mean DNA base composition (**46.1–48.0 mol% G + C**) and genome size (1.8–2.4 Gdal), and they are all **euryhaline**. They are all capable of **photoheterotrophic growth**, most using glucose as the organic carbon substrate; PCC 6906 utilizes glycerol, but not glucose, distinguishing this strain from all other members of the form-genus *Synechocystis*. The temperature maximum for growth is similar to that of strains of cluster 1. The strains examined (PCC 6702, PCC 6714, PCC 6803, PCC 6805, and PCC 6806) differ from those of cluster 1 in having a **high content of polyunsaturated fatty acids**. Seven strains (PCC 6803, PCC 6902, PCC 6905, PCC 7201, PCC 8906, PCC 8915, and PCC 8962) exhibit gliding motility. Five (PCC 6803, PCC 8906, PCC 8912, PCC 8915, and PCC 8962) have been shown to be close genetic relatives, as evidenced by DNA–DNA hybridization and by DNA restriction patterns revealed by hybridization to a 16S ribosomal DNA probe (R. Rippka, T. Coursin and M. Herdman, unpublished data). Four additional strains (PCC 7338, PCC 7339, PCC 7903, and PCC 8932), of undetermined mean DNA base composition, are assigned to cluster 2.1 on the basis of their close relationship to the other members, revealed by RFLP analysis of the ITS region (I. Iteman and J. Filée, unpublished data).

Reference strain   PCC 6803 (co-identity ATCC 27184, for other identities see Rippka and Herdman, 1992); isolated from freshwater, California, United States, 1968 by R. Kunisawa (Stanier et al., 1971; Rippka et al., 1979); mol% G + C is 47.5; genome size is 2.1 Gdal. Strain PCC 6803 replaces PCC 6714 as reference strain for cluster 2.1, since the former has been extensively studied and the sequence of the entire genome is now available (Kaneko et al., 1996).

Cluster 2.2   The two strains of this subcluster, PCC 8938 and PCC 8974, have a mean DNA base composition (**47.5 and 46.6 mol% G + C**, respectively) and genome size (2.2 and 2.0 Gdal) similar to members of subcluster 2.1 and, like the latter, are **euryhaline**. However, they differ in two properties: their cells are slightly larger (**3–4 μm** in diameter) and both are **obligate photoautotrophs**. Their separation from members of cluster 2.1 is justified by the low level of genetic similarity revealed by DNA–DNA hybridization (Rippka and Herdman, unpublished data).

Reference strain   PCC 8938; isolated from salt marsh, Salicornia habitat, Lafarge factory, Fos/mer, Mediterranean coast, France, 1988 by I. Thiery (Thiery et al., 1991); mol% G + C is 47.5; genome size is 2.2 Gdal.

Cluster 3   This cluster was part of the "high GC cluster" in the last edition. It comprises four strains (PCC 6902, PCC 7008, PCC 7509, and PCC 7511) of known mean DNA base composition (range **42.1–44.9 mol% G + C**), and one (PCC 7921) that is closely related to the above on the basis of RFLP analysis of the ITS region of the rRNA operon (I. Iteman and J. Filée, unpublished data). The members of this cluster are **intermediate between clusters 1 and 2 in mean DNA base composition**. They do not contain C-phycoerythrin. Two strains, PCC 6902 and PCC 7008, isolated from brackish water and freshwater, respectively; are euryhaline and are obligate photoautotrophs, but differ in mean DNA base composition (42.1 and 44.9 mol% G + C, respectively). PCC 6902 is the only member of cluster 3 that exhibits gliding motility. Strains PCC 7509 and PCC 7511 were isolated from calcareous rocks in different locations in the Swiss Alps and may represent a single species. They share similar cell dimensions

(3–3.5 µm in diameter), mean DNA base composition (42.5 and 43.2 mol% G + C, respectively), and genome size (3.5 and 3.3 Gdal); both have a low salt tolerance and are facultative photoheterotrophs. Furthermore, they are the only members of the form-genus *Synechocystis* capable of using not only glucose, but also sucrose, for photoheterotrophic growth. Both strains seem to divide in three planes, as judged from their almost cubical colonies on solid medium.

Reference strain PCC 7509 (co-identity ATCC 29235); isolated from rock, Schöllenen, below Teufelsbrücke, Switzerland, 1972 by R. Rippka (Rippka et al., 1979); mol% G + C is 42.5; genome size is 3.5 Gdal.

TAXONOMIC COMMENTS

The form-genus *Synechocystis* is more restricted here than in the last edition, where the "group" included the "*Microcystis* cluster". Members of the latter are similar in cell size and mean DNA base composition to strains of cluster 3. However, they were elevated to generic status by Rippka and Herdman (1992), and are thus removed from the form-genus *Synechocystis*. Two other strains (WH 8501 and WH 8502), previously included in the *Synechocystis* "group" as the "marine cluster", do not correspond to the definitions of any of the genera treated here and are listed as an addition to the form-genus *Cyanothece*. Strain PCC 7511, previously (Rippka et al., 1979) misidentified as a member of the form-genus *Synechococcus*, has been included. A new cluster, containing strains of intermediate mean DNA base composition, has been revealed by RFLP analysis of the 16S rRNA–23S rRNA ITS region (I. Iteman and J. Filée, unpublished data).

In addition to *Microcystis*, four botanical genera for unicellular cyanobacteria that divide in two or three planes have been described: *Synechocystis* Sauvageau 1892, *Aphanocapsa* Nägeli 1849, *Merismopedia* Meyen 1839, and *Eucapsis* Clements and Shantz 1909. Many of the strains here included in the form-genus *Synechocystis* were previously identified as *Aphanocapsa* (Stanier et al., 1971), the members of which differ from those of *Synechocystis* only by producing mucilaginous colonies. This distinction was not recognized by Rippka et al. (1979), a decision which we maintain here. None of the strains seems to correspond to the type species, *Synechocystis aquatilis*. However, from their size and widespread occurrence in nature, members of cluster 2.1 could correspond to *Synechocystis salina* (see Komárek, 1976).

Merismopedia and *Eucapsis* are traditionally distinguished from *Synechocystis* and *Aphanocapsa* by their characteristic plate-like or cubical colonial aggregates, resulting from two or three synchronous cell divisions in successive planes, respectively. Species of *Merismopedia* form rectangular tablets composed of cells that occur in multiples of two (from a minimum of 4–8 and up to 64 cells per colony). All strains assigned here to cluster 1 of the form-genus *Synechocystis* form rectangular colonies, either in old and undisturbed liquid cultures exposed to a diurnal cycle of low light (Stanier et al., 1971), or in slide cultures on a solid substrate (R. Rippka, unpublished data), and would conform to the description of the genus *Merismopedia*. In contrast, plate-like colonies have never been observed in members of cluster 2. However, phylogenetic analyses showed that three strains (strain SAG B1448–1 and two new isolates) identified as *Merismopedia* on the basis of their characteristic colony formation, and having identical 16S rRNA gene sequences, unexpectedly do not cluster closely with *Synechocystis* PCC 6308 (a member of cluster 1) but rather are closely related to strains PCC 6906 (Palinska et al., 1996) and PCC 6803 (see Wilmotte and Herdman, this volume), which are here assigned to cluster 2 on the basis of cell size and mean DNA base composition. Furthermore, the postdivisional cell size of the three *Merismopedia* isolates was reported to vary from 1.9 to 4.9 µm (Palinska et al., 1996), thus spanning the entire size range of strains here assigned to the form-genus *Synechocystis*. Separation of *Merismopedia* and *Synechocystis* must therefore await further investigations. Strain PCC 6906 was isolated as a species of *Eucapsis* from a hypersaline lake, but we have never seen any evidence for cell division in three planes, even after growth on solid medium. Two strains (PCC 7509 and PCC 7511) of cluster 3 of the form-genus *Synechocystis* do form more or less cubical aggregates on solid medium but, in the absence of supporting evidence, it is premature to assign them to the genus *Eucapsis*.

Subsection II. (Formerly **Pleurocapsales** Geitler 1925, emend. Waterbury and Stanier 1978)

ROSMARIE RIPPKA, JOHN B. WATERBURY, MICHAEL HERDMAN AND RICHARD W. CASTENHOLZ

FURTHER DESCRIPTIVE INFORMATION

Cyanobacteria assigned to this Subsection reproduce either exclusively, or at some stages of their life cycle, by the formation of small, spherical cells (baeocytes). These reproductive cells arise from multiple fissions of a parental cell ("vegetative" cell) and are released after rupture of its fibrous outer wall layer. The number of baeocytes released from a parental cell ranges from 4 to over 1000. Enlargement of vegetative cells is accompanied by progressive thickening of the fibrous outer wall layer. Members of some genera divide exclusively by multiple fission. In others, the baeocytes give rise to cell aggregates or pseudofilaments which reproduce by binary fission ("vegetative division"). Subsequently, some or all of the cells composing the aggregates undergo multiple fission and release new baeocytes.

Multiple fission is a phenotypic property that distinguishes members of this subsection from all other cyanobacteria. Three primary features characterize the different developmental patterns in this group: the relative contributions of multiple and binary fission, the planes of successive vegetative divisions, and the changes in the structure of the cell envelope that occur during the cell cycle. In spite of certain reproductive similarities in this group of cyanobacteria, 16S rDNA sequence data do not support the separation of this group as a natural phylogenetic clade (see "Phylogenetic relationships amongst the *Cyanobacteria*," p. 487). However, given the restricted number of baeocyte-forming isolates presently in culture, the total lack of DNA–DNA hybridization data necessary for confirmation of generic and specific validity, and the limited number of available 16S rRNA sequences, it is premature to reclassify the members of subsection

II at the "divisional" level. Thus, the assemblage of baeocyte-forming cyanobacteria in Subsection II, corresponding, with minor modifications, to the botanical order *Pleurocapsales* Geitler 1925 emend. Waterbury and Stanier 1978, emphasizes their common developmental properties, but does not imply a monophyletic origin.

The process of multiple fission is initiated by the rapid successive cleavage of a vegetative cell into at least four (and up to ~1000) spherical daughter cells called baeocytes (Greek, small cells), a term first introduced by Waterbury and Stanier (1978). During the course of multiple fission there is no significant increase in cell volume following each successive cleavage. Consequently, the volume of each daughter cell is a quarter or less than that of each parental cell. Binary fission, on the other hand, is always followed by cell growth. The marked diminution in cell size that accompanies multiple fission distinguishes it from a series of binary fissions during which each round of division is followed by cell growth. This represents a real distinction between the two processes.

In the older botanical literature, the small vegetative cells generated by multiple fission are termed "endospores" or "nanocytes". The term "endospore" was widely used and has a precise definition in the botanical literature, but it is confusing when used in a procaryotic context, since the "pleurocapsalean endospore" differs extensively in its mode of formation, structure, and development from the extremely resistant bacterial endospore. The term "nanocyte" has never been precisely defined, but has been used in the botanical literature to describe unusually small cells produced by cyanobacteria (Geitler, 1932; Fritsch, 1945; Bourrelly, 1970, 1985; Komárek and Anagnostidis, 1986). However, the name was not restricted to "pleurocapsalean" members, but was also applied to describe small cells in cyanobacterial groups that never undergo multiple fission, notably in some genera of subsection I ("*Chroococcales*" Geitler, 1932, 1942, 1960; Fritsch, 1945; Komárek and Anagnostidis, 1986).

In unicellular members of subsection II, cell division occurs after release solely by multiple fission. In these genera, the baeocyte enlarges into a spherical, or ovoid to pyriform, vegetative cell that subsequently undergoes multiple fission and releases new baeocytes. Following enlargement, the baeocytes will reinitiate multiple fission. This developmental pattern is maintained over successive generations, although the cell diameter of individual parental cells at the onset of baeocyte release (and consequently the number of baeocytes) varies within a clonal population. In some members, baeocyte enlargement is followed by binary fission to produce small aggregates of vegetative cells. In others, binary fission is more extensive, resulting in relatively large cell aggregates. The characteristic morphologies of these

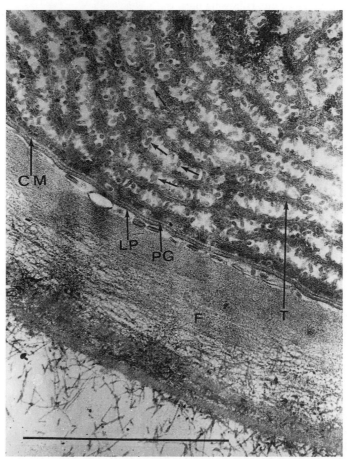

FIGURE B10.28. Electron micrograph of a thin section of part of a large vegetative cell of *Stanieria* (PCC 7302) (formerly *Dermocarpa*). *F*, fibrous outer wall layer, 750 nm thick and of variable density; *LP*, lipopolysaccharide layer (i.e., outer membrane or OM); *PG*, peptidoglycan wall layer; *CM*, cytoplasmic membrane; *T*, thylakoids; *arrows*, glycogen granules. Bar = 1.0 µm. (Reproduced with permission from J.B. Waterbury and R.Y. Stanier, Microbiological Reviews *42*: 2–44, 1978, ©American Society for Microbiology.)

aggregates are determined by the degree of variability in the planes of successive binary fissions. If divisions in one plane predominate during successive binary fissions in several planes, as in the *Pleurocapsa*-group, the mature vegetative cell aggregates will bear numerous pseudofilamentous, branched or unbranched, extensions. Some or all of the cells that compose these vegetative cell aggregates eventually undergo multiple fissions and release baeocytes.

In addition to the two outer cell wall layers typical of Gram-negative bacteria, the vegetative cells of "pleurocapsalean" representatives are always surrounded by a third external layer that structurally resembles the sheath of other cyanobacteria. Electron micrographs of thin sections show that this wall layer has a fibrous structure of varying thickness and density and is usually closely appressed to the outer membrane (Fig. B10.28). During growth and enlargement of vegetative cells, the peptidoglycan and outer membrane layers of the cell wall maintain a constant thickness. In contrast, the fibrous outer wall layer increases progressively throughout the vegetative cell cycle and may reach a thickness of 1 μm. Thus, a pseudofilament is here distinguished from a filament by the intercalation of fibrous material between adjacent cells in the former but not in the latter.

During either binary or multiple fission, the fibrous outer wall layers do not participate in cell cleavage. Thus, in binary fission the transverse wall is formed by ingrowth of the layers of the Gram-negative cell envelope, i.e., the cytoplasmic membrane, the peptidoglycan layer, and the outer membrane. Following the completion of transverse wall formation, synthesis of new fibrous outer wall material is initiated over the entire surfaces of the two daughter cells. Daughter cells are thus pushed apart by the fibrous outer wall material, which continues to thicken until onset of the next fission. Successive layers of fibrous outer wall material serve to hold vegetative cell aggregates together and to maintain their characteristic topology (Fig. B10.29).

The synthesis of fibrous outer wall material during multiple fission follows two distinct patterns. In some "pleurocapsalean" genera, the cleavage products of consecutive multiple fission events immediately initiate synthesis of fibrous outer wall material, which is deposited externally to their Gram-negative envelope. Such baeocytes consequently exhibit a Gram-negative envelope as well as a thin fibrous outer wall layer at the time of their release. Thus, cell wall synthesis during multiple fission in these representatives follows the same pattern as during binary fission. The resulting baeocytes can be distinguished from vegetative cells by their size, but not by wall structure.

However, in the majority of genera, synthesis of the fibrous outer wall material is completely suppressed during the course of multiple fission, resulting in baeocytes that at the time of their release are surrounded only by a Gram-negative envelope.

In all members of Subsection II, baeocytes are released shortly after the completion of multiple fission. As they initiate vegetative growth, the increase in cell volume of the parental cell causes the fibrous outer wall layer to rupture, presumably as the result of physical stress (Fig. B10.30).

The vegetative cells of all "pleurocapsalean" cyanobacteria are permanently nonmotile. In cases where synthesis of the fibrous outer wall layer is suppressed during multiple fission, the baeocytes may be capable of gliding motility for a short period (6–24 h) immediately following their release. Motile baeocytes react phototactically in a light gradient, moving toward or away from the light source, presumably depending on the intensity (Fig. B10.31). Loss of motility appears to coincide with onset of syn-

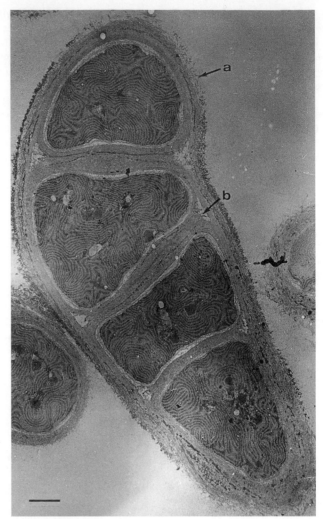

FIGURE B10.29. Electron micrograph of a thin section of *Pleurocapsa* (PCC 7314), grown in an agar overlay. Each cell is enclosed by individual peptidoglycan, outer membrane, and fibrous outer wall layers. The group of cells is also enclosed by additional fibrous outer wall layers synthesized during previous generations. As cell volume increases, the outermost layers of fibrous wall material become either stretched (*a*) or torn (*b*). Bar = 1.0 μm. (Reproduced with permission from J.B. Waterbury and R.Y. Stanier, Microbiological Reviews *42*: 2–44, 1978, ©American Society for Microbiology.)

thesis of the fibrous outer cell wall material that occurs as the baeocytes begin to enlarge into vegetative cells.

In nature, members of this group grow attached to solid substrates. Observations made on laboratory cultures suggest that attachment occurs at the baeocyte stage. The firmness of attachment appears to be enhanced during subsequent vegetative growth, as synthesis of fibrous outer wall material causes this layer of the cell wall to flare out around the base of the developing vegetative cell, enlarging the area in contact with the substrate.

Ecological Information The taxa included in this subsection vary greatly with respect to ecology and physiology. Many are aquatic (freshwater or marine) where they generally grow attached to inorganic or organic substrates; none are known to be planktonic. The marine intertidal zone contains a particularly abundant and diverse population of these organisms, where they may grow as epiliths or as epiphytes on algae and on the shells

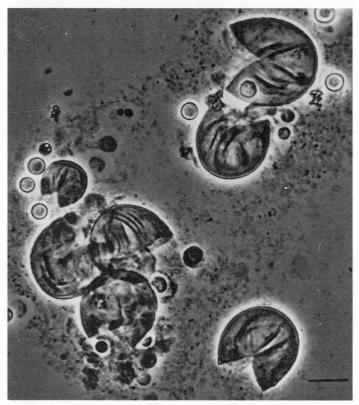

FIGURE B10.30. Fibrous outer wall layers of parental cells, emptied of their baeocytes, in a liquid culture of *Stanieria* (PCC 7437) (formerly *Dermocarpa*). Phase contrast. Bar = 10.0 μm. (Reproduced with permission from J.B. Waterbury and R.Y. Stanier, Microbiological Reviews *42:* 2–44, 1978, ©American Society for Microbiology.)

of marine invertebrates. Many are endoliths that in nature are capable of penetrating calcareous substrates, where they develop in microscopic tunnels formed through dissolution of calcium carbonate. Others (e.g. *Chroococcidiopsis*) are terrestrial, often growing as constituents of epilithic crusts or as cryptoendoliths within sandstones in both cold and hot deserts (see Vincent, 1988). *Chroococcidiopsis* also occurs as the cyanobiont in some lichens (Büdel et al., 1997).

Culturing Methods for the study of growth and development of "pleurocapsalean" cyanobacteria in culture are given in Waterbury and Stanier (1978), Rippka (1988a), and Rippka et al. (1981). The entire developmental cycle, however, may take several weeks to be observed, as a result of both slow growth and the complexity and size of the cell aggregates attained before the onset of multiple fission. For solid media, replacement of Difco agar by agarose, may prove useful for obtaining a higher viability of single cells or small cell aggregates.

The Cooper dish culture technique employed by Waterbury and Stanier (1978) was designed to sustain growth and permit semicontinuous observations of development over long periods of time. The technique has proven extremely useful for elucidation of the developmental patterns typical of the various "pleurocapsalean" genera but does not have sufficient resolution for the characterization of division patterns in small cyanobacteria (diameter <5 μm). The optical limitations are imposed by the thickness of the agar layer employed, the thickness of the plastic in the Cooper dishes, and the relatively low power of the objective (× 16) necessary to achieve a sufficiently long working distance.

Unfortunately, Cooper dishes are no longer available commercially.

TAXONOMIC COMMENTS

With minor changes, the generic definitions employed here are mainly those used by Waterbury (1989). They are based on developmental patterns determined on axenic cultures by use of light and electron microscopy (Waterbury and Stanier, 1978). Semicontinuous light microscopic observations of development were made throughout entire developmental cycles, starting with individual baeocytes. These observations were supplemented by electron microscopic examination of thin sections of cells fixed at various growth stages. The resulting developmental groups were correlated as closely as possible with existing generic descriptions from the botanical "Geitlerian" system of classification. In some instances, the botanical definitions were modified to incorporate new properties and the reinterpretation of features from the botanical descriptive literature documented during the study of developmental cycles in culture. The authors concluded that all cyanobacteria that reproduce by multiple fission with the production of baeocytes could be placed in a single order, the *Pleurocapsales.*

Most botanical taxonomic treatments of "pleurocapsalean" cyanobacteria (Fritsch, 1945; Desikachary, 1959; Bourrelly, 1985) follow the system of classification originally proposed by Geitler (1925, 1942), although at least two other systems exist (Drouet and Daily, 1956; Komárek and Anagnostidis, 1998).

Approximately 25 genera of "pleurocapsalean" cyanobacteria have been described in the "Geitlerian system", based on the

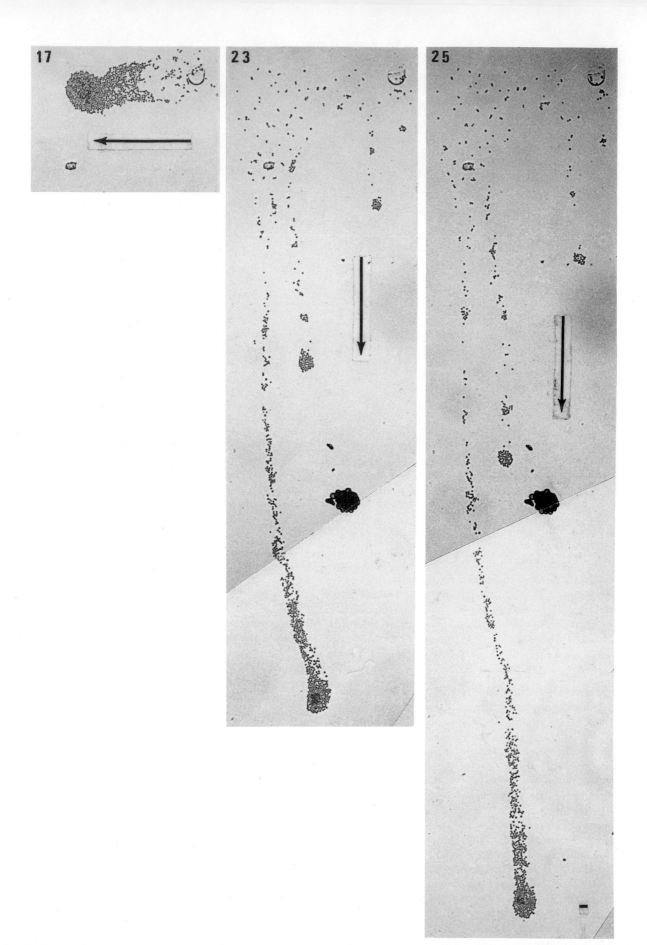

FIGURE B10.31. Phototactic response of baeocytes liberated from a parental cell of *Stanieria* (PCC 7302) (formerly *Dermocarpa*), which had been placed in a Cooper dish culture and illuminated laterally with unidirectional light. Successive light micrographs of the same field, taken 17, 23, and 25 h after the preparation was made. *Arrows* indicate the direction of illumination, which was changed at 17 h. Note remnants of the parental fibrous outer layer in the upper right-hand corner of each micrograph. Bar = 10.0 μm. (Reproduced with permission from J.B. Waterbury and R.Y. Stanier, Microbiological Reviews *42*: 2–44, 1978, ©American Society for Microbiology.)

examination of field material. Of these, five genera plus the morphologically heterogeneous *Pleurocapsa*-group have been recognized and redefined on the basis of differences in structure and development established through studies on pure cultures (Waterbury and Stanier, 1978; Waterbury, 1989). Of the remaining "Geitlerian genera", some have not yet been successfully cultured, while others will almost certainly be combined as the extent of environmentally induced variation in morphology is documented for some members of the *Pleurocapsa*-group (Waterbury, 1989).

Komárek and Anagnostidis (1998) have transferred all genera here treated as members of Subsection II into the order *Chroococcales* (equivalent to Subsection I), the botanical order *Pleurocapsales* no longer being recognized. This combination (with the order *Chroococcales* now comprising 13 families) was justified by

the authors' argument that a distinction should not be made between binary fission and multiple fission. Although still limited, phylogenetic analyses based on 16S rRNA gene sequences (Wilmotte and Herdman, this volume; Turner, 1997; Rudi et al., 1997; Garcia-Pichel et al., 1998) support the elimination of a separate higher taxon for baeocyte-forming cyanobacteria.

Included for the first time in this edition is the genus *Cyanocystis* Borzi 1982 (ex *Dermocarpa* Crouan 1858), described on the basis of a cultured isolate (Hua et al., 1989). This strain exhibits developmental properties in common with *Xenococcus* (*sensu* Waterbury 1989), but displays apical-basal polarity of mature vegetative cells and sometimes of baeocytes as well. The genus *Dermocarpa* (*sensu* Waterbury 1989) has been renamed *Stanieria* (*sensu* Komárek and Anagnostidis 1986) for reasons outlined below.

Key to the form-genera of Subsection II

I. Reproduction by multiple fissions only, or in combination with limited (1–3) binary fissions.
    A. Vegetative cells with apical–basal polarity (ovoid to pyriform cells).
        1. Baeocytes immotile, with fibrous wall layer at time of release.
            Form-genus *Cyanocystis*, p. 520
        2. Baeocytes without fibrous wall layer at time of release, may be motile; undergo 1–3 binary fissions resulting in single "apical" cell forming 6–120 baeocytes.
            Form-genus *Dermocarpella*, p. 520
    B. Vegetative cells not heteropolar (spherical cells); up to 1000 baeocytes.
        1. Baeocytes devoid of outer fibrous wall when released; may or may not be motile.
            Form-genus *Stanieria*, p. 523
        2. Baeocytes with fibrous wall layer at time of release; immotile.
            Form-genus *Xenococcus*, p. 524
II. Reproduction by extensive binary fissions, followed by multiple fissions.
    A. Vegetative cells undergo repeated binary fissions in three planes to form cubical or irregular cell clusters; usually all of the cells undergo multiple fissions to form a few to numerous baeocytes.
        1. Baeocytes with fibrous wall layer when released; immotile.
            Form-genus *Chroococcidiopsis*, p. 528
        2. Baeocytes devoid of fibrous wall layer at time of release, motile or immotile.
            Form-genus *Myxosarcina*, p. 531
    B. Binary fissions in several planes resulting in irregular aggregates of cells from which branching or non-branching pseudofilaments may arise due to intermittent divisions in one plane; only some cells undergo multiple fissions to form baeocytes that often are motile.
        Pleurocapsa-"group" (including *Pleurocapsa, Hyella, Solentia*), p. 533

FURTHER READING

Geitler, L. 1932 (Reprinted 1971). Cyanophyceae. *In* Rabenhorst (Editor), Kryptogamenflora von Deutschland, Österreich und der Schweiz, Vol. 14, Akademische Verlags, Leipzig. (Johnson Reprint Co., New York).

Holt, J.G., N.R. Krieg, P.H.A. Sneath, J.T. Staley and S.T. Williams. 1994. Bergey's Manual of Determinative Bacteriology, 9th Ed., The Williams & Wilkins Co., Baltimore.

Komárek, J. and K. Anagnostidis. 1986. Modern approach to the classification system of cyanophytes: 2. *Chroococcales*. Arch. Hydrobiol. Suppl. *73*: 157–226.

Komárek, J. and K. Anagnostidis. 1998. Cyanoprokaryota 1. Teil *Chroococcales*, G. Fischer, Jena.

Rippka, R., J. Deruelles, J.B. Waterbury, M. Herdman and R.Y. Stanier. 1979. Generic assignments, strain histories and properties of pure cultures of cyanobacteria. J. Gen. Microbiol. *111*: 1–61.

Rippka, R., J.B. Waterbury and R.Y. Stanier. 1981. Isolation and purification of cyanobacteria: some general principles. *In* Starr, Stolp, Trüper, Balows and Schlegel (Editors), The Prokaryotes. A Handbook on Habitats, Isolation, and Identification of Bacteria, Springer-Verlag, Berlin. 212–220.

Waterbury, J.B. 1989. Subsection II. Order *Pleurocapsales*. *In* Staley, Bryant, Pfennig and Holt (Editors), Bergey's Manual of Systematic Bacteriology,1st Ed., Vol. 3, The Williams & Wilkins Co., Baltimore. pp. 1746–1770.

Waterbury, J.B. 1992. The cyanobacteria - isolation, purification, and identification. *In* Balows, Trüper, Dworkin, Harder and Schleifer (Editors), The Prokaryotes. A Handbook of Bacteria: Ecophysiology, Isolation, Identification, Applications,2nd Ed., Vol. 2, Springer-Verlag, Berlin. 2058–2078.

Waterbury, J.B. and R.Y. Stanier. 1978. Patterns of growth and development in pleurocapsalean cyanobacteria. Microbiol. Rev. *42*: 2–44.

Genera Reproducing By Multiple Fissions Only, Or In Combination With Limited (1–3) Binary Fissions

Form-genus I. **Cyanocystis** Borzi 1882 (ex **Dermocarpa** Crouan 1858)

ROSMARIE RIPPKA, JOHN B. WATERBURY, MICHAEL HERDMAN AND RICHARD W. CASTENHOLZ

Cy.a.no.cyst' is. Gr. f. n. *cyanos* blue-green color; Gr. n. *cystis* bladder, bag; M.L. fem. n. blue-green bladder.

Unicellular cyanobacteria that have apical-basal polarity (i.e., ovoid to pyriform in shape), and that reproduce only by multiple fission (i.e., do not exhibit binary fission). Baeocytes are immotile, possessing an outer fibrous wall layer at the time of release from the parental cell.

FURTHER DESCRIPTIVE INFORMATION

The description of this genus is based on a single marine isolate (strain CCMEE 407), proposed by Hua et al. (1989) as the neotype for *Cyanocystis violacea* (Crouan) Komárek and Anagnostidis 1986. This strain has recently been rendered axenic in the PCC (R. Rippka, unpublished) and carries the number PCC 9504. The spherical to ovoid baeocytes (about 3 μm in diameter) at time of release possess a fibrous outer wall layer and thus do not exhibit transient motility. They give rise to ovoid to pyriform cells (up to 13 μm wide and 23 μm long) that at their narrow base are attached to the substrate by fibrous adhesive material. Multiple fission generally is synchronous and comprises the entire cytoplasm, resulting in baeocytes ranging in number from 12 to ~220 per individual cell. In a small fraction (about 1% of the cell population), however, multiple fission is retarded either at the base, at the apex, or in the central part of the cytoplasm. Subsequent completion of multiple fission in the undivided parts

will give rise to 2–4 baeocyte-forming cell clusters within the original parental cell. Depending on the developmental stages of these infrequent events, the cells may appear as dividing by both binary fission and multiple fission and may be confused with either members of the genus *Dermocarpella* (*sensu* Waterbury and Stanier 1978) or *Xenococcus* (*sensu* Waterbury and Stanier 1978).

Reference strain PCC 9504 (co-identity CCMEE 407), collected from marine benthic rock at Marseille, Veyre by T. Le-Campion. Isolated by E.I. Friedmann (Hua et al., 1989). This isolate has a high C-PE content (red color), but the capacity for complementary chromatic adaptation has not yet been examined; immotile baeocytes (3 μm in diameter) give rise to vegetative cells with apical-basal polarity (width ~9 μm, length ~17 μm). Medium: ASNIII, or a mixture of ASNIII and BG11.

TAXONOMIC COMMENTS

This genus is considered valid at this time only because of the distinct heteropolarity of the mature cells. Komárek and Anagnostidis (1998), using morphological, ecological, and pigment criteria of field material, propose that there are eight species of *Cyanocystis*. However, this can only be confirmed after more representatives of this genus have been isolated and analyzed genetically.

Form-genus II. **Dermocarpella** Lemmermann 1907

ROSMARIE RIPPKA, JOHN B. WATERBURY, MICHAEL HERDMAN AND RICHARD W. CASTENHOLZ

Der.mo.car.pel' la. L. suffix *-ella* diminutive; M.L. fem. n. *Dermocarpella* small *Dermocarpa*.

Cyanobacteria that undergo binary fission to form an ovoid aggregate consisting of a large apical cell and from one to three smaller basal cells. The cell aggregate possesses an intrinsic apical-basal polarity. Multiple fission occurs initially in the apical part of the small cell aggregate, followed by the release of baeocytes that do not possess an outer fibrous layer, and which may exhibit gliding motility. The baeocytes enlarge asymmetrically into ovoid vegetative cells before initiating binary fission. After the initial round of multiple fission, the remaining basal cells enlarge asymmetrically and undergo binary fission to form apical reproductive cells and smaller basal cells; multiple fission is reinitiated in the new apical cells.

FURTHER DESCRIPTIVE INFORMATION

The description of *Dermocarpella* (*sensu* Waterbury and Stanier 1978) is based on a single axenic representative, strain PCC 7326 (Waterbury and Stanier, 1978; Waterbury, 1989), whose devel-

opmental cycle incorporates division by both binary and multiple fissions. Characteristic developmental stages of this strain are shown in the photomicrographs of Fig. B10.32.

The distinctions between binary and multiple fission in "pleurocapsalean" cyanobacteria are revealed with particular clarity in the electron micrographs of thin sections illustrating the division stages of *Dermocarpella* (Figs. B10.33 and B10.34). The ovoid parental cell first undergoes transverse binary fission, producing a small basal vegetative cell and a much larger apical reproductive cell (Fig. B10.33A). The outer membrane of the basal vegetative cell remains closely appressed to the fibrous outer wall layer of the parental cell, both during and after cell division. In contrast, the envelope of the apical reproductive cell begins to retract from the fibrous outer wall layer of the parental cell even before complete separation from the basal vegetative cell. Immediately after the completion of transverse division, the basal vegetative cell begins to synthesize new fibrous outer wall material over the entire surface. If the basal vegetative cell undergoes further bi-

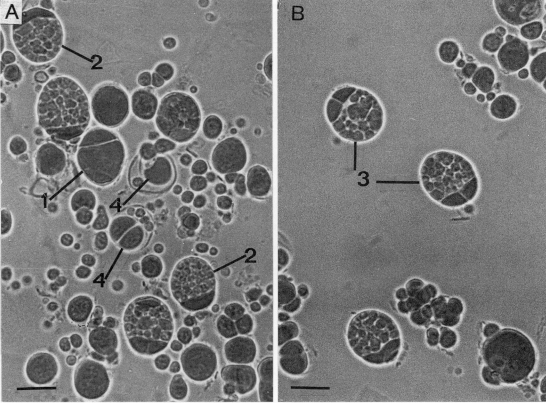

FIGURE B10.32. *A* and *B*, light micrographs of a liquid culture of *Dermocarpella* (PCC 7326). These micrographs show various division stages which cannot be resolved in Cooper dish culture: *1*, an individual which has just undergone binary transverse fission to form a small basal and a large apical cell; *2*, individuals containing a single basal cell and an apical cell which has completed multiple fission; *3*, individuals containing a pair of basal cells and an apical cell which has completed multiple fission; and *4*, individuals in which baeocytes have been released from the apical cell, revealing the outline of a parental fibrous outer wall layer which also encloses the basal cell or cells. Bars = 10 μm. (Reproduced with permission from J.B. Waterbury and R.Y. Stanier, Microbiological Reviews *42:* 2–44, 1978, ©American Society for Microbiology.)

nary fissions (Figs. B10.33B and B10.34), new fibrous outer wall material is intercalated between the septa and continues to thicken throughout the vegetative cell cycle. Multiple fission of the large reproductive cell occurs through rapid successive cleavage of the entire cell contents (Figs. B10.33B and B10.34A), producing a large number of baeocytes (Fig. B10.34B). Synthesis of fibrous outer wall material is suppressed during the cleavage process, resulting in potentially motile baeocytes.

As in *Cyanocystis*, the number of baeocytes produced per individual cell in *Dermocarpella* is not greatly influenced by environmental factors. The relative constancy in size of the ovoid parental cell at the onset of division restricts the number of baeocytes produced per reproductive cell; typically 60–120 baeocytes are formed.

Dermocarpella and *Cyanocystis* are unique among the "pleurocapsalean" cyanobacteria now in culture because of their intrinsic heteropolarity. In both genera this polarity results from the asymmetric enlargement of the baeocyte to produce an ovoid vegetative cell. This development is not determined by the site of attachment of baeocytes to their substrates. Electron micrographs of thin sections of *Dermocarpella*, developed from baeocytes attached to a sheet of dialysis membrane deposited on the surface of an agar plate, show that the planes of elongation and division are completely unrelated to the plane of the underlying substrate (Waterbury and Stanier, 1978).

Reference strain   PCC 7326 (co-identity ATCC 29376, SAG 29.84); isolated from snail shell, intertidal marine, Puerto Penasco, Mexico, 1971, by J.B. Waterbury (Waterbury and Stanier, 1978). It most closely resembles the types species *Dermocarpella incrassata* Lemmermann 1907 (Waterbury, 1989) (but see Taxonomic Comments). Maximal diameter of vegetative cells 10–13 μm; transiently motile baeocytes (diameter, 3 μm); C-PE present, complementary chromatic adaptation type III (Bryant, 1982). No

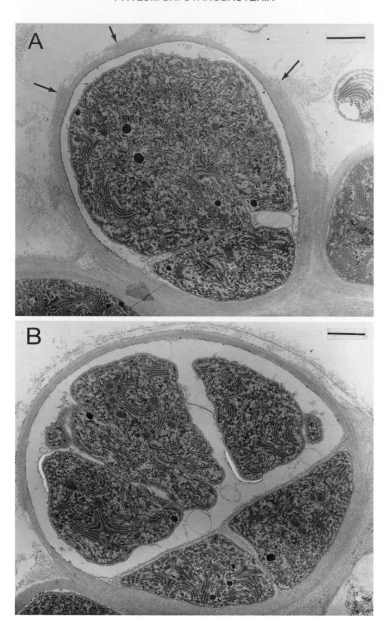

FIGURE B10.33. Electron micrographs of thin sections of *Dermocarpella* (PCC 7326) grown on agar plates. *A*, an individual which has just completed the binary fission which separates the apical cell from the basal cell. *B*, an individual which has undergone a second binary fission to form two basal cells. The apical cell has begun to undergo multiple fission, generating cleavage products which are initially not of equal size. The basal cells have begun to synthesize new fibrous outer wall layer material over their entire surfaces, but no new synthesis of fibrous outer wall layer material has occurred around the dividing cells of the apical portion. Bars = 1.0 μm. (Reproduced with permission from J.B. Waterbury and R.Y. Stanier, Microbiological Reviews *42*: 2–44, 1978, ©American Society for Microbiology.)

N₂-fixation under anoxic conditions. Capable of chemohetero-trophic growth. Mol% G + C = 45.1; genome size, 3.33 Gdal (Herdman et al., 1979a, b). Medium: MN or ASN-III.

TAXONOMIC COMMENTS

The definition of *Dermocarpella* is identical to that given in the previous edition of this *Manual*, but differs from that of Komárek and Anagnostidis (1986, 1998). The latter authors placed mem-bers of *Dermocarpella* (*sensu* Waterbury and Stanier 1978) into a

new genus, *Chamaecalyx* Komárek and Anagnostidis 1986, on the grounds that the illustration of the botanical type species *Der-mocarpella incrassata* Lemmermann 1907 showed two to three par-allel binary fissions at right angle to the long axis of the basal part of the mature cells, a mode of division never observed by Waterbury and Stanier (1978). However, in the absence of iso-lates that entirely conform to the latter description, this nomen-clatural change has not been accepted in this edition.

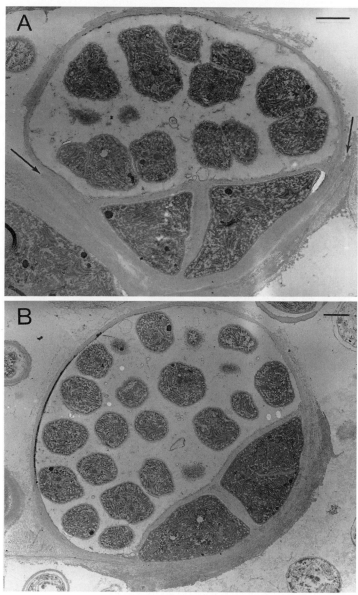

FIGURE B10.34. Electron micrographs of thin sections of *Dermocarpella* (PCC 7326) from the same preparation as the cells illustrated in Figure 10.33. *A*, an individual nearing completion of multiple fission. The fibrous outer wall layer surrounding the basal cells is markedly thicker than in the earlier developmental stage shown in Figure 10.33. No fibrous outer wall layer material surrounds the cleavage products in the apical cell. *B*, an individual which has completed multiple fission and is filled with baeocytes. Bars = 1.0 μm. (Reproduced with permission from J.B. Waterbury and R.Y. Stanier, Microbiological Reviews *42*: 2–44, 1978, ©American Society for Microbiology.)

Form-genus III. **Stanieria** *Komárek and Anagnostidis 1986*

ROSMARIE RIPPKA, JOHN B. WATERBURY, MICHAEL HERDMAN AND RICHARD W. CASTENHOLZ

Stan'ier.ia. M.L fem. n. named after R.Y. Stanier, microbiologist.

This genus encompasses unicellular cyanobacteria that are generally spherical in shape, not exhibiting intrinsic heteropolarity. Reproduction occurs exclusively by multiple fission. The baeocytes lack a fibrous outer wall layer and therefore are generally motile.

FURTHER DESCRIPTIVE INFORMATION

Strains included in this genus have previously been assigned to *Dermocarpa* (*sensu* Waterbury and Stanier 1978). Among the five axenic strains presently carried in the PCC, the mature spherical vegetative cells may vary in diameter from 20 to 30 µm, and the fibrous external wall layer may reach thicknesses of >1 µm. The number of motile baeocytes (1.5–4 µm in diameter) produced per mother cell varies greatly (from 16 to >1000), depending on the size of the mother cell prior to division. The PCC strains have been isolated from both freshwater and marine environments and vary with respect to physiological properties and mean DNA base composition (range 41–44 mol% G + C) (Herdman et al., 1979b; Waterbury, 1989). They can be subdivided into three clusters.

Cluster 1 This cluster contains two freshwater isolates from Cuba, PCC 7437 and PCC 7438, that can be grown in medium BG11. Both exhibit relatively large baeocytes (3–4 µm in diameter), but the maximum size of the vegetative cells differs (30 and 20 µm, respectively). Both have a relatively low mean DNA base composition (40.7 and 38.3 mol% G + C, respectively), produce C-PE, and undergo complementary chromatic adaptation of type III (Bryant, 1982). Both are facultative photoheterotrophs, but do not synthesize nitrogenase anaerobically. As might be expected from the tropical habitat, the upper temperature limit for growth is relatively high (44°C) (see Fig. B10.35).

Reference strain PCC 7437 (co-identity ATCC 29371), collected by J. Komárek from pool in botanical garden, Havana, Cuba in 1965; mol% G + C: 40.7; a.k.a. *Chroococcidiopsis cyanosphaera*, *Dermocarpa* sp.

Cluster 2 The single member of this "cluster", PCC 7301, is a marine isolate from an aquarium in California, having an obligate requirement for both vitamin B$_{12}$ and elevated concentrations of salt typical of true marine cyanobacteria (Waterbury, 1989). The baeocytes (3–4 µm in diameter) give rise to vegetative cells with a maximum diameter of 30 µm. This strain is distinguished from the freshwater isolates of cluster 1 by a slightly higher mean DNA base composition (44 mol% G + C). It produces C-PE constitutively (i.e., no complementary chromatic adaptation) (Bryant, 1982), is a facultative photoheterotroph, and synthesizes nitrogenase anaerobically. The upper temperature limit for growth is 37°C.

Reference strain PCC 7301 (co-identity ATCC 29367, CCAP 1416/1, UTEX 1635) collected from marine aquarium, Scripps Institution of Oceanography, California in 1964 by R.A. Lewin; mol% G + C: 44; genome size 3.07 Gdal; a.k.a. *Dermocarpa violacea*.

Cluster 3 The two true marine strains of this cluster (PCC 7302 and 7304) are distinguished from strain PCC 7301 (cluster 2) by smaller baeocytes (1.5–2.0 µm in diameter) and by not synthesizing nitrogenase anaerobically. The maximum diameter of the vegetative cells is 30 and 20 µm, respectively. Both synthesize C-PE and undergo type III complementary chromatic adaptation (Bryant, 1982). The two strains differ from each other in that PCC 7302 is an obligate photoautotroph and PCC 7304 is a facultative photoheterotroph. The upper temperature limit for growth (35–37°C) and the mean DNA base composition (42.9 and 44 mol% G + C, respectively) are similar to strain PCC 7301 (see Figs. B10.36 and B10.37).

Reference strain PCC 7302 (co-identity ATCC 29368, SAG 27.84) collected by J.B. Waterbury from sea water tank, Arizona Marine Station, Puerto Penasco, Mexico in 1971; mol% G + C: 42.9; a.k.a. *Dermocarpa* sp.

TAXONOMIC COMMENTS

The definition of *Stanieria* given here corresponds entirely to that of the genus *Dermocarpa* (*sensu* Waterbury and Stanier 1978). However, the nomenclatural change was made based on a proposition by Komárek and Anagnostidis (1986). These authors proposed to invalidate the genus name *Dermocarpa* on the grounds that the original description of *Dermocarpa violacea* Crouan 1858 was doubtful (see also Hua et al., 1989), and created a new genus, *Stanieria* Komárek and Anagnostidis 1986, to accommodate baeocyte-forming species exhibiting the structural and developmental properties of *Dermocarpa* (*sensu* Waterbury and Stanier 1978). The generic nomen *Stanieria* was created to honor Roger Y. Stanier, who pioneered the creation of a cyanobacterial taxonomy based on a polyphasic approach using pure cultures.

Form-genus IV. **Xenococcus** *Thuret 1880, emend. Waterbury and Stanier 1978*

ROSMARIE RIPPKA, JOHN B. WATERBURY, MICHAEL HERDMAN AND RICHARD W. CASTENHOLZ

Xe.no.coc′cus. Gr. n. *xenos* stranger, foreigner; Gr. n. *kokkos* a grain or kernel; M.L. masc. n. *Xenococcus* strange (spherical) cell.

Unicellular cyanobacteria that reproduce exclusively by multiple fission, leading to the formation and release of baeocytes that possess a fibrous outer wall and are therefore nonmotile. The baeocyte enlarges into a spherical vegetative cell. The successive cleavage stages of multiple fission are readily detectable by light microscopy.

FURTHER DESCRIPTIVE INFORMATION

The developmental pattern of members assigned to this genus is similar to that of *Stanieria*, differing primarily by one feature: in *Xenococcus*, the synthesis of fibrous outer wall material occurs during the course of multiple fission, being initiated immediately after each cleavage process. Consequently, the baeocytes are immotile.

Differentiating *Xenococcus* from *Stanieria* Superficially, mass cultures of *Xenococcus* appear very similar to those of *Stanieria* when examined by light microscopy. Photomicrographs comparing strains of both genera (Fig. B10.35, parts A and B) show spherical, internally homogeneous cells ranging from 2 to 30 µm in diameter. Some of the larger cells have undergone multiple fission and are filled with baeocytes, particularly visible in *Stanieria* (Fig. B10.35A). However, *Xenococcus* differs strikingly from *Stanieria* in exhibiting cells that appear to be dividing by binary fission (Fig. B10.35B) but that are, in fact, in the process of undergoing multiple fission. This difference is caused by the refractile nature of the fibrous outer wall material, which in *Xenococcus* is synthesized during the cleavage process (cf. Fig. B10.36,

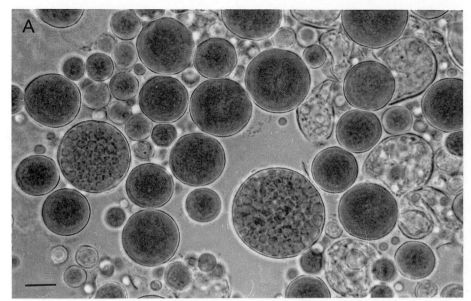
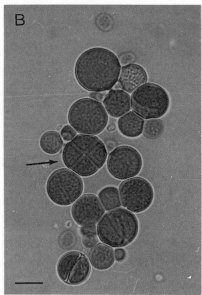

FIGURE B10.35. *A*, light micrograph of a mass culture of *Stanieria* (PCC 7437) (ex *Dermocarpa*), revealing spherical cells of varying size that appear either homogeneous or filled with baeocytes. *B*, light micrograph of a mass culture of *Xenococcus*, showing that during multiple fission the cleavage stages are visible by light microscopy (*arrow*). (Reproduced with permission from J.B. Waterbury and R.Y. Stanier, Microbiological Reviews *42:* 2–44, 1978, ©American Society for Microbiology.)

parts A and B), making the successive division stages readily evident by light microscopy.

Three axenic strains are presently assigned to *Xenococcus* (see Waterbury, 1989). All of them are true marine cyanobacteria, growing only at elevated concentrations of salt (media MN or ASN III), but none of them have an obligate requirement for vitamin B_{12}. They all produce baeocytes of similar diameter (2–3 µm), but the maximum size of the vegetative cells differs (PCC 7305,15 µm; PCC 7306, 25 µm; PCC 7307, 10–15 µm). All produce C-PE and undergo type III complementary chromatic adaptation (Bryant, 1982). Only strain PCC 7307 is capable of facultative photoheterotrophic growth. The mean DNA base composition (only determined for strains PCC 7305 and PCC 7307) is identical at ~40 mol% G + C (Herdman et al., 1979b). Strain PCC 7305 is capable of N_2 fixation under anoxic conditions.

Reference strain PCC 7305 (co-identity ATCC 29373); isolated from marine aquarium, Scripps Institute of Oceanography, La Jolla, California, U.S.A., 1971, by R.A. Lewin as *Dermocarpa* sp.; mol% G + C of the DNA: 44.2.

TAXONOMIC COMMENTS

The definition of *Xenococcus* is that of Waterbury and Stanier (1978) (Waterbury, 1989), but differs from that of Komárek and Anagnostidis (1986, 1998) in that the latter authors interpret the reproductive stages observed in field material as involving binary fission rather than multiple fission.

Further taxonomic and descriptive comments on *Dermocarpella*, *Cyanocystis*, *Stanieria* (ex *Dermocarpa*), and *Xenococcus* *Stanieria* and *Xenococcus* are here distinguished from *Cyanocystis* and *Dermocarpella* by lacking bipolarity. These genera can be further

distinguished from the each other by the delay (*Stanieria* and *Dermocarpella*) or early onset (*Xenococcus* and *Cyanocystis*) of the synthesis of fibrous wall material in the course of baeocyte formation. The developmental patterns of *Xenococcus* and *Stanieria* are indistinguishable at the level of resolution obtained using the Cooper dish culture technique (Fig. B10.37). The baeocytes enlarge symmetrically into spherical vegetative cells that increase in size until the onset of multiple fission. In both genera, the baeocytes of individual strains are nearly uniform in size but may differ in diameter between strains. In contrast, the size of the vegetative cells at the onset of multiple fission can vary widely within each strain. Since the size of the vegetative cells of individual strains at the onset of multiple fission varies, while the size of the baeocytes at the time of their release is constant, it follows that the number of baeocytes produced from a single parental cell is variable. It is not unusual for a strain to produce 10–1000 baeocytes per parental cell. If a culture in the stationary phase of growth is transferred to fresh medium, a large fraction of the vegetative cells, irrespective of size, will cleave and release baeocytes within 12–24 h, suggesting that environmental factors strongly affect the onset of multiple fission. If Cooper dish cultures are followed for more than one generation, more baeocytes are usually produced per parental cell in each successive generation as the cells become crowded and the growth conditions less favorable, implying that favorable growth conditions lead to early baeocyte release.

In contrast to *Dermocarpella* and *Cyanocystis*, which exhibit intrinsic heteropolarity, the vegetative cells in all cultured strains of *Stanieria* and *Xenococcus* lack heteropolarity and are generally spherical (Figs. B10.35, parts A and B, and B10.37). However,

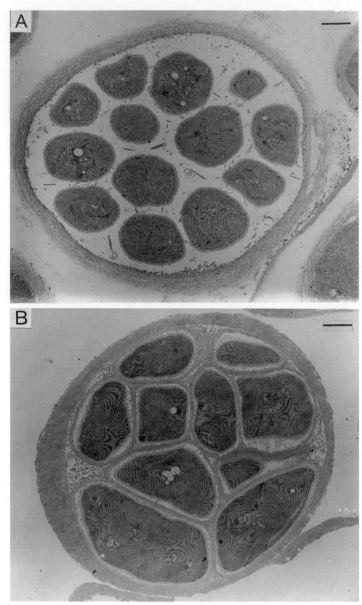

FIGURE B10.36. Electron micrographs of thin sections of parental cells that have completed multiple fission and are filled with baeocytes. *A, Stanieria* (PCC 7304) (formerly *Dermocarpa*). *B, Xenococcus* (PCC 7307). Bars = 1.0 µm. (Reproduced with permission from J.B. Waterbury and R.Y. Stanier, Microbiological Reviews *42:* 2–44, 1978, ©American Society for Microbiology.)

the shape of vegetative cells can be markedly modified by environmental factors. When populations of baeocytes of the latter two genera attach in dense clusters to a common substrate (e.g., the glass wall of a culture vessel), the enlarging vegetative cells are subjected to mutual compression and assume a pyriform or clavate shape. Cells of this form are frequently observed in natural populations and may be identified incorrectly.

Currently, there is considerable disagreement and confusion within the botanical literature concerning the structural and developmental differences among these four genera. This is due

in part to the inherent difficulty of inferring developmental cycles from isolated observations on field material or mass cultures and is exacerbated by the practice of describing new taxa rather than amending existing ones to incorporate new or reinterpreted properties. Furthermore, some variation in division patterns may occur among strains of the same genus and even within a single isolate. For example, in the heteropolar vegetative cells of *Cyanocystis violacea*, strain CCMEE 407 (PCC 9504), an occasional partial delay of multiple fission (or incomplete multiple fission) occurs, resulting in divisional stages that resemble one or more

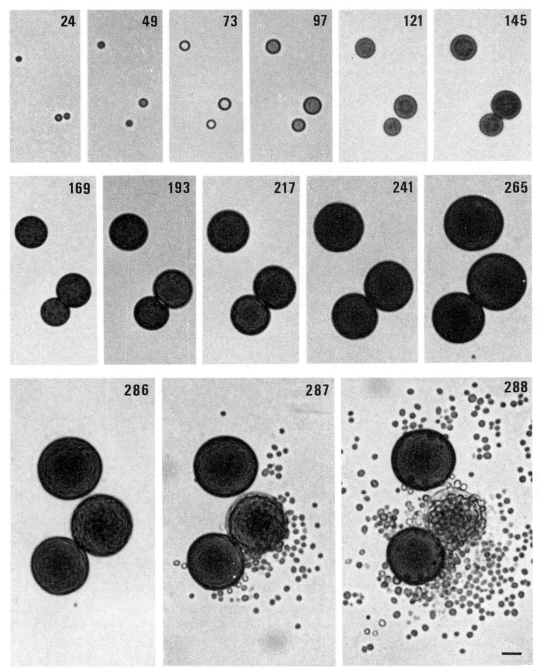

FIGURE B10.37. Development of *Stanieria* (PCC 7302) (formerly *Dermocarpa*) in a Cooper dish culture. Numbers in the upper right-hand corners in the micrographs of Cooper dish cultures indicate the elapsed time, in hours, following the initial observation. Bar = 10 μm. (Reproduced with permission from J.B. Waterbury and R.Y. Stanier, *Microbiological Reviews 42:* 2–44, 1978, ©American Society for Microbiology.)

binary fissions (Hua et al., 1989). This phenomenon has also been (though less frequently) observed in strains of *Stanieria* (ex *Dermocarpa sensu* Waterbury and Stanier 1978), where it was interpreted as a lethal error in division, since the undivided portion was never observed to retain viability.

The major problems in correlating cultured strains of the four genera provisionally accepted here with those of "species" described in the botanical literature (mainly based on field material) result from different interpretations of the types of division that may occur and the emphasis given to the presence or absence of cell polarity. Komárek and Anagnostidis (1998) divided

Cyanocystis (*sensu* Bourrelly 1970) and reserved *Cyanocystis* Borzi 1882 ("*sensu stricto*") for "species" displaying apical-basal polarity. Those exhibiting spherical cells (thus lacking bipolarity) were placed into *Stanieria* Komárek and Anagnostidis 1986, a new genus now replacing *Dermocarpa* (*sensu* Waterbury and Stanier 1978). *Dermocarpella* Lemmermann 1907, supported by Ginsburg-André (1966), corresponds with minor differences to the definition of Waterbury and Stanier (1978) and Waterbury (1989). However, "species" conforming to the latter definition have recently been transferred into a new genus, *Chamaecalyx* Komárek and Anagnostidis 1986 (Komárek and Anagnostidis, 1986, 1998),

Dermocarpella Lemmermann 1907 having been reserved for as yet uncultured members. This nomenclatural change has not been accepted here (see comments on the genus *Dermocarpella*).

Although in this edition we retain four of the traditional botanical genera for unicellular baeocyte-forming cyanobacteria, it is likely that many variations of division patterns may occur in members of this group, without them necessarily being phylogenetically very distant. This seems to be suggested by 16S rRNA gene sequence analyses (albeit still very limited), which reveal that *Dermocarpella* PCC 7326, two strains of *Stanieria* (PCC 7301 and PCC 7437), *Xenococcus* PCC 7305, *Myxosarcina* PCC 7312, and three strains of the *Pleurocapsa*-group (PCC 7321, PCC 7327, and PCC 7516) cluster in a clade that also includes unicellular and filamentous cyanobacteria that do not reproduce by multiple fission (see Fig. B10.38) (see also Wilmotte and Herdman, this volume; Turner, 1997; Rudi et al., 1997; Turner et al., 1999). Illustrating the importance of ecological distinctions, the freshwater isolate of *Stanieria* (strain PCC 7437) is located on a different branch from that of the marine strain of this genus (PCC 7301), the latter being most closely related to one of the marine strains of the *Pleurocapsa*-group (PCC 7516). Thus it can be anticipated that in the future many (if not all) of the generic arrangements described here may need to be substantially altered. Given the lack of DNA–DNA hybridization studies needed to validate the generic distinctions made here, it seems wise to abstain from any further nomenclatural changes. Consequently, unless a nearly perfect match with an original description can be made and the generic identification supported by appropriate genetic evidence, it is here suggested that attaching new generic epithets to material closely resembling these genera be done with caution.

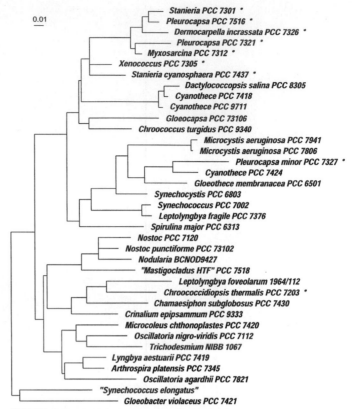

FIGURE B10.38. 16S rRNA subtree showing the phylogenetic relationships of cyanobacteria of Subsection II (indicated by *). Analyses were performed by maximum likelihood as described by Wilmotte and Herdman (this volume.)

Genera In Which Extensive Vegetative Binary Fission Precedes Multiple Fission

Form-genus I. **Chroococcidiopsis** *Geitler 1933, emend. Waterbury and Stanier 1978*

ROSMARIE RIPPKA, JOHN B. WATERBURY, MICHAEL HERDMAN AND RICHARD W. CASTENHOLZ

Chro.o.coc' ci.di.op' sis. Gr. fem. n. *chroa* skin; Gr. n. *kokkidios* presumed diminutive of *kokkos* small seed or grain; Gr. fem. n. *opsis* appearance; M.L. fem. n. *Chroococcidiopsis* small cells appearing in skin or coating.

Cyanobacteria that undergo repeated binary fission in three planes to produce more or less regular cubical cell aggregates. Multiple fission occurs simultaneously in most cells of the aggregate and is followed by the release of nonmotile baeocytes which possess an outer fibrous wall layer. The baeocytes differ little in diameter from mature parental cells and enlarge symmetrically into a spherical vegetative cells that, just before the onset of binary fission, attain dimensions characteristic and constant for any given strain.

FURTHER DESCRIPTIVE INFORMATION

In this genus binary fission plays a dominant role in the developmental pattern. Photomicrographs of mass cultures typically show cubical or somewhat irregularly shaped cell aggregates that vary considerably in size (Fig. B10.39). The developmental stages, examined at the level of resolution of Cooper dish cultures (Fig. B10.40), are virtually indistinguishable from those typical of the

genus *Myxosarcina*. The spherical baeocyte enlarges symmetrically into a vegetative cell that differs little in size from the mature vegetative cells. Because of this similarity, the distinction between baeocytes and vegetative cells, as revealed by either light or electron microscopy, is particularly difficult to make in this genus and rests entirely on the relatively small size differences between the two cell types (Fig. B10.39). The cleavage products of multiple fission are individually surrounded by fibrous outer wall material (Fig. B10.41), in accordance with the observation that free baeocytes in *Chroococcidiopsis* are never motile. Although baeocyte counts for this genus are difficult to determine, the strains examined appear to produce a constant but low number (about 4) of baeocytes per vegetative cell, in agreement with the small differences in size compared to the mature vegetative cells.

Environmental factors may influence the developmental pattern of *Chroococcidiopsis* by affecting the number of vegetative binary fissions that occur prior to the onset of multiple fission. Thus, the size of the cell aggregates, rather than the number of

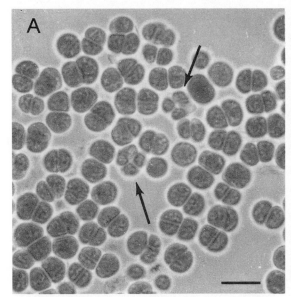
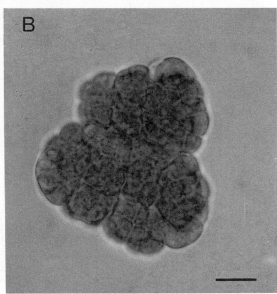

FIGURE B10.39. Light micrographs of mass cultures of *Chroococcidiopsis* grown on agar plates. *A*, dispersed cells from a crushed cell aggregate of strain PCC 6712, showing the regularity of cell size. Some of the cells (*arrows*) have undergone multiple fission, producing four baeocytes per parental cell. *B*, characteristic cell aggregates of strain PCC 7431. Bars = 10 μm. (Reproduced with permission from J.B. Waterbury and R.Y. Stanier, Microbiological Reviews *42*: 2–44, 1978, ©American Society for Microbiology.)

baeocytes produced per vegetative cell, is affected. Multiple fission, when it occurs, seems to be relatively synchronous, with almost all cells in the aggregates undergoing cleavage simultaneously.

Strains of *Chroococcidiopsis* in the PCC

Cluster 1 This cluster comprises 7 strains (PCC 7203, PCC 7431, PCC 7432, PCC 7433, PCC 7434, PCC 7436, and PCC 7439), all of which (excluding strain PCC 7434) show sufficient developmental, morphological, and physiological similarity to suggest that they may be representatives of a single species (Waterbury, 1989). All are freshwater or soil isolates, growing in medium BG11. Their vegetative cells (~5 μm in diameter) each give rise to 4 baeocytes (~3 μm in diameter). The range of mean DNA base composition is extremely narrow (45.8–46.4 mol% G + C) (Herdman et al., 1979b). All (except strain PCC 7334) synthesize phycoerythrocyanin (PEC), are facultative photo and chemoheterotrophs, and synthesize nitrogenase under anoxic conditions. Strain PCC 7434 differs by synthesizing neither PEC nor C-PE and will not grow chemoheterotrophically.

REFERENCE STRAIN PCC 7203 (co-identity ATCC 27900, SAG 42.79, CCAP 1451/1), isolated from soil sample near Greifswald, DDR, 1962 (Komárek, 1972; Waterbury, 1989). Mol% G + C: 45.8; genome size 4.06 Gdal (Herdman et al., 1979a, b).

Cluster 2 The single strain representative of this "cluster" (PCC 6712) differs from those of cluster 1 by producing larger vegetative cells (6.3 μm diameter) and baeocytes (~4 μm). Although identical to strains of cluster 1 with respect to growth requirements (medium BG11), heterotrophy, and N₂-fixation, this isolate lacks PEC but produces C-phycoerythrin and exhibits complementary chromatic adaptation of type III (Bryant, 1982). The lower mean DNA base composition (mol% G + C = 40.2) (Herdman et al., 1979b) suggests that this strain may represent a different species, if not a separate genus.

REFERENCE STRAIN PCC 6712 (co-identity ATCC 27176, CCAP 1411/2), isolated from freshwater reservoir, Marin County, California, 1967. Mol% G + C: 40.2; genome size 3.31 Gdal (Herdman et al. 1979a, b).

TAXONOMIC COMMENTS

The definition of *Chroococcidiopsis* given here is the same as in the previous edition of this *Manual* (Waterbury, 1989) and agrees well with current botanical descriptions (Komárek and Anagnostidis, 1986, 1998). However, earlier diagnoses of this genus, based both on observations of field material and on development inferred from mass cultures, incorporated many of the characteristic developmental stages of *Chroococcidiopsis* but were unclear about how these stages were linked together to form the complete developmental cycle (Geitler, 1933; Friedmann, 1961; Komárek, 1972; Komárek and Hindak, 1975).

It should also be mentioned that in the absence of detailed developmental studies and electron microscopy, members of this genus may be easily confused with *Myxosarcina* or even with unicellular forms such as members of the genus *Gloeocapsa* (*sensu* Rippka et al., 1979), which possess extracellular sheath layers, divide by binary fission in three planes, and produce cell aggregates that may resemble those typical of both *Chroococcidiopsis* and *Myxosarcina*. An example of these difficulties is strain PCC 6712, now recognized as a *Chroococcidiopsis* (cluster 2), which for many years was thought to divide solely by binary fission and was consequently placed in the order *Chroococcales* (Stanier et al., 1971). The ability of this strain to reproduce by multiple fission was not recognized until its development was followed in Cooper dish cultures (Waterbury and Stanier, 1978).

Chroococcidiopsis is distinguished from *Myxosarcina* by lack of baeocyte motility, associated with the early synthesis of fibrous wall material around the cleavage products in the course of multiple fission (Waterbury and Stanier, 1978; Waterbury, 1989). This feature was not incorporated into earlier botanical diagnoses of

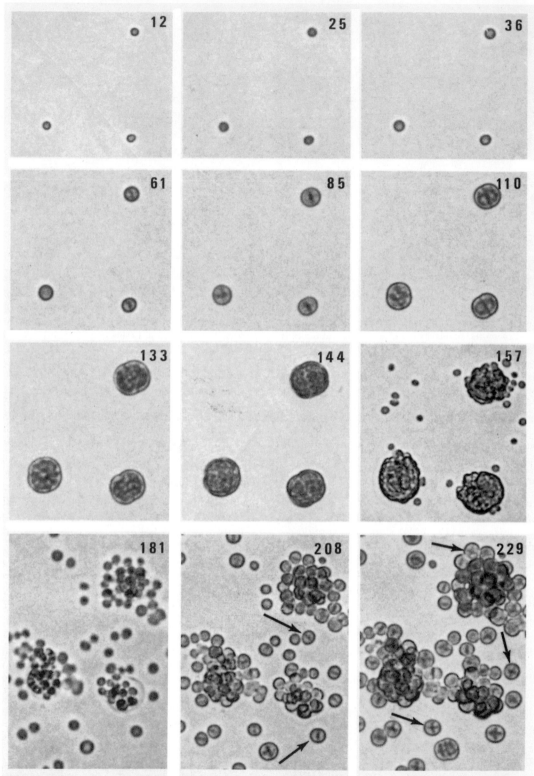

FIGURE B10.40. Development of three baeocytes of *Chroococcidiopsis* (PCC 7432) on medium BG-11 supplemented with 0.3% sucrose in a Cooper dish culture. *Arrows* indicate baeocytes that have enlarged and begun to divide by binary fission (×500). (Reproduced with permission from J.B. Waterbury and R.Y. Stanier, *Microbiological Reviews 42*: 2–44, 1978 ©American Society for Microbiology.)

these genera; consequently, the assignment of cultured strains to botanically described species is largely arbitrary.

A single strain of *Chroococcidiopsis* (strain PCC 7203) has been analyzed phylogenetically and was shown to be positioned in the 16S rRNA tree in a clade distant from all other baeocyte-forming cyanobacteria, including *Myxosarcina* PCC 7312, considered to be the closest morphological counterpart of the genus *Chroococcidiopsis* (see Fig. B10.38; Wilmotte and Herdman, this volume).

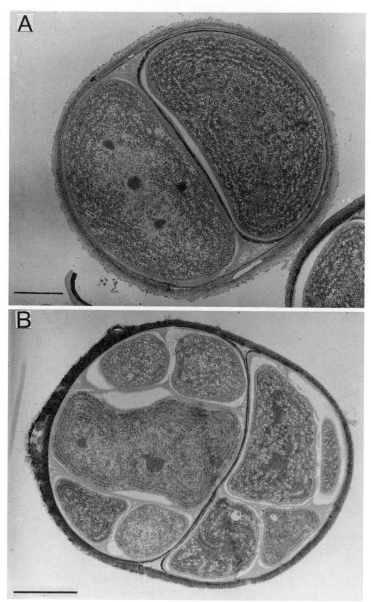

FIGURE B10.41. Electron micrograph of thin sections of *Chroococcidiopsis* (PCC 7436) grown on an agar plate. *A*, two vegetative cells produced by binary fission. *B*, a pair of cells that have undergone multiple fission after one binary fission; note the small size of the cells and the thinness of their individual fibrous outer cell wall layers. Bars = 10 μm. (Reproduced with permission from J.B. Waterbury and R.Y. Stanier, *Microbiological Reviews 42:* 2–44, 1978, ©American Society for Microbiology.)

Form-genus II. **Myxosarcina** *Printz 1921, emend. Waterbury and Stanier 1978*

ROSMARIE RIPPKA, ISABELLE ITEMAN, JOHN B. WATERBURY, MICHAEL HERDMAN, AND RICHARD W. CASTENHOLZ

Myx.o.sar′ ci.na. Gr. fem. n. *myxa* slime, mucus; L. n. *sarcina* bundle or package; M.L. fem. n. *Myxosarcina* slime bundle or packet of cells in slime.

Cyanobacteria that undergo repeated binary fission in three planes to form more or less cubical aggregates of cells. Multiple fission occurs simultaneously in most vegetative cells, resulting in the liberation of a restricted number (about four per vegetative cell) of baeocytes that lack an outer fibrous wall layer and are often capable of gliding motility. The baeocytes enlarge symmetrically into spherical vegetative cells that, just prior to the onset of binary fission, attain a size that is characteristic and constant for any given strain.

FURTHER DESCRIPTIVE INFORMATION

The morphology of cell aggregates and the developmental pattern in members of this genus are very similar to those of the

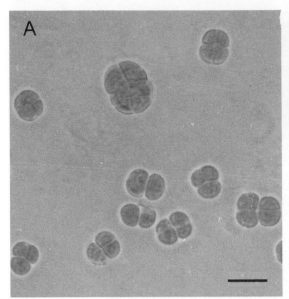
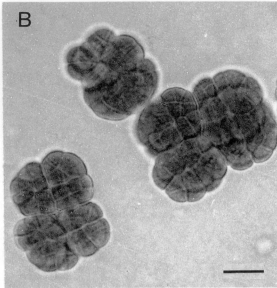

FIGURE B10.42. Light micrograph of *Myxosarcina* (PCC 7312) grown on an agar plate. *A*, young aggregates showing the uniformity of cell size and the regularity of the planes of successive divisions. *B*, older aggregates showing the maintenance of fairly regular cubical packets of cells. Bars = 10 μm. (Reproduced with permission from J.B. Waterbury and R.Y. Stanier, Microbiological Reviews *42:* 2–44, 1978, ©American Society for Microbiology.)

genus *Chroococcidiopsis*. Photomicrographs of young and older cell aggregates of *Myxosarcina* PCC 7312 are shown in Fig. B10.42 and can be compared with *Chroococcidiopsis* PCC 6712 (Fig. B10.39). As in *Chroococcidiopsis*, the spherical baeocytes enlarge symmetrically into vegetative cells and divide by binary fission over many generations to give rise to large cubical aggregates before undergoing sudden baeocyte liberation that seems to occur simultaneously in all cells (Fig. B10.40). The principal difference between *Myxosarcina* and *Chroococcidiopsis* is the lack of a fibrous wall layer surrounding the baeocytes at the time of release. Consequently, the two *Myxosarcina* strains presently in axenic culture (PCC 7312 and PCC 7325) have been observed to exhibit gliding motility.

Although isolated at the same time from the same snail shell collected in the intertidal zone (Puerto Penasoco, Mexico) and requiring elevated salt concentrations typical of marine forms (media MN or ASNIII), the two axenic strains currently assigned to this genus differ from each other in many ways. Strain PCC 7312 has smaller vegetative cells (5 μm in diameter), releases only 4 baeocytes (2 μm in diameter) per vegetative cell, and fixes N_2 anaerobically. Strain PCC 7325 exhibits slightly larger dimensions (vegetative cells, 8–10 μm in diameter; baeocytes, 3 μm in diameter), releases more baeocytes (8–16), and does not synthesize nitrogenase under anoxic conditions. Both strains are facultative photoheterotrophs, produce C-PE (Bryant, 1982), and undergo type III complementary chromatic adaptation (Tandeau de Marsac, 1977). The mean DNA base composition is similar for both isolates (mol% G + C 44.0 and 42.9, respectively) (Herdman et al., 1979b).

Reference strains PCC 7325 (co-identity ATCC 29378) isolated from snail shell, intertidal marine, Puerto Penasco, Mexico, 1971, by J.B. Waterbury; mol% G + C = 2.7. Note: this strain replaces the previous reference strain PCC 7312 (co-identity ATCC 29377), which is no longer dispatched from the PCC because of doubts concerning its generic identity (see Taxonomic Comments below).

TAXONOMIC COMMENTS

Myxosarcina is here defined as previously (Waterbury and Stanier, 1978; Waterbury, 1989) and conforms well to botanical descriptions of the genus (Printz, 1921; Geitler, 1932; Desikachary, 1959; Bourrelly, 1985; Komárek and Anagnostidis, 1986, 1998). However, the legitimacy of this genus will need to be carefully examined for several reasons:

1. Although 16S rRNA sequence analyses have clearly demonstrated that *Myxosarcina* PCC 7312 can easily be distinguished from *Chroococcidiopsis* PCC 7203 by its position in a different and rather distant clade of the phylogenetic tree, the same studies also revealed that this isolate clusters relatively closely with members of the *Pleurocapsa*-group (see Fig. B10.38) (see also Turner, 1997; Turner et al., 1999);

2. Recent examinations of strain PCC 7312 by light microscopy revealed short "pseudofilamentous extensions" of the cell aggregates, particularly in old cultures, a morphological property more typical of the *Pleurocapsa*-group;

3. Restriction enzyme analyses (RFLP) of PCR amplicons of the intergenic transcribed spacer region (ITS) of the ribosomal operon of all pleurocapsalean isolates presently available in the PCC, suggest that strain *Myxosarcina* PCC 7312 is extremely closely related, if not identical, to strains PCC 7314 and PCC 7315 of the *Pleurocapsa*-group. *Myxosarcina* PCC 7325 has no close relatives in the ITS tree, but forms a distinct branch among other *Pleurocapsa*-group isolates (K. Romari and I. Iteman, unpublished data). Nevertheless, 16S rRNA gene sequence comparisons of many more strains and DNA–DNA hybridization data are required before any definitive conclusions concerning appropriate generic reassignments can be made. Given the uncertainty associated with identification of *Myxosarcina* PCC 7312, strain PCC 7325 has here been designated as reference strain, in contrast to the last edition of the *Manual* (Waterbury, 1989).

Pleurocapsa-group *sensu Waterbury and Stanier 1978, Waterbury 1989*

ROSMARIE RIPPKA, JOHN B. WATERBURY, MICHAEL HERDMAN,
AND RICHARD W. CASTENHOLZ

Pleu.ro.cap' sa. Gr. n. *pleyra* rib or side; L. n. *capsa* box; M.L. fem. n. *Pleurocapsa* box
with ribs.

The *Pleurocapsa*-group is a provisional assemblage which can loosely be defined as cyanobacteria that undergo repeated binary fissions in many different planes to produce cell aggregates that are diverse in size and form. They range from smaller or larger cell aggregates of variable compactness to very complex structures, all of which exhibit at the periphery pseudofilamentous appendages of varying length. Multiple fission occurs in some cells of the aggregates, resulting in baeocytes which may be motile because they lack a fibrous wall layer at the time of release.

FURTHER DESCRIPTIVE INFORMATION

For the present, *Pleurocapsa* is regarded as a "super-genus" or "form-genus", within which several different genera may eventually be recognized. Cyanobacteria placed into this assemblage, the "*Pleurocapsa*-group" of Waterbury and Stanier (1978), comprise a number of seemingly diverse strains. The diversity in size of cell aggregates and arrangements typical of this group is illustrated by light micrographs of mass cultures of representative strains (Figs. B10.43, B10.44, and B10.45).

Two modes of baeocyte enlargement have been recognized by Waterbury and Stanier (1978) and are illustrated in Figs. B10.45 and B10.46. Fig. B10.46 shows the developmental cycle of strain PCC 7319 in a Cooper dish culture. In this strain (a member of cluster I; Waterbury, 1989), the baeocyte initially enlarges symmetrically, then divides in many irregular planes to produce a complex cell aggregate with prominent pseudofilamentous appendages. Multiple fission then occurs in some of the cells in the aggregate, resulting in the release of motile baeocytes. Characteristic growth stages of strain PCC 7516 (cluster II; Waterbury, 1989) are shown in Fig. B10.45. The baeocyte elongates and enlarges asymmetrically before the onset of binary fission (Fig. B10.45, parts A and B). Binary fission in irregular planes results in a cell aggregate with pseudofilamentous appendages. Aggregate formation is followed by multiple fission (Fig. B10.45D) of some of the vegetative cells to form motile baeocytes.

Environmental factors affect the developmental patterns of strains assigned to the *Pleurocapsa*-group in two ways. As in *Myxosarcina* and *Chroococcidiopsis*, the number of binary fissions that precede the onset of multiple fission is not fixed. Multiple fission occurs earlier under favorable growth conditions. Indeed, some strains that normally produce pseudofilamentous appendages may undergo multiple fission after only one or two binary fissions (Fig. B10.43, C and D).

In contrast to *Myxosarcina* and *Chroococcidiopsis*, the planes of successive binary fissions that determine the three-dimensional configuration of vegetative cell aggregates may be altered by environmental factors in most strains of the *Pleurocapsa*-group. As a result, the range of morphological and developmental variation within individual strains of this group can be extensive, as illustrated by the following examples.

The differences in developmental expression caused by variable growth conditions are illustrated for strain PCC 7322 (Fig. B10.47). Under photoheterotrophic conditions, cubical cell aggregates that can be confused with those of *Myxosarcina* are produced (Fig. B10.47B), whereas under photoautotrophic conditions, cell aggregates exhibit pseudofilamentous appendages (Fig. B10.47A) (Waterbury, 1989).

In strain PCC 7319, the addition of sucrose to the mineral growth medium does not change the gross structure of the aggregate. Instead, the pseudofilamentous extensions, which are uniseriate at their tips when grown photoautotrophically, become multiseriate throughout their entire length when grown photoheterotrophically (Fig. B10.44, A and B) (Waterbury, 1989).

Strain PCC 7324, like the other strains of cluster I (Waterbury, 1989), is characterized by symmetrical baeocyte enlargement but is unusual in several other respects. Binary fission occurs predominantly in two planes, forming relatively flat cell aggregates which can reach considerable size before undergoing multiple fission. Nearly all the cells in the aggregate undergo multiple fission simultaneously, a property not possessed by any other strain assigned to the *Pleurocapsa*-group. In addition, when growing under favorable conditions, cells of this strain occasionally become covered by an amorphous precipitate, giving them an appearance characteristic of natural populations of *Entophysalis*, a genus currently placed in the *Chroococcales* in the botanical literature (Komárek and Anagnostidis, 1998).

Based on the phenotypic properties of axenic strains presently available in the PCC, the *Pleurocapsa*-group has been divided into three clusters.

Cluster 1 This cluster corresponds to subgroup I of Waterbury and Stanier (1978) (cluster I of Waterbury, 1989) and accommodates five strains (PCC 7317, 7319, 7320, 7322, and 7324) in which the baeocytes enlarge symmetrically to produce a spherical vegetative cell. The cultured strains were isolated from surfaces of solid substrates collected in the marine intertidal zone. They all have requirements for elevated concentrations of salts typical of true marine strains, being cultivated in medium MN or ASNIII. Strain PCC 7317 has an obligate requirement for vitamin B_{12}. All strains synthesize C-PE (Bryant, 1982), but only strains PCC 7319, 7320, and 7322 undergo complementary chromatic adaptation of type III (Bryant, 1982). All are facultative photoheterotrophs with the exception of strain PCC 7324. The baeocytes of all strains are similar in size (2–3 μm) and vegetative cells are also relatively similar in dimensions (5–8 μm). However, these strains differ from one another in several other characteristics (e.g., the formation of pseudofilamentous appendages, the maximum temperature for growth, and the ability to synthesize nitrogenase anaerobically (see Waterbury, 1989; Rippka and Herdman, 1992). The mean DNA base composition is in the range of 43.0–45.4 mol% G + C (Herdman et al., 1979b).

Reference strain PCC 7319 (co-identity ATCC 29388), isolated from a snail shell, intertidal marine, Puerto Penasco, Mexico, 1971, by J.B. Waterbury (Waterbury and Stanier, 1978). Mol% G + C = 43.2.

Cluster 2 This cluster corresponds to subgroup II of Waterbury and Stanier (1978) (cluster II, Waterbury, 1989) and includes three strains (PCC 7314, 7321, and 7516) in which the baeocytes are characterized by asymmetric enlargement to form elongated vegetative cells. As in cluster 1, the strains are truly

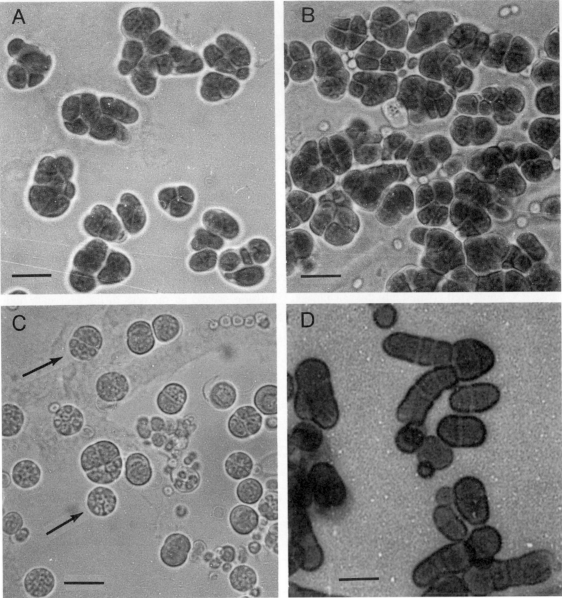

FIGURE B10.43. Light micrographs of mass cultures of members of the *Pleurocapsa*-group grown on agar plates. *A* and *B*, small cell aggregates of strain PCC 7317. *C*, strain PCC 7320 showing cells (*arrows*) that have undergone multiple fission at the one- or two-celled stage. *D*, small branched cell aggregates typical of strain PCC 7320. Bars = 10 μm. (*A* is reproduced with permission from J.B. Waterbury and R.Y. Stanier, *Microbiological Reviews 42:* 2–44, 1978, ©American Society for Microbiology.)

marine, epilithic on mollusc shells or rocks, show requirements for elevated salt concentrations (media MN or ASNIII), and have similar baeocyte dimensions (2.5–3 μm in diameter); the size of the vegetative cells vary among the isolates (8–16 μm in diameter). All produce C-PE, synthesize nitrogenase under anaerobic conditions, are facultative photoheterotrophs, and undergo complementary chromatic adaptation of type III (with the exception of PCC 7321, which synthesizes C-PE constitutively). The mean DNA base composition is in the range of 41.0–43.3 mol% G + C (Herdman et al., 1979b) (for detailed properties, see Waterbury, 1989).

Reference strain PCC 7516 (co-identity ATCC 29396), isolated from a rock chip, marine intertidal zone, I'lle Riou, Marseille, France, 1974, by T. Le Campion-Alsumard (Waterbury and Stanier, 1978). Mol% G + C = 41.0.

Cluster 3 Strain PCC 7327 assigned to this "cluster" was originally included in subgroup I of Waterbury and Stanier (1978) and thus exhibits the structural and developmental properties described for cluster 1 above. However, this strain differs physiologically and ecologically from cluster 1 strains and was placed into a separate cluster (cluster 3) by Rippka and Herdman (1992). It was isolated from a freshwater hot spring at 50°C (Castenholz, 1970) and grows in freshwater medium BG11. It synthesizes C-PE constitutively (Bryant, 1982), is a facultative photoheterotroph, and is capable of N₂ fixation under anoxic conditions. The mean DNA base composition of strain PCC 7327 is slightly higher (46.5 mol% G + C) than that of members of clusters 1 and 2 (Herdman et al., 1979b).

Reference strain PCC 7327 (co-identity ATCC 29393): collected by R.W. Castenholz from 50°C, Hunter's Hot Springs,

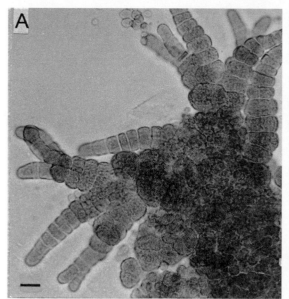
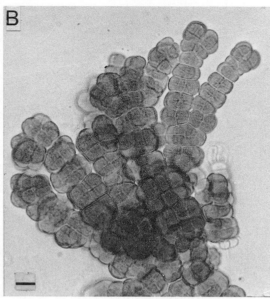
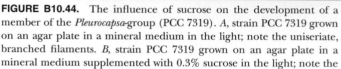

FIGURE B10.44. The influence of sucrose on the development of a member of the *Pleurocapsa*-group (PCC 7319). *A*, strain PCC 7319 grown on an agar plate in a mineral medium in the light; note the uniseriate, branched filaments. *B*, strain PCC 7319 grown on an agar plate in a mineral medium supplemented with 0.3% sucrose in the light; note the multiseriate structure of the filaments. Bars = 10 μm. (Reproduced with permission from J.B. Waterbury. 1979. *In* Parish (Editor), Developmental Biology of Prokaryotes, Vol. 1: Studies in Microbiology. ©University of California Press.)

Oregon in 1969 (Waterbury and Stanier, 1978). Mol% G + C = 46.5.

The *Pleurocapsa*-group Compared to the Genera *Hyella* and *Solentia* *Pleurocapsa*, *Hyella*, and *Solentia*, three of the principal genera that largely correspond to the *Pleurocapsa*-group described in the botanical literature, have been distinguished from each other by their growth habits in the natural environment and on the basis of non-axenic cultures (Fig. B10.48, Fig. B10.49). Members of *Pleurocapsa* are epiphytes or epiliths, whereas members of *Hyella* and *Solentia* are endoliths that penetrate calcareous substrates. It is generally believed that the ability to penetrate calcareous substrates is an inherent property possessed by some cyanobacteria, and that the mechanism of penetration, while not known, follows a pattern suggestive of chemical dissolution of the substrate.

Several marine strains of the *Pleurocapsa*-group in the PCC were isolated from mollusc shells or rock chips and could, in principle, have been either epiliths or endoliths. Except for strain PCC 7516, observed to be endolithic in its natural habitat, the relations of these strains to their natural substrates were not determined at the time of their isolation. Following purification, several of these strains (PCC 7314, PCC 7319, PCC 7321, PCC 7322, and PCC 7324) were grown on oyster shell chips that were transferred frequently in a liquid growth medium for a period of 2 years (Waterbury and Stanier, 1978; Waterbury, 1989). All grew profusely on the surface of the oyster shell chips, but none penetrated the surface, indicating that all five strains were epiliths. Some experiments suggested that penetration of calcareous substrates might depend on the activity of associated acid-producing bacteria. This hypothesis was tested by isolating acid-producing, aerobic, chemoheterotrophic bacteria from an impure culture of strain PCC 7506 (a *Pleurocapsa*-group isolate that was later lost; Waterbury, 1989). One of the strongest acid-producing bacterial isolates was grown for two months on oyster shells to-

gether with either strain PCC 7440 or PCC 7516 (*Pleurocapsa*-group clusters I and II, respectively), the latter of which was known to exhibit endolithic growth. In these two-membered cultures, maintained in a mineral medium and incubated in the light, both strains penetrated the oyster shell chips, whereas controls lacking the acid-producing bacteria (i.e., axenic cultures of either strain) showed no penetration of the substrate. These results indicate that associated chemoheterotrophic bacteria may play a role in facilitating the penetration of calcareous substrates by members of the *Pleurocapsa*-group. Thus, the effects of chemoheterotrophic bacteria on expression of the endolithic habit by cyanobacteria, whether obligate or facultative, limits the utility of endolithy as a major taxonomic character for the differentiation of genera within the *Pleurocapsa*-group.

Nevertheless, Le Campanion-Alsumard et al. (1996), using a non-axenic strain of *Solentia*, claim that no chemoheterotrophic bacteria could be detected anywhere close to the areas of carbonate dissolution. They remain convinced that the apical cells typical of the latter genus (and presumably of members of *Hyella*) are capable of carbonate dissolution.

Another example of the variability in morphological expression and the difficulties associated with identifying genera of the *Pleurocapsa*-group is provided by the following observations on strain PCC 7324. In several-months-old liquid cultures of this strain, some cells of the population were observed to be embedded in a mass of extracellular material, which had the form of dichotomously branched threads with a diameter slightly less than that of the connected cells (Waterbury, 1989). Many of the cells associated with this material were widely separated from one another and typically located at the apices of the threads. Such forms, present only in old and presumably senescent cultures of this strain, are striking because they display the principal diagnostic feature that distinguishes members of the genus *Solentia* from *Pleurocapsa* and *Hyella* (Fig. B10.50) (Golubic et al., 1996; Komárek and Anagnostidis, 1998).

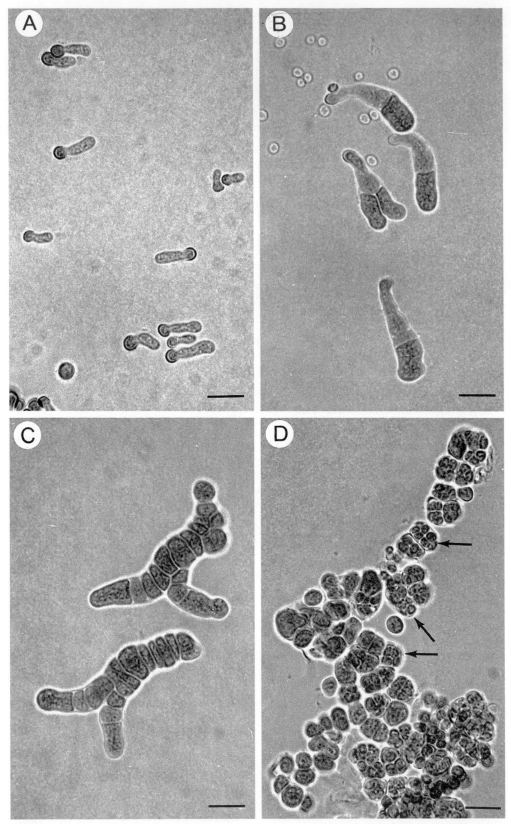

FIGURE B10.45. Light micrographs of wet mounts of *Pleurocapsa*-group strain PCC 7516, prepared by sampling an agar plate culture at various intervals after inoculation. *A*, initial outgrowth of baeocytes. *B*, asymmetric enlargement and initial vegetative divisions in young aggregates. *C*, pseudofilamentous cell aggregate. *D*, large cell aggregate containing cells that have undergone multiple fission and are filled with baeocytes. Bars = 10 μm.

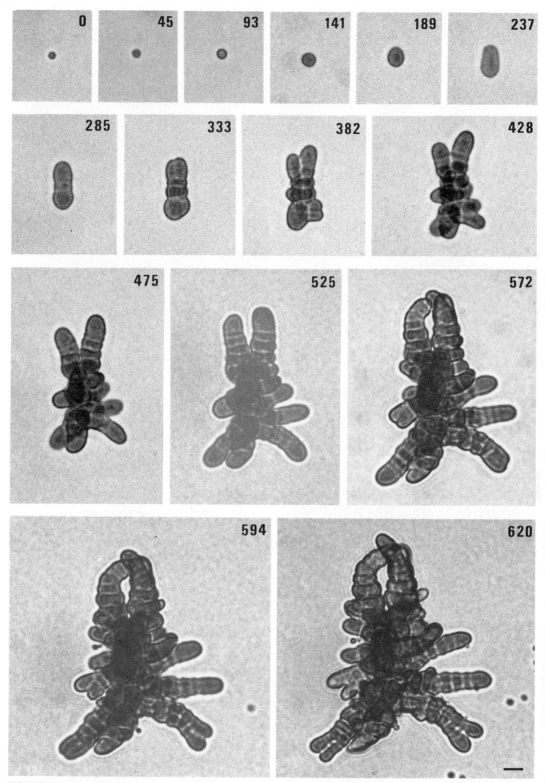

FIGURE B10.46. Development of PCC 7319, a member of cluster 1 (subgroup I) of the *Pleurocapsa*-group, in a Cooper dish culture. Numbers in the upper right-hand corner of each micrograph indicate hours following the initial observation. Bar = 10 μm. (Reproduced with permission from J.B. Waterbury and R.Y. Stanier, Microbiological Reviews *42:* 2–44, 1978, ©American Society for Microbiology.)

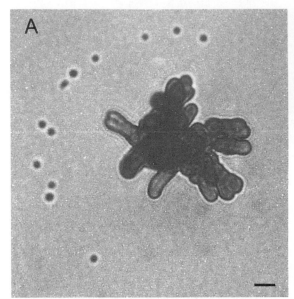

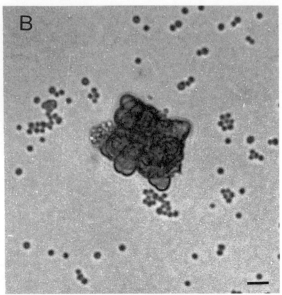

FIGURE B10.47. Cell aggregates of *Pleurocapsa*-group (PCC 7322). Some cells in the aggregate have undergone multiple fission and released baeocytes. Grown in the light, on an agar plate in mineral medium (*A*) and on an agar plate in a mineral medium supplemented with sucrose (*B*).

Bars = 10 µm. (Reproduced with permission from J.B. Waterbury. 1979. *In* Parish [ed.], Developmental Biology of Prokaryotes, Vol. 1: Studies in Microbiology. ©University of California Press.)

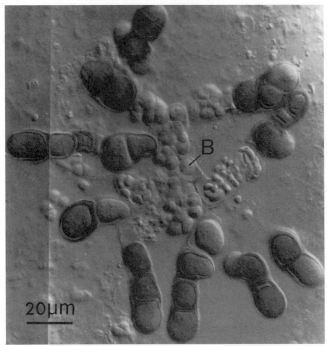

FIGURE B10.48. *Hyella gigas* (*Pleurocapsa*-group) colony from a decalcified ooid. Most of the coccoid cells have divided to produce clusters of baeocytes (*B*), which will be released from the colony by rupturing of the cell walls and dissolution of the gel sheaths. Nomarski interference contrast microscopy (Reproduced with permission of S. Golubic from Lukas and Golubic, Journal of Phycology *19*: 129–136, 1983,© Allen Press.)

TAXONOMIC COMMENTS

Cyanobacteria that display the general developmental features described for the *Pleurocapsa*-group are placed in the botanical literature in a variety of genera, including: *Pleurocapsa, Hyella,*

Myxohyella, Radaisia, Radaisiella, Solentia, Hormathonema, and *Cyanostylon.* The distinctions between these genera are made on the basis of characteristics not easily determinable with cultures or characteristics that merely reflect physical and biological responses of a given cyanobacterium to the environment. Waterbury and Stanier (1978) therefore concluded that the characterization of this group was insufficient to permit even generic boundaries to be established for the strains in pure culture.

The 16S rRNA gene sequences for three strains of the *Pleurocapsa*-group (PCC 7321, PCC 7327, and PCC 7516), that differ with respect to baeocyte development and ecology, have so far been determined (see Turner, 1997; Turner et al., 1999). Comparative analyses (see Fig. B10.38) reveal that the marine strain PCC 7321 (cluster 1), exhibiting symmetric baeocyte enlargement, is located on a different branch of the phylogenetic tree than the marine strain PCC 7516 (cluster 2), in which baeocytes enlarge asymmetrically. The former is most closely related to the marine *Myxosarcina* PCC 7312, whereas the latter clusters most tightly with strain PCC 7301, a marine isolate of *Stanieria* (ex *Dermocarpa sensu* Waterbury 1989). Further demonstrating the need to distinguish taxonomic entities on the basis of ecological aspects, the hot spring isolate strain PCC 7327 (cluster 3) of the *Pleurocapsa*-group is located in a different clade of the 16S rRNA tree and clusters most closely with *Cyanothece* strain PCC 7424, a unicellular rod-shaped cyanobacterium dividing exclusively by binary fission (for more details on this genus, see description in Subsection I).

The relatedness of strains PCC 7312 and strain PCC 7321 may not be very surprising, since the primary phenotypic characteristic that distinguishes *Myxosarcina* from the *Pleurocapsa*-group is the regularity of successive planes of binary fission and lack of pseudofilamentous appendages. Given that within members of the *Pleurocapsa*-group the degree of pseudofilamentous extensions may vary with growth conditions, it is reasonable to assume that misidentifications could occur if this discriminatory property is not expressed (see also Taxonomic Comments to the genus

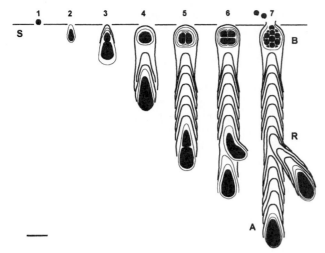

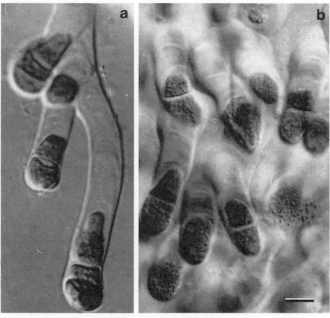

FIGURE B10.49. Schematic presentation of the life cycle of a *Solentia* (*Pleurocapsa*-group): An initially motile, gliding baeocyte settles on the carbonate substrate (*1*), excretes polysaccharide envelope and penetrates into the substrate (*2*); this cell increases in size, assumes an asymmetric clavate shape, and divides transversally into a larger cell distal, and a smaller cell proximal, to the substrate surface (*S*); its envelope bursts open at the distal end (*3*); both cells increase in size but only the distal cell continues to bore, becoming the apical cell of a pseudofilament composed of cup-shaped downward opening envelopes (*4*); the clavate apical cell continues to divide transversally, while the isodiametric proximal cell changes the plane of division (*5*); the subapical cell forms its own envelope, slips out of the alignment (*6*) forming a false branch (*R*). While the apical cells (*A*) continue to bore and divide, the proximal cell undergoes multiple fission, producing and releasing baeocytes (*B*) at the substrate surface (*7*). Bar = 10 mm. (Reproduced with permission of T. Le Campion-Alsumard from T. Le Campion-Alsumard et al., Algological Studies *83*: 107–127, 1996, ©Schweizerbart'sche.)

FIGURE B10.50. *Solentia sanguinea* (*Pleurocapsa*-group) cultured on agar maintains its natural growth habit: *a*, a pseudofilament with multiple ramifications. All apical cells are in the process of transversal cell fission. *b*, growth of several *Solentia sanguinea* pseudofilaments on agar. Nomarski interference contrast. Bar = 10 μm. (Reproduced with permission of S. Golubic from Golubic et al., 1996, Algological Studies *83*: 291–301, 1996, ©Schweizerbart'sche.)

Myxosarcina). In contrast, the close relationship between *Pleurocapsa*-group strain PCC 7516 and *Stanieria* PCC 7301 is more difficult to understand, since the developmental patterns in these two baeocyte-forming isolates differ extensively, the latter strain being easily distinguished from the former both by its unicellular mode of growth and by reproducing exclusively by multiple fissions. Consequently, as mentioned before, the definitive setting of boundaries for genera and species among baeocyte-forming cyanobacteria will require many more carefully characterized isolates, and extensive further studies at the genotypic level.

Subsection III. (Formerly **Oscillatoriales** Elenkin 1934)

RICHARD W. CASTENHOLZ, ROSMARIE RIPPKA, MICHAEL HERDMAN, AND ANNICK WILMOTTE

FURTHER DESCRIPTIVE INFORMATION

It is now apparent from multiple 16S rDNA sequence analyses that this subsection is not phylogenetically coherent. Some of the genera of this former order and even named species of the genus *Oscillatoria*, for example, are polyphyletic (see Turner, 1997; Phylogenetic relationships amongst the *Cyanobacteria*, p. 487). It is premature to reclassify this "oscillatorian" group of morphotypes on the basis of the still incompletely determined genetic relationships. The present treatment, therefore, will follow the older, artificial grouping of Subsection III, although the degree of genetic similarities of some genera (on the basis of nucleotide and other sequence similarities) are mentioned in the text description of each genus and also in the discussion in the chapter on phylogeny.

Phenotypic Characteristics of Subsection III. Subsection III (*"Oscillatoriales"* in the traditional sense) includes almost all filamentous cyanobacteria that undergo **binary fission only in a single plane at right angles to the long axis and are, therefore, uniseriate**. They produce only "vegetative" cells, with the exception, in some cases, of patterned "necridial" cells (biconcave "separation discs") which result from cell lysis phenomena and provoke transcellular trichome fragmentation. Trichome breakage may also result from intercellular separation. The terminal cell of a trichome may be tapered, with or without a cap or calyptra, but in a few forms the taper may also be carried through several subterminal cells and sometimes the end of the trichome is distinctly hooked. **Heterocysts and akinetes do not occur**. Trichome diameters may range from about 1 to (rarely) >100 μm. Generally the trichome diameter within a strain varies little (<10%)

in contrast to some other subsections of cyanobacteria described here. Trichomes of small diameter (e.g., 1–3 μm) are generally composed of isodiametric or elongate cells, whereas trichomes of large diameter (>7–8 μm) possess discoid cells that are shorter than they are broad. As a rule, all cells retain the ability to divide, although the terminal cell in some strains does not. Trichomes may be flexible or semirigid. In some cases, the entire trichome is wound into a loose or tight helix; in others, only terminal portions of the trichome may be openly spiraled.

A visible sheath may be present in some genera, but even species without an apparent sheath leave behind at least a very thin, transparent collapsed sheath when moving by a gliding mechanism. Those species that commonly produce a thickened (sometimes laminated) sheath may move only slightly and occasionally within this confinement. Some ensheathed members also produce **false branches**, which result from breaks in the sheath with protrusion of the growing trichome as a loop which eventually breaks (geminate false branches), or from the breakage of the trichome before protrusion (thus, a single false branch). In both cases, no change in the division plane occurs. When short fragments of a few cells separate from the remainder of the trichome near the free open end of a sheath, these units (termed hormogonia) may move out, elongate by cell division, and eventually form new sheaths. In contrast to some members of Subsections IV and V (see Rippka et al. and Hoffman and Castenholz, this volume), these hormogonia are not structurally different from vegetative trichomes. In species without apparent sheaths, a distinct hormogonial phase is not easily detectable, since both short and longer trichome fragments (except for those composed of only a few cells) are either permanently motile or, in some members, permanently immotile.

Gliding motility, which occurs in many members, takes place only when trichomes are in contact with a solid to semisolid substrate. Movement forward or backward may or may not be accompanied by a right- or left-handed rotation of the trichome.

In the "Geitlerian" system (Geitler, 1932), generic distinctions in the family *Oscillatoriaceae* (equivalent to Subsection III) are based primarily on sheath properties and other morphological characters, some of which may be trivial. These characters are still used as the main diagnostic characters in the present edition, but increased knowledge of physiology and biochemistry, as well as nucleotide sequence analyses of various genes (mainly 16S rDNA), has already demonstrated the need for drastic revisions in the future (see Wilmotte and Herdman, this volume).

Cyanobacteria included in Subsection III occur in an enormous diversity of habitats: freshwater and marine, as plankton, benthic mats, and periphyton. Terrestrial crusts, mats, and turfs are also common. Hot spring mats of some Subsection III cyanobacteria develop at temperatures of up to ~62°C. Both Antarctic and Arctic freshwater ponds are usually dominated by "oscillatorian" cyanobacteria. Intimate symbiotic relationships are unknown among these filamentous forms.

The physiology of some species of "oscillatorian" cyanobacteria, including several of the planktonic forms (e.g., *Limnothrix* and *Planktothrix*), has been studied (e.g., Van Liere and Walsby, 1982; Mur, 1983, Ahlgren, 1985). Most species are obligate photoautotrophs; some are facultative photoheterotrophs (Rippka et al., 1979; Rippka and Herdman, 1992). Among the latter, a few are also capable of very slow growth under chemoheterotrophic conditions (Rippka, unpublished), either by aerobic or fermentative metabolism (see Stal, 1995). Many strains are capable of fixing N_2, but generally this process is oxygen sensitive. Thus N_2 fixation occurs only when the medium is purged of O_2 or during darkness, when lack of photosystem II activity and occurrence of enhanced respiration leads to a reduction in ambient O_2 (Stal, 1995). A few strains of *Oscillatoria* and an isolate of *Symploca* are able to fix N_2 in the presence of atmospheric concentrations of oxygen, even if incubated under a continuous light regime (Rippka, unpublished). The same is true for marine planktonic forms. *Trichodesmium* species that grow in bundles and fix N_2 in nature, even in O_2-rich sea water (Bergman et al., 1997). Sulfide-dependent anoxygenic photosynthesis is known in several species of oscillatorian cyanobacteria (see Cohen et al., 1975; 1986; Castenholz and Utkilen, 1984). In thermophilic *Leptolyngbya* (ex *Oscillatoria*) cf. *amphigranulata* only reduced nitrogen compounds can be used as a source of nutritional nitrogen (Castenholz, unpublished).

TAXONOMIC COMMENTS

The cyanobacteria of Subsection III are here grouped in a context similar to Geitler's usage of the family *Oscillatoriaceae* (Geitler, 1932) and as used in the 1989 edition of *Bergey's Manual of Systematic Bacteriology* (Castenholz, 1989a). Recent "botanical" revisions of this group by Anagnostidis and Komárek (1988) and Anagnostidis (1989) have defined new genera, some of which are adopted here. However, in most cases the primary distinguishing features between genera are clearly insufficient to accommodate species (or strains) of widely differing ecology (fresh water, soil, hot springs, salt marshes, etc). Furthermore, separation is particularly difficult and arbitrary when gradations in cell size and sheath development exist. Some generic names, therefore, merely emphasize "pinnacles" in what often appears to be a continuum of phenotypic properties. It follows that all generic categories included or excluded here are provisional and, in many cases, artificial with respect to phylogenetic relationships (see Wilmotte and Herdman, this volume). Thus, they may be referred to as form-genera or "genera pro tem" that ultimately will have to be redefined on the basis of the morphological and genetic properties of cultivated or noncultivated type species.

Additional comments on the taxonomy of the traditional grouping of these morphotypes are included in the first edition of *Bergey's Manual of Systematic Bacteriology* (Castenholz, 1989a).

Key to the form-genera of Subsection III

I. Trichomes not cylindrical (flattened or triradiate in cross-section).
    A. Trichomes flattened, i.e., elliptical in cross-section.
                Form-genus *Crinalium*, p. 543
    B. Trichomes triradiate in cross-section.
                Form-genus *Starria*, p. 559
II. Trichomes cylindrical.
    A. Trichomes helically coiled (open or closed), usually for entire length; motility by rotation.

1. Trichome helix usually nearly closed (i.e., spring-like coil); cross walls thin and usually invisible with light microscope; cell width typically 2–4 µm, permanently motile by rotation.

Form-genus *Spirulina*, p. 557

2. Trichome helix usually open (i.e., as stretched spring); cross walls visible with light microscopy; cell width typically 6–12 µm; gas vesicles generally present.

Form-genus *Arthrospira*, p. 542

B. Trichomes straight or, if sinuous, for only a portion of length.

1. Trichomes more or less ensheathed and, if motile, do not seem to rotate; some members produce short hormogonia that migrate out of the sheath; in others a distinct hormogonial phase cannot be recognized.

a. Sheath persistent and thick (usually >1 µm); single trichome per sheath; trichome width typically >8 µm; motility restricted to hormogonia.

Form-genus *Lyngbya*, p. 547

b. Sheath persistent but thin (usually <0.2 µm), sometimes only visible when cells are missing and gap is produced; trichome width typically <6 µm.

i. In nature, common sheath regularly harbors 2 or more parallel trichomes; in culture trichomes generally only united by mucilage; filaments exhibit slow motility; hormogonial phase not distinct.

Form-genus *Microcoleus*, p. 548

ii. Single trichome per sheath (trichome <6 µm wide); slow or transient motility. Filaments with upright aerial growth, forming tufts; trichome width 4–6 µm, aerobic N_2 fixation; motility restricted to hormogonia.

Form-genus *Symploca*, p. 559

iii. Filaments not as above; slow or transient motility; trichome diameter <3 µm.

Form-genus *Leptolyngbya*, p. 544

2. Trichomes permanently motile or nonmotile; motility may be by rotation or by "creeping"; a hormogonial phase not readily evident; no persistent sheath (although gliding trichomes may leave nearly transparent, mucous or sheath-like trail).

a. Cells longer than broad.

i. Trichomes with slow (<1 µm/s) or no motility. Constriction at cross-walls present, but minor (1/5 to <1/8 of cell diameter); 1–3 µm broad; gas vacuoles usually present, both in polar and intracellular positions. Contains phycobilin pigments in addition to chlorophyll *a*.

Form-genus *Limnothrix*, p. 546

ii. Trichomes with slow (<1 µm/s) or no motility. Constriction at cross-walls present, but minor (1/5 to <1/8 of cell diameter; 1–3 µm broad; gas vacuoles usually present, both in polar and central positions. Lacks phycobilin pigments, but contains chlorophyll *b* in addition to chlorophyll *a*.

Form-genus *Prochlorothrix*, p. 554

iii. Trichomes with slow (<1 µm/s) or no motility. Constrictions at cross-walls pronounced (exceeding 1/5 of trichome diameter); polar gas vacuoles generally present.

Form-genus *Pseudanabaena*, p. 554

iv. Trichomes with high rate of motility (>1 µm/s); cell width 1–4 µm.

Form-genus *Geitlerinema*, p. 544

b. Cells shorter than broad or isodiametric (4 to >80 µm broad).

i. Trichomes consistently short (2–30 cells); constrictions at cross-walls prominent; motility slow or lacking.

Form-genus *Borzia*, p. 543

ii. Trichomes not consistently short, except under conditions of frequent breakage, generally without pronounced constrictions at cross-walls. 4–100 µm broad trichome, lacking conspicuous gas-vesicles; permanently motile by rotation.

Form-genus *Oscillatoria*, p. 550

iii. Trichomes not consistently short, except under conditions of frequent breakage, generally without pronounced constrictions at cross-walls. 3.5–6 µm wide, abundant gas vesicles; fresh water, planktonic.

Form-genus *Planktothrix*, p. 553

iv. Trichomes not consistently short, except under conditions of frequent breakage, generally without pronounced constrictions at cross-walls. Cells isodiametric or longer than broad; marine planktonic, PUB-containing, size from 6 to 22 μm wide.

Form-genus *Trichodesmium*, p. 560

v. Trichomes not consistently short, except under conditions of frequent breakage, generally without pronounced constrictions at cross-walls. Cells contain a reticulated network of cytoplasmic strands containing the thylakoids; freshwater or marine.

Form-genus *Tychonema*, p. 561

Form-genus I. **Arthrospira** Stizenberger 1852

RICHARD W. CASTENHOLZ, ROSMARIE RIPPKA, MICHAEL HERDMAN AND ANNICK WILMOTTE

Ar.thro.spi′ra. Gr. n. *arthron* a joint; L. n. *spira* a coil; M.L. fem. n. *Arthrospira* jointed coil.

The entire trichome is arranged as an **open helix**, in which cross-walls may be seen via light microscopy, but not too easily when gas vacuoles are abundant (Fig. B10.51). However, the vesicles can be easily collapsed by centrifugation or by exerting pressure on a suspension in a plastic syringe by pushing against a rubber stopper. Cells are generally shorter than broad or isodiametric. Slight constrictions at crosswalls may be present or absent. In some strains the terminal cells may acquire a cap-like thickening of the outer cell wall, termed a calyptra. A single circle of junctional pores (i.e., near crosswalls) seems to be typical for members of this genus, and fimbriae are closely appressed along the trichome (E.M.) (Guglielmi and Cohen-Bazire, 1982a, b). Persistent sheaths are not produced. Gliding motility is evident in most strains. Thylakoids have a radial arrangement in trichome cross-section. Trichome width among the numerous isolates in culture varies from about 5–12 μm, although smaller forms (3 μm) have been described from nature. The helix is an open spiral with wave amplitudes ranging from 30 to 60 μm, with interhelical spacing of similar dimensions (Ciferri, 1983; Ciferri and Tiboni, 1985). On solid medium, the helix undergoes a transition to a "flat spiral". Considerable variation in degree of helix pitch occurs within some strains, and culture variants that are nearly straight are also found (Jeeji-Bai, 1985; R. Rippka, unpublished).

FURTHER DESCRIPTIVE INFORMATION

The physiology of some strains has been studied extensively (see Ciferri, 1983; Vonshak, 1997). PCC strain 7345 is a gas-vacuolate euryhaline organism. It is an obligate photoautotroph, as are all other known strains, and is unable to synthesize nitrogenase

aerobically or anaerobically. It contains C-PC and APC, but no C-PE or PEC. This isolate was identified by R. Lewin as *A. (ex Spirulina) platensis* (Nordst.) Gomont, but morphologically similar strains have also been assigned to *Arthrospira (ex Spirulina) maxima* Setchell and Gardner. Much attention has been paid to members of these two "species" as sources of human protein (Ciferri, 1983). Given the close genetic relationships between these strains revealed in a recent study by Scheldeman et al. (1999), it is likely that strains assigned to either *A. platensis* or *A. maxima*, and to some other additional "species", may represent one major genotype with only minor internal distinctions, but with a wide range of morphological variability. The "life cycle" of *Arthrospira* in laboratory culture involves the breaking up of trichomes at the sites of a necridium (lysing cell) at intervals of every 4–6 cells (Ciferri, 1983). The resulting short and uncoiled hormogonia form a migratory phase. Each hormogonium then undergoes cell division, growing into a new helical trichome.

This group with spirally formed trichomes and visible crosswalls, known interchangeably as *Spirulina* or *Arthrospira*, has been found mainly in marine, brackish water, and saline lake environments of tropical and semitropical regions, but some representatives are also thought to occur in freshwater (Geitler, 1932). Many isolates have been used in aquaculture. Most forms are planktonic, producing gas vesicle clusters that are dispersed throughout the cells. However, synthesis of these buoyant structures may vary with growth conditions or may be lost due to mutation. Some species that lack gas vesicles have been described, but these reports have not been confirmed by studies in culture. Cyanobacteria of the genus *Arthrospira* often dominate the plankton of warm lakes with high carbonate/bicarbonate levels and pH levels as high as 11. The biology of *Arthrospira platensis* is the subject of a comprehensive book (Vonshak, 1997). The use of the nomen *Spirulina* for the strains used as food supplements is inappropriate.

Culturing Isolation is achieved by micromanipulation of single gliding filaments; various media (e.g., "*Spirulina*" medium: Schlösser, 1994) may be used, including the common medium BG-11, supplemented with either bicarbonate (10 mM) alone, or in combination with, various amounts of additional salts (NaCl, $MgSO_4$, $MgCl_2$, KCl, and $CaCl_2$) for samples originating from more saline sources.

Reference strain PCC 7345 (UTEX 1928), as *Arthrospira platensis*, isolated from a saline marsh, Del Mar Slough, California, 1969 (Rippka et al., 1979). Mol% G + C = 44.3 (Herdman et al., 1979b). The trichome diameter is 8–9 μm.

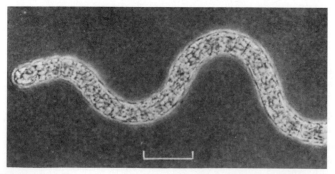

FIGURE B10.51. *Arthrospira* (*Spirulina*) PCC 7345. Bar = 20 μm. (Reproduced with permission from R. Rippka et al., Journal of General Microbiology *111:* 1–61, 1979, ©Society for General Microbiology.)

TAXONOMIC COMMENTS

Arthrospira has frequently been submerged in the genus *Spirulina* (Geitler, 1932) or in the genus *Oscillatoria* along with *Spirulina* (Bourrelly, 1985). On the basis of ultrastructural properties (Guglielmi and Cohen-Bazire, 1982a, b) and major differences in mean DNA base composition (Herdman et al., 1979b; Herdman, unpublished) it is clear that the smaller, coiled *Spirulina* strains lacking crosswalls discernible by light microscopy constitute an independent taxonomic unit. This is also supported by 16S rDNA sequence data, which show that the two *Arthrospira* strains examined (PCC 7345 and PCC 8005) are more closely related to a strain of *Lyngbya* (PCC 7419) and some species of the genus *Oscillatoria* than to the tightly or less tightly coiled, much smaller *Spirulina* strains (Nelissen et al., 1994; Wilmotte and Herdman, this volume). In agreement with phylogenetic affiliations, the patterns of junctional pores and fimbriae in *Arthrospira* show great structural similarities to strains of *Oscillatoria* (*sensu* Rippka

et al., 1979) and an isolate of *Lyngbya* (as defined here) (Guglielmi and Cohen-Bazire, 1982a, b). However, given the large morphological and genetic diversity among members of the genus *Oscillatoria*, some of which may eventually attain independent generic status (see Wilmotte and Herdman, this volume), it seems preferable to preserve the genus *Arthrospira*. All members of *Arthrospira* in culture are highly related, as shown in a recent study by Scheldeman et al. (1999). These authors compared the Internally Transcribed Spacer (ITS) region between the 16S and 23S rDNA, after amplification and hydrolysis by various restriction enzymes, of about fifty strains of *Arthrospira* from four continents. Only two main clusters, without correlation to geographic origin or different specific assignments, were revealed by this analysis. Given that one strain from each cluster (PCC 7345 and PCC 8005) have been shown to share 99.7% sequence similarity (Nelissen et al., 1994), it is highly likely that all *Arthrospira* strains are representatives of a single nomen species.

Form-genus II. **Borzia** Cohn ex Gomont 1892

RICHARD W. CASTENHOLZ, ROSMARIE RIPPKA AND MICHAEL HERDMAN

Bor'zi.a. M.L. fem. n. *Borzi* named after A. Borzi, Italian cyanobacteriologist.

This genus is characterized by **very short trichomes of 2–8 (16) cells,** with conspicuous constrictions at crosswalls, of various diameters; motility slight or lacking; **cells usually shorter than broad.** This last characteristic seems to be the main feature separating this genus from *Pseudanabaena* (as represented in this volume), since many strains of *Pseudanabaena* consistently exhibit comparably short trichomes. Brevitrichomy may also occur in other filamentous cyanobacteria (Pinevich et al., 1997), and mutants with short trichomes have been isolated from heterocystous

forms such as *Anabaena* (Bauer et al., 1995). Thus this property may be trivial, which could be demonstrated by genetic analyses on typical representatives of this "genus". *Hormoscilla* Anagnostidis and Komárek, 1988 is very similar to *Borzia* and will not be considered here. A species isolated by E.G. Pringsheim and conforming to the description of *Borzia trilocularis* (Pringsheim, 1968) exists in culture strain SAG B 1407–1, but has been assigned to *Hormoscilla pringsheimii* (Schlösser, 1994).

Form-genus III. **Crinalium** Crow 1927

RICHARD W. CASTENHOLZ, ROSMARIE RIPPKA AND MICHAEL HERDMAN

Cri.na' li.um. L. adj. *crinalis* of hair; Gr. suff. *-ion* diminutive; M.L. neut. n. *Crinalium* small hairlike (filament).

Members of this genus share structural similarities with *Oscillatoria*. **The trichome is flattened to elliptical in cross-section** rather than cylindrical (or triradiate as in *Starria*). The only representative in axenic culture (strain PCC 9333), described as *Crinalium epipsammum* (SAG 22.89), has short cells (1–1.5 μm) that are 2–2.5 μm in narrow width and 5–7 μm in breadth (De Winder et al., 1990). The terminal cells are undifferentiated. Binary fission as in *Oscillatoria*. Trichomes are nonmotile. As seen in transmission electron microscopy, thylakoids are mainly parallel to the long axis and close to the cell periphery. The cell wall peptidoglycan layer is thick (80–100 nm). Reported to have a mean DNA base composition of 33.9 mol% G + C and to contain high levels of cellulose II, associated with the cell envelope (de Winder et al., 1990). A thin sheath may exist. No PE; β-carotene and echinenone are the major carotenoids. There is no N_2 fixation under anoxic or oxic conditions. Apparently common in terrestrial sandy areas as surface crusts and is desiccation-resistant. Two other species from field collections have been characterized poorly (see Desikachary, 1959).

FURTHER DESCRIPTIVE INFORMATION

Culturing Isolation by micromanipulation on agar plates. The optimum growth temperature is ~25°C. Maintained in medium BG-11 containing 2 mM $NaNO_3$, or in medium Z8, under low light intensity.

Reference strains PCC 9333, an axenic isolate of *Crinalium epipsammum* (SAG 22.89), isolated from sandy crusts in the Netherlands.

TAXONOMIC COMMENTS

Although immotile and distinctive in morphology, *Crinalium* has many characteristics in common with strains of *Oscillatoria* (as defined here), including a thick cell wall peptidoglycan layer and junctional pores. However, phylogenetic analysis indicates that strain PCC 9333 is not closely related to members of the latter genus (Wilmotte and Herdman, this volume).

Form-genus IV. **Geitlerinema** *stat. nov. (Anagnostidis and Komárek 1988) Anagnostidis 1989*

RICHARD W. CASTENHOLZ, ROSMARIE RIPPKA AND MICHAEL HERDMAN

Geit' ler.i.nema. Geitler Lothar Geitler, Austrian cyanobacterial taxonomist, 1899–1990; Gr. n. *nema* thread; M.L. fem. n. *Geitlerinema* Geitler's thread or filament.

Filamentous cyanobacteria which form straight to slightly sinuous cylindrical trichomes that divide by binary fission; trichome breakage transcellular. **Trichomes less than 5–6 μm in diameter; cells longer than broad or isodiametric, never disc- or coin-like; constrictions between adjacent cells are absent or very shallow;** apical cells are rounded, conical or distinctly pointed, tapered, and often bent. Polar gas vesicles rarely present; large cyanophycin granules near crosswalls in some species. Ultrastructural studies of strain PCC 7105 demonstrated a single circular row of pores near the crosswalls (Guglielmi and Cohen-Bazire, 1982a); whether this holds for the other strains assigned to this genus is unknown. Trichomes of freshly isolated cultures or observed in field material show active gliding motility accompanied by trichome rotation, sometimes leaving behind an extremely thin trail (or almost invisible sheath). Copious amounts of gel-like matter may be produced by some strains, particularly in axenic liquid culture.

The mol% G + C of the DNA is: 53–67 (Rippka and Herdman, 1992).

FURTHER DESCRIPTIVE INFORMATION

The strains presently in culture were obtained from a variety of hypersaline, marine, and freshwater sources, including hot spring habitats with temperatures as high as 55°C. Given this ecological diversity, it is obvious that *Geitlerinema* is a genus "pro tem" (a form-genus) that may require more than one generic subdivision once phylogenetic analyses are performed for a large number of representatives.

Culturing *Geitlerinema* encompasses a large number of ecospecies, but since all possess active gliding motility, the best method for isolation is self-isolation on agar (Castenholz, 1988a, b). Many types of media are acceptable, depending on the source of the sample (see General Characteristics of the Cyanobacteria).

Reference strains PCC 7105, a true marine strain; PCC 7407, a freshwater strain, but capable of growth in marine media (Rippka and Herdman, 1992; Rippka, unpublished); PCC 9228 (formerly *Oscillatoria limnetica*), isolated from a hypersaline lake (Cohen et al., 1975). About 12 additional strains in the PCC, including a marine isolate previously identified as *Microcoleus* 10mfx (Wilmotte, 1994), have also been tentatively assigned to this genus. Several thermophilic strains of *Oscillatoria* cf. *terebriformis* Agardh 1827 in the University of Oregon Culture Collection (e.g., CCMEE OH-93-Ot, OH-80-Ot, SA-Alwagh-Ot) should also be referred to the genus *Geitlerinema* (Fig. B10.52).

TAXONOMIC COMMENTS

This form-genus was created from the morass of the LPP-B "oscillatorian" group (*sensu* Rippka et al., 1979) almost simultaneously as a subgenus of *Phormidium* (*sensu* Anagnostidis and Komárek, 1988) and as *"Oscillothrix"* by Rippka (1988b). Anagnostidis (1989) raised the subgenus *Geitlerinema* to genus level, a status and name later adopted in the PCC (Rippka and Herdman, 1992). The representatives of this genus differ from *Leptolyngbya* by very active gliding motility and lack of sheath envelopes. They may be distinguished from *Oscillatoria* by cell shape (longer than broad) and, with some exceptions, by the typically narrow trichomes (usually <5 μm). Previously, most cyanobacteria possessing the characteristics of *Geitlerinema* were considered species of *Oscillatoria*, and in some cases, species of *Microcoleus* or *Phormidium*. Phylogenetically, *Geitlerinema* PCC 9452 (*Microcoleus* strain 10 mfx) clusters fairly closely with *Geitlerinema* strain PCC 7105, in agreement with their similar trichome morphology and cell size (2–3 μm), but not with marine isolates typical of *Microcoleus chthonoplastes* such as strain PCC 7420 (Garcia-Pichel et al., 1996; Turner, 1997). In contrast, the halophile *Geitlerinema* PCC 9228 (Solar Lake isolate) is located distantly from both the marine strains of *Geitlerinema* (PCC 7105 and PCC 9454) and the *Microcoleus* cluster (see also Wilmotte and Herdman, this volume).

FIGURE B10.52. *Geitlerinema* (*Oscillatoria*) cf. *terebriformis* (strain University of Oregon OH-51–Ot, Clone 1) from hot springs. Nomarski interference contrast optics. Normal trichome tip (*arrow*). Photograph by D.C. Nelson. Bar = 10 μm. (Reproduced with permission from R.W. Castenholz, Mitteilungen Internationale Vereinigung für Limnologie *21:* 296–315, 1978, ©Schweizebart'sche Verlag.)

Form-genus V. **Leptolyngbya** *Anagnostidis and Komárek 1988*

RICHARD W. CASTENHOLZ, ROSMARIE RIPPKA, MICHAEL HERDMAN AND ANNICK WILMOTTE

Lep.to.lyng' by.a. Gr. adj. *leptos* delicate or thin; *lyngbya* named after H.C. Lyngbye, Danish botanist, 1782–1837; *Lyngbya* another genus of cyanobacteria; M.L. fem. n. *Leptolyngbya* delicate or thin Lyngbya.

Cylindrical trichomes less than 3 μm in diameter; cells isodiametric or longer than wide; constrictions between cells are gen- erally absent or shallow, seldom exceeding 1/8 of the trichome diameter; trichome breakage may be trans- or inter-cellular; apical cells rounded (or exceptionally conical) (Fig. B10.53); **persistent sheaths may be produced** but are not of firm consistency (and not laminated); in some members, several trichomes may be surrounded by a "mucous-like" matrix. **Trichomes nonmotile (or inconspicuously motile as seen by light microscopy, <0.1 μm/ s), except for transient motile hormogonia when free of any sheath or coalescent "mucous" material.** Some strains or natural specimens may undergo false branching, which may depend mainly on sheath strength.

FURTHER DESCRIPTIVE INFORMATION

The form-genus *Leptolyngbya* harbors a large number of ecologically diverse filamentous cyanobacteria, whose common traits are primarily a **narrow trichome diameter** and **lack of conspicuous motility**, the latter property distinguishing them from the very similar form-genus *Geitlerinema*. The strains in culture have been isolated from a variety of hypersaline, marine, and freshwater sources, including hot spring habitats with temperatures as high as 62–63°C.

Leptolyngbya is a genus "pro tem" (a form-genus) that may require generic subdivisions once more extensive phylogenetic analyses are performed. Not surprisingly, phylogenetic analyses have already demonstrated a high degree of internal genetic diversity (Wilmotte, 1994; Wilmotte and Herdman, this volume).

Culturing Since this form-genus includes forms from a great variety of habitats, many methods of isolation and maintenance may be appropriate. However, since gliding motility is absent or slow, micromanipulation from agar plates or from liquid in wells may be most successful. Outgrowth on agar may also allow clonal isolation (see Castenholz, 1988a, b; Rippka, 1988a).

Reference strains

Cluster 1 PCC 6306 (formerly of LPP group B), freshwater, source unknown; collected before 1952; *Plectonema notatum*; host to cyanophage LPP-1.

Cluster 3 PCC 7104 (formerly of LPP group B), coastal freshwater or brackish, N.E. coast USA, 1960; BG-11 medium.

Cluster 4 PCC 7375 (formerly LPP-B), PE-rich, marine plankton, N.E. coast USA, 1952; ASN-III B12 medium (obligate B_{12} requirement); *Phormidium ectocarpi* (from R.A. Lewin). Another well-known strain that would currently fit into this assemblage is the thermophilic isolate UO-CCMEE NZ-Concert-Oa 1, referred to in the literature as "*Oscillatoria* cf. *amphigranulata*", isolated in 1971 from a New Zealand hot spring (Fig. B10.54).

TAXONOMIC COMMENTS

There are over 50 cultured strains currently assigned to this form-genus in the PCC. They differ with respect to phycobiliprotein composition, most lacking phycoerythrinoid pigments, but some synthesizing PE and undergoing chromatic adaptation. A few marine strains constitutively produce high levels of PE and are bright red in color. Some fix nitrogen anaerobically; many are hosts to cyanophage LPP-1 (Rippka and Herdman, 1992). Most of these strains would have traditionally been identified as species of *Lyngbya* Agardh 1824, *Phormidium* Kützing 1843, *Plectonema* Thuret 1875, (if false branching is frequent), or even *Oscillatoria* Vaucher 1803. Rippka et al. (1979) grouped the PCC strains that were then available into the LPP group, the name of which is based on the initials of the first three genera. Anagnostidis and Komárek (1988) and Anagnostidis (1989) drastically revised the order *Oscillatoriales* and created several new genera, including *Leptolyngbya*, which was to accommodate almost all species of small trichome diameter previously assigned to one or other of the three "classical" genera mentioned above. For now, the form-genus *Leptolyngbya* simply serves as a repository for small-celled "oscillatorian" cyanobacteria without evident motility that are difficult to classify. The accumulation of genetic evidence will eventually result in further generic subdivisions, since several discreet phylogenetic lineages of *Leptolyngbya* seem to be emerging (Turner, 1997; Wilmotte and Herdman, this volume).

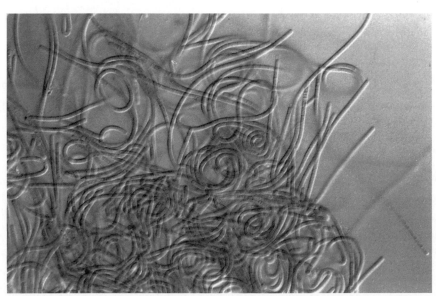

FIGURE B10.53. *Leptolyngbya* sp. (CCMEE Ant-Brack-2-O). Trichome width 2.5–2.8 μm. From slightly saline Brack Pond, McMurdo Ice Shelf (near Bratina Island), Antarctica, 1998.

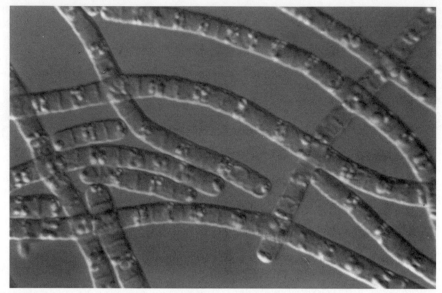

FIGURE B10.54. *Leptolyngbya* (*Oscillatoria* cf. *amphigranulata*) sp. (CCMEE NZ-Concert-"Oa", Cl 1). Trichome width ~2.1 μm. The granules centrally located in most cells are cyanophycin. From hot spring in Rotorua, New Zealand, 1971.

Form-genus VI. **Limnothrix** *Meffert 1988*

RICHARD W. CASTENHOLZ, ROSMARIE RIPPKA, MICHAEL HERDMAN AND ANNICK WILMOTTE

Lim' no.thrix. Gr. n. *lime* lake; Gr. n. *thrix, trichos* thread or hair; M.L. fem. n. *Limnothrix* lake hair or lake thread.

The members of this genus, although morphologically similar and apparently phylogenetically closely related to *Pseudanabaena* spp., deserve separate mention because they are typical constituents of the freshwater plankton (Meffert, 1987). Trichomes composed of elongate cells (although sometimes isodiametric), typically 1–3 μm in diameter (exceptionally up to 6 μm); constrictions between cells indistinct or shallow. Apical cells rounded. **Generally synthesis of conspicuous gas vacuoles near both cell poles and in the center of predivisional cells, but sometimes dispersed throughout the cells** (Fig. B10.55). **Motility lacking or very slight**; reproduction by random intercellular trichome breakage. With T.E.M., thylakoids are mainly peripheral.

FURTHER DESCRIPTIVE INFORMATION

Although some members resemble *Prochlorothrix* morphologically, the genus can be distinguished by the lack of chlorophyll *b* and possession of APC and PC in *Limnothrix* strains. In addition, some synthesize PE and may or may not undergo complementary chromatic adaptation.

Culturing Isolation by removal of trichomes from agar plates after a series of washings, filtrations with membrane filters of 8 μm pore size, and dilutions (see Meffert and Chang, 1978 for details); Medium Z8, or BG11₀ containing sodium nitrate (2 mM) (Rippka, 1988a).

Reference strains PCC 9416 (SAG 3.89), a PE-rich axenic culture of *L. redekei* strain Meffert N8 7206E, isolated from plankton (Edebergsee, E. Holstein, Germany; Meffert and Chang, 1978).

TAXONOMIC COMMENTS

The taxon *Limnothrix* as defined here is based mainly on the studies by Meffert (Meffert and Oberhäuser, 1982; Meffert, 1987, 1988) and includes isolates (or descriptions) of species previously assigned to *Oscillatoria* (e.g., *O. redekei* van Goor, *O. amphigranulata* van Goor, *O. planktonica* Wolosz., *O. rosea* Utermöhl). The only strain of *Limnothrix* analyzed (Meffert 6705) clusters in the 16S rRNA tree with strains of *Pseudanabaena* (Turner, 1997; Wilmotte and Herdman, this volume). Unfortunately, this isolate is no longer available (Meffert, personal communication). Therefore, until additional isolates and sequences have been compared, it seems preferable to retain *Limnothrix* as a separate genus.

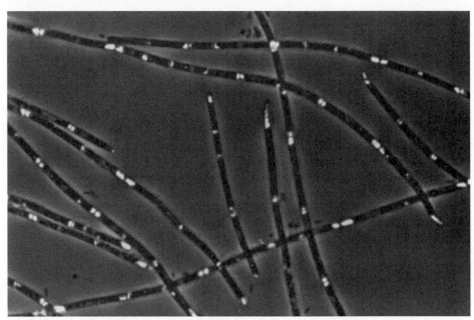

FIGURE B10.55. *Limnothrix* (*Oscillatoria*) sp. Trichome width ~1.5 μm; white areas are gas vacuoles; plankton from lake in western England. (Reprinted with permission from H. Canter-Lund and John W.G. Lund. 1995. Freshwater Algae: Their Microscopic World Explored. Biopress Ltd., Bristol, ©H. Canter-Lund.)

Form-genus VII. **Lyngbya** *Agardh 1824 (sensu* Anagnostidis and Komárek 1988)

RICHARD W. CASTENHOLZ, ROSMARIE RIPPKA AND MICHAEL HERDMAN

Lyng'by.a. M.L. fem. n. *Lyngbye* named after H.C. Lyngbye, Danish botanist, 1782–1837.

Filamentous organisms that share the entire range of cellular types with *Oscillatoria* (as here defined), but which produce a **distinct, persistent, and firm sheath** (Fig. B10.56). The sheath may be thin but can be seen with phase contrast optics, particularly where it extends beyond the terminal cell of the trichome (Fig. B10.56). The trichome diameters range from about 6 to ~80 μm. Sheaths in some species may accrete to several μm in thickness and display laminations. **Cells composing the trichomes are disk shaped (shorter than broad)**, a feature also characteristic of *Oscillatoria* (as here defined).

FURTHER DESCRIPTIVE INFORMATION

Trichomes are usually nonmotile within the sheath, but short sections of trichomes (hormogonia) sometimes move slowly when placed on new agar-solidified medium, glide free of the sheaths, and resemble *Oscillatoria* until new sheath production again immobilizes them. In some cases, rapid growth extends a portion of the trichome out of old sheaths, and these portions are sheathless for a while.

The sheaths of some strains, including reference strain PCC 7419, are quite prominent and strong, so that an entire entangled mass of filaments in liquid culture will hold together if attempts are made to remove only a small part with forceps. Laminated sheaths occur commonly in the large diameter species, in both freshwater and marine forms. A yellow UV-screening pigment (scytonemin) commonly occurs in the sheaths of some marine mat-forming species, giving the whole filament a yellow to brown color. The occurrence of this pigment and its effectiveness in preventing damage by UV radiation is reviewed by Castenholz and Garcia-Pichel (2000). In some filamentous species (i.e., *Porphyrosiphon* Kützing 1849), intense purple to red pigments occur in multilayered sheaths. Species of *Lyngbya*, described in the literature, may or may not synthesize PE. The reference strain has neither PE nor PEC, fixes N_2 under anoxic conditions, and has photoheterotrophic capability. *Lyngbya*, as here defined, has a worldwide distribution, species occurring in benthic marine and freshwater habitats. One easily identified morphotype (*Lyngbya* cf. *aestuarii* Liebman 1841) forms extensive mats in many intertidal mud flats and marshes (e.g., Javor and Castenholz, 1981). *Lyngbya* spp. also occur extensively as attached elongate tufts in subtidal, protected marine waters. The morphotype *Lyngbya* cf. *majuscula* Harvey ex Gomont 1892 has trichomes up to 80 μm in diameter with sheaths as thick as 11 μm and occurs in fresh and marine waters.

Culturing   In general, the more heavily ensheathed species of *Lyngbya* are more difficult to isolate in axenic culture than are species of *Oscillatoria*. On agar, the production of motile hormogonia is not always assured, and a large number of various contaminants are often attached to the thick sheath envelopes.

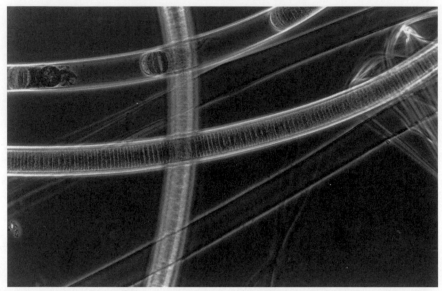

FIGURE B10.56. *Lyngbya* cf. *aestuarii.* University of Oregon strain WH 82-L (lost). Filament width 17–18 μm; some empty or partly empty sheaths shown; from marine microbial mat near Woods Hole, Massachusetts, U.S.A.

Thus, micromanipulation is generally required to obtain axenic cultures, often with the aid of antibiotic treatments (Rippka, 1988a); initial isolation and partial purification may be effected by dragging a single filament through or over agar with a watch-maker's forceps.

Reference strains PCC 7419, isolated from salt marsh, Woods Hole, Massachusetts, 1974 (Rippka et al., 1979). PCC 7419 was separated by Rippka et al. (1979) as the single member of the LPP-A group. It is characterized by a thick and persistent sheath, wide trichome (15–16 μm) and short disk-like cells. The mol% G % C content is 43.3 and the genome size is 4.58×10^9 daltons (Herdman et al., 1979a, b). It appears to be equivalent to the morphotype *L. aestuarii* Liebman 1841. The PCC has four strains, all marine, that have been tentatively assigned to this genus.

TAXONOMIC COMMENTS

The definition used here is based on the revision by Anagnostidis and Komárek (1988). They restricted this previously poorly de-fined genus to include only species with cell diameters exceeding 6 μm and possessing disc-like cells, thus corresponding to the LPP-group A (Rippka et al., 1979). This was in contrast to the treatments of Geitler (1932) and Desikachary (1959), who used much broader limits of cell diameter and cell shape. Although *Lyngbya* strain PCC 7419 is structurally very similar to large spe-cies of *Oscillatoria*, such as strain PCC 7515 (cf. *O. sancta*), the two isolates do not cluster tightly in the 16S rRNA tree, the closest relatives of the former, surprisingly, being two strains of *Arthro-spira* (Wilmotte, 1994; Wilmotte and Herdman, this volume).

Form-genus VIII. **Microcoleus** Desmazieres 1823

RICHARD W. CASTENHOLZ, ROSMARIE RIPPKA AND MICHAEL HERDMAN

Mi.cro.co'le.us. Gr. adj. *mikros* small (or thin); Gr. n. *koleos* sheath; M.L. masc. n. *Microcoleus* small or thin sheath.

In members assigned to this form-genus, **two to several trichomes oriented parallel**, often spirally and tightly interwoven, are **en-closed by a common homogeneous sheath** (Fig. B10.57). Cells (diameter 3–6 μm) composing the trichomes are **longer than wide** (or isodiametric); **visible constrictions at the cross-walls** (but usually much less than 1/4 of the cell diameter); **mature end cells are conical**; the bundles of filaments **generally exhibit slow gliding motility, not involving rotation**. Based on analysis of a limited number of isolates, a **radial arrangement of thylakoids** (in cross-section, T.E.M.) seems to be **typical**.

FURTHER DESCRIPTIVE INFORMATION

Several species from marine and freshwater have been described (Geitler, 1932), but the restricted definition of this genus is based on a single species (*M. chthonoplastes* Thuret), extensively studied using field specimens and an important component of intertidal

mats, including mats of hypersaline waters (up to ~100 g/l) (Javor and Castenholz, 1981). Although cultures (including axenic strains) have been available in the past (Rippka et al., 1979; Javor and Castenholz, 1981), the morphotype, *M. chthon-oplastes*, has only recently been characterized in more detail (Gar-cia-Pichel et al., 1996), by examining strain PCC 7420 and 13 new independent isolates from saline and hypersaline habitats of widespread geographical regions. As observed in the original field samples, all cultured strains exhibited conical ("bullet-shaped") **end cells** (diameter 3–6 μm) and formed typical **en-sheathed trichome bundles**. With unidirectional illumination the trichomes align themselves parallel to the direction of the in-coming light. This trait was conserved even in strains rendered axenic and no longer forming ensheathed bundles (i.e., strains PCC 7420, MPI-ND1, and MPI-NCR-1) (Garcia-Pichel et al., 1996). Other common properties were: **gliding motility** (speed

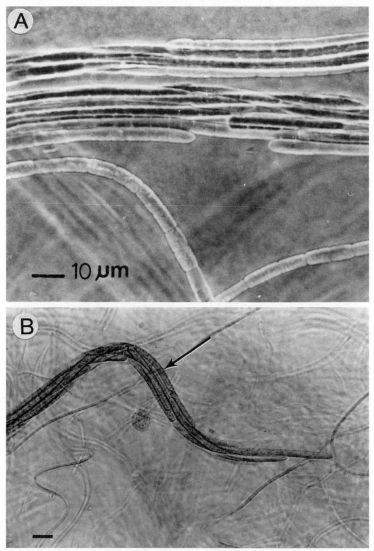

FIGURE B10.57. *Microcoleus* cf. *chthonoplastes. A,* culture showing bundles of trichomes. (Reprinted with permission from B.J. Javor and R.W. Castenholz, Geomicrobiology Journal *2:* 237–273, 1981, ©Taylor & Francis.) *B,* collection showing bundle (*arrow*) within common sheath. Bar ~20 μm. Both *A* and *B* are from microbial mats in hypersaline lagoons at Guerrero Negro, Baja California, Mexico.

0.3–0.6 μm/s) **without rotation**; synthesis of **PEC** (in addition to APC and PC) and **MAA** (mycosporine-like amino acids which absorb UVB/A radiation). Although variable in response to salinity (Karsten, 1996) and exhibiting slight but consistent differences in carotenoid and MAA composition (Karsten and Garcia-Pichel, 1996), all isolates, including the PCC reference strain (PCC 7420), were shown to be virtually identical at the 16S rDNA level (Garcia-Pichel et al., 1996). Thus, in spite of geographical diversity, this group of strains corresponds to a tightly delimited morphological and genetic cluster.

Culturing Enrichment cultures are not useful, since opportunistic *Geitlerinema*-like filaments tend to overgrow the *Microcoleus* bundles. The latter may be pulled out of collected material on agar plates with watchmaker's forceps and dragged through the agar for cleaning and isolation (Prufert-Bebout and Garcia-Pichel, 1994). F/2 medium, prepared with filter-sterilized sea water, was successfully used for isolation of *M. chthonoplastes.*

Reference strains Strain PCC 7420, collected as *Microcoleus chthonoplastes* from a salt marsh at Woods Hole, Massachusetts in 1974 (Rippka et al., 1979; formerly designated a member of the LPP-B group). The mol% G + C is 45.5 with a genome size of 4.08×10^9 daltons (Herdman et al., 1979a, b); strains of the same species, maintained in the collection of the Max Planck Institute of Marine Microbiology (MPI), are: MEL-1 Mellum (North Sea Germany), MPI EBD-1 (Ebro Delta, Spain), and MPI GN5-1 (hypersaline pond, Guerrero Negro, Mexico) (Garcia-Pichel et al., 1996).

TAXONOMIC COMMENTS

The genus *Microcoleus* as defined traditionally (Geitler, 1932) harbors species that differ more in trichome morphology and cell size than described in the rather restricted definition given here. Furthermore, some species were described as occurring in freshwater habitats as well as in terrestrial crusts on desert soils. However, as exemplified by the marine strains "*Microcoleus* 10 mfx" and SAG 3192, or the freshwater isolate *"Microcoleus vaginatus"* (*Oscillatoria* PCC 6304), that differ from *Microcoleus chthonoplastes* in having terminal cells (rounded in 10mfx and SAG 3192, "hooklike" in PCC 6304) and are either practically immotile (10 mfx) or glide rapidly by rotation (SAG 3192 and PCC 6304), the trait of a common sheath envelope is insufficient alone to allow correct generic assignment. None of these three strains is phylogenetically related to the *Microcoleus chthonoplastes* cluster (Garcia-Pichel et al., 1996; Wilmotte and Herdman, this volume). Whether some of the freshwater isolates, such as strain PCC 7113 or PCC 8701, that exhibit gliding motility without rotation and

are morphologically very similar to *Microcoleus chthonoplastes*, are closer phylogenetic relatives remains to be determined. A few other traditional "genera" have been described on the basis of a common sheath envelope, but their validity should be confirmed using cultured or uncultured representatives. Such morphotypes are: *Schizothrix* Kützing 1843, in which the common sheath is often branched toward the termini and commonly laminated; *Hydrocoleum* Kützing 1843, *Sirocoleus* Kützing 1849, *Polychlamydum* West and West 1879, and *Dasygloea* Thwaites 1848, all described on the basis of variations in sheath consistency and trichome number (see Geitler, 1932; Desikachary, 1959).

Form-genus IX. **Oscillatoria** *Vaucher 1803 (sensu Rippka, Deruelles, Waterbury, Herdman and Stanier 1979)*

RICHARD W. CASTENHOLZ, ROSMARIE RIPPKA AND MICHAEL HERDMAN

Os.cil.la.to' ri.a. L. v. *oscillare* to swing; L. adj. suffix *-torius* belonging to; M.L. fem. n. *Oscillatoria* (filament) that swings.

Filamentous organisms that divide exclusively by binary fission and in one plane. The cylindrical trichomes are straight (except for the terminal region in some cases); flexible or semirigid; **over ~4 µm in diameter; disc-like cells wider than long**. The trichome diameter of cultured representatives is 4–16 µm, but species with trichomes up to >100 µm in diameter have been described in the botanical literature. Invariably, in larger trichomes (>10 µm in diameter) the cells are much shorter than broad. Cross-walls are generally visible by light microscopy (Figs. B10.58 and B10.59). Constrictions at cross-walls are lacking or slight, but the total indentation never exceeds 1/8 of the trichome diameter. Generally, the cross-wall is thinner than the longitudinal wall. **The initiation of cross-wall synthesis by centripetal growth precedes the completion of previously initiated cross-walls** (Fig. B10.60). During fission, the cytoplasmic membrane invaginates, with a thinner peptidoglycan layer following and separating the new membranes of the daughter cells. This characteristic also applies to the genera *Spirulina*, *Arthrospira*, and *Lyngbya*. **Repro-**

duction of trichomes is **by transcellular breakage** involving sacrificial cell death (i.e., "necridial" cells).

FURTHER DESCRIPTIVE INFORMATION

The terminal cell of many species of *Oscillatoria* is differentiated to a shape distinct from the simple bulging of an unattended cross-wall. Shapes include round, blunt, truncate, conical, prolonged-attenuate, and capitate. In addition, some terminal cells acquire a cap-like thickening of the outer cell wall termed the calyptra. Trichomes may be attenuated as well, but only for the length of a few cells near the terminus. The terminal cell of recently fragmented trichomes will generally appear undifferentiated at the breakage site, but will redifferentiate as growth continues. The terminus of many species (whether tapered or not) may be bent, and this may extend for the length of several cells. Levels of photosynthetic pigments may be reduced and cell division may be arrested in the fully differentiated terminal cell, or even in a few of the subterminal cells.

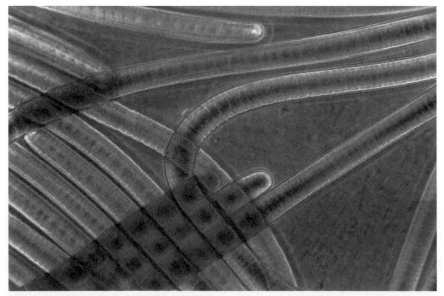

FIGURE B10.58. *Oscillatoria* cf. *margaritifera.* (CCMEE D-89-Om Cl 6) Trichome width 18–19 µm. Isolated from marine Limfjord, Denmark, 1989.

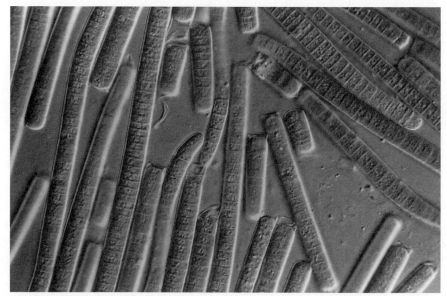

FIGURE B10.59. *Oscillatoria* sp. (CCMEE Ant-G-17-O). Trichome width 7–8 μm. Isolated from Gentle Pond G-17, near Bratina Island, McMurdo Ice Shelf, Antarctica, 1990.

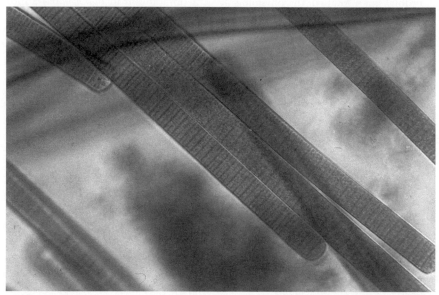

FIGURE B10.60. *Oscillatoria* sp. Trichome width ~6–7 μm. Collection from mud flats, Fern Ridge Lake, Lane County, Oregon.

The trichome exhibits **gliding motility by rotation** in a left- or right-handed manner with respect to the direction of movement (species or strain specific). If terminal regions are not in contact with a solid substrate, the free end appears to oscillate as the trichome rotates, particularly if the free end is curved. Rates of movement range from ~1 to about 11 μm/s (Halfen and Castenholz, 1971). Usually, **sheaths are nearly invisible** and consist of extremely thin casings that are shed as flattened trails when the trichomes move on solid substrates. Occasionally, a more visible sheath may build up around trichomes, particularly on old agar plates or after prolonged incubation in unshaken liquid cultures (Chang, 1977; Rippka, unpublished). Trichomes are sol-

itary or, if clustered in mats, they are not surrounded by a common sheath, diffluent or otherwise.

The color is variable, ranging from bright blue-green to deep red. Some species, olive or almost black in color, synthesize C-PE constitutively, in addition to containing APC and PC; complementary chromatic adaptation (Tandeau de Marsac, 1977) is relatively rare. Phycoerythrin with an absorption maximum at 493–495 nm, in addition to that at 543–546 nm, occurs in some marine forms, including shade-inhabiting species (R.W. Castenholz, unpublished), a strain of *O.* cf. *corallinae* CJ1 (Hoffmann et al., 1990), and the marine planktonic *Trichodesmium* (considered as *Oscillatoria* by some authors) (Fujita and Shimura, 1974).

Absorption at the shorter wavelengths is characteristic of phycourobilin-containing phycoerythrin (PUB), instead of the more common cyanobacterial C-PE (absorption maximum around 545 nm) which contains phycoerythrobilin (PEB) as the chromophore (Glazer, 1988, 1989).

Guglielmi and Cohen-Bazire (1982a) found that the junctional pores occurred in a single circular row near each crosswall in four strains assigned to *Oscillatoria* (PCC 6304, PCC 6407, PCC 7112, and PCC 8008), though the same pattern was also observed in *Arthrospira* (PCC 7345). Mid-cell and cross-wall perforations varied among strains. In addition, *Oscillatoria* and *Arthrospira* produced fimbriae that appeared closely appressed along the trichome, almost forming an envelope (Guglielmi and Cohen-Bazire, 1982b).

The mean DNA base composition of strains included in *Oscillatoria* (*sensu* Rippka and Herdman, 1992) at the Pasteur Culture Collection varies from 40 to 50 mol% G + C, and the genome sizes range from 2.50 to 4.38 $\times$ 10^9 daltons (Herdman et al, 1979a, b).

The species of *Oscillatoria* (even as narrowly defined here) have a wide distribution in freshwater, marine, and brackish waters. They also occur in inland saline lakes, and a few species tolerate temperatures as high as 45°C in some hot springs. Some are known as mat-formers in streams. Others are known as motile (gliding) components of microbial mats. Many of the physiological studies of *Oscillatoria* have used strains or populations that, as defined here, would fit into other genera, such as *Geitlerinema*.

Culturing   Isolation can usually be carried out by "self-isolation" on agar due to rapid gliding rates (Castenholz, 1988a; Rippka, 1988a).

Reference strains

Cluster 1   PCC 7515, isolated from a greenhouse water tank, Stockholm, Sweden, 1972 (Rippka et al., 1979) is closest to the morphotype *Oscillatoria sancta* Gomont 1892. Cells of this strain are 15–16 μm in diameter; mol% G + C is 40.1; genome size is 3.63 $\times$ 10^9 daltons. It is euryhaline and has photoheterotrophic ability; capable of aerobic N$_2$ fixation; C-PE synthesis constitutive.

Cluster 2   PCC 7112, isolated from greenhouse soil in San Francisco, 1970; it appears close to the morphotype *Oscillatoria nigro-viridis* Thwaites 1846 ex Gomont 1892. Cells of this strain is 7–8 μm in diameter; mol% G + C is 49.5; obligate photoautotroph; no N$_2$ fixation; synthesizes C-PE constitutively.

Cluster 3   PCC 6304, freshwater, originally identified as *Microcoleus vaginatus*; cells are 4–5 μm in diameter; mol% G + C is 49; genome size 4.38 $\times$10^9 daltons; obligate photoautotroph; N$_2$ fixation; neither PE nor PEC; it resembles morphotype *Oscillatoria acuminata* Gomont 1892.

Cluster 4   PCC 6407, isolated from freshwater in California, 1964; cells are 4–5 μm in diameter; mol% G + C is 45.4; anaerobic N$_2$ fixation; photoheterotrophic ability; neither C-PE nor PEC; small gas vesicles may occasionally be observed; close to morphotype *Oscillatoria formosa* Bory 1827 ex Gomont 1892.

Cluster 5   (Rippka and Herdman, 1992) corresponds to *Oscillatoria agardhii* Gomont 1892, which in this volume is referred to the form-genus *Planktothrix* Anagnostidis and Komárek 1988.

All of the above reference strains, and morphological relatives within each cluster, can be found in Rippka and Herdman (1992). Over 50 strains assigned to *Oscillatoria* are currently carried in the PCC.

TAXONOMIC COMMENTS

The genus *Oscillatoria* is now considered in the restrictive sense used by Rippka et al. (1979). The much broader definition of *Oscillatoria* formulated in the 1989 edition of *Bergey's Manual of Systematic Bacteriology* (and by others earlier) included strains or species of rather different trichome morphologies and cell dimensions (such as members now placed into *Geitlerinema*). However, phylogenetic analyses of some representatives demonstrated the validity of exclusion (Turner, 1997; Wilmotte and Herdman, this volume). However, in agreement with a morphology-based classification into five clusters, such comparative studies also suggest that further generic subdivisions may eventually be necessary within the current genus *Oscillatoria*. This approach has been initiated here by exclusion of strains assigned to *Oscillatoria agardhii* Gomont 1892 (equivalent to *Oscillatoria* cluster 5), now assigned to *Planktothrix* (*sensu* Anagnostidis and Komárek 1988).

Oscillatoria cluster 1, represented by reference strain PCC 7515, best fits the description of *O. sancta* Gomont 1892, but is also similar to *Oscillatoria princeps* Vaucher 1803, the botanical type species. Thus if generic subdivisions and nomenclatural changes are made, the name *Oscillatoria* should be retained for members of cluster 1. In their revision of the order *Oscillatoriales*, Anagnostidis and Komárek (1988) created a number of new genera (in addition to *Planktothrix*) appropriate for some members of *Oscillatoria* (as defined here), some of which (such as *Gomontinema* or *Tychonema*) may be suitable for some of the other morphological and genetic clusters in the PCC. *Oscillatoria* cluster 2 conforms to the newly created genus *Gomontinema* (*sensu* Anagnostidis and Komárek 1988), but clusters genetically relatively closely with strains identified as *Tychonema* (*sensu* Anagnostidis and Komárek 1988) (see Wilmotte and Herdman, this volume). Thus, further analyses are required prior to adopting a new generic name. Isolates of *Oscillatoria* cluster 3, with distinctly pointed or "hooked" terminal cells, conform to some species included by Anagnostidis and Komárek (1988) in the genus *Geitlerinema*. However, the reference strain of this cluster (PCC 6304) shows a high degree of genetic relatedness neither to members of *Geitlerinema*, nor to any representatives of the other four subclusters of *Oscillatoria*. Consequently, this genetic entity may require a new generic name. Members of *Oscillatoria* cluster 4 have narrow trichomes (4–6 μm) and are extremely common in freshwater, brackish water, or soil. Apart from lacking, or having low numbers of, gas vesicles, they resemble *Planktothrix* morphologically but would also fit the description of the genus *Phormidium* (as redefined by Anagnostidis and Komárek, 1988). However, the genus *Phormidium* Kützing 1843 has in the past been used to encompass widely diverse morphotypes and has been associated with more than 100 years of taxonomic confusion. We therefore suggest that this nomen be considered invalid or be relegated to some of the genetic clusters of the current genus *Leptolyngbya*, on the grounds that some of these (such as *"Phormidium ectocarpi"* or *"Phormidium fragile"*) are well entrenched in the literature and conform at least in part to the traditional generic epithet. *Oscillatoria* Vaucher 1803 and *Lyngbya* Agardh 1824, as defined in this volume, are very similar, with only the presence of a firm sheath in *Lyngbya* separating the two genera. However, based on sequence analyses of the PCC reference strain for *Lyngbya* (PCC 7419), marine species having trichome properties similar to those of the morphotype *Lyngbya aestuarii* Liebman 1841, seem to be genetically distinct from all isolates assigned to *Oscillatoria* (see

Wilmotte and Herdman, this volume). *Trichodesmium*, although phylogenetically closest to *Oscillatoria* PCC 7515 (see Turner, 1997; Wilmotte and Herdman, this volume), will be here treated as a separate genus because of some rather distinctive features (planktonic, oceanic habitat, and importance in global N_2 fixation).

Form-genus X. **Planktothrix** *Anagnostidis and Komárek 1988*

RICHARD W. CASTENHOLZ, ROSMARIE RIPPKA, MICHAEL HERDMAN AND ANNICK WILMOTTE

Plank' to.thrix. Gr. n. *planktos* wandering; Gr. n. *thrix* thread or hair; M.L. fem. n. *Planktothrix* wandering (or planktonic) hair.

This genus (Anagnostidis and Komárek, 1988) corresponds to Cluster 5 of *Oscillatoria* (*sensu* Rippka and Herdman, 1992). It is based primarily on the **structural and ecological features of** *O. agardhii* and *O. rubescens*, which are well known planktonic filamentous cyanobacteria **with abundant gas vacuoles** (Fig. B10.61). Typical of eutrophic or hypereutrophic freshwater lakes or reservoirs, the trichomes of these organisms often form clusters or fascicles and often represent a major constituent of "water blooms". Like *Oscillatoria* (*sensu stricto*) the **cells are shorter than broad or isodiametric**, and the **trichomes exhibit gliding motility** by rotation; prominent sheath layers are not produced; the cell diameter of all representatives presently in culture ranges from 3.5 to 10 μm (Skulberg and Skulberg, 1985; Rippka, unpublished). C-PE may be present or absent. Some strains were reported to produce the secondary metabolite 7-methylheptadecane (Jüttner, 1991).

FURTHER DESCRIPTIVE INFORMATION

Culturing Isolation may be accomplished by micromanipulation in a series of multiple depression slides with liquid medium. The medium may be varied (Castenholz, 1988a; Rippka, 1988a); Z8 was used by Skulberg and Skulberg (1985). Single gliding filaments have been isolated and purified using medium BG11$_o$, containing 2 mM NaNO$_3$ and 10 mM NaHCO$_3$, and solidified with washed agar (0.8%, w/v) (Rippka, unpublished).

Reference strains PCC 7805, corresponding to the classical description of *Oscillatoria agardhii* Gomont 1892 (Rippka and Herdman, 1992). This strain is co-identic with CCAP 1459/27 and NIVA CYA 68, has a mol% G + C content of 40.9 and a genome size of 2.5×10^9 daltons; neither PEC nor PE is produced. It is grown in BG-11 liquid medium at 20–25°C. Additional axenic isolates (7), including a PE-rich strain (PCC 7821 = NIVA CYA-18) which conforms to *Oscillatoria rubescens* D.C. ex Gomont 1825, are presently available in the PCC.

TAXONOMIC COMMENTS

Although similar in many respects to members of the genus *Oscillatoria*, numerous strains corresponding to *O. agardhii* or *O. rubescens* have been shown by phylogenetic analyses to form a separate and tight genetic cluster (Wilmotte and Herdman, this volume). Amplification and restriction enzyme analysis of the ITS region of the *rrn* operon, an approach permitting discrimination of strains in which 16S rDNA sequences are almost identical (Scheldeman et al., 1999), demonstrated that, in spite of its different pigment composition (red color due to high levels of PE), strain PCC 7821 is highly related to some of the blue-green isolates lacking PE (I. Iteman and L. Via-Ordorika, unpublished results). Thus, the distinction between *Planktothrix rubescens* (ex *Oscillatoria rubescens*) and *Planktothrix agardhii* (ex *O. agardhii*) is valid, if at all, only at the subspecies level.

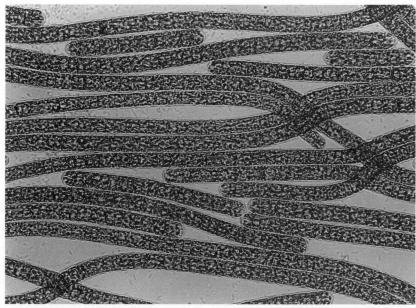

FIGURE B10.61. *Planktothrix* (*Oscillatoria*) sp. Trichome width ~6 μm. (Reproduced with permission from H. Canter-Lund and John W.G. Lund. 1995. Freshwater Algae: Their Microscopic World Explored. Biopress Ltd., Bristol, ©H. Canter-Lund.)

Form-genus XI. **Prochlorothrix** Burger-Wiersma, Stal and Mur 1989

RICHARD W. CASTENHOLZ, ROSMARIE RIPPKA AND MICHAEL HERDMAN

Pro' chlor' o.thrix. Gr. pref. *protos* first (primordial); Gr. adj. *chloros* green; Gr. n. *thrix* hair or thread; M.L. fem. n. *Prochlorothrix* primordial green hair.

Prochlorothrix is a filamentous organism that divides exclusively by binary fission in one plane and has slight to prominent constrictions at the cross-walls. Cells are longer (3–10 µm) than broad (0.5–1.5 µm in diameter). However, under salt stress (i.e., 25 mM NaCl) cells may enlarge to reach 15 µm in length and 3 µm in width. Thylakoids arranged in layers parallel to the long axis at the periphery of the cytoplasm. **Phycobilisomes absent**. Motility absent; **polar gas vacuoles present**; sheaths absent or poorly developed; apical cells undifferentiated (rounded). Lacks N$_2$-fixing ability. **Phycobiliproteins (i.e., CPE, C-PC, and APC) absent; chlorophyll *a* and chlorophyll *b* present**. Zeaxanthin and β-carotene are the major carotenoids.

FURTHER DESCRIPTIVE INFORMATION

The description of this genus is based on a living type strain and type species, *P. hollandica* (see Burger-Wiersma et al., 1989), later rendered axenic in the PCC (strain PCC 9006). The organism was discovered as a constituent of "blooms" in shallow, freshwater, eutrophic lakes which originated from peat mining. However, it is possible that this and related, as yet unknown, species of this genus may be common elsewhere. A second 16S rDNA sequence has recently become available (see Wilmotte and Herdman, this volume).

Culturing Isolation by micromanipulation (Castenholz, 1988a); FPG medium (see Burger-Wiersma et al., 1989) or BG-11o containing 2 mM NaNO$_3$ (Schyns et al., 1997). Optimal growth temperature 20–30°C; pH optimum 8.4.

Reference strain PCC 9006, type strain of the genus and species *Prochlorothrix hollandica* Burger-Wiersma et al. 1989, isolated from a shallow, highly eutrophic freshwater lake (Lake Loosdrecht) in the Netherlands in 1984. The non-axenic strain was deposited in the CCAP (CCAP 1490/1). The mean DNA base composition reported by Burger-Wiersma et al. (1989) was

53 mol% G + C, but a slightly different percentage (51.6 mol% G + C) was later determined for the axenic strain (Schyns et al., 1997).

TAXONOMIC COMMENTS

This organism, along with *Prochloron* Lewin 1977 and *Prochlorococcus* Chisholm et al. 1992 was until recently, considered a member of a "new" and separate order (*Prochlorales*) and phylum (*Prochlorophyta*); hence the separate treatment of these organisms (except for *Prochlorococcus* which had not yet been discovered) in the last edition of this *Manual* (see Burger-Wiersma and Mur, 1989; Lewin, 1989). This classification was based on the marked difference in pigment composition and thylakoid structure (including lack of phycobilisomes) between these "green" oxyphotobacteria and the more typical phycobiliprotein-containing cyanobacteria. The pigment complement and thylakoid stacking of this organism resemble the photosynthetic machinery encountered in the chloroplasts of green algae and higher plants. It was therefore proposed that members of this group represent the living descendents of an ancient lineage that, in the course of evolution involving an endosymbiotic event, gave rise to the chloroplasts, hence the *Prochlorophyta*. However, recent phylogenetic analyses have demonstrated that representatives of these three "prochloralian" genera do not represent a coherent cluster particularly affiliated with chloroplasts. On the contrary, examination of these data lead to the conclusion that chloroplasts evolved in multiple and independent events within the cyanobacterial radiation (see Palenik and Swift, 1996; Turner, 1997; Urbach et al., 1998; see also Wilmotte and Herdman, this volume). Consequently, all three "prochloralian" members are now regarded as cyanobacteria (Urbach et al., 1992) and the phylum *Prochlorophyta* and order *Prochlorales* are rejected. For descriptions of *Prochloron* and *Prochlorococcus*, the reader may consult Subsection I (this *Manual*).

Form-genus XII. **Pseudanabaena** Lauterborn 1916

RICHARD W. CASTENHOLZ, ROSMARIE RIPPKA, MICHAEL HERDMAN AND ANNICK WILMOTTE

Pseud' an.a.baen' a. Gr. adj. *pseud* false; *anabaena* (cf. genus *Anabaena*); M.L. fem. n. *Pseudanabaena* false *Anabaena*.

Trichomes that divide exclusively by binary fission in one plane and have **conspicuous constrictions at the crosswalls** that may exceed half the cell diameter (Fig. B10.62). In some strains, the total constriction is less but is still more than 1/8 the cell diameter (Guglielmi and Cohen-Bazire, 1984a). **Cells are longer than broad to isodiametric and are often barrel-shaped**. Trichomes of strains characterized in culture range from ~1 to 3 µm in diameter. The formation of the transverse septum involves a partial centripetal ingrowth of all wall layers. In some cases, the remaining connection appears quite narrow, as if the cells were

strung as beads. The structural (peptidoglycan) layer of the crosswall is 3–6 times thicker than the layer surrounding the rest of the cell (Guglielmi and Cohen-Bazire, 1984a) (Fig. B10.63). Another characteristic observed through T.E.M. is the consistency of **peripheral thylakoids, parallel to the cell walls**. Near the crosswalls, the continuity of thylakoids may be interrupted by gas vesicle clusters. The trichomes are usually straight and, depending on the morphotype, may be quite short (consisting of less than 3 to ~10 cells) or may be longer (>10 cells). Reproduction by **intercellular trichome breakage**. Single detached cells are fre-

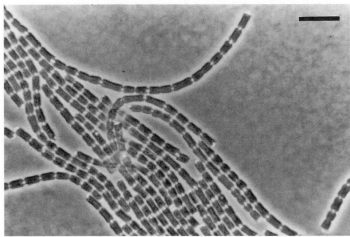

FIGURE B10.62. *Pseudanabaena* cf. *galeata* (University of Oregon OL-75-Ps; culture lost). Gas vacuoles at cell poles (*light areas*). Bar = 10 μm.

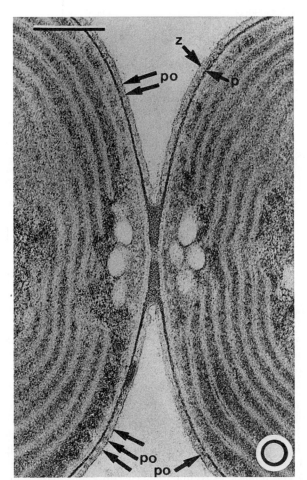

FIGURE B10.63. Transmission electron micrograph of longitudinal section of *Pseudanabaena* sp. (PCC 7367) at point of constricted cross-wall. *p*, peptidoglycan layer of wall; *po*, multiple pores restricted to cell pole region. Gas vesicles at both cell poles. Bar = 200 nm. (Reprinted with permission from G. Guglielmi and G. Cohen-Bazire, Protistologica *18*: 167–177, 1982, ©Editions due CNRS.)

quently found in cultured populations and may be confused with unicellular cyanobacteria (Castenholz, unpublished). Gliding motility, probably without rotation, occurs at rates usually <1 μm/s. Gliding motility has been lost in a few cultured strains (Guglielmi and Cohen-Bazire, 1984a).

Over 35 strains of *Pseudanabaena* are maintained at the PCC.

FURTHER DESCRIPTIVE INFORMATION

In one morphotype, *Pseudanabaena* cf. *galeata* (CCMEE strain OL-75-Ps), gliding aggregates of many trichomes develop in the form of "comet"-shaped bundles and usually move at rates greater than those of individual trichomes (Castenholz, 1982) (Fig. B10.64).

Many strains have **polar gas vacuoles**, but this characteristic is not expressed under all culture conditions and in some clones seems to have been lost permanently. Although vegetative trichomes are very similar in appearance to hormogonia of some *Nostoc* species (see Subsection IV), as in all members of Subsection III, *Pseudanabaena* is incapable of producing heterocysts. Terminal cells of trichomes are not differentiated with respect to shape, except in some strains where they may be slightly more pointed than end cells created by recent intercellular trichome breakage. Guglielmi and Cohen-Bazire (1984b) found that all strains then assigned to *Pseudanabaena* possessed rings (60–500 nm wide) of irregularly distributed pores (150–500 per ring) in the longitudinal peptidoglycan wall near the cell poles, in contrast to the single ring of pores found in strains of *Oscillatoria* (also see Guglielmi and Cohen-Bazire, 1984a) (Fig. B10.63). However, the same pore pattern typical of *Pseudanabaena* was also observed in some strains now assigned to *Leptolyngbya* that are genetically remote from *Pseudanabaena*.

The color of isolates of *Pseudanabaena* may be blue-green, olive, or red, depending on the presence or absence of C-PE; complementary chromatic adaptation occurs quite frequently. Some strains are capable of N_2-fixation under anoxic conditions (Rippka and Herdman, 1992).

The mol% G + C of the DNA ranges from 42 to 48 (Herdman et al., 1979a; Guglielmi and Cohen-Bazire, 1984a; Rippka and Herdman, 1992), and the genome size ranges from 2.14 to 5.19 $\times 10^9$ daltons, demonstrating the great genetic diversity of this "genus".

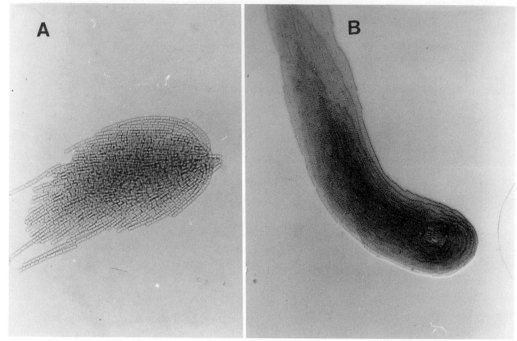

FIGURE B10.64. *Pseudanabaena* cf. *galeata* (University of Oregon OL-75-Ps; culture lost). *A*, "young" migrating aggregate ("comet"). Individual trichomes may be seen. *B*, complex "older" "comet" with several tiers of cells. Width of "comet" A is ~60 μm; width of "comet" B is ~120 μm.

The distribution of the morphotype *Pseudanabaena* is worldwide. It is seen in collections from hot springs (usually not above 55°C) and from marine and freshwater muds. It is particularly common in anoxic sulfide-containing sediments. Some forms are known to occur in freshwater plankton and include those found in the mucilage of other planktonic cyanobacteria. All known strains are obligate photoautotrophs.

Culturing Isolation can usually be performed using gliding self-isolation on agar; streak dilution plating may be preferred for the rapid establishment of axenic cultures, since the resulting flat and slightly spreading *Pseudanabaena* colonies can be isolated more easily from the bacterial contaminants; the medium will depend on source of the sample (see Castenholz, 1988a, b; Rippka, 1988a).

Reference strains Rippka and Herdman (1992) have divided the partially characterized strains of this genus into five clusters, with one reference strain for each.

Cluster 1 PCC 7429, conforms to *Pseudanabaena biceps* and produces relatively long trichomes (>10 cells). However, three other strains with shorter trichomes (PCC 6901, PCC 6903, and PCC 7402) are also included in this cluster. All were collected from fresh or brackish water in widespread locations from 1969 to 1972. The cell diameter of PCC 6901 is 1.3–1.8 μm. The others are ~2.5 μm. All have polar gas vacuoles, are motile, lack C-PE and PEC, are obligate photoautotrophs, and lack N_2-fixing ability. The mol% G + C content of these four strains ranges from 43.8 to 46.6; the genome size ranges from 2.14 to 3.38 × 10^9 daltons.

Cluster 2 PCC 7367 harbors a single representative collected from an intertidal zone in Mexico in 1971. Mol% G + C is 47.7; genome size is 2.95 × 10^9 daltons. It is similar in most characteristics to Cluster 1 strains, except that it contains C-PE, but does not undergo complementary chromatic adaptation; however, it is a marine form with requirements for elevated salt concentrations.

Cluster 3 PCC 7408 was isolated in 1940 from the river Thames in England and identified by the isolator (E.G. Pringsheim) as *Pseudanabaena catenata*. The second strain in this cluster (PCC 7409) was isolated from an *Azolla filiculoides* sample collected in California in 1971. The mean DNA base composition is 41.7 and 43.6 mol% G + C for strains PCC 7408 and PCC 7409, respectively. Genome size is 5.19 (PCC 7408) and 4.63 (PCC 7409) × 10^9 daltons. Both strains produce C-PE and exhibit complementary chromatic adaptation. These two strains do not produce gas vacuoles visible by light microscopy. However, PCC 7409 has been shown to contain the gene *gvp*A, which encodes the major structural protein of gas vesicles (Damerval et al., 1989); this information is still lacking for strain PCC 7408.

Cluster 4 PCC 7403 is the only representative and was collected from a bog in Switzerland in 1972. Mol% G + C is 46.3; genome size is 3.17 × 10^9 daltons. It produces short (3–6 cells) trichomes and was originally motile, but this ability was lost in culture. Will fix N_2 under anoxic conditions; PE and PEC absent.

Cluster 5 PCC 6802 is the only representative and was collected in 1968 from a freshwater pond in California. Mol% G + C is 45.9; genome size is 3.43 × 10^9 daltons. It forms single cells or very short chains of cells (2–3) and has lost motility in culture; is capable of N_2 fixation under anoxic conditions. It synthesizes C-PE and undergoes complementary chromatic adaptation.

TAXONOMIC COMMENTS

Although most specimens from nature, or freshly isolated strains, typical of *Pseudanabaena* show gliding motility and possess polar gas vesicle clusters, the definition of this genus as formulated here is not quite as restrictive as that proposed by Rippka et al. (1979). Gas vacuoles may be lost in culture, and some strains of unicellular *Synechococcus* from hot springs often possess polar gas vacuoles (R.W. Castenholz, unpublished). In addition, gliding motility cannot be used to distinguish *Pseudanabaena*, since some strains of *Pseudanabaena* lose their gliding ability, particularly if maintained in liquid cultures (Rippka, unpublished), and some strains of *Synechococcus* are known to move by gliding (Ramsing et al., 1997). For these reasons, correct assignments at the genus level are best made on the basis of the characteristic ultrastructural features established by Guglielmi and Cohen-Bazire (1984a, b). A similar definition of the genus *Pseudanabaena* to that formulated here is also given by Anagnostidis and Komárek (1988), although the latter authors divided the genus into three subgenera based on some minor morphological distinctions. The complexity of identifying the more atypical members of this genus, however, is demonstrated by the case of three strains: PCC 7409 lacks gas vesicles (and was therefore placed into the LPP-group B (Rippka et al., 1979). This strain (together with *Lim-nothrix redekei* 6705), clusters in the phylogenetic tree with typical gas vacuole-containing isolates of *Pseudanabaena*, whereas two strains (PCC 6406 and PCC 7376) assigned to the latter genus on the basis of ultrastructural similarities (Guglielmi and Cohen-Bazire, 1984a, b) are remote both from each other and from the monophyletic "*Pseudanabaena* cluster" (Turner, 1997; Wilmotte and Herdman, this volume), each occupying a position close to, or within, two different "*Leptolyngbya*" clades. Two biochemical markers may correlate with the relative uniqueness of strain PCC 6406: it is the only "*Pseudanabaena*" strain synthesizing PEC (Bryant, 1982), and it has a significantly higher mean DNA base composition (52.2 mol% G + C) than all other strains (42–48 mol% G + C) assigned to this genus (Herdman et al., 1979b). For these reasons, this "borderline *Pseudanabaena*" (exhibiting only very shallow constrictions between the cells) was placed by Rippka and Herdman (1992) into Cluster 3 of *Leptolyngbya*. In conclusion, as is also the case for the genera *Leptolyngbya* and *Limnothrix*, unambiguous identification of the representatives of *Pseudanabaena* will require supporting genetic data.

Form-genus XIII. **Spirulina** *Turpin 1829 ex Gomont 1892*

RICHARD W. CASTENHOLZ, ROSMARIE RIPPKA, MICHAEL HERDMAN AND ANNICK WILMOTTE

Spi.ru.li' na. L. n. *spira* a coil; L. n. *linea* a line; M.L. fem. n. *Spirulina* coiled filament.

Filamentous organisms that divide exclusively by binary fission and in one plane but that grow in the form of a **tight to nearly tight, coiled right- or left-handed helix**, although this helix may become partly unwound in some trichomes of cultured populations. More loosely coiled representatives often form partial double helices composed of two individual filaments that rotated into each other. The **cross-walls are thin and barely visible even if observed by light microscopy using phase contrast objectives**, a characteristic that distinguishes this genus from *Arthrospira* (Holmgren et al., 1971) (Fig. B10.65). Originally *Spirulina* was described as being a long unicellular thread. **No sheath** is visible under the light microscope, and "healthy" trichomes are in constant motion. Gliding motility consists of a "turning of the screw", thus with great transverse movement and little forward motion. **Motility is by rotation around the outer surface of the helix**. Free ends not in contact with a solid support may oscillate wildly as the coil turns. The terminus of the trichome is either blunt or pointed. In different species the diameter of the trichome can range from <1 μm to ~5 μm in size. In the latter case, the width of the whole helix may be as great as 12 μm. Color is variable, blue-green to red; the latter being typical of some marine representatives that synthesize C-PE as the major light-harvesting pigment, but contain relatively little PC and APC. Although C-PE content may vary inversely with light intensity, complementary chromatic adaptation has not been observed (Tomaselli et al., 1995). *Spirulina* isolates seem to be stable with respect to trichome structure. Straight variants, as found often in *Arthrospira* (Jeeji-Bai and Seshadri, 1980; Lewin, 1980; Rippka, unpublished), have not been observed, even after maintenance in culture for more than 30 years (Rippka, unpublished).

FURTHER DESCRIPTIVE INFORMATION

In four strains, Guglielmi and Cohen-Bazire (1982b) found patches of pores at cross-walls but only on the inner concave surfaces of the helix (Fig. B10.66). This may be a general diagnostic feature for this taxonomic unit, since very similar pat-

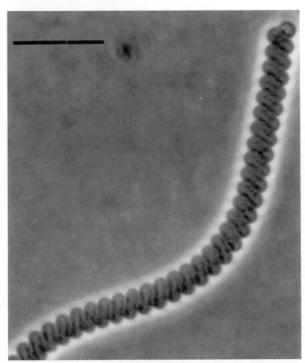

FIGURE B10.65. Photomicrograph (phase contrast) of *Spirulina* cf. *labyrinthiformis*, a moderately thermophilic strain isolated from Mammoth Hot Springs, Yellowstone National Park, Wyoming, U.S.A. Bar = 10 μm. (Reprinted with permission from R.W. Castenholz, *Microbial Ecology* 3: 79–105, 1977, ©Springer-Verlag.)

terns were also observed in a more recent study of two isolates from a microbial mat in the North Sea (Palínska et al., 1998).

The members of this genus are known to have a worldwide distribution in freshwater, marine, and brackish waters. *Spirulina* species are also common in inland saline lakes and in some hot

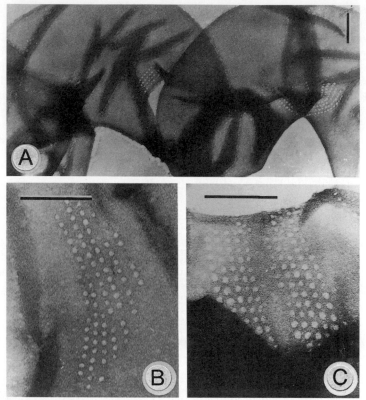

FIGURE B10.66. Electron micrograph of sacculus of *Spirulina* sp. (PCC 6313) treated with sulfuric acid. *A*, spiral envelope showing distribution of pores. *B*, pores marking an incomplete transverse septum. *C*, pores disposed on both sides of a complete transverse septum. Bar = 200 nm. (Reprinted with permission from G. Guglielmi and G. Cohen-Bazire, Protistologica *18:* 151–165, 1982, ©Editions due CNRS.)

springs at temperatures as high as 50°C (Castenholz, 1977, 1978). Some species or ecotypes are tolerant to a wide range of salinity (Javor, 1989; Rippka, unpublished). The are often found in habitats with high concentrations of free sulfide, and some have been shown to be capable of using sulfide in anoxygenic photosynthesis (Castenholz, 1977). Some red (shade-adapted) marine strains require "oligotrophic" media and grow very slowly (R.W. Castenholz, unpublished).

Culturing The best isolation procedures involve self-isolation by gliding motility. However, because of poor forward progress on agar relative to the spread of heterotrophic bacteria, isolation by successive dilution in drops of liquid medium on slides with multiple depressions, followed by a final manual picking of a single filament with drawn-out capillary tubes, is often more successful (Castenholz, 1988a). One tightly coiled strain was isolated and rendered axenic by repeated picking of single gliding filaments from the edge of the inoculum and subsequent transfer onto plates prepared with medium ASNIII, supplemented with $NaHCO_3$ (10 mM) and solidified with washed agar (Rippka, unpublished).

Reference strains PCC 6313, isolated from brackish water, Berkeley, California, 1963. The mol% G + C is 53.4, and the genome size is 2.53×10^9 daltons (Herdman et al., 1979a, b). No C-PE or PEC; relatively loosely coiled; no gas vacuoles; obligate photoautotroph; no N_2 fixation; grows in BG11 and ASNIII (i.e., euryhaline) media. Corresponds to morphotype *Spirulina* cf. *major* Kützing ex Gomont 1892. PCC 9445, isolated from

crude sample (provided by A. Gambacorta, Napoli), originating from a brackish lake with hydrothermal activity and rich in H_2S (Lago di Venere, Pantellaria Island, Italy, 1994). Filaments have the same trichome diameter as in strain PCC 6313, but appear larger due to very tight coiling. Corresponds to morphotype *Spirulina* cf. *subsalsa*; growth in BG11/ASNIII (v/v) or ASNIII (i.e., slightly halophilic); no PE; mean DNA base composition 47 mol% G + C (Herdman, unpublished). This isolate is not included in the PCC Catalog (Rippka and Herdman, 1992).

TAXONOMIC COMMENTS

In the traditional literature (e.g., Geitler, 1932), the genus *Spirulina* was used to accommodate morphotypes conforming to the current definition, but also harbored the larger coiled members assigned to *Arthrospira*, a concept maintained (though with some reservations) as late as 1979 (Rippka et al., 1979). Bourrelly (1985), considering that the degree of coiling represented only gradual transitions from one extreme to the other, included both *Spirulina* and *Arthrospira* in the genus *Oscillatoria*. This point of view may be justified for the morphotype *Arthrospira*, since 16S rDNA analyses revealed that two representatives of this genus cluster more or less tightly with *Lyngbya* (PCC 7419) and certain isolates of *Oscillatoria* (Turner, 1997; Wilmotte and Herdman, this volume). In contrast, both morphotypes of *Spirulina* (*S. major* and *S. subsalsa*), though not closely related, group together (but are not closely related) and are genetically remote from the "*Oscillatoria*" and "*Arthrospira*" clades (Turner, 1997; Wilmotte and Herdman, this volume).

In a recent study of the physiology, morphology, and phylogeny of 11 strains of *Spirulina*, all morphological forms that would normally be assigned to this genus clustered phylogenetically far from *Arthrospira*, but still exhibited considerable genetic diversity (U. Nübel, personal communication). Three strains that were highly halotolerant formed a separate cluster. Thus the genera *Spirulina* Turpin 1829 ex Gomont 1892 and *Arthrospira* Stizen-

berger 1852 ex Gomont 1892 constitute, beyond doubt, separate taxonomic units. On these grounds, the cyanobacteria referred to as *Spirulina* by Jeeji-Bai (1985) and Lewin (1980), and the commercially available "*Spirulina*", used as a protein supplement, should be considered species (or varieties) of the genus *Arthrospira*.

Form-genus XIV. **Starria** *Lang 1977*

RICHARD W. CASTENHOLZ, ROSMARIE RIPPKA AND MICHAEL HERDMAN

Star' ri.a. M.L. fem. n. *Starria* Starr, named after R.C. Starr, U.S. phycologist (1924–1998).

This nonbranching filamentous cyanobacterium is unique in that the short **trichomes are triradiate in cross-section** (Lang, 1977). The **immotile** trichomes are straight to helically twisted. The **cells are significantly shorter** (length 1–2 µm) **than wide** (diameter ~15 µm). The triradiate form usually has broad, arm-like projections 120 degrees apart, separated by U-shaped depressions (Fig. B10.67). The pigmentation is concentrated in the arms. A thin sheath (3 µm thick) covers the trichome. Cross-walls are as found in the genus *Oscillatoria*, with an invagination of the peptidoglycan layer but not of the outer membrane. The relatively thick peptidoglycan layer of longitudinal walls is characterized

by evenly distributed pits 70 nm in diameter, as in some species of *Oscillatoria* (e.g., *O. princeps*) (Halfen and Castenholz, 1971; Guglielmi and Cohen-Bazire, 1982b). Little is known about the physiology or biochemistry of this organism, but its very unusual structure has been documented by light and electron microscopy (Lang, 1977). This organism is known from a single strain isolated from a terrestrial sample in Zimbabwe. The scarce record of *Starria* may not imply that it is rare, since cyanobacteria of the soil have not been studied extensively.

FURTHER DESCRIPTIVE INFORMATION

Culturing Growth occurs in freshwater media with soil extract added (Lang, 1977). Low light intensity is recommended for incubation. Due to its large size, the organism can be easily isolated by picking single filaments under the dissection scope, but due to a lack of gliding motility and poor growth on agar of less contaminated (washed) trichomes, no axenic cultures are presently available (Rippka, unpublished).

Reference strain UTEX 1754, isolated from a depression in granite rock, Zimbabwe, 1969.

TAXONOMIC COMMENTS

Although morphological variants different from the wild type may occur, most share a distinct resemblance from possessing the triradiate feature. However, some of the less triradiate variants resemble in structure species described in the literature as *Crinalium* Crow 1927 (Rippka, unpublished), a genus described earlier. Both *Starria* and *Crinalium* are also to some extent similar to *Gomontiella* Teodoresco 1901 (see Bourrelly, 1985). However, for lack of cultured representatives (or phylogenetic data on uncultured members), the latter genus is not discussed in this *Manual*.

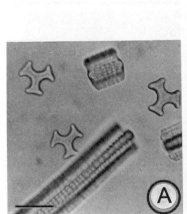 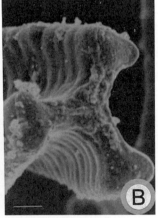

FIGURE B10.67. *Starria zimbabweënsis* in culture. *A*, longitudinal and cross-sectional views, with the latter showing projecting arms with slightly broadened extremities. Bar = 20 µm. *B*, enlarged cross-section. Bar = 2 µm. (Both *A* and *B* are reproduced with permission from N.J. Lang, ©Journal of Phycology *13*: 288–296, 1977.)

Form-genus XV. **Symploca** *Kützing 1843*

RICHARD W. CASTENHOLZ, ROSMARIE RIPPKA AND MICHAEL HERDMAN

Sym.plo' ca. Gr. adj. *sym* together; Gr. n. *plokos* lock of hair; M.L. fem. n. *Symploca* clustered hair.

Trichomes that divide by binary fission in one plane only, with a prominent and persistent sheath (**a single trichome per sheath, often with frequent false branching**). In contrast to *Lyngbya* (as treated here), the **cells are isodiametric, or slightly longer than broad, and less than 6 µm in width. The constrictions between cells are shallow.** The sheathed filaments are **arranged in bundles, often forming tufts or turfs** in moist terrestrial or intertidal habitats. Geitler (1932) recognized 22 species. This form-genus is very tentative, since many types of ensheathed filamentous

cyanobacteria may correspond to this poor definition and some strains of the form-genus *Leptolyngbya*, and of *Lyngbya* as defined here, exhibit aerial growth on solid substrates or form pinnacles or tufts under moist culture conditions. However, an approach such as has been taken for the genus *Microcoleus*, which under the current definition is primarily based on a single morpho- and ecotype (*Microcoleus chthonoplastes*, see earlier), may be necessary for a more precise concept of *Symploca*. Strain PCC 8002, although originally described as *Microcoleus chthonoplastes* (Pearson et al., 1979), does not share the structural features of this species, but rather conforms more to the description of *Symploca atlantica* Gomont 1892 or *Symploca funicularis* Setchell and Gardner 1918. The former species has been extensively described by Gomont (1893) on the basis of field collections from intertidal mats in France (where it was found in association with *Microcoleus chthonoplastes*), whereas the description of *S. funicularis*, possibly a synonymous species, is based on field material from the coast in California (Setchell and Gardner, 1918). Typically for these two botanical species, the trichomes of strain PCC 8002 are of medium size (**5–6 μm in diameter**) and are **heavily ensheathed**. In contrast to *Microcoleus chthonoplastes*, the **terminal cells** are not conical but **rounded and carry a small "hat-like" flat-pointed to convex calyptra; 1–10 subterminal cells are consistently less pigmented** (yellowish), irrespective of the source of nitrogen; on solid media the trichomes exhibit upright aerial growth; although isolated from a marine environment, this strain is euryhaline (Rippka, unpublished). It synthesizes neither **PE nor PEC, and MAA** (mycosporine-like amino acids) were not detected (Karsten and Garcia-Pichel, 1996). Most importantly, strain PCC 8002 is one of the rare nonheterocystous filamentous cyanobacteria **capable of fixing N$_2$ under oxic conditions** (Pearson et al., 1979), even if incubated under a continuous light regime (Rippka, unpublished).

FURTHER DESCRIPTIVE INFORMATION

Culturing Purification by single filament isolation of less ensheathed and slightly motile "hormogonia" from the edges of an inoculum placed on solidified medium. This euryhaline isolate grows in freshwater medium BG11 or in marine media (both MN or ASNIII). To avoid loss of the capacity to fix N$_2$, or selection of salt resistant or salt sensitive mutants, the strain is is presently maintained in a mixture of media BG11$_o$ and ASoIII (v/v) (Rippka, unpublished).

Reference strains PCC 8002, isolated as *Microcoleus chthonoplastes* from stabilized mud, high intertidal zone, Menai Straits, North Wales, UK (Pearson et al., 1979). The properties of this strain are described above.

TAXONOMIC COMMENTS

As it is based on a single representative, The description of this form-genus is very tentative. However, phylogenetic analyses, comparing a partial 16S rDNA sequence available in GenBank (accession number AF 013026), suggest that strain PCC 8002 clusters with *Microcoleus chthonoplastes* PCC 7420 but is sufficiently remote to merit independent taxonomic status (Herdman, unpublished).

Form-genus XVI. **Trichodesmium** *Ehrenberg 1830 ex Gomont 1892*

RICHARD W. CASTENHOLZ, ROSMARIE RIPPKA AND MICHAEL HERDMAN

Trich' o.des' mium. Gr. n. *trichos* hair; Gr. n. *desmos* bond; M.L. neut. n. *Trichodesmium* bonded filaments.

Filamentous cyanobacteria that divide exclusively by binary fission and in one plane. Trichomes of different species or strains range from 6 to 22 μm in diameter. Cells are isodiametric or slightly longer than broad; constrictions at cross-walls are shallow. **Trichomes are arranged in parallel fascicles or form radiate macroscopic clusters**; *Trichodesmium* species **often occur as major constituents of the plankton in tropical to semi-tropical oceans**; they are **reddish to orange in color, containing PE with phycourobilin (PUB) chromophores**, exhibiting an absorption peak at ~493–495 nm (Fujita and Shimura, 1974; Glazer, 1989). All known members contain **abundant gas vacuoles** (Fig. B10.68), **generally dispersed throughout the cells, and particularly resistant to collapse**, even at hydrostatic pressures of 12–27 bars (Walsby, 1994).

FURTHER DESCRIPTIVE INFORMATION

Trichomes are capable of slow gliding motility on agar. Most attempts to establish cultures in the past have failed, but have proved successful more recently (Ohki et al., 1986; Waterbury and Willey, 1988; Prufert-Bebout et al., 1993). Nitrogen fixation under oxic conditions was first reported on field collected material more than 30 years ago (Dugdale et al., 1961), but has only recently been confirmed on cultured species (Ohki et al., 1991, 1992; Prufert-Bebout et al., 1993). The mechanisms for protecting the oxygen-sensitive enzyme nitrogenase from inactivation by atmospheric O$_2$ seem to vary among the different species, involving the confinement of the nitrogenase either to certain cells in a trichome or to a limited number of filaments within a colony. However, it is assumed that all species of *Trichodesmium* fix N$_2$ in the light, without temporally separating oxygenic pho-

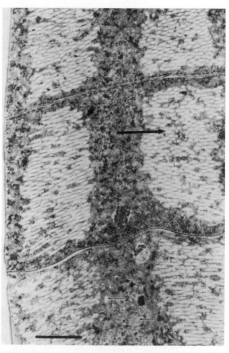

FIGURE B10.68. Transmission electron micrograph of a median longitudinal section of a trichome of *Trichodesmium* sp. showing the concentric arrangement of gas vesicles (*arrow*). Bar = 1.0 μm. (Reprinted with permission from C. Van Baalan and R.M. Brown, Jr., Archiv für Mikrobiologie *69*: 79–91, 1969, ©Springer-Vergag.)

tosynthesis from the seemingly incompatible process of dinitrogen fixation (Carpenter et al., 1990; Prufert-Bebout et al., 1993; Bergman et al., 1997).

Culturing For the isolation and culturing of *Trichodesmium*, two different media have been successfully employed: a modification of the defined medium "Aquil", for species from Japan (see Ohki et al., 1986), and a seawater based medium described for isolates from North Carolina (Prufert-Bebout et al., 1993; Chen et al., 1996).

Reference strains NIBB 1067 (Ohki et al., 1986) and IMS 101 (Prufert-Bebout et al., 1993).

TAXONOMIC COMMENTS

The 16S rDNA sequences of several different *Trichodesmium* morphotypes (*T. contortum, T. erythraeum, T. hildenbrandtii, T. tenue,* and *T. thiebautii*) show a high degree of similarity (Turner, 1997; see also Wilmotte and Herdman, this volume), demonstrating the validity of assigning all morphotypes to a single genus (in spite of major size differences exhibited by some of the "species"). Their closest relative outside the genus *Trichodesmium* is *Oscillatoria sancta* PCC 7515 (see *Oscillatoria* Cluster 1). However, separation of the genus *Trichodesmium* from *Oscillatoria* (*sensu stricto*), and even more so from other "*Oscillatoria*" clusters, seems to be justified, as the genetic distances between these groups is almost as high as those determined between *Arthrospira* and *Lyngbya.*

Genus XVII. **Tychonema** *Anagnostidis and Komárek 1988 emend.*

RICHARD W. CASTENHOLZ, ROSMARIE RIPPKA AND MICHAEL HERDMAN

Ty'cho.nem'a. Gr. n. *tychos* luck, accident; Gr. n. *nema* thread; M.L. fem. n. *Tychonema* accidental thread.

This genus was created by Anagnostidis and Komárek (1988) and includes filamentous cyanobacteria that resemble in most respects strains typical of *Oscillatoria* (as defined here). The only characteristics of diagnostic value are thought to be the presence of a sap vacuole (as in eucaryotic algae) and **a reticulated network of cytoplasmic strands which contain the thylakoids** (Figs. B10.69 and B10.70). Trichomes of freshwater morphotypes range in diameter from 2 to 16 μm and **gas vacuoles are lacking** (Anagnostidis and Komárek, 1988). However, cultured isolates of the botanical morphotypes *Oscillatoria bornetii* Zukal 1894 ex Geitler 1932 and *O. bornetii* f. *tenuis* Skuja 1930, later assigned to *Tychonema* (Anagnostidis and Komárek, 1988), exhibit a more narrow range in size (diameters 12–16 μm and 4–8 μm, respectively) (Skulberg and Skulberg, 1985). These isolates all contain PE and undergo complementary chromatic adaptation; myxoxantho-

phyll and oscillaxanthin seem to be the characteristic carotenoids (Skulberg and Skulberg, 1985). Another representative of this "genus" that has been cultured but subsequently lost, is a much larger species (diameter up to 60 μm) that occurs commonly in tychoplankton of protected marine waters of Bermuda (R.W. Castenholz, unpublished; Fig. B10.74).

FURTHER DESCRIPTIVE INFORMATION

Culturing Strains assigned to *Tychonema* ("*Oscillatoria bornetii* group") have been isolated and maintained in modified medium Z8 (Skulberg and Skulberg, 1985)

Reference strains NIVA CYA 60, conforming to *O. bornetii,* and NIVA CYA 33/1, assignable to *O. bornetii* cf *tenuis* 1985; both were isolated from Lake Mjøsa, Norway (Skulberg and Skulberg, 1985).

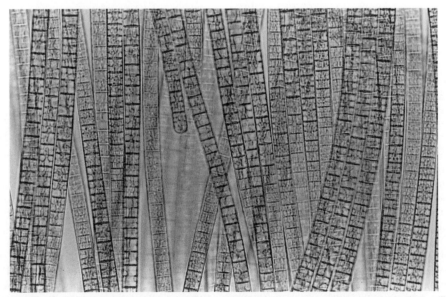

FIGURE B10.69. *Tychonema* (*Oscillatoria*) sp. Trichome width is 5–7 μm. From freshwater plankton. (Reprinted with permission from H. Canter-Lund and John W.G. Lund. 1995. Freshwater Algae: Their Microscopic World Explored. Biopress Ltd., Bristol, ©H. Canter-Lund.)

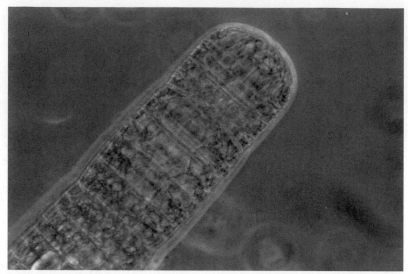

FIGURE B10.70. *Tychonema* (*Oscillatoria*) sp. A large diameter marine species. Trichome width is ~50 µm. Collected from marine lagoon in Bermuda. Dark reticulations are cytoplasmic strands which contain the thylakoids.

TAXONOMIC COMMENTS

Although the definition of this genus is far from satisfactory, it seems clear from phylogenetic analyses that *Tychonema bornetii* NIVA CYA-60 (ex *Oscillatoria bornetii* Zukal 1894 ex Geitler 1932) and several strains (NIVA CYA 33/1, CYA-33/6, CYA-58, and CYA 70) of *Tychonema bourrellyi* (ex *O. bornetii* f. *tenuis* Skuja 1930) are closely related, their 16S rDNA sequences being practically identical (see Wilmotte and Herdman, this volume). Their closest relatives outside this "genus" are two strains identified as "*Phormidium*" sp. and strain PCC 7112, a member of *Oscillatoria* Cluster 2 (cf. "*Oscillatoria nigro-viridis*", see earlier). Further comparative ultrastructural, biochemical, and genetic studies (DNA–DNA hybridization) between all members of related group of strains are required in order to propose a more definitive nomenclature.

ADDENDUM TO SUBSECTION III

A number of traditional genera, some with numerous "species", have not been included here because their distinctions are relatively subtle (*Schizothrix*, for example). Others lack representatives are lacking or have not been studied in detail (see Geitler, 1932; Desikachary, 1959; Bourrelly, 1985; Anagnostidis and Komárek, 1988 for additional generic names). Given the ease and power of molecular analyses, the validity of such genera can now be examined even using uncultured or impure specimens.

Subsection IV. (Formerly **Nostocales** Castenholz 1989b sensu Rippka, Deruelles, Waterbury, Herdman and Stanier 1979)

ROSMARIE RIPPKA, RICHARD W. CASTENHOLZ AND MICHAEL HERDMAN

Members of Subsection IV are **filamentous cyanobacteria that divide exclusively by binary fission in one plane**, perpendicularly to the long axis of the trichomes. Cell division occurs at similar rates throughout the trichome or, in some genera, may be more rapid in "meristematic" zones of the filaments (Castenholz, 1989b; Komárek and Anagnostidis, 1989; Whitton, 1989). With the exception of some representatives of the genus *Nostoc*, the trichomes are always clearly uniseriate. The **ensheathed filaments** of some genera (e.g., *Calothrix* or *Scytonema*) **may exhibit false branching**, a developmental process distinct from true branching that occurs only in members of Subsection V (see Hoffman and Castenholz, this volume) and involves cellular division in more than one plane. **Several genera** are defined by a basal–apical polarity of the filaments (i.e., **tapering** trichomes), the apical extremity often terminating in thin and colorless cells, termed "hair" cells (Komárek and Anagnostidis, 1989; Whitton, 1989). The trichome diameter among members of Subsection IV presently in culture varies greatly, ranging from about 2 µm (*Cylin-*

drospermopsis) to 15 µm (i.e., members of *Scytonema*, *Calothrix*, and *Rivularia*).

In the absence of combined nitrogen (ammonium or nitrate), 5–10% of the vegetative cells differentiate into **heterocysts**. These specialized cells are the sites of **aerobic nitrogen fixation** and may occupy **terminal** or **intercalary positions** in the trichomes. Heterocysts are characterized by a thick envelope usually composed of an inner laminated layer, an intermediate homogeneous layer, and an outer fibrous layer that successively are deposited externally to the existing cell walls. Heterocysts generally have a low content of phycobiliproteins and lack both a functional PS II and the enzymes of the reductive pentose phosphate pathway. Together with other metabolic changes, these properties enable them to fix atmospheric nitrogen and to **protect the O₂-labile nitrogenase complex from inactivation by oxygen** (Haselkorn, 1978; Adams and Carr, 1981; Wolk, 1982; Fay, 1992; Wolk et al., 1994).

Some members of Subsection IV **produce akinetes that are**

resistant to cold and to desiccation. These thick-walled cells, generally yellowish to brownish in appearance, are formed as the result of light or nutrient limitation (Nichols and Carr, 1978; Sutherland et al., 1979; Adams and Carr, 1981; Herdman, 1987, 1988). However, with some exceptions (*Cylindrospermum*), their synthesis is not repressed by combined nitrogen. **Akinetes** are located either **adjacent to or distant from the heterocysts**. In the latter case they often occur in chains that, in the absence of combined nitrogen, are initiated at a site that is equidistant from two intercalary or terminal heterocysts.

Many representatives undergo a distinct developmental cycle involving the formation of hormogonia (sensu Rippka et al., 1979). Their differentiation is best detected after transfer of aged cultures to fresh medium. Most hormogonia are distinguished from the parental trichomes by **transient gliding motility, reduced trichome length**, and **smaller cells**. They are produced by rapid successive cell divisions in the absence of growth. Hormogonial cells may also **differ in shape and/or may contain gas vesicle clusters**. In "species" characterized by heavily ensheathed mature filaments, hormogonia are sheathless or thinly sheathed. In some axenic isolates of *Nostoc* and *Tolypothrix* both the mature filaments and the hormogonia are immotile and the differences of their cell size are relatively minor (Rippka, 1988b). In such cases, correct generic assignments have been made only on the basis of DNA–DNA hybridization studies (see Lachance, 1981). Irrespective of the genus, freshly released **hormogonia never exhibit heterocysts** (Rippka et al., 1979; Herdman and Rippka, 1988), even if produced in media lacking combined nitrogen. Consequently, they do not fix N_2 (Campbell and Meeks, 1989). In the natural environment, hormogonia serve as a means of dispersal (Herdman and Rippka, 1988; Castenholz, 1989b; Tandeau de Marsac, 1994).

Some genera **do not exhibit a distinct developmental cycle**, the trichomes being either **motile at all stages of growth** or, as in most planktonic representatives, being **permanently immotile** (Rippka, 1988b; R. Rippka, unpublished data).

Further descriptive information

Heterocystous cyanobacteria are found in a great diversity of habitats, as a result of their universal ability to fix atmospheric nitrogen under oxic conditions. Several "species" of *Anabaena*, *Aphanizomenon*, and *Gloeotrichia* are planktonic and are major constituents of water blooms in temperate or tropical freshwater lakes or reservoirs. *Anabaenopsis*, *Cyanospira*, and *Nodularia* are more typically found in blooms of saline lakes and brackish waters such as the Baltic Sea, though *Anabaena* and *Aphanizomenon* may also be found in the latter habitat (Castenholz, 1989b). Most planktonic members from freshwater environments produce gas vesicles (Walsby, 1981). In marine habitats planktonic heterocystous cyanobacteria are rare, though *Richelia* often occurs as an endosymbiont of centric diatoms (Geitler, 1932; Waterbury and Willey, 1988). Representatives of many other genera occur as epiphytes or epiliths in both freshwater and marine habitats. "Species" of *Calothrix*, *Gloeotrichia*, and *Rivularia* (or *Dichothrix* and *Gardnerula*, only known from field descriptions) form gelatinous or firm cushions on solid substrates, some of which may be encrusted by $CaCO_3$ deposition. Some members of *Nostoc* form firm spherical to amorphous gelatinous colonies that range considerably in size, being either microscopic or attaining diameters equivalent to bowling balls (Castenholz, 1989b). *Calothrix*, *Tolypothrix*, and *Scytonema* and possibly other morphologically related

"genera", are typical of subaerial habitats, particularly in moist or tropical climates.

Many representatives are capable of surviving cold temperatures and severe conditions of desiccation (Castenholz, 1989b). However, none occurs in hot springs exceeding 52–54°C and the only member of Subsection IV known to persist at such temperatures is a member of the genus *Calothrix* (Wickstrom and Castenholz, 1978).

Some members of this Subsection are endosymbionts or exosymbionts, supplying the host mainly with fixed nitrogen, though fixed carbon may also be provided (Stewart, 1980; Stewart et al., 1980, 1983). Almost all cyanobionts are representatives of the genus *Nostoc* (Stewart et al., 1980). The most common symbiotic hosts are ascomycetes (to form lichens), liverworts, hornworts, ferns and cycads, and an angiosperm, *Gunnera* (Stewart, 1980; Stewart et al., 1980, 1983; Enderlin and Meeks, 1983; Meeks, 1988).

Gliding motility (if exhibited) is commonly an order of magnitude slower (i.e., <1.0 µm/s) than that of some of the "*Oscillatoriales*" (see Subsection III) and the hormogonia or mature trichomes (some representatives) do not rotate as do members of the genus *Oscillatoria* (Nultsch and Wenderoth, 1983). The mean DNA base composition of all strains of Subsection IV examined to date (about 50) ranges only from 38.3 to 46.7 mol% G + C (Herdman et al., 1979b; Lachance, 1981; M. Herdman, unpublished data). The genome sizes, however, are quite different, ranging from 3.17 to 8.58 Gdal (Herdman et al., 1979a).

Taxonomic Comments

The separation of members of Subsection IV ("*Nostocales*" sensu Castenholz 1989b) from those of Subsection III (cf. "*Oscillatoriales*", see Castenholz et al., this volume) is fully supported by 16S rRNA sequence data, all heterocystous cyanobacteria forming a coherent genetic cluster distinct from nonheterocystous representatives (Giovannoni et al., 1988; Wilmotte, 1994; Turner, 1997; Turner et al., 1999; Honda et al., 1999; see also Willmotte and Herdman, this volume). In contrast, these data also demonstrate that the traditional ordinal distinctions between genera of Subsections IV (cf. "*Nostocales*") and V (cf. "*Stigonematales*", see Hoffman and Castenholz, this volume) are not justified, the genetic distances spanning "nostocalean" genera being equal to or greater than those that separate them from the "stigonematalean" representatives. However, until a single higher taxon for members of these two botanical orders has officially been proposed and legalized, we will retain in this edition the morphologically informative separation of Subsections IV and V.

In the last edition, Castenholz (1989b) divided Subsection IV (cf. "*Nostocales*") into three families: *Nostocaceae*, *Scytonemataceae*, and *Rivulariaceae*. However, phylogenetic analyses do not support these higher taxa, since members of *Tolypothrix*, a "scytonematacean" genus, and one strain of *Calothrix*, a member of the "*Rivulariaceae*", cluster with "nostocacean" representatives (see Wilmotte and Herdman, this volume). Therefore, we have here departed from the traditional family concept and describe Subsection IV in two parts (IV.I and IV.II), based on differences in trichome morphology. However, these subdivisions are merely made for ease of description and do not reflect a phylogenetic grouping.

Subsection IV.I corresponds mainly to the "*Nostocaceae*" in the last edition (Castenholz, 1989b), but also includes the genus *Scytonema*, traditionally assigned to the "*Scytonemataceae*" (Geitler, 1932). Subsection IV.II harbors all heterocystous cyanobacteria previously assigned to *Calothrix* (sensu Rippka et al., 1979), a genus of *"Rivulariaceae"*. However, this "genus" is here subdivided into five genera on the basis of the degree of tapering of mature trichomes and other morphological or ecological features. Two genera (*Tolypothrix* and *Microchaete*) were traditionally treated as members of the botanical families *"Scytonemataceae"* and *"Microchaetaceae"*, respectively (Geitler, 1932). The separation of *Tolypothrix* (exhibiting a low degree of tapering) and of marine *Calothrix* strains (now assigned to *Rivularia*) from *Calothrix* (sensu Rippka et al., 1979) is largely based on the DNA–DNA hybridization data of Lachance (1981), but is also supported by 16S rRNA gene sequence analyses (see Wilmotte and Herdman, this volume). However, only a limited number of strains have been analyzed genetically and no data are yet available for representatives of *Microchaete* or *Gloeotrichia*, for example. Furthermore, a high degree of genetic diversity is already evident within some "genera". Thus the internal organization of Subsection IV remains largely arbitrary.

Some genera of the *"Nostocales"* have been described on the basis of field studies as lacking heterocysts, but so far no genuine representatives of the latter have been cultured. Their correct assignment to Subsection IV will require confirmation by genetic evidence once such representatives become available.

For the strain descriptions below, unless otherwise stated, phenotypic data are taken from Rippka et al,. 1979; Bryant, 1982; Rippka, 1988b; Rippka and Herdman, 1992; Iteman et al., 1999; and R. Rippka, unpublished data. Genotypic data are taken from Herdman et al., 1979a, b; Lachance, 1981; Iteman et al., 1999; and M. Herdman, unpublished data. Co-identities of strains in addition to those given in this chapter may be found in Rippka and Herdman, 1992. Growth media cited can be found in Rippka, 1988a and Castenholz and Waterbury, 1989.

Key to the form-genera of Subsection IV

I. Trichomes do not exhibit basal-apical polarity (cf. *"Nostocaceae"* and *"Scytonemataceae"*, in part).
   A. Developmental cycle does not involve formation of structurally distinct hormogonia.
      1. Trichome in a loose spiral coil.
         a. Trichome very short (1/2 to 1 coil); permanently immotile; mature heterocyst in terminal positions; *de novo* heterocyst differentiation in the center of the trichome, obligatorily in pairs; akinetes distant from the heterocysts, predominantly single; gas vesicles throughout life cycle.

                  Form-genus *Anabaenopsis*, p. 568
         b. Trichome consisting of several coils; permanently immotile; heterocysts predominantly intercalary, not occurring in pairs; akinetes in chains, initiated distant from heterocysts; gas vesicles throughout life cycle.

                  Form-genus *Cyanospira*, p. 570
      2. Trichome straight.
         a. Trichome composed of cylindrical cells; motile or immotile; constrictions between cells shallow; trichome ends slightly attenuated; heterocysts intercalary; akinetes distant from, or (rarely) adjacent to, heterocysts; gas vesicles throughout life cycle. Fasicle-like bundles of trichomes are typical of material from nature.

                  Form-genus *Aphanizomenon*, p. 569
         b. Trichome composed of discoid cells, often surrounded by a thin sheath; motile or immotile; heterocysts intercalary and terminal; akinetes in chains, usually initiated distant from heterocysts; gas vesicles in some members.

                  Form-genus *Nodularia*, p. 574
         c. Trichome composed of cylindrical cells; motile or immotile, end cells conical; exclusively terminal heterocysts; gas vesicles lacking; akinetes adjacent to heterocysts.

                  Form-genus *Cylindrospermum*, p. 572
         d. Trichome composed of cylindrical cells; permanently immotile, end cells conical; exclusively terminal heterocysts; permanent gas vesicle formation; akinetes adjacent to, or distant from, terminal heterocysts.

                  Form-genus *Cylindrospermopsis*, p. 571
         e. Trichome composed of barrel-shaped or cylindrical cells; motile or immotile; end cells conical (gliding species) or undifferentiated (planktonic members); heterocysts exclusively intercalary, or both intercalary and terminal; akinetes adjacent to, or remote from, heterocysts; gas vesicles in some members.

                  Form-genus *Anabaena*, p. 566
   B. Developmental cycle involves structurally distinct hormogonia.
      1. Straight, motile or immotile hormogonia differentiate terminal heterocysts at both extremities; growth leads to wave-like twisted or tightly coiled, immotile mature trichomes with additional intercalary heterocysts; akinetes in chains, initiated distant from heterocysts; transient gas vesicle formation in hormogonia of some members.
         a. Developmental cycle readily evident; hormogonia motile.

                  Form-genus *Nostoc* (clusters 1 and 2), p. 577

b. Developmental cycle not readily evident due to lack of motility or low degree of trichome differentiation.

Form-genus *Nostoc* (clusters 3–5), p. 578

2. Thinly sheathed, motile or immotile hormogonia differentiate a single terminal heterocyst; growth leads to heavily ensheathed mature trichomes with predominantly intercalary heterocysts; false branches, single or geminate, frequent; aerial growth on solid substrates; gas vesicles absent.

Form-genus *Scytonema*, p. 580

II. Trichomes exhibit basal-apical polarity (cf. *"Rivulariaceae"* and *"Scytonemataceae"*, in part).

A. Developmental cycle involves structurally distinct hormogonia.

1. Sheathless or thinly sheathed, motile hormogonia of uniform cell size differentiate a single terminal heterocyst upon maturation; the mature trichome is generally heavily ensheathed and exhibits very pronounced tapering (the ratio of basal to apical cell width being 3–5); false branching frequent; akinetes (if produced) occur singly or in short chains, initiated adjacent to the basal heterocyst.

a. Akinetes lacking. No requirement for elevated concentrations of Na^+, Mg^{2+}, and Ca^{2+}.

Form-genus *Calothrix*, p. 582

b. Akinetes lacking. Obligate requirement for elevated salt concentrations.

Form-genus *Rivularia*, p. 586

c. Large subterminal akinetes produced; most members known from freshwater habitats.

Form-genus *Gloeotrichia*.

2. Properties as II.A.1, but tapering not distinct (the ratio of basal to apical cell width being 2 or less).

a. Apical region of trichome of even width; transient gas vesicle formation (hormogonia) in most members; akinetes not produced.

Form-genus *Tolypothrix*, p. 587

b. Apical region of trichome exhibits slightly enlarged terminal cells; akinetes in chains, adjacent to, or remote from, intercalary and basal heterocysts.

Form-genus *Microchaete*.

FURTHER READING

Bourrelly, P. 1970. Les Algues d'eau Douce. III. Les Algues Bleues et Rouges, les Eugleniens, Peridiniens et Cryptomonadines, N. Boubée & Cie, Paris.

Bourrelly, P. 1985. Les Algues d'eau Douce. III. Les Algues Bleues et Rouges, les Eugleniens, Peridiniens et Cryptomonadines, 2nd ed., N. Boubée & Cie, Paris.

Castenholz, R.W. 1989. Subsection IV. Order *Nostocales. In* Staley, Bryant, Pfennig and Holt (Editors), Bergey's Manual of Systematic Bacteriology, 1st ed., Vol. 3, The Williams & Wilkins Co., Baltimore. pp. 1780–1793.

Desikachary, T.V. 1959. Cyanophyta, Indian Council of Agricultural Research, New Delhi.

Geitler, L. 1932. Cyanophyceae. *In* Kolkwitz (Editor), Rabenhorst's Kryptogamenflora von Deutschland, Österreich und der Schweiz, Vol. 14, Akademische Verlags, Leipzig.

Giovannoni, S.J., S. Turner, G.J. Olsen, S. Barnes, D.J. Lane and N.R. Pace. 1988. Evolutionary relationships among cyanobacteria and green chloroplasts. J. Bacteriol. *170*: 3584–3592.

Komárek, J. and K. Anagnostidis. 1989. Modern approach to the classification system of cyanophytes, 4. *Nostocales.* Arch. Hydrobiol. Suppl. *82*: 247–345.

Turner, S. 1997. Molecular systematics of oxygenic photosynthetic bacteria. Plant Syst. Evol. Suppl. *11*: 13–52.

Turner, S., K.M. Pryer, V.P.W. Miao and J.D. Palmer. 1999. Investigating deep phylogenetic relationships among cyanobacteria and plastids by small subunit rRNA sequence analysis. J. Eukaryot. Microbiol. *46*: 327–338.

Subsection IV.I

In cyanobacteria of subsection IV.I the **trichomes never exhibit basal–apical polarity**, all **vegetative cells** being of **uniform size**. With the exception of *Scytonema*, cell divisions seem to occur at similar rates throughout the trichomes. Members of **some genera undergo a distinct developmental cycle**, involving hormogonia formation. Although some representatives may produce ensheathed trichomes, **false branching is extremely rare**, occurring frequently only in *Scytonema*. Under nitrogen-limiting conditions, some members differentiate **heterocysts exclusively in terminal or intercalary positions**; others exhibit both terminal and intercalary heterocysts. **Akinetes** (if present) may be formed **singly or in short chains** (two to three akinetes) exclusively adjacent to, or very close to, heterocysts. In some members, they are initiated equidistant between two heterocysts, successive differentiation on either side of the first akinete often leading to very long chains of akinetes, **often comprising the majority of the cell population**. Some botanical genera (e.g., *Anabaenopsis* and *Cylindrospermum*) are easily identified by the location and/or spacing of heterocysts or akinetes.

The botanical family *"Nostocaceae"* (comprising all genera of

Subsection IV.I, except *Scytonema*) was subdivided by Komárek and Anagnostidis (1989) into two subfamilies, *"Anabaenoideae"* and *"Nostocoideae"* (Komárek and Anagnostidis, 1989). The former included all genera in which akinete formation is initiated adjacent (or very close) to heterocysts, whereas the latter subfamily was restricted to members that initiate akinetes equidistant between two heterocysts. However, as demonstrated by morphological studies and DNA–DNA hybridization for *Anabaena*, both types of akinete pattern may occur in different "species" of the same genus (Stulp and Stam, 1982, 1984a).

Form-genus I. **Anabaena** *Bory de St. Vincent 1822 sensu Rippka, Deruelles, Waterbury, Herdman and Stanier 1979*

ROSMARIE RIPPKA, RICHARD W. CASTENHOLZ, ISABELLE ITEMAN AND MICHAEL HERDMAN

A.na.bae'na. Gr. adj. *ana* up again; Gr. v. *baino* to walk, to step; M.L. fem. n. *Anabaena* to step up again, repeating line of units (cells).

Trichomes straight or slightly sinuous, vegetative cells spherical, cylindrical, or barrel shaped, and separated by conspicuous constrictions at the crosswalls. Heterocysts predominantly intercalary, though terminal heterocysts may also occur. Akinetes located, singly or in groups (2–3), adjacent to heterocysts or occurring in chains, initiated distant from heterocysts. A firm sheath is absent, but mucilage production may be abundant. False branching generally not observed. Reproduction by random trichome fragmentation. Hormogonia, distinct in morphology from the parental trichomes, not produced.

FURTHER DESCRIPTIVE INFORMATION

The cell width of botanical "species" of *Anabaena* ranges from 2 to 15 μm (Geitler, 1932). However, most representatives presently in culture have cylindrical or barrel-shaped cells of relatively small dimensions (diameter 3–6 μm). The terminal cells often differ from intercalary cells by their conical shape. Some representatives may exhibit a slight taper toward both termini of the filament. Heterocysts are typically separated by 10–20 vegetative cells. Intercalary heterocysts are similar in shape to vegetative cells, but terminal heterocysts may be conical in members forming differentiated end cells. The trichomes typically exhibit slow motility (<1 μm/s) throughout the growth cycle and discrete colonies are not formed, although planktonic "species" that synthesize gas vesicles seem to be permanently immotile.

Many representatives of this genus are known worldwide as major components of the freshwater plankton or tychoplankton and there are at least 15 gas-vacuolate botanical "species" described from such habitats, *A. spiroides*, *A. circinalis*, and *A. flos-aquae* being the most common. However, *Anabaena flos-aquae* may also occur in brackish environments such as the low-salinity regions of the Baltic Sea (Castenholz, 1989b). Although one marine strain, *Anabaena* CA, has been brought into axenic culture (Bottomley et al., 1979), members of this genus have never been reported as an important constituent of marine plankton. Some "species" have been described from moderate hot springs or terrestrial environments (Geitler, 1932), or as endosymbionts of the water fern *Azolla* or cycad species (e.g., *Anabaena azollae* and *Anabaena cycadeae*), but the latter may more appropriately be assigned to the genus *Nostoc*, given that members established in culture produce motile hormogonia (Stewart et al., 1983; Rippka, 1988b).

The physiology and biochemistry of *Anabaena cylindrica* Lemmerman 1896, a typical member of this genus, has been extensively studied on isolates co-identic with strains PCC 7122 or PCC 73105 and served to elucidate the role of heterocysts for aerobic N_2 fixation (Fogg, 1949, 1951; Stewart et al., 1969). In contrast, *"Anabaena"* sp. strain PCC 7120, a model organism for genetic studies on nitrogen fixation and cell differentiation (Haselkorn, 1978, 1986, 1992, 1995; Buikema et al., 1993), differs extensively from the former species and was shown to exhibit a very high degree (more than 70% relative binding) of DNA–DNA similarity to some representatives of the genus *Nostoc* (Lachance, 1981). For this reason, the latter strain and its relatives have been transferred to *Nostoc* (Rippka and Herdman, 1992) and will be discussed in more detail below.

Only three axenic strains (PCC 6309, PCC 7108, and PCC 7122) of *Anabaena* have been characterized in detail and were assigned to two of the three clusters below (Rippka and Herdman, 1992).

Cluster 1　This cluster is represented by two strains, PCC 6309 and PCC 7122, that conform to the description of *Anabaena cylindrica* Lemmerman 1896. The trichomes are composed of cylindrical cells (diameter 4–5 μm) that are separated by relatively deep constrictions; mature terminal cells are conical; heterocysts are differentiated both in intercalary and terminal positions (see Fig. B10.71A); cylindrical akinetes (2–3 times the length of vegetative cells) are formed singly, or in pairs, on one or both sides of the heterocysts (see Fig. B10.71B). Both are freshwater isolates from Europe and originally exhibited gliding motility, though strain PCC 6309 has become immotile (R. Rippka, unpublished data). Both strains are obligate photoautotrophs and synthesize phycoerythrocyanin (Bryant, 1982). In spite of a relatively important difference in mean DNA base composition (Herdman et al., 1979b; Lachance, 1981), strains PCC 6309 (38.3 mol% G + C) and PCC 7122 (43.9 mol% G + C) exhibit a high degree of DNA–DNA similarity (relative binding of 83%; $\Delta T_{m(e)}$ 0°C) and are thus representatives of the same nomenspecies (Lachance, 1981). The genome sizes are similar (3.6 and 3.2 Gdal, respectively) and relatively small compared to some members of Subsection IV (Herdman et al., 1979a).

Based on strain history records (Rippka et al., 1979; Rippka and Herdman, 1992), strain PCC 7122 is co-identic with strains UTEX 629, UTEX 1609, UTEX 1611, and CCAP 1403/2a, all being subisolates of *Anabaena cylindrica* Lemmerman 1896, originating from pond water (Cambridge, U.K.) and first described by Fogg (1942). The latter four strains were included in the studies of Stulp (1983) and Stulp and Stam (1982; 1984a, b; 1985) and, in agreement with their history, were shown to exhibit ~100% DNA–DNA similarity (or $\Delta T_{m(e)}$ values below 2°C) (Stulp and Stam, 1984a; see also Castenholz, 1989b). In contrast, the conflicting hybridization results for strains PCC 6309 and PCC 7122 of Lachance (1981) compared to those of Stulp and Stam (1984a) on supposedly co-identic strains (PCC 6309 = UTEX 377 and CCAP 1403/4b; PCC 7122 = UTEX 629, UTEX 1609, UTEX 1611, and CCAP 1403/2a), strongly suggest that strain

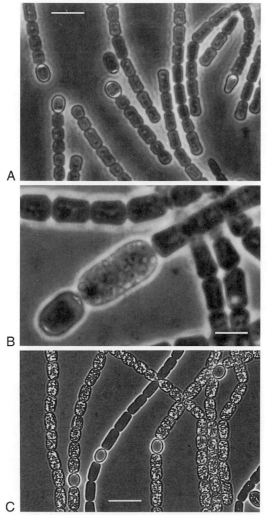

FIGURE B10.71. *Anabaena* PCC 6309 (*A,B*) and *Anabaena flos-aquae* PCC 9332 (*C*). Phase contrast. Bar = 10 μm (*A,C*) and 5 μm (*B*).

PCC 6309 is not the same as its counterparts in UTEX and CCAP collections (for more details, see Rippka, 1988b).

Reference strain PCC 7122 (co-identity ATCC 27899), isolated in 1939 by S.P. Chu from pond water, Cambridge, England (Fogg, 1942; Rippka et al., 1979). Proposed as type strain of *Anabaena cylindrica* (Rippka and Herdman, 1992). The mol% G + C is 43.9; genome size is 3.2 Gdal. Note that this strain is also co-identic with CCAP 1403/2a, UTEX 629, UTEX 1609, and UTEX 1611).

Cluster 2 This "cluster" harbors a single representative, *Anabaena* PCC 7108, isolated from the intertidal zone in California, but maintained in standard freshwater medium BG11₀. The trichome properties are very similar to members of cluster 1, but ellipsoidal akinetes are differentiated in short chains mainly distant from heterocysts, a property more typical of *Anabaena variabilis* Kützing 1843. As members of cluster 1, strain PCC 7108 is an obligate photoautotroph and produces phycoerythrocyanin (PEC) (Bryant, 1982). It also has a similar mean DNA base composition (42.3 mol% G + C) and genome size (3.7 Gdal), but is genetically sufficiently remote from members of cluster 1 (only 32% relative binding with strain PCC 7122) to represent a different species (Lachance, 1981). Based on its morphological

properties, strain PCC 7108 may be a relative of the *Anabaena variabilis* strains studied by Stulp and Stam (1982; 1984a, b; and 1985), but this needs to be confirmed by DNA–DNA hybridization studies. However, the latter isolates should not be confused with the well-known "Wolk strain" of *Anabaena variabilis*, ATCC 29413 (co-identity *Nostoc* PCC 7937), that has become popular as one of the model organisms in molecular biology (see Tandeau de Marsac and Houmard, 1993 and Wolk et al., 1994 for reviews) but that evidently was misidentified. In contrast to strains PCC 7108 and the isolates studied by Stulp and Stam, the trichomes of the "Wolk strain" have rounded end cells, form chains of rather inconspicuous akinetes and produce gas vacuolated hormogonia (R. Rippka, unpublished data).

Reference strain PCC 7108, isolated in 1970 by J.B. Waterbury from intertidal zone, Moss beach, California (Rippka et al., 1979). PCC reference strain of *Anabaena* "cluster 2" (Rippka and Herdman, 1992). The mol% G + C is 42.3; genome size is 3.7 Gdal.

Cluster 3 This cluster (new in this edition) is represented by three planktonic isolates (PCC 9302, PCC 9332, and PCC 9349), only recently rendered axenic (R. Rippka, unpublished data). Two of them originate from freshwater lakes in Canada (PCC 9302 and PCC 9349, co-identic with *Anabaena flos-aquae* strains 525–17-b1-c and A113) and produce anatoxin a (s) and anatoxin (c) (Mahmod and Carmichael, 1986, and W.W. Carmichael, personal communication, respectively). Strain PCC 9332 corresponds to *Anabaena flos-aquae*, strain CCAP 1403/13f, first described by Booker and Walsby (1979). All three strains exhibit similar cell size (diameter 4–5 μm), are immotile, and produce abundant gas vesicle clusters that are dispersed throughout the spherical vegetative cells (see Fig. B10.71C). The terminal cells do not differ from intercalary cells. In strain PCC 9332, heterocysts are produced predominantly in intercalary positions, whereas this property has been lost in the other two strains due to prolonged maintenance in media containing combined nitrogen prior to entry in the PCC. Akinete formation has never been observed in any of the strains (R. Rippka, unpublished data). 16S rRNA gene sequence data (Iteman et al., 1999) demonstrated that strain PCC 9302 is relatively distant from *Anabaena cylindrica* strain PCC 7122 and clusters most closely with *Aphanizomenon flos-aquae* strain PCC 7905 (see Wilmotte and Herdman, this volume). However, in the absence of DNA–DNA hybridization studies, no conclusions concerning generic or specific reassignment can be made.

Reference strain PCC 9332 (co-identity CCAP 1403/13F), isolated in 1964 by J.W.G. Lund from a water sample of Lake Windermere, Cumbria, England; (see Booker and Walsby, 1979). Proposed as type strain of *Anabaena flos-aquae* (this edition). The mol% G + C is 39.1; genome size is 3.3 Gdal.

ENRICHMENT AND ISOLATION PROCEDURES

Isolation and purification of most *Anabaena* species can be easily performed on solidified freshwater media (BG11₀ or Z8₀, supplemented with 1–10 mM NaHCO₃) and micromanipulation of gliding filaments (Castenholz, 1988a; Rippka, 1988a). For planktonic species that generally are immotile, dilution streaking as applicable for unicellular strains and incubation at slightly lower temperatures (20–23°C) proved successful to isolate axenic colonies (R. Rippka, unpublished data). A washing step prior to plating may also improve the chances of obtaining axenic isolates (Carmichael and Gorham, 1974; Rippka, 1988a). As for all heterocystous cyanobacteria, it is prudent to maintain *Anabaena* strains in media devoid of combined nitrogen in order to prevent selection of mutants that have lost the capacity to differentiate functional heterocysts.

TAXONOMIC COMMENTS

The definition of this genus is essentially the same as given previously (Rippka et al., 1979; Castenholz, 1989b), except that in this edition *"Anabaena spiroides"* is not treated as a member of this taxon. The latter botanical "species" exhibits trichome properties very similar to *Cyanospira* but, as a bloom-forming freshwater representative, differs in ecology (*Cyanospira* having a higher salt requirement). However, in the absence of phylogenetic studies confirming correct generic assignment of *"Anabaena spiroides"*, the latter coiled member has temporarily been excluded from both *Anabaena* and *Cyanospira*.

Minor modifications have also been made with respect to the revision of Komárek and Anagnostidis (1989). According to the latter authors, heterocyst differentiation in *Anabaena* always occurs in an intercalary position. This, however, is not a general rule: the conical end cells, typically found in *Anabaena cylindrica*, give rise to similarly shaped heterocysts (see Fig. B10.71A); these are easily distinguished from the more or less cylindrical heterocysts that differentiate in intercalary positions but may become "terminal" after trichome breakage.

Furthermore, we have included in this genus representatives that form akinetes distant from heterocysts, such as *Anabaena variabilis*, a "species" that was placed by Komárek and Anagnostidis (1989) into the genus *Trichormus* (*Trichormus variabilis*). Given the genetic diversity already evident for some strains of *Anabaena* (see Wilmotte and Herdman, this volume), it is likely that some of the distinctions proposed by Komárek and Anagnostidis (1989) may eventually prove useful for further generic subdivisions and nomenclatural changes.

Form-genus II. **Anabaenopsis** *(Woloszynska) Miller 1923*

ROSMARIE RIPPKA, RICHARD W. CASTENHOLZ, ISABELLE ITEMAN AND MICHAEL HERDMAN

A.na.bae' n.op' sis. Anabaena genus of cyanobacteria; Gr. *opsis* appearance; M.L. fem. n. *Anabaenopsis* *Anabaena*-like in appearance.

Short, immotile trichomes form a loose half circular to a full circular coil composed of 8–16 ellipsoidal, cylindrical, or barrel-shaped cells that contain gas vesicle clusters dispersed throughout the vegetative cells. Sheath layers are lacking. Terminal heterocysts often present at both ends of a trichome; new heterocyst differentiate in pairs in the center of the trichome. Akinetes, singly or in pairs, are formed distant from the heterocysts. Structurally distinct hormogonia not produced.

FURTHER DESCRIPTIVE INFORMATION

The cells composing the short coiled trichomes are separated by relatively deep constrictions and end cells do not differ in morphology from intercalary cells. Following trichome elongation to approximately twice the original length, **asymmetric division of two intercalary cells in the center of the trichome** results in the formation of two adjacent small vegetative cells that give rise to **transiently paired intercalary heterocysts**; subsequent trichome fragmentation between the latter results in two short trichomes that carry a heterocyst at both termini. **Large cylindrical akinetes** occur predominantly singly and differentiate one to two cells away from the paired heterocysts or, occasionally, form in place of the paired heterocysts at the center of the trichome.

Members of this genus are mainly known from soda lakes in Siberia and East Africa, but have also been found as constituents of blooms in lakes of Europe, Africa, India, and North and South America (Desikachary, 1959; Jeeji-Bai et al., 1977). Four axenic isolates are presently available in the PCC. Two of them (PCC 9215 and PCC 9216) were isolated from a coastal lagoon in Spain by H. Rodriguez (see Rippka and Herdman, 1992), the others originated from the alkaline Lake Nakuru in Kenya (PCC 9420) and from a dam in Sweden (PCC 9608) (Iteman et al., 1999). Strain PCC 9420 (SAG 252.80) entered the PCC as *Anabaenopsis elenkinii* and, as strain PCC 9608, has slightly larger cells (7–8 µm in diameter) than the two isolates from Spain (5–7 µm in diameter). A photomicrograph of strain PCC 9215 is shown in Fig. B10.72. In contrast to strains PCC 9215, PCC 9216, and PCC

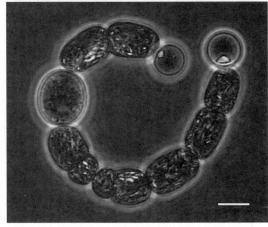

FIGURE B10.72. *Anabaenopsis* PCC 9215. Phase contrast. Bar = 5 µm.

9608, which grow in medium BG11$_0$ supplemented with NaHCO$_3$ (10 mM), strain PCC 9420 requires higher concentrations of NaHCO$_3$ (50–100 mM). As gas vesicle clusters, the coiled trichome structure of *Anabaenopsis* isolates is easily lost in culture, as in strain PCC 9216, which now exhibits exclusively straight trichomes (R. Rippka, unpublished data). However, the asymmetric cell division prior to de novo heterocyst differentiation is maintained even in the spontaneous mutant phenotype.

Reference strains

PCC 9215, isolated by H. Rodriguez from a coastal lagoon, Albufera de Valencia, Valencia, Spain. Carried in the PCC as *Anabaenopsis* sp. (Rippka and Herdman, 1992). Mol% G + C is 43.2.

PCC 9420, isolated by E. Hegewald from Lake Nakuru, Kenya, and identified as *Anabaenopsis elenkinii* (co-identic with SAG 252.80) (Schlösser, 1984). Mol% G + C is 43.1.

ENRICHMENT AND ISOLATION PROCEDURES

Isolation has been successful by Jeeji-Bai et al. (1977) in a medium described by Uherkovich (1969). Purification of three strains carried in the PCC was performed on plates prepared with medium BG11$_o$, supplemented with filter sterilized NaHCO$_3$ (10 mM) and solidified with 0.8% (w/v) Sigma washed agar (A 8678). For strain PCC 9420, which requires a higher Na$^+$ concentration, the medium was supplemented with 65 mM NaHCO$_3$ and 15 mM Na$_2$CO$_3$.

TAXONOMIC COMMENTS

In the last edition, Castenholz (1989b) combined this genus with *Anabaena*, but suggested that it may warrant independent generic status. Recent 16S rRNA gene sequence data (Iteman et al., 1999) demonstrated that *Anabaenopsis* PCC 9215 is distant in the phylogenetic tree from both *Anabaena cylindrica* PCC 7122 and *Anabaena flos-aquae* PCC 9302. Thus its generic separation is justified, even though isolates of *Anabaena flos-aquae* have also been reported to produce helical coiled variants in culture (Booker and Walsby, 1979; see also Castenholz, 1989b). Iteman et al. (1999) also showed that *Anabaenopsis* PCC 9215 clusters relatively closely with *Cyanospira rippkae* PCC 9501 (see Wilmotte and Herdman, this volume). Unfortunately, no DNA–DNA hybridization data are available to correlate total genomic relatedness, necessary to validate generic distinctions, with the observed degree of 16S rRNA gene sequence divergence between the latter two strains. *Anabaenopsis* as the older genus has legal priority over *Cyanospira* Florenzano et al., 1985. Thus, if these two "genera" would eventually need to be combined, only *Cyanospira* would lose validity.

Form-genus III. **Aphanizomenon** *Morren 1838*

ROSMARIE RIPPKA, RICHARD W. CASTENHOLZ, ISABELLE ITEMAN AND MICHAEL HERDMAN

A.pha.ni.zo'me.non. Gr. v. *aphanizo* to disappear; Gr. neut. pass. part. *aphanizomenon;* M.L. neut. n. *Aphanizomenon* obscure thing.

Trichomes are straight and generally exhibit a slight attenuation in size toward their termini; a firm sheath is never produced. Intercalary vegetative cells are cylindrical, significantly longer than wide, and contain abundant gas vesicle clusters. The terminal cells are rounded to subconical and generally lack gas vesicles. Constrictions between individual cells are relatively shallow. Heterocysts differentiate exclusively in intercalary positions. Akinetes, singly or in pairs, are formed adjacent to, or distant from, heterocysts. Structurally distinct hormogonia are not produced.

FURTHER DESCRIPTIVE INFORMATION

This genus is defined on the basis of *Aphanizomenon flos-aquae* (L.) Ralfs ex Bornet and Flahault 1888, the lectotype proposed by Geitler (1942). This "species" was first described by Linnaeus in 1753 as *Byssus flos-aquae* (see Baker, 1981; Anagnostidis et al., 1988). **All members of this genus are planktonic** (Geitler, 1932) and **in their natural environment** occur **singly or form feathery, flake-like, or spindle-shaped bundles of trichomes**. *Aphanizomenon flos-aquae* and *Aphanizomenon gracile* Lemm. occur in many mesotrophic and eutrophic lakes in temperate climates, being most abundant in late summer and fall. *Aphanizomenon* also occurs in **brackish waters** of the Baltic Sea together with *Nodularia spumigena* (Castenholz, 1989b). The **end cells** of the trichomes may be extremely long (up to 10 times the length of intercalary cells) and appear **less pigmented** ("hyaline"), though these traits may not be observed in cultured isolates. Heterocysts are similar in shape and dimensions to the vegetative cells. Akinetes are generally slightly broader and significantly longer (2–3 fold) than the vegetative cells. A representative of this genus is shown in Fig. B10.73.

A single axenic isolate (PCC 7905) conforming to the description of this genus is presently available in the PCC, although two other axenic representatives (strains TR 183 and 202, from the Baltic Sea and a freshwater lake in Finland, respectively) were included in a genetic study by Lyra et al. (1997). Strain PCC 7905 was isolated as a bloom-forming cyanobacterium from a Dutch lake and was originally identified as *Aphanizomenon flos-aquae* (W. Zevenboom, personal communication). However, although the terminal cells often exhibit reduced content of gas vesicle clusters, the cell elongation typical for the latter species in nature has never been observed in culture. Furthermore, strain PCC 7905 differentiates numerous intercalary heterocysts at fairly regular intervals in the relatively long trichomes. This is in contrast to the description of *Aphanizomenon flos-aquae* by Komárek and Anagnostidis (1989), who define this species as typically producing only one to two heterocysts in a "subsymmetric" pattern along the trichome. In addition, akinetes are lacking in strain PCC 7905, or their formation may require special nutritional stimuli. This isolate grows in even suspension in liquid medium and does not produce flake-like aggregates. However, O'Flaherty and Phinney (1970) observed on unicyanobacterial nonaxenic cultures of *Aphanizomenon flos-aquae* that typical flake or bundle formation could be maintained if sufficient iron was

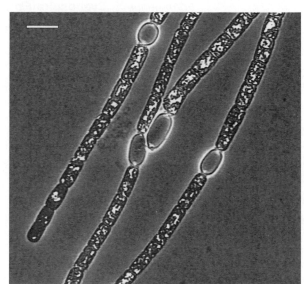

FIGURE B10.73. *Aphanizomenon flos-aquae* PCC 7905. Phase contrast. Bar = 10 μm.

available in the medium. Thus the influence of this element on characteristic colony morphology may be worth reexamining. It should also be mentioned that (as with most planktonic isolates) care has to be taken not to lose the wild-type phenotype, since strain PCC 7905 generates at relatively high frequencies non-gas-vacuolated mutants such as strain PCC 7905/1 (Rippka and Herdman, 1992). Strain PCC 7905 produces PEC (Bryant, 1982), but neither the mean DNA base composition nor its genome size have been determined.

Reference strains PCC 7905 (co-identity DCC D0654), isolated by F.I. Kappers from lake water (Brielse Meer), The Netherlands, and identified as *Aphanizomenon flos-aquae* (Zevenboom et al., 1981). PCC reference strain for *Aphanizomenon* (Rippka and Herdman, 1992).

MAINTENANCE PROCEDURES

Strain PCC 7905 and the mutant derivative, strain PCC 7905/1, are maintained in liquid medium BG11$_o$, supplemented with NaHCO$_3$ (5 mM) at 20–23°C; growth on solid medium proved difficult. Isolation conditions and other media appropriate for members of this genus are described by McLachlan et al. (1963), Gentile and Maloney (1969), O'Flaherty and Phinney (1970), Heaney and Jaworski (1977), and Rouhiainen et al. (1995).

TAXONOMIC COMMENTS

The definition of this genus is identical to that of Castenholz (1989b). However, it should also be mentioned that certain planktonic species described for the genus *Anabaena* differ from those of *Aphanizomenon* by such subtle differences that identification can be very problematic, not only at the specific but also at the generic level (see Komárek and Anagnostidis, 1989). This is evidenced by some recent phylogenetic studies that showed that a number of planktonic isolates identified as either *Anabaena* or *Aphanizomenon* cluster very tightly in the 16S rRNA tree (Iteman et al., 1999), or may not even be distinguished (Lyra et al., 1997). However, in the absence of DNA–DNA hybridization studies demonstrating the necessity to combine the planktonic members of these two "genera", *Aphanizomenon* has been retained in this edition.

Form-genus IV. **Cyanospira** *Florenzano, Sili, Pelosi and Vincenzini 1985, 305*

ROSMARIE RIPPKA, RICHARD W. CASTENHOLZ, ISABELLE ITEMAN AND MICHAEL HERDMAN

Cy'an.o.spi'ra. Gr. adj. *kuanos* blue (color); L. n. *spira* a coil; M.L. fem. n. *Cyanospira* blue-(green) coils.

Helically coiled, immotile, trichomes with intercalary and terminal heterocysts; the vegetative cells are spherical to ovoid and are separated by relatively deep constrictions; gas vesicle clusters are dispersed throughout the vegetative cells. Defined sheaths are lacking. Asymmetric cell division prior to de novo heterocyst differentiation does not take place. Akinetes differentiate distant from heterocysts, often forming chains. Hormogonia distinct in morphology from mature trichomes not produced.

FURTHER DESCRIPTIVE INFORMATION

This genus is based on the description of the type species *Cyanospira rippkae* Florenzano et al. 1985, **strain Mag II 702** (PCC 9501) having been **designated as the holotype**. This isolate originated from a soda lake (Lake Magadi) in the Eastern Rift Valley (Kenya), a habitat also typical of members of the genus *Anabaenopsis* and of the filamentous nonheterocystous representatives of *Arthrospira platensis* (see Subsection III, this volume). The pH in the alkaline lakes of the Eastern rift Valley ranges from 9.5 to 10.2, sodium, bicarbonate, and carbonate being the major ions (Florenzano et al., 1985). As shown in Fig. B10.74 strain Mag II 702 (PCC 9501) has ovoid vegetative cells 4–5 μm in diameter, the spherical heterocysts and akinetes exhibiting slightly larger dimensions (diameter 7–8 μm and 9–9.5 μm, respectively).

A second isolate, strain **Mag I 504** (PCC 9502) from the same habitat was assigned to a second species, *Cyanospira capsulata* (Florenzano et al., 1985). It is slightly larger (diameter of vegetative cells 6–7 μm; heterocysts 8–8.5 μm; akinetes 9–10 μm) and differs from the former strain (Mag II 702) by its spherical cells, lower content of gas vesicle clusters, and abundant mucilage production.

In both strains, reproduction by random trichome breakage results in short coiled chains of cells. Both isolates are obligate photoautotrophs, synthesize only phycocyanin and allophycocyanin and were reported to have a low mean DNA base compo-

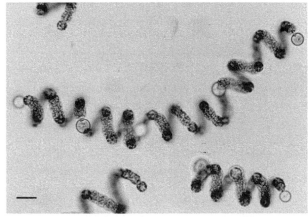

FIGURE B10.74. *Cyanospira rippkae* PCC 9501 (strain Mag II 702). Courtesy of C. Sili. Bright field. Bar = 10 μm.

sition (35 and 36 mol% G + C, respectively) (Florenzano et al., 1985). However, higher values (42.8 and 42.3 mol% G + C) were determined for the corresponding axenic strains PCC 9501 and PCC 9502 (M. Herdman, unpublished data).

Type strains

Cyanospira rippkae Florenzano et al. 1985, strain Mag II 702, isolated in 1981/1982 from Lake Magadi, Kenya (Florenzano et al., 1985). Corresponds to the axenic strain PCC 9501 (Iteman et al., 1999). The mol% G + C is 42.8 (M. Herdman, unpublished).

Cyanospira capsulata Florenzano et al. 1985, strain Mag I 504, isolated in 1981/1982 from Lake Magadi, Kenya (Florenzano et al., 1985). Corresponds to the axenic strain PCC 9502 (Iteman et al., 1999). The mol% G + C is 42.3 (M. Herdman, unpublished).

ENRICHMENT AND ISOLATION PROCEDURES

The strains *Cyanospira* Mag II 702 and Mag I 504 were isolated on plates prepared with medium $BG11_o$, supplemented with $NaHCO_3$ (about 120 mM) and solidified with Difco Bacto agar (1% w/v) (Florenzano et al., 1985). Final purification from highly motile contaminants, resulting in strains PCC 9501 and PCC 9502, respectively, was achieved in the same medium, solidified with Sigma washed agar (A 8678) and supplemented with 65 mM $NaHCO_3$ and 15 mM Na_2CO_3 (Rippka, unpublished). Maintenance of the axenic strains is performed in liquid medium.

TAXONOMIC COMMENTS

Cyanospira is one of the rare genera whose description was made under the rules of the *Bacteriological Code of Nomenclature*, and both species are based on cultured representatives. Cyanobacteria assigned to this new genus resemble both *Anabaena spiroides*

and some "species" traditionally assigned to *Anabaenopsis*. However, they can be distinguished from the former (a typical freshwater inhabitant) by having a higher salt requirement and from the latter by some morphological features (some of which may be more trivial than others): the coiled filaments are longer; uneven cell division leading to two transiently paired heterocysts is never observed; akinetes are spherical and occur in chains, often comprising a significant part of the trichomes.

In agreement with their similar ecology, *Cyanospira* PCC 9501 and *Anabaenopsis* PCC 9215 cluster relatively tightly in the 16S rRNA tree (Iteman et al., 1999; see also Wilmotte and Herdman, this volume). However, for lack of DNA–DNA hybridization studies, it is presently unknown whether these two genera will eventually need to be combined (in which case *Anabaenopsis* would have legal priority) or whether both can be justified. Since no 16S rRNA sequence data are available for *Anabaena spiroides*, the relationship between the latter and *Cyanospira* remains unknown.

Form-genus V. **Cylindrospermopsis** *Seenayya and Subba-Raju 1972*

ROSMARIE RIPPKA, RICHARD W. CASTENHOLZ AND MICHAEL HERDMAN

Cy.lin.dro.sper'm.op'sis. *Cylindrospermum* genus of cyanobacteria; Gr. *opsis* appearance; M.L. fem. n. *Cylindrospermopsis* *Cylindrospermum*-like in appearance.

The thin trichomes (<4 µm in diameter) are straight or loosely coiled; vegetative cells are cylindrical and exhibit randomly distributed gas vesicle clusters. Heterocysts occupy exclusively terminal positions, at one or both ends of the trichomes. As the vegetative cells that give rise to them, heterocysts are conical to spear-shaped. Cylindrical akinetes, singly, in pairs or short chains, are formed adjacent to, or slightly distant from, the heterocysts. Sheath layers and hormogonia that differ structurally from the mature filaments are not produced.

FURTHER DESCRIPTIVE INFORMATION

This genus is best known for the type species of the genus, *Cylindrospermopsis raciborskii* (Woloszynska) Seenayya and Subba-Raju 1972, first described from the plankton of lakes in Indonesia (as a species of *Anabaena*) by Woloszynska (1912). However, the same or similar representatives have been found, often as constituents of water blooms, in ponds or lakes in many tropical parts of the world (East Africa, Central and South America, Australia) (see Komárková, 1998), but occur also in temperate water bodies of North America and Central and Southern Europe (Padisák, 1996; Couté et al., 1997). Based on some relatively subtle morphological differences at least seven botanical "species" have been recognized (see Komárková, 1998). The potential toxicity associated with members of this genus has only recently been discovered (Hawkins et al., 1985; Ohtani et al., 1992). These authors reported the presence of a new type of hepatotoxic alkaloid (cylindrospermopsin) in blooms and isolates of *Cylindrospermopsis raciborskii*.

No axenic representatives of *Cylindrospermopsis* are presently available. However, a photomicrograph of a typical representative from nature is shown in Fig. B10.75A and B.

Reference strain AWT 205 (not axenic), as *Cylindrospermopsis raciborskii* from Solomon dam (Palm Island, a tropical continental island about 28 km from the northeastern coast of Australia)

(Hawkins et al., 1985) (B. Neilan, personal communication to R. Rippka).

MAINTENANCE PROCEDURES

Growth conditions for an isolate of *Cylindrospermopsis raciborskii* from a domestic water supply reservoir close to the northeastern coast of Australia were described by Hawkins et al. (1985). The nonaxenic clonal strain AWT 205 derived from the latter habitat, grows in medium BG11 (B. Neilan, personal communication). However, purification of this strain has not yet been achieved, in spite of varying the media and the solidifying agents (R. Rippka, unpublished data).

TAXONOMIC COMMENTS

The botanical type species of this genus, *Cylindrospermopsis raciborskii*, has long been considered as a member of the genus *Anabaenopsis* (see Geitler, 1932; Hawkins et al., 1985; Couté et al., 1997; Komárková, 1998). However, as first critically analyzed by Seenayya and Subba-Raju (1972), it differs from typical members of the latter genus (such as *A. elenkinii* or related "species") primarily by its different pattern of heterocyst differentiation. Consequently, the latter authors proposed the creation of a new genus, *Cylindrospermopsis*, designating *C. raciborskii* (Woloszynska) Seenayya and Subba-Raju 1972 as the type species. In *Anabaenopsis* de novo heterocyst differentiation always occurs after two asymmetric cell divisions in the center of the filaments, resulting temporarily in paired heterocysts (for more details, see description of the genus *Anabaenopsis*, above). This is in contrast to representatives now assigned to *Cylindrospermopsis*. Although it has been proposed that uneven cell division of the terminal cells may give rise to the conical to spear-like terminal heterocysts typical of the latter genus (Komárková, 1998), this theory needs to be confirmed by more careful developmental studies. "Species" described as *Cylindrospermopsis* resemble superficially in morphology members of *Aphanizomenon* or *Cylindrospermum*. However, they

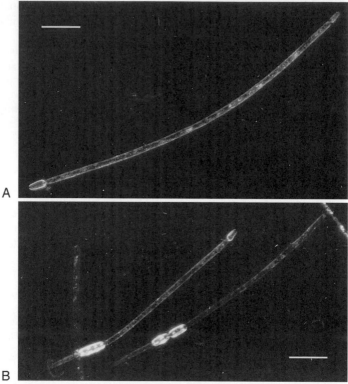

FIGURE B10.75. *Cylindrospermopsis.* Dark field photomicrograph of a natural sample. (*A*) Filament with pointed terminal heterocyst typical of this genus; (*B*) filament showing the position of the akinetes. Bar = 5 μm. Courtesy of A. Couté, J.-F. Briand, and C. Bernard.

can be distinguished from the former by the pointed terminal heterocysts (lacking in *Aphanizomenon*) and from the latter by the presence of conspicuous gas vesicle clusters (never observed in *Cylindrospermum*). Another interesting morphological trait of *Cylindrospermopsis* that distinguishes this genus from practically all other heterocystous cyanobacteria is the small trichome diameter (0.8–4 μm) typical of most members (Komárek and Kling, 1991; Komárková, 1998).

This genus is new in this edition of the *Manual.* Although no axenic representatives are yet available, it has been included here on the basis of 16S rRNA sequence data (see Wilmotte and Herdman, this volume). The long phylogenetic distances that separate representatives of the genera *Anabaena, Anabaenopsis, Aphanizomenon,* and *Cylindrospermum* from those of *Cylindrospermopsis* clearly indicate the validity of the latter as an independent genus.

Form-genus VI. **Cylindrospermum** *Kützing 1843*

ROSMARIE RIPPKA, RICHARD W. CASTENHOLZ AND MICHAEL HERDMAN

Cy.lin.dro.sper' mum. Gr. n. kylindros roller, cylinder; Gr. n. *sperma* seed; M.L. neut. n. *Cylindrospermum* (filament) cylinder of seedlike cells.

Trichomes, generally motile, are untapered and differentiate exclusively terminal heterocysts at one or both ends. The vegetative cells are cylindrical in shape and do not contain gas vesicle clusters. A single large cylindrical to ovoid akinete is formed adjacent to one or both heterocysts in the trichome. Sheath envelopes and structurally distinct hormogonia are not produced.

FURTHER DESCRIPTIVE INFORMATION

Cylindrospermum is best known as nonplanktonic, i.e., as a part of the tychoplankton or periphyton of freshwaters. Some "species" also occur in moist subaerial (terrestrial) habitats. The trichomes are relatively long, and are composed of cells that are of equal length and width, or are longer than wide; constrictions at the crosswalls are conspicuous but shallow. A confluent mu-

cilage uniting many individual trichomes is relatively common. In the absence of combined nitrogen, members of this genus **differentiate initially a terminal heterocyst** at only one end of the trichome; further trichome elongation by binary fission in a single plane eventually leads to heterocyst formation at the opposite end. Trichome breakage is random, but often occurs at a point toward the middle of the filament, thus producing daughter trichomes each possessing only a single terminal heterocyst. **Mature akinetes in *Cylindrospermum* are rather conspicuous in size** and when mature generally exhibit a brown pigmentation (see Geitler, 1932). Although some "species" were reported to produce multiple akinetes (see Geitler, 1932; Hirosawa and Wolk, 1979a, b), these specialized cells are more often solitary and **adjacent to the terminal heterocysts.** Fully mature akinetes, together with the adjacent heterocysts, detach easily from the trichomes and are liberated into the surrounding medium. Under conditions where heterocyst differentiation is completely repressed (i.e., in the presence of ammonium), akinetes have never been observed. A typical representative of this genus is shown in Fig. B10.76.

Only three axenic strains have been characterized (PCC 73101, PCC 7417, and PCC 7604) and have been assigned to two clusters (Rippka and Herdman, 1992).

Cluster 1 This cluster includes strains PCC 73101 and PCC 7417. When first rendered axenic, both strains exhibited cylindrical akinetes with dimensions (12–14 × 30–35 μm) conforming to the description of *Cylindrospermum stagnale* Kützing ex Bornet and Flahault 1888. However, strain PCC 73101 seems to have lost the capacity to form akinetes (R. Rippka, unpublished data). Both strains produce phycoerythrocyanin (PEC) (Bryant, 1982), but differ with respect to physiological properties, strain PCC 7417 (but not PCC 73101) using fructose and sucrose for photo- and chemoheterotrophic growth. The mean DNA base composition was reported to be 46.7 and 42.1 mol% G + C, respectively (Herdman et al., 1979b), but a slightly lower value (45 mol% G + C) for the former strain was determined by Lachance (1981). Strains PCC 73101 and PCC 7417 have a relatively large genome (5.7 and 6.2 Gdal, respectively). DNA–DNA hybridization performed between these strains resulted in 100% relative binding (Lachance, 1981). Thus, despite nutritional differences and the slightly divergent mean DNA base compositions, these two strains are undoubtedly members of the same nomenspecies, *C. stagnale*.

Reference strain PCC 7417 (co-identity ATCC 29204), isolated in 1972 by A. Neilson from soil in a greenhouse, Stockholm, Sweden (Rippka et al., 1979). Proposed as type strain of the type species *Cylindrospermum stagnale* (Rippka and Herdman, 1992). The mol% G + C is 42.1; genome size is 6.2 Gdal.

Cluster 2 Strain PCC 7604, the only member of this "cluster", produces ellipsoidal akinetes that are similar in width (12–14 μm) to those of strains of cluster 1, but may reach a length up to 50 μm. This strain was originally isolated and identified by J. Komárek as *Cylindrospermum majus* Kützing ex Bornet and Flahault 1888 (see strain history in Rippka and Herdman, 1992). The latter assignment seems acceptable, even though the maximal length of the akinetes observed for this strain is slightly more than that described for *C. majus*. Strain PCC 7604 is an obligate photoautotroph, produces PEC (Bryant, 1982), and has a mean DNA base composition of 42.9 mol% G + C (Herdman et al., 1979b). On the basis of DNA–DNA hybridizations performed with strain PCC 73101 (cluster 1), strain PCC 7604 is only distantly related to the former (59% relative binding) (Lachance, 1981). Thus, in agreement with its different specific epithet at the time of entering the PCC, strain PCC 7604 represents a second distinct species, *C. majus*.

Reference strain PCC 7604 (co-identity CCAP 1415/2), isolated by J. Komárek, source unknown. Proposed as type strain of the species *Cylindrospermum majus* (Rippka and Herdman, 1992). The mol% G + C is 42.9.

It should be noted that one of the traditional discriminatory characters used to separate *C. stagnale* from *C. majus* seems to be invalid: the papillate outer wall layers supposed to be typical of mature akinetes of the latter species (Bornet and Flahault, 1888) may also be observed in strains here assigned to *C. stagnale*.

MAINTENANCE PROCEDURES

All strains of this genus presently in culture grow well in medium BG-11$_o$, but prefer liquid media rather than slants for maintenance.

TAXONOMIC COMMENTS

The definition of *Cylindrospermum* is as given previously (Castenholz, 1989b) and is in full agreement with classical (Geitler, 1932)

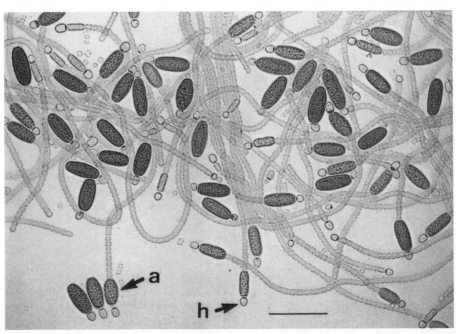

FIGURE B10.76. *Cylindrospermum* PCC 73101. *h*, heterocyst; *a*, akinete. Bar = 50 μm. (Reprinted with permission from R. Rippka et al., Journal of General Microbiology *111*: 1–61, 1979, ©Society for General Microbiology.)

and modern botanical taxonomic treatises (Komárek and Anagnostidis, 1989). This may be attributed to the fact that members of this genus are easily identified, providing that environmental or culture conditions are suitable for the expression of heterocyst and akinete development.

Comparative analyses of the 16S rRNA gene sequence of *Cylindrospermum stagnale* PCC 7417 with those of other heterocystous cyanobacteria, allow the following conclusions: this strain is member of a clade that includes *Nostoc* PCC 7120, several strains of *Tolypothrix*, and, as the closest relative, *"Calothrix"* PCC 7507 (see Wilmotte and Herdman, this volume). The latter isolate differs from other members of *Calothrix* by its low degree tapering, but interestingly, is the only "rivularian" member that, like *Cylindrospermum*, forms akinetes adjacent to the basal heterocysts. However, the genetic distance separating these two strains is larger than those between *Anabaenopsis* PCC 9215 and *Cyanospira* PCC 9501, or *Aphanizomenon* PCC 7905 and *Anabaena flos-aquae* PCC 9302 (see Willmotte and Herdman, this volume). Thus it would be premature to conclude that *Cylindrospermum* PCC 7417 and *"Calothrix"* PCC 7507 should be placed into the same genus.

Form-genus VII. **Nodularia** *Mertens 1822*

ROSMARIE RIPPKA, RICHARD W. CASTENHOLZ, ISABELLE ITEMAN AND MICHAEL HERDMAN

Nod.u.la′ri.a. L. adj. *nodulus* diminutive of nodus, small and knotty, knobby; L. n. *arium* place; M.L. fem. n. *Nodularia* knobby microbial filament.

The trichomes, motile or immotile, are composed of vegetative cells that are shorter than broad ("discoid" or "disk-like") and may be enclosed in a thin sheath layer. Terminal cells do not differ from intercalary cells. Heterocysts and akinetes are similar in shape to the vegetative cells but often appear compressed. Differentiation of heterocysts takes place predominantly in intercalary positions. Akinetes are initiated usually distant from heterocysts and may occur in short chains. Structurally distinct hormogonia not produced. Planktonic members exhibit gas vesicle clusters dispersed throughout the vegetative cells.

FURTHER DESCRIPTIVE INFORMATION

Members of this genus are benthic or planktonic, inhabiting brackish coastal waters and inland alkaline ponds or lakes in geographical diverse areas, but may also occur on the surface of soil (Öström, 1976; Nordin and Stein, 1980; Komárek et al., 1993; Blackburn et al., 1996; Hayes and Barker, 1997; Bolch et al., 1999). The aquatic habitats typically have pH values of 8–11 and salinities in the range of 3–67%, but the range for optimal growth of cultured isolates is lower (5–20%) (Nordin and Stein, 1980). A *Nodularia spumigena* variety producing abundant gas vesicle clusters is best known for forming toxic water blooms in the Baltic Sea and in brackish coastal lakes in Australia (Sivonen et al., 1989a, b; Bolch et al., 1999). A typical representative of *Nodularia* is shown in Fig. B10.77.

Clusters 1 and 2   Among the strains presently carried in the PCC, only two (PCC 73104 and PCC 7804) have been characterized in more detail. The former was received as *Nodularia spumigena* and is co-identic with UTEX 2091, originating from soil, near an alkaline lake in Canada (Nordin and Stein, 1980). The vegetative cells of this isolate are 7–8 μm × 3–6 μm. Strain PCC 7804 was isolated by P. Pourriot from a thermal spring in Southern France and was identified as *Nodularia harveyana*. However, the cell shape and dimensions of the latter are very similar to strain PCC 73104 (see Bolch et al., 1999), and neither strain produces gas vesicles or synthesizes a phycoerythrinoid pigment (Bryant, 1982). Strain PCC 73104 is capable of photoheterotrophic growth at the expense of glucose, fructose, and sucrose, whereas strain PCC 7804 is an obligate photoautotroph and produces the hepatotoxin nodularin (G. Codd, personal communication; Bolch et al., 1999). The strains also differ with respect to mean DNA base composition (40.5 and 44.7 mol% G + C, respectively) (Herdman et al., 1979b; Lachance, 1981). DNA–

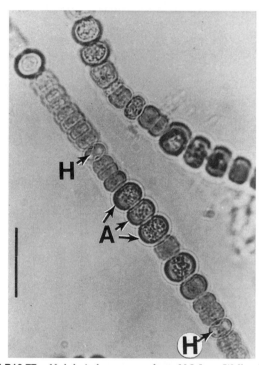

FIGURE B10.77. *Nodularia harveyana* culture N-8 from Wallender Lake, British Columbia. Cultured at 25°C, pH 9, and 1.6% NaCl. *H*, heterocyst; *A*, akinete. Bar = 10 μm. (Reprinted with permission from R.N. Nordin and J.R. Stein, Canadian Journal of Botany *58*: 1211–1224, 1980, ©National Research Council of Canada.)

DNA hybridizations revealed a relative binding of 65% and a $\Delta T_{m(e)}$ of 6°C (Lachance, 1981). Thus the two strains may be considered members of two different species of the same genus.

Reference strains

CLUSTER 1   PCC 73104 (co-identity UTEX 2091), isolated by R.N. Nordin from alkaline soil, Spotted Lake, British Columbia, Canada (Nordin and Stein, 1980). Accidentally quoted as *Nodularia harveyana* (co-identity UTEX 2093) by Castenholz (1989b). Proposed as type strain of the botanical lectotype species, *Nodularia spumigena* Mertens ex Bornet and Flahault emend. Nordin and Stein 1980 (Rippka and Herdman, 1992, and this edition). The mol% G + C is 40.5; genome size is 3.3 Gdal.

CLUSTER 2 PCC 7804 (co-identic with SAG 50.79 and CCALA 114), isolated in 1966 by P. Pourriot from a thermal spring, Dax, France; identified as *Nodularia harveyana*. Reference strain of *Nodularia* cluster 2 (Rippka and Herdman, 1992, and this edition). The mol% G + C is 44.7. This strain produces the hepatotoxin nodularin.

Cluster 3 Strain PCC 9350 was isolated as strain AV2 from a toxic bloom in the Baltic Sea (Martin et al., 1990; K. Sivonen, personal communication). This isolate has cell dimensions identical to strains PCC 73104 and PCC 7804, but the vegetative cells contain gas vesicle clusters, though their content is significantly lower than originally observed in the impure culture (R. Rippka, unpublished data). Using traditional means of identification this isolate corresponds best to *Nodularia spumigena* var. *vacuolata* (Geitler, 1932). The mean DNA base composition of this strain (42.1 mol% G + C) lies within the range of the former two isolates (M. Herdman, unpublished data). The genome complexity of strain PCC 7804 has not yet been determined, but both strains PCC 73104 and the Baltic isolate PCC 9350 have a relatively small genome (3.3 and 3.06 Gdal, respectively) compared to some genera of Subsection IV (Herdman et al., 1979a; M. Herdman, unpublished data). Based on 16S rRNA sequence data (Iteman et al., 1999), strain PCC 9350 represents a species distinct from strain PCC 73104. Its relationship to strain PCC 7804 remains to be examined.

Reference strain PCC 9350, isolated by K. Sivonen from a toxic bloom in the Baltic Sea (Martin et al., 1990; K Sivonen, personal communication). Reference strain of *Nodularia* cluster 3 (this edition). The mol% G + C is 42.1; genome size is 3.1 Gdal. This strain produces the hepatotoxin nodularin.

MAINTENANCE PROCEDURES

Strains PCC 73104 and PCC 7804 are presently maintained in medium $2N_{10}$ (BG-11$_o$, supplemented with 2 mM NaNO$_3$ and 10 mM NaHCO$_3$), since growth in the unsupplemented medium was poor and growth in medium BG11 selected for spontaneous mutants lacking heterocysts. Although containing a source of combined nitrogen, heterocyst differentiation is not repressed in medium $2N_{10}$ (R. Rippka, unpublished data). Strain PCC 9350 is cultured in medium BG-11$_o$, supplemented with artificial seawater (20% v/v) prepared according to the instructions given for Turks Island Salts (Merck Index 9954), and 5 mM NaHCO$_3$ (R. Rippka, unpublished data). Other media appropriate for isolates of *Nodularia* are reported by Blackburn et al. (1996), Hayes and Barker (1997), and Bolch et al. (1999). A photon flux density of 10–30 μmol m^{-2} s^{-1}, a light dark cycle (12–16 h light), and relatively narrow temperature range (20–25°C) seem advisable, particularly for the cultivation of gas vesicle containing members (Bolch et al., 1999; R. Rippka, unpublished data).

TAXONOMIC COMMENTS

The definition of this genus is the same as that given in the previous edition and agrees with Komárek and Anagnostidis (1989). However, the present "species" concept is rather confusing and warrants some comments. Nordin and Stein (1980), on the basis of carefully studied unicyanobacterial cultures, freshly collected field material and reexamination of herbarium specimens, judged that only two species of *Nodularia* described in the botanical literature deserve to be maintained, *N. spumigena* Mertens ex Bornet and Flahault 1886 and *N. harveyana* (Thwaites) Thuret 1875. Rippka and Herdman (1992) accepted this point of view and designated strain PCC 73104 (UTEX 2091) as the type strain of the botanical lectotype species, *Nodularia spumigena* Mertens ex Bornet and Flahault emend. Nordin and Stein 1980. However, the practice of creating new species or revising existing ones (Komárek et al., 1993), together with an error in quoting the correct co-identity of strain PCC 73104 (see reference strains, above), has led to the present situation where genetic distinctions between isolates are easily made (Hayes and Barker, 1997; Bolch et al., 1999) but the nomenclature of "species" remains highly controversial (Bolch et al., 1999). As an example, in some publications strains PCC 73104 (UTEX 2091) and UTEX 2093 (PCC 9336, but not described here) are cited as *"Nodularia sphaerocarpa"* and *Nodularia* sp., respectively (see Komárek et al., 1993; Bolch et al., 1999), though the latter had previously been designated as the type strain of *Nodularia harveyana* (Thwaites) Thuret 1875 (Nordin and Stein, 1980). Until these more recent taxonomic problems have been solved, we maintain the assignment of Nordin and Stein (1980) and consider strain PCC 73104 the type strain of *Nodularia spumigena*.

The 16S rRNA gene sequences of three strains of *Nodularia* (PCC 73104, PCC 9350, and BCNOD 9427) cluster tightly together in a clade mainly composed of planktonic heterocystous cyanobacteria (see Wilmotte and Herdman, this volume). The genetic distances between the strains suggest that they may be assignable to three different species. However, this needs to be confirmed by DNA–DNA hybridization studies. The phylogenetic distinction of PCC 73104 and BCNOD 9427 is in good agreement with their separation into different clusters based on sequence analyses of the intergenic spacer (IGS) of the *cpc* operon (Bolch et al., 1999). Strain PCC 9350 was not included in the latter study.

Form-genus VIII. **Nostoc** *Vaucher 1803*

MICHAEL HERDMAN, RICHARD W. CASTENHOLZ AND ROSMARIE RIPPKA

Nos′toc. M.L. n. *Nostoc* origin uncertain, supposedly invented by Paracelsus from the old English word *Nosthryl* (nostril) and the corresponding German word *Nasenloch = Nostoch.*

Mature trichomes are nonmotile, wave-like or tightly coiled, and composed of spherical or ovoid vegetative cells. In typical members, hormogonia are straight, distinct in cell morphology, and motile; some may contain gas vesicle clusters. Constrictions at cross-walls are conspicuous both in hormogonia and mature trichomes. Terminal heterocysts are differentiated at both ends of the maturing hormogonia; subsequent development gives rise to trichomes containing predominantly intercalary heterocysts.

Mucilage production or sheath formation is typical of many members. Akinetes are initiated distant from heterocysts, often occurring in long chains.

FURTHER DESCRIPTIVE INFORMATION

Cyanobacteria assigned to this genus exhibit structural properties that are similar to members of *Anabaena*. Reproduction occurs by binary fission in a single plane and the trichomes are of equal diameter throughout their entire length. However, with a few exceptions, members of *Nostoc* are **characterized by a developmental cycle**, in which **hormogonia are produced** that are clearly distinct from the mature parental trichomes. In most representatives, hormogonia possess relatively small cylindrical intercalary cells and slightly pointed to conical end cells; during maturation hormogonial cells increase in size and eventually assume the spherical to ellipsoidal shape typical of those of mature filaments (see Fig.B10.78). **Newly liberated hormogonia lack heterocysts** and **generally exhibit transient gliding motility** ("migrating phase"); some synthesize gas vesicle clusters that are dispersed throughout the cells or are located on either side of the crosswalls. In the latter case, the hormogonia resemble microscopically the trichomes of *Pseudanabaena* (compare Fig.B10.78A with Fig. B10.62). Under nitrogen limiting conditions, hormogonia

give rise to **young trichomes that differentiate heterocysts at both termini**. Gas vesicles, if present, are lost during maturation. After further increase in cell size, often associated with changes in cell shape, intercalary heterocysts are produced, motility is arrested, and mucilage or sheath formation may occur. Due to concomitant shifts of cell orientation, the trichomes may display a loose to tightly coiled configuration ("aseriate" stage) (Lazaroff and Vishniac, 1961; Lazaroff, 1972, 1973). Some "species" of *Nostoc* were thought to divide in a plane parallel to the long axis in the late stages of development (Bornet and Flahault, 1888; Geitler, 1932; Kantz and Bold, 1969; Lazaroff, 1972, 1973). However, as evidenced by the lack of lateral heterocysts, a true change in the plane of division, such as is typical of members of Subsection V (see Hoffman and Castenholz, this volume), does not occur. **Akinete formation** in members of *Nostoc* **is always initiated equidistant between two heterocysts**; successive differentiation on either side of these cold and desiccation-resistant cells results in chains of akinetes, which in old cultures may comprise almost the totality of the vegetative trichomes. They may also detach to produce a cell population composed almost entirely of single akinetes.

Some members included in this genus are less typical, since the hormogonia may lack gliding motility and be similar in size and shape to the mature trichomes. Some could only be distinguished by transient gas vesicle formation (Rippka, 1988b). Correct assignment of such representatives requires confirmation by genetic evidence.

Nostoc, as typically observed in nature, is characterized by a confluent gelatinous matrix embedding numerous trichomes. Such aggregates may be spherical, ovoid, or hair-like in shape. Some colonies may be encountered as flattened disks or large sheets, or they may be soft and amorphous with no distinct boundaries. In others, a high density of trichomes is observed in the firm outer layer of a colony while the interior is of softer consistency and contains trichomes that may be radially arranged (see Fig. 19.69 in Castenholz, 1989b). The size of colonies range from microscopic (representing the progeny of a single hormogonium) to over 20 cm in diameter (Castenholz, 1989b).

Various "species" of *Nostoc* are known from benthic habitats in freshwater lakes or streams. They may either grow firmly attached on solid substrates, or the semispherical colonies rest only loosely on more or less consolidated sediments. Many occur as amorphous sheets or aggregates of mucilage-bound trichomes in freshwater or on moist soil. *Nostoc* is also known to form thick, soft mats in polar regions where members of this genus are the major N_2-fixing microorganisms. Although in the past sometimes identified as *Anabaena*, representatives of *Nostoc* are the most commonly encountered cyanobacterial partners in exo- or endo-symbiotic associations with ascomycetes (to form lichens), bryophytes, pteridophytes, cycads, and *Gunnera* (a genus of angiosperms), in which they benefit the hosts mainly by the supply of fixed nitrogen (see Stewart, 1980; Stewart et al., 1980, 1983; Enderlin and Meeks, 1983; Meeks, 1988).

The desiccation tolerance typical of many members of this genus, as well as the effects of rehydration, have been studied in detail by Potts and collaborators (Scherer and Potts, 1989; Hill et al., 1994; Xie et al., 1995; see also Potts, 1994, 1996).

McGuire (1984), using numerical taxonomy, has distinguished several "species" of *Nostoc* on the basis of 30 morphological characteristics. The etymology of the name *Nostoc* has been proposed by Potts (1997).

Among the numerous strains (about 100) of *Nostoc* carried in the PCC (Rippka and Herdman, 1992; R. Rippka, unpublished

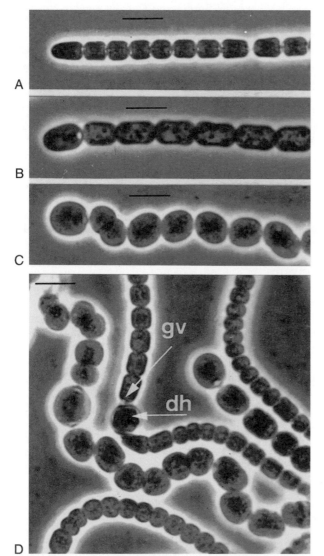

FIGURE B10.78. *Nostoc* PCC 6720. (*A* to *C*) successive stages of the development of a hormogonium into a mature trichome. (*D*) a degenerating heterocyst (*dh*) from which a rapidly dividing, newly formed hormogonium will subsequently detach; gas vacuoles (*gv*) are already present in the cells adjacent to the heterocyst. Phase contrast. Bar = 5 μm. (Reproduced with permission Rippka et al., Journal of General Microbiology *111*: 1–61,1979, ©Society for General Microbiology.)

data), only 17 have been characterized in more detail (Rippka et al., 1979; Herdman et al., 1979a, b). Four additional isolates were characterized for pigment composition (Bryant, 1982) and examined for genetic relatedness by DNA–DNA hybridization (Lachance, 1981). The range of mean DNA base composition for all strains is relatively narrow, 39–47.2 mol% G + C (Herdman et al., 1979b; Lachance, 1981). The genome size (determined only for some representatives) is 3.7–6.4 Gdal (Herdman et al., 1979a). Except for one tight genetic cluster composed of six strains (Cluster 3.1), the other 15 strains analyzed represent many different species, but seemed to be sufficiently related to warrant assignment to a single genus (Lachance, 1981). On the basis of pigment content and DNA similarities the strains can be assigned to several clusters.

Cluster 1 This cluster comprises six strains from freshwater habitats or soil (PCC 6302, PCC 6310, PCC 7121, PCC 7706, PCC 7803, and PCC 7807) and two (PCC 73102, PCC 7422) that were isolated as endosymbionts from cycad species. All produce C-phycoerythrin (C-PE), but differ in their response to light quality. Strains PCC 6302, PCC 7121, PCC 7422, and PCC 7807 synthesize C-PE constitutively, whereas the others undergo complementary chromatic adaptation of type II (Bryant, 1982). Except for strain PCC 7422, an obligate photoautotroph, all are facultative photoheterotrophs; two of the latter (PCC 6310 and PCC 73102) also exhibit growth under chemoheterotrophic conditions (Rippka and Herdman, 1992). The degree of trichome coiling in the advanced stages of the developmental cycle is highly variable among these isolates, being most pronounced in strain PCC 73102 (see Fig. B10.79A-C), the reference strain of cluster 1 (Rippka and Herdman, 1992, and this edition). Strains PCC 6310 and PCC 7807 are atypical representatives of this genus, since their hormogonia differ little from mature trichomes and exhibit respectively little or no gliding motility. Strains PCC 6302 and PCC 7121 are spontaneous mutants that no longer form functional heterocysts and thus are unable to fix N_2 under aerobic conditions. In addition, the latter strain produces very short immotile trichomes, composed of two to four cells. However, in

agreement with the original description of this isolate (Allison et al., 1937), a replacement culture of strain PCC 7121 (PCC 7906, co-identic with UTEX 486) still exhibits all the features typical of *Nostoc*, including motile hormogonia.

DNA–DNA hybridizations of these eight strains (Lachance, 1981) revealed only a moderate degree of similarity (49–64% relative binding, $\Delta T_{m(e)}$ of 7–16°C), when using PCC 6302 and PCC 73102 as sources of reference DNA. Thus, although clearly being representatives of the same genus, strains of cluster 1 are genetically heterogeneous (see Fig. B10.80), none of them being sufficiently related as to be considered members of the same nomenspecies.

Reference strain PCC 73102 (co-identity ATCC 29133), isolated in 1973 by R. Rippka from a root section of the cycad *Macrozamia* (Australia) (Rippka et al., 1979). Proposed as type strain of the species *Nostoc punctiforme* (Rippka and Herdman, 1992, and this edition). The mol% G + C is 45.2; genome size is 4.9 Gdal.

Cluster 2 This cluster is represented by three strains (PCC 6720, PCC 7107, and PCC 7416), isolated from soil or shallow ponds. Strain PCC 6720 was identified prior to entering the PCC as *Anabaenopsis circularis* (Rippka et al., 1979), which evidently was an incorrect identification (see description of *Anabaenopsis*, above). All three strains synthesize PEC (Bryant, 1982), are facultative photo- and chemoheterotrophs, and exhibit a developmental cycle involving readily identifiable motile hormogonia and loosely coiled mature filaments. Strain PCC 6720 produces hormogonia with polar gas vesicle clusters (Fig. B10.78), whereas strains PCC 7107 and PCC 7416 seem to lack these structures. Based on DNA–DNA hybridization studies by Lachance (1981), they can be divided into two subclusters.

Cluster 2.1 Strains PCC 7107 and PCC 7416 were both isolated from shallow ponds in California, exhibited a high degree of DNA–DNA similarity (93% relative binding, $\Delta T_{m(e)}$ 0°C) and can be considered members of the same nomenspecies (Lachance, 1981).

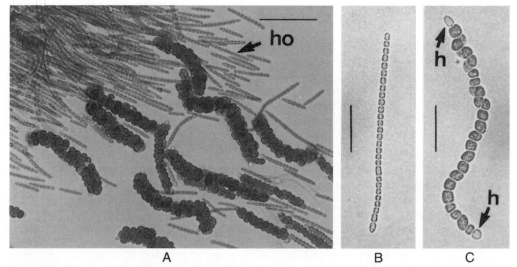

FIGURE B10.79. *Nostoc* PCC 73102. (A) coiled vegetative filaments together with hormogonia (*ho*). (B) enlarged view of a hormogonium. (C) a mature, but still uncoiled, vegetative trichome with terminal heterocysts (*h*). Bright field. Bar = 50 μm (A) and 20 μm (B,C). (Reproduced with permission Rippka et al., Journal of General Microbiology *111*: 1–61,1979, ©Society for General Microbiology.)

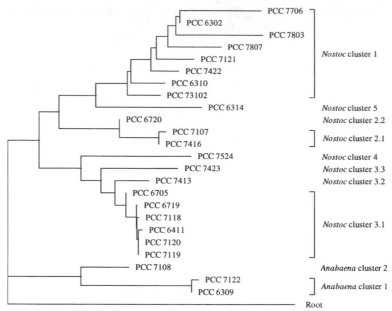

FIGURE B10.80. Dendrogram showing relationships between representative strains of the genera *Anabaena* and *Nostoc*, deduced from the DNA–DNA hybridization results of Lachance (1981) and M.-A. Lachance and R.Y. Stanier (personal communication). Relative binding values were converted to dissimilarity, incorporated into a distance matrix, and analyzed by the neighbor-joining method in TREECON (Van de Peer and de Watcher, 1994). The different clusters of each genus are indicated on the right. The tree was rooted by the use of an arbitrary value ("root") representing 15% relative binding.

Reference strain PCC 7107 (co-identity ATCC 29150), isolated in 1970 by A. Neilson from a shallow pond, Point Reyes Peninsula, California (Rippka et al., 1979). Reference strain of *Nostoc* cluster 2 (Rippka and Herdman, 1992). Proposed as reference strain of cluster 2.1 (this edition). The mol% G + C is 41.8.

Cluster 2.2 Strain PCC 6720, isolated from soil in Indonesia (Rippka et al., 1979), is relatively closely related to the strains of cluster 2.1 (69% relative binding, $\Delta T_{m(e)}$ of 4°C, using strain PCC 7107 as the source of reference DNA), but represents a different species (Lachance, 1981).

All three isolates of cluster 2 are only distantly related (23–42% relative binding) to representatives of *Nostoc* clusters 1, 3, and 4 (see Fig. B10.80).

Cluster 3 The nine strains included in cluster 3 are relatively atypical members of the genus *Nostoc*, since none of them exhibits a very distinct developmental cycle under standard growth conditions, except strain PCC 6705, which produces motile gas vesicle-containing hormogonia. Based on DNA–DNA hybridization studies by Lachance (1981), they can be divided into several subclusters.

Cluster 3.1 The six strains included in this cluster (PCC 6411, PCC 6705, PCC 6719, PCC 7118, PCC 7119, and PCC 7120) have trichomes composed of barrel-shaped cells that are very similar in dimensions (3–3.5 μm in width). Although for many of them the original habitat is unknown (Rippka et al, 1979), they are most likely freshwater or soil isolates from North America. All are obligate photoautotrophs and synthesize PEC (Bryant, 1982). Except for strain PCC 6705, all are permanently immotile. Upon transfer of the cultures to fresh medium all, except strain PCC 7120, produce hormogonia that exhibit gas vesicle clusters dispersed throughout the cells. However, in contrast to strain PCC 6705, this transient developmental feature is only observed if carryover of old medium (containing an inhibitor of hormogonial differentiation) is avoided (Rippka, 1988b; Herdman and Rippka, 1988). Strain PCC 7120 may carry a spontaneous mutation that is responsible for the lack of gas vesicle-containing hormogonia, since Damerval et al. (1989) showed that all strains here assigned to cluster 3.1 contain the genes *gvp*A and *gvp*C, which encode the structural proteins of gas vesicles. With the exception of PCC 6705 and PCC 6719, all strains are hosts to the lytic cyanophage N-1 (Adolph and Haselkorn, 1971; Rippka et al., 1979; R. Rippka, unpublished data). Like strain PCC 7118, which was unable to form heterocysts and to fix nitrogen under aerobic conditions upon receipt in the PCC, strain PCC 6411 has lost these properties (due to prolonged maintenance in nitrate-containing liquid culture). However, revertants of both these strains, capable of differentiating functional heterocysts, have been isolated (strains PCC 6411/1, PCC 7118/1 and PCC 7118/2) (Rippka and Herdman, 1992).

DNA–DNA hybridizations of the six strains revealed a high degree of similarity (80–102% relative binding, $\Delta T_{m(e)}$ of 0–1°C) using both *Nostoc* PCC 6705 and PCC 7120 as sources of reference DNA (Lachance, 1981). Consequently, they can be considered members of the same nomenspecies. In contrast, only a low degree of similarity (25–28% relative binding, $\Delta T_{m(e)}$ of 12–16°C) was observed between strain PCC 7120 and representatives of *Nostoc* clusters 1 and 2 (Lachance, 1981; M.-A. Lachance and R.Y. Stanier, personal communication) (see also Fig. B10.80). Strain PCC 6705 was proposed as reference strain of *Nostoc* cluster 3 (Rippka and Herdman, 1992), but is here designated the reference strain of cluster 3.1. However, strain PCC 7120, known in the literature as *Anabaena* sp., has become a model organism for genetic studies on the regulation of N₂-fixation and heterocyst development (see Haselkorn, 1992, 1995; Buikema and Haselkorn, 1993; Tandeau de Marsac and Houmard, 1993;

Wolk et al., 1994; Wolk, 1996). Consequently, it may in the future replace strain PCC 6705 as the reference strain of this cluster.

Reference strain PCC 6705 (co-identity ATCC 29131), isolated in 1967 by G. Cohen-Bazire from botanical garden, Berkeley, California (Rippka et al., 1979). Reference strain of *Nostoc* cluster 3 (Rippka and Herdman, 1992). Proposed as reference strain of cluster 3.1 (this edition). The mol% G + C is 43.4; genome size is 5.6 Gdal.

Clusters 3.2 and 3.3 These clusters harbor two representatives (PCC 7413, and PCC 7423), isolated from surface garden soil in England and dry soil in Senegal, respectively (Rippka et al., 1979). They are similar in morphology to members of cluster 3.1, but none of them is host to cyanophage N-1, and gas vesicle-containing hormogonia are not produced (R. Rippka, unpublished data). Strain PCC 7413 is an obligate photoautotroph, whereas strain PCC 7423 is a facultative photoheterotroph (Rippka et al., 1979). Strains PCC 7413 and PCC 7423 synthesize neither PEC nor C-PE (Bryant, 1982). Although strain PCC 7413 was identified prior to entering the PCC as *Cylindrospermum licheniforme*, large subterminal akinetes typical of the latter genus have never been observed (R. Rippka, unpublished data). Strain PCC 7423 was originally identified as *Anabaena* sp. (for strain history, see Rippka and Herdman, 1992).

DNA–DNA hybridization studies by Lachance (1981) showed that strains PCC 7413 and PCC 7423 are different species (48% relative binding using strain PCC 7413 as the source of reference DNA). Both are also distant from members of cluster 3.1, though strain PCC 7413 exhibited a higher degree of DNA similarity to the latter (65–69% relative binding, $\Delta T_{m(e)}$ 6°C using DNA of PCC 7413 as the reference) than strain PCC 7423 (41–50% relative binding, $\Delta T_{m(e)}$ 11–12°C with reference DNA of strains PCC 6705 and PCC 7120, respectively). Thus strains of cluster 3.1, 3.2, and 3.3 represent three different species (see also Fig. B10.80).

Cluster 4 This "cluster" harbors a single representative, strain PCC 7524, isolated from a moderate hot spring in Sri Lanka. It synthesizes PEC (Bryant, 1982) and is a facultative photoheterotroph (Rippka et al., 1979).

Although exhibiting similar properties to members of cluster 3.2 and being distantly related to strain PCC 7413 (43% relative binding, $\Delta T_{m(e)}$ 14°C) (Lachance, 1981), this strain is genetically distinct. Not being more closely related to any other strains of *Nostoc* examined (Lachance, 1981), it either represents a very remote relative of members of cluster 3, or warrants independent generic status (see also Fig. B10.80).

Reference strain PCC 7524 (co-identity ATCC 29411), isolated in 1973 by A. Neilson from a moderate hot spring, Maya Oya, Sri Lanka (Rippka et al., 1979). Proposed as reference strain of cluster 4 (this edition). The mol% G + C is 39; genome size is 5.6 Gdal.

Cluster 5 This "cluster" includes a single strain PCC 6314, isolated from crude algal material in North America. It is an obligate photoautotroph and synthesizes PEC (Bryant, 1982). Strain PCC 6314 no longer forms functional heterocysts and thus does not fix nitrogen under aerobic conditions. However, revertants capable of differentiating functional heterocysts have been isolated (strains PCC 6314/1 and PCC 6314/2) (Rippka and Herdman, 1992).

DNA–DNA hybridization studies revealed that strain PCC 6314 is not closely related to any of the *Nostoc* strains examined

(M.A. Lachance and R.Y. Stanier, personal communication), the highest degree of similarity (36% relative binding, $\Delta T_{m(e)}$ 18°C) having been observed with DNA of two strains of cluster 1, PCC 73102 and PCC 6302 (Lachance, 1981) (see also Fig. B10.80). Consequently, strain PCC 6314 may represent a very distant species, loosely affiliated with members of cluster 1, or may warrant independent generic status.

Reference strain PCC 6314 (co-identity ATCC 27904), isolated in 1963 by M.M. Allen from crude algal material (Rippka et al., 1979). Previously assigned to *Nostoc* cluster 4 (Rippka and Herdman, 1992). Proposed as reference strain of cluster 5 (this edition). The mol% G + C is 43.9; genome size is 5.1 Gdal.

Note Strains PCC 6314, PCC 7413, PCC 7423, and PCC 7524 were previously assigned to *Nostoc* cluster 4 (Rippka and Herdman, 1992), an assemblage created for strains of relatively uncertain genetic affiliations. However, more careful analyses of the results of Lachance (1981) and of unpublished data (M.A. Lachance and R.Y. Stanier, personal communication) revealed that two of these strains (PCC 7413 and PCC 7423) may be considered distant relatives of members of cluster 3.1 (see Fig. B10.80). Consequently, the latter have here been transferred from cluster 4 to cluster 3. Strain PCC 7524 has been maintained as *Nostoc* cluster 4, whereas strain PCC 6314 has been transferred to cluster 5.

Enrichment and Isolation Procedures

Members of *Nostoc* are generally grown in media BG11$_0$ or Z8$_0$, unless they have lost the capacity to produce functional heterocysts and to fix N$_2$. In the latter case, the media need to be supplemented with NaNO$_3$ (2–20 mM). Isolation and purification of representatives with actively gliding hormogonia is generally easy; the freshly collected material is placed onto the center of an agar (or agarose) plate and the hormogonia, or young filaments, are recovered by micromanipulation after having moved away from the zone of deposit (Castenholz, 1988a; Rippka, 1988a). Less typical members (i.e., those that form immotile hormogonia) need to be isolated and purified by dilution streaking to obtain single colonies as typically performed for unicellular cyanobacteria. If mucilage or sheath production is abundant, an antibiotic treatment followed by washing of the trichomes by differential centrifugation or by filtration through polycarbonate membranes of appropriate pore size (8 µm) may be necessary, prior to plating, to speed up the purification process (Castenholz, 1988a; Rippka, 1988a).

Taxonomic Comments

The developmental cycle typical of most members of the genus *Nostoc* was already described in great detail in the last century (Thuret, 1844; Janczewski, 1874; Sauvageau, 1897). These early descriptions were confirmed and extended by later authors (Harder, 1917; Lazaroff and Vishniac, 1961, 1964; Kantz and Bold, 1969; Mollenhauer, 1970, 1986a, b; Lazaroff, 1972, 1973). However, in traditional taxonomic treatises (Bornet and Flahault, 1887; Geitler, 1932; Desikachary, 1959; Bourrelly, 1970), this developmental property was never used as a discriminatory character to separate this genus from *Anabaena*. Instead, *Nostoc* was primarily defined by the shape and consistency of the gelatinous colonies, as observed in nature. This characteristic is undoubtedly influenced by environmental factors and thus may not be expressed, or may be lost, in laboratory cultures. For this reason Rippka et al. (1979), in overall agreement with the taxonomic conclusions of Kantz and Bold (1969), defined the genus *Nostoc* on the basis of the developmental cycle, which in typical members is easily observed in culture even when sheath or mucilage envelopes are lacking. This emendation to the traditional definition of the genus *Nostoc* has been accepted by Castenholz (1989b) and by Komárek and Anagnostidis (1989). However, as mentioned above, hormogonium formation in some isolates may be difficult to detect or (due to the selection of spontaneous mutants through prolonged maintenance in liquid media?) may not be observed at all. In such instances appropriate generic assignments can be made only on the basis of genetic evidence.

Strains assigned to cluster 1 were identified prior to entering the PCC under various generic and specific designations such as *Anabaena spiroides* (PCC 6310), *Anabaena* sp. (PCC 7422), and *Nostoc muscorum* (PCC 6302, PCC 7121, and PCC 7906) (see Rippka and Herdman, 1992). The relatively close relatedness of strains PCC 6310 and PCC 7422 to *Nostoc* PCC 73102, a very typical representative of *Nostoc*, excludes their assignment to *Anabaena*. This is also confirmed by the low degree of hybridization observed between members of *Nostoc* cluster 1 with *Anabaena cylindrica* PCC 7122, the reference strain of the genus *Anabaena* (see Fig. B10.80).

Strain PCC 73102, isolated as an endosymbiont from a root section of *Macrozamia* sp., was previously designated as the reference strain of *Nostoc* without carrying a specific epithet (Rippka et al., 1979). On the basis of its habitat and the very characteristic tightly coiled filaments in the mature stage of development, it fits best the description of *Nostoc punctiforme* (Kützing) Hariot 1891, a "species" typically found in the angiosperm *Gunnera* (see Geitler, 1932). Given that the host specificity of cyanobacterial endosymbionts does not seem to be very stringent, since strain PCC 73102 has been shown to enter symbiosis even with the bryophyte *Anthoceros* (Enderlin and Meeks, 1983; Meeks, 1988), strain PCC 73102 was proposed as the type strain of *Nostoc punctiforme* (Rippka and Herdman, 1992).

Strain PCC 7803 was isolated from a dry habitat (the surface of a sand dune in Scotland), thought to be typical of *Nostoc commune* (Vaucher) ex Bornet and Flahault 1886, the botanical type species of the genus *Nostoc* (Geitler, 1932). In agreement with the description of the latter "species", the colonies as observed in the feral sample exhibited both coiled trichomes and balloon-like sheath envelopes (R. Rippka, unpublished data). Consequently, it would seem appropriate to designate strain PCC 7803 both as the type strain of the latter species and as the type strain of the genus. However, Rippka and Herdman (1992) decided to postpone this nomenclatural proposal until genetic data become available on other isolates that fit the description of *Nostoc commune*. It should also be mentioned that the relation-

ships of members of cluster 1 have mainly been deduced on the basis of hybridizations with strains PCC 73102 and PCC 6302 as sources of reference DNA. Additional cross-hybridizations (or other methods of genetic analysis) may reveal more refined genetic affiliations and are required to confirm more precisely the number of species that constitute this cluster.

Strains of cluster 3.1 (this edition) were identified prior to their deposition in the PCC as *Anabaena variabilis* (PCC 7118), *Anabaena* sp. (PCC 6411 and PCC 6705), or *Nostoc muscorum* (PCC 6719, PCC 7119, and PCC 7120). Four of them (PCC 6411, PCC 7118, PCC 7119, and PCC 7120) were subsequently assigned to *Anabaena*, whereas both PCC 6705 and PCC 6719 were considered members of the genus *Nostoc* (Rippka et al., 1979). These nomenclatural changes were made because PCC 6705 and PCC 6719 were known to produce hormogonia containing gas vesicle clusters (Rippka et al., 1979; Armstrong et al., 1983), whereas such a developmental cycle had not been observed in the other four strains. However, as shown in Fig. B10.80, DNA–DNA hybridization studies revealed that these six strains represent a single species, loosely affiliated with all other strains of *Nostoc* (Lachance, 1981) but much more remote (12–21% relative binding) from the reference strain (PCC 7122) of the genus *Anabaena* (Lachance, 1981; M.-A. Lachance and R.Y. Stanier, personal communication). Consequently, strains PCC 6411, PCC 7118, PCC 7119, and PCC 7120 were later transferred to the genus *Nostoc* (Rippka, 1988b; Rippka and Herdman, 1992), and all (except strain PCC 7120) were shown to produce gas vesicle-containing hormogonia after washing the filaments in fresh medium prior to establishing new cultures.

Analyses of 16S rRNA gene sequences clearly support the generic distinction between *Anabaena* PCC 7122 and *Nostoc* PCC 73102, the reference strain of cluster 1 (see Wilmotte and Herdman, this volume). They also agree with the relatively low degree of DNA–DNA similarity observed by Lachance (1981) between *Nostoc* PCC 73102 (cluster 1) and *Nostoc* PCC 7120 (cluster 3.1). However, given that the former clusters with two strains of *Cylindrospermopsis*, whereas the latter is most closely related to three strains assigned to *Tolypothrix* (see Wilmotte and Herdman, this volume), it is highly likely that members of "*Nostoc*" clusters 1, 2, and 3 may in fact represent different genera rather than distantly related species of the same genus. Consequently, strains PCC 7524 and PCC 6314 (clusters 4 and 5, respectively) may equally warrant independent generic status. However, any nomenclatural changes need to await further genetic studies. If new genera are created to reflect the high internal genetic diversity of this "genus", the name *Nostoc* should be conserved for members of cluster 1, which harbors the most typical representatives of this botanical taxon.

Form-genus IX. **Scytonema** *Agardh 1824*

RICHARD W. CASTENHOLZ, MICHAEL HERDMAN AND ROSMARIE RIPPKA

Scy.to.ne'ma. Gr. n. *skytos* leather; Gr. n. *nema* thread; M.L. neut. n. *Scytonema* leather thread.

The mature trichomes are immotile, heavily ensheathed, and form numerous false branches that exhibit upright aerial mode of growth. Constrictions between the disk-shaped to cylindrical cells are shallow. Heterocysts generally occupy intercalary posi-

tions, although some false branches may carry at their base a terminal heterocyst. Hormogonia are less ensheathed, exhibit no or only slow motility, and differ little in cell morphology from the mature filaments. Akinetes are not produced.

FURTHER DESCRIPTIVE INFORMATION

Strains assigned to *Scytonema* exhibit the morphological properties resembling those of *Scytonema hofmanni* Agardh 1824 ex Bornet and Flahault 1888, the type species of this genus. Reproduction occurs by binary fission in only one plane, **intracellular trichome breakage involving sacrificial cell death (necridium formation)**, and hormogonia. **False branching, singly or geminate**, may be observed **in either the presence or absence of combined nitrogen** and thus is not intimately linked to heterocyst differentiation. The formation of geminate false branches is often preceded by a loop-like extrusion of a trichome through the sheath envelope (see Fig. 59 in Rippka et al., 1979). Under N_2-fixing conditions, the false branches generally occur equidistant between two intercalary heterocysts and may themselves carry one or more heterocysts in intercalary positions (Fig. B10.2) (see also Geitler, 1932; Rippka et al., 1979). However, de novo differentiation of a true terminal heterocyst may also occur on either side of the breakage point between the vegetative cells. Perforation of the growing ends of such trichome segments through the sheath layer will, as in rivularian genera (see Subsection IV.II), result in the formation of single false branches that subtend a terminal heterocyst (see Fig. 60 in Rippka et al., 1979).

Rapid cell division in "meristematic" regions of the ensheathed mature filaments or at the tips of the false branches leads to hormogonia formation. The latter reproductive structures differ little from the mature filaments and can mainly be distinguished by their slightly shorter cells, thinner sheath envelopes, and shorter trichome length. The representatives presently in axenic culture produce **hormogonia** that exhibit little gliding motility and, in the absence of combined nitrogen, **differentiate a single terminal heterocyst**. However, upon maturation they **give rise to long ensheathed trichomes of even width, in which heterocyst differentiation occurs predominantly in intercalary positions**.

Typical members of this genus have filaments that are sufficiently large (diameter >10 μm) to be visible by eye. Most "species" have been described from freshwater, soil, terrestrial rocks, and hot springs, but a few are known from marine environments (Geitler, 1932). In aquatic habitats, they grow suspended as irregular star-like bundles of filaments, whereas on solid substrates they form velvet- to crust-like covers, depending on the degree of humidity. The characteristic upright aerial mode of growth, typical of many feral samples, is a stable feature maintained even in laboratory cultures. The conditions that may influence the degree of false branching in nonaxenic, unicyanobacterial cultures of *Scytonema stuposum* were described by Jeeji-Bai (1976).

The yellow-brown lipid-soluble pigment scytonemin, a constituent uniquely found in cyanobacterial sheath layers, owes its name to a member of this genus (Garcia-Pichel and Castenholz, 1991). This compound, however, is synthesized by representatives of many different genera in response to high light intensities and was shown to protect against the damaging effects of exposure to UV-A (320–400 nm) and UV-C (190–280 nm) irradiance (Garcia-Pichel and Castenholz, 1991; Dillon and Castenholz, 1999).

Only two axenic strains (PCC 7110 and PCC 7814) are presently available in the PCC. They were isolated from a limestone wall of a cave in Bermuda and a lichen (*Heppia*) collected in France, respectively. Both strains synthesize phycoerythrocyanin (Bryant, 1982) and are facultative photoheterotrophs. Strain PCC 7110 also grows on glucose, fructose, or sucrose under che-

moheterotrophic conditions (Rippka et al., 1979; Rippka and Herdman, 1992). Only strain PCC 7110 has been characterized in more detail. It has a mean DNA base composition of 44 mol% G + C (Herdman et al., 1979b) and a high degree of genetic complexity (genome size 7.4 Gdal) (Herdman et al., 1979a). DNA–DNA hybridization studies of Lachance (1981) have shown that strain PCC 7110 is not significantly related to any of the PCC reference strains examined. Justification to maintain *Scytonema* as a distinct genus has also been provided by 16S rRNA sequence data, which showed that strain PCC 7110 is remote from all other heterocystous strains so far examined (see Wilmotte and Herdman, this volume).

Reference strain PCC 7110 (co-identity ATCC 29171), isolated in 1971 by J.B. Waterbury from limestone, Crystal Cave, Bermuda (Rippka et al., 1979). Proposed as type strain of the type species *Scytonema hofmanni* (Rippka and Herdman, 1992, and this edition). The mol% G + C is 44.4; genome size is 7.4 Gdal.

ENRICHMENT AND ISOLATION PROCEDURES

Strains PCC 7110 and PCC 7814 were isolated and purified on medium $BG11_0$. Purification of strain PCC 7814 was achieved by ampicillin treatment as described by Rippka (1988a).

TAXONOMIC COMMENTS

The definition of the genus *Scytonema* is similar to that given previously (Rippka et al., 1979; Rippka, 1988b), but differs from earlier taxonomic treatises (Geitler, 1932; Desikachary, 1959). In the latter, *Scytonema* was defined on the basis of geminate false branches, distinguishing this genus from *Tolypothrix*, for which single false branches were the diagnostic trait. However, the formation of single or geminate false branches is variable, and may be observed even within a single isolate assignable to either of these two genera (Jaag, 1945; Hoffmann and Demoulin, 1985; R. Rippka, unpublished data). For this reason, Rippka and Herdman (1992), in overall agreement with Castenholz (1989b) and Komárek and Anagnostidis (1989), adopted a minor amendment: *Tolypothrix* was distinguished from *Scytonema* by its basal–apical trichome polarity as evidenced by a low degree of tapering and the differentiation of predominantly terminal heterocysts. This redefinition is supported by traditional descriptions and drawings of *Scytonema* "species" (Geitler, 1932; Desikachary, 1959), which strongly suggest that in most representatives of this taxon trichome polarity is lacking and heterocysts in mature filaments occur almost exclusively in intercalary positions.

Scytonema has traditionally been placed into the botanical family *"Scytonemataceae"*, which included (among others) the supposedly nonheterocystous genus *Plectonema* and the genus *Tolypothrix* (Geitler, 1932; Desikachary, 1959; Bourrelly, 1970). Members of the former vary greatly in trichome width (diameter 0.5–22 μm). The smaller representatives (<3 μm in diameter) have recently been placed into the genus *Leptolyngbya* (Anagnostidis and Komárek, 1988) (see Subsection III, this volume), and their genuine lack of heterocysts has been demonstrated in laboratory cultures (Stewart and Lex, 1970; Rippka and Waterbury, 1977; Rippka et al., 1979). Their position among nonheterocystous genera is also supported by DNA–DNA hybridization studies (Stam and Venema, 1977; Stam, 1980) and 16S rRNA gene sequence data (see Wilmotte and Herdman, this volume). In contrast, the larger representatives, including the type species *Plectonema tomasianum* Bornet 1889, have not yet been brought into culture, but may in fact be members of *Scytonema* that in the natural samples did not exhibit heterocysts (see also Hoffmann and Demoulin, 1985). The inclusion of *Scytonema* and *Tolypothrix* at the level of a family is not supported by phylogenetic studies, since the representatives of the latter genus are more closely related to "nostocacean" genera than to *Scytonema* PCC 7110 (see Wilmotte and Herdman, this volume). A few additional genera traditionally included in the *Scytonemataceae*, but not yet represented in culture, are described by Castenholz (1989b) in the last edition of this *Manual.*

Subsection IV.II

This Subsection includes two genera, *Calothrix* and *Rivularia*, traditionally assigned to the botanical family "*Rivulariaceae*", and two genera, *Tolypothrix* and *Microchaete*, members of the "*Scytonemataceae*" and "*Microchaetaceae*", respectively (see Geitler, 1932). Representatives of the former two genera exhibit **mature trichomes that have a distinct basal–apical polarity and display a high degree of tapering**, the ratios of cell width of the basal to apical cells varying between 3 and 5. Freshly released **hormogonia** do not display basal–apical polarity, **exhibit transient gliding motility, and differ distinctly in cell size and morphology** from the mature trichomes. Members of the genera *Tolypothrix* and *Microchaete* share most structural and developmental properties with the "rivulariacean" genera, but the **mature trichomes**, even if grown in the absence of combined nitrogen, **exhibit only a very low degree of tapering**, the ratios of cell width of the basal to apical cells being less than 2. Other distinguishing traits are described in more detail below.

A large number of other "genera" have been included in the botanical families *Rivulariaceae*, *Scytonemataceae*, and *Microchaetaceae*, most of which are only known from field populations. For this reason, they will not be discussed here. However, the reader may consult Castenholz (1989b) and Komárek and Anagnostidis (1989) for more comprehensive reviews.

Form-genus I. **Calothrix** *Agardh 1824*

ROSMARIE RIPPKA, RICHARD W. CASTENHOLZ AND MICHAEL HERDMAN

Ca'lo.thrix. Gr. adj. *kalos* beautiful; Gr. n. *thrix* hair; M.L. fem. n. *Calothrix* beautiful filament.

Mature ensheathed trichomes exhibit a pronounced degree of tapering. The motile hormogonia are generally very short, exhibit a significantly reduced cell diameter, and are sheathless (or less ensheathed). They give rise to trichomes that differentiate a basal, one-pored heterocyst. False branching frequent. Akinetes lacking in most members. All are freshwater, soil, or hot spring inhabitants.

FURTHER DESCRIPTIVE INFORMATION

Members assigned to this genus are filamentous heterocystous cyanobacteria from various **freshwater or terrestrial habitats**, not requiring elevated concentrations of Na^+, Mg^{2+}, and Ca^{2+}, typical of marine cyanobacteria. Constrictions at the cross-walls are shallow; reproduction occurs by binary fission in a single plane and by **transcellular trichome breakage**. The basal–apical polarity is most readily evident in the absence of combined nitrogen from the medium (Whitton, 1987, 1989; R. Rippka, unpublished data). The **mature trichomes**, composed of disk-shaped, isodiametric, or short cylindrical cells, are **generally relatively large** and the basal vegetative cells may be **up to 24 μm in diameter**, but a few "species" have been described from field populations as being smaller (maximal diameter, 4–6 μm) (Geitler, 1932). Depending on the "species" and the phase of the developmental cycle, the sheath material varies considerably in thickness and consistency. Most members produce **single false branches**, but loop-like extrusion of the growing part of the trichome through the sheath envelope may occasionally also lead to geminate false branching. Some representatives may exhibit coiling of the trichomes within the sheath envelopes; in others, the sheath layers may enclose more than one trichome. These phenomena are most likely the result of crowding effects, caused by localized rapid growth, and are undoubtedly influenced by the consistency and constraints exerted by the sheath envelopes.

Hormogonia are released from "meristematic" zones at the apical to subapical end of the mature trichome, and are therefore **composed of small cells (~10) that exhibit no tapering**. As in "nostocacean" genera, they lack heterocysts, even if produced in the absence of combined nitrogen, and generally exhibit active gliding motility. Some representatives may produce hormogonia that exhibit transient gas vesicle formation (Whitton, 1989; R. Rippka, unpublished data). Sheath layers of hormogonia are thin or seem to be lacking, even when the mature trichome is heavily ensheathed. As hormogonia mature into **young filaments**, they **differentiate a single terminal heterocyst**, the size and shape (rounded or conical) of which is determined by the morphological characteristics of the vegetative cell from which it derives. Presumably as a result of less frequent cell divisions (Geitler, 1932), the vegetative cells in the proximity of the first terminal ("basal") heterocyst enlarge more than the distal cells. This leads to tapering trichomes, the degree of which is indicative of the maturation process, but may also vary depending on the "species". Heterocysts have only a limited functional lifetime and, when senescent, may detach from the filaments. A nonfunctional heterocyst is replaced by a new heterocyst that differentiates from its adjacent vegetative cell. Successive rounds of differentiation will lead to heterocysts of increasing size, as a consequence of the enlargement of the basal vegetative cells in the course of trichome maturation. Strains in which the senescent heterocysts remain attached thus exhibit (at only one end of the trichome) a chain of heterocysts of different diameter, illustrating clearly the hormogonial maturation process and the succession of heterocyst differentiation (see Fig. B10.81). It should be noted that

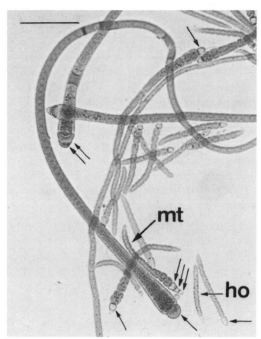

FIGURE B10.81. *Calothrix* PCC 7103. Note taper of trichomes with basal heterocyst or heterocysts (*arrows*). Hormogonia (*ho*) do not bear a heterocyst, but one is differentiated early in further development (*lowest right-hand arrow*). *mt*, maturing trichome. Bar = 50 μm. (Reprinted with permission from R. Rippka et al., Journal of General Microbiology *111*: 1-61, ©Society for General Microbiology.)

these developmental characteristics are observable in standard media (BG11$_0$ or Z8$_0$) (Rippka et al., 1979; R. Rippka, unpublished data), contrary to Whitton (1989), who suggested that multiple basal heterocysts are mainly produced under conditions of iron and molybdenum limitation.

In more mature filaments, particularly if enclosed by relatively firm sheaths, "terminal" heterocysts (connected only at one pole to a chain of vegetative cells) may be formed rather frequently in what appear to be intercalary positions. This is the case when the continuity of a long trichome has been interrupted at one or several intervals by necridium formation. Since de novo heterocyst differentiation will occur at either end of the separated chains of cells, one often observes a "terminal" heterocyst on both sides of a necridium (see Fig. 62 in Rippka et al., 1979). Further development will lead to basal–apical trichome polarity in opposite directions. If such fragments elongate and perforate the sheaths with their apical part, they will lead to false branches that subtend a terminal heterocyst. However, some members may produce true intercalary heterocysts, once the filaments reach a certain maturity.

Only a few "species" of *Calothrix* that form akinetes have been described in the botanical literature (see Geitler, 1932). In agreement with their apparent rarity in nature, only 1 (PCC 7507) of the 20 axenic strains presently carried in the PCC produces these specialized cells (R. Rippka, unpublished data). However, in contrast to traditional "species" that were observed to form mainly one or two subterminal akinetes (see Geitler, 1932), strain PCC 7507 has been shown to produce **chains of akinetes that are initiated adjacent to the basal heterocysts** (see Fig. 69 in Rippka et al., 1979).

In nature, members of *Calothrix* grow either suspended in water as individual filaments or irregularly shaped trichome bun-

dles, or may be attached to solid substrates forming flat strata, "tufts", or crustlike aggregates. Many "species" observed in feral samples have been reported to exhibit **hair formation**, a differentiation process rarely observed under standard growth conditions in the laboratory. However, in at least some laboratory strains, these nonreproductive apical trichome extensions, composed of narrow, colorless, and "vacuolated" cells, may be **induced (or derepressed) in response to deprivation of phosphate or iron**, although the former condition is more effective in promoting their expression (Sinclair and Whitton, 1977b; Whitton, 1989). According to these authors, hair formation does not involve de novo cell divisions, but merely results from narrowing of existing apical cells that thus increase in length and cell surface area. "Vacuolation" in mature hair cells seems to be associated with cellular degeneration and the formation of one or more intrathylakoidal spaces. Readdition of inorganic or organic phosphate to such cultures leads to rapid production of polyphosphate bodies in the vegetative cells, these inclusions often being visible by light microscopy, particularly in the basal cells, in less than 5 min. Eventually, hormogonium differentiation is reinitiated below the hair, and the hairs are shed (Sinclair and Whitton, 1977b; Whitton, 1989).

The light to dark brown pigmentation of the sheath envelopes, often observed in natural samples, has traditionally been considered a taxonomic trait (Whitton, 1989). However, as for other genera, this characteristic is due to the relative abundance of scytonemin, a sheath constituent synthesized in response to high light and UV irradiation (Garcia-Pichel and Castenholz, 1991), though it may also be influenced by nutrient deprivation (Whitton, 1989).

If heterocyst differentiation is repressed by combined nitrogen (nitrate or ammonia), trichome tapering may not be observed at all, may be significantly reduced, or may occur from the center of the filaments toward both extremities (Sinclair and Whitton, 1977a; R. Rippka, unpublished data). Such filaments may resemble the nonheterocystous genera *Homoeothrix* (Thuret) Kirchner 1898 (tapering in one direction) or *Hammatoidea* West and West 1897 (tapering in both directions) described from field observations (see Geitler, 1932).

As for other heterocystous representatives that synthesize C-phycoerythrin, this pigment is lacking in the heterocysts of strains of *Calothrix*, giving them a blue to yellow-green appearance (depending on their age and prior to losing pigmentation completely) in contrast to the vegetative cells which may be olive to red in color. This is the case for strains that synthesize C-PE constitutively, or if chromatically adapting members are cultivated in white or green light (R. Rippka, unpublished data; see also Whitton, 1987, 1989).

A correlation of laboratory cultures with traditional botanical "species" of *Calothrix* is confounded by similar problems as those encountered for other genera in which cell width varies in the course of trichome maturation. Furthermore, the degree of tapering (and thus the range of cell size) varies with culture conditions. Consequently, although it is possible to distinguish laboratory strains on the basis of overall cell dimensions, pigmentation, sheath coloration, or by the presence of akinetes (Whitton, 1989; R. Rippka, unpublished data), specific assignments are largely arbitrary, since many of the botanical "species" were described by observing a particular stage of development and/or on the basis of variable environmental factors.

Among the representatives here assigned to *Calothrix*, some (PCC 6303, PCC 7102, PCC 7103, and PCC 7507) were described in detail by Rippka et al., (1979). The others (PCC 7709, PCC 7711, PCC 7713, PCC 7714, PCC 7715, and PCC 7716) were characterized with respect to pigment composition (Bryant, 1982) and, like the former strains, were included in the DNA–DNA hybridization studies of Lachance (1981). Together with seven additional strains (all isolated from marine habitats), they were previously included in the *Calothrix* group that, based on structural, ecological, and genetic differences, was divided into three clusters (Rippka and Herdman, 1992). Members of cluster 2 (the marine isolates) have been transferred to the genus *Rivularia*. The representatives of clusters 1 and 3, provisionally retained as clusters 1 and 2 of *Calothrix*, have a narrow range of mean DNA base composition (40–43 mol% G + C), though strain PCC 7713 has not yet been examined. The genome size determined for three of these strains (PCC 6303, PCC 7103, and PCC 7507) is similar (5.1–5.5 Gdal), whereas strain PCC 7102 was reported to have a larger genome (8.6 Gdal) (Herdman et al., 1979a). Based on differences in pigment composition and genetic relatedness (Lachance, 1981), the strains retained in *Calothrix* can be subdivided as described below.

Cluster 1   The eight strains included in this cluster (PCC 6303, PCC 7102, PCC 7103, PCC 7709, PCC 7713, PCC 7714, PCC 7715, and PCC 7716) are of widely different geographical origin and were isolated from terrestrial or freshwater habitats (lakes, ponds, thermal springs, desert sand, or tree bark). None is capable of differentiating akinetes. All except strains PCC 6303 and PCC 7103 were included in the studies on the influence of nutrient deficiency on hair formation by Sinclair and Whitton (1977b). With the exception of strain PCC 7714, hair formation was favored by phosphate limitation in all strains examined.

With the exception of PCC 6303 (which due to poor quality DNA was thought to give inconclusive results, but see cluster 1.5), all members of cluster 1 were shown to share a relatively high degree of genetic relatedness (53–86% relative binding, $\Delta T_{m(e)}$ 0–10°C) using strains PCC 7102, PCC 7103, or PCC 7709 as the sources of labeled reference DNA (Lachance, 1981; M.-A. Lachance and R.Y. Stanier, personal communication) (see also Fig. B10.82). This level of DNA similarity confirms that all of them are assignable to a single genus. Based on the degree of genetic relationships, some further subdivisions are possible.

Cluster 1.1   This cluster includes three strains (PCC 7709, PCC 7715, and PCC 7716), isolated respectively from samples in Cuba, France, and India, and carried in the Durham Culture Collection (DCC) as three different species: *Calothrix vigueri* (D0253), *Calothrix thermalis* (D0266), and *Calothrix fusca* (D0269) (Sinclair and Whitton, 1977b; B. Whitton, personal communication). They all synthesize C-PE and undergo chromatic adaptation of type II (Bryant, 1982) and are facultative photoheterotrophs. In spite of their different specific assignments, they were shown to exhibit 85–86% relative binding ($\Delta T_{m(e)}$ 0–1°C), and thus can be considered members of the same nomenspecies.

Reference strain   PCC 7709 (co-identity DCC D0253), isolated in 1964 by J. Komárek from tree bark, Las Villas, Cuba (Sinclair and Whitton, 1977b; B. Whitton, personal communication). Member of cluster 1 of the *Calothrix* group (Rippka and Herdman, 1992). Reference strain of *Calothrix* cluster 1.1 (this edition). The mol% G + C is 41.8.

Cluster 1.2   The single strain, PCC 7102, assigned to this "cluster" was isolated from an extreme environment (arid soil in Chile) and described as *Calothrix desertica* (Schwabe, 1960b). As

strains of cluster 1.1, it is a facultative photoheterotroph, but produces neither C-PE nor PEC (Bryant, 1982). It was shown to exhibit 74% relative binding ($\Delta T_{m(e)}$ 4°C) with strain PCC 7709 and thus represents the same, or a very closely related, species as members of cluster 1.1. Strain PCC 7102 was previously designated, without a specific epithet, as the reference strain of *Calothrix* (Rippka et al., 1979). However, Rippka and Herdman (1992) accepted the specific nomen "*desertica*", with which this strain entered the PCC, given the fact that the new species created by Schwabe (1960b) was based on a cultured isolate (Schwabe, 1960b), is still available and has been characterized genetically. Hence, they proposed strain PCC 7102 as the type strain of *Calothrix desertica* Schwabe 1960b. This assignment is maintained here.

Reference strain   PCC 7102 (co-identity ATCC 27901), isolated in 1958 by G.H. Schwabe from fine desert sand in Antofagasta, Chile (Schwabe, 1960b; Rippka et al., 1979). Proposed as type strain of the species *Calothrix desertica* (Rippka and Herdman, 1992, and this edition). The mol% G + C is 39.8; genome size is 8.6 Gdal.

Cluster 1.3   This cluster includes two strains, PCC 7103, of unknown origin and originally incorrectly identified as *Nodularia sphaerocarpa* (see strain history in Rippka and Herdman, 1992), and PCC 7713 (co-identic with *Calothrix* sp. strain DCC D0184; B. Whitton, personal communication), isolated from a laboratory tank in the UK. Both are facultative photoheterotrophs, synthesize C-PE and undergo chromatic adaptation of type III or type II, respectively (Bryant, 1982). These two strains are highly related (80% relative binding, $\Delta T_{m(e)}$ 1°C, using PCC 7103 as the source of reference DNA) and can be considered independent isolates of the same nomenspecies (Lachance, 1981). The DNA homologies of strain PCC 7103 with representatives of clusters 1.1 and 1.2 were significantly lower (62–65% relative binding, $\Delta T_{m(e)}$ 5°C, using DNA of strain PCC 7103 as the reference) (Lachance, 1981). However, this level of relatedness confirms that strains PCC 7103 and PCC 7713 are assignable to the same genus, though representing a species that is distinct from members of clusters 1.1 and 1.2.

Reference strain   PCC 7103 (co-identity ATCC 27905), of unknown origin. Member of cluster 1 of the *Calothrix* group (Rippka and Herdman, 1992). Reference strain of *Calothrix* cluster 1.3 (this edition). The mol% G + C is 39.8; genome size is 5.2 Gdal.

Cluster 1.4   Strain PCC 7714, received as *Calothrix marchica* (DCC D0202; B. Whitton, personal communication), was isolated from a pool in India. It is a facultative photoheterotroph, synthesizes C-PE and undergoes chromatic adaptation of type II (Bryant, 1982). This strain showed only moderate DNA similarity to three representatives of the former three clusters (53%, 37%, and 58% relative binding, $\Delta T_{m(e)}$ of 10–11°C with DNA of strains PCC 7709, PCC 7102, and PCC 7103, respectively), and thus is a representative of the same genus, but is a distinct species.

Reference strain   PCC 7714 (co-identity DCC D0202), isolated from pool in India (Sinclair and Whitton, 1977b; B. Whitton, personal communication). Member of cluster 1 of the *Calothrix* group (Rippka and Herdman, 1992). Reference strain of *Calothrix* "cluster" 1.4 (this edition). The mol% G + C is 41.6.

Cluster 1.5   This "cluster" contains a single strain, PCC 6303, isolated in 1948 by G.C. Gerloff from lake water in Wisconsin (United States) and identified as *Calothrix parietina*. It is a fac-

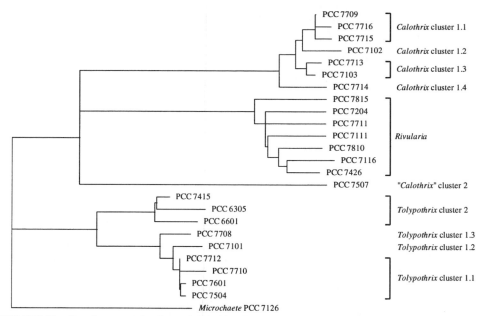

PCC 7709
PCC 7716] *Calothrix* cluster 1.1
PCC 7715
PCC 7102   *Calothrix* cluster 1.2
PCC 7713] *Calothrix* cluster 1.3
PCC 7103
PCC 7714   *Calothrix* cluster 1.4
PCC 7815
PCC 7204
PCC 7711
PCC 7111   *Rivularia*
PCC 7810
PCC 7116
PCC 7426
PCC 7507   "*Calothrix*" cluster 2
PCC 7415
PCC 6305] *Tolypothrix* cluster 2
PCC 6601
PCC 7708   *Tolypothrix* cluster 1.3
PCC 7101   *Tolypothrix* cluster 1.2
PCC 7712
PCC 7710   *Tolypothrix* cluster 1.1
PCC 7601
PCC 7504
Microchaete PCC 7126

FIGURE B10.82. Dendrogram showing relationships between strains of the genera *Calothrix, Rivularia,* and *Tolypothrix,* deduced from the DNA–DNA hybridization results of Lachance (1981) and M.-A. Lachance and R.Y. Stanier (personal communication). Relative binding values were analyzed as in Figure B10.80. The different clusters of the genera *Calothrix* and *Tolypothrix* are indicated on the right. The tree was rooted with *Microchaete* PCC 7126.

ultative photohetcrotroph and, like most members of cluster 1, synthesizes C-PE and undergoes chromatic adaptation of type II (Bryant, 1982). Strain PCC 6303 exhibited only very low DNA–DNA similarity with strain PCC 7102, a representative of cluster 1.2, but the results were thought to be inconclusive due to poor DNA quality of the former isolate (Lachance, 1981). However, the relatively long phylogenetic distance that separates this strain from two other strains of *Calothrix* in the 16S rRNA tree (see Wilmotte and Herdman, this volume) suggests that this strain is indeed very remote (for more details, see taxonomic comments to *Calothrix* and *Rivularia,* below).

A second strain (D0550) identified as *Calothrix parietina,* carried in the Durham Culture Collection, was previously proposed as the reference strain of this genus (Whitton, 1989). Its genetic relatedness to strain PCC 6303, or to any other members here included in *Calothrix,* remains to be examined.

Reference strain PCC 6303 (co-identity ATCC 29156), isolated in 1948 by G.C. Gerloff from lake water, Wisconsin (United States); identified as *Calothrix parietina* (Gerloff et al., 1950; Rippka et al., 1979). Member of cluster 1 of the *Calothrix* group (Rippka and Herdman, 1992). Reference strain of *Calothrix* "cluster" 1.5 (this edition). The mol% G + C is 41.8; genome size is 5.1 Gdal.

Cluster 2 This "cluster" is represented by a single strain, PCC 7507. It was isolated from a sphagnum bog in Switzerland and is the only strain of *Calothrix* that is capable of forming akinetes. In the absence of combined nitrogen, these specialized cells are differentiated singly or in short chains adjacent to the basal heterocysts. Among the "species" described from Europe, it conforms best, though not perfectly, to the botanical species *Calothrix stagnalis* Gomont 1895 (see Geitler, 1932). Unlike most members of cluster 1, strain PCC 7507 does not synthesize C-phycoerythrin,

but produces phycoerythrocyanin constitutively (Bryant, 1982). The trichomes are enclosed in a relatively thin sheath and may form true intercalary heterocysts. Strain PCC 7507 grows photoheterotrophically with fructose as the sole carbon source but not at the expense of other sugars (Rippka et al., 1979). As reported by Lachance (1981) and shown in Fig. B10.82, strain PCC 7507 is genetically very remote from strain PCC 7102 (cluster 1.2) (8% relative binding) and, in agreement with 16S rRNA sequence data, requires generic separation. For lack of a suitable traditional botanical genus, it is here designated as the reference strain of cluster 2, without proposing an alternative generic nomen.

Reference strain PCC 7507 (co-identity ATCC 29112), isolated in 1972 by R. Rippka from a sphagnum bog, near Vierwaldstättersee, Switzerland (Rippka et al., 1979). Previously reference strain of *Calothrix* group cluster 3 (Rippka and Herdman, 1992). Reference strain of *Calothrix* cluster 2 (this edition). The mol% G + C is 42.8; genome size is 5.5 Gdal.

ENRICHMENT AND ISOLATION PROCEDURES

Members of *Calothrix* are generally easily isolated and purified on solidified freshwater media BG11₀ or Z8₀, supplemented with 5–10 mM NaHCO₃, the motile hormogonia or young filaments being recovered by micromanipulation as described for members of the genus *Nostoc.* For expression of hair formation, and to prevent loss of this trait in laboratory cultures, Whitton (1989) recommends growth and maintenance at low concentrations (maximum 10 μM) of inorganic or organic sources of phosphorus.

TAXONOMIC COMMENTS

Taxonomic comments on *Calothrix* are included with those for *Rivularia.*

Form-genus II. **Rivularia** Agardh 1824

ROSMARIE RIPPKA, RICHARD W. CASTENHOLZ AND MICHAEL HERDMAN

Ri.vu.la' ri.a. L. adj. *rivularius* pertaining to a small creek; M.L. fem. n. *Rivularia* the one of a small creek.

As in *Calothrix* Agardh 1824, the developmental cycle involves structurally distinct, motile hormogonia and the mature trichomes exhibit a high degree of tapering. Akinetes are lacking. They differ from members of the former genus by having higher salt requirements or tolerance, reflecting their ecology in nature.

FURTHER DESCRIPTIVE INFORMATION

The seven strains here assigned to *Rivularia* were **previously included in the genus** *Calothrix* (Rippka et al., 1979), or were assigned to cluster 2 of the *Calothrix* group (Rippka and Herdman, 1992). However, Lachance (1981) demonstrated the importance of ecological characteristics, since all *"Calothrix"* strains of marine origin proved to be genetically distinct. One of these marine strains (PCC 7116, see Fig. B10.83) was originally identified by R.A. Lewin as *Rivularia* sp. In need of generic distinction and accepting the identification of the latter author, a botanical expert, all *"Calothrix"* strains that showed a high degree of DNA similarity to strain PCC 7116 have here been transferred to the genus *Rivularia*.

Three of the isolates assigned to *Rivularia* are from California (PCC 7111, PCC 7116, and PCC 7204), one originated in Scotland (PCC 7711), and three (PCC 7426, PCC 7810, and PCC 7815) were collected in France. **They were all isolated from saline environments** (intertidal zone, supralittoral shore, seawater aquarium), and five of them are truly marine, having an obligate requirement for elevated concentrations of Na^+, Mg^{2+}, and Ca^{2+}; two (PCC 7204 and PCC 7711) are merely euryhaline. None synthesize C-phycoerythrin, but all (except strains PCC 7111 and PCC 7711) synthesize phycoerythrocyanin (Bryant, 1982). Strain PCC 7711 is an obligate photoautotroph, whereas the others are facultative photoheterotrophs; some also grow under chemoheterotrophic conditions (see Rippka and Herdman, 1992). The mean DNA base composition (only determined for strains PCC 7111, PCC 7116, PCC 7204, and PCC 7426) ranges from 40.5 to 44.4 mol% G + C (Herdman et al., 1979b). The genome complexity for three of the latter isolates is similar (5.3–5.4 Gdal), but strain PCC 7426 has a larger genome (8.2 Gdal) (Herdman et al., 1979a).

Although originally identified as members of several different genera, *Rivularia* (PCC 7111 and PCC 7116), *Isactis* (PCC 7426), and various species of *Calothrix* (see Rippka and Herdman, 1992), it is clear that these generic distinctions are unjustified; using strains PCC 7116 and PCC 7810 as sources of the labeled reference DNA, variable but significant degrees of DNA similarities (34–60% relative binding, $\Delta T_{m(e)}$ of 9–18°C) were observed for the seven strains of marine origin (Lachance, 1981). Thus they are all assignable to a single genus, but each strain represents a different species (see Fig. B10.82). In support of the proposed recognition of *Rivularia* for marine strains, the DNA similarities observed between strain PCC 7116 and the freshwater or soil isolates here assigned to *Calothrix* were significantly lower (9–19% relative binding) (Lachance, 1981).

Reference strain PCC 7116 (co-identity ATCC 29111), isolated in 1968 by R.A. Lewin from a sample at La Paz, Baja, California (United States) (Rippka et al., 1979). PCC reference strain of cluster 2 of the *Calothrix* group (Rippka and Herdman, 1992). Reference strain of the genus *Rivularia* (this edition). The mol% G + C is 40.8; genome size is 5.4 Gdal.

ENRICHMENT AND ISOLATION PROCEDURES

Five of the strains here assigned to *Rivularia* were isolated by use of, and are maintained in, media M_o or AS_oIII. The two euryhaline strains (PCC 7204 and PCC 7711) can be cultured in medium $BG11_o$.

TAXONOMIC COMMENTS

Rippka et al. (1979) included all heterocystous cyanobacteria with a low and high degree of tapering, and irrespective of their ecology, in the genus *Calothrix* on the grounds that colony appearance observed in feral samples seemed to be a doubtful taxonomic trait. The latter argument was largely proven to be justified, since marine strains identified as *Calothrix* Agardh 1824, *Isactis* Thuret 1875, or *Rivularia* (Roth) Agardh 1824, were shown to share sufficient DNA similarity to be assignable to a single genus (Lachance, 1981). However, DNA–DNA hybridization studies (Lachance, 1981) also showed that freshwater and soil isolates of *Calothrix* (sensu Rippka et al., 1979) are unrelated to the marine representatives, and form three major genetic clusters that can be correlated with the degree of tapering, or the formation of akinetes. Consequently, Rippka and Herdman (1992) transferred a number of freshwater *"Calothrix"* strains that exhibit a low degree of tapering to the genus *Tolypothrix* and assigned the single akinete-forming strain (PCC 7507) to "cluster" 3 of the *Calothrix* group. The marine members were assigned to cluster 2 of the *Calothrix* group, but have here been assigned to

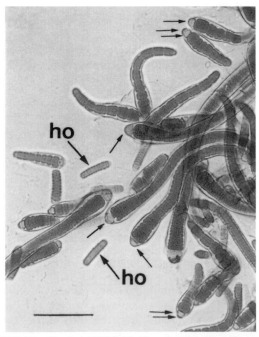

FIGURE B10.83. *Rivularia* PCC 7116. Note the sheathless aheterocystous hormogonia (*ho*) and the heavily ensheathed tapering trichomes that bear terminal heterocysts (*arrows*). Bright field. Bar = 50 μm. (Reprinted with permission from R. Rippka et al., Journal of General Microbiology *111*: 1-61, ©Society for General Microbiology.)

Rivularia. The latter genus is now distinguished from *Calothrix* on both an ecological and genetic basis. Consequently, the definitions given here for both genera depart significantly from those of traditional and modern taxonomic treatises (Geitler, 1932; Desikachary, 1959; Komárek and Anagnostidis, 1989).

Phylogenetic analyses of 16S rRNA gene sequence data (see Wilmotte and Herdman, this volume) have shown that *Calothrix* PCC 6303, PCC 7102, and PCC 7709 (clusters 1.5, 1.2, and 1.1, respectively) are positioned in the same clade, though PCC 6303 is genetically more remote and may warrant independent generic status. As expected from the results of Lachance (1981), the akinete-forming *"Calothrix"* strain PCC 7507 (cluster 2) is located in a different clade, where it shares a branch with *Cylindrospermum stagnale* PCC 7417. Consequently, this strain warrants a new generic nomen. *Rivularia* PCC 7116 groups with the three *Calothrix* strains of cluster 1, but the long genetic distance that separates this isolate from the latter fully supports its generic distinction.

Members of the genus *Gloeotrichia* Agardh 1842 share many structural and developmental properties with *Calothrix*. These planktonic "rivulariacean" representatives typically inhabit freshwater lakes and reservoirs and form spherical colonies composed of radially arranged trichomes that exhibit gas vesicle formation throughout the life cycle. They are further characterized by very large subterminal cylindrical akinetes. Consequently, they are easily identified in natural samples and most likely represent a valid genus. However, no detailed description of *Gloeotrichia* has been given here, since both axenic representatives and 16S rRNA sequence data are still lacking.

Form-genus III. **Tolypothrix** *Kützing 1843* (*sensu Rippka and Herdman 1992*)

MICHAEL HERDMAN, RICHARD W. CASTENHOLZ AND ROSMARIE RIPPKA

*To.ly.po' thrix.*Gr. n. *tolypa* bundle, tangle; Gr. fem. n. *thrix* hair; M.L. fem. n. *Tolypothrix* bundle/tangle of hairs.

Mature trichomes exhibit a low degree of tapering, the ratio of the width of basal to apical cells not exceeding 1.5. False branches are frequent, often subtending a "basal" heterocyst. Hormogonia, generally motile and often containing gas vesicle clusters dispersed throughout the cells, give rise to young trichomes that differentiate a single terminal heterocyst. All are freshwater or soil inhabitants. Akinetes are lacking.

FURTHER DESCRIPTIVE INFORMATION

The genus *Tolypothrix* was used by Rippka and Herdman (1992) as a repository for **freshwater or soil isolates** that exhibit a very **low degree of tapering, do not form akinetes**, and exhibit a developmental cycle similar to that of members of the genus *Calothrix*. This generic assignment implies some minor amendments to the traditional definition of *Tolypothrix*, a genus primarily based on the trait of single false branch formation (Geitler, 1932). However, both the ecological and structural amendments reported here are in overall agreement with the botanical descriptions of the various "species" of this genus; none has been described from saline environments, akinetes have not been reported, and the trichomes of many "species" show a low degree of basal to apical tapering, or at least exhibit a certain heteropolarity, as evidenced by shorter cells at one end of the trichome or at the tips of the false branches (see Geitler, 1932).

Among the 12 strains included in *Tolypothrix*, three (PCC 7101, PCC 7504, and PCC 7415) were characterized by Rippka et al. (1979). Six additional isolates (PCC 6305, PCC 6601, PCC 7601, PCC 7708, PCC 7710, and PCC 7712) were examined for pigment composition (Bryant, 1982) and were included in the DNA–DNA hybridization studies of Lachance (1981). The range of mean DNA base composition for the strains examined (all except PCC 7708) is narrow (41–46 mol% G + C). The genome size for three strains (PCC 6305, PCC 6601, and PCC 7415) is similar (ranging from 4.1 to 5.4 Gdal), but that of strain PCC 7101 is higher (7.8 Gdal). On the basis of their genetic relationship, members of this genus can be divided into two major clusters.

Cluster 1 The six strains included in this cluster (PCC 7101, PCC 7504, PCC 7601, PCC 7708, PCC 7710, and PCC 7712) produce relatively long hormogonia that are liberated from the ends of the trichomes and contain gas vesicle clusters dispersed throughout their cells (see Fig. B10.84A and B). They generally break up to yield shorter fragments, prior to losing the gas vesicles and initiating heterocyst differentiation at one terminus. All strains are facultative photoheterotrophs, synthesize C-phycoerythrin, and undergo complementary chromatic adaptation of type II (PCC 7710) or type III (all other strains) (Bryant, 1982). Strain PCC 7601 entered the PCC as a mutant strain, no longer capable of producing functional heterocysts. However, a revertant (PCC 7601/1), forming mature heterocysts and fixing N_2 aerobically, has been isolated (Rippka and Herdman, 1992). Strain PCC 7712 was previously identified as *Gloeotrichia* sp. (see strain history in Rippka and Herdman, 1992), which for lack of large subterminal akinetes, was most likely a misidentification (R. Rippka, unpublished data).

As shown by Lachance (1981) the six strains of cluster 1 share a relatively high degree of genetic relatedness (54–93% relative binding, $\Delta T_{m(e)}$ 0–6°C, using strains PCC 7101 and PCC 7504 as the sources of labeled reference DNA) and are clearly members of a single genus, although further subdivisions are possible.

Cluster 1.1 This cluster includes four strains that were isolated from a freshwater aquarium in Sweden (PCC 7504), freshwater samples in Yale (United States) (PCC 7601 and PCC 7710), and from a soil sample, New York (United States) (PCC 7712). The genetic relatedness determined for these strains (Lachance, 1981) is relatively high (72–93% relative binding, $\Delta T_{m(e)}$ 0–1°C, using PCC 7504 as the source of reference DNA), demonstrating that they are members of the same nomenspecies (see also Fig. B10.82). It should be noted that due to a printing mistake two lines were omitted from Table 8 of Lachance (1981) and thus some data such as the hybridization of strain PCC 7504 with DNA of strain PCC 7712 (relative binding 93%, $\Delta T_{m(e)}$ 0°C) (M.-A. Lachance and R.Y. Stanier, personal communication) are missing.

Reference strain PCC 7504 (co-identity ATCC 29158), isolated in 1972 by A. Neilson from freshwater aquarium, Stockholm, Sweden; included in the genus *Calothrix* (Rippka et al., 1979). Member of cluster 1 of *Tolypothrix* (Rippka and Herdman, 1992). Reference strain of *Tolypothrix* cluster 1.1 (this edition). The mol% G + C is 41.5.

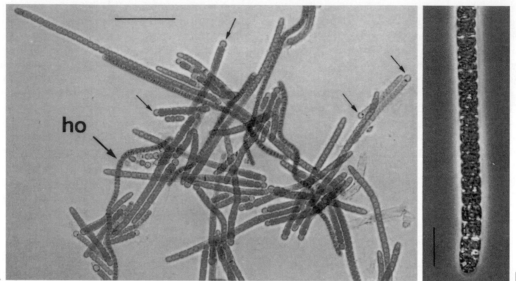

FIGURE B10.84. *Tolypothrix* PCC 7101. (*A*) note the low degree of tapering of the mature trichomes, the exclusively terminal heterocysts (*arrows*), and the long gas-vacuolated hormogonia (*ho*). Bright field. Bar = 50 μm. (*B*) enlargement of a hormogonium, showing the irregularly distributed gas vacuoles typical of this strain. Phase contrast. Bar = 10 μm. (Reprinted with permission from R. Rippka et al., Journal of General Microbiology *111:* 1-61, ©Society for General Microbiology.)

Cluster 1.2 This "cluster" includes a single strain, PCC 7101, isolated from a soil sample in Borneo and identified as *Tolypothrix tenuis* (Watanabe, 1959). If used as the source of reference DNA, relative binding to DNA of all strains of cluster 1.1 was in the range of 54–74%, with $\Delta T_{m(e)}$ values of 4–5°C (M.-A. Lachance and R.Y. Stanier, personal communication, see comments about missing data above). In a reciprocal cross, using strain PCC 7504 (cluster 1.1) as reference DNA, strain PCC 7101 gave 67% relative binding and a $\Delta T_{m(e)}$ of 3°C (Lachance, 1981). Thus strain PCC 7101, though relatively close to members of cluster 1.1, represents a different species. The generic and specific epithets originally given by Watanabe (1959) to strain PCC 7101 can largely be defended on the basis of its structural and developmental characteristics. Consequently, Rippka and Herdman (1992) proposed the latter isolate as the type strain of the species *Tolypothrix tenuis* Kützing 1843. These assignments are maintained here.

Reference strain PCC 7101 (co-identity ATCC 27914), isolated in 1950 by A. Watanabe from soil, Borneo (Watanabe, 1959). Included in the genus *Calothrix* (Rippka et al., 1979). Proposed as type strain of the species *Tolypothrix tenuis* (Rippka and Herdman, 1992, and this edition). The mol% G + C is 41.2; genome size is 7.8 Gdal.

Cluster 1.3 This "cluster" is represented by strain PCC 7708, isolated from Indian soil and identified as *Calothrix membranacea* (Sinclair and Whitton, 1977b; Rippka and Herdman, 1992). Hybridizations with DNA of strains PCC 7504 (cluster 1.1) and PCC 7101 (cluster 1.2) gave relative binding values of 57% and 56%, respectively ($\Delta T_{m(e)}$ 5°C and 6°C) (Lachance, 1981). Thus strain PCC 7708 most likely represents a third species.

Reference strain PCC 7708 (co-identity DCC D0179), isolated by E.G. Pringsheim (prior to 1950) from Indian soil sample; identified as *Calothrix membranacea* (Sinclair and Whitton, 1977b; B. Whitton, personal communication). Included in *Tolypothrix* cluster 1 (Rippka and Herdman, 1992). Reference strain of *Tolypothrix* cluster 1.3 (this edition). The mol% G + C is not determined.

Cluster 2 The three strains included in this cluster (PCC 6305, PCC 6601, and PCC 7415) differ from members of cluster 1 by hormogonia that do not contain gas vesicle clusters. Strains PCC 6305 and PCC 6601 have lost the capacity to produce functional heterocysts and are permanently immotile (Rippka and Herdman, 1992; R. Rippka, unpublished data). All are facultative photoheterotrophs (Rippka et al., 1979), synthesize C-phycoerythrin and undergo complementary chromatic adaptation of type III (Bryant, 1982).

With strain PCC 7415 as the source of reference DNA only relatively low homologies (61% relative binding, $\Delta T_{m(e)}$ 5°C and 7°C) were observed with strains PCC 6601 and PCC 6305, respectively, suggesting that they represent at least two different species (Lachance, 1981; see also Fig. B10.82). Using the same reference DNA, lower homologies (33–40% relative binding, $\Delta T_{m(e)}$ of 12°C) were determined with all representatives of cluster 1 (Lachance, 1981). Thus strains of cluster 2 are heterogeneous and remote from members of cluster 1, but show sufficient DNA homologies to be included in the same genus.

Reference strain PCC 7415 (co-identity ATCC 29157), isolated in 1972 by A. Neilson from soil, greenhouse, Stockholm, Sweden. Included in the genus *Calothrix* (Rippka et al., 1979). Reference strain of *Tolypothrix* cluster 2 (Rippka and Herdman, 1992, and this edition). The mol% G + C is 41.8; genome size is 5.4 Gdal.

MAINTENANCE PROCEDURES

All strains of this genus have been purified on, and are maintained in, freshwater medium BG11$_o$; for isolates that no longer form functional heterocysts, medium BG11 is employed.

TAXONOMIC COMMENTS

Members here assigned to *Tolypothrix* were previously included in *Calothrix* (Rippka et al. 1979) since, apart from a low degree of tapering, they share all of the structural and developmental properties that define the latter genus. However, the low level of DNA similarity observed (Lachance, 1981) with typical members of *Calothrix* provided the justification to separate these strains at the generic level (Rippka and Herdman, 1992).

Strains PCC 7601 and PCC 7710 (cluster 1.1) are co-identic (see strain histories in Rippka and Herdman, 1992), though the former does not form functional heterocysts (Wyatt et al., 1973). Strain PCC 7710 is carried in the DCC as strain *Calothrix* sp. D0255 (Sinclair and Whitton, 1977b; B. Whitton, personal communication). Co-identic subisolates of these two strains have also been deposited in other collections (UTEX 481, CCAP 1429/1, and SAG 1429–1a) under the generic and specific epithets *"Fremyella diplosiphon"*, according to Wyatt et al. (1973). Consequently, this assignment is now widely associated with strain PCC 7601 and its "relatives", some of which have become important model organisms in genetic studies on complementary chromatic adaptation and hormogonium development (see Tandeau de Marsac and Houmard, 1993). However, being a later synonym of the genus *Microchaete* Thuret 1875, *"Fremyella"* is illegitimate under the Botanical Code of Nomenclature (see Bourrelly, 1970). Furthermore, the "species" *Microchaete diplosiphon* Gomont 1895 (synonym *"Fremyella diplosiphon"*) forms akinetes in rows (see Geitler, 1932), which by definition is never observed in cultures of PCC 7601 and PCC 7710. Thus the latter generic and specific epithets are invalid, the correct generic nomen being *Tolypothrix*.

In agreement with their assignment to a single genus, the three strains of *Tolypothrix* examined (PCC 7101, PCC 7601, and PCC 7415) cluster tightly phylogenetically (see Wilmotte and Herdman, this volume). Although the genus *Tolypothrix* was traditionally placed with *Scytonema* (among other genera) in the botanical family *Scytonemataceae* (Geitler, 1932), a close phylogenetic relationship between members of these two genera is not supported by 16S rRNA gene sequence data, since *Scytonema hofmanni* PCC 7110 is located in a different clade, remote from the three strains of *Tolypothrix*.

ADDENDUM TO SUBSECTION IV

The genus *Microchaete* Thuret 1875 was traditionally reserved for filamentous heterocystous cyanobacteria that exhibit a **low degree of tapering** and do not (or only rarely) form false branches. In contrast to the somewhat similar genus *Tolypothrix*, many "species" of *Microchaete*, including the type species *Microchaete tenera* Thuret 1875, have been described as **producing akinetes that occur in chains**, located either adjacent to basal heterocysts or some distance away from them.

Strain PCC 7126 seems to conforms to the latter "species" but has not yet been characterized in detail. This strain can be distinguished from members of *Tolypothrix* by motile hormogonia that may differentiate heterocysts at both termini, thus giving rise to filaments that exhibit a basal to apical tapering towards the thinner center of the trichomes. Most importantly, this strain forms akinetes as described for *Microchaete tenera* (Geitler, 1932; Schwabe, 1967).

Strain PCC 7126 was isolated in 1971 by J.B. Waterbury from a freshwater aquarium. This strain synthesizes C-phycoerythrin and undergoes complementary chromatic adaptation of type III (Bryant, 1982). DNA–DNA hybridizations with *Tolypothrix* strains PCC 7101, PCC 7415, and PCC 7504 as sources of reference DNA gave very low relative binding values (16–22%), and hybridization to DNA of *Calothrix* strain PCC 7102 as the reference resulted in even lower binding (5%) (M.-A. Lachance and R.Y. Stanier, personal communication) (see also Fig. B10.82). Thus, strain PCC 7126 is genetically unrelated to members of *Tolypothrix* and *Calothrix*. The highest degree of relatedness (35% relative binding) was observed with *Nodularia* strain PCC 73104 (Lachance, 1981). However, in the absence of 16S rRNA sequence data, at this low level of DNA similarity the taxonomic position of strain PCC 7126 remains uncertain. Although strain PCC 7126 was proposed as the type species, *Microchaete tenera* Thuret 1875 (Rippka and Herdman, 1992), support for this assignment requires further studies on this and additional isolates (or representatives from nature).

Subsection V. (Formerly **Stigonematales** Geitler 1925)

LUCIEN HOFFMANN AND RICHARD W. CASTENHOLZ

Genera of this subsection exhibit the highest degree of morphological complexity and differentiation within the cyanobacteria; the few available 16S rDNA sequence data analyses suggest that it is a monophyletic group, arising from within an assemblage of filamentous heterocystous cyanobacteria that divide in a single plane (Turner, 1997; "Phylogenetic relationships amongst the *Cyanobacteria*," p. 487). Longitudinal and oblique cell divisions occur in addition to transverse cell divisions. This results in periodic true branching in all genera and in multiseriate trichomes (two or more rows of cells) in some genera. Three major types of true branchings, named "T", "V", and "Y", can be distinguished (Golubic et al., 1996). T-branching is a lateral, nearly perpendicular branching originating from the change of division plane from transverse to longitudinal (Fig. B10.85), V-branching is a dichotomous or pseudodichotomous bifurcation originating

from a change in division plane at, or close to, the trichome tip (Fig. B10.86), and Y-branching arises from the displacement of an intercalary branch-point cell by surrounding meristematic growth (Fig. B10.87). False branching also occurs in some genera. Pit-like synapses or pore channels occur between the cells of trichomes in some of these genera. Heterocysts are both intercalary and terminal. Although reproduction occurs by random breakage of trichomes, hormogonia are also formed in most strains. Akinetes or akinete-like cells are produced by some genera.

In most cases, the width of trichomes varies greatly even within a single clone, since narrower secondary trichomes, arising as branches of thickened primary trichomes, occur as a regular phenomenon, even in uniseriate species.

The mol% G + C of the DNA ranges only from 41.9 to 44.4

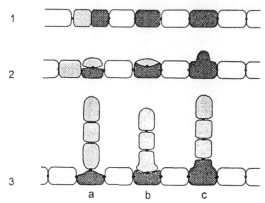

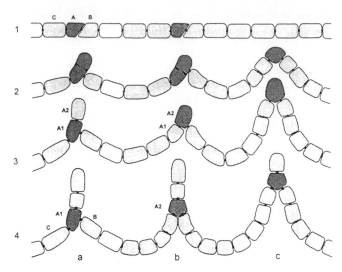

FIGURE B10.85. Schematic presentation of variations of true lateral T-branching through three developmental stages: *a*, change in division plane by longitudinal septation immediately following a transversal division; *b*, branch formation starting with a lens-shaped septation; *c*, branching with development of a lateral protuberance or bulge, which is subsequently cut off by longitudinal septation. (Reprinted with permission from S. Golubic et al., Algological Studies *83:* 303–329, 1996, ©E. Schweizerbart'sche Verlag).

FIGURE B10.87. Schematic presentation of reverse Y-branching types: *a*, initiation of Y-branch by oblique division of an intercalary cell (*stage 1*), followed by a second oblique division perpendicular to the previous one and positioned above the synapse left by it (*stage 2*). The septum separates the branch-generating cell (*A2*) from the branch-point cell (*A1*) (*stages 3–4*). A third cell (*C*) participates in the lateral displacement of the branch point; *b*, a variant of the former with the second oblique division intercepting below the synapse of the previous stage one (*stage 2*), thus separating the branch-point cell (*A2*) from a trichome cell (*A1*) below (*stage 3*). Branch generation starts with the next division of the branch-point cell (*stage 4*); *c*, formation of a Y-branch by inversed timing, in which a trichome loop is generated by an intercalary meristem (*stages 1–3*), followed by a change in division plane in the cell at the top of the loop (*stage 4*). (Reprinted with permission from S. Golubic et al., Algological Studies *83:* 303-329, 1996, ©E. Schweizerbart'sche Verlag).

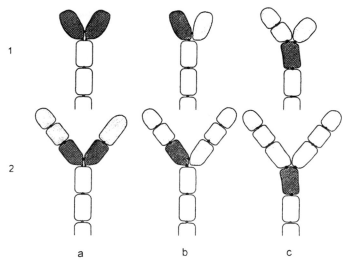

FIGURE B10.86. Schematic presentation of dichotomous and pseudo-dichotomous V-branching: *a*, true dichotomy with median longitudinal division of the apical cell that passed through the synapse of the preceding division, resulting in two equal branch-point cells; *b*, apical pseudodichotomy with longitudinal septum through the apical cell passing outside the synapse of the preceding division; *c*, lateral (subapical) pseudodichotomy originating from a change in division plane and angled septation of an intercalary cell. (Reprinted with permission from S. Golubic et al., Algological Studies *83:* 303–329, 1996, ©E. Schweizerbart'sche Verlag).

for the strains examined. However, these strains represent only two genera and probably only two species as well. The genome size of the same two genera (six strains in all) ranges from 3.2 to 5.4 × 10⁹ daltons (Rippka and Herdman, 1992).

The best known aquatic forms in culture are the various isolates of *Fischerella* (*Mastigocladus* cf. *laminosus*), which are typically found in the flowing waters of hot springs, at least below tem-

peratures of 57–58°C. However, a large number of uncultured or poorly studied stigonematalean cyanobacteria are known from slightly acidic oligotrophic lakes and from fast-flowing streams. An equally large number occur in moist subaerial environments, whereas the order is poorly represented in marine habitats. Some members of Subsection V (e.g., *Stigonema, Chlorogloeopsis*) are capable of producing a UV-absorbing protective pigment (scytonemin) in the extracellular sheaths (see Garcia-Pichel and Castenholz, 1991).

Some genera (e.g., *Chlorogloeopsis, Fischerella, Iyengariella, Herpyzonema, Mastigocladopsis, Nostochopsis, Stigonema, Westiella*) have been studied in uni-cyanobacterial culture, but only two, *Fischerella* and *Chlorogloeopsis*, have been recharacterized using axenic culture strains. However, it is certain from field observations that several other genera exist and have not yet been cultured. Geitler (1932) included seven families and 29 genera. Bourrelly (1985) retained six families and 35 freshwater genera within the *Stigonematales*. Anagnostidis and Komárek (1990) reorganized the order at the family level and recognized eight families with 48 genera. Because of the large number of genera and species, few of which have been obtained in culture, only 6 genera are described here. The following simplified key includes few of the genera considered by Bourrelly (1985) and Anagnostidis and Komárek (1990), but gives an overview of the morphological diversity of the group.

Key to the form-genera of Subsection V

1. True branching, dichotomous or subdichotomous.
   A. Calcareous deposits surrounding filaments; absence of terminal whips (hairs).
      1. Heterocysts present.
         Form-genus *Loriella* (not described in text).
      2. Heterocysts absent.
         Form-genus *Geitleria*, p. 595
   B. Non-calcareous; presence of terminal whips.
         Form-genus *Iyengariella*, p. 598
2. True branching lateral, irregular, or filamentous habit indistinct.
   A. Heterocysts as a distinct lateral cell, or terminal on short branches, rarely intercalary.
      1. Presence of Y-branching.
         Form-genus *Mastigocladopsis* (not described in text).
      2. Absence of Y-branching.
        a. Marine, endolithic in carbonates.
          Form-genus *Mastigocoleus* (not described in text).
        b. Freshwater or terrestrial, nonendolithic.
          Form-genus *Nostochopsis*, p. 598
   B. Heterocysts intercalary, sometimes terminal.
      1. Produces hormocysts and hormogonia; trichomes uniseriate.
         Form-genus *Westiella* (not described in text).
      2. Hormogonia may be produced but hormocysts absent; trichomes uniseriate or multiseriate or semi- amorphous.
        a. Mature trichomes fragment into irregular *Gloeocapsa*-like aggregates.
          Form-genus *Chlorogloeopsis*, p. 591
        b. Mature trichomes do not fragment. Trichomes multiseriate, at least in part.
          Form-genus *Stigonema*, p. 599
        c. Mature trichomes do not fragment. Trichomes uniseriate or multiseriate in small parts only. Heterocysts absent.
          Form-genus *Doliocatella* (not described in text).
        d. Mature trichomes do not fragment. Trichomes uniseriate or multiseriate in small parts only. Heterocysts present.
        Form-genus *Fischerella* Gomont 1895 (*Fischerella* as designated also includes *Mastigocladus* Cohn 1862 and *Hapalosiphon* Nägeli in Kützing 1849), p. 593

Form-genus I. **Chlorogloeopsis** Mitra and Pandey 1966

LUCIEN HOFFMANN AND RICHARD W. CASTENHOLZ

Chlor'ogloe.op'sis. Gr. adj. *chloros* green; Gr. adj. *gloios* sticky; Gr. n. *opsis* appearance, hence resemblance; M.L. fem. n. *Chlorogloeopsis* (cyanobacterium) resembling *Chlorogloea* Mitra.

The filamentous nature of this organism is often unclear, except in the hormogonia. Hormogonia are composed of short chains of cylindrical or barrel-shaped cells which, after ceasing motility, enlarge to become spherical cells (Fig. B10.88). Extracellular sheath is present. Heterocysts develop in both intercalary and terminal positions when levels of combined nitrogen are low. Growth continues, with cell divisions in more than one plane (Fig. B10.88), so that multiseriate trichomes develop. The filamentous nature of the organisms is usually lost, however, since the growing mass of cells commonly fragments into clusters or amorphous aggregates of cells, generally within a mucilaginous sheath (Rippka et al., 1979) (Fig. B10.88). Hormogonia arise from such aggregates. Uneven (asymmetric) divisions of vegetative cells occur; small heterocysts occur when strains are deprived of combined nitrogen (Foulds and Carr, 1981). Masses of vegetative cells may also enlarge to form thicker-walled cells (akinetes or akinete-like cells). Akinete germination takes place with division in several planes and the shedding of extra wall layers

(Rippka et al., 1979). Synaptic "pore channels", common in many other genera in Subsection V, are not present.

FURTHER DESCRIPTIVE INFORMATION

The mol% G + C of the DNA of three studied strains vary between 42.1 and 43.5 and the genome sizes range from 4.7 to 5.4×10^9 daltons (Rippka and Herdman, 1992). The two strains studied so far (Rippka et al., 1979) are facultative aerobic chemoheterotrophs utilizing sucrose best, but also glucose, fructose, and ribose. Phycoerythrocyanin is synthesized by both strains.

Culturing *Chlorogloeopsis* is one of the cyanobacteria most easily grown as a chemoheterotroph. A commonly used growth medium is that of Kratz and Myers (1955) supplemented with $NaHCO_3$ (0.1%) (Table B10.3). Sucrose at 10 mM concentration is included when necessary (Evans et al., 1976). Under photoautotrophic conditions (after inoculation), there is a progression from short trichomes or hormogonia (Fig. B10.88) to multiseriate trichomes, to aseriate clusters of aggregates. Motile hor-

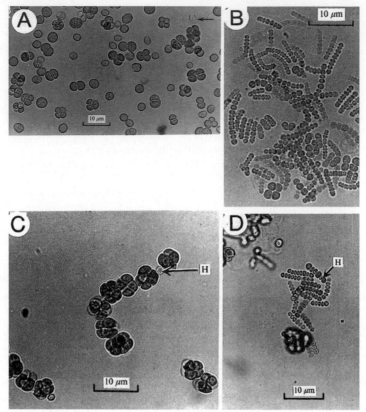

FIGURE B10.88. *Chlorogloeopsis fritschii* (CCAP 1411/1). *A*, photoheterotrophic growth after growth in solid medium, *arrow*, discarded sheath material. *B*, photoautotrophic growth, after transfer from photoheterotrophic conditions. Many hormogonia are present. *C*, photoheterotrophic growth in liquid medium lacking combined nitrogen. *H*, heterocyst. *D*, earlier stage of growth as depicted in *C*, with heterocysts (*H*) sometimes differentiating in hormogonia. (Reprinted with permission from E.H. Evans et al., Journal of General Microbiology *92*: 147–155, 1976, ©Society for General Microbiology).

mogonia may self-isolate, and clones can be established by mechanical removal of the hormogonium-containing agar block with watchmaker's forceps. Growth in the light in the presence of sucrose, however, does not allow the development of trichomes. Unicells or small groups of cells (Fig. B10.88A) give rise to larger spherical clusters of cells and eventually to large aseriate aggregates (Fig. B10.88). Dark chemoheterotrophic growth on sucrose maintains aseriate clusters of cells and again no trichome formation was evident (Evans et al., 1976).

Reference strain   PCC 6912 (ATCC 27193, CCAP 1411/1). This culture was first isolated from soil in India by A.K. Mitra as *Chlorogloea fritschii*.

DIFFERENTIATION OF THE GENUS *CHLOROGLOEOPSIS* FROM OTHER GENERA

There is a close resemblance between *Chlorogloeopsis* and some species of *Nostoc* in phases of the developmental cycle. The aseriate stages of *Nostoc* are hardly distinguishable from similar phases of *Chlorogloeopsis*. The motile hormogonia are also similar. However, in *Chlorogloeopsis* there are very clearly divisions in more than one plane, in hormogonia that have come to rest and in the "unicellular" and cluster phases (see Fig. B10.88). In the aseriate stages of *Nostoc*, cells become detached and disoriented, but two or more planes of division are not discernible.

TAXONOMIC COMMENTS

Although there are few available taxonomic criteria besides morphological characteristics, the thermophilic cyanobacterium referred to as "high temperature form (HTF) *Mastigocladus*" and later as "HTF *Chlorogloeopsis*" appears to be closely related to *Chlorogloeopsis* in most respects (Castenholz, 1978, 1996), although it may represent a distinct genus (Wilmotte et al., 1993; Wilmotte, 1994). It occurs in hot springs worldwide at temperatures up to 63–64°C. Similar hormogonia are formed by both aseriate and essentially unicellular cultures (Fig. B10.89).

The mol% G + C of the DNA in the three studied strains of "HTF *Mastigocladus*" varies between 43.2 and 43.8 (Rippka and Herdman, 1992).

Earlier authors, observing forms in nature similar to "HTF *Mastigocladus*", have simply referred to these as forms or varieties of *Mastigocladus (Hapalosiphon) laminosus* (e.g., Frémy, 1936; Schwabe, 1960a). *Mastigocladus laminosus* (*sensu stricto*), which displays true branches, is here included in the genus *Fischerella* as in Rippka and Herdman (1992).

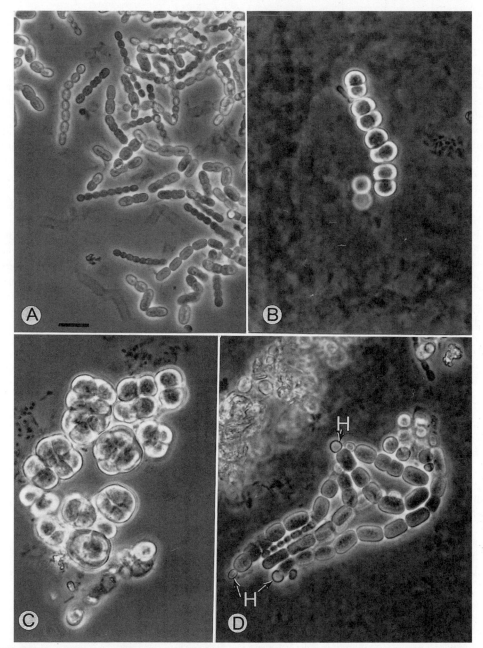

FIGURE B10.89. HTF *Chlorogloeopsis* (strain CCMEE I-15-HTF, clone 1). *A*, motile hormogonia in nitrate-replete medium (45 °C). *B*, later stage under the same conditions. *C*, later amorphous phase. *D*, trichomes in medium lacking combined nitrogen. *H*, heterocysts. Bar = 10 μm.

Form-genus II. **Fischerella** Gomont 1895

LUCIEN HOFFMANN AND RICHARD W. CASTENHOLZ

Fisch.er.el' la. M.L. fem. n. *Fischerella* Fischer, probably named for B. Fischer, botanical bacteriologist, 1852–1915.

The true branches of these organisms are uniseriate and composed of cells, particularly those distal from the base, that are generally longer than broad. The axis (primary trichome) from which they arise is also mainly uniseriate, but may become multiseriate in part, with divisions in more than one plane. The axis in these regions, however, is seldom more than 2–3 cells in thickness. In addition, the cells of the axes become enlarged, often semispherical in shape. The older cells of a main axis may be-

come separated from each other by sheath material and may act as akinetes (Martin and Wyatt, 1974). Most of the widened cells of the axial trichome, however, display a true filamentous nature, with only a peptidoglycan septum separating cells (Nierzwicki et al., 1982; Balkwill et al., 1984). The main axis forms when a hormogonium comes to rest, cells enlarge, and cell divisions that are parallel to the long axis or diagonal (oblique) begin; some of the resulting cells elongate to form branches (secondary trichomes) (Balkwill et al., 1984; Nierzwicki-Bauer et al., 1984) (Fig. B10.90). Extracellular sheaths are absent.

FURTHER DESCRIPTIVE INFORMATION

These narrow secondary trichomes (which may taper) become progressively longer with cell elongation and transverse divisions. In field populations of *Fischerella* (*Mastigocladus* cf. *laminosus*) in flowing hot springs, almost the entire mass is composed of tufts or streamers of secondary trichomes several centimeters in length, and no branching is seen except in the prostrate attached mass of primary trichomes (main axes). In cultures, at least, secondary trichomes of thermophilic *Fischerella* may eventually differentiate into series of spherical, thick-walled cells that are akinete-like. Typical cyanobacterial akinetes are not easily recognizable. In some cases, the widened cells may divide in a second plane and form new branches (i.e., secondary trichomes).

The hormogonium is a gliding trichome composed of few (~11–16), narrow, morphologically uniform cells that are cylindrical or slightly barrel-shaped (Hernández Muñiz and Stevens, 1987) (Fig. B10.91). They are formed as the distal portion of branches. No heterocysts differentiate until after hormogonia have ceased motility. Heterocysts of *Fischerella* (*Mastigocladus* cf. *laminosus*) are elongate, spherical, or even compressed (shorter than broad), and are lateral, terminal, or intercalary. Heterocysts of thermophilic strains are different from typical heterocysts of the *Nostocales*, in that they possess only one additional wall layer (homogeneous type) and have densely stacked lamellar membranes (Nierzwicki-Bauer et al., 1984).

Most of the information on *Fischerella* in culture has been obtained from the thermophilic species, which are common in neutral pH and alkaline hot springs throughout the world (Castenholz, 1978, 1996). However, Martin and Wyatt (1974) have compared the physiology of six strains of stigonematalean cyanobacteria, including five strains of three species of *Fischerella*, none of which were thermophilic. Besides the true branching habit, the great diversity of form as described by Frémy (1936), Schwabe (1960a), Rippka et al. (1979), and Balkwill et al. (1984) for *Fischerella* (*Mastigocladus* cf. *laminosus*) also applies to these nonthermophilic representatives. In the nonthermophilic strains, hormogonia were not always formed under conditions favorable for their production, and sometimes the multiseriate condition of the axis was rare or lacking.

Thurston and Ingram (1971) have studied the morphology and fine structure of another nonthermophilic species of *Fischerella*, in which they have shown details of synaptic connections, with microplasmodesmata joining adjacent cells in the branch trichomes (Fig. B10.92).

All strains of *Fischerella* examined by Rippka and Herdman (1992) synthesize phycoerythrocyanin (see also Füglistaller et al., 1981). All strains studied so far are capable of photoheterotrophic and dark chemoheterotrophic growth, utilizing glucose, fructose, or sucrose and, in two cases, ribose.

The mol% G + C of the DNA of the studied strains range from 41.9–44.4 and the genome size ranges from 4.0 to 4.7 × 10^9 daltons (Rippka and Herdman, 1992).

Although *Fischerella* (*Mastigocladus laminosus*) forms dominant, almost monotypic, populations in many hot springs, species of nonthermophilic *Fischerella* are generally not as conspicuous and often occur in moist subaerial habitats. Marine forms are rare or lacking.

Culturing   The several strains used for the description of *Fischerella* by Rippka et al. (1979) were all thermophiles ascribed to the genus *Mastigocladus*.

Cultures are easily maintained in a variety of media with or without combined nitrogen, e.g., BG-11 (Rippka et al., 1979), medium B (Stevens et al., 1973), and D medium or D medium lacking NO_3 (ND) (Castenholz, 1981) (Table B10.3). When ma-

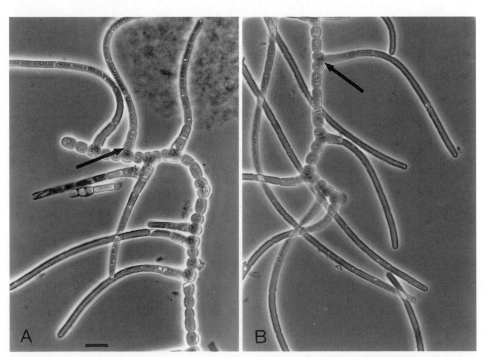

FIGURE B10.90. *Fischerella* (*Mastigocladus* cf. *laminosus*) strain University of Oregon I-S5-M, clone 5 (culture lost), growing in replete medium at 45°C. *A*, showing true branching (*arrow*) from uniseriate primary trichome (main axis). Bar = 10 μm. *B*, showing Y-branching (arrow) (see Castenholz, 1972). For *B*, the same scale as in *A* applies.

FIGURE B10.91. *Fischerella* (*Mastigocladus laminosus*) culture NZ-69-M, clone 1, showing the circling of motile hormogonia on agar surface at 45°C. Bar = 10 μm.

ture cultures are inoculated on agar-solidified medium, motile hormogonia are readily formed and easily isolated after migration on agar. Although thermophilic forms are the cosmopolitan "weeds" of hot springs, medium with a high content of ammonium (i.e., >2 mM) is usually inhibitory. Nonthermophilic species of *Fischerella* have been grown in CG-10 medium (Thurston and Ingram, 1971) and ASM-1 medium (Gorham et al., 1964; Martin and Wyatt, 1974).

Reference strains PCC 7115 (ATCC 27929), isolated from Tassajara Hot Springs, California (Rippka et al., 1979) as *Mas-*

tigocladus species H2; and PCC 7522 (ATCC 29539, CCMEE NZ-86-m), isolated from a hot spring, Whakarewarewa, New Zealand (Rippka et al., 1979) as *Mastigocladus laminosus*.

DIFFERENTIATION OF THE GENUS *FISCHERELLA* FROM OTHER GENERA

The complex variety of forms or developmental stages exhibited by *Fischerella*, which may include primary and secondary trichomes, hormogonia, unicells and amorphous cell aggregates, makes identifying any single stage difficult. *Fischerella*, however, is most easily confused with the genera *Stigonema* and *Hapalosiphon*. In *Stigonema*, the main axis and the branches become multiseriate or multicellular; however, the growing tips are usually uniseriate, and this condition may extend for several cells. *Fischerella*, on the other hand, has secondary trichomes (branches) with more elongate and narrower cells than those of the primary trichome (main axis). The main axis may be multiseriate in part.

Herpyzonema, known from terrestrial (cave) and marine habitats, is similar to *Fischerella* and *Hapalosiphon*, but has typical Y-branchings which originate after the formation of lateral loops (Hernández-Mariné and Canals, 1994).

Doliocatella Geitler 1935 is a benthic freshwater genus mostly known from tropical waters, where it forms thalli up to 5 mm high. It is characterized by the absence of heterocysts and, often, unilateral branches.

TAXONOMIC COMMENTS

For the present, there appears to be insufficient reason to separate the genera *Mastigocladus* and *Hapalosiphon* from *Fischerella*. The boundaries separating these three genera are not clear for all described species (see Geitler, 1932; Desikachary, 1959; Bourrelly, 1985). Furthermore, strains variously identified by their isolators as members of these three genera have indeed been shown by DNA–DNA hybridization to be closely related and are probable members of the same genus (Rippka and Herdman, 1992). The original description of *Mastigocladus* Cohn 1862 gave insufficient information, and subsequent redefinitions have attempted to use the possession of Y-branches, in addition to true and false branches, to distinguish *Mastigocladus* from *Fischerella* (Schwabe, 1960a) or to establish *Mastigocladus laminosus* as *Hapalosiphon laminosus* (Frémy, 1936). Although such Y-branches do occur in some strains (see Fig. B10.90B), this characteristic does not warrant separation of *Mastigocladus* as a separate genus until further study is completed. The main difference between *Mastigocladus* and *Fischerella* on the one hand, and *Hapalosiphon* on the other hand, is the presence of branches similar to the main axis in *Hapalosiphon*, whereas they differ in the other two genera. Rippka and Herdman (1992) distinguish two clusters within the genus *Fischerella*, for which they propose the names *F. muscicola* and *F. thermalis*.

Form-genus III. **Geitleria** *Friedmann 1955*

LUCIEN HOFFMANN AND RICHARD W. CASTENHOLZ

Geit.ler' i.a. M.L. fem. n. *Geitleria* Geitler, named after L. Geitler, Austrian botanist and specialist in cyanobacteria, 1899–1990.

Branching is pseudodichotomous, dichotomous, and lateral (Fig. B10.93). The sheath becomes heavily calcified, and only cells near the tips of trichomes are able to give rise to a lateral branch or, when an apical cell undergoes an oblique division followed by further divisions of both cells, to a false dichotomy (Friedmann, 1955). The ultrastructure of decalcified specimens, in-

cluding pore channels (i.e. "pit connections"), has been studied by Couté (1982, 1985) (Fig. B10.93B). No cultures of this genus have been established.

Geitleria calcarea, together with *Scytonema julianum*, is a calcified cyanobacterium of limestone caves (Friedmann, 1955; Couté, 1985).

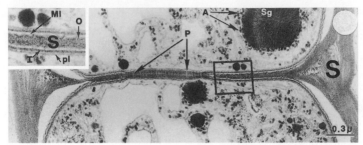

FIGURE B10.92. Electron micrograph of a stage of cell division and sheath formation in a branch filament of *Fischerella*. Cell division is almost complete. *P*, pores (microplasmodesmata) in the remaining septum. *A*, alpha granules (glycogen granules); *Sg*, structured granule (cyanophycin in granule); *S*, sheath. *Insert*: enlargement of septum showing two walls separated by sheath(s). *Pl*, plasmalemma; *MI*, middle layer; *I*, inner layer (peptidoglycan layer); *O*, outer membrane; *S*, sheath. (Reprinted with permission from E.L. Thurston and L.O. Ingram, ©Journal of Phycology *7*: 203–210, 1971.)

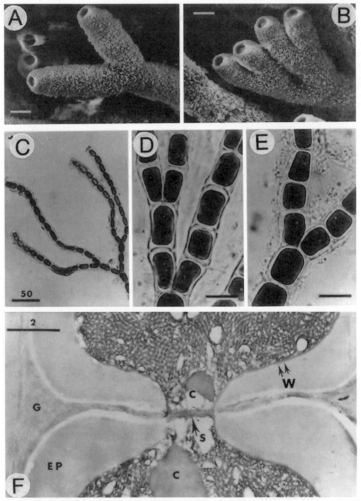

FIGURE B10.93. *Geitleria calcarea*. A and B, scanning electron micrographs of apical ramifications of filaments with CaCO$_3$-encrusted sheath. Bars = 10 μm. C, D and E, photomicrograph of material decalcified with the aid of EDTA. Various views of branching system. Bars = 10 μm except for C. F, electron micrograph of section of decalcified material, showing pore channel ("pit connection") between two cells of a trichome. *S*, synapses which contains microplasmodesmata; *C*, cyanophycin granules; *W*, cell wall; *EP*, parietal (additional) wall; *G*, sheath. Bar = 2 μm. Patterns of phycobilisomes on thylakoids can be seen, particularly in the upper cell. (Reprinted with permission from A. Couté, Hydrobiologia *97*: 255–274, 1982, ©Kluwer Academic Publishers.)

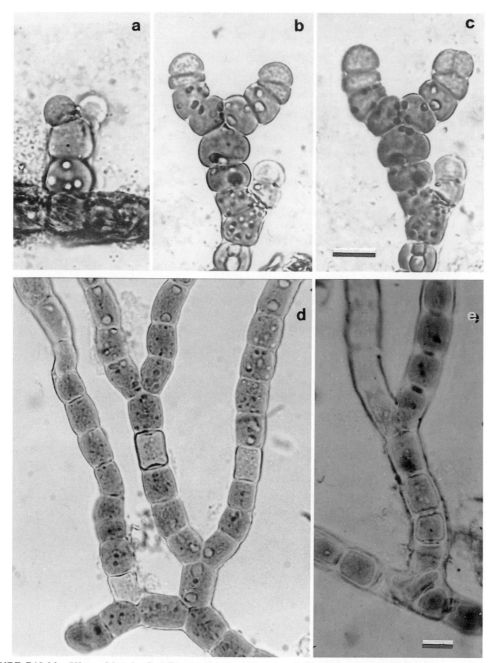

FIGURE B10.94. V-branching in *Loriella osteophila* in culture. *a*, initial stage of a true dichotomy on a lateral branch; *b*, symmetrical true dichotomy following a lateral pseudodichotomy; *c*, the same branch at a later stage, starting another true dichotomy on one of the branches. (Reprinted with permission from S. Golubic et al., Algological Studies *83:* 303–329, 1996, ©E. Schweizerbart'sche Verlag).

DIFFERENTIATION OF THE GENUS *GEITLERIA* FROM OTHER GENERA

Geitleria is morphologically and ultrastructurally (Hernández-Mariné et al., 1999) similar to the genus *Loriella* Borzi 1892 (Fig. B10.94), another aerophytic cyanobacterium generally found in low light environments (Hoffmann, 1990; Hernández- Mariné et al., 1999); the main difference between the two genera is the presence of heterocysts in *Loriella*. The ultrastructure of another dichotomously branching stigonematalean genus, *Pulvinularia* Borzi 1916, with macroscopic cartilaginous thalli formed by radially arranged dichotomous branches, has been studied by Rott and Hernández-Mariné (1994).

Form-genus IV. **Iyengariella** Desikachary 1953

LUCIEN HOFFMANN

I.yen.gar.iel'la. M.L. fem. n. *Iyengariella* Iyengar, named after M.O.P. Iyengar, Indian phycologist.

Besides true and pseudodichotomous branchings, false branchings are also found; false branches occur after trichome breakage as a result of the formation of a necridium. True dichotomous branches originate from the longitudinal division of the apical cell and pseudodichotomous true branches originate from the oblique division of an intercalary cell. Filaments taper, the trichomes ending in long whips of the rivulariacean type. Heterocysts and akinetes are absent. Reproduction is through hormogonia- formation.

Iyengariella is a freshwater genus of which one species, *Iyengariella endolithica*, is a carbonate-boring species (Seeler and Golubic, 1991). This species, when grown on agar, exhibits differentiated growth with prostrate filaments on the agar surface and erect, aerotactic filaments; it shows the ability to undergo complementary chromatic adaptation.

Form-genus V. **Nostochopsis** Wood 1869

LUCIEN HOFFMANN

Nost' och.op' sis. Old English n. *Nost'hryl* nostril; German n. *Nasen'loch;* Gr. n. opsis appearance, hence resemblance; M.L. fem. n. *Nostochopsis* (cyanobacterium) resembling *Nostoc* Vaucher, "excrement from the nostrils of some rheumatick planet" (Paracelsus 1493–1541) referring to the rapid appearance of *Nostoc commune* colonies after thundershowers.

The uniseriate trichomes form lateral T-branchings of two types, one long and many celled, the other one of limited growth (1–3 cells) with a heterocyst at the end; besides these pedicellate heterocysts, heterocysts are intercalary or lateral and sessile. A

culture of this genus has been established (Weber et al., 1996), but has not been studied taxonomically.

Nostochopsis is mainly a genus of tropical to warm temperate freshwater, where it forms benthic, macroscopic, gelatinous *Ri-*

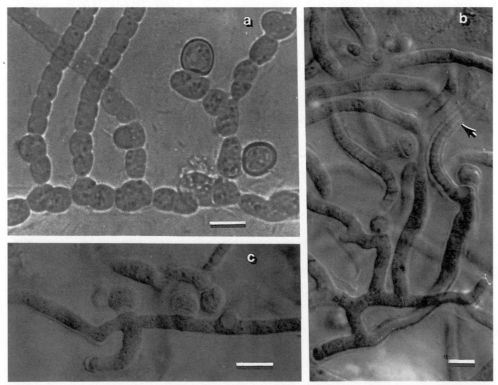

FIGURE B10.95. Lateral T-branching and terminal heterocyst differentiation: *a, Mastigocladopsis,* developing lateral branches with "unlimited" and "limited" growth. The latter differentiate immediately into a terminal unipolar heterocyst which appears laterally attached to the trichome; *b, Mastigocoleus testarum,* with a dense display of true lateral branches (long branches penetrate calcium carbonate or differentiate into cellular whips (*arrow*); the short, "limited" ones differentiate into terminal heterocysts); *c, Mastigocoleus testarum,* with opposing branches. Bars = 10 µm. (Reprinted with permission from S. Golubic et al., Algological Studies *83:* 303–329, 1996, ©E. Schweizerbart'sche Verlag.)

vularia-like colonies, although it has also been found growing cryptoendolithically in sandstone (Weber et al., 1996).

DIFFERENTIATION OF THE GENUS *NOSTOCHOPSIS* FROM OTHER GENERA

Two other genera with lateral heterocysts exist. *Mastigocladopsis* Iyengar and Desikachary 1946 (Fig. B10.95), characterized by Y-branchings, is found in the soil or forms macroscopic benthic colonies in running freshwaters. Its ultrastructure, including its pit-like synapses, and morphological variability in culture have been studied by Hernández-Mariné et al. (1992) and Hoffmann (1994). *Mastigocoleus* forms only lateral T-branchings and is probably a cosmopolitan endolith in marine carbonate substrates. Its ultrastructure has been studied by Pantazidou (1991).

Form-genus VI. **Stigonema** Agardh 1824

LUCIEN HOFFMANN AND RICHARD W. CASTENHOLZ

Sti.go.ne′ma. Gr. n. *stigeus* one who marks; Gr. n. *nema* thread; M.L. fem. n. *Stigonema* segmented filament.

Many species of this genus are extreme examples of multicellular complexity in the cyanobacteria. Main axes with numerous branches may reach thicknesses of 1 mm (Fig. B10.96). Extracellular sheaths are present. Pore channels (synapses of microplasmodesmata) occur between all cells derived by division (Fig. B10.93). Branches, although initially simpler than the main axis, eventually become as complex. Tips of branches may be uniseriate for a considerable distance.

Stigonema cf. *ocellatum* and *S.* cf. *minutum* have been isolated in culture, but taxonomic studies have not been made (Zehnder, 1985).

Stigonema is a freshwater or subaerial (terrestrial) genus found commonly on moist rocks or soil or in oligotrophic, slightly acidic lakes and some streams. Conspicuous tufted benthic mats of *Stigonema* occur in some oligotrophic lakes. Some of the species are also photobionts in lichen symbioses.

Three strains are held in the Pasteur Culture Collection.

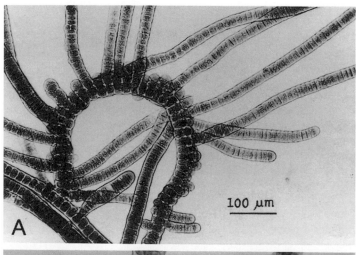

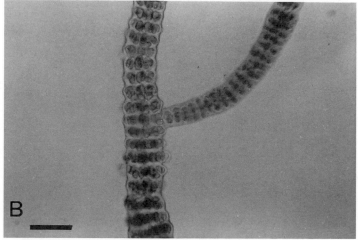

FIGURE B10.96. *A, Stigonema minutum.* Young culture showing multiseriate axis and primarily uniseriate branches. (Reprinted with permission from A. Zehnder, Archiv für Hydrobiologie Supplement *71:* 281–289, 1985, ©E. Schweizerbart'sche Verlag.) *B, Stigonema* species with multiseriate axis and branch. Bar = 20 μm.

Phylum BXI. Chlorobi *phy. nov.*

GEORGE M. GARRITY AND JOHN G. HOLT

Chlo.ro' bi. M.L. neut. n. *Chlorobium* genus of the phylum; dropping ending to denote a phylum; M.L. fem. pl. n. *Chlorobi* phylum of the genus *Chlorobium*.

The *Chlorobi* share a common root with *Bacteroidetes* in both of the major reference trees. At present, the phylum contains a single class, order, and family. Gram-negative, spherical, ovoid, straight, or curved rod-shaped *Bacteria*. Strictly anaerobic, obligately phototrophic. Cells grow preferentially by photoassimilation of simple organic compounds. Some species may utilize sulfide or thiosulfate as an electron donor for CO_2 accumulation.

Sulfur globules accumulate on the outside of the cells when grown in the presence of sulfide and light, and sulfur is rarely oxidized further to sulfate. Ammonia and dinitrogen used as the nitrogen source. Most genera require one or more growth factors; the most common are biotin, thiamine, niacin, and *p*-aminobenzoic acid.

Class I. *"Chlorobia"*

Subclass I.

Order I. *"Chlorobiales"*

Family I. *"Chlorobiaceae"*

Green Sulfur Bacteria

JÖRG OVERMANN

Introduction The green sulfur bacteria are predominately aquatic bacteria that grow photosynthetically under anoxic conditions. Similar to *Chromatiaceae* (purple sulfur bacteria) and purple nonsulfur bacteria (*Proteobacteria*), the photosynthetic metabolism of green sulfur bacteria differs from that of *Cyanobacteria*, algae, and green plants in that water cannot serve as an electron-donating substrate and molecular oxygen is not generated. The external electron donors for photosynthetic CO_2 assimilation have low redox potentials and comprise reduced sulfur compounds and molecular hydrogen. Green sulfur bacteria can be distinguished from the *Chromatiaceae* by their distinct bacterio-chlorophylls (BChls), the presence of special light-harvesting structures, and their exclusively photolithotrophic metabolism.

The Gram-negative cells are spherical, ovoid, straight, or curved rod shaped; they multiply by binary fission or binary plus ternary fission. In one genus (*Chloroherpeton*), cells are long unicellular filaments, highly flexible, and motile by gliding; all other genera are nonmotile. Six of the fourteen species possess gas

vesicles. Photosynthetic pigments are located in the cytoplasmic membrane and in the light-harvesting chlorosomes (formerly chlorobium vesicles; Cohen-Bazire et al., 1964; Staehelin et al., 1978). The latter are ovoid structures (70–180 nm long and 30–60 nm wide) which underlie and are attached to the cytoplasmic membrane (Fig. B11.1). The chlorosome envelope (width, 2–3 nm) consists of monogalactosyl diglyceride and several proteins. Chlorosomes are also found in multicellular filamentous gliding bacteria of the genus *Chloroflexus* and *Oscillochloris*, which are distantly related phylogenetically.

The existing species exhibit one of two clearly distinguishable colors: cell suspensions are either green (grass green) or brown (chocolate brown). All strains of the green species contain BChl *c* or *d* (Table B11.1) as the major component and small amounts of BChl *a*. The major carotenoid of the green species is chlorobactene, in addition to OH-chlorobactene and γ-carotene. All strains of the brown species contain BChl *e* as the major component, as well as small amounts of BChl *a* (Gloe et al., 1975);

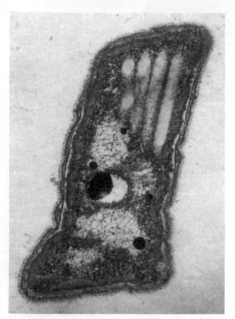

FIGURE B11.1. Fine structure of green sulfur bacterium *Pelodictyon clathratiforme* strain 1831. Note the chlorosomes (*dark gray area*) underlying and attached to the cytoplasmic membrane, the bundle of gas vesicles (*light gray areas with pointed ends*) in the upper part of the section, and the broadened end of the cell at the site of ternary fission. Electron micrograph (×105,000). (Courtesy of G. Cohen-Bazire.)

TABLE B11.1. Nomenclature of bacteriochlorophylls

Bacterio-chlorophyll	Designation by	Characteristic absorption maxima in living cells (nm)
a	Jensen et al. (1964)	375, 590, 805
c	Jensen et al. (1964)	335, 460, 745–755
d	Jensen et al. (1964)	325, 450, 710–740
e	Gloe et al. (1975)	345, 450–460, 700–710

the carotenoids of the brown species are isorenieratene and β-isorenieratene (Schmidt, 1978). In the genus *Chloroherpeton*, γ-carotene is the main carotenoid.

Antenna BChls are mainly farnesol esters. In each species, several BChl homologs are present which differ with respect to the aliphatic side chains attached to the tetrapyrrole (methyl, ethyl, *n*-propyl, isobutyl, or neopentyl). A high degree of alkylation occurs in cells grown at low light intensities (Huster and Smith, 1990). A minor fraction of the BChls contain a different esterifying alcohol. Overall, up to 15 different BChl homologs can be found in one strain (Borrego and Garcia-Gil, 1994).

Each chlorosome contains about 10,000 molecules of BChl *c*, *d*, or *e*, small amounts of BChl *a*, and three isoprenoid quinones (1'-oxomenaquinone-7, a derivative thereof, and menaquinone-7; Frigaard et al., 1997) and is energetically connected to 5–10 reaction centers (Amesz, 1991). This size of the photosynthetic antennae of green sulfur bacteria (and as a result their specific BChl content of up to 915 μg/mg of protein) exceeds that of *Chromatiaceae* by one order of magnitude.

Green sulfur bacteria are strictly anaerobic, obligately phototrophic, and capable of photolithoautotrophic growth with sulfide, sulfur, or sometimes thiosulfate as an electron donor. If sulfide is oxidized, highly refractile globules of zero-valence sulfur are formed outside the cells; the sulfur is further oxidized

to sulfate by most species. All genera are potentially mixotrophic and photoassimilate a number of simple organic substrates in the presence of both sulfide and bicarbonate. Acetate is incorporated by reductive carboxylation, which requires the reducing equivalents from sulfide and involves a reduced ferredoxin. Most strains reach light saturation of growth at 10 μmol·m^{-2}·s^{-1} (daylight fluorescent tubes), while light inhibition has been observed at intensities above 200 μmol·m^{-2}·s^{-1}.

Basically different from all purple bacteria, the green sulfur bacteria carry out autotrophic CO_2 assimilation by the reductive tricarboxylic acid cycle (Arnon cycle; Fuchs et al., 1980), which requires less ATP per molecule of CO_2 fixed than the Calvin cycle. This may partially explain why green sulfur bacteria exhibit much lower maintenance energy requirements than *Chromatiaceae* (Veldhuis and van Gemerden, 1986; Overmann et al., 1992a). The lowest light intensity supporting growth of a phototrophic organism (< 0.25 μmol·m^{-2}·s^{-1}) was determined for *Chlorobium phaeobacteroides* MN1, isolated from the chemocline of the Black Sea (Overmann et al., 1992a). In addition to reduced sulfur species, molecular hydrogen is an electron donor for many green sulfur bacteria. However, due to the lack of assimilatory sulfate reduction, a reduced sulfur source is required under these growth conditions (Lippert and Pfennig, 1969). The affinity for sulfide of green sulfur bacteria is one order of magnitude higher than that of purple sulfur bacteria, but the affinity for acetate is one order of magnitude lower (van Gemerden and Mas, 1995). In the dark, cultures of *Chlorobium limicola* f. sp. *thiosulfatophilum* oxidize polyglucose and the degradation products acetate, propionate, caproate, and succinate are excreted (Sirevåg and Ormerod, 1977). Vitamin B_{12} is required by most strains of all species (Pfennig and Lippert, 1966). In its absence, the specific BChl content is drastically reduced and cells are incapable of forming chlorosomes (Fuhrmann et al., 1993).

On the basis of the presence or absence of gas vacuoles and the selective advantage in enrichment cultures, two physiological-ecological subgroups can be differentiated among the green sulfur bacteria. The first group includes species of the genera *Chlorobium* and *Prosthecochloris*, which do not contain gas vacuoles and have a selective advantage in liquid cultures at high sulfide concentrations (4–8 mM) and light intensities (100 μmol·m^{-2}·s^{-1}). The second subgroup includes the gas vacuole-containing species, which compete successfully only at very low sulfide concentrations (0.4–2 mM), low light intensities (5–20 μmol·m^{-2}·s^{-1}), and low incubation temperatures (10–20°C). *Pelodictyon clathratiforme*, *P. phaeoclathratiforme*, and *Ancalochloris perfilievii* are examples of this second subgroup.

The mol% G + C of the DNA is: 47.8–58.1 (Bd).

Type genus: *Chlorobium* Nadson 1906, 190.

The major differentiating characteristics of the genera of the green sulfur bacteria are presented in Table B11.2. The consortium-forming symbiotic aggregates of green bacteria with colorless bacteria are given under "Addendum to the Green Sulfur Bacteria". This appears justified in that the green or brown components of the consortia contain chlorosomes and are morphologically similar to known species of the green sulfur bacteria.

Ecology

Habitats In nature, green sulfur bacteria occur where light reaches anoxic water layers or sediments containing reduced sulfur compounds. Dense accumulations are mostly found in planktonic habitats, like thermally stratified or meromictic lakes and

TABLE B11.2. Differential characteristics of the genera of the Green Sulfur Bacteria, a subcategory of Green Bacteria[a]

Characteristic	Chlorobium	Prosthecochloris	Pelodictyon	Ancalochloris	Chlorherpeton
Cultures and cells green	+	+	+	+	+
Cultures and cells brown	+	+	+	−	−
Gas vacuoles	−	−	+	+	+
Nonmotile	+	+	+	+	−
Gliding motility	−	−	−	−	+
Cells unicellular, filamentous, flexible	−	−	−	−	+
Cells starlike, with extrusions, prosthecae	−	+	−	+	−
Cells spherical, ovoid or rod-shaped	+	−	+	−	−
Cells curved, vibrioid-shaped, may form spirals	+	−	−	−	−

[a]Symbols: +, positive or present; −, negative or absent.

brackish lagoons (van Gemerden and Mas, 1995). Unlike *Chromatiaceae*, green sulfur bacteria rarely develop profusely in marine sediments, where they have sometimes been found in multilayered microbial mats ("Farbstreifensandwatt") (Nicholson et al., 1987). Dense populations of green sulfur bacteria also develop as 3-mm-thick, unlaminated mats on the bottom substrates of New Zealand sulfur springs at sulfide concentrations above 50 μM, pH values between 6.0 and 4.5, and temperatures between 55 and 45°C (Castenholz et al., 1990). Macroscopically visible accumulations have only rarely been described for other sediment systems.

Meanwhile, 16S rRNA probes which permit the detection of green sulfur bacteria *in situ* have become available (Tuschak et al., 1999).

Autecology In the natural environment, growth of green sulfur bacteria is limited to the narrow zone of overlap between the opposing gradients of light and sulfide. As a consequence, layers of green sulfur bacteria are generally present at considerable depth below the lake or sediment surface and anoxygenic photosynthesis is generally limited by light penetrating to the bacterial layer (van Gemerden and Mas, 1995). The depth, vertical extension, and cell density of the layers of green sulfur bacteria are correlated. In lakes, they occur at depths of 2–20 m, have a vertical extension of 0.5–2 m, and reach a biomass density of about 600 μg BChl/l. The largest and deepest extant population of green sulfur bacteria was found in the chemocline of the Black Sea at 68–98 m depth and had a biomass density of only 0.94 ng BChl/l (Repeta et al., 1989). By comparison, benthic populations of *Prosthecochloris aestuarii* form extremely thin layers only 1 mm thick but reach much higher biomass densities (120,000 μg BChl/l; Pierson et al., 1987).

Species capable of gas vesicle formation usually dominate in pelagic microbial communities. In *Pelodictyon phaeoclathratiforme* DSMZ 5477[T], formation of gas vesicles is induced exclusively at low light intensities (<5 μmol·m^{-2}·s^{-1}; Overmann et al., 1991a). So far, no diurnal vertical migrations of gas-vacuolated green sulfur bacteria have been observed (Parkin and Brock, 1981), but gas vesicles confer a selective advantage on the cells by reducing their buoyant density and thus preventing their sedimentation into dark bottom layers.

Competition among anoxygenic phototrophic bacteria The species composition in layers of green sulfur bacteria is determined by the spectrum of the underwater light reaching the bacterial cells. In productive lakes, and in lakes where the oxic–anoxic interface is located at greater depths (≥ 9 m), selective absorption of light by phytoplankton occurs in upper water layers and only light in the green wavelength range (500–550 nm) reaches the anoxic bottom waters. Brown-colored species of green sulfur

bacteria usually dominate the anoxygenic phototrophic community in these ecosystems. These species contain the carotenoids isorenieratene and β-isorenieratene (absorption maximum, 505–520 nm); their specific carotenoid content is up to four times higher than that of their green-colored counterparts (Montesinos et al., 1983).

Green and purple sulfur bacteria are similar with respect to their nutritional requirements. Unlike their purple counterparts, however, green sulfur bacteria assimilate only very few organic carbon compounds and cannot grow chemotrophically with molecular oxygen as the terminal electron acceptor. However, the superior light-harvesting potential of green sulfur bacteria compensates for their limited physiological flexibility. As a result, populations of green sulfur bacteria are often located below those of purple sulfur bacteria in many stratified lakes (Caldwell and Tiedje, 1975a; Kuznetsov, 1971; Overmann et al., 1998). Where underwater light is filtered by populations of purple sulfur bacteria, the remaining light has its maximum intensity at 420–450 nm. Because the short-wavelength absorption maxima of BChl *c* and *d in vivo* coincide with this wavelength range (Table B11.1), green-colored species of green sulfur bacteria dominate under these conditions.

Syntrophy Cocultures of *C. limicola* f. sp. *thiosulfatophilum* and *Chromatium vinosum* have been established with sulfide as the sole substrate. Sulfide is oxidized by the green sulfur bacterium to zero-valence sulfur, and polysulfides form abiotically. Polysulfides are used as electron donors by *Chromatium* but cannot be exploited by *Chlorobium* in the presence of hydrogen sulfide (van Gemerden and Mas, 1995). Thus, the higher sulfide affinity of *Chlorobium* is compensated for by the instantaneous polysulfide utilization of *Chromatium*. This type of syntrophic interaction, together with the capacity of the *Chromatiaceae* to assimilate a wider variety of carbon compounds, may partially explain the coexistence of purple and green sulfur bacteria that is frequently observed in lakes.

Green sulfur bacteria form stable syntrophic associations with sulfur- and sulfate-reducing bacteria (Pfennig, 1980; Warthmann et al., 1992). A closed sulfur cycle is the physiological basis for the syntrophic growth of both partners: sulfur or sulfate is reduced by the chemotrophic component, and the sulfide generated is reoxidized photosynthetically by the green sulfur bacteria. Syntrophic growth is fueled by the oxidation of organic carbon substrates, while only traces of reduced sulfur compounds are required because of the extensive cycling of sulfur compounds.

In many pelagic ecosystems, conspicuous associations of green sulfur bacteria with colorless bacteria ("*Chlorochromatium aggregatum*", "*Pelochromatium roseum*", and others) are present. The physiological basis and selective value of these phototrophic consortia remain unclear (see the Appendix) but it has been suggested that the second bacterium belongs to the group of sulfate-reducing bacteria (Kuznetsov, 1977; Pfennig, 1980).

Antagonism   Only one case of an antagonistic relationship with other bacteria has been reported for green sulfur bacteria. The Gram-negative bacterium *Stenotrophomonas maltophilia* causes growth inhibition zones on cell lawns of several *Chlorobium* species (Nogales et al., 1997). Cells of the green sulfur bacteria are lysed and form ghosts. This interaction is not obligatory and is not limited to green sulfur bacteria as hosts.

Ecological significance   Dense populations are present in some lakes (up to 4000 μg of BChl d/l; 4.6×10^7 cells/ml), and anoxygenic photosynthetic carbon fixation can amount to 83% of the total primary production in lakes (Culver and Brunskill, 1969). With the exception of sulfide springs, anoxygenic photosynthesis is ultimately and indirectly fueled by carbon that has already been fixed by oxygenic phototrophs within or outside the ecosystem. Therefore, the production of green sulfur bacteria represents a net input of easily degradable biomass only if it is based on the anaerobic degradation of allochthonous organic matter (Overmann, 1997). In contrast, the CO_2 present in hydrothermal fluids of sulfide springs is often of geological origin. Consequently, carbon fixation by anoxygenic photosynthesis represents autochthonous primary production in these special cases.

In laboratory experiments, cladocera (*Daphnia magna* and *Ceriodaphnia reticulata*) and copepods (*Mesocyclops leukartii*) are able to ingest *C. phaeobacteroides* and grow at its expense (Gophen, 1977; Gophen et al., 1974). Because (1) the numbers of zooplankters in unproductive lakes are unexpectedly high, (2) herbivorous zooplankton accumulates directly above the layers of green sulfur bacteria, and (3) colored bacterial cells have been observed in the intestinal tracts of the zooplankters, it has been concluded that populations of green sulfur bacteria are also heavily grazed in the natural environment (Culver and Brunskill, 1969). The gross production of green sulfur bacteria frequently is similar in magnitude to their net biomass increase, and consequently losses through grazing must be insignificant (Parkin and Brock, 1981). In addition, analysis of stable carbon and sulfur isotopes has revealed that the biomass of green sulfur bacteria does not enter the food web in the oxic parts of lakes (Fry, 1986). This discrepancy between laboratory and field data has been explained by the inhibition of grazers due to the toxicity of the sulfide present in natural populations of green sulfur bacteria (van Gemerden and Mas, 1995).

Phylogenetic considerations   So far, green sulfur bacteria have been divided into five genera and 14 species. The species *Clathrochloris sulfurica* was eliminated in the First Edition of the *Manual* (Pfennig, 1989) because its morphology (trellis-shaped cell aggregates and formation of gas vesicles) closely resembled that of *Pelodictyon luteolum*. Because their physiological capacities are too similar to permit a differentiation between the taxa, the six genera *Ancalochloris, Chlorobium, Chloroherpeton, Clathrochloris, Pelodictyon,* and *Prosthecochloris* have been classified according to their cell morphologies, motilities, and abilities to form gas vesicles. The different species were distinguished according to their morphologies and pigment compositions.

A phylogenetic tree of 13 strains of green sulfur bacteria (representing 10 different species and four different genera) has recently been constructed based on their 16S rDNA sequences

(Overmann and Tuschak, 1997). Similarity values of >90.1% and K_{nuc} values of <0.11 indicate close phylogenetic relatedness among all strains. The 16S rDNA sequence of *Chloroherpeton thalassium* was recently determined (John Stolz, personal communication), and it has been included in the phylogenetic tree presented in Fig. B11.2. Within the green sulfur bacterial radiation, *C. thalassium* clearly is the most isolated branch. The green sulfur bacteria represent an isolated group within the eubacterial radiation, confirming earlier results of ribosomal oligonucleotide cataloguing (Gibson et al., 1985a; Woese et al., 1985b). Consequently, the family *Chlorobiaceae* Copeland 1956, 31 (Trüper and Pfennig, 1971) should be maintained as a coherent systematic group.

Strains which have been combined in the species *C. limicola* or *C. vibrioforme* clearly represent different taxa. This confirms the genetic heterogeneity of *C. limicola*, in which the mol% G + C varies between 49–58.1 and total genome size ranges from 1435 to 3342 kb (Méndez-Alvarez et al., 1996). Subsets of closely related strains of *C. limicola* have been described, however (Figueras et al., 1997). In addition, the genus *Pelodictyon* will have to be redefined.

In Fig. B11.2, the phylogenetic relatedness among the different species is compared with their biochemical and physiological properties. The mol% G + C values confirm the polyphyletic origin of *C. vibrioforme* DSMZ 260^T and DSMZ 262 and the monophyly of strains DSMZ 262 and *P. luteolum* DSMZ 273^T. Only a few phenotypic characteristics are of taxonomic significance. Closely related strains have similar salt requirements for growth, and although only few data are available, the fatty acid composition may similarly reflect the phylogenetic relationships. Of the morphological characteristics, only ternary fission (present in *P. clathratiforme* and *P. phaeoclathratiforme*) may be of taxonomic value. In contrast, pigment composition and the formation of gas vesicles are cytological traits without any phylogenetic relevance. Because the current classification scheme of green sulfur bacteria is mainly based on the latter two characteristics, it is not congruent with the phylogenetic tree.

As mentioned above, "*Clathrochloris sulfurica*" was eliminated as a separate species in the First Edition of *Bergey's Manual of Systematic Bacteriology* (Pfennig, 1989) because its morphology closely resembles that of *Pelodictyon* spp. (trellis-shaped cell aggregates and formation of gas vesicles). One strain, "*C. sulfurica*" 1, has been sequenced (Witt et al., 1989) and phylogenetically represents a clearly distinct species within the green sulfur bacterial radiation (Fig. B11.2).

Given the phylogenetic heterogeneity and the limited number of strains analyzed so far, it appears premature to establish a new classification system for green sulfur bacteria at this point. Additional nucleic acid analyses are required in order to avoid further rearrangements. The previous arrangement of genera and species of the green sulfur bacteria, which was based on simple phenotypic characteristics (Pfennig, 1989), has therefore been kept for the present edition of the *Manual*.

Further Comments   Strains of the green sulfur bacteria with very similar or identical 16S rDNA sequences can be distinguished rapidly by using banding patterns of PCR-amplified dispersed repetitive DNA sequences (Overmann and Tuschak, 1997).

C. limicola 8327 exhibits a natural competence for intrastrain genetic transformation (Ormerod, 1988). However, DNA extracts of this strain did not transform other *Chlorobium* strains.

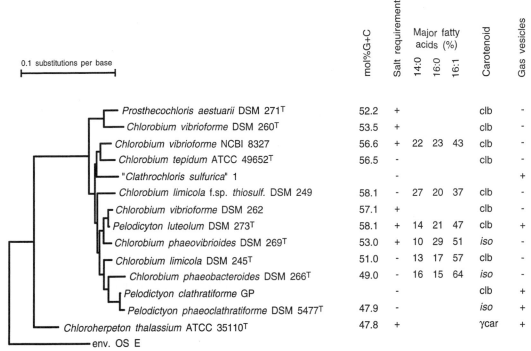

	mol%G+C	Salt requirement	Major fatty acids (%)			Carotenoid	Gas vesicles
			14:0	16:0	16:1		
Prosthecochloris aestuarii DSM 271T	52.2	+				clb	-
Chlorobium vibrioforme DSM 260T	53.5	+				clb	-
Chlorobium vibrioforme NCBI 8327	56.6	+	22	23	43	clb	-
Chlorobium tepidum ATCC 49652T	56.5	-				clb	-
"*Clathrochloris sulfurica*" 1		-					+
Chlorobium limicola f.sp. *thiosulf.* DSM 249	58.1	-	27	20	37	clb	-
Chlorobium vibrioforme DSM 262	57.1	+				clb	-
Pelodicyton luteolum DSM 273T	58.1	+	14	21	47	clb	+
Chlorobium phaeovibrioides DSM 269T	53.0	+	10	29	51	iso	-
Chlorobium limicola DSM 245T	51.0	-	13	17	57	clb	-
Chlorobium phaeobacteroides DSM 266T	49.0	-	16	15	64	iso	-
Pelodictyon clathratiforme GP		-				clb	+
Pelodictyon phaeoclathratiforme DSM 5477T	47.9	-				iso	+
Chloroherpeton thalassium ATCC 35110T	47.8	+				γcar	+
env. OS E							

0.1 substitutions per base

FIGURE B11.2. Phylogenetic tree of the green sulfur bacteria compared to five different biochemical or cytological traits (data from Kenyon and Gray, 1974; Knudsen et al., 1982; Imhoff, 1988, and Pfennig, 1989). Construction of the tree was done as described by Overmann and Tuschak (1997), using the most closely related 16S rRNA gene sequence available as the outgroup (OS E is the sequence of an environmental clone isolated from Octopus Spring, Yellowstone National Park (Ward et al., 1990), and of all entries in the RDP database is the one most closely related to the green sulfur bacteria).

Genus I. **Chlorobium** *Nadson 1906, 190*[AL]

NORBERT PFENNIG AND JÖRG OVERMANN

Chlo.ro' bi.um. Gr. adj. *chloros* green, yellowish green; Gr. n. *bios* life; M.L. neut. n. *Chlorobium* green life.

Cells spherical, ovoid, straight, or curved rod shaped, 0.3–1.1 µm wide and 0.4–3 µm long (sometimes much longer). **Cells are often united in chains resembling streptococci or filaments; curved rod-shaped strains may form long spirals.** Multiplication by binary fission. **Nonmotile,** Gram negative. The existing species exhibit one of two clearly distinguishable colors: **culture or cell material is either green (grass green) or brown (chocolate brown).** Significantly, these colors can also be recognized under the light microscope with bright-field illumination. **Photosynthetic pigments are located in the cytoplasmic membrane and the chlorosomes** (Staehelin et al., 1978), which underlie and are attached to the cytoplasmic membrane. Bacteriochlorophylls (BChl) *c, d,* or *e* occur as major photosynthetic pigments, in addition to small amounts of BChl *a*. Chlorobactene or isorenieratene are the major carotenoid components (Schmidt and Schiburr, 1970).

Obligately anaerobic and phototrophic. Photolithoautotrophic growth with sulfide or sulfur as electron donor. **During sulfide oxidation, globules of S⁰ are formed outside the cells; the sulfur may be further oxidized to sulfate.** In sulfide-reduced media, thiosulfate may be used as electron donor substrate. In the presence of reduced sulfur compounds and bicarbonate, a number of simple organic substrates can be photoassimilated. Ammonia is used as nitrogen source; molecular nitrogen is fixed by many strains. Growth temperature: 5–40°C (optimum: 20–

32°C) with the exception of *C. tepidum* (temperature range: 32–52°C). Storage materials are polyphosphate and polysaccharides (Sirevåg and Ormerod, 1977). Habitat: hydrogen sulfide-containing mud and water of freshwater, brackish-water, and marine environments.

The mol% G + C of the DNA is: 49.0–58.1 (Mandel et al., 1971).

Type species: **Chlorobium limicola** Nadson 1906, 190.

FURTHER DESCRIPTIVE INFORMATION

The green or brown strains of the green sulfur bacteria included in the genus *Chlorobium* show essentially two types of cell morphology. One type is represented by strains with straight to slightly curved rod-shaped cells. In undisturbed stationary cultures, these strains may form long filaments of rod-shaped cells or, in the stationary phase of growth, chains of almost spherical cells resembling streptococci. Green, rod-shaped *Chlorobium* strains belonging to the species *C. limicola* can be isolated regularly from the sulfide-containing mud or sediment of all kinds of freshwater habitats. The brown, rod-shaped *Chlorobium* strains of the species *C. phaeobacteroides* are also typical freshwater bacteria; however, they thrive preferentially as planktonic forms in the sulfide-containing hypolimnia of lakes.

The second morphological type of the *Chlorobium* species is represented by strains with curved rod-shaped or C-shaped cells or vibrioid-shaped to ring-like cells. Characteristically, the cell diameters of all strains of this morphological type are smaller than those of the rod-shaped strains. Depending on the culture conditions, the curved rod- or vibrioid-shaped cells may form C-shaped or ring-like chains of cells that may extend into coils or helical filaments. Both the green and brown strains of the curved rod-shaped cell type are isolated preferentially if anoxic water or mud samples from estuarine, brackish-water, or marine habitats are used as inocula.

Depending on the culture conditions, the strains of all *Chlorobium* species tend to form soluble extracellular slime that turns the culture medium more or less viscous. This feature is particularly pronounced in the strains of *C. chlorovibrioides*.

Most strains of the *Chlorobium* species are able to use molecular hydrogen as an electron donor for CO_2 assimilation if reduced sulfur compounds are provided as a source of cell sulfur. Molecular nitrogen can be assimilated by most strains. Photosynthetic pigments of the green strains of all *Chlorobium* species are either BChl *c* or *d*, together with the carotenoid chlorobactene. In the brown strains of all species, the pigment is BChl *e*, together with the carotenoid isorenieratene. Small amounts of BChl *a* are present in all species.

An outstanding characteristic of all green sulfur bacteria, including the *Chlorobium* species, is their capacity to grow reasonably well at light intensities of $0.5–1$ $\mu mol \cdot m^{-2} \cdot s^{-1}$, which is too dim to allow multiplication of any other phototrophic organisms, including the purple bacteria (Biebl and Pfennig, 1978). Therefore, this feature can be used for the selective enrichment culture of *Chlorobium* strains from anoxic mud samples. In good agreement with this, *Chlorobium* is often found below the layers of other phototrophic organisms in muddy or sandy freshwater or marine sediments.

The pH range for growth of *Chlorobium* is fairly narrow, between 6.5 and 7.2, with an optimum of 6.8. Most strains require vitamin B_{12} for growth.

ENRICHMENT AND ISOLATION PROCEDURES

Chlorobium species may be enriched and isolated from any anoxic water-and-mud samples of freshwater, brackish-water, marine, and saline habitats. The source and species composition of the inoculum is of primary significance for the results of enrichment experiments. In liquid enrichment cultures, however, strains or species that were not directly detected in the original sample may eventually be obtained. A *Chlorobium* species which is the dominant type in a given natural habitat may be entirely outgrown by a species that occurs only incidentally in the habitat. When certain color or cell types that were detected microscopically in a given sample are to be isolated, agar shake dilution series should be prepared directly from the sample without prior use of liquid enrichments. The incubation conditions for the agar cultures should closely resemble the conditions to be used for liquid enrichment cultures of the particular species. The defined medium* has been found to be relatively nonspecific and useful

for the cultivation of all green sulfur bacteria currently in pure culture. Other culture media useful for *Chlorobium* species were published by Larsen (1952), Sirevåg and Ormerod (1977), and Pfennig and Trüper (1981). For enrichment culture of *Chlorobium* species in the presence of large numbers of purple sulfur bacteria, advantage can be taken of the ability of *Chlorobium* to compete successfully with other phototrophic bacteria at growth-limiting intensities of white fluorescent light between 0.5 and 5 $\mu mol \cdot m^{-2} \cdot s^{-1}$. *Chlorobium* species may be selectively enriched from other genera of the green sulfur bacteria by the application of high sulfide concentrations (4–6 mM) and high light intensities (50–100 $\mu mol \cdot m^{-2} \cdot s^{-1}$) at incubation temperatures of about 30°C.

The sulfide concentration that is initially provided in the fresh culture medium does not support much growth. To achieve reasonably high population densities in enrichment or pure cultures, repeated additions of neutralized sodium sulfide solution are necessary. Additions are made if the previously added sulfide is consumed and the transiently formed S^0 is largely oxidized. The sulfide solution for feeding of cultures is prepared by neutralizing (pH ~7.2) a stirred sodium sulfide solution (60 mM) with sterile 2 M sulfuric acid. The neutralized solution has to be applied to the culture medium immediately. Alternatively, a special device for preparation and storage of sterile neutral sulfide solution may be used (Siefert and Pfennig, 1984).

In agar shake dilution cultures, growth may be enhanced by the addition of 3.0–5.0 mM acetate to the defined medium. Pure cultures are obtained by repeated application of the agar shake dilution method. For this method, water agar is prepared with 3.0% (w/v) repeatedly washed agar in distilled water. Depending on the salinity of the sample from nature or the enrichment culture medium, corresponding amounts of NaCl and $MgCl_2 \cdot 6H_2O$ are added per liter of water agar. The agar solution is dispensed in 3 ml amounts into test tubes, which are stoppered with cotton plugs and autoclaved. The agar tubes are kept molten in a water bath at 55°C. Prepared defined culture medium is prewarmed to 40°C, and 6 ml amounts are added to the tubes of liquefied agar; exposure to air is minimized by dipping the tip of the pipette into the agar medium. Starting with a few drops of a sample from nature or an enrichment culture as an inoculum and using 6–8 tubes, serial dilutions are made. All tubes are then hardened in cold water and immediately sealed with a sterile overlay consisting of one part paraffin wax and three parts mineral oil; the overlay should be 2 cm thick. The tubes are finally flushed with N_2/CO_2 (90%:10%) and sealed with butyl rubber stoppers. The tubes are kept in the dark for 12 h and subse-

---

*Defined medium for green sulfur bacteria is prepared in an Erlenmeyer flask with an outlet near the bottom at one side. Connected to the outlet is a silicon rubber tube with a pinchcock and a bell for aseptic distribution of the medium into bottles or tubes. The flask is closed by a silicon rubber stopper with an inlet and outlet for gas and a screw-cap glass tube, through which additions or samplings can be made.

The defined basal medium has the following composition (per liter of distilled water): KH_2PO_4, 0.30 g; NH_4Cl, 0.34 g; KCl, 0.34 g; $CaCl_2 \cdot 2H_2O$, 0.15 g; NaCl (only

for seawater medium), 20.0 g; $MgSO_4 \cdot 7H_2O$, 0.5 g (for seawater medium, 3.0 g); and trace element solution (see below), 1.0 ml. After the medium has been autoclaved and cooled under an atmosphere of N_2/CO_2 (90%:10%), the following components (per liter of medium) are added aseptically from sterile stock solutions while access of air is prevented by continuous flushing with the gas mixture of N_2/CO_2 (90%:10%): 15 ml of a 10% (w/v) solution of $NaHCO_3$ (saturated with CO_2 and autoclaved under a CO_2 atmosphere), 6 ml of a 10% (w/v) solution of $Na_2S \cdot 9H_2O$ (autoclaved under an N_2 atmosphere), and 1 ml of a vitamin B_{12} solution containing 2 mg of vitamin B_{12} in 100 ml of distilled water. The pH of the medium is adjusted to 6.7 with a sterile 2 M H_2SO_4 or 2 M Na_2CO_3 solution while the medium is magnetically stirred. The medium is then dispensed aseptically into sterile 100-ml bottles with metal screw tops containing autoclavable rubber seals. A small air bubble is left in each bottle to accommodate possible pressure changes.

Each liter of trace element solution contains the following: $FeSO_4 \cdot 7H_2O$, 2.0 g; $CoCl_2 \cdot 6H_2O$, 190 mg; $MnCl_2 \cdot 4H_2O$, 100 mg; $ZnCl_2$, 70 mg; $Na_2MoO_4 \cdot 2H_2O$, 36 mg; $NiCl_2 \cdot 6H_2O$, 24 mg; H_3BO_3, 6 mg; and $CuCl_2 2 \cdot H_2O$, 2 mg. $FeSO_4$ is dissolved in 10 ml HCl (25% v/v), distilled water is then added, followed by the other components.

quently incubated at the desired light intensity and temperature. During the first two days of incubation, the paraffin overlay is gently reheated to achieve a complete sealing effect.

Well-separated yellowish green or brown colonies that occur in the higher dilutions can be removed with sterile Pasteur pipettes and without breaking the tubes. The cells are suspended in 0.5–1.0 ml of anoxic medium and used as the inoculum for subsequent agar shake cultures. The process is repeated until a pure culture is obtained.

When pure agar cultures are obtained, individual colonies are isolated and inoculated into liquid medium. It is advisable to start with small bottles or screw-cap tubes (10 or 25 ml) and then to scale up to the regularly used sizes. Purity is checked by use of both a microscope and growth on A-C medium (Difco) adjusted to the respective salinities.

MAINTENANCE PROCEDURES

Pure cultures of all species of the phototrophic green sulfur bacteria can be maintained in the defined mineral medium; 100-ml screw-cap bottles are preferentially used as culture vessels. The freshly grown cultures should have just used up the transiently formed S^0. At this stage, the stock cultures can be stored at 4°C for 3–4 months. After this time, the cultures keep well if they are supplemented with neutralized sulfide solution and put back in dim light at room temperature. After consumption of sulfide and sulfur, the cultures may be stored for another 3–4 months. Thereafter, the cultures are transferred to fresh culture medium.

Long-term preservation of green sulfur bacteria is successfully carried out by storage in liquid nitrogen. For this purpose, heavy cell suspensions of liquid cultures are supplemented with DMSO as a cryoprotectant to a final concentration of 5%. Plastic ampoules (2 ml) are filled with such suspensions, sealed, and stored frozen.

Some *Chlorobium* strains have been successfully preserved by lyophilization.

DIFFERENTIATION OF THE GENUS CHLOROBIUM FROM OTHER GENERA

The differentiation of the *Chlorobium* species from the other genera of the green sulfur bacteria presents no problems. The genus *Chlorobium* comprises all green or brown strains with spherical to straight or curved rod-shaped cells that do not form gas vacuoles under any conditions of culture or storage at 4°C. The cells of *Chlorobium* species never carry prosthecae, which are characteristic of the genera *Prosthecochloris* and *Ancalochloris*. In order to detect the prosthecae, high-resolution phase-contrast microscope objectives (magnification: ×100; oil) should be used. *Chlorobium* strains are nonmotile under all conditions and thus are different from the unicellular filamentous gliding green sulfur bacterium *Chloroherpeton*.

TAXONOMIC COMMENTS

The formation of species within the genus *Chlorobium* is entirely based on simple morphological and biochemical characteristics (photosynthetic pigments). In the two species *C. limicola* and *C. vibrioforme*, the *forma specialis thiosulfatophilum* was established for the thiosulfate-utilizing strains (Pfennig and Trüper, 1971c). These strains contain an additional cytochrome c_{551} which is absent in the non-thiosulfate-using strains of the two species (Meyer et al., 1968; Steinmetz and Fischer, 1982) and which is supposed to be specifically involved in thiosulfate metabolism. Although in the non-thiosulfate-using strains the first oxidation product of sulfide is S^0, the *thiosulfatophilum* strains form thiosulfate as the first product and release it into the medium. In a second step, thiosulfate is then further oxidized to S^0. These results indicate that the thiosulfate-using strains of the two species differ significantly from the other strains in their metabolism of sulfide. The wide variation of the DNA base ratios both within the genus and within single species indicates a marked genetic heterogeneity of the physiologically very homogeneous group. Nucleic acid studies, such as rRNA sequences and DNA–DNA hybridization are required to establish a genetically meaningful classification.

DIFFERENTIATION OF THE SPECIES OF THE GENUS CHLOROBIUM

The phenotypic traits that are currently used to differentiate the species of the genus *Chlorobium* are listed in Table B11.3.

List of species of the genus Chlorobium

1. **Chlorobium limicola** Nadson 1906, 190.[AL]

   li.mi' co.la. L. n. *limus* mud; L. suff., n. *cola* dweller; M.L. masc. n. *limicola* the mud dweller.

   Cells straight or slightly curved rod-shaped, 0.7–1.1 μm wide and 0.9–1.5 μm long (sometimes much longer) (Fig. B11.3). Cells often united in chains resembling streptococci. Depending on culture conditions, strains may produce slime. Color of individual cells, light green; color of cell suspensions, green. Major photosynthetic pigments are BChl *c* (occasionally *d*) and carotenoid chlorobactene.

   Photoautotrophic growth occurs with sulfide and sulfur; molecular hydrogen may be used as additional electron donor. In the presence of sulfide and bicarbonate, acetate or propionate is photoassimilated; some strains may assimilate pyruvate, glutamate, and fructose. Not utilized: thiosulfate, ethanol, succinate, butyrate, or higher fatty acids. Nitrogen source: ammonium salts. Vitamin B_{12} may be required for growth. pH range, 6.5–7.0; optimum pH, 6.8. Growth temperature: 25–35°C. Habitat: mud and water of ditches, ponds and lakes with stagnant freshwater containing hydrogen sulfide and exposed to light.

   The mol% G + C of the DNA is: 51.0 (Bd).

   Type strain: 6330, DSMZ 245.

   GenBank accession number (16S rRNA): Y10640.

   Additional Remarks: Strain 6330 is a neotype strain, isolated from Gilroy Hot Spring.

2. **Chlorobium limicola f. sp. thiosulfatophilum** (Larsen 1952) Pfennig and Trüper 1971c, 14.

   thi.o.sul.fa.to' phi.lum. M.L. n. *thiosulfatum* thiosulfate; Gr. adj. *philos* loving; M.L. neut. adj. *thiosulfatophilum* thiosulfate loving.

   Description is the same as for *C. limicola*, except that thiosulfate is utilized as photosynthetic electron donor.

   The mol% G + C of the DNA is: 58.1 (Bd).

   Type strain: 6230, DSMZ 249.

   GenBank accession number (16S rRNA): Y08102.

TABLE B11.3. Characteristics of the species of the genus *Chlorobium*[a]

Characteristic	1. C. limicola	2. C. limicola f. sp. thiosulfatophilum	3. C. chlorovibrioides	4. C. phaeobacteroides	5. C. phaeovibrioides	6. C. vibrioforme	7. C. vibrioforme f. sp. thiosulfatophilum
Cells or cell suspensions:							
Green	+	+	+	−	−	+	+
Brown	−	−	−	+	+	−	−
Cell shape:							
Rod-shaped to spherical	+	+	−	+	(+)	−	−
Curved rod- and vibrioid-shaped to ringlike	−	−	+	−	+	+	+
Cell diameter, μm:							
0.7–1.1	+	+	−	−	−	−	−
0.5–0.8	−	−	−	+	−	+	+
0.3–0.4	−	−	+	−	+	−	−
Thiosulfate used	−	+	−	−	−	−	+
NaCl requirement (%)	0	0	2–3	0	2	2	2
Mol% G + C	51–52	52.5–58	54	49–50	52–53	52–57	53.5

[a]Symbols: +, positive, − negative.

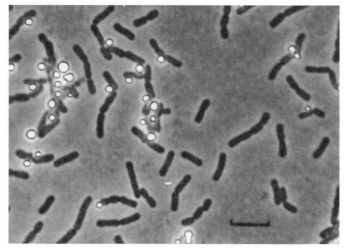

FIGURE B11.3. *C. limicola* f. sp. *thiosulfatophilum* DSMZ 249 cultured photoautotrophically with sulfide. Globules of elemental sulfur are outside the cells. Phase-contract micrograph. Bar = 5 μm.

Additional Remarks: Strain 6230 is a neotype strain, isolated from Tassajara Springs, California

3. **Chlorobium chlorovibrioides** Gorlenko, Chebotarev and Kachalkin 1974, 908[AL] (See also Puchkova and Gorlenko 1982.)
chlo.ro.vi.bri.o.i' des. Gr. adj. *chlorus* green, greenish yellow; L. v. *vibro* vibrate; M.L. masc. n. *vibrio* that which vibrates, a generic name; Gr. suffix *-oides*; M.L. neut. adj. *chlorovibrioides* green, vibrio shaped.

Cells curved rod shaped or vibrioid shaped, 0.3–0.4 μm wide and 0.4–0.8 μm long; in stationary phase, short rod shaped to spherical. Color of cell suspensions, bright green.

Major photosynthetic pigments are BChl *d* and carotenoid chlorobactene.

Photoautotrophic growth occurs with sulfide and sulfur. In the presence of sulfide and bicarbonate, formate, acetate, and propionate are photoassimilated. Not utilized: thiosulfate, methanol, ethanol, and glutamate. Ammonium salts are used as nitrogen source. NaCl, in the range of 2–3% is required for optimal growth. Vitamin B_{12} is required. Optimum pH: 6.8. Growth temperature: 20–30°C. Habitat: thrives planktonically in sulfide-containing stagnant water of meromictic Lake Repnoye, Lake Pomyaretskoye, and others (Russia).

The mol% G + C of the DNA is: 54 ± 0.3 (Bd).

Type strain: PM-1, DSMZ 1377.

Additional Remarks: Strain PM-1 was isolated from Pomyaretskoye, Ukraine. This strain has been lost.

4. **Chlorobium phaeobacteroides** Pfennig 1968, 225.[AL]
phae.o.bac.te.ro.i' des. Gr. adj. *phaeus* brown; Gr. neut. n. *bakterion* rod; Gr. suffix *-oides*; M.L. adj. *phaeobacteroides* brown, rod shaped.

Cells straight or slightly curved rod shaped, 0.6–0.8 μm wide and 1.3–2.7 μm long (sometimes longer). Color of individual cells, light yellowish brown; color of cell suspensions, yellowish to reddish brown or chocolate brown. Major photosynthetic pigments are BChl *e* and carotenoid isorenieratene (Schmidt, 1978).

Photoautotrophic growth occurs with sulfide and sulfur; molecular hydrogen may be used as additional electron donor. In the presence of sulfide and bicarbonate, acetate or fructose are photoassimilated. Not utilized: thiosulfate, alcohols, pyruvate, propionate, and higher fatty acids. Nitrogen source: ammonium salts. Strains from brackish water habitats may require at least 1% NaCl. Vitamin B_{12} required for growth. pH range: 6.5–7.3. Growth temperature: 20–30°C. Habitat: hydrogen sulfide-containing stagnant water of ponds and lakes, frequently occurring in the upper hypolimnia of freshwater or meromictic lakes (Trüper and Genovese, 1968).

The mol% G + C of the DNA is: 49.0 (Bd).

Type strain: 2430, DSMZ 266.

GenBank accession number (16S rRNA): Y08104.

Additional Remarks: Isolated from Blankvann, Oslo, Norway

5. **Chlorobium phaeovibrioides** Pfennig 1968, 226.[AL]

phae.o.vi.bri.o.i' des. Gr. adj. *phaeus* brown; L. v. *vibro* vibrate; M.L. masc. n. *vibrio* that which vibrates, a generic name; Gr. suffix *-oides*; M.L. adj. *phaeovibrioides* brown, vibrio shaped.

Cells curved rod shaped to vibrioid shaped, 0.3–0.4 μm wide and 0.7–1.4 μm long (Fig. B11.4). Under certain conditions, cells remain attached and grow into partially wound coils and spirals. Color of cells, light yellowish brown; color of cell suspensions, yellowish to reddish brown or chocolate brown. Major photosynthetic pigments are BChl *e* and carotenoid isorenieratene.

Photoautotrophic growth occurs with sulfide and sulfur; in the presence of sulfide and bicarbonate, acetate and propionate are photoassimilated. Not utilized: thiosulfate. Ammonium salts used as nitrogen source. Brackish-water and marine strains require at least 1% NaCl. Vitamin B$_{12}$ required for growth. pH range: 6.5–7.3. Growth temperature: 20–30°C. Habitat: mud and stagnant water of anoxic marine and brackish-water environments with hydrogen sulfide; hypolimnia of meromictic lakes.

The mol% G + C of the DNA is: 53.0 (Bd).

Type strain: 2631, DSMZ 269.

GenBank accession number (16S rRNA): Y08105.

Additional Remarks: Isolated from Langvikvann, Norway.

6. **Chlorobium tepidum** Wahlund, Woese, Castenholz and Madigan 1996, 1189[VP] (Effective publication: Wahlund, Woese, Castenholz and Madigan 1991, 88.)

te' pi.dum. L. neut. adj. *tepidum* lukewarm.

Cells of type strain straight or slightly curved rod shaped, 0.6–0.8 μm wide and 1.3–2.6 μm long (Fig. B11.5). Cells of other strains can reach 10–25 μm in length. Color of cell suspensions, dark green. Major photosynthetic pigments are BChl *c* and the carotenoid chlorobactene.

Photoautotrophic growth occurs with sulfide and thiosulfate. Sulfide alone supports only very poor growth, but a minimum of 1 mM is required. High cell yields obtained with thiosulfate. In the presence of sulfide and bicarbonate, acetate or pyruvate is photoassimilated. Other organic or fatty acids (including propionate) and sugars not utilized. Nitrogen source: ammonium salts, glutamine, or N$_2$. Vitamin B$_{12}$ required for growth. Optimum pH: 6.8–7.0. Optimum growth temperature: 47–48°C with generation times of 2 h. Maximum temperature, 51–52°C; minimum, 32°C. NaCl not required for growth. Habitat: high-sulfide acidic (pH 4.5–6) thermal springs at temperatures to 55°C.

The mol% G + C of the DNA is: 56.5 (T_m).

Type strain: TLS, ATCC 49652.

GenBank accession number (16S rRNA): M58468.

Additional Remarks: Isolated from "Travelodge Stream," Rotorua, New Zealand.

7. **Chlorobium vibrioforme** Pelsh 1936, 63.[AL]

vi.bri.o.for' me. L. v. *vibro* vibrate; M.L. n. *vibrio* that which vibrates, a generic name; L. adj. suff. *-formis* like, of the shape of; M.L. neut. adj. *vibrioforme* of vibrio shape.

Cells curved rod shaped, C shaped, or vibrioid shaped, 0.5–0.7 μm wide and 1–2 μm long. In natural habitats and under unfavorable culture conditions, ring-like chains of cells occur which may extend into coils or helical filaments (Fig. B11.6). Color of individual cells, light green; color of cell suspensions, green. Major photosynthetic pigments are BChl *d* (occasionally *c*) and carotenoid chlorobactene. Besides BChl *d*, cells of *C. vibrioforme* strain DSMZ 260 contain significant amounts of BChl *c* (36%; Otte et al., 1993).

Photoautotrophic growth occurs with sulfide and sulfur; molecular hydrogen may be used as additional electron donor. In the presence of sulfide and bicarbonate, acetate or propionate may be photoassimilated. Not utilized: thiosulfate, ethanol, succinate, or higher fatty acids. Nitrogen sources: ammonium salts. Most strains require vitamin B$_{12}$

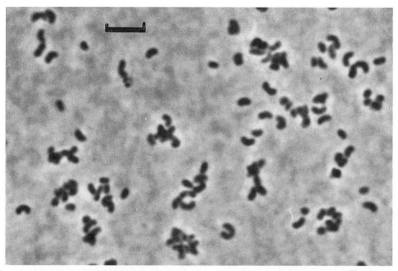

FIGURE B11.4. *C. phaeovibrioides* DSMZ 269 cultured photoautotrophically with sulfide. Vibrioid- and half-circle-shaped cells are present. Phase contract micrograph. Bar = 3 μm.

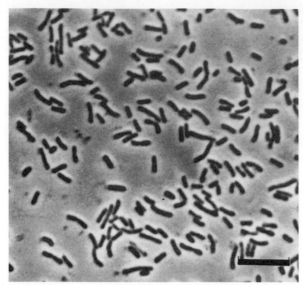

FIGURE B11.5. *C. tepidum* strain TLS. Phase-contrast micrograph. Bar = 5 μm. (Courtesy of M.T. Madigan).

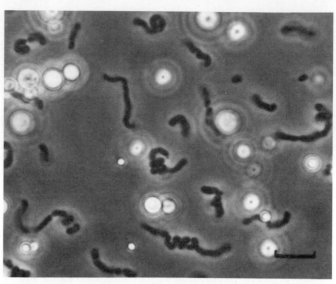

FIGURE B11.6. *C. vibrioforme* DSMZ 260 cultured photoautotrophically at high sulfide concentration (5 mM). The bacterium formed more or less coiled chains of cells; large extracellular sulfur globules are also seen. Phase-contrast micrograph. Bar = 5 μm.

for growth and at least 1% NaCl. pH range: 6.5–7.3. Growth temperature: 25–35°C. Habitat: mud and stagnant water of anoxic marine and brackish-water environments with hydrogen sulfide.

> *The mol% G + C of the DNA is*: 53.5 (Bd).
> *Type strain*: strain 6030, DSMZ 260.
> *GenBank accession number (16S rRNA)*: M62791.
> *Additional Remarks*: Isolated from Moss Landing, California.

8. **Chlorobium vibrioforme f. sp. thiosulfatophilum** (Larsen 1952) Pfennig and Trüper 1971c, 14.

   thi.o.sul.fa.to' phi.lum. M.L. n. *thiosulfatum* thiosulfate; Gr. adj. *philos* loving; M.L. adj. *thiosulfatophilum* thiosulfate loving.

Description is the same as for *C. vibrioforme*, except that thiosulfate is utilized as photosynthetic electron donor. An almost complete replacement of BChl *d* by BChl *c* occurred in a culture of this species that had been kept in the laboratory for a number of years (Smith and Bobe, 1987). The same effect could be induced by reduced light intensity during growth.

> *The mol% G + C of the DNA is*: 53.5 (Bd).
> *Type strain*: strain 1930, DSMZ 265.
> *Additional Remarks*: Isolated from Sehestedt, Northern Germany.

Genus II. Ancalochloris *Gorlenko and Lebedeva 1971, 1038*[AL]

VLADIMIR M. GORLENKO

An.ca' lo.chlo' ris Gr. masc. n. *ancalos* arm; Gr. adj. *chloros* green; M.L. neut. n. *Ancalochloris* arm (-producing) green (microbe)

Bacteria of irregular shape, forming prosthecae of irregular length that are wide at the base and pointed at the end (Fig. B11.7). The length of prosthecae can exceed the cell diameter. These bacteria **multiply by unequal fission** and form irregular chains of cells and typically form perforated microcolonies. **Nonmotile. Gram negative. Contain bacteriochlorophylls and carotenoid pigments.** The photosynthetic apparatus includes antenna structures of the **chlorosome** type, ovoid vesicles underlying and attached to the cytoplasmic membrane (Figs. B11.7 and B11.8). **Cells contain gas vacuoles.**

Anaerobic. Capable of photosynthesis in the presence of hydrogen sulfide. **Sulfur is deposited extracellularly.** Cell material and cultures appear **green or yellowish green** due to the photosynthetic pigments

> *The mol% G + C of the DNA is:* unknown.
> *Type species*: **Ancalochloris perfilievii** Gorlenko and Lebedeva 1971, 1038.

FURTHER DESCRIPTIVE INFORMATION

Cells are 0.5–1.0 μm wide, of irregular starlike shape due to the formation of several appendages (prosthecae) up to 2 μm long with tapering ends. Prosthecae are often as wide as the cell at the base but can also be of smaller diameter. Gas vacuoles are present in the central parts of cells and consist of elongated gas vesicles measuring 40–80 × 230–360 nm.

These bacteria occur in nature as typical conglomerates consisting of 10–30 cells; less frequently, they have been observed as short chains of cells with long prosthecae. The length of the prosthecae depends on the growth conditions. The longer prosthecae are observed at low light intensity in the habitat of *Ancalochloris*.

Multiplication is unequal, and division proceeds by constriction. Prosthecae between cells have not been observed. Prosthecae are formed by local polar growth of the cell wall together

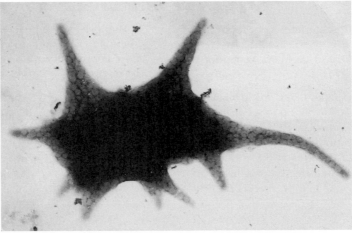

FIGURE B11.7. *A. perfilievii* from a water sample of the hypolimnion of meromictic Lake Bol'shoi, Kichier, Russia. Electron micrograph of a single cell stained with phosphotungstic acid ($\times 30,000$).

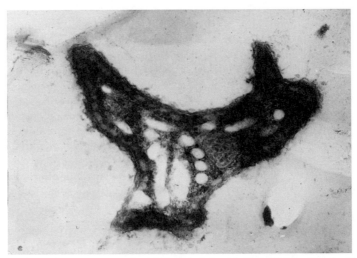

FIGURE B11.8 . *A. perfilievii* from water samples of dimictic Lake Glukhaya Lamba, Russia. Electron micrograph of ultrathin section ($\times 50,000$).

with the cytoplasmic membrane. Additional chlorosomes are synthesized in emerging prosthecae. The process of prostheca formation resembles budding, but ultrathin sections of prosthecae do not reveal DNA filaments. Prosthecae may branch.

Ancalochloris is widely distributed in the anoxic hypolimnia of dimictic and meromictic freshwater lakes with low H_2S content (1–60 mg/l) at low light intensity at depths down to 10 m.

ENRICHMENT AND ISOLATION PROCEDURES

Enrichment cultures can be obtained by using the same culture medium described for *Chlorobium* with a low H_2S content (1.26 mM $Na_2S \cdot 9H_2O$; pH 6.8–7.0). Samples are incubated at low light intensity (10–20 μmol $\cdot m^{-2} \cdot s^{-1}$) and at temperatures between 18–20°C. Vitamin B_{12} (20 μg/l) should be added. Pure culture isolation should be attempted by dilution in liquid medium, since *Ancalochloris* does not grow in agar shake dilution media.

MAINTENANCE PROCEDURES

Ancalochloris can be maintained in liquid cultures if transferred every 2–3 months. Grown cultures are stored at 4°C and very low light intensity.

DIFFERENTIATION OF THE GENUS *ANCALOCHLORIS* FROM OTHER GENERA

The genus *Ancalochloris* is characterized by the starlike shape of the cells and the possession of gas vesicles. The prosthecae differ from those of *Prosthecochloris* by their greater length and tapering ends.

TAXONOMIC COMMENTS

The use of pure cultures is required to clarify the systematic position of *Ancalochloris* and its relationship to other genera of the green sulfur bacteria.

List of species of the genus Ancalochloris

1. **Ancalochloris perfilievii** Gorlenko and Lebedeva 1971, 1038.[AL]

per.fi' lie.vi.i. M.L. gen. n. *perfilievii* of Perfil'ev; named for B. V. Perfil'ev, a Russian microbiologist.

Cells of irregular starlike shape, 0.5–1 µm in diameter without the prosthecae. One cell forms up to six prosthecae of varying lengths up to 2 µm long (Fig. B11.7). The lengths of prosthecae vary with the conditions of incubation. Prosthecae have tapering ends, 0.1 µm wide at the end and up to 0.7 µm wide at the base; sometimes the prosthecae are branched. Gas vacuoles are located in the centers of the cells.

Multiplication is by unequal fission. The direction of fission appears random. Microcolonies are of irregular shape and comprise up to 30 cells. The peripheral cells of a microcolony may have longer prosthecae than do cells in the center.

The chlorosomes of the photosynthetic apparatus are 70–85 nm long and 45–80 nm wide; they may occupy a large part of the prosthecae (Figs. B11.7 and B11.8).

Habitat: the anoxic metalimnia or hypolimnia of dimictic and meromictic lakes with low H_2S content.

The mol% G + C of the DNA is: unknown.

Type strain: No type material available.

Genus III. **Chloroherpeton** *Gibson, Pfennig and Waterbury 1985b, 223[VP] (Effective publication: Gibson, Pfennig and Waterbury 1984, 100)*

JANE GIBSON

Chlo.ro.her' pe.ton. Gr. adj. *chloros* green; Gr. n. *herpeton* a creeping organism; M.L. neut. n. *Chloroherpeton* green creeping organism.

Long rods, 8–20 × 1 µm. Cells separate promptly after division, and no septa are seen in units (Figs. B11.9 and B11.10). Cells tend to grow in clumps and produce some extracellular slime. Gram negative. **Gliding motility,** 10 µm/min at 20°C. **Cells flex up to 180°.**

Cultures and cell material are green. **Photosynthetic pigments are located in the cytoplasmic membrane and chlorosomes,** which underlie and are attached to the cytoplasmic membrane

(Fig. B11.11); bacteriochlorophyll (BChl) *c* and γ–carotene are the main photosynthetic pigments.

Strictly anaerobic and obligately phototrophic. Sulfide and CO_2 (bicarbonate) are required for growth. Growth yield slightly increased by acetate, propionate, malate, succinate, or glutamate but not by glucose, fructose, or Casamino Acids. During sulfide oxidation, S^0 **is formed outside the cells** and only slowly oxidized further to sulfate. Optimum temperature, 25°C; maximum temperature, about 30°C. Optimum pH: 6.8–7.2.

The mol% G + C of the DNA is: 45.0–48.2.

Type species: **Chloroherpeton thalassium** Gibson, Pfennig and Waterbury 1985, 223 (Effective publication: Gibson, Pfennig and Waterbury 1984, 100.)

FURTHER DESCRIPTIVE INFORMATION

The characteristic morphology of *Chloroherpeton* permits its identification in natural samples. It has been observed regularly on decaying vegetation and mud surfaces from littoral pools and channels on Cape Cod, Massachusetts. Isolation has been achieved by repeated shake culture in medium containing 0.8% agar, which allows some cell gliding, so that the light green, fluffy colonies are clearly distinguishable from the smaller, darker, and compact colonies of *Chlorobium* species. Specific enrichment conditions have not been found. *Chloroherpeton thalassium* isolates require 1–2% NaCl in the medium and do not tolerate more than 3 mM Na_2S; repeated sulfide additions must be made to obtain dense cultures in liquid medium. Vitamin B_{12} is required for growth, and the culture appears yellow-orange instead of deep green after a single transfer without B_{12} addition. Hydrogen can be used as an electron donor, provided that small additions of sulfide are also made to serve as a S source. Nitrogen fixation is indicated both by growth in the absence of ammonium and by acetylene reduction in suspensions incubated in light in the absence of NH_4^+. The cells contain gas vesicles, which confer a characteristic banded appearance under phase-contrast optics. Electron micrographs of cross-sections or sagittal sections show chlorosomes of typical dimensions underlying the cytoplasmic membrane. The type species and four other isolates contain con-

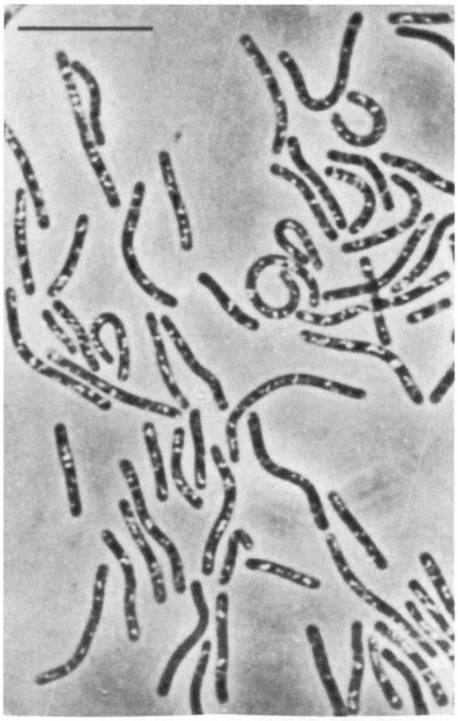

FIGURE B11.9. *C. thalassium* showing cell flexibility and refractive areas within the cells resulting from gas vacuoles. Phase-contrast photomicrograph. *Bar*, 10 µm. (Reprinted with permission from J. Gibson et al., *Archives of Microbiology 138:* 96–101, 1984, ©Springer-Verlag.)

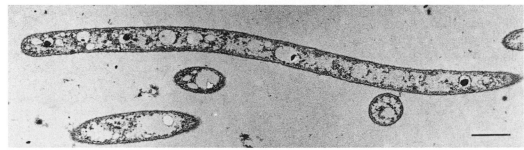

FIGURE B11.10. *C. thalassium* showing absence of cross-walls. Electron micrograph of a longitudinal thin section of a single cell. Bar = 1 μm. (Reprinted with permission from J. Gibson et al., Archives of Microbiology *138:* 96–101, 1984, ©Springer-Verlag.)

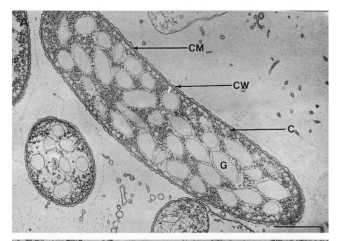

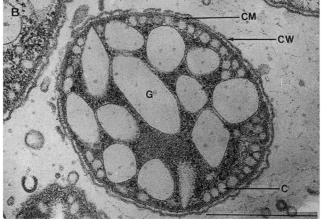

FIGURE B11.11. *C. thalassium.* Electron micrographs of thin cross-sections. **G**, vesicles; **C**, *chlorosomes; CW*, cell wall; *CM*, cytoplasmic membrane. Bar = 0.5 μm. (Reproduced with permission from J. Gibson et al., Archives of Microbiology *138:* 96–101, 1984, ©Springer-Verlag.)

ical-ended gas vesicles; a single isolate (mol% G + C of 45.0) has never produced gas vesicles in culture.

The current isolates all contain BChl *c* as the main light-harvesting pigment, together with a small quantity of BChl *a* (*c:a* = 50:1). More than 80% of the carotenoid is γ-carotene in the type species.

Acetate carbon appears in all cell fractions in proportions similar to those found in *Chlorobium*. Ribulosebisphosphate carboxylase activity is not detectable (D.G. Nelson, personal communication), suggesting that the main carbon fixation reactions

also correspond to those found in *Chlorobium* (Evans et al., 1966; Fuchs et al., 1980) rather than to those in *Chloroflexus* (Sirevåg and Castenholz, 1979).

In addition to *C. thalassium*, a second type of unicellular filamentous green sulfur bacteriaum exhibiting gliding motility has been isolated from a highly eutrophic freshwater pond (Eichler and Pfennig, 1988). The cells were smaller (0.8 μm) than *C. thalassium* and, in contrast to the latter species, were inhibited by NaCl concentrations of ≥0.2%. Because of its different morphology and physiology, the isolate was considered a new species and provisionally named "*Chloroherpeton limnophilum*". Unfortunately, it was lost before further studies could be made.

ENRICHMENT AND ISOLATION PROCEDURES

Samples of littoral marine muds containing large quantities of decaying vegetation will often develop green as well as red or purple patches on the container walls when brought into the laboratory and incubated in glass containers in dim light at 20–25°C. *Chloroherpeton* may form a significant component of the microbial population of such green areas, especially those which appear first, and can be identified by its characteristic morphology. Isolation can be achieved by repeated shake cultures in medium consisting of the following: NaCl, 170 mM; $MgCl_2$, 15 mM; KCl, 4 mM; $CaCl_2$, 1 mM; NH_4Cl, 5 mM; KH_2PO_4, 0.1 mM; $NaHCO_3$, 30 mM; Na_2S, 2.5 mM; trace element mixture (Widdel and Pfennig, 1981); vitamin B_{12}, 0.05 μg/100 ml; and 0.8% washed agar, pH 6.9. The inoculated tubes are sealed with a Vaseline/paraffin oil mixture (1:1) and incubated 50–100 cm from a 40W incandescent lamp. Light green, fluffy colonies should appear within a week and are picked when they are 1–2 mm in diameter.

MAINTENANCE PROCEDURES

The cultures may be maintained illuminated in liquid culture as described above by additions of CO_2-neutralized Na_2S (pH 6.5–6.8) to a final concentration of 2 mM. The best growth is obtained if the cells are kept in suspension by magnetic stirring. Cultures should be transferred to fresh liquid medium once a month. Stab cultures in the agar medium used for isolation do not survive as well as liquid cultures. Long-term preservation in liquid nitrogen is carried out with freshly grown cultures after addition of 5% DMSO.

DIFFERENTIATION OF THE GENUS *CHLOROHERPETON* FROM OTHER GENERA

Chloroherpeton differs from the multicellular filamentous green bacteria *Chloronema* (Dubinina and Gorlenko, 1975), *Oscillochloris*

(Gorlenko and Pivovarova, 1977), and *Chloroflexus* (Pierson and Castenholz, 1974a) in cell dimensions, in being unicellular rather than filamentous, and in photosynthetic pigment composition. The differences among the other genera of the green sulfur bacteria are listed in Table B11.2.

TAXONOMIC COMMENTS

The physiology, pigmentation, and fine structure of *Chloroherpeton* are similar to those of the green species of *Chlorobium*. A major

difference lies in the flexing and gliding motility, which might suggest affinities with *Chloroflexus*. The 16S rRNA catalog, however, indicates that *Chloroherpeton* belongs with the green sulfur bacteria (*Chlorobiaceae*) (Gibson, 1980; Gibson et al., 1985b), indicating that motility and flexing are of lesser taxonomic importance than are the other characteristics. The possibility that more than one species has been isolated requires determination of DNA mol% G + C and DNA–DNA similarity, and of 16S rRNA gene sequence data of more strains.

List of species of the genus Chloroherpeton

1. **Chloroherpeton thalassium** Gibson, Pfennig and Waterbury 1985b, 223^{VP} (Effective publication: Gibson, Pfennig and Waterbury 1984, 100.)

   tha.las′si.um. Gr. n. *thalassa* the sea; Gr. adj. *thalassios* pertaining to the sea, marine; M.L. neut. adj. *thalassium* marine.

   The characteristics are the same as those described for

the genus. Habitat: anoxic sulfide-containing marine littoral sediments, salt marshes, and tidal inlets rich in rotting plant material.

   The mol% G + C of the DNA is: 47.8 (T_m).

   Type strain: GB-78, ATCC 35110.

   GenBank accession number (16S rRNA): AF170103.

Genus IV. *Pelodictyon* Lauterborn 1913, 98^{AL}

NORBERT PFENNIG AND JÖRG OVERMANN

Pe.lo.dic′ ty.on. Gr. adj. *pelos* dark-colored; Gr. n. *diktyon* net; M.L. neut. n. *Pelodictyon* dark-colored net.

Cells rod shaped to ovoid, occurring singly or in netlike or more or less spherical aggregates. Multiplication is by binary fission. **Branching may occur as a result of ternary fission. Nonmotile. Gram negative. Cells regularly contain gas vacuoles. Cultures or cell material is either green or brown. Photosynthetic pigments are located in the cytoplasmic membrane and the chlorosomes,** which underlie and are attached to the cytoplasmic membrane. Bacteriochlorophyll (BChl) *c, d,* or *e* occurs as the major photosynthetic pigment, in addition to small amounts of BChl *a*. Chlorobactene or isorenieratene is the major carotenoid component (Schmidt and Schiburr, 1970; Gorlenko, 1972).

Obligately anaerobic and phototrophic. Photoautotrophic growth with sulfide or sulfur as electron donor. During sulfide oxidation, globules of S⁰ are formed outside the cells; the sulfur may be further oxidized to sulfate. In the presence of reduced sulfur compounds and bicarbonate, a number of simple organic substrates may be photoassimilated. Growth temperature: 15–30°C. Habitat: hydrogen sulfide-containing water and mud of freshwater, brackish-water, and marine environments

   The mol% G + C of the DNA is: 48.5–58.1 (Mandel et al., 1971).

   Type species: **Pelodictyon clathratiforme** (Szafer 1911) Lauterborn 1913, 98 (*Aphanothece clathratiformis* Szafer 1911, 162.)

FURTHER DESCRIPTIVE INFORMATION

The genus *Pelodictyon* was introduced by Lauterborn as a green sulfur bacterium that may form distinctive cell aggregate structures reminiscent of those found in the purple sulfur bacterium *Thiodictyon* and the green alga *Hydrodictyon*. In *Pelodictyon clathratiforme*, the rod-shaped cells are united into more or less large three-dimensional nets. Characteristically, the cells contain gas vacuoles, which confer buoyancy on the cell aggregates. Two green strains of *P. clathratiforme*, free of other phototrophic bacteria, were obtained in laboratory cultures and characterized in detail (Pfennig and Cohen-Bazire, 1967). The microscopic study of various anoxic water samples collected from blooms of phototrophic sulfur bacteria in sulfide-containing ponds and lakes revealed the existence of a brown species with the same mor-

phology as the green *P. clathratiforme* (Pfennig, 1977). The brown species was studied in pure culture and named *P. phaeoclathratiforme* (Overmann and Pfennig, 1989).

Another gas vacuole-containing green sulfur bacterium which may form spherical or irregular round colonies (*Schmidlea luteola* Lauterborn 1913) was classified in the genus *Pelodictyon* (*P. luteolum*) by Pfennig and Trüper (1971c). The third species of the genus *Pelodictyon* is a gas-vacuolate, brown sulfur bacterium, *P. phaeum* (Gorlenko, 1972). Thus, the genus *Pelodictyon* comprises the straight or curved rod-shaped, gas vacuole-containing green and brown species of the green sulfur bacteria group.

Although the species of the genera *Chlorobium* and *Pelodictyon* are physiologically and biochemically fairly similar, the possession of gas vacuoles by the latter genus is ecologically rather significant. The gas-vacuolated *Pelodictyon* species are buoyant because of their gas vesicles. Consequently, these species occur predominantly planktonically in sulfide-containing stagnant bodies of water, in which they may exhibit a selective advantage over *Chlorobium* species.

ENRICHMENT AND ISOLATION PROCEDURES

Enrichments for green or brown *Pelodictyon* species may be unsuccessful when liquid culture media are applied. Under laboratory conditions, such enrichments are readily outgrown by *Chlorobium* species. Isolations are more successful, therefore, if agar shake cultures are prepared directly from the natural sample in which cells of this genus were detected microscopically. Some species (e.g., *P. clathratiforme* and *P. phaeoclathratiforme*) can only be isolated in media reduced by the addition of 200 μM sodium dithionite (Overmann and Pfennig, 1989). Because of the presence of gas vacuoles, the colonies of *Pelodictyon* species maintain a chalky appearance (compared with that of *Chlorobium* colonies) even if the intermediately formed S⁰ is consumed. The liquid culture medium and agar shake dilution methods are carried out in the same way as described for the isolation of *Chlorobium* species. Not more than 1.5–2 mM sulfide and 3 mM acetate are added to the medium; the light intensity should be limited to 5-

20 μmol·m^{-2}·s^{-1} from a tungsten lamp. The cultures are incubated at about 20°C (Pfennig and Cohen-Bazire, 1967). Before second agar shake dilution tubes are inoculated with suspensions of green or brown colonies isolated from the first dilution series, the cells are microscopically checked for the presence of gas vacuoles. Only cell suspensions of gas-vacuolated cells are cultivated further. The three-dimensional cell aggregates characteristic of some species may develop only in liquid culture media and depending on the culture conditions.

MAINTENANCE PROCEDURES

The maintenance procedures are the same as described for the genus *Chlorobium* (see page 605).

DIFFERENTIATION OF THE GENUS *PELODICTYON* FROM OTHER GENERA

The phenotypic traits currently used to differentiate the species of the genus *Pelodictyon* are listed in Table B11.4.

List of species of the genus Pelodictyon

1. **Pelodictyon clathratiforme** (Szafer 1911) Lauterborn 1913, 98[AL] (*Aphanothece clathratiformis* Szafer 1911, 162.)
clath.ra.ti.for′ me. L. part. adj. *clathratus* latticed; L. fem. n. *forma* shape, form; M.L. neut. adj. *clathratiforme* lattice-like.

Cells are rod shaped, 0.7–1.2 μm wide and 1.5–2.5 μm long, although elongated cells up to 7 μm long may occur. Cells are characteristically united in three-dimensional nets, which are formed as follows (Fig. B11.12): Successive binary fissions result in the formation of chains of cells. Occasionally, two adjacent cells in such a chain change their mode of growth. The contiguous poles start to branch simultaneously, resulting in the formation, in the middle of the chain, of two Y-shaped cells, in opposition at the ends of both arms of the Y. If these two cells do not separate, the ring structure between them enlarges, by subsequent cell elongation and binary fission, into a typical many-celled mesh. The arrangement of cells at the branch points of fully formed nets implies that these cells eventually undergo ternary fission to yield three daughter cells, all in apposition at one pole. The colonial structure of *P. clathratiforme* is therefore caused by its ability to perform both binary and ternary fissions (Pfennig and Cohen-Bazire, 1967).

Color of individual cells, light green; color of cell suspensions, green. Major photosynthetic pigments are BChl *c* and carotenoids.

Photoautotrophic growth occurs with sulfide and sulfur. In their presence, acetate is photoassimilated. Thiosulfate, pyruvate, succinate, and higher fatty acids not utilized. Ammonium salts serve as nitrogen source. pH range: 6.5–7.0. Growth temperature: 15–25°C. Only partially purified cultures exist.

The mol% G + C of the DNA is: unknown.

Type strain: A pure culture is not available.

2. **Pelodictyon phaeoclathratiforme** Overmann and Pfennig 1990, 212[VP] (Effective publication: Overmann and Pfennig 1989, 405.)
phae.o.clath.ra.ti.for′ me. Gr. adj. *phaeus* brown; L. part. adj. *clathratus* latticed; L. fem. n. *forma* shape, form; M.L. neut. adj. *phaeoclathratiforme* brown latticelike.

Individual cells are rod shaped, 0.8–1.1 μm wide and 1.5–3.0 μm long. Elongated cells up to 6.0 μm long occur in cultures grown at 5 or 10°C. Similar to *P. clathratiforme*, some of the cells become trapezoid at one end and develop into Y-shaped cells by ternary fission. Large three-dimensional netlike microcolonies may be formed (Fig. B11.13). At low light intensities, intracellular gas vesicles are formed (Fig. B11.13) and the thickness of extracellular slime layers increases significantly.

Color of individual cells, brown; color of cell suspensions, dark brown. In transparent light, cultures exhibit a rose-red shine. Major photosynthetic pigments are BChl *e*, small amounts of BChl *a*, and the carotenoids isorenieratene and β-isorenieratene.

Photoautotrophic growth occurs with sulfide, sulfur, and thiosulfate; in their presence, acetate and propionate are photoassimilated. Not utilized: glucose, fructose, mannitol, alcohols, pyruvate, lactate, gluconate, succinate, fumarate, malate, tartrate, glycollate, 2-oxoglutarate, citrate, acetoin, benzoate, L(+)-arginine, L(+)-glutamate, formate, butyrate, and higher fatty acids. Optimum sulfide concentration for growth, 2–3 mM. After transfer of sulfide-grown cells to medium containing thiosulfate as an electron donor, a lag phase of 48 h is observed. Ammonium salts serve as nitrogen source. Optimum pH: 6.6. Optimum growth temperature: 20–25°C. No growth at or above NaCl concentrations of 0.5%. Vitamin B$_{12}$ required for growth.

The mol% G + C of the DNA is: 47.9 (T_m, Bd).

Type strain: BU 1, DSMZ 5477.

GenBank accession number (16S rRNA): Y08108.

Additional Remarks: Isolated from Buchensee, South Germany

3. **Pelodictyon luteolum** (Schmidle 1901) Pfennig and Trüper 1971c, 13[AL] (*Aphanothece luteola* Schmidle 1901, 179.)
lu.te′ o.lum. L. masc. adj. *luteus* yellow; L. masc. dim. adj. *luteolus* yellowish, somewhat yellow.

Cells straight or curved short rod shaped, 0.6–0.9 μm wide and 1.2–2.0 μm long. Strains of larger cells (1.0–1.2 μm wide and 2–4 μm long) do occur. Pure cultures may contain free

TABLE B11.4. Characteristics of the species of the genus *Pelodictyon*[a]

Characteristic	1. *P. clathratiforme*	2. *P. luteolum*	3. *P. phaeoclathratiforme*	4. *P. phaeum*
Gas vacuoles	+	+	+	+
Cells or cell suspensions green	+	+	−	−
Cells or cell suspensions brown	−	−	+	+
Cells rod-shaped, binary and ternary fission, cells in netlike structures	+	−	+	−
Cells rod-shaped or ovoid, singly or in round colonies	−	+	−	+
Cells straight or curved rod-shaped, singly or in chains or aggregates	−	−	−	+

[a]Symbols: +, positive; −, negative.

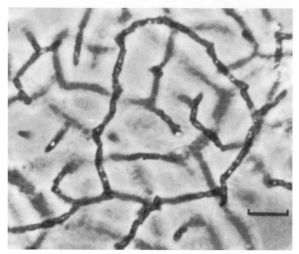

FIGURE B11.12. *P. clathratiforme* strain 1831 grown photoautotrophically with sulfide. Gas vacuole-containing cells are united in three-dimensional netlike structures. Phase-contrast micrograph. Bar = 5 μm.

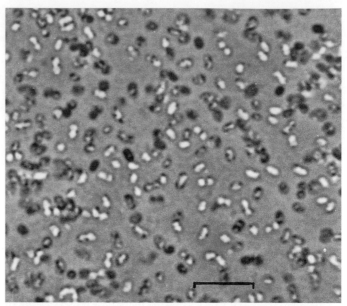

FIGURE B11.14. *P. luteolum* strain 2532 grown photoautotrophically with sulfide. The light areas inside the cells are the gas vacuoles. Phase-contrast micrograph. Bar = 5 μm.

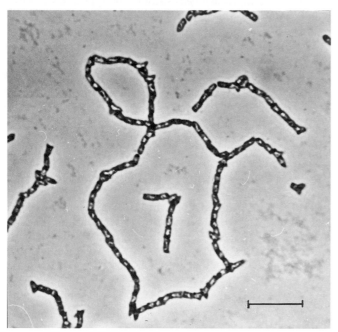

FIGURE B11.13. *P. phaeoclathratiforme* strain BU-1. Cells with gas vacuoles and ternary fission form rings and netlike colonies. Phase-contrast micrograph. Bar = 10 μm.

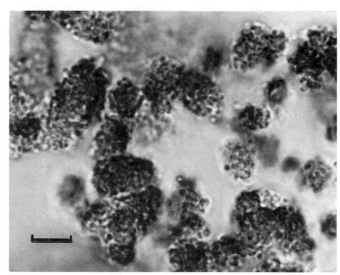

FIGURE B11.15. *P. luteolum* strain 2532 grown photoautotrophically with sulfide. Cells are united in irregular round colonies. Phase-contrast micrograph. Bar = 5 μm.

individual cells (Fig. B11.14); under certain conditions, the cells produce slime and are united into hollow spherical or irregular round colonies with the cells in a single layer (Fig. B11.15). The latter growth forms are occasionally observed in mud samples or enrichment cultures. Color of cells or cell suspensions, green. Major photosynthetic pigments are BChl *c* or *d* and the carotenoid chlorobactene (Schmidt and Schiburr, 1970).

Photoautotrophic growth with sulfide and sulfur. In the presence of sulfide and bicarbonate, acetate and propionate are photoassimilated. Thiosulfate, pyruvate, succinate, higher fatty acids, or peptone are not utilized. Reduced sulfur compounds are required as a source of cell sulfur. Ammonium

salts are used as nitrogen source. Strains from habitats with seawater require at least 1% NaCl in the medium. pH range: 6.5–7.0. Growth temperature: 15–25°C.

The mol% G + C of the DNA is: 58.1 (Bd).

Type strain: 2530, DSMZ 273.

GenBank accession number (16S rRNA): Y08107.

Additional Remarks: Strain 2530 is a neotype strain, isolated from Lake Polden, Norway.

4. **Pelodictyon phaeum** Gorlenko 1972, 370.[AL]

phae' um. Gr. adj. *phaeus* brown; M.L. neut. adj. *phaeum* brown.

Cells are short, straight or curved and rod shaped, 0.6–0.9 μm wide and 1.0–2.0 μm long, occurring either as single cells or as coiled chains of cells or irregularly shaped cell aggregates embedded in slime. Color of cell suspensions, chocolate brown. Major photosynthetic pigments are BChl *e* and carotenoid isorenieratene.

Photoautotrophic growth occurs with sulfide and sulfur; in their presence, acetate is photoassimilated. Thiosulfate.not utilized. Ammonium salts serve as nitrogen source. Vitamin

B_{12} required for growth. Optimum NaCl concentration: 3%. Optimum pH: 7.0. Growth temperature: 25°C.

The mol% G + C of the DNA is: unknown.

Type strain: WS-6, DSMZ 728.

Additional Remarks: Isolated from Lake Veisovo, Donetsk Region, Ukraine. This strain has been lost.

Genus V. **Prosthecochloris** Gorlenko 1970, 148[AL]

VLADIMIR M. GORLENKO

Pros.the′ co.chlo.ris. Gr. n. *prostheca* appendage; Gr. adj. *chloros* green; M.L. fem. n. *Prosthecochloris* green (organism) with appendages.

Spherical to ovoid bacteria forming nonbranching prosthecae and multiplying by **binary fission** in various directions. When separation is incomplete, cells form groups and branched chains, the configuration of which depends on the direction of fission. **Nonmotile.** Gram negative. Cell suspensions appear green or chocolate-brown. Cells **contain bacteriochlorophyll** (BChl) *c, d,* or *e* as the major bacteriochlorophyll component, **and carotenoids.** The photosynthetic apparatus includes antenna structures and **chlorosomes,** i.e., elongated ovoid vesicles underlying and attached to the cytoplasmic membrane. Cells do not contain gas vacuoles.

Anaerobic. Capable of photosynthesis in the presence of hydrogen sulfide, during which they produce and deposit, as an intermediate oxidation product, S^0 **in the form of globules outside the cells** in the medium.

The mol% G + C of the DNA is: 50.0–56.1.

Type species: **Prosthecochloris aestuarii** Gorlenko 1970, 148.

FURTHER DESCRIPTIVE INFORMATION

Two known species, *P. aestuarii* (Gorlenko, 1968, 1970) and "*P. phaeoasteroidea*" (Puchkova and Gorlenko, 1976), are identical morphologically (Figs. B11.16, B11.17 and B11.18). Cells are rigid and ovoid, 0.3–0.6 × 0.5–0.8 µm; cells form 10–20 prosthecae with rounded, occasionally thickened ends. Prosthecae are 0.1–0.16 µm wide and 0.07–0.3 µm long. Cells are enclosed in thin slimy capsules. Considerable slime formation was reported

in *P. aestuarii* (Gorlenko and Zhilina, 1968). Multiplication by binary fission. The daughter cell is sometimes considerably smaller than the mother cell. Mother and daughter cells remain connected for some time by one or two thin filaments, which subsequently form prosthecae of new cells. The direction of fission appears random. When separation is incomplete, cells form branched chains. The length of prosthecae in "*P. phaeoasteroidea*" was shown to depend on the light conditions of incubation; longer prosthecae are formed under low light intensities.

Ultrastructure of cells is typical of green sulfur bacteria (Gorlenko and Zhilina, 1968). The cell wall is of the Gram-negative type. The outer layer of the cell wall in *P. aestuarii* is assumed to take part in the excretion and holding of S^0 near the cell. The chlorosomes underlie the cytoplasmic membrane and fill the prosthecae (Fig. B11.19). Gas vacuoles have never been observed.

In *P. aestuarii*, the cell suspension appears green and the major pigment is BChl *c* and is located predominantly in the chlorosomes (*in vivo* absorption maximum: 740–750 nm). BChl *a* (*in vivo* absorption maximum: 805 nm) is present in the reaction centers that are located in the cytoplasmic membrane. These centers also contain BChl *c* and unidentified pigment P-665, which appears to be an unknown pheophytin. Since bacteriopheophytin *a* was not found in *P. aestuarii* (as distinct from *Chlorobium limicola*; Swarthof et al., 1982), P-665 instead of bacteriopheophytin *c* is assumed to serve as an acceptor in the photoreactions

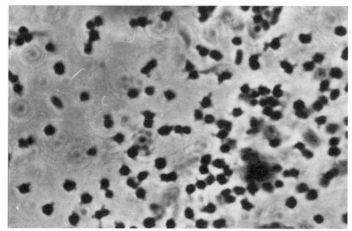

FIGURE B11.16. *P. aestuarii.* Phase-contrast micrograph (×3500). (Reprinted with permission from N.N. Puchkova and V.M. Gorlenko, Mikrobiologiya *45:* 656–660, 1976.)

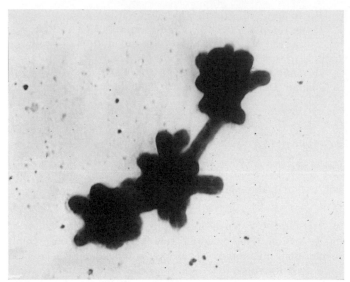

FIGURE B11.17.   *P. aestuarii* strain SK-413. Electron micrograph of cells stained with phosphotungstic acid (×20,000). (Reprinted with permission from N.N. Puchkova and V.M. Gorlenko, Mikrobiologiya *45:* 656–660, 1976.)

Green *P. aestuarii* strains contain two major carotenoids, chlorobactene (or its hydroxyl derivative) and rhodopin or lycopene (or its hydroxyl derivative). Three minor carotenoids are also present.

Brown *"P. phaeoasteroidea"* strains contain the major pigment BChl *e* (*in vivo* absorption maximum: 713 nm) and the minor pigment BChl *a* (*in vivo* absorption maximum: 805 nm). The major carotenoid is isorenieratene (*in vivo* absorption maximum: 523 nm; Puchkova and Gorlenko, 1976).

Species of the genus *Prosthecochloris* are strictly anaerobic and photolithotrophic. H_2S and S^0 are utilized as electron donors for photosynthesis. H_2S is oxidized to S^0, which is deposited extracellularly and then partly or fully oxidized to sulfate. Thiosulfate cannot be utilized as an electron donor during photosynthesis. These cell cultures are capable of autotrophic growth with CO_2 (bicarbonate) as the sole carbon source and require vitamin B_{12}. In the presence of H_2S and CO_2, a number of organic compounds are assimilated; acetate and pyruvate are utilized by the majority of strains.

Since these cells are moderately halophilic, growth is possible with NaCl concentrations ranging from 0.2 to 10% NaCl (optimum salinity: 0.5–2% NaCl). Some strains of *P. aestuarii* are able to develop in the presence of up to 20% NaCl, with an optimum at 4% NaCl (Puchkova, 1984).

Subsurface colonies in 0.8% agar medium appear as uneven lumps, easily distinguishable from colonies of other species of green sulfur bacteria.

Ammonium salts are used as a nitrogen source. Brown *"P. phaeoasteroidea"* also utilizes glutamate or casein hydrolysate; nitrate and asparagine are not utilized.

The salt tolerance of *Prosthecochloris* strains correlates with their tolerance for sulfide (maximum, up to 8 mM; optimum, 2–3 mM $Na_2S \cdot 9H_2O$) and their ability to live under high light intensities (up to 1,000 $\mu mol \cdot m^{-2} \cdot s^{-1}$), i.e., they are eurysulfidophilic and euryphotophilic (Matheron and Baulaigue, 1972; Puchkova, 1984).

The two species occupy different ecological niches. Among the green *P. aestuarii*, benthic forms are typical inhabitants of shallow marine environments, whereas these forms are found less frequently in meromictic saline bodies of water. The cells occur in microzones below purple sulfur bacteria under strictly anoxic conditions up to 3.8 mM H_2S, pH 7.0–7.8, E_h of 185 mV, 1.5–18% NaCl, and 15–40°C. *P. aestuarii* has been isolated from different parts of the globe: a region of the White Sea, the Arkhangel'sk region, saline waters off the Crimean Peninsula, and coastal areas of France, and also from the chemocline of a meromictic salt lake (Overmann et al., 1991b).

Among the brown *"P. phaeoasteroidea"*, the planktonic form develops in the metalimnia-hypolimnia of saline meromictic lakes of marine origin (e.g., Lake Mogil'noe, Kil'din Island, Barents

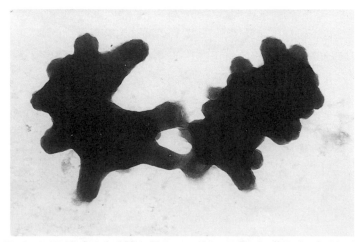

FIGURE B11.18.   *"P. phaeoasteroidea"* strain MG-1. Electron micrograph of cells stained with phosphotungstic acid (×22,000). (Reprinted with permission from N.N. Puchkova and V.M. Gorlenko, Mikrobiologiya *45:* 656–660, 1976.)

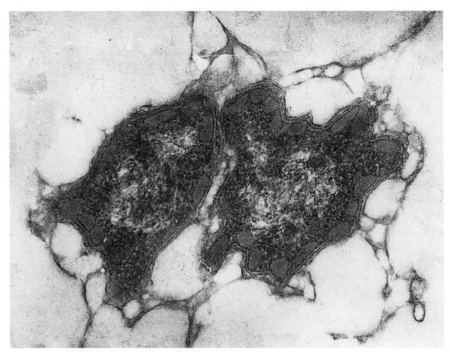

FIGURE B11.19. *P. aestuarii* strain SK-413. Electron micrograph of ultrathin section of cells fixed with OsO$_4$ (×100,000). (Reprinted with permission from N.N. Puchkova, *Mikrobiologiya 53*: 324–328, 1984.)

Sea; Lake Faro, Sicily; and the Mediterranean Sea) (Gorlenko et al., 1976). Parameters of habitat: 0.13–0.5 mM H$_2$S; pH 6.5–7.5; E$_h$, 50–240 mV; 33.5% NaCl; 10–15°C; and a depth of occurrence of 10–13 m.

ENRICHMENT AND ISOLATION PROCEDURES

P. aestuarii often occurs in the hydrogen sulfide-containing mud of shallow saline bodies of water (Matheron and Baulaigue, 1972; Puchkova, 1984). It can be isolated by direct inoculation of a mud sample with dilutions into 0.8% agar medium of Pfennig (1965) (or the medium for *Chlorobium*, see p. 605) containing the following: NaCl, 2%; Na$_2$S·9H$_2$O, 700–1,500 mg/l; NH$_4$-acetate, 400 mg/l (instead of NH$_4$Cl, 330 mg/l); and vitamin B$_{12}$, 20 μg/l. Green colonies of *P. aestuarii* are lumpy, typically olive green, with an uneven surface and are easily distinguishable from (a) smaller diffuse colonies of *Chlorobium chlorovibrioides* and (b) brighter, flat, dense, often lens-shaped colonies of *Chlorobium vibrioforme*. Colonies of *P. luteolum* are distinguished from those of *C. vibrioforme* by their greater transparency and emerald color.

Enrichment of *P. aestuarii* occurs when mud columns are exposed to diffused daylight. After growth of green bacteria in water or on the mud surface becomes evident, a sample is inoculated into agar medium of Pfennig. Isolated colonies are purified by repeated application of the agar shake dilution method. When a pure culture is obtained, a single colony is transferred into liquid medium of Pfennig and incubated in screw-cap glass bottles in a luminostat at 25–35°C and a light intensity of 50–200 μmol·m^{-2}·s^{-1}.

Brown *"P. phaeoasteroidea"* are enriched and purified in the same way as described for *P. aestuarii*.

In enrichment cultures containing brown strains of several species of green sulfur bacteria during prolonged storage in the dark, *Chlorobium phaeovibrioides* and *Pelodictyon phaeum* die out first, and thus a predominance of *"P. phaeoasteroidea"* can be obtained. Colonies of the latter in 0.8% agar are similar to those of *P. aestuarii* but appear dark brown. They are easily distinguishable by shape and color from those of other brown strains of green sulfur bacteria.

MAINTENANCE PROCEDURES

Prosthecochloris can be maintained in liquid cultures if transferred every 2–3 months. Grown cultures are stored at 4°C and very low light intensity.

DIFFERENTIATION OF THE GENUS *PROSTHECOCHLORIS* FROM OTHER GENERA

The genus *Prosthecochloris* differs morphologically from other genera of green sulfur bacteria. The presence of short prosthecae with thickened ends and the absence of gas vacuoles differentiate it from the genera *Chlorobium*, *Pelodictyon*, and *Ancalochloris*. The latter genus also possesses prosthecae, but the cells contain gas vacuoles. The shape and the number of prosthecae are different in *Ancalochloris* than in representatives of *Prosthecochloris*.

TAXONOMIC COMMENTS

Based on the analysis of 16S rRNA gene sequences (Fig. B11.2), *P. aestuarii* DSMZ 271[T] is phylogenetically closely related to the type strain of *C. vibrioforme* DSMZ 260[T] and therefore does not represent an isolated branch within the radiation of green sulfur bacteria. No further studies on the genetic relationships of the species of *Prosthecochloris* with other green sulfur bacteria have been conducted to date.

List of species of the genus Prosthecochloris

1. **Prosthecochloris aestuarii** Gorlenko 1970, 148.[AL]

    aes.tu.a' rii. L. gen. n. *aestuarii* of the estuary.

    Cells appear spherical after division and elongated before division, measuring 0.5–0.7 × 1.0–1.2 μm, and form 10–20 prosthecae/cell (Figs. B11.16 and B11.17). Ends of prosthecae are rounded and slightly thickened. Prosthecae are 0.1– 0.17 μm wide; their length rarely exceeds the cell diameter and corresponds to 0.1– 0.25 μm. Cells are enclosed in slimy microcapsules.

    Slimy strands are observed between divided cells. Multiplication is by binary fission; unequal division is possible. The direction of fission is random. Sister cells can be connected by one or more filaments, which subsequently form new prosthecae. Incomplete separation results in formation of branched chains.

    Strictly anaerobic; obligately phototrophic. Sulfide and S^0 are utilized as electron donors for photosynthesis. Thiosulfate is not utilized. Cells photoassimilate simple organic compounds, acetate, or pyruvate only in the presence of sulfide and CO_2. Optimum pH: 6.7–7.0. Cells grow in 1–8% NaCl (optimum salinity: 2–5% NaCl). Color of liquid culture varies from intensive green to grayish green. Following precipitation of S^0, the medium becomes milky turbid.

    Vitamin B_{12} is required for growth. Sulfide is required as a sulfur source for biosynthesis.

    The major photosynthetic pigment is BChl *c*, the minor photosynthetic pigment is BChl *a*. The carotenoids are chlorobactene, rhodopin, or lycopene or their hydroxyl derivatives. Ammonium salts are utilized as a nitrogen source.

    Habitat: hydrogen sulfide-containing mud of shallow bodies of water with up to 18% NaCl.

    The mol% G + C of the DNA is: 52.0–56.1 (T_m, Bd).

Type strain: SK-413, INMI, DSM271.

GenBank accession number (16S rRNA): Y07837.

2. **"Prosthecochloris phaeoasteroidea"** Puchkova and Gorlenko 1976, 658.

    phae' o.as.te.ro.i.de' a. Gr. adj. *phaeus* brown; Gr. adj. *asteroides* star shaped; M.L. fem. adj. *phaeoasteroidea* brown, star shaped.

    Morphology of the cells is similar to that of *P. aestuarii*; size of the cell is 0.5–0.6 × 0.5–1.2 μm; the number of prosthecae is 10–20, and the prosthecae are 0.13–0.16 μm wide and 0.07– 0.3 μm long. Slimy strands are absent (Fig. B11.18).

    Color of cell suspensions varies from dark brown to chocolate brown. Colonies in 0.8% agar are dark brown, lumpy, and uneven. The major pigment component is BChl *e*; BChl *a* is present in small amounts. The major carotenoid is isorenieratene.

    Photoassimilated: acetate, pyruvate, lactate, malate, fumarate, fructose, mannitol, glutamate, and casein hydrolyzate in the presence of H_2S and CO_2. Not utilized: citrate, succinate, propionate, valerate, methanol, ethanol, glycerol, glucose, malonate, glycolate, and asparagine. Optimum pH for growth: 7.0. pH range: 6.0–7.5. Salinity range, 0.2–7.0% NaCl; optimum salinity, 0.5–2% NaCl. Ammonium salts, casein hydrolysate, or glutamate can be used as a nitrogen source. Nitrate and asparagine are not utilized.

    Habitat: the anoxic monimolimnia of meromictic saline lakes of marine origin. Depth of occurrence: 10–13 m. NaCl concentration: 3–3.5%. H_2S concentration: 10–40 mg/l.

    The mol% G + C of the DNA is: 52.2 ± 0.8 (chemical analysis).

    Deposited strain: MG-1, INMI.

    Additional Remarks: Isolated from Lake Mogil'noe, Kil'din Island, Barents Sea, Russia. This strain has been lost.

Addendum to the Green Sulfur Bacteria
Phototrophic green sulfur bacteria living in consortia with other microorganisms.

JÖRG OVERMANN

Phototrophic consortia are structural associations between green- or brown-colored and colorless bacteria. The regular structure of these consortia is unique in the microbial world (Schink, 1991). Colored cells contain chlorosomes (e.g., Caldwell and Tiedje, 1975b) and hybridize with 16S rDNA oligonucleotide probes specific for green sulfur bacteria (Tuschak et. al., 1999). This indicates that the colored cells belong to the green sulfur bacteria. So far neither the phototrophic nor the chemotrophic partner are available in pure culture. The isolation of a green sulfur bacterium from the *"Chlorochromatium"* consortium by Mechsner (1957) was the first step in this direction. Unfortunately, this green sulfur bacterium was lost before detailed metabolic and taxonomic studies were made. Not one of the consortia has so far been isolated in its intact form.

Phototrophic consortia represent a significant fraction of total bacterial biomass in some lakes and can amount to two-thirds of total bacterial biomass (Gasol et al., 1995). Typically, consortia are found in the chemocline of stratified lakes to depths of 22 m and seem to be adapted to a very narrow regime of light intensities (<5% light transmission, irradiance <5 μmol quanta·m^{-2}·s^{-1}), and sulfide concentrations (50–500 μM) (compilation of data in Overmann et al., 1998).

It is possible that phototrophic consortia represent fortuitous combinations whose success depend on environmental factors. However, the numbers of phototrophic epibionts in *"Chlorochromatium aggregatum"* and *"Pelochromatium roseum"* show a nonrandom frequency distribution (Overmann et al., 1998), indicating that under natural conditions, the cell division of all epibionts proceeds synchronously and parallel to that of the central bacterium.

The type of interaction in the consortia is still unknown, but it has been speculated that the colorless bacterium reduces sulfur or sulfate to sulfide which is reoxidized phototrophically by the green sulfur bacteria (Kuznetsov, 1977; Pfennig, 1980). Recent experiments with artificial electron acceptors have revealed that the green sulfur bacteria of the consortium *"Chlorochromatium aggregatum"* indeed use sulfide as the electron donor in the light (Fröstl and Overmann, 1998). Based on the observation that the whole consortium exhibits chemotaxis towards sulfide, it has been concluded that a rapid interspecies signal transfer occurs between the nonmotile, green epibiont and the flagellated, colorless central bacterium (Fröstl and Overmann, 1998). Even more significantly, whole consortia accumulate scotophobically in microcuvettes at a wavelength of 740 nm. This wavelength corresponds to the position of the absorption maximum of BChls *c* or *d*. Because only the chemotrophic partner seems to be motile, the scotophobic response provides strong evidence for signal transduction between the cells within the phototrophic consortia.

The use of generic and species designations for the apparently stable complexes composed of two different microorganisms has been questioned (Buder, 1914; Van Niel, 1957). According to the rules of bacterial nomenclature, the generic designations are not validly published (Trüper and Pfennig, 1971). Therefore, the generic and species designations with the addition of the term consortium are used here as laboratory names without taxonomic significance.

Only those consortia are included here that were repeatedly observed and described to thrive in water and mud samples collected in nature. A schematic presentation is provided in Fig. B11.20.

1. **"Chlorochromatium aggregatum" consortium** Lauterborn 1906, 197 (*Chloronium mirabile* Buder 1914, 80) (See also Pfennig 1980.)

Barrel-shaped aggregates consisting of a rather large, colorless, polar flagellated bacterium as the center, which is surrounded by green bacteria arranged in 5–6 rows, ordinarily 2–4 cells long (Fig. B11.20). The entire consortium behaves as a unit, is motile and is scotophobically active. Multiplication occurs by the more or less simultaneous fission of the component cells; growth is dependent on both sulfide and light.

Cells of the green component are 0.5–1.0 μm × 1.0–2.5 μm. Morphologically, the green cells resemble those of *Chlorobium limicola*. The central bacterium is usually surrounded by 12–24 green cells (Fig. B11.21). The size of the entire motile barrel-shaped consortium is variable, generally 2.5–4.0 μm × 4–10 μm.

"Chlorochromatium aggregatum" exhibits chemotaxis toward 2-oxoglutarate, citrate, sulfide, and thiosulfate, but not sulfite (Fröstl and Overmann, 1998). No chemotaxis occurs toward acetate, propionate, or pyruvate, which are typical carbon substrates of green sulfur bacteria. Using 2-oxoglutarate and low concentrations of sulfide (300 μM), this type of consortium could be enriched from sediment samples of a eutrophic freshwater lake and maintained at high numbers in anoxic sulfide-reduced medium (Fröstl and Overmann, 1998). Growth of intact consortia strictly dependent on light and 2-oxoglutarate. No growth occurs with short chain fatty acids, alcohols, pyruvate, lactate, malate, fumarate, succinate, citrate and isocitrate. Maximum growth rates (t_d = 1 d) are reached at light intensities ~5 μmol quanta·m^{-2}·s^{-1}.

The consortia usually occur together with other green and purple sulfur bacteria in the chemocline and hypolimnion of stratified freshwater lakes or mud and sulfide-containing water of stagnant freshwater ponds exposed to light.

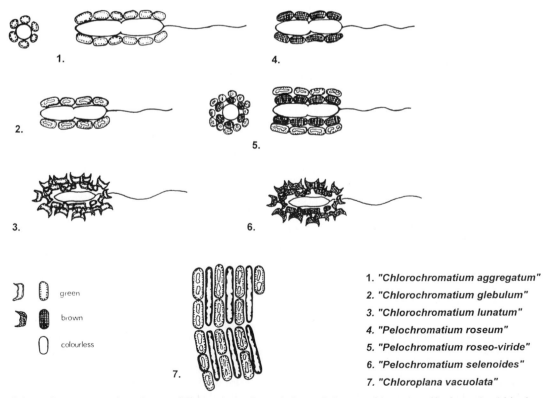

1. *"Chlorochromatium aggregatum"*
2. *"Chlorochromatium glebulum"*
3. *"Chlorochromatium lunatum"*
4. *"Pelochromatium roseum"*
5. *"Pelochromatium roseo-viride"*
6. *"Pelochromatium selenoides"*
7. *"Chloroplana vacuolata"*

green
brown
colourless

FIGURE B11.20. Schematic representation of seven different types of associations of phototrophic green sulfur bacteria with unknown chemotrophic bacteria. *1–6*, longitudinal and cross-sectional views through different consortia; *7*, top view of *"Chloroplana"* (modified after Pfennig, 1980; and Abella et al., 1998).

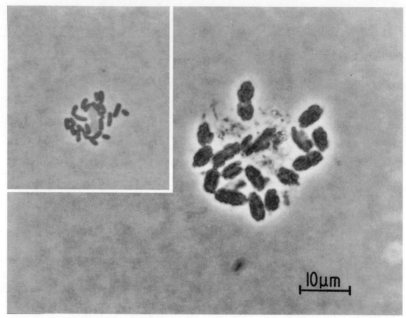

FIGURE B11.21. *"Chlorochromatium aggregatum"*. Intact consortia are shown, some of them dividing. *Insert*, squeeze preparation of one consortium exposing the central rod-shaped bacterium. Phase-contrast micrograph.

2. **"Chlorochromatium glebulum" (sic) consortium** Skuja 1956, 36.

Cells of the green component are 0.5–0.6 μm × 0.7–1.0 μm and contain gas vacuoles (Fig. B11.20). Morphologically, the green cells resemble those of *Pelodictyon luteolum*. The curved central bacterium may be surrounded by 7–40 individual green cells. The size of the barrel-shaped motile consortium is 3–4 μm wide and 4–8 μm long. This kind of consortium was described as occurring in the hypolimnion of Swedish lakes.

3. **"Chlorochromatium lunatum" consortium** Abella, Cristina, Martinez, Pibernat and Vila 1998, 452.

Similar to the other consortia described above, a central spindle-shaped bacterium is surrounded by an array of green-colored cells (Fig. B11.20). Based on the presence of chlorosomes the outer cells were identified as members of the green sulfur bacteria. Epibionts are 0.5–0.7 μm ×1–1.5 μm and have a half-moon shaped morphology (Fig. B11.22) which distinguishes them from all other unicellular green sulfur bacteria described so far. The mean range of epibionts per consortium is 49–69 and thus more than twice as high as that of most other motile consortia. Predominant pigments are bacteriochlorophyll *d* and the carotenoid chlorobactene. *"Chlorochromatium lunatum"* was found in the hypolimnia of several freshwater lakes in Wisconsin and Michigan (USA) at depths of 2–7 m in lakes containing low sulfide concentrations (below <60 μM).

4. **"Pelochromatium roseum" consortium** Lauterborn 1913, 99 (See also Utermöhl 1924; Gorlenko and Kuznetsov 1972; Pfennig 1980.)

The morphology, multiplication and behavior of the *"Pelochromatium"* consortium (Fig. B11.20) resemble those of the *"Chlorochromatium"* consortium (Fig. B11.23). Different from the latter, the phototrophic sulfur bacteria surrounding the

colorless motile central bacterium morphologically resemble those of the brown *Chlorobium phaeobacteroides*.

Cells of the pinkish-brown component are straight or slightly curved rod-shaped, 0.6–1.0 μm × 1.2–2.5 μm. The rod-shaped central bacterium is slightly bigger and has tapered ends; it is surrounded by 10–20 individual brown cells. In natural samples the whole consortium is often surrounded by an extracellular slime layer of varying thickness (Overmann et al., 1998). The size of the entire barrel-shaped motile consortium is variable, mostly 2.5–4.0 μm × 4–8 μm long.

In lake water cultures, bacterial cells of the consortium *"Pelochromatium roseum"* remain in a stable association only if incubated at low light intensities (5–10 μmol quanta·m^{-2}·s^{-1}).

The brown consortia have been observed most often in the sulfide-containing hypolimnion of stratified freshwater lakes. In these habitats, the consortia were found together with other green and purple sulfur bacteria. Generally, brown consortia are found at greater depths in the water column as compared to their green counterparts.

5. **"Pelochromatium roseo-viride" consortium** Gorlenko and Kuznetsov 1972, 7.

The consortium represents a particular variation of the *"Pelochromatium roseum"* consortium (Fig. B11.20). Instead of being surrounded by only one layer of a brown strain of green sulfur bacteria, the motile central bacterium of *"P. roseo-viride"* is covered by two layers of phototrophic bacteria. The inner layer consists of brown cells, as in *"P. roseum"*. The second, outer layer consists of rod-shaped green cells with gas vacuoles. The green cells are of about the same size as the brown cells and resemble *Pelodictyon luteolum*. The length of the entire motile barrel-shaped consortium is approximately 4 μm after division and 8 μm before division.

The brown plus green consortia consisting of three different kinds of bacteria occur in Lake Kononjer in the Mariji

FIGURE B11.22. *"Chlorochromatium lunatum"*. Electron micrograph of intact consortia. Bar = 1 μm.

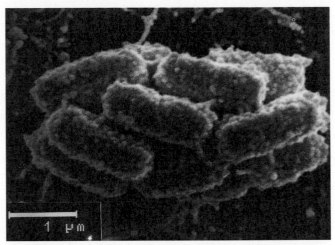

FIGURE B11.23. *"Pelochromatium roseum"*. Electron micrograph of intact consortia.

Republic of Russia. The consortia were found to thrive together with *"Pelochromatium roseum"* in the sulfide-containing hypolimnion at a depth of 11.5 m.

6. **"Pelochromatium selenoides" consortium** Abella, Cristina, Martinez, Pibernat and Vila 1998, 452.

The morphology resembles that of *"Chlorochromatium lunatum"*, but the half- moon-shaped epibionts are brown in color. Main photosynthetic pigments are bacteriochlorophyll *e* and isorenieratene/β-isorenieratene. The mean number of epibionts per consortium is 36–45. *"Pelochromatium selenoides"* was found in the hypolimnia of several freshwater lakes in Wisconsin and Michigan (USA) at greater depths (8–16 m) than its green counterpart *"Chlorochromatium lunatum"*.

7. **"Chloroplana vacuolata" consortium** Dubinina and Kusnezov 1976, 8 (See also Pfennig, 1980).

Flat sheaths, lamellae or platelets composed of parallel rows of alternating green and colorless bacteria with gas vacuoles. The platelets are variable in size and may contain from a few to 300–400 green cells; consortia are nonmotile. Cells of the green bacteria are 0.6– 0.8 μm × 1.2–2.0 μm and resemble the gas-vacuolated green *Pelodictyon luteolum*. Cells of the gas vacuole-containing colorless rod-shaped bacterium are 0.35–0.45 μm × 3.0–6.0 μm. By electron microscopy, the green cells were shown to contain chlorosomes, the typical light-harvesting structures of the green bacteria. The presence of BChl *c* or *d* was inferred from the green fluorescence in reflected light.

The *"Chloroplana"* consortia were found in large number together with other green and purple sulfur bacteria in the anoxic sulfide-containing water of small forest lakes, Lesnaya Lamba, in southern Karelia, Russia. The occurrence of these consortia in stagnant freshwater ponds has been reported repeatedly.

Phylum BXIII. Firmicutes

Gibbons and Murray 1978, 5[AL] emend. Garrity and Holt

Class I. *"Clostridia"*

Order I. **"Clostridiales"**

Family VI. **"Heliobacteriaceae"** Beer-Romero and Gest 1987, 113

MICHAEL T. MADIGAN

He.li.o.bac.te.ri.a' ce.ae. M.L. neut. n. *Heliobacterium* type genus of the family; *-aceae* ending to denote a family; M.L. fem pl. n. *Heliobacteriaceae* the *Heliobacterium family.*

Rod-shaped to short filaments or spirillum-shaped cells; multiplication by binary fission. **Anoxygenic phototrophs**. Rod-shaped cells are frequently curved slightly and in one genus (*Heliophilum*), cells are straight and tapered and group together to form bundles that are motile as a unit. **Cells stain Gram-negative, although the phylogenetic position of the group is Gram-positive** (Woese et al., 1985a; Madigan and Ormerod, 1995; Ormerod et al., 1996a). Motility is by gliding or flagellar means; if the latter then by polar, subpolar, or peritrichous flagella. **Contain bacteriochlorophyll *g* as major bacteriochlorophyll** (Brockman and Lipinski, 1983; Michalski et al., 1987) and diaponeurosporene (Taikaichi et al., 1997) as the major carotenoid. Small amounts of 8′-OH-chlorophyll *a* is also present (Amesz, 1995). Bacteriochlorophyll *g* and 8′-OH-chlorophyll *a* are esterified with the C-15 alcohol farnesol rather than the C-20 alcohol phytol. **Color of phototrophically grown cultures is brownish green** although cultures turn a more emerald green in color in stationary phase. Extensively developed **intracytoplasmic membranes, of the kind observed in phototrophic purple bacteria or chlorosomes characteristic of the green sulfur or green nonsulfur bacteria, are not observed**. Gas vesicles are not present. Metabolism is strictly anaerobic; cells grow under anoxic conditions in the light as photoheterotrophs on a limited number of organic carbon sources or chemoorganotrophically (in darkness) by pyruvate fermentation (pyruvate fermentation has not been observed in *Heliorestis*) (Kimble et al., 1994). Aerobic or microaerobic dark growth does not occur. **Photoautotrophic growth on CO_2/H_2 or CO_2/H_2S has not been observed**, although if sulfide is added to culture media, it is frequently oxidized to S^0. Optimal pH is ~7 with a pH growth range of 5.5–8, depending on the species; *Heliorestis* is alkaliphilic, having a pH optimum of 9. Optimal temperature is 37– 42°C (52°C for *Heliobacterium modesticaldum* and 30°C for *Heliorestis daurensis*). Ammonium salts and gluta-

mine are used as nitrogen sources and certain species can use other amino acids as well. **Nitrogen fixation is a property of all heliobacteria. Biotin is required as a growth factor by all heliobacteria** and some species also require a reduced sulfur compound (sulfide, thiosulfate, or cysteine) as a sulfur source for biosynthesis. Poly-β-hydroxybutyrate has not been observed as a storage material. True endospores containing dipicolinic acid and elevated levels of Ca^{2+} have been observed in some species and, considering the phylogenetic position of all members of the family, this property may well be universal among the group; the loss of ability to form endospores in laboratory culture has been a common observation with several isolates. *Heliobacteriaceae* occur in nature primarily in soil and in this regard differ significantly in their ecology from purple and green anoxygenic phototrophs.

The mol% G + C of the DNA is: 44.9–55.

Type genus: **Heliobacterium** Gest and Favinger 1985, 223 (Effective publication: Gest and Favinger 1983, 15.)

FURTHER DESCRIPTIVE INFORMATION

The family *Heliobacteriaceae* contains all of the anoxygenic phototrophic bacteria that produce bacteriochlorophyll *g*. This structurally and spectrally unique bacteriochlorophyll distinguishes the heliobacteria from the purple bacteria, which contain bacteriochlorophyll *a* or *b*, and the green sulfur and green nonsulfur (*Chloroflexus*) bacteria, which contain bacteriochlorophylls *c*, *d*, or *e*, or bacteriochlorophyll *c$_s$*, respectively. *In vivo* absorption spectra (performed anoxically) of heliobacteria show a major peak at 786–792 nm (depending on the species) due to absorption by bacteriochlorophyll *g*, and comparatively small peaks at about 575 and 670 nm; the latter is due to a small amount of 8′-OH-chlorophyll *a* present in the photosynthetic reaction center. Exposure of cultures of heliobacteria to air causes the apparently irreversible oxidation of bacteriochlorophyll *g* to chlo-

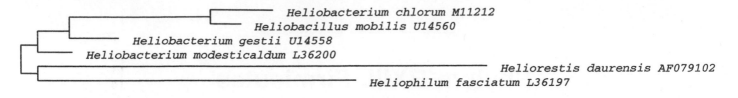

5% difference

FIGURE 1. Evolutionary distance tree of the family *"Heliobacteriaceae"* based on comparative 16S rRNA sequencing. The tree shown was computed from evolutionary distances by the algorithm of De Soete using the Jukes and Cantor correction as detailed in Wahlund et al. (1991). Organisms and their GenBank accession numbers include: *Heliobacterium chlorum* (M11212); *Heliobacterium gestii* (strain Chainat, U14558); *Heliobacterium modesticaldum* (strain Ice1, L36200); *Heliobacillus mobilis* (U14560); *Heliophilum fasciatum* (strain Tanzania, L36197); *Heliorestis daurensis* (strain BT-H1, AF079102).

rophyll *a*, a major increase in 670 nm absorbance coupled to a corresponding loss of 788 nm absorbance, and loss of cell viability. Fermentatively (dark)-grown cells of heliobacteria also produce photopigments.

The cell wall of heliobacteria species is rather fragile and stationary phase cells of some species tend to form spheroplasts. The peptidoglycan, which is present in small amounts, contains L,L-diaminopimelic acid instead of *meso*-diaminopimelic acid as muramic acid crosslinks. Cell walls also contain a considerable amount of lipid, although this lipid is not in the form of lipopolysaccharide (Beck et al., 1990). Heliobacteria are unusually sensitive to penicillin—growth of *Heliobacterium chlorum* is inhibited by 2 ng/ml of penicillin G, a level 1000-fold lower than the inhibitory level for *E. coli* (Beer-Romero et al., 1988).

From a phylogenetic standpoint, heliobacteria belong to the low mol% G + C Gram-positive phylum (*Firmicutes*) (Fig. 1), that includes chemotrophic endospore-forming bacteria. Currently, four genera with a total of six species are recognized.

Enrichment, Isolation, and Ecology Enrichment of heliobacteria takes advantage of their ability to form endospores and their presence in soils; rice soils are particular good sources of heliobacteria (Stevenson et al., 1997). A soil sample is pasteurized (80°C for 10 minutes) in a stoppered tube of anoxically prepared growth media (for enrichment, a 0.25% [w/v] yeast extract medium at pH 7 works well), sealed under an atmosphere of N_2/CO_2 (95:5 v/v), and incubated in incandescent light (40 $\mu E \cdot m^{-2} \cdot s^{-1}$) at 38–40°C. It should be emphasized that all heliobacteria isolated to date are strict anaerobes, such that extreme care should be taken to ensure that media and growth conditions for the isolation of heliobacteria are strictly anoxic. Successful

enrichments usually show a green film of cells atop the soil in the tube. This material can be removed with a sterile Pasteur pipette and plated onto enrichment medium (or other media; see Kimble and Madigan, 1992; Kimble et al., 1995; Stevenson et al., 1997; Bryantseva et al., 1999a). Plating should be done inside an anoxic chamber. Streaked plates are incubated inside anoxic jars (e.g., Gas-Pak® jars) placed in the light and green colonies with irregular edges form within 2–3 d. Pure cultures can usually be obtained by repeated restreaking. Alternatively, isolated colonies of heliobacteria can be obtained by using the agar "shake culture" technique (Trüper and Pfennig, 1982). If the latter is used without benefit of an anoxic chamber, it is advisable to add 1–2 mM sulfide to the medium to ensure reducing conditions.

In a survey of soils for the presence of heliobacteria it was found that these anoxyphototrophs are common in rice (paddy) soils in both traditional rice-growing regions such as Southeast Asia and also in modern cultivated rice fields such as those in the southern United States (Stevenson et al., 1997). Other agricultural or garden soils were not reliable sources of heliobacteria, although occasional isolates from these soils were obtained. Aquatic habitats, with the exception of neutral to alkaline hot springs ≤60°C, did not yield heliobacteria (Stevenson et al., 1997). The close connection between heliobacteria and rice plants suggest that a specific plant-bacterium association may exist, with the photoheterotrophic heliobacteria assimilating organic compounds excreted by rice plant roots and the plants benefiting from fixed nitrogen excreted by the nitrogen-fixing heliobacteria; all species of heliobacteria have been shown to be strong nitrogen fixers (Kimble and Madigan, 1992; Madigan, 1995).

Genus I. **Heliobacterium** *Gest and Favinger 1985, 223^(VP) (Effective publication: Gest and Favinger 1983, 15)*

MICHAEL T. MADIGAN

He.li.o.bac.ter'i.um. Gr. n. *helios* sun; Gr. neut. n. *bakterion,* a small rod; M.L. neut. n. *Heliobacterium* sun bacterium.

Rod-shaped cells, slightly curved, 0.8–1 × 2.5–9 µm (Fig. 2), **or spirilla,** 1–1.2 µm in width and of similar length to rod-shaped cells (Fig. 3). From an ultrastructural standpoint, cells of heliobacteria lack intracytoplasmic membranes or chlorosomes,

and thus electron micrographs of the cytoplasm give no indication that the cells are phototrophic (Fig. 4). Colonies on a yeast extract/salts medium (medium PYE, see Kimble et al.,1995) frequently have irregular margins because of wavy protrusions

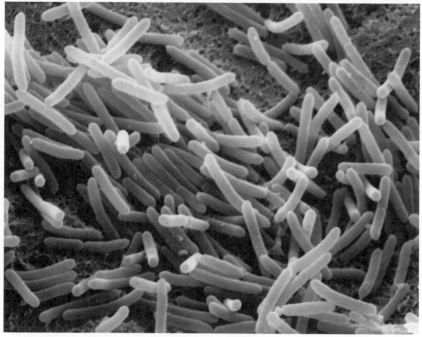

FIGURE 2. Scanning electron micrograph of cells of *Heliobacterium chlorum* ATCC 35205. (×5000) (Courtesy of F. Rudy Turner, Indiana University).

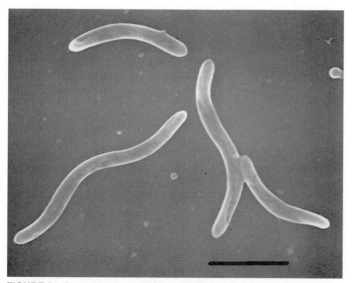

FIGURE 3. Scanning electron micrograph of cells of *Heliobacterium gestii*. Bar = 5 µm.

of cell masses in palisade formations, similar to what is seen in gliding bacteria. Motility variable but usually by polar or sub-polar flagella; *H. chlorum* may show slow gliding movement. **Some species,** particularly *H. chlorum*, **lyse readily at stationary phase in rich or defined media.** One species is thermophilic.

Type species: **Heliobacterium chlorum** Gest and Favinger 1985, 223 (Effective publication: Gest and Favinger 1983, 15.)

DIFFERENTIATION OF THE GENUS *HELIOBACTERIUM* FROM OTHER GENERA

The genus *Heliobacterium* is differentiated from other genera of heliobacteria by aspects of motility (gliding or sub-polar flagella in species of *Heliobacterium* versus peritrichous flagella in *Heliobacillus*) and by the fact that cells exist as single entities rather than as cell bundles that move as a unit in culture (genus *Heliophilum*). The genus *Heliobacterium* is differentiated from the genus *Heliorestis* by an inability to grow at pH 9 and an optimal growth temperature near 40°C instead of 30°C. As a group, heliobacteria are differentiated from all other anoxygenic phototrophs (and all other procaryotes, for that matter) by virtue of their production of bacteriochlorophyll *g*.

DIFFERENTIATION OF THE SPECIES OF THE GENUS *HELIOBACTERIUM*

The major differentiating properties for species of *Heliobacterium* are listed in Table 1.

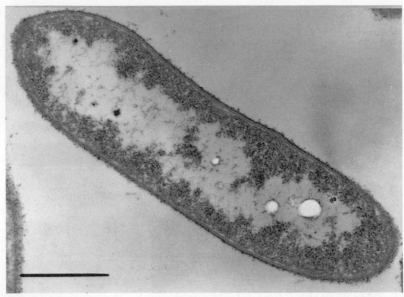

FIGURE 4. Thin section transmission electron micrograph of a cell of the thermophilic heliobacterium species, *Heliobacterium modesticaldum* strain Ice1 (Reprinted with permission from Kimble et al., Archives of Microbiology *163:* 259–267, 1995. Bar = 1μm.)

List of species of the genus Heliobacterium

1. **Heliobacterium chlorum** Gest and Favinger 1985, 223[VP] (Effective publication: Gest and Favinger 1983, 15.)

   chlo' rum. M.L. neut. adj. *chlorum* from the Gr. adj. *chloros* green.

   Cells rod-shaped; other characteristics are the same as those described for the genus. Isolated from garden soil on the campus of Indiana University, Bloomington (USA).

   The mol% G + C of the DNA is: 52 (T_m).

   Type strain: ATCC 35205, DSMZ 3682.

   GenBank accession number (16S rRNA): M11212.

2. **Heliobacterium gestii** Ormerod, Kimble, Nesbakken, Torgersen, Woese and Madigan 1996b, 1189[VP] (Effective publi-

cation: Ormerod, Kimble, Nesbakken, Torgersen, Woese and Madigan 1996a, 233.)

gest.i.i. M.L. gen. n. *gestii* named for Howard Gest, an American microbiologist.

Cells spirilloid (Fig. 3), motile by polar or subpolar flagella. Subterminal cylindrical endospores (Fig. 5) formed in freshly isolated strains. Optimal pH 7. Incapable of assimilatory sulfate reduction. NaCl not required and growth inhibited at concentrations above 1%. Only species of heliobacteria known to photocatabolize sugars.

Strain Chainat was isolated from a Thailand rice soil.

The mol% G + C of the DNA is: 55.1 (T_m).

Type strain: Chainat, ATCC 43375.

GenBank accession number (16S rRNA): L36198, U14558.

3. **Heliobacterium modesticaldum** Kimble, Mandelco, Woese and Madigan 1996, 1189[VP] (Effective publication: Kimble, Mandelco, Woese and Madigan 1995, 266.)

mo.des.ti.cal' dum. L. neut. adj. *modesticum;* L. neut. adj. *caldum;* M.L. neut. adj. *modesticaldum* moderately hot; named for its thermophilic character.

Cells rod-shaped or slightly curved, 0.8–1 × 2.5–9 μm (Fig. 4). Motile by flagella or nonmotile. Form cylindrical subterminal endospores. Thermophilic; optimum growth temperature 52°C (temperature range 30–56°C). NaCl not required for growth and growth inhibited by 1% NaCl. Optimal pH 6.5. Fixes N_2 up to its maximum growth temperature. Isolated from neutral to alkaline hot spring microbial mats and volcanic soils.

Strain Ice1 was isolated from soil obtained near alkaline hot springs, Reykjanes, Iceland.

The mol% G + C of the DNA is: 54.6–55 (T_m).

Type strain: Ice1, ATCC 51547.

GenBank accession number (16S rRNA): L36200, U14559.

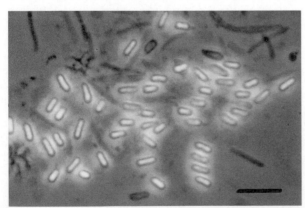

FIGURE 5. Phase-contrast photomicrograph of endospores produced in a culture of *Heliobacterium gestii* (Reprinted with permission from Ormerod et al., Archives of Microbiology *165:* 226–234, 1996. Bar = 5μm.)

TABLE 1. Summary of properties of heliobacteria[a]

Property	Heliobacterium chlorum	Heliobacterium gestii	Heliobacterium modesticaldum	Heliobacillus mobilis	Heliophilum fasciatum	Heliorestis daurensis
Habitat	Temperate soil	Tropical paddy soil	Neutral/alkaline hot springs and volcanic soils	Tropical paddy soil	Tropical paddy soil	Microbial mats of brackish soda lakes
Morphology	Rod	Spirillum	Rod/curved rod	Rod	Straight rods with tapered ends grouped in bundles of two to many cells	Coiled to bent rod or short filament
Dimensions (μm)	1 × 7–9 (see Fig. 2)	1 × 7–10 (see Fig. 3)	0.8–1 × 2.5–9 (see Fig. 4)	1 × 7–10	0.8–1 × 5–8	0.8–1.2 × 10–20
Motility	Gliding	Multiple subpolar flagella	Flagellar or none	Peritrichous flagella	Polar to subpolar flagella; cell bundles move as a unit	Peritrichous flagella
Carbon sources photo-metabolized	Pyruvate, lactate, yeast extract	Pyruvate, lactate, acetate, butyrate, ethanol (+ CO_2), yeast extract	Pyruvate, lactate, acetate, yeast extract	Pyruvate, lactate, acetate, butyrate (+ CO_2); yeast extract	Pyruvate, lactate, acetate, butyrate (+ CO_2)	Pyruvate, acetate, yeast extract
Biosynthetic sulfur sources	SO_4^{2-}, $S_2O_3^{2-}$, methionine, or cysteine	$S_2O_3^{2-}$, methionine, or cysteine	$S_2O_3^{2-}$, sulfide, methionine, or cysteine	SO_4^{2-}, $S_2O_3^{2-}$ methionine, or cysteine	SO_4^{2-}, $S_2O_3^{2-}$	Sulfide
Endospores produced	None observed	Yes (see Fig. 5)	Yes	None observed	Yes	Yes
Optimum temperature, °C	37–42	37–42	50–52	38–42	37–40	30
Optimum pH	7	7	6–7	6.5–7	6.5–7	9
Mol% G + C content	52	55.1	54.6–55	50.3	51.8	44.9

[a]Data obtained from Gest and Favinger (1983), Beer-Romero and Gest (1987), Kimble et al. (1995), Ormerod et al. (1996a), Stevenson et al. (1997), and Bryantseva et al. (1999a).

Genus II. **Heliobacillus** Beer-Romero and Gest 1998, 627[VP] (Effective publication: Beer-Romero and Gest 1987, 113)

MICHAEL T. MADIGAN

He.li.o.ba.cil' lus. Gr. n. helios sun; L. dim.n. bacillum a small rod; M.L. masc. n. Heliobacillus sun rod.

Highly motile cells containing peritrichous flagella. For other properties, see Table 1.

Type species: **Heliobacillus mobilis** Beer-Romero and Gest 1998, 627 (Effective publication: Beer-Romero and Gest 1987, 113.)

List of species of the genus Heliobacillus

1. **Heliobacillus mobilis** Beer-Romero and Gest 1998, 627[VP] (Effective publication: Beer-Romero and Gest 1987, 113.)
   mo' bil.is. L. masc. adj. mobilis movement, named for its rapid motility.

   Cells rod-shaped 1 × 7–10 μm, highly motile in young cultures. Contain peritrichous flagella. Optimal pH 7. NaCl not required. Capable of assimilatory sulfate reduction. Isolated from dry soil from Thailand.

   The mol% G + C of the DNA is: 50.3 (T_m).
   Type strain: ATCC 43427, DSMZ 6151.
   GenBank accession number (16S rRNA): L36199, U14560.

Genus III. **Heliophilum** Ormerod, Kimble, Nesbakken, Torgersen, Woese and Madigan 1996b, 1189[VP] (Effective publication: Ormerod, Kimble, Nesbakken, Torgersen, Woese and Madigan 1996a, 233)

MICHAEL T. MADIGAN

He.li.o.phi' lum. Gr. n. helios sun; philos loving; M.L. neut. n. Heliophilum sun lover.

Cells straight rods with tapered ends. Form bundles of several cells that are motile as a unit. Entire bundles show scotophotophobic response. Heat-resistant endospores produced in newly isolated cultures. Contain bacteriochlorophyll g. Capable of assimilatory sulfate reduction; growth inhibited by sulfide above 0.1 mM. Biotin required. NaCl not required and growth inhibitory at concentrations 1%. Other properties are the same as those described for the family *Heliobacteriaceae* (see also Table 1).

Type species: **Heliophilum fasciatum** Ormerod, Kimble, Nesbakken, Torgersen, Woese and Madigan 1996b, 1189 (Effective publication: Ormerod, Kimble, Nesbakken, Torgersen, Woese and Madigan 1996a, 233.)

List of species of the genus Heliophilum

1. **Heliophilum fasciatum** Ormerod, Kimble, Nesbakken, Torgersen, Woese and Madigan 1996b, 1189[VP] (Effective publication: Ormerod, Kimble, Nesbakken, Torgersen, Woese and Madigan 1996a, 233.)

fas.ci.a' tum. L. n. *fasicle* bundle, named for the fact that cells form into bundles that move as a unit.

The characteristics are the same as those described for the genus. Isolated from rice soil from Tanzania, continent of Africa.

The mol% G + C of the DNA is: 51.8 (T_m).

Type strain: ATCC 51790.

GenBank accession number (16S rRNA): L36197, U14557.

Genus IV. **Heliorestis** Bryantseva, Gorlenko, Kompantseva, Achenbach and Madigan 2000, 949[VP] (Effective publication: Bryantseva, Gorlenko, Kompantseva, Achenbach and Madigan 1999a, 173)

MICHAEL T. MADIGAN

He.li.o.res' tis. Gr. n. *helios* sun; L. fem. n. *restis* a rope; *Heliorestis* sun rope.

Cells in coils or bent rods to short filaments. Alkaliphilic. NaCl not required and growth inhibitory above 1%. Sulfide tolerant, growing in media containing up to 10 mM sulfide. Optimal growth temperature 30°C with no growth above 40°C. Biotin required and yeast extract at 0.025–0.05% [w/v] is highly growth stimulatory. **Lowest G + C content of all known heliobacteria**. Other properties are the same as those described for the family *"Heliobacteriaceae"* (see also Table 1). The genus *Heliorestis* currently contains one species.*

Type species: **Heliorestis daurensis** Bryantseva, Gorlenko, Kompantseva, Achenbach and Madigan 2000, 949 (Effective publication: Bryantseva, Gorlenko, Kompantseva, Achenbach and Madigan 1999a, 173.).

List of species of the genus Heliorestis

1. **Heliorestis daurensis** Bryantseva, Gorlenko, Kompantseva, Achenbach and Madigan 2000, 949[VP] (Effective publication: Bryantseva, Gorlenko, Kompantseva, Achenbach and Madigan 1999a, 173.).

dau.ren' sis. L. adj. *daurensis* the name of the geographic region Dauria (Daur Steppe, Russia), from which the type strain was isolated.

Cells 0.8–1.2 µm wide forming a coil or bent filament with a length up to 20 µm. Motile by peritrichous flagella. Optimal growth pH, 9; pH range from 7.5 to 10.5. Other characteristics are as described for the genus. Isolated from Lake Barun Torey located on the Daur Steppe (South Chita region, Southeast Siberia, Russia).

The mol% G + C of the DNA is: 44.9 (T_m).

Type strain: BT-H1, ATCC 700798.

GenBank accession number (16S rRNA): AF079102.

---

*As this volume was going to press, a second species of *Heliorestis*, the rod-shaped *H. baculata*, was described, along with an emendation of the genus *Heliorestis* (Bryantseva, I.A., V.M. Gorlenko, E.I. Kompantseva, T.P. Tourova, B.B. Kuznetsov and G.A. Osipov. 2000. Alkaliphilic heliobacterium *Heliorestis baculata* sp. nov. and emended description of the genus *Heliorestis*. Arch. Microbiol. 174: 283–291).

The Anoxygenic Phototrophic Purple Bacteria

Johannes F. Imhoff

INTRODUCTION

The anoxygenic phototrophic purple bacteria are *Proteobacteria*, a new class proposed for all phototrophic purple bacteria and their purely chemotrophic relatives (Stackebrandt et al., 1988) and elevated to the status of a phylum in this edition of *Bergey's Manual of Systematic Bacteriology*, which had been called the "Purple Bacteria" by C.R. Woese (Woese et al., 1984a, b, 1985c; Woese, 1987). Phototrophic purple bacteria are representatives of the *"Alphaproteobacteria"* (*"Rhodospirillum* and relatives", aerobic bacteriochlorophyll-containing bacteria), the *"Betaproteobacteria"* (*"Rhodocyclus* and relatives"), and the *"Gammaproteobacteria"* (*Chromatiaceae* and *Ectothiorhodospiraceae*). The taxonomy of these bacteria, including determinative keys and description of the taxa, will be covered in detail in Volume 2 of this *Manual* and only a short overview is given here.

The anoxygenic phototrophic purple bacteria represent an assemblage of predominantly aquatic bacteria that are able to grow under anoxic conditions by photosynthesis without oxygen production. The most striking and common property of these bacteria is their ability to carry out light-dependent bacteriochlorophyll mediated energy transfer processes, a property shared with cyanobacteria, *Chlorobiaceae*, *"Chloroflexaceae"*, and heliobacteria. The various photosynthetic pigments give the cell cultures a distinct coloration from green, yellowish-green, brownish- green, brown, brownish-red, red, pink, purple, and purple-violet to even blue (carotenoid-less mutants of certain species containing bacteriochlorophyll *a*) depending on the pigment content. The major pigments are bacteriochlorophyll *a* or *b* and various carotenoids of the spirilloxanthin, rhodopinal, spheroidene (alternative spirilloxanthin), and okenone series (Schmidt, 1978). In all purple bacteria the photosynthetic pigments and the structures of the photosynthetic apparatus are located within a more or less extended system of internal membranes that is considered as originating from and being continuous with the cytoplasmic membrane. These intracellular membranes consist of small fingerlike intrusions, vesicles, tubules, or lamellae parallel to or at an angle to the cytoplasmic membrane and they carry the photosynthetic apparatus, the reaction centers, and light-harvesting pigment-protein complexes surrounding the reaction center in the plane of the membrane (see Drews and Imhoff, 1991).

Photosynthesis in anoxygenic phototrophic purple bacteria in general requires oxygen-deficient conditions, because synthesis of the photosynthetic pigments is repressed by oxygen. These bacteria are unable to use water as an electron donor, but need more reduced compounds. Most characteristically, sulfide and other reduced sulfur compounds, but also hydrogen and a number of small organic molecules, are used as photosynthetic electron donors. Recently even growth with reduced iron as electron donor has been demonstrated in some phototrophic purple bacteria (Widdel et al., 1993; Ehrenreich and Widdel, 1994). The group of the "aerobic bacteriochlorophyll-containing *Proteobacteria*" such as *Erythrobacter longus* and others (see below) is exceptional in that oxygen does not repress synthesis of photosynthetic pigments in these bacteria.

HISTORICAL ASPECTS OF THE TAXONOMY OF PURPLE BACTERIA

Molisch was the first to consider the pigmentation of purple sulfur and purple nonsulfur bacteria (and as a consequence their ability to perform photosynthesis) as a criterion to combine these groups taxonomically in a new order *Rhodobacteria* Molisch 1907. The two groups were recognized as two families. All purple bacteria that were able to form globules of elemental sulfur inside the cells were arranged in the *Thiorhodaceae*, and all purple bacteria which lack this ability were placed into the *Athiorhodaceae*. This order and these two families were later renamed to *Rhodospirillales* Pfennig and Trüper 1971b, 17[AL], *Chromatiaceae* Bavendamm 1924, 125[AL] [*Nom. cons.* Pfennig and Trüper 1971a, 16; emended description Imhoff 1984b, 339] and *Rhodospirillaceae* Pfennig and Trüper 1971b, 17[AL], respectively. Newly recognized purple sulfur bacteria that accumulate elemental sulfur outside their cells were classified in the new genus *Ectothiorhodospira* (Pelsh, 1937; Trüper, 1968). They were included into the *Chromatiaceae* (Trüper, 1968; Pfennig and Trüper, 1971b) and this classification was presented in the Eigth Edition of *Bergey's Manual of Determinative Bacteriology* (Pfennig and Trüper, 1974).

With the first available information on their genetic relatedness (Fowler et al., 1984; Stackebrandt et al., 1984; Woese et al., 1985c), the phylogenetic distinctness of *Ectothiorhodospira* from other *Chromatiaceae* became apparent. With this background and on the basis of fundamental phenotypic differences both groups were recognized as separate families (Imhoff, 1984b). The *Ectothiorhodospiraceae* Imhoff 1984b are purple sulfur bacteria that form sulfur globules outside the cells, while the *Chromatiaceae* exclusively comprise those phototrophic sulfur bacteria able to deposit elemental sulfur inside their cells (Imhoff, 1984b), which is in agreement with Molisch's (1907) definition of the *Thiorhodaceae*. Quite interestingly, Pelsh (1937) had already differentiated the *"Ectothiorhodaceae"* from the *"Endothiorhodaceae"* on a family level. However, because of their illegitimacy these family names had no standing in nomenclature.

Molisch's *Athiorhodaceae*, the *Rhodospirillaceae* Pfennig and

Trüper 1971b, had always been recognized as a heterogeneous group of bacteria, first on the basis of cell morphology (shape, internal membrane system) and metabolic versatility (Van Niel, 1944; Pfennig, 1967, 1977; Drews and Imhoff, 1991), and later, by genetic relatedness expressed as similarity of 16S rRNA oligonucleotide catalogs and 16S rDNA gene sequences. In the classification of these bacteria according to morphological and physiological properties three genera were recognized in the Eighth Edition of *Bergey's Manual of Determinative Bacteriology*: the spiral- shaped *Rhodospirillum*, the rod-shaped *Rhodopseudomonas*, and the stalk-forming *Rhodomicrobium* (Pfennig and Trüper, 1974). Later, the genus *Rhodocyclus* was described on the basis of a new species that characteristically possessed cells that formed a half-circle to a full-circle resembling *Microcyclus* species (Pfennig, 1978).

In particular for the purple nonsulfur bacteria, an increasing amount of information on sequences of macromolecules, suitable for phylogenetic delineations, became available with the establishment of sequencing techniques for proteins and nucleic acids. These included cytochrome *c* (Ambler et al., 1979; Dickerson, 1980), oligonucleotide catalogs obtained from 16S rRNA after digestion with specific endonucleases (Gibson et al., 1979; Fowler et al., 1984; Stackebrandt et al., 1984), and later almost complete sequences of the 16S rRNA gene (e.g., Kawasaki et al., 1993b; Hiraishi, 1994; Hiraishi and Ueda, 1994a, b; Imhoff et al., 1998a). A close relationship between purple nonsulfur bacteria and chemotrophic representatives of the genera *Paracoccus* and *Pseudomonas* had already been delineated from cytochrome *c* sequences and a first simple phylogenetic tree based on these data was presented (Almassy and Dickerson, 1978). The analysis of 16S rRNA oligonucleotide catalogs and 16S rDNA sequences revealed deep branches within the *Rhodospirillaceae* as well as the close relationship of these phototrophic bacteria to purely chemotrophic bacteria in numerous cases (e.g., Gibson et al., 1979; Woese et al., 1984a, b; Woese, 1987; Kawasaki et al., 1993b; Hiraishi and Ueda, 1994a, b).

Rhodospirillaceae species, despite their common physiological traits, were found in the *"Alphaproteobacteria"* as well as in the *"Betaproteobacteria"* (Woese et al., 1984a, b; Woese, 1987). With our current knowledge, the classification of the diverse assemblage of *"Alphaproteobacteria"* (*"Rhodospirillum rubrum"* and relatives") and *"Betaproteobacteria"* (*"Rhodocyclus purpureus"* and relatives") within one family, the *Rhodospirillaceae* Pfennig and Trüper 1971b, was certainly the most significant "phylogenetic misclassification" within the phototrophic purple bacteria. A first consequence of this recognition was the proposal to abandon the use of a true family name (*Rhodospirillaceae*) and to use the popular name Purple Nonsulfur Bacteria for these bacteria until a major decision in regard to the higher ranks of taxa within the *Proteobacteria* was made (Imhoff et al., 1984; Imhoff and Trüper, 1989).

A brief overview of the groups of phototrophic purple bacteria and their genera is given in Table 1 and a list of the currently recognized species together with previous species designations in Table 2.

THE *"GAMMAPROTEOBACTERIA"*

Ectothiorhodospiraceae *Ectothiorhodospiraceae* represent a group of haloalkaliphilic purple sulfur bacteria that form a separate line of phylogenetic descent related to the *Chromatiaceae*, as was first demonstrated by analysis of their rRNA oligonucleotide catalogs (Stackebrandt et al., 1984). On the basis of physiological and available molecular information the genus *Ectothiorhodospira* was

removed from the *Chromatiaceae* into a new family, the *Ectothiorhodospiraceae* (Imhoff, 1984b). In a phylogenetic tree based on 16S rRNA/DNA data both families form distinct but related groups within the *"Gammaproteobacteria"* (Fowler et al., 1984; Stackebrandt et al., 1984; Woese et al., 1985c; Imhoff and Süling, 1996). Only a single, purely chemotrophic bacterium which clusters within the closely related species of the *Ectothiorhodospiraceae*, namely *Arhodomonas aquaeolei* (Adkins et al., 1993), has been found so far. A new genus and species of the *Ectothiorhodospiraceae*, *Thiorhodospira sibirica* (Bryantseva et al., 1999b), that also deposits elemental sulfur outside the cells, has been described only recently. Part of the sulfur globules remain attached to the cells or are located in the periplasm, and they occur in such a way that microsocopically the impression is given that they are located inside the cells.

The *Ectothiorhodospiraceae* Imhoff 1984b include phototrophic bacteria (in addition to a branch of purely chemotrophic bacteria) that, during oxidation of sulfide, deposit elemental sulfur outside their cells. Furthermore, they are distinguished from the *Chromatiaceae* by lamellar intracellular membrane structures, by significant differences in polar lipid composition (Imhoff et al., 1982; Imhoff and Bias-Imhoff, 1995), and by the dependence on saline and alkaline growth conditions (Imhoff, 1989). *Halorhodospira halophila* is the most halophilic eubacterium known and even grows in saturated salt solutions. *Ectothiorhodospiraceae* are clearly separated from the *Chromatiaceae* by sequence dissimilarity and signature sequences of their 16S rDNA (Imhoff and Süling, 1996; Imhoff et al., 1998b). Recently, sequences of the 16S rDNA have been determined from all type strains of the recognized *Ectothiorhodospira* species and a number of additional strains (Imhoff and Süling, 1996). These data resolved the phylogenetic relationships of the whole family, confirmed the established species, and improved the classification of strains of uncertain affiliation. On the basis of sequence similarities and a number of characteristic signature sequences, two major phylogenetic groups could be recognized and classified as separate genera. The existence of two groups of species within *Ectothiorhodospiraceae* had been noted earlier on the basis of several phenotypic properties (Tindall, 1980; Imhoff, 1984b; Stackebrandt et al., 1984). The extremely halophilic species were removed from the genus *Ectothiorhodospira* and reassigned to the new genus *Halorhodospira*, resulting in the species *Halorhodospira halophila*, *Halorhodospira halochloris*, and *Halorhodospira abdelmalekii*. Among the slightly halophilic species, the classification of strains belonging to *Ectothiorhodospira mobilis* and *Ectothiorhodospira shaposhnikovii* was improved and the close relationship between *Ectothiorhodospira shaposhnikovii* and *Ectothiorhodospira vacuolata* was demonstrated. *Ectothiorhodospira marismortui* was also confirmed as a distinct species on a genetic basis. Several strains that had previously been tentatively identified as *Ectothiorhodospira mobilis* formed a separate cluster based on 16S rDNA sequence analysis and were recognized as two new species: *Ectothiorhodospira haloalkaliphila*, which includes the most alkaliphilic strains originating from strongly alkaline soda lakes and *Ectothiorhodospira marina*, describing isolates from the marine environment (Imhoff and Süling, 1996).

The taxonomy of bacteria belonging to the *Ectothiorhodospiraceae* is now in accord with their genetic relationship based upon 16S rDNA sequence comparison. In addition, species and genera are well distinguished by a number of phenotypic properties.

Chromatiaceae The *Chromatiaceae* Bavendamm 1924 [emended description Imhoff 1984b] comprise those photo-

TABLE 1. Major groups of phototrophic purple bacteria

Characteristic	Chromatiaceae	Ectothiorhodospiraceae	Purple Nonsulfur Bacteria	Purple Nonsulfur Bacteria	Aerobic Purple Bacteria[a]
Phylogenetic affiliation	"Gammaproteobacteria"	"Gammaproteobacteria"	"Betaproteobacteria"	"Alphaproteobacteria"	"Alphaproteobacteria"
Major Bchl	Bchl a or Bchl b	Bchl a or Bchl b	Bchl a	Bchl a or Bchl b	Bchl a
Type of cytochrome c	c_{551}	c_{551}	c_{551}	Lc_2/Sc_2	c_{551}[b]
Major quinones	MK-8/Q-8	MK-8/Q-8, MK-7/Q-7 or Q-8	Q-8, MK-8	Q-10, MK-10, RQ-10, Q-9, MK-9	Q-10[b]
Major fatty acids	$C_{16:0}$, $C_{16:1}$ and $C_{18:1}$ (32–45%)	$C_{18:1}$ (60–75%)	$C_{16:0}$ and $C_{16:1}$ (5–25% $C_{18:1}$)	$C_{18:1}$ (68–94%)	$C_{18:1}$
List of genera	Chromatium	Ectothiorhodospira	Rhodocyclus	Rhodospirillum	Acidiphilium
	Allochromatium	Halorhodospira	Rhodoferax	Blastochloris	Erythrobacter
	Halochromatium	Thiorhodospira	Rubrivivax	Phaeospirillum	Erythromicrobium
	Isochromatium			Rhodobacter	Erythromonas
	Lamprobacter			Rhodobium	Methylobacterium
	Lamprocystis			Rhodocista	"Photorhizobium"
	Marichromatium			Rhodomicrobium	Porphyrobacter
	Rhabdochromatium			Rhodopila	Roseobacter
	Thermochromatium			Rhodoplanes	Roseococcus
	Thiocapsa			Rhodopseudomonas	Sandaracinobacter
	Thiococcus			Rhodospira	Acidisphaera
	Thiocystis			Rhodothalassium	Craurococcus
	Thiodictyon			Rhodovibrio	Paracraurococcus
	Thiohalocapsa			Rhodovulum	Citromicrobium
	Thiolamprovum			Roseospira	Rubrimonas
	Thiopedia			Roseospirillum	Roseovarius
	Thiorhodococcus				Roseivivax
	Thiorhodovibrio				
	Thiospirillum				

[a]A betaproteobacterium that has physiological properties similar to the aerobic bacteriochlorophyll-containing bacteria has been described as *Roseateles depolymerans* (Suyama et al., 1999).

[b]These properties have been determined only for a limited number of species and therefore may not be representative for the whole group.

trophic purple sulfur bacteria that, under the proper growth conditions, deposit globules of elemental sulfur inside their cells (Imhoff, 1984b). This definition agrees with that of the *Thiorhodaceae* by Molisch (1907). The family represents a quite coherent group of species, based on physiological properties, on the similarity of oligonucleotide catalogs and sequences of 16S rRNA/DNA (Fowler et al., 1984; Guyoneaud et al., 1998; Imhoff et al., 1998b), and on chemotaxonomic markers such as fatty acid and quinone composition (Imhoff and Bias-Imhoff, 1995) and lipopolysaccharide structures (Meissner et al., 1988b; Weckesser et al., 1995).

At present, all but one of the recognized species (*Thiococcus pfennigii*) have a vesicular type of intracellular membranes. Many species are motile by means of flagella, some have gas vesicles, and others completely lack motility. All species are able to grow photoautotrophically under anaerobic conditions in the light using sulfide or elemental sulfur as an electron donor. Several species are also able to grow under photoheterotrophic conditions, some even as chemoautotrophs, and a few species can also grow chemoheterotrophically (Gorlenko, 1974; Kondratieva et al., 1976; Kämpf and Pfennig, 1980).

Until recently, classification of the *Chromatiaceae* was based upon traditional morphological and physiological properties and a basic definition of the genera was given according to cell form, motility by flagella and presence of gas vesicles (see Imhoff and Trüper, 1982; Pfennig and Trüper, 1989). More recently new genera such as *Thiorhodovibrio* (Overmann et al., 1992b), *Rhabdochromatium* (Dilling et al., 1995), and *Thiorhodococcus* (Guyoneaud et al., 1997) have been described without clear reference to the outlined classical scheme that still was the basis for the

treatment of the *Chromatiaceae* in the last edition of this *Manual* (Pfennig and Trüper, 1989) and in part used the available sequence information of rRNA genes as supporting evidence.

Until recently, sequences of 16S rDNA of *Chromatiaceae* have been analysed to a much lesser extent than those of purple nonsulfur bacteria. Initial surveys of these bacteria were made by comparison of oligonucleotide catalogs (Fowler et al., 1984). The first complete 16S rDNA sequences were obtained for *Chromatium vinosum* (DeWeerd et al., 1990) and *Chromatium tepidum* (Wahlund et al., 1991). With the description of the new species and genera *Rhabdochromatium marinum* (Dilling et al., 1995), *Chromatium glycolicum* (Caumette et al., 1997), and *Thiorhodococcus minus* [correct name: *Thiorhodococcus minor*] (Guyoneaud et al., 1997), more 16S rDNA sequences became available. Complete sequences are now available for most *Chromatiaceae* species (Guyoneaud et al., 1998; Imhoff et al., 1998b). Despite the fact that a much higher resolution is obtained and many more species have been included in 16S rDNA sequence analyses, the early results obtained by comparison of oligonucleotide catalogs (Fowler et al., 1984) by and large have been confirmed, at least in principal, by the more recent work. The phylogenetic analyses revealed the existence of major groups of species that did not correlate well with their classification. Therefore, a reclassification of these bacteria on the basis of their genetic relationship and supported by diagnostic phenotypic properties was proposed (Guyoneaud et al., 1998; Imhoff et al., 1998b).

Most significantly, the sequence comparison revealed a genetic divergence between *Chromatium* species originating from freshwater sources and those of truly marine and halophilic nature. Major phylogenetic branches of the *Chromatiaceae* contain

TABLE 2. Species of phototrophic purple bacteria and aerobic bacteriochlorophyll-containing bacteria[a]

Species name	Previous names	Reference
"Gammaproteobacteria": Chromatiaceae		
Allochromatium minutissimum	*Chromatium minutissimum*	Imhoff et al., 1998b
Allochromatium vinosum	*Chromatium vinosum*	Imhoff et al., 1998b
Allochromatium warmingii	*Chromatium warmingii*	Imhoff et al., 1998b
Chromatium okenii		
Chromatium weissei		
Halochromatium glycolicum	*Chromatium glycolicum*	Imhoff et al., 1998b
Halochromatium salexigens	*Chromatium salexigens*	Imhoff et al., 1998b
Isochromatium buderi	*Chromatium buderi*	Imhoff et al., 1998b
Lamprobacter modestohalophilus		
Lamprocystis purpurea	*Amoebobacter purpureus*	see this *Manual*, Volume 2
Lamprocystis roseopersicina		
Marichromatium gracile	*Chromatium gracile*	Imhoff et al., 1998b
"Marichromatium marinum"	*"Rhodobacter marinus"*	Imhoff et al., 1998b
Marichromatium purpuratum	*Chromatium purpuratum*	Imhoff et al., 1998b
Rhabdochromatium marinum	new species	Dilling et al., 1995
Thermochromatium tepidum	*Chromatium tepidum*	Imhoff et al., 1998b
Thiocapsa pendens	*Amoebobacter pendens*	Guyoneaud et al., 1998
Thiocapsa rosea	*Amoebobacter roseus*	Guyoneaud et al., 1998
Thiocapsa roseopersicina		
Thiococcus pfennigii	*Thiocapsa pfennigii*	Imhoff et al., 1998b
Thiocystis gelatinosa		
Thiocystis minor	*Chromatium minus*	Imhoff et al., 1998b
Thiocystis violacea		
Thiocystis violascens	*Chromatium violascens*	Imhoff et al., 1998b
Thiodictyon bacillosum		
Thiodictyon elegans		
Thiohalocapsa halophila	*Thiocapsa halophila*	Imhoff et al., 1998b
Thiolamprovum pedioforme	*Amoebobacter pedioforme*	Guyoneaud et al., 1998
Thiopedia rosea		
Thiorhodococcus minor	new species	Guyoneaud et al., 1997
Thiorhodovibrio winogradskyi	new species	Overmann et al., 1992b
Thiospirillum jenense		
"Gammaproteobacteria": Ectothiorhodospiraceae		
Ectothiorhodospira haloalkaliphila	new species	Imhoff and Süling, 1996
Ectothiorhodospira marina	new species	Imhoff and Süling, 1996
Ectothiorhodospira marismortui	new species	Oren et al., 1989
Ectothiorhodospira mobilis		
Ectothiorhodospira shaposhnikovii		
Ectothiorhodospira vacuolata		
Halorhodospira abdelmalekii	*Ectothiorhodospira abdelmalekii*	Imhoff and Süling, 1996
Halorhodospira halochloris	*Ectothiorhodospira halochloris*	Imhoff and Süling, 1996
Halorhodospira halophila	*Ectothiorhodospira halophila*	Imhoff and Süling, 1996
Thiorhodospira sibirica	new species	Bryantseva et al., 1999b
"Betaproteobacteria": Purple Nonsulfur Bacteria		
Rhodocyclus purpureus		
Rhodocyclus tenuis	(*Rhodospirillum tenue*)	
Rhodoferax fermentans	new species	Hiraishi et al., 1991
Rubrivivax gelatinosus	*Rhodocyclus gelatinosus*	
	(*Rhodopseudomonas gelatinosa*)	Willems et al., 1991b

(continued)

i) marine and halophilic species, ii) freshwater *Chromatium* species together with *Thiocystis* species, and iii) species of the genera *Thiocapsa* and *Amoebobacter* as reclassified by Guyoneaud et al. (1998), namely *Thiocapsa roseopersicina*, *Thiocapsa pendens* (formerly *Amoebobacter pendens*), *Thiocapsa rosea*, (formerly *Amoebobacter roseus*), *Amoebobacter purpureus* (this bacterium will be assigned to the genus *Lamprocystis* as *Lamprocystis purpurea*; see Volume 2 of the *Manual*), and *Thiolamprovum pedioformis* (formerly *Amoebobacter pedioformis*). Apparently, ecological aspects and adaptation of bacteria to specific factors of their habitat are of importance in a phylogenetically oriented taxonomy, because evolution takes place in particular natural environments. Separate phylogenetic lines can be expected to result from bacterial evolution in distinct habitats that, due to their peculiar properties, allow only the development of specifically adapted bacteria. Such properties may be salt concentration, temperature, and pH. In fact, the taxonomic relevance of the salt requirement has been recognized already in the Purple Nonsulfur Bacteria, *Ectothiorhodospiraceae* and *Chromatiaceae*. It proved to be of relevance in distinguishing phototrophic bacteria on the family level (*Ectothiorhodospiraceae* and *Chromatiaceae*; Imhoff, 1984b) and on the genus level (e.g., *Rhodovulum* and *Rhodobacter*, *Halorhodospira* and *Ectothiorhodospira*; Hiraishi and Ueda, 1994a; Imhoff and Süling, 1996). However, morphological and physiological properties used so far in the classification of these bacteria apparently have minor relevance in a genetically oriented classification system.

TABLE 2. *(continued)*

Species name	Previous names	Reference
"Alphaproteobacteria": Purple Nonsulfur Bacteria		
Blastochloris sulfoviridis	*Rhodopseudomonas sulfoviridis*	Hiraishi, 1997
Blastochloris viridis	*Rhodopseudomonas viridis*	Hiraishi, 1997
Phaeospirillum fulvum	*Rhodospirillum fulvum*	Imhoff et al., 1998a
Phaeospirillum molischianum	*Rhodospirillum molischianum*	Imhoff et al., 1998a
Rhodobacter azotoformans	new species	Hiraishi et al., 1996
Rhodobacter blasticus	*Rhodopseudomonas blastica*	Kawasaki et al., 1993a
Rhodobacter capsulatus	*Rhodobacter capsulatus*	Imhoff et al., 1984
Rhodobacter sphaeroides	*Rhodobacter sphaeroides*	Imhoff et al., 1984
Rhodobacter veldkampii		
Rhodobium marinum	*Rhodopseudomonas marina*	Hiraishi et al., 1995
Rhodobium orientis	new species	Hiraishi et al., 1995
Rhodocista centenaria	*Rhodospirillum centenum*	Kawasaki et al., 1992
Rhodomicrobium vannielii		
Rhodopila globiformis	*Rhodopseudomonas globiformis*	Imhoff et al., 1984
Rhodoplanes elegans	new species	Hiraishi and Ueda, 1994b
Rhodoplanes roseus	*Rhodopseudomonas rosea*	Hiraishi and Ueda, 1994b
Rhodopseudomonas acidophila		
Rhodopseudomonas cryptolactis		
Rhodopseudomonas julia		
Rhodopseudomonas palustris		
Rhodopseudomonas rutila		
Rhodospira trueperi	new species	Pfennig et al., 1997
Rhodospirillum photometricum		
Rhodospirillum rubrum		
Rhodothalassium salexigens	*Rhodospirillum salexigens*	Imhoff et al., 1998a
Rhodovibrio salinarum	*Rhodospirillum salinarum*	Imhoff et al., 1998a
Rhodovibrio sodomensis	*Rhodospirillum sodomense*	Imhoff et al., 1998a
Rhodovulum adriaticum	*Rhodobacter adriaticus*	
	(*Rhodopseudomonas adriatica*)	Hiraishi and Ueda, 1994a
Rhodovulum euryhalinum	*Rhodobacter euryhalinus*	Hiraishi and Ueda, 1994a
Rhodovulum strictum	new species	Hiraishi and Ueda, 1995
Rhodovulum sulfidophilum	*Rhodobacter sulfidophilus*	
	(*Rhodopseudomonas sulfidophila*)	Hiraishi and Ueda, 1994a
Roseospira mediosalina	*Rhodospirillum mediosalinum*	Imhoff et al., 1998a
Roseospirillum parvrum	new species	Glaeser and Overmann, 1999
"Alphaproteobacteria": Aerobic bacteriochlorophyll-containing bacteria		
(without methylotrophic species and "rhizobia")		
Acidiphilium cryptum		
Acidiphilium acidophilum	*Thiobacillus acidophilus*	Hiraishi et al., 1998
Acidiphilium angustum	new species	Wichlacz et al., 1986
Acidiphilium multivorum	new species	Wakao et al., 1994
Acidiphilium organovorum	new species	Lobos et al., 1986
Acidiphilium rubrum	new species	Wichlacz et al., 1986
Acidisphaera rubrifaciens	new species	Hiraishi et al., in press
Citromicrobium bathyomarinum	new species	Yurkov et al., 1999
Craurococcus roseus	new species	Saitoh et al., 1998
Erythrobacter litoralis	new species	Yurkov et al., 1994
Erythrobacter longus		
Porphyrobacter neustonensis	new species	Fuerst et al., 1993
Porphyrobacter tepidarius	new species	Hanada et al., 1997
Erythromonas ursincola	*Erythromicrobium ursincola*	Yurkov et al., 1997
Erythromicrobium ramosum	new species	Yurkov et al., 1994
Erythromicrobium hydrolyticum	new species	Yurkov and Gorlenko, 1992
Erythromicrobium ezovicum	new species	Yurkov and Gorlenko, 1992
Paracraurococcus ruber	new species	Saitoh et al., 1998
Roseivivax haldurans	new species	Suzuki et al., 1999a
Roseivivax halotolerans	new species	Suzuki et al., 1999a
Roseobacter litoralis	new species	Shiba, 1991
Roseobacter denitrificans	new species	Shiba, 1991
Roseococcus thiosulfatophilus	new species	Yurkov et al., 1994
Roseovarius tolerans	new species	Labrenz et al., 1999
Rubrimonas cliftonensis	new species	Suzuki et al., 1999b
Sandaracinobacter sibiricus	*Erythromicrobium sibiricum*	Yurkov et al., 1997

aNew taxa described since this chapter has been completed will be included in the corresponding chapters of Volume 2 of this *Manual.*

PURPLE NONSULFUR BACTERIA

The Purple Nonsulfur Bacteria [*Rhodospirillaceae*, Pfennig and Trüper 1971b] represent by far the most diverse group of the phototrophic purple bacteria (Imhoff and Trüper, 1989). The high diversity of these bacteria is reflected in greatly varying morphology, internal membrane structure, carotenoid composition, utilization of carbon sources and electron donors, among other features. Furthermore, this high diversity is now well documented by a number of chemotaxonomic observations, such as cytochrome c structures, lipid composition, quinone composition, lipopolysaccharide structure, and fatty acid composition (Ambler et al., 1979; Weckesser et al., 1979, 1995; Dickerson, 1980; Hiraishi et al., 1984; Imhoff, 1984a, 1991, 1995; Imhoff et al., 1984; Imhoff and Bias-Imhoff, 1995). It was the recognition of the close genetic relationship between phototrophic purple bacteria and chemotrophic bacteria on the basis of 16S rRNA oligonucleotide catalogs and 16S rDNA sequences, respectively, that led C.R. Woese to call them the "Purple Bacteria and their relatives" and to discuss the role of phototrophic purple nonsulfur bacteria as ancestors of numerous chemotrophic representatives of these *Proteobacteria* groups (Woese et al., 1984a, b, 1985c; Woese, 1987).

Most of the purple nonsulfur bacteria are representatives of the *"Alphaproteobacteria"*, but a distinct group of species belongs to the *"Betaproteobacteria"*.

THE *ALPHAPROTEOBACTERIA*

The *"Alphaproteobacteria"* contain species that have traditionally been assigned to the genera *Rhodospirillum*, *Rhodopseudomonas*, and *Rhodomicrobium* (Pfennig and Trüper, 1974). As a first consequence of early phylogenetic studies (similarity of oligonucleotide catalogs obtained from 16S rRNA analyses, Gibson et al., 1979) and chemotaxonomy, *Rhodopseudomonas globiformis* was transferred to a new genus as *Rhodopila globiformis*. Also, "nonbudding" species of *Rhodopseudomonas* with spheroidene as major carotenoid and vesicular type of intracellular membranes were removed and considered as species of the new genus *Rhodobacter* (Imhoff et al., 1984). Later, based on complete 16S rDNA sequence comparison, *Rhodobacter* species were divided into two genera of marine and freshwater species. The marine species were transferred to the new genus *Rhodovulum* (Hiraishi and Ueda, 1994a).

In accordance with their genetic distance to the type species of the genus *Rhodopseudomonas* (*Rhodopseudomonas palustris*), several additional species were transferred to new genera. *Rhodopseudomonas marina* was transferred to *Rhodobium marinum* and *Rhodobium orientis* described as a new species of this genus (Hiraishi et al., 1995). Together with the new species *Rhodoplanes elegans*, *Rhodopseudomonas rosea* was reclassified into the new genus *Rhodoplanes* as *Rhodoplanes roseus* (Hiraishi and Ueda, 1994b). The two species that contain bacteriochlorophyll *b*, *Rhodopseudomonas viridis* and *Rhodopseudomonas sulfoviridis*, were transferred to the new genus *Blastochloris* as *Blastochloris viridis* and *Blastochloris sulfoviridis* (Hiraishi, 1997).

Like the rod-shaped purple nonsulfur bacteria (*Rhodopseudomonas* Pfennig and Trüper 1974), spiral-shaped purple nonsulfur bacteria, known as *Rhodospirillum* species, are extremely heterogeneous phylogenetically and intermixed with chemoheterotrophic non-phototrophic bacteria. Until recently, the genus *Rhodospirillum* was comprised of eight species, with *Rhodospirillum rubrum* as the type species. In addition, *Rhodocista centenaria*, originally described as a species of the genus *Rhodospirillum* (*Rhodospirillum centenum*, Favinger et al., 1989), was subsequently recognized as a species of the new genus *Rhodocista*, primarily on the basis of significant differences in the rRNA gene sequences from *Rhodospirillum rubrum* (Kawasaki et al., 1992). Another new spiral-shaped species is *Rhodospira trueperi*, which was described recently as a new species and a new genus on the

basis of phenotypic and genotypic properties (Pfennig et al., 1997).

It soon became obvious from the work of Kawasaki et al. (1993b) that, based on 16S rDNA sequences, the recognized species of the genus *Rhodospirillum* of the *"Alphaproteobacteria"* are phylogenetically very distantly related to each other and do not warrant classification in one and the same genus. On the basis of the distinct phenotypic properties and 16S rDNA sequence similarities of all recognized spiral-shaped purple nonsulfur *"Alphaproteobacteria"*, a reclassification of these bacteria was proposed (Imhoff et al., 1998a). Phylogenetic relationships determined by 16S rDNA sequence analyses of these bacteria are in good agreement with differences in major quinone and fatty acid composition and also in their growth requirement for NaCl or sea salt. This is in agreement with separate phylogenetic lines containing freshwater and salt water strains. Therefore, these properties were considered of primary importance in defining and differentiating these genera. Four of these genera are defined as salt-dependent and three as freshwater bacteria. Only *Rhodospirillum rubrum* and *Rhodospirillum photometricum* were maintained as species of the genus *Rhodospirillum*. Other species were transferred to new genera:

1. *Rhodospirillum fulvum* and *Rhodospirillum molischianum* to the genus *Phaeospirillum* gen. nov. as the new combinations *Phaeospirillum fulvum* comb. nov. and *Phaeospirillum molischianum* comb. nov.
2. *Rhodospirillum salinarum* and *Rhodospirillum sodomense* to the new genus *Rhodovibrio* gen. nov. as *Rhodovibrio salinarum* comb. nov. and *Rhodovibrio sodomensis* comb. nov.
3. *Rhodospirillum salexigens* to the new genus *Rhodothalassium* gen. nov. as *Rhodothalassium salexigens* comb. nov.
4. *Rhodospirillum mediosalinum* to the new genus *Roseospira* gen. nov. as *Roseospira mediosalina* comb. nov.

The aerobic bacteriochlorophyll-containing bacteria A quite remarkable group of bacteria, containing bacteriochlorophyll and belonging to the *"Alphaproteobacteria"*, but unable to grow phototrophically under anaerobic conditions, is represented by a number of Gram-negative aerobic bacteria (Sato, 1978; Shiba et al., 1979; Nishimura et al., 1981; Trüper, 1989). These bacteria are physiologically and also genetically different from the anoxygenic phototrophic bacteria *sensu-stricto* discussed above. In contrast to all previously known phototrophic purple bacteria, synthesis of bacteriochlorophyll *a* and carotenoids is stimulated by oxygen (Harashima et al., 1980). The best studied of these bacteria are *Erythrobacter longus* (Shiba and Simidu, 1982) and *Roseobacter denitrificans* (Shiba, 1991), both of which can synthesize bacteriochlorophyll *a*, form intracellular membranes, and have reaction center complexes similar to those of other purple bacteria (Harashima et al., 1980; Shimada et al., 1985; Iba et al., 1988). The number of aerobic bacteria known to contain bacteriochlorophyll has steadily increased during the last few years. These include *Erythrobacter longus* and *Erythrobacter litoralis* (Shiba and Simidu, 1982; Yurkov et al., 1994), *Roseobacter litoralis* and *Roseobacter denitrificans* (Shiba, 1991), *Erythromicrobium ramosum*, *Erythromicrobium ezovicum*, and *Erythromicrobium hydrolyticum* (Yurkov and Gorlenko, 1992; Yurkov et al., 1994), *Porphyrobacter neustonensis* and *Porphyrobacter tepidarius* (Fuerst et al., 1993; Hanada et al., 1997), *Roseococcus thiosulfatophilus* (Yurkov et al., 1994), *Acidiphilium cryptum*, *Acidiphilium angustum*, *Acidiphilium organovorum*, and *Acidiphilium rubrum* (Kishimoto et al., 1995), *Acidiphilium acidophilum* (Hiraishi et al., 1998) and *Acidiphilium mul-*

tivorum (Wakao et al., 1994), *Sandaracinobacter sibiricus* (Yurkov et al., 1997), and *Erythromonas ursincola* (Yurkov et al., 1997). A complete list of species is given in Table 2. Properties of these taxa will be described in Volume 2 of this *Manual.*

THE *"BETAPROTEOBACTERIA"*

The *"Betaproteobacteria"* contain species that have been known as *Rhodopseudomonas* and *Rhodospirillum* species (Pfennig and Trü per, 1974) as well as *Rhodocyclus purpureus* (Pfennig, 1978). With growing knowledge based on chemotaxonomic data and genetic sequence informations (Ambler et al., 1979; Dickerson, 1980; Hiraishi et al., 1984; Imhoff, 1984a, 1991; Imhoff and Bias-Imhoff, 1995), the need to separate these bacteria from the purple nonsulfur *"Alphaproteobacteria"* was recognized. It became obvious that *Rhodocyclus purpureus* (Pfennig, 1978), together with *Rhodopseudomonas gelatinosa* and *Rhodospirillum tenue* (Pfennig, 1969), formed a group of species separate from all other purple nonsulfur bacteria on the basis of fatty acid and quinone composition as well as cytochrome c sequences. As a consequence, *Rhodospi-*

rillum tenue (Pfennig, 1969) was transferred to *Rhodocyclus tenuis* (Imhoff et al., 1984). *Rhodopseudomonas gelatinosa* was also transferred to this genus as *Rhodocyclus gelatinosus* (Imhoff et al., 1984) and later to *Rubrivivax gelatinosus* (Willems et al., 1991b), because of its significant phylogenetic distance to *Rhodocyclus purpureus.* 16S rDNA sequence analysis clearly indicates these bacteria as members of the *"Betaproteobacteria"* (Hiraishi, 1994; Maidak et al., 1994).

Additional new bacteria that are also members of the *"Betaproteobacteria"* have been isolated and described as a new species and genus *Rhodoferax fermentans* (Hiraishi and Kitamura, 1984; Hiraishi et al., 1991). Although physiological properties and structures of the photosynthetic apparatus are not suitable for distinguishing between the two groups of purple nonsulfur bacteria, the *"Betaproteobacteria"* can be easily differentiated from the *"Alphaproteobacteria"* on the basis of differences in cytochrome *c* size and sequence (Ambler et al., 1979; Dickerson, 1980) as well as fatty acid, quinone, and lipid composition (Table 1; Imhoff and Bias- Imhoff, 1995).

Bibliography

Abella, C.A., X.P. Cristina, A. Martinez, I. Pibernat and X. Vila. 1998. Two new motile phototrophic consortia: *"Chlorochromatium lunatum"* and *"Pelochromatium selenoides"*. Arch. Microbiol. *169*: 452–459.

Abella, C.A. and L.J. Garcia-Gil. 1992. Microbial ecology of planktonic filamentous phototrophic bacteria in holomictic freshwater lakes. Hydrobiologia *243–244*: 79–86.

Acca, M., M. Bocchetta, E. Ceccarelli, R. Creti, K.O. Stetter and P. Cammarano. 1994. Updating mass and composition of archaeal and bacterial ribosomes. Archaeal-like features of ribosomes from the deep-branching bacterium *Aquifex pyrophilus*. Syst. Appl. Microbiol. *16*: 629–637.

Achenbach-Richter, L., R. Gupta, K.O. Stetter and C.R. Woese. 1987. Were the original eubacteria thermophiles? Syst. Appl. Microbiol. *9*: 34–39.

Adams, D.G., D. Ashworth and B. Nelmes. 1999. Fibrillar array in the cell wall of a gliding filamentous cyanobacterium. J. Bacteriol. *181*: 884–892.

Adams, D.G. and N.G. Carr. 1981. The developmental biology of heterocyst and akinete formation in cyanobacteria. CRC Crit. Rev. Microbiol. *9*: 45–100.

Adams, M.W. and A. Kletzin. 1996. Oxidoreductase-type enzymes and redox proteins involved in fermentative metabolisms of hyperthermophilic archaea. *In* Adams (Editor), Advances in Protein Chemistry, Vol. 48: Enzymes and Proteins from Hyperthermophilic Microorganisms, Academic Press, San Diego. pp. 101–180.

Adkins, J.P., M.T. Madigan, L. Mandelco, C.R. Woese and R.S. Tanner. 1993. *Arhodomonas aquaeolei* gen. nov., sp. nov., an aerobic, halophilic bacterium isolated from a subterranean brine. Int. J. Syst. Bacteriol. *43*: 514–520.

Adnan, S., N. Li, H. Miura, Y. Hashimoto, H. Yamamoto and T. Ezaki. 1993. Covalently immobilized DNA plate for luminometric DNA–DNA hybridization to identify viridans streptococci in under 2 hours. FEMS Microbiol. Lett. *106*: 139–142.

Ado, Y., T. Kawamoto, I. Masunaga, K. Takayama, S. Takasawa and K. Kimura. 1982. Production of 1-malic acid with immobilized thermophilic bacterium *Thermus rubens* sp. nov. Enzyme Eng. *6*: 303–310.

Adolph, K.W. and R. Haselkorn. 1971. Isolation and characterization of a virus infecting the blue-green alga *Nostoc muscorum*. Virology *46*: 200–208.

Aeckersberg, F., F.A. Rainey and F. Widdel. 1998. Growth, natural relationships, cellular fatty acids and metabolic adaptation of sulfate-reducing bacteria that utilize long-chain alkanes under anoxic conditions. Arch. Microbiol. *170*: 361–369.

Agardh, C.A. 1824. Systema Algarum, Litteris Berlingianis, Lund, Sweden.

Ahlgren, G. 1985. Growth of *Oscillatoria agardhii* in chemostat culture: 3. Simultaneous limitation of nitrogen and phosphorus. Br. Phycol. J. *20*: 249–262.

Ahmann, D., A.L. Roberts, L.R. Krumholz and F.M. Morel. 1994. Microbe grows by reducing arsenic. Nature *371*: 750.

Ahring, B.K., F. Alatriste-Mondragon, P. Westermann and R.A. Mah. 1991a. Effects of cations on *Methanosarcina thermophila* TM-1 growing on moderate concentrations of acetate: production of single cells. Appl. Microbiol. Biotechnol. *35*: 686–689.

Ahring, B.K., P. Westermann and R.A. Mah. 1991b. Hydrogen inhibition of acetate metabolism and kinetics of hydrogen consumption by *Methanosarcina thermophila* TM-1. Arch. Microbiol. *157*: 38–42.

Ainsworth, G.C. and P.H.A. Sheath. 1962. Microbial Classification: Appendix I. Symp. Soc. Gen. Microbiol. *12*: 456–463.

Alam, M. and D. Oesterhelt. 1984. Morphology, function and isolation of halobacterial flagella. J. Mol. Biol. *176*: 459–476.

Albertano, P. and L. Kovacik. 1994. Is the genus *Leptolyngbya* (Cyanophyte) a homogeneous taxon? Arch. Hydrobiol. Suppl. *105*: 37–51.

Aldrich, H.C., D.B. Beimborn and P. Schönheit. 1987. Creation of artifactual internal membranes during fixation of *Methanobacterium thermoautotrophicum*. Can. J. Microbiol. *33*: 844–849.

Alfredsson, G.A., S. Baldursson and J.K. Kristjánsson. 1985. Nutritional diversity among *Thermus* spp. isolated from Icelandic hot springs. Syst. Appl. Microbiol. *6*: 308–311.

Alfredsson, G.A. and J.K. Kristjánsson. 1995. Ecology, distribution, and isolation of *Thermus*. *In* Sharp and Williams (Editors), *Thermus* Species, Plenum Press, New York. pp. 43–66.

Alfreider, A., J. Pernthaler, R.I. Amann, B. Sattler, F.O. Glöckner, A. Wille and R. Psenner. 1996. Community analysis of the bacterial assemblages in the winter cover and pelagic layers of a high Mountain Lake by *in situ* hybridization. Appl. Environ. Microbiol. *62*: 2138–2144.

Allen, M.B. 1959. Studies with *Cyanidium caldarium* an anomalously pigmented chlorophyte. Arch. Mikrobiol. *32*: 270–277.

Allison, F.E., S.R. Hoover and H.J. Morris. 1937. Physiological studies with the nitrogen-fixing alga *Nostoc muscorum*. Bot. Gaz. *98*: 433–463.

Allison, M.J., W.R. Mayberry, C.S. McSweeney and D.A. Stahl. 1992. *Synergistes jonesii*, gen. nov., sp. nov.: a rumen bacterium that degrades toxic pyridinediols. System. Appl. Microbiol. *15*: 522–529.

Allison, M.J., W.R. Mayberry, C.S. McSweeney and D.A. Stahl. 1993. *In* Validation of the publication of new names and new combinations previously effectively published outside the IJSB. List No. 45. Int. J. Syst. Bacteriol. *43*: 398.

Alm, E.W., D.B. Oerther, N. Larsen, D.A. Stahl and L. Raskin. 1996. The oligonucleotide probe database. Appl. Environ. Microbiol. *62*: 3557–3559.

Almassy, R.J. and R.E. Dickerson. 1978. *Pseudomonas* cytochrome c_{551} at 2.0 Å resolution: enlargement of the cytochrome *c* family. Proc. Natl. Acad. Sci. U.S.A. *75*: 2674–2678.

Altekar, W. and R. Rajagopalan. 1990. Ribulose bisphosphate carboxylase activity in halophilic Archaebacteria. Arch. Microbiol. *153*: 169–174.

Amann, R.I. 1995a. Fluorescently labeled, rRNA-targeted oligonucleotide probes in the study of microbial ecology. Mol. Ecol. *4*: 543–554.

Amann, R.I. 1995b. *In situ* identification of micro-organisms by whole cell hybridization with rRNA-targeted nucleic acid probes. *In* Akkermans, van Elsas and de Bruijn (Editors), Molecular Microbial Ecology Manual, Vol. 3.3.6, Kluwer Academic Publishers, Dordrecht. pp. 1–15.

Amann, R.I., B.J. Binder, R.J. Olson, S.W. Chisholm, R. Devereux and

D.A. Stahl. 1990a. Combination of 16S rRNA-targeted oligonucleotide probes with flow cytometry for analyzing mixed microbial populations. Appl. Environ. Microbiol. *56*: 1919–1925.

Amann, R.I., L. Krumholz and D.A. Stahl. 1990b. Fluorescent-oligonucleotide probing of whole cells for determinative, phylogenetic, and environmental studies in microbiology. J. Bacteriol. *172*: 762–770.

Amann, R.I. and M. Kühl. 1998. *In situ* methods for assessment of microorganisms and their activities. Curr. Opin. Microbiol. *1*: 352–358.

Amann, R.I., W. Ludwig and K.H. Schleifer. 1994. Identification of uncultured bacteria: a challenging task for molecular taxonomists. ASM News *60*: 360–365.

Amann, R.I., W. Ludwig and K.H. Schleifer. 1995. Phylogenetic identification and *in situ* detection of individual microbial cells without cultivation. Microbiol. Rev. *59*: 143–169.

Amann, R.I., W. Ludwig, R. Schulze, S. Spring, E. Moore and K.H. Schleifer. 1996a. rRNA-targeted oligonucleotide probes for the identification of genuine and former pseudomonads. Syst. Appl. Microbiol. *19*: 501–509.

Amann, R.I., J. Snaidr, M. Wagner, W. Ludwig and K.H. Schleifer. 1996b. *In situ* visualization of high genetic diversity in a natural microbial community. J. Bacteriol. *178*: 3496–3500.

Amann, R.I., N. Springer, W. Ludwig, H.-D. Görtz and K.H. Schleifer. 1991. Identification *in situ* and phylogeny of uncultured bacterial endosymbionts. Nature *351*: 161–164.

Amann, R.I., J. Stromley, R. Devereux, R. Key and D.A. Stahl. 1992a. Molecular and microscopic identification of sulfate-reducing bacteria in multispecies biofilms. Appl. Environ. Microbiol. *58*: 614–623.

Amann, R.I., B. Zarda, D.A. Stahl and K.H. Schleifer. 1992b. Identification of individual prokaryotic cells by using enzyme-labeled, rRNA- targeted oligonucleotide probes. Appl. Environ. Microbiol. *58*: 3007–3011.

Ambler, R.P., M. Daniel, J. Hermoso, T.E. Meyer, R.G. Bartsch and M.D. Kamen. 1979. Cytochrome c_2 sequence variation among the recognised species of purple nonsulphur photosynthetic bacteria. Nature *278*: 659–660.

Amesz, J. 1991. Green photosynthetic bacteria and heliobacteria. *In* Barton and Shively (Editors), Variations in Autotrophic Life, Academic Press, London. pp. 99–119.

Amesz, J. 1995. The antenna-reaction center of heliobacteria. *In* Blankenship, Madigan and Bauer (Editors), Anoxygenic Photosynthetic Bacteria, Kluwer Academic Publishers, Dordrecht. pp. 687–697.

Amy, P.S. and H.D. Hiatt. 1989. Survival and detection of bacteria in an aquatic environment. Appl. Environ. Microbiol. *55*: 788–793.

Anagnostidis, K. 1989. *Geitlerinema*, new genus new status of oscillatorialean cyanophytes. Plant Syst. Evol. *164*: 33–46.

Anagnostidis, K., A. Economou-Amilli and T. Tafas. 1988. *Aphanizomenon* sp. from Lake Trichonis, Hellas (Greece)—a taxonomic consideration in relation to morphological and ecological parameters. Arch. Hydrobiol. Suppl. *80*: 529–543.

Anagnostidis, K. and J. Komárek. 1985. Modern approach to the classification system of cyanophytes. 1. Introduction. Arch. Hydrobiol. Suppl. *71*: 291–302.

Anagnostidis, K. and J. Komárek. 1988. Modern approach to the classification system of cyanophytes. III. *Oscillatoriales*. Arch. Hydrobiol. Suppl. *80*: 327–472.

Anagnostidis, K. and J. Komárek. 1990. Modern approach to the classification system of cyanophytes: V. *Stigonematales*. Arch. Hydrobiol. Suppl. *86*: 1–74.

Anderson, A.W., H.C. Nordan, R.F. Cain, G. Parrish and D. Duggan. 1956. Studies on a radio-resistant micrococcus. I. Isolation, morphology, cultural characteristics, and resistance to gamma radiation. Food Technol. *10*: 575–578.

Anderson, K.L., T.A. Tayne and D.M. Ward. 1987. Formation and fate of fermentation products in hot spring cyanobacterial mats. Appl. Environ. Microbiol. *53*: 2343–2352.

Anderson, R. and K. Hansen. 1985. Structure of a novel phosphoglycolipid from *Deinococcus radiodurans*. J. Biol. Chem. *260*: 12219–12223.

Andrä, S., G. Frey, M. Nitsch, W. Baumeister and K.O. Stetter. 1996.

Purification and structural characterization of the thermosome from the hyperthermophilic archaeum *Methanopyrus kandleri*. FEBS Lett. *379*: 127–131.

Andrews, K.T. and B.K.C. Patel. 1996. *Fervidobacterium gondwanense* sp. nov., a new thermophilic anaerobic bacterium isolated from nonvolcanically heated geothermal waters of the Great Artesian Basin of Australia. Int. J. Syst. Bacteriol. *46*: 265–269.

Angelidaki, I., S.P. Petersen and B.K. Ahring. 1990. Effects of lipids on thermophilic anaerobic digestion and reduction of lipid inhibition upon addition of bentonite. Appl. Microbiol. Biotechnol. *33*: 469–472.

Angert, E.R., K.D. Clements and N.R. Pace. 1993. The largest bacterium. Nature *362*: 239–241.

Antoine, E., V. Cilia, J.R. Meunier, J. Guezennec, F. Lesongeur and G. Barbier. 1997. *Thermosipho melanesiensis* sp. nov., a new thermophilic anaerobic bacterium belonging to the order *Thermotogales*, isolated from deep-sea hydrothermal vents in the Southwestern Pacific Ocean. Int. J. Syst. Bacteriol. *47*: 1118–1123.

Anton, J., R. Amils, C.L. Smith and P. López-García. 1995. Comparative restriction maps of the archaeal megaplasmid pHM300 in different *Haloferax mediterranei* strains. Syst. Appl. Microbiol. *18*: 439–447.

Apolinario, E.A. and K.R. Sowers. 1996. Plate colonization of *Methanococcus maripaludis* and *Methanosarcina thermophila* in a modified canning jar. FEMS Microbiol. Lett. *145*: 131–137.

Arahal, D.R., F.E. Dewhirst, B.J. Paster, B.E. Volcani and A. Ventosa. 1996. Phylogenetic analyses of some extremely halophilic archaea isolated from Dead Sea water, determined on the basis of their 16S rRNA sequences. Appl. Environ. Microbiol. *62*: 3779–3786.

Aranki, A. and R. Freter. 1972. Use of anaerobic glove boxes for the cultivation of strictly anaerobic bacteria. Am. J. Clin. Nutr. *25*: 1329–1334.

Archer, D.B. and N.R. King. 1983. A novel ultrastructural feature of a gas-vacuolated *Methanosarcina*. FEMS Microbiol. Lett. *16*: 217–223.

Archer, D.B. and N.R. King. 1984. Isolation of gas vesicles from *Methanosarcina barkeri*. J. Gen. Microbiol. *130*: 167–172.

Armstrong, R.E., P.K. Hayes and A.E. Walsby. 1983. Gas vacuole formation in hormogonia of *Nostoc muscorum*. J. Gen. Microbiol. *129*: 263–270.

Asakawa, S., H. Morii, M. Akagawa Matsushita, Y. Koga and K. Hayano. 1993. Characterization of *Methanobrevibacter arboriphilicus* SA isolated from a paddy field soil and DNA–DNA hybridization among *M. arboriphilicus* strains. Int. J. Syst. Bacteriol. *43*: 683–686.

Ash, C., J.A.E. Farrow, S. Wallbanks and M.D. Collins. 1991. Phylogenetic heterogeneity of the genus *Bacillus* revealed by comparative-analysis of small subunit ribosomal RNA sequences. Lett. Appl. Microbiol. *13*: 202–206.

Assmus, B., P. Hutzler, G. Kirchhof, R.I. Amann, J.R. Lawrence and A. Hartmann. 1995. *In situ* localization of *Azospirillum brasilense* in the rhizosphere of wheat with fluorescently labeled, rRNA-targeted oligonucleotide probes and scanning confocal laser microscopy. Appl. Environ. Microbiol. *61*: 1013–1019.

Atlas, R.M. and R. Bartha. 1993. Microbial Ecology — Fundamentals and Applications, Benjamin-Cummings Publishing Co., Redwood City.

Baas-Becking, A. 1931. Historical notes on salt and salt manufacture. Sci. Monthly *32*: 354–360.

Bachleitner, M., W. Ludwig, K.O. Stetter and K.H. Schleifer. 1989. Nucleotide sequence of the gene coding for the elongation factor Tu from the extremely thermophilic eubacterium *Thermotoga maritima*. FEMS Microbiol. Lett. *48*: 115–120.

Bak, F. and N. Pfennig. 1987. Chemolithotrophic growth of *Desulfovibrio sulfodismutans*, new species by disproportionation of inorganic sulfur compounds. Arch. Microbiol. *147*: 184–189.

Bakeeva, L.E., K.M. Chumakov, A.L. Drachev, A.L. Metlina and V.P. Skulachev. 1986. The sodium cycle: III. *Vibrio alginolyticus* resembles *Vibrio cholerae* and some other vibrios by flagellar motor and ribosomal 5S-RNA structures. Biochim. Biophys. Acta *850*: 466–472.

Baker, K.K. 1981. Ecology and taxonomy of five natural populations of the genus *Aphanizomenon* (Cyanophyceae). Arch. Hydrobiol. *92*: 222–251.

Balashova, V.V., I.Y. Vedenina, G.E. Markosyan and G.A. Zavarzin. 1974. The auxotrophic growth of *Leptospirillum ferrooxidans*. Mikrobiologiya *43*: 491–494.

Balch, W.E., G.E. Fox, L.J. Magrum, C.R. Woese and R.S. Wolfe. 1979. Methanogens: reevaluation of a unique biological group. Microbiol. Rev. *43*: 260–296.

Balch, W.E. and R.S. Wolfe. 1976. New approach to the cultivation of methanogenic bacteria: 2-mercaptoethanesulfonic acid (HS-CoM)-dependent growth of *Methanobacterium ruminantium* in a pressurized atmosphere. Appl. Environ. Microbiol. *32*: 781–791.

Balch, W.E. and R.S. Wolfe. 1979a. Specificity and biological distribution of coenzyme M (2-mercaptoethanesulfonic acid). J. Bacteriol. *137*: 256–263.

Balch, W.E. and R.S. Wolfe. 1979b. Transport of coenzyme M (2-mercaptoethanesulfonic acid) in *Methanobacterium ruminantium*. J. Bacteriol. *137*: 264–273.

Balch, W.E. and R.S. Wolfe. 1981. *In* Validation of the publication of new names and new combinations previously effectively published outside the IJSB. List No. 6. Int. J. Syst. Bacteriol. *31*: 215–218.

Baldauf, S.L., J.D. Palmer and W.F. Doolittle. 1996. The root of the universal tree and the origin of eukaryotes based on elongation factor phylogeny. Proc. Natl. Acad. Sci. U.S.A. *93*: 7749–7754.

Balkwill, D.L., S.A. Nierzwicki-Bauer and S.E. Stevens. 1984. Modes of cell division and branch formation in the morphogenesis of the cyanobacterium, *Mastigocladus laminosus*. J. Gen. Microbiol. *130*: 2079–2088.

Bank, S., B. Yan and T.L. Miller. 1996. Solid ^{13}C CPMAS NMR spectroscopy studies of biosynthesis in whole cells of *Methanosphaera stadtmanae*. Solid State Nucl. Magn. Reson 7: 253–261.

Barbier, G., A. Godfroy, J.R. Meunier, J. Quérellou, M.A. Cambon, F. Lesongeur, P.A.D. Grimont and G. Raguénès. 1999. *Pyrococcus glycovorans* sp. nov., a hyperthermophilic archaeon isolated from the East Pacific Rise. Int. J. Syst. Bacteriol. *49*: 1829–1837.

Barkay, T., D.L. Fouts and B.H. Olson. 1985. Preparation of a DNA gene probe for detection of mercury resistance genes in Gram-negative bacterial communities. Appl. Environ. Microbiol. *49*: 686–692.

Barker, H.A. 1936. Studies upon the methane-producing bacteria. Arch. Mikrobiol. 7: 420–438.

Barker, H.A. 1956. Bacterial Fermentations, John Wiley and Sons, New York.

Barnes, L.D. and E. Stellwagen. 1973. Enolase from the thermophile *Thermus* X-1. Biochemistry *12*: 1559–1565.

Barnes, S.M., C.R. Delwiche, J.D. Palmer and N.R. Pace. 1996. Perspectives on archaeal diversity, thermophily and monophyly from environmental rRNA sequences. Proc. Natl. Acad. Sci. U.S.A. *93*: 9188–9193.

Barns, S.M., C.F. Delwiche, J.D. Palmer and N.R. Pace. 1996. Perspectives on archeal diversity, thermophily and monophyly from environmental rRNA sequences. Proc. Natl. Acad. Sci. U.S.A. *93*: 9188–9193.

Barns, S.M., R.E. Fundyga, M.W. Jeffries and N.R. Pace. 1994. Remarkable archaeal diversity detected in a Yellowstone National Park hot spring environment. Proc. Natl. Acad. Sci. U.S.A. *91*: 1609–1613.

Barrett, D.M., D.O. Faigel, D.C. Metz, K. Montone and E.E. Furth. 1997. *In situ* hybridization for *Helicobacter pylori* in gastric mucosal biopsy specimens: quantitative evaluation of test performance in comparison with the CLO test and thiazine stain. J. Clin. Lab. Anal. *11*: 374–379.

Bartosch, S., I. Wolgast, E. Spieck and E. Bock. 1999. Identification of nitrite-oxidizing bacteria with monoclonal antibodies recognizing the nitrite oxidoreductase. Appl. Environ. Microbiol. *65*: 4126–4133.

Bateson, M.M., K.J. Thibault and D.M. Ward. 1990. Comparative analysis of 16S ribosomal RNA sequences of *Thermus* species. Syst. Appl. Microbiol. *13*: 8–13.

Bateson, M.M. and D.M. Ward. 1988. Photoexcretion and fate of glycolate in a hot spring cyanobacterial mat. Appl. Environ. Microbiol. *54*: 1738–1743.

Battaglia, F., D. Morin, J.-L. Garcia and P. Ollivier. 1994. Isolation and study of two strains of *Leptospirillum*-like bacteria from a natural mixed

population cultured on a cobaltiferous pyrite substrate. Antonie Leeuwenhoek *66*: 295–302.

Baudet, C., G.D. Sprott and G.B. Patel. 1988. Adsorption and uptake of nickel in *Methanothrix concilii*. Arch. Microbiol. *150*: 338–342.

Bauer, C.C., W.J. Buikema, K. Black and R. Haselkorn. 1995. A short-filament mutant of *Anabaena* sp. strain PCC 7120 that fragments in nitrogen-deficient medium. J. Bacteriol. *177*: 1520–1526.

Bauld, J. 1981. Occurrence of benthic microbial mats in saline lakes. Hydrobiologia *81*: 87–111.

Baumeister, W., O. Kubler and H.P. Zingsheim. 1981. The structure of the cell envelope of *Micrococcus radiodurans* as revealed by metal shadowing and decoration. J. Ultrastruct. Res. *75*: 60–71.

Baumeister, W. and G. Lembcke. 1992. Structural features of archaebacterial cell envelopes. J. Bioenerg. Biomembr. *24*: 567–575.

Bavendamm, W. 1924. Die farblosen und roten Schwefelbakterien des Süss- und Salzwassers, Fischer Verlag, Jena.

Beck, H., G.D. Hegeman and D. White. 1990. Fatty acid and lipopolysaccharide analyses of three *Heliobacterium* spp. FEMS Microbiol. Lett. *69*: 229–232.

Beck, P. and R. Huber. 1997. Detection of cell viability in cultures of hyperthermophiles. FEMS Microbiol. Lett. *147*: 11–14.

Becker, R.J. and M.J. Starzyk. 1984. Morphology and rotund body formation in *Thermus aquaticus*. Microbios *41*: 115–129.

Beeder, J., T. Torsvik and T. Lien. 1995. *Thermodesulforhabdus norvegicus* gen. nov., sp. nov., a novel thermophilic sulfate-reducing bacterium from oil field water. Arch. Microbiol. *164*: 331–336.

Beer-Romero, P., J.L. Favinger and H. Gest. 1988. Distinctive features of bacilliform photosynthetic heliobacteria. FEMS Microbiol. Lett. *49*: 451–454.

Beer-Romero, P. and H. Gest. 1987. *Heliobacillus mobilis*, a peritrichously flagellated anoxyphototroph containing bacteriochlorophyll *g*. FEMS Microbiol. Lett. *41*: 109–114.

Beer-Romero, P. and H. Gest. 1998. *In* Validation of the publication of new names and new combinations previously effectively published outside the IJSB. List No. 65. Int. J. Syst. Bacteriol. *48*: 627.

Beffa, T., M. Blanc, P.F. Lyon, G. Vogt, M. Marchiani, J.L. Fischer and M. Aragno. 1996. Isolation of *Thermus* strains from hot composts (60 to 80°C). Appl. Environ. Microbiol. *62*: 1723–1727.

Beh, M., G. Strauss, R. Huber, K.O. Stetter and G. Fuchs. 1993. Enzymes of the reductive citric acid cycle in the autotrophic eubacterium *Aquifex pyrophilus* and in the archaebacterium *Thermoproteus neutrophilus*. Arch. Microbiol. *160*: 306–311.

Behammer, W., Z. Shao, W. Mages, R. Rachel, K.O. Stetter and R. Schmitt. 1995. Flagellar structure and hyperthermophily: analysis of a single flagellin gene and its product in *Aquifex pyrophilus*. J. Bacteriol. *177*: 6630–6637.

Beimfohr, C., W. Ludwig and K.H. Schleifer. 1997. Rapid genotypic differentiation of *Lactococcus lactis* subspecies and biovar. Syst. Appl. Microbiol. *20*: 216–221.

Belay, N., R. Johnson, B.S. Rajagopal, E. Conway de Macario and L. Daniels. 1988a. Methanogenic bacteria from human dental plaque. Appl. Environ. Microbiol. *54*: 600–603.

Belay, N., K.-Y. Jung, B.S. Rajagopal, J.D. Kremer and L. Daniels. 1990. Nitrate as the sole nitrogen source for *Methanococcus thermolithotrophicus* and its effect on the growth of several methanogenic bacteria. Curr. Microbiol. *21*: 193–198.

Belay, N., R. Sparling, B.B. Choi, M. Roberts, J.E. Roberts and L. Daniels. 1988b. Physiological and ^{15}N-NMR analysis of molecular nitrogen fixation by *Methanococcus thermolithotrophicus*, *Methanobacterium bryantii* and *Methanospirillum hungatei*. Biochim. Biophys. Acta *971*: 233–245.

Belay, N., R. Sparling and L. Daniels. 1984. Dinitrogen fixation by a thermophilic methanogenic bacterium. Nature *312*: 286–288.

Belkin, S. and E. Padan. 1978. Hydrogen metabolism in the facultative anoxygenic cyanobacteria (blue-green algae) *Oscillatoria limnetica* and *Aphanothece halophytica*. Arch. Microbiol. *116*: 109–111.

Belkin, S., C.O. Wirsen and H.W. Jannasch. 1986. A new sulfur-reducing, extremely thermophilic eubacterium from a submarine thermal vent. Appl. Environ. Microbiol. *51*: 1180–1185.

Belly, R.T., B.B. Bohlool and T.D. Brock. 1973. The genus *Thermoplasma.* Ann. N. Y. Acad. Sci. *225:* 94–107.

Belly, R.T. and T.D. Brock. 1972. Cellular stability of a thermophilic, acidophilic mycoplasma. J. Gen. Microbiol. *73:* 465–469.

Belyaev, S.S., A.Y. Obraztsova, K.S. Laurinavichius and L.V. Bezrukova. 1987. Characteristics of rod-shaped methane-producing bacteria isolated from an oil pool and the description of *Methanobacterium ivanovii,* sp. nov. Mikrobiologiya *55:* 1014–1020.

Belyaev, S.S., R. Wolkin, W.R. Kenealy, M.J. DeNiro, S. Epstein and J.G. Zeikus. 1983. Methanogenic bacteria from the Bondyuzhskoe oil field (USSR): general characterization and analysis of stable-carbon isotopic fractionation. Appl. Environ. Microbiol. *45:* 691–697.

Benlloch, S., A.J. Martínez-Murcia and F. Rodríguez-Valera. 1995. Sequencing of bacterial and archaeal 16S rRNA genes directly amplified from a hypersaline environment. Syst. Appl. Microbiol. *18:* 574–581.

Benstead, J., D.B. Archer and D. Lloyd. 1991. Formate utilization by members of the genus *Methanobacterium.* Arch. Microbiol. *156:* 34–37.

Berenguer, J., M.L. Faraldo and M.A. de Pedro. 1988. Ca^{2+}-stabilized oligomeric protein complexes are major components of the cell envelope of *"Thermus thermophilus"* HB8. J. Bacteriol. *170:* 2441–2447.

Bergey, D.H., R.S. Breed, B.W. Hammer, F.M. Huntoon, E.G.D. Murray and F.C. Harrison (Editors). 1934. Bergey's Manual of Determinative Bacteriology, 4th Ed., The Williams & Wilkins Co., Baltimore.

Bergey, D.H., R.S. Breed, E.G.D. Murray and A.P. Hitchens (Editors). 1939. Bergey's Manual of Determinative Bacteriology, 5th Ed., The Williams & Wilkins Co., Baltimore.

Bergey, D.H., F.C. Harrison, R.S. Breed, B.W. Hammer and F.M. Huntoon (Editors). 1923. Bergey's Manual of Determinative Bacteriology, 1st Ed., The Williams & Wilkins Co., Baltimore. pp. 1–442.

Bergey, D.H., F.C. Harrison, R.S. Breed, B.W. Hammer and F.M. Huntoon (Editors). 1930. Bergey's Manual of Determinative Bacteriology, 3rd Ed., The Williams & Wilkins Co., Baltimore.

Bergman, B., J.R. Gallon, A.N. Rai and L.J. Stal. 1997. N_2 fixation by non-heterocystous cyanobacteria. FEMS Microbiol. Lett. *19:* 139–185.

Bergman, B., A.N. Rai, C. Johansson and E. Söderbäck. 1992. Cyano-bacterial-Plant Symbioses. Symbiosis *14:* 61–81.

Bergthorsson, U. and H. Ochman. 1995. Heterogeneity of genome sizes among natural isolates of *Escherichia coli.* J. Bacteriol. *177:* 5784–5789.

Bernhardt, G., R. Jaenicke, H.-D. Lüdemann, H. König and K.O. Stetter. 1988. High pressure enhances the growth rate of the thermophilic archaebacterium *Methanococcus thermolithotrophicus* without extending its temperature range. Appl. Environ. Microbiol. *54:* 1258–1261.

Bertram, P.A., R.A. Schmitz, D. Linder and R.K. Thauer. 1994. Tungstate can substitute for molybdate in sustaining growth of *Methanobacterium thermoautotrophicum.* Identification and characterization of a tungsten isoenzyme of formylmethanofuran dehydrogenase. Arch. Microbiol. *161:* 220–228.

Betti, J.A., R.E. Blankenship, L.V. Natarajan, L.C. Dickinson and R.C. Fuller. 1982. Antenna organization and evidence for the function of a new antenna pigment species in the green photosynthetic bacterium, *Chloroflexus aurantiacus.* Biochim. Biophys. Acta *680:* 194–201.

Betzl, D., W. Ludwig and K.H. Schleifer. 1990. Identification of lactococci and enterococci by colony hybridization with 23S rRNA-targeted oligonucleotide probes. Appl. Environ. Microbiol. *56:* 2927–2929.

Beveridge, T.J., B.J. Harris, G.B. Patel and G.D. Sprott. 1986a. Cell division and filament splitting in *Methanothrix concilii.* Can. J. Microbiol. *32:* 779–786.

Beveridge, T.J., G.B. Patel, B.J. Harris and G.D. Sprott. 1986b. The ultrastructure of *Methanothrix concilii,* a mesophilic aceticlastic methanogen. Can. J. Microbiol. *32:* 703–710.

Beveridge, T.J., G. Southam, M.H. Jericho and B.L. Blackford. 1990. High-resolution topography of the S-layer sheath of the archaebacterium *Methanospirillum hungatei* provided by scanning tunneling microscopy. J. Bacteriol. *172:* 6589–6595.

Beveridge, T.J., G.D. Sprott and P. Whippey. 1991. Ultrastructure, inferred porosity, and Gram-staining character of *Methanospirillum hungatei* fil-

ament termini describe a unique cell permeability for this archaeo-bacterium. J. Bacteriol. *173:* 130–140.

Bezrukova, L.V., A.Y. Obraztsova and T.N. Zhilina. 1989. Immunological studies on a group of methane-producing bacterial strains. Mikrobiologiya *58:* 92–98.

Bhatnagar, L., M.K. Jain, J.P. Aubert and J.G. Zeikus. 1984. Comparison of assimilatory organic nitrogen, sulfur and carbon sources for growth of *Methanobacterium* species. Appl. Environ. Microbiol. *48:* 785–790.

Bhatnagar, L., J.G. Zeikus and J.P. Aubert. 1986. Purification and characterization of glutamine synthetase from the archaebacterium *Methanobacterium ivanovi.* J. Bacteriol. *165:* 638–643.

Bhattacharya, D. and H.A. Schmidt. 1997. Division *Glaucocystophyta.* Plant Syst. Evol. Suppl. *11:* 139–148.

Biavati, B., M. Vasta and J.G. Ferry. 1988. Isolation and characterization of *Methanosphaera cuniculi* sp. nov. Appl. Environ. Microbiol. *54:* 768–771.

Biavati, B., M. Vasta and J.G. Ferry. 1990. *In* Validation of the publication of new names and new combinations previously effectively published outside the IJSB. List No. 35. Int. J. Syst. Bacteriol. *40:* 470–471.

Bibel, M., C. Brettl, U. Gosslar, G. Kriegshäuser and W. Liebl. 1998. Isolation and analysis of genes for amylolytic enzymes of the hyper-thermophilic bacterium *Thermotoga maritima.* FEMS Microbiol. Lett. *158:* 9–15.

Bidnenko, E., C. Mercier, J. Tremblay, P. Tailliez and S. Kulakauskas. 1998. Estimation of the state of the bacterial cell wall by fluorescent *in situ* hybridization. Appl. Environ. Microbiol. *64:* 3059–3062.

Biebl, H. and N. Pfennig. 1978. Growth yields of green sulfur bacteria in mixed cultures with sulfur and sulfate reducing bacteria. Arch. Microbiol. *117:* 9–16.

Birk, I.M., R. Dierstein, I. Kaiser, U. Matern, W.A. König, R. Krebber and J. Weckesser. 1989. Nontoxic and toxic oligopeptides with D-amino acids and unusual residues in *Microcystis aeruginosa* PCC 7806. Arch. Microbiol. *151:* 411–415.

Bizio, B. 1823. Lettera di Bartolomeo Bizio al chiarissimo canonico Angelo Bellani sopra il fenomeno della polenta porporina. Biblioteca Italiana o sia Giornale di Letteratura Scienze e Arti (Anno VIII) *30:* 275–295.

Bjourson, A.J., C.E. Stone and J.E. Cooper. 1992. Combined subtraction hybridization and polymerase chain reaction amplification procedure for isolation of strain-specific *Rhizobium* DNA sequences. Appl. Environ. Microbiol. *58:* 2296–2301.

Black, F.T., E.A. Freundt, O. Vinther and C. Christiansen. 1979. Flagellation and swimming motility of *Thermoplasma acidophilum.* J. Bacteriol. *137:* 456–460.

Blackburn, S.I., M.A. McCausland, C.J.S. Bolch, S.J. Newman and G.J. Jones. 1996. Effect of salinity on growth and toxin production in cultures of the bloom-forming cyanobacterium *Nodularia spumigena* from Australian waters. Phycologia *35:* 511–522.

Blackford, B.L., W. Xu, M.H. Jericho, P.J. Mulhern, M. Firtel and T.J. Beveridge. 1994. Direct observation by scanning tunneling microscopy of the two-dimensional lattice structure of the S-layer sheath of the archaebacterium *Methanospirillum hungatei* GP1. Scanning Microsc. *8:* 507–512 .

Blake, R.C., II and E.A. Shute. 1997. Purification and characterization of a novel cytochrome from *Leptospirillum ferrooxidans.* International Biohydrometallurgy Symposium IBS97/Biomine 97, Glenside, SA, Australia. American Mineral Foundation.

Blamey, J.M., S. Mukund and M.W. Adams. 1994. Properties of a thermostable 4Fe-ferredoxin from the hyperthermophilic bacterium *Thermotoga maritima.* FEMS Microbiol. Lett. *121:* 165–169.

Blankenship, R.E., J.M. Olson and M. Miller. 1995. Antenna complexes from green photosynthetic bacteria. *In* Blankenship, Madigan and Bauer (Editors), Anoxygenic Photosynthetic Bacteria, Kluwer Academic Publishers, Dordrecht. pp. 399–435.

Blattner, F.R., G. Plunkett, III, C.A. Bloch, N.T. Perna, V. Burland, M. Riley, J. Collado Vides, J.D. Glasner, C.K. Rode, G.F. Mayhew, J. Gregor, N.W. Davis, H.A. Kirkpatrick, M.A. Goeden, D.J. Rose, B. Mau and Y. Shao.

1997. The complete genome sequence of *Escherichia coli* K-12. Science *277*: 1453–1462.

Bleicher, K. and J. Winter. 1991. Purification and properties of F$_{420}$- and NADP$^+$-dependent alcohol dehydrogenases of *Methanogenium liminatans* and *Methanobacterium palustre*, specific for secondary alcohols. Eur. J. Biochem. *200*: 43–51.

Bleicher, K., G. Zellner and J. Winter. 1989. Growth of methanogens on cyclopentanol/CO$_2$ and specificity of alcohol dehydrogenase. FEMS Microbiol. Lett. *59*: 307–312.

Blöchl, E., R. Rachel, S. Burggraf, D. Hafenbradl, H.W. Jannasch and K.O. Stetter. 1997. *Pyrolobus fumarii*, gen. and sp. nov., represents a novel group of archaea, extending the upper temperature limit for life to 113°C. Extremophiles *1*: 14–21.

Blöchl, E., R. Rachel, S. Burggraf, D. Hafenbradl, H.W. Jannasch and K.O. Stetter. 1999. *In* Validation of the publication of new names and new combinations previously effectively published outside the IJSB. List No. 71. Int. J. Syst. Bacteriol. *49*: 1325–1326.

Blotevogel, K.H. and U. Fischer. 1985. Isolation and characterization of a new thermophilic and autotrophic methane producing bacterium: *Methanobacterium thermoaggregans*, sp. nov. Arch. Microbiol. *142*: 218–222.

Blotevogel, K.H. and U. Fischer. 1989. Transfer of *Methanococcus frisius* to the genus *Methanosarcina* as *Methanosarcina frisia* comb. nov. Int. J. Syst. Bacteriol. *39*: 91–92.

Blotevogel, K.H., U. Fischer and K.H. Lüpkes. 1986. *Methanococcus frisius*, sp. nov., a new methylotrophic marine methanogen. Can. J. Microbiol. *32*: 127–131.

Blotevogel, K.H., R. Gahl-Janssen, S. Janssen, U. Fischer, F. Pilz, G. Auling, A.J.L. Macario and B.J. Tindall. 1991. Isolation and characterization of a novel mesophilic, freshwater methanogen from river sediment *Methanoculleus oldenburgensis* sp. nov. Arch. Microbiol. *157*: 54–59.

Blotevogel, K.H., R. Gahl-Janssen, S. Janssen, U. Fischer, F. Pilz, G. Auling, A.J.L. Macario and B.J. Tindall. 1998. *In* Validation of the publication of new names and new combinations previously effectively published outside the IJSB. List No. 64. Int. J. Syst. Bacteriol. *48*: 327–328.

Blotevogel, K.H. and J.L. Macario. 1989. Antigenic relationship of *Methanococcus frisius*. Syst. Appl. Microbiol. *11*: 148–150.

Bocchetta, M., E. Ceccarelli, R. Creti, A.M. Sanangelantoni, O. Tiboni and P. Cammarano. 1995. Arrangement and nucleotide sequence of the gene (*fus*) encoding elongation factor G (EF-G) from the hyperthermophilic bacterium *Aquifex pyrophilus*: phylogenetic depth of hyperthermophilic bacteria inferred from analysis of the EF-G/*fus* sequences. J. Mol. Evol. *41*: 803–812.

Böck, A. and O. Kandler. 1985. Antibiotic sensitivity of archaebacteria. *In* Woese and Wolfe (Editors), The Bacteria, Vol. VIII, Academic Press, New York. pp. 525–544.

Böck, A.K., A. Priegerkraft and P. Schönheit. 1994. Pyruvate: a novel substrate for growth and methane formation in *Methanosarcina barkeri*. Arch. Microbiol. *161*: 33–46.

Böck, A.K. and P. Schönheit. 1995. Growth of *Methanosarcina barkeri* (Fusaro) under nonmethanogenic conditions by the fermentation of pyruvate to acetate: ATP synthesis via the mechanism of substrate level phosphorylation. J. Bacteriol. *177*: 2002–2007.

Bodnar, I.V., T.N. Zhilina and G.A. Zavarzin. 1987. Hydrogen production from methylamines by halophilic methanogenic bacteria. Mikrobiologiya *56*: 501–503.

Bogosian, G., P.J.L. Morris and J.P. O'Neil. 1998. A mixed culture recovery method indicates that enteric bacteria do not enter the viable but nonculturable state. Appl. Environ. Microbiol. *64*: 1736–1742.

Bohlool, B.B. and T.D. Brock. 1974. Immunodiffusion analysis of membranes of *Thermoplasma acidophilum*. Infect. Immun. *10*: 280–281.

Bolch, C.J.S., P.T. Orr, G.J. Jones and S.I. Blackburn. 1999. Genetic, morphological, and toxicological variation among globally distributed strains of *Nodularia* (Cyanobacteria). J. Phycol. *35*: 339–355.

Bonch-Osmolovskaya, E.A., M.L. Miroshnichenko, N.A. Kostrikina, N.A. Chernych and G.A. Zavarzin. 1990. *Thermoproteus uzoniensis* sp. nov., a new extremely thermophilic archaebacterium from Kamchatka continental hot springs. Arch. Microbiol. *154*: 556–559.

Bonch-Osmolovskaya, E.A., A.I. Slesarev, M.L. Miroshnichenko, T.P. Svetlichnaya and V.A. Alekseev. 1988. Characteristics of *Desulfurococcus amylolyticus* sp. nov.—a new extremely thermophilic archaebacterium isolated from thermal springs of Kamchatka and Kunashir Island. Mikrobiologiya *57*: 94–101.

Bonelo, G., A. Ventosa, M. Megias and F. Ruiz-Berraquero. 1984. The sensitivity of halobacteria to antibiotics. FEMS Microbiol. Lett. *21*: 341–345.

Booker, M.J. and A.E. Walsby. 1979. The relative form resistance of straight and helical blue-green algal filaments. Br. Phycol. J. *14*: 141–150.

Boone, D.R. 1987. Replacement of the type strain of *Methanobacterium formicicum*: request for an opinion and reinstatement of *Methanobacterium bryantii* (ex Balch and Wolfe, 1981) with M.o.H. (=DSM 863) as the type strain. Int. J. Syst. Bacteriol. *37*: 172–173.

Boone, D.R. 1991. Strain GP6 is proposed as the neotype strain of *Methanothrix soehngenii*^{VP} pro synon. *Methanothrix concilii*^{VP} and *Methanosaeta concilii*^{VP}. Request for an opinion. Int. J. Syst. Bacteriol. *41*: 588–589.

Boone, D.R. 1995. Short- and long-term maintenance of methanogenic stock cultures. *In* Sowers and Schreier (Editors), Methanogens, Cold Spring Harbor Laboratory Press, Plainview. pp. 79–83.

Boone, D.R. and M.P. Bryant. 1980. Propionate-degrading bacterium, *Syntrophobacter wolinii* sp. nov. gen. nov., from methanogenic ecosystems. Appl. Environ. Microbiol. *40*: 626–632.

Boone, D.R., R.L. Johnson and Y. Liu. 1989. Diffusion of the interspecies electron carriers H$_2$ and formate in methanogenic ecosystems and its implications in the measurement of K_m for H$_2$ or formate uptake. Appl. Environ. Microbiol. *55*: 1735–1741.

Boone, D.R. and Y. Kamagata. 1998. Rejection of the species *Methanothrix soehngenii*^{VP} and the genus *Methanothrix*^{VP} as nomina confusa, and transfer of *Methanothrix thermophila*^{VP} to the genus *Methanosaeta*^{VP} as *Methanosaeta thermophila* comb. nov. Request for an opinion. Int. J. Syst. Bacteriol. *48*: 1079–1080.

Boone, D.R. and R.A. Mah. 1987. Effects of calcium, magnesium, pH, and extent of growth on the morphology of *Methanosarcina mazei* S-6. Appl. Environ. Microbiol. *53*: 1699–1700.

Boone, D.R. and R.A. Mah. 1989. Genus I. *Methanobacterium*. *In* Staley, Bryant, Pfennig and Holt (Editors), Bergey's Manual of Systematic Bacteriology, 1st Ed., Vol. 3, The Williams & Wilkins Co., Baltimore. pp. 2175–2177.

Boone, D.R., I.M. Mathrani, Y. Liu, J.A.G.F. Menaia, R.A. Mah and J.E. Boone. 1993a. Isolation and characterization of *Methanohalophilus portucalensis*, sp. nov. and DNA reassociation study of the genus *Methanohalophilus*. Int. J. Syst. Bacteriol. *43*: 430–437.

Boone, D.R., J.A.G.F. Menaia, J.E. Boone and R.A. Mah. 1987. Effects of hydrogen pressure during growth and effects of pregrowth with hydrogen on acetate degradation by *Methanosarcina* species. Appl. Environ. Microbiol. *53*: 83–87.

Boone, D.R., W.B. Whitman and P.E. Rouvière. 1993b. Diversity and taxonomy of methanogens. *In* Ferry (Editor), Methanogenesis: Ecology, Physiology, Biochemistry, and Genetics, Chapman & Hall, New York. pp. 35–80.

Boone, D.R., S. Worakit, I.M. Mathrani and R.A. Mah. 1986. Alkaliphilic methanogens from high-pH lake sediments. Syst. Appl. Microbiol. *7*: 230–234.

Boopathy, R. 1996. Methanogenic transformation of methylfurfural compounds to furfural. Appl. Environ. Microbiol. *62*: 3483–3485.

Boopathy, R. and L. Daniels. 1991. Pattern of organotin inhibition of methanogenic bacteria. Appl. Environ. Microbiol. *57*: 1189–1193.

Boopathy, R. and L. Daniels. 1992. Isolation and characterization of a marine methanogenic bacterium from the biofilm of a ship hull in Los Angeles harbor. Curr. Microbiol. *25*: 157–164.

Borges, K.M. and P.L. Bergquist. 1993. Genomic restriction map of the extremely thermophilic bacterium *Thermus thermophilus* HB8. J. Bacteriol. *175*: 103–110.

Borneman, J. and E.W. Triplett. 1997. Molecular microbial diversity in soils from eastern Amazonia: evidence for unusual microorganisms and microbial population shifts associated with deforestation. Appl. Environ. Microbiol. *63*: 2647–2653.

Bornet, E. 1892. Les algues de P.-K.-A. Schousboe, récoltées au Maroc et dans la Mediterranée de 1815 à 1829. Mem. Soc. Nat. Sci. *28*: 175.

Bornet, E. and C. Flahault. 1886. Revision des Nosocacées hétérocystées. Ann. Sci. Nat. Bot. *3*: 323–380.

Bornet, E. and C. Flahault. 1887. Revision des Nosocacées hétérocystées. Ann. Sci. Nat. Bot. *5*: 51–129.

Bornet, E. and C. Flahault. 1888. Revision des Nosocacées hétérocystées. Ann. Sci. Nat. Bot. *7*: 177–262.

Borowitzka, L.J. 1986. Osmoregulation in blue-green algae. *In* Round and Chapman (Editors), Progress in Phycological Research, Vol. 4, Biopress Ltd., Great Britain. pp. 243–256.

Borrego, C.M. and L.J. Garcia-Gil. 1994. Separation of bacteriochlorophyll homologs from green photosynthetic sulfur bacteria by reversed-phase HPLC. Photosynth. Res. *41*: 157–163.

Bory de St. Vincent, J.B. 1822. *Anabaena. In* Dictionnaire Classique d'Histoire Naturelle, Vol. 1, Rey and Gravier, Paris. p. 307.

Bott, M., B. Eikmanns and R.K. Thauer. 1985. Defective formation and/or utilization of carbon monoxide in H_2/CO_2 fermenting methanogens dependent on acetate as carbon source. Arch. Microbiol. *143*: 266–269.

Bott, M. and R.K. Thauer. 1989. Proton translocation coupled to the oxidation of carbon monoxide to CO_2 and H_2 in *Methanosarcina barkeri*. Eur. J. Biochem. *179*: 469–472.

Bottomley, P.J., J.F. Grillo, C. van Baalen and F.R. Tabita. 1979. Synthesis of nitrogenase and heterocysts by *Anabaena* sp. CA in the presence of high levels of ammonia. J. Bacteriol. *140*: 938–943.

Bourrelly, P. 1970. Les Algues d'Eau Douce. III. Les Algues Bleues et Rouges, les Eugléniens, Peridiniens et Cryptomonadines, N. Boubée & Cie, Paris.

Bourrelly, P. 1985. Les Algues d'Eau Douce. III. Les Algues Bleues et Rouges, les Eugléniens, Peridiniens et Cryptomonadines, 2nd Ed., N. Boubée & Cie, Paris.

Bouvet, P.J. and S. Jeanjean. 1989. Delineation of new proteolytic genomic species in the genus *Acinetobacter*. Res. Microbiol. *140*: 291–299.

Bradford, D., P. Hugenholtz, E.M. Seviour, M.A. Cunningham, H. Stratton, R.J. Seviour and L.L. Blackall. 1996. 16S rRNA analysis of isolates obtained from Gram-negative, filamentous bacteria micromanipulated from activated sludge. Syst. Appl. Microbiol. *19*: 334–343.

Braks, I.J., M. Hoppert, S. Roge and F. Mayer. 1994. Structural aspects and immunolocalization of the F_{420}-reducing and non-F_{420}-reducing hydrogenases from *Methanobacterium thermoautotrophicum* Marburg. J. Bacteriol. *176*: 7677–7687.

Brandis, A., R.K. Thauer and K.O. Stetter. 1981. Relatedness of strains ΔH and Marburg of *Methanobacterium thermoautotrophicum*. Zentbl. Bakteriol. Mikrobiol. Hyg. 1 Abt Orig. C *2*: 311–317.

Branton, D., S. Bullivant, N.G. Gilula, M.J. Karnovsky, H. Moor, K. Muhlethaler, D.H. Northcote, L. Packer, P. Sater, V. Speth, L.A. Staehlin, R.L. Steere and R.S. Weinstein. 1975. Freeze-etching nomenclature. Science *190*: 54–56.

Braun, A. and A. Grunow. 1865. *In* Rabenhorst's flora Europaea algarum aquae dulcis et submarinae, section II, Eduard Kummer, Leipzig. pp. 148–149.

Breed, R.S., E.G.D. Murray and A.P. Hitchens (Editors). 1948a. Bergey's Manual of Determinative Bacteriology, 6th Ed., Williams & Wilkins Co., Baltimore.

Breed, R.S., E.G.D. Murray and A.P. Hitchens (Editors). 1948b. Bergey's Manual of Determinative Bacteriology, abridged 6th Ed., Biotech Publications, Geneva.

Breed, R.S., E.G.D. Murray and N.R. Smith (Editors). 1957. Bergey's Manual of Determinative Bacteriology, 7th Ed., Williams & Wilkins Co., Baltimore.

Brenner, D.J. 1984. Family I. *Enterobacteriaceae. In* Krieg and Holt (Editors), Bergey's Manual of Systematic Bacteriology, 1st Ed., Vol. 1, The Williams & Wilkins Co., Baltimore. pp. 408–420.

Brenner, D.J., A.C. McWhorter, J.K. Leete Knutson and A.G. Steigerwalt.

1982. *Escherichia vulneris*, a new species of *Enterobacteriaceae* associated with human wounds. J. Clin. Microbiol. *15*: 1133–1140.

Brenner, M.C., L. Ma, M.K. Johnson and R.A. Scott. 1992. Spectroscopic characterization of the alternate form of S-methylcoenzyme M reductase from *Methanobacterium thermoautotrophicum* (strain ΔH). Biochim. Biophys. Acta *1120*: 160–166.

Breuil, C. and G.B. Patel. 1980. Viability and depletion of cell constituents of *Methanospirillum hungatii* GP1 during starvation. Can. J. Microbiol. *26*: 887–892.

Breznak, J.A. and R.N. Costilow. 1994. Physiochemical factors in growth. *In* Gerhardt, Murray, Wood and Krieg (Editors), Methods for General and Molecular Bacteriology, American Society for Microbiology, Washington, DC. pp. 137–154.

Bridges, B.A., M.J. Ashwood-Smith and R.J. Munson. 1969a. Correlation of bacterial sensitivities to ionizing radiation and mild heating. J. Gen. Microbiol. *58*: 115–124.

Bridges, B.A., M.J. Ashwood-Smith and R.J. Munson. 1969b. Susceptibility of mild thermal and of ionizing radiation damage to the same recovery mechanisms in *Escherichia coli*. Biochem. Biophys. Res. Commun. *35*: 193–196.

Brinkmann, H., P. Martinez, F. Quigley, W. Martin and R. Cerff. 1987. Endosymbiotic origin and codon bias of the nuclear gene for chloroplast glyceraldehyde-3-phosphate dehydrogenase from maize. J. Mol. Evol. *26*: 320–328.

Broch-Due, M. and J.G. Ormerod. 1978. Isolation of Bchl *c* mutant from *Chlorobium* with Bchl *d* by cultivation at low light intensities. FEMS Microbiol. Lett. *3*: 305–308.

Brock, T.D. 1978. Thermophilic Microorganisms and Life at High Temperatures, Springer–Verlag, New York.

Brock, T.D. 1984. Genus *Thermus. In* Krieg and Holt (Editors), Bergey's Manual of Systematic Bacteriology, 1st Ed., Vol. 1, The Williams & Wilkins Co., Baltimore. pp. 333–337.

Brock, T.D. 1995. The road to Yellowstone – and beyond. Annu. Rev. Microbiol. *49*: 1–28.

Brock, T.D. and K.L. Boylen. 1973. Presence of thermophilic bacteria in laundry and domestic hot-water heaters. Appl. Microbiol. *25*: 72–76.

Brock, T.D., K.M. Brock, R.T. Belly and R.L. Weiss. 1972. *Sulfolobus*: a new genus of sulfur-oxidizing bacteria living at low pH and high temperature. Arch. Mikrobiol. *84*: 54–68.

Brock, T.D. and M.R. Edwards. 1970. Fine structure of *Thermus aquaticus*, an extreme thermophile. J. Bacteriol. *104*: 509–517.

Brock, T.D. and H. Freeze. 1969. *Thermus aquaticus* gen.n. and sp.n., a nonsporulating extreme thermophile. J. Bacteriol. *98*: 289–297.

Brock, T.D. and S. Petersen. 1976. Some effects of light on the viability of rhodopsin-containing halobacteria. Arch. Microbiol. *109*: 199–200.

Brock, T.D. and I. Yoder. 1971. Thermal pollution of a small river by a large university: bacteriological studies. Proc. Indiana Acad. Sci. *80*: 183–188.

Brockmann, E., B.L. Jacobsen, C. Hertel, W. Ludwig and K.H. Schleifer. 1996. Monitoring of genetically modified *Lactococcus lactis* in gnotobiotic and conventional rats by using antibiotic resistance markers and specific probe or primer based methods. Syst. Appl. Microbiol. *19*: 203–212.

Brockmann, H., Jr. and A. Lipinski. 1983. Bacteriochlorophyll *g*: a new bacteriochlorophyll from *Heliobacterium chlorum*. Arch. Microbiol. *136*: 17–19.

Brooks, B.W. and R.G.E. Murray. 1981. Nomenclature for "*Micrococcus radiodurans*" and other radiation-resistant cocci: *Deinococcaceae* fam. nov. and *Deinococcus* gen. nov., including five species. Int. J. Syst. Bacteriol. *31*: 353–360.

Brooks, B.W., R.G.E. Murray, J.L. Johnson, E. Stackebrandt, C.R. Woese and G.E. Fox. 1980. Red-pigmented micrococci: a basis for taxonomy. Int. J. Syst. Bacteriol. *30*: 627–646.

Brown, J.R. and W.F. Doolittle. 1995. Root of the universal tree of life based on ancient aminoacyl-tRNA synthetase gene duplications. Proc. Natl. Acad. Sci. U.S.A. *92*: 2441–2445.

Brown, J.R. and W.F. Doolittle. 1997. Archaea and the prokaryote-to-eukaryote transition. Microbiol. Mol. Biol. Rev. *61*: 456–502.

Bruce, B.D., R.C. Fuller and R.E. Blankenship. 1982. Primary photochemistry in the facultatively aerobic green photosynthetic bacterium *Chloroflexus aurantiacus*. Proc. Natl. Acad. Sci. U.S.A. *79*: 6532–6536.

Bryant, D.A. 1982. Phycoerythrocyanin and phycoerythrin: properties and occurrence in cyanobacteria. J. Gen. Microbiol. *128*: 835–844.

Bryant, D.A. 1994. The Molecular Biology of Cyanobacteria, Kluwer Academic Publishers, Dordrecht.

Bryant, D.A., G. Cohen-Bazire and A.N. Glazer. 1981. Characterization of the biliproteins of *Gloeobacter violaceus*: chromophore content of a cyanobacterial phycoerythrin carrying phycourobilin chromophore. Arch. Microbiol. *129*: 190–198.

Bryant, M.P. 1965. Rumen methanogenic bacteria. *In* Dougherty, Allen, Burroughs, Jacobson and McGilliard (Editors), Physiology of Digestion in the Ruminant, Buttersworth Publishing Inc., Washington, DC. pp. 411–418.

Bryant, M.P. 1974. Methane-producing bacteria. *In* Buchanan and Gibbons (Editors), Bergey's Manual of Determinative Bacteriology, 8th Ed, The Williams & Wilkins Co., Baltimore. pp. 472–477.

Bryant, M.P. and D.R. Boone. 1987a. Emended description of strain MS^T (DSM 800^T), the type strain of *Methanosarcina barkeri*. Int. J. Syst. Bacteriol. *37*: 169–170.

Bryant, M.P. and D.R. Boone. 1987b. Isolation and characterization of *Methanobacterium formicicum* MF. Int. J. Syst. Bacteriol. *37*: 171.

Bryant, M.P. and I.M. Robinson. 1961. An improved nonselective culture medium for ruminal bacteria and its use in determining diurnal variation in number of bacteria in the rumen. J. Dairy Sci. *44*: 1446–1456.

Bryant, M.P., S.F. Tzeng, I.M. Robinson and A.E. Joyner. 1971. Nutrient requirements of methanogenic bacteria. *In* Gould (Editor), Anaerobic Biological Treatment Processes. Advances in Chemistry Series 105, American Chemical Society, Washington, DC. pp. 23–40.

Bryant, M.P., E.A. Wolin, M.J. Wolin and R.S. Wolfe. 1967. *Methanobacillus omelianskii*, a symbiotic association of two species of bacteria. Arch. Mikrobiol. *59*: 20–31.

Bryantseva, I.A., V.M. Gorlenko, E.I. Kompantseva, L.A. Achenbach and M.T. Madigan. 1999a. *Heliorestis daurensis*, gen. nov., sp. nov., an alkaliphilic rod-to-coiled-shaped phototrophic heliobacterium from a Siberan soda lake. Arch. Microbiol. *172*: 167–174.

Bryantseva, I.A., V.M. Gorlenko, E.I. Kompantseva, L.A. Achenbach and M.T. Madigan. 2000. *In* Validation of the publication of new names and new combinations previously effectively published outside the IJSEM. List No. 74. Int. J. Syst. Evol. Microbiol. *50*: 949–950.

Bryantseva, I., V.M. Gorlenko, E.I. Kompantseva, J.F. Imhoff, J. Süling and L. Mityushina. 1999b. *Thiorhodospira sibirica* gen. nov., sp. nov., a new alkaliphilic purple sulfur bacterium from a Siberian soda lake. Int. J. Syst. Bacteriol. *49*: 697–703.

Bryniok, D. and W. Trosch. 1989. Taxonomy of methanogens by ELISA techniques. Appl. Microbiol. Biotechnol. *32*: 243–247.

Bsat, N. and C.A. Batt. 1993. A combined modified reverse dot-blot and nested PCR assay for the specific non-radioactive detection of *Listeria monocytogenes*. Mol. Cell. Probes 7: 199–207.

Buchanan, R.E. 1916. Studies in the nomenclature and classification of the bacteria.I. The problem of bacterial nomenclature. J. Bacteriol. *1*: 591–596.

Buchanan, R.E. 1917a. Studies in the nomenclature and classification of the bacteria. II. The primary subdivisions of the *Schizomycetes*. J. Bacteriol. *2*: 155–164.

Buchanan, R.E. 1917b. Studies in the nomenclature and classification of the bacteria. III. The families of the *Eubacteriales*. J. Bacteriol. *2*: 347–350.

Buchanan, R.E. 1917c. Studies in the nomenclature and classification of the bacteria. IV. Subgroups and genera of the *Coccaceae*. J. Bacteriol. *2*: 603–617.

Buchanan, R.E. 1918a. Studies in the nomenclature and classification of the bacteria.V. Subgroups and genera of the *Bacteriaceae*. J. Bacteriol. *3*: 27–61.

Buchanan, R.E. 1918b. Studies in the nomenclature and classification of

the bacteria. VI. Subdivisions and genera of the *Spirillaceae* and *Nitrobacteriaceae*. J. Bacteriol. *3*: 175–181.

Buchanan, R.E. 1918c. Studies in the nomenclature and classification of the bacteria. VII. The subgroups and genera of the *Chlamydobacteriales*. J. Bacteriol. *3*: 301–306.

Buchanan, R.E. 1918d. Studies in the nomenclature and classification of the bacteria. VIII. The subgroups and genera of the *Actinomycetales*. J. Bacteriol. *3*: 403–406.

Buchanan, R.E. 1918e. Studies in the nomenclature and classification of the bacteria. IX. The subgroups and genera of the *Thiobacteriales*. J. Bacteriol. *3*: 461–474.

Buchanan, R.E. 1918f. Studies in the nomenclature and classification of the bacteria.X. Subgroups and genera of the *Myxobacteriales* and *Spirochaetales*. J. Bacteriol. *3*: 541–545.

Buchanan, R.E. 1925. General systematic bacteriology, The Williams & Wilkins Co., Baltimore.

Buchanan, R.E. 1948. How bacteria are named and identified. *In* Breed, Murray and Hitchens (Editors), Bergey's Manual of Determinative Bacteriology, 6th Ed., The Williams & Wilkins Co., Baltimore. pp. 39–48.

Buchanan, R.E. 1994. Chemical terminology and microbiological nomenclature. Int. J. Syst. Bacteriol. *44*: 588–590.

Buchanan, R.E. and N.E. Gibbons (Editors). 1974. Bergey's Manual of Determinative Bacteriology, 8th Ed., The Williams & Wilkins Co., Baltimore.

Buchanan, R.E., J.G. Holt and E.F.J. Lessel (Editors). 1966. Index Bergeyana, The Williams & Wilkins Co., Baltimore.

Buchanan, R.E., R. St. John-Brooks and R.S. Breed. 1948. International Bacteriological Code of Nomenclature. J. Bacteriol. *55*: 287–306.

Büdel, B., U. Becker, S. Porembski and W. Barthlott. 1997. Cyanobacteria and cyanobacterial lichens from inselbergs of the Ivory Coast, Africa. Bot. Acta *110*: 458–465.

Buder, J. 1914. *Chloronium mirabile*. Ber. Dtsch. Bot. Ges. *31*: 80–97.

Buikema, W.J. and R. Haselkorn. 1993. Molecular genetics of cyanobacterial development. Annu. Rev. Plant Physiol. Plant Mol. Biol. *44*: 33–52.

Burger-Wiersma, T. and L.R. Mur. 1989. Genus "*Prochlorothrix*". *In* Staley, Bryant, Pfennig and Holt (Editors), Bergey's Manual of Systematic Bacteriology, 1st Ed., Vol. 3, The Williams & Wilkins Co., Baltimore. pp. 1805–1806.

Burger-Wiersma, T., L.J. Stal and L.R. Mur. 1989. *Prochlorothrix hollandica* gen. nov., sp. nov., a filamentous oxygenic photoautotrophic procaryote containing chlorophylls *a* and *b*: assignment to *Prochlorotrichaceae* fam. nov. and order *Prochlorales* Florenzano, Balloni, and Materassi 1986, with emendation of the ordinal description. Int. J. Syst. Bacteriol. *39*: 250–257.

Burggraf, S., H. Fricke, A. Neuner, J. Kristjansson, P.E. Rouvière, L. Mandelco, C.R. Woese and K.O. Stetter. 1990a. *Methanococcus igneus* sp. nov., a novel hyperthermophilic methanogen from a shallow submarine hydrothermal system. Syst. Appl. Microbiol. *13*: 263–269.

Burggraf, S., P. Heyder and N. Eis. 1997a. A pivotal Archaea group. Nature *385*: 780.

Burggraf, S., H. Huber and K.O. Stetter. 1997b. Reclassification of the crenarchaeal orders and families in accordance with 16S rRNA sequence data. Int. J. Syst. Bacteriol. *47*: 657–660.

Burggraf, S., H.W. Jannasch, B. Nicolaus and K.O. Stetter. 1990b. *Archaeoglobus profundus*, sp. nov., represents a new species within the sulfate-reducing archaebacteria. Syst. Appl. Microbiol. *13*: 24–28.

Burggraf, S., H.W. Jannasch, B. Nicolaus and K.O. Stetter. 1990c. *In* Validation of the publication of new names and new combinations previously effectively published outside the IJSB. List No. 34. Int. J. Syst. Bacteriol. *40*: 320–321.

Burggraf, S., N. Larsen, C.R. Woese and K.O. Stetter. 1993. An intron within the 16S ribosomal RNA gene of the archaeon *Pyrobaculum aerophilum*. Proc. Natl. Acad. Sci. U.S.A. *90*: 2547–2550.

Burggraf, S., G.J. Olsen, K.O. Stetter and C.R. Woese. 1992. A phylogenetic analysis of *Aquifex pyrophilus*. Syst. Appl. Microbiol. *15*: 352–356.

Burggraf, S., K.O. Stetter, P.E. Rouvière and C.R. Woese. 1991. *Methano-*

pyrus kandleri: an archaeal methanogen unrelated to all other known methanogens. Syst. Appl. Microbiol. *14*: 346–351.

Button, D.K., B.R. Robertson, P.W. Lepp and T.M. Schmidt. 1998. A small, dilute-cytoplasm, high-affinity, novel bacterium isolated by extinction culture and having kinetic constants compatible with growth at ambient concentrations of dissolved nutrients in seawater. Appl. Environ. Microbiol. *64*: 4467–4476.

Button, D.K., F. Schut, P. Quang, R. Martin and B.R. Robertson. 1993. Viability and isolation of marine bacteria by dilution culture: theory, procedures, and initial results. Appl. Environ. Microbiol. *59*: 881–891.

Caccavo, F., Jr., J.D. Coates, R.A. Rossello-Mora, W. Ludwig, K.H. Schleifer, D.R. Lovley and M.J. McInerney. 1996. *Geovibrio ferrireducens*, a phylogenetically distinct dissimilatory Fe(III)-reducing bacterium. Arch. Microbiol. *165*: 370–376.

Caldwell, D.E. 1994. End of the pure culture era? ASM News *60*: 231–232.

Caldwell, D.E. and J.M. Tiedje. 1975a. A morphological study of anaerobic bacteria from the hypolimnia of two Michigan lakes. Can. J. Microbiol. *21*: 362–376.

Caldwell, D.E. and J.M. Tiedje. 1975b. The structure of anaerobic bacterial communities in the hypolimnia of several Michigan lakes. Can. J. Microbiol. *21*: 377–385.

Campbell, E.L. and J.C. Meeks. 1989. Characteristics of hormogonia formation by symbiotic *Nostoc* spp. in response to the presence of *Anthoceros punctatus* or its extracellular products. Appl. Environ. Microbiol. *55*: 125–131.

Canganella, F., W.J. Jones, A. Gambacorta and G. Antranikian. 1998. *Thermococcus guaymasensis* sp. nov. and *Thermococcus aggregans* sp. nov., two novel thermophilic archaea isolated from the Guaymas Basin hydrothermal vent site. Int. J. Syst. Bacteriol. *48*: 1181–1185.

Canhos, V.P., G.P. Manfio and D.A.L. Canhos. 1996. Networking the microbial diversity information. J. Ind. Microbiol. *17*: 498–504.

Cann, I.K.O., S. Ishino, N. Nomura, Y. Sako and Y. Ishino. 1999. Two family B DNA polymerases from *Aeropyrum pernix*, an aerobic hyperthermophilic crenarchaeote. J. Bacteriol. *181*: 5984–5992.

Carballeira, N.M., M. Reyes, A. Sostre, H. Huang, M.F. Verhagen and M.W. Adams. 1997. Unusual fatty acid compositions of the hyperthermophilic archaeon *Pyrococcus furiosus* and the bacterium *Thermotoga maritima*. J. Bacteriol. *179*: 2766–2768.

Carbon, P., C. Ehresmann, B. Ehresmann and J.P. Ebel. 1979. The complete nucleotide sequence of the ribosomal 16S RNA from *Escherichia coli*. Experimental details and cistron heterogeneities. Eur. J. Biochem. *100*: 399–410.

Carbonneau, M.A., A.M. Melin, A. Perromat and M. Clerc. 1989. The action of free radicals on *Deinococcus radiodurans* carotenoids. Arch. Biochem. Biophys. *275*: 244–251.

Carmichael, W.W. 1988. Toxins of freshwater algae. *In* Tu (Editor), Handbook of Natural Toxins, Vol. 4, Marcel Dekker, New York. pp. 121–147.

Carmichael, W.W. and P.R. Gorham. 1974. An improved method for obtaining axenic clones of planktonic blue-green algae. J. Phycol. *10*: 238–240.

Carpenter, E.J., J. Chang, M. Cottrell, J. Schubauer, H.W. Paerl, B.M. Bebout and D.G. Capone. 1990. Re-evaluation of nitrogenase oxygen-protective mechanism in the planktonic marine cyanobacterium *Trichodesmium*. Mar. Ecol. Prog. Ser. *65*: 151–158.

Carr, N.G. and B.A. Whitton (Editors). 1982. The Biology of Cyanobacteria, University of California Press, Berkeley.

Carreto, L., R. Wait, M.F. Nobre and M.S. Da Costa. 1996. Determination of the structure of a novel glycolipid from *Thermus aquaticus* 15004 and demonstration that hydroxy fatty acids are amide linked to glycolipids in *Thermus* spp. J. Bacteriol. *178*: 6479–6486.

Castenholz, R.W. 1969. Thermophilic blue-green algae and the thermal environment. Bacteriol. Rev. *33*: 476–504.

Castenholz, R.W. 1970. Laboratory culture of thermophilic cyanophytes. Schweiz. Z. Hydrol. *32*: 538–551.

Castenholz, R.W. 1972. The occurrence of the thermophilic blue-green

alga, *Mastigocladus laminosus*, on Surtsey in 1970. Surtsey Progr. Rep. Reykjavik *6*: 1–6.

Castenholz, R.W. 1973. The possible photosynthetic use of sulfide by the filamentous phototrophic bacteria of hot springs. Limnol. Oceanogr. *18*: 863–876.

Castenholz, R.W. 1976. The effect of sulfide on the blue-green algae of hot springs.I. New Zealand and Iceland. J. Phycol. *12*: 54–68.

Castenholz, R.W. 1977. The effect of sulfide on the blue-green algae of hot springs. II. Yellowstone National Park. Microb. Ecol. *3*: 79–105.

Castenholz, R.W. 1978. The biogeography of hot spring algae through enrichment cultures. Mitt. Int. Ver. Theor. Angew. Limnol. *21*: 296–315.

Castenholz, R.W. 1981. Isolation and cultivation of thermophilic cyanobacteria. *In* Starr, Stolp, Trüper, Balows and Schlegel (Editors), The Prokaryotes: A Handbook on Habitats, Isolation, and Identification of Bacteria, Springer-Verlag, Berlin. pp. 236–246.

Castenholz, R.W. 1982. Motility and taxes. *In* Carr and Whitton (Editors), The Biology of Cyanobacteria, Blackwell, Oxford, and University of California Press, Berkeley. pp. 413–439.

Castenholz, R.W. 1984a. Composition of hot spring microbial mats: a summary. *In* Cohen, Castenholz and Halvorson (Editors), Microbial Mats: Stromatolites, Alan R. Liss, New York. pp. 101–119.

Castenholz, R.W. 1984b. Habitats of *Chloroflexus* and related organisms. *In* Klug and Reddy (Editors), Current Perspectives in Microbial Ecology, ASM Publications, Washington DC. pp. 196–200.

Castenholz, R.W. 1988a. Culturing methods for cyanobacteria. Methods Enzymol. *167*: 68–93.

Castenholz, R.W. 1988b. Thermophilic cyanobacteria: special problems. Methods Enzymol. *167*: 96–100.

Castenholz, R.W. 1989a. Subsection III. Order *Oscillatoriales*. *In* Staley, Bryant, Pfennig and Holt (Editors), Bergey's Manual of Systematic Bacteriology, 1st Ed., Vol. 3, The Williams & Wilkins Co., Baltimore. pp. 1771–1780.

Castenholz, R.W. 1989b. Subsection IV. Order *Nostocales*. *In* Staley, Bryant, Pfennig and Holt (Editors), Bergey's Manual of Systematic Bacteriology, 1st Ed., Vol. 3, The Williams & Wilkins Co., Baltimore. pp. 1780–1793.

Castenholz, R.W. 1996. Endemism and biodiversity of thermophilic cyanobacteria. Nova Hedwigia Beih. *112*: 33–47.

Castenholz, R.W., J. Bauld and B.B. Jørgensen. 1990. Anoxygenic microbial mats of hot springs: thermophilic *Chlorobium* sp. FEMS Microbiol. Ecol. *74*: 325–336.

Castenholz, R.W. and F. Garcia-Pichel. 2000. Cyanobacterial responses to UV-radiation. *In* Whitton and Potts (Editors), Ecology of Cyanobacteria: Their Diversity in Time and Space, Kluwer Academic Publishing, Dordrecht. pp. 591–611.

Castenholz, R.W. and B.K. Pierson. 1995. Ecology of thermophilic anoxygenic phototrophs. *In* Blankenship, Madigan and Bauer (Editors), Anoxygenic Photosynthetic Bacteria, Kluwer Academic Publishers, Dordrecht. pp. 87–103.

Castenholz, R.W. and H.C. Utkilen. 1984. Physiology of sulfide tolerance in a thermophilic *Oscillatoria*. Arch. Microbiol. *138*: 299–305.

Castenholz, R.W. and J.B. Waterbury. 1989. Cyanobacteria. *In* Staley, Bryant, Pfennig and Holt (Editors), Bergey's Manual of Systematic Bacteriology, 1st Ed., Vol. 3, The Williams & Wilkins Co., Baltimore. pp. 1710–1727.

Castón, J.R., J.L. Carrascosa, M.A. de Pedro and J. Berenguer. 1988. Identification of a crystalline surface layer on the cell envelope of the thermophilic eubacterium *Thermus thermophilus*. FEMS Microbiol. Lett. *51*: 225–230.

Caumette, P., J.F. Imhoff, J. Süling and R. Matheron. 1997. *Chromatium glycolicum* sp. nov., a moderately halophilic purple sulfur bacterium that uses glycolate as substrate. Arch. Microbiol. *167*: 11–18.

Chang, T.P. 1977. Sheath formation in *Oscillatoria agardhii* Gomont. Schweiz. Z. Hydrol. *39*: 178–181.

Charlebois, R.L. 1995a. Physical and genetic map of the genome of *Halobacterium* sp. GRB. *In* Robb, Sowers, DasSarma, Place, Schreier and

Fleischmann (Editors), Archaea: A Laboratory Manual, Cold Spring Harbor Laboratory Press, Plainview. pp. 237–241.

Charlebois, R.L. 1995b. Physical and genetic map of the genome of *Haloferax volcanii* DS2. *In* Robb, Sowers, DasSarma, Place, Schreier and Fleischmann (Editors), Archaea: A Laboratory Manual, Cold Spring Harbor Laboratory Press, Plainview. pp. 231–236.

Charlebois, R.L. and W.F. Doolittle. 1989. Transposable elements and genome structure in halobacteria. *In* Berg and Howe (Editors), Mobile DNA, American Society for Microbiology, Washington, DC. pp. 297– 307.

Charlebois, R.L., L.C. Schalkwyk, J.D. Hofman and W.F. Doolittle. 1991. Detailed physical map and set of overlapping clones covering the genome of the archaebacterium *Haloferax volcanii* DS2. J. Mol. Biol. *222*: 509–524.

Chatzinotas, A., R.A. Sandaa, W. Schönhuber, R.I. Amann, F.L. Daae, V. Torsvik, J. Zeyer and D. Hahn. 1998. Analysis of broad-scale differences in microbial community composition of two pristine forest soils. Syst. Appl. Microbiol. *21*: 579–587.

Chen, M. and M.J. Wolin. 1979. Effect of monensin and lasalocid-sodium on the growth of methanogenic and rumen saccharolytic bacteria. Appl. Environ. Microbiol. *38*: 72–77.

Chen, Y.B., B. Dominic, M.T. Mellon and J.P. Zehr. 1998. Circadian rhythm of nitrogenase gene expression in the diazotrophic filamentous nonheterocystous cyanobacterium *Trichodesmium* sp. strain IMS 101. J. Bacteriol. *180*: 3598–3605.

Chen, Y.B., J.P. Zehr and M. Mellon. 1996. Growth and nitrogen fixation of the diazotrophic filamentous nonheterocystous cyanobacterium *Trichodesmium* sp. IMS 101 in defined media: evidence for a circadian rhythm. J. Phycol. *32*: 916–923.

Chester, F.D. 1897. Report of the mycologist: bacteriological work. Del. Agr. Exp. Stn. Bull. *9*: 38–145.

Chester, F.D. 1898. Report of the mycologist: bacteriological work. Del. Agr. Exp. Stn. Bull. *10*: 47–137.

Chester, F.D. 1901. A Manual of Determinative Bacteriology, The Macmillan Co., New York.

Cheung, J., K.J. Danna, E.M. O'Connor, L.B. Price and R.F. Shand. 1997. Isolation, sequence, and expression of the gene encoding halocin H4, a bacteriocin from the halophilic archaeon *Haloferax mediterranei* R4. J. Bacteriol. *179*: 548–551.

Chien, Y.T. and S.H. Zinder. 1994. Cloning, DNA sequencing, and characterization of a *nif* D-homologous gene from the archaeon *Methanosarcina barkeri* 227 which resembles *nif* D1 from the eubacterium *Clostridium pasteurianum*. J. Bacteriol. *176*: 6590–6598.

Childers, S.E., M. Vargas and K.M. Noll. 1992. Improved methods for cultivation of the extremely thermophilic bacterium *Thermotoga neapolitana*. Appl. Environ. Microbiol. *58*: 3949–3953.

Chisholm, S.W., S.L. Frankel, R. Goericke, R.J. Olson, B. Palenik, J. Waterbury, L. West-Johnrud and E.R. Zettler. 1992. *Prochlorococcus marinus* nov. gen. sp.: an oxyphototrophic prokaryote containing divinyl chlorophyll *a* and *b*. Arch. Microbiol. *157*: 297–300.

Chistoserdova, L., J.A. Vorholt, R.K. Thauer and M.E. Lidstrom. 1998. C_1 transfer enzymes and coenzymes linking methylotrophic bacteria and methanogenic archaea. Science *281*: 99–102.

Choi, I.G., S.S. Kim, J.R. Ryu, Y.S. Han, W.G. Bang, S.H. Kim and Y.G. Yu. 1997. Random sequence analysis of genomic DNA of a hyperthermophile: *Aquifex pyrophilus*. Extremophiles *1*: 125–134.

Choquet, C.G., J.C. Richards, G.B. Patel and G.D. Sprott. 1994a. Purine and pyrimidine biosynthesis in methanogenic bacteria. Arch. Microbiol. *161*: 471–480.

Choquet, C.G., J.C. Richards, G.B. Patel and G.D. Sprott. 1994b. Ribose biosynthesis in methanogenic bacteria. Arch. Microbiol. *161*: 481–488.

Chorus, I. and J. Bartram (Editors). 1999. Toxic Cyanobacteria in Water, St. Edmundsbury Press, Suffolk.

Christensen, B., T. Torsvik and T. Lien. 1992. Immunomagnetically captured thermophilic sulfate-reducing bacteria from North Sea oil field waters. Appl. Environ. Microbiol. *58*: 1244–1248.

Christensen, E.A. and H. Kristensen. 1981. Radiation-resistance of microorganisms from air in clean premises. Acta. Pathol. Microbiol. Scand. B. *89*: 293–301.

Christiansen, C., E.A. Freundt and F.T. Black. 1975. Genome size and deoxyribonucleic acid base composition of *Thermoplasma acidophilum*. Int. J. Syst. Bacteriol. *25*: 99–101.

Christiansen, C., E.A. Freundt and O. Vinther. 1981. Lack of deoxyribonucleic acid-deoxyribonucleic acid homology between *Thermoplasma acidophilum* and *Sulfolobus acidocaldarius*. Int. J. Syst. Bacteriol. *31*: 346–347.

Chumakov, K.M. 1987. Evolution of nucleotide sequences. Sov. Sci. Rev. Sect. D Biol. Rev. *7*: 51–94.

Chumakov, K.M., T.N. Zhilina, I.S. Zvyagintseva, A.L. Tarasov and G.A. Zavarzin. 1987. 5S rRNA in archaebacteria. Zh. Obshch. Biol. *48*: 167–181.

Chung, A.P., F. Rainey, M.F. Nobre, J. Burghardt and M.S. Da Costa. 1997. *Meiothermus cerbereus* sp. nov., a new slightly thermophilic species with high levels of 3-hydroxy fatty acids. Int. J. Syst. Bacteriol. *47*: 1225–1230.

Ciferri, O. 1983. *Spirulina*, the edible microorganism. Microbiol. Rev. *47*: 551–578.

Ciferri, O. and O. Tiboni. 1985. The biochemistry and industrial potential of *Spirulina*. Annu. Rev. Microbiol. *39*: 503–526.

Ciulla, R.A., S. Burggraf, K.O. Stetter and M.F. Roberts. 1994a. Occurrence and role of di- *myo*-inositol-1,1'-phosphate in *Methanococcus igneus*. Appl. Environ. Microbiol. *60*: 3660–3664.

Ciulla, R.A., C. Clougherty, N. Belay, S. Krishnan, C. Zhou, D. Byrd and M.F. Roberts. 1994b. Halotolerance of *Methanobacterium thermoautotrophicum* ΔH and Marburg. J. Bacteriol. *176*: 3177–3187.

Clarens, M. and R. Moletta. 1990. Kinetic studies of acetate fermentation by *Methanosarcina* sp. MSTA-1. Appl. Microbiol. Biotechnol. *33*: 239–244.

Clark-Curtiss, J.E. and M.A. Docherty. 1989. A species-specific repetitive sequence in *Mycobacterium leprae* DNA. J. Infect. Dis. *159*: 7–15.

Clayton, R.A., G. Sutton, P.S. Hinkle, Jr, C. Bult and C. Fields. 1995. Intraspecific variation in small-subunit rRNA sequences in GenBank: why single sequences may not adequately represent prokaryotic taxa. Int. J. Syst. Bacteriol. *45*: 595–599.

Clements, A.P. and J.G. Ferry. 1992. Cloning, nucleotide sequence, and transcriptional analyses of the gene encoding a ferredoxin from *Methanosarcina thermophila*. J. Bacteriol. *174*: 5244–5250.

Clements, A.P., L. Kilpatrick, W.P. Lu, S.W. Ragsdale and J.G. Ferry. 1994. Characterization of the iron-sulfur clusters in ferredoxin from acetate-grown *Methanosarcina thermophila*. J. Bacteriol. *176*: 2689–2693.

Clements, F.E. and H.L. Shantz. 1909. A new genus of blue-green algae. Minn. Bot. Stud. *4*: 133–135.

Cline, S.W. and W.F. Doolittle. 1992. Transformation of members of the genus *Haloarcula* with shuttle vectors based on *Halobacterium halobium* and *Haloferax volcanii* plasmid replicons. J. Bacteriol. *174*: 1076–1080.

Codd, G.A. and W.W. Carmichael. 1982. Toxicity of a cloned isolate of the cyanobacterium *Microcystis aeruginosa* from Great Britain. FEMS Microbiol. Lett. *13*: 409–411.

Cohen, Y. 1984. The Solar Lake cyanobacterial mats: strategies of photosynthetic life under sulfide. *In* Cohen, Castenholz and Halvorson (Editors), Microbial Mats: Stromatolites, Alan R. Liss, New York. pp. 133–148.

Cohen, Y., B.B. Jørgensen, N.P. Revsbech and R. Poplawski. 1986. Adaptation to hydrogen sulfide of oxygenic and anoxygenic photosynthesis among cyanobacteria. Appl. Environ. Microbiol. *51*: 398–407.

Cohen, Y., E. Padan and M. Shilo. 1975. Facultative anoxygenic photosynthesis in the cyanobacterium *Oscillatoria limnetica*. J. Bacteriol. *123*: 855–861.

Cohen-Bazire, G. and D.A. Bryant. 1982. Phycobilisomes: composition and structure. *In* Carr and Whitton (Editors), The Biology of Cyanobacteria, University of California Press, Berkeley. pp. 143–190.

Cohen-Bazire, G., N. Pfennig and R. Kunisawa. 1964. The fine structure of green bacteria. J. Cell Biol. *22*: 207–225.

Cohn, F. 1872. Untersuchungen über Bakterien. Beitr. Biol. Pflanz. *1875 1 (Heft 2)*: 127–224.

Cohn, F. 1875. Untersuchungen über Bakterien II. Beitr. Biol. Pflanz. *1875 1 (Heft 3)*: 141–207.

Cole, S.T., R. Brosch, J. Parkhill, T. Garnier, C. Churcher, D. Harris, S.V. Gordon, K. Eiglmeier, S. Gas, C.E. Barry, III, F. Tekaia, K. Badcock, D. Basham, D. Brown, T. Chillingworth, R. Connor, R. Davies, K. Devlin, T. Feltwell, S. Gentles, N. Hamlin, S. Holroyd, T. Hornsby, K. Jagels, A. Krogh, J. McLeah, S. Moule, L. Murphy, K. Oliver, J. Osborne, M.A. Quail, M.A. Rajandream, J. Rogers, S. Rutter, K. Seeger, J. Skelton, R. Squares, S. Squares, J.E. Sulston, K. Taylor, S. Whitehead and B.G. Barrett. 1998. Deciphering the biology of *Mycobacterium tuberculosis* from the complete genome sequence. Nature *393*: 537–544.

Collins, M.D. and D. Jones. 1981. Distribution of isoprenoid quinone structural types in bacteria and their taxonomic implication. Microbiol. Rev. *45*: 316–354.

Collins, M.D., H.N.M. Ross, B.J. Tindall and W.D. Grant. 1981. Distribution of isoprenoid quinones in halophilic bacteria. J. Appl. Bacteriol. *50*: 559–565.

Collins, M.D. and B.J. Tindall. 1987. Occurences of menaquinones and some novel methylated menaquinones in the alkaliphilic, halophilic archaebacterium *Natronobacterium gregoryi*. FEMS Microbiol. Lett. *43*: 307–312.

Collins, M.D. and F. Widdel. 1986. Respiratory quinones of sulphate-reducing and sulphur-reducing bacteria: a systematic investigation. Syst. Appl. Microbiol. *8*: 8–18.

Colwell, R.R. 1970. Polyphasic taxonomy of the genus *Vibrio*: numerical taxonomy of *Vibrio cholerae*, *Vibrio parahaemolyticus*, and related *Vibrio* species. J. Bacteriol. *104*: 410–433.

Colwell, R.R. 1973. Genetic and phenetic classification of bacteria. Adv. Appl. Microbiol. *16*: 137–175.

Colwell, R.R. 1993. Nonculturable but still viable and potentially pathogenic. Int. J. Med. Microbiol. Virol. Parasitol. Infect. Dis. *279*: 154–156.

Colwell, R.R., C.D. Litchfield, R.H. Vreeland, L.A. Kiefer and N.E. Gibbons. 1979. Taxonomic studies of red halophilic bacteria. Int. J. Syst. Bacteriol. *29*: 379–399.

Cometta, S., B. Sonnleiter, W. Sidler and A. Fiechter. 1982. Population distribution of aerobic extremely thermophilic microorganisms from an Icelandic natural hot spring. Eur. J. Appl. Microbiol. Biotechnol. *16*: 151–156.

Consalvi, V., R. Chiaraluce, L. Politi, R. Vaccaro, M. De Rosa and R. Scandurra. 1991. Extremely thermostable glutamate dehydrogenase from the hyperthermophilic archaebacterium *Pyrococcus furiosus*. Eur. J. Biochem. *202*: 1189–1196.

Convention on Biological Diversity Secretariat. 1992. Conference of the Parties Documentation on COP I. UNEP, 15, chemin des Anemones, CP 356, CH 1219, Geneva, Switzerland.

Conway de Macario, E., H. König and A.J.L. Macario. 1986. Immunologic distinctiveness of archaebacteria that grow in high salt. J. Bacteriol. *168*: 425–427.

Conway de Macario, E. and A.J.L. Macario. 1986. Immunology of archaebacteria: identification, antigenic relationships and immunochemistry of surface structures. Syst. Appl. Microbiol. *7*: 320–324.

Conway de Macario, E., A.J.L. Macario and O. Kandler. 1982a. Monoclonal antibodies for immunochemical analysis of methanogenic bacteria. J. Immunol. *129*: 1670–1674.

Conway de Macario, E., A.J.L. Macario, T. Mok and T.J. Beveridge. 1993. Immunochemistry and localization of the enzyme disaggregatase in *Methanosarcina mazei*. J. Bacteriol. *175*: 3115–3120.

Conway de Macario, E., A.J.L. Macario and M.J. Wolin. 1982b. Antigenic analysis of *Methanomicrobiales* and *Methanobrevibacter arboriphilus*. J. Bacteriol. *152*: 762–764.

Conway de Macario, E., M.J. Wolin and A.J.L. Macario. 1981. Immunology of archaebacteria that produce methane gas. Science *214*: 74–75.

Conway de Macario, E., M.J. Wolin and A.J.L. Macario. 1982c. Antibody analysis of relationships among methanogenic bacteria. J. Bacteriol. *149*: 316–319.

Cooper, I.P. 1997. Biotechnology and the Law, Clark Boardman Callaghan Co., New York.

Copeland, H.F. 1956. The Classification of Lower Organisms, Pacific Book, Palo Alto.

Corder, R.E., L.A. Hook, J.M. Larkin and J.I. Frea. 1983. Isolation and characterization of two new methane-producing cocci: *Methanogenium olentangyi*, sp. nov. and *Methanococcus deltae*, sp. nov. Arch. Microbiol. *134*: 28–32.

Corder, R.E., L.A. Hook, J.M. Larkin and J.I. Frea. 1988. *In* Validation of the publication of new names and new combinations previously effectively published outside the IJSB. List No. 25. Int. J. Syst. Bacteriol. *38*: 220–222.

Cord-Ruwisch, R., W. Kleinitz and F. Widdel. 1986. Sulfatreduzierende Bakterien in einem Erdölfeld—Arten und Wachstumsbedingungen. Erdöl Erdgas Kohle *102*: 281–289.

Cote, R.J. and R.L. Gherna. 1994. Nutrition and media. *In* Gerhardt, Murray, Wood and Krieg (Editors), Methods for General and Molecular Bacteriology, American Society for Microbiology, Washington, DC. pp. 155–178.

Counsell, T.J. and R.G.E. Murray. 1986. Polar lipid profiles of the genus *Deinococcus*. Int. J. Syst. Bacteriol. *36*: 202–206.

Couté, A. 1982. Ultrastructure d'une cyanophycée aérienne calcifée cavernicole: *Geitleria calcarea* Friedmann. Hydrobiologia *97*: 255–274.

Couté, A. 1985. Essai préliminaire de comparaison de deux Cyanophycées cavernicoles calcifiées: *Geitleria calcarea* Friedmann et *Scytonema julianum* Meneghini. Arch. Hydrobiol. Suppl. *71*: 91–98.

Couté, A., M. Leitao and C. Martin. 1997. Première observation du genre *Cylindrospermopsis* (*Cyanophyceae*, *Nostocales*) en France. Cryptogam. Algol. *18*: 57–70.

Cowan, D.A. and R.M. Daniel. 1982a. Purification and some properties of an extracellular protease (caldolysin) from an extreme thermophile. Biochim. Biophys. Acta *705*: 293–305.

Cowan, D.A. and R.M. Daniel. 1982b. The properties of immobilized Caldolysin, a thermostable protease from an extreme thermophile. Biotechnol. Bioeng. *24*: 2053–2062.

Cowan, S.T. 1965. Principles and practice of bacterial taxonomy—a forward look. J. Gen. Microbiol. *37*: 143–153.

Cowan, S.T. 1971. Sense and nonsense in bacterial taxonomy. J. Gen. Microbiol. *67*: 1–8.

Cowan, S.T. 1974. Cowan and Steel's Manual for the Identification of Medical Bacteria, 2nd Ed., Cambridge University Press, Cambridge.

Cowan, S.T. 1978. *In* Hill (Editor), A Dictionary of Microbial Taxonomy, Cambridge University Press, Cambridge.

Crosa, J., D.J. Brenner and S. Falkow. 1973. Use of a single-strand specific nuclease for the analysis of bacterial and plasmid DNA homo- and heteroduplexes. J. Bacteriol. *115*: 904–911.

Cruden, D., R. Sparling and A.J. Markovetz. 1989. Isolation and ultrastructure of the flagella of *Methanococcus thermolithotrophicus* and *Methanospirillum hungatei*. Appl. Environ. Microbiol. *55*: 1414–1419.

Culver, D.A. and G.J. Brunskill. 1969. Fayetteville Green Lake, New York. V. Studies of primary production and zooplankton in a meromictic marl lake. Limnol. Oceanogr. *14*: 862–873.

Curnow, A.W., D.L. Tumbula, J.T. Pelaschier, B. Min and D. Soll. 1998. Glutamyl-tRNA$^{(Gln)}$ amidotransferase in *Deinococcus radiodurans* may be confined to asparagine biosynthesis. Proc. Natl. Acad. Sci. U.S.A. *95*: 12838–12843.

Dalgaard, J.Z. and R.A. Garrett. 1992. Protein-coding introns from the 23S rRNA-encoding gene form stable circles in the hyperthermophilic archaeon *Pyrobaculum organotrophum*. Gene *121*: 103–110.

D'Amelio, E.D.A., Y. Cohen and D.J. Des Marais. 1989. Comparative functional ultrastructure of two hypersaline submerged cyanobacterial mats: Guerrero Negro, Baja California Sur, Mexico, and Solar Lake, Sinai, Egypt. *In* Cohen and Rosenberg (Editors), Microbial Mats: Physiological Ecology of Benthic Microbial Communities, ASM Publications, Washington, DC. pp. 97–113.

Damerval, T., A.M. Castets, G. Guglielmi, J. Houmard and N. Tandeau de Marsac. 1989. Occurrence and distribution of gas vesicle genes among cyanobacteria. J. Bacteriol. *171*: 1445–1452.

Dangel, W., H. Schulz, G. Diekert, H. König and G. Fuchs. 1987. Occurrence of corrinoid-containing membrane proteins in anaerobic bacteria. Arch. Microbiol. *148*: 52–56.

Daniels, L., N. Belay and B.S. Rajagopal. 1986. Assimilatory reduction of sulfate and sulfite by methanogenic bacteria. Appl. Environ. Microbiol. *51*: 703–709.

Daniels, L., G. Fuchs, R.K. Thauer and J.G. Zeikus. 1977. Carbon monoxide oxidation by methanogenic bacteria. J. Bacteriol. *132*: 118–126.

Daniels, L. and J.G. Zeikus. 1978. One-carbon metabolism in methanogenic bacteria: analysis of short-term fixation products of $^{14}CO_2$ and $^{14}CH_3OH$ incorporated into whole cells. J. Bacteriol. *136*: 75–84.

Danon, A. and S.R. Caplan. 1977. CO_2 fixation by *Halobacterium halobium*. FEBS Lett. *74*: 255–258.

Danson, M.J. 1988. Archaebacteria: the comparative enzymology of their central metabolic pathways. Adv. Microb. Physiol. *29*: 165–231.

Danson, M.J. and D.W. Hough. 1992. The enzymology of archaebacterial pathways of central metabolism. Biochem. Soc. Symp. *58*: 7–21.

Darimont, B. and R. Sterner. 1994. Sequence, assembly and evolution of a primordial ferredoxin from *Thermotoga maritima*. EMBO (Eur. Mol. Biol. Organ.) J. *13*: 1772–1781.

Darland, G., T.D. Brock, W. Samsonoff and S.F. Conti. 1970. A thermophilic, acidophilic mycoplasma isolated from a coal refuse pile. Science *170*: 1416–1418.

Das, S.K., A.K. Mishra, B.J. Tindall, F.A. Rainey and E. Stackebrandt. 1996. Oxidation of thiosulfate by a new bacterium, *Bosea thiooxidans* (strain BI-42) gen. nov., sp. nov.: analysis of phylogeny based on chemotaxonomy and 16S ribosomal DNA. Int. J. Syst. Bacteriol. *46*: 981–987.

DasSarma, S. 1993. Identification and analysis of the gas vesicle gene cluster on an unstable plasmid of *Halobacterium halobium*. Experientia (Basel) *49*: 482–486.

DasSarma, S. 1995. Natural plasmids and plasmid vectors of halophiles. *In* Robb, Sowers, DasSarma, Place, Schreier and Fleischmann (Editors), Archaea: A Laboratory Manual, Cold Spring Harbor Laboratory Press, Plainview. pp. 241–250.

DasSarma, S. and P. Arora. 1997. Genetic analysis of the gas vesicle gene cluster in haloarchaea. FEMS Microbiol. Lett. *153*: 1–10.

DasSarma, S. and E.M. Fleischmann. 1995. Archaea: A Laboratory Manual, Cold Spring Harbor Laboratory Press, Plainview.

DasSarma, S., E.M. Fleischmann and F. Rodríguez-Valera. 1995. Media for halophiles. *In* DasSarma and Fleischmann (Editors), Archaea: A Laboratory Manual, Cold Spring Harbor Laboratory Press, Plainview. pp. 225–230.

DasSarma, S. and P. Stolt. 1995. Halophages. *In* Robb, Sowers, DasSarma, Place, Schreier and Fleischmann (Editors), Archaea: A Laboratory Manual, Cold Spring Harbor Laboratory Press, Plainview. pp. 251–252.

Davey, M., W.A. Wood, R. Key, K. Nakamura and D.A. Stahl. 1993a. Isolation of three species of *Geotoga* and *Petrotoga*: two new genera, representing a new lineage in the bacterial line of descent distantly related to the "*Thermotogales*". Syst. Appl. Microbiol. *16*: 191–200.

Davey, M.E., W.A. Wood, R. Key, K. Nakamura and D.A. Stahl. 1993b. *In* Validation of the publication of new names and new combinations previously effectively published outside the IJSB. List No. 47. Int. J. Syst. Bacteriol. *43*: 864–865.

Davies, D.G., M.R. Parsek, J.P. Pearson, B.H. Iglewski, J.W. Costerton and E.P. Greenberg. 1998. The involvement of cell-to-cell signals in the development of a bacterial biofilm. Science *280*: 295–298.

Davis, N.S., G.J. Silverman and E.B. Masurovsky. 1963. Radiation-resistant, pigmented coccus isolated from Haddock tissue. J. Bacteriol. *86*: 294–298.

Deckert, G., P.V. Warren, T. Gaasterland, W.G. Young, A.L. Lenox, D.E. Graham, R. Overbeek, M.A. Snead, M. Keller, M. Aujay, R. Huber, R.A. Feldman, J.M. Short, G.J. Olsen and R.V. Swanson. 1998. The complete genome of the hyperthermophilic bacterium *Aquifex aeolicus*. Nature *392*: 353–358.

Degryse, E. and N. Glansdorff. 1976. Metabolic function of the glyoxylic shunt in an extreme thermophilic strain of the genus *Thermus*. Arch. Int. Physiol. Biochim. *84*: 598–599.

Degryse, E., N. Glansdorff and A. Piérard. 1978. A comparative analysis

of extreme thermophilic bacteria belonging to the genus *Thermus*. Arch. Microbiol. *117*: 189–196.

DeLang, R.J., G.R. Green and D.G. Searcy. 1981. A histone-like protein (HTa) from *Thermoplasma acidophilum*. I. Purification and properties. J. Biol. Chem. *256*: 900–904.

De Ley, J. 1992. The *Proteobacteria*: ribosomal RNA cistron similarities and bacterial taxonomy. *In* Balows, Trüper, Dworkin, Harder and Schleifer (Editors), The Prokaryotes: A Handbook on the Biology of Bacteria, Ecophysiology, Isolation, Identification, Applications, 2nd Ed., Vol. 2, Springer-Verlag, Berlin. pp. 2111–2140.

De Ley, J. and J. De Smedt. 1975. Improvement of the membrane filter method for DNA–rRNA hybridization. Antonie Leeuwenhoek *41*: 287–307.

DeLong, E.F. 1992. *Archaea* in coastal marine environments. Proc. Natl. Acad. Sci. U.S.A. *89*: 5685–5689.

DeLong, E.F., G.S. Wickham and N.R. Pace. 1989. Phylogenetic stains: ribosomal RNA-based probes for the identification of single cells. Science *243*: 1360–1363.

DeLong, E.F., K.Y. Wu, B.B. Prezelin and R.V.M. Jovine. 1994. High abundance of *Archaea* in Antarctic marine picoplankton. Nature *371*: 695–697.

DeMoll, E. and L. Tsai. 1986. Utilization of purines or pyrimidines as the sole nitrogen source by *Methanococcus vannielii*. J. Bacteriol. *167*: 681–684.

Denhardt, D.T. 1966. A membrane-filter technique for the detection of complementary DNA. Biochem. Biophys. Res. Commun. *23*: 641–646.

Denman, S., K. Hampson and B.K. Patel. 1991. Isolation of strains of *Thermus aquaticus* from the Australian artesian basin and a simple and rapid procedure for the preparation of their plasmids. FEMS Microbiol. Lett. *66*: 73–78.

Denner, E.B.M., T.J. McGenity, H.J. Busse, W.D. Grant, G. Wanner and H. Stan-Lotter. 1994. *Halococcus salifodinae* sp. nov., an archaeal isolate from an Austrian salt mine. Int. J. Syst. Bacteriol. *44*: 774–780.

Dennis, P.P., S. Ziesche and S. Mylvaganam. 1998. Transcription analysis of two disparate rRNA operons in the halophilic archaeon *Haloarcula marismortui*. J. Bacteriol. *180*: 4804–4813.

Deppenmeier, U., M. Blaut and G. Gottschalk. 1989. Dependence on membrane components of methanogenesis from methyl-CoM with formaldehyde or molecular hydrogen as electron donors. Eur. J. Biochem. *186*: 317–323.

de Rijk, P., E. Robbrecht, S. de Hoog, A. Caers, Y. van de Peer and R. de Wachter. 1999. Database on the structure of large subunit ribosomal RNA. Nucleic Acids Res. *27*: 174–178.

de Rijk, P., Y. van de Peer, I. van den Broeck and R. de Wachter. 1995. Evolution according to large ribosomal subunit RNA. J. Mol. Evol. *41*: 366–375.

De Rosa, M., A. Gambacorta, R. Huber, V. Lanzotti, B. Nicolaus, K.O. Stetter and A. Trincone. 1988. A new 15,16-dimethyl-30-glyceroloxytriacontanoic acid from lipids of *Thermotoga maritima*. J. Chem. Soc. Chem. Commun. *19*: 1300–1301.

De Rosa, M., A. Gambacorta, B. Nicolaus and W.D. Grant. 1983. A $C_{25}C_{25}$ diether core lipid from archaebacterial haloalkaliphiles. J. Gen. Microbiol. *129*: 2333–2338.

De Rosa, M., A. Gambacorta, B. Nicolaus, H.N.M. Ross, W.D. Grant and J.D. Bu'lock. 1982. An asymmetric archaebacterial diether lipid from alkaliphilic halophiles. J. Gen. Microbiol. *128*: 343–348.

Desikachary, T.V. 1959. Cyanophyta, Indian Council of Agricultural Research, New Delhi.

Devereux, R., S.-H. He, C.L. Doyle, S. Orkland, D.A. Stahl, J. LeGall and W.B. Whitman. 1990. Diversity and origin of *Desulfovibrio* species: phylogenetic definition of a family. J. Bacteriol. *172*: 3609–3619.

Devereux, R., M.D. Kane, J. Winfrey and D.A. Stahl. 1992. Genus- and group-specific hybridization probes for determinative and environmental studies of sulfate-reducing bacteria. Syst. Appl. Microbiol. *15*: 601–609.

DeWeerd, K.A., L. Mandelco, R.S. Tanner, C.R. Woese and J.M. Suflita. 1990. *Desulfomonile tiedjei* gen. nov. and sp. nov., a novel anaerobic, dehalogenating, sulfate-reducing bacterium. Arch. Microbiol. *154*: 23– 30.

de Winder, B., L.J. Stal and L.R. Mur. 1990. *Crinalium epipsammum* sp. nov.: a filamentous cyanobacterium with trichomes composed of elliptical cells and containing poly-β-(1,4) glucan (cellulose). J. Gen. Microbiol. *136*: 1645–1653.

De Wulf-Durand, P., L.J. Bryant and L.I. Sly. 1997. PCR-mediated detection of acidophilic, bioleaching-associated bacteria. Appl. Environ. Microbiol. *63*: 2944–2948.

Dhar, N.M. and W. Altekar. 1986. Distribution of class I and class II fructose bisphosphate aldolases in halophilic archaebacteria. FEMS Microbiol. Lett. *35*: 177–181.

Dickerson, R.E. 1980. Evolution and gene transfer in purple photosynthetic bacteria. Nature *283*: 210–212.

Diekert, G., U. Konheiser, K. Piechulla and R.K. Thauer. 1981. Nickel requirement and factor F_{430} content of methanogenic bacteria. J. Bacteriol. *148*: 459–464.

Dilling, W., W. Liesack and N. Pfenning. 1995. *Rhabdochromatium marinum* gen. nom. rev., sp. nov., a purple sulfur bacterium from a salt marsh microbial mat. Arch. Microbiol. *164*: 125–131.

Dillon, J.G. and R.W. Castenholz. 1999. Scytonemin, a cyanobacterial sheath pigment, protects against UVC radiation: implications for early photosynthetic life. J. Phycol. *35*: 673–681.

DiMarco, A.A., T.A. Bobik and R.S. Wolfe. 1990. Unusual coenzymes of methanogenesis. Annu. Rev. Biochem. *59*: 355–394.

Dirmeier, R., M. Keller, G. Frey, H. Huber and K.O. Stetter. 1998. Purification and properties of an extremely thermostable membrane-bound sulfur-reducing complex from the hyperthermophilic *Pyrodictium abyssi*. Eur. J. Biochem. *252*: 486–491.

Doddema, H.J., J.W.M. Derksen and G.D. Vogels. 1979. Fimbriae and flagella of methanogenic bacteria. FEMS Microbiol. Lett. *5*: 135–138.

Doi, R.H. and R.T. Igarashi. 1965. Conservation of ribosomal and messenger ribonucleic acid cistrons and *Bacillus* species. J. Bacteriol. *90*: 384–390.

Dolganov, N.A., D. Bhaya and A.R. Grossman. 1995. Cyanobacterial protein with similarity to the chlorophyll *a/b* binding proteins of higher plants: evolution and regulation. Proc. Natl. Acad. Sci. U.S.A. *92*: 636–640.

Donato, M.M., E.A. Seleiro and M.S. Da Costa. 1990. Polar lipid and fatty acid composition of strains of the genus *Thermus*. Syst. Appl. Microbiol. *13*: 234–239.

Donato, M.M., E.A. Seleiro and M.S. Da Costa. 1991. Polar lipid and fatty acid composition of strains of *Thermus ruber*. Syst. Appl. Microbiol. *14*: 235–239.

Doolittle, W.F. and J.M. Logsdon, Jr. 1998. Archaeal genomics: do archaea have a mixed heritage? Curr. Biol. *8*: R209–R211.

Douglas, C., A. Achatz and A. Böck. 1980. Electrophoretic characterization of ribosomal proteins from methanogenic bacteria. Zentbl. Bakteriol. Mikrobiol. Hyg. 1 Abt. Orig. C *1*: 1–11.

Drews, G. and J.F. Imhoff. 1991. Phototrophic purple bacteria. *In* Shively and Barton (Editors), Variations in Autotrophic Life, Academic Press, London. pp. 51–97.

Drouet, F. and W.A. Daily. 1956. Revision of the coccoid *Myxophyceae*. Butler Univ. Bot. Stud. *10*: 1–218.

Dubinina, G.A. and V.M. Gorlenko. 1975. New filamentous photosynthetic green bacteria containing gas vacuoles. Mikrobiologiya *44*: 511–517.

Dubinina, G.A. and S.I. Kusnetzov. 1976. The ecological and morphological characteristics of microorganisms in Lesnaya Lamba (Karelia). Int. Rev. Gesamten Hydrobiol. *61*: 1–19.

Dubnau, D., I. Smith, P. Morel and J. Marmur. 1965. Gene conservation in *Bacillus* species. 1. Conserved genetic and nucleic acid base sequence homologies. Proc. Natl. Acad. Sci. U.S.A. *54*: 491–498.

Ducharme, L., A.T. Matheson, M. Yaguchi and L.P. Visentin. 1972. Utilization of amino acids by *Halobacterium cutirubrum* in chemically defined medium. Can. J. Microbiol. *18*: 1349–1351.

Duckworth, A.W., W.D. Grant, B.E. Jones and R. van Steenbergen. 1996.

Phylogenetic diversity of soda lake alkaliphiles. FEMS Microbiol. Ecol. *19*: 181–191.

Dugdale, R.C., D.W. Menzel and J.H. Ryther. 1961. Nitrogen fixing in the Sargasso Sea. Deep Sea Res. *7*: 297–299.

Dunn, G. and B.S. Everitt. 1982. An Introduction to Mathematical Taxonomy, Cambridge University Press, Cambridge.

Dyall-Smith, M. and W.F. Doolittle. 1994. Construction of composite transposons for halophilic archaea. Can. J. Microbiol. *40*: 922–929.

Dyer, D.L. and R.D. Gafford. 1961. Some characteristics of a thermophilic blue-green alga. Science *134*: 616–617.

Dyksterhouse, S.E., J.P. Gray, R.P. Herwig, J.C. Lara and J.T. Staley. 1995. *Cycloclasticus pugetii* gen. nov., sp. nov., an aromatic hydrocarbon-degrading bacterium from marine sediments. Int. J. Syst. Bacteriol. *45*: 116–123.

Dym, O., M. Mevarech and J.L. Sussmann. 1995. Structural features that stabilise halophilic malate dehydrogenase from an archaebacterium. Science *267*: 1344–1346.

Eberhard, M.D., C. Vasquez, P. Valenzuela, R. Vicuña and A. Yudelevich. 1981. Physical characterization of a plasmid (pTT1) isolated from *Thermus thermophilus*. Plasmid *6*: 1–6.

Ebert, K. and W. Goebel. 1985. Conserved and variable regions in the chromosomal and extrachromosomal DNA of halobacteria. Mol. Gen. Genet. *200*: 96–102.

Ebert, K., W. Goebel and F. Pfeifer. 1984. Homologies between heterogeneous extrachromosomal DNA populations of *Halobacterium halobium* and 4 new halobacterial isolates. Mol. Gen. Genet. *194*: 91–97.

Edelman, M., D. Swinton, J.A. Schiff, H.T. Epstein and B. Zeldin. 1967. Deoxyribonucleic acid of the blue-green algae (Cyanophyta). Bacteriol. Rev. *31*: 315–331.

Edgell, D.R., N.M. Fast and W.F. Doolittle. 1996. Selfish DNA: the best defense is a good offense. Curr. Biol. *6*: 385–388.

Edwards, K.J., P.T. Bond, T.M. Gihring and J.F. Banfield. 2000. An archaeal iron-oxidizing extreme acidophile important in acid mine drainage. Science *287*: 1796–1799.

Edwards, K.J., T.M. Gihring and J.F. Banfield. 1999. Seasonal variations in microbial populations and environmental conditions in an extreme acid mine drainage environment. Appl. Environ. Microbiol. *65*: 3627–3632.

Edwards, P.R. and W.H. Ewing. 1962. Identification of *Enterobacteriaceae*, 2nd Ed., Burgess Publishing Company, Minneapolis.

Edwards, T. and B.C. McBride. 1975. New method for the isolation and identification of methanogenic bacteria. Appl. Microbiol. *29*: 540–545.

Egas, M.C.V., M.S. Da Costa, D.A. Cowan and E.M.V. Pires. 1998. Extracellular alpha-amylase from *Thermus filiformis* Ork A2: purification and biochemical characterization. Extremophiles *2*: 23–32.

Eggen, R.I.L., A.C.M. Geerling, P.W.J. de Groot, W. Ludwig and W.M. de Vos. 1992. Methanogenic bacterium Göl1: an acetoclastic methanogen that is closely related to *Methanosarcina frisia*. Syst. Appl. Microbiol. *15*: 582–586.

Ehling-Schulz, M., W. Bilger and S. Scherer. 1997. UV-B-induced synthesis of photoprotective pigments and extracellular polysaccharides in the terrestrial cyanobacterim *Nostoc commune*. J. Bacteriol. *179*: 1940–1945.

Ehrenberg, C.G. 1838. Die Infusionthierchen als vollkommene Organismen: ein Blick in das tiefere organische Leben der Natur, L. Voss, Leipzig. pp. i–xvii; 1–547.

Ehrenreich, A. and F. Widdel. 1994. Anaerobic oxidation of ferrous iron by purple bacteria, a new type of phototrophic metabolism. Appl. Environ. Microbiol. *60*: 4517–4526.

Ehrich, S., D. Behrens, E. Lebedeva, W. Ludwig and E. Bock. 1995. A new obligately chemolithoautotrophic, nitrite-oxidizing bacterium, *Nitrospira moscoviensis* sp. nov. and its phylogenetic relationship. Arch. Microbiol. *164*: 16–23.

Ehrmann, M., W. Ludwig and K.H. Schleifer. 1994. Reverse dot blot hybridization: a useful method for the direct identification of lactic acid bacteria in fermented food. FEMS Microbiol. Lett. *117*: 143–150.

Eichler, B. and N. Pfennig. 1988. A new green sulfur bacterium from a freshwater pond. *In* Olson, Stackebrandt and Trüper (Editors), Green Photosynthetic Bacteria, Plenum Publishing Corporation, New York. pp. 233–235.

Eichler, B. and N. Pfennig. 1990. Seasonal development of anoxygenic phototrophic bacteria in a holomictic drumlin lake (Schleinsee, Germany). Arch. Hydrobiol. *119*: 369–392.

Eisen, J.A. 1995. The RecA protein as a model molecule for molecular systematic studies of bacteria: comparison of trees of RecAs and 16S rRNAs from the same species. J. Mol. Evol. *41*: 1105–1023.

Eisenreich, W., G. Strauss, U. Werz, G. Fuchs and A. Bacher. 1993. Retrobiosynthetic analysis of carbon fixation in the phototrophic eubacterium *Chloroflexus aurantiacus*. Eur. J. Biochem. *215*: 619–632.

Ekiel, I., I.C.P. Smith and G.D. Sprott. 1983. Biosynthetic pathways in *Methanospirillum hungatei* as determined by ^{13}C NMR. J. Bacteriol. *156*: 316–326.

Ekiel, I., G.D. Sprott and G.B. Patel. 1985. Acetate and CO_2 assimilation by *Methanothrix concilii*. J. Bacteriol. *162*: 905–908.

Elazari-Volcani, B. 1940. Studies on the microflora of the Dead Sea, Doctoral thesis, Hebrew University, Jerusalem.

Elazari-Volcani, B. 1957. Genus XII. *Halobacterium*. *In* Breed, Murray and Smith (Editors), Bergey's Manual of Determinative Bacteriology, 7th Ed., The Williams & Wilkins Co., Baltimore. pp. 207–212.

Elhardt, D. and A. Böck. 1982. An *in vitro* polypeptide synthesizing system from methanogenic bacteria: sensitivity to antibiotics. Mol. Gen. Genet. *188*: 128–134.

Eloff, J.N. 1981. Autecological studies on *Microcystis*. *In* Carmichael (Editor), The Water Environment—Algal Toxins and Health, Plenum Press, New York. pp. 71–96.

Embley, T.M., A.G. O'Donnell, R. Wait and J. Rostron. 1987. Lipid and cell wall amino acid composition in the classification of members of the genus *Deinococcus*. Syst. Appl. Microbiol. *10*: 20–27.

Embley, T.M., R.H. Thomas and R.A.D. Williams. 1993. Reduced thermophilic bias in the 16S rDNA sequence from *Thermus ruber* provides further support for a relationship between *Thermus* and *Deinococcus*. Syst. Appl. Microbiol. *16*: 25–29.

Emerson, D., S. Chauhan, P. Oriel and J.A. Breznak. 1994. *Haloferax* sp. D1227, a halophilic archaeon capable of growth on aromatic compounds. Arch. Microbiol. *161*: 445–452.

Enderlin, C.S. and J.C. Meeks. 1983. Pure culture and reconstitution of the *Anthoceros-Nostoc* symbiotic association. Planta *158*: 157–165.

Engel, A.M., M. Brunen and W. Baumeister. 1993. The functional properties of Ompβ, the regularly arrayed porin of the hyperthermophilic bacterium *Thermotoga maritima*. FEMS Microbiol. Lett. *109*: 231–236.

Engel, A.M., Z. Cejka, A. Lupas, F. Lottspeich and W. Baumeister. 1992. Isolation and cloning of Ompα, a coiled-coil protein spanning the periplasmic space of the ancestral eubacterium *Thermotoga maritima*. EMBO (Eur. Mol. Biol. Organ.) J. *11*: 4369–4378.

Engel, M. 1999. Untersuchungen zur Sequenzheterogenität multipler rRNS-Operone bei Vertretern verschiedener Entwicklungslinien der Bacteria, Thesis, Technical University, Munich.

Erauso, G., A.L. Reysenbach, A. Godfroy, J.R. Meunier, B. Crump, F. Partensky, J.A. Baross, V. Marteinsson, G. Barbier, N.R. Pace and D. Prieur. 1993. *Pyrococcus abyssi* sp. nov., a new hyperthermophilic archaeon isolated from a deep-sea hydrothermal vent. Arch. Microbiol. *160*: 338–349.

Erdmann, V.A., J. Wolters, E. Huysmans, A. Vandenberghe and R. de Wachter. 1984. Collection of published 5S and 5.8S ribosomal RNA sequences. Nucleic Acids Res. *12*: r133–r166.

Erhart, R., D. Bradford, R.J. Seviour, R.I. Amann and L.L. Blackall. 1997. Development and use of fluorescent *in situ* hybridization probes for the detection and identification of *"Microthrix parvicella"* in activated sludge. Syst. Appl. Microbiol. *20*: 310–318.

Ermler, U., M. Merckel, R. Thauer and S. Shima. 1997. Formylmethanofuran: tetrahydromethanopterin formyltransferase from *Methanopyrus kandleri*—new insights into salt-dependence and thermostability. Structure (Lond.) *5*: 635–646.

Euzéby, J.P. 1997. Revised nomenclature of specific or subspecific epithets

that do not agree in gender with generic names that end in -bacter. Int. J. Syst. Bacteriol. *47*: 585.

Evans, E.H., I. Foulds and N.G. Carr. 1976. Environmental conditions and morphological variation in the blue-green alga *Chlorogloea fritschii*. J. Gen. Microbiol. *92*: 147–155.

Evans, M.C.W., B.B. Buchanan and D.I. Arnon. 1966. A new ferredoxin dependent carbon reduction cycle in a photosynthetic bacterium. Proc. Natl. Acad. Sci. U.S.A. *55*: 928–934.

Evans, P.J., D.T. Mang, K.S. Kim and L.Y. Young. 1991. Anaerobic degradation of toluene by a denitrifying bacterium. Appl. Environ. Microbiol. *57*: 1139–1145.

Evans, R.W., S.C. Kushwaha and M. Kates. 1980. The lipids of *Halobacterium marismortui*, an extremely halophilic bacterium in the Dead Sea. Biochim. Biophys. Acta *619*: 533–544.

Everett, K.D.E., R.M. Bush and A.A. Andersen. 1999. Emended description of the order *Chlamydiales*, proposal of *Parachlamydiaceae* fam. nov. and *Simkaniaceae* fam. nov., each containing one monotypic genus, revised taxonomy of the family *Chlamydiaceae*, including a new genus and five new species, and standards for the identification of organisms. Int. J. Syst. Bacteriol. *49*: 415–440.

Ewing, H.W. 1986. Edwards and Ewing's Identification of *Enterobacteriaceae*, 4th Ed., Elsevier, New York.

Ezaki, T., Y. Hashimoto and E. Yabuuchi. 1989. Fluorometric deoxyribonucleic acid-deoxyribonucleic acid hybridization in microdilution wells as an alternative to membrane filter hybridization in which radioisotopes are used to determine genetic relatedness among bacterial strains. Int. J. Syst. Bacteriol. *39*: 224–229.

Faguy, D.M., D.P. Bayley, A.S. Kostyukova, N.A. Thomas and K.F. Jarrell. 1996. Isolation and characterization of flagella and flagellin proteins from the thermoacidophilic archaea *Thermoplasma volcanium* and *Sulfolobus shibatae*. J. Bacteriol. *178*: 902–905.

Faguy, D.M., S.F. Koval and K.F. Jarrell. 1993. Effect of changes in mineral composition and growth temperature on filament length and flagellation in the archaeon *Methanospirillum hungatei*. Arch. Microbiol. *159*: 512–520.

Faguy, D.M., S.F. Koval and K.F. Jarrell. 1994. Physical characterization of the flagella and flagellins from *Methanospirillum hungatei*. J. Bacteriol. *176*: 7491–7498.

Faraldo, M.L.M., M.A. de Pedro and J. Berenguer. 1988. Purification, composition, and Ca^{2+}-binding properties of the monomeric protein of the S-layer of *Thermus thermophilus*. FEBS Lett. *235*: 117–121.

Faraldo, M.L.M., M.A. de Pedro and J. Berenguer. 1992. Sequence of the S-layer gene of *Thermus thermophilus* HB8 and functionality of its promoter in *Escherichia coli*. J. Bacteriol. *174*: 7458–7462.

Fardeau, M.-L., B. Ollivier, B.K.C. Patel, M. Magot, P. Thomas, A. Rimbault, F. Rocchiccioli and J.-L. Garcia. 1997. *Thermotoga hypogea* sp. nov., a xylanolytic, thermophilic bacterium from an oil-producing well. Int. J. Syst. Bacteriol. *47*: 1013–1019.

Farlow, W.G. 1880. On the nature of the peculiar reddening of salted codfish during the summer season, U.S. Commission of Fish and Fisheries. 969–974.

Farlow, W.G. 1886. Vegetable parasites of codfish. Bull. U.S. Fish Commission *6*: 1–4.

Fauque, G., A.R. Lino, M. Czechowski, L. Kang, D.V. DerVartanian, J.J. Moura, J. LeGall and I. Moura. 1990. Purification and characterization of bisulfite reductase (desulfofuscidin) from *Desulfovibrio thermophilus* and its complexes with exogenous ligands. Biochim. Biophys. Acta *1040*: 112–118.

Favinger, J., R. Stadtwald and H. Gest. 1989. *Rhodospirillum centenum*, sp. nov., a thermotolerant cyst-forming anoxygenic photosynthetic bacterium. Antonie Leeuwenhoek *55*: 291–296.

Fay, P. 1983. The Blue-greens (Cyanophyta-Cyanobacteria), Vol. 160, Edward Arnold, London.

Fay, P. 1992. Oxygen relations of nitrogen fixation in cyanobacteria. Microbiol. Rev. *56*: 340–373.

Fay, P. and C. van Baalen. 1987. The Cyanobacteria, Elsevier Science Publishers B.V. (Biomedical Division), Amsterdam.

Feick, R., A. Ertlmaier and U. Ermler. 1996. Crystallization and x-ray analysis of the reaction center from the thermophilic green bacterium *Chloroflexus aurantiacus*. FEBS Lett. *396*: 161–164.

Feick, R., J.A. Shiozawa and A. Ertlmaier. 1995. Biochemical and spectroscopic properties of the reaction center of the green filamentous bacterium, *Chloroflexus aurantiacus*. *In* Blankenship, Madigan and Bauer (Editors), Anoxygenic Photosynthetic Bacteria, Kluwer Academic Publishers, Dordrecht. pp. 699–708.

Felsenstein, J. 1982. Numerical methods for inferring evolutionary trees. Q. Rev. Biol. *27*: 44–57.

Felsenstein, J. 1992. PHYLIP (Phylogeny Inference Package), version 3.5c, Department of Genetics, University of Washington, Seattle, WA.

Felske, A., H. Rheims, A. Wolterink, E. Stackebrandt and A.D.L. Akkermans. 1997. Ribosome analysis reveals prominent activity of an uncultured member of the class Actinobacteria in grassland soils. Microbiology (Reading) *143*: 2983–2989.

Feltham, R.K., A.K. Power, P.A. Pell and P.A. Sneath. 1978. A simple method for storage of bacteria at −76°C. J. Appl. Bacteriol. *44*: 313–316.

Feltham, R.K.A., P.A. Wood and P.H.A. Sneath. 1984. A general-purpose system for characterizing medically important bacteria to genus level. J. Appl. Bacteriol. *57*: 279–290.

Ferguson, T.J. and R.A. Mah. 1983. Isolation and characterization of an H₂-oxidizing thermophilic methanogen. Appl. Environ. Microbiol. *45*: 265–274.

Fernández-Herrero, L.A., G. Olabarría and J. Berenguer. 1997. Surface proteins and a novel transcription factor regulate the expression of the S-layer gene in *Thermus thermophilus* HB8. Mol. Microbiol. *24*: 61–72.

Fernández-Herrero, L.A., G. Olabarría, J.R. Castón, I. Lasa and J. Berenguer. 1995. Horizontal transference of S-layer genes within *Thermus thermophilus*. J. Bacteriol. *177*: 5460–5466.

Ferrante, G., J.R. Brisson, G.B. Patel, I. Ekiel and G.D. Sprott. 1989. Structures of minor ether lipids isolated from the aceticlastic methanogen, *Methanothrix concilii*. J. Lipid Res. *30*: 1601–1610.

Ferrante, G., I. Ekiel, G.B. Patel and G.D. Sprott. 1988a. A novel core lipid isolated from the aceticlastic methanogen, *Methanothrix concilii* GP6. Biochim. Biophys. Acta *963*: 173–182.

Ferrante, G., I. Ekiel, G.B. Patel and G.D. Sprott. 1988b. Structure of the major polar lipids isolated from the aceticlastic methanogen, *Methanothrix concilii* GP6. Biochim. Biophys. Acta *963*: 162–172.

Ferrari, A., T. Brusa, A. Rutili, E. Canzi and B. Biavati. 1994. Isolation and characterization of *Methanobrevibacter oralis* sp. nov. Curr. Microbiol. *29*: 7–12.

Ferrari, A., T. Brusa, A. Rutili, E. Canzi and B. Biavati. 1995. *In* Validation of the publication of new names and new combinations previously effectively published outside the IJSB. List No. 55. Int. J. Syst. Bacteriol. *45*: 879–880.

Ferraz, A.S., L. Carreto, S. Tenreio, M.F. Nobre and M.S. Da Costa. 1994. Polar lipids and fatty acid composition of *Thermus* strains from New Zealand. Antonie Leeuwenhoek *66*: 357–363.

Ferreira, A.C., M.F. Nobre, F.A. Rainey, M.T. Silva, R. Wait, J. Burghardt, A.P. Chung and M.S. Da Costa. 1997. *Deinococcus geothermalis* sp. nov. and *Deinococcus murrayi* sp. nov., two extremely radiation-resistant and slightly thermophilic species from hot springs. Int. J. Syst. Bacteriol. *47*: 939–947.

Ferreira, A.M., R. Wait, M.F. Nobre and M.S. Da Costa. 1999. Characterization of glycolipids from *Meiothermus* spp. Microbiology (Reading) *145*: 1191–1199.

Ferris, M.J., S.C. Nold, N.P. Revsbech and D.M. Ward. 1997. Population structure and physiological changes within a hot spring microbial mat community following disturbance. Appl. Environ. Microbiol. *63*: 1367–1374.

Ferris, M.J., A.L. Ruff Roberts, E.D. Kopczynski, M.M. Bateson and D.M. Ward. 1996. Enrichment culture and microscopy conceal diverse thermophilic *Synechococcus* populations in a single hot spring microbial mat habitat. Appl. Environ. Microbiol. *62*: 1045–1050.

Ferris, M.J. and D.M. Ward. 1997. Seasonal distributions of dominant 16S rRNA-defined populations in a hot spring microbial mat exam-

ined by denaturing gradient gel electrophoresis. Appl. Environ. Microbiol. *63*: 1375–1381.

Ferry, J.G., R.D. Sherod, H.D. Peck and L.G. Ljungdahl. 1976. Autotrophic fixation of CO₂ via tetrahydrofolate intermediates by *Methanobacterium thermoautotrophicum*. *In* Schlegel, Gottschalk and Decker (Editors), Symposium on Microbial Production and Utilization of Gases (H₂, CH₄, and CO), Goltz, Göttingen. pp. 173–180.

Ferry, J.G., P.H. Smith and R.S. Wolfe. 1974. *Methanospirillum*, a new genus of methanogenic bacteria, and characterization of *Methanospirillum hungatii* sp. nov. Int. J. Syst. Bacteriol. *24*: 465–469.

Ferry, J.G. and R.S. Wolfe. 1977. Nutritional and biochemical characterization of *Methanospirillum hungatii*. Appl. Environ. Microbiol. *34*: 371–376.

Festl, H., W. Ludwig and K.H. Schleifer. 1986. DNA hybridization probe for the *Pseudomonas fluorescens* group. Appl. Environ. Microbiol. *52*: 1190–1194.

Fiala, G. and K.O. Stetter. 1986a. *Pyrococcus furiosus*, sp. nov. represents a novel genus of marine heterotrophic archaebacteria growing optimally at 100°C. Arch. Microbiol. *145*: 56–61.

Fiala, G. and K.O. Stetter. 1986b. *In* Validation of the publication of new names and new combinations previously effectively published outside the IJSB. List No. 22. Int. J. Syst. Bacteriol. *36*: 573–576.

Fiala, G., K.O. Stetter, H.W. Jannasch, T.A. Langworthy and J. Madon. 1986. *Staphylothermus marinus* sp.nov. represents a novel genus of extremely thermophilic submarine heterotrophic archaebacteria growing up to 98°C. Syst. Appl. Microbiol. *8*: 106–113.

Fiala, G., C.R. Woese, T.A. Langworthy and K.O. Stetter. 1990. *Flexistipes sinusarabici*, a novel genus and species of eubacteria occurring in the Atlantis II Deep brines of the Red Sea. Arch. Microbiol. *154*: 120–126.

Figueras, J.B., L.J. Garcia-Gil and C.A. Abella. 1997. Phylogeny of the genus *Chlorobium* based on 16S rDNA sequence. FEMS Microbiol. Lett. *152*: 31–36.

Firtel, M., G. Southam, G. Harauz and T.J. Beveridge. 1993. Characterization of the cell wall of the sheathed methanogen *Methanospirillum hungatei* GP1 as an S layer. J. Bacteriol. *175*: 7550–7560.

Firtel, M., G. Southam, G. Harauz and T.J. Beveridge. 1994. The organization of the paracrystalline multilayered spacer-plugs of *Methanospirillum hungatei*. J. Struct. Biol. *112*: 160–171.

Fischer, A. 1895. Untersuchungen über Bakterien. J. Wiss. Bot. *27*: 1–163.

Fitch, W.M. and E. Margoliash. 1967. Construction of phylogenetic trees: a method based on mutational distances as estimated from cytochrome *c* sequences of general applicability. Science *155*: 279–284.

Fitz-Gibbon, S., A.J. Choi, J.H. Miller, K.O. Stetter, M.I. Simon, R. Swanson and U.J. Kim. 1997. A fosmid-based genomic map and identification of 474 genes of the hyperthermophilic archaeon *Pyrobaculum aerophilum*. Extremophiles *1*: 36–51.

Flärdh, K., P.S. Cohen and S. Kjelleberg. 1992. Ribosomes exist in large excess over the apparent demand for protein synthesis during carbon starvation in marine *Vibrio* sp. strain CCUG 15956. J. Bacteriol. *174*: 6780–6788.

Fleischmann, R.D., M.D. Adams, O. White, R.A. Clayton, E.F. Kirkness, A.R. Kerlavage, C.J. Bult, J.F. Tomb, B.A. Dougherty, J.M. Merrick, K. McKenney, G. Sutton, W. Fitzhugh, C. Fields, J.D. Gocayne, J. Scott, R. Shirley, L.I. Liu, A. Glodek, J.M. Kelley, J.F. Weidman, C.A. Phillips, T. Spriggs, E. Hedblom, M.D. Cotton, T.R. Utterback, M.C. Hanna, D.T. Nguyen, D.M. Saudek, R.C. Brandon, L.D. Fine, J.L. Fritchman, J.L. Fuhrmann, N.S.M. Geoghagen, C.L. Gnehm, L.A. McDonald, K.V. Small, C.M. Fraser, H.O. Smith and J.C. Venter. 1995. Whole-genome random sequencing and assembly of *Haemophilus influenzae*. Science *269*: 496–512.

Florenzano, G., C. Sili, E. Pelosi and M. Vincenzini. 1985. *Cyanospira rippkae* and *Cyanospira capsulata* (gen. nov. and spp. nov.): new filamentous heterocystous cyanobacteria from Magadi Lake (Kenya). Arch. Microbiol. *140*: 301–306.

Flügge, C. 1886. Die Mikroorganismen, F.C.W. Vogel, Leipzig.

Fodor, S.P.A., J.L. Read, M.C. Pirrung, L. Stryer, A.T. Lu and D. Solas.

1991. Light-directed, spatially addressable parallel chemical synthesis. Science *251*: 767–773.

Fogg, G.E. 1942. Studies on nitrogen fixation by blue-green algae.I. Nitrogen fixation by *Anabaena cylindrica* Lemm. J. Exp. Biol. *19*: 78– 87.

Fogg, G.E. 1949. Growth and heterocyst production in *Anabaena cylindrica* Lemm. II. In relation to carbon and nitrogen metabolism. Ann. Bot. N.S. *13*: 241–259.

Fogg, G.E. 1951. Growth and heterocyst production in *Anabaena cylindrica* Lemm. III. The cytology of heterocysts. Ann. Bot. N.S. *15*: 23–35.

Formisano, M. 1962. Richerche sul "calore rosso" della pell salate: II. Isolamento e caratterizzazione degli agenti microbici causanti l'alterazione. Bolletini della Stazione Spermentale per l'Industria delle Pelli e delle Materie Concianti *38*: 183–213.

Forster, A.C., J.L. McInnes, D.C. Skingle and R.H. Symons. 1985. Non-radioactive hybridization probes prepared by the chemical labeling of DNA and RNA with a novel reagent, photobiotin. Nucleic Acids Res. *13*: 745–761.

Fossing, H., V.A. Gallardo, B.B. Jørgensen, M. Huttel, L.P. Nielsen, H. Schulz, D.E. Canfield, S. Forster, R.N. Glud, J.K. Gundersen, J. Kuver, N.B. Ramsing, A. Teske, B. Thamdrup and O. Ulloa. 1995. Concentration and transport of nitrate by the mat-forming sulphur bacterium *Thioploca*. Nature *374*: 713–715.

Foulds, I.J. and N.G. Carr. 1981. Unequal cell division preceding heterocyst development in *Chlorogloeopsis fritschii*. FEMS Microbiol. Lett. *10*: 223–226.

Fournier, D., R. Lemieux and D. Couillard. 1998. Genetic evidence for highly diversified bacterial populations in wastewater sludge during biological leaching of metals. Biotechnol. Lett. *20*: 27–31.

Fowler, V.J., N. Pfennig, W. Schubert and E. Stackebrandt. 1984. Towards a phylogeny of phototrophic purple sulfur bacteria: 16S ribosomal RNA oligonucleotide cataloging of 11 species of *Chromatiaceae*. Arch. Microbiol. *139*: 382–387.

Fox, G.E., K.R. Pechman and C.R. Woese. 1977. Comparative cataloging of 16S ribosomal ribonucleic acid: molecular approach to procaryotic systematics. Int. J. Syst. Bacteriol. *27*: 44–57.

Fox, G.E. and E. Stackebrandt. 1987. The application of 16S rRNA cataloging and 5S rRNA sequencing in bacterial systematics. Methods Microbiol. *19*: 405–458.

Fox, G.E., E. Stackebrandt, R.B. Hespell, J. Gibson, J. Maniloff, T.A. Dyer, R.S. Wolfe, W.E. Balch, R.S. Tanner, L.J. Magrum, L.B. Zablen, R. Blakemore, R. Gupta, L. Bonen, B.J. Lewis, D.A. Stahl, K.R. Luehrsen, K.N. Chen and C.R. Woese. 1980. The phylogeny of prokaryotes. Science *209*: 457–463.

Fox, G.E., J.D. Wisotzkey and P. Jurtshuk, Jr. 1992. How close is close: 16S rRNA sequence identity may not be sufficient to guarantee species identity. Int. J. Syst. Bacteriol. *41*: 166–170.

Franzmann, P.D., Y. Liu, D.I. Balkwill, H.C. Aldrich, E. Conway de Macario and D.R. Boone. 1997. *Methanogenium frigidum* sp. nov., a psychrophilic, H₂-using methanogen from Ace Lake, Antarctica. Int. J. Syst. Bacteriol. *47*: 1068–1072.

Franzmann, P.D., N. Springer, W. Ludwig, E. Conway de Macario and M. Rohde. 1992. A methanogenic archaeon from Ace Lake, Antarctica: *Methanococcoides burtonii*, sp. nov. Syst. Appl. Microbiol. *15*: 573– 581.

Franzmann, P.D., N. Springer, W. Ludwig, E. Conway de Macario and M. Rhode. 1993. *In* Validation of the publication of new names and new combinations previously effectively published outside the IJSB. List No. 45. Int. J. Syst. Bacteriol. *43*: 398–399.

Franzmann, P.D., E. Stackebrandt, K. Sanderson, J.K. Volkman, D.E. Cameron, P.L. Stevenson, T.A. McMeekin and H.R. Burton. 1988. *Halobacterium lacusprofundi*, sp. nov., a halophilic bacterium isolated from Deep Lake, Antarctica. Syst. Appl. Microbiol. *11*: 20–27.

Fredrickson, H.L., W.I.C. Rijpstra, A.C. Tas, J. van der Greef, G.F. LaVos and J.W. de Leew. 1989. Chemical characterization of benthic microbial assemblages. *In* Cohen and Rosenberg (Editors), Microbial Mats: Physiological Ecology of Benthic Microbial Communities, American Society for Microbiology, Washington, DC. pp. 455–468.

Freeze, H. and T.D. Brock. 1970. Thermostable aldolase from *Thermus aquaticus*. J. Bacteriol. *101*: 541–550.

Frémy, P. 1936. Remarques sur la morphologie et la biologie de l'*Hapalosiphon laminosus* Hanog. Ann. Protistol. *5*: 175–200.

Friedmann, I. 1955. *Geitleria calcarea* n. gen. et n. sp., a new atmophytic lime-incrusting blue-green alga. Bot. Not. *5*: 439–455.

Friedmann, I. 1961. *Chroococcidiopsis kashaii* sp.n. and the genus *Chroococcidiopsis* (studies on cave algae from Israel III). Österr. Bot. Z. *108*: 354–367.

Friedrich, A.B. and G. Antranikian. 1996. Keratin degradation by *Fervidobacterium pennavorans*, a novel thermophilic anaerobic species of the order *Thermotogales*. Appl. Environ. Microbiol. *62*: 2875–2882.

Friedrich, A.B. and G. Antranikian. 1999. *In* Validation of the publication of new names and new combinations previously effectively published outside the IJSB. List No. 68. Int. J. Syst. Bacteriol. *49*: 1–3.

Frigaard, N.-U., S. Takaichi, M. Hirota, K. Shimada and K. Matsuura. 1997. Quinones in chlorosomes of green sulfur bacteria and their role in the redox-dependent fluorescence studied in chlorosome-like bacteriochlorophyll *c* aggregates. Arch. Microbiol. *167*: 343–349.

Frischer, M.E., P.J. Floriani and S.A. Nierzwicki Bauer. 1996. Differential sensitivity of 16S rRNA targeted oligonucleotide probes used for fluorescence *in situ* hybridization is a result of ribosomal higher order structure. Can. J. Microbiol. *42*: 1061–1071.

Fritsch, F.E. 1945. The Structure and Reproduction of Algae, Cambridge University Press, Cambridge.

Fröstl, J.M. and J. Overmann. 1998. Physiology and tactic response of the phototrophic consortium "*Chlorochromatium aggregatum*". Arch. Microbiol. *169*: 129–135.

Fry, B. 1986. Sources of carbon and sulfur nutrition for consumers in three meromictic lakes of New York State. Limnol. Oceanogr. *31*: 79–88.

Fu, W. and P. Oriel. 1998. Gentisate 1,2-dioxygenase from *Haloferax* sp. D1227. Extremophiles 2: 439–446.

Fu, W. and P. Oriel. 1999. Degradation of 3-phenylpropionic acid by *Haloferax* sp. D1227. Extremophiles *3*: 45–53.

Fuchs, B.M., G. Wallner, W. Beisker, I. Schwippl, W. Ludwig and R.I. Amann. 1998. Flow cytometric analysis of the *in situ* accessibility of *Escherichia coli* 16S rRNA for fluorescently labeled oligonucleotide probes. Appl. Environ. Microbiol. *64*: 4973–4982.

Fuchs, G., E. Stupperich and G. Eden. 1980. Autotrophic CO₂ fixation *Chlorobium limicola*. Evidence for the operation of a tricarboxylic acid cycle in growing cells. Arch. Microbiol. *128*: 64–71.

Fuchs, T., H. Huber, S. Burggraf and K.O. Stetter. 1996a. 16S rDNA-based phylogeny of the archaeal order *Sulfolobales* and reclassification of *Desulfurolobus ambivalens* as *Acidianus ambivalens* comb. nov. Syst. Appl. Microbiol. *19*: 56–60.

Fuchs, T., H. Huber, S. Burggraf and K.O. Stetter. 1996b. *In* Validation of the publication of new names and new combinations previously effectively published outside the IJSB. List No. 58. Int. J. Syst. Bacteriol. *46*: 836–837.

Fuchs, T., H. Huber, T. Teiner, S. Burggraf and K.O. Stetter. 1995. *Metallosphaera prunae*, sp. nov., a novel metal-mobilizing, thermoacidophilic archaeum, isolated from a uranium mine in Germany. Syst. Appl. Microbiol. *18*: 560–566.

Fuchs, T., H. Huber, K. Teiner, S. Burggraf and K.O. Stetter. 1996c. *In* Validation of the publication of new names and new combinations previously effectively published outside the IJSB. List No. 57. Int. J. Syst. Bacteriol. *46*: 625.

Fuerst, J.A., J.A. Hawkins, A. Holmes, L.I. Sly, C.J. Moore and E. Stackebrandt. 1993. *Porphyrobacter neustonensis* gen. nov., sp. nov., an aerobic bacteriochlorophyll-synthesizing budding bacterium from fresh water. Int. J. Syst. Bacteriol. *43*: 125–134.

Füglistaller, P., H. Widmer, W. Sidler, G. Frank and H. Zuber. 1981. Isolation and characterization of phycoerythrocyanin and chromatic adaptation of the thermophilic cyanobacterium *Mastigocladus laminosus*. Arch. Microbiol. *129*: 268–274.

Fuhrman, J.A., K. McCallum and A.A. Davis. 1992. Novel major archaebacterial group from marine plankton. Nature *356*: 148–149.

Fuhrman, J.A. and C.C. Ouverney. 1998. Marine microbial diversity studied via 16S rRNA sequences: cloning results from coastal waters and counting of native archaea with fluorescent single cell probes. Aquat. Ecol. *32*: 3–15.

Fuhrmann, S., J. Overmann, N. Pfennig and U. Fischer. 1993. Influence of vitamin B_{12} and light on the formation of chlorosomes in green- and brown-colored *Chlorobium* species. Arch. Microbiol. *160*: 193–198.

Fujita, S.C., T. Oshima and K. Imahori. 1976. Purification and properties of D-glyceraldehyde-3-phosphate dehydrogenase from an extreme thermophile, *Thermus thermophilus* strain HB8. Eur. J. Biochem. *64*: 57–68.

Fujita, Y. and S. Shimura. 1974. Phycoerythrin of the marine blue-green alga *Trichodesmium thiebautii*. Plant Cell Physiol. *15*: 939–942.

Fukuzaki, S., N. Nishio and S. Nagai. 1990. Kinetics of the methanogenic fermentation of acetate. Appl. Environ. Microbiol. *56*: 3158–3163.

Fuqua, C. and E.P. Greenberg. 1998. Cell-to-cell communication in *Escherichia coli* and *Salmonella typhimurium*: they may be talking, but who's listening? Proc. Natl. Acad. Sci. U.S.A. *95*: 6571–6572.

Fuqua, W.C., S.C. Winans and E.P. Greenberg. 1994. Quorum sensing in bacteria: the *LuxR-LuxI* family of cell density-responsive transcriptional regulators. J. Bacteriol. *176*: 269–275.

Gaasterland, T. and M.A. Ragan. 1999. Microbial genescapes: phyletic and functional patterns of ORF distribution among prokaryotes. Microbiol. Comp. Genomics *3*: 199–217.

Galindo, I., R. Rangel-Aldao and J.L. Ramirez. 1993. A combined polymerase chain reaction-colour development hybridization assay in a microtitre format for the detection of *Clostridium* spp. Appl. Microbiol. Biotechnol. *39*: 553–557.

Galperin, M.Y., K.M. Noll and A.H. Romano. 1996. The glucose transport system of the hyperthermophilic anaerobic bacterium *Thermotoga neapolitana*. Appl. Environ. Microbiol. *62*: 2915–2918.

Garcia-Pichel, F. and J. Belnap. 1996. Microenvironments and microscale productivity of cyanobacterial desert crusts. J. Phycol. *32*: 774–782.

Garcia-Pichel, F. and R.W. Castenholz. 1991. Characterization and biological implications of scytonemin, a cyanobacterial sheath pigment. J. Phycol. *27*: 395–409.

Garcia-Pichel, F. and R.W. Castenholz. 1993. Occurrence of UV-absorbing, mycosporine-like compounds among cyanobacterial isolates and an estimate of their screening capacity. Appl. Environ. Microbiol. *59*: 163–169.

Garcia-Pichel, F., U. Nübel and G. Muyzer. 1998. The phylogeny of unicellular, extremely halotolerant cyanobacteria. Arch. Microbiol. *169*: 469–482.

Garcia-Pichel, F., L. Prufert-Bebout and G. Muyzer. 1996. Phenotypic and phylogenetic analyses show *Microcoleus chthonoplastes* to be a cosmopolitan cyanobacterium. Appl. Environ. Microbiol. *62*: 3284–3291.

Garlick, S., A. Oren and E. Padan. 1977. Occurrence of facultative anoxygenic photosynthesis among filamentous and unicellular cyanobacteria. J. Bacteriol. *129*: 623–629.

Garza, D.R. and C.A. Suttle. 1998. The effect of cyanophages on the mortality of *Synechococcus* spp. and selection for UV resistant viral communities. Microb. Ecol. *36*: 281–292.

Gasol, J.M., K. Jürgens, R. Massana, J.I. Calderon-Paz and C. Pedros-Alio. 1995. Mass development of *Daphnia pulex* in a sulfide-rich pond (Lake Ciso). Arch. Hydrobiol. *132*: 279–296.

Gast, D.A., U. Jenal, A. Wasserfallen and T. Leisinger. 1994. Regulation of tryptophan biosynthesis in *Methanobacterium thermoautotrophicum* Marburg. J. Bacteriol. *176*: 4590–4596.

Gehrke, T., R. Hallmann and W. Sand. 1995. Importance of exopolymers from *Thiobacillus ferrooxidans* and *Leptospirillum ferrooxidans* for bioleaching. *In* Jerez, Vargas, Toledo and Weirtz (Editors), Biohydrometallurgical Processing I, University of Chile, Santiago. pp. 1–11.

Geitler, L. 1925. Synoptische Darstellung der Cyanophyceen in morphologischer und systematischer Hinsicht. Beih. Bot. Zentralbl. *41*: 163–294.

Geitler, L. 1927. Neue Blaualgen aus Lunz. Arch. Protistenkd. *60*: 440–448.

Geitler, L. 1932 (Reprinted 1971). Cyanophyceae. *In* Kolkwitz (Editor), Rabenhorst's Kryptogamenflora von Deutschland, Österreich und der Schweiz, Vol. 14, Akademische Verlag, Leipzig. (Johnson Reprint Co., New York).

Geitler, L. 1933. Diagnosen neuer Blaualgen von den Sunda-Inseln. Arch. Hydrobiol. Suppl. *12*: 622–634.

Geitler, L. 1935. Kleine Mitteilungen über neue oder wenig bekannte Blaualgen. Österr. Bot. Z. *84*: 287.

Geitler, L. 1942. Schizophyceae. *In* Engler and Prantl (Editors), Die Natürlichen Pflanzenfamilien, 2nd Ed., Duncker and Humboldt, Berlin. pp. 1–232.

Geitler, L. 1960. Schizophyceen. *In* Zimmermann and Ozenda (Editors), Linsbauers Handbuch der Pflanzenanatomie, Band. VI, Borntraeger, Berlin. pp. 1–131.

Gelwicks, J.T., J.B. Risatti and J.M. Hayes. 1994. Carbon isotope effects associated with aceticlastic methanogenesis. Appl. Environ. Microbiol. *60*: 467–472.

Gemmell, R.T., T.J. McGenity and W.D. Grant. 1998. Use of molecular techniques to investigate possible long-term dormancy of halobacteria in evaporite deposits. Ancient Biomol. *2*: 125–133.

Gentile, J.H. and T.E. Maloney. 1969. Toxicity and environmental requirements of a strain of *Aphanizomenon flos-aquae* (L.) Ralfs. Can. J. Microbiol. *15*: 165–173.

Georganta, G., K.E. Smith and R.A.D. Williams. 1993. DNA:DNA homology and cellular components of *Thermus filiformis* and other strains of *Thermus* from New Zealand hot springs. FEMS Microbiol. Lett. *107*: 145–150.

Gerloff, G.C., G.P. Fitzgerald and F. Skoog. 1950. The isolation, purification and nutrient solution requirements of blue-green algae. *In* Proceedings of the Symposium on the Culturing of Algae, Charles F. Kettering Foundation, Dayton, Ohio. pp. 27–44.

Gest, H. 1999. Gest's postulates. ASM News *65*: 123.

Gest, H. and J.L. Favinger. 1983. *Heliobacterium chlorum* gen. nov. sp. nov., an anoxygenic brownish-green photosynthetic bacterium containing a new form of bacteriochlorophyll. Arch. Microbiol. *136*: 11–16.

Gest, H. and J.L. Favinger. 1985. *In* Validation of the publication of new names and new combinations previously effectively published outside the IJSB. List No. 17. Int. J. Syst. Bacteriol. *35*: 223–225.

Gherna, R.L. 1994. Culture preservation. *In* Gerhardt (Editor), Methods for General and Molecular Bacteriology, 2nd Ed., American Society for Microbiology, Washington, DC. pp. 278–292.

Ghiorse, W.C., D.N. Miller, R.L. Sandoli and P.L. Siering. 1996. Applications of laser scanning microscopy for analysis of aquatic microhabitats. Microsc. Res. Tech. *33*: 73–86.

Ghosh, M. and H.M. Sonawat. 1998. Kreb's TCA cycle in *Halobacterium salinarum* investigated by ^{13}C nuclear magnetic resonance. Extremophiles *2*: 427–434.

Gibbons, N.E. 1974a. Family V. *Halobacteriaceae*. *In* Buchanan and Gibbons (Editors), Bergey's Manual of Determinative Bacteriology, 8th Ed., The Williams & Wilkins Co., Baltimore. pp. 269–273.

Gibbons, N.E. 1974b. Reference collections of bacteria—the need and requirements for type and neotype strains. *In* Buchanan and Gibbons (Editors), Bergey's Manual of Determinative Bacteriology, 8th Ed., The Williams & Wilkins Co., Baltimore. pp. 14–17.

Gibbons, N.E. and R.G.E. Murray. 1978. Proposals concerning the higher taxa of bacteria. Int. J. Syst. Bacteriol. *28*: 1–6.

Gibbons, N.E., K.B. Pattee and J.G. Holt (Editors). 1981. Supplement to Index Bergeyana, The Williams & Wilkins Co., Baltimore.

Gibson, C.E. and R.V. Smith. 1982. Freshwater plankton. *In* Carr and Whitton (Editors), The Biology of Cyanobacteria, University of California Press, Berkeley. pp. 463–489.

Gibson, J. 1980. Phylogenetic analysis of photosynthetic bacteria based on comparison of 16S ribosomal RNA catalogs. *In* Halvorson and van Holde (Editors), The Origins of Life and Evolution, Alan R. Liss, New York. 97–102.

Gibson, J., W. Ludwig, E. Stackebrandt and C.R. Woese. 1985a. The phylogeny of the green photosynthetic bacteria: absence of a close relationship between *Chlorobium* and *Chloroflexus*. Syst. Appl. Microbiol. *6*: 152–156.

Gibson, J., N. Pfennig and J.B. Waterbury. 1984. *Chloroherpeton thalassium* gen. nov. et spec. nov., a non-filamentous, flexing and gliding green sulfur bacterium. Arch. Microbiol. *138*: 96–101.

Gibson, J., N. Pfennig and J.B. Waterbury. 1985b. *In* Validation of the publication of new names and new combinations previously effectively published outside the IJSB. List No. 17. Int. J. Syst. Bacteriol. *35*: 223–225.

Gibson, J., E. Stackebrandt, L.B. Zablen, R. Gupta and C.R. Woese. 1979. A phylogenetic analysis of the purple photosynthetic bacteria. Curr. Microbiol. *3*: 59–64.

Gicklhorn, J. 1921. Über den Blauglanz zweier neuer Oscillatorien. Österr. Bot. Z. *70*: 1–11.

Ginsburg-Ardré, F. 1966. *Dermocarpa, Xenococcus, Dermocarpella* (Cyanophycées): nouvelles observations. Österr. Bot. Z. *113*: 362–367.

Ginzburg, M. 1978. Ion metabolism in whole cells of *Halobacterium halobium* and *Halobacterium marismortui*. *In* Caplan and Ginzburg (Editors), Energetics and Structure of Halophilic Microorganisms, Elsevier, Amsterdam. pp. 561–577.

Giovannoni, S.J. 1991. The polymerase chain reaction. *In* Stackebrandt and Goodfellow (Editors), Nucleic Acid Techniques in Bacterial Systematics, John Wiley and Sons, Chichester. pp. 177–203.

Giovannoni, S.J., T.B. Britschgi, C.L. Moyer and K.G. Field. 1990. Genetic diversity in Sargasso Sea bacterioplankton. Nature *345*: 60–63.

Giovannoni, S.J., M.S. Rappé, K.L. Vergin and N.L. Adair. 1996. 16S rRNA genes reveal stratified open ocean bacterioplankton populations related to the green non-sulfur bacteria. Proc. Natl. Acad. Sci. U.S.A. *93*: 7979–7984.

Giovannoni, S.J., N.P. Revsbech, D.M. Ward and R.W. Castenholz. 1987a. Obligately phototrophic *Chloroflexus*: primary production in anaerobic hot spring microbial mats. Arch. Microbiol. *147*: 80–87.

Giovannoni, S.J., E. Schabtach and R.W. Castenholz. 1987b. *Isophaera pallida*, gen and comb. nov., a gliding, budding eubacterium from hot springs. Arch. Microbiol. *147*: 276–284.

Giovannoni, S.J., S. Turner, G.J. Olsen, S. Barns, D.J. Lane and N.R. Pace. 1988. Evolutionary relationships among cyanobacteria and green chloroplasts. J. Bacteriol. *170*: 3584–3592.

Glaeser, J. and J. Overmann. 1999. Selective enrichment and characterisation of *Roseospirillum parvum*, gen. nov., a new purple nonsulfur bacterium with unusual light absorption properties. Arch. Microbiol. *171*: 405–416.

Glazer, A.N. 1988. Phycobilisomes. Methods Enzymol. *167*: 304–312.

Glazer, A.N. 1989. Light guides. Directional energy transfer in a photosynthetic antenna. J. Biol. Chem. *264*: 1–4.

Gleason, F.K. and C.A. Baxa. 1986. Activity of the natural algicide, cyanobacterin, on eukaryotic microorganisms. FEMS Microbiol. Lett. *33*: 85–88.

Glöckner, F.O., R.I. Amann, A. Alfreider, J. Pernthaler, R. Psenner, K. Trebesius and K.H. Schleifer. 1996. An *in situ* hybridization protocol for detection and identification of planktonic bacteria. Syst. Appl. Microbiol. *19*: 403–406.

Gloe, A., N. Pfennig, H. Brockmann, Jr. and W. Trowitzsch. 1975. A new bacteriochlorophyll from brown-colored *Chlorobiaceae*. Arch. Microbiol. *102*: 103–109.

Gloe, A. and N. Risch. 1978. Bacteriochlorophyll c_s, a new bacteriochlorophyll from *Chloroflexus aurantiacus*. Arch. Microbiol. *118*: 153–156.

Glover, H.E. 1986. The physiology and ecology of the marine cyanobacterial genus *Synechococcus*. *In* Jannasch and Williams (Editors), Advances in Aquatic Microbiology, Vol. 3, Academic Press, Inc., London. pp. 49–107.

Gluch, M.F., D. Typke and W. Baumeister. 1995. Motility and thermotactic responses of *Thermotoga maritima*. J. Bacteriol. *177*: 5473–5479.

Godfroy, A., F. Lesongeur, G. Raguénès, J. Quérellou, E. Antoine, J.-R. Meunier, J. Guezennec and G. Barbier. 1997. *Thermococcus hydrothermalis* sp. nov., a new hyperthermophilic archaeon isolated from a deep-sea hydrothermal vent. Int. J. Syst. Bacteriol. *47*: 622–626.

Godfroy, A., J.-R. Meunier, J. Guezennec, F. Lesongeur, G. Raguénès, A. Rimbault and G. Barbier. 1996. *Thermococcus fumicolans* sp. nov., a new

hyperthermophilic archaeon isolated from a deep-sea hydrothermal vent in the North Fiji Basin. Int. J. Syst. Bacteriol. *46*: 1113–1119.

Goebel, B.M. 1997. Prokaryotic diversity in acidic, metal-leaching habitats, Doctoral thesis, University of Queensland, Brisbane.

Goebel, B.M. and E. Stackebrandt. 1995. Molecular analysis of the microbial biodiversity in a natural acidic environment. *In* Jerez, Vargas, Toledo and Weirtz (Editors), Biohydrometallurgical Processing II, University of Chile, Santiago. pp. 43–52.

Goericke, R. and D.J. Repeta. 1993. Chlorophylls *a* and *b* and divinyl chlorophylls *a* and *b* in the open subtropical North Atlantic Ocean. Mar. Ecol. Prog. Ser. *101*: 307–313.

Gokhale, J.U., H.C. Aldrich, L. Bhatnagar and J.G. Zeikus. 1993. Localization of carbon monoxide dehydrogenase in acetate-adapted *Methanosarcina barkeri*. Can. J. Microbiol. *39*: 223–226.

Golovacheva, R.S., O.V. Golyshina, G.I. Karavaiko, A.G. Dorofeev, T.A. Pivovarova and N.A. Chernykh. 1992. A new iron-oxidizing bacterium, *Leptospirillum thermoferrooxidans* sp. nov. Mikrobiologiya *61*: 1056–1065.

Golovacheva, R.S., K.M. Val'ekho-Roman and A.V. Troitskii. 1987. *Sulfurococcus mirabilis* gen. nov., sp. nov., a new thermophilic archaebacterium with the ability to oxidize sulfur. Mikrobiologiya *56*: 100–107.

Golovacheva, R.S., K.M. Val'ekho-Roman and A.V. Troitskii. 1995. *In* Validation of the publication of new names and new combinations previously effectively published outside the IJSB. List No. 55. Int. J. Syst. Bacteriol. *45*: 879–880.

Golovacheva, R.S., N.V. Zhilina, L.O. Severina and A.G. Dorofeev. 1991. Effect of external conditions on the behavior of *Sulfurococcus* B. Mikrobiologiya *60*: 628–636.

Golubic, S., M. Hernández-Mariné and L. Hoffmann. 1996. Developmental aspects of branching in filamentous Cyanophyta/Cyanobacteria. Algol. Stud. *83*: 303–329.

Gomont, M. 1892–93. Monographie des Oscillariées. Ann. Sci. Nat. Ser. Bot. *15–16*: 263–368; 91–264.

Gonzalez, C., C. Gutiérrez and C. Ramirez. 1978. *Halobacterium vallismortis* sp. nov., an amylolytic and carbohydrate-metabolizing, extremely halophilic bacterium. Can. J. Microbiol. *24*: 710–715.

González, J.M., C. Kato and K. Horikoshi. 1995. *Thermococcus peptonophilus* sp. nov., a fast-growing, extremely thermophilic archaebacterium isolated from deep-sea hydrothermal vents. Arch. Microbiol. *164*: 159–164.

González, J.M., C. Kato and K. Horikoshi. 1996. *In* Validation of the publication of new names and new combinations previously effectively published outside the IJSB. List No. 57. Int. J. Syst. Bacteriol. *46*: 625–626.

González, J.M., Y. Masuchi, F.T. Robb, J.W. Ammerman, D.L. Maeder, M. Yanagibayashi, J. Tamaoka and C. Kato. 1998. *Pyrococcus horikoshii* sp. nov., a hyperthermophilic archaeon isolated from a hydrothermal vent at the Okinawa Trough. Extremophiles *2*: 123–130.

González, J.M., F.T. Robb and C. Kato. 1999. *In* Validation of the publication of new names and new combinations previously effectively published outside the IJSB. List No. 71. Int. J. Syst. Bacteriol. *49*: 1325–1326.

Gophen, M. 1977. Feeding of *Daphnia* on *Chlamydomonas* and *Chlorobium*. Nature *265*: 271–273.

Gophen, M., B.Z. Cavari and T. Berman. 1974. Zooplankton feeding on differently labelled algae and bacteria. Nature *247*: 393–395.

Gordon, R.E. 1967. The taxonomy of soil bacteria. *In* Gray and Parkinson (Editors), The Ecology of Soil Bacteria. An International Symposium, University of Toronto Press, Toronto. pp. 293–321.

Gorham, P.R., J.S. McLachlan, U.T. Hammer and W.K. Kim. 1964. Isolation and culture of toxic strains of *Anabaena flos-aquae* (Lyng.) de Bréb. Verh. Int. Ver. Theor. Angew. Limnol. *15*: 796–804.

Gorkovenko, A., M.F. Roberts and R.H. White. 1994. Identification, biosynthesis, and function of 1,3,4,6-hexanetetracarboxylic acid in *Methanobacterium thermoautotrophicum* ΔH. Appl. Environ. Microbiol. *60*: 1249–1253.

Gorlenko, V.M. 1968. A new species of the green sulfur bacteria. Dokl. Akad. Nauk. SSSR. *179*: 1229–1231.

Gorlenko, V.M. 1970. A new phototrophic green sulfur bacterium —

Prosthechloris aestuarii nov. gen. nov. sp. Z. Allg. Mikrobiol. *10*: 147–149.

Gorlenko, V.M. 1972. A new species of phototropic brown sulfur bacteria *Pelodictyon phaeum* nov. spec. Mikrobiologiya *41*: 370–371.

Gorlenko, V.M. 1974. Oxidation of thiosulfate by *Amoebobacter roseus* in the darkness under microaerobic conditions. Mikrobiologiya *43*: 729–731.

Gorlenko, V.M. 1975. Characteristics of filamentous phototrophic bacteria from freshwater lakes. Mikrobiologiya *44*: 756–758.

Gorlenko, V.M., E.N. Chebotarev and V.I. Kachalkin. 1974. Participation of microorganisms in sulfur turnover in Pomiaretzkoe Lake. Mikrobiologiya *43*: 908–914.

Gorlenko, V.M. and S.A. Korotkov. 1979. Morphological and physiological features of the new filamentous gliding green bacterium *Oscillochloris trichoides* nov. comb.. Izv. Akad. Nauk Uzb. S.S.S.R. Ser. Biol. *5*: 848–857.

Gorlenko, V.M. and S.A. Korotkov. 1989. *In* Validation of the publication of new names and new combinations previously effectively published outside the IJSB. List No. 31. Int. J. Syst. Bacteriol. *39*: 495–497.

Gorlenko, V.M. and S.I. Kuznetsov. 1971. Vertical distribution of photosynthetic bacteria in lake Konon'er in Mari ASSR. Mikrobiologiya *40*: 746–747.

Gorlenko, V.M. and S.I. Kusnetsov. 1972. Über die photosynthetisierenden Bakterien des Kononjer Sees. Arch. Hydrobiol. *70*: 1–13.

Gorlenko, V.M. and E.V. Lebedeva. 1971. New green bacteria with outgrowths. Mikrobiologiya *40*: 1035–1039.

Gorlenko, V.M. and S.I. Lokk. 1979. Vertical distribution and characteristics of the species composition of microorganisms from some stratified Estonian lakes. Mikrobiologiya *48*: 351–359.

Gorlenko, V.M. and T.A. Pivovarova. 1977. On the assignment of the blue-green alga *Oscillatoria coerulescens* Gicklhorn, 1921 to the new genus of chlorobacteria *Oscillochloris* nov. gen. Izv. Akad. Nauk Uzb. SSR Ser. Biol. *3*: 396–409.

Gorlenko, V.M. and T.A. Pivovarova. 1989. *In* Validation of the publication of new names and new combinations previously effectively published outside the IJSB. List No. 31. Int. J. Syst. Bacteriol. *39*: 495–497.

Gorlenko, V.M., M.B. Vainshtein and V.I. Kachalkin. 1976. Microbial characteristic of Lake Mogil'noe. Arch. Hydrobiol. *81*: 475–492.

Gorlenko, V.M. and T.N. Zhilina. 1968. A study on a fine structure of green sulfur bacteria, strain SK-413. Mikrobiologiya *37*: 1052–1056.

Gosink, J.J., C.R. Woese and J.T. Staley. 1998. *Polaribacter* gen. nov., with three new species, *P. irgensii* sp. nov., *P. franzmannii* sp. nov. and *P. filamentus* sp. nov., gas vacuolate polar marine bacteria of the *Cytophaga-Flavobacterium-Bacteroides* group and reclassification of '*Flectobacillus glomeratus*' as *Polaribacter glomeratus* comb. nov. Int. J. Syst. Bacteriol. *48*: 223–235.

Gould, S.J. 1996. Full house: the spread of excellence from Plato to Darwin, Harmony Books, New York.

Grant, I.R. and M.F. Patterson. 1989. A novel radiation-resistant *Deinobacter* sp. isolated from irradiated pork. Lett. Appl. Microbiol. *8*: 21–24.

Grant, S., W.D. Grant, B.E. Jones, C. Cato and L. Li. 1999. Novel archaeal phylotypes from an East African saltern. Extremophiles *3*: 139–146.

Grant, W.D., R.T. Gemmell and T.J. McGenity. 1998a. Halobacteria: the evidence for longevity. Extremophiles *2*: 279–287.

Grant, W.D. and H. Larsen. 1989a. Extremely halophilic archaeobacteria. *In* Staley, Bryant, Pfennig and Holt (Editors), Bergey's Manual of Systematic Bacteriology, 1st Ed., Vol. 3, The Williams & Wilkins Co., Baltimore. pp. 2216–2219.

Grant, W.D. and H. Larsen. 1989b. *In* Validation of the publication of new names and new combinations previously effectively published outside the IJSB. List No. 31. Int. J. Syst. Bacteriol. *39*: 495–497.

Grant, W.D., A. Oren and A. Ventosa. 1998b. Proposal of strain NCIMB 13488 as neotype of *Halorubrum trapanicum*. Request for an opinion. Int. J. Syst. Bacteriol. *48*: 1077–1078.

Grant, W.D., G. Pinch, J.E. Harris, M. DeRosa and A. Gambacorta. 1985. Polar lipids in methanogen taxonomy. J. Gen. Microbiol. *131*: 3277–3286.

Grant, W.D. and H.N.M. Ross. 1986. The ecology and taxonomy of halobacteria. FEMS Microbiol. Rev. *39*: 9–15.

Grassia, G.S., K.M. McLean, P. Glénat, J. Bauld and A.J. Sheehy. 1996. A systematic survey for thermophilic fermentative bacteria and archaea in high temperature petroleum reservoirs. FEMS Microbiol. Ecol. *21*: 47–58.

Greene, A.C., B.K.C. Patel and A.J. Sheehy. 1997. *Deferribacter thermophilus* gen. nov., sp. nov., a novel thermophilic manganese- and iron-reducing baterium isolated from a petroleum reservoir. Int. J. Syst. Bacteriol. *47*: 505–509.

Greuter, W., D.L. Hawksworth, J. McNeill, M.A. Mayo, A. Minelli, P.H.A. Sneath, B.J. Tindall, P. Trehane and P. Tubbs. 1998. Draft Biocode (1997): the prospective international rules for the scientific naming of organisms. Taxon *47*: 129–150.

Griffiths, D.J. and L.-V. Thinh. 1989. Current status of the taxonomy of the genus *Prochloron. In* Staley, Bryant, Pfennig and Holt (Editors), Bergey's Manual of Systematic Bacteriology, 1st Ed., Vol. 3, The Williams & Wilkins Co., Baltimore. pp. 1802–1805.

Grimm, D., H. Merkert, W. Ludwig, K.H. Schleifer, J. Hacker and B.C. Brand. 1998. Specific detection of *Legionella pneumophila*: construction of a new 16S rRNA-targeted oligonucleotide probe. Appl. Environ. Microbiol. *64*: 2686–2690.

Grimont, P.A.D., F. Grimont, N. Desplaces and P. Tchen. 1985. DNA probe specific for *Legionella pneumophila*. J. Clin. Microbiol. *21*: 431– 437.

Grimont, P.A.D., M.Y. Popoff, F. Grimont, C. Coynault and M. Lemelin. 1980. Reproducibility and correlation study of three deoxyribonucleic acid hybridization procedures. Curr. Microbiol. *4*: 325–330.

Grimsley, J.K., C.I. Masters, E.P. Clark and K.W. Minton. 1991. Analysis by pulsed-field gel electrophoresis of DNA double-strand breakage and repair in *Deinococcus radiodurans* and a radiosensitive mutant. Int. J. Radiat. Biol. *60*: 613–626.

Grobbelaar, N., T.C. Huang, H.Y. Lin and T.J. Chow. 1986. Dinitrogen-fixing endogenous rhythm in *Synechococcus* RF-1. FEMS Microbiol. Lett. *37*: 173–177.

Grogan, D., P. Palm and W. Zillig. 1990. Isolate B12, which harbors a virus-like element, represents a new species of the archaebacterial genus *Sulfolobus, Sulfolobus shibatae*, sp. nov. Arch. Microbiol. *154*: 594–599.

Grogan, D., P. Palm and W. Zillig. 1991. *In* Validation of the publication of new names and new combinations previously effectively published outside the IJSB. List No. 38. Int. J. Syst. Bacteriol. *41*: 456–457.

Groot Obbink, D.J., L.J. Ritchie, F.H. Cameron, J.S. Mattick and V.P. Ackerman. 1985. Construction of a gentamicin resistance gene probe for epidemiological studies. Antimicrob. Agents Chemother. *28*: 96–102.

Grossman, A.R., M.R. Schaefer, G.G. Chiang and J.L. Collier. 1993. The phycobilisome, a light-harvesting complex responsive to environmental conditions. Microbiol. Rev. *57*: 725–749.

Grunstein, M. and D.S. Hogness. 1975. Colony hybridization: a method for the isolation of cloned DNAs that contain a specific gene. Proc. Natl. Acad. Sci. U.S.A. *72*: 3961–3965.

Guglielmi, G. and G. Cohen-Bazire. 1982a. Comparative study of the structure and distribution of the extracellular filaments (fimbriae) in some cyanobacteria. Protistologica *18*: 167–178.

Guglielmi, G. and G. Cohen-Bazire. 1982b. Structure and distribution of pores and perforations of the peptidoglycan wall layer in some cyanobacteria. Protistologica *18*: 151–166.

Guglielmi, G. and G. Cohen-Bazire. 1984a. Taxonomic study of a cyanobacterial genus belonging to the *Oscillatoriaceae*: the genus *Pseudanabaena*: 1. Ultrastructure. Protistologica *20*: 377–392.

Guglielmi, G. and G. Cohen-Bazire. 1984b. Taxonomic study of a cyanobacterial genus belonging to the *Oscillatoriaceae*: the genus *Pseudanabaena*: 2. Molecular composition and structure of the phycobilisomes. Protistologica *20*: 393–414.

Guigliarelli, B., P. Bertrand, C. More, P. Papavassiliou, E.C. Hatchikian and J.P. Gayda. 1985. Interconversions between the 3Fe and 4Fe forms of the iron-sulfur clusters in the ferredoxin from *Thermodesulfobacter-*

ium commune. EPR characterization and potentiometric titration. Biochim. Biophys. Acta *810:* 319–324.

Gunsalus, R.P. and R.S. Wolfe. 1978. ATP activation and properties of the methyl coenzyme M reductase system in *Methanobacterium thermoautotrophicum.* J. Bacteriol. *135:* 851–857.

Gupta, R.S. 1996. Evolutionary relationships of chaperonins. *In* Ellis (Editor), The Chaperonins, Academic Press, New York. pp. 27–64.

Gupta, R.S. 1998. Protein phylogenies and signature sequences: a reappraisal of evolutionary relationships among archaebacteria, eubacteria, and eukaryotes. Microbiol. Mol. Biol. Rev. *62:* 1435–1491.

Gupta, R.S., B. Kevin, F. Mizied and S. Davindra. 1997. Sequencing of heat shock protein 70 (DnaK) homologs from *Deinococcus proteolyticus* and *Thermomicrobium roseum* and their integration in a protein-based phylogeny of prokaryotes. J. Bacteriol. *179:* 345–357.

Gupta, R.S. and C.R. Woese. 1980. Unusual modification patterns in the transfer ribonucleic acids of archaebacteria. Curr. Microbiol. *4:* 245–249.

Guschin, D.Y., B.K. Mobarry, D. Proudnikov, D.A. Stahl, B.E. Rittmann and A.D. Mirzabekov. 1997. Oligonucleotide microchips as genosensors for determinative and environmental studies in microbiology. Appl. Environ. Microbiol. *63:* 2397–2402.

Gutiérrez, M.C., M.T. García, A. Ventosa, J.J. Nieto and F. Ruiz-Berraquero. 1986. Occurrence of megaplasmids in halobacteria. J. Appl. Bacteriol. *61:* 67–72.

Gutiérrez, M.C., A. Ventosa and F. Ruiz-Berraquero. 1989. DNA–DNA homology studies among strains of *Haloferax* and other halobacteria. Curr. Microbiol. *18:* 253–256.

Gutiérrez, M.C., A. Ventosa and F. Ruiz-Berraquero. 1990. Deoxyribonucleic acid relatedness among species of *Haloarcula* and other halobacteria. Biochem. Cell Biol. *68:* 106–110.

Guyoneaud, R., R. Matheron, W. Liesack, J.F. Imhoff and P. Caumette. 1997. *Thiorhodococcus minus,* gen. nov., sp. nov., a new purple sulfur bacterium isolated from coastal lagoon sediments. Arch. Microbiol. *168:* 16–23.

Guyoneaud, R., J. Süling, R. Petri, R. Matheron, P. Caumette, N. Pfennig and J.F. Imhoff. 1998. Taxonomic rearrangements of the genera *Thiocapsa* and *Amoebobacter* on the basis of 16S rDNA sequence analyses, and description of *Thiolamprovum* gen. nov. Int. J. Syst. Bacteriol. *48:* 957–964.

Haas, E.S. and J.W. Brown. 1998. Evolutionary variation in bacterial RNase P RNAs. Nucleic Acids Res. *26:* 4093–4099.

Hackett, N.R., Y. Bobovnikova and N. Heyrovska. 1994. Conservation of chromosomal arrangement among three strains of the genetically unstable archaeon *Halobacterium salinarium.* J. Bacteriol. *176:* 7711–7718.

Häder, D.P. 1987. Photosensory behavior in procaryotes. Microbiol. Rev. *51:* 1–21.

Hafenbradl, D., M. Keller, R. Dirmeier, R. Rachel, P. Rossnagel, S. Burggraf, H. Huber and K.O. Stetter. 1996. *Ferroglobus placidus* gen. nov., sp. nov., a novel hyperthermophilic archaeum that oxidized Fe^{2+} at neutral pH under anoxic conditions. Arch. Microbiol. *166:* 308–314.

Hafenbradl, D., M. Keller, R. Dirmeier, R. Rachel, P. Rossnagel, S. Burggraf, H. Huber and K.O. Stetter. 1997. *In* Validation of the publication of new names and new combinations previously effectively published outside the IJSB. List No. 61. Int. J. Syst. Bacteriol. *47:* 601–602.

Hafenbradl, D., M. Keller, R. Thiericke and K.O. Stetter. 1993. A novel unsaturated archaeal ether core lipid from the hyperthermophile *Methanopyrus kandleri.* Syst. Appl. Microbiol. *16:* 165–169.

Hahn, D., R.I. Amann, W. Ludwig, A.D.L. Akkermans and K.H. Schleifer. 1992. Detection of microorganisms in soil after *in situ* hybridization with rRNA-targeted, fluorescently labelled oligonucleotides. J. Gen. Microbiol. *138:* 879–887.

Hahn, D., R.I. Amann and J. Zeyer. 1993. Detection of mRNA in *Streptomyces* cells by whole-cell hybridization with digoxigenin-labeled probes. Appl. Environ. Microbiol. *59:* 2753–2757.

Halfen, L.N. and R.W. Castenholz. 1971. Gliding motility in the blue-green alga, *Oscillatoria princeps.* J. Phycol. *7:* 133–145.

Halfen, L.N., B.K. Pierson and G.W. Francis. 1972. Carotenoids of a gliding organism containing bacteriochlorophylls. Arch. Mikrobiol. *82:* 240–246.

Hall, B.D. and S. Spiegelman. 1961. Sequence complementarity of T2-DNA and T2-specific RNA. Proc. Natl. Acad. Sci. U.S.A. *47:* 137–146.

Hall, I.C. 1927. Some fallacious tendencies in bacteriological taxonomy. J. Bacteriol. *13:* 245–253.

Hallmann, R., A. Friedrich, H.P. Koops, A. Pommerening-Roeser, K. Rohde, C. Zenneck and W. Sand. 1992. Physiological characteristics of *Thiobacillus ferrooxidans* and *Leptospirillum ferrooxidans* and physicochemical factors influence microbial metal leaching. Geomicrobiol. J. *10:* 193–206.

Hamana, K., H. Hamana, M. Niitsu, K. Samejima and T. Itoh. 1996. Distribution of long linear and branched polyamines in thermophilic eubacteria and hyperthermophilic archaebacteria. Microbios *85:* 19–33.

Hamana, K., S. Matsuzaki, M. Niitsu and K. Samejima. 1990. Pentaamines and hexamine are present in a thermophilic eubacterium *Thermomicrobium roseum.* FEMS Microbiol. Lett. *68:* 31–34.

Hamana, K., M. Niitsu, K. Samejima, T. Itoh, H. Hamana and T. Shinozawa. 1998. Polyamines of the thermophilic eubacteria belonging to the genera *Thermotoga, Thermodesulfovibrio, Thermoleophilum, Thermus, Rhodothermus* and *Meiothermus,* and the thermophilic archaebacteria belonging to the genera *Aeropyrum, Picrophilus, Methanobacterium* and *Methanococcus.* Microbios *94:* 7–21.

Hammond, A.C., M.J. Allison, M.J. Williams, G.M. Prine and D.B. Bates. 1989. Prevention of leucaena toxicosis of cattle in Florida (USA) by ruminal inoculation with 3-hydroxy-4-(1H)-pyridone-degrading bacteria. Am. J. Vet. Res. *50:* 2176–2180.

Hanada, S., A. Hiraishi, K. Shimada and K. Matsuura. 1995a. *Chloroflexus aggregans* sp. nov., a filamentous phototrophic bacterium which forms dense cell aggregates by active gliding movement. Int. J. Syst. Bacteriol. *45:* 676–681.

Hanada, S., A. Hiraishi, K. Shimada and K. Matsuura. 1995b. Isolation of *Chloroflexus aurantiacus* and related thermophilic phototrophic bacteria from Japanese hot springs using an improved isolation procedure. J. Gen. Appl. Microbiol. *41:* 119–130.

Hanada, S., Y. Kawase, A. Hiraishi, S. Takaichi, K. Matsuura, K. Shimada and K.V.P. Nagashima. 1997. *Porphyrobacter tepidarius* sp. nov., a moderately thermophilic aerobic photosynthetic bacteium isolated from a hot spring. Int. J. Syst. Bacteriol. *47:* 408–413.

Hansen, M.T. 1978. Multiplicity of genome equivalents in the radiation-resistant bacterium *Micrococcus radiodurans.* J. Bacteriol. *134:* 71–75.

Hansgirg, A. 1888. Synopsis generum subgenerumque Myxophycearum (Cyanophycearum), huiusque cognitorum cum descriptione generis nov. *Dactylococcopsis.* Notarisia *3:* 548–590.

Harashima, K., J.-I. Hayashi, T. Ikari and T. Shiba. 1980. O$_2$-stimulated synthesis of bacteriochlorophyll and carotenoids in marine bacteria. Plant Cell Physiol. *21:* 1283–1294.

Harder, R. 1917. Ernährungsphysiologische Untersuchungen an Cyanophyceen, hauptsächlich dem endophytischen *Nostoc punctiforme.* Z. Bot. *9:* 149–242.

Harmsen, H.J.M., A.D.L. Akkermans, A.J.M. Stams and W.M. de Vos. 1996a. Population dynamics of propionate-oxidizing bacteria under methanogenic and sulfidogenic conditions in anaerobic granular sludge. Appl. Environ. Microbiol. *62:* 2163–2168.

Harmsen, H.J.M., H.M.P. Kengen, A.D.L. Akkermans, A.J.M. Stams and W.M. de Vos. 1996b. Detection and localization of syntrophic propionate-oxidizing bacteria in granular sludge by *in situ* hybridization using 16S rRNA-based oligonucleotide probes. Appl. Environ. Microbiol. *62:* 1656–1663.

Harmsen, H.J.M., D. Prieur and C. Jeanthon. 1997. Group-specific 16S rRNA-targeted oligonucleotide probes to identify thermophilic bacteria in marine hydrothermal vents. Appl. Environ. Microbiol. *63:* 4061–4068.

Harriott, O.T., R. Huber, K.O. Stetter, P.W. Betts and K.M. Noll. 1994. A cryptic miniplasmid from the hyperthermophilic bacterium *Thermotoga* sp. strain RQ7. J. Bacteriol. *176:* 2759–2762.

Harris, J.E., P.A. Pinn and R.P. Davis. 1984. Isolation and characterization of a novel thermophilic, freshwater methanogen. Appl. Environ. Microbiol. *48*: 1123–1128.

Harris, J.E., P.A. Pinn and R.P. Davis. 1996. *In* Validation of the publication of new names and new combinations previously effectively published outside the IJSB. List No. 57. Int. J. Syst. Bacteriol. *46*: 625– 626.

Harris, J.I., J.D. Hocking, M.J. Runswick, K. Suzuki and J.E. Walker. 1980. D-glyceraldehyde-3-phosphate dehydrogenase. The purification and characterisation of the enzyme from the thermophiles *Bacillus stearothermophilus* and *Thermus aquaticus*. Eur. J. Biochem. *108*: 535–547.

Harrison, A.P. and P.R. Norris. 1985. *Leptospirillum ferrooxidans* and similar bacteria: characteristics and genomic diversity. FEMS Microbiol. Lett. *30*: 99–102.

Harrison, F.C. and M.E. Kennedy. 1922. The red discolouration of cured codfish. Proc. Trans. R. Soc. Can. Sect. V. *16*: 101–152.

Harsojo, S. Kitayama and A. Matsuyama. 1981. Genome multiplicity and radiation resistance in *Micrococcus radiodurans*. J. Biochem. (Tokyo) *90*: 877–880.

Hartmann, R., H.-D. Sickinger and D. Oesterhelt. 1980. Anaerobic growth of halobacteria. Proc. Natl. Acad. Sci. U.S.A. *77*: 3821–3825.

Hartmann, R.K., J. Wolters, B. Kröger, S. Schultze, T. Specht and V.A. Erdmann. 1989. Does *Thermus* represent another deep eubacterial branching? Syst. Appl. Microbiol. *11*: 243–249.

Haselkorn, R. 1978. Heterocysts. Annu. Rev. Plant Physiol. *29*: 319–344.

Haselkorn, R. 1986. Organization of the genes for nitrogen fixation in photosynthetic bacteria and cyanobacteria. Annu. Rev. Microbiol. *40*: 525–547.

Haselkorn, R. 1992. Developmentally regulated gene rearrangements in prokaryotes. Annu. Rev. Genet. *26*: 113–130.

Haselkorn, R. 1995. Molecular genetics of nitrogen fixation in photosynthetic prokaryotes. *In* Provorov, Romanov, Newton and Tikhonovich (Editors), Nitrogen Fixation: Fundamentals and Applications, Kluwer Academic Publishers, Dordrecht. pp. 29–36.

Hastings, J.W. and E.P. Greenberg. 1999. Quorum sensing: the explanation of a curious phenomenon reveals a common characteristic of bacteria. J. Bacteriol. *181*: 2667–2668.

Hatchikian, E.C. 1994. Desulfofuscidin: dissimilatory, high-spin sulfite reductase of thermophilic, sulfate-reducing bacteria. Methods Enzymol. *243*: 276–295.

Hatchikian, E.C., P. Papavassiliou, P. Bianco and J. Haladjian. 1984. Characterization of cytochrome c_3 from the thermophilic sulfate reducer *Thermodesulfobacterium commune*. J. Bacteriol. *159*: 1040–1046.

Hatchikian, E.C. and J.G. Zeikus. 1983. Characterization of a new type of dissimilatory sulfite reductase present in *Thermodesulfobacterium commune*. J. Bacteriol. *153*: 1211–1220.

Hauser, N.C., M. Vingron, M. Scheidler, B. Krems, K. Hellmuth, K.D. Entian and J.D. Hoheisel. 1998. Transcriptional profiling on all open reading frames of *Saccharomyces cerevisiae*. Yeast *14*: 1209–1221.

Hawkins, P.R., M.T.C. Runnegar, A.R.B. Jackson and I.R. Falconer. 1985. Severe hepatotoxicity caused by the tropical cyanobacterium (blue-green alga) *Cylindrospermopsis raciborskii* (Woloszynoska) Seenaya and Subba Raju isolated from a domestic water supply reservoir. Appl. Environ. Microbiol. *50*: 1292–1295.

Hawksworth, D.L. and J. McNeill. 1998. The International Committee on Bionomenclature (ICB), the draft BioCode (1997), and the IUBS resolution on bionomenclature. Taxon *47*: 123–136.

Hayes, P.K. and G.L.A. Barker. 1997. Genetic diversity within Baltic Sea populations of *Nodularia* (Cyanobacteria). J. Phycol. *33*: 919–923.

Heaney, S.I. and G.H.M. Jaworski. 1977. A simple separation technique for purifying micro-algae. Br. Phycol. *12*: 171–174.

Hedlund, B.P., J.J. Gosink and J.T. Staley. 1996. Phylogeny of *Prosthecobacter*, the fusiform caulobacters: members of a recently discovered division of the Bacteria. Int. J. Syst. Bacteriol. *46*: 960–966.

Hedlund, B.P., J.J. Gosink and J.T. Staley. 1997. *Verrucomicrobia* div. nov., a new division of the *Bacteria* containing three new species of *Prosthecobacter*. Antonie Leeuwenhoek *72*: 29–38.

Hedrick, D.B., J.B. Guckert and D.C. White. 1991. Archaebacterial ether lipid diversity analyzed by supercritical fluid chromatography: integration with a bacterial lipid protocol. J. Lipid Res. *32*: 659–666.

Hefter, J., H.H. Richnow, U. Fischer, J.M. Trendel and W. Michaelis. 1993. (–)-Verrucosan-2β-ol from the phototrophic bacterium *Chloroflexus aurantiacus*: first report of a verrucosane-type diterpenoid from a prokaryote. J. Gen. Microbiol. *139*: 2757–2761.

Hendriksen, H.V. and B.K. Ahring. 1991. Effects of ammonia on growth and morphology of thermophilic hydrogen-oxidizing methanogenic bacteria. FEMS Microbiol. Lett. *85*: 241–246.

Henry, E.A., R. Devereux, J.S. Maki, C.C. Gilmour, C.R. Woese, L. Mandelco, R. Schauder, C.C. Remsen and R. Mitchell. 1994a. Characterization of a new thermophilic sulfate-reducing bacterium *Thermodesulfovibrio yellowstonii*, gen. nov. and sp. nov.: its phylogenetic relationship to *Thermodesulfobacterium commune* and their origins deep within the bacterial domain. Arch. Microbiol. *161*: 62–69.

Henry, E.A., R. Devereux, J.S. Maki, C.C. Gilmour, C.R. Woese, R. Mandelco, R. Schauder, C.C. Remsen and R. Mitchell. 1994b. *In* Validation of the publication of new names and new combinations previously effectively published outside the IJSB. List No. 50. Int. J. Syst. Bacteriol. *44*: 595.

Hensel, R., W. Demharter, O. Kandler, R.M. Kroppenstedt and E. Stackebrandt. 1986. Chemotaxonomic and molecular-genetic studies of the genus *Thermus*: evidence for a phylogenetic relationship of *Thermus aquaticus* and *Thermus ruber* to the genus *Deinococcus*. Int. J. Syst. Bacteriol. *36*: 444–453.

Hensel, R., K. Matussek, K. Michalke, L. Tacke, B.J. Tindall, M. Kohlhoff, B. Siebers and J. Dielenschneider. 1997a. *Sulfophobococcus zilligii* gen. nov., spec. nov., a novel hyperthermophilic archaeum isolated from hot alkaline springs of Iceland. Syst. Appl. Microbiol. *20*: 102–110.

Hensel, R., K. Matussek, K. Michalke, L. Tacke, B.J. Tindall, M. Kohlhoff, B. Siebers and J. Dielenschneider. 1997b. *In* Validation of the publication of new names and new combinations previously effectively published outside the IJSB. List No. 62. Int. J. Syst. Bacteriol. *47*: 915–916.

Henze, K., A. Badr, M. Wettern, R. Cerff and W. Martin. 1995. A nuclear gene of eubacterial origin in *Euglena gracilis* reflects cryptic endosymbioses during protist evolution. Proc. Natl. Acad. Sci. U.S.A. *92*: 9122–9126.

Herdman, M. 1987. Akinetes: structure and function. *In* Fay and van Baalen (Editors), The Cyanobacteria, Elsevier, New York. pp. 227–250.

Herdman, M. 1988. Cellular differentiation: Akinetes. Methods Enzymol. *167*: 222–232.

Herdman, M., M. Janvier, R. Rippka and R.Y. Stanier. 1979a. Genome size of cyanobacteria. J. Gen. Microbiol. *111*: 73–85.

Herdman, M., M. Janvier, J.B. Waterbury, R. Rippka, R.Y. Stanier and M. Mandel. 1979b. Deoxyribonucleic acid base composition of cyanobacteria. J. Gen. Microbiol. *111*: 63–71.

Herdman, M. and R. Rippka. 1988. Cellular differentiation: hormogonia and baeocytes. Methods Enzymol. *167*: 232–242.

Hernández-Mariné, M., A. Asencio-Martinez, A. Canals, X. Arino, M. Aboal and L. Hoffmann. 1999. Discovery of populations of the lime-incrusting *Loriella* (*Stigonematales*) in Spanish caves. Algol. Stud. *94*: 121–138.

Hernández-Mariné, M. and T. Canals. 1994. *Herpyzonema pulverulentum* (*Mastigocladaceae*), a new cavernicolous atmophytic and lime-encrusted cyanophyte. Arch. Hydrobiol. Suppl. *105*: 123–136.

Hernández-Mariné, M., M. Fernandez and V. Meriño. 1992. *Mastigocladopsis repens*, new species (*Nostochopsaceae*), a new cyanophyte from Spanish soils. Cryptogam. Algol. *13*: 113–120.

Hernández-Muñiz, W. and S.E.J. Stevens. 1987. Characterization of the motile hormogonia of *Mastigocladus laminosus*. J. Bacteriol. *169*: 218–223.

Hertel, C., W. Ludwig, M. Obst, R.F. Vogel, W.P. Hammes and K.H. Schleifer. 1991. 23S ribosomal RNA-targeted oligonucleotide probes for the rapid identification of meat lactobacilli. Syst. Appl. Microbiol. *14*: 173–177.

Hess, W.R., F. Partensky, G.W. van der Staay, J.M. Garcia Fernandez, T. Borner and D. Vaulot. 1996. Coexistence of phycoerythrin and a chlorophyll *a/b* antenna in a marine prokaryote. Proc. Natl. Acad. Sci. U.S.A. *93*: 11126–11130.

Hess, W.R., C. Steglich, C. Lichtlé and F. Partensky. 1999. Phycoerythrins of the oxyphotobacterium *Prochlorococcus marinus* are associated with the thylakoid membranes and are encoded by a single large gene cluster. Plant Mol. Biol. *40*: 507–521.

Heuer, H., M. Krsek, P. Baker, K. Smalla and E.M.H. Wellington. 1997. Analysis of actinomycete communities by specific amplification of genes encoding 16S rRNA and gel-electrophoretic separation in denaturing gradients. Appl. Environ. Microbiol. *63*: 3233–3241.

Hicks, R.E., R.I. Amann and D.A. Stahl. 1992. Dual staining of natural bacterioplankton with 4′,6-diamidino-2-phenylindole and fluorescent oligonucleotide probes targeting kingdom-level 16S rRNA sequences. Appl. Environ. Microbiol. *58*: 2158–2163.

Hilario, E. and J.P. Gogarten. 1993. Horizontal transfer of ATPase genes — the tree of life becomes a net of life. Biosystems *31*: 111–119.

Hilario, E. and J.P. Gogarten. 1998. The prokaryote-to-eukaryote transition reflected in the evolution of the V/F/A-ATPase catalytic and proteolipid subunits. J. Mol. Evol. *46*: 703–715.

Hill, D.R., S.L. Hladun, S. Scherer and M. Potts. 1994. Water stress proteins of *Nostoc commune* (Cyanobacteria) are secreted with UV-A/B-absorbing pigments and associate with 1,4-β-D-xylanxylanohydrolase activity. J. Biol. Chem. *269*: 7726–7734.

Hilpert, R., J. Winter, W. Hammes and O. Kandler. 1981. The sensitivity of archaebacteria to antibiotics. Zentbl. Bakteriol. Mikrobiol. Hyg. 1 Abt Orig. C *2*: 11–20.

Hippe, H. 1984. Maintenance of methanogenic bacteria. *In* Kirsop and Snell (Editors), Maintenance of Microorganisms: a Manual of Laboratory Methods, Academic Press, London. pp. 69–81.

Hippe, H. 1991. Maintenance of methanogenic bacteria. *In* Kirsop and Doyle (Editors), Maintenance of Microorganisms and Cultured Cells, 2nd Ed., Academic Press, London. pp. 101–113.

Hippe, H. 2000. *Leptospirillum* gen. nov. (ex Markosyan 1972), nom. rev., including *Leptospirillum ferrooxidans* sp. nov. (ex Markosyan 1972), nom. rev. and *Leptospirillum thermoferrooxidans* sp. nov. (Golovacheva et al. 1992). Int. J. Syst. Evol. Microbiol. *50*: 501–503.

Hippe, H., D. Caspari, K. Fiebig and G. Gottschalk. 1979. Utilization of trimethylamine and other *N*-methyl compounds for growth and methane formation by *Methanosarcina barkeri*. Proc. Natl. Acad. Sci. U.S.A. *76*: 494–498.

Hiraishi, A. 1994. Phylogenetic affiliations of *Rhodoferax fermentans* and related species of phototrophic bacteria as determined by automated 16S rDNA sequencing. Curr. Microbiol. *28*: 25–29.

Hiraishi, A. 1997. Transfer of the bacteriochlorophyll *b*-containing phototrophic bacteria *Rhodopseudomonas viridis* and *Rhodopseudomonas sulfoviridis* to the genus *Blastochloris* gen. nov. Int. J. Syst. Bacteriol. *47*: 217–219.

Hiraishi, A., Y. Hoshino and H. Kitamura. 1984. Isoprenoid quinone composition in the classification of *Rhodospirillaceae*. J. Gen. Appl. Microbiol. *30*: 197–210.

Hiraishi, A., Y. Hoshino and T. Satoh. 1991. *Rhodoferax fermentans*, gen. nov., sp. nov., a phototrophic purple nonsulfur bacterium previously referred to as the "*Rhodocyclus gelatinosus*-like" group. Arch. Microbiol. *155*: 330–336.

Hiraishi, A. and H. Kitamura. 1984. Distribution of phototrophic purple nonsulfur bacteria in activated sludge systems and other aquatic environments. Bull. Jpn. Soc. Sci. Fish. *50*: 1929–1937.

Hiraishi, A., Y. Matsuzawa, T. Kanbe and N. Wakao. 2000. *Acidisphaera rubrifaciens* gen. nov., sp. nov., an aerobic phototrophic bacterium isolated from acidic environments. Int. J. Syst. Evol. Microbiol. *50*: in press.

Hiraishi, A., K. Muramatsu and Y. Ueda. 1996. Molecular genetic analyses of *Rhodobacter azotoformans* sp. nov. and related species of phototrophic bacteria. Syst. Appl. Microbiol. *19*: 168–177.

Hiraishi, A., K.V.P. Nagashima, K. Matsuura, K. Shimada, S. Takaichi, N. Wakao and Y. Katayama. 1998. Phylogeny and photosynthetic features of *Thiobacillus acidophilus* and related acidophilic bacteria: its transfer

to the genus *Acidiphilum* as *Acidiphilum acidophilum* comb. nov. Int. J. Syst. Bacteriol. *48*: 1389–1398.

Hiraishi, A. and Y. Ueda. 1994a. Intrageneric structure of the genus *Rhodobacter*: transfer of *Rhodobacter sulfidophilus* and related marine species to the genus *Rhodovulum* gen. nov. Int. J. Syst. Bacteriol. *44*: 15–23.

Hiraishi, A. and Y. Ueda. 1994b. *Rhodoplanes* gen. nov., a new genus of phototrophic bacteria including *Rhodopseudomonas rosea* as *Rhodoplanes roseus* comb. nov. and *Rhodoplanes elegans* sp. nov. Int. J. Syst. Bacteriol. *44*: 665–673.

Hiraishi, A. and Y. Ueda. 1995. Isolation and characterization of *Rhodovulum strictum* sp. nov. and some other purple nonsulfur bacteria from colored blooms in tidal and seawater pools. Int. J. Syst. Bacteriol. *45*: 319–326.

Hiraishi, A., K. Urata and T. Satoh. 1995. A new genus of marine budding phototrophic bacteria, *Rhodobium* gen. nov., which includes *Rhodobium orientis* sp. nov. and *Rhodobium marinum* comb. nov. Int. J. Syst. Bacteriol. *45*: 226–234.

Hirosawa, T. and C.P. Wolk. 1979a. Factors controlling the formation of akinetes adjacent to heterocysts in the cyanobacterium *Cylindrospermum licheniforme* Kütz. J. Gen. Microbiol. *114*: 423–432.

Hirosawa, T. and C.P. Wolk. 1979b. Isolation and characterization of a substance which stimulates the formation of akinetes in the cyanobacterium *Cylindrospermum licheniforme* Kütz. J. Gen. Microbiol. *114*: 433–441.

Hishinuma, F., T. Tanaka and K. Sakaguchi. 1978. Isolation of extrachromosomal deoxyribonucleic acids from extremely thermophilic bacteria. J. Gen. Microbiol. *104*: 193–199.

Hobbie, J.E., R.J. Daley and S. Jasper. 1977. Use of nuclepore filters for counting bacteria by fluorescence microscopy. Appl. Environ. Microbiol. *33*: 1225–1228.

Hoffmann, L. 1990. Rediscovery of *Loriella osteophila* (Cyanophyceae). Br. Phycol. J. *25*: 391–395.

Hoffmann, L. 1994. Characterization of *Mastigocladopsis jogensis* (*Cyanophyceae, Stigonematales*) in culture. Arch. Hydrobiol. Suppl. *103*: 43–55.

Hoffmann, L. and V. Demoulin. 1985. Morphological variability of some species of *Scytonemataceae* (Cyanophyceae) under different culture conditions. Bull. Soc.R. Bot. Belg. *118*: 189–197.

Hoffmann, L., L. Talarico and A. Wilmotte. 1990. Presence of CU-phycoerythrin in the marine benthic blue-green alga *Oscillatoria* cf. *corallinae*. Phycologia *29*: 19–26.

Höfle, M.G. 1990. Transfer RNA as genotypic fingerprints of eubacteria. Arch. Microbiol. *153*: 299–304.

Höfle, M.G. 1991. Rapid genotyping of pseudomonads by using low-molecular-weight RNA profiles. *In* Galli, Silver and Witholt (Editors), *Pseudomonas* Molecular Biology and Biotechnology, American Society for Microbiology, Washington, DC. pp. 116–126.

Hoiczyk, E. and W. Baumeister. 1998. The junctional pore complex, a prokaryotic secretion organelle, is the molecular motor underlying gliding motility in cyanobacteria. Curr. Biol. *8*: 1161–1168.

Holben, W.E. and D. Harris. 1995. DNA-based monitoring of total bacterial community structure in environmental samples. Mol. Ecol. *4*: 627–631.

Holländer, R. 1978. The cytochromes of *Thermoplasma acidophilum*. J. Gen. Microbiol. *108*: 165–167.

Holliger, C., S.W.M. Kengen, G. Schraa, A.J.M. Stams and A.J.B. Zehnder. 1992a. Methyl-coenzyme-M reductase of *Methanobacterium thermoautotrophicum* ΔH catalyzes the reductive dechlorination of 1,2-dichloroethane to ethylene and chloroethane. J. Bacteriol. *174*: 4435–4443.

Holliger, C., G. Schraa, E. Stupperich, A.J.M. Stams and A.J.B. Zehnder. 1992b. Evidence for the involvement of corrinoids and factor F_{430} in the reductive dechlorination of 1,2-dichloroethane by *Methanosarcina barkeri*. J. Bacteriol. *174*: 4427–4434.

Holmes, M.L. and M.L. Dyall-Smith. 1990. A plasmid vector with a selectable marker for halophilic archaebacteria. J. Bacteriol. *172*: 756–761.

Holmes, M.L. and M.L. Dyall-Smith. 1991. Mutations in DNA gyrase result in novobiocin resistance in halophilic archaebacteria. J. Bacteriol. *173*: 642–648.

Holmes, M.L. and M.L. Dyall-Smith. 1999. Cloning, sequence and heterologous expression of *bgaH*, a β-galactosidase gene of *"Haloferax alicantei"*. *In* Oren (Editor), Microbiology and Biogeochemistry of Hypersaline Environments, CRC Press, Boca Raton. pp. 265–271.

Holmes, M.L., S.D. Nuttall and M.L. Dyall-Smith. 1991. Construction and use of halobacterial shuttle vectors and further studies on *Haloferax* DNA gyrase. J. Bacteriol. *173*: 3807–3813.

Holmes, M.L., F. Pfeifer and M.L. Dyall-Smith. 1994. Improved shuttle vectors for *Haloferax volcanii* including a dual-resistance plasmid. Gene *146*: 117–121.

Holmes, M.L., R.K. Scopes, R.L. Moritz, R.J. Simpson, C. Englert, F. Pfeifer and M.L. Dyall-Smith. 1997. Purification and analysis of an extremely halophilic β-galactosidase from *Haloferax alicantei*. Biochim. Biophys. Acta *1337*: 276–286.

Holmgren, P.R., H.P. Hostetter and V.E. Scholes. 1971. Ultrastructural observation of crosswalls in the blue-green alga *Spirulina major*. J. Phycol. *7*: 309–311.

Holo, H. 1989. *Chloroflexus aurantiacus* secretes 3-hydroxypropionate, a possible intermediate in the assimilation of CO_2 and acetate. Arch. Microbiol. *151*: 252–256.

Holt, J.G. (Editor). 1977. Shorter Bergey's Manual of Determinative Bacteriology, The Williams & Wilkins Co., Baltimore.

Holt, J.G. (Editor). 1984–1989. Bergey's Manual of Systematic Bacteriology, 1st Ed., Vols. 1–4, The Williams & Wilkins Co., Baltimore.

Holt, J.G., M.A. Bruns, B.J. Caldwell and C.D. Pease (Editors). 1992. Stedman's Bergey's Bacteria Words, The Williams & Wilkins Co., Baltimore.

Holt, J.G., N.R. Krieg, P.H.A. Sneath, J.T. Staley and S.T. Williams (Editors). 1994. Bergey's Manual of Determinative Bacteriology, 9th Ed., The Williams & Wilkins Co., Baltimore.

Holt, J.G. and R.A. Lewin. 1968. *Herpetosiphon aurantiacus*, gen. nov., sp. nov., a new filamentous gliding organism. J. Bacteriol. *95*: 2407–2408.

Honda, D., A. Yokota and J. Sugiyama. 1999. Detection of seven major evolutionary lineages in cyanobacteria based on the 16S rRNA gene sequence analysis with new sequences of five marine *Synechococcus* strains. J. Mol. Evol. *48*: 723–739.

Hönerlage, W., D. Hahn and J. Zeyer. 1995. Detection of mRNA of *nprM* in *Bacillus megaterium* ATCC 14581 grown in soil by whole-cell hybridization. Arch. Microbiol. *163*: 235–241.

Horn, C., B. Paulmann, G. Kerlen, N. Junker and H. Huber. 1999. *In vivo* observation of cell division of anaerobic hyperthermophiles by using a high-intensity dark-field microscope. J. Bacteriol. *181*: 5114–5118.

Hoshino, T., Y. Yoshino, E.D. Guevarra, S. Ishida, T. Hiruta, R. Fujii and T. Nakahara. 1994. Isolation and partial characterization of carotenoid underproducing and overproducing mutants from an extremely thermophilic *Thermus thermophilus* HB27. J. Ferment. Bioeng. *77*: 131–136.

Hovanec, T.A., L.T. Taylor, A. Blakis and E.F. Delong. 1998. *Nitrospira*-like bacteria associated with nitrite oxidation in freshwater aquaria. Appl. Environ. Microbiol. *64*: 258–264.

Hsung, J.C. and A. Haug. 1975. Intracellular pH of *Thermoplasma acidophila*. Biochim. Biophys. Acta *389*: 477–482.

Hua, M., E.I. Friedmann, R. Ocampo-Friedmann and S.B. Campbell. 1989. Heteropolarity in unicellular cyanobacteria: structure and development of *Cyanocystis violacea*. Plant Syst. Evol. *164*: 17–26.

Huang, T.C. and T.J. Chow. 1988. New type of N_2-fixing cyanobacterium (blue-green algae). FEMS Microbiol. Lett. *36*: 109–110.

Huang, Y. and R. Anderson. 1989. Structure of a novel glucosamine-containing phosphoglycolipid from *Deinococcus radiodurans*. J. Biol. Chem. *264*: 18667–18672.

Huang, Y. and R. Anderson. 1991. Phosphatidylglyceroylalkylamine, a novel phosphoglycolipid precursor in *Deinococcus radiodurans*. J. Bacteriol. *173*: 457–462.

Huber, G. 1987. Isolierung, Charakterisierung und taxonomische Einordnung neuer thermophiler, metallmobilisierender Archaebakterien, Thesis, University of Regensburg, Germany.

Huber, G., E. Drobner, H. Huber and K.O. Stetter. 1992a. Growth by aerobic oxidation of molecular hydrogen in Archaea—a metabolic property so far unknown for this domain. Syst. Appl. Microbiol. *15*: 502–504.

Huber, G., C. Spinnler, A. Gambacorta and K.O. Stetter. 1989a. *Metallosphaera sedula* gen. and sp. nov. represents a new genus of aerobic, metal-mobilizing, thermoacidophilic archaebacteria. Syst. Appl. Microbiol. *12*: 38–47.

Huber, G., C. Spinnler, A. Gambacorta and K.O. Stetter. 1989b. *In* Validation of the publication of new names and new combinations previously effectively published outside the IJSB. List No. 31. Int. J. Syst. Bacteriol. *39*: 495–497.

Huber, G. and K.O. Stetter. 1991. *Sulfolobus metallicus*, sp. nov., a novel strictly chemolithoautotrophic thermophilic archaeal species of metal-mobilizers. Syst. Appl. Microbiol. *14*: 372–378.

Huber, G. and K.O. Stetter. 1992a. *In* Validation of the publication of new names and new combinations previously effectively published outside the IJSB. List. No. 40. Int. J. Syst. Bacteriol. *42*: 191–192.

Huber, H., S. Burggraf, T. Mayer, I. Wyschkony, M. Biebl, R. Rachel and K.O. Stetter. 2000. *Ignicoccus* gen. nov., a novel genus of hyperthermophilic, chemolithoautotrophic *Archaea*, represented by two new species *Ignicoccus islandicus* sp. nov. and *Ignicoccus pacificus* sp. nov. Int. J. Syst. Evol. Microbiol. *50*: 2093–2100.

Huber, H., H. Jannasch, R. Rachel, T. Fuchs and K.O. Stetter. 1997. *Archaeoglobus veneficus* sp. nov., a novel facultative chemolithoautotrophic hyperthermophilic sulfite reducer, isolated from abyssal black smokers. Syst. Appl. Microbiol. *20*: 374–380.

Huber, H., H. Jannasch, R. Rachel, T. Fuchs and K.O. Stetter. 1998a. *In* Validation of the publication of new names and new combinations previously effectively published outside the IJSB. List No. 64. Int. J. Syst. Bacteriol. *48*: 327–328.

Huber, H., M. Thomm, H. König, G. Thies and K.O. Stetter. 1982. *Methanococcus thermolithotrophicus* a novel thermophilic lithotrophic methanogen. Arch. Microbiol. *132*: 47–50.

Huber, R., S. Burggraf, T. Mayer, S.M. Barns, P. Rossnagel and K.O. Stetter. 1995a. Isolation of a hyperthermophilic archaeum predicted by *in situ* RNA analysis. Nature *376*: 57–58.

Huber, R., D. Dyba, H. Huber, S. Burggraf and R. Rachel. 1998b. Sulfur-inhibited *Thermosphaera aggregans* sp. nov., a new genus of hyperthermophilic archaea isolated after its prediction from environmentally derived 16S rRNA sequences. Int. J. Syst. Bacteriol. *48*: 31–38.

Huber, R., W. Eder, S. Heldwein, G. Wanner, H. Huber, R. Rachel and K.O. Stetter. 1998c. *Thermocrinis ruber* gen. nov., sp. nov., a pink-filament-forming hyperthermophilic bacterium isolated from Yellowstone National Park. Appl. Environ. Microbiol. *64*: 3576–3583.

Huber, R., W. Eder, S. Heldwein, G. Wanner, H. Huber, R. Rachel and K.O. Stetter. 1999. *In* Validation of the publication of new names and new combinations previously effectively published outside the IJSB. List No. 69. Int. J. Syst. Bacteriol. *49*: 341–342.

Huber, R., G. Huber, A. Segerer, J. Seger and K.O. Stetter. 1987a. Aerobic and anaerobic extremely thermophilic autotrophs. *In* van Verseveld and Duine (Editors), Microbial Growth on C1 Compounds, Martinus Nijhoff Publisher, Dordrecht. pp. 44–51.

Huber, R., J.K. Kristjansson and K.O. Stetter. 1987b. *Pyrobaculum* gen. nov., a new genus of neutrophilic, rod-shaped archaebacteria from continental solfataras growing optimally at 100°C. Arch. Microbiol. *149*: 95–101.

Huber, R., M. Kurr, H.W. Jannasch and K.O. Stetter. 1989c. A novel group of abyssal methanogenic archaebacteria (*Methanopyrus*) growing at 110°C. Nature *342*: 833–834.

Huber, R., T.A. Langworthy, H. König, M. Thomm, C.R. Woese, U.B. Sleytr and K.O. Stetter. 1986. *Thermotoga maritima* sp. nov. represents a new genus of unique extremely thermophilic eubacteria growing up to 90°C. Arch. Microbiol. *144*: 324–333.

Huber, R. and K.O. Stetter. 1992b. The order *Thermoproteales*. *In* Balows, Trüper, Dworkin, Harder and Schleifer (Editors), The Prokaryotes.

A handbook of Bacteria: Ecophysiology, Isolation, Identification, Applications, 2nd Ed., Vol. I, Springer-Verlag, New York. pp. 677–683.

Huber, R. and K.O. Stetter. 1992c. The order *Thermotogales*. *In* Balows, Trüper, Dworkin, Harder and Schleifer (Editors), The Prokaryotes: A Handbook of Bacteria: Ecophysiology, Isolation, Identification, Applications, 2nd Ed., Springer-Verlag, New York. pp. 3809–3815.

Huber, R. and K.O. Stetter. 1992d. The *Thermotogales*: hyperthermophilic and extremely thermophilic bacteria. *In* Kristjansson (Editor), Thermophilic Bacteria, CRC Press, Boca Raton. pp. 185–194.

Huber, R. and K.O. Stetter. 1992e. *In* Validation of the publication of new names and new combinations previously effectively published outside the IJSB. List No. 43. Int. J. Syst. Bacteriol. *42*: 656–657.

Huber, R. and K.O. Stetter. 1996. *In* Validation of the publication of new names and new combinations previously effectively published outside the IJSB. List No. 58. Int. J. Syst. Bacteriol. *46*: 836–837.

Huber, R., J. Stöhr, S. Hohenhaus, R. Rachel, S. Burggraf, H.W. Jannasch and K.O. Stetter. 1995b. *Thermococcus chitonophagus* sp. nov., a novel chitin-degrading, hyperthermophilic archaeum from a deep-sea hydrothermal vent environment. Arch. Microbiol. *164*: 255–264.

Huber, R., T. Wilharm, D. Huber, A. Trincone, S. Burggraf, H. König, R. Rachel, I. Rockinger, H. Fricke and K.O. Stetter. 1992b. *Aquifex pyrophilus* gen. nov. sp. nov., represents a novel group of marine hyperthermophilic hydrogen-oxidizing bacteria. Syst. Appl. Microbiol. *15*: 340–351.

Huber, R., C.R. Woese, T.A. Langworthy, H. Fricke and K.O. Stetter. 1989d. *Thermosipho africanus* gen. nov., represents a new genus of thermophilic eubacteria within the *"Thermotogales"*. Syst. Appl. Microbiol. *12*: 32–37.

Huber, R., C.R. Woese, T.A. Langworthy, H. Fricke and K.O. Stetter. 1989e. *In* Validation of the publication of new names and new combinations previously effectively published outside the IJSB. List No. 31. Int. J. Syst. Bacteriol. *39*: 495–497.

Huber, R., C.R. Woese, T.A. Langworthy, J.K. Kristjansson and K.O. Stetter. 1990. *Fervidobacterium islandicum* sp. nov., a new extremely thermophilic eubacterium belonging to the *"Thermotogales"*. Arch. Microbiol. *154*: 105–111.

Huber, R., C.R. Woese, T.A. Langworthy, J.K. Kristjansson and K.O. Stetter. 1991. *In* Validation of the publication of new names and new combinations previously effectively published outside the IJSB. Int. J. Syst. Bacteriol. *41*: 178–179.

Hudson, J.A., H.W. Morgan and R.M. Daniel. 1986. A numerical classification of some *Thermus* isolates. J. Gen. Microbiol. *132*: 531–540.

Hudson, J.A., H.W. Morgan and R.M. Daniel. 1987a. Numerical classification of some *Thermus* isolates from Icelandic hot springs. Syst. Appl. Microbiol. *9*: 218–223.

Hudson, J.A., H.W. Morgan and R.M. Daniel. 1987b. *Thermus filiformis* sp. nov., a filamentous caldoactive bacterium. Int. J. Syst. Bacteriol. *37*: 431–436.

Hudson, J.A., H.W. Morgan and R.M. Daniel. 1989. Numerical classification of *Thermus* isolates from globally distributed hot springs. Syst. Appl. Microbiol. *11*: 250–256.

Hugenholtz, P., B.M. Goebel and N.R. Pace. 1998a. Impact of culture-independent studies on the emerging phylogenetic view of bacterial diversity. J. Bacteriol. *180*: 4765–4774.

Hugenholtz, P., C. Pitulle, K.L. Hershberger and N.R. Pace. 1998b. Novel division level bacterial diversity in a Yellowstone hot spring. J. Bacteriol. *180*: 366–376.

Hughes, E.O., P.R. Gorham and A. Zehnder. 1958. Toxicity of a unialgal culture of *Microcystis aeruginosa*. Can. J. Microbiol. *4*: 225–236.

Hungate, R.E. 1950. The anaerobic mesophilic cellulolytic bacteria. Bacteriol. Rev. *14*: 1–49.

Hungate, R.E. 1969. A roll tube method for cultivation of strict anaerobes. *In* Norris and Ribbons (Editors), Methods in Microbiology, Vol. 3B, Academic Press, London. pp. 117–132.

Hungate, R.E. 1979. Evolution of a microbial ecologist. Annu. Rev. Microbiol. *33*: 1–20.

Huser, B.A., K. Wuhrmann and A.J.B. Zehnder. 1982. *Methanothrix soehn-genii*, gen. nov. sp. nov., a new acetotrophic non-hydrogen-oxidizing methane bacterium. Arch. Microbiol. *132*: 1–9.

Huster, M.S. and K.M. Smith. 1990. Biosynthetic studies of substituent homologation in bacteriochlorophylls *c* and *d*. Biochemistry *29*: 4348–4355.

Hyypiä, T., A. Jalava, S.H. Larsen, P. Terho and V. Hukkanen. 1985. Detection of *Chlamydia trachomatis* in clinical specimens by nucleic acid spot hybridization. J. Gen. Microbiol. *131*: 975–978.

Iba, K., K.-I. Takamiya, Y. Toh and M. Nishimura. 1988. Roles of bacteriochlorophyll and carotenoid synthesis in formation of intracytoplasmic membrane systems and pigment-protein complexes in an aerobic photosynthetic bacterium, *Erythrobacter* sp. strain OCH114. J. Bacteriol. *170*: 1843–1847.

Ihara, K., S. Watanabe and T. Tamura. 1997. *Haloarcula argentinensis* sp. nov. and *Haloarcula mukohataei* sp. nov., two new extremely halophilic archaea collected in Argentina. Int. J. Syst. Bacteriol. *47*: 73–77.

Ikuta, S., K. Takagi, R.B. Wallace and K. Itakura. 1987. Dissociation kinetics of 19 base paired oligonucleotide-DNA duplexes containing different single mismatched base pairs. Nucleic Acids Res. *15*: 797–811.

Imhoff, J.F. 1984a. Quinones of phototrophic purple bacteria. FEMS Microbiol. Lett. *256*: 85–89.

Imhoff, J.F. 1984b. Reassignment of the genus *Ectothiorhodospira* Pelsh 1936 to a new family, *Ectothiorhodospiraceae* fam. nov., and emended description of the *Chromatiaceae* Bavendamm 1924. Int. J. Syst. Bacteriol. *34*: 338–339.

Imhoff, J.F. 1988. Lipids, fatty acids and quinones in taxonomy and phylogeny of anoxygenic phototrophic bacteria. *In* Olson, Ormerod, Amesz, Stackebrandt and Trüper (Editors), Green Photosynthetic Bacteria, Plenum Publishing Corporation, New York. pp. 223–232.

Imhoff, J.F. 1989. Family *Ectothiorhodospiraceae*. *In* Staley, Bryant, Pfennig and Holt (Editors), Bergey's Manual of Systematic Bacteriology, 1st Ed., Vol. 3, The Williams & Wilkins Co., Baltimore. pp. 1654–1658.

Imhoff, J.F. 1991. Polar lipids and fatty acids in the genus *Rhodobacter*. Syst. Appl. Microbiol. *14*: 228–234.

Imhoff, J.F. 1995. Taxonomy and physiology of phototrophic purple bacteria and green sulfur bacteria. *In* Blankenship, Madigan and Bauer (Editors), Anoxygenic Photosynthetic Bacteria, Kluwer Academic Publishing, Dordrecht. pp. 1–15.

Imhoff, J.F. and U. Bias-Imhoff. 1995. Lipids, quinones and fatty acids of anoxygenic phototrophic bacteria. *In* Blankenship, Madigan and Bauer (Editors), Anoxygenic Photosynthetic Bacteria, Kluwer Academic Publishing, Dordrecht. pp. 179–205.

Imhoff, J.F., D.J. Kushner, S.C. Kushwaha and M. Kates. 1982. Polar lipids in phototrophic bacteria of the *Rhodospirillaceae* and *Chromatiaceae* families. J. Bacteriol. *150*: 1192–1201.

Imhoff, J.F., R. Petri and J. Süling. 1998a. Reclassification of species of the spiral-shaped phototrophic purple non-sulfur bacteria of the α-*Proteobacteria*: description of the new genera *Phaeospirillum* gen. nov., *Rhodovibrio* gen. nov., *Rhodothalassium* gen. nov. and *Roseospira* gen. nov. as well as transfer of *Rhodospirillum fulvum* to *Phaeospirillum fulvum* comb. nov., of *Rhodospirillum molischianum* to *Phaeospirillum molischianum* comb. nov., of *Rhodospirillum salinarum* to *Rhodovibrio salexigens*. Int. J. Syst. Bacteriol. *48*: 793–798.

Imhoff, J.F. and J. Süling. 1996. The phylogenetic relationship among *Ectothiorhodospiraceae*: a reevaluation of their taxonomy on the basis of 16S rDNA analyses. Arch. Microbiol. *165*: 106–113.

Imhoff, J.F., J. Süling and R. Petri. 1998b. Phylogenetic relationships among the *Chromatiaceae*, their taxonomic reclassification and description of the new genera *Allochromatium*, *Halochromatium*, *Isochromatium*, *Marichromatium*, *Thiococcus*, *Thiohalocapsa* and *Thermochromatium*. Int. J. Syst. Bacteriol. *48*: 1129–1143.

Imhoff, J.F. and H.G. Trüper. 1982. Taxonomic classification of photosynthetic bacteria (anoxyphotobacteria, phototrophic bacteria). *In* Mitsui and Black (Editors), Handbook of Biosolar Resources, Vol. 1. Basic Principles. Part 1, CRC Press, Boca Raton. pp. 513–522.

Imhoff, J.F. and H.G. Trüper. 1989. The purple nonsulfur bacteria. *In* Staley, Bryant, Pfennig and Holt (Editors), Bergey's Manual of Systematic Bacteriology, 1st Ed., Vol. 3, The Williams & Wilkins Co., Baltimore. pp. 1658–1661.

Imhoff, J.F., H.G. Trüper and N. Pfennig. 1984. Rearrangement of the species and genera of the phototrophic "purple nonsulfur bacteria". Int. J. Syst. Bacteriol. *34*: 340–343.

Inatomi, K.I., Y. Kamagata and K. Nakamura. 1993. Membrane ATPase from the aceticlastic methanogen *Methanothrix thermophila*. J. Bacteriol. *175*: 80–84.

International Committee on Systematic Bacteriology. 1997. VIIth International Congress of Microbiology and Applied Bacteriology. Minutes of the Meetings, 17, 18, and 22 August 1996, Jerusalem, Israel. Int. J. Syst. Bacteriol. *47*: 597–600.

International Committee on Systematic Bacteriology Subcommittee on the Taxonomy of *Mollicutes*. 1979. Proposal of miminal standards for descriptions of new species of the class *Mollicutes*. Int. J. Syst. Bacteriol. *29*: 172–180.

Ishida, T., A. Yokota and J. Sugiyama. 1997. Phylogenetic relationships of filamentous cyanobacterial taxa inferred from 16S rRNA sequence divergence. J. Gen. Appl. Microbiol. *43*: 237–241.

Ishiguro, E.E. and R.S. Wolfe. 1970. Control of morphogenesis in *Geodermatophilus*: ultrastructural studies. J. Bacteriol. *104*: 566–580.

Ishii, M., Y. Igarashi and T. Kodama. 1987. Colony formation of *Hydrogenobacter thermophilus* on a plate solidified with GELRITE. Agric. Biol. Chem *51*: 3139–3141.

Ishii, M., Y. Igarashi and T. Kodama. 1989. Purification and characterization of ATP:citrate lyase from *Hydrogenobacter thermophilus* TK-6. J. Bacteriol. *171*: 1788–1792.

Ishii, M., T. Omori, Y. Igarashi, O. Adachi, M. Ameyama and T. Kodama. 1991. Methionaquinone is a direct natural electron acceptor for the membrane-bound hydrogenase in *Hydrogenobacter thermophilus* strain TK-6. Agric. Biol. Chem. *55*: 3011–3016.

Ishii, M., Y. Ueda, K.S. Yoon, Y. Igarashi and T. Kodama. 1996. Purification and characterization of ferredoxin from *Hydrogenobacter thermophilus* strain TK-6. Biosci. Biotechnol. Biochem. *60*: 1513–1515.

Istock, C.A., J.A. Bell, N. Ferguson and N.L. Istock. 1996. Bacterial species and evolution: theoretical and practical perspectives. J. Ind. Microbiol. *17*: 137–150.

Iteman, I., R. Rippka, N. Tandeau de Marsac and M. Herdman. 1999. Use of molecular tools for the study of genetic relationships of heterocystous cyanobacteria. *In* Charpy and Larkum (Editors), Marine Cyanobacteria, Bulletin de l'Institut Océanographique Monaco, special issue 19, Institut Océanographique, Monaco. pp. 13–20.

Ito, H. 1977. Isolation of *Micrococcus radiodurans* occurring in radurized sawdust culture media of mushroom. Agric. Biol. Chem. *41*: 35–41.

Ito, H., H. Watanabe, M. Takeshia and H. Iizuka. 1983. Isolation and identification of radiation-resistant cocci belonging to the genus *Deinococcus* from sewage sludges and animal feeds. Agric. Biol. Chem. *47*: 1239–1247.

Itoh, T., K. Suzuki and T. Nakase. 1998. *Thermocladium modesticus* gen. nov., sp. nov., a new genus of rod-shaped, extremely thermophilic crenarchaeote. Int. J. Syst. Bacteriol. *48*: 879–887.

Itoh, T., K. Suzuki, P.C. Sanchez and T. Nakase. 1999. *Caldivirga maquilingensis* gen. nov., sp. nov., a new genus of rod-shaped crenarchaeote isolated from a hot spring in the Philippines. Int. J. Syst. Bacteriol. *49*: 1157–1163.

Ivancich, A., R. Feick, A. Ertlmaier and T.A. Mattioli. 1996. Structure and protein binding interactions of the primary donor of the *Chloroflexus aurantiacus* reaction center. Biochemistry *35*: 6126–6135.

Ivanovsky, R.N., Y.I. Fal, I.A. Berg, N.V. Ugolkova, E.N. Krasilnikova, O.I. Keppen, L.M. Zakharchuc and A.M. Zyakun. 1999. Evidence for the presence of the reductive pentose phosphate cycle in a filamentous anoxygenic photosynthetic bacterium, *Oscillochloris trichoides* strain DG-6. Microbiology (Reading) *145*: 1743–1748.

Ivanovsky, R.N., E.N. Krasilnikova and Y.I. Fal. 1993. A pathway of the autotrophic CO_2 fixation in *Chloroflexus aurantiacus*. Arch. Microbiol. *159*: 257–264.

Iwabe, N., K. Kuma, M. Hasegawa, S. Osawa and T. Miyata. 1989. Evolutionary relationship of archaebacteria, eubacteria, and eukaryotes inferred from phylogenetic trees of duplicated genes. Proc. Natl. Acad. Sci. U.S.A. *86*: 9355–9359.

Jaag, O. 1945. Experimentelle Untersuchungen über die Variabilität einer Blaualge unter dem Einfluss verschieden starker Belichtung. Verh. Naturforsch. Ges. Basel *56*: 28–40.

Jablonski, E., E.W. Moomaw, R.H. Tullis and J.L. Ruth. 1986. Preparation of oligodeoxynucleotide-alkaline phosphatase conjugates and their use as hybridization probes. Nucleic Acids Res. *14*: 6115–6128.

Jablonski, P.E. and J.G. Ferry. 1992. Reductive dechlorination of trichloroethylene by the carbon monoxide-reduced carbon monoxide dehydrogenase enzyme complex from *Methanosarcina thermophila*. FEMS Microbiol. Lett. *96*: 55–59.

Jackson, T.J., R.F. Ramaley and W.G. Meinschein. 1973. *Thermomicrobium*, a new genus of extremely thermophilic bacteria. Int. J. Syst. Bacteriol. *23*: 28–36.

Jaenicke, R., H. Schurig, N. Beaucamp and R. Ostendorp. 1996. Structure and stability of hyperstable proteins: glycolytic enzymes from hyperthermophilic bacterium *Thermotoga maritima*. Adv. Protein Chem. *48*: 181–269.

Jain, M.K., T.E. Thompson, E. Conway de Macario and J.G. Zeikus. 1987a. Speciation of *Methanobacterium* strain Ivanov as *Methanobacterium ivanovii*, sp. nov. Syst. Appl. Microbiol. *9*: 77–82.

Jain, M.K., T.E. Thompson, E. Conway de Macario and J.G. Zeikus. 1988a. *In* Validation of the publication of new names and new combinations previously effectively published outside the IJSB. List No. 24. Int. J. Syst. Bacteriol. *38*: 136–137.

Jain, R.K., R.S. Burlage and G.S. Sayler. 1988b. Methods for detecting recombinant DNA in the environment. Crit. Rev. Biotechnol. *8*: 33–84.

Jain, R.K., G.S. Sayler, J.T. Wilson, L. Houston and D. Pacia. 1987b. Maintenance and stability of introduced genotypes in groundwater aquifer material. Appl. Environ. Microbiol. *53*: 996–1002.

Jan, R.L., J. Wu, S.M. Chaw, C.W. Tsai and S.D. Tsen. 1999. A novel species of thermoacidophilic archaeon, *Sulfolobus yangmingensis* sp. nov. Int. J. Syst. Bacteriol. *49*: 1809–1816.

Janczewski, E. 1874. La reproduction de quelques Nostochacées. Ann. Sci. Nat. Bot. *19*: 119–130.

Janekovic, D., S. Wunderl, I. Holz, W. Zillig, A. Gierl and H. Neumann. 1983. TTV1, TTV2 and TTV3, a family of viruses of the extremely thermophilic, anaerobic, sulfur reducing archaebacterium *Thermoproteus tenax*. Mol. Gen. Genet. *192*: 39–45.

Jannasch, H.W., R. Huber, S. Belkin and K.O. Stetter. 1988. *Thermotoga neapolitana* sp. nov. of the extremely thermophilic, eubacterial genus *Thermotoga*. Arch. Microbiol. *150*: 103–104.

Jannasch, H.W., R. Huber, S. Belkin and K.O. Stetter. 1989. *In* Validation of the publication of new names and new combinations previously effectively published outside the IJSB List No. 28. Int. J. Syst. Bacteriol. *39*: 93–94.

Janssen, P.H., L.E. Parker and H.W. Morgan. 1991. Filament formation in *Thermus* species in the presence of some D-amino acids or glycine. Antonie Leeuwenhoek *59*: 147–154.

Jarrell, K.F. and S.F. Koval. 1989. Ultrastructure and biochemistry of *Methanococcus voltae*. Crit. Rev. Microbiol. *17*: 53–87.

Javor, B.J. 1988. CO_2 fixation in halobacteria. Arch. Microbiol. *149*: 433–440.

Javor, B.J. 1989. Hypersaline Environments: Microbiology and Biochemistry, Springer-Verlag, Berlin.

Javor, B.J. and R.W. Castenholz. 1981. Laminated microbial mats, Laguna Guerrero Negro, Mexico. Geomicrobiol. J. *2*: 237–274.

Javor, B.J., C. Requadt and W. Stoeckenius. 1982. Box-shaped halophilic bacteria. J. Bacteriol. *151*: 1532–1542.

Jeanthon, C., S. L'Haridon, N. Pradel and D. Prieur. 1999a. Rapid identification of hyperthermophilic methanococci isolated from deep-sea hydrothermal vents. Int. J. Syst. Bacteriol. *49*: 591–594.

Jeanthon, C., S. L'Haridon, A.-L. Reysenbach, E. Corre, M. Vernet, P. Messner, U.B. Sleytr and D. Prieur. 1999b. *Methanococcus vulcanius* sp. nov., a novel hyperthermophilic methanogen isolated from East Pacific Rise, and identification of *Methanococcus* spp. DSM 4213^T as *Methanococcus fervens* sp. nov. Int. J. Syst. Bacteriol. *49*: 583–598.

Jeanthon, C., S. L'Haridon, A.L. Reysenbach, M. Vernet, P. Messner, U.B. Sleytr and D. Prieur. 1998. *Methanococcus infernus* sp. nov., a novel hyperthermophilic lithotrophic methanogen isolated from a deep-sea hydrothermal vent. Int. J. Syst. Bacteriol. *48*: 913–919.

Jeanthon, C., A.L. Reysenbach, S. L'Haridon, A. Gambacorta, N.R. Pace, P. Glénat and D. Prieur. 1995. *Thermotoga subterranea* sp. nov., a new thermophilic bacterium isolated from a continental oil reservoir. Arch. Microbiol. *164*: 91–97.

Jeeji-Bai, N. 1976. Morphological variation of certain blue-green algae in culture: *Scytonema stuposum* (Kütz) Born. Schweiz. Z. Hydrol. *38*: 55–62.

Jeeji-Bai, N. 1985. Competitive exclusion or morphological transformation? A case study with *Spirulina fusiformis*. Z. Hydrol. Suppl. *71*: 191–199.

Jeeji-Bai, N., E. Hegewald and C.J. Soeder. 1977. Revision and taxonomic analysis of the genus *Anabaenopsis*. Arch. Hydrobiol. Suppl. *51*: 3–24.

Jeeji-Bai, N. and C.V. Seshadri. 1980. Coiling and uncoiling of trichomes in the genus *Spirulina*. Arch. Hydrobiol. Suppl. *60*: 32–47.

Jeffrey, C. 1977. Biological Nomenclature, 2nd Ed., Arnold, London.

Jensen, A.O., O. Aasmundrud and K.E. Eimhjellen. 1964. Chlorophylls of photosynthetic bacteria. Biochim. Biophys. Acta *88*: 466–479.

Jensen, M.T., J. Knudsen and J.M. Olson. 1991. A novel aminoglycosphingolipid found in *Chlorobium limicola* f. *thiosulfatophilum* 6230. Arch. Microbiol. *156*: 248–254.

Jochimsen, B., S. Peinemann-Simon, H. Völker, D. Stüben, R. Botz, P. Stoffers, P.R. Dando and M. Thomm. 1997. *Stetteria hydrogenophila*, gen, nov. and sp. nov., a novel mixotrophic sulfur-dependent crenarchaeote isolated from Milos, Greece. Extremophiles *1*: 67–73.

Jochimsen, B., S. Peinemann-Simon, H. Völker, D. Stüben, R. Botz, P. Stoffers, P.R. Dando and M. Thomm. 1998. *In* Validation of the publication of new names and new combinations previously effectively published outside the IJSB. List. No. 64. Int. J. Syst. Bacteriol. *48*: 327– 328.

Johnson, D.B. 1995. Selective solid media for isolating and enumerating acidophilic bacteria. J. Microbiol. Methods *23*: 205–218.

Johnson, J.L. 1984. Nucleic acids in bacterial classification. *In* Krieg and Holt (Editors), Bergey's Manual of Systematic Bacteriology, 1st Ed., Vol. 1, The Williams & Wilkins Co., Baltimore. pp. 8–11.

Johnson, J.L. 1985. DNA reassociation and RNA hybridization of bacterial nucleic acids. *In* Gottschalk (Editor), Methods in Microbiology, Vol. 18, Academic Press, New York. pp. 33–74.

Joklik, W.K. (Editor). 1999. Microbiology: A Centenary Perspective, American Society for Microbiology, Washington, DC.

Jones, B.E., W.D. Grant, A.W. Duckworth and G.G. Owenson. 1998. Microbial diversity of soda lakes. Extremophiles *2*: 191–200.

Jones, J.B., B. Bowers and T.C. Stadtman. 1977. *Methanococcus vannielii*: ultrastructure and sensitivity to detergents and antibiotics. J. Bacteriol. *130*: 1357–1363.

Jones, J.B. and T.C. Stadtman. 1977. *Methanococcus vannielii*: culture and effects of selenium and tungsten on growth. J. Bacteriol. *130*: 1404–1406.

Jones, R.J. and R.G. Megarrity. 1986. Successful transfer of dihydroxy-pyridine-degrading bacteria from Hawaiian (USA) goats to Australian ruminants to overcome the toxicity of *Leucaena*. Aust. Vet. J. *63*: 259–262.

Jones, W.J., M.I. Donnelly and R.S. Wolfe. 1985. Evidence of a common pathway of carbon dioxide reduction to methane in methanogens. J. Bacteriol. *163*: 126–131.

Jones, W.J. and G.U. Holzer. 1991. The polar and neutral lipid composition of *Methanosphaera stadtmanae*. Syst. Appl. Microbiol. *14*: 130–134.

Jones, W.J., J.A. Leigh, F. Mayer, C.R. Woese and R.S. Wolfe. 1983a. *Methanococcus jannaschii* sp. nov., an extremely thermophilic methanogen from a submarine hydrothermal vent. Arch. Microbiol. *136*: 254–261.

Jones, W.J., M.J.B. Paynter and R. Gupta. 1983b. Characterization of *Me-*

thanococcus maripaludis sp. nov., a new methanogen isolated from salt marsh sediment. Arch. Microbiol. *135*: 91–97.

Jones, W.J., M.J.B. Paynter and R. Gupta. 1984. *In* Validation of the publication of new names and new combinations previously effectively published outside the IJSB. List No. 14. Int. J. Syst. Bacteriol. *34*: 270–271.

Jones, W.J., W.B. Whitman, R.D. Fields and R.S. Wolfe. 1983c. Growth and plating efficiency of methanococci on agar media. Appl. Environ. Microbiol. *46*: 220–226.

Jørgensen, B.B. and D.C. Nelson. 1988. Bacterial zonation, photosynthesis, and spectral light distribution in hot spring microbial mats of Iceland. Microb. Ecol. *16*: 133–148.

Judicial Commission. 1986a. Opinion 62. Transfer of the type species of the genus *Methanococcus* to the genus *Methanosarcina* as *Methanosarcina mazei* (Barker 1936) comb. nov. et emend. Mah and Kuhn 1984 and conservation of the genus *Methanococcus* (Approved Lists, 1980) emend. Mah and Kuhn 1984 with *Methanococcus vannielii* (Approved Lists, 1980) as the type species. Int. J. Syst. Bacteriol. *36*: 491.

Judicial Commission. 1986b. Opinion 63. Rejection of the type species *Methanosarcina methanica* and conservation of the genus *Methanosarcina* emend. Mah and Kuhn 1984 with *Methanosarcina barkeri* as the type species. Int. J. Syst. Bacteriol. *36*: 492.

Juez, G., F. Rodriguez-Valera, A. Ventosa and D.J. Kushner. 1986a. *Haloarcula hispanica*, spec. nov. and *Haloferax gibbonsii* spec. nov., two new species of extremely halophilic archaebacteria. Syst. Appl. Microbiol. *8*: 75–79.

Juez, G., F. Rodriguez-Valera, A. Ventosa and D.J. Kushner. 1986b. *In* Validation of the publication of new names and new combinations previously effectively published outside the IJSB. List No. 22. Int. J. Syst. Bacteriol. *36*: 573–576.

Jukes, T.H. and R.R. Cantor. 1969. Evolution of protein molecules. *In* Munzo (Editor), Mammalian Protein Metabolism, Academic Press, New York. pp. 21–132.

Juretschko, S., G. Timmermann, M. Schmid, K.H. Schleifer, A. Pommerening-Röser, H.P. Koops and M. Wagner. 1998. Combined molecular and conventional analyses of nitrifying bacterium diversity in activated sludge: *Nitrosococcus mobilis* and *Nitrospira*-like bacteria as dominant populations. Appl. Environ. Microbiol. *64*: 3042–3051.

Jurgens, G., K. Lindstrom and A. Saano. 1997. Novel group within the kingdom *Crenarchaeota* from boreal forest soil. Appl. Environ. Microbiol. *63*: 803–805.

Jürgens, K., J. Pernthaler, S. Schalla and R.I. Amann. 1999. Morphological and compositional changes in a planktonic bacterial community in response to protozoal grazing. Appl. Environ. Microbiol. *65*: 1241–1250.

Jürgens, U.J., J. Meissner, U. Fischer, W.A. König and J. Weckesser. 1987. Ornithine as a constituent of the peptidoglycan of *Chloroflexus aurantiacus*, diaminopimelic acid in that of *Chlorobium vibrioforme* f. *thiosulfatophilum*. Arch. Microbiol. *148*: 72–76.

Jussofie, A., F. Mayer and G. Gottschalk. 1986. Methane formation from methanol and molecular hydrogen by protoplasts of new methanogenic isolates and inhibition by dicyclohexylcarbodiimide. Arch. Microbiol. *146*: 245–249.

Jüttner, F. 1984. Characterization of *Microcystis* strains by alkyl sulfides and β-cyclocitral. Z. Naturforsch. *39*: 867–871.

Jüttner, F. 1991. Taxonomic characterization of *Limnothrix* and *Planktothrix* using secondary metabolism (hydrocarbons). Algol. Stud. *64*: 261– 266.

Kadam, P.C. and D.R. Boone. 1995. Physiological characterization and emended description of *Methanolobus vulcani*. Int. J. Syst. Bacteriol. *45*: 400–402.

Kadam, P.C. and D.R. Boone. 1996. Influence of pH on ammonia accumulation and toxicity in halophilic, methylotrophic methanogens. Appl. Environ. Microbiol. *62*: 4486–4492.

Kadam, P.C., D.R. Ranade, L. Mandelco and D.R. Boone. 1994. Isolation and characterization of *Methanolobus bombayensis* sp. nov., a methylotrophic methanogen that requires high concentrations of divalent cations. Int. J. Syst. Bacteriol. *44*: 603–607.

Kafatos, F.C., C.W. Jones and A. Efstratiadis. 1979. Determination of nucleic acid sequence homologies and relative concentrations by a dot hybridization procedure. Nucleic Acids Res. *7*: 1541–1552.

Kageyama, A., Y. Benno and T. Nakase. 1999. Phylogenic and phenotypic evidence for the transfer of *Eubacterium fossor* to the genus *Atopobium* as *Atopobium fossor* comb. nov.. Microbiol. Immunol. *43*: 389–395.

Kakinuma, Y., K. Igarashi, K. Konishi and I. Yamato. 1991. Primary structure of the alpha-subunit of vacuolar-type Na$^{(+)}$-ATPase in *Enterococcus hirae*. Amplification of a 1000-bp fragment by polymerase chain reaction. FEBS Lett. *292*: 64–68.

Kallas, T., T. Coursin and R. Rippka. 1985. Different organization of *nif* genes in nonheterocystous and heterocystous cyanobacteria. Plant Mol. Biol. *5*: 321–329.

Kalmbach, S., W. Manz and U. Szewzyk. 1997. Dynamics of biofilm formation in drinking water: phylogenetic affiliation and metabolic potential of single cells assessed by formazan reduction and *in situ* hybridization. FEMS Microbiol. Ecol. *22*: 265–279.

Kamagata, Y., H. Kawasaki, H. Oyaizu, K. Nakamura, E. Mikami, G. Endo, Y. Koga and K. Yamasato. 1992. Characterization of three thermophilic strains of *Methanothrix* (*"Methanosaeta"*) *thermophila*, sp. nov., and rejection of *Methanothrix* (*"Methanosaeta"*) *thermoacetophila*. Int. J. Syst. Bacteriol. *42*: 463–468.

Kamagata, Y. and E. Mikami. 1991. Isolation and characterization of a novel thermophilic *Methanosaeta* strain. Int. J. Syst. Bacteriol. *41*: 191–196.

Kamekura, M. 1998a. Diversity of extremely halophilic bacteria. Extremophiles *2*: 289–296.

Kamekura, M. 1998b. Diversity of members of the family *Halobacteriaceae*. *In* Oren (Editor), Microbiology and Biogeochemistry of Hypersaline Environments, CRC Press, Boca Raton. pp. 13–25.

Kamekura, M. and M.L. Dyall-Smith. 1995. Taxonomy of the family *Halobacteriaceae* and the description of two new genera *Halorubrobacterium* and *Natrialba*. J. Gen. Appl. Microbiol. *41*: 333–350.

Kamekura, M. and M.L. Dyall-Smith. 1996. *In* Validation of the publication of new names and new combinations previously effectively published outside the IJSB. List No. 57. Int. J. Syst. Bacteriol. *46*: 625–626.

Kamekura, M., M.L. Dyall-Smith, V. Upasani, A. Ventosa and M. Kates. 1997. Diversity of alkaliphilic halobacteria: proposals for transfer of *Natronobacterium vacuolatum*, *Natronobacterium magadii*, and *Natronobacterium pharaonis* to *Halorubrum*, *Natrialba*, and *Natronomonas* gen. nov., respectively, as *Halorubrum vacuolatum* comb. nov., *Natrialba magadii* comb. nov., and *Natronomonas pharaonis* comb. nov., respectively. Int. J. Syst. Bacteriol. *47*: 853–857.

Kamekura, M., D. Oesterhelt, R. Wallace, P. Anderson and D.J. Kushner. 1988. Lysis of halobacteria in Bacto-peptone by bile acids. Appl. Environ. Microbiol. *54*: 990–995.

Kamekura, M. and Y. Seno. 1989. Lysis of halobacteria with bile acids and proteolytic enzymes of halophilic archaebacteria. *In* Rodriguez-Valera (Editor), General and Applied Aspects of Halophilic Microorganisms, Plenum Press, New York. 359–365.

Kamekura, M., Y. Seno, M.L. Holmes and M.L. Dyall-Smith. 1992. Molecular cloning and sequencing of the gene for a halophilic alkaline serine protease (halolysin) from an unidentified halophilic archaea strain (172P1) and expression of the gene in *Haloferax volcanii*. J. Bacteriol. *174*: 736–742.

Kamekura, M., Y. Seno and H. Tomioka. 1998. Detection and expression of a gene encoding a new bacteriorhodopsin from an extreme halophile strain HT (JCM 9743) which does not possess bacteriorhodopsin activity. Extremophiles *2*: 33–39.

Kamlage, B. and M. Blaut. 1992. Characterization of cytochromes from *Methanosarcina* strain Gö1 and their involvement in electron transport during growth on methanol. J. Bacteriol. *174*: 3921–3927.

Kämpf, C. and N. Pfennig. 1980. Capacity of *Chromatiaceae* for chemotrophic growth. Specific respiration rates of *Thiocystis violacea* and *Chromatium vinosum*. Arch. Microbiol. *127*: 125–135.

Kämpfer, P., R. Erhart, C. Beimfohr, J. Böhringer, M. Wagner and R.I. Amann. 1996. Characterization of bacterial communities from acti-

vated sludge: culture-dependent numerical identification versus *in situ* identification using group- and genus-specific rRNA-targeted oligonucleotide probes. Microb. Ecol. *32*: 101–121.

Kanai, H., T. Kobayashi, R. Aono and T. Kudo. 1995. *Natronococcus amylolyticus* sp. nov., a haloalkaliphilic archaeon. Int. J. Syst. Bacteriol. *45*: 762–766.

Kandler, O. (Editor). 1982a. Archaebacteria, Gustav Fischer Verlag, Stuttgart.

Kandler, O. 1982b. Cell wall structures and their phylogenetic implications. Zentbl. Bakteriol. Mikrobiol. Hyg. 1 Abt Orig. C *3*: 149–160.

Kandler, O. 1994. Cell wall biochemistry and three-domain concept of life. Syst. Appl. Microbiol. *16*: 501–509.

Kandler, O. and H. Hippe. 1977. Lack of peptidoglycan in the cell walls of *Methanosarcina barkeri*. Arch. Microbiol. *113*: 57–60.

Kandler, O. and H. König. 1978. Chemical composition of the peptidoglycan-free cell walls of methanogenic bacteria. Arch. Microbiol. *118*: 141–152.

Kandler, O. and H. König. 1985. Cell envelopes of archaebacteria. *In* Woese and Wolfe (Editors), The Bacteria, Vol. VII, Academic Press, New York. pp. 413–457.

Kandler, O. and W. Zillig (Editors). 1986. Archaebacteria '85, Gustav Fischer Verlag, Stuttgart.

Kane, M.D., L.K. Poulsen and D.A. Stahl. 1993. Monitoring the enrichment and isolation of sulfate-reducing bacteria by using oligonucleotide hybridization probes designed from environmentally derived 16S rRNA sequences. Appl. Environ. Microbiol. *59*: 682–686.

Kaneko, T., S. Sato, H. Kotani, A. Tanaka, E. Asamizu, Y. Nakamura, N. Miyajima, M. Hirosawa, M. Sugiura, S. Sasamoto, T. Kimura, T. Hosouchi, A. Matsuno, A. Muraki, N. Nakazaki, K. Naruo, S. Okumura, S. Shimpo, C. Takeuchi, T. Wada, A. Watanabe, M. Yamada, M. Yasuda and S. Tabata. 1996. Sequence analysis of the genome of the unicellular cyanobacterium *Synechocystis* sp. strain PCC 6803. II. Sequence determination of the entire genome and assignment of potential protein-coding regions. DNA Res. *3*: 109–136.

Kann, E. 1972. Zur Systematik und Ökologie der Gattung *Chamaesiphon* (Cyanophyceae). I. Systematik. Arch. Hydrobiol. Suppl. 41, Algol. Stud. *7*: 117–171.

Kantz, T. and H.C. Bold. 1969. Phycological Studies. Morphological and Taxonomic Investigations of *Nostoc* and *Anabaena* in Culture. Publication no. 6924, University of Texas, Austin.

Karavaiko, G.I., O.V. Golyshina, A.V. Troitskii, K.M. Val'ekho-Roman, R.S. Golovacheva and T.A. Pivovarova. 1994. *Sulfurococcus yellowstonii* sp. nov., a new species of iron- and sulfur-oxidizing thermoacidophilic archaebacteria. Microbiology *63*: 379–387.

Karavaiko, G.I., O.V. Golyshina, A.V. Troitskii, K.M. Val'ekho-Roman, R.S. Golovacheva and T.A. Pivovarova. 1995. *In* Validation of the publication of new names and new combinations previously effectively published outside the IJSB. List No. 55. Int. J. Syst. Bacteriol. *45*: 879–880.

Karl, D.M. 1980. Cellular nucleotide measurements and applications in microbial ecology. Microbiol. Rev. *44*: 739–796.

Karlin, S., G.M. Weinstock and V. Brendel. 1995. Bacterial classifications derived from *recA* protein sequence comparisons. J. Bacteriol. *177*: 6881–6893.

Karrasch, M., M. Bott and R.K. Thauer. 1989. Carbonic anhydrase activity in acetate grown *Methanosarcina barkeri*. Arch. Microbiol. *151*: 137–142.

Karsten, U. 1996. Growth and organic osmolytes of geographically different isolates of *Microcoleus chthonoplastes* (Cyanobacteria) from benthic microbial mats: response to salinity change. J. Phycol. *32*: 501–506.

Karsten, U. and F. Garcia-Pichel. 1996. Carotenoids and mycosporine-like amino acid compounds in members of the genus *Microcoleus* (cyanobacteria): a chemosystematic study. Syst. Appl. Microbiol. *19*: 285–294.

Kates, M. 1978. The phytanyl ether-linked polar lipids and isoprenoid lipids of extremely halophilic bacteria. Prog. Chem. Fats Other Lipids *15*: 301–342.

Kates, M. 1993. Membrane lipids of extreme halophiles: biosynthesis, function and evolutionary significance. Experientia (Basel) *49*: 1027–1036.

Kawarabayashi, Y., Y. Hino, H. Horikawa, S. Yamazaki, Y. Haikawa, K. Jinno, M. Takahashi, M. Sekine, S.A. Baba, A., H. Kosugi, A. Hosoyama, S. Fukui, Y. Nagai, K. Nishijima, H. Nakazawa, M. Takamiya, S. Masuda, T. Funahashi, T. Tanaka, Y. Kudoh, J. Yamazaki, N. Kushida, A. Oguchi, K. Aoki, K. Kubota, Y. Nakamura, N. Nomura, Y. Sako and H. Kikuchi. 1999. Complete genome sequence of an aerobic hyper-thermophilic crenarchaeon, *Aeropyrum pernix* K1. DNA Res. *6*: 83–101.

Kawarabayasi, Y., M. Sawada, H. Horikawa, Y. Haikawa, Y. Hino, S. Yamamoto, M. Sekine, S. Baba, H. Kosugi, A. Hosoyama, Y. Nagai, M. Sakai, K. Ogura, R. Otsuka, H. Nakazawa, M. Takamiya, Y. Ohfuku, T. Funahashi, T. Tanaka, Y. Kudoh, J. Yamazaki, N. Kushida, A. Oguchi, K. Aoki and H. Kikuchi. 1998. Complete sequence and gene organization of the genome of a hyper-thermophilic archaebacterium, *Pyrococcus horikoshii* OT3. DNA Res. *5*: 55–76.

Kawasaki, H., Y. Hoshino, A. Hirata and K. Yamasato. 1993a. Is intracytoplasmic membrane structure a generic criterion? It does not coincide with phylogenetic interrelationships among phototrophic purple nonsulfur bacteria. Arch. Microbiol. *160*: 358–362.

Kawasaki, H., Y. Hoshino, H. Kuraishi and K. Yamasato. 1992. *Rhodocista centenaria* gen. nov., sp. nov., a cyst-forming anoxygenic photosynthetic bacterium and its phylogenetic postion in the *Proteobacteria* alpha group. J. Gen. Appl. Microbiol. *38*: 541–551.

Kawasaki, H., Y. Hoshino and K. Yamasato. 1993b. Phylogenetic diversity of phototrophic purple non-sulfur bacteria in the *proteobacteria* alpha group. FEMS Microbiol. Lett. *112*: 61–66.

Kawasumi, T., Y. Igarashi, T. Kodama and Y. Minoda. 1980. Isolation of strictly thermophilic and obligately autotrophic hydrogen bacteria. Agric. Biol. Chem. *44*: 1985–1986.

Kawasumi, T., Y. Igarashi, T. Kodama and Y. Minoda. 1984. *Hydrogenobacter thermophilus* gen. nov. sp. nov., an extremely thermophilic, aerobic, hydrogen-oxidizing bacterium. Int. J. Syst. Bacteriol. *34*: 5–10.

Kayton, I. 1982. Copyright in living genetically engineered works. Geo. Wash.L. Rev. *50*: 191–218.

Keightley, J.A., B.H. Zimmermann, M.W. Mather, P. Springer, A. Pastuszyn, D.M. Lawrence and J.A. Fee. 1995. Molecular genetic and protein chemical characterization of the cytochrome ba_3 from *Thermus thermophilus* HB8. J. Biol. Chem. *270*: 20345–20358.

Kellenberger, E., A. Ryter and J. Sechaud. 1958. Electron microscope study of DNA-containing plasmids. II. Vegetative and mature phage DNA as compared with normal bacterial nucleoids in different physiological states. J. Biophys. Biochem. Cytol. *4*: 671–678.

Keller, M., F.-J. Braun, R. Dirmeier, D. Hafenbradl, S. Burggraf and K.O. Stetter. 1995. *Thermococcus alcaliphilus* sp. nov., a new hyperthermophilic archaeum growing on polysulfide at alkaline pH. Arch. Microbiol. *164*: 390–395.

Keller, M., F.-J. Braun, R. Dirmeier, D. Hafenbradl, S. Burggraf and K.O. Stetter. 1997. *In* Validation of the publication of new names and new combinations previously effectively published outside the IJSB. List No. 61. Int. J. Syst. Bacteriol. *47*: 601–602.

Kellerman, K.F. 1915. Micrococci causing red deterioration of salted codfish. Zentbl. Bakteriol. Parasitenkd. Infektkrankh. Hyg. Abt. II *34*: 398–494.

Kelly, D.P. and A.P. Wood. 2000. Reclassification of some species of *Thiobacillus* to the newly designated genera *Acidithiobacillus* gen. nov., *Halothiobacillus* gen. nov. and *Thermithiobacillus* gen. nov. Int. J. Syst. Evol. Microbiol. *50*: 511–516.

Keltjens, J.T., M.J. Huberts, W.H. Laarhoven and G.D. Vogels. 1983. Structural elements of methanopterin, a novel pterin present in *Methanobacterium thermoautotrophicum*. Eur. J. Biochem. *130*: 537–544.

Keltjens, J.T. and G.D. Vogels. 1988. Methanopterin and methanogenic bacteria. Biofactors *1*: 95–103.

Kengen, S.W.M. and A.J.M. Stams. 1994. Formation of L-alanine as a reduced end product in carbohydrate fermentation by the hyperthermophilic archaeon *Pyrococcus furiosus*. Arch. Microbiol. *161*: 168–175.

Kengen, S.W.M., A.J.M. Stams and W.M. de Vos. 1996. Sugar metabolism of hyperthermophiles. FEMS Microbiol. Rev. *18*: 119–137.

Kenyon, C.N. 1972. Fatty acid composition of unicellular strains of blue-green algae. J. Bacteriol. *109*: 827–834.

Kenyon, C.N. and A.M. Gray. 1974. Preliminary analysis of lipids and fatty acids of green bacteria and *Chloroflexus aurantiacus*. J. Bacteriol. *120*: 131–138.

Keppen, O.I., O.I. Baulina and E.N. Kondratieva. 1994. *Oscillochloris trichoides* neotype strain DG-6. Photosynth. Res. *41*: 29–33.

Keppen, O.I., O.I. Baulina, A.M. Lysenko and E.N. Kondratieva. 1993. A new filamentous green bacterium belonging to the *Chloroflexaceae* family. Mikrobiologiya *62*: 267–275.

Kersters, K. and J. De Ley. 1984. Genus III. *Agrobacterium. In* Krieg and Holt (Editors), Bergey's Manual of Systematic Bacteriology, 1st. Ed., Vol. 1, The Williams & Wilkins Co., Baltimore. pp. 244–254.

Kersters, K. , W. Ludwig, M. Vancanneyt, P. de Vos, M. Gillis and K.H. Schleifer. 1996. Recent changes in the classification of the pseudomonads: an overview. Syst. Appl. Microbiol. *19*: 465–476.

Kessel, M. and Y. Cohen. 1982. Ultrastructure of square bacteria from a brine pool in Southern Sinai. J. Bacteriol. *150*: 851–860.

Kessel, M., Y. Cohen and A.E. Walsby. 1985. Structure and physiology of square-shaped and other halophilic bacteria from the Gavish Sabkha. *In* Friedman and Krumbein (Editors), Ecological Studies, Vol. 53, Hypersaline Ecosystems, Springer-Verlag, Berlin. pp. 268–287.

Kessel, M. and F. Klink. 1982. Identification and comparison of eighteen archaebacteria by means of the diphtheria toxin reaction. Zentbl. Bakteriol. Mikrobiol. Hyg. 1 Abt Orig. C *3*: 140–148.

Kessler, C. 1991. The digoxigenin:anti-digoxigenin (DIG) technology - a survey on the concept and realization of a novel bioanalytical indicator system. Mol. Cell. Probes *5*: 161–205.

Kessler, C. 1994. Non-radioactive analysis of biomolecules. J. Biotechnol. *35*: 165–189.

Keswani, J., S. Orkand, U. Premachandran, L. Mandelco, M.J. Franklin and W.B. Whitman. 1996. Phylogeny and taxonomy of mesophilic *Methanococcus* spp. and comparison of rRNA, DNA hybridization, and phenotypic methods. Int. J. Syst. Bacteriol. *46*: 727–735.

Kevbrin, V.V., A.M. Lysenko and T.N. Zhilina. 1997. Physiology of the alkaliphilic methanogen Z-7936, a new strain of *Methanosalsus zhilinaeae* isolated from Lake Magadi. Microbiology *66*: 261–266.

Kiene, R.P., R.S. Oremland, A. Catena, L.G. Miller and D.G. Capone. 1986. Metabolism of reduced methylated sulfur compounds in anaerobic sediments and by a pure culture of an estuarine methanogen. Appl. Environ. Microbiol. *52*: 1037–1045.

Kim, C.W., P. Markiewicz, J.J. Lee, C.F. Schierle and J.H. Miller. 1993. Studies of the hyperthermophile *Thermotoga maritima* by random sequencing of cDNA and genomic libraries. J. Mol. Biol. *231*: 960–981.

Kimble, L.K. and M.T. Madigan. 1992. Nitrogen fixation and nitrogen metabolism in heliobacteria. Arch. Microbiol. *158*: 155–161.

Kimble, L.K., L. Mandelco, C.R. Woese and M.T. Madigan. 1995. *Heliobacterium modesticaldum*, sp. nov., a thermophilic heliobacterium of hot springs and volcanic soils. Arch. Microbiol. *163*: 259–267.

Kimble, L.K., L. Mandelco, C.R. Woese and M.T. Madigan. 1996. *In* Validation of the publication of new names and new combinations previously effectively published outside the IJSB. List No. 59. Int. J. Syst. Bacteriol. *46*: 1189–1190.

Kimble, L.K., A.K. Stevenson and M.T. Madigan. 1994. Chemotrophic growth of heliobacteria in darkness. FEMS Microbiol. Lett *115*: 51–56.

Kirsop, B.E. 1996. The Convention on Biological Diversity: some implications for microbiology and microbial culture collections. J. Ind. Microbiol. Biotechnol. *17*: 505–511.

Kishimoto, N., Y. Kosako, N. Wakao, T. Tano and A. Hiraishi. 1995. Transfer of *Acidiphilium facilis* and *Acidiphilium aminolytica* to the genus *Acidocella* gen. nov., and emendation of the genus *Acidiphilium*. Syst. Appl. Microbiol. *18*: 85–91.

Kishino, H. and M. Hasegawa. 1989. Evaluation of the maximum likelihood estimate of the evolutionary tree topologies from DNA sequence data, and the branching order in *Hominoidea*. J. Mol. Evol. *29*: 170–179.

Kjems, J., N. Larsen, J.Z. Dalgaard, R.A. Garrett and K.O. Stetter. 1992. Phylogenetic relationships amongst the hyperthermophilic archaea determined from partial 23S rRNA gene sequences. Syst. Appl. Microbiol. *15*: 203–208.

Klebahn, H. 1919. Die Schädlinge des Klippfisches. Ein Beitrag zur Kenntnis der salzliebenden Organismen. Mitt. Inst. Allg. Bot. Hambg. *4*: 11–69.

Klein, A.R., J. Breitung, D. Linder, K.O. Stetter and R.K. Thauer. 1993. N^5,N^{10}-methenyltetrahydromethanopterin cyclohydrolase from the extremely thermophilic sulfate reducing *Archaeoglobus fulgidus*: comparison of its properties with those of the cyclohydrolase from the extremely thermophilic *Methanopyrus kandleri*. Arch. Microbiol. *159*: 213–219.

Kleinig, H. and H. Reichenbach. 1977. Carotenoid glucosides and menaquinones from the gliding bacterium *Herpetosiphon giganteus* Hpa2. Arch. Microbiol. *112*: 307–310.

Klenk, H.-P., R.A. Clayton, J.-F. Tomb, O. White, K.E. Nelson, K.A. Ketchum, R.J. Dodson, M. Gwinn, E.K. Hickey, J.D. Peterson, D.L. Richardson, A.R. Kerlavage, D.E. Graham, N.C. Kyrpides, R.D. Fleischmann, J. Quackenbush, N.H. Lee, G.G. Sutton, S. Gill, E.F. Kirkness, B.A. Dougherty, K. McKenny, M.D. Adams, B. Loftus, S. Peterson, C.I. Reich, L.K. McNeil, J.H. Badger, A. Glodek, L. Zhou, R. Overbeek, J.D. Gocayne, J.F. Weidman, L. McDonald, T. Utterback, M.D. Cotton, T. Spriggs, P. Artiach, B.P. Kaine, S.M. Sykes, P.W. Sadow, K.P. D'Andrea, C. Bowman, C. Fujii, S.A. Garland, T.M. Mason, G.J. Olsen, C.M. Fraser, H.O. Smith, C.R. Woese and J.C. Venter. 1997a. The complete genome sequence of the hyperthermophilic, sulphate-reducing archaeon *Archaeoglobus fulgidus*. Nature *390*: 364–370.

Klenk, H.-P., T.D. Meier, P. Durovic, V. Schwass, F. Lottspeich, P.P. Dennis and W. Zillig. 1999. RNA polymerase of *Aquifex pyrophilus*: implications for the evolution of the bacterial *rpoBC*-operon and the extreme thermophilic bacteria. J. Mol. Evol. *48*: 528–541.

Klenk, H.-P., P. Palm and W. Zillig. 1994. DNA-dependent RNA polymerases as phylogenetic marker molecules. Syst. Appl. Microbiol. *16*: 638–647.

Klenk, H.-P., L. Zhou and J.C. Venter. 1997b. Understanding life on this planet in the age of genomics. Proceedings of SPIE. *3111*: 306–317.

Kluyver, A.J. and C.G.T.P. Schnellen. 1947. On the fermentation of carbon monoxide by pure cultures of methanogenic bacteria. Arch. Biochem. *14*: 57–70.

Kluyver, A.J. and C.B. Van Niel. 1936. Prospects for a natural system of classification of bacteria. Zentbl. Bakteriol. Parasitenkd. Infektkrankh. Hyg. Abt. II *94*: 369–403.

Kneifel, H., K.O. Stetter, J.R. Andreesen, J. Wiegel, H. König and S.M. Schoberth. 1986. Distribution of polyamines in representative species of archaebacteria. Syst. Appl. Microbiol. *7*: 241–245.

Knoll, A.H. 1985. The distribution and evolution of microbial life in the late Proterozoic era. Ann. Rev. Microbiol. *39*: 391–417.

Knudsen, E., E. Jantzen, K. Bryn, J.G. Ormerod and R. Sirevåg. 1982. Quantitative and structural characteristics of lipids in *Chlorobium* and *Chloroflexus*. Arch. Microbiol. *132*: 149–154.

Kobatake, M., S. Tanabe and S. Hasegawa. 1973. Nouveau *Micrococcus* radiorésistant à pigment rouge, isolé de féces de *Llama glama*, et son utilisation comme indicateur microbiologique de la radiostérilisation. Comtes Rendus des Séances de la Societé de Biologie *167*: 1506–1510.

Kobayashi, T., S. Higuchi, K. Kimura, T. Kudo and K. Horikoshi. 1995. Properties of glutamate dehydrogenase and its involvement in alanine production in a hyperthermophilic archaeon, *Thermococcus profundus*. J. Biochem. (Tokyo) *118*: 587–592.

Kobayashi, T. and K. Horikoshi. 1995. *In* Validation of the publication of new names and new combinations previously effectively published outside the IJSB. List No. 53. Int. J. Syst. Bacteriol. *45*: 418–419.

Kobayashi, T., H. Kanai, R. Aono, K. Horikoshi and T. Kudo. 1994a. Cloning, expression, and nucleotide sequence of the α-amylase gene from the haloalkaliphilic archaeon *Natronococcus* sp. strain Ah-36. J. Bacteriol. *176*: 5131–5134.

Kobayashi, T., H. Kanai, T. Hayashi, T. Akiba, R. Akaboshi and K. Horikoshi. 1992. Haloalkaliphilic maltotriose-forming α-amylase from the archaebacterium *Natronococcus* sp. strain Ah-36. J. Bacteriol. *174*: 3439–3444.

Kobayashi, T., Y.S. Kwak, T. Akiba, T. Kudo and K. Horikoshi. 1994b. *Thermococcus profundus* sp. nov., a new hyperthermophilic archaeon isolated from a deep-sea hydrothermal vent. Syst. Appl. Microbiol. *17*: 232–236.

Kocur, M. and W. Hodgkiss. 1973. Taxonomic status of the genus *Halococcus* Schoop. Int. J. Syst. Bacteriol. *23*: 151–156.

Koga, Y., H. Morii, M. Akagawa-Matsushita and M. Ohga. 1998. Correlation of polar lipid composition with 16S rRNA phylogeny in methanogens: further analysis of lipid component parts. Biosci. Biotechnol. Biochem. *62*: 230–236.

Koga, Y., M. Nishihara, H. Morii and M. Akagawa-Matsushita. 1993. Ether polar lipids of methanogenic bacteria: structures, comparative aspects, and biosyntheses. Microbiol. Rev. *57*: 164–182.

Kogure, K., U. Simidu and N. Taga. 1979. A tentative direct microscopic method for counting living marine bacteria. Can. J. Microbiol. *25*: 415–420.

Kogure, K., U. Simidu and N. Taga. 1984. An improved direct viable count method for aquatic bacteria. Arch. Hydrobiol. *102*: 117–122.

Komárek, J. 1969. On the validity of the genus *Dactylococcopsis* (Cyanophyceae). Österr. Bot. Z. *117*: 248–257.

Komárek, J. 1970. Generic identity of the *"Anacystis nidulans"* strain Kratz-Allen/Bloom 625 with *Synechococcus* Näg. 1849. Arch. Protistenkd. *112*: 343–364.

Komárek, J. 1972. Reproduction process and taxonomy of unicellular endosporine blue-green algae. Proceedings of the Symposium on Taxonomy and Biology of Blue-green Algae, University of Madras. pp. 41–47.

Komárek, J. 1976. Taxonomic review of the genera *Synechocystis* Sauv. 1892, *Synechococcus* Näg, 1849, and *Cyanothece* gen. nov. (Cyanophyceae). Arch. Protistenkd. *118*: 119–179.

Komárek, J. 1983. *Rhabdogloea*, the correct name of cyanophycean *Dactylococcopsis* sensu Auctt, non Hansgirg (1888). Taxon *32*: 464–466.

Komárek, J. 1994. Current trends and species delimitation in the cyanoprokaryote taxonomy. Arch. Hydrobiol. Suppl. *105*: 11–29.

Komárek, J. and K. Anagnostidis. 1986. Modern approach to the classification system of cyanophytes: 2. *Chroococcales*. Arch. Hydrobiol. Suppl. *73*: 157–226.

Komárek, J. and K. Anagnostidis. 1989. Modern approach to the classification system of cyanophytes, 4. *Nostocales*. Arch. Hydrobiol. Suppl. *82*: 247–345.

Komárek, J. and K. Anagnostidis. 1998. Cyanoprokaryota 1. Teil *Chroococcales*, G. Fischer, Jena.

Komárek, J. and F. Hindak. 1975. Taxonomy of the new isolated strains of *Chroococcidiopsis* (Cyanophyceae). Arch. Hydrobiol. Suppl. *46*: 311–329.

Komárek, J., M. Hübel, H. Hübel and J. Šmarda. 1993. The *Nodularia* studies 2. Taxonomy. Arch. Hydrobiol. Suppl. *96*: 1–4.

Komárek, J. and H. Kling. 1991. Variation in six planktonic cyanophyte genera in Lake Victoria (East Africa). Algol. Stud. *61*: 21–45.

Komárek, J., J. Kopecký and V. Cepák. 1999. Generic characters of the simplest cyanoprokaryotes *Cyanobium*, *Cyanobacterium* and *Synechococcus*. Cryptogam. Algol. *20*: 209–222.

Komárková, J. 1998. The tropical genus *Cylindrospermopsis* (Cyanophytes, Cyanobacteria). *In* de Paiva Azevedo (Editor), Anais do IV Congresso Latino-Americano, II Reunião Ibero-Americano, VII Reunião Brasileira de Ficologia. Biodiversity Conservation and New Technologies: Promises and Perils, Vol. 1, Sociedade Ficológica da América Latina e Caribe, São Paulo. pp. 327–340.

Kondo, T. and M. Ishiura. 1994. Circadian rhythms of cyanobacteria: monitoring the biological clocks of individual colonies by bioluminescence. J. Bacteriol. *176*: 1881–1885.

Kondo, T., T. Mori, N.V. Lebedeva, S. Aoki, M. Ishiura and S.S. Golden.

1997. Circadian rhythms in rapidly dividing cyanobacteria. Science *275*: 224–227.

Kondratieva, E.N., V.G. Zhukov, R.N. Ivanovsky, U.P. Petushkova and E.Z. Monosov. 1976. The capacity of phototrophic sulfur bacterium *Thiocapsa roseopersicina* for chemosynthesis. Arch. Microbiol. *108*: 287–292.

König, H. 1984. Isolation and characterization of *Methanobacterium uliginosum*, new species from a marshy soil. Can. J. Microbiol. *30*: 1477–1481.

König, H. 1985. *In* Validation of the publication of new names and new combinations previously effectively published outside the IJSB. List No. 18. Int. J. Syst. Bacteriol. *35*: 375–376.

König, H. 1986. Chemical composition of cell envelopes of methanogenic bacteria isolated from human and animal feces. Syst. Appl. Microbiol. *8*: 159–162.

König, H., R. Kralik and O. Kandler. 1982. Structure and modifications of pseudomurein in *Methanobacteriales*. Zentbl. Bakteriol. Mikrobiol. Hyg. 1 Abt Orig. C *3*: 179–191.

König, H., P. Messner and K.O. Stetter. 1988. The fine structure of the fibers of *Pyrodictium occultum*. FEMS Microbiol. Lett. *49*: 207–212.

König, H., E. Nusser and K.O. Stetter. 1985. Glycogen in *Methanolobus* and *Methanococcus*. FEMS Microbiol. Lett. *28*: 265–269.

König, H. and K.O. Stetter. 1982. Isolation and characterization of *Methanolobus tindarius*, sp. nov., a coccoid methanogen growing only on methanol and methylamines. Zentbl. Bakteriol. Mikrobiol. Hyg. 1 Abt Orig. C *3*: 478–490.

König, H. and K.O. Stetter. 1983. *In* Validation of the publication of new names and new combinations previously effectively published outside the IJSB. List No. 10. Int. J. Syst. Bacteriol. *33*: 438–440.

König, H. and K.O. Stetter. 1986. Studies on archaebacterial S-layers. Syst. Appl. Microbiol. *7*: 300–309.

Koonin, E.V., R.L. Tatusov and M.Y. Galperin. 1998. Beyond complete genomes: from sequence to structure and function. Curr. Opin. Struct. Biol. *8*: 355–363.

Korolik, V., P.J. Coloe and V. Krishnapillai. 1988. A specific DNA probe for the identification of *Campylobacter jejuni*. J. Gen. Microbiol. *134*: 521–530.

Kosko, B. 1994. Fuzzy Thinking, Harper Collins Publishers, London.

Kostrikina, N.A., I.S. Zvyagintseva and V.I. Duda. 1991. Cytological peculiarities of some extremely halophilic soil archaeobacteria. Arch. Microbiol. *156*: 344–349.

Kotelnikova, S.V., A.J.L. Macario and K. Pedersen. 1998. *Methanobacterium subterranium* sp. nov., a new alkaliphilic, eurythermic and halotolerant methanogen isolated from deep granitic groundwater. Int. J. Syst. Bacteriol. *48*: 357–367.

Kotelnikova, S.V., A.Y. Obraztsova, K.H. Blotevogel and I.N. Popov. 1993a. Taxonomic analysis of thermophilic strains of the genus *Methanobacterium*: reclassification of *Methanobacterium thermoalcaliphilum* as a synonym of *Methanobacterium thermoautotrophicum*. Int. J. Syst. Bacteriol. *43*: 591–596.

Kotelnikova, S.V., A.Y. Obraztsova, G.M. Gongadze and K.S. Laurinavichius. 1993b. *Methanobacterium thermoflexum* sp. nov. and *Methanobacterium defluvii* sp. nov.: thermophilic rod-shaped methanogens isolated from anaerobic digestor sludge. Syst. Appl. Microbiol. *16*: 427–435.

Koyama, Y. and K. Furukawa. 1990. Cloning and sequence analysis of tryptophan synthetase genes of an extreme thermophile, *Thermus thermophilus* HB27: plasmid transfer from replica-plated *Escherichia coli* recombinant colonies to competent *T. thermophilus* cells. J. Bacteriol. *172*: 3490–3495.

Koyama, Y., T. Hoshino, N. Tomizuka and K. Furukawa. 1986. Genetic transformation of the extreme thermophile *Thermus thermophilus* and of other *Thermus* spp. J. Bacteriol. *166*: 338–340.

Koyama, Y., S. Okamoto and K. Furukawa. 1989. Development of host-vector systems in the extreme thermophile *Thermus thermophilus*. *In* da Costa, Duarte and Williams (Editors), Microbiology of Extreme Environments and Its Potential for Biotechnology, Elsevier, London. pp. 103–105.

Koyama, Y., S. Okamoto and K. Furukawa. 1990. Cloning of α- and β-galactosidase genes from an extreme thermophile, *Thermus* strain T2,

and their expression in *Thermus thermophilus* HB27. Appl. Environ. Microbiol. *56*: 2251–2254.

Krabbenhoft, K.L., A.W. Anderson and P.R. Elliker. 1965. Ecology of *Micrococcus radiodurans*. Appl. Microbiol. *13*: 1030–1037.

Kraepelin, G. and H.U. Gravenstein. 1980. Experimental induction of rotund bodies in *Thermus equaticus*. Z. Allg. Mikrobiol. *20*: 33–45.

Krafft, T. and J.M. Macy. 1998. Purification and characterization of the respiratory arsenate reductase of *Chrysiogenes arsenatis*. Eur. J. Biochem. *255*: 647–653.

Kratz, W.A. and J. Myers. 1955. Nutrition and growth of several blue-green algae. Am. J. Bot. *42*: 282–287.

Kreisl, P. and O. Kandler. 1986. Chemical structure of the cell wall polymer of *Methanosarcina*. Syst. Appl. Microbiol. *7*: 293–299.

Kremer, D.R., M. Veenhuis, G. Fauque, H.D. Peck, Jr, J. Legall, J. Lampreia, J.J.G. Moura and T.A. Hansen. 1988. Immunocytochemical localization of APS reductase and bisulfite reductase in three *Desulfovibrio* species. Arch. Microbiol. *150*: 296– 301.

Krichevsky, M.I. and L.M. Norton. 1974. Storage and manipulation of data by computers for determinative bacteriology. Int. J. Syst. Bacteriol. *24*: 525–531.

Krieg, N.R. and P. Gerhardt. 1981. Solid Culture. *In* Gerhardt, Murray, Costilow, Nester, Wood, Krieg and Phillips (Editors), Manual of Methods for General Bacteriology, American Society for Microbiology, Washington, D.C. pp. 143–150.

Krieg, N.R. and J.G. Holt (Editors). 1984. Bergey's Manual of Systematic Bacteriology, 1st Ed., Vol. 1, The Williams & Wilkins Co., Baltimore.

Kristensen, H. and E.A. Christensen. 1981. Radiation-resistant micro-organisms isolated from textiles. Acta Pathol. Microbiol. Scand. B *89*: 303–309.

Kristjánsson, J.K. and G.A. Alfredsson. 1983. Distribution of *Thermus* spp. in Icelandic hot springs and a thermal gradient. Appl. Environ. Microbiol. *45*: 1785–1789.

Kristjánsson, J.K., S. Hjörleifsdóttir, V.T. Marteinsson and G.A. Alfredsson. 1994. *Thermus scotoductus*, sp. nov., a pigment-producing thermophilic bacterium from hot tap water in Iceland and including *Thermus* sp. X-1. Syst. Appl. Microbiol. *17*: 44–50.

Kristjánsson, J.K., S. Hjörleifsdóttir, V.T. Marteinsson and G.A. Alfredsson. 1995. *In* Validation of the publication of new names and new combinations previously effectively published outside the IJSB. List No. 53. Int. J. Syst. Bacteriol. *45*: 418–419.

Kristjánsson, J.K., G.O. Hreggvidsson and G.A. Alfredsson. 1986. Isolation of halotolerant *Thermus scotoductus* spp. from submarine hot springs in Iceland. Appl. Environ. Microbiol. *52*: 1313–1316.

Krone, U.E., K. Laufer, R.K. Thauer and H.P.C. Hogenkamp. 1989. Coenzyme F_{430} as a possible catalyst for the reductive dehalogenation of chlorinated C_1 hydrocarbons in methanogenic bacteria. Biochemistry *28*: 10061–10065.

Krone, U.E. and R.K. Thauer. 1992. Dehalogenation of trichlorofluoromethane (CFC-11) by *Methanosarcina barkeri*. FEMS Microbiol. Lett. *90*: 201–204.

Krüger, K. and F. Pfeifer. 1996. Transcript analysis of the c-vac region and differential synthesis of the two regulatory gas vesicle proteins GvpD and GvpE in *Halobacterium salinarium* PHH4. J. Bacteriol. *178*: 4012–4019.

Krumholz, L.R., J.P. McKinley, G.A. Ulrich and J.M. Suflita. 1997. Confined subsurface microbial communities in Cretaceous rock. Nature *386*: 64–66.

Kruschel, C. and R.W. Castenholz. 1998. The effect of solar UV and visible irradiance on the vertical movements of cyanobacteria in microbial mass of hypersaline waters. FEMS Microbiol. Ecol. *27*: 53–72.

Kryukov, V.R., N.D. Savel'eva and M.A. Pusheva. 1983. *Calderobacterium hydrogenophilum*, nov. gen. nov. sp., an extreme thermophilic hydrogen bacterium, and its hydrogenase activity. Mikrobiologiya *52*: 781–788.

Kryukov, V.R., N.D. Savel'eva and M.A. Pusheva. 1984. *In* Validation of the publication of new names and new combinations previously effectively published outside the IJSB. List No. 14. Int. J. Syst. Bacteriol. *34*: 270–271.

Krzycki, J. and J.G. Zeikus. 1980. Quantification of corrinoids in meth-anogenic bacteria. Curr. Microbiol. *3*: 243–245.

Kubler, O. and W. Baumeister. 1978. The structure of a periodic cell wall component (HPI-layer of *Micrococcus radiodurans*). Cytobiologie *17*: 1–9.

Kuchino, Y., M. Ihara, Y. Yabusaki and S. Nishimura. 1982. Initiator tRNAs from archaebacteria show common unique sequence characteristics. Nature *298*: 684–685.

Kuhner, C.H., S.S. Smith, K.M. Noll, R.S. Tanner and R.S. Wolfe. 1991. 7-Mercaptoheptanoylthreonine phosphate substitutes for heat-stable factor (mobile factor) for growth of *Methanomicrobium mobile*. Appl. Environ. Microbiol. *57*: 2891–2895.

Kujo, C. and T. Ohshima. 1998. Enzymological characteristics of the hy-perthermostable NAD-dependent glutamate dehydrogenase from the archaeon *Pyrobaculum islandicum* and effects of denaturants and or-ganic solvents. Appl. Environ. Microbiol. *64*: 2152–2157.

Kumazawa, Y., T. Fujiwara, Y. Fukumori, Y. Koga and T. Yamanaka. 1994. Cytochrome *bc* purified from the methanogen *Methanosarcina barkeri*. Curr. Microbiol. *29*: 53–56.

Kurosawa, N., Y.H. Itoh, T. Iwae, A. Sugai, I. Uda, N. Kimura, T. Horiuchi and T. Itoh. 1998. *Sulfurisphaera ohwakuensis* gen. nov., sp. nov., a novel extremely thermophilic acidophile of the order *Sulfolobales*. Int. J. Syst. Bacteriol. *48*: 451–456.

Kurr, M., R. Huber, H. König, H.W. Jannasch, H. Fricke, A. Trincone, J.K. Kristjansson and K.O. Stetter. 1991. *Methanopyrus kandleri*, gen. and sp. nov. represents a novel group of hyperthermophilic meth-anogens, growing at 110°C. Arch. Microbiol. *156*: 239–247.

Kurr, M., R. Huber, H. König, H.W. Jannasch, H. Fricke, A. Trincone, J.K. Kristjansson and K.O. Stetter. 1992. *In* Validation of the publi-cation of new names and new combinations previously effectively pub-lished outside the IJSB List No. 41. Int. J. Syst. Bacteriol. *42*: 327–329.

Kurtzman, C.P. 1986. The ARS Culture Collection: present status and new directions. Enzyme Microb. Technol. *8*: 328–333.

Kushner, D.J. 1985. The *Halobacteriaceae*. *In* Woese and Wolfe (Editors), The Bacteria: A Treatise on Structure and Function, Vol. VIII, The Archaebacteria, Academic Press, New York. pp. 171–214.

Kushner, D.J. 1993. Growth and nutrition of halophilic bacteria. *In* Vree-land and Hochstein (Editors), The Biology of Halophilic Bacteria, CRC Press, Boca Raton. pp. 87–103.

Kushwaha, S.C., M.B. Gochnauer, D.J. Kushner and M. Kates. 1974. Pig-ments and isoprenoid compounds in extremely and moderately halo-philic bacteria. Can. J. Microbiol. *20*: 241–245.

Kushwaha, S.C., M. Kates, G.D. Sprott and I.C.P. Smith. 1981. Novel polar lipids from the methanogen, *Methanospirillum hungatii* GP1. Biochim. Biophys. Acta *664*: 156–173.

Kützing, F.T. 1843. Phycologia Generalis, oder Anatomie, Physiologie und Systemkunde der Tange, F.A. Brockhaus, Leipzig.

Kuznetsov, S.I. 1977. Trends in the development of ecological microbi-ology. Adv. Aquat. Microbiol. *1*: 1–48.

Labrenz, M., M.D. Collins, P.A. Lawson, B.J. Tindall, P. Schumann and P. Hirsch. 1999. *Roseovarius tolerans* gen. nov., sp. nov., a budding bacterium with variable bacteriochlorophyll a production from hy-persaline Ekho Lake. Int. J. Syst. Bacteriol. *49*: 137–147.

Lachance, M.A. 1981. Genetic relatedness of heterocystous cyanobacteria by DNA–DNA reassociation. Int. J. Syst. Bacteriol. *31*: 139–147.

Lah, M.S., M.M. Dixon, K.A. Pattridge, W.C. Stallings, J.A. Fee and M.L. Ludwig. 1995. Structure-function in *Escherichia coli* iron superoxide dismutase: comparisons with the manganese enzyme from *Thermus thermophilus*. Biochemistry *34*: 1646–1660.

Lai, M.C., K.R. Sowers, D.E. Robertson, M.F. Roberts and R.P. Gunsalus. 1991. Distribution of compatible solutes in the halophilic methano-genic archaebacteria. J. Bacteriol. *173*: 5352–5358.

Lake, J.A., R. Jain and M.C. Rivera. 1999. Mix and match in the tree of life. Science *283*: 2027–2028.

Lane, D.J. 1991. 16S/23S rRNA sequencing. *In* Stackebrandt and Good-fellow (Editors), Nucleic Acid Techniques in Bacterial Systematics, John Wiley & Sons, Chichester. pp. 115–175.

Lane, D.J., A.P. Harrison, Jr., D.A. Stahl, B. Pace, S.J. Giovannoni, G.J. Olsen and N.R. Pace. 1992. Evolutionary relationships among sulfur- and iron-oxidizing eubacteria. J. Bacteriol. *174*: 269–278.

Lane, D.J., B. Pace, G.J. Olsen, D.A. Stahl, M.L. Sogin and N.R. Pace. 1985. Rapid determination of 16S ribosomal RNA sequences for phy-logenetic analyses. Proc. Natl. Acad. Sci. U.S.A. *82*: 6955–6959.

Lang, N.J. 1977. *Starria zimbabweënsis* (*Cyanophyceae*) gen. nov. et sp. nov.: a filament triradiate in transverse section. J. Phycol. *13*: 288–296.

Langenberg, K.F., M.P. Bryant and R.S. Wolfe. 1968. Hydrogen-oxidizing methane bacteria. II. Electron microscopy. J. Bacteriol. *95*: 1124–1129.

Langer, P.R., A.A. Waldrop and D.C. Ward. 1981. Enzymatic synthesis of biotin-labeled polynucleotides: novel nucleic acid affinity probes. Proc. Natl. Acad. Sci. U.S.A. *78*: 6633–6637.

Langworthy, T.A. 1977. Long-chain diglycerol tetraethers from *Thermo-plasma acidophilum*. Biochim. Biophys. Acta *487*: 37–50.

Langworthy, T.A. 1979. Special features of Thermoplasmas. *In* Barile and Razin (Editors), The Mycoplasmas I: Cell Biology, Academic Press, Inc., New York. pp. 495–513.

Langworthy, T.A. 1985. Lipids of Archaebacteria. *In* Woese and Wolfe (Editors), The Bacteria, Vol. 8, Academic Press, New York. pp. 459–497.

Langworthy, T.A., G. Holzer, J.G. Zeikus and T.G. Tornabene. 1983. Iso-and anteiso-branched glycerol diethers of the thermophilic anaerobe *Thermodesulfotobacterium commune*. Syst. Appl. Microbiol. *4*: 1–17.

Langworthy, T.A. and J.L. Pond. 1986a. Archaebacterial ether lipids and chemotaxonomy. Syst. Appl. Microbiol. *7*: 253–275.

Langworthy, T.A. and J.L. Pond. 1986b. Membranes and lipids of ther-mophiles. *In* Brock (Editor), Thermophiles: General, Molecular, and Applied Microbiology, John Wiley & Sons, New York. pp. 107–135.

Langworthy, T.A. and P.F. Smith. 1989. Group IV. Cell wall-less archaeo-bacteria. *In* Staley, Bryant, Pfennig and Holt (Editors), Bergey's Man-ual of Systematic Bacteriology, 1st Ed., Vol. 3, The Williams & Wilkins Co., Baltimore. pp. 2233–2236.

Langworthy, T.A., T.G. Tornabene and G. Holzer. 1982. Lipids of ar-chaebacteria. Zentbl. Bakteriol. Mikrobiol. Hyg. 1 Abt Orig. C *3*: 228–244.

Lanyi, J.K. 1991. Mechanism of chloride transport in halophilic archae-bacteria. *In* Rodriguez-Valera (Editor), General and Applied Aspects of Halophilic Microorganisms, Plenum Press, New York. pp. 73–80.

Lanyi, J.K. 1993. Pathways of proton transfer in the light-driven proton pump bacteriorhodopsin. Experientia (Basel) *49*: 514–517.

Lanyi, J.K. 1995. Bacteriorhodopsin as a model for proton pumps. Nature *375*: 461–463.

Lanzotti, V., B. Nicolaus, A. Trincone, M. De Rosa, W.D. Grant and A. Gambacorta. 1989. A complex lipid with a cyclic phosphate from the archaebacterium *Natronococcus occultus*. Biochim. Biophys. Acta *1001*: 31–34.

Lanzotti, V., B. Nicolaus, A. Trincone and W.D. Grant. 1988. The gly-colipid of *Halobacterium saccharovorum*. FEMS Microbiol. Lett. *55*: 223–228.

Lapage, S.P., S. Bascomb, W.R. Willcox and M.A. Curtis. 1973. Identifi-cation of bacteria by computer. I. General aspects and perspectives. J. Gen. Microbiol. *77*: 273–290.

Lapage, S.P., P.H.A. Sneath, E.F. Lessel, Jr., V.B.D. Skerman, H.P.R. See-liger and W.A. Clark (Editors). 1975. International Code of Nomen-clature of Bacteria, 1976 revision, American Society for Microbiology, Washington, DC.

Lapage, S.P., P.H.A. Sneath, E.F. Lessel, Jr., V.B.D. Skerman, H.P.R. See-liger and W.A. Clark (Editors). 1992. International Code of Nomen-clature of Bacteria, (1990) Revision. Bacteriological Code, American Society for Microbiology, Washington, DC.

Larsen, H. 1952. On the culture and general physiology of the green sulfur bacteria. J. Bacteriol. *94*: 889–895.

Larsen, H. 1967. Biochemical aspects of extreme halophilism. Adv. Mi-crob. Physiol. *1*: 97–132.

Larsen, H. 1981. The family *Halobacteriaceae*. *In* Starr, Stolp, Trüper, Bal-ows and Schlegel (Editors), The Prokaryotes: a Handbook on Habitats, Isolation and Identification of Bacteria, Springer-Verlag, Berlin. pp. 985–994.

Larsen, H. 1984. *Halobacteriaceae. In* Staley, Bryant, Pfennig and Holt (Editors), Bergey's Manual of Systematic Bacteriology, 1st Ed., Vol. 3, The Williams & Wilkins Co., Baltimore. pp. 2261–2267.

Larsen, H. 1989. Genus VI. *Halococcus. In* Staley, Bryant, Pfennig and Holt (Editors), Bergey's Manual of Systematic Bacteriology, 1st Ed., Vol. 3, The Williams & Wilkins Co., Baltimore. pp. 2228–2230.

Larsen, H. and W.D. Grant. 1989. Genus I. *Halobacterium. In* Staley, Bryant, Pfennig and Holt (Editors), Bergey's Manual of Systematic Bacteriology, 1st Ed., Vol. 3, The Williams & Wilkins Co., Baltimore. pp. 2219–2224.

Larsen, N., H. Leffers, J. Kjems and R.A. Garrett. 1986. Evolutionary divergence between the ribosomal RNA operons of *Halococcus morrhuae* and *Desulfurococcus mobilis.* Syst. Appl. Microbiol. *7:* 49–57.

Larsen, N., R. Overbeek, S. Pramanik, T.M. Schmidt, E.E. Selkov, O. Strunk, J.M. Tiedje and J.W. Urbance. 1997. Towards microbial data integration. J. Ind. Microbiol. Biotechnol. *18:* 68–72.

Lasa, I., J.R. Castón, L.A. Fernández-Herrero, M.A. de Pedro and J. Berenguer. 1992a. Insertional mutagenesis in the extreme thermophilic eubacteria *Thermus thermophilus* HB8. Mol. Microbiol. *6:* 1555–1564.

Lasa, I., M. de Grado, M.A. de Pedro and J. Berenguer. 1992b. Development of *Thermus* and *Escherichia* shuttle vectors and their use for expression of the *Clostridium thermocellum* celA gene in *Thermus thermophilus.* J. Bacteriol. *174:* 6424–6431.

Lathe, R. 1985. Synthetic oligonucleotide probes deduced from amino acid sequence data: theoretical and practical considerations. J. Mol. Biol. *183:* 1–12.

Lauerer, G., J.K. Kristjansson, T.A. Langworthy, H. König and K.O. Stetter. 1986. *Methanothermus sociabilis,* sp. nov., a second species within the *Methanothermaceae* growing at 97°C. Syst. Appl. Microbiol. *8:* 100–105.

Laurinavichus, K.S., S.V. Kotelnikova and A.Y. Obraztsova. 1987. *Methanobacterium thermophilum,* a new species thermophilic methane-forming bacterium. Mikrobiologiya *57:* 1035–1041.

Lauterborn, R. 1906. Zur Kenntnis der sapropelischen Flora. Allg. Bot. Z. *12:* 196–197.

Lauterborn, R. 1913. Zur Kenntnis einiger sapropelischer Schizomyceten. Allg. Bot. Z. *19:* 97–100.

Lauterborn, R. 1915. Die sapropelische Lebewelt. Verh. Naturh.-Med. Ver. Heidelb. *13:* 437–438.

Laverman, A.M., J.S. Blum, J.K. Schaefer, E.J.P. Phillips, D.R. Lovley and R.S. Oremland. 1995. Growth of strain SES-3 with arsenate and other diverse electron acceptors. Appl. Environ. Microbiol. *61:* 3556–3561.

Lawrence, J.G. and H. Ochman. 1998. Molecular archaeology of the *Escherichia coli* genome. Proc. Natl. Acad. Sci. U.S.A. *95:* 9413–9417.

Layton, A.C., C.A. Lajoie, J.P. Easter, R. Jernigan, J. Sanseverino and G.S. Sayler. 1994. Molecular diagnostics and chemical analysis for assessing biodegradation of polychlorinated biphenyls in contaminated soils. J. Industrial Microbiol. Biotechnol. *13:* 392–401.

Lazaroff, N. 1972. Experimental control of Nostocacean development. *In* Desikachary (Editor), Taxonomy and Biology of Blue-green Algae, University of Madras, Madras. pp. 521–544.

Lazaroff, N. 1973. Photomorphogenesis and nostocacean development. *In* Carr and Whitton (Editors), The Biology of Blue-green Algae, Blackwell Scientific Publications, Oxford. pp. 279–319.

Lazaroff, N. and W. Vishniac. 1961. The effect of light on the develpmental cycle of *Nostoc muscorum,* a filamentous blue-green alga. J. Gen. Microbiol. *25:* 365–374.

Lazaroff, N. and W. Vishniac. 1964. The relationship of cellular differentiation to colonial morphogenesis of the blue-green alga, *Nostoc muscorum* A. J. Gen. Microbiol. *35:* 447–457.

Leadbetter, J.R. and J.A. Breznak. 1996. Physiological ecology of *Methanobrevibacter cuticularis* sp. nov. and *Methanobrevibacter curvatus* sp. nov., isolated from the hindgut of the termite *Reticulitermes flavipes.* Appl. Environ. Microbiol. *62:* 3620–3631.

Leadbetter, J.R. and J.A. Breznak. 1997. *In* Validation of the publication of new names and new combinations previously effectively published outside the IJSB, List No. 61. Int. J. Syst. Bacteriol. *47:* 601–602.

Leadbetter, J.R., L.D. Crosby and J.A. Breznak. 1998a. *Methanobrevibacter filiformis* sp. nov., a filamentous methanogen from termite hindguts. Arch. Microbiol. *169:* 287–292.

Leadbetter, J.R., L.D. Crosby and J.A. Breznak. 1998b. *In* Validation of the publication of new names and new combinations previously effectively published outside the IJSB, List No. 67. Int. J. Syst. Bacteriol. *47:* 1083–1084.

Le Campion-Alsumard, T., S. Golubic and A. Pantazidou. 1996. On the endolithic genus *Solentia* Ercegovic (Cyanophyta/Cyanobacteria). Algol. Stud. *83:* 107–127.

Lee, M.J. and S.H. Zinder. 1988. Isolation and characterization of a thermophilic bacterium which oxidizes acetate in syntrophic association with a methanogen and which grows acetogenically on H_2–CO_2. Appl. Environ. Microbiol. *54:* 124–129.

Lee, S., C. Malone and P.F. Kemp. 1993. Use of multiple 16S rRNA-targeted fluorescent probes to increase the signal strength and measure cellular RNA from natural planktonic bacteria. Mar. Ecol. Prog. Ser. *101:* 193–201.

Lehmann, K.B. and R. Neuman. 1896. Atlas und Grundriss der Bakteriologie und Lehrbuch der speciellen bakteriologischen Diagnostik, 1st Ed., J.F. Lehmann, München.

Lehmann, R.P., R.A. Brunisholz and H. Zuber. 1994. Structural differences in chlorosomes from *Chloroflexus aurantiacus* grown under different conditions support the BChl c-binding function of the 5.7 kDa polypeptide. FEBS Lett. *342:* 319–324.

Lemmermann, E. 1896. Ber. Biol. Stat. Plön. *4:* 186.

Lemmermann, E. 1907. Die Algenflora der Chatham Islands. Bot. Jahrb. Syst. Pflanzengeschich. Pflanzengeogr. *38:* 343–382.

Leuschner, C. and G. Antranikian. 1995. Heat-stable enzymes from extremely thermophilic and hyperthermophilic microorganisms. World J. Microbiol. Biotechnol. *11:* 95–114.

Lewin, R.A. 1970. A new *Herpetosiphon* species (*Flexibacteriales*). Can. J. Microbiol. *16:* 517–520.

Lewin, R.A. 1977. *Prochloron,* type genus of the Prochlorophyta. Phycologia *16:* 216.

Lewin, R.A. 1980. Uncoiled variants of *Spirulina platensis* (*Cyanophyceae: Oscillatoriaceae*). Arch. Hydrobiol. Suppl. *60:* 48–52.

Lewin, R.A. 1989. Genus *Prochloron. In* Staley, Bryant, Pfennig and Holt (Editors), Bergey's Manual of Systematic Bacteriology, 1st Ed., Vol. 3, The Williams & Wilkins Co., Baltimore. pp. 1800–1802.

Lewin, R.A., L. Cheng and R.S. Alberte. 1984. *Prochloron*-ascidian symbioses: photosynthetic potential and productivity. Micronesia *19:* 165–170.

Lewis, N.F. 1971. Studies on a radio-resistant coccus isolated from Bombay duck (*Harpodon nehereus*). J. Gen. Microbiol. *66:* 29–35.

L'Haridon, S., V. Cilia, P. Messner, B. Raguénès, A. Gambacorta, U.W. Sleytr, D. Prieur and C. Jeanthon. 1998. *Desulfurobacterium thermolithotrophum* gen. nov., sp. nov., a novel autotrophic, sulfur-reducing bacterium isolated form a deep-sea hydrothermal vent. Int. J. Syst. Bacteriol. *48:* 701–711.

L'Haridon, S., A.L. Reysenbach, P. Glénat, D. Prieur and C. Jeanthon. 1995. Hot subterranean biosphere in a continental oil reservoir. Nature *377:* 223–224.

Lien, T., M. Madsen, F.A. Rainey and N.K. Birkeland. 1998. *Petrotoga mobilis* sp. nov., from a North Sea oil-production well. Int. J. Syst. Bacteriol. *48:* 1007–1013.

Lin, C. and T.L. Miller. 1998. Phylogenetic analysis of *Methanobrevibacter* isolated from feces of humans and other animals. Arch. Microbiol. *169:* 397–403.

Lin, X.L. and R.H. White. 1986. Occurrence of coenzyme F_{420} and its γ-monoglutamyl derivative in nonmethanogenic archaebacteria. J. Bacteriol. *168:* 444–448.

Link, H.F. 1809. Observationes in ordines plantarum naturales. Dissertatio prima complectens anandrarum ordines Epiphytas, Mucedines, Gastromycos et Fungos. Ges. Nat. *3:* 3–42.

Linnaeus, C. 1753. Species plantarum, exhibentes plantas rite cognitas, and genera relatas, cum differentüs specificis, nominibus trivialibus, synonymis selectis, locis natalibus, secundum systema sexuale digestas. *In* Tomus II, Stockholm. pp. 561–1200.

Lippert, K.D. and N. Pfennig. 1969. Utilisation of molecular hydrogen by *Chlorobium thiosulfatophilum*. Growth and CO₂-fixation. Arch. Mikrobiol. *65*: 29–47.

Liston, J., W. Weibe and R.R. Colwell. 1963. Quantitative approach to the study of bacterial species. J. Bacteriol. *85*: 1061–1070.

Liu, W.-T., T.L. Marsh, H. Cheng and L.J. Forney. 1997. Characterization of microbial diversity by determining terminal restriction fragment length polymorphisms of genes encoding 16S rRNA. Appl. Environ. Microbiol. *63*: 4516–4522.

Liu, Y., D.R. Boone and C. Choy. 1990. *Methanohalophilus oregonense*, sp. nov., a methylotrophic methanogen from an alkaline, saline aquifer. Int. J. Syst. Bacteriol. *40*: 111–116.

Liu, Y., D.R. Boone, R. Sleat and R.A. Mah. 1985. *Methanosarcina mazei* LYC, a new methanogenic isolate which produces a disaggregating enzyme. Appl. Environ. Microbiol. *49*: 608–613.

Llobet-Brossa, E., R. Rossello-Mora and R.I. Amann. 1998. Microbial community composition of Wadden Sea sediments as revealed by fluorescence *in situ* hybridization. Appl. Environ. Microbiol. *64*: 2691–2696.

Lobo, A.L. and S.H. Zinder. 1988. Diazotrophy and nitrogenase activity in the archaebacterium *Methanosarcina barkeri* 227. Appl. Environ. Microbiol. *54*: 1656–1661.

Lobo, A.L. and S.H. Zinder. 1990. Nitrogenase in the archaebacterium *Methanosarcina barkeri* 227. J. Bacteriol. *172*: 6789–6796.

Lobos, J.H., T.E. Chisolm, L.H. Bopp and D.S. Holmes. 1986. *Acidiphilium organovorum*, new species, an acidophilic heterotroph isolated from a *Thiobacillus ferrooxidans* culture. Int. J. Syst. Bacteriol. *36*: 139–144.

Lobyreva, L.B., R.S. Fel'dman and V.K. Plakunov. 1987. The influence of aromatic amino acids on the growth of *Halobacterium salinarium* and the uptake of ¹⁴C phenylalanine. Mikrobiologiya *56*: 16–20.

Lochhead, A.G. 1934. Bacteriological studies on the red discolouration of salted hides. Can. J. Res. *10*: 275–286.

Lockhart, W.R. and J. Liston. 1970. Methods for Numerical Taxonomy, American Society for Microbiology, Washington, DC.

Lodwick, D., H.N.M. Ross, J.E. Harris, J.W. Almond and W.D. Grant. 1986. *dam* Methylation in the archaebacteria. J. Gen. Microbiol. *132*: 3055–3060.

Logan, N.A. 1994. Bacterial Systematics, Blackwell Scientific Publications, Oxford.

Loginova, L.G. and L.A. Egorova. 1975. An obligately thermophilic bacterium, *Thermus ruber*, from hot springs of Kamchatka. Mikrobiologiya *44*: 661–665.

Loginova, L.G., L.A. Egorova, R.S. Golovacheva and L.M. Seregina. 1984. *Thermus ruber* sp. nov., nom. rev. Int. J. Syst. Bacteriol. *34*: 498–499.

Loginova, L.G., G.I. Khraptsova, T.I. Bogdanova, L.A. Egorova and L.M. Seregina. 1978. A thermophilic bacterium *Thermus ruber* producing a bright orange pigment. Mikrobiologiya *47*: 561–562.

Løken, O. and R. Sirevåg. 1982. Evidence for the presence of the glyoxylate cycle in *Chloroflexus aurantiacus*. Arch. Microbiol. *132*: 276–279.

Londei, P., S. Altamura, R. Huber, K.O. Stetter and P. Cammarano. 1988. Ribosomes of the extremely thermophilic eubacterium *Thermotoga maritima* are uniquely insensitive to the miscoding-inducing action of aminoglycoside antibiotics. J. Bacteriol. *170*: 4353–4360.

Lopez-Cortez, A. 1990. Microbial mats in tidal channels at San Carlos, Baja California Sur, Mexico. Geomicrobiol. J. *8*: 69–86.

López-García, P., J.P. Abad, C. Smith and R. Amils. 1992. Genomic organization of the halophilic archaeon *Haloferax mediterranei*: physical map of the chromosome. Nucleic Acids Res. *20*: 2459–2464.

López-García, P., A. St. Jean, R. Amils and R.L. Charlebois. 1995. Genomic stability in the archaea *Haloferax volcanii* and *Haloferax mediterranei*. J. Bacteriol. *177*: 1405–1408.

Lovley, D.R., R.C. Greening and J.G. Ferry. 1984. Rapidly growing rumen methanogenic organism that synthesizes coenzyme M and has a high affinity for formate. Appl. Environ. Microbiol. *48*: 81–87.

Lübben, M. and K. Morand. 1994. Novel prenylated hemes as cofactors of cytochrome oxidases. Archaea have modified hemes A and O. J. Biol. Chem. *269*: 21473–21479.

Ludwig, W. 1995. Sequence databases. *In* Akkermans, van Elsas and de Bruijn (Editors), Molecular Microbial Ecology Manual, Vol. 3.3.5, Kluwer Academic Publishers, Dordrecht. pp. 1–22.

Ludwig, W., R.I. Amann, E. Martinez-Romero, W. Schönhuber, S. Bauer, A. Neef and K.H. Schleifer. 1998a. rRNA based identification and detection systems for rhizobia and other bacteria. Plant Soil *204*: 1–19.

Ludwig, W., S.H. Bauer, M. Bauer, I. Held, G. Kirchhof, R. Schulze, I. Huber, S. Spring, A. Hartmann and K.H. Schleifer. 1997. Detection and *in situ* identification of representatives of a widely distributed new bacterial phylum. FEMS Microbiol. Lett. *153*: 181–190.

Ludwig, W., J. Neumaier, N. Klugbauer, E. Brockmann, C. Roller, S. Jilg, K. Reetz, I. Schachtner, A. Ludvigsen, M. Bachleitner, U. Fischer and K.-H. Schleifer. 1993. Phylogenetic relationships of *Bacteria* based on comparative sequence analysis of elongation factor Tu and ATP-synthase β-subunit genes. Antonie Leeuwenhoek *64*: 285–305.

Ludwig, W., R. Rossello Mora, R. Aznar, S. Klugbauer, S. Spring, K. Reetz, C. Beimfohr, E. Brockmann, G. Kirchhof, S. Dorn, M. Bachleitner, N. Klugbauer, N. Springer, D. Lane, R. Nietupsky, M. Weiznegger and K.H. Schleifer. 1995. Comparative sequence analysis of 23S rRNA from Proteobacteria. Syst. Appl. Microbiol. *18*: 164–188.

Ludwig, W. and K.H. Schleifer. 1994. Bacterial phylogeny based on 16S and 23S rRNA sequence analysis. FEMS Microbiol. Rev. *15*: 155–173.

Ludwig, W., O. Strunk, S. Klugbauer, N. Klugbauer, M. Weizenegger, J. Neumaier, M. Bachleitner and K.H. Schleifer. 1998b. Bacterial phylogeny based on comparative sequence analysis. Electrophoresis *19*: 554–568.

Luehrsen, K., G.E. Fox, M.W. Kilpatrick, R.T. Walker, H. Domdey, G. Krupp and H.J. Gross. 1981. The nucleotide sequence of the 5S rRNA from the archaebacterium *Thermoplasma acidophilum*. Nucleic Acids Res. *9*: 965–970.

Lundbäck, K.M.O., K.T. Klasson, E.C. Clausen and J.L. Gaddy. 1990. Kinetics of growth and hydrogen uptake by *Methanobacterium formicicum*. Biotechnol. Lett. *12*: 857–860.

Lupas, A., S. Muller, K. Goldie, A.M. Engel, A. Engel and W. Baumeister. 1995. Model structure of the Ompα rod, a parallel four-stranded coiled coil from the hyperthermophilic eubacterium *Thermotoga maritima*. J. Mol. Biol. *248*: 180–189.

Lyra, C., J. Hantula, E. Vainio, J. Rapala, L. Rouhiainen and K. Sivonen. 1997. Characterization of cyanobacteria by SDS-PAGE of whole-cell proteins and PCR/RFLP of the 16S rRNA gene. Arch. Microbiol. *168*: 176–184.

Lysenko, A.M. and T.N. Zhilina. 1985. Taxonomic position of *Methanosarcina vacuolata* and *Methanococcus halophilus* determined by the technique of DNA–DNA hybridization. Mikrobiologiya *54*: 501–502.

Ma, Y.-Z., R.P. Cox, T. Gillbro and M. Miller. 1996. Bacteriochlorophyll organization and energy transfer kinetics in chlorosomes from *Chloroflexus aurantiacus* depend on the light regime during growth. Photosynth. Res. *47*: 157–165.

MacAdoo, T.O. 1993. Nomenclatural literacy. *In* Goodfellow and O'Donnell (Editors), Handbook of New Bacterial Systematics, Academic Press, London. pp. 339–360.

Macario, A.J.L. and E. Conway de Macario. 1982. The immunology of methanogens: a new development in microbial biotechnology. Immunol. Today *3*: 279–284.

Macario, A.J.L. and E. Conway de Macario. 1983. Antigenic fingerprinting of methanogenic bacteria with polyclonal antibody probes. Syst. Appl. Microbiol. *4*: 451–458.

Macario, A.J.L. and E. Conway de Macario. 1985. Monoclonal antibodies of predefined molecular specificity for identification and classification of methanogens and for probing their ecological niches. *In* Macario and Conway de Macario (Editors), Monoclonal Antibodies Against Bacteria, Vol. II, Academic Press, Orlando. pp. 213–247.

Macario, A.J.L. and E. Conway de Macario. 1987. Antigenic distinctiveness, heterogeneity, and relationships of *Methanothrix* spp. J. Bacteriol. *169*: 4099–4103.

Macario, A.J.L., M.W. Peck, E. Conway de Macario and D.P. Chynoweth. 1991a. Unusual methanogenic flora of a wood-fermenting anaerobic bioreactor. J. Appl. Bacteriol. *71*: 31–37.

Macario, A.J.L., F.A. Visser, J.B. van Lier and E. Conway de Macario. 1991b. Topography of methanogenic subpopulations in a microbial consortium adapting to thermophilic conditions. J. Gen. Microbiol. *137*: 2179–2190.

MacGregor, B.J., D.P. Moser, E.W. Alm, K.H. Nealson and D.A. Stahl. 1997. Crenarchaeota in Lake Michigan sediment. Appl. Environ. Microbiol. *63*: 1178–1181.

Machida, M., H. Matsuzawa and T. Ohta. 1985. Fructose 1,6-bisphosphate-dependent L-lactate dehydrogenase from *Thermus aquaticus* YT- 1, an extreme thermophile: activation by citrate and modification reagents and comparison with *Thermus caldophilus* GK24 L-lactate dehydrogenase. J. Biochem. (Tokyo) *97*: 899–910.

Mack, E.E. and B.K. Pierson. 1988. Preliminary characterization of a temperate marine member of the Chloroflexaceae. *In* Olson, Ormerod, Amesz, Stackebrandt and Trüper (Editors), Green Photosynthetic Bacteria, Plenum Press, New York. pp. 237–241.

MacKay, M.W., G.H. al Bakri and B.E. Moseley. 1985. The plasmids of *Deinococcus* spp. and the cloning and restriction mapping of the *D. radiophilus* plasmid pUE1. Arch. Microbiol. *141*: 91–94.

Macy, J.M., K. Nunan, K.D. Hagen, D.R. Dixon, P.J. Harbour, M. Cahill and L.I. Sly. 1996. *Chrysiogenes arsenatis* gen. nov., sp. nov., a new arsenate-respiring bacterium isolated from gold mine wastewater. Int. J. Syst. Bacteriol. *46*: 1153–1157.

Madern, D., C. Pfister and G. Zaccai. 1995. Mutation at a single amino acid enhances the halophilic behaviour of malate dehydrogenase from *Haloarcula marismortui*. Eur. J. Biochem. *30*: 1088–1095.

Madigan, M.T. 1976. Studies on the physiological ecology of *Chloroflexus aurantiacus*, a filamentous photosynthetic bacterium, Doctoral thesis, University of Wisconsin, Madison.

Madigan, M.T. 1995. Nitrogen fixation by photosynthetic bacteria. *In* Blankenship, Madigan and Bauer (Editors), Anoxygenic Photosynthetic Bacteria, Kluwer Academic Publishers, Dordrecht. pp. 915–928.

Madigan, M.T. and T.D. Brock. 1975. Photosynthetic sulfide oxidation by *Chloroflexus aurantiacus*, a filamentous, photosynthetic, gliding bacterium. J. Bacteriol. *122*: 782–784.

Madigan, M.T. and T.D. Brock. 1977. CO_2 fixation in photosynthetically grown *Chloroflexus aurantiacus*. FEMS Microbiol. Lett. *1*: 301–304.

Madigan, M.T. and J.G. Ormerod. 1995. Taxonomy, physiology, and ecology of heliobacteria. *In* Blankenship, Madigan and Bauer (Editors), Anoxygenic Photosynthetic Bacteria, Kluwer Academic Publishers, Dordrecht. pp. 17–30.

Madigan, M.T., S.R. Peterson and T.D. Brock. 1974. Nutritional studies on *Chloroflexus*, a filamentous photosynthetic gliding bacterium. Arch. Microbiol. *100*: 97–103.

Maestrojuán, G.M. and D.R. Boone. 1991. Characterization of *Methanosarcina barkeri* MS^T and 227, *Methanosarcina mazei* S-6^T, and *Methanosarcina vacuolata* Z-761^T. Int. J. Syst. Bacteriol. *41*: 267–274.

Maestrojuán, G.M., D.R. Boone, L. Xun, R.A. Mah and L. Zhang. 1990. Transfer of *Methanogenium bourgense*, *Methanogenium marisnigri*, *Methanogenium olentangyi*, and *Methanogenium thermophilicum* to the genus *Methanoculleus* gen. nov., emendation of *Methanoculleus marisnigri* and *Methanogenium*, and description of new strains of *Methanoculleus bourgense* and *Methanoculleus marisnigri*. Int. J. Syst. Bacteriol. *40*: 117–122.

Maestrojuán, G.M., J.E. Boone, R.A. Mah, J.A.G.F. Menaia, M.S. Sachs and D.R. Boone. 1992. Taxonomy and halotolerance of mesophilic *Methanosarcina* strains, assignment of strains to species, and synonymy of *Methanosarcina mazei* and *Methanosarcina frisi*. Int. J. Syst. Bacteriol. *42*: 561–567.

Magingo, F.S.S. and C.K. Stumm. 1991. Nitrogen fixation by *Methanobacterium formicicum*. FEMS Microbiol. Lett. *81*: 273–278.

Magot, M., O. Possot, N. Souillard, M. Henriquet and L. Sibold. 1986. Structure and expression of *nif* (nitrogen fixation) genes in methanogens. *In* Dubourguier, Albagnac, Montreuil, Romond, Sautière and Guillaume (Editors), Biology of Anaerobic Bacteria, Elsevier Science Publishers, Amsterdam. pp. 193–199.

Mah, R.A. 1980. Isolation and characterization of *Methanococcus mazei*. Curr. Microbiol. *3*: 321–326.

Mah, R.A. and D.A. Kuhn. 1984a. Rejection of the type species *Methanosarcina methanica*, conservation of the genus *Methanosarcina* with *Methanosarcina barkeri* as the type species, and emendation of the genus *Methanosarcina*. Int. J. Syst. Bacteriol. *34*: 266–267.

Mah, R.A. and D.A. Kuhn. 1984b. Transfer of the type species of the genus *Methanococcus* to the genus *Methanosarcina*, naming it *Methanosarcina mazei* (Barker 1936) comb. nov. et emend. and conservation of the genus *Methanococcus* (Approved Lists 1980) with *Methanococcus vannielii* (Approved Lists 1980) as the type species. Request for an opinion. Int. J. Syst. Bacteriol. *34*: 263–265.

Mah, R.A., M.R. Smith and L. Baresi. 1978. Studies on an acetate-fermenting strain of *Methanosarcina*. Appl. Environ. Microbiol. *35*: 1174–1184.

Mah, R.A., M.R. Smith, T. Ferguson and S.H. Zinder. 1981. Methanogenesis from H_2–CO_2, methanol, and acetate by *Methanosarcina*. *In* Dalton (Editor), Microbial Growth on C_1 Compounds, Heyden & Sons, Philadelphia. pp. 131–142.

Mah, R.A., D.M. Ward, L. Baresi and T.L. Glass. 1977. Biogenesis of methane. Annu. Rev. Microbiol. *31*: 309–341.

Mahmood, N.A. and W.W. Carmichael. 1986. The pharmacology of anatoxin a(s), a neurotoxin produced by the freshwater cyanobacterium *Anabaena flos-aquae* NRC 525–17. Toxicon. *24*: 425–434.

Mai, B., G. Frey, R.V. Swanson, E.J. Mathur and K.O. Stetter. 1998. Molecular cloning and functional expression of protein-serine/threonine phosphatase from the hyperthermophilic archaeon *Pyrodictium abyssi* TAG11. J. Bacteriol. *180*: 4030–4035.

Maidak, B.L., J.R. Cole, C.T. Parker, Jr., G.M. Garrity, N. Larsen, B. Li, T.G. Lilburn, M.J. McCaughey, G.J. Olsen, R. Overbeek, S. Pramanik, T.M. Schmidt, J.M. Tiedje and C.R. Woese. 1999. A new version of the RDP (Ribosomal Database Project). Nucleic Acids Res. *27*: 171–173.

Maidak, B.L., N. Larsen, J. McCaughey, R. Overbeek, G.J. Olsen, K. Fogel, J. Blandy and C.R. Woese. 1994. The Ribosomal Database project. Nucleic Acids Res. *22*: 3483–3487.

Maidak, B.L., G.J. Olsen, N. Larsen, R. Overbeek, M.J. McCaughey and C.R. Woese. 1997. The RDP (Ribosomal Database Project). Nucleic Acids Res. *25*: 109–111.

Malik, K.A. 1996a. A convenient method for maintaining *Chloroflexus* for long time periods as slow growing liquid cultures. J. Microbiol. Methods *27*: 151–155.

Malik, K.A. 1996b. A modified medium and method for the cultivation of *Chloroflexus*. J. Microbiol. Methods *27*: 147–150.

Malik, K.A. and D. Claus. 1987. Bacterial culture collections: their importance to biotechnology and microbiology. *In* Russell (Editor), Biotechnology and Genetic Engineering Reviews, Vol. 5, Intercept, Ltd., Dorset. pp. 137–198.

Manaia, C.M. and M.S. Da Costa. 1991. Characterization of halotolerant *Thermus* isolates from shallow marine hot springs on S. Miguel, Azores. J. Gen. Microbiol. *137*: 2643–2648.

Manaia, C.M., B. Hoste, M.C. Gutiérrez, M. Gillis, A. Ventosa, K. Kersters and M.S. Da Costa. 1994. Halotolerant *Thermus* strains from marine and terrestrial hot springs belong to *Thermus thermophilus* (ex Oshima and Imahori, 1974) nom. rev. emend. Syst. Appl. Microbiol. *17*: 526–532.

Manaia, C.M., B. Hoste, M.C. Gutiérrez, M. Gillis, A. Ventosa, K. Kersters and M.S. Da Costa. 1995. *In* Validation of the publication of new names and new combinations previously effectively published outside the IJSB. List No. 54. Int. J. Syst. Bacteriol. *45*: 619–620.

Manca, M.C., B. Nicolaus, V. Lanzotti, A. Trincone, A. Gambacorta, J. Peter-Katalinic, H. Egge, R. Huber and K.O. Stetter. 1992. Glycolipids from *Thermotoga maritima*, a hyperthermophilic microorganism belonging to Bacteria domain. Biochim. Biophys. Acta *1124*: 249–252.

Mandel, M. 1969. New approaches to bacterial taxonomy: perspective and prospects. Annu. Rev. Microbiol. *23*: 239–274.

Mandel, M., E.R. Leadbetter, N. Pfennig and H.G. Trüper. 1971. Deoxyribonucleic acid base composition of phototrophic bacteria. Int. J. Syst. Bacteriol. *21*: 222–230.

Manz, W., R.I. Amann, W. Ludwig, M. Wagner and K.H. Schleifer. 1992. Phylogenetic oligodeoxynucleotide probes for the major subclasses of proteobacteria: problems and solutions. Syst. Appl. Microbiol. *15*: 593–600.

Manz, W., R.I. Amann, R. Szewzyk, U. Szewzyk, T.A. Stenstrom, P. Hutzler and K.H. Schleifer. 1995. *In situ* identification of *Legionellaceae* using 16S rRNA-targeted oligonucleotide probes and confocal laser scanning microscopy. Microbiology (Reading) *141*: 29–39.

Manz, W., U. Szewzyk, P. Ericsson, R.I. Amann, K.H. Schleifer and T.A. Stenström. 1993. *In situ* identification of bacteria in drinking water and adjoining biofilms by hybridization with 16S and 23S rRNA-directed fluorescent oligonucleotide probes. Appl. Environ. Microbiol. *59*: 2293–2298.

Manz, W., M. Wagner, R.I. Amann and K.H. Schleifer. 1994. *In situ* characterization of the microbial consortia active in two wastewater treatment plants. Water Res. *28*: 1715–1723.

Marchesi, J.R., T. Sato, A.J. Weightman, T.A. Martin, J.C. Fry, S.J. Hiom and W.G. Wade. 1998. Design and evaluation of useful bacterium-specific PCR primers that amplify genes coding for bacterial 16S rRNA. Appl. Environ. Microbiol. *64*: 795–799.

Mardia, K.V., J.T. Kent and J.M. Bibby. 1979. Multivariate Analysis, Academic Press, London.

Markosyan, G.E. 1972. A new iron-oxidizing bacterium, *Leptospirillum ferrooxidans* gen. et sp. nov. Biol. Zh. Arm. *25*: 26.

Marmur, J. and P. Doty. 1961. Thermal renaturation of DNA. J. Mol. Biol. *3*: 584–594.

Marmur, J. and D. Lane. 1960. Strand separation and specific recombination in deoxyribonucleic acids: biological studies. Proc. Natl. Acad. Sci. U.S.A. *46*: 453–461.

Marteinsson, V.T., J.L. Birrien, J.K. Kristjánsson and D. Prieur. 1995. First isolation of thermophilic aerobic non-sporulating heterotrophic bacteria from deep-sea hydrothermal vents. FEMS Microbiol. Ecol. *18*: 163–174.

Marteinsson, V.T., J.L. Birrien, G. Raguenes, M.S. da Costa and D. Prieur. 1999. Isolation and characterization of *Thermus thermophilus* Gy1211 from a deep-sea hydrothermal vent. Extremophiles *3*: 247–251.

Marteinsson, V.T., J.-L. Birrien, A.-L. Reysenbach, M. Vernet, D. Marie, A. Gambacorta, P. Messner, U.B. Sleytr and D. Prieur. 1999. *Thermococcus barophilus* sp. nov., a new barophilic and hyperthermophilic archaeon isolated under high hydrostatic pressure from a deep-sea hydrothermal vent. Int. J. Syst. Bacteriol. *49*: 351–359.

Martin, C., K. Sivonen, U. Matern, R. Dierstein and J. Weckesser. 1990. Rapid purification of the peptide toxins microcystin-LR and nodularin. FEMS Microbiol. Lett. *56*: 1–6.

Martin, S.M. and V.B.D. Skerman (Editors). 1972. World Directory of Collections of Cultures of Microorganisms, Wiley-Interscience, New York.

Martin, T.C. and J.T. Wyatt. 1974. Comparative physiology and morphology of six strains of stigonematacean blue-green algae. J. Phycol. *10*: 57–65.

Martin, W., H. Brinkmann, C. Savonna and R. Cerff. 1993. Evidence for a chimeric nature of nuclear genomes: eubacterial origin of eukaryotic glyceraldehyde-3-phosphate dehydrogenase genes. Proc. Natl. Acad. Sci. U.S.A. *90*: 8692–8696.

Martin, W. and R. Cerff. 1986. Prokaryotic features of a nucleus-encoded enzyme. cDNA sequences for chloroplast and cytosolic glyceraldehyde-3-phosphate dehydrogenases from mustard (*Sinapis alba*). Eur. J. Biochem. *159*: 323–331.

Martinec, T., M. Kocur and I. Habetova. 1966–1967. František Král, founder of the first collection of microorganisms. Publ. Fac. Sci. Univ. J.E. Purkyně Brno. no. 475 *K38*: 261–265.

Martins, L.O., L.S. Carreto, M.S. Da Costa and H. Santos. 1996. New compatible solutes related to Di-*myo*-inositol-phosphate in members of the order *Thermotogales*. J. Bacteriol. *178*: 5644–5651.

Masters, C.I., R.G.E. Murray, B.E.B. Moseley and K.W. Minton. 1991. DNA

polymorphisms in new isolates of '*Deinococcus radiopugnans*'. J. Gen. Microbiol. *137*: 1459–1469.

Mather, M.W. and J.A. Fee. 1990. Plasmid-associated aggregation in *Thermus thermophilus* HB8. Plasmid *24*: 45–56.

Mather, M.W. and J.A. Fee. 1992. Development of plasmid cloning vectors for *Thermus thermophilus* HB8: expression of a heterologous, plasmid-borne kanamycin nucleotidyltransferase gene. Appl. Environ. Microbiol. *58*: 421–425.

Matheron, R. and R. Baulaigue. 1972. Bactéries photosynthétiques sulfureuses marines. Arch. Mikrobiol. *86*: 291–304.

Matheson, A.T., G.D. Sprott, I.J. McDonald and H. Tessier. 1976. Some properties of an unidentified halophile: growth characteristics, internal salt concentration, and morphology. Can. J. Microbiol. *22*: 780–786.

Mathrani, I.M. and D.R. Boone. 1985. Isolation and characterization of a moderately halophilic methanogen from a solar saltern. Appl. Environ. Microbiol. *50*: 140–143.

Mathrani, I.M., D.R. Boone, R.A. Mah, G.E. Fox and P.P. Lau. 1988. *Methanohalophilus zhilinae*, sp. nov. an alkaliphilic, halophilic, methylotrophic methanogen. Int. J. Syst. Bacteriol. *38*: 139–142.

Matsubara, T., N. Iida-Tanaka, M. Kamekura, N. Moldoveanu, I. Ishizuka, H. Onishi, A. Hayashi and M. Kates. 1994. Polar lipids of a non-alkaliphilic extremely halophilic archaebacterium strain 172: a novel bis-sulfated glycolipid. Biochim. Biophys. Acta *1214*: 97–108.

Matsuzawa, H., K. Tokugawa, M. Hamaoki, M. Mizoguchi, H. Taguchi, I. Terada, S.T. Kwon and T. Ohta. 1988. Purification and characterization of Aqualysin I (a thermophilic alkaline serine protease) produced by *Thermus aquaticus* YT-1. Eur. J. Biochem. *171*: 441–448.

Mattimore, V. and J.R. Battista. 1996. Radioresistance of *Deinococcus radiodurans*: functions necessary to survive ionizing radiation are also necessary to survive prolonged desiccation. J. Bacteriol. *178*: 633–637.

Mattimore, V., K.S. Udupa, G.A. Berne and J.R. Battista. 1995. Genetic characterization of forty ionizing radiation-sensitive strains of *Deinococcus radiodurans*: linkage information from transformation. J. Bacteriol. *177*: 5232–5237.

Maxam, A.M. and W. Gilbert. 1980. Sequencing end-labeled DNA with base-specific chemical cleavages. Methods Enzymol. *65*: 499–560.

Mayberry-Carson, K.J., I.L. Roth, J.L. Harris and P.F. Smith. 1974. Scanning electron microscopy of *Thermoplasma acidophilum*. J. Bacteriol. *120*: 1472–1475.

Mayerhofer, L.E., A.J.L. Macario and E. Conway de Macario. 1992. Lamina, a novel multicellular form of *Methanosarcina mazei* S-6. J. Bacteriol. *174*: 309–314.

Maymó-Gatell, X., Y.T. Chien, J.M. Gossett and S.H. Zinder. 1997. Isolation of a bacterium that reductively dechlorinates tetrachloroethene to ethene. Science *276*: 1568–1571.

Maynard Smith, J. 1995. Do bacteria have population genetics? *In* Baumberg, Young, Wellington and Saunders (Editors), Population Genetics of Bacteria, Cambridge University Press, Cambridge. 1–12.

Mayr, A. and F. Pfeifer. 1997. The characterization of the *nv-gvp*ACNOFGH gene cluster involved in gas vesicle formation in *Natronobacterium vacuolatum*. Arch. Microbiol. *168*: 24–32.

Mayr, E. 1998. Two empires or three? Proc. Natl. Acad. Sci. U.S.A. *95*: 9720–9723.

Mazé, P. 1903. Sur la fermentation formènique et le ferment qui la produit. C.R. Hebd. Séanc. Acad. Sci. *137*: 887–889.

McGenity, T.J., R.T. Gemmell and W.D. Grant. 1998. Proposal of a new halobacterial genus *Natrinema* gen. nov., with two species *Natrinema pellirubrum* nom. nov. and *Natrinema pallidum* nom. nov. Int. J. Syst. Bacteriol. *48*: 1187–1196.

McGenity, T.J. and W.D. Grant. 1995. Transfer of *Halobacterium saccharovorum*, *Halobacterium sodomense*, *Halobacterium trapanicum* NRC 34021 and *Halobacterium lacusprofundi* to the genus *Halorubrum* gen. nov., as *Halorubrum saccharovorum* comb. nov., *Halorubrum sodomense* comb. nov., *Halorubrum trapanicum* comb. nov., and *Halorubrum lacusprofundi* comb. nov. Syst. Appl. Microbiol. *18*: 237–243.

McGenity, T.J. and W.D. Grant. 1996. *In* Validation of the publication of new names and new combinations previously effectively published outside the IJSB. List No. 56. Int. J. Syst. Bacteriol. *46*: 362–363.

McGowan, V.F. and V.B.D. Skerman (Editors). 1982. World Directory of Collections of Cultures of Microorganisms, 2nd Ed., World Data Center for Microorganisms, Brisbane.

McGuire, R.F. 1984. A numerical taxonomic study of *Nostoc* and *Anabaena*. J. Phycol. *20*: 454–460.

McInerney, M.J., M.P. Bryant, R.B. Hespell and J.W. Costerton. 1981a. *Syntrophomonas wolfei* gen. nov. sp. nov., an anaerobic syntrophic, fatty acid-oxidizing bacterium. Appl. Environ. Microbiol. *41*: 1029–1039.

McInerney, M.J., R.I. Mackie and M.P. Bryant. 1981b. Syntrophic association of a butyrate-degrading bacterium and *Methanosarcina* enriched from bovine rumen fluid. Appl. Environ. Microbiol. *41*: 826–828.

McLachlan, J. 1973. Growth media-marine. *In* Stein (Editor), Handbook of Phycological Methods, Culture Methods and Growth Measurements, Cambridge University Press, Cambridge. pp. 25–51.

McLachalan, J.L., U.T. Hammer and P.R. Gorham. 1963. Observations on the growth and colony habits of ten strains of *Aphanizomenon flos-aquae*. Phycologia *2*: 157–168.

McLean, R.J.C., M. Whiteley, D.J. Stickler and W.C. Fuqua. 1997. Evidence of autoinducer activity in naturally occurring biofilms. FEMS Microbiol. Lett. *154*: 259–263.

McSweeney, C.S., M.J. Allison and R.L. Mackie. 1993a. Amino acid utilization by the ruminal bacterium *Synergistes jonesii* strain 78–1. Arch. Microbiol. *159*: 131–135.

McSweeney, C.S., R.I. Mackie, A.A. Odenyo and D.A. Stahl. 1993b. Development of an oligonucleotide probe targeting 16S rRNA and its application for detection and quantitation of the ruminal bacterium *Synergistes jonesii* in a mixed-population chemostat. Appl. Environ. Microbiol. *59*: 1607–1612.

Mechsner, K. 1957. Physiologische und morphologische Untersuchungen an Chlorobakterien. Arch. Mikrobiol. *26*: 32–51.

Meeks, J.C. 1988. Symbiotic associations. Methods Enzymol. *167*: 113–121.

Meffert, M.E. 1987. Planktic unsheathed filaments (*Cyanophyceae*) with polar and central gas vacuoles: I. Their morphology and taxonomy. Arch. Hydrobiol. Suppl. *76*: 315–346.

Meffert, M.E. 1988. *Limnothrix* Meffert nov. gen., the unsheathed planktic cyanophycean filaments with polar and central gas vacuoles. Arch. Hydrobiol. Suppl. *80*: 269–276.

Meffert, M.E. and T.P. Chang. 1978. The isolation of planktonic blue-green algae (*Oscillatoria* species). Arch. Hydrobiol. *82*: 231–239.

Meffert, M.E. and R. Oberhäuser. 1982. Polar and central gas vacuoles in planktonic *Oscillatoria* spp. (Cyanophyta). Arch. Hydrobiol. *95*: 235–248.

Meile, L., P. Abendschein and T. Leisinger. 1990. Transduction in the archaebacterium *Methanobacterium thermoautotrophicum* Marburg. J. Bacteriol. *172*: 3507–3508.

Meile, L., K. Fischer and T. Leisinger. 1995. Characterization of the superoxide dismutase gene and its upstream region from *Methanobacterium thermoautotrophicum* Marburg. FEMS Microbiol. Lett. *128*: 247–253.

Meile, L., U. Jenal, D. Studer, M. Jordan and T. Leisinger. 1989. Characterization of ψM1, a virulent phage of *Methanobacterium thermoautotrophicum* Marburg. Arch. Microbiol. *152*: 105–110.

Meissner, J., J.H. Krauss, U.J. Jürgens and J. Weckesser. 1988a. Absence of a characteristic cell wall lipopolysaccharide in the phototrophic bacterium *Chloroflexus aurantiacus*. J. Bacteriol. *170*: 3213–3216.

Meissner, J., N. Pfennig, J.H. Krauss, H. Mayer and J. Weckesser. 1988b. Lipopolysaccharides of the *Chromatiaceae* species *Thiocystis violacea*, *Thiocapsa pfennigii*, and *Chromatium tepidum*. J. Bacteriol. *170*: 3267–3272.

Melin, A.M., M.A. Carbonneau and N. Rebeyrotte. 1986. Fatty acids and carbohydrate-containing lipids in four *Micrococcaceae* strains. Biochimie *68*: 1201–1209.

Méndez-Alvarez, S., I. Esteve, R. Guerrero and N. Gaju. 1996. Genomic analysis of different *Chlorobium* strains by pulsed-field gel electrophoresis and ribotyping. Int. J. Syst. Bacteriol. *46*: 1177–1179.

Meredith, R. 1997. Winning the race to invent. Nat. Biotechnol. *15*: 283–284.

Merkel, G.J., D.R. Durham and J.J. Perry. 1980. The atypical cell wall composition of *Thermomicrobium roseum*. Can. J. Microbiol. *26*: 556–559.

Merkel, G.J., S.S. Stapleton and J.J. Perry. 1978a. Isolation and peptidoglycan of gram-negative hydrocarbon-utilizing thermophilic bacteria. J. Gen. Microbiol. *109*: 141–148.

Merkel, G.J., W.C. Underwood and J.J. Perry. 1978b. Isolation of thermophilic bacteria capable of growth solely on long-chain hydrocarbons. FEMS Microbiol. Lett. *3*: 81–83.

Mertens, K.H. 1822. *In* Jürgens (Editor), Algae Auquatica, Decas XV, No. 4.

Mescher, M.F. and J.L. Strominger. 1976. Purification and characterization of a prokaryotic glycoprotein from the cell envelope of *Halobacterium salinarium*. J. Biol. Chem. *251*: 2005–2014.

Meseguer, I. and F. Rodriguez-Valera. 1985. Production and purification of halocin H4. FEMS Microbiol. Lett. *28*: 177–182.

Meseguer, I. and F. Rodriguez-Valera. 1986. Effect of halocin H4 on cells of *Halobacterium halobium*. J. Gen. Microbiol. *132*: 3061–3068.

Meseguer, I., F. Rodriguez-Valera and A. Ventosa. 1986. Antagonistic interactions among halobacteria due to halocin production. FEMS Microbiol. Lett. *36*: 177–182.

Meseguer, I., M. Torreblanca and T. Konishi. 1995. Specific inhibition of the halobacterial Na^+/H^+ antiporter by halocin H6. J. Biol. Chem. *270*: 6450–6455.

Meyen, F.J.F. 1839. Neues System der Pflanzenphysiologie, 3 Bände, 1827–1839, Haude und Spenersche Buchhandlung, Berlin.

Meyer, T.E., R.G. Bartsch, M.A. Cusanovich and J.H. Mathewson. 1968. The cytochromes of *Chlorobium thiosulfatophilum*. Biochim. Biophys. Acta *153*: 854–861.

Michalski, T.J., J.E. Hunt, M.K. Bowman, U. Smith, K. Bardeen, H. Gest, J.R. Norris and J.J. Katz. 1987. Bacteriopheophytin *g*: properties and some speculations on a possible primary role for bacteriochlorophylls *b* and *g* in the biosynthesis of chlorophylls. Proc. Natl. Acad. Sci. U.S.A. *84*: 2570–2574.

Migas, J., K.L. Anderson, D.L. Cruden and A.J. Markovetz. 1989. Chemotaxis in *Methanospirillum hungatei*. Appl. Environ. Microbiol. *55*: 264–265.

Migula, W. 1890. Bakterienkunde für Landwirte, Berlin.

Migula, W. 1894. Über ein neues System der Bakterien. Arb. Bakteriol. Inst. Karlsruhe *1*: 235–238.

Migula, W. 1895. Schizomycetes (Bacteria, Bakterien). *In* Engler and Prantl (Editors), Pflanzenfamilien, Tiel I, Abt. 1a, W. Englemann, Leipzig. pp. 1–44.

Migula, W. 1897. System der Bakterien, Vol. 1, Gustav Fischer, Jena.

Migula, W. 1900. System der Bakterien, Vol. 2, Gustav Fischer, Jena.

Mikesell, M.D. and S.A. Boyd. 1990. Dechlorination of chloroform by *Methanosarcina* strains. Appl. Environ. Microbiol. *56*: 1198–1201.

Miller, J.M. and C.M. O'Hara. 1995. Substrate utilization systems for the identifcation of bacteria and yeasts. *In* Murray, Baron, Pfaller, Tenover and Yolken (Editors), Manual of Clinical Microbiology, 6th Ed., American Society for Microbiology, Washington, D.C. pp. 103–109.

Miller, T.L., X. Chen, B. Yan and S. Bank. 1995. Solution ^{13}C nuclear magnetic resonance spectroscopic analysis of the amino acids of *Methanosphaera stadtmanae*: biosynthesis and origin of one-carbon units from acetate and carbon dioxide. Appl. Environ. Microbiol. *61*: 1180–1186.

Miller, T.L. and M.J. Wolin. 1974. A serum bottle modification of the Hungate technique for cultivating obligate anaerobes. Appl. Microbiol. *27*: 985–987.

Miller, T.L. and M.J. Wolin. 1982. Enumeration of *Methanobrevibacter smithii* in human feces. Arch. Microbiol. *131*: 14–18.

Miller, T.L. and M.J. Wolin. 1983. Oxidation of hydrogen and reduction of methanol to methane is the sole energy source for a methanogen isolated from human feces. J. Bacteriol. *153*: 1051–1055.

Miller, T.L. and M.J. Wolin. 1985a. *Methanosphaera stadtmaniae*, gen. nov., sp. nov.: a species that forms methane by reducing methanol with hydrogen. Arch. Microbiol. *141*: 116–122.

Miller, T.L. and M.J. Wolin. 1985b. *In* Validation of the publication of new names and new combinations previously effectively published outside the IJSB. List No. 19. Int. J. Syst. Bacteriol. *35*: 535.

Miller, T.L. and M.J. Wolin. 1986. Methanogens in human and animal intestinal tracts. Syst. Appl. Microbiol. *7*: 223–229.

Miller, T.L., M.J. Wolin, E. Conway de Macario and A.J.L. Macario. 1982. Isolation of *Methanobrevibacter smithii* from human feces. Appl. Environ. Microbiol. *43*: 227–232.

Miller, T.L., M.J. Wolin, Z. Hongue and M.P. Bryant. 1986a. Characteristics of methanogens isolated from bovine rumen. Appl. Environ. Microbiol. *51*: 201–202.

Miller, T.L., M.J. Wolin and E.A. Kusel. 1986b. Isolation and characterization of methanogens from animal feces. Syst. Appl. Microbiol. *8*: 234–238.

Miller, V. 1923. Zum Systematik der Gattung *Anabaena* Bory. Arch. Protistol. *2*: 116–126.

Min, H. and S.H. Zinder. 1989. Kinetics of acetate utilization by two thermophilic acetotrophic methanogens: *Methanosarcina* sp., strain CALS-1 and *Methanothrix* sp. strain CALS-1. Appl. Environ. Microbiol. *55*: 488–491.

Mink, R.W. and P.R. Dugan. 1977. Tentative identification of methanogenic bacteria by fluorescence microscopy. Appl. Environ. Microbiol. *33*: 713–717.

Minuth, T., M. Henn, K. Rutkat, S. Andrä, G. Frey, R. Rachel, K.O. Stetter and R. Jaenicke. 1999. The recombinant thermosome from the hyperthermophilic archaeon *Methanopyrus kandleri*: *in vitro* analysis of its chaperone activity. Biol. Chem. *380*: 55–62.

Miroshnichenko, M.L. 1990. *In* Validation of the publication of new names and new combinations previously effectively published outside the IJSB. List No. 34. Int. J. Syst. Bacteriol. *40*: 320–321.

Miroshnichenko, M.L., E.A. Bonch-Osmolovskaya, A. Neuner, N.A. Kostrikina, N.A. Chernych and V.A. Alekseev. 1989. *Thermococcus stetteri* sp. nov., a new extremely thermophilic marine sulfur-metabolizing archaebacterium. Syst. Appl. Microbiol. *12*: 257–262.

Miroshnichenko, M.L., G.A. Gongadze, F.A. Rainey, A.S. Kostyukova, A.M. Lysenko, N.A. Chernyh and E.A. Bonch-Osmolovskaya. 1998. *Thermococcus gorgonarius* sp. nov. and *Thermococcus pacificus* sp. nov.: heterotrophic extremely thermophilic archaea from New Zealand submarine hot vents. Int. J. Syst. Bacteriol. *48*: 23–29.

Mirzabekov, A.D. 1994. DNA sequencing by hybridization: a megasequencing method and a diagnostic tool? Trends Biotechnol. *12*: 27–32.

Miyashita, H., K. Adachi, N. Kurano, H. Ikemoto, M. Chihara and S. Miyachi. 1997. Pigment compostion of a novel oxygenic photosynthetic prokaryote containing chlorophyll *d* as the major chlorophyll. Plant Cell Physiol. *38*: 274–281.

Miyazaki, K. 1996. Isocitrate dehydrogenase from *Thermus aquaticus* YT1: purification of the enzyme and cloning, sequencing, and expression of the gene. Appl. Environ. Microbiol. *62*: 4627–4631.

Mobarry, B.K., M. Wagner, V. Urbain, B.E. Rittmann and D.A. Stahl. 1996. Phylogenetic probes for analyzing abundance and spatial organization of nitrifying bacteria. Appl. Environ. Microbiol. *62*: 2156– 2162.

Mohn, W.W. and J.M. Tiedje. 1992. Microbial reductive dehalogenation. Microbiol. Rev. *56*: 482–507.

Mojica, F.J., C. Ferrer, G. Juez and F. Rodríguez-Valera. 1995. Long stretches of short tandem repeats are present in the largest replicons of the archaea *Haloferax mediterranei* and *Haloferax volcanii* and could be involved in replicon partitioning. Mol. Microbiol. *17*: 85–93.

Molisch, H. 1907. Die Purpurbakterien nach neuen Untersuchungen, G. Fischer, Jena.

Molitor, M., C. Dahl, I. Molitor, U. Schäfer, N. Speich, R. Huber, R. Deutzmann and H.G. Truper. 1998. A dissimilatory sirohaem-sulfite-reductase-type protein from the hyperthermophilic archaeon *Pyrobaculum islandicum*. Microbiology (Reading) *144*: 529–541.

Mollenhauer, D. 1970. Beiträge zur Kenntnis der Gattung *Nostoc*. Abh. Senckenb. Naturforsch. Ges. *524*: 1–80.

Mollenhauer, D. 1986a. Blaualgen der Gattung Nostoc—ihre Rolle in Forschung und Wissenschaftsgeschichte, III. Nat. Mus. *116*: 43–64.

Mollenhauer, D. 1986b. Blaualgen der Gattung Nostoc—ihre Rolle in Forschung und Wissenschaftsgeschichte, IV. Nat. Mus. *116*: 104–120.

Montalvo-Rodríguez, R. 1996. Taxonomic studies on extremely halophilic archaebacteria isolated from the solar salterns of Cabo Rojo, Puerto Rico, Doctoral thesis, University of Puerto Rico.

Montalvo-Rodríguez, R., R.H. Vreeland, A. Oren, M. Kessel, C. Betancourt and J. López-Garriga. 1998. *Halogeometricum borinquense* gen. nov., sp. nov., a novel halophilic archaeon from Puerto Rico. Int. J. Syst. Bacteriol. *48* : 1305–1312.

Montero, C.G., H.P. Klenk, J.J. Nieto and A. Ventosa. 1993. DNA–rRNA hybridization studies on *Halococcus saccharolyticus* and other halobacteria. FEMS Microbiol. Lett. *111*: 68–72.

Montero, C.G., A. Ventosa, F. Rodriguez-Valera, M. Kates, N. Moldoveanu and F. Ruiz-Berraquero. 1989. *Halococcus saccharolyticus* sp. nov., a new species of extremely halophilic non-alkaliphilic cocci. Syst. Appl. Microbiol. *12*: 167–171.

Montero, C.G., A. Ventosa, F. Rodriguez-Valera, M. Kates, N. Moldoveanu and F. Ruiz-Berraquero. 1990. *In* Validation of the publication of new names and new combinations previously effectively published outside the IJSB. List No. 32. Int. J. Syst. Bacteriol. *40*: 105–106.

Montero, C.G., A. Ventosa, F. Rodriguez-Valera and F. Ruiz-Berraquero. 1988. Taxonomic study of non-alkaliphilic halococci. J. Gen. Microbiol. *134*: 725–732.

Montesinos, E., R. Guerrero, C. Abella and I. Esteve. 1983. Ecology and physiology of the competition for light between *Chlorobium limicola* and *Chlorobium phaeobacteroides* in natural habitats. Appl. Environ. Microbiol. *46*: 1007–1016.

Montgomery, L., B. Flesher and D. Stahl. 1988. Transfer of *Bacteroides succinogenes* (Hungate) to *Fibrobacter* gen. nov. as *Fibrobacter succinogenes* comb. nov. and description of *Fibrobacter intestinalis* sp. nov. Int. J. Syst. Bacteriol. *38*: 430–435.

Moore, R.L. and B.J. McCarthy. 1969. Base sequence homology and renaturation studies of the deoxyribonucleic acid of extremely halophilic bacteria. J. Bacteriol. *99*: 255–262.

Moore, W.E.C. and Moore, L.V.H. (Editors). 1989. Index of the Bacterial and Yeast Nomenclatural Changes Published in the *International Journal of Systematic Bacteriology* since the 1980 Approved Lists of Bacterial Names (January 1, 1980 to January 1, 1989), American Society for Microbiology, Washington, DC.

Moreira, L.M., M.S. da Costa and I. Sá-Correia. 1995. Plasmid RFLP profiling and DNA homology in *Thermus* isolated from hot springs of different geographical areas. Arch. Microbiol. *164*: 7–15.

Moreira, L.M., M.S. da Costa and I. Sá-Correia. 1997. Comparative genomic analysis of isolates belonging to the six species of the genus *Thermus* using pulsed-field gel electrophoresis and ribotyping. Arch. Microbiol. *168*: 92–101.

Morgan, R.M., T.D. Pihl, J. Nölling and J.N. Reeve. 1997. Hydrogen regulation of growth, growth yields, and methane gene transcription in *Methanobacterium thermoautotrophicum* ΔH. J. Bacteriol. *179*: 889– 898.

Morii, H., M. Nishihara and Y. Koga. 1983. Isolation, characterization and physiology of a new formate-assimilable methanogenic strain (A2) of *Methanobrevibacter arboriphilus*. Agric. Biol. Chem. *47*: 2781–2790.

Morii, H., M. Nishihara and Y. Koga. 1988. Composition of polar lipids of *Methanobrevibacter arboriphilicus* and structure determination of the signature phosphoglycolipid of *Methanobacteriaceae*. Agric. Biol. Chem *52*: 3149–3156.

Morii, H., H. Yagi, H. Akatsu, N. Nomura, Y. Sako and Y. Koga. 1999. A novel phosphoglycolipid archaetidyl (glucosyl) inositol with two sesterterpanyl chains from the aerobic hyperthermophilic archaeon *Aeropyrum pernix* K1. Biochim. Biophys. Acta *1436*: 426–436.

Morren, C. 1838. Mém. Acad. R. Belg. *11*: 5–20.

Morth, S. and B.J. Tindall. 1985a. Evidence that changes in the growth conditions affect the relative distribution of diether lipids in haloalkaliphilic bacteria. FEMS Microbiol. Lett. *29*: 285–288.

Morth, S. and B.J. Tindall. 1985b. Variation of polar lipid composition within haloalkaliphilic archaebacteria. Syst. Appl. Microbiol. 6: 247–250.

Moseley, B.E. and A. Mattingly. 1971. Repair of irradiation transforming deoxyribonucleic acid in wild type and a radiation-sensitive mutant of Micrococcus radiodurans. J. Bacteriol. 105: 976–983.

Moseley, S.L., P. Echeverria, J. Seriwatana, C. Tirapat, W. Chaicumpa, T. Sakuldaipeara and S. Falkow. 1982. Identification of enterotoxigenic Escherichia coli by colony hybridization using three enterotoxin gene probes. J. Infect. Dis. 145: 863–869.

Mountfort, D.O. and M.P. Bryant. 1982. Isolation and characterization of an anaerobic syntrophic benzoate-degrading bacterium from sewage sludge. Arch. Microbiol. 133: 249–256.

Mrozek, C. 1990. Physical and chemical factors influencing the phototactic steering response in Lyngbya sp. (U-Sara.-L), Master's thesis, University of Oregon, Eugene.

Mukamolova, G.V., A.S. Kaprelyants, D.I. Young, M. Young and D.B. Kell. 1998. A bacterial cytokine. Proc. Natl. Acad. Sci. U.S.A. 95: 8916–8921.

Mukhopadhyay, B., E. Purwantini and L. Daniels. 1993. Effect of methanogenic substrates on coenzyme F_{420}-dependent N^5,N^{10}-methylene-H_4MPT dehydrogenase, N^5,N^{10}-methenyl-H_4MPT cyclohydrolase and F_{420}-reducing hydrogenase activities in Methanosarcina barkeri. Arch. Microbiol. 159: 141–146.

Mukohata, Y. 1994. Comparative studies on ion pumps of the bacterial rhodopsin family. Biophys. Chem. 50: 191–201.

Mullakhanbhai, M.F. and H. Larsen. 1975. Halobacterium volcanii spec. nov., a Dead Sea halobacterium with a moderate salt requirement. Arch. Microbiol. 104: 207–214.

Müller, O.F. 1773. Vermium Terrestrium et Fluviatilium, seu Animalium Infusoriorum, Helminthicorum et Testaceorum, non Marinorum. Succincta Historia 1: 1–135.

Müller, O.F. 1786. Animalcula Infusoria Fluviatilia et Marina, quae Detexit, Systematice Descripsit et ad Vivum Delineari Curavit.

Müller, V., M. Blaut and G. Gottschalk. 1986. Utilization of methanol plus hydrogen by Methanosarcina barkeri for methanogenesis and growth. Appl. Environ. Microbiol. 52: 269–274.

Munson, M.A., D.B. Nedwell and T.M. Embley. 1997. Phylogenetic diversity of Archaea in sediment samples from a coastal salt marsh. Appl. Environ. Microbiol. 63: 4729–4733.

Munster, M.J., A.P. Munster and R.J. Sharp. 1985. Incidence of plasmids in Thermus spp. isolated in Yellowstone National Park (USA). Appl. Environ. Microbiol. 50: 1325–1327.

Munster, M.J., A.P. Munster, J.R. Woodrow and R.J. Sharp. 1986. Isolation and preliminary taxonomic studies of Thermus strains isolated from Yellowstone National Park, USA. J. Gen. Microbiol. 132: 1677–1684.

Mur, L.R. 1983. Some aspects of the ecophysiology of cyanobacteria. Ann. Microbiol. (Paris) 134b: 61–72.

Murray, P.A. and S.H. Zinder. 1984. Nitrogen fixation by a methanogenic archaebacterium. Nature 312: 284–285.

Murray, P.A. and S.H. Zinder. 1985. Nutritional requirements of Methanosarcina sp. strain TM-1. Appl. Environ. Microbiol. 50: 49–55.

Murray, R.G.E. 1992. The family Deinococcaceae. In Ballows, Trüper, Dworkin, Harder and Schleifer (Editors), The Prokaryotes: A Handbook of Bacteria: Ecophysiology, Isolation, Identification, Applications, 2nd Ed., Vol. 4, Springer-Verlag, New York. pp. 3732–3744.

Murray, R.G.E. and B.W. Brooks. 1986. Genus I. Deinococcus. In Sneath, Mair, Sharpe and Holt (Editors), Bergey's Manual of Systematic Bacteriology, 1st Ed., Vol. 2, The Williams & Wilkins Co., Baltimore. pp. 1035–1043.

Murray, R.G.E., M. Hall and B.G. Thompson. 1983. Cell division in Deinococcus radiodurans and a method for displaying septa. Can. J. Microbiol. 29: 1412–1423.

Murray, R.G.E. and K.-H. Schleifer. 1994. Taxonomic notes: a proposal for recording the properties of putative taxa of Procaryotes. Int. J. Syst. Bacteriol. 44: 174–176.

Murray, R.G.E. and E. Stackebrandt. 1995. Taxonomic note: implementation of the provisional status Candidatus for incompletely described procaryotes. Int. J. Syst. Bacteriol. 45: 186–187.

Muyzer, G. and K. Smalla. 1998. Application of denaturing gradient gel electrophoresis (DGGE) and temperature gradient gel electrophoresis (TGGE) in microbial ecology. Antonie Leeuwenhoek 73: 127–141.

Mwatha, W.E. and W.D. Grant. 1993. Natronobacterium vacuolata, sp. nov., a haloalkaliphilic archaeon isolated from Lake Magadi, Kenya. Int. J. Syst. Bacteriol. 43: 401–404.

Mylroie, R.L. and R.E. Hungate. 1954. Experiments on the methane bacteria in sludge. Can. J. Microbiol. 1: 55–64.

Mylvaganam, S. and P.P. Dennis. 1992. Sequence heterogeneity between the two genes encoding 16S rRNA from the halophilic archaebacterium Haloarcula marismortui. Genetics 130: 399–410.

Nadson, G.A. 1906. The morphology of inferior algae. III. Chlorobium limicola Nads., the green chlorophyll bearing microbe. Bull. Jard. Bot. St. Petersb. 6: 190.

Nägeli, C. 1849. Gattungen einzelliger Algen, physiologisch und systematisch bearbeitet. Neue Denkschriften der allgemeinen schweizerischen Gesellschaft für die gesamten Naturwissenschaften 8: 44–60.

Namikoshi, M. and K.L. Rihehart. 1996. Bioactive compounds produced by cyanobacteria. J. Ind. Microbiol. Biotechnol. 17: 373–384.

Nash, T.H. (Editor). 1996. Lichen Biology, Cambridge University Press, Cambridge.

Neef, A., A. Zaglauer, H. Meier, R.I. Amann, H. Lemmer and K.H. Schleifer. 1996. Population analysis in a denitrifying sand filter: conventional and in situ identification of Paracoccus spp. in methanol-fed biofilms. Appl. Environ. Microbiol. 62: 4329–4339.

Neilan, B.A., J. Jacobs, T. Del Dot, L.L. Blackall, P.R. Hawkins, P.T. Cox and A.E. Goodman. 1997a. rRNA seqeunces and evolutionary relationships among toxic and nontoxic cyanobacteria of the genus Microcystis. Int. J. Syst. Bacteriol. 47: 693–697.

Neilan, B.A., J.L. Stuart, A.E. Goodman, P.T. Cox and P.R. Hawkins. 1997b. Specific amplification and restriction polymorphisms of the cyanobacterial rRNA operon spacer region. Syst. Appl. Microbiol. 20: 612–621.

Nelissen, B., R. de Baere, A. Wilmotte and R. de Wachter. 1996. Phylogenetic relationships of nonaxenic filamentous cyanobacterial strains based on 16S rRNA sequence analysis. J. Mol. Evol. 42: 194–200.

Nelissen, B., A. Wilmotte, J.-M. Neefs and R. de Wachter. 1994. Phylogenetic relationships among filamentous helical cyanobacteria investigated on the basis of 16S ribosomal RNA gene sequence analysis. Syst. Appl. Microbiol. 17: 206–210.

Nelson, D.C., J.B. Waterbury and H.W. Jannasch. 1982. Nitrogen fixation and nitrate utilization by marine and freshwater Beggiatoa. Arch. Microbiol. 133: 172–177.

Nelson, K.E., R.A. Clayton, S.R. Gill, M.L. Gwinn, R.J. Dodson, D.H. Haft, E.K. Hickey, J.D. Peterson, W.C. Nelson, K.A. Ketchum, L. McDonald, T.R. Utterback, J.A. Malek, K.D. Linher, M.M. Garrett, A.M. Stewart, M.D. Cotton, M.S. Pratt, C.A. Phillips, D. Richardson, J. Heidelberg, G.G. Sutton, R.D. Fleischmann, J.A. Eisen, O. White, S.L. Salzberg, H.O. Smith, J.C. Venter and C.M. Fraser. 1999. Evidence for lateral gene transfer between Archaea and Bacteria from genome sequence of Thermotoga maritima. Nature 399: 323–329.

Neumaier, J. 1996. Gene der katalytischen Untereinheit der V-Typ- und der F_1F_0-ATPase bei Bakterien, Thesis, Technical University, Munich.

Neuner, A., H.W. Jannasch, S. Belkin and K.O. Stetter. 1990. Thermococcus litoralis, sp. nov.: a new species of extremely thermophilic marine archaebacteria. Arch. Microbiol. 153: 205–207.

Newman, D.K., E.K. Kennedy, J.D. Coates, D. Ahmann, D.J. Ellis, D.R. Lovley and F.M.M. Morel. 1997. Dissimilatory arsenate and sulfate reduction in Desulfotomaculum auripigmentum sp. nov. Arch. Microbiol. 168: 380–388.

Ng., W.V, S.P. Kennedy, G.G. Mahaira, B. Berquist, M. Pan, H.D. Shulka, S.R. Lasky, N.S. Baliga, V. Thorsson, J. Sbrogna, S. Swartzell, D. Weir, J. Hall, T.A. Dahl, R. Welti, Y. Ah Goo, B. Leithauser, K. Keller, R. Cruz, .M.J. Danson, D.W. Hough, D.G. Maddocks, P.E. Jablonski, M.P. Krebs, C.M. Angevine, H. Dale, T.A. Isenbarger, R.F. Peck, M. Pohlschroder, J.L. Spudich, K.-H. Jung, M. Alam, T. Freitas, S. Hou, C.J. Daniels, P.P. Dennis, A.D. Omer, H. Ebhardt, T.M. Lowe, P. Liang, M.

Riley, L. Hood and S. DasSarma. 2000. Genome sequence of *Halobacterium* species NRC-1. Proc. Natl. Acad. Sci. U.S.A. *97*: 12176–12181.

Ni, S. and D.R. Boone. 1991. Isolation and characterization of a dimethyl sulfide-degrading methanogen, *Methanolobus siciliae* HI350, from an oil well, characterization of *M. siciliae* T4/MT, and emendation of *M. siciliae*. Int. J. Syst. Bacteriol. *41*: 410–416.

Ni, S., J.E. Boone and D.R. Boone. 1994a. Potassium extrusion by the moderately halophilic and alkaliphilic methanogen *Methanolobus taylorii* GS-16 and homeostasis of cytosolic pH. J. Bacteriol. *176*: 7274–7279.

Ni, S., C.R. Woese, H.C. Aldrich and D.R. Boone. 1994b. Transfer of *Methanolobus siciliae* to the genus *Methanosarcina*, naming it *Methanosarcina siciliae*, and emendation of the genus *Methanosarcina*. Int. J. Syst. Bacteriol. *44*: 357–359.

Nichols, J.M. and N.G. Carr. 1978. Akinetes of cyanobacteria. *In* Chambliss and Vary (Editors), Spores VII, American Society for Microbiology, Washington, D.C. pp. 335–343.

Nicholson, J.A.M., J.F. Stolz and B.K. Pierson. 1987. Structure of a microbial mat at Great Sippewissett Marsh, Cape Cod, Massachusetts. FEMS Microbiol. Ecol. *45*: 343–364.

Nielsen, P.H., K. Andreasen, M. Wagner, L.L. Blackall, H. Lemmer and R. Seviour. 1997. Variability of type 021N in activated sludge as determined by *in situ* substrate uptake pattern and *in situ* hybridization with fluorescent rRNA targeted probes. 2nd International Conference on Microorganisms in Activated Sludge and Biofilm Processes, Berkeley, California. IAWQ. pp. 255–262.

Niemetz, R., U. Kärcher, O. Kandler, Tindall, B.J. and H. König. 1997. The cell wall polymer of the extremely halophilic archaeon *Natronococcus occultus*. Eur. J. Biochem. *249*: 905–911.

Nierzwicki, S.A., D. Maratea, D.L. Balkwill, L.P. Hardie, V.B. Mehta and S.E. Stevens, Jr. 1982. Ultrastructure of the cyanobacterium, *Mastigocladus laminosus*. Arch. Microbiol. *133*: 11–19.

Nierzwicki-Bauer, S.A., D.L. Balkwill and S.E. Stevens, Jr. 1984. Morphology and ultrastructure of the cyanobacterium *Mastigocladus laminosus* growing under nitrogen-fixing conditions. Arch. Microbiol. *137*: 97–103.

Nieto, J.J., A. Ventosa and F. Ruiz-Berraquero. 1987. Susceptibility of halobacteria to heavy metals. Appl. Environ. Microbiol. *53*: 1199–1202.

Nilsen, R.K. and T. Torsvik. 1996. *Methanococcus thermolithotrophicus* isolated from North Sea oil field reservoir water. Appl. Environ. Microbiol. *62*: 728–731.

Nishihara, H., Y. Igarashi and T. Kodama. 1990. A new isolate of *Hydrogenobacter*, an obligately chemolithoautotrophic, thermophilic, halophilic and aerobic hydrogen-oxidizing bacterium from seaside saline hot spring. Arch. Microbiol. *153*: 294–298.

Nishihara, M., H. Morii and Y. Koga. 1987. Structure determination of a quartet of novel tetraether lipids from *Methanobacterium thermoautotrophicum*. J. Biochem. (Tokyo) *101*: 1007–1016 .

Nishihara, M., H. Morii and Y. Koga. 1989. Heptads of polar ether lipids of an archaebacterium, *Methanobacterium thermoautotrophicum*: structure and biosynthetic relationship. Biochemistry *28*: 95–102.

Nishimura, Y., M. Kano, T. Ino, H. Iizuka, Y. Kosako and T. Kaneko. 1987. Deoxyribonucleic acid relationship among the radiation-resistant *Acinetobacter* and other *Acinetobacter*. J. Gen. Appl. Microbiol. *33*: 371–376.

Nishimura, Y., M. Shimizu and H. Iizuka. 1981. Bacteriochlorophyll formation in radiation-resistant *Pseudomonas radiora*. J. Gen. Appl. Microbiol. *27*: 427–430.

Nishiyama, M., N. Matsubara, K. Yamamoto, S. Iijima, T. Uozumi and T. Beppu. 1986. Nucleotide sequence of the malate dehydrogenase gene of *Thermus flavus* and its mutation directing an increase in enzyme activity. J. Biol. Chem. *261*: 14178–14183.

Nishiyama, Y., T. Takashina, W.D. Grant and K. Horikoshi. 1992. Ultrastructure of the cell wall of the triangular halophilic archaebacterium *Haloarcula japonica* strain TR-1. FEMS Microbiol. Lett. *99*: 43–47.

Nitsch, M., M. Klumpp, A. Lupas and W. Baumeister. 1997. The thermosome: alternating alpha and beta-subunits within the chaperonin of the archaeon *Thermoplasma acidophilum*. J. Mol. Biol. *267*: 142–149.

Nobre, M.F., L. Carreto, R. Wait, S. Tenreiro, O. Fernandes, R.J. Sharp and M.S. da Costa. 1996a. Fatty acid composition of the species of the genera *Thermus* and *Meiothermus*. Syst. Appl. Microbiol. *19*: 303–311.

Nobre, M.F., H.G. Trüper and M.S. da Costa. 1996b. Transfer of *Thermus ruber* (Loginova et al. 1984), *Thermus silvanus* (Tenreiro et al. 1995), and *Thermus chliarophilus* (Tenreiro et al. 1995) to *Meiothermus* gen. nov. as *Meiothermus ruber* comb. nov., *Meiothermus silvanus* comb. nov., and *Meiothermus chliarophilus* comb. nov., respectively, and emendation of the genus *Thermus*. Int. J. Syst. Bacteriol. *46*: 604–606.

Nogales, B., R. Guerrero and I. Esteve. 1997. A heterotrophic bacterium inhibits growth of several species of the genus *Chlorobium*. Arch. Microbiol. *167*: 396–399.

Nold, S.C. and D.M. Ward. 1995. Diverse *Thermus* species inhabit a single hot spring microbial mat. Syst. Appl. Microbiol. *18*: 274–278.

Noll, K.M. 1989. Chromosome map of the thermophilic archaebacterium *Thermococcus celer*. J. Bacteriol. *171*: 6720–6725.

Nölling, J. and W.M. de Vos. 1992. Identification of the CTAG-recognizing restriction-modification systems *Mth*ZI and *Mth*FI from *Methanobacterium thermoformicicum* and characterization of the plasmid-encoded *mth*ZIM gene. Nucleic Acids Res. *20*: 5047–5052.

Nölling, J., A. Elfner, J.R. Palmer, V.J. Steigerwald, T.D. Pihl, J.A. Lake and J.N. Reeve. 1996. The phylogeny of *Methanopyrus kandleri* based on methyl coenzyme M reductase operons. Int. J. Syst. Bacteriol. *46*: 1170–1173.

Nölling, J., M. Frijlink and W.M. de Vos. 1991. Isolation and characterization of plasmids from different strains of *Methanothermobacterium thermoformicicum*. J. Gen. Microbiol. *137*: 1981–1986.

Nölling, J., A. Groffen and W.M. de Vos. 1993a. φF1 and φF3, two novel virulent, archaeal phages infecting different thermophilic strains of the genus *Methanobacterium*. J. Gen. Microbiol. *139*: 2511–2516.

Nölling, J., D. Hahn, W. Ludwig and W.M. de Vos. 1993b. Phylogenetic analysis of thermophilic *Methanobacterium* sp.: evidence for a formate-utilizing ancestor. Syst. Appl. Microbiol. *16*: 208–215.

Nölling, J., F.J.M. van Eeden, R.I.L. Eggen and W.M. de Vos. 1992. Modular organization of related archaeal plasmids encoding different restriction–modification systems in *Methanobacterium thermoformicicum*. Nucleic Acids Res. *20*: 6501–6507.

Nomura, N., Y. Sako and A. Uchida. 1998. Molecular characterization and postselecting fate of three introns within the single rRNA operon of the hyperthermophilic archaeon *Aeropyrum pernix* K1. J. Bacteriol. *180*: 3635–3643.

Nordin, R.N. and J.R. Stein. 1980. Taxonomic revision of *Nodularia* (Cyanophyceae/Cyanobacteria). Can. J. Bot. *58*: 1211–1224.

Norris, P.R. 1989. Mineral-oxidizing bacteria: metal-organism interactions. *In* Poole and Gadd (Editors), Metal-Microbe Interactions, IRL Press, Oxford. pp. 99–107.

Norris, P.R., D.W. Barr and D. Hinson. 1988. Iron and mineral oxidation by acidophilic bacteria: affinities for iron and attachment to pyrite. Biohydrometallurgy: Proceedings of the International Symposium, Warwick. Science and Technology Letters, Kew. pp. 43–59.

Norris, P.R., J.C. Murrell and D. Hinson. 1995. The potential for diazotrophy in iron- and sulfur-oxidizing acidophilic bacteria. Arch. Microbiol. *164*: 294–300.

Norton, C.F., T.J. McGenity and W.D. Grant. 1993. Archaeal halophiles (halobacteria) from two British salt mines. J. Gen. Microbiol. *139*: 1077–1081.

Nottingham, P.M. and R.E. Hungate. 1968. Isolation of methanogenic bacteria from feces of man. J. Bacteriol. *96*: 2178–2179.

Nozhevnikova, A.N. and V.I. Chudina. 1985. Morphology of the thermophilic acetate methane bacterium *Methanothrix thermoacetophila* sp. nov. Microbiology *53*: 618–624.

Nübel, U., B. Engelen, A. Felske, J. Snaidr, A. Weishuber, R.I. Amann, W. Ludwig and H. Backhaus. 1996. Sequence heterogeneities of genes encoding 16S rRNAs in *Paenibacillus polymyxa* detected by temperature gradient gel electrophoresis. J. Bacteriol. *178*: 5636–5643.

Nultsch, W. and K. Wenderoth. 1983. Partial irradiation experiments with *Anabaena variabilis* (Kütz). Z. Pflanzenphysiol. *111*: 1–7.

Nunes, O.C., M.M. Donato and M.S. da Costa. 1992. Isolation and characterization of *Rhodothermus* strains from Sao Miguel, Azores. Syst. Appl. Microbiol. *15*: 92–97.

Nüsslein, K. and J.M. Tiedje. 1998. Characterization of the dominant and rare members of a young Hawaiian soil bacterial community with small-subunit ribosomal DNA amplified from DNA fractionated on the basis of its guanine and cytosine composition. Appl. Environ. Microbiol. *64*: 1283–1289.

Nuttall, S.D. and M.L. Dyall-Smith. 1993a. Ch2, a novel halophilic archaeon from an Australian solar saltern. Int. J. Syst. Bacteriol. *43*: 729– 734.

Nuttall, S.D. and M.L. Dyall-Smith. 1993b. HF1 and HF2: novel bacteriophages of halophilic archaea. Virology *197*: 678–684.

Nuttall, S.D. and M.L. Dyall-Smith. 1995. Halophage HF2: genome organization and replication strategy. J. Virol. *69*: 2322–2327.

Ochi, K. 1995. Comparative ribosomal protein sequence analyses of a phylogenetically defined genus, *Pseudomonas*, and its relatives. Int. J. Syst. Bacteriol. *45*: 268–273.

Oelze, J. 1992. Light and oxygen regulation of the synthesis of bacteriochlorophylls *a* and *c* in *Chloroflexus aurantiacus*. J. Bacteriol. *174*: 5021–5026.

Oelze, J. and R.C. Fuller. 1983. Temperature dependence of growth and membrane-bound activities of *Chloroflexus aurantiacus* energy metabolism. J. Bacteriol. *155*: 90–96.

Oertling, W.A., K.K. Surerus, Ó. Einarsdóttir, J.A. Fee, R.B. Dyer and W.H. Woodruff. 1994. Spectroscopic characterization of cytochrome ba_3, a terminal oxidase from *Thermus thermophilus*: comparison of the a_3/CuB site to that of bovine cytochrome aa_3. Biochemistry *33*: 3128–3141.

O'Flaherty, L.M. and H.K. Phinney. 1970. Requirements for the maintenance and growth of *Aphanizomenon flos-aquae* in culture. J. Phycol. *6*: 95–97.

Ohki, K., J.G. Rueter and Y. Fujita. 1986. Cultures of the pelagic cyanophytes *Trichodesmium erythraeum* and *Trichodesmium thiebautii* in synthetic medium. Mar. Biol. *91*: 9–14.

Ohki, K., J.P. Zehr, P.G. Falkowski and Y. Fujita. 1991. Regulation of nitrogen-fixation by different nitrogen sources in the marine non-heterocystous cyanobacterium *Trichodesmium* sp. NIBB1067. Arch. Microbiol. *156*: 335–337.

Ohki, K., J.P. Zehr and Y. Fijita. 1992. *Trichodesmium*: establishment of culture and characteristics of N_2-fixation. *In* Carpenter, Capone and Rueter (Editors), Marine Pelagic Cyanobacteria: *Trichodesmium* and Other Diazotrophs, Kluwer, Dordrecht. pp. 307–318.

Ohtani, I., R.E. Moore and M.T.C. Runnegar. 1992. Cylindrospermopsin: a potent hepatotoxin from the blue-green alga *Cylindrospermopsis raciborskii*. J. Am. Chem. Soc. *114*: 7941–7942.

Olabarría, G., J.L. Carrascosa, M.A. de Pedro and J. Berenguer. 1996a. A conserved motif in S-layer proteins is involved in peptidoglycan binding in *Thermus thermophilus*. J. Bacteriol. *178*: 4765–4772.

Olabarría, G., L.A. Fernández-Herrero, J.L. Carrascosa and J. Berenguer. 1996b. *slpM*, a gene coding for an "S-layer-like array" overexpressed in S-layer mutants of *Thermus thermophilus* HB8. J. Bacteriol. *178*: 357–365.

Ollivier, B.M., J.L. Cayol, B.K.C. Patel, M. Magot, M.L. Fardeau and J.L. Garcia. 1997. *Methanoplanus petrolearius* sp. nov., a novel methanogenic bacterium from an oil-producing well. FEMS Microbiol. Lett. *147*: 51–56.

Ollivier, B.M., J.L. Cayol, B.K.C. Patel, M. Magot, M.L. Fardeau and J.L. Garcia. 1998a. *In* Validation of the publication of new names and new combinations previously effectively published outside the IJSB. List No. 67. Int. J. Syst. Bacteriol. *48*: 1083–1084.

Ollivier, B.M., M.-L. Fardeau, J.-L. Cayol, M. Magot, B.K.C. Patel, G. Prensier and J.-L. Garcia. 1998b. *Methanocalculus halotolerans* gen. nov., sp. nov., isolated from an oil-producing well. Int. J. Syst. Bacteriol. *48*: 821– 828.

Ollivier, B.M., A. Lombardo and J.L. Garcia. 1984. Isolation and char-

acterization of a new thermophilic *Methanosarcina* strain (strain MP). Ann. Microbiol. *2*: 187–198.

Ollivier, B.M., R.A. Mah, J.L. Garcia. and D.R. Boone. 1986. Isolation and characterization of *Methanogenium bourgense*, sp. nov. Int. J. Syst. Bacteriol. *36*: 297–301.

Ollivier, B.M., R.A. Mah, J.L. Garcia and R. Robinson. 1985. Isolation and characterization of *Methanogenium aggregans* sp. nov. Int. J. Syst. Bacteriol. *35*: 127–130.

Olsen, G.J., H. Matsuda, R. Hagstrom and R. Overbeek. 1994a. FastDNAml: a tool for construction of phylogenetic trees of DNA sequences using maximum likelihood. Comput. Appl. Biosci. *10*: 41–48.

Olsen, G.J. and C.R. Woese. 1993. Ribosomal RNA: a key to phylogeny. FASEB J. *7*: 113–123.

Olsen, G.J. and C.R. Woese. 1997. Archaeal genomics: an overview. Cell *89*: 991–994.

Olsen, G.J., C.R. Woese and R. Overbeek. 1994b. The winds of (evolutionary) change: breathing new life into microbiology. J. Bacteriol. *176*: 1–6.

Olson, J.M. 1998. Chlorophyll organization and function in green photosynthetic bacteria. Photochem. Photobiol. *67*: 61–75.

Olson, K.D., C.W. McMahon and R.S. Wolfe. 1991. Light sensitivity of methanogenic archaebacteria . Appl. Environ. Microbiol. *57*: 2683–2686.

Onderdonk, A.B. and M. Sasser. 1995. Gas-liquid and high-performance chromatographic methods for the identification of micoorganisms. *In* Murray, Baron, Pfaller, Tenover and Yolken (Editors), Manual of Clinical Microbiology, 6th Ed., American Society for Microbiology, Washington, D.C.. pp. 123–129.

Ong, L. and A. Glazer. 1984. An unusual phycoerythrin from a marine cyanobacterium. Science *224*: 80–83.

Onishi, H., T. Kobayashi, S. Awao and M. Kamekura. 1985. Archaebacterial diether lipids in a non-pigmented extremely halophilic bacterium. Agric. Biol. Chem. *49*: 3053–3056.

Onishi, H., M.E. McCance and N.E. Gibbons. 1965. A synthetic medium for extremely halophilic bacteria. Can. J. Microbiol. *11*: 365–373.

Oremland, R.S., J.S. Blum, C.W. Culbertson, P.T. Visscher, L.G. Miller, P. Dowdle and F.E. Strohmaier. 1994. Isolation, growth, and metabolism of an obligately anaerobic, selenate-respiring bacterium, strain SES-3. Appl. Environ. Microbiol. *60*: 3011–3019.

Oremland, R.S. and D.R. Boone. 1994. *Methanolobus taylorii* sp. nov., a new methylotrophic, estuarine methanogen. Int. J. Syst. Bacteriol. *44*: 573–575.

Oremland, R.S., R.P. Kiene, I. Mathrani, M.J. Whiticar and D.R. Boone. 1989. Description of an estuarine methylotrophic methanogen which grows on dimethyl sulfide. Appl. Environ. Microbiol. *55*: 994–1002.

Oren, A. 1983. *Halobacterium sodomense*, sp. nov., a Dead Sea halobacterium with an extremely high magnesium requirement. Int. J. Syst. Bacteriol. *33*: 381–386.

Oren, A. 1990. Starch counteracts the inhibitory action of Bacto-peptone and bile salts in media for the growth of halobacteria. Can. J. Microbiol. *36*: 299–301.

Oren, A. 1991. Anaerobic growth of halophilic archaeobacteria by reduction of fumarate. J. Gen. Microbiol. *137*: 1387–1390.

Oren, A. 1993. Ecology of extremely halophilic microorganisms. *In* Vreeland and Hochstein (Editors), The Biology of Halophilic Bacteria, CRC Press, Boca Raton. pp. 25–53.

Oren, A. 1994. The ecology of the extremely halophilic archaea. FEMS Microbiol. Rev. *13*: 415–440.

Oren, A. 1999. The enigma of square and triangular halophilic bacteria. *In* Seckbach (Editor), Enigmatic Microorganisms and Life in Extreme Environments, Kluwer Academic Publishers, Dordrecht. pp. 339–355.

Oren, A., S. Duker and S. Ritter. 1996. The polar lipid composition of Walsby's square bacterium. FEMS Microbiol. Lett. *138*: 135–140.

Oren, A., M. Ginzburg, B.-Z. Ginzburg, L.I. Hochstein and B.E. Volcani. 1990. *Haloarcula marismortui* (Volcani) sp. nov., nom. rev., an extremely halophilic bacterium from the Dead Sea. Int. J. Syst. Bacteriol. *40*: 209–210.

Oren, A. and P. Gurevich. 1993. Characterization of the dominant halophilic archaea in a bacterial bloom in the Dead Sea. FEMS Microbiol. Ecol. *12*: 249–256.

Oren, A., P. Gurevich, R.T. Gemmell and A. Teske. 1995. *Halobaculum gomorrense* gen. nov., sp. nov., a novel extremely halophilic archaeon from the Dead Sea. Int. J. Syst. Bacteriol. *45*: 747–754.

Oren, A., M. Kamekura and A. Ventosa. 1997a. Confirmation of strain VKM B-1733 as the type strain of *Halorubrum distributum*. Int. J. Syst. Bacteriol. *47*: 231–232.

Oren, A., M. Kessel and E. Stackebrandt. 1989. *Ectothiorhodospira marismortui*, sp. nov., an obligately anaerobic, moderately halophilic purple sulfur bacterium from a hypersaline sulfur spring on the shore of the Dead Sea. Arch. Microbiol. *151*: 524–529.

Oren, A. and C.D. Litchfield. 1999. A procedure for the enrichment and isolation of *Halobacterium*. FEMS Microbiol. Lett. *73*: 353–358.

Oren, A. and H.G. Trüper. 1990. Anaerobic growth of halophilic archaeobacteria by reduction of dimethylsulfoxide and trimethylamine *N*-oxide. FEMS Microbiol. Lett. *7*: 33–36.

Oren, A. and A. Ventosa. 1996. A proposal for the transfer of *Halorubrobacterium distributum* and *Halorbrobacterium coriense* to the genus *Halorubrum* as *Halorubrum distributum* comb. nov. and *Halorubrum coriense* comb. nov., respectively. Int. J. Syst. Bacteriol. *46*: 1180.

Oren, A., A. Ventosa and W.D. Grant. 1997b. Proposed minimal standards for description of new taxa in the order *Halobacteriales*. Int. J. Syst. Bacteriol. *47*: 233–238.

Oren, A., A. Ventosa, M.C. Gutiérrez and M. Kamekura. 1999. *Haloarcula quadrata* sp. nov., a square, motile archaeon isolated from a brine pool in Sinai (Egypt). Int. J. Syst. Bacteriol. *49*: 1149–1155.

Orla-Jensen, S. 1909. Die Hauptlinien des natürlichen Bakterien-systems. Zentbl. Bakteriol. Parasitenkd. Infektkrankh. Hyg. Abt. II *22*: 97–98 and 305–346.

Orla-Jensen, S. 1919. The Lactic Acid Bacteria, Høst, Copenhagen.

Ormerod, J.G. 1988. Natural genetic transformation in *Chlorobium*. *In* Olson, Stackebrandt and Trüper (Editors), Green Photosynthetic Bacteria, Plenum Publishing Corporation, New York. pp. 315–319.

Ormerod, J.G., L.K. Kimble, T. Nesbakken, Y.A. Torgersen, C.R. Woese and M.T. Madigan. 1996a. *Heliophilum fasciatum* gen. nov. sp. nov. and *Heliobacterium gestii* sp. nov.: endospore-forming heliobacteria from rice field soils. Arch. Microbiol. *165*: 226–234.

Ormerod, J.G., L.K. Kimble, T. Nesbakken, Y.A. Torgersen, C.R. Woese and M.T. Madigan. 1996b. *In* Validation of the publication of new names and new combinations previously effectively published outside the IJSB. List No. 59. Int. J. Syst. Bacteriol. *46*: 1189–1190.

Oshima, M. and T. Yamakawa. 1974. Chemical structure of a novel glycolipid from an extreme thermophile, *Flavobacterium thermophilum*. Biochemistry *13*: 1140–1146.

Oshima, T. and K. Imahori. 1974. Description of *Thermus thermophilus* (Yoshida and Oshima) comb. nov., a nonsporulating thermophilic bacterium from a Japanese thermal spa. Int. J. Syst. Bacteriol. *24*: 102– 112.

Osipov, G.A., E.A. Shabanova, O.V. Morozov, G.I. El'Registan, A.N. Kozlova and T.N. Zhilina. 1984. Lipids of *Methanosarcina vacuolata* and *Methanococcus halophilus*. Mikrobiologiya *53*: 633–638.

Osipov, G.A., T.N. Zhilina, A.N. Koslova and G.I. El'Registan. 1988. Pattern recognition of archaebacteria by lipid profile. *In* Zavarzin (Editor), Archaebacteria, The Scientific Center for Biological Research, Pustchino. pp. 126–134.

Öström, B. 1976. Fertilization of the Baltic by nitrogen fixation in the blue-green alga *Nodularia spumigena*. Remote Sens. Environ *4*: 305–310.

Otsuka, S., S. Suda, R. Li, M. Watanabe, H. Oyaizu, S. Matsumoto and M.M. Watanabe. 1999. Phylogenetic relationships between toxic and nontoxic strains of the genus *Microcystis* based on 16S to 23S internal transcribed spacer sequence. FEMS Microbiol. Lett. *172*: 15–21.

Otte, S.C.M., E.J. van de Meent, P.A. van Veelen, A.S. Pundsnes and J. Amesz. 1993. Identification of the major chlorosomal bacteriochlorophylls of the green sulfur bacteria *Chlorobium vibrioforme* and *Chlorobium phaeovibrioides*: their function in lateral energy transfer. Photosynth. Res. *35*: 159–169.

Ouverney, C.C. and J.A. Fuhrman. 1997. Increase in fluorescence intensity of 16S rRNA *in situ* hybridization in natural samples treated with chloramphenicol. Appl. Environ. Microbiol. *63*: 2735–2740.

Overmann, J. 1997. Mahoney Lake: a case study of the ecological significance of phototrophic sulfur bacteria. Adv. Microb. Ecol. *15*: 251–288.

Overmann, J., J.T. Beatty, K.J. Hall, N. Pfennig and T.G. Northcote. 1991a. Characterization of a dense, purple sulfur bacterial layer in a meromictic salt lake. Limnol. Oceanogr. *36*: 846–859.

Overmann, J., H. Cypionka and N. Pfennig. 1992a. An extremely low-light-adapted phototrophic sulfur bacterium from the Black Sea. Limnol. Oceanogr. *37*: 150–155.

Overmann, J., U. Fischer and N. Pfennig. 1992b. A new purple sulfur bacterium from saline littoral sediments, *Thiorhodovibrio winogradskyi* gen. nov. and sp. nov. Arch. Microbiol. *157*: 329–335.

Overmann, J., S. Lehmann and N. Pfennig. 1991b. Gas vesicle formation and buoyancy regulation in *Pelodictyon phaeoclathratiforme* (green sulfur bacteria). Arch. Microbiol. *157*: 29–37.

Overmann, J. and N. Pfennig. 1989. *Pelodictyon phaeoclathratiforme* sp. nov., a new brown-colored member of the *Chlorobiaceae* forming net-like colonies. Arch. Microbiol. *152*: 401–406.

Overmann, J. and N. Pfennig. 1990. *In* Validation of the publication of new names and new combinations previously effectively published outside the IJSB. List No. 33. Int. J. Syst. Bacteriol. *40*: 212.

Overmann, J. and C. Tuschak. 1997. Phylogeny and molecular fingerprinting of green sulfur bacteria. Arch. Microbiol. *167*: 302–309.

Overmann, J., C. Tuschak, J. Fröstl and H. Sass. 1998. The ecological niche of the consortium *"Pelochromatium roseum"*. Arch. Microbiol. *169*: 120–128.

Oyaizu, H., E. Stackebrandt, K.H. Schleifer, W. Ludwig, H. Pohla, H. Ito, A. Hirata, Y. Oyaizu and K. Komagata. 1987. A radiation-resistant rod-shaped bacterium, *Deinobacter grandis* gen. nov., sp. nov., with peptidoglycan containing ornithine. Int. J. Syst. Bacteriol. *37*: 62–67.

Pace, N.R. 1997. A molecular view of microbial diversity and the biosphere. Science *276*: 734–740.

Pace, N.R., D.A. Stahl, D.L. Lane and G.J. Olsen. 1985. Analyzing natural microbial populations by rRNA sequences. ASM News *51*: 4–12.

Packer, B. and A.N. Glazer (Editors). 1988. Cyanobacteria, Vol. 167: Methods in Enzymology, Academic Press, San Diego.

Padisák, J. 1996. Occurrence of *Anabaenopsis raciborskii* Wolosz. in the pond of Tômalom near Sopron, Hungary. Act. Bot. Acad. Sci. Hung. *36*: 163–165.

Padmaja, T.D. and T.V. Desikachary. 1968. Studies on coccoid blue-green algae—I. *Synechococcus elongatus* and *Anacystis nidulans*. Phykos 7: 62–89.

Palenik, B. and H. Swift. 1996. Cyanobacterial evolution and prochlorophyte diversity as seen in DNA-dependent RNA polymerase gene sequences. J. Phycol. *32*: 638–646.

Palinska, K.A., W.E. Krumbein and U. Schlemminger. 1998. Ultramorphological studies on *Spirulina* sp. Bot. Mar. *41*: 349–355.

Palinska, K.A., W. Liesack, E. Rhiel and W.E. Krumbein. 1996. Phenotype variability of identical genotypes: the need for a combined approach in cyanobacterial taxonomy demonstrated on *Merismopedia*-like isolates. Arch. Microbiol. *166*: 224–233.

Palleroni, N.J., R. Kunisawa, R. Contopoulou and M. Doudoroff. 1973. Nucleic acid homologies in the genus *Pseudomonas*. Int. J. Syst. Bacteriol. *23*: 333–339.

Palm, P., C. Schleper, I. Arnold-Ammer, I. Holz, T. Meier, F. Lottspeich and W. Zillig. 1993. The DNA-dependent RNA-polymerase of *Thermotoga maritima*; characterisation of the enzyme and the DNA-sequence of the genes for the large subunits. Nucleic Acids Res. *21*: 4904–4908.

Palmer, J.R., T. Baltrus, J.N. Reeve and C.J. Daniels. 1992. Transfer RNA genes from the hyperthermophilic archaeon, *Methanopyrus kandleri*. Biochim. Biophys. Acta *1132*: 315–318.

Palmisano, A.C., S.E. Cronin and D.J. Des Marais. 1988. Analysis of lipophilic pigments from a phototrophic microbial mat community by high performance liquid chromatography. J. Microbiol. Methods 8: 209–217.

Pantazidou, A.I. 1991. Studies on marine euendolithic cyanophytes from marine coastal carbonate sediments of Greece, University of Athens, Athens. pp. 298.

Papavassiliou, P. and E.C. Hatchikian. 1985. Isolation and characterization of a rubredoxin and a two-(4Fe-4S) ferredoxin from *Thermodesulfobacterium commune*. Biochim. Biophys. Acta *810*: 1–11.

Parkin, T.B. and T.D. Brock. 1981. Photosynthetic bacterial production and carbon mineralization in a meromictic lake. Arch. Hydrobiol. *91*: 366–382.

Partensky, F., W.R. Hess and D. Vaulot. 1999. *Prochlorococcus*, a marine photosynthetic prokaryote of global significance. Microbiol. Mol. Biol. Rev. *63*: 106–127.

Partensky, F., N. Hoepffner, W.K.W. Li, O. Ulloa and D. Vaulot. 1993. Photoacclimation of *Prochlorococcus* sp. (Prochlorophyta) strains isolated from the north Atlantic and the Mediterranean sea. Plant Physiol. *101*: 285–296.

Pask-Hughes, R.A. and N. Shaw. 1982. Glycolipids from some extreme thermophilic bacteria belonging to the genus *Thermus*. J. Bacteriol. *149*: 54–58.

Pask-Hughes, R.A. and R.A.D. Williams. 1975. Extremely thermophilic gram-negative bacteria from hot tap water. J. Gen. Microbiol. *88*: 321–328.

Pask-Hughes, R.A. and R.A.D. Williams. 1977. Yellow-pigmented strains of *Thermus* spp. from Icelandic hot springs. J. Gen. Microbiol. *107*: 375–383.

Pask-Hughes, R.A. and R.A.D. Williams. 1978. Cell envelope components of strains belonging to the genus *Thermus*. J. Gen. Microbiol. *107*: 65–72.

Patel, B.K.C., H.W. Morgan and R.M. Daniel. 1985a. *Fervidobacterium nodosum* gen. nov. and spec. nov., a new chemoorganotrophic, caldoactive, anaerobic bacterium. Arch. Microbiol. *141*: 63–69.

Patel, B.K.C., H.W. Morgan and R.M. Daniel. 1985b. *In* Validation of the publication of new names and new combinations previously effectively published outside the IJSB. List No. 19. Int. J. Syst. Bacteriol. *35*: 535.

Patel, B.K.C., J.H. Skerratt and P.D. Nichols. 1991a. The phospholipid ester-linked fatty acid composition of thermophilic bacteria. Syst. Appl. Microbiol. *14*: 311–316.

Patel, G.B. 1984. Characterization and nutritional properties of *Methanothrix concilii*, sp. nov., a mesophilic, aceticlastic methanogen. Can. J. Microbiol. *30*: 1383–1396.

Patel, G.B. 1985. *In* Validation of the publication of new names and new combinations previously effectively published outside the IJSB. List No. 17. Int. J. Syst. Bacteriol. *35*: 223–225.

Patel, G.B. 1992. A contrary view of the proposal to assign a neotype strain for *Methanothrix soehngenii*. Int. J. Syst. Bacteriol. *42*: 324–326.

Patel, G.B., B.J. Agnew and C.J. Dicaire. 1991b. Inhibition of pure cultures of methanogens by benzene ring compounds. Appl. Environ. Microbiol. *57*: 2969–2974.

Patel, G.B., C. Baudet and B.J. Agnew. 1988. Nutritional requirements for growth of *Methanothrix concilii*. Can. J. Microbiol. *34*: 73–77.

Patel, G.B., L.A. Roth, L. van den Berg and D.S. Clark. 1976. Characterization of a strain of *Methanospirillum hungatii*. Can. J. Microbiol. *22*: 1404–1410.

Patel, G.B. and G.D. Sprott. 1990. *Methanosaeta concilii* gen. nov., sp. nov. (*"Methanothrix concilii"*) and *Methanosaeta thermoacetophila* nom. rev., comb. nov. Int. J. Syst. Bacteriol. *40*: 79–82.

Patel, G.B. and G.D. Sprott. 1991. Cobalt and sodium requirements for methanogenesis in washed cells of *Methanosaeta concilii*. Can. J. Microbiol. *37*: 110–115.

Patel, G.B., G.D. Sprott and J.E. Fein. 1990. Isolation and characterization of *Methanobacterium espanolae*, sp. nov., a mesophilic, moderately acidophilic methanogen. Int. J. Syst. Bacteriol. *40*: 12–18.

Patel, G.B., G.D. Sprott, R.W. Humphrey and T.J. Beveridge. 1986. Comparative analyses of the sheath structures of *Methanothrix concilii* GP6

and *Methanospirillum hungatei* strains GP1 and JF1. Can. J. Microbiol. *32*: 623–631.

Paterek, J.R. and P.H. Smith. 1985. Isolation and characterization of a halophilic methanogen from Great Salt Lake. Appl. Environ. Microbiol. *50*: 877–881.

Paterek, J.R. and P.H. Smith. 1988. *Methanohalophilus mahii* gen. nov., sp. nov., a methylotrophic halophilic methanogen. Int. J. Syst. Bacteriol. *38*: 122–123.

Paulsrud, P., J. Rikkinen and P. Lindblad. 1998. Cyanobiont specificity in some *Nostoc*-containing lichens and in a Peltigera aphthosa photosymbiodeme. New Phytol. *139*: 517–524.

Paynter, M.J.B. and R.E. Hungate. 1968. Characterization of *Methanobacterium mobilis*, sp.n., isolated from the bovine rumen. J. Bacteriol. *95*: 1943–1951.

Pearion, C.T. and P.E. Jablonski. 1999. High level, intrinsic resistance of *Natronococcus occultus* to potassium tellurite. FEMS Microbiol. Lett. *174*: 19–23.

Pearson, H.W., R. Howsley, C.K. Kjeldsen and A.E. Walsby. 1979. Aerobic nitrogenase activity associated with a non-heterocystous filamentous cyanobacterium. FEMS Microbiol. Lett. *5*: 163–167.

Pease, A.C., D. Solas, E.J. Sullivan, M.T. Cronin, C.P. Holmes and S.P.A. Fodor. 1994. Light-generated oligonucleotide arrays for rapid DNA sequence analysis. Proc. Natl. Acad. Sci. U.S.A. *91*: 5022–5026.

Pecher, T. and A. Böck. 1981. *In vivo* susceptibility of halophilic and methanogenic organisms to protein synthesis inhibitors. FEMS Microbiol. Lett. *10*: 295–297.

Peck, M.W. 1989. Changes in concentrations of coenzyme F_{420} analogs during batch growth of *Methanosarcina barkeri* and *Methanosarcina mazei*. Appl. Environ. Microbiol. *55*: 940–945.

Peek, K., R.M. Daniel, C. Monk, L. Parker and T. Coolbear. 1992. Purification and characterization of a thermostable proteinase isolated from *Thermus* sp. strain Rt41A. Eur. J. Biochem. *207*: 1035–1044.

Peer, C.W., M.H. Painter, M.E. Rasche and J.G. Ferry. 1994. Characterization of a CO:heterodisulfide oxidoreductase system from acetate-grown *Methanosarcina thermophila*. J. Bacteriol. *176*: 6974–6979.

Pelsh, A.D. 1936. Hydrobiology of Karabugaz Bay of the Caspian Sea. Tr. Vses. Nauchno-Issled. Inst. Galurgii Leningrad *5*: 49–126.

Pelsh, A.D. 1937. Photosynthetic sulfur bacteria of the eastern reservoir of Lake Sakskoe. Mikrobiologiya *6*: 1090–1100.

Pennisi, E. 1998. Genome data shake tree of life. Science *280*: 672–674.

Pentecost, A. 1995. The microbial ecology of some Italian hot-spring travertines. Microbios *81*: 45–58.

Pepper, A.E. 1998. Molecular Evolution: old branches on the phytochrome family tree. Curr. Biol. *8*: R117–R120.

Pernthaler, J., F.-O. Glöckner, S. Unterholzner, A. Alfreider, R. Psenner and R.I. Amann. 1998. Seasonal community and population dynamics of pelagic bacteria and archaea in a high mountain lake. Appl. Environ. Microbiol. *64*: 4299–4306.

Pernthaler, J., T. Posch, K. Simek, J. Vrba, R.I. Amann and R. Psenner. 1997. Contrasting bacterial strategies to coexist with a flagellate predator in an experimental microbial assemblage. Appl. Environ. Microbiol. *63*: 596–601.

Perski, H.J., P. Schönheit and R.K. Thauer. 1982. Sodium dependence of methane formation in methanogenic bacteria. FEBS Lett. *143*: 323–326.

Peters, J., M. Nitsch, B. Kühlmorgen, R. Golbik, A. Lupas, J. Kellermann, H. Engelhardt, J.-P. Pfander, S. Müller, K. Goldie, A. Engel, K.O. Stetter and W. Baumeister. 1995. Tetrabrachion: a filamentous archaebacterial surface protein assembly of unusual structure and extreme stability. J. Mol. Biol. *245*: 385–401.

Petter, H.F.M. 1931. On bacteria of salted fish. Proc. K. Ned. Akad. Wet. Amsterdam *34*: 1417–1423.

Petter, H.F.M. 1932. Over roode en andere bacterien van gezouten visch, Doctoral thesis, Rijks-Universiteit te Utrecht, Utrecht.

Pfeifer, F. 1988. Genetics of halobacteria. *In* Rodriguez-Valera (Editor), Halophilic Bacteria, Vol. 2, CRC Press, Boca Raton. pp. 105–133.

Pfeifer, F. 1995. Fractionation of halophilic archaeal DNA into FI and FII using affinity chromatography on malachite green bisacrylamide.

In DasSarma and Fleischmann (Editors), Archaea: A Laboratory Manual, Cold Spring Harbor Laboratory Press, Plainview. pp. 189–192.

Pfeifer, F., U. Blaseio and M. Horne. 1989. Genome structure of *Halobacterium halobium*: plasmid dynamics in gas vacuole-deficient mutants. Can. J. Microbiol. *35*: 96–100.

Pfeifer, F., K. Krüger, R. Röder, A. Mayr, S. Ziesche and S. Offner. 1997. Gas vesicle formation in halophilic Archaea. Arch. Microbiol. *167*: 259–268.

Pfeifer, F., S. Offner, K. Krüger, P. Ghahraman and C. Englert. 1994. Transformation of halophilic archaea and investigation of gas vesicle synthesis. Syst. Appl. Microbiol. *16*: 569–577.

Pfeifer, F., G. Weidinger and W. Goebel. 1981. Characterization of plasmids in halobacteria. J. Bacteriol. *145*: 369–374.

Pfennig, N. 1965. Anreicherungskulturen für rote und grüne Schwefelbakterien. Zentbl. Bakteriol. Parasitenkd. Infektkrankh. Hyg. Abt. I Orig. *Suppl. I*: 179–189, 503–505.

Pfennig, N. 1967. Photosynthetic bacteria. Annu. Rev. Microbiol. *21*: 285–324.

Pfennig, N. 1968. *Chlorobium phaeobacteroides* nov. spec. und *Chlorobium phaeovibrioides* nov. spec., zwei neue Arten der grünen Schwefelbakterien. Arch. Mikrobiol. *63*: 224–226.

Pfennig, N. 1969. *Rhodospirillum tenue* sp.n., a new species of the purple nonsulfur bacteria. J. Bacteriol. *99*: 619–620.

Pfennig, N. 1977. Phototrophic green and purple bacteria: a comparative, systematic survey. Annu. Rev. Microbiol. *31*: 275–290.

Pfennig, N. 1978. *Rhodocyclus purpureus* gen. nov. and sp. nov. a ring-shaped, vitamin B_{12}-requiring member of the family *Rhodospirillaceae*. Int. J. Syst. Bacteriol. *28*: 283–288.

Pfennig, N. 1980. Syntrophic mixed cultures and symbiotic consortia with phototrophic bacteria: a review. *In* Gottschalk, Pfennig and Werner (Editors), Anaerobes and Anaerobic Infections, Gustav Fischer Verlag, Stuttgart. pp. 127–131.

Pfennig, N. 1989. Green sulfur bacteria. *In* Staley, Bryant, Pfennig and Holt (Editors), Bergey's Manual of Systematic Bacteriology, 1st Ed., Vol. 3, The Williams & Wilkins Co., Baltimore. pp. 1682–1697.

Pfennig, N. and G. Cohen-Bazire. 1967. Some properties of the green bacterium *Pelodictyon clathratiforme*. Arch. Microbiol. *59*: 226–236.

Pfennig, N. and K.D. Lippert. 1966. Über das Vitamin B_{12}-Bedürfnis phototropher Schwefelbakterien. Arch. Mikrobiol. *55*: 245–256.

Pfennig, N., H. Lünsdorf, J. Süling and J.F. Imhoff. 1997. *Rhodospira trueperi* gen. nov., sp. nov., a new phototrophic proteobacterium of the alpha group. Arch. Microbiol. *168*: 39–45.

Pfennig, N. and H.G. Trüper. 1971a. Conservation of the family name *Chromatiaceae* Bavendamm 1924 with the type genus *Chromatium* Perty 1852. Int. J. Syst. Bacteriol. *21*: 15–16.

Pfennig, N. and H.G. Trüper. 1971b. Higher taxa of the phototrophic bacteria. Int. J. Syst. Bacteriol. *21*: 17–18.

Pfennig, N. and H.G. Trüper. 1971c. New nomenclatural combinations in the phototrophic sulfur bacteria. Int. J. Syst. Bacteriol. *21*: 11–14.

Pfennig, N. and H.G. Trüper. 1974. The phototrophic bacteria. *In* Buchanan and Gibbons (Editors), Bergey's Manual of Determinative Bacteriology, 8th Ed., The Williams & Wilkins Co., Baltimore. pp. 24–60.

Pfennig, N. and H.G. Trüper. 1981. Isolation of members of the families *Chromatiaceae* and *Chlorobiaceae*. *In* Starr, Stolp, Trüper, Balows and Schlegel (Editors), The Prokaryotes: a Handbook on Habitats, Isolation and Identification of Bacteria, Springer-Verlag, Berlin. pp. 279–289.

Pfennig, N. and H.G. Trüper. 1989. Family I. *Chromatiaceae*. *In* Staley, Bryant, Pfennig and Holt (Editors), Bergey's Manual of Systematic Bacteriology, 1st Ed., Vol. 3, The Williams & Wilkins Co., Baltimore. pp. 1637–1653.

Pfennig, N., F. Widdel and H.G. Trüper. 1981. The dissimilatory sulfate-reducing bacteria. *In* Starr, Stolp, Trüper, Balows and Schlegel (Editors), The Prokaryotes: a Handbook on Habitats, Isolation and Identification of Bacteria, Springer-Verlag, Berlin. pp. 926–940.

Phillips, W.E. and J.J. Perry. 1976. *Thermomicrobium fosteri* sp. nov., a hydrocarbon-utilizing obligate thermophile. Int. J. Syst. Bacteriol. *26*: 220–225.

Phipps, B.M., H. Engelhardt, R. Huber and W. Baumeister. 1990. Three-dimensional structure of the crystalline protein envelope layer of the hyperthermophilic archaebacterium *Pyrobaculum islandicum*. J. Struct. Biol. *103*: 152–163.

Phipps, B.M., A. Hoffmann, K.O. Stetter and W. Baumeister. 1991a. A novel ATPase complex selectively accumulated upon heat shock is a major cellular component of thermophilic archaebacteria. EMBO (Eur. Mol. Biol. Organ.) J. *10*: 1711–1722.

Phipps, B.M., R. Huber and W. Baumeister. 1991b. The cell envelope of the hyperthermophilic archaebacterium *Pyrobaculum organotrophum* consists of two regularly arrayed protein layers: three-dimensional structure of the outer layer. Mol. Microbiol. *5*: 253–266.

Phipps, B.M., D. Typke, R. Hegerl, S. Volker, A. Hoffmann, K.O. Stetter and W. Baumeister. 1993. Structure of a molecular chaperone from a thermophilic archaebacterium. Nature *361*: 475–477.

Pickup, R.W. and J.R. Saunders (Editors). 1996. Molecular approaches to environmental microbiology, Ellis Horwood, London.

Pieper, U., G. Kapadia, M. Mevarech and O. Herzberg. 1998. Structural features of halophilicity derived from the crystal structure of dehydrofolate dehydrogenase from the Dead Sea archaeon *Haloferax volcanii*. Structure *6*: 75–88.

Pierson, B.K. 1994. The emergence, diversification, and role of photosynthetic eubacteria. *In* Bengston (Editor), Early Life on Earth, Nobel Symposium No. 84, Columbia University Press, New York. pp. 161–180.

Pierson, B.K. and R.W. Castenholz. 1971. Bacteriochlorophylls in gliding filamentous prokaryotes from hot springs. Nat. New Biol. *233*: 25–27.

Pierson, B.K. and R.W. Castenholz. 1974a. A phototrophic gliding filamentous bacterium of hot springs, *Chloroflexus aurantiacus*, gen. and sp. nov. Arch. Microbiol. *100*: 5–24.

Pierson, B.K. and R.W. Castenholz. 1974b. Studies of pigments and growth in *Chloroflexus aurantiacus*, a phototrophic filamentous bacteria. Arch. Microbiol. *100*: 283–305.

Pierson, B.K. and R.W. Castenholz. 1992. The Family *Chloroflexaceae*. *In* Balows, Trüper, Dworkin, Harder and Schleifer (Editors), The Prokaryotes. A handbook of Bacteria: Ecophysiology, Isolation, Identification, Applications, 2nd Ed., Springer-Verlag, New York. pp. 3754–3774.

Pierson, B.K. and R.W. Castenholz. 1995. Taxonomy and physiology of filamentous anoxygenic phototrophs. *In* Blankenship, Madigan and Bauer (Editors), Anoxygenic Photosynthetic Bacteria, Kluwer Academic Publishers, Dordrecht. pp. 31–47.

Pierson, B.K., S.J. Giovannoni and R.W. Castenholz. 1984a. Physiological ecology of a gliding bacterium containing bacteriochlorophyll *a*. Appl. Environ. Microbiol. *47*: 576–584.

Pierson, B.K., S.J. Giovannoni, D.A. Stahl and R.W. Castenholz. 1985. *Heliothrix oregonensis*, gen. nov., sp. nov., a phototrophic filamentous gliding bacterium containing bacteriochlorophyll *a*. Arch. Microbiol. *142*: 164–167.

Pierson, B.K., S.J. Giovannoni, D.A. Stahl and R.W. Castenholz. 1986. *In* Validation of the publication of new names and new combinations previously effectively published outside the IJSB. List No. 20. Int. J. Syst. Bacteriol. *36*: 354–356.

Pierson, B.K., L.M. Keith and J.G. Leovy. 1984b. Isolation of pigmentation mutants of the green filamentous photosynthetic bacterium *Chloroflexus aurantiacus*. J. Bacteriol. *159*: 222–227.

Pierson, B.K., H.K. Mitchell and A.L. Ruff-Roberts. 1993. *Chloroflexus aurantiacus* and ultraviolet radiation: implications for archean shallow-water stromatolites. Origins Life Evol. Biosph. *23*: 243–260.

Pierson, B.K., A. Oesterle and G.L. Murphy. 1987. Pigments, light penetration, and photosynthetic activity in the multi-layered microbial mats of Great Sippewissett Salt Marsh, Massachusetts. FEMS Microbiol. Ecol. *45*: 365–376.

Pierson, B.K. and J.P. Thornber. 1983. Isolation and spectral characterization of photochemical reaction centers from the thermophilic green bacterium *Chloroflexus aurantiacus* strain J-10-fl. Proc. Natl. Acad. Sci. U.S.A. *80*: 80–84.

Pierson, B.K., J.P. Thornber and R.E.B. Seftor. 1983. Partial purification, subunit structure and thermal stability of the photochemical reaction center of the thermophilic green bacterium *Chloroflexus aurantiacus.* Biochim. Biophys. Acta *723*: 322–326.

Pierson, B.K., D. Valdez, M. Larsen, E. Morgan and E.E. Mack. 1994. *Chloroflexus*-like organisms from marine and hypersaline environments: distribution and diversity. Photosynth. Res. *41*: 35–52.

Pihl, T.D., S. Sharma and J.N. Reeve. 1994. Growth phase-dependent transcription of the genes that encode the two methyl coenzyme M reductase isoenzymes and N^5-methyltetrahydromethanopterin: coenzyme M methyltransferase in *Methanobacterium thermoautotrophicum* ΔH. J. Bacteriol. *176*: 6384–6391.

Pinevich, A.V., S.G. Averina and O.V. Gavrilova. 1997. *Pseudanabaena* sp. with short trichomes: a comment on the occurrence and taxonomic implications of brevitrichomy in oscillatorian cyanophytes. Algol. Stud. *86*: 1–9.

Pitulle, C., Y. Yang, M. Marchiani, E.R.B. Moore, J.L. Siefert, M. Aragno, P. Jurtshuk, Jr. and G.E. Fox. 1994. Phylogenetic position of the genus *Hydrogenobacter.* Int. J. Syst. Bacteriol. *44*: 620–626.

Pivovarova, T.A. and V.M. Gorlenko. 1977. Fine structure of *Chloroflexus aurantiacus* var. *mesophilus* (nom. prof.) grown in the light under aerobic and anaerobic conditions. Mikrobiologiya *46*: 329–334.

Pivovarova, T.A., G.E. Markosyan and G.I. Karavaiko. 1981. Morphogenesis and fine structure of *Leptospirillum ferrooxidans.* Mikrobiologiya *50*: 482–486.

Platt, T. and W.K.W. Li (Editors). 1986. Photosynthetic Picoplankton, Vol. 214: Canadian Bulletin of Fisheries and Aquatic Sciences, Department of Fisheries and Oceans, Ottawa.

Pley, U., J. Schipka, A. Gambacorta, H.W. Jannasch, H. Fricke, R. Rachel and K.O. Stetter. 1991. *Pyrodictium abyssi*, sp. nov. represents a novel heterotrophic marine archaeal hyperthermophile growing at 110°C. Syst. Appl. Microbiol. *14*: 245–253.

Pley, U. and K.O. Stetter. 1991. *In* Validation of the publication of new names and new combinations previously effectively published outside the IJSB. List No. 39. Int. J. Syst. Bacteriol. *41*: 580–581.

Polz, M.F. and C.M. Cavanaugh. 1998. Bias in template-to-product ratios in multitemplate PCR. Appl. Environ. Microbiol. *64*: 3724–3730.

Pond, J.L. and T.A. Langworthy. 1987. Effect of growth temperature on the long-chain diols and fatty acids of *Thermomicrobium roseum.* J. Bacteriol. *169*: 1328–1330.

Pond, J.L., T.A. Langworthy and G. Holzer. 1986. Long-chain diols: a new class of membrane lipids from a thermophilic bacterium (*Thermomicrobium roseum*). Science *231*: 1134–1136.

Porta, D., R. Rippka and M. Hernández-Mariné. 2000. Unusual ultrastructural features in three strains of *Cyanothece* (cyanobacteria). Arch. Microbiol. *173*: 154–163.

Porter, J.R. 1976. The world view of culture collections. *In* Colwell (Editor), The Role of Culture Collections in the Era of Molecular Biology, American Society for Microbiology, Washington, D.C. pp. 62–72.

Porter, R.W. and Y.S. Feig. 1980. The use of DAPI for identifying and counting aquatic microflora. Limnol. Oceanogr. *25*: 943–948.

Post, F.J. 1977. The microbial ecology of the Great Salt Lake. Microb. Ecol. *3*: 143–165.

Post, F.J. and F.A. Al-Harjan. 1988. Surface activity of halobacteria and potential use in microbially enhanced oil recovery. Syst. Appl. Microbiol. *11*: 97–101.

Postgate, J.R. 1963. Versatile medium for the enumeration of sulfate-reducing bacteria. Appl. Microbiol. *11*: 265–267.

Postgate, J.R. 1969. Media for sulphur bacteria: some amendments. Lab. Pract. *18*: 286–294.

Potts, M. 1994. Desiccation tolerance of prokaryotes. Microbiol. Rev. *58*: 755–805.

Potts, M. 1996. The anhydrobiotic cyanobacterial cell. Physiol. Plant. *97*: 788–794.

Potts, M. 1997. Etymology of the genus name *Nostoc* (Cyanobacteria). Int. J. Syst. Bact. *47*: 584.

Potts, M., R. Ocampo-Friedmann, M.A. Bowman and B. Tözün. 1983. *Chroococcus* S24 and *Chroococcus* N41 (cyanobacteria): morphological, biochemical and genetic characterization and effects of water stress on ultrastructure. Arch. Microbiol. *135*: 81–90.

Poulsen, L.K., G. Ballard and D.A. Stahl. 1993. Use of rRNA fluorescence *in situ* hybridization for measuring the activity of single cells in young and established biofilms. Appl. Environ. Microbiol. *59*: 1354–1360.

Poulsen, V.A. 1879. Om nogle mikroskopiske planteorganismer. Vidensk. Medd. Dan. Nathist. Foren. *1879–1880*: 231–254.

Prado, A., M.S. da Costa and V.M.C. Madeira. 1988. Effect of growth temperature on the lipid composition of two strains of *Thermus* sp. J. Gen. Microbiol. *134*: 1653–1660.

Preston, C.M., K.Y. Wu, T.F. Molinski and E.F. DeLong. 1996. A psychrophilic crenarchaeon inhabits a marine sponge: *Crenarchaeum symbiosum* gen. nov., sp. nov. Proc. Natl. Acad. Sci. U.S.A. *93*: 6241–6246.

Priest, F. and B. Austin. 1993. Modern Bacterial Taxonomy, 2nd Ed., Chapman and Hall, London.

Pringsheim, E.G. 1968. Cyanophyceen-Probleme. Planta *79*: 1–9.

Printz, H. 1921. Subaerial algae from South Africa. K. Nor. Vidensk. Selsk. Skr. *1*: 35–36.

Prufert-Bebout, L. and F. Garcia-Pichel. 1994. Field and cultivated *Microcoelus chthonoplastes*: the search for clues to its prevalence in marine microbial mats. *In* Stal and Caumette (Editors), Microbial Mats: Structure, Development and Environmental Significance, NATO ASI Series G, Vol. 35, Springer-Verlag, Berlin. pp. 111–116.

Prufert-Bebout, L., H.W. Paerl and C. Lassen. 1993. Growth, nitrogen fixation, and spectral attenuation in cultivated *Trichodesmium* species. Appl. Environ. Microbiol. *59*: 1367–1375.

Puchkova, N.N. 1984. Green sulfur bacteria as a component of the "sulfureta" of shallow saline waters of the Crimea and nothern Caucasus. Mikrobiologiya *53*: 324–328.

Puchkova, N.N. and V.M. Gorlenko. 1976. New brown chlorobacteria *Prosthechloris phaeoasteroidea* nov. sp. Mikrobiologiya *45*: 656–660.

Puchkova, N.N. and V.M. Gorlenko. 1982. *Chlorobacterium chlorovibrioides* nov. sp., a new green sulfur bacterium. Mikrobiologiya *51*: 118–124.

Pugh, E.L. and M. Kates. 1994. Acylation of proteins of the archaebacteria *Halobacterium cutirubrum* and *Methanobacterium thermoautotrophicum.* Biochim. Biophys. Acta *1196*: 38–44.

Pum, D., P. Messner and U.B. Sleytr. 1991. Role of the S layer in morphogenesis and cell division of the archaebacterium *Methanocorpusculum sinense.* J. Bacteriol. *173*: 6865–6873.

Quesada, E., A. Ventosa, F. Rodriguez-Valera and A. Ramos-Cormenzana. 1982. Types and properties of some bacteria isolated from hypersaline soils. J. Appl. Bacteriol. *53*: 155–162.

Quintela, J.C., F. Garcia del Portillo, E. Pittenauer, G. Allmaier and M.A. de Pedro. 1999. Peptidoglycan fine structure of the radiotolerant bacterium *Deinococcus radiodurans* Sark. J. Bacteriol. *181*: 334–337.

Quintela, J.C., E. Pittenauer, G. Allmaier, V. Arán and M.A. de Pedro. 1995. Structure of peptidoglycan from *Thermus thermophilus* HB8. J. Bacteriol. *177*: 4947–4962.

Rachel, R. 1999. Fine structure of hyperthermophilic procaryotes. *In* Seckbach (Editor), Enigmatic Microorganisms and Life in Extreme Environments, Kluwer Academic Publishers, Dordrecht. 277–289.

Rachel, R., A. Engel, R. Huber, K.O. Stetter and W. Baumeister. 1990. A porin-type protein is the main constituent of the cell envelope of the ancestral eubacterium *Thermotoga maritima.* FEBS Lett. *262*: 64–68.

Rachel, R., J. Wildhaber, K.O. Stetter and W. Baumeister. 1988. The structure of the surface protein of *Thermotoga maritima. In* Sleytr, Messner, Pum and Sara (Editors), Proceedings of the Second International Workshop on S-layers in Procaryotes, Springer-Verlag, Berlin. pp. 83–86.

Radax, C., O. Sigurdsson, G.O. Hreggvidsson, N. Aichinger, C. Gruber, J.K. Kristjánsson and H. Stan-Lotter. 1998. F- and V-ATPases in the genus *Thermus* and related species. Syst. Appl. Microbiol. *21*: 12–22.

Raemakers-Franken, P.C., F.G.J. Voncken, J. Korteland, J.T. Keltjens, C. van der Drift and G.D. Vogels. 1989. Structural characterization of tatiopterin, a novel pterin isolated from *Methanogenium tationis.* Biofactors *2*: 117–122.

Rainey, F.A., M.F. Nobre, P. Schumann, E. Stackebrandt and M.S. da Costa. 1997. Phylogenetic diversity of the deinococci as determined by 16S ribosomal DNA sequence comparison. Int. J. Syst. Bacteriol. *47*: 510–514.

Rainey, F.A., N.L. Ward-Rainey, P.H. Janssen, H. Hippe and E. Stackebrandt. 1996. *Clostridium paradoxum* DSM 7308^T contains multiple 16S rRNA genes with heterogeneous intervening sequences. Microbiology (Reading) *142*: 2087–2095.

Raj, H.D., F.L. Duryee, A.M. Deeny, C.H. Wang, A.W. Anderson and P.R. Elliker. 1960. Utilization of carbohydrates and amino acids by *Micrococcus radiodurans*. Can. J. Microbiol. *6*: 289–298.

Rajagopal, B.S. and L. Daniels. 1986. Investigation of mercaptans, organic sulfides, and inorganic sulfur compounds as sulfur sources for the growth of methanogenic bacteria. Curr. Microbiol. *14*: 137–144.

Rajagopalan, R. and W. Altekar. 1991. Products of non-reductive CO_2 fixation in the halophilic archaebacterium *Haloferax mediterranei*. Indian J. Biochem. Biophys. *28*: 65–67.

Rajagopalan, R. and W. Altekar. 1994. Characterization and purification of ribulose-bisphosphate carboxylase from heterotrophically grown halophilic archaebacterium *Haloferax mediterranei*. Eur. J. Biochem. *221*: 863–869.

Ramaley, R.F. and K. Bitzinger. 1975. Types and distribution of obligate thermophilic bacteria in man-made and natural thermal gradients. Appl. Microbiol. *30*: 152–155.

Ramaley, R.F. and J. Hixson. 1970. Isolation of a nonpigmented, thermophilic bacterium similar to *Thermus aquaticus*. J. Bacteriol. *103*: 526– 528.

Ramalay, R.F., F.R. Turner, L.E. Malick and R.B. Wilson. 1978. The morphology and surface structure of some extremely thermophilic bacteria found in slightly alkaline hot springs. *In* Friedman (Editor), Biochemistry of Thermophily, Academic Press, New York.

Ramsing, N.B., M.J. Ferris and D.M. Ward. 1997. Light-induced motility of thermophilic *Synechococcus* isolates from Octopus Spring, Yellowstone National Park. Appl. Environ. Microbiol. *63*: 2347–2354.

Ramsing, N.B., H. Fossing, T.G. Ferdelman, F. Andersen and B. Thamdrup. 1996. Distribution of bacterial populations in a stratified Fjord (Mariager Fjord, Denmark) quantified by *in situ* hybridization and related to chemical gradients in the water column. Appl. Environ. Microbiol. *62*: 1391–1404.

Ramsing, N.B., M. Kühl and B.B. Jørgensen. 1993. Distribution of sulfate-reducing bacteria, O_2, and H_2S in photosynthetic biofilms determined by oligonucleotide probes and microelectrodes. Appl. Environ. Microbiol. *59*: 3840–3849.

Raskin, L., L.K. Poulsen, D.R. Noguera, B.E. Rittmann and D.A. Stahl. 1994a. Quantification of methanogenic groups in anaerobic biological reactors by oligonucleotide probe hybridization. Appl. Environ. Microbiol. *60*: 1241–1248.

Raskin, L., J.M. Stromley, B.E. Rittmann and D.A. Stahl. 1994b. Group-specific 16S rRNA hybridization probes to describe natural communities of methanogens. Appl. Environ. Microbiol. *60*: 1232–1240.

Ratanakhanokchai, K., J. Kaneko, Y. Kamio and K. Izaki. 1992. Purification and properties of a maltotetraose- and maltotriose-producing amylase from *Chloroflexus aurantiacus*. Appl. Environ. Microbiol. *58*: 2490– 2494.

Ratcliff, R. 1981. Terminal deoxynucleotidyltransferase. *In* Boyer (Editor), The Enzymes, Vol. 14, Academic Press, New York. pp. 105–118.

Raven, N.D.H. 1995. Genetics of *Thermus*: plasmids, bacteriophage, potential vectors, gene transfer systems. *In* Sharp and Williams (Editors), *Thermus* Species, Plenum Press, New York. pp. 157–184.

Ravin, A.W. 1963. Experimental approaches to the study of bacterial phylogeny. Am. Natur. *97*: 307–318.

Ravot, G., M. Magot, M.L. Fardeau, B.K.C. Patel, G. Prensier, A. Egan, J.L. Garcia and B. Ollivier. 1995a. *Thermotoga elfii* sp. nov., a novel thermophilic bacterium from an African oil-producing well. Int. J. Syst. Bacteriol. *45*: 308–314.

Ravot, G., B. Ollivier, M.L. Fardeau, B.K. Patel, K.T. Andrews, M. Magot and J.L. Garcia. 1996a. L-Alanine production from glucose fermentation by hyperthermophilic members of the domains bacteria and

Archaea: a remnant of an ancestral metabolism? Appl. Environ. Microbiol. *62*: 2657–2659.

Ravot, G., B. Ollivier, M. Magot, B.K.C. Patel, J.L. Corlet, M.L. Fardeau and J.L. Garcia. 1995b. Thiosulfate reduction, an important physiological feature shared by members of the order *Thermotogales*. Appl. Environ. Microbiol. *61*: 2053–2055.

Ravot, G., B. Ollivier, B.K.C. Patel, M. Magot and J.-L. Garcia. 1996b. Emended description of *Thermosipho africanus* as a carbohydrate-fermenting species using thiosulfate as an electron acceptor. Int. J. Syst. Bacteriol. *46*: 321–323.

Rawlings, D.E., H. Tributsch and G.S. Hansford. 1999. Reasons why *"Leptospirillum"*-like species rather than *Thiobacillus ferrooxidans* are the dominant iron-oxidizing bacteria in many commercial processes for the biooxidation of pyrite and related ores. Microbiology *145*: 5–13.

Ray, P.H., D.C. White and T.D. Brock. 1971. Effect of temperature on the fatty acid composition of *Thermus aquaticus*. J. Bacteriol. *106*: 25–30.

Reddy, D.M., P.F. Crain, C.G. Edmonds, R. Gupta, T. Hashizume, K.O. Stetter, F. Widdel and J.A. McCloskey. 1992. Structure determination of two new amino acid-containing derivatives of adenosine from tRNA of thermophilic bacteria and archaea. Nucleic Acids Res. *20*: 5607–5615.

Rees, G.N., G.S. Grassia, A.J. Sheehy, P.P. Dwivedi and B.K.C. Patel. 1995. *Desulfacinum infernum* gen. nov., sp. nov., a thermophilic sulfate-reducing bacterium from a petroleum reservoir. Int. J. Syst. Bacteriol. *45*: 85–89.

Reichenbach, H. 1992. The genus *Herpetosiphon*. *In* Balows, Trüper, Dworkin, Harder and Schleifer (Editors), The Prokaryotes. A Handbook of Bacteria: Ecophysiology, Isolation, Identification, Applications, 2nd Ed., Springer-Verlag, New York. pp. 3785–3805.

Reichenbach, H. and J.R. Golecki. 1975. The fine structure of *Herpetosiphon*, and a note on the taxonomy of the genus. Arch. Microbiol. *102*: 281–291.

Reistad, R. 1970. On the composition and nature of the bulk protein of extremely halophilic bacteria. Arch. Mikrobiol. *71*: 353–360.

Reistad, R. 1975. Amino sugar and amino acid constituents of the cell wall of the extremely halophilic cocci. Arch. Mikrobiol. *102*: 71–73.

Relman, D.A. 1999. The search for unrecognized pathogens. Science *284*: 1308–1310.

Repeta, D.J., D.J. Simpson, B.B. Jørgensen and H.W. Jannasch. 1989. Evidence for anoxygenic photosynthesis from the distribution of bacteriochlorophylls in the Black Sea. Nature *342*: 69–72.

Revsbech, N.P. and D.M. Ward. 1984. Microprofiles of dissolved substances and photosynthesis in microbial mats measured with microelectrodes. *In* Cohen, Castenholz and Halvorson (Editors), Microbial Mats: Stromatolites, Alan R. Liss, New York. pp. 171–188.

Reysenbach, A.L., G.S. Wickham and N.R. Pace. 1994. Phylogenetic analysis of the hyperthermophilic pink filament community in Octopus Spring, Yellowstone National Park. Appl. Environ. Microbiol. *60*: 2113– 2119.

Rieger, G., K. Müller, R. Hermann, K.O. Stetter and R. Rachel. 1997. Cultivation of hyperthermophilic archaea in capillary tubes resulting in improved preservation of fine structures. Arch. Microbiol. *168*: 373– 379.

Rieger, G., R. Rachel, R. Hermann and K.O. Stetter. 1995. Ultrastructure of the hyperthermophilic archaeon *Pyrodictium abyssi*. J. Struct. Biol. *115*: 78–87.

Rippka, R. 1988a. Isolation and purification of cyanobacteria. Methods Enzymol. *167*: 3–27.

Rippka, R. 1988b. Recognition and identification of cyanobacteria. Methods Enzymol. *167*: 28–67.

Rippka, R. and G. Cohen-Bazire. 1983. The *Cyanobacteriales*: a legitimate order based on the type strain *Cyanobacterium stanieri*? Ann. Microbiol. (Institut Pasteur) *134B*: 21–36.

Rippka, R., T. Coursin, W. Hess, C. Lichtlé, D.J. Scanlan, K.A. Palinska, I. Iteman, F. Partensky, J. Houmard and M. Herdman. 2000. *Prochlorococcus marinus* Chisholm et al. 1992, subsp. *pastoris* subsp. nov. strain

PCC 9511, the first axenic chlorophyll a_2/b_2-containing cyanobacterium (*Oxyphotobacteria*). Int. J. Syst. Evol. Microbiol. *50*: 1833– 1847.

Rippka, R., J. Deruelles, J.B. Waterbury, M. Herdman and R.Y. Stanier. 1979. Generic assignments, strain histories and properties of pure cultures of cyanobacteria. J. Gen. Microbiol. *111*: 1–61.

Rippka, R. and M. Herdman. 1992. Pasteur Culture Collection of Cyanobacteria Catalog and Taxonomic Handbook: Catalog of Strains, Institut Pasteur, Paris.

Rippka, R. and J.B. Waterbury. 1977. The synthesis of nitrogenase by nonheterocystous cyanobacteria. FEMS Microbiol. Lett. *2*: 83–86.

Rippka, R., J.B. Waterbury and G. Cohen-Bazire. 1974. A cyanobacterium which lacks thylakoids. Arch. Microbiol. *100*: 419–436.

Rippka, R., J.B. Waterbury and R.Y. Stanier. 1981. Provisional generic assignments for cyanobacteria in pure culture. *In* Starr, Stolp, Trüper, Balows and Schlegel (Editors), The Prokaryotes: A Handbook on Habits, Isolation and Identification of Bacteria, Vol. 1, Springer-Verlag, Berlin. pp. 247–256.

Risatti, J.B., W.C. Capman and D.A. Stahl. 1994. Community structure of a microbial mat: the phylogenetic dimension. Proc. Natl. Acad. Sci. U.S.A. *91*: 10173–10177.

Rivard, C.J., J.M. Henson, M.V. Thomas and P.H. Smith. 1983. Isolation and characterization of *Methanomicrobium paynteri*, sp. nov., a mesophilic methanogen isolated from marine sediments. Appl. Environ. Microbiol. *46*: 484–490.

Rivard, C.J. and D.H. Smith. 1982. Isolation and characterization of a thermophilic marine methanogenic bacterium, *Methanogenium thermophilicum* sp. nov. Int. J. Syst. Bacteriol. *32*: 430–436.

Robertson, D.E., M.C. Lai, R.P. Gunsalus and M.F. Roberts. 1992a. Composition, variation, and dynamics of major osmotic solutes in *Methanohalophilus* strain FDF-1. Appl. Environ. Microbiol. *58*: 2438–2443.

Robertson, D.E., D. Noll and M.F. Roberts. 1992b. Free amino acid dynamics in marine methanogens. β-amino acids as compatible solutes. J. Biol. Chem. *267*: 14893–14901.

Robertson, D.E., D. Noll, M.F. Roberts, J.A.G.F. Menaia and D.R. Boone. 1990a. Detection of the osmoregulator betaine in methanogens. Appl. Environ. Microbiol. *56*: 563–565.

Robertson, D.E., M.F. Roberts, N. Belay, K.O. Stetter and D.R. Boone. 1990b. Occurrence of β-glutamate, a novel osmolyte, in marine methanogenic bacteria. Appl. Environ. Microbiol. *56*: 1504–1508.

Robinson, I.M. and M.J. Allison. 1969. Isoleucine biosynthesis from 2-methylbutyric acid by anaerobic bacteria from the rumen. J. Bacteriol. *97*: 1220–1226.

Robinson, R.W. 1986. Life cycles in the methanogenic archaebacterium *Methanosarcina mazei*. Appl. Environ. Microbiol. *52*: 17–27.

Robinson, R.W., H.C. Aldrich, S.F. Hurst and A.S. Bleiweis. 1985. Role of the cell surface of *Methanosarcina mazei* in cell aggregation. Appl. Environ. Microbiol. *49*: 321–327.

Rodrigo, A.G., K.M. Borges and P.L. Bergquist. 1994. Pulsed-field gel electrophoresis of genomic digests of *Thermus* strains and its implications for taxonomic and evolutionary studies. Int. J. Syst. Bacteriol. *44*: 547–552.

Rodriguez-Valera, F., G. Juez and D.J. Kushner. 1982. Halocins: salt-dependent bacteriocins produced by extremely halophilic rods. Can. J. Microbiol. *28*: 151–154.

Rodriguez-Valera, F., G. Juez and D.J. Kushner. 1983a. *Halobacterium mediterranei* sp. nov., a new carbohydrate-utilizing extreme halophile. Syst. Appl. Microbiol. *4*: 369–381.

Rodriguez-Valera, F., J.J. Nieto and F. Ruiz-Berraquero. 1983b. Light as an energy source in continuous cultures of bacteriorhodopsin-containing halobacteria. Appl. Environ. Microbiol. *45*: 868–871.

Rodriguez-Valera, F., F. Ruiz-Berraquero and A. Ramos-Cormenzana. 1979. Isolation of extreme halophiles from sea water. Appl. Environ. Microbiol. *38* : 164–165.

Rodriguez-Valera, F., F. Ruiz-Berraquero and A. Ramos-Cormenzana. 1980. Isolation of extremely halophilic bacteria able to grow in defined inorganic media with single carbon sources. J. Gen. Microbiol. *119*: 535–538.

Rodriguez-Valera, F., F. Ruiz-Berraquero and A. Ramos-Cormenzana.

1981. Characteristics of the heterotrophic bacterial populations in hypersaline environments of differing salinities. Microb. Ecol. *7*: 235–243.

Rodriguez-Valera, F., A. Ventosa, G. Juez and J.F. Imhoff. 1985. Variation of environmental features and microbial populations with salt concentrations in a multi-pond saltern. Microb. Ecol. *11*: 107–116.

Roller, C., M. Wagner, R.I. Amann, W. Ludwig and K.H. Schleifer. 1994. *In situ* probing of Gram-positive bacteria with high DNA G + C content using 23S rRNA-targeted oligonucleotides: environmental application of nucleic acid hybridization. Microbiology (Reading) *140*: 2849–2858.

Rolstad, A.K., E. Howland and R. Sirevåg. 1988. Malate dehydrogenase from the thermophilic green bacterium *Chloroflexus aurantiacus*: purification, molecular weight, amino acid composition, and partial amino acid sequence. J. Bacteriol. *170*: 2947–2953.

Romesser, J.A. and R.S. Wolfe. 1982. Coupling of methyl coenzyme M reduction with carbon dioxide activation in extracts of *Methanobacterium thermoautotrophicum*. J. Bacteriol. *152*: 840–847.

Romesser, J.A., R.S. Wolfe, F. Mayer, E. Spiess and A. Walther-Mauruschat. 1979. *Methanogenium*, a new genus of marine methanogenic bacteria, and characterization of *Methanogenium cariaci* sp. nov. and *Methanogenium marisnigri* sp. nov. Arch. Microbiol. *121*: 147–153.

Romesser, J.A., R.S. Wolfe, F. Mayer, E. Spiess and A. Walther-Mauruschat. 1981. *In* Validation of the publication of new names and new combinations previously effectively published outside the IJSB. List No. 6. Int. J. Syst. Bacteriol. *31*: 215–218.

Rondon, M.R., S.J. Raffel, R.M. Goodman and J. Handelsman. 1999. Toward functional genomics in bacteria: analysis of gene expression in *Escherichia coli* from a bacterial artificial chromosome library of *Bacillus cereus*. Proc. Natl. Acad. Sci. U.S.A. *96*: 6451–6455.

Ronimus, R.S., A.-L. Reysenbach, D.R. Musgrave and H.W. Morgan. 1997. The phylogenetic position of the *Thermococcus* isolate AN1 based on 16S rRNA gene sequence analysis: a proposal that AN1 represents a new species, *Thermococcus zilligii* sp. nov.. Arch. Microbiol. *168*: 245–248.

Ronimus, R.S., A.-L. Reysenbach, D.R. Musgrave and H.W. Morgan. 1999. *In* Validation of publication of new names and new combinations previously effectively published outside the IJSB. List. No. 70. Int. J. Syst. Bacteriol. *49*: 935–936.

Rooney-Varga, J.N., R. Devereux, R.S. Evans and M.E. Hines. 1997. Seasonal changes in the relative abundance of uncultivated sulfate-reducing bacteria in a salt marsh sediment and in the rhizosphere of *Spartina alterniflora*. Appl. Environ. Microbiol. *63*: 3895–3901.

Rospert, S., D. Linder, J. Ellermann and R.K. Thauer. 1990. Two genetically distinct methyl coenzyme M reductases in *Methanobacterium thermoautotrophicum* strain Marburg and ΔH. Eur. J. Biochem. *194*: 871–878.

Ross, H.N.M. and W.D. Grant. 1985. Nucleic acid studies on halophilic archaebacteria. J. Gen. Microbiol. *131*: 165–173.

Ross, H.N.M., W.D. Grant and J.E. Harris. 1985. Lipids in archaebacterial taxonomy. *In* Goodfellow and Minnikin (Editors), Chemical Methods in Bacterial Systematics, Academic Press, New York. pp. 289–299.

Rossello-Mora, R.A., B. Thamdrup, H. Schäfer, R. Weller and R.I. Amann. 1999. Marine sediment microbial community response to organic carbon amendment under anaerobic conditions. Syst. Appl. Microbiol. *22*: 237–248.

Rossello-Mora, R.A., M. Wagner, R.I. Amann and K.H. Schleifer. 1995. The abundance of *Zoogloea ramigera* in sewage treatment plants. Appl. Environ. Microbiol. *6*: 702–707.

Roth, R., R. Duft, A. Binder and R. Bachofen. 1986. Isolation and characterization of a soluble ATPase from *Methanobacterium thermoautotrophicum*. Syst. Appl. Microbiol. *7*: 346–348.

Rott, E. and M. Hernández-Mariné. 1994. *Pulvinularia suecica*, a rare stigonematalean cyanophyte. Arch. Hydrobiol. Suppl. *105*: 313–322.

Rouhiainen, L., K. Sivonen, W.J. Buikema and R. Haselkorn. 1995. Characterization of toxin-producing cyanobacteria by using an oligonucleotide probe containing a tandemly repeated heptamer. J. Bacteriol. *177*: 6021–6026.

Rouvière, P.E., L. Mandelco, S. Winker and C.R. Woese. 1992. A detailed phylogeny for the *Methanomicrobiales*. Syst. Appl. Microbiol. *15*: 363–371.

Rouvière, P.E. and R.S. Wolfe. 1987. Use of subunits of the methylreductase protein for taxonomy of methanogenic bacteria. Arch. Microbiol. *148*: 253–259.

Rozanova, E.P. and A.I. Khudyakova. 1974. A new nonspore-forming thermophilic sulfate-reducing organism, *Desulfovibrio thermophilus* nov. sp. Mikrobiologiya *43*: 1069–1075.

Rozanova, E.P. and T.A. Pivovarova. 1988. Reclassification of *Desulfovibrio thermophilus* (Rozanova, Khudyakova, 1974). Mikrobiologiya *57*: 102–106.

Rozanova, E.P. and T.A. Pivovarova. 1991. *In* Validation of the publication of new names and new combinations previously effectively published outside the IJSB. List No. 36. Int. J. Syst. Bacteriol. *41*: 178–179.

Rudi, K., O.M. Skulberg, F. Larsen and K.S. Jakobsen. 1997. Strain characterization and classification of oxyphotobacteria in clone cultures on the basis of 16S rRNA sequences from the variable regions V6, V7, and V8. Appl. Environ. Microbiol. *63*: 2593–2599.

Rudnick, H., S. Hendrich, U. Pilatus and K.H. Blotevogel. 1990. Phosphate accumulation and the occurrence of polyphosphates and cyclic 2,3-diphosphoglycerate in *Methanosarcina frisia*. Arch. Microbiol. *154*: 584–588.

Ruffett, M., S. Hammond, R.A.D. Williams and R.J. Sharp. 1992. A taxonomic study of red pigmented gram negative thermophiles. Thermophiles: Science and Technology IceTec, Reykjavik. 74.

Ruff-Roberts, A.L., J.G. Kuenen and D.M. Ward. 1994. Distribution of cultivated and uncultivated cyanobacteria *Chloroflexus*-like bacteria in hot spring microbial mats. Appl. Environ. Microbiol. *60*: 697–704.

Saenger, W. 1984. Principles of Nucleic Acid Structure, Springer-Verlag, Berlin.

Safferman, R.S., I.R. Schneider, R.L. Steere, M.F. Morris and T.O. Diener. 1969. Phycovirus SM-1: a virus infecting unicellular blue-green algae. Virology *37*: 386–395.

Sahm, K., B.J. MacGregor, B.B. Jørgensen and D.A. Stahl. 1999. Sulphate reduction and vertical distribution of sulphate-reducing bacteria quantified by rRNA slot-blot hybridization in a coastal marine sediment. Environ. Microbiol. *1*: 65–74.

Saiki, R.K., D.H. Gelfand, S. Stoffel, S.J. Scharf, R. Higuchi, G.T. Horn, K.B. Mullis and H.A. Erlich. 1988. Primer-directed enzymatic amplification of DNA with a thermostable DNA polymerase. Science *239*: 487–491.

Saiki, R.K., P.S. Walsh, C.H. Levenson and H.A. Erlich. 1989. Genetic analysis of amplified DNA with immobilized sequence-specific oligonucleotide probes. Proc. Natl. Acad. Sci. U.S.A. *86*: 6230–6234.

Saiki, T., R. Kimura and K. Arima. 1972. Isolation and characterization of extremely thermophilic bacteria from hot springs. Agric. Biol. Chem. *36*: 2357–2366.

Saitoh, S., T. Suzuki and Y. Nishimura. 1998. Proposal of *Craurococcus roseus* gen. nov., sp. nov. and *Paracraurococcus ruber* gen. nov., sp. nov., novel aerobic bacteriochlorophyll *a*-containing bacteria from soil. Int. J. Syst. Bacteriol. *48*: 1043–1047.

Saitou, N. and M. Nei. 1987. The neighbor-joining method: a new method for reconstructing phylogenetic trees. Mol. Biol. Evol. *4*: 406–425.

Sakaki, Y. and T. Oshima. 1975. Isolation and characterization of a bacteriophage infectious to an extreme thermophile, *Thermus thermophilus* HB8. J. Virol. *15*: 1449–1453.

Sakane, T., I. Fukuda, T. Itoh and A. Yokota. 1992. Long-term preservation of halophilic archaebacteria and thermoacidophilic archaebacteria by liquid drying. J. Microbiol. Methods *16*: 281–287.

Sako, Y., N. Nomura, A. Uchida, Y. Ishida, H. Morii, Y. Koga, T. Hoaki and T. Maruyama. 1996. *Aeropyrum pernix* gen. nov., sp. nov., a novel aerobic hyperthermophilic archaeon growing at temperatures up to 100°C. Int. J. Syst. Bacteriol. *46*: 1070–1077.

Sambrook, J., E.F. Fritsch and T. Maniatis. 1989. Molecular cloning: a laboratory manual, Cold Spring Harbor Press, Cold Spring Harbor.

Samsonoff, W.A., T. Hashimoto and S.F. Conti. 1970. Ultrastructural

changes associated with germination and outgrowth of an appendage-bearing clostridial spore. J. Bacteriol. *101*: 1038–1045.

Sand, W., T. Gehrke, R. Hallmann and A. Schippers. 1995. Sulfur chemistry, biofilm, and the (in)direct attack mechanism: a critical evaluation of bacterial leaching. Appl. Microbiol Biotechnol. *43*: 961–966.

Sanders, S.W. and R.B. Maxcy. 1979. Isolation of radiation-resistant bacteria without exposure to irradiation. Appl. Environ. Microbiol. *38*: 436–439.

Santos, M.A., R.A.D. Williams and M.S. da Costa. 1989. Numerical taxonomy of *Thermus* isolates from hot springs in Portugal. Syst. Appl. Microbiol. *12*: 310–315.

Sanz, J.L., I. Marín, L. Ramirez, J.P. Abad, C.L. Smith and R. Amils. 1988. Variable rRNA gene copies in extreme halobacteria. Nucleic Acids Res. *16*: 7827–7832.

Sanz, J.L., I. Marín, D. Ureña and R. Amils. 1993. Functional analysis of seven ribosomal systems from extremely halophilic archaea. Can. J. Microbiol. *39*: 311–317.

Sapienza, C. and W.F. Doolittle. 1982. Unusual physical organization of the *Halobacterium* genome. Nature *295*: 384–389.

Saravani, G.A., D.A. Cowan, R.M. Daniel and H.W. Morgan. 1989. Caldolase, a chelator-insensitive extracellular serine proteinase from a *Thermus* spp. Biochem. J. *262*: 409–416.

Sato, K. 1978. Bacteriochlorophyll formation by facultative methylotrophs, *Protaminobacter ruber* and *Pseudomonas* AM 1. FEBS Lett. *85*: 207–210.

Saul, D.J., A.G. Rodrigo, R.A. Reeves, L.C. Williams, K.M. Borges, H.W. Morgan and P.L. Bergquist. 1993. Phylogeny of twenty *Thermus* isolates constructed from 16S rRNA gene sequence data. Int. J. Syst. Bacteriol. *43*: 754–760.

Sauvageau, C. 1892. Sur les algues d'eau douce récoltées en Algérie. Bull. Soc. Bot. France *39*: 115–117.

Sauvageau, C. 1897. Sur le *Nostoc punctiforme*. Ann. Sci. Nat. Bot. *8*: 366–378.

Savel'eva, N.D., V.R. Kryukov and M.A. Pusheva. 1982. An obligate thermophilic hydrogen bacterium. Mikrobiologiya *51*: 765–769.

Sayler, G.S. and A.C. Layton. 1990. Environmental application of nucleic acid hybridization. Annu. Rev. Microbiol. *44*: 625–628.

Sayler, G.S., M.S. Shields, E.T. Tedford, A. Breen, S.W. Hooper, K.M. Sirotkin and J.W. Davis. 1985. Application of DNA–DNA colony hybridization to the detection of catabolic genotypes in environmental samples. Appl. Environ. Microbiol. *49*: 1295–1303.

Schaechter, M.O., O. Maaløe and N.O. Kjeldgaard. 1958. Dependency on medium and temperature of cell size and chemical composition during balanced growth of *Salmonella typhimurium*. J. Gen. Microbiol. *19*: 592–606.

Schäfer, S., C. Barkowski and G. Fuchs. 1986. Carbon assimilation by the autotrophic thermophilic archaebacterium *Thermoproteus neutrophilus*. Arch. Microbiol. *146*: 301–308.

Schalkwyk, L.C. 1993. Halobacterial genes and genomes. *In* Kates (Editor), The Biochemistry of Archaea (Archaebacteria), Elsevier Science Publishers, Amsterdam. pp. 467–496.

Schauer, N.L., D.P. Brown and J.G. Ferry. 1982. Kinetics of formate metabolism in *Methanobacterium formicicum* and *Methanospirillum hungatei*. Appl. Environ. Microbiol. *44*: 549–554.

Schauer, N.L. and J.G. Ferry. 1980. Metabolism of formate in *Methanobacterium formicicum*. J. Bacteriol. *142*: 800–807.

Scheldeman, P., D. Baurain, D. Bouhy, M. Scott, M. Mühling, B.A. Whitton, A. Belay and A. Wilmotte. 1999. *Arthrospira* ("*Spirulina*") strains from four continents are resolved into only two clusters, based on amplified ribosomal DNA restriction analysis of the internally transcribed spacer. FEMS Microbiol. Lett. *172*: 213–222.

Scherer, P.A. 1989. Vanadium and molybdenum requirement for the fixation of molecular nitrogen by two *Methanosarcina* strains. Arch. Microbiol. *151*: 44–48.

Scherer, P.A. and H.P. Bochem. 1983. Ultrastructural investigation of 12 methanosarcinae and related species grown on methanol for occurrence of polyphosphate-like inclusions. Can. J. Microbiol. *29*: 1190–1199.

Scherer, P.A. and H. Kneifel. 1983. Distribution of polyamines in methanogenic bacteria. J. Bacteriol. *154*: 1315–1322.

Scherer, S. and M. Potts. 1989. Novel water stress protein from a desiccation-tolerant cyanobacterium: purification and partial characterization. J. Biol. Chem. *264*: 12546–12553.

Schink, B. 1991. Syntrophism among prokaryotes. *In* Balows, Trüper, Dworkin, Harder and Schleifer (Editors), The Prokaryotes. A Handbook of Bacteria: Ecophysiology, Isolation, Identification, Applications, 2nd Ed., Springer-Verlag, New York. pp. 276–299.

Schink, B. 1997. Energetics of syntrophic cooperation in methanogenic degradation. Microbiol. Mol. Biol. Rev. *61*: 262–280.

Schleifer, K.H., M. Ehrmann, C. Beimfohr, E. Brockmann, W. Ludwig and R.I. Amann. 1995. Application of molecular methods for the classification and identification of lactic acid bacteria. Int. Dairy J. *5*: 1081–1094.

Schleifer, K.H. and O. Kandler. 1972. Peptidoglycan types of bacterial cell walls and their taxonomic implications. Bacteriol. Rev. *36*: 407–477.

Schleifer, K.H. and W. Ludwig. 1989. Phylogenetic relationships among bacteria. *In* Fernholm, Bremer and Joernvall (Editors), The Hierarchy of Life: Molecules and Morphology in Phylogenetic Analysis, Elsevier Science Publishing Co., Amsterdam. pp. 103–117.

Schleifer, K.H., W. Ludwig and R.I. Amann. 1993. Nucleic acid probes. *In* Goodfellow and O'Donnell (Editors), Handbook of New Bacterial Systematics, Academic Press Ltd., London. pp. 464–512.

Schleifer, K.H. and E. Stackebrandt. 1983. Molecular systematics of prokaryotes. Annu. Rev. Microbiol. *37*: 143–187.

Schleifer, K.H., J. Steber and H. Mayer. 1982. Chemical composition and structure of the cell wall of *Halococcus morrhuae*. Zentbl. Bakteriol. Mikrobiol. Hyg. 1 Abt Orig. C *3*: 171–178.

Schleper, C., G. Pühler, I. Holz, A. Gambacorta, D. Janekovic, U. Santarius, H.P. Klenk and W. Zillig. 1995. *Picrophilus* gen. nov., fam. nov.: a novel aerobic, heterotrophic, thermoacidophilic genus and family comprising archaea capable of growth around pH 0. J. Bacteriol. *177*: 7050–7059.

Schleper, C., G. Puhler, H.-P. Klenk and W. Zillig. 1996. *Picrophilus oshimae* and *Picrophilus torridus* fam. nov., gen. nov., sp. nov., two species of hyperacidophilic, thermophilic, heterotrophic, aerobic archaea. Int. J. Syst. Bacteriol. *46*: 814–816.

Schlesner, H. and E. Stackebrandt. 1986. Assignment of the genera *Planctomyces* and *Pirella* to a new family *Planctomycetaceae* fam. nov. and description of the order *Planctomycetales* ord. nov. Syst. Appl. Microbiol. *8*: 174–176.

Schlösser, U.G. 1984. Sammlung von Algenkulturen Göttingen: additions to the collection since 1982. Ber. Dtsch. Bot. Ges. *97*: 465–475.

Schlösser, U.G. 1994. SAG-Sammlung von Algenkulturen at the University of Göttingen. Catalog of strains. Bot. Acta *107*: 113–186.

Schmid, K., M. Thomm, A. Laminet, F.G. Laue, C. Kessler, K.O. Stetter and R. Schmitt. 1984. Three new restriction endonucleases MaeI, MaeII and MaeIII from *Methanococcus aeolicus*. Nucleic Acids Res. *12*: 2619–2628.

Schmidhuber, S., W. Ludwig and K.H. Schleifer. 1988. Construction of a DNA probe for the specific identification of *Streptococcus oralis*. J. Clin. Microbiol. *26*: 1042–1044.

Schmidle, W. 1901. Neue Algen aus dem Gebiete des Oberrheins. Beih. Bot. Zentralbl. *10*: 179–180.

Schmidle, W. 1904. Hedwigia *44*: 415.

Schmidt, K. 1978. Biosynthesis of carotenoids. *In* Clayton and Sistrom (Editors), The Photosynthetic Bacteria, Plenum Press, New York. pp. 729–750.

Schmidt, K., M. Maarzahl and F. Mayer. 1980. Development and pigmentation of chlorosomes in *Chloroflexus aurantiacus* strain OK-70-fl. Arch. Microbiol. *127*: 87–98.

Schmidt, K. and R. Schiburr. 1970. Die Carotinoide der grünen Schwefelbakterien: Carotinoidzusammensetzung in 18 Stämmen. Arch. Mikrobiol. *74*: 350–355.

Schmitz, R.A., S.P.J. Albracht and R.K. Thauer. 1992a. A molybdenum and a tungsten isoenzyme of formylmethanofuran dehydrogenase in

the thermophilic archaeon *Methanobacterium wolfei*. Eur. J. Biochem. *209*: 1013–1018.

Schmitz, R.A., P.A. Bertram and R.K. Thauer. 1994. Tungstate does not support synthesis of active formylmethanofuran dehydrogenase in *Methanosarcina barkeri*. Arch. Microbiol. *161*: 528–530.

Schmitz, R.A., M. Richter, D. Linder and R.K. Thauer. 1992b. A tungsten-containing active formylmethanofuran dehydrogenase in the thermophilic archaeon *Methanobacterium wolfei*. Eur. J. Biochem. *207*: 559–565.

Schnellen, C.G.T.P. 1947. Onderzoekingen over de methaangisting, Thesis, Technische Hoogeschool Delft, Drukkerij "De Maasstad," Rotterdam, The Netherlands.

Schönheit, P. and T. Schäfer. 1995. Metabolism of hyperthermophiles. World J. Microbiol. Biotechnol. *11*: 26–57.

Schönhuber, W., B. Fuchs, S. Juretschko and R.I. Amann. 1997. Improved sensitivity of whole-cell hybridization by the combination of horseradish peroxidase-labeled oligonucleotides and tyramide signal amplification. Appl. Environ. Microbiol. *63*: 3268–3273.

Schook, L.B. and R.S. Berk. 1978. Nutritional studies with *Pseudomonas aeruginosa* grown on inorganic sulfur sources. J. Bacteriol. *133*: 1378–1382.

Schoop, G. 1935a. *Halococcus litoralis*, ein obligat halophiler Farbstofflbildner. Deut. Tierärztl. Wochenschr. *43*: 817–820.

Schoop, G. 1935b. Obligat halophile Mikroben. Zentbl. Bakteriol. Parasitenkd. Infektkrankh. Hyg. Abt. I Orig. *134*: 14–26.

Schopf, W.J. 1996. Cyanobacteria: pioneers of the early Earth. Nova Hedwigia Beih. *112*: 13–32.

Schramm, A., D. de Beer, M. Wagner and R. Amann. 1998. Identification and activities *in situ* of *Nitrosospira* and *Nitrospira* spp. as dominant populations in a nitrifying fluidized bed reactor. Appl. Environ. Microbiol. *64*: 3480–3485.

Schramm, A., L.H. Larsen, N.P. Revsbech, N.B. Ramsing, R.I. Amann and K.H. Schleifer. 1996. Structure and function of a nitrifying biofilm as determined by *in situ* hybridization and the use of microelectrodes. Appl. Environ. Microbiol. *62*: 4641–4647.

Schroeter, J. 1886. Schizomycetes. *In* Cohn (Editor), Kryptogamenflora von Schlesien, Bd. 3, Heft 3, Pilze, J.U. Kern's Verlag, Breslau. pp. 1–814.

Schumann, J., A. Wrba, R. Jaenicke and K.O. Stetter. 1991. Topographical and enzymatic characterization of amylases from the extremely thermophilic eubacterium *Thermotoga maritima*. FEBS Lett. *282*: 122–126.

Schwabe, G.H. 1960a. Über den thermobionten Kosmopoliten *Mastigocladus laminosus* Cohn. Z. Hydrol. *22*: 759–792.

Schwabe, G.H. 1960b. Zur autotrophen Vegetation in ariden Böden. Blaualgen und Lebensraum IV. Österr. Bot. Z. *107*: 281–309.

Schwabe, G.H. 1967. *Microchaete tenera* Thur. als Vertreter der pleistomorphen Schicht des Cyanophytenstamms. Nova Hedwigia *13*: 423–448.

Schwörer, B. and R.K. Thauer. 1991. Activities of formylmethanofuran dehydrogenase, methylenetetrahydromethanopterin dehydrogenase, methylenetetrahydromethanopterin reductase, and heterodisulfide reductase in methanogenic bacteria. Arch. Microbiol. *155*: 459–465.

Schyns, G., R. Rippka, A. Namane, D. Campbell, M. Herdman and J. Houmard. 1997. *Prochlorothrix hollandica* PCC 9006: genomic properties of an axenic representative of the chlorophyll *a*/*b*-containing oxyphotobacteria. Res. Microbiol. *148*: 345–354.

Scott, O.T., R.W. Castenholz and H.T. Bonnet. 1984. Evidence for a peptidoglycan envelope in the cyanelles of *Glaucocystis nostochinearum* Itzigsohn. Arch. Microbiol. *139*: 130–138.

Searcy, D.G. 1976. *Thermoplasma acidophilum*: intracellular pH and potassium concentration. Biochim. Biophys. Acta *451*: 278–286.

Searcy, D.G. and E.K. Doyle. 1975. Characterization of *Thermoplasma acidophilum* deoxyribonucleic acid. Int. J. Syst. Bacteriol. *25*: 286–289.

Searcy, K.B. and D.G. Searcy. 1981. Superoxide dismutase from the archaebacterium *Thermoplasma acidophilum*. Biochim. Biophys. Acta *670*: 39–46.

Seeler, J.-S. and S. Golubic. 1991. *Iyengariella endolithica* sp. nova, a carbonate boring stigonematalean cyanobacterium from a warm spring-fed lake: nature to culture. Algol. Stud. *64*: 399–410.

Seenayya, G. and N. Subba-Raju. 1972. On the ecology and systematic position of the alga known as *Anabaenopsis raciborskii* (Wolosz.) Elenk. and a critical evaluation of the forms described under the gunus *Anabaenopsis. In* Desikachary (Editor), Taxonomy and Biology of Blue-green Algae, University of Madras, Madras. pp. 52–57.

Segerer, A., T.A. Langworthy and K.O. Stetter. 1988a. *Thermoplasma acidophilum* and *Thermoplasma volcanium,* new species from solfatara fields. Syst. Appl. Microbiol. *10*: 161–171.

Segerer, A., T.A. Langworthy and K.O. Stetter. 1988b. *In* Validation of the publication of new names and new combinations previously effectively published outside the IJSB. List No. 26. Int. J. Syst. Bacteriol. *38*: 328–329.

Segerer, A., A. Neuner, J.K. Kristjansson and K.O. Stetter. 1986. *Acidianus infernus* gen. nov., sp. nov., and *Acidianus brierleyi* comb. nov.: facultatively aerobic, extremely acidophilic thermophilic sulfur-metabolizing archaebacteria. Int. J. Syst. Bacteriol. *36*: 559–564.

Segerer, A. and K.O. Stetter. 1989a. Genus I. *Sulfolobus. In* Staley, Bryant, Pfennig and Holt (Editors), Bergey's Manual of Systematic Bacteriology, 1st Ed., Vol. 3, The Williams & Wilkins Co., Baltimore. pp. 2250–2251.

Segerer, A. and K.O. Stetter. 1989b. Genus II. *Acidianus. In* Staley, Bryant, Pfennig and Holt (Editors), Bergey's Manual of Systematic Bacteriology, 1st Ed., Vol. 3, The Williams & Wilkins Co., Baltimore. pp. 2251–2253.

Segerer, A.H., A. Trincone, M. Gahrtz and K.O. Stetter. 1991. *Stygiolobus azoricus* gen. nov., sp. nov., represents a novel genus of anaerobic, extremely thermoacidophilic archaebacteria of the order *Sulfolobales.* Int. J. Syst. Bacteriol. *41*: 495–501.

Sehgal, S.N. and N.E. Gibbons. 1960. Effect of some metal ions on the growth of *Halobacterium cutirubrum.* Can. J. Microbiol. *6*: 165–169.

Selig, M. and P. Schönheit. 1994. Oxidation of organic compounds to CO_2 with sulfur or thiosulfate as electron acceptor in the anaerobic hyperthermophilic archaea *Thermoproteus tenax* and *Pyrobaculum islandicum* proceeds via the citric acid cycle. Arch. Microbiol. *162*: 286–294.

Selig, M., K.B. Xavier, H. Santos and P. Schonheit. 1997. Comparative analysis of Embden-Meyerhof and Entner-Doudoroff glycolytic pathways in hyperthermophilic archaea and the bacterium *Thermotoga.* Arch. Microbiol. *167*: 217–232.

Serebryakova, L.T., N.A. Zorin, I.N. Gogotov and O.I. Keppen. 1989. Hydrogenase activity of the thermophilic green bacterium *Chloroflexus aurantiacus.* Mikrobiologiya *58*: 539–543.

Setchell, W.A. and N.L. Gardner. 1918. *Symploca funicularis. In* Gardner (Editor), New Pacific Coast Algae III, University of California Press, Berkeley. p. 469.

Shand, R.F. and A.M. Perez. 1999. Haloarchaeal growth physiology. *In* Seckbach (Editor), Enigmatic Microorganisms and Life in Extreme Environments, Kluwer Academic Publishing, Dordrecht. 411–424.

Shand, R.F. , L.B. Price and E.M. O'Connor. 1999. Halocins: protein antibiotics from hypersaline environments. *In* Oren (Editor), Microbiology and Biogeochemistry of Hypersaline Environments, CRC Press, Boca Raton. pp. 295–306.

Shao, Z., W. Mages and R. Schmitt. 1994. A physical map of the hyperthermophilic bacterium *Aquifex pyrophilus* chromosome. J. Bacteriol. *176*: 6776–6780.

Shapiro, A., D. DiLello, M.C. Loudis, D.E. Keller and S.H. Hutner. 1977. Minimal requirements in defined media for improved growth of some radio-resistant pink tetracocci. Appl. Environ. Microbiol. *33*: 1129–1133.

Shapiro, J. 1997. The role of carbon dioxide in the initiation and maintenance of blue-green dominance in lakes. Freshw. Biol. *37*: 307–323.

Sharak Genthner, B.R., C.L. Davis and M.P. Bryant. 1981. Features of rumen and sewage sludge strains of *Eubacterium limosum,* a methanol- and H_2-CO_2-utilizing species. Appl. Environ. Microbiol. *42*: 12–19.

Sharp, R., D. Cossar and R. Williams. 1995. Physiology and metabolism of *Thermus. In* Sharp and Williams (Editors), *Thermus* Species, Plenum Press, New York. pp. 67–91.

Sharp, R.J. and R.A.D. Williams. 1988. Properties of *Thermus ruber* strains isolated from Icelandic hot springs and DNA:DNA homology of *Thermus ruber* and *Thermus aquaticus.* Appl. Environ. Microbiol. *54*: 2049–2053.

Sherman, L.A. and M. Connelly. 1976. Isolation and characterization of a cyanophage infecting the unicellular blue-green algae *A. nidulans* and *S. cedrorum.* Virology *72*: 540–544.

Shiba, T. 1991. *Roseobacter litoralis* gen. nov., sp. nov., and *Roseobacter denitrificans* sp. nov., aerobic pink-pigmented bacteria which contain bacteriochlorophyll *a.* Syst. Appl. Microbiol. *14*: 140–145.

Shiba, T. and U. Simidu. 1982. *Erythrobacter longus,* gen. nov., sp. nov., an aerobic bacterium which contains bacteriochlorophyll *a.* Int. J. Syst. Bacteriol. *32*: 211–217.

Shiba, T., U. Simidu and N. Taga. 1979. Distribution of aerobic bacteria which contain bacteriochlorophyll *a.* Appl. Environ. Microbiol. *38*: 43–45.

Shima, S., D.A. Herault, A. Berkessel and R.K. Thauer. 1998a. Activation and thermostabilization effects of cyclic 2,3-diphosphoglycerate on enzymes from the hyperthermophilic *Methanopyrus kandleri.* Arch. Microbiol. *170*: 469–472.

Shima, S. and K.-I. Suzuki. 1993. *Hydrogenobacter acidophilus* sp. nov., a thermoacidophilic, aerobic, hydrogen-oxidizing bacterium requiring elemental sulfur for growth. Int. J. Syst. Bacteriol. *43*: 703–708.

Shima, S., C. Tziatzios, D. Schubert, H. Fukada, K. Takahashi, U. Ermler and R.K. Thauer. 1998b. Lyotropic-salt-induced changes in monomer/dimer/tetramer association equilibrium of formyltransferase from the hyperthermophilic *Methanopyrus kandleri* in relation to the activity and thermostability of the enzyme. Eur. J. Biochem. *258*: 85–92.

Shimada, K., H. Hayashi and M. Tasumi. 1985. Bacteriochlorophyll-protein complexes of aerobic bacteria, *Erythrobacter longus* and *Erythrobacter* species OCH 114. Arch. Microbiol. *143*: 244–247.

Shiozawa, J.A. 1995. A foundation for the genetic analysis of green sulfur, green filamentous and heliobacteria. *In* Blankenship, Madigan and Bauer (Editors), Anoxygenic Photosynthetic Bacteria, Kluwer Academic Publishers, Dordrecht. pp. 1159–1173.

Shiozawa, J.A., F. Lottspeichi and R. Feick. 1987. The photochemical reaction center of *Chloroflexus aurantiacus* is composed of two structurally similar polypeptides. Eur. J. Biochem. *167*: 595–600.

Shiozawa, J.A., F. Lottspeich, D. Oesterhelt and R. Feick. 1989. The primary structure of the *Chloroflexus aurantiacus* reaction-center polypeptides. Eur. J. Biochem. *180*: 75–84.

Shporer, M. and M.M. Civan. 1977. Pulsed nuclear magnetic resonance study of ^{39}K within halobacteria. J. Membr. Biol. *33*: 385–400.

Sibley, C.G. and J.E. Ahlquist. 1987. DNA hybridization evidence of hominoid phylogeny: results from an expanded data set. J. Mol. Evol. *26*: 99–121.

Sibley, C.G., J.A. Comstock and J.E. Ahlquist. 1990. DNA hybridization evidence of hominoid phylogeny: a reanalysis of the data. J. Mol. Evol. *30*: 202–236.

Siefert, E. and N. Pfennig. 1984. Convenient method to prepare neutral sulfide solution for cultivation of phototrophic sulfur bacteria. Arch. Microbiol. *139*: 100–101.

Silver, S. 1996. Bacterial resistances to toxic metal ions: a review. Gene *179*: 9–19.

Silvestri, L., M. Turri, L.R. Hill and E. Gilardi. 1962. A quantitative approach to the systematics of *Actinomycetales* based on overall similarity. Microbial Classification, Symp. Soc. Gen. Microbiol., *12*: 333–360.

Simpson, P.G. and W.B. Whitman. 1993. Anabolic pathways in methanogens. *In* Ferry (Editor), Methanogenesis: Ecology, Physiology, Biochemistry, and Genetics, Chapman & Hall, New York. pp. 445–472.

Sinclair, C. and B.A. Whitton. 1977a. Influence of nitrogen source on morphology of *Rivulariaceae* (Cyanophyta). J. Phycol. *13*: 335–340.

Sinclair, C. and B.A. Whitton. 1977b. Influence of nutrient deficiency on hair formation in the *Rivulariaceae.* Br. Phycol. *12*: 297–313.

Singh, P.K. 1973. Nitrogen fixation by the unicellular blue-green alga *Aphanothece.* Arch. Mikrobiol. *92*: 59–62.

Sirevåg, R. 1995. Carbon metabolism in green bacteria. *In* Blankenship, Madigan and Bauer (Editors), Anoxygenic Photosynthetic Bacteria, Kluwer Academic Publishers, Dordrecht. pp. 871–883.

Sirevåg, R. and R.W. Castenholz. 1979. Aspects of carbon metabolism in *Chloroflexus.* Arch. Microbiol. *120*: 151–153.

Sirevåg, R. and J.G. Ormerod. 1977. Synthesis, storage and degradation of polyglucose in *Chlorobium thiosulfatophilum.* Arch. Microbiol. *111*: 239–244.

Sivonen, K., K. Kononen, W.W. Carmichael, A.M. Dahlem, K.L. Rinehart, J. Kiviranta and S.I. Niemelä. 1989a. Occurrence of the hepatotoxic cyanobacterium *Nodularia spumigena* in the Baltic Sea and structure of the toxin. Appl. Environ. Microbiol. *55*: 1990–1995.

Sivonen, K., K. Kononen, A.L. Esala and S.I. Niemelä. 1989b. Toxicity and isolation of the cyanobacterium *Nodularia spumigena* from the southern Baltic Sea in 1986. Hydrobiologia *185*: 3–8.

Skerman, V.B.D. 1967. A Guide for the Identification of the Genera of Bacteria, 2nd Ed., The Williams & Wilkins Co., Baltimore.

Skerman, V.B.D., V. McGowan and P.H.A. Sneath. 1980. Approved lists of bacterial names. Int. J. Syst. Bacteriol. *30*: 225–420.

Skinner, F.E., R.C.T. Jones and J.E.A. Mollison. 1952. Comparison of a direct- and a plate-counting technique for the quantitative estimation of soil microorganisms. J. Gen. Microbiol. *32*: 261–271.

Skuja, H. 1956. Taxonomische und biologische Studien über das Phytoplankton schwedischer Binnengewässer. Nova Acta Reg. Soc. Sci. Ups. Ser. IV *16*: 1–404.

Skulberg, O.M., G.A. Codd and W.W. Carmichael. 1984. Toxic blue-green algal blooms in Europe: a growing problem. Ambio. *13*: 224–247.

Skulberg, O.M. and R. Skulberg. 1985. Planktic species of *Oscillatoria* (*Cyanophyceae*) from Norway: characterization and classification. Algol. Stud. *38/39*: 157–174.

Slesarev, A.I., J.A. Lake, K.O. Stetter, M. Gellert and S.A. Kozyavkin. 1994. Purification and characterization of DNA topoisomerase V. An enzyme from the hyperthermophilic prokaryote *Methanopyrus kandleri* that resembles eukaryotic topoisomerase I. J. Biol. Chem. *269*: 3295–3303.

Slesarev, A.I., K.O. Stetter, J.A. Lake, M. Gellert, R. Krah and S.A. Kozyavkin. 1993. DNA topoisomerase V is a relative of eukaryotic topoisomerase I from a hyperthermophilic prokaryote. Nature *364*: 735–737.

Sleytr, U.B. and A.M. Glauert. 1982. Bacterial cell walls and membranes. *In* Harris (Editor), Electron Microscopy of Proteins, Vol. 3, Academic Press, Inc., London. pp. 41–76.

Sleytr, U.B. and P. Messner. 1983. Crystalline surface layers on bacteria. Annu. Rev. Microbiol. *37*: 311–339.

Sly, L.I., M. Taghavi and M. Fegan. 1998. Phylogenetic heterogeniety within the genus *Herpetosiphon*: transfer of the marine species *Herpetosiphon cohaerens, Herpetosiphon nigricans* and *Herpetosiphon persicus* to the genus *Lewinella* gen. nov. in the *Flexibacter-Bacteroides-Cytophaga* phylum. Int. J. Syst. Bacteriol. *48*: 731–737.

Smibert, R.M. and N.R. Krieg. 1995. Phenotypic characterization. *In* Gerhardt, Murray, Wood and Krieg (Editors), Methods for General and Molecular Bacteriology, American Society for Microbiology, Washington, D.C. pp. 607–654.

Smith, E., P. Leeflang and K. Wernars. 1997. Detection of shifts in microbial community structure and diversity in soil caused by copper contamination using amplified ribosomal DNA restriction analysis. FEMS Microbiol. Ecol. *23*: 249–261.

Smith, K.M. and F.W. Bobe. 1987. Light adaptation of bacteriochlorophyll-d producing bacteria by enzymatic methylation of their antenna pigments. J. Chem. Soc. Chem. Commun. *1987*: 276–277.

Smith, M.R. and R.A. Mah. 1978. Growth and methanogenesis by *Methanosarcina* strain 227 on acetate and methanol. Appl. Environ. Microbiol. *36*: 870–879.

Smith, P.F. 1980. Sequence and glycoside bond arrangement of sugars in lipopolysaccharide from *Thermoplasma acidophilum.* Biochim. Biophys. Acta *619*: 367–373.

Smith, P.F., T.A. Langworthy , W.R. Mayberry and A.E. Hougland. 1973. Characterization of the membranes of *Thermoplasma acidophilum.* J. Bacteriol. *116*: 1019–1028.

Smith, P.F., T.A. Langworthy and M.R. Smith. 1975. Polypeptide nature of growth requirement in yeast extract for *Thermoplasma acidophilum.* J. Bacteriol. *124*: 884–892.

Smith, P.H. 1961. Studies on the methanogenic bacteria of domestic sewage sludge. Bacteriological Proceedings, p. 60.

Smith, P.H. 1966. The microbial ecology of sludge methanogenesis. Dev. Ind. Microbiol. *7*: 155–161.

Smith, P.H. and R.E. Hungate. 1958. Isolation and characterization of *Methanobacterium ruminantium* n. sp. J. Bacteriol. *75*: 713–718.

Snaidr, J., R.I. Amann, I. Huber, W. Ludwig and K.H. Schleifer. 1997. Phylogenetic analysis and *in situ* identification of bacteria in activated sludge. Appl. Environ. Microbiol. *63*: 2884–2896.

Snaidr, J., B.M. Fuchs, G. Wallner, M. Wagner and K.H. Schleifer. 1999. Phylogeny and *in situ* identification of a morphologically conspicuous bacterium, *Candidatus Magnospira bakii*, present in very low frequency in activated sludge. Environ. Microbiol. *1*: 125–136.

Sneath, P.H.A. 1972. Computer taxonomy. *In* Norris and Ribbons (Editors), Methods in Microbiology, Academic Press, London. 29–98.

Sneath, P.H.A. 1974. Phylogeny of microorganisms. Symp. Soc. Gen. Microbiol. *24*: 1–39.

Sneath, P.H.A. 1977. A method for testing the distinctness of clusters: a test for the disjunction of two clusters in euclidean space as measured by their overlap. J. Int. Assoc. Math. Geol. *9*: 123–143.

Sneath, P.H.A. 1978a. Classification of microorganisms. *In* Norris and Richmond (Editors), Essays in Microbiology, John Wiley, Chichester. 9/1–9/31.

Sneath, P.H.A. 1978b. Identification of microorganisms. *In* Norris and Richmond (Editors), Essays in Microbiology, John Wiley, Chichester. 10/1–10/32.

Sneath, P.H.A. 1979a. BASIC program for a significance test for clusters in UPGMA dendrograms obtained from square euclidean distances. Comput. Geosci. *5*: 127–137.

Sneath, P.H.A. 1979b. BASIC program for a significance test for two clusters in euclidean space as measured by their overlap. Comput. Geosci. *5*: 143–155.

Sneath, P.H.A. 1992. International Code of Nomenclature of Bacteria (1990 Revision), American Society for Microbiology, Washington, DC.

Sneath, P.H.A., N.S. Mair, M.E. Sharpe and J.G. Holt (Editors). 1986. Bergey's Manual of Systematic Bacteriology, 1st Ed., Vol. 2, The Williams & Wilkins Co., Baltimore.

Sneath, P.H.A. and R.R. Sokal. 1973. Numerical Taxonomy. The Principles and Practice of Numerical Classification, W.H. Freeman, San Francisco.

Sokal, R.R. and P.H.A. Sneath. 1963. Principles of Numerical Taxonomy, W.H. Freeman and Co., San Francisco.

Soliman, G.S.H. and H.G. Trüper. 1982. *Halobacterium pharaonis* sp. nov., a new extremely haloalkaliphilic archaebacterium with a low magnesium requirement. Zentbl. Bakteriol. Mikrobiol. Hyg. 1 Abt Orig. C *3*: 318–329.

Sonawat, H.M., S. Srivastava, S. Swaminathan and G. Govil. 1990. Glycolysis and Entner–Doudoroff pathways in *Halobacterium halobium*: some new observations based on ^{13}C-NMR spectroscopy. Biochem. Biophys. Res. Commun. *173*: 358–362.

Sonne-Hansen, J. and B.K. Ahring. 1997. Anaerobic microbiology of an alkaline Icelandic hot spring. FEMS Microbiol. Ecol. *23*: 31–38.

Sonne-Hansen, J. and B.K. Ahring. 1999. *Thermodesulfobacterium hveragerdense* sp. nov., and *Thermodesulfovibrio islandicus* sp. nov., two thermophilic sulfate reducing bacteria isolated form a Icelandic hot spring. Syst. Appl. Microbiol. *22*: 559–564.

Soppa, J. and D. Oesterhelt. 1989. *Halobacterium* sp. GRB: a species to work with!? Can. J. Microbiol. *35*: 205–209.

Sørheim, R., V.L. Torsvik and J. Goksøyr. 1989. Phenotypical divergences between populations of soil bacteria isolated on different media. Microb. Ecol. *17*: 181–192.

Soriano, S. 1973. Flexibacteria. Annu. Rev. Microbiol. *27*: 155–170.

Souillard, N. and L. Sibold. 1986. Primary structure and expression of a gene homologous to *nif*H (nitrogenase Fe protein) from the archaebacterium *Methanococcus voltae*. Mol. Gen. Genet. *203*: 21–28.

Soulimane, T., M. Von Walter, P. Hof, M.E. Than, R. Huber and G. Buse. 1997. Cytochrome-c_{552} from *Thermus thermophilus*: a functional and crystallographic investigation. Biochem. Biophys. Res. Commun. *237*: 572–576.

Southam, G. and T.J. Beveridge. 1991. Dissolution and immunochemical analysis of the sheath of the archaeobacterium *Methanospirillum hungatei* GP1. J. Bacteriol. *173*: 6213–6222.

Southam, G. and T.J. Beveridge. 1992a. Characterization of novel, phenol-soluble polypeptides which confer rigidity to the sheath of *Methanospirillum hungatei* GP1. J. Bacteriol. *174*: 935–946.

Southam, G. and T.J. Beveridge. 1992b. Detection of growth sites in the protomer pools for the sheath of *Methanospirillum hungatei* GP1 by use of constituent organosulfur and immunogold labeling. J. Bacteriol. *174*: 6460–6470.

Southam, G., M. Firtel, B.L. Blackford, M.H. Jericho, W. Xu, P.J. Mulhern and T.J. Beveridge. 1993. Transmission electron microscopy, scanning tunneling microscopy, and atomic force microscopy of the cell envelope layers of the archaeobacterium *Methanospirillum hungatei* GP1. J. Bacteriol. *175*: 1946–1955.

Southam, G., M.L. Kalmokoff, K.F. Jarrell, S.F. Koval and T.J. Beveridge. 1990. Isolation, characterization, and cellular insertion of the flagella from 2 strains of the archaebacterium *Methanospirillum hungatei*. J. Bacteriol. *172*: 3221–3228.

Sowers, K.R., S.F. Baron and J.G. Ferry. 1984a. *Methanosarcina acetivorans*, sp. nov., an acetotrophic methane-producing bacterium isolated from marine sediments. Appl. Environ. Microbiol. *47*: 971–978.

Sowers, K.R., S.F. Baron and J.G. Ferry. 1986. *In* Validation of the publication of new names and new combinations previously effectively published outside the IJSB. List No. 20. Int. J. Syst. Bacteriol. *36*: 354–356.

Sowers, K.R., J.E. Boone and R.P. Gunsalus. 1993. Disaggregation of *Methanosarcina* spp. and growth as single cells at elevated osmolarity. Appl. Environ. Microbiol. *59*: 3832–3839.

Sowers, K.R. and J.G. Ferry. 1983. Isolation and characterization of a methylotrophic marine methanogen, *Methanococcoides methylutens*, gen. nov., sp. nov. Appl. Environ. Microbiol. *45*: 684–690.

Sowers, K.R. and J.G. Ferry. 1985a. Trace metal and vitamin requirements of *Methanococcoides methylutens* grown with trimethylamine. Arch. Microbiol. *142*: 148–151.

Sowers, K.R. and J.G. Ferry. 1985b. *In* Validation of the publication of new names and new combinations previously effectively published outside the IJSB. List No. 17. Int. J. Syst. Bacteriol. *35*: 223–225.

Sowers, K.R. and R.P. Gunsalus. 1988a. Adaptation for growth at various saline concentrations by the archaebacterium *Methanosarcina thermophila*. J. Bacteriol. *170*: 998–1002.

Sowers, K.R. and R.P. Gunsalus. 1988b. Plasmid DNA from the acetotrophic methanogen *Methanosarcina acetivorans*. J. Bacteriol. *170*: 4979–4982.

Sowers, K.R. and R.P. Gunsalus. 1995. Halotolerance in *Methanosarcina* spp.: role of N^ε-acetyl-β-lysine, α-glutamate, glycine, betaine, and K^+ as compatible solutes for osmotic adaptation. Appl. Environ. Microbiol. *61*: 4382–4388.

Sowers, K.R., J.L. Johnson and J.G. Ferry. 1984b. Phylogenetic relationships among the methylotrophic methane-producing bacteria and emendation of the family *Methanosarcinaceae*. Int. J. Syst. Bacteriol. *34*: 444–450.

Sowers, K.R. and K.M. Noll. 1995. Techniques for anaerobic growth. *In* DasSarma and Fleischmann (Editors), Archaea: A Laboratory Manual, Cold Spring Harbor Laboratory Press, Plainview. pp. 15–47.

Sowers, K.R., D.E. Robertson, D. Noll, R.P. Gunsalus and M.F. Roberts. 1990. N^ε-acetyl-β-lysine: an osmolyte synthesized by methanogenic archaebacteria. Proc. Natl. Acad. Sci. U.S.A. *87*: 9083–9087.

Sparling, R., M. Blaut and G. Gottschalk. 1993a. Bioenergetic studies of *Methanosphaera stadtmanae*, an obligate hydrogen–methanol utilising methanogen. Can. J. Microbiol. *39*: 742–748.

Sparling, R., L.T. Holth and Z. Lin. 1993b. Sodium ion dependent active transport of leucine in *Methanosphaera stadtmanae*. Can. J. Microbiol. *39*: 749–753.

Spieck, E., S. Ehrich, J. Aamand and E. Bock. 1998. Isolation and immunocytochemical location of the nitrite-oxidizing system in *Nitrospira moscoviensis*. Arch. Microbiol. *169*: 225–230.

Spring, S., R.I. Amann, W. Ludwig, K.H. Schleifer, H. van Gemerden and N. Petersen. 1993. Dominating role of an unusual magnetotactic bacterium in the microaerobic zone of a freshwater sediment. Appl. Environ. Microbiol. *59*: 2397–2403.

Springer, E., M.S. Sachs, C.R. Woese and D.R. Boone. 1995. Partial gene sequences for the A subunit of methyl-coenzyme M reductase (*mcr*I) as a phylogenetic tool for the family *Methanosarcinaceae*. Int. J. Syst. Bacteriol. *45*: 554–559.

Sprott, G.D., T.J. Beveridge, G.B. Patel and G. Ferrrante. 1986. Sheath disassembly in *Methanospirillum hungatei* strain GP1. Can. J. Microbiol. *32*: 847–854.

Sprott, G.D., J.R. Colvin and R.C. McKellar. 1979. Spheroplasts of *Methanospirillum hungatii* formed upon treatment with dithiothreitol. Can. J. Microbiol. *25*: 730–738.

Sprott, G.D., I. Ekiel and G.B. Patel. 1993. Metabolic pathways in *Methanococcus jannaschii* and other methanogenic bacteria. Appl. Environ. Microbiol. *59*: 1092–1098.

Sprott, G.D. and K.F. Jarrell. 1982. Sensitivity of methanogenic bacteria to dicyclohexylcarbodiimide. Can. J. Microbiol. *28*: 982–986.

Sprott, G.D. and R.C. McKellar. 1980. Composition and properties of the cell wall of *Methanospirillum hungatii*. Can. J. Microbiol. *26*: 115–120.

Sprott, G.D., M. Meloche and J.C. Richards. 1991. Proportions of diether, macrocyclic diether, and tetraether lipids in *Methanococcus jannaschii* grown at different temperatures. J. Bacteriol. *173*: 3907–3910.

Sprott, G.D. and G.B. Patel. 1986. Ammonia toxicity in pure cultures of methanogenic bacteria. Syst. Appl. Microbiol. *7*: 358–363.

Sprott, G.D., K.M. Shaw and K.F. Jarrell. 1983. Isolation and chemical composition of the cytoplasmic membrane of the archaebacterium *Methanospirillum hungatei*. J. Biol. Chem. *258*: 4026–4031.

Spudich, J.L., R.A. Zacks and R.A. Bogomolni. 1995. Microbial sensory rhodopsins: photochemistry and function. Isr. J. Chem. *35*: 495–513.

Stackebrandt, E., V.J. Fowler, W. Schubert and J.F. Imhoff. 1984. Toward a phylogeny of phototrophic purple sulfur bacteria: the genus *Ectothiorhodospira*. Arch. Microbiol. *137*: 366–370.

Stackebrandt, E. and B.M. Goebel. 1994. Taxonomic note: a place for DNA–DNA reassociation and 16S rRNA sequence analysis in the present species definition in bacteriology. Int. J. Syst. Bacteriol. *44*: 846–849.

Stackebrandt, E., R.G.E. Murray and H.G. Trüper. 1988. *Proteobacteria* classis nov., a name for the phylogenetic taxon that includes the "purple bacteria and their relatives". Int. J. Syst. Bacteriol. *38*: 321–325.

Stackebrandt, E., F.A. Rainey and N. Ward-Rainey. 1996. Anoxygenic phototrophy across the phylogenetic spectrum: current understanding and future perspectives. Arch. Microbiol. *166*: 211–223.

Stackebrandt, E., F.A. Rainey and N.L. Ward-Rainey. 1997. Proposal for a new hierarchic classification system, *Actinobacteria* classis nov. Int. J. Syst. Bacteriol. *47*: 479–491.

Stackebrandt, E. and C.R. Woese. 1981. The evolution of the prokaryotes. *In* Carlile, Collins and Moseley (Editors), Molecular and Cellular Aspects of Microbial Evolution, Cambridge University Press, Cambridge. pp. 1–31.

Stadtman, T.C. and H.A. Barker. 1951a. Studies on the methane fermentation. IX. The origin of methane in the acetate and methanol fermentations by *Methanosarcina*. J. Bacteriol. *61*: 81–86.

Stadtman, T.C. and H.A. Barker. 1951b. Studies on the methane fermentation. X. A new formate-decomposing bacterium, *Methanococcus vannielii*. J. Bacteriol. *62*: 269–280.

Staehelin, L.A., J.R. Golecki, R.C. Fuller and G. Drews. 1978. Visualization of the supramolecular architecture of chlorosomes (*Chlorobium*-type vesicles) in freeze-fractured cells of *Chloroflexus aurantiacus*. Arch. Microbiol. *119*: 269–277.

Stahl, D.A. 1986. Evolution, ecology, and diagnosis: unity in variety. Bio/Technol. *4*: 623–628.

Stahl, D.A. and R.I. Amann. 1991. Development and application of nucleic acid probes in bacterial systematics. *In* Stackebrandt and Goodfellow (Editors), Nucleic Acid Techniques in Bacterial Systematics, John Wiley & Sons, Chichester. pp. 205–248.

Stahl, D.A., B. Flesher, H.R. Mansfield and L. Montgomery. 1988. Use of phylogenetically based hybridization probes for studies of ruminal microbial ecology. Appl. Environ. Microbiol. *54*: 1079–1084.

Stal, L. 1995. Physiological ecology of cyanobacteria in microbial mats and other communities. New Phytol. *131*: 1–32.

Staley, J.T. 1997. Biodiversity: are microbial species threatened? Curr. Opin. Biotechnol. *8*: 340–345.

Staley, J.T. 1999. Bacterial biodiversity: a time for place. ASM News *65*: 681–687.

Staley, J.T., M.P. Bryant, N. Pfennig and J.G. Holt (Editors). 1989. Bergey's Manual of Systematic Bacteriology, 1st Ed., Vol. 3, The Williams & Wilkins Co., Baltimore.

Staley, J.T. and A. Konopka. 1985. Measurement of *in situ* activities of nonphotosynthetic microorganisms in aquatic and terrestrial habitats. Annu. Rev. Microbiol. *39*: 321–346.

Stallings, W.C., K.A. Pattridge, R.K. Strong and M.L. Ludwig. 1985. The structure of manganese superoxide dismutase from *Thermus thermophilus* HB8 at 2.4-Å resolution. J. Biol. Chem. *260*: 16424–16432.

Stam, W.T. 1980. Relationships between a number of filamentous blue-green algal strains (Cyanophyceae) revealed by DNA–DNA hybridization. Arch. Hydrobiol. Suppl. *56*: 351–374.

Stam, W.T. and H.C. Holleman. 1979. Cultures of *Phormidium, Plectonema, Lyngbya* and *Synechococcus* (Cyanophyceae) under different conditions: their growth and morphological variability. Acta Bot. Neerl. *28*: 45–66.

Stam, W.T. and G. Venema. 1977. The use of DNA–DNA hybridization for determination of the relationship between some blue-green algae (Cyanophyceae). Acta Bot. Neerl. *26*: 327–342.

Stanier, R.Y. and G. Cohen-Bazire. 1977. Phototrophic prokaryotes: The cyanobacteria. Annu. Rev. Microbiol. *31*: 225–274.

Stanier, R.Y., R. Kunisawa, M. Mandel and G. Cohen-Bazire. 1971. Purification and properties of unicellular blue-green algae (order *Chroococcales*). Bacteriol. Rev. *35*: 171–205.

Stanier, R.Y. and C.B. Van Niel. 1962. The concept of a bacterium. Arch. Mikrobiol. *42*: 17–35.

Starr, R.C. 1978. The culture collection of algae at the University of Texas at Austin. J. Phycol. Suppl. *14*: 47–100.

Steenbergen, C.L.M. and H.J. Korthals. 1982. Distribution of phototrophic microorganisms in the anaerobic and microaerophilic strata of Lake Vechten (The Netherlands): pigment analysis and role in primary production. Limnol. Oceanogr. *27*: 883–895.

Steenbergen, C.L.M., H.J. Korthals, A.L. Baker and C.J. Watras. 1989. Microscale vertical distribution of algal and bacterial plankton in Lake Vechten (The Netherlands). FEMS Microbiol. Ecol. *62*: 209–220.

Steensland, H. and H. Larsen. 1969. A study of the cell envelope of the halobacteria. J. Gen. Microbiol. *55*: 325–336.

Steinberger, B., N. Petersen, H. Petermann and D.G. Weiss. 1994. Movement of magnetic bacteria in time-varying magnetic fields. J. Fluid Mech. *273*: 189–211.

Steinmetz, M.A. and U. Fischer. 1982. Cytochromes of green sulfur bacterium *Chlorobium vibrioforme* f. sp. *thiosulfatophilum*. Purification, characterization and sulfur metabolism. Arch. Microbiol. *131*: 19–26.

Stetter, K.O. 1982a. Ultrathin mycelia-forming organisms from submarine volcanic areas having an optimum growth temperature of 105°C. Nature *300*: 258–260.

Stetter, K.O. 1982b. *In* Validation of the publication of new names and new combinations previously effectively published outside the IJSB. List No. 8. Int. J. Syst. Bacteriol. *32*: 266–268.

Stetter, K.O. 1986a. Diversity of extremely thermophilic archaebacteria. *In* Brock (Editor), Thermophiles: General, Molecular and Applied Microbiology, John Wiley & Sons, New York. pp. 39–74.

Stetter, K.O. 1986b. *In* Validation of the publication of new names and

new combinations previously effectively published outside the IJSB. List No. 22. Int. J. Syst. Bacteriol. *36*: 573–576.

Stetter, K.O. 1988a. *Archaeoglobus fulgidus* gen. nov., sp. nov.: a new taxon of extremely thermophilic archaebacteria. Syst. Appl. Microbiol. *10*: 172–173.

Stetter, K.O. 1988b. *In* Validation of the publication of new names and new combinations previously effectively published outside the IJSB. List No. 26. Int. J. Syst. Bacteriol. *38*: 328–329.

Stetter, K.O. 1989a. Genus II. *Methanolobus*. *In* Staley, Bryant, Pfennig and Holt (Editors), Bergey's Manual of Systematic Bacteriology, 1st Ed., Vol. 3, The Williams & Wilkins Co., Baltimore. pp. 2205–2297.

Stetter, K.O. 1989b. Group II. Archaeobacterial sulfate reducers. Order *"Archaeoglobales"*. *In* Staley, Bryant, Pfennig and Holt (Editors), Bergey's Manual of Systematic Bacteriology, 1st Ed., Vol. 3, The Williams & Wilkins Co., Baltimore. p. 2216.

Stetter, K.O. 1989c. Order III. *Sulfolobales*. *In* Staley, Bryant, Pfennig and Holt (Editors), Bergey's Manual of Systematic Bacteriology, 1st Ed., Vol. 3, The Williams and Wilkins Co., Baltimore. pp. 2250.

Stetter, K.O. 1989d. *In* Validation of the publication of new names and new combinations previously effectively published outside the IJSB. List No. 31. Int. J. Syst. Bacteriol. *39*: 495–497.

Stetter, K.O. 1992. The genus *Archaeoglobus*. *In* Balows, Trüper, Dworkin, Harder and Schleifer (Editors), The Prokaryotes. A Handbook of Bacteria: Ecophysiology, Isolation, Identification, Applications, 2nd Ed., Vol. 1, Springer-Verlag, Berlin. pp. 707–711.

Stetter, K.O. 1995. Microbial life in hyperthermal environments. ASM News *61*: 285–290.

Stetter, K.O. and G. Fiala. 1986. *In* Validation of the publication of new names and new combinations previously effectively published outside the IJSB. List No. 22. Int. J. Syst. Bacteriol. *36*: 573–576.

Stetter, K.O. and G. Gaag. 1983. Reduction of molecular sulphur by methanogenic bacteria. Nature *305*: 309–311.

Stetter, K.O. and R. Huber. 1986. *In* Validation of the publication of new names and new combinations previously effectively published outside the IJSB. List No. 22. Int. J. Syst. Bacteriol. *36*: 573–576.

Stetter, K.O., R. Huber, E. Blöchl, M. Kurr, R.D. Eden, M. Fiedler, H. Cash and I. Vance. 1993. Hyperthermophilic archaea are thriving in deep North Sea and Alaskan oil reservoirs. Nature *365*: 743–745.

Stetter, K.O., R. Huber and J.K. Kristjansson. 1988. *In* Validation of the publication of new names and new combinations previously effectively published outside the IJSB. List No. 25. Int. J. Syst. Bacteriol. *38*: 220–222.

Stetter, K.O., H. König and E. Stackebrandt. 1983. *Pyrodictium* gen. nov., a new genus of submarine disc-shaped sulphur reducing archaebacteria growing optimally at 105°C. Syst. Appl. Microbiol. *4*: 535–551.

Stetter, K.O., H. König and E. Stackebrandt. 1984. *In* Validation of the publication of new names and new combinations previously effectively published outside the IJSB. List No. 14. Int. J. Syst. Bacteriol. *34*: 270–271.

Stetter, K.O., H. König and M. Thomm. 1989. *In* Validation of the publication of new names and new combinations previously effectively published outside the IJSB. List No. 31. Int. J. Syst. Bacteriol. *39*: 495–497.

Stetter, K.O., G. Lauerer, M. Thomm and A. Neuner. 1987. Isolation of extremely thermophilic sulfate reducers: evidence for a novel branch of archaebacteria. Science *236*: 822–824.

Stetter, K.O., M. Thomm, J. Winter, G. Wildgruber, H. Huber, W. Zillig, D. Janecovic, H. König, P. Palm and S. Wunderl. 1981. *Methanothermus fervidus*, sp. nov., a novel extremely thermophilic methanogen isolated from an icelandic hot spring. Zentbl. Bakteriol. Mikrobiol. Hyg. *C2*: 166–178.

Stetter, K.O., J. Winter and R. Hartlieb. 1980. DNA-dependent RNA polymerase of the archaebacterium *Methanobacterium thermoautotrophicum*. Zentbl. Bakteriol. Mikrobiol. Hyg. 1 Abt Orig. *C1*: 201–214.

Stetter, K.O. and W. Zillig. 1985. *Thermoplasma* and thermophilic sulfur-dependent archaebacteria. *In* Woese and Wolfe (Editors), The Bacteria, Vol. 8, Academic Press, New York. pp. 85–170.

Stettler, R. and T. Leisinger. 1992. Physical map of the *Methanobacterium thermoautotrophicum* Marburg chromosome. J. Bacteriol. *174*: 7227–7234.

Stettler, R., P. Pfister and T. Leisinger. 1994. Characterization of a plasmid carried by *Methanobacterium thermoautotrophicum* ZH3, a methanogen closely related to *Methanobacterium thermoautotrophicum* Marburg. Syst. Appl. Microbiol. *17*: 484–491.

Stettler, R., C. Thurner, D. Stax, L. Meile and T. Leisinger. 1995. Evidence for a defective prophage on the chromosome of *Methanobacterium wolfei*. FEMS Microbiol. Lett. *132*: 85–89.

Stevens, S.E., Jr., C.O.P. Patterson and J. Myers. 1973. The production of hydrogen peroxide by blue-green algae: a survey. J. Phycol. *9*: 427–430.

Stevenson, A.K., L.K. Kimble, C.R. Woese and M.T. Madigan. 1997. Characterization of new phototrophic heliobacteria and their habitats. Photosynth. Res. *53*: 1–11.

Stewart, W.D. 1980. Some aspects of structure and function in N₂-fixing cyanobacteria. Annu. Rev. Microbiol. *34*: 497–536.

Stewart, W.D., A. Haystead and H.W. Pearson. 1969. Nitrogenase activity in heterocysts of blue-green algae. Nature *224*: 226–228.

Stewart, W.D. and M. Lex. 1970. Nitrogenase activity in the blue-green alga *Plectonema boryanum* strain 594. Arch. Mikrobiol. *73*: 250–260.

Stewart, W.D., P. Rowell and A.N. Rai. 1980. Symbiotic nitrogen-fixing cyanobacteria. *In* Stewart and Gallon (Editors), Proceedings of the International Symposium on Nitrogen Fixation, Academic Press, London. pp. 239–277.

Stewart, W.D., P. Rowell and A.N. Rai. 1983. Cyanobacteria-eukaryotic plant symbioses. Ann. Microbiol. (Paris) *134b*: 205–228.

St. Jean, A., B.A. Trieselmann and R.L. Charlebois. 1994. Physical map and set of overlapping cosmid clones representing the genome of the archaeon *Halobacterium* sp. GRB. Nucleic Acids Res. *22*: 1476–1483.

Stoeckenius, W. 1981. Walsby's square bacterium: fine structure of an orthogonal procaryote. J. Bacteriol. *148*: 352–360.

Stoeckenius, W. and W.H. Kunau. 1968. Further characterization of particulate fractions from lysed cell envelopes of *Halobacterium halobium* and isolation of gas vacuole membranes. J. Cell Biol. *38*: 337–357.

Stolt, P. and W. Zillig. 1994. Gene regulation in halophage φH; more than promoters. Syst. Appl. Microbiol. *16*: 591–596.

Stolz, J.F. 1983. Fine structure of the stratified microbial community at Laguna Figueroa, Baja California, Mexico. I. Methods of *in situ* study of the laminated sediments. Precambrian Res. *20*: 479–492.

Stolz, J.F. 1984. Fine structure of the stratified microbial community at Laguna Figueroa, Baja California, Mexico. II. Transmission electron microscopy as a diagnostic tool in studying microbial communities *in situ*. *In* Cohen, Castenholz and Halvorson (Editors), Microbial Mats: Stromatolites, Alan R. Liss, New York. pp. 23–28.

Stolz, J.F. 1990. Distribution of phototrophic microbes in the flat laminated microbial mat at Laguna Figueroa, Baja California, Mexico. Biosystems *23*: 345–358.

Stolz, J.F., D.J. Ellis, J.S. Blum, D. Ahmann, D.R. Lovley and R.S. Oremland. 1999. *Sulfurospirillum barnesii* sp. nov. and *Sulfurospirillum arsenophilum* sp. nov., new members of the *Sulfurospirillum* clade of the ε *Proteobacteria*. Int. J. Syst. Bacteriol. *49*: 1177–1180.

Stramer, S.L. and M.J. Starzyk. 1981. The occurrence and survival of *Thermus aquaticus*. Microbios *32*: 99–110.

Strauss, G., W. Eisenreich, A. Bacher and G. Fuchs. 1992. ¹³-C-NMR study of autotrophic carbon dioxide fixation pathways in the sulfur-reducing archaebacterium *Thermoproteus neutrophilus* and in the phototrophic eubacterium *Chloroflexus aurantiacus*. Eur. J. Biochem. *205*: 853–866.

Strauss, G. and G. Fuchs. 1993. Enzymes of a novel autotrophic CO₂ fixation pathway in the phototrophic bacterium *Chloroflexus aurantiacus*, the 3-hydroxypropionate cycle. Eur. J. Biochem. *215*: 633–643.

Stulp, B.K. 1983. Morphological and molecular approaches to the taxonomy of the genus *Anabaena* (*Cyanophyceae*, Cyanobacteria), Drukkerij van Denderen, Groningen.

Stulp, B.K. and W.T. Stam. 1982. General morphology and akinete germination of *Anabaena* strains (*Cyanophyceae*) in culture. Arch. Hydrobiol. Suppl. *63*: 35–52.

Stulp, B.K. and W.T. Stam. 1984a. Genotypic relationships between strains of *Anabaena* (*Cyanophyceae*) and their correlation with morphological affinities. Br. Phycol. J. *19*: 287–301.

Stulp, B.K. and W.T. Stam. 1984b. Growth and morphology of *Anabaena* strains (*Cyanophyceae*, Cyanobacteria) in cultures under different salinities. Br. Phycol. J. *19*: 281–286.

Stulp, B.K. and W.T. Stam. 1985. Taxonomy of the genus *Anabaena* (*Cyanophyceae*) based on morphological and genotypic criteria. Arch. Hydrobiol. Suppl. *71*: 257–268.

Stupperich, E. and H.J. Eisinger. 1989. Biosynthesis of *p*-cresolyl cobamide in *Sporomusa ovata*. Arch. Microbiol. *151*: 372–377.

Stupperich, E. and G. Fuchs. 1981. Products of CO₂ fixation and ¹⁴C labeling pattern of alanine in *Methanobacterium thermoautotrophicum* pulse-labeled with ¹⁴CO₂. Arch. Microbiol. *130*: 294–300.

Sturm, S., U. Schönefeld, W. Zillig, D. Janekovic and K.O. Stetter. 1980. Structure and function of the DNA dependent RNA polymerase of the archaebacterium *Thermoplasma acidophilum*. Zentbl. Bakteriol. Parasitenkd. Infektkrankh. Hyg. Abt. I Orig. *1*: 12–25.

Sugawara, H., S. Miyazaki, J. Shimura and Y. Ichiyanagi. 1996. Bioinformatics tools for the study of microbial diversity. J. Ind. Microbiol. *17*: 490–497.

Suggs, S.V., T. Hirose, T. Miyake, E.H. Kawashima, M.J. Johnson, K. Itakura and R.B. Wallace. 1981. Use of synthetic oligonucleotides for the isolation of specific cloned DNA sequences. *In* Brown and Fox (Editors), Developmental Biology Using Purified Genes, Academic Press, New York. pp. 683–693.

Sugiyama, Y., N. Yamada and Y. Mukohata. 1994. The light-driven proton pump, cruxrhodopsin-2 in *Haloarcula* sp. arg-2 (bR⁺, hR⁻), and its coupled ATP formation. Biochim. Biophys. Acta *1188*: 287–292.

Sundaram, T.K. and G.P. Bridger. 1979. Regulatory characteristics of phosphoenolpyruvate carboxylase from the extreme thermophile, *Thermus aquaticus*. Biochim. Biophys. Acta *570*: 406–410.

Sutherland, J.M., M. Herdman and W.D.P. Stewart. 1979. Akinetes of the cyanobacterium *Nostoc* PCC 7524: Macromolecular composition, structure and control of differentiation. J. Gen. Microbiol. *115*: 273–287.

Suyama, T., T. Shigematsu, S. Takaichi, Y. Nodasaka, S. Fujikawa, H. Hosoya, Y. Tokowa, T. Kanagawa and S. Hanada. 1999. *Roseateles depolymerans* gen. nov., sp. nov. a new bacteriochlorophyll *a*-containing obligate aerobe belonging to the β-subclass of the *Proteobacteria*. Int. J. Syst. Bacteriol. *49*: 449–457.

Suzuki, M., M.S. Rappe and S.J. Giovannoni. 1998. Kinetic bias in estimates of coastal picoplankton community structure obtained by measurements of small-subunit rRNA gene PCR amplicon length heterogeneity. Appl. Environ. Microbiol. *64*: 4522–4529.

Suzuki, T., Y. Muroga, M. Takahama and Y. Nishimura. 1999a. *Roseivivax halodurans* gen. nov., sp. nov. and *Roseivivax halotolerans* sp. nov., aerobic bacteriochlorophyll-containing bacteria isolated from a saline lake.. Int. J. Syst. Bacteriol. *49*: 629–634.

Suzuki, T., Y. Muroga, M. Takahama, T. Shiba and Y. Nishimura. 1999b. *Rubrimonas cliftonensis* gen. nov., sp. nov., an aerobic bacteriochlorophyll-containing bacterium isolated from a saline lake. Int. J. Syst. Bacteriol. *49*: 201–205.

Swanson, K.L. and K.M. Smith. 1990. Biosynthesis of bacteriochlorophyll-*c* via the glutamate C-5 pathway in *Chloroflexus aurantiacus*. J. Chem. Soc. Chem. Commun. *23*: 1696–1697.

Swanson, R.V., M.G. Sanna and M.I. Simon. 1996. Thermostable chemotaxis proteins from the hyperthermophilic bacterium *Thermotoga maritima*. J. Bacteriol. *178*: 484–489.

Swarthof, T., H.J.M. Kramer and J. Amesz. 1982. Thin-layer chromatography of pigments of the green photosynthetic bacterium *Prosthecochloris aestuarii*. Biochim. Biophys. Acta *681*: 354–358.

Swofford, D.L., G.J. Olsen, P.J. Waddell and D.M. Hillis. 1996. Phylogenetic inference. *In* Hillis, Moritz and Mable (Editors), Molecular Systematics, 2nd Ed., Sinauer Associates, Inc., Sunderland. pp. 407–514.

Synstad, B., O. Emmerhoff and R. Sirevåg. 1996. Malate dehydrogenase from the green gliding bacterium *Chloroflexus aurantiacus* is phylogenetically related to lactic dehydrogenases. Arch. Microbiol. *165*: 346–353.

Szafer, W. 1911. Zur Kenntnis der Schwefelflora im der Umgebung von Lemberg. Bull. Int. Acad. Sci. Ser. V. Cracovie. 160–167.

Szostak, J.W., J.I. Stiles, B.K. Tye, P. Chiu, F. Sherman and R. Wu. 1979. Hybridization with synthetic oligonucleotides. Methods Enzymol. *68*: 419–428.

Tabata, K., S. Ishida, T. Nakahara and T. Hoshino. 1994. A carotenogenic gene cluster exists on a large plasmid in *Thermus thermophilus*. FEBS Lett. *341*: 251–255.

Tabata, K., T. Kosuge, T. Nakahara and T. Hoshino. 1993. Physical map of the extremely thermophilic bacterium *Thermus thermophilus* HB27 chromosome. FEBS Lett. *331*: 81–85.

Taguchi, H., M. Yamashita, H. Matsuzawa and T. Ohta. 1982. Heat-stable and fructose 1,6-bisphosphate-activated L-lactate dehydrogenase (EC 1.1.1.27) from an extremely thermophilic bacterium. J. Biochem. (Tokyo) *91*: 1343–1348.

Takaichi, S., K. Inoue, M. Akaike, M. Kobayashi, H. Oh-oka and M.T. Madigan. 1997. The major carotenoid in all known species of heliobacteria is the C_{30} carotenoid 4,4'-diaponeurosporene, not neurosporene. Arch. Microbiol. *168*: 277–281.

Takaichi, S., K. Tsuji, K. Matsuura and K. Shimada. 1995. A monocyclic carotenoid glucoside ester is a major carotenoid in the green filamentous bacterium *Chloroflexus aurantiacus*. Plant Cell Physiol. *36*: 773–778.

Takashina, T., T. Hamamoto, K. Otozai, W.D. Grant and K. Horikoshi. 1990. *Haloarcula japonica* sp. nov., a new triangular halophilic archaebacterium. Syst. Appl. Microbiol. *13*: 177–181.

Takashina, T., T. Hamamoto, K. Otozai, W.D. Grant and K. Horikoshi. 1991. *In* Validation of the publication of new names and new combinations previously effectively published outside the IJSB. List No. 36. Int. J. Syst. Bacteriol. *41*: 178–179.

Takayanagi, S., H. Kawasaki, K. Sugimori, T. Yamada, A. Sugai, T. Ito, K. Yamasato and M. Shioda. 1996. *Sulfolobus hakonensis* sp. nov., a novel species of acidothermophilic archaeon. Int. J. Syst. Bacteriol. *46*: 377–382.

Tamakoshi, M., M. Uchida, K. Tanabe, S. Fukuyama, A. Yamagishi and T. Oshima. 1997. A new *Thermus-Escherichia coli* shuttle integration vector system. J. Bacteriol. *179*: 4811–4814.

Tanaka, T., N. Kawano and T. Oshima. 1981. Cloning of 3-isopropylmalate dehydrogenase (EC 1.1.1.85) gene of an extreme thermophile and partial purification of the gene product. J. Biochem. (Tokyo) *89*: 677–682.

Tandeau de Marsac, N. 1977. Occurrence and nature of chromatic adaptation in cyanobacteria. J. Bacteriol. *130*: 82–91.

Tandeau de Marsac, N. 1994. Differentiation of hormogonia and relationships with other biological processes. *In* Bryant (Editor), The Molecular Biology of Cyanobacteria, Kluwer Academic Publishers, Dordrecht. pp. 825–842.

Tandeau de Marsac, N. and J. Houmard. 1993. Adaptation of cyanobacteria to environmental stimuli: new steps towards molecular mechanisms. FEMS Microbiol. Rev. *104*: 119–190.

Tang, E.P.Y., R. Tremblay and W.F. Vincent. 1997. Cyanobacterial dominance of polar freshwater ecosystems: are high-latitude mat-formers adapted to low temperature? J. Phycol. *33*: 171–181.

Tanner, J.J., R.M. Hecht and K.L. Krause. 1996. Determinants of enzyme thermostability observed in the molecular structure of *Thermus aquaticus* D-glyceraldehyde-3-phosphate dehydrogenase at 2.5 Å resolution. Biochemistry *35*: 2597–2609.

Tanner, R.S. 1982. Novel compounds from methanogens: characterization of component B of the methylreductase system and mobile factor, Doctoral thesis, University of Illinois at Urbana-Champaign.

Tanner, R.S. and R.S. Wolfe. 1988. Nutritional requirements of *Methanomicrobium mobile*. Appl. Environ. Microbiol. *54*: 625–628.

Tao, T.S., Y.Y. Yue and C.X. Fang. 1996. Irregularities in the validation of the genus *Thermodesulfobacterium* and its species. Request for an opinion. Int. J. Syst. Bacteriol. *46*: 622.

Tarasov, A.L., I.S. Zvyagintseva and A.M. Lysenko. 1987. Nucleotide composition and homology of DNA in new extremely halophilic soil archaebacteria. Mikrobiologiya *56*: 938–942.

Tarasov, A.L., I.S. Zvyagintseva and V.K. Plakunov. 1989. Regulation of succinate utilisation by extreme halophilic archaebacteria. Mikrobiologiya *58*: 15–19.

Tateno, M., K. Ihara and Y. Mukohata. 1994a. The novel ion pump rhodopsins from *Haloarcula* form a family independent from both the bacteriorhodopsin and archaerhodopsin families/tribes. Arch. Biochem. Biophys. *315*: 127–132.

Tateno, Y., N. Takezaki and M. Nei. 1994b. Relative efficiencies of the maximum-likelihood, neighbor joining, and maximum-parsimony methods when substitution rate varies with site. Mol. Biol. Evol. *11*: 261–277.

Tatton, M.J., D.B. Archer, G.E. Powell and M.L. Parker. 1989. Methanogenesis from ethanol by defined mixed continuous cultures. Appl. Environ. Microbiol. *55*: 440–445.

Tatusov, R.L., E.V. Koonin and D.J. Lipman. 1997. A genomic perspective on protein families. Science *278*: 631–637.

Taylor, C.D., B.C. McBride, R.S. Wolfe and M.P. Bryant. 1974. Coenzyme M, essential for growth of a rumen strain of *Methanobacterium ruminantium*. J. Bacteriol. *120*: 974–975.

Taylor, G.T., D.P. Kelly and S.J. Pirt. 1976. Intermediary metabolism in methanogenic bacteria (*Methanobacterium*). *In* Schlegel, Gottschalk and Decker (Editors), Symposium on Microbial Production and Utilization of Gases (H_2, CH_4, and CO), Goltz, Göttingen. pp. 173–180.

Tayne, T.A., J.E. Cutler and D.M. Ward. 1987. Use of *Chloroflexus*-specific antiserum to evaluate filamentous bacteria of a hot spring microbial mat. Appl. Environ. Microbiol. *53*: 1962–1964.

Tchelet, R. and M. Mevarech. 1994. Interspecies genetic transfer in halophilic archaebacteria. Syst. Appl. Microbiol. *16*: 578–581.

Tenreiro, S., M.F. Nobre and M.S. da Costa. 1995a. *Thermus silvanus* sp. nov. and *Thermus chliarophilus* sp. nov., two new species related to *Thermus ruber* but with lower growth temperatures. Int. J. Syst. Bacteriol. *45*: 633–639.

Tenreiro, S., M.F. Nobre, B. Hoste, M. Gillis, J.K. Kristjánsson and M.S. da Costa. 1995b. DNA:DNA hybridization and chemotaxonomic studies of *Thermus scotoductus*. Res. Microbiol. *146*: 315–324.

Tenreiro, S., M.F. Nobre, F.A. Rainey, C. Miguel and M.S. da Costa. 1997. *Thermonema rossianum* sp. nov., a new thermophilic and slightly halophilic species from saline hot springs in Naples, Italy. Int. J. Syst. Bacteriol. *47*: 122–126.

Teske, A., P. Sigalevich, Y. Cohen and G. Muyzer. 1996. Molecular identification of bacteria from a coculture by denaturing gradient gel electrophoresis of 16S ribosomal DNA fragments as a tool for isolation in pure cultures. Appl. Environ. Microbiol. *62*: 4210–4215.

Thauer, R.K., R. Hedderich and R. Fischer. 1993. Reactions and enzymes involved in methanogenesis from CO_2 and H_2. *In* Ferry (Editor), Methanogenesis, Chapman and Hall, New York. pp. 209–252.

Thiery, I., L. Nicholas, R. Rippka and N. Tandeau de Marsac. 1991. Selection of cyanobacteria isolated from mosquito breeding sites as a potential food source for mosquito larvae. Appl. Environ. Microbiol. *57*: 1354–1359.

Thomas, I., D. Verrier, H.C. Dubourguier, N. Hanoune and C. Langrand. 1986. Numerical analysis of whole-cell protein patterns of methanogenesis. *In* Dubourguier, Albagnac, Montreuil, Romond, Sautière and Guillaume (Editors), Biology of Anaerobic Bacteria, Elsevier Science Publishers, Amsterdam. pp. 245–253.

Thomm, M., J. Altenbuchner and K.O. Stetter. 1983. Evidence for a plasmid in a methanogenic bacterium. J. Bacteriol. *153*: 1060–1062.

Thompson, B.G., R. Anderson and R.G. Murray. 1980. Unusual polar lipids of *Micrococcus radiodurans* strain Sark. Can. J. Microbiol. *26*: 1408–1411.

Thompson, B.G. and R.G. Murray. 1981. Isolation and characterization of the plasma membrane and the outer membrane of *Deinococcus radiodurans* strain Sark. Can. J. Microbiol. *27*: 729–734.

Thompson, B.G. and R.G.E. Murray. 1982a. The association of the surface array and the outer membrane of *Deinococcus radiodurans*. Can. J. Microbiol. *28*: 1081–1088.

Thompson, B.G. and R.G.E. Murray. 1982b. The fenestrated peptidoglycan layer of *Deinococcus radiodurans*. Can. J. Microbiol. *28*: 522–525.

Thongthai, C., T.J. McGenity, P. Suntinanalert and W.D. Grant. 1992. Isolation and characterization of an extremely halophilic archaeobacterium from traditionally fermented Thai fish sauce (nam pla). Lett. Appl. Microbiol. *14*: 111–114.

Thornley, M.J. 1975. Cell envelopes with regularly arranged surface subunits in *Acinetobacter* and related bacteria. Crit. Rev. Microbiol. *4*: 65–100.

Thornley, M.J., R.W. Horne and A.M. Glauert. 1965. The fine structure of *Micrococcus radiodurans*. Arch. Mikrobiol. *51*: 267–289.

Thuret, G. 1844. Note sur le mode de reproduction du *Nostoc verrucosum*. Ann. Sci. Nat. Bot. Biol. Vég. *2*: 319–323.

Thuret, G. 1875. Essai de classification des Nostochinées. Ann. Sci. Nat. Bot. VI *1*: 372–382.

Thurston, E.L. and L.O. Ingram. 1971. Morphology and fine structure of *Fischerella ambigua*. J. Phycol. *7*: 203–210.

Tiboni, O., R. Cantoni, R. Creti, P. Cammarano and A.M. Sanangelantoni. 1991. Phylogenetic depth of *Thermotoga maritima* inferred from analysis of the *fus* gene: amino acid sequence of elongation factor G and organization of the *Thermotoga str* operon. J. Mol. Evol. *33*: 142–151.

Tindall, B.J. 1980. Phototrophic bacteria from Kenyan soda lakes, Doctoral thesis, University of Leicester, Leicester, England.

Tindall, B.J. 1985. Qualitative and quantitative distribution of diether lipids in haloalkaliphilic archaebacteria. Syst. Appl. Microbiol. *6*: 243–246.

Tindall, B.J. 1986. The natronobacteria, alkaliphilic members of the aerobic, halophilic archaebacteria. *In* Kandler and Zillig (Editors), Archaebacteria, Gustav Fischer Verlag, Stuttgart. p. 410.

Tindall, B.J. 1988. Prokaryotic life in the alkaline, saline, athalassic environment. *In* Rodriguez-Valera (Editor), Halophilic Bacteria, Vol. 1, CRC Press, Boca Raton, FL. 31–67.

Tindall, B.J. 1989. Fully saturated menaquinones in the archaebacterium *Pyrobaculum islandicum*. FEMS Microbiol. Lett. *60*: 251–254.

Tindall, B.J. 1990. A comparative study of the lipid composition of *Halobacterium saccharovorum* from various sources. Syst. Appl. Microbiol. *13*: 128–130.

Tindall, B.J. 1991. Cultivation and preservation of members of the family *Halobacteriaceae*. World J. Microbiol. Biotechnol. *7*: 95–98.

Tindall, B.J. 1992. The family *Halobacteriaceae*. *In* Balows, Trüper, Dworkin, Harder and Schleifer (Editors), The Prokaryotes. A Handbook of Bacteria: Ecophysiology, Isolation, Identification, Applications, 2nd Ed., Vol. 1, Springer-Verlag, New York. pp. 768–808.

Tindall, B.J., B. Amendt and C. Dahl. 1991a. Variations in the lipid composition of aerobic, halphilic archaeobacteria. *In* Rodriguez-Valera (Editor), General and Applied Aspects of Halophilic Microorganisms, NATO ASI Series A: Life Sciences, Vol. 201, Plenum Press, New York. pp. 199–205.

Tindall, B.J., A.A. Mills and W.D. Grant. 1980. An alkalophilic red halophilic bacterium with a low magnesium requirement from a Kenyan soda lake. J. Gen. Microbiol. *116*: 257–260.

Tindall, B.J., H.N.M. Ross and W.D. Grant. 1984a. *Natronobacterium*, gen. nov. and *Natronococcus*, gen. nov., two new genera of haloalkaliphilic archaebacteria. Syst. Appl. Microbiol. *5*: 41–57.

Tindall, B.J., H.N.M. Ross and W.D. Grant. 1984b. *In* Validation of the publication of new names and new combinations previously effectively published outside the IJSB. List No. 15. Int. J. Syst. Bacteriol. *34*: 355–356.

Tindall, B.J., G.A. Tomlinson and L.I. Hochstein. 1989. Transfer of *Halobacterium denitrificans* (Tomlinson, Jahnke, and Hochstein) to the genus *Haloferax* as *Haloferax denitrificans* comb. nov. Int. J. Syst. Bacteriol. *39*: 359–360.

Tindall, B.J., V. Wray, R. Huber and M.D. Collins. 1991b. A novel, fully saturated cyclic menaquinone in the archaebacterium *Pyrobaculum organotrophum*. Syst. Appl. Microbiol. *14*: 218–221.

Tomaselli, L., M.C. Margheri and A. Sacchi. 1995. Effects of light on pigments and photosynthetic activity in a phycoerythrin-rich strain of *Spirulina subsalsa*. Aquat. Microb. Ecol. *9*: 27–31.

Tomioka, H. and H. Sasabe. 1995. Isolation of photochemically active archaebacterial photoreceptor, pharaonis phoborhodopsin from *Natronobacterium pharaonis*. Biochim. Biophys. Acta *1234*: 261–267.

Tomlinson, G.A. and L.I. Hochstein. 1976. *Halobacterium saccharovorum* sp. nov., a carbohydrate-metabolizing, extremely halophilic bacterium. Can. J. Microbiol. *22*: 587–591.

Tomlinson, G.A., L.L. Jahnke and L.I. Hochstein. 1986. *Halobacterium denitrificans*, sp. nov., an extremely halophilic denitrifying bacterium. Int. J. Syst. Bacteriol. *36*: 66–70.

Topley, W.W.C. and G.S. Wilson. 1929. The Principles of Bacteriology and Immunity, Edward Arnold and Co., London.

Topley, W.W.C. and G.S. Wilson. 1964. Topley and Wilson's Principles of Bacteriology and Immunity, The Williams & Wilkins Co., Baltimore.

Tornebene, T.G. and T.A. Langworthy. 1979. Diphytanyl and dibiphytanyl glycerol ether lipids of methanogenic archaebacteria. Science *203*: 51–53.

Tornabene, T.G., T.A. Langworthy, G. Holzer and J. Oro. 1979. Squalenes, phytanes and other isoprenoids as major neutral lipids of methanogenic and thermoacidophilic archaebacteria. J. Mol. Evol. *13*: 73–83.

Torreblanca, M., I. Meseguer and F. Rodriguez-Valera. 1989. Halocin H6, a bacteriocin from *Haloferax gibbonsii*. J. Gen. Microbiol. *135*: 2655–2662.

Torreblanca, M., I. Meseguer and A. Ventosa. 1994. Production of halocin is a practically universal feature of archaeal halophilic rods. Lett. Appl. Microbiol. *19*: 201–205.

Torreblanca, M., F. Rodriguez-Valera, G. Juez, A. Ventosa, M. Kamekura and M. Kates. 1986a. Classification of non-alkaliphilic halobacteria based on numerical taxonomy and polar lipid composition, and description of *Haloarcula* gen. nov. and *Haloferax* gen. nov. Syst. Appl. Microbiol. *8*: 89–99.

Torreblanca, M., F. Rodriguez-Valera, G. Juez, A. Ventosa, M. Kamekura and M. Kates. 1986b. *In* Validation of the publication of new names and new combinations previously effectively published outside the IJSB. List No. 22. Int. J. Syst. Bacteriol. *36*: 573–576.

Torreblanca, M., F. Rodriguez-Valera, G. Juez, A. Ventosa, M. Kamekura and M. Kates. 1987. *In* Validation of the publication of new names and new combinations previously effectively published outside the IJSB. List No. 23. Int. J. Syst. Bacteriol. *37*: 179–180.

Torsvik, V., J. Goksøyr and F.L. Daae. 1990a. High diversity in DNA of soil bacteria. Appl. Environ. Microbiol. *56*: 782–787.

Torsvik, V., K. Salte, R. Sørheim and J. Goksøyr. 1990b. Comparison of phenotypic diversity and DNA heterogeneity in a population of soil bacteria. Appl. Environ. Microbiol. *56*: 776–781.

Totten, P.A., K.K. Holmes, H.H. Handsfield, J.S. Knapp, P.L. Perine and S. Falkow. 1983. DNA hybridization technique for the detection of *Neisseria gonorrhoeae* in men with urethritis. J. Infect. Dis. *148*: 462–471.

Touzel, J.P. and G. Albagnac. 1983. Isolation and characterization of *Methanococcus mazei* strain MC₃. FEMS Microbiol. Lett. *16*: 241–245.

Touzel, J.P., E. Conway de Macario, J. Nölling, W.M. de Vos, T. Zhilina and A.M. Lysenko. 1992. DNA relatedness among some thermophilic members of the genus *Methanobacterium*: emendation of the species *Methanobacterium thermoautotrophicum* and rejection of *Methanobacterium thermoformicicum* as a synonym of *Methanobacterium thermoautotrophicum*. Int. J. Syst. Bacteriol. *42*: 408–411.

Touzel, J.P., D. Petroff and G. Albagnac. 1985. Isolation and characterization of a new thermophilic *Methanosarcina*, strain CHTI-55. Syst. Appl. Microbiol. *6*: 66–71.

Touzel, J.P., G. Prensier, J.L. Roustan, I. Thomas, H.C. Dubourguier and G. Albagnac. 1988. Description of a new strain of *Methanothrix soehngenii* and rejection of *Methanothrix concilii* as a synonym of *Methanothrix soehngenii*. Int. J. Syst. Bacteriol. *38*: 30–36.

Trench, R.K. 1982. Physiology, biochemistry, and ultrastructure of cyanelles. *In* Round and Chapman (Editors), Progress in Phycological Research, Vol. 1, Elsevier Biomed. Press (B.V.), Amsterdam. pp. 257–288.

Trevisan, V. 1887. Sul micrococco della rabbia e sulla possibilità di riconoscere durante il periode d'incubazione, dall'esame del sangue della persona moricata, se ha contratta l'infezione rabbica. Rend. Ist. Lombardo (Ser. 2) *20*: 88–105.

Trevisan, V. 1889. I generi e le specie delle Batteriacee, Zanaboni and Gabuzzi, Milan.

Trincone, A., B. Nicolaus, L. Lama, M. De Rosa, A. Gambacorta and W.D. Grant. 1990. The glycolipid of *Halobacterium sodomense*. J. Gen. Microbiol. *136*: 2327–2331.

Trincone, A., B. Nicolaus, G. Palmieri, M. De Rosa, R. Huber, G. Huber, K.O. Stetter and A. Gambacorta. 1992. Distribution of complex and core lipids within new hyperthermophilic members of the Archaea domain. Syst. Appl. Microbiol. *15*: 11–17.

Trincone, A., E. Trivellone, B. Nicolaus, L. Lama, E. Pagnotta, W.D. Grant and A. Gambacorta. 1993. The glycolipid of *Halobacterium trapanicum*. Biochim. Biophys. Acta *1210*: 35–40.

Trost, J.T., J.D. McManus, J.C. Freeman, B.L. Ramakrishna and R.E. Blankenship. 1988. Auracyanin, a blue copper protein from the green photosynthetic bacterium *Chloroflexus aurantiacus*. Biochemistry *27*: 7858–7863.

Trüper, H.G. 1968. *Ectothiorhodospira mobilis* Pelsh, a photosynthetic sulfur bacterium depositing sulfur outside the cells. J. Bacteriol. *95*: 1910–1920.

Trüper, H.G. 1976. Higher taxa of the phototrophic bacteria: *Chloroflexaceae* fam. nov., a family for the gliding filamentous, phototrophic "green" bacteria. Int. J. Syst. Bacteriol. *26*: 74–75.

Trüper, H.G. 1987. Phototrophic bacteria (an incoherent group of prokaryotes). A taxonomic versus phylogenetic survey. Microbiol. SEM. *3*: 71–89.

Trüper, H.G. 1989. Genus *Erythrobacter*. *In* Staley, Bryant, Pfennig and Holt (Editors), Bergey's Manual of Systematic Bacteriology, 1st Ed., Vol. 3, The Williams & Wilkins Co., Baltimore. pp. 1708–1709.

Trüper, H.G. 1992. Prokaryotes: an overview with respect to biodiversity and environmental importance. Biodivers. Conserv. *1*: 227–236.

Trüper, H.G. 1996. Help! Latin! How to avoid the most common mistakes while giving Latin names to newly discovered prokaryotes. Microbiol. SEM *12*: 473–475.

Trüper, H.G. and L. de'Clari. 1997. Taxonomic note: necessary correction of specific epithets formed as substantives (nouns) "in apposition". Int. J. Syst. Bacteriol. *47*: 908–909.

Trüper, H.G. and L. de'Clari. 1998. Taxonomic note: erratum and correction of further specific epithets formed as substantives (nouns) "in apposition". Int. J. Syst. Bacteriol. *48*: 615.

Trüper, H.G. and S. Genovese. 1968. Characterization of photosynthetic sulfur bacteria causing red water in Lake Faro (Messina, Sicily). Limnol. Oceanogr. *13*: 225–232.

Trüper, H.G. and N. Pfennig. 1971. Family of phototrophic green bacteria: *Chlorobiaceae* Copeland, the correct family name; rejection of *Chlorobacterium* Lauterborn; and the taxonomic situation of the consortium-forming species. Int. J. Syst. Bacteriol. *21*: 8–10.

Trüper, H.G. and N. Pfennig. 1982. Characterization and identification of anoxygenic phototrophic bacteria. *In* Starr, Stolp, Trüper, Balows and Schlegel (Editors), The Prokaryotes, Springer-Verlag, Berlin. pp. 299–312.

Tsutsumi, S., K. Denda, K. Yokoyama, T. Oshima, T. Date and M. Yoshida. 1991. Molecular cloning of genes encoding two major subunits of a eubacterial V-type ATPase from *Thermus thermophilus*. Biochim. Biophys. Acta *1098*: 13–20.

Tu, J., D. Prangishvilli, H. Huber, G. Wildgruber, W. Zillig and K.O. Stetter. 1982. Taxonomic relations between archaebacteria including 6 novel genera examined by cross hybridization of DNA and 16S ribosomal RNA. J. Mol. Evol. *18*: 109–114.

Tumbula, D.L., J. Keswani, J. Shieh and W.B. Whitman. 1995. Long-term maintenance of methanogenic stock cultures in glycerol. *In* DasSarma and Fleischmann (Editors), Archaea: A Laboratory Manual, Cold Spring Harbor Laboratory Press, Plainview. pp. 85–87.

Turner, S. 1997. Molecular systematics of oxygenic photosynthetic bacteria. Plant Syst. Evol. Suppl. *11*: 13–52.

Turner, S., K.M. Pryer, V.P.W. Miao and J.D. Palmer. 1999. Investigating deep phylogenetic relationships among cyanobacteria and plastids by small subunit rRNA sequence analysis. J. Eukaryot. Microbiol. *46*: 327–338.

Tuschak, C., J. Glaeser and J. Overmann. 1999. Specific detection of green sulfur bacteria by *in situ*-hybridization with fluorescently labeled oligonucleotide probes. Arch. Microbiol. *171*: 265–272.

Tzeng, S.F., M.P. Bryant and R.S. Wolfe. 1975a. Factor 420-dependent pyridine nucleotide-linked formate metabolism of *Methanobacterium ruminantium*. J. Bacteriol. *121*: 192–196.

Tzeng, S.F., R.S. Wolfe and M.P. Bryant. 1975b. Factor 420-dependent pyridine nucleotide-linked hydrogenase system of *Methanobacterium ruminantium*. J. Bacteriol. *121*: 184–191.

Uherkovich, G. 1969. Beiträge zur Kenntnis der Algenvegetation der Natron - bzw. Soda - (Szik) Gewässer Ungarns. II. Über die Algen des Teichs Öszeszek. Hydrobiologia *33*: 250–288.

Upasani, V.N. and S.G. Desai. 1990. Sambhar Salt Lake (Sambhar, Rajasthan, India): chemical composition of the brines and studies on haloalkaliphilic archaebacteria. Arch. Microbiol. *154*: 589–593.

Upasani, V.N., S.G. Desai, N. Moldoveanu and M. Kates. 1994. Lipids of extremely halophilic archaeobacteria from saline environments in India: a novel glycolipid in *Natronobacterium* strains. Microbiology (Reading) *140*: 1959–1966.

Urbach, E., D.L. Robertson and S.W. Chisholm. 1992. Multiple evolutionary origins of prochlorophytes within the cyanobacterial radiation. Nature *355*: 267–270.

Urbach, E., D.J. Scanlan, D.L. Distel, J.B. Waterbury and S.W. Chisholm. 1998. Rapid diversification of marine picophytoplankton with dissimilar light-harvesting structures inferred from sequence of *Prochlorococcus* and *Synechococcus* (Cyanobacteria). J. Mol. Evol. *46*: 188–201.

Urdea, M.S., B.D. Warner, J.A. Running, M. Stempien, J. Clyne and T. Horn. 1988. A comparison of non-radioisotopic hybridization assay methods using fluorescent, chemiluminescent and enzyme labeled synthetic oligodeoxyribonucleotide probes. Nucleic Acids Res. *16*: 4937–4956.

Utermöhl, H. 1924. Phaeobakterien (Bakterien mit braunen Farbstoffen). Biol. Zentralbl. *43*: 605–610.

Vaisanen, O.M., A. Weber, A. Bennasar, F.A. Rainey, H.J. Busse and M.S. Salkinoja Salonen. 1998. Microbial communities of printing paper machines. J. Appl. Microbiol. *84*: 1069–1084.

van Baalen, C. 1962. Studies on marine blue-green algae. Bot. Mar. *4*: 129–139.

van Bruggen, J.J.A., K.B. Zwart, J.G.F. Hermans, E.M. van Hove, C.K. Stumm and G.D. Vogels. 1986a. Isolation and characterization of *Methanoplanus endosymbiosus* sp. nov., an endosymbiont of the marine sapropelic ciliate *Metopus contortus* Quennerstedt. Arch. Microbiol. *144*: 367–374.

van Bruggen, J.J.A., K.B. Zwart, J.G.F. Herman, E.M. van Hove, C.K. Stumm and G.D. Vogels. 1986b. *In* Validation of the publication of new names and new combinations previously effectively published outside the IJSB. List No. 22. Int. J. Syst. Bacteriol. *36*: 573–576.

van Bruggen, J.J.A., K.B. Zwart, R.M. van Assema, C.K. Stumm and G.D. Vogels. 1984. *Methanobacterium formicicum*, an endosymbiont of the anaerobic ciliate *Metopus striatus* McMurrich. Arch. Microbiol. *139*: 1– 7.

Vandamme, P. 1998. Speciation. *In* Williams, Ketley and Salmond (Editors), Methods in Microbiology, Vol. 27. Bacterial Pathogenesis, Academic Press, London. pp. 51–56.

Vandamme, P., E. Falsen, R. Rossau, B. Hoste, P. Segers, R. Tytgat and J. De Ley. 1991. Revision of *Campylobacter*, *Helicobacter*, and *Wolinella* taxonomy: emendation of generic descriptions and proposal of *Arcobacter* gen. nov. Int. J. Syst. Bacteriol. *41*: 88–103.

Vandamme, P., B. Pot, M. Gillis, P. de Vos, K. Kersters and J. Swings. 1996a. Polyphasic taxonomy, a consensus approach to bacterial systematics. Microbiol. Rev. *60*: 407–438.

Vandamme, P., M. Vancanneyt, A. Van Belkum, P. Segers, W.G.V. Quint, K. Kersters, B.J. Paster and F.E. Dewhirst. 1996b. Polyphasic analysis of strains of the genus *Capnocytophaga* and Centers for Disease Control group DF-3. Int. J. Syst. Bacteriol. *46*: 782–791.

van de Casteele, M., P. Chen, M. Roovers, C. Legrain and N. Glansdorff.

1997. Structure and expression of a pyrimidine gene cluster from the extreme thermophile *Thermus* strain ZO5. J. Bacteriol. *179*: 3470–3481.

van de Casteele, M., M. Demarez, C. Legrain, N. Glansdorff and A. Piérard. 1990. Pathways of arginine biosynthesis in extreme thermophilic archaeo- and eubacteria. J. Gen. Microbiol. *136*: 1177–1183.

Vandenberghe, A., A. Wassink, P. Raeymaekers, R. de Baere, E. Huysmans and R. de Wachter. 1985. Nucleotide sequence, secondary structure and evolution of the 5S ribosomal RNA from five bacterial species. Eur. J. Biochem. *149*: 537–542.

van den Eynde, H., Y. van de Peer, J. Perry and R. de Wachter. 1990. 5S ribosomal RNA sequences of representatives of the genera *Chlorobium, Prosthecochloris, Thermomicrobium, Cytophaga, Flavobacterium, Flexibacter,* and *Saprospira* and a discussion of the evolution of eubacteria in general. J. Gen. Microbiol. *136*: 11–18.

van de Peer, Y. and R. de Wachter. 1994. TREECON for windows: a software package for the construction and drawing of evolutionary trees for the Microsoft Windows environment. Comput. Appl. Biosci. *10*: 569–570.

van de Peer, Y., J.M. Neefs, P. de Rijk, P. de Vos and R. de Wachter. 1994. About the order of divergence of the major bacterial taxa during evolution. Syst. Appl. Microbiol. *17*: 32–38.

van de Peer, Y., J.M. Neefs, P. de Rijk and R. de Wachter. 1993. Reconstructing evolution from eukaryotic small ribosomal subunit RNA sequences: calibration of the molecular clock. J. Mol. Evol. *37*: 221–232.

van de Peer, Y., E. Robbrecht, S. de Hoog, A. Caers, P. de Rijk and R. de Wachter. 1999. Database on the structure of small subunit ribosomal RNA. Nucleic Acids Res. *27*: 179–183.

van de Wijngaard, W.M.H., J. Creemers, G.D. Vogels and C. van der Drift. 1991. Methanogenic pathways in *Methanosphaera stadtmanae.* FEMS Microbiol. Lett. *80*: 207–211.

van Dok, W. and B.T. Hart. 1996. Akinete differentiation in *Anabaena circinalis* (Cyanophyta). J. Phycol. *32*: 557–565.

van Dok, W. and B.T. Hart. 1997. Akinete germination in *Anabaena circinalis* (Cyanophyta). J. Phycol. *33*: 12–17.

van Gemerden, H. and J. Mas. 1995. Ecology of phototrophic sulfur bacteria. *In* Blankenship, Madigan and Bauer (Editors), Anoxygenic Photosynthetic Bacteria, Kluwer Academic Publishers, Dordrecht. pp. 49–85.

van Liere, L. and A.E. Walsby. 1982. Interactions of cyanobacteria with light. *In* Carr and Whitton (Editors), The Biology of Cyanobacteria, Blackwell, Oxford, and University of California Press, Berkeley. pp. 9–45.

van Niel, C.B. 1944. The culture, general physiology, morphology, and classification of the non-sulfur purple and brown bacteria. Bacteriol. Rev. *8*: 1–118.

van Niel, C.B. 1957. The photosynthetic bacteria. Suborder I. *Rhodobacteriineae. In* Breed, Murray and Smith (Editors), Bergey's Manual of Determinative Bacteriology, 7th Ed., The Williams & Wilkins Co., Baltimore. pp. 35–67.

Vargas, M. and K.M. Noll. 1994. Isolation of auxotrophic and antimetabolite-resistant mutants of the hyperthermophilic bacterium *Thermotoga neapolitana.* Arch. Microbiol. *162*: 357–361.

Vasquez, C., A. Venegas and R. Vicuña. 1981. Characterization and cloning of a plasmid isolated from the extreme thermophile *Thermus flavus* AT-62. Biochem. Int. *3*: 291–299.

Vasquez, C., J. Villanueva and R. Vicuña. 1983. Plasmid curing in *Thermus thermophilus* and *Thermus flavus.* FEBS Lett. *158*: 339–342.

Vaucher, J.P. 1803. Histoire des conferves d'eau douce, contenant leurs different modes de reproduction, et la description de leurs principales espèces, J. Paschoud, Genève.

Vauterin, L., B. Hoste, K. Kersters and J. Swings. 1995. Reclassification of *Xanthomonas.* Int. J. Syst. Bacteriol. *45*: 472–489.

Veldhuis, M.J.W. and H. van Gemerden. 1986. Competition between purple and brown phototrophic bacteria in stratified lakes: sulfide, acetate, and light as limiting factors. FEMS Microbiol. Ecol. *38*: 31–38.

Venebles, W.N. and B.D. Ripley. 1994. Modern Applied Statistics with S-Plus, Springer-Verlag, New York.

Venetskaya, S.L. and L.M. Gerasimenko. 1988. Electron-microscopic study of microorganisms in a halophilic cyanobacterial community. Mikrobiologiya *57*: 450–457.

Ventosa, A., D.R. Arahal and B.E. Volcani. 1999a. Studies on the microbiota of the Dead Sea—50 years later. *In* Oren (Editor), Microbiology and Biogeochemistry of Hypersaline Environments, CRC Press, Boca Raton. pp. 139–147.

Ventosa, A., M.C. Gutiérrez, M. Kamekura and M.L. Dyall-Smith. 1999b. Proposal for the transfer of *Halococcus turkmenicus, Halobacterium trapanicum* JCM 9743 and strain GSL 11 to *Haloterrigena turkmenica* gen. nov., comb. nov. Int. J. Syst. Bacteriol. *49*: 131–136.

Ventosa, A. and A. Oren. 1996. *Halobacterium salinarum* nom. corrig., a name to replace *Halobacterium salinarium* (Elazari-Volcani) and to include *Halobacterium halobium* and *Halobacterium cutirubrum.* Int. J. Syst. Bacteriol. *46*: 347.

Veríssimo, A., G. Marrao, F.G.D. Silva and M.S. da Costa. 1991. Distribution of *Legionella* spp. in hydrothermal areas in continental Portugal and the island of São Miguel, Azores. Appl. Environ. Microbiol. *57*: 2921–2927.

Viale, A.M., A. Arakaki, F.C. Soncini and R.G. Ferreyra. 1994. Evolutionary relationships among eubacterial groups as inferred from GroEL (chaperonin) sequence comparisons. Int. J. Syst. Bacteriol. *44*: 527–533.

Vijgenboom, E., L.P. Woudt, P.W. Heinstra, K. Rietveld, J. van Haarlem, G.P. van Wezel, S. Shochat and L. Bosch. 1994. Three *tuf*-like genes in the kirromycin producer *Streptomyces ramocissimus.* Microbiology *140*: 983–998.

Villanueva, J., J.O. Grimalt, R. De Wit, B.J. Keely and J.R. Maxwell. 1994. Chlorophyll and carotenoid pigments in solar saltern microbial mats. Geochim. Cosmochim. Acta *58*: 4703–4715.

Vincent, W.F. 1987. Special issue: dominance of bloom-forming cyanobacteria (blue-green algae). N.Z. J. Mar. Freshw. Res. *21*: 361–542.

Vincent, W.F. 1988. Microbial Ecosystems of Antarctica, Cambridge University Press, Cambridge.

Völkl, P., R. Huber, E. Drobner, R. Rachel, S. Burggraf, A. Trincone and K.O. Stetter. 1993. *Pyrobaculum aerophilum* sp. nov., a novel nitrate-reducing hyperthermophilic archaeum. Appl. Environ. Microbiol. *59*: 2918–2926.

Völkl, P., R. Huber and K.O. Stetter. 1996. *In* Validation of the publication of new names and new combinations previously effectively published outside the IJSB. List No. 58. Int. J. Syst. Bacteriol. *46*: 836–837.

Völkl, P., P. Markiewicz, K.O. Stetter and J.H. Miller. 1994. The sequence of a subtilisin-type protease (aerolysin) from the hyperthermophilic archaeum *Pyrobaculum aerophilum* reveals sites important to thermostability. Protein Sci. *3*: 1329–1340.

Vonshak, A. (Editor). 1997. *Spirulina platensis* (*Arthrospira*): Physiology, Cell-biology, and Biotechnology, Taylor & Francis, London.

von Wintzingerode, F., U.B. Gobel and E. Stackebrandt. 1997. Determination of microbial diversity in environmental samples: pitfalls of PCR-based rRNA analysis. FEMS Microbiol. Rev. *21*: 213–229.

Voordouw, G., J.K. Voordouw, T.R. Jack, J. Foght, P.M. Fedorak and D.W.S. Westlake. 1992. Identification of distinct communities of sulfate-reducing bacteria in oil fields by reverse sample genome probing. Appl. Environ. Microbiol. *58*: 3542–3552.

Voordouw, G., J.K. Voordouw, R.R. Karkoff-Schweizer, P.M. Fedorak and D.W.S. Westlake. 1991. Reverse sample genome probing, a new technique for identification of bacteria in environmental samples by DNA hybridization, and its application to the identification of sulfate-reducing bacteria in oil field samples. Appl. Environ. Microbiol. *57*: 3070–3078.

Vorholt, J.A., D. Hafenbradl, K.O. Stetter and R.K. Thauer. 1997. Pathways of autotrophic CO$_2$ fixation and of dissimilatory nitrate reduction to N$_2$O in *Ferroglobus placidus.* Arch. Microbiol. *167*: 19–23.

Vorholt, J.A., J. Kunow, K.O. Stetter and R.K. Thauer. 1995. Enzymes and coenzymes of the carbon monoxide dehydrogenase pathway for autotrophic CO$_2$ fixation in *Archaeoglobus lithotrophicus* and the lack of carbon monoxide dehydrogenase in the heterotrophic *A. profundus*. Arch. Microbiol. *163*: 112–118.

Wade, W.G., J. Downes, D. Dymock, S.J. Hiom, A.J. Weightman, F.E.P. Dewhirst, B.J., N. Tzellas and B. Coleman. 1999. The family *Coriobacteriaceae*: reclassification of *Eubacterium exiguum* (Poco et al. 1996) and *Peptostreptococcus heliotrinreducens* (Lanigan 1976) as *Slackia exigua* gen. nov., comb. nov. and *Slackia heliotrinireducens* gen. nov., comb. nov., and *Eubacterium lentum* (Prevot 1938) as *Eggerthella lenta* gen. nov., comb. nov. Int. J. Syst. Bacteriol. *49*: 595–600.

Wagner, M., R.I. Amann, H. Lemmer and K.H. Schleifer. 1993. Probing activated sludge with oligonucleotides specific for proteobacteria: inadequacy of culture-dependent methods for describing microbial community structure. Appl. Environ. Microbiol. *59*: 1520–1525.

Wagner, M., R. Erhart, W. Manz, R.I. Amann, H. Lemmer, D. Wedi and K.H. Schleifer. 1994. Development of an rRNA-targeted oligonucleotide probe specific for the genus *Acinetobacter* and its application for *in situ* monitoring in activated sludge. Appl. Environ. Microbiol. *60*: 792–800.

Wagner, M., G. Rath, R.I. Amann, H.P. Koops and K.H. Schleifer. 1995. *In situ* identification of ammonia-oxidizing bacteria. Syst. Appl. Microbiol. *18*: 251–264.

Wagner, M., G. Rath, H.P. Koops, J. Flood and R.I. Amann. 1996. *In situ* analysis of nitrifying bacteria in sewage treatment plants. Water Sci. Technol. *34*: 237–244.

Wagner, M., M. Schmid, S. Juretschko, T.K. Trebesius, A. Bubert, W. Goebel and K.H. Schleifer. 1998. *In situ* detection of a virulence factor mRNA and 16S rRNA in *Listeria monocytogenes*. FEMS Microbiol. Lett. *160*: 159–168.

Wahl, G.M., S.L. Berger and A.R. Kimmel. 1987. Molecular hybridization of immobilized nucleic acids: theoretical concepts and practical considerations. Methods Enzymol. *152*: 399–407.

Wahlund, T.M., C.R. Woese, R.W. Castenholz and M.T. Madigan. 1991. A thermophilic green sulfur bacterium from New Zealand hot springs, *Chlorobium tepidum* sp. nov. Arch. Microbiol. *156*: 81–90.

Wahlund, T.M., C.R. Woese, R.W. Castenholz and M.T. Madigan. 1996. *In* Validation of the publication of new names and new combinations previously effectively published outside the IJSB. List No. 59. Int. J. Syst. Bacteriol. *46*: 1189.

Wais, A.C. 1988. Recovery of halophilic archaebacteria from natural environments. FEMS Microbiol. Ecol. *53*: 211–216.

Wait, R., L. Carreto, M.F. Nobre, A.M. Ferreira and M.S. da Costa. 1997. Characterization of novel long-chain 1,2-diols in *Thermus* species and demonstration that *Thermus* strains contain both glycerol-linked and diol-linked glycolipids. J. Bacteriol. *179*: 6154–6162.

Wakao, N., N. Nagasawa, T. Matsuura, H. Matsukura, T. Matsumoto, A. Hiraishi, Y. Sakurai and H. Shiota. 1994. *Acidiphilium multivorum* sp. nov., an acidophilic chemoorganotrophic bacterium from pyritic acid mine drainage. J. Gen. Appl. Microbiol. *40*: 143–159.

Wallner, G., R.I. Amann and W. Beisker. 1993. Optimizing fluorescent *in situ* hybridization with rRNA-targeted oligonucleotide probes for flow cytometric identification of microorganisms. Cytometry *14*: 136–143.

Wallner, G., R. Erhart and R.I. Amann. 1995. Flow cytometric analysis of activated sludge with rRNA-targeted probes. Appl. Environ. Microbiol. *61*: 1859–1866.

Wallner, G., B. Fuchs, S. Spring, W. Beisker and R.I. Amann. 1997. Flow cytometric sorting of microorganisms for molecular analysis. Appl. Environ. Microbiol. *63*: 4223–4231.

Walsby, A.E. 1980. A square bacterium. Nature *283*: 69–71.

Walsby, A.E. 1981. Cyanobacteria: planktonic gas-vacuolate forms. *In* Starr, Stolp, Trüper, Balows and Schlegel (Editors), The Prokaryotes. A Handbook on Habitats, Isolation, and Identification of Bacteria, Springer-Verlag, Berlin. pp. 224–235.

Walsby, A.E. 1994. Gas vesicles. Microbiol. Rev. *58*: 94–144.

Walsby, A.E., J. van Rijn and Y. Cohen. 1983. The biology of a new gas vacuolate cyanobacterium *Dactylococcopsis salina*, a new species in Solar Lake, Sinai, Egypt. Proc.R. Soc. Lond. Ser.B. Biol. Sci. *217*: 417–448.

Wang, D. and Q. Tang. 1989. *Natronobacterium* from soda lakes of China. *In* Hattori, Ishida, Maruyama, Morita and Uchida (Editors), Recent Advances in Microbial Ecology, Japan Scientific Societies Press, Tokyo. pp. 68–72.

Ward, D.M., E. Beck, N.P. Revsbech, K.A. Sandbeck and M.R. Winfrey. 1984. Decomposition of hot spring microbial mats. *In* Cohen, Castenholz and Halvorson (Editors), Microbial Mats: Stromatolites, Alan R. Liss, New York. pp. 191–214.

Ward, D.M., M.J. Ferris, S.C. Nold and M.M. Bateson. 1998. A natural view of microbial biodiversity within hot spring cyanobacterial mat communities. Microbiol. Mol. Biol. Rev. *62*: 1353–1370.

Ward, D.M., R. Weller and M.M. Bateson. 1990. 16S rRNA sequences reveal numerous uncultured microorganisms in a natural community. Nature *345*: 63–65.

Ward, D.M., R. Weller, J. Shiea, R.W. Castenholz and Y. Cohen. 1989. Hot spring microbial mats: anoxygenic and oxygenic mats of possible evolutionary significance. *In* Cohen and Rosenberg (Editors), Microbial Mats: Physiological Ecology of Benthic Microbial Communities, American Society for Microbiology, Washington, DC. pp. 3–15.

Ward, J.M. 1970. The microbial ecology of estuarine methanogenesis, Master's thesis, University of Florida.

Waring, G.A. 1965. Thermal springs of the United States and other countries of the world; a summary, US Govt. Print. Office, Washington DC.

Warthmann, R., H. Cypionka and N. Pfennig. 1992. Photoproduction of hydrogen from acetate by syntrophic cocultures of green sulfur bacteria and sulfur-reducing bacteria. Arch. Microbiol. *157*: 343–348.

Wasserfallen, A., K. Huber and T. Leisinger. 1995. Purification and structural characterization of a flavoprotein induced by iron limitation in *Methanobacterium thermoautotrophicum* Marburg. J. Bacteriol. *177*: 2436–2441.

Wasserfallen, A., J. Nölling, P. Pfister, J.N. Reeve and E. Conway de Macario. 2000. Phylogenetic analysis of 18 thermophilic *Methanobacterium* isolates supports the proposals to create a new genus, *Methanothermobacter* gen. nov., and to reclassify several isolates in three species, *Methanothermobacter thermautotrophicus*, comb. nov., *Methanothermobacter wolfeii*, comb. nov., and *Methanothermobacter marburgense*, sp. nov. Int. J. Syst. Evol. Microbiol. *50*: 43–53.

Wassill, L., W. Ludwig and K.H. Schleifer. 1998. Development of a modified subtraction hybridization technique and its application for the design of strain specific PCR systems for lactococci. FEMS Microbiol. Lett. *166*: 63–70.

Watanabe, A. 1959. Distribution of nitrogen-fixing blue-green algae in various areas of South and East Asia. J. Gen. Appl. Microbiol. *5*: 21–29.

Watanabe, Y., R.G. Feick and J.A. Shiozawa. 1995. Cloning and sequencing of the genes encoding the light-harvesting B806–866 polypeptides and initial studies on the transcriptional organization of *puf2B*, *puf2A* and *puf2C* in *Chloroflexus aurantiacus*. Arch. Microbiol. *163*: 124–130.

Waterbury, J.B. 1989. Subsection II. Order *Pleurocapsales*. *In* Staley, Bryant, Pfennig and Holt (Editors), Bergey's Manual of Systematic Bacteriology, 1st Ed., Vol. 3, The Williams & Wilkins Co., Baltimore. pp. 1746–1770.

Waterbury, J.B. and R. Rippka. 1989. Subsection I. Order *Chroococcales*. *In* Staley, Bryant, Pfennig and Holt (Editors), Bergey's Manual of Systematic Bacteriology, 1st Ed., Vol. 3, The Williams & Wilkins Co., Baltimore. pp. 1728–1746.

Waterbury, J.B. and R.Y. Stanier. 1977. Two unicellular cyanobacteria which reproduce by budding. Arch. Microbiol. *115*: 249–257.

Waterbury, J.B. and R.Y. Stanier. 1978. Patterns of growth and development in pleurocapsalean cyanobacteria. Microbiol. Rev. *42*: 2–44.

Waterbury, J.B. and R.Y. Stanier. 1981. Isolation and growth of Cyanobacteria from marine and hypersaline environments. *In* Starr, Stolp, Trüper, Balows and Schlegel (Editors), The Prokaryotes: A Handbook on Habitats, Isolation, and Identification of Bacteria, Vol. 1, Springer-Verlag, Berlin. pp. 221–223.

Waterbury, J.B., S.W. Watson, F.W. Valois and D.G. Franks. 1986. Biological and ecological characterization of the marine unicellular cyanobacterium *Synechococcus*. Can. Bull. Fish. Aquat. Sci. *214*: 71–120.

Waterbury, J.B. and J.M. Willey. 1988. Isolation and growth of marine planktonic cyanobacteria. Methods Enzymol. *167*: 100–105.

Waterbury, J.B., J.M. Willey, D.G. Franks, F.W. Valois and S.W. Watson. 1985. A cyanobacterium capable of swimming motility. Science *230*: 74–76.

Watson, S.W., E. Bock, H. Harms, H.P. Koops and A.B. Hooper. 1989. Nitrifying bacteria. *In* Staley, Bryant, Pfennig and Holt (Editors), Bergey's Manual of Systematic Bacteriology, 1st Ed., Vol. 3, The Williams & Wilkins Co., Baltimore. pp. 1808–1834.

Watson, S.W., E. Bock, F.W. Valois, J.B. Waterbury and U. Schlosser. 1986a. *Nitrospira marina*, gen. nov. sp. nov.: a chemolithotrophic nitrite-oxidizing bacterium. Arch. Microbiol. *144*: 1–7.

Watson, S.W., E. Bock, F.W. Valois, J.B. Waterbury and U. Schlosser. 1986b. *In* Validation of the publication of new names and new combinations previously effectively published outside the IJSB. List No. 21. Int. J. Syst. Bacteriol. *36*: 489.

Wayne, J. and S.Y. Xu. 1997. Identification of a thermophilic plasmid origin and its cloning within a new *Thermus-E. coli* shuttle vector. Gene *195*: 321–328.

Wayne, L.G. 1994. Actions of the Judicial Committee on Systematic Bacteriology on Requests for Opinions published between January 1985 and July 1993. Int. J. Syst. Bacteriol. *44*: 177–178.

Wayne, L.G., D.J. Brenner, R.R. Colwell, P.A.D. Grimont, O. Kandler, M.I. Krichevsky, L.H. Moore, W.E.C. Moore, R.G.E. Murray, E. Stackebrandt, M.P. Starr and H.G. Trüper. 1987. Report of the ad hoc committee on reconciliation of approaches to bacterial systematics. Int. J. Syst. Bacteriol. *37*: 463–464.

Weaver, G.A., J.A. Krause, T.L. Miller and M.J. Wolin. 1986. Incidence of methanogenic bacteria in a sigmoidoscopy population: an association of methanogenic bacteria and diverticulosis. Gut *27*: 699–704.

Weber, B., D.C.J. Wessels and B. Büdel. 1996. Biology and ecology of cryptoendolithic cyanobacteria of a sandstone outcrop in the Northern Province, South Africa. Algol. Stud. *83*: 565–579.

Weber, J.M., S.P. Johnson, V. Vonstein, M.J. Casadaban and D.C. Demirjian. 1995. A chromosome integration system for stable gene transfer into *Thermus flavus*. Biotechnology *13*: 271–275.

Wechsler, T., R.A. Brunisholz, G. Frank, F. Suter and H. Zuber. 1987. The complete amino acid sequence of the antenna polypeptide B806-866-β from the cytoplasmic membrane of the green bacterium *Chloroflexus aurantiacus*. FEBS Lett. *210*: 189–194.

Wechsler, T., R.A. Brunisholz, F. Suter, R.C. Fuller and H. Zuber. 1985. The complete amino acid sequence of a bacteriochlorophyll *a* binding polypeptide isolated from the cytoplasmic membrane of the green photosynthetic bacterium *Chloroflexus aurantiacus*. FEBS Lett. *191*: 34–38.

Weckesser, J., G. Drews and H. Mayer. 1979. Lipopolysaccharides of photosynthetic prokaryotes. Annu. Rev. Microbiol. *33*: 215–239.

Weckesser, J., H. Mayer and G. Shulz. 1995. Anoxygenic phototrophic bacteria: model organisms for studies on cell wall macromolecules. *In* Blankenship, Madigan and Bauer (Editors), Anoxygenic Photosynthetic Bacteria, Kluwer Academic Publishing, Dordrecht. pp. 207–230.

Weimer, P.J. and J.G. Zeikus. 1978. One carbon metabolism in methanogenic bacteria; cellular characterization and growth of *Methanosarcina barkeri*. Arch. Microbiol. *119*: 49–57.

Weisburg, W.G., S.J. Giovannoni and C.R. Woese. 1989. The *Deinococcus-Thermus* phylum and the effect of ribosomal RNA composition on phylogenetic tree construction. Syst. Appl. Microbiol. *11*: 128–134.

Weller, R., M.M. Bateson, B.K. Heimbuch, E.D. Kopczynski and D.M. Ward. 1992. Uncultivated cyanobacteria, *Chloroflexus*-like inhabitants and spirochete-like inhabitants of a hot spring microbial mat. Appl. Environ. Microbiol. *58*: 3964–3969.

Wen, A., M. Fegan, C. Hayward, S. Chakraborty and L.I. Sly. 1999. Phylogenetic relationships among members of the *Comamonadaceae*, and description of *Delftia acidovorans* (den Dooren de Jong 1926 and Tamaoka et al. 1987) gen. nov., comb. nov. Int. J. Syst. Bacteriol. *49*: 567–576.

Werber, M.M., J.L. Sussmann and H.L. Eisenberg. 1986. Molecular basis for the special properties of proteins and enzymes from *Halobacterium marismortui*. FEMS Microbiol Rev. *39*: 129–135.

Werner, F.C. (Editor). 1972. Wortelemente Lateinisch-Griechischer Fachausdrücke in den Biologischen Wissenschaften, 3rd Ed., Suhrkamp Taschenbuch Verlag, Berlin.

West, W. and G.S. West. 1897. J. Bot. London *35*: 1–300.

Westermann, P., B.K. Ahring and R.A. Mah. 1989a. Acetate production by methanogenic bacteria. Appl. Environ. Microbiol. 55: 2257–2261.

Westermann, P., B.K. Ahring and R.A. Mah. 1989b. Threshold acetate concentrations for acetate catabolism by aceticlastic methanogenic bacteria. Appl. Environ. Microbiol. 55: 514–515.

Wetmur, J.G., D.M. Wong, B. Ortiz, J. Tong, F. Reichert and D.H. Gelfand. 1994. Cloning, sequencing, and expression of RecA proteins from three distantly related thermophilic eubacteria. J. Biol. Chem. *269*: 25928–25935.

Wettstein, F.V. and von Westersheim, R. 1924. Handbuch der Systematischen Botanik, 3rd Ed., F. Deuticke, Leipzig and Wien.

Weyant, R.S., C.W. Moss, R.E. Weaver, D.G. Hollis, J.G. Jordan, E.C. Cook and M.I. Daneshvar (Editors). 1996. Identification of Unusual Pathogenic Gram-negative Aerobic and Facultatively Anaerobic Bacteria, 2nd Ed., The Williams & Wilkins Co., Baltimore.

White, O., J.A. Eisen, J.F. Heidelberg, E.K. Hickey, J.D. Peterson, R.J. Dodson, D.H. Haft, M.L. Gwinn, W.C. Nelson, D.L. Richardson, K.S. Moffat, H. Qin, L. Jiang, W. Pamphile, M. Crosby, M. Shen, J.J. Vamathevan, P. Lam, L. McDonald, T. Utterback, C. Zalewski, K.S. Makarova, L. Aravind, M.J. Daly, K.W. Minton, R.D. Fleischmann, K.A. Ketchum, K.E. Nelson, S. Salzberg, H.O. Smith, J.C. Venter and C.M. Fraser. 1999. Genome sequence of the radioresistant bacterium *Deinococcus radiodurans* R1. Science *286*: 1571–1577.

Whitman, W.B. 1989. Order II. *Methanococcales*. *In* Staley, Bryant, Pfennig and Holt (Editors), Bergey's Manual of Systematic Bacteriology, 1st Ed., Vol. 3, The Williams & Wilkins Co., Baltimore. pp. 2185–2190.

Whitman, W.B., E. Ankwanda and R.S. Wolfe. 1982. Nutrition and carbon metabolism of *Methanococcus voltae*. J. Bacteriol. *149*: 852–863.

Whitman, W.B., D.C. Coleman and W.J. Wiebe. 1998. Prokaryotes: the unseen majority. Proc. Natl. Acad. Sci. U.S.A. *95*: 6578–6583.

Whitman, W.B., J. Shieh, S. Sohn, D.S. Caras and U. Premachandran. 1986. Isolation and characterization of 22 mesophilic methanococci. Syst. Appl. Microbiol. *7*: 235–240.

Whitman, W.B., S. Sohn, S. Kuk and R. Xing. 1987. Role of amino acids and vitamins in nutrition of mesophilic *Methanococcus* spp. Appl. Environ. Microbiol. *53* : 2373–2378.

Whitton, B.A. 1987. The biology of *Rivulariaceae*. *In* Fay and van Baalen (Editors), The Cyanobacteria, Elsevier Science Publishers, Amsterdam.

Whitton, B.A. 1989. Genus I. *Calothrix*. *In* Staley, Bryant, Pfennig and Holt (Editors), Bergey's Manual of Systematic Bacteriology, 1st Ed., Vol. 3, The Williams & Wilkins Co., Baltimore. pp. 1791–1793.

Whitton, B.A. and M. Potts (Editors). 2000. Ecology of Cyanobacteria: Their Diversity in Time and Space, Kluwer Academic Publishers, Dordrecht.

Wichlacz, P.L., R.F. Unz and T.A. Langworthy. 1986. *Acidiphilium angustum*, sp. nov., *Acidiphilium facilis*, sp. nov. and *Acidiphilium rubrum*, sp. nov.: acidophilic heterotrophic bacteria isolated from acidic coal mine drainage. Int. J. Syst. Bacteriol. *36*: 197–201.

Wickstrom, C.E. and R.W. Castenholz. 1978. Association of *Pleurocapsa* and *Calothrix* (Cyanophyta) in a thermal stream. J. Phycol. *14*: 84–88.

Widdel, F. 1980. Anaerober Abbau von Fettsäuren und Benzoesäuren durch neu isolierte Arten Sulfat-reduzierender Bakterien, Doctoral thesis, Universität Göttingen, Göttingen..

Widdel, F. 1986. Growth of methanogenic bacteria in pure cultures with 2-propanol and other alcohols as hydrogen donors. Appl. Environ. Microbiol. *51*: 1056–1062.

Widdel, F. 1992. The genus *Thermodesulfobacterium*. *In* Balows, Trüper, Dworkin, Harder and Schleifer (Editors), The Prokaryotes. A Handbook of Bacteria: Ecophysiology, Isolation, Identification, Applications, 2nd Ed., Springer-Verlag, New York. pp. 3390–3392.

Widdel, F. and F. Bak. 1992. Gram-negative mesophilic sulfate-reducing bacteria. *In* Balows, Trüper, Dworkin, Harder and Schleifer (Editors), The Prokaryotes. A Handbook of Bacteria: Ecophysiology, Isolation, Identification, Applications, 2nd Ed., Vol. IV, Springer-Verlag, New York. 3352–3378.

Widdel, F., G.-W. Kohring and F. Mayer. 1983. Studies on dissimilatory sulfate-reducing bacteria that decompose fatty acids III. Characterization of the filamentous gliding *Desulfonema limnicola* gen. nov. sp. nov., and *Desulfonema magnum* sp. nov. Arch. Microbiol. *134*: 286–294.

Widdel, F. and N. Pfennig. 1981. Studies on dissimilatory sulfate-reducing bacteria that decompose fatty acids I. Isolation of new sulfate-reducing bacteria with acetate from saline environments. Description of *Desulfobacter postgatei* gen. nov., sp. nov. Arch. Microbiol. *129*: 395–400.

Widdel, F. and N. Pfennig. 1984. Dissimilatory sulfate- or sulfur-reducing bacteria. *In* Krieg and Holt (Editors), Bergey's Manual of Systematic Bacteriology, 1st Ed., Vol. 1, The Williams & Wilkins Co., Baltimore. pp. 663–679.

Widdel, F., P.E. Rouvière and R.S. Wolfe. 1988. Classification of secondary alcohol-utilizing methanogens including a new thermophilic isolate. Arch. Microbiol. *150*: 477–481.

Widdel, F., P.E. Rouvière and R.S. Wolfe. 1989. *In* Validation of the publication of new names and new combinations previously effectively published outside the IJSB. List No. 28. Int. J. Syst. Bacteriol. *39*: 93–94.

Widdel, F., S. Schnell, S. Heising, A. Ehrenreich, B. Assmus and B. Schink. 1993. Ferrous iron oxidation by anoxygenic phototrophic bacteria. Nature *362*: 834–836.

Wieland, F., J. Lechner and M. Sumper. 1982. The cell wall glycoprotein of the cell wall of halobacteria: structural, functional, and biosynthetic aspects. Zentbl. Bakteriol. Mikrobiol. Hyg. 1 Abt Orig. C *3*: 161–170.

Wieland, F., G. Paul and M. Sumper. 1985. Halobacterial flagellins are sulfated glycoproteins. J. Biol. Chem. *260*: 15180–15185.

Wildgruber, G., M. Thomm, H. König, K. Ober, T. Ricchiuto and K.O. Stetter. 1982. *Methanoplanus limicola*, a plate-shaped methanogen representing a novel family, the *Methanoplanaceae*. Arch. Microbiol. *132*: 31–36.

Wildgruber, G., M. Thomm and K.O. Stetter. 1984. *In* Validation of the publication of new names and new combinations previously effectively published outside the IJSB. List No. 14. Int. J. Syst. Bacteriol. *34*: 270–271.

Wilharm, T., T.N. Zhilina and P. Hummel. 1991. DNA–DNA hybridization of methylotrophic halophilic methanogenic bacteria and transfer of *Methanococcus halophilus*[VP] to the genus *Methanohalophilus* as *Methanohalophilus halophilus*, comb. nov. Int. J. Syst. Bacteriol. *41*: 558–562.

Willcox, W.R., S.P. Lapage and B. Holmes. 1980. A review of numerical methods in bacterial identification. Antonie Leeuwenhoek *46*: 233–299.

Willems, A., J. De Ley, M. Gillis and K. Kersters. 1991a. *Comamonadaceae*, a new family encompassing the acidovorans ribosomal RNA complex, including *Variovorax paradoxus*, gen. nov., comb. nov., for *Alcaligenes paradoxus* (Davis 1969). Int. J. Syst. Bacteriol. *41*: 445–450.

Willems, A., M. Gillis and J. de Ley. 1991b. Transfer of *Rhodocyclus gelatinosus* to *Rubrivivax gelatinosus* gen. nov., comb. nov., and phylogenetic relationships with *Leptothrix*, *Sphaerotilus natans*, *Pseudomonas saccharophila*, and *Alcaligenes latus*. Int. J. Syst. Bacteriol. *41*: 65–73.

Willems, A., B. Pot, E. Falsen, P. Vandamme, M. Gillis, K. Kersters and J. de Ley. 1991c. Polyphasic taxonomic study of the emended genus *Comamonas*: relationship to *Aquaspirillum aquaticum*, E. Falsen group 10, and other clinical isolates. Int. J. Syst. Bacteriol. *41*: 427–444.

Willén, T. and R. Mattsson. 1997. Water-blooming and toxin-producing cyanobacteria in Swedish fresh and brackish waters, 1918–1995. Hydrobiologia *353*: 181–192.

Williams, R.A.D. 1989. Biochemical taxonomy of the genus *Thermus*. *In* da Costa, Duarte and Williams (Editors), Microbiology of Extreme Environments and Its Potential for Biotechnology, Elsevier, London. pp. 82–97.

Williams, R.A.D. and M.S. da Costa. 1992. The genus *Thermus* and related microorganisms. *In* Balows, Trüper, Dworkin, Harder and Schleifer (Editors), The Prokaryotes: A Handbook of Bacteria: Ecophysiology, Isolation, Identification, Applications, 2nd Ed., Springer-Verlag, New York. pp. 3745–3753.

Williams, R.A.D., K.E. Smith, S.G. Welch and J. Micallef. 1996. *Thermus oshimai* sp. nov., isolated from hot springs in Portugal, Iceland, and the Azores, and comment on the concept of a limited geographical distribution of *Thermus* species. Int. J. Syst. Bacteriol. *46*: 403–408.

Williams, R.A.D., K.E. Smith, S.G. Welch, J. Micallef and R.J. Sharp. 1995. DNA relatedness of *Thermus* strains, description of *Thermus brockianus* sp. nov., and proposal to reestablish *Thermus thermophilus* (Oshima and Imahori). Int. J. Syst. Bacteriol. *45*: 495–499.

Williams, S.T., M.E. Sharpe and J.G. Holt (Editors). 1989. Bergey's Manual of Systematic Bacteriology, 1st Ed., Vol. 4, The Williams & Wilkins Co., Baltimore.

Wilmotte, A. 1991. Taxonomic study of marine oscillatoriacean strains (*Cyanophyceae*, Cyanobacteria) with narrow trichomes. I. Morphological variability and autecological features. Arch. Hydrobiol. Suppl. *92*: 215–248.

Wilmotte, A. 1994. Molecular evolution and taxonomy of the cyanobacteria. *In* Bryant (Editor), The Molecular Biology of Cyanobacteria, Kluwer Academic Publishers, Dordrecht. pp. 1–25.

Wilmotte, A.M.R. and W.T. Stam. 1984. Genetic relationships among cyanobacterial strains originally designated as '*Anacystis nidulans*' and some other *Synechococcus* strains. J. Gen. Microbiol. *130*: 2737–2740.

Wilmotte, A., W. Stam and V. Demoulin. 1997. Taxonomic study of marine oscillatoriacean strains (*Cyanophyceae*, Cyanobacteria) with narrow trichomes. III. DNA–DNA hybridization studies and taxonomic conclusions. Arch. Hydrobiol. Suppl. *122*: 11–28.

Wilmotte, A., S. Turner, Y. van de Peer and N.R. Pace. 1992. Taxonomic study of marine oscillatoriacean strains (Cyanobacteria) with narrow trichomes: II. Nucleotide sequence analysis of the 16S ribosomal RNA. J. Phycol. *28*: 828–838.

Wilmotte, A., G. van der Auwera and R. de Wachter. 1993. Structure of the 16S ribosomal RNA of the thermophilic cyanobacterium *Chlorogloeopsis* HTF ("*Mastigocladus laminosus* HTF") strain PCC 7518, and phylogenetic analysis. FEBS Lett. *317*: 96–100.

Wilson, E.O. 1994. Naturalist, Island Press, Washington, DC.

Wilson, I.G. 1997. Inhibition and facilitation of nucleic acid amplification. Appl. Environ. Microbiol. *63*: 3741–3751.

Windberger, E., R. Huber and K.O. Stetter. 1992. *In* Validation of the publication of new names and new combinations previously effectively published outside the IJSB. List No. 41. Int. J. Syst. Bacteriol. *42*: 327–329.

Windberger, E., R. Huber, A. Trincone, H. Fricke and K.O. Stetter. 1989. *Thermotoga thermarum* sp. nov. and *Thermotoga neapolitana* occurring in African continental solfataric springs. Arch. Microbiol. *151*: 506–512.

Winslow, C.-E.A., J. Broadhurst, R.E. Buchanan, C.J. Krumwiede, L.A. Rogers and G.H. Smith. 1917. The families and genera of the bacteria. Preliminary report of the Committee of the Society of American Bacteriologists on characterization and classification of bacterial types. J. Bacteriol. *2*: 506–566.

Winslow, C.-E.A., J. Broadhurst, R.E. Buchanan, C.J. Krumwiede, L.A. Rogers and G.H. Smith. 1920. The families and genera of the bacteria. Final report of the Committee of the Society of American Bacteriologists on characterization and classification of bacterial types. J. Bacteriol. *5*: 191–229.

Winslow, C.-E.A. and A. Winslow. 1908. The Systematic Relationships of the *Coccaceae*, John Wiley and Sons, New York.

Winter, J. 1983. Maintenance of stock cultures of methanogens in the laboratory. Syst. Appl. Microbiol. *4*: 558–563.

Winter, J., C. Lerp, H.P. Zabel, F.X. Wildenauer, H. König and F. Schindler. 1984. *Methanobacterium wolfei*, sp. nov., a new tungsten-requiring, thermophilic, autotrophic methanogen. Syst. Appl. Microbiol. *5*: 457–466.

Wise, M.G., J.V. McArthur and L.J. Shimkets. 1997. Bacterial diversity of a Carolina bay as determined by 16S rRNA gene analysis: confirmation of novel taxa. Appl. Environ. Microbiol. *63*: 1505–1514.

Witt, D., T.B.-B. Dan and E. Stackebrandt. 1989. Nucleotide sequence of 16S rRNA and phylogenetic position of the green sulfur bacterium *Clathrochloris sulfurica*. Arch. Microbiol. *152*: 206–208.

Woese, C.R. 1987. Bacterial evolution. Microbiol. Rev. *51*: 221–271.

Woese, C.R., L. Achenbach, P.E. Rouvière and L. Mandelco. 1991. Archaeal phylogeny: reexamination of the phylogenetic position of *Archaeoglobus fulgidus* in light of certain composition-induced artifacts. Syst. Appl. Microbiol. *14*: 364–371.

Woese, C.R., B.A. Debrunner-Vossbrinck, H. Oyaizu, E. Stackebrandt and W. Ludwig. 1985a. Gram-positive bacteria: possible photosynthetic ancestry. Science *229*: 762–765.

Woese, C.R., O. Kandler and M.L. Wheelis. 1990. Towards a natural system of organisms: proposal for the domains *Archaea, Bacteria,* and *Eucarya.* Proc. Natl. Acad. Sci. U.S.A. *87*: 4576–4579.

Woese, C.R., J. Maniloff and L.B. Zablen. 1980. Phylogenetic analysis of the mycoplasmas. Proc. Natl. Acad. Sci. U.S.A. *77*: 494–498.

Woese, C.R., M. Sogin, D. Stahl, B.J. Lewis and L. Bonen. 1976. A comparison of the 16S ribosomal RNAs from mesophilic and thermophilic bacilli: some modifications in the Sanger method for RNA sequencing. J. Mol. Evol. *7*: 197–213.

Woese, C.R., E. Stackebrandt, T.J. Macke and G.E. Fox. 1985b. The phylogenetic definition of the major eubacterial taxa. Syst. Appl. Microbiol. *6*: 143–151.

Woese, C.R., E. Stackebrandt, W. Weisburg, B.J. Paster, M.T. Madigan, V.J. Fowler, C.M. Hahn, P. Blanz, R. Gupta, K.H. Nealson and G.E. Fox. 1984a. The phylogeny of purple bacteria: the alpha subdivision. Syst. Appl. Microbiol. *5*: 315–326.

Woese, C.R., W.G. Weisburg, C.M. Hahn, B.J. Paster, L.B. Zablen, B.J. Lewis, T.J. Macke, W. Ludwig and E. Stackebrandt. 1985c. The phylogeny of purple bacteria: the gamma subdivision. Syst. Appl. Microbiol. *6*: 25–33.

Woese, C.R., W.G. Weisburg, B.J. Paster, C.M. Hahn, R.S. Tanner, N.R. Krieg, H.P. Koops, H. Harms and E. Stackebrandt. 1984b. The phylogeny of purple bacteria: the beta subdivision. Syst. Appl. Microbiol. *5*: 327–336.

Wolk, C.P. 1982. Heterocysts. *In* Carr and Whitton (Editors), The Biology of Cyanobacteria, Blackwell Scientific Publications, Oxford. pp. 359–386.

Wolk, C.P. 1996. Heterocyst formation. Annu. Rev. Genet. *30*: 59–78.

Wolk, C.P., A. Ernst and J. Elhai. 1994. Heterocyst metabolism and development. *In* Bryant (Editor), The Molecular Biology of Cyanobacteria, Kluwer Academic Publishers, Dordrecht. pp. 769–823.

Woloszynska, J. 1912. O glonach planktonowych niektôrych jezior jawanskich, z uwzglednieniem glônow Sawy. Das Phytoplankton einiger javanischer Seen mit Berücksichtigung des Sawa-Planktons. Bull. Int. Acad. Sci. Cracovie *6B*: 649–709.

Wood, A.M. 1985. Adaptation of photosynthetic apparatus of marine ultraphytoplankton to natural light fields. Nature *316*: 253–255.

Worakit, S., D.R. Boone, R.A. Mah, M.E. Abdel-Samie and M.M. El-Halwagi. 1986. *Methanobacterium alcaliphilum,* sp. nov., an H$_2$-utilizing methanogen that grows at high pH values. Int. J. Syst. Bacteriol. *36*: 380–382.

Work, E. and H. Griffiths. 1968. Morphology and chemistry of cell walls of *Micrococcus radiodurans.* J. Bacteriol. *95*: 641–657.

Wu, L.C., K.C. Chow and K.K. Mark. 1983. The role of pigments in *Halobacterium cutirubrum* against UV irradiation. Microbios Lett. *24*: 85–90.

Wu, W.M., R.F. Hickey, M.K. Jain and J.G. Zeikus. 1993. Energetics and regulations of formate and hydrogen metabolism by *Methanobacterium formicicum.* Arch. Microbiol. *159*: 57–65.

Wyatt, J.T., T.C. Martin and J.W. Jackson. 1973. An examination of three strains of the blue-green algal genus, *Fremyella.* Phycologia *12*: 153–161.

Xie, W.Q., D. Tice and M. Potts. 1995. Cell-water deficit regulates expression of *rpoC1C2* (RNA polymerase) at the level of mRNA in desiccation-tolerant *Nostoc commune* UTEX 584 (Cyanobacteria). FEMS Microbiol. Lett. *126*: 159–164.

Xu, W., P.J. Mulhern, B.L. Blackford, M.H. Jericho, M. Firtel and T.J. Beveridge.

1996. Modeling and measuring the elastic properties of an archaeal surface, the sheath of *Methanospirillum hungatei,* and the implication for methane production. J. Bacteriol. *178*: 3106–3112.

Xu, Y., P. Zhou and X. Tian. 1999. Characterization of two novel halo-alkaliphilic archaea *Natronorubrum bangense* gen. nov., sp. nov. and *Natronorubrum tibetense* gen. nov., sp. nov. Int. J. Syst. Bacteriol. *49*: 261–266.

Xun, L., D.R. Boone and R.A. Mah. 1989. Deoxyribonucleic acid hybridization study of *Methanogenium* and *Methanocorpusculum* species, emendation of the genus *Methanocorpusculum,* and transfer of *Methanogenium aggregans* to the genus *Methanocorpusculum* as *Methanocorpusculum aggregans* comb. nov. Int. J. Syst. Bacteriol. *39*: 109–111.

Xun, L.Y., R.A. Mah and D.R. Boone. 1990. Isolation and characterization of disaggregatase from *Methanosarcina mazei* LYC. Appl. Environ. Microbiol. *56*: 3693–3698.

Yamada, Y. 1983. *Acetobacter xylinus* sp. nov., nom. rev., for the cellulose-forming and cellulose-less acetate-oxidising acetic acid bacteria with the Q-10 system. J. Gen. Appl. Microbiol. *29*: 417–420.

Yamada, Y., H. Takinami, Y. Tahara and K. Kondo. 1977. The menaquinone system in the classification of radiation-resistant micrococci. J. Gen. Appl. Microbiol. *23*: 105–108.

Yang, L.L. and A. Haug. 1979. Purification and partial characterization of a procaryotic glycoprotein from the plasma membrane of *Thermoplasma acidophilum.* Biochim. Biophys. Acta *556*: 265–277.

Yano, T., S.S. Chu, V.D. Sled, T. Ohnishi and T. Yagi. 1997. The proton-translocating NADH-quinone oxidoreductase (NDH-1) of thermophilic bacterium *Thermus thermophilus* HB-8: complete DNA sequence of the gene cluster and thermostable properties of the expressed NQO2 subunit. J. Biol. Chem. *272*: 4201–4211.

Yanyushin, M.F. 1988. Isolation and characterization of F$_1$-ATPase from green nonsulfur photosynthesizing bacterium *Chloroflexus aurantiacus.* Biokhimiya *53*: 1288–1295.

Yeh, K.C., S.H. Wu, J.T. Murphy and J.C. Lagarias. 1997. A cyanobacterial phytochrome two-component light sensory system. Science *277*: 1505–1508.

Yeh, M.F. and J.M. Trela. 1976. Purification and characterization of a repressible alkaline phosphatase from *Thermus aquaticus.* J. Biol. Chem. *251*: 3134–3139.

Yershov, G., V. Barsky, A. Belgovsky, E. Kirillov, E. Kreindlin, I. Ivanov, S. Parinov, D. Guschin, A. Drobishev, S. Dubiley and A. Mirzabekov. 1996. DNA analysis and diagnostics on oligonucleotide microchips. Proc. Natl. Acad. Sci. U.S.A. *93*: 4913–4918.

Yokoyama, A., G. Sandmann, T. Hoshino, K. Adachi, M. Sakai and Y. Shizuri. 1995. Thermozeaxanthins, new carotenoid-glycoside-esters from thermophilic eubacterium *Thermus thermophilus.* Tetrahedron Lett. *36*: 4901–4904.

Yokoyama, A., Y. Shizuri, T. Hoshino and G. Sandmann. 1996. Thermocryptoxanthins: novel intermediates in the carotenoid biosynthetic pathway of *Thermus thermophilus.* Arch. Microbiol. *165*: 342–345.

Yokoyama, K., T. Oshima and M. Yoshida. 1990. *Thermus thermophilus* membrane-associated ATPase: indication of a eubacterial V-type ATPase. J. Biol. Chem. *265*: 21946–21950.

Yoon, H.-S. and J.W. Golden. 1998. Heterocyst pattern formation controlled by a diffusible peptide. Science *282*: 935–938.

Yoon, K.S., M. Ishii, Y. Igarashi and T. Kodama. 1996a. Purification and characterization of 2-oxoglutarate: ferredoxin oxidoreductase from a thermophilic, obligately chemolithoautotrophic bacterium, *Hydrogenobacter thermophilus* TK-6. J. Bacteriol. *178*: 3365–3368.

Yoon, K.S., M. Ishii, T. Kodama and Y. Igarashi. 1997. Purification and characterization of pyruvate:ferredoxin oxidoreductase from *Hydrogenobacter thermophilus* TK-6. Arch. Microbiol. *167*: 275–279.

Yoon, K.S., Y. Ueda, M. Ishii, Y. Igarashi and T. Kodama. 1996b. NADH:Ferredoxin reductase and NAD-reducing hydrogenase activities in *Hydrogenobacter thermophilus* strain TK-6. FEMS Microbiol. Lett. *139*: 139–142.

Yoshida, M. 1972. Allosteric nature of thermostable phosphofructokinase from an extreme thermophilic bacterium. Biochemistry *11*: 1087–1093.

Yoshida, M. and T. Oshima. 1971. The thermostable allosteric nature of fructose 1,6-diphosphatase from an extreme thermophile. Biochem. Biophys. Res. Commun. *45*: 495–500.

Yoshida, M., T. Oshima and K. Imahori. 1971. The thermostable allosteric enzyme: phosphofructokinase from an extreme thermophile. Biochem. Biophys. Res. Commun. *43*: 36–39.

Young, J.P.W. and K.E. Haukka. 1996. Diversity and phylogeny of rhizobia. New Phytol. *133*: 87–94.

Yu, I.K. and F. Kawamura. 1988. *In* Validation of the publication of new names and new combinations previously effectively published outside the IJSB. List No. 26. Int. J. Syst. Bacteriol. *38*: 328–329.

Yu, J.P., J. Ladapo and W.B. Whitman. 1994. Pathway of glycogen metabolism in *Methanococcus maripaludis*. J. Bacteriol. *176*: 325–332.

Yu, J.S. and K.M. Noll. 1997. Plasmid pRQ7 from the hyperthermophilic bacterium *Thermotoga* species strain RQ7 replicates by the rolling-circle mechanism. J. Bacteriol. *179*: 7161–7164.

Yurkov, V. and V.M. Gorlenko. 1992. New species of aerobic bacteria from the genus *Erythromicrobium* containing bacteriochlorophyll *a*. Mikrobiologiya *61*: 248–255.

Yurkov, V., V.M. Gorlenko, L.L. Mityushina and D.A. Starynin. 1991. The effect of limiting factors on the structure of phototrophic associations in thermal springs of the river Bolshaya. Mikrobiologiya *60*: 129–138.

Yurkov, V.V., S. Krieger, E. Stackebrandt and T. Beatty. 1999. *Citromicrobium bathyomarinum*, a novel aerobic bacterium isolated from deep-sea hydrothermal vent plume waters that contains photosynthetic pigment-protein complexes. J. Bacteriol. *181*: 4517–4525.

Yurkov, V., E. Stackebrandt, O. Buss, A. Vermeglio, V. Gorlenko and J.T. Beatty. 1997. Reorganization of the genus *Erythromicrobium*: description of "*Erythromicrobium sibiricum*" as *Sandaracinobacter sibiricus* gen. nov., sp. nov., and of "*Erythromicrobium ursincola*" as *Erythromonas ursincola* gen. nov., sp. nov. Int. J. Syst. Bacteriol. *47*: 1172–1178.

Yurkov, V., E. Stackebrandt, A. Holmes, J.A. Fuerst, P. Hugenholtz, J. Golecki, N. Gad'on, V.M. Gorlenko, E.I. Kompantseva and G. Drews. 1994. Phylogenetic positions of novel aerobic bacteriochlorophyll *a*-containing bacteria and description of *Roseococcus thiosulfatophilus* gen. nov., sp. nov., *Erthyromicrobium ramosum* gen. nov., sp. nov., and *Erythrobacter litoralis* sp. nov. Int. J. Syst. Bacteriol. *44*: 427–434.

Zabel, H.P., H. König and J. Winter. 1984. Isolation and characterization of a new coccoid methanogen, *Methanogenium tatii*, spec. nov. from a solfataric field on Mount Tatio. Arch. Microbiol. *137*: 308–315.

Zabel, H.P., H. König and J. Winter. 1985. Emended description of *Methanogenium thermophilicum* and assignment of new isolates to this species. Syst. Appl. Microbiol. *6*: 72–78.

Zannoni, D. 1995. Aerobic and anaerobic electron transport chains in anoxygenic phototrophic bacteria. *In* Blankenship, Madigan and Bauer (Editors), Anoxygenic Photosynthetic Bacteria, Kluwer Academic Publishers, Dordrecht. pp. 949–971.

Zarda, B., D. Hahn, A. Chatzinotas, W. Schönhuber, A. Neef, R.I. Amann and J. Zeyer. 1997. Analysis of bacterial community structure in bulk soil by *in situ* hybridization. Arch. Microbiol. *168*: 185–192.

Zarilla, K.A. and J.J. Perry. 1984. *Thermoleophilum album* gen. nov. and sp. nov., a bacterium obligate for thermophily and *n*-alkane substrates. Arch. Microbiol. *137*: 286–290.

Zehnder, A.J.B. 1985. Isolation and cultivation of large cyanophytes for taxonomic purposes. Arch. Hydrobiol. Suppl. *71*: 281–289.

Zehnder, A.J.B. 1989. Genus III. *Methanothrix*. *In* Staley, Bryant, Pfennig and Holt (Editors), Bergey's Manual of Systematic Bacteriology, 1st Ed., Vol. 3, The Williams & Wilkins Co., Baltimore. pp. 2207–2209.

Zehnder, A.J.B. and P.R. Gorham. 1960. Factors influencing the growth of *Microcystis aeruginosa* Kütz emend. Elenkin. Can. J. Microbiol. *6*: 645–660.

Zehnder, A.J.B., B.A. Huser, T.D. Brock and K. Wuhrmann. 1980. Characterization of an acetate-decarboxylating, non-hydrogen-oxidizing methane bacterium. Arch. Microbiol. *124*: 1–11.

Zehnder, A.J.B. and K. Wuhrmann. 1977. Physiology of a *Methanobacterium* strain AZ. Arch. Microbiol. *111*: 199–205.

Zehr, J.P., K. Ohki, Y. Fujita and D. Landry. 1991. Unique modification

of adenine in genomic DNA of the marine cyanobacterium *Trichodesmium* sp. strain NIBB 1067. J. Bacteriol. *173*: 7059–7062.

Zeikus, J.G., A. Ben-Bassat and P.W. Hegge. 1980. Microbiology of methanogenesis in thermal, volcanic environments. J. Bacteriol. *143*: 432–440.

Zeikus, J.G. and V.G. Bowen. 1975a. Comparative ultrastructure of methanogenic bacteria. Can. J. Microbiol. *21*: 121–129.

Zeikus, J.G. and V.G. Bowen. 1975b. Fine structure of *Methanospirillum hungatei*. J. Bacteriol. *121*: 373–380.

Zeikus, J.G., M.A. Dawson, T.E. Thompson, K. Ingvorsen and E.C. Hatchikian. 1983. Microbial ecology of volcanic sulphidogenesis: isolation and characterization of *Thermodesulfobacterium commune* gen. nov. and sp. nov. J. Gen. Microbiol. *129*: 1159–1169.

Zeikus, J.G., M.A. Dawson, T.E. Thompson, K. Ingvorsen and E.C. Hatchikian. 1995. *In* Validation of the publication of new names and new combinations previously effectively published outside the IJSB. List No. 52. Int. J. Syst. Bacteriol. *45*: 197–198.

Zeikus, J.G. and D.L. Henning. 1975. *Methanobacterium arbophilicum* sp.nov. an obligate anaerobe isolated from wetwood of living trees. Antonie Leeuwenhoek *41*: 543–552.

Zeikus, J.G. and M.R. Winfrey. 1976. Temperature limitation of methanogenesis in aquatic sediments. Appl. Environ. Microbiol. *31*: 99–107.

Zeikus, J.G. and R.S. Wolfe. 1972. *Methanobacterium thermoautotrophicus* sp. n., an anaerobic, autotrophic, extreme thermophile. J. Bacteriol. *10*: 707–713.

Zellner, G., C. Alten, E. Stackebrandt, E. Conway de Macario and J. Winter. 1987. Isolation and characterization of *Methanocorpusculum parvum* gen. nov., spec. nov., a new tungsten requiring, coccoid methanogen. Arch. Microbiol. *147*: 13–20.

Zellner, G., C. Alten, E. Stackebrandt, E. Conway de Macario and J. Winter. 1988. *In* Validation of the publication of new names and new combinations previously effectively published outside the IJSB. List No. 24. Int. J. Syst. Bacteriol. *38*: 136–137.

Zellner, G., K. Bleicher, E. Braun, H. Kneifel, B.J. Tindall, E. Conway de Macario and J. Winter. 1989a. Characterization of a new mesophilic, secondary alcohol-utilizing methanogen, *Methanobacterium palustre*, spec. nov. from a peat bog. Arch. Microbiol. *151*: 1–9.

Zellner, G., K. Bleicher, E. Braun, H. Kneifel, B.J. Tindall, E. Conway de Macario and J. Winter. 1990a. *In* Validation of the publication of new names and new combinations previously effectively published outside the IJSB. List No. 35. Int. J. Syst. Bacteriol. *40*: 470–471.

Zellner, G., D.R. Boone, J. Keswani, W.B. Whitman, C.R. Woese, A. Hagelstein, B.J. Tindall and E. Stackebrandt. 1999. Reclassification of *Methanogenium tationis* and *Methanogenium liminatans* as *Methanofollis tationis* gen. nov., comb. nov. and *Methanofollis liminatans* comb. nov. and description of a new strain of *Methanofollis liminatans*. Int. J. Syst. Bacteriol. *49*: 247–255.

Zellner, G., M. Geveke, E. Conway de Macario and H. Diekmann. 1991. Population dynamics of biofilm development during start-up of a butyrate-degrading fluidized-bed reactor. Appl. Microbiol. Biotechnol. *36*: 404–409.

Zellner, G., A.J.L. Macario and E. Conway de Macario. 1996. Microbial subpopulations in the biofilm attached to the substratum and in the free flocs of a fixed-bed anaerobic bioreactor. Appl. Microbiol. Biotechnol. *46*: 443–449.

Zellner, G., A.J.L. Macario and E. Conway de Macario. 1997. A study of three anaerobic methanogenic bioreactors reveals that syntrophs are diverse and different from reference organisms. FEMS Microbiol. Ecol. *22*: 295–301.

Zellner, G., P. Messner, H. Kneifel, B.J. Tindall, J. Winter and E. Stackebrandt. 1989b. *Methanolacinia* gen. nov., incorporating *Methanomicrobium paynteri* as *Methanolacinia paynteri* comb. nov. J. Gen. Appl. Microbiol. *35*: 185–202.

Zellner, G., P. Messner, H. Kneifel, B.J. Tindall, J. Winter and E. Stackebrandt. 1990b. *In* Validation of the publication of new names and new combinations previously effectively published outside the IJSB. List No. 35. Int. J. Syst. Bacteriol. *40*: 470–471.

Zellner, G., P. Messner, J. Winter and E. Stackebrandt. 1998. *Methano-*

culleus palmolei sp. nov., an irregularly coccoid methanogen from an anaerobic digester treating wastewater of a palm oil plant in North-Sumatra, Indonesia. Int. J. Syst. Bacteriol. *48*: 1111–1117.

Zellner, G., U.B. Sleytr, P. Messner, H. Kneifel and J. Winter. 1990c. *Methanogenium liminatans*, spec. nov., a new coccoid, mesophilic methanogen able to oxidize secondary alcohols. Arch. Microbiol. *153*: 287–293.

Zellner, G., E. Stackebrandt, P. Messner, B.J. Tindall, E. Conway de Macario, H. Kneifel, U.B. Sleytr and J. Winter. 1989c. *Methanocorpusculaceae* fam. nov., represented by *Methanocorpusculum parvum, Methanocorpusculum sinense* spec. nov. and *Methanocorpusculum bavaricum* spec. nov. Arch. Microbiol. *151*: 381–390.

Zellner, G., E. Stackebrandt, P. Messner, B.J. Tindall, E. Conway de Macario, H. Kneifel, U.B. Sleytr and J. Winter. 1989d. *In* Validation of the publication of new names and new combinations previously effectively published outside the IJSB. List No. 30. Int. J. Syst. Bacteriol. *39*: 371.

Zellner, G. and J. Winter. 1987a. Growth promoting effect of tungsten on methanogens and incorporation of tungsten-185 into cells. FEMS Microbiol. Lett. *40*: 81–87.

Zellner, G. and J. Winter. 1987b. Secondary alcohols as hydrogen donors for CO_2-reduction by methanogens. FEMS Microbiol. Lett. *44*: 323–328.

Zevenboom, W., J. van der Does, K. Bruning and L.R. Mur. 1981. A non-heterocystous mutant of *Aphanizomenon flos-aquae*, selected by competition in light-limited continuous culture. FEMS Microbiol. Lett. *10*: 11–16.

Zhao, H., A.G. Wood, F. Widdel and M.P. Bryant. 1988. An extremely thermophilic *Methanococcus* from a deep sea hydrothermal vent and its plasmid. Arch. Microbiol. *150*: 178–183.

Zhao, Y., D.R. Boone, R.A. Mah, J.E. Boone and L. Xun. 1989. Isolation and characterization of *Methanocorpusculum labreanum*, a new species from the LaBrea Tar Pits (California, USA). Int. J. Syst. Bacteriol. *39*: 10–13.

Zhao, Y., H. Zhang, D.R. Boone and R.A. Mah. 1986. Isolation and characterization of a fast-growing, thermophilic *Methanobacterium* species. Appl. Environ. Microbiol. *52*: 1227–1229.

Zhilina, T.N. 1971. The fine structure of *Methanosarcina*. Mikrobiologiya *40*: 674–680.

Zhilina, T.N. 1976. Biotypes of *Methanosarcina*. Mikrobiologiya *45*: 481–489.

Zhilina, T.N. 1978. Growth of a pure *Methanosarcina* culture, biotype 2, on acetate. Mikrobiologiya *47*: 396–399.

Zhilina, T.N. 1983. A new obligate halophilic methane-producing bacterium. Mikrobiologiya *52*: 375–382.

Zhilina, T.N. 1986. Methanogenic bacteria from hypersaline environments. Syst. Appl. Microbiol. *7*: 216–222.

Zhilina, T.N. and S.A. Ilarionov. 1984. Isolation and comparative characteristics of methanogenic bacteria assimilating formate with the description of *Methanobacterium thermoformicicum*, sp. nov. Mikrobiologiya *53*: 785–790.

Zhilina, T.N. and T.P. Svetlichnaya. 1989. The ultrafine structure of *Methanohalobium evestigatus*, an extremely halophilic methanogenic bacterium. Mikrobiologiya *58*: 312–318.

Zhilina, T.N. and G.A. Zavarzin. 1979a. Comparative cytology of methanosarcinea and a description of *Methanosarcina vacuolata* sp. nova. Mikrobiologiya *48*: 279–285.

Zhilina, T.N. and G.A. Zavarzin. 1979b. Cyst formation by *Methanosarcina*. Mikrobiologiya *48*: 451–456.

Zhilina, T.N. and G.A. Zavarzin. 1987a. *Methanohalobium evestigatus*, gen. nov. sp. nov., the extremely halophilic methanogenic archaebacterium. Dokl. Akad. Nauk. SSSR. *293*: 464–468.

Zhilina, T.N. and G.A. Zavarzin. 1987b. *Methanosarcina vacuolata* sp. nov., a vacuolated methanosarcina. Int. J. Syst. Bacteriol. *37*: 281–283.

Zhilina, T.N. and G.A. Zavarzin. 1988. *In* Validation of the publication of new names and new combinations previously effectively published outside the IJSB List No. 24. Int. J. Syst. Bacteriol. *38*: 136–137.

Zhilina, T.N. and G.A. Zavarzin. 1990. Extremely halophilic, methylotrophic, anaerobic bacteria. FEMS Microbiol. Rev. *87*: 315–321.

Zhou, P., Y. Xu, C. Xiao , Y. Ma and H. Liu. 1994. New species of *Haloarcula*. Acta Microbiol. Sin. *34*: 89–95.

Zillig, W. 1983. *In* Validation of the publication of new names and new combinations previously effectively published outside the IJSB. List No. 11. Int. J. Syst. Bacteriol. *33*: 672–674.

Zillig, W. 1988. *In* Validation of the publication of new names and new combinations previously effectively published outside the IJSB. List No. 24. Int. J. Syst. Bacteriol. *38*: 136–137.

Zillig, W. 1989a. Genus I. *Thermoproteus. In* Staley, Bryant, Pfennig and Holt (Editors), Bergey's Manual of Systematic Bacteriology, 1st Ed., Vol. 3, The Williams & Wilkins Co., Baltimore. pp. 2241.

Zillig, W. 1989b. *In* Validation of the publication of new names and new combinations previously effectively published outside the IJSB. List No. 31. Int. J. Syst. Bacteriol. *39*: 495–497.

Zillig, W. and A. Gierl. 1983. *In* Validation of the publication of new names and new combinations previously effectively published outside the IJSB. List No. 11. Int. J. Syst. Bacteriol. *33*: 672–674.

Zillig, W., A. Gierl, G. Schreiber, S. Wunderl, D. Janekovic, K.O. Stetter and H.P. Klenk. 1983a. The archaebacterium *Thermofilum pendens* represents a novel genus of the thermophilic, anaerobic sulfur respiring *Thermoproteales*. Syst. Appl. Microbiol. *4*: 79–87.

Zillig, W., F. Gropp, A. Henschen, H. Neumann, P. Palm, W.-D. Reiter, M. Rettenberger, H. Schnabel and S. Yeats. 1986a. Archaebacterial virus host systems. Syst. Appl. Microbiol. *7*: 58–66.

Zillig, W., I. Holz, D. Janekovic, W. Schäfer and W.D. Reiter. 1983b. The archaebacterium *Thermococcus celer*, represents a novel genus within the thermophilic branch of the archaebacteria. Syst. Appl. Microbiol. *4*: 88–94.

Zillig, W., I. Holz, H.-P. Klenk, J. Trent, S. Wunderl, D. Janekovic, E. Imsel and B. Haas. 1987. *Pyrococcus woesei*, sp. nov., an ultra-thermophilic marine archaebacterium, representing a novel order, *Thermococcales.* Syst. Appl. Microbiol. *9*: 62–70.

Zillig, W., I. Holz and S. Wunderl. 1991. *Hyperthermus butylicus* gen. nov., sp. nov., a hyperthermophilic anaerobic, peptide-fermenting, facultatively H_2S-generating archaebacterium. Int. J. Syst. Bacteriol. *41*: 169– 170.

Zillig, W., H.-P. Klenk, P. Palm, G. Pühler, F. Gropp, R.A. Garrett and H. Leffers. 1989. The phylogenetic relations of DNA-dependent RNA polymerases of archaebacteria, eukaryotes, and eubacteria. Can. J. Microbiol. *35*: 73–80.

Zillig, W., A. Kletzin, C. Schleper, I. Holz, D. Janekovic, J. Hain, M. Lanzendörfer and J.K. Kristjansson. 1994. Screening for *Sulfolobales*, their plasmids and their viruses in Icelandic solfataras. Syst. Appl. Microbiol. *16*: 609–628.

Zillig, W., W.D. Reiter, P. Palm, F. Gropp, H. Neumann and M. Rettenberger. 1998. Viruses of Archaebacteria. *In* Calender (Editor), The Bacteriophages, Vol. 1, Plenum Press, New York. pp. 517–568.

Zillig, W. and K.O. Stetter. 1982. *In* Validation of the publication of new names and new combinations previously effectively published outside the IJSB. List No. 8. Int. J. Syst. Bacteriol. *32*: 266–268.

Zillig, W. and K.O. Stetter. 1983. *In* Validation of the publication of new names and new combinations previously effectively published outside the IJSB. List No. 10. Int. J. Syst. Bacteriol. *33*: 438–440.

Zillig, W., K.O. Stetter, D. Prangishvilli, W. Schäfer, S. Wunderl, D. Janekovic, I. Holz and P. Palm. 1982. *Desulfurococcaceae*: the second family of the extremely thermophilic, anaerobic, sulfur-respiring *Thermoproteales*. Zentbl. Bakteriol. Mikrobiol. Hyg. 1 Abt Orig. C *3*: 304– 317.

Zillig, W., K.O. Stetter, W. Schäfer, D. Janekovik, S. Wunderl, I. Holz and P. Palm. 1981. *Thermoproteales*: a novel type of extremely thermoacidophilic anaerobic archaebacteria isolated from Icelandic solfataras. Zentbl. Bakteriol. Mikrobiol. Hyg. 1 Abt Orig. C *2*: 205–227.

Zillig, W., K.O. Stetter, S. Wunderl, W. Schulz, H. Priess and I. Scholz. 1980a. The *Sulfolobus-*"*Caldariella*" group: taxonomy on the basis of the structure of DNA-dependent RNA polymerases. Arch. Microbiol. *125*: 259–269.

Zillig, W., K.O. Stetter, S. Wunderl, W. Schulz, H. Priess and I. Scholz.

1980b. *In* Validation of the publication of new names and new combinations previously effectively published outside the IJSB. List No. 5. Int. J. Syst. Bacteriol. *30*: 676–677.

Zillig, W., S. Yeats, I. Holz, A. Böck, M. Rettenberger, F. Gropp and G. Simon. 1986b. *Desulfurolobus ambivalens*, gen. nov., sp. nov., an autotrophic archaebacterium facultatively oxidizing or reducing sulfur. Syst. Appl. Microbiol. *8*: 197–203.

Zimmermann, B.H., C.I. Nitsche, J.A. Fee, F. Rusnak and E. Munck. 1988. Properties of a copper-containing cytochrome ba_3: a second terminal oxidase from the extreme thermophile *Thermus thermophilus*. Proc. Natl. Acad. Sci. U.S.A. *85*: 5779–5783.

Zinder, S.H. and T. Anguish. 1992. Carbon monoxide, hydrogen, and formate metabolism during methanogenesis from acetate by thermophilic cultures of *Methanosarcina* and *Methanothrix* strains. Appl. Environ. Microbiol. *58*: 3323–3329.

Zinder, S.H., T. Anguish and A.L. Lobo. 1987. Isolation and characterization of a thermophilic acetotrophic strain of *Methanothrix*. Arch. Microbiol. *146*: 315–322.

Zinder, S.H. and R.A. Mah. 1979. Isolation and characterization of a thermophilic strain of *Methanosarcina* unable to use H_2–CO_2 for methanogenesis. Appl. Environ. Microbiol. *38*: 996–1008.

Zinder, S.H., K.R. Sowers and J.G. Ferry. 1985. *Methanosarcina thermophila* sp. nov., a thermophilic, acetotrophic, methane-producing bacterium. Int. J. Syst. Bacteriol. *35*: 522–523.

ZoBell, C.E. 1941. Studies on marine bacteria.I. The cultural requirements of heterotrophic anaerobes. J. Mar. Res. *4*: 42–75.

Zopf, W. 1885. Die Spaltpilze, 3rd Ed., Edward Trewendt, Breslau.

Zuckerkandl, E. and L. Pauling. 1965. Molecules as documents of evolutionary history. J. Theor. Biol. *8*: 357–366.

Zvyagintseva, I.S., E.B. Kudryashova and E.S. Bulygina. 1996. Proposal of a new type strain of *Halobacterium distributum*. Mikrobiologiya 65: 399–402.

Zvyagintseva, I.S. and A.L. Tarasov. 1987. Extreme halophilic bacteria from saline soils. Mikrobiologiya 56: 839–844.

Zwart, G., W.D. Hiorns, B.A. Methe, M.P. van Agterveld, R. Huismans, S.C. Nold, J.P. Zehr and H.J. Laanbroek. 1998. Nearly identical 16S rRNA sequences recovered from lakes in North America and Europe indicate the existence of clades of globally distributed freshwater bacteria. Syst. Appl. Microbiol. *21*: 546–556.

Zweifel, U.L. and A. Hagström. 1995. Total counts of marine bacteria include a large fraction of non-nucleoid-containing bacteria (ghosts). Appl. Environ. Microbiol. *61*: 2180–2185.

Index of Scientific Names of *Archaea* and *Bacteria*

Key to the fonts and symbols used in this index:

Nomenclature 　Lower case, Roman	Genera, species, and subspecies of bacteria. Every bacterial name mentioned in the *Manual* is listed in the index. Specific epithets are listed individually and also under the genus.*
CAPITALS, ROMAN:	Names of taxa higher than genus (tribes, families, orders, classes, divisions, kingdoms).
Pagination 　Roman:	Pages on which taxa are mentioned.
Boldface:	Indicates page on which the description of a taxon is given.†

---

* Infrasubspecific names, such as serovars, biovars, and pathovars, are not listed in the index.

† A description may not necessarily be given in the *Manual* for a taxon that is considered as *incertae sedis* or that is listed in an addendum or note added in proof; however, the page on which the complete citation of such a taxon is given is indicated in boldface type.

ISBN 0-387-98771-1

EAN

9 780387 987712 >